4th International Conference on Engineering, Applied Sciences and Technology (ICEAST 2018)

Exploring Innovative Solutions for Smart Society

MATEC Web of Conferences Volume 192 (2018)

Phuket, Thailand
4-7 July 2018

Editor:

C. Hamontree

ISBN: 978-1-5108-6872-4

Printed from e-media with permission by:

Curran Associates, Inc.
57 Morehouse Lane
Red Hook, NY 12571

Some format issues inherent in the e-media version may also appear in this print version.

Printed by Curran Associates, Inc. (2018)

For additional information, please contact EDP Sciences – Web of Conferences at the address below.

EDP Sciences – Web of Conferences
17, Avenue du Hoggar
Parc d'Activité de Courtabœuf
BP 112
F-91944 Les Ulis Cedex A
France

Phone: +33 (0) 1 69 18 75 75
Fax: +33 (0) 1 69 28 84 91

contact-edps@webofconferences.org

Additional copies of this publication are available from:

Curran Associates, Inc.
57 Morehouse Lane
Red Hook, NY 12571 USA
Phone: 845-758-0400
Fax: 845-758-2633
Email: curran@proceedings.com
Web: www.proceedings.com

TABLE OF CONTENTS

https://doi.org/10.1051/matecconf/201819200001

Preface

The 4th International Conference on Engineering, Applied Sciences and Technology (ICEAST), was held on July 4-7, 2018 in Phuket, Thailand. The ICEAST conference was inaugurated in 2007 to provide opportunities for enhancing collaboration and exploring current interdisciplinary topics. The conference aims to provide forum for presentations and discussions of the most recent advancements in all fields of engineering. The topics covered in the conference (published in MATEC Web of Conference) are categorized into three tracks: Track 1- Industrial Engineering, Materials and Manufacturing; Track 2- Mechanical, Mechatronics and Civil Engineering; and Track 3- Food, Chemical and Agricultural Engineering.

We would like to express our gratitude toward the plenary speakers, invited speakers, reviewers and participants. We also would like to extend our congratulations to the authors and presenters. This conference would not have been successful without the support from our sponsors and we greatly appreciate their contributions.

Editors of ICEAST 2018

Message from the Honorary Chairman

On behalf of King Mongkut's Institute of Technology Ladkrabang (KMITL), I am honored and delighted to welcome all the delegates and their guests to the 4th International Conference on Engineering, Applied Sciences and Technology (ICEAST) in the beautiful city of Phuket, Thailand.

The theme of the conference this year is "Exploring Innovative Solutions for Smart Society" which is organized in four different tracks to provide opportunities for enhancing interdisciplinary interactions, dialogue, and collaborations among all engineers, scientists, technologists and educators in various topics of current interest.

I would like to thank the organizing committees who have worked hard to plan and organize this important conference. I also would like to thank the co-sponsors for their contributions. Hopefully, their efforts will be rewarded and the conference will meet everyone's highest expectations. Lastly, I wish you all a successful conference and an enjoyable experience in Phuket, Thailand.

Prof. Dr. Suchatvee Suwansawat

President

King Mongkut's Institute of Technology Ladkrabang

Message from the General Chairman

I am delighted to see many colleagues, professors, researchers, and participants from around the world, all in one place. It is an honor and pleasure for King Mongkut's Institute of Technology Ladkrabang (KMITL) to greet and welcome you to the 4th International Conference on Engineering, Applied Sciences and Technology (ICEAST).

We organize ICEAST 2018 as a platform to gather the latest knowledge in engineering with the spectrum of topics that are current and broad. This is so that academicians, researchers, scientists, and practitioners can share and discuss new innovations or findings in engineering. Moreover, we hope that this event will be the starting point of collaborative partnership creation and innovation community building. I would like to take this opportunity to express my sincere gratitude to you for being here today.

Lastly, I hope that you enjoy your participation in ICEAST 2018 and have a memorable time in Phuket. On behalf of KMITL and ICEAST 2018 committee, I am very pleased to have you as our guests, thank you.

Assoc. Prof. Dr. Komsan Maleesee

Dean of Faculty of Engineering

King Mongkut's Institute of Technology Ladkrabang

Message from the General Secretary

On behalf of the organizing committee, I am pleased to welcome you to the 4th International Conference on Engineering, Applied Sciences and Technology (ICEAST). The ICEAST conference was inaugurated in 2007 to provide opportunities for enhancing collaboration and exploring current interdisciplinary topics. The conference aims to provide forum for presentations and discussions of the most recent advancements in all fields of engineering. In addition to the technical program, participants will have the opportunity to enjoy the charms of Phuket which is one of the world's top tourism destinations.

I would like to express my gratitude toward the plenary speakers, invited speakers and participants. I also would like to extend my congratulations to the authors and presenters. This conference would not have been successful without the support from our sponsors and we greatly appreciate their contributions. I hope that this conference will provide not only outstanding technical knowledge but also enjoyable experience for the participants. In conclusion, I am delighted to give a warm welcome to all you.

Dr. Chaowalit Hamontree

Head of Department of Industrial Engineering

King Mongkut's Institute of Technology Ladkrabang

iCEAST 2018

The 4th International Conference on Engineering, Applied Sciences and Technology

"Exploring Innovative Solutions for Smart Society"

July 4-7, 2018

Swissôtel Resort Phuket Patong Beach, Phuket, Thailand

The International Conference on Engineering, Applied Sciences and Technology (ICEAST) was inaugurated in 2007 to provide opportunities for enhancing collaboration and exploring current interdisciplinary topics. The ICEAST conferences were held in the past in 2007, 2012 and 2013. To continue the conference mission, the fourth ICEAST conference will be held at Swissôtel Resort Phuket Patong Beach in Phuket, Thailand during July 4-7, 2018. The accepted conference papers will be published in MATEC Web of Conference (indexed by SCOPUS) and the proceedings volume in IEEE Xplore Digital Library (indexed by EI Compendex and SCOPUS).

In addition to the technical program, participants will have the opportunity to enjoy the many charms of Phuket which is one of the World's top tourism destinations. The Organizing Committee hopes that this conference will provide not only outstanding technical knowledge but also enjoyable experience for the participants.

Honorary Chair
- Suchatvee Suwansawat *(KMITL, Thailand)*

General Chair
- Komsan Maleesee *(KMITL, Thailand)*

Co-Chairs
- Somsak Choomchuay *(KMITL, Thailand)*
- Chuwong Phongcharoenpanich *(KMITL, Thailand)*

International Committee
- Tadashi Ariga *(Tokai University, Japan)*
- Pannapa Changpetch *(Bentley University, USA)*
- Hsun-Yi Chen *(National Taiwan University, Taiwan)*
- Chen-Yang Cheng *(National Taipei University of Technology, Taiwan)*
- Hector Gabriel Allende Cid *(Pontificia Universidad Catolica de Valparaiso, Chile)*
- Giuseppe Di Fatta *(University of Reading, UK)*
- Tomotaka Homae *(National Institute of Technology, Toyama College, Japan)*
- Kwei-Long Huang *(National University of Taiwan, Taiwan)*
- Nobutaka Ito *(Chiang Mai University, Thailand)*
- Namhun Kim *(Ulsan National Institute of Science and Technology, South Korea)*
- Mettaya Kitiwan *(Tohoku University, Japan)*
- Yoshikazu Miyanaga *(Hokkaido University, Japan)*
- Yasuyuki Miyazawa *(Tokai University, Japan)*
- Baorong Ni *(Fukuoka Institute of Technology, Japan)*
- Takao Onoye *(Osaka University, Japan)*
- Jun Sato *(National Institute of Technology, Tsuruoka College, Japan)*
- Satoru Seto *(National Institute of Technology, Ishikawa College, Japan)*
- M. Shamsuzzaman *(University of Sharja, UAE)*
- Yoshiro Takeichi *(National Institute of Technology, Tsuruoka College, Japan)*
- Munehiro Tanaka *(Saga University, Japan)*
- I Gede Agus Widyadana *(Petra Christian University, Indonesia)*
- Kuniaki Yajima *(National Institute of Technology, Sendai College, Japan)*
- Nerda Zura Zaibidi *(Universiti Utara, Malaysia)*

Steering Committee
- Somsak Choomchuay *(KMITL, Thailand)*
- Vanchai Riewruja *(KMITL, Thailand)*
- Monai Krairiksh *(KMITL, Thailand)*
- Chuwong Phongcharoenpanich *(KMITL, Thailand)*
- Uma Seeboonruang *(KMITL, Thailand)*
- Pornchai Supnithi *(KMITL, Thailand)*
- Supan Tungjitkusolmun *(KMITL, Thailand)*

Technical Committee
- Sompob Polmai *(KMITL, Thailand)*
- Surapan Airphaiboon *(KMITL, Thailand)*
- Yutthapong Rangsanseri *(KMITL, Thailand)*
- Watchara Chatwiriya *(KMITL, Thailand)*
- Jarruwat Charoensuk *(KMITL, Thailand)*
- Apinan Namkanisorn *(KMITL, Thailand)*
- Sawai Pongswatd *(KMITL, Thailand)*
- Arthit Petchsasithon *(KMITL, Thailand)*
- Ekkapong Cheevitsopon *(KMITL, Thailand)*
- Panmanas Sirisomboon *(KMITL, Thailand)*
- Ranon Jientrakul *(KMITL, Thailand)*
- Sorawat Chivapreecha *(KMITL, Thailand)*
- Pearawut Yutthagowith *(KMITL, Thailand)*
- Vanvisa Chutchavon *(KMITL, Thailand)*
- Montree Siriprachyanun *(KMUTNB, Thailand)*
- Nantakrit Yodpijit *(KMUTNB, Thailand)*
- Sukanya Praesomboon *(KMUTNB, Thailand)*
- Banhan Lila *(BUU, Thailand)*
- Supphachan Rajsiri *(MU, Thailand)*
- Anupun Terdwongworakul *(KU, Thailand)*
- Anirut Chaijaruwanich *(CMU, Thailand)*
- Wassanai Wattanutchariya *(CMU, Thailand)*
- Wasawat Nakkiew *(CMU, Thailand)*
- Chusit Pradabpet *(RU, Thailand)*

Technical Program Chairs
Track 1
- Kannachai Kanlayasiri *(KMITL, Thailand)*
- Tossapol Kiatcharoenpol *(KMITL, Thailand)*

Track 2
- Worapong Tangsriat *(KMITL, Thailand)*
- Monsak Pimsarn *(KMITL, Thailand)*

Track 3
- Vasu Udompetaikul *(KMITL, Thailand)*
- Jedsada Chaishome *(KMITL, Thailand)*

Track 4
- Somsak Choomchuay *(KMITL, Thailand)*
- Wiboon Prompharich *(KMITL, Thailand)*

General Secretary
- Chaowalit Hamontree *(KMITL, Thailand)*

Important Dates

Paper Submission Deadline	February 28, 2018
Notification of Acceptance	April 30, 2018
Final Paper Submission Deadline	May 15, 2018
Early Registration Deadline	May 15, 2018
Conference Date	July 4-7, 2018

Call for Papers
Track 1 - Industrial Engineering, Materials and Manufacturing
Track 2 - Mechanical, Mechatronics and Civil Engineering
Track 3 - Food, Chemical and Agricultural Engineering
Track 4 - Telecommunications, Computer Science, Instrumentation and Control, Electrical and Electronic Engineering

Publications
Accepted papers will be published in the Proceedings of ICEAST 2018 and
- **(Track 1-3)** MATEC Web of Conference (indexed by SCOPUS).
- **(Track 4)** Proceedings volume to be included into IEEE Xplore Digital Library. *(The IEEE provides SCOPUS with all of the IEEE Xplore Digital Library content so that the bibliographic information (what is seen on an abstract page in IEEE Xplore, including the abstract) can be made visible.)*

Registration Rate (THB)

	Early Registration	Standard Registration
Regular Paper	12,000	14,000
Student Paper	8,000	9,000
Additional Paper	8,000	9,000
Participant	5,000	6,000

iceast2018@kmitl.ac.th
www.iceast.org

Statement of Peer review

In submitting conference proceedings to *Web of Conferences*, the editors of the proceedings certify to the Publisher that

1. They adhere to its **Policy on Publishing Integrity** in order to safeguard good scientific practice in publishing.
2. All articles have been subjected to peer review administered by the proceedings editors.
3. Reviews have been conducted by expert referees, who have been requested to provide unbiased and constructive comments aimed, whenever possible, at improving the work.
4. Proceedings editors have taken all reasonable steps to ensure the quality of the materials they publish and their decision to accept or reject a paper for publication has been based only on the merits of the work and the relevance to the journal.

Title, date and place of the conference

The title of the conference: The 4th International Conference on Engineering, Applied Sciences and Technology

Date of conference:July 4-7, 2018

Location: Location: Swissôtel Resort Phuket Patong Beach, Phuket, Thailand

Proceedings editor(s):

Chaowalit Hamontree, PhD

Date and editor's signature

July 27, 2018

Numerical simulation of two tool turning process

Kalidasan Rathinam[1*] *and Sandeep* Kumar[2]

[1]School of Mechanical Engineering, Lovely Professional University, Punjab, India
[2]Department of Mechanical Engineering, Indian Institute of Technology Ropar, Punjab, India

Abstract Double tool turning process is used to improve productivity. A 2D finite element model was developed using commercially available finite element analysis software Abaqus 6.13. The workpiece and the cutting tool materials are modelled as elasto-plastic and elastic material respectively. Johnson-Cook damage criterion was used for chip separation. The friction between the cutting tool and the workpiece is modelled based on penalty contact approach. The coefficient of friction between the chip and the first and second cutting tool was taken as 0.8 and 0.6 respectively. In this numerical investigation the effect tool separation distance over the cutting force, feed force and cutting temperature were studied. Three different tool separation distances were considered. The simulation result shows that cutting force and feed force of the front cutting tool and the rear cutting tool do not change appreciably with the variation of the tool separation distance. It was revealed that the temperature rise of the work material due to machining by two cutting tool is well below the recrystallization temperature. Hence the forces on front and rear cutting tool remain same for various tool separation distances. It was also observed that the cutting temperatures remained unchanged for the various tool separation distances.

1 Introduction

Numerical simulation of double tool turning process helps in increasing the productivity by conserving the resources. It also provides a detailed insight of the turning process. Many researchers had focused their attention on turning simulation, since it is a fundamental machining process. Baker [1] developed a generic two dimensional model using finite element method to study the influence of cutting speed on cutting force. It was revealed that the cutting force decreased when the cutting speed is increased. In addition the morphology of chip also changed from continuous to segmented chip. Dogu et al. [2] determined the temperature distribution in orthogonal cutting by a numerical model. The maximum cutting temperature was found for various cutting conditions. The simulation results agreed well with the literature. Duan et al. [3] simulated the chip formation process of AISI 1045 hardened steel. The influence of rake angle on cutting forces and chip morphology was studied. It was reported that finite element simulation could accurately determine the cutting force and chip features agreeing with the experimental results. Davim et al. [4] employed a Lagrangian FE model to predict the plastic strain distribution, cutting forces and cutting temperature. It was observed that the difference between simulated and experimental cutting force was 20% where us for cutting temperature it was 15%. Akbar et al. [5] used a fully coupled thermo-mechanical FE model to study the fraction of heat passed into the cutting tool. Commercially available software Abaqus 6.6 was used. It was shown that the tool-chip interface temperature, tool-chip contact length and von Mises stresses depends on the heat partition between the chip and tool. The model was also validated through experiments. Ucun and Aslantas [6] carried out numerical simulation of AISI 4340 steel using single layer and multi-layer coated carbide tools. Johnson-Cook material model was used for elastic-plastic work deformation. The effect of coating on cutting forces, tool stresses and cutting temperatures was evaluated. It was observed that the shear angle was maximum for coated tool and minimum for uncoated tool. Calamaz et al. [7] analyzed shear localization process in titanium alloy. The influence of thermal and strain softening and cutting parameters on the cutting temperature was studied. It was noted that the coefficient of friction affected the tool-chip interface temperature and shear band frequency. Hu and Huang [8] performed computational analysis of AISI 4340 steel with a ceramic tool for various turning speeds. Deform 3D software was employed to compute the effects of cutting speed on tool wear, cutting forces and temperature. It was noted that the diffusion wear becomes dominant when the cutting speed is increased, and the cutting forces decreases due to the softening of work material caused by its temperature rise.

It is observed that lot of research on numerical simulation has been performed on traditional turning process. In case of two tool turning process, two cutting tools are simultaneous used to machine the workpiece resulting in higher productivity. Few analytical and experimental works on double tool turning are reported as follows.

Budak and Ozturk [9] analysed the dynamic stability while turning with two tools simultaneously. The use of two tools created a shock absorber effect, thereby increasing the stability of turning process. Ozturk et al. [10] formulated the stability of parallel turning process in frequency domain and time domain for two different conditions. It was observed that best process stability was achieved when the natural frequencies of the two tools are slightly different. Azvar and Budak [11] presented a multi-dimensional model to

* Corresponding author: kalidasan.22180@lpu.co.in

analyse chatter stability during parallel double tool turning. The increase in nose radii and side cutting edge angle of the cutting tools increased the stability of turning process. Kalidasan et al. [12] investigated experimentally the effect of turning parameters on cutting forces and temperatures in double tool turning of cast iron and steel with a carbide tool. It was reported that, the separation distance between the two cutting tool did not have a significant effect on the cutting force component. Another experimental work of Kalidasan et al. [13] dealt about tool wear, chip morphology and machining accuracy. It was found that the front cutting tool exhibited predominant abrasive wear while the rear cutting tool showed both adhesive and diffusion wear.

It is observed from the literature that no computational work has been performed on double tool turning process even though it offers a lot of advantages such as savings in time and cost, increased tool life and improved surface finish, reduced cutting forces and temperature. The present work aims to fulfil that research gap. The primary objective of this research work is to develop a numerical model of double tool turning process and to determine the effect of tool separation distance on cutting force components and cutting temperature.

2 Finite Element Model

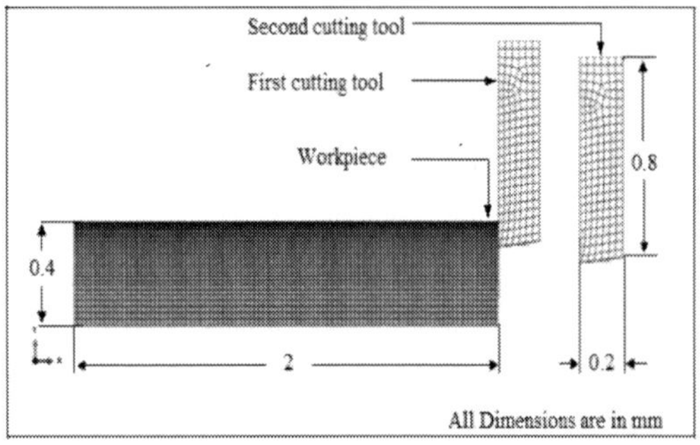

Fig. 1. Meshed FE model of double tool turning process

Fig. 1 shows the meshed finite element model of double tool turning process. The geometric model consists of two 2D deformable shell elements. The cutting tool is provided with neutral rake angle and with a clearance angle of 7°. The values are taken from the literature [5].

2.1 Material model

Johnson-Cook material model is considered for the AISI 4140 work material. This model is very widely used in machining process simulation because it takes into account the effect of strain, strain rate and temperature on the flow stress behavior of the material. The flow stress is given by

$$\bar{\sigma} = (A + B\bar{\varepsilon}^n)\left[1 + C\ln\left(\frac{\dot{\bar{\varepsilon}}}{\dot{\bar{\varepsilon}}_0}\right)\right]\left[1 - \left(\frac{\theta - \theta_{room}}{\theta_{melting} - \theta_{room}}\right)^m\right] \quad (1)$$

where A is the initial yield stress, B is the hardening modulus, C is the strain rate dependency coefficient, n is the work hardening coefficient, m is the thermal softening , [redacted] is the plastic strain rate, $\bar{\varepsilon}$ is the equivalent plastic strain and [redacted] is the reference strain rate, $\theta_{melting}$ is the melting temperature of the workpiece, θ is the process temperature θ_{room} is the ambient temperature. An equivalent plastic strain criterion is adopted to formulate the chip formation. Material failure depends upon the critical value of equivalent plastic strain. This cumulative law is stated as

$$D = \sum\left(\frac{\Delta\bar{\varepsilon}}{\bar{\varepsilon}_f}\right) \quad (2)$$

where D is the damage parameter, $\bar{\varepsilon}_f$ is the equivalent strain at failure $\Delta\bar{\varepsilon}$ is the increment of equivalent plastic strain. Equivalent plastic strain is given by

$$\bar{\varepsilon}_f = \left[D_1 + D_2\exp\left(D_3\frac{P}{\bar{\sigma}}\right)\right]\left[1 + D_4\ln\left(\frac{\dot{\bar{\varepsilon}}}{\dot{\bar{\varepsilon}}_0}\right)\right]\left[1 + D_5\left(\frac{\theta - \theta_{room}}{\theta_{melt} - \theta_{room}}\right)\right] \quad (3)$$

where P is the hydrostatic pressure. D_1, D_2, D_3, D_4 and D_5 are the experimentally obtained failure constants. Numerical model of Pantale et al. [14] provides the Johnson-Cook and damage parameters as shown in Table 1. The thermo-mechanical properties of work material and cutting tool are referred from the work of Akbar et al. [5]. Material stiffness start degrading after element satisfies the damage initiation criterion. Failure is assumed to occur when D value exceeds 1. After the element is fully degraded, element deletion equal to yes is specified in Abaqus software to completely remove the element.

Table 1. Johnson-Cook parameters of AISI 4140

A (MPa)	*B* (MPa)	n	*C*	m
595	580	0.133	0.023	1.03
D_1	D_2	D_3	D_4	D_5
1.5	3.44	-2.12	0.002	0.1

2.2 Explicit dynamic analysis

Formulation based on explicit dynamic analysis is generally used for highly non-linear problems like machining which involves large amount of deformations. Explicit dynamic is an integration technique which uses the central difference method to move the solution further in time. Explicit analysis uses the information already it has for obtaining the unknown information. It does not require any iteration or convergence check for obtaining the solution. It is designed to study the highly non-linear, discontinuous and high speed dynamic problems. It requires lesser disk space as compare to standard analysis. In this simulation work temperature displacement coupled explicit dynamic analysis is used.

2.3 Contact model

A kinematic contact algorithm has been used to establish the contact between the tool rake surface and the chip. Penalty contact with a constant coefficient of friction is used to model the contact between the chip and cutting tool. The experimental work of Kalidasan et al. [12] revealed the coefficient of friction between the chip and the first and second cutting tool as 0.8 and 0.6 respectively. Frictional forces are taken into consideration with the help of tangential behavior effect.

2.4 Meshing and Boundary condition

CPE4RT four node plane strain bilinear displacement and temperature quadrilateral elements are used. Additionally reduced integration scheme and hourglass control are used for both the cutting tool and workpiece. Workpiece consist of 10010 elements and 10332 nodes while the tool consists of 168 elements and 200 nodes. In coupled temperature displacement elements stress analysis is coupled with the heat transfer. The selected elements have both temperature and displacement degree of freedom. Elements can use linear or parabolic interpolation for displacement but can use only the linear interpolation for the temperature in coupled temperature displacement type element. Machining generates a lot of heat which affects the mechanical properties of the work piece hence coupled analysis is preferred. Workpiece is kept fixed. All the degrees of freedom at the base of workpiece are arrested to prevent displacement. Workpiece dimension are kept such that it will attain a steady state condition while machining. Cutting tools are given velocity in the negative X direction and it is constrained in the Y direction. In this work, a cutting velocity of 395 m/min and depth of cut of 0.1 mm is taken with 0.4, 0.8 and 1.2 mm as separation distances between the cutting tools.

3 Numerical results and discussion

The distance between the two cutting tools is an additional cutting parameter in double tool turning process. Therefore numerical simulation was performed to evaluate the effect of the tool separation distance on the cutting force and cutting temperature. It is to be noted that first and second cutting tool are referred as front and rear cutting tool.

3.1 Effect of tool separation distance on cutting force components

The cutting forces generated by two cutting tools during the numerical simulation process are shown in Fig. 2.

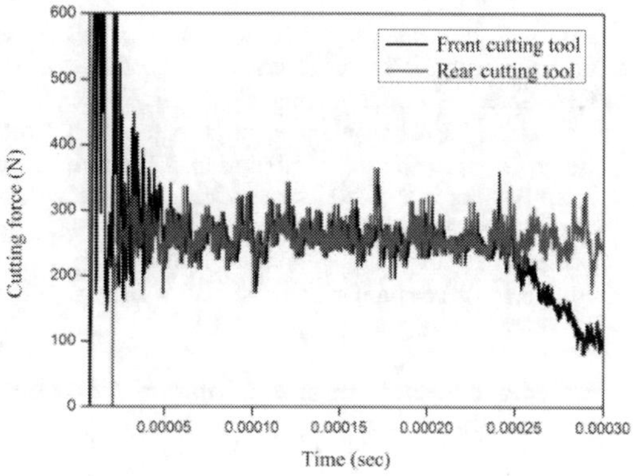

Fig. 2. Simulated cutting forces of front and rear cutting tool

The cutting conditions are already mentioned in section 2.4. It was observed that a transient region existed for about 0.05 milliseconds for the front cutting tool and thereafter a steady cutting force of 260 N was generated. On contrary for the rear cutting tool the transient region is almost non-existence as it is machining the already machined surface. It is to be noted that the mean cutting force of the rear cutting tool was 265 N. Thus the difference between the main cutting forces of front and rear cutting tool is less than 2%. Hence it can be said the front and rear cutting generated same magnitude of cutting force for similar cutting conditions. On similar lines the feed force generated by the front and rear tools are 108 N and 96 N. This accounts for 42% and 36% of the cutting forces of front and rear cutting tool respectively. The lesser value of feed forces of the rear cutting tool is due to lower coefficient of friction. Chisholm [15] reported that in orthogonal cutting the friction coefficient value lies in the range of 0.5 to 1. For the present simulation the friction coefficient values are mentioned in section 2.3. Lower friction coefficient of 0.6 between the rear cutting tool and chip may be attributed to reduced feed force, compared with front cutting tool.

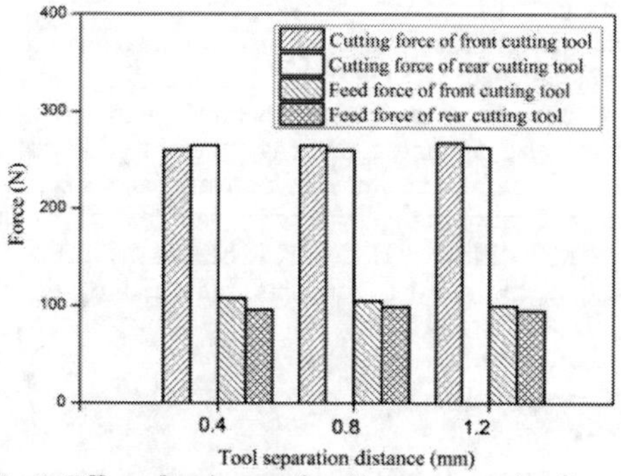

Fig. 3. Effect of tool separation distance on cutting force and feed force of front cutting tool and rear cutting tool

Fig. 3 shows the influence of tool separation distance on cutting and feed forces of both the cutting tools. It can be seen that when the tool separation distance is increased from 0.4 mm to 0.8 mm the variation in cutting force and feed force of front cutting tools are 2% and 3% respectively. For the rear tool it is 3% for both forces. Again on increasing from 0.8 mm to 1.2 mm the change in cutting force and feed force of front and rear cutting tools is 1%, 5%, 1% and 4% respectively. Thus it is evident that the cutting and feed forces of both the cutting tools does not vary appreciably on changing the tool separation distance.

3.1 Effect of tool separation distance on cutting temperature

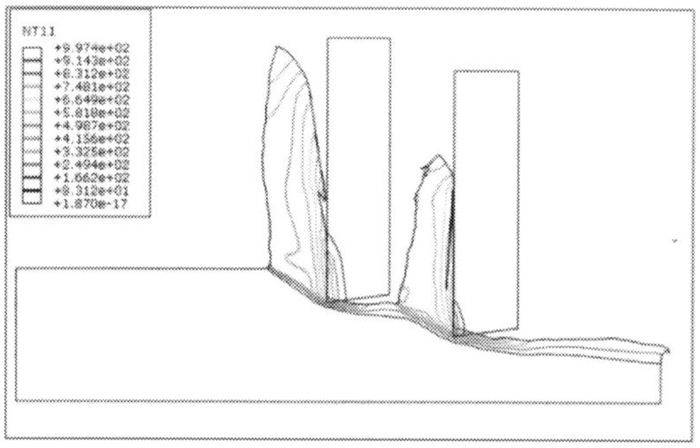

Fig. 4. Cutting temperature distribution for 0.4 mm tool separation distance

Fig. 4 shows the cutting temperature distribution on the chip, cutting tool and workpiece for a cutting velocity of 395 m/min, 0.1 mm depth of cut with 0.4 mm as distance between the two tools. The maximum cutting temperature at secondary deformation zone was 914°C where us at the tertiary deformation zone the temperature was 83°C. The bulk mean temperature of the workpiece is well below the recrystallization temperature of the work material. Kalidasan et al.[16] experimentally measured the cutting temperature while turning gray cast iron with two carbide cutting tools. It was reported that the bulk mean temperature of 56°C was attained by the work material for certain cutting condition. For the tool separation distance of 0.8 mm and 1.2 mm the maximum cutting temperatures are 908°C and 912°C. On the other hand the work material temperature remained 83°C under all cutting conditions. The work material properties remain unaltered as the workpiece is below its recrystallization temperatures in all the three tool separation distances. Hence it can be said that the cutting temperatures remains constant immaterial of the tool separation distances.

4 Conclusions

FE model was developed and numerical simulation was performed for the double tool turning process. Following conclusions are arrived based on the numerical simulation process.

- The cutting and feed forces generated by the front and rear cutting remain almost same for the simulated cutting conditions. The average cutting force and feed force are 264 N and 101 N with variation less than 7%.
- The magnitude of cutting and feed force is not influenced by tool separation distances. Therefore cutting and feed forces remained constant immaterial of tool separation distance for the selected cutting conditions.
- The maximum cutting temperature reached was above 900°C but the bulk mean temperature of the workpiece was 83°C. The bulk work material temperature is well below its recrystallization temperature. Variation of tool separation distance did not cause any change in the mean temperature of the work material.

References

1. M. Baker, J. Mater. Process. Technol., **176**, 117–126 (2006)
2. Y. Dogu, E. Aslan, N. Camuscu, J. Mater. Process. Technol., **171**, 1–9 (2006)
3. C.Z. Duan, T. Dou, Y.J. Cai, Int. J. Recent trends Eng., **1(5)**, 46–50 (2009)
4. J. Paulo Davim, C. Maranhão, P. Faria, A. Abrão, J. C. Rubio, Leonardo R. Silva, Int J Adv Manuf Technol., **42**, 842–849 (2009)
5. F. Akbar, P.T. Mativenga, M.A. Sheikh, Int J Adv Manuf Technol., **46**, 491–507 (2010)
6. I. Ucun, K. Aslantas, Int J Adv Manuf Technol., **54**, 899–910 (2011)
7. M. Calamaz, D. Coupard, M. Nouari, F. Girot, Int J Adv Manuf Technol., **52**, 887–895 (2011)
8. H. Hu, W. Huang, Int J Adv Manuf Technol., **67** (1–4), 907–915 (2013)
9. E. Budak, E. Ozturk, CIRP Ann Manuf Technol., **60**, 383–386 (2011)
10. E. Ozturk, A. Comak, E. Budak, J. Sound Vibration, **360**, 17–30 (2016)
11. M. Azvar, E. Budak, Int J Mach Tools Manuf, **123**, 116–128 (2017)
12. R. Kalidasan, M. Yatin, D.K. Sarma, S. Senthilvelan, U.S. Dixit. IJMMM, **18 (5/6)**, 540–551 (2016)
13. R. Kalidasan, S. Senthilvelan, U.S. Dixit, J. Vaibhav, Int. J. Precision Technology, **6(2)**, 142–157 (2016)
14. O. Pantale, J.L. Bacaria, O. Dalverny, R. Rakotomalala, S. Caperaa, Comput. Methods Appl. Mech. Engrg., **193**, 4383–4399 (2004)
15. A.J. Chisholm, Proc. Mech Eng., **165**, 14–21 (1951)
16. R. Kalidasan, M. Yatin, D.K. Sarma, S. Senthilvelan, Appl Mech Mater., **592–594**, 211–215 (2014)

Effect of forming condition on compressive strength of hydroxyapatite-bioactive glass compact rod

Wassanai Wattanutchariya [1,*] and *Teerawat* Sangkas[2]

[1]Advanced Manufacturing Technology Research Center, Department of Industrial Engineering, Faculty of Engineering, Chiang Mai University, Chiang Mai, Thailand
[2]Department of Tools and Die Engineering, Faculty of Engineering, Rajamangala University of Technology Lanna, Chiang Mai, Thailand

Abstract. With rising concerns regarding alloy implants, alternative biomaterials are currently studied, to avoid the adverse effects of metal on the human body such as irritation and inflammation. Hydroxyapatite (HA) and Bioactive glass (BG) are two bio-ceramics, which have been implemented in medical applications such as bone implants and fixation parts due to their biocompatibility and close resemblance to the mineralized phase of human bone. Furthermore, these materials can be synthesized from natural sources. In this study, M8 screws rod which are commonly implemented for bone fixation was selected as a case study to investigate an effect of forming conditions on the mechanical property of the composite structure. HA and BG were synthesized and formed into a composite specimen using a hydraulic pressing machine. Full factorial experimental design was employed to solve for an optimal forming condition. 3 factors consisting of mixing ratio, pressure and holding time for pressing were investigated for their impacts on the specimen's compressive strength. The result revealed that the BG ratio and pressure have a significant effect on the structural strength. The maximum compressive strength of 32.20 MPa was obtained from the specimen with 7.5 wt% of BG, 120 kg/cm^2 of pressure and 30 second of holding time.

1 Introduction

Nowadays, fixation devices or artificial human organ implants often contain metal parts, due to its superior mechanical properties. However, much literature has illustrated the adverse effects of metal implants on the human body such as irritation and inflammation [1-2]. Thus, investigation on biomaterials for medical part fabrication is a relevant area of research [3-4]. HA and BG are biomaterials that have been widely investigated for their potential use in medical implant and fixation parts since their composition and biocompatibility are close to the mineralized phase of human bones [5-7]. HA is currently used in many medical and dental applications such as orthopedic implants and implant coatings because HA has high biocompatibility and osteoconductivitiy [8-11]. Nonetheless, the disadvantage of HA is its low elasticity, which becomes a problem in load-bearing applications [12]. Therefore, reinforced biomaterials are being investigated; HA-composites may be developed for medical application to improve its characteristics in terms of mechanical strength, as well as other relevant physical and biological properties. BG is one of alternative reinforced material which improves on the mechanical performance of HA [13-15]. In this study, the investigation is focused on the development of a BG-HA composite bone fixation screw because this screw is frequently used as a basis for qualitative comparison [16-19]. Thus, the purpose of this research is to investigate the effect of forming conditions on the compressive property of BG-HA compact rod using factorial design as a technique for statistical evaluation and optimization of these parameters.

2 Methodology

2.1. Preparation of materials

For this study, 2 types of biomaterials were synthesis in the research center. HA was synthesized from cattle bone, while BG was partially synthesized from mollusc shell. For preparation of HA powder, a cattle bone was cut and cleaned by hydrogen peroxide (H_2O_2) solution to eliminate organic substances and collagen. Then, its moisture was reduced by drying in an oven at 120 °C for 7 days. Finally, it was calcined at 800 °C for 3 hr, and ground into powder with an average size of 37 μm. In the meantime, BG with the composition of 45 wt% SiO_2, 24.5 wt% CaO, 24.5 wt% Na_2O, and 6 wt% P_2O_5 [20], which CaO was prepared using mollusc shell as the raw material. This CaO was thoroughly mixed with other chemical reagents and poured into a crucible. Then, it was heated at 1,100 °C for 2.5 hr to generate glass phase [21]. Finally, the glass was ground into powder with an average size of 17.58 μm.

* Corresponding author: wassanai@eng.cmu.ac.th

https://doi.org/10.1051/matecconf/201819201002

2.2. Fabrication of HA/BG compact specimen

To investigate the effect of mixing ratio, various HA and BG mixtures were prepared, and then uniaxially pressed on a hydraulic pressing machine, as shown in Figure 1(a). The specimen dimension was conformed to ASTM F1044-05 (2011). Then, a specimen with 8-mm diameter and 50-mm long compact rods (Figure 1(b)) were sintered in an electrical furnace at 1,200 °C for 3 hr.

(a) (b)

Fig. 1. (a) Hydraulic pressing machine (b) compact specimen

2.3 Experimental Setup

The effects of mixing ratio and pressing condition were investigated using full factorial design with center point. Table 1 describes the experimental parameters and boundary conditions, consisting of BG mixing ratio, pressure and time, for a total of 12 experimental conditions (2 replications). The specimen for each condition will be further characterized to investigate the relationships among these parameters based on statistical evaluation.

Table 1. Experimental parameters

Factors	**Level**		**Unit**
	Low (-)	**High (+)**	
1.BG mixing ratio	5.0	7.5	wt%
2.Pressure	100	120	kg/cm^2
3.Time	30	60	Second

2.4 Evaluation of Mechanical Property

The compressive strength of specimen was tested on a Instron 5566 universal testing machine with a setup of crosshead speed at 0.5 mm/min and 10 kN load cell. The test was evaluated based on the load to failure of the cylindrical specimens, with dimensions of 8 mm. and 5 mm. in height, as shown in Figure 2.

Fig. 2. Compression test on universal testing machine

3 Results and Discussion

The compressive strength results of the HA-BG specimen based on the experimental design is illustrated in Table 2. The full factorial design with center point was generated by MINITAB 16 software to investigate the effect of 3 forming conditions on the mechanical property of the specimen. Statistical evaluation in terms of the significant effect of each parameter and its interaction was performed using Analysis of Variance (ANOVA) at 95% confidence interval, as shown in Table 3. In addition, R-squared and adjusted R-squared statistics were analysed to specify the extent to which fitted models could explain the variability of the results.

Table 2. Experimental results

No.	Factors			Compressive Strength (MPa)
	BG (wt%)	Pressure (Kg/cm^2)	Time (Sec)	
1	5	100	30	14.02
2	7.5	100	30	25.99
3	5	120	30	12.43
4	7.5	120	30	29.81
5	5	100	60	14.97
6	7.5	100	60	27.18
7	5	120	60	19.21
8	7.5	120	60	31.08
9	5	100	30	14.50
10	7.5	100	30	27.09
11	5	120	30	19.51
12	7.5	120	30	28.93
13	5	100	60	16.82
14	7.5	100	60	26.21
15	5	120	60	18.02
16	7.5	120	60	27.82
17	6.25	110	45	22.65
18	6.25	110	45	25.00

Table 3. Regression coefficients and ANOVA

Term	Coefficients	T	P
Constant	22.1000	43.53	0.000
BG	5.9150	11.65	0.000
PRESS	1.2516	2.47	0.036
TIME	0.5655	1.11	0.294
BG*PRESS	0.1454	0.29	0.781
BG*TIME	-0.5059	-1.00	0.345
PRESS*TIME	0.1156	0.23	0.825
BG*PRESS*TIME	-0.1355	-0.27	0.796
R-Sq = 94.18% R-Sq (adj.) = 89.00%			

According to the ANOVA, the parameters that influence the compressive strength of the HA-BG specimen were the BG ratio and forming pressure and. Hence, the regression model of rod specimen compressive strength (Y) based on significant factors can be described as equation 1.

$$Y = 22.1+5.915(BG)+1.2516(PRESS) \quad (1)$$

The validation of experimental results in terms of compressive strength's residual plot is shown in Figure 3. This normal probability plot was evaluated to evaluate the fitted distribution, estimate percentiles, and correlate different sample distributions. Since the fitted distribution, the relation of the theoretical and sample percentiles, form a straight line, it can be concluded that our experimental results were normally distributed. In the meantime, the surface and contour plot, as illustrated in Figure 4 and 5 show the effect of both forming parameters on the compressive strength of the HA-BG specimen. After analyzing with the response optimizer function, the optimal conditions within the experimental boundary was found at 120 kg/cm^2 of pressure, 7.5 wt% of BG ratio which yields the highest compressive strength of 29.95 MPa (as shown in Figure 6). It is worth mention that pressing time and parameter interaction have no significant effect on the compressive strength. Consequently, the obtained optimal condition was validated by performing another set of experimental setup (with 10 replications). The average compressive strength from confirmation tests was 32.20 MPa, which found to be not significantly different from the previous evaluation (p-value = 0.108).

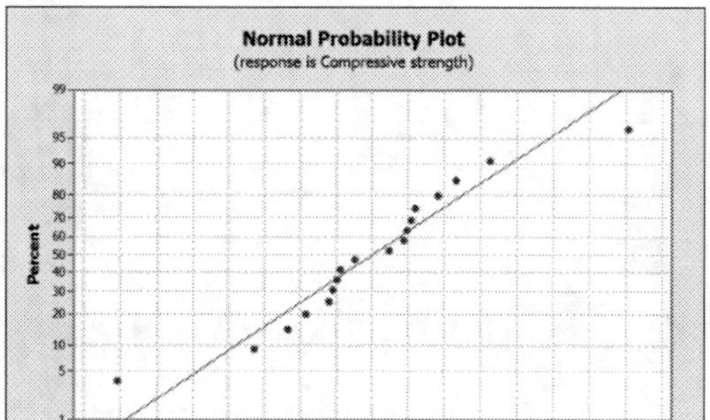

Fig. 3. Normal probability plot of compressive strength

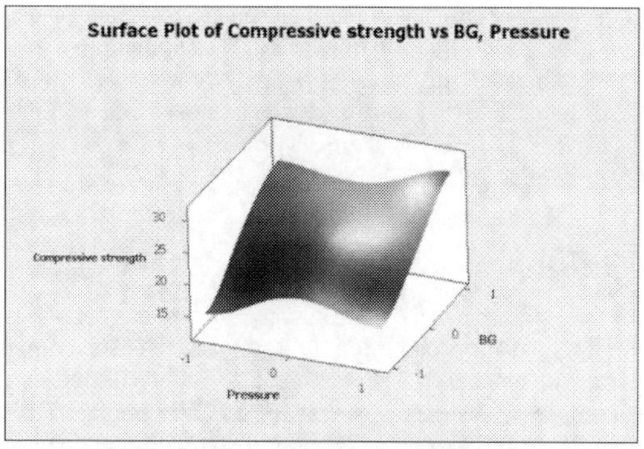

Fig. 4. Response surface of compressive strength

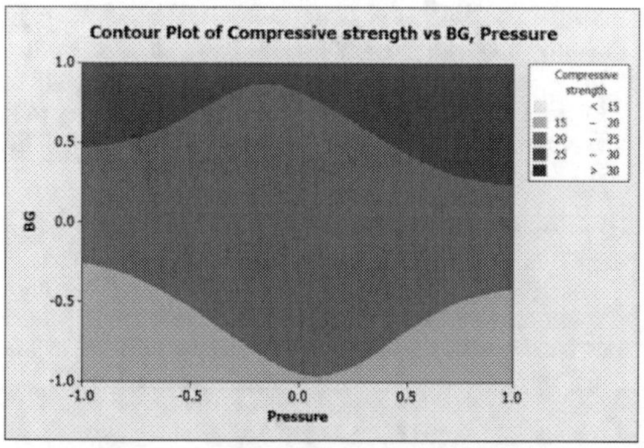

Fig. 5. Contour plot of compressive strength

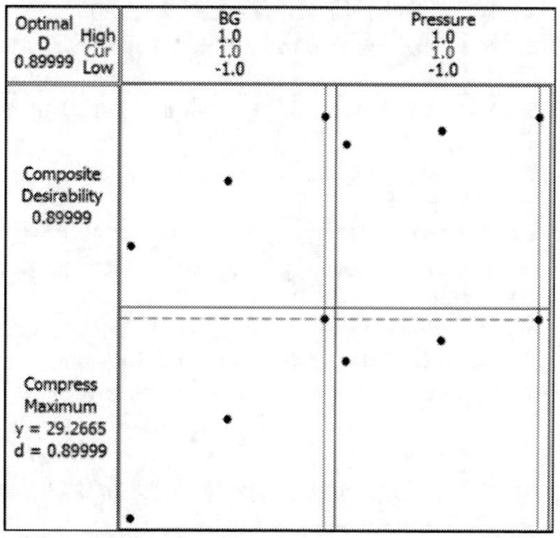

Fig. 6. Optimal condition of compressive strength for experimental boundary

According to the experimental results, the compressive strength of this rod-type specimen was greater at the upper boundary conditions, which could relate to the formation of crystalline phase [22]. Furthermore, this study revealed that the compressive strength of the rod-type structure is significantly lower than bar-like specimen at the similar experimental setup of the former investigation [15]. Because during fabrication of the rod-type specimen, the friction between the side of model and the mold was proportional to the pressure applied during forming process. Too much forming pressure could lead to difficulty in specimen removal. Nonetheless, this present study shows that rod-type structure fabricated by compression molding technique still has superior mechanical properties than 3D printed techniques [11].

4 Conclusion

The formation condition of rod-type HA-BG compact was investigated. The effect of forming conditions on compressive strength was analysed and optimized using full factorial experimental design. The optimal condition of 32.20 MPA compressive strength was found at 7.5 wt% BG ration, with a pressure of 120 kg/cm^2, while holding time has no significant effect on the strength. Further study will focus on other mechanical and biological properties both *in vitro* and *in vivo* prior to the development of bio-based artificial screw for medical application.

The authors would like to acknowledge the financial support from Chiang Mai University (Department of Industrial Engineering and Biomedical Engineering Center, Faculty of Engineering), as well as the technical support from Department of Tools and Die Engineering and Machine Tool Technology, Rajamangala University of Technology Lanna (RMUTL).

References

1. D. C. Hansen., Elec. Soc. Int. **17**, 2 (2008)
2. L. Chenglong, Y. Dazhi, L. Guoqiang, Q. Min, Mat. Let. **59**, (2005)
3. S. Bauer, P. Schmuki, K. von der Mark, J. Park., Prog- in Mat- Sc-, **58**, 3, '2013(
4. S. R. Paital, N. B. Dahotre, Mat- Sc- & Eng- R66, '2009(
5. M. N. Rahaman, D. E. Day, B. S. Bal, Q. Fu, S.B. Jung, L.F. Bonewald, A. P. Tomsia, Acta bio, **7**+6, '2011(.
6. P. D. Silva, M. H., Lemos, A. F. Ferreira, J. M. F. and J. D. Santos, Mat- Res-+6, 3+' 2003(
7. O. Bretcanu, X. Chatzistavrou, K. Paraskevopoulos, R. Conradt, I. Thompson, A. R. Boccaccini, J- Euro- Cer-Soc-, **29**, (2009)
8. S. Pal, *Design of Artificial Human Joint & Organs*, (Springer, 2013)
9. Y. W. Gu, N. H. Loh, K. A. Khor, S. B. Tor and P. Cheang, Biomat., **23**, (2002)
10. W. Wattanutchariya and W. Changkowchai, Lec. Notes Eng. & Com. Sc., 2210, (2014)
11. L. H. Hernández-Gómez, A. I. Rangel-Elizalde, J. A. Beltrán-Fernández, A. González-Rebatú, N. Corro-Valdez, G. Urriolagoitia-Calderón, R. Rodríguez-Martínez. App. of Com. Tool. in Biosc. and Med. Eng., 43-59, (2015)
12. H. B. Guo, X. Miao, Y. Chen, P. Cheang and K. A. Khor, Mat. Let., **58**, (2004)
13. R. Ravarian, F. Moztarzadeh, M. S. Hashjin, S. M. Rabiee, P. Khoshakhlagh and M. Tahriri, Cer. Int., **36**, (2010)
14. W. Wattanutchariya and P. Yenbut, Adv. Mat. Res., 931-932, (2014)
15. W. Wattanutchariya, J. Ruennareenard, P. Suttakul, Eng. J., **20**, (2016)
16. A. Gupta, C. Lattermann, M. Busam, A. Riff, Amer. J. of Sp. Med., **37**, (2009)
17. R. M. Felfel, I. Ahmed, A.J. Parsons, C.D. Rudd, J. Me. Beh. of Bio. Mat., **18**, (2013)
18. L. Handolin, T. Pohjonen, E.K. Partio, I. Arnala., P. Törmäläb, P. Rokkanena, Biomat., **23**, (2002)
19. J. A. Rano, R.T. Savoy-Moore, L.M. Fallat, J. Foot & Ank. Sur., **41**, (2002)
20. L. L. Hench and J. Wilson, *An Introduction to bioceramics*, (World Scientific, 1993)
21. X. Chatzistavrou, T. Zorba, E. Kontonasaki, K. Chrissafis and P. Koidis, Phy. St. Sol.(a), **201**, (2004)
22. Q. Z. Chen, J. L. Xu, L. G. Yu, X. Y. Fang and K. A. Khor, Mat. Sc. & Eng., **C32**, (2012)

Preparation and characterization of oyster shell powder-treated rice husk ash adsorbent pellet for As(III) removal

Pichnipa Khownpurk and *Walairat* Chandra-ambhorn*

Department of Chemical Engineering, Faculty of Engineering, King Mongkut's Institute of Technology Ladkrabang, 1 Soi Chalongkrung 1, Ladkrabang, Bangkok 10520, Thailand

Abstract. The adsorbent pellet for As(III) removal was prepared from ground oyster shell and rice husk ash. The effects of particle size of oyster shell powder (OS) and the ratio between the OS and Treated rice husk ash (TRHA) on the stability of the adsorbent pellet were studied. The adsorbent pellet was characterized by XRD, XRF and SEM. The solubility and As(III) adsorption tests were performed. The results showed that the adsorbent pellet prepared from OS size <106 μm with OS:TRHA ratio of 0.7:0.3 could provide As(III) maximum adsorption capacity of approximately 26.20 mg/g. Furthermore, the XRD and SEM results indicated that the adsorbent pellet could consist of two parts i.e. CaO that could adsorb As(III) in the form of Ca-As-O and the $CaSiO_3$ and C-S-H compounds which behaved as a binder to bind the precursor powders to be stable adsorbent pellet without cracking.

1. Introduction

Arsenic is a hazardous element that contaminated in soil and water. It was reported that the toxicity of arsenic could cause the human health problem such as cancer, pigmentation changes, neurological disorder etc.[1-4]. In Thailand, more than 25 provinces are affected by the arsenic pollution such as Suphanburi province and Ubon Ratchathani [5, 6]. Furthermore, almost of arsenic contaminated in water exists in two forms; arsenite (As(III)) and arsenate (As(V)). It was also reported that the As(III) was much more toxic and more difficult to remove than As(V) [7].

Adsorption is an effective method for removing arsenic from contaminated water. It was reported in the previous work that the calcine oyster shell powder (COS) could be used as the adsorbent to remove the As(III) from contaminated water with the maximum As(III) adsorption capacity approximately 195.5 mg/g (pH 11) [8]. However, fine powder of COS is difficult to separate from treated water.

The industrial crops in Thailand caused a lot of waste rice husk ash (RHA). It was reported that RHA was mainly contained SiO_2 [9] which could react with $Ca(OH)_2$ and form calcium silicate hydrate compound (C-S-H) called the pozzolanic reaction. The C-S-H compound could improve the strength of concreate [10, 11].

This research aims to develop the As(III) adsorbent pellet in order to be easily separated from treated water after used by using oyster shells and RHA as raw materials. In this work, the oyster shell powder (OS) was blended with treated rice husk ash (TRHA) and pelletized by using the metal mould. The effects of OS particle size and ratio between OS and TRHA on the stability of the adsorbent pellet and adsorption performance were investigated.

2. Materials and methods

2.1 Adsorbent pellet preparation

The adsorbent pellets were prepared from oyster shells and rice husk ash (RHA) obtained from Chanthaburi province and a power plant in Thailand, respectively. Firstly, the treated rice husk ash (TRHA) was prepared by calcining RHA at 600 °C in air for 13 h. Oyster shells were washed, dried, ground and sieved to obtain the oyster shell powder (OS) with the particle sizes <106 μm, 106-150 μm and 150-250 μm. Secondly, OS and TRHA were weighed and mixed in various ratios. Then, distilled water was added into the mixture powder to provide plasticity. Then, the mixture paste was pressed in a pellet metal mould. After that, the pellets were calcined at 700°C in air for 8 h. The X-ray fluorescence spectrometer (Bruker model S8 Tiger) was used to evaluate the elemental composition of OS and TRHA. The crystalline phase of the adsorbent pellets was determined by an x-ray diffractometer (XRD 6100, SHIMADZU, Japan) using Cu-Kα radiation, at 40 kV and 2θ range from 5-80° with the scan speed of 2.0 °/min. The morphologies of the adsorbent pellets before

* Corresponding author: walairat.ch@kmitl.ac.th (Walairat Chandra-ambhorn)

and after As(III) adsorption were observed by a scanning electron microscope (SEM, Carl Zeiss EVO®MA10).

2.2 Solubility tests

The solubility tests of COS and the adsorbent pellet were performed in Erlenmeyer flasks with rubber cork. Firstly, pH of distilled water was adjusted to 7 by 0.1 M of NaOH and/or HCl. Next, COS and the adsorbent pellet were weighed and individually added into the solutions and kept for 24 h. The pH of solutions were measured by a pH meter (Hanna instrument, HI 98107).

2.3 As(III) adsorption test

As(III) adsorption on the adsorbent pellet was performed via batch test. The stock solution of As(III) prepared from As_2O_3 was diluted to 100 mg/L with distilled water. The pH of the solution was adjusted to 7 by using 0.1 M of NaOH and/or HCl. Then, the adsorbent pellets were individually added into the Erlenmeyer flasks which contained 100 ml of As (III) solution and kept for 24 h. After that, the adsorbent pellet was removed from the As(III) solution. The concentration of As(III) remaining in the solution was determined by titration with $KMnO_4$. The As(III) adsorption capacity of the adsorbent pellet was calculated by using Equation (1).

$$\textit{As(III) adsorption capacity } (q_t) = \frac{(C_0 - C_e)V}{W} \quad (1)$$

where C_0 and C_e are concentrations of As(III) at initial and at equilibrium (mg/L), respectively, while V is volume of the solution (L) and W is mass of the adsorbent (g).

3. Results and discussion

3.1 XRF and XRD analysis

The chemical compositions of OS and TRHA were evaluated by XRF and XRD. The results indicated that $CaCO_3$ (95.99%) was the main component of OS while SiO_2 (87.00%) was the major compound of TRHA. Moreover, the XRD result of TRHA showed the broad smooth hump peak at 2Θ approximately 22° (the XRD pattern is not shown in this paper) which indicates that SiO_2 in TRHA should exist in amorphous form. Habeeb [12] and Srinivasreddy [13] reported that amorphous SiO_2 could react with $Ca(OH)_2$ and form calcium silicate hydrate compound (C-S-H) which could improve the compressive strength in the materials.

The crystalline phases of the adsorbent pellet before and after As(III) adsorption test were determined by using XRD technique. The XRD results [14] showed that the adsorbent pellet before As(III) adsorption composed of calcium silicate ($CaSiO_3$), C-S-H compounds, calcium oxide (CaO) and $Ca(OH)_2$. This indicated that after mixing OS with TRHA and water, some of $Ca(OH)_2$, the product produced from the reaction between $CaCO_3$ and water, in OS should react with SiO_2 in the TRHA and water and formed H_4SiO_4. During the calcination at 700°C for 8 h, H_4SiO_4 transformed into $CaSiO_3$ and C-S-H insoluble compounds [15] while the remaining $CaCO_3$ in the adsorbent pellet converted into CaO. After the adsorption test, the peaks of CaO disappeared while the peaks of Ca-As-O were observed. This implies that after the adsorbent pellet was added into As(III) solution, the CaO changed into $Ca(OH)_2$ and further reacted with As(III) and formed Ca-As-O. However, the intensities of the XRD peaks of $CaSiO_3$ and C-S-H compounds insignificantly changed. This implies that the $CaSiO_3$ and C-S-H compounds should not react with As(III).

3.2 The solubility test

The amounts of dissolved COS and adsorbent pellet reported in terms of dissolved CaO are presented in Fig 1. It is noted that COS <106, 106-150 and 150-250 in Fig.1 referred to COS prepared from OS particle size ranges of <106 μm, 106-150 μm and 150-250 μm, respectively while pellet <106, 106-150 and 150-250 referred to adsorbent pellet prepared from OS:TRHA ratio of 0.7:0.3 with OS sizes <106 μm, 106-150 μm and 150-250 μm, respectively.

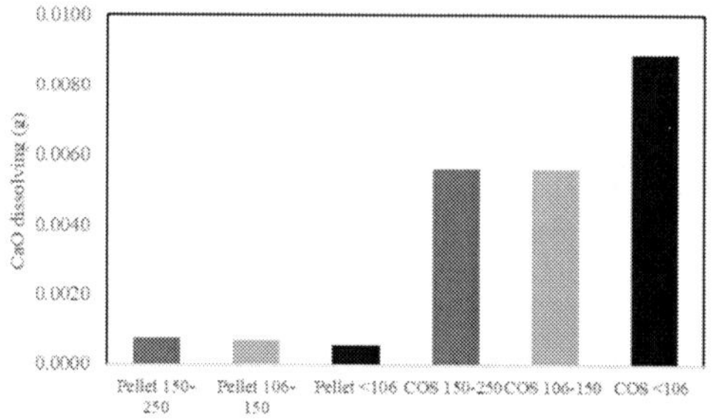

Fig. 1. The amount of CaO in COS and in adsorbent pellets dissolved in the solutions.

The solubility tests were performed twice, the error was in the range of 0.00 – 21.51%. As seen in Fig. 1, the amount of dissolved CaO in adsorbent pellets were much less than that of COS in all size ranges. This should be because some amount of $CaCO_3$ in the adsorbent pellet precursor transformed into $CaSiO_3$ and C-S-H insoluble compounds causing less CaO to be produced after calcination and thus less CaO dissolving in the solution. The other reason is because surface area of the adsorbent pellet is much less than that of COS resulting in less contact of CaO with water in the solution and therefore less dissolution can occur. As seen in Fig.1, COS prepared from OS size <106 μm had the greatest amount of CaO dissolved in the solution due to the largest contact area.

3.3 Adsorption capacity and pellet stability

The stability and As(III) adsorption tests of adsorbent pellets were performed in order to find the most appropriate condition for the adsorbent pellet preparation. The results are showed in Table 1.

Table 1. The stability and As(III) adsorption capacity of adsorbent pellets with different conditions.

OS size (μm)	OS:TRHA	Stability	Adsorption capacity (mg/g)
	0:1.0		1.38
<106	1.0:0		241.67
150-250	0.8:0.2	Crack	42.50±0.09
150-250	0.7:0.3	Crack	22.38±2.85
150-250	0.6:0.4	Crack	24.26±24.23
106-150	0.8:0.2	Crack	35.64±7.47
106-150	0.7:0.3	Crack	14.02±0.65
106-150	0.6:0.4	Crack	34.08±10.65
<106	0.8:0.2	Crack	33.10±0.41
<106	0.7:0.3	Not Crack	26.20±2.02
<106	0.6:0.4	Not Crack	19.50±2.83

The adsorption tests were performed twice and the As(III) adsorption capacities were reported as the average values. The errors of As(III) adsorption capacity of cracked pellets were in the range of 0.21% - 62.37% while un-cracked pellets had the errors of As(III) adsorption in the range of 0.88%-15.08%. The greater error of the cracked pellets was because of the uncontrolled cracking patterns of the cracked pellets causing in different area of cracked pieces exposed to the solution. The stability tests of adsorbent pellets showed that the crack of adsorbent pellets prepared from OS sizes 106-150 μm and 150-250 μm with OS:TRHA ratios between 0.8:0.2 and 0.6:0.4 was observed within 30 min after the adsorbent pellets had been added into the solution while the adsorbent pellets prepared from OS size < 106 μm with OS:TRHA ratios of 0.7:0.3 and 0.6:0.4 did not crack even after finishing the adsorption experiment (24 h). There are two reasons that could explain why the adsorbent pellets prepared from OS size < 106 μm with ratio 0.7:0.3 and 0.6:0.4 did not crack. (1) The OS size <106 μm contained the particles with the sizes that small enough to be well compacted and better sintered than the larger ones. After the adsorbent pellets were added into solution, the solution could less diffuse into the denser adsorbent pellets causing lower CaO to dissolve. (2) The OS size <106 μm had larger surface area that allows the pozzolanic reaction to occur and form the binder substances i.e. $CaSiO_3$ and C-S-H compounds more than the other OS sizes. The $CaSiO_3$ and C-S-H compounds could bind the precursor powder to be stable without cracking. In addition, the crack of the adsorbent pellet prepared from OS size <106 μm with ratio of 0.8:0.2 may occur due to too little TRHA contained in the adsorbent pellet. Then, the adsorbent pellet contained insufficient $CaSiO_3$ and C-S-H compounds to bind the precursor particles in the pellet.

As also seen in Table 1, the As(III) adsorption tests showed that As(III) adsorption capacity of COS<106 was extremely more than that of TRHA. This indicates that only CaO should be adsorption sites in the adsorbent pellet. The As(III) adsorption capacity of un-cracked adsorbent pellets were further compared. It can be seen that As(III) adsorption capacity of adsorbent pellet prepared from OS size < 106 μm with ratio of 0.7:0.3 was greater than that of the pellet prepared from OS size <106 μm with the ratio of 0.6:0.4. The adsorption result confirmed the previous statement as the amount of CaO contained in the adsorbent pellet prepared from OS< 106 μm with ratio of 0.7:0.3 was more than that of the adsorbent pellet prepared from OS size < 106 μm with ratio of 0.6:0.4. The results of stability test and As(III) adsorption capacity indicate that the adsorbent pellet prepared from OS <106 μm with ratio of 0.7:0.3 is the most appropriate condition for the adsorbent pellet preparation for As(III) removal.

3.4 Morphology

The surface morphologies of the adsorbent pellets prepared from OS sizes <106 μm, 106-150 μm and 150-250 μm with OS:TRHA ratio of 0.7:0.3 before and after As(III) adsorption tests were observed by using SEM. The SEM images are illustrated in Fig 2.

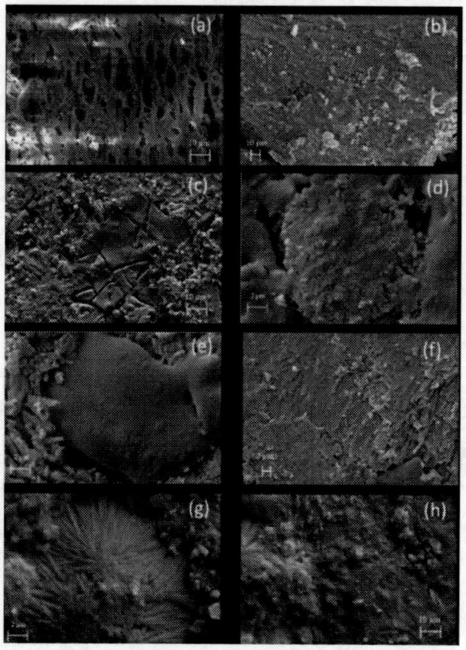

Fig. 2. The SEM images of the (a) TRHA (b) COS before adsorption test, the adsorbent pellet before adsorbed As(III) which prepared from (c) OS<106 μm, (d)-(f) OS106-150 μm and 150-250 μm and (g)-(h) the adsorbent pellet after adsorbed As(III).

The SEM image in Fig.2(c) shows that the surface morphologies of the adsorbent pellet prepared from OS size <106 μm before As(III) adsorption test could be divided into two types ; the rough one and the plain one. The EDX result (not shown in this paper) indicated that the plain surface morphology should consist of $CaSiO_3$ and C-S-H compounds while the surface morphologies before As(III) adsorption test of the adsorbent pellets prepared from OS sizes 106-150 μm and 150-250 μm were observed in three forms as shown in Figs.2(d) - 2(f). The rough surface (Fig.2(d)) and the plain surface (Fig.2(e)) look similar to the surface morphologies of the adsorbent pellet prepared from OS size <106 μm before As(III) adsorption test. However, the third form looks like the layers of porous bars (Fig 2(f)) which is similar to the morphology of COS before As(III) adsorption test (Fig. 2(b)). This implies that some parts of the adsorbent pellet prepared from OS sizes 106-150 μm and 150-250 μm were still in the porous bars form of COS. As mentioned before, the crack of adsorbent pellets prepared from OS sizes 106-150 μm and 150-250 μm with ratios between 0.8:0.2 and 0.6:0.4 resulted from the lowering compaction of the precursor powders. During the calcination, $CaCO_3$ changed into CaO and released CO_2 caused the porous structure on the adsorbent surface. Furthermore, the volatile impurity removal and loss of humidity during the calcination caused porosity in the adsorbent pellets. After the adsorbent pellet was added into the solution, the solution could better diffuse into the pores of porous structure in the adsorbent pellets and contact with CaO inside the adsorbent pellet. The dissolution of CaO inside the pellet might cause the crack of the adsorbent pellet.

After As(III) adsorption test, the morphologies of adsorbent pellets could be classified into two forms. The first one looks like the needle-like shape as shown in Fig. 2(g). The latter is a plain surface as shown in Fig. 2(h) which looks similar to the plain surface of the adsorbent pellets before As(III) adsorption test (Fig.2 (e)). The needle-like shape morphology on the adsorbent surface looks like the morphology of COS after As(III) adsorption test in the previous work [8]. In addition, Phenrat et al. [16] studied the iron-arsenic sludge immobilized by using cement and lime and found that the needle-like shape was one form of calcium-arsenic compound. The EDX result of needle-like shape morphology (not shown in this paper) showed Ca, O and As elements which should refer to Ca-As-O. The SEM result also agrees with the XRD result which indicated that the As(III) should be adsorbed on CaO adsorption sites on the surface of adsorbent pellet in the form of Ca-As-O. In the case of the plain surface found both before and after adsorption test, this plain surface should be the surface of $CaSiO_3$ and C-S-H compounds which were inactive with As(III) adsorption.

4. Conclusions

The adsorbent pellet for As(III) removal was prepared from OS and TRHA. The most appropriate condition to prepare the adsorbent pellet was OS size <106 μm with OS:TRHA ratio of 0.7:0.3 which provided the As(III) maximum capacity of 26.20 mg/g. The adsorbent pellet could be classified into two parts. The first one was the CaO which could adsorb As(III) in the form of Ca-As-O. The other one was $CaSiO_3$ and C-S-H compounds which could not adsorb As(III) but could bind the precursor powder to be the stable adsorbent pellet without cracking.

This work is supported by King Mongkut's Institute of Technology Ladkrabang (Grant number KREF015905).

References

1. Y. Tian, M. Wu, X. Lin, P. Huang,Y. Huang, J. Hazard. Mater.,**193**, 6, (2011).
2. V.K. Sharma, M. Sohn, Environment International, **35**(4), 16, (2009).
3. D. Mohan and C.U. Pittman Jr., J. Hazard. Mater., **142**,53 (2007).
4. Y. Du, Q. Lu, H.Chen,Y. Du and D.Du, Journal of Water Process Engineering, **12**, 5, (2016).
5. S. Pansamut and G. Wattayakorn, J. Environ. Res. Develop., **4**, 8, (2010).
6. P. Pattanapipitpaisal and P. Suraruk, J. Environ. Sci. Eng. Technol., **1**, 8, (2012).
7. S.K. Maji, Y.H. Kao and C.W. Liu, Desalination., **280**, 7, (2011).
8. P. Khownpurk, W. Wongpromrat, W. Chandra-Ambhorn, Key Eng. Mater., **728**,7, (2017).
9. J. Prasara-A and S.H. Gheewala, J. Clean. Prod., **167**, 8, (2016).
10. S.A. Saad, M.F. Nuruddin, N. Shafiq and M. Ali, Appl. Mech. Mater., **773-774**, 4, (2015).
11. A.A. Ramezanianpour, M. Mahdi khani and G. Ahmadibeni, Int. J. Civ. Eng., **7**, 8, (2009).
12. G.A. Habeeb and H.B. Mahmud, Mat. Res., **13**, 5, (2010).
13. A.B. Srinivasreddy, T.J. McCarthy and E. Lume, *6th Biennial Concrete Institute of Australia's National Conference (Concrete 2013), Australia,* 10, (2013).
14. P. Khownpurk and W. Chandra-Ambhorn, JCIE (submitted)
15. A.Seidell, *Solubility of inorganic compounds. A compilation of quantitative solubility data from the periodical literature. D. Van Nostrand Company, New York* (1919)
16. T. Phenrat, T.F. Marhaba and M. Rachakornkij, J. Hazard. Mater., **118**, 10, (2005).

Designing number of Kanban in glove box assembly

Chonthicha Limpayanak[1,*] and *Oran* Kittithreerapronchai[1,*]
[1]Industrial Engineering Department, Faculty of Engineering, Chulalongkorn University, Bangkok, Thailand

Abstract. For many decades, Just-In-Time manufacturing and lean manufacturing have adopted kanban system as a mean to control inventory and production. A number of tagged kanban cards represents inventory of the system, whereas untagged kanban cards equate to production orders. The ideal conditions are a consistent product mixed with a small production lot size and minimum setup time. Such conditions are difficult to obtain even in many automotive assembly lines similar to a case study. The case study assembly line is a glove box assembly production line that assembles two plastic parts and wields them together with a vibration machine after positioning in a mold designed for different products. Because of storage equipment, each kanban card represents 60 identical glove boxes. Therefore, a production engineer re-schedules the production planning to minimize the numbers of changeover without any effect on downstream. This ad hoc production scheduling leads to urgent production orders and interruption of lot size. As a result, the factory embeds historical data into a computer simulation that accounts for current molds, receiving time, and numbers of pending kanban cards, as well as number of storage equipment. The results of this simulation reveals that scheduling criteria and suggest a suitable lot size if the factory decides to change a storage equipment.

1 Introduction

Pioneered by Toyota, Just-In-Time (JIT) and lean manufacturing systems have become synonymous with a pull system and adopted kanban system as a mean to control inventory and production. Many researchers and practitioners view kanban system as a controlling mechanism that monitors a constant work-in-process or base stock system [1].

Particularly, a number of tagged kanban cards represents inventory of the system, whereas untagged kanban cards equate to production orders. In the ideal conditions, each individual part, including nuts and bolts, should have their own kanban cards and a consistent product mixed with a small production lot size and a minimal setup time. Nevertheless, such conditions are difficult, if not impossible, to obtain even in many automotive assembly lines because the trade-off between the efficiency in batch production and/or batch transportation and the responsiveness in production changeover. Hence, the goal of this article is to address and analyse the trade-off using the production of a case study assembly line as an example.

Similar to many manufacturing plant, the case study company embraces a pull system with kanban system to manufacture plastic components used in the automotive industry. The case study assembly line is a glove box assembly, as shown in Figure 1.

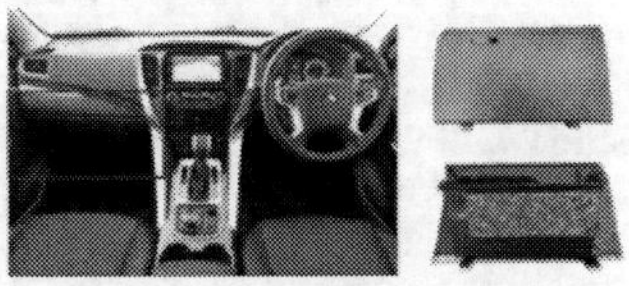

Fig. 1. Example of glove boxes manufactured in a case study company.

At the time of writing, the company manufactures eleven models that consist of two different molds for different passing lane, particularly left-hand-side (LHS) and right-hand-side (RHS) driving. Because of the capacity of a mobile rack used as storage and transportation equipment, each kanban cards equivalents to 60 glove boxes with an identical model. Therefore, a production engineer must re-schedule the production planning to address that the trade-off between minimizing the numbers of changeover and possible product shortage downstream. Before evaluating the production scheduling criteria and design scheduling and suggesting a suitable lot size if the company decides to change a mobile rack, it is important review relevant literature.

2 Literature Review

Sequencing the glove box production planning in an automotive assembly line has related to two interdependent streams of research JIT Manufacturing System and Number of kanban.

* Corresponding author: climpayanak@gmail.com

2.1 JIT Manufacturing System

As a control production and inventory control mechanism in JIT manufacturing system, a kanban card is used to transmit inventory and production by tagged or untagged with items, respectively. Managing a number of kanban cards is one of key successes of JIT manufacturing system as it must ensure production synchronization between different department while maintains minimum inventory. In addition to analyze the formulation used Toyota to determine kanban cards, Monden [2] applied the Little's law by determining an appropriated safety factor, demand rate, lead time, and container capacity. Since then literature has described and proposed both optimal one and/or heuristic processes for different situation. For example, Philipoom *et al* [3] proposed an Integer Linear Programming to determine an optimal number of kanban cards for multi-stage and multi-product assembly line. Few researchers have addressed operational issues as majority of literature assumed that the variation of processing time is small as well as ignored setup time and sequencing of kanban cards. One empirical study [5] found that four factors affect number of kanban cards are throughput rate, variation in processing time, utilization of machine, and continuity of processing time. An interesting reader can consult [6] for the design of kanban card and operational issues.

Based on these literature, JIT System in the glove box assembly line can be classified as a single kanban card part-type and queue-size blocking mechanism with batch production system similar to Tayfus [7] that addressed multiple production type. Rather assuming Poisson distribution, this article explores empirical distribution of parts. Having reviewed and established the connection with the related literature, the remaining sections are organized as follows. Section 3 highlights and analyzes the problem in the glove box assembly line. After understanding nature of the assembly line, a discrete event time simulation is introduced in Section 4 whereas the result is reported and discussed in Section 5.

3 Case Study Company

Located in the eastern of Thailand, this company is the first tier manufacturer that has produced automotive plastic parts. Relied on kanban system, a case study company has used four different kanban based on different party, as shown in Figure 2.

Depending on a design shape and size, the main manufacturing process is the plastic injection process. In general, the company divides its plastic products into different categories based on its functions and installation locations inside an automotive vehicle. Located underneath a vehicle dashboard, a glove box is viewed as an instrument panel. Designed to storage documents or small items, a typical globe box design consists of two plastic parts and wields them together at the glove box assembly line.

3.1 Glove Box Assembly line

The glove box assembly line comprises of two operators and single vibration machine. The purpose of the assembly line is to prepare and assemble the plastic parts with the vibration machine after positioning them in a mold designed for different product models, as shown in Figure 3.

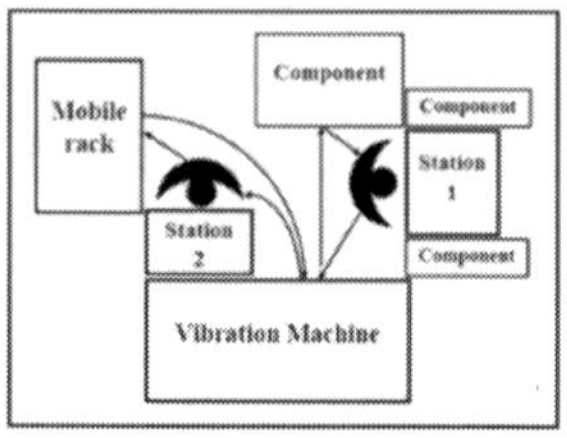

Fig. 2. Vibration machine used in the glove box assembly line.

Each operator has different roles. Before transferring the plastics parts to the vibration machine, the first operator manually assembles components onto glove box and positions the inner and outer parts of a glove box into their respective molds inside the vibration machine for welding. After the vibration welding completed, the second operator removes welded glove box and assemble additional components before packing in to a mobile rack for storage and transportation. In the change-over event, an addition operator is required to assist the assembly-line operators to remove and replace mold because of their size and weight. Without the changing over, this assembly line is a simple tendum line system with a single machine and two operators to load and unload a component to a machine. Hence, the trivial solution to maximize the utilization and throughput is to minimize the chance-over. Nevertheless, the mixed glove box requirement of downstream process as well as its JIT environment prohibits such solution. Therefore, it is interesting to analyze the change-over.

3.2 Change-Over Analysis

To improve the efficiency, the production engineer of the assembly line re-schedules the production planning to minimize the numbers of changeover, while observing three factors that indicate the downstream demands:

Duration time is the time in which a kanban card arrives can indicates consumption rate of downstream production. Assuming a single model assembly, the production engineer can estimate the consumption rate of recent arrival kanban. It is important to note that the interarrival time of a kanban card also help to validate this assumption.

The remaining kanban is the difference between available and total kanban cards. Because of a constant planning cycle, a number of kanban cards in each model every remains constant for ten days before revising in the next cycle. Therefore, a number of available kanban cards of each model with respected to its total reveals the inventory at store department and downstream production line.

Change-Over Sequence: In additional of an individual model, the relationship between current model

* Corresponding author: climpayanak@gmail.com

and potential change-over model is also an important factor as some models may require complete change-over of mold, while the others may require only cleaning and additional processing time.
Based on these factors and personal experience, the engineer schedule production to balance between utilization and change-over downtime, as depicted in Figure 4.

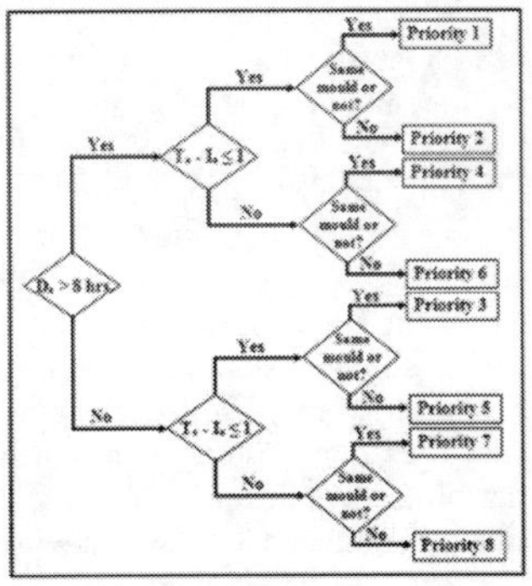

Fig. 4. Decision tree diagram of the current scheduling practice.

The practice of a simple decision tree diagram that ignores sequence of kanban cards has raised questions and concerns among department and led to the analysis of production sequence, as shown in Figure 5.

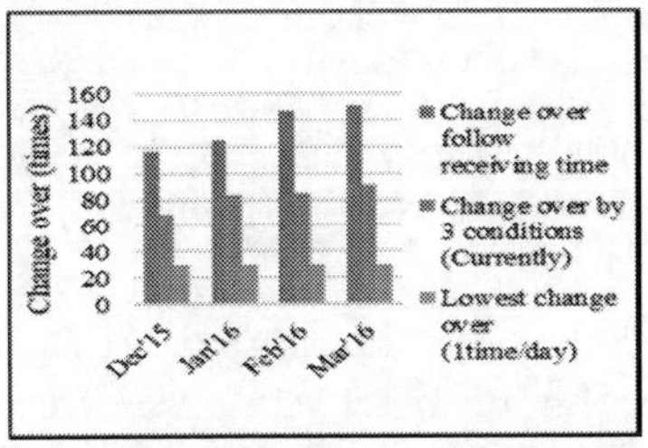

Fig. 5. Comparison of monthly change-over times between current practice and kanban sequence

Figure 5 reveals the advantage of the current practice. Particularly, scheduling by three factors could significantly reduce a number of change-over. This practice, however, lacks of theatrical ground as it relies heavily on the experience of an engineer without a useful insight on important factors. Furthermore, the past urgent request was usually done without any record. As a result, the case study company wish to analyze the most important factors including a capacity of mobile rack as well as to develop a suitable decision criteria. To answer these questions, a discrete event-time simulation model, as discussed in the next section.

3.3 Processing Time

To Before developing a simulation model, it is important to analyze the processing time of each operator as well as welding machine, as shown in Figure 6.

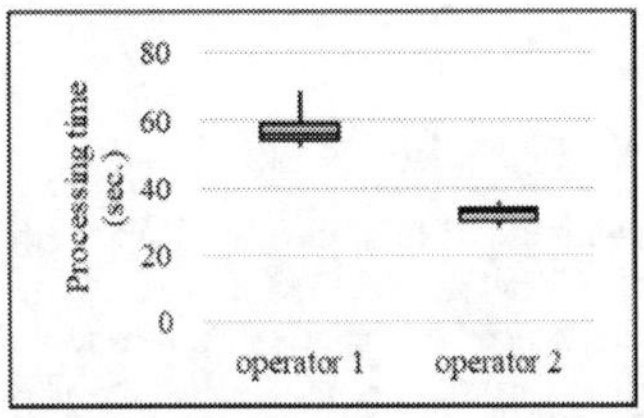

Fig. 6. Distribution of processing time of each operator

Figure 6 results from actual time and work of each operation denoted by his station. During the study, 30 samples of activity are randomly corrected and analyzed. The processing times of both operators are normally distributed with mean 57.0 and 35.0 seconds and variance 2.5 and 3.5 seconds, respectively. The processing time of vibration machine is 15 second regardless mold type and model.

4 Glove Box Assembly Simulation

The purpose of glove box assembly simulation is to determine effects and interaction of possible factors by simulating a manufacturing one production cycle or ten days with empirical schedule. Based on the decision diagram in Figure 4, three factors are introduced, partially.

The levels of each factor as well as the scenarios of experiments are based on current operation and extreme values of each factor as shown in Table 1.

Table 1. Scenarios and levels of factors in experiment

Scenario	R (piece)	D (hours)	K (cards)	Note
A0	60	8	1	Current
A1	30	8	1	
A2	90	8	1	
A3	60	∞	∞	FIFO
A4	60	4	1	
A5	60	6	1	
A6	60	10	1	
A7	60	16	1	
A8	60	∞	1	
A9	60	8	0	
A10	60	8	2	
A11	60	8	∞	

Having defined factors and contemplated on values of each factor, each scenario is modeled as a discrete event time simulation model.

The operation flow of model begins with creating entities or kanbans with status associated score representing mold side, duration time, and available kanban. The model re-evaluates scores of every kanban time when a kanban is assigned to different operator, including created or exited the system. If the first operator is available, the simulation assigns the highest score kanban to the operator who assembles components into glove boxes and loads them into the vibration

* Corresponding author: climpayanak@gmail.com

machine. The particular kanban is occupied until completed its batch size based on rack capacity.

4.1 Result of Simulation

The simulation is embedded in Rockwell Arena 16.5 with 100 replications. The model was verified by number of kanban comparing between a kanban that generated from simulation. The results of each scenario in terms of system performance and operator performance are shown in Table 2 and Table 3, respective.

With exception of a higher machine utilization, the performance of the current re-scheduling kanban (A0) dominates the performance of FIFO kanban (A3). This suggests a merit of re-scheduling kanban and leads to analysis of each factor.

Table 2. System preformance of each scenario

Scenario	number of mold changed	Glove box completed (pcs)	Ratio completed kanban
A0	49.7 ± 1.2	8,583.5	0.974
A1	73.3 ± 1.6	8,516.5	0.974
A2	35.4 ± 1.0	8,611.8	0.992
A3	68.3 ± 1.5	8,519.9	0.993
A4	53.9 ± 1.2	8,545.9	0.975
A5	51.4 ± 1.2	8,531.2	0.974
A6	48.6 ± 1.2	8,551.2	0.977
A7	46.5 ± 1.2	8,586.4	0.975
A8	46.2 ± 1.2	8,581.9	0.993
A9	49.7 ± 1.2	8,583.6	0.974
A10	49.7 ± 1.2	8,583.6	0.974
A11	49.7 ± 1.2	8,583.6	0.974

Table 3. Operator preformance of each scenario

Scenario	Machine Utilization	Cycle Time (hours)	
		Avg	Max
A0	0.22	2.24 ± 0.21	5.74
A1	0.25	2.79 ± 0.21	6.28
A2	0.20	1.90 ± 0.20	6.69
A3	0.24	2.69 ± 0.27	9.45
A4	0.22	2.27 ± 0.22	8.87
A5	0.22	2.16 ± 0.19	6.04
A6	0.21	2.22 ± 0.22	8.00
A7	0.21	2.15 ± 0.18	5.09
A8	0.21	2.11 ± 0.17	4.95
A9	0.21	2.24 ± 0.21	5.74
A10	0.21	2.24 ± 0.21	5.74
A11	0.21	2.24 ± 0.21	5.74

The results of Scenarios A0 to A2 suggest that changing rack capacity greatly affects performance of system and operators. As the capacity increases, productivity increase as number of molds and average cycle time are reduced. Nevertheless, there is little impact to overall utilization as the utilization of vibration machine increase slightly. It is important to note that utilizations of the first operator and the second operator remain unchanged at 0.71 and 0.44 throughout the experiment.

Changing duration time required to process kanban leads to increase number of mold changed as it reduces probability of using the same mold especially LHS mold which has fewer models that RHS mold. Interestingly, the benefit of duration is marginal decreasing and other performance indicators remain unchanged.

Based on Scenarios A0, A9, A10, and A11, the simulation reveals that remaining kanban has little or no effect to any performance indicators.

5 Summary

The article addresses and analyzes the trade-off between minimizing the numbers of changeover and possible product shortage down-stream in a glove box assembly. Using the production. Based on current decision tree to re-schedule and another suggested factor, the operation and re-scheduling procedure were modeled as a discretetime event simulation. The result suggests that changing rack storage capacity would greatly affect performance, whereas increasing duration before process kanban could reduce mold change-over with little effect on other performance indicators. The result also reveals that counting remaining kanban in system has no benefit for re-scheduling.

References

1 Karmarkar U., Push, pull and hybrid control schemes, Tijdschrift voor Economie en Management 26, 345-363 (1991)
2 Monden Y., Toyota Production System, Norcross, Georgia Industrial Engineering and Management Press, 215-216 (1983)
3 Philipoom P., Rees L., Taylor III B. and Huang P., A mathematical programming approach for determining workcentre lotsizes in a just-in-time system with signal Kanbans, 1-15 (1990)
4 Philipoom P., Rees L., Taylor III B. and Huang P., A mathematical programming approach for determining workcentre lotsizes in a just-in-time system with signal Kanbans, 1-15 (1990)
5 Philipoom P., Rees L. and Taylor III B. W., Simultaneously Determining the Number of Kanbans, Container Sizes, and the Final-Assembly Sequence of Products in a Just-in-Time Shop, International Journal of Production Research, Vol.34, 51–69 (1996)
6 Akturk M. S. and Erhun F., An overview of design and operational issues of kanban systems, International Journal of Production Research, Vol.37, 3859-3881 (1999)
7 Tayfur A., and Goang A. S., Pull-type manufacturing systems with multiple product types, IIE Transactions journal, Vol.32, 115-124 (2000)

* Corresponding author: climpayanak@gmail.com

A study on the run length properties of the side sensitive group runs double sampling (SSGRDS) control chart

Nger Ling Chong[1,*], *Michael Boon Chong* Khoo[1], *Zhi Lin* Chong[2] and *Wei Lin* Teoh[3]

[1]School of Mathematical Sciences, Universiti Sains Malaysia, 11800 Minden, Penang, Malaysia
[2]Department of Physical and Mathematical Science, Faculty of Science, Universiti Tunku Abdul Rahman, 31900 Kampar, Perak, Malaysia
[3]School of Mathematical and Computer Sciences, Heriot-Watt University Malaysia, 62200 Putrajaya, Malaysia

Abstract. The side sensitive group runs double sampling (SSGRDS) chart incorporates the control charting concepts of the side sensitive group runs (SSGR) and double sampling (DS) charts. The SSGRDS chart which combines the efficiency of its basic charts is an effective approach to increase the speed of mean shift detection. The performance of the SSGRDS chart, based on the average number of observations to signal (ANOS), median number of observations to signal (MNOS) and percentiles of the number of observations to signal (PNOS) is investigated in this paper. Based on the results obtained, it is found that the SSGRDS chart becomes more sensitive in detecting mean shifts with an increase in the size of the process mean shift. With the use of MNOS and PNOS to measure the performance of the SSGRDS chart, the entire run length distribution is considered and this leads to a more complete understanding of the performance of the chart. The findings in this paper will provide a clearer picture on the run length properties of the SSGRDS chart which will facilitate practitioners in using the chart.

1 Introduction

Companies strive to manufacture products of high quality standards to meet the requirements of customers. The ability of a company to produce high quality products leads to a competitive advantage over other companies and is a primary factor in achieving customer satisfaction. Statistical Process Control (SPC) refers to the employment of statistical techniques to enhance the quality of products manufactured. Control chart is an SPC tool which is extensively adopted to monitor how a process changes with time in industries and non-manufacturing sectors. The Shewhart $\bar{X}$ chart possesses the ability to identify large process mean shifts quickly. However, the main drawback of the Shewhart $\bar{X}$ chart is its lack of sensitivity towards small and moderate process mean shifts.

Among the strategies used to enhance the performance of the Shewhart $\bar{X}$ chart is the double sampling (DS) $\bar{X}$ control charting approach. The DS $\bar{X}$ chart was introduced by [1]. A modified version of the DS scheme that makes an out-of-control decision at the combined-sample stage with regard to information obtained from both samples was proposed by [2]. It was found that the DS procedure provides superior statistical efficiency without increasing sampling in comparison to the Shewhart $\bar{X}$ chart [2]. On the other hand, the DS procedure can also minimize sampling while maintaining the statistical efficiency. Meanwhile, [3] investigated the performance of the DS $\bar{X}$ chart under non-normality and found that the DS $\bar{X}$ chart surpasses the Shewhart $\bar{X}$ chart in detecting small mean shifts. Some of the literature on the DS $\bar{X}$ chart can be seen in [4-7].

[8] developed the group runs (GR) chart which is an integration of the Shewhart $\bar{X}$ chart and an extended version of the conforming run length (CRL) chart. In comparison to the synthetic and Shewhart $\bar{X}$ charts, the GR chart has a considerably lower out-of-control average time to signal (ATS_1). Additionally, [9] proposed a side sensitive group runs (SSGR) control chart by combining the GR chart with the side sensitivity approach. The SSGR chart outperforms the Shewhart $\bar{X}$, GR and synthetic charts. Other related studies on the SSGR chart can be found in [10-12].

Past research has been done with the purpose of developing alternative control charts that are superior in terms of efficiency in detecting shifts. In fact, a control chart with better sensitivity to shifts and has a faster detection speed will lead to more savings in cost. Hence, [13] introduced the side sensitive group runs double sampling (SSGRDS) chart which integrates the efficiency of both the DS and SSGR charts. Merging of control charting methods is an approach to produce an enhanced chart that surpasses the performances of its individual counterparts. In this paper, the SSGRDS chart is studied in terms of the average number of observations to signal (ANOS), median number of observations to signal (MNOS) and percentiles of the number of observations to signal (PNOS). The rest of this paper is as follows: Section 2 outlines the implementation of the SSGRDS

* Corresponding author: ngerlingc@gmail.com

chart while Section 3 focuses on the performance measures. In Section 4, the results are shown. The run length properties are discussed in Section 5 and conclusions are deduced in Section 6.

2 Implementation of the SSGRDS chart

The SSGRDS chart is comprised of the DS sub-chart and an extended version of the CRL sub-chart. CRL is the number of sampling stages between two successive non-conforming sampling stages, which includes the last non-conforming sampling stage. The first and second samples are represented by n_1 and n_2, respectively. For the DS sub-chart, L and L_2 denote the control limits of the first and second sampling stages while L_1 is the warning limit of the first stage. Let $I_1 = [-L_1, L_1]$, $I_2 = [-L, -L_1) \cup (L_1, L]$, $I_3 = (-\infty, -L) \cup (L, +\infty)$ and $I_4 = [-L_2, L_2]$. Fig. 1 shows the schematic representation of the DS sub-chart. Meanwhile, L_3 is the lower limit of the CRL sub-chart. The implementation of the SSGRDS chart is as shown below [13]:

1. Compute the charting parameters n_1, n_2, L, L_1, L_2 and L_3.
2. A sample with n_1 observations is taken from a population and the sample mean $\bar{X}_{1,i} = \frac{\sum_{j=1}^{n_1} X_{1j}}{n_1}$ is computed, where X_{1j}, for $j = 1,2,\ldots,n_1$, are the observations in the first sample of sampling stage i. This is followed by the computation of $Z_{1,i} = \frac{\bar{X}_{1,i}-\mu_0}{\sigma/\sqrt{n_1}}$.
3. Sampling stage i is conforming if $Z_{1,i} \in I_1$. Then return to Step 2.
4. Sampling stage i is non-conforming if $Z_{1,i} \in I_3$. Then, proceed to Step 9.
5. A second sample of size n_2 is taken from the same population as that of the first sample if $Z_{1,i} \in I_2$. Subsequently, $\bar{X}_{2,i} = \frac{\sum_{j=1}^{n_2} X_{2j}}{n_2}$ is computed, where X_{2j}, for $j = 1,2,\ldots,n_2$, are the observations in the second sample of sampling stage i.
6. The overall sample mean, $\bar{X}_i = \frac{n_1\bar{X}_{1,i}+n_2\bar{X}_{2,i}}{n_1+n_2}$ and the Z statistic, $Z_i = \frac{\bar{X}_i-\mu_0}{\sigma/\sqrt{n_1+n_2}}$ are computed for sampling stage i.
7. The sampling stage i is conforming if $Z_i \in I_4$ and return to Step 2. Otherwise, the process is declared as non-conforming and proceed to Step 8.
8. Calculate the K_m value, which is the amount of sampling stages between the current (included in the count) and ending (excluded in the count) non-conforming sampling stages.
9. The process is out-of-control if $K_1 \le L_3$; or for $m > 1$, $K_m \le L_3$ and $K_{m+1} \le L_3$ for the first time and the value of the Z statistic corresponding to the mth and $(m+1)$th non-conforming sampling stages lies on the similar side of the target mean μ_0.
10. An out-of-control point is signaled and corrective actions are taken to investigate the process and eliminate the assignable cause(s).

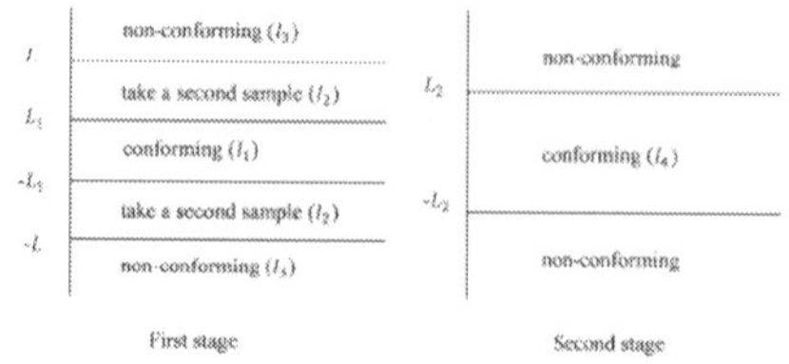

Fig. 1. Schematic representation of the DS sub-chart

3 Performance Measures

In this paper, ANOS is used to study the performance of the SSGRDS chart instead of the average run length (ARL) as the chart's sample size is not constant at each sampling stage (the sample size can be either n_1 or $n_1 + n_2$). In fact, ARL which represents the expected number of sampling stages required to signal an out-of-control condition is not as informative as ANOS that measures the actual number of observations needed by the chart to signal an out-of-control condition. MNOS refers to the median number of observations taken from the start of process monitoring until a signal is given by the chart. On the other hand, PNOS measures the percentiles of the number of observations required since the beginning of process monitoring until the chart signals an out-of-control condition. Let the probability that the process is declared in-control be $P_a = P_{a1} + P_{a2}$ where P_{aq}, $q = 1,2$ is the probability that the process is in-control at the qth stage. It is given by [2] that

$$P_{a1} = \Pr(Z_1 \in I_1)$$
$$= \Phi(L_1 + \delta\sqrt{n_1}) - \Phi(-L_1 + \delta\sqrt{n_1}) \quad (1)$$

where δ is the mean shift and $\Phi(.)$ is the standard normal cumulative distribution function. Meanwhile [2],

$$P_{a2} = \Pr(Z \in I_4 \text{ and } Z_1 \in I_2) = \int_{z\in I_2^*} P_4 \phi(z)dz \quad (2)$$

such that $\phi(.)$ is the standard normal probability density function. Note that $I_2^* = [-L + \delta\sqrt{n_1}, -L_1 + \delta\sqrt{n_1}) \cup (L_1 + \delta\sqrt{n_1}, L + \delta\sqrt{n_1}]$ and

$$P_4 = \Pr(Z \in I_4 | Z_1 = z) \quad (3)$$

Then ANOS can be computed as follows [13]:

$$\text{ANOS} = \frac{[1-\alpha(1-\alpha)A^2](n_1+n_2P_2)}{A^2(1-P_a)[1+\alpha(1-\alpha)(A-2)]} \quad (4)$$

where

$$P_2 = \Pr(Z_1 \in I_2), \quad (5)$$

$$A = \Pr(K_m \le L_3) = 1 - (1-P)^{L_3} \quad (6)$$

such that

$$P = 1 - \Phi(y_{ssgr} - \delta\sqrt{n}) + \Phi(-y_{ssgr} - \delta\sqrt{n}), \quad (7)$$

and

$$\alpha = \Pr(Z_1 > L \text{ or } \{Z > L_2 \text{ and } Z_1 \in (L_1, L)\} \mid \text{non-conforming sampling stage})$$

$$= \frac{1 - \Phi(L_1 + \delta\sqrt{n_1}) - \int_{z \in I_2'} \Phi\left(cL_2 + rc\delta - z\sqrt{\frac{n_1}{n_2}}\right)\phi(z)dz}{1 - P_a} \quad (8)$$

such that $I_2' = (L_1 + \delta\sqrt{n_1}, L + \delta\sqrt{n_1})$, $r = \sqrt{n_1 + n_2}$ and $c = \sqrt{\frac{n_1+n_2}{n_2}}$. Note that y_{ssgr} is the width constant of the SSGR chart.

4 Research Findings

This section presents the findings of the performance measures ANOS, MNOS and PNOS of the SSGRDS chart. The in-control ANOS ($ANOS_0$) is fixed as 370. In other words, it is assumed that when the process is in-control, an average of 370 observations are needed before the SSGRDS chart signals a false alarm. It is worth noting that the 50th percentile is equivalent to the MNOS. Simulation studies are conducted using the Statistical Analytical System (SAS) software to compute the out-of-control ANOS, MNOS and PNOS denoted as $ANOS_1$, $MNOS_1$ and $PNOS_1$, respectively. The various combinations of the optimal charting parameters $n_1, n_2, L, L_1, L_2, L_3$ are shown in Table 1 with the in-control average sample sizes, ASS_0 = 5 and 10. The $ANOS_1$ and $MNOS_1$ values are shown in Table 2 while the $PNOS_1$ values are enumerated in Table 3. Fig. 2 shows the comparison of the out-of-control ANOS and MNOS values when ASS_0 = 5.

Table 1. Optimal charting parameters of the SSGRDS chart when $ANOS_0$ = 370 for ASS_0 = 5 and 10 (italicized)

δ	n_1	n_2	L	L_1	L_2	L_3
0.2	3	10	3.2384	1.2781	1.5400	22
	5	*15*	*3.0355*	*0.9626*	*1.3730*	*12*
0.6	3	4	3.2340	0.6726	1.4710	5
	7	*4*	*3.0301*	*0.3154*	*1.2840*	*3*
1.0	4	2	3.2340	0.6726	1.3870	3
	7	*4*	*3.0301*	*0.3154*	*1.1350*	*2*
1.4	4	2	3.2340	0.6726	1.2270	2
	6	*5*	*3.0298*	*0.2502*	*1.1280*	*2*
1.8	3	3	3.2332	0.4290	1.2250	2
	5	*6*	*3.0295*	*0.2073*	*1.1240*	*2*
2.0	3	3	3.2332	0.4290	1.2250	2
	4	*7*	*3.0294*	*0.1769*	*1.1230*	*2*
2.6	3	3	3.2332	0.4290	1.2250	2
	3	*8*	*3.0293*	*0.1542*	*1.2670*	*3*
3.0	2	4	3.2329	0.3170	1.3630	3
	2	*9*	*3.0292*	*0.1366*	*1.2680*	*3*

Table 2. $ANOS_1$ and $MNOS_1$ of the SSGRDS chart when ASS_0=5 and 10

δ	$ANOS_1$		$MNOS_1$	
	ASS_0 = 5	ASS_0 = 10	ASS_0 = 5	ASS_0 = 10
0.2	85.70	75.33	45	45
0.6	13.48	14.45	10	11
1.0	6.79	9.76	6	11
1.4	5.44	7.77	6	6
1.8	4.64	5.98	6	5
2.0	4.23	5.17	3	4
2.6	3.30	3.57	3	3
3.0	2.64	3.03	2	2

Table 3. $PNOS_1$ of the SSGRDS chart when ASS_0 = 5 and 10

δ	$PNOS_1$							
	ASS_0 = 5				ASS_0 = 10			
	20th	40th	60th	80th	20th	40th	60th	80th
0.2	19	35	56	94	25	35	55	90
0.6	8	11	11	17	11	11	11	18
1.0	6	6	6	6	7	11	11	11
1.4	4	6	6	7	6	6	6	11
1.8	3	3	6	6	5	5	5	5
2.0	3	3	6	6	4	4	4	4
2.6	3	3	3	3	3	3	3	3
3.0	2	2	2	2	2	2	2	2

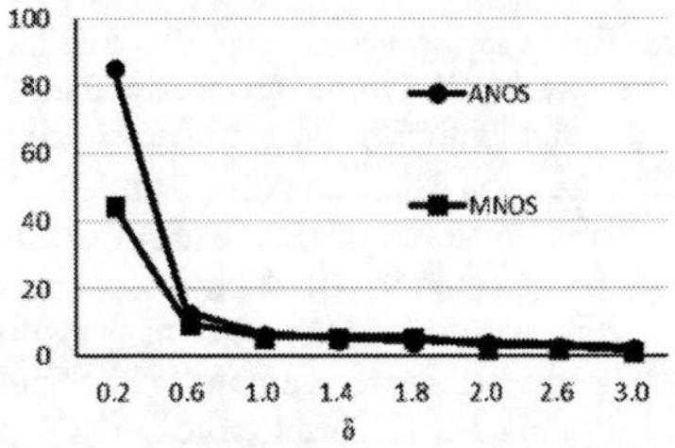

Fig. 2. Comparison of the out-of-control ANOS and MNOS values when ASS_0 = 5.

5 Run Length Properties

In this paper, MNOS and PNOS are used besides ANOS to measure the performance of the SSGRDS chart. Based on Table 2 and Fig. 2, most of the $ANOS_1$ values are greater than the $MNOS_1$ values, indicating that the distribution of the number of observations is skewed to the right. However, the skewness level decreases as δ increases. Due to the skewness of the distribution of the number of observations, the interpretation of the SSGRDS chart's performance based on solely ANOS can be misleading. With MNOS and PNOS, the entire run length distribution is considered which provides more information on the performance of the SSGRDS chart. Additionally, median is less impacted by outliers compared to average, hence making it a more credible value to interpret a chart's performance. In Table 3, the percentiles depict the confidence level of the occurrence of an out-of-control signal. As an illustration, when ASS_0 = 10 and δ = 0.2, we are 80% confident that an out-of-control condition is signalled by the SSGRDS chart at the 90th observation. It is also observed from Table 3 that the out-of-control condition is signaled earlier as δ increases and the out-of-control condition is signaled at the same observation for all percentiles when δ is large ($\delta \geq 2.6$).

It can be seen from Tables 2 and 3 that $ANOS_1$, $MNOS_1$ and $PNOS_1$ decrease as δ increases for $ASS_0 = 5$ and 10. To illustrate, for $ASS_0 = 5$, the $ANOS_1(MNOS_1)$ decrease from 85.70 (45) when $\delta = 0.2$ to 2.64 (2) when $\delta = 3.0$. Similarly, for the 60th percentile of $ASS_0 = 5$, the $PNOS_1$ decreases from 56 when $\delta = 0.2$ to 2 when $\delta = 3.0$. For an out-of-control process, it is desirable to have small ANOS, MNOS and PNOS values so that the process mean shift is detected quickly. Thus, the SSGRDS chart becomes more sensitive in detecting mean shifts as δ increases. Control charts are more likely to detect large process mean shifts swiftly compared to small process mean shifts as large shifts lead to a significant loss of quality.

6 Conclusions

This paper studies the performance of the SSGRDS chart in terms of the ANOS, MNOS and PNOS. Owing to the sample size of the SSGRDS chart that is not constant at each sampling stage, ANOS which is more informative as it evaluates the actual number of observations needed to signal an out-of-control condition is used instead of the typical performance measure ARL. Additionally, MNOS and PNOS are used as performance measures in this paper to assess the overall performance of the SSGRDS chart. As most of the $ANOS_1$ values are greater than the $MNOS_1$ values, the distribution of the number of observations is skewed to the right. Thus, the MNOS and PNOS are more suitable representations of the central tendency and are better indicators of the SSGRDS chart's performance. The results of this paper also show that the sensitivity of the SSGRDS chart increases with the shift size, which is justified by the need to detect large shifts that lead to a considerable loss of quality quickly.

This research is supported by the USM Research University Grant, Grant number: 1001.PMATHS.8011039 and funded by USM School of Mathematical Sciences (KPI MYRA PTJ), Grant number: 1001.PMATHS.AUPS001

References

1. R. Croasdale, Int. J. Prod. Res. **12**, 585-592 (1974).
2. J.J. Daudin, J. Qual. Technol. **24**, 78-87 (1992).
3. C.C. Torng and P.H. Lee. Commun. Stat. Simul. Comput. **38**, 541-557 (2009).
4. L.F. Hsu, Int. J. Prod. Res. **42**, 1043-1047 (2004).
5. M.B.C. Khoo, W.L. Teoh, P. Castagliola, and M.H. Lee, Int. J. Prod. Econ. **144**, 345-357 (2013).
6. W.L. Teoh, M.B.C. Khoo., P. Castagliola, and S. Chakraborti, Comput. Ind. Eng. **67**, 104-115 (2014).
7. M.B.C. Khoo, H.C. Lee, Z. Wu, C.H. Chen, and P. Castagliola, IIE Trans. **43**, 23-38 (2010).
8. M.P. Gadre and R.N. Rattihalli, Econ. Qual. Contr. **19**, 29-43 (2004).
9. M.P. Gadre and R.N. Rattihalli, Stat. Method Appl. **16**, 27-37 (2007).
10. H.W. You, M.B.C. Khoo, P. Castagliola, and Y. Ou, Comput. Stat. **30**, 1245-1278 (2015).
11. S.L. Lim, M.B.C. Khoo, W.C. Yeong, and M.H. Lee, Comput. Ind. Eng. **90**, 314-325 (2015).
12. M.P. Gadre, K.A. Joshi, and R.N. Rattihalli, J. Appl. Stat. **37**, 2073-2087 (2010).
13. M.B.C. Khoo, E.K. Tan, Z.L. Chong, and S. Haridy, Int. J. Prod. Res. **53**, 4735-4753 (2015).

Intelligent robot of inclined assembly sequence planning in Industrial 4.0

Yu Cheng Chiang[1], *Chumpol* Yuangyai[2], and *Chen yang* Cheng[1*]

[1]Industrial Engineering Department, National Taipei University of Technology

[2]Department of Industrial Engineering, Faculty of Engineering, King Mongkut's Institute of Technology Ladkrabang, Bangkok 10520,Thailand

[1*]cycheng@ntut.edu.tw

Abstract. In the industry 4.0, the Cyber-physical system (CPS) is one of the most important core which makes the manufacturing process more intelligent. Intelligent assembly operation is an important key in intelligent manufacturing of CPS. To complete the intelligent assembly operation, the cooperation between assembly robotic arm and assembly sequence planning (ASP) is necessary. However, the ASP and writing robotic codes manually is time consuming and requires professional knowledge and experience. Because the Local Coordinate System (LCS) is often ignored when checking for interference. If product have inclined interference and without considering LCS and causing and infeasible ASP. Therefore, this paper proposes a LCCPIAS (Local Coordinate Cyber-Physical Intelligent Assembly System) system to achieve three objective functions. First, this paper presents a dual-projected-based interference analysis approach (DPIAA) that analyzes the relations between components. Second, this paper generates optimal assembly sequence automatically to let the assembly sequence more suitable for the robotic arm to perform the assembly operation. The last one is LCS can recognize inclined interference between components and generate feasible ASP. Furthermore, this paper uses CAD model to verify that the DPIAA is faster and consider LCS interference can solve inclined interference problem. In the future assembly factory, the proposed method can help to realize intelligent manufacturing.

1 Introduction

In the manufacturing process, assembly work consume a lot of cost and time of the whole production. The product assembly cost accounts for more than 40% of the total cost in the whole product manufacturing process [2], and the assembly time of the product accounts for the total manufacturing time 20-70%, so intelligent assembly operation is an important key to the CPS intelligent manufacturing. In order to complete the smart assembly, the cooperation between the robot and the ASP is necessary.

If the best assembly sequence of products is automatically generated, it will not only reduce manufacturing time and costs but also customers' rapidly changing needs can be met immediately [4, 8]. In order to generate the best assembly sequence, several heuristic algorithms are proposed to solve the problem. The three most important ones are Genetic Algorithm (GA) [1] , Particle Swarm Optimization (PSO) [3] and ACO [9] . However, the input data of these algorithms such as the contact and interference between parts must be input manually. In addition if the assembly product has inclined assembly and without considering the LCS coordinates of the part itself only considering GCS, it will may generate wrong ASPs and lead to assembly process errors. Therefore some methods to automatically obtain the part information through CAD are proposed [5-7] .

Therefore, this paper purpose LCCPIAS to solve problem. Firstly, DPIAA method is proposed in this paper. The relationship between the parts in the product is analyzed. According to the output of DPIAA, the ACO algorithm is used to automatically generate Good assembly sequence, so that the assembly sequence is more suitable for assembly of the robot arm operation.

2 Literature discussion

There are probably many assembly sequences for products made up of dozens of parts, so how to find out the best or even the best assembly sequence has been discussed for a long time. There should be no possible collision of assembly sequences, so it is necessary to analyze the interference between the parts. Interference analysis is a common method of calculating the assembly direction of parts[5]. The final pattern of assembly sequence is the optimal assembly sequence, and although the possible assembly sequence allows the assembly operation to be performed without any interference, some assembly sequences still have many repetitive motions during the assembly process and many literature studies Find out the best assembly sequence by different algorithms and different objective functions [1, 9]

3 Research methods

In this paper, the LCCPIAS method was proposed to let the product designer design the product in the CAD software. The designer provided the CAD file to the service provider. The service provider imported the CAD file into the LCCPIAS then it will output assembly sequence automatically to robot.

The method is divided into two parts as shown in Figure 1. Firstly, DPIAA method is used to analyze the contact and interference between parts. After the analysis is completed, the ACO algorithm is used to generate the best assembly sequence. Because ACO uses the assembly robot's objective function calculates each feasible assembly sequence, so assembly can be done without human intervention.

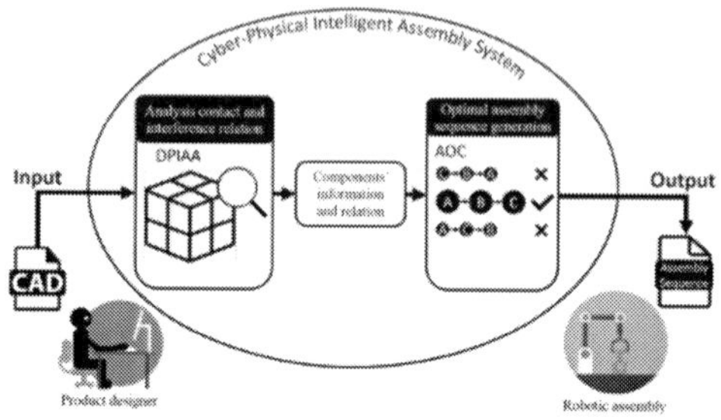

Figure 1 LCCPIAS Method flow chart

3.1 Dual Projection-Based Interferometric Analysis Method: IM is generated by collision detection. Using the proposed DPIAA and combining AABB can reduce the collision detection time.

- Symbol definition:

d_k : Cartesian axis in the direction of k。

e_s : Parts "s" to be assembled。

n : The total number of parts

CM_{st} : The contact matrix representing the contact relationship between $e_s$and e_t

DCM_{st}^k : The contact matrix representing the contact relationship between $e_s$and e_t in the direction d_k

PM_{st}^k : Projection matrix representing the projection of $e_s$and $e_t$in the$d_k$direction.

IM_{st}^k : The interference matrix representing the interference between $e_s$and $e_t$in the d_k direction.

LCS_{st}^k: The interference matrix representing the interference between $e_s$and $e_t$in the itself direction.

- DPIAA Steps

Step 1: Interfere detection and generation of CM and DCM

In the contact matrix (CM), $CMst$ denotes the contact between e_s and e_t. CM has two kinds of relations: if there is no contact between $e_s$and e_t, $CMst$ value will be 0,otherwise 1 .Directional

Contact Matrix (DCM), where DCM_{st}^{k} indicates the contact between e_s and e_t in a certain direction. There are two types of relationship in DCM. If $e_s$and e_s have no contact relationship in the d_k direction, DCM_{st}^{k} value will be 0,otherwise 1.

Step 2 Double projection inspection and generation of PM

Projection Matrix (PM) is generated using dual projection detection. PM_{st}^{k} represent projection relationship between e_s and e_t. PM used to represent e_s and e_t AABB projection onto the plane is overlapped.

The second projection is the projection of AABB to the moving direction d_k axis. As shown in Figure 2(a), if any part e_u overlaps parts e_s and e_t ,e_u will between two parts (e_s, e_t). It's means e_s and $e_t$is not collision in d_k axis. The projection matrix is modified to 0, otherwise 2,

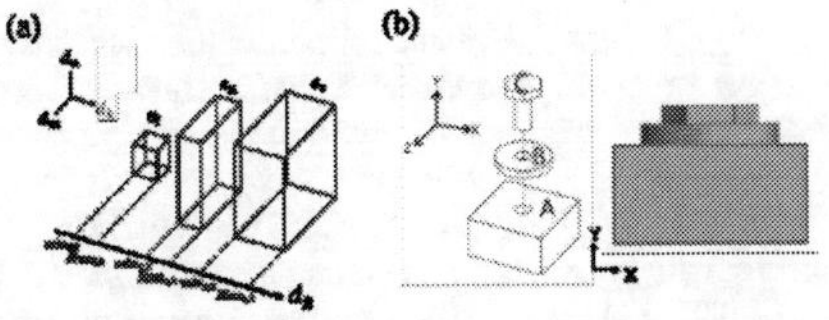

Figure 2 Project AABB to the direction of movement

Step 3 Collision detect

Collision detection is the last method used to determine the collision between parts. If the double projection determines that there is a collision between the parts e_s and e_t, the DPIAA will perform collision detection to verify that the parts e_s and e_t interfere in the direction d_k. If there is interference between e_s and e_t, it will modify the interference matrix value to 1.

In order to generate LCS, collision detection must be performed on all parts of the LCS itself. Considering the order of collision, if e_s and e_t interfere with each other in the LCS direction, If there is any interference with other parts, if interference occurs, test the collision of multiple parts by one part. If there is interference, modify the value of 1 in LCS_{st}^{k}.

Step 4 Generate interference matrix

After the DPIAA process is completed, the interference matrix is automatically generated.

3.2 Ant colony algorithm to solve ASP

By simulating the behavior of ants and using the pheromone message to find the best path, the ACO proposed an ACO algorithm for finding the best assembly order. The target of the ACO algorithm is minimum rotation angle and the highest stability. The input assembly required for ACO is CM, DCM, and IM from DPIAA.

Step 1 Initialize the basic data

The first step in this paper is to initialize the basic ACO data. The information to be initialized includes control parameters (such as rotation angle, stability, weight of pheromones, total number of ants and number of iterations), and assembly data (such as the total number of parts n, CM, DCM and IM), the total number of ants m is set by the number of can be disassembled parts.

Step 2 Initialize ant data

Each ant is assigned to a set of *IM, CM*, and *DCM* that are automatically generated from DPIAA.

Step3 Find the first moveable part

In each iteration, each ant starts browsing to find out the possible assembly order. Every ant can get all the parts that can be removed by IM. At the beginning of the first iteration, all the movable parts must be located and assigned to different ants, and all the movable parts can be selected according to equation(1).

$$Ant_{k}^{start} = P_{feasible} \quad (1)$$

Ant_{k}^{start} represent the ant first choice of parts, ants select the first part of the probability is based on equation(2).

$$P_{k}^{i}(0) = \sum_{j=0}^{n-1} \tau ij \,/ \sum_{s\in allowed_k} \sum_{j=0}^{n-1} \tau sj \quad (2)$$

Step 4. Look for detachable parts

After selecting the first detachable part, the ant selects the next part according to equation (3) .

$$P_k^i(t) = \tau_{ij}(t)^{\alpha} * \eta_{ij}{}^{\beta} \Big/ \sum_{s \in allowed_k} \tau_{is}(t)^{\alpha} * \eta_{is}{}^{\beta} \quad (3)$$

$P_k^i(t)$ represents the probability of ant k from i to j at time t, α and β are the weights and heuristic functions of pheromone, and the heuristic function $\eta_{ij}j$ is composed of the following two criteria:

1. Rotation angle：The rotation angle indicates the combined direction difference between the assembly directions of the parts i and j as equation (4),

$$\begin{cases} 0.1, d_i - d_j = 180^o \\ 1, d_i - d_j = 0^o \\ 0.5, other\ angle \end{cases} \quad (4)$$

2. Stability：Stability indicates that the part does not fall or displace during disassembly. Heuristic functions for stability (S) are shown in equations(5), (6) and (7)

$$S_{ij} = S_j^1 + S_{ij}^2 (j = 1, \ldots, n-1)(i = j-1) \quad (5)$$

$$\begin{cases} 4 \\ 0.1 \\ \frac{CCD_{max} - CCD_j}{CCD_{max}} + \frac{DCCD_{G,max} - DCCD_{G,j}}{DCCD_{G,max}}, \end{cases} \begin{cases} \sum_{k=0}^{n} IM_{j,k}^{+Y} \\ \sum_{k=0}^{n} IM_{j,k}^{-Y} \\ other \end{cases} \quad (6)$$

$$S_{ij}^2 = \begin{cases} 0.5, DCM_{ij}^{-Y} = 0 \\ 0, other \end{cases} \quad (7)$$

In this paper, we modify the definition of the direction of gravity and increase the stability of the previous component. The multi-objective heuristic function is given by equations(8) .S_{ij} represents the stability between i and j, and $w1$ and $w2$ represent the weight of each target.

$$\eta_{ij} = w_1 * D_{ij} + w_2 * S_{ij} \quad (8)$$

Step5 Pseudo-random scaling rules

After calculating the probability of all moving parts, the ants in the ACO use the pseudorandom scaling rule of equation (9) to select the parts from i to j

$$s = \begin{cases} arg \max_{k \in allowed_k} \tau_{ij}(t)^{\alpha} * \eta_{ik}^{\beta}, & if\ q \leq Q \\ S & other \end{cases} \quad (9)$$

CM, DCM, and IM are updated from the data of each ant, and if there are still parts that have not been selected yet, it will go back to Step 4. Otherwise, when all the ants have finished searching in the current iteration, go to Step 6; otherwise, go to Step 3.

Step 6 Update Pheromone matrix

After all the ants finishes searching, the pheromone matrix is updated with the global update rule given by equation(10) and (11).

$$\tau_{ij}(t+1) = (1-p)\tau_{ij}(t) + \sum_{k=1}^{m} \Delta\tau_{ij}^k(t) \quad (10)$$

$$\Delta\tau_{ij}^k(t) \begin{cases} (F_{gb})^{-1} \\ (F_{ib})^{-1}, as \\ 0 \end{cases} \begin{cases} Mod(t,5) = 0\ \&(i,j) \in & Global\ best\ solution \\ Mod(t,5) \neq 0\ \&(i,j) \in & Iterated\ optimal\ solution \\ other \end{cases} \quad (11)$$

References

1. Gunji, B., et al., *Hybridized genetic-immune based strategy to obtain optimal feasible assembly sequences.* International Journal of Industrial Engineering Computations, 2017. **8**(3): p. 333-346.
2. Hongbo, S., et al., *Genetic simulated annealing algorithm-based assembly sequence planning.* 2006.
3. Li, M., et al., *A hybrid assembly sequence planning approach based on discrete particle swarm optimization and evolutionary direction operation.* The International Journal of Advanced Manufacturing Technology, 2013. **68**(1-4): p. 617-630.
4. Li, X., et al., *Assembly sequence planning based on an improved harmony search algorithm.* The International Journal of Advanced Manufacturing Technology, 2016. **84**(9-12): p. 2367-2380.
5. Ou, L.-M. and X. Xu, *Relationship matrix based automatic assembly sequence generation from a CAD model.* Computer-Aided Design, 2013. **45**(7): p. 1053-1067.
6. Pan, C., *Integrating CAD files and automatic assembly sequence planning.* 2005.
7. Pintzos, G., et al., *Assembly precedence diagram generation through assembly tiers determination.* International Journal of Computer Integrated Manufacturing, 2016. **29**(10): p. 1045-1057.
8. Yin, Z., et al., *A connector-based hierarchical approach to assembly sequence planning for mechanical assemblies.* Computer-Aided Design, 2003. **35**(1): p. 37-56.
9. Yu, J. and C. Wang, *A max–min ant colony system for assembly sequence planning.* The International Journal of Advanced Manufacturing Technology, 2013. **67**(9-12): p. 2819-2835.

Selection of electrode tips for the resistance spot welding of dissimilar stainless steels

Teerawut Khuenkaew and *Kannachai* Kanlayasiri *

Industrial Engineering Department, Faculty of Engineering, King Mongkut's Institute of Technology Ladkrabang, Bangkok, Thailand

Abstract. This research sought to resolve the issue of electrode tip selection used in the resistance spot welding for two different stainless steel grades under otherwise identical welding parameters: SUS316 and SUS425. The study used a full fractional design in order to perform the analysis of the variables involved, while the quality of the weld was assessed by considering the indentation depth, penetration, and nugget diameter. The results of the study described the nugget diameter and penetration for all welds with the various different electrode tips which met the specified requirements of the customers. A single type of upper R30 and lower R30 electrodes achieved acceptable indentation depth. Ideally, good weld quality would demand a large nugget diameter, shallow indentation, and good penetration, all contribution to an acceptable overall appearance. Both R30 electrode tips are able to achieve the desired quality requirements in the spot welding process.

1 Introduction

Within the automotive and manufacturing sectors, Resistance spot welding (RSW) is a common approach when joining sheet metal. The creation of joint exhaust pipe parts has made significant use of both ferritic and austenitic stainless steel grades because they have an attractive appearance, can be readily welded, and resist corrosion [1]. When RSW techniques are applied to the dissimilar stainless steel grades 316 and 425, problems can arise when the nuggets are formed because the high temperatures in the molten metal along with high current density distribution can lead to cracking [2]. The outcome can lead to problems for the vehicle parts in the form of excessive noise and vibration, as well as compromised strength. This particular spot welding problem has its influence upon the nugget formation in RSW. In addition, the geometric qualities of the electrode tip are also able to influence the way in which weld nugget formation occurs. This also has a critical effect upon current conduction [3].

The electrode geometry has already been studied for resistance spot welding parameters in terms of nugget growth. Y. Li et al. presented the effect of the cone angle of truncated electrodes in terms of heat and mass transfer in RSW. The result recommended electrode cone angles of 30° and 45° in vehicle body production [2]. W. Zhang et al. reported that a planar circular tip diameter of 10 mm and spherical tip diameter of 70 mm were the optimal tip electrodes morphology for dissimilar RSW of galvanised high strength steel and aluminium alloy, respectively [4]. Y. Luo et al. studied the calculation of dynamic resistance of nugget growing in RSW. The nugget growth in RSW can be divided into three stages, which were the initial stage, growing stage, and stable stage [5]. T. Khuenkaew and K. Kanlayasiri investigated the use of the Box-Behnken experiment design method with 6 factors for quality characteristics, which are penetration, nugget diameter and nugget area, on the RSW of dissimilar stainless steel grade 316 and grade 425 [6].

The objective of this paper was to determine the electrode tip for weldment quality characteristics, which are penetration (t1, t2), indentation depth (h1, h2) and nugget diameter (di) on the RSW of dissimilar stainless steel grade 316 and grade 425. The variables used in this study relating to the upper electrode tip shape (UT) and lower electrode tip shape (LT) were flatness and radius, which consider the appearance and macrostructure of weld nuggets in the spot welding process. The full factorial design method and ANOVA were used to investigate the RSW parameter factors on the nugget geometry of RSW.

2 Materials and Methodologies

The base materials used were austenitic stainless steel 316 (SUS316) and ferritic stainless steel 425 (SUS425) sheets. Both materials were divided into plates of 45 × 45 × 1.5 mm. Spot welding was performed using a 50 kVA AC device, 50 Hz RSW machine. The spot welding parameters were applied for experimentation as shown in Table 1. The 13 mm outer diameter of the electrode tips made from CrCuZr alloy was applied. The spots were taken for appearance analysis. The characteristics of the weld nuggets were measured for all samples of metallographic cross-sections of the weld zone, as shown in Fig. 1 (a). The metallographic samples were ground

* Corresponding author: kannachai.ka@kmitl.ac.th

and polished, and were then electrolytically etched by 30% nitric acid solution at 6 V for 10 - 15s at room temperature. The macrostructures of all samples were assessed using an optical microscope (OM), as shown in Fig. 1 (b). The full factional experiment design was used as the experiment technique (27 runs).

Table 1. Resistance spot welding parameters used in the study.

Spot welding parameters	Value
Electrode force (kN)	3.0
Welding time (cycle)	30
Welding current (amperes)	8,800

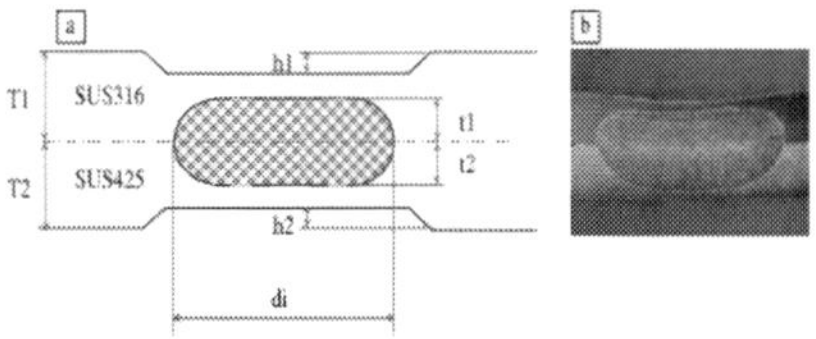

Fig. 1. (a) Schematic illustration of the weld zone
(b) Cross-sectional macrostructure of the weld joint from the electrode UT R30, LT flat

3 Results and Discussion

3.1 The appearance of the weldment

The appearance of the weldment of SUS316 and SUS425 using the flat type on the upper electrode and a different electrode tip type on the lower electrode is shown. Fig. 2 (a) and Fig. 3 (a) present the appearance with similar results; the results appeared to splash at the edge around the periphery of the contact surface. Fig. 2 (b) and (c) show that the splash of the weld joint appeared in a circular shape. Obviously for each examination, the indentation depth has not occurred. In Fig. 3 (b) and (c) the indentation depth is shown. The indentation depth of LT R16 is higher than LT R30. It can be interpreted that the indentation depth increases with a smaller contact area and higher cone angle of the electrode tip.

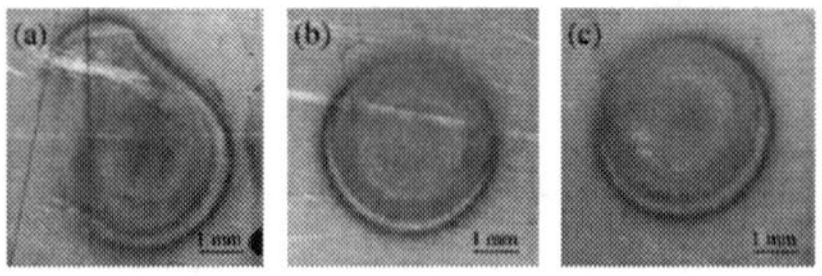

Fig. 2 Appearances of the welded joints of SUS316 and SUS425 by all of UT flat and LT different electrode shapes on SUS316 side (a) UT flat – LT flat, (b) UT flat – LT R 16, (c) UT flat – LT R 30

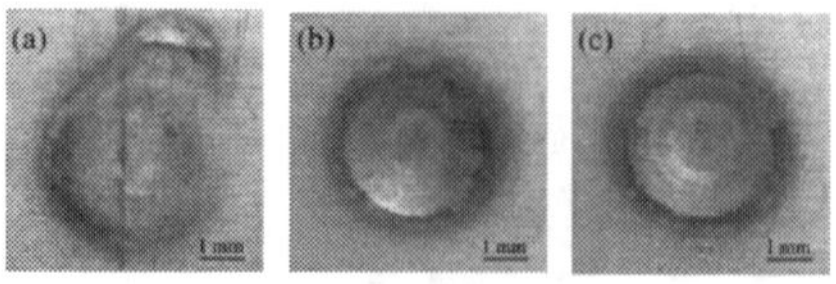

Fig. 3 Appearances of the welded joints of SUS316 and SUS425 by all of UT flat and LT different electrode shapes on SUS425 side (a) UT flat – LT flat, (b) UT flat – LT R 16, (c) UT flat – LT R 30

The appearance of the weldment of SUS316 and SUS425 using R16 type on the upper electrode and different electrode tip types on the lower electrode is shown. Fig. 4 (a-c) shows the appearance of severe expulsion on the SUS316 surface of the welded joint which occurred during welding. This is because of the high current density in a small contact surface, which may generate excessive heat in the small area. The expulsion reaches to the formation of diffused indentation depth. Fig. 5 (a-c) shows the value of indentation depth for LT R16 is the highest, with LT R30, flat, respectively. The expulsion was shown only for the LT R16 type, and clearly UT R16 and LT R16 induced expulsion.

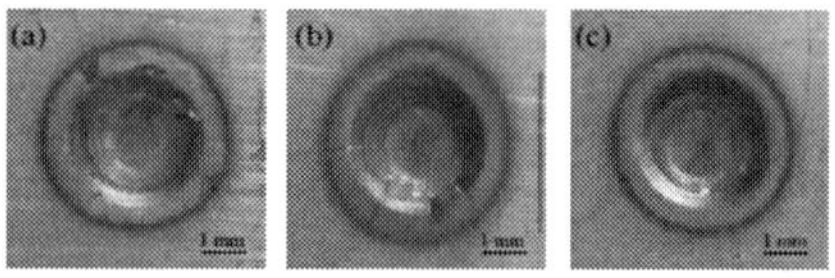

Fig. 4 Appearances of the welded joints of SUS316 and SUS425 by all of UT R16 and LT different electrode shapes on SUS316 side (a) UT R16 – LT flat, (b) UT R16 – LT R 16, (c) UT R16 – LT R 30

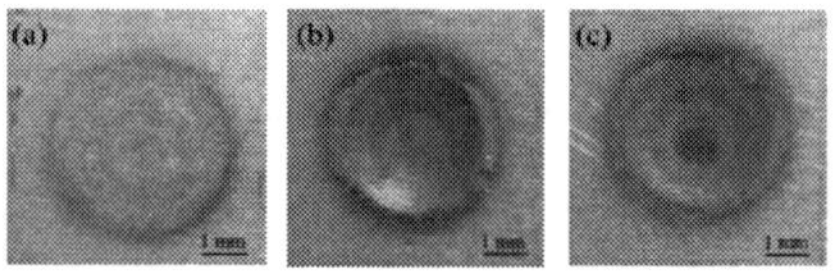

Fig. 5 Appearances of the welded joints of SUS316 and SUS425 by all of UT R16 and LT different electrode shapes on SUS425 side (a) UT R16 – LT flat, (b) UT R16 – LT R 16, (c) UT R16 – LT R 30

The appearance of the weldment of SUS316 and SUS425 using the R30 type on the upper electrode and different electrode tip types on the lower electrode is shown. Fig. 6 (a-c) and Fig. 7 (a-c) clearly show the appearance of the weld nugget. When the electrode tips at each side of the joint are the same, the force delivered is symmetrical, as is the distribution of the current density. This allows the weld nuggets to develop symmetrically within the welded sheets, as Fig 6 (c) and Fig. 7 (c) show. However, during the heating and cooling cycles the heat distribution is asymmetrical.

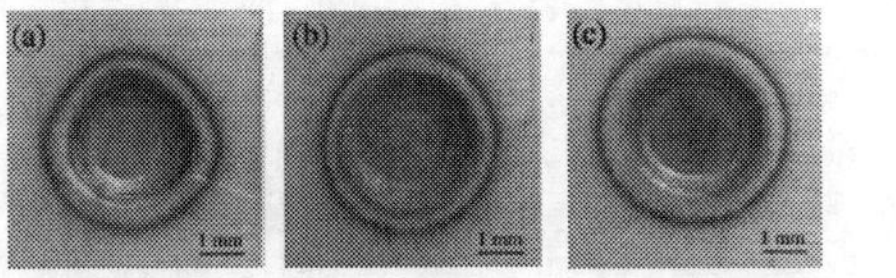

Fig. 6 Appearances of the welded joints of SUS316 and SUS425 by all of UT R30 and LT different electrode shapes on SUS316 side (a) UT R30 – LT flat, (b) UT R30 – LT R 16, (c) UT R30 – LT R 30

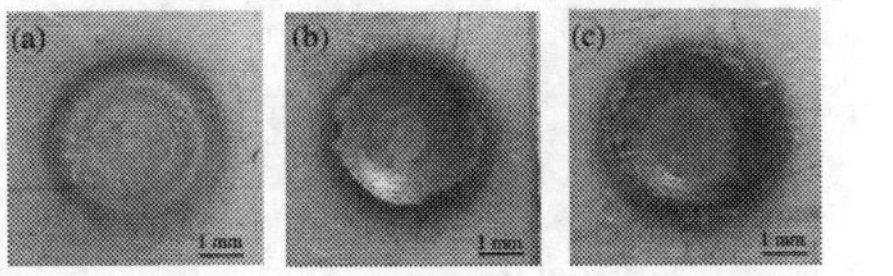

Fig. 7 Appearances of the welded joints of SUS316 and SUS425 by all of UT R30 and LT different electrode shapes on SUS425 side (a) UT R30 – LT flat, (b) UT R30 – LT R 16, (c) UT R30 – LT R 30

In this study, good symmetry was developed in the penetration when using the same electrode tips for each side of the joint, as shown in Fig. 8: R16 (UL R16 – LT R16) and R30 (UL R30 – LT R30).

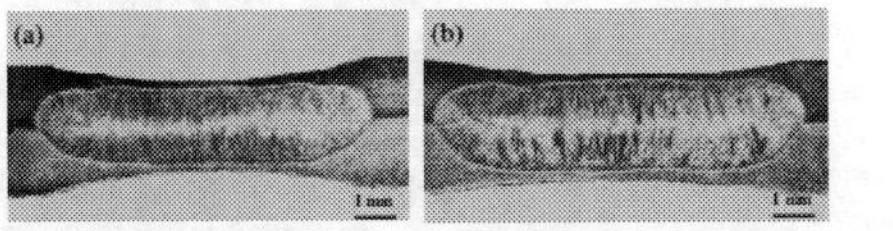

Fig. 8. Cross-sectional macrostructure of the weld joint from symmetry of the electrode tip geometry (a) UT R16 – UT R16, (b) UT R30 – LT R 30

Comparisons of the appearances were made for the two different types. When the same R30 electrode tip is used on both sides, the overall appearance is improved due to good indentation depth, increased nugget diameter, and lower expulsion. The parts for vehicle exhaust pipes are predominantly welded on their curved surfaces, and therefore the radius type of electrode geometry would be the best choice for spot welding in these circumstances.

3.2 Weldment quality characteristics

MINITAB statistical software was used to create and analyse the experiment data. The significance for the welding quality (t1, t2, h1, h2, di) was evaluated using the p-value. In this case the p-value was smaller than the confidence level of 0.05, indicating the statistical significance of the model. The R^2 values of this analysis were 99.83%, 99.81%, 99.00%, 97.19% and 99.83%, respectively, while the adjusted R^2 values of the model were 99.76%, 99.72%, 98.55%, 95.94% and 99.75%, respectively. These values were relatively moderate to high.

3.2.1 Penetration

The effects of different electrode tips on the penetration of the spot welded joints were illustrated in Fig. 9 (a) and (b). Penetration should be kept greater than 20% (>0.3 mm) of the thinnest thickness of the material. Fig. 9 (a) shows the penetration 1. The penetration 1 of all results was higher than 0.3 mm. The maximum and minimum values of penetration 1 used UT R30 – LT flat (1.18 mm), UT R16 – LT R30 (0.66 mm), respectively. Fig. 9 (b) shows the penetration 2. The maximum and minimum values of penetration 2 used UT R16 – LT flat (1.26 mm), UT flat – LT R16 (0.75 mm), respectively. The penetration of the SUS425 side was higher than that of the SUS316 side because of the difference in the mechanical and physical properties.

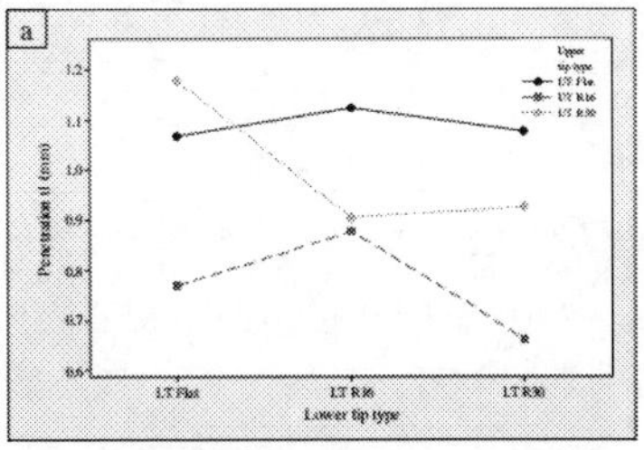

Fig. 9a. Interaction plot for penetration t1

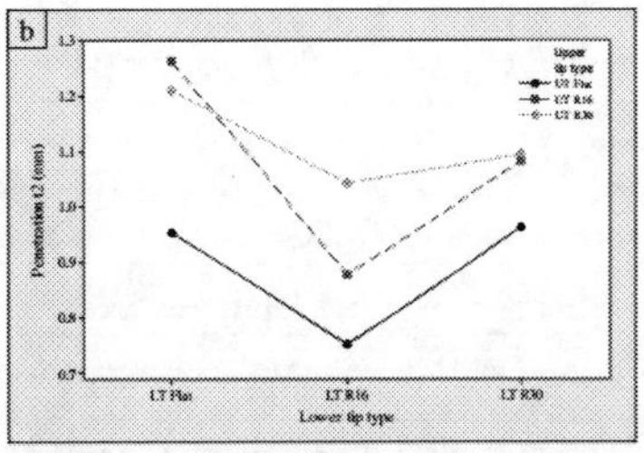

Fig. 9b. Interaction plot for penetration t2

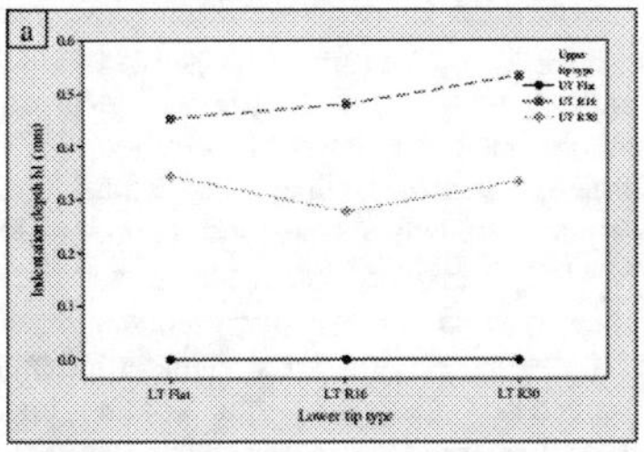

Fig. 10a. Interaction plot for indentation depth h1

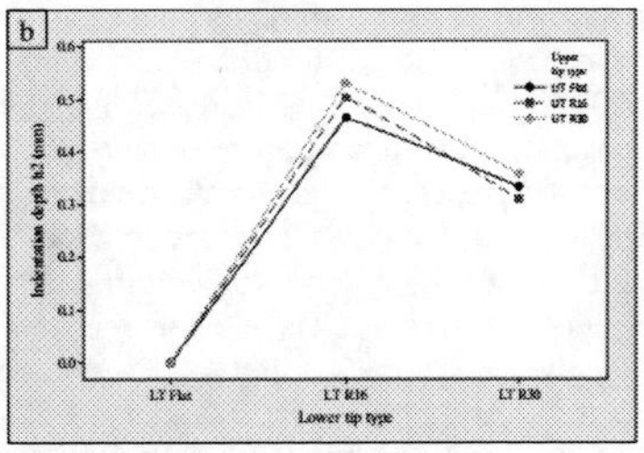

Fig. 10b. Interaction plot for indentation depth h2

 https://doi.org/10.1051/matecconf/201819201007

3.2.2 Indentation depth

In the general designs, the indentation depth was less than 30% (<0.45 mm) of the thickness of the sheet. Fig. 10 (a) and (b) shows the indentation depth of the spot welded joints. For both the UT and LT of flat electrode tips there was no indentation depth. The electrode flat shape was used to improve the weld surface quality, conversely the indentation depth of UT R16 and LT R30 electrode tips reached the maximum value (0.53 mm) in all experiments because the cone electrode R16 offers a higher angle and smaller contact surface, which could not improve the cooling ability of the electrode. Obviously for each examination, the expulsion occurs. The indentation depth of UT 30 and LT R30 achieved the target (0.33 mm).

3.2.3 Nugget diameter

The nugget diameter was larger than 4√t (> 5.0 mm), in which t represents the thinnest part of the thickness of the sheet. Fig. 11 shows the nugget diameter of the spot welded joints.

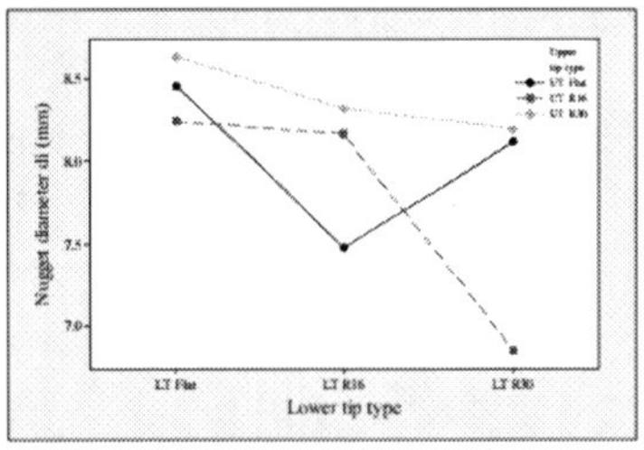

Fig. 11. Interaction plot for nugget diameter

The nugget diameter of all results was larger than 5.0 mm. The maximum and minimum values of nugget diameter used UT R30 – LT flat (8.63 mm), UT R16 – LT R30 (6.85 mm), respectively. The customer requirements demanded good weldment quality through criteria such as good penetration, a large nugget diameter, shallow indentation depth, lower expulsion, adequate strength and a good overall weld appearance. The RSW process used electrode tips of differing geometry, while the welding involved the same materials with the same thickness, under identical welding parameters. Therefore, the main influence upon differences in nugget formation would be electrode geometry. The current density determines the contact surface between the workpiece and the electrodes, and the use of flat type electrodes to increase the contact surface can serve to increase the nugget shape. Concentric electrodes present one concern because of a lack of homogeneity in the distribution of the current density when flat type electrodes are used. This results in the greatest current densities being found at the edges of the contact surface. The outcome is that when weld nuggets are formed using a larger contact surface electrode on the face side, whether UT flat or R30, the nugget edges will protrude from the electrode at the smaller contact surface [7]. Expulsion occurs when the upper or lower electrode is of the R16 type, because of the increased current density within a small contact area, since this can create excessively high temperatures. The expulsion leads to the creation of diffuse indentation [8].

4. Conclusions

In this study, electrode tips used in the RSW process were selected for use with varying stainless steel grades in order to vary the quality of the nuggets formed. Furthermore, the morphology of the electrode has to be chosen on the basis of the task to be performed. In this case, the welding of vehicle exhaust pipe parts must be carried out on a curved surface, and therefore the radius electrode geometry should be selected for the spot welding process. In such circumstances, R30 can be used for both the upper and lower electrode tip geometry in order to achieve suitable weldment quality in the RSW process.

Acknowledgements

The researchers are grateful to Yokoyama Kogyo Thailand Co., Ltd for providing access to the RSW machine. The authors would also like to express their thanks to the Faculty of Engineering of King Mongkut's Institute of Technology Ladkrabang who offered support in covering the costs incurred in conducting this research.

References

1. D. Kianersi, A. Mostafaei, A. Ali Amadeh, Mater. Des. **61** 251-263 (2014).
2. Y. Li, Z. Wei W, Y. Li, Q. Shen, Z. Lin, Int. J. Heat Mass Tranf. **65** 400-408 (2013).
3. D.Min, L.Shi-sheng, H. Hong, P. Tao, Z. Pei-Lei, Mater. Des. **52** 353-358 (2013).
4. W. Zhang, D. Sun, L. Han, Y. Li, Mater. Des. **85** 461-470 (2015).
5. Y. Luo, R. Wan, Z. Yang, X. Xie, Measuement. **87** 18-28 (2016).
6. T. Khuenkaew, K. Kanlayasiri, ICMM, *International Conference on Mechatronics and Manufacturing (2018).*
7. W. Marzur, A. Kyriakopoulos, N. Bott, D. West, J. Manuf. Process. **22** 60-73 (2016).
8. H. Zhang, X. Qiu, Y. Bai, F. Xing, H. Yu, Y. shi, Mater. Des. **63** 151-158 (2014).

MATEC Web of Conferences **192**, 01008 (2018) https://doi.org/10.1051/matecconf/201819201008
ICEAST 2018

Pluronic lecithin organogel with d-limonene as a transdermal delivery system for *Kaempferia parviflora* extract

Worranan Rangsimawong, *Paisit* Wattanasri, *Prasert* Akkaramongkolporn, *Prasopchai* Tonglairoum, *Tanasait* Ngawhirunpat and *Praneet* Opanasopit*

Department of Pharmaceutical Technology, Faculty of Pharmacy, Silpakorn University, Nakhon Pathom 73000, Thailand

Abstract. *Kaempferia parviflora* (KP) extract has been used in the Thai medicinal plant recipe, which the methoxyflavones are the main active compound. These compounds have low water solubility, high lipophilicity, and low bioavailability. The aim of this study was to develop the pluronic lecithin organogel (PLO) and PLO with d-limonene (PLO-L) for enhancing transdermal delivery of KP extract. These formulations were prepared and their physicochemical properties, stability, and *in vitro* skin permeation were evaluated. For the result, all formulations exhibited good physicochemical properties and stable under storage condition for 3 months. The permeation of KP extract-loaded PLO-L and PLO formulation showed significantly higher total methoxyflavones permeated through the skin than KP extract in water, which PLO-L provided the highest permeated flux of total methoxyflavones. This result suggested that d-limonene play a role as skin permeation enhancer. Organogel consisting of poloxamer 407 and lecithin also increased the skin permeation of KP extract. In conclusion, PLO-L could be a potential transdermal delivery system for KP extract.

1 Introduction

Kaempferia parviflora (KP) that is commonly known as "krachaidum", or "black ginger", was previously demonstrated to have a number of physiological effects, including male sexual enhancement, antimicrobial, aphrodisiac effect, anti-gastric ulcer, antidepressant, anticholinesterase activity, anti-obesity effects, vasodilator and antioxidant effects [1]. The main active ingredient of KP extract is the methoxyflovone derivative, which 5,7-dimethoxyflavone (DMF), 5,7,4'-trimethoxyflavone (TMF), and 3,5,7,3', 4'-pentamethoxyflavone (PMF) are commonly found. These compounds have low water solubility, high lipophilicity, and low bioavailability (about 1-4%) [2-4]. Therefore, a potential transdermal delivery system should be developed to increase KP into the body and improve its therapeutic effects.

Organogel is a thermodynamically stable semi-solid system, which an organic liquid phase is immobilized by a three-dimensional network of a gelator [5]. It has also been used as a potential transdermal drug delivery system. In recent years, organogel consisting of an aqueous solution of pluronic F127™ (poloxamer 407) and lecithin have been reported to facilitate hydrophilic and lipophilic drugs through the skin [6, 7].

The aim of the study was to develop the pluronic lecithin organogel formulations for enhancing transdermal delivery of KP extract. These formulations were prepared and their physicochemical properties, stability, and *in vitro* skin permeation were investigated.

2 Materials and methods

2.1. Materials

Isopropyl palmitate (IPM) was purchased from Nikkol, Singapore. Oleic acid and d-limonene were obtained from Sigma Aldrich, MO, USA. Lecithin (Emulmetik® 900) was purchased from Lucas Meyer Cosmetics, France. Poloxamer 407 (Kolliphor® P 407) was obtained from BASF Corporation, Germany.

For KP extract, the dried rhizomes were purchased from Loei province, Thailand. The percolation technique with ethanol was employed to extract the methoxy flavones from KP.

2.2 Preparation of KP extract-loaded pluronic lecithin organogel (PLO)

PLO formulation containing oil phase (5%w/w IPM, and 5%w/w oleic acid dissolved in 10%w/w lecithin) and aqueous phase (20%w/w Poloxamer 407 dissolved in cold water) was prepared. For PLO-L formulation, 5%w/w d-limonene was added into 10%w/w KP extract. Then, KP extract, oil phase, and the aqueous phase were mixed at ambient temperature.

2.3 Physicochemical properties

The pH value was evaluated using a pH meter (Metler Toledo, Seven compact S220) at room

* Corresponding author: opanasopit p@su.ac.th

 https://doi.org/10.1051/matecconf/201819201008

temperature. For viscosity measurement, the samples were measured using a Brookfield viscometer (DV-III ultra, Brookfield Engineering Laboratories, USA) at 25 °C. The stability of the formulations were monitored for 3 months under long-term (at room temperature) and accelerated (40 ± 2 °C, 75 ± 5 %RH) conditions. The quantities of total methoxyflavones remaining in the formulations were determined at the beginning and after 3 months. Each experiment was determined in triplicate.

2.4 In vitro skin permeation study

Abdominal skin was obtained from naturally dead newborn pig from a farm in Ratchaburi province, Thailand. The subcutaneous layer was separated by medical scissors and surgical blades (thickness 0.6-0.7 mm). Each skin was placed between donor and receiver compartment of Franz-type diffusion cells. Approximately 6 ml of phosphate buffer saline (PBS, pH 7.4) was filled in the receiver part and stirred by a magnetic bar at 32 °C. The 1 g of each formulation was added into the donor compartment. At 15, 30 min, 1, 2, 4, 8 and 24 h, 0.8 mL of the receiver fluid was withdrawn, and fresh PBS was added to keep the constant volume of the receiver medium. The KP content was analyzed using a UPLC.

The cumulative amounts of total methoxyflavones permeated across the skin vs time were plotted, and the slope of each line represents a permeated flux. The permeability coefficients (K_p) were calculated from the cumulative skin permeation profile using the flux, and the donor concentration (C_d) of the formulations by equations (1).

$$K_p = Flux/C_d \quad (1)$$

2.5 UPLC analysis

Three methoxyflavones (DMF, TMF and PMF) were analyzed using an ACQUITY UPLC Core system (Waters Corporation, Milford, USA) binary solvent management and dual switching solvent/Degas/ ACQUITY TUV Detector and column heater. ACQUITY UPLC BEC C18 column (2.1 mm x 100 mm, 1.7 μm, Waters®) was used. The mobile phase consisted of acetonitrile and 0.5% formic acid in water (15:85 to 85:15), with the flow rate of 0.5 ml/min and 2.0 μL injection volume. The operating temperature of column and samples was controlled at 38 and 25 °C, respectively. The detection was operated using a UV detector at 335 nm. The linearity range was 0.001–35 μg/ml and the R^2 of the mean calibration plot was 0.9999.

2.6 Data analysis

Each experiment was measured in triplicate. The results were reported as mean ± standard deviation (SD). The statistically significant differences between formulations were determined using analysis of variance (ANOVA) with an LSD post hoc test. The significance level was set at $p < 0.05$.

3 Results and discussion

3.1 Physicochemical properties

The amounts of DMF, PMF and TMF in KP extract were 10.39, 9.18, and 10.13%, respectively. All formulations containing 10%w/w of KP extract were successfully prepared and exhibited brown opaque gel with pH 4.66. The percent total methoxyflavones content of PLO and PLO-L were 99.44 ± 1.28 % and 99.37 ± 2.07 %, respectively. The viscosity of PLO and PLO-L were 42,256.96 ± 771.47 cP and 35,513.29 ± 561.49 cP, respectively.

As shown in Fig. 1, the physicochemical properties of PLO-L formulation under storage conditions for 3 months exhibited a little change from the initial day of preparation, representing good formulation stability.

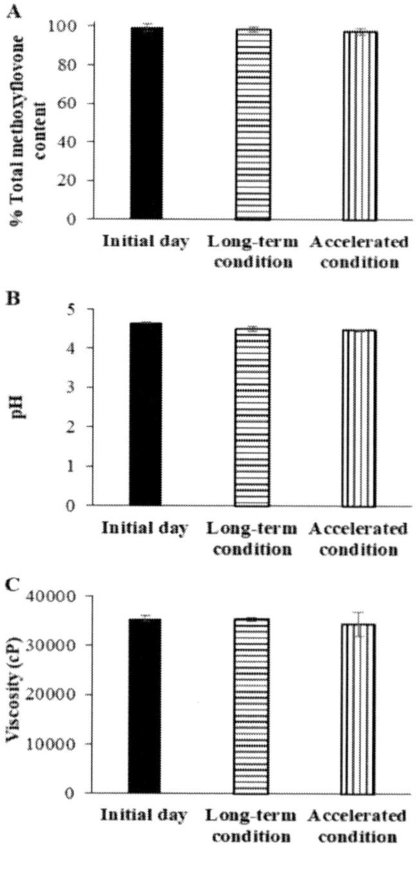

Fig. 1. Physicochemical properties (total methoxyflavone content (A), pH (B), and viscosity (C)) of PLO-L formulations at initial day and 3 months. Each value represents the mean ± SD (n=3).

3.2 *In vitro* skin permeation study

As shown in Fig. 2, the permeation of KP extract-loaded PLO-L and PLO formulation showed significantly higher total methoxyflavones permeated through the skin than KP extract in water, which the permeated flux and permeability coefficient were in the order: PLO-L > PLO > KP extract in water (Table 1). PLO-L provided the highest permeated flux of total methoxyflavones that was 2.00-fold enhancement from KP extract. This result indicated that d-limonene play a role as skin permeation enhancer and intercalate into stratum corneum intercellular lipids, leading to increase lipid lamellae fluidity and enhance skin permeability of drug via an intercellular pathway [8].

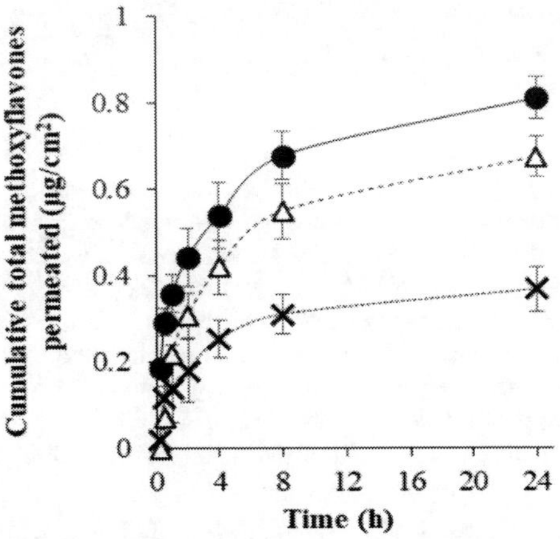

Fig. 2. Skin permeation profile of PLO (•) and PLO-L (Δ) compared with KP extract in water (x). Each value represents the mean ± SD (n=3).

Table 1. The skin permeation flux and permeability coefficient (K_p) of total methoxyflavones from KP extract in water, KP extract-loaded PLO, and KP extract-loaded PLO-L. Each value represents the mean ± SD (n=3). *indicates significant different from KP extract in water.

Formulations	Flux ($\mu g.cm^{-2}.h^{-1}$)	Kp ($cm.h^{-1}$) ($x10^{7}$)
KP extract in water	0.013 ± 0.001	4.377 ± 0.337
PLO	0.023 ± 0.002*	7.744 ± 0.673*
PLO-L	0.026 ± 0.002*	8.754 ± 0.673*

For PLO formulation that also enhanced total methoxyflavones permeability. Lecithin organogels might be efficient carriers for transdermal delivery, because lecithin promoted the skin permeation and partitioning of the drug into the skin layers. In addition, pluronic lecithin organogel is known as a new lecithin organogel. These systems have been developed for improving the skin permeation and stabilizing the formulation [9, 10]. Poloxamer 407 is a copolymer of polyoxyethylene and polyoxypropylene, which it can form a thermoreversible gel [11] and disrupts lipid layers of the skin stratum corneum [12]. Moreover, the PLO system has both oil and aqueous phases, so this system might be a potential transdermal delivery system of hydrophilic as well as lipophilic drugs [6, 7].

4 Conclusion

In this study, pluronic lecithin organogel consisting d-limonene could be a potential transdermal delivery system for KP extract. This system provided the highest skin permeability of total methoxylflavone and suitable physicochemical properties with good stability under long-term and accelerated condition.

The authors would like to thank Faculty of Pharmacy, Silpakorn University for financial supports, and Bangkok Lab & Cosmetic Co., Ltd., for *Kaempferia parviflora* (KP) extract and standard methoxyflavones.

References

1. A. Chuthaputti, Journal Thai traditional & Alternative Medicine **11**, 1(2013)
2. K. Sutthanut, B. Sripanidkulchai, C. Yenjai, M. Jay. J. Chromatogr. A **1143**, 1-2 (2007)
3. K. Sutthanut, X. Lu, B. Sripanidkulchai, C. Yenjai, M. Jay, J. Biomed. Nanotechnol **5**, 2(2009).
4. C. Mekjaruskul, M. Jay, B. Sripanidkulchai, Drug. Metab. Dispos **40**, 12(2012).
5. A. Vintiloiu, J. C. Leroux, J. Control. Release **125**, 3 (2008).
6. B. L. Bramwell, L. V. A. Williams, Int. J. Pharm. Compd. **16**, 1(2012).
7. H. Almeida, M. H. Amaral, P. Lobão, J. M. S. Lobo, J. Pharm. Pharm. Sci **15**, 4(2012).
8. S. A. Ibrahim, S. K. Li, Int. J. Pharm **383**, 1–2(2010).
9. R. Kumar, O. P. Katare, AAPS PharmSciTech **6**, 2 (2005).
10. S. Raut, S. S. Bhadoriya, V. Uplanchiwar, V. Mishra, A. Gahane, S. K. Jain, Acta. Pharm. Sin. B **2**, 1(2012).
11. G. Dumortier, J. L. Grossiord, F. Agnely, J. C. Chaumeil, Pharm. Res **23**, 12 (2006).
12. V. Agrawal, V. Gupta, S. Ramteke, P. Trivedi, AAPS PharmSciTech **11**, 4 (2010).

Single machine scheduling with consideration of preventive maintenance and machine health

Kwei-Long Huang [1], *Jakey* Blue [2], *Hao-Chen* Weng [1] and *Shu-Han* Liu [1,*]

[1] Institute of Industrial Engineering, National Taiwan University, IYC Building, No 1, Roosevelt Road, Sec 4, 10617 Taipei, Taiwan

[2] Manufacturing Sciences and Logistics, École des Mines de Saint-Étienne, Center of Microelectronics in Provence - site Georges Charpak, 880 Route de Mimet, 13541 Gardanne, France

Abstract. Because of Industry 4.0 and Internet of Things, it is easier to collect data from machines through sensors that are embedded inside machines. Once the status change of a machine is detected, production on that machine may need to be adjusted accordingly. In this research, we focus on single machine scheduling with considering the Preventive Maintenance (PM) and machine health index. Machine health index is categorized into three states: good, fair, and breakdown. When the machine moves from one state to another, the processing time of jobs will change as well as the machine failure rate. We develop a model to determine an optimal interval of performing PM and production sequence of jobs. A two-phase heuristic method is proposed to solve a large-size problem. Through different parameter settings, such as the machine failure rate, number of jobs, repair and maintenance cost, we show that the two-phase heuristic can obtain a solution with high quality.

1 Introduction

As the advance of Internet of Things, more and more production data are collected in a manufacturing factory such that the machine health index can be more truthfully characterized. Well utilizing these data can further enhance and improve production scheduling. For a traditional scheduling problem, the processing time of a job may be assumed independent of the status of machines. That is, a job always requires the same amount of time to be finished on a machine regardless of whether the machine is in a good condition or not. In a practical system, the processing time of a job can vary according to the machine status. When the machine is in different conditions, its capacity or processing speed changes accordingly, e.g., the job processing time becomes longer or shorter. In this study, we consider this characteristic to better reflect the real production environment. In addition, when the machine status changes, actions of maintenance or repair are taken to restore the machine to a better condition. Therefore, we develop a model that determines the production sequence of jobs on one machine with the minimal makespan and the timing of maintenance or repair with the least cost.

Several research topics related to our work such as maintenance, PM, machine health, and hazard rate function are introduced below:

1.1 Maintenance

Machine will deteriorate or breakdown through planning horizon. In order to avoid machine breakdown, we need to do maintenance effectively. In this research, we adopt so-called perfect repair policy or called "As Good As New (AGAN)" repair [1]. In recent research, Peng and Houtum use joint optimization model to consider maintenance [2]. Manzini, Accorsi, Cennerazzo, Maranesi relaxed the constraints with consideration of cost, reliability and resource respectively to solve the problem [3]. Contrary to them, we aim on the problem that remains all objective and constraints without compromising the soundness and completeness of the problem when solving it.

1.2 Preventive Maintenance (PM)

As the information technology advances, it is more and more viable to estimate the machine health [4]. Machine health status affects the machine processing time or workload. When the machine health starts to deteriorate, the machine failure rate may increase and the yield rate will reduce. Through perform the PM, machine health can be significantly improved [5], and the machine failure rate and the repair cost will be efficiently reduced [6, 7]. In this research, we assume that the machine health will become as good as new after performing PM. We also use the objective of minimizing unit time repair cost to determine the optimal interval of PM [8].

1.3 Machine health

* Corresponding author: Shu-Han Liu r05546023@ntu.edu.tw

https://doi.org/10.1051/matecconf/201819201009

In the past, machine status was only considered as two states: available and breakdown. Due to the installation of more and more sensors, machine health statuses can be specified more truthfully [9]. Conventionally, machine health status can be measured by hazard rate [10] and reliability [11]. Hazard rate is used to model the transition of machine health status [12] and is also called the failure rate, defined as the instantaneous rate of being irreparable given the machine has functioned for time *t*. In this research, we use hazard function as the transition rate among machine health statuses [8, 13].

2 Problem description

We try to minimize the total maintenance cost and repair cost through considering the machine health and sequencing the jobs on one machine, and to find the optimal sequence of jobs and the best timing to perform a maintenance.

2.1. Problem assumption

• There are three states for machine health: Good (denoted as 2), Fair (1), and Breakdown (0). In the good state, the processing time of each job follows original settings; in the fair state, the processing time of each job will increase; in the breakdown state, the machine cannot process jobs.

• Machine health status will not change until the machine completes a job

• Differ from transition probability of a stochastic problem, we assume the machine health transition (from 2 to 1) follows Hazard Rate Function with Exponential distribution. Machine health transition (from 1 to 0) follows Hazard Rate Function with Weibull distribution. The transition is shown in Figure 1.

• When the machine breakdowns, the minimal repair policy is used to fix the machine, that is, the machine is back to Fair state (1).

• PM is assumed to be perfectly performed, which means the machine health status will become as good as new after the maintenance. Also, PM can only be scheduled at the time when the machine is idle.

• Non-Preemptive is assumed, i.e., a job cannot be interrupted in the middle of the processing.

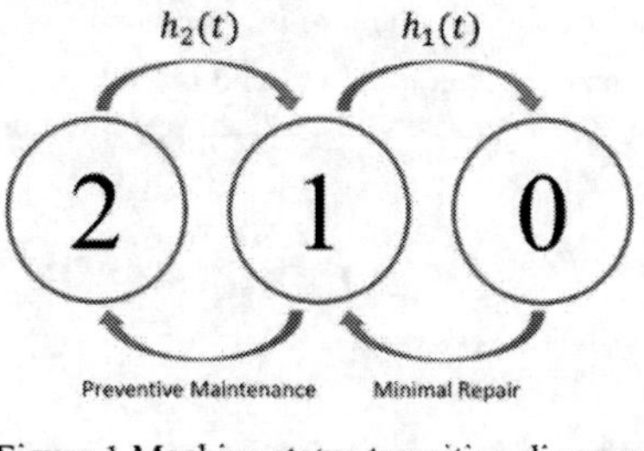

Figure 1 Machine status transition diagram

2.2 Model

Tables 1 and 2 summarize the decision variables and parameters used in the model.

Table 1. Decision variables.

Symbols	Description
y_{iak}	Job *i* is scheduled in *ath* preventive maintenance on *kth* position.
x_{ia}	Job *i* is scheduled in *ath* period preventive maintenance.
y_{ik}	Job *i* is scheduled on *kth* position.
C_a	Repair cost during *ath* period of preventive maintenance.

Table 2. Parameters

Symbols	Description
p_{ij}	Processing time of job *i* when machine is in state *j*.
C_{PM}	Cost of preventive maintenance.
C_r	Cost of minimal repair.
T_{PM}	Time duration of preventive maintenance.
T_r	Time duration of minimal repair.
N	Number of jobs.
J	Upper bound of number of jobs in one interval of preventive maintenance.
$h_2(t_2)$	Hazard function (from 2 to 1).
$h_1(t_1)$	Hazard function (from 1 to 0).
α, β	Weibull distribution, scale parameter α, shape parameter β
λ	Exponential distribution, parameter λ
T^*	Optimal interval of preventive maintenance.

Before modeling, we have to find out the optimal interval PM time T^*. We can see that Hazard function is defined as $h(t) = \frac{f(t)}{1-F(t)}$, and $UC = \frac{C_{PM}+C_r[\int_0^T h_2(t_2)\int_x^T h_1(t_1)dt_1dt_2]}{T+T_{PM}+T_r[\int_0^T h_2(t_2)\int_x^T h_1(t_1)dt_1dt_2]}$, represents machine repair cost and maintenance cost in one unit of time. In order to obtain T^* which minimizes unit time repair and maintenance cost, we take first derivative of $\frac{dUC}{dT} = 0$ and the repair cost vs time is shown in Figure 2.

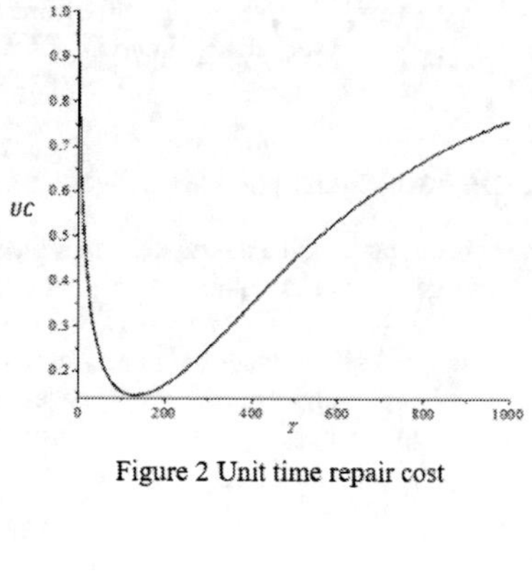

Figure 2 Unit time repair cost

After finding the PM duration T^*, we have to define the time duration of state one (T_1^*) and two (T_2^*) between two maintenances according to the distribution parameter λ. Then we can use this for the model to find the optimal schedule of the problem.

$$\text{Min } \sum_{a=1}^{N_g} C_a + N_g \cdot C_{PM} \quad (1)$$

Subject to:

$$C_a = C_r \cdot \sum_{z_a=1}^{J} \{[\int_{\sum_{i=1}^{N}\sum_{k=1}^{z_a-1} p_{i2} \cdot y_{iak}}^{\sum_{i=1}^{N}\sum_{k=1}^{z_a} p_{i2} \cdot y_{iak}} h_2(x)dx] \cdot$$

$$\left[\int_{\sum_{i=1}^{N}\sum_{k=1}^{z_a} p_{i2} \cdot y_{iak}}^{\sum_{i=1}^{N}\sum_{k=1}^{z_a} p_{i2} \cdot y_{iak} + \sum_{i=1}^{N}\sum_{k=z_a}^{J} p_{i1} \cdot y_{iak}} \mathrm{h}_1(t_1)dt_1\right]\}, \forall\, \mathrm{a} \quad (2)$$

$$N_g = \sum_{a=1}^{N} G_a \quad (3)$$

$$T_2^* = T^* \cdot \frac{\lambda - T^*}{\lambda} \quad (4)$$

$$T_1^* = T^* \cdot \frac{T^*}{\lambda} \quad (5)$$

$$\sum_{i=1}^{N}\sum_{k=1}^{N} p_{i2} \cdot y_{iak} \leq T_2^* \cdot G_a,\ \forall\, a \quad (6)$$

$$\sum_{i=1}^{N}\sum_{k=1}^{N} (p_{i1} - p_{i2}) \cdot y_{iak} \leq T_1^*, \forall\, a \quad (7)$$

$$\sum_{k=1}^{N}\sum_{i=1}^{N} y_{iak} \leq J,\ \forall a \quad (8)$$

$$\sum_{i=1}^{N} y_{iak} \leq 1,\ \forall k,\ \forall a \quad (9)$$

$$\sum_{a=1}^{N}\sum_{k=1}^{J} y_{iak} = 1,\ \forall i \quad (10)$$

$$y_{iak},\ G_a \in \{0,\ 1\} \quad (11)$$

Objective (1) is to minimize total expected preventive maintenance cost ($N_g \cdot C_{PM}$) and repair cost ($\sum_{a=1}^{N_g} C_a$), N_g represents the number of preventive maintenances, C_{PM} represents the PM cost for once. Constraint (2) represents the expected repair cost in one interval of PM. Constraint (3) represents the total number of preventive maintenance. Constraint (4) represents time duration of a machine in good state between two PMs. Constraint (5) represents time duration of machine in fair state between two PMs. Constraint (6) means that total the processing time of jobs cannot be longer than the optimal interval of two PMs when the machine is in good state. Constraint (7) means that the total increase of processing time cannot be longer than T_1^* when the machine is in fair state. Constraint (8) means that total number of jobs in the period of PM cannot exceed J. Constraint (9) means that one job can be only processed on one machine. Constraint (10) means that one job can only be processed in one interval between two PMs.

3 Two-phase Heuristic

It is time consuming to solve the model when the number of jobs is large, so we develop a two-phase heuristic method to solve the problem efficiently. The model in Section 2.2 is try to schedule all the jobs simultaneously. Because of the complexity, our two-phase heuristic method tried to divide jobs into groups to form a few small problems. In Phase 1, we separate jobs into different intervals of preventive maintenance. In Phase 2, we sequence the jobs in each interval between preventive maintenance. The idea of the proposed algorithm is shown in Figure 3.

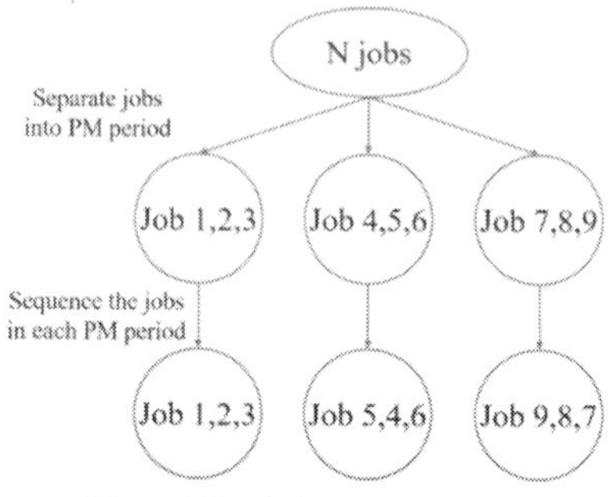

Figure 3 Heuristic method process

• Phase 1: Separating jobs into groups

In this phase, jobs are assigned to different preventive maintenance interval.

$$Min\ N_g = \sum_{a=1}^{N} G_a \quad (12)$$

Subject to:

$$T_2^* = T^* \cdot \frac{\lambda - T^*}{\lambda} \quad (13)$$

$$T_1^* = T^* \cdot \frac{T^*}{\lambda} \quad (14)$$

$$\sum_{i=1}^{N} p_{i2} \cdot x_{ia} \leq T_c^* \cdot G_a,\ \forall\, a \quad (15)$$

$$\sum_{i=1}^{N}\sum_{k=1}^{N} (p_{i1} - p_{i2}) \cdot y_{iak} \leq T_1^*, \forall\, a \quad (16)$$

$$\sum_{a=1}^{N} x_{ia} = 1, \forall\, i \quad (17)$$

$$G_a,\ x_{ia} \in \{0,1\} \quad (18)$$

Objective (12) is to minimize the total number of PMs. Constraint (13) represents time duration of the machine in good state during the interval of preventive maintenance. Constraint (14) represents time duration of the machine in fair state in the interval of preventive maintenance. Constraints (15) means that total processing time of jobs with machine good state cannot be longer than the interval of preventive maintenance. Constraint (16) means that total increase of processing time with machine fair state cannot be longer than T_1^*. Constraint (17) means that each job will be processed in only one period of preventive maintenance. Constraint (18) is a binary constraint.

• Phase 2: Sequence the jobs in a group

In this phase, given jobs assigned in each interval of the PM, the job sequence will be determined to minimize the repair cost:

$$Min\ C_r \cdot \sum_{z_a=1}^{n} \{[\int_{\sum_{i=1}^{n}\sum_{k=1}^{z_a-1} p_{i2} \cdot y_{ik}}^{\sum_{i=1}^{n}\sum_{k=1}^{z_a} p_{i2} \cdot y_{ik}} h_2(t_2)dt_2] \cdot$$

$$\left[\int_{\sum_{i=1}^{n}\sum_{k=1}^{z_a} p_{i2} \cdot y_{ik}}^{\sum_{i=1}^{n}\sum_{k=1}^{z_a} p_{i2} \cdot y_{ik} + \sum_{i=1}^{n}\sum_{k=z_a}^{n} p_{i1} \cdot y_{ik}} \mathrm{h}_1(t_1)dt_1\right]\} \quad (19)$$

Subject to:

$$\sum_{i=1}^{n} y_{ik} = 1,\ \forall k \quad (20)$$

$$\sum_{k=1}^{n} y_{ik} = 1,\ \forall i \quad (21)$$

$$y_{ik} \in \{0,\ 1\} \tag{22}$$

Objective (19) is to minimize repair cost in one interval of preventive maintenance. Constraint (20) means that one position on the machine can only process one job. Constraint (21) means that one job can only be processed on one position on the machine. Constraint (22) is a binary constraint.

4 Results and Conclusion

We consider different settings of parameters to examine the effectiveness of the heuristic method.

• Parameter settings

Two types of machines are considered, one with a higher failure rate and the other with a lower failure rate. Machine conditions are shown in Table 3.

Table 3. Machine type condition.

Machine type	C_{PM}	C_r	T_{PM}	T_r	λ	(α, β)
1	20000	30000	60	80	1/900	(100,2)
2	20000	30000	60	80	1/900	(200,2)

When the machine is in different health status, the processing time of jobs will be different. Processing time becomes longer in fair state. There are 14 jobs and the processing time of each job in states 2 and 1 is shown in Table 4.

Table 4. Processing time of each job.

Job		1	2	3	4	5	6	7
Processing time	State 2	30	28	36	40	44	48	38
	State 1	38	33	45	49	54	55	46
Job		8	9	10	11	12	13	14
Processing time	State 2	26	33	50	39	42	20	32
	State 1	31	42	66	47	52	27	41

• Parameter settings

According to the machine health conditions, we find out the optimal interval of preventive maintenance from the model of Section 2.2, T^* of machine type 1 is 162, and type 2 is 259.

Based on the results shown in Table 5 and Table 6, we can find out that the proposed two-phase heuristic can efficiently solve the problem, and the quality of solution remains good.

Table 5. Number of jobs in two-phase heuristic (type 1).

# of jobs	Repair and maintenance cost			Computation time (sec)	
	Optimal model	Two-phase heuristic	Gap	Optimal model	Two-phase heuristic
8	65127	65696	99%	75	<1
10	72248	72292	99.9%	86	<1
12	91838	92073	99.7%	882	<1
14	95938	96094	99.8%	3633	<1

Table 6. Number of jobs in two-phase heuristic (type 2).

# of jobs	Repair and maintenance cost			Computation time (sec)	
	Optimal model	Two-phase heuristic	Gap	Optimal model	Two-phase heuristic
8	43383	43649	99.3%	23	<1
10	63105	65605	96%	67	<1
12	65900	66089	99.7%	376	<1
14	67770	68548	98.8%	1418	<1

In the study, we consider different processing times of a job with respect to different machine states, which is known as the machine health index. An optimal model is constructed to determine the job sequence that minimizes the cost of preventive maintenance and repair. For a large -scale problem, a two-phase heuristic is proposed to find a good solution in a short computational time.

References

1. Dinh. T. N, Yann. D, Mitra. F, European Journal of Operational Research, **256(2)**, 439-453. (2017).
2. Peng. H, Houtum. G, European Journal of Operational Research, **253(1)**, 94-107. (2016).
3. Manzini. R, Accorsi. R, Cennerazzo. T, Maranesi. F, Computers & Industrial Engineering, **87**, 561-568. (2015).
4. Lee. J, Wu. F, Zhao. W, Ghaffari. M, Liao. L, & Siegel. D, Mechanical Systems and Signal Processing, **42(1)**, 314-334. (2014).
5. Kazaz. B, & Sloan. T. W, European Journal of Operational Research, **227(1)**, 88-100. (2013).
6. Mobley. R.K, *An introduction to predictive maintenance: Butterworth-Heinemann*. (2002).
7. Ruiz-Castro. J. E., & Li. Q-L, European Journal of Operational Research, **211(1)**, 97-111. (2011).
8. Xia. T, Jin. X, Xi. L, & Ni. J, European Journal of Operational Research, **240(3)**, 781-790. (2015).
9. Pham. H. T., & Yang. B-S, Mechanical Systems and Signal Processing, **24(2)**, 546-558. (2010).
10. Lim. J-H, & Park. D. H, Asia-Pacific Journal of Operational Research, **24(01)**, 111-124. (2007).
11. Liao. W, Pan. E, & Xi. L, Journal of Intelligent Manufacturing, **21(6)**, 875-884. (2010).
12. Wang. S, & Liu. M, International Journal of Production Research, **51(3)**, 847-868. (2013).
13. Yong. J, *Condition-based hazard rate estimation and optimal maintenance scheduling for electrical transmission system*. (2006).

Effects of brazing parameters on the microstructure and tensile shear force of copper sheets using amorphous filler metal

Prajak Jattakul and *Kannachai* Kanlayasiri*

Industrial Engineering Department, Faculty of Engineering, King Mongkut's Institute of Technology Ladkrabang, Bangkok, Thailand

Abstract. In this research, the objective was to investigate and measure the microstructure, tensile shear force, and fracture surface behavior when copper sheets are brazed using Cu-Ni-Sn-P amorphous filler metal. In order to examine the microstructure and its properties, assessment of the Cu-Ni-Sn-P/copper brazed joints was conducted after furnace brazing under an argon atmosphere using specific parameters relating to temperature, holding time, and loading pressure. In order to assess the tensile shear force, tensile testing was carried out. All the joints exhibited sound bonding without voids or cracks, when brazing temperatures reached 680 °C, this resulted in a maximum value for tensile shear force of 696.325 N. The microstructure consisted of both Cu-rich solid solution and a $(Cu, Ni)_3P$ phase as a eutectic structure formed in the brazing joint, and then the Cu-rich solid solution was produced as a matrix region around the eutectic structure. Fracture analysis was conducted for brazed joints which showed the ductile fractures after the shear test.

1 Introduction

Today there are numerous industrial applications for devices which employ both micro-channel process technology (MTP) and laminated micro-channel array (LMA). These include the conversion of natural gas to liquid form, the recovery of waste heat, and thermal management of laser diodes. While it is possible for large MPT devices to be constructed, with dimensions measured in meters, they include microchannel components in sizes measuring as little as 100 μm [1,2] . The most notable advantage offered by MPT devices over the conventional technologies is the large surface area to volume ratio, which permits much faster transfers of heat and mass inside the microchannels on account of the shortened distances over which diffusion takes place. The consequence is that the microchannels are able to generate energy and chemical systems which are both smaller and lighter [3].

Typically, there are three key stages in the process of micro-lamination. These comprise laminate registration, patterning, and bonding. The bonding process is usually conducted via soldering, diffusion brazing, diffusion bonding, or fusion joinings [4,5]. In the soldering process this method requires a low joining temperature and short joining time, however suffers from a weaker joint and fluid stage emerges in the joining process and is unsatisfactory for high-temperature applications [6,7]. In the case of diffusion bonding and diffusion brazing, these techniques create a strong joint but require a high welding temperature and extended welding time, giving rise to the high production cost [8]. The ultrasonic welding process, requires a short joining time but yields low joint strength and is unsuited for high-temperature applications [9–12]. In the case of fusion weldings, such as electron beam welding and tungsten inert gas welding, the sheets are joined together only at the edges, such that the microchannels are not tight to each other leading to mixing of the medium fluid [13]. Amongst them, brazing has been taken into consideration to be one of the optimal joining techniques due to its high joint strength, simplicity, low expenses, as well as a vast variety of filler metals [14,15].

This study performs the brazing of copper sheets using Cu-Ni-Sn-P amorphous filler metal, whereby the particular point of interest lies in the characteristics of the microstructure and the tensile shear force of the joints when different brazing temperatures are used.

2 Experiment process

2.1 Materials preparation

The base material employed in the brazing process was commercially pure copper sheet, with thickness of 0.350 mm. The samples were divided into rectangular plates with 10×26.25×0.35 mm dimensions. The filler metal was Cu-Ni-Sn-P amorphous ribbon and its thickness was about 0.025 mm. In the experiment, rectangular plates of the filler metal were cut to a size of 2×10×0.025 mm. Furnace brazing was carried out using a CARBOLITE No. 20-600024 model. The brazing process was performed under an argon atmosphere to protect the specimens from oxidation.

*Corresponding author: kannachai.ka@kmitl.ac.th

2.2 Brazing process

The joints of the foil and the copper sheets were then affixed between two copper sheets using a brazing fixture. The loading pressure was then set to 30.433 kPa and an argon atmosphere used for the furnace brazing at a variety of brazing temperatures. Initially, the heating of the brazing couple was carried out in order to reach the brazing temperature, with the increase in temperature progressing at 10 °C/min. The various brazing temperatures used were 600 °C, 640 °C, 660 °C, and 680 °C while the holding time was 5 minutes. At last, the furnace was cooled down to room temperature. In this study, each brazing condition was performed twice to ensure the result.

2.3 Microstructure observation and mechanical testing

Upon completion of the brazing, the joints were cut along their length, before grinding using an abrasive paper and polishing using alumina powder at 1 µm and 0.3 µm. Following the brazing process, the microstructure was examined and the element composition of the brazing joints was assessed using SEM (JEOL model JSM-6610LV, Japan). An EDS (Inca, Oxford, UK) was also employed for the analysis. The test to measure the tensile shear force at the brazed joints was performed at regular room temperature using a Shimadzu AG-100 universal testing machine with a crosshead speed of 0.1 mm per minute.

3 Results and discussion

3.1 Microstructure analysis of brazed joints

The first experiment involved the brazing at 680 °C for 5 minutes of amorphous Cu-Ni-Sn-P BFM, while the loading pressure was set at 30.433 kPa. Fig. 1(a-b) presents the backscattered electron image showing the cross section of the Cu/Cu-Ni-Sn-P/Cu joint. It is clear that there are no cracks or voids to be seen at the joint, indicating that the mechanical properties should suffer no weaknesses. The brazed zone has a width exceeding that of the original joint clearance, measuring around 40 µm, since the interaction between the base material and BFM occurs via the base material solution containing the liquid filler alloy, while diffusion occurs as both liquid and solid. The brazed zone can be separated into two separate reaction zones (I and II: Fig. 1(a)) on the basis of the chemical composition and the microstructural morphology. There are two phases in the brazed zone, marked as A and B (Fig. 1(b-d)). Phase A is shown in light gray while phase B is dark gray. Phase A is the major phase and is located adjacent to the base alloy, while phase B can be seen within the intergranular area of phase A at the middle of the brazed zone. The microstructure of Cu/Cu-Ni-Sn-P/Cu joint mainly comprised the eutectic phase on the brazed zone, as shown in Fig. 2(a-f) presenting the backscattering electron image and quantitative phase compositions of the eutectic phase in the brazed seam. It can be seen that there were $(Cu, Ni)_3P$ and Cu-rich phases in the eutectic structure labeled by point A and B. Element analysis of the brazed zone by EDS and x-ray mapping showed that the major element found in the two phases in Cu, as indicated in Fig. 2(a-f) and Table 1.

Table 1. The compositions of the marked zones from Fig. 1(b).

Position	Eutectic composition (at. %)				Possible phase
	Cu	Ni	Sn	P	
A	95.06	-	4.94	-	Cu Rich
B	36.05	37.69	0.31	25.96	$(Cu, Ni)_3P$

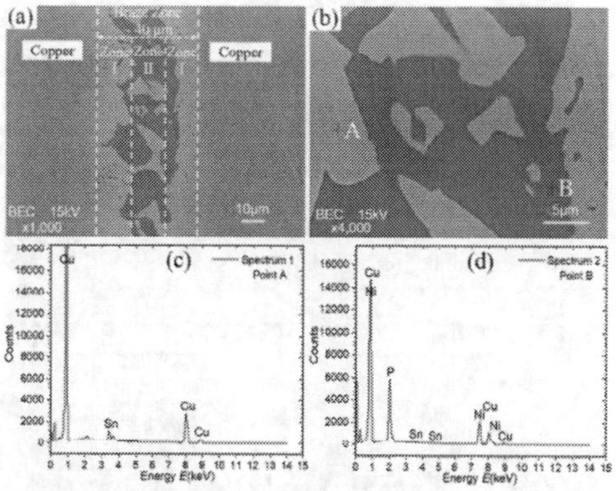

Fig. 1. (a-b) Brazed joint morphology of a brazed joint under conditions of 680°C, 5 minutes, and 30.433 kPa, (c-d) EDS spectrum of point A and B, respectively.

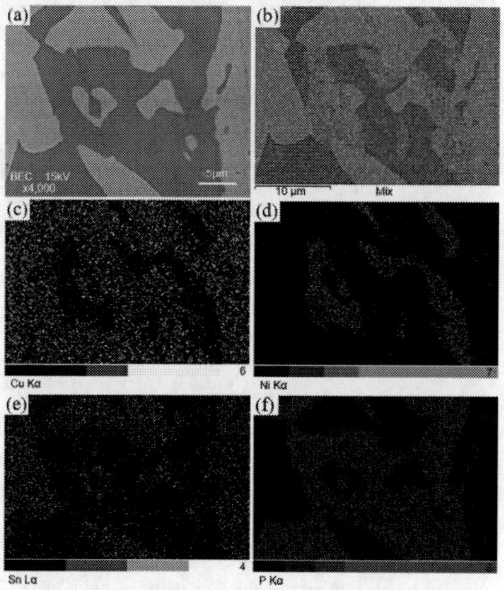

Fig. 2. (a) The interfacial microstructure of Cu/Cu-Ni-Sn-P/Cu, (b) the x-ray mapping for Cu, Ni, Sn and P (c) Cu Kα mapping, (d) Ni Kα mapping, (e) Sn Lα mapping, (f) P Kα mapping.

The EDS findings are shown in Fig. 2 and Table 1 as compositional maps illustrating the brazed joint. In this format, it is possible to determine the spatial distribution of the elements; it is thus clear that within the joint there is a continuous Cu-rich phase along with a (Cu, Ni)$_3$P phase. By considering the atomic ratio of the microstructure of the brazed joint microstructure in addition to the binary Cu-P phase diagram, it may be inferred that there exist both Cu_3P and CuP_2 phases in the reaction layer [16]. This study did not, however, record any sign of a CuP_2 phase being formed at the brazed joint. The binary phase diagram of Cu-Ni and Cu-Sn (at Sn 4.94 at. %) indicates the absence of an intermetallic phase, thereby revealing that a solid solution was produced by Cu-Ni and Cu-Sn. This demonstrates conclusively that it is possible for the brazing process to be completed without generating an intermetallic phase. When producing (Cu, Ni)$_3$P, this was achieved using filler metal containing nickel and phosphorus atoms diffused in the direction of the interface, while the base material contained copper atoms which also diffused in the direction of the interface, thereby creating the observed (Cu, Ni)$_3$P compound.

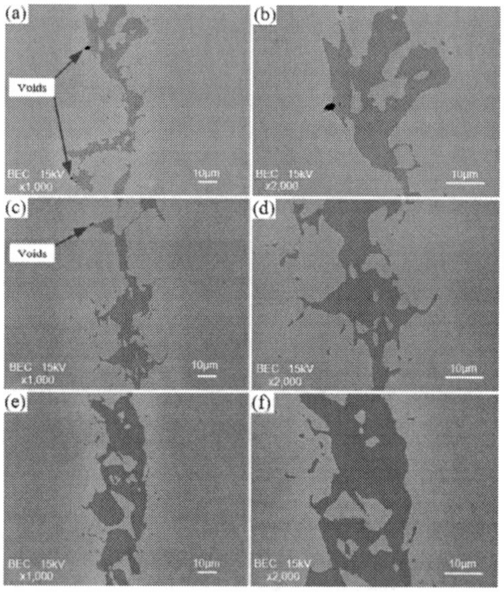

Fig. 3. The backscattered electron image showing the morphology of the cross-section of the Cu/Cu-Ni-Sn-P/Cu brazed joint using 5 minutes holding time and a loading pressure of 30.433 kPa with amorphous BFM at varying brazing temperatures of: (a-b) 640 °C, (c-d) 660 °C and (e-f) 680 °C.

Fig. 3(a-f) indicated the microstructure morphologies of the microstructure of Cu/Cu-Ni-Sn-P/Cu joints at both low and high magnification when brazed at varying temperatures of 640 °C, 660 °C, and 680 °C in an argon atmosphere furnace for 5 minutes at 30.433 kPa. Fig. 3(b, d, f) shows the microstructure of the brazed zone in high magnification and also indicates the eutectic phase for the various different brazing temperatures. The difference in image contrast shows that the brazed zone is a mixed region comprising two phases. Examination revealed that there were micro-voids to be found between the filler metal and the copper substrate, leading to doubts about the efficiency of the joint. Fig. 3(a-b) depicts this particular issue. The reason for this shortcoming is that at relatively low brazing temperatures (640 °C for 5 min and 30.433 kPa), the filler is unable to completely react with the copper sample in order to form a continuous reaction layer. This issue can, however, be resolved by increasing the temperature (660 °C for 5 min and 30.433 kPa), whereupon the voids are minimized and a continuous reaction layer can be clearly identified. Fig. 3(c-d) shows this variation. Finally, upon further increasing the temperature (680 °C for 5 min and 30.433 kPa), no cracks at all can be observed, as indicated in Fig. 3(e-f).

3.2 Mechanical properties brazed joint

Fig. 4 indicates how the different brazing temperatures influence the joint strength after brazing with Cu-Ni-Sn-P amorphous BFM. The findings revealed that when brazing takes place for 5 minutes at 600 °C at 30.433 kPa, the maximum joint strength recorded was 352.526 N. Increasing the temperature to 640 °C, 660 °C, and finally 680 °C caused further increases in the joint strength to 696.325 N, with the improvement showing a similar pattern to that observed in the tensile shear force test. Furthermore, Fig. 4 also confirms that when the brazing temperature is insufficient, a strong joint with good bonding cannot be achieved. At the maximum temperature for this study, of 680 °C, a strong joint could be formed.

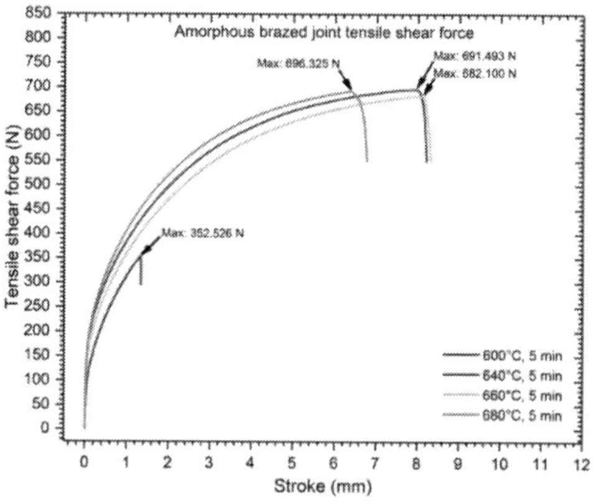

Fig. 4. Effect of brazing temperature on joint strength.

3.3 Fracture characteristics of brazed joints

Fracture analysis was carried out using SEM to establish the fracture mode. The findings are shown in Fig. 5(a-b) which presents an overview along with magnified images of the specimen fracture surfaces following brazing at 600 °C upon completion of the shear tests. Cleavage steps are observed on the fracture surface (Fig. 5(a-b)), suggesting the dimples fracture mode of the joint, as denoted in the

https://doi.org/10.1051/matecconf/201819201010

mark line in Fig. 5(a). The formation of the dimples implies plastic deformation in the micro-regions, which is sometimes taken as a positive indication of enhancement of the mechanical properties of the joint. According to these results, the improvement of the Cu/Cu-Ni-Sn-P/Cu brazed joint was achieved because the formation of ductile dimple fracture resulted from the strong bonding and fatigue resistance, and when the brittle fracture formation occurred on the brazing joints, the cracks preferred to propagate at the brittle intermetallic compounds, resulting in an easy propagation of the cracks, which leads to a decrease in the strength and fatigue life of the joints [17]. Fig. 5(b), shows a large number of dimples densely distributed on the fracture surface, which is a typical feature of ductile failure. According to these results, it is suggested that the fracture mode of the joint is a ductile fracture mode.

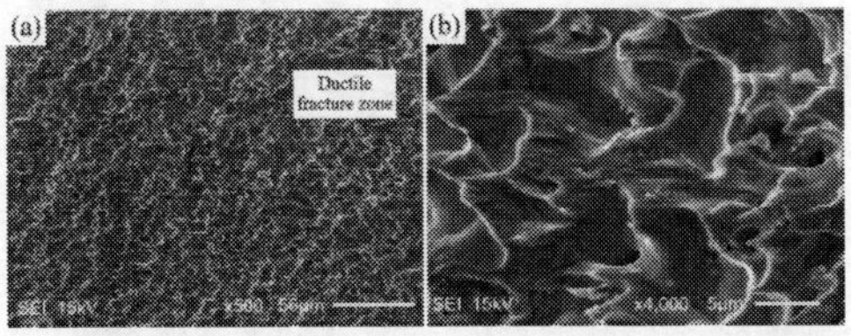

Fig. 5. Fracture morphologies of the Cu/Cu-Ni-Sn-P/Cu brazed joint after shear tests: (a) overall view; (b) magnified views of the ductile fracture zone.

4 Conclusions

The current study assessed the properties and microstructure of the brazed joints between copper sheets and Cu-Ni-Sn-P amorphous filler metal. The major conclusions are summarized as follows:

(1) Solid joints without micro-pores or any defects in the brazed joint were obtained when the brazing temperature increased to 680 °C, the highest value achieved for the tensile shear force was 696.325 N.

(2) The of Cu/Cu-Ni-Sn-P/Cu brazed joint consisted of Cu-rich phase and the eutectic structure formed the phases in eutectic region were Cu-rich solid solution and a $(Cu, Ni)_3P$ phase.

(3) Analysis of all the fractured surfaces of the brazed joints revealed ductile fracture characteristic upon completion of the shear test.

Acknowledgements

The researchers are grateful to King Mongkut's Institute of Technology Ladkrabang in acknowledgement of the financial support offered for this study.

References

1. K. Kanlayasiri, B.K. Paul, J. Manuf. Process. **6** 72–80 (2004).
2. J. Li, S. Wang, W. Cai, W. Zhang, Appl. Therm. Eng. **30** 2738–2745 (2010).
3. P. Paulraj, B.K. Paul, R.B. Peterson, Appl. Therm. Eng. **48** 194–201 (2012).
4. R. Eluri, B. Paul, J. Manuf. Process. **14** 471–477 (2012).
5. T. Watanabe, H. Sakuyama, A. Yanagisawa, J. Mater. Process. Technol. **209** 5475–5480 (2009).
6. B.K. Paul, N. Sharma, T. Doolen, J. Manuf. Process. **7** 174–181 (2005).
7. P. Paulraj, B.K. Paul, J. Manuf. Process. **13** 85–95 (2011).
8. R. Eluri, B.K. Paul, Mater. Des. **36** 13–23 (2012).
9. M. Shakil, N.H. Tariq, M. Ahmad, M.A. Choudhary, J.I. Akhter, S.S. Babu, Mater. Des. **55** 263–273 (2014).
10. S. S. Lee, T.H. Kim, S.J. Hu, W.W. Cai, J.A. Abell, J. Manuf. Sci. Eng. **137** 31016-1–8 (2015).
11. J.S. Seo, H.S. Jang, D.S. Park, Mater. Manuf. Process. **30** 1069–1073 (2015).
12. M.P. Satpathy, B.R. Moharana, S. Dewangan, S.K. Sahoo, Eng. Sci. Technol. Int. J. **18** 634–647 (2015).
13. W.W. Basuki, O. Kraft, J. Aktaa, Mater. Sci. Eng. A **538** 340–348 (2012).
14. J. Zou, Z. Jiang, Q. Zhao, Z. Chen, Mater. Sci. Eng. A **507** 155–160 (2009).
15. P. Jattakul, K. Kanlayasiri, *ICMM, International Conference on Mechatronics and Manufacturing* (2018)
16. H. Okamoto, P.R. Subramanian, L. Kacpzak, T.B. Massalski, *Binary Alloy Phase Diagram* (The Materials Information Society, 2001).
17. E. Ganjeh, H. Sarkhosh, M.E. Bajgholi, H. Khorsand, M. Ghaffari, Mater. Charact. **71** 31–40 (2012).

The effect of cell boundary on 2D foam

Atef AbdelKader[1*]

[1]Department of Mathematics and Science, Ajman University, UAE.

Abstract. We have studied the effect of cell boundary on 2D foam, with particular attention to perfect arrays of identical bubbles, and those containing only a single defect with time. We have also examined the effect of the wetness of the foam, observing the stability of two-dimensional foam comprising bubble rafts constrained to a fixed area of liquid surface. Perfectly six-fold coordinated foam appear to be unstable against loss of cohesion, but the lifetime to breakage of the perfect foam increases systematically with changing the cell boundaries. Foams containing a single defect are stable against such breakage due to the elastic stress fields around it.

1 Introduction

Relatively ordered foam exhibits an initial transient in its evolution [1], [2], [3] which has been interpreted in terms of the growth of disorder from individual topological defects. The past decades has witnessed increasing interests in the study of such growing disorder. Considerable attention was addressed to the scaling state by many authors and it has been studied experimentally and numerically [4], [5], and [6]. Levitan investigated the evolution of an isolated defect in an otherwise ideal hexagonal froth [7] concentrating on the cluster of disordered bubbles around the initial defect (the set of bubbles having at least one nonhexagonal neighbour). As cluster of transformed bubbles grow; the time dependence of the number of bubbles in this cluster in investigated and the distribution of the topological classes in the evolving part of the system is calculated. The results of the latter study challenged the common wisdom that the scaling state dynamics does not depend on the initial condition, suggesting as they did, that the long-time topological distribution function, while of stable form, differed from that for generic initial conditions (random 2D froth). Not in standing with, considerable controversy has been initiated [8], [9] and subsequent computer simulations have been carried out suggesting that more conventional ideas are more likely correct [10], [11] ,[12], [13]. In the present work, we report an experimental study of such perfectly triangular lattices, as well as ones containing a dislocation defect.

2 Experimental method

The simulations discussed above concern dry 2D froth. While it is possible to make 2D froth which is nearly dry [14], it appears essentially impossible to create perfectly ordered froth. Two-dimensional arrays of bubbles on the surface of soap solutions have been exploited as useful models of various aspects of condensed matter physics, following the pioneering work of Bragg [15], [16]. We have adapted these bubble rafts to permit formation of perfectly ordered 2D foam. Fortes et al. [17], have demonstrated that bubble rafts trapped between the surface of a soap solution and a glass cover plate endure essentially indefinitely, temporal evolution being restricted to that due to coarsening. Restricting such 2D foam to a hexagonal shape helps enforce the desired symmetry. The glass cover is supported just 2 mm above the soap solution on a metal plate which contains a hexagonal hole (typically 6 cm on a side). The plate extends into the solution and 2D foam is formed by bubbling N2 into the solution below this hexagonal cell via a long hypodermic needle. By systematically sweeping the tip of the needle to and fro as the lines of bubbles form it is possible, with practice, to create perfectly 6-fold

* Corresponding author: a.abdelkader@ajman.ac.ae

coordinated lattices comprising several thousand bubbles about 2 mm in diameter. A dislocation is introduced by interrupting the process when the lattice is part made, and then completing the lattice as shown in Fig. 1. This process has been repeated several times using hexagonal frame with elastic boundaries, less wet bubbles, and also by changing the depth between the glass cover and the soap solution. The 2D foams produced as above are of course wet, but we may hope that their behavior may reflect at least some of the generic aspects of the evolution of 2D froth as revealed in the simulations [10], [11].

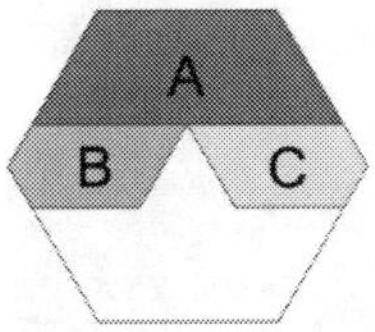

Fig. 1. A sketch outlining the procedure for creating a defect. A crystalline lattice is first made in region A, then regions B and C are filled with bubbles in register with those in A, leaving a gap. This fills in as the remaining space is filled with bubbles, creating a dislocation.

3 Results and discussion

Most of the data shown are bubbles in the order 2 mm in diameter in a cell 6 cm on a side. We mainly consider the effects of boundaries of the system as well as the effect of the wetness of the foam.

The ideal 6-fold coordinated 2D foam shows no great change for the stability of the foam a period of order 25 minutes and then breaks, creating an area that lacking any bubbles as shown in Fig. 2.

(a) (b)

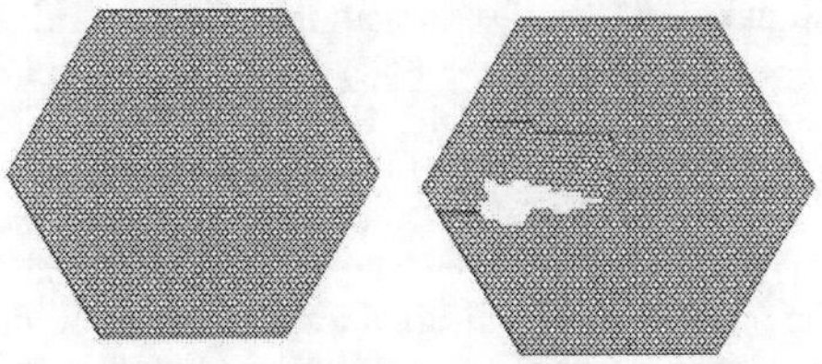

Fig. 2. Pictures of ideal 2D foam. (a) The 2D foam as formed (t = 0) and (b) t = 25 minutes after creation.

During the rupture, the number of bubbles is conserved. The system is clearly under tension and the breakage consists of a crack appear between a few bubbles which then spread rapidly to form a large and ultimately stable hole in the foam. When the boundary of the cell is replaced by more-or-less elastic one, the lifetime of the cell has increased up to 100 minutes as shown in Fig. 3. When the dry bubbles are used, a relatively long lifetime is indicated. Experiments with hexagonal cells that are 3 cm and 10 cm on a side indicate that the system size has no effect upon the phenomena.

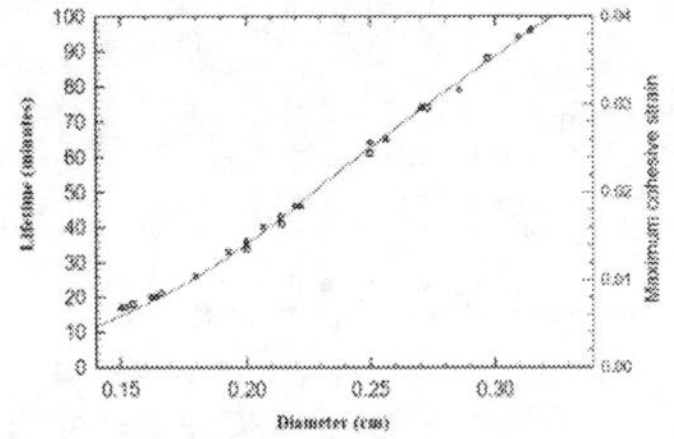

Fig. 3. Variation of lifetime of perfect foam with bubble diameter in hexagonal cells of different sizes using elastic boundary. The line represents the variation of the cohesive strain with the bubble diameter.

However, the lifetime to breakage of the perfect foam increases systematically with bubble size comparing with that obtained by using a hard boundary and wet foam as shown in Fig. 4.

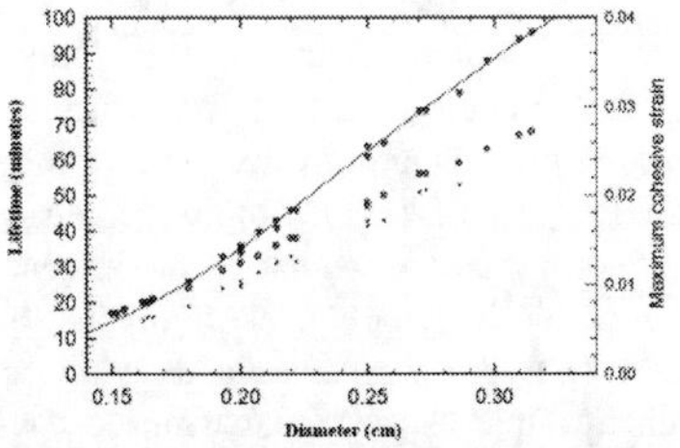

Fig. 4. Variation of lifetime of perfect foam with bubble diameter in hexagonal cells of different sizes for elastic and hard boundary. The line represent the variation of the cohesive strain with the bubble diameter.

Optional introduction of a dislocation into 2D foam as shown in Fig. 5, (either located close to a boundary or more centrally), drastically alters the system's evolution. These differences in behaviour were completely unexpected. At this stage of manuscript preparation, authors are not aware of any other experimental studies of the long-time behaviour of ideal 2D foam including defects. It is suggested that stabilization arises from the deformation of the bubbles in response to the stress field surrounding the defect. The quadratic dependence of the surface energy of the bubbles on the deformation leads to a contribution to the system energy which increases as the system incorporating a defect attempts to approach equilibrium packing. Turning to the 2D foam containing a single defect, the results appear to be independent of system size, at least over the entire range indicated as above. A typical example of the evolution of a single defect in an otherwise regular hexagonal lattice is shown in Fig. 5.

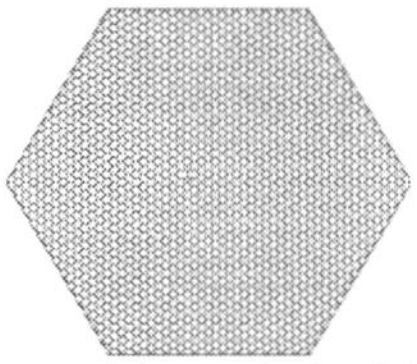

Fig. 5. 2D foam containing a single dislocation as made where no breakage has occurred.

The initial cluster grows as the disorder increases around the initial defect due to coarsening. Essentially no observable changes occur in the system for 12-15 hours, about which time the first increase in disorder around the defect appears.

We study the evolution of a 'cluster', defined as all bubbles having at least one neighbour, which is not 6fold coordinated [7], [10]. We state here that in the first stage, the whole bubbles in the cluster are having six-coordinates. The stability of the foam has been followed by investigating the evolution of certain statistics of the cluster. These include the distribution of the so-called topological classes (bubble coordination numbers) and its second moment. The tail of the topological distribution function, P(n), was extended towards larger values of n as time progressed, while the peak of the distribution is kept at n = 6 (Fig. 6). Turning into quantitative measures, the second moment of the topological class distribution of the cluster (μ2) grew monotonically with time (Fig. 7), reaching rather larger values when less wet bubbles is used as well as a hard boundaries. Here we compare current data with the data obtained when we used wet bubbles and an elastic boundary. We found that (μ2) dropped to lower value when we used soft boundaries. On the other hand, higher values of (μ2) are reported when we use less wet bubbles as expected. Paying attention to the perfect foam, an answer to the ordinal question of why the diameter and the wetness of the bubbles affected the lifetime should be provided. In that respect, we consider our ideal foam as perfect in the sense that all bubbles are 6-fold coordinated even though it should be noted that these bubbles must also have small variations in both size and separation. The latter lead to small readjustments of the positions of the bubbles in which the system locally seeks to reach equilibrium at r = r0 (Fig. 8). The bubbles thus become closer in some places and more separated in others. In that sense, the global energy of the system must reduce. As this process continues, the local stresses accumulate at various points, of which few are becoming weak. As the local strain reaches a critical level, the cracks appear in this area just one or two minutes before the breakage of the foam. This critical strain is just the strain at which the stress (force/bubble) is maximum

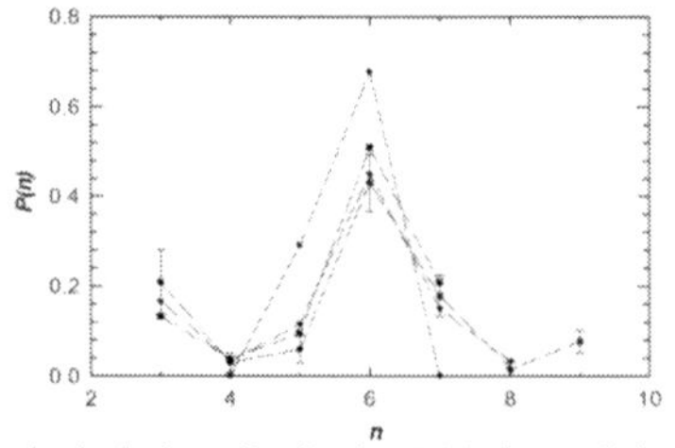

Fig. 6. Topological class distribution P(n) for evolving foam containing an isolated defect in several experiments.

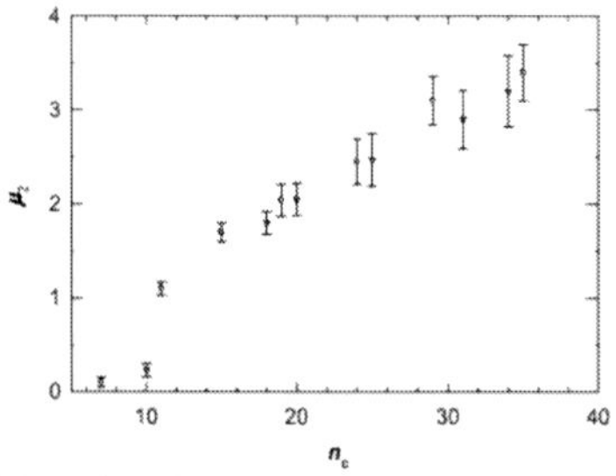

Fig. 7. Variation of μ_2, the second moment of the cluster with nc as different foams (different symbols) containing a dislocation evolve.

If the internal tension exceeds this value, the lattice of bubbles will rupture. This corresponds to the cohesive strain of the system (defined as the ratio of the separation between r0 and that r corresponding to maximum resistive force (i.e. maximum δV(r)/δr) and the bubble diameter), which increases monotonically with bubble size [19]. The cohesive strain between two bubbles was defined as the ratio of the critical separation distance, between the position of initial equilibrium and the position of maximum resistive force on the pair force-displacement law, to the diameter of the free bubble. Here we consider the dependence of the lifetime of the ideal foam upon the bubble size. We mentioned that the lifetime of the ideal foam increases systematically with the bubble size, in qualitative agreement with the variation of the cohesive strain. When the bubbles become larger, the breakage of the foam requires a large strain. A long time is needed for the local strain to accumulate to such a large value, so the lifetime increases. It is difficult to compute the cohesive strain in the present case quiet precisely. In addition, to our knowledge, the separation between the bubbles in the initial foam remain unjustified however it should rather exceeds ro. By far, the potential which is computed by Nicolson et. al [18] and Lomer et.al [20] is for a bubble raft with a free surface, where the top of each bubble being a spherical cap, whereas in our case the bubbles are pushed up against the flat cover glass, which the soap films must meet at 90°. The boundary conditions are thus not those used by previous workers. However, for the present approximate calculation we use Nicolson's potential. Lomer's more correct form [20] differs only in the repulsive contribution, which is not very significant at r rather above r0, and is less simple to use. For a range of bubble sizes, we computed r0 and that r corresponding to maximum inter-bubble force, to calculate the cohesive strain. The general trend of the plot of the cohesive strain against the diameter of the bubbles shows very good general agreement with the variation of our lifetime data (see Fig. 3, 4). The displacement between the theoretical line and some of the experiment data may be due to the differences of our system from that treated by Nicolson discussed above. The lifetime is defined as the time required to accumulate strain to the maximum cohesive strain. In our result, the latter is not found to depend on the thickness of the air gap between soap solution and the cover glass. However, the violence of the foam breakage is observed to increase when the gap is reduced. Moreover, we deduce that the force on the bubbles depends on the air-gap, while the maximum cohesive strain does not. Stated-other-wise, the meniscus must meet the cover plate at 90° and become horizontal at large distances. The inter-bubble potential depends on the height of the meniscus. Thus, the force proportional to dV(r)/dr, must increase accounting for the more curved meniscus for the thinner air gap.

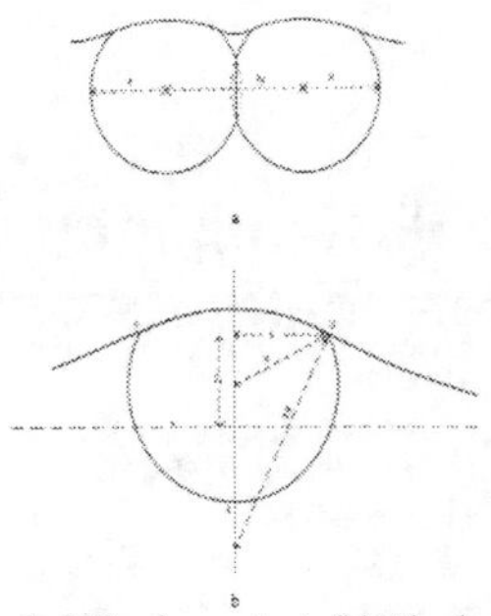

Fig. 8. (a) Two bubbles in contact. (b) Single bubble floating on liquid PQ is the ring of contact of radius b. The free surface is indicated by the broken line. (From [18]).

4 CONCLUSIONS

We show that arrays of bubbles constrained within a rigid cell are metastable with respect to breakage (loss of cohesion) if they are initially ideally six-fold coordinated. However, when the rigid cell replaced by more elastic one and when the foam is less wet, the lifetime increases relatively. The significant increasing in the lifetime is related to the elastic stress fields in the cell. Such foams are stabilized by the introduction of a single point defect, the temporal evolution that associated with disproportionate of gas between small and large bubbles. The evolution of a single defect in an isotropic hexagonal lattice has been investigated experimentally for comparison with previous study. In general, our results qualitatively support results from previous work.

References

1. Stavans and J.A. Glazier et al., Phys. Rev. Lett. 62. 1318 (1989).
2. J.A. Glazier, S.P. Gross, and J. Stavans, Phys. Rev. A 36, 306 (1987).
3. Stavans and J.A. Glazier et al., Phys. Rev. Lett. 62. 1318 (1989).
4. H. Flyvbjerg and C. Jeppesen, Phys. Scr. T38, 49 (1991).
5. V.E. Fradkov, L.S. Shvidlerman, and D.G. Udler, Philos. Mag. Lett. 55, 289 (1987).
6. B. Levitan, M.Sc. thesis (1992).
7. B. Levitan, Phys. Rev. Lett. Volume 72, No. 26, 4 057 (1994).
8. Weaire, D., Phys. Rev. Lett. 74: 3710 (1994).
9. Sire, C. Phys. Rev. Lett. 74: (1995) 3708.
10. Jiang, Y., Mombach, M., & Glazier, J. A., Phys. Rev. E 52: (1995) 3333-3338.
11. Ruskin, H. J., Feng, Y., J. Phys. Condens. Matter 7: (1995) L553-562.
12. Levitan, L., Domany, E., Phys. Rev. E 54: (1996) 27662771.
13. Chae, J. J., Tabor, M., Phys. Rev. E 55: (1997) 598-610. T38, 49 (1991).
14. Stavans, J., Phys. Rev. A 42: (1990) 5049-5053.
15. Bragg, W. L., Proc. Phys. Soc., 52: (1940) 54-56.
16. Bragg, W. L. J. Sci. Instrum., 19: (1942) 148-150.
17. Fatima Vaz, M., Fortes, M. A. J. Phys.: Condense Matter 9: (1997) 8921-8930.
18. Nicolson, Mathematical proceedings of the Cambridge Philosophical Society, Math. phys. Sci., 45: (1949) 288-293.
19. Shi, L. T., Argon, A. S., Philosophical Magazine A 46: (1982) 275-286.
20. Lomer, W., Proc. Royal Soc. London Series A, 196: (1949) 182-194.

Variable sampling interval run sum $\bar{X}$ chart with estimated process parameters

Sajal Saha[1,*], *Michael* Boon Chong Khoo[1], *Peh* Sang Ng[2] and *Mahfuza* Khatun[1]

[1] School of Mathematical Sciences, Universiti Sains Malaysia, 11800 Penang, Malaysia
[2] Department of Physical and Mathematical Science, Faculty of Science, Universiti Tunku Abdul Rahman, 31900 Kampar, Perak, Malaysia

Abstract. The $\bar{X}$ type control chart is often evaluated by assuming the process parameters are known. However, the exact values of process parameters are hardly known and thus Phase-I dataset is needed to estimate them. In this paper, the performance of the variable sampling interval run sum $\bar{X}$ chart with estimated process parameters is evaluated by using the performance measure of the average of the average time to signal (AATS) and the optimal design of the proposed chart in minimizing the out-of-control AATS is developed. The performance measure of the standard deviation of the average time to signal (SDATS) is then used to identify the number of Phase-I samples (w) needed to have an in-control AATS performance close to its known process parameter case. Results show that large w is needed to minimize the performance gap between known and unknown process parameters cases of the VSI RS $\bar{X}$ chart.

1 Introduction

The $\bar{X}$ chart is the first control chart invented by Shewhart in 1924 for process monitoring [1]. However, the $\bar{X}$ chart is not sensitive in signaling an out-of-control, when small and moderate mean shifts are detected. Thus, the more advancement control chart such as run sum (RS-$\bar{X}$) chart was proposed.

RS-$\bar{X}$ chart was developed by [2]. It is also well known as a zone chart, where the control chart is segregated into many regions according to the preference of the practitioners. The run sum chart is shown to have better performance than the Shewhart charts with and with-out run-rules [3]. In addition, the performance of the run sum chart is comparable with the performance of the exponentially weighted moving average (EWMA) and the cumulative sum (CUSUM) charts when the number of regions of the chart is increased [4]. To enhance the performance as well as to simplify the design of the run sum chart, [5] integrated the fast initial response (FIR) feature into the run sum chart with one parameter. They pointed out that the proposed chart is competitive with the CUSUM, FIR CUSUM and Shewhart FIR CUSUM charts. To have a quick detection of the changes of the process mean, [6] employed the variable sampling interval (VSI) schemes on the RS-$\bar{X}$ control charts. They showed that their proposed charts outperform the RS-$\bar{X}$ chart for all of the shift sizes under steady state and zero state modes, respectively. However, all these mentioned run sum-type charts assumed the process parameters are known in monitoring process. However, without such assumption, an in-control Phase-I dataset is needed by the practitioners to estimate the process parameters. Thus, numerous research studied the impact of the estimated process parameters on the performance of control charts.

Some of the contemporary works of estimated parameters based control charts are discussed. [7] evaluated the impact of estimated control limits on the adaptive $\bar{X}$ chart, where the results showed that the run length performance are significantly impacted by the parameter estimation. [8] investigated the process parameter estimation on the double sampling S^2 chart. To address the variability in run length performance and practitioner to practitioner variation in estimating process parameters, [9] employed the performance measures of AARL and SDARL to evaluate the RS-$\bar{X}$ chart with estimated process parameters.

To the best of our knowledge, the study on the impact of estimated process parameters on the variable sampling interval run sum $\bar{X}$ chart (VSI RS-$\bar{X}$) chart has not been proposed in the literature. Therefore, this paper proposes the VSI RS-$\bar{X}$ chart with estimated process parameters-based on the criteria of average of the average time to signal (AATS) and standard deviation of the average time to signal (SDATS). The rest of the paper is organized as follows: Section 2 demonstrates the optimal design of the VSI RS-$\bar{X}$ chart with estimated process parameters in minimizing the out-of-control average of the average time to signal (AATS_1). The optimal charting parameters and the performance of the chart under known and unknown process parameters are studied in Section 3. Lastly, conclusions and recommendations are summarized in Section 4.

* Corresponding author: sajal.saha@iubat.edu

2 VSI RS-$\bar{X}$ chart with estimated process parameters

When a chart consists of the unknown process parameters such as the mean (μ_0) and standard deviation (σ_0), an in-control Phase-I dataset is needed to estimate them. By assuming a Phase-I dataset consists of w samples, where each of the sample consists of size n such as $\{X_{i,1}, X_{i,2}, \ldots, X_{i,n}\}$ for $i = 1, 2, \ldots, w$. By further considering the Phase-I data are normally distributed with μ_0 and σ_0, i.e. $X_{i,j} \sim N(\mu_0, \sigma_0^2)$ and there is independence within and between samples. The commonly used estimator for μ_0 and σ_0 are defined as ([10])

$$\hat{\mu}_0 = \frac{1}{wn} \sum_{i=1}^{w} \sum_{j=1}^{n} X_{i,j} \tag{1}$$

and

$$\hat{\sigma}_0 = \sqrt{\frac{1}{w(n-1)} \sum_{i=1}^{w} \sum_{j=1}^{n} \left(X_{i,j} - \bar{X}_i \right)^2}, \tag{2}$$

respectively, where $\bar{X}_i = \frac{1}{n} \sum_{j=1}^{n} X_{i,j}$ is the sample mean of ith sample.

In the Phase-II process, let $Y_{i,1}, Y_{i,2}, \ldots, Y_{i,n}$ be the n observations of sample i and we further assume that $Y_{i,j}$ follows a normal $N(\mu_0 + \delta\sigma_0, \sigma_0^2)$ distribution, where $j = 1, 2, \ldots, n$. Note that, $\delta = 0$ and $\delta \neq 0$ indicate the process is statistically in-control and out-of-control, respectively.

Fig. 1 depicts the M regions above the centre line (CL), i.e. $\hat{R}_1, \hat{R}_2, \ldots, \hat{R}_{(M-1)}, \hat{R}_M$ and M regions below the CL, i.e. $\hat{R}_{1b}, \hat{R}_{2b}, \ldots, \hat{R}_{(Mb-1)}, \hat{R}_{Mb}$ of the VSI RS-$\bar{X}$ chart with estimated process parameters. The M regions above the CL are allocated with positive integer scores, $0 \le +S_1 \le +S_2 \le \ldots \le +S_K$, while the M regions below the CL are allocated with negative integer scores, $-S_M \le \ldots \le -S_2 \le -S_1 \le 0$. The interval of the upper regions ($\hat{R}_H$) and lower regions ($\hat{R}_{Hb}$) of the CL for the chart are shown as

$$\hat{R}_H = [\hat{\mu}_0 + A_{H-1}\hat{\sigma}_0, \hat{\mu}_0 + A_H\hat{\sigma}_0) \tag{3a}$$

and

$$\hat{R}_{Hb} = [\hat{\mu}_0 - A_H\hat{\sigma}_0, \hat{\mu}_0 - A_{H-1}\hat{\sigma}_0) \tag{3b}$$

where, $A_0 = 0$, $A_M = \infty$ and $A_H = \frac{k}{\sqrt{n}} \left[\frac{3H}{q-1} \right]$ for H = 1, 2,..., $M-1$, and k is the constant chosen to set the value of in-control average of the average time to signal (AATS_0).

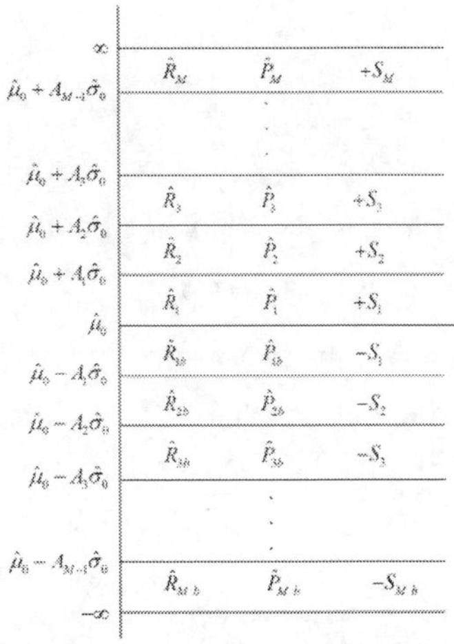

Fig. 1. The regions (M) with the corresponding scores and probabilities of the VSI RS-$\bar{X}$ chart with estimated process parameters.

The estimated process parameter-based VSI RS-$\bar{X}$ chart reacts based on the upper cumulative scores (U_i) and lower cumulative scores (L_i), defined as

$$U_i = \begin{cases} 0 & \text{if } \bar{Y}_i < CL \\ U_{i-1} + S(\bar{Y}_i) & \text{if } \bar{Y}_i \ge CL \end{cases}, \tag{4a}$$

$$L_i = \begin{cases} 0 & \text{if } \bar{Y}_i \ge CL \\ L_{i-1} + S(\bar{Y}_i) & \text{if } \bar{Y}_i < CL \end{cases}, \tag{4b}$$

where $i = 1, 2, \ldots$ and $H = 1, 2, \ldots M$, while

$$S(\bar{Y}_i) = \begin{cases} S_H & \text{if } \bar{Y}_i \in \hat{R}_H \\ -S_H & \text{if } \bar{Y}_i \in \hat{R}_{Hb} \end{cases} \tag{5}$$

The chart signals an out-of-control signal if $U_i \ge +S_M$ or $L_i \le -S_M$, where, $+S_M$ and $-S_M$ act as the triggering scores for the M regions above CL and M regions below CL, respectively.

For simplicity, we consider the initial value $U_0 = L_0 = 0$ for the estimated parameters-based VSI RS-$\bar{X}$ chart. Let t_i be the sampling interval of selecting the next sample of the estimated VSI RS-$\bar{X}$ chart based on the cumulative scores values, (i.e. U_i and L_i), then t_i is defined as:

$$t_i = \begin{cases} d_1 & \text{if } S_M/D \le U_i < S_M \text{ or } -S_M < L_i \le -S_M/D \\ d_2 & \text{if } 0 \le U_i < S_M/D \text{ or } -S_M/D < L_i \le 0 \end{cases}, \tag{6}$$

where D is the positive integer decided by user to control the threshold limit, while d_1 and d_2 denote the short and long sampling intervals, respectively, Note that $d_2 > d_1 > 0$. The probability of $\bar{Y}_i$ falls in the region $\hat{R}_H$ is calculated as

$$\hat{P}_H = \Pr\left(\bar{Y}_i \in \hat{R}_H\right)$$
$$= \Phi\left(\left(\hat{\mu}_0 - \mu_0\right)\frac{\sqrt{n}}{\sigma_0} + A_H\frac{\hat{\sigma}_0\sqrt{n}}{\sigma_0} - \delta\sqrt{n}\right) - \Phi\left(\left(\hat{\mu}_0 - \mu_0\right)\frac{\sqrt{n}}{\sigma_0} + A_{H-1}\frac{\hat{\sigma}_0\sqrt{n}}{\sigma_0} - \delta\sqrt{n}\right) \quad (7)$$

By letting $U = (\hat{\mu}_0 - \mu_0)(\sqrt{n}/\sigma_0)$ and $V = (\hat{\sigma}_0\sqrt{n}/\sigma_0)$, $\hat{P}_H$ is simplified as

$$\hat{P}_H = \Phi\left(U + A_H V - \delta\sqrt{n}\right) - \Phi\left(U + A_{H-1}V - \delta\sqrt{n}\right) \quad (7a)$$

Similarly, the probability of $\bar{Y}_i$ falls in the region $\hat{R}_{Hb}$ can be computed as

$$\hat{P}_{Hb} = \Phi\left(U - A_{H-1}V - \delta\sqrt{n}\right) - \Phi\left(U - A_H V - \delta\sqrt{n}\right) \quad (7b)$$

Here, the probability density functions (pdf) of U and V can be obtained from [9].

The performance measures of AATS and SDATS are used to evaluate the performance of the estimated process parameter-based VSI RS-$\bar{X}$ chart. The AATS of the chart is computed as ([11])

$$\text{AATS}(\delta) = \int_{-\infty}^{+\infty}\int_{0}^{+\infty} \widehat{\text{ATS}} f_{(U)}(u \mid w) f_V(v \mid w, n)\, dv\, du, \quad (8)$$

where $\widehat{\text{ATS}}$ denotes the average time to signal of the estimated case, and it is defined as ([6])

$$\widehat{\text{ATS}} = \boldsymbol{q}^T\left(\boldsymbol{I} - \hat{\boldsymbol{Q}}\right)^{-1}\boldsymbol{t} - \boldsymbol{q}^T\boldsymbol{t} \quad (9)$$

Note that $\hat{\boldsymbol{Q}}$ is a $p \times p$ transition probability matrix of estimated process parameters-based VSI RS-$\bar{X}$ chart. where, the value of p depends on the scores combinations and [9] described the procedure of finding $\hat{\boldsymbol{Q}}$. The column vector $\boldsymbol{t}$ consists of the elements t_i, for $i = 1, 2, ..p$, determined by Eqs (6), while, $\boldsymbol{q}^T = (1, 0, 0, ..., 0)$ is a initially probability vector.

The SDATS is defined as ([11])

$$\text{SDATS}(\delta) = \sqrt{E\left(\widehat{\text{ATS}}^2\right) - \left[E\left(\widehat{\text{ATS}}\right)\right]^2}, \quad (10)$$

Where

$$E[\widehat{\text{ATS}}^2] = \int_{-\infty}^{+\infty}\int_{0}^{+\infty} \widehat{\text{ATS}}^2 f_{(U)}(u \mid w) f_V(v \mid w, n)\, dv\, du.$$

Consequently, to minimize the out-of-control AATS (AATS_1) of the estimated process parameter-based VSI RS-$\bar{X}$ chart, the associated optimal algorithm is mathematically expressed as

$$\underset{(k, d_2, S_1, S_2, ..., S_M)}{\text{Minimize}} \ \text{AATS}_1(\delta) \quad (11)$$

subject to the constraints of in-control AATS = AATS_0 and in-control average sampling interval $\text{ASI}_0 = t_0$.

3 Numerical analysis

In this section, the methodology presented in Section 2 is used to evaluate the performance of the VSI RS-$\bar{X}$ chart when the process parameters are estimated from Phase-I samples. Table 1 shows the optimal charting parameters $\{k, d_2(S_1, S_2, ..., S_M)\}$ in minimizing the AATS_1 with respect to the constraints of $\text{AATS}_0 = 370$ and $\text{ASI}_0 = 1$, by considering the combinations of $n = 5$, $d_1 = 0.01$, $M = 4$, $D = 4$, $\delta \in \{0.2, 0.4, 0.6, 0.8, 1, 1.5, 2\}$ and $w \in \{20, 40, 80\}$.

Table 2 provides the optimal AATS_1 and the associated out-of-control SDATS (SDATS_1) values of the estimated VSI RS-$\bar{X}$ chart for $w \in \{20, 40, 80\}$ corresponding to the optimal parameters given in Table 1 and the out-of-control ATS (ATS_1) values for known parameters case (i.e. $w = \infty$). The values of the AATS_1 and SDATS_1 are computed using Eqs. (8) and (10). For the chart performance to be at acceptable level, [12] suggested that the SDATS value should not more than 10% of the ATS value. The details can refer to [9]. From Table 2, we observe most of the SDATS_1 values are greater than its corresponding 10% of the ATS_1 values. Furthermore, the chart's performance based on the estimated process parameters is significantly different compared to its known parameters counterpart, especially when the values of δ and w are small.

Table 1. Values of optimal parameters of the 4 regions estimated process parameter-based VSI RS-$\bar{X}$ chart when $\text{AATS}_0 = 370$, $d_1 = 0.01$ and $D = 4$.

		$w = 20$						$w = 40$						$w = 80$					
n	δ	k	d_2	S_1	S_2	S_3	S_4	k	d_2	S_1	S_2	S_3	S_4	k	d_2	S_1	S_2	S_3	S_4
5	0.2	1.2336	1.5089	0	2	5	8	1.2196	1.5496	0	2	3	7	1.2103	1.5695	0	2	3	7
	0.4	1.2336	1.5089	0	2	4	7	1.2256	1.5418	0	2	4	7	1.217	1.5605	0	2	4	7
	0.6	1.2336	1.5089	0	2	5	8	1.2256	1.5418	0	2	5	8	1.217	1.5605	0	2	4	7
	0.8	1.2336	1.5089	0	2	4	7	1.2256	1.5418	0	2	5	8	1.217	1.5605	0	2	4	7
	1.0	1.1665	1.6008	0	2	3	8	1.154	1.6385	0	2	3	8	1.1432	1.6616	0	2	3	8
	1.5	1.1665	1.6008	0	2	3	8	1.154	1.6385	0	2	3	8	1.1432	1.6616	0	2	3	8
	2.0	1.1593	1.6131	0	2	2	7	1.1458	1.6524	0	2	2	7	1.1344	1.6763	0	2	2	7

Table 2. $AATS_1$ and $SDATS_1$ values of the 4 regions estimated process parameter-based VSI RS-$\bar{X}$ chart when $AATS_0 = 370$, $d_1 = 0.01$ and $D = 4$.

n	δ	$w = 20$		$w = 40$		$w = 80$		$w = \infty$
		$AATS_1$	$SDATS_1$	$AATS_1$	$SDATS_1$	$AATS_1$	$SDATS_1$	ATS_1
5	0.2	132.705	222.402	95.851	104.934	76.323	54.059	58.350
	0.4	16.157	30.004	11.871	9.729	10.313	4.954	8.949
	0.6	3.131	2.882	2.7395	1.370	2.565	0.829	2.387
	0.8	0.100	0.598	0.934	0.357	0.898	0.234	0.856
	1.0	0.384	0.204	0.362	0.130	0.350	0.087	0.335
	1.5	0.037	0.021	0.034	0.014	0.033	0.009	0.031
	2.0	0.003	0.002	0.003	0.001	0.003	0.001	0.002

Table 3 presents the values of $AATS_0$ and in-control SDATS ($SDATS_0$) for different number of w with the optimal combinations $\{k, d_2, (S_1, S_2, S_3, S_4)\}$ ={1.1121, 1.7095, (0, 1, 2, 4)} and {1.1932, 1.5964, (0, 2, 3, 7)} for ATS_0=200 and 370, respectively. The purpose of this analysis is to identify the number of w needed to generate the $AATS_0$ values that close to its known process parameters case ($w = \infty$), where the corresponded $SDATS_0$ should not exceed 10% of the ATS_0. For example, at least $w = 1000$ Phase-I samples are needed to attain $SDATS_0$ within 10% of the associated ATS_0, when $AATS_0$ performance of the known process parameter-based VSI RS-$\bar{X}$ is 200 (see $w = \infty$). The bolded values in Table 3 are the values of $SDATS_0$ that fall within 10% of the associated ATS_0 values, and the corresponding w are suggested, so that the run length performance between the known and unknown process parameters cases is acceptable.

Table 3. $AATS_0$ and $SDATS_0$ values of the 4 regions estimated process parameter-based VSI RS-$\bar{X}$ chart for different number of Phase-I samples w, $ATS_0 \in \{200, 370\}$, $d_1 = 0.01$, $D = 4$ and $n = 5$.

	$ATS_0 = 200$		$ATS_0 = 370$	
w	$AATS_0$	$SDATS_0$	$AATS_0$	$SDATS_0$
100	185.080	62.532	336.256	124.893
300	193.913	35.690	355.630	71.301
500	196.186	27.481	360.825	54.831
700	197.233	23.145	363.249	46.133
900	197.835	20.366	364.655	40.564
1000	198.050	**19.305**	365.157	38.438
1100	198.227	18.393	365.572	**36.614**
1150	198.304	17.983	365.754	35.793
1200	198.375	17.600	365.921	35.025
$w = \infty$	200	0	370	0

4 Conclusions

The impacts of the process parameters estimation on the AATS and SDATS of the VSI RS-$\bar{X}$ chart are investigated in this paper. As shown in numerical analysis, the $AATS_0$ and $AATS_1$ performances of the estimated process parameter-based VSI RS-$\bar{X}$ chart tend to similar to its known process parameters case when large w is considered. Thus, based on the performance of SDATS, the appropriate number of Phase-I samples required for the parameter estimation is recommended. While, this work focuses on univariate VSI RS chart, in future, the multivariate charts can be investigated under the assumption of estimated process parameters.

This research is supported by the Universiti Sains Malaysia (USM) Fellowship and funded by USM School of Mathematical Sciences (KPI MYRA PTJ), Grant number : 1001.PMATHS.AUPS001

References

1. D.C. Montgomery, *Introduction to Statistical Quality Control* (New York : John Wiley & Sons, 2012)
2. S.W. Roberts, Dent Tech., **8**, 3 (1996)
3. R.B. Davis, A. Homer, W.H. Woodall, Commun. Stat -Theory Methods, **19**, 5 (1990)
4. C.W. Champ, S.E. Rigdon, J. Qual. Technol., **29**, 4 (1997)
5. C.A. Acosta-Mejia, L. Rincon, Commun. Stat -Theory Methods, **43**, 20 (2014)
6. X.Y. Chew, M.B.C. Khoo, S.Y. Teh, P. Castagliola, Comput. Ind. Eng., **90**, (2015)
7. M. Zhou, Oper. Res., **17**, 1 (2017)
8. P. Castagliola, P.C. Oprime, M.B. Khoo, Commun. Stat-Theory Methods, **46**, 7 (2017)
9. S. Saha, M.B.C. Khoo, W.L. Teoh, M.H. Lee, Qual. Reliab. Eng. Int., **33**, 8 (2017)
10. Y. Zhang, P. Castagliola, Z. Wu, M.B.C. Khoo, Qual. Reliab. Eng. Int. **28**, 1 (2012)
11. N.A.Saleh, M.A. Mahmoud, M.J. Keefe, W.H. Woodall, J. Qual. Technol. **47**, 2 (2015)
12. M. Zhang, F.M. Megahed, W.H. Woodall, Qual. Reliab. Eng. Int. **30**, 2 (2014)

The research of hidden Markov models for overall equipment effectiveness analysis in smart manufacturing system

Chia-Yu Hung* and *Chen*-Yang Cheng

Industrial Engineering of Management, National Taipei University of Technology, Taipei, Taiwan

Abstract. In the manufacturing industry, excellent product quality and increased production flexibility can be achieved by eliminating waste and improving production efficiency. In the past, the manufacturing industry used manual records of production information, but this method is characterized by low efficiency and high error rates. Even if a programmable logic controller and radio-frequency identification are employed, problems still occur because of constraints such as different machine types and high costs. The use of a cyber–physical system and information visualization requires the collection of manufacturing information in order to facilitate the analysis of manufacturing data. Monitoring the machining status. This study proposes an approach for segmenting machine-processed signals. With plug-and-play noninvasive current-sensing equipment to collect machine production information, this approach can immediately determine the state of the manufacturing process and calculate the machine utilization, machine production cycle, and production quantity. The goal is to enable the use of this method with this equipment, improve machine utilization, instantly identify the production quantity, and reduce equipment idle time to reduce manufacturing waste, thus rendering production management more convenient and faster.

1 Introduction

The market demand for customized products, which requires high-quality, low-cost products and fast delivery, the manufacturing industry must ensure that the operating time of machines exhibits more flexibility, adaptability, and intelligent decision-making. Features of Industry 4.0 must be adapted in the manufacturing industry, one of which is cyber–physical systems (CPSs). A CPS combines the network information world with the real world of physics. Therefore, a CPS must collect information to effectively integrate an actual situation, and then through information visualization (infovis), it can transform abstract data into an intuitive presentation. Infovis can transform the data into valuable information. Using infovis in factory management indicators and applying it in plant management can help the plant to effectively analyze manufacturing problems. This will increase flexibility and adaptability and thus facilitate achieving high-quality, low-cost products and fast delivery. In factory management, overall equipment efficiency (OEE) is the most commonly used indicator to measure the operation of production equipment. It simplifies complex problems in manufacturing and reformulates them through simple indicators. Using OEE to measure equipment operation is the most effective method of analyzing the efficiency of a manufacturing system. In this approach, the overall efficiency of a factory is analyzed in terms of equipment availability, performance, and quality.

The manufacturing process usually involves monitoring the state of a machine and the progress of production to improve efficiency. In the past, this information was obtained orally or was manually copied onto paper, and these methods are inefficient and have a high error rate. With technological progress, the demand for automatic production monitoring systems, such as the widely used programmable logic controllers (PLCs). However, a PLC is limited to different types of machines and different operating systems. The processing of information in a production line requires a considerable amount of money and time. Consequently, many plant managers use radio-frequency identification (RFID). However, this method is excessively expensive, because it requires the use of a large number of tags. In addition, the message between the tag and the reader can be easily truncated and has a limited usage range and distance.

In general, factories use an electricity consumption-based monitoring approach, because production equipment typically accounts for the majority of electricity consumption in factories. In this approach, current-sensing devices are used to noninvasively monitor production equipment on the basis of current signals; hence, this approach is not limited by the aforementioned constraints.

This paper presents an approach for analyzing machine status signals, and it can obtain machine production information using noninvasive current-sensing devices. This approach can distinguish between the signals produced during processing and can yield information on different statuses, in addition to calculating equipment

* Corresponding author: claire830110@gmail.com

availability, processing period, and production quantity. The objective of the proposed approach is to achieve a "plug-and-play" function, improve machine utilization, identify production quantity immediately, reduce equipment idle time, and manage production progress instantly, thus reducing resource wastage in the manufacturing process. This approach increases the speed and convenience of production management.

2 Related work

OEE is a simple and practical production management tool used in the manufacturing industry to measure actual and theoretical production capacity levels by analyzing equipment availability, performance, and quality. OEE facilitates identifying the cause of equipment-related problems. Ljungberg [1] stated that in total productive maintenance (TPM), the severity of the loss of different pieces of production equipment must be assessed to optimize resource allocation. In the manufacturing industry, collecting production site data is essential for the real-time monitoring of production conditions. With technological progress, the demand for automatic production monitoring systems, such as the widely used PLCs, has increased. Lin, Lin [2] developed a PLC system for automatically monitoring the quality of eggshells; this system can identify abnormal eggs with a success rate of >80%. In addition to PLCs, some manufacturing systems employ RFID. A tag is attached to each product, and an RFID reader is subsequently used to receive the tag information; thus, information on the product can be obtained immediately to collect production data. Wang, Cao [3] used RFID to obtain accurate production data for monitoring manufacturing systems and to obtain data from RFID tags and readers in order to analyze production information.

Signals can be analyzed through various methods; the most common methods are described herein. A wavelet transform is a commonly practiced time–frequency analysis method. This approach entails using a series of fundamental wavelet transforms to simultaneously analyze the time domain of the original signal, as well as the frequency domain [4]. The fundamental wave of the wavelet transform exhibits the characteristics of elastic time and frequency; this wave widens at low frequencies and yields time-domain information. Conversely, the fundamental wave narrows at high frequencies and yields frequency-domain information. Gritli, Zarri [5] used a wavelet transform-based diagnostic method to detect faults in three-phase wound rotor motors. Through the FT, the signal is decomposed into several sine wave frequencies. In other words, the FT is the sum of many sine functions. Therefore, the FT is often used for spectral analysis [6]. Peng, Chu [7] applied the fast FT (FFT) and wavelet transform to vibration signals and reported that the time–frequency characteristic of the vibration signals can be used to obtain fault characteristics.

Fault signals contain noise and other interference, rendering the fault characteristics difficult to identify. Therefore, effective noise filtering is crucial in diagnosing machine problems. Mechanical failure-related signals are usually nonlinear, and the information is nonstationary. Huang, Shen [8] proposed a novel adaptive signal analysis method called empirical mode decomposition (EMD), which decomposes complex vibration signals into several inherent modal functions (IMFs). Each IMF represents a single shock signal component. After decomposition, each IMF has only one frequency at any given moment; therefore, the time–frequency distribution of the complex signals can be clearly defined and presented. A different instantaneous frequency calculation method is used to obtain each IMF instantaneous frequency and instantaneous amplitude to construct the time–frequency energy distribution. The resulting spectrum is known as the Hilbert time spectrum. This spectrum is integrated with time to obtain the frequency–energy relationship plot, called the marginal Hilbert spectrum. In the marginal Hilbert spectrum, different IMFs represent different signal characteristics, and faults can be identified by analyzing the different signal characteristics. Musaruddin and Zivanovic [9] proposed an EMD method for decomposing faulty power signals and distinguishing between the power failure signals of different sections by observing the anomalous segments on the Hilbert time–frequency diagram. DTW is one of the methods utilized for exploring time series data. It is a dynamic programming method and can determine the similarity between two groups of the same length or different length time series. DTW is applied to time series to measure similarity. Muda, Begam [10] analyzed voice messages by using the mel frequency cepstral coefficient (MFCC). The analysis results were uploaded to a database, and DTW was applied to match voice signals from an external environment with the voice signals in the database. Anguera, Macrae [11] proposed an unbounded DTW (U-DTW) method, which improved the shortcomings of DTW. The rule of the U-DTW method is to determine the possible match between two sequences by using a fixed range for forward and backward calculations and using the minimum length L_{min}. Thus, calculating unnecessary similarities can be avoided, consequently reducing the calculation time. U-DTW can commence from any point of comparison, and the average of each step is used to meet the conditions to determine the path; therefore, calculating all the values is not required. Miro and Macrae [12] applied U-DTW to speech matching and compared it with other DTW methods. The results revealed that U-DTW is fast and accurate. HMMs are used to determine hidden and unknown parameters from parameters that can be observed. HMMs can be applied in various situations. Mon and Tun [13] used the MFCC to decompose the speech signals of different words, and they subsequently entered these signals into different HMMs. The results revealed that the higher the number of HMMs, the higher the accuracy of the model.

3 Research Method

Figure 1 presents a flowchart of the proposed approach. In this approach, the signal is preprocessed to facilitate its analysis. Next, the signal is divided into numerous segments. These cut signals are analyzed to determine their status. The status of each signal segment can be determined by the characteristics of each signal segment. The availability of the machine can be calculated by determining the processing section and nonprocessing section of the time. Next, every state is trained separately, and then the model is established and subsequently compared with the signal status. Finally, the processing status of the signal is analyzed, and the production quantity is calculated.

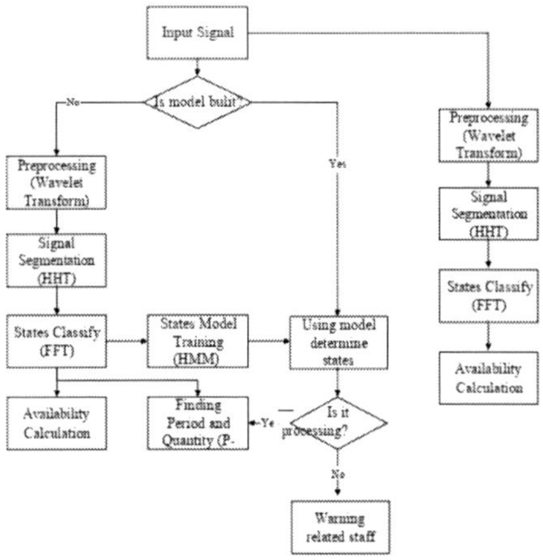

Fig. 1. Flowchart of research method

4 Experimental results

We propose a method for ascertaining current processing status, production cycle time, production quantities, and equipment availability through signal analysis and using noninvasive current-sensing hardware based on current signals to connect to equipment to collect current signal information. In the proposed approach, the signal is first segmented and classified, and the segmentation and classification results are then used to calculate equipment availability and the production cycle time and quantity. Finally, an updated classification model is established to ensure the same processing states can be quickly resolved and analyzed. Figure 2 provides the experimental data used in this study. The experimental scenario included equipment standby, refueling, processing product A, breakdowns, stoppages, and processing product B. Six different states were considered. The experiment results show in table 1~3 and figure 3.

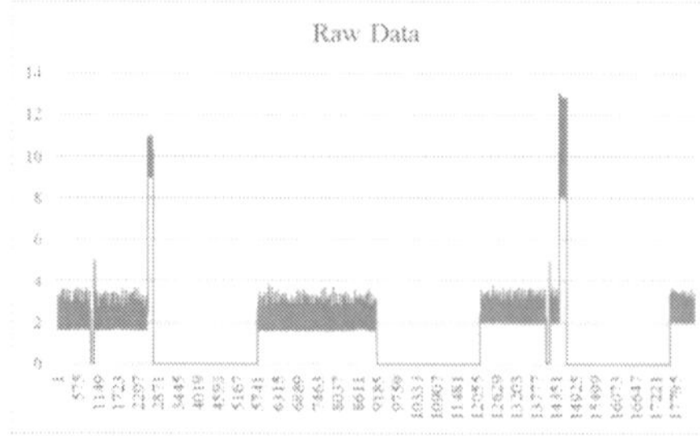

Fig. 2. Experimental data

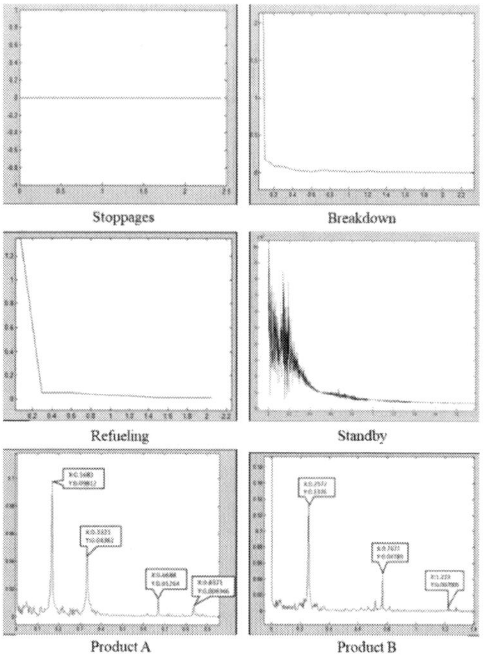

Fig. 3. FFT comparison in different states

Table 1. HHT segment error rate.

	Standby	Refueling	Breakdown	Stoppages	Product A	Product B
HHT data length	4969	144	218	2999	10905	4420
Actual data length	5538	149	219	2930	10384	4435
error	569	5	1	69	521	15
Error rate	10.3%	3.4%	0.5%	2.4%	5.0%	0.3%

Table 2. Actual availability and calculated availability

Actual availability	Calculated availability	Error
71.5%	74.2%	2.7%

Table 3. Results of processing cycle time and quantity

	Cycle time (sec)	Actual quantity	Calculated quantity	Error	Error rate(%)
Product A	5.8	345	348	3	0.87%
Product B	4	226	222	4	1.77%

5 Conclusion

In recent years, the concept of Industry 4.0 has been widely applied in the manufacturing industry. Typically, PLCs and RFID processes are implemented to collect processing information and monitor the production site in real time. Therefore, this paper presents an approach for analyzing machine status signals, and this approach can be used to obtain machine production information by using noninvasive current-sensing devices. The proposed algorithm involves four stages: noise elimination, signal cutting and segmentation, signal state classification, machine availability calculation, and production quantity calculation. Finally, to enable managers to quickly ascertain the processing state, the classified states are trained to construct a model.

To prove the accuracy and validity of the proposed state segmentation method, an experiment was conducted on six states: standby, refueling, failure, shutdown, product A, and product B. The purpose of this study was to identify the various states in the manufacturing process and to calculate equipment availability and the production quantity. After data analysis, the average error associated the determination of states in this study was less than 4%, the error associated with the determination of equipment availability was less than 3%, and the error associated with the determination of production quantity was less than 2%. Therefore, the proposed method is highly accurate in determining the status, availability, and production quantity. Applying this method to manage machine status and availability can assist managers to master the state of real-time processing and reduce machine idle time.In the proposed approach, the real-time calculation of the quantity of production can reduce the number of human errors caused by missing data. Moreover, the approach can enable managers to gain immediate knowledge of the progress of production and improve machine scheduling. Finally, training the classified states to construct a model expedites the determination of production states. Therefore, the proposed approach can be applied to factories.

References

1. Ljungberg, Õ ., Measurement of overall equipment effectiveness as a basis for TPM activities. International Journal of Operations & Production Management, 1998. 18(5): p. 495-507.
2. Lin, J., et al. An automatic system for eggshell quality monitoring. in 2001 ASAE Annual Meeting. 1998. American Society of Agricultural and Biological Engineers.
3. Wang, B., et al., Fundamental technology for RFID-based supervisory control of shop floor production system. The International Journal of Advanced Manufacturing Technology, 2011. 57(9): p. 1123-1141.
4. Mallat, S., A wavelet tour of signal processing. 1999: Academic press.
5. Gritli, Y., et al., Advanced diagnosis of electrical faults in wound-rotor induction machines. IEEE Transactions on Industrial Electronics, 2013. 60(9): p. 4012-4024.
6. Sedláček, M. and M. Titěra, Interpolations in frequency and time domains used in FFT spectrum analysis. Measurement, 1998. 23(3): p. 185-193.
7. Peng, Z., F. Chu, and Y. He, Vibration signal analysis and feature extraction based on reassigned wavelet scalogram. Journal of Sound and Vibration, 2002. 253(5): p. 1087-1100.
8. Huang, N.E., et al. The empirical mode decomposition and the Hilbert spectrum for nonlinear and non-stationary time series analysis. in Proceedings of the Royal Society of London A: mathematical, physical and engineering sciences. 1998. The Royal Society.
9. Musaruddin, M. and R. Zivanovic. Signal segmentation of fault records based on Empirical Mode Decomposition. in TENCON 2011-2011 IEEE Region 10 Conference. 2011. IEEE.
10. Muda, L., M. Begam, and I. Elamvazuthi, Voice recognition algorithms using mel frequency cepstral coefficient (MFCC) and dynamic time warping (DTW) techniques. arXiv preprint arXiv:1003.4083, 2010.
11. Anguera, X., R. Macrae, and N. Oliver. Partial sequence matching using an unbounded dynamic time warping algorithm. in Acoustics Speech and Signal Processing (ICASSP), 2010 IEEE International Conference on. 2010. IEEE.
12. Miro, X.A. and R. Macrae, Fast partial pattern matching system and method. 2014, Google Patents.
13. Mon, S.M. and H.M. Tun, Speech-To-Text Conversion (STT) System Using Hidden Markov Model (HMM). International Journal of Scientific & Technology Research, 2015. 4(6): p. 349-352.

Reliability analysis of the aircrafts normal braking system

Agung Premono[1,*] and *Martin Danang* Laksono[1]

[1]Mechanical Engineering Department, Faculty of Engineering, State University of Jakarta, Jakarta, Indonesia

Abstract. An identified component of Airbus A-330 aircraft which should be maintained properly is the normal braking system. The objectives of this study are to: (1) identify the critical component of the normal braking system; and (2) determine the optimal replacement time interval of replacement of the normal braking system component. This study is conducted in Garuda Maintenance Facility Aero Asia company. Primary data in this study are the components replacement data of normal braking system in period January 2015 - January 2017. The life distribution such as Mean Time to Failure (MTTF), reliability, probability density function, cumulative distribution function, and failure rate are obtained by the Weibull++ software. Results show that the critical component of the normal braking system is the brake unit assembly. Meanwhile, the life distribution analysis show that the reliability of importance value of the normal breaking system IR is 74.07 percent which occurs at MMTF and system reliability are 4365.5 hours and 55.58 percent, respectively. These values indicate that optimal replacement time interval of the Brake Unit Assembly is 5599.7 hours.

1. INTRODUCTION

Aeroplane safety in flight is under the influence of many factors. One of these factors is braking safety while the aircraft is landed. Several plane crashes are caused by failure of the braking system[1]. In order to reduce failure of the braking system, monitoring and checking of the system are required. These activities involve analysis of the pilot report, delay report and maintenance report. Pilot report is the data from the pilot while an error occurs in flight such as taxing, take off, flight, and landing. Maintenance report obtained by maintenance operator while the aircraft is repaired. Delay report is the incident flight report which causes airplane delay. Therefore, these data are obtained from the real flight conditions.

Research of the aircrafts braking system has been carried out by previous researcher in limited quantities. Aircraft braking system has explained by Shruti and Shreya N[2]. Meanwhile, reliability analysis of the aircraft braking system was studied by Al-Garni et.al[3] and performed in Boeing 737 series aircraft. This research has not been determined the optimum time for critical component replacement. In this present research, reliability of the normal braking system of the Airbus A-330 series aircraft is investigated. Furthermore, optimum time for critical component replacement is determined. This research is performed based on the damage data of the braking system during January 2015 – January 2017 in Garuda Maintenance Facility AeroAsia. In this period, the damage is dominated by the normal breaking system.

2. LITERATURE REVIEW

2.1. The Norml Braking System in The Aircraft

Airbus A-330 series has main landing gear with eight wheels. Braking system in this aircraft is located in each main landing gear in each wheel. The brake is made of carbon multi-gear which is equipped with 14 piston cylinders. This piston is driven by two hydraulic systems, hydraulic green and blue. Flight by wire control system technology is adopted in this aircraft. Main component of the braking system consists of: (1) brake unit assembly, brake pedal transmitter, Brakes and Steering Control Unit (BSCU), automatic selector valve, normal brake selector, servo valve, return accumulators, and green manifold assembly.

Work principle of normal braking is started from the first input by stepping the brake pedal or by pressing the landing gear push button at the cockpit panel. Signal from the input is passed to BSCU. Furthermore, the signal is ordering to open the normal selector valve. This valve distributes the hydraulic fluid from the manifold to automatic selector valve. Fluid flows to normal servo valve, then pushing the piston to activate the brake[4].

2.2. Reliability of The Normal Braking System

Reliability is the probability of success or the probability that the system will perform its intended function under specified design limits[5]. Reliability distribution function is required to estimate the reliability value of component. Several methods are used to determine the reliability distribution function. These are: (1) Weibull distribution; (2) Normal distribution; (3) Lognormal distribution; and (4) Exponential distribution.

* Corresponding author: agung-premono@unj.ac.id

All methods have similarity parameters -- probability density function, cumulative distribution function, reliability function, failure rate function, and MTTF -- to determine the reliability function. In this paper, only Weibull distribution is presented.

The probability density function is[6]:

$$f(t)=\frac{\beta(t-\gamma)^{\beta-1}}{\theta^{\beta}}e^{-\left(\frac{t-\gamma}{\theta}\right)^{\beta}} \quad t\geq\gamma\geq 0 \qquad (1)$$

where θ and β are known as the scale and shape parameters, respectively, and γ is known as the location parameter.

The reliability function R(t) is

$$R(t)=e^{-\left(\frac{t-\gamma}{\theta}\right)^{\beta}} \text{ for } t>\gamma>0, \beta>0, \theta>0 \qquad (2)$$

Hence, the failure rate function (h) is

$$h(t)=\frac{\beta(t-\gamma)^{\beta-1}}{\theta^{\beta}} \quad t>\gamma>0, \beta>0, \theta>0 \qquad (3)$$

Mean Time to Failuer (MTTF) is

$$MTTF=\theta\Gamma\left(\frac{1+\beta}{\beta}\right) \qquad (4)$$

where value of $\Gamma\left(\frac{1+\beta}{\beta}\right)$ is obtained from table of $\Gamma(x)$ Gamma function.

Reliability of Importance (I_i^B) of component i in the component system n is predicted by Birnbaum equation as follow:

$$I_i^B=\frac{\partial R_s[R(t)]}{\partial R_i(t)} \qquad (5)$$

where Rs is reliability value of system and Ri is reliability value of component

3. METHODOLOGY

The damage data of the braking system during January 2015 – January 2017 for Airbus A-330 series are obtained in Garuda Maintenance Facility AeroAsia Co. Data is obtained from the component removal data which is caused by unscheduled maintenance. These data are collected from pilot report, maintenance report, and delay report. Fault tree analysis is used to analyse the data. The life distribution such as Mean Time to Failure (MTTF), reliability, probability density function, cumulative distribution function, and failure rate are obtained by the Weibull++ software. Several steps is conducted sequentially to obtained the results. There are: (1) obtaining the Time to Failure (TTF) value from the replacement component data; (2) Estimating the Maximum Likelihood Estimate (MLE) for each component.; (3) Calculating MTTF of each component in the normal breaking system; (4) Calculating reliability of each component in the normal breaking system; (5) Creating a Reliability Block Diagram (RBD); (6) Determining Reliability Importance; and (7) Determining the optimal interval schedule of critical component.

4. RESULTS AND DISCUSSION

Unscheduled maintenance data which is obtained from the problem beyond the schedule maintenance is recorded as the replacement component data in this research. Distribution component of the normal braking system is determined by Reliasoft Weibull ++ 9.0 software using Maximum Likelihood Estimation (MLE) method[7]. Weibull, Normal, Lognormal, and Exponential are calculated. Results show suitable method to evaluate the reliability function and parameter distribution for each component with appropriate with TTF distribution data. Suitable method and parameter for this calculation are summarized in Table 1.

Table 1 Suitable Method and Parameter for Each Component of Normal Braking System

Component	Suitable Method	Parameter	
Brake Unit Assembly	2P- Weibull	b = 3,5287	θ = 6216,95
Return Accumulator , Normal Brake	2P- Weibull	b =1.580395	θ = 12410.21
Servo valve, Normal Brake	Normal	μ = 7832,918	σ = 3586,578
Brake & Steering Control Unit (BSCU)	2P-Weibull	b= 2,019235	θ= 12327,96
Automatic Selector Valve	Normal	μ = 9638.4	σ = 2709.16
Transmitter , Pedal Brake	2P-Weibull	b= 5.247662	θ= 9759.277

Table 1 show that the brake unit assembly, return accumulator, BSCU, and transmitter pedal brake are several critical components in the normal braking system. These data are used to calculate MTTF which refer to Eq. (4). MTTF value of each component in the normal braking system is shown in table 2.

Table 2. MTTF value and distribution function of each component in the normal braking system

Component	t=MTTF				MTTF (hours)
	f(t)	F(t)	R(t)	h(t)	
Brake Unit Assembly	*0.00021*	0.49	50.1%	0.000435	5599,73
Return Accumulator	5.147E-05	0.56	43.0%	0.000119	11139.65
Servo valve	0.00011	0.50	50%	0.000222	7832.918
Brake Steering Control Unit (BSCU)	6.61E-05	0.54	45.7%	0.000145	10921.96
Automatic Selector Valve	0.000147	0.5	50%	0.000294	9638.4
Pedal Transmitter Brake	6.998E-05	0.41	58.9%	0.000118	8985.63

Table 2 show that the lowest of MTTF value occurs in the Brake Unit Assembly. Reliability importance is used to carried out the most critical component of the normal breaking system. The reliability importance value of each component is determined by Eq. (2). Reliability importance value of each component in the normal braking system is summarized in Table 3.

Table 3 Reliability importance value of each component in the normal breaking system

Component	Suitable Method	t=MTTFs R(t)
Brake Unit Assembly	2P-Weibull	75.04%
Return Accumulator	2P-Weibull	82.55%
Servo valve	Normal	83.32%
Brake Steering Control Unit (BSCU)	2P-Weibull	88.43%
Automatic Selector Valve	Normal	97.42%
Pedal Transmitter	2P-Weibull	98.54%

The reliability of the normal braking system which has calculated by Reliasoft Weibull ++ 9.0 is 55.58 %. Therefore, the reliability importance index for each component have summarized in Table 4. These values are calculated by Eq. (5). Table 4 shows that the most critical component in the normal braking system is the brake unit assembly. This conclusion refers to the biggest of the reliability importance index. Hence, the most priority component that should be prioritized on the preventive maintenance is the Brake Unit Assembly. The optimal replacement time interval of replacement of the normal braking system component -- the brake unit assembly – is 5599.7 hours. This value refers to table 2.

Table 4 Reliability importance index of each component in the normal breaking system

Component	Suitable Method	I Ri = Rs/R(t)
Brake Unit Assembly	2P-Weibull	74.07
Return Accumulator	2P-Weibull	67,33
Servo valve	Normal	66.71
Brake Steering Control Unit (BSCU)	2P-Weibull	62.85
Automatic Selector Valve	Normal	57.05
Pedal Transmitter Brake	2P-Weibull	56.40

5. CONCLUSION

From the calculation results, it may be concluded that the most critical component of the aircraft normal braking system is the brake unit assembly. This conclusion is based on the reliability importance index 74.07 which is obtained at yaitu β = 3,528781, θ = 6216,95, R(t) 75.04 %, and R (s) is 55.58 %. The optimal replacement time interval of replacement of the this component is 5599.7 hours.

References

[1]http://www.telegraph.co.uk/news/worldnews/europe/russia/9771664/Faulty-brake-system-blamed-for-Moscow-plane-crash.html

[2]Shruti N and Shreya N, Aircraft Braking System, IJRMET Vol. 4, Issue 1, Nov 2013 – April 2014.

[3]Al-Garni A.Z, et.al, Reliability Analysis of Aeroplane Brakes, Qual. Reliab. Engng. Int. 15: 143-150 (1999)

[4]Airbus, Aircraft Maintenance Manual type A330 halaman 18 Figure 32-42-27-991-00900-00-B A

[5]Elsayed, A. Elsayed, Reliability Engineering 2nd Edition, New Jersey, John Wiley & Sons, INC (2012)

[6]Modarres, M., Kaminsky, M. & Krivtsov,V., Reliability Engineering and Risk Analysis : A Partical Guide 2nd Edition, Boca Raton: CRC Press Taylor & Francis Group, (2010)

[7]Reliasoft Weibul++ Life Data Analysis Software Tool, http://www.reliasoft.com/Weibull/index.htm.

A hybrid particle swarm optimization for the generalized assignment problem with time window

Phattaraphorn Phudphad[1], *Kanchana* Sethanan[1,*] and *Thitipong* Jamrus[1]

[1]Research Unit on System Modeling for Industry, Department of Industrial Engineering, Faculty of Engineering, Khon Kaen University, Khon Kaen, Thailand

Abstract. This study focuses on the inbound logistics of the sugarcane industry, which has three main procedures consisting of cultivation, harvest and transportation. Generally, small-scale growers cannot manage all of the procedures effectively, because of their lack of bargaining power and inadequate equipment. For this reason a resource-sharing policy, such as harvester and truck sharing, is used by factories to reduce the cost of the sugarcane harvest, and increase harvester and truck utilization. To solve the generalized assignment problem (GAP) with time window, thus minimizing the total cost from the assignment of the third-party logistics providers to service small-scale growers under capacity and time limitations, a mathematical model has been developed for small-sized problems. For large-scale problems, particle swarm optimization (PSO) is applied and improved by the hybridization of PSO with k-cyclic moves algorithm (PSOK). The results demonstrate that the proposed metaheuristics can solve the problem efficiently since the results are equal to, or close to, the optimal solutions in which the averaged performances of PSO and PSOK are 99.61% and 99.64%, respectively and the averaged relative improvement is 0.1519%.

1 Introduction

The sugarcane industry is important agricultural industries in Thailand. Currently, Thailand is the second-largest sugar exporter in the world after Brazil, with exporting approximately 8 million tons of sugar per year [1]. Thus it certainly affects the potential economic growth of the country. However, certain problems prevail in this industry related to limited resources and intense competition. As a result of resource constraints, the operating costs have increased greatly, especially the inbound logistics process that involves planting, harvesting and transportation, all of which need to be reduced. Approximately 80% of Thai sugarcane growers are small-scale growers [2] who have a small area of less than 30 rai [3] or about 4.8 hectares. In practice, these growers are not able to manage all procedures effectively because of their lack of bargaining power and inadequate resources. Therefore, some of the small-scale growers have shifted to other economic crops. To reduce costs for the small-scale growers, the mill factories play an important role liaising between the small-scale growers and third-party logistics providers [2], in terms of resource allocation, scheduling and guaranteeing fair prices. Thus, this research considers the allocation of resources for harvesting (sugarcane harvesters) and transportation (trucks), which are mostly required by small-scale growers [4], from third-party logistics providers i (i = 1...m) to service small-scale growers j (j = 1...n), which is the generalized assignment problem (GAP). Each small-scale grower has to be processed by exactly one third-party logistics provider [5]. This is subject to the restrictions of the capacity and time, so that the problem becomes the GAP with time window (GAPTW) to minimize the total cost of the assignment.

The GAP has been extended in many ways. For example, the study of [6], [7], and [8]. Additionally, the GAP is known as an NP-hard problem [9] Large size problems cannot be resolved by the exact method; thus, many researchers have had to develop heuristics for their solutions (see [10], [11], and [12]). However, there are no publications of the GAPTW. And there is to date no research that has previously applied particle swarm optimization (PSO) to solve the GAP, which is the simple algorithm, observed to be performing optimization, so it became popular. For example, [13] and [14] applied PSO to solve their problems. Due to the attractive features of PSO, this research was focused on implementation of this methodology to solve the GAPTW with the expectation to minimize the total cost. Moreover, the k-cyclic moves algorithm from [15], a local search technique which prevents trapping of the local optima, is combined with PSO called PSOK for improving the solutions. Thus, this study promotes the sugar mill as an intermediary, with assistive equipment for making decisions using metaheuristics. PSOK algorithms are applied in order to obtain approximate solutions with measurement of the performance achieved compared to the mathematical model, so that the third-party logistics providers are able to manage their logistics activities more effectively and the small-scale grower can be serviced efficiently at a fair price.

* Corresponding author: ksethanan@gmail.com

2 Materials and methods

2.1 Mathematical model

A mathematical model, which is a mixed integer programming model with a concise explanation of each constraint, was developed for solving GAPTW with minimizing the total cost by assignment of the third-party logistics providers i to service small-scale growers j under capacity and time limitations. The parameters and decision variables used in this model are defined as follows:

Assumptions

1. The total number of trucks is sufficient for transportation in the harvesting process, regardless of deficiency of trucks or waiting trucks and problems during operation.
2. The implementation of third-party logistics providers from harvesting sugarcane to small-scale growers can be performed simultaneously.
3. After the sugarcane harvest process has been completed, the sugarcane harvesters and the trucks will return to the base location of the third-party logistics provider.
4. Service time includes the harvest and transportation times.

Indices

i : Index of third-party logistics provider; i = 1, 2, 3, …, m

j : Index of small-scale grower; j = 1, 2, 3, …, n

Parameters

C_{ij} : Fuel cost of of third-party logistics provider i to serviced small-scale grower j

DC_{ij} : Driver cost of of third-party logistics provider i to serviced small-scale grower j

D_{ij} : Demand from third-party logistics provider i to serviced small-scale grower j

P_{ij} : The capacity of third party logistics provider i

S_{ij} : Service time of third-party logistics provider i to serviced small-scale grower j

S_i^P : Period time of third-party logistics provider i to serviced small-scale grower j

T_j^{gs} : Start time of requirement of small-scale grower j

T_j^g : Ending time of requirement of small-scale grower j

T_i^{ps} : Start time of third-party logistics provider i for service

T_i^p : Ending time of third-party logistics provider i for service

M : Large constant number

Decision Variable

X_{ij} : = 1 if third-party logistics provider i is assigned to service small-scale grower j, Otherwise 0

Objective function

$$\text{Minimize Cost} \quad \sum_{i=1}^{m}\sum_{j=1}^{n} X_{ij}(C_{ij}+DC_{ij}) \tag{1}$$

$$\text{Subject to:} \quad \sum_{j=1}^{n} X_{ij}D_{ij} \le P_{ij} \quad \forall_i \tag{2}$$

$$\sum_{i=1}^{m} X_{ij} = 1 \quad \forall_j \tag{3}$$

$$\sum_{j=1}^{n} X_{ij}S_{ij} \le S_i^P \quad \forall_i \tag{4}$$

$$(T_j^g - T_i^{ps}) + M(1 - X_{ij}) \ge S_{ij} \quad \forall_{i,j} \tag{5}$$

$$(T_i^p - T_j^{gs}) + M(1 - X_{ij}) \ge S_{ij} \quad \forall_{i,j} \tag{6}$$

$$T_j^{gs}, T_j^g, T_i^{ps}, T_i^p, S_i^P \text{ and } S_{ij} \ge 0 \quad \forall_{i,j} \tag{7}$$

$$X_{ij} \in \{0,1\} \quad \forall_{i,j} \tag{8}$$

Equation (1) is an objective function which purposes to minimize the total cost of the assignment. Constraint (2) ensures that the total resource requirement of the small-scale grower does not exceed the capacity. Constraint (3) ensures that each small-scale grower is assigned to exactly one third-party logistics provider. Constraint (4) ensures that service time to be carried out is less than the available time of third-party logistics provider. Constraints (5) - (6) ensure that the service time must not exceed time window of small-scale growers and third-party logistics. And constraints (7) - (8) are the basic restrictions on the parameters of time and a binary variable.

A mathematical model was validated by comparing the optimal solution obtained from the small-sized problem. The instance of input data used in the test for 3 third-party logistics providers and 5 small-scale growers are shown in Table 1-3.

Table 1 The input data of $C_{ij} + DC_{ij}$ and D_{ij}

$C_{ij} + DC_{ij}$ (Baht)	1	2	3	4	5
1	1,080	1,080	1,944	1,656	2,736
2	2,088	936	2,448	1,800	2,088
3	3,384	576	2,448	1,440	1,512
D_{ij} (ton)	240	260	216	264	80

Table 2 The input data of P_{ij}, S_i^P, T_i^{ps} and T_i^p

Third party logistics i	P_{ij} (Ton)	S_i^P (Day)	T_i^{ps}	T_i^p
1	6,500	50	9	59
2	1,560	12	10	22
3	2,850	19	80	99

Table 3 The input data of S_{ij}, T_j^{gs} and T_j^g

S_{ij} (Day)	1	2	3	4	5
1	1.33	1.44	1.20	1.47	0.44
2	1.85	2.00	1.66	2.03	0.62
3	1.60	1.73	1.44	1.76	0.53
T_j^{gs}	50	39	35	46	1
T_j^g	81	100	100	58	89

Table 4 The cost of the allocation of the third-party logistics provider *i* to serve the small-scale grower *j*

Third party logistics *i*	Small-scale grower *j*	Cost (Baht)
1	1	1,080
1	3	1,944
1	4	1,656
3	2	576
3	5	1,512
Total cost (Baht)		6,768

The allocation to meet the lowest cost as shown in Table 4 show that third-party logistics provider 1 was allocated to small-scale grower 1, 3 and 4, and the third-party logistics provider 3 was allocated to the small-scale grower 2 and 5. The total cost was 6,768 baht.

From the example, the mathematical model executed by LINGO13 gives the optimal solution because the result is equal to the solution, which has been validated and verified by Microsoft Excel 2013.

2.2 Metaheuristic development

The mathematical model cannot solve large problems; thus, a metaheuristic was developed by applying PSO to solve this problem executed by MATLAB (R2017a).

2.2.1 Particle swarm optimization

The PSO is originally attributed to [16], which is an optimization technique, inspired by the collective behavior of movement of organisms in a bird flock or fish school which is the advantages of having several optimization parameters that govern its behavior and efficiency, especially in the behavior of particles while searching, the influence of control parameters on performance, and the convergence properties of the algorithm [17].

2.2.2 K-cyclic moves algorithm

K-cyclic moves algorithm, which is local search, is applied to prevent trapping of the local optima. The notation and procedure of the k-cyclic moves algorithm are presented below.

Notation:

k : Maximum number of moves
nr : The number of moves required
z : Position of particles

1. The value of nr is randomly selected from [2, k].
2. Randomize the unique position of z which is equal to a number of nr.
3. Cyclic move selected positions.
4. Re-do steps 1.–3. for all particles.

2.2.3 PSOK algorithm

The notation and PSOK algorithm is given as follows:

Notation:

t : Iteration t; $t = 1, 2, 3, \ldots, T$
l : Particle l; $l = 1, 2, 3, \ldots, L$
$x_l\,(t)$: The position of the particle l in iteration t
$v_l\,(t)$: The velocity of the particle l in iteration t
w : The inertia weight
c_p : The weight coefficient for the personal best
c_g : The weight coefficient for global best
r_1, r_2 : Random numbers generated from a uniform distribution in [0, 1]
$pBest_l$: The personal best of the particle l
$gBest$: The global best

Procedure: PSOK for the GAPTW
Input: $r_1, r_2, x_i, v_i, c_p = 2, c_g = 2$ and $w = 0.9$ [18]
Output: Best solution
Begin $t \leftarrow 1$;
  Initialize L particles as a swarm
  While $t \leq$ T; looping
    Decode particles into solutions
    Evaluate the particles
    Do k-cyclic moves algorithm
    Update $pbest_l$ and $gbest$
    Update velocity by equation (9)

$$v_l\,(t+1) = w \cdot v_l\,(t) + c_p \cdot r_1\,(pBest_l - x_l\,(t)) + c_g \cdot r_2\,(gBest - x_l\,(t)) \quad (9)$$

    Update position by equation (10)

$$x_l\,(t+1) = x_l\,(t) + v_l\,(t) \quad (10)$$

    $t \leftarrow t + 1$;
  End while
  Output best solution
End

Factorial design of the experiment and statistical test by SPSS19 were performed which demonstrated that L = 30 particles and T = 500 iterations are proper to obtain the best solution for this problem.

3 Computational results

The heuristic performance (HP %) between the mathematical model and the proposed algorithm (PSO and PSOK) were calculated. In Table 5 illustrate that the results of both metaheuristics are close and equal to the optimal solution with average the performance of PSO and PSOK is 99.61% and 99.64%, respectively. An experimental design of the factors taken under consideration is 1) the number of third-party logistics providers, 2) the number of small-scale growers, and 3) the time of the third-party logistics provider. Each factor has 3 levels; thus, the total number of trials is $3^3 = 27$ combinations and the results of experiments with the relative improvement (RI%) of PSOK demonstrate that the averaged relative improvement of the PSOK is 0.1389% for the best solution and 0.1519% for the averaged solution. The statistical tests based on SPSS19 showed that Factors 1) and 2) significantly affect the

Table 5 The results of the experiment and the computation time

No.	*i*	*j*	Optimal (Baht)	Time (s)	PSO (Baht)	Time (s)	*HP%*	PSOK (Baht)	Time (s)	*HP%*
1	5	50	40,080	0.10	40,080	5.94	100	40,080	10.50	100
2	10	100	61,708	0.10	61,708	8.96	100	61,708	18.03	100
3	15	500	265,824	0.10	265,824	39.63	100	265,824	79.97	100
4	20	700	385,488	0.10	385,542	73.04	99.99	385,493	90.57	100
5	30	1,000	527,832	28.00	536,549	100.72	98.38	536,208	141.45	98.44
6	40	1,300	688,920	61.39	699,307	145.96	98.51	697,930	181.25	98.71
7	50	1,500	N/A	-	843,073	172.60	-	841,980	344.82	-
8	70	2,000	N/A	-	1,083,788	282.80	-	1,082,918	562.20	-

*N/A = Out of memory

experimental response at the confidence interval of 95%. In other words, these factors affect improving PSOK.

4 Discussion and conclusion

This paper proposes a mathematical model and metaheuristics to solve the GAPTW. The objective is to minimize the total assignment cost for third-party logistics providers to service small-scale growers under capacity and time limitations, which made the steps more complex than the original GAP. From our literature database, there are no research studies using a metaheuristic algorithm that applies PSO, so the PSO has been used in this research. Furthermore, PSO has been improved by the hybridization of PSO with the k-cyclic moves algorithm (PSOK) to prevent the local trap and develop the solution. A comparison of the performance between the mathematical model and both metaheuristics demonstrate that for small-sized problems, both meta-heuristics can give the optimal solutions. When the problem size was increased, the mathematical model could not find the solutions, whereas both metaheuristics could obtain results either equal to or approximately equal to the optimal solutions with the averaged performances of 99.61% and 99.64%, respectively. The results of improving the performance of PSOK demonstrate that the relative improvement is 0.1389% for the best solution and 0.1519% for averaged the best solution, which indicated that the proposed hybrid PSO outperforms the original. The solutions found in this research lead to efficient cost savings, which is a major help for third-party logistics providers to serve the small-scale growers at a more standardized and fair price. Therefore, it mitigates the risk of sugarcane growers in changing to other competing economic crops. Thus, the sugar mills would have more sugarcane to the sugarcane supply system, which would achieve a sugar productivity increased to meet the demands of the present global marketing.

Acknowledgement. The authors wish to express their grateful and sincere thanks to the anonymous reviewers for their helpful comments. In addition, this work was supported by the Research Unit on System Modelling for Industry and Department of Industrial Engineering, Faculty of Engineering, Khon Kaen University, Thailand.

References

1. Statista, "Thailand's sugar exports from 2009/2010 to 2017/2018", Available: https://www.statista.com/statistics/249653/exported-amount-of-sugar-from-thailand/, 2018, [Accessed on 9 May 2018]
2. W. Neungmatcha, *A Model for Integrating Inbound Logistics of a Mill Region* (Ph.D. Dissertation, Khon Kaen Univ., Thailand, 2014)
3. Office of the Cane and Sugar Fund, "Cost and productivity of sugarcane small-scale growers", Available: http://www.ocsf.or.th/Research/rese-43.doc, n.d., [Accessed on 16 Fab 2016]
4. C. Kusoncum, *Sugarcane Harvest and Transportation Management Model for Small Growers: A Case Study of Mitr Kalasin Sugar Mill* (M.S. Thesis, Khon Kaen Univ., Thailand, 2013)
5. I. H. Osman, Oper. Res. spektrum, **17**, 211–225 (1995)
6. C. Rainwater, J. Geunes, H. Edwin Romeijn, Discret. Appl. Math., **157**, 49–67 (2009)
7. S. Mitrović-mini, ć A. P. Punnen, Discret. Optim., **6**, 370–377 (2009)
8. I. Litvinchev, M. Mata, S. Rangel, J. Saucedo, Comput. Math. with Appl., **60**, 1115–1123 (2010)
9. L. A. N. Lorena, M. G. Narciso, J. E. Beasley, Evol. Optim., **5**, 1–19 (2002)
10. P. C. Chu, J. E. Beasley, Comput. Oper. Res., **24**, 17–23 (1997)
11. N. Palasak, S. Tharmlikit, N. Indra-Payoong, Proceedings of the 6th Int. Joint Conf. on Comput. Sci. and Software Eng., **2**, 93-101 (2009)
12. K. Sethanan, R. Pitakaso, Comput. Electron. Agric., **121**, 245–259 (2016)
13. W. Wisittipanich, C. Kasemset, Chiang Mai Univ. J. Nat. Sci., **14**, 361–378 (2015)
14. K. Sethanan, W. Neungmatcha, Eur. J. Oper. Res., **252**, 969–984 (2016)
15. K. Sethanan, R. Pitakaso, Comput. Electron. Agric., **121**, 245–259 (2016)
16. J. Kennedy, R. Eberhart, Proceedings of the IEEE int. conf. on neural networks, **4**, 1942–1948 (1995)
17. F. Marini, B. Walczak, Chemom. Intell. Lab. Syst., **149**, 153–165 (2015)
18. Y. Shi, R. Eberhart, 1998 IEEE Int. Conf. Evol. Comput. Proceedings. IEEE World Congr. Comput. Intell., **24**, 69–73 (1998)

Lipid-based nanocarriers to enhance skin permeation and antioxidant activity of Centella asiatica extract

Boonnada Pamornpathomkul[1], *Worranan* Rangsimawong[1], *Theerasak* Rojanarata[1], *Praneet* Opanasopit[1], *Chuleerath* Chaiyodsilp[2] and *Tanasait* Ngawhirunpat[1,*]

[1]Pharmaceutical Development of Green Innovations Group (PDGIG), Faculty of Pharmacy, Silpakorn University, Nakhon Pathom, Thailand.

[2]Thai Herb for Hair Co., Ltd. Nakhon Pathom, Thailand

Abstract. The purpose of this study was to evaluate the use of different formulations, including solution, gel, liposome and niosome for in vitro skin permeation and antioxidant activity of *Centella asiatica* (CA) extract. The liposomes and niosomes loaded with CA were characterized to observe the physicochemical properties i.e., particle size, zeta potential, percentage of entrapment efficiency (%EE) and percentage of loading efficiency (%LE). In vitro skin permeation studies revealed that liposome formulations had a superior enhancing effect on skin permeation compared to niosome, gel and solution formulation. Upon applied niosome formulations for the delivery of CA extract at 24 hours (h), the antioxidant activity was higher than liposome, gel and solution formulation, as evidenced by the increased in percent inhibition using 2,2-diphenyl-1-picrylhydrazyl (DPPH) assay. However, there was no significant difference in antioxidant activity between niosome and liposome formulations. Accordingly, both the liposome and noisome formulations are promising approaches for transdermal delivery of CA extract for promoting successful antioxidant activity.

1 Introduction

Centella asiatica (CA) is a tropical medicinal plant from the Apiaceae family native to Southeast Asian countries. It is commonly known as "bua-bok" in Thailand. It has been widely used as a traditional herbal medicine due to biological activities from asiaticoside as a major constituent of CA, including being an antioxidant [1], anti-aging, antipsoriatic [2] and anti-inflammatory [3]. CA is the chief herb for treating hair loss and promoting hair re-growth, which strengthens the hair blood vessels and expands hair follicles due to the entry of more nutrients and oxygen. In addition, antioxidant agents can interrupt radical chain processes, protect against oxidative damage and help to repair hair systems [4].

It is challenging to deliver molecules into the skin because of the foremost barrier from the stratum corneum of skin [5]. Therefore, improved delivery strategies are necessary to solve this issue. Lipid-based nanocarriers are attractive formulations for topical and transdermal drug delivery. Among the lipid-based nanocarriers systems, liposomes have been widely investigated as potentially efficacious lipid nanocarriers. Liposomes are small, sphere-shaped vesicles. They can be created from cholesterol and natural nontoxic phospholipids resulting in being consisted of amphipathic phospholipids arranged in one or more phospholipid bilayer(s). Hydrophilic drugs can be entrapped within the inner aqueous core and hydrophobic drugs can be intercalated in the lipid bilayers. Due to the small size vesicle, hydrophobic and hydrophilic character, liposomes have the ability to interact with lipids in skin and decreased systemic absorption [6].

Niosomes are also widely used as lipid-based nanocarriers for drug delivery. Although liposomes can encapsulate various types of drugs and are advantageous, their formulation is expensive and there is a limited shelf life. Niosomes have the functional and physicochemical properties similar to a liposome vesicle. However, niosomes differ from liposomes in their chemical structure resulting in greater stability than liposomes [7]. Niosomes are created from cholesterol and non-ionic surfactants resulting in bilayer structures being formed. They are composed of amphiphilic molecules surrounded by an aqueous compartment that contain both hydrophobic and hydrophilic groups [8]. Niosome nanocarriers have been used as potential transdermal drug delivery systems due to enhanced drug permeation and localized depots for sustained drug release [9].

In this study, CA extract was selected as our model compound because it is a hydrophilic compound with many biological activities, especially, antioxidant activity and promotion of hair growth. Physicochemical characterizations were investigated to compare the different lipid nanocarriers. Moreover, the effect of different formulations, i.e., CA solution, CA gel, CA liposome and CA niosome on skin permeation and antioxidant activity were investigated.

2 Materials and methods

2.1. Materials

* Corresponding author: ngawhirunpat_t@su.ac.th

Egg phosphatidylcholine (PC) was purchased from Lipoid GmbH (Ludwigshafen, Germany). Cholesterol was obtained from Carlo Erba Reagent (Ronado, Italy). Span 20, DPPH and d-limonene were purchased from Sigma-Aldrich (MO, U.S.A.). All chemicals used were of analytical reagent grade.

2.2 Preparation of CA extract

Fresh leaves of CA were dried at 50°C, powdered and sifted. Powder was refluxed with 36% ethanol at 60°C for 1 h. The extract was then filtered, mixed with 80% propylene glycol and stored at 4°C [10]. To control the concentration of extract, high-performance liquid chromatography (HPLC) was used to control the initial concentration of CA extract (4 %w/v).

2.3 Preparation of formulations

2.3.1 CA gel preparation

The gel formulation was prepared by dispersing 4% of Carbopol® 940 polymer into the deionized water. Then, 4% w/v of CA extract was added and mixed until homogeneous. Triethanolamine was then added with gentle sweeping agitation.

2.3.2 CA liposome preparation

Liposome formulations were prepared by thin film hydration method. Briefly, 250 μl of PC (200 mM) and 500 μl of cholesterol (20 mM) were dissolved with 2:1 v/v of chloroform/methanol. The solvent was then evaporated using stream of nitrogen. The thin film was placed in a desiccator until completely dry (6 h). The 4% CA extract was dissolved in phosphate-buffered saline (PBS; pH 7.4) and added to the lipid film to give hydrated liposome vesicles. The dispersion was mixed with Tween 20 and d-limonene. The formulation was sonicated for 30 min to reduce the size of the liposomes. Remaining materials were removed by centrifuge at 15,000 rpm for 15 min. The supernatant was collected.

2.3.3 CA niosome preparation

The niosome formulation was formulated similarly with the method used for the liposomes. Briefly, 125 μl of Span 20 (200 mM) and 125 μl of cholesterol (200 mM) were dissolved with chloroform/ethanol (1:1). The solvent was evaporated with a stream of nitrogen and then placed in a desiccator for 6 h. The 4% CA extract was dissolved in PBS (pH 7.4) 5 ml and added to the lipid film to give hydrated niosome. The dispersion was mixed with Tween 20 and d-limonene. The formulation was sonicated for 30 min. Remaining materials were also removed. The supernatant was collected.

2.4 Characterization of nanocarrier formulations

2.4.1 Particle size and zeta potential analysis

The particle size and surface charge of the nanocarrier formulations were determined at 25°C by photon correlation spectroscopy using a Zetasizer Nano ZS (Malvern Instrument, Worcestershire, U.K.).

2.4.2 Percentage of entrapment efficiency (%EE) and loading efficiency (%LE)

Each lipid nanocarrier formulation (0.5 mL) was placed in a Microcon YM-3 ultrafiltration (Minipore, Billerica, U.S.A.) and centrifuged at 10000 ×g at 4°C for 60 min. The filtrate was discarded, and 0.25 mL of PBS was then added to the retentate before further centrifugation at 4°C at 10000 ×g for 40 min [11]. The collected CA-loaded nanocarriers in the retentate were then disrupted with 0.1%w/v TritonX-100 and centrifuged at 4°C at 10000×g for 10 min. The CA contents of the supernatants were determined using HPLC. The drug %EE and %LE were calculated with equations 1 and 2:

$$\%EE = (C / C_i) \times 100 \quad (1)$$

$$\%LE = (C / \text{Lipid composition}) \times 100 \quad (2)$$

where C is the concentration of CA extract in the formulation, and C_i is the initial concentration of CA extract.

2.5 *In vitro* skin permeation studies

A Strat-M™ membrane was used as a human skin model because it has multiple layers with varied diffusivity as it correlates more closely to human skin than animal skin models. The Strat-M™ membranes were mounted between the donor and the receptor compartments of Franz diffusion cells with the epidermis side facing up towards the donor compartment, and the apparatus was affixed using clamps. The receptor compartment was filled with PBS and continuously stirred at 400 rpm. To evaluate the effect of different formulations, including CA solution, gel, liposome and niosome, the concentration of CA extract placed into the donor compartment was 4 %w/v. Two milliliters of different CA formulations were placed onto the skin. The top of the donor compartment was then covered with Parafilm® for occlusion. The temperature was maintained at 32 ± 1 °C. To investigate the cumulative permeation profiles, a 500 μl aliquot of receptor was withdrawn at the time intervals of 2, 4, 6, 8 and 24 h and replaced with equal volumes of fresh PBS. Each sample was divided for measuring the concentration of asiaticoside using HPLC and antioxidant activity by DPPH assay.

After the concentration of asiaticoside was measured using HPLC, the cumulative amount of asiaticoside permeated into the skin samples was plotted against time, and the steady-state flux was determined as the slope of the linear portion of the plot. The lag time was obtained by extrapolating the linear portion of the penetration profile to the abscissa. The enhancement ratio (ER) was calculated with equation 3:

$$ER = \frac{\text{Flux of nanocarrier formulations}}{\text{Flux of 4\% CA solution}}$$ [redacted]

2.6 Quantitative analysis of asiaticoside

The amount of asiaticoside permeated through Strat-M® membrane was quantized using HPLC 1260 infinity Agilent Technologies® (California, U.S.A.) with a UV detector. Chromatographic separation was performed with a Zobrax Eclipse XDB-C18® column (4.6×250 mm I.D.; particle size: 5 μm) (Agilent Technologies, California, U.S.A.). Mobile phase was a gradient of acetonitrile/water as shown in Table 1, a flow rate of 1 ml/min and detection at 206 nm.

Table 1. Gradient conditions for HPLC

Time (min)	Pump A, water (%)	Pump B, acetronitrile (%)
0	80	20
15	65	35
30	35	65
35	20	80
40	20	80

2.7 Antioxidant activity

The DPPH assay method was used in this study. This method was based on the reduction of DPPH, a stable free radical. The free radical DPPH with an odd electron gives a maximum absorption at 517 nm (purple color). Upon reduction, the solution color fades; the reaction progress is monitored by a spectrophotometer. One-hundred microliters of samples were added into each well of 96-well plates. A 100 μl of DPPH solution (200 μM) was then added, mixed and left at 25 °C for 30 min. The absorbance was then measured at 550 nm using a microplate reader. The results of the assay were expressed as %inhibition (equation 4).

$$\%\text{inhibition} = (1 - A_{sample}/A_{control}) \times 100 \quad (4)$$

Where A_{sample} and $A_{control}$ are the absorbance in the presence of the different formulations and control, respectively.

2.8 Data analysis

All data were statistically analyzed by one-way analysis of variance (ANOVA). In all cases, a value $p < 0.05$ was considered statistically significant.

3 Results and Discussions

3.1 The physiochemical characterization of lipid formulations

3.1.1 Particle size and zeta potential

The physiochemical characteristics of the different lipid formulations are shown in Table 2. The lipid nanocarrier formulations showed the nanoscale range of particle size with a negative charge. The average particle size of the liposome formulation was smaller than the niosome formulation. Moreover, the liposome formulation showed a narrow size distribution (polydispersity index; PDI<0.3), while the niosome formulation has large variability in the particle size. An increase in the particle size of niosomes might be due to the interaction of drug with surfactant head groups of niosomes resulting in increasing the charge and repulsion force of the surfactant bilayers [12].

3.1.2 %Entrapment efficiency and %loading efficiency

As shown in table 2, the %EE of the liposome formulation was higher than that of the niosome formulation, suggesting that the niosome structure was more rigid and decreased drug entrapment. The drug entrapment efficiency of niosomes was lower than liposomes. An increase in cholesterol content of the niosome formulation resulted in a decrease in the fluidity of the niosome bilayers but an increase in the drug loading efficiency. Moreover, the thickness of the niosome membrane structure and solubility of the drug in water affected the drug loading efficiency [13]. In case of the liposome formulation, the drug loading efficiency depends on drug-lipid interactions, the solubility of the drug in the liposome membrane and the ability of liposomes to encapsulate aqueous drugs during vesicle formation [6, 14]

Table 2. Characterization parameters of the lipid nanocarrier formulations

Parameters	CA liposome	CA niosome
Particle size (nm)	42.33 ± 0.47	55.73 ± 2.52
PDI	0.28 ± 0.03	0.43 ± 0.01
Zeta potential (mV)	-13.30 ± 0.64	-3.50 ± 0.33
% EE	94.31 ± 3.84	72.55 ± 2.74
%LE	16.93 ± 0.69	28.94 ± 1.26

3.2 *In vitro* skin permeation studies

The *in vitro* skin permeation profiles of asiaticoside from the different formulations are illustrated in Fig. 1A and 1B. The permeation parameters of asiaticoside from the different formulations are shown in Table 3. The cumulative amount of asiaticoside at 24 h after applying various formulations could be ranked as follows: CA liposome > CA niosome > CA gel > CA solution. The results showed that the CA liposome formulation was significantly increased in the skin permeation of asiaticoside when compared with other formulations. Although CA niosome formulation slightly enhanced the skin permeation of asiaticoside in comparison with the gel formulation, there was no significant difference in cumulative amounts of asiaticoside at 24 h. The most enhancing effect of liposome formulation might be caused from the lipid composition of liposome structures, as it was similar to the epidermis layer of the skin which enables them to penetrate the epidermal barrier to a greater extent compared to other formulations [15]. Moreover, the liposome formulation forms a barrier around their drugs, which is resistant to oxidation and degradation [6]. Compared with the liposome formulation, the skin permeation of the niosome formulation was significantly lower. The particle size of the liposome formulation was smaller and the %EE of the liposome formulation was higher than the niosome formulation. In addition, CF loaded into the liposomes with higher %EE could present higher skin permeation than those with lower %EE [16]. Therefore, it could be assumed that the particle size and %EE substantially affect the transdermal drug delivery of lipid-based nanocarriers.

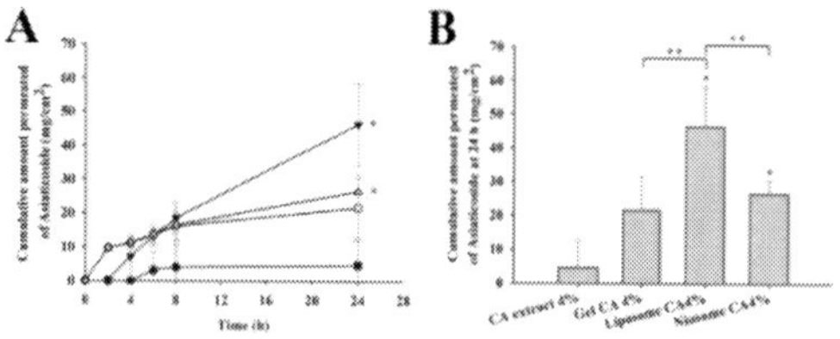

Fig. 1. *In vitro* permeation profiles of asiaticoside (A) The cumulative amount profiles of asiaticoside in (●) CA solution, (○) CA gel, ([redacted]CA niosome and ([redacted]CA liposome (B) The cumulative amount at 24 h of all groups. * indicates significant difference compared to CA solution (p<0.05). **indicates significant difference between two groups (p<0.05).

Table 3. The permeation parameters of asiaticoside from different formulations

Formu-lations	Flux (mg/cm^2/h)	ER	Lag time (h)	K_p (×10^{-3}) (cm/h)
solution	0.2036±0.35	-	0.46±0.80	0.10±0.18
gel	0.7084±0.41	3.48	6.77±0.55	0.36±0.21
liposome	1.9846±0.49*,**	9.56	0.05±0.08	1.01±0.25*,**
niosome	0.9222±0.17*,**	4.53	6.14±0.45	0.47±0.08*,**

3.3 Antioxidant activity

The niosome formulation had a superior antioxidant activity after 24 h of being topically applied compared to the liposome, solution and gel formulation, respectively (Fig.2). Antioxidant levels were found to increase significantly in both the liposome and niosome formulations after 24 h of being topically applied. Although the niosome formulation slightly improved antioxidant activity in comparison with liposome formulation, there was no significant difference between percentage inhibitions at 24 h. It can be concluded that both the liposomes and niosomes are very suitable formulations for transdermal delivery of CA extract.

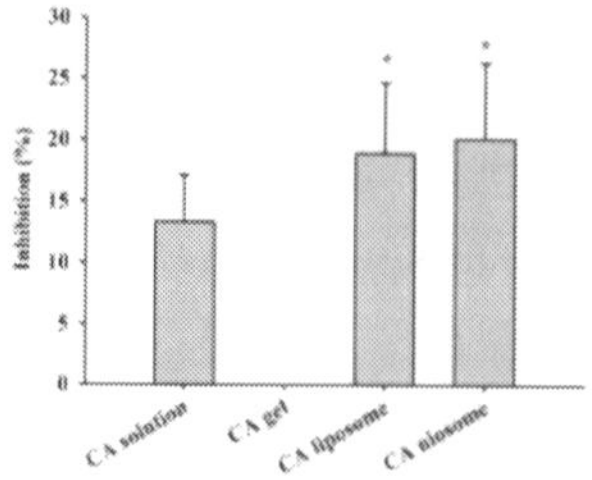

Fig. 2. Antioxidant activity of DPPH assay of different CA formulations

4 Conclusion

The purpose of this study was to evaluate the different lipid formulations with CA extract on skin permeation and antioxidant activity. The particle size of lipid-based nanocarrier formulations was in the nanoscale range. Both liposomes and niosomes revealed a high cumulative amount at 24 h and antioxidant activity of CA extract. The ability of liposomes and niosomes to enhance drug delivery and promoting antioxidant activity promotes the two lipid-based nanocarriers to be useful for topical formulations for CA extract.

5 Acknowledment

We gratefully acknowledge the financial support provided by the National Innovation Agency, Thailand.

References

1. M. Bajpai, A. Pande, S.K. Tewari, D. Prakash, *Int J Food Sci Nutr*, **56** (2005).
2. J.H. Sampson, A. Raman, G. Karlsen, H. Navsaria, I.M. Leigh, Phytomedicine, **8** (2001).
3. M. George, L. Joseph, Ramaswamy, African Journal of Traditional, Complementary, and Alternative Medicines, **6** (2009).
4. M.P. Lupo, Clin Dermatol, **19** (2001).
5. O.G. Jepps, Y. Dancik, Y.G. Anissimov, M.S. Roberts, Adv Drug Deliv Rev, **65** (2013).
6. A. Akbarzadeh, R. Rezaei-Sadabady, S. Davaran, S.W. Joo, N. Zarghami, Y. Hanifehpour, M. Samiei, M. Kouhi, K. Nejati-Koshki, Nanoscale Res Lett, **8** (2013).
7. N.B. Mahale, P.D. Thakkar, R.G. Mali, D.R. Walunj, S.R. Chaudhari, Advances in Colloid and Interface Science, **183–184** (2012).
8. I.F. Uchegbu, A.T. Florence, Advances in Colloid and Interface Science, **58** (1995).
9. R. Muzzalupo, L. Tavano, Research and Reports in Transdermal Drug Deliver, **4** (2015).
10. P. Hashim, H. Sidek, M.H. Helan, A. Sabery, U.D. Palanisamy, M. Ilham, Molecules, **16** (2011).
11. W. Rangsimawong, P. Opanasopit, T. Rojanarata, S. Duangjit, T. Ngawhirunpat, Biol Pharm Bull, **39** (2016).
12. K.M. Kazi, A.S. Mandal, N. Biswas, A. Guha, S. Chatterjee, M. Behera, K. Kuotsu, Journal of Advanced Pharmaceutical Technology & Research, **1** (2010).
13. B. Shi, C. Fang, Y. Pei, J Pharm Sci, **95** (2006).
14. S. Hua, Frontiers in Pharmacology, **6** (2015).
15. J. Dua, A. Rana, A. Bhandari, Int J Pharm Stud Res, **3** (2012).
16. D.D. Verma, S. Verma, G. Blume, A. Fahr, Int J Pharm, **258** (2003).

https://doi.org/10.1051/matecconf/201819201017

Estimation of residual life of a cutting tool used in a machining process

Sourath Ghosh[1], *Sukanta* Kumar Naskar[2] and *Nirmal* Kumar Mandal[3]

[1] Manufacturing Technology Department, National Institute of Technical Teachers' Training & Research, Kolkata, India
[2] Education and Management Department, National Institute of Technical Teachers' Training & Research, Kolkata, India
[3] Mechanical Engineering Department, National Institute of Technical Teachers' Training & Research, Kolkata, India

Abstract A significant part of cost of machining is associated with non-optimum use of cutting tool. Moreover cutting tool failure is responsible for almost 20% of the machining downtime. Thus, having knowledge of residual life of cutting tool is highly recommended so as to maximise the availability time and reduce the machining cost. The aim of this work is to find out residual life of a worn cutting tool which has been used for turning of Ti-6Al-4V alloy under constant cutting condition. The lognormal distribution is used to model the cutting tool life data. Remaining useful life of cutting tool is estimated using Mean Remaining Life (MRL) function. The results obtained from model are compared with the experimental results and it shows good agreement.

1 Introduction

Aim of modern manufacturing industry is to produce economical and reliable product along with good quality. Quality of machining product is generally referred as surface finish and accuracy in dimensions, which are highly dependent on the condition of cutting tool. Besides product quality, tool failure puts high impact on machining system and productivity. Moreover, in the total cost of machining, tooling cost accounts for a significant part (approximately 25%) [1]. Most of the time cutting tools are either overused or underused with respect to optimum tool life. Study suggests that, cutting tools are typically utilised only up to 70–80% of its optimum life [2]. Hence, the prediction of remaining useful life (RUL) of cutting tool is important when a significant amount of cost is involved in a machining process.

Till now many researchers have developed various models for estimating cutting tool life. Devor, et al [3] concluded that the life span of cutting tool is a probabilistic phenomenon. Based on this fact, till now different reliability model has been proposed [4-5]. Rodriguez, et al [6] suggested a strategy of optimum tool replenishment based on analysis of cutting tool reliability. Xiao-Sheng, et al [7] presented brief review on recent developments of various models for estimating the RUL. Benkedjouh, et al [8] proposed a methodology for estimating the remaining useful life of cutting inserts which is based on the support vector regression. Gebraeel, et al [9] used artificial neural network approach for predicting remaining life of bearing. Aramesh, et al [10] introduced a model for predicting the remaining tool life in variable machining conditions using proportion hazard model. In this paper, the weibull distribution was used for modelling baseline hazard function of PHM.

The aim of this study is to present a probabilistic approach to assess useful remaining life of a worn cutting tool which has been used for turning of Ti-6Al-4V alloy under constant cutting condition. The lognormal distribution is used to fit the tool life data. Finally, Mean Remaining Life (MRL) equation is used to estimate expected residual life of cutting tool.

2 Tool Wear

Failure of cutting tool mainly occurs due to gradual blunting of cutting edge. Tool wear which takes place on the rake surface of the tool is called crater wear and which occurs on the flank surface called flank wear. Flank wear being the most predominant wear mechanism, maximum length of flank wear (VB_{max}) is generally used as failure criteria for determining tool life. However, depending on the desired application crater wear could also be used as failure criteria [11].

[1] Sourath Ghosh: sourath93@gmail.com

As shown in Fig. 1, a typical tool flank wear curve involves three different states. At the initial state, the rate of tool wear is high which is known as rapid initial wear state. Next, at the steady state, wear rate remains almost constant followed by the accelerating wear state where tools wear out very rapidly.

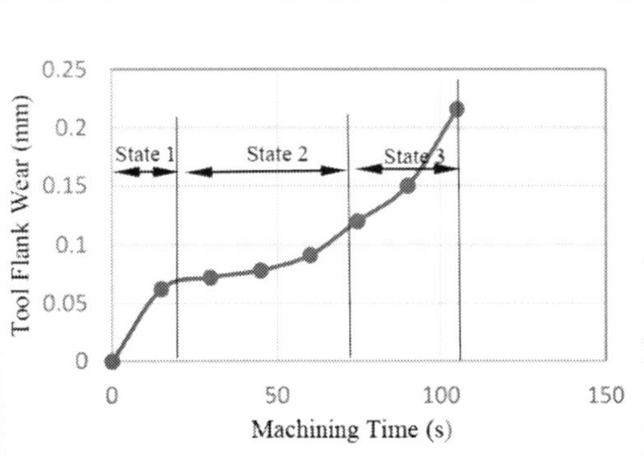

Fig. 1. Different states of flank wear

3 Experiment Setup

Dry turning operations were performed on MAXTURN Plus CNC turning centre. Fig. 2 shows the experimental setup used in this study. Cylindrical bar made up of Ti-6Al-4V alloy was used as work piece material. Turning tool chosen for this study was TiN coated tungsten carbide insert having specification CNMG 120408 as shown in Fig. 3.

Fig. 2. The experimental setup

Experiments were performed using total eight number of new tool inserts. Turning operations were carried out for each tool under constant cutting parameters. Cutting parameters are chosen based on the supplier's recommendations which are listed in Table 1.

Table 1. Machining parameters

Parameters	Speed (m/min)	Feed Rate (mm/rev)	Depth of Cut (mm)
Value	50	0.15	0.2

While each test, sequential observations of flank wear were performed using Upright Materials Microscope Leica DM2700 M as shown in Fig.5. For example, the experimental data of sequential wear measurements for tool-3 is shown in Table 2. The same procedure is repeated for each new tools.

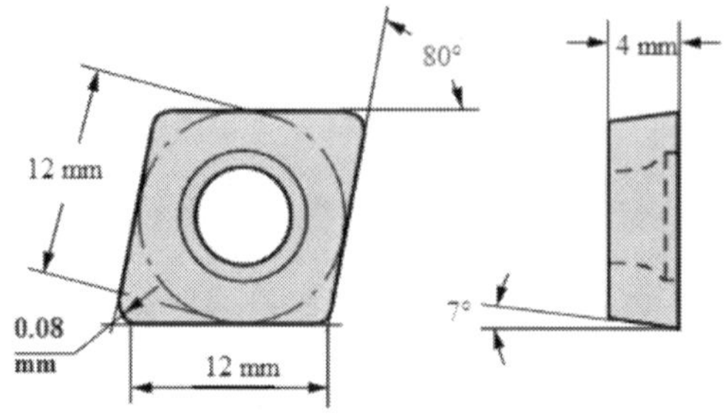

Fig. 3. Geometry of insert type tool (CNMG 120408) [16]

Table 2. Progressive wear of tool-3

Sequential measurement no.	Machining Time (s)	Tool Wear (mm)
1	0	0
2	150	0.064
3	300	0.079
4	450	0.103
5	600	0.121
6	750	0.133
7	900	0.146
8	1050	0.153

It is assumed that nature of the progression of tool wear is linear [12] between two measurement points as shown in Fig. 4. Time-to-failure (t) corresponding to the threshold limit (VB_{max} = 0.15 mm) is calculated by interpolating between last two observations (i_{th} *and* $(i+1)_{th}$) by using equation (1). This procedure is performed for the each tools and the obtained results are presented in Table 3.

$$\frac{t-t_i}{t_{i+1}-t_i} = \frac{VB_{max}-VB_i}{VB_{i+1}-VB_i} \qquad (1)$$

Table 3. Tool life data for eight inserts

Tool Id	Tool Life (s)
1	912
2	873
3	982
4	1018
5	958
6	1007
7	854
8	898

4 Methodology

Eight number of cutting tool inserts are used for this study. While machining, sequential observations (measurements of tool flank wear) are performed until critical tool wear level is reached. Same procedure is carried out for eight inserts.

Time to failure data is calculated for each insert by the regression process between last two observation points.

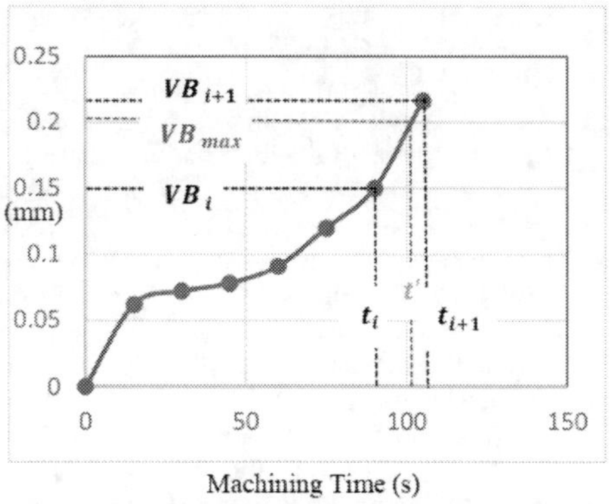

Fig. 4. Linear variation of tool flank wear between two points

Tool wear threshold limit is chosen based upon the economic aspect of machining. As, once the tool gets worn out significantly, it effects the surface finish and dimensional accuracy of machining product. In this study, maximum tool flank wear, $VB_{max} = 0.15$ mm, is taken as failure criteria. Which means, machining time corresponding to tool flank wear of 0.15 mm is considered as useful tool life.

In this study, lognormal distribution is used to fit time-to-failure data of eight inserts. From previous literature, it can be stated that lognormal distribution describes tool life accurately over other distributions [13]. The probability density function $f(t)$, for lognormal distribution is represented as follow:

$$f(t) = \frac{1}{\sigma t\sqrt{2\,\pi}}\, e^{\left[-\frac{(\ln t-\mu)^2}{2\,\sigma^2}\right]} \qquad (2)$$

Where, σ and μ denote the standard deviations and mean respectively and t represents tool life as continuous variable. The corresponding cumulative distribution function F(t), for lognormal distribution can be derived using the following relationship:

$$F(t) = \int_0^t f(t)\,dt$$

$$= \int_0^t \frac{1}{\sigma t\sqrt{2\,\pi}}\, e^{\left[-\frac{(\ln t-\mu)^2}{2\sigma^2}\right]}\,dt \qquad (3)$$

Consequently, the reliability function $R(t)$, for lognormal distribution can be obtained from following relation:

$$R\,(t) = 1 - F(t)$$

$$= 1 - \int_0^t \frac{1}{\sigma\, t\,\sqrt{2\pi}}\, e^{\left[-\frac{(\ln t-\mu)^2}{2\sigma^2}\right]}\,dt \qquad (4)$$

After constructing probability density function and reliability function, next step is to estimate the useful residual life of cutting tool using mean residual life function (MRL) [14]. Given that a tool has used for machining up to a certain time t_o, the expected residual life of the tool from that observation point can be calculated by the following equation:

$$\mathrm{MRL}\,(t_o) = \frac{\int_{t_o}^{\infty} t\,.\,f(t)dt}{R(t_o)} - t_o \qquad (5)$$

5 Results and discussion

Firstly, turning operation is performed for each tool and tool life data are collected as shown in Table 3. Maximum likelihood estimation (MLE) method is used to calculate unknown parameters (μ and σ) of Lognormal distribution using experimental data set [15]. This MLE is performed using the statistical software R. The estimated parameters of lognormal distribution are shown in Table4 and corresponding probability density function is shown in

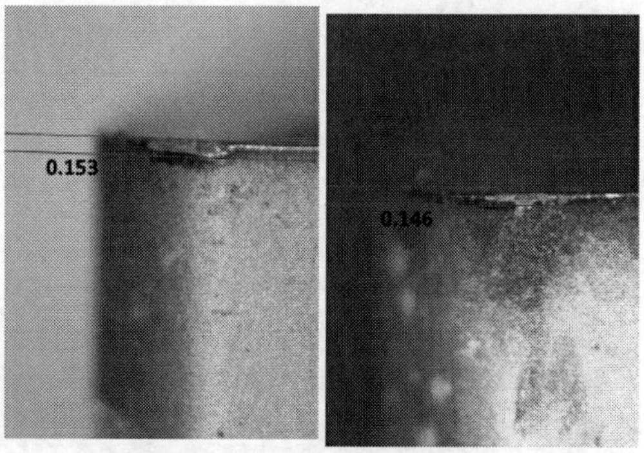

Fig. 5. Worn-out cutting edge after machining

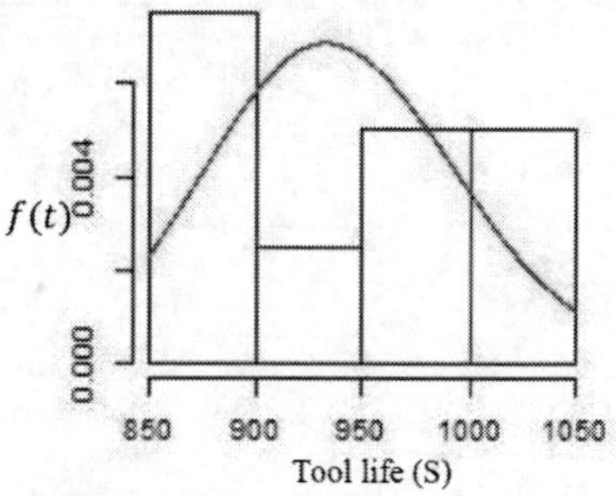

Fig. 6. Probability density function

After estimating distribution parameters and constructing reliability model, the next step is to calculate expected residual life. This calculation is done by the help of Matlab software.

Table 4. Parameters of Lognormal distribution

Parameters	μ	σ
Estimated value	6.842	0.062

Let a cutting tool insert which already used t_r= 120 (s) while operating under the same constant cutting parameters, then it's Mean Residual Life will be given by the Eq. 6 as below:

$$\text{MRL}(120) = \frac{\int_{120}^{\infty} t.f(t)dt}{R(120)} - 120 \quad (6)$$

Three Confirmation tests were conducted on three cutting tools and the estimated results of residual life from the statistical model are presented in Table 5. In order to check the accuracy of the estimated result, the residual life was observed by further machining with the tools and noted down as presented in Table 5.

Table 5. Comparison between estimated and experimental result

Sl. no	Consumed life (s)	Residual life from model (s)	Residual life from experiment (s)	Perce nt-age error (%)
1	120	818.961	837.000	2.15
2	600	338.961	322.000	5.26
3	840	103.977	114.000	8.79

It can be seen in table 5, that the probabilistic results obtained from the model are quite close to the actual experimental results.

6 Conclusions

In this study, a methodology for predicting residual life of cutting tool is introduced which is intended for optimize the cost of machining and maximise the availability time of machine tool. Turning of Ti-6Al-4V is performed on a CNC turning centre and experimental data are collected. Statistical model is constructed based on these tool life data. Once the reliability model is developed, it can estimate expected useful life of a worn out cutting tool. Experimental results show that the estimated residual life is in good agreement with the experimental one. It can be concluded that, this approach of predicting remaining tool life is helpful for mass production of parts, made up of same workpiece material and with same type of cutting tool.

References

1. G. N. Sakharov, V. Ilinykh, V. Yu. Konyukhov, Soviet engineering research, *Improvement of fastening elements in an assembled cutting tool,* **10(11)**, 102-103 (1990)
2. H. Wiklund, Quality and Reliability Engineering International, *Bayesian and regression approaches to on-line prediction of residual tool life,* **14(5)**, 303-309 (1998)
3. R. E. DeVor, D. L. Anderson, W. J. Zdeblick, Journal of Engineering for Industry Transactions of the ASME, *Tool Life Variation and its Influence on the Development of Tool Life Models,* **99**, 578–584 (Sec B) (1977)
4. K. Hitomi, N. Nakamura, S. Inoue, Manufacturing Science and Engineering, *Reliability analysis of cutting tools,* **101(2)**, 185–190 (1979)
5. H. Negishi, K. Aoki, Precis *Machining Investigations on reliability of carbide cutting tools,* **42(6)**, 578–589 (1976)
6. C. E. P. Rodriguez, Souza de, Reliability Engineering & System Safety, *Reliability concepts applied to cutting tool change time,* **95(8)**, 866–873 (2010)
7. X. ShengSi, W. Wang, C. HuaHu, D. HuaZhou, European Journal of Operational Research, *Remaining useful life estimation – A review on the statistical data driven approaches,* **213**, 1-14 (2011)
8. T. Benkedjouh, K. Medjaher, N. Zerhouni, and S. Rechak, Journal of Intelligent Manufacturing, *Health assessment and life prediction of cutting tools based on support vector regression,* **26(2)**, 1-11 (2013)
9. N. Gebraeel, M. Lawley, R. Liu, V. Prameswaran, IEEE Transactions on Industrial Electronics, *Residual life predictions from vibration based degradation signals: a neural network approach,* **51(3)**, 694–700 (2004)
10. M. Aramesh , M. H. Attia, H. A. Kishawy, M. Balazinski, CIRP Journal of Manufacturing Science and Technology, *Estimating the remaining useful tool life of worn tools under different cutting parameters: A survival life analysis during turning of titanium metal matrix composites (Ti-MMCs),* **12**, 35–43 (2016)
11. H. A. Kishawy, S. Kannan, M. Balazinski, CIRP Annals – Manufacturing Technology, *Analytical modeling of tool wear progression during turning particulate reinforced metal matrix composites,* **54(1)**, 55–58, (2005)
12. Y. Shaban, M. Aramesh, S. Yacout, M. Balazinski, H. Attia, H. Kishawy, Journal on Engineering Manufacture, *Optimal replacement times for machining tool during turning titanium metal matrix composites under variable machining conditions,* **231(6)**, 924-932 (2017)
13. F. Ding, Z. He, International Journal of Advanced Manufacturing Technology, *Cutting tool wear monitoring for reliability analysis using proportional hazards model,* **57**, 565-574 (2011)
14. R. J. Cook, J. F. Lawless, *Springer, The statistical analysis of recurrent events,* (2007)
15. D. Banjevic, A. Jardine, V. Makis, M. Ennis, INFOR-OTTAWA, *A control-limit policy and software for condition based maintenance optimization,* **39**, 32-50 (2001)
16. https://www.sandvik.coromant.com/engb/products/t-max_p/Pages/default.aspx , (2018)

https://doi.org/10.1051/matecconf/201819201018

Bendability and failure mechanisms of dual phase steel under air-bending at elevated temperatures

Natthasak Pornputsiri and *Kannachai* Kanlayasiri

Industrial Engineering Department, Faculty of Engineering, King Mongkut's Institute of Technology Ladkrabang, Bangkok, Thailand

Abstract. In this study, process parameters such as the bending angle and bending temperature of the metal sheets for pure bending of advanced high strength steel sheet DP980 were investigated using the air-bending method. This investigation focused on bendability and failure mechanisms. Experimental studies were carried out with various bending temperatures and the bending angle parameters to identify effect on springback angle, initial crack, and the evolution of the microstructure. Prior to testing, the DP steel sheets were heat-treated at different temperatures (room temperature, 200 °C, 400 °C, and 600 °C) with a heating rate of 2 °C/S and 5 minutes of holding time in an electric furnace. As a result of the experiments performed at different temperature values and bending angle values, it was determined that an increase in the bending angle reduces the size of the springback angle. It was also determined that an increase in air-bending-temperature reduces the amount of springback. Air-bending-temperature also affected the microstructure evolution and slowed the cracking of the bent surface.

1 Introduction

High strength steel, dual-phase (DP) steels type, have been commonly used in the motor manufacturing sector due to their high strength and formability, which makes them ideal for body construction and for use in the chassis [1]. The typical characteristics of DP steels, in terms of mechanical properties, include relatively low-yield strength, high-strain hardening, and continuous yielding behavior. Due to the relatively high-hardening rate of DP steels, a reasonable combination of formability and strength is provided [2]. These strong DP steels earned their name on account of the two phases from which they are composed; a ferrite matrix in combination with a martensitic phase. The benefits to the manufacturer include good formability along with increased strength, thereby allowing lighter cars and trucks to be built while offering an environmental advantage due to the lower cost offered by reducing the overall weight. The only drawback is that such steels can fail unexpectedly during bending by even the smallest radius, and the bending surface will frequently indicate signs of failure below the forming limit strain. One of the problems that needs to be solved is the springback issue [3].

Bendability is an indicator that shows the extent of formability for a sheet metal sample. One important manufacturing technique is air-bending, whereby the sheet metal components are produced using pure bending stress and severe forming. Therefore, it is necessary to fully understand the bendability properties of sheet metals in order to produce high-quality components. The sheet metal bending process for high-strength steel has extensive industrial applications, but it is difficult to complete the forming or bending process at room temperature due to the significant degree of deformation involved, as well as the high-flow stresses that affect the materials [4,5]. If the temperature is increased for the forming process, the flow stresses are lowered, residual stresses are also reduced, and deformation becomes easier, hence improving formability [6,7]. The lower flow stresses and heightened formability make it possible to achieve greater stretching, and permits deeper drawing in creating the final product [8]. Sheet metal bending at elevated temperatures that are below recrystallization temperature is considered warm forming. Materials can be formed into complex shapes without much difficulty at this temperature. While high temperature air-bending has not yet found effective commercial use, it is likely that it will become much more important in the future as the technique develops.

Good bendability is absolutely vital when making metal components which take complex forms. Therefore, it is important to develop more effective ways for AHSS to perform in industrial applications. Ability of the bending is could be evaluated as the lowest ratio of the inside bending radius (r) to the sheet thickness (t) that the sheet metal can accept without the manifestation of surface defects during bending to an angle of 90º in pure bending [9]. Generally, when the sheet metal stronger is affected to decreasing of the bending ability. The damaging effect of stretched inclusions is well known, especially in advanced high strength steel (AHSS), on mechanical properties and bendability in conventional structural steels.

While the use of high temperature metal forming processes has not yet become widespread in industrial

* Corresponding author: kkkannac@kmitl.ac.th

applications, it is expected that this will change. Therefore, one study that investigated the formability of pure titanium and titanium alloy at high temperatures is of particular interest, primarily because its findings present the effects of temperature when applied with high-strength metals [10]. However, studies involving high-strength steels are rather limited, meaning there is a gap that must be addressed because the applications are potentially valuable. The required temperatures for the formation of steel sheets are particularly high. The problems this presents may be responsible for the fact that few experiments have been conducted to date.

This research seeks to carry out an investigation of the effect of bending temperature on bendability and the damage mechanism in dual phase steel sheets (DP980), which would be examined with air-bending tests conducted at varying temperatures (room temperature to 600 °C) in order to assess the suitability and limitations of the technique along with the results for springback angle, microstructure evolution, and failure mechanisms by optical microscope (OM), scanning electron microscope (SEM), and crystal structure by XRD.

2 Experimental Procedures

2.1 Materials

This research experimented with AHSS sheets of 1 mm thickness. For the dual phase steel (DP980) sheets used in this study, Table 1 tabulates their tensile mechanical properties, which were obtained on the Zwick universal testing machine according to ASTM E 8M at room temperature. The microstructures of the DP steel sheets were investigated via 50X Optical Microscope (OM), scanning electron microscopy (SEM), while the XRD measurements of the DPS microstructures were also investigated. Specimens of each heat treatment were metallographically prepared with a final polishing step of 0.1 µm. The macro-scale microstructure (OM) and SEM study was made after etching with 2% Nital using the Optical Microscope (OLYMPUS, BX 60 M) and JEOL JSM-6510 LV FRG-SEM (field emission gun scanning electron microscope) working at 15kV. The characterized microstructure of each heat treatment before the bending is shown in Fig. 1, where the ferrite (α) appears as light brown and the martensite (α') as a whitish color.

Table 1. Mechanical properties at room temperature

Material	Ys (MPa)	UTS (MPa)	TE (%)
DP980	740	1020	16

2.2 Mechanical conformation tests

Figure 2 illustrates the pure-bending, plane-strain experimental setup at different deformation temperatures, which was used to investigate the bendability and failure mechanisms of the DP980 steel sheets using three-point air-bending.

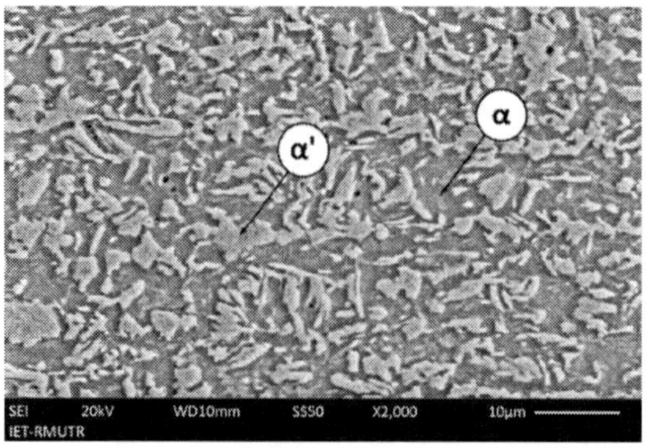

Fig 1. Microstructure of dual phase steel (DP980) observed via SEM image after 2% Nital etching

During experimentation, the specimen is subjected to a punch with a cylindrical body. The experiment was devised in accordance with the ISO 7438:2010 standard. The punch (R_p) and roller (R_d) radii were 0.25mm and 10mm, respectively. The metal sheets were 12 mm in width (b), 30mm in length (l), and 1.0mm in thickness (t). In the operation, the punch pressed downward while the support set remained stationary. The experiments were carried out by varying the bending angles in 15° increments from 0° to 75°. Uni-axial tensile tests were conducted with a Zwick Z20 universal testing machine at a constant bending speed of 60 mm/min.

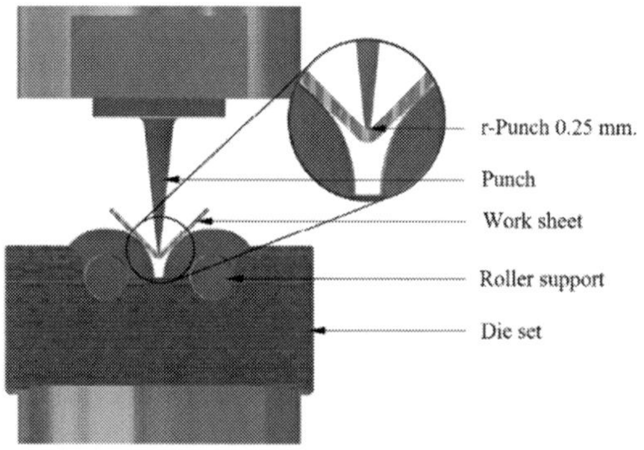

Fig 2. Air-bending tool ISO 7438:2010 standard

The tests were completed at varying temperatures including room temperature(RT), 200°C, 400°C, and 600°C and without lubrication. Regulation of the temperature was accomplished through the use of a thermostatic chamber with sensitivity of ± 2 °C under warm and hot conditions. A thermocouple was set up at the position of the specimen, where the temperature was controlled precisely. The test pattern is shown in Fig. 3. First, the specimen was heated to the target temperature at a heat increase rate of 2 °C /s using a thermostatic chamber, after which it was soaked for 5 min to assure complete austenitization [7]. The item being tested was then bent to accomplish the target bending angle utilizing the punch stroke and subsequently cooled in the air to room temperature.

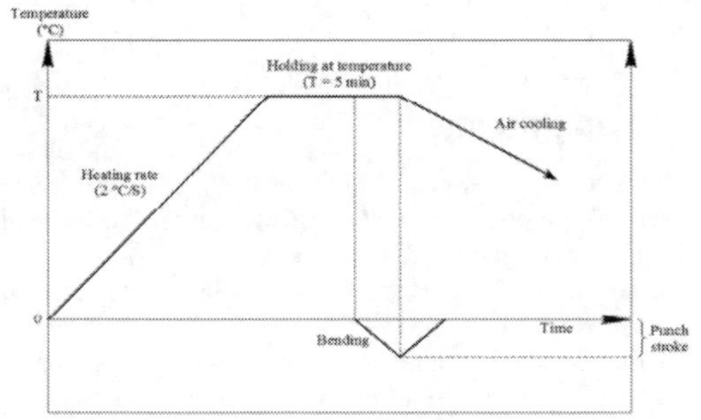

Fig 3. Scheme of the elevated temperature air-bending test procedure, which consists of four main steps: heating, holding, air-bending to the testing level, and air-cooling.

3 Results and Discussion

3.1 Springback

In the sheet metal forming process, the product geometry must be within dimensional tolerances. However, dimensional errors due to springback occur following the plastic deformation, which is are known to be among the most important problems encountered in the sheet metal forming process [11].

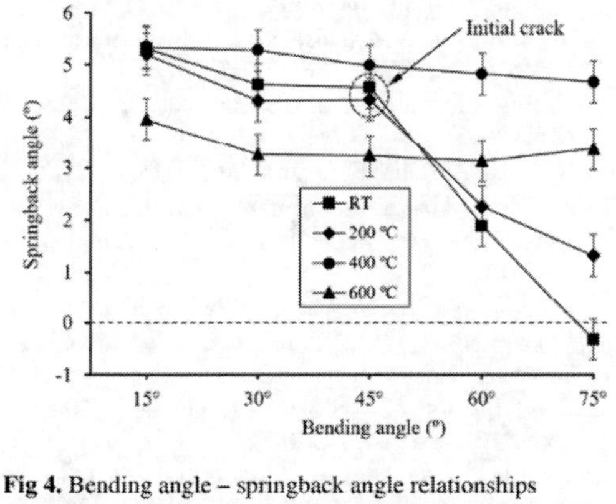

Fig 4. Bending angle – springback angle relationships

The springback results after bending in this study are shown in Fig. 4, which illustrates the influence of the bending angle on the springback angle. It was determined that an increase in bending angle decreased the springback angle at all temperatures. It was also clear that the angle of the springback increased for the temperature increase to 400 °C, for bending at room temperature, and at 200 °C. For bending at 45°, the springback decreased continuously, with the initial crack appearing on the convex bending surface, while the springback angle was likely to decrease as temperatures increased higher than 400 °C. This was likely due to the yield strength and strain hardening decrease [12]. This leads to a reduction in the amount of springback occurring after the forming process [13]. In conclusion, springback is likely to decrease markedly for temperature increases above 400 °C, while the microstructures of this steel are in the ferrite-pearlite phase at warm temperatures [7]. According to previous literature [14], recovery and recrystallization of partial ferrite occurs in specimens at this temperature. Thus, the bending strength and hardness becomes lower as the temperature rises higher than 400 °C.

3.2 Failure mechanisms

The experimental process was continued up to the point when the convex surface began to show signs of fracture. These cracks tend to appear first in the central areas of the specimen because the plane strain is higher in these regions than at the edges, as confirmed by the ASM Handbooks (2000) [15].

Figure 5 shows the condition of the various bent specimens in differing bending states. The convex surface can be examined for cracks using 20X magnification. For the purposes of this study, minor wrinkling effects or an orange peel appearance is not taken to be a significant type of defect.

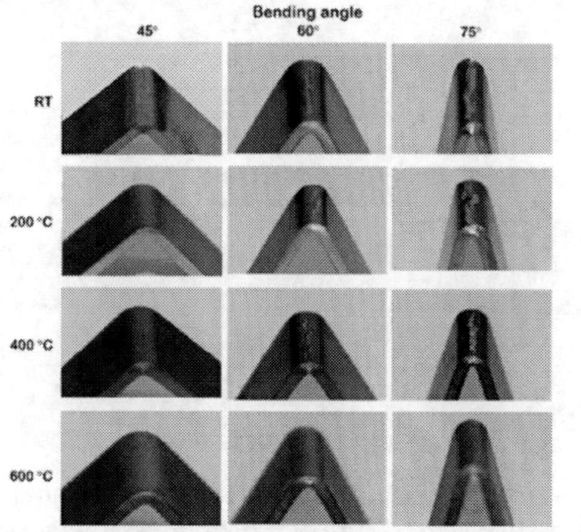

Fig 5. Various bent specimens shown at different temperatures and bending angles

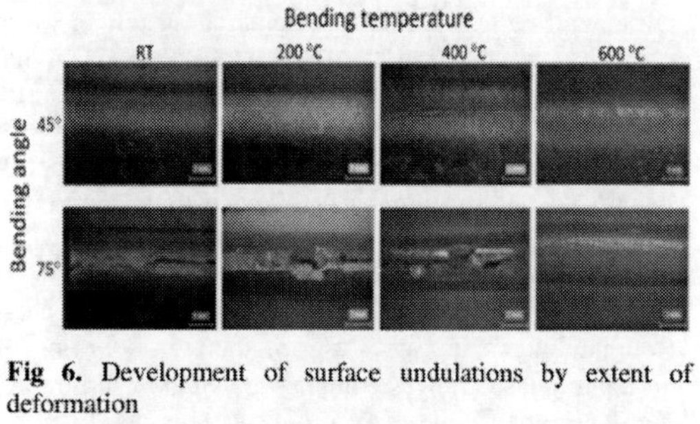

Fig 6. Development of surface undulations by extent of deformation

The different stages of a developing crack are shown in Fig. 6. The orange peel appearance becomes visible along with a wavy effect upon the surface. This undulation appears on the convex part of the bending region. As explained by Dao and Lie [16], the grooves that are formed cause the strain to be increased, leading to a lack of homogeneity in the plastic flow measured in micrometers. However, the defects are limited to narrow slip bands.

When these slip bands intersect with a free surface, they can become the source of both intrusions and extrusions. Cracks tend to originate from intrusions, as would be expected on the basis of the findings related to bending, whereby the intruding undulations are where cracks are observed to start.

3.3 Relationship between microstructures and crystal structures

The microstructures of dual phase steel sheets produced by different temperatures show no changes in the ferrite phase following heat treatment that affects the austenite and ferrite regions. As the temperature rises, the fraction by volume of martensite declines.

The results in Fig. 7 reveal the usual DP980 XRD patterns as indicated by the XRD phase analysis performed at room temperature, 200 °C, 400 °C, and 600 °C using Cu Kα radiation. This technique offers accuracy of around 5%, while a step of 2Θ occurs from 40 to 100.

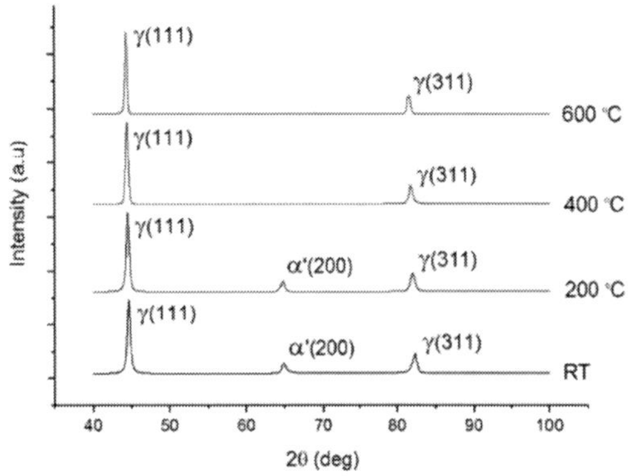

Fig 7. The result of the XRD scan, which applied Cu Kα radiation with the DP980 microstructures, the γ austenite phase and the α'-martensite phase

Figure 7 also clearly shows the body-centered tetragonal martensite and face-centered austenite, which comprise each separate diffracting plane. When experimenting at room temperature and 200 °C, the presence of the planes can be readily observed, along with the existence of the α'-martensite phase, which appears before the deformation occurs. The material has lower formability and bending at this temperature range, which can easily cause cracking, as shown in Fig. 6. If the temperature is increased to 400 °C, the martensite phase (α') no longer appears. Instead, there is a prominent austenite phase (γ) by the high stress yield, resulting in a high springback angle. According to previous literature [17], a higher martensite volume fraction was obtained with a higher work hardening, but it led to a decreased amount of springback. Therefore, the XRD result shows that the martensite phase (α') no longer appears at 400 °C, resulting in springback angle increase. Increasing the temperature to 600 °C leads to an increase in ductility and results in a lack of fractures on the convex bending surface. Other authors have reported similar findings [18]. However, the findings in this study show the dominance of the austenite phase in terms of the eventual condition of the microstructure.

4 Conclusions

Concerning the fundamentals of the experimental study carried out and exhibited in this research, the following conclusions can be drawn:

- The amount of springback occurring in the sheet matal following the pure-bending is significantly reduced as the bending angle increases, and the amount of springback is significantly affected by temperature increased at about 400 °C.
- The initial crack occurring on the convex surface of the specimen occurs significantly more slowly as the bending temperature increases.
- Intermediate heat temperature of dual phase steel resulted in a product that comprised a ferrite-plus-martensite grain structure. The lower volume fraction of α'-martensite declined as the temperature increased to intermediate and high levels. The quantity of austentite forms was shown to be higher than that of martensite within the DPS microstructure.

References

1. N. Pathak, C. Butcher, M.J. Worswick, E. Bellhouse, J. Gao, Materials. **10**, 4(2017).
2. T. Sirinakorn, V. Uthaisanksuk, S.Srimanosaowapak, Procedia Eng. **81**, (2014).
3. E. Hug, M. Martinez, J. Chottin, Mater. Sci. Eng., A. **626**, (2015).
4. S. Toros, A. Polat, F. Ozturk., Mater. Des., **41**, (2012).
5. *N. Pornputsiri, K. Kanlayasiri, ICMM, International Conference on Mechatronics and Manufacturing (2018).*
6. L. Luo, L. Wei, L. Wang, S. Zhou, X. Jin, Mater. Sci. Eng., A. **682**, (2017).
7. T. Yong, L. Zhuang, J. Iron Steel Res. Int. **19**, 6(2012).
8. S.M. Hussaini, S.K. Singh, A.K. Gupta, J. King Saud Univ., Eng. Sci. **26**, 2(2014).
9. J. Heikkala, A. Väisänen, Proc. 11th Bienn. Conf Eng. Syst. Des. Anal, (2012).
10. M.H. Shipton and W.T. Roberts, Mat. Sci. & Tech. **7**, (1991).
11. O. Çavuşoğlu, S. Toros, H. Gürün, A. Güral, Ironmaking & Steelmaking, (2017).
12. S. Kurukuri, A. Miroux, M. Ghosh, A.H. Boogaard, Int. J. Mater. Form. **2**, (2009).
13. H. Laurent, J. Coër, P.Y. Manach, M.C. Oliveira, L.F. Menezes, Int. J. Mech. Sci. **93**, (2015)
14. J. Yanagimoto, K. Oyamada, Annals CIRP. **54**, (2005).
15. ASM Handbooks, **2**, (2000).
16. M. Dao, M. Lie, Phil. Mag. **81**, (2001).
17. H. Fathi, H.R.M. Semnani, E. Emadoddin, B.M. Sadeghi, Met. Mater. Int. **22**, 406 (2016).
18. S. K. Akay, M. Yazici, A. Avinc, Proceeding of Romanian. Acd. A. **10**, 1(2009).

Indirect test methods for the mechanical characterization of building stones

Giulia Forestieri

Facultad de Ingeniería, Universidad de la Sabana, Campus Universitario del Puente del Común, km 7 Autopista norte de Bogotá, Chía, Cundinamarca, Colombia

Abstract. The main objective of this study is to evaluate the usefulness of indirect methods to estimate the uniaxial compressive strength (UCS) of building stones. For this purpose, the results of the UCS test on five types of stones from southern Italy, one igneous and four sedimentary stones are firstly correlated with the corresponding results from Schmidt hammer, point load and UCS direct tests. Then, derived correlations are compared with the equations obtained by different researchers in the mechanical stone characterization.

1 Introduction

The uniaxial compressive strength (UCS) is widely employed in civil engineering to define mechanical resistance of building materials. Current testing procedures especially refer to the most employed worldwide standards of the International Society for Rock Mechanics (ISRM) and the American Society for Testing and Materials (ASTM). The advantage of performing direct tests consists in obtaining results very close to the mechanical behaviour of the tested materials but, on the contrary, these methods suppose large costs to the sample preparation according to the above standards. Thus, indirect methods are always more frequently preferred thanks to their less cost and to their ease, especially for in situ tests.

The aim of this research consists in predicting UCS using indirect tests, in particular Schmidt hammer test and Point load test and to compare results with direct UCS in order to establish correlations.

2 Methods and materials

2.1. Sampling

Five building stone types are sampled and tested in this study. Stone blocks are collected from active quarries and natural outcrops in the Calabria area of Italy (southern Italy). Each block sample is inspected for macroscopic defects so that it would provide test specimens free from fractures, partings or alteration zones. Selected stone types are:

1. San Giovanni in Fiore granite (GF) from San Giovanni in Fiore (39°15′N and 16°41′E), belonging to the Sila Batholith Unit of Upper Paleozoic and classified as granite [1];
2. Grisolia limestone (DG) from Grisolia (39°43′N and 15°51′E), from the Verbicaro Unit of the Upper Triassic and classified as crystalline carbonate [2];
3. San Lucido calcarenite (CS) from San Lucido (39°18′N and 16°03′E), geologically belonging to the Miocenic sedimentary Successions of the Tyrrhenian Coastal Range and petrographically known as biocalcarenite [3];
4. Fuscaldo sandstone (AF) from Fuscaldo (39°24′N and 16°01′E), belonging to the same Miocenic Succession of CS and known as graywacke [3];
5. Trebisacce limestone (TL) from Trebisacce (39°52′N and 16°32′E), belonging to the sedimentary successions of Holocene to Middle-Upper Pleistocene of the Albidona Formation and classified as limestone [4].

2.2 Experimental procedure

The Schmidt hammer test (SHT) provides a quick and inexpensive measure of surface hardness and it is widely used for estimating the mechanical properties of stone materials in the field [5]. L-type SHT are conducted in situ directly in blocks or outcrops surfaces. All tests are carried out with the hammer held vertically downwards and at right angles to horizontal faces of large stone blocks. 60 readings are obtained for each analyzed block. Readings are rejected if any individual impact test results in cracking or any other visible damage. The average value is recorded as the SHT rebound value according to the ASTM standards [6]. Equations correlating the compressive strength to SHT number are given in Table 1.

The point load test (PLT) is often employed as an indirect test method in order to evaluate the compressive or tensile strength of rock [7]. Table 2 lists the equations correlating compressive strength to PLT employed in this study. According to the ISRM standards [8], the ratio

*Corresponding author: giulia.forestieri@unisabana.edu.co

between compressive strength and PLT varies between 20 and 25 [9]. 32 cubic specimens for each stone type of 100 ± 5 mm edge are tested through a portable PLT machine connected to a barometer to register the maximum pressure. Results are corrected to a specimen diameter of 50 mm and the average value is recorded as the PLT strength.

Table 1. Equations correlating the estimated compressive strength ($UCS_{(SHT)}$) to the Schmidt hammer rebound (SHT).

Authors	Equation
Katz et al. (2000)	$UCS_{(SHT)}= 2.208e^{0.067SHT}$
Yilmaz and Sendir (2002)	$UCS_{(SHT)}= 2.27e^{0.054SHT}$
Fener et al. (2005)	$UCS_{(SHT)}= 4.24e^{0.059SHT}$

The direct uniaxial compressive strength (UCS) is evaluated through the Maschinen fabric Liezen system (MFL) testing machine at a constant speed rate of 1 mm/min with a maximum load capacity of 3000 kg. The test is repeated 10 times for each cubic stone specimen (50 ± 5 mm edge) and the average value is recorded as the UCS mean value.

Table 2. Equations correlating the indirect compressive strength ($UCS_{(PLT)}$) to the point load index (PLT).

Authors	Equation
D'Andrea et al. (1964)	$UCS_{(PLT)}= 15.3$ PLT + 16.3
Deere and Miller (1966)	$UCS_{(PLT)}= 20.7$ PLT + 29.6
Singh (1981)	$UCS_{(PLT)}= 18.7$ PLT - 13.2
Gunsallus and Kulhawy (1984)	$UCS_{(PLT)}= 16.5$ PLT + 51.0
Cargill and Shakoor (1990)	$UCS_{(PLT)}= 23.0$ PLT + 13.0

3 Results and discussion

Test results are given in Table 3 and are analysed using the method of least squares regression. Indirect tests values are correlated with the corresponding direct UCS values. The equation of the best-fit line, and the correlation coefficient are determined for each regression.

Table 3. Uniaxial compressive strength (UCS), Schmidt hammer rebound values (SHT) and point load strengths (PLT) for each investigated stone type.

Samples	UCS (MPa)	SHT (rebound)	PLT (MPa)
GF	104,1 ± 30,5	51,0 ± 2,1	2,1 ± 0,1
DG	68,4 ± 14,2	33,0 ± 2,0	1,2 ± 0,2
CS	33,8 ± 5,9	17,0 ± 0,9	0,8 ± 0,2
AF	31,0 ± 11,0	16,0 ± 0,8	0,5 ± 0,1
TL	82,8 ± 7,8	49,0 ± 3,0	1,6 ± 0,3

An exponential relation between SHT and UCS is found (Fig. 1) and the equation of the curve is:

$$UCS = 20.08e^{0.0316SHT} \quad r=0.95 \qquad (1)$$

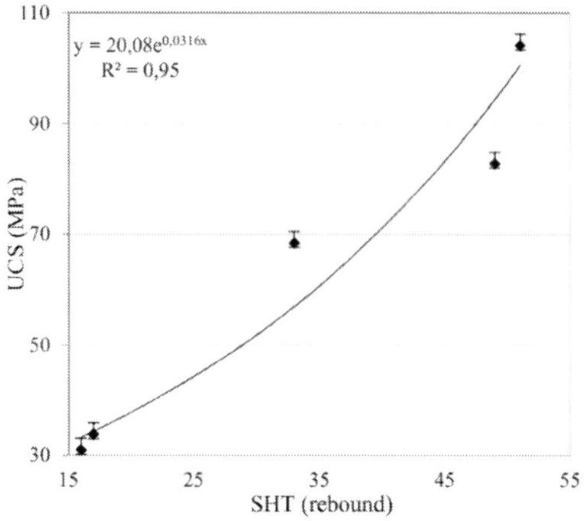

Fig. 1. Schmidt rebound value (SHT) vs. uniaxial compressive strength (UCS).

A linear relation between PLT and UCS is reported in Fig. 2. The equation of the line is:

$$UCS = 48.875PLT + 3.4154 \quad r=0.97 \qquad (2)$$

To check the estimation capability of the derived equations (1) and (2), the estimated UCS values are plotted against the direct UCS values for each equation, respectively (Figs. 3–4).

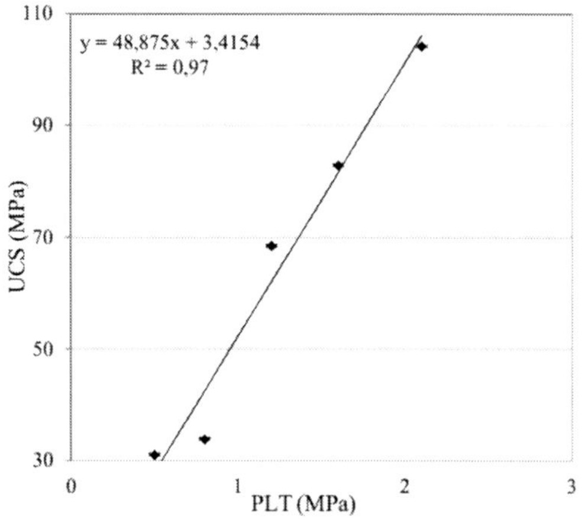

Fig. 2. Point load strength (PLT) vs. uniaxial compressive strength (UCS).

The error in the estimated value is represented by the distance that each data point plots from the line. A point lying close to the line indicates a better estimation than those points that are far than the interpolation line. As it is shown in Figs. 3–4, the points are scattered uniformly about the diagonal line, suggesting that the obtained models are reasonable.

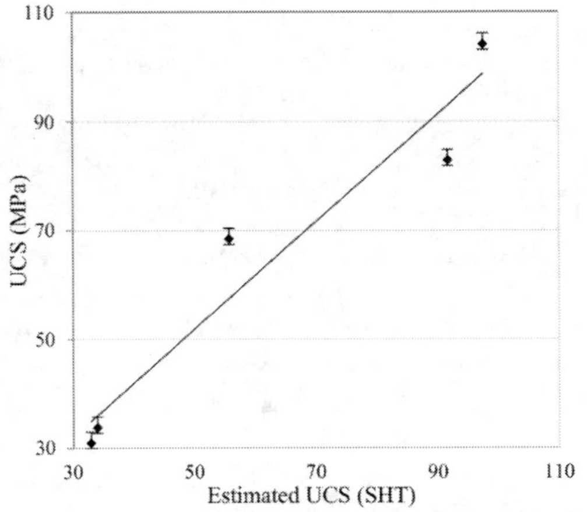

Fig. 3. Estimated UCS from Schmidt hammer rebound (SHT) vs. uniaxial compressive strength (UCS) using the derived equation 1.

Researchers used both linear and exponential equations to correlate SHT with UCS. Some researchers multiplied SHT by density to improve the correlation or by the porosity of the stone. In this study, exponential function gives the highest correlation coefficient. The relation found in this study is similar to the relations of Katz et al. (2000), Yilmaz and Sendir (2002), and Fener et al. (2005) [9].

As it is shown in the Fig. 1, the higher the stone strength the more scattered the data points are. The data points fall closer to the line at low strength values but become more scattered at higher strength values. This suggests that the ability to estimate the UCS of stones using SHT is better at low strength values, and is less reliable at higher strength values.

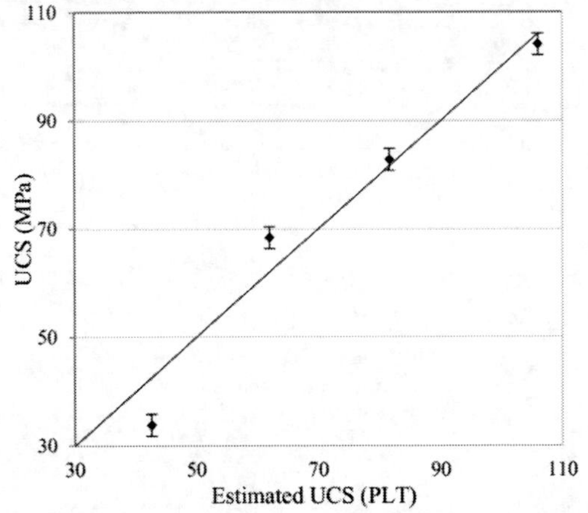

Fig. 4. Estimated UCS from point load strength (PLT) vs. uniaxial compressive strength (UCS) using the derived equation 2.

The comparison of the obtained equation between SHT and UCS and those derived from other researchers is shown in the Fig. 5. As shown in the graph, the derived equation is very similar to the equation developed by Fener et al. (2005) [9]. On the other hand, the most distant equation from the derived model of this study is represented by the equation of Ylmax and Sendir (2002) at first, and by the equation of Katz et al. (2000) [9].

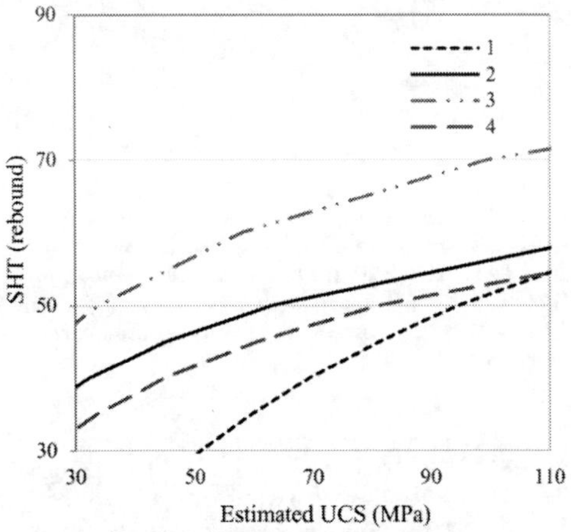

Fig. 5. Comparison to predict UCS from SHT of the derived equation from this study (1) with the equations [9] of: (2) Katz et al. (2000); (3) Ylmaz and Sendir (2002); (4) Fener et al. (2005).

Although, ISRM [8] stated that the ratio between UCS and PLT varies between 20 and 25, many researchers found different ratios, lower and higher than the above interval [9]. While some equations conform to y=ax form, the others conform to y=ax+b form as well as derived in this study. The derived equation of PLT vs. UCS is compared with the equations conforming to y=ax+b form from literature. As shown in the Fig. 6, the obtained equation in this research shows a trend that is similar to some of the other equations. In particular it is similar to the equations of Deere and Miller (1966), Gunsallus and Kulhawy (1984) and Cargill and Shakoor (1990) [9] due to the proximity of the estimated values from this study to those obtained by the above researchers. Among them, the relationship of Cargill and Shakoor (1990) crosses the estimated equation and shows a more horizontal trend, quite parallel to the equation of this study. On the contrary, the equations of Grasso et al. (1992), Kahraman (2001) and Fener et al. (2005) [9] can be rejected because their trend does not respect the trend shown by the derived equation (in particular the lines are not parallel to the line of the estimated equation from this study) and values obtained by those researchers are very distant from those calculated experimentally in this study.

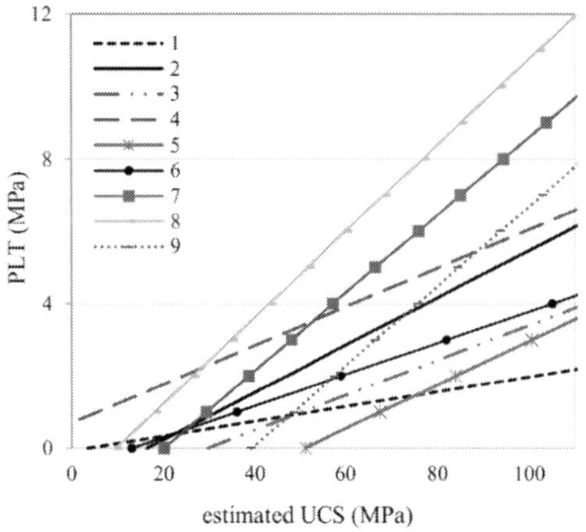

Fig. 6. Comparison to predict UCS from PLT of the derived equation from this study (1) with the equations [9] of: (2) D'Andrea et al. (1964); (3) Deere and Miller (1966); (4) Singh (1981); (5) Gunsallus and Kulhawy (1984); (6) Cargill and Shakoor (1990); (7) Grasso et al. (1992); (8) Kahraman (2001); (9) Fener et al. (2005).

Thanks to the Department of Civil Engineering of the University of Calabria (Italy) and to the IPG Laboratory of Castrolibero (Italy) where mechanical tests were performed. Thanks to the Universidad la Sabana (Colombia) for financing the research.

Conclusions

The uniaxial compression test, Schmidt hammer test and point load test are carried out on five Italian building stones (igneous and sedimentary) in order to develop predictive equations for uniaxial compressive strength. The usefulness of the indirect test methods is assured through test results. The obtained linear and exponential equations are compared to the relationships previously obtained by other researchers. It is found that there is no agreement between the equations suggested by different researchers. While some equations exhibit the same trend, the others differ. This is probably due to the differentiation of stone types and test conditions. Correlations show that the prediction of uniaxial compressive strength on the basis of different relationship gives in some cases inaccurate estimates and in others a good prediction. The most reliable results, especially in the case of the Schmidt hammer test, are obtained for stones with lower strength rather than for those that present higher uniaxial compressive strength. The most fitted relationship between other researches and this study is obtained by the equation of Fener et al. In the case of the point load test, the equation of Cargill and Shakoor shows the most similar trend to the estimated equation from this study. The application of the suggested equations must be practiced with care probably taking into account detailed stone features like the porosity or density.

References

1. G. Forestieri, D.M. Freire-Lista, A.M. De Francesco, M. Ponte, R. Fort, Int. J. Archit. Herit. **8** (2017), pp. 1-13 http://dx.doi.org/10.1080/15583058.2017.1354096
2. G. Forestieri, A. Tedesco, M. Ponte, Ge-Conservación (2017), pp. 102-109
3. G. Forestieri, A. Campolongo, M. Ponte, History of Engineering, **1** (2016) pp. 213-222
4. G. Forestieri, M. Ponte, Rend. Online Soc. Geol. It. **38** (2016), pp. 47-50, doi: 10.3301/ROL.2016.14
5. A. Aydin, A. Basu, Eng. Geol., **81** (2005), pp. 1-14
6. ASTM D5873, ASTM International, 2001.
7. Z.T. Bieniawski, Eng. Geol. **9**, 1 (1975).
8. ISRM, Int. J. Rock Mech. Min. Sci. Geomech. Abstr. **22**, 2 (1985)
9. M. Fener, S. Kahraman, A. Bilgil, O. Gunaydin, Rock Mech. Rock Eng. **38,** 4 (2005), pp. 329-343

MATEC Web of Conferences **192**, 01020 (2018) https://doi.org/10.1051/matecconf/201819201020
ICEAST 2018

Synthesis of *N*-vinylpyrrolidone/Acrylic acid nanoparticles for drug delivery: Method optimization

Chaiyakarn Pornpitchanarong, *Theerasak* Rojanarata, *Praneet* Opanasopit, *Prasopchai* Patrojanasophon and *Tanasait* Ngawhirunpat*

Department of Pharmaceutical Technology, Faculty of Pharmacy, Silpakorn University, Nakhon Pathom, 73000 Thailand

Abstract. There are various approaches to deliver therapeutic agents to the preferred target. Polymeric nanoparticles were found to have pleasing suitability as a drug carrier. The goal of this research was to optimize the synthesis method to obtain the desirable %yield and particle properties of the new biocompatible polymer-based nanoparticles. The non-toxic polymer, *N*-vinyl pyrrolidone (NVP) and a widely used hydrophilic biocompatible acrylic acid (AA) monomer were used to form the drug nanocarriers. The synthesis method was optimized by changing the types of initiator (KPS or V50) and the monomers molar ratio (NVP:AA). It was found that by varying both the monomer molar ratio and the type of reaction initiator, did not have significant effect on the physicochemical characteristics of the nanocarriers. The FT-IR spectra of all products exhibited the peaks of carboxylic acid, carbonyl, and tertiary amine functional group vibration. The particle size of the nanocarriers was in the range of 173.6 ± 18.4 to 201.4 ± 17.1 nm with negative surface charge. However, the yield obtained increased as the initiator was altered from KPS to V50, and when the acrylic acid molar ratio was increased from 1:1 to 1:3. In conclusion, changing the initiator and monomer molar ratio may affect the physicochemical properties of the nanocarriers and the %yield of the nanocarrier product. Further investigations are essential to obtain the favorable drug nanocarriers for drug delivery.

1 Introduction

Ideal drug carriers offer several advantages such as good stability, high drug loading capacity, be able to incorporate various drugs, varies in administration route, and be capable of control drug release.[1] Recently, polymer-based nanoparticles (PNPs) are attracting lots of interest from researchers due to their marvelous characteristics.[2] These particles demonstrate very small particle size ranging between 10 – 100 nm, and they can be generated from both natural and synthetic polymers. Nevertheless, the polymers used to prepare a nanocarrier are required to be nontoxic, biocompatible, and biodegradable.[3] In the synthesis of a polymer, chain-growth polymerization reactions can be used to polymerize one or more type of monomer with the aid of an initiator.[4]

The addition polymerization technique requires decomposition of an initiator to generate active species of free radicals, cations, or anions to polymerize the unsaturated monomers.[5] Free radical initiators (such as benzoyl peroxide, potassium persulfate (KPS), and 2,2'-azobis(isobutyronitrile) (AIBN)) propagate the polymer chain with a carbon radical. Cationic initiator such as Boron trifluoride (BF_3) and 2,2'-azobis(2-methylpropionamidine) dihydrochloride (V50) generates carbocation, while alkyl lithium anionic initiator grew up the chain with carbanion. To reach the process of initiation, the initiators must be solubilized and decomposed at a certain temperature while monomers remain stable. The selection of suitable initiator brings a worthwhile synthesis.[6-9]

The molar ratio of monomers plays an important role in polymer synthesis. Different monomer ratios may lead to the difference in the polymers' properties. Also, non-reacted or exceeded monomers can be classified as synthesis impurities and needed to be removed after the reaction terminated.[10] To gain a consistence PNPs in each synthesis, the molar ratio is needed to be justified.

NVP is a safe hydrophillic monomer and is widely used as PNPs stabilizers. When formed into a polymer, it helps to prevent aggregation by having steric hindrance structure and act as polymers or NPs dispersant; moreover, it is remarkably stable and inert.[11]

AA can be classified as a hydrophilic and biocompatible monomer. With its carboxylic acid functional group, polymers containing acrylic acid can be extensively modified in adjustment of their properties to gain PNPs of plentiful advantages. [12, 13]

Therefore, this work aims to optimize the synthesis polymerization reaction between NVP and AA through

*Corresponding author: tonglairoum_p@su.ac.th; Tel.: +66-08-5191-2848

the variation of monomer molar ratios and initiator concentrations. While considering the NVP/AA nanoparticle as the product, factors that are needed to be evaluated include %yielding, particle size, charge, and structural characteristics.

2 Materials and Methods

2.1 Materials

Acrylic acid (AA), *N*-vinylpyrrolidone (NVP), Potassium persulfate (KPS), 2,2'-Azobis(2-methylpropionamidine) dihydrochloride (V50), and *N,N'*-methylenebisacrylamide (MBA) were purchased from Sigma-Chemical Co. (St. Louis, MO, USA). All chemicals were of analytical reagent grade.

2.2 Preparation of NVP/AA nanoparticles using polymerization reaction

2.2.1 Effect of initiator

Nanoparticles were synthesized using surfactant-free free radical polymerization reaction. In brief, 50 mL of deionized water in a clean round-bottom flask was heated to 65-70 °C before the initiator (0.1 %w/w V50 or 0.2 %w/w KPS) was added. The reaction mixture was purged with nitrogen gas for 20 min to create an inert environment. In a 5 mL glass vial, NVP and AA were prepared at the weight ratio of 1:1. After that 10 wt% MBA, which was used as a crosslinking agent, was added to the glass vial and mixed. The mixture was then dispersed in chloroform (5 mL) before adding dropwise to the flask under vigorous stirring. The reaction environment was kept inert throughout 18 h of the reaction period. After that, the synthesized product was purified by dialysis against deionized water for 3 days before being freeze dried and lyophilized.

2.2.2 Effect of the NVP/AA molar ratio

The synthesis process mentioned earlier was repeated. V50 was selected to be used as the initiator, while the molar ratio of NVP/AA was varied from 1:1 to 1:3.

2.3 Nanoparticle characterizations

2.3.1 Fourier transform infrared spectroscopy (FT-IR)

The chemical structure and functional groups of the synthesized PNPs was elucidated using FT-IR (Spectrum 100, Perkin Elmer) The synthesized polymers were ground and pressed into KBr disks prior to the investigation.

2.3.2 Particle size and surface charge

The particle size and surface charge of the nanoparticles were evaluated using a Zetasizer Nano ZS (Malvern Instruments, Malvern, UK) at 25°C. Samples were dispersed, diluted to 1:100 and place into a zeta cell (1 cm^3) prior to the measurement. The measurement was performed in triplicate.

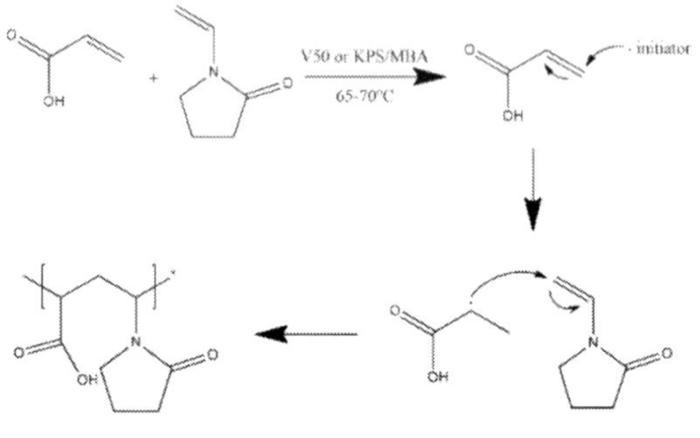

Fig. 1. Synthesis reaction of NVP/AA nanoparticle

3 Results and Discussion

3.1 NVP/AA nanoparticles: influence of initiator

The synthesis reaction was performed as shown in Fig. 1. Once the monomers were initiated with free radical, polymerization occurred while electrons transfer to one another and cross-linked to form the particle. The reaction was left for 18 h to complete the synthesis process prior to purification and drying. After the synthesis, the nanoparticles were characterized using FT-IR. The spectrum of all formulations were similar. The peaks at 3419, 1647-1725, 1174, and 1114 cm^{-1} were assigned to O-H, C=O, C-O, and C-N stretching vibration, respectively, which are the characteristics peaks of carboxylic acid (from AA) and tertiary amine (from NVP) functional groups. The appearance of these peaks are greatly suggestive of the successful polymerization of NVP and AA.

The %yield and nanoparticle characteristics of the nanoparticles obtained from the synthesis reaction using different initiators and monomer ratios are listed in Table 1. As it can be seen from the results, the %yield of the synthesized product increased as the initiator was changed to V50. However, this did not affect the size and surface charge of the obtained nanoparticles. The increase in the %yield may be because the different decomposition rate (k_d) of the initiators. The k_d of KPS at 70 °C is 2.83×10^{-5} sec^{-1}, while the rate of V50 is 1.21×10^{-4} sec^{-1}[14]. Therefore, it is obvious that at a higher rate of initiator decomposition, V50 may propagate the chain at a greater rate compared to KPS. Due to the increment of %yield and the smaller size of the particles, V50 was selected as the initiator for the molar ratio variation.

Table 1. Summary of method optimization results.

Initiator	Molar ratio	%yield	Size (nm)	Zeta Potential (mV)
KPS	1:1	12.3	201.4 ± 17.1	-42.7 ± 0.8
V50	1:1	24.0	173.6 ± 18.4	-40.0 ± 3.2
V50	1:3	42.3	178.9 ± 9.5	-22.3 ± 1.0

3.2 NVP/AA nanoparticles: influence of monomer molar ratio

To investigate the effect of monomer ratios on the %yield and the properties of the nanoparticles, V50 was selected as an initiator. The NVP:AA ratio was varied from 1:1 to 1:3. The FT-IR spectra of the nanoparticles prepared from different monomer ratios are presented in Fig. 2. The FT-IR spectra of the nanoparticles synthesized from different monomer ratios showed identical characteristic peaks of carboxylic acid and tertiary amine. As it can be observed from Table 1, at the monomer ratio of 1:3, the % yield was almost twice greater compared that obtained at the monomer ratio of 1:1 (42.29 vs 23.95%). In addition, changing the monomer molar ratio (NVP:AA) from 1:1 to 1:3 led to the more negative surface charge of the particles. This may be because the greater amount of acrylic acid was used to form the nanoparticles. However, this did not affect the particle size.

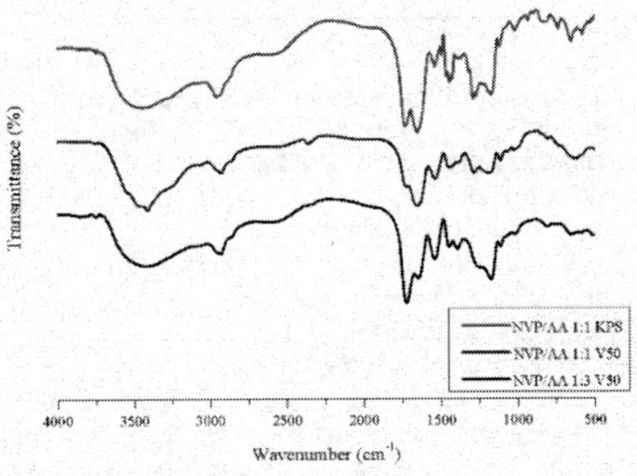

Fig 2. FT-IR spectra of the synthesized nanoparticles

4 Conclusions

There are various factors affecting polymer synthesis, including type of initiator and monomers molar ratio. This study demonstrated the influence of these elements on the final product obtained. It was found that while other reaction factors including monomer stoichiometry remains unchanged, percent yielding obtain form the reaction using KPS as reaction initiator was not comparable to V50. However, the PNPs obtained from the both reaction exhibited similar FT-IR peak characteristics, particle size and charge. Although the yield might differ, type of initiator did not affect the PNPs attributes. The molar ratio of the monomers also has an impact on the product yield. Stoichiometric equivalent ratio may not practically bring about the highest amount of product. Excess monomer input can cause greater product impurities, but also generates a more promising productive synthesis. Altering the monomer ratio could lead to the alteration in the product yield and particle attributes. Nevertheless, other synthesis reaction factors such as synthesis method, time, temperature, initiator concentration and cross-linker ratio may also affect the properties of the synthesis products.

The authors acknowledge the Faculty of Pharmacy, Silpakorn University and The Commission of Higher Education (Thailand) and the Thailand Research Funds through the Golden Jubilee Ph.D. Program (Grant No.PHD/0021/2560) for all financial supports.

References

1. A. Z. Wilczewska, K. Niemirowicz, K. H. MarkiewiczH. Car, Pharmacological Reports, **64**, 1020-1037, (2012)
2. D. M. Eckmann, R. J. Composto, A. TsourkasV. R. Muzykantov, J Mater Chem B, **2**, 8085-8097, (2014)
3. W. H. De JongP. J. A. Borm, International Journal of Nanomedicine, **3**, 133-149, (2008)
4. B. Kostova, E. Kamenska, G. Momekov, D. Rachev, G. GeorgievK. Balashev, European Polymer Journal, **49**, 637-645, (2013)
5. S. Nimesh, Gene Therapy, 13-42, (2013)
6. H.-R. Lin, European Polymer Journal, **37**, 1507-1510, (2001)
7. D. Hunkeler, Macromolecules, **24**, 2160-2171, (1991)
8. C. Costa, V. H. S. Santos, P. H. H. Araujo, C. Sayer, A. F. Santos, C. DarivaM. Fortuny, Journal of Applied Polymer Science, n/a-n/a, (2010)
9. G. Chen, X. Zhu, Z. Cheng, W. XuJ. Lu, Radiation Physics and Chemistry, **69**, 129-135, (2004)
10. S. Ilic-Stojanovic, L. Nikolic, V. Nikolic, M. Stankovic, J. Stamenkovic, I. Mladenovic-RanisavljevicS. Petrovic, Chemical Industry and Chemical Engineering Quarterly, **18**, 1-9, (2012)
11. K. M. Koczkur, S. Mourdikoudis, L. PolavarapuS. E. Skrabalak, Dalton Trans, **44**, 17883-17905, (2015)
12. Y. Hu, X. Jiang, Y. Ding, H. Ge, Y. YuanC. Yang, Biomaterials, **23**, 3193-3201, (2002)
13. R. Melinda Molnar, M. Bodnar, J. F. HartmannJ. Borbely, Colloid and Polymer Science, **287**, 739-744, (2009)
14. C. Costa, V. H. S. Santos, P. H. H. Araujo, C. Sayer, A. F. Santos, C. Dariva, M. Fortuny, J Appl Polym Sci, **118**, 1421-1429, (2010)

Environmental performance assessment using the evidential reasoning approach: The case of logistics service providers

Panitas Sureeyatanapas[1,*] and *Korrapat Tawwan*[1]

[1]Department of Industrial Engineering, Faculty of Engineering, Khon Kaen University, 123 Moo 16, Mittraphap Road, Khon Kaen 40002, Thailand

Abstract. The demand for environmental performance assessment is increasing among business practitioners, and it has nowadays become one of the key factors for a company's self-improvement as well as for selecting suppliers and logistics providers. The assessment is, in essence, a multiple criteria decision analysis (MCDA) problem comprised of many quantitative and qualitative criteria. Frequently, the assessment data of some criterion is inevitably imprecise and/or incomplete since the nature of environmental assessment relies heavily on professional and complex methods which might not be fully available for every company. Also, qualitative criteria can only be assessed based upon human judgment. This paper, therefore, proposes an application of the evidential reasoning (ER) approach to the assessment of environmental performance for logistics service providers. The lists of criteria and indicators are adapted from ISO 14031. The ER approach is able to logically aggregate all assessment information, although different forms of data (precise or imprecise; complete or incomplete) are obtained. For this paper, assessment data from two logistics companies were gathered and analysed to illustrate the implementation process. The results are in the form of aggregated belief distributions on a unified set of evaluation grades, and the company can use this information for performance improvement and benchmarking.

1 Introduction

Due to growing public awareness of environmental issues and stricter laws and regulations in many countries, product manufacturers nowadays intensively focus on auditing and assessing environmental performance of their suppliers, including logistics partners, in order to improve the overall supply chain performance [1]. Since transportation activities have been claimed as one of the major sources of global warming and the depletion of non-renewable energy, one of the emerging criteria for selecting and auditing suppliers is their abilities to preserve the environment [2-4]. The logistics industry, therefore, has slightly shifted from pricing competition to value-adding services in terms of green and sustainability capabilities [2, 5].

It is generally recognised that performance measurement is crucial for business success. The assessment helps promote continuous improvement, presents achievements of corporate goals of environmental protection, provides data for internal decision making, and ensures that firm's environmental performance meets or goes beyond legislative requirements, as such risks of legal fines and penalties are reduced [4, 6, 7]. ISO 14031 is one of the most well-known standards for environmental management that specifically focuses on performance evaluation. The growing popularity of ISO 14031 can be explained by a number of reasons. First of all, it can be simply adopted by every single organisation without any limitations in terms of specific sizes and types of industry. Secondly, since it is a kind of standardised framework, comparison with others is possible. It also, thirdly, facilitates the communication of environmental performances in a systematic way as well as provides comprehensive information for policy decision-making [7, 8]. However, the application of ISO 14031 to the logistics industry is still limited among the literature.

Although ISO 14031 could be a choice for logistics firms to start assessing and reporting their environmental performance, a number of criticisms have been made. Firstly, most aspects of performance evaluation suggested in ISO 14031 are still multi-dimensional and qualitative, such as "Wastes" or "Community relations". That means the evaluation still relies on subjectivity and may vary according to personal interpretation. Secondly, since a large number of indicators are exemplified within the standard, it is likely that monitoring too many indicators simultaneously can distract practitioners from the main focuses of their organisations [1]. From this, a method to aggregate various indicators is needed in order for firms to be able to monitor the overall picture of their environmental performance. The obtained aggregated number is also open for business benchmarking in a credible way.

Another issue is that ISO 14031 does not provide a solution to deal with uncertainty and/or incompleteness of the assessment information. A number of studies in

* Corresponding author: panisu@kku.ac.th

green supply management implied that many firms feel uncomfortable when reporting information relating to environmental issues [9, 10]. This might be because such information relates to pollution and negative impacts upon their operations release into the environment. Furthermore, some firms may lack the capability to conduct a self-assessment in relation to some indicators, particularly the ones that require specific technology, laboratory, or skilled personnel. Moreover, firm's performances according to some indicators may fluctuate over time or vary by its branches or sites. In these cases, the average may not always be a good choice since it possibly causes the loss of meaningful information. For some qualitative aspects, in addition, the assessor may lack confidence in giving a precise judgement due to the unavailability or incompleteness of evidence. Although a set of evaluation grades or rating scales can be used to transform subjective opinions into numerical data, each assessor may interpret the meaning of each grade or scale differently. Also, they might not be confident when concluding that a situation being considered matches a particular grade, and one or more grades may better suit the actual practice. Lastly, among the various aspects or criteria in ISO 14031, practitioners within the same organisation, or within the same chain of benchmarking partners, may perceive the importance levels of those aspects differently. This may cause disagreement with regard to the aggregated score obtained. These issues support that the method used to determine the overall environmental performance needs the ability to solve the issues of incompleteness and uncertainty of the information in a sensible way. Furthermore, in the logistics industry, the lack of studies which focus on managers' attitudes toward the importance or the weight of each criterion still represents a gap in the literature.

The aim of this study is therefore to propose a logical method to assess and analyse environmental performance for logistics operations. The tool must be able to perform even when uncertainty and/or incompleteness of the assessment information, as well as uncertainty of the criteria weights, exist. To respond to these, the evidential reasoning approach (ER) is employed. ER can theoretically cope with the limitations of other MCDA methods generally used to determine a composite index. For example, the simple additive weighting (SAW) method which relies on the additive approach assumes that all indicators are preferentially independent [11, 12]. This assumption may not be realistic in conducting an environmental performance assessment. The aggregation based on the weighted geometric mean (WGM) method [13], on the other hand, does not allow for any compensation among indicators. That means the overall performance will be zero when at least one indicator has zero value. This assumption may not always be acceptable in business performance evaluation since a company's weakness in one indicator may be compensated by a very strong point in another, which helps to build a competitive advantage. For the data envelopment analysis (DEA) [14] and the technique for order preference by similarity to an ideal solution (TOPSIS) [15], the composite score or the overall performance for a company is determined by using a comparison of the performance of other companies. They might be therefore applicable only for ranking alternatives but not for a single company's self-assessment [16]. SAW, WGM, DEA, or TOPSIS are also not designed to deal with qualitative indicators, nor any uncertainty of the data.

This paper is organised as follows: Following the introduction, ISO 14031 is briefly described in Section 2. Section 3 then explains the algorithms of the ER approach. Section 4 gives details of the research methodology. The applicability of the ER approach is then demonstrated using data from two logistics companies. The results and discussion are given in Section 5, and Section 6 is the conclusion.

2 ISO 14031

ISO 14031 is an international standard that provides a guideline for assessing and monitoring an organisation's environmental performance [1, 17]. It describes two dimensions of performance indicators: (i) Management performance indicators (MPIs) and (ii) Operational performance indicators (OPIs). MPIs indicate management capability and efforts to improve an organisation's environmental performances while OPIs provide direct measurements about such performances (e.g., air emission; the use of energy and water). MPIs are also linked to economic and social dimensions, such as indicators that show an organisation's capabilities to educate or encourage staff to follow environmental protection guidelines; costs and investments spent on this matter. MPIs are divided into four criteria: (i) Implementation of environmental policies and programmes, (ii) Regulatory compliance, (iii) Financial performance associated with environmental performance, and (iv) Community relations. OPIs, on the other hand, are broken down into nine criteria: (i) Materials, (ii) Energy, (iii) Services supporting organisation's operations, (iv) Physical facilities and equipment, (v) Supply and delivery, (vi) Products, (vii) Services provided by the organisation, (viii) Wastes, and (ix) Emissions [18]. For this study, however, the initial set of OPIs was adapted to suit the characteristics of the logistics industry. Finally, only seven criteria remain to represent OPIs, including (i) Materials, (ii) Energy, (iii) Services supporting organisation's operations, (iv) Physical facilities and equipment, (v) Products and services, (vi) Wastes, and (vii) Emissions.

As suggested by ISO 14031, it is not necessary for practitioners to include all criteria of MPIs and OPIs in their evaluation system. Each firm normally has its own interests or concerns that relate to the environment. To implement ISO 14031, for each firm, suitable indicators (and measurement units) under each criterion should be identified [18].

3 The evidential reasoning approach

The ER approach is one MCDA technique that has been applied to many areas of performance assessment and decision making, such as the self-assessment following the European Foundation for Quality Management (EFQM) Model [19], the assessment of new product design [20], the assessment based upon the balanced scorecard [21], the evaluation and selection of research projects [22], and the prioritisation of energy sources [23].

The ER approach was firstly proposed by Yang and Singh [24] as a procedure to combine multiple criteria based upon the evidence theory that employs a belief distribution to describe the assessment. The algorithm developed in their article is based on the recursive approach that requires pairwise aggregations. Wang, et al. [25] then proposed an analytical ER algorithm to present an explicit aggregation function by combining all indicators simultaneously. The assessment of each indicator, e_i, is conducted towards a set of grades H, $H = \{H_1, H_2, ..., H_n, ..., H_N\}$, where H_{n+1} is preferred to H_n (H_1 and H_N denote the worst and the best grades, respectively). All grades must be defined to be mutually exclusive and collectively exhaustive [26].

Suppose that the total performance could be assessed towards L indicators, or a set of indicators e_i ($i=1,...,L$), the assessment result (S) for a company M in terms of each indicator e_i can be expressed in the form of a belief distribution, as shown below.

$$S(e_i(M)) = \{(H_n, \beta_{n,i}(M)), n = 1, ..., N; (H, \beta_{H,i}(M))\}, \quad i = 1, ..., L \tag{1}$$

$\beta_{n,i}(M)$ represents a belief degree in which a company M is assessed to a grade H_n when an indicator e_i is considered, where $\beta_{n,i}(M) \geq 0$ and $\sum_{n=1}^{N} \beta_{n,i}(M) \leq 1$. $B_{H,i}(M)$ is the degree of global ignorance or the portion of belief unassigned to subsets of H, $\beta_{H,i}(M) \geq 0$ and $\sum_{n=1}^{N} \beta_{n,i}(M) + \beta_{H,i}(M) = 1$ [26].

Wang, et al. [27] extended the ability of the ER approach to handle interval belief degree as a result of the inclusion of interval data from the assessment. Guo, et al. [28] then demonstrated a case in which interval weights of criteria were inserted into the ER algorithm. When such kinds of uncertainty exist, the belief distribution of company M towards the indicator e_i is also expressed in the interval form, as shown below.

$$S(e_i(M)) = \{(H_n, \beta_{n,i}(M) \in [\beta_{n,i}^-(M), \beta_{n,i}^+(M)]), \ n = 1, ..., N; \ (H, \beta_{H,i}(M) \in [\beta_{H,i}^-(M), \beta_{H,i}^+(M)])\}, \ i = 1, ..., L \tag{2}$$

In response to a case in which interval belief degrees and interval weights are involved, the analytical ER algorithm is handled through the optimisation functions, as shown below. The next explanations of the ER algorithm follow Wang, et al. [27], Guo, et al. [28], and Sureeyatanapas, et al. [16]. Note that $n = 1,..., N$ and $i = 1,...,L$.

Objective functions:

$$\text{Max/Min } \beta_n(M) = \frac{m_n}{1 - \bar{m}_H} \tag{3}$$

$$\text{Max/Min } \beta_H(M) = \frac{\tilde{m}_H}{1 - \bar{m}_H} \tag{4}$$

Subject to (s.t.):

$$m_n = k\left[\prod_{i=1}^{L}(m_{n,i} + \bar{m}_{H,i} + \tilde{m}_{H,i}) - \prod_{i=1}^{L}(\bar{m}_{H,i} + \tilde{m}_{H,i})\right] \tag{5}$$

$$\tilde{m}_H = k\left[\prod_{i=1}^{L}(\bar{m}_{H,i} + \tilde{m}_{H,i}) - \prod_{i=1}^{L}\bar{m}_{H,i}\right] \tag{6}$$

$$\bar{m}_H = k\left[\prod_{i=1}^{L}\bar{m}_{H,i}\right] \tag{7}$$

$$k = \left[\sum_{n=1}^{N}\prod_{i=1}^{L}(m_{n,i} + \bar{m}_{H,i} + \tilde{m}_{H,i}) - (N - 1)\prod_{i=1}^{L}(\bar{m}_{H,i} + \tilde{m}_{H,i})\right]^{-1} \tag{8}$$

$$m_{n,i} = w_i\beta_{n,i}(M) \tag{9}$$

$$\tilde{m}_{H,i} = w_i\beta_{H,i}(M) \tag{10}$$

$$\bar{m}_{H,i} = 1 - w_i \tag{11}$$

$$w_i^- \leq w_i \leq w_i^+ \tag{12}$$

$$\sum_{i=1}^{L} w_i = 1 \tag{13}$$

$$\beta_{n,i}^-(M) \leq \beta_{n,i}(M) \leq \beta_{n,i}^+(M) \tag{14}$$

$$\beta_{H,i}^-(M) \leq \beta_{H,i}(M) \leq \beta_{H,i}^+(M) \tag{15}$$

$$\sum_{n=1}^{N} \beta_{n,i}(M) + \beta_{H,i}(M) = 1 \tag{16}$$

$$\sum_{n=1}^{N} m_{n,i} + \bar{m}_{H,i} + \tilde{m}_{H,i} = 1 \tag{17}$$

where w_i represents the relative weight of the indicator e_i. The weights of all indicators must be normalised to sum to one; $m_{n,i}$ is the weighted belief which supports the assessment of M to the grade H_n based on the indicator e_i; The probability mass unassigned to any grade after all N grades have been considered is a combination of two parts: $\bar{m}_{H,i}$ and $\tilde{m}_{H,i}$. $\bar{m}_{H,i}$ is the remaining probability mass caused by the fact that an indicator e_i only partly contributes to the assessment relative to its weight. $\bar{m}_{H,i}$ is equal to zero if e_i dominates the assessment or its relative weight is equal to one. $\tilde{m}_{H,i}$ is an unassigned probability mass resulting from the incompleteness of the assessment $S(e_i)$. $\tilde{m}_{H,i}$ is zero if the assessment is complete, or $\sum_{n=1}^{N} \beta_{n,i} = 1$.

For this model, two sets of variables, including the intervals for (i) criteria weights and (ii) belief degrees, are the inputs of the optimisation process. When the precise assessment is given for some indicators, the minimum and maximum values of the interval are equal to that precise value. After combining all L indicators, the total belief degrees, β_n and β_H, can be generated. β_H denotes the degree of belief unassigned to any single grade, or it reflects the total degree of incompleteness in the assessment. The aggregated result for each company M, or $S(y(M))$, can be described as follows.

$$S(y(M)) = \{(H_n, \beta_n(M) \in [\beta_n^-(M), \beta_n^+(M)]); (H, \beta_H(M) \in [\beta_H^-(M), \beta_H^+(M)])\}$$
$$\text{with } \sum_{n=1}^{N} \beta_n(a_l) + \beta_H(a_l) = 1 \tag{18}$$

The number of nonlinear optimisation models required to aggregate belief degrees of each company is equal to $2(N+1)$ for one combination process. The obtained result shows a belief distribution presenting a panoramic view of the overall performance. However, if the aim of the assessment is for ranking or comparing alternatives, the aggregated belief distribution cannot straightforwardly support those purposes. The determination of the expected utilities is then suggested. Utility reflects the assessor's preference for the value of the indicator being considered [12]. The utility of a grade H_n, or $u(H_n)$, could be between zero to one, and $u(H_{n+1}) > u(H_n)$. The expected utility of M can be determined, without any kind of uncertainty, through equation (19).

$$u\big(S(y(M))\big) = \sum_{n=1}^{N} \beta_n(a_i)u(H_n) \tag{19}$$

Nevertheless, the expected utility of M will be in the form of an interval if the assessment involves incomplete or uncertain information. The interval is determined by simply assuming that the unassigned belief degree, (β_H), is transferred to the best and the worst grades to represent the upper and lower bounds of the interval, respectively. To determine the maximum and minimum expected utilities, denoted by $U_{max}(M)$ and $U_{min}(M)$, the following pair of optimisation models are employed. The comparison of alternatives can be conducted based on the average utility, or $U_{avg}(M)$, which can be defined as the midpoint between $U_{max}(M)$ and $U_{min}(M)$. The input variables for this model are the aggregated degrees of belief and the utilities of each grade.

Objective functions:

$$\text{Max } U_{max}(M) = \sum_{n=1}^{N-1} u(H_n)\beta_n(M) + u(H_N)\big(\beta_N(M) + \beta_H(M)\big) \tag{20}$$

$$\text{Min } U_{min}(M) = u(H_1)(\beta_1(M) + \beta_H(M)) + \sum_{n=2}^{N} u(H_n)\beta_n(M) \tag{21}$$

s.t.

$$\beta_n^-(M) \leq \beta_n(M) \leq \beta_n^+(M) \tag{22}$$

$$\beta_H^-(M) \leq \beta_H(M) \leq \beta_H^+(M) \tag{23}$$

$$\sum_{n=1}^{N} \beta_n(M) + \beta_H(M) = 1 \tag{24}$$

4 Research methodology

The assessment process starts by investigating the importance (or the weights) of each criterion in respect of the logistics industry. It is considered that the weights should not be individually determined by a single organisation, although the assessment results are only monitored internally for that firm. This is because, for each firm, the image of 'being green' or sustainability does not depend only on their own perception, but it is also judged by the external community. Taking into account how people within the same industry perceive which criteria significantly contribute to the overall environmental image can help firms improve their own performance in a more efficient way. For this study, the weights were elicited from 10 industry experts (hereafter referred to as the "decision makers", or "DMs") based on interviews and the direct rating technique. The 10 DMs include two academic lecturers in logistics management and eight senior managers from different truck fleet logistics companies in Thailand. There are several product types delivered by the eight companies, including metals and steel, cement, concrete, parcels, agricultural products, frozen foods, beverages, etc.

The direct rating is one of the most straightforward weighting methods available. It is highly recommended when the performance evaluation relies on a large number of criteria and indicators, and when a respondent does not feel comfortable using complex weighting methods [29, 30]. The direct rating only requires a DM to directly assign the weight to each criterion through a simple scale, such as 0-10 points ranging from 'extremely unimportant' to 'extremely important', respectively [31]. For this study, each DM was asked to assign the weights to each MPI criterion, followed by each criterion under OPIs. Finally, the weights were assigned to the overall picture of MPIs and then to that of OPIs. The DMs were also asked to provide supporting reasons. The weights of criteria belonging to the same group were then normalised to be relative to each other (they were sum to one). Indicators being under the same criterion were assumed to take equal weight in the aggregation processes. This is because the list of indicators under each criterion is likely to be changeable depending on a firm's current focuses and strategies.

To demonstrate the applicability of the ER approach to the ISO 14031-based environmental assessment, indicators representing each criterion were identified. Each criterion was able to be assessed through either quantitative or qualitative indicators, or both, and the numbers of indicators were not necessary to be equal among criteria. A measurement unit was specified for each quantitative indicator, and a set of evaluation grades was determined for each qualitative one. A clear definition referring to objective evidence or feasible situations was attached to each grade in order to standardise the assessment or to minimise subjective bias. For each qualitative indicator, the set of grades should cover all feasible practices. The number of grades might be different among different indicators. This depends on how many distinct levels of performances can be classified for each indicator [32]. Then, actual data sets from two logistics companies were employed to demonstrate the analysis. The data was gathered through the interviews of the company managers. They were two of the eight managers giving the weight information. The interviews were conducted during January 2018. For each quantitative indicator, the managers were asked to provide the data of the year 2017. They, however, were informed that they could give interval values if they were uncertain about the answers, or if the data fluctuated over a year. Another option was to skip some indicators if they completely lacked information about such indicators, or if they did not want to disclose such information. In terms of qualitative assessment, the managers were asked to describe the extent of their practices regarding each particular indicator, without showing them the evaluation grades. Belief degree was then assigned to each grade by the researchers based upon the evidence or the stories provided by the managers. Since the aggregation of indicators according to the ER approach requires that all indicators must be assessed towards the same scale, five general grades (None, Poor, Fair, Good, and Excellent) were employed for this study. To unlock such condition, the rule-based transformation technique, introduced by Yang [26] and further developed by Wang, et al. [27] and Guo, et al. [28], was employed to transform various forms of original assessment data into belief distributions attached to the general grades. Due to the word limit, the transformation processes cannot be fully presented in this paper. After the transformation, the results of all indicators were arranged into a form that was applicable for the ER to derive the overall performance for each company, using the weight ranges determined by the 10 industry experts. Note that the nonlinear optimisation here was performed through LINGO software.

5 Results and discussion

5.1 The importance of the criteria (the weights)

Table 1 shows the relative weights of the criteria elicited from the 10 experts. The weights are presented in the form of the intervals to reflect different attitudes towards the importance or the contribution of each criterion to the overall environmental performance. The averages are also presented.

Table 1. Relative weights of the criteria.

Criteria	Weight ranges	Weight average
MPI1 - Implementation of environmental policies and programmes	0.21-0.30	0.25
MPI2 - Regulatory compliance	0.26-0.50	0.30
MPI3 - Financial performance associated with environmental performance	0.10-0.27	0.23
MPI4 - Community relations	0.10-0.24	0.22
OPI1 - Materials	0.00-0.15	0.09
OPI2 - Energy	0.15-0.30	0.19
OPI3 - Services supporting organisation's operations	0.10-0.16	0.12
OPI4 - Physical facilities and equipment	0.10-0.25	0.15
OPI5 - Products and services	0.00-0.17	0.10
OPI6 - Wastes	0.13-0.20	0.16
OPI7 - Emissions	0.14-0.30	0.19
MPIs	0.30-0.56	0.46
OPIs	0.44-0.70	0.54

For MPIs, in general, most DMs assigned a high weight to MPI2 indicating that they greatly realised the impact of legal fines and penalties on the image of their environmental performance. Many of them stressed that, for a firm to sustain itself in the logistics business, regulations relating to freight and transportation must be strictly considered. A bad image regarding this point would destroy customers' loyalty and decrease prospective sales. MPI1, based on an average basis, seemed to be the second-most important management criterion. Many DMs stated that, to be successful in green programs, environmental concerns must be embedded into a company's policies and strategies.

Energy consumption (OPI2) and Emissions (OPI7) became the two most important operational criteria for the logistics industry. From the interviews, this sector consumed a large amount of diesel, and this led to the emission of various greenhouse gases. Most of the green logistics practices, therefore, focused on minimising fuel consumption and air pollution. The cost of diesel also represented the main cost of a transportation service, which needed to be minimised. Wastes (OPI6) also received a high weight from many DMs due to a strong association between logistics operations and many kinds of solid wastes, such as used tyres, pallets, paper boxes.

Regarding the two main aspects of environmental performance, based on an average basis, the weight of OPIs seemed to be only slightly higher than that of MPIs. Many DMs believed that actual outputs and outcomes better reflected a company's successful implementation of green initiatives than merely displaying management strategies. On the other hand, a number of DMs expressed a contrastive opinion that management should be the most important since it is the driving mechanism for a company to reach an expected outcome. However, the weight ranges of MPIs and OPIs greatly overlapped each other, indicating that the ranking based on the averages was not generalised. The intervals of the weights were then inputted into the optimisation model in order to analyse the environmental performance of the two companies.

5.2 Indicators and the transformation into belief distributions

To demonstrate the applicability of the ER approach, 20 indicators (belonging to the four MPIs and the seven OPIs) were generated, as shown in Table 2. They were adapted from the examples of indicators listed in ISO 14031 [18]. A measurement unit was specified for each quantitative indicator, and a set of evaluation grades (with clear definitions attached to each grade) was determined for each qualitative one. Equivalent rules for the transformation were also established for each indicator by referring to the industry's best and worst cases, based mainly on the review of the literature and the interviews. Unfortunately, these kinds of information were not displayed in this article due to the word limit.

For this paper, MPI3.1 and MPI1.3 were used to respectively demonstrate the transformation of assessment data into belief distributions for quantitative and qualitative indicators. For MPI3.1, its measurement unit was a percentage (%). Figure 1 shows the equivalent rule used for the transformation. The value 20% was considered here to be equivalent to the grade 'Excellent', while 'None' was assigned to 0%. The concept of linear utility function was then applied to the remaining grades.

Table 2. List of indicators under each criterion.

Criteria	Indicators
MPI1	MPI1.1: The number of environmental initiatives per year (number)
	MPI1.2: Percentage of employees participating in environmental initiatives (%)
	MPI1.3: Environmental performance assessment (evaluation grades *A-D*)*
	MPI1.4: Monitoring of fuel consumption (evaluation grades *A-D*)*
MPI2	MPI2.1: Number of non-compliance with laws and regulations per year (number)
MPI3	MPI3.1: Percentage of total expenditure on environmental research and initiatives relative to the net profit per year (%)
MPI4	MPI4.1: The number of complaints from the local community per year (number)
	MPI4.2: Progress on local remediation and social development activities (evaluation grades *A-E*)*
	MPI4.3: Progress on a social survey about corporate community relations (evaluation grades *A-E*)*
OPI1	OPI1.1: The use of recycled or reused materials (evaluation grades *A-E*)*
OPI2	OPI2.1: Diesel used relative to a kilometre of transportation per year (litres per kilometre)
OPI3	OPI3.1: Waste management related to supporting operations (evaluation grades *A-D*)*
OPI4	OPI4.1: Percentage of vehicles in fleet with pollution-abatement technology (%)
	OPI4.2: Progress on vehicle preventive maintenance program (evaluation grades *A-C*)*
OPI5	OPI5.1: Ability to launch environmentally-friendly products/services (evaluation grades *A-D*)*
OPI6	OPI6.1: Management of used tyres (evaluation grades *A-C*)*
	OPI6.2: Management of used lubricating oil (evaluation grades *A-C*)*
	OPI6.3: The weight of solid waste disposed of per year (tonnes)
OPI7	OPI7.1: Carbon dioxide (CO_2) emission per a tonne and a kilometre of transport (kilogram of CO_2 equivalent/tonne-kilometre)
	OPI7.2: Emission of particulate matter under 10 microns diameter per a kilometre of transport (grams/kilometre)

*Qualitative evaluation

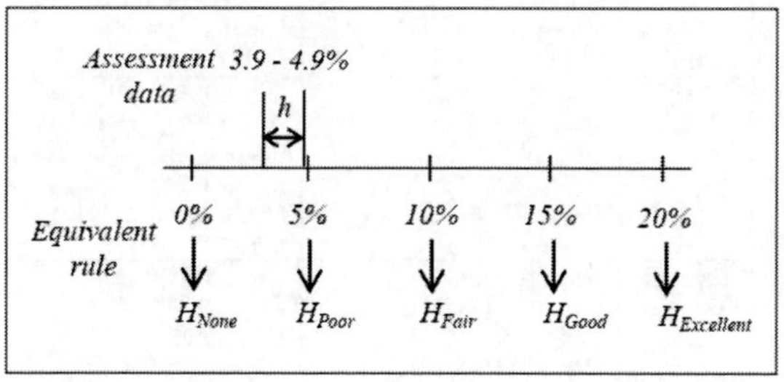

Fig. 1. Assessment data of company *M* on MPI3.1 and the equivalent rule.

Given h is the assessment data of MPI3.1 of company *M*, and $h \in [3.9, 4.9]$, which is an interval value covered by two adjacent grades (H_{None} and H_{Poor}), as shown in Figure 1. Following Wang, et al. [27], the degrees of belief assigned to H_{None} and H_{Poor} are then derived as the intervals, $[\beta^-_{n,i}, \beta^+_{n,i}]$ and $[\beta^-_{n+1,i}, \beta^+_{n+1,i}]$, as follows:

$$\beta^-_{None} = \frac{5 - 4.9}{5 - 0} = 0.02, \qquad \beta^+_{None} = \frac{5 - 3.9}{5 - 0} = 0.22$$

$$\beta^-_{Poor} = \frac{3.9 - 0}{5 - 0} = 0.78, \qquad \beta^+_{Poor} = \frac{4.9 - 0}{5 - 0} = 0.98$$

The data was finally transformed into the general form: *S(MPI3.1(M))* = $\{(H_{None}, [0.02, 0.22]); (H_{Poor}, [0.78, 0.98])\}$.

For MPI1.3, an example of qualitative indicators, it was assessed through four grades (A-D). The equivalent rule is displayed by Figure 2.

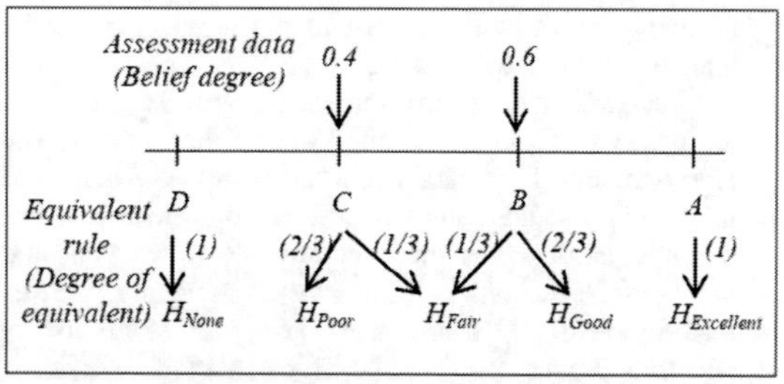

Fig. 2. Assessment data (belief degree) of company *N* on MPI1.3 and the equivalent rule.

Figure 2 shows that the local grade '*C*' was equivalent to the general grade 'Poor' for only around 0.666 degree of equivalent, while another 0.334 degree falls within the grade 'Fair'. The same concept is also applied to the grade '*B*'. According to the story given by the manager of company *N* and the definitions attached to the assessment grades *A-D* for MPI1.3, the company's performance towards this indicator falls within the grade '*B*' for 60% and another 40% is in accordance with grade '*C*'. Then, following Yang [26], the assessment data was transformed into the general form: *S(MPI1.3(N))* = $\{(H_{Poor}, 0.27); (H_{Fair}, 0.33); (H_{Good}, 0.4)\}$. At the end, the transformed belief distributions are summarised in Table 3. Note that $(H_H, 1)$ means that there was a lack of information due to sufficient data not being available or completely unknown, or the company did not want to disclose such information.

5.3 Application of the ER approach

The data from Table 3 was then used to analyse the environmental performance of each company under uncertain and incomplete information through the ER nonlinear optimisation model, or Equations (3)-(17). For each company, the aggregation of all indicators under the same criterion was firstly performed. This was followed by the aggregation of all MPIs. Then, all OPIs were combined. Finally, the overall performance was determined by aggregating the two main aspects. Table 4 shows the aggregated interval degrees of belief for the overall MPI, OPI, and the environmental performance of the two companies. The intervals of the expected utilities were derived by Equations (20)-(24). The results are shown in Table 5. For this study, the utilities of the five general grades; $u(H_{None})$, $u(H_{Poor})$, $u(H_{Fair})$, $u(H_{Good})$, and $u(H_{Excellent})$; were set to be 0, 0.25, 0.5, 0.75, and 1, respectively.

Table 5 shows that company M receives the highest average utility in terms of the overall environmental performance. However, when considering the ranges of the utilities, the maximum utility of company N (0.74) is greater than the minimum of company M (0.42). This means M does not absolutely dominate N. If all information is precisely known, it is possible that the N's performance might be greater than M's.

Based on the average basis, M is ranked first in both MPIs and OPIs. When considering MPIs, the expected utility of M completely dominates that of N (the minimum of M is greater than the maximum of N). This is mainly due to the fact that M performs much better than N in both MPI1 and MPI2 which receive high weights from the industry representatives. For M, there are plenty of programmes and projects that relate to the environment. The manager of M also stressed, during the interview, that the company strictly conformed to legal requirements, and, as such, the amount of fines and the number of sanctions/warnings were considerably lower than those of N. There are a few indicators that N performs better than M, such as MPI4.2 which is assessed towards a linguistic scale (grades A-E). Company N is able to show evidence of their efforts to improve the quality of life of the community, such as sponsoring local religious activities and contributing to public infrastructure. CSR activities have been embedded in the company N's policies whereas they have only been conducted unofficially by company M. The weight of MPI4, unfortunately, is not high enough to push N to be greater than M in terms of the overall management performance.

For OPIs, the ranges of the expected utilities of the two companies greatly overlap each other, and the average of M (0.67) is just slightly higher than that of N (0.50). In general, company M seems to perform better than N, particularly in OPI3 and OPI4. M has a policy of waste segregation. Food waste is used to produce organic fertilisers. Recyclable waste is sold to external parties. There is also a campaign to replace the use of printed documents with email. M, moreover, has a greater percentage of trucks that utilise liquefied petroleum gas (LPG), a kind of low-carbon energy source, when compared to N. On the other hand, when considering OPI2, one of the OPI criteria receiving the highest weight, the manager of company N gave information indicating that the average rate of companies N's fuel consumption (diesel) fluctuated between 0.167-0.25 litres/kilometre while the manager of M only gave a single number (around 0.24 litres/kilometre). This implies that M performs slightly poorer than N in terms of energy consumption.

Table 3. Distribution assessment matrix of the two companies.

Indica-tors	Company M	Company N
MPI1.1	$(H_{Fair},1)$	$(H_{Fair},0.5);(H_{Good},0.5)$
MPI1.2	$(H_{Excellent},1)$	$(H_{Excellent},1)$
MPI1.3	$(H_{Fair},0.33);(H_{Good},0.67)$	$(H_{Poor},0.27);(H_{Fair},0.33);$ $(H_{Good},0.4)$
MPI1.4	$(H_{Fair},0.33);(H_{Good},0.67)$	$(H_{Fair},0.33);(H_{Good},0.67)$
MPI2.1	$(H_{Good},0.2);(H_{Excellent},0.8)$	$(H_{None},1)$
MPI3.1	$(H_{None},[0.02,0.22]);$ $(H_{Poor},[0.78,0.98])$	$(H_{Poor},[0.32,0.78]);$ $(H_{Fair},[0.22,0.68])$
MPI4.1	$(H_{Good},0.6);(H_{Excellent},0.4)$	$(H_{Good},1)$
MPI4.2	$(H_{Fair},0.5);(H_{Good},0.5)$	$(H_{Good},1)$
MPI4.3	$(H_{None},1)$	$(H_{None},0.9);(H_{Poor},0.1)$
OPI1.1	$(H_{Good},0.7);(H_{Excellent},0.3)$	$(H_{Fair},1)$
OPI2.1	$(H_{Fair},0.4);(H_{Good},0.6)$	$(H_{Fair},[0,0.5]);(H_{Good},[0,1]);$ $(H_{Excellent},[0,0.33])$
OPI3.1	$(H_{Excellent},1)$	$(H_{None},0.7);(H_{Poor},0.2);$ $(H_{Fair},0.1)$
OPI4.1	$(H_{Fair},0.4);(H_{Good},0.6)$	$(H_{None},0.9);(H_{Poor},0.1)$
OPI4.2	$(H_{Poor},0.065);(H_{Fair},0.07);$ $(H_{Good},0.065);(H_{Excellent},0.8)$	$(H_{Poor},0.33);(H_{Fair},0.34);$ $(H_{Good},0.33)$
OPI5.1	$(H_{None},1)$	$(H_{None},0.5);(H_{Poor},0.33);$ $(H_{Fair},0.17)$
OPI6.1	$(H_{Excellent},1)$	$(H_{Excellent},1)$
OPI6.2	$(H_{Excellent},1)$	$(H_{Excellent},1)$
OPI6.3	$(H_H,1)$	$(H_H,1)$
OPI7.1	$(H_H,1)$	$(H_H,1)$
OPI7.2	$(H_H,1)$	$(H_H,1)$

Table 4. Aggregated interval belief degrees of the two companies.

Perfor-mances	Company M	Company N
MPI1	$(H_{Fair},0.44);(H_{Good},0.33);$ $(H_{Excellent},0.23)$	$(H_{Poor},0.06);(H_{Fair},0.30);$ $(H_{Good},0.42);(H_{Excellent},0.22)$
MPI2	$(H_{Good},0.2);(H_{Excellent},0.8)$	$(H_{None},1)$
MPI3	$(H_{None},[0.02,0.22]);$ $(H_{Poor},[0.78,0.98])$	$(H_{Poor},[0.32,0.78]);$ $(H_{Fair},[0.22,0.68])$
MPI4	$(H_{None},0.32);(H_{Fair},0.16);$ $(H_{Good},0.4);(H_{Excellent},0.12)$	$(H_{None},0.26);(H_{Poor},0.03);$ $(H_{Good},0.71)$
MPIs	$(H_{None},[0.02,0.14]);$ $(H_{Poor},[0.05,0.26]);$ $(H_{Fair},[0.08,0.17]);$ $(H_{Good},[0.2,0.28]);$ $(H_{Excellent},[0.3,0.56]);$	$(H_{None},[0.29,0.66]);$ $(H_{Poor},[0.03,0.23]);$ $(H_{Fair},[0.06,0.29]);$ $(H_{Good},[0.12,0.31]);$ $(H_{Excellent},[0.04,0.07]);$
OPI1	$(H_{Good},0.7);(H_{Excellent},0.3)$	$(H_{Fair},1)$
OPI2	$(H_{Fair},0.4);(H_{Good},0.6)$	$(H_{Fair},[0,0.5]);(H_{Good},[0,1]);$ $(H_{Excellent},[0,0.33])$
OPI3	$(H_{Excellent},1)$	$(H_{None},0.7);(H_{Poor},0.2);$ $(H_{Fair},0.1)$
OPI4	$(H_{Poor},0.03);(H_{Fair},0.24);$ $(H_{Good},0.34);(H_{Excellent},0.39)$	$(H_{None},0.45);(H_{Poor},0.23);$ $(H_{Fair},0.16);(H_{Good},0.16)$
OPI5	$(H_{None},1)$	$(H_{None},0.5);(H_{Poor},0.33);$ $(H_{Fair},0.17)$
OPI6	$(H_{Excellent},0.79);(H_H,0.21)$	$(H_{Excellent},0.79);(H_H,0.21)$
OPI7	$(H_H,1)$	$(H_H,1)$
OPIs	$(H_{None},[0,0.18]);$ $(H_{Poor},[0,0.01]);$ $(H_{Fair},[0.08,0.20]);$ $(H_{Good},[0.12,0.39]);$ $(H_{Excellent},[0.23,0.49]);$ $(H_H,[0.12,0.3])$	$(H_{None},[0.1,0.34]);$ $(H_{Poor},[0.04,0.15]);$ $(H_{Fair},[0.02,0.39]);$ $(H_{Good},[0.04,0.4]);$ $(H_{Excellent},[0.09,0.29]);$ $(H_H,[0.13,0.3])$
Overall environ-mental perfor-mance	$(H_{None},[0.01,0.17]);$ $(H_{Poor},[0.01,0.17]);$ $(H_{Fair},[0.07,0.20]);$ $(H_{Good},[0.13,0.38]);$ $(H_{Excellent},[0.25,0.58]);$ $(H_H,[0.04,0.22])$	$(H_{None},[0.14,0.59]);$ $(H_{Poor},[0.03,0.21]);$ $(H_{Fair},[0.03,0.38]);$ $(H_{Good},[0.05,0.40]);$ $(H_{Excellent},[0.05,0.25]);$ $(H_H,[0.04,0.22])$

Table 5. Expected utilities of the two sample companies $(U_{min}, \boldsymbol{U_{avg}}, U_{max})$.

Performance aspects	Company *M*	Company *N*
MPIs	(0.56,**0.70**,0.83)	(0.19,**0.32**,0.46)
OPIs	(0.41,**0.67**,0.93)	(0.19,**0.50**,0.80)
Overall environmental performance	(0.42,**0.67**,0.92)	(0.12,**0.43**,0.74)

5 Conclusion

This paper presents a unique application of the ER approach to the assessment of environmental performance towards ISO 14031 for logistics operations. It shows that the ER algorithm is able to facilitate the combination of various criteria and indicators even when uncertainties in criteria weights and belief assignments (assessment data) are involved. Uncertainties and incompleteness in the assessment data are still preserved in the aggregated results. This enhances transparency and minimises any inaccuracies of the results that might be caused by too much effort being put into the process of converting imprecise or uncertain data into crisp numbers. The ER approach also suggests the design of evaluation grades in the assessment of qualitative criteria by linking the grades to pieces of evidence or feasible practices in order to minimise subjectivity. The relative weights of criteria are employed in the ER algorithm, and this indicates a compensatory approach among multiple criteria which is consistent with the nature of business performance evaluation and decision-making.

This study, also, is the first effort of its kind to elicit the weights or the importance of ISO 14031 criteria perceived by the logistics industry. Although it is difficult to reach a complete consensus within the industry regarding this matter, the obtained ranges of the weights are able to reveal some general viewpoints with major agreement. Overall, in terms of the management performance (MPIs), the industry tends to praise high contributions of conformance to legal regulations and the establishment of official environmental policies and programmes to the image of environmental management. In contrast, focusing only on building community relations tends to have only a small impact. In terms of operational performance (OPIs), fuel consumption and air emissions seem to be at the top of the most important criteria used to reflect the performance of a logistics operator. Understanding major concerns and perceptions of the industry could help create better guidelines for logistics companies to initiate their environmental improvement programmes in a way that is powerful and well-recognised by the community.

The ER approach is effective not only for performance comparison across different companies but a single company can also employ such method for their own self-assessment. The obtained aggregated degrees of belief facilitate the companies to establish improvement plans. For example, regarding the two sample companies, company *M* may choose to focus on indicator MPI1.1 (due to its high weight and the fact that the company itself still has only a 'Fair' performance) by establishing a team to investigate the feasibilities of conducting more environmental initiatives. Company *N*, on the other hand, still has a low performance in many indicators. For this case, high-weight indicators, such as MPI2.1 and OPI7, should be firstly considered. For MPI2.1, the company should place emphasis on educating their staff and drivers of the updated laws and regulations that relate to their operations in order to avoid being subject to legal fines and penalties. In terms of OPI7, both companies completely lack the capability of properly estimating their carbon footprints of transportation services, which, nowadays, has become one of the key aspects considered by shippers or product manufacturers in their selection-process of logistics service providers. For unknown information, the ER approach assumes that their actual performance could be any value within the entire range of possible cases, and this lowers the minimums of their expected utilities.

This study does not aim to standardise the list of indicators under each ISO 14031 criterion for the logistics industry. The list of indicators shown in Table 2 only represents the examples for demonstrating the implementation process of the ER approach. Subsequent studies may consider investigating practical indicators that are generally recognised and implemented among logistics service providers in order to promote better performance comparison within the industry, using a general set of indicators.

Acknowledgements

This work was supported by the Thailand Research Fund and the Office of the Higher Education Commission [grant numbers MRG6080084].

References

1. L.M.D.F. Ferreira, C. Silva, and S.G. Azevedo, An environmental balanced scorecard for supply chain performance measurement (Env_BSC_4_SCPM), Benchmarking: An International Journal. **23(6)**, 1398-1422 (2016)
2. C.-N. Wang, H.-X.T. Ho, S.-H. Luo, and T.-F. Lin, An integrated approach to evaluating and selecting green logistics providers for sustainable development, Sustainability. **9(2)**, 218 (2017)
3. P. Oberhofer and M. Dieplinger, Sustainability in the transport and logistics sector: lacking environmental measures, Business Strategy and the Environment. **23(4)**, 236-253 (2014)
4. M. Björklund and H. Forslund, The purpose and focus of environmental performance measurement systems in logistics, International Journal of Productivity and Performance Management. **62(3)**, 230-249 (2013)
5. M.I. Piecyk and M. Björklund, Logistics service providers and corporate social responsibility: sustainability reporting in the logistics industry, International Journal of Physical Distribution & Logistics Management. **45(5)**, 459-485 (2015)

6. J.F. Henri and M. Journeault, Environmental performance indicators: an empirical study of Canadian manufacturing firms, Journal of Environmental Management. **87**, 165-176 (2008)
7. A. Scipioni, A. Mazzi, F. Zuliani, and M. Mason, The ISO 14031 standard to guide the urban sustainability measurement process: an Italian experience, Journal of Cleaner Production. **16**, 1247-1257 (2008)
8. C. Jasch, Environmental performance evaluation and indicators, Journal of Cleaner Production. **8(1)**, 79-88 (2000)
9. H. Walker, L. Di Sisto, and D. McBain, Drivers and barriers to environmental supply chain management practices: lessons from the public and private sectors, Journal of Purchasing and Supply Management. **14(1)**, 69–85 (2008)
10. D. Nawrocka, Environmental supply chain management, ISO 14001 and RoHS. How are small companies in the electronics sector managing?, Corporate Social Responsibility and Environmental Management. **15(6)**, 349–360 (2008)
11. R.L. Keeney and H. Raiffa, *Decision with multiple objectives: preferences and value tradeoffs* (John Wiley & Sons, New York, 1976)
12. W.L. Winston, *Operations research: applications and algorithms* (Thomson Learning, Toronto, 2004)
13. P. Zhou, B.W. Ang, and K.L. Poh, Comparing aggregating methods for constructing the composite environmental index: an objective measure, Ecological Economics. **59(3)**, 305-311 (2006)
14. W.W. Cooper, L.M. Seiford, and K. Tone, *Data Envelopment Analysis: a comprehensive text with models, applications, reference and DEA-solver software* (Springer, New York, 2007)
15. K.P. Yoon and C.L. Hwang, *Multiple attribute decision making: an introduction* (SAGE Publications, Thousand Oaks, California, 1995)
16. P. Sureeyatanapas, J.B. Yang, and D. Bamford, Evaluation of corporate sustainability, Frontiers of Engineering Management. **1(2)**, 176-194 (2014)
17. C.M. Tam, V.W.Y. Tam, and S.X. Zeng, Environmental performance evaluation (EPE) for construction, Building Research & Information. **30(5)**, 349-361 (2002)
18. International Standard ISO 14031: Environmental management - Environmental performance evaluation - Guidelines (International Organization for Standardization, 2013)
19. J.B. Yang, B.G. Dale, and C.H.R. Siow, Self-assessment of excellence: an application of the evidential reasoning approach, International Journal of Production Research. **39(16)**, 3789-3812 (2001)
20. K.S. Chin, J.B. Yang, M. Guo, and J.P.K. Lam, An evidential-reasoning-interval-based method for new product design assessment, IEEE Transactions on Engineering Management. **56(1)**, 142-156 (2009)
21. C. Fu and S. Yang, The combination of dependence-based interval-valued evidential reasoning approach with balanced scorecard for performance assessment, Expert Systems with Applications. **39(3)**, 3717-3730 (2012)
22. F. Liu, W.-d. Zhu, Y.-w. Chen, D.-l. Xu, and J.-b. Yang, Evaluation, ranking and selection of R&D projects by multiple experts: an evidential reasoning rule based approach, Scientometrics. **111(3)**, 1501–1519 (2017)
23. H. Sellak, B. Ouhbi, and B. Frikh, Energy planning under uncertain decision-making environment: an evidential reasoning approach to prioritize renewable energy sources, Inteligencia Artificial. **20(59)**, 21-31 (2017)
24. J.B. Yang and M.G. Singh, An evidential reasoning approach for multiple attribute decision making with uncertainty, IEEE Transactions on Systems, Man, and Cybernetics. **24(1)**, 1-18 (1994)
25. Y.M. Wang, J.B. Yang, and D.L. Xu, Environmental impact assessment using the evidential reasoning approach, European Journal of Operational Research. **174(3)**, 1885-1913 (2006)
26. J.B. Yang, Rule and utility based evidential reasoning approach for multiple attribute decision analysis under uncertainty, European Journal of Operational Research. **131(1)**, 31-61 (2001)
27. Y.M. Wang, J.B. Yang, D.L. Xu, and K.S. Chin, The evidential reasoning approach for multiple attribute decision analysis using interval belief degrees, European Journal of Operational Research. **175(1)**, 35-66 (2006)
28. M. Guo, J.B. Yang, K.S. Chin, and H.W. Wang, Evidential reasoning based preference programming for multiple attribute decision analysis under uncertainty, European Journal of Operational Research. **182(3)**, 1294-1312 (2007)
29. P.A. Bottomley, J.R. Doyle, and R.H. Green, Testing the reliability of weight elicitation methods: direct rating versus point allocation, Journal of Marketing Research. **37(4)**, 508-513 (2000)
30. P. Sureeyatanapas and S. Pathumnakul, Impacts of using relative weights in multiple criteria decision making: a comparative study between independent- and overlapping-criteria decision problems, International Journal of Applied Decision Sciences. **10(2)**, 101-117 (2017)
31. P. Sen and J.B. Yang, *Multiple criteria decision support in engineering design* (Springer, London, 1998)
32. P. Sureeyatanapas, J.B. Yang, and D. Bamford, The sweet spot in sustainability: a framework for corporate assessment in sugar manufacturing, Production Planning & Control. **26(13)**, 1128-1144 (2015)

Application of element decomposing method for solving integer problems

Ekkaphon Jaiyen[1,*] *and Komgrit* Leksakul[2]

[1]Ph.D.'s Degree Program in Industrial Engineering, Department of Industrial Engineering, Faculty of Engineering, Chiang Mai University, Chiang Mai 50200, Thailand
[2]Department of Industrial Engineering, Faculty of Engineering, Chiang Mai University, Chiang Mai 50200, Thailand

Abstract. This research objective is to solve the integer problems by using an algorithm that applied to the element decomposing method (EDCM). Integer problem is a NP-hard when the problem is large-size, more time needed to solve the problem. The EDCM cuts a structure into several elements and reconnects elements at "nodes". This process can disconnect other nodes that not connected with the element. Moreover, it appropriately need in optimization solutions for applications, in which can give answers faster by cutting the nodes in finding the answers. There's 2 phases of step on this research. The first phase is input data and simplex method, while the second phase is creating and developing the algorithm from EDCM application. The comparison results show how two methods are carried out between EDCM and B&B Method. The results from two methods are focused on the value and solution step by step. According to the problem, it can be solved within the number of variables 2-15. The value of the difference in the answer is on average at 0.00% and solution step used for solving problem is less on average at 40.89 %.

1 Introduction

Operation Research (OR) is using of the model in mathematics, statistics and algorithm to help in decision making. This will bring to use much in the business, the industry, the government sector etc. The function to search for the practice that will give the best result (Search for Optimality) will be the main idea. The operation research has come into existence during the period of the World War II, the military administration of British has given the scientist team to study research about strategy and tactic in protecting the country for both on the road and the air [1].

Integer linear programming (ILP) is main part of mathematical programing model that is NP-Hard problem. Integer liner programming is optimal solution in linear programming to the integer number, in which the objective function and constraint functions are linear. ILP consist production planning problem, capital budgeting, fixed-change problem, scheduling problem, telecommunications network problem etc.

Algorithm is the sequence of the procedure in solving the problem to solution to find the best answer or this is closed to the complexity and the difficulty. Algorithm can be used to solve the integer linear programming that is consist of 2 types, the exact method and the heuristic method. For both of 2 types, this will be different in the viewpoint of the quality of the answer and the solution time. If this has used the exact method in solving the problem, it would have made to get the best answer but it would have used the long time period. Heuristic techniques are powerful and flexible search methodologies that have successfully tackled practical difficult problems. Heuristic and meta-heuristic algorithms are used to produce good-quality solutions in reasonable amount of computation times and acceptable for practical purposes.

In solving the integer problem, it consists of 2 methods, which is a popular method that are consist of the branch and bound (B&B) method and the cutting plane method. B&B method is widely popular method, which uses tools for solving large-scale (NP-hard) combinatorial optimization problems. The B&B algorithm operates according to two principles: (1) It recursively splits the search space into smaller spaces, thus minimizing the function on those smaller spaces; the splitting is called branching. (2) Branching alone would amount to brute-force enumeration of candidate solutions and the testing of them all. In other part, cutting plane method is an alternative to the branch and bound method, which also can be used to solve integer problems. The fundamental idea behind cutting planes is to add constraints to linear program until the feasible optimal basic solution takes on integer values. On contrary, this method has to be carefully performed as it has constraints, and one would not want to change the problem because of the addition of the constraints. This method will add a special type of constraint called a cut. [2, 3] In both methods, the constraints added will eliminate the portions of relaxed solution space, but never any of the feasible integer points. Neither of the two methods can be claimed to be uniformly more

* Corresponding author: bang0285@gmail.com

effective in solving ILP's. Nevertheless, B&B methods are computationally far more successful than the cutting-plane methods. For this reason, most of commercial codes are based on the use of the B&B procedure. However, the explicit enumeration is normally impossible due to the exponentially increasing number of potential solutions. The use of bounds for optimizing the function, combined with the value of the current best solution enables the algorithm to search parts of the solution space implicitly [4]. Therefore, when there are many variables in the solution, the time and number of branches required are more. This method takes longer time to process the solution.

Finite element method (FEM) is a numerical technique to finds approximate solutions to a boundary value problem for partial differential equations, which now comprehensively used in solid and structural mechanics [5, 6]. The method of The FEM is a mathematical method that brought to be applied in software application creation, used for calculation to solve the engineering problem. The Finite element method is done by filling the shape of the product with the small piece that has geometric form. In FEM, the structural system is modelled by a set of appropriate finite elements interconnected at a point called "nodes". The elements might have physical properties, such as thickness, density, Young's modulus, shear modulus, Poisson's ratio and etc.

Currently, from the problem in using the algorithm that has applied to solve the mathematic problem. This has still the limitation and this will use the time period in solution the answer for a long time. This will have the idea in applying to use the method in the operation research in creating and developing the new algorithm for solving the problem to will be faster further.

2 Research Methodology

The research study aims to solve an optimization problem using an algorithm that applies the decomposed element method, as shown in Figure 1.

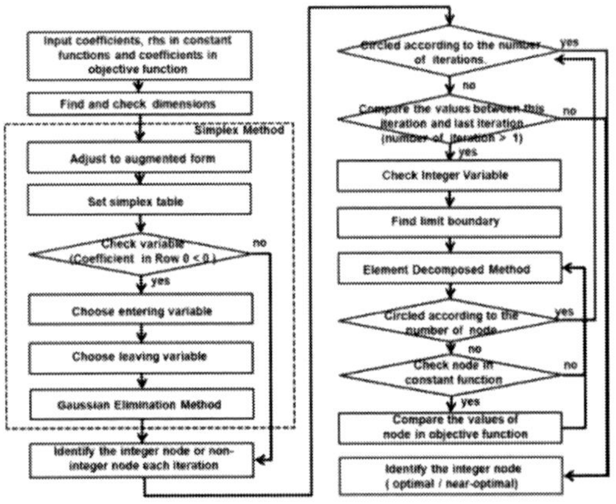

Figure 1. Steps of the research methodology.

This research study has two phases. The first phase consists of the input data and simplex method. The second phase is about creating and developing the algorithm via application of the element decomposing method (EDCM).

Phase 1; consists of the input data and simplex method, which is performed as follows.

Step 1.1; The input coefficients, the right-hand side (RHS) in constant functions and coefficients in the objective function follow the mathematical equation form.

Example; Integer problem in 2 variables solved by graph method that is shown figure 2.

Objective function : $\max z = 4x_1 + 3x_2$
Constant function : $15x_1 + 9x_2 \leq 85$
$3x_1 + 4x_2 \leq 24$
$x_1 \geq 0,\ x_2 \geq 0$

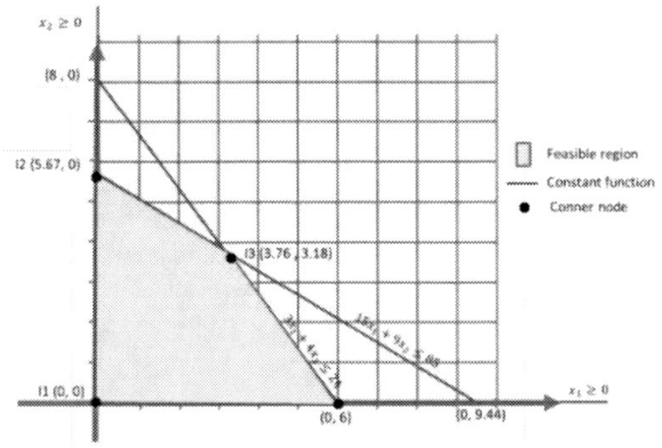

Figure 2. Feasible region of example from graph method

Coefficient of objective function = [4 3]
Coefficient of constant functions = [15 9 ; 3 4]
Rhs. of constant functions = [85 ; 24]

Step 1.2; Finding examined according to the following conditions.

1. number of column in coefficients of objective function and coefficients of constant functions is equal.
2. number of row in coefficients of objective function and Rhs. of constant functions is equal.

Subsequently, write all the values of the coefficient of objective function, coefficient of the constant functions, and the right-hand side of the constant functions in a simplex table adjust to the augmented form for prepare to simplex method.

Step 1.3; Solving the problem using simplex method which is shown the node each iteration in Table 1.

Table 1. Node of Example Problem from Simplex Method.

Iteration	X_1	X_2	Z
Iteration 1 (i1)	0	0	0
Iteration 2 (i2)	5.67	0	22.67
Iteration 3 (i3)	3.76	3.18	24.58

Phase 2; consists of creating and developing the algorithm via application of the element decomposed method (EDCM) which is performed as follows.

Step 2.1; Using nodes from phase 1 to check the integer number in each iteration that starts from last iteration (Iteration 3; i3).

According to Example, the answer is that the optimal value (z) = 24.58 is at node i3 which is (x1, x2) = (3.76, 3.18). However, if it is a pure integer problem, it is necessary to use the element decomposing method (EDCM) for solving the problem. In this method, the feasible region is divided into small sub-areas such as those of the integer unit, or it is the feasible region which has the Conner node as the integer node, as shown in Figure 3.

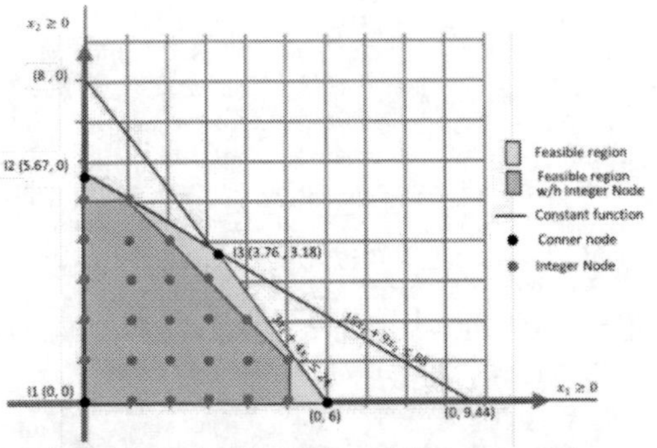

Figure 3. The feasible region with the integer problem

Step 2.2; Using the nodes in each iteration by making them adjust to integer node by using the element decomposing method (EDCM) which has three cases, namely Case 1 which is the complete integer node, Case 2 which is the non-complete integer node and Case 3 which is the non-integer node. In Case 2, the non-complete integer node (Node form iteration 2) and Case 3, the non-integer node (Node form iteration 3) adjust to the integer node by using the element decomposing method (EDCM), which is shown in Figure 4. The number of integer nodes is four, 2^number of variable.

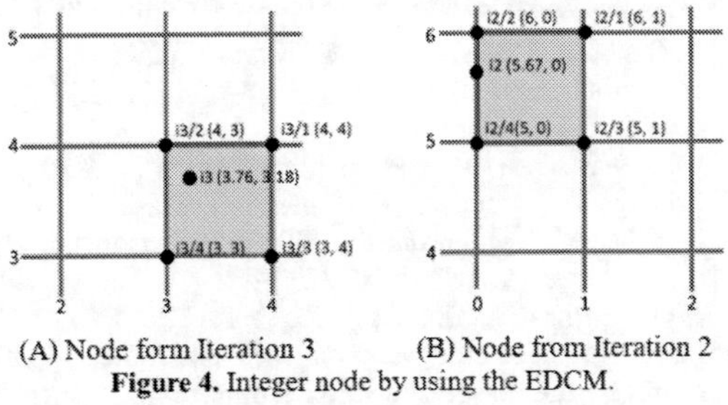

(A) Node form Iteration 3 (B) Node from Iteration 2

Figure 4. Integer node by using the EDCM.

Table 2. Integer Node by Using EDCM.

Iteration	Original Node	Node 1	Node 2	Node 3	Node 4
Iteration 2 (i2)	i2 (5.67, 0)	i2/1 (6, 1)	i2/2 (6, 0)	i2/3 (5, 1)	i2/4 (5, 0)
Iteration 3 (i3)	i1 (3.76, 3.18)	i1/1 (4, 4)	i1/2 (4, 3)	i1/2 (3, 4)	i1/4 (3, 3)

Step 2.3; Checking the integer node in each of the constant functions. This starts from the node arranged in Step 2.2, as shown in Figure 5.

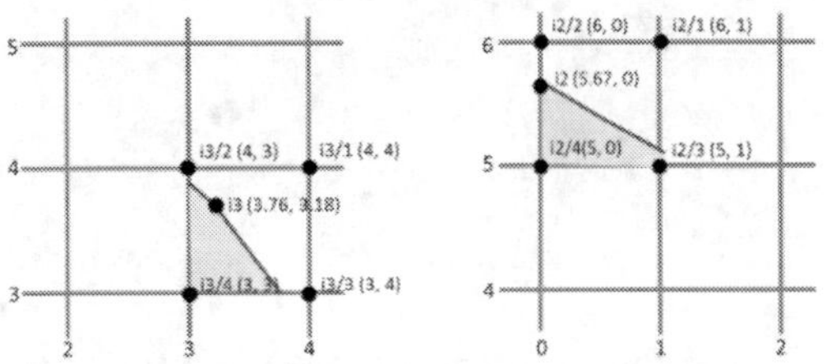

(A) Node form Iteration 3 (B) Node from Iteration 2

Figure 5. The integer node after checking constant function.

In iteration 3 has node i3/1, node i3/2 and node i3/3 which cannot pass the condition of the constraint function but node i3/4 pass the conditions of both of the constraint functions. In iteration 2 has node i2/1 and node i2/2 which cannot pass the condition of the constraint function but node i2/3 and node i2/4 pass the conditions of both of the constraint functions.

Step 2.4; Finding the value in the objective function using the integer node after checking the constant function (from Step 2.3).

Checking all the points may need a lot of time, so there are steps for checking and finding the answers in order to reduce the length of time, as shown in Figure 6.

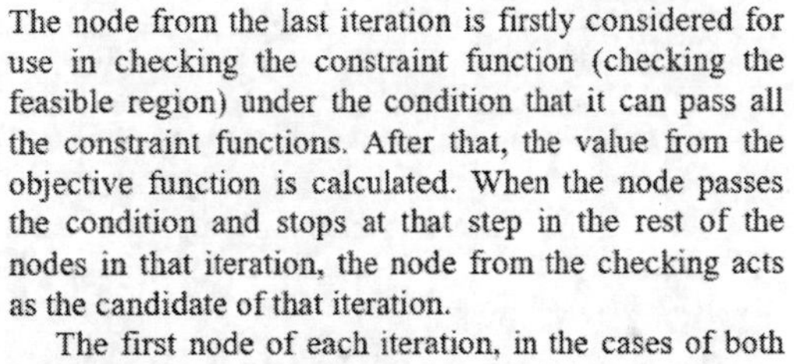

Figure 6. Step for checking node each iteration

The node from the last iteration is firstly considered for use in checking the constraint function (checking the feasible region) under the condition that it can pass all the constraint functions. After that, the value from the objective function is calculated. When the node passes the condition and stops at that step in the rest of the nodes in that iteration, the node from the checking acts as the candidate of that iteration.

The first node of each iteration, in the cases of both the non-integer node and the non-complete integer node, is out of the feasible region, or it cannot pass the constraint function of at least one equation, such as Node i3 and Node i2.

After that, a comparison is made between the value in one iteration and that in the next iteration. If the result shows the value to be greater than that of the Maximum Problem or less than that of the Minimum Problem, examining for the answer to this problem must be stopped. If the result does not show, it is necessary to get the candidate of the new node in the next iteration of checking, as shown in Figure 7.

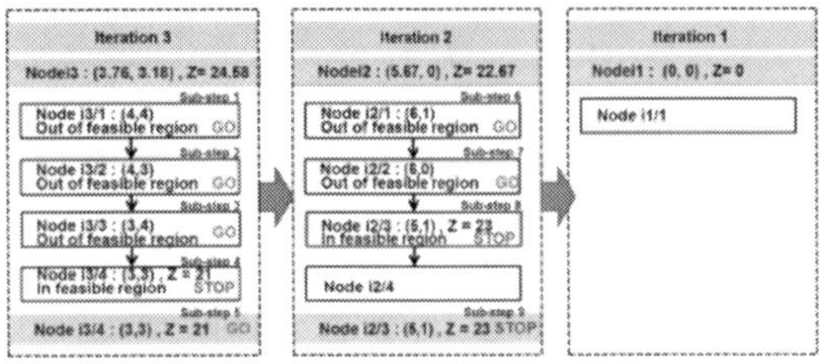

Figure 7. Step for checking the node and the finding value.

In example problem and Figure 7, it starts from Nodei3/1 (Iteration 3) as (x1, x2) = (4, 4).

Sub-step 1, it starts from Nodei3/1 as (x1, x2) = (4, 4) which cannot pass the condition of the constraint function no. 1 15*(4) + 9*(4) = 96 > 85. (Go to consider node i3/2).

Sub-step 2, consider Nodei3/2 as (x1, x2) = (4, 3) which cannot pass the condition of the constraint function no. 1. (Go to consider node i3/3).

Sub-step 3, consider Nodei3/3 as (x1, x2) = (3, 4) which cannot pass the condition of the constraint function no. 2. (Go to consider node i3/4).

Sub-step 4, consider Node3/4 as (x1, x2) = (3, 3) which can pass the conditions of both of the constraint functions and possess the value (z) = 21. After that, stop the examination at iteration 3, which would make Node3/4 become the candidate of the iteration 3. (Stop consider in iteration 3).

Sub-step 5, consider the value from Node i3/4 in comparison to that from Node i2 and determine that Node i3/4 is less than Node i2 (21 < 22.67) into consider the new node in the next iteration. (Go to consider iteration 2).

Sub-step 6, it starts from Nodei2/1 (Iteration 2) as (x1, x2) = (6, 1) which cannot pass the condition of the constraint function no. 1. (Go)

Sub-step 7, consider Nodei2/2 as (x1, x2) = (6, 0) which cannot pass the condition of the constraint function no. 1. (Go)

Sub-step 8, consider Node2/3 as (x1, x2) = (5, 1) which can pass the conditions of both of the constraint functions and possess the value (z) = 23. After that, stop the examination at iteration 2, which would make Node2/3 become the candidate of the iteration 2. (Stop in iteration 2)

Sub-step 9, consider the value from Node i2/3 in comparison to that from Node i1 and determine that Node i2/3 is higher than Node i1 (23 > 0). Then, stop the examination for the answer to the problem when the answer is at Node i2/3 as (x1, x2) = (5,1) and value (z) = 23. (Stop this solution)

3 Result and Discussion

The comparison of results is carried out between the results from algorithm obtained by the application of the element decomposed method (EDCM) and the results from exact solution (Branch and bound method; B&B).

The results from two methods are focusing on values and solution step, as shown in table 3. EDCM count number of node from decomposing and B&B count number of branch. According to the problem, it can be solved with the number of variables that are 2, 3, 6 and 9. These results are comparing by percentages of gap in terms of value and solution step, as shown in table 3.

Table 3. Values and solution step

Problem (NO. of variable)	Value		Solution Step		
	B&B	EDCM	B&B	EDCM	Percentages of gap
Example (2)	23	23	13	7	- 46.15 %
Telfa Problem (2)	40	40	7	3	- 57.14 %
Dorian Auto (3)	6000	6000	11	4	- 63.64 %
Food Choices (6)	57	57	15	14	- 6.67 %
S.Confederation (9)	633250	633250	15	6	- 60.00 %
Stein (15)	9	9	17	15	- 11.76 %

Based on Table 3, the results are solved by B&B method and EDCM that can resolve problems for 2 to 15 variables and shows the percentages of difference in terms of solution step and solution time when a comparison is made between the B&B method and EDCM. Value of the difference in the answer is at an average of 0.00 % (no difference) and in the solution step used for solving the problem is at an average of 40.89 %. B&B and EDCM can solve small problem less than 1 second. From the result of the small problem, this will show the difference of the solution step in solving the integer problem clearly. This will lead to the large size problem more.

The authors world like to gratefully acknowledge Department of Industrial Engineering, Chiang Mai University, for the supporting of this research work. Financial support from the Thailand Research Fund through the Research and Researchers for Industries Project (Grant NO. PHD59I0002) to Mr. Ekkaphon Jaiyen and Assoc. Prof. Dr. Komgrit Leksakul.

References

[1] Joseph F. M., (1987) OR Forum—British Operational Research in World War II. Operations Research 35(3):453-470.
[2] Michael, A. T., (2014), Cutting Plane Techniques, available at: http://mat.gsia.cmu.edu /orclass/integer /node14.html. (accessed 8 April 2016).
[3] Lewis, F. D., (1999), Cutting Plane Techniques,available at: http://www.cs.uky.edu/~lewis/ cs-heuristic/text/integer/cutting.html. (accessed 8 April 2016).
[4] Jens, C., (1999), Branch and Bound Algorithms - Principles and Examples, available at: http://janders.eecg.toronto.edu/1387/readings/b_and_b.pdf. (accessed 18 January 2016).
[5] Thomas, J. H., Wing K., and Alec B., (1979), Finite Element Analysis of Incompressible Viscous Flows by the Penalty Function Formulation, JOURNAL OF COMPUTATIONAL PHYSICS, Vol 30, pp.1-60.
[6] Charles, E. K., The Finite Element Method in Mechanical Design, Virginia Polytechnic Institute and State University, Boston: PWS-KENT Publishing Company, pp. 1-7.

A comparison of latin hypercube sampling techniques for a supply chain network design problem

Kantapit Kaewsuwan[1], *Chumpol* Yuangyai[1,*], *Udom* Janjarassuk[1] and *Kanokporn* Rienkhemaniyom[2]

[1]Industrial Engineering Department, Faculty of Engineering, King Mongkut's Institute of Technology Ladkrabang, Bangkok, Thailand
[2]Graduate School of Management and Innovation, King Mongkut's University of Technology Thonburi, Bangkok, Thailand

Abstract. Currently, supply chain network design becomes more complex. In designing a supply chain network to withstand changing events, it is necessary to consider the uncertainties and risks that cause network disruptions from unexpected events. The current research related to the designing problem considers network disruptions using Monte Carlo Sampling (MCS) or Latin Hypercube Sampling (LHS) techniques. Both have a disadvantage that sample points or disruption locations are not scattered entirely sample space leading to high variation in objective function values. The purpose of this study is to apply a modified LHS or Improved Distributed Hypercube Sampling (IHS) techniques to reduce the variation. The results show that IHS techniques provide smaller standard deviation than that of the LHS technique. In addition, IHS can reduce not only the number of sample size but also and the computational time.

1 Introduction

Supply chain network design is a critical decision-making process that affects the efficiency of enterprise management. Especially in high volatility business environment. This is due to the risk and the uncertainty of the raw materials of suppliers. A disaster occurs in one country will affect the global supply chain, resulting in disruptions, such as in 2011, a major flood in Thailand, one of global supply chain disruptions [1].

Therefore, the proactive supply chain network design is important for all organizations. It is necessary to consider the risks that occur in countries where the suppliers, manufacturers, and warehouses are located. And also, the ability to recover from disruptions. In many supply chain network design studies.

In this paper, we use a two-stage stochastic programming model[2] for supply chain design. Our supply chain network consists of suppliers, manufacturing, warehouses, and retailers. A single product is considered. The objective function is to maximize profits under disruptions.

The two stages involve 1) to find an initial solution by a heuristic method. Then, 2) evaluation of those solutions from step 1 with difference disruptions by using Monte Carlo Sampling (MCS)

In general, MCS techniques are commonly used. However, it requires many sample size to increase the precision of the solution leading longer computation time. Latin Hypercube Sampling (LHS) and its variants can reduce the number of sample sizes and use less time to evaluate a solution. In this paper, we compare two type of LHS to improve the efficiency of the two-stage stochastic programming.

2 Literature review

Since our main is on the sample techniques for stochastic programming, therefore only MCS, LHS, and its variants are discussed.

MCS techniques[3] is generated by using a random number that independent of each other and uniformly distributed on the interval [0, 1]. MCS is a matrix (M) with dimensions N x S where N is the number of samples and S is the number of independent variables

McKay et al., 1979 proposed one of the sampling techniques used for computer-aided design, The LHS techniques provide a more uniform random sample distribution. LHS is a matrix (L) with dimensions N x S The design steps is shown below:

<u>Step 1</u> The matrix $P(N, S)$ is consists of random shuffles of integers ranging from 1 to N.

<u>Step 2</u> The matrix $M(N, S)$ is generated from a random number in Monte Carlo method, where $M(N, S) \in U[0, 1]$ which are independent of each other.

<u>Step 3</u> The matrix $L(N, S)$ is constructed from equation (1), and this matrix is used to simulate the event.

$$L(N,S) = \frac{1}{N}\left[P(N,S) - M(N,S)\right] \qquad (1)$$

Brian K. B., 2002 [4][5] developed Improved Distributed Hypercube Sampling (IHS) from the LHS techniques, which adds conditions for finding the distance between points by using the Euclidean distances. The design process of the IHS techniques is similar to the design of the LHS techniques. It is different to construct the $P(N, S)$ matrix with the distance between the points that are described in Section 3.3.2.

*Corresponding author: chumpol.yu@kmitl.ac.th

3. Supply Chain Network Model Description

3.1 Supply chain network design

In this paper, we consider a network with four echelons that consists of suppliers, manufacturing, warehouses, and retailers. As shown in Figure 1, the location of supplier, manufacturing, and warehouses depends on the probability of occurrence of disruptions within the supply chain network. The probability in each location of suppliers, manufacturing, and warehouses are varying. The goal is to decide the location of the supplier, and warehouses while maximizing profit.

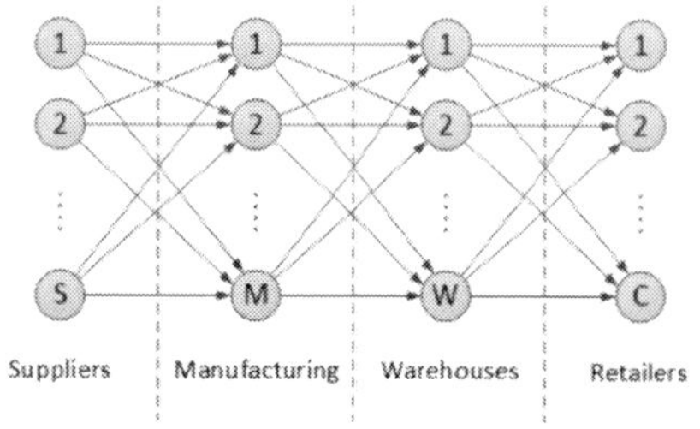

Figure 1. Supply chain network [6]

3.2 Mathematical model and solution approach

Our mathematical model is developed from Kanokporn et al., 2015 [6]. The model is shown below:

Sets: *S:* Set of suppliers; *M:* Set of manufacturing plants; *W:* Set of warehouses; *C:* Set of retailers; *L:*Set of warehouse capacities; *K:* Set of disruptive scenarios

Indices: *s:* Index of suppliers $s \in S$; *m*: Index of manufacturing plants $m \in M$; *w*: Index of warehouses $w \in W$; *c*: Index of retailer $c \in C$; *l*: Index of warehouse capacities $l \in L$; *k*: Index of disruptive scenarios $k \in K$

Parameters

cap_m Production capacity at manufacturing plant m
cap_s Capacity at supplier s
cap^l_w Capacity at warehouse w of size l
d_c Demand for products at retailer c
msm Minimum transportation quantity from supplier s to manufacturer m
f^l_w Fixed cost of opening a warehouse w of capacity l
pm_{sm} Purchasing cost of material from supplier s by plant m
tr_{sm} Transportation cost per unit from plant m to warehouse w
tr_{wc} Transportation cost per unit from warehouse w to retailer c
pc_m Production cost for a product at plant m
np Price of a product
ls_c Lost sales cost at retailer c

Random Parameters

$$\alpha_{sk} = \begin{cases} 1 & \text{if supplier } s \text{ is operated in scenario } k \\ 0 & \text{if a disruption occurs at supplier } s \text{ in scenario } k \end{cases}$$

$$\delta_{wk} = \begin{cases} 1 & \text{if warehouse } w \text{ is operated in scenario } k \\ 0 & \text{if a disruption occurs at warehouse } w \text{ in scenario } k \end{cases}$$

$$\beta_{mk} = \begin{cases} 1 & \text{if plant } m \text{ is operated in scenario } k \\ 0 & \text{if a disruption occurs at plant } m \text{ in scenario } k \end{cases}$$

p_k The probability of disruption occurs in scenario k

First-stage Decision Variables

$$x^l_w = \begin{cases} 1 & \text{if warehouse } w \text{ is operated with size } l \\ 0 & \text{otherwise} \end{cases}$$

$$y_s = \begin{cases} 1 & \text{if supplier } s \text{ is selected} \\ 0 & \text{otherwise} \end{cases}$$

Second-stage Decision Variables

QSM_{smk} Quantity of raw material purchased from supplier s by plant m in scenario k
QMW_{mwk} Quantity of products shipped from plant m to warehouse w in scenario k
QWC_{wck} Quantity of products shipped from warehouse w to retailer c in scenario k
LD_{ck} Quantity of sales lost at retailer c in scenario k

3.2.1 Objective function

The objective is to maximize the expected supply chain profit (Z) which is the difference between the total cost and the expected revenue.

$$\text{Maximize } Z = -\sum_{w \in W}\sum_{l \in L} f^l_w x^l_w + \sum_{k \in K} p_k \Bigg[np\Big(\sum_{w \in W}\sum_{c \in C} QWC_{wck}\Big) - \Big(\sum_{s \in S}\sum_{m \in M} pm_{sm} QSM_{smk}\Big) - \Big(\sum_{m \in M}\sum_{w \in W} tr_{mw} QMW_{mwk} + \sum_{w \in W}\sum_{c \in C} tr_{wc} QWC_{wck}\Big) - \sum_{m \in M} pc_m \Big(\sum_{w \in W} QMW_{mwk}\Big) - \sum_{c \in C} ls_c LD_{ck} \Bigg]$$

3.2.2 Constraints

Supplier capacity

$$\sum_{m \in M} QSM_{smk} \le cap_s \alpha_{sk} y_s, \quad \forall s \in S, k \in K$$

Inter-stage flow

$$msm \cdot \alpha_{sk} y_s \le QSM_{smk} \le cap_s \alpha_{sk} y_s, \quad \forall s \in S, k \in K$$

Production capacity

$$\sum_{w \in W} QMW_{mwk} \le cap_m \beta_{mk}, \quad \forall m \in M, k \in K$$

Material flow between suppliers and plants

$$\sum_{s \in S} QSM_{smk} = \sum_{w \in W} QMW_{mwk}, \quad \forall m \in M, k \in K$$

Warehouse capacity

$$\sum_{m \in M} QMW_{mwk} \le \sum_{l \in L} cap_w \delta_{wk} x^l_w, \quad \forall w \in W, k \in K$$

$$\sum_{l \in L} x^l_w \le 1, \quad \forall w \in W$$

Product flow between warehouse and retailers

$$\sum_{m \in M} QMW_{mwk} = \sum_{C \in C} QWC_{wck}, \quad \forall w \in W, k \in K$$

Demand requirement

$$\sum_{w \in W} QWC_{wck} + LD_{ck} = d_c, \quad \forall c \in C, k \in K$$

Non-negativity

$$QSM_{smk}, QMW_{mwk}, QWC_{wck}, LD_{ck} \geq 0$$

3.3 Solution Methodology

We use two-stage stochastic programming which is composed of 2 Stages: 1) Simulated Annealing algorithm (SA) which is used to find the initial solution. Then, 2) disruptions are generated to evaluate the solution's objective function. In the first stage, we used an SA to find the solution which includes 2 steps. The first step is to set the model's initial parameters. The second step is to find the initial solution, which is the fixed cost of suppliers' location, warehouse's location and the size of the warehouse. In the second stage is to generate disruptive scenarios by using two types of LHS techniques. We generate a sample average problem instead of finding the answer to all possible events. This technique is called sample average approximation (SAA)[3]. Then we find the value of the objective function, which is the mean of the maximum profit that we obtained under disruptive scenarios by using the CPLEX program. The structure of finding the solution as shown in Figure 2.

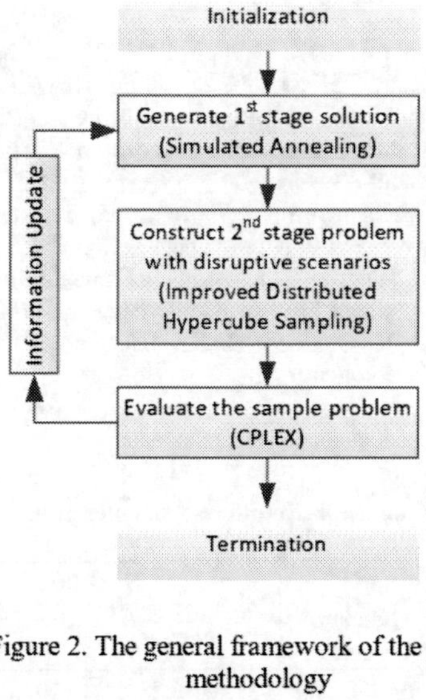

Figure 2. The general framework of the proposed methodology

3.3.1 Simulated Annealing

The Simulated Annealing method [7] is used to find the near optimal solution.

3.3.2 Disruptive Scenarios

In this study, we use IHS to generate disruptive scenarios to ensure that all disruptive scenarios will occur in each input variables. We expect to obtain solutions that have a smaller standard deviation than the LHS techniques.

Euclidean distances that use in IHS can be calculated from the equation 2, should be close to or equal to d_{opt}, which can be calculated from equation 3. This techniques has a distribution of sample points covering the sample space and have low coefficients of variation. It is one of the most popular sampling techniques used in improving the performance of a solution. This research utilizes the sampling techniques by the hypercube distribution method to randomly sample the disruptions to ensure that random events can represent all events.

$$d(x_i, x_j) = d_{ij} = \left[\sum_{k=1}^{S} \left| x_{ik} - x_{jk} \right|^2 \right]^{1/2}, \quad (2)$$

$$i \in \{1,..,(N-1)\}, j \in \{(i+1),..,N\}$$

$$d_{opt} = \frac{N}{\sqrt[S]{N}} \quad (3)$$

Where

x_{ik} Integer at scenarios i of variable k

x_{jk} Integer at scenarios j of variable k

$d(x_i, x_j)$ Distance between x_{ik} and x_{jk}, $i \neq j$

d_{opt} Optimal distance

IHS is a matrix (I) with dimensions N x S where N is the number of samples and S is the number of independent variables. The design steps are as follows.

Step 1: Define initial parameters (N, S, D) where D is the number of sets of numbers generated.

Step 2: Create the final $P(N, S)$ matrix from the design approach.

Step 3: Define the initial parameter $r: = N\text{-}1$

Step 4: When the required number of cycles ($r \leq 2$) is not reached

For the answer matrix. Complete the following steps.

1. Create a matrix $A(c, S)$ by $c = r.(D\text{-}1) + 1, ..., rd$ and $d = 1, ..., D$. The random number of integers is 1 to N and is not chosen from the matrix $P(N, S)$.

2. Find the distance between the points of $P(r + 1, S)$ and $A(l, S)$ from equation 2, where $l = 1, ..., r.d.$

3. Put an integer in matrix A where l is the distance between the nearest point and d_{opt} from equation 3. Put in the matrix $P(r, S)$

Step 5 Take the integer value from 1 to N and not be chosen from the final matrix $P(N, S)$. Put in the matrix $P(1, S)$

Step 6 Create the Matrix $I(N, S)$ from Equation 1.

Then, Compare the distribution of random numbers obtained from the LHS techniques and IHS techniques. As shown in Figure 3.

Figure 3. shows a scatter plot of LHS techniques. The number of samples size (N) is 40 with two input variables (S). When dividing the sample space into 5x5 in each row and each column it contains eight points, which shows that the sample points are uniformly distributed in the corresponding row and column. It shows that each input variables has all portions of its range. But the distribution of sample points is also adjacent, not spreading the sample space. Because of the design, there are no conditions to find the distance between sample points. While Figure 4. shows that the dispersion of the sample points has spread over the sample area.

In this paper, the sampling techniques by LHS and IHS method is used as a set of disruptive scenarios in the evaluation of the objective function.

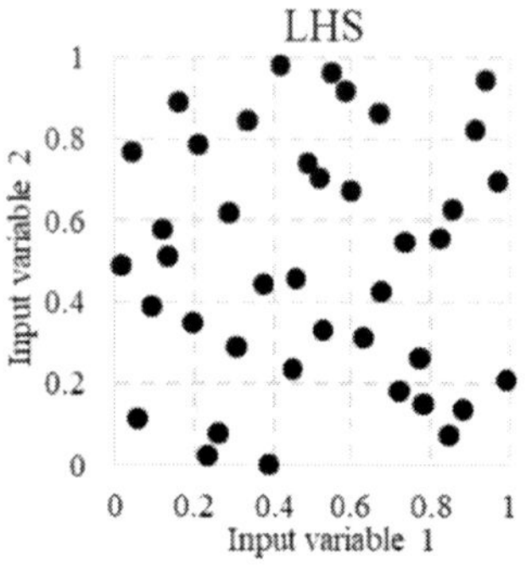

Figure 3. Scatter plot of samples generated by LHS

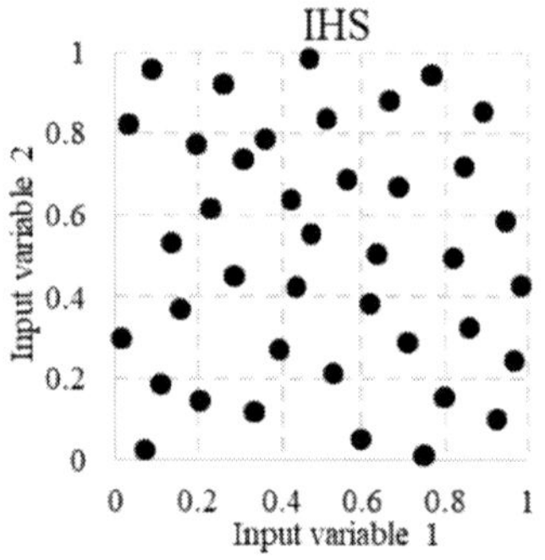

Figure 4. Scatter plot of samples generated by IHS

4 Result and discussions

We compared the two sampling techniques of LHS and IHS, which are used to randomize disruption events in designing supply chain network problem. The sample size (N) was varied from 20, 50, and 100 respectively. Each sample size run was repeated 10 times and calculate their average and standard deviation, as shown in Table 1.

Based on the comparison of the standard deviation, IHS outperforms LHS. It reduces the standard deviation by 22.45%, 15.54%, and 21.99%, for all three sample sizes. Furthermore, smaller sample size with IHS provides the similar or lower standard deviation of those of larger sample size with LHS.

5 Conclusion

This study presents a comparison of two sampling techniques Latin Hypercube Sampling and Improve Distributed Hypercube Sampling. The IHS is used with two-stage stochastic programming to solve a supply chain network design. IHS provides a lower standard deviation, smaller sample size, and better computation efficiency. In addition, a modification of LHS or Optimal Latin Hypercube Sampling will be further studied to improve computational efficiency.

References

1. Haraguchi M, Lall U (2015) Flood risks and impacts: A case study of Thailand's floods in 2011 and research questions for supply chain decision making. Int J Disaster Risk Reduct 14:256–272
2. Rienkhemaniyom K, Yuangyai C, Janjarassuk U A two-stage stochastic program for supply chain design under disruption
3. Shapiro A, Ruszczynski A (2003) Monte Carlo sampling methods. Handbooks Oper Res Manag 10:353–425
4. Beachkofski BK, Grandhi R V (2002) Improved Distributed Hypercube Sampling. AIAA Pap 1–8
5. Randall D. Manteufel (2001) Distributed Hypercube Sampling Algorithm. Am Inst Aeronaut Astronaut
6. Rienkhemaniyom K (2015) Resilience improvement in supply chain network design. In: Asia Pacific Industrial Engineering and Management Systems Conference. pp 475–482
7. Sastry K, Goldberg D, Kendall G (2005) Search Methodologies

Table 1: Comparison of objective values between two sampling techniques with different samples sizes

Sampling techniques	N = 20		N = 50		N = 100	
	Average	Standard deviation	Average	Standard deviation	Average	Standard deviation
Latin Hypercube	12,486,781.44	244,354.22	11,876,213.93	197,586.00	11,680,288.82	194,683.05
Improve Distributed Hypercube	12,228,495.99	199,558.44	11,943,134.72	171,003.85	11,676,678.56	159,589.15
Difference	-	-22.45%	-	-15.54%	-	-21.99%

Influences of soldering time on wettability and intermetallic phase between Sn-3.0Cu solder and copper substrate

Niwat Mookam and *Kannachai* Kanlayasiri*

Industrial Engineering Department, Faculty of Engineering, King Mongkut's Institute of Technology Ladkrabang, Bangkok, Thailand

Abstract. In this paper, the influences of soldering time on the wettability and intermetallic phase between Sn-3.0Cu lead-free solder and copper substrate were investigated. Reflow soldering was performed at 350 °C under variable soldering times of 10, 20, 40, 60, 120, 240 and 480 s. The results indicated that the wettability and intermetallic growth depend on the soldering time. In addition, the Cu_6Sn_5 and Cu_3Sn intermetallic phases with a hexagonal crystal structure were found between the lead-free solder and the copper substrate. The growth of intermetallic phases increased with soldering time, and the growth of intermetallic phases remarkably depended on grain boundary diffusion and was volume diffusion-controlled for Cu_6Sn_5 and Cu_3Sn, respectively.

1 Introduction

In the electronic packaging field there is a commitment to quality, and the continuous improvement in developing electronic products also extends to protecting the environment. This often involve which prohibits the use of lead in electronic parts, highlights the requirement of lead-free solder alloys [1]. With the various lead-free solders, Sn-Cu alloy is exposed to the best providing excellent mechanical properties and low cost. Normally, the high-temperature solders are widely used within automotive and energy production industries [2-3]. Sn-3.0Cu alloy is lead-free solder within the high-temperature family suitable for electronics devices.

In the soldering process, a molten solder was contacts with a solid substrate resulting in wetting of the substrate surface. This physical spreading of the solder has several interrelated phenomena such as surface energy, surface tension reduction, interfacial, and chemical reactions [4, 5]. A review of literature also revealed the relationship between the wettability and intermetallic phase formation. The intermetallic phase formation of the interface provides strong wettability [6]. The faster formation rate of intermetallic phase at the interface zone can explain the shorter wetting time [7]. In the work of Wang et al. [8], reported that the intermetallic phase formation can supply driving force for wettability because of the energy relaxation at the interface zone, which promotes the wettability of the solder. However, the phase transformation from Cu_6Sn_5 to Cu_3Sn is a process to discharge energy, which may lead to the increase of the wettability due to the surface tensions of the liquid solder changes according to the change of intermetallic compound (IMC) in the soldering process [9]. Arenas and Acoff [10] reported that the correlation between the intermetallic formation and wettability was no explicit interpretation. Thus, limitation of both prior studies has not been shown relationship of wetting and formation of intermetallic phase, especially with Sn-3.0Cu solder.

This current research investigates the influences of the soldering time on the wettability in term contact angle and the intermetallic phase between Sn-3.0Cu solder and copper substrate. Reflow soldering was then carried out at 350°C under variable soldering times of 10, 20, 40, 60, 120 and 480 s.

2 Experiment

High-temperature Sn-3.0Cu lead-free solder was used in this research. The solidus and liquidus temperatures of this solder are 227 and 309°C, respectively. The solder bar was formed into a cylindrical shape with a diameter of 6.5 mm and thickness of 1.24 mm in accordance with JIS Z3198-3. The substrate was pure copper (99.99%, grade C1100), according to the JIS H3100 standard with thickness 0.2 mm and cut to 25x 30 mm. A confocal laser scanning microscope (CLMS, OLYMPUS, OLS5000) was used to examine microstructures and surface roughness. Substrate grain size was measured following the line-intercept method using OLYMPUS Stream software. The copper surface was cleaned in an HCl solution before fluxing the copper substrate with RC-15SH RMA (15%). A reflow soldering process was followed throughout this research. Copper foil and cylindrical solder rods were soldered using a hot plate at 350°C for times of 10s-480s. After the soldering process, samples were rinsed with ethanol to remove the flux. Wettability of the solders on the foil sheet was tested as the expression of contact angle. Optical microscopy (OM, OLYMPUS, BX 53) was employed to examine the shape of liquid solder spreading over the solid substrate and the contact angle was determined using OLYMPUS Stream software. Microstructure of each specimen after reflow soldering was cut and mounted. Specimens were polished using an argon (Ar) ion beam (IBM, HITACHI, IM 4000

* Corresponding author: kkkannac@kmitl.ac.th

PLUS). Scanning electron microscopy (SEM, JEOL, JSM-5800LV) was then used to verify the microstructure and intermetallic phases, followed by energy dispersive spectroscopy (EDS, OXFORD INSTRUMENTS, X-Max) to determine chemical compositions of the intermetallic phases. SEM images obtained were used to measure the thickness of the intermetallic phases. X-ray diffraction (XRD, Bruker, D8-Discover) was also carried out to analyze the intermetallic compound structures.

3 Results and discussion

3.1 Wettability and contact angle of solder

Surface characteristics and microstructure of the copper substrate play a major role regarding wetting behavior of molten solder. A rough substrate surface improves the wettability in terms of contact angle [11-12]. In addition, grain size (i.e. grain boundary) affects the wetting behavior of molten solder on the substrate. Atomic migrations occur more easily at grain boundaries and increase the free energy of the system which may lead to an increase in wetting kinetics [13]. Fig. 1. demonstrates surface microstructure of the copper substrate before soldering. To examine the grain size number of copper, intercept counting via the test pattern was conducted.

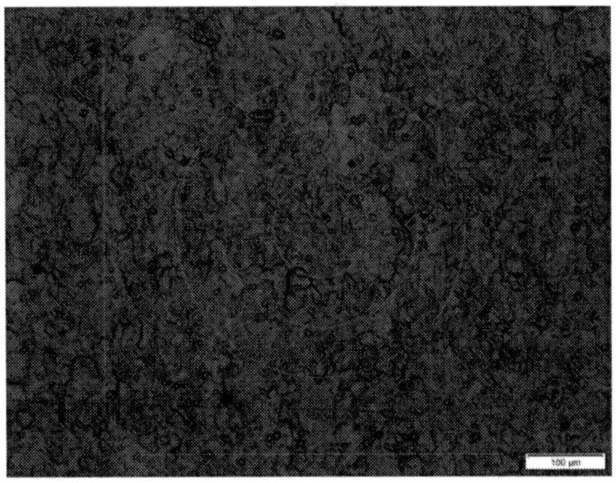

Fig. 1. Surface microstructure of copper substrate.

The grain size number based on the ASTM E112-10 standard was calculated using Equation (1). The original copper substrate exhibited a grain size number of 8.36 for the surface microstructure and was classified as fine grain structure.

$$G = (6.643856 \log \overline{N_L}) - 3.288 \qquad (1)$$

Where G is the grain size number and $\overline{N_L}$ is the number of grain intersections per unit length.

Surface roughness was measured at the center of the copper substrate. Surface area was 258.571 µm x 257 µm. Regarding orientation (or direction) analysis, a single-direction was present in the copper substrate due to the rolling process. Examination length was 258.571 µm with arithmetic mean roughness obtained for the copper substrate at 0.165 µm. After soldering, liquid solder in contact with copper in a solid state is depicted in 2D and 3D as a solder joint in Fig. 2. Contact angle of the solder at different soldering times is shown in Fig. 3. Results indicated that the contact angle decreased with reflow time. Zang et al. [14] defined "good wettability" as a contact angle less than 40° and a contact angle between 40 and 50° as "fair wettability". Average contact angles were in the range 8.26°-30.35° and exhibited good wetting behavior of the solder with soldering times ranging from 20s to 240s. In addition, fair wettability was achieved at less than 10s soldering time with contact angle of 48.83°.

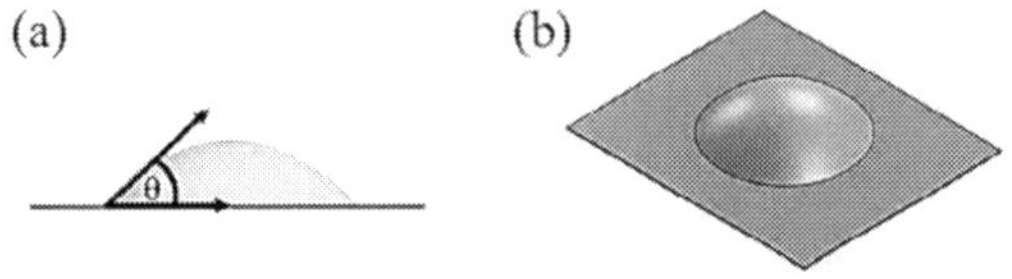

Fig. 2. Schematic of a solder joint; (a) 2D and (b) 3D.

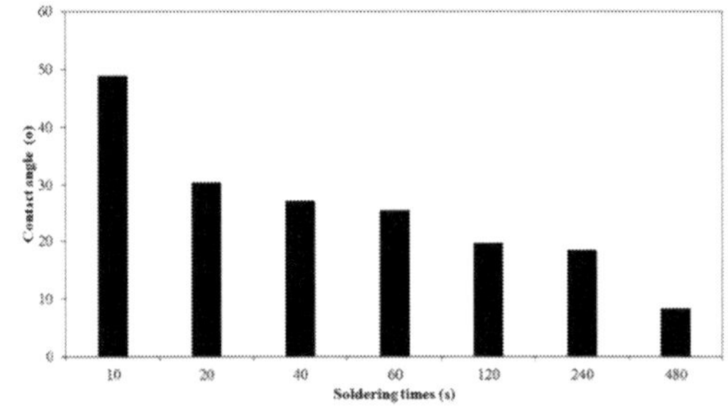

Fig. 3. Contact angle of solder at different soldering times.

3.2 Intermetallic phases

The microstructure and thickness plot of intermetallic phases at different soldering conditions was also examined, as demonstrated in Figures 4 and 5, respectively. To observe the grains of the copper substrate effectively, microstructures were viewed with low magnification and thickness of the intermetallic phases was measured using high magnification. For a given copper substrate, no significant change in grain size of copper substrate was found as soldering time increased. The metal phases found in this experiment were Cu_6Sn_5 and Cu_3Sn. Under soldering conditions of 240s and 480s, a Cu_3Sn phase formed between the copper substrate and the Cu_6Sn_5 phase ($Cu/Cu_3Sn/Cu_6Sn_5$). Generally, increasing thickness of Cu_6Sn_5 was observed with time. This thickness reduced at 240s, with transformation of Cu_6Sn_5 to the Cu_3Sn phase by depletion of Sn which diffused to the Cu side to form a thin phase of Cu_3Sn. In addition, Cu_6Sn_5 phase growth was restricted by the Cu_3Sn phase as fewer Cu atoms diffused from the substrate to the Cu_6Sn_5 phase [15]. After soldering time of 480s, both Cu_6Sn_5 and Cu_3Sn phases showed high growth rates. This occurred as a result of either adding Cu or subtracting Sn from Cu_6Sn_5. Meanwhile, addition of Cu by Cu atom refers to the dissolved substrate [16-17]. Therefore, Cu_3Sn phase grows rapidly. Elemental compositions of Cu_6Sn_5 and Cu_3Sn were determined by EDS analysis as illustrated in Table 1, and results were consistent with Zuozhu et al. [18]. Crystal structure of the intermetallic phases was hexagonal. The intermetallic phases found in this study were in good agreement with those reported in previous literature [19-20].

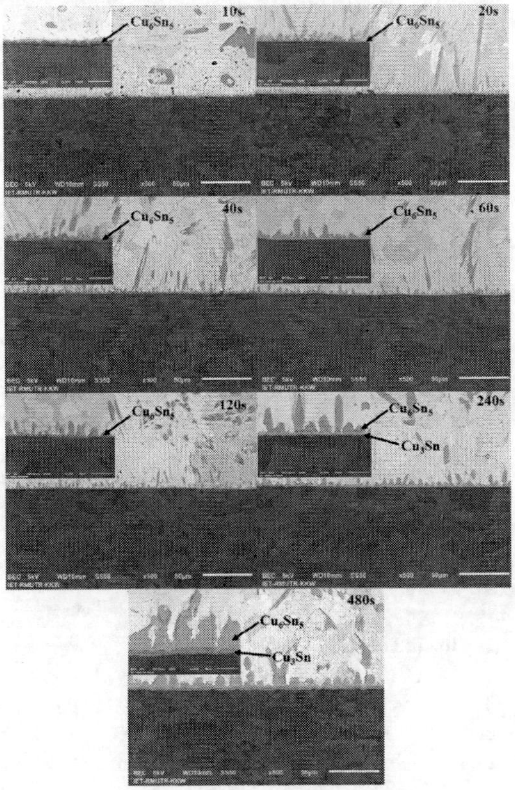

Fig. 4. SEM microstructure with soldering time.

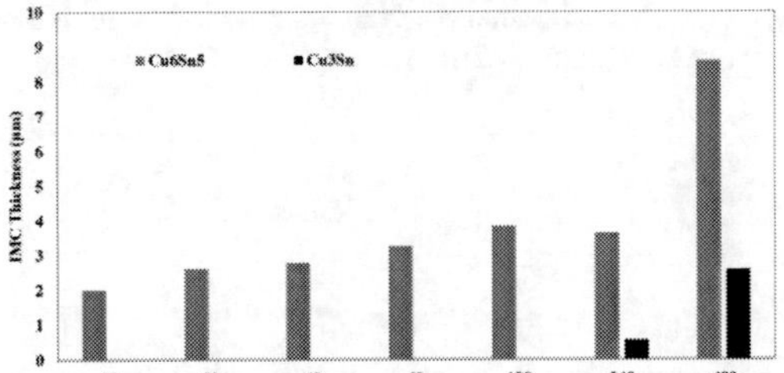

Fig. 5. IMC thickness with soldering time.

Fig. 6 shows the phase fraction of Cu_6Sn_5 in the Sn matrix determined by advanced phase analysis, with results expressed as phase fraction area. Fig. 7 illustrates the percentage of Cu_6Sn_5 phase in the solder for all time points. Volume fraction of Cu_6Sn_5 phase in the solder matrix increased with soldering time due to increased nucleation during solder solidification.

Table 1. Composition of intermetallic phases.

Phase	Cu (at.%)	Sn (at.%)	Ref.
Cu_6Sn_5	51.11-52.77	Balanced	This study
Cu_3Sn	70.74-73.38	Balanced	This study
Cu_6Sn_5	53.30	44.70	[18]
Cu_3Sn	74.90	25.1	[18]

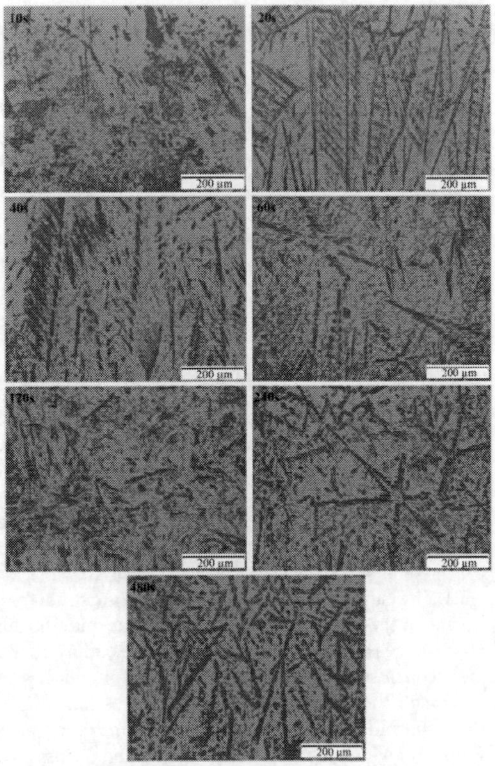

Fig. 6. Phase fraction of Cu_6Sn_5 with soldering time.

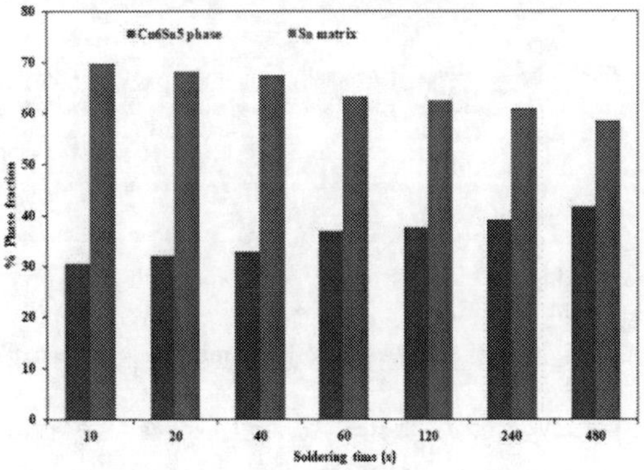

Fig. 7. Percentage of Cu_6Sn_5 phase in Sn matrix.

Thickness of the intermetallic phase was elucidated by equations (2) and (3).

$$Y = kt^n \quad (2)$$

$$Log\ Y = log\ k + n\ log\ t \quad (3)$$

where Y is the thickness of the intermetallic phase, k is the intermetallic growth rate constant, t is the reflow soldering time, and n is the time exponent.

The value of n was determined from the slope of plot between log Y with log t. It is well known that volume of n has a significant effect on the growth rate of the intermetallic phase. When n=1/3, the intermetallic phase growth follows the grain boundary diffusion-controlled; when n=1/2, the intermetallic phase growth follows the volume diffusion-controlled; when n=1, the intermetallic phase growth follows the reaction-controlled [21-23]. The time exponent values for Cu_6Sn_5 and Cu_3Sn phases were 0.083 and 0.682, respectively. This phenomenon can be explained in terms of the growth rate controlled by grain boundary diffusion and volume diffusion-controlled for Cu_6Sn_5 and Cu_3Sn phases, respectively. In summary, relationship between the contact angle and intermetallic phase formation resulted from increase of the intermetallic phase during soldering in response to decreasing contact angle.

4. Conclusion

Influence of soldering time on contact angle and intermetallic phase formation during soldering were investigated. Insignificant change in grain size of copper substrate was recorded as soldering time increased. However, increasing soldering time significantly altered contact angle of the solder and intermetallic phases. Cu_6Sn_5 and Cu_3Sn phases were found at the interface zone. The Cu_3Sn phase only formed when the reflow time became very long. Total thickness of the intermetallic phase increased with post-soldering time, implying that the mechanisms were controlled via grain boundary and volume diffusion. Thus, soldering time was fundamental for controlling wettability and intermetallic phase formation within the soldering process, and necessary to determine a comprehensive approach to develop overall performance of high-temperature solders.

Acknowledgements:

This research was financially supported by Faculty of Engineering, King Mongkut's Institute of Technology Ladkrabang (KMITL).

References

1. D.R. Reinosa, *IEEE International Conference on Test* (2005)
2. S. Prabhu, K.N, Adv. Colloid. Interface. Sci. **166**, 87–118 (2011)
3. J. Watson, G. Castro. Analog Dialogue 46-04 (2012)
4. J. Liang, N. Dariavach, P. Callahan, D. Shangguan, Mater Trans. **47**, 317–325 (2006)
5. C. Lei, H. Huang, Y. Huang, Z. Cheng, S. Tang, Y. Du, Powder Tech. **301**, 356–359 (2016)
6. L. Zang, Z. Yuan, H. Zhao, X. Zhang, Mater Lett. **63**, 2067–2069 (2009)
7. Q. Zeng, J. Guo1, X. Gu, X. Zhao, X. Liu, J. Mater. Sci. Technol. **26**, 156–162 (2010)
8. H. Wang, F. Wang, F. Gao, X. Ma, Y. Qian, J. Alloy Comp. **433**, 302–305 (2007)
9. D.Q. Yu, H.P. Xie, L. Wang, J. Alloy Comp. **385**, 119–125 (2004)
10. M. F. Arenas, V. L. Acoff, J. Electron Mater. **33**, 1452–1458 (2004)
11. T. Novak, J. Stary, F. Steiner, P. Stejskal, *Electroscope, Conference EDS* (2009)
12. Y.Y. Chen, J.G. Duh, J. Mater Sci. **11**, 279–283 (2000)
13. L. Lin, S. Hui, G. Lu, S-L. Wang, X-D. Wang, D.J. Lee, Chem Eng Sci. **174**, 127–135 (2017)
14. L. Zang, Z. Yuan, H. Xu, B. Xu, Appl Surf Sci. **257**, 4877–4884 (2011)
15. X. Wang, J. Han, F. Guo, L. Ma, Y. Wang, *International Conference on Electronic Packaging Technology* (2017)
16. C. Yu, H. Lu, S. Li, J. Alloy Comp. **460**, 594–598 (2008)
17. Y. Wang, D.T. Chu, K.N. Tu, *IEEE Electronic Components and Technology Conference*, 439–446 (2016)
18. Z. Yin, F. Sun, M. Guo, Mater Lett. **215**, 207–210 (2018)
19. S. Fürtauer, D. Li, D. Cupid, H. Flandorfer, Intermetallics, **34**, 142–147 (2013)
20. N. Mookam, K. Kanlayasiri, *ICMM, international Conference on Mechatronics and Manufacturing* (2018)
21. M. Schaefer, R.A. Fournelle, J. Liang. J. Electron Mater. **27**, 1167–1176 (1998)
22. X. Deng, G. Piotrowski, J.J. Williams, N. Chawlai, J. Electron Mater. **32**, 1403–1413 (2003)
23. Y. Yao, J. Zhou, F. Xue, X. Chen, J. Alloy Comp. **682**, 627–633 (2016)

Influence of serum on DNA protection ability and transfection efficiency of cationic lipid-based nanoparticles for gene delivery

Supusson Pengnam[1], *Lalita* Leksantikul[1], *Prasopchai* Tonglairoum[1], *Praneet* Opanasopit[1], *Nattisa* Ni-yomtham[2], *Boon-Ek* Yingyongnarongkul[2], *Samarwadee* Plianwong[3,*]

[1]Faculty of Pharmacy, Silpakorn University, Nakhon Pathom, 73000 Thailand
[2]Department of Chemistry and Center of Excellence for Innovation in Chemistry, Faculty of Science, Ramkhamhaeng University, Bangkok, 10240 Thailand
[3]Faculty of Pharmacy, Burapha University, Chonburi, 20130 Thailand

Abstract. Cationic lipid-based nanoparticulate systems are delivery systems that has been widely used in pharmaceutical field including gene delivery. There are many barriers obstructing genetic materials and their delivery systems to reach the target. Serum is one of the imperative factor that should be investigated. Therefore, the aim of this study was to examine the effect of serum on DNA protection ability of spermine-liposomes and niosomes by evaluating the percentage of transfection efficiency in Hela cell and observing the DNA degradation band using agarose gel electrophoresis in the presence of serum. The results showed that the percentage of transfection efficiency of spermine-liposomes was dramatically decreased when serum is presented ($p < 0.05$). In contrast, whether or not the serum is presented, the spermine-niosomes showed no significant difference in transfection efficiency. Concisely, liposomes could slightly protect DNA from DNase in the serum, whereas, niosomes had potential ability to protect DNA from the enzymes in serum. This result revealed an advantage of the cationic niosomes system as a gene carrier over the cationic liposomes.

1 Introduction

Numerous researches have been using cationic lipid-based nanoparticles (cLNPs) such as cationic liposomes and cationic niosomes for gene delivery which can be applied in various applications such as infectious diseases, metabolic diseases, cancers and regenerative medicines [1,3]. Liposomes are composed of cholesterol and non-toxic phospholipids which self-assembled as small artificial vesicles of spherical shape. Niosomes consist of non-ionic surfactant and cholesterol as excipients. Besides liposomes, niosomes show some significant advantages, such as low cost and storage stability [1]. Some literature indicates that genetic material has a short half-life in the blood circulation because of the rapid degradation caused by nucleases [2,3]. Otherwise cationic lipids are included in both carriers to generate positive charge in their systems and facilitate transfection by electrostatic interaction with DNA and cellular membrane. In the previous study, spermine-liposomes and niosomes exhibited a potential gene delivery and high transfection efficiency [4, 5]. Generally, hypothesis of serum incompatibility with cLNPs begin through interaction between negative charge of serum proteins onto the positive charge of cationic systems affected electrostatic interaction with proteins on the cell surface and their internalization [8,9]. Therefore, the objective of this research is to explore the effect of serum on the DNA protection ability of liposomes and niosomes by evaluating the percentage of transfection efficiency in Hela cell and observing the DNA protection ability by agarose gel electrophoresis in the presence of serum.

2 Materials and Methods

2.1 Materials

Phosphatidylcholine (PC) (Phospholipon® 90G) was purchased from GmbH, Nordrhein-Westfalen, Germany. Span20 (Sp20) were obtained from Sigma-Aldrich, MO. Cholesterol (Chol) and DOPE were purchased from Carlo Erba Reagenti, MI, Italy. Spermine-based cationic lipids (shown in Fig. 1) were synthesized via solid-phase synthesis in the previous

* Corresponding author: samarwadee.pl@go.buu.ac.th

 https://doi.org/10.1051/matecconf/201819201025

report [5]. HeLa cells; the human cervical cancer cell line was obtained from American Type Culture Collection (ATCC, Rockville, MD). The pEGFP-C2 plasmid DNA, encoding green fluorescent protein (GFP), was obtained from Clontech, CA. The Hind III were obtained from Promega, CA. Lipofectamine® 3000 was purchased from Invitrogen, NY. All other reagents were employed of cell culture and molecular biology quality.

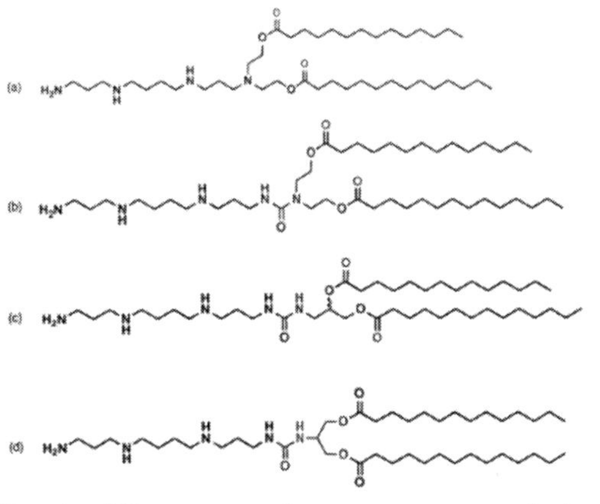

Fig. 1. different central core structures of the spermine-based cationic; di(oxyethyl)amino (a), di(oxyethyl)amino carboxy (b), 3-amino-1,2-dioxypropyl (c) and 2-amino-1,3-dioxypropyl (d).

2.2 Preparation of the cLNPs and the cLNPs/pDNA complexes

The cLNPs were prepared using thin film hydration in ethanol:chloroform (1:1) with sonication method. This study performed 4 cationic lipids with different central core structure; di(oxyethyl)amino (a), di(oxyethyl)amino carboxy (b), 3-amino-1,2-dioxypropyl (c) and 2-amino-1,3-dioxypropyl (d) that were showed in fig. 1 to prepared liposomes; Li-a, Li-b, Li-c, Li-d and niosomes; Ni-a, Ni-b, Ni-c, Ni-d, respectively. The optimal molar ratio of each formulation and the weight ratio of cLNPs to pDNA were obtained in the previous studies [4, 5]. In brief, for cationic niosomes, the molar ratio of Sp20:Chol:cationic lipid was fixed at 2.5:2.5:0.5 mM. For cationic liposomes, the molar ratio of PC:DOPE:cationic lipid was fixed at 2.5:2.5:1.5 mM. Thin film was obtained after evaporated the solvent in a fume hood under N_2 gas flow and left in a desiccator overnight. Then hydrated thin film with pH 7.4 Tris-buffer. After hydration, the particle size was reduced by bath sonicator for 30 min followed by a probe sonicator for 30 min with two cycles in ice bath. The prepared cationic niosomes were kept at 4 °C.

Before the experimentations, pDNA and cLNPs were mixed to allow complexation, and incubated for 30 min at room temperature. The weight ratio of cLNPs to pDNA was fixed at 2.5:1 and 5:1 for niosomes and liposomes, respectively.

2.3 Effect of serum on the transfection in HeLa cells

HeLa cells were cultured in MEM with 10 % FBS and cultured at 37 °C, 5 % CO_2 for 24 h in 48-well plate by seeding density of 1 x 10^4 cells/well. The cLNPs/pDNA complexes were separated into 2 groups that were diluted with two different transfection mediums. One was diluted with MEM with 10 % FBS, the other was diluted with serum-free MEM. After 6 h, the medium was replaced with MEM with 10 % FBS and the cells were incubated at 37 °C under 5% CO_2 for 48 h. The transfected cell (TF) was observed under fluorescence microscopy and calculated in term of percentage of transfection efficiency according to equation 1. Naked pDNA and Lipofectamine® 3000/pDNA complexes were incubated as a negative control and a positive control, respectively.

$$\%\ \text{transfection efficiency} = \frac{[TF_{(cLNPs)} - TF_{(negative)}] \times 100}{[TF_{(positive)} - TF_{(negative)}]} \quad (1)$$

2.3 Serum stability of pDNA in the cLNPs/pDNA complexes

The cLNPs/pDNA complexes with the optimal weight ratio was incubated with serum at 37 °C for 6 h. Then, heated at 70 °C for 15 min to inactivate the serum enzymes. After that, 0.5% sodium dodecyl sulfate (SDS) solution was added to the sample to extract pDNA from the cLNPs and further incubated at room temperature for 10 min. The extracted pDNA was analyzed using 1% agarose gel electrophoresis for 45 min at 100 V. The DNA bands were visualized under a UV transilluminator using a GelDoc system. A naked pDNA was tested under the same conditions as control.

2.4 Statistical analysis

All experiments were performed in triplicate measurements and the data was represent as mean ± standard deviation (SD). Statistical analysis of the data was examined by F-test and t-test (Microsoft® Excel® 2016 MSO for Windows) with $p < 0.05$.

3 Results and discussion

The optimal formula of the niosomes and liposomes in the previous studied were selected to compare the effect of serum on the transfection ability between liposomes and niosomes [4, 5]. All experiments were tested along with lipofectamine® 3000 to obtain a percentage of transfection efficiency. From the results in Fig. 2, serum affected to the transfection efficiency of all liposome formulations. Although Li-d showed high transfection (104.17 %), the percentage of transfection efficiency was dramatically decreased to 59.49 % with the presence of serum ($p < 0.05$). In contrast, when the cationic lipids were prepared as niosomes, no significant effect of serum on the percentage of transfection efficiency has been observed ($p > 0.05$).

The DNA protection ability was investigated by gel agarose electrophoresis after complexing with cLNPs. DNA band could not be observed when pDNA was contacted with serum (positive control, lane 3 in Fig. 3a, 3b). The percentage of transfection efficiency of spermine-liposomes in the presence of serum was displayed in Fig. 3a (lane 4-8). DNA bands of the complexes was faded compared with the DNA band which was not contacted with serum (lane 4, negative control), indicating that DNA degradation may occur in the presence of serum. However, the DNA bands were not clearly disappeared. Therefore, liposome formulations may slightly protect DNA from degradation. On the other hand, the niosome formulations exhibited ability to protect DNA for at least 6 h as it can be seen from the DNA bands (Fig. 3b; lane 4-8). This result was consistent with the transfection efficiency result. However, niosomes which are composed of non-ionic surfactants might be more stable than liposome formulation which consisted of phospholipids. The phospholipids contained in liposomes may be subjected to oxidation and degradation by phospholipases in the blood circulation [6, 7]. Therefore, liposomes might be degraded by serum before internalization. Serum is a one important factor that should be considered to break through the delivery barrier.

The effect of different central core structure of cationic lipids to transfection efficiency was reported in the previous study [4, 5]. In addition, the effect of different central core structure of cationic lipids with or without urea functional group on DNA protection ability in presence of serum was showed in fig. 3b. Niosomes with cationic lipid a exhibited high intensity of released DNA band. Notice that cationic lipid a without urea in their structure provides slightly higher DNA protection ability than the other cationic lipids with urea (b, c, d). The intensity of released DNA band from liposomes in fig. 3a did not show the difference of intensity in each cationic lipid. The stability of phospholipid in serum that contained in liposomes might affect DNA protection ability more than the effect of central core structure. However, structure-activity relationships of changing central core structure was enigmatical [10].

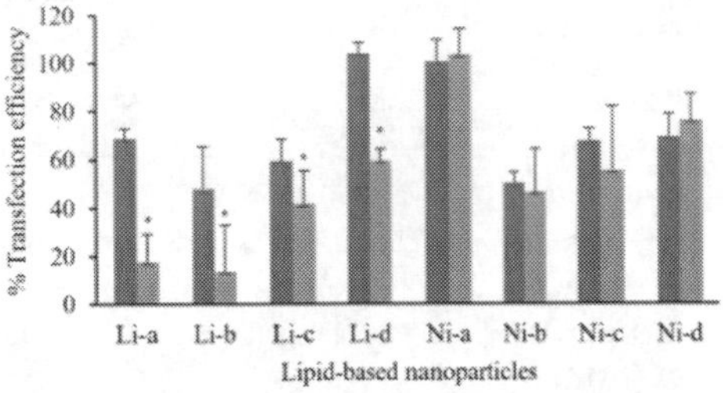

Fig. 2. Percentage of transfection efficiency of spermine-liposomes (Li-a, Li-b, Li-c, Li-d), and spermine-niosomes (Ni-a, Ni-b, Ni-c, Ni-d) (■) without serum and (■) with serum. Data was represented as mean ± standard deviation in triplicate measurements. *Significant different from without serum ($p<0.05$)

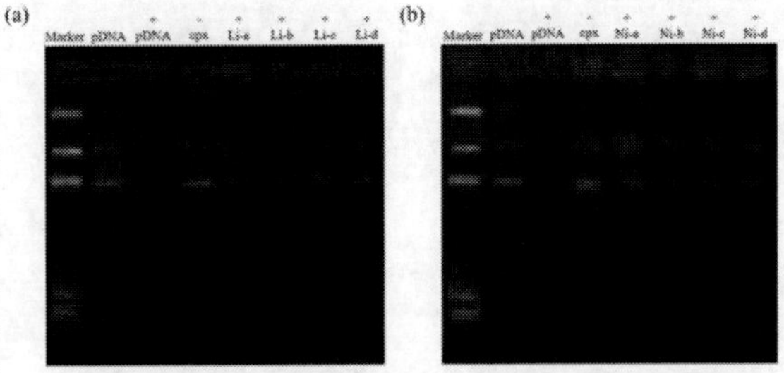

Fig. 3. Agarose gel images of (a) liposome formulations and (b) niosome formulations under the (+) presence of serum for 6 h and (-) absence of serum.

4 Conclusion

Serum contains various enzymes that may affect the transfection efficiency and DNA protection ability of a gene delivery system. In order to improve the transfection efficiency, the delivery systems need to have the ability to protect the DNA from being destroyed by DNase in the serum. From this study, spermine-liposomes could slightly protect the DNA from DNase in the serum. On the other hand, spermine-niosomes exhibited potential ability to protect the DNA whether or not the serum was presented. Therefore, this is the another strength of a cationic niosomes as gene carrier over cationic liposomes. Otherwise spermine-based cationic lipid with no urea provides slightly higher DNA protection ability than cationic lipids with urea in central core structure. However, structure-activity relationships of changing central core structure was enigmatical. It was necessary for study new compounds containing different central core structures, which might be useful for the development of cationic lipid-based gene delivery.

References

1. K.M. Kazi, A.S. Mandal, N. Biswas, A. Guha, S. Chatterjee, M. Behera, K. Kuotsu, J Adv Pharm Technol Res. **1(4)**, 374–380 (2010)
2. M. Saffari, H.R. Moghimi, C.R. Dassc, Iran J Pharm Res. 15(Suppl), 3–17 (2016)
3. X. Cheng, R.J. Lee. Adv. Drug Deliv. Rev. 99, 129-137 (2016)
4. O. Paecharoenchai, N. Niyomtham ,L. Leksantikul, et al. AAPS PharmSciTech. **15(3)**, 722-30 (2014)
5. Niyomtham N, Apiratikul N, Suksen K, et al. Bioorg Med Chem Lett. **25**, 496–503 (2015)
6. M.N. Jones, A.R. Nicholas. Biochim Biophys Acta. **1065(2)**, 145-52 (1991)
7. K. Agarwal, A. Bali, C.M. Gupta. Biochim Biophys Acta. **856(1)**, 36-40 (1986)
8. T. Sato, T. Ishii, Y. Okahata. Biomaterials. **22(15)**, 2075-80 (2001)
9. S. Li, W.C. Tseng, D.B. Stolz, et al. Gene Ther. **6(4)**, 585-94 (1999)
10. D. Zhi, Y. chao, B.J. Yang, et al. Adv. Colloid Interface Sci. **253**, 117-40 (2018)

Allocation strategy for an ambulance base under traffic congestion

Suriyaphong Nilsang[1], *Chumpol* Yuangyai[1*], *Sirisuda* Buatongkue[2], and *Chen-Yang* Cheng[3]

[1]Department of Industrial Engineering, Faculty of Engineering, King Mongkut's Institute of Technology Ladkrabang, Bangkok, Thailand
[2]Department of Information technology, Faculty of Science and Arts, Burapha University, Chanthaburi, Thailand
[3]Department of Industrial Engineering & Management National Taipei University of Technology (Taipei Tech), Taipei, Taiwan

Abstract. One of crucial issues for emergency medical service (EMS) is to reduce response time. However, in metropolis city, a traffic congestion is an obstacle for an ambulance to responsively reach at the scene, then patient mortality and disability rates increase. Traffic congestion is considered as a complex spatial–temporal situation. It is often triggered by repeating factors, such as car lane capacity, weather, and unexpected events. Therefore, a real-time traffic condition is required to effectively determine the location of an ambulance. The current ambulance base allocation strategy model considers only demand point, resulting inability to handle high traffic congestion. This paper proposed a covering model based on traffic congestion (using Google map API) to allocate ambulance bases that covering all demand point, while minimizing the number of the ambulance. In addition, our model was applied to the case study of Bangkok EMS.

1 Introduction

One of crucial issues for emergency medical service (EMS) in metropolis city is traffic congestion. Especially, the ambulance services are highly affected. Their response time increases leading to the causes of death and disabilities of patients in crisis, which impose severe socio-economic costs across the world. [1]. The main objective of emergency ambulance service is to reach the scene within the effective target time for treatment. Typically, EMS responds to 90% of emergency calls within 8 min [2].

For example, the traffic condition for Bangkok is considered as the second-most congested city in the world, it has 61% congestion level (overall travel time when compared uncongested), which is 64 minutes of extra travel time (TomTom Traffic 2016). Extra travel time of traffic congestion affect to ambulance and survival rate of patients in the scene [3], therefore allocation strategy of ambulance has been affected by this traffic congestion. However, it is impossible to place an ambulance all possible bases. EMS managers and administrators have difficulties of locating a limited number of ambulances to cover all accidents or demands points.

Allocation strategy for ambulance have been studied for over 40 years, mostly called "covering model". Researchers have discussed two major models for the allocation of ambulance bases, such as a location set covering problem (LSCP) [4] and a maximal covering location problem (MCLP) [5], The latter is useful in solving strategic problems, in which the MCLP model lacks the flexibility required to solve operational problems because real-time demands vary spatially, temporally. For general covering model consider only a demand point covered and a number of ambulances to reduce response time. In practice, traffic congestion, patient conditions, public events, and geographical location of calls/ bases are also important. Specifically, this paper considers a covering model based on real time traffic congestion using information form Google Map API.

This paper is organized as follows: detailed description of traffic information are presented in Section 2. The overview of integrated model with traffic condition is discussed in Section 3. The Bangkok case is discussed in Section 4. The results, conclusion, and future work are in Section 5.

2 Traffic Congestion Problem

Traffic congestion is considered as a complex spatial–temporal situation. It is often triggered by repeating factors, such as car lane capacity, weather, and unexpected events (e.g. accidents, lane closures, and public events) [6]. Understanding those factors that cause congestion has become a critical issue in traffic management.

The causes of traffic congestion can be measured by physical and psychological factors. Physical causes measure the traffic volume, speed, and street density [7]. In June 2005 Google officially released their Google Maps Application Programming Interface (API), which enables users to mix Google streamed base data with other spatially referenced material that used Floating Car Data (FCD) or Floating Phone Data (FPD) that using the Global Positioning System (GPS). These data can then

* Corresponding author: chumpol.yuangyai@gmail.com

be served as another applications through the Google map interface.

Therefore, this study bring traffic information (traffic speed) from Google map API with ambulance deployment on allocation strategic (covering model) to determine ambulance location that considering environment of traffic. Moreover, we considering number of ambulance on each bases that reduce response time and satisfied traffic speed in this area.

3 Our Proposed Model

The main purpose of this study was to use traffic information from Google map for the redeployment of the ambulances. An overview of the system in this study is shown in Figure 1. When high the traffic congestion occurs, traffic information and images at the scene shared to social media such as Google map, which publishes traffic information. We used this information to analyze traffic speed in data per unit period by filtering based on spatial-term, and temporal-term using floating phone data (FPD) of Google traffic for calculate traffic speed each main road and evaluate traffic congestion; moreover, we updated the ambulance base locations in next period.

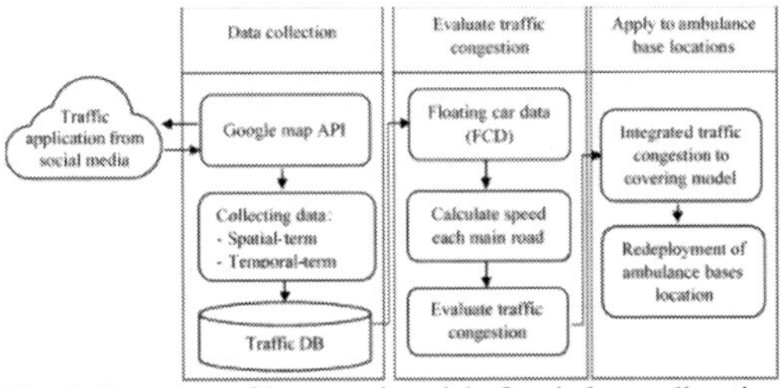

Fig. 1. Overview of integrated model of ambulance allocation with traffic condition from Google map.

3.1 Data collection of traffic congestion from Google map

The proposed framework of integrated traffic information for allocating ambulance bases is shown in Figure 1. First of all, we collected a traffic information database from Google map by using Google map API, that publish traffic information to study in this area information, such as main road, and level of traffic congestion (color on road) and average of car speed, that is shown in Figure 2. We then filtered spatial terms and temporal terms from the collected data in assigned area. In assigned area, there are more than one road and each line has a different road length, thus this study use average speed of car on this area to identity traffic condition. After got information, We compute the traffic speed on each area for allocate ambulance base by using the ratio of Traffic speed of road and a number of road in the area.

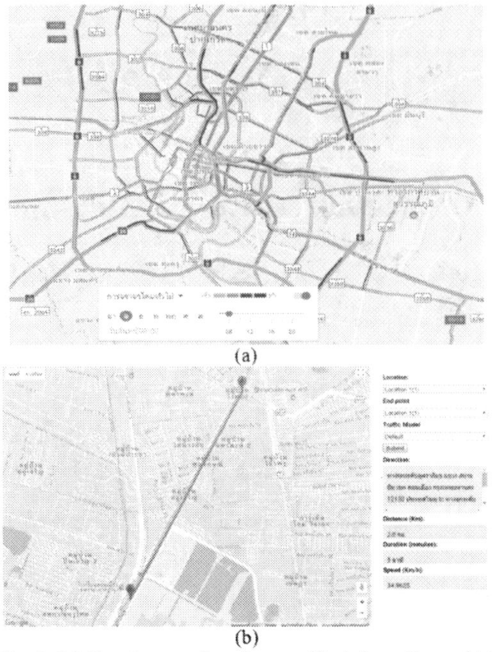

Fig. 2. (a) Google map that show traffic information, which indicate by color on road network. (b) Interface of google map API, which used for collect data of traffic information.

3.2 Mathematical model with traffic condition

We allocate both number of ambulances and base locations by using the covering model. In this study, we consider traffic congestion information into the model by adding a constraint to balance between average car speed and acceptable speed in a location. We named the proposed model as "Traffic Speed of Location Set Covering Problem for Traffic Speed" (LSCP-TS). Our model was under the following assumptions: traffic speed are deterministic, the base locations must be available in advance and all location can be ambulance base.

3.2.1 Notation

The model description is given as:

V = Set of demand points

W = Set of potential base locations

A_{ij} = Set of potential base locations that cover demand point i within the distance standard.

i = Index of demand points, $i \in V$

j = Index of potential base locations, $j \in W$

P = Total potential base location

C_i = Average of speed in location i

S_i = Acceptable car speed in location i

$$B_j = \begin{cases} 1 & \text{if base location is location - forced j} \\ 0 & \text{otherwise} \end{cases}$$

$$x_j = \begin{cases} 1 & \text{if base location is selected at base j} \\ 0 & \text{otherwise} \end{cases}$$

$$y_{ik} = \begin{cases} 1 & \text{if demand point i is covered at k} \\ 0 & \text{otherwise} \end{cases}$$

3.2.2 Mathematical model

This mathematical model is extended from modified covering model of Nilsang et al. [8] in location-forced constraint to determine impotent location to model. This can be formulated as follows:

$$\text{Minimize} \quad \sum_{j \in W} x_j \qquad (1)$$

$$\text{Subject to} \quad \sum_{j \in W} A_{ij} x_j \geq \sum_{k=1}^{P} y_{ik} \qquad i \in V \qquad (2)$$

$$\sum_{k=1}^{P} y_{ik} \geq 1 \qquad i \in V \qquad (3)$$

$$\beta_j \leq x_j \qquad j \in W \qquad (4)$$

$$\sum_{i=V} C_i \sum_{k=1}^{P} y_{ik} \geq S_i \theta \qquad i \in V \qquad (5)$$

$$y_{ik} \in \{0,1\} \qquad i \in V, k = 1,\ldots,P \qquad (6)$$

$$x_j \text{ Integer} \qquad j \in W \qquad (7)$$

The objective of the model is to minimize number of ambulance base that cover all demand points while observing traffic congestion in all location. Traffic congestion occurred, ambulance vehicle may not be able to arrival to the scene if there is only one ambulance. Therefore, if the scene is covered by multiple bases and covered more than one ambulance, the scene will be taken care within target time. The objective function is given in (1). Constraint (2) and (3) ensures that ambulance base location cover demand point within the distance standard least than once. Sometime, allocation of ambulance base must assigned specific locations (e.g., public events) that shows in constraint (4). And our proposed constraint that consider traffic condition with average of traffic speed in this location must more than acceptable speed per one ambulance in constraint (5). Otherwise, that is variable constraints x_j is integer and y_{ik} is binary variable.

3.3 Traffic information constraint into allocation strategy

The integrated model, which is covering model with traffic condition and traffic data from Google map information can be used to determine ambulance base location in the future and reduce response time each periods of daily. Moreover, we added location-forced constraints that can be used when abnormalities (increasing accidents or public events) are observed on social media information. Both constraints traffic constraint and location-forced constraint are more useful on the strategic and operational levels.

For traffic speed constraint, we used Google Maps API for evaluate average traffic speed in real-time. Then we added traffic condition to our covering model (LSCP-TS) to determine new ambulance base locations that use minimize ambulance base location and number of ambulance. To accept traffic speed that ambulance can arrive to the scene under target response time. For the new ambulance base, we set a new position of an ambulance base by using a geographic information system to set positions on a real map and we checked the covered demand points.

4 A Case Study in Bangkok

Bangkok is the capital and the most populous city in Thailand. Bangkok has a population of more than 14 million. Traffic congestion is major problem in Thailand, which has a tremendous impact on the economy and public health. In 2016 The INRIX Global Traffic Scorecard rated Bangkok the 12th most congested of all cities rated (1st most congested of Asia), considerably worse than 30th in 2015, with a traffic scorecard rating of 11, down from 20 in 2015 (INRIX 2017). Bangkok drivers spent an average 64.1 hours a year in traffic jams, according to the scorecard 23% of overall time and an average 33% of their time during peak hours. Moreover,

For historical traffic status in Bangkok area (google map 2017), it shows in Figure 3. We divided the area into 56 locations (6 kilometers by 6 kilometers) include 44 full of area and 12 some part of area and have 3 levels of traffic speed; (i) green area, average traffic speed more than 50 km/h, (ii) orange area, average traffic speed lower than 50 km/h and more than 30 km/h, and finally (iii) rad area, average traffic speed lower than 30 km/h. that indicated most area in Bangkok have traffic speed lower than 50 km/h. At 8 minutes of response time, ambulance must travel to the scene by traffic speed 75 kilometers per hour.

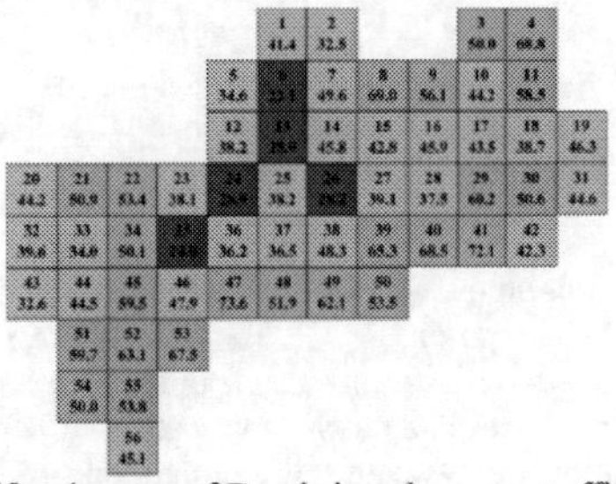

Fig. 3. Fifty-six areas of Bangkok and average traffic speed 3 level that show traffic problem in Bangkok.

5 Results

For results, we compare difference 3 periods, which difference traffic speed at each location. We collect traffic speed data from Google map API at 3 periods, such as daytime in weekend (assume all location traffic speed 80 Km/h), peck hour in normally day and daytime in normally day to analyze traffic condition of each locations. Afterthought, we setting parameters follow that; potential ambulance base (P) is 30 bases, traffic speed acceptance (S_i) is 75 Km/h (Basically smooth) and location-forced (B_j) is location No 27 for determine ambulance base locations by LSCP-TS model.

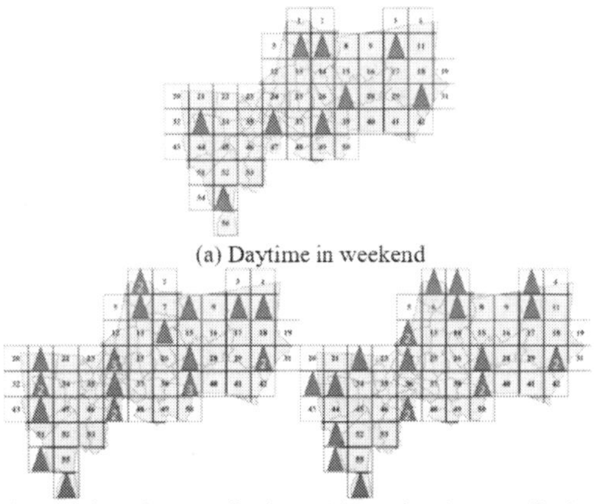

(a) Daytime in weekend

(b) Peck hour in normally day (c) Daytime in normally day

Fig. 4. Fifty-six areas of Bangkok and average traffic speed 3 types that show traffic problem in Bangkok.

After using the LSCP-TS model, the number of locations was 9 and 9 number of ambulances for daytime in weekend, which we used result as a base line to compare with another results. Two more results, daytime in weekday and peak hour in weekday, the number of locations is 17 bases, but difference location and number of ambulance (21 ambulances and 25 ambulances, respectively) to accept traffic speed, which show in Figure 4. It can be seen that LSCP-TS modal provide ambulance bases for daytime more than those of peak hour, because of a low traffic speed in peak hour so that model must add ambulance base to system. Although LSCP-TS model determine ambulance bases and ambulance vehicles a lot, but it can alleviate the effect of traffic congestion.

6 Conclusions and Discussion

In this study, we proposed an integrated framework of traffic condition, LSCP-TS constraints, to update ambulance base locations in response to traffic problem in urban. We explained the application of social media information for evaluate traffic condition, in the context of the EMS, and proposed an allocation model that does not require periodic relocation. We also validated the model by applying it to Bangkok EMS. Result showed that the locations of ambulance bases obtained that difference period from the proposed approach can cater to acceptable traffic speed and minimize the response times of ambulances under traffic conditions.

In future research, the relationship between uncertainty in risk management and the allocation of ambulance bases from social media information, will be considered. In general covering model for ambulance, we used only demand that not enough for ambulance allocation strategy. Therefore, we will focus on additional factors that related with ambulance management such as population density, disaster area, and public event etc.

References

1. Anand, J. and T.A. Flora, *Emergency Traffic Management for Ambulance using Wireless Communication*. IPASJ International Journal of Electronics & Communication (IIJEC), 2014. **2**(7): p. 43-52.
2. De Maio, V.J., et al., *Optimal defibrillation response intervals for maximum out-of-hospital cardiac arrest survival rates*. Annals of emergency medicine, 2003. **42**(2): p. 242-250.
3. Karaman, M., *A Genetic Algorithm for the Multi-Level Maximal Covering Ambulance Location Problem*. Master of Science Thesis, 2008.
4. Toregas, C., et al., *The location of emergency service facilities*. Operations research, 1971. **19**(6): p. 1363-1373.
5. Church, R. and C. ReVelle. *The maximal covering location problem*. in *Papers of the Regional Science Association*. 1974. Springer.
6. Kwon, J., M. Mauch, and P. Varaiya, *Components of congestion: Delay from incidents, special events, lane closures, weather, potential ramp metering gain, and excess demand*. Transportation Research Record: Journal of the Transportation Research Board, 2006(1959): p. 84-91.
7. Lu, J. and L. Cao. *Congestion evaluation from traffic flow information based on fuzzy logic*. in *Intelligent Transportation Systems, 2003. Proceedings. 2003 IEEE*. 2003. IEEE.
8. Nilsang, S., Yuangyai, C., Janjarassuk, U., and Cheng, C.Y. Locating an Ambulance Base by Using Social Media: A Case Study in Bangkok. Annals of Operations Research, (In-process of publish).

Critical path analysis programming method without network diagram

Paing Hein Soe[1,*] and *Thein* Min Htike[1]

[1]Department of Mechanical Engineering, Yangon Technological University, Yangon, Myanmar

Abstract. Project scheduling is an important task in project management which monitors the project duration, resource utilization and affects the benefit of project for all kinds of project such as construction, IT, renewable energy, product design and manufacturing etc. Among many methods for project management, critical path method (CPM) is one of the better-known planning and control techniques in project scheduling. In this study, CPM-based computerized program was developed together with modeling of step by step calculations of parameters of interest such as earliest starts and finishes, slack and float times of project activities resulting in determination of critical path for the project. One advantage of computer implementation of CPM using MATLAB in this study is that unlike conventional method, it eliminates necessity to conventionally draw network diagram to perform critical path analysis.

I. Introduction

A project is any series of activities and tasks having a specific objective to be completed within some specifications and limitations such as a defined start and end date, funding, human resources and equipment [11,12]. Project management is the management of project activities applying knowledge, skills, tools and techniques to meet the project requirements [1,11]. Project management is a key function for successful project. The term "Project management" is firstly used in the US defense aerospace sector in 1953 [11,13]. Project scheduling also plays an important role in project planning under project management. Scheduling issue is one of the main functions in project planning. It is a frequent task at the very early stage of project to control many systems, for examples, in construction projects, construction and careful planning [2]. There is a clear correlation between the project completion time and its profitability in construction undertakings [10].

Project schedule network diagram is one of the methods to assist project scheduling issue. To assist project planning, many techniques and tools such as Guntt Chart, Critical Path Method (CPM) and Program Evaluation and Review Technique (PERT) have been developed and these are seriously used by a large majority of project managers to identify critical activities and to calculate minimum time required for project completion [2,3,4,5]. PERT and CPM are most popular diagramming methods among others used in project management [20]. In planning and controlling both large and small projects of all types including construction, research, IT, manufacturing, defense, development and many others, network-based procedures of PERT and CPM are well known and widely used to assist managers, and these two techniques give considerable benefit to decision makers. These two methods are also the first major computerized project management decision support systems (DSS) [21,22,23,24,25]. PERT and CPM were outgrowths of the new discipline of project management, in which PERT, developed in the late 1950's, was first initiated by the US Navy to plan and control the Polaris missile program to have shortest possible completion time, and the first attempt for CPM was in DuPont company in 1956 [11,14,15,16,17,18].

In this study, computer program for determining critical path based on CPM is developed. CPM provides parameters of activities including earliest starting time, latest starting time, earliest finish time, latest finish time, maximum available time and slack time [2,8,9]. Since CPM and PERT are traditional methods, they have some disadvantages such as the need to draw network diagram to determine the minimum project duration. To overcome these, several models such as vertical method, linear scheduling method and time scheduling method have been introduced for scheduling of construction projects [18,19]. CPM and PERT techniques can also be blended with fuzzy logic and earned value technique [20]. One of the main differences in CPM and PERT is a probabilistic model with uncertainty in activity duration, but CPM is a deterministic model. The most important benefit of the CPM deterministic assumption may be facilitating the use of optimization models for trading off time and cost [24] and for sequencing [21,26,27]. In agricultural projects, network models also have been used as planning and project controlling techniques [2,6,7]. This study

* Corresponding author: paingheinsoe.pht@gmail.com

focuses on the development of computer program which can find critical path for the project without the need to draw network diagram by using MATLAB program.

II. Methodology and Terms

CPM mainly includes forward pass and backward pass calculation to get parameters of activities including earliest, latest time and flexibility of path. Earliest start and earliest finish time for each activity within the network are usually first computed by forward pass and latest finish and start time are by backward pass [2]. To perform this, it generally needs to have network diagrams to show logical relationships between activities usually represented by arrow and node formats. In this study, calculation methods based on CPM are based and programming approach without network diagram to be followed is presented in detailed calculations.

Here is an example project in design and producing component to consider logical sequence of activities: finishing component development (A), design marketing programme (B), design production system (C), selecting advertising media (D), initial production run (E) and releasing component to market (F). In these six activities, it is relevant that activity A should be the first step which is the earliest starting activity. Activity A is the predecessor activity for B and C. B is the follower activity of D. E should be after C and F should be before D and E.

Duration of each activity is known or estimated, and project duration is unknow to be solved for and it has to cover all these activities. Interested parameters are the critical path and its duration, minimum or longest duration to complete the entire project, and knowledge of these will help to check the actual progress against the scheduled duration of the project. The activities in critical path are critical activities, and if there is any delay in its start, further delay will result in project completion time. So, these are in critical conditions. Some activities can be delayed for some time without causing any delay in completion the entire project. These activities are called non-critical activities. The allowed interval of delay time is known from earliest and latest time of start and end for each activity. Therefore, these parameters of information are useful for project management in (1) avoiding any delay in critical activities or tasks to prevent delay in project completion, and (2) shifting start time for any cause in some non-critical activities without affecting entire project completion time. This sample sequence of activities will be determined by using CPM by logically sequencing them in MATLAB script in order to automatically execute and display the result based on the activity sequence and normal time of each activity.

III. Calculation Steps Procedure

The following procedure is applied to determine the critical path:

1. Normal time consumed by each task or activity involved in a project are estimated or assigned.
2. Follower activities for each activity are listed based on the logical sequence of problem or project.
3. All paths including respective activities to complete the project are listed based on the data from step (2).
4. Earliest finish time for each activity is calculated.
5. Earliest start time is acquired by earliest finish time and normal time of each activity.
6. Latest finish time for each activity is calculated.
7. Latest start time is achieved by latest finish time and normal time of each activity.
8. Total float is calculated and critical path is decided.

IV. Detailed Procedure for Each Step

Six activities A to J in sample problem are logically arranged as follows:

A < D,E ; B,D < F ; C < G ; B,G < H ; F,G < I

where, the sign < means "precedes".

IV.I Step 1: Assigning Normal Time

Normal time for each activity (A to J) is a known parameter based on the experience of performing task before, in actual. Here, normal time (in days) is assigned as follows in Table 1.

Table 1. Normal Time

Task	A	B	C	D	E	F	G	H	I	J
Normal time	23	8	20	16	24	18	19	4	10	15

IV.II Step 2: Logical Sequence of Activities

Logical sequence is now arranged as in Table 2 based on the data given in this example.

Table 2. Logical Sequence

Activity	Followers
A	D , E
B	F , H
C	G
D	F
E	-----
F	I
G	H , I
H	-----
I	-----
J	-----

IV.III Step 3: All Paths to Complete Project

Firstly, it needs to find the activities of initial start which are not listed in the followers by the aid of result in step (2). In this case, activities A, B, C and J are not included in the followers list, so they are initial starters. Some starters alone may be one of the paths which means that some of them do not have followers and become paths involved itself alone. Activities which do not have followers are E, H, I and J. So, among initial starters (A, B, C and J), J becomes first end path containing only one activity itself. Then, other three starters remain to be continued. For starter

A, it forms instant two groups of path A-D and A-E (D and E are followers of A). For first group of A-D path, activity F which is a follower of last activity D, transforms one new path A-D-F. Then, for A-D-F path, F has follower I which gives A-D-F-I path. For A-D-F-I path, last activity I has no follower so it becomes second end path A-D-F-I.

Then for second group of path AE, E has no follower activity, so it terminates and becomes third end path. Like for starter A, B and C are traced by logical sequence. Finally, after all starter activities, all paths to be performed to complete entire project are achieved. These results are shown in Table 3.

Table 3. All Paths and Duration

No.	Path	Path duration (days)
1.	J	15 = 15
2.	A-D-F-I	23+16+18+10 = 67
3.	A-E	23+24 = 47
4.	B-F-I	8+18+10 = 36
5.	B-H	8+ 4 = 12
6.	C-G-H	20+19+ 4 = 43
7.	C-G-I	20+19+10 = 49

IV.IV Step 4: Earliest Finish Time (EF_j)

Earliest finish time is the time for an activity which should be finished as the earliest. Earliest finish time for each activity is calculated based on forward calculation. For each activity, the paths containing specific activity are extracted. Then, sum of normal time for the activities from start to till specific activity (including itself) is calculated for each extracted path. Then, maximum sum of normal time for extracted activities becomes earliest finish time EF for specific activity j.

As an example, for activity F, there are two paths contained F: A-D-F-I and B-F-I. Here, first finish time is 57 days [time(A)+time(D)+time(F)] and second finish time is 26 days [time(B)+time(F)]. Then earliest finish time for activity F is 57 days [max (57,26)] and shown in Table 4.

Table 4. Earliest Finish Time in Detail

Activity	Extracted path	Extracted activity	Sum of normal time	EF_j
F	A-D-F-I	A-D-F	23+16+18=57	57
	B-F-I	B-F	8+18=26	

IV.V Step 5: Earliest Start Time (ES_j)

Earliest start time is the time for an activity which should be started as the earliest. The earliest start time for each activity is just the difference between EF_j and normal time of specific activity j.

IV.VI Step 6: Latest Finish Time (LF_j)

Latest finish time is the time for an activity which should be finished as the latest. Latest finish time calculation for each activity is based on backward calculation. Firstly, maximum path duration from step 3 is determined. 67 days will be maximum path duration for this project.

Then, for each activity, the paths contained specific activity are extracted like in EF_j calculation. Unlike this, starting from activity beyond specific activity to the last activity of each path, sum of normal time of these activities is calculated for each path. Among them, maximum sum time is differenced from maximum path duration. This difference becomes the latest finish time LF_j for specific activity j and activity A, as an example, is processed in Table 5.

Table 5. Latest Finish Time in Detail

Activity	Extracted path	Extracted activity	Sum of normal time	EF_j
A	A-D-F-I	D-F-I	16+18+10=44	67-44 =23
	A-E	E	24	

IV.VII Step 7: Latest Start Time (LS_j)

Latest start time is the time for an activity which should be started as the latest. The latest start time for each activity is just the difference between LF_j and normal time of specific activity j.

IV.VIII Step 8: Total Float (TF_j) and Critical Path

Total float of an activity expresses the amount of time by which an activity can be delayed without delay in completing the project. It is the difference between latest start time (ESj) and earliest start time (ESj) or between Latest finish time (LFj) and Earliest finish time (EFj) for specific activity j.

V. Programming By MATLAB

All the steps shown in above are programmed using MATLAB in order to automatically get the results depending upon the change in input of logical sequence and normal duration of activities.

In this program, activities A,B,C,D,E,F,G,H, I are inputs as 1,2,3,4,5,6,7,8,9 respectively in order to be directly and easily processed in program. For example, follower(1).p = [4,5] means that immediate followers of activity A are activities 4 and 5.

VI. Results

The results in Table and Excel sheet achieved by the program are shown in following Figures 1 and 2.

```
There are 1 critical path(s).
The critical path processed by Program is ADFI.
Completion time for this path is 67 units.

The minimum total time to complete the whole prject is 67 units.

ans =
```

Activity	Time	E_Start	E_Finish	L_Start	L_Finish	T_Float
A	23	0	23	0	23	0
B	8	0	8	31	39	31
C	20	0	20	18	38	18
D	16	23	39	23	39	0
E	24	23	47	43	67	20
F	18	39	57	39	57	0
G	19	20	39	38	57	18
H	4	39	43	63	67	24
I	10	57	67	57	67	0
J	15	0	15	52	67	52

Fig.1. Results in Table on command window

In Figure 1, it can be seen that A-D-F-I is the critical path and the result is also validated by the conventional method with network diagram. Delay of any activity in this critical path will cause further delay in project completion time. In addition, free float and independent float for each activity can also be computed by considering its influence on delay time of another activity of sequence. Both free and independent float are the portion of total float and represent the time in which an activity can be manipulated without affecting the float of subsequent activities and can be delayed for start without affecting floats of the preceding activities respectively.

1	Activity	Time	Earliest Start Time	Earliest Finish Time	Latest Start Time	Latest Finish Time	Total Float	Path	Activity Sequence	Path Duration
2	A	23	0	23	0	23	0	Path 1	J	15
3	B	8	0	8	31	39	31	Path 2	AE	47
4	C	20	0	20	18	38	18	Path 3	BH	12
5	D	16	23	39	23	39	0	Path 4	BFI	36
6	E	24	23	47	43	67	20	Path 5	CGH	43
7	F	18	39	57	39	57	0	Path 6	CGI	49
8	G	19	20	39	38	57	18	Path 7	ADFI	67
9	H	4	39	43	63	67	24			
10	I	10	57	67	57	67	0			
11	J	15	0	15	52	67	52			
12										

Fig.2. Results in Excel Sheet

Figure 2 shows the results in Excel sheet and also includes all paths to complete the project together with each path duration. There are total 7 paths and activity sequence for each path is also presented in the result. Out of all paths, critical path A-D-F-I has the longest path duration of 67 weeks.

Conclusion

Project scheduling and control is a key factor in successful project management. If delayed or exceeded the deadline of the project, there will have negative financial consequence and will degrade the benefits of the project. The results obtained by CPM will assist the user in an effective manner realizing the relationships between the activities in the project and having the idea on the progress along the project duration. The delay in any activity on the critical path will delay overall project duration. The possible delayed time for other activities will also be obtained from the results by CPM. In this study, MATLAB-based computer program was developed to determine what task or job or activity should be delayed or not and when they should be started as the earliest, latest and should be finished as the earliest, latest. It was also suggested to use the developed program to perform critical path analysis without needing to draw diagram of network. For this purpose, step-by-step procedures for calculation were also presented. Future research should also focus on generating network diagram and presenting critical path with graphical presentation such as MATLAB GUI. This would further highlight the significance of this proposed method in realizing its application in managing projects and finding possible flexibility particularly construction, manufacturing and renewable energy implementation projects.

REFERENCES

[1] Project Management Institute, *A Guide to the Project Management Body of Knowledge (PMBOK® Guide)* – Fifth Edition (2013)
[2] Samira Zareei, Project Scheduling for Constructing Biogas Plant Using Critical Path Method. Renewable and Sustainable Energy Reviews **81**, (2018), p. 756-759
[3] Yang JB, How the Critical Chain Scheduling Method is Working for Construction, Cost Eng **49**(4), (2007), p. 25–32
[4] Fahimifard SM, Kehkha AA, Application of Project Scheduling in Agriculture (Case Study: Grape Garden Stabilization), Am-Eurasia J Agric Environ Sci **5**(3), (2009), p. 313–21
[5] Pollack JB. Project Management Software Usage Patterns and Suggested Research Directions for Future Development. Proj Manag J **29**, (1998), p. 19–29
[6] Monjezi N, Sheikhdavoodi MJ, Basirzadeh H, Zakidizaji H., Analysis and Evaluation of Mechanized Greenhouse Construction Project Using CPM Methods, Res Appl Sci Eng Technol **4**(18), (2012), p. 3267–73
[7] Abdi R, Ghasemzadeh HR, Abdollahpour S, Sabzeparvar M, Dabbag Mohammadi, Nasab A., Modeling and Analysis of Mechanization Projects of Wheat Production by GERT Networks, Agric Sci China **9**(7), (2010), p. 1078–83.
[8] Smith P. *Agricultural Project Management: Monitoring and Control of Implementation*, Elsevier Applied Science Publisher; (1984). p. 190.
[9] Zargar M., *Project Control*, Parham, Tehran, Iran; (2004)
[10] Krystyna A., *Application of Critical Chain Management in Construction Projects Schedules in a Multi-Project Environment: a Case Study*. Procedia Engineering **182**, (2017), p. 33– 41.
[11] H.A. Hornstein. The Integration of Project Management and Organizational Change Management iss Now A Necessity, Int. J. Proj. Manag., (2014)
[12] Kerzner, H., *Project Management: A Systems Approach to Planning, Scheduling, and Controlling*, John Wiley & Sons, Hoboken, N. J, 11th edition, (2013)
[13] Johnson, S.B., *The Secret of Apollo: Systems Management in American and European Space Programs*, The Johns Hopkins University Press, Baltimore, (2002).
[14] Morris, P.W.G., Pinto, J.K., Soderlund, J. (Eds.), *The Oxford Handbook of Project Management*, Oxford University Press, Oxford, UK, (2012)
[15] Ma Jun-jie, Qi Jian-xun, Study on Critical Path Method with Fixed Time Parameter in Network Planning Technology, Third International Symposium on Information Science and Engineering, **96**, (2010)
[16] S. E. Elmaghraby, On criticality and sentivity in activity networks, European Journal of Operational Research, **127**, (2000), p. 220-238
[17] J. E. Kelley Jr. and M. R. Walker, Critical path planning and scheduling, Proceedings of the Eastern Joint Computational Conference, **16**, (1959), p. 160-172
[18] Amy H. I. Lee, He-Yau Kang, Tzu-Ting Huang, Project Management Model for Constructing A Renewable Energy Plant, Procedia Engineering, **174**, (2017), p. 145–154
[19] K.A. Chrysafis, B.K. Papadopoulos, Approaching Activity Duration in PERT By Means of Fuzzy Sets Theory And Statistics, J Intell Fuzzy Syst **26**(2), (2014), p. 577-587
[20] Mete MAZLUM, Ali Fuat GÜNERİ. CPM, PERT and Project Management with Fuzzy Logic Technique and Implementation on a Business, Procedia - Social and Behavioral Sciences **210**, (2015) , p. 348 – 357
[21] Dan Trietsch, Kenneth R. Baker, PERT 21: Fitting PERT/CPM for Use in the 21st Century, International Journal of Project Management **30**, (2012), p. 490–502.
[22] Clark, C.E., The PERT Model for The Distribution of an Activity Time, Operations Research **10**, (1962), p. 405–406.
[23] Malcolm, D.G., Roseboom, J.H., Clark, C.E., Fazar, W., Application of a Technique for a Research and Development Program Evaluation, Operations Research **7**, (1959), p. 646–669.
[24] Kelley, J.E., Critical-Path Planning and Scheduling: Mathematical Basis, Operations Research **9** (3), (1961), p. 296–320.
[25] Kelley, J.E., Walker, M.R., Critical Path Planning and Scheduling, Proceedings of the Eastern Joint Computer Conference, Boston, (1959), p. 160–173
[26] Kelley, J.E., The Critical Path Method: Resources Planning and Scheduling, Chapter 21, Muth, J., Thompson, G.L. (Eds.), *Industrial Scheduling*, Prentice-Hall, (1963), p. 347–365
[27] Wiest, J.D., Some Properties of Schedules for Large Projects with Limited Resources, Operations Research **12**, (1964), p. 395–418.

Fourier-spectral-method implementation of deformation in the phase-field crystal model

Nirand Pisutha-Arnond[1,*]

[1]Department of Industrial Engineering, Faculty of Engineering, King Mongkut's Institute of Technology Ladkrabang, Bangkok 10520, Thailand

Abstract. The phase-field crystal (PFC) method is a promising computational model with atomistic resolution and diffusive time scale. In this work, the Fourier-spectral-method (FSM) scheme was developed for evaluating the PFC free energy of a system subjected homogeneous deformation. This scheme addresses the complication where, in numerical implementation of FSM using discrete Fourier transform (DFT), the discretized data may no longer lie along the directions of the Cartesian basis due to deformation. In this scheme, the real-space coordinate transformation is employed so that the (continuous) Fourier transform is performed on the function of the undeformed coordinates. This transformation allows straightforward DFT implementation because the sampling at the undeformed configuration is unaffected by the deformation. This scheme is also shown to be applicable to both the original PFC model and a "CDFT"-type PFC model containing a two-body correlation function.

1 Introduction

The phase-field crystal (PFC) model [1] is a promising model for simulating atomistic phenomena on diffusive time scale. The PFC model is based on a free energy functional that can be minimized by uniform or periodic order-parameter functions. The latter, in particular, can be used to represent atomic density in a crystal lattice, which gives the PFC model an atomic resolution. The time-evolution of the order parameter is driven by dissipative dynamics which leads to the (diffusive) time scale that is not bounded by prohibitively small atomic-vibration time scale. These features make the PFC model appealing for simulating nano- to meso-scale materials phenomena over the time scale that is relevant to experimental observation. (The interested reader may consult Ref. [2] and the references therein for comprehensive information of the PFC model.)

Since the order-parameter profiles in the PFC model are typically smooth, periodic, and contained in a geometrically simple computational domain, the Fourier spectral method (FSM) provides an efficient and accurate algorithm for evaluating the PFC free energy and solving the dynamic equations. The reason is that, in Fourier space, spatial derivative becomes multiplication, which can be computed efficiently. Also, the FSM is a global approach which yields high accuracy (known as spectral accuracy) unmatched by local methods such as finite difference. These advantages make the FSM a robust numerical method for PFC calculations.

In numerical implementation of FSM, the discrete Fourier transform (DFT) is used to approximate the continuous Fourier transform. For the DFT, it is assumed that the discrete data are sampled along the directions of the Cartesian basis; in two dimensions, for example, one can visualize the data forming a rectangular grid. However, the complication arises when the system is deformed and the grid is no longer rectangular. How does one implement the DFT, and FSM in general, in this situation?

In this work, I presented the FSM scheme for calculating the PFC free energy of a system subjected to homogeneous deformation. This scheme utilizes real-space coordinate transformation so that the (continuous) Fourier transform is performed on the function of the undeformed coordinates and the effect of the deformation instead manifests in the transformation of the Fourier wave vector. The advantage of this scheme is the straightforward DFT implementation because the function of the undeformed configuration and its discretized data are not affected by the deformation. This technique is, in part, similar to that used in imaging processing to obtain the Fourier representation of affine-transformed image data [3]. However, in the context of the PFC model, the spatial derivative needs to be included in the derivation and the application of this scheme to the PFC free energy that contains a two-body correlation function (referred to as the "CDFT"-type PFC [4]) is discussed. The scheme developed in this work not only applies to the PFC model, but also to other models that can take advantages of the efficiency and accuracy of the FSM.

The paper is organized as follows. In Section 2, the background information on the PFC model, the (continuous) Fourier transform, and the measure of deformation are reviewed. The FSM scheme to evaluate the PFC free energy of the deformed system is presented in Section 3

*e-mail: nirand.pi@kmitl.ac.th

and the extension of the scheme to the CDFT-type PFC is presented in Section 4. The numerical verification of the proposed scheme is shown in Section 5, followed by the summary in Section 6.

2 Background

2.1 Phase-Field Crystal Model

The original form of the PFC free energy functional, F_{pfc}, is [1, 5]

$$F_{\text{pfc}} = \int_V f_{\text{pfc}}(\mathbf{x})d\mathbf{x}, \tag{1}$$

$$f_{\text{pfc}}(\mathbf{x}) = \frac{\phi(\mathbf{x})}{2}\left[-\epsilon + \left(\nabla_{\mathbf{x}}^2 + 1\right)^2\right]\phi(\mathbf{x}) + \frac{\phi(\mathbf{x})^4}{4}, \tag{2}$$

where the (field) variable f_{pfc} is the free energy density (FED), $\phi(\mathbf{x})$ is the order parameter representing the atomic density field, ϵ is the model parameter, and V is the volume of the system. The variable, $\mathbf{x} = x_i\mathbf{e}_i$ is a position vector expressed in terms of the Cartesian basis vectors, $\mathbf{e}_i$, where the Einstein summation convention is used throughout the paper and the subscript i varies from 1 to 3 in three dimensions. The operatior $\nabla_{\mathbf{x}}^2$ is the Laplacian written in terms of $\mathbf{x}$, or $\nabla_{\mathbf{x}}^2 = \partial^2/\partial x_i \partial x_i$.

2.2 Fourier Transform

The (continuous) Fourier transform used in this work is defined as

$$\hat{h}(\mathbf{s}) = \int_{-\infty}^{\infty} h(\mathbf{x})e^{-i\mathbf{s}\cdot\mathbf{x}}d\mathbf{x}, \tag{3}$$

where $\mathbf{s}$ is the Fourier wave vector and $\hat{h}(\mathbf{s})$ is the Fourier representation of the function $h(\mathbf{x})$. The corresponding inverse Fourier transform is defined as

$$h(\mathbf{x}) = \frac{1}{(2\pi)^d}\int_{-\infty}^{\infty} h(\mathbf{s})e^{i\mathbf{s}\cdot\mathbf{x}}d\mathbf{s}, \tag{4}$$

where d is the dimensionality of the problem.

2.3 Measure of Deformation

To represent deformation, one can define $\mathbf{R} = R_i\mathbf{e}_i$ and $\mathbf{r} = r_i\mathbf{e}_i$ to be the position vectors that are attached to material points in the undeformed and deformed configurations, respectively. A transformation that maps $\mathbf{R}$ to $\mathbf{r}$ is described by a deformation-gradient tensor, $\boldsymbol{\alpha}$, whose elements are defined as $\alpha_{ij} = \partial r_i/\partial R_j$. For homogeneous deformation, $\boldsymbol{\alpha}$ is spatially independent and one can write the transformation in tensor notations: $\mathbf{r} = \boldsymbol{\alpha}\mathbf{R}$ and $\mathbf{R} = \boldsymbol{\alpha}^{-1}\mathbf{r}$. Also, $J = \det(\boldsymbol{\alpha})$ is the volume ratio $V/\mathcal{V}$, where V and $\mathcal{V}$ is the deformed and undeforemd volumes respectively.

The deformation type used for numerical verification is characterized by $r_1 = \eta R_1 - \xi R_2$, $r_2 = \eta R_2$, and $r_3 = \eta R_3$, where $\eta = 1 + \xi$. This gives $J = \eta^3$ and

$$\boldsymbol{\alpha} = \begin{bmatrix} \eta & -\xi & 0 \\ 0 & \eta & 0 \\ 0 & 0 & \eta \end{bmatrix}, \quad \boldsymbol{\alpha}^{-1} = \frac{1}{\eta}\begin{bmatrix} 1 & \xi/\eta & 0 \\ 0 & 1 & 0 \\ 0 & 0 & 1 \end{bmatrix}. \tag{5}$$

3 FSM Scheme to Evaluate PFC Free Energy of a Deformed System

To calculate the free energy of the deformed system [6], one expresses the order parameter with the coordinates $\mathbf{R}(\mathbf{r}) = \boldsymbol{\alpha}^{-1}\mathbf{r}$, which yields $\phi(\mathbf{R}(\mathbf{r}))$. This functional representation relates the functional values in the deformed configuration to those in the undeformed configuration. Writing the integral (Eq. (1)) and the Laplacian (Eq. (2)) in terms of $\mathbf{r}$, one arrives at

$$F_{\text{def}} = \int_V f_{\text{def}}(\mathbf{r})d\mathbf{r} = J\int_{\mathcal{V}} f_{\text{def}}(\mathbf{r}(\mathbf{R}))d\mathbf{R}, \tag{6}$$

$$f_{\text{def}}(\mathbf{r}) = \frac{\phi(\mathbf{R}(\mathbf{r}))}{2}\left[-\epsilon + \left(\nabla_{\mathbf{r}}^2 + 1\right)^2\right]\phi(\mathbf{R}(\mathbf{r})) + \frac{\phi(\mathbf{R}(\mathbf{r}))^4}{4}, \tag{7}$$

where $\nabla_{\mathbf{r}}^2 = \partial^2/\partial r_i \partial r_i$. Since the effective use of the FSM is on evaluation of the terms in Eq. (7) with Laplacian (and the higher order Laplacian), I will only consider a simplified example of

$$f_{\text{ex1}}(\mathbf{r}) = \nabla_{\mathbf{r}}^2\phi\left(\mathbf{R}(\mathbf{r})\right) \tag{8}$$

to demonstrate the FSM scheme.

3.1 Forward Transformation

By using Eq. (3), the continuous Fourier transform of $f_{\text{ex1}}(\mathbf{r})$ is written as

$$\hat{f}_{\text{ex1}}(\mathbf{w}) = \int_{-\infty}^{\infty} f_{\text{ex1}}(\mathbf{r})e^{-i\mathbf{w}\cdot\mathbf{r}}d\mathbf{r} = \int_{-\infty}^{\infty} \nabla_{\mathbf{r}}^2\phi\left(\mathbf{R}(\mathbf{r})\right)e^{-i\mathbf{w}\cdot\mathbf{r}}d\mathbf{r}. \tag{9}$$

The goal here is to transform the coordinates from $\mathbf{r}$ to $\mathbf{R}$. First, the volume integration is rewritten as

$$\int_{-\infty}^{\infty} d\mathbf{r} = J\int_{-\infty}^{\infty} d\mathbf{R}. \tag{10}$$

Second, the Fourier basis is re-expressed as

$$\begin{aligned} \exp(-i\mathbf{w}\cdot\mathbf{r}) &= \exp\left[-i\mathbf{w}\cdot(\boldsymbol{\alpha}\mathbf{R})\right] \\ &= \exp\left[-i\left(\boldsymbol{\alpha}^T\mathbf{w}\right)\cdot\mathbf{R}\right] \\ &= \exp\left[-i\mathbf{W}\cdot\mathbf{R}\right], \end{aligned} \tag{11}$$

where $\mathbf{W} = \boldsymbol{\alpha}^T\mathbf{w}$ is the transformed Fourier wave vector. This transformation is how the effect of deformation is "carried" into the Fourier space. Finally, the Laplacian can be transformed using the chain rule $\partial/\partial r_m = (\partial R_i/\partial r_m)\partial/\partial R_i = \alpha_{im}^{-1}\partial/\partial R_i$. This yields

$$\nabla_{\mathbf{r}}^2 = \alpha_{im}^{-1}\alpha_{jm}^{-1}\frac{\partial}{\partial R_i}\frac{\partial}{\partial R_j} \equiv \nabla_{\mathbf{r}(\mathbf{R})}^2. \tag{12}$$

Using Eqs. (10), (11), and (12), Eq. (9) can be rewritten as

$$\hat{f}_{\text{ex1}}(\mathbf{w}(\mathbf{W})) = J\int_{-\infty}^{\infty}\left(\alpha_{im}^{-1}\alpha_{jm}^{-1}\frac{\partial}{\partial R_i}\frac{\partial}{\partial R_j}\right)\phi(\mathbf{R})e^{-i\mathbf{W}\cdot\mathbf{R}}d\mathbf{R}, \tag{13}$$

where $\mathbf{w}(\mathbf{W}) \equiv \alpha^{-T}\mathbf{W}$. Next, the partial derivative $\partial/\partial R_m$ can be eliminated by performing integration by parts; as it turns out, each operation $\partial/\partial R_m$ results in a factor of $-iW_m$. Therefore, the above equation can be expressed as

$$\begin{aligned}\hat{f}_{\text{ex1}}(\mathbf{w}(\mathbf{W})) &= J\left(-\alpha_{im}^{-1}\alpha_{jm}^{-1}W_iW_j\int_{-\infty}^{\infty}\phi(\mathbf{R})e^{-i\mathbf{W}\cdot\mathbf{R}}d\mathbf{R}\right)\\ &= J\left(-\alpha_{im}^{-1}\alpha_{jm}^{-1}W_iW_j\hat{\phi}(\mathbf{W})\right)\\ &\equiv J\hat{\zeta}_{\text{ex1}}(\mathbf{W}),\end{aligned} \tag{14}$$

where $\hat{\phi}(\mathbf{W})$ is the Fourier representation of $\phi(\mathbf{R})$ and $\hat{\zeta}_{\text{ex1}}(\mathbf{W})$ is the Fourier representation of $\zeta_{\text{ex1}}(\mathbf{R}) \equiv f_{\text{ex1}}(\mathbf{r})|_{\mathbf{r}=\mathbf{R}}$. The above equation indicates that $\hat{f}_{\text{ex1}}(\mathbf{w}(\mathbf{W}))$, or $\hat{\zeta}_{\text{ex1}}(\mathbf{W})$, can be calculated from the (continuous) Fourier transform of $\phi(\mathbf{R})$. Since $\phi(\mathbf{R})$ is expressed with $\mathbf{R}$, the DFT can be implemented on the discrete $\phi(\mathbf{R})$ data since one can always choose to sample $\phi(\mathbf{R})$ along the directions of $\mathbf{e}_i$. Furthermore, it can be easily shown that for $f_{\text{ex1}} = \nabla_{\mathbf{r}}^{2n}\phi(\mathbf{R}(\mathbf{r}))$, where n is a positive integer, one will have $\hat{\zeta}_{\text{ex1}}(\mathbf{W}) = \left(-\alpha_{im}^{-1}\alpha_{jm}^{-1}W_iW_j\right)^n\hat{\phi}(\mathbf{W})$.

3.2 Inverse Transformation

Here I derive the scheme for the inverse Fourier transform to obtain $f_{\text{ex1}}(\mathbf{r})$. By using Eqs. (4) and (14), one writes

$$\begin{aligned}f_{\text{ex1}}(\mathbf{r}) &= \frac{1}{(2\pi)^d}\int_{-\infty}^{\infty}\hat{f}_{\text{ex1}}(\mathbf{w}(\mathbf{W}))e^{i\mathbf{w}\cdot\mathbf{r}}d\mathbf{w}\\ &= \frac{J}{(2\pi)^d}\int_{-\infty}^{\infty}\hat{\zeta}_{\text{ex1}}(\mathbf{W})e^{i\mathbf{w}\cdot\mathbf{r}}d\mathbf{w},\end{aligned} \tag{15}$$

where the Fourier basis $e^{i\mathbf{w}\cdot\mathbf{r}}$ is used to maintain consistency with that used in Eq. (9). Next, one rewrites the integral with

$$\int_{-\infty}^{\infty}d\mathbf{w} = \frac{1}{J}\int_{-\infty}^{\infty}d\mathbf{W}, \tag{16}$$

and modifies the Fourier basis as

$$\begin{aligned}\exp(i\mathbf{w}\cdot\mathbf{r}) &= \exp\left[i\left(\alpha^{-T}\mathbf{W}\right)\cdot\mathbf{r}\right]\\ &= \exp\left[i\mathbf{W}\cdot\left(\alpha^{-1}\mathbf{r}\right)\right]\\ &= \exp\left[i\mathbf{W}\cdot\mathbf{R}\right].\end{aligned} \tag{17}$$

From Eqs. (16) and (17), Eq. (15) becomes

$$\begin{aligned}f_{\text{ex1}}(\mathbf{r}(\mathbf{R})) &= \frac{1}{(2\pi)^d}\int_{-\infty}^{\infty}\hat{\zeta}_{\text{ex1}}(\mathbf{W})e^{i\mathbf{W}\cdot\mathbf{R}}d\mathbf{W}\\ &= \zeta_{\text{ex1}}(\mathbf{R}).\end{aligned} \tag{18}$$

Equation (18) indicates that when one performs the inverse DFT on the discrete $\hat{\zeta}_{\text{ex1}}(\mathbf{W})$ data obtained from Section 3.1, the result is the discrete values of $\zeta_{\text{ex1}}(\mathbf{R})$. Then one simply associates these data with the corresponding deformed coordinates $\mathbf{r}(\mathbf{R})$ to obtain the discretized $f_{\text{ex1}}(\mathbf{r})$ data.

4 Application to CDFT-Type PFC Free Energy

A more general form of the PFC FED is that of the CDFT-type PFC model [4]:

$$f_{\text{pfc}} = \frac{\phi(\mathbf{x})}{2}\int_{-\infty}^{\infty}C(\mathbf{x}-\mathbf{x}')\phi(\mathbf{x}')d\mathbf{x}' + \mathcal{L}(\phi(\mathbf{x})), \tag{19}$$

where $\mathcal{L}(\phi(\mathbf{x}))$ is a function of terms with no derivative. The other part is the "non-local" term where the (non-local) operation is expressed in terms of a convolution of a two-body correlation function $C(\mathbf{x}-\mathbf{x}')$ and ϕ. In the case of the FED in Eq. (2), one can see that $\mathcal{L}(\phi(\mathbf{x})) = \phi(\mathbf{x})^4/4$ and $C(\mathbf{x}-\mathbf{x}') = \left[-\epsilon + \left(\nabla^2+1\right)^2\right]\delta(\mathbf{x}-\mathbf{x}')$. The latter gives $\hat{C}(\mathbf{s})$ (Fourier representation of $C(\mathbf{x}-\mathbf{x}')$) in the form of

$$\hat{C}(\mathbf{s}) = -\epsilon + \left(-|\mathbf{s}|^2+1\right)^2. \tag{20}$$

Similar to Section 3, one can calculate the free energy of a deformed system from Eq. (6) where $f_{\text{def}}(\mathbf{r})$ is

$$\begin{aligned}f_{\text{def}}(\mathbf{r}) =& \frac{\phi\left(\alpha^{-1}\mathbf{r}\right)}{2}\int_{-\infty}^{\infty}C(\mathbf{r}-\mathbf{r}')\phi\left(\alpha^{-1}\mathbf{r}'\right)d\mathbf{r}'\\ &+\mathcal{L}\left[\phi\left(\alpha^{-1}\mathbf{r}\right)\right].\end{aligned} \tag{21}$$

To derive the FSM scheme, I consider a simplified example of

$$f_{\text{ex2}}(\mathbf{r}) = \int_{-\infty}^{\infty}C(\mathbf{r}-\mathbf{r}')\phi\left(\alpha^{-1}\mathbf{r}'\right)d\mathbf{r}'. \tag{22}$$

As shown in Appendix A, the (continuous) Fourier transform of $f_{\text{ex2}}(\mathbf{r})$ is

$$\hat{f}_{\text{ex2}}(\mathbf{w}) = \hat{C}(\mathbf{w})\int_{-\infty}^{\infty}e^{-i\mathbf{w}\cdot\mathbf{r}}\phi\left(\alpha^{-1}\mathbf{r}\right)d\mathbf{r}. \tag{23}$$

Using Eqs. (10) and (11), one arrives at

$$\begin{aligned}\hat{f}_{\text{ex2}}(\mathbf{w}(\mathbf{W})) &= J\left[\hat{C}\left(\alpha^{-T}\mathbf{W}\right)\int_{-\infty}^{\infty}e^{-i\mathbf{W}\cdot\mathbf{R}}\phi(\mathbf{R})d\mathbf{R}\right]\\ &= J\left[\hat{C}\left(\alpha^{-T}\mathbf{W}\right)\hat{\phi}(\mathbf{W})\right]\\ &= J\hat{\zeta}_{\text{ex2}}(\mathbf{W}).\end{aligned} \tag{24}$$

The above equation indicates that one can obtain $\hat{f}_{\text{ex2}}(\mathbf{w}(\mathbf{W}))$, and also $\hat{\zeta}_{\text{ex2}}(\mathbf{W})$, from the product of $\hat{\phi}(\mathbf{W})$ and $\hat{C}\left(\alpha^{-T}\mathbf{W}\right)$. Since the two-body correlation function is typically specified in Fourier space, $\hat{C}\left(\alpha^{-T}\mathbf{W}\right)$ is simply obtained from functional evaluation at the (transformed) Fourier wave vector $\alpha^{-T}\mathbf{W}$. From the above equation, the result from Section 3.1 can be obtained by recognizing that $\hat{C}(\mathbf{w}) = -|\mathbf{w}|^2 = -\mathbf{w}\cdot\mathbf{w}$ and, therefore, $\hat{C}\left(\alpha^{-T}\mathbf{W}\right) = -\left[\alpha^{-T}\mathbf{W}\right]_m\left[\alpha^{-T}\mathbf{W}\right]_m = -\alpha_{im}^{-1}W_i\alpha_{jm}^{-1}W_j$. This leads to $\hat{\zeta}_{\text{ex2}}(\mathbf{W}) = -\alpha_{im}^{-1}\alpha_{jm}^{-1}W_iW_j\hat{\phi}(\mathbf{W}) = \hat{\zeta}_{\text{ex1}}(\mathbf{W})$, as expected. The procedure for the inverse Fourier transform is similar to that described in Section 3.2.

5 Numerical Verification

The FSM scheme developed in this work will be verified by calculating F_{def} (Eq. (6)) resulting from the deformation specified in Eq. (5). The order parameter profile is that of a body-centered cubic structure, $\phi(\mathbf{R}) = \bar{\phi} + A[\cos(qR_1)\cos(qR_2) + \cos(qR_1)\cos(qR_3) + \cos(qR_2)\cos(qR_3)]$, where A is the amplitude, $\bar{\phi}$ is the average value and q is the parameter that sets the scale of the structure. The parameters used are $A = 1$, $\epsilon = 0.15$, $\bar{\phi} = -0.23$, and $q = 1/\sqrt{2}$; the latter sets the lattice parameter $L = 2\sqrt{2}\pi$. The coordinates R_i is discretized into $N = 32$ intervals with the spacing $\Delta R_i = L/N$ and the coordinate values are $R_i = \Delta R_i[0, 1, 2, ..., N-1]$. The Fourier coordinates $\mathbf{W}$ will have the spacing $\Delta W_i = 2\pi/L$ and, since Matlab software is used, $W_i = \Delta W_i[0, 1, ..., N/2, -N/2 + 1, -N/2 + 2, ..., -1]$. The DFT and the inverse DFT is performed using the "fftn" and "ifftn" functions, respectively, which employ a fast Fourier transform algorithm.

Figure 1 shows three values of F_{def}/V as functions of ξ. The first result, labeled "Analytical", is the analytical solution obtained from symbolic computation. The second result, denoted "Laplacian", is the numerical calculation where the terms with derivatives in f_{def} (Eqs. (7)) are calculated using the FSM scheme developed in Section 3. The third result, labeled "Correlation", is the numerical calculation using the FED from Eq. (19) with the two-body correlation function from Eq. (20). From the figure, one can see that the results from the three calculations agree with one another; the differences among these values are negligibly small. These results validate the numerical implementation of the FSM scheme proposed for both the original and CDFT-type PFC models.

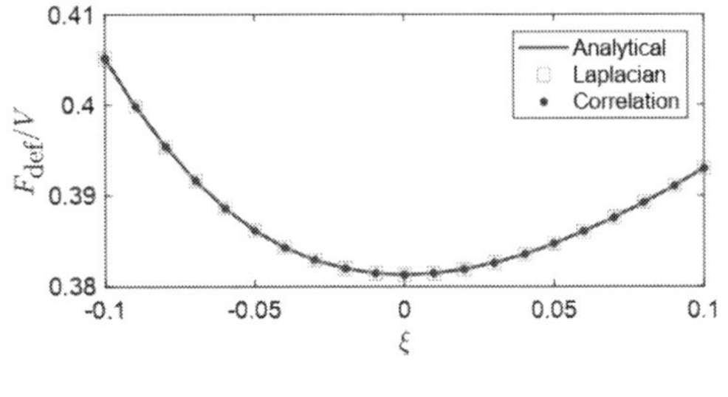

Figure 1. Plot of F_{def}/V as a function of ξ. (See text for explanation.)

6 Summary

The FSM scheme for evaluating the PFC free energy of a system subjected to homogeneous deformation was developed. This scheme employs real-space coordinate transformation so that the (continuous) Fourier transform is performed on the function of the undeformed coordinates and the effect of the deformation is realized in the transformation of the Fourier wave vector. The advantage of this scheme is that it allows straightforward DFT implementation because the sampling at the undeformed configuration can be chosen to be along the directions of $\mathbf{e}_i$. This scheme can be used to efficiently and accurately evaluate the derivative of an order parameter and the convolution of an order-parameter and a two-body correlation function. In addition to the PFC model, this scheme applies to other models that can take advantages of the efficiency and accuracy of the FSM.

This work was supported by King Mongkut's Institute of Technology Ladkrabang Research Fund under Grant No. KREF046103, Thailand Research Fund (TRF) under Grant No. TRG5880008, and High Performance Computing Services from (Thailand) National Electronics and Computer Technology Center (NECTEC). The author would like to thank Dr. Victor W. L. Chan for helpful discussion and comments.

A Fourier Transform of Convolution Integral

The (continuous) Fourier transform of f_{ex2} is derived in this section. Denoting $\Delta = \mathbf{r} - \mathbf{r}'$, one can write

$$\hat{f}_{\text{ex2}} = \int_{-\infty}^{\infty} e^{-i\mathbf{w}\cdot\mathbf{r}} \left[\int_{-\infty}^{\infty} C(\Delta)\phi\left(\alpha^{-1}\mathbf{r}'\right) d\mathbf{r}' \right] d\mathbf{r}. \quad (25)$$

Using the fact that $e^{-i\mathbf{w}\cdot\mathbf{r}} = e^{-i\mathbf{w}\cdot\Delta} e^{-i\mathbf{w}\cdot\mathbf{r}'}$ and rearranging the order of integral, one arrives at

$$\hat{f}_{\text{ex2}} = \int_{-\infty}^{\infty} e^{-i\mathbf{w}\cdot\mathbf{r}'}\phi\left(\alpha^{-1}\mathbf{r}'\right) \left[\int_{-\infty}^{\infty} e^{-i\mathbf{w}\cdot\Delta} C(\Delta) d\mathbf{r} \right] d\mathbf{r}'. \quad (26)$$

Since $\mathbf{r}'$ is constant for the inner integral, $d\mathbf{r} = d\Delta$ and one can write

$$\begin{aligned} \hat{f}_{\text{ex2}} &= \int_{-\infty}^{\infty} e^{-i\mathbf{w}\cdot\mathbf{r}'}\phi\left(\alpha^{-1}\mathbf{r}'\right) \left[\int_{-\infty}^{\infty} e^{-i\mathbf{w}\cdot\Delta} C(\Delta) d\Delta \right] d\mathbf{r}' \\ &= \int_{-\infty}^{\infty} e^{-i\mathbf{w}\cdot\mathbf{r}'}\phi\left(\alpha^{-1}\mathbf{r}'\right) \hat{C}(\mathbf{w}) d\mathbf{r}'. \end{aligned} \quad (27)$$

Changing the coordinates from $\mathbf{r}'$ to $\mathbf{r}$, one obtains $\hat{f}_{\text{ex2}} = \hat{C}(\mathbf{w}) \int_{-\infty}^{\infty} e^{-i\mathbf{w}\cdot\mathbf{r}}\phi\left(\alpha^{-1}\mathbf{r}\right) d\mathbf{r}$, which is Eq. (23).

References

[1] K.R. Elder, M. Katakowski, M. Haataja, M. Grant, Phys. Rev. Lett. **88**, 245701 (2002)

[2] H. Emmerich, H. Löwen, R. Wittkowski, T. Gruhn, G.I. Tóth, G. Tegze, L. Gránásy, Adv. Phys. **61**, 665 (2012)

[3] G. Shechter, Ph.D. thesis, Johns Hopkins University (2004)

[4] K.R. Elder, N. Provatas, J. Berry, P. Stefanovic, M. Grant, Phys. Rev. B **75**, 064107 (2007)

[5] K.R. Elder, M. Grant, Phys. Rev. E **70**, 051605 (2004)

[6] K.A. Wu, A. Adland, A. Karma, Phys. Rev. E **81**, 061601 (2010)

An investigation on dimensional accuracy of EDM deep hole using multi-hole interior flushing electrode

Suppawat Chuvaree and *Kannachai* Kanlayasiri*

Industrial Engineering Department, Faculty of Engineering, King Mongkut's Institute of Technology Ladkrabang, Bangkok, Thailand

Abstract. Improper flushing of dielectric fluid adversely affects the dimensional accuracy produced in EDM deep hole process. This paper investigates the dimensional accuracy (e.g. roundness error and gap clearance) in EDM deep hole process of AISI P20 tool steel. The inner flushing electrode is newly proposed in this study, using the multi-hole interior flushing technique coupled with electrode rotation. The results showed that the multi-hole interior flushing with electrode rotation improves the deviation of roundness error due to improved flushing. In addition, using multi-hole interior flushing with electrode rotation provides the cylindrical shape in gap clearance. Finally, a more powerful flushing technique is achieved with newly purposed multi-hole interior flushing technique on roundness error and gap clearance, respectively.

1 Introduction

Electrical discharge machining (EDM) is finding a lot of applications for machining certain materials which traditionally have been more difficult to machine. Such developments further enable the growing demand of higher precision parts, for advancements in automotive and other industries [1]. The dimensions and accuracies resultant on machined a hole by EDM process plays a crucial role when close tolerance components are required for critical applications like plastic mould and die part, automotive part etc. During EDM of a deep hole, machining conditions can be optimized leading to high process accuracy and greater machining productivity. However, the flushing stage has been identified as the most important factor in the determining the dimensional accuracy [2-3]. In an Electrical discharge machining operation, the flushing function is employed to remove the debris in the machining gap and to sustain the dielectric temperature well below its flash point. In addition, the influence of electrode rotation coupled with the flushing technique can increase the material removal rate (MRR), assisting the removal of the debris and assuring that the hole is round [4].

During the EDM process both the workpiece and tool electrode are eroded simultaneously. When the machining depth is increased, the viscous resistance within the machining gap impedes in removal of debris and bubbles from the eroding area resulting in reduced extensive performance and dimensional accuracy. This debris activation and secondary spark appearance on the wall, result in a machined hole that is not ideally cylindrical but concaved and tapered, which is undesirable in the precision machining processes.

The study of the surface and accuracies produced during EDM with rotating electrode have been reported in [5]. It found that the electrode rotation improves the surface finish and roundness. Diver et al. [6] developed a new flushing technique to produce reverse tapered micro-holes. Excellent hole formation with hole diameter variations within 3 μm were reported.

Recently the authors [7] developed a new technique of flushing for deep hole by using a multi-hole interior flushing electrode. We found that flushing through the tool electrode and multi-hole interior flushing are more preferred than side flushing. In addition, the effects of improved flushing increases material removal rate and reduces machining time. Therefore, in this paper, an attempt has been made to investigate dimensional accuracy characteristics of deep holes produced by EDM using multi-hole interior flushing technique like roundness error and gap clearance, respectively.

2 Experiment

In this research, the electrode was designed with regard to improved flushing ability to reduce the error of hole dimensions caused by EDM condition and improper flushing. As shown in Figure 1, the copper electrode of diameter 12 mm, and 8.4 mm. in flushing hole. In addition, the electrode rotation helps in flushing away debris, preventing secondary spark phenomena and enhancing the performance of EDM process.

Figures 2a shows the machined workpieces after deep holes machining. To investigate the influence of flushing condition on roundness error and gap clearance, we considered an electrode rotation at 105 rpm and a fixed machining depth of 50 mm

*Corresponding author: kkkannac@kmitl.ac.th

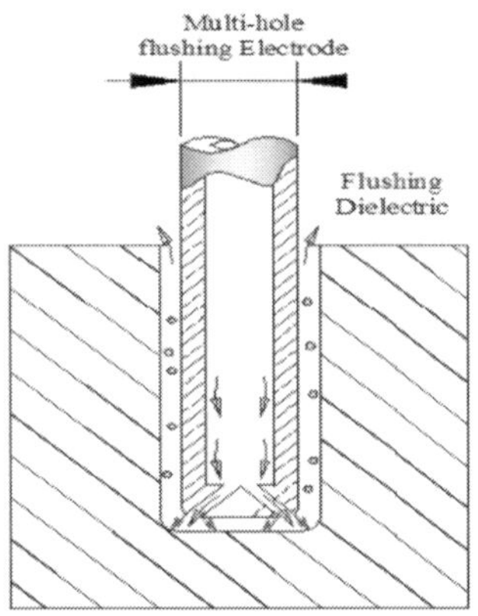

Fig. 1. Tool electrode and flushing system.

The dimensional accuracy was tested at six locations (1, 10, 20, 30, 40, and 50 mm) through the machining direction as shown in figure 2b. The precision measurements of roundness error of the hole were analyzed from 3D surface roughness, Olympus model LEXT OLS5000 laser scanning confocal microscope noncontact. The measurement of roundness error of the hole profile was defined by maximum inscribed circle (LSC) method according to GD&T methodology. Whereby, least square circle, it is a circle which separates the roundness profile of an object by separating the sum of total areas of the inside and outside it in equal amounts. The roundness error then can be estimated difference between the maximum and minimum distance from this reference circle. So, the roundness error is defined as equation (1).

$$\text{Roundness} = R_{max} - R_{min} \quad (1)$$

The gap clearance profiles were examined using a laser scanning model Creaform REV scan 3D handy scanner. The technical specifications are given in Table 1.

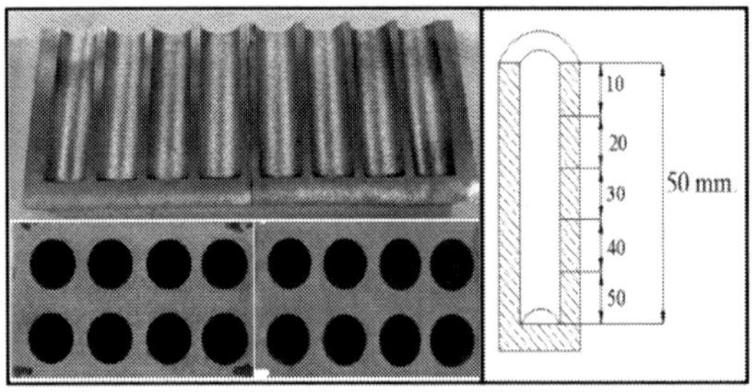

a) b)

Fig. 2 a) AISI P20 tool steel workpieces after deep hole machining.
b) Dimensional accuracy measurement on the test workpiece.

Table 1. Experimental conditions.

Process parameters	Values
Pulse on-time Pulse off-time	150 μs. 20 μs.
Peak current	15 A.
Electrode diameter	12 mm.
Electrode polarity	+
Machining depth	50 mm.
Dielectric	EDM oil-TOTAL DIET MS 700
Flushing pressure	1kg/cm^2
Workpiece Material	AISI P20

3 Results and Discussion

3.1. Roundness

In this experiment, the electrode rotation and flushing mode using a multi-hole interior flushing technique are used to investigate the effect of dimensional accuracy in terms of roundness error of the hole. In the roundness measurement operation, the results of the deviation of roundness error when using conventional flushing and multi-hole interior flushing electrode are shown in Table 2 and 3 respectively.

Table 2. The deviation of roundness error when using conventional flushing with electrode rotation.

- Z position	Diameter of hoe (mm.)	Max. Deviation (mm.)	Min. Deviation (mm.)	Range (mm.)
-1	12.327	0.026	-0.035	0.061
-10	12.543	0.040	-0.057	0.097
-20	12.545	0.048	-0.061	0.109
-30	12.530	0.061	-0.059	0.120
-40	12.440	0.076	-0.098	0.174
-50	12.263	0.096	-0.151	0.247

Table 3. The deviation of roundness error when using multi-hole interior flushing electrode with electrode rotation.

- Z position	Diameter of hoe (mm.)	Max. Deviation (mm.)	Min. Deviation (mm.)	Range (mm.)
-1	12.220	0.057	-0.066	0.123
-10	12.350	0.063	-0.070	0.133
-20	12.416	0.074	-0.047	0.121
-30	12.490	0.048	-0.061	0.109
-40	12.409	0.059	-0.048	0.107
-50	12.152	0.061	-0.061	0.122

In Figure 3, a comparison of the deviation of roundness error produced under the different flushing conditions across the six depths were plotted.

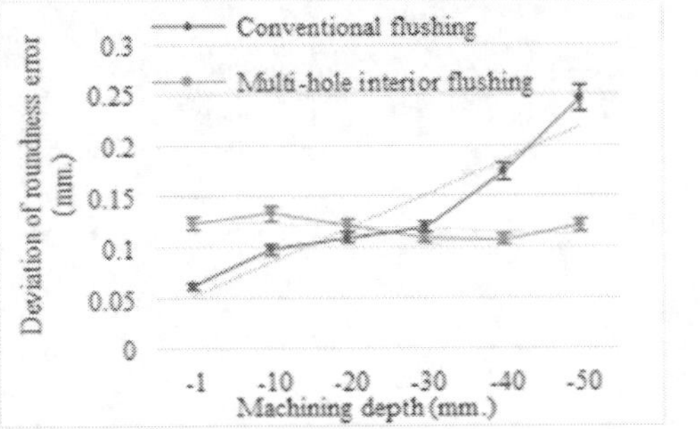

Fig. 3. The comparison of the deviation of roundness error produced under the different flushing conditions against machining depth.

It is observed that with conventional flushing the deviation of the roundness error increases continuously with the machining depth up to 50 mm. In conventional flushing coupled with electrode rotation, the debris particles of material are moving out slowly from the machining gap and formation of secondary spark phenomena between the debris particles and side wall of the hole. As machining depth is increased, long-term effects of the secondary spark phenomena result in roundness error of the hole in higher depth [1]. In addition, it might be due to improper flushing at higher depths, as the debris particles are irregularly distributed on the side of wall of the hole leading to roundness error along the depth of hole. However, when using the multi-hole interior flushing electrode, a more stable roundness was observed. It is clear that the deviation of roundness error is slightly decreased, a cylindricity deviation only occurred due to the flushing as the multi-hole interior electrode is flowed out of small gap between electrode and side of wall of the hole to remove debris particles continuously. In addition, the electrode rotation during EDM deep hole process promotes cleaning of the debris in the machining gap. Therefore, the deviation of roundness error is uniformly changed. The 3D view measurement of roundness from 3D laser scanning surface roughness is shown in Figure 4.

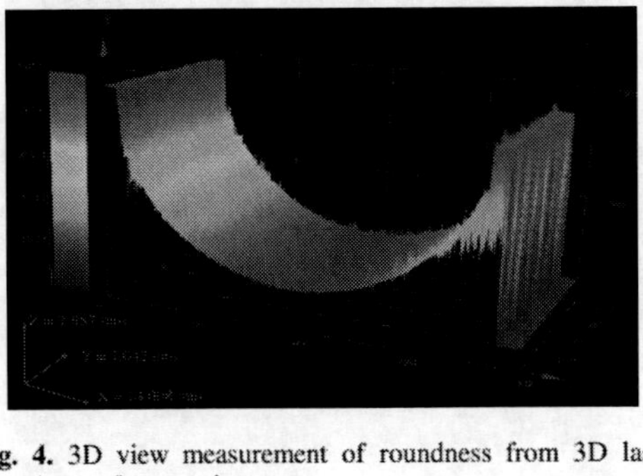

Fig. 4. 3D view measurement of roundness from 3D laser scanning surface roughness.

3.2. Gap clearance

For investigation of dimensional accuracy, the gap clearance is another important quality. Figure 5 shows the values of gap clearance with different flushing conditions. The electrode rotation effect at 105 rpm was investigated. It is clear that for each condition the gap clearance increases towards the half-way depth and then narrows towards the finish. As such, the characteristic shape of the machined hole was concave. The figure also demonstrates that when using conventional electrode by side flushing the gap clearance is increased and that the gap clearance increase becomes pronounced with the depth of hole. However, the electrode rotation effect coupled with side flushing causes a decrease in gap clearance. Moreover, using the multi-hole interior flushing with rotational electrode causes a cylindrical shape in gap clearance. It is explained through a reduction in debris build up within the machining gap, and by a more uniform distribution of debris from the multi-hole flushing with rotational electrode during the EDM deep hole process [7]. This means that a more powerful flushing technique is achieved with the newly purposed multi-hole interior flushing technique on dimensional accuracy like roundness error and gap clearance, respectively. The measurement of gap clearance from 3D handy laser scanning is shown in Figure 6.

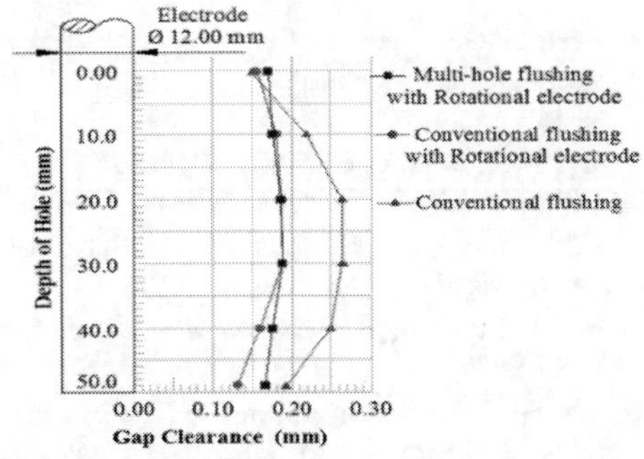

Fig. 5 Effect of flushing condition on gap clearance characteristic.

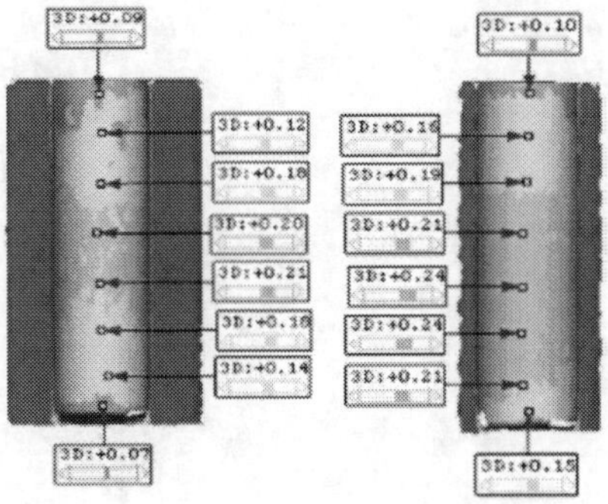

a) b)

Fig. 6. Measurement of gap clearance profiles was examined using 3D handy laser scanning. a) Multi-hole flushing with rotational electrode b) Conventional flushing.

4 Conclusion

This paper investigates the dimensional accuracy characteristics of EDM deep holes produced by a newly purposed multi-hole interior flushing technique on AISI P20 tool steel like roundness error and gap clearance, respectively. This study shows that:

1. The multi-hole interior flushing with electrode rotation improved the deviation of roundness error due to improved flushing.
2. When using multi-hole interior flushing with rotational electrode causes the cylindrical shape in gap clearance.
3. A more powerful flushing technique is achieved with a newly purposed multi-hole interior flushing technique on dimensional accuracy like roundness error and gap clearance, respectively.

Acknowledgment

The authors acknowledge the financial support provided by Faculty of Engineering, King Mongkut's Institute of Technology Ladkrabang and Rajamangala University of Krungthep were thanked for providing facilities for testing.

References

1. V. K. Meena, M. S. Azed, S. Mitra, Int. J. Machining and Machinability of Material. V **12**, 4 (2012)
2. M. Risto, R. Haas, M. Munz, *Proceeding CIRP of ISEM XVIII.* V **42,** (2016), 127-134
3. M. N. Islam, N. H. Rafai, S. S. Subramanian, *Proceeding of the WCE.* V **III**, (2010)
4. M. M. Makenzi, B. W. Ikua, *Proceeding of the MECSRI 2012.* V **4**, (2012)
5. J. S. Soni, G. Chakraverti, Ind. J. Eng. Mat. Sci. V **1**, pp. 127-134 (1994)
6. C. Diver, J. Atkinson, H. J. Helml, L. Li, J. Mater. Process. Technol. V **169** (2005) 427-436
7. S. Chuvaree, K. Kanlayasiri, *Proceeding of International conference on ICMM 2018.*

MATEC Web of Conferences **192**, 01030 (2018)
https://doi.org/10.1051/matecconf/201819201030
ICEAST 2018

A design of HDPE flexible spline of harmonics gear

Pholchai Chotiprayanakul, *Nattakul* Khamsri and *Neeracha* Kumjaroen

Industrial Engineering Department, Faculty of Engineering,
King Mongkut's Institute of Technology Ladkrabang, Bangkok, Thailand

Abstract. In a design of a small robot, selecting a servo-motor is the most important method in designing process. Common miniature servo motors for small robots normally use conventional gear train set or planetary gear set. Gear train and planetary gear give very low gear ratio whereas its weight is too heavy. On the other hand, Harmonic gear system, which is developed from strain wave drive gearing, gives better the highest gear ratio per weight than those two gear systems. In this paper, a plastic flexible spline gear is presented in order to replace a thin metal spline. The plastic flexible spline gear is designed under gear physical requirement and servo-motor properties. The spline thicknesses are varied in a range of 1 to 4 millimetre and simulations on strength, fatigue, and torque requirement are made to verify the designs.

1. Introduction

There are many brands of servo-motors for a small robot, all of them contain with gear train sets. A robot joint normally needs a servo-motor provide speed about 30 RPM and torque depending on the size of the robot. For DC motor running at 1,000 to 3,000 RPM, a gear set must have gear ratio of 1:33 to 1:100. In 1957, C.W. Musser introduced a strain wave gear system, which is now known as harmonic gear system. The harmonic gear set provides high gear ratio and low weight and it is generally used in industrial robot today. Analysis and Design of Harmonics Gear presented by Lih-Min Hsia [1] shows a guide line to consider the geometry of the harmonic gear. Sensor monitoring of torque and force in the harmonics gear is presented in [2], forces detect and formulate a force function. Many novel types of the harmonics [3-4] were introduced for industrial use. In a household robot, servo-motor may not be need as strong as the servo-motor in the industrial robot thus a plastic flexible spline for harmonic gear set in servo-motor is considered to replace a metal spline. The harmonic gear designing method can be separated into 4 steps: first gear geometry design and then second strength consideration. After spline expansion force is calculated, torque to roll the flexible spline is formulated in this third step. At final step, the flexible spline fatigue is simulated.

2. Harmonic Gear Geometry

A harmonic gear set consists of 3 parts; 1) Circular Spline, 2) Flexible Spline, and 3) Wave Generator and the geometry of harmonic gear includes number of gear teeth, the gear tooth shape, diameter of circular spline, and thickness of the flexible spline.

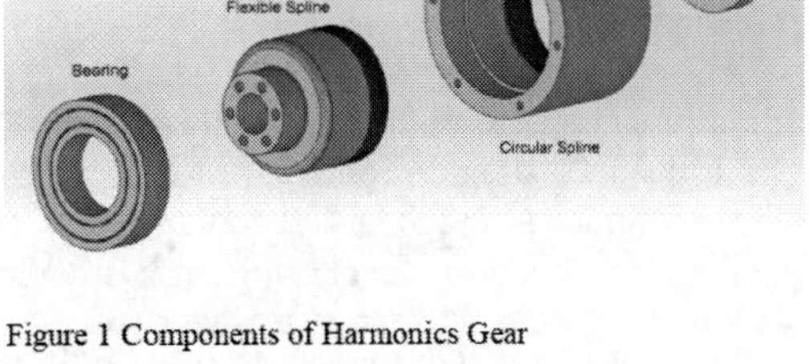

Figure 1 Components of Harmonics Gear

Harmonics gear reduces rotation speed by the uneven number of the teeth between the flexible spline and the circular spline. By number of teeth, diameter of the flexible spline is slightly smaller than the circular spline. The wave generation rotor will stretch the flexible

* Pholchai Chotiprayanakul: kcpholch@kmitl.ac.th

spline shell along a diameter line until teeth of both splines attach. When the wave generation rotor rolls, contact points of the attached teeth will move around. After the wave generation rotor rotates a cycle, the flexible spline will shift couple teeth depending on different number of teeth between flexible and circular spline. With 1:100 gear ratio configuration, the number of gear teeth is considered from strain wave gear ratio equation as show in (1) where CST is number of circular spline teeth and FST is number of flexible spline teeth.

$$\textbf{\textit{Gear ratio}} = \frac{CST - FST}{CST} \quad (1)$$

The smallest number of teeth will be 99 teeth of flexible spline and 100 teeth of circular spline but this configuration has only a contact point. Single contact point, the wave generator touches only a point on the flexible spline makes the wave generator is unbalanced and shakes the whole body of gear. Avoiding balancing issue, 2 contact points flexible spline is considered, the gear will have 198 teeth of flexible spline and 200 teeth of circular spline.

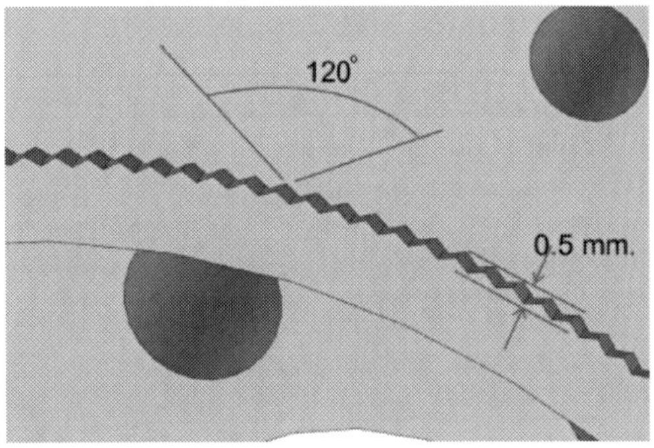

Figure 2 Harmonics Gear's Teeth Geometry

Simply triangle tooth shape is chosen with 32 teeth per inch thread. Tooth's vertex is at 120 degree and its height is 0.5 mm. For 10 millimetres of a tooth length, the shear strength is about 23 MPa and ultimate shear force per tooth will be 398N. Thus the payload that the harmonic gear set is able to support will be 19.8 N.m.

Figure 3 Simulation of Teeth Strength and Torque Load

This tooth shape configuration at 198/200 teeth spline will give the inside diameter of circular spline at 50.53 mm. and the outside diameter of flexible spline at 50.03 mm. thus the stretching distance of flexible spline is 0.5 mm. The thickness of flexible spline will define the stretching force, which lead to the loss of motor's torque. For prelim study, researchers define for 4 thicknesses as 1.3 to 4.3 mm. of the flexible spline shell to find out the optimal thickness which gives minimum torque loss and maximum life cycle.

3. Stretching Force And Life Cycle Estimation

The force to stretch the flexible spline is considered in this section. The size of force depends on thickness of the flexible spline varied from 1.3 to 4.3 mm. and what material the flexible spline made of. High-Density PolyethylEne (HDPE) is selected to make the flexible spline. HDPE tensile strength is around 32 MPa and flexural modulus at 1.25 GPa. Stretching force is simulated on a CAD-software which the flexible spline expansion distance at 0.5 mm. is a control parameter. The result of simulation shows the stretching force is increasing when the flexible spline shell is more thicken.

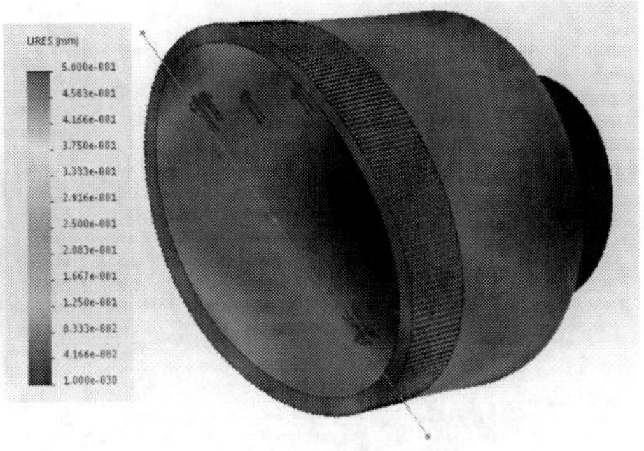

Figure 4 Simulation of Stretching Force at 0.5 mm. Displacement

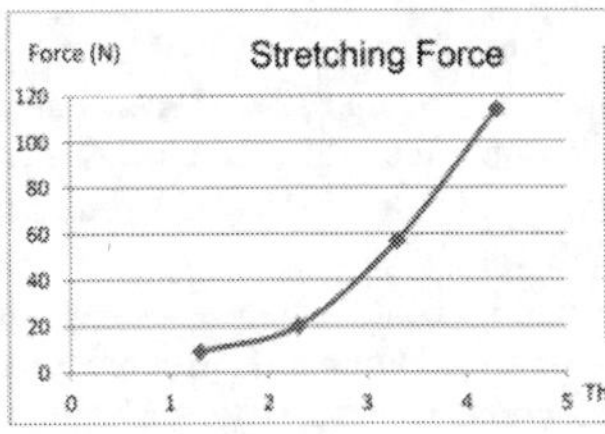

Thickness (mm.)	Force (N)
1.3	9
2.3	20
3.3	57
4.3	114

Figure 5 Stretching Force and Flexible Spline Thickness

The harmonic wave generator rolls inside the flexible spline and creates a repeating stretching force on the flexible spline. This cycle load may conduct fatigue on the flexible spline. From [5], researchers provide a S-N formula to estimate the life cycle for HDPE as shown in equation 2.

$$S = 32.24 \times N^{-0.068} \qquad (2)$$

Where S is stress on workpiece and N is the life cycle. We ran a simulation with thickness series from 1 mm. to 5.5 mm. step 0.5 mm. of the flexible spline. The maximum stresses of the stretching flexible splines are collected from simulation tests and shows in Figure 6.

The result shows relation between life cycle and thickness of flexible spline. The stress of stretching flexible spline depends on the thickness. However, the stress at 3 - 3.5 mm. of the flexible spline thickness is on critical condition that is the stress closed to the yield strength of the HDPE material. The cup shape of the flexible spline has started to deform in the plastic mode and may not full recover spring back. Thus, we do only focus on 1.3 mm. thickness that provides a low stretching force and thinnest flexible spline's shell which can be built. The life cycle of this thickness will be over 296 billion cycles.

Thickness (mm.)	Stress (MPa)	Life Cycle
1	5.28	3.58E+11
1.5	5.35	2.96E+11
2	5.52	1.87E+11
2.5	7.05	5.13E+09
3	8.52	3.15E+08
3.5	17.33	9.22E+03
4	19.29	1.91E+03
4.5	21.72	3.33E+02
5	24.25	6.59E+01
5.5	26.46	1.83E+01

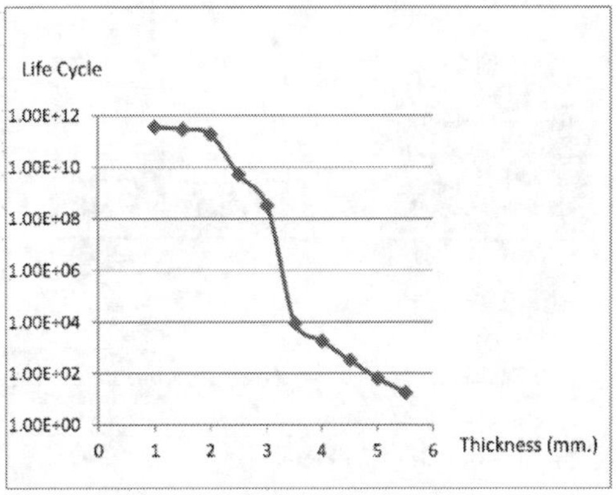

Figure 6 Thickness of Flexible Spline and Stress vs. Life Cycle

4. Stall Torque

When harmonic gear is used in a servo-motor set, the stretching flexible spline will resist the motor to roll. The resistance is known as stall torque load. Stall torque related on the stretching force and the radius of the flexible spline.

$$T_s = 2\pi R F_s \qquad (3)$$

where T_s is stall torque, R is radius of the flexible spline and F_s is the stretching force. Stall torque would be 2.8 N.m. thus the servo-motor must has torque greater than stall torque (2.8 N.m.) but it should not excess over the payload torque (19.8 N.m. from topic 2) multiply by gear ratio (1/100). The servo-motor torque will be around 3 N.m.

5. Prototyping

After all configurations of the flexible spline is studied and selected as shown topics above. All 3 components, which include circular spline, flexible spline, and wave generator rotor, are made for a prototype of harmonics gear system. Circular spline is made with steel for the first prototype but we plan to replace it with a plastic circular spline in the future work. The wave generator rotor is made by 3D printing process and assembly with 2 roller bearings and an aluminium shaft. The flexible spline is made by the injection molding process thus a mold that has a cavity of flexible spline shape was design and built before injecting melt HDPE into the mold.

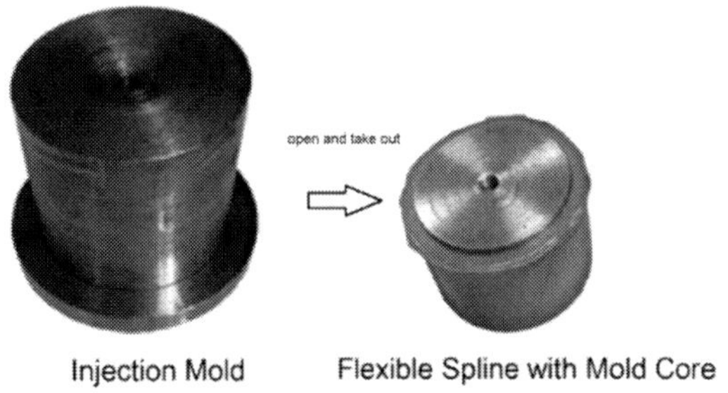

Figure 7 Injection Mold for Flexible Spline

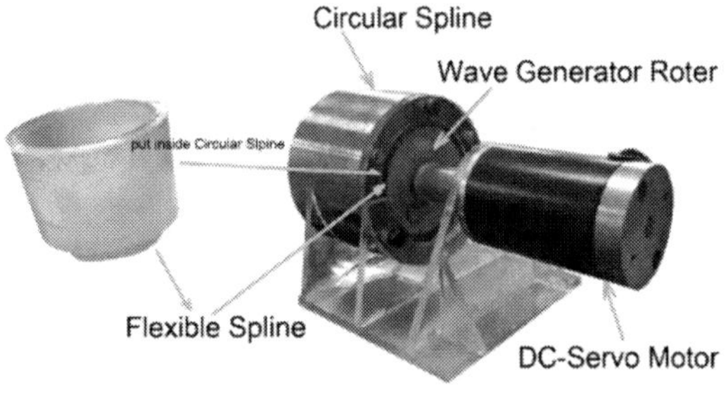

Figure 8 The 1:100 Harmonics Gear System Prototype

6. Conclusion

This paper shows a design of HDPE plastic flexible spline of harmonic gear. Strength of gear tooth, forces, fatigue and life cycle are considered with simulation tests. A prototype of the harmonics gear system is built to verify our design. This research needs to extend more experiment to find actual gear ratio, actual stall torque, and the life of the flexible spline.

7. Reference

1. Lih-Min Hsia, The Analysis and Design Of Harmonic Gear Drives (1988), Proceedings of the 1988 IEEE International Conference on Systems, Man, and Cybernetics, 8-12 Aug. 1988.
2. I. Godler, M. Hashimoto, Torque control of harmonic drive gears with built-in sensing (1998), Proceedings of the 24th Annual Conference of the IEEE Industrial Electronics Society, vol: 3, pp.1818-1823, Aachen, Germany 31 Aug-4 Sep 1998
3. Jan Rens, Kais Atallah, Stuart D. Calverley, and David Howe, A Novel Magnetic Harmonic Gear (2010), IEEE Transactions On Industry Applications, Vol. 46, No. 1, January/February2010
4. Jonathon W. Sensinger and James H. Lipsey, Cycloid vs. Harmonic Drives for use in High Ratio, Single Stage Robotic Transmissions (2012), IEEE International Conference on Robotics and Automation, River Centre, Saint Paul, Minnesota, USA May 14-18, 2012
5. A. Aid1; M. Bendouba, A. Talha, N. Benseddiq, M. Benguediab, and S. Zengah , Uniaxial Fatigue of HDPE-100 Pipe. Experimental Analysis (2014), Directory of Open Access Journals in Engineering, Volume 4, Number 2, pp. 600-604(5), 2014

Conversion of metal-organic halide perovskite from PbI_2 precursor films grown by hot-wall method

Satoru Seto*, *Rintaro* Shimizu and *Makoto* Tokuda

National Institute of Technology, Ishikawa College, Tsubata, Kahoku, Ishikawa 929-0392, Japan

Abstract. We report on metal-organic halide perovskite $CH_3NH_3PbI_3$ films converted from PbI_2 precursors for planar heterojunction perovskite solar cells. PbI_2 films as a precursor were deposited by hot-wall method and conventional vacuum evaporation. The conversion to perovskite phase from the PbI_2 films were performed by annealing in methyl ammonium iodine (MAI) vapour at 120-150 °C. We confirmed that no residual PbI_2 phase can be detected in the converted perovskite films by x-ray diffraction measurements. The surface morphology of the perovskite films was measured by AFM. Roughness *Ra* of the films is 17.8 nm, which is comparable value to the reported ones. Using the converted perovskite films we fabricated tentative perovskite solar cells with a device architecture of ITO/PEDOT:PSS/Perovskite/C_{60}/Ag. The power conversion efficiencies of the fabricated solar cells from a conventional evaporation and the hot-wall method exhibited 2.22 and 2.33%, respectively.

1 INTRODUCTION

Metal-organic hybrid perovskite solar cells have been extensively studied by many researchers all over the world and its power conversion efficiency (PCE) reaches over 20%, recently [1-3]. The fabrication methods of perovskite film are classified mainly into two method: one is a solution growth method from precursor solutions of PbI_2 and MAI or their mixed solution by a spin-coating technique, and another is a vapour growth method by conversion of perovskite phase from PbI_2 films by annealing under methyl ammonium iodide (MAI) vapour. The PbI_2 films as precursor are deposited by a spin-coating or a conventional vacuum evaporation. These perovskite films have been deposited on porous TiO_2 scaffolds. Recently, perovskite solar cells with flat substrates such as compact TiO_2 [4], PEDOT:PSS/ITO [5, 6] have extensively been studied. The perovskite solar cells using these flat substrates possess planar heterojunctions. In planar heterojunction perovskite solar cells, the qualities of perovskite film itself such as surface morphology (roughness and pin-hole free) and crystallinity is key factors to fabricate high-efficiency perovskite solar cells. It is also reported that the crystallinity of PbI_2 as a precursor affects the film quality of converted perovskite layer by thermal annealing under a MAI vapour atmosphere [7]. PbI_2 films deposited from vapour phase are usually used by a conventional vacuum evaporation. On the other hand, the hot-wall method is known as a high-quality deposition technique for inorganic and organic semiconductor films [8-11]. The hot-wall method is one of the vapour deposition techniques and can be deposited under a quasi-thermal equilibrium condition [12]. The deposited films are therefore high quality, compared to those by a conventional evaporation technique.

In this paper, we have studied conversions of $CH_3NH_3Pb_3$ perovskite films from PbI_2 films deposited by the hot-wall method as well as a conventional evaporation method. We compared crystalline quality of the converted perovskite films by x-ray diffraction and surface morphology by atomic force microscope (AFM). In addition, we have tentatively fabricated perovskite solar cells using the converted perovskite films from PbI_2 deposited by the hot-wall method. The fabricated perovskite solar cells were measured current-density voltage characteristics under an AM1.5G light irradiation using a solar simulator.

2 Experimental

Perovskite films were fabricated by a conversion from PbI_2 film by an annealing under a MAI vapour at 120-150 °C. The PbI_2 films were deposited by two kinds of method: a conventional vacuum evaporation and a hot-wall method. The hot-wall evaporation system equipped with a hot-wall furnace was a home made one. The hot-wall furnace installed into a vacuum chamber is shown in Fig. 1. The quartz tube is surrounded by three zones with separated heaters. Each zone consists of wall zone, source zone and reservoir zone. Their temperatures were controlled independently by PID temperature controllers. PbI_2 source materials (99.99%) are placed in the source zone of a quartz tube. The reservoir zone was not used in this study. The substrate temperature was kept at 100 °C during deposition of PbI_2 film by the hot-wall method. In the case of the depositions of PbI_2 films by a conventional vacuum evaporation, on the other hand, no

* Corresponding author: seto@ishikawa-nct.ac.jp

substrate heating was performed. The source and wall temperatures are 200 and 180 °C, respectively. The base pressure of the vacuum chamber is less than 5×10^{-6} Torr.

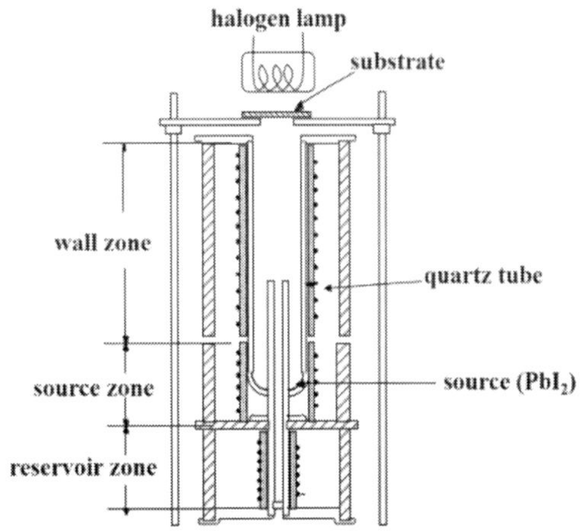

Fig. 1. Hot-wall furnace in a vacuum chamber.

The device structure and band diagram of perovskite solar cells are shown in Fig. 2. The device architecture as shown in Fig. 2(a) is the so-called inverted planar heterostructure: ITO/PEDOT:PSS/Perovskite/C60/Ag. PEDOT:PSS film was spin-coated on patterned ITO glass followed by annealed for 30 min at 150 C. C60 and Ag films were deposited by a conventional vacuum evaporation. It is pointed out here that the sample transfers from one process to a subsequent process were performed in an ambient atmosphere with humidity. On the other hand, the reported perovskite films in solar cells with high PCEs were fabricated in a glove box without oxygen and water vapour.

Surface morphology of the perovskite films was observed by an atomic force microscope (AFM). Crystalline phase of the converted perovskite films was confirmed by x-ray diffraction (XRD) measurements. Current density-voltage characteristics was measured with a source meter (Keithley 2612B) under AM1.5G light irradiation using a solar simulator.

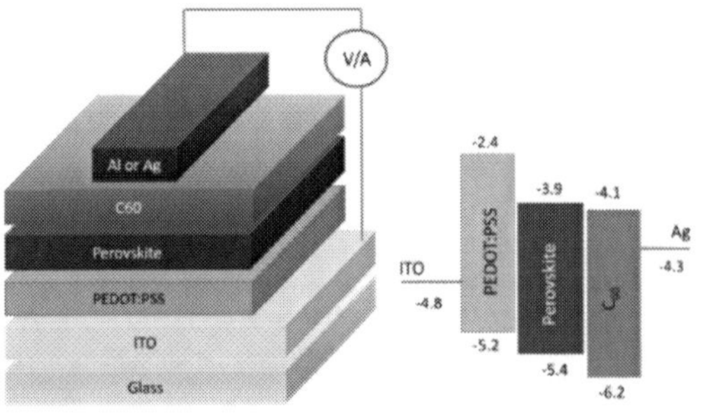

Fig. 2. The device architecture (left) and energy diagram (right) of the fabricated planar heterojunction solar cells.

3 Results and discussion

Figure 3 shows morphologies of the converted perovskite films from PbI_2 precursor deposited by a conventional evaporation (left) and the hot-wall method (right). The arithmetical mean roughness R_a of the films are 19.3 and 17.8 nm, respectively. Although, the perovskite film converted from PbI_2 deposited by a vacuum evaporation exhibits uniform surface, the roughness is still high for planar heterojunction solar cells. It is noticed that flatter areas can be seen on the surface of the perovskite film converted from the hot-wall method. This result shows that the hot-wall method is a potential method to fabricate flatter perovskite film compared to a vacuum evaporation. We believe that more flat perovskite films for planar heterojunction solar cells can be achieved by adjusting the growth condition like deposition rate and substrate temperature.

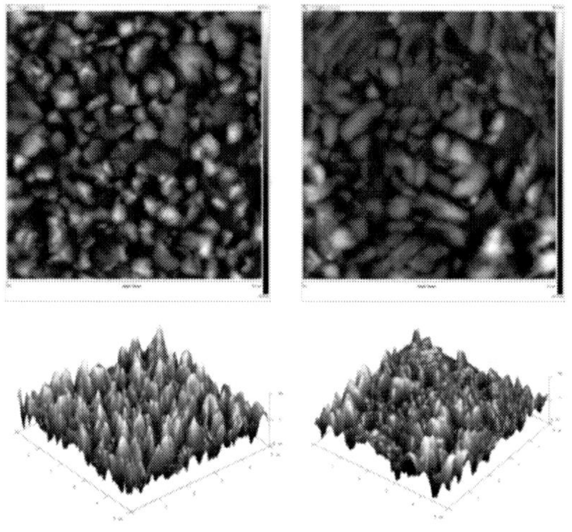

Fig. 3. AFM images (5x5 μm) of the perovskite films converted fromPbI_2 films deposited by a conventional vacuum evaporation (left) and a hot-wall method (right).

In order to confirm a perovskite phase of the converted films from PbI_2 precursor by the thermal annealing under a MAI vapour, we measured XRD. The measured XRD patterns are shown in Fig. 4. Both of the XRD patterns of the converted perovskite films shows relatively strong XRD peaks at 14.1°, 24.5°, 28.4°, 31.9° and 43.2° exhibiting (110), (211), (220), (310) and (330) planes from a perovskite phase, respectively. No additional XRD peaks related to PbI_2 phase (e.g. 12.7°) can be seen in the converted perovskite films in Fig. 4. This imply that the PbI_2 films as precursor converted completely to a perovskite phase by the thermal annealing under MAI vapour. Furthermore, high intensity peaks from (110) and (220) planes indicates that the converted perovskite films are highly oriented with respect to the substrate.

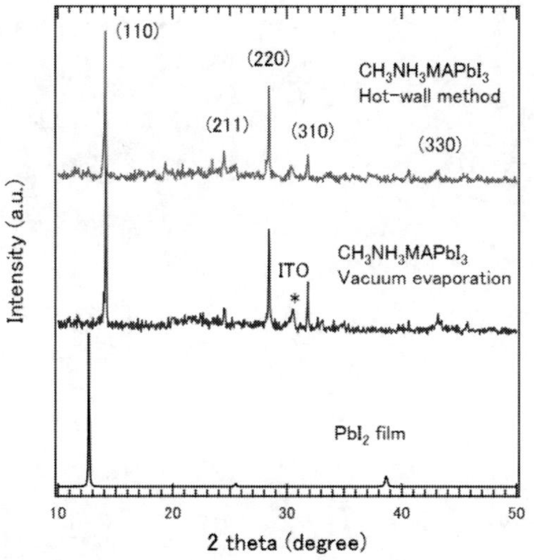

Fig. 4. X-ray diffraction patterns of the converted perovskite films from PbI_2 precursor deposited by a conventional vacuum evaporation and hot-wall method. X-ray diffraction pattern of a bare PbI_2 film is also shown.

Finally, we fabricated planar heterojunction perovskite solar cells by using the converted perovskite films discussed above. We named here each solar cell #1 and #2. The solar cell #1 is a solar cell fabricated from PbI_2 precursor deposited by a conventional vacuum evaporation, the solar cell #2 is a solar cell fabricated from the hot-wall method. The current density-voltage (*J-V*) characteristics is shown in Fig. 5, and the device performance parameters, short-circuit current density (J_{SC}), open-circuit voltage (V_{OC}), fill factor (FF) and PCE are summarized in Table 1. Although the *J-V* characteristics of the fabricated perovskite solar cells vary from sample to sample in this study, no significant differences can be found in the two kinds of solar cells, as can be seen in Fig. 5. The J_{SC} of the perovskite solar cell #1 is slightly higher than the solar cell #2. On the contrary, the V_{OC} shows higher value in the solar cell #2, which was fabricated from the hot-wall deposited PbI_2 precursor film, than solar cell #1, which was fabricated from a vacuum evaporated one. The cell performance parameters FF and PCE of two solar cell #1 and #2 are comparable. However, these performance parameters is considerably lower than the reported ones. We think that the main reasons of the low performances are most probably due to a fabrication process under an ambient environment with humidity, pin-holes in the perovskite film and flatness of the film. Several researchers reported efficient perovskite solar cells prepared in ambient air with humidity [13, 14]. Therefore we expect that high efficient perovskite solar cells could also be fabricated by our process. However, pin-holes and flatness of perovskite film is serious issues to achieve higher performance of planar perovskite solar cells. In particular, low FF is caused mainly by low shunt resistance arising from pin-holes in perovskite film. As mentioned before, the perovskite film prepared by the hot-wall deposited PbI_2 precursor has flatter surface. If the flatter surface expanded to an entire surface, the pin-holes could significantly be reduced. Further experiments using the hot-wall method are needed to achieve the depositions of high quality flat perovskite films by optimizing the growth and anneal conditions of PbI_2 films.

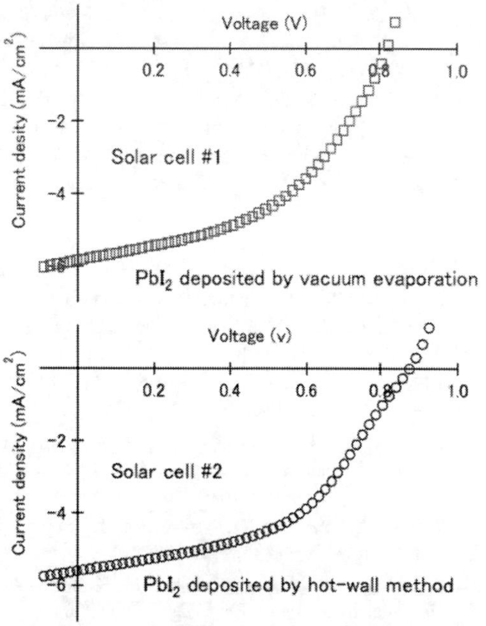

Fig. 5. Current density-voltage characteristics of the perovskite solar cells fabricated by two kinds of converted perovskite films.

Table 1. Performance parameters of fabricated planar heterojunction solar cells.

Solar cell	J_{SC} **(mA/cm²)**	V_{OC} **(V)**	**FF**	**PCE (%)**
#1	5.82	0.822	0.463	2.22
#2	5.60	0.873	0.479	2.33

4 Summary

We developed a method to convert PbI_2 precursor films into perovskite phase. The PbI_2 films were prepared with two kinds of deposition method: a conventional evaporation and the hot wall method. The conversion of PbI_2 films into perovskite phase was completed by a thermal annealing under MAI vapour. We fabricated solar cells with a planar heterojunctions using these perovskite films. The PCE was as low as 2.33%. We speculated that the low PCEs were mainly due to pin-holes and flatness in the perovskite films. We found that the hot-wall method for PbI_2 precursor film was a potential method to fabricate pin-hole free and flat perovskite films.

Acknowledgments

The authors thank Dr. S. Yamada for x-ray diffraction measurements. This study was supported in part by the Iwatani Naoji Foundation.

References

1. D. Bi, C. Yi, J. Luo, J-D. Décoppet, F. Zhang, S. M. Zakeeruddin, X. Li, A. Hagfeldt, M. Grätzel, Nat Energy **1**, 16142 (2016).
2. W.S. Yang, J.H. Noh, N.J. Jeon, Y.C. Kim, S. Ryu, J. Seo, S. Seok, Science **348**, 1234 (2015).
3. H. Back, G. Kim, J. Kim, J Kong, T.K. Kim, H. Kang, H. Kim, J. Lee, S. Lee, K. Lee, Energy Environ. Sci. **9**, 1258 (2016).
4. M. Liu. M.B. Johnston, H.J. Snaith, Nature **501**, 395 (2013).
5. J. You, Z. Hong, Y. Yang, Q. Chen, M. Cai, T-B, Song, C-C. Chen, S. Lu, Y. Liu, H. Zhou, Y. Yang, ACS NANO **8**, 1674 (2014).
6. J.H. Heo, H.J. Han, D. Kim, T.K. Ahn, S.H. Im, Energy & Environ. Sci. **8**, 1602 (2015).
7. H.A. Abbas, R. Kottokkaran, B. Canapathy, M. Samiee, L. Zhang, A. Kitahara, M. Noack, V.L. Dalal, AIP Materials **3**, 016105 (2015).
8. S. Seto, S. Yamada, T. Miyakawa, K. Suzuki, J. Cryst. Growth **237-239**, 1585 (2002).
9. S. Seto, Y. Nosho, T. Kousho. H. Kitani and S. Yamada, Jpn. J. Appl. Phys.**42**, L1123 (2003).
10. S. Seto, S. Yamada, A. Kitazaki, K. Sebald, I. Rückmann, J. Gutowski, J. Cryst. Growth **318**, 1067 (2011).
11. S. Seto, Molecular Crystals & Liquid Crystals **471**, 245 (2007).
12. A. Lopez-Otero, Thin Solid Films **49**, 3 (1978).
13. H-S. Ko, K-W. Lee, N-G. Park, J. Mater. Chem. A **3**, 8808 (2015).
14. Q. Tai, P. You, H. Sang, Z. Liu, C. Hu, H.L.W. Chan, F. Yau, Nature Commun. **7**, 11105 (2016).

Dielectric elastomer generator with diaphragm configuration

Zhen-Qiang Song[1], *Sriyuttakrai* Sathin[2], *Wei* Li[1], *Kazuhiro* Ohyama[2] and *ShiJie* Zhu[1*]
[1]Department of Intelligent Mechanical Engineering, Fukuoka Institute of Technology, Japan
[2]Department of Electrical Engineering, Fukuoka Institute of Technology, Japan

Abstract. The dielectric elastomer generator (VHB 4905, 3M) with diaphragm configuration was investigated with the constant-voltage harvesting scheme in order to investigate its energy harvesting ability. The maximum energy density and energy conversion efficiency is measured to be 65 J/kg and 5.7%, respectively. The relatively low efficiency indicates that higher energy conversion efficiency is impeded by the viscosity of the acrylic elastomer, suggesting that higher conversion efficiency with new low-viscosity elastomer should be available.

1. Introduction

As one kind of electroactive polymers, the dielectric elastomer (DE) possesses some unique properties, such as extraordinarily large strain, excellent flexibility and high response speed [1]. Thus, the DE shows great promise as electromechanical transducer, which is able to convert energy between the mechanical energy and the electrical energy [2, 3]. When applied as power generators, the DE is able to harvest electricity from various environmental resources, such as the ocean waves, wind, and human activity, which provides a new strategy to acquire clean and renewable source of energy [4, 5]. Compared with other competing power generation technologies, the energy density of dielectric elastomer generator (DEG) shows great advantages, which is more than two orders of magnitude higher than electromagnetics and piezoelectric ceramics [1].

For decades, numerous prototypes adopting different DE materials and various harvesting circuits and schemes have been developed. Huang et al. [6] reported that the energy density of acrylic dielectric elastomer within a single harvesting cycle can be as high as 560 J/kg adopting the equi-biaxial stretching method with the constant-voltage energy harvesting strategy. Later, Shian et al. [7] optimized the electromechanical cycle of DEG with a "triangle" harvesting strategy, which boosted the energy density of dielectric elastomer to 780 J/kg. Even though the maximum energy density is expected to be achieved with equi-biaxial loading strategy in the laboratory, the complex mechanical structure confined its large-scale applications. In contrast, the diaphragm configuration is widely adopted in real applications due to its simple structure and easy installation. However, the research on the diaphragm configuration of DEG with constant-voltage scheme is very few and far between.

In this work, the performance of DEG with diaphragm configuration was investigated adopting the constant-voltage scheme. Then, the energy density and energy conversion efficiency of DEG were analyzed. Finally, strategies for optimizing the performance of diaphragm DEG was discussed.

2. Experimental methods

2.1 Specimen preparation

The acrylic elastomer VHB 4905 (3M) with a thickness of 0.5 mm was used for the investigation, and carbon conductive grease was painted on both sides to serve as compliant electrodes. The initial diameter D_0 of the VHB 4905 film was 12 cm, and then the film was equibiaxially

* Corresponding author: zhu@fit.ac.jp

pre-stretched by a ratio (D_1/D_0) of 2 before attached to the rigid plastic plates. The schematic of the diaphragm specimen is shown in Fig. 1.

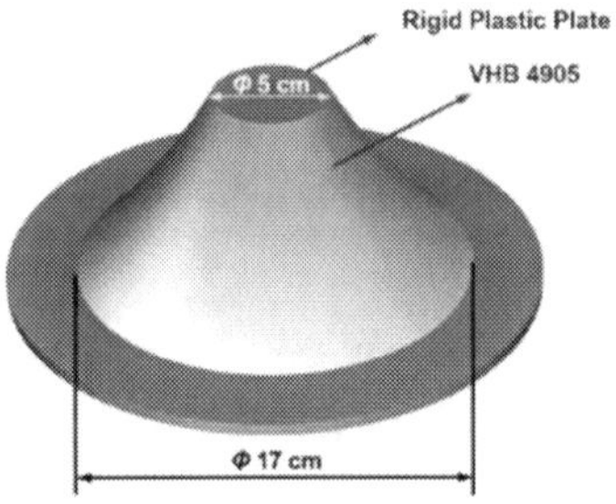

Fig.1 Schematic of the diaphragm specimen.

2.2 The energy harvesting scheme

The constant-voltage scheme was adopted in this work. The operational cycles were schematically illustrated on the voltage-charge work-conjugate plane in Fig. 2, and the corresponding electric circuit diagrams is shown in Fig. 3. Twenty Zener diodes with a reverse bias breakdown voltage of 200 V were connected in series to build a constant harvesting voltage of 4 kV.

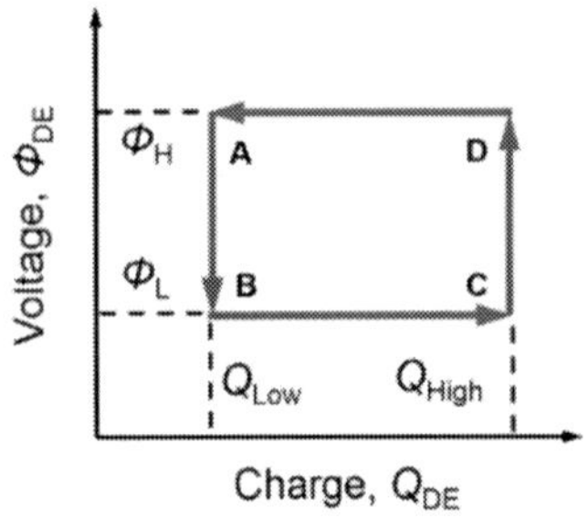

Fig.2 The schematics show the energy-harvesting principle of dielectric elastomer generator.

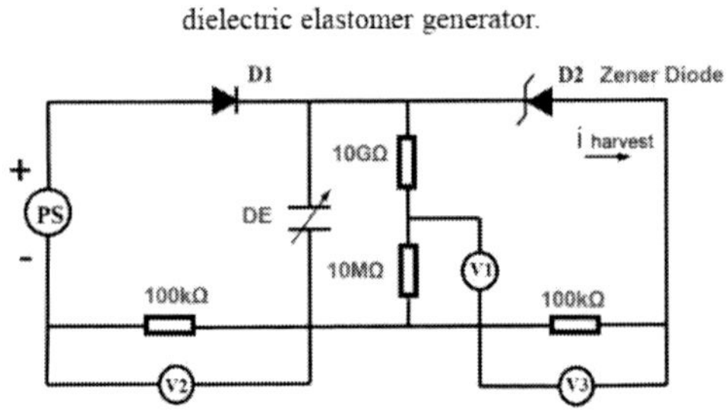

Fig.3 Electrical circuits for energy harvesting.

2.3 Mechanical loading configuration

The tests were operated on a self-made testing bench. The stretching of the DE film was accomplished by the motion of a linear servo-motor (GLM20AP, THK) installed with a force sensor (Minebea Co., Ltd.). A multichannel voltmeter (National Instruments) was used to measure the voltages and currents in the circuit. All the data acquisition and the motion of the servo-motors were accomplished with the LabVIEW software.

3. Results and discussions

The servo-motor position (S), the force applied by the linear motor (F), the DE voltage (U_{DE}), the charging current (i_{charge}), and the harvesting current ($i_{harvest}$) *versus* time (t) were plotted in Fig. 4. The maximum stretch, the input voltage and the harvest voltage were experimentally chosen to be 20 cm, 2 kV and 4 kV, respectively, which maximized the energy density within a single harvesting cycle while avoiding premature failure of DE, such as mechanical fracture, electric breakdown as well as loss of tension. It can be seen that charge flowed from the power source to the dielectric elastomer membrane when the film was stretched at a constant voltage of 2 kV, and then the charge on the dielectric elastomer was harvested when the DE recovered to the initial state at a constant voltage of 4 kV.

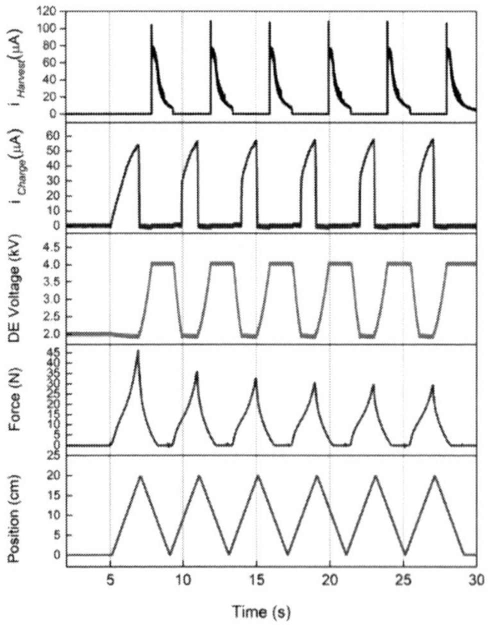

Fig.4 The plot of servo-motor position (S), the force applied by the linear motor (F), the DE voltage (U_{DE}), the charging current (i_{charge}) and harvesting current ($i_{harvest}$) *versus* time (t).

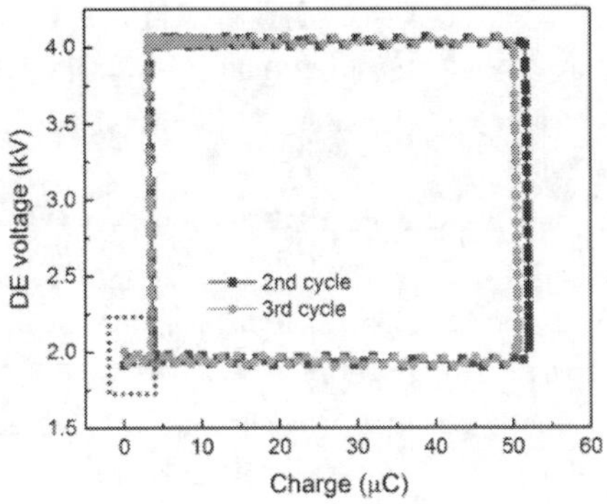

Fig. 5 Voltage of DE as a function of the charge.

The experimental energy harvesting plot, Fig. 5, have the similar basic characteristics as the theoretical scheme on the work-conjugate plane in Fig. 2. The slight distinction is that the charge flowed from the power sources cannot completely harvested during the DE relaxation, as marked by the dotted box in Fig. 5, which is due to the charge leakage within each cycle. The area enclosed by the curve is the net energy gained in a single cycle, ΔE_{Net}, and the energy density is $E_{Density}=\Delta E_{Net}/M$, where M is the mass of the electromechanical active region of the DE membrane coated with carbon grease.

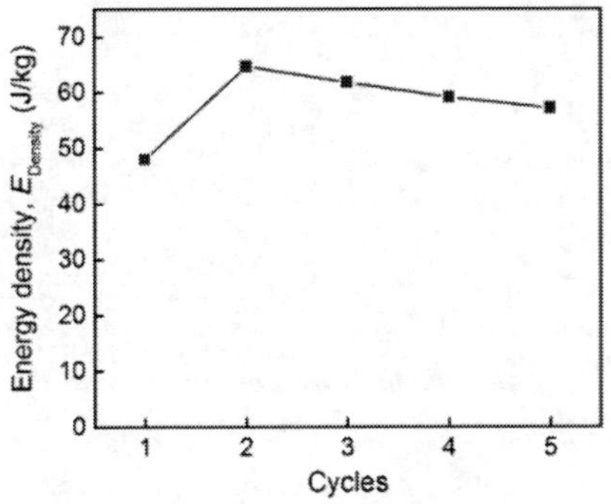

Fig. 6 The energy density within each loading cycle.

The energy density of each cycle is shown in Fig. 6. It can be seen that the average energy density is 58±6 J/kg, and the maximum energy density is 65 J/kg. The corresponding average power and the maximum power is calculated to be 14.5±1.5 W/kg and 16.3 W/kg, respectively. It should be noted that the power of the DEG should be further improved when the moving velocity of the motor increases. It is found that the energy density decreased after two harvesting cycles, which is due to the loss of tension of DE. It should be noted that in this work, the DE was stretched to approaching its strain limit in order to investigate its maximum energy harvesting ability, while in the actual application with long-term operation, the stretch ratio of the DE was much smaller than its strain limit, and the energy conversion efficiency will not be significantly affected by the increase of the harvesting cycle.

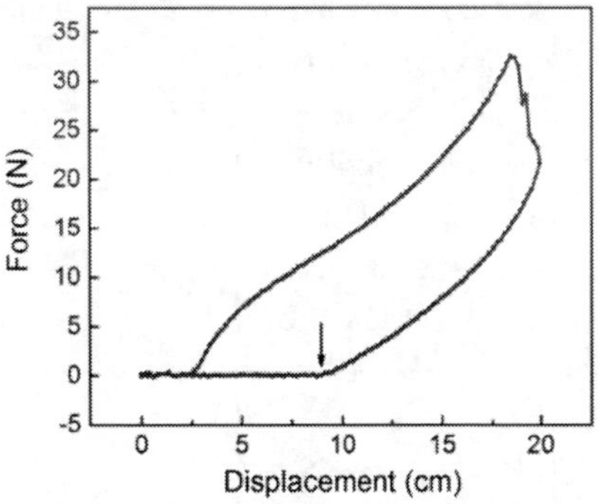

Fig. 7 The force applied by the linear motor versus the servo-motor position.

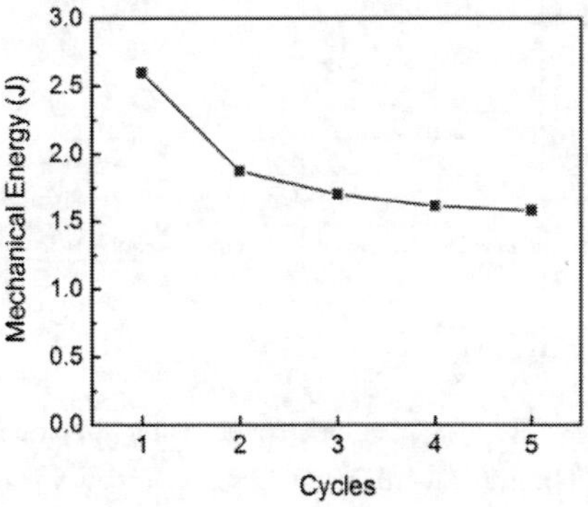

Fig. 8 The mechanical energy dissipation in each cycle.

The relationship between the force applied to stretch dielectric elastomer film and the displacement of linear motor is shown in Fig. 7. It can be seen that during the relaxation process of the DE, the force decreased to be zero when the motor is at the position of 9 cm. Thus, the DE can not relax to the initial state due to the loss of tension within a small motor displacement range, which is caused by the viscosity of the acrylic elastomer. This effect is also evident in Fig. 4, where a delay can be detected between harvesting electric current shut-off and the start of the following harvesting cycle. The mechanical energy dissipation, ΔE_{mech}, can be calculated based on the area enclosed by the force-displacement plot in Fig. 7. The mechanical dissipation in each cycle is shown in Fig. 8. It can be seen that as the loading cycle increased, the value of ΔE_{mech} decreased from 2.6 J to 1.6 J.

The mechanical-electrical energy conversion efficiency of the DE diaphragm is calculated in terms of the ratio of the net electric energy to the mechanical energy consumed within a single cycle, $\eta=\Delta E_{Net}/\Delta E_{mech}$. The value of η in each cycle is shown in Fig. 9. It can be seen that the average energy conversion efficiency is 4.8±1.2 % with the maximum value of 5.7%, which indicated that a large proportion of the mechanical work was consumed via viscous dissipation of acrylic elastomer instead of being transformed into the electrical energy. Therefore, the energy conversion efficiency is expected to be further improved through development of new elastomer materials with low viscosity.

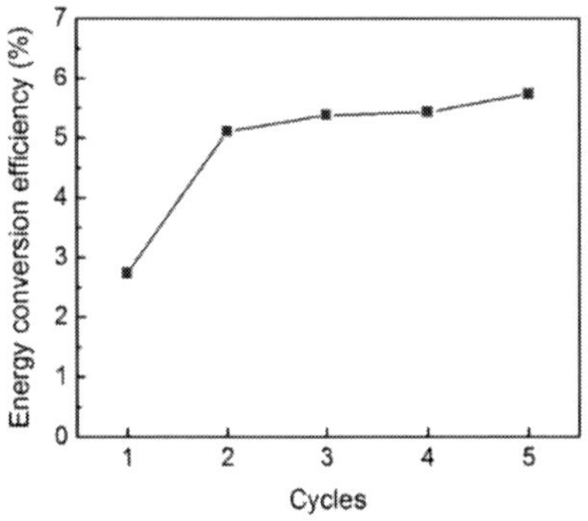

Fig. 9 The energy conversion efficiency in each cycle.

4. Conclusions

The dielectric elastomer generator with diaphragm configuration was investigated with the constant-voltage harvesting scheme. The maximum energy density and energy conversion efficiency is measured to be 65 J/kg and 5.7%, respectively. After two harvesting cycles, the energy density in a single cycle decreased with the loading cycles, while the energy conversion efficiency increased slightly. The relatively low conversion efficiency indicates that a majority of mechanical energy was consumed by the viscosity of the acrylic dielectric elastomer.

This work was supported by the Strategic Research Foundation Grant-Aided Projects for Private University funded by Japanese Ministry of Education, Culture, Sports, Science and Technology (MEXT).

References

1. R.D. Kornbluh, R. Pelrine, H. Prahlad, A. Wong-Foy, B. McCoy, S. Kim, et al., MRS Bull. **37**, 246-53(2012)
2. R. Pelrine, R. Kornbluh, Q. Pei, J. Joseph, Science **287**, 836-9(2000)
3. S. Shian, K. Bertoldi, D.R. Clarke, Adv. Mater. **27**, 6814-9(2015)
4. S. Chiba, M. Waki, R. Kornbluh, R. Pelrine, Smart Materials and Structures **20**, 124006(2011)
5. K. Ahnert, M. Abel, M. Kollosche, P.J. Jørgensen, G. Kofod, J. Mater. Chem. **21**, 14492(2011)
6. J. Huang, S. Shian, Z. Suo, D.R. Clarke, Adv. Funct. Mater. **23**, 5056-61(2013)
7. S. Shian, J. Huang, S. Zhu, D.R. Clarke, Adv. Mater. **26**, 6617-21(2014)

MATEC Web of Conferences **192**, 01033 (2018) https://doi.org/10.1051/matecconf/201819201033
ICEAST 2018

Selection of multi-criteria plant layout design by combining AHP and DEA methodologies

Sirawadee Arunyanart* and *Surangkana* Pruekthaisong

Department of Industrial Engineering, Faculty of Engineering, Khon Kaen University, Khon Kaen, Thailand

Abstract. This paper deals with the problem of finding the optimal plant layout design. Decision-making methodologies based on analytic hierarchy process (AHP) and data envelopment analysis (DEA) approach are applied in the selection of the best plant layout. New layouts are developed, in an efficient manner, based on the systematic layout planning (SLP). By using the multiple criteria, AHP is applied to weigh the qualitative performance measures. DEA is then used to determine the suitable layout design by measuring layouts' efficiency, using the information of AHP and combining with the quantitative data. The utilization of the proposed procedure is applied to a real data set of a machining precision parts manufacturing company.

1 Introduction

Plant layout is one of the most important components that significantly has an impact on productivity, profitability, and performance of any manufacturing industries, regardless of the size of plant or the type of operation. The necessity of designing a plant layout arises not only when a new plant is about to be elected but also during production process due to various reasons. The selection of an appropriate layout design is a multi-criteria decision making (MCDM) which requires joint consideration of multiple alternatives, which are various designed layouts and several conflicting evaluation criteria that influence the layout decision. Moreover, in the evaluation process, both of the qualitative and quantitative criteria are necessary to be taken into consideration. Consequently, it is challenging and time consuming for layout generation and evaluation.

This paper explores the problem of finding the optimal plant layout design by applying a decision-making methodologies based on an integration of analytic hierarchy process (AHP) and data envelopment analysis (DEA). Empirical illustration from a practical case study of a machining precision parts manufacturing company illustrates the effectiveness of the proposed methodology. Firstly, an existing plant layout is improved by adopting systematic layout planning (SLP) methodology in order to generate new layouts as well as to collect quantitative performance data of each alternative, namely distance and cost. In the next step, an expert judgment is applied to perform the individual pairwise comparisons of the qualitative performance data related to flexibility, safety, and space utilization in AHP. Finally, DEA is used to solve the problem of layout design selection by simultaneously considering both quantitative and qualitative performance criteria which leads to the identification of the more optimal plant layout design alternative.

2 Literature review

One of the effective and frequently used methods for layout design or redesign is SLP. The method presents layout planning step that can be easily used in developing a new layout or improving an existing layout [1]. It is also a proven tool in providing layout design guidelines and it is widely used in the manufacturing industry for facility layout planning [2].

Pitil and Kuber [3] improved productivity of Auto Ancillary company by using SLP. Hossain et al. [4] developed plant layout of jute industry based on SLP pattern theory. The new plant layout showed that the distance and overall cost of material flow from stores to dispatch area were significantly decreased. Sutari and Rao [5] applied SLP to design plant layout of a nacelle production. The new plant layout could significantly decrease the distance of material and work flow travel and resulted in increasing the productivity. Ojaghi et al. [6] minimized travel distance, material handling and losses of company producing meatball and soup paste by using SLP to generate new layouts. Chen et al. [7] adopted SLP approach to enhance the performance on optimizing the layout design of a large auto parts manufacturing enterprise. In this study, AHP was adopted to evaluate optimal layout.

AHP has been accepted as one of robust and flexible multi-criteria decision making tools for dealing with complex problems [8,9]. The method requires the decision maker to provide judgements about the relative importance of each criterion which is then used in specifying a preference for each decision alternative. The

* Corresponding author: sirawadee@kku.ac.th

output of AHP is in a form of weights of each alternative based on the overall preferences evaluated by the decision maker [10-13]. The methodology of AHP has been applied to support decision of facility layout design selection in various cases.

Yang and Deuse [14] made use of Matlab-based to generate three new layout designs. Afterwards, the weights for each of six decision criteria was obtained by AHP calculation. Each layout alternative was then evaluated using PROMETHEE to get final ranking. Yang et al. [15] generated ten new layout alternatives of a paint department by a Pareto-based multi-objective optimization approach. Rough set theory was integrated with AHP to determine the weights for each of five criterion of alternatives. The final alternative facility layout design was obtained by using TOPSIS method. Hadi-Vencheh and Mohamadghasemi [16] adopted a computer-aided layout planning tool, Spiral, to generate the facility layout patterns. Six criteria were taken into account. AHP was applied to determine weights of qualitative criteria and nonlinear programming model was then used to solve the problem of selecting one out of 18 alternatives by considering both quantitative and qualitative data. Shokri et al. [17] tackled the same facility layout design problem as Hadi-Vencheh and Mohamadghasemi, by employing AHP to weigh the qualitative performance measures. VIKOR was then applied to find a compromise priority ranking of alternatives according to the selected criteria. Bacudio et al. [18] generated three layout alternatives of a metal manufacturing company by applying SLP. Considering four criteria, AHP was applied to determine the priority weights and the best layout.

DEA is one of the most popular tools in management analysis. The technique is applied to evaluate performance or efficiency of a homogeneous set of entities which is referred as decision making units (DMUs) by considering multiple input and multiple output factors [19]. The DMUs could represent any businesses, operations, or entities under an evaluation that convert multiple inputs into multiple outputs. The maximum ratio of weighted sum of outputs to weighted sum of inputs is defined as the efficiency, which is calculated for each DMU. Then the efficiency of observed DMU is evaluated relatively by other DMUs [20]. The goal of DEA is to determine the productive efficiency of DMUs by comparing how well the DMU converts inputs into outputs [21].

Liu [22] applied fuzzy DEA with an assurance region approach for selecting a flexible manufacturing system. Twelve alternatives were compared using two inputs and four outputs performance measures. Yand and Kuo [23] used commercial software called Spiral for generation of 17 layouts of IC packaging company. Qualitative measures were weighted by AHP. The DEA was adopted to select the final design. Ertay et al. [24] utilized VisFactory, which is a computer-aided layout-planning tool, to create 18 layout alternatives of a plastic production company. The performance measures on qualitative criteria were developed using AHP. Ranking of the alternatives was later obtained by considering six criteria using DEA.

3 Methodology

The proposed framework used in solving the layout design problem is illustrated in Fig.1.

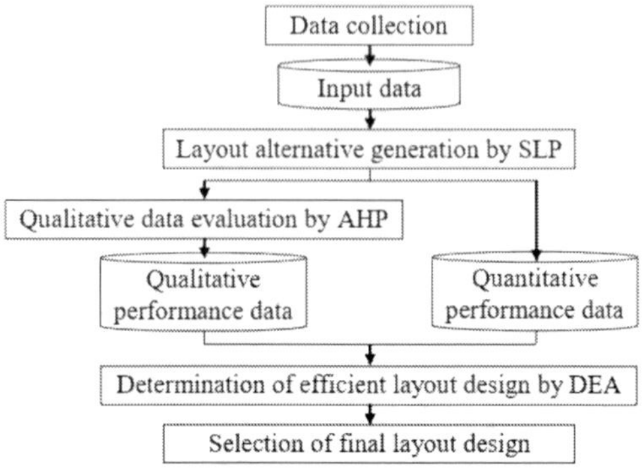

Fig. 1. Framework for plant layout selection.

3.1 Data collection

An important data of the existing plant including flow of materials, procedures and activities, space available and requirements have to be collected in order to assure the validity of input data at the layout design stage. Information related to quantitative and qualitative criteria also have to be assembled.

3.2 SLP for layout alternative generation

SLP is a procedural layout design approach which its process is relatively straightforward but yet is proven a powerful tool in providing the layout design guideline [25]. A step-by-step planning procedure of SLP uses to determine a reasonable production system layout scheme and develop a new plant layout with an improved process flow and an effective utilization of space. A workplace in a plant is arranged by locating two areas that contain high frequency and logical relationships close to each other. Readers are suggested reference [25] for detailed discussion and explanation of the SLP technique.

3.3 AHP for qualitative data evaluation

Since qualitative data which is part of criteria to be evaluated is usually complicated and conflict, AHP are applied in order to provide the weights that indicate the relative importance of the layout alternatives for each criterion. The management are asked to determine a comparison matrix by comparing any pairs of qualitative criteria one at a time. Scale of importance according to Saaty 1-9 scale for the pairwise comparison is used in the comparison process. In order to avoid potential comparative inconsistency between any pairs of criteria in pairwise comparison matrix, a consistency ratio (CR), which is an index for consistency, is then calculated to assure the appropriateness of the comparisons. For the details of AHP methodology and CR development, readers are referred to look further at Saaty [26].

3.4 DEA for optimal layout design determination

DEA uses the linear programming to assess efficiencies of DMUs with the goal to determine the productive efficiency of DMUs by comparing how well the DMU converts inputs into outputs [21]. Assume that there are n DMUs to be evaluated, where each DMU consumes varying amounts of inputs to produce different amounts of outputs. x_{ij} and y_{rj} are respectively amounts of input ith and output rth of DMU$_j$, which are positive number. v_i and u_r are decision variables and are respectively called input and output multipliers or input and output weights. The efficiency score, h_O, of a particular DMU being evaluated, DMU$_O$, is obtained by solving the following model:

$$\max h_0(u,v) = \frac{\sum_r u_r y_{r0}}{\sum_i v_i x_{i0}} \tag{1}$$

$$\text{s.t.:} \quad \frac{\sum_r u_r y_{rj}}{\sum_i v_i x_{ij}} \le 1 \qquad \forall j \tag{2}$$

$$u_r, v_i \ge 0 \qquad \forall r,i \tag{3}$$

The objective function is to maximize the ratio of weighted outputs to weighted inputs or efficiency value for a particular DMU, subject to the constraints that the ratio of weighted outputs to weighted inputs is less than or equal to one. The model can be converted to a linear programming model, which is called CCR model as following:

$$\max h_0 = \sum_{r=1}^{s} u_r y_{r0} \tag{4}$$

$$\text{s.t.:} \quad \sum_{i=1}^{m} v_i x_{io} = 1 \tag{5}$$

$$\sum_{r=1}^{s} u_r y_{rj} - \sum_{i=1}^{m} v_i x_{ij} \le 0 \quad \forall j \tag{6}$$

$$u_r, v_i \ge 0 \qquad \forall r,i \tag{7}$$

The model is run n times to identify the efficiency scores of all DMUs. DEA will identify the DMU(s) that produces the largest amounts of outputs by consuming the least amounts of inputs and then allocate efficiency score equal to one to the DMU(s). Other inefficient DMUs will be given efficiency scores relatively to the efficient DMU(s) which is less than one.

4 Performance evaluation for selection of plant layout design

A manufacturing company considered in this study focuses on manufacturing and design of high precision and high complexity mechanical components, including design, test, production, distribution and return/repair services. The company located in Thailand offers its products worldwide, especially in Europe and the USA. Due to a business expansion plan, the company has to double its machines in order to be able to handle the plastics precision line of product separately. Therefore, to ensure the effectiveness of the production line, the new plant layout needs to be function on supporting the production's upcoming activities.

4.1 Data collection

Fundamental data of the company such as product data, manufacturing process data, flow of process, plant layout pattern, and manufacturing facilities are collected. The current facility layout and the size of each area are shown in Fig. 2 and Table 1. The criteria (or the performance measures) for the selection of this plant layout design are determined by reviewing literatures as general layout guidelines and by discussing with the company's management. As a result, those criteria are: distance, cost, flexibility, safety, and utilization.

The *distance* is an average transit distance of any materials that are carried into and out of four department areas throughout the production process—turning, milling, quality control, and washing. The *cost* involves an investment cost in a renovation of the plant for the new layout design. The *flexibility* involves accessibility, maintenance, and ability to perform a variety of tacks for the future expansion. The *safety* involves both of personal and working environment that is safe from danger or harm in manufacturing workplace. The *utilization* is a proper utilization of the floor space.

4.2 Generation of plant layout alternatives

In order to support the expansion of the company's production line, the management decides to move several working areas from the current layout to the building nearby. The office area which includes meeting

Table 1. Size of facility's areas.

Available floor space = 18 m (Width) × 45 m (Length)

No.	Area	W×L	Size (m^2)
1	Turning	14×15	210
2	Milling	14×21	294
3	Finish goods storage	9×14	126
4	Quality control (QC room)	4×6	24
5	Washing room	4×6	24
6	Office	4×25	100
7	WC (East)	4×3	12
8	WC (West)	4×5	20

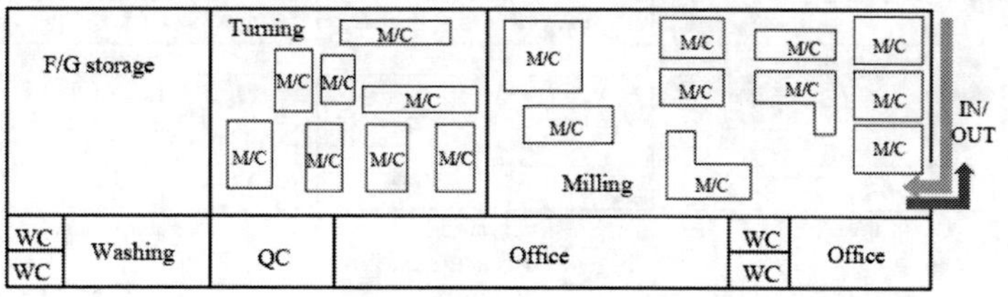

Fig. 2. Current facility layout.

room, CAD/CAM and planning rooms, and finish good storage area will be moved. The washing and quality control process, these working areas previously set separately but next to each other, are required during the manufacturing of turning and milling process. Since both processes are linked, the new layout will group them together for gaining advantage of the area extension.

By applying SLP technique, the relationships between departments in each process from material storage area to final product storeroom is collected. Accordingly, a relationship chart is drawn. As shown in Fig. 3, the closeness values are defined as following: A as absolutely necessary, E as an especially important, I as an important, O as ordinary closeness, and U as unimportant. Furthermore, the indicator of this relationship are defined as following: 1 as continuous operation, 2 as convenience, 3 as noise, 4 as lighting, 5 as shared workers, and 6 as vibration.

Three alternative plant layouts are developed based on modifying of the current layout. Since the management is deciding if the restrooms are staying in the new layout or not, six more choices of alternative layouts are generated, including three layouts with two WCs and other three layouts without any. All nine alternative plant layouts are displayed in Fig. 4 for further evaluation. The quantitative performance data related with each alternative layout, which will be used as the parameters of the DMU for DEA methodology, are shown in Table 2.

4.3 Evaluation of qualitative data

The qualitative performance data are obtained by AHP methodology and solved by a spreadsheet. The eigenvector or relative weight of each three qualitative criteria namely flexibility, safety, and utilization are 0.293, 0.078, and 0.629 respectively. The CR value is 0 (zero). Since CR is smaller than the commonly critical value of 0.1, there is no evidence of inconsistency.

The resulting relative weights of each criterion for each of nine alternatives are shown in Table 3. These weights then become parameters for the DMUs in DEA calculation. The resulting CR values for the comparison of alternatives are shown in Table 4. The pairwise comparison is considered to be consistent enough since all values are smaller than 0.1.

4.4 Determination of optimal layout design

The inputs and outputs of DMUs are necessary information for the computation of the DEA model. The criteria that are to be minimized are considered as inputs whereas the criteria to be maximized are viewed as outputs. For this case study of the plant layout designs, all output variables—flexibility, safety, and utilization—of DEA are from AHP analysis, and all inputs—distance and cost—are from layout design.

The DEA equations (4) to (7) are applied to the data set of the nine facility layout alternatives. Table 5 shows the efficiency scores obtained by using DEA. The results show that alternatives 1 and 7 are relatively efficient. They are equally viable candidates for the final layout design.

5 Conclusions

This paper addresses the evaluation of the facility layout design by developing framework based on AHP and DEA methodologies. SLP is adopted to develop the layout alternatives as well as to gather the quantitative performance data. The qualitative performance data are obtained by applying AHP to assess weights expressing

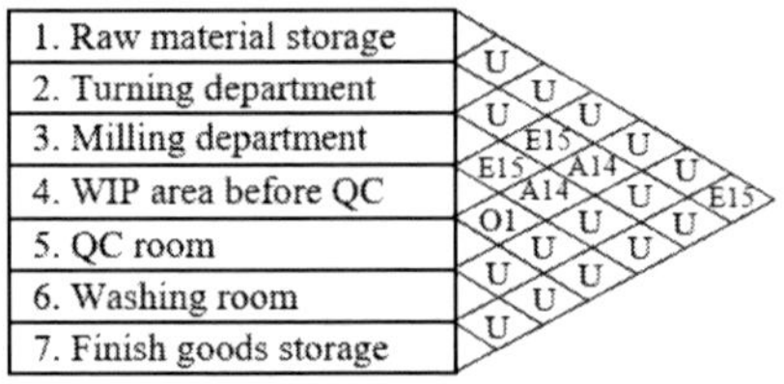

Fig. 3. Relationship chart.

Table 2. Quantitative measures for layout alternatives.

DMU no.	Distance (m)	Cost (thousand THB)
1	117.75	10
2	103.00	10
3	112.00	10
4	117.75	15
5	120.00	15
6	93.50	15
7	115.50	20
8	128.50	20
9	112.00	20

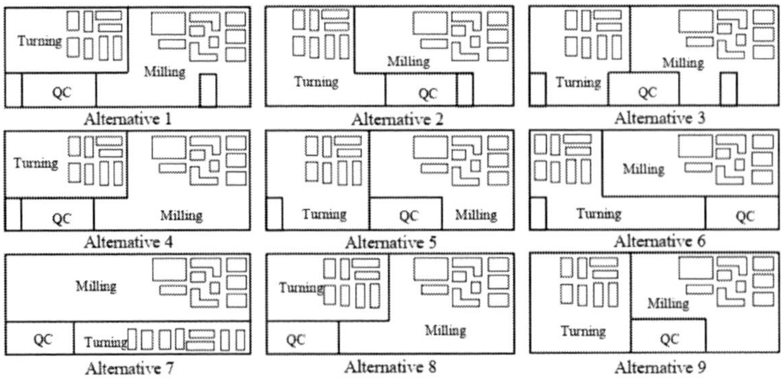

Fig. 4. Layout alternatives.

Table 3. Qualitative measures for layout alternatives.

DMU no.	Flexibility	Safety	Utilization
1	0.0739	0.0197	0.0552
2	0.0276	0.0067	0.0369
3	0.0215	0.0058	0.0323
4	0.0522	0.0141	0.0287
5	0.0232	0.0046	0.0244
6	0.0134	0.0039	0.0202
7	0.0429	0.0117	0.1672
8	0.0276	0.0074	0.1414
9	0.0107	0.0041	0.1227

Table 4. CR values.

Criteria	Flexibility	Safety	Utilization
CR	0.0652	0.0642	0.0656

Table 5. Final efficiency scores.

DMU no.	Efficiency score
1	1.0000
2	0.5436
3	0.4619
4	0.7143
5	0.3282
6	0.2807
7	1.0000
8	0.8457
9	0.7567

the relative importance of the layout alternatives for each criterion. DEA model is then applied to identify the efficiency scores by considering both the quantitative and qualitative performance data. The result indicates that DEA model is useful for finding global priorities among the layout designs. The proposed decision support framework offers a systematic guideline for the decision makers in planning the facility layout which could contribute to a successful final decision.

The framework is applied to a real data set of the manufacturing company which consists of both the quantitative and qualitative data for evaluating nine alternative layouts. Since the optimize outcome turns out to be more than one option, the final decision still needs to be made on which layout to be selected. However, this case study has successfully illustrated the effectiveness of this proposed framework, as a result. Moreover, the framework presented in this paper can easily be implemented without the need of using any software tools.

Acknowledgements

This work is supported by the Supply Chain and Logistics Systems Research Unit, Department of Industrial Engineering, Faculty of Engineering, Khon Kaen University, Khon Kaen, Thailand.

References

1. M.M.D. Hassan, J. of Manuf. Tech. Manage. **18**, 3, 292-303 (2007)
2. T. Yang, C.T. Su, Y.R. Hsu, Int. J. Oper. & Prod. Manage. **20**, 11, 1360-1372 (2000)
3. S.B. Patil, S.S. Kuber, Int. J. Res. in Eng. & Tech. **3**, 4, 770-775 (2014)
4. R. Hossain, K. Rasel, S. Talapatra, Glob. J. Res. in Eng. J **14**, 7, 71-75 (2014)
5. O. Sutari, S. RAO, Int. J. Mech. & Prod. Eng. **2**, 8, 63-66 (2014)
6. Y. Ojaghi, A. Khademi, N.M. Yusof, N.G. Renani, S.A.H.B.S. Hassan, Procedia CIRP **26** 247-251 (2015)
7. W. Chen, C. Liu, X. Huang, H. Lai, B. Li, Sci. J. of Busi. & Manag. **4**, 5, 172-180 (2016)
8. N. Bhushan, K. Rai, *Strategic decision making: Applying the Analytic Hierarchy Process* (Springer, London, 2004)
9. T.L. Saaty, Eur. J. Oper. Res. **74**, 426-447 (1994)
10. S.J. Chen, C.L. Hwang, *Fuzzy multi attribute decision making-methods and applications* (Springer-Verlag, 1992)
11. J. Dyer, Manage. Sci. **36**, 3, 249-258 (1990)
12. T.L. Saaty, L.G. Vargas, *Fundamentals of decision-making and priority theory with AHP* (RWS Publications, Pittsburgh, PA, 2000)
13. T.L. Saaty, *Decision making for leaders: The Analytic Hierarchy Process for decision in a complex world* (RWS Publications, Pittsburgh, PA, 2012)
14. L. Yang, J. Deuse, Procedia CIRP **3**, 149-154 (2012)
15. L. Yang, J. Deuse, P. Jiang, Int. J. Adv. Manuf. Tech. **66**, 795-807 (2013)
16. A. Hadi-Vencheh, A. Mohamadghasemi, J. of Manuf. Syst. **32**, 40-45 (2013)
17. H. Shokri, B. Ashjari, M. Saberi, J.H. Yoon, Ind. Eng. & Manage. Syst. **12**, 4, 389-405 (2013)
18. L. Bacudio, D.M. Sy, M.A. Promentilla, *Proceedings of the DLSU Research Congress*, **3** (2015)
19. R.D. Banker, A. Charns, W.W. Cooper, Manage. Sci. **30**, 1078-92 (1984)
20. E. Thanassoulis, *Introduction to the theory and application of Data Envelopment Analysis: A foundation text with integrated software* (Kluwer Academic Publishers, Boston, 2001)
21. A. Charnes, W.W. Cooper, E.L. Rhodes, Eur. J. Oper. Res. **2**, 429-444 (1978)
22. S.T. Liu, Comp. & Ind. Eng. **54**, 66-76 (2008)
23. T. Yang, C.A. Kuo, Eur. J. Oper. Res. **147**, 128-136 (2003)
24. T. Ertay, D. Ruan, U.R. Tuzkaya, Info. Sci. **176**, 3, 237-262 (2006)
25. R. Muther, *Systematic layout planning (SLP)* (Cahners Books, Boston, 1974)
26. T.L Saaty, *The Analytic Hierarchy Process* (McGraw-Hill, New York, 1980)

Patient waiting time and satisfaction in GP clinic at a tertiary hospital in Thailand

Apichart Boonma*, *Kanchana* Sethanan, *Sukangkana* Talangkun and *Teerawat* Laonapakul

Industrial Engineering Department, Faculty of Engineering, Khon Kaen University, Khon Kaen, Thailand

Abstract. This study aims to investigate patient waiting time in General Practice Clinic and identify the main cause of the problem through patient satisfaction survey. 9,232 Patients were randomly observed and quantitative data such as arrival time and waiting time were recorded. Other 1,200 patients were randomly selected for satisfaction survey which is considered as qualitative data. The result indicated that average waiting time of patients was 175.87 minutes with standard deviation of 68.66 minutes. Waiting time at GP clinic appeared to be the longest with an average of 92.9 minutes and standard deviation of 43.4 minutes. Also, based on the obtained data, patients remained positive until their waiting time exceeded 2 hours.

1 Introduction

Recently, there has been increasing attention to quality of care in hospitals and healthcare providers around the globe. Deteriorating quality in health services is a major concern in many countries as they are challenged by problems such as the approaching of an aging society, significant rise in healthcare costs, and etc. Patient satisfaction was reported to have an impact on health service quality [1]. Several studies have investigated the factors affecting patient dissatisfaction and addressed that patient waiting time was one of the major causes [2, 3, 4, 5].

Waiting time is the amount of time a patient spent on being seen or treated in a healthcare institution. It is one of the six key quality indexes in US healthcare system and widely used to evaluate performance and efficiency of health services [6, 7]. Long waiting time does not only slow down the medical treatment, it also increases stress, dissatisfaction, healthcare cost, and may even cause fatality to patients.

Outpatient department (OPD) is where patients are examined and given medications without staying overnight. Among all OPDs, General Practice Clinic (GP) is one of the most crowed departments. It is also the foremost clinic that most patients need to visit. GP doctors typically help patients with different conditions such as acute medical and surgical problems, long-term health problems (e.g. diabetes and asthma), ENT (ear, nose and throat), and dermatology (skin). The OPD patients therefore often experience long waiting time, long length of stay, and dissatisfaction.

The objective of this study is to explore patient waiting time and its cause based on data extracted from patient satisfaction survey at GP clinic of a tertiary hospital in Thailand.

2 Materials and methods

The case study hospital is a tertiary hospital and a teaching school for medical students in the Northeast of Thailand. At the time of study, the hospital had about 1,466 beds available for inpatients. Hospital record indicated that average annual outpatient and inpatient visits increased from 718,527 in 2010 to 828,931 in 2013 (~15% increment). With this large number of patient visits, waiting time in the hospital has attracted more attention recently.

This study was conducted during Jan – March 2014 at a GP clinic of the case study hospital located in the Northeastern of Thailand. Patients were randomly observed without prior notification. All hospital staffs not involved in the study were not informed about the existence of the study. Observers were placed at main service stations to record important statistics. Two different types of dataset were gathered including quantitative and qualitative data.

The quantitative data consist of arrival, waiting, and leaving time of patients who visited main service stations (triage, medical record, financial support, x-ray, lab test, medical examination, cashier, and pharmacy). Flow chart demonstrating outpatient flow in the hospital is shown in Fig 1. Prior to data collection, sample size was calculated using equation (1).

$$S = [(1/e^2)z^2p(1-p)]/[1+(z^2p(1-p)/(e^2N))] \quad (1)$$

Where S is the sampling size, N is population size, e is margin of error, and z is z-score. In this study, N is the average number of patients who daily visit GP clinic which is about 120 patients per day. The prevalence of the factor under study (p) was set to 0.8. The margin of

* Corresponding author: apicbo@kku.ac.th

error and z-score were set to 0.05 and 1.96 respectively. The z-score of 1.96 represents the confident level of 95%. The result suggested that the sample size should be 80.64 or 81 patients.

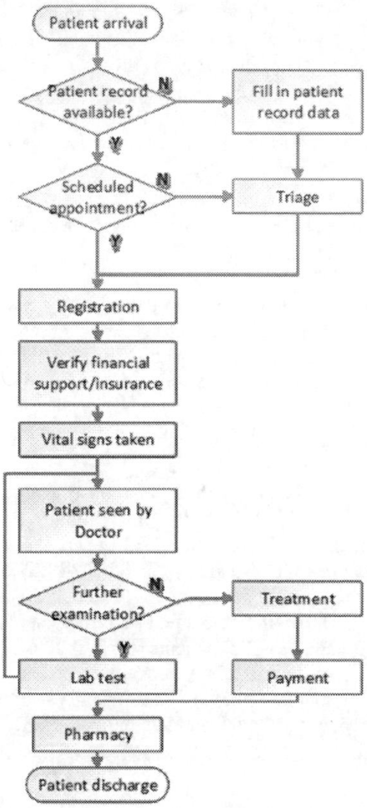

Fig 1. Flow chart illustrating outpatient activities after entering the hospital.

The qualitative data are developed from patient satisfaction survey containing a set of questions. Patients were asked to give their opinion on a scale of 1 to 10. The survey was conducted at the following service stations; triage, medical record, financial support, x-ray, lab test, medical examination, cashier, and pharmacy. All collected data were analyzed using Analysis ToolPak add-in for Microsoft Excel and Rockwell's Arena Input Analyzer.

3 Results

3.1 Patient arrival

Based on the observation (Monday – Friday), total number of patients being observed for analyzing arriving pattern was 9,232 patients. Patient arrival times at each station were recorded and then analyzed using Rockwell Arena Input Analyzer with an attempt to identify patient arrival distribution. Based on mean square error (MSE), best fitted distribution can be illustrated as shown in Table 1. The data indicated that patient inter-arrival time was ranging between 0.5 – 3 minutes.

Table 1. Best fitted statistical distribution of patient arrival

Section	Expression	MSE
Triage	-0.5+LOGN(2.76,2.73)	0.007080
Registration	-0.5+LOGN(2.04,2.15)	0.004021
Insurance	-0.5+LOGN(1.28,1.17)	0.003281
Laboratory	-0.5+LOGN(1.46,1.31)	0.007911
Xray	-0.001+EXPO(2.78)	0.000289
GP clinic	-0.001+WEIB(1.78,0.469)	0.002852
Cashier	-0.5+ERLA(0.709,2)	0.015499
Pharmacy	-0.5+LOGN(0.989,0.784)	0.003613

Moreover, the pattern of patient arrival shown in Fig 2 indicates the time each service station was busy. The busiest time for Triage, Medical Record/Insurance, Xray, GP, Cashier, and Pharmacy were 7.00-8.00, 8.00-9.00, 9.00-11.00, 9.00-12.00, and 10.00-12.00 respectively. After 10.00, Pharmacy and Cashier had largest number of patient visits. Since doctors often start working at OPD around 9.00, patients who have been diagnosed and given medication or prescription arrive at the stations around some time after that. Laboratory, on the contrary, was very busy early morning (6.00 – 10.00). This is due to the fact that patients are required to have laboratory tests (e.g. blood, urine sampling, etc) before they can be seen by doctors.

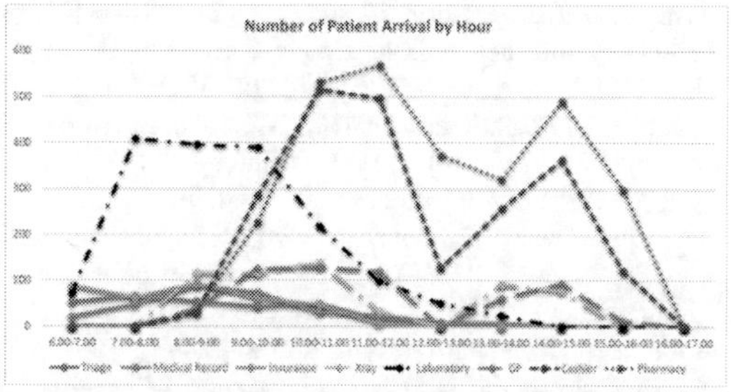

Fig 2. Patient arrival at each service station by hour

3.2 Waiting time

Waiting time or the time patients have to wait idly before they can get serviced is one of the most favorite key performance index to evaluate the performance of healthcare units. In our study, waiting time was determined as the time a patient checked in at a service station and waited until he/she received service or treatment.

When patients arrive at the hospital, they need to go to triage first for initial interview. There were 2-3 nurses stationed in triage. A total of 284 patients were observed at triage. The data showed that most patients (85.92%) received service as soon as they arrived. The remaining (14.08%), however, waited up to 13 minutes. The average waiting time was about 3.53 minutes with standard deviation of 2.95.

Next, patients need to complete registration process. There was an average of 3.5 staffs posed at this station and total 309 patients were monitored. One hundred and fifty-nine patients (51.46%) waited less than 5 minutes. Seventy-four patients (23.95%) waited between 5 and 10 minutes. Forty-nine patients (15.86%) waited for 10 to 20 minutes. Twenty-one (6.8%) and six patients (1.94%) waited for 20-30 minutes and more than 30 minutes respectively. The average waiting time of 6.7 minutes was obtained with standard deviation of 7.61.

Prior to being seen by physician, patients need to inform hospital staffs regarding their health insurance and payment method. There were 4-7 hospital staffs providing support at this station (5.5 staffs on average). Total 81 out of 358 observed patients (22.63%) waited between 5 and 10 minutes. Nine patients (2.51%) waited longer than 10 minutes. The average waiting time at this process was 3.65 minutes with standard deviation of 2.47.

In case blood sample or other lab tests are required, patients need to visit clinical laboratory and go through blood sampling procedure. An average of 4.04 staffs providing service at this station. Total of 2,661 patients were observed and their waiting time are listed as shown by the following 1) less than 5 minutes – 1,803 patients (67.76%), 2) 5-10 minutes – 294 patients (11.05%), 3) 10-20 minutes – 227 patients (8.53%), 4) 20-30 minutes – 122 patients (4.58%), 5) longer than 40 minutes – 71 patients (2.67%). The average waiting time was 7.56 minutes with standard deviation of 10.91.

At Xray, an average of 2.79 rooms were available with 2 staffs per room. Total of 1,212 patients were observed and their waiting time are listed as follows 1) less than 5 minutes – 375 patients (30.94%), 2) 5-10 minutes – 49 patients (4.04%), 3) 10-20 minutes – 85 patients (7.01%), 4) 20-30 minutes – 193 patients (15.92%), 5) 30-40 minutes – 216 patients (17.82%), 6) 40-50 minutes – 120 patients (9.90%), 7) 50-60 minutes – 123 patients (10.15%), and 8) more than 60 minutes – 51 patients (4.21%). The average waiting time was 25.63 minutes with standard deviation of 22.25.

Services at GP clinic were divided into 2 stages, 1) Patient Interview and 2) Medical Examination. Patients arrived at GP clinic were first interviewed by a nurse and then they waited for their turn to get examined by a doctor. For patient interview where nurses asked patients for their information, number of patients and their waiting time are listed as follows 1) less than 5 minutes – 68 patients, 2) 5-10 minutes – 20 patients (3.83%), 3) 10-20 minutes – 29 patients (5.56%), 4) 20-30 minutes – 42 patients (8.05%), 5) 30-40 minutes – 52 patients (9.96%), 6) 40-50 minutes – 40 patients (7.66%), 7) 50-60 minutes – 43 patients (8.24%), 8) 60-120 minutes – 153 patients (29.31%), and 9) longer than 120 minutes – 75 patients (14.37%). The average waiting time at patient interview was 63.70 minutes with standard deviation of 51.92.

For medical examination, there were 2-13 doctors providing service depending on the day of the week and the time of the day (5.74 doctors on average). Number of patients and their waiting time are listed as follows, 1) less than 5 minutes – 66 patients (11.40%), 2) 5-10 minutes – 69 patients (11.92%), 3) 10-20 minutes – 126 patients (21.76%), 4) 20-30 minutes – 92 patients (15.89%), 5) 30-40 minutes – 80 patients (13.82%), 6) 40-50 minutes – 50 patients (8.64%), 7) 50-60 minutes – 44 patients (7.60%), 8) 60-120 minutes – 47 patients (7.60%) and 9) more than 120 minutes – 5 patients (0.86%). The average waiting time at medical examination was 29.20 minutes with standard deviation of 24.27. Considering both patient interview and medical examination as single process, the average waiting time became 92.9 minutes with standard deviation of 43.40.

After consulting with a doctor, patients make payment at Cashier. There were 2-5 staffs available at this station (3.37 staffs on average). 1981 out of 2039 patients (97.16%) received service within 5 minutes. Fifty-two patients (2.55%) had to wait between 5 to 10 minutes before they could get serviced. Only 6 patients (0.29%) received service after waiting longer than 10 minutes. The average waiting time was about 1 minute with standard deviation of 1.54.

Lastly, patients can pick up medication at Pharmacy. There were 2-7 staffs posed at this station (average of 4.58). Number of patients and their waiting times are listed as follows 1) less than 5 minutes - 42 patients (1.26%), 2) 5-10 minutes - 212 patients (6.34%), 3) 10-20 minutes – 864 patients (25.85%), 4) 20-30 minutes – 719 patients (21.51%), 5) 30-40 minutes – 533 patients (21.51%), 6) 40-50 minutes – 347 patients (10.38%), 7) 50-60 minutes – 229 patients (6.85%), 8) 60-120 minutes – 341 patients (10.20%), and 9) longer than 120 minutes – 56 patients (1.68%). The average waiting time at Pharmacy was 34.90 minutes with standard deviation of 27.21.

3.3 Patient satisfaction

Patient satisfaction is considered qualitative data in our study. Total number of patients who participated in the survey was 1,200 patients. The questionnaires were evenly distributed to the patients receiving services at four different departments including Medical Record, GP clinic, Cashier, and Pharmacy. There were 10 questions in the questionnaire, but, there was only a single question directly relating to patient waiting time. The results of satisfaction survey can be shown by follows.

At Registration, patients participating in the survey consisted of 122 males and 178 females. Patients rated how fast they received service as 6.03 on a scale of 10. Most patients thought waiting time was the result of a large number of patients and insufficient staffs.

At GP clinic, total 140 males and 160 females were interviewed. Their response to service speed was 4.73 on a scale of 10. They also mentioned that in their opinion the delay was caused by a great number of early registered patients.

Three hundred patients consisting of 141 males and 159 females at Cashier agreed to take part in the survey. Their response to the service at this station was 5.57 out of 10. One of the main cause was insufficient staffs and service desks.

Lastly, at Pharmacy, total 110 males and 190 females gave a score of 6.30 out of 10 for the services they received. The main cause of delay was reported similar to that found at GP clinic.

4 Discussion

Based on the data shown in Fig 3, the longest waiting time was observed at GP clinic (92.9 minutes). Pharmacy comes second (34.9 minutes), and the third and fourth are medical exam (29.2 minutes) and Xray (25.63 minutes) respectively. Early patient arrival and limited number of hospital staffs are probably the primary causes of the problem. The similar finding was found in another secondary hospital in Thailand where the bottleneck processes are patient interview and medical exam.

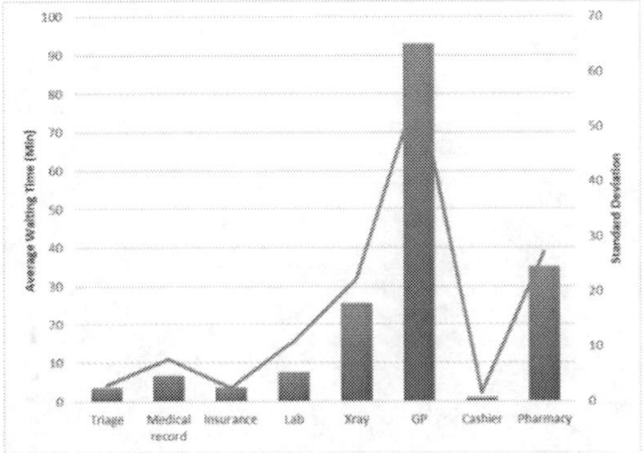

Fig 3. Average waiting time at each service station

The data also indicated total waiting time that patients required in addition to processing time to complete all services at the case study hospital was approximately 175.87 minutes with standard deviation of 68.66 on the average. This amount of time falls within the range of time that patients in most developing countries spent in outpatient department (2-4 hours). It is also in close agreement to total waiting time observed in Michigan (USA), Nigeria, and Trinidad and Tobago where 188, 173, and 160 minutes were recorded [8, 9, 10]. Shorter waiting time was reported elsewhere such as Atlanta (USA – 60 minutes), and Toronto (Canada 126.79 minutes) [1]. These data are shown in Fig 4.

Considering the satisfaction survey, the minimum satisfaction score of 4.73 was discovered at GP clinic. It is the only station that received a score less than 5 and thus its satisfaction level is considered as "below average". This is probably due to the fact that about 50% of patients had to wait at least 2 hours before they could be seen by a doctor. This is four times longer than the recommended waiting time of the Institute of Medicine [11].

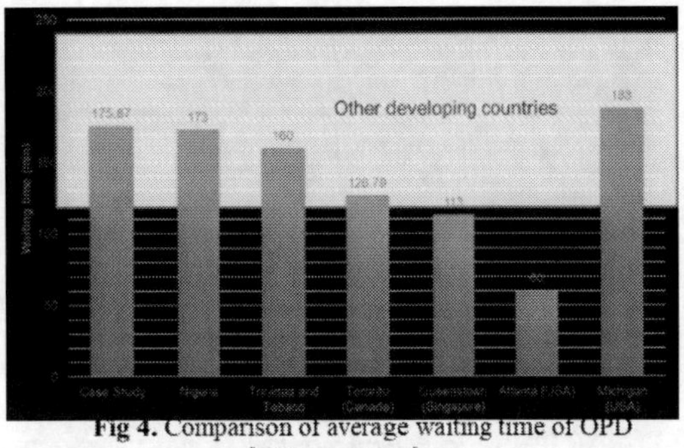

Fig 4. Comparison of average waiting time of OPD between countries

Also, from the questionnaires, the primary causes of long waiting time perceived by patients include large number of patients receiving service, insufficient staffs, and early arrival of patients. As Thailand is changing into aging society, an increase in patients needing health service has become more common. Moreover, since the case study hospital is a famous teaching hospital in the region, many patients requiring advance diagnosis and treatment are transferred here. In Thailand, although the ratio of doctor and nurse to patient of 2.28 per 1000 population is higher than that recommended by World Health Organization of 1 per 1000 population [12], allocating manpower to cope with incoming patients is still a challenge. Many patients still believe that early bird registration would minimize waiting time and they therefore come to the hospital extremely early. This does not only prolong the average waiting time of every patient, but it also applies pressure to hospital staff and eventually decrease their performance and efficiency. To make the problem worse, most patients in the case study hospital are walk-in patients who come to the hospital without making prior arrangements. This is different from hospital in developed countries where patients are required to make appointment before their arrival.

5 Conclusion

Our findings indicated that patient waiting time of 175.87 minutes or 2.93 hours at the case study hospital is in good agreement with other hospital in other developing countries. Also, patient dissatisfaction was discovered once patients need to wait longer than 2 hours. To prepare for increasing number of patients due to aging society, there is a need for GP clinic to develop necessary strategies in order to reducing waiting time. Staff allocation should be planned according to patient incoming pattern, while patients should be familiar and comfortable with making appointment before arriving at the hospital. Moreover, efficient appointment system should be used in order to properly distributing patients across the hospital service hours. Further research should focus on measuring related factors contributing to waiting time in more details, develop, and evaluate possible solutions.

Acknowledgements

This research was partially funded by Faculty of Engineering, Khon Kaen University. The author would also like to thank the managements and staffs of the case study hospital for their support, cooperation and suggestion.

References

1. K. Syed, D. Parente, S. Johnson, J. Davies. Factors determining wait-time and patient satisfaction at post-operative orthopaedic follow-up. OJMP **2**(1), 47-53 (2013).
2. X. M. Huang. Patient attitude towards waiting in an outpatient clinic and its applications. Health Serv Manage Res **7**(1), 2-8 (1994).
3. L.P. Fenny, T. Smith, J. Dwyer. Client flow analysis: A practical management technique for outpatient clinic settings. Int J Qual Health Care **6**, 179-186 (1994).
4. K. C. Chung, J. B. Hamil, H. M. Kim, M. R. Walters, E. G. Wilkins. Predictors of patient satisfaction in an outpatient plastic surgery clinic. Ann Plast Surg **42**(1), 56-60 (1999).
5. J. Levesque, E. R. Bogoch, B. Cooney, B. Johnson, J. G. Wright. Improving patient satisfaction with time spent in an orthopedic outpatient clinic. Can J Surg **43**(6), 431-436 (2000)
6. R. J. Maxwell. Quality assessment in health. Br Med J (Clin Res Ed) **288**(6428), 1470-1472 (1984).
7. Committee on Quality of Health Care in America; Institute of Medicine, Crossing the Quality Chasm: A New Health System for the 21st Century (National Academies Press, 2001).
8. L. M. dos Santos, G. Stewart, N. M. Rosenberg. Pediatric emergency department walk-outs. Pediatr Emerg Care **10**(2), 76-78 (1994).
9. A. N. Ofilli, C. E. Ofowye. What patients want: A content analysis of key qualities that influence patient satisfaction. J Med Pract Manage **22**, 255-261 (2007).
10. A. N. Ofilli, C. E. Ofowye. Patient's assessment of efficiency of services at a teaching hospital in a developing country. Ann Afr Med **4**, 150-153 (2005).
11. M. S. O'Malley, S. W. Fletcher, R. H. Fletcher, J. A. Earp, J Ambul Care Manage **6**, 20-27 (1983).
12. M. O. Oche, H. Adamu. Determinants of patient waiting time in the General Outpatient Department of a tertiary health institution in North Western Nigeria. Ann Med Health Sci Res **3**(4), 588-592 (2013).

Bio-degradable material realization for antenna fabrication in wireless communications

Thanakorn Homsai[1], *Chuwong* Phongcharoenpanich[2], and *Wipoo* Sriseubsai[1*]

[1]Industrial Engineering Department, Faculty of Engineering, King Mongkut's Institute of Technology Ladkrabang, Bangkok, Thailand
[2]Telecommunication Engineering Department, Faculty of Engineering, King Mongkut's Institute of Technology Ladkrabang, Bangkok, Thailand

Abstract. The bio-degradable PBS material is presented to be applied for microwave device design (antenna). The PBS is forming by using compression molding. The PBS thickness is 1 mm with the flexible characteristic. In this paper, the antenna design for the operating frequency band of 2400-2500 MHz is design using the bio-degradable material with the dielectric constant of 3.2. The prototype was fabricated. The characteristics of the proposed device were presented. The effected of the curved antenna and substrate are described in this work.

1 Introduction

The assumption of the technology product has been rapidly grown because nowadays the technology become a part of human daily life. Eevery day many electronic and microwave devices are improved in both of the model and characteristics to support the modern and advance technology [1]. This creates the plenty of electronic waste (e-waste) which is included the e-waste from the end-of-life electronic equipment. Therefore, the e-wastes become a considerable problem and needs immediately reaction nowadays [2]-[3]. Many techniques are presented to decrease e-wastes such as recycling, using environmentally friendly electronics [4]

No one can be denied that one of an important technology is a wireless communication technology because it can work efficient without wire-connection having more flexibility, infrastructure cost saving, improve customer service, enhanced collaboration and others. To communication by wireless technology, three basically components namely transmitter, receiver and medium are need. Mostly common wireless communication uses radio wave; therefore, an antenna is important to convert transmitted electrical signal to electromagnetic wave (EM wave) propagated in directing direction [5]. At the receiver, the antenna coverts the EM wave to electrical signal.

As same as modern technology, the changing of trend of technology and model of wireless device, the changing of antenna and printed circuit board is relatively important. Normally, the antenna for modern wireless communication are located on a printed circuit board as a substrate of the antenna. The substrate properties are effect to size and characteristics of the antenna. For the antenna design, the antenna substrate can improve or decrease electrical and mechanical stability of a device. There are a lot of commercial substrate such as epoxy, FR-4, PET, PE and many others. However, they cannot decay or hardly degradable by nature. Hence, the bio-degradable substrate is one of the way to decrease e-waste. Using biological materials and products are a possibility for decreasing amounts and recycling of waste.

The bio-gradable material is a material that can be broken down by natural process and it can be returned to environment or reused [6]. The bio-polymer becomes an interesting choice to be the antenna substrate. In this paper, the polybutylene succinate (PBS) which is a bio-degradable material is provided for microwave device (antenna). The PBS has a highly flexible, heat resistant, easy to compression. PBS is used in a large scale of production such as eco-smart food packaging, agriculture, pharmaceutical, automotive part, electronic structure part and others [7].

The applied of PBS for antenna design is studied and designed to ensure that the presented bio-degradable based device is proper for wireless communication technology. The compression process and the characteristics of the proposed device such as impedance, reflected power, radiation pattern and gain are described in this work. An introduction provides the detail about the basic theory, importance, and the detail of this work. In section II, the compression process is expressed. The design of the device shows in section III.

* Corresponding author: wipoo.sr@kmitl.ac.th

The characteristics of the bio-martial based device are presented in section IV. Final section provides the conclusion of this work.

2 PBS compression process

For preparation process of PBS forming, the devices consist of 1. Polybutylene succinate (FZ91PM) 2. LAB-companion oven 3. Weighing scales 4 . Mold with the size of 18 cm x 18 cm with the thickness of 1 mm. To determine the volume for PBS forming, the volume can be calculated from $D = m/v$. Where, m = Mass (Gram), D = density of polybutylene succinate (Gram/ cubic centimetre) and v = volume (cubic centimetres). The density of polybutylene succinate is 1.26 ram/ Cubic centimetre. The forming process is described in figure 1.

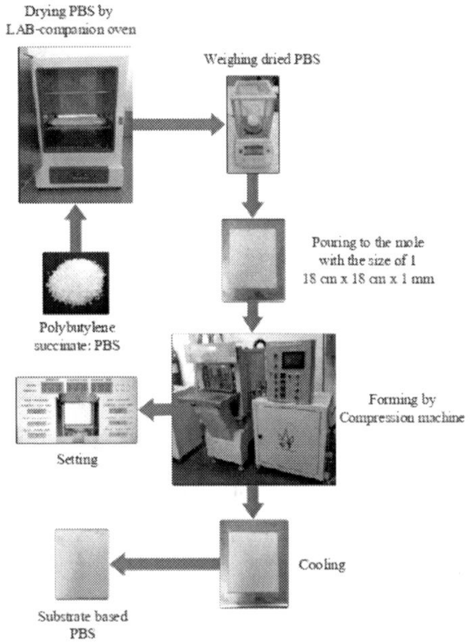

Fig. 1. Forming process of PBS substrate

One of the important process is drying process. The PBS (FZ91PM) is dried by heating at 80 degrees with 8 hours to decrease humidity in the material. The humidity of material causes air bubble making non-homogeneous in a texture of material. The non-homogeneous can decrease antenna efficiency. The dried PBS is weighed to find the mass for a compression process. The dielectric constant is measured by 85070E Dielectric Probe Kit together with Network Analyzer. The measured dielectric of fabricated PBS is 3.2.

3 Apply of material on antenna design

The PBS is applied to be an antenna substrate with the thickness of 1 mm. The radiated patch used a stick copper patch. The designed antenna is based on a printed circular monopole fed by coplanar waveguide. A size of the circular monopole is related with the wavelength of the operating frequency. The wavelength (λ) is the ratio of frequency (f) divided by speed of light (C). The initial printed monopole size can be calculated via $L=(L_d+2r)$. Where L_d = length of monopole, r = radius of monopole and L = length of feeding monopole. The size of an initial parameter (structure A) tabulated in table 1 and the parameter of developed structure (structure B) is tabulated in table II. Table III shows the parameter of the proposed antenna. From the initial stature, the parameter is adjusted to achieve the desired characteristics. The improvement of reflection power from each structure is provided in figure 2. The structure and the photograph of the proposed antenna are shown in figure 3.

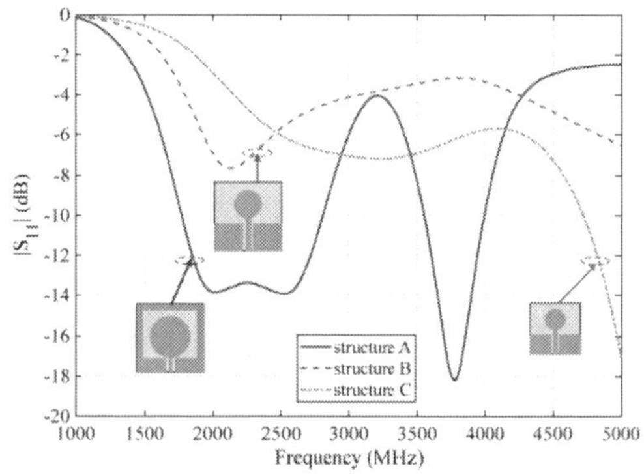

Fig. 2. The developed structure of the PBS-based antenna

Table 1. Initial antenna parameters

Parameters	Physical size (mm)	Parameters	Physical size (mm)
L	37	W_f	3.23
W	37	S	1.00
H	1	L_d	3.00
L_g	15	r	8.00

Table 2. Developed antenna parameters

Parameters	Physical size (mm)	Parameters	Physical size (mm)
L	50	W_f	3.23
W	50	S	1.00
H	1	L_d	0.58
L_g	11	r	15.5

The initial parameters are calculated based on printed circular monopole fed by coplanar waveguide. Owing to integrating of antenna with a SMA connector, the characteristic of the antenna changes as shown in figure 2. It is obvious that the antenna resonates at the relatively high frequency. Then, the parameters are developed by changing the antenna size. From improving of the antenna size, the trend of the operating frequency is shifted to the lower frequency. However, the reflected power is quite higher than that of desired reflection power (less than -10 dB). Therefore, to meet the requirement, the antenna structure is revised. The proposed antenna that meet the requirement is presented in figure 4.

Table 3. Antenna parameters

Parameters	Physical size (mm)	Parameters	Physical size (mm)
L	50	W_f	2.00
W	50	S	1.00
H	1	L_d	0.51
L_g	11	r	15.5
W_m	3.5	L_m	4.00

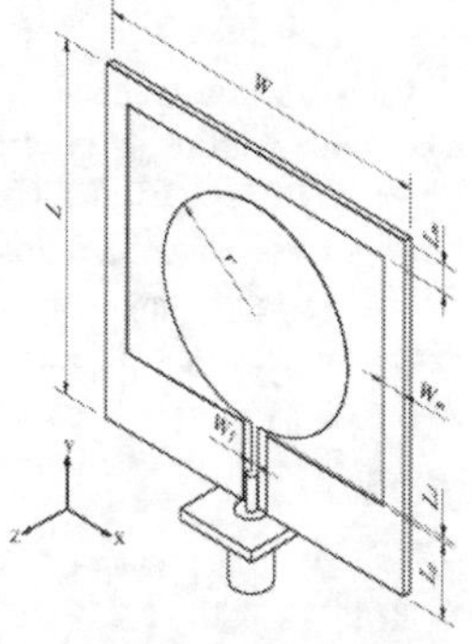

(a) Structure

(b) Photograph

Fig. 3. The PBS-based antenna

4 Characteristic of the Proposed Device

To achieve a maximum power transfers, the antenna and transmission line should be matched impedance. A normal antenna always designs to match with 50 ohms of transmission line system along the proposed operating frequency. Hence, the antenna should have the impedance close to 50 ohms. In figure 4, it is illustrated that the antenna impedance is average nearly 50 ohms having around 49.2 ohms from the operating frequency between 2400 to 2500 MHz.

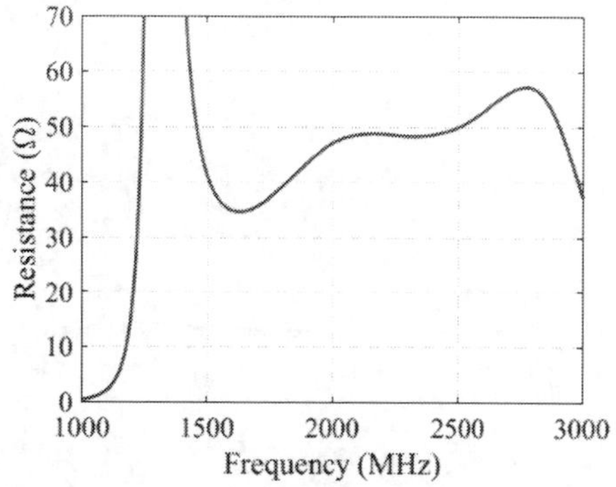

(a) Resistance

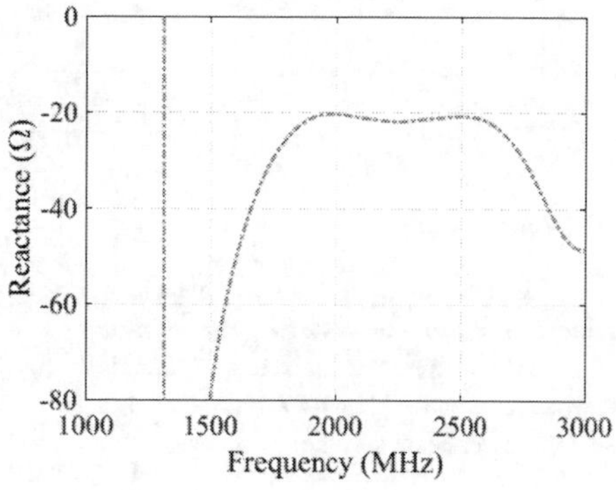

(b) Reactance

Fig. 4. Impedance of the proposed antenna

The impedance bandwidth is defined as the range of the operating frequency that the reflection power is acceptable with the transmission line. The reflected power should have a less value normally should be less than -10 dB for the desired operating frequency resulting in the limitation of operating frequency of the designed antenna. In this work, impedance bandwidth is 1020 MHz (1780 to 2800 MHz) covering the proposed operating frequency (2400 to 2500 MHz).

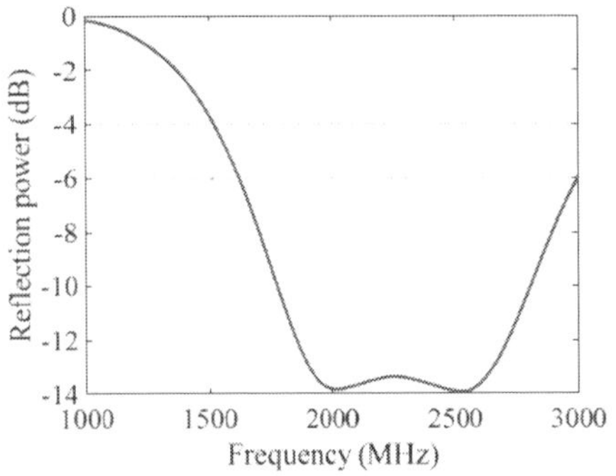

Fig. 5. Reflection power of the proposed antenna

The antenna gain is one of an important parameter that effect to the communication range. Gain of the antenna is normally describes in decibel (dB). The proposed antenna gain is 3.12 dBi as presented in figure 6.

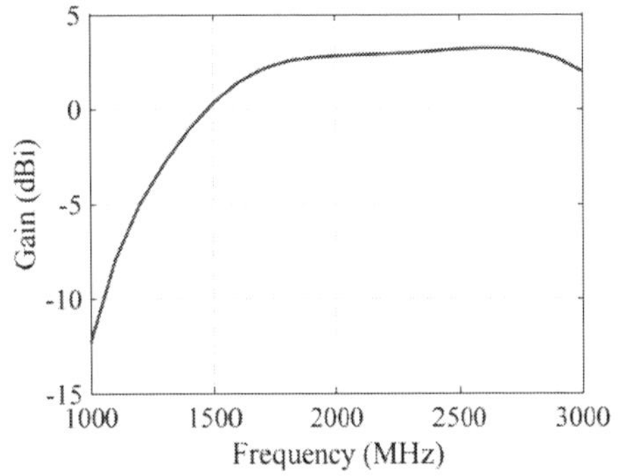

Fig. 6. Antenna gain

The radiation pattern of the antenna presents in figure 7. The radiation pattern is shown in both xz- and yz-planes. The proposed antenna is ommi-directional like radiation pattern having a symmetrical pattern with the half-power beam width of 77.8 degree.

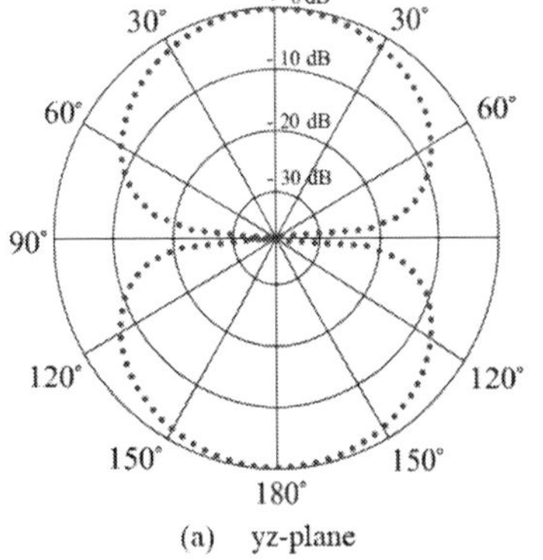

(a) yz-plane

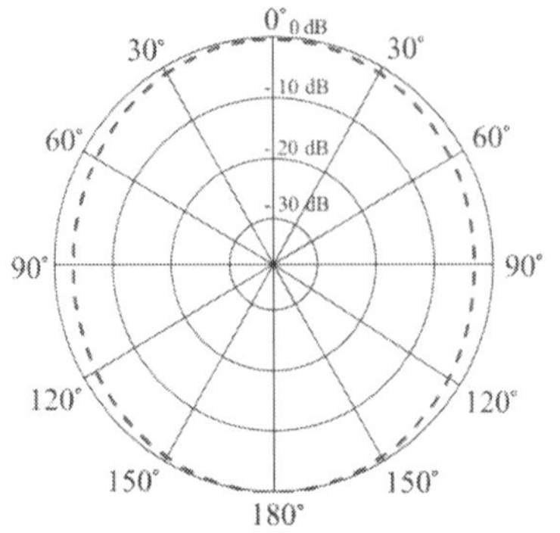

(b) xz-plane

Fig. 7. 2D-Radiation pattern

5. Conclusion

The bio-degradable material namely PBS is used to be an antenna substrate. The proposed operating frequency band is from 2400 MHz to 2500 MHz. The forming process for PBS substrate is a compression process. The substrate has the thickness of 1 mm with the dielectric constant of 3.2. The antenna size is 50 mm x50 mm.

The impedance of the antenna is around 49.2 ohms. The impedance bandwidth covers the frequency from 1780 to 2800 which is covered the desired operating frequency. The radiation pattern of the antenna is omnidirectional like with the gain of 3.12 dBi. From the results, the PBS-based antenna can be a good candidate for wireless communication system and the PBS is efficient enough to be a material for microwave device in the part of the antenna design.

References

1. G. Kiemel, IEEE Electromagn. Compat **1**, 4, (2012)
2. H-Y.Kang, J.M.Schoenung, Resour Conserv Recycl **45**, 4 (2005)
3. L. M. Hilty, Environ Impact Asses **25**, 5 (2005)
4. F. F. Rochman, W. S. Ashton, M. G.M. Wiharjo, Dev **24**, 1-8(2017)
5. C. A. Balanis, *Antenna Theory – Analysis and Design* (John Wileys & Sons, Inc., 2005)
6. P. Ducheyne, *Comprehensive Biomaterials II* (Elsevier, Oxford, 2017)
7. M.I. Vladu, Chem Soc Rev **43**, 2, ·2014

Effect of cooling rates of production process of Al-3%B-3%Sr master alloy on grain refinement and eutectic modification efficiency in cast A356 alloy

Krittee eidhed[1,*] and *Phisith* muangnoy[1]

[1] Department of Materials and Production Technology Engineering, Faculty of Engineering, King Mongkut's University of Technology North Bangkok, Bangkok, Thailand

Abstract. In this paper, size and morphology of the grain refiner and modifier particles in the Al-3%B-3%Sr master alloy production by using different cooling rates were investigated. Two Al-3%B-3%Sr master alloys were produced with 0.2 and 10°C/s, respectively. The grain refinement and eutectic modification efficiency of the Al-3%B-3%Sr master alloy were tested in casting process of A356 alloy by addition of 4wt.% and holding times for 10-120 min. The experimental result showed that microstructure of the M1 alloy (Slow cooling) consisted of larger solidified particles of AlB_2, SrB_6 and Al_4Sr in the matrix of α-Al compared to the M2 alloy (Rapid cooling). The addition of the M1 alloy in cast A356 alloy, it was found that small grain size and fully modify eutectic silicon were obtained from the holding time in a range of 10-60 min. While the addition of M2 alloy, a small grain size was achieved in shorter holding time in a range of 10-30 min but the eutectic silicon was partly modify. From the thermal analyzed result, solidification of un-modified A356 alloy was changed after addition of Al-3%B-3%Sr master alloy. It was clearly observed that both the undercooling of nucleation and eutectic reaction was reduced and the solidification time was shifted to longer.

1 Introduction

Al-Si-Mg/Al-Si-Cu based alloys are widely used for cast complex part because of an excellent castability, resistance to hot tearing in casting, enhance mechanical properties by using the precipitation strengthening by controlling the precipitated particle of Mg_2Si or Al_2Cu [1]. There are two melt treatment processes in casting of hypoeutectic Al-Si based alloy. One is the grain refinement process by the addition of grain refiner such as Al-Ti, Al-Ti-C, Al-Ti-B, Al-B master alloys [2-4]. A fine α-Al grain is achieved by producing of heterogeneous nucleation site in the melt during it solidified in the mold [5-6]. Second, the eutectic modification process by addition of modifier such as Al-Sr alloy [7]. The eutectic modification process can be improvement mechanical properties, the die filling, controlling the porosity distribution [7-8] and reducing the solution heat treatment in the age hardening process [9]. Moreover, other elements such as Cr can refine the eutectic silicon however the sludge phase can easily form with high Cr content and slow cooling rate [10]. Upon addition with Al-Sr alloy, it was dissolve into the melt and form the $SrAl_2Si_2$ compound [11] which increases the impurity twining for changed eutectic Si from acicular into a fibrous morphology.

The combination between the grain refiner and modifier of master alloys can enhance quality and mechanical properties of casting part [12]. Recently, there are many grain refiner and modifier were developed such Al-Ti-B-Sr and Al-B-Sr master alloys by producing various types of particles such as TiB_2, $TiAl_3$, AlB_2 and SrB_6 as the heterogeneous nucleation site and modification in the melt treatment process, simultaneously. The Al_4Sr particle is found in the Al-Sr or Al-B-Sr master alloys was dissolved when it was added into the melt during casting [13]. Then, Sr will affect to the eutectic reaction temperature (ERT) and modify microstructure of eutectic phase will be obtained [14].

Form many research papers, the size and morphology of various phases in the master alloy production depend on the melting temperature, cooling rate, chemical compositions and casting method. Ding reported that the blocky $TiAl_3$ phase was fully refined grain size compared to combined the needle-like $TiAl_3$ phase [15]. Liao reported that the size and morphology of Al_4Sr phase from different processes affects to the modification efficiency [13]. Wang found that the morphologies of $TiAl_3$, TiB_2, AlB_2 and AlB_{12} phases form different synthesis temperature of the Al-3%Ti-3%B master alloy were effect to reduce the α-Al phase in Al-Si alloys [16]. Cui, reported SrB_6 phase of Al-3B-5Sr master alloy can refine the α-Al of A356 alloy but have not investigated the size and morphology on grain refinement and eutectic modification efficiencies [17].

This paper studied the effect of cooling rates on size and morphology of AlB_2 and SrB_6 particles in production of Al-3%B-3%Sr master alloy. The comparison of grain refinement and eutectic modification efficiency in cast A356 alloy were discussed.

2 Experimental procedures

In order to develop and produce the Al-3%B-3%Sr master alloy, the Al-6%B alloy was melted by using a low frequency induction furnace at 900°C. Then, the

* Corresponding author: weh@kmutnb.ac.th

melt was added of the Al-10%Sr master alloy and held for 30min. After that the melt was poured into a molds. There were two different cooling rates of casting for 0.2 and 10°C/s, respectively. The difference wall thickness of the molds between a stainless-steel cup and cylinder steel molds were used to control the cooling rate. The solidification curve of casting was detected by chromel-alumel thermocouple with 0.5°C/s in a stainless-steel cup by using the thermal analysis technique. The specimens were polished and etched by the Keller's reagent. The microstructure was analyzed by optical microscope. The grain refiner and modifier phase were identified by the scanning electron microscopy with energy dispersive spectroscopy (SEM-EDS) technique. The phase fraction of each phase was analyzed by the image analyzer as shown in table 1.

Table 1. Cooling and phase fraction of master alloys

Master alloys	Cooling rate	Phase fraction (%)		
		AlB_2	SrB_6	Al_4Sr
Al-3%B-3%Sr (M1 alloy)	0.2°C/s (Slow cooling)	0.34	3.88	0.38
Al-3%B-3%Sr (M2 alloy)	10°C/s (Rapid cooling)	0.32	3.43	0.27

In the grain refinement and eutectic silicon modification efficiency tests, the Al-7%Si-0.3%Mg alloy was melted at 900°C by using the low frequency induction furnace. The Al-3%B-3%Sr master alloy was added into the melt with 4wt.% and holding for 10-120 min. The covering flux was used to prevent the hydrogen gas diffuse into the melt during casting. The Ar gas was used to remove hydrogen gas and various inclusions with flow rate 10L/min. The melt was poured at 720°C into a stainless-steel cup with control cooling rate for 0.2°C/s. The specimen preparations of macrostructure and microstructure were grinded, polished and etched by the Keller's and Tucker's reagents, respectively. The grain size was measured by the linear intercept method according to ASTM E112.

3 Result and discussion

3.1 Microstructure of Al-3%B-3%Sr master alloy

Microstructures of Al-3%B-3%Sr master alloy with resulted from different cooling rates of the production were shown in Fig.1 and Fig.2.

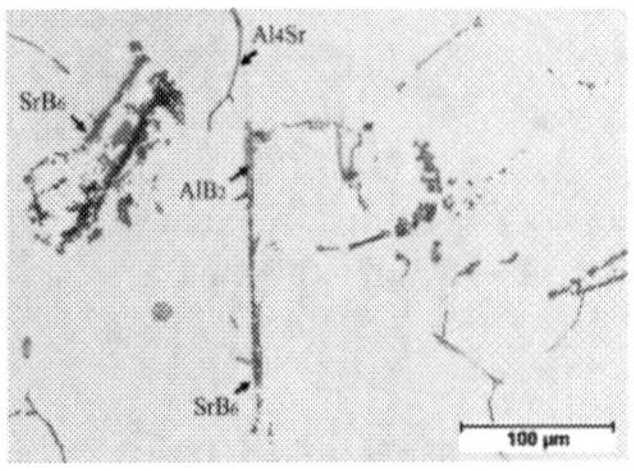

Fig. 1. Microstructure of Al-3%B-3%Sr master alloy with cooling rate of 0.2°C/s

Each phase in the microstructures were identified by the SEM-EDS in order to clarify the composition. The microstructure of the M1 alloy (Slow cooling) consisted of solidified phase of AlB_2, SrB_6 and Al_4Sr in the matrix of α-Al as shown in Fig. 1 It can be seen that the SrB_6 (gray-black) was blocky morphology. The AlB_2 was surrounded by the SrB_6 particle while Al_4Sr was the plate-like morphology.

Microstructure of the M2 alloy (Rapid cooling) consisted of the smaller particles of AlB_2, blocky SrB_6 and Al_4Sr. It can be seen that the SrB_6 was found in cluster, while AlB_2 was uniform distribution as shown in Fig. 2. From phase fraction analyzed as shown in Table.1, it was found that the M1 alloy have higher amount of phase fraction than the M2 alloy.

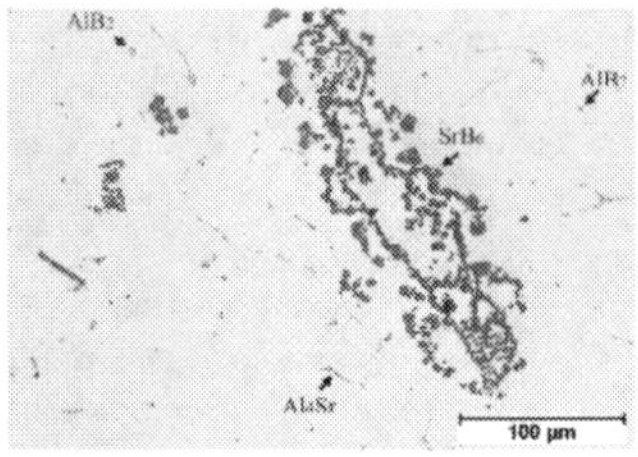

Fig. 2. Microstructure of Al-3%B-3%Sr master alloy with cooling rate of 10°C/s.

3.2 Microstructure of un-modified A356 alloy

Fig. 3 (a-b) show the macrostructure and microstructure of un-modified A356 alloy which consisted of coarse grain size (Fig.3 (a)) and acicular eutectic Si (Fig.3(b)).

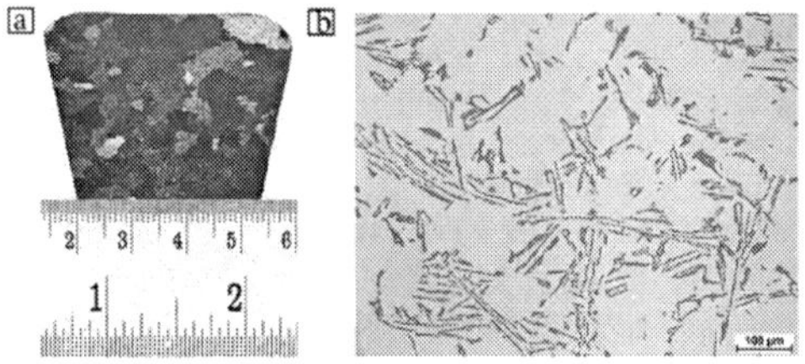

Fig. 3. Showed (a) macrostructure (b) microstructure of un-modified A356 alloy.

3.3 Solidification and microstructure of modified A356 with M1 alloy

Fig. 4 shows comparison of cooling curves of modified A356 by addition of M1 alloy with various holding times. The solidification behavior as observed from cooling curve, it was found smaller peaks of undercooling of α-Al in comparison to cooling curve of un-modified A356. This is because AlB_2 and SrB_6 particles in master alloy act as the heterogeneous nucleation site. When comparison effect of holding time in furnace, it found that holding time in a range of 10-60min specimens have smaller grain size compared to 120min specimen as shown in Fig.5.

The eutectic reaction temperature (ERT) of un-modified A356 is 578°C after cooling time in the mold for 110s. When addition of M1 alloy after holding in furnace for 10-30min, it was found that the ERT occurred at lower temperature of 572°C and increase cooling time to longer for 140s and 153s, respectively. When holding in furnace for prolong time for 120min, the ERT occurred at 576°C with cooling time for 120s. These solidification behaviors can be explained by the Sr in master alloy affect to ETR and solidification time as shown Fig. 6.

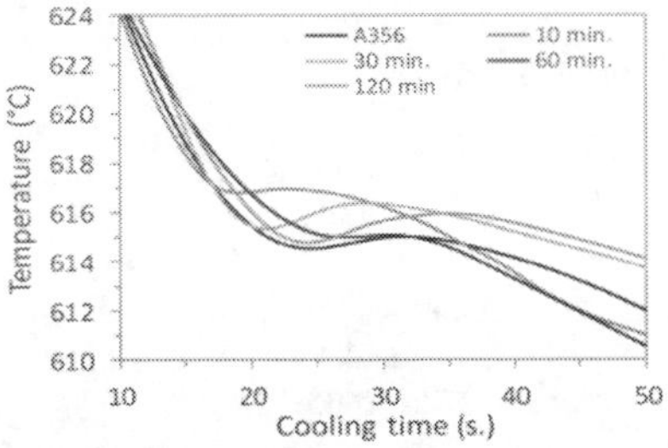

Fig. 4. Comparison of cooling curves at α-Al formation with various holding time for 10 to 120 min.

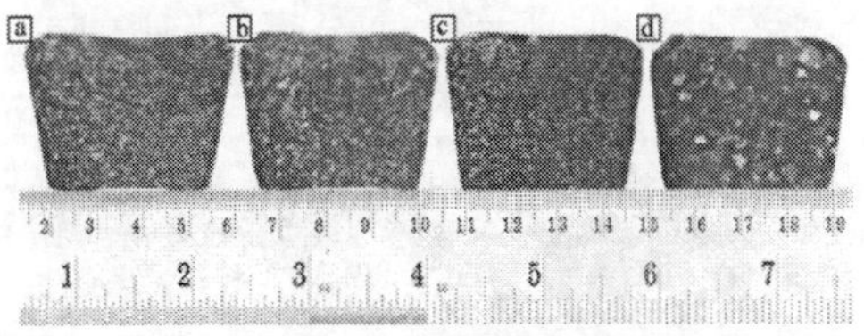

Fig. 5. Macrostructures of modified A356 by added M1 alloy (a) 10 (b) 30 (c) 60 (d) 120 min, respectively.

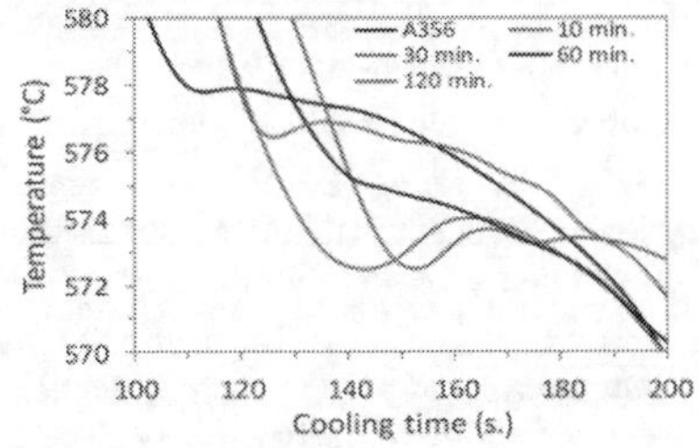

Fig. 6. Comparison of cooling curves at ERT with various holding time for 10-120 min.

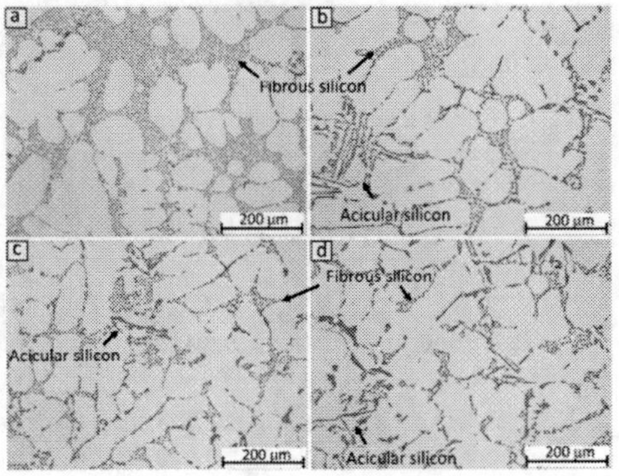

Fig. 7. Microstructure of modified A356 by addition of M1 alloy (a) 10 (b) 30 (c) 60 (d) 120 min, respectively.

Fig. 7 (a-d) show microstructure of various holding times. It is clearly seen that the 10min specimen (Fig. 7 (a)) has finer fibrous Si compared to other specimens (Fig 7 (b-e)). Thus, when prolong holding time the modification efficiency was reduced.

3.4 Solidification and microstructure of modified A356 with M2 alloy

Fig. 8 shows comparison of cooling curves of modified A356 by added M2 alloy with various holding times. The solidification behavior, it was found smaller peaks of undercooling when the melt was added M2 alloy compared to un-modified A356. When comparison effect of holding time, it was found that holding time in a range 10-30min specimens have smaller grain size compared to 60-120min specimen as shown in Fig. 9.

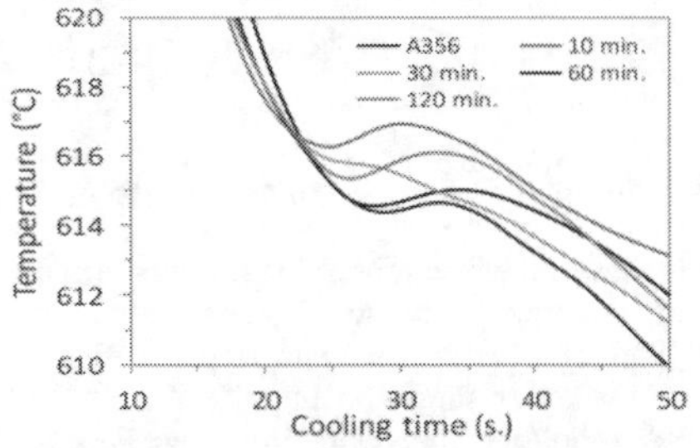

Fig. 8. Comparison of cooling curve at α-Al formation with various holding time for 10 to 120 min.

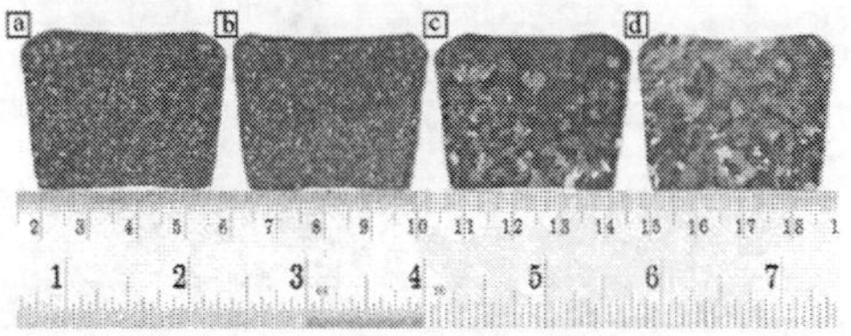

Fig. 9. Macrostructures of modified A356 by added M2 alloy (a) 10, (b) 30, (c) 60, (d) 120 min, respectively.

When addition of M2 alloy after holding in furnace for 10-30min, it was found that the ERT occurred at lower temperature of 573°C and longer cooling time for 140s and 145s, respectively. When holding in furnace for prolong time for 120min, the ERT occurred at 575.8°C and cooling time for 130s as shown Fig. 10.

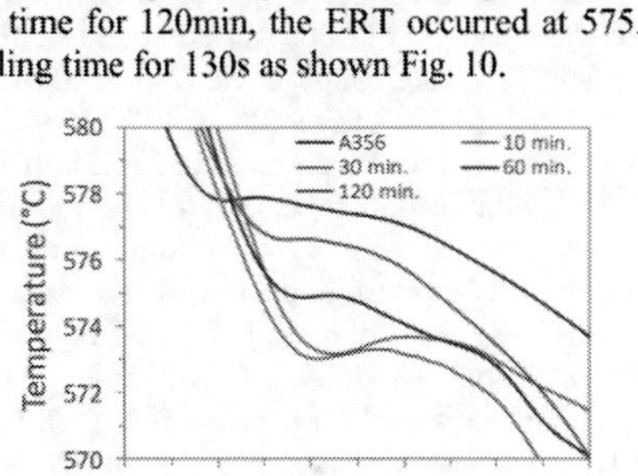

Fig. 10. Comparison of cooling curves at ERT with various holding time for 10-120 min.

Fig. 11 (a-d) show microstructures of various holding times. It is clearly seen that the 60min specimen has finer fibrous Si compared to 10-30min specimens (Fig. 11 (a-b)). When holding for prolong time the modification efficiency was reduced (Fig. 11 (d)).

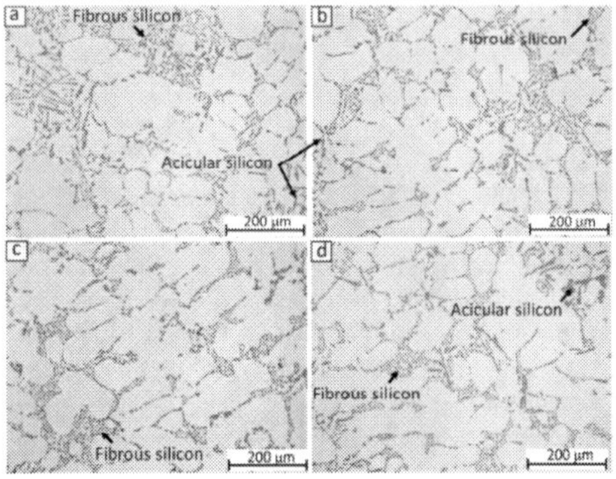

Fig. 11. Microstructure of alloy with M1 alloy (a) 10 (b) 30 (c) 60 (d) 120 min, respectively.

4.4 Effect of size and morphology of AlB_2 and SrB_6

Form these experimental results, it can be discussed that addition of the M1 alloy with higher amount of AlB_2 and SrB_6 particles has higher grain refining efficiency than the M2 alloy as shown in Fig. 12. The M2 alloy has shorter efficiency time with in 30min because a fine particles is easy dissolved into the melt when prolong holding time.

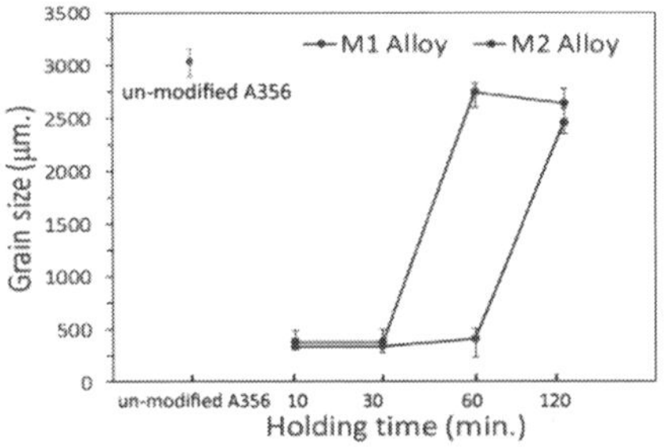

Fig. 12. Comparison of grain size with various holding time.

Moreover, the M1 alloy has higher modification efficiency than the M2 alloy in all period holding times, especially for 10min specimen (Fig. 7 (a-b)). This is because the ERT of the modified A356 is become lower 5-6°C compared to the un-modified A356. It also found that the solidification time (ST) of the modified A356 was expended to 110s to 140s, simultaneously. Thus, when the ERT reduced and ST shifted to longer (shift to right of Al-Si phase diagram) it should be achieved fine microstructure. Therefore, from these experimental results, the cooling rate of production of Al-3%B-3%Sr master alloys is very important parameter in order to achieve high efficiency of grain refinement and modification.

5 Conclusions

1. The cooling rate of master alloy production affected to the size and morphology. The slow cooling have larger particle than rapid cooling.
2. The larger particle of AlB_2 and SrB_6 was higher efficiency than smaller particles.
3. The plate-like Al_4Sr particle was better modifying the eutectic silicon compared to smaller particle.
4. The addition of M1 and M2 alloys were reduced of ERT and shifted ST to longer.

References

1. R. Lumley, *Fundamentals of Aluminium Metallurgy, Production, Processing and Applications* (Woodhead Publishing Limited, Great Abington, 2011)
2. P. Li, E.G. Kandalova, V.I. Nikitin, Materials Letters **59,** 723 (2005)
3. Z. Chen, T. Wang, L. Gao, H. Fu, T. Li, Materials Science and Engineering A **553**, 32 (2012)
4. C. Limmaneevichitr, W. Eidhed, Materials Science and Engineering A **355**, 174 (2003)
5. P. S. Mohanty, J. E. Gruzleski, Acta mater **44**, 3749 (1996)
6. V.P. Patel, H.R. Prajapati, International Journal of Engineering Research and Applications (IJERA) **2**, 147 (2012)
7. J. Gilbert Kaufman, Elwin L. Rooy, *Aluminum Alloy Castings: Properties, Processes, and Applications*, (ASM International, OH, 2004)
8. S. Hegde, K. Narayan Prabhu, J Mater Sci **43**, 3009 (2008)
9. Eidhed, W, Journal of Materials Science & Technology **24**, 29 (2008)
10. Eidhed, W, Journal of Materials Science & Technology, **24**, 45 (2008)
11. J. Barrirero , J. Li , M. Engstler , N. Ghafoor , P. Schumacher, M. Odén , F. Mücklich, Scripta Materialia **117**, 16 (2016)
12. G. L. Liu, N. C. Si, Applied Mechanics and Materials 312, 396 (2013)
13. C. Liao, J. Chen, Y. Li, H. Chen, C. Pan, Progress in Natural Science: Materials International **24**, 87 (2014)
14. G. Eisaabadi B., M. NouriR Beygi, M. Zarezadeh Mehrizi, A. Nouri, M. Ebrahimi, Shape Casting: 6th International Symposium (to be published)
15. W. Ding, T. Xia, W. Zhao, Materials **7**, 3663 (2014)
16. T. Wang, H. Fu, Z. Chen, J. Xu, J. Zhu, F. Cao, T. Li, Journal of Alloys and Compounds **511**, 45 (2012)
17. X.L. Cui, Y.Y. Wu, T Gao, X.F. Liu, Journal of Alloys and Compounds **615,** 906 (2014)

Machine components clustering with DSM and repeating method : case study of a soil mixing machine

Tanongsak Kongsin* and *Sakon* Klongboonjit

Industrial Engineering Department, Faculty of Engineering, King Mongkut's Institute of Technology Ladkrabang, Bangkok, Thailand

Abstract. In this study, components of the machine are analyzed to group all components into modular groups with a case study of a soil mixing machine. The study begins by creating a design structure matrix of all components. Next, the design structure matrix is transferred into a distance matrix of all components with Jaccard method. After that, the equation of complete linkage must be applied to change the distance matrix to a tree dendrogram for showing the relationship of machine components and dependent coefficient. With this tree dendrogram, six clusters are arranged:- the 1st cluster has 8 modules at the lowest dependent coefficient, the 2nd cluster has 7 modules, the 3rd cluster has 6 modules, the 4th cluster has 5 modules, the 5th cluster has 4 modules, and the 6th cluster has 2 modules at the highest dependent coefficient. Finally, the 1st cluster with 8 modules is considered to be the most proper cluster for this soil mixing machine by applying the repeating method to analyze all six clusters.

1. Introduction

Modular design has continuously been developed with a base on multi-resolution modeling to serve with a flexible manufacturing system (FMS). With FMS, machines in production processes must easily change machine functions in response to unstable marketing demand without buying new machines. To design changeable machines, Design Structure Matrix (DSM) should be applied to design a process of the machine to develop new generation machine which consists of many modules and this type of machine can change its function by changing its modules. Many research studies of the DSM show decomposition and integration analysis whether the system is a product or a process. [1] They include the designs assembly system layout for mass customization. [2] This approach could assist selecting model resolution more flexibility, based on needs or understanding degree for the complex system [3].

In this paper, an application of DSM with Jaccard Method and repeating method is introduced to be a new idea to design a modular machine [4] that can easily change its function, its components and maintenance.

2. Purpose

The main purpose of this study is to apply the clustering with DSM and repeating method for modular design of the soil mixing machine [10]. The main expected result is to group machine components into independent modules for easily developing new generation machine, disassembling its components, and setting a maintenance plan.

3. Study Methodology

In this study, Design Structure Matrix (DSM) and Repeating Method are applied for clustering machine components into independent modules [5]. To obtain the expected results, the study processes consist of 5 steps as shown in Fig. 1.

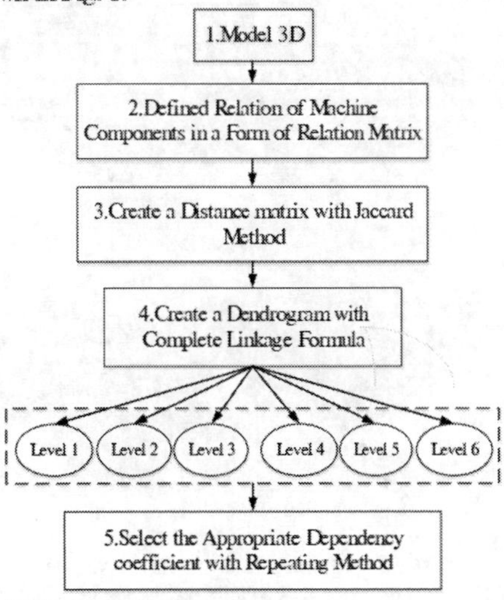

Fig. 1. The Clustering flow diagram.

* Corresponding author: tanongsak9797@gmail.com, sklonboonjit@hotmail.com

4. Results

4.1 Model 3D

The soil mixing machine is designed by CAD software to create model 3D of this machine as shown in Fig. 2.

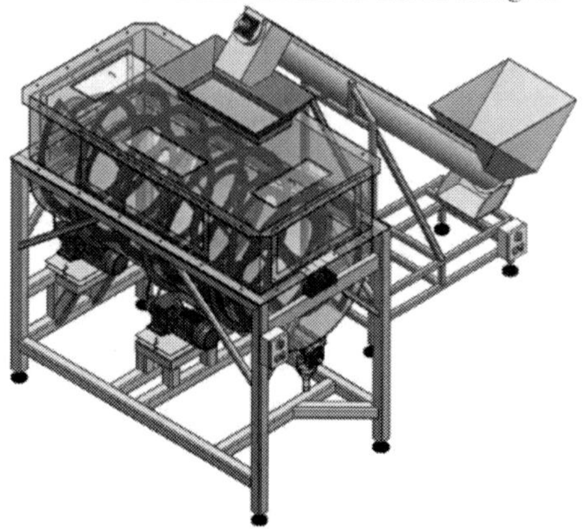

Fig. 2. 3D model for soil mixing machine

4.2 Defined Relation of Machine Components in a Form of Relation Matrix

To define the relation between each machine component and its adjacent components [6-7], CAD software is a main tool to check this relation with formula (Equation 1): Relation = 1 means that considered component surface is contacted with its adjacent components and Relation = 0 means that considered component surface is not contacted with its adjacent components [8-10].

$$a_{ij} = \begin{cases} 0 & (i=j \text{ or } a_j \nrightarrow a_i) \\ 1 & (a_j \rightarrow a_i) \end{cases} \quad (1)$$

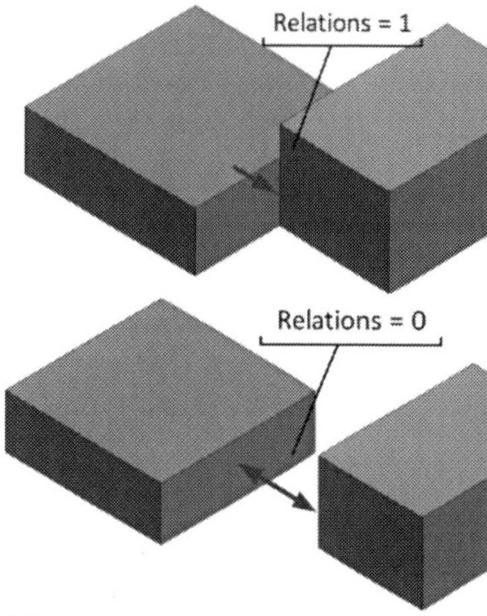

Fig. 3 Show how to define relation value.

With CAD software, this machine can be exploded into 75 components and relation of each compont and its adjacent components are defined and shown in Fig. 4 in form of a design matrix.

Fig. 4. Relation matrix of soil mixing machine components.

4.3 Create a Distance matrix with Jaccard Method

Jaccard index formula (Equation 2 and 3) is applied to a relation matrix [9] in Fig. 4. Finally, a relation matrix shall be transferred into a distance matrix as shown in Fig. 5.

$$\mathrm{Jsim}(A,B) = \frac{A \cap B}{A \cup B} = \frac{a}{a+b+c} \quad (2)$$

$$\mathrm{Jdist}(A,B) = 1 - \mathrm{Jsim}(A,B) \quad (3)$$

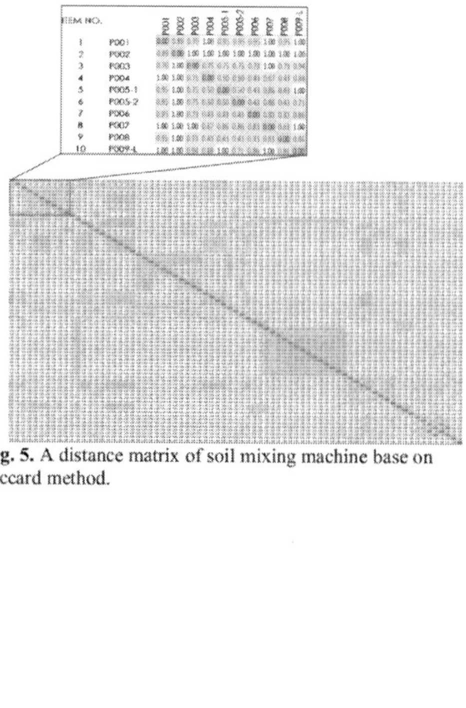

Fig. 5. A distance matrix of soil mixing machine base on Jaccard method.

https://doi.org/10.1051/matecconf/201819201037

4.4 Create a Dendrogram with Complete Linkage Formula

In this step, complete linkage formula (Equation 4) is applied to a distance matrix [8] in Fig. 5 to create a dendrogram of this machine as shown in Fig. 6.

$$D((C_i \cup C_j), C_k) = \max_{x \notin c_i, y \notin c_j} \{d(x_i, x_k), d(x_j, x_k)\} \quad (4)$$

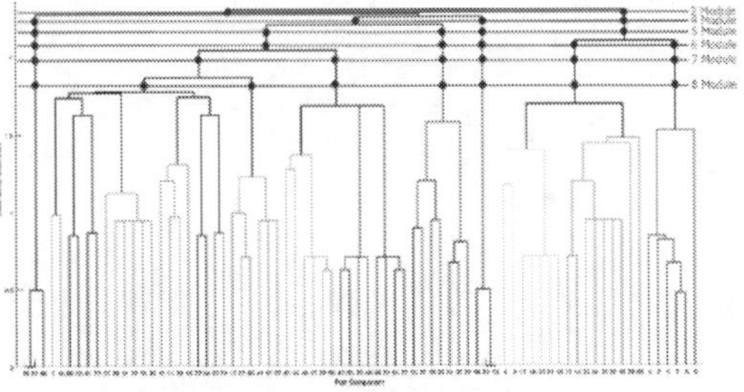

Fig. 6. A dendrogram of soil mixing machine.

Fig. 6 shows that the numbers of cluster are depend on the value of dependency coefficient. In this study, six dependency coefficient values from 2.4 to 1.7 are defined so that 1) There are 2 modules for the dependency coefficient = 2.4, 2) There are 4 modules for the dependency coefficient = 2.35, 3) There are 5 modules for the dependency coefficient = 2.3, 4) There are 6 modules for the dependency coefficient = 2.2, 5) There are 7 modules for the dependency coefficient = 1.9 and 6) There are 8 modules for the dependency coefficient = 1.7

4.5 Select the Appropriate Dependency coefficient with Repeating Method

Among six dependency coefficient values, the most appropriate value for this machine is selected with repeating method Fig. 7. And Fig. 8. shows that.

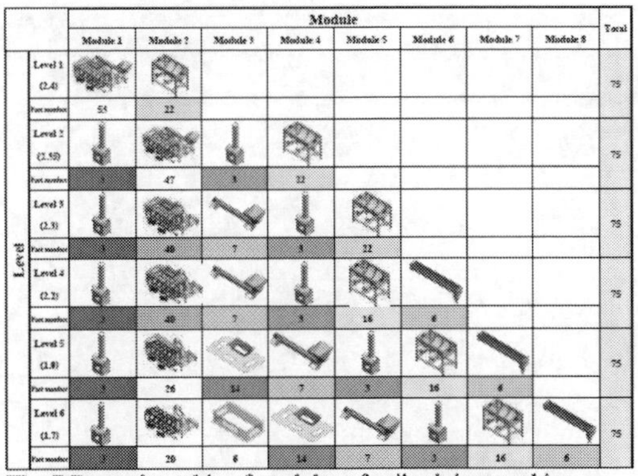

Fig. 7 Repeating table of modules of soil mixing machine.

Fig.7. Show the repeating of module soil mixing machine by searching a module that has been repeating most frequently in the repeating table.

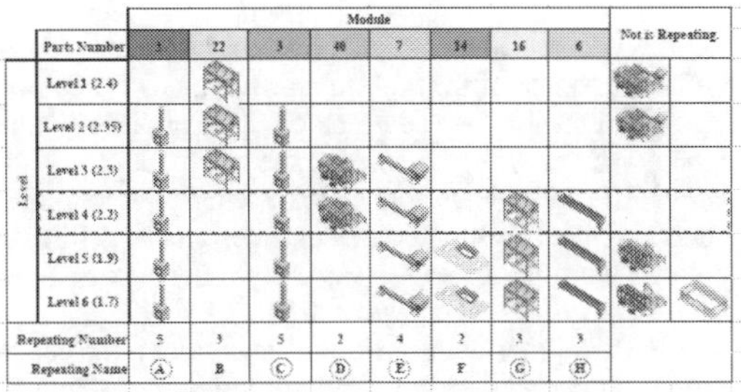

Fig. 8. Select of Repeating Module.

Fig. 8. Shows the selection with repeating of module from Repeating Table in Fig. 7., indicate clearly the module of soil mixing machine as shown in Fig. 9.

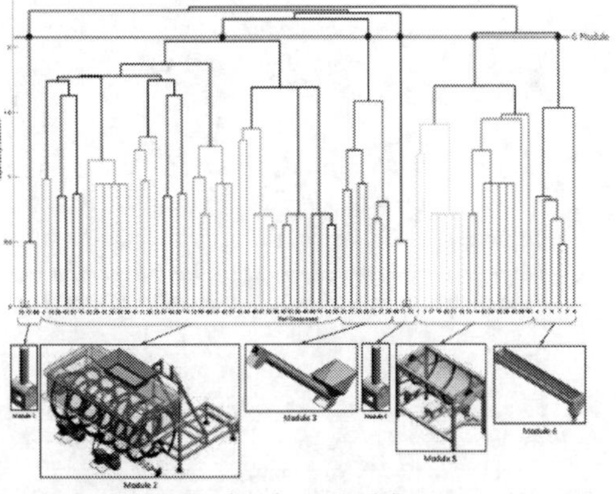

Fig. 9 The dendrogram has shown module.

From the case studies in this research can explain the clustered components of soil mixing machine, as in the Fig. 10.

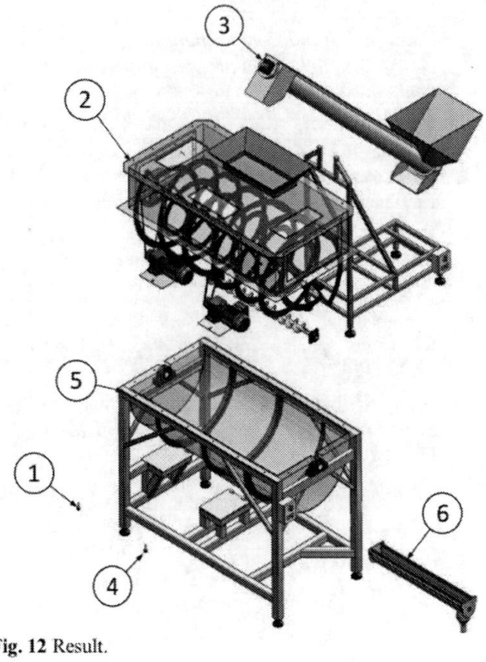

Fig. 12 Result.

5. Conclusion

In this paper, we have studied how to cluster components for the module of the soil mixing machine. By defining the relationships between each component and repeating method by creating a module of the machine.

The approach is applied to the clustering with DSM and repeating method of modular design of soil mixing machine are results so the value dependent coefficient = 2.2 and modules of machine = 6 Modules.

The cluster with 6 modules is considered to be the most proper cluster for this soil mixing machine by applying the repeating method to analyze six clusters.

References

1. Tyson R. Browing, *Applying the Design Structure Matrix to System Decomposition and Integration Problems: A Review and New Directions,* IEEE Transactions on Engineering Management, Vol 48, No.3, August 2001
2. Xuejian Gong, Titao Liu, Roger, J. Jiao, *Variety-Driven Assembly Layout Design by Design Structure Matrix Clustering Analysis*, CIRP Conference on Manufacturing, 63 (2017)
3. Tarek ALGeddawy, *A DSM Cladistics module for product family architecture design*, CIRP Design Conference, 21 (2014)
4. Vladimir Maslak, Dmitry Boytsov, *Modular design of subway facilities*, International scientific conference "Underground Urderanisation as a Perequisite for Sustainable Development", 165 (2016) 782-787
5. Simon Li, Mehrnaz Mirhosseini, *A matrix-based modularization apporting secure collaboration in parametric design*, Computers in Industry, 63 (2012)
6. Xiaoguang Deng, Greg Huet, Suo Tan, Clement Fortin, *Product decomposition using design structure matrix for intellectual property protection in supply chain outsourcing*, Computers in Industry, 63 (2012)
7. Robert Helmer, Ali Yassine and Christoph Meier, *Systematic module and interface definition using component design structure matrix*, Journal of Engineering Design, Vol. 21, No. 6, December 2010
8. Brian Everitt, Sabine Landau, Morven Leese, Daniel Stahl, *Cluster Analysis* (John Wiley & Sone, Ltd, 2011)
9. Guojun Gan, Chaoqun Ma, Jianhong Wu, *Data Clustering Thery, Algorithms and Applications*, (SIAM, 2007)
10. Steven D. Eppinger and Trson R. Browning, *Design Structure Matrix Methods and Application*, (MIT Press, 2012)

MATEC Web of Conferences **192**, 01038 (2018) https://doi.org/10.1051/matecconf/201819201038
ICEAST 2018

Integrating Spatial-Temporal Risk Factors for an Ambulance Allocation Strategy: A Case Study in Bangkok

Pakinai Chaicharoenwut[1], *Jarotwan* Koiwanit[1], *Pannapa* Changpetch[2], *Sirisuda* Buatongkue[3], *Chumpol* Yuangyai[1*]

[1]Industrial Engineering Department, Faculty of Engineering, King Mongkut's Institute of Technology Ladkrabang, Bangkok, Thailand
[2]Mathematical Science Department, Bentley University, Massachusetts, United States
[3]Information Technology Department, Faculty of Science and Arts, Burapha University, Chanthaburi, Thailand

Abstract. Dedicated emergency medical services (EMS) are important to patients' chances of survival. In particular , the quicker such services arrive at the scene of an incident, the higher the survival rate. Therefore, the management of ambulance bases is an essential aspect of emergency medical services. Further, the locations of ambulance bases are determined based on patient demand. However, in practice, many elements should be taken into account in a risk assessment of given areas within a locale. Specifically, each area should be assessed for the number and severity of accidents that ordinarily take place there, the number and size of the public events it hosts, its population density, and the number of elderly people resident. In this study, we use a spatial-temporal approach to integrate those factors into a risk assessment of areas relative to each other in a locale. Based on this risk assessment, we determine the optimal locations for ambulance bases in order to minimize response time. We validate our approach using Bangkok as a case study.

1 Introduction

Emergency medical services (EMS) are crucial to health management [1]. Their primary purpose is to provide care to injured people at the scene of an accident or other incident requiring immediate care by providing primary care and then getting them to a hospital as fast as possible. Most EMS system planners draw on experience and information pertaining to the historical accidents per area to determine where to locate ambulance bases. However, this traditional approach is not optimal in terms of minimizing expenses, i.e., more bases than are necessary for a given area are built. Nor is this approach optimal in terms of minimizing response time, i.e., injured people wait longer than they would under another design.

For example, in 2014–2017 in Bangkok, Thailand, the EMS took longer than 10 minutes to reach the scene in 99% of incidents (Emergency Medical Information System: ITEMS 2559), which exceeds the standard time of 8 minutes [2]. A delay of this nature may have an adverse effect on victims' chances of surviving.

The allocation problem in regard to emergency base location has generally been considered using a covering model. Introduced in 1971 by Toregas et al. [3], the first covering model referred to as the Location Set Covering Problem (LSCP) had the goal of establishing the minimum number of ambulance bases needed to cover a service area. Variants include the Location Set Covering Problem (LSCP) [3] and the Maximal Covering Location Problem (MCLP) [4]. Most of the models used to determine where to locate bases take only patient demand into account. Yet, several kinds of information should be considered concurrently. In this study, therefore, we integrate multiple kinds of spatial-temporal information such as the number and the severity of accidents per area, population density, the number of elderly living in each area, and the number and size of the public events that take place there.

Designed to consider factors involving space and time simultaneously, spatial-temporal analysis is usually applied to a geographic information system (GIS) in order to define areas of relative risk. The higher the risk area, the greater the number of accidents that take place there. On this basis, we undertake accessibility risk identification in order to identify high-, medium-, and low-risk areas in order to allocate the number of ambulances per area more effectively than is possible under the standard approach.

2 Literature Review

In this section, we present accounts of the research relevant to our study, the coverage models, a method for accessibility risk identification, and a spatial-temporal analysis of accessibility risk.

2.1 A covering models

Determining the ambulance bases for an emergency medical service system is very important. In studies over published over the last 40 years, researchers have proposed several methods of management. In particular, the covering model is often used to determine ambulance bases, and mathematical versions of the covering model are used in public service planning to cover most areas of service [5]. Overall, the covering model has been applied in many kinds of organizations, such as transport planning, fire stations, and emergency medical services including allocating ambulances. In terms of ambulance allocation, the goal is to meet (cover) the needs of victims

* Corresponding author: chumpol.yuangyai@gmail.com

at the scene by ensuring that ambulances reach them quickly and thereby optimizing chances of survival. The Location Set Covering Problem (LSCP) model [3] was used to determine the minimum number of ambulance bases needed for a given service area. However, in this basic model there are no restrictions on the resources available to establish and operate bases. In 1974, Church and ReVelle [4] modified this basic covering model in their Maximal Covering Location Problem (MCLP) by adding demand (i.e., meeting patients' needs) to the covering model. In this model, the goal is to cover most of the demand using a limited number of ambulance base. Each of these models has been applied at both the operational level and the strategic planning level of ambulance base allocation.

In most of the relevant research, patient demand constitutes the principal or even the only information taken into account in determining where to locate ambulance bases. In practice, numerous factors affect ambulance allocation, such as the number and severity of accidents in an area, population density, the number of elderly, public events, and whether or not the area is likely to suffer a natural disaster. Thus, in the present research, we integrate these risk factors into the covering model proposed by Nilsang et al. [6] with the overall purpose of improving response time.

2.2 Determining risk factors to account for in determining ambulance base allocation

In this section, we discuss the spatial-temporal risk factors that should be taken into account in determining the locations of ambulance bases. These factors include:

- Number of accidents: This aspect, which refers to the demand for emergency medical services, is the principal factor considered in the covering model [7].
- Severity of accidents: Accidents vary greatly in terms of severity, such that emergency medical services should be capable of adapting accordingly. [8].
- Population density: When the population density of an area is high, there is likely to be a correspondingly high need for emergency services. It is, therefore, necessary to locate an ambulance base or bases in such areas.
- Number of elderly: Elderly people have a high need for medical services. Therefore, it is necessary to locate an ambulance base or bases in areas where a high number of elderly people live [9].
- Number and size of public events: Public events that attract a large number of people, whether national sporting events, University graduation ceremonies, or music festivals, present risk such that emergency medical services should be readily available. Therefore, an ambulance base or bases should be located in areas where such events take place [10].

We evaluate all 5 of these risk factors in our model, and we describe how we assess risk level in Section 2.3.

2.3 Risk assessment and analysis of spatial-temporal risk factors

The analysis of spatial-temporal risk factors is an aspect of the Geographic Information System (GIS) in which location and time are considered concurrently. This approach has been used in several areas: anomaly detection 46% on disaster management, 27% in traffic management 14% inference place 13% health management 5% which information from Twitter [11].

In this study, we assess multiple spatial-temporal risk factors to determine the level of risk (risk values) in each area. We use risk values to determine the optimal places to locate ambulance bases. As we will show, our model is designed to take all risks affecting ambulance base allocation into account.

3 Overview of Our Integrated Model

In this section, we present our new covering model in which risk values instead of patient demand values are used to determine the locations of ambulance bases.

3.1 Assessment of spatial-temporal risk factors

Our assessment of spatial risk factors includes the number of accidents in the area, the severity of the accidents, the population density, the number of elderly people living in the area, and the number and size of public events that take place there. The criteria for risk assessment are presented in Appendix 1. The assessment of the risk value of each factor is presented in Table 1.

To determine the spatial risk of each factor, the spatial risk of each factor used to obtain the spatial risk totally applied to a covering model (Equation 1):

$$R_i = \prod_{h=1}^{5} (L_{ih} \times I_{ih}) \quad (1)$$

The model is described as follows: R_i is the risk value for area i; L_{ih} is the likelihood of risk factor h for area i; I_{ih} is the impact of risk factor h for area i; i is the index of the area; and h is the index of the risk factor.

In a covering model, the total risk of each area is used to determine the ambulance bases needed to cover the maximum risk and set the range of the total risk values for each area. The areas with the lowest risk are shown in purple, and the highest-risk areas are shown in red.

3.2 Covering model with risk factors

Next, the risk assessment of each study area is used with a mathematical model to determine the ambulance base allocation. We integrate spatial-temporal risk values into the MCLP model. In addition, we incorporate the concept of location-forced constraints as proposed by Nilsang et al. [6] into our model. An MCLP model relies on the following equations (2–7):

$$\textbf{Maximize} \quad \sum_{i=1} R_i y_i \tag{2}$$

Constraints

$$\sum_{j \in W_i} x_j \geq y_i \,, \; i \in V \tag{3}$$

$$\sum_{j \in W_i} x_j \leq P \tag{4}$$

$$\beta_j \leq x_j \quad j \in W \tag{5}$$

$$x_j \in \{0,1\} \,, \; j \in W \tag{6}$$

$$y_i \in \{0,1\} \,, \; i \in V \tag{7}$$

Notations:

V = Set of area points
W = Set of potential base locations
i = Index of area points, $i \in V$
j = Index of potential base locations, $j \in W$
R_i = Risk value in area point i
P = Total potential base location

$$\beta_j = \begin{cases} 1 \text{ if base location is location-forced } j \\ 0 \text{ otherwise} \end{cases}$$

$$x_j = \begin{cases} 1 \text{ if base location is selected at site } j \\ 0 \text{ otherwise} \end{cases}$$

$$y_i = \begin{cases} 1 \text{ if area point i is covered at least onec} \\ 0 \text{ otherwise} \end{cases}$$

Objective (2) is to maximize the risk value coverage. Constraint (3) guarantees that an area is covered only if at least one ambulance base is located within the distance standard. Constraint (4) guarantees that the total number of potential base locations is less than P. Constraint (5) guarantees that the control location j is selected. Constraints (6) and (7) x_j and y_i together constitute the binary decision variable.

4 Case Study

We demonstrate our proposed approach to ambulance base allocation and spatial risk assessment using the EMS in Bangkok, Thailand, as a case study. The data collection and parameters are described next.

4.1 Description of the EMS in Bangkok

The capital of Thailand, Bangkok has an area of 1,568.74 kilometers^2 and a population of 5,686,646, which means that it has the highest population density in the country. There are 43,959 elderly people living in each area of Bangkok, who are at a high risk of needing emergency medical services. As the country's capital, Bangkok has a large convention center and a national stadium such that the city hosts many public events. When there is an emergency situation at events in particular, it is important that extensive and appropriate medical services be readily available.

4.2 Data collection and risk assessment

For this case study, we collected data pertaining to the number of severity of accidents in the Bangkok city area in January 2018. We obtained data on population density and the number of elderly people living in the city from the Ministry of the Interior, and public event information from the Internet. Next, we divided Bangkok into 56 blocks of 36 km^2 (6×6 km) to use as a basis for decision making. We used this information to assess the risk of each area in our covering model to determine emergency base allocation. The color temperature is divided into 10 levels according to the risk used in decision making, parking ambulance. The purple area represents low risk ($R_i \leq 99$), and the red area represents high risk ($R_i \geq 1{,}000$). We used the proposed coverage model to find the ambulance location.

5 Results

We compared the results of the MCLP-Demand and MCLP-Risk values in the MCLP model for Bangkok, which, as stated, we divided into 56 blocks for which we represented risk using a color scheme (Figure 1). A black circle symbol is used to represent the ambulance bases. The result is that the MCLP-Risk value determines the number of ambulance bases which cover more than MCLP-demand shown in Figure 1.

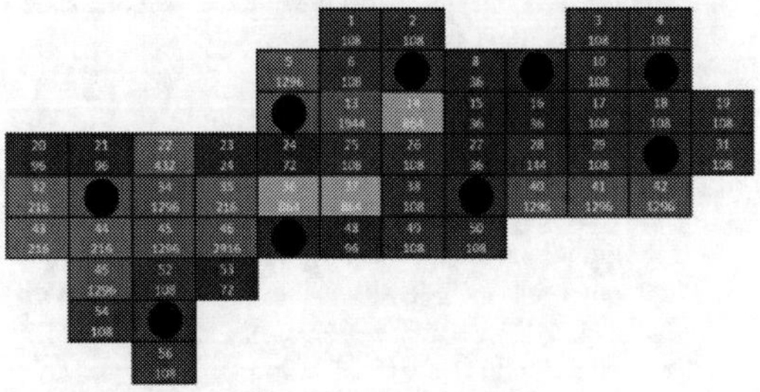

(a) MCLP – Risk Value

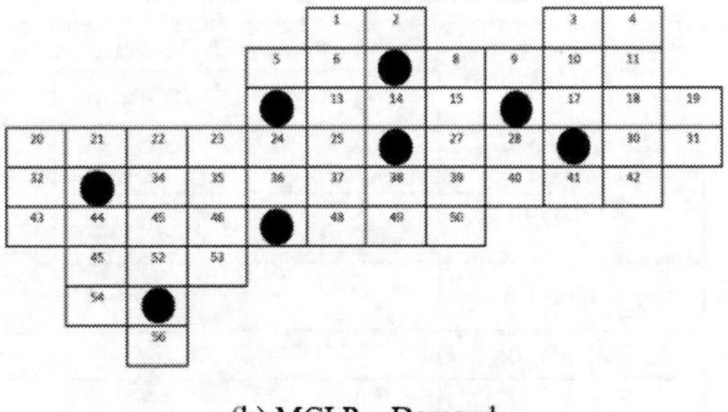

(b) MCLP – Demand

Fig. 1. Ambulance base locations determined using the MCLP model for 56 areas of Bangkok.

Our analysis shows a difference in the results in the MCLP model. The MCLP-Risk value model assigns 9 ambulance bases to cover the risk in the Bangkok area, whereas the MCLP-Demand model assigns 8 ambulance, which do not cover the entire Bangkok area. Our model relies on these 5 risk factors in a covering model together with the color scheme for decision making. On the basis of our model compared with the standard model, the ambulance base allocation has a greater coverage area and victims are admitted to hospital more quickly.

6 Conclusions and Discussion

We have presented a spatial-temporal model for identifying risk factors that affect the effectiveness of emergency medical systems. We found that 5 factors should be considered in allocating ambulance base locations: the number and severity of the accidents per area, its population density, the number of elderly people living there, and the number and size of the public events held there. We bring all of these risk factors to bear on assessing the risk-value of each area in terms of providing emergency medical services. We defined the criteria for determining the risk value of each area and presented this information using a color scheme to indicate risk levels in a given area. Next, we integrated the risk values into the standard MCLP model to allocate ambulance bases in order to cover all the high-risk areas. We also validated our model by applying it to Bangkok.

In future research, we plan to include information pertaining to the number and size of public events from social media. Further, we plan to add other spatial-temporal risk factors such as the amount of rainfall and conflagration incidents in determining where to allocate ambulance bases in order to improve the performance of emergency medical systems.

References

1. Lim CS, Mamat R, Braunl T (2011) Impact of ambulance dispatch policies on performance of emergency medical services. IEEE Trans Intell Transp Syst 12:624–632. doi: 10.1109/TITS.2010.2101063
2. De Maio VJ, Stiell IG, Wells GA, Spaite DW (2003) Optimal defibrillation response intervals for maximum out-of-hospital cardiac arrest survival rates. Ann Emerg Med 42:242–250. doi: 10.1067/mem.2003.266
3. Toregas, C., et al., The location of emergency service facilities. Operations research, 1971. 19(6): p. 1363-1373.
4. Church, R. and C. ReVelle. The maximal covering location problem. in Papers of the Regional Science Association. 1974. Springer.
5. Farahani RZ, Asgari N, Heidari N, et al. (2012) Covering problems in facility location: A review. Comput Ind Eng 62:368–407. doi: 10.1016/j.cie.2011.08.020
6. Nilsang S, Yuangyai C, Janjarassuk, U, Cheng CY, Locating an ambulance base by using social media: A case study in Bangkok. Annals of Operations. Research In process of public.
7. Li X, Zhao Z, Zhu X, Wyatt T (2011) Covering models and optimization techniques for emergency response facility location and planning: A review. Math Methods Oper Res 74:281–310. doi: 10.1007/s00186-011-0363-4
8. Ratanavaraha V, Suangka S (2014) Impacts of accident severity factors and loss values of crashes on expressways in Thailand. IATSS Res 37:130–136. doi: 10.1016/j.iatssr.2013.07.001
9. Peacock PJ, Peacock JL (2006) Emergency call work-load, deprivation and population density: An investigation into ambulance services across England. J Public Health (Bangkok) 28:111–115. doi: 10.1093/pubmed/fdi079
10. Laylavi F, Rajabifard A, Kalantari M (2017) Event relatedness assessment of Twitter messages for emergency response. Inf Process Manag 53:266–280. doi: 10.1016/j.ipm.2016.09.002
11. Steiger E, de Albuquerque JP, Zipf A (2015) An advanced systematic literature review on spatiotemporal analyses of Twitter data. Trans GIS 19:809–834. doi: 10.1111/tgis.12132

Appendix 1. Assessment of Risk Values

Factor	Likelihood			Impact		
	1 (Low)	2 (Medium)	3 (High)	1 (Low)	2 (Medium)	3 (High)
Frequency of accidents	0–2 days/week	3–5 days/week	5–7 days/week	< 3 times/day	3–6 times/day	> 6 times/day
Severity of accidents	< 3 times/day	3–5 times/day	> 5 times/day	Small accidents	Medium accidents	Big accidents
Population density	assign = Low			< 50,000 units	50,000–100,000 units	> 100,000 units
Number of elderly	assign = Low			50–500 units	501–1,000 units	> 1,000 units
Number of Public events	0–2 days/week	3–5 days/week	5–7 days/week	< 2,000 units	2,001-5,000 units	> 5,000 units

Polymer based microneedle patch fabricated using microinjection moulding

Pattanaphong Janphuang[1,*], *Mongkhol* Laebua[2], *Chanwut* Sriphung[1], *Phatsakon* Taweewat[2], *Anan* Sirichalarmkul[2], *Kasiphisan* Sukjantha[2], *Napatporn* Promsawat[1], *Pimchanok* Leuasoongnoen[1], *Somphong* Suphachiaraphan[1], *Kantapon* Phimol[1], *Don* Isarakorn[3] and *Somchai* Tancharakorn[1]

[1]Synchrotron Light Research Institute (Public Organization), 111 Moo 6, University Avenue, Muang District, Nakhon Ratchasima, 30000, Thailand
[2]Thai-German Institute, 700/1 Moo 1 Amata Nakorn Industrial Estate, Bangna Trad Road, Klongtamru, Muang District, Chonburi, 20000, Thailand
[3]Department of Instrumentation and Control Engineering, Faculty of Engineering, King Mongkut's Institute of Technology Ladkrabang, Bangkok, 10520, Thailand

Abstract. This paper presents the development of a polymer based microneedle patch for transdermal drug delivery application using plastic microinjection moulding. Design and analysis of the microneedle cavities and mould insert used in the injection moulding process were carried out using Computer-Aided Engineering (CAE) software. A mould insert with low surface roughness was fabricated using Micro Electrical Discharge Machining (μ-EDM). The injection moulding parameters including clamping force, temperature, injection pressure and velocity were characterized in order to obtain the optimum reproducibility. Solid truncated cone microneedles, made of biocompatible polymethyl methacrylate (PMMA), with a round tip radius of 50 μm and 500 μm in height have been realized by microinjection moulding process demonstrating the potential of a low cost, high production efficiency, and suitable for mass production. In addition, a mould insert of cylindrical microneedles fabricated using X-ray LIGA has been proposed.

1 Introduction

Currently, transdermal drug delivery (TDD) is one of a convenient way to deliver medicine into the subcutaneous tissues without drawbacks such as pain and digestive system disorders [1-2]. To achieve this, the drug is injected through the outermost layer of skin, Stratum Corneum (SC), from where it diffuses into the blood vessels of the deeper dermal skin layer through microneedle conduits. Since there are no nerves in SC layer, thus, this method is pain-free in comparison to hypodermic needle [3].

Microneedles (MNs) are micro-scale needles having thorn-like mechanical structure with at least one dimension of their shank smaller than 500 μm [4]. The MNs are typically designed in arrays in order to improve the surface contact with the skin. Microneedle array technology is an emerging technique combining the ease of transdermal patch and the capability of hypodermic needle to facilitate penetration of actives into the skin. Various types of MNs yield have been developed since late of 1990s, including coated MNs enable drug on the surface of the needles, hollow MNs for drug infusion, and dissolving MNs encapsulated drug and release after insertion [5]. Previously, MNs were made of silicon, metals, and organic materials fabricated using laser ablation, wet etching, and metal electrodeposition processes [1]. In recent years, polymers based MNs is of great interest. Most of polymers have high toughness and biocompatible enabling polymers based MNs to avoid crack and hardly induce side effect during their insertion into the skin. Some polymers are water-soluble or biodegradable. Thus, drugs can be encapsulated in dissolvable MNs. Moreover, due to the relatively low melting temperature, microinjection moulding can be used in preparation of polymer MNs. Injection molding is a process in which melt materials, mostly polymer, is forced to flow under high pressure through a plunger into a mold cavity, where it solidifies and then ejected to produce a replicated structure on the surface. Thus, mould based fabrication technique could further reduce the manufacturing cost and is suitable for mass production.

Here, we present the design and fabrication method for an inexpensive microneedle patch made of biocompatible polymer, PMMA, by means of plastic microinjection moulding process. A detailed description on the flow analysis of molten polymer into the cavity, driven by the pressure applied by the plunger was studied through simulation in Moldex3D and μ-EDM was used to fabricate the mould insert. By optimizing the injection moulding parameters, results show that low surface roughness polymer based microneedle patch can be produced. In addition, a mould insert of microneedle patch fabricated using X-ray LIGA are proposed.

[*] Corresponding author: pattanaphong@slri.or.th

2 Design and fabrication

2.1 Microneedle design

The feature of MNs should meet several requirements in order to achieve painless penetration into skin. The MNs should be sharp and strong enough to handling induced mechanical failure during the insertion of MNs into skin. The MNs should have proper length and diameter to avoid nerve contact and the short-shot during the microinjection moulding process. As illustrated in [6], considering the skin deformation, the MNs shank length is limited to 300 – 400 µm. However, it can be assumed that the entire needle shank will not penetrate into the skin since the human skin is very flexible. Thus, the length of the MNs in this work is defined to be 500 µm. Figure 1 shows the schematic diagram and dimensions of the designed microneedle. The material properties are given in table 1.

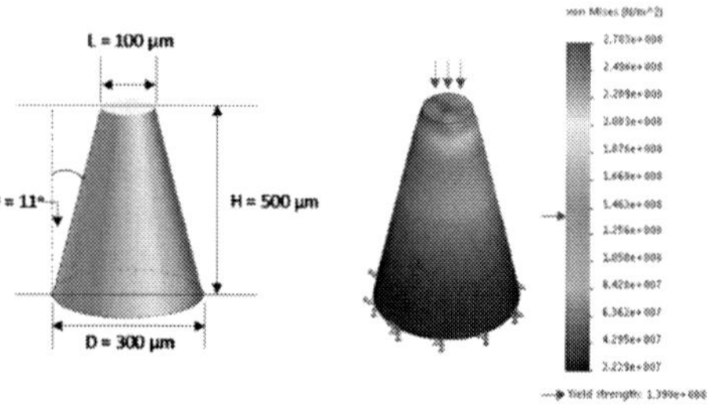

(a) (b)

Fig. 1. (a) Geometric shape of a microneedle (MN); (b) Simulated stress and deformation of the MN under applied pressure of 300 MPa on the tip.

The designed MN in this study is solid truncated cone structure, as illustrated in Fig. 1(a), in order to minimize the bending stress and detrimental buckling failure during the insertion of MN into the skin compared with other high aspect ratio or column-like structures [1]. It is obvious that sharpening tip, by reducing the tip diameter (L) and increasing the bevel angle (Θ), can effectively diminish the insertion force. However, it may result a tip damage during the penetration. Thus, a trade-off between sharpness and mechanical strength of the tip should be taken into account.

Table 1. PMMA material properties used in simulation [7].

Properties	
Elastic modulus	8.30 GPa
Poisson ratio	0.28
Shear modulus	3.20 GPa
Density	1400 kg m^{-3}
Tensile strength	1.43 GPa
Yield strength	1.39 GPa

The designed MN was then computer simulated in SolidWorks in order to determine the stress distribution and deformed shape of the MN. Fig. 1(b) illustrates the stress developed on the MN under applied pressure of 300 MPa to the tip of the MN. It is clearly seen that the tip could endure the applied pressure but the MN is deformed at this point. As described in [7], a pressure of 15 MPa is sufficient for the penetration of human skin by liquid jet, diameter of 0.1-0.5 mm. The penetration of microneedle requires less pressure than high-speed liquid jet. Consequently, we can estimate that the designed MN should not deformed in the human skin during the insertion. To evaluate the capability of microinjection moulding process in this work, 400 MNs made of PMMA with 1 mm pitch between needles will be fabricated.

2.2 Mould design

The CAD model of a mould insert (or cavity insert) is illustrated in Fig. 2. It consists of 400 microneedle cavities (Fig. 2(b)) occupying approximately 500 mm^2. The thickness of the base portion is designed to be 2 mm. In order to successfully fabricate MNs by microinjection moulding process, guidelines for the design of a runner system and a detailed description on the flow analysis of molten polymer into the cavity were carried out by FE simulation in Moldex3D.

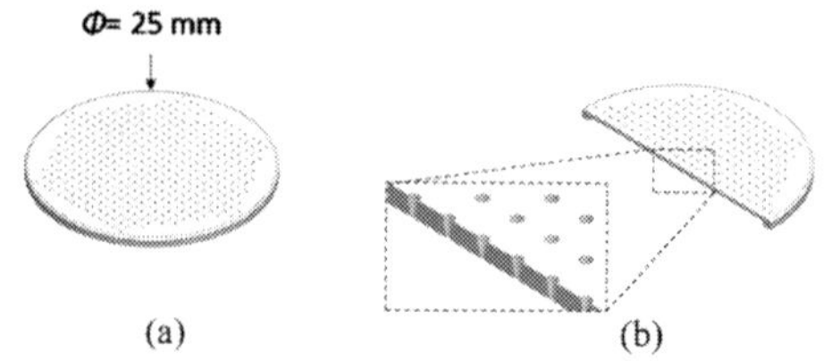

(a) (b)

Fig. 2. (a) Schematic diagram of the mould insert designed for plastic injection of microneedle patch; (b) Cross-sectional view of the mould insert.

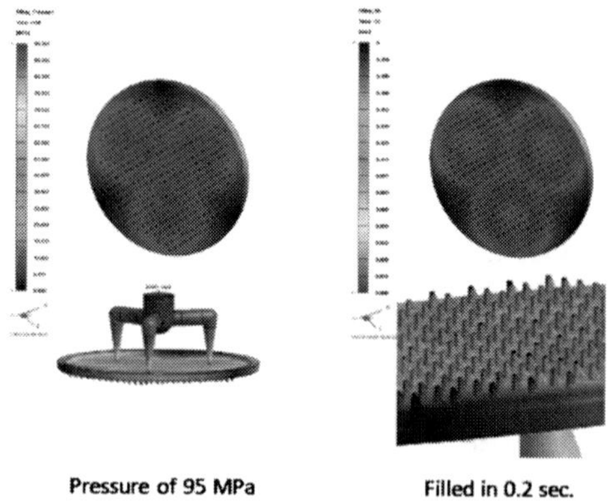

(a) (b)

Fig. 3. Simulated results of (a) injection pressure loaded to the designed runners and the mould insert and (b) molten PMMA filled in the cavity.

The runner optimization analysis is first performed in order to define the size and the number of runners, and the injection pressure applied to the cavity. To assure symmetric filling, pin point gate type was designed. In simulation, polymer melt temperature (T_p = 230 °C), mould temperature (T_m = 50 °C), holding pressure (P_h = 60 MPa), and injection flow (V_i = 9 ccm/s) were kept constant (to the values recommended by the manufacturer). The simulated results revealed that three-point gate, with diameter of 5 mm, were sufficient for the injection of molten PMMA, as shown in Fig. 3. More than 90% of MNs cavities were fulfilled by the molten PMMA in 0.2 s using injection pressure of 95 MPa and gate freeze time of 0.38 s, as illustrated in Fig. 3(b). Both fill and demoulding of high aspect ratio structure like microneedle are likely to be difficult. Air could be trapped inside the cavities blocking the filling process. The vacuum condition could be introduced to the system in order to improve the injection moulding capability.

2.3 Fabrication of the mould insert

The MN cavities in the mould insert were fabricated using µ-EDM (SARIX SX-100, Switzerland). The whole MN mould was mainly made of high temperature alloy (SKD11) in order to enhance the corrosion resistance and wear resistance with less thermal expansion stress. The basic working principle of µ-EDM process is described in Fig. 4.

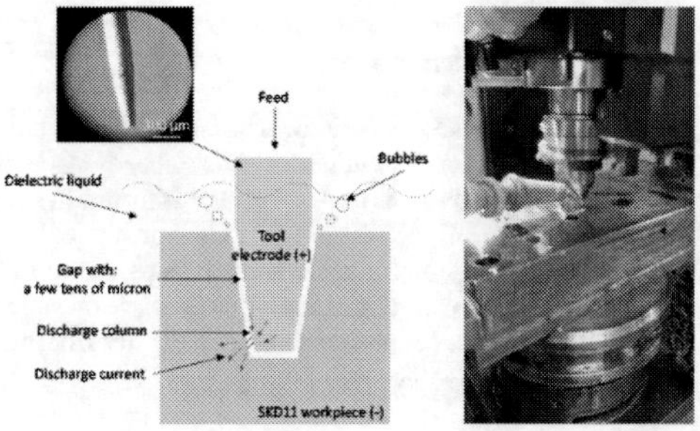

(a) (b)

Fig. 4. (a) Schematic diagram of µ-EDM process; (b) Machining of mould insert by µ-EDM.

To produce the MN cavities, a submerging of the anode-cathode system into a dielectric liquid is required (Fig 4(a)). In μ-EDM process, the designed MN structure was firstly created on a brass tool electrode using 100 µm solid carbide end milling tools from SARIX. The material was then removed from the workpiece electrode by current discharges between two electrodes subjected to a voltage. When the voltage was increased, the electric field in the volume between two electrodes become larger than the strength of the dielectric, resulting in the breakdown of the dielectric allowing current to flow between electrodes. Consequently, the workpiece was then eroded and formed a replica of the tool electrode on the workpiece by a thermal process under a controlled electric spark [8]. During the fabrication, the open gap voltage between the electrodes was maintained at 100 V until the dielectric broke down. Figure 5 shows optical images of the fabricate mould insert fabricated by μ-EDM. It should be noted that mould insert surfaces in which molten polymer undergoes have to be very smooth in order to avoid sticking of polymer during the demoulding process.

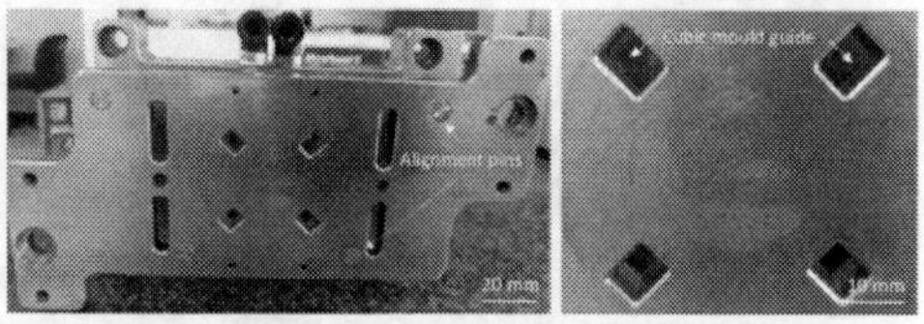

Fig. 5. Optical images of the MNs mould insert made of SKD11 alloy fabricated using precision μ-EDM.

3 Microinjection moulding

The microinjection mould insert was carefully aligned inside the mould base through the alignment pins and the mould guide to ensure proper matching of the cavities. The mould insert and mould based were then installed into the MircroPower-5t microinjection moulding machine (Wittmann Battenfeld, Austria), as shown in Fig. 6, to investigate the capability of microinjection moulding process for the production of microstructure.

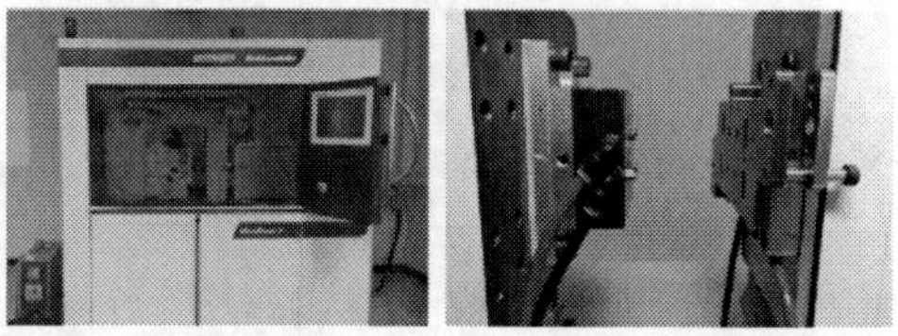

(a) (b)

Fig. 6. Optical images of (a) MicroPower-5t microinjection moulding machine and (b) the installed microneedle mould.

The plunger diameter of the machine is set to be 5 mm. During the microinjection moulding process, injection moulding parameters such as shot size, temperature (melt and mould), injection flow, clamping force, injection and holding pressure have been characterized in order to optimize the moulding results and reproducibility. Details of microinjection moulding parameters are described in table 2.

Table 2. Microinjection moulding parameters.

Parameters	
Shot size	1 ccm
Melt temperature (T_p)	230 °C (max. 250 °C)
Mold temperature (T_m)	50 °C (max. 70 °C)
Injection flow (V_i)	9 ccm/s
Injection pressure (P_i)	120 MPa
Holding pressure (P_h)	60 MPa
Clamping force	5000 kg
Cooling duration	9 s

The machine temperature was set at 230 °C to maintain the PMMA in the molten state. The shot size of 1 ccm was selected. Using large shot size (greater than 1 ccm), flush will occur. The clamping force was 5 tons (5000 kg) and the injection pressure was found to be 120 MPa. Both of these parameters are close to the maximum capacity of the MicroPower-5t. Complete polymer based microneedle patch fabricated using microinjection moulding is illustrated in Fig. 7.

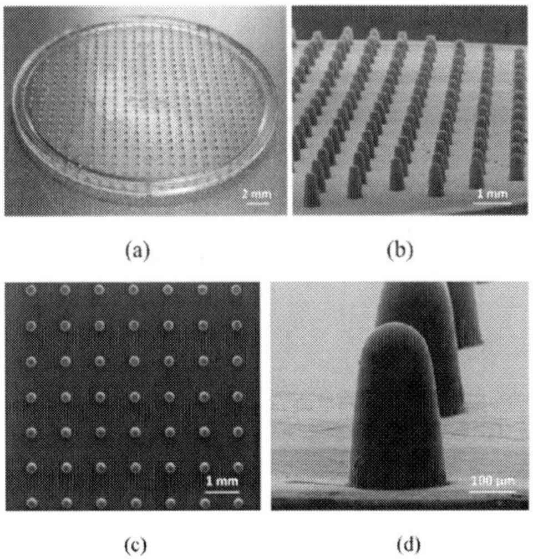

Fig. 7. (a) Optical images of a fabricated microneedle patch, and SEM images of (b) MNs array, (c) top view of the MNs array, and (d) a close-up of a MN.

4 Fabrication of microneedle mould insert using X-Ray LIGA

LIGA, a German acronym for Lithographie, Galvanoformung, and Abformung) process utilizes deep x-ray exposure for the fabrication of high aspect ratio microstructure. Using high-energy photons, which is deep x-ray, microneedle shank can be fabricated. However, LIGA process is a planar exposure technique to construct 3D microstructure. Thus, a vertical sidewall microstructure could be achieved. In this section, we present a fabrication of MN mould insert using x-ray LIGA and the potential of using x-ray exposure for the fabrication of sharp tipped microneedle is proposed.

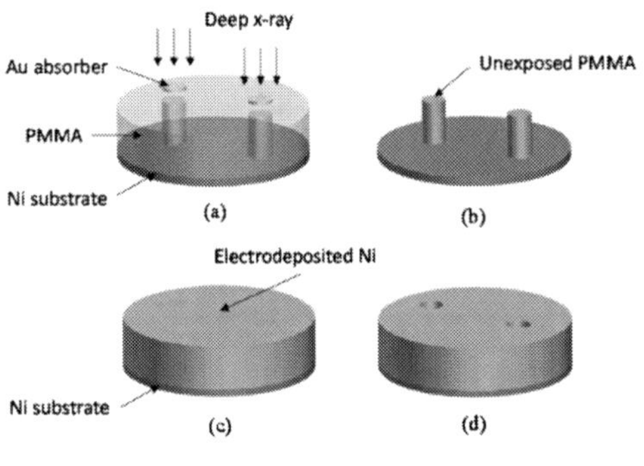

Fig. 8. Schematic diagram of the microfabrication process for the MNs mould insert using x-ray LIGA. (a) Vertical deep x-ray exposure; (b) Development of column-like microneedles; (c) Electrodeposition of Ni; and (d) PMMA stripping.

As shown in Fig. 8, the unexposed PMMA area blocked by gold (Au) absorber (Fig. 8(a)) making high aspect ratio column-like (cylindrical) MNs using deep x-ray produced by synchrotron radiation (Fig. 8(b)). 500 µm-thick Nickel (Ni) is then electrodeposited directly on top of Ni substrate (Fig. 8(c)). After stripping PMMA, the Ni mould insert is achieved (Fig. 8(d)). Figure 9 shows Ni the mould insert fabricated using x-ray LIGA.

Fig. 9. Optical image of (a) cylindrical MNs based on PMMA exposed by deep x-ray and (b) Ni mould insert fabricated using x-ray LIGA.

Using several combinations of deep x-ray exposure conditions including deep x-ray exposure angle (exposure of an inclined plane), and a gap between the gold absorber mask and PPMA, sharp tipped MNs made of PMMA can be realized.

5 Conclusions

Polymer based microneedle patch fabricated using microinjection moulding has been successfully demonstrated in this paper. Using this technique, the mass production of microneedle patch can be realized. The design of the microneedle architecture to avoid the bed-of-nail effect, and penetration testing experiments are ongoing for the optimization of the transdermal drug delivery using microneedle.

References

1. G. Ma, Ch. Wu. J Control Release **251**, 11-23 (2017)
2. M.R. Prausnitz, R. Langer. Nat. Biotechnol **26**, 1261-1268 (2008)
3. S. Indermun, R. Luttge, Y.E. Choonara, P. Kumar, L.C. du Toit, G. Modi, V. Pillay. J. Control. Release **185**, 130-138 (2014)
4. B. Stoeber, D. Liepmann. JMEMS **14**, 472-479 (2005)
5. E. Larraneta, R.E.M Lutton, A.D. Woolfson, R.F. Donnelly. Mater. Sci. Eng. R. Rep **104**, 1-32 (2016)
6. S.J. Moon, S.S. Lee. J. Micromech. Microeng **15**, 903-911 (2005)
7. K.L. Yung, Ch. Kang, H. Liu, K.F. Tam, S.M. Ko, F.Y. Kwan, T.M.H. Lee. J. Micromech. Microeng **22**, 015016 (2012)
8. K. Maghsoudi, R. Jafari, G. Momen, M. Farzaneh. Mater. Today. Commun. **13**, 126-143 (2017)

Wear characteristics of die surface in deep drawing of stainless steel 304 by using finite element simulation

Numchoak Sabangban [1,*], *Warit* Sirikun [1], *Krittatee* Sangounsak[1] and *Sedthawatt* Sucharitpwatskul [2]
[1]Department of Industrial Engineering, Faculty of Engineering, Nakhon Phanom University 48000
[2]National Metal and Materials Technology Center (MTEC), Thailand Science Park, Pathum Thani 12120

Abstract. The characteristics of adhesive wear of TiN coating in drawing process by using finite element method were carried out to investigate the effects of drawing angles and sliding velocities of punch to explained wear index. Bending radius of die was used to increase the contact pressure between stainless sheet and TiN coating surface. Three sliding velocities (0.5, 1 and 5 mm/s) were studied each bending angle between (0° to 90°) by keeping the hardness of material coating to 24 GPa (HV.0.05). The results showed that the wear index were decreased with drawing angle but sliding speed of punch was not affected in slightly speed.

1. INTRODUCTION

The demand for stainless steel has significantly increased for automotive industries due to their effective performance in mass production of safety parts and specific assembly in engine parts. [1] In the automotive sector the several of stainless steel was essential application in components parts such as fuel system, head gasket, windscreen wipers and safety part. These were tested to demonstrate and improved their suitability of car making from mechanical properties, corrosion resistance, and forming ability to other material parts. On the one hand, that their strength increases as they are cold formed into the shape of the component by conventional or advanced forming methods. In stamping process, the galling is the major issue to reduced tool life and terribly appearance of final product parts. The contact pressures (being bending) of sheet and die was increased and undesirable to prevent galling. As a result, adhesive wear of the interaction between die-sheet in the drawing process of stainless steel must be understood and develop to increase tool life quality of parts.

[2] was verify the FEM deep-drawing model with experiments variety conditions such as sliding times and hardness of TiNi. The comparison results of punch forces were shown characteristic of pseudoelasticity is greater with increased beneficial wear resistance of TiNi alloy. The major failure from FEM model was shown wrinkling point deepened on die temperature, sheet thickness and die temperature.

[3] introduction concept of a device for deep drawing in extreme conditions, represents drawing process in vacuum by high temperatures and molybdenum sheets forming. The functional of structural, economical, technological, operational, safety, environmental, legal and design requirements was investigated with description of possible variants of the mechanism structure. The initial blank metal sheet was deformed to complicated geometry shape by punch force in deep drawing process. Wrinkling and deep drawing parameters under different Blank-Holder Pressure were studied the behaviour by 3- Dimensional elasto-plastic Finite Element (FE) model for Mild Steel IS 2062 [4]. Deep drawing is the geometry process parameters that effected on deformation characteristic. [5] used 3-D experiment blank-holder pressure to identified coefficient of friction (COF) and improved quality of aluminium alloys.

According to interaction surfaces between blank and die in deep drawing process, [6] frictional condition, surface asperities on the micro deep drawing differ from that in normal scale. A Finite Element (FE) model representing surface morphology of SUS304 sheets was built for micro deep drawing simulation. [7] The local stress in SUS304 was measured during tensile deformation. Tensile stress was stopped at engineering strains of 0.33%, 0.78% and 0.43%, respectively.

However, stamping process are permanently used multi tools that affected unpredictable wear phenomena due to interaction surface friction between sheet metal (which will be deformed) and the tool. [8] identifying wear mechanism in the main surface on Tin coated steel sheet. In additional two conditions of advanced PVD coatings were studied and experimented to improve wear characteristics resistant on punch and die surface with the friction coefficient of this sheet against some selected coatings under stamping process. Results confirm show that the Titanium Nitride (TiN) covered steel sheet to die and punch ensuring a longer life of these parts. Due to [9] shown the structural of Diamond like carbon (DLC) coatings have high wear resistance and low coefficient of friction. On the other hand the wear rate of Titanium Nitride and Titanium Carbon Nitride (TICN) TiN/TICN were increase at the high load values.

[10] The sliding friction characteristics and wear mechanism of the TiAlN/TiN was calculated in the sliding test and observed by Environmental scanning electron microscope (ESEM) and an Energy Dispersive X-Ray Fluorescence Spectrometer (EDX) by investigated the combine action of the peeling off, chipping off, adhesion and mechanical scratch of the Titanium Aluminium Nitride (TiAlN); TiAlN/TiN coated tool.

2. Finite Element Simulation Method

A schematic geometry of the one-sided deep drawing test was shown in Fig. 1(a). The sheet having an initial state on the drawing die in Fig. 1(b). During the deep drawing process the punch was moving down vertically direction and pulling the sheet pass drawing die to U-shape bending, which surface of stainless sheet was contact with drawing die.

The deep drawing simulation set-up is displayed in Fig. 1. The FE 2D nonlinear plane strain, static structural has been analysed by Marc software. The geometry of punch has been selected to rigid body, the die and sheet are set to be deformable bodies in Table 1. In addition the material properties of sheet and die coating are stainless steel 304 (SS304) and TiN were based on [11] and [12] respectively. The assumption of Coulomb arctangent (velocity) is applied to all of contact body area with coefficient of friction in dry contact surface is 0.3 and moreover the boundary conditions of FE setup was shown in Fig 2. The boundaries of right surface nodes has been fixed displacement in horizontal x direction in order to present the haft of symmetry in geometric body. The relative sliding velocities between sheet and die were started from punch velocities under investigation by 0.5, 1 and 5 mm/s respectively in vertical –y axis and the die angles start from 0 to 90 degree are shown in Table 2. The displacement of punch travel is 20 mm from the beginning point. The number of elements of sheet and die are show in Table 1. All element types analysed by using Plain Strain full integral by keeping Explicit Large Strain.

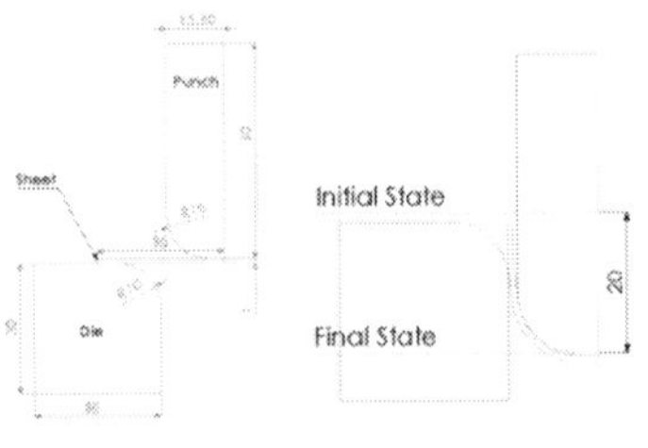

(a) (b)

Fig. 1. (a) Geometries of the one-sided deep drawing and (b) geometries and final state of deep drawing.

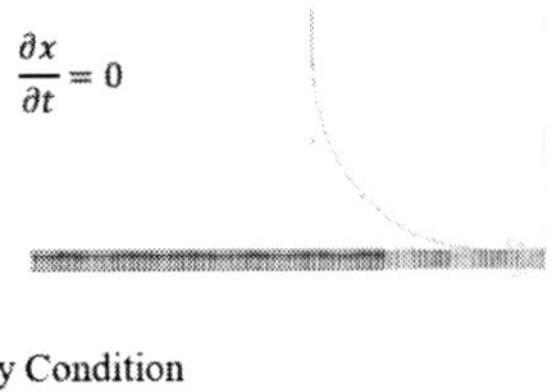

Fig 2. Boundary Condition

Table 1 Type of contact body and number of elements

Contact body	Contact Type (Number of Elements)
Die	Deformable (845)
Sheet	Deformable (749)
Punch	Rigid (-)

Table 2 Parameter Condition

Contact Body	Type of parameters	Conditions
Die	Radius (mm)	10
	Angle (Degree)	0-90
Punch	Velocity (mm/s)	0.5,1 and 5
Sheet	Thickness (mm)	1

2.1. Material Properties

The Finite Element Analysis (FEA) studied parameters are shown in Table 3. obtained from [12] experimental TiN wear tests and Fig. 3. [13] shows true stress of SUS 304 tensile tests. Those results leading to the relative sliding velocity and normal stress values can be exacted from the FE simulations results.

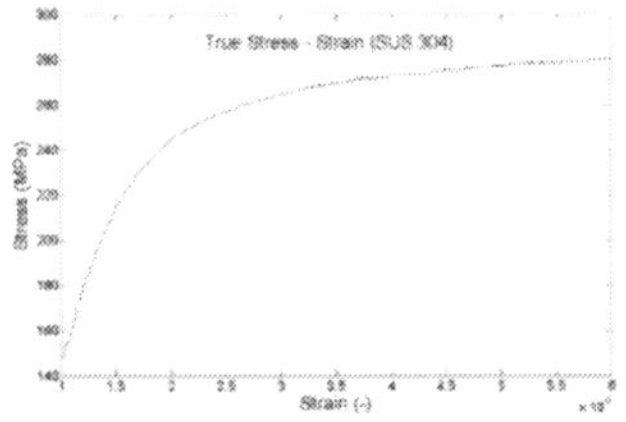

Fig 3. True stress of SUS 304 tensile tests

2.2. Wear Model

The following equation wear rate of die surface can be established by Archard's wear model [14] in Eq. (1)

$$\dot{w} = \frac{K}{H}\sigma_n V_{rel} \qquad (1)$$

Where $\dot{w}$ is the wear rate per times of drawing is, K is wear coefficient, H is the indentation hardness in order to shown the Vicker Hardness (Hv) of materials, σ_n is normal stress at identified contact area and V_{rel} is the relative velocities of contact body.

During the interaction contact, sheet was moving down to surface radius of the die. Difference surface angles provide difference value of normal stress and relative sliding speed values was shown in Fig 4.

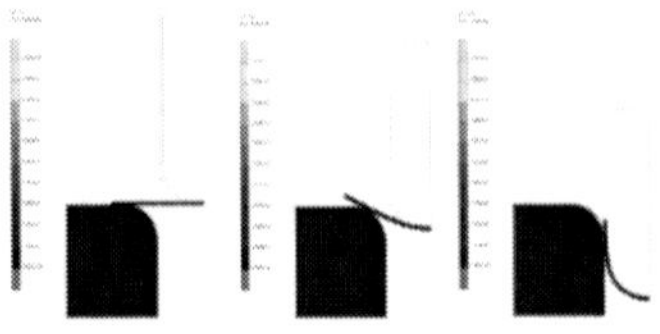

(a) (b) (c)

Fig. 4. (a) Initaial state (increment 0), (b) increment 50 and (c) Final state increment 100

Table 3 Values of experimental TiN and SUS304

Properties	Symbols	Unit	TiN	SUS304
Young's modulus	E	GPa	341.9	196.12
coefficient of Poisson	v	-	0.19	0.275
Hardness	H	GPa (HV)	25.6	2.1
Wear Coefficient	K	μN^{-1}	0.15	-
Yield Stress	σ_y	MPa	-	271.2
Friction coefficient	μ	-	0.3	0.3

To a great extent Eq. (1) in FEM, the relative of velocities directions and normal stress can be estimated by Cauchy's stress formula [15] in Eq. (2).

$$t(N) = -t(-N) = -t(n) \quad (2)$$

And

$$\sigma_n = t(n) \cdot n \quad (3)$$

Where t(N) is the traction on a body at its surface element with the normal direction N, and n is the normal vector inward.

For determining the synopsis of relative velocity vector of die and sheet it can be obtain by Eq. (4).

$$V_{rel} = V_{ns} - V_{nd} \quad (4)$$

The V_{ns} is velocity of node inside element of sheet body and V_{nd} velocity of node inside element of die body. In this studied the die element was set to zero, thrust the velocity of die element was neglected. Leading to only parameter of V_{ns} can be explained by Eq. (5).

$$V_{rel} = V_{ns}\cos(\beta) \quad (5)$$

Where cos β is the drawing angle of V_n is shown in Fig. 5(a)

In the FE simulations, the wear rate from Eq. (1) can be calculated to predict die surface wear, Wear Index (WI) after each increment steps. Leading to Eq. (6) can be calculated by

$$WI = \sum_{i=1}^{N} \dot{w} \quad (6)$$

Where N is the total number of increment and i is the initial increment.

The arc curve of die surface was investigated by determined number of nodes in the arc length of die surface is shown in Fig. 5(b). is 20 nodes (between node number 911 to 930).

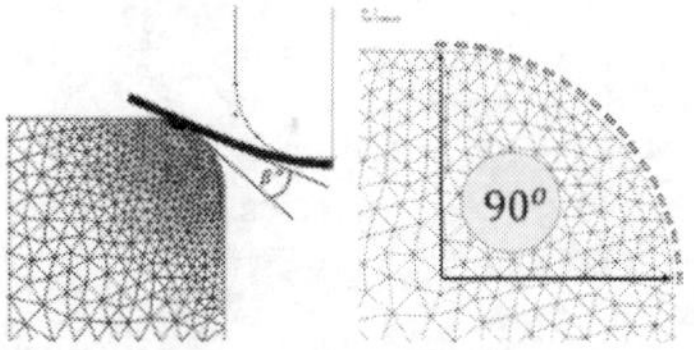

Fig 5. (a) Drawing angle and (b) nodes in the arc length of die surface

2.3 Scientifically prof and Verification

This FE model was analyzed under thin thickness (X-Y plane) was less than sheet geometry depth (Z direction), the total deformation could be assumed to plane strain. On the other hand velocities of punch are 0.1, 1 and 5 mm/s, respectively. Fig. 6 was shown the plane die force verification with mesh resolution results.

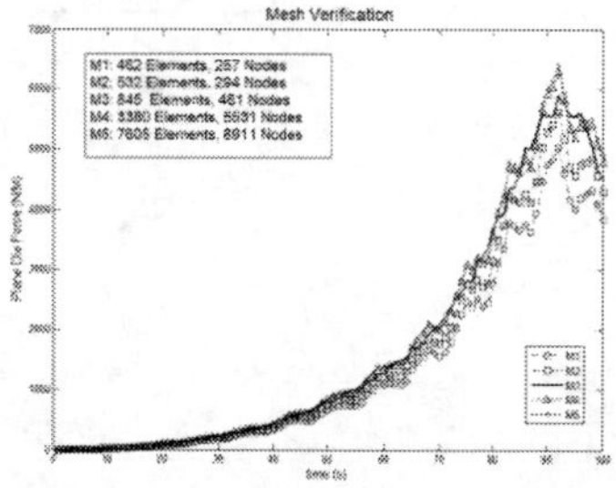

Fig 6. Plane die force verification with mesh resolution results.

The M1, M2, M3, M4 and M5 are represent mesh resolution models. The verification in this study revealed the followings comparison of plane die force average less than 5 percent error are shown in Table 5. It could be explained the M3 was suitable model in FE.

Table 5 The comparison model and percent error.

Comparison Model	Percent Error (%)
M1 Vs M2	12.527
M2 Vs M3	11.274
M3 Vs M4	4.807
M4 Vs M5	4.721

3. Results and Discussion

The change in mechanical properties base on FE simulations can be rationalized from normal stress in relative velocity. It could be observed the punch speed is not significantly values in Static Structural theory because the relative velocity in governing equation was slightly values.

Since the first increment sheet was moving down, the interactions between die areas node to finally state 20 mm distance .The intensity of normal stress was deepened on the edge of arc curve. Leading to FE simulation can be calculated the wear index with the effect of drawing angle, which could be explained in Fig. 7. The results were decrease by increasing sliding node angle along arc length of drawing die surface. This cloud also be explained by the fact that the direction of material flow between node of sheet and die respectively.

The polynomial equation (Order 5) can be explained and predicted the characteristics effect of wear index in function of drawing angle by Eq. (7).

$$y = A(\beta^5) + B(\beta^4) - C(\beta^3) + D(\beta^2) - E(\beta) + F \quad (7)$$

Table 6 The constants values of Eq. (7)

Constants	Values	Unit
y	Wear Index (WI)	m
β	Drawing Angle	(-°)
A	1.3 (10^{-16})	m
B	3.8 (10^{-14})	m
C	4.4 (10^{-12})	m
D	2.6 (10^{-10})	m
E	8.1 (10^{-9})	m
F	1.3 (10^{-7})	m

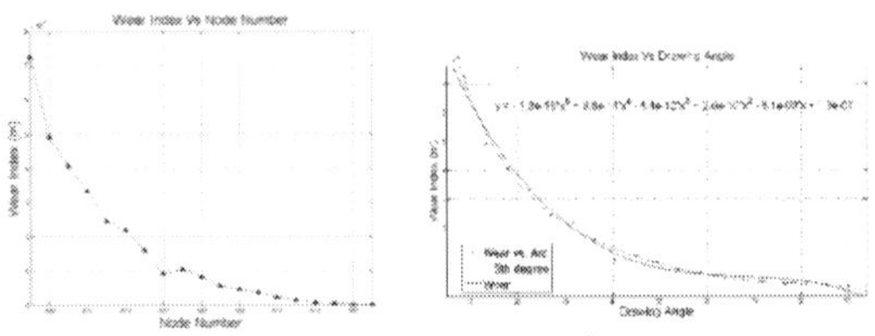

(a) (b)

Fig 7. (a) The Wear Index along the arc length nodes and (b) The polynomial fitting equation of wear index depended on drawing angle.

4. Conclusions

This model is the simple method utilized to evaluate the characteristics of tool wear in die geometry. For additional the range of TiN coating thickness was between 1 to 5μm, the resolving of galling on die surface was depended on various parameter in stamping process such as sampling per minute (SPM), lubrication, material geometry shape, surface roughness and cleanness etc.

5. Acknowledgements

The authors would like to acknowledge the technical support of National Metal and Materials Technology Center (MTEC) and Department of Industrial Engineering, Faculty of Engineering, Nakhon Phanom University. In addition, the authors would like to sincerely thank Dr.Numpon Mahayotsanun (Khon Kaen University) and Mr. Sedthawatt Sucharitpwatskul (MTEC) for their suggestions and advice.

6. References

1. Automotive, Railways, Transport, "Next generation vehicle Stainless steel in structural automotive applications," 2009, vol. 2009, no. 10, p. 53.
2. M. Wallmeier, E. Linvill, M. Hauptmann, J.-P. Majschak, and S. Östlund, "Explicit FEM analysis of the deep drawing of paperboard," Mech. Mater., vol. 89, pp. 202–215, Oct. 2015.
3. M. Tropp, M. Tomasikova, R. Bastovansky, L. Krzywonos, and F. Brumercik, "Concept of Deep Drawing Mechatronic System Working in Extreme Conditions," Procedia Eng., vol. 192, pp. 893–898, Jan. 2017.
4. A. K. Choubey, G. Agnihotri, and C. Sasikumar, "Numerical Validation of Experimental Result in Deep-Drawing," Mater. Today Proc., vol. 2, no. 4, pp. 1951–1958, Jan. 2015.
5. R. Dwivedi and G. Agnihotri, "Study of Deep Drawing Process Parameters," Mater. Today Proc., vol. 4, no. 2, Part A, pp. 820–826, Jan. 2017.
6. L. Luo, Z. Jiang, and D. Wei, "Influences of micro-friction on surface finish in micro deep drawing of SUS304 cups," Wear, vol. 374–375, pp. 36–45, Mar. 2017.
7. M. Chen et al., "Local Stress Evaluation During Deformation in SUS304 Austenitic Stainless Steel," Mater. Today Proc., vol. 2, pp. S937–S940, Jan. 2015.
8. L. Fernandes, F. J. G. Silva, M. F. Andrade, R. Alexandre, A. P. M. Baptista, and C. Rodrigues, "Improving the punch and die wear behavior in tin coated steel stamping process," Surf. Coat. Technol., vol. 332, pp. 174–189, Dec. 2017.
9. H. Cicek, "Wear behaviors of TiN/TiCN/DLC composite coatings in different environments," Ceram. Int., vol. 44, no. 5, pp. 4853–4858, Apr. 2018.
10. G. Zheng, G. Zhao, X. Cheng, R. Xu, J. Zhao, and H. Zhang, "Frictional and wear performance of TiAlN/TiN coated tool against high-strength steel," Ceram. Int., vol. 44, no. 6, pp. 6878–6885, Apr. 2018.
11. R. K. Blandford, D. K. Morton, S. D. Snow, and T. E. Rahl, "Tensile Stress-Strain Results for 304L and 316L Stainless Steel Plate at Temperature," pp. 617–628, Jan. 2007.
12. R. C. Cozza, "A study on friction coefficient and wear coefficient of coated systems submitted to micro-scale abrasion tests," Surf. Coat. Technol., vol. 215, pp. 224–233, Jan. 2013.
13. X. Li, J. Chen, L. Ye, W. Ding, and P. Song, "Influence of strain rate on tensile characteristics of SUS304 metastable austenitic stainless steel," Acta Metall. Sin. Engl. Lett., vol. 26, no. 6, pp. 657–662, Dec. 2013.
14. J. Chang, H. Chenyuan, and J. Xiaoxiong, "FE simulation of tire wear with complicated tread pattern," Procedia Eng., vol. 15, pp. 5015–5019, Jan. 2011.
15. M. E. Gurtin, V. J. Mizel, and W. O. Williams, "A note on Cauchy's stress theorem," J. Math. Anal. Appl., vol. 22, no. 2, pp. 398–401, May 1968.

Comparative study of dissimilar tailor-welded blanks between DP590 and DP980 dual phase steels

Rittichai Phaoniam[1,*], *Jesada* Kaewwichit[1] and *Komgrit* Lawanwong[2]

[1] Department of Mechanical and Industrial Engineering, Faculty of Engineering, Rajamangala University of Technology Krungthep, Bangkok, Thailand
[2]Department of Production Engineering Technology, Faculty of Industrial and Technology, Rajamangala University of Technology Rattanakosin, Prachubkirikhan, Thailand

Abstract. This research aims to study the dissimilar tailor-welded blanks between DP590 and DP980 dual phase steels using the autonomous GTAW process. The summarized results are as follows. It was found that utilized welding parameters in DP590 and DP980 steel butt joint were able to achieve complete penetration joint and there were not any defects. Furthermore, the dissimilar weld zone produced the hardness profile value between the DP590 and DP980 base metal. In particularly, DP590-HAZ region resulted in harden zone. Meanwhile, DP980-HAZ region induced a soften zone. By the way, the harden DP590-HAZ exhibited fine columnar martensite with a ferrite matrix. On the other hand, the soften DP980-HAZ represented a tempered martensite structures. Comparing the tensile test was carried out in order to investigate dissimilar welded joint strength in the different rolling directions. It was suggested that the specimen welded along to rolling direction was evidently larger tensile strength (1092 MPa) than the welding transverse to rolling direction (638 MPa). Moreover, the weakest fracture occurred apparently on the region of DP590-BM rather than the HAZ region.

1 Introduction

In recent years, automotive industry is concerned with energy saving, to reduce in the vehicle weight as a focus concern in the automobile industry. This was the concept of Tailor-welded blanks (TWBs) has been applied to the various materials. TWBs consist of two or more sheets of material with different properties such as thickness, material, strength, surface coating or uncoating and different formability properties. [1] In addition, Dual phase (DP) steel is now widely used for fabricating automotive components such as frames, wheels, and bumpers as they offer a good combination of high strength and ductility.[2] Furthermore, from the literature review a research related on the DP steels welded with other steels and the dissimilar DP steels or the similar DP steels such as [3-5] laser welding DP780 and DP980 high strength steel joints, Friction Stir Welding DP590 steel, resistance spot welding of DP590 steel and other researches. However, such works have not siginificantly reported in the combined joints between DP590 and DP980 steels. Consequently, the purpose of this study aims to investigate the dissimilar tailor-welded blanks between DP590 and DP980 dual phase steels. This work is beneficial for the development of future automotive technology.

2 Materials and methods

In this work, dissimilar joints between DP590 and DP980 dual phase steels were carried out with gas tungsten arc welding process (GTAW). Chemical compositions of used materials at room temperature shown in Table 1

Table 1. Chemical compositions of experimental (wt%)

Materials	C	Si	Mn	P	S	Al	V	B	Cu	Cr	Fe
DP590	0.12	0.40	1.60	0.025	0.010	0.015	0.2	0.005	0.2	-	Bal.
DP980	0.177	1.426	1.386	0.019	-	-	-	-	-	0.031	Bal.

* Corresponding author: rittichai.p@mail.rmutk.ac.th

As for the welding method, GTAW process was employed in order to autogenously butt weld joints between dual phase steels. Welding procedure was divided into two cases, namely, 1) a specimen welded along to rolling direction, and 2) a specimen welded transverse to rolling direction as illustrated in Fig. 1 a) and b) respectively. Unique device with backing argon gas, backing cupper plates, and clamping locks was used during welding in order to accomplish appropriate weld quality as shown in Fig 1 c). Before welding, these specimens were polished to eliminate undesired oxides on surfaces. Then, acetone was applied. During welding, specimens were clamped to prevent distortion. Specimen size and utilized welding conditions were set as Table 2. After joining, welded specimens were cut through electrical discharge machining (wire-cut EDM) for determining mechanical properties. Welded specimen geometry for tensile test was given as Fig.2 a) the welding along to rolling direction, and Fig.2 b) the welding transverse to rolling direction. Shimadzu Auto Graph AG-X plus Tensile tester

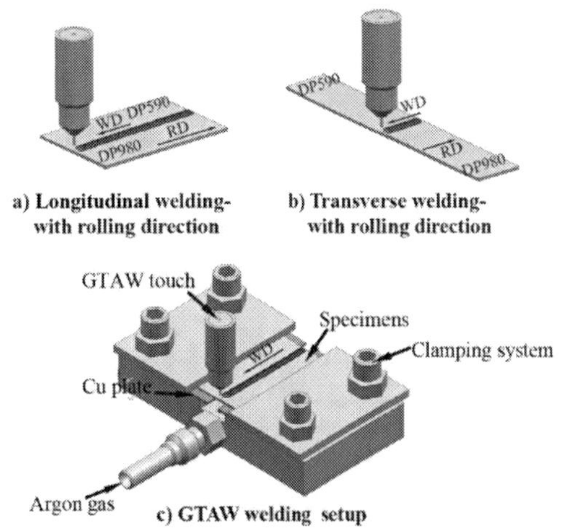

Fig. 1. Schematic representation of the GTAW setup

Table 2. Utilized welding conditions

Base metal	DP590 (60 mm x 20 mm x *t*1 mm)
	DP980 (60 mm x 20 mm x *t*1.2 mm)
Welding current	50 A
Shielding gas	Pure argon
Gas flow rate	15 L/min
Electrode extension	5 mm
Arc length	2 mm
Welding speed	6 mm/sec
Ceramic nozzle	Ø 6 mm

In addition, hardness profile along the cross-section weld from DP590 to DP980 was measured with micro-vickers hardness tester (Matsuzawa MMT-X3). Measuring point interval was given as indent spacing of 0.1 mm. with applied load of 100 grams, as well as hold time of 10 seconds. In microstructure examination according to ASTM E407 standards, the cross-sections of the welds were polished using an automatic polishing machine and then etched with 2% nitric acid solution for about 5-10 sec. Then, three different regions of the welded specimen were comparatively observed for metallurgical microstructures with optical microscopic (Leica DM-2500M), namely the base metals, the dissimilar weld metal, and the heat affected zone (HAZ).

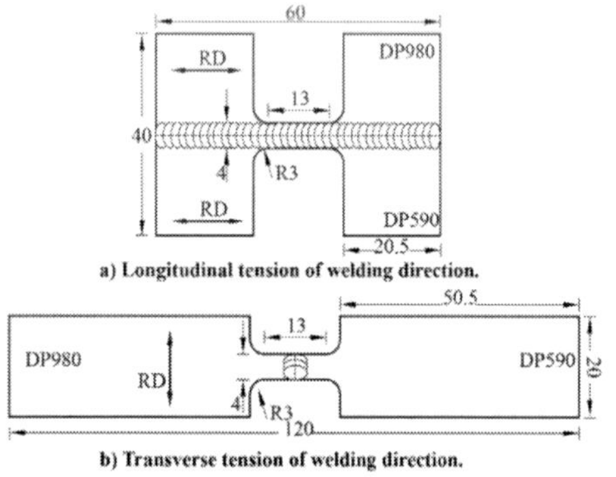

Fig. 2. Welded specimen dimension of tensile test,
a) Welding along to rolling direction,
b) Welding transverse to rolling direction. (Unit: mm)

3 Results and discussion

3.1 Weld surface observation

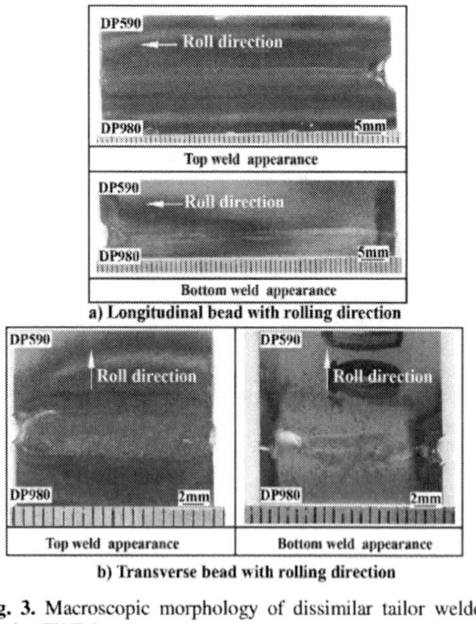

Fig. 3. Macroscopic morphology of dissimilar tailor welded blanks (TWBs)

After experimented with GTAW, the weld surface appearances of top and bottom side are exhibited as Fig. 3 a) the welding along to rolling direction and Fig. 3 b)

the welding transverse to rolling direction. As a result, it revealed that the weld was full penetration. The weld surface neither have undesired oxide films nor any defects. These evidences testified to appropriate welding conditions and pursuing gas system as well.

3.2 Mechanical properties

3.2.1 Hardness

In order to investigate the evolution of mechanical properties in the dissimilar tailor welded blanks between DP590 and DP980 steels, Fig. 4 exhibited hardness profile along the cross-section weld. It was found that the hardness value of the dissimilar weld was approximately 300-400 HV. Such hardness value was between the base metal of DP590 (217 HV) and DP980 (426 HV). However, particularly in HAZ, it evidently revealed that DP590-HAZ region induced the hardness value more than its base metal. And then, gradually decreasing approach to its base metal. This increased hardness is probably caused by critical cooling rate nearby the dissimilar weld during welding.

In contrast, DP980-HAZ region exhibited the soften zone with the hardness value lower than its base metal. This decreased hardness was probably due to the tempering treatment during welding [6, 7]. As a result, it strongly indicated fluctuating characteristics of dissimilar tailor welded blanks, especially in HAZ.

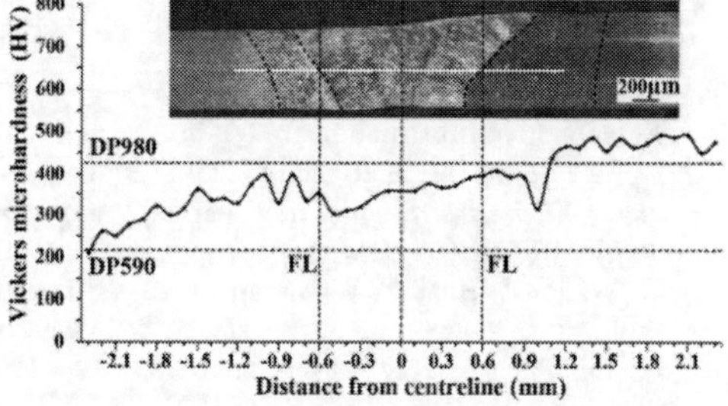

Fig. 4. Distribution profile of hardness value of dissimilar tailor-welded blanks.

3.2.2 Tensile strength.

A comparative study between the specimens welded along and transverse to rolling direction was carried out through tensile test. Engineering stress-strain curves are represented in each case as Fig. 5. Consequently, it clearly revealed that the specimens welded along to rolling direction induced a superior tensile strength (1092 MPa), and yield point (692 MPa) as well. On the other hand, it is noticed that such tensile strength is close to DP980 properties. Whilst, the specimens welded transverse to rolling direction encouraged slightly more elongation. This welding case contributed to tensile strength of 638 MPa and yield point of 438 MPa. In addition, the fracture characteristics were substantially considered. It was found that, in case of welding along to rolling direction, the fracture location occurred at the central dissimilar weld. This result was able to prove the strength of longitudinal dissimilar weld. On the other hand, welding transverse to rolling direction, the fracture was apparently located on the region of DP590 base metal rather than HAZ or dissimilar weld. It was remarkably noticed welding with a given conditions could not influence on the strength of HAZ regions.

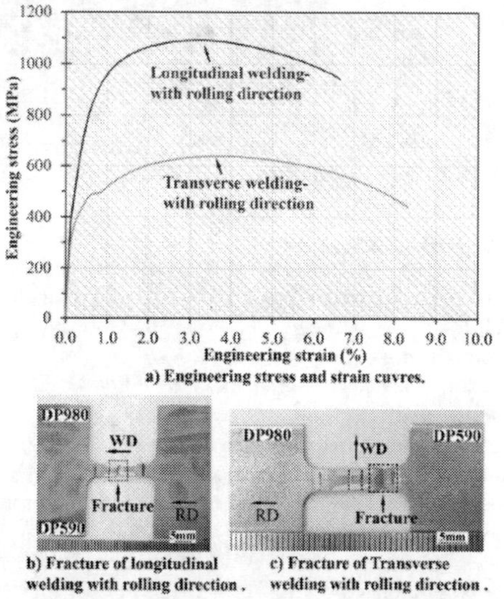

Fig. 5. Comparison of engineering stress and strain curves.
a) Engineering stress and strain curves,
b) Fracture of longitudinal welding with rolling direction.
c) Fracture of transverse welding with rolling direction.

3.2.2 Microstructure morphology.

Figure 6 exhibits typically cross-sectional weld in the case of welding transverse to rolling direction. Microstructures on each region were compared to each other, namely dissimilar weld metal (WM), DP590-HAZ, DP590 base metal (DP590-BM), DP980-HAZ, and DP980 base metal (DP980-BM). As a result, welding with a given conditions was able to attain complete penetration weld. There were no imperfections like porosity or impurity inclusion as represented in Fig. 6 a) Whilst, Fig. 6 b) and c) represents typical morphology of DP590 and DP980 base metal which consist of a ferrite matrix with martensite particles. It can be observed that the matrix structure in DP980 was coarser than DP590 base metal. In the dissimilar weld, this region was re-melted in order to combine between DP590-BM and DP980-BM. Therefore, microstructure characteristics in the fusion zone were radically evolved. As Fig. 6 c), WM exhibits columnar grained martensite along the ferrite matrix in re-melted region. This was the reason why there was the hardness transfiguration in the weld as mentioned before. Aside, DP980-HAZ exhibited a tempered martensite structures. For that reason, lower hardness in DP980-HAZ than DP980-BM was significantly induced. On the contrary, fine columnar martensite with a ferrite matrix was found in DP590-HAZ. These evidences encouraged to increasing hardness more than DP590-BM. As a consequence, it was particularly noticed that thermal distribution during welding effects on HAZ in different ways.

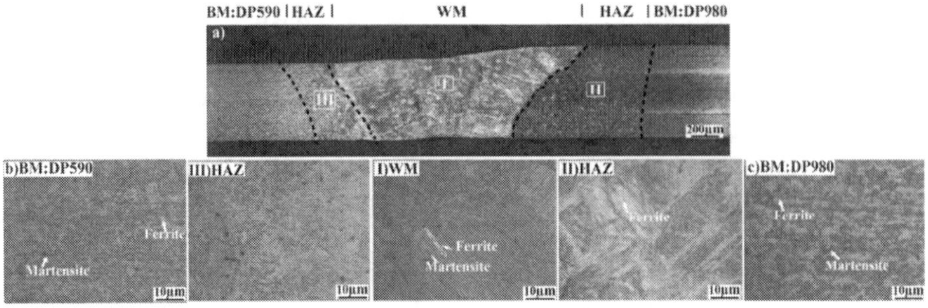

Fig. 6. Overview of microstructural weld joint and each position of weld metal a) dissimilar weld joint b) BM DP590 c) BM DP980

4 Conclusions

This paper describes the study of dissimilar tailor-welded blanks between DP590 and DP980 dual phase steels using GTAW process. The remarkable investigations can be summarized as follows.

1) According to the utilized GTAW condition and pursuing gas system, it revealed that DP590 and DP980 steels butt joint was able to achieve complete penetration joint. There were not any defects and undesired oxide films. Weld bead appearance was smoothly blended.
2) In hardness profile, dissimilar tailor-welded blanks exhibited strongly fluctuating characteristics. In other words, the dissimilar weld hardness value was between the DP590 and DP980 base metal. In particularly, DP590-HAZ region increased hardness value while DP980-HAZ region induced a soften zone.
3) The harden DP590-HAZ was testified through fine columnar martensite with a ferrite matrix. On the other hand, the soften DP980-HAZ represented a tempered martensite structures. Microstructure evolution was strongly caused by thermal distribution during welding and distinguished material characteristics.
4) Tensile test was carried out in order to investigate dissimilar welded joint strength in different rolling directions. It revealed that the welding along to rolling direction produced virtually greater tensile strength of the welded specimen than the welding transverse to rolling direction. Furthermore, the weakest fracture occurred apparently on the region of DP590-BM rather than HAZ.
In the future, additional investigations should be conducted with computational welding simulation in order to demonstrably understand thermal distribution in HAZ which influences harden or soften zone.

Acknowledgments

The authors thank the research and development institute of Rajamangala University of Technology Krungthep for project funding and support.

References

[1] A. Zadpoor, J. Sinke, and R. Benedictus, *Mechanics of Tailor Welded Blanks: An Overview* vol. 344, 2007.
[2] J. H. Lee, S. H. Park, H. S. Kwon, G. S. Kim, and C. S. Lee, "Laser, tungsten inert gas, and metal active gas welding of DP780 steel: Comparison of hardness, tensile properties and fatigue resistance," *Materials & Design,* vol. 64, pp. 559-565, 2014/12/01/ 2014.
[3] S. Gao, Y. Li, L. Yang, and W. Qiu, "Microstructure and mechanical properties of laser-welded dissimilar DP780 and DP980 high-strength steel joints," *Materials Science and Engineering: A,* vol. 720, pp. 117-129, 2018/03/21/ 2018.
[4] H. S. Di, Q. Sun, X. K. Nie, X. N. Wang, and X. M. Chen, "Mirostructure and properties of laser welded joints of dual phase and press-hardened steels," *Procedia Engineering,* vol. 207, pp. 1665-1670, 2017/01/01/ 2017.
[5] B. Wang, L. Hua, X. Wang, and J. Li, "Effects of multi-pulse tempering on resistance spot welding of DP590 steel," *The International Journal of Advanced Manufacturing Technology,* vol. 86, pp. 2927-2935, 2016/10/01 2016.
[6] K. Bandyopadhyay, S. K. Panda, P. Saha, and G. Padmanabham, "Limiting drawing ratio and deep drawing behavior of dual phase steel tailor welded blanks: FE simulation and experimental validation," *Journal of Materials Processing Technology,* vol. 217, pp. 48-64, 2015/03/01/ 2015.
[7] W. Xu, D. Westerbaan, S. S. Nayak, D. L. Chen, F. Goodwin, and Y. Zhou, "Tensile and fatigue properties of fiber laser welded high strength low alloy and DP980 dual-phase steel joints," *Materials & Design,* vol. 43, pp. 373-383, 2013/01/01/ 2013.

In-service operating conditions affecting on weld HAZ hardness for API5L Gr.B pipe steel maintenance.

Supolchai Kwankaew[1], *Rittichai* Paoniam[2*], *Bovornchok* Poopat[1] and *Sermsak* Srisutraporn[1]

[1]Production Engineering Department, Faculty of Engineering, King Mongkut's University of Technology Thonburi, Bangkok, Thailand
[2]Mechanical and Industrial Engineering Department, Faculty of Engineering, Rajamangala University of Technology Krungthep Bangkok, Thailand

Abstract. This study aims to investigate the influences of in-service welding conditions on maximum HAZ hardness, namely heat input, pipe wall thickness, and medium flow rate. The research consists of two parts; computationally thermal analysis and actual welding experiments. In conclusion, it suggested that increasing heat input resulted in decreasing maximum HAZ hardness, as well as, the range of run pipe thickness in the experiment was insignificantly influenced on HAZ hardness. Additionally, the higher flow rate induced more HAZ hardness with a certain flow rate level. HAZ hardness between the sleeve plate and the run pipe was insignificantly different. Nevertheless, it indicated that HAZ hardness prediction via computational analysis was more conservative than experiments.

1 Introduction

The In-service welding is widely accepted solution for pipeline repairing that can be applied onto pressurized pipeline and non-stopping natural gas transportation. Especially, when the system cannot be isolated or shut down. However, welding onto lived pipeline can cause Hydrogen Induced Cracking (HIC)[1-3] due to the flowing fluid within the system. In practice, avoiding hydrogen induced cracking is undertake through reducing HAZ hardness. However, there is unclear main effect of individual welding conditions on HAZ hardness. The limited welding parameters are capable to select to meet the desire conditions.

As relevant studies [3-9] have reported involving thermal analysis models and welding procedure qualifications. Fischer[3] developed a calculation method using finite difference to determine the heat transfer coefficient in In-service welding this model can use to calibrate the medium conditions which are used during welding. Nolan[4] developed two hardness prediction models for using to select welding parameters to achieve a satisfied hardness level to avoid the risk of cracking in In-service welding. One of factors is considered as an indicator to evaluate the susceptibility of Hydrogen Induce Cracking is HAZ hardness[1]. Bailey[5] investigated HAZ cracking by the use of SMAW wires with different hydrogen level. The result showed that the cracking did not occur when the critical hardness was not more than 350 HVN. Gravell[6] advices 350 is a conservative hardness limit for using to asses HIC in In-service welding. Cassie[7] studied the effect of material composition on weld hardness In-service welding with CE varied from 0.25 to 0.48. Cassie found that the material composition is directly related to weld hardness. Weld hardness raised with increasing carbon equivalent. DeHertogh[8] conducted welding with artificial flow to study effect of preheat temperature on hardness by simulating four different preheat levels and found that the size of HAZ decreased as decreasing preheat temperature. However, these studies are unclear regarding the effects of in-service welding conditions, especially in API5L Gr.B pipe steel applications.

Therefore, in order to find out properly welding procedures based on severely in-service operating conditions, this research is aimed to investigate the influence of three (3) welding parameters on the maximum HAZ hardness on API 5L GR. B pipe steel, namely welding heat input, run pipe thickness, and medium flow rate. These operating conditions needs to appropriate parameters for in-service piping welding in the future.

2 Experimental Procedure

In this study, operating conditions during in-service pipeline welding were investigated in order to determine maximum HAZ hardness. There are three (3) factors were studied, namely welding heat input, run pipe thickness, and medium flow rate.

2.1 Tested Materials

Pipe steel API5L Gr.B standard was used as tested material. Pipe size of 219.8 mm in diameter and 8.18 mm thick was employed. Chemical compositions of tested material was examined with optical emission

* Corresponding author: rittichai.p@mail.rmutk.ac.th

spectrometer (OES) as shown in Table 1. Carbon equivalent (CE_{IIW}) was calculated as 0.37.

Table 1. Chemical compositions of tested materials (wt%)

C	Mn	Cu	Ni	Cr	Mo	V
0.134	0.969	0.181	0.099	0.223	0.049	0.003

Note. $CE_{IIW} = C + Mn/6 + (Cu+Ni)/15 + (Cr+Mo+V)/5$

GTAW process was performed with a filler metal of ER70S-G according to AWS A5.18 standard, as well as, EWTH-2 ttungsten electrode of 2.4 mm in diameter. Argon shielding gas of 99.9% was used as flow rate of 15 L/min. DCEN current polarity was given.

2.2 Methods

This work was divided into two parts. Firstly, computational simulation, commercial PRCI program (Thermal analysis model for hot tap welding V4.2) was applied. In-service operating conditions during GTAW were considered for calculation, namely welding heat input, chemical compositions of tested material, run pipe thickness, and amount of medium flow rate. Such thermal analysis program was beneficial to estimate maximum HAZ hardness value depending on in-service operating conditions. PRCI program computed thermally based on finite element method (FEM) and HAZ hardness prediction by means of Yurioka's Model[9]. Schematic of PRCI-Thermal analysis during welding is illustrated as Fig.1. Utilized parameters for simulations and experiments were given as Table 2 respectively.

Secondly, actual experiments were carried out as in-service operating conditions which were suggested through PRCI thermal analysis. The experimental set up as shown in Figure 2 was accomplished according to API 1104 Annex B. This means that a sleeve plate is attached to a main run pipe to form a fillet weld lap joint. The welding was performed with GTAW(KEPPI MasterTIG) in 5G uphill position. In this experiment, water was circulated instead of methane as the medium for safety reason. The water flow rate during welding was calculated in order to obtain the convective heat transfer be equivalent to in-service natural gas (Methane) pipelines[10]. Water flow rate were measured and monitored with ultrasonic flow meter (LONGRUN LRF-200 series). In order to regulate the certain experiment conditions, at both inlet and outlet of the main run pipe, temperature gauges were installed for monitoring difference between the outlet and the inlet. And also a pressure gauge was set to measure inline pressure during welding.

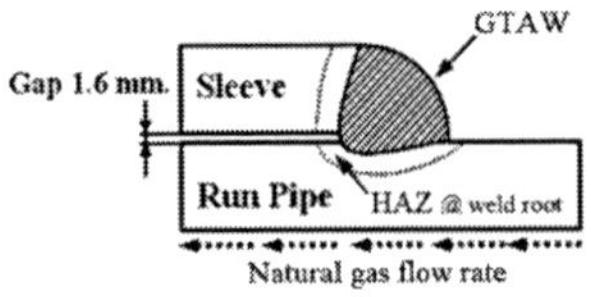

Fig. 1. Schematic of PRCI-Thermal analysis during welding

Table 2. Utilized parameters for experiments

Case	Heat Input (HI) kJ/mm (kJ/inch)	Pipe Thickness (mm)	PRCI (Methane) Flow rate (mmscfd)	Experiment (Water) Flow rate (m^3/hr.)
A	1.2(30)	8	42	19.1
B	2.0(50)	8	42	19.1
C	2.8(70)	8	42	19.1
D	2.8(70)	4	37	16.9
E	2.8(70)	6	37	16.9
F	2.8(70)	8	37	16.9
G	2.8(70)	8	0	0
H	2.8(70)	8	44	19.7

Note. mmscfd: Million standard cubic feet per day

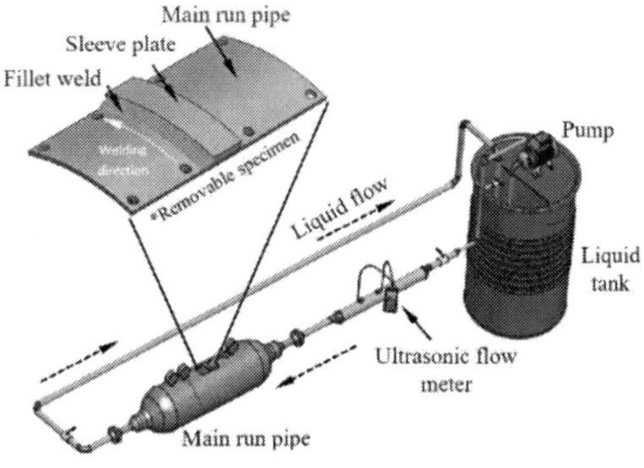

(a) Schematic of in-service pipeline welding

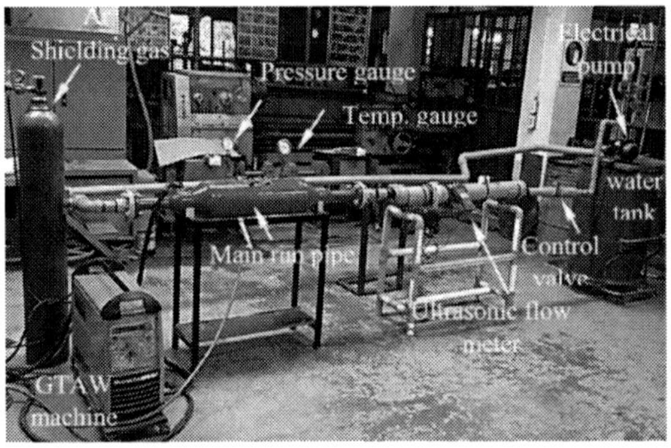

(b) Experimental setup of in-service pipeline welding

Fig. 2. Experimental Set Up

Generally, the weld root has the potential for hydrogen induced cracking because of the stress concentration due to the gap between sleeve and pipe. Then, each welded specimen was cut into 3 test pieces. Later on, those test pieces were grinded, polished and etched with 2% Nital acid. Optical microscopic (Leica DM-2500M) was used so as to examine a sound weld. In addition, maximum hardness in the HAZ area of coarse-grained region at the root weld pass was averagely measured with micro-vickers hardness tester (Matsuzawa MMT-X3). Measuring procedure was operated in accordance with ASTM E384-11 as shown in Fig 3. That is measuring

https://doi.org/10.1051/matecconf/201819201042

point interval was given as 0.2 mm from a fusion line, an indent spacing of 0.5 mm with applied load of 500 gram, as well as hold time of 10 seconds. The hardness value for evaluation was obtained from average maximum hardness of 3 specimens.

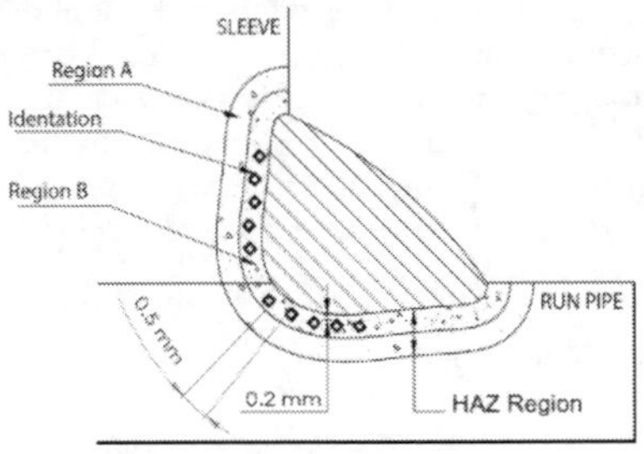

Figure 3. HAZ Hardness Measuring Procedure

3 Results and discussion

3.1 Effect of welding heat input

As completed welding experiments, Figure 3 shows the cross sectional area of the welded specimen in different heat inputs. The weld profile of entire specimens welded with GTAW were concave shape. The HAZ dimension in the pipe was wider than in the sleeve. As the results, heat input of 1.2, 2.0, and 2.8 kJ/mm revealed that the depth of penetration was not different, but the HAZ of the run pipe was wider than the sleeve. On the top of that as shown in Fig. 4, the PRCI result indicated that reducing maximum HAZ hardness at the weld root was influenced with increasing heat input. Similarly to the hardness of run pipe, HAZ hardness reduce significantly after increasing heat input. On the other hand, the hardness of sleeve did not change although heat input increased.

3.2 Effect of run pipe thickness

In this case, the run pipe thickness of 4, 6, and 8 mm was studied. Figure 5 shows the weld cross sections of different run pipe thickness. From the results, PRCI calculations exhibited that maximum HAZ hardness was almost unchanged when increasing the pipe thickness. In addition to PRCI calculations, experimental results were slightly difference and not agree to computational prediction as presented in Figure 6.Comparing HAZ hardness between run pipe and sleeve, the HAZ hardness of the run pipe and the sleeve of 4 and 6 mm were insignificantly difference. But the HAZ hardness is different when the run pipe thickness reach 8 mm. It may affect from the different gap between the run pipe and sleeve.

Welding Heat Input

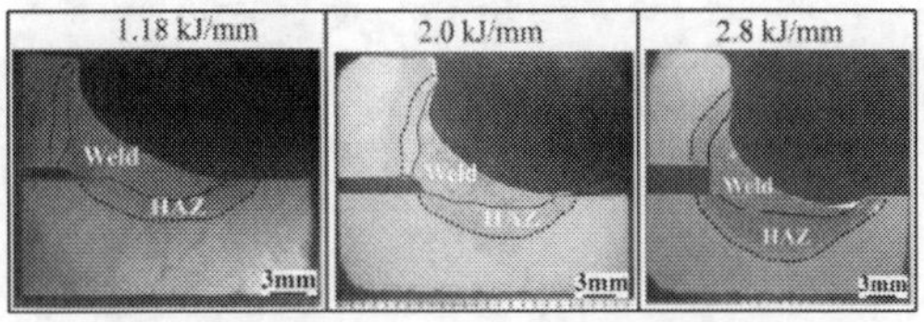

Fig. 3. The weld cross sections due to heat input
(the main pipe of 8 mm thick, flow rate of 42 mmscfd)

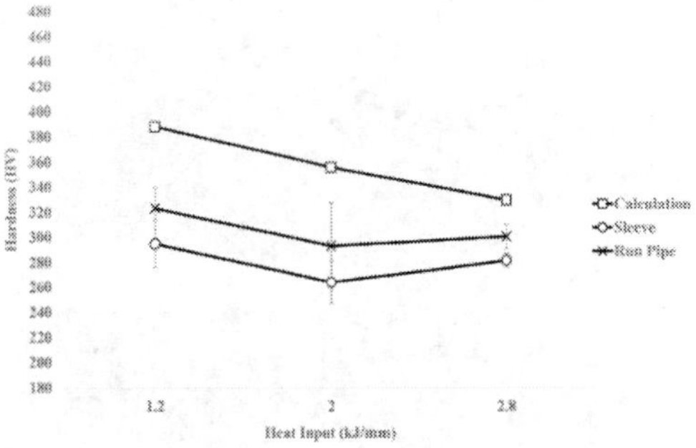

Fig. 4. Max. HAZ hardness and Heat Inputs
(the main pipe of 8 mm thick, flow rate of 42 mmscfd)

Run Pipe Thickness

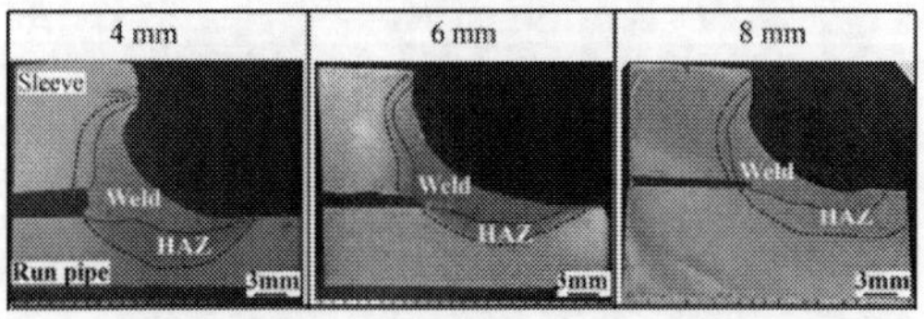

Fig. 5. The weld cross sections due to different pipe thickness
(heat input of 2.8 kJ/mm, flow rate of 37 mmscfd)

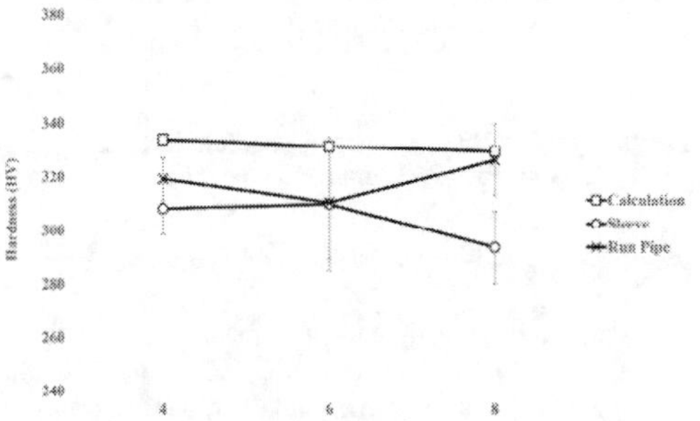

Fig. 6. Max. HAZ hardness and the run pipe thickness
(heat input of 2.8 kJ/mm, flow rate of 37 mmscfd)

3.3 Effect of medium flow rate

The influence of flow rates on maximum HAZ hardness is presented in Fig.7 and 8. Both the results of the experiments and the PRCI calculations indicated that when the flow rate increased from 0 to 37 mmscfd, maximum HAZ Hardness raised up accordingly. However, when the flow rate was at the range of flow rate between 37 and 44 mmscfd, the maximum HAZ hardness was constantly remained. This is mean that, increasing flow rate has a tendency to promote higher maximum HAZ hardness but there is reaching a certain level.

Volumetric Flow Rate

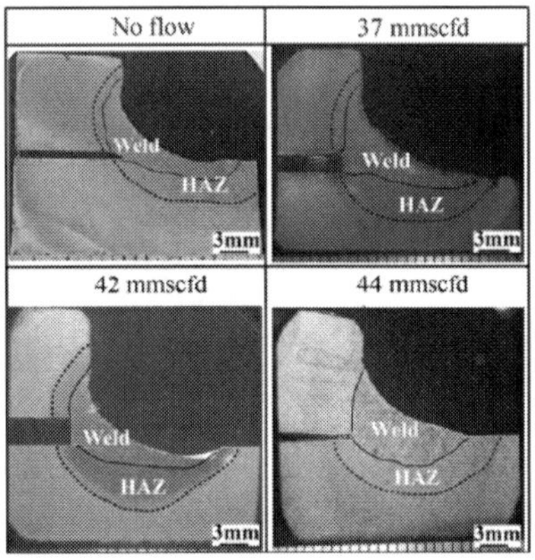

Fig. 7. The weld cross sections due to different flow rate (heat input of 2.76 kJ/mm, run pipe thickness of 8 mm)

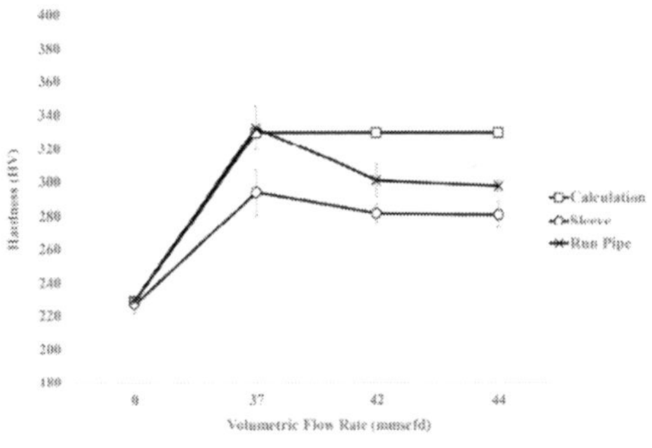

Fig. 8. Max. HAZ hardness and the medium flow rate (heat input of 2.76 kJ/mm, run pipe thickness of 8 mm)

5. Conclusions

This research investigated the influences of the in-service operating conditions on HAZ hardness, namely welding heat input, run pipe thickness, and medium flow rate. The results can be concluded that;

1) Welding heat input was significant effect on maximum HAZ hardness at the weld root. That is an increasing heat input resulting in decreasing maximum HAZ hardness.
2) In this case, HAZ hardness was insignificantly influenced on the range of run pipe thickness from 4 to 8 mm. This was suggested through PRCI calculations and the actual experiments.
3) As for medium flow rate, maximum HAZ hardness increased with rising up the flow rate. However, when increasing the flow rate to reach a certain level, the hardness was no longer changed significantly.
4) From the results, HAZ hardness prediction using PRCI thermal analysis was significantly higher than experiments indicating more conservative. Whilst HAZ hardness between the sleeve plate and the run pipe was slightly different.

Acknowledgments

The authors thank to the research and development institute of Rajamangala University of Technology Krungthep, King Mongkut's University of Technology Thonburi and Four Wheelers Company for project funding and support.

References

[1] W. A. Bruce, "Welding Onto In-Service Pipelines: A Review, paper presented at Pipeline Welding," International Symposium on Pipeline Welding, 1998.
[2] Phelps B Cassie B. A & Evans N. H, "Welding Onto Live Natural Gas Pipelines, Metal Construction," August 1976.
[3] K. J. F. W. G. R. Fischer R. D., "User Manual for Modell & Model 2 Computer Programs for the Predicting Critical Cooling Rates and Temperatures During Repair and Hot Tap Welding on Pressurised Pipelines," Battelle Memorial Institute ReportJune 1981 1981.
[4] Z. S. a. D. D. D. Nolan, "Modelling of HAZ Hardness in C - Mn Pipeline Steels Subjected t o In - Service Welding Procedures "."
[5] Bailey N, "Welding Procedures for Low Alloy Steels," ed. The Welding Institute Cambridge England, 1970.
[6] Graville B. A. & Read J. A, "Optimization of Filletweldsizes," Welding journal, vol. 53, pp. 161s-169s, 1974.
[7] Cassie B. A, "The Welding of Hot Tap Connections to High Pressure Gas Pipelines," paper presented Pipeline Industries Guild J. W. Jones Memorial Lecture, 1974.
[8] DeHertogh J. & Illeghems H, "Welding Natural Gas Filled Pipelines," Metal Construction & British Welding Journal, 1972.
[9] N. Yurioka S. Ohsita andH Tamehiro, "Study on Carbon Equivalents to Assess Cold Cracking Tendency and Hardness in Steel Welding," Proceedings of The Specialist Symposium on Pipeline Welding in the 80s.
[10] I.-W. BANG Y.-P. SON K. H. OH Y.-P. KIM AND W.-S. KIM, "Numerical simulation of sleeve repair welding of in-service gas pipelines," Welding Journal, vol. 81, pp. 273S-282S, 2002.

Production of low impurity aluminium rotor for motor efficiency enhancement

Savitri Payakkapol[1], *Nattapon* Chayopitak[2], *Panya* Kunsuwan[3], *Naoto* Ohtake[4] and *Sompong* Srimanosaowapak[5,*]

[1]International College, King Mongkut's Institute of Technology Ladkrabang, Bangkok 10520, Thailand
[2]National Electronics and Computer Technology Center (NECTEC), National Science and Technology Development Agency (NSTDA), Klong Luang, Pathumthani 12120, Thailand
[3]Faculty of Mechanical Engineering, King Mongkut's Institute of Technology Ladkrabang, Bangkok 10520, Thailand
[4]Department of Mechanical Engineering, Tokyo Institute of Technology, Tokyo, Japan
[5]National Metal and Materials Technology Center (MTEC), National Science and Technology Development Agency (NSTDA), Klong Luang, Pathumthani 12120, Thailand

Abstract. Induction motors have been widely used in various electrical applications. However, their efficiency are still limited due to losses occurring during operation,especially for smaller ratings. Higher the electrical conductivity of motor rotor by lower its impurity content is one of practical and simple way to improve motor efficiency. The aim of this work is to reduce the impurity content of aluminium rotor by addition of boron in the form of Al-5wt%B master alloy and $Na_2B_4O_7$-NaCl-KCl flux, and then sedimention of their precipitated particles. The purer melt was then cast as rotors for single-phase induction motors. With the amount of addition of boron in aluminium melt in this work, boron in Al-5wt%B master alloy was more pronounced in removing impurities in aluminium melt than that in $Na_2B_4O_7$. It was found that the rotors made of aluminium melt with lower impurity contents and hence higher electrical conductivity resulted in more enhancement of motor efficiency.

1 Introduction

Induction motors with aluminium squirrel-cage rotors have been widely used in various electrical applications. However, their efficiency are still limited due to losses associated with iron core, stray load, stator, rotor resistance, windage losses and friction, resulting in the low efficiency of around 70 % for smaller rating [1-2]. Consequently, improvement of motor efficiency is a subject which has recently received increasing attention [3-4]. Practical and simple method to improve motor efficiency is to enhance conductivity of existing aluminium squirrel-cage rotor, by increasing purity of the aluminium used. In general, it is more economical to use lower grade aluminium in the production process and enhance its purity during melting and rotor casting process, rather than using higher aluminium grade. This work aims to increase the conductivity of aluminium rotor by precipitation and sedimentation method for 750 W shaded-pole induction motor used in small refrigeration applications.

The experimental procedures consist of two parts. First, the impurity removal methods in aluminium using Boron and $Na_2B_4O_7$ treatments were investigated for aluminium sample with 99.91wt% in laboratory setup. Second, the more convenient method is then chosen to prototype the rotor of induction motor for efficiency test with industrial grade alunimium with 99.85wt% in real manufacturing production setup.

2 Experimental Procedures

2.1 Impurity removal methods using Boron and $Na_2B_4O_7$ treatments

Aluminium sample with 99.91 wt% purity, whose main chemical compositions measured using optical emission spectroscopy (OES) are given in Table 1, was used as a material to remove its impurity using Al-5wt%B (Boron) and $Na_2B_4O_7$ addition. Al-5wt%B master alloy in the form of rod produced by AMG Aluminum was used. The amount of B required for chromium, titanium, vanadium and zirconium removal as insoluble borides is determined using the below stoichiometric equation.

$$B = (\frac{M}{P}) * \frac{Cr + Ti + V + Zr}{2} \qquad (1)$$

where B is the total weight of the Al-B master alloy required (kg), Cr, Ti, V, Zr are the concentration of impurity elements (wt%), M is the weight of aluminium to be treated (kg), and P is the concentration of B in Al-B master alloy (wt.%) [5]. For the impurity removal using $Na_2B_4O_7$ addition, 0.9wt% $Na_2B_4O_7$, which gave high iron removal in commercial purity aluminium [6],

* Corresponding author: sompongs@mtec.or.th

Table 1. Chemical compositions of aluminium sample with 99.91wt% purity used in laboratory setup (wt%).

Fe	Si	Zn	Ti	V	Cr	Al
0.055	0.01	0.001	0.003	0.012	0.001	99.91

Table 2. Weight ratio of $Na_2B_4O_7$-NaCl-KCl flux.

NaCl	KCl	$Na_2B_4O_7$
10	10	9

Table 3. Chemical compositions of industrial grade aluminium 99.85wt% purity used for rotor prototype in manufacturing production (wt%).

Fe	Si	Zn	Ti	V	Zr	Al
0.076	0.036	0.002	0.002	0.017	0.001	99.85

was added to aluminium melt in the form of flux having been mixed with NaCl and KCl using a ball mill with the ratio given in Table 2.

Al-5wt%B master alloy and $Na_2B_4O_7$-NaCl-KCl flux were each vigorously stirred for 1 minute after addition in each approximately 3 kg aluminium melt at 720 ± 2°C. Each melt including the one without any addition was held for 15 hrs to precipitate and sediment of boride and other particles. The melt was sampled for chemical analysis prior to boron addition and at intervals during the holding period. At the end of the experiments, each remaining melt was left to cool inside the furnace, after which the observation of boride particles at the bottom of the remaining metal was performed using Hitachi SU 5000 scanning electron microscopy with energy dispersive spectroscopy (EDS) system.

2.2 Aluminium rotor prototype using Boron treatment

Industrial grade aluminium with 99.85wt% purity, which was boron treated using Al-5wt%B master alloy, was used for rotor production at 720 ± 2°C using a 350 ton high pressure die casting machine. Chemical compositions of the aluminium prior to impurity removal are given in Table 3. After degassing and 9.5g Al-5wt%B master alloy addition, melt was sampled for chemical analysis and taken for rotor casting each of about 0.38 kg at 5-minute intervals up to 1 hr. Biscuit and runner of each produced rotor casting were then sectioned and measured their porosity using Archimedes method [7]. Selected rotor castings having low porosities of their biscuit and runner were machined and assembled with a single phase shaded-pole induction motor rating 750 W, 1,300 rpm, 220 V AC, as shown for example in Figure 1. Efficiency of motors was then tested using a MAGTROL Hysteresis dynamometer at the speed of 1,300 rpm.

3 Results and discussion

3.1 Impurity removal using Boron and $Na_2B_4O_7$ treatments

Fig 1. A prototype of the single phase shaded-pole induction motor rating 750 W, 1,300 rpm, 220 V AC.

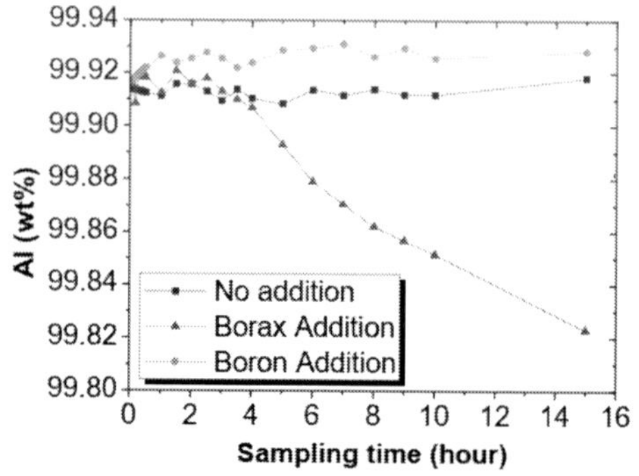

Fig. 2. Al content in 99.91wt% purity Al melts which had experienced different purification processes and holding times.

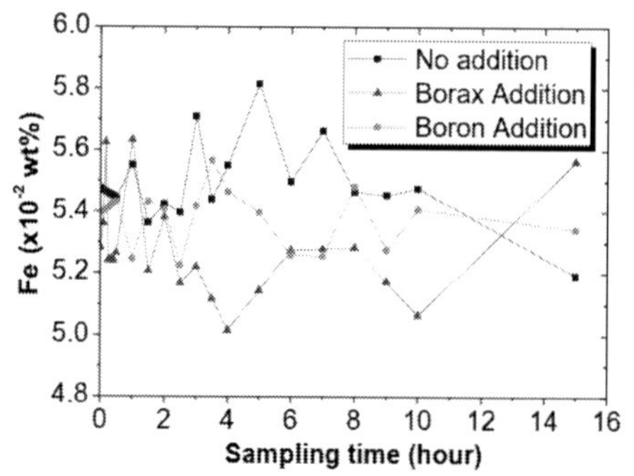

Fig. 3. Fe content in 99.91wt% purity Al melts which had experienced different purification processes and holding times.

In laboratory setup, after holding the melts with Al-5wt%B master alloy addition, $Na_2B_4O_7$-NaCl-KCl flux addition and without any addition for 15 hrs, the chemical compositions of the melts, except for the one without addition, changed from their initial values, where the results are shown in Figures 2 - 5. The melts were found to be purer after the addition of Al-5wt%B master alloy and $Na_2B_4O_7$-NaCl-KCl flux during the whole and only initial period of holding, respectively, as shown in Figure 2. It should be noted that the addition of Al-5wt%B master alloy results in purer melt than the addition of $Na_2B_4O_7$-NaCl-KCl flux. However, Fe content in the melt with $Na_2B_4O_7$-NaCl-KCl flux addition decreased more noticeably than the melt with Al-5wt%B master alloy addition, as shown in Figure 3. Al-5wt%B master alloy addition, on the other hands, was more effective in removing V and Ti than $Na_2B_4O_7$-NaCl-KCl flux addition, as shown in Figures 4 and 5. Samples of the sedimented layer in the melts after the precipitation-sedimentation process showed the presence of particles. Figure 6 shows, for example, the backscattered electron image of sedimented layer in the melt with Al-5wt%B master alloy addition after precipitation and holding at 720°C for 15 hrs. Most particles are borides and the majority of them contained V and Ti, which were identified as $(V, Ti)B_2$. Evidence and discussion of transition metal borides precipitating after boron additions in aluminium melts have been reported [6, 8-10]. This suggests that transition metal borides are stable in molten aluminium which paves the way for simple and practical removal of transition metal impurities and hence enhancement of electrical conductivity in aluminium.

3.2 Aluminium rotor prototype using Boron treatment

In manufacturing production setup, only the Boron treatment with Al-5wt%B master alloy is chosen for effectiveness and practicality because it can reduce Ti, V and Zr, whereas the application of $Na_2B_4O_7$-NaCl-KCl flux can only reduce Fe significantly. Also it is more convenient in existing manufacturing procedure with simple addition of Al-5wt%B master alloy in the melt, whereas the application of $Na_2B_4O_7$-NaCl-KCl flux requires more steps in preparation. The 99.85wt% purity aluminium melt after the Al-5wt%B master alloy addition was sampled for chemical analysis and taken for rotor casting every 5-minute intervals. According to the chemical analysis result of the samples, it was found that the content of V, Ti and Zr decreased with increasing holding time as shown in Figure 7(a), whereas Fe and Si remained unchanged as shown in Figure 7(b). Figure 8 shows an example of aluminium rotor casting. Biscuit and runner of each rotor castings were taken for porosity measurement, whose result is shown in Figure 9. It can be noted from this figure that the majority of biscuits and runners contained porosity less than 0.75%. However, some castings contained higher porosities which may occur from the poor handling or pouring of the melt, which were carried out manually. Only the rotors produced from the reasonably low porosities of biscuits and runners were therefore used for prototyping 750 W single phase induction motors. Efficiency of the motors having the selected low porosity rotors is shown in Figure 10. It can be seen from this figure that, for the samples considered, the motor efficiency increased as the holding time of melt increased, although more test data are needed to establish a more defined trend. The enhance of motor efficiency with increasing holding time seems to be associated with a decrease in impurity quantity to a greater or lesser extent. Based on the relation of resistivity on impurity contents given in [11], the resistivity of aluminium rotors, which were taken to make-up the motors, in this work is estimated to be 3.151, 3.149, 3.139, 3.145, 3.132 μΩ.cm for the rotors holding time after the Al-5wt%B master alloy addition. Hence, with the increase of aluminium purity, resulting in decreasing resistivity of aluminium casting in the rotor, this procedure helps reduceing the power losses in the rotor that eventually improves the overall efficiency of the motor.

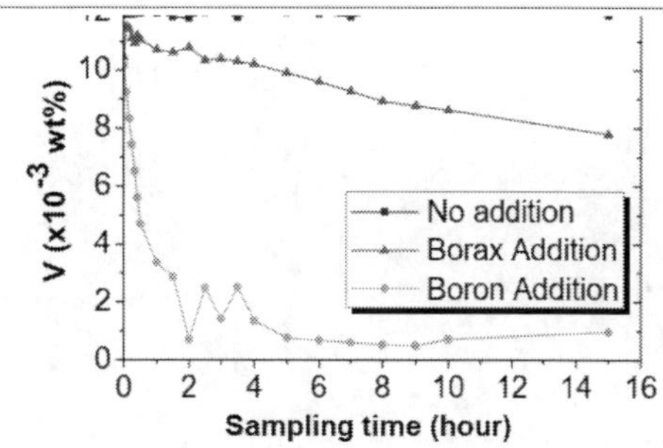

Fig. 4. V content in 99.91wt% purity Al melts which had experienced different purification processes and holding times.

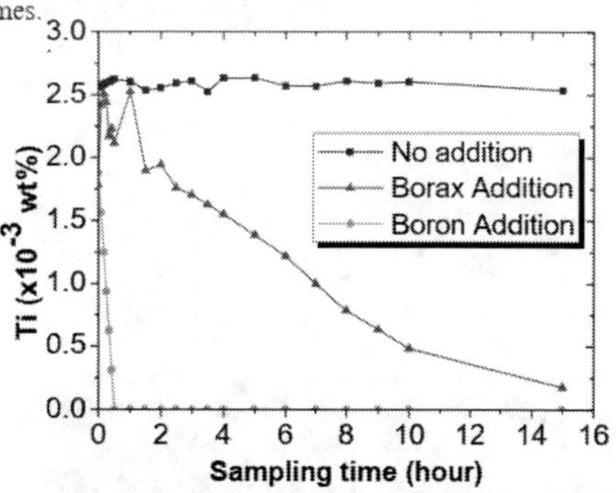

Fig. 5. Ti content in 99.91wt% purity Al melts which had experienced different purification processes and holding times.

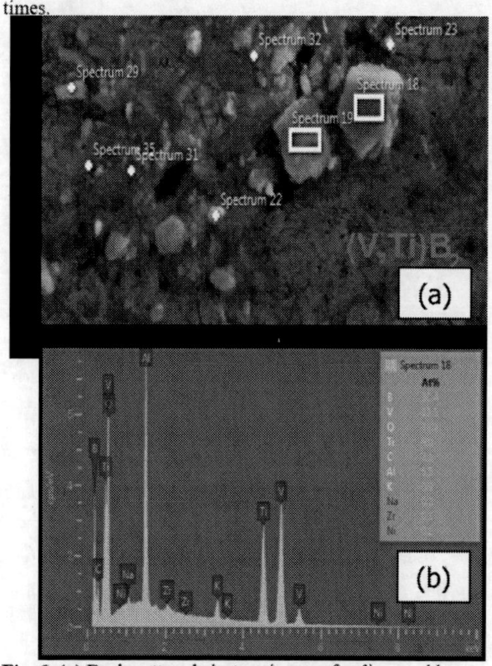

Fig. 6. (a) Backscattered electron image of sedimented layer and **(b)** chemical spectrum showing the presence of borides.

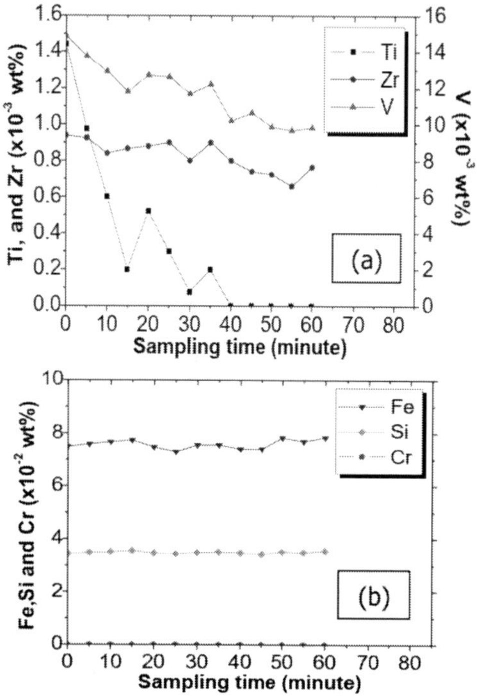

Fig. 7. Content of main impurities: (a) Ti, Zr, V and (b) Fe, Si, Cr, in 99.85wt% purity Al with Al-5wt%B master alloy held for sedimentation at different times.

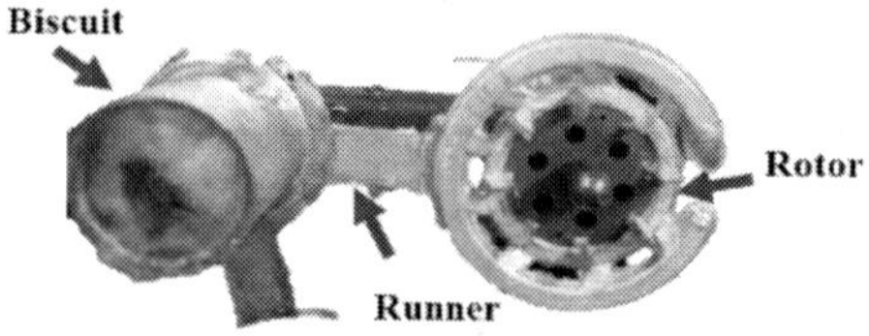

Fig. 8. Example of an aluminium rotor casting.

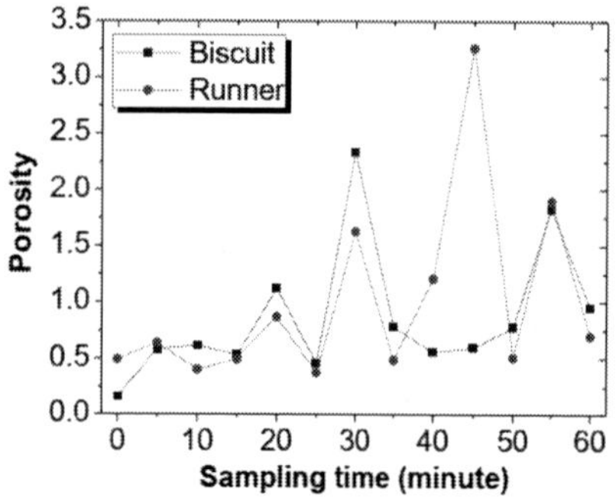

Fig. 9. Porosity of biscuit and runner of aluminium rotor castin

4 Conclusions

The precipitation and sedimentation of borides of transition metals in aluminium melt after addition of Al-5wt%B master alloy and $Na_2B_4O_7$-NaCl-KCl flux resulted in purer aluminium melt. With the amount of addition of boron in aluminium melt in this work, boron in Al-5wt%B master alloy was more pronounced in removing impurities in aluminium melt than that in

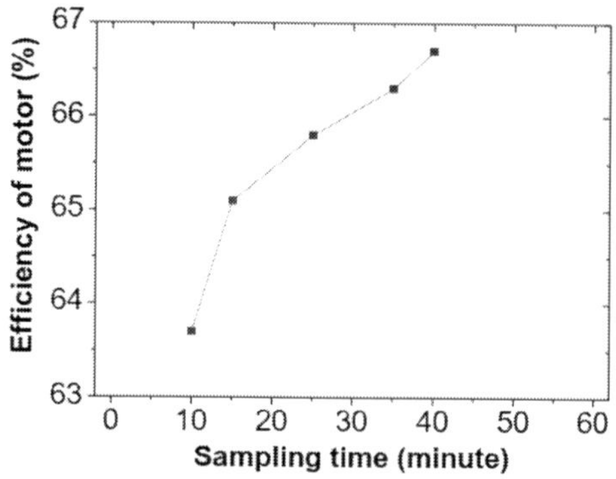

Fig. 10. Efficiency of single phase induction motors (750 W and 1,300 rpm) having rotors made of aluminium melt with Al-5wt%B master alloy held for sedimentation at different times.

$Na_2B_4O_7$. Although the addition of $Na_2B_4O_7$-NaCl-KCl flux can reduce Fe significantly, the addition of Al-5wt%B master alloy can reduce Ti, V and Zr. It was found that the rotors made of aluminium melt of industrial grade aluminium with lower impurity contents resulted in more enhancement of motor efficiency. It is expected that the motor efficiency can be improved further with the increase of the holding time of melt of the proposed method. The possible future work is to find the practical methods for transitioning more impurity metal to improve the efficiency of the motor.

The authors would like to express the appreciation to Kulthorn Electric Co.,Ltd Company for supplying materials and supporting with rotor casting production of motor prototypes and efficiency test.

References

1. Technical note: IEC 60034-30-1 standard on efficiency classes for low volttage AC motors, ABB, 1-2 (2014)
2. R. Tiwari, A.K. Bhardwaj, Inter. J. Innov. Res. Elec. Electron. Instr. Contr. Eng., **2**, 1552-1558 (2014)
3. S. Bauer, W. Braun, P. Orban, K. Schleede, P. Smith, Cs. Walczer, J. Mach. Manuf., **XLIX**, 82-83 (2009)
4. H. Su, H. Wen, X. Zheng, J. Su, Procedia Eng., **174**, 1221-1228 (2017)
5. S. Karabay, I. Uzman, J. Mater. Process. Tech., **160**, 174–182 (2005)
6. J. W. Gao, D. Shu, J. Wang, B.D. Sun, Mater. Sci. Technol., **25**, 619–624 (2009)
7. V.D. Tsoukalas, Mater. Des., **29**, 2027-2033 (2008)
8. P. Cooper, M. Kearns, *Alminium Alloys: Their Physical and Mechanical Properties*, Pts 1-3, **217**, 141-146 (1996)
9. A. Khaliq, M.A. Rhamdhani, G.A. Brooks, J. Grandfield, Metall. Mater. Trans. B, **45**, 784-794 (2014)
10. T.X. Ping, ZK. Hong, S.D. Fu, Z.X. Ming, Chinese J. Nonferrous Met., **24**, 1401-1407 (2014)
11. W.A. Dean, Aluminium, **1**, 174 (1967)

Application of artificial neural network for design: a case of paper helicopter

Tossapol Kiatcharoenpol and *Tanaporn* Klangpetch

Industrial Engineering Department, Faculty of Engineering, King Mongkut's Institute of Technology Ladkrabang, Bangkok, Thailand

Abstract. The design engineering is one of essential work in modern manufacturing environment. The optimization is principal technique to be used widely for searching the solution. However, primary process of optimization is to know the relation between design input parameters and target output. In this work, an artificial neural network (ANN) approach as an intelligent algorithm is proposed to construct the relation and also provides it in form of mathematic modeling. Even though the ANN modeling is so call a back-block due to difficulty to understand complicated equations, it is simply constructed by automate iteration process. A case of paper helicopter is used as an example of the application. The classical 2k Factorial design is used to provide an experiment plan to create training and testing data. 93 experiments are carried out. The architecture of ANN is set according to lowest Mean square error (MSE) of training and testing procedure. The result of 5-10-1 architecture has shown ability to accurately predict output, landing time, with MSE of 0.012. With such a highly quantitative accuracy of results, the developed model using the neural network approach can be used for finding the suitable input parameters to achieve a desired target output. In this case, the design of dimension (A) Depth of cut wing is 1.3 cm., (B) Length of wing is 12.9 cm., (C) Length of body is 9.0, (D) Width of body is 2.0 cm., and (E) Depth of cut body is 0 cm. yield the lowest area of a paper helicopter that can meet the target landing time, 2.85 $\pm$ 5% second.

1 Introduction

Application of artificial neural networks (ANN) has been reported by a number of researchers [1-3]. In this study, a Backpropagation neural network has been proposed to map relation between inputs and outputs of a paper helicopter case. The inputs are the dimensions of parts of a helicopter and output are landing times. The optimization based on the ANN model constructed can to be used for designing the suitable dimensions of the helicopter body. To ensure the practical usefulness, a knowledge base for training neural networks was stemmed from experiments carried out over a comprehensive working range of conditions according to full factorial design concept [4]. At the beginning, a brief introduction of neural networks is initially discussed, before the development of neural network models for the intelligent modelling for the estimation of landing time is presented. The suitable features of a paper helicopter that yields the lowest area will be set to achieving the target landing time.

2 Literature Review

While the specified literature provides adequate theory on the neural network models studied in this paper, it is useful here to consider the basic theory associated with each of these neural networks with an understanding of the industrial application applied. The architecture and algorithms used by each network to achieve this are significantly different. A brief note on the Backpropagation networks is discussed below.

The standard back propagation network [5-6] comprises 3 layers of processing elements, fully feed-forward connected (Fig.1). With the sigmoid on the hidden layer, only the basic equations are :

$$y_k = \sum_{j=1}^{H} u_{kj} z_j \tag{1}$$

$$z_j = \frac{1}{1+\exp(-\theta_j)} \tag{2}$$

$$\theta_j = \sum_{i=0}^{N} w_{ji} x_i \tag{3}$$

The Least Mean Square error

$$E = \frac{1}{2}\sum_{k=1}^{M}\left(y_k - t_k\right)^2 \tag{4}$$

All the data are scaled between 0..1 but it can be scaled between -1..1 to standardise all the inputs with various dimensions. The weights w_{ji} and u_{kj} are assigned random numbers in the range -1..1, and a random pair of

* Corresponding author: tossapol_k@yahoo.com

input / output vectors are picked from the training set. The input vector is fed through the network to get an output vector (feed-forward process); this is then compared with the output vector and an error is found. This error is then passed back through the neural network (back propagation process) to modify the weights using the following equations

$$u_{kj}^{new} = u_{kj}^{old} + \Delta u_{kj} \qquad (5)$$

$$w_{ji}^{new} = w_{ji}^{old} + \Delta w_{ji} \qquad (6)$$

The gradient descent optimization technique is used to calculate the change in each weight. This is then repeated by picking another random pair of input / output vectors and continuing until the error is at a minimum. This was done 1.5 million times which was sufficient for the network to reach a suitable minimum.

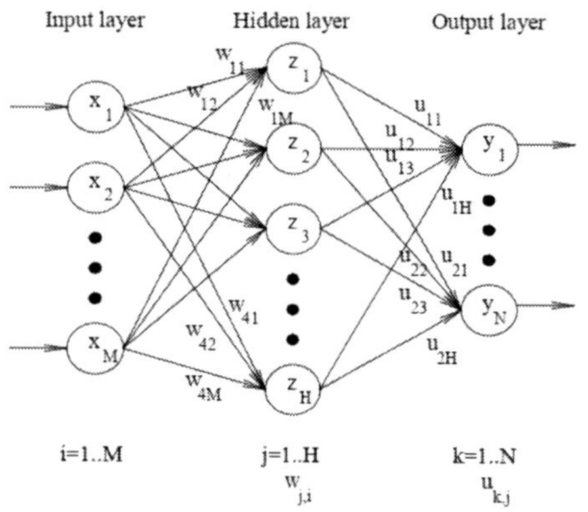

Fig. 1. Multi-layered Back Propagation Neural Network

Momentum can be used to decrease times in training and the chance of the network getting stuck in a shallow minimum. This is done by accelerating the convergence of the error but is not applied in this situation.

3 Develop of Database

The ANN is applied to build up a model of a paper helicopter case. There are five inputs: (A) Depth of cut wing, (B) Length of wing, (C) Length of body, (D) Width of body, and (E) Depth of cut body in Figure 2. The output is landing time in second where a paper helicopter is released from eight feet height.

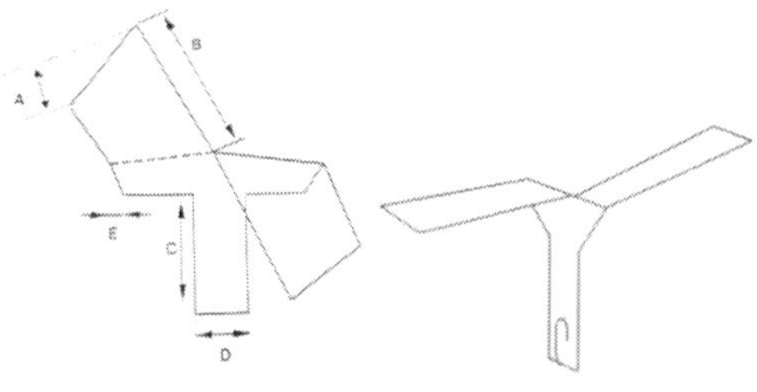

Fig. 2. Factors of A Paper Helicopter

Full Factorial 2k is used to design experiment plan to construct the database. Five inputs or factors are conditioned at two levels of values as shown in Table 1.

Table 1. Factors and Levels

Input or Factor	Level		unit
	Low (-)	High (+)	
(A) Depth of cut wing	1.0	2.0	cm.
(B) Length of wing	11.0	13.0	cm.
(C) Length of body	8.0	9.0	cm.
(D) Width of body	2.0	3.0	cm.
(E) Depth of cut body	0.0	1.0	cm.

The experiments of 32 various conditions with three replications were carried out. The overall 96 data are obtained to be the database for training and testing ANN.

4 Training and Testing ANN

Back propagation ANN with one hidden layer is used for modeling. The number of neurons in input layer is set to five neurons according to the number of inputs and for the output Layer, the number of neurons are set to one. The number of neurons in hidden layer is basically determined, which based on the minimum error of testing data. The mean square error (MSE) is used in trail of neurons from 7 to 13 neurons. The results are shown in Table 2.

The Backpropagation ANN of 5-10-1 yielding the minimum of error is used in this case of paper helicopter. The relation of output or landing time and inputs of dimensions are modeling even though the mathematical model is hardly depicted. The prediction of landing time could be estimated based on the ANN black box.

Table 2. Factors and Levels

No.	Number of Neurons			MSE
	Input	Hidden	Output	
1	5	7	1	0.031065
2	5	8	1	0.029954
3	5	9	1	0.023788
4	**5**	**10**	**1**	**0.012283**
5	5	11	1	0.019383
6	5	12	1	0.029194
7	5	13	1	0.021212

5 Design of Suitable Condition

5.1 Selection of Suitable Condition

The target of Landing Time is set at 2.90 ±0.05. Then the conditions of the five inputs have to be searched in order to meet the target. Since five inputs are the dimension of a paper helicopter, the condition of five inputs providing the lowest area is suitable condition. The 307,461 combination of five inputs are put in to ANN Model constructed as shown in Fig. 3. The results are presented in Table 3.

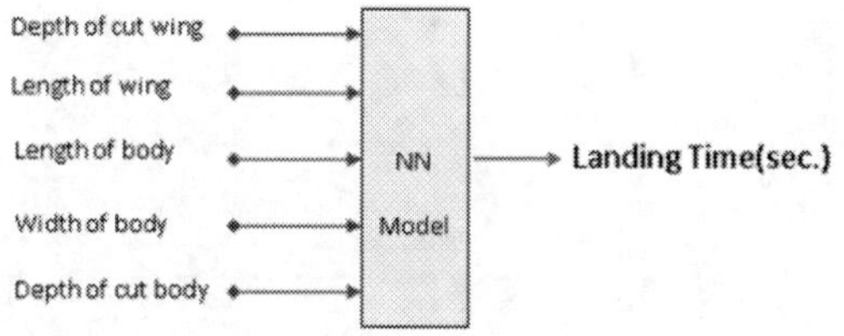

Fig. 3. ANN model for Estimating Landing Time

Table 3. Results of Searching Suitable Conditions

No.	A	B	C	D	E	Time	Area
1	1.1	12.9	9.0	2.0	0.0	2.852	110.1
2	1.2	12.9	9.0	2.0	0.0	2.852	109.8
3	**1.3**	**12.9**	**9.0**	**2.0**	**0.0**	**2.850**	**109.5**
4	1.0	13.0	9.0	2.0	0.0	2.854	111.0
5	1.1	13.0	9.0	2.0	0.0	2.856	110.7
6	1.2	13.0	9.0	2.0	0.0	2.855	110.4
7	1.3	13.0	9.0	2.0	0.0	2.853	110.1
8	1.0	13.0	9.0	2.1	0.0	2.851	111.9

The suitable condition should be the number 3; (A) Depth of cut wing is 1.3 cm., (B) Length of wing is 12.9 cm., (C) Length of body is 9.0, (D) Width of body is 2.0 cm., and (E) Depth of cut body is 0 cm. The lowest area of a paper helicopter is 109.5 cm2.

5.2 Verification of Result

The suitable condition of design inputs are tested in order to confirm that could yield the output, landing time 2.85 second within ± 5%. The twenty experiments were carried out and the experimental results are shown in Fig. 4.

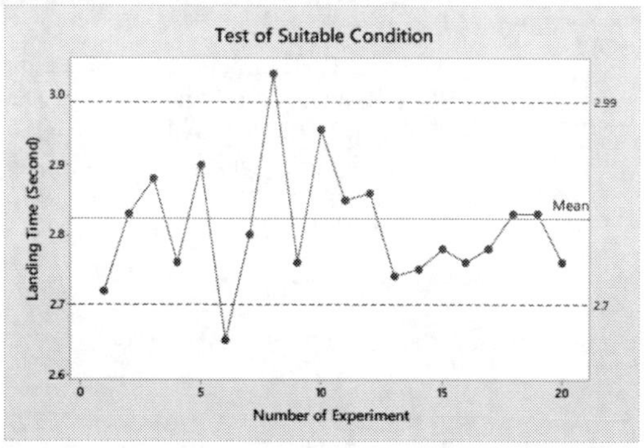

Fig. 4. Experimental Results of Suitable Condition

In suitable condition, A =1.3, B= 12.9, C = 9.0, D=2.0 and E =0, almost of twenty data are between 2.85 ± 5% (2.70-2.99) second. The arithmatic mean is 2.81second when compare to 2.85 second from the prediction of the ANN. The error is about 1.4%, which support the accuracy of developed ANN model.

6 Conclusion

A Backpropagation ANN is studied to apply in modeling a case of paper helicopter. The landing time as an output is targeted in order to find suitable dimensions of a paper helicopter. Five dimensions are chosen as inputs in this design. 2k Full factorial design is used to provide the experimental plan to build a database for training and testing ANN. The 32 experimental conditions with three replicates are carried out to create 96 databases. The architecture of ANN is determined based on providing the lowest error of test data set. After training and testing procedure, the ANN (5-10-1) with one hidden layer is set and used to model relation between the output and inputs. The design of dimension (A) Depth of cut wing is 1.3 cm., (B) Length of wing is 12.9 cm., (C) Length of body is 9.0, (D) Width of body is 2.0 cm., and (E) Depth of cut body is 0 cm. yield the lowest area of a paper helicopter that can meet the target landing time, 2.85 ± 5% second. This ANN for modeling in this case would be useful to mapping input and output for selecting suitable condition.

References

1. V. Karri, and T. Kiatcharoenpol, A monitoring system of drill wear states using a hybrid neural network, *Materials Science Forum* (2004)
2. V. Karri, and T. Kiatcharoenpol, Tool condition monitoring in drilling using artificial neural networks, *Lecture Notes in Computer Science* (including subseries Lecture Notes in Artificial Intelligence and Lecture Notes in Bioinformatics) (2003)
3. V. Karri, and T. Kiatcharoenpol, Prediction of Internal Surface Roughness in Drilling Using Three Feedforward Neural Networks - A Comparison, 9th International Conference On Neural Information Processing, Singapore, (2002)
4. C. Montgomery Douglas, Design and analysis of experiments. 8.th ed. Hoboken, NJ : Wiley, (2013)
5. F. Ortega-Zamorano, J. M. Jerez, D. Urda Muñoz, R. M. Luque-Baena, L. Franco, Efficient Implementation of the Backpropagation Algorithm in FPGAs and Microcontrollers *IEEE Transactions on Neural Networks and Learning Systems* (2016)
6. N. V. Irukulapati, D. Marsella, P. Johannisson, E. Agrell, M. Secondini, H. Wymeersch, Stochastic Digital Backpropagation With Residual Memory Compensation, *Journal of Lightwave Technology* (2016)

Automatic virgin coconut oil (VCO) extractor

Fabian R. Tagle

BS Industrial Technology Department, Faculty of Polytechnic School, Isabela State University Cauayan Campus, Isabela Philippines

Abstract. Virgin coconut oil (VCO) is a vegetable oil extracted from coconut milk that undergone either of the following extraction method: natural fermentation method with heat or without heat, expelling method or centrifugation method. Research showed that the extraction of VCO using expelling method had the highest percent oil recovery with 88.35% and yield of 30-31% followed by centrifugation method with oil recovery of 86.62% and yield of 31% then natural fermentation method with 65.95% oil recovery and yield of 16.5-19%. Even with low percent oil recovery and yield, VCO producers here in the Philippines particularly in Quezon province still employs the natural fermentation among other extraction method of VCO due to its cost-effectivity. The natural fermentation method involves several manual scooping activities in removing the VCO from the other component of the coconut milk which also takes time of waiting for about 24 to 48 hours for the VCO to be fractioned from the coconut milk mixture. This research therefore, focused in improving the natural fermentation method by developing a machine that automatically extracts the VCO from the coconut milk with higher percent oil recovery and yield. The designed machine was evaluated based on its oil recovery, and yield with respect to the current method of extraction. Furthermore, the effects of temperature and maturity of coconut kernel to the machine's capability of extracting the VCO were carried out. The tests conducted showed that the Automatic Virgin Coconut Oil (VCO) Extractor had an oil recovery of 89.84%. The study also showed that the yield using the automatic extractor is 31.27%. It was also concluded that it is better to use the Automatic Virgin Coconut Extractor in the area with temperature of 35-37 °C and preferably good coconut kernel should be used for the extraction of VCO.

1 Introduction

Virgin coconut oil productions were still considered as one of the promising industries in the country, since Philippines is one of the largest coconut producing countries in the world. It is simply because of the benefits that the oil could give such as: it could be used as raw materials for beauty enhancing products or even for products treating minor illnesses [1]. But what is virgin coconut oil? According to niulifekokonut pacific of Australia, the definition of virgin coconut oil depends on its processing method. They define virgin coconut oil as the oil produced from fresh coconut meat without the addition of enzyme/bacteria/additive and made without heat. And Mansor et al defined VCO as the oil from a coconut kernel through natural and mechanical means, either with or without the application of heat provided that there is no transformation of the oil. Generally, VCO could be extracted using the following processes: natural fermentation method with heat or without heat, expeller method, and centrifuge method [1,2]. And the common method of producing VCO in Quezon province is by natural fermentation method wherein the recovery of the oil employs series of manual scooping activities [3].

And according to studies, the VCO extraction was evaluated based on either yield or oil recovery. For the centrifugation method the oil recovery was 86.62%, for the expelling method it was 88.35% and for the fermentation method, its oil recovery was 65.95%.[1]. And the yield for centrifugation, expelling and fermentation method was 31%, 30-31%, and 16.5 – 19% respectively [4].

Since the natural fermentation employs the manual separation of the oil by scooping it from the mixture, this study aimed to improve the oil recovery and yield of the fermentation method of producing VCO through automatic extraction of the oil from the other components of the coconut milk.

Specifically, the study sought to achieve the following objectives:

1. To design a machine that automatically extract the oil from the other components of the coconut milk.
2. To determine the effectiveness of the machine in recovering the VCO with respect to the manual scooping process.
3. To determine the effect of temperature in extracting VCO using the Automatic VCO Extractor.

*E-mail: frtagle@yahoo.com

https://doi.org/10.1051/matecconf/201819201045

4. To determine the effect of the maturity of coconut in the extraction of VCO using the Automatic VCO Extractor.

2 Methodology

2.1. Machine Design

The Automatic Virgin Coconut Oil (VCO) Extractor was divided into two parts: the container design and the controller design. In designing the container, several factors were considered: the amount of coconut milk to be processed, the size of the container, and the materials to be used for the container. The size of the container was based from the amount of coconut milk that the container could handle. The material used for the container was stainless steel which is not chemically reactive to the virgin coconut oil. A glass strip was also inserted in order for the light of the sensor to pass through the VCO once it was fractioned in the mixture. The finished designed container undergone several testing to determine where the solenoid valve and through beam sensors should be placed

2.2 Evaluation

The Automatic Virgin Coconut Oil Extractor performance was evaluated based on: (1) the volume of oil that can be separated automatically (yield) and comparing the result with the yield of the manual scooping;(2) volume of VCO extracted at ambient temperature and at 35-37 °C temperature; (3) and volume of VCO extracted using good coconut kernel, green coconut kernel and matured coconut kernel.

The yield was computed based from the formula used by Y.C.Wong et. al. in their research Virgin Coconut Oil Production by Centrifugation Method.

$$\text{Yield} = \left(\frac{\text{Volume of VCO extracted}}{\text{Volume of coconut milk}}\right) x\ 100\% \qquad (1)\ [5]$$

The first activity was conducted in 5 trials and the model used was shown in Figure 1 as guide for the testing. One batch of coconut milk was brought to the Philippine Coconut Authority for laboratory test. And the VCO extracted from this same batch using both the manual scooping and automatic extractor were subjected to laboratory test at the Philippine Coconut Authority to determine if the machine contributes contaminants to the machine's output.

The extracted coconut milk was mixed with water with a ratio of 1:1 [4]. The first group undergone the automatic extraction and the other group passed on the manual scooping. Both method of extraction was conducted in a controlled room with temperature ranging from 35-37 °C [4]. And after 48 hours of natural fermentation, the volume of VCO for both group were measured using graduated cylinder. The average extracted VCO was computed using the the given formula.

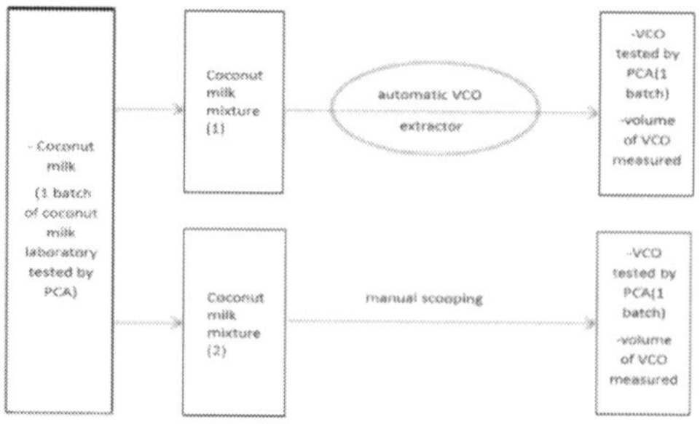

Figure 1.Automatic VCO Extrcator and Manual Scooping Test Process

$$Average\ volume\ of\ VCO = \frac{\sum Voil\ in\ cubic\ centimeters}{5} \qquad (2)$$

For the oil recovery, the formula used to compute for the available oil in the mixture (Voil) was based from the container measurements. This principle of determing the available oil was based from the study of Nour et.al. [6].The formula was given as

$$Voil = \pi r^2(h_{oil}) - 2(7.15)(h_{oil}) \qquad (3)$$

where: r – radius of the container

h_{oil} – height or thickness of the oil

And the percent oil recovery was computed by dividing the extracted oil by the available oil (Voil) times 100 %.

For the second activity, the machine was evaluated on the effect of temperature in extracting the VCO. The machine was tested in two scenarios; ambient (28 °C) temperature and the 35-37 °C temperature (measured temperature range in the testing area). The required temperature for maximum yield of VCO is 35-40 °C[1] [4]. The activity was conducted in 5 trials for both temperature setting.

And for the third activity, the performance of the machine was tested using different types of coconut based on maturity; the good coconut kernel, green coconut kernel and matured coconut kernel. Good coconut kernels are those coconuts with brown husk, and with half full coconut water while green coconut kernels are coconuts with green husk and with full coconut water and matured coconut kernel are those coconuts with brown husk, almost empty coconut water and with haustorium (coconut shoot). The temperature in the testing area was 35-37 °C. And 5 trials were conducted each for green coconut kernel and matured coconut kernel. And the data for good coconut kernel was already gathered in the previous activity.

3 Results

3.1 VCO Settlement in the Container

With the designed container, the location where the solenoid valve be placed as well as the through beam sensor were identified. From Table 1, the approximate volume of oil that could be extracted from the coconut milk mixture was shown. The average volume of VCO that could be separated from an input of 1900 ml coconut milk is 661.43 ml.

Table 1. VCO Settlement in the Container

Trials	Thickness of the oil in cm	V_{oil} in cubic centimeters
1	10.8	660.20
2	11.1	678.54
3	10.8	660.20
4	10.7	654.09
5	10.7	654.09
Mean ($\overline{X}$)		661.43
Standard Deviation (SD)		10.04

3.2 Performance Evaluation of the Automatic Virgin Coconut Oil Extractor

Automatic extraction
After selecting the good coconut kernel, cracking, grating and coconut milk extraction, 1900ml of water was added to the 1900ml extracted coconut milk. Good coconut kernel are those coconuts with brown coconut shell, good liquid sound when shake and no leak in any part of its shell. Shown in Table 2 is the summary of results for the good coconut kernel testing with fermentation temperature for 48 hours ranging from 35-37 °C.

Table 2. Volume of VCO and Yield Using the Automatic and Manual Extraction

Trial	Automatic Extraction (ml)	Yield	Manual Extraction (ml)	Yield
1	595	31.32%	512	26.95%
2	595	31.32%	530	27.90%
3	605	31.84%	519	27.32%
4	586	30.84%	510	26.84%
5	590	31.05%	521	27.42%
$\overline{X}$	594.2	31.27%	518.4	27.28%
SD	7.12		7.96	

Table 2 shows the yield results of the automatic extraction for five (5) trials. The average yield was 31.27%. The data also revealed that it is normally distributed within the first sigma of the mean of volume extracted both for automatic and manual extraction.

Table 3 shows the results of the tests for the VCO percent oil recovery. The resulting Oil recovery of the automatic extraction of VCO with the fermentation method was 89.84%. And the oil recovery of the fermentation method was 65.95% [1]. From these values, the Automatic VCO Extractor has improved the percent oil recovery of the VCO significantly.

Table 3. VCO Percent Oil Recovery

Trials	VCO Height (cm)	Volume of VCO Automatically Extracted (ml)	Percent Oil Recovery
1	10.8	595	90.12%
2	11	595	88.49%
3	11	605	89.97%
4	10.7	586	89.59%
5	10.6	590	91.05%
Mean		594.2	89.84%
Standard Deviation		7.12	

Manual Scooping
After selecting the good coconut kernel, cracking, grating and coconut milk extraction, 1900ml of water was added to the 1900ml extracted coconut milk. Shown in Table 2 is the summary of results for the good coconut kernel testing with fermentation temperature for 48 hours ranging from 35-37 °C. In this method of extraction, the curd above the VCO layer was set aside so that the VCO could be easily scooped. Careful and controlled scooping should be observed so that the curds and skim milk will not be scooped with the VCO.

Test 1 of Figure 2 shows the comparison between automatic extraction and manual scooping when it comes to volume of VCO extracted. It is noticeable that the automatic extraction has a significant difference in the amount of extracted VCO compared to manual scooping. In automatic extraction the mean of the volume of oil recovered was 594.2 while in manual scooping, it was 518.4ml.

For the second test. the coconut milk mixture is subjected to ferment at ambient temperature and compare it to VCO extracted at the required temperature of 35-37 °C. The data for the automatic extraction at the required temperature of 35-37 °C was already gathered and stated on Test 2 of Figure 2 wherein the average volume of VCO extracted was 594.2 ml. While, the average volume of VCO extracted at ambient temperature of 28 °C was 483.2 ml. And the computed yield for the extraction at ambient temperature is 25.43% This results verified that temperature affect the VCO extraction.

For the third test, the Automatic Extractor was used to determine the volume of VCO that can be extracted if the coconut milk came from good coconut kernel, green coconut kernel and matured coconut kernel. The data for the automatic extraction of VCO with coconut milk from good coconut kernel was already gathered and indicated on Test 2 of Figure 2 wherein the average volume of VCO extracted was 594.2 ml. And with coconut milk coming from green coconut kernel, the VCO extracted was 384.2 ml while from matured coconut kernel, the VCO extracted

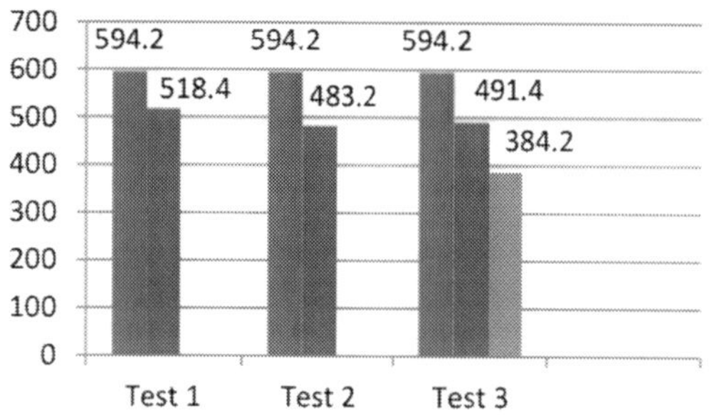

Figure 2. Graph of the Three Tests Conducted

was 491.4 ml. Among the three types of coconut kernel the green coconut kernel had the least amount of VCO extracted because it is young and the VCO is not yet fully grown at its kernel. The matured coconut kernel which already had a haustorium had a lesser amount of VCO extracted than the good coconut kernel because some of the VCO are already distributed to the other parts of the kernel (haustorium). And based from the VCO extracted and observations, the VCO from good coconut kernel generated higher yield than the green coconut kernel (20.21%) and matured coconut kernel (25.86%).

4 Conclusions

1. The Gizduino microcontroller together with relay circuit, through beam sensors, and solenoid valve comprises the design in extracting the virgin coconut oil automatically.
2. The results of testing revealed that the Automatic Virgin Coconut Oil (VCO) Extractor had an oil recovery of 89.84% and yield of 31.27% compared to the manual scooping with 65.95% and 27.28% respectively.
3. More VCO were extracted when the coconut milk mixture was fermented in the area with the required temperature of 35-37 ℃ than the area with ambient temperature of 28 ℃, The resulting volume of VCO are 594.2 ml and 483.2 ml respectively. And the yields are 31.27 % for the automatic extraction at the required temperature and 25.43 % for the ambient temperature.
4. The good coconut kernel could produce more VCO (594.2 ml) with a yield of 31.27 % compared to green coconut kernel (384.2 ml) with a yield of 20.22 % and matured coconut kernel (491.4 ml) with a yield of 25.86 %.

References

1. Mansor, T.S.T., Che Man, Y.E., Shuhaimi, M, Abdul Afiq, M.J., & Ku Nurul, F.K.M (2012). Physicochemical properties of virgin coconut oil extracted from different processing methods. Int. Fd. Rsrch. Jour. 19(3):837-845. Retrieved on November 16, 2015 from http://www.ifrj.upm.edu.my/19%20(03)%202012/(8)%20IFRJ%2019%20(03)%202012%20Che%20Man.pdf. (2012)
2. Santos, J. E. R., Villarino, B.N., Zosa A.R. &Dayrit F.M. Analysis of Volatile Organic Compounds in Virgin Coconut Oil and their Sensory Attibutes. Phil. Jour. of Sci. ISSN 0031- 7683. Retrieved on December 10, 2015 from http://philjournal.dost.gov.ph/vol140no2/pdfs/Analysis%20of%20volatile%20organic%20compounds%20in%20VCO5.pdf. (2009)
3. "Coco Corporation formed in Lopez, Quezon". Retrieved on November 15, 2015 from http://www.prrm.org/news/coco-corporation-formed-in-lopez-quezon.html
4. Bawalan, Divina ; Processing Manual for Virgin Coconut Oil, its Products and By-products for Pacific Island Countries and Territories. Retrived on November 12, 2015 from http://lrd.spc.int/publication/doc_download/1535-processing-manual-for-vco-its-products-and-by-products-for-picts--chapters-1-6. (2011).
5. Wong Y.C. & Hartina Virgin Coconut Oil Production by Centrifugation Method. Retrieved on December 2, 2015 on http://www.orientjchem.org/pdf/vol30no1/OJCV030I01P237-245.pdf. (2014).
6. Nour,A. H., Mohammed, F.S., Yunus, R.M., and Arman, A. Demulsification of Virgin Coconut Oil by Centrifugation Method: A Feasibility Study. Retrieved on January 20, 2016 fromhttp://scialert.net/qredirect.php?doi=ijct.2009.59.64&linkid=pdf. (2009).
7. Alicay, Calixto B. *Research Methods and Techniques*. (2014)

Effect of tempered bead techniques on maximum HAZ hardness for in-service pipeline welding

Sermsak Srisutraporn[1], *Rittichai* Paoniam[2*], *Bovornchok* Poopat[1] and *Supolchai* Kwankaew[1]

[1]Production Engineering Department, Faculty of Engineering, King Mongkut's University of Technology Thonburi, Bangkok, Thailand
[2]Mechanical and Industrial Engineering Department, Faculty of Engineering, Rajamangala University of Technology Krungthep Bangkok, Thailand

Abstract. This research intends to investigate the main factors of tempered bead techniques affecting on maximum HAZ hardness for in-service pipeline welding. Tempering parameters to be considered are the overlap ratio, weld bead sequences, and subsequent welding processes. This research consists of two parts of experimental procedure. Firstly, critical HAZ hardness (> 350 HV) in the first weld bead was estimated using computational simulation. Secondly, welding experiments were conducted with tempered techniques. Experimental setup included the used material of API 5L Gr. B pipe steel with nominal size of DN 200, wall thickness of 8.18 mm, and water piping flow of 18.77 m^3/hr. As a results, it suggested that the overlap weld ratio of 50% and 75%, weld bead sequences, as well as subsequent SMAW processes, were proficient of reducing significantly maximum HAZ hardness at the weld root. Nevertheless, in the case that the weld root was built up, maximum HAZ hardness was slightly changed with different weld bead sequences.

1 Introduction

In the oil and gas industry, a number of pipelines are widely used to transport natural gas or related products. Generally, welding is substantially employed for construction or maintainance. In this case, in-service welding plays an important role because this approach do not need to stop natural gas transportation in a main run pipe. Therefore, it is more benefit of saving production rate and income. However, there is issue to be noticeably considered such as the occurrence of hydrogen induced cracking (HIC). This crack typically occurs in a heat affected zone (HAZ). In practice, avoiding hydrogen induced cracking is undertake through reducing HAZ hardness. When HAZ hardness is lower than the critical value of 350 HV [1, 2]. Preheat or post weld heat treatment are alternative method to decrease HAZ hardness. Nevertheless, these methods are more difficult for in-service pipeline welding due to limitation of installation when the pipeline system is underground. As a result, tempered bead welding is practised to alleviate HAZ hardness.

As relevant researches[1-4], many studies have reported involving welding techniques to improve mechanical properties of the In-service welds such as Aloraier reported the change of metallurgical after post-weld heat treatment in repair welding using the temper bead technique. Mark Keeler had developed procedure to reduce the risk of burn through, blowout and HAZ Cracking. Likewise, Nicholas had studied and set Procedure qualification record (PQR) which used in In-service welding. However, there is no such woks focused on welding procedure such as covering area of next weld bead ,weld bead pattern and different welding processes for next bead.

Consequently, this research intends to investigate the main factors of tempered bead techniques affecting maximum on HAZ hardness for in-service pipeline welding.

2 Experimental Works

In this work, In- service welding with tempered bead techniques was carried out. There were three (3) essential factors on maximum HAZ hardness to be investigated, namely overlap ratio, weld bead sequences, and subsequent welding processes.

2.1 Used Materials

Fillet welding was performed on a main pipe and a sleeve plate according to API5L Grade B seamless pipe material standard. Nominal pipe size of specimen was DN 200 Sch. 40 (219.8 mm outside diameter, 8.18 mm thickness). The sleeve plate of 8.18 mm thickness was employed. Chemical compositions of specimen was examined through optical emission spectrometer (OES) as shown in Table 1.

Table 1. Chemical composition of used material (wt%)

C	Mn	Cu	Ni	Cr	Mo	V
0.134	0.969	0.181	0.099	0.223	0.049	0.003

* Corresponding author: rittichai.p@mail.rmutk.ac.th

Two (2) different welding processes were conducted, namely gas tungsten arc welding (GTAW) and shield metal arc welding (SMAW). As for GTAW, specimens were welded with a filler metal of ER70S-G according to AWS A 5.18, EWTH-2 tungsten electrode size of 2.4 mm in diameter, DCEN electrical polarity, argon shielding gas of 99.9%, as well as gas flow rate of 15 l/min. Meanwhile, SMAW was utilized with E7018 according to AWS A 5.5, electrode size of 3.2 mm in diameter, DCEP electrical polarity.

2.2 Methods

This work was divided into two steps. Firstly, heat input for a first weld bead with GTAW was estimated through computational simulation. Commercial PRCI program (Thermal Analysis Model for Hot Tap Welding V4.2) was utilized. Such heat input was used in order to achieve critical HAZ hardness value (over 350 HV). This could possibly lead to hydrogen induced cracking.

Secondly, actual welding experiments were carried out. Tempered bead techniques were employed to reduce the maximum HAZ hardness value of the first weld bead. Three (3) factors of tempered bead welding were investigated as follows; ratio of overlap weld, weld bead sequences, and subsequent welding processes. Utilized welding parameters were as given in Table 2.

The experimental set up as shown in Figure.1 was accomplished in accordance with API 1104 Annex B. This means that a sleeve plate is attached to a main run pipe to form a circumferential fillet weld lap joint, as well as 5G uphill position as illustrated in Figure.1(a). Meanwhile, water flows in the main run pipe during welding. The water flow rate of 18.8 m^3/hr. (Methane flow rate of 41.5 MMSCFD) was set up in order to obtain the convective heat transfer same as in-service natural gas pipelines[5].

2.2.1 Ratio of Overlap Weld

After experimental setup as mentioned earlier, GTAW process was performed. The influence of overlap weld ratios as illustrated in Figure.2 on HAZ hardness was investigated. That is to say, 1st weld bead was overlay welded by 2nd weld bead in the different percentages of overlap weld, namely 25%, 50%, and 75%. Welding parameters was employed as given in Table 2. Then, maximum HAZ hardness at the root weld pass was examined.

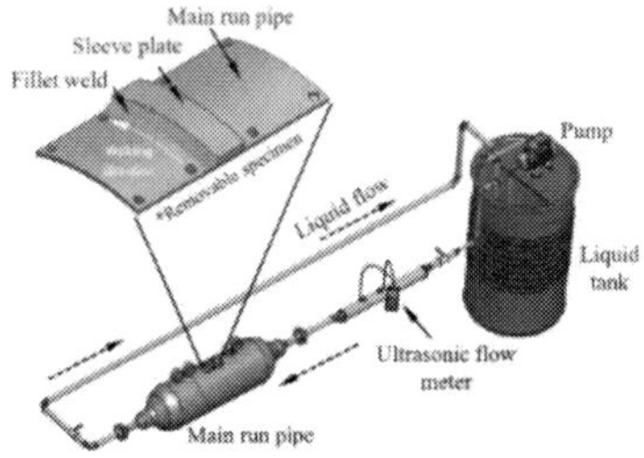

(a) Schematic of in-service pipeline welding

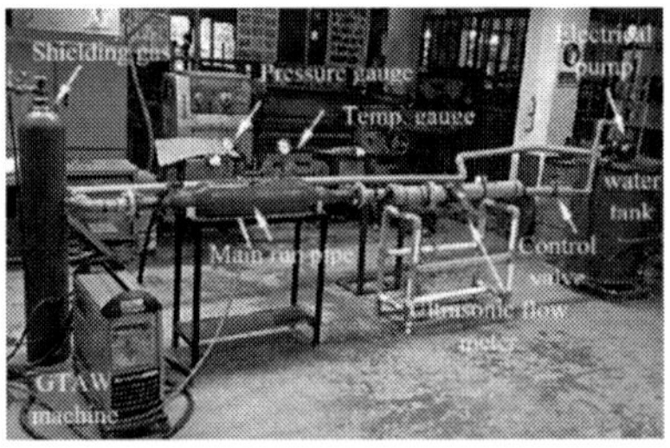

(b) Experimental setup of in-service pipeline welding

Fig 1 Experimental Set Up

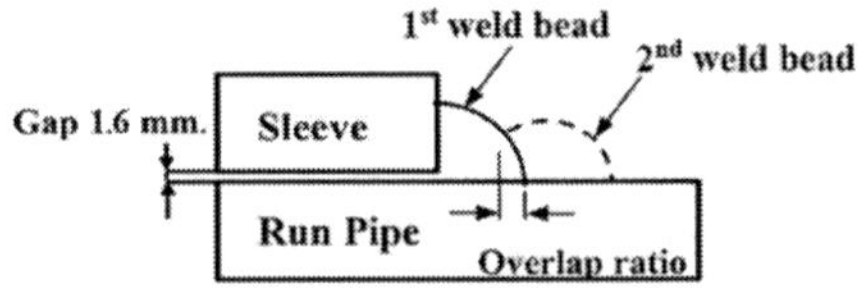

Fig2. Overlap Weld Ratio

Table2. Utilized welding parameters for experiments

Conditions / Variables	Overlap Weld		Weld Bead Sequences			Subsequent Welding Process			
	1st bead	2nd bead	1st bead	2nd bead	3rd bead	1st bead	2nd bead		3rd bead
Process	GTAW		GTAW			GTAW	GTAW	SMAW	SMAW
Voltage (Volts)	150	200	150	200		150	200	150	
Current (Amps)	10.4	12.4	10.4	12.4		10.4	12.4	26.6	
Travel Speed	7.75	7.53	7.75	7.53		7.75	7.53	12.19	
Heat Input (kJ/mm)	1.2	2.0	1.2	2.0		1.2	2.0		

Note. Heat Input(kJ/mm) = (Current x Voltage x 60)/(Travel Speed x 1000

2.2.2 Weld Bead Sequences

HAZ hardness at the root weld pass was influenced through the thermal energy during welding. Therefore, two (2) different formations of the weld sequences were conducted in order to obviously determine maximum HAZ hardness. The weld sequence of type A and type B were illustrated as Figure. 3. GTAW process was employed in entire procedure. In addition, severe HAZ hardness was considered that located on whether the sleeve plate or the run pipe.

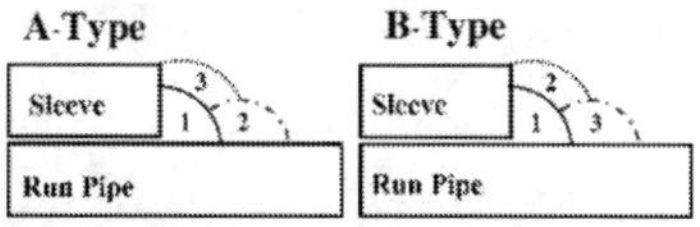

Figure 3. Types of Weld Bead Sequence

2.2.3 Subsequent Welding Processes

In this case, the effect of different welding procedures on HAZ hardness was studied. Two (2) patterns of subsequently different welding processes were carried out, namely 1st weld bead was built-up with GTAW and then followed with SMAW as exhibited in Figure 4.

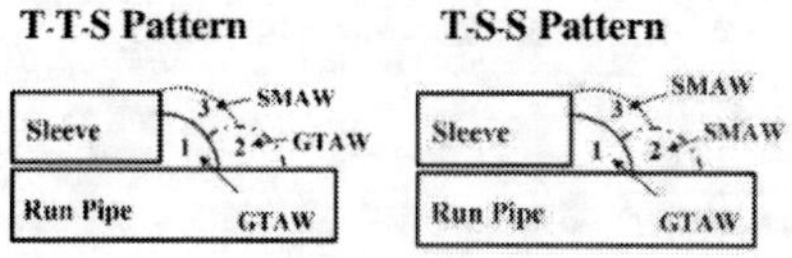

Figure 4. Different welding procedures

2.2.3 Hardness Measurement

After completed welding, each welded specimen was cut into 3 test pieces. Later on, those test pieces were grinded, polished and etched with Nital Acid 2%. Optical microscopic (Leica DM-2500M) was used so as to examine a sound weld. Besides, Maximum hardness in the HAZ area of coarse-grained region at the root weld pass was measured with micro-vickers hardness tester (Matsuzawa MMT-X3). Measuring procedure was operated in accordance with ASTM E384-11 as shown in Fig 5. Five indentations were pressed on individual area of HAZ. Measuring interval point was given as 0.2 mm from a fusion line, an indent spacing of 0.5 mm with applied load of 500 gram, as well as hold time of 10 seconds.

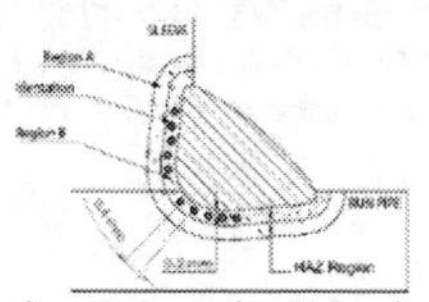

Figure 5. HAZ Hardness Measuring Procedure

*Corresponding author: rittichai.p@mail.rmutk.ac.th

3 Results and Discussion

3.1 Heat input to induce critical HAZ hardness

First step, welding parameter like a heat input for 1st weld bead was approximately calculated by commercial PRCI program. As the results, it exhibited that the critical heat input of 1.18 kJ/mm (30 kJ/in) was suggested in order to obtain critical HAZ hardness of 388.1 HV. However, maximum HAZ hardness of 323.1 HV at the run pipe was attained in the actual welding. This welding condition was utilized for 1st weld bead in entire experiments.

3.2 Effect of Overlap Weld Ratio

The effect of tempering welds was considered through overlap ratio. Fig. 6 represents the cross section appearance of different overlap weld ratios. It exhibited entire sound welds. Furthermore, Fig. 7 presents the comparison of maximum HAZ hardness in each overlap ratio between 25%, 50%, 75%. It revealed that HAZ hardness at the root weld pass decreased when using overlap ratio of 50% and 75%. However, it was found that the hardness was not significantly different.

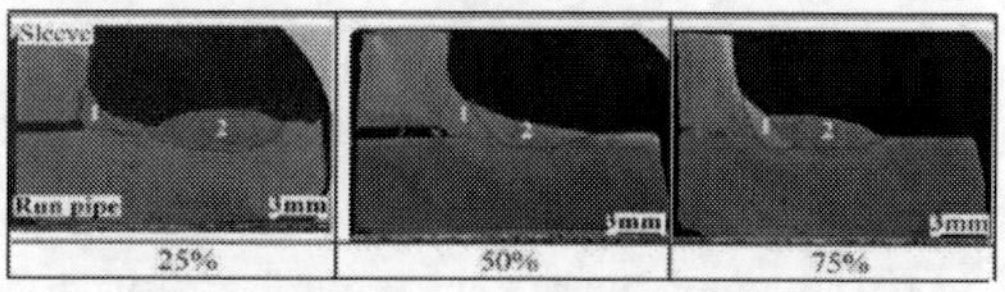

Figure 6. The weld cross sections due to the different overlap ratios

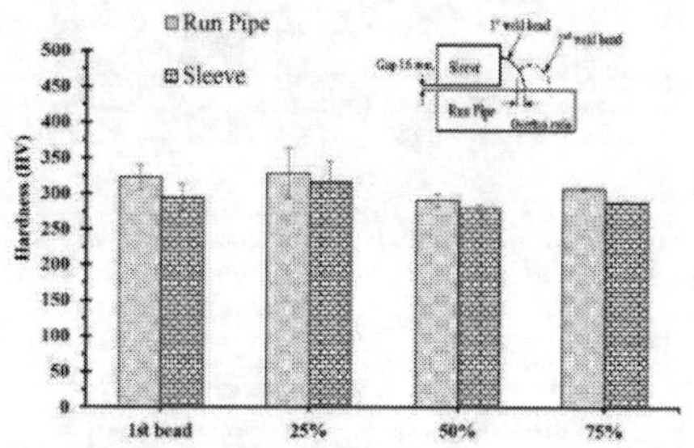

Figure 7. Max. HAZ hardness in each overlap ratio

Moreover ,HAZ hardness between Run pipe and sleeve was not different.

3.3 Effect of Weld Bead Sequences

In order to investigate the different formations of the weld sequences affecting HAZ hardness, Fig. 8 represents the weld cross sections of type A and type B weld sequence. As the results, it exhibited that maximum HAZ hardness reduced consistently with increasing the number of weld beads, namely reducing from 323 HV to 260 HV (run pipe A-Type) and reduce to 265 HV (run pipe B-Type). And also the maximum HAZ hardness occurred at the run pipe in all cases. However, comparing between the weld bead sequence of type A and type B indicate that these tempered techniques were not obviously different as shown in Fig.9.

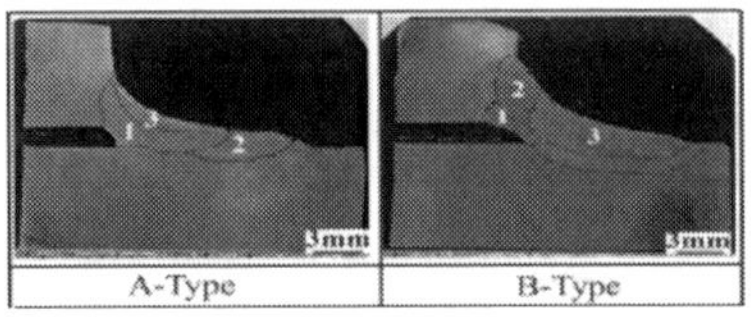

Figure 8. Cross-Sectional Welds Due to Bead Sequences

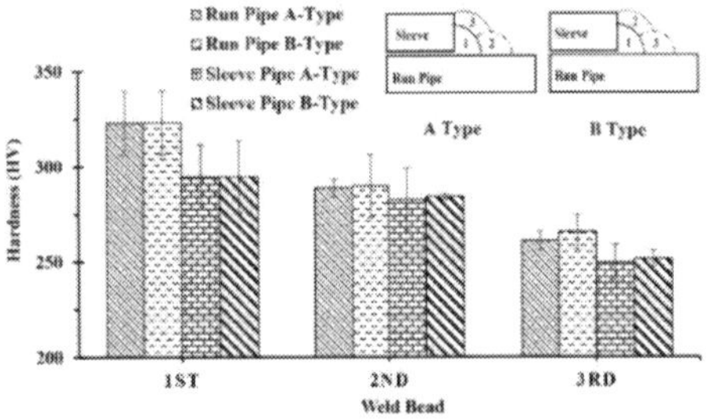

Figure 9. Max. HAZ Hardness in Different Weld Sequences

3.4 Effect of Subsequent Welding Processes

In this case, critical HAZ hardness at 1st weld bead was thermally tempered by different welding processes. Fig. 10 represents cross-sectional welds due to T-T-S pattern and T-S-S pattern procedure.

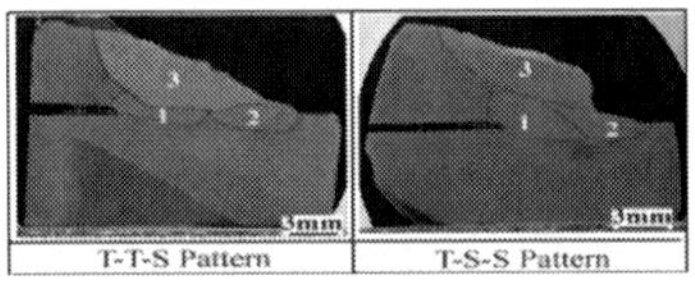

Figure 10. Cross-sectional welds due to different welding processes

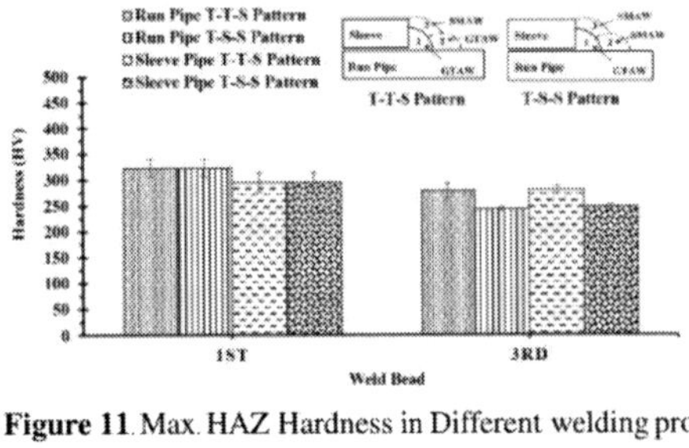

Figure 11. Max. HAZ Hardness in Different welding procedures

It was found that subsequent SMAW process was able to produce larger fusion zone. Especially, critical HAZ hardness of 1st weld bead dropped remarkably with T-S-S pattern. Maximum HAZ hardness reduced from 323 HV to 279 HV (run pipe T-T-S Pattern) and reduced to 243 HV (run pipe T-S-S Pattern)

4 Conclusion

This research describes the effect of tempered bead techniques on maximum HAZ hardness for in-service pipeline welding. The outstanding investigations can be summarized as follows.

1) As the experimental results, it was found that the overlap weld ratios of 50 %and 75% was able to significantly reduce critical HAZ hardness.

2) In the case that critical HAZ hardness reduced with increasing the number of weld beads, the different types of the weld bead sequences was insignificantly influenced.

3) As for subsequent welding process, it was suggested that SMAW process should be subsequently employed to built-up on 1st weld bead. This procedure was able to obviously reduce the maximum HAZ hardness at the root weld pass.

4) When comparing maximum HAZ hardness between the run pipe and the sleeve plate, maximum HAZ hardness always occurred at the root weld pass on the run pipe.

Acknowledgments

The authors thank to the research and development institute of RMUTK, KMUTT and 4WL Company for project funding and support.

5 References

[1] A. S. Aloraier, R. N. Ibrahim, and J. Ghojel, "Eliminating post-weld heat treatment in repair welding by temper bead technique: role bead sequence in metallurgical changes," *Journal of Materials Processing Technology*, vol. 153-154, pp. 392-400, 2004/11/10/ 2004.

[2] Mark keeler, "The step needed to avoid failure during in-service welding on a live gas line " Master of science, School of Applied Science, Cranesfield University, Mark Keeler,, 2009-2011.

[3] T. L. Joanna Nicholas, "Development and Qualification of Welding Procedures and Inspection Methodology for In-service Welding," presented at the Evaluation, Rehabilitation and Repair of Pipelines Conference, Berlin, Germany, 2010.

[4] K. Meszaros, C. Vrolyk, J. Pepin, M. Yarmuch, and T. Mah-Paulson, "The Effects of Temper Bead Welding Technique on Weld Integrity for In-Service Welding of Carbon Steels," pp. 563-570, 2012.

[5] I. W. Bang, Y. P. Son, K. H. Oh, Y. P. Kim, and W. S. Kim, *Numerical simulation of sleeve repair welding of in-service gas pipelines - A model was developed to predict conditions for sucessful sleeve repair of in-service gas pipelines* vol. 81, 2002.

Broadband dielectric characterization of novel organic substrates for microwave applications

Muhammad Yusof Ismail[1,*], *Hasan Ijaz* Malik[1], *Sharmiza* Adnan[2] *and Shaiful Rizal* Masrol[1]

[1]Universiti Tun Hussein Onn Malaysia (UTHM), Parit Raja 86400, Batu Pahat, Johor, Malaysia
[2] Forest Research Institute Malaysia (FRIM), Jalan Frim, Kepong, 52109 Kuala Lumpur, Selangor, Malaysia

Abstract. This paper presents three novel substrate materials with efficient dielectric characteristics. Different types of innovative substrate materials used for microwave electronics have been discussed. It has been demonstrated that the repeatability of results for paper substrate based electronics can be maintained using the proposed substrate materials. Three different substrate materials derived from recycled sources such as newspaper, copier paper and carton paper and banana pulp have been presented. The paper substrates were manufactured and processed to reduce the moisture content that might degrade the performance. A broadband dielectric material characterization of the proposed substrate shows a stable electromagnetic behaviour over the X – band operation with relative permittivities of 1.81, 1.63 and 1.84 along with loss tangents of 0.053, 0.046 and 0.057 respectively.

1 Introduction

The introduction of innovative materials with potential applications for millimetre and submillimetre frequencies have revolutionized the communication technologies [1]–[3]. Low – loss dielectric substrate materials with stable electromagnetic behaviour are of great interest for millimetre wave frequencies. Conventional dielectric substrate materials such as epoxy and rubber based substrates include the FR – 4 and Roger Duroid substrate materials. They have been widely used in the communication industry for microwave electronic circuit designs. The substrate materials can be grouped in two basic categories depending upon the dielectric permittivity they present i.e. high permittivity material and low permittivity materials. High permittivity materials find applications in electronic circuit designing where no spurious radiations are desired. Low permittivity material are preferred for antenna applications where fringing field are required for radiation purposes.

Sustainable dielectric materials derived from organic sources such as plant sources demonstrate efficient capabilities for antenna applications [4]–[8]. They provide advantages in term of low cost, stable and excellent properties in dry conditions. Different paper substrate materials have been presented for antenna applications. They include commercially available paper materials such as the Mitsubishi Electric photo-paper [4] and the usual paper [5], [6], [9]–[11]. The characterization results of conventional paper differs a lot between the individual runs due to difference of the paper ingredients. The composition concern is of utter importance since the paper used in the stated works might be from different manufacturers and may consist of different additives. Thus an accurate material characterisation of dielectric is always recommended that could ensure realization of high performance antennas on paper substrate materials.

A substrate material based on aromatic organic material Benzocyclobutene have been proposed for millimetre and sub-millimetre wave frequencies [12]. It has been proposed to address the large dielectric losses of conventional substrate materials at sub-millimetre wave frequencies. The proposed dielectric substrate material has been characterized over a wide frequency range of 11 – 65 GHz, where it presents a stable dielectric constant of 2.65 and a loss tangent of 0.001 – 0.009 [13].

Dielectric substrates based on textile materials for wearable electronics are also being discussed in [14]–[17]. These materials provide low dielectric constants between 1-2 due to porous structure and the presence of air. Textile substrates are always having a dynamic equilibrium with humidity and temperature of air of their surroundings. This makes textile materials vulnerable to unstable properties due to changing moisture content. The moisture content absorbed in the textile fibre increases the dielectric constant and loss tangent and it decreases the impedance bandwidth of the antenna [15]. A successful integration of a waveguide antenna on a textile substrate material has been demonstrated at a measured gain of 3.9 dBi for 2.45 GHz [14]. Thus a further improvement of textile materials in terms of designs and optimization will help enhance the performance and behaviour of textile antennas.

* Corresponding author: yusofi@uthm.edu.my

This article provides an innovative substrate material based on recycled materials derived from different organic sources such as copier paper, newspaper and the carton paper. It has been demonstrated that the controlled composition of the of different ingredients lead to diverse material properties. The proposed dielectric material is thoroughly characterized over X – band frequency range and the results of material characterization are presented and compared for analysis.

2 Proposed Dielectric Materials

Recycled materials derived from sources newspaper, carton paper and the copier paper was processed to manufacture dielectric substrates. The constituents of the proposed paper substrate are depicted in Fig. 1. The proposed dielectric material is based on the combination of two different constituents: the main ingredient and the binding ingredient. Banana fibre is used as a binding ingredient for the paper and it adds tensile strength properties to the paper substrate. The amount of banana pulp depends on the particle size of the main constituent. The composition ratios of the material are presented in Table 1.

Table 1. Physical composition of proposed dielectric materials

Substrate	Banana fibre	News-paper	Carton paper	Copier paper
RCP50	50	--	--	50
RCR75	25	--	75	--
RNP50	50	50	--	--

A paper material is a suspension of different size particles suspended in a bonding medium. The ratio between the bonding medium and the particles depends upon the individual size characteristics of the particles. In the manufacturing of the paper substrate the banana fibre was added with controlled compositions according to the size of main ingredient. As shown in Table 1 the percentage of banana fibre is 50% for copier paper (RCP50) and newspaper (RNP50) based substrate materials and it is 25 % for carton based paper substrate (RCR75). Initially the substrate materials were produced in the form of thin sheets which are not suitable for microwave antenna applications. Thus multiple sheets were combined together to increase the substrate thickness. The thickness of the substrate material plays an important role in the accuracy of dielectric material characterization results. The substrate materials were then passed thoroughly through rolling and pressing stages at temperatures of 30 – 35°C to remove any moisture content. Higher temperatures were avoided during pressing and rolling stages so that paper properties might not be affected.

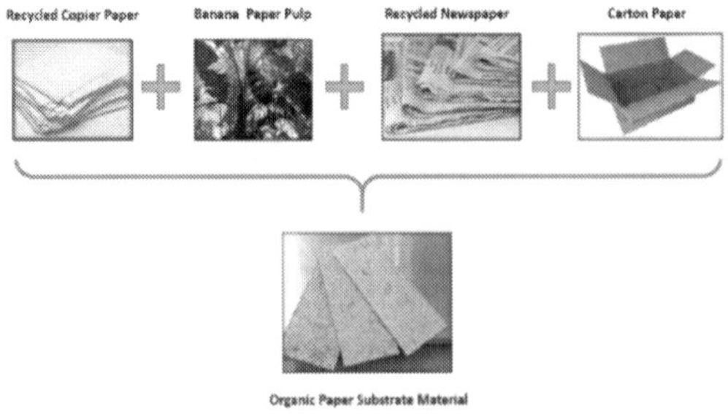

Fig. 1. Different recycled ingredients for paper substrate material

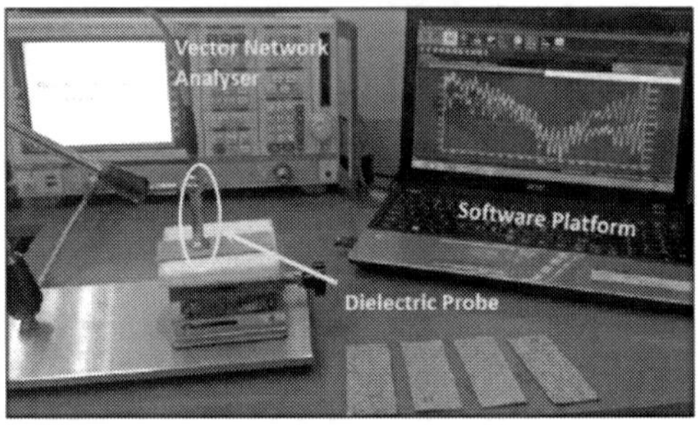

Fig. 2. Dielectric material characterization set-up

3 Dielectric Material Characterization

The electrical behaviour and the performance of proposed paper substrates was determined using a broadband characterization technique. A dielectric probe was used for this purpose. The principle of the dielectric probe is based on an open ended coaxial feed line. The fringing fields at the open end of the probe are utilized to estimate the properties of the material held in close proximity. The method is well suited for flat solids, liquids and semi-solids. The material under test is held in close contact with the face of the dielectric probe where the fringing fields pass through the material. The velocity of a wave in a medium is dependent to on the dielectric constant of material, using this principle the dielectric behaviour of material under test is estimated. The process can be used for determining the relative dielectric constant and the electrical loss factors of materials.

The complete set-up used for the dielectric material characterization is presented in Fig. 2. The dielectric probe was a DAK 3.5 mm probe with a broadband frequency analysis range of 200 – 20 GHz. The complete set-up shown in Fig. 1 consists of the probe, a Rodhe & Schwarz 14 GHz vector network analyser and a software platform installed on a PC. The probe is operated by the software platform via network analyser. Before the material testing the probe is thoroughly calibrated for any errors and discrepancies using air, copper strip and water for open, short and load calibrations. Load calibration was performed using water as load due to universal properties of water. The temperature of water was specially taken in to account due its temperature dependent properties.

https://doi.org/10.1051/matecconf/201819201047

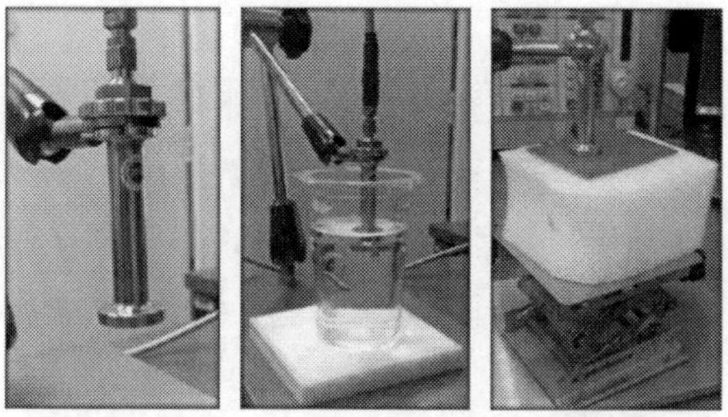

(a) (b) (c)

Fig. 3. Material characterization (a) Dielectric probe (b) Load calibration (c) Material under test

The dielectric probe and load calibrations are presented in Fig. 3. The probe is completely submerged in the water for load calibration as shown in Fig. 3(b). A vertically moveable fixture was used to maintain a firm contact between the probe face and the material under test as shown in Fig. 3(c). A firm contact is necessary for reliable results since any air gap will lead to measurement discrepancies.

Multiple samples were measured to ensure repeatability and reliability of the results. The results of the dielectric material characterization in term of relative dielectric permittivity and the loss tangent are presented in Fig. 4 and Fig. 5. Dielectric permittivity results presented in Fig. 4 show a broadband analysis of proposed paper substrate material over the X – band frequency range. The presented results show a gentle behaviour of relative permittivity of the material over the entire band of interest with minor deviations. The results show the lowest dielectric permittivity for RCR75 substrate material while highest permittivity results are shown by RNP50 substrate material. Fig. 5 shows the loss tangent results for the proposed substrate materials. The presented curves show a close trend of loss tangent for RNP50 and RCP50 substrate materials. The trend shows a decrease in loss from 8 – 10 GHz while a constant loss is shown for the range of 10 – 12 GHz. The finding from the Fig. 4 and Fig. 5 are tabulated in Table 2.

Table 2. Dielectric material characterization results

Substrate	ε_r	tanδ	Height (mm)
RCP50	1.81	0.053	1.45
RCR75	1.63	0.046	1.62
RNP50	1.84	0.057	1.12

The substrate heights of samples for the material used for characterization are also presented in Table 2. The presented results are the mean values over the X – band frequency range. The results show a mean relative permittivity of 1.81, 1.63 and 1.84 for RCP50, RCR75 and RNP50 substrate materials. The loss tangent results show a maximum loss tangent of 0.057 for RNP50 substrate

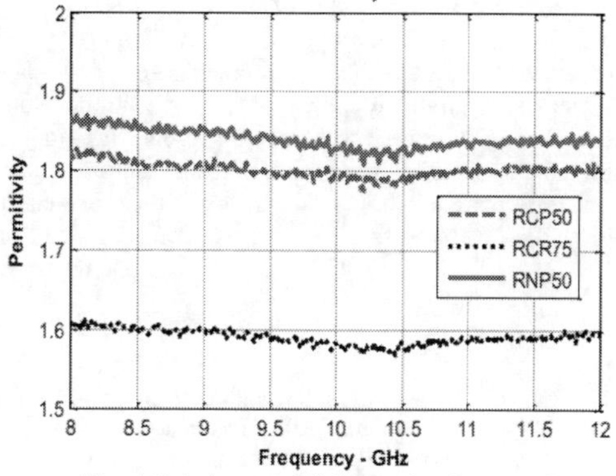

Fig. 4. Relative permittivity results of proposed substrate materials

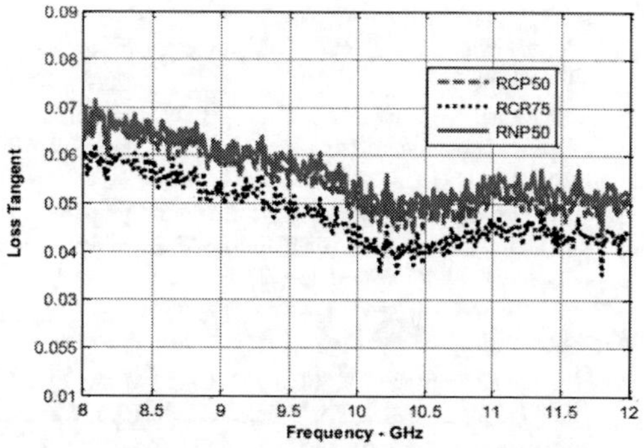

Fig. 5. Loss tangent measurement results for proposed paper substrates

material, while a minimum loss tangent of 0.046 is shown by RCR75. An intermediate loss tangent of 0.053 is presented by RCP50 substrate material.

Material characterization of the proposed substrate material shows that a diverse trend in dielectric properties can be achieve by controlling the composition of organic substrate materials. The proposed substrate materials will provide excellent electromagnetic properties for microstrip patch antenna design. The low permittivity of the material will help to enhance the fringing field effects required for microstrip patch antennas [18]. This will improve the radiation efficiency of the antenna. In order to protect the paper substrate from harsh environmental conditions water proof laminations should be used for microstrip antennas. Since the moisture content may contaminate the substrate resulting in degradation of material properties.

Conclusion

Dielectric materials find a wide variety of applications in different components such as capacitors, PCBs and microstrip antennas. Researchers have introduced a lot of different types of substrate materials with diverse material properties. The use of paper substrates for low cost, quick printed of electronic components have revolutionized the circuit printing industry. However due to difference in the composition and additives of paper from different manufacturers, performance of paper substrate based electronics lack repeatability of results. In order to address the issue this article proposes three different types of paper substrate materials. The composition of the paper substrate is carefully controlled in order to develop a relationship between the material properties and the ingredient of paper. A thorough characterization of paper substrate material show efficient low dielectric material properties with relative permittivities of 1.81, 1.63 and 1.84 along with dissipation factors of 0.053, 0.046 and 0.057 respectively.

The work was funded by GPPS Grant (VOT466), RACE Grant (VOT1119) and Research and Development (R&D) Fund awarded by Ministry of Higher Education Malaysia.

References

[1] Y. Malallah et al., 16th Mediterranean Microw. Symp. (MMS), (2016).

[2] K. Nadaud, R. Gillard, E. Fourn, H. W. Gundel, and C. Borderon, Loughborough Ant. & Propa. Conf., LAPC, November, 214–217, (2014).

[3] F. L. Martinez et al., Proc. 21st IEEE Int. Symp. on Appl. of Ferroelect. & 11th IEEE Euro. Conf. on Appl. Polar Dielec. IEEE PFM, 1–4. (2012)

[4] F. Alimenti et al., 2015 IEEE MTT-S Int. Micr. Symp., 1–4. (2015)

[5] T. Leng, X. Huang, K. Chang, J. Chen, M. A. Abdalla, and Z. Hu, IEEE Ant. Wirel. Propag. Lett. **15**, 1565–1568, (2016)

[6] B. S. Cook and A. Shamim, IEEE Ant. Wirel. Propag. Lett., **12**, 76–79. (2013)

[7] Aastha, A. Kaur, A. S. Dhillon, and E. Sidhu, Int. Conf. on Wirel. Comm., Sig. Proces. and Netw. (WiSPNET), 2090–2093. (2016)

[8] L. Xu and J. L. W. Li, 10th Int. Symp. on Ant., Prop. and EM Theory, 109–112. (2012)

[9] H. AbuTarboush and a. Shamim, IEEE Ant. Wirel. Propa. Lett., **11**, 1-1. 2012.

[10] A. Shamim and H. F. Abutarboush, IET Micro. Ant. Propa., **8**, 52–56. (2014)

[11] M. Kanagasabai and J. Kizhekke, IET Micro. Ant. Propag. **9**. (2015)

[12] S. Costanzo, F. Venneri, A. Borgia, I. Venneri, and G. Di Massa, IET Sci. Meas. Technol., **5**, 134, (2011)

[13] S. Costanzo, I. Venneri, G. Massa, and A. Borgia, J. Infr., Mill., Terahertz Waves, **31**, 66, (2009)

[14] S. Agneessens, M. Bozzi, R. Moro, and H. Rogier, Electron. Lett., **48**, 985–987, (2012)

[15] R. Salvado, C. Loss, Gon, and P. Pinho, Sens., **12**, 15841–15857, (2012)

[16] F. Declercq, H. Rogier, and C. Hertleer, IEEE Trans. Ant. Propag., **58**. (2008)

[17] T. Blecha, R. Linhart, and J. Reboun, Proc. 5th Electron. Syst. Technol. Conf. ESTC, 14–17, (2014)

[18] C. A. Balanis, 4 ed. Wiley Pub. **4.** (2016)

MATEC Web of Conferences 192, 01048 (2018) https://doi.org/10.1051/matecconf/201819201048
ICEAST 2018

Improve inspection process in tire manufacturing by using computer simulation

Sathitthep Sangthong*, *Pattara* Woraphu *and Kanokbhorn* Arayikanon

Industrial Engineering Department, Faculty of Engineering, Rajamangala University of Technology Rattanakosin, Nakorn Pathom, Thailand

Abstract. The purity of this research was to study the feasibility of improving the quality inspection in the manufacturing of tire by focusing on the utilization process. The current plant production is continuous and there are four working groups which can be divided into 3 shifts with 24 hours produced a day (Average production capacity is 350 days per year). Improved quality inspection processes will inevitably result in unnecessary production stops. Therefore, the researchers have proposed a feasibility study on how to improve the quality of the test procedures using engineering and technology techniques. These create models that have the same functionality as the current working condition. The result was found that Model 2 could reduce unnecessary processes and also four workers per shift if the worker's wage was calculated. This model shows a cost reduction of approximately 1.68 million baht per year, as well as utilization process

1 Introduction

This factory case study is a big tire production company which has make-to-order manufacturing processes. Job flow is a flow shop which is the same direction from start to finish. The production is carried out by the production department and it employs around 2,000 people. The Maximum tire production capacity is about 35,000 units per day. Work scope is categorized into 4 groups of 3 running cycles for 24 hours excluding 10-15 holidays. In case of engine break down or any improvements will affect the production line greatly especially in high volume. Hence, the machine maintenance should be avoided in the peak period. If the machine needs to be repaired, it has to be done after production is finished. As for the problem mentioned above, the management team recognizes the importance of increasing productivity but not stop working. Due to limitations and conditions of the factory, Researchers propose a feasibility study on the improvement of the performance for the tire quality inspection department by using various tools in industrial engineering, problem solving and simulation to study the feasibility of improvement and find alternative ways for better decision making. To use simulation, each models can be configured to study the consequences of each model. There is efficiency measure for the existing and new working scenarios. To select an effective improvement scenario before starting the actual plan can help the business to be successful.

2 Theories and Research

Situational modelling is one of the best way to solve process problems. At present, the design and development of the system often rely on Modelling for analysis prior to applying to the real system. There are simulation software which can simulate the work system. Computer simulation is in the form of computer programs the principle is to create a systematic decision-making approach troubleshoot the system or improve the existing system is good. It is not without interference in the real process. [1] Simulation is a way to analyse performance of system which is based on interaction the intersection of several random processes. These random processes can be identified by the probability model [2]. The system we are interested in may really exist. For example, if we want to improve the existing system "better" (the criteria for measuring "good") or a situation that does not exist yet. For example, changing the storage layout of a warehouse will use a rack rather than a place on the pallet. The current simulation cannot measure performance directly. In the case the numerical method cannot be determined estimation error. [3] Situational modelling is a model of the real process to learn the behavior of the real process under the conditions laid down in the analysis and evaluation the implementation of the system before the actual. [4] Yonlada Shomya and Udom Jantarajatsuk [5]. present to improved performance of the queuing process by using simulation out-patient services of Lum look Ka Hospital in Pratumthanee province, by modelling the situation and taking into account the dynamic behavior of the patient's arrival time and validation of the model with statistical

* Corresponding author: sathitthep.san@rmutr.ac.th

validation techniques. Then presented the revised model in 3 approaches which can be concluded to be the best choices. The result showed that there should have an additional channel for drug dispensing and this would lead a reduction in the average number of service hours per patient by an average of 20.90% and reduced outpatient services by an average of 29.09%.

3 Operation

3.1 Work Study

To study the work process of the audit department (Inspection Department) for inspection of tire quality is to understand the correct work and simulation model which should be very close to the original process. The study found that there are 5 steps in the process.

- Trimming rubber (Trimming)
- Mark and weigh (Balance Section)
- Quality Inspection 100% (100% Section)
- Check the balance (Uniformity Section)
- Quality Inspection 200% (200% Section)

The procedure of the inspection quality department is shown in Figure 1.

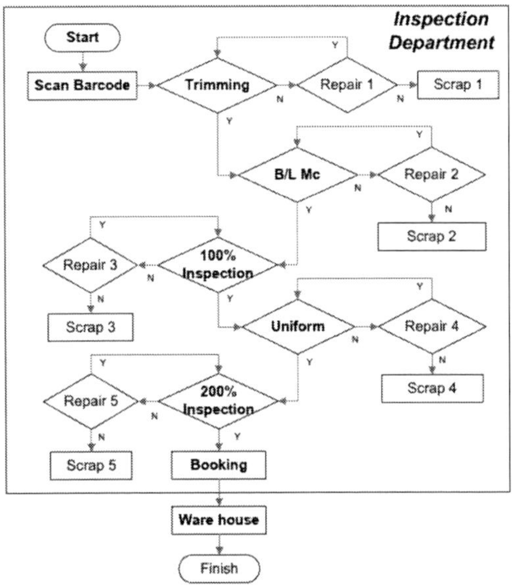

Fig. 1. Procedure of the inspection quality.

3.2 Data collection and Analysis

Tire Quality Inspection Department will use data from working time Monday to Friday in the morning 8 a.m. to 4 p.m. (day shift) with no break time. In day time, the data will have an interruption break period for 15 days. Most of the information is from the monitoring department's timekeeping. Quality and information generated by the machine. In order to provide statistical analysis, every data sets need to have an outlier test. The outlier will be removed before input into statistical analysis in the next step.

$$n = [Z\alpha/2x\ S/d]2 \tag{1}$$

d = Margin of error in estimating mean (setting α = 0.05)

The appropriate sample size is 22.67, but the data collection must be an integer, so the appropriate amount of Section 100% is 23.

3.3 Modelling the situation

When data is collected as required. The next step is to find the probability distribution of the data by using Chi-square test [6] for one-way discriminant analysis. (Goodness of Fit Test)

$$\chi^2 = \sum_{i=1}^{k} \frac{(o_i - E_i)^2}{E_i} \tag{2}$$

Table 1. Data distribution model of tire quality inspection process.

Work station	Amount	Pattern of Data
Cut Tentacle	2	Binomial (57,0.948)
Mark and weigh	2	Uniform (26,31)
Quality Inspection100%	2	Binomial (71,0.923)
Check Equation	2	Uniform (27,32)
Quality Inspection200%	2	Binomial (68,0.725)

Situational modelling needs to be closest to the real system. In order to be able to describe the various functions correctly from the analysis. Statistically, the distribution pattern of all data process simulation Modelling [7,8] is used to model the position of the tire quality inspection process.

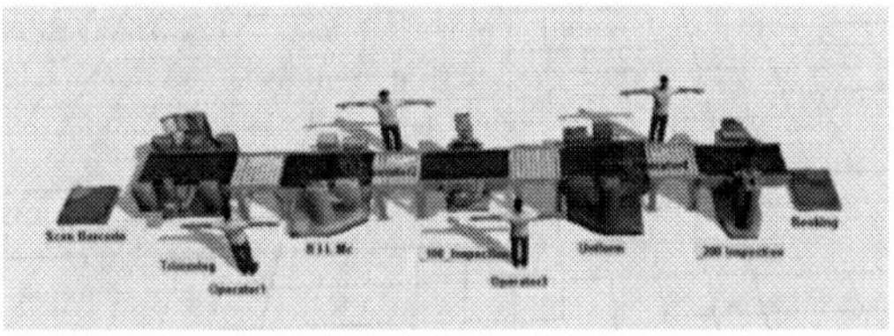

Fig. 2. Current Model Simulation

3.4 Analysis of the accuracy of the situation model

The process of analyzing the accuracy of the procedure is a necessary. It also requires those who understand the system thoroughly to assist in the investigation. If the generated model does not work properly with the diagram design. The improvement will result in errors. Therefore, the generated model must have a valid process flow based on the actual process and statistical tests. Comparison of the results is obtained by Welch Test Confidence Interval, the simulation model of the tire inspection department was created is close to the real system average at the 95% confidence level, and this model can be used instead of the real system.

3.5 Rounding out the simulation

System design for modelling situations requires multiple tests for accuracy and data variance. This can be calculated by the following equation

$$R \cong R_0 \frac{{h_0}^2}{h^2} \qquad (3)$$

R_0 = Number of preliminary simulation rounds

h_0 = Half width from the simulation cycle.

Introduction at the 95% confidence level,

h = Acceptable 5% half width

The experimental results of the tire quality testing department were 240 hours per cycle in 8 replication experiments.

4 Overall Result

4.1 Solutions of problems

The concept of work improvement for waste reduction in the manufacturing process in the tire quality inspection department is trying to eliminate and reduce unnecessary work. There are many outstanding works from production that is redundant of working procedures. The problem is solved by 5w1H technique to find solutions and then design the production process using ECRS technique. In order to reduce waste from production and increase efficiency, it is necessary to consider tackling quality control process to 100% and quality control process up to 200% quality control process in this section. The analysis can propose ways to improve tire quality inspection system in 2 ways as follows.

4.2 Model Simulation 1

Apply ECRS to improve performance by removing the 100% Inspection section. The newly created models are shown in Figures 3 and 4. Figure 4 shows the procedure for inspecting the tire quality. (Simulation 1)

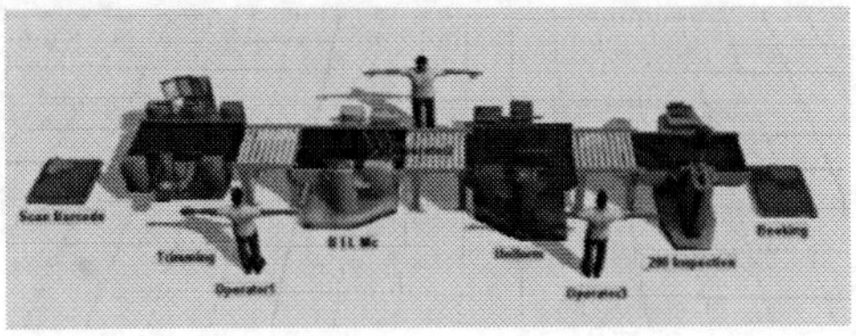

Fig. 3. Model Simulation 1

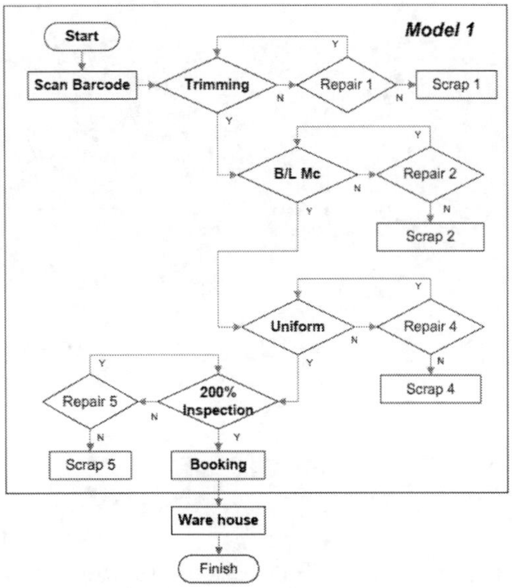

Fig. 4. Procedure for inspecting the tire quality. (Model Simulation 1)

From Tire Quality Inspection Models after improvement, the first test used 240 hours with 8 replications. The average all work sections are over 80% except for rubber cutter which has machine efficiency 75.02%

4.3 Model Simulation 2

Using ECRS principles to improve the original functionality by eliminating the 200% Inspection section. The Model Simulation 2 is shown in Figures 5 and 6.

Fig. 5. Model Simulation 2

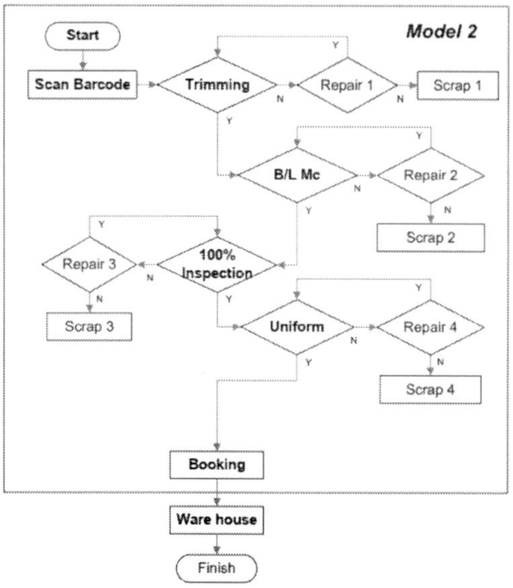

Fig. 6. Procedure for inspecting the tire quality. (Model Simulation 2)

The Model 2 was test by 240 hours of 8 replications. The machine efficiency of all work stations have average percentage more than 80% except for rubber cutter which has machine efficiency 74.89% as show on Table 2.

Table 2. Average performance of each workstation classified by Model type.

Work Station	Machine efficiency (%)		
	Cur	M.1	M.2
Cut Tentacle	74.99	75.02	74.89
Mark and weigh	83.51	83.56	84.25
Quality Inspection 100%	94.71	-	91.63
Check Equation	85.13	84.64	85.11
Quality Inspection 200%	91.74	91.24	-
Average	86.02	83.62	83.97

After defining guidelines for improvement and the results are summarized in Table 3.

Table 3. Result of each simulation model.

Situation simulation	Machine efficiency	Number of employees	Amount of waste
Current model Simulation	86.02	6	253
Model Simulation 1	83.62	4	288
Model Simulation 2	83.97	4	279

5 Summary and recommendations

Based on a feasibility study, the Quality Assurance Department found that the Product Inspection Division had a redundant work process between the 100% Inspection Section and the 200% Inspection Section which will be increasing confidence for quality assurance of customers and the quality of the product slightly. If you want to improve the inspection process, the Model Simulation 2, which can cut 200% Inspection Section out, can be a very good candidate. Using the ECRS principle and simulation model to improve performance is a guideline by creating a model to provide an improved system which cannot be tested on real situation. If we improve the inspection process based on the proposed model, it will be possible to reduce 4 workers per shift (cost reduction is about 1.68 million baht per year). Performance is closed to work situation before improvement. The Factory plant layout is very small and worth the investment.

Acknowledgments

The author would like to thank the Institute of Research and Development Rajamangala University of Technology Rattanakosin, Thailand

References

1. R.E. Rhanon, "System Simulation: The Art and Science," Prentice-hall Inc., Englewood Cliff., 1975.
2. Nelson, B. L. 2013. Foundations and methods of stochastic simulation. USA: Springer.
3. Juta Phichitlumken "Random-Simulation Basics Model to apply to solve the real problem." Department of Engineering
4. Industrial Engineering Faculty Kasetsart University Publisher of Kasetsart University.
5. Kelton, D.W., Sadowski, R.P. and Sturrock D.T., 2003, "Simulation with Arena"-3 rd ed., International Edition, McGraw-Hill, The McGraw-Hill Company, Inc.
6. Yonlada Shomya and Udom Jantarajatsuk "Improving queuing performance using situation simulation Case Study: Outpatient Services System Lamlukka Hospital" Industrial Engineering Faculty of Engineering King Mongkut's Institute of Technology Ladkrabang Ladkrabang Engineering Year 30 No. 1 March 2013
7. Rungrat Phisashpen, "Modeling Guide with Arena Program Revised edition," 1st edition, SE-EDUCATION Co., Ltd., Bangkok, 2008
8. Maria, A., 1997, "Introduction to model and simulation", Proceeding of the 1997 Winter simulation Conference ed. S. Andradottir, K.J. Healy, D.H. Withers, and B.L.Nelson.
9. Charles Harrell, Biman K.Ghosh, Royce Bowden. "Simulation using ProModel" Boston: McGraw-Hill/Higher Education, c2004

Effect of forming conditions of poly-lactic acid/hydroxyapatite to tensile strength of canine bone fixation plate using full factorial experimental design

Pornpailin Watcharaprapapong, *Wasawat* Nakkiew*, *Wassanai* Wattanuchariya, *Anirut* chaijaruwanit and *Siwasit* Pitjamit

Advanced Manufacturing Technology Research Center (AMTech), Department of Industrial Engineering, Faculty of Engineering, Chiang Mai University, Chiang Mai 50200, Thailand.

Abstract. Problems of using metallic bone fixation plates for canine lead to the idea of the replacement with biocomposite materials. In this research, polylactic acid (PLA) blends with hydroxyapatite (HA) powder 5-15% was used as biocomposite material. The specimens were formed by hot compression molding and the experiment in investigating suitable forming condition was based on the 2^3 full factorials with the center point design of experiment. There are many types of research attempting to develop a prototype of bone fixation plates from a substitute material including forming conditions, in order to obtain the effective bone fixation plate to use. This paper presents the tensile strength for the proposed biocomposite bone fixation plates. After the mechanical testing on various conditions, the tensile strength results ranged from 40 to 60 MPa and it showed that the higher HA ratio had significant effect to the decrease of the tensile strength. This preliminary experiment reveals the tensile strength of forming conditions of PLA/HA composite, which may indicate a direction for improving the better mechanical properties of the bone fixation plates.

1 Introduction

The metallic implants have been widely used in orthopedic and dental applications. The advantages of these materials such as good mechanical strength, easy forming and biocompatibility. However, many problems such as like stress shielding or stress protection atrophy, ionic release and allergic or toxic reactions have been introduce [1]. These risks related to the use of implants have led to the requirement of an additional operation to remove the implant after fracture healing [2]. To overcome these problems such as using noble metals coated with bioinert or bioactive materials. Because of these disadvantages, research into other materials became very significant. These disadvantages show obviously a need for non-metallic biodegradable devices for osteosynthesis of sufficient strength during the healing period and without causing side effects [3]. These would take away the necessity for an unpopular second operation to remove plates and screws used for internal fixation of fractures or after orthognathic surgery [4]. Furthermore, there is also significant importance in developing biodegradable replacements for metallic implants to eliminate the requisite to remove the implant after the bone has healed. Human bone is a highly ordered and complex structure consisted in organic part is hydroxyapatite (HA) along with collagen type II. Properties of HA are bioactive and good mechanical properties using for coated-metallic implant and bone repair [5]. Nevertheless, mechanical properties of HA is insufficient load bearing applications. Hence, synthetic, as well as some natural polymers have been being studied as appropriate materials for biomimicking the fibrous component in natural bone. There is a very short list of synthetic biodegradable polymers that are components in medical device implants. Polylactic acid (PLA) was the primary reported resorbable polymer to be used for fracture fixation [7]. Currently, PLA is still subjects of ongoing research into the development of biodegradable orthopedic biomaterials.

In a current animal study, PLA/HA plate for fixations of mandibular fractures was tested in comparison with stainless-steel fixation plate. These fixation plates were used to stabilize transverse mandible shaft fractures in canine.

*Corresponding author: Email address: wasawat@eng.cmu.ac.th

Therefore, this work aims to investigate and identify appropriate forming conditions for a PLA/HA composite using the experimental design based on the full factorial method. The optimal conditions were evaluated based on tensile strength. Finally, the actual mechanical properties of the composite specimen were used to apply to the forming composite bone fixation plate for the future.

2 Materials and methods

2.1 Material

2.1.1 Preparation of poly-L-lactic acid (PLA)

PLA was synthesized from lactic acid monomer through ring opening polymerization at 120 °C for 72 h using Tn (II). butoxide 0.01% mol per 500 g lactic acid to PLA powders (Molecular weight = 100,000–150,000 and viscosity = 2.17 ± 0.059 cP)[6, 8, 9].

2.1.2 Preparation of hydroxyapatite (HA)

HA powder prepared from removing tissue from the bovine bone and boiled the bones. Then, soaked the bones in oxidizing solution for 72 h and calcined at 950 °C for 12 h [8]. Finally, milling and sieving to be powders [6].

2.1.3 Preparation of biocomposite material

Biocomposite materials can be prepared from mixing the poly-L-lactic acid (PLA) and hydroxyapatite (HA) 5-15% of PLA using alumina balls to shaker about 15 min in the cylindrical Teflon mold to homogeneous material [10].

2.2 Specimens forming

Add biocomposite materials to hot compression mold to form the specimens using a custom-made mold. The applied molding temperature at 140-160 °C [11], molding pressure at 3-7 MPa and set holding time for 30 second [12]. every condition. Then, specimens are tested mechanical properties and statistical.

2.3 The experimental design and optimization

The experimental design for optimization study compression conditions of biocomposite materials according to 2^3 full-factorial design with the center point using Minitab16 software. The three factors include HA ratio, molding temperature and molding pressure as shown in table 1. Total of 20 experimental consists of 8 factorial designs, 2 center points and all of the design experiments were 2 replicates. The experimental design considers responses as tensile strength and bending strength of specimens.

2.4 Mechanical testing

The specimens are tested tensile strength from ASTM D 368 specimens by the universal testing machine (Instron5566). The cross-head speed was set at 0.1 mm/min.

Table 1. Parameters for experimental design.

Factors	Variation Levels		
	Low (-1)	Medium (0)	High (+1)
HA (%)	5	10	15
Molding Temperature (°C)	140	150	160
Molding Pressure (MPa)	3	5	7

3 Results and discussion

3.1 Microstructural characterization

3.1.1 Particle size and x-ray diffraction (XRD) analysis of materials

The particle size analysis of PLA and HA powders were characterized by laser diffraction method in ethanol. The mean of particle size of PLA and HA were 231.29 µm and 8.11 µm, respectively. The HA powder was verified by x-ray diffraction analysis because the synthesized HA must be confirmed their chemical composition with JCPDS-ICDD No.09-0432, the comparison of synthesis HA and standard HA composition pattern show in Fig. 1.

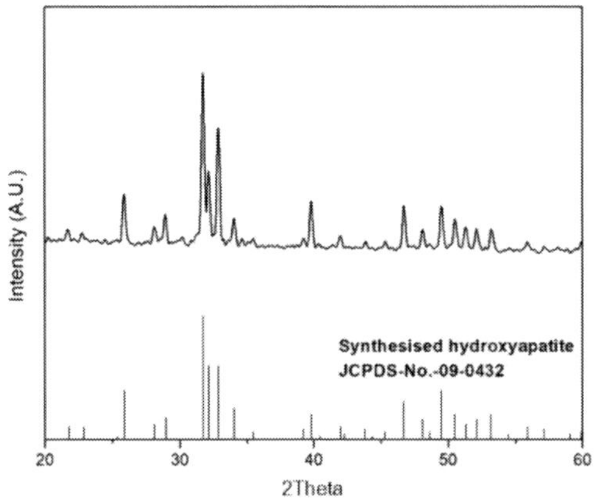

Fig. 1. XRD pattern of synthesized HA powder.

3.1.2 Differential scanning calorimetry (DSC) analysis of PLA

From the DSC curves of PLA showed the glass transition temperature (Tg), onset, endset and peak of endothermic temperature were 55 °C, 145 °C, 160 °C and 152 °C respectively. Because the DSC information describes the thermal behavior of matrix material, these data lead to control the range of forming temperature in experimental at 140-160 °C.

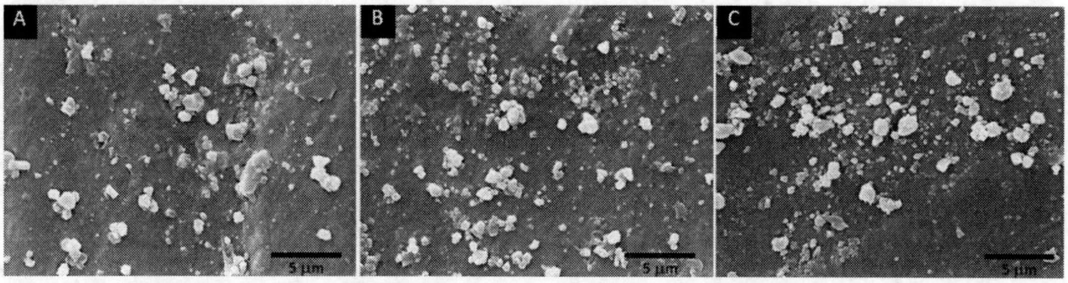

Fig. 2. SEM images of biocomposite material at 5 (A), 10 (B) and 15 wt% HA (C).

3.1.3 Scanning electron micrographs (SEM) of biocomposite material

The Scanning electron micrograph of biocomposite material at 5000x magnification for 5-10 wt% HA shows in Fig. 2. It shows the HA particle distribution was found to be successfully integrated into PLA matrix.

3.2 Mechanical properties

The results of mechanical testing for different molding condition specimens. All three factors are HA ratio, molding temperature and molding pressure. The values obtained from the mechanical tested statistically analyzed using Minitb16 software. Table 2 shows the R-Sq and R-Sq (adj) in the experiment are 88.03% and 79.32%, respectively. In addition, it presents the HA ratio had a significant effect on tensile strength because p-value of HA ratio is less than 0.05. The highest tensile strength of specimen was found at 5 wt% HA for molding temperature and molding pressure were 160 °C and 7 MPa, respectively. The tensile strength of various ratios of PLA/HA are varied from 40-60 MPa. Hence, this tensile strength value could be compared with cortical bone that shows in table 3. The result in table 4 shows the tensile strength dropped when HA particle contains higher than 5 wt% HA. Probably more loading HA particle that effect to ductility of PLA matrix. Consequently, brittle of the composite with 10 and 15 wt% HA will be increased and decreased tensile strength as well. Nevertheless, both of molding temperature and molding pressure had no significant effect to tensile strength. Due to molding temperature was not exceed the melting point of PLA, there, the micro structure of PLA matrix was not difference with similar cooling rate. In addition, the molding pressure used in the experimental at each level are not sufficiently different to affect to tensile strength because of the limitation of molding machine. Finally, Fig. 3 shows that the residual plots for tensile strength to indicating normally distribution data and normal probability plot, it can be predicted the trend of data to occur.

Table 2. The estimated effects and coefficients tensile strength.

Term	Coef	SE Coef	T	P
Constant	49.443	0.6670	74.12	0.000
HA ratio	-5.469	0.6670	-8.20	0.000
Temperature	1.394	0.6670	2.09	0.061
Pressure	1.334	0.6670	2.00	0.071
HA ratio*Temp	-0.745	0.6670	-1.12	0.288
HA ratio*Pressure	-0.330	0.6670	-0.49	0.631
Temp*Pressure	0.650	0.6670	0.97	0.351
HA *Temp*Pressure	-1.111	0.6670	-1.67	0.124
Ct Pt	-0.408	1.4916	-0.27	0.790
R-Sq = 88.03%		R-Sq (adj) = 79.32%		

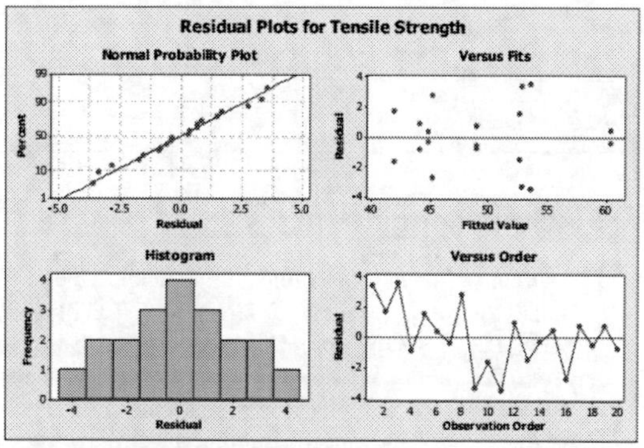

Fig. 3. The residual plots for tensile strength.

Table 3. The tensile strength of cortical bone and materials.

	Tensile Strength (MPa)
Cortical Bone (transverse direction)	52 [13]
Stainless Steel (316L)	586 [13]
Polylactic Acid	50 [14]
Hydroxyapatite	40 [15]

4 Conclusions

The study of the forming conditions of composite PLA/HA by hot compression molding using 2^3 full-factorial design with the center point. All factor in this study consist of the HA ratio, molding temperature, and molding pressure, but the HA ratio only had a significant effect on tensile strength. The range of tensile strength at 5-15 wt% HA about 40-60 MPa on average but, this value decreases when increasing of HA ratio. The highest tensile strength is using 5 wt% HA for molding temperature at 160 °C and molding pressure at 7 MPa, this condition gives the tensile strength about 60 MPa. For future work, the mechanical values are input into the finite element analysis to determine the probability of actual use on transverse mandible shaft fractures in canine.

Table 4. Tensile strength of the specimens.

Orders	HA ratio (wt%)	Temperature (°C)	Pressure (MPa)	Tensile Strength (MPa)
1	5	140	3	56.23
2	15	140	3	43.57
3	5	160	3	57.19
4	15	160	3	43.18
5	5	140	7	54.20
6	15	140	7	45.11
7	5	160	7	60.09
8	15	160	7	47.97
9	5	140	3	49.51
10	15	140	3	40.15
11	5	160	3	50.06
12	15	160	3	44.98
13	5	140	7	51.15
14	15	140	7	44.47
15	5	160	7	60.86
16	15	160	7	42.36
17	10	150	5	49.73
18	10	150	5	48.45
19	10	150	5	49.73
20	10	150	5	48.23

5 Acknowledgement

The authors of this study would like to acknowledge the National Research Council of Thailand (NRCT) under Project ID 270376 for financial sponsorship of research.The authors are also grateful to Advanced Manufacturing Technology Research Center (AMTech), Department of Industrial Engineering, Faculty of Engineering, Small Animal Hospital, Faculty of Veterinary Medicine, Bioplastic Production Laboratory for Medical Applications, Faculty of Science and Graduate School, Chiang Mai University for their facility support.

6 References

1. Van der Elst, M., Klein, C. P. A. T., de Blieck-Hogervorst, J. M., Patka, P., & Haarman, H. T. M. (1999). Bone tissue response to biodegradable polymers used for intra medullary fracture fixation: a long-term in vivo study in sheep femora. *Biomaterials*, *20*(2), 121-128.
2. Friedman, R. J., Black, J., Galante, J. O., Jacobs, J. J., & Skinner, H. B. (1993). Current concepts in orthopaedic biomaterials and implant fixation. *JBJS*, *75*(7), 1086-1109.
3. Eglin, D. and M. Alini, *Degradable polymeric materials for osteosynthesis: Tutorial.* European Cells and Materials, 2008. **16**: p. 80-91.
4. Rauso, R., et al., *Plates removal in orthognathic surgery and facial fractures: when and why.* J Craniofac Surg, 2011. **22**(1): p. 252-4.
5. Nuss, K. M., & von Rechenberg, B. (2008). Biocompatibility issues with modern implants in bone-a review for clinical orthopedics. *The open orthopaedics journal*, *2*, 66.
6. Catledge, S.A., V. Thomas, and Y.K. Vohra, *Nanostructured diamond coatings for orthopaedic applications.* Woodhead Publ Ser Biomater, 2013. **2013**: p. 105-150.
7. P Pawar, R., U Tekale, S., U Shisodia, S., T Totre, J., & J Domb, A. (2014). Biomedical applications of poly (lactic acid). *Recent Patents on Regenerative Medicine*, *4*(1), 40-51.
8. Girdthep, S., et al., *Formulation and characterization of compatibilized poly(lactic acid)-based blends and their nanocomposites with silver-loaded kaolinite.* Polymer International, 2015. **64**(2): p. 203-211.
9. Girdthep, S., et al., *Biodegradable nanocomposite blown films based on poly(lactic acid) containing silver-loaded kaolinite: A route to controlling moisture barrier property and silver ion release with a prediction of extended shelf life of dried longan.* Polymer, 2014. **55**(26): p. 6776-6788.
10. Russias, J., et al., *Fabrication and mechanical properties of PLA/HA composites: A study of in vitro degradation.* Mater Sci Eng C Biomim Supramol Syst, 2006. **26**(8): p. 1289-1295.
11. Ochi, S., *Mechanical properties of kenaf fibers and kenaf/PLA composites.* Mechanics of Materials, 2008. **40**(4-5): p. 446-452.
12. Kandar, M.I.M. and H.M. Akil, *Application of Design of Experiment (DoE) for Parameters Optimization in Compression Moulding for Flax Reinforced Biocomposites.* Procedia Chemistry, 2016. **19**: p. 433-440.
13. Kharazi, A.Z., M.H. Fathi, and F. Bahmany, *Design of a textile composite bone plate using 3D-finite element method.* Materials & Design, 2010. **31**(3): p. 1468-1474.
14. Hartmann, M. H. (1998). High molecular weight polylactic acid polymers. In *Biopolymers from renewable resources* (pp. 367-411). Springer, Berlin, Heidelberg.
15. Chen, Q., Zhu, C., & Thouas, G. A.)2012(. Progress and challenges in biomaterials used for bone tissue engineering: bioactive glasses and elastomeric composites. *Progress in Biomaterials*, *1*)1(, 2.

Development of natural acne patch from local materials using quality function deployment technique

Nattawadee Kapao and *Wassanai* Wattanutchariya*

Advanced Manufacturing Technology Research Center (AMTech), Department of Industrial Engineering, Faculty of Engineering, Chiang Mai University, Chiang Mai 50200, Thailand.

Abstract. This research aims to develop an acne patch based on local material extracted from silk and mulberry and the implementation of quality tools. The development started with a consumer needs' survey about the acne patch product. Technical requirements were evaluated based on quality function deployment (QFD) technique phase 1 or house of quality (HoQ). The results show that the highest score of consumer requirements is standard, followed by acne pus absorption and acne bacteria inhibition property, respectively. After transforming the consumer requirements into technical requirements, it was found that the highest score of technical requirements is qualified to anti-acne bacteria test, followed by swelling and antioxidant property, respectively. According to QFD result, the product prototype was developed using local materials extract and then evaluated based on consumer satisfaction. The most consumer satisfying feature of the natural acne patch product was the acne treatment and no irritation properties. However, the consumer made suggestion on the strength and stickiness of the product for further improvement.

1. Introduction

The cosmetic industry is one industry that is growing and expanding continuously. Cosmeceuticals are another option for people who care for themselves, whether male and female consumers. Cosmeceuticals combines the properties of cosmetics and drugs together. [1] It is a new product category in cosmetic industry. According to statistics, the market for cosmeceutical in Thailand has a market value of up to 2,000 million baht per year and a high growth rate of 30 percent per year, making the cosmeceutical product sector grow rapidly. Most cosmeceuticals are skin care and skin repair product. For example, they help increase radiance, add moisture, eliminate dullness and reduce wrinkles. They are also used to relieve the acne problem. [1-3] Acne treatment is available in many forms, such as drugs, comedone extractor and Intralesional Kenacort. Acne patches are another method to treat acne. It can conceal and treat acne at the same time. Acne patches are popular among consumers because it is portable and easy to buy. Generally, acne patches in the market are composed of hydrocolloid a synthetic substance. If there is a natural acne patch product, then there are no synthetic substances. It may be another option for consumers who are interested in natural products. However, those natural products need to have properties that meet the requirement for cosmeceutical. Silk and mulberry are local materials that can be easily produced or grew in Thailand. [4] Silk consists of fibroin and sericin. Fibroin can enhance the growth rate of fibroblast which can quickly treat acne wounds. [5] Meanwhile, mulberry extract has substances which can inhibit and kill acne bacteria. [6] Furthermore, fibroin and mulberry extract have antioxidant properties. [7, 8] Because of the range of cosmeceutical products in market, consumers are able to choose products that can satisfy their needs. Therefore, manufacturer must make their product to be interesting and be able to response to consumer's expectation. Thus, the purpose of this research is to develop an acne patch made from silk extract and mulberry extract. Principle of product design and development tools along with the house of quality

* Corresponding author: wassanai@eng.cmu.ac.th

technique were employed as tools to develop the acne patches from local materials, as well as to satisfy the consumer's needs and be able to compete in the cosmeceutical market.

2. Materials and methods

The research methodology started with preparing the fibroin extract, mulberry extract and chitosan. Then, the QFD was employed to develop the acne patch prototype, which then was evaluated based on consumers satisfaction.

2.1 Local material preparation

2.1.1 Silk Extract preparation

Silk cocoons were cleaned and cut into small pieces. Then, the pieces were boiled in solution of 0.02 mol Na_2CO_3 for 30 minutes. After that, the solution was washed with distilled water for 10 minutes. The extraction procedure was repeated two times before drying in hot air oven at 37°C for 12 hours. Next, the degummed silk was dissolved in 3 solutions of $CaCl_2$: CH_3CH_2OH: DI water at 70°C, followed by soaking the extracts packed in a cellulose membrane in distilled water for 3 days. Next, they were centrifuged at 2000 rpm. for 10 minutes and stored at -20°C and -80°C for further processing by lyophilization. [7, 9]

2.1.2 Mulberry Extract preparation

Mulberry leaf was cleaned before extraction with water at 45°C for 3 hours. The extracts were spin by centrifuge, with the supernatant liquid collected and filtered by filter paper. Then, rotary evaporation under reduced pressure at 45°C for sublimation was performed until the water was removed. [8]

2.1.3 Chitosan preparation

Commercial chitosan that produced by squid pen was purchased from a local supplier. Chitosan was prepared in lactic acid for 3 days. Then, it was stirred until homogeneous before adding into the mixture. [5, 9]

2.2 Consumer needs evaluation

Total of 20 consumers who had used acne patch before, were recruited and interviewed in order to develop the questionnaire for QFD implementation. After gathering raw data from the interview answers, it was converted to consumer needs. Then, consumer needs were categorized by tree diagram for easy conversion to the features of acne patches and state on the questionnaire. Then, this questionnaire was distributed online to target consumer which the related question focused on the voice of customer towards acne patch products in the market, criteria in buying decisions and product satisfaction.

2.3 QFD matrix development

Data from online questionnaires were applied in a HOQ matrix. HOQ was thus created from finding the relationship between consumer requirements and technical requirements. Including to, itself relationship of technical requirements and define target values of technical requirements.

2.4 Prototype development

The technical requirements that have a high score of at least 2 from HOQ was chosen to develop a product prototype before analyzing and assessment.

2.5 Consumer satisfaction assessment

The prototype was assessed by 20 assessors using a sensory evaluation and questionnaire by focus group. The result was analyzed and can tell how much the consumers are satisfied with the acne patch product that was developed.

3. Results and discussion

3.1 Local material characterization

The local material was tested for cytotoxicity by MTT reduction assay, which found that all of extraction and preparation of substrate were not toxic to cells. Silk extract was verified by Fourier transforms infrared spectroscopy (FTIR), which confirmed the standard property of general silk extraction.

3.2 Evaluation of consumer's response

The result from consumers responses show that the general acne patch can fix the acne problem by absorbing pus in the acne and concealing acne. But they cannot treat acne or kill the acne bacteria, so this acne patch should treat and fix the acne problem at the same time. Moreover, if the acne patch contains natural ingredients it can enhance the consumer satisfaction. By analyzing and summarizing the consumers' needs, a tree diagram was employed to show the consumer's needs in to proper category, as shown in figure 1. The importance rating of consumers' needs is presented on the left side of the HoQ. The result of consumer survey on acne patch development is also shown in figure 2. This score showed that the priority of consumer's requirement is the product safety, with a score of 4.46. In second place is the ability to absorb acne pus, with a score of 4.39, following by acne bacteria inhibition and good adherence to the face with score of 4.38 and 4.28, respectively. In fact, most consumers are interested

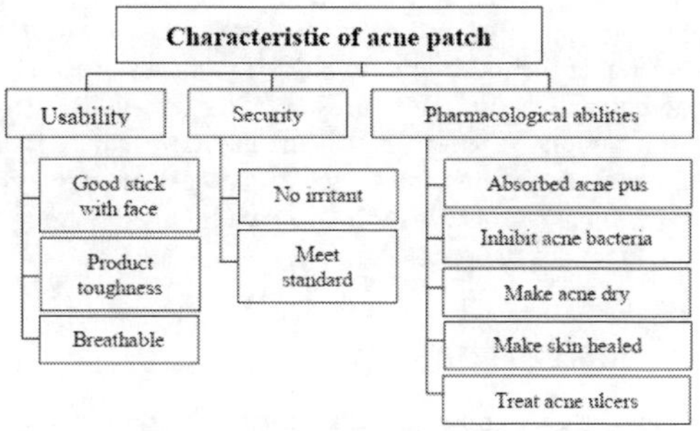

Fig. 1. Tree diagram of consumer needs on acne patch

about acne patch made from natural, which is safer than synthetic substance and do not cause allergic reactions.

3.3 Application of HoQ for acne patch

HoQ is a tool to transform customer needs into product specification, which then is converted into product prototype which can be used to evaluate consumer satisfaction as well as to seek for further improvement.

Symbol	Meaning	Score
	No relationship	0
Δ	Possible	1
O	Moderate relationship	3
●	Strong relationship	9

Symbol	Meaning
	No impact
++	Strong positive
+	Moderately positive
-	Moderately negative
--	Strong negative

Movement of Target Value
↑ Large the better
O Target the best
↓ Smaller the better

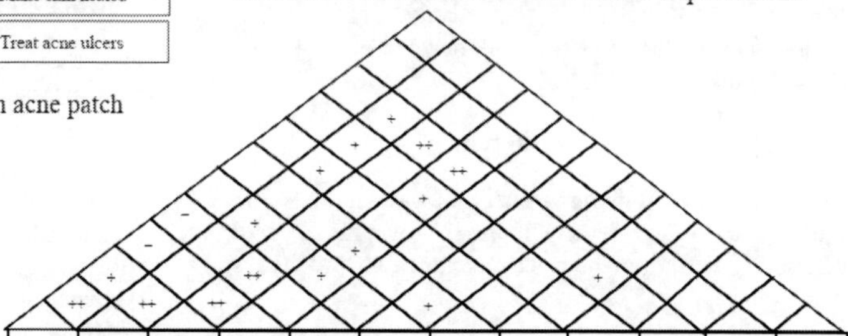

			Physical characteristics					Pharmacological characteristics					Safety	Cost	
			O	↑	↑	↑	↑	↑	↑	↑	↑	↑	O	↓	
Demand factors that affect product attributes		Importance score	Stickiness of surface	Ability to seize with skin	Efficacy of liquid suction	Ability to release substance	Porous	Skin lightening	Antioxidant	Heal lesion from acne	Add moisture	Treat acne	Certified	Cost of product	
Physical abilities	Product toughness	3.49	●	O	O	O							Δ	Δ	
	Good stick with face	4.28	●	●	●					O	Δ	O	Δ	Δ	
	Breathable	2.91		Δ			●			O	O			Δ	
Safety standards	Meet standard	4.46								O		O	●	O	
Pharmacological abilities	Inhibit acne bacteria	4.38				O			Δ			●	O		
	Make skin healed	3.99		O	O	Δ			O	●	O	Δ			
	Absorb acne pus	4.39			●	O	●					●			
	Treat acne ulcers	3.64		O		O		O	●	O	Δ				
Target Value			Passed sensory evaluation	Passed sensory evaluation	Passed the swelling test	Passed chromatography	Passed the porosity test	Tested by DPPH. assay	Tested by DPPH. assay	Tested by DPPH. assay	Passed sensory evaluation	Passed anti-acne bacteria test	Regulated by FDA.	20Baht/pack (6sheets)	Total score
Relative Technical Requirement			69.93	74.79	100.4	51.69	65.7	10.92	49.11	81.78	28.62	109.1	61.05	24.06	727.26
Absolute Technical Requirement			9.62	10.28	13.81	7.11	9.03	1.50	6.75	11.24	3.94	15	8.40	3.31	100

Fig. 2. House of Quality (HOQ) for natural acne patch product

The goal of this technique is ensuring that the product can meet the consumer's expectation. [10, 11] In defining relationships, symbols were used to show the type of relationship between consumer requirements and technical requirements. The symbols replaced the scores were that calculated as relative technical scores. HoQ was used to find the most important technical requirements of the acne patch product that shown in figure 2. The technical requirements that should be developed in the acne patch product include proven to treat acne with a score of 109.1, efficacy of liquid suction and healing lesion from acne with a score 100.4 and 81.78 respectively. Therefore, all three are importance attribute in the development of acne patch prototype.

3.4 Prototype of natural acne patch

The prototype was fabricated based on local materials and lyophilization is shown in figure 3. An initial cost evaluation of the prototype natural acne patch product is 3.33 baht/piece.

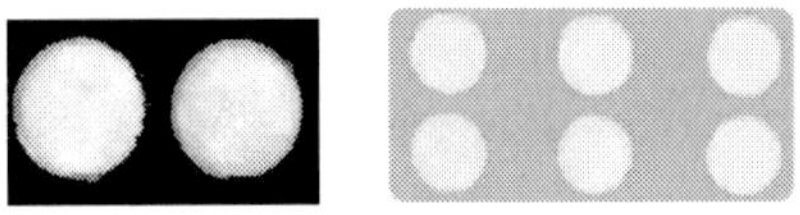

Fig. 3. Prototype of natural acne patch

3.5 Evaluation of consumer satisfaction

Focus group technique was employ for consumer's satisfaction evaluation. The result shows that most consumers had good level of satisfaction with the product's prototype. The consumers satisfied with the safety of the natural product. Moreover, it can treat the acne problem. Cost is another mentioned factor, which is found to be in an acceptable range. However, customers showed concern about the solidity of the patch's structure, as well as the stickiness of the product.

4. Conclusion

Three types of local material: silk extract, mulberry extract and chitosan were synthesized and fabricated into a natural acne patch product. These materials have been assessed to do no harm to cells in preliminary test and pass the property for acne treatment. Quality function deployment was implemented to evaluate the consumer needs and found that the important requirement is meeting the safety standard, followed by absorbance of acne pus and inhibition of acne bacteria, respectively. After transforming into technical requirements, it was found that the first important technical requirements is the efficiency of acne treatment, followed by efficacy of liquid suction and healing lesion from acne, respectively. Therefore, these specifications were than used to help develop a prototype. Afterward, this prototype was evaluated for consumer satisfaction, which found to be in good agreement with HoQ result. The main criteria of satisfaction are safety as it is the natural product, followed by acne treatment, color, smell, and cost of product. The recommendations for further improvement are strength and stickiness of the product.

5. Acknowledgement

The authors appreciate the financial support from the Graduate School of Chiang Mai University.

6. References

1. Nanjwade, B.K., *Development of Cosmeceuticals.* World Journal of Pharmacy and Pharmaceutical Sciences, 2017: p. 643-691.
2. Manela-Azulay, M. and E. Bagatin, *Cosmeceuticals vitamins.* Clin Dermatol, 2009. **27**(5): p. 469-74.
3. Gao, X.H., et al., *Efficacy and safety of innovative cosmeceuticals.* Clin Dermatol, 2008. **26**(4): p. 367-74.
4. Vepari, C. and D.L. Kaplan, *Silk as a Biomaterial.* Prog Polym Sci, 2007. **32**(8-9): p. 991-1007.
5. Zhang, X., et al., *Silk fibroin microfibers and chitosan modified poly (glycerol sebacate) composite scaffolds for skin tissue engineering.* Polymer Testing, 2017. **62**: p. 88-95.
6. Emniyet, A.A., et al., *Antioxidant and antimicrobial activities with GC/MS analysis of the Morus alba L. Leaves.* Hittite Journal of Science & Engineering, 2015. **1**(1): p. 37-41.
7. Thunsiri, K., S. Udomsom, and W. Wattanutchariya, *Characteristic of Pore Structure and Cells Growth on the Various Ratio of Silk Fibroin and Hydroxyapatite in Chitosan Base Scaffold.* Key Engineering Materials, 2016. **675-676**: p. 459-462.
8. Arabshahi-Delouee, S. and A. Urooj, *Antioxidant properties of various solvent extracts of mulberry (Morus indica L.) leaves.* Food Chemistry, 2007. **102**(4): p. 1233-1240.
9. Sionkowska, A. and A. Płanecka, *Preparation and characterization of silk fibroin/chitosan composite sponges for tissue engineering.* Journal of Molecular Liquids, 2013. **178**: p. 5-14.
10. Wattanutchariya, W. and P. Krutasood. *Purple rice dessert product development by Quality Function Deployment technique and Mood Consumption concept.* in Industrial Engineering and Engineering Management (IEEM), 2015.
11. Kasemset, C., N. Kumpong, and W. Wattanutchariya, *Processed Parboiled-Rice Product Development Using HoQ Technique.* Advanced Materials Research, 2014. 931-932: p. 1642-1646.

The performance improvement of look-alike sound-alike drug pre-dispensing

Manutchanok Jongprasithporn [1,*], *Wichai* Sunkarat [1], *Thanakorn* Homsai [1], *Teppakorn* Sittiwanchai [2] and *Nantakrit* Yodpijit [2]

[1]Department of Industrial Engineering, Faculty of Engineering, King Mongkut's Institute of Technology Ladkrabang, Bangkok, Thailand
[2]Center for Innovation in Human Factors Engineering and Ergonomics (CIHFE^2), Department of Industrial Engineering, Faculty of Engineering, King Mongkut's University of Technology North Bangkok, Bangkok, Thailand

Abstract. Medical dispensing errors are common and classic human errors in the hospitals. Look-alike, Sound-alike (LASA) drug is one of the main issues in pre-dispensing and dispensing processes. The data in this study were observed from one middle size hospital in Thailand. In 2017, there were two months that the medication errors KPI (0.065 and 0.061) were greater than the accepted error threshold (0.06). The SBAR (Situation, Background, Assessment, and Recommendation) solution and HOSxP software were applied for three months in the pre-dispensing process to reduce the medical dispensing errors. The results showed that the medication errors KPI (0.04, 0.03, and 0.02, respectively) were reduced dramatically. Therefore, implementation of the SBAR and HOSxP software is recommended as a fast and easy solution for medical dispensing errors.
Keywords - LASA; Human errors; Pre-dispensing process

1 Introduction

Pre-dispensing error and dispensing error are defined as a misunderstanding of dispensed processing between prescription and pharmacist [1]. The errors of pharmacist are incorrectly dispensing dose as specified in the prescriptions, such as medical type, dosage form, dosage, medical size, administration, number of prescriptions. The errors are also included dispensing without prescriptions such as dilution/ incorrect mixture using improper container, incorrect labeling, having wrong patient name, drug name and so on [2].

Look-alike sound-alike (LASA) drug is medications that could make people confuse because of their appeal or name. LASA can cause a confusing and medication error. In sound–alike, medications have common names or trade names with similar written, spelling, phonetics. In look-alike, medication products have similar color and size of tablets, bottles, labels or packaging to other drugs or similar to the same drug with different strength [3].

In 2007, world health organization (WHO) published "Nine patient safety solutions". LASA is one of the nine topics in the Nine patient safety solutions document [4]. It provided guidance about the prevention of LASA problems to ensure the safety of patients. WHO also distributed a guideline to managing LASA problems in member countries. In Thailand, the ministry of public health announced the national patient safety policy during 2007-2008 [5]. There are two main issues that are focusing on preventing infection from medical treatment and drug safety. In drug safety issue, there are three points that need to concern such as high alert drug, drugs that cause severe adverse reactions and repeatedly allergic drug. The pharmacist polls indicated that 82.46% of problems in hospitals came from LASA drug. LASA is the first rank causing of the medication errors [6]. In 2009, the drug system situation was reported that the drug safety related to LASA drug. A survey of 25 hospitals found the first rank problem of LASA drug was caused by a medicine labels and drug bottles from the same company. The second rank problem was the drug tablets that were looked alike and had similar brand name to other drugs.

Several studies [7] suggested that solving LASA drug problem in hospitals is continually processing by not buying look-alike and sound-alike drug. Putting the caution sign on the drug packet is another method to remind for the patients. However, medication error problem from LASA drug still cannot avoid because of physical limitations of human such as visual problem which could effect on choosing the wrong type of drug [8]. Therefore, the purpose of this study is to apply human factors engineering and human information processing to solve LASA problem in a pharmacy department in the hospital.

The objective of this paper is focused on the transcribing process of a pharmacist. The transcribing

* Corresponding author: kjmanutc@kmitl.ac.th

process is a part of the pre-dispensing process and it is one of the risk processes causing the medical errors, directly related to human. In this paper, the SBAR (Situation, Background, Assessment, and Recommendation) solution [9] is used along with HOSxp program. The HOSxP is a software providing to connect the medicine data in the hospital database in order to label the assigned medicine labels.

2 The Study of Transcribing Error and Objective

As described earlier, the scope of this study was to focus on a part of the transcribing process of pharmacists. The transcribing process is directly related to the model of human information processing (HIP) [10]. HIP model shows that the working memory is involved in decision-making from sight and hearing of human. Attention also play an important role to perception, working memory, decision making and response execution while one is dealing with look-alike, sound-alike medications.

The FDA U.S journal indicated that "miscommunication of drug orders can involve poor handwriting, confusion between drugs with similar names, poor packaging design, and confusion of metric or other dosing units". The content in the journal was specified that the poor handwriting of a doctor was affected by the transcribing error of pharmacists. Moreover, it can lead to the pre-description errors. The issue about unavoidable of LASA drug from one of the private hospital in Thailand was studied in this study.

Figure 1 shows the example of Look-Alike, Sound-Alike medication that can cause transcribing error. Both medications have the same ending name, which is CILLIN. They also have similar bottle and same amount 500 mg. Key Performance Indicator (KPI) is one of the techniques, used for a performance appraisal in order to achieve the expected value. In this case, the expected KPI should be less than 0.06 that means the errors should be less than 60 per 1,000 numbers of visits. The data was collected in 2017 from a sampling hospital. The blue line is the occurred errors and the red line is the hospital error threshold (0.06).

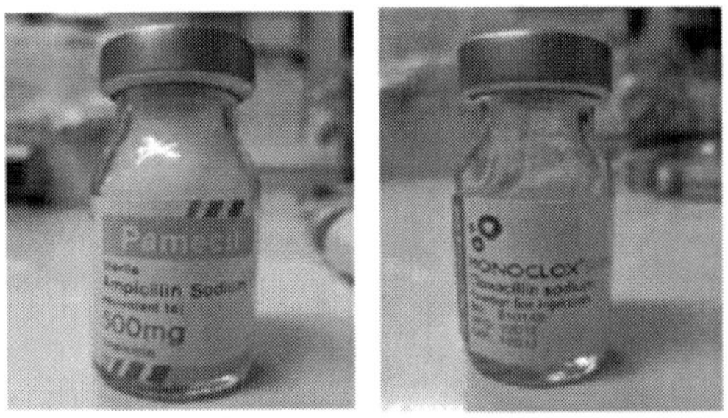

(a) (b)

Fig. 1. Look-Alike, Sound-Alike Medication
(a) Ampicillin 500 mg (b) Cloxacillin 500 mg [11]

In 2017, a hospital's KPI (Figure 2) showed that the occurred medication errors in February and July (0.065 and 0.061, respectively) were greater than the hospital error threshold. In March and May, there were the occurred medication errors, equalling to hospital error threshold. When KPI value was equal or greater the error threshold, it could indicate that patients might have high risks to get effect by medication errors. Patients who receive wrong dose and/or wrong medicine could be harmful to their life. Hence, the methods to preventing LASA medication errors had been an important issue for the hospitals.

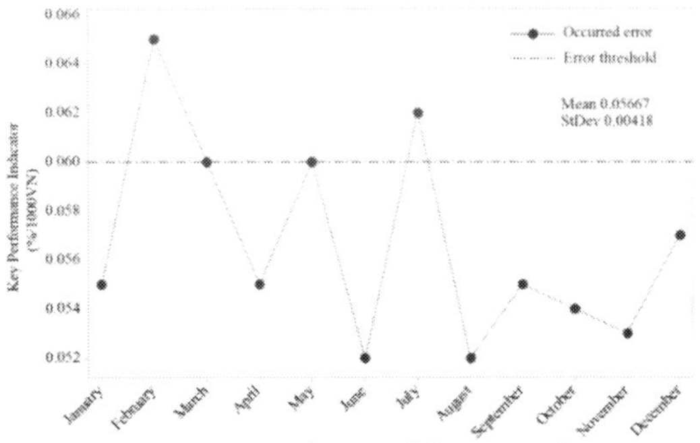

Fig. 2. Occurred transcribing error of a sample hospital in 2017

To improve LASA, this study is focusing on the SBAR (Situation, Background, Assessment, and Recommendation) and HOSxP software. The SBAR is a tool for rechecking the medical prescription and it is used for practicing in taking and giving report among nursing students [12]. In Situation (S), healthcare givers have to describe the specific situation including the patient's name, consultant, patient location, code status, and vital signs. In Background (B), healthcare givers have to inform patient's background including admitting diagnosis, date of admission, prior procedures, current medications, allergies, pertinent laboratory results and other relevant diagnostic results. In Assessment (A), healthcare givers have to think critically about the reason for your patient's condition including laboratory results. In Recommendation (R), healthcare givers have to inform about what you would like to happen by the end of the conversation with the physician and clarify your expectations.

The HOSpital eXPerience (HOSxP) is a free software, developed by Thai programmer [13]. It has been used by 150 health centers and mid-size hospitals in Thailand. The advantage of the HOSxP program is that the front of the medical label can be adjusted such as resize, making bold letter and changing lower case to upper case and so on. The writing of drug's name in upper case letter can help distinguish LASA drug.

3 Research Methodologies

There are two main methodology parts in this study, including observing of a pre-dispensing procedure and improving process.

The pre-dispensing procedure was summarized as shown in Figure 3. The pre-dispensing procedure was including five steps. First, the prescription was written by doctor's handwriting. Second, the lists of the medicine were transcribed by a pharmacist. Third, the lists of medicine were prepared in the pre-dispensing process. Fourth, the medicines of the patients were checked in the validity checking process. In the fourth step, if the pharmacist found incorrect prescription, the lists of medicine will be sent to transcribing process again. The lists of the medicine were sent to the dispensing process when the prescription was correct. The depended medicine was recorded and reported to the doctors.

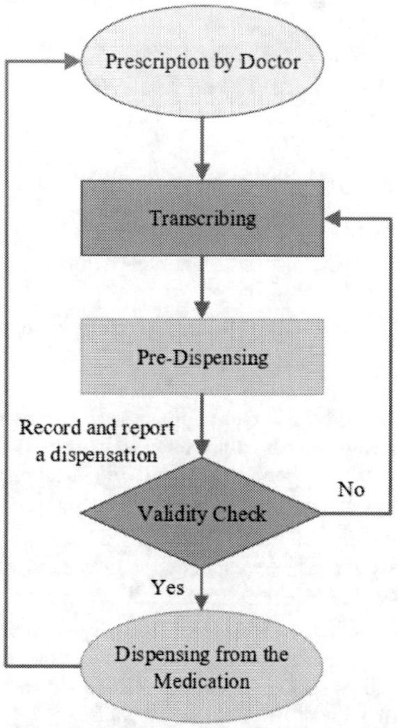

Fig. 3. Flow-chart of pre-dispensing procedure without the proposed solution

The problem can be divided into three main issues. The first problem was identified as a communication issue. Miscommunication between doctors and pharmacists was one of the most important LASA issues. The problem occurred when pharmacists could not recognize handwriting of doctor. The second issue was lacking of the pre-dispensing process skill of new pharmacists. This issue could happen when the pharmacy department admitted new pharmacists who had less experience. The third issue was related to a software. Normally, the efficient pharmacy software for medicine management is very costly to hospitals. Therefore, reducing medication errors were limited by the software cost. For example, using of a normally the labelled software the LASA medicine label cannot be labelled in a Tallman letter. The Tallman letter is the using of upper/lower case letter or different letter size to prevent the medical error.

A flow chart of pre-dispensing procedure with SBAR and HOSxP was described in Figure 4. To solve the transcribing errors, the SBAR (Situation, Background, Assessment, and Recommendation) and HOSxP software were applied in the pre-dispensing procedure for three months from January to March, 2018. The SBAR and HOSxP software were added between the prescription by doctors and transcription by pharmacists. The prescriptions have been written by the doctor's handwriting. The SBAR could help enhance communication between doctors and pharmacists and ensure the correct lists of medication before it was sending to the pre-dispensing process. In the SBAR process, an unreadable prescription was checked with the previous hospital database of the patients. The prescription was rechecked by a pharmacist in transcribing process. Then, the checked prescription was sent to the doctor again for authentication. In the HOSxP process, the authorized prescription was labeled by HOSxP software and sent out to pre-dispensing process. The redundant visually check was implemented in the improved process to increase patient's safety. Although SBAR and HOSxP were added to reduce human errors, the lists of medicine still needed to check in the validity checking process again. Then the lists of the medicine were sent to the dispensing process.

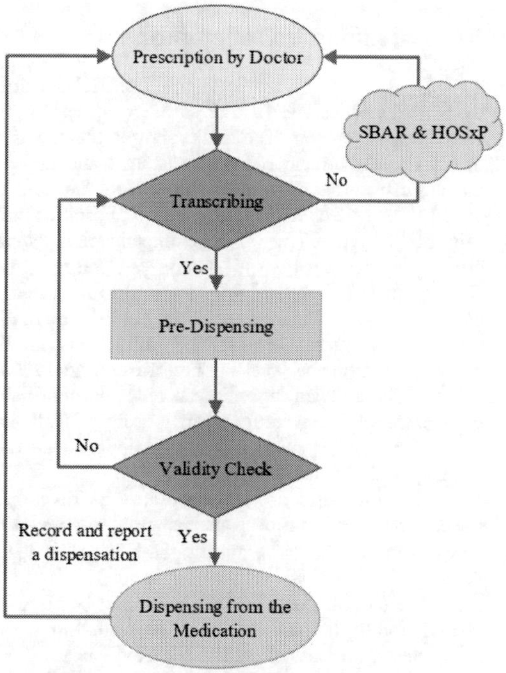

Fig. 4. Flow-chart of pre-dispensing procedure with the SBAR and HOSx

4 Results

After implementing SBAR and HOSxP, the KPI of the transcribing error was presented from January to March, 2018 (0.04, 0.03, and 0.02, respective). The data was compared between before and after applying SBAR solution with HOSxP software between the prescription by doctors and transcription by pharmacists. The average KPI of the transcribing errors were compared as shown in Figure 5. It is obvious that the average occurred error in 2018 is less than that of the occurred error in 2017. The expected KPI is that the 0.06. However, before applying the proposed techniques in February and July, the KPI are higher than 0.06. After using the proposed techniques to improve the transcribing process, the KPI is less than 0.04 that means the error from transcribing process is less than 40 visit numbers per 1,000 visit number. It indicated that the proposed solution can decrease the transcribing error efficiently.

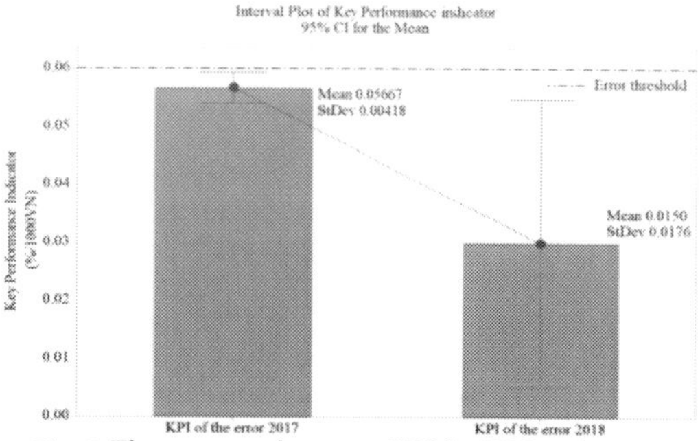

Fig. 5. The compared avearge KPI in 2017 and 2018

5 Discussions & Conclusions

To decrease the medical error from LASA drug, the SBAR solution and HOSxP software are used in the pre-describing process. The results showed that the using of SBAR solution and HOSxP software could improve the pre-dispensing process and decreased the medical errors from LASA drug. Hence, the proposed solution can solve the problems of miscommunication, human errors, and inexperienced pharmacist issues. The HOSxP, an efficient free software is recommended to use in pre-describing process to solve the LASA problem. The advantage of HOSxP software is that the letter in the medicine label can be formatted in Tallman letters. The Tallman lettering can reduce a confusable drug name of LASA drug because using of Tall man lettering in a part of medicine name in upper case letters can help to distinguish LASA drugs.

The limitations of this study are including three issues. The first limitation is the difficulty of connecting between the HOSxP program and the hospital database. Although the HOSxP software is a free software, it requires a specific program to help connect between the HOSxP and the hospital database. The second limitation is lacking pharmacists' skill and experience in pre-dispensing process. The pharmacists' skill and experience is one of the most important factors in order to solve LASA problem. The pharmacists who is lacking of skill and working experience could cause the transcribing error. The last limitation is the issue about the hospital confidential information. Therefore, it is difficult to provide some results in the public.

In the future, the doctor should type prescriptions instant of writing in the prescription process. Moreover, the hospitals should have their own prescription program, which enhances data communication between doctors and pharmacists.

References

1. W. T. Basco; M. Ebeling; T.C. Hulsey; K. Simpson, "Using Pharmacy Data to Screen for Look-Alike, Sound-Alike Substitution Errors in Pediatric Prescriptions", Academic pediatrics, vol 10, issue 4, pp. 233 – 237, 2010
2. T. Nagata, M. Kimura, F. Tsuchiya, "Similarity Index for Sound-Alikeness of Drug Names with Pitch Accents", Procedia Computer Science, vol 35, pp 1519-1528, 2014
3. N. Tuohy, S. Paparella, "Look-Alike and Sound-Alike Drugs: Errors Just Waiting to Happen", Journal of Emergency Nursing, vol 31, Issue 6, pp. 569 – 571, 2005
4. World Health Organization. WHO launches 'Nine patient safety solutions' [online]. 2007 [cited Oct 7,2016]. Available from: www.who.int/mediacentre/news/releases/2007/pr22/en/ index.html.
5. Department of Health Service Support. National patient safety goal. Nonthaburi: Ministry of Public Health: 2007-2008.
6. Chumchit C. Development of medication safety management system for Look-Alike Sound-Alike drugs in public hospital [dissertation]. Nakorn Pratom: Silpakorn University; 2013.
7. Nursing Students' perspectives of the Cause of Medication errors. Nurse Education Today 34 (2014) 434 – 440, April 16,2013.
8. Academic Centers for Drug System Surveillance and Development. Drug system situation report [online]. 2009 [cited Oct 7,2016]. Available from: www.thaidrugwatch.org/download/ Other/print/otherprint01.pdf
9. Thomas CM, Bertram E, Johnson D. The SBAR communication technique: teaching nursing students professional communication skills. Nurse Educ. 2009;34(4):176–80. www.ncbi.nlm.nih.gov/pub med/19574858
10. Wickens,C.D. (1984). Processing resources in attention. In R. Parasuraman & D.R. Davies(Eds.), Varieties of attention(pp.63-102). New York: Academic Press
11. U.S. Food and Drug Administration, [online] Available in https://www.fda.gov/ForConsumers/ConsumerUpda tes /ucm048644.htm
12. R. Charuwanno, J. Vongchanglaw, T. Pongsananurak, "Effects of Teaching by Using SBAR Technique on Knowledge Attitude and Practice in Taking and Giving Report Among Nursing Students" Vol 15, No 3, pp 690-397, 2014
13. HOSxP, [online] Available: http://hosxp.net/.

 https://doi.org/10.1051/matecconf/201819201052

Variation minimization in tele-sandblasting system: the effect of human-arm movement error

Ranon Jientrakul[1], *Pornsak Attavanish*[2], *Pholchai* Chotiprayanakul[1], *Sunpasit* Limnararat[1] and *Chumpol* Yuangyai[1,*]

[1]Department of Industrial Engineering, Faculty of Engineering, King Mongkut's Institute of Technology Ladkrabang, Bangkok, Thailand
[2]Faculty of Science and Technology, Southeast Bangkok College, Bangkok, Thailand

Abstract. In tele-sandblasting task, human arm movement is a critical source of producing variation in position of sandblasting nozzle resulting in high operating cost and low productivity. Each operator behaves differently leading to unpredictable movements. Skilled operators are able to reduce the variation; however, developing skills requires a training period. In this paper, we proposed a new approach which is the use of a novel operator's arm movement pattern incorporated with a Kalman filter to reduce the effect of human-arm movement error. A virtual tele-sandblasting system is used to validate our approach. The experimental results verify that our proposed approach is able to significantly reduce the effect of human arm movement error. The approach helps operators to perform the task more comfortably and takes short training time.

1 Introduction

Teleoperation system has been an active research area and used in diverse applications as it provides several benefits such as preventing operators from hazardous environment and accident, enabling small-scale or large-scale tasks, reaching inaccessible spots. Specifically, a tele-sandblasting system employs teleoperation system for complex structure maintenance and cleaning purpose [1].

Despite many benefits, teleoperation system inherently associates with problems causing variation in the system. The system consists of master controller stage and slave device stage. Recently, much research has contributed on variation occurred at the slave device which is caused by robot movement error and time delay [2][3]. However, the study of the variation occurred at the master controller which is mainly caused by human arm movement has been rarely found [4]. To the best of our knowledge, we realized that variation in master controller position in two-dimensional space caused by human arm movement has not been studied. This problem affects total variation of sandblasting nozzle position which is explained by system's accuracy and precision which eventually results in higher operating cost and lower productivity [5]. Skilled operators are able to reduce the variation; however, developing skills generally requires training period.

In this paper, we describe tele-sandblasting as two-stage system and we propose a novel arm movement pattern incorporated with Kalman filter to withstand human skill differences so that the system total variation is reduced. We then validate our approach by an experiment.

2 System Description

Tele-sandblasting system mainly consists of two stages, master controller and slave device as shown in Fig. 1. Master controller is the part where the operator interacts with the system to create a command for the slave device. Generally, master controller is a passive manipulator. Its movement is controlled by the operator. The slave device is an active manipulator which is sent into a working environment. It moves following the command received from the master controller. A sandblasting nozzle is mounted at the end-effector. At the working area, a camera is placed to capture the movement of the slave device for the operator.

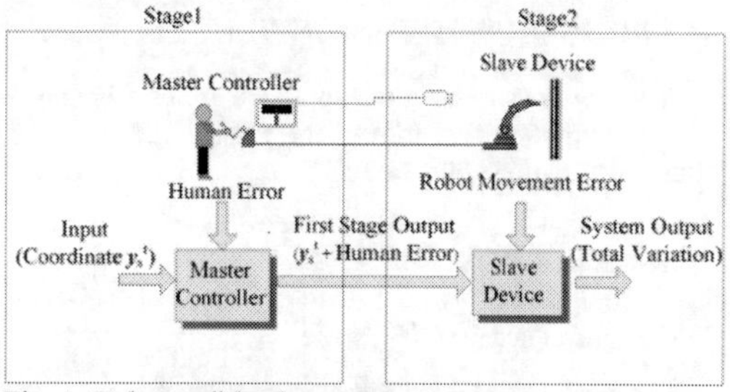

Fig. 1: Tele-sandblasting system

The preferable outcomes of the task is to ensure that the surface is completely clean as fast as possible. To do so, the width of the blasting path is set to be equivalent to or slightly smaller than the blasting spot size to achieve optimal operating time and cost [6], shown in Fig. 2. With this setup, navigating the nozzle inaccurately and imprecisely will result in poor surface quality. Therefore,

* Corresponding author: chumpol.yu@kmitl.ac.th

pathwidth was frequently set to be much smaller than the blasting spot size to cover variation caused by human-arm movement error resulting in the longer time required to complete the task as the total traveling distance increases. Note that, the output of interest is the variation in Y axis of the sandblasting nozzle position.

Recently, many variation reduction approaches were proposed to minimize robot movement error [3]. Liu and Zhang, (2014), proposed an adaptive ANFIS model to control human arm movement velocity in gas tungsten arc welding task [5]. Popp et al.,(2016) pointed out the problem of physical interface between human and computer which requires an approach to assist human operator in performing a hand-movement task more comfortable [7]. However, the aforementioned research did not consider the variation in the position caused by human.

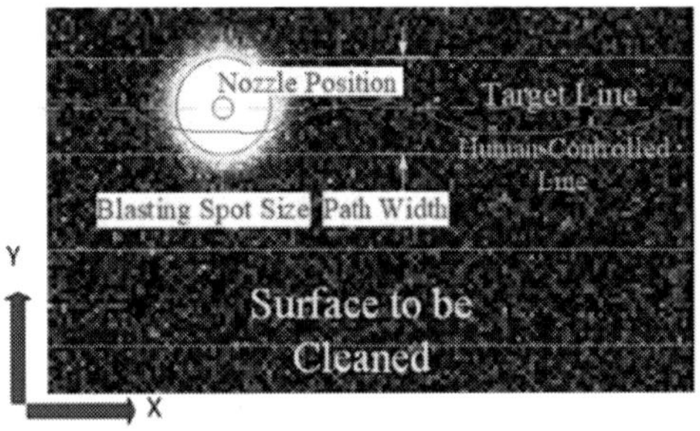

Fig. 2: A surface having tele-sandblasting operation where the sandblasting spot is controlled to blasting along target line to blast the path.

Variation caused by human-arm movement error can be divided into two portions: accuracy and precision. A blasting line created by human contains both of them which produce multiple types of possible variation behaviors. Also, for an operator, the behavior was not consistent. Therefore, it is difficult to define a model to estimate the variation for a specific person. Therefore, we proposed a new approach to cope with the problem.

3 Our Proposed Approach

There are two steps in our approach as described in Fig. 3: 1) navigating arm movement pattern and 2) using a linear estimation with Kalman filter

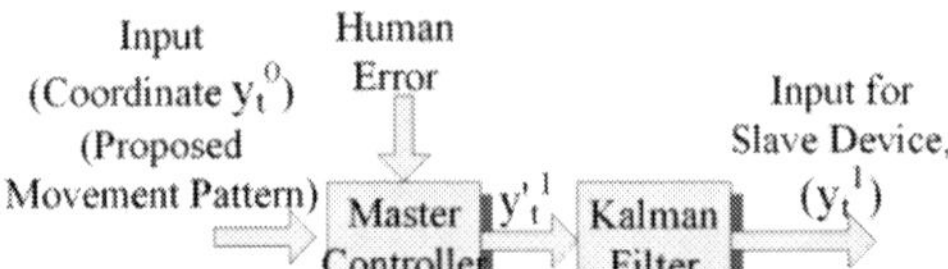

Fig. 3 : diagram describing our proposed approach

3.1 Proposed a navigating arm movement pattern

This step is motivated by noise filtering in electrical signal processing which uses linear estimators or filtering techniques to smooth out the jitter signal. We employ this concept to solve the variation in position of the master controller in tele-sandblasting application.

The existing approach requires highly skilled operator and takes very long time to train to move along the target line. The reason is that, in the existing approach, a specific guidance is not given to operator. They are allowed to move freely leading to a high variation. Therefore a certain movement must be specified in advance. Following electrical noise filtering, existing arm movement pattern is changed from "straight line following the target" to "jitter line across the target" to imitates the noise in electrical signal, depicted in Fig. 4. After this step, we have a jitter line ready for Kalman filter. Then, Kalman filter is applied to the line to filter the jitter line into a straight line.

To perform the task with our proposed arm movement pattern, during training the operators will be trained by moving the cursor along the guideline to be familiar with the new arm movement pattern before operating the task. Mathematical model to construct the guideline in two dimensional space is expressed in (1). The guideline is shown in Fig. 4(a).

$$\begin{bmatrix} x_t^G \\ y_t^G \end{bmatrix} = \begin{bmatrix} v_x \cdot i \cdot t \\ v_y \cdot i \cdot t \end{bmatrix} + \begin{bmatrix} \frac{v_y}{v_x+v_y} \\ \frac{v_x}{v_x+v_y} \end{bmatrix} (-1)^{\lceil \frac{t}{k} \rceil^-} \cdot (\frac{H}{2} + \frac{H}{k} \cdot (t \bmod j)) \quad (1)$$

Where, x_t^G and y_t^G are the positions in x axis and y axis, respectively, at time t. v_x and v_y are the desired blasting velocities of the sandblasting nozzle in x and y directions, respectively (mm./sec.). H is the amplitude (height of the wave). i is the amount of time per interval (second). j is the number of time interval in a half wavelength, depicted in Fig. 4(b). W is the wavelength which can be calculated by using (2).

$$W = 2 \cdot j \cdot t \cdot (||v_x + v_y||) \quad (2)$$

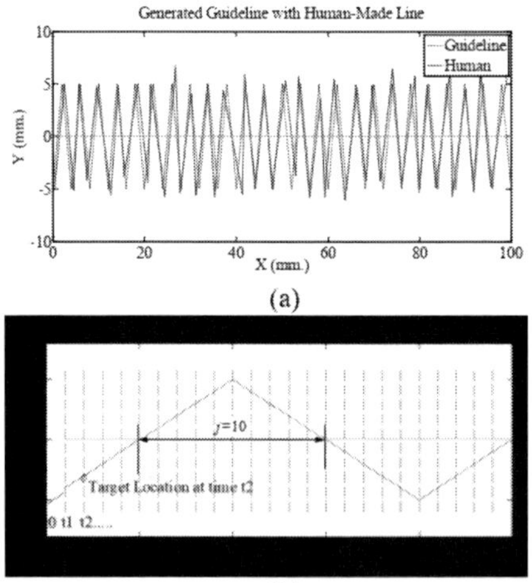

(a)

(b)

Fig. 4: (a) An example of a guideline (dash) with a human-made line (solid). (b)The guideline is design to have 10 mm. of amplitude (Y = -5:5), 4 mm. of wavelength. Time interval is 0.2 second. The generated guideline truncated by time interval lines, t_n's.

The shape of the guideline is a connection of multiple waves with equivalent size of amplitude. As human operated the task, the effect of human-arm movement error caused the difference in the sizes of amplitudes and wavelengths so that the positions form a bell-shape histogram which imitates the random noise behavior in electrical signal.

3.2 Using linear estimation with Kalman filter

Kalman filter is an approach used to minimize output variation by predicting the output based on previous values. Then, update the output based on the error between measured output and predicted output. Practically, Kalman filter works by predicting the current state first. The predicting equation is given in (3).

$$\hat{\boldsymbol{y}}_t^- = \boldsymbol{D}_t\hat{\boldsymbol{y}}_{t-1} + \boldsymbol{E}_t\boldsymbol{s}_t \quad (3)$$

Where $\hat{\boldsymbol{y}}_k^-$ is the predicted current state. $\hat{\boldsymbol{y}}_{t-1}$ is the state vector containing the sandblasting nozzle positions at time t. $\boldsymbol{D}_t$ is the state transition matrix which applies the effect of system state parameter(s) at time t-1on the system state at time t. $\boldsymbol{E}_t$ is the control input matrix which applies the effect of control input parameter(s) in the vector $\boldsymbol{s}_t$ on the state vector. Then, the confidence of the prediction, $\boldsymbol{P}_k^-$, is estimated by (4).

$$\boldsymbol{P}_t^- = \boldsymbol{D}\boldsymbol{P}_{t-1}\boldsymbol{D}^T + \boldsymbol{Q} \quad (4)$$

Where $\boldsymbol{Q}$ is the covariance of the process noise which is assumed to normally distribute with zero mean. After the current state has been predicted by (3) and (4), we then check how good the prediction is using updating equations as follows:

$$\boldsymbol{K}_t = \boldsymbol{P}_t^-\boldsymbol{F}^T(\boldsymbol{F}\boldsymbol{P}_t^-\boldsymbol{F}^T + \boldsymbol{R})^{-1} \quad (5)$$

$$\hat{\boldsymbol{y}}_t = \hat{\boldsymbol{y}}_t^- + \boldsymbol{K}_t(\boldsymbol{z}_t - \boldsymbol{F}\hat{\boldsymbol{y}}_t^-) \quad (6)$$

$$\boldsymbol{P}_t = (\boldsymbol{I} - \boldsymbol{K}_t\boldsymbol{F})\boldsymbol{P}_t^- \quad (7)$$

Where $\boldsymbol{K}_t$ is Kalman gain used in the updating process. $\boldsymbol{F}_t$ is the observation matrix used to map the true state space into the observed space (unit of measurement). $\boldsymbol{z}_t$ is the observed state matrix of $\hat{\boldsymbol{y}}_t$ which is the updated state. The measurement noise is assumed to normally distribute with zero mean and covariance $\boldsymbol{R}$.

According to Fig. 3, we define that $\boldsymbol{y'}_t^1 = \boldsymbol{z}_t$ in (6) and $\boldsymbol{y'}_t^1 = \hat{\boldsymbol{y}}_t$ in (6).

4 Validation of Our Approach

In this section, an experiment is set up to validate our proposed. Then, the results are shown and discussed.

4.1 Experiment Setup

Fig. 5: A virtual tele-sandblasting workstation.

The experiment is conducted on a virtual tele-sandblasting system, depicted in Fig. 5, [8]. The surface for blasting is virtually created with a size of 350 mm. x 25 mm. placed horizontally along the X axis. Pathwidth is 10 mm. wide. The target line is at the middle of the path. The blasting operation is illustrated in Fig. 6.

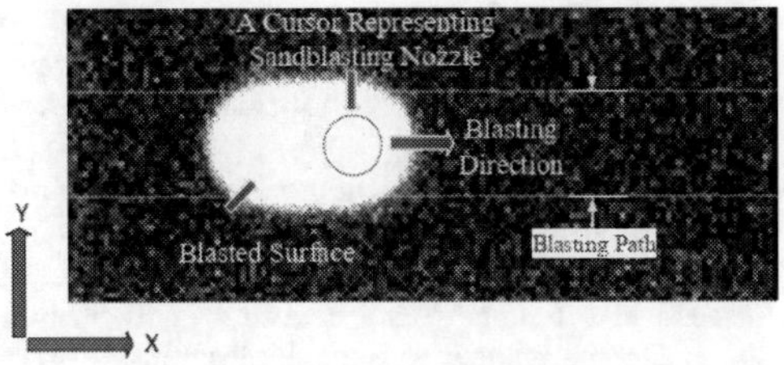

Fig. 6: A surface to be blasted in the experiment.

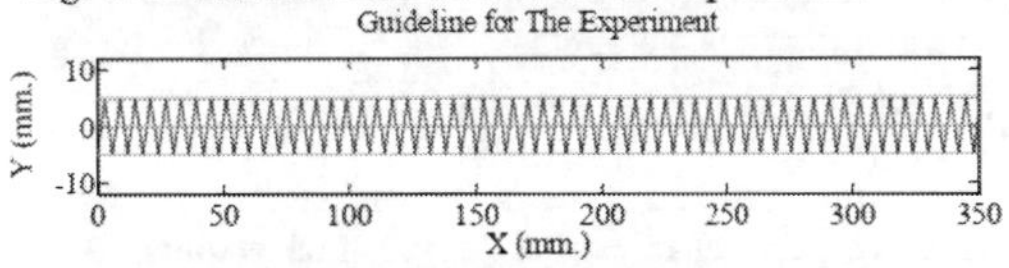

Fig. 7: A guideline for the experiment.

The guideline for training is designed to have the amplitude, *H*, of 10 mm. The wavelength, *W*, is 6 mm. The amplitude and wavelength are set according to the numbers obtained from our sub experiment to find values that the participant was comfortable with in the movements of wrist and finger during the operation. The proper operating time for our virtual tele-sandblasting system is expected to be in a range of 35-40 seconds. The guideline is shown in Fig. 7.

4.2 Experimental Results and Discussion

In this section, the results obtained from the experiment are discussed. First, the collected data is plotted to see how the proposed arm movement pattern reforms the variation in the output. The plots of the master controller positions over time and the histograms of its position in Y axis of the existing approach and our proposed approach from a participant are shown in Fig. 8.

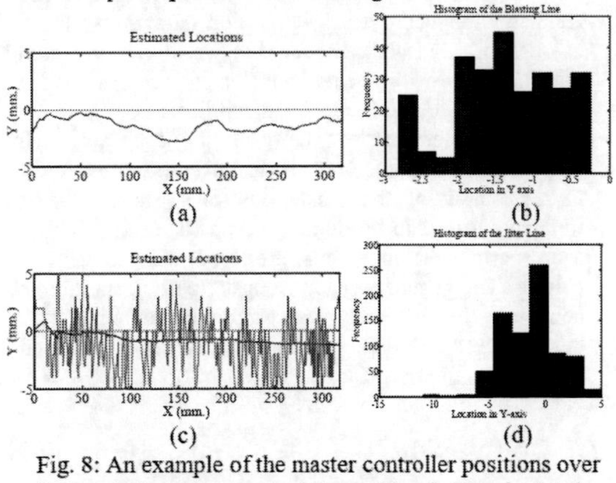

Fig. 8: An example of the master controller positions over time and the histograms of its position in Y axis. (a) the existing approach and (b) its corresponding histogram. (c) our proposed approach and (d) its corresponding histograms of the positions before filtering.

In Fig. 8 (a), with existing approach, we can see both inaccuracy and imprecision throughout the blasting line. The lack of accuracy was caused by the lack of an ability to guess the target accurately, while with the proper velocity, the participants struggled to maintain the precision due to the ability in arm control which resulted in high variance in the line. In Fig. 8 (b), a histogram clearly shows low accuracy (the peaks are far away from zero) and imprecision (multiple high frequency bins), suggesting that the existing approach (straight movement) is not consistent and difficult to manage to achieve low variation in the output.

Considering our proposed approach, the plots of master controller positions and the filtered output, shown in Fig. 8(c), are represented by red lines and black lines, respectively. It can be seen that when the participant tried to imitate the guideline under a constraint i.e., proper velocity, the effect of human-arm movement error exists as variation in wavelengths and amplitudes. Focusing on the upper and lower boundaries, it can be seen that the participant failed to navigate the master controller to touch the guideline which is the nature of human. However, this characteristic of the line is suitable with Kalman filter because the error causes the data to form a bell-shape histogram (imitating random noise in electrical signal), shown in Fig. 8(d), which centers closer to zero comparing to that of the existing approach suggesting that the accuracy is reduced. The reason is that although the operator does not know the target position exactly, when he moves the cursor to approximately touch both side of the path alternately, there is a great chance that the cursor frequently crosses the target line so that the frequency of the positions will be high at the value close to zero. Consequently, the filtered output, represented by the black line is straighter and closer to zero than that of the existing approach resulting in lower total variation.

It can be seen that our proposed arm movement pattern incorporating with Kalman filter significantly improves precision and accuracy in the output. This is because moving the master controller across instead of parallel to the target line helps reduce the chance of missing the target line so that the accuracy is maintained. Therefore, the error in accuracy is minimized (close to target). Then, Kalman filter improves the precision (straightness). An example of the results can be seen in Fig. 8(c). The filtered blasting line provides better precision and accuracy which is as impressive as other Kalman applications, [9][10].

We have demonstrated how our proposed approach helps reduce variation due to human-arm movement error in the output of the master-controller stage. During training period, short training time was taken for training each participant with our proposed arm movement pattern. The given feedback from the participant can be concluded that they feel more comfortable and relax with our proposed arm movement pattern and it yet provided better system output as a result.

5. Conclusion

We have demonstrated a new approach which is composed of two steps: navigating arm movement pattern and using a linear estimation with Kalman filter. Our suggested jitter pattern minimizes the error in accuracy while Kalman filter helps improving the precision. Consequently, the total variation is significantly reduced. The approach is then validated by a virtual tele-sandblasting system. However, in the future, an experiment will be conducted with more number of participants (with different ages and genders) to prove that our approach is applicable for multiple ranges of users.

Our approach can be an efficient alternative for tele-sandblasting task and other similar tasks to assist operators especially for those who prone to make high variation sandblasting nozzle position. Additionally, all participants stated that our proposed approach is more comfortable to perform than the existing approach leading to less stress and fatigue during the operation. We strongly believe that our approach can be applied to other similar applications.

Acknowledgment

This research is supported by the funding from faculty of engineering, KMITL. The authors would like to thank Mr. Kamtorn Sukpimay for laboratory cooperation.

References

[1] P. Chotiprayanakul, D.K. Liu, G. Dissanayake, Human–robot–environment interaction interface for robotic grit-blasting of complex steel bridges, In Automation in Construction, Volume 27, 2012, Pages 11-23, ISSN 0926-5805
[2] T. Li, D. Wang, S. Zhang, Y. Zhang and C. Yu, "Speed-accuracy tradeoff of controlling absolute magnitude of fingertip force," 2015 IEEE World Haptics Conference (WHC), Evanston, IL, 2015, pp. 408-414.
[3] S. J. Chen, N. Huang, Y. K. Liu, Y. M. Zhang; "Machine assisted manual torch operation: system design, response modeling, and speed control", Journal of Intelligence Manufacturing, Volume 28, Issue 6, pp 1249–1258, August 2017.
[4] Y.K. Liu, Y.M. Zhang, "Control of human arm movement in machine-human cooperative welding process", In Control Engineering Practice, Volume 32, 2014, Pages 161-171, ISSN 0967-0661,
[5] R. Jientrakul and C. Yuangyai, "Minimizing multi-stage system variation using Kalman filter for a tele-sandblasting system," *2016 IEEE International Conference on Management of Innovation and Technology (ICMIT)*, Bangkok, 2016, pp. 133-138.
[6] R. Jientrakul, C. Yuangyai, C.Y. Cheng, P. Chotiprayanakul and S. Limnararat, "Investigating multiple human performance measures in teleoperation task: A translation task in a tele-sandblasting maintenance system.", Hum Factors Man. 2017:1–13.
[7] W. L. Popp, O. Lambercy, C. Müller, R. Gassert, "Effect of handle design on movement dynamics and muscle co-activation in a wrist flexion task", In International Journal of Industrial Ergonomics, Volume 56, 2016, Pages 170-180.
[8] R. Jientrakul, S. Limnararat, C. Yuangyai, W. Srisuebsai, P. Chotiprayanakul, "Investigating of Human-Robot Collaboration in a Mixed-Reality Tele-Sandblasting Maintenance System", The 16th Asia Pacific Industrial Engineering and Management Systems Conference, Ho Chi Minh City, Vietnam, December 8-11, 2015
[9] G. Chang, Kalman filter with both adaptivity and robustness, In Journal of Process Control, Volume 24, Issue 3, 2014, Pages 81-87, ISSN 0959-1524
[10] M. Toivanen, An advanced Kalman filter for gaze tracking signal, In Biomedical Signal Processing and Control, Volume 25, 2016, Pages 150-158, ISSN 1746-8094

The simulation of volume shrinkage for double-injection molding with PC and ABS

Pichai Leklong[1] and *Wipoo* sriseubsai[1,*]

[1]Industrial Engineering Department, Faculty of Engineering, King Mongkut's Institute of Technology Ladkrabang, Bangkok, Thailand.

Abstract. The objective of this paper was to study and analyze the plastic injection molding parameters to reduce the volume shrinkage of double injection molded part. The specimen was molded with Acrylonitrile-butadiene-styrene (ABS) after Polycarbonate was molded as a half part. The weld line occurred at the haft of the molded part. The simulation with Moldex 3D R13 and the design of experiment with Taguchi method was used to perform the experiments and analyze the data to get the optimum of volume shrinkage. The results showed that melt temperature and packing pressure was significant to the volume shrinkage. When using low melt temperature and high packing pressure, the shrinkage was low and related to the thermal expansion of the material. It would be one of the parameters to the design of double injection molding.

1 Introduction

Plastic injection molding is one of the most popular plastic forming processes. It can produce the complex plastic parts from the mold which is designed to make a plastic part. Over the decades, the high demand of production rates is required because of the increasing industries competition. There are many special tools and manufacturing technique are developed to improve the injection molding process and to reduce the molding problems. One of them is the reducing volume shrinkage when the injection molded part starts to cool down [1].

Volume shrinkage is a high important issue in designing complex plastic part and when it is designed as an assembly part. The part quality is related to the volume shrinkage which is caused by the injection factors such as pressure and temperature during the filling phase. The PvT diagram of the material was used and analyzed in the simulation program to reduce the volume shrinkage problem. [2-4]

It is difficult to analyze and make a troubleshooting for the plastic molded part problem. The simulation program was developed to solve the complex mathematic model with the numerical method such as finite element and finite volume. It is helpful for the designers and operators such as to reduce cost, cycle time and shrinkage of the injection molding [5-6].

This research aims to apply the design of experiment (DOE) method with Taguchi model, statistical and engineering techniques for double injection molding. The objective is the optimum of volume shrinkage for injection of two types of materials, polycarbonate (PC) and acrylonitrile butadiene styrene (ABS). Modex 3D was used to simulate the plastic injection molding process to reduce the defect and trial time which allowed the design process to be faster [7]

Taguchi uses the quadratic equation to determine the loss function as the following:

$$L(x) = k(x\text{-}N)^2 \qquad (1)$$

Where L(x) = Loss function.

K = C/d^2 Constant of proportionality,

where C - Loss associated with sp limit.

d - Deviation of specification from target.

value.

x = Quality features of selected product.

N = Nominal value of the product.

(x-N) = Tolerance.

2 Material and methodology.

The simulation process was studied with the melt temperature, mold temperature, injection pressure, packing time, cooling time, and packing pressure as the parameters of DOE. The plastic PC and ABS were manufactured by PC WANDERLITEPC-122 CHI-MEI manufacturer and ABS POLYLAC PA777B CHI-MEI manufacturer respectively which were in the database of the simulation software. Normally, simulation software has the large material and machinery database [8]. They were used to perform the simulation. The injection process was simulated by 2 steps for injected 2 plastic materials to get the plastic part as shown in Fig. 1. The first step PC was injected to fill the half of the part. Then, the second step, ABS, was injected to fill the remaining part. The mold was designed to make a part

* Corresponding author: wipoo.sr@kmitl.ac.th

following the ASTM D638 specimen as shown in Fig.2 with cooling line. The most important step was to specify, select the parameters and levels [9]. The injection molding parameters with 2 levels were set and shown in Table 1 to determine the relationship between the factors that affect the process and objective of the process [10]. The Taguchi method with L8, 2^6 orthogonal array was selected to perform the experiment.

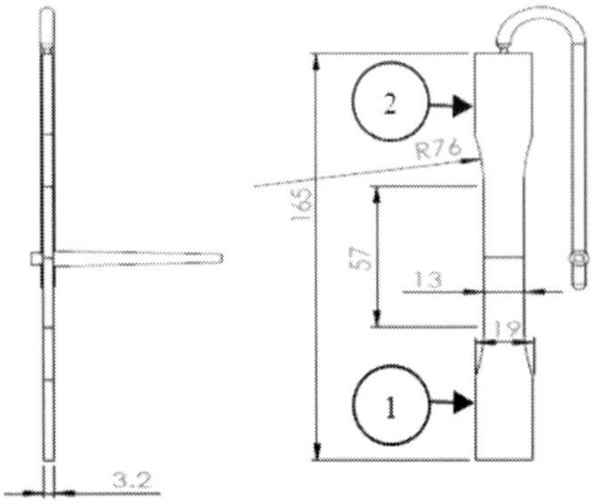

Fig 1. Part of two-material the 1 is PC plastic and the 2 is ABS Plastic

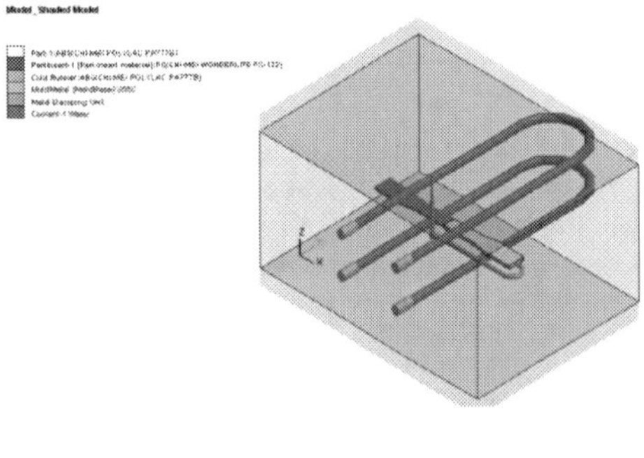

Fig 2. Standard tensile test specimen ASTM D638

Table 1. Process condition parameter in the DOE.

No.	Control factor	Unit	Level 1	Level 2
1	Melt temperature	(°C)	220	250
2	Mold temperature	(°C)	30	70
3	Injection pressure	(Mpa.)	180	210
4	Packing time	(sec.)	7	9
5	Cooling time	(sec.)	18	21
6	Packing pressure	(Mpa.)	180	210

3. Results and discussion

The example of the simulation result for volume shrinkage was shown in Fig.3. The chart showed the distribution and the average of the shrinkage of the molded part. The specimen showed two different color due to different material. The experiments were performed following the Taguchi model with L8, 2^6 orthogonal array. The average volume shrinkage results for each injection molding conditions were shown in Table 2. The results showed that the smallest volume shrinkage was occurred when using the melt temperature of 250°C, the mold temperature of 70°C, injection pressure of 180 MPa, packing time of 7 sec., cooling time of 21 sec and packing pressure of 210°C. While using the melt temperature of 250°C, the mold temperature of 30°C, injection pressure of 210 MPa, packing time of 7 sec., cooling time of 21 sec and packing pressure of 180°C, the volume shrinkage was the highest.

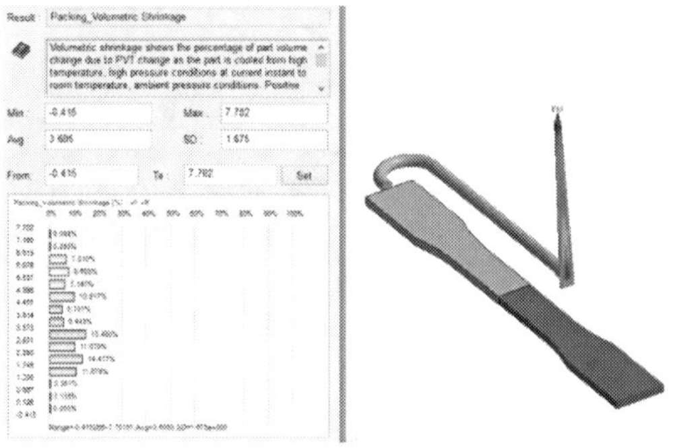

Fig 3. The volume shrinkage of the simulation result.□

Table 2. The results of volume shrinkage in the simulation.

No.	Tm	Tmo	Ip	Pt	Ct	Pp	Volume shrinkage
1	220	30	180	7	18	180	3.249
2	220	30	180	9	21	210	2.986
3	220	70	210	7	18	210	2.993
4	220	70	210	9	21	180	3.228
5	250	30	210	7	21	180	3.582
6	250	30	210	9	18	210	3.574
7	250	70	180	7	21	210	3.703
8	250	70	180	9	18	180	3.585

When considering the main effect plot as shown in Fig.4, those six injection parameters were significant for injection molded part. The molded part had less volume shrinkage when injection condition of melt temperature, injection pressure, packing time was low but mold temperature, cooling time and packing pressure was high. Moreover, melt temperature and packing pressure was effect to volume shrinkage than other factors. The less shrinkage occurred when injection molded part with low melt temperature because of less thermal expansion. While the volume shrinkage was known and defined as equation 1, the volume shrinkage was related to the temperature as shown in equation 2.

$$V_s = \frac{V_i - V_d}{V_d} x100 \qquad (2)$$

Where, V_s = Volume shrinkage (%)

V_i = Initial test mass of solid (cm^3)

V_d = Final volume of solid (cm^3)

$$a_v = \frac{1}{V}\frac{\Delta V}{\Delta T} \qquad (3)$$

Where, a_v = Volume thermal expansion coefficient

V = Area of the object (m^3)

ΔV = Change in volume (m^3)

ΔT = Change in temperature (K)

Many materials, the thermal expansion is related to the melting temperature of that material. The coefficient of thermal expansion is related to the volume changing and as well represented in the PvT diagram shown in Fig.5. The diagram also presented that using high pressure, especially packing pressure during molding, the volume shrinkage was reduced.

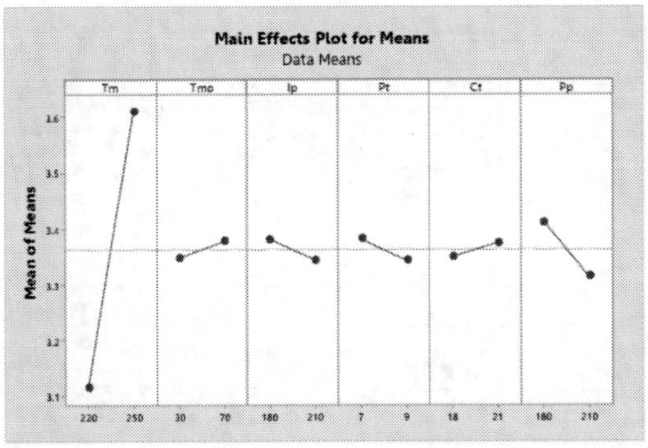

Fig 4. The relationship between injection parameter and volume shrinkage.

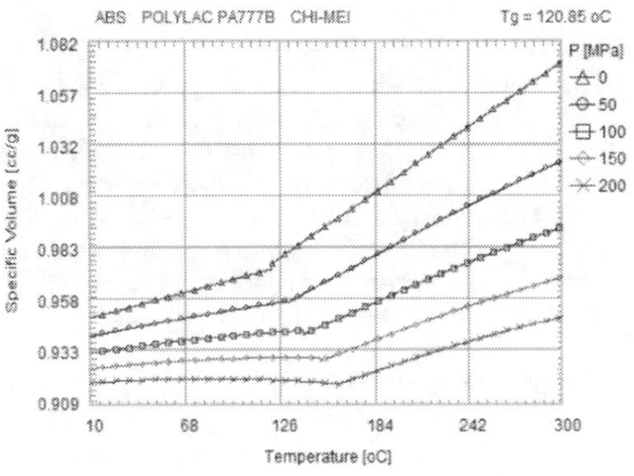

Fig 5. PvT diagram of plastic ABS POLYLAC PA777B CHI-MEI manufacturer.

Table 3. Response Table for Means.

Level	Tm	Tmo	Ip	Pt	Ct	Pp
1	3.114	3.348	3.381	3.382	3.350	3.411
2	3.611	3.377	3.344	3.343	3.375	3.314
Delta	0.497	0.030	0.036	0.038	0.024	0.097
Rank	1	5	4	3	6	2

Table 3 showed an analysis to find out the effect of injection parameters to reduce volume shrinkage. Based on Taguchi method technique and data analysis by configuration (α=0.05). The results showed that the factor of Tmo, Ip, Pt, and Ct were no significant. Opposite, Tm and Pp were significant.

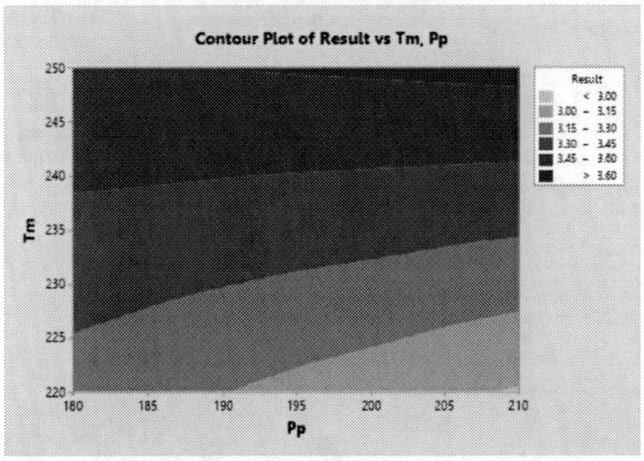

Fig 6. Contour plot of melt temperature with packing pressure.

Fig 6 showed the contour plot of melt temperature with packing pressure. The results showed that the less shrinkage occurred when using Tm of 220 °C and Pp of 210 Mpa.

5. Conclusions

The simulation results showed that the ABS molded part had the less shrinkage when using low melt temperature and high packing pressure. The mold temperature, injection pressure, packing time and cooling time was less significant than those parameters which can be explained by the thermal expansion and PvT diagram. The results of this experiment can be the case study for double injection molding and can be applied to a commercial product such as a button air car and an automotive light bulb in the future.

Acknowledgment

The authors would like to thank the Department of Industrial Engineering, Faculty of Engineering, King Mongkut's Institute of Technology Ladkrabang, Bangkok, Thailand, and Core Tech System Co., Ltd. for supporting to access to their software simulation and research facilities.

References

1. X.Yan, J.Jianbo, Comput. Ind CD **(2010)**

2. S.S.Deng, C.Y.Ho,S.J.Hwang, D.Y.Hong 76-82,IEEE TRANSACTIONS ON COMPONENTS **(2011)**

3. L.-C. Hong and S.-J. Hwang, IEEE Trans. Compon. Packag. Technol. 27, no. 2, pp. 291–295, Jun. **(2004)**

4. Y.-S. Chang, Mech. Eng., *Propulsion* (2005).

5. P.K.Kennedy, Propulsion physics (EDP Sciences, TUe, **(2008)**

6. C.Yen, J. C.Lin, W.Li, M. F. Huang, J. Mater. Process. Technol. **(2006)**

7 J. G.Kovács, B. Sikló, *Polym. Test.*(2010).

8. Moldex3D Reference Manual R13, www.moldex3d.com

9. C.Grolleau, C.Coutanceau, F Pierre, J. M.Leger, J. Power Sources **(2010)**.

10. J. B.Tranter,; P.Refalo,; Rochman, A. Towards, *J. Manuf. Process.* **(2017)**.

Surface morphology and wettability for thin films of beta-iron disilicide produced through direct-current sputtering utilizing a pair of facing targets

Peerasil Charoenyuenyao[1], *Nathaporn* Promros[1,*], *Rawiwan* Chaleawpong[1] and *Tsuyoshi* Yoshitake[2]

[1]Department of Physics, Faculty of Science, King Mongkul's Institute of Technology Ladkrabang, Bangkok 10520
[2]Thailand Department of Applied Science for Electronic and Materials, Kyushu University, Fukuoka 816-8580, Japan

Abstract. In the current work, beta-$FeSi_2$ films were epitaxially produced onto Si(111) wafer substrates via usage of facing-targets direct-current sputtering (FTDCS). The temperature for substrate heating was maintained at 600 °C and the sputtering pressure was set at 1.33×10^{-1} Pa. The surface morphology and contact angles of the beta-$FeSi_2$ films were explored consistently in this research. Images of three-dimensional AFM and FESEM for the beta-$FeSi_2$ film surface revealed a smooth surface with a root mean square roughness of 1.31 nm and a porous area. The average contact angle between the dropped water and beta-$FeSi_2$ film surface was found to be 98.7°, establishing that the surface of the beta-$FeSi_2$ films was hydrophobic. The acquired experimental results revealed the commencement of the hydrophobic surface feature of the beta-$FeSi_2$ films produced via FTDCS approach.

1 Introduction

Over the past few years, iron disilicide ($FeSi_2$) possessing an orthorhombic semiconducting phase (beta) has garnered considerable attention from numerous researchers as a result of its suitable semiconducting features [1, 2]. It has been experimentally supported that it is possible to be epitaxially produced onto Si layers. The lattice mismatches for epitaxial production between films of beta-$FeSi_2$ and Si layers were from 2 to 5 percent [3, 4]. Further, the compound beta-$FeSi_2$ was produced from silicon and iron, which are environmentally friendly, non-toxic, and plentiful in nature [5]. Beta-$FeSi_2$ holds direct (0.85 electron volt) and indirect (0.76 electron volt) optical band gaps as well as an absorption coefficient > 10^5 cm^{-1} at 1.2 electron volt [6-8]. It has been evidentially confirmed by previous research that the beta-$FeSi_2$ film surface was quite smooth [9]. These features of beta-$FeSi_2$ are appropriate for usage in optoelectronic devices and superhydrophobic surfaces. Employing facing-targets direct-current sputtering (FTDCS), the production of beta-$FeSi_2$ films onto Si(111) wafers was achieved at low temperatures for substrate heating in the previous study [9]. The FTDCS approach consists of a pair of targets set in a parallel direction. Particularly, the substrate is situated outside of the plasma area and vertical with the pair of targets. In comparison to common radio-frequency sputtering, this approach has the following advantages: (a) sputtering at low pressure operation, (b) higher density of plasma created from sputtering, (c) less increment in substrate temperature, (d) less damage and (e) obtained films with low difference stoichiometry in comparison with a target as a result of the substrate being free of plasma [4, 9]. Additionally, the energy of particles that reached the substrate surface was higher compared to utilization of different approaches because of sputtering at low pressure operation [4, 9]. Based on the merits of FTDCS, the epitaxial production of beta-$FeSi_2$ thin films onto the Si(111) was accomplished when the temperature for substrate heating was 600 °C. The thin films of beta-$FeSi_2$ created in this work presented n-type property as well as carrier density of approximately 5×10^{17} cm^{-3} [9]. The produced hetero-structure between p-type Si wafers and n-type beta-$FeSi_2$ thin films can be utilized in photovoltaic and photodiode applications [10, 11]. Despite the face beta-$FeSi_2$ possesses desirable properties for hydrophobic surface applications, there have been few studies or reports on the assessment of wettability properties of beta-$FeSi_2$ films produced via FTDCS for employment in such applications. Consequently, the current research proposes study of their surface morphology as well as wettability behavior. It was anticipated that the average contact angle between the dropped water and the beta-$FeSi_2$ film surface would be greater than 90° (hydrophobic surface). According to all available knowledge and information, this research comprises the first exploration and report concerning wettability based on beta-$FeSi_2$ thin films produced through FTDCS.

*Corresponding author: nathaporn promros@kyudai.jp

2 Experimental details

Based on the usage of FTDCS, a 300 nm beta-$FeSi_2$ layer was produced onto a wafer of p-type Si(111) by employing a pair of $FeSi_2$ alloy targets. The temperature for substrate heating was fixed at 600 °C during film production. Before epitaxial production, the organic and oil layers were eliminated from the p-type Si(111) wafer by rinsing in acetone, methanol, and deionized water for five minutes in each step. Next, a diluted hydrofluoric (HF) acid solution (concentration of 1 percent) was applied to eradicate the existing oxide layer. The HF residue was purged instantaneously from the p-type Si wafer in deionized water. After cleaning, the prepared Si wafer was attached to the substrate holder inside the FTDCS chamber, where the space between the Si wafer substrate and $FeSi_2$ targets was set at 7.5 cm. The pressure inside of the chamber was evacuated to 3×10^{-5} Pa using both a turbomolecular pump and rotary pump. The sputtering process was operated in argon gas (purity: 6N). The flow rate of the argon gas and pressure of sputtering were maintained at 15 sccm and 1.33×10^{-1} Pa, respectively. The applied voltage for sputtering was assigned to 1 kV, while the generated sputtering current was about 1.5 mA. Thorough study of X-ray diffraction (XRD) for the beta-$FeSi_2$ layer epitaxially produced on Si(111) wafers was performed in a Rigaku RINT 2000/PC. The surface morphology view of beta-$FeSi_2$ films was explored using FESEM (Carl Zeiss Auriga Field Emission Scanning Electron Microscope). The roughness of the film surface was inspected by means of an XE-120 Atomic Force Microscope (AFM). The wettability of the film surface was determined using an OCA 20 measuring instrument.

3 Results and discussion

Figure 1(a) presents the scanned XRD pattern of the produced thin films of beta-$FeSi_2$ onto the layer of Si(111) wafers. Measurement was executed utilizing a mode of 2θ-θ scan. From the appearance pattern, weak 404/440 and intense 202/220 peaks could be found. These peaks are essential for beta-$FeSi_2$ films, which can be epitaxially produced on Si(111). In order to measure the pole figure, a specific crystallographic plane of the 2θ-θ scan should be fixed, which determines the Miller indices of the films. From the XRD pattern, the beta-202/220 peak was close to the strong peak of the Si-111 plane. Namely, the obtained pole figure image, which was measured in this plane, might show the pole figure of the beta-202/220 peak with interaction from the substrate plane. Therefore, the diffraction peak of beta-404/440 was chosen as the main scanning plane in this XRD pole figure measurement. In the measurement process, the pole figure was performed by varying two geometrical parameters, such as the tilt angle (Ψ) from beta-$FeSi_2$ film surface normal direction and the rotation angle (ϕ) around the surface of the films in the normal direction. Figure 1(b) depicts the acquired pattern of pole-figure concerning the peak of beta-404/440 for the beta-$FeSi_2$ films produced onto Si(111) wafers. This result reveals the occurrence of 3 kinds of epitaxial variants. The rotation angle was 120 degrees to each other [3].

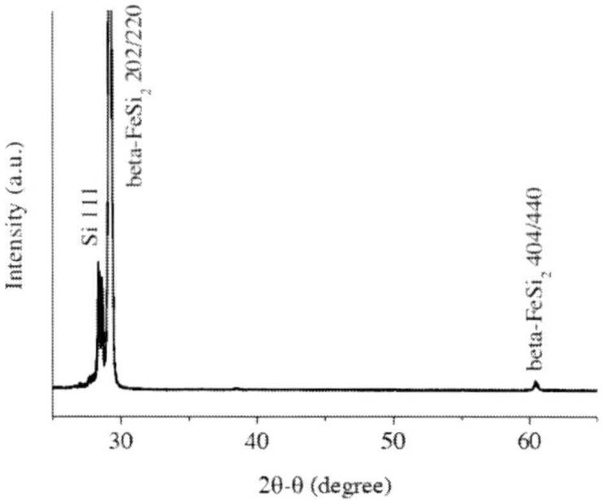

(a)

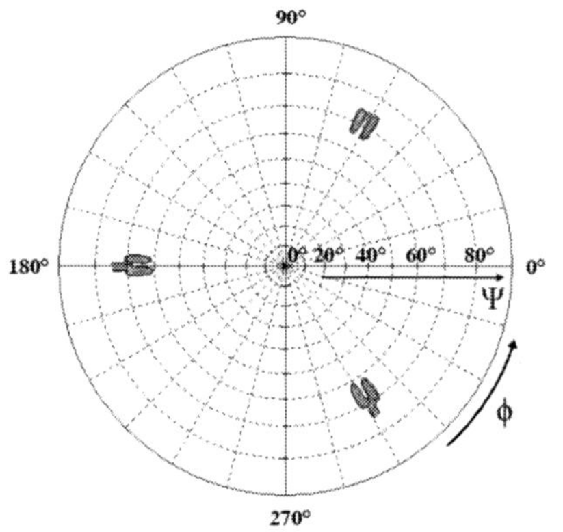

(b)

Fig. 1. (a) The XRD pattern measured in a mode of 2θ-θ scan and (b) the pole-figure pattern of beta-$FeSi_2$ films produced onto Si(111) wafers.

Figure 2(a) illustrates a FESEM micrograph with 50k magnitude in plain surface view of the produced beta-$FeSi_2$ film. It was observed that a large amount of crystallites with small sizes covered the film surface. Further, the produced beta-$FeSi_2$ films contained many grain boundaries and a porous area that existed at the edge of the grain boundary. In Figure 2(b), a cross-sectional FESEM micrograph is presented showing a layer of beta-$FeSi_2$ thin film coated onto a layer of Si(111) wafer. In the cross-section micrograph, the beta-$FeSi_2$ film surface produced through FTDCS was level and appeared non-fractured along the path.

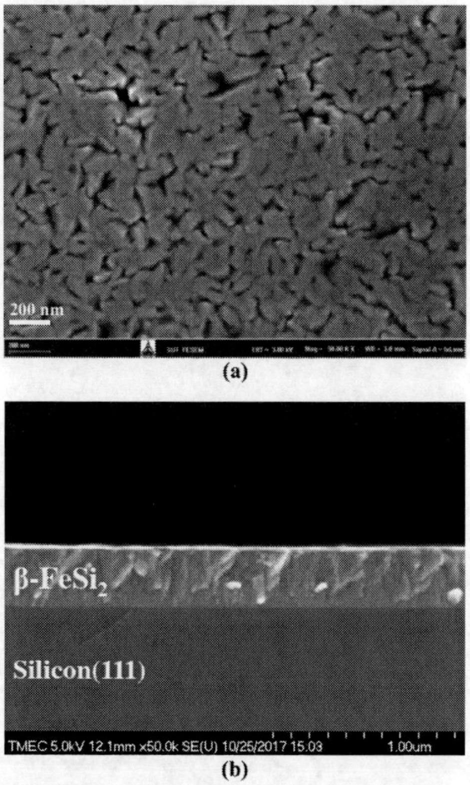

Fig. 2. FESEM micrograph in (a) plain view and (b) cross-sectional view for the layer of beta-$FeSi_2$ films produced onto the layer of Si (111) wafer.

A three-dimensional AFM image of beta-$FeSi_2$ films, scanning in non-contact mode, is illustrated in Figure. 3. The scanning result demonstrates the rather smooth surface of the beta-$FeSi_2$ film. The estimated root mean square roughness (R_{rms}) of the film surface was 1.31 nm. Some pinholes are observable within the scanning area of 500 × 500 nm, which conform to FESEM surface images. This can be represented by the fact that, in the production process of beta-$FeSi_2$ films by usage of FTDCS, the Si substrate is located on a substrate holder far away from the plasma that was generated from sputtering [4, 9]. Furthermore, the temperature of the produced beta-$FeSi_2$ layer surface did not increase as a result of the Si substrate being arranged far from the plasma region. Based on this situation, the surface of the produced beta-$FeSi_2$ layer suffered minimal damage from the destruction of sputtering plasma [4, 9].

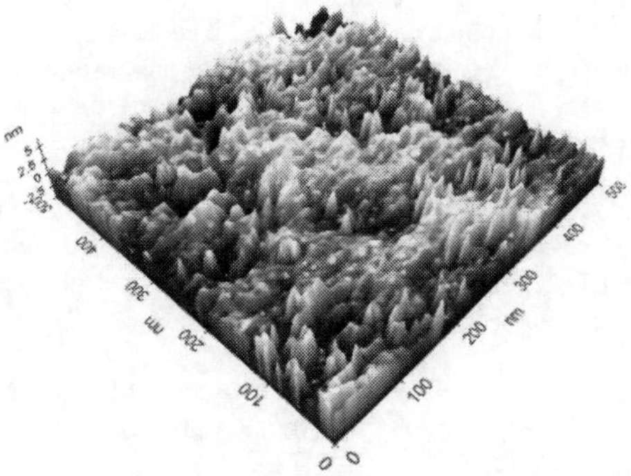

Fig. 3. Three-dimensional AFM image of beta-$FeSi_2$ films produced by FTDCS.

In order to expose the wettability behavior of the beta-$FeSi_2$ thin films produced by means of FTDCS in this study, the contact angle between the beta-$FeSi_2$ film surface and the dropped water was measured. Figure 4 presents an image of the measured contact angle between film surface and water droplet. The average contact angle between the water droplet and film layer was 98.7° (99.0° and 98.4° for the left and right sides of contact angle, respectively). From the obtained results, the surface of the beta-$FeSi_2$ films exhibited hydrophobic properties. Generally, a hydrophobic surface is labeled as having contact angles in the angle range of $90° < \theta < 150°$ [12-14]. According to the FESEM image, the beta-$FeSi_2$ film surface has an occurrence of porous territory. This can be explained by the wetting model of Cassie-Baxter, which describes the wetting phenomena of the porous area underneath liquid drops. The porous area is an act to the air pocket, where air can infiltrate into the grooves and support the lifting of dripped liquid, resulting in a higher contact angle than for a flat surface [15-17].

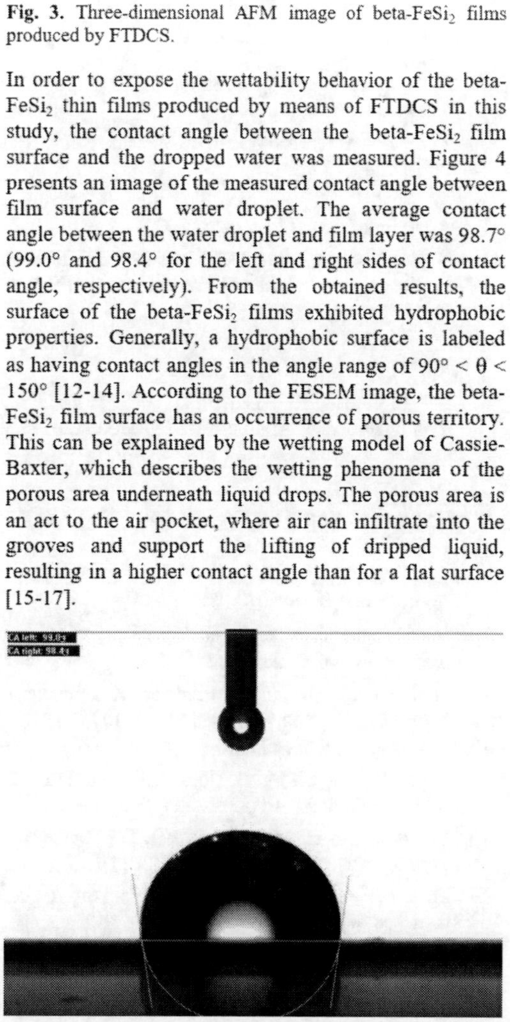

Fig. 4. The contact angle between the water droplet and plane of beta-$FeSi_2$ layer produced on Si(111) wafer via FTDCS.

4 Conclusions

In the present work, a layer of beta-$FeSi_2$ was produced onto a Si(111) wafer via FTDCS. The temperature for substrate heating and the sputtering pressure were maintained at 600 °C and 1.33 × 10^{-1} Pa, respectively. Surface morphology for the produced beta-$FeSi_2$ films was explored using FESEM and three-dimensional AFM images, which demonstrated the existence of a porous area (pinholes) and R_{rms} of 1.31 nm. The average contact angle of the produced beta-$FeSi_2$ film surface was 98.7°. This proved the hydrophobic behavior of the film surface. From the acquired results, it was revealed that beta-$FeSi_2$ films produced via FTDCS could potentially be applied to hydrophobic surfaces.

The present study was supported by a research grant from the AUN/SEED-Net as a CRA project as well as by the Faculty of Science (KMITL).

References

1. D. Leong, M. Harry, K.J. Reeson, K.P. Homewood, Nature **387**, 686 (1997)
2. T. Sunohara, K. Kobayashi, T. Suemasu, Thin Solid Films **508**, 371 (2006)
3. M. Shaban, K. Nomoto, S. Izumi, T. Yoshitake Appl. Phys. Lett. **94**, 222113 (2009)
4. N. Promros, R. Baba, M. Takahara, T.M. Mostafa, P. Sittimart, M. Shaban, T. Yoshitake, Jpn. J. Appl. Phys. **55**, 06HC03 (2016)
5. M. Shaban, S. Izumi, K. Nomoto, T. Yoshitake, Appl. Phys. Lett. **95**, 162102 (2009)
6. N. Promros, K. Yamashita, R. Iwasaki, T. Yoshitake, Jpn. J. Appl. Phys. **51**, 108006 (2012)
7. M.C. Bost, J.E. Mahan, J. Appl. Phys. **64**, 2034 (1988)
8. B. Tatar, K. Kutlu, M. Ürgen, Thin Solid Films **516**, 13 (2007)
9. T. Yoshitake, Y. Inokuchi, A. Yuri, K. Nagayama, Appl. Phys. Lett. **88**, 182104 (2006)
10. M. Shaban, K. Nakashima, T. Yoshitake, Jpn. J. Appl. Phys. **46**, 7708 (2007)
11. S. Izumi, M. Shaban, N. Promros, K. Nomoto, T. Yoshitake, Appl. Phys. Lett. **102**, 032107 (2013)
12. A. Kumar, G.K. Dalapati, H. Hidayat, F. Law, H.R. Tan, P.I. Windenborg, B. Hoex, C.C. Tan, D.Z. Chi, A.G. Aberle, RSC Adv. **3**, 7733 (2013)
13. J. Drelich, E. Chibowski, D.D. Meng, K. Terpilowski, Soft Matter **7**, 9804 (2011)
14. M. Krasowska, J. Zawala, K. Malysa, Adv. Colloid Interface Sci. **147-148**, 155 (2009)
15. A. Lafuma, D. Quéré, Nat. Mater. **2**, 457 (2003)
16. V. Mortazavi, M.M. Khonsari, Wear **372-373**, 145 (2017)
17. A.M.A. Mohamed, A.M. Abdullah, N.A. Younan, Arab. J. Chem. **8**, 749 (2015)

Design and simulate of LQR-Fuzzy controller for unicycle robot with double flywheels

Surachat Chantarachit*

Department of Mechatronics Engineering, Rajamangala University of Technology Thanyaburi, Thailand

Abstract. This research is focus on design and simulate unicycle robot with double flywheels model with LQR-Fuzzy controller. Roll balancing torque is generated by gyroscopic effect. Pitch balancing torque is applied by inverted pendulum concept. To control the heading of the robot, the angular momentum from both flywheel is applied to control this. The robot model is based on Euler-Lagrange equations. The non-linear model is linearization by Taylor series expansion. The simulation results conducted by MATLAB/Simulink. LQR-Fuzzy is combination algorithm between LQR and Fuzzy controller. The main structure control is the LQR controller and use the Fuzzy controller to adjust the close loop controller gain. The simulation results is simulated and compared with conventional LQR.

1. Introduction

Unicycle robot is a mobile robot which has only one contact point on ground. The unicycle robot dynamic model is very challenging and interesting in research topic in dynamic control. Since, model is non-linear, unstable and difficult to control. A. Schoowinkel [1] is the first one who derives unicycle dynamic model by using the turntable mechanism concept. Z. Sheng and K. Yamafuji [2], who are successful to balance the unicycle robot. The next generation of unicycle balancing is gyroscope force (flywheel). S. Majima et al [3] proposed a single-flywheel technique for balancing a unicycle robot. The produced torque from this technique is higher than the turntable technique. To generate greater torque, the double flywheel is introduced in this paper [4]. There are many control algorithms which applied to control unicycle robot. LQR is well-known and widely used controller in robot control. This paper presents the combination of LQR and Fuzzy controller. The performance of LQR-Fuzzy [5] is evaluated and compared in this paper.

2. Concept Design

Unicycle robot can be falling in lateral and longitudinal direction. In lateral direction, the gyro scope is the concept to generate torque. There are two flywheels with produce the rolling torque. Both of them are rotated in difference direction with high speed in order to generate angular momentum. By opposite rotating of both flywheels, there are some component of angular momentum effect from rolling torque, it can be applied to control heading of the robot.

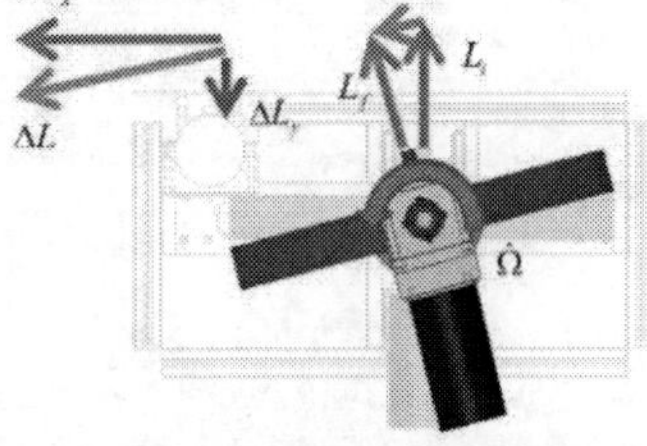

Fig. 1 Angular momentum of flywheel

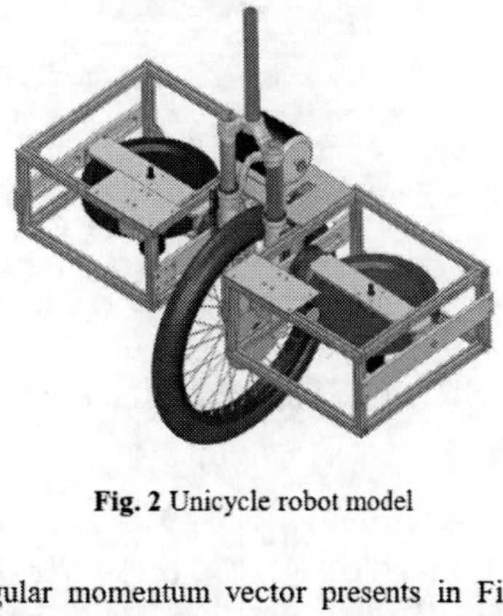

Fig. 2 Unicycle robot model

The angular momentum vector presents in Fig.1. The green vector is the rolling torque component and purple vector is the heading torque component. Last balancing axis is the longitudinal direction, the invert pendulum concept is applied to generate torque. The unicycle robot model presents in Fig.2.

Corresponding author: surachat_c@rmutt.ac.th

3. Robot Dynamic Model

The dynamic model of the unicycle robot is derived into two main parts: wheel and body are presented in Fig.3. The model is referred to Chantarachit, S. [4] model. Unicycle robot model is coupling model between each axis (x y and z).

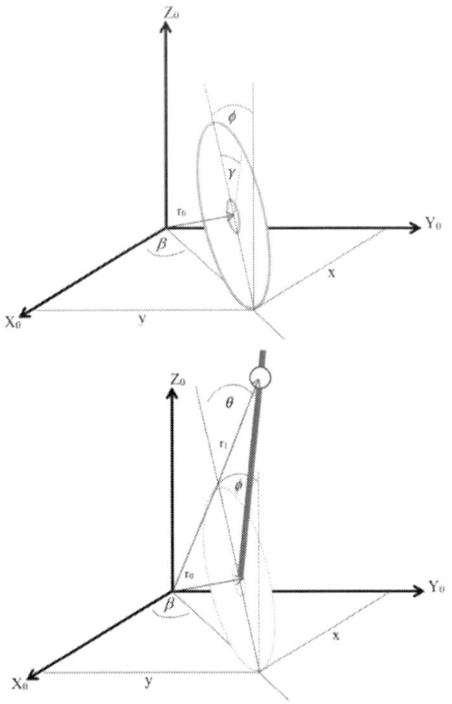

Fig. 3 Wheel and Body model

The Dynamic model is derived by using Euler-Lagrange (Eq.1). Friction in the system is small and negligible.

$\frac{d}{dt}\Big(\frac{\partial L}{\partial}$ ∂L [redacted] f (1)

L is the difference between total kinetic energy and potential energy. From (Eq.1), the nonlinear equation will be presented in 4 terms and rearranged in to non-linear dynamic equation. The equation of the unicycle robot is shown in (Eq. 2). M is inertia matrix. C is Coriolis matrix. G is gravity matrix, and D is disturbance matrix. Since the robot is operated at low speed, viscous friction is small and negligible (D=0).

$M(q)$ [redacted] $+D=F$ (2)

Where,

$$M=\begin{bmatrix} M11 & M12 & M13 & M14 \\ M21 & M22 & M23 & M24 \\ M31 & M32 & M33 & M34 \\ M41 & M42 & M43 & M44 \end{bmatrix},\ C=\begin{bmatrix} C1 \\ C2 \\ C3 \\ C4 \end{bmatrix},\ G=\begin{bmatrix} G1 \\ G2 \\ G3 \\ G4 \end{bmatrix}, F=\begin{bmatrix} \tau_\phi \\ \tau_\theta \\ -\tau_\gamma \\ \tau_\gamma \end{bmatrix}$$

The details of M C and G matrices which are derived from unicycle robot model are shown below

$$M11=i_{F1}\mathrm{Cos}\theta^2+i_{F3}\mathrm{Sin}\theta^2+i_{w1}+rm_F\left(r+\mathrm{Cos}\theta l_F\right)+ m_F l_F\mathrm{Cos}\theta\left(r+\mathrm{Cos}\theta l_F\right)+r^2m_w$$

$$M12=-i_{F1}\mathrm{Cos}\theta\mathrm{Cos}\phi\mathrm{Sin}\theta+i_{F3}\mathrm{Cos}\theta\mathrm{Cos}\phi\mathrm{Sin}\theta- l_F m_F\mathrm{Cos}\phi\mathrm{Sin}\theta\left(r+\mathrm{Cos}\theta l_F\right)$$

$$M13,M14=0\ ,\ M21=M12$$

$$\begin{aligned} M22=\ &\mathrm{Cos}\phi^2\mathrm{Sin}\theta^2 i_{F1}+\mathrm{Sin}\phi^2 i_{F2}+\mathrm{Cos}\theta\mathrm{Cos}\phi^2 i_{F3}+ \\ &\mathrm{Sin}\phi^2 i_{w2}+\mathrm{Cos}\phi^2 i_{w3}+0.5r^2m_F-0.5m_F r^2\mathrm{Cos}2\phi+ \\ &r\mathrm{Cos}\theta l_F m_F-0.5r\mathrm{Cos}[\theta-2\phi]l_F m_F- \\ &0.5r\mathrm{Cos}[\theta+2\phi]l_F m_F+0.75l_F^2m_F- \\ &0.25\mathrm{Cos}2\theta l_F^2m_F-0.125\mathrm{Cos}[2(\theta-\phi)]l_F^2m_F- \\ &0.25\mathrm{Cos}2\phi l_F^2m_F-0.125\mathrm{Cos}[2(\theta+\phi)]l_F^2m_F+ \\ &r^2\mathrm{Sin}\phi^2m_w \end{aligned}$$

$$M23=\mathrm{Sin}\phi i_{F2}+r\mathrm{Cos}\theta\mathrm{Sin}\phi l_F m_F+\mathrm{Sin}\phi l_F^2m_F$$

$$\begin{aligned} M24=\ &\mathrm{Sin}\phi i_{w2}+r^2\mathrm{Cos}\varphi^2\mathrm{Sin}\phi m_F+r^2\mathrm{Sin}\phi\mathrm{Sin}\varphi^2m_F+ \\ &r\mathrm{Cos}\theta\mathrm{Cos}\varphi^2\mathrm{Sin}\phi l_F m_F+r\mathrm{Cos}\theta\mathrm{Sin}\phi\mathrm{Sin}\varphi^2 l_F m_F+ \\ &r^2\mathrm{Cos}\varphi^2\mathrm{Sin}\phi m_w+r^2\mathrm{Sin}\phi\mathrm{Sin}\varphi^2m_w \end{aligned}$$

$$M31=M13\ ,\ M32=M23$$

$$M33=i_{F2}+l_F^2m_F$$

$$M34=rl_F m_F\mathrm{Cos}\theta$$

$$M41=M14\ ,\ M42=M24\ ,\ M43=M34$$

$$M44=i_{w2}+r^2m_F+r^2m_w$$

$$\begin{aligned} C1=\ &-2\mathrm{Cos}\theta\mathrm{Sin}\theta i_{F1}\theta'\phi'+2\mathrm{Cos}\theta\mathrm{Sin}\theta i_{F3}\theta'\phi'-\mathrm{Cos}\phi i_{w2}\gamma'\varphi'- \\ &\mathrm{Cos}\theta^2\mathrm{Cos}\phi i_{F1}\theta'\varphi'+\mathrm{Cos}\phi\mathrm{Sin}\theta^2 i_{F1}\theta'\varphi'-\mathrm{Cos}\phi i_{F2}\theta'\varphi'+ \\ &\mathrm{Cos}\theta^2\mathrm{Cos}\phi i_{F3}\theta'\varphi'-\mathrm{Cos}\phi\mathrm{Sin}\theta^2 i_{F3}\theta'\varphi'+\mathrm{Cos}\phi\mathrm{Sin}\theta^2\mathrm{Sin}\phi i_{F1}\varphi'^2- \\ &\mathrm{Cos}\phi\mathrm{Sin}\phi i_{F2}\varphi'^2+\mathrm{Cos}\theta^2\mathrm{Cos}\phi\mathrm{Sin}\phi i_{F3}\varphi'^2-\mathrm{Cos}\phi\mathrm{Sin}\phi i_{w2}\varphi'^2+ \\ &\mathrm{Cos}\phi\mathrm{Sin}\phi i_{w3}\varphi'^2-rm_w\left(r\mathrm{Cos}\phi\gamma'\varphi'+r\mathrm{Cos}\phi\mathrm{Sin}\phi\varphi'^2\right)- \\ &\left(r+\mathrm{Cos}\theta l_F\right)m_F\begin{pmatrix} r\mathrm{Cos}\phi\gamma'\varphi'+r\mathrm{Cos}\phi\mathrm{Sin}\phi\varphi'^2+ \\ l_F\begin{pmatrix} \mathrm{Cos}\theta\mathrm{Cos}\phi\mathrm{Sin}\phi\varphi'^2+ \\ 2\theta'\left(\mathrm{Sin}\theta\phi'+\mathrm{Cos}\theta\mathrm{Cos}\phi\varphi'\right)\end{pmatrix}\end{pmatrix} \end{aligned}$$

$$\begin{aligned} C2=\ &-\mathrm{Cos}\theta^2\mathrm{Cos}\phi i_{F1}\theta'\phi'+\mathrm{Cos}\phi\mathrm{Sin}\theta^2 i_{F1}\theta'\phi'+\mathrm{Cos}\phi i_{F2}\theta'\phi'+ \\ &\mathrm{Cos}\theta^2\mathrm{Cos}\phi i_{F3}\theta'\phi'-\mathrm{Cos}\phi\mathrm{Sin}\theta^2 i_{F3}\theta'\phi'+\mathrm{Cos}\theta\mathrm{Sin}\theta\mathrm{Sin}\phi i_{F1}\phi'^2- \\ &\mathrm{Cos}\theta\mathrm{Sin}\theta\mathrm{Sin}\phi i_{F3}\phi'^2+2\mathrm{Cos}\theta\mathrm{Cos}\phi^2\mathrm{Sin}\theta i_{F1}\theta'\varphi'- \\ &2\mathrm{Cos}\theta\mathrm{Cos}\phi^2\mathrm{Sin}\theta i_{F3}\theta'\varphi'-2\mathrm{Cos}\phi\mathrm{Sin}\theta^2\mathrm{Sin}\phi i_{F1}\phi'\varphi'+ \\ &2\mathrm{Cos}\phi\mathrm{Sin}\phi i_{F2}\phi'\varphi'-2\mathrm{Cos}\theta^2\mathrm{Cos}\phi\mathrm{Sin}\phi i_{F3}\phi'\varphi'- \\ &2\mathrm{Cos}\phi\mathrm{Sin}\phi i_{w3}\phi'\varphi'+r^2\mathrm{Sin}2\phi m_F\phi'\varphi'+2r^2\mathrm{Cos}\phi\mathrm{Sin}\phi m_w\phi'\varphi'+ \\ &i_{w2}\left(\mathrm{Cos}\phi\gamma'\phi'+2\mathrm{Cos}\phi\mathrm{Sin}\phi\phi'\varphi'\right)+ \\ &rl_F m_F\begin{pmatrix} -\mathrm{Sin}\theta\mathrm{Sin}\phi\theta'^2+\mathrm{Sin}\theta\mathrm{Sin}\phi\phi'^2+\mathrm{Cos}\varphi^2\mathrm{Sin}\theta\gamma'\varphi'+ \\ \mathrm{Sin}\theta\mathrm{Sin}\varphi^2\gamma'\varphi'+(-\mathrm{Sin}\theta+0.5\mathrm{Sin}[\theta-2\phi]+ \\ 0.5\mathrm{Sin}[\theta+2\phi])\theta'\varphi'+(-\mathrm{Sin}[\theta-2\phi]+ \\ \mathrm{Sin}[\theta+2\phi])\phi'\varphi' \end{pmatrix}+ \\ &l_F^2m_F\begin{pmatrix} (0.25\mathrm{Cos}[2\theta-\phi]-0.25\mathrm{Cos}[2\theta+\phi])\phi'^2+ \\ (-0.25\mathrm{Sin}[2(\theta-\phi)]+0.5\mathrm{Sin}[2\phi]+ \\ \theta'\begin{pmatrix} (-0.5\mathrm{Cos}[2\theta-\phi]+\mathrm{Cos}\phi- \\ 0.5\mathrm{Cos}[2\theta+\phi])\phi'+(0.5\mathrm{Sin}2\theta+ \\ 0.25\mathrm{Sin}2(\theta-\phi)+ \\ 0.25\mathrm{Sin}2(\theta+\phi))\varphi' \end{pmatrix}+ \\ 0.25\mathrm{Sin}[2(\theta+\phi)])\phi'\varphi' \end{pmatrix} \end{aligned}$$

https://doi.org/10.1051/matecconf/201819202001

$$C3 = -\frac{K_e K_t \gamma'}{RG_{ratio}} - \text{Cos}\theta\text{Sin}\theta i_{F3}\phi'^2 + \text{Cos}\phi i_{F2}\phi'\varphi' - \text{Cos}\theta^2\text{Cos}\phi i_{F3}\phi'\varphi' +$$
$$\text{Cos}\phi\text{Sin}\theta^2 i_{F3}\phi'\varphi' + \text{Cos}\theta\text{Cos}\phi^2\text{Sin}\theta i_{F3}\varphi'^2 +$$
$$i_{F1}\begin{pmatrix}\text{Cos}\theta\text{Sin}\theta\phi'^2 + \text{Cos}\phi\left(\text{Cos}\theta^2 - \text{Sin}\theta^2\right)\phi'\varphi' - \\ \text{Cos}\theta\text{Cos}\phi^2\text{Sin}\theta\varphi'^2\end{pmatrix} +$$
$$l_F m_F\begin{pmatrix} r\text{Sin}\theta\phi'^2 + r\text{Sin}\theta\text{Sin}\phi\gamma'\varphi' + 2r\text{Cos}\theta\text{Cos}\phi\phi'\varphi' + \\ 0.5r\text{Sin}\theta\varphi'^2 - 0.5r\text{Cos}\phi^2\text{Sin}\theta\varphi'^2 + 0.5r\text{Sin}\theta\text{Sin}\phi^2\varphi'^2\end{pmatrix} +$$
$$l_F^2 m_F\begin{pmatrix} 0.5\text{Sin}2\theta\phi'^2 + (0.5\text{Cos}[2\theta - \phi] + \text{Cos}\phi + \\ 0.5\text{Cos}[2\theta + \phi])\phi'\varphi' + \\ (-0.25\text{Sin}2\theta - 0.125\text{Sin}[2(\theta - \phi)] - \\ 0.125\text{Sin}[2(\theta + \phi)])\varphi'^2 \end{pmatrix}$$

$$C4 = -\frac{K_e K_t \gamma'}{RG_{ratio}} + \text{Cos}\phi i_{w2}\phi'\varphi' + 2r^2\text{Cos}\phi m_F\phi'\varphi' + 2r^2\text{Cos}\phi m_w\phi'\varphi' +$$
$$rl_F m_F\begin{pmatrix} -\text{Sin}\theta\theta'^2 - 2\text{Sin}\theta\text{Sin}\phi\theta'\varphi' + 2\text{Cos}\theta\text{Cos}\phi\phi'\varphi' - \\ \text{Cos}\varphi^2\text{Sin}\theta\varphi'^2 - \text{Sin}\theta\text{Sin}\varphi^2\varphi'^2 \end{pmatrix}$$

$$G1 = -m_F g\text{Sin}\phi\left(r + \text{Cos}\theta l_F\right) - gm_w r\text{Sin}\phi$$
$$G2 = 0\ ,\ G4 = 0$$
$$G3 = -g\text{Cos}\phi\text{Sin}\theta l_F m_F$$

The unicycle robot equation of motion is in non-linear form. Taylor series expansion method is applied to linearize robot model and rearranged to state space form as shows in (Eq.3)

[redacted] u (3)

where:

$$-A = \begin{bmatrix} \begin{matrix} 0 & 0 & 0 & 0 & 1 & 0 & 0 & 0 \\ 0 & 0 & 0 & 0 & 0 & 1 & 0 & 0 \\ 0 & 0 & 0 & 0 & 0 & 0 & 1 & 0 \\ 0 & 0 & 0 & 0 & 0 & 0 & 0 & 1 \end{matrix} \\ M^{-1} * \begin{bmatrix} \Delta_1 & 0 & \Delta_7 + \Delta_3 & 0 & 0 & \Delta_4 & i_{F2} + 2\Delta_3 + \Delta_7 & i_{w2} + \Delta_8 \\ \Delta_2 & 0 & -\Delta_3 - 2\Delta_7 & 0 & \Delta_4 & 0 & -i_{F2} + \Delta_7 + l_F^2 m_F & 0 \\ 0 & 0 & gl_F m_F & 0 & \Delta_5 & \Delta_5 & 0 & 0 \\ 0 & 0 & 2rl_F m_F & 0 & i_{w2} - 2\Delta_6 & -i_{w2} - 2\Delta_6 & 0 & 0 \end{bmatrix} \end{bmatrix}$$

$$-B = \begin{bmatrix} 0 & 0 & 0 \\ 0 & 0 & 0 \\ 0 & 0 & 0 \\ 0 & 0 & 0 \\ \left[M_*^{-1} * \begin{bmatrix} 1 \\ 0 \\ 0 \\ 0 \end{bmatrix}\right] & \left[M_*^{-1} * \begin{bmatrix} 0 \\ 1 \\ 0 \\ 0 \end{bmatrix}\right] & \left[M_*^{-1} * \begin{bmatrix} 0 \\ 0 \\ -1 \\ 1 \end{bmatrix}\right] \end{bmatrix},\ -u = \begin{bmatrix} \tau_\phi \\ \tau_p \\ \tau_\gamma \end{bmatrix}$$

where:

$-\Delta_1 = g\left(\left(r + l_F\right)m_F + rm_w\right)$ $\quad -\Delta_2 = -2\left(\Delta_4 + rl_F m_F + l_F^2 m_F\right)$

$-\Delta_3 = rl_F m_F + l_F^2 m_F$ $\quad -\Delta_4 = i_{F1} - i_{F2} - i_{F3} - i_{w2}$

$-\Delta_5 = -i_{F1} - i_{F2} + i_{F3} - 2rl_F m_F - 2l_F^2 m_F$

$-\Delta_6 = r^2 m_F + rl_F m_F + r^2 m_w$ $\quad -\Delta_7 = i_{F1} - i_{F3}$

M_* is the symmetrical positive definite matrix of the robot inertia and evaluated at the upright position.

4. Control Algorithm

This section is presented the implemented control algorithm. By combination LQR with Fuzzy controller, the details presents below.

4.1 LQR Controller

The conventional LQR controller is the optimal controller that optimally determines the gains by compromising the state and control-input cost. The LQR cost function is expressed by

$$J = \int_0^\infty \left(x^T Qx + u^T Ru\right)dt \tag{4}$$

The Q and R matrices are the state and control weighting matrices. The control signal follows (Eq. 5)

$$u = -Kx \tag{5}$$

The optimal gain matrix is solved by the algebraic Riccati equation (Eqs. 6-7).

$$A^T P + PA + Q - PBR^{-1}B^T P = 0 \tag{6}$$

$$K = R^{-1}B^T P \tag{7}$$

The metrics Q and R are the weighting metrics which define depend on significant state. The Q matrix is combined of many states of the unicycle robot. The balancing states (roll and pitch) of the unicycle robot are set to highest priority than other states.

5.2 LQR-Fuzzy Controller

The LQR-Fuzzy applies both advantages from LQR and Fuzzy controller. LQR controller gain is optimal gain which designed by the cost function (4). It can easily satisfy the response of the system by set the priority of the Q matrix. And Fuzzy controller can handle the non-linear term of unicycle robot system. Therefore, the combination between LQR and Fuzzy controller can improve the performance of the robot. The block diagrams shows in Fig.4.

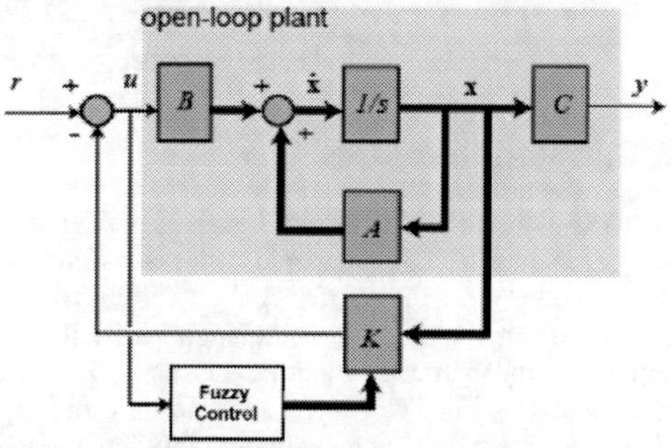

Fig. 4 LQR-Fuzzy Block diagrams

LQR-Fuzzy algorithm is designed for decreasing overshoot and increasing performance of the system. The input of Fuzzy control block in Fig.4 is position state error of the unicycle robot: roll yaw pitch and wheel position. There are four inputs and one output. The output of the fuzzy controller is torque ratio. By This torque ratio, the LQR controller gains are adjusted depend on environment of each state input. This technique can produce good dynamic and static performance.

The input membership function is Gaussian type. There are two numbers of membership function of each input. For example, the membership function of the roll angle is combined with two Gaussian sets. The first set is normal period angle, and another one is aggressive angle as show in Fig.5. All member ship functions are set similar to roll angle.

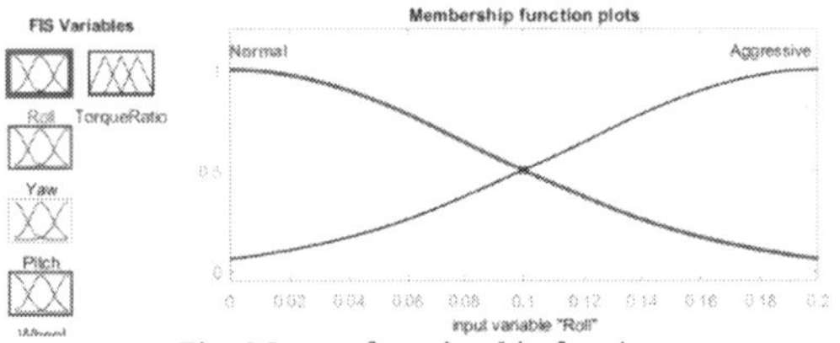

Fig. 5 Input of membership function

The membership function of the output is combined with three Gaussian sets: low normal and aggressive. The defuzzification output is solved by centroid concept.

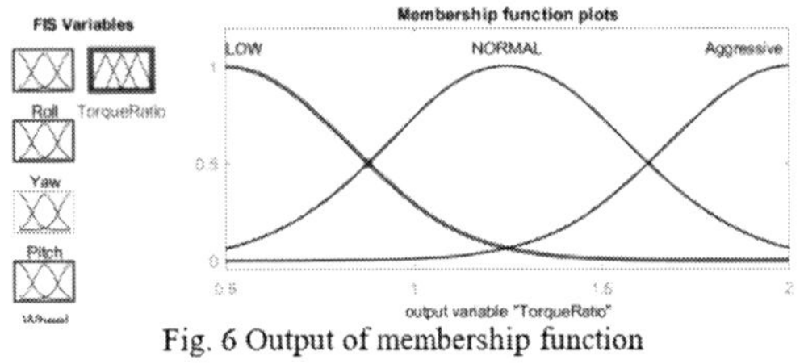

Fig. 6 Output of membership function

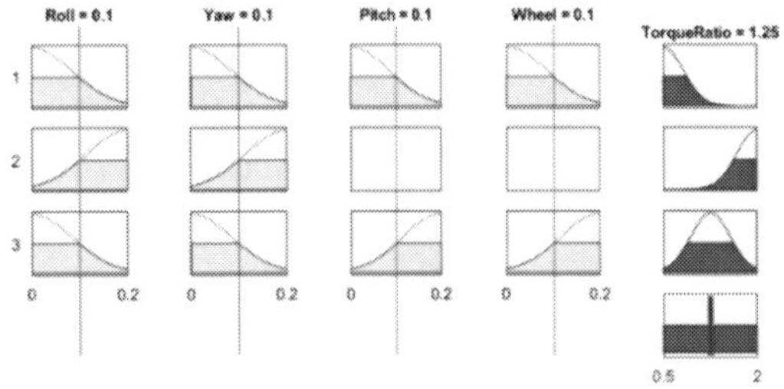

Fig. 7 Rule base of Fuzzy controller

5. Simulation results

In this section, the balancing performances of the unicycle robot are simulated and compared with convention LQR. The robot simulation is simulated by MATLAB/Simulink. In the simulation parts, there are initial conditions in each states of the robot. The results show that the robot can reject disturbances and try to convert each states to set point (zero). The comparison of LQR-Fuzzy and LQR are simulate, presents in Fig.8.

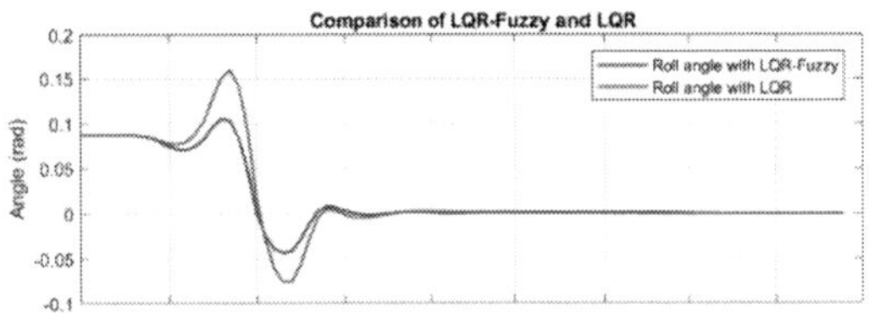

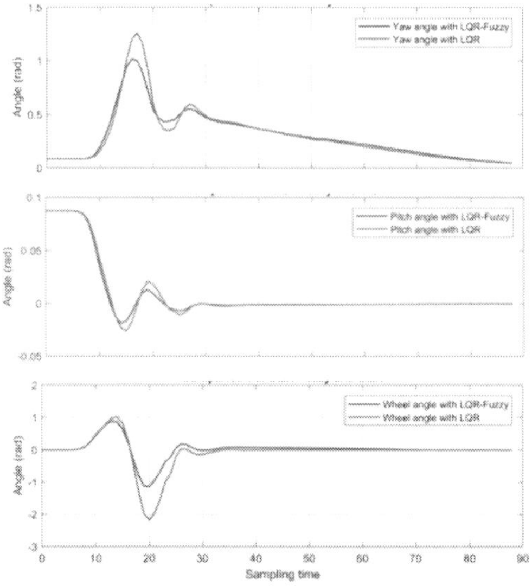

Fig. 8 Simulation results of Roll Yaw Pitch and Wheel Position

From the simulation results, the LQR-Fuzzy is designed which meets requirements by decreasing the overshoot and increasing the performance of the unicycle robot. The LQR-Fuzzy can decease the overshoot about fifty percent compares with conventional LQR.

6. Conclusion

In this paper, the unicycle robot model is simulated by LQR-Fuzzy controller. LQR controller is optimal control for multivariable feedback and known the good performance for linear model. Fuzzy controller ability is to deal with non-linearity of the system. The combination of both controller can increase the performance of robot.

References

1. Schoonwinkel, A. (1988). Design and test of a computer-stabilized unicycle.
2. Sheng, Z., & Yamafuji, K. (1995, May). Realization of a human riding a unicycle by a robot. In Robotics and Automation, 1995. Proceedings., 1995 IEEE International Conference on (Vol. 2, pp. 1319-1326). IEEE.
3. Majima, S., Kasai, T., & Kadohara, T. (2006, November). A design of a control method for changing yaw direction of an underacuatted unicycle robot. In TENCON 2006. 2006 IEEE Region 10 Conference (pp. 1-4). IEEE.
4. Chantarachit, S., & Parnichkun, M. (2016). Development and control of a unicycle robot with double flywheels. Mechatronics, 40, 28-40.
5. Yazdanpanah, R., Mahjoob, M. J., & Abbasi, E. (2013). Fuzzy LQR controller for heading control of an unmanned surface vessel. In International Conference in Electrical and Electronics Engineering (pp. 73-78).

Seismic performance of a high-rise residential building model in Purwokerto, Indonesia

Yanuar Haryanto[1,*], *Buntara* Sthenly Gan[2], *Nanang* Gunawan Wariyatno[1], and *Eva* Wahyu Indriyati[1]

[1]Jenderal Soedirman University, Department of Civil Engineering, Jl. Mayjend. Sungkono KM 5, Blater, Purbalingga, 53371, Indonesia
[2]Nihon University, Department of Architecture, Koriyama 963-8642, Japan

Abstract. We evaluated the performance of a high-rise residential building model in Purwokerto, Indonesia due to the seismic load. The evaluation was performed based on seismic loads given in the 2002 and 2012 Indonesian National Standard (SNI) using linear static analysis, dynamic response analysis and pushover analysis. Based on the linear static analysis, the drift ratio decreased by an average of 34.42 and 32.61% for the X and Y directions respectively. Meanwhile, based on the dynamic response analysis, the drift ratio also decreased by an average of 30.74 and 27.33% for the X and Y directions respectively. In addition, the pushover analysis indicates that the performance of this high-rise residential building model is still at Immediate Occupancy (IO) level. The post-earthquake damage state in which the building remains safe to occupy, essentially retaining the pre-earthquake design strength and stiffness of the structure. The risk of life-threatening injury as a result of structural damage is very low. Although some minor structural repairs may be appropriate, these would generally not be required prior to re-occupancy.

1 Introduction

These days, the demand for housing has increased because of the economic development and increase in population. One approach to fully use a limited size of land is by vertically developing a high-rise residential building. There are numerous perspectives to consider in constructing a high-rise residential building such as safety, effectiveness and feasibility components. Safety is the major factor to be considered in developing a tall housing. The high-rise residential building is designed in such a way to maintain the vertical gravitational forces and withstand natural disaster such as an earthquake.

Because of its seismicity, Indonesia has made major changes on the earthquake of Indonesian National Standards: SNI 03-1726-2002 [1] to SNI 1726:2012 [2]. These changes will change the building's structural behaviors. Purwokerto has a site-particular estimation of spectral acceleration response for short-period (S_s) is 0.5g and one-second (S_1) is 0.2g. Table 1 demonstrates that a locale with a site-particular estimation of spectral acceleration response for S_s larger than or equivalent to 0.5g but smaller than 1.0g and S_1 larger than or equivalent to 0.2g but smaller than 0.4g can be considered as a region with modestly high seismicity potential [3]. In this manner, Purwokerto is a locale with modestly high seismicity potential which needs to follow earthquake-resistant structural framework in every housing structure built up in Purwokerto. The purpose of the current study is to assess the performance of the high-rise residential building model in Purwokerto, Indonesia because of the seismic load. This will create awareness in the society or any parties.

Table 1. Typical seismicity region [3].

Seismicity Region	Spectral Acceleration Response, S_s (short-period, or 0.2 seconds)	Spectral Acceleration Response, S_1 (long-period, or 1.0 second)
Low	less than 0.250g	less than 0.100g
Moderate	greater than or equal to 0.250g but less than 0.500g	greater than or equal to 0.100g but less than 0.200g
Moderately High	greater than or equal to 0.500g but less than 1.000g	greater than or equal to 0.200g but less than 0.400g
High	greater than or equal to 1.000g but less than 1.500g	greater than or equal to 0.400g but less than 0.600g
Very High	greater than or equal to 1.500g	greater than or equal to 0.600g

The seismic hazard relies on the danger level as well as the susceptibility to know plausible seismic impact [4]. The hazard level cannot diminish since it is an environmental occurrence. Thusly, the seismic hazard can be diminished by bringing down the susceptibility. Earthquake engineering comprises two critical terms: performance-based design and performance-based evaluation. The performance-based design focuses on structural design [5-8] while performance-based evaluation focuses on a constructed structural analysis [9-11].

* Corresponding author: yanuar.haryanto@unsoed.ac.id

2 Methodology

2.1. Building model

The building model was based on the horizontally irregular structures. The dimension of the X and Y axes varied. The soil condition was determined in the moderate state. An improvement was done in the design as well. For instance, the stiffness levels of the positive and negative X and Y axes were made equivalently. The dead loads were the structures' self-weight, comprised columns, floor plate profile, beams and walls. It was utilized as a distributed load on the beam to involve the load-bearing walls. The plate loads were formed based on the plate dimensional features. The live loads for a housing building type of 250 kg/m^2 were formed as equal loads distributed on the plates. The assessed three-dimensional (3D) building model was displayed in Figure 1 and the resulting structure plan was displayed in Figure 2.

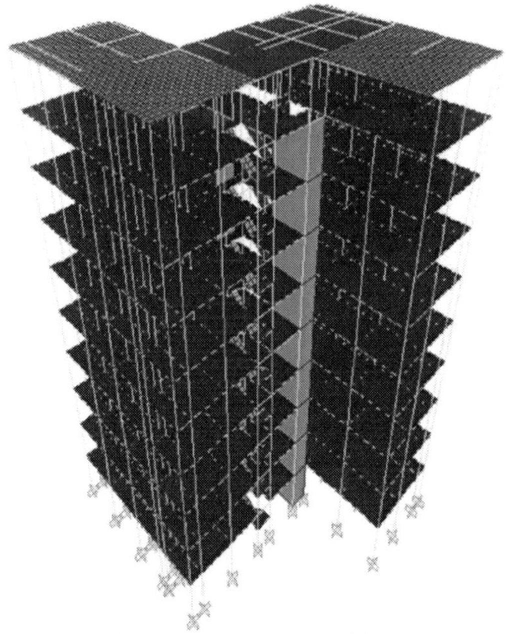

Fig. 1. Three-dimensional (3D) building model.

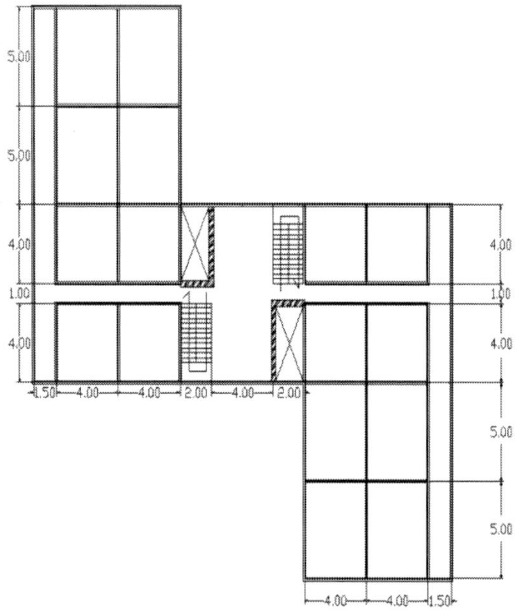

Fig. 2. The structure plan.

2.2 Structure details

The beams utilized profile I of 600 x 300 x 12 x 17 with 64 shear connectors. Every pair of shear connectors was attached to a floor with a separation of 18.75 cm between them on floors 1-9. In the meantime, the columns utilized profile I of 600 x 300 x 12 x 17 for floors 1-5 and profile I of 250 x 250 x 11 x 11 for floors 6-10. The roof utilized profile I of 500 x 300 x 11 x 15 with 52 shear connectors. Every pair of shear connectors was attached 23.10 cm separated.

2.3 Analysis of seismic evaluation

The static linear analysis was done by using the equivalent static load analysis based on the Non-Masonry Model. The base shear *(V)* happening at the structural basic level was computed based on equations in SNI 1726:2012 and SNI 03-1726-2002. The base shear *(V)* was then allocated to all the building structures to form the equivalent static seismic load (F_i) through all the joints on every floor.

The spectrum dynamic response seismic loads were utilized in dynamic response analysis by altering the location categorization and seismicity level in Purwokerto, Indonesia. The spectrum response was modeled using SAP2000 software while dynamic response analysis utilizes the damping of 0.05 [12]. The function of the response spectrum was defined clearly in *Spectrum Response Function Definition*. Next, data were inputted into the spectral response accelerations (S_a) and fundamental period of the structure *(T)*. In order to simulate the earthquake arbitrary load towards the building structure model, the effect of the seismic loading on the main directions were set to be 100% effective. This is because the effect of the seismic loading is in the direction perpendicular to the main direction (30% effective only).

Pushover analysis [13, 14] is a static and nonlinear procedure whereby the value of the structural loading was increased based on the predefined pattern. The relationship between the weak links and failure modes of the structure were found after increasing the magnitude of the loading. The loading is categorized to be monotonic. The effects of the cyclic behavior and load reversals were estimated by using the altered monotonic force-deformation criterion with damping approximations [15].

* Corresponding author: yanuar.haryanto@unsoed.ac.id

3 Result and discussions

3.1. Linear static analysis

The base shear *(V)* on the moderate soil type occurring in the structure's basic level calculated base on SNI 03-1726-2002 is 4929.53 kN in the X direction and 6433.04 kN in the Y direction. Meanwhile, the base shear referring to SNI 1726: 2012 is 3426.77 kN in the X direction and 4471.94 kN in the Y direction. The base shear is reduced by 30.48%. The drift ratio experiences average decreases of 34.42 and 32.61% respectively in the X and Y directions, as displayed in Figure 5.

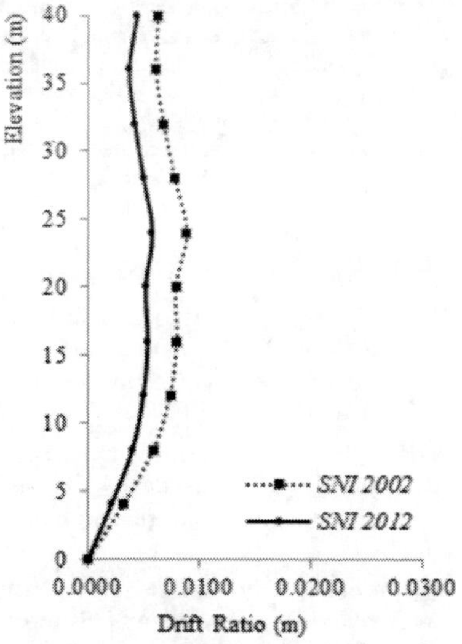

(a) X direction

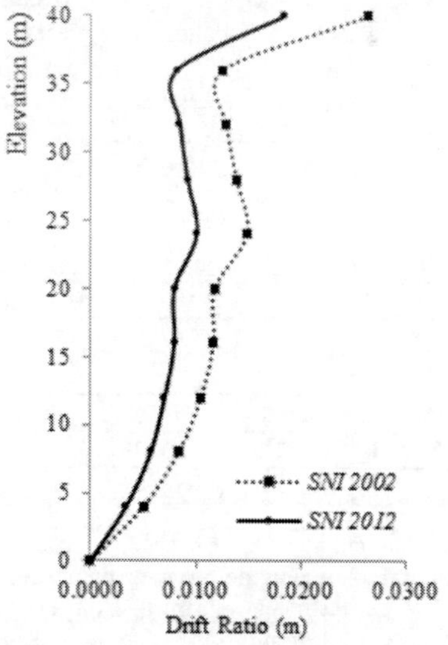

(b) Y direction

Fig. 5. Drift ratio resulting from the linear static analysis

3.2. Dynamic response analysis

In general, the response spectrum based on SNI 03-1726-2002 was larger than SNI 1726:2012. In other words, the high-rise residential building model in Purwokerto, Indonesia still met the requirements of SNI 1726:2012 even SNI 03-1726-2002 was used. The drift ratio reduced by 30.74 and 27.33% in the X and Y direction respectively as displayed in Figure 6.

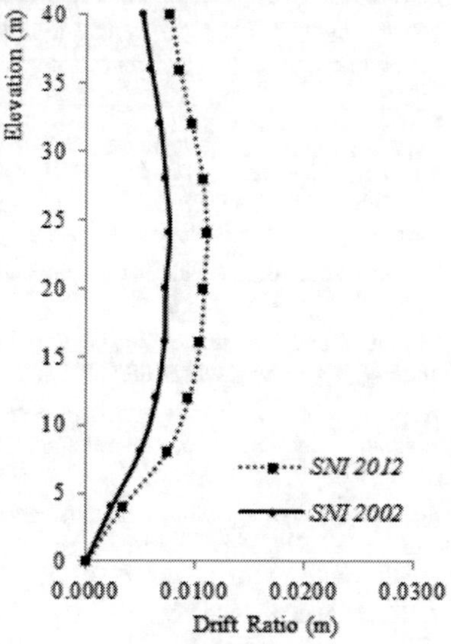

(a) X direction

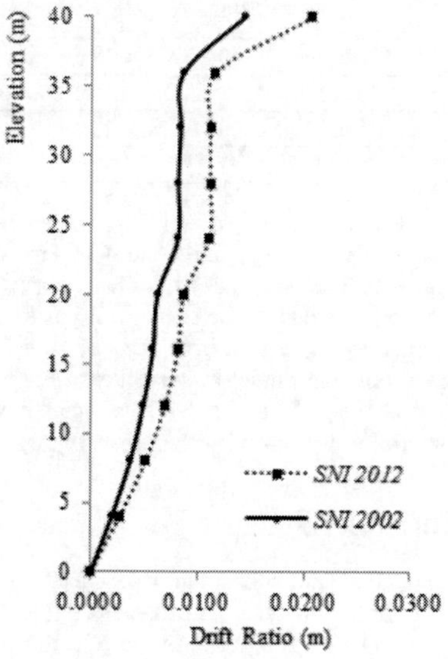

(b) Y direction

Fig. 6. Drift ratio resulting from the dynamic response analysis

* Corresponding author: yanuar.haryanto@unsoed.ac.id

3.2. Pushover analysis

The results of the pushover analysis displayed the magnitude of displacement *(D)* and basic shear *(V)* when the high-rise residential building model nearly collapsed. As illustrated in Table 2, the base shear reduced by 30.57 and 18.27% for X and Y direction due to the high-rise residential building model in Purwokerto, Indonesia when the performance point was reached. The ductility of the structure was acquired by associating both the yield displacement (δ_y) and ultimate displacement (δ_u) as displayed in Table 3.

Table 2. Comparison of pushover analytical results.

Parameter	SNI			
	2002		2012	
	Dir. X	Dir. Y	Dir. X	Dir. Y
V (kg)	7571.40	7420.61	5256.80	6064.94
D (m)	0.077	0.061	0.053	0.050

Table 3. Ductility of the structure.

Direction	δ_y (m)	δ_u (m)	Ductility
X	0.21229	0.21714	1.02
Y	0.19704	0.64398	3.27

The structural performance of the high-rise residential building model was distinguished by drift ratio. The pushover analysis results displayed that the drift ratio was 0.193 and 0.153% for X and Y direction respectively for the seismic load of SNI 03-1726-2002. On the other hand, the drift ratio was 0.133 and 0.125% for X and Y direction respectively for the seismic load of SNI 1726:2012. The structural performance was displayed in Table 4.

Table 4. Structural performance.

SNI	Dir.	D_t (m)	Elevation (m)	Drift Ratio (%)	Performance level
2002	X	0.0770	40	0.193	*IO*
	Y	0.0610	40	0.153	*IO*
2012	X	0.0530	40	0.133	*IO*
	Y	0.0500	40	0.125	*IO*

The analyzed drift of this high-rise residential building model in the Y direction was larger than X direction as X direction was stiffer than Y direction. This phenomenon was due to the shear wall position and the stair structure. The performance of the high-rise residential building model in Purwokerto, Indonesia still remained at the level of Immediate Occupancy (IO) as the value of the drift ratio was less than 1%.

4 Conclusions

About 30.48% reduction in the base shear *(V)* based on the static linear analysis. Besides that, the drift ratio reduced to 34.42 and 32.61% in the X and Y direction, respectively. However, the drift ratio reduced to 30.74 and 27.33% in the X and Y direction respectively if based on the dynamic response analysis. The results of the pushover analysis displayed that the performance of this high-rise residential building model still at Immediate Occupancy (IO) level. In other words, the building still remained safe to occupy as the pre-earthquake design strength and stiffness of the structure did not alter greatly. The risk of getting injury due to minor structural damage was low and safe to stay.

Acknowledgements The authors express their gratitude and appreciation to the Research and Public Services Institution (LPPM) of Jenderal Soedirman University (Unsoed), Indonesia, for funding this study. The authors also thank Society Coffee House, Purwokerto, with its great music playlist, for providing a cozy place that allowed the writing process of this manuscript such an enjoyable duty.

References

1. The National Standardization Agency of Indonesia, *SNI 03-1726-2002 Tata cara perencanaan ketahanan gempa untuk bangunan gedung* (2002)
2. The National Standardization Agency of Indonesia Badan Standarisasi Nasional Indonesia, *SNI 1726: 2012 Tata cara perencanaan ketahanan gempa untuk struktur bangunan gedung dan non gedung* (2012)
3. Federal Emergency Management Agency, *FEMA P-154 Rapid visual screening of building for potential seismic hazards: a handbook* (2015)
4. Y. Haryanto, G.H. Sudibyo, F.C.A. Effendi, Procedia Eng. **171**, 1025 (2017)
5. A.V. Bergami, X. Liu, C. Nuti, J. Civil Eng. Architect. Res. **2**, 842 (2015)
6. J. Seo, J.W. Hu, B. Davaajamts, Sustainability **7**, 14287 (2015)
7. S. Bhagat, S. Kono, A.C. Wijeyewickrema, K. Kusunoki, S. Tajiri, H. Kashiwa, *New technologies for urban safety of mega cities in asia* (2015)
8. S.Tesfamariam, K. Goda, Front. Built Environ. **1**, 1 (2015)
9. Y. Haryanto, B.S. Gan, N.G. Wariyatno, E.W. Indriyari, ARPN JEAS, **12,** 4858 (2017)
10. M. Midorikawa, I. Okawa, M. Iiba, M. Teshigawara, Earthquake Eng. Eng. Seis. **4**, 15 (2002)
11. I. Hajirasouliha, K. Pilakoutas, J. Earthquake Eng. **16**, 1 (2012)
12. Computer and Structures, *Structural and earthquake engineering software SAP2000* (2016)
13. Applied Technology Council. 1996. Seismic evaluation and retrofit of concrete buildings, Vol. 1 (ATC-40)
14. Federal Emergency Management Agency, *FEMA 273 NEHRP guidelines for the seismic rehabilitation of buildings* (1997)
15. A. Habibullah, S. Pyle, *Structure Magazine* (1998)

* Corresponding author: yanuar.haryanto@unsoed.ac.id

Shear model with shear-flexure interaction for non-linear analysis of reinforced concrete frame element

Worathep Sae-Long[1,*], and *Suchart* Limkatanyu[1]

[1]Department of Civil Engineering, Faculty of Engineering, Prince of Songkla University, Hat Yai, Songkhla, Thailand

Abstract. This paper presents the shear constitutive model for the reinforced concrete (R/C) frame structures analysis under monotonic and cyclic loading. The proposed model is adopted and modified from Mergos and Koppos model [1] that accounts the shear stiffness degradation effect by the shear-flexure interaction in the plastic hinge region. Firstly, the proposed shear model starts from the primary curve without the damages due to the shear-flexure interaction effect. Then, the shear-flexure interaction effect is taken into consideration at the locations of plastic hinges and this effect leads to the degradation of the shear strength and shear stiffness on the undamaged primary curve that is replaced with the damaged primary curve. To determine the sectional shear stiffness with the shear-flexure interaction, an alternative way of the iterative procedure is proposed here. Finally, a numerical example is used to verify the characteristics and behavior of the R/C frame system and confirm accuracy and computational efficiency of the proposed model among the experimental data.

1 Introduction

The reinforced concrete structures (R/C-Structure) have been used extensively in engineering works, for example, the building and bridge structures etc. From the past earthquakes, the R/C members before the proposed seismic codes were often collapsed and heavily damaged due to the member without ductile design requirement. The most dangerous damage mechanism was related to shear that occurred suddenly because of the insufficient shear strength. However, the characteristics of R/C member failure didn't account only the shear failure but also included the flexural failure and the flexure-shear failure [2].

The flexure-shear failure might be significant in some cases such as the lightly R/C columns. To represent the accurate responses of those members, there were several researchers [2-3] who developed the analytical procedures to include the shear-flexure interaction into the assessment of element by using the different strategies. Although these procedures were attractive, these methods were not successful with the degradation of shear strength in the plastic hinge region. Later, Mergos and Kappos [1, 4] presented the shear model and the shear-flexure interaction procedure that was able to capture the degradation of the shear strength in the plastic hinges region. However, although these methods were attractive, it didn't seem reasonable in the real analysis because their method approximated the shear stiffness with the shear-flexure interaction which based on a constant stiffness in every step.

In the light of this reason, this paper proposes an alternative of a shear model with the modified technique for the degraded shear stiffness estimation due to the shear-flexure interaction effect. Moreover, a numerical example is used to confirm the accuracy and efficiency of the proposed model. The example presents the member with failing in shear after the flexural yielding that is defined to the shear-flexure critical member. All the responses of the model in this paper are implemented in the general-purpose finite element platform FEAP [5].

2 Shear model

The shear model represents the shear force-shear strain relation of the R/C member. The original shear model was proposed by Mergos and Kappos [1] and their model covered all the behaviors of R/C member prior and after concrete cracking, the flexural yielding and the transverse reinforcement yielding or failure as indicated in the primary curve for this paper.

2.1 Primary curve of shear model

The primary curve presents the initial shear force-shear strain relation that is first derived without considering the damage due to the shear-flexure interaction effect. As shown in Fig. 1, it comprises four branches with three different slopes. The first branch oa with uncracked slope (GA_0) shows the elastic behavior of uncracked section in shear. The origin point o connects to the cracking point a (V_{cr}, γ_{cr}) at which the nominal principal tensile stress becomes larger than the nominal tensile strength of concrete. The cracking shear force (V_{cr}) [4],

* Corresponding author: s.worathep@yahoo.com

the uncracked slope (GA_0) and the cracking shear strain (γ_{cr}) can be written as:

$$V_{cr} = \frac{f_t'}{L_a / h}\sqrt{1+\frac{N}{f_t' A_g}}\,0.80A_g \quad (1)$$

$$GA_0 = 0.80GA_g \quad (2)$$

$$\gamma_{cr} = V_{cr} / GA_0 \quad (3)$$

Where f_t' is the mean concrete tensile strength, A_g is the gross area of the concrete section, L_a is the shear span, h is the section height, N is the compressive axial load and G is the shear modulus respectively.

The second branch ab and third branch bc have the same slope of GA_1. The second branch ab links the cracking point a (V_{cr} , γ_{cr}) to the flexural-yielding point b (V_y , γ_y) at which the longitudinal reinforcement yields for the first time. The third branch bc joins the flexural-yielding point b (V_y , γ_y) to the shear-yielding point c (V_{u0} , γ_{st}) at which shear strain corresponds to the onset of the transverse reinforcement yielding (γ_{st}) while shear force reaches its ultimate value (V_{u0}).

To estimate the shear strength (V_u) on the primary curve, the shear strength model, proposed by Priestly et al. [6], is used in this study. The model is able to account for the shear strength degradation with increasing curvature ductility demand that can be defined as:

$$V_u = k_\varphi \sqrt{f_c'}\left(0.80A_g\right) + \frac{A_v f_{yv} D' \cot 30^o}{s} + \frac{d-c}{2L_a}N \quad (4)$$

Where k_φ is a parameter associated the curvature ductility demand μ_φ [1], f_c' is concrete strength , A_v is the area of the transverse reinforcement, f_{yv} is the yield strength of transverse reinforcement, D' is the distance measured parallel to the applied shear between the centres of transverse reinforcement, s is the spacing of transverse reinforcement, d is the section depth and c is the neutral axis depth respectively.

The slope of the second and third branch (GA_1) can be expressed as:

$$GA_1 = \frac{V_{u0} - V_{cr}}{\gamma_{st} - \gamma_{cr}} \quad (5)$$

Where V_{u0} is the non-degraded shear strength that can determine from Eq. (4) by setting the curvature ductility demand which becomes lesser than 3.

The shear strain at the onset of transverse reinforcement yielding (γ_{st}) can simply be determined by using the truss analogy. However, it is not corresponding well with experimental results due to the axial load and aspect ratio effects. Mergos and Kappos [1] suggested two modification factors (κ and λ) to improve the accuracy of these equations. The value of γ_{st} was given by

$$\gamma_{truss} = \frac{V_{cr}}{GA_0} + \frac{A_v f_{yv}\left(\sin^4 45^o + \frac{E_s}{E_c}p_v\right)}{sE_s b p_v \sin^4 45^o \cot 45^o} \quad (6)$$

$$\kappa = 1-1.07v \text{ and } \lambda = 5.37 - 1.59min(2.5, L_a / h) \quad (7)$$

$$\gamma_{st} = \kappa\lambda\gamma_{truss} \quad (8)$$

Where γ_{truss} is the shear associated with the yielding of transverse reinforcement based on truss anology, E_s is the elastic modulus of steel, E_c is the elastic modulus of concrete, p_v is the volumetric ratio of transverse reinforcement, b is the section width and $v = N/(f_c' A_g)$ is the normalized compressive axial load.

Finally, the last branch cd represents the plastic behavior of shear and connects the shear-yielding point c (V_{u0} , γ_{st}) to the ultimate point d (V_{u0} , γ_u) which associates with the onset of shear failure. The value of γ_u is based on regression analysis of experimental results and suggested by Mergos and Kappos [1]:

$$\lambda_1 = 1.0 - 2.5\min\left(0.40, v\right) \text{ and } \lambda_2 = \min(2.5, L_a / h)^{2.0}$$
$$\lambda_3 = 0.31 + 17.8\min\left(\left(A_v f_{yv}\right)/\left(bsf_c'\right), 0.08\right) \quad (9)$$

$$\gamma_u = \lambda_1\lambda_2\lambda_3\gamma_{st} \geq \gamma_{st} \quad (10)$$

Where λ_1, λ_2 and λ_3 are the parameter acounting for axial load, aspect ratio and amount of transverse reinforcement respectively. It is important to note that the Mergos and Kappos formula for γ_{st} and γ_u estimation suggested the criteria: $1.11 \leq L_a / h \leq 3.91$, $0 \leq v \leq 0.61$ and $0.47\% \leq \left(A_v f_{yv}\right)/\left(bsf_c'\right) \leq 8.13\%$ [1].

2.2 The modified Mergos-Kappos shear-flexure interaction procedure (MMKSFIP)

According to the primary curve, there are two phenomena results due to the shear-flexure interaction effect on the undamaged primary curve. The first one is the degradation of the shear strength because of the increasing curvature ductility demand as defined by the shear strength model in Eq. (4). From this result, the last branch cd is degraded from the undamaged primary curve to the new portion that associates with the shear strength at the start of load increment step. Moreover, the reduction of the shear strength also leads to the degradation of the shear stiffness as defined in the second phenomena result. The authors refer to the degraded shear stiffness as "the effective shear stiffness $(GA)_{eff}$ ". As shown in Fig. 1, there are three cases to be confronted when the shear-flexure interaction appears after the flexural reinforcement yielding.

The first case refers to the case of the non-degradation of the shear strength, the sectional curvature ductility is lesser than 3 that based on the shear strength model in Eq. (4). In this case (Fig.1), the sectional shear force at the start and end of load increment step are located on the undamaged primary curve with the cracked sectional shear stiffness (GA_1).

The second case presents the case of the degradation of the shear strength for the first time, the sectional curvature ductility is larger than 3 for the first time. From this case, the sectional shear force at the start of load increment step lies on the undamaged primary curve but the sectional shear force at the end of load increment step is situated on the damage primary curve due to the shear-flexure interaction effect with the effective shear stiffness ($(GA)_{eff}$) as shown in Fig.1.

The last case shows the case after the degradation of the shear strength for the first time. The sectional shear force at the start and end of load increment step are located on the damaged primary curve with the effective shear stiffness ($(GA)_{eff}$).

According to the effective shear stiffness estimation, the original procedure proposed by Mergos and Kappos [1] is adopted and modified herein. The authors refer to this procedure as "modified Mergos-Kappos shear-flexure interaction procedure (MMKSFIP)". From all three possible cases, The incremental sectional shear force (ΔV) and the effective shear stiffness ($(GA)_{eff}$) can determine form a given incremental sectional shear strain ($\Delta\gamma$) by using the MMKSFIP through the proposed reference slope that links the sectional shear force at the start of load increment step to the sectional shear force at the end of load increment step on the undamaged primary curve as shown in Fig. 1. In this procedure, the so-called reference sectional shear stiffness ($\left(GA_{ref}\right)_i^k$) can define as:

$$\left(GA_{ref}\right)_i^k = \frac{V_{0,i}^{k+1} - V^k}{\Delta\gamma_i^k} \tag{11}$$

Where $V_{0,i}^{k+1}$ is the non-degradation sectional shear force that correponds to the sectional shear strain $\gamma_i^{k+1} = \gamma_i^k + \Delta\gamma_i^k$ and can be written as:

$$V_{0,i}^{k+1} = V_{cr} + GA_1\left(\gamma_i^{k+1} - \gamma_{cr}\right) \tag{12}$$

It is worth noting that, the reference sectional shear stiffness ($\left(GA_{ref}\right)_i^k$) in the first and second cases is simply equal to the cracked sectional shear stiffness (GA_1) as shown in the Fig. 1. Considering the relation between incremental sectional shear force (ΔV) and shear strain ($\Delta\gamma$) in Fig. 1 leads to the following expression:

$$\Delta\gamma_i^k = \frac{\Delta V_i^k}{\left(GA_{eff}\right)_i^k} = \frac{\Delta V_i^k + \left(\Delta V_c^{deg}\right)_i^k}{\left(GA_{ref}\right)_i^k} \tag{13}$$

Where $\left(\Delta V_c^{deg}\right)_i^k$ is the degradation of the concrete shear resisting mechanism can calculate by:

$$\left(\Delta V_c^{deg}\right)_i^k = \left(GA_{ref}\right)_i^k \Delta\gamma_i^k - \frac{\left(V_{u,i}^k - V_i^k\right)}{\left(\gamma_{st} - \gamma_i^k\right)}\Delta\gamma_i^k \tag{14}$$

The effective shear stiffness ($\left(GA_{eff}\right)_i^k$) can calculate by solving Eq. (13). The following equation is obtained.

$$\left(GA_{eff}\right)_i^k = \frac{\Delta V_i^k}{\Delta V_i^k + \left(\Delta V_c^{deg}\right)_i^k}\left(GA_{ref}\right)_i^k \tag{15}$$

It is observed from Eq. (15) that the effective shear stiffness ($\left(GA_{eff}\right)_i^k$) is a function of the incremental sectional shear force (ΔV_i^k) but both quantities are unknown. To overcome this problem, an iterative analytical scheme is applied within the iterative step i of the load increment k. It needs to note that the following parameters $\Delta\gamma_i^k$, $\left(GA_{ref}\right)_i^k$ and $\left(\Delta V_c^{deg}\right)_i^k$ aren't changed during an iterative scheme except $\left(GA_{eff}\right)_i^k$ and ΔV_i^k. For this reason, the additional subscript " j " is added into $\left(GA_{eff}\right)_i^k$ and ΔV_i^k to define the iterative step of the shear-flexural interaction process. The iterative process within the shear-flexural interaction procedure in this study is adopted from Newton-Raphson iterative technique. Firstly, an initial effective shear stiffness ($\left(GA_{eff}\right)_{i,j=1}^k$) is assumed. Shear force increment ($\Delta V_{i,j}^k$) is calculated from $\Delta V_{i,j}^k = \left(GA_{eff}\right)_{i,j}^k \Delta\gamma_i^k$ in the second step and Eq.(15) is enforced to a residual function by moving $\left(GA_{eff}\right)_{i,j}^k$ into the opposite side in the third step. The fourth step is the derivative of residual function determination. Next, the new value of $\left(GA_{eff}\right)_{i,j+1}^k$ is updated from the residual function and its derivative. Finally, the iterative procedure terminates, when the value of residual function converges with the specified tolerance.

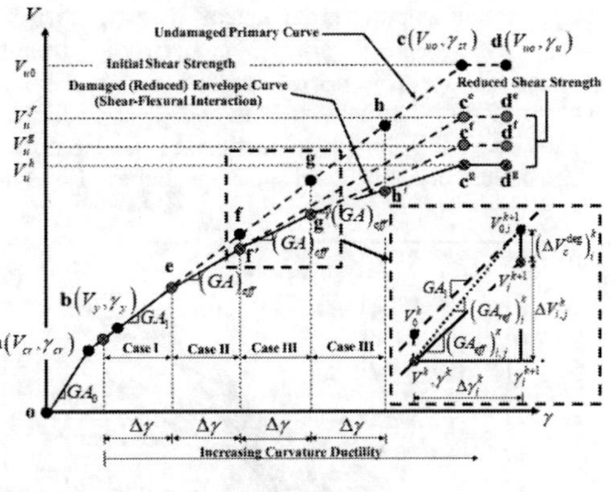

Fig. 1. Primary curve of the shear model

2.3 The hysteretic shear behavior

In this study, the hysteretic relation between the shear force and shear strain is based on the uniaxial behaviors and expressed in term of shear strain. Moreover, the hysteretic law is adapted from the hysteretic moment-rotation relation with pinching and damage, proposed by Filippou et. al. [7]. It comprises of three parts. The first one is the monotonic envelop curve (Primary curve). The second and third parts are unloading and reloading curves respectively.

3 Numerical example

To verify the accuracy and efficiency of the proposed model, a numerical example is presented herein. The example is presented by using 16 elements with 7 Gauss-Lobatto integration points under the displacement-based method that are enough for the convergent responses. Moreover, the frame section is discretized into 40 fibers (Layers) to represent the nonlinear responses under the fiber-section model.

The specimen 3SMD12 was one of the full-scale R/C column tests by Lynn [8]. The column was a double curvature specimen under a constant axial load and cyclic lateral displacements at the end of the column and collapsed in shear after the longitudinal reinforcement bars yielding. It's so-called "Flexure-shear failure". The initial shear strength can predict from Eq. (4) about 597.93 kN by using the properties in Table 1.

Table 1. Properties of specimen 3SMD12

Properties	Value	Properties	Value
f_c'	25.5 (MPa)	s	305 (mm)
b and h	457.2 (mm)	p_v	0.0017
f_{yl}	331 (MPa)	L	2,946 (mm)
f_{yv}	400 (MPa)	N	1,512 (kN)

Fig. 2 shows the lateral load vs. lateral displacement response as derived by the proposed model and the experimental data. It can be seen that the proposed model is able to represent the complex behavior of the experimental results very well. These include the member capacity, stiffness degradation with increasing displacement amplitude, amount of dissipated hysteretic energy and general shape of hysteretic response. Moreover, the proposed model can capture the degradation of the shear strength and shear stiffness due to the increasing curvature ductility demand before the shear failure occurs at the displacement about 17.12 mm.

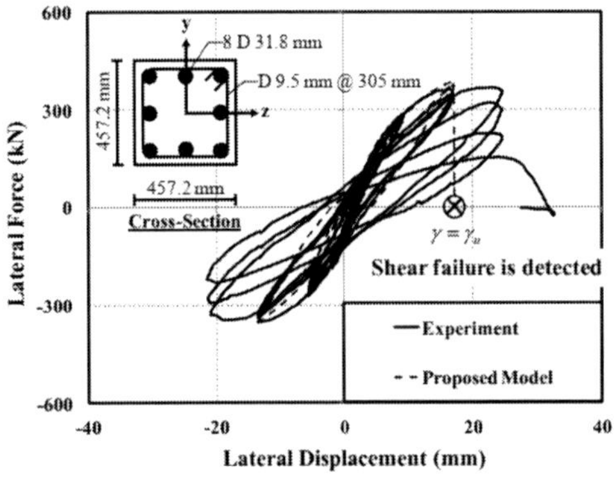

Fig. 2. The lateral load and total displacement relationship between the proposed model and experimental data

Fig. 3. presents the shear force vs. shear strain hysteretic relationship of a specimen 3SMD12 at the end of a column (Inside plastic hinge regions). It can be observed that after the curvature ductility demand value exceeds 3, the shear strain increases immediately and shear stiffness is degraded from the undamaged primary curve into the damaged primary curve. Finally, the models can predict shear failure when the shear strain reaches γ_u, that is found to be 3.33 x10^{-3}.

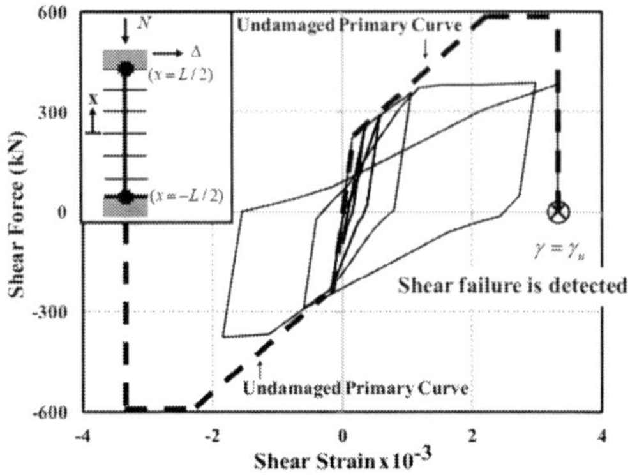

Fig. 3. Shear force vs. shear strain hysteresis loops at the end of a column (inside plastic hinges)

4 Conclusions

This paper presents an alternative of the shear model for the R/C frame structures analysis under monotonic and cyclic loading. The degradation of the shear strength and shear stiffness due to the shear-flexure interaction effect is included in this model. Moreover, the modified Mergos-Kappos shear-flexure interaction procedure (MMKSFIP) is proposed here and used to estimate the effective shear stiffness. Finally, the response analysis of a numerical example confirms the accuracy and computational efficiency of the proposed model.

Funding

This study was partially supported by Faculty of engineering, Prince of Sonkla University.

References

1. P.E. Mergos, A.J. Kappos, Eng. Struct., **44**, 94-106 (2012)
2. P. Ceresa, L. Petrini, R. Pinho, *IUSS*, Pavia, Italy (2008)
3. F.J. Vecchio, M.P. Collins, ACI Struct. J., **83**, 219-231 (1986)
4. P.E. Mergos, A.J. Kappos, Earthquake Eng. Struct. Dyn., **37**, 1349-1370 (2008)
5. R.L. Taylor, *User manual: version 7.3* (Department of Civil and Environmental Engineering, UC, Berkeley, 2000)
6. M.J.N. Priestley, F. Seible, R. Verma,Y. Xiao, *Report no SSRP-93/06* (USD, Calif., 1993)
7. F.C. Filippou, E.P. Popov, V.V. Bertero, *EERC Report 83-19* (EERC, UC, Berkeley, 1983)
8. A. Lynn, *Ph.D. thesis* (UCB, Berkeley, Calif., 2001)

Construction claim problems in Malaysia: from the contractors perspective

Bakhary Nor Azmi[1], Adnan Hamimah[2], and Ibrahim Azmi[3]

[1,3] Faculty of Civil Engineering, [2] Faculty of Architecture, Planning and Surveying Universiti Teknologi MARA, Malaysia

Abstract. There is high frequency of disputes arising from construction contract claims. Even with the most expert understanding of construction contract and the most reasonable risk-allocation system, claims will continue to present problems if they are poorly managed in practice. The significantly increasing number of construction claims indicates the need for the implementation of an effective construction claim management. This research aimed at surveying the most common problems experienced by Malaysian contractors in construction project. The paper presents the survey results and main findings, which indicate that the lack of site staff awareness to proactively detect claims, inaccessibility or unavailability of relevant documents, and conflicts arising during owner/contractor negotiation are all critical problems associated with the process of claim management. The problems observed from this research could be used to solve or improve the contractors' claim management system.

1 Introduction

Construction claims are found in almost every construction project. They have significant effect to project cost and time [1]. Claims management is the process of employing and co-ordinating resources to progress a claim from identification and analysis through preparation, and presentation, to negotiation and settlement [2]. The key objective of the claim management process is to resolve a certain problem in an effective and efficient way. Avoiding litigation and arbitration in claim settlement is a good practice that the successful contractors must keep in mind [3]. The title is set in bold 16-point Arial, justified. The first letter of the title should be capitalised with the rest in lower case. You should leave 35 mm of space above the title and 6 mm after the title. Generally, there are 6 stages of claim process. It starts with identification followed by notification, examination, documentation, presentation and negotiation of claims [6]. Construction claim identification involves timely and accurate recognition of a change. It is the first and critically important step followed by notification to the other party of a potential problem. Time limit requirements are very crucial and critical [7]. Normally, the contract specifies such duties to both parties. Establishing legal and factual ground on which the claim is to be based is done during examination stage. The fourth stage is the documentation. It plays a very important role in the settlement of claims. All the supporting documents including drawings, specification, written instruction, cost breakdown, measurement records and many more should be compile together. However, the importance of record management is not realized as much as it should be [8]. The entire completed document then will be submitted and presented to client for assessment. According to PWD form, the claim should be submitted not later than 90 days after practical completion of the works. Upon receiving the official claim, client will assess and decide the outcome. They should act fast and avoid procrastination. The final stage is the negotiation. This process concerns the process of negotiation claim to the owner, and mutual resolution of such claim [9]. If an agreement cannot be reached and either party believes his position is correct, he should propose an alternative dispute resolution method. If this fails, the remaining choice is to take the matter to court.

2 Results

After 3 months period of inquiry, only 130 completed questionnaires returned. 110 respondents are G7 contractors registered with CIDB, 16 are G6 and 4 are G5. In order for a contractor to register with G7 grade, they need to have paid up capital of RM700, 000 and can enjoy the privilege of unlimited tendering capacity. G6 contractors need to have RM500, 000 paid up capital and the tendering capacity should not exceed RM10 million. RM250, 000 paid up capital needed for G5 contractors with the maximum tendering capacity of RM5 million. From the 130 respondents, 47 of them are Project Manager, 42 are Quantity Surveyor, 40 are engineer and 1 is an Architect. PWD 203A 2010 is the most common contract used followed by PAM 2006 and CIDB 2000. PWD 203A 2010 contract is mainly used in public project whereas PAM 2006 contact is preferred in private project. Newly introduced CIDB 2000 contract is not favoured because many contractors still unfamiliar with the contract.

2.1 Current Issues in Claim Management

Table 1 shows the most common types of claim experienced by Malaysian contractors. Top of the table is claim due to "extra work" followed by "changes in the scope of work" and claim due to the "extension of time".

Table 1: Common types of claim in construction.

Rank	Common Types of Claim	Mean
1	Extra works	4.05
2	Changes	3.98
3	Extension of time	3.82
4	Measurement and payments	3.74
5	Delay	3.69
6	Contract ambiguity	3.28
7	Error and omission	3.21
8	Site condition	3.04
9	Acceleration	2.92
10	Damages	2.90
11	Non performance	2.50
12	Suspension of works	2.25
13	Termination of contract	1.88

Table 2 shows the main reason for construction claims. The most popular reason for construction claims is due to the "design changes being introduced at the post-tender stage". Second is due to the "project being implemented in unduly short time periods with inadequate site investigation, design work, tender and contract documentation" and the third is due to "changes in client's requirement during the construction stage".

Table 2: Common reasons of claim in construction.

Rank	Common Reasons of Claim.	Mean
1	Design changes	3.63
2	Short time period	3.48
3	Requirement changes	3.42
4	Inadequate specification	3.42
5	Incomplete design	3.40
6	Lack of clarity	3.28
7	Competitive contract	3.16
8	Unclear tender offer	3.16
9	Rights awareness	3.02
10	Authority changes	2.99
11	Contract failure	2.87
12	Deterioration of standards	2.73
13	Claim consultant	2.70
14	Philosophy shift	2.60
15	Economic swing	2.58
16	Political factors	2.18

2.2 Problems in Claim Process

2.2.1 Identification stage

As illustrated in Table 3, the major problems experienced by the contractors are: "lack of awareness of site staff to detect a claim", "insufficient skilled personnel for detecting a claim", and "insufficient time to identifying a claim due to high workload". It is unsurprising that matters relating to awareness and skills of staff are the most crucial problem in the claim identification process. Site staffs are expected to fully understand contract provisions as agreed and signed by the contracting parties. The lack of awareness, skills and knowledge of site personnel caused loss of chances of the right to declare a claim. They are also burden with heavy workload and this contributes to the slow identification of claim in construction.

Table 3: Problems in Claim Identification

Rank	Problems in Claim Identification	Mean
1	Lack awareness	3.42
2	Insufficient knowledge	3.36
3	Insufficient time	3.35
4	High workload	3.34
5	Insufficient skilled person	3.34
6	Poor communication	3.29
7	Document inaccessibility	3.02
8	Ambiguous responsibility	2.92
9	Ambiguous procedures	2.88

2.2.2 Notification stage

As shown in Table 4 the order ranks suggests that the most serious problems during claim notification are "inaccessibility of documents to be attached along with the notice", "ambiguous procedures in notice preparation", and "poor communication/instruction to proceed with submitting the notice". Time limit is a crucial aspect in notifying a valid claim. Having accurate site records is also another critical issue facing the Malaysian contractors. It is directly contributes from the factors of employing inexperienced and low skilled staff. Procedures of notice preparation need to be spelled out clearly and understand by parties involved. In addition, site staffs need to have good communication skill so that proper instruction can be given and received.

Table 4: Problems in Claim Notification

Rank	Problems in Claim Notification	Mean
1	Inaccessibility of documents	3.33
2	Ambiguous procedures	3.12
3	Poor communication	3.10
4	Insufficient time	3.08
5	Queries	3.06

6	Ambiguous responsibility	3.00
7	Prescribed time	2.98
8	No standard form	2.84

2.2.3 Examination stage

Claim evaluation is considered as a critical factor in winning a claim. The analysis of the contractors' responses was summarized in Table 5. For the problems associated with claim examination, the most familiar problems are "unavailability of record used to analyze and estimate the potential recovery", "inadequate time to thoroughly perform examination due to high workload", and "poor communication for gathering the required information to analyze a claim". Respondents indicated that record availability is very important in analyzing and estimating the expenses of claims. In this process, the contractors need to be aware of the need to check the required files and to estimate their claim's cost by presenting accurate documents to the owner. Therefore, it is advisable for the contractors to examine claims recovery and support it by providing precise related documents. Insufficient time and poor communication still contribute to the problems in this stage.

Table 5: Problems in Claim Examination

Rank	Problems in Claim Examination	Mean
1	Record unavailability	3.28
2	Insufficient time	3.17
3	Poor communication	3.03
4	Lack of legal	2.98
5	No standard formula	2.96
6	Ambiguous procedures	2.95
7	Ambiguous responsibility	2.85
8	Unrealistic formula	2.72
9	Insufficient computerised machine	2.43

2.2.4 Documentation stage

An effective documentation will prove the alleged claim and improve the chances of winning such a claim. The results shown in Table 6 suggest that not all information and instructions generated during project execution are kept in writing. Construction industry in Malaysia appears to be famous for consistently not documenting procedures and transactions properly. Verbal instructions given by the Superintending Officer and the inaccurate recorded information are ranked in the first and second positions, respectively. Furthermore ineffective record keeping will worsen the documentation process. The contractor needs to ensure that the owner gives their instructions in the form of written requests to the contractors. This would provide support and evidence (if needed) to the contractor's claims. In general, the problem of claim documentation starts from the lack of appreciation by site staff on the importance of record keeping. This contributes to poor quality of site records which in turn impacts on the contractor's ability to recover the compensation when making the claims.

Table 6: Problems in Claim Documentation

Ran	Problems in Claim	Mean
1	Verbal instruction	3.95
2	Inaccurate Information	3.78
3	Ineffective record	3.45
4	Inaccurate record	3.39
5	Document inaccessibility	3.31
6	Overdue	3.12
7	No standard form	2.98
8	Documentation system	2.86
9	High cost	2.50

2.2.5 Presentation stage

Table 7 suggests that the most crucial problems in preparing a claim submission are: "inaccessibility of relevant documents to submit along with the claim", "insufficient skilled staff for preparing a claim submission", and "insufficient time to thoroughly prepare due to high workload". Documentation is still the key problem in claim presentation requiring particular effective means to resolve. The respondents agreed that some documents are still ineffectively kept because on-site staffs do not realize the importance of recording such documents. Most respondents agreed that it is impossible to submit the relevant document during the claims presentation, especially where the contractors need to support their claims with accurate formal documentation. Presenting a claim requires a knowledgeable, skilled and experienced person to effectively organize the full detailed claim submission. Contractors face difficulties in identifying and hiring experienced staff in preparing a claim submission. This issue affects the claim presentation process, possibly causing the client to reject the claim. Due to heavy workload, they have insufficient time to fully concentrate on the preparation of claim presentation.

Table 7: Problems in Claim Presentation

Ran	Problems in Claim	Mean
1	Document inaccessibility	3.15
2	Insufficient skilled staff	3.12
3	Insufficient time	3.08
4	Poor communication	3.02
5	Ambiguous responsibility	2.98
6	Ambiguous procedures	2.87
7	No standard format	2.78

2.2.6 Negotiation stage

According to Table 8, the main problems associated with claim negotiation are "disagreement arising during negotiation", "unsatisfactory evidence to convince other parties and "poor negotiation skills". It is believed that these three problems are somehow linked. Lack of negotiation skills will give impact on the ability of the contractor to convince the owner, and hence conflicts may arise. Strong evidence is very much needed for presentation during this stage in order to support the claim and to convince the owner. In many cases, the contractors lose their claim as a result of the lack of documentation evidence therefore affecting the contractor's position during claim negotiation. In some cases, the negotiation process could jeopardise the relationship between the contractor and the owner as a result of unsatisfactory claims preparation and negotiation. The contractors becomes the weakest party because the owner, because of the adversarial relationship, may reject the claim with very little consideration given to the contractor's situation.

Table 8: Problems in Claim Negotiation.

Rank	Problems in Claim Negotiation	Means
1	Disagreement	3.74
2	Unsatisfactory evidence	3.52
3	Poor negotiation	3.28
4	Adversarial relationship	3.12
5	Inadequate time	2.95
6	Difficult	2.74

In the construction industry, the problems associated with not having experienced construction personnel with high level of negotiation skills is predictable. Most contracting companies do not have a dedicated unit or individual who is responsible for managing the claims. Normally, the duty to manage the claim process is assigned to the project manager or site engineer, who may not be skilled negotiators. Clearly, there is a need to enhance staff negotiation skills to avoid from facing difficulties and potential conflicts during negotiation with the client. Failing to do so may result in losing the claim.

3 Conclusion

This research aimed to explore the problems related to the process of claim from contractor's point of view by analyzing common procedure categories, distributed into six stages (claim identification, claim notifications, claim examination, claim documentation, claim presentation and claim negotiation). The study findings emphasize on the need for a good documentation system with a capable site staff that can identify a claim during project execution. These competencies must be joined so that it would allow easy accessibility to project records when needed in order to support the submitted claim. Thus, keeping relevant project information should be viewed as an essential part of project monitoring and reporting. It also appears that a standard and transparent procedure should be put in place whereby contractor's staff would be able to follow to properly prepare the claim. This procedure should highlight the importance of linking the claim to cost recovery so as not to disadvantage the contractor. The contractors can increase their claim management efficiencies by deliberately scrutinizing the contract provisions, enhancing the cooperative atmosphere among all parties, establishing the good claim management data collection and documentation, promptly notifying to the client the event that change their statuses, and signing another separate contract for the large-amounted variation order. The government can help the contractors by improving the relevant laws, regulations and standards and establishing the claim settlement organization.

References

1. Hossam H., Ahmed H.I. & Asmaa A.S. (2014). Reducing construction disputes through effective claims management. American Journal of Civil Engineering and Architecture, 2014, Vol. 2, No. 6, 186-196
2. Surawongsin, P. (2002). The Implementation of Construction Claims Management in the Thai Construction Industry. Master Thesis School of Civil Engineering Asian Institute of Technology (AIT), Pathumthani.
3. Mohamed, M., & Tarek, I.E. (2014). Analyzing delay causes in Egyptian construction projects. Journal of Advanced Research. 5. 49–55.
4. Hassanein, A.A.G., & El-Nemr, W. (2008). Claims management in the Egyptian industrial construction sector: A contractor's perspective. Engineering, Construction, and Architectural Management, 15(3), 246-259.
5. Abdul-Rahman H, Berawi M.A., Berawi A.R., Mohamed O, Othman M, & Yahya I.A. (2006). Delay mitigation in the Malaysian construction industry Journal of Construction Engineering and Management, 132 (2):125-33.
6. Enshassi, A. & Mohamed, S. (2009). Problems Associated with the Process of Claim Management in Palestine. Journal of Engineering Construction and Architectural Management Vol 16, 61-72.
7. Abowitz, D.A & Toole, T.M. (2010). Mixed method research: fundamental issues of design, validity, and reliability in construction research, Journal of Construction Engineering and Management, 136(1), 108-116.
8. Chan, T.K. (2009). Measuring performance of the Malaysian construction industry. Construction Management & Economics. 27((12), 1231-1244.
9. Kamal, E.M. (2012). The critical review on the Malaysian construction industry. Journal of Economics and Sustainable Development, 3(13), 81-87

A traffic micro-simulation model to estimate four-lane highways' capacity in Thailand

Wuttikrai Chaipanha[1], *Ladda* Tanwanichkul[1,*], and *Jumrus* Pitaksringkarn[2]

[1]Department of Civil Engineering, Faculty of Engineering, Khon Kaen University, Khon Kaen, Thailand
[2]Department of Civil Engineering, Faculty of Engineering, King Mongkut's Institute of Technology Ladkrabang, Bangkok, Thailand

Abstract. This article purposed to present the maximum capacity and to develop the equation in the capacity estimation of 4 types of four-lane highways by using the micro-simulation model. Regarding the analysis, the factors affecting the capacity include access-point, heavy vehicles and median u-turn. According to the study, it was found that the maximum capacity of the four-lane highways in type 1 is 2194 passenger car/hour/lane. In the descending orders, the maximum capacity of the highways with four lanes in type 2, 3, and 4 are 2161, 2094 and 2017 passenger car/hour/lane, respectively. At the same time, the maximum capacity of the prevailing condition in the study is 1300-1600 vehicles/hour/lane, which is different from the HCM2010 method for 20-30%, due to the Thai's highway characteristics directly affect the traffic and driving behavior. Median u-turn affects the four-lane highways in type 1 and 2 the most, while access-point factor has the most influence towards type 3 and 4. In addition, the author has developed the equation models for capacity estimation, which the result derived from the relationship between the capacities and affecting factors. It was aimed at using as the guidelines in the capacity assessment of Thai's four-lane highways in the future.

1 Introduction

Highway capacity is one of the most important applications of traffic theory, with analysis and design of transportation facilities leading to competent decision-making to improve road networks [1-3]. The United States Highway Capacity Manual (US-HCM) is regarded as the global standard [4]. However, this was developed specifically for the United States and may not be relevant in other countries which have unique traffic characteristics. Some nations have attempted to modify the US-HCM and develop highway capacity manuals specifically suited to their individual requirements as the Indian, Swedish, Indonesian, Korea and German.

Estimation of the highways' capacity in Thailand was also based on the US-HCM; however, the physical characteristics of Thai highways are different from the USA [5]. Therefore, it is an interesting gap to consider the development of a highway capacity estimation method designed especially for Thai's highways.

Methods of estimating roadway capacity can be divided into two categories as Direct Empirical Methods, based on observed traffic flow characteristics and Indirect Empirical Methods, based on guidelines and simulation models [6]. However, a direct empirical method generally requires a survey to be conducted under capacity conditions, and it also is relatively uncommon for a capacity situation to occur on a regular weekday. A micro-simulation model is often used in indirect empirical analyses to reduce the limitations of the survey and complex phenomena which are difficult to analyze by mathematical techniques. Several previous studies have used traffic micro-simulation models to identify factors affecting roadway capacity [7-12].

This research investigated the application of a micro-simulation model to estimate the capacity of four-lane highways in Thailand, which is important and account for about one-third of the total highway distance [13]. Maximum capacity for base and prevailing conditions was identified and equation models were developed to estimate the capacity of four types of four-lane highways. Consider the key factors negatively affecting the capacity included (1) access-points, (2) heavy vehicles, and (3) median u-turn (Fig. 1).

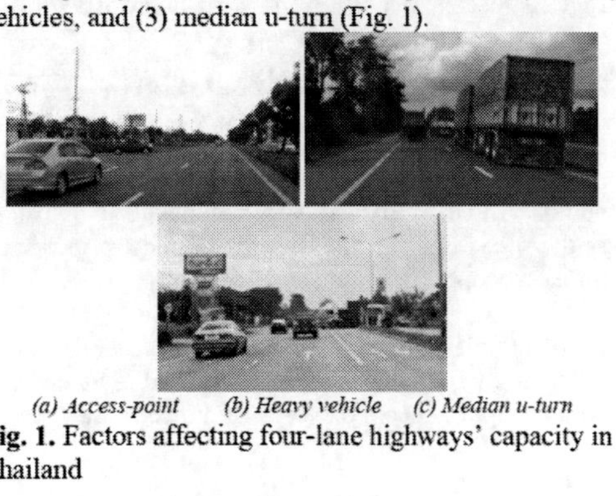

(a) Access-point (b) Heavy vehicle (c) Median u-turn
Fig. 1. Factors affecting four-lane highways' capacity in Thailand

* Corresponding author: ladpit@kku.ac.th

2 Methodology

2.1 Site selection

Four-lane highways represent multi-lane rural road networks in Thailand and many characteristics and factors differ from the US-HCM analysis conditions. The study was divided into four types of cross-sectional design in accordance with the standard of the Department of Highways [14] as shown in (Table 1).

Table 1. Type of four-lane highways in Thailand

Type no.	Divided/ Undivided	Median type	Median width (m.)
1	Divided	Depressed median	varies
2		Raised median	4.6
3	Undivided	Painted median*	2.0
4		Undivided	-

**Driving behavior on the four-lane highways with painted median is no differences from type 4, and was identified for undivided.*

Highways in the northeast region of Thailand were examined. The criterion of site selection followed the HCM 2010 base condition and represented the prevailing conditions of four-lane highways in Thailand as follows:

- Consist of physical factors that affect the capacity (access-point and median u-turn) to be used to calibrate the driving behavior in the model.
- Each selected highway type shall be designed in accordance with the standard of the Department of Highways, and the highest traffic volume will be compared to the same type [15].

Based on these conditions, the representative four-lane highways sections were selected (Fig. 2).

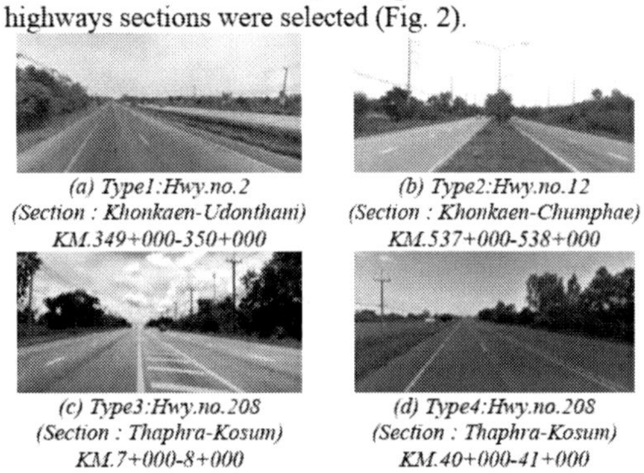

(a) Type1:Hwy.no.2 (Section : Khonkaen-Udonthani) KM.349+000-350+000
(b) Type2:Hwy.no.12 (Section : Khonkaen-Chumphae) KM.537+000-538+000
(c) Type3:Hwy.no.208 (Section : Thaphra-Kosum) KM.7+000-8+000
(d) Type4:Hwy.no.208 (Section : Thaphra-Kosum) KM.40+000-41+000

Fig. 2. Selected sections of four-lane highways

2.2 Data Collection and Preparation

Data were categorized into four groups as (1) Geometric data: lane width, median width, median type, and aerial photos, (2) Demand data: entry volume, (3) Control data as the speed limits, and (4) Calibration data: spot speed, travel speed, traffic count, and headway.

Entry volume was measured at two intervals of 1 hour 30 minutes each during the morning and evening peak hours. Spot speed surveys followed the desired distributions in the micro-simulation model as a significant variable which affected the capacity [16]. Radar speed gun surveys were conducted during non-peak hours by vehicle type and traffic lane. In the car following model, the time gap (Headway) that has a determining influence on capacity [12, 16]. Video data were collected and applied in calibration of the micro-simulation model.

2.3 Base Model Development

PTV Vissim software was applied to develop a traffic micro-simulation model (Fig. 3.) divided into five steps as (1) Base data for simulation, (2) Traffic networks, (3) Evaluation and configuration, (4) Simulation, and (5) Error checking.

Fig. 3. Example of the base model

2.4 Model Calibration and Validation

Traffic volume, and travel speed were used as measures of consistency between the model and local conditions. The calibration and validation target criteria were based on the Design Manual for Roads and Bridges (DMRB) and other acceptable standards [17]. The traffic count comparison has a GEH statistics value of less than 5 and speed of travel not different from the observed over 5%. In order to calibrate and validate the model, the configuration and adjustment of the values for the VISSIM parameters CC0, CC1 and CC2 were the most influential variables for cars following model and capacity [18]. From the model calibration and validation, all the GEH values were ranged from 0.1 to 0.4 for the traffic count comparison, while the average percentage differences of the travel speed for each type at 4.2, 0.8, 1.3 and 2.5, respectively.

2.5 Model Application

The traffic micro-simulation model was applied as a tool to estimate the four-lane highways' capacity, considering various factors including number of access-points per km (0, 2, 4 and 6), different proportions of heavy vehicles (0%, 5%, 15% and 25%), and number of median u-turns per km (0, 1 and 2) [19].

2.6 Capacity Estimation

A calibrated and validated micro-simulation model was used in the capacity estimation, where speed data were estimated under different traffic volumes by increasing the flow rate in increments of 500 veh/hr, from an initial value till highway capacity. When the simulated volume reached capacity level, increments in the input traffic volumes did not equate to increase in the exit volume and resulted in a decrease in the rate of traffic flow [20]. Highway capacity estimation was performed using speed-volume relationships by plotting speed on the graph ordinate and volume on the abscissa.

3 Results and discussion

3.1 Estimation of capacity under base condition and prevailing condition

For the base condition of no access-points, all vehicles as passenger cars, and no median u-turns, the Type 1 four-lane highways had the maximum capacity at 2194 pc/hr/ln. With higher physical and standard design, drivers were more likely to travel faster with greater capacity. Four-lane highways Types 2, 3 and 4 recorded 2161, 2094 and 2071 pc/hr/ln, respectively. All results slightly different from the HCM2010 calculation by 0.4% - 4.5% because this was not impacted by any factors.

The traffic micro-simulation model estimated the capacity under prevailing condition (surveyed state) for each type at 1574, 1524, 1433 and 1368 veh/hr/ln, respectively. The capacity results, which represented the highways in each type, was significantly different from the HCM2010 method due to the factors that affected the capacity to disturb the traffic flow (Table 2).

Table 2. Capacity estimation under prevailing condition compared with the HCM2010 calculation

Type no.	Median type	Capacity (Veh/hr/ln)		Diff. (%)
		Micro-Simulation	HCM 2010	
1	Depressed median	1574	1963	-19.8
2	Raised median	1524	1992	-23.5
3	Painted median	1433	1999	-28.3
4	Undivided	1368	1991	-31.3

3.2 Impact of factors on capacity

3.2.1 Impact of access-point on capacity

Impact of the access-point on four-lane highway capacity for Types 1 and 2 decreased by 1.87 and 1.90%, respectively, compared to the base condition. However, in the case of four-lane highways for Types 3 and 4, access-points had a significant effect on capacity, which reduced by 47.18 and 46.16%, respectively, compared to the base condition. Access-points involved turning through the traffic resulting in disruption in the flow and reduction in both speed and capacity (Fig. 4).

3.2.2 Impact of heavy vehicle on capacity

The impact of heavy vehicles on traffic flow between 0% and 25% resulted in a decrease in the capacity of all highways. There was a relatively stable trend of 0.7-0.8% decrease in the proportion of heavy vehicles, with a 1% increase from the base condition. Heavy vehicles need more space between the vehicles in the stream and overtaking or changing lanes require longer distances and cause speed and capacity reduction (Fig. 5).

3.2.3 Impact of median u-turn on capacity

Results for four-lane capacity at 0 - 2 points/km for Type 1 and 2 highways showed a decrease in capacity by 26.2% and 29.5%, respectively, compared to the base condition. Reduced capacity was due to disturbing the traffic in the opposite direction and obstructing the traffic in the same direction as the traffic volume increased (Fig. 6).

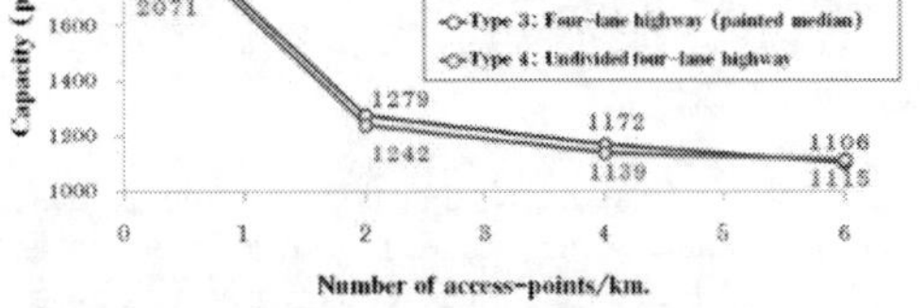

Fig. 4. Impact of access-points on capacity

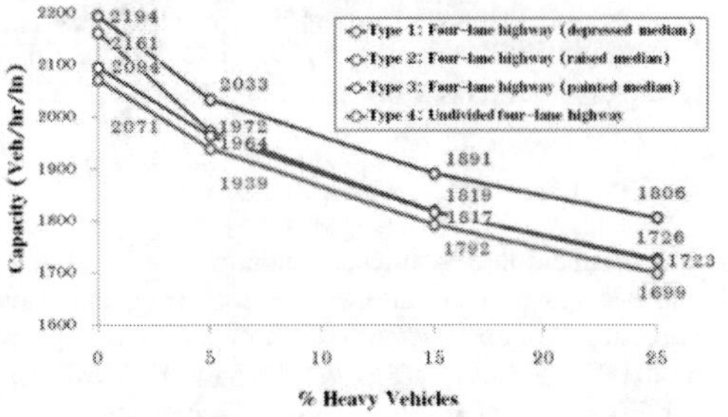

Fig. 5. Impact of heavy vehicles on capacity

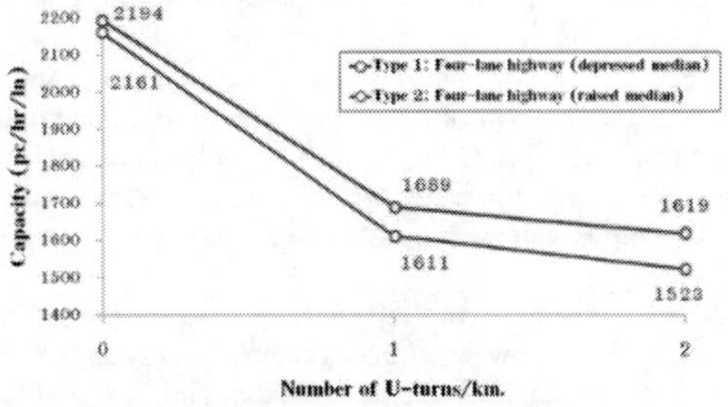

* *Types 3 and 4 were assumed as unaffected by median u-turns*
Fig. 6. Impact of median u-turns on capacity

3.3 Capacity model

The relationships between capacity as a dependent variable and access-point, heavy vehicle, and median u-turn as independent variables were investigated. Simple regression analysis was used to check the correlation coefficient (r) between the independent variables, and non-collinear variables are induced into the multiple linear regression models. A multiple linear regression model was generated from the micro-simulation results which estimated four-lane highways' capacity on prevailing conditions as follows:

$$C_{Type1} = 2018.466 - 9.167_A - 6.932_H - 211.063_U \quad (1)$$

$$C_{Type2} = 2002.649 - 13.846_A - 7.977_H - 262.438_U \quad (2)$$

$$C_{Type3} = 1794.303 - 130.125_A - 6.749_H \quad (3)$$

$$C_{Type4} = 1755.045 - 123.288_A - 5.955_H \quad (4)$$

Where C is the capacity for each four-lane highway type (veh/hr/ln), A is the number of access-points per kilometer (in the range of 0-6), H is the percentage of heavy vehicles in the traffic stream (in the range of 0-25%), and U is the number of median u-turns per kilometer (in the range of 0-2). From statistical analysis of the model, the coefficients of determination (R^2) at 0.781, 0.875, 0.750 and 0.712 reflected a high goodness of fit. Significance at the 95% confidence level, with significance of the F statistic at <0.001. A negative sign indicated that as the effect of the factors increased, the capacity decreased.

The standardized coefficients (β) show the influence of each variable. The median u-turn was the most influential factor for four-lane highways with depressed median (Type 1) and raised median (Type 2). The access-point factor had the most influence toward four-lane highways with painted median (Type 3) and undivided four-lane highways (Type 4).

4. Conclusion

The application of a micro-simulation model is another approach to estimate highways' capacity. In addition to reducing the limitations of survey data, micro-simulation models can also effectively emulate local traffic conditions, resulting in more accurate estimations of highways' capacity. In this study, four types of four-lane highways were selected as representative in Thailand; these differed from the conditions of the HCM analysis. Highways with physical and higher design standards have greater capacity due to fewer traffic distractions and higher speeds. However, capacity reduces when affected by various factors and comparisons with the HCM approach were significantly different.

Interactions and conflicts between vehicles interfered with traffic flow on the carriageway and significantly reduced stream speed and capacity. The availability of median u-turns affected four-lane highways with median as Types 1 and 2, while access-points has the highest negative influence on four-lane highways without median as Types 3 and 4. The capacity estimation model can serve as a guideline for transport and traffic operation, and improve design and planning based on the standard of Thailand's Department of Highways (DOH) with assumptions of the level terrain. Further studies should analyze other conditions and roadways that fall under the responsibility of other departments as an interesting approach in the future.

We would like to acknowledge the Department of Civil Engineering, Faculty of Engineering, Khon Kaen University for the research funding.

References

1. J. Banks, J.S. Carson, B.L.Nelson, D.M. Nicol, *Discrete-event system simulation*, (2004)
2. V.T. Arasan, R. Koshy, J Transp Eng, **131(7)**: 544-551. (2005)
3. Transport Research Board. *Highway capacity manual 2010*, (2010)
4. Central Road Research Institute. *Development of Indian highway capacity manual*, (2014)
5. L. Tanwanichkul, J. Pitaksringkarn, P. Thongkrew. KKU Enj, **39(3)**:241-248. (2012)
6. M.M. Minderhoud, H. Botma, P.H.L. Bovy. Transp Res Rec, **1572;1997**, pp.59-67. (1997)
7. J.H. Kim. *A thesis in civil engineering*, Pennsylvania state university. (2006)
8. S.S. Arkatkar. IJEE, **volume 04**, no 06, 2011, pp. 375-380. (2011)
9. Madhu, S. Velmurugan. IJSTER, **vol. 1(6)**, (2011)
10. M.S. Bains, A. Bhardwaj, S. Arkatkar, S. Velmurugan. Procedia Soc Behav Sci, **104**. 458 – 467, (2013)
11. A. Mehar, S. Chandra, S. Velmurugan., KSCE **Volume 18**, pp 639–645. (2014)
12. S.Chandra, A.Mehar, S.Velmurugan. KSCE, **Volume 20**, pp 2033–2040. (2015)
13. Department of Highways [Internet]. [cited 2017]. Available from: : http://maintenance.doh.go.th/website/download/distance2513-2558.xls. (2014).
14. Department of Highways. *Annual Average Daily Traffic on Highways 2015*. (2015)
15. Department of Highway. *Standard drawings for highway design and construction rev. 2015*. (2016)
16. PTV. *Vissim user manual*. (2011)
17. R. Dowling, A. Skabardonis, V. Alexiadis. *Traffic analysis toolbox volume III: guidelines for applying traffic microsimulation modeling software*, (2004)
18. G. Gomes, A. May, R. Horowitz. TRR Journal, **Vol.1876**, pp 71-81, (2004)
19. Department of Highway. *Suggestions to consider opening the median of the highway construction*. (2005)
20. P.S.Praveen, V.T.Arasan. IJTTE, **3(3):** 302 – 330. (2013)

Impact of climate change on the groundwater sustainability in the lower Chao Phraya basin, Thailand

Pinit Tanachaichoksirikun[1,*], *Uma* Seeboonruang[2], and *Phayom* Saraphirom[3]

[1] PhD. Student, Civil Engineering Department, Faculty of Engineering, King Mongkut's Institute of Technology Ladkrabang, Bangkok, Thailand

[2] Associate Professor, Civil Engineering Department, Faculty of Engineering, King Mongkut's Institute of Technology Ladkrabang, Bangkok, Thailand

[3] Lecturer, Agricultural Engineering Department, Faculty of Engineering, Khon Kean University, Thailand

Abstract. This research investigates the impact of climate change on the hydraulic heads of Thailand's Lower Chao Phraya basin. The research also determines the sustainability of groundwater as the result from climate change. In the study, the climatic scenario **(IPSL-CM5A-MR)** of the **R**epresentative Concentration **P**athways **(RCP)** between 2.6, 4.5 and 8.5 were considered, and the simulations were carried out using the three-dimensional groundwater flow model (i.e. **MODFLOW-2000)** predicting the groundwater behavior between 2017 and 2036. The findings revealed that the impact of climate change on the hydraulic head fluctuation was positively correlated. Specifically, under the IPSL-CM5A-MR RCP 4.5 that has the highest average precipitation, the average hydraulic head increased. In conclusion, the sustainability of groundwater in the Lower Chao Phraya basin was sufficient during the simulated time. However, the groundwater budget was lower than the average groundwater budget during 2009 – 2014 indicating, the groundwater storage was continuously decreased. Specifically, the 2nd, and 3rd (Phra Pradeang and Nakorn Luang) aquifers may be facing the groundwater shortage in the future.

1 Introduction

Groundwater is an important resource and a secondary source of water when facing critical droughts and disasters. In addition, climate change is probably one of the most challenging pressures facing hydrological systems and water resources. Consequently, there are several researches subject to the effect of climate change on the groundwater resource such as [1], [2], [3], [4], [5], [6]. [2] investigated the impact of climate change on two small aquifers in western Canada and the United States and revealed the impact from climate change on recharge and groundwater levels at both study areas. In addition, [1] examined the impact of climate change on a chalky groundwater basin in Belgium and showed two models had the groundwater levels decreased, while the other showed no significantly changed. Moreover, [7] carried out the study of climate change on groundwater levels in Lansing, Michigan and found that the simulated steady-state groundwater levels were generally predicted to increase or decrease depending on climate change scenarios.

Recent research investigated the effect of climate change on the groundwater In Thailand, for example, [4] performed climate change and groundwater resources in Thailand and explained the climate change effect on the groundwater quality and quantity and the groundwater policy. Moreover, [8] determined impact of climate change on groundwater and vulnerability to drought of areas in eastern Thailand and reported the risky hotspots and prioritized zones threatened by persistent droughts. In addition, [9] applied an empirical decomposition to study deep groundwater and possible link to climate variability in the Lower Chao Phraya basin and revealed that the groundwater significantly related to the groundwater withdraw than the climate variability.

Even though, there are several studies that focus on the impact of climate change on the groundwater resource, little research focused on the groundwater sustainability. Thus, this study aims to predict the impact of climate change on the hydraulic head in the Lower Chao Phraya (LCP) basin, Thailand by using numerical simulations. In addition, the groundwater sustainability was determined by two variables: groundwater recharge and groundwater demand.

2 Study area

2.1 General and topography

Fig. 1 illustrates the Lower Chao Phraya (LCP) basin, Thailand. The area covers 43,317 km^2, 21 provinces including Bangkok, which is the capital city. The LCP basin located in the central Thailand. The western part is the Tenasserim hills range, and the northern part is a small hill which separates the Upper Chao Phraya basin. The southern part connects to the Gulf of Thailand. The basin has slope from the north to the south. The LCP

* Corresponding author: 57601070@kmitl.ac.th

basin contains four constituent rivers (i.e. the Chao Phraya, Mae Klong, Pa Sak and Tha Chin rivers).

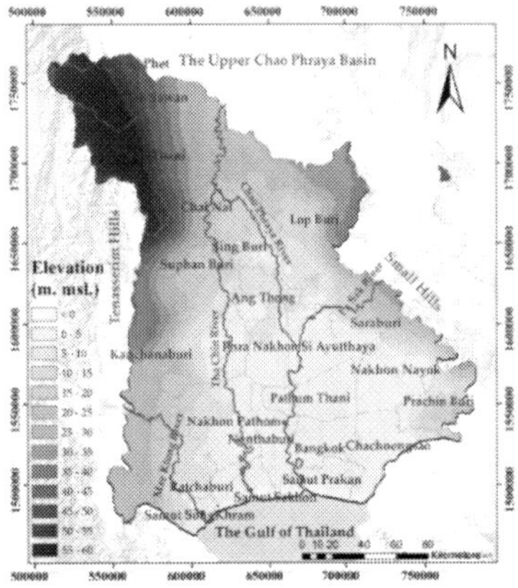

Fig. 1 The geographical map of the Lower Chao Phraya basin

2.2 Hydrogeology

The hydrogeology of the LCP basin is the groundwater in unconsolidated materials and accounts for the largest proportion of groundwater in the basin [10]. In addition, sediments in the basin are predominantly of old flood-plain deposits aquifer. The LCP basin has eight aquifer layers, which are Bangkok (BK ~50 m), Phra Pradeang (PD ~100 m), Nakorn Luang (NL ~150 m), Nonthaburi (NB ~200 m), Sam Khok (SK ~ 250 m), Phayathai (PT ~300 m), Thonburi (TB ~350 m) and Pak Nam (PN ~400 m) aquifers.

3 Methodology

3.1 Model design and boundary conditions

In this research, the LCP basin was divided into nine layers, consisting of the eight aquifers and the top layer was Bangkok clay (BKC).

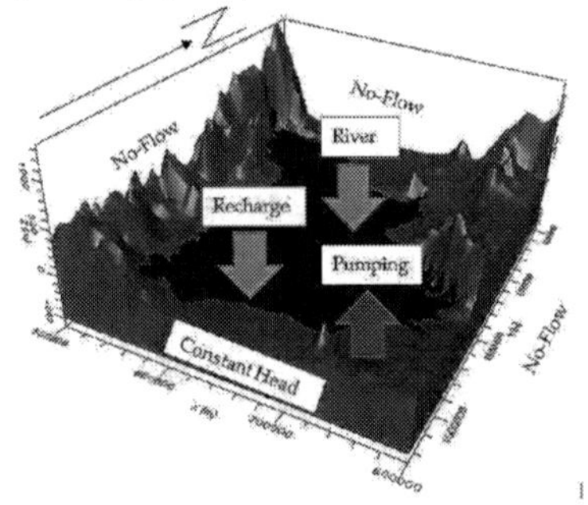

Fig. 2 The 3D model of the LCP basin and the boundaries

Fig.2 illustrated the three-dimensional model of groundwater and the boundaries in the LCP basin. The model composes of nine layers which individually discretized into 350 x 300 (row x column) grid sizes of 1 x 1 km^2 each to efficiently determine the hydraulic heads. The model generation was carried out using MODFLOW-2000 [11]. The underside was treated as a no-flow boundary of impermeable bedrock; the west side as other no-flow boundary of the Tenasserim hills range; and the north as another no-flow boundary of the small hills separating the upper and lower Chao Phraya basins. The four constituent rivers (the Chao Phraya, Mae Klong, Pa Sak and Tha Chin rivers) were treated as the river boundary, which was constantly gaining and losing between the surface water and groundwater. On the south side, the interface between the aquifers layers and the Gulf of Thailand were treated as the constant-head boundary, where the groundwater-seawater connected. The top of the model was treated as the recharge boundary from the rainfall. The aquifers in the LCP basin were pumped for consumption, agriculture and industry.

3.2 Transient calibration and validation

Fig. 3(a) illustrated the calibration of the groundwater model, which was performed during the period 2009 – 2014 based on 271 observation wells. The groundwater levels were compared to the actual groundwater levels. The results of the transient calibrated model reveal absolute residual mean of 1.74 m and root mean squared (RMS) error of 2.35 m with normalized RMS of 2.38%.

Fig. 3(b) showed the validation of the groundwater model, which was analyzed between 2007 and 2008 based on 235 observation wells. The results of the validated model show the absolute residual mean of 1.58 m and root mean squared (RMS) error of 2.15 m with normalized RMS of 2.16%.

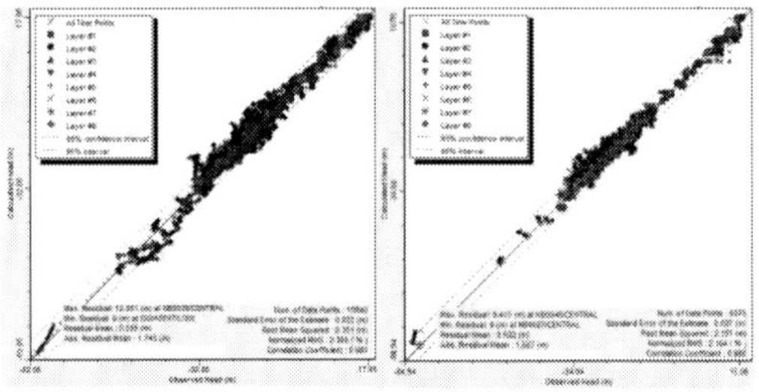

Fig. 3 The relationship between simulated and observed hydraulic head (a) calibrated model (b) validated model

3.3 Selected and simulated climate scenarios

The climatic model was selected and prepared by [12] and the model was conducted based on six criteria (i.e., average temperature, annual rainfall, humidity, pressure, evaporation and water discharge). IPSL-CM5A-MR (The Institute Pierre Simon Laplace) which is the one of climatic scenarios of the Coupled Model International research group (CMIP5) was selected, giving preference to scenarios offering the minimum bias and root mean square error on the annual precipitation in the Chao Phraya watershed [13]. Moreover, the **R**epresentative **C**oncentration **P**athways (RCP) 2.6, 4.5 and 8.5 were chosen to give the most contrasted climate changes.

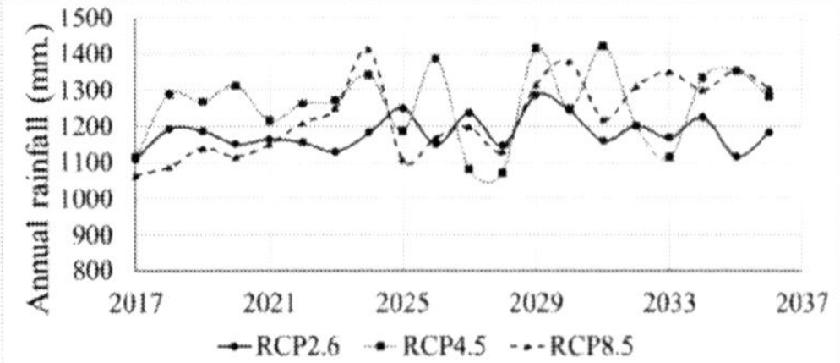

Fig. 4 Annual rainfall of the IPSL-CM5A-MR was varied the RCP 2.6, 4.5 and 8.5 during period 2017 – 2036

Fig. 4 illustrated the annual rainfall of the climatic scenarios of RCP 2.6, 4.5 and 8.5 during period 2017 –

2036. The average annual rainfall under RCP 2.6, 4.5 and 8.5 were 1,182, 1,258 and 1,227 mm, respectively, while the trend of the annual rainfall increased by 1.85, 2.64 and 12.78 mm, respectively.

3.4 Simulation scenarios

In the model, the period during 2009 – 2014 was used as the base case to determine the impact of climate change on the hydraulic head in the LCP basin. In addition, the research investigated the sustainability of groundwater in the specific basin. In the climatic scenario of RCP 2.6, 4.5 and 8.5, monthly precipitation was generated in the future between 2017 and 2036. The assessment was carried out with regard to the impact of climate change on the hydraulic head and the sustainability of groundwater in the area-specific.

4 Results and discussion

4.1 Impact of climate change on the hydraulic heads

Figs.5 illustrated the simulated hydraulic head fluctuation associated with the PD, NL and NB aquifers under the base case (2009 – 2014) and IPSL-CM5A-MR with the variable RCPs (2.6, 4.5 and 8.5). In RCP 2.6 during the period 2017 – 2036, the average hydraulic head in PD, NL and NB increased by 4.94, 4.70 and 3.93 m, respectively vis-à-vis the base case, under RCP 4.5 the average hydraulic head in PD, NL and NB increased by 5.23, 4.87 and 4.03 m, respectively vis-à-vis the base case, under RCP 8.5 the average hydraulic head in PD, NL and NB increased by 5.09, 4.76 and 3.96 m, respectively vis-à-vis the base case. Essentially, the high precipitation gives rise to the hydraulic head increase.

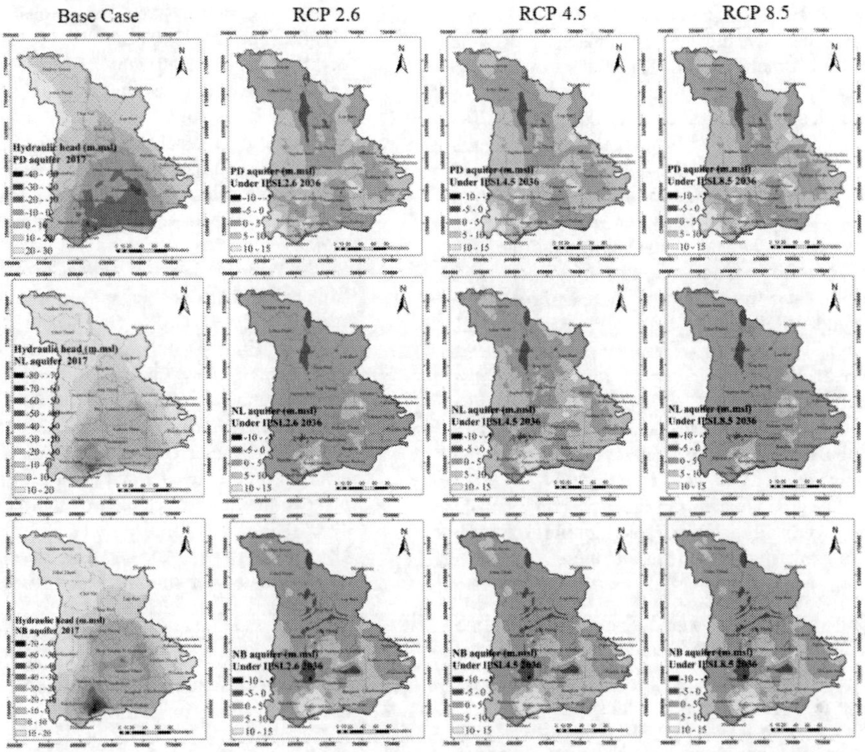

Fig. 5 The simulated hydraulic head in the PD, NL and NB aquifers under the base case and IPSL-CM5A-MR with RCP 2.6, 4.5 and 8.5 between the period 2017-2036

4.2 Sustainability of groundwater

Figs. 6(a-c) respectively illustrated groundwater budget in dry and rainy season and groundwater demand in the PD, NL and NB aquifers under IPSL-CM5A-MR RCP of 2.6, 4.5 and 8.5. The influence of seasonal fluctuates in PD aquifer was shown in Fig. 6(a). Under RCP 2.6, the average rate of groundwater budget in dry and rainy season were 1.65 and 2.11 million m^3/day, respectively. Under RCP 4.5, the average rate of groundwater budget in dry and rainy season were 1.76 and 2.29 million m^3/day, respectively. Under RCP 8.5, the average rate of groundwater budget in dry and rainy season were 1.71 and 2.19 million m^3/day, respectively. While the average groundwater demand is 0.30 million m3/day and the rate of groundwater demand continuously increased by 20 m^3/day. Fig. 6(b), under RCP 2.6, the average rate of groundwater budget in dry and rainy season were 1.24 and 1.27 million m^3/day, respectively. Under RCP 4.5, the average rate of groundwater budget in dry and rainy season were 1.30 and 1.33 million m^3/day, respectively. Under RCP 8.5, the average rate of groundwater budget in dry and rainy season were 1.26 and 1.30 million m^3/day, respectively. While the average groundwater demand is 0.1 million m^3/day and the rate of groundwater demand continuously increased by 9 m^3/day. Fig. 6(c), under RCP 2.6, the average rate of groundwater budget in dry and rainy season were 1.12 and 1.10 million m^3/day, respectively. Under RCP

4.5, the average rate of groundwater budget in dry and rainy season were 1.16 and 1.15 million m^3/day, respectively. Under RCP 8.5, the average rate of groundwater budget in dry and rainy season were 1.13 and 1.12 million m^3/day, respectively. While the average groundwater demand is 0.05 million m^3/day and the rate of groundwater demand continuously increased by 1 m^3/day. The high precipitation rendered the great groundwater recharge then the higher groundwater is stored.

The groundwater budget during period 2017 – 2036 was greater than the groundwater extraction. Consequently, the sustainability of groundwater in the LCP basin was sufficient. However, the groundwater budget was lower than the average groundwater budget during period 2009 – 2014 (baseline) that means groundwater storage continuously decreased. Therefore, PD and NL aquifers may be facing the groundwater shortage in the future.

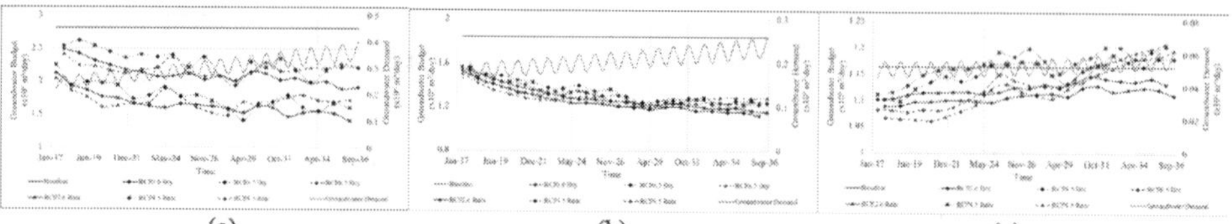

(a) (b) (c)

Fig. 6 The rate of groundwater budget and groundwater demand in dry and rainy season (a) PD, (b) NL, (c) NB aquifers

5 Conclusions

This research has investigated the impact of variable climate on the hydraulic head in Thailand's Lower Chao Phraya (LCP) basin, and the sustainability of groundwater in the area-specific. In the study, the period during 2009 – 2014 was used as the base case. Using the climatic scenario (IPSL-CM5A-MR) was varied between RCP 2.6, 4.5 and 8.5 during periods 2017 and 2036. The simulations were carried out using MODFLOW-2000. The findings revealed that the impact of climate change on the hydraulic head fluctuation was positively correlated. Specifically, the simulation under IPSL-CM5A-MR RCP 4.5 that had the highest average precipitation resulted in the greatest average hydraulic head increase. In addition, the sustainability of groundwater of the Lower Chao Phraya (LCP) basin was obtained during period 2017 – 2036. However, the groundwater budget was lower than the average groundwater budget during 2009 – 2014 indicating, the groundwater storage was continuously decreased. Specifically, PD and NL aquifers may be facing the groundwater shortage in the future.

Acknowledgements

The authors would like to extend deep gratitude to the Thailand Research Fund through the Royal Golden Jubilee PhD Program (Grant No. PHD/0189/2556) for the financial sponsorship. Sincere appreciation goes to the Land Development Department, the Thai Meteorological Department and the Thailand Department of Groundwater Resources for the data.

References

1. S. Brouyère, G. Carabin, A. Dassargues. Climate change impacts on groundwater resources: modelled deficits in a chalky aquifer, Geer basin, Belgium. Hydrogeol. J., **12**, 123-134 (2004)
2. D. M. Allen, D. C. Mackie, M. Wei. Groundwater and climate change: a sensitivity analysis for the Grand Forks aquifer, southern British Columbia, Canada. Hydrogeol. J., **12**, 270-290 (2004)
3. M. I. Jyrkama, J. F. Sykes. The impact of climate change on spatially varying groundwater recharge in the Grand River watershed (Ontario). J. Hydrol., **338**, 237-250 (2007)
4. K. Srisuk, T. Nettasana. Climate change and groundwater resources in Thailand. *J.* Gw. Sc. and E., **5**, 67-75 (2016)
5. F. Bouraoui, G. Vachaud, L. Z. X. Li, H. Le Treut, T. Chen. Evaluation of the impact of climate changes on water storage and groundwater recharge at the watershed scale. Clim. Dyn., **15**, 153-161 (1999)
6. S. T. Woldeamlak, O. Batelaan, F. De Smedt. Effects of climate change on the groundwater system in the Grote-Nete catchment, Belgium. Hydrogeol. J., **15**, 891-901 (2007)
7. T. E. Croley, C. L. Luukkonen. Potential effects of climate change on ground water in Lansing, Michigan. JAWRA, **39**, 149-163 (2003)
8. U. Seeboonruang. Impact assessment of climate change on groundwater and vulnerability to drought of areas in Eastern Thailand. Envir. Earth Sc., **75**, 42 (2016)
9. U. Seeboonruang. An empirical decomposition of deep groundwater time series and possible link to climate variability. GNEST J., **16**, 87-103 (2014)
10. Thailand Department of Groundwater Resources *Groundwater situation, Thailand 2015 Quarter 1,* Ministry of Natural Resources and Environment, Bangkok. (2015) (in Thai)
11. A. W. Harbaugh, E. R. Banta, M. C. Hill, M. G. McDonald. *MODFLOW-2000, The U. S. Geological Survey Modular Ground-Water Model-User Guide to Modularization Concepts and the Ground-Water Flow Process,* Open-file Report. USGS **92**, 134 (2000)
12. J. Wattanasetpong, P. Charoenvaravut, W. Laosinwattana. *Downscaling climate models in Thailand by artificial neural network method.* Thesis of civil engineering, King Mongkut's Institute of Technology Ladkrabang, Bangkok, Thailand. (2015) (in Thai)
13. P. Ruangrassamee, A. Khamkong, P. Chuenchum. Assessment of precipitation simulations from CMIP5 climate models in Thailand. ICWRE3, 591-599 (2015)

Cluster and regression analysis for predicting salinity in groundwater

Phiraphat Aphiphan[1,*], *Uma* Seeboonruang [2], and *Somyot* Kaitwanidvilai[3]

[1]Master degree student, Environmental and energy engineering for sustainability, Department of Civil Engineering, Faculty of Engineering, King Mongkut's Institute of Technology Ladkrabang, Thailand, 10520.

[2,3]Assoc.Prof.Dr, Faculty of Engineering, King Mongkut's Institute of Technology Ladkrabang, Thailand, 10520.

Abstract. Groundwater salinity is a major problem particularly in the northeastern region of Thailand. Saline groundwater can cause widespread saline soil problem resulting in reducing agricultural productivity as in the Lower Nam Kam River Basin. In order to better manage the salinity problem, it is important to be able to predict the groundwater salinity. The objective of this research was to create a cluster-regression model for predicting the groundwater salinity. The indicator of groundwater salinity in this study was electrical conductivity because it was simple to measure in field. Ninety-eight parameters were measured including precipitation, surface water levels, groundwater levels and electrical conductivity. In this study, the highest groundwater salinity at 3 wells was predicted using the combined cluster and multiple linear regression analysis. Cross correlation and cluster analysis were applied in order to reduce the number of parameters to effectively predict the quality. After the parameter selection, multiple linear regression was applied and the modeling results obtained were R^2 of 0.888, 0.918, and 0.692, respectively. This linear regression model technique can be applied elsewhere in the similar situation.

1 Introduction

Groundwater is an important natural resource for ecosystem, organism and human living. The groundwater salinity problem was important to study at the risk areas. Groundwater quality monitoring information was required in the interested area. However, continuous water quality measurement spent a lot of time and investment.

The salinity of groundwater may be derived from natural salts and human actions. The important issue of this research is groundwater salinity due to rapid population, rapid industrialization growth and the use of enormous chemicals in agriculture because of poor management. The flow of water in groundwater will increase the saline and the evaporation process will bring the saline to surface cause salinity problems. In the long term, salinity problem will severely affect the environment such as salty soil, water salinity and a shortage of fresh water. Electrical conductivity was a useful indicator of saline in this research because it was easy to measure in fieldwork. A few numbers of literatures were used regression equation for groundwater quality prediction data in different areas. Modeling used time series techniques was an alternative tool that can be used to determine the relationship between water quality and index variables for unknown parameters [1,6,8,9]. So predicting the fluctuation of groundwater quality is important for proper water The title is set in bold 16-point Arial, justified. The first letter management and land use [5,9]. Ground water samples from different area have been analyzed for correlation between electrical conductivity and parameters. An attempt has been made to develop linear regression for predicting the concentration of water quality constituents having significant correlation coefficients with electrical conductivity. Water quality indexes prediction was applied by multiple linear regression modeling [2,7]. Cluster analysis was used to assess the water quality and it useful to manage, control pollution and protect water quality [12].

This research is necessary to predict the salinity of groundwater for better planning or proper management about salinity area in the future because it is the one of the main causes of salinity problems. So the purpose of this study was to create a cluster-regression model for predicting the groundwater salinity and finds the relationships of variables.

2 Methodology

2.1 Data Preparation

First step in this research, the lower basin data was focused area for this study. It was sub-basin of the Mekong River basin located in Nakhon Phanom province in the northeastern of Thailand [11]. Fig. 1 shows well locations. In this area, groundwater faces salinity problem because this area may have salt pits or

* Corresponding author: jack_ozaka@hotmail.com

rock salt which dissolves into high conductivity value. The data was collected into monthly from December 2010 to October 2012. The data of the groundwater were from 27 wells, each of well has water level, pH and EC data.There were data of 12 water gates and rainfall data from 5 areas. Data obtained from previous research was incomplete. It was necessary to convert data with average data, remove the abnormal values and interpolation data (piecewise linear interpolation method) by the software, PAST (Paleontological Statistics Software Package for Education and Data Analysis). Fig. 2 shows incomplete data that we have presented by plotting graph and it shows the missing data and this is not suitable for next process. Fig. 3 shows the complete graph and this graph has no missing data because the missing data has fulfilled by interpolation and outlier to the data. This data is a sample of the conductivity data of BLK_Y, BDD_P, BDD, BDD_C and BDY_S wells. We had the descriptive statistics of incomplete and complete data, we have compared the incomplete data with the complete data to observe the variance and standard deviation of the data.

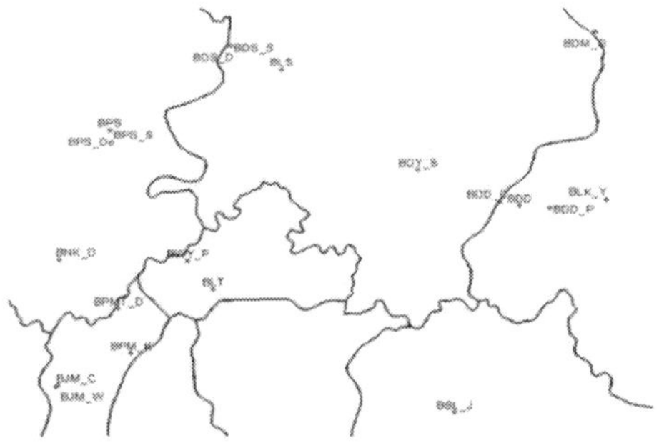

Fig. 1. Well locations

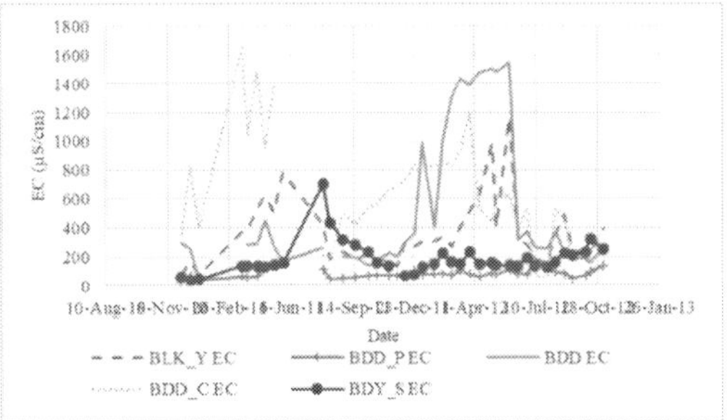

Fig. 2. Samples of incomplete data

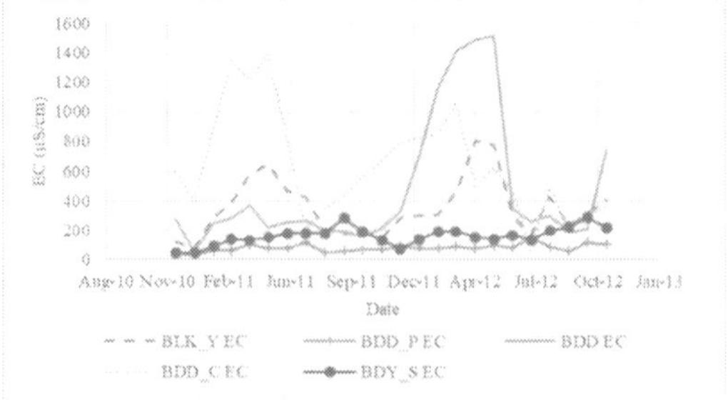

Fig. 3. Samples of complete data

2.2 Cross correlation analysis

After the data preparation step, the data has been set. The variables used in this research were 98 variables, which included conductivity, pH, groundwater level of well, precipitation and surface water level data. We would like to create 3 regression models. So we created three experiments. We used 3 highest electrical conductivity as the main variables in 3 experiments which in each experiment had 1 main variable be dependent variable and remaining 97 variables be independent variables. These dependent variables were Y_1, Y_2 and Y_3 and they were the electrical conductivity of BDS_S, BDS_D and BPS_Do wells, they had highest electrical conductivity were 27365.58 µS/cm, 134402.334 µS/cm, 95527.84 µS/cm, which their conductivity value higher than conductivity of seawater. The remaining 95 variables were defined as independent variables X_1 to X_{95}. In cross correlation step, we need to create 3 experiments by each experiment had 98 variables and we used 1 main variable (dependent variable) matching one by one with 98 independent variables. Cross correlation is the finding relationship between variable and variable of lag time. In general, Cross correlation is given by Eqn. (1)

$$r_k(x_t,y_t) = \frac{\sum_{t=k+1}^{n}(Y_t-\bar{Y})(X_{t-k}-\bar{X})}{\sqrt{\sum_{t=1}^{n}(X_t-\bar{X})^2\sum_{t=1}^{n}(Y_t-\bar{Y})^2}} \quad \text{and}$$

$$r_{-k}(x_t,y_t) = \frac{\sum_{t=k+1}^{n}(Y_{t-k}-\bar{Y})(X_t-\bar{X})}{\sqrt{\sum_{t=1}^{n}(X_t-\bar{X})^2\sum_{t=1}^{n}(Y_t-\bar{Y})^2}} \quad (1)$$

When r is coefficient cross correlation, y is dependent variable, x is independent variable. Cross correlation is useful for finding the relationships of two variables and it can explain about lag time.

2.3 Cluster analysis

In this step, the data from cross correlation analysis will be classified. If any variable had no relationship from cross correlation, it will be cut off. Correlated variables were grouped by cluster analysis by the software, SPSS (Statistical Package for the Social Sciences). Representative data selection by Selected the most cross correlated value data of each group to represent the group. This step is important. This step was the process of grouping the variables and finding the representatives of the groups to simplify the regression analysis. The variables in this research were quite numerous. It was necessary to find the representatives of them in order to make predictions and the grouping will be in the dendrogram. Example of dendrogram is shown in Fig. 4.

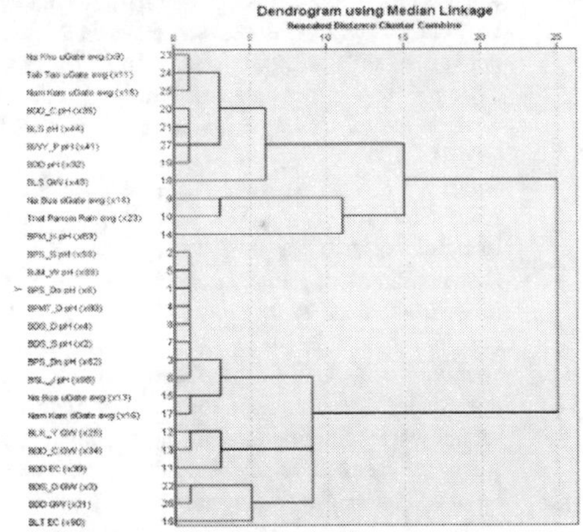

Fig. 4. Example Dendrogram of cluster Y3

2.4 Multiple linear regression

In final step, after grouping each group's representative data was used for create multiple linear regression models. A multiple linear regression model is applied to study any linear relationship between one dependent variable and several of independent variables. The multiple regression model is given by Eqn. (2)

$$Y = b_0 + b_1 X_1 + b_2 X_2 + \dots + b_m X_m \qquad (2)$$

When Y is the dependent variable, b_o is the intercept, b_i is the regression coefficients or slope in linear regression. Regression model is the equation for find the relationship of variables and predicting interested variable. This model is very useful in statistics and time series. The regression model can be used to predict the accuracy of forecasting.

3 Result and Discussion

3.1 Measured and interpolated data comparison

According to research, In the process of preparing the complete data. It had made the missing data range more complete and result in lower standard deviation and variance. This made the data suitable for data entry for cross correlation analysis.

3.2 Parameter elimination by cross correlation

In the cross correlation process, we had 3 main variables which in each main variable had matching with other variables and it shows in Table 1.

Table 1. The number of cross correlation between dependent variable with independent variable

Dependent variable	Number of independent variables
Y_1	69
Y_2	71
Y_3	27

3.3 Grouping variables by cluster analysis

In the clustering process, we had 3 cluster models by importing cross correlated data into grouping and in each cluster models were many groups in itself. So we used the variances and standard deviations to decide how many groups in each models and the number of appropriate groups for each models is shown in Table 2.

Table 2. Group of cluster model

Cluster Model	groups
Y_1	8
Y_2	4
Y_3	10

3.4 Multiple linear regression

For the regression process, Table 3 shows the model of Y_1 Y_2 Y_3 had R^2 of 0.888, 0.918 and 0.692 respectively. The statistical significance of models is 95%, the regression models of Y_1 Y_2 Y_3 can be described by regression equations with R^2 of 88.8%, 91.8%, 69.2%, respectively, indicated the regression models appropriated for predicting the dependent variables, which were electrical conductivity at 3 wells. Regression model of Y_1, Y_2 and Y_3 were

$$Y_1 = 26489.345 + 20522.46X_{76} - 21411.258X_{77}$$

$$Y_2 = 88786.137 + 0.871X_{60} - 36632.942X_3$$

$$Y_3 = -199546.303 + 33901.825X_{62} + 25275.749X_{35} - 15236.37X_{83}$$

Fig. 5 shows an example of graph plotting between observed data (Y_{obs}) and computed data (Y_{fc}) for Y_1 model. This graph shows these lines were very similar and the mean absolute percent error of observed data (Y_{obs}) and computed data (Y_{fc}) for Y_1 is 6.01%. It can be explained that the computed data was close to the observed data. It possible to explain the computed data, it can use to be a substitute for observed data.

Table 3. Models summary

Model	R	R Square	Adjusted R Square	Std. Error of the Estimate
Y_1	.942	.888	.877	2010.42372
Y_2	.958	.918	.910	11245.21347
Y_3	.832	.692	.646	12133.25304

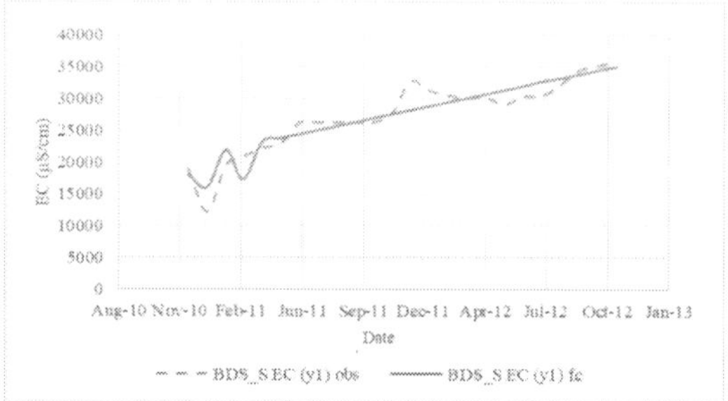

Fig. 5. Example of graph plotting between observed data (Yobs) and computed data (Yfc)

4 Conclusion

Groundwater is very important natural resource. Groundwater salinity is a huge problem for agriculture and environment. Predicting groundwater salinity is vital for appropriate planning and management. In this research, electrical conductivity at 3 observation wells was predicted by using the new cluster – regression models. Cross correlation and cluster analysis techniques were applied in order to eliminate unnecessary variables for prediction. Then, multiple regression equations were formed for the 3 locations. The models can predict the measured groundwater salinity very well. This model can be applied elsewhere to predict other variables.

References

1. Seeboonruang U., A Multiple Regression Analysis for Predicting Salinity in Shallow Groundwater. King Mongkut's Institute of Technology Ladkrabang. *IMEC 2017*.
2. Joarder M.A.M., Raihan F., Alam J.B. and Hasanuzzaman S., Regression analysis of ground water quality data of Sunamganj District, Bangladesh, *Int. J. Environ. Res. Vol* **2**(3), pp. 291-296 (2008).
3. Aflatooni M. and Mardaneh M., Time series analysis of ground water table fluctuations due to temperature and rainfall change in Shiraz plain. *International Journal of Water Resources and Environmental Engineering Vol.* **3**(9), pp. 176-188 (2011).
4. Hajigholizadeh M. and Assefa M., Melesse. Assortment and spatiotemporal analysis of surface water quality using cluster and discriminant analyses. *Catena 151* p.247–258 (2017).
5. Anderson M.P. and Woessner W.W., Applied Groundwater Modeling.Academic Press, San Diego 1992.
6. Maiti S. and Tiwari R.K., A comparative study of artificial neural networks, Bayesian neural networks and adaptive neuro-fuzzy inference system in groundwater level prediction. *Environmental Earth Sciences Vol* **71** pp.3147 (2014).
7. Agarwal M. and Agarwal A., Linear Regression And Correlation Analysis Of Water Quality Parameters: A Case Study Of River Kosi at District Rampur, India. *International Journal of Innovative Research in Science, Engineering and Technology Vol.* **2**, 2013.
8. Seeboonruang U., An application of time-lag regression technique for assessment of groundwater fluctuations in a regulated river basin: a case study in Northeastern Thailand. *Environmental Earth Sciences Vol* **73** pp 6511–6523 (2015).
9. Jain C.K. and Sharma M.K., Relationship among Water Quality Parameters of Groundwater of Jammu District. *Pollution Research,* **16** (4), 241-246 (1997).
10. Hayashi M., *Temperature-Electrical Conductivity relation of water for environmental monitoring and geophysical data.* University of Calgary (2003).
11. Thailand Mineral Resource Department (TMRD). *The study of salt rock in the Lower Namkam Basin Irrigation Project-Nakhon Panom Province.* Final Report, Thailand (1998).

Causes of construction delay from contractors and suppliers in Thailand's oil and gas platform projects

Pittayaporn Gomarn[1,*], and *Jakrapong* Pongpeng[1]

[1]Civil Engineering Department, Faculty of Engineering, King Mongkut's Institute of Technology Ladkrabang, Bangkok, Thailand

Abstract. Construction project delays caused by contractors and suppliers are the top problems in Thailand's oil and gas industries. Recognizing this importance and their relationships between factors can help reduce the risk of delays in construction projects. Therefore, this study set out to confirm factors and inspect relationships between delay factors of contractors and suppliers. A 16 item questionnaire survey was distributed to 134 managers, engineers, and supervisors in oil and gas platform construction projects in Thailand. A confirmatory factor analysis (CFA) was performed by the use of Amos Version 20 software program. The analysis results showed that delays caused by the contractors and suppliers had high relationships, due to high regression weighs. The delays caused by the contractors included seven factors which included poor site management and supervision (17%), lack of safety rules and regulations (16%), poor communication and coordination with others (15%), poor procurement system management (15%), defective components and mistakes during construction (14%), supplier payments lateness (13%), and poor planning and scheduling (10%). The delays caused by the suppliers included six factors which included the supply of unqualified and unskilled personnel (22%), supply of low efficiency equipment (20%), late delivery of materials and equipment (20%), supply of low quality materials (16%), late supply of workers (16%), and price escalation (5%).

1 Introduction

Over the years, construction project delays cast a dark shadow over many construction projects, with the most important being caused by the stakeholders. Confirmation of this has come from Fallahnejad, which reported that in many countries (including Iran), delays caused by contractors and suppliers are in the top 10 reasons for delay in the oil and gas industry [1].

Ruqaishi and Bashir [2] also indicated that poor site management and contractor supervision (such as ineffective planning and scheduling) are the main causes of the oil and gas construction projects delays in Oman. In Malaysia, the most important causes are reported to also be planning and scheduling, and contractor site management [3].

In Jordan construction projects, evidence shows that contractor financial difficulties are common causes of project delays [4]. Assaf and Al-Hejji [5] conducted a site survey of 23 contractors in Saudi Arabia to determine the causes of delays, and found that 76% of the most common causes come from contractor difficulties in financing, poor site management, poor communication, poor planning and scheduling, and improper construction methods.

This is supported by Aziz [6], which surveyed post-revolution construction project delays in private, public, and local general construction firms. Results showed that significant contributing factors include ineffective contractor project planning and scheduling (83.9%), poor site management and supervision (83.8%), poor site financial control (82.3%), and reworks due to errors (82.1%). Another survey result from Jordan indicated that construction projects delays were often caused by financing and payments [7].

Doloi and others [8] studied 45 key factors affecting project delays in the Indian construction industry. Results indicated that the most critical factors were lack of commitment (11.6%), which included a lack of safety measures and the use of improper or obsolete construction methods. Second was inefficient site management (10.9%), which included poor site management and supervision. Third, was improper planning (10.8%), which included lateness in materials procurement), and finally, lack of communication (8.6%).

In Thailand, construction delays often mean time and cost overruns. Ogunlana and Promkuntong [9] explored delays of high-rise building projects in Bangkok, and found that most delays could be tracked back to problems caused by contractors and suppliers. The main reasons stated for these delays by contractors were materials management problems (75%), coordination deficiencies (67%), and planning and scheduling problems (58%). Delays by suppliers were related to late delivery (41%), price escalation (16%), and supply of low quality materials (8%).

* Corresponding author: gomarnp@gmail.com

Given both the financial importance caused by delays and the fact that project delays can contribute to site safety issues. However, the oil and gas platform construction projects and general construction projects have a different contractual project delivery system. Therefore, researchers wanted to survey and study delay causes and factors in Thailand's oil and gas platform construction projects caused by contractors and suppliers. The results of this study can be applied in the risk assessment analysis, and help with the prevention and reduction of project delays.

2 Research methods

2.1 The conceptual framework

From a review of the literature, Table 1 is used to present the primary reasons for construction project delays by contractors and suppliers.

Table 1. Cause of project delay from contractor and supplier factors.

	[1]*	[2]*	[3]*	[4]*	[5]*	[6]*	[7]*	[8]*	[9]*
C1**		x	x	x	x	x	x		x
C2**	x		x		x	x	x	x	x
C3**	x				x	x		x	
C4**		x	x	x	x	x	x	x	
C5**	x			x		x		x	
C6**	x	x	x		x	x	x	x	
C7**				x	x	x		x	
S1**								x	x
S2**		x		x	x	x		x	x
S3**			x			x	x		x
S4**		x		x	x	x			
S5**			x		x	x	x		
S6**	x	x			x	x		x	

*References and **contractor (C) and supplier (S) factors

2.2 Panel of experts

This research was discussed with three experts to define and finalize factors from the literature review, from which, Table 1 was created as a summary of the related factors. Additionally, the observed factors were finalized into two classification groups related to contractors (C) and suppliers (S).

2.3 The research instrument

To analyse the observed factors and inspect the relationships between the latent factors, a 5-level, Likert type agreement survey was used for data collection.

Part 1 of the questionnaire contained items related to the individual's position level, experience and project value. Part 2 of the survey asked the individual to evaluate the importance level of the cause of project delay from contractors and suppliers on a Likert type scale from 1 to 5, with "1" indicating the importance level of the item as very low, whereas a "5" rated the importance level of the item as very high [10].

2.4 Content and construct validity

2.4.1 Content validity

The questionnaire was discussed with three experts (the experts are working as suppliers and contractors in the offshore oil and gas industry with experiences more than fifteen years) to ensure content validity.

2.4.2 Construct validity

The Spearman Rank Correlation was used to ensure construct validity of the questionnaire. The correlation coefficient is a measure of the relationship between factors from a -1 to a +1 [11]. In this study, the observed factors were determined to be correlated.

Cronbach's alpha was also used to test the internal consistency (reliability) of the questionnaire. The criteria value of Cronbach's alpha is suggested to be above 0.7 [12]. In this study, the Cronbach's alpha value of 0.866 is determined.

3 Results

3.1 Respondent's characteristics

A total of 134 questionnaires were returned from the respondents having experienced in oil and gas platform construction projects in Gulf of Thailand. Of these 4.5% were managers, 67.2% were engineers, and 28.4% were supervisors (Table 2). Experience level responses indicated that 41.8% had 5-10 years' experience. Also, 36.6% worked in enterprises valued at over 1 billion Thai baht ($31.4 million).

3.2 Confirmatory factor analysis (CFA)

A CFA was used to analyze and confirm the sets of observed factors and relationships between the latent factors by using the IBM® SPSS® Amos™ 20 software program. Fit of the CFA model included the indices Chi-squares's p-value, relative chi-square ($\chi 2 = df$), goodness-of-fit index (GFI), and the root mean square error of approximation (RMSEA). [13, 14] The criteria and index values of the goodness-of-fit (GOF) are presented in Table 3.

Table 2. Demographic characteristic

Position level	**Quantity**	**Percent**
Manager level	6	4.5%
Engineer level	90	67.2%
Supervisor level	38	28.4%
Experience	**Quantity**	**Percent**
Less than 5 years	28	20.9%
5–10 years	56	41.8%
11–15 years	30	22.4%
More than 15 years	20	14.9%
Project value	**Quantity**	**Percent**
Less than 100 million THB	27	20.1%
100 – 500 million THB	17	12.7%
501 – 1000 million THB	41	30.6%
More than 1 billion THB	49	36.6%

3.3 Data analysis

Table 3 shows the results of the goodness-of-fit (GOF) criteria and recommended index values.

Table 3. Goodness-of-fit criteria and indices.

GOF recommended index values	**CFA***		
	Contractor	**Supplier**	**Relationship between contractor and supplier**
$0.05 < p \leq 1.00$	0.055	0.050	0.077
$0 \leq \chi^2/df \leq 2$	1.668	1.936	1.280
$0.09 \leq GFI \leq 1.00$	0.952	0.964	0.930
$0 \leq RMSEA \leq 0.08$	0.071	0.084	0.046

*Remarks: The delay factor of price escalation (S1) was removed from the supplier factors, because S1 does not significantly results delays from the supplier.

Fig.1 shows the CFA analysis results of the factors and their relationships between contractor and supplier delay causes. The results indicated that the delay causes from contractors has a high correlation with delay causes from suppliers, as the regression weight was 0.78. Information suggests that this is due to the majority of the suppliers and subcontractors have to work under the main contractors due to financial capability and production capacity of the project's requirements. Therefore, any supplier mistakes and delays directly affect contractors. These include items such as the supply of low quality materials, low efficiency of equipment, and unqualified and unskilled personnel to the contractor. Meanwhile, any delays caused from contractors affect suppliers also. These include poor supplier communication and coordination, and contractor late payments to suppliers.

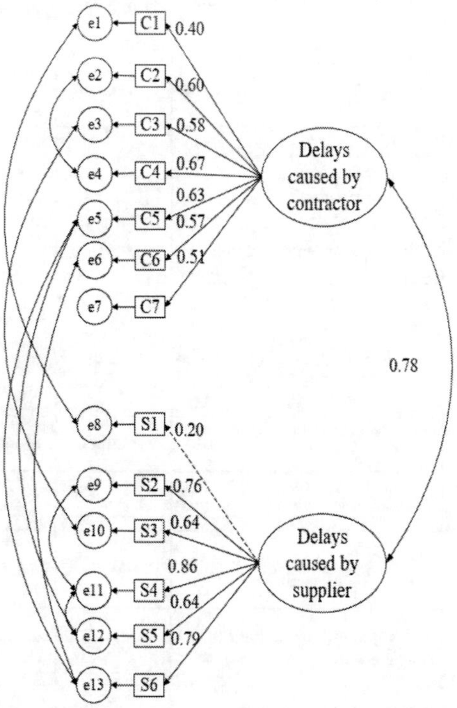

Fig. 1. CFA results.

Table 4 shows the factor loadings for delays caused by construction project contractor, with poor site management and supervision being the leading cause for delay.

Table 5 shows the factor loadings for delays caused by the construction project supplier, with the supply of unqualified and unskilled personnel being the number one reason for delay.

Normally, in construction projects, a key element is management's knowledge and experience. If however key personnel within the organization lack the required project management knowledge, experience, and supervisory skills, projects will have a much higher probability of experience a delay and possibility even a project's failure. Also, if the suppliers assign and supply unqualified and unskilled personnel to the project organization and team, those projects also have a high chance of delay and failure also.

Table 4. Factor loading of delays causes by the contractor.

Project delay causes	Regression weights	Factors loading
C1: Poor planning and scheduling	0.40	10%
C2: Poor communication and coordination with others	0.60	15%
C3: Poor procurement system management	0.58	15%
C4: Poor site management and supervision	0.67	17%
C5: Lack of safety rules and regulations	0.63	16%
C6: Defective material and mistakes during construction	0.57	14%
C7: Late payments to suppliers	0.51	13%

Table 5. Factor loading of delays causes by the supplier.

Project delay causes	Regression weights	Factors loading
S1: Price escalation	0.20	5%
S2: Late delivery of materials and equipment	0.76	20%
S3: Supply of low quality materials	0.64	16%
S4 :Supply of unqualified and unskilled personnel	0.86	22%
S5: Late supply of workers	0.64	16%
S6: Supply of low efficiency equipment	0.79	20%

Remarks: S1 does not significantly indicate delays from the supplier (p-value < 0.05).

4 Conclusion

From the use and analysis of a CFA on contractor and supplier delay factors, it was determined that there is a significant relationship between the sets of factors.

The major reasons for delays from contractors in construction projects are poor site management and supervision, lack of safety rules and regulations, poor communication and coordination with others, poor procurement system management, defective components and mistakes during construction, late payments to suppliers, and poor planning and scheduling, respectively.

The delays caused by suppliers in construction projects are the supply of unqualified and unskilled personnel, supply of low efficiency equipment, late delivery of materials and equipment, supply of materials of low quality, late supply of workers, respectively. In addition, price escalation had a low impact on project delay from supplier's causes.

Finally, this analysis points to the key factors that have been confirmed and the relationship between the delays caused by the contractors and the suppliers. These results may help to reduce and prevent delay risk in ongoing and future oil and gas platform construction projects. Furthermore, the project controller should take a stronger role in evaluating contractors and suppliers in the project's prequalification phase and additionally monitor projects during the execution phase.

The results of this study should be defined with a limitations as the conceptual framework was developed based on the experts experience and an international literature review. Therefore, further inspection of the project types is required.

References

1. M.H. Fallahnejad, Int. J. Proj. Manag. **31**(1), 136–146 (2013)
2. M. Ruqaishi, H.A. Bashir, J. Manag. Eng. **31**(3), 1–8 (2015)
3. M. Sambasivan, Y.W. Soon, Int. J. Proj. Manag. **25** (5), 517–526 (2007)
4. G. Sweis, R. Sweis, A. Abu Hammad, A. Shboul, Int. J. Proj. Manag. **26** (6), 665–674 (2008)
5. S.A. Assaf, S. Al-Hejji, Int. J. Proj. Manag. **24** (4), 349–357 (2006)
6. B.R. Aziz, Alexandria Eng. J. **52**, 387–406 (2013)
7. A.M. Odeh, H.T. Battaineh, Int. J. Proj. Manag. **20** (1), 67–73 (2002)
8. H. Doloi, A. Sawhney, S. Rentala, Int. J. Proj. Manag. **30** (4), 479–489 (2012)
9. S.O. Ogunlana, K. Promkuntong, V. Jearkjirm, Int. J. Proj. Manag. **14** (1), 37–45 (1996)
10. R. Likert, *Social psychology: Experimentation, theory, research* (Intext, Scranton, PA, 1972)
11. SPSS Training Department, *Survey research using SPSS.* (SPSS, Inc., Chicago, 1998)
12. J. Nunnally, *Psychometric theory* (McGraw-Hill. New York, 1967)
13. J.L. Arbuckle, *IBM SPSS amoss 20 user's guide* (IBM Corporation. U.S.A., 2011)
14. K. Schermelleh-Engell, H. Moosbrugger, Meths. Psych. Resea. **8** (2), 23-74 (2003)

Construct and control of feet gait mechanisms for walking training

Warawut Suwalai*

Mechatronics Engineering Department, Faculty of Technical Education, Rajamangala University of Technology Thanyaburi, Pathumthani, Thailand

Abstract. Gait training or gait rehabilitation is one of the major physiotherapy for stroke patients. Evidently, the robot-assisted gait training, as one part of medical technology innovation breakthrough, has important role in the rehabilitation process. The robot effectively improves treatment outcomes, fast recovery time and better convenience from traditional treatment. Typically, stroke patients are trained to walk on treadmill while a physiotherapist carefully supports and re-correct the gait pattern of the patient. For repetitive and long-time rehabilitation, it is so difficult that the physiotherapist can maintain the quality of treatment consistently. To solve such difficulties, the robotic platform is proposed for automatic guiding the gait pattern for ankle rehabilitation. The robot consists of left and right sides. Each of them is actuated by two linear and one rotational actuator. PID algorithm is implemented for position control of each joint. The experimental results show the tracking error in non-loaded and loaded cases.

1 Introduction

Nowadays, people suffers from the brain injury and nervous disorder. There exists some study of Thai Epidemiological Stroke [1] that discovers 1.88 percentage of the over 45-age people become stroked. Also, the number of strokes significantly increase every year.

Normally, stroke patients have problem with muscle weakness and no ability to control their legs. To recover the muscle strength and nervous system, they are properly trained by physiotherapy program until they can move by themselves or recover motor function. Typically, stroke patients are guided to walk on treadmill while a physiotherapist carefully supports and re-correct the gait pattern of the patient. However, in long-time and repetitive procedures such in physiotherapy, the quality of training is quite difficult to be maintained by human ability. To improve the quality of the physiotherapy, the robot technology is introduced to efficiently assist these procedures.

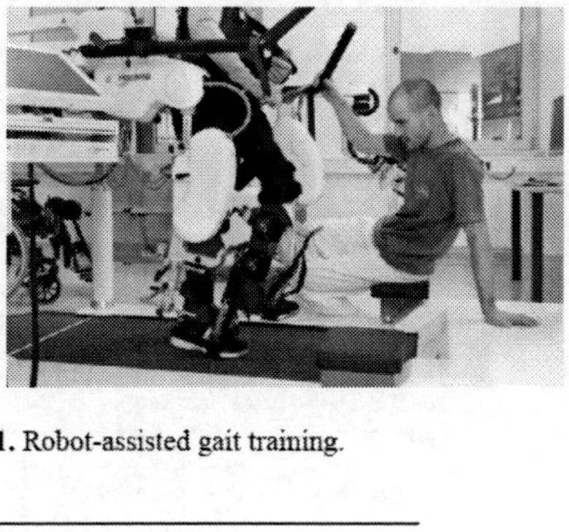

Fig. 1. Robot-assisted gait training.

Jungwon et al.[2] proposed a 6-DOF gait rehabilitation robot. The robot has footplate on a sliding device for each foot. This allows creation of the different pattern for gait training. Also, during gait training, that used the virtual environments(VEs) to guideline patient for walking procedure. Hesse et al.[3] designed a novel gait robot(G-EO-System) for repetitive practice of stroke patients based on the end-effector principle. This concept can modify the paths of the foot such as floor walking or stair climbing up and down.

In this research, the rehabilitation machine is developed for automatically assisting the stroke rehabilitation. Especially, the robot focuses on rehabilitation of the feet positions and ankle joints.

2 Implementation

The methodology for research compose of three topics. There are hardware design, communication system design, and programing design. The details of each topics present in this section.

2.1. Hardware design

2.1.1 Design pattern

In this research, the feet gait mechanism comprises two sides. There are left and right feet. This mechanism has the coordinate system, that use to the design for comprehensive of walking motion. The reference frame of this mechanism has x and y direction on the plane and z direction perpendicular to x-y plane, respectively. Each

* Corresponding author: warawut_s@rmutt.ac.th

side compose the two linear motion on x-z planar and one rotational motion on y-axis, that can show a three-dimensional model of feet gait mechanism as figure 1.

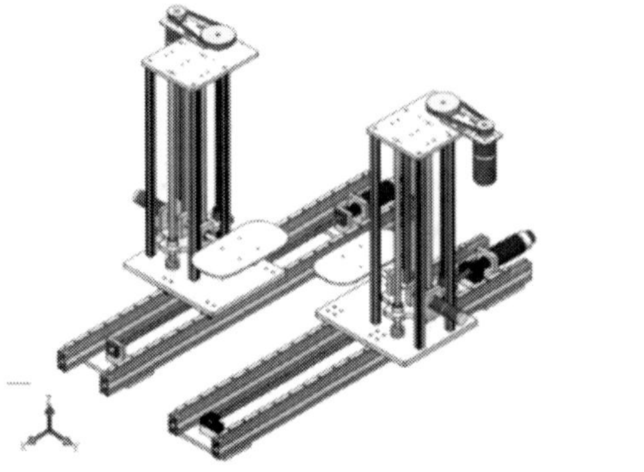

Fig. 2. 3D model and the reference frame of Feet Gait Mechanisms for Walking Training.

2.1.2 Simulation of torque and power consumption

Before the machine is built, the selection of mechanical parts and actuators to suit for application. It is very important. At present, we have a lot of software for designing and simulating. It is very convenient to use. Therefore, this research uses the SOLIDWORKS software to simulate torque and power consumption of actuators like a motor.

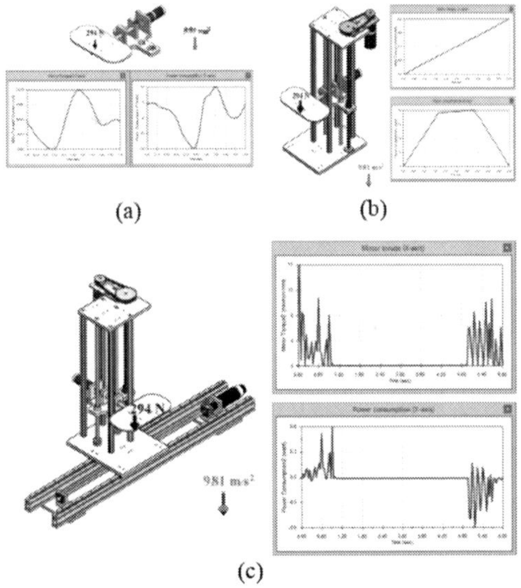

Fig. 3. Torque and power consumption results are simulated by software: rotational motion(a) linear motion(b), (c).

Firstly, we start simulation the rotational motion. Supposing situation, a weight of 30 kilograms act on the footrest. The distance of force from the rotating center is 0.035 meter. These two parameters including gravity force are set into the software. Then, the simulation results are represented by graphs. Each graph shows minimum and maximum values relative to time. The result of rotational motion simulation can be show as figure 3(a).

Secondly, we simulate the vertical linear motion. This motion include weight from rotational motion also. Total weight is sum of load and mechanical weight. The result of vertical linear motion simulation can be present as figure 3(b).

Lastly, we simulate the horizontal linear motion. This motion support total weight from rotational and vertical linear motion. The result of horizontal linear motion simulation can be show as figure 3(c). All parameters setting of two linear motion simulations are duplicated as rotational motion.

2.1.3 Physical model

All of feet gait mechanisms for walking training, that consists of two main parts. The first part is mechanism for movement left and right sides of foot. This mechanism force ankle position following the gait trajectory by electrical control. Mechanism are made up like 3D design that show as previous section. The second part is electrical control cabinet, that compose of electrical and electronics device such as microcontroller, motor driver, and the other electrical devices. All of actuators and sensors are connected to this cabinet. The physical model shown as figure 4.

Fig. 4. Hardware of feet gait mechanisms for walking training.

2.2 Communication system design

The operating and monitoring of machine need to using computer. The computer communicates with the controller for transfer data between each other by using serial communication. This approach used the serial communication converter because the logic level of two devices are not completable. The logic of computer is not transistor-transistor logic(TTL) level, but logic level of microcontroller is TTL. Communication system diagram is expressed as figure 5.

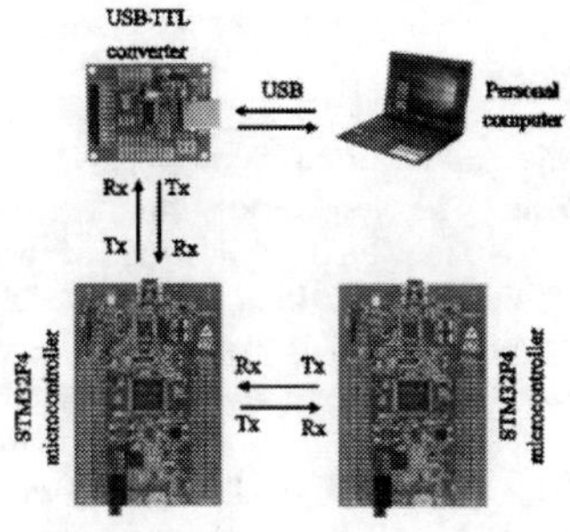

Fig. 5. Communication system connection.

2.3 Programming design

Software design for operation this machine, include three section program that consists of main control, user interface, and data loggers. The design procedure software preliminary used flowchart of program, that allows to guideline programming process. Flowchart of each program function expressed as figure 6.

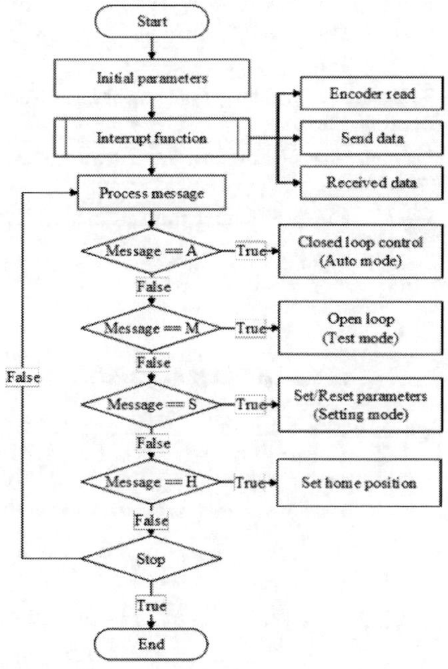

Fig. 6. Flowchart of main function in main processor.

3 Controller Design

For automatic control of any machine, the controller is very important for control it. The accuracy and precision of motion depend on designing the control algorithm. There are many control algorithms. In this research, The PID control algorithm is selected to control the position of each motion. The PID controller need to adjustable the gain of controller. These parameters should be suitable for system. This research consists of control the ankle joint and foot position. Therefore, every joint of the machine must be tune the gain of PID controller.

The PID controller have many methods for tuning the gain. One of methods is Ziegler-Nichols method, which is extensively used. This method is a simple and it is a trial and error tuning method based on continuous oscillations.

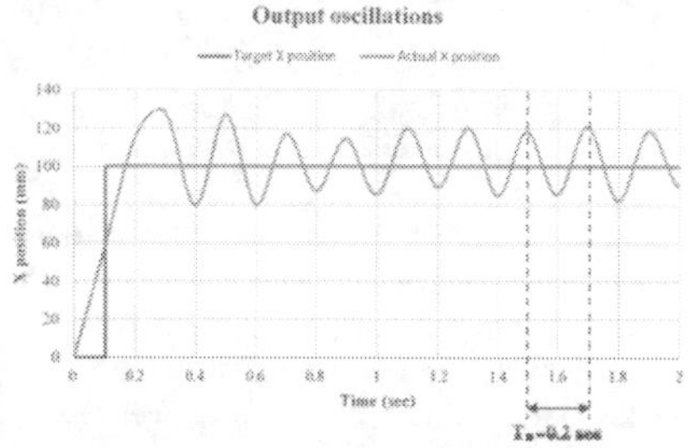

Fig. 7. Consistent oscillation and the oscillation period

For procedure to get the gain, substitution parameter which is received from figure 7 fill up into Ziegler-Nichols tuning rule table. The results are given as table 1. The classical PID controller was selected in this research.

Table 1. Gain values from Ziegler-Nichols tuning method.

Controller	k_c	τ_I	τ_D
P	75	-	-
PI	67.5	0.167	-
PD	120	-	0.025
classic PID	90	0.1	0.025

4 Experimental results

This section presents the experimental results of simulation of walking training. PID control algorithm is used in this task. The experiment used two scenarios, no load and weight, respectively. To investigate, that machine still perform in every suppose scenarios.

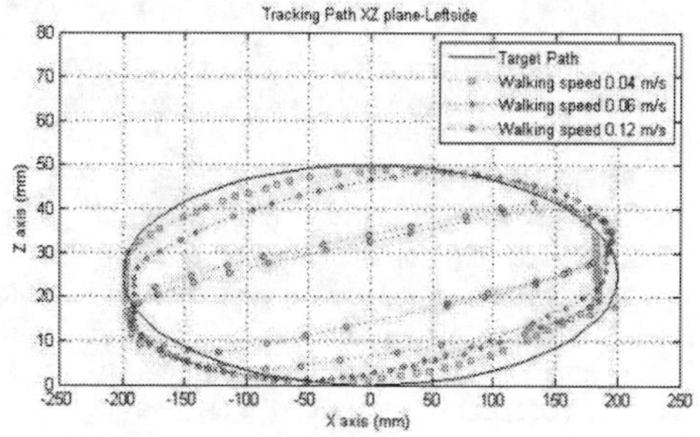

Fig. 8. Experimental results of without weight of left side foot position on x-z plane at different walking speed.

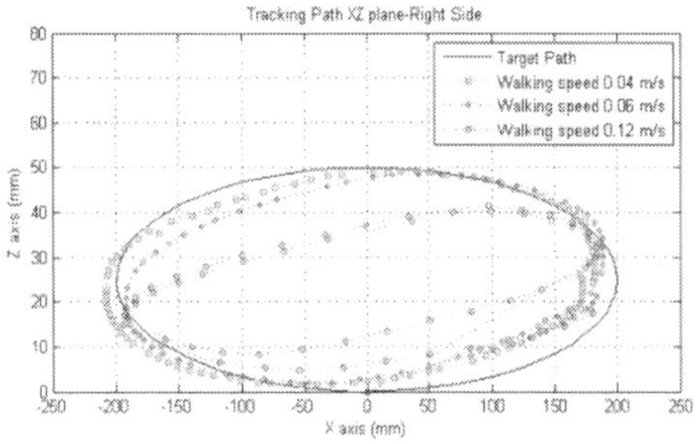

Fig. 9. Experimental results of without weight of right side foot position on x-z plane at different walking speed.

First test result as figure 8 and 9 that experiments are without weight by using walking speed 0.04 m/s up to 0.12 m/s. The results of tracking path at low speed are close to the path. Increasing the speed make some error from the path, but the positions of feet are still in the pattern.

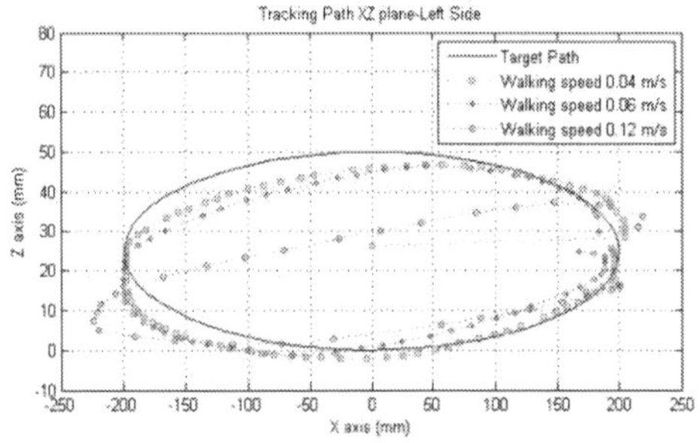

Fig. 10. Experimental results of weighted 30-kilograms of left side foot position on x-z plane at different walking speed.

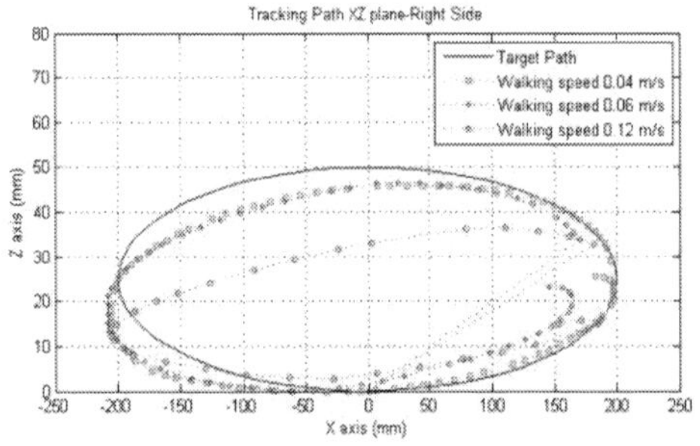

Fig. 11. Experimental results of weighted 30-kilograms of right side foot position on x-z plane at different walking speed.

Second Experimental result as figure 10 and 11 that experiments are with carry load 30-kilograms by using walking speed 0.04 m/s up to 0.12 m/s also. The results of tracking path at low speed are close to the without weight experimentation. Increasing the speed gets large error than without weight experimentation. Some positions of the feet are in some of path form.

5 Conclusions and recommendations

5.1 Conclusions

To summarize, all the experimental results are obtained in this research. The test results are satisfactory. The mechanism can movement following the example gait pattern. Although, the position of feet is slight accuracy from the pattern when increase the speed of walking but the movement form still resembles to simulation target path. In this paper, the experimental results show only the test of feet position because the control of ankle joints is developing phase.

5.2 Recommendations

To Improve the position accuracy, the PID controller gain should be fine adjustment or use higher performance controller such as LQR, LQG.

References

1. NC. Suwanwela, J. Storke., 1-7 (2014)
2. J. Yoon, B. Novandy, C.H. Yoon, K.J. Park., IEEE/ASME Trans. Mechatronics., **15 no. 2**, 201-215 (2010)
3. S. Hesse, A. Waldner, C. Tomelleri, J. Neuroeng. Rehabil., (2010)
4. J. Yoon, J. Ryu, G. Burdea, *HAPTICS 2003*, 374-381 (2003)
5. W. H. Chang, Y.H. Kim, J. Storke., **15 no. 3**, 174-181 (2013)

Parametric Study for Nonlinear Dynamic Behaviors using Integrated Cable Elongation Model of Cable Driven Parallel Robot

Sung-Hyun Choi and Kyoung-Su Park

Abstract—**Polymer is one of the most widely used cable materials in the CDPR system due to its light weight and low inertia. However, it is not easily to achieve high accuracy because of the complicated response of polymer cable while CDPR is operated. In our previous study, the integrated dynamics model of cable was derived with the visco-elastic model. With the integrated dynamics model of cable, the parametric studies were carried out for the cable length, the applied force and the tensile rate for various nonlinear cable elongation behaviors. At first, dynamic creep behavior was defined and investigated in CDPR system. Dynamic creep was saturated with one value as the cyclic load/unload is continuously carried out. The saturation number was inversely proportional to the cable length because of the residual stress. And as the tensile rate was increased, the dynamic creep was decreased linearly. In long-term recovery, the long-term recovery is finished more quickly for the case of a shorter cable.**

Index Terms—**Integrated non-linear dynamic model, dynamic creep, hysteresis, short- and long- term recovery, hardening effect**

I. Introduction

In recent years, cable-driven parallel robots (CDPRs) have become popular due to their advantages over conventional link parallel robots, such as high sensitivity to movement, ease of maintenance and large workspace [1, 2]. To achieve the high sensitivity and large workspace, they have used 3 to 8 cable members. The polymer cable is one of the most widely used materials in the CDPR system due to its light weight and inertia. However, the complex responses of polymer cables make it difficult to operate CDPR systems with a high precision [3, 4]. Polymer cables have several important characteristics: hardening effects, creep, short- and long-term recovery, hysteresis and so on. When combined, these characteristics have a significant impact on position control. Even though many studies have tried to compensate these characteristics in static condition, they were not effective [5, 6].

In this study, we conducted the experimental studies and simulation to analyze the elongation characteristics on polymer cable for various cable lengths and tensile rates. In the first and second section of this paper, we describe the experiment and integrated non-linear dynamic model. Next, we report the results of our parametric study for the design parameters which we mentioned above. Finally, the effect of parameters for the nonlinear cable elongation is evaluated for CDPR system.

S.H. Choi is with the Gachon University, 1342 Seongnamdaero, Sujeong-gu, Seongnam-si, Gyeonggi-do, 461-701, Korea (e-mail: csh5974@gc.gachon.ac.kr)

K.S. Park is with the Gachon University, 1342 Seongnamdaero, Sujeong-gu, Seongnam-si, Gyeonggi-do, 461-701, Korea (e-mail: pks6348@gachon.ac.kr)

II. Experimental Setup and Characteristics of the Polymer Cable Used in CDPR

A. Experimental setup

The parametric studies were based on tensile loading tests. We investigated the dynamic characteristics of a Dyneema polyethylene SK78 cable using a tensile tester. We applied forces of 100 N to the cables. As the end-effector pose is changed, the length of cable is changed. Thus, it is necessary to investigate the length effect of cable. Therefore, we used the two kinds of cable lengths (100 mm and 300 mm). And, constant tensile rates corresponding to CDPR speed are based on the 3 mm/min and 50 mm/min. Also, we measured a long-term recovery after rest times of 24, 48 and 72 hours

III. Dynamic Characteristics of Polymer Cable

The polymer cables used in CDPR have complex dynamic characteristics, including nonlinear elongation, hysteresis, creep, and short- and long-term recovery. Each characteristic interacts with each other while operating CDPRs. Hysteresis is the result of this interaction. In our previous research [7], the dynamics model was derived based on the visco-elastic model for time-dependent applied force not static force.

$$\varepsilon(t)=\left(\frac{F_f}{E_{1,c}A}\right)^{h(\varepsilon_{i-1})}-\frac{F_f}{E_{1,r}A} + \frac{\int_0^{t_{total}}F(t)dt}{t_{total}}\left(\frac{1}{E_{2,c}A}\left(1-e^{-t_{total}/t_c}\right)+\frac{1}{\eta_0 A}-\frac{1}{E_{2,r}A}\left(1-e^{-t_{total}/t_{r,1}}\right)\right) \quad (1)$$

$$\varepsilon(t)=\int_0^{t_{total}}\hat{\varepsilon}(t)dt\approx\hat{\varepsilon}t_{total}\,(for\ constant\ strain\ rate) \quad (2)$$

Each parameter was defined in the [7]. Equation (1) is the strain equation for constant strain rate from 0 to t_{total} of loading/unloading time. The first term of equation (1) is a nonlinear elongation. And the second term is the time dependent dynamic creep and short time recovery. However, it is difficult to analyze the non-linear dynamic behavior for time-dependent applied force. Therefore, we were carried out the cyclic test and numerical simulation for case of constant

https://doi.org/10.1051/matecconf/201819202010

strain rate. For the case, equation (1) can be solved by calculating the loading and unloading time. To numerically resolve this problem, Newton-Raphson method is applied based on the condition that equation (1) is equal to (2) for constant strain rate. After then, the dynamic strains (Eqs. (3) and (4)) are derived using the predicted loading and unloading time for loading and unloading cases as following:

$$\varepsilon_{i-loading} = \left(\frac{F_f}{E_1 A}\right)^{h(\varepsilon_{ncl})} - \frac{\int_0^{t_{total}} F(t)dt}{t_{total}} \left(\frac{1}{E_2 A}\left(1-e^{-t_{total}/\tau_s}\right) + \frac{1}{\eta_0 A}\right) \tag{3}$$

$$\varepsilon_{i-unloading} = \left(\frac{F_f}{E_1 A}\right)^{h(\varepsilon_{ncl})} - \frac{F_f}{E_{1,f} A} + \frac{\int_0^{t_{total}} F(t)dt}{t_{total}} \left(\frac{1}{E_2 A}\left(1-e^{-t_{total}/\tau_s}\right) + \frac{1}{\eta_0 A} - \frac{1}{E_{2,f} A}\left(1-e^{-t_{total}/\tau_{s,1}}\right)\right) - \frac{\Delta F}{E_{3,f} A}\left(1-e^{-t_{total}/\tau_{s,2}}\right) \tag{4}$$

The equations contain the non-linear elongation as mentioned above. The model is divided into two cases, loading and unloading. Figure 1 shows the qualitative dynamic behavior of the cable based on this dynamic model. Sections in Fig. 1 are divided into five parts. Section ① shows the behavior of static elongation and dynamic creep. In section ②, cable has been hardened because of a residual stress. So, the gradient becomes stiff. Section ③ shows the effect of time dependent-dynamic creep while progressing the loading cycle. Section ④ shows the unloading process. When the cable is unloaded, the short-term recovery and dynamic creep occurs simultaneously. And then, the cable remains unloading state. The long term recovery dominantly occurred like ⑤ for a long time. Because of elongation difference between the loading and unloading, the hysteresis is generated in cyclic loading/unloading process.

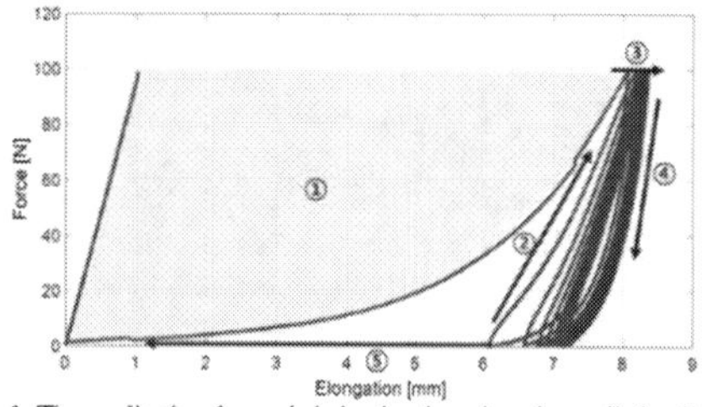

Fig. 1. The qualitative dynamic behaviors based on the cyclic loading and unloading process.

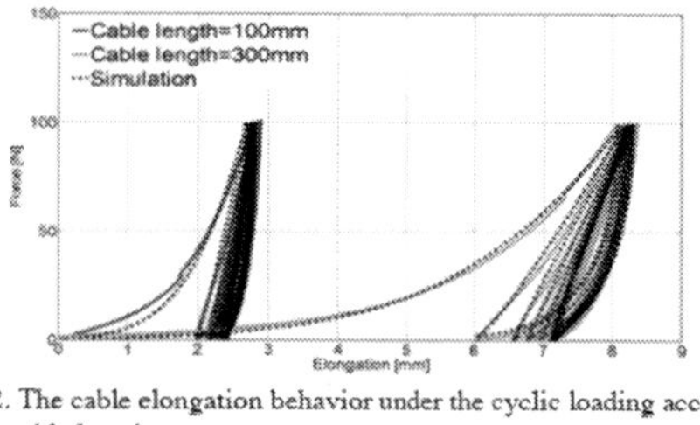

Fig. 2. The cable elongation behavior under the cyclic loading according to the cable length

IV. Parametric studies of dynamic behavior

A. *The investigation of dynamic creep*

The dynamic creep is dependent on the loading time. To investigate the dynamic creep, the simulation is performed based on the 10 times of cyclic loading/unloading process as shown in Fig. 2. The tensile rate is a 3 mm/min in this simulation. In Fig. 2, the each colored line means the experiment result for a different cable length. The black dot line means the simulation result. The analysis of dynamic creep, hardening, short-term recovery, and hysteresis are based on the simulation results as shown in Fig. 2. Figure 3 shows the dynamic creep according to the cycle number for a different cable length. In overall, the dynamic creep is saturated as the loading cycle is processed. And the saturation is delayed as cable length is longer. This is because the longer the cable length is, the larger amount of residual stress the cable molecules have. Therefore, it takes a long time for the long cable to be saturated by creep. Another factor that affects the dynamic creep is the tensile rate. Figure 4 shows the dynamic creep with the tensile rate. As the tensile rate increases, the dynamic creep decreases linearly. In the case of high strain rate, since the loading time to approaching at the same final applied force is shorter than that of low strain rate, the effect of time-dependent creep is smaller. As the result, less dynamic creep is occurred at high strain rate.

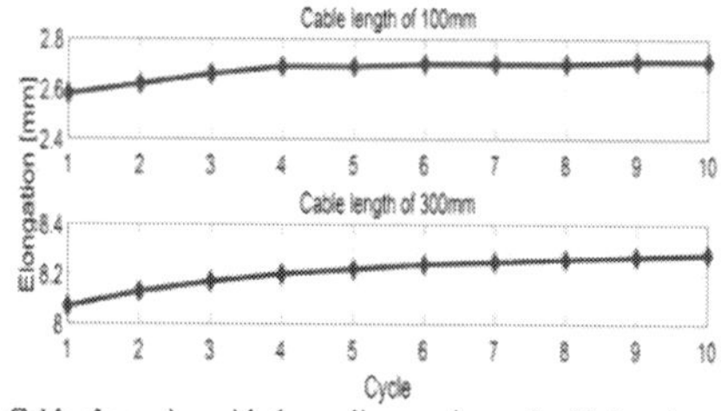

Fig. 3. Cable elongation with the cycling number and cable length

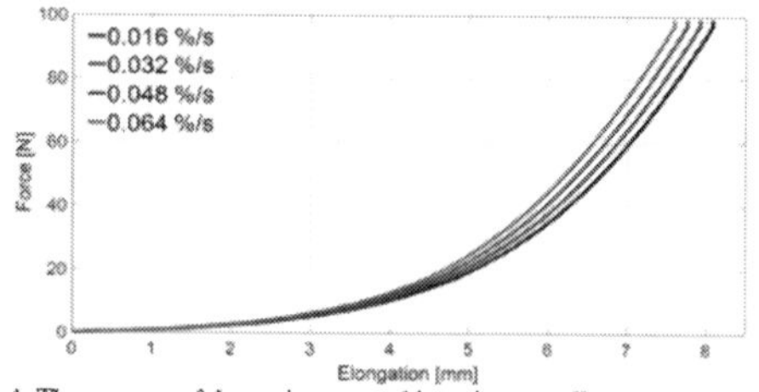

Fig. 4. The amount of dynamic creep with various tensile rates (length:300 mm)

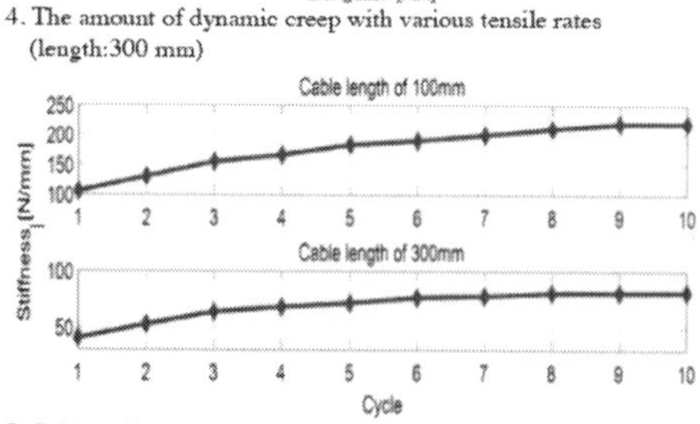

Fig. 5. Cable stiffness with the cycling number and cable length

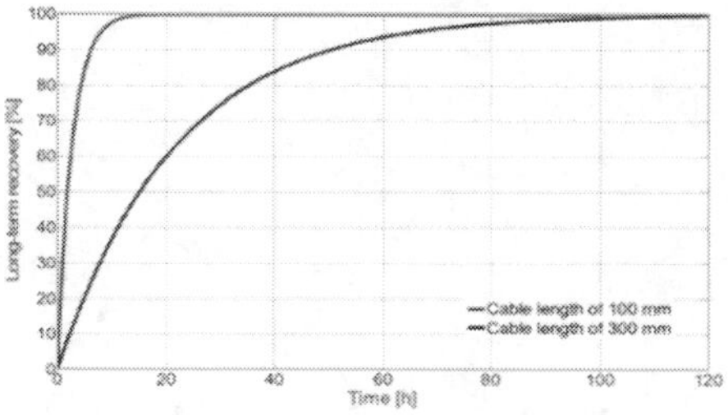

Fig. 6. The long-term recovery with various cable lengths

B. *The investigation of hardening effect*

The hardening is dependent on the strain of the cable. Figure 5 shows the stiffness with loading cycle and cable length. As the cycle is carried out, the stiffness increases and then saturates. This means the cable is no longer stretched because it is sufficiently hardened. In the CDPR system, it is very important to consider the hardening effect because the variation of the stiffness is about two times as large as that of the first operation. So, the warming up process might be effective in order to improve the control accuracy of CDPR.

C. *The investigation of short- and long-term recovery*

The short-term recovery occurs as soon as the unloading begins in the CDPR system. Table I summarizes the amount of short-term recovery for the cycle and cable length. The short-term recovery tends to decrease as the cycle is carried out, and finally saturated. Thus, the short term recovery has the opposite tendency to dynamic creep and hardening effect. And Fig. 6 shows a long- term recovery behavior for the time and cable length. At the cable length of 100 mm and 300mm, the cable length is completely recovered after 20, 120 hours. In cast that the cable length is long, the long-term recovery time increases exponentially. The result is very important to operate the CDPR system because it is continuously operated and rated sometimes.

D. *The investigation of hysteresis characteristics*

Hysteresis is result of dynamic behavior interactions as shown in Fig. 2. The loading part of hysteresis is caused by dynamic creep and hardening effect, the unloading part is caused by dynamic creep and short-term recovery. To investigate hysteresis, the area of hysteresis is used (Table II). The area is decreased and then saturated by the combined dynamic characteristics as the loading cycles are progressed. It is because the hysteresis is result of the combined dynamic behaviors. Also, it critically affects CDPR's accuracy because the elongated length for the loading step is largely different from that for the unloading step at the same applied force

TABLE I. THE SHORT-TERM RECOVERY

Cycle		1	2	3	4	5
Short-term recovery [mm]	100 mm	0.64	0.51	0.46	0.42	0.38.
	300 mm	2.03	1.56	1.22	1.14	1.13
Cycle		6	7	8	9	10
Short-term recovery [mm]	100 mm	0.36	0.36	0.36	0.36	0.36
	300 mm	1.13	1.13	1.12	1.12	1.11

TABLE II. AREA OF THE HYSTERESIS

Cycle		1	2	3	4	5
Hysteresis energy dissipation[J]	100mm	0.250	0.092	0.084	0.076	0.070
	300mm	0.777	0.226	0.184	0.170	0.150
Cycle		6	7	8	9	10
Hysteresis energy dissipation [J]	100mm	0.067	0.063	0.062	0.060	0.060
	300mm	0.149	0.141	0.138	0.138	0.138

V. CONCLUSIONS

In this paper, we investigated the non-linear dynamics behavior of cable according to the various effective parameters using the derived model. It contained a hardening effect, a dynamic creep, a short and a long- term recovery and hysteresis. To investigate the effect of the effective parameters, we focused on the individual characteristics based on the cyclic loading test. As the loading cycle was increased, dynamic creep and hardening effect are increased and then, saturated. In contrary, short-term recovery is decreased and saturated. The longer the cable length, the longer-term recovery time is increased. Finally the hysteresis is decreased and then saturated by the combined dynamic characteristics as the loading cycles are progressed. As a result, slow CDPR system should be considered the cable dynamics because it makes the complicated errors.

ACKNOWLEDGMENT

This research was supported by the Leading Foreign Research Institute Recruitment Program through the National Research Foundation of Korea (NRF), funded by the Ministry of Science and ICT (MSIT) (2012K1A4A3026740) and by Development of Space Core Technology Program through the National Research Foundation of Korea(NRF) funded by the Ministry of Science, ICT (MSIT)(2017M1A3A3A02016340) and by the Korea Institute of Energy Technology Evaluation and Planning(KETEP) and the Ministry of Trade, Industry & Energy(MOTIE) of the Republic of Korea (2017030201530)

REFERENCES

[1] R. Dekker, A. Khajepour, and S. Behzadipour, "Design and testing of an ultra-high-speed cable robot," Int J Robot Autom., vol. 21,no. 1, pp. 25-34 2006

[2] P. Miermeister, A. Pott, and A. Verl, "Dynamic modeling and hardware-in-the-loop simulation for the cable-driven parallel robot IPAnema," in international symposium on robotics and german conference on robotics, Munich, Germany, 2011, pp. 1-8

[3] Maeda K, Tadokoro S, Takamori T, et al. "On design of a redundant wire-driven parallel robot WARP manipulator," In: Proceedings of the IEEE

international conference robotics and automation, Detroit, pp.895–900. New York: IEEE, 1999
[4] Gouttefarde M and Gosselin CM. Analysis of the wrench-closure workspace of planar parallel cable driven mechanisms. IEEE T Robot, vol 22, pp.434–445, 2006.
[5] M. Miyasaka, M. Haghighipanah, and Y. Li "Hysteresis model of longitudinally loaded cable for cable driven robots and identification of the parameters," in IEEE international conference on robotics and automation (ICRA), Stockholm, Sweden, 2016, pp. 4051-4057
[6] V. L. Schmidt, and A. Pott "Increase of position accuracy forb cable-driven parallel robots using a model for elongation of plastic fiber ropes," in new trends in mechanism and machine science, Switzerland, 2017, pp. 335-343
[7] S.H. Choi, and K.S. Park "Integrated and nonlinear dynamic model of a polymer cable for low-speed cable-driven parallel robot" Mycro syst tech, 2017

A study on dynamic response of functionally graded sandwich beams under different dynamic loadings

Wachirawit Songsuwan[1*], *Monsak* Pimsarn[1], and *Nuttawit* Wattanasakulpong[2]

[1]Mechanical Engineering Department, Faculty of Engineering, King Mongkut's Institute of Technology Ladkrabang, Bangkok, Thailand
[2]Mechanical Engineering Department, Faculty of Engineering, Mahanakorn University of Technology, Nongchok, Bangkok, Thailand

Abstract. In this research, free and forced vibration of functionally graded sandwich beams is considered using Timoshenko beam theory which takes into account the significant effects of transverse shear deformation and rotary inertia. The governing equations of motion are formulated from Lagrange's equations and they are solved by using The Ritz and Newmark methods. The results are presented in both tabular and graphical forms to show the effects of layer thickness ratios, boundary conditions, length to height ratios, etc. on natural frequencies and dynamic deflections of the beams. According to the numerical results, all parametric studies considered in this research have significant impact on free and forced behaviour of the beams; for example, the frequency is low and the dynamic deflection is large for the beams which are hinged at both ends.

1 Introduction

Recently, functionally graded (FG) sandwich structures play a vital role in modern structural engineering with excellent properties in high strength-to-weight ratio. The FG sandwich structures have no problems of de-bonding and delaminating modes of failure between layers which often occur in conventional sandwich structures due to the mismatch of materials at the interface. In the literature, buckling and vibration of FG sandwich beams with the combination of general and non-general boundary conditions were considered by Tossapanon and Wattanasakulpong [1] and Trinh et al. [2]. Moreover, in the study of Nguyen et al. [3], they also provided the solutions of natural frequencies and critical buckling load of such beams. A quasi-3D theory was employed to deal with static bending, buckling and vibration problems of FG sandwich beams in Refs. [4-5]. Based on the literature, most of previous studies presented only the static and free vibration analysis of FG sandwich beams. In the study of Vo et al. [6], the development of finite element model was presented for vibration and buckling of FG sandwich beams under various general boundary conditions. The relationship between fundamental natural frequency and in-plane loading was also presented in the study.

This investigation aims to extend research work to deal with dynamic behaviour of FG sandwich beams under various kinds of different dynamic loadings. Ritz method incorporated with Newmark time integration procedure are used to solve equations of motion based on Timoshenko beam theory. Many important parametric studies such as boundary condition, layer thickness ratio, material volume fraction index, etc. are investigated.

2 FG Sandwich beams

A geometry of FG sandwich beam including three layers of FG face sheets and metal homogenous core is shown in Fig. 1. The FG face sheets are the composite materials with the graded mix of ceramic and metal phases. The three numeric notations are used to represent the layer thickness ratio. For example, 1-1-1 indicates that the beam is composed of equal thickness in each layer.

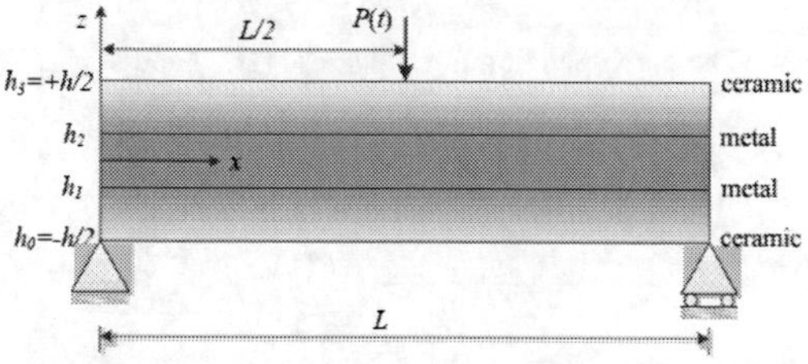

Fig. 1 Geometry and coordinate of FG sandwich beam.

The equations for estimating the effective material properties of the beam can be expressed as follows [1]:

$$E^{(i)}(z)=(E_b-E_t)V_b^{(i)}(z)+E_t, \quad (1)$$

$$\rho^{(i)}(z)=(\rho_b-\rho_t)V_b^{(i)}(z)+\rho_t, \quad (2)$$

where $E^{(i)}(z)$ and $\rho^{(i)}(z)$ are the Young's modulus and material density in each layer. The positions of material

* Corresponding author: w.songsuwan@gmail.com

properties at the faces and the core are denoted by the subscripts t and b respectively. The constant value of Poisson's ratio (ν) is used in this investigation. The material volume fraction, ($V_b^{(i)}$), related to the power law distribution can be expressed as follows [1]:

$$\begin{cases} V_b^{(1)}(z) = \left(\dfrac{z-h_0}{h_1-h_0}\right)^n & z\in[h_0,h_1] \\ V_b^{(2)}(z) = 1 & z\in[h_1,h_2] \\ V_b^{(3)}(z) = \left(\dfrac{z-h_3}{h_2-h_3}\right)^n & z\in[h_2,h_3] \end{cases} \quad (3)$$

where n is the material volume fraction index or power law index, $0 \le n \le \infty$ For the beam in Fig. 1, we use ($E_b = E_m$, $\rho_b = \rho_m$) and ($E_t = E_c$, $\rho_t = \rho_c$). It is also noted that the subscripts c and m are used to define the material properties of ceramic and metal phases, respectively.

3 Mathematical Formulations

The dynamic models of FG sandwich beams are formulated in accordance with Timoshenko beam theory [1]. The displacements of any point of the beams along the x- and z-axes, which are denoted by $u(x,z,t)$, $w(x,z,t)$, respectively, can be written below:

$$u(x,z,t) = u_0(x,t) + z\psi(x,t), \quad (4)$$

$$w(x,z,t) = w_0(x,t), \quad (5)$$

where u_0 and w_0 are axial and transverse displacements in the middle plane (z=0), respectively, ψ is the rotation of the beam cross-section and t is time.
The strain-displacement relations in terms of normal strain (ε_{xx}) and shear strain (γ_{xz}) are given by

$$\varepsilon_{xx} = \frac{\partial u}{\partial x} = \frac{\partial u_0}{\partial x} + z\frac{\partial \psi}{\partial x}, \quad (6)$$

$$\gamma_{xz} = \frac{\partial u}{\partial z} + \frac{\partial w}{\partial x} = \frac{\partial w_0}{\partial x} + \psi. \quad (7)$$

The corresponding normal stress (σ_{xx}) and shear stress (γ_{xz}) can be obtained from the elastic constitutive law as

$$\sigma_{xx} = E(z)\varepsilon_{xx}, \quad (8)$$

$$\tau_{xz} = G(z)\gamma_{xz} = \frac{E(z)}{2(1+\nu)}\gamma_{xz}. \quad (9)$$

The strain energy (U) of the FG sandwich beams at any instant can be defined in form of normal stress (σ_{xx}), normal strain (ε_{xx}), shear stress (τ_{xz}) and shear strain (γ_{xz}) as

$$U = \frac{b}{2}\int_0^L \int_{-h/2}^{h/2} (\sigma_{xx}\varepsilon_{xx} + \tau_{xz}\gamma_{xz})dzdx. \quad (10)$$

For the kinetic energy (K), it is expressed in form of displacements of u and w as

$$K = \frac{b}{2}\int_0^L \int_{-h/2}^{h/2} \rho(z)\left[(\dot{u})^2 + (\dot{w})^2\right]dzdx. \quad (11)$$

The work done (V) by the dynamic load (P) is $V = P(t)w(L/2)$ and the total energy of the system is

$$\Pi = K + V - U. \quad (12)$$

The total energy can be solved by using the Ritz and Newmark method [7]. The implementation of such methods to free and forced vibration of the beams was described elsewhere in details [8]. An example of the trial displacement functions in form of algebraic polynomials for a case of beam clamped at both ends is:

$$\begin{cases} u_0(x,t) = \sum_{j=1}^{N} A_j(t)\left(\dfrac{x}{L}\right)^j\left(1-\dfrac{x}{L}\right) \\ w_0(x,t) = \sum_{j=1}^{N} B_j(t)\left(\dfrac{x}{L}\right)^j\left(1-\dfrac{x}{L}\right) \\ \psi(x,t) = \sum_{j=1}^{N} C_j(t)\left(\dfrac{x}{L}\right)^j\left(1-\dfrac{x}{L}\right) \end{cases} \quad (13)$$

4 Results

In this research, FG sandwich beams having thickness (h=1 m), breadth (b=0.5 m) and length (L=20 m) are made of a mixture of Alumina (Al_2O_3) as ceramic phases and Aluminum (Al) as metal phases whose material properties such as Young's modulus (E), Poisson's ratio (ν) and material density (ρ) are:

For Alumina (Al_2O_3)

$$E_c = 380\,\text{GPa}, \nu_c = 0.3, \rho_c = 3960\,\text{kg/m}^3$$

For Aluminum (Al)

$$E_m = 70\,\text{GPa}, \nu_m = 0.3, \rho_m = 2702\,\text{kg/m}^3$$

Table 1 shows the convergence study and validation of dimensionless frequencies of the beams with hinged at both ends (H-H). Accuracy is confirmed by the result of Ref. [6] for the case of the beam with material volume fraction index (n=0.5). N is the polynomial number of terms in the Ritz method.

Table 1 Convergence study and validation for free vibration of 1-1-1 FG sandwich beams.

N	ω_1	ω_2	ω_3
4	4.6405	22.9665	57.4058
6	4.6391	18.3272	40.6471
8	4.6391	18.2530	40.0224
9	4.6391	18.2527	40.0223
10	4.6391	18.2527	40.0223
[6]	4.6294	-	-

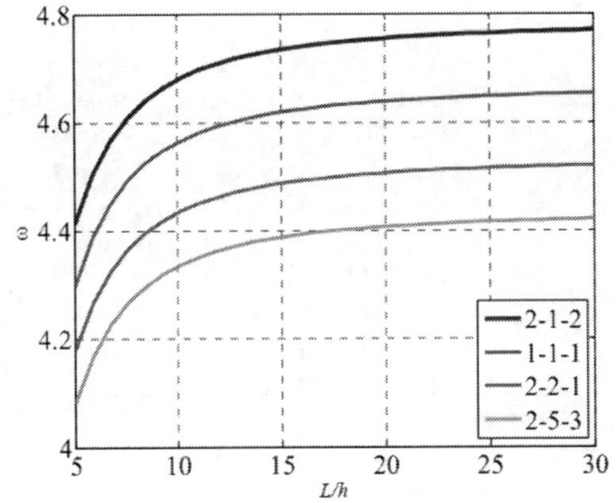

Fig. 2 Dimensionless fundamental frequency of FG sandwich beams.

To study an influence of the beam thickness ratio (L/h) on fundamental frequency of FG sandwich beams (n=0.5, H-H), Fig. 2 shows the plot of the 1st frequency of the beams with various layer thickness ratios. It can be observed that the frequency increases significantly as the increase of L/h ratio and the frequency of 2-1-2 is higher than that of other beams considered in this research.

In the following investigations, we consider the forced vibration of FG sandwich beams under Heaviside step loading and harmonic loading as shown in Fig. 3.

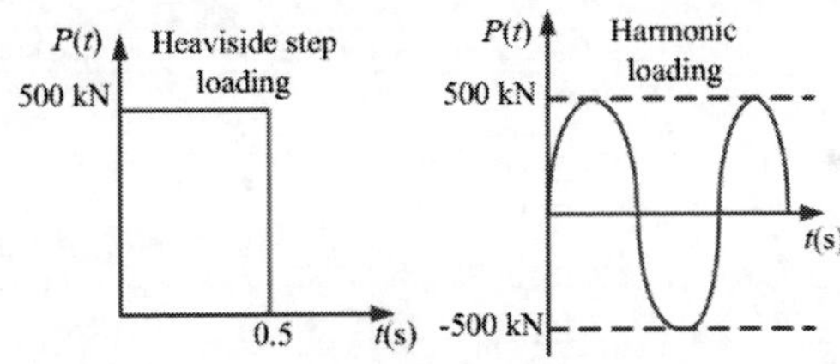

Fig. 3 Heaviside step and harmonic loadings.

All dynamic deflections are measured at the middle position of the beams. Fig. 4 and Fig. 5 show the effects of the index n and boundary condition on dynamic deflections of 1-1-1 FG sandwich beams under the Heaviside step loading, respectively. The dynamic deflection in this research is presented in the non-dimensional form of $w(x,t)/w_s$ where $w(x,t)$ is dynamic deflection and w_s is static deflection, $w_s = PL^3/48E_m I$ where $I = bh^3/12$. As can be seen, the beam with low value of n, $n<1.0$, (percentage of metal more than that of ceramic) is soft with larger dynamic deflection throughout the time. $n=1.0$ means that the percentage of metal is equal to the ceramic; while $n>1.0$, the beam is composed of much more ceramic than metal. For the effect of boundary condition, it is observed that the beam hinged at both ends (H-H) has larger deflection than that of the beam which is clamped at one end and hinged at another one defined as C-H beam. And the beam clamped at both ends (C-C) shows its lowest deflection through the considered time. Moreover, when the beams subjected to the Heaviside step loading, we can obtain the dynamic deflections for both areas which are the area of forced vibration ($t = 0 \rightarrow 0.5$ s) and free vibration ($t = 0.5 \rightarrow 1.0$ s).

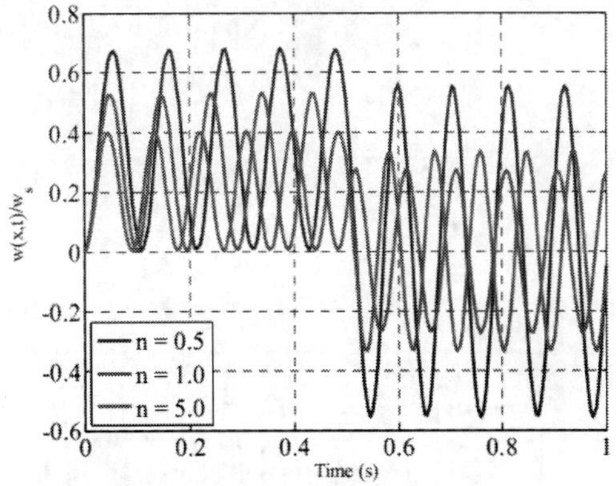

Fig. 4 Dynamic deflections of FG sandwich beams supported by H-H under Heaviside step loading: Effect of n.

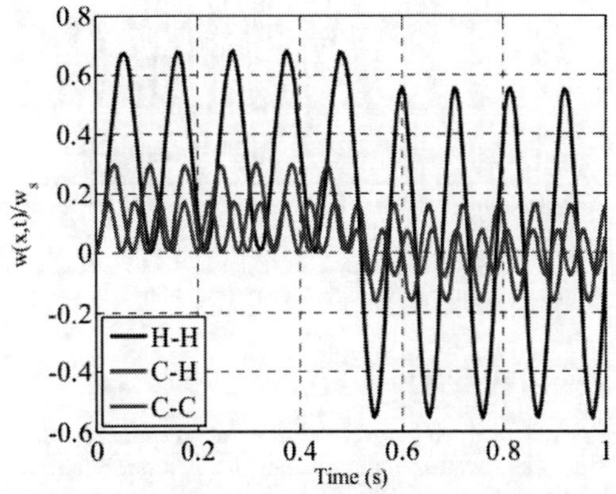

Fig. 5 Dynamic deflections of FG sandwich beams with n=0.5 under Heaviside step loading: Effect of boundary condition.

In Fig.6, the beams with various layer thickness ratios are investigated to find out their dynamic deflections under the action of harmonic loading. The boundary condition of the beams is H-H and the material volume fraction is n=0.5. From this figure, the deflection of the beam with 2-5-3 ratio is larger than that of other beams

To consider the significant effect of boundary condition on dynamic response of FG sandwich beams under the harmonic loading, Fig. 7 shows the dynamic deflections of beams with different supports. The beams having 1-1-1 and n=0.5 are considered. It is clearly seen that the beam, which is clamped at both ends (C-C), is very strong and gives less deflection throughout the considered time as compared to the deflections of C-H and H-H beams, respectively.

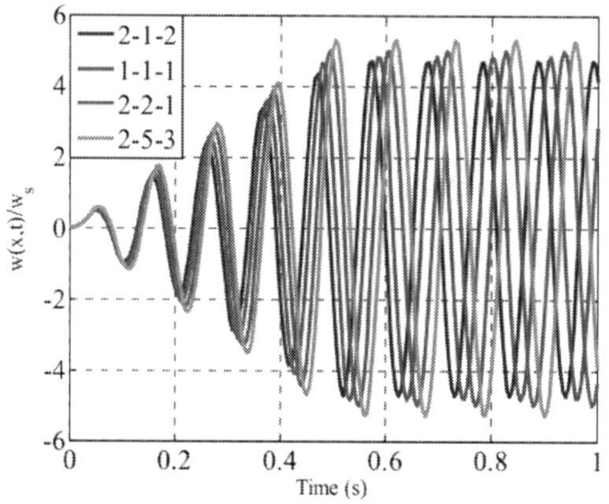

Fig. 6 Dynamic deflections of FG sandwich beams under harmonic loading: Effect of layer thickness ratio.

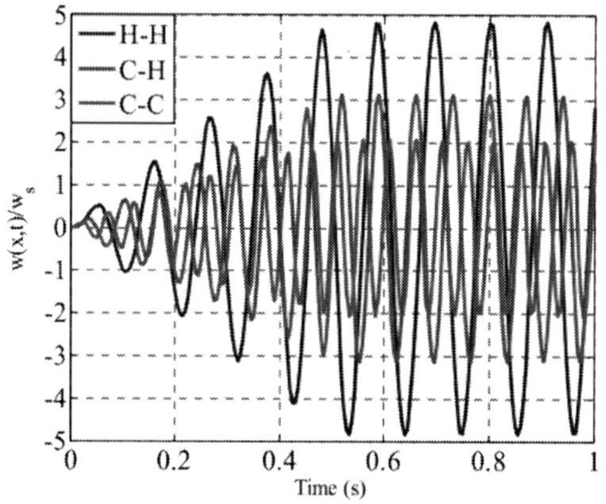

Fig. 7 Dynamic deflections of FG sandwich beams under harmonic loading: Effect of boundary condition.

5 Conclusions

The free and forced vibration responses of FG sandwich beams under Heaviside step and harmonic loadings are investigated using the Ritz and Newmark methods. Within the framework of Timoshenko beam theory including the effects of shear deformation and rotary inertia, the equation of motion is established for the beams with different boundary conditions. The influences of parametric studies such as layer thickness ratio, boundary condition, length to height ratio, material volume fraction index of the beams are presented and discussed in detail. The present modelling is validated and accurate with the comparison between the obtained solutions and the previous ones in the literature. Numerical results reveal that the beams which are clamped at both ends (C-C) have higher natural frequencies and lower dynamic deflections than those of C-H and H-H beams, respectively. Additionally, the dynamic deflection decreases as the increase of parameter n. It is due to material properties of the beams are dependent on the parameter.

References

1. P. Tossapanon, N. Wattanasakulpong, *Compos Struc.* **142**, 215-225 (2016)
2. L.C. Trinh, T.P. Vo, A.I. Osofero, J. Lee, *Compos Struc.* **156**, 263-275 (2016)
3. T.K. Nguyen, T.T.P. Nguyen, T.P. Vo, H.T. Thai, *Compos Part B.* **76**, 237-285 (2015)
4. T.P. Vo, H.T. Thai, T.K. Nguyen, F. Inam, J. Lee, *Compos Struc* **119**, 1-12 (2015)
5. T.P. Vo, H.T. Thai, T.K. Nguyen, F. Inam, J. Lee, *Compos Part B* **68**, 59-74 (2015)
6. T.P. Vo, H.T. Thai, T.K. Nguyen, A. Maheri, J. Lee, *Eng Struc* **64**, 12-22 (2014)
7. D. Chen, J. Yang, S. Kitipornchai, *Int J Mech Sci* **108-109**, 14-22 (2016)
8. W. Songsuwan, M. Pimsarn, N. Wattanasakulpong, *In. J. Struct. Stab. Dyn.* **18**(9), 1850112 (2018)

Study of fatigue crack initiation location of wheel and rail under rolling contact using finite element method

Apichai Jaifu[1,*], *Suthep* Raeon[2] and *Monsak* Pimsarn[3]

[1,3]Mechanical Engineering Department, Faculty of Engineering, King Mongkut's Institute of Technology Ladkrabang, Bangkok, Thailand

[2]Mass Rapid Transit Authority of Thailand, 175 Rama IX Rd., Huay Khwang, Bangkok 10310

Abstract The rail transit system is widely used for freight and passenger transportation. Due to the fact that its economic worthiness and high safety mode. Maintenance and damage prevention of wheel and rail are important factors affecting the safety of the system. The previous studies show that the most damage of wheel and rail is fatigue cracking, which is caused by the contact stress resulting from wheel and rail interaction. This article presents the study of the fatigue crack initiation location of wheel and rail under rolling contact at the wheel speed of 80 km/h using Finite Element Method (FEM). The three dimensional finite element models were created using the UIC60E1 wheel profile and BS100 rail profile. The Dang Van criteria was applied to analyse the fatigue crack initiation location in case of the wheel's position was changed along the rail lateral direction while the rail inclination angle was also varied at 0, 1/40, 1/30 and 1/20, respectively. The analysing results show that the fatigue crack initiation, determined by the Dang Van stress ratio, tends to increase when the wheel is moved from gauge side to field side. Additionally, the fatigue crack damage is likely to decrease when the rail inclination increases up to the inclination of 1/30 and the fatigue crack initiation locations were found underneath the wheel and rail surfaces. The obtained result can be a primary guideline for maintenance planning.

1 Introduction

Rolling contact fatigue is a damage generally appearing on contact surfaces between wheel and rail. This fatigue occurs from the repeated load on a surface by millions of intensive wheel and rail contact cycles. However, the previous studies[1] reported that the several accidents of train derailment resulted from damages of wheel/rail, such as, spalling, fatigue crack, shelling, rolling fatigue and thermal crack [2]. Rolling fatigue crack is very common to all kind of trains and can be found in both wheel and rail. Thus, maintenance of wheel/rail must be the prime concern and necessary to be steadily improve. Toumi et. al. [3] employed finite element method to study rolling contact stress between wheel and rail with increased spin effect. The FEM solution is compared to the solution of the CONTACT software [4]. This paper reported that the creep force characteristic from FEM solution is slightly lower than the result obtained by the CONTACT software. Martin Pletz et al. [5] purposed a full scale test rig of rolling contact fatigue and performed the simulation of wheel and rail loading of test-rig by using FEM combined with CONTACT software. From this research, it was found that there are two contacts, one of them is near on the top and the other is on the gauge corner of its rail. The contact area on the top of the rail is bigger than the other at the gauge corner. This might be due to the lower lateral load.

However, this article aims to employ finite element method and Dang Van criterion to predict the rolling fatigue crack initiation in the wheel/rail when the contact area is varied from field side to gauge side and the wheel is rolling under constant axle load and constant speed. Additionally, the effect of rail inclination angle will be numerically investigated.

2. Dang Van high cycle fatigue criterion

In 1993, Dang Van proposed a multiaxial fatigue criterion to predict crack initiation when the two solids are in a rolling contact. This criterion is based on the elastic shakedown principle at the mesoscopic scale. The Dang Van criterion predicts that the fatigue crack of the material will occur on the grain boundary. In the macroscopic scale, the Dang Van criterion considers this to happen when the stress inside the material reaches critical value and can be expressed by

$$\tau_{\max}(t)+\alpha_{DV}\sigma_H(t)=\tau_W \qquad (1)$$

When α_{DV} is a constant to be determined, τ_w is the fatigue limit of pure torsion. $\sigma_H(t)$ is the instantaneous hydrostatic component of stress tensor and $\tau_{\max}(t)$ is the

* Corresponding author: apichai.jaifu@gmail.com

value of the Tresca shear stress which can be calculated by[6] :

$$\tau_{\max}(t) = \left(\frac{\sigma_1(t)-\sigma_3(t)}{2}\right) \quad (2)$$

When $\sigma_1(t)$ and $\sigma_3(t)$ are maximum and minimum principal microscopic deviatoric stress at each instant t. The constant of α_{DV} is evaluated by [7];

$$\alpha_{DV} = 3\left(\frac{\tau_w}{\sigma_w}-\frac{1}{2}\right) \quad (3)$$

In the relationship of ratio from maximum shear stress $\tau_{\max}(t)$ and fatigue limit of bending σ_W and fatigue limit of pure torsion τ_w and hydrostatic pressure. Therefore, in order to predict crack initiation, the maximum Dan Van stress ratio at any time t can be defined as

$$DV = Max\left[\frac{\tau(t)}{\tau_w-\alpha_{DV}\sigma_H(t)}\right] \quad (4)$$

In Eq.(4), if DV value is more than one, the fatigue failure will be expected to occur and the position that has the maximum DV is prone to damage faster.

3 Simulation of wheel rolling on rail using Finite Element Method

In this article, the wheel and rail geometry are based on the data of purple line project, Bangsue to Bangyai, Bangkok, Thailand. The wheel diameter is 860 mm and the rail is UIC60E1[1]. The material properties of wheel and rail are shown in Table 1.

Table 1. Material properties of wheel and rail

Properties	Wheel	Rail
Material type	SSW-Q3S	R260
Standard	JIS E 5402	EN 13674
Young's modulus, E	210 x 10^3 MPa	210 x 10^3 MPa
Yield stress, σ_y	500 MPa	528 MPa
Poisson's ratio, ν	0.25	0.25

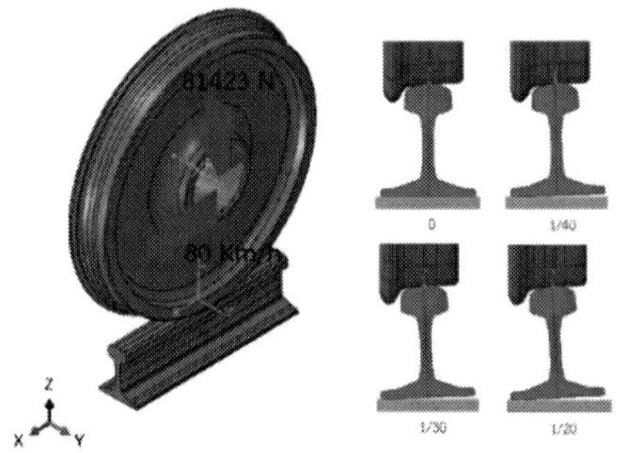

Fig. 1 Three-dimensional solid model of wheel and rail

Fig. 1 shows that the axle loading of 81,420 N is applied to the wheel and the wheel speed is 80 km/h. Also, rail inclination is varied at 0, 1/20, 1/30 and 1/40, respectively. Contact condition between wheel and rail is frictional and the friction coefficient is assumed to be 0.3. The bottom surface of rail foot is fixed in all directions.

Fig. 2 shows the assembly FEM model of wheel and rail. In FEM model, the hexahedron and tetrahedron 3D elements are employed and the model consists of 210,541 elements. Additionally, in the contact zone, the hexahedron element size of 1 mm is employed to capture the correct contact stress distribution.

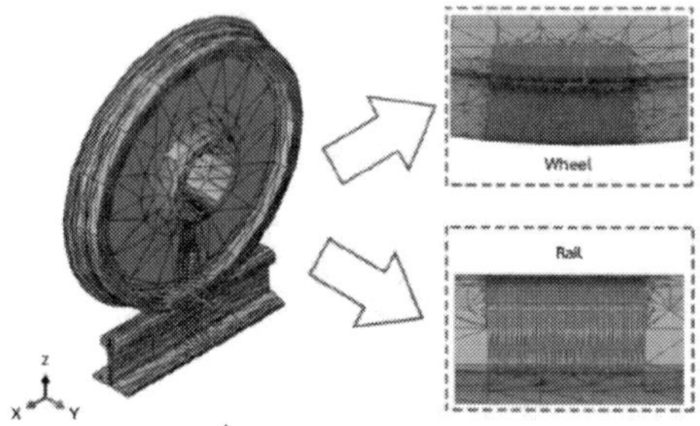

Fig. 2 Finite element model of wheel and rail

4 Results and discussions

4.1 Validation of finite element model

In order to confirm that the FEM model of Fig. 2 is valid, the Hertz's contact theory[8,9] is employed. Fig. 3 shows the stress variation underneath the contact area along z-axis obtained from FEM and Hertz's contact theory. The results obtained from both methods are in good agreement and the maximum discrepancy is approximately 1%.

Moreover, the comparison of contact dimensions and the peak contact pressure were also investigated. Table 2 shows the comparison of contact dimensions and the peak contact pressure. It was found that the finite element results are reasonable in a good agreement. Therefore, the finite element model of wheel and rail developed in this article can be accepted.

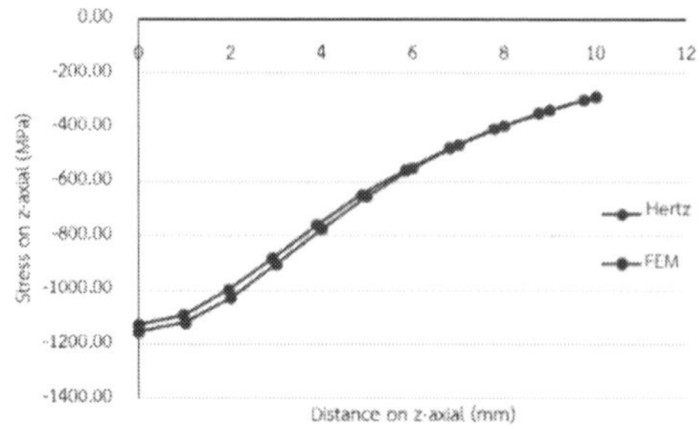

Fig. 3 Stress underneath contact area distribution along z-direction

Table 2. Comparison of the maximum contact pressure and contact dimensions

	Hertz	FEM	Discrepancy
Half of width contact a	6.53 mm	6.09 mm	6.81 %
Half of width contact b	5.13 mm	5.11 mm	0.30 %
Peak contact pressure P_0	1152.53 MPa	1153.55 MPa	0.09 %

4.2 Peak contact pressure and Von-Mises stress at rail inclination of 1/40

In this article, the peak contact pressure and area occurring between wheel and rail with the rail inclination of 1/40 were firstly investigated when a wheel lateral displacement is varied from -4 mm to 4 mm. The wheel nominal position is at 0 mm displacement. If the wheel displacement is less than 0 mm, it is considered as gauge side of rail. On the other hand, the field side of rail is a wheel displacement that is from 0 mm to 4 mm. Fig.4 shows the variation of peak contact pressure. According to the graph, it shows that the peak contact pressure is gradually increased from gauge side to field side. Due to the fact that the contact are is reduced, as shown in Fig.5. The maximum contact area is 142.31 mm^2 as the wheel lateral displacement is -4 mm and, at this position, the peak contact pressure is minimum, 920.55 MPa. On the other hand, with the wheel lateral displacement of 4 mm, the contact area is minimum, 88.73 mm^2, and the peak contact pressure is maximum, 1408.78 MPa.

Fig.6 shows the variation of maximum Von-Mises stress underneath the wheel and rail contact surfaces when the wheel lateral position is varied. It is found that the Von-Mises stress is higher in the rail. The Von-Mises stress tends to increase when the wheel lateral position is moved from gauge side to the field side, in the same manner as the peak contact pressure. Moreover, the maximum Von-Mises stresses of wheel and rail are 518.18 MPa and 549.45 MPa, respectively, which are higher than the material yield stress. Therefore, it is expected that plastic deformation will occur.

4.3 Effect of rail inclination on peak contact pressure and contact area

In this section, the rail inclination effect on peak pressure and contact area had been studied. Fig.7 shows the variation of peak contact pressure as the rail inclination angles are 0, 1/40, 1/30 and 1/20. The maximum contact pressure is found as rail inclination angle is zero. The peak contact pressure tends to increase if the wheel lateral position is varied from gauge side to field side. However, this trend is opposite to the case of 1/20 rail inclination. On the other hand, as depicted in Fig.8, the minimum contact area is found as rail inclination is zero.

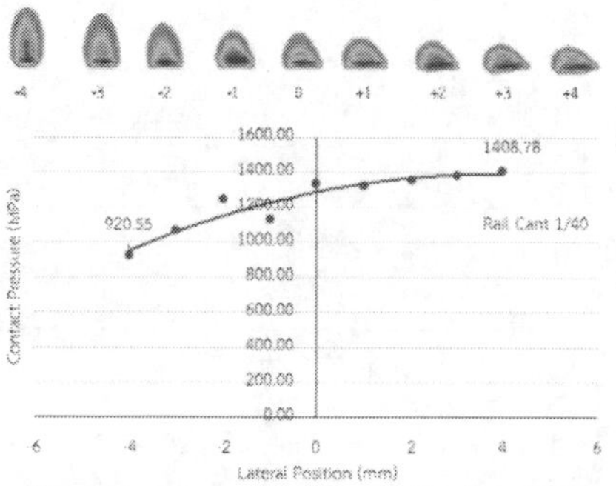

Fig. 4 Contact pressure at rail inclination of 1/40

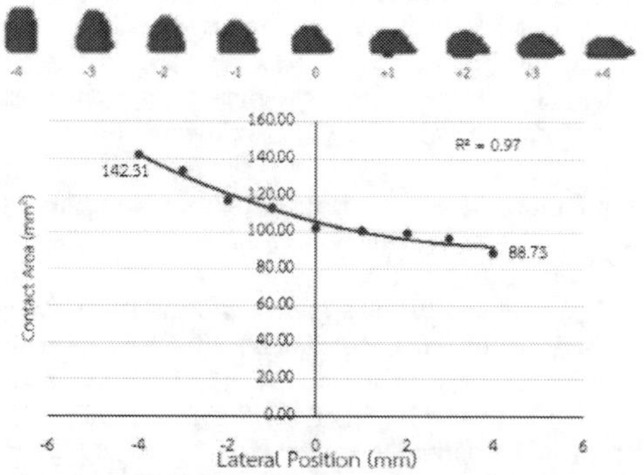

Fig. 5 Contact area at rail inclination of 1/40

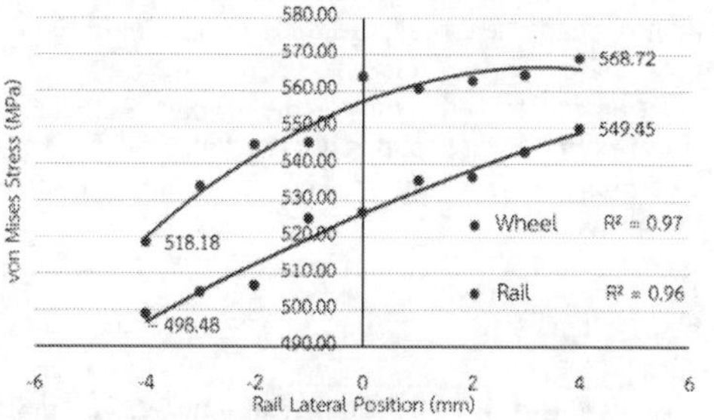

Fig. 6 Maximum Von-Mises stress variation of wheel and rail at rail inclination of 1/40

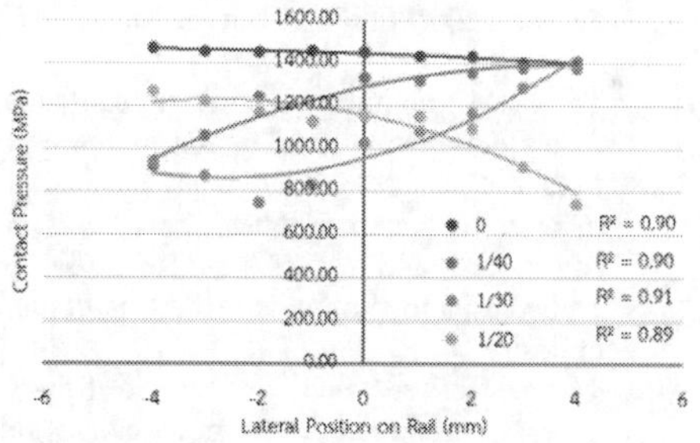

Fig. 7 Peak contact pressure at rail inclinataion of 0, 1/40, 1/30, 1/20

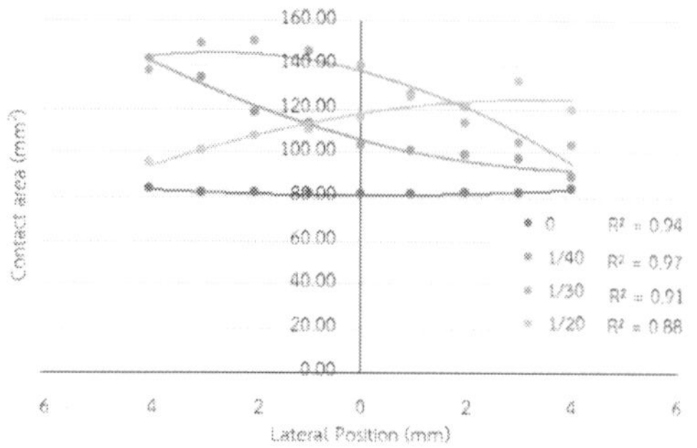

Fig. 8 Contact area at rail inclinaton of 0, 1/40, 1/30, 1/20

Also, the contact area is not much affected by the wheel lateral position. In the case of the rail inclination of 1/40 and 1/30, the contact area is gradually decreased if the wheel lateral position is varied from gauge side to field side. This posture is completely different from the rail inclination, which is 1/20, as shown in Fig.8.

4.4 Rail inclination effect on Dang Van Stress Ratio

Fig. 9 and 10 show the Dang-Van(DV) stress ratio, as defined in eq.(4), distribution on the wheel and rail. Maximum DV stress ratio is found as rail inclination is zero and its value is higher on the rail. Therefore, fatigue crack initiation on the rail is expected to happen faster. The DV stress ratio tends to decrease as the wheel lateral position is moved from gauge side to field side in all cases. However, this is not the case for 1/20 rail inclination. Therefore, the fatigue crack initiation is potential to happen on the field side of the rail.

From the FE results, the maximum Dang-Van stress ratio of wheel and rail were found at the depth of 2.45 mm and 5.65 mm from wheel and rail surfaces, respectively.

5 Conclusions

In this article, the numerical simulations of wheel and rail under rolling contact using finite element method were carried out by varying the rail inclination angles and the wheel lateral position, from gauge side to field side. The major findings can be summarized as follows:

(1) The peak contact pressure is potential to increase as the rail inclination increases up to 1/30 and the peak contact pressure gradually increases if the wheel lateral position is moved from gauge side to field side. However, for the 1/20 rail inclination, this posture is completely opposite.

(2) The contact area tends to decrease if the wheel lateral position is varied from gauge side to field side in most cases. However, this is not valid for the 1/20 rail inclination.

(3) The Dang Van stress ratios of wheel and rail were influenced by the rail inclination and the wheel lateral position. Its value is potential to be higher if the rail inclination increases up to 1/30 and the wheel lateral position is on the field side. Therefore, the fatigue crack initiation location is prone to occur earlier on the field side. However, for the 1/20 rail inclination, this posture is completely different. Moreover, the FE results reveal that the maximum Dang Van stress ratio was found below the wheel and rail surfaces, at the depth of 2.45 mm and 5.65 mm, respectively.

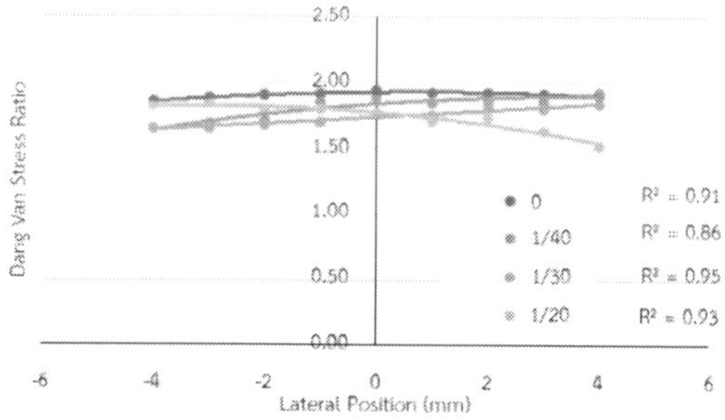

Fig. 9 Dang-Van stress ratio of wheel on various rail inclinations

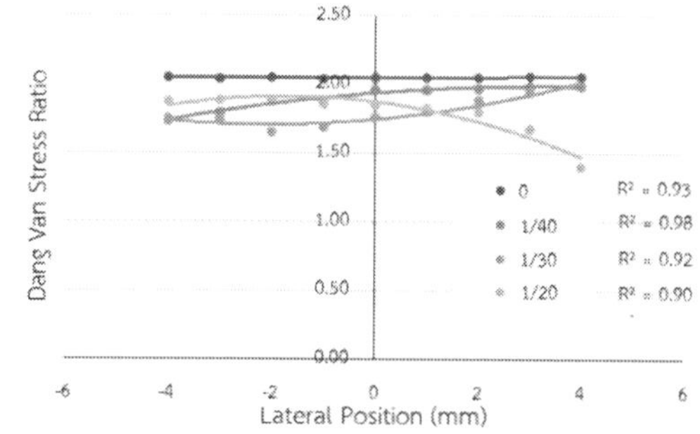

Fig. 10 Dang-Van stress ratio of rail on various rail inclinations

References

1. S. Reaon, M. Pimsarn, *Thailand Rail Academic Symposium* (TRAS2016)
2. Y.wu, Y.Wei, Y.Liu, Z.Duan, L.Wang, *Applied Thermal Engineering*, 115 (2017)
3. M.Toumi, H.Chollet, H.Yin, *Wear 366-367*, (2016) 157-166
4. E.A.H.Vollebregt, User Guide for CONTACT, Vollebregt & Kaller's Rolling cantact and sliding Contact model, 2013
5. M.Pletz, W.Daves, W.Yao, W.Kubin, S.Scheriau, *Tribology International 80*, (2014) 147-155
6. J.Alexander, E.Mamiya, R.Castro, 17th *International Congress of Mechanical Engineering*, 1224(2003)
7. M.Cerullo, *13th International conference on fracture*, 2013
8. W. Yan, F.D. Fischer, *Archive of Applied Mechanics*, 70 (2000)
9. M. Busquat, L. Baillet, C. Bordreuil, Y. Berthier, *Engineering Science*, 258 (2005)

MATEC Web of Conferences **192**, 02013 (2018) https://doi.org/10.1051/matecconf/201819202013
ICEAST 2018

CO_2 concentration from turbocharged common rail diesel engine dually fueled with compressed biomethane gas controlled at optimum ratio

Niti Kammuang-lue[1,*], and *Matas* Bhudtiyatanee[1]

[1]Department of Mechanical Engineering, Faculty of Engineering, Chiang Mai University, Chiang Mai 50200, Thailand

Abstract. The objectives of this study are to investigate the carbon dioxide (CO_2) concentration from the compressed biomethane gas (CBG) and diesel dual-fueled diesel engine and to compare the CO_2 concentration produced from the dual-fueled and the diesel-fueled engines. The duration of CBG injection was controlled by following the optimum ratio of the CBG obtained from the previous study. During the test, the engine speed was varied from 1,000 to 4,000 rpm and the engine torque was maintained to be 25, 50, 75 and 100% of the maximum engine torque. Experiment was divided into two parts consisting of the dual-fueled and the diesel-fueled modes. From the dual-fueled mode, when the engine speed increased, the CO_2 concentration decreased. Because the optimum ratio of the CBG and the volumetric efficiency decrease during the high engine speed range, the proportion of the diesel increases, the incomplete combustion occurs. The unburned carbon oxidizes to be the CO in higher proportion than the CO_2, thus, the CO_2 consequently decreases. From the CO_2 comparison, the dual-fuel mode produced the CO_2 nearly the same as that of the diesel-fuel mode during the low engine torque. On contrary, the dual-fuel mode had higher CO_2 concentration during the high engine torque.

1 Introduction

A diesel engine is one type of an internal combustion engine that is classified into a compression-ignition engine or CI engine. Advantages of diesel engines over gasoline engines are: (i) Diesel engines have higher compression ratio, the thermal efficiency is higher than that of gasoline engines. (ii) Combustion in diesel engines can be directly controlled by adjusting of injected fuel amount. And (iii) overall cost of operation of diesel engines is lower than that of gasoline engines [1-2]. These are reasons causing number of diesel powered vehicles to increase continuously, especially, in a case of the light commercial vehicles that ratio of the vehicles installed with diesel engine is obviously higher than that of with gasoline engine [3].

According to the oil's price crisis and rapid decreasing in amount of reserved crude oil, people and researchers involved with diesel-powered vehicles are interested to find alternative fuel for used in diesel engines to minimize the fuel cost. Various past studies have reported that diesel engines can run on a mixture of alternative fuel and diesel. The alternative fuel, which is in liquid or gas phase, is injected into intake air and they will completely mix as homogeneous air and alternative fuel mixture. Diesel is consequently injected into the combustion chamber and become the pilot ignition that ignites the surrounding alternative fuel [4]. The engine that utilizes two types of fuel simultaneously is defined as a "dual-fueled engine" and so called the mixture of fuels as "dual fuel".

Biogas was become well-known alternative fuel for a few decades. Biogas is the raw product obtained from the biological process in many diversities of resource, such as, sewage treatment plants, industrial site, landfills, digestion plants of agricultural organic waste, etc. Raw biogas primarily consists of methane (CH_4, 40-75%) and carbon dioxide (CO_2, 15-60%) [5, 6]. Since proportion of the methane being the main composition in the biogas is lower than that of in the CNG, the biogas is taken into an upgrading process to enhance proportion of the methane and purity and the upgraded biogas is called as "biomethane" [6]. And the pressurized biomethane is named as "compressed biomethane gas" (CBG).

It has been clearly known that if commercial carriers need to decrease the fuel cost of dual-fueled engines, the dual fuel mixing ratio must be adjusted to have the maximum proportion of the CBG as possible while engines can produce efficient performance. However, the acceptable increase in the ratio of the CBG is not only an increase the ratio of the CBG to be the maximum, it is important to concern with safety condition of engines and country regulations. In the light of this reason, the previous studies on the optimum ratio of CBG and diesel dual fuel used in diesel engine have defined the limitation factors for increasing the ratio of the CBG that consisted of: (i) The exhaust gas temperature did not exceed 580 °C. (ii) The voltage measured from a knock

* Corresponding author: niti@eng.cmu.ac.th

sensor did not exceed 1 V. And (iii) the opacity of the smoke in the exhaust gas from the diesel engine must not exceed 35% measured by a partial-flow opacity-type smoke tester, which has been defined in the national regulation of The Pollution Control Department, Ministry of Natural Resources and Environment, Thailand. Since these limitation factors were defined from the experimental investigation on the diesel-fueled engine, if the optimum ratio of the CBG was adjusted by following these factors, the duel-fueled engine could operate without failure and the emission did not exceed the maximum value defined by the regulation. And the engine torque obtained from the duel-fueled engine with the optimum ratio of the CBG was identical to a case of the diesel-fueled engine [7, 8].

Although the dual fuel has been adjusted to be in line with the optimum ratio of the CBG obtained from the previous studies and the soot in the exhaust gas has already corresponded to the national regulation, concentration of other emissions occurring from combustion in the CBG and diesel dual-fueled engines is interesting issue. These became background and significance of this study with objectives as: (i) To investigate the carbon dioxide (CO_2) concentration in the exhaust gas from the CBG and diesel dual-fueled turbocharged common rail diesel engine (TCRD engine) with the optimum ratio of the CBG. And (ii) to compare the CO_2 concentration in the exhaust gas from the dual-fueled and the diesel-fueled engines at the identical operating condition and engine torque. Knowledge obtained from this study will be usefulness for researchers and automotive engineers involved in adjusting the optimum ratio of the CBG and diesel dual fuel. Obtained data can be used as a guideline to adjust the ratio for effective decreasing in the emission from the combustion.

2 Experimental setup and procedure

The engine used in this study was the common rail, direct injection diesel engine (Toyota, model 2KD-FTV, four-cylinder, four-stroke, total displacement 2,494 cm^3) that was directly taken from the market-sold vehicle. The engine had the maximum power of 75 kW at 3,600 rpm and the maximum torque of 200 N-m at 1,400–3,200 rpm. The compression ratio was 18.5:1. The bore and the stroke was 92.0 and 93.8 mm, respectively. The turbocharger was installed onto the engine without intercooler as default arrangement from manufacturer. It should be noted that the primary hardware of the engine was not modified except the installation of gas electronic control unit (Gas ECU) and the gas supplying system. The CBG with the CH_4 proportion of 80.0% by volume and the diesel with the Cetane index of 55 were used. The diesel injection timing and duration were controlled by a factory engine's electronic control unit (ECU) in which none of the modification on the programmed values. The factory sensors located on the engine were completely wired to the ECU as the manufacturer default. The solenoid diesel injectors (Denso, type 415F, 6 holes) were located on the factory fuel common rail. The fuel pressure was 160 MPa. The injected timing and duration of the CBG was controlled by the gas ECU (Autogaz, model Stag Diesel-4, for 4 injectors) which could be programmed via a computer. The CBG supplying equipments were installed on the engine to activate the dual-fuel mode, as shown in Figure 1. The configuration of the gas and the diesel systems are described as follows. The pressure of the CBG was reduced from 200 bar_g in the tanks to 2 bar_g. The inlet gas pressure and the inlet water temperature probes were attached on the gas reducer and wired to the gas ECU. Another set of the gas pressure and temperature sensors (Autogaz, PS-02, maximum temperature 125 °C, maximum pressure 6.75 bar) were installed on the outlet tube of the gas reducer for measuring the pressure and the temperature of the CBG supplied to the gas injectors. The gas injectors (Hana, maximum gas flow rate 0.13±0.0015 m^3/min, maximum temperature 120 °C, maximum pressure 3.5 bar), which were directly controlled by the gas ECU, were installed on the gas rail to inject the CBG into the intake runners of the engine. In addition, the knock sensor (Bosch, 3-wire, frequency range 1–20 kHz) was installed on the outer surface of the cylinder block. The exhaust gas temperature sensor (Termo-Precyzja, Type-K, temperature range -100–1,300 °C) was installed on an outlet pipe of the turbocharger's turbine side.

A hydraulic engine dynamometer (HPA) was applied to load the engine and to maintain its speed. The CO_2 concentration was measured by the gas analyzer (Testo, model 340) with an accuracy of ±0.2%v/v for CO_2. In addition, the smoke optical absorption coefficient was measured by the opacimeter-type smoke tester (Hamann, DO 285, range 0–9.99 m^{-1}). The smoke opacity could be obtained in real-time as an advantage of using the opacimeter. Probes of the gas analyzer and the smoke meter were attached into the exhaust pipe behind the turbine side of the turbocharger.

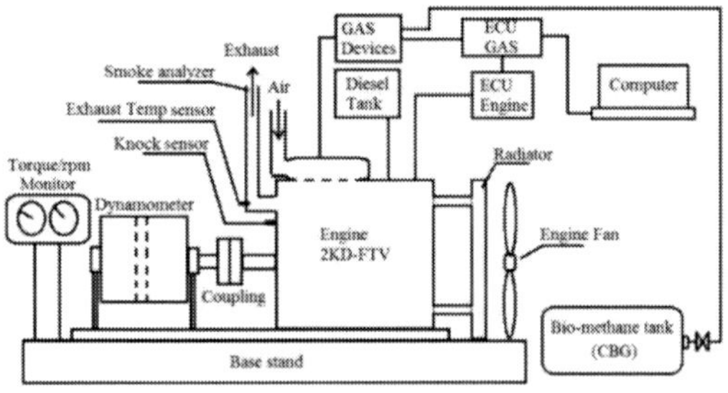

Fig. 1. Schematic diagram of experimental setup.

The experimental procedure was divided into two main sections as follows:

(i) First section, the test on the diesel-fueled engine: The engine speeds in the test were defined to be varied from 1,000 to 4,000 rpm in an increment of 200 rpm. At each defined engine speed, the load from the dynamometer applying on the engine and the throttling valve of the engine were simultaneously adjusted to maintain the engine torque to correspond with the reference engine torque curves at 25, 50, 75 and 100% of the maximum engine torque obtained from the previous study [8] as depicted in Figure 2. This procedure was conducted in order to control the engine torque to be identical in each of the engine speed to simulate the same operating condition of the engine. When the engine speed was steady, the engine speed, the engine torque, emissions in the exhaust gas, and data from all sensors were monitored and recorded by the interval of 180 s. Then, the engine speed and the engine torque were increased step-by-step until they reached the maximum values as scoped in this study.

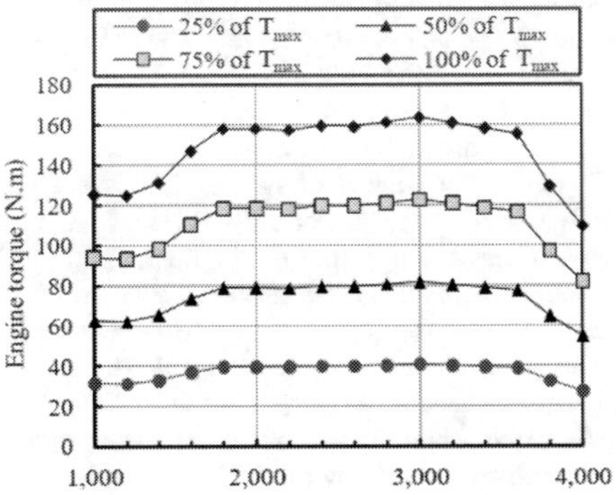

Fig. 2. Reference engine torque curve obtained from [8].

(ii) Second section, the test on the dual-fueled engine: The gas injection duration in each engine speed and torque was initially programmed onto the gas ECU, thus, the CBG was injected within the optimum ratio corresponding to the results obtained from the previous study [8]. Due to the optimum ratio of the CBG, the engine did not operate in the condition that exceeded the limitation factors as described in the introduction. The experimental procedure in this section was identical to the previous section except the CBG and diesel dual fuel was used instead of the diesel.

Finally, the measured CO_2 concentrations were plotted against the engine speed and the variations were analyzed further in the next section.

3 Results and discussions

Carbon dioxide (CO_2) is the emission occurring in the complete and incomplete combustions. For the complete combustion, the CO_2 is produced without the CO but the CO_2 and CO are both produced in the incomplete combustion. It was found that when the engine speed increased from 1,000 to 4,000 rpm, the CO_2 concentration was nearly constant. After it reached the high engine speed range, the CO_2 concentration obviously decreased. The trend of the variation in the CO_2 concentration corresponded in all reference engine torques. Moreover, It could be noticed that when the engine torque increased, the engine speed, where the CO_2 concentration began to decrease, continuously decreased as shown in Figure 3.

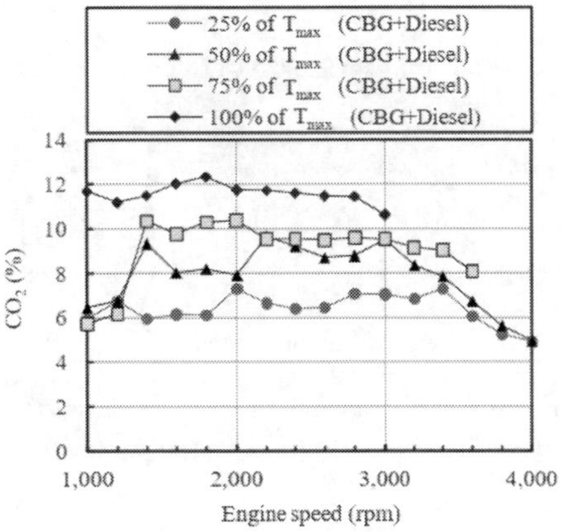

Fig. 3. CO_2 concentration from dual-fueled engine.

The variations in the CO_2 concentration were investigated in these trends because the CBG and diesel dual fuel are completely burnt during the low-to-medium engine speed. The combustion is rather complete. Only the CO_2 or the CO_2 with small proportion of the CO was produced from the combustion. Therefore, the CO_2 concentration during the low-to-medium engine speed range is higher than that of the high engine speed range. In a case of the high engine speed range, the CBG ratio is decreasingly supplied due to the limitation factors of the exhaust gas temperature and the engine knock on the optimum ratio of the CBG as obtained from the previous study [8]. This causes the ratio of the diesel, which has higher carbon atom than the CBG, to increase. The possibility that the carbon atom in diesel will not be burned consequently increases. The incomplete combustion subsequently occurs. In addition, the volumetric efficiency of the reciprocating engines decreases due to an increase in the engine speed. Mass of the air flowing into cylinders decreases. This promotes the occurrence of the incomplete combustion. The exhaust gas, thus, contains both of the CO_2 and the CO. In the light of this reason, when a proportion of the unburned carbon oxidizing to be the CO increases, the remaining proportion that will change into the CO_2 consequently decreases. These cause the CO_2 concentration during high engine speed range to decrease and vice versa for the CO concentration. This corresponds to the results obtained from the past study [9] that the complete combustion produced higher CO_2 and lower CO concentrations than that of the incomplete one. Relation between the CO_2 and the CO concentrations can be described by examples of the chemical combustion equations of the CBG and diesel dual fuel as follow. The reactions of the CBG and diesel dual fuel are considered in a case that the mass ratio between the CBG and the diesel is 80:20. The total mass of the dual fuel is 1 kg. In a case of the complete combustion with 100% of the air, the combustion equation can be found as in Equation (1). For a case of the incomplete combustion with 70% of the air, the reaction is expressed as in Equation (2). It could be seen that, in a case of the complete combustion, the mole fraction of the CO_2 is 2.85 without the CO. After the incomplete combustion takes place, in turn, during the high engine speed range, the mole fraction of the CO_2 decreases to be 0.34 while the CO increases to be 3.34.

$$0.2C_{14.4}H_{24.9} + 0.8(0.80CH_4 + 0.20CO_2) + 5.405O_2 + 20.33N_2 \rightarrow 3.68CO_2 + 3.77H_2O + 20.33N_2 \quad (1)$$

$$0.2C_{14.4}H_{24.9} + 0.8(0.80CH_4 + 0.20CO_2) + 3.735O_2 + 14.05N_2 \rightarrow 0.34CO_2 + 3.34CO + 3.77H_2O + 14.05N_2 \quad (2)$$

The CO_2 concentration releasing from the CBG and diesel dual fuel and the diesel fuel could be compared as shown in Figure 4. It could be seen that the dual fuel and the diesel had similar CO_2 concentration during the engine torque of 25% of the maximum torque. Since the optimum ratio of the CBG is relatively high during the low engine torque, the CBG being in the gas phase replaces the volume of the intake air flowing into the cylinders. The O_2 in the combustion process decreases and then the rich-fuel mixing ratio tends to occur. The unburned carbon oxidizes to be the CO in higher proportion. The CO_2 concentration consequently decreases as above mentioned relation between the CO_2 and the CO. However, during the engine torque of 100% of the maximum torque, the dual fuel produced higher CO_2 concentration than the diesel. Since the optimum ratio of the CBG is quite low during the high engine torque, the unburned carbon changes to be the CO_2 in higher proportion and vice versa for the CO as previous described. These discussions were well agreed with the one in the past study [10]. It was discussed that during the part engine load, the CO concentration sharply increases as an increase in the amount of gaseous fuel. This could be implied to be the reason why the CO_2 concentration was low as a case of the diesel fuel. In addition, it was reported for the high engine load that the CO concentration increased with an increase in gas mass ratio. Since the optimum ratio of the CBG was relatively low during the high load, the CO concentration decreased and the CO_2 concentration consequently increased to be higher than that of the diesel fuel.

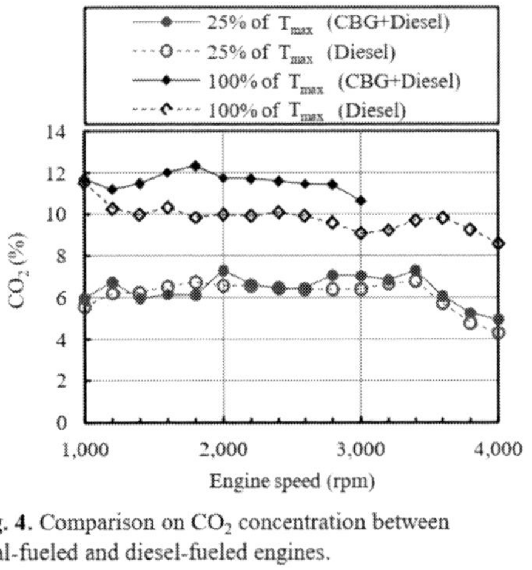

Fig. 4. Comparison on CO_2 concentration between dual-fueled and diesel-fueled engines.

4 Conclusions

The CO_2 concentration in the exhaust gas from the CBG and diesel dual-fueled turbocharged common rail diesel engine has been thoroughly investigated following the torque curves and the optimum ratio of the CBG obtained from the previous study [8]. It could be concluded in a case of the CBG and diesel dual fuel that when the engine speed increased, the CO_2 concentration decreased. Because the optimum ratio of the CBG decreases during the high engine speed range, the proportion of the diesel increases. In conjunction with a decrease in the mass of the air flowing into the cylinders due to a decrease in the volumetric efficiency, the incomplete combustion occurs. The unburned carbon oxidizes to be the CO in higher proportion than the CO_2, thus, the CO_2 consequently decreases. In addition, it was found that the CBG and diesel dual fuel produced the CO_2 nearly the same as of the diesel fuel for a case of the low engine torque. On contrary, the dual fuel had higher CO_2 concentration during the high engine torque. These variations depend on the relation between the CO_2 and the CO concentrations following the chemical combustion equation. Experimental studies on other emissions, such as, CO, NO_x, SO_x, unburned hydrocarbon, particulate matter, etc., from turbocharged common rail diesel engine dually fueled with CBG are necessary and suggested to be conducted in the future.

This study was supported by the Department of Mechanical Engineering, Faculty of Engineering, Chiang Mai University. The authors would like to express their sincere appreciation for all of the support provided.

References

1. T.K. Garrett, *Automotive Fuels and Fuel Systems Volume 2: Diesel*, 1-12 (1994).
2. K.K. Jain, R.B. Asthana, *Automobile Engineering*, 16-24 (2004).
3. *European Vehicle Market Statistics Pocketbook 2015/16*, 47-100 (2016).
4. M.I. Khan, T. Yasmin, A. Shakoor, *Renew. Sust. Energ. Rev.* **51**, 785-797 (2015).
5. O. Bordelanne, M. Montero, F. Bravin, A. Prieur-Vernat, O. Oliveti-Selmi, H. Pierre, M. Papadopoulo, T. Muller, *J. Nat. Gas. Sci. Eng.* **3**, 617-624 (2011).
6. E. Ryckebosch, M. Drouillon, H. Vervaeren, *Biomass Bioenergy* **35**, 1633-1645 (2011).
7. M. Bhudtiyatanee, N. Kammuang-lue, P. Aggarangsi, T. Sucharitakul, *Proc. Bangkok Intl. Conf. on Eng. and Appl. Sci.* 94-102 (2015).
8. N. Kammuang-lue, M. Bhudtiyatanee, *J. Energy Inst.* (in press)
9. B.J. Bora, U.K. Saha, *Appl. Therm. Eng.* **92**, 111-121 (2016).
10. R.G. Papagiannakis, D.T. Hountalas, *Appl. Therm. Eng.* **23**, 353-365 (2003).

Causes of delay on highway construction projects in Thailand

Pornsirichotirat Thapanont[1,2*], *Charoenpornpattana* Santi[1], and *Xinghatiraj* Pruethipong [2]

[1] Department of Civil Engineering, Faculty of Engineering, King Mongkut's University of Technology Thonburi, Bangkok, Thailand
[2]Bureau of Research and Development, Department of Highways, Ministry of Transports, Bangkok, Thailand

Abstract. Delay in construction project is an important issue in construction management. This is because delay will cause negative impacts on both the owner and the contractor. For owner, it will lead to the late opening of new road. For contractor, it will cause extra operating cost. In this paper, therefore, the causes of delay in road construction projects are researched. This paper studied a list of road construction delay causes gathered from literature having different countries. Internationally, twenty-six factors that might cause delays of road construction projects are identified in this research. Among 26 factors, considering in Thailand, eight key factors are typically found as the cause of delays in road construction projects. These 8 factors are then prioritized to identify the most important causes of delays in road construction projects in Thailand. A questionnaire and personal interviews through project engineer have formed the basis of this paper. The study indicates that the 5 top factors affecting road construction delays are: incomplete drawings, lack of equipment efficiency or financial status of contractors, delay in relocating existing infrastructure structures, less of project engineer experiences, and delay in relieving environmental impact, respectively

1.Introduction

Background

Transportation infrastructure development is an important role to promoting the economic development of the country. Currently, the department of Highways develop highways' network development plan. This plan has objective to support national logistic networks, ensure the highway network to support ASEAN economic, and improve and maintain highway network throughout Thailand. Presently, the Department of Highways operate highways over 51,504 kilometres across all regions of Thailand, divided as 1,992 km of RC paved roads, 49,253 km of AC paved roads, and 258 km of non-paved roads. The travel patterns in Thailand are classified as: 89.60% by personal vehicle, 7.50% by bus, 1.80% by train, and 1.10% by airplane. From data in 2014, the main pattern of goods and products transportation is by land, about 81.20%. This information provides the necessity of high investment in road improvement and maintenance to provide the effective highway networks for support land transportation.

In highway construction, the delay in construction is one of the major problems that brought the department of highways concern. This caused the essential of research on construction delays in this study. The interest on finding causes of the highway construction delay can locate approaches to prevent or surround the delay problem. This research emphasizes on the analysis of causes on highways construction delays. Therefore, the main objectives of this study are:

1. Study the causes of delay in highways construction projects in Thailand;
2. Prioritizing delay causes in highways construction projects in Thailand.

2.Literature Review

[1, 2, 3, 4] stated that construction period is the key factor to determine the construction expenses. The construction delay are important for contractors as it will cause extra expenses associated with the problems or delays and raise some unexpected damage costs.

There are a number of papers cited on delay causes in construction which can be summarized as the followings.

[2] The research in West Bank, Palestine, found that there are 52 delay causes in road construction project. The most important 5 factors are 1) political situation, 2) limited movement between area of west bank segmentation, 3) award project to lowest bid price, 4) progress payment delay by owner and 5) shortage of equipment that should consideration to improve project performance.

[3] Carried out the analysis of delay factor for Hospital Projects in Vietnam. Thirty-three have found and ranked in their study. The research reveals that the 5 top reasons of the delay in hospital projects in Vietnam are:

* Corresponding author: thapanont.por@mail.kmutt.ac.th

financial difficulties to owner, supervisor's responsibility problem, owner design change, incompetence of contractor, and insufficient contractor experience.

[4] Performed the study on factors affecting schedule delay, cost overrun, and quality level in public construction projects. This study is based on a questionnaire with 26 factors which sent to the full population of publicly employed project managers. The analysis found the most influential factor for time is lack of project funding, for cost is error in consultant material, and for quality is construction work error.

[5] Examined the causes of construction delays on large scale construction project in Saudi Arabia. The research investigation based on data collected from project stakeholders, e.g. owner, construction company, construction consultant. The results indicated that there are 73 causes of construction delays which could be classified as 9 groups of causes. The research shows that change order is the prime reason of construction delay.

[6] The research found significant factors that causing delay in Malaysia building construction project. The study indicated that the prime cause of delay is financial problems and seconded by the lack of cooperation of stakeholder.

[7] The delay and cost overruns that studied in Vietnam construction focus on large projects. This paper examined the delay causes by employing a questionnaire survey from 87 experts from Vietnamese construction. The 21 delay causes and cost overruns was applied to categorize with Factor analysis technique and found the 7 yielded factors: i)Lack of Constraint and cumbrous, ii) incapable, iii) Design, iv) Market state and Estimate, v) Financial Capability, vi) Government, vii) Labor or Worker

[8] The research in Benin, the study focus on the delay factors which affect in construction completion. The 35 factors were identified and used to make the questionnaires and sent to the respondent like construction managers, contractor, owner, consultant and architects. The outcome show that the 10 most significant delay factors for example: contractor financial capability, financial difficulties from owner, subcontractor performance insufficiently.

[9] Stated that for the lights of enhancing highway construction proficiency in Thailand, the Department of highways performs the quality evaluation of highway construction work. From the review of this reports, the authors found that eight factors are the top most cause of construction delay are: i) incomplete drawing, ii) lack of equipment efficiency or financial status of contractors, iii) delay in relocating old infrastructure structures, iv) less of project engineer experiences, v) delay in relieving environmental impact, vi) lack of traffic safety during construction, vii) lack of training of management to monitor the construction operations, and viii) shortage of materials.

3. Research Methodology

3.1 Review Causes of Delay

From the literature reviews, there are 26 causes of highway construction delay that are quite found in many researches as summarized in Table 1. The items are:

1. Difficulties on Financial
2. Material shortage on construction site
3. Poor construction site management
4. Mistakes and defective of work
5. Delay in delivery of materials on construction site
6. Problems on coordination with others
7. Site labour Shortage
8. Low productivity of labour
9. Poor skills and experience of labour
10. Lack of skill of subcontractor
11. Lack of contractor's staff on site
12. Shortage of equipment and tool on construction site
13. Inadequate planning and scheduling
14. Financial problem of owner
15. Slowness on making decisions
16. Contract modifications during construction
17. Lack of owner's coordination with contractor
18. Lack of material on the market
19. Poor construction site condition
20. Lack of equipment and tools on the market
21. Poor conditions of weather
22. Delays of Transportation
23. External work due to public agency
24. Poor conditions of economic
25. Laws and regulations change
26. Rise in price of the material

From these 26 major causes of delay, this research investigated the top rank reasons of delay based on data collected through the interview of highway construction management project manager from Bureau of Highways Construction, Department of Highways. The 8 major causes of highway construction delay are:

i) Incomplete drawing,
ii) Lack of equipment efficiency or financial status of contractors,
iii) Delay in relocating existing utilities structures,
iv) Less of project engineer experiences,
v) Delay in relieving environmental impact,
vi) Lack of traffic safety during construction,
vii) Poor site management, and
viii) Shortage of materials.

Table 1. Review for delay causes in former research

Causes of delay item	1	2	3	4	5	6	7	8	9	10	11	12	13	14	15	16	17	18	19	20	21	22	23	24	25	26
Sadi A. Assaf et al. (2006)	√		√	√	√	√	√	√		√	√	√	√	√	√	√	√	√	√			√	√		√	
Wa'el Alaghbari et al. (2007)	√	√	√	√	√	√	√	√	√	√	√	√		√	√	√	√	√	√	√	√	√	√	√	√	
Long Le Hoai et al. (2008)	√		√	√					√	√	√			√				√	√		√				√	√
Ibrahim Mahamid et al. (2012)	√			√		√		√	√		√	√	√	√	√	√	√	√			√			√	√	
Romuald-Kokou T.M. Akogbe et al.(2013)	√		√	√					√	√		√	√	√	√	√			√		√	√			√	√
Jesper Kranker Larsen et al. (2016)		√									√		√		√			√	√		√		√			
Soo-Yong Kim et al. (2016)	√		√	√						√	√	√		√		√					√			√	√	

3.2 Questionnaire Design

The eight causes of delay stated in section 3.1 were used in the questionnaire. The structure of questionnaire was designed into 3 parts, Part I: Personal Information of the Respondent, Part II: ranking the severity of causes of construction delay which have 5 level; level 5 = most severe, 4 = very severe, 3= severe, 2= quite severe, 1= less severe, Part III ranking the Frequency of Occurrence of causes of construction delay which have 5 levels; 5 = most frequent, 4 = very frequent, 3 = frequent, 2 quite frequent, and 1 = less frequent.

4. Data Analysis

4.1 Respondent Personal Information

All of project manager; 10 persons from Bureau of Highways Construction Department of Highways. The respondent average age are 51 year and average Experian are 24.20 year

4.2 Ranking causes of delay

In this study, the technique of mean square was used to rank the cause of delay, as shown in Equation 1.

$$MS = \left(\frac{\sum f \times s}{N}\right) \quad (1)$$

Where

MS = Mean Square,
f = Frequency of Occurrence,
s = Severity of Occurrence (5 levels), and
N = Number of respondents (Max.10 respondents)

Table 2. Ranking cause of delay

Causes	Mean	SD	Rank
Incomplete drawing	16.00	5.9	1
Less of project engineer experiences	11.80	3.5	4
Lack of equipment efficiency or financial status of contractors,	13.90	6.0	2
Delay in relieving environmental impact	10.20	4.0	5
Shortage of materials.	7.20	4.0	8
Delay in relocating existing utilities structures	12.20	3.4	3
Lack of traffic safety during construction	10.00	3.8	6
Poor site management	9.30	3.9	7

5. Discussion

The results from Table 3 indicated that the cause of delays in Thailand can be ranked by the severity of occurrence as followings.

i) Incomplete drawing
ii) Lack of equipment efficiency or financial status of contractors,
iii) Delay in relocating existing utilities structures,
iv) Less of project engineer experiences,
v) Delay in relieving environmental impact,
vi) Lack of traffic safety during construction,
vii) Poor site management, and
viii) Shortage of materials.

Furthermore, in this research, the comparison of causes of road construction delay between Thailand and international are performed. The 26 causes of construction delay from reviewing of various international literatures which were mentioned on section 3.1 are used as international causes of delay. The details of comparison describes on Table 4. It can be concluded that 4 causes of delay in Thailand are commonly found in International environment, namely 1. Difficulties on Financial, 2. External work due to public agency, 3. Poor site management and 4. Lack of material on the market. However, the results indicated that 4 reasons of construction delays are found only in Thailand environment, i.e. 1.Incomplete drawing, 2. Less of project engineer experiences, 3. Delay in relieving environmental impact, and 4. Lack of traffic safety during construction.

Table 3. Comparison of Causes of Delay

Item	Causes of Delay in Thailand	Cause of Delay in Literature Review (Section 3.1)
1	Incomplete drawing	-
2	Lack of equipment efficiency or financial status of contractors,	Item 1.Difficulties on Financial
3	Delay in relocating existing utilities structures	Item 23. External work due to public agency
4	Less of project engineer experiences	-
5	Delay in relieving environmental impact	-
6	Lack of traffic safety during construction	-
7	Poor site management	Item 3. Poor construction site management
8	Shortage of materials.	Item 18. Lack of material on the market

6. Conclusion

In this paper, the causes of delay in road construction projects are researched. A questionnaire and personal interviews through project engineer have formed the basis of this paper. The study reveals that the 5 top factors affecting road construction delays are:

1. Incomplete drawings,
2. Lack of equipment efficiency or financial status of contractors,
3. Delay in relocating existing utilities structures,
4. Less of project engineer experiences, and
5. Delay in relieving environmental impact, respectively.

The benefit of this study will educate project engineer the important of surrounding these delay factors to provide the completion of Highways construction project on time. However this study that has a population constraint then in the future work could be study in wider scope.

References

1. José Ramón San Cristbal, *Time, Cost, and Quality in a Road Building Project*, Journal of Construction Engineering and Management, **Vol.135**, 1271-1274 (2009)
2. Ibrahim Mahamid, Amund Bruland and Nabil Dmaidi, *Cause of Delay in Road Construction Projects*, Journal of Management in Engineering, **Vol.28**, 300 – 310 (2012)
3. Soo-Yong Kim, Kiet Nguyen Tuan and Van Truong Luu, *Delay Factor Analysis for Hospital Projects in Vietnam*, KSCE Journal of Civil Engineering **20**, 519 – 529 (2016)
4. Jesper Kranker Larsen, Geoffrey Qiping Shen, Soren Munch Lindhard and Thomas Ditlev Bronoe, *Factors Affecting Schedule Delay, Cost Overrun, and Quality Level in Public Construction Projects*, Journal of Management in Engineering ,**32** (2016)
5. Sadi A. Assaf, Sadiq Al-Hejji, *Causes of delay in large construction projects*, International Journal of Project Management, **24**, 349-357 (2006)
6. Wa'el Alaghbari, Mohd. Razali A. Kadir, Azizah Salim and Ernawati, *The significant factors causing delay of building construction projects in Malaysia*, **Vol.14**,192 – 206 (2007)
7. Long Le Hoai, Young Dai Lee and Jun Yong Lee, *Delay and Cost Overruns in Vietnam Large Construction Projects :A Comparison with Other Selected Countries*, KSCE Journal of Civil Engineering, **12**,367–377 (2008)
8. Romuald-Kokou T.M. Akogbe, Xin Feng, and Jing Zhou, *Importance and Ranking Evaluation of Delay Factors for Development Construction Projects in benin*, KSCE Journal of Civil Engineering **17**, 1213 – 1222 (2013)
9. Report on Department of Highways, *Highways Construction Performance Evaluation in Fiscal Year 2014 and 2015*, Department of Highways, Ministry of Transports, Bangkok, Thailand

Improvement of dynamic characteristics of manipulator driven by a gas-liquid phase-change actuator using an antagonistic drive

Tomonori Kato[1,*], *Kenya* Higashijima[1], *Yusuke* Kuradome[1], *Kohei* Noguchi[1], and *Manabu* Ono[2]

[1]Department of Intelligent Mechanical Engineering, Faculty of Engineering, Fukuoka Institute of Technology, Fukuoka, Japan
[2]Monozukuri Department, Tokyo Metropolitan College of Industrial Technology, Tokyo, Japan

Abstract. The goal of this research is to improve the dynamic characteristics of a manipulator composed of pneumatic artificial rubber muscles driven by gas-liquid phase change. Pneumatic actuators, such as pneumatic artificial rubber muscle (PARM) or rubber bellows, have been widely used in many industrial and research fields. They have merits of being compact and lightweight. However, the large size of the compressor driving the actuator is a problem. To overcome this, the authors researched soft actuators driven by the gas-liquid phase change (GLPC) of fluorocarbon. Fluorocarbon ($C_5F_{11}NO$) is a substance with a relatively low boiling point (50 °C) and a low heat of evaporation (104.65 kJ/kg). The heat of evaporation of water is 2260 kJ/kg. This paper presents the overview of an actuator driven by GLPC. Then, fabrication of a manipulator using the GLPC driven PARM, and details of experiments conducted to determine manipulator characteristics are given. To improve the dynamic characteristics of the manipulator, a force control method using the antagonistic drive of two PARMs is proposed, and experiments are conducted to validate the effectiveness of the proposed method.

1 Introduction

A pneumatic artificial rubber muscle (PARM) is a pneumatically driven actuator that imitates muscle contraction. As shown in Fig. 1, when the inside is pressurized, PARM expands in a radial direction, shrinking and generating a pulling force in the axial direction. Several studies have applied PARMs to robots that contact humans, because PARMs have beneficial properties, such as light weight and flexibility [1] [2]. However, a PARM requires many cumbersome peripheral devices, such as a compressor, pressure regulators, servo valves, and piping to drive it. (Fig.2).

To solve this problem and to realize a compact PARM driving system, the authors researched PARM drives using gas-liquid phase-change of fluorocarbon (GLPC) instead of a compressor and other peripheral devices. At the initial stage of our research (2000–2010), it took more than 200 s to generate the adequate pressure to drive an actuator with GLPC [3].

Therefore, in recent years, we studied the dynamic response of GLPC-driven PARM by modifying the structure of the actuator and applying a feedback control scheme. Thus, the pressure response was improved [4]. Particularly, when the inside pressure was raised, the time constant of the pressure of less than 1 s was realized [5]. Thus, PARMs, driven by GLPCs, can apply to many practical problems. However, when the inside pressure is reduced, the time constant tends to be larger, because cooling the fluorocarbon takes longer than heating it.

In this paper, the overview of the GLPC-driven actuator is first explained. Second, design and fabrication of the manipulator is explained, and the result of the gripping experiment is shown. Third, to compensate the dynamic characteristics of the manipulator, the force control method using antagonistic drive of two PARMs is proposed and experiments are conducted to validate the effectiveness of the proposed method.

2 Manipulator driven by GLPC

2.1 Concept of a GLPC-driven actuator

GLPC is a phenomenon in which a substance transitions from liquid to gas or from gas to liquid. When a liquid is heated, it begins to boil and transitions to a gas. This transition causes its volume to expand and the pressure in the container to increase. When the substance is removed from the heat source, it loses its energy and returns to the liquid phase because of the heat transfer from inside the container to the ambient environment. Consequently, the substance volume contracts and the pressure decreases. Fig. 3 illustrates the concept of an actuator driven by GLPCs.

The actuator takes advantage of volume expansion during the liquid-to-gas phase change. A working fluid is added to an actuator, such as a PARM, and an electric heater (e.g., constantan heater) is installed. When the

* Corresponding author: t-kato@fit.ac.jp

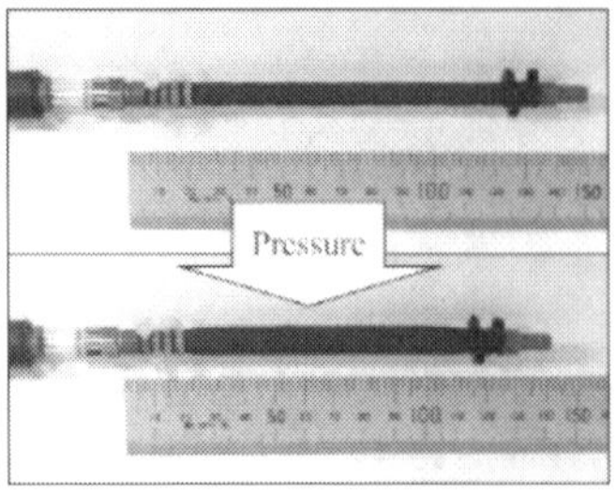

Fig. 1 Pneumatic artificial rubber muscle.

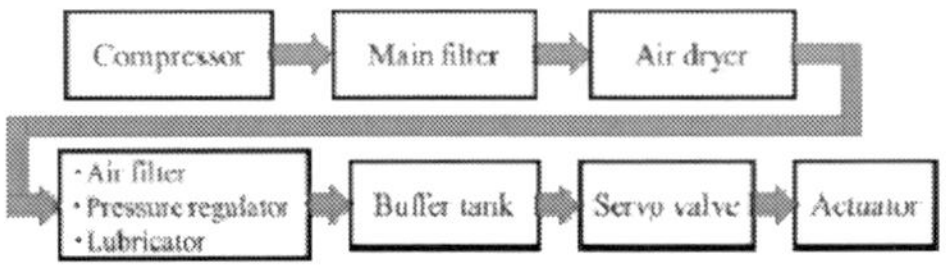

Fig. 2 Pneumatic components driving PARM.

heater in the actuator is powered on, the heated liquid expands and boils into gas. The actuator is driven by the pressure generated by the volume expansion because of the thermal expansion of the liquid and the following GLPC. When the heat source is removed, the generated gas returns to liquid and the volume contracts. Because the GLPC actuator does not require a compressor or other peripheral pneumatic components, it is possible to miniaturize the entire apparatus for driving the actuator.

In this study, fluorocarbon ($C_5F_{11}NO$) is used as the working fluid. Its characteristics are compared to those of water, and presented in Table 1. Because this fluorocarbon has a low boiling point of 50 °C at atmospheric pressure, and a vaporization heat of 104.65 kJ/kg, which is 1/22 that of water (2260 kJ/kg), a small thermal energy supply can induce the liquid-to-gas phase change. The coefficient of thermal expansion of the fluorocarbon at 20 °C is 0.00154 $°C^{-1}$, which is approximately seven times that of water. Moreover, it is non-poisonous and incombustible, and its properties are not changed by heating, according to its product manual issued by 3M. Thus, this working fluid is suitable for use in the GLPC actuator.

2.2 Robotic manipulator driven by PARM using GLPC

A manipulator whose driving force is generated by a PARM using GLPCs is designed and fabricated. The image and that of the controlling devices of the fabricated manipulator are shown in Figs. 4 and 5. The PARM used in this research is FESTO MXAM-5-AA. Beneath the PARM, a fixed container with a volume of 3.93 cm^3 is installed. The PARM and the container are filled with the fluorocarbon working fluid, which is heated and boiled by powering the constantan heater (Tokyo Wire Works, Ltd., diameter: 0.231 mm, resistance per length: 16.02 Ω/m, total resistance: 20.6 Ω) at the bottom of the container. Additionally, a pressure sensor (SMC PSE510-R06) is installed between the PARM and the container. The control signal (voltage: 0–10 V) is generated by the digital signal processor (DSP) (i.e., MTT s-BOX). By inputting the control signal into the power source, a voltage of 0–35 V is applied to the heater. The working fluid in the container is heated by the heater to induce the GLPC. According to the kinematic calculation, when the pulling force generated by the PARM is 20 N, the hand can generate a gripping force of about 2.57 N.

A PI feedback control was utilized to control the temperature as shown in Fig. 6. The PI control gains were set as follows. Proportional gain was set to 1000 V/Pa, and integral gain was set to 3 V/(Pa·s). The reference pressure, P_{ref}, was initially set to 0.3 MPa (i.e., gauge). At 40 s, P_{ref} was increased to 0.35 MPa. Then, at 80 s, it was decreased back to 0.3 MPa. The gripping force was measured by a load cell (Tokyo Sokki, TCLZ-20NA, Dynamic Strain Meter DA-18A). The generated pressure P, gripping force F, and voltage to the heater E, were measured by the data logger.

The experimental results are shown in Fig. 7. The time constants were 0.18 s when the pressure was increased from 0.3 to 0.35 MPa and 1.35 s when the pressure was decreased from 0.35 to 0.3 MPa. During 40–80 s, the pressure P was kept nearly steady at 0.35 MPa. Therefore, the gripping force was kept at 2.57 N.

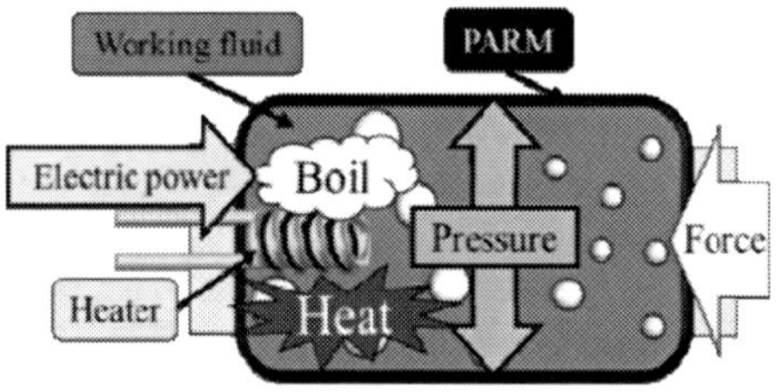

Fig. 3 Concept of the GLPC-driven actuator

Table 1. Characteristics of working fluid compared to water.

Working fluid (Chemical formula)	Boiling point (1 atm) [°C]	Heat of vaporization [kJ/kg]	Coefficient of thermal expansion [$°C^{-1}$]
Fluorocarbon ($C_5F_{11}NO$)	50	104.65	0.00154
Water (H_2O)	100	2257	0.00021

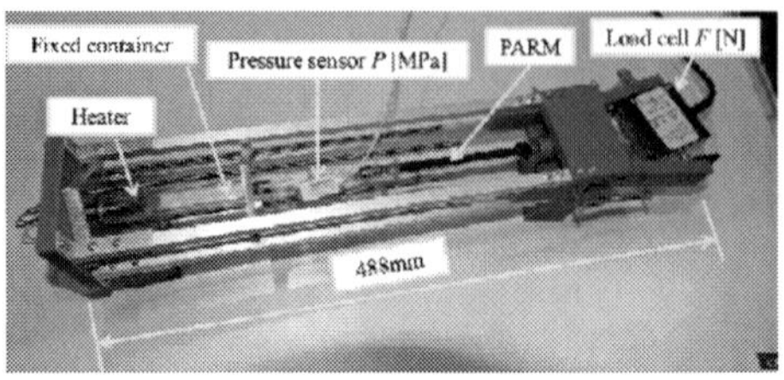

Fig. 4 Fabricated manipulator.

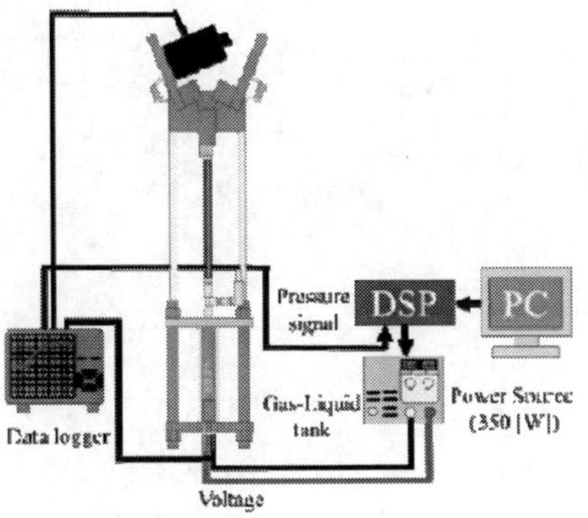

Fig. 5 Controlling devices of the manipulator.

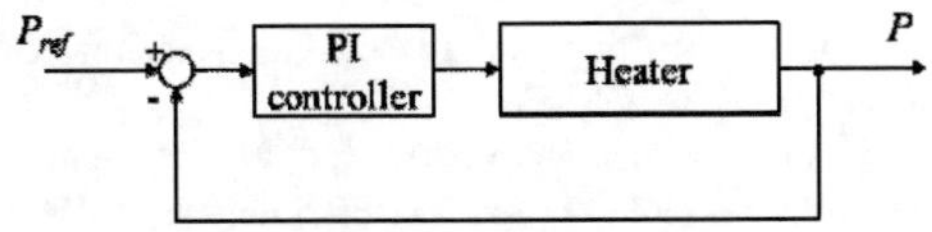

Fig. 6 Block diagram of PI feedback control.

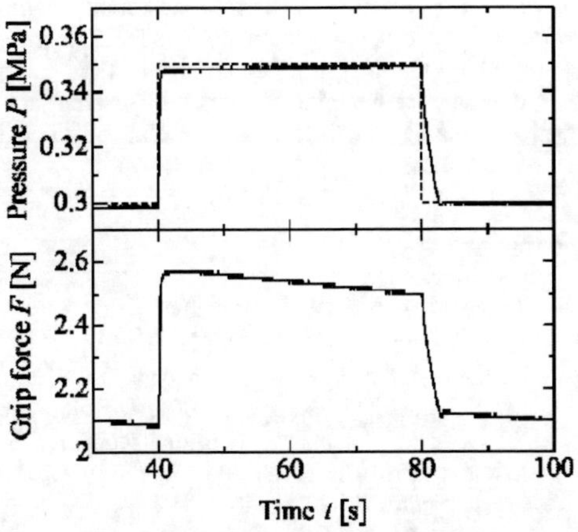

Fig. 7 Results of the manipulator gripping experiment.

3 Compensation of dynamic characteristics

3.1 Force control of antagonistic drive of 2 PARMs

Response of the PARM driven by GLPC is slow, particularly when the inside is depressurized. To solve this problem, an antagonistic drive system shown in Fig. 8 is designed and fabricated. The device is composed of two PARMs, a gear-wheel, a chain, and a rod. At the tip of the rod, a load cell is set. Because the two PARMs are antagonistically driven, when PARM A is contracted (i.e., pressurized), PARM B is not contracted, and torque τ [N m] is generated. Therefore, pushing force F' [N] is generated at the tip of the rod. The antagonistic drive system is especially beneficial when repeating force change is required, because PARM A and PARM B can alternately generate pulling force.

The fabricated device is shown in Fig. 9. The length of the rod and the diameter of the pully are indicated in the figure. With both PARM A and PARM B, PI pressure feedback control is conducted as in the previous section.

3.2 Experimental procedures and results

By using the experimental device shown in Fig. 9, repeating force control was performed to validate the effectiveness of the proposed antagonistic drive. The timing chart of the pressure reference values of both PARMs is shown in Fig. 10. In the experiment, both reference values were set at 0.3 MPa (i.e., gauge). Then, at 40 s, the reference value with PARM A was raised to 0.35 MPa. By geometric calculation, this should generate a pushing force of F' =0.65 N. Third, at 80 s, the reference value of PARM B was raised to 0.36 MPa, whereas the value for PARM A was kept at 0.35 MPa. In the experiment, the pressure values in both PARM A and B, and the pushing force F' [N] were measured.

The experimental results are shown in Fig. 11, where the dark black line indicates the pushing force, F' [N], measured at the tip of the rod. Between 40–80 s, the F' was kept at around 0.63–0.65 N. Before 40 s and after 80 s, the value was almost 0 N. The time constants for the force F' [N] were 0.78 s when increasing the force (soon after 40 s) and 0.31 s when decreasing the force (soon after 80 s).

A comparative experiment was conducted using the pressure reference values shown in Fig. 12. Because the value for PARM B is unchanged throughout the experiment, this condition is close to a manipulator driving experiment using a single PARM, as explained in the last chapter. With the reference values shown in Fig. 13, the time constant when decreasing the force was 1.08 s. The comparison of the experimental results with the proposed driving method and that of the comparative experiment are shown in Fig.13.

Therefore, the proposed antagonistic driving method can compensate the weak point in the dynamic characteristics of GLPC-driven PARM.

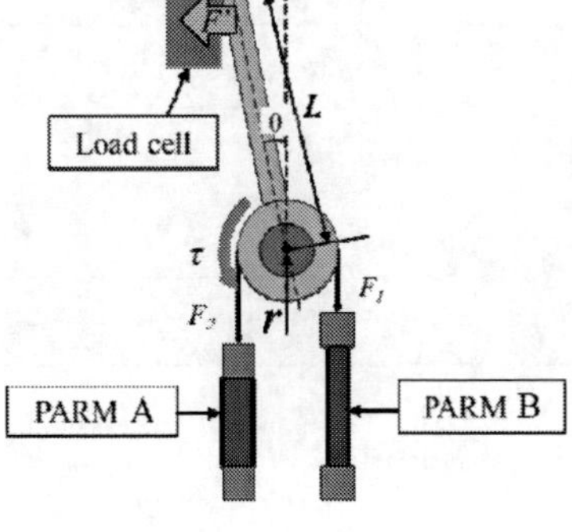

Fig. 8 Antagonistic drive system.

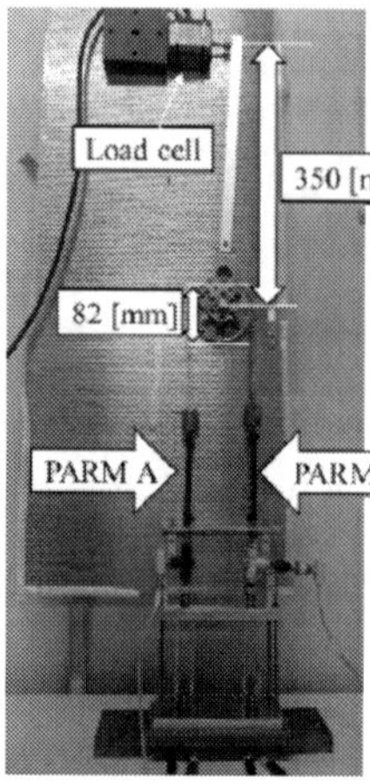

Fig. 9 Experimental device of the antagonistic drive.

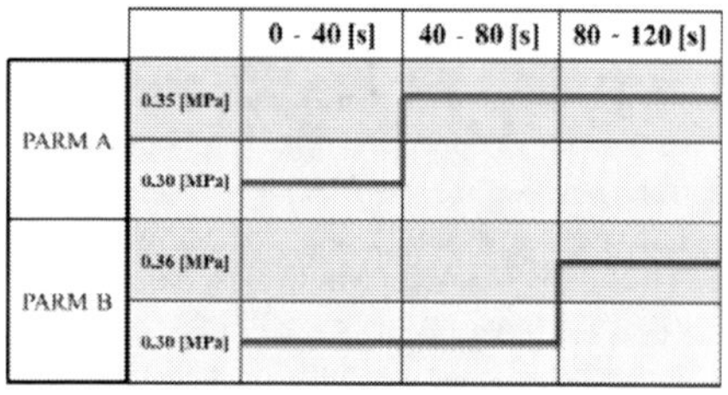

Fig. 10 Timing chart of the antagonistic drive system.

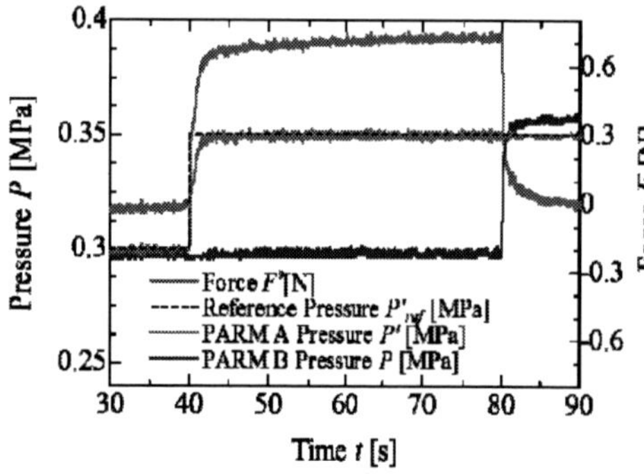

Fig.11 Experimental results of antagonistic driving device.

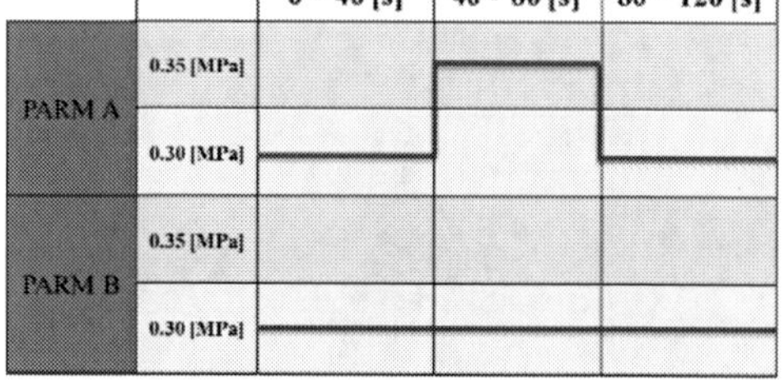

Fig. 12 Timing chart of antagonistic drive system for the comparative experiment.

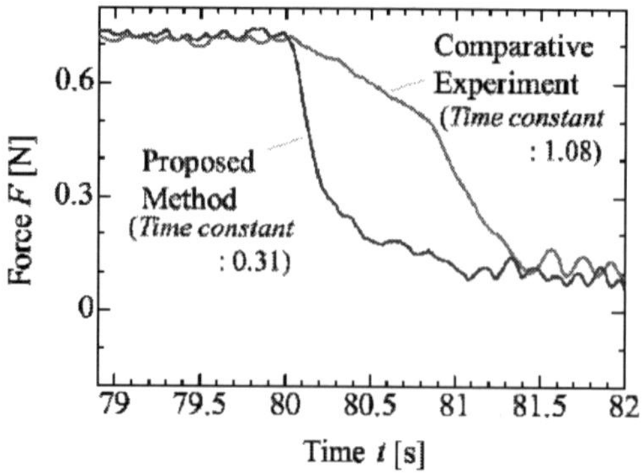

Fig. 13 Comparison of experimental results.

4 Conclusions

In this paper, first, we presented an overview of the actuator driven by GLPC.

Second, design and fabrication of a manipulator was explained, and the result of gripping experiment was presented.

Third, to compensate the dynamic characteristics of the manipulator, force control method using antagonistic drive of two PARMs was proposed, and experiments were conducted using a fabricated device. Experimental results show that the proposed control method can make the dynamic characteristics of GLPC-driven actuator faster, reducing the time constant value of force from 1.08 s to 0.31 s.

References

1. T. Noritsugu. 1997. Pneumatic Actuators, *J. Robotics Soc., Japan*, **15**(3), pp. 355-359.
2. K. Kawashima, T. Sasaki, T. Miyata, N. Nakamura, M. Sekiguchi, T. Kagawa. 2004. Development of Robot Using Pneumatic Artificial Rubber Muscles to Operate Construction Machinery, *J. Robotics, Mechatronics*, **16**(1), pp. 8-15.
3. S. Matsuda, M. Ono, T. Izumi, S. Kato. 2005. Metal Bellows Type Gas-liquid Phase-change Actuator: Applications for Endoscopic Surgery Forceps, *J. Japan Soc. for Precis. Engin.*, **71**(6), pp. 723-728.
4. T. Kato, S. Honda, M. Cheng, K. Sakuragi, M. Ono. 2016. Fabrication of a Miniature Rubber Muscle Actuator Driven by Gas-Liquid Phase Change, *J. Fluid Power Sys.*, **9**(1), pp. 11-17.
5. K. Higashijima, T. Kato, K. Sakuragi, T. Sato, M. Ono. 2017. Development of Manipulator using a Gas-Liquid Phase-Change Actuator. *10th JFPS Intl. Symp. Fluid Power 2017*, 2B04.

Numerical simulation of porous media combustion for high temperature heat exchanger

Panu Iamsakulpanich[1], *Kittipass* Wasinarom[2], *Thanathon* Sesuk[3], *Jarruwat* Charoensuk[2], *Katsunori* Hanamura[4], *Preecha* Karin[1], and *Visarn* Lilavivat[3*]

[1]Automotive engineering, International College, King Mongkut's Institute of Technology Ladkrabang, Bangkok, Thailand
[2]Department of Mechanical Engineering, Faculty of Engineering, King Mongkut's Institute of Technology Ladkrabang, Bangkok, Thailand
[3]Materials for energy research unit, National metal and materials technology center (MTEC), Patumthani, Thailand
[4]Tokyo Institute of Technology, Tokyo, Japan

Abstract. The purpose of this work is developing the numerical 1D model of porous media combustion for investigating porous media burner systems. The software is used to solve energy, mass transfer and chemical reaction equation of the combustion. The operating condition and property parameters, which mainly affect the functions and quality of the industrial burner design, such as the inlet velocity of the reactants, the equivalence ratio, the extinction coefficient and the thermal conductivity of porous media, will be investigated and validated with experimental data. For developing the procedure of experiment, three diameter sizes of porous media materials (5 mm, 10 mm, and 15 mm.) were used. As a result, the developed model will be used as a tool to explore temperature distribution of heat exchange to improve thermal performance and overall efficiency system. Moreover, this knowledge can be applied to design porous media burner systems for uniform temperature distribution operation.

1 Introduction

In the present, environmental concerns and global warming problems have an impact on our environment. Although, the sustainable energy such as biomass, hydropower, geothermal, solar, wind and marine energies has been progressively used, it still cannot supply enough for the total world energy. And the nuclear energy has a problem about safety. Furthermore, the world considers reducing the consumption of coal and petroleum to reduce greenhouse gases. Under these conditions, improve the efficiency of in the present combustion has important.

Currently, the heat exchanger is widely used in commercial industry such as power stations, chemical plants and petrochemical plants etc. Operating of the heat exchanger in commercial industry is transfer heat from combustion to use. high temperature heat exchanger is an improved efficiency of the heat exchanger in high temperature operating condition but they have problems about meltdown of material, how to design heat transfer in heat exchanger etc.

Accordingly, porous media materials in the combustion of heat exchanger have attracted attention for improving heat exchanger efficiency in this work because of their advantages such as higher burning rates than free flame, extended lean flammability limits, low emissions of pollutants, high radiant output, and increased power dynamic range.

This paper presents the development of the combustion modeling in porous media. The developing code is based on "One-dimensional CFD combustion modeling in porous media" written by Wasinarom[1]. The model is governed by conservation of mass, fuel species, energy in fluid pore space, energy in a solid porous matrix. According to the 2 phases of Heat transfer between solid and fluid phase, local thermal non-equilibrium was been modeling via energy source term of the solid and fluid energy equation. Arrhenius law is used to calculate fuel consumption and heat transfer from oxidation. The code is developed in one dimension due to prevent fluid dynamics complication at this stage. And, velocity field was merely calculated by ideal gas law as the calorific equation of state with 1D models. One dimensional code is recommended at the early stage of the complex porous media with combustion development.

2 Methodology

2.1 Porous media combustion modeling (PMC modeling)

The porous media combustion modeling trends up to 1994[2] are mainly described 1D models only. Takeno, Echigo, and their coworkers[3] are the pioneers in PMC modeling to study the effects of mass flow rate and heat transfer coefficient on flame characteristics in excess enthalpy flames and suggested inserting a porous, highly conductive solid into the flame to conduct heat from the solid to the reactants. They found that increasing the mass flow rate above the laminar burning rate increased the heat

* Corresponding author: visarn.lil@mtec.or.th

release rate and the reaction zone became more concentrated.

Generally, all 1D models assume 1D flow conditions and no radial heat losses[4]. These two assumptions may become inaccurate if the porous media combustor is a commercial burner prototype with complex geometry. In such a situation 1D relations will no longer be valid and multidimensional models are imperative. It is, therefore, necessary to predict 3D combustion and emissions in complex geometrical burner configurations to help the design of commercial inert porous burners[5].

Mohamad et al.[6,7] modeled a PM burner with embedded coolant tubes. The 2D continuity, momentum, energy and fuel mass fraction equations were solved and the combustion was described as a one-step reaction. A numerical code has been developed by Bidi et al.[8] to evaluate the effects of different parameters of combustion in porous media. The Navier-Stokes, the solid and gas energy and the chemical species transport equations were solved using a multi-step reduced kinetic mechanism. The discrete ordinates method was used to solve the radiative transfer equation and a finite volume method (FVM) based on SIMPLE method was applied to discretize the conservation equations. Moraga et al.[9] who studied the convective heat transfer within a cylindrical inert porous media combustor have used a 2D, two temperature mathematical model, based on fluid mechanics, energy and chemical species governing equations The FVM was used to solve the discrete model for methane combustion with air.

Hayashi and coworkers[10] had introduced the 3D modeling of a two-layer burner. The first layer was a perforated plate made of an insulating material (Al_2O_3) with the purpose of avoiding flash-back, while the second layer, a thin plate made of SiC foam to act as the reaction layer. They claimed that the proposed 3D model could facilitate a detailed study of the flow at the interface of the two solid layers, which is not possible utilizing one- or two-dimensional models, owing to the complex flow structure originated by the 3D jets from the perforated plate into the SiC foam.

For most of the cases, a 1D model with single-step reaction kinetics could yield the results with reasonable accuracy. In fact, the realistic prediction of pollutants formation necessitated detailed reaction kinetics to be incorporated in the model. And in 3D simulation, the inclusion of detailed reaction chemistry is still lacking, or 3D simulation is too small[11].

2.2 Porous media burner

Porous media combustion has three modes of heat transfer conduction, convection, and radiation. Advantages of Porous media burner[12] is a better homogenization of temperature across the porous media and the significant amount of radiation helps to preheat the incoming air-fuel mixture at upstream. The technique of premixed combustion within porous media has been studied and applied to steady combustion with great success. The porous media combustion has proved to be one of the applicable options to solve the problems to a remarkable extent in both technical and economic perspectives. This technique has[12] been used for both gaseous and liquid fuels in steady or unsteady combustion. Flame stability in porous media with lean and rich mixtures, a significant reduction in pollutants and increasing combustion efficiency, was proven.

The material of porous media burner[13], aluminum oxide (Al_2O_3), silicon carbide (SiC), and zirconium dioxide (ZrO_2) proposed as suitable materials for application. Aluminum oxide and zirconium dioxide were recognized as high temperature resistant materials. SiC shows good thermal shock resistance, mechanical strength, and conductive heat transport. Silicon carbide also has a high melting point (3260 K), against cyclic thermal stress and strength retention at the peak regenerator temperature (1673 K), and excellent oxidation resistance. Metallic materials were found less suitable for porous media because of their inadequate thermal stability and high thermal inertia.

A comparison of the data for materials relevant for use in porous media burners is shown in Table 1.

Table 1 Most important material data for Al_2O_3, SiC and ZrO_2[14]

Property	Unit	Al_2O_3	SiC	ZrO_2
Maximum use temperature in air	°C	1900	1650	1800
Thermal expansion coefficient (20-1000 °C)	10^{-6} K^{-1}	8	4-5	10-13
Thermal conductivity at 20 °C	W m^{-1} K^{-1}	20-30	80-150	2-5
Thermal conductivity at 1000 °C	W m^{-1} K^{-1}	5-6	20-50	2-4
Specific thermal capacity	$J\,g^{-1}$ K^{-1}	0.9-1	0.7-0.8	0.5-0.6
Thermal stress resistance parameter, hard shock, R (σ/Eα)	K	100	230	230
Thermal stress resistance parameter, mild thermal shock, R' (Rλ)	10^{-3} W m^{-1}	3	23	1

3 Experiment section

3.1 Experiment setup

Fig. 1 shows the schematic of the experimental setup. This system is composed of methane gas and air supply systems, a mixing chamber, four sizes of porous media, and a measurement system. Methane gas and air were controlled by MFC and then mixed in the mixing chamber. The methane/air mixture flowed through flashback protection and after that combustion in a quartz tube 75 mm. in diameter and 170 mm. long.

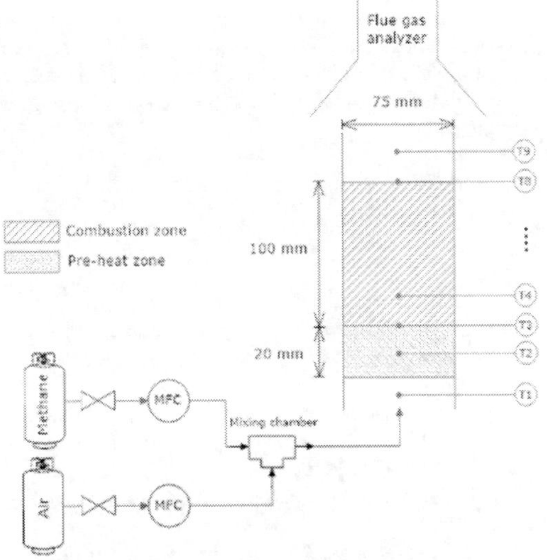

Fig. 1 Schematic of the experimental setup.

In the burner, a quartz tube was separated into two layers of porous media. The first layer was a preheat zone that was 20 mm. long and filled with Al_2O_3 pellets diameter size 3.0 mm. The second layer was the flame region that was 100 mm. long and filled with Al2O3 pellets diameter size 5.0 mm. in the 1st experiment, 10 mm. in the 2nd experiment, and 15 mm. in the 3rd experiment.

Thermocouples were set up as No.1 in the premixing zone and No.2 is in center of flashback protection zone. The No.3 is on the border between flashback protection zone and combustion zone. Thermocouples No.4-8 begin to start apart from No.3 in every 20mm until No.9 set depart of the combustion zone. The recorded data is temperature data and emission of the combustion exhausts such as CO, O_2, NO, C_xH_y and NO_2 at the burner outlet.

The air flow rate at inlet burner of methane/air mixture is kept constant between 0.09-0.15 m/s. And, equivalent ratio is developed between 0.5-1.0 that was suggested by Marker J. et al.[15].

3.2 Modeling equations

For simplify the problem, Model was followed under the below assumptions.

1. The porous media was setup as gray homogeneous media.
2. The wall is no slip, adiabatic, and radiative gray.
3. In high temperature, Potential catalytic effects of the solid are ignored.
4. Gas radiation and Body forces is neglected.
5. Reactants and Products setup as incompressible ideal gases.

A set of differential equations is in the following form.

Continuity equation,

$$\frac{\partial \varepsilon \rho_g}{\partial t} + \nabla \cdot \left(\varepsilon \rho_g u\right) = 0 \tag{1}$$

Where

ρ_g is the density of methane/air mixture

ε is the porosity, u is the velocity

Momentum equation,

$$\varepsilon \rho_g \left(\frac{\partial u}{\partial t} + u \cdot \nabla u \right) = -\varepsilon \nabla p + \mu \nabla^2 u + \frac{1}{3} \mu \nabla \left(\nabla \cdot u \right) \tag{2}$$

Where

μ is the viscosity of gas.

Species conservation equation,

$$\frac{\partial \varepsilon \rho_g Y_i}{\partial t} + \nabla \cdot \left(\varepsilon \rho_g Y_i\right) = -\nabla \cdot \left(\varepsilon \rho_g Y_i V_i\right) + \varepsilon \text{[redacted]} \tag{3}$$

Where

Y_i is the molar faction of substance

V_i is the diffuse velocity of substance

[redacted] is the reaction rate of substance

W_i is the molar mass of substance

Energy equation,

For gas

$$\frac{\partial \left(\varepsilon \rho_g C_g T_g\right)}{\partial t} + \nabla \cdot \left(\varepsilon u \left(\rho_g C_g T_g + p\right)\right) = h_v \left(T_s - T_g\right) + \nabla \cdot \left(\varepsilon k_g\right) \nabla T_g + \varepsilon \text{[redacted]} \tag{4}$$

For solid

$$\frac{\partial \left[\left(1-\varepsilon\right)\rho_s C_s T_s\right]}{\partial t} = \nabla \cdot \left(k_s \nabla T_s\right) + h_v \left(T_s - T_g\right) \tag{5}$$

Where

[redacted] is the heat from the chemical reaction

h_v is the convection of heat transfer coefficient

k_s is the heat transfer coefficient for solids

The inlet velocity is defined by

$$u = \frac{\text{[redacted]}}{A} \text{[redacted]} \frac{}{\pi D^2} \text{[redacted]} \frac{}{\pi D^2} \tag{6}$$

Where

[redacted] is the gas flow rate of methane/air mixture

A is the sectional area of the flow path

D is the diameter of the flow path

[redacted] is the gas flow rates of methane

[redacted] is the gas flow rates of air

The porosity is defined as

$$\text{Porosity}(\varepsilon) = \frac{V_{\text{cavity}}}{V_{\text{total}}} \times 100 \tag{7}$$

Where

ε is the ratio of the void volume within a porous media to the total bulk volume of the porous media

V_{Cavity} is a volume of cavity of porous media pack bed

V_{total} is a volume of cylindrical quartz tube

4 Results

In this study, the combustion position occurs at the rapidly changed temperature of the fluid. And the location of the combustion is depended upon the porosity. The porosity. At the porosity of 0.30 or 5 mm. of diameter size of porous media has the combustion position at approximately 1.9 cm., and porosity of 0.35 (diameter size 10mm) has the combustion position at 2.2 cm., and porosity of 0.40 (diameter size 15mm.) has the combustion position at 2.5 cm., respectively

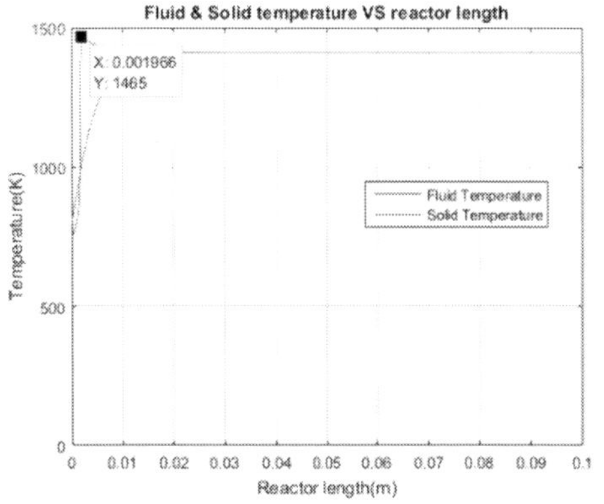

(a)

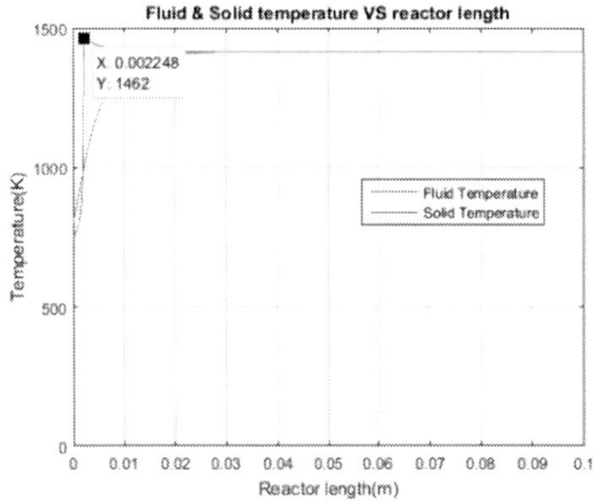

(b)

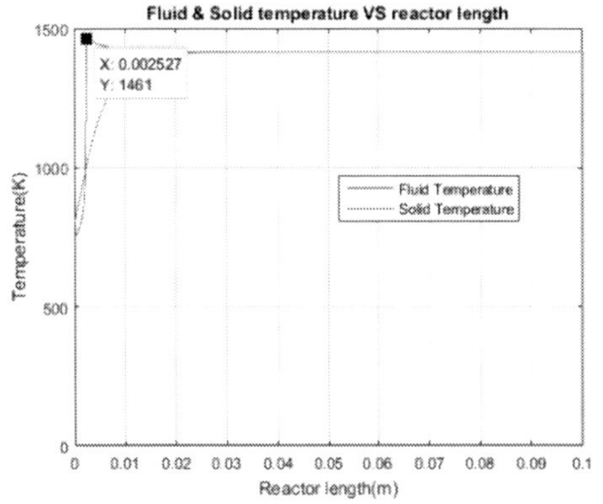

(c)

Fig. 2 Temperature distribution along the axial distance of modeling at porosity (a) 0.3, (b) 0.35, and (c) 0.4

5 Conclusions

The comparison three diameter size of porous media model results, the porosity of porous media affects the position of combustion that occurs in the model. The lower value of porosity shows the faster combustion will occurred.

This work was supported by National metal and materials technology center (MTEC).

References

1. K. Wasinarom, K. Onthong, D. Boonchauy, J. Charoensuk, *TSME ICOME DECEMBER2016.*, (2016).
2. M. A. Mujeebu, M. Z. Abdullah, A. A. Mohamad, and M. Z. A. Bakar, *Prog. Energy Combust. Sci.*, **36**, 627–650 (2010).
3. T. Takeno and K. Sato, *Combust. Sci. Technol.* (1979).
4. A. Loukou et al., *Int. J. Hydrogen Energy*, **42**, 652–663 (2017)
5. I. MALICO, X. Y. ZHOU, J. C. F. PEREIRA, and J. C. F. Pereira, *Combust. Sci. Technol.*, **152**, 57–79 (2000)
6. A. A. MOHAMAD et al., *Combust. Sci. Technol.*, **96**, 387–407 (1994)
7. A. A. Mohamad, S. Ramadhyani, and R. Viskanta, *Int. J. Heat Mass Transf.* **37**, 1181–1191 (1994).
8. M. Bidi, M. Saffar Avval, and M. Heyrani Nobari, in *Eurotherm Seminar n° 81 Reactive Heat Transfer in Porous Media*, (2007).
9. N. O. Moraga, C. E. Rosas, V. I. Bubnovich, and N. A. Solari, *Int. J. Heat Mass Transf.* (2008).
10. T. C. Hayashi, I. Malico, and J. C. F. Pereira, in *Computers and Structures*, (2004).
11. V. Vandadi and C. Park, *Energy* (2016).
12. Z. Masoud and M. Arash, *Numerical Simulation - From Theory to Industry*, chapter 22, *Intech* (2012).
13. P. Chitmongkol, thesis, RMUTT (2014).
14. F. Avdic, Thesis, Universität Erlangen-Nürnberg, (2004).
15. T. D. B. Mark J. Khinkis, Hamid A. Abbasi, PATENT, US5476375A (1994)

MATEC Web of Conferences **192**, 02017 (2018) https://doi.org/10.1051/matecconf/201819202017
ICEAST 2018

Assessment of land cover on soil erosion in Lam Phra Phloeng watershed by USLE model

Jatuwat Wattanasetpong[1,*], *Uma* Seeboonruang[2], *Uba* Sirikaew[3] , *Walter* Chen[4]

[1]Master's degree. Student, Civil Engineering Department, Faculty of Engineering, King Mongkut's Institute of Technology Ladkrabang, Bangkok, Thailand

[2]Associate Professor, Civil Engineering Department, Faculty of Engineering, King Mongkut's Institute of Technology Ladkrabang, Bangkok, Thailand

[3]Lecturer, Civil Engineering Department, Faculty of Engineering, King Mongkut's Institute of Technology Ladkrabang, Bangkok, Thailand

[4]Professor, Department of Civil Engineering, College of Engineering, National Taipei University of Technology, Taipei, Taiwan

Abstract. Soil loss due to surface erosion has been a global problem not just for developing countries but also for developed countries. One of the factors that have greatest impact on soil erosion is land cover. The purpose of this study is to estimate the long-term average annual soil erosion in the Lam Phra Phloeng watershed, Nakhon Ratchasima, Thailand with different source of land cover by using the Universal Soil Loss Equation (USLE) and GIS (30 m grid cells) to calculate the six erosion factors (R, K, L, S, C, and P) of USLE. Land use data are from Land Development Department (LDD) and ESA Climate Change Initiative (ESA/CCI) in 2015. The result of this study show that mean soil erosion by using land cover from ESA/CCI is less than LDD (29.16 and 64.29 ton/ha/year respectively) because soil erosion mostly occurred in the agricultural field and LDD is a local department that survey land use in Thailand thus land cover data from this department have more details than ESA/CCI.

1 Introduction

Nowadays many global problems are caused from climate change and human activities. Soil erosion problem is one of them. For example, the total soil loss to the European Union is estimated to be 970 million tons annually [1], which is a major threat to the ecosystem, crop production, and drinking water. Soil erosion is the displacement of the upper layer of soil, one form of soil degradation. There are several variables that affect soil erosion such as water, soil type, human activity and so on. A low rate of soil erosion has been occurring in every land on the earth and high rate of soil erosion is mostly happening on areas with high steepness slope and deforestation. The effect of soil erosion has led to sedimentation in streams and rivers; when sediment has delivered through the river into the bottom of reservoir. The sedimentation increases loads on the dams and gates, damages mechanical equipment and creates a wide range of environmental impacts [2]. In the past three decades, rapid increases in human developments have caused some significant environmental problems, such as deforestation and slope mass movement in every place in Thailand. There is a department in Thailand called Land Development Department (LDD) that studies about soil erosion [3]. They published a table and data for evaluating soil erosion with Universal Soil Loss Equation (USLE) [4] for every region of Thailand [3] because they recognized the importance of soil erosion.

Lam Phra Phloeng reservoir is currently being used for agriculture and irrigation in Thailand. Severe erosion of the areas becomes a big problem consequently [5] the reservoir is one of the most seriously affected by soil erosion related sedimentation in Thailand. Most of the area in upper watershed is cultivated by sugarcane and cassava. After the crop has been harvested the land is tilled and becomes sensitive to sheet erosion [6]. Due to this area has many high slope on the mountain near the boundary of watershed, agriculture area and urban, this area is facing with many problem about soil erosion.

As mentioned previously about LDD, there exists no update of overviews of spatial soil erosion. It's necessary to evaluate soil erosion in this specific area. Lam Phra Phloeng watershed has enough reason to be used in the study. One of the methods that has been used widely to evaluate soil erosion is Universal Soil Loss Equation (USLE)[4]. It is designed to evaluate long-term average annual soil loss from field slopes under a specific land use and management system, based on the product of rainfall-runoff erosivity (R), soil erodibility (K),slope length and steepness (LS), cover and management (C) and support practice factor (P). This model can work with geographic information system (GIS). One of the most important factors is the coverage and management (C) thus this study aims to evaluate soil erosion in Lam Phra Phloeng watershed by USLE with different sources of land cover data; one is from LDD and another one is from ESA Climate Change Initiative (ESA/CCI).

* .Corresponding author: 59601202@kmitl.ac.th

2 Study area

Lam Phra Phloeng watershed is located in northeast part of Thailand from 14°18'30" - 14°38'30"N, 101°30'00" - 101°54'00"E in WGS 84 coordinate system as shown in Fig. 1 with 815.2 km^2 (81,520 ha) of total area. Lam Phra Phloeng reservoir was constructed since 1963. Total capacity of reservoir is 155,000,000 m^3 and average annual rainfall 1,135.8 mm/year. Lam Phra Phloeng River is a branch of Mun River in Nakhonratchasima province. Upper watershed area is close to Khao Yai National Park. This area has many high steep slope mountains. The land cover types in this area include paddy field, crop field, shrubland, residential area and forests. Normally this area has low rainfall rate but in rainy season it has too much rainfall. This causes flooding every year. To solve this problem Royal Irrigation Department has created the Lam Phra Phloeng reservoir to manage the flooding, irrigation and water supply. Construction of reservoir was completed in 1963, and reservoir has suffered from very high levels of sedimentation.

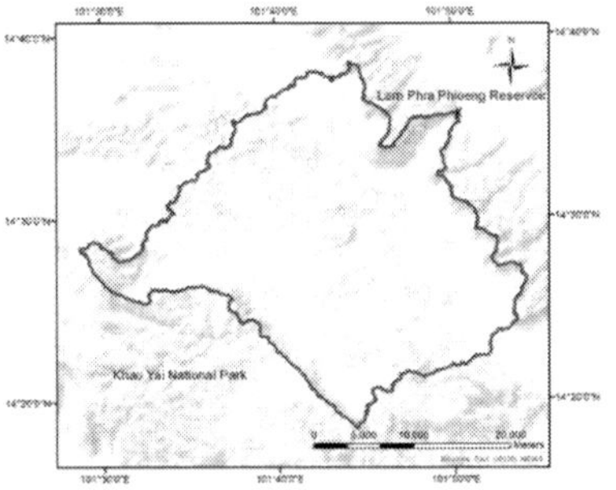

Fig. 1 The geographical map of Lam Phra Phloeng watershed

3 Methodology

3.1 Universal Soil Loss Equation

The Universal Soil Loss Equation (USLE) can evaluate an average rate of soil erosion for each feasible alternative combination of crop system and management practices in association with a specified soil type, rainfall pattern, and topography [4]. The equation of USLE is as follows [7].

$$A_m = R_m \times K_m \times L \times S \times C \times P \quad (1)$$

A_m : soil loss (ton/ha/year)
R_m : rainfall factor (ton/ha/year)
K_m : soil erodibility factor (ton-hour/106joule/mm)
L : slope length factor
S : steepness factor
C : vegetation cover factor
P : conservation practice factor

This equation is widely accepted worldwide for erosion prediction. Each variable is considered as a layer in the GIS database to be used in the modeling process.

3.1.1 *R_m: rainfall factor layer*

$$R_m = 0.4669X - 12.1415 \quad (2)$$

where:
R_m – rainfall and runoff erosivity (Mg/ha/year)
X – average annual rainfall (mm/year)

The average annual rainfall of the study area was calculated based on the 10-year period (from 2005 to 2015) rainfall by 9 stations in the watershed from Royal Irrigation Department (RID). Thiessen Polygons method was applied to create R_m factor layer.

3.1.2 *K_m: soil erodibility factor layer*

K_m factor layer was based on soil group map from LDD which the value identified by LDD [8]. Soil group map was provided by LDD at a scale of 1:50000

3.1.3 *L,S-factor layer*

These layers were topographic factors which were extracted from 30 m × 30 m digital elevation model (DEM) by using spatial analysis on ArcGIS 10.5. This DEM derived from Global Land Cover (USGS).

(a) L factor was calculated by following equations.

$$L = (\lambda/22.13)^m \quad (3)$$

where, λ is field slope length(m); m is a variable slope-length exponent that depends on slope described by [3], [4], [8], and [9].

m = 0.2 for gradients 0-1.0%
m = 0.3 for gradients 1.1-3.0%
m = 0.4 for gradients 3.1-5.0%
m = 0.5 for gradients >5.0

(b) S factor is the ratio of soil loss from the field slope gradient to a 9 percent slope. It was calculated by following equations.

$$S = 0.065 + 0.045s + 0.0065s^2 \quad (4)$$

where: s – slope (%)[4]

3.1.4 *C vegetation cover factor layer*

The cover and management factor, is the ratio of soil loss from an area with specified cover and management to that from an identical area in tilled continuous fallow [4]. To evaluate the C-factor layer, the value for each class of the land cover was assigned as identified by LDD [3] in Table. 1. In this study use 2015 land cover map from LDD and ESA/CCI to compare soil loss with different C value as shown in Fig. 2.

3.1.5 *P conservation practice factor layer*

In this study use land cover map from LDD and ESA/CCI to evaluate P factor layer. This P factor layer was established by LDD [3] in Table. 1.

Table. 1 Vegetation cover factor (C) and Conservation practice factor (P) for land use land cover classes (LULC) [3].

LULC Class	C Value	P Value
Mixed crops	0.255	1
Paddy field	0.280	0.10
Field crops	0.525	1
Perennial trees	0.150	1
Orchards	0.300	1
Horticulture crops	0.600	1
Grassland	0.100	1
Shifting cultivation	0.250	1
Evergreen forest	0.003	1
Deciduous forest	0.048	1
Forest plantation	0.088	1
Agro forestry	0.088	1
Natural grassland	0.015	1
Water body	0	0
Urban	0	0

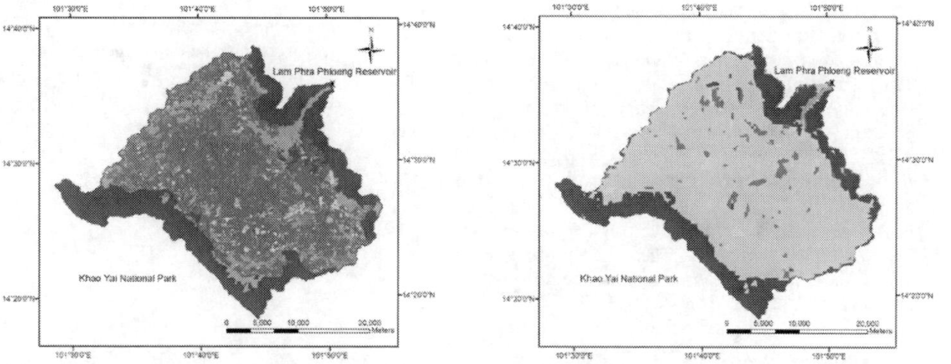

Fig. 2 Vegetation cover factor map (a) Land cover from LDD (b) Land cover from ESA/CCI

4 Results and discussion

4.1 Assessment of land use on soil erosion in Lam Phra Phloeng watershed by USLE model

The soil erosion map resulting from spatial overlay of the factor layers with the USLE model in Lam Phra Ploeng watershed is presented in Fig. 3 and Table. 2.

As mentioned before, this study evaluated soil erosion with USLE model by using land cover from different sources, (a) land cover data from LDD has 6.96% high risk soil erosion, 16.90% very high risk soil erosion and the average of soil erosion was 64.29 ton/ha/year as shown in Table.2.

In Fig. 3(a) show a high risk of soil erosion in the central of watershed and near Khao Yai National Park because the most agriculture in this area is field crop, paddy field and high steepness slope that affect high value of soil erosion.

(b) land cover data from ESA/CCI has 2.05% high risk soil erosion, 3.30% very high risk soil erosion and the average of soil erosion was 29.16 ton/ha/year as shown in Table.2

In Fig. 3(b) show a small risk of soil erosion because land use data from ESA/CCI is a global land cover data then it has less detailed than LDD.

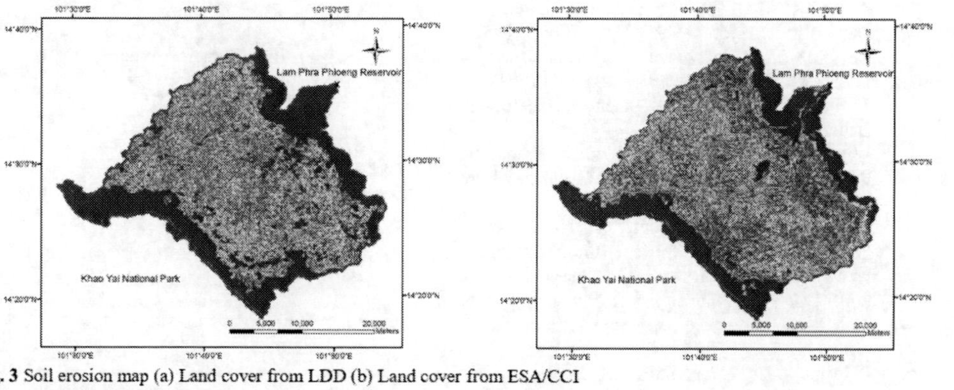

Fig. 3 Soil erosion map (a) Land cover from LDD (b) Land cover from ESA/CCI

Table. 2 Soil erosion risk categories

	Land cover data from LDD		Land cover data from ESA/CCI	
Category	Area(ha)	%	Area(ha)	%
Low risk (0-12.5 ton/ha/year)	32134	39.42	34181	41.93
Slightly low risk (12.5-31.25 ton/ha/year)	9354	11.47	25146	30.85
Moderately risk (31.25-93.75 ton/ha/year)	20582	25.25	17827	21.87
High risk (93.75-125 ton/ha/year)	5674	6.96	1676	2.05
Very high risk (>125 ton/ha/year)	13776	16.90	2690	3.30

5 Conclusions

This research has evaluated a soil erosion value by using USLE model. This model is widely acceptable for land cover management and organization that work about environment. In this study focus on different land cover sources to see an effect of soil loss with different land covers. Land cover data from LDD mostly has field crops, horticulture crops and paddy field. These land cover has huge impact to soil erosion as shown in Table. 1 but Land cover data from ESA/CCI mostly has grassland that has less impact than land cover from LDD. Therefore, the land cover management should emphasize on agriculture area to prevent and reduce a soil erosion rate.

Acknowledgements

This research was funded by King Mongkut's Institute of Technology Ladkrabang. for the financial sponsorship. Sincere appreciation goes to the Land Development Department, Royal Irrigation Department Global Land Cover(USGS) and ESA Climate change initiative for the data.

References

1. Panagos, P., Borrelli, P., Meusburger, K., Alewell, C., Lugato, E., & Montanarella, L. Estimating the soil erosion cover-management factor at the European scale. Land Use Policy, **48**, 38-50. (2015)
2. Schellenberg, G., Donnelly, C. R., Holder, C., & Ahsan, R. Dealing with Sediment: Effects on Dams and Hydropower Generation. Hydro Review Worldwide. (2017)
3. Land Development Department. Soil Erosion in Thailand. Land Development Department, Bangkok. (2000)
4. Wischmeier, W.H. & Smith, D.D. Predicting Rainfall Erosion Losses. USDA Agriculture Handbook No.537. (1978)
5. Phetprayoon, T., Sarapirome, S., Navanugraha, C., & Wonprasaid, S. Surface runoff estimation using grid-based curve number method in the upper Lam Phra Phloeng Watershed, Thailand. In 30th Asian conference on remote sensing, 18-23. (2009)
6. Polpanich, O., Nongpho, K., Boonwan, K., Chandraphabha, C., Hakari1, M. & Samarakoon, L. Integration of RS/GIS for Estimation of Soil Erosion in Lam Phra Phloeng Reservior, Nakhon Ratchasima Province, Thailand. Asian Association on Remote Sensing. (2010)
7. Gray, D. H. & Sotir, R. B.,. Biotechnical and Soil Bioengineering Slope Stabilization: A Practical Guide for Erosion Control, John Wiley & Sons, Inc., (1996)
8. McCool, D.K., Foster, G.R., Mutchler, C.K., Meyer, L.D. Revised Slope Length Factor for The Universal Soil Loss Equation. Trans. ASAE, **32**, 1571-1576. (1989)
9. Toxopeus, A. G. Spatial and temporal modelling for sustainable management of semi-arid rangelands: the wildlife versus livestock issue in the Amboseli ecosystems, Southern Kenya. In Proceedings of the conference on geo-information for sustainable land management, ITC. (1997).

Hybrid artificial intelligence-based bond strength model of CFRP-lightweight concrete composite

Emmanuel Macabenta Lazo[1], and *John Pepard Mendoza* Rinchon[1,*]

[1]Civil Engineering Department, College of Engineering, Technological Institute of the Philippines, Quezon City, Philippines 1109

Abstract. Different retrofitting techniques are commonly used to sustain the design life of heavy damage and deteriorated concrete structures, whilst epoxy-bonded carbon fiber reinforced polymer (CFRP) has emerged as a widely known retrofitting method. Consequently, a sound understanding of the bond strength between structural lightweight concrete (LWC) and CFRP based on influential factors is essential in safety and economic requirements. In this study, a hybrid bond strength model using the artificial neural network (ANN) and genetic algorithm (GA) was developed to furtherly understand the bond of a CFRP strengthened LWC structure. ANN was able to establish under satisfactory performance the relationship between the maximum bond load and the following influential parameters: width of CFRP (b_{frp}), total CFRP bond length (L_{frp}), CFRP thickness (t_{frp}), and CFRP angle of orientation (θ_{frp}). Furthermore, GA was able to derive the optimal configuration of the influential parameters resulted in high bond performance. Moreover, the optimization results also validated the sensitivity of each parameter on the interfacial bond behavior between LWC and CFRP.

1 Introduction

Rapid deterioration, excessive seismic damage, and outdated design codes and provisions of concrete infrastructure become the principal challenges of all nations worldwide. In response to these challenges, repair and strengthening of existing concrete structures became the solutions in order to still meet the required ultimate load carrying capacity and serviceability. Consequently, carbon fiber reinforced polymer (CFRP) has emerged as one of the notable accepted retrofitting materials of the civil engineering community in recent years [1]. Several studies account the success of CFRP as reinforcement in a wide variety of applications for different materials such as aluminum [2, 3], steel [4, 5], and concrete [6-8]. The utilization of CFRP as retrofitting material provides technically sound and cost-efficient due to its lightweight, resistance to corrosion, and ease of handling and installation [9-11]. CFRP can be in the form of laminates, sheets, or strips as shown in Fig. 1. Furthermore, CFRP has very advantageous mechanical properties of high specific strength and high specific stiffness compare to other strengthening materials [12].

On the other hand, interfacial debonding is concluded to be the most common mode of failure of CFRP strengthened concrete structures. According to Pellegrino et al. [13], the bond mechanism for effective stress transfer is vital in the performance of the CFRP-concrete composite. Sayed-Ahmed et al. [14] conducts an extensive investigation to review the different analytical models of bond strength between CFRP and concrete considering different parameters and assumptions. Furthermore, several studies have developed different prediction models on the bond strength of CFRP and concrete under different considerations [15-20]. Different modeling techniques are also utilized in developing the maximum bond load and bond strength prediction models [21-24].

However, previous studies focus only on the performance of CFRP on normal weight concrete even there is an apparent growing demand for the use of lightweight concrete (LWC) in construction projects over the past decades. Still, less attention is given in the investigation of bond performance between LWC and CFRP [9]. Thus, this present study aimed to model the underlying behavior of the bond strength performance between LWC and CFRP using a hybrid artificial intelligence (AI) model wherein recent studies [25-28] show its promising application in the field of civil engineering and construction materials. The study specifically aimed to develop a prediction model of the bond strength between LWC and CFRP using the artificial neural network (ANN); and to utilize genetic algorithm (GA) optimization in producing high bond strength performance between LWC and CFRP.

2 Materials and Methods

2.1 Double Lap Shear (DLS) Test Datasets

In this study, a total of fifty-five datasets from the double lap shear (DLS) test were retrieved from the paper published by Al-Allaf et al. [9] (Refer to table 1). These

* Corresponding author: jprinchon27@gmail.com

datasets were summarized and utilized as the primary input in the development of the hybrid model. The following factors considered by Al-Allaf et al. [9] in the DLS test such as the width of CFRP (b_{frp}), total CFRP bond length (L_{frp}), CFRP thickness (t_f), and CFRP angle of orientation (θ_{frp}) were also used in this study as the influential parameters for the prediction of bond strength between LWC and CFRP.

2.2 Hybrid Neural Network-Genetic Algorithm

The development of hybrid bond strength model was composed of two stages namely the ANN modeling and the GA optimization. Generally, ANN was utilized to develop a prediction model that established the relationship between the maximum bond load (P_{max}) and the influential parameters. The neural network for the maximum bond load between LWC and CFRP was developed using the simplest and widely used ANN model known as the feedforward multilayered supervised neural network with error back-propagation algorithm. The network was structured based on the following internal parameters such as the training algorithm, transfer function, number of hidden layers, and number of neurons per hidden layer. The Levernberg-Marquardt (LM) algorithm and hyperbolic tangent sigmoid function or *tansig* function were utilized as training algorithm and transfer function respectively. The number of the hidden layer was set at one and the number of neurons per hidden layer was based on the value proposed by Asteris et al. [29]. Moreover, the predicting performance of the neural network was assessed based on Pearson correlation coefficient (R) and mean square error (MSE) while the topology of the bond strength neural network model was constructed and run using MatLab® R2015a software.

The GA was deployed for the purpose of optimizing the independent variables in Equation 1 that will yield a high bond load performance. The objective function and the corresponding constraints for the GA optimization are expressed as follows:

$$\text{Maximize: } P_{max} = f(b_{frp}, L_{frp}, t_f, \theta_{frp}) \quad (1)$$

$$50 \leq b_{frp} \leq 150 \quad (2)$$

$$100 \leq L_{frp} \leq 200 \quad (3)$$

$$0.1178 \leq t_f \leq 0.2356 \quad (4)$$

$$0 \leq \theta_{frp} \leq 90 \quad (5)$$

The P_{max} is the prediction model developed during the first stage of the hybrid model. Equations 2 to 5 were used as constraints for the algorithm intended to search for the global and realistic solution of the optimization problem. Furthermore, the genetic algorithm was formed based on the procedure originally proposed by Goldberg [30]. This algorithm is generally composed of three genetic operators namely the selection, mutation, and crossover. In this study, the stochastic uniform method was used as the selection operator while the adaptive feasible model was utilized as mutation operator. Moreover, scattered and one-point models were explored as crossover operators.

3 Results and Discussions

After several simulations, the final topology of the bond strength prediction model of the LWC-CFRP composite was structured. The final model was composed of three neurons with one hidden layer. The biases and weights for the input layer of the final ANN architecture are given by Equations 6 to 7 respectively, while for the hidden layer, Equations 8 and 9 represent the biases and weights respectively.

(a) (b) (c)

Fig. 1. Forms of CFRP ©2017 Google Image: (a) sheets, (b) strips, and (c) laminates

Table 1. Summary of ANN input and output parameters. Retrieved from Al-Allaf [9]

Variable	Details of Variable			
	Minimum	*Maximum*	*Range*	*Mean*
Input				
CFRP width, b_{fr} (mm)	50	150	100	100
Total CFRP bond length, L_{frp} (mm)	100	200	100	130
CFRP thickness, t_f(mm)	0.1178	0.2356	0.1178	0.1392
CFRP angle of orientation, θ_{frp} (°)	0	90	90	22.91
Output				
Test failure load, P_{test} (kN)	0.47	29.69	29.22	18.87

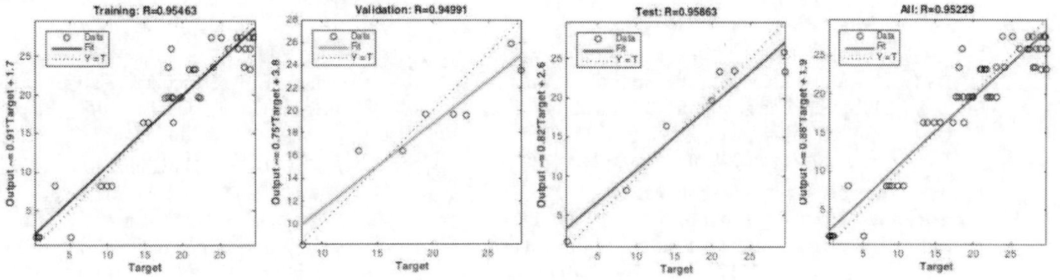

Fig. 2. Pearson correlation coefficient performance of the bond strength neural network

$$b_I = \begin{bmatrix} 3.31225 \\ 1.96781 \\ -3.86852 \end{bmatrix} \quad (6)$$

$$W_I = \begin{bmatrix} 2.75362 & 3.78588 & -0.034438 & -1.92568 \\ -1.84853 & 1.472366 & 2.019534 & -0.55476 \\ -6.91359 & 0.52008 & 0.58746 & -4.14345 \end{bmatrix} \quad (7)$$

$$b_H = [-0.42735] \quad (8)$$

$$W_H = [0.72335 \quad 0.63921 \quad -0.83450] \quad (9)$$

The performance of the final ANN bond strength prediction model was remarkably high at 0.95 correlation coefficient. This was also evident based on the correlation coefficients of the training, validation and testing stages with values greater than 0.95 as shown in Fig. 2. Additionally, LM Algorithm and *Tansig* Function were still proven to be high performing training algorithm and transfer function respectively in the present study.

For GA optimization, the optimal combination of the influential parameters (i.e. $b_{frp} = 91.5$ mm, $L_{frp} = 200$ mm, $t_f = 0.2356$, and $\theta_{frp} = 90°$) for high bond load performance was determined for over 200 generations as shown in Fig. 3. Based on this combination, the optimization procedure was able to yield a maximum bond load of $P_{max} = 40$ kN. Furthermore, it is observed that larger values of the CFRP's total bond length, thickness, and angle of orientation increase the bond load while increasing the width of the CFRP sheet provides limited benefit on the bond load capacity.

4 Conclusion

Further investigation of the interfacial bond behavior of LWC with externally bonded CFRP using previously proposed hybrid model was studied in this research. At this point, the hybrid neural network-genetic algorithm was able to provide satisfactory results in terms of developing the bond strength prediction model and optimization of the considered influential parameters. Satisfactory establishment of the relationship between the influential parameters and the bond load was completely achieved based on justifiable Pearson correlation coefficient. Moreover, the optimization results showed that each influential parameter has a different effect on the bond quality between LWC and CFRP, wherein the orientation of the CFRP sheet (θ_{frp}) greatly affects the shear distribution of load to the fiber of CFRP sheet.

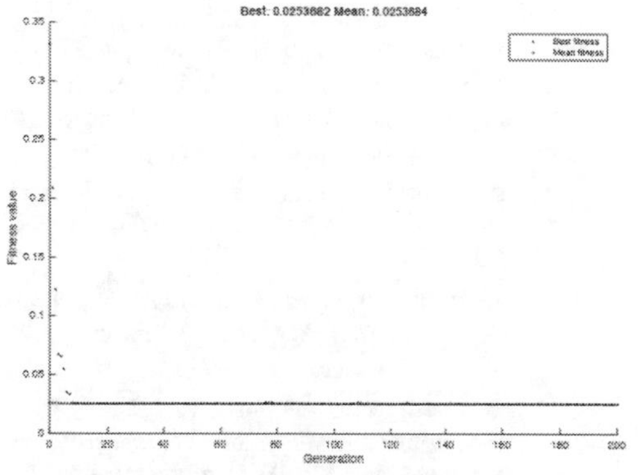

Fig. 3. Results of genetic algorithm optimization

References

[1] P. Sarker, M. Begum, and S. Nasrin, "Fibre reinforced polymers for structural retrofitting: A review," *J Civil Eng*, vol. 39, no. 1, pp. 49-57, 2011.

[2] U. Tamilarasan, L. Karunamoorthy, and K. Palanikumar, "Mechanical Properties Evaluation of the Carbon Fibre Reinforced Aluminium Sandwich Composites," *Materials Research*, vol. 18, no. 5, pp. 1029-1037, 2015.

[3] J. Wang *et al.*, "Bond strength between carbon fiber–reinforced plastic tubes and aluminum joints for racing car suspension," *Advances in Mechanical Engineering*, vol. 8, no. 10, p. 1687814016674627, 2016.

[4] M. Kamruzzaman, M. Z. Jumaat, N. Ramli Sulong, and A. Islam, "A review on strengthening steel beams using FRP under fatigue," *The Scientific World Journal*, vol. 2014, 2014.

[5] A. Shaat, D. Schnerch, A. Fam, and S. Rizkalla, "Retrofit of steel structures using fiber-reinforced polymers (FRP): State-of-the-art," in *Transportation research board (TRB) annual meeting. CD-ROM (04-4063)*, 2004.

[6] A. Belarbi and B. Acun, "FRP systems in shear strengthening of reinforced concrete structures," *Procedia Engineering*, vol. 57, pp. 2-8, 2013.

[7] A. Belarbi, M. Reda, P. Poudel, H. Tahsiri, M. Dawood, and B. Gencturk, "Prestressing Concrete with CFRP Composites for Sustainability and Corrosion-Free Applications," in *MATEC Web of Conferences*, 2018, vol. 149, p. 01010: EDP Sciences.

[8] A. S. Karzad, S. Al Toubat, M. Maalej, and P. Estephane, "Repair of reinforced concrete beams using carbon fiber reinforced polymer," in *MATEC Web of Conferences*, 2017, vol. 120, p. 01008: EDP Sciences.

[9] M. H. Al-Allaf, L. Weekes, L. Augusthus-Nelson, and P. Leach, "An experimental investigation into the bond-slip behaviour between CFRP composite and lightweight concrete," *Construction and Building Materials*, vol. 113, pp. 15-27, 2016.

[10] S. E. Günaslan, A. Karaşin, and M. E. Öncü, "Properties of FRP Materials for Strengthening," *International Journal of Innovative Science, Engineering & Technology*, vol. 1, no. 9, 2014.

[11] S. Rizkalla and T. Hassan, "Various FRP strengthening techniques for retrofitting concrete structures," in *CICE 2001 Conference proceedings*, 2001.

[12] M. R. Piggott, *Load bearing fibre composites*, 2nd ed. Boston: Kluwer Academic Publishers, 2002, pp. xii, 475 p.

[13] C. Pellegrino, D. Tinazzi, and C. Modena, "Experimental study on bond behavior between concrete and FRP reinforcement," *Journal of Composites for Construction*, vol. 12, no. 2, pp. 180-189, 2008.

[14] E. Sayed-Ahmed, R. Bakay, and N. Shrive, "Bond strength of FRP laminates to concrete: state-of-the-art review," *Electronic Journal of structural engineering*, vol. 9, no. 1, pp. 45-61, 2009.

[15] C. Arya and N. Farmer, "Design guidelines for flexural strengthening of concrete members using FRP composites," in *FRPRCS-5: Fibre-reinforced plastics for reinforced concrete structures Volume 1: Proceedings of the fifth international conference on fibre-reinforced plastics for reinforced concrete structures, Cambridge, UK, 16–18 July 2001*, 2001, pp. 167-176: Thomas Telford Publishing.

[16] F. Externally Bonded, "Reinforcement for RC Structures," *fib Bulletin*, no. 14, pp. 51-58, 2001.

[17] M. A. Issa, M. Rahman, and R. Alrousan, "Prediction of Ultimate Load at the CFRP-Concrete Interface under Pure Shear Mode," *International Journal of Composite Materials*, vol. 6, no. 6, pp. 183-195, 2016.

[18] R. Seracino, M. Raizal Saifulnaz, and D. Oehlers, "Generic debonding resistance of EB and NSM plate-to-concrete joints," *Journal of Composites for Construction*, vol. 11, no. 1, pp. 62-70, 2007.

[19] A. Serbescu, M. Guadagnini, and K. Pilakoutas, "Standardised double-shear test for determining bond of FRP to concrete and corresponding model development," *Composites Part B: Engineering*, vol. 55, pp. 277-297, 2013.

[20] C. Tuakta and O. Büyüköztürk, "Conceptual model for prediction of FRP-concrete bond strength under moisture cycles," *Journal of Composites for Construction*, vol. 15, no. 5, pp. 743-756, 2011.

[21] J. A. Abdalla, R. Hawileh, and A. Al-Tamimi, "Prediction of FRP-concrete ultimate bond strength using Artificial Neural Network," in *Modeling, Simulation and Applied Optimization (ICMSAO), 2011 4th International Conference on*, 2011, pp. 1-4: IEEE.

[22] J. Dai, T. Ueda, and Y. Sato, "Development of the nonlinear bond stress–slip model of fiber reinforced plastics sheet–concrete interfaces with a simple method," *Journal of Composites for Construction*, vol. 9, no. 1, pp. 52-62, 2005.

[23] J. L. Huang, P. Y. Huang, Z. W. Li, and X. H. Zheng, "Prediction of bond strength of FRP-concrete based on multiple linear regression method," in *Advanced Materials Research*, 2011, vol. 163, pp. 3623-3628: Trans Tech Publ.

[24] G. S. Tong and S. S. Chen, "Prediction Bond Strength between FRP and Concrete Interface by LEFM Method," in *Advanced Materials Research*, 2014, vol. 988, pp. 195-200: Trans Tech Publ.

[25] N. C. Concha and E. P. Dadios, "Optimization of the rheological properties of self compacting concrete using neural network and genetic algorithm," in *Humanoid, Nanotechnology, Information Technology, Communication and Control, Environment and Management (HNICEM), 2015 International Conference on*, 2015, pp. 1-6: IEEE.

[26] N. C. Concha, "Rheological Optimization of Self Compacting Concrete with Sodium Lignosulfate Based Accelerant Using Hybrid Neural Network-Genetic Algorithm," in *Materials Science Forum*, 2016, vol. 866, pp. 9-13: Trans Tech Publ.

[27] J. P. M. Rinchon, N. C. Concha, and M. G. V. Calilung, "Reinforced concrete ultimate bond strength model using hybrid neural network-genetic algorithm," in *Humanoid, Nanotechnology, Information Technology, Communication and Control, Environment and Management (HNICEM), 2017 IEEE 9th International Conference on*, 2017, pp. 1-6: IEEE.

[28] J. P. M. Rinchon, "Strength Durability-Based Design Mix of Self-Compacting Concrete with Cementitious Blend using Hybrid Neural Network-Genetic Algorithm," *IPTEK Proceedings Series*, no. 6, 2017.

[29] P. Asteris, K. Kolovos, M. Douvika, and K. Roinos, "Prediction of self-compacting concrete strength using artificial neural networks," *European Journal of Environmental and Civil Engineering*, vol. 20, no. sup1, pp. s102-s122, 2016.

[30] D. E. Goldberg, "Genetic algorithms in search, optimization and machine learning 'addison-wesley, 1989," *Reading, MA*, 1989.

PID electro-hydraulic cylinder force tracking system with friction compensation

Weerapong Chanbua[1]*, and *Unnat* Pinsopon[1]

[1]Department of Mechanical Engineering, Faculty of Engineering, King Mongkut's Institute of TechnologyLadkrabang, Bangkok, Thailand

Abstract. Friction can be found in all mechanical systems, and is the cause of control tracking error. In this article, LuGre friction model for symmetric hydraulic cylinder was studied and obtained experimentally. The estimated friction force was applied for reference force compensation. Force tracking performances of PID force control systems with and without friction compensation were tested and compared. Control system with friction compensation outperformed one without in all cases of tracking tests. The best result was found in the square force tracking tests, with an average error at maximum compression force of 86.105 N in the case of with friction compensation compared with error of 511.996 N in the case without. However, the estimated friction achieved in the study was noisy. This is due to the use of noisy numerically differentiated velocity signal in the friction estimation procedure.

1 INTRODUCTION

Electro-hydraulic control systems could be applied for position control, velocity control or force control. Electro-hydraulic force control systems are used various industrial applications such as press machine, earth moving equipment [1] and flight simulator [2]. Hydraulic cylinder is the main actuator used in force control applications. Friction occurs in cylinder at the contact areas between piston seal and cylinder barrel, and between piston rod seal and wiper. Friction depends on surface contact, temperature, lubrication and contact velocities. Friction always occurs in mechanical systems, and leads to control tracking error. Friction compensation is one of most popular method to solve the problem. Various friction models were reviewed in great details in [3] and [4]. The prediction ability of four friction models were compared in experiments [4], and LuGre friction model was found to be the all-round best predictor.

In this paper, an electro-hydraulic cylinder force control system was constructed. The force feedback signal used in the control scheme was calculated using the value of oil pressures at both sides of cylinder. LuGre friction model was used to predict cylinder friction force. The predicted friction would be used as force reference compensation in the PID control scheme. Force tracking performances of the PID controller with and without friction compensation would be tested and compared.

2 FRICTION MODEL

General steady state friction characteristics are presented in figure 1, whereas friction is shown as a velocity function. The Coulomb and the stiction friction forces are constant and independent of velocity. The viscous friction force is caused by the viscosity of lubricants, and increases with velocity. The Stribeck effect is the phenomenon at low velocity region that friction decreases with the increasing of velocity.

LuGre friction model is a widely used dynamic friction model which describes the microscopic deflection between the contact surfaces in terms of a damping-spring behavior [3]. The LuGre model described the deflection of elastic bristles which represents the asperities between two contacting surfaces. The average deflection of the bristles is denoted by z and the deflection rate of z is modeled by equation 1. The average deflection of the bristles is related to the friction force as shown in equation 2.

$$\frac{dz}{dt} = v - \frac{\sigma_0}{g(v)}|v| \tag{1}$$

$$F_{friction} = \sigma_0 z + \sigma_1 \frac{dz}{dt} + \sigma_2 v \tag{2}$$

Where v is velocity between contact surfaces, σ_0 is the average stiffness of the bristles, σ_1 is the damping coefficient and σ_2 is viscous friction coefficient. At steady state, $\frac{dz}{dt}$ is zero, and $\sigma_0 z$ is equal to $g(v)$. Therefore the steady state friction is calculated by

$$F_{friction\ ss} = g(v) + \sigma_2 v \tag{3}$$

* Corresponding author : weerapong.kmitl@gmail.com

$$g(v) = F_{coulomb} + (F_{stiction} - F_{coulomb})\, e^{-\left|\frac{v}{v_{stribeck}}\right|^{n}} \quad (4)$$

Where $F_{coulomb}$ is the coulomb friction force, $F_{stiction}$ is the stiction friction force, $v_{stribeck}$ is the stribeck velocity, n is the exponent and

$F_{coulomb} + (F_{stiction} - F_{coulomb})\, e^{-\left|\frac{v}{v_{stribeck}}\right|^{n}}$ is the stribeck friction. The values of these parameters are to be determined experimentally.

The steady state friction can be expressed by

$$F_{friction\ ss} = F_{coulomb} + (F_{stiction} - F_{coulomb})\, e^{-\left|\frac{v}{v_{stribeck}}\right|^{n}} + \sigma_2 v \quad (5)$$

The steady state friction force calculated by equation 5 could well describe the general steady state friction characteristics (figure 1).

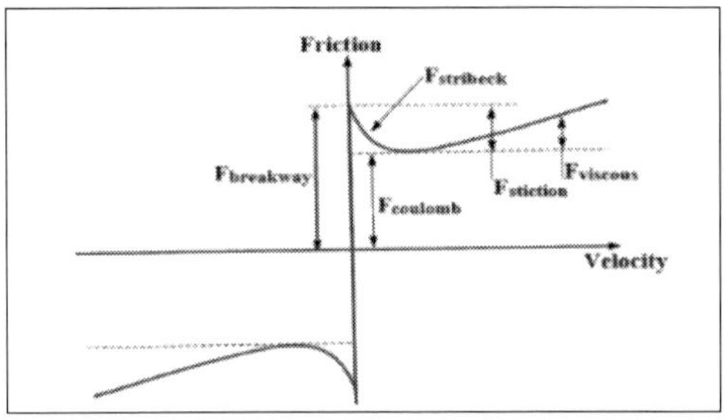

Fig. 1. Steady state friction characteristics.

3 EXPERIMENTAL APPARATUS AND METHODS

3.1 Experimental apparatus

Figure 2 shows the electro-hydraulic force control experimental apparatus. The cylinder used is the symmetric double-rod type with 0.04 m bore diameter, 0.028 m rod diameter and 0.50 m stroke. The motion of the cylinder was controlled by a proportional valve which receives oil flow from a gear pump. Pressure transducers were installed at both sides of the cylinder. Since direct force measurements are usually impractical in actual industrial applications, feedback cylinder force was calculated based on the cylinder pressures (equation 6). A load cell is also installed at the end of the cylinder rod in the setup. However, the force signal measured by the load cell is used as reference only. A draw wire potentiometer was used to measure piston displacement which would be numerically differentiated to obtain cylinder velocity. A compression spring was used as the simulated cylinder load.

$$F_{cylinder} = (P_1 - P_2)A \quad (6)$$

Where $F_{cylinder}$ is cylinder force, P_1 and P_2 are oil pressures in hydraulic cylinder and A is piston area.

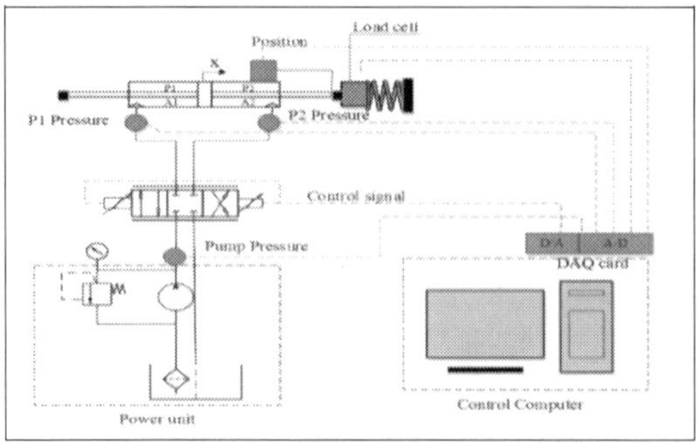

Fig. 2. Electro-hydraulic force control experimental apparatus.

3.2 Experimental methods

The experiments are divided in two sections. First, the friction model parameters would be identified, and the steady state LuGre friction model according to equation 5 will be constructed. Second, PID force tracking controls with and without friction compensation would be tested and compared.

3.2.1 Friction parameters identification

Friction force in hydraulic cylinder is obtained from force balance of the cylinder piston.

$$F_{friction} = (P_1 - P_2)A - m\ddot{x} \quad (7)$$

Where $F_{friction}$ is friction force, m is piston mass, $\ddot{x}$ is piston acceleration. Various constant commands were sent to the proportional valve, frictions at various velocities were then calculated by equation 6 and recorded. Generalized reduced gradient nonlinear (GRG) algorithm was used for fitting the recorded frictions into the LuGre steady state friction model (equation 5) in order to obtain the values of friction parameters.

3.2.2 Force tracking control

Figure 3 shows the diagram of PID force control system. In addition to the cylinder force feedback, velocity signal was also fed back to LuGre friction model. The output from LuGre friction model, the estimated friction, would be used as compensation to the reference force signal. The tracking performances of PID force tracking with and without friction compensation would be tested and compared.

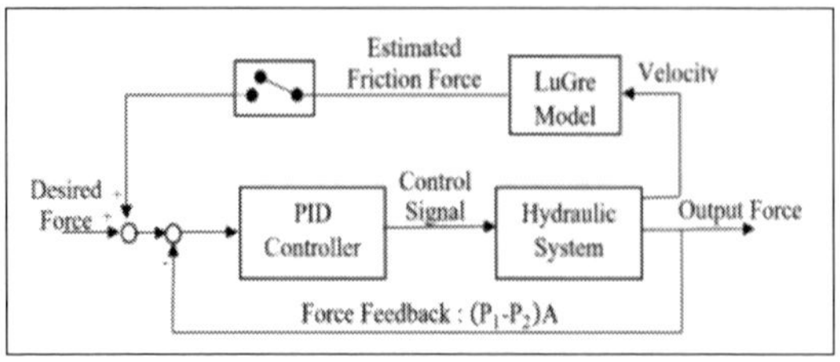

Fig. 3. Block diagram of PID force control with friction compensation.

https://doi.org/10.1051/matecconf/201819202019

4 EXPERIMENTAL RESULT

4.1 Friction parameters identification

Figure 4 shows the comparison between the measured steady state friction (obtained from equation 7) and the estimated friction (obtained from equation 5). The LuGre model parameters were obtained by GRG algorithm. The friction parameters were obtained separately according to the direction of motion. Equation 8 described the friction for the positive (right hand side) motion, while equation 9 described the friction for the negative motion. Equations 8 and 9 were coded in the friction model of the control scheme.

$$F_{friction\ ss,extend} = 32.37 + 440.1163\, e^{-\left|\frac{v}{0.000000111}\right|^{0.089938}} + 1068.905\, v \quad (8)$$

$$F_{friction\ ss,retract} = (-204.351) + (-307.096)\, e^{-\left|\frac{v}{-0.0128}\right|^{1.215129}} + 581.397\, v \quad (9)$$

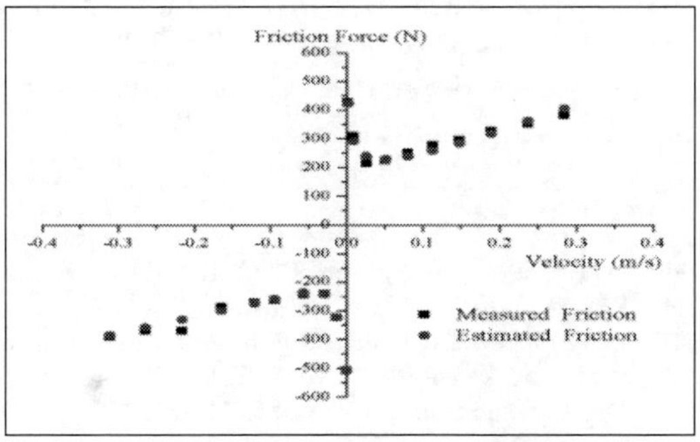

Fig. 4. Comparison of measured frinction and estimated friction

4.2 Experimental Results and Discussion

The force tracking performances were done with two type force reference signals, sinusoidal signal and square signal. The same PID controller gains were used in the control schemes with and without friction compensation. The PID gains were first tuned with Ziegler-Nichols method, and finely tuned later in experiments. The desired sinusoidal and square force signals had minimum and maximum amplitudes of 500N and 3,500 N, both at a frequency of 0.2 Hz.

Figure 5 shows the tracking performances of the PID controller systems with sinusoidal force reference signal. As mentioned earlier, the feedback force signals used in both control schemes were calculated using the cylinder pressures. In order to realize the actual forces on the load, the tracking forces of both control schemes shown in figure 6 were directly obtained from measurement of the load cell. It can be seen that the control system with friction compensation performed better when the piston moved in the positive direction. The average tracking errors at maximum compression force are 264.826 N and 569.342 N for the controllers with and without friction compensation, respectively. Figure 6 shows the estimated friction force and the calculated friction force obtained from the sinusoidal force tracking test. The estimated friction force signal was noisy in zones a and b (figure 6). The cylinder velocity signal was obtained by numerically differentiation of position signal, and was noisy at the low velocity regions of zones a and b. This leads to noisy estimated friction force according to equation 5. The control actions of the controllers with and without friction compensation are shown in figure 7. The control actions of two cases resembled each other for almost all of the time, except at the times about 4, 9, 13, and 18 seconds (figure 7). The control action of the controller with friction compensation surged at higher values at those instants time. Those are the times that cylinder switched the direction of motion from negative to positive direction. The surged control action would give the cylinder a kick start at the instant of direction switch, and overcame the breakaway friction. This resulted in smaller tracking error at maximum compression force. The RMS errors of sinusoidal force tracking are 371.386 N and 457.737 N for the controllers with and without friction compensation, respectively. The RMS errors as well as average errors at maximum compression forces are also shown in in table 1.

Figures 8 to 10 show the experimental results from square force tracking tests. The controller with friction compensation outperformed the controller without friction compensation, with an average error at maximum compression force of 86.105 N compared with an average error of 511.996 N in the case of without friction compensation. The control action of the controller with friction compensation surged at higher values compared with the controller without friction compensation at the times 5, 10 and 15 seconds (figure 10). This would give the cylinder a kick start at the motion starting instant, the same action that could be observed in the sinusoidal force tracking tests. The cylinder velocity in square force tracking tests was low about zero at almost all of the time. Therefore, the numerically differentiated velocity signal was noisy almost all of the time, and so was the estimated friction force (figure 9). The RMS tracking errors in the cases of with and without friction compensation are 809.743 N and 976.401 N, respectively. For these square force tracking tests, the maximum compression lasted only half of the period of the reference command. Had the maximum compression lasted longer, the difference in RMS errors between two cases would be larger.

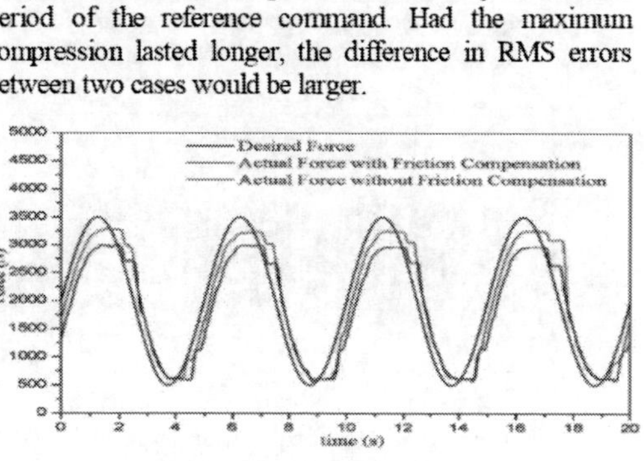

Fig. 5. Force tracking control performances of sinusoidal force tracking tests.

 https://doi.org/10.1051/matecconf/201819202019

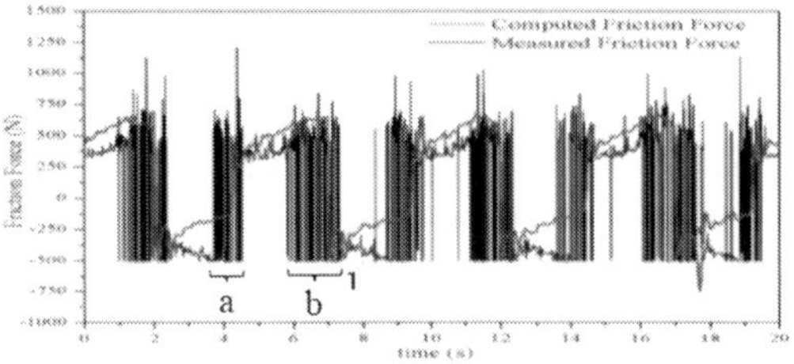

Fig. 6. Comparison of measured and computed friction forces in sinusoidal force tracking tests.

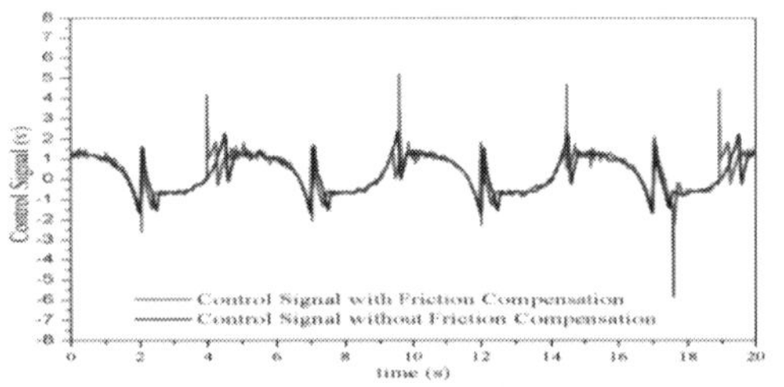

Fig. 7. Comparison of control signals in sinusoidal force tracking tests.

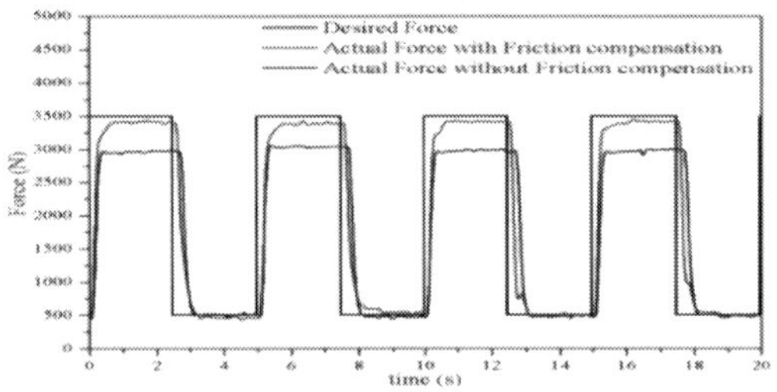

Fig. 8. Force tracking control performances of square force signal tracking tests.

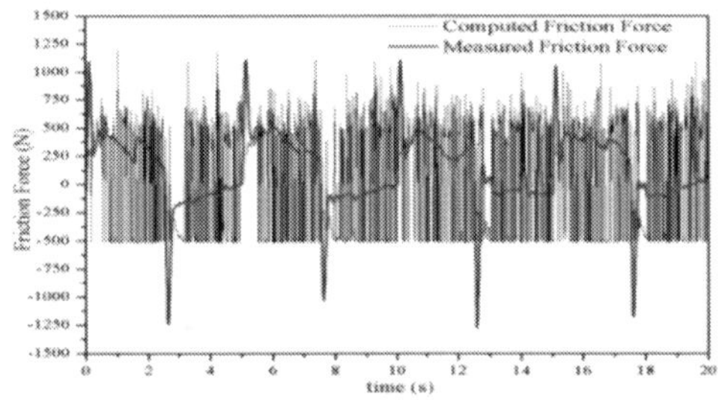

Fig. 9. Comparison of measured and computed friction forces in square force tracking tests.

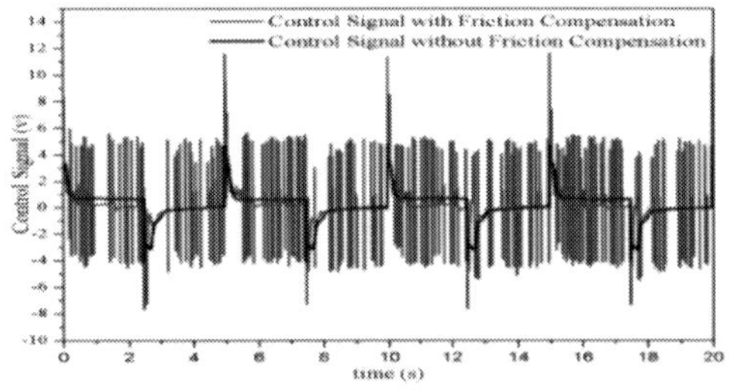

Fig. 10. Comparison of control signals in square force tracking tests.

Table 1. RMS error and error at maximum force

Reference Command	Without friction compensation		With friction compensation	
	RMS error (N)	Error at maximum force (N)	RMS error (N)	Error at maximum force (N)
Sinusoidal	457.737	569.342	371.386	264.826
Square	976.401	511.996	809.743	86.105

5 CONCLUSIONS

In this paper, LuGre friction model in symmetric hydraulic cylinder was experimentally constructed. The estimated friction force was applied as compensation to the reference force signal. Tracking performances of PID force controllers with and without friction estimation were tested and compared. Feedback force signal used in both control schemes were calculated using the cylinder pressures.The control system friction compensation achieved better tracking performances. Friction compensation could give cylinder a kick start and overcame the breakaway friction force. This resulted in much lower tracking errors at maximum compression force in both sinusoidal and square force tracking. The estimated friction force obtained in this study was noisy due to the use of numerical differentiated velocity signal. Had a higher performance velocity filter been used, better estimated friction force would had been obtained, better tracking performance could had been achieved.

Reference

1. Q. P. Ha, Q. H. Nguyen, D. C. Rye,H. F. Durrant-Whyte, *Force and Position Control for Electrohydraulic systems of a Robotic Excavator*, Proceedings of the 39th IEEE Conference on Decision and Control, **Vol. 5**, 5224-5229 (2000)
2. D. Q. Truong, K. K. Ahn, *Force Control for Hydraulic Load Simulator using Self-Tuning Grey Predictor – fuzzy PID*, International Federation of Automatic Control, Mechatronics 19, 233–246 (2009)
3. H.Olsson, K.J.Åström, Canudas de Wit M.Gäfvert, P.Lischinsky, *Friction Models and Friction Compensation*, European Journal of Control, **Vol. 4**, Issue 3, 176-195 (1998)
4. Y. Sun, Tao Chen, C. Q. Wu, C. Shafai, *Comparison of Four Friction Models: Feature Prediction*, Journal of Computational and Nonlinear Dynamics, **Vol. 11**, 1-10 (2016)

Bio-hydrogen production from waste materials: A review

Vinod Singh Yadav[1,*], *Vinoth* R[2], *Dharmesh* Yadav[2]

[1]Mechanical Engineering, NIT, Uttarakhand, Srinagar, India
[2]Automobile Engineering, SAMM, Manipal University, Jaipur, India

Abstract. When hydrogen burns in air, it produces nothing but water vapour. It is therefore the cleanest possible, totally non-polluting fuel. This fact has led some people to propose an energy economy based entirely on hydrogen, in which hydrogen would replace gasoline, oil, natural gas, coal, and nuclear power. Hydrogen is a clean energy source. Therefore, in recent years, demand on hydrogen production has increased considerably. Electrolysis of water, steam reforming of hydrocarbons and auto-thermal processes are well-known methods for hydrogen gas production, but not cost-effective due to high energy requirements. As compare to chemical methods, biological production of hydrogen gas has significant advantages such as bio-photolysis of water by algae, dark and photo-fermentation of organic materials, usually carbohydrates by bacteria. New approach for bio-hydrogen production is dark and photo-fermentation process but with some major problems like dark and photo-fermentative hydrogen production is the raw material cost. By using suitable bio-process technologies hydrogen can be produced through carbohydrate rich, nitrogen deficient solid wastes such as cellulose and starch containing agricultural and food industry wastes and some food industry wastewaters such as cheese whey, olive mill and baker's yeast industry wastewaters. Utilization of aforementioned wastes for hydrogen production provides inexpensive energy generation with simultaneous waste treatment. This review article summarizes bio-hydrogen production from some waste materials with recent developments and relative advantages.

1. Introduction

The energy need has been increasing exponentially all over world-wide, the reserves of fossil fuels are going to deplete very soon in near future, and serious negative effects of the combustion of fossil fuels are spreading throughout because of CO_2 emission is growing day by day in the atmospheric air. For such types of reasons, motivating the researchers to find some kind of new sustainable energy sources those should be capable to substitute the conventional energy sources with less emissions.

The major demerits in utilization of hydrogen as a gaseous fuel is its availability in nature is nil or can say not easy and expensive production methods are required. It has been reported that 50 million tonnes of hydrogen are traded annually worldwide with a growth rate of nearly 10% per year for the time being [1]. Based on the National Hydrogen program of the United States, the contribution of hydrogen to total energy market will be 8–10% by 2025 [2]. It was reported by the US Department of Energy (US-DOE) that H2 power and transport systems will be available in all regions of the United States by the year 2040 [3]. Due to this level of increasing need for hydrogen energy, in recent years there is urgent need to pay attention on finding the cost-effective development and efficient hydrogen production technologies.

Some methods are developed to improve the existing technologies like, membrane processes, selective oxidation of methane and oxidative dehydrogenation [2]. Biomass and water can be used as renewable resources for hydrogen gas production. Utilization of wide variety of gaseous, liquid and solid carbonaceous wastes was investigated by Kim [4] as renewable sources for formation of hydrogen gas by steam reforming. Despite the low cost of waste materials used, high temperature requirement (T = 1200 °C) is still the major limitation for this process. Electrolysis of water may be the cleanest technology for hydrogen gas production. However, electrolysis should be used in areas where electricity is inexpensive since electricity costs account for 80% of the operating cost of H_2 production. In addition, feed water has to be demineralized to avoid deposits on the electrodes and corrosion [2].

Photosynthetic processes include algae which use CO_2 and H_2O for hydrogen gas production. Some photo-heterotrophic bacteria utilize organic acids such as acetic, lactic and butyric acids to produce H_2 and CO_2. However, the rate of H_2 production is low and the technology for this process needs further development [5].

With the use of sustainable technologies production of clean energy source and utilization of waste materials make biological hydrogen production a novel and promising approach to meet the increasing energy needs.

2. Types of waste materials

Availability, cost, carbohydrate content and biodegradability are the major concern for the selection of waste materials from which bio-hydrogen is going to be produced. Pure carbohydrate sources are expensive raw materials.

2.1. Carbohydrate industrial wastewaters

Some biodegradable carbohydrate containing and non-toxic industrial effluents such as dairy industry, olive

* Corresponding author: vinodsingh.yadav@jaipur.manipal.edu

mill, baker's yeast and brewery wastewaters can be used as raw material for bio-hydrogen production [6]. To remove undesirable components and for nutritional balancing, pre-treatment of waste waters may require. Fig. 1 shows schematic diagram for bio-hydrogen production from food industry wastewaters by two stage anaerobic dark and photo-fermentations.

2.2 Waste sludge of WTP

For production of methane or hydrogen gas, large quantities of carbohydrate and proteins are useful which can be available from waste sludge generated in wastewater treatment plants.

2.3. Agricultural or food industry wastes (Starch and cellulose)

Biodegradability may be affected badly due to the complex nature of these types of wastes. Cellulose containing agricultural wastes requires further pre-treatment. Agricultural wastes should be ground and then de-lignified by mechanical or chemical means before fermentation.

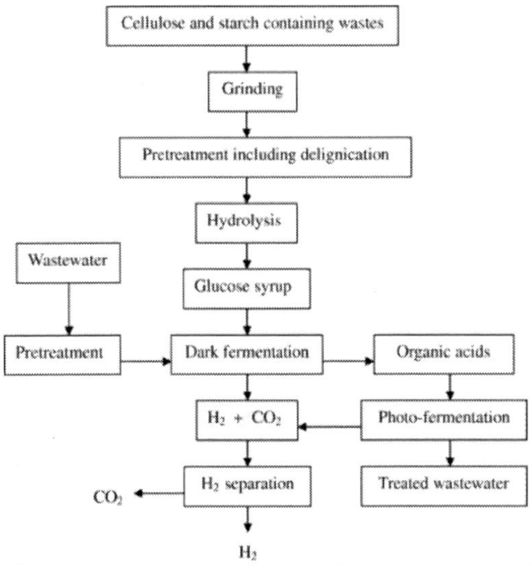

Fig. 1: Bio-hydrogen production from agriculture and food industry wastewaters by two stage anaerobic dark and photo-fermentations

3. Bio-processes for hydrogen gas production

For hydrogen gas production, three major categories can be classified:

1. Bio-photolysis of water through algae.
2. Dark/photo-fermentative (two stage) hydrogen production.
3. Dark-fermentative hydrogen production during acidogenic phase of anaerobic digestion of organic matter.

3.1. Bio-photolysis of water through algae

Generally, the work of algae is division of water molecules into hydrogen ion and oxygen via photosynthesis process. By the use of hydrogenase enzyme, hydrogen ions can be converted in hydrogen gas. Chlamydomonas reinhardtii is one of the well-known hydrogen producing algae [7,8]. Hydrogenase activity has been detected in green algae, Scenedesmus obliquus [9], in marine green algae Chlorococcum littorale [10,11], Playtmonas subcordiformis [12] and in Chlorella fusca [13]. The hydrogenase activity of different algae species was compared by Winkler et al. [14] and it was reported that enzyme activity of the Scenedesmus sp. (150 nmol/_g Chl a.h) is lower than C. reindhartii (200 nmol/_g Chl a.h). It was reported that Anabaena sp. have higher hydrogen evolution potential over the other cyanobacter species [15]. Four to 18% CO_2 concentrations were reported to increase cell density during growth phase resulting in higher hydrogen evolution in the later stage [16].

3.2. Hydrogen gas production through photo-fermentations (two stage)

3.2.1. Types of organisms with conditions

Hydrogen gas production capabilities of some purple photosynthetic bacteria such as Rhodobacter spheroides [17,18-27], Rhodobacter capsulatus [29], Rhodovulum sulfidophilum W-1S and Rhodopseudomonas palustris have been investigated. Photoproduction of hydrogen from CO or other organic acids by carbon-monoxide dependent dehydrogenase (CODH) enzyme containing cultures such as Rhodospirillum rubrum and Rodopseudomonos palsutris P4 has also been investigated.

Therefore, the process requires ammonium limited and oxygen free conditions. Hydrogen gas production was lower in the presence of ammonia salts, while proteins such as albumin, glutamate and yeast extract as a nitrogen source enhanced the production. The metabolism shifts to utilization of organic substance for cell synthesis rather than hydrogen production in the presence of high nitrogen concentrations resulting in excess biomass growth and reduction in light diffusion [27]. It was reported that presence of carbonate enhanced ammonia removal and stimulated hydrogen production. Light intensity is one of the parameters which affects the performance of photofermentation. Increasing light intensity has a stimulatory effect on hydrogen yield and production rate, but has an adverse effect on the light conversion efficiency [29]. Light intensity might also affect the consumption rates of organic acids. Similarly, Wakayama reported that hydrogen production rate during 30 min dark/light cycle was 22 L/m2 d which was twice as much as that obtained by illuminated culture during a 12 h cycle under the same condition.

3.2.2. Substrates

Utilization of industrial effluents for hydrogen gas production by photosynthetic bacteria is possible although, these cultures prefer organic acids as carbon sources. Color of wastewaters is the one of the effluents, which could reduce the light penetration. Hydrogen gas production from potato starch, sugar cane, juice and whey by using Rhodopseudomonas sp. Was investigated by Singh Among the three substrates sugar cane juice yielded the maximum level of hydrogen production (45 mL/mg DWh) as compared to potato (30 mL/mg DWh) and whey (25 mL/mg DWh)[29].

3.2.3. Bio-hydrogen production via Photo-bioreactors

Tubular, flat panel and bubble column reactors are the major types of photo-bioreactors developed for hydrogen production. A mutant type photosynthetic bacteria have been developed to increase the light conversion efficiency and hence hydrogen production rate.
El-Shishtawy reported 9.23% maximum light conversion efficiency by using light-induced and diffused photo-bioreactor (IDPBR) at 300 W/m^2 light intensity. The width of the culture significantly affected the productivity which reached to 7577 mL H_2/m^2 h hydrogen production rate with 1 cm culture width was investigated by Nakada in a photo-bioreactor composed of four compartments aligned along the light penetration axis. In addition, a delay of 2–4 h was observed in maximum hydrogen production rate (3.4 L H_2/m^2 h) after the highest light intensity at noon with an average light conversion efficiency of 1.4%.

3.3. Dark-fermentative hydrogen production during acidogenic phase of anaerobic digestion of organic matter

3.3.1. Type of organisms with conditions

Hydrogen can be produced through many anaerobic organisms from carbohydrate containing organic wastes. Investigations on microbial diversity of a mesophilic hydrogen producing sludge indicated the presence of Clostridia species as 64.6%. By mixed acid or the 2–3 butanediol fermentation, CO_2 and H_2 are produced from formic acid in addition to ethanol and the 2–3 butanediol. Recently, researchers identified hydrogen producing aerobic cultures such as Aeromonos spp., Pseudomonos spp. and Vibrio spp. Under anaerobic conditions when the cultures were cultivated, the hydrogen yield varied between 1 and 1.2 mmol/mol glucose. Hydrogen production by Thermotogales species and Bacillus sp. were detected in mesophilic acidogenic cultures.
Medium pH affects biogas content, hydrogen production yield, the specific hydrogen production rate, and type of the organic acids produced. pH range for the maximum hydrogen yield or specific hydrogen production rate is reported between pH 5.0 and 6.0. However, the optimum pH range is reported by some investigators in between 6.8 and 8.0. Most of the studies indicated that final pH in anaerobic hydrogen production is around 4.0–4.8 regardless of initial pH. Due to production of organic acids which depletes the buffering capacity of the medium results in decrease in pH. Gradual decreases in pH inhibit hydrogen production since pH affects the activity of iron containing hydrogenase enzyme.
When lactose and molasses (sucrose) were used as the substrates, formation of lactic acid was observed. pH also affects the type of organic acids produced. Concentration of acetate and butyrate could be almost equal at pH 6.5–7.0. Ethanol production was observed depending on the environmental conditions. Methane was not detected in most of the hydrogen production studies because of elimination of methane producers by heat digestion of sludge. However, long retention times may cause methane formation by the mesophilic cultures. Methane production was also observed when sewage sludge was used as the substrate.

3.3.2. Substrates

3.3.2.1. Use of simple sugars

An easily biodegradable carbon source is Glucose which is present in most of the industrial effluents and can be obtained abundantly from agricultural wastes. The highest hydrogen yield obtained from glucose is around 2.0–2.4 mol/mol. Though, even when more than 95% glucose was degraded, the yield could be less than 1.7 mol H_2/mol glucose. Therefore, utilization of substrate as an energy source for bacterial growth is the main reason for obtaining the yields lower than theoretical estimations. Similarly, cumulative hydrogen production from sucrose was 300mL while it was only 140mL from starch. These studies showed that the higher hydrogen yields could be obtained from sucrose compared to other simple sugars.

3.3.2.2. Use of starch containing wastes.

As a carbohydrate source for hydrogen production, starch containing materials are abundant in nature and have great potential to be used. As per the reaction stoichiometry, a maximum of 553mL hydrogen gas is produced from one gram of starch with acetate as a by-product.

3.3.2.3. Use of cellulose containing wastes.

The major constitute of plant biomass and highly available in agricultural wastes and industrial effluents such as pulp/paper and food industry is Cellulose. If cellulose concentration increases, it resulted in lower yields with the maximum value of 2.18 mol H_2/mol cellulose with 12.5 g/L cellulose concentration.
During an 81 h period of stationary culture, the organisms consumed 0.92 mmol glucose/h and produced 4.10 mmol H_2/h. The hydrogen yield from the hydrolysate was higher than that of carbohydrates as 19.6 and 18.6 mmol H_2 per gram of substrate consumed, respectively.

3.3.2.4. Use of food industry wastes and wastewater.

The conventional approaches for the solid waste management are landfilling, composting and incineration. The variations in carbohydrate and protein types and concentrations in the mixture is the problem with the food waste. Each component requires different environmental and bio-processing conditions for hydrogen gas production.

Addition of sewage sludge onto food waste as a rich protein source did not improve the production rate. Similarly, hydrogen production potential of carbohydrate rich high solid organic waste (HSOW) was 20 times larger than those of fat rich HSOW and protein rich HSOW because of the consumption of hydrogen gas to form ammonium using nitrogen generated from biodegradation of protein rich solid wastes.

3.3.2.5. Use of waste sludge.

Biosolids (sludges) from waste water treatment plants contain large amounts of polysaccharides and proteins. Hydrogen yields of 1.2 mg H_2/g COD and 0.6 mol/kg CODi were reported when sludge was used as the raw material. However, higher hydrogen yields (15 mg H_2/g COD) were obtained from the filtrate.

3.4. Hydrogen gas production by sequential dark and photo-fermentation

A new approach in biological hydrogen gas production is sequential dark and photo-fermentation. The limitation by the organic acid availability would be removed. Higher hydrogen production yields can be attained when two systems are combined. Further, utilization of organic acids by photo-fermentative bacteria could offer better effluent quality in terms of COD. The ammonia concentration and C/N ratio in the effluent of anaerobic fermentation should not be at the inhibitory level for the photosynthetic bacteria. Dilution and neutralization of dark fermentation effluents are required before photo-fermentation to adjust the organic acid concentration and the pH7 for the optimal performance of photosynthetic bacteria.

4. Conclusions

Hydrogen is considered as the 'energy for future' since it is a clean energy source with high energy content as compared to hydrocarbon fuels. Therefore, new processes need to be developed for cost-effective production of hydrogen. Chemical methods such as steam reforming of hydrocarbons and partial oxidation of fossil fuels operate at high temperatures, and therefore are energy intensive and expensive.

Biological methods offer distinct advantages for hydrogen production such as operation under mild conditions and specific conversions. However, raw material cost is one of the major limitations for bio-hydrogen production. Utilization of some carbohydrate rich, starch or cellulose containing solid wastes and/or some food industry wastewaters is an attractive approach for bio-hydrogen production.

Among the various methods used for bio-hydrogen production are: (a) water splitting by photosynthetic algae, (b) dark fermentation of carbohydrate rich wastes and (c) photofermentation of organic acid rich wastewaters. Organic acids produced during the dark fermentation of carbohydrate rich wastes may be converted to hydrogen and CO_2 by photoheterotrophic bacteria. The process requires special organisms, light and strict control of the environmental conditions. Sequential or combined bio-processes of dark and photo-fermentations seem to be the most attractive approach resulting in high hydrogen yields for hydrogen production from carbohydrate rich wastes.

The major problems in bio-hydrogen production from wastes are the low rates and yields of hydrogen formation. Large reactor volumes are required for bio-hydrogen production due to low hydrogen production rates. Low yields and the rates of hydrogen formation may be overcome by selecting and using more effective organisms or mixed cultures, developing more efficient processing schemes, optimizing the environmental conditions, improving the light utilization efficiency and developing more efficient photo-bioreactors. Considerable research and development studies are needed to improve the 'state of the art' in bio-hydrogen production.

References

[1] Winter CJ. Into the hydrogen energy economy-milestones. International Journal of Hydrogen Energy 2005;30:681–5.

[2] Armor JN. The multiple roles for catalysis in the production of H2. Appl Catal A: Gen 1999;176:159–76.

[3] United States Department of Energy. An integrated research, development and demonstration plan. Hydrogen Pasteur Plan, 2004.

[4] Kim HY. A low cost production of hydrogen from carbonaceous wastes. International Journal of Hydrogen Energy 2003;28:1179–86.

[5] Levin DB, Pitt L, Love M. Biohydrogen production: prospects and limitations to practical application. International Journal of Hydrogen Energy 2004;29:173–85.

[6] De Vrije T, De Haas GG, Tan GB, Keijsers ERP, Claassen PAM. Pretreatment of Miscanthus for hydrogen production by Thermotoga elfii. International Journal of Hydrogen Energy 2002;27:1381–90.

[7] Ghirardi ML, Zhang L, Lee JW, Flynn T, Seibert M, Greenbaum E, et al. Microalgae: a green source of renewable H2. Tibtech 2000;18:506–11.

[8] Melis A. Green alga hydrogen production: progress, challenges and prospects. International Journal of Hydrogen Energy 2002;27:1217–28.

[9] Florin L, Tsokoglou A, Happe T. A novel type of iron hydrogenase in the green alga Scenedesmus obliquus is linked to the photosynthetic electron transport chain. J Biol Chem 2001;276:6125–32.

[10] Ueno Y, Kurano N, Miyachi S. Purification and characterization of hydrogenase from the marine green alga, Chlorococcum littorale. FEBS Lett 1999;443:144–8.

[11] Schnackenberg J, Ikemoto H, Miyachi S. Photosynthesis and hydrogen evolution under stress

conditions in a CO2-tolerant marine green alga, Chlorococcum littorale. J Photochem Photobiol B: Biol 1996;34:59–62.
[12] Guan Y, Deng M, Yu X, Zang W. Two stage photo-production of hydrogen by marine green algae Platymonas subcordiformis. Biochem Eng J 2004;19:69–73.
[13] Winkler M, Heil B, Heil B, Happe T. Isolation and molecular characterization of the [Fe]-hydrogenase from the unicellular green alga Chlorella fusca. Biochim Biophys Acta (BBA): Gene Struct Express 2002;1576:330–4.
[14] Winkler M, Hemschemeier A, Gotor C, Melis A, Happer T. [Fe]- hydrogenases in green algae: photo-fermentation and hydrogen evolution under sulfur deprivation. International Journal of Hydrogen Energy 2002;27:1431–9.
[15] Pinto FAL, Troshina O, Lindbald P. A brief look at three decades of research on cyanobacterial hydrogen evolution. International Journal of Hydrogen Energy 2002;27:1209–15.
[16] Yoon JH, Sim SJ, Kim M, Park TH. High cell density culture of Anabaena variabilis using repeated injection of carbon dioxide for the production of hydrogen. International Journal of Hydrogen Energy 2002;27:1265–70.
[17] Koku H, Ero˘glu ˙I, G¨und¨uz U, Y¨ucel M, T¨urker L. Aspects of metabolism of hydrogen production by Rhodobacter sphaeroides. International Journal of Hydrogen Energy 2002;27:1315–29.
[18] Federov AS, Tsygankov AA, Rao KK, Hall DO. Hydrogen photoproduction by Rhodobacter sphaeroides immobilized on polyurethane foam. Biotechnol Lett 1998;20:1007–9.
[19] Ero˘glu ˙I, Aslan K, G¨und¨uz U, Y¨ucel M, T¨urker L. Substrate consumption rate for hydrogen production by Rhodobacter sphaeroides in a column photobioreactor. J Biotechnol 1999;70:103–13.
[20] Koku H, Ero˘glu ˙I, G¨und¨uz U, Y¨ucel M, T¨urker L. Kinetics of biohydrogen production by the photosynthetic bacterium Rhodobacter spheroids O.U. 001. International Journal of Hydrogen Energy 2003;28:381–8.
[21] Ero˘glu E, G¨und¨uz U, Y¨ucel M, T¨urker L, Ero˘glu ˙I. Photobiological hydrogen production from olive mill wastewater as sole substrate sources. International Journal of Hydrogen Energy 2004;29:163–71.
[22] Yetis M, G¨und¨uz U, Ero˘glu ˙I, Y¨ucel M, T¨urker L. Photoproduction of hydrogen from sugar refinery wastewater by Rhodobacter sphaeroides O.U.001. International Journal of Hydrogen Energy 2000;25:1035–41.
[23] Kondo T, Arakawa M, Hiral T, Wakayama T, Hara M, Miyake J. Enhancement of hydrogen production by a photosynthetic bacterium mutant with reduced pigment. J Biosci Bioeng 2002;93:145–50.
[24] Kim MS, Baek JS, Lee JK. Comparison of H2 accumulation by Rhodobacter sphaeroides KD131 and its uptake hydrogenase and PHB synthase deficient mutant. International Journal of Hydrogen Energy 2006;31:121–7.
[25] Zhu H, Ueda S, Asada Y, Miyake J. Hydrogen production as a novel process of wastewater treatment—studies on tofu wastewater with entrapped R. sphaeroides and mutagenesis. International Journal of Hydrogen Energy 2002;27:1349–57.
[26] Zhu H, Suzuki T, Tsygankov AA, Asada Y. Miyake J. hydrogen production from tofu wastewater by Rhodobacter sphaeroides immobilized agar gels. International Journal of Hydrogen Energy 1999;24:305–10.
[27] Fascetti E, Todini O. Rhodobacter sphaeroids RV cultivation and hydrogen production in a one and two stage chemostat. Appl Microbial Biotechnol 1995;44:300–5.
[28] He D, Bultel Y, Magnin JP, Roux C, Willison JC. Hydrogen photosynthesis by Rhodobacter capsulatus and its coupling to PEM fuel cell. J Power Sources 2005;141:19–23.
[29] Fang HHP, Liu H, Zhang T. Phototrophic hydrogen production from acetate and butyrate in wastewater. International Journal of Hydrogen Energy 2005;30:785–93.

Project failure Indicators: perceptions of Thai and Malaysian engineers

Pittayaporn Gomarn[1,*], and *Jakrapong* Pongpeng[1]

[1]Civil Engineering Department, Faculty of Engineering, King Mongkut's Institute of Technology Ladkrabang, Bangkok, Thailand

Abstract. This research aims to compare the rank and importance level of failure indicators of the construction projects from the perception of Thai and Malaysian engineers. The questionnaire was used to survey the opinions of the engineers who supervised construction projects in Bangkok, Thailand and Kuala Lumpur, Malaysia. Analysis of the data was undertaken with the use of SPSS software which included a comparison of the importance level of construction project failure indicators from the perception of Thai and Malaysian engineers. Additionally, the Mann Whitney U test was used to compare the similarities and differences of construction project failure indicators from the perceptions of both Thai and Malaysian engineers. The results showed that construction project failure indicators from both Thai and Malaysian engineers were similar. Failure indicators ranked in importance were: 1) negative effects on safety, health and environment (18.83%), 2) cost overrun (18.44%), 3) time overruns (17.41%), 4) quality defects (16.43%), 5) stakeholders' dissatisfaction (15.36%), and 6) dispute and litigation (13.54%). The outcome of the research is therefore useful in assisting supervisory staff better understand project failure indicators. In addition, the research can be used as a guideline for analyzing the risk of failures in construction projects.

1 Introduction

According to the ASEAN Construction Federation (ACF) Conference in 2017, in 2016 the Association of Southeast Asian Nations (ASEAN), had a combined GDP of $2.4 trillion. The two member nations of Malaysia and Thailand contributes 29% to the total and are currently growing at a 3-5% growth rate. Malaysia plans to invest about US $30 billion in construction projects, while Thailand plans on spending approximately $35 billion [1].

This shows that the construction industry is very important for the economic and social development of the two countries. However, there is great competition within the sector that creates many obstacles and performance pressures from many factors. These factors can include but are not limited to; project time and cost overruns, as well as safety, health and environmental issues. Therefore, in addition to understanding the key success indicators, engineering managers should understand project failure indicators as well. These can help guide the analysis and prevention of future project failure.

2 Literature review

Over the years, researchers have studied and presented indicators of success and failure of construction projects in Thailand and Malaysia. Ogunlana et al. [2] collected data from 12 high-rise construction projects in Bangkok and found that time and cost overruns were important indicators of the potential for a projects failure.

In Malaysia, Sambasivan and Soon [3] collected responses from 150 contractors and managers and found that there were six main factors affecting a project's completion. These included time overruns, cost overruns, disputes, arbitration, litigation, and total abandonment. Toor and Ogunlana [4] inspected the perception of the key performance indicators (KPIs) in the context of large government projects in Thailand. It was found that the traditional indicator of the steel triangle (on-time, under-budget and according to specifications) are not sufficient for measuring project performance. Other performance indicators such as safety, efficient use of resources, effectiveness, satisfaction of stakeholders, and reduced conflicts and disputes are progressively becoming significant. They were suggests to considered the new measure of performance on large construction projects.

Mirza et al. [5] determined that a major contribution to unsuccessful projects is the lack of understanding or defining project and product scope at the beginning of the project. A properly defined and managed scope results in delivering a quality product, in agreed cost and within planned schedules to the stakeholders.

Nguyen et al. [6] evaluated indexes of construction project success, which included classic indicators such as project cost, project time, and project quality and safety. In addition, projects can also be evaluated by assessing technical performance, functionality, productivity, construction productivity, project stakeholders'

* Corresponding author: gomarnp@gmail.com

satisfaction, environment, communication, and litigation evaluations, and disputes that occur during construction.

Based on the above literature, the researchers defined the concept and meaning of construction project failures as the inability to achieve time, cost, quality, and other goals as required [7]. That is to say, whenever a project does not meet the expectation of the stakeholders, the project is considered a failure [8]. Therefore, the researchers have determined that few studies have presented construction projects key failure indicators, and for those that have, the indicators are not consistent. Therefore, this research aims to analyze the indicator importance level and compare how they are different or similar as perceived by both Thai and Malaysian engineers.

3 Research methodology

The research instrument used for the study was a questionnaire which examined the perceptions of key failure indicators by both Thai and Malaysian engineers. The research methodology was designed as follows:

- The literature and theory were reviewed on construction project failures. These indicators are summarised in Table 1.
- Then, a detailed conceptual framework of indictors was developed from the literature review.
- From the conceptual framework, a questionnaire was designed which included six key indicators. In Part 1 of the questionnaire, there were four items related to the respondents profile regarding the country where they were employed, their position and level, experience, and gross yearly revenue of their organization (Table 2). In Part 2 of the survey, six items were used to evaluate the importance level of the respondents' opinions related to a construction project's failure. For this purpose, a five-level, Likert type agreement scale was used [9], with "1" indicating "the importance of that factor is very low" to "5" indicating "the importance of that factor is very high" as the anchor points, with "3" representing the response "the importance of that factor is moderate". To ensure variable and content validity, the questionnaire was evaluated by four experts in the Thai and Malaysian construction industry.
- To test the construct validity of the questionnaire, the *Spearman's rank* correlation coefficient was used to measure the relationship between factors. The testing results showed that all the variable were correlated indicating the construct validity of these variables [10].
- To test the reliability of the questionnaire scale, the Cronbach's alpha was used with a value from 0 to 1, where 0 indicates the lowest reliability and 1 indicates the highest reliability. For the study, questionnaire scale reliability was calculated as 0.794, indicating satisfactory reliability [11].
- A total of 60 questionnaires were returned from a total of 70 questionnaires sent to both Thai and Malaysian engineers, giving a highly acceptable survey response rate of 85.7% [12].

Table 1. Construction project failure indicators

Project failure indicators / References and country	Time overrun	Cost overrun	Quality defects	Stakeholders' dissatisfaction	Negative effects on safety, health and environment	Dispute and litigation
[2] Thailand	√	√				
[3] Malaysia	√	√				√
[4] Thailand	√	√	√	√	√	√
[5] Malaysia	√	√	√	√		
[6] Thailand	√	√	√	√	√	√
[14] Malaysia	√	√	√		√	

- Data analysis and hypotheses testing was accomplished as follows:

1) Analysis of the importance level of the indictors comes from the following equation (1).

$$\textit{Importance level} = \frac{\textit{Mean}}{\textit{Standard Deviation}} \quad (1)$$

2) Comparison of the perceptions between Thai and Malaysian engineers on construction project's failure indicators are presented in hypotheses H_0 and H_1:
   - H_0: Mean of perceptions between Thai and Malaysian engineers on construction project's failure indicators are not different.
   - H_1: Mean of perceptions between Thai and Malaysian engineers on construction project's failure indicators are different.
3) A significant level (α) of 0.05 was used to test the hypotheses by comparing p-value obtained by the Mann Whitney U Test and the significant level as follows [13]:
   - If p-value < 0.05, H_0 is rejected.
   - If p-value ≥ 0.05, H_0 is accepted.

Table 2. Respondents profile.

Respondents	Thai engineer	Malaysia engineer
Sent	32	38
Return	30	30
Percentage return	93.8%	78.9%
Position		
Senior engineer	40.0%	40.0%
Engineer	60.0%	60.0%
Experience		
6–10 years	56.7%	60.0%
11–15 years	26.7%	36.7%
More than 16 years	16.7%	3.3%
Average value of organization per year		
Less than USD 10 million	16.7%	0.0%
USD 10 million to USD 20 million	33.3%	40.0%
USD 20 million to USD 30 million	23.3%	43.3%
More than USD 30 million	26.7%	16.7%

4 Data analysis results

4.1 Ranking of construction project's failure indicators

The analysed ranking results of perceptions between Thai and Malaysian engineers on construction project's failure indicators are shown in Table 3 and Table 4.

Table 3 and Table 4 show that the ranking indicators influencing the failure of the construction project were the same for both Thai and Malaysian engineers except for ranked second and third (time and cost overruns importance level indices of 8.54 and 9.75 by Thai engineers, and 9.24 and 9.07 by Malaysian engineers). In fact, both of the indicators have the same linear relationship: when working time increases, cost overrun also increases.

The greatest importance in both countries was judged to be the negative effects on safety, health and the environment (importance level indices of 9.92 and 9.45 by Thai and Malaysian engineers, respectively). A possible explanation is perhaps because the operation of construction projects can be stopped if they do not comply with safety requirements, and environmental laws. Therefore, the negative effects on safety, health and the environment possibly affecting project cost, time, quality, satisfaction of stakeholders, including dispute and litigation, that most likely importance effect to the project success or failure.

Table 3. Ranking of construction projects failure indicators obtained from Thai engineers.

Indicator	Mean	Standard deviation	Importance level	Ranking
Negative effects on safety, health and environment	4.80	0.484	9.92	1
Cost overrun	4.67	0.479	9.75	2
Time overrun	4.67	0.547	8.54	3
Quality defects	4.73	0.583	8.11	4
Stakeholders' dissatisfaction	4.57	0.568	8.05	5
Dispute and litigation	4.33	0.606	7.15	6

Table 4. Ranking of construction projects failure indicators obtained from Malaysian engineers.

Indicator	Mean	Standard deviation	Importance level	Ranking
Negative effects on safety, health and environment	4.63	0.490	9.45	1
Time overrun	4.60	0.498	9.24	2
Cost overrun	4.57	0.504	9.07	3
Quality defects	4.53	0.507	8.93	4
Stakeholders' dissatisfaction	4.43	0.504	8.79	5
Dispute and litigation	4.30	0.651	6.61	6

4.2 Mann Whitney Testing

Table 5 presents the results of the perceptions between Thai and Malaysian engineers on construction projects' failure indicators.

Table 5. Hypotheses testing results.

Indicator	p-value*	Hypotheses testing results
Time overrun	0.312	Not different
Cost overrun	0.217	Not different
Quality defects	0.081	Not different
Stakeholders' dissatisfaction	0.170	Not different
Negative effects on safety, health and environment	0.095	Not different
Dispute and litigation	0.419	Not different

*When p-value ≥ 0.05, H_0 is accepted.

Table 5 also showed the hypotheses testing results after p-value analysis. These results indicated that the six indicators influencing the perceived failure of Thai and Malaysian engineers were significantly similar. The possible reason why all indicators in perceived both Thai and Malaysian engineers are similarly is that all those indicators are considered as important on project performance in the construction industry of both countries. In addition, over the past years the stakeholders in the construction industry of both countries are almost focused on safety, health and environment as a first priority along with cost, time, quality and other indicators. Therefore, a combined analysis was also done to find the weight of the indicators. These are shown in Table 6.

Table 6. Importance weights of construction project's failure indicators after a combined analysis.

Indicator	Importance weights (%)
Time overrun	17.41%
Cost overrun	18.44%
Quality defects	16.43%
Stakeholders' dissatisfaction	15.36%
Negative effects on safety, health and environment	18.83%
Dispute and litigation	13.54%

5 Conclusion

The purpose of this research was to compare the ranking and significance of failure indicators with the Thai and Malaysian construction industry. From the questionnaire and the subsequent analysis, the results were ranked as follows (most important to least important as weights of percentage): 1) negative effects on safety, health and environment, 2) cost overrun, 3) time overrun, 4) quality defects, 5) stakeholders' dissatisfaction, and 6) dispute and litigation.

Analysis by use of the Mann Whitney U Test also determined that the perception differences between engineers in their respective countries was also small. However, the negative effects on safety, health and environment had the highest priority (9.92) on Thai engineer perceptions, while with Malaysian engineers it was slightly smaller at 9.45. This is probable because project construction safety, health and environmental issues directly and indirectly affect other project failure indicators. For example, when an accident occurs, the project is most likely to experience a delay in the schedule, budget overruns, stakeholders' dissatisfaction, and possibly dispute and litigation at the end.

Therefore, to avoid the project failure, "safety should be job one". Also, there should be a heightened awareness amongst all parties and managers about the indicators and how to avoid them if possible. In the future research may studies in different countries and regions to compare the rank and importance level of failure indicators.

References

1. Prachachat Business News, Retrieved at: https://tinyurl.com/y8bdaxol (2018 Feb 1)
2. S.O. Ogunlana, K. Promkuntong, V. Jearkjirm, Int. J. Proj. Manag. **14** (1), 37–45 (1996)
3. M. Sambasivan, Y.W. Soon, Int. J. Proj. Manag. **25** (5), 517–526 (2007)
4. S. Toor, S.O. Ogunlana, Int. J. Proj. Manag. **28** (3), 228-236 (2010)
5. M.N. Mirza, Z. Pourzolfaghar, N. Shahnazari, Procedia Technology. **9**, 722-729 (2013)
6. T.A. Nguyen, V. Chovichien, S. Takano, Int. J. Const. Eng. Manag. **2** (3), 70-84 (2013)
7. A.O. Gwaya, S.M. Masu, G. Wanyona, Int. J. Soft. Comp. Eng. **4** (1), 64-69 (2014)
8. O.P. Kharbanda, J.K. Pinto, *What Made Gertie Gallop: Learning from Project Failure* (London: Chapman and Hall, 1996)
9. R. Likert, *Social psychology: Experimentation, theory, research* (Intext, Scranton, PA, 1972)
10. J. Nunnally, *Psychometric theory* (McGraw-Hill. New York, 1967)
11. SPSS Training Department, *Survey research using SPSS.* (SPSS, Inc., Chicago, 1998)
12. E. Babbie, *The practice of social research, 5th Ed.*(Wadsworth Publishing, Belmont, CA, 1989)
13. T. Harris, J.W. Hardin, StataJournal **13**, 337–343 (2013)
14. B. Xiong, M. Skitmore, B. Xia, M.A. Masrom, K. Ye, A. Bridge, Int. J. Proj. Manag. **32**, 482–491 (2014)

MATEC Web of Conferences **192**, 02022 (2018)
ICEAST 2018
https://doi.org/10.1051/matecconf/201819202022

Track-quality index and degradation of railway track structure: The construction track doubling project of northeast line from thanon chira junction to khon kaen station, Thailand

Thawatchai Phanyakit[1,*] and *Thaned* Satiennam[2]

[1]Department of Civil Engineering, Faculty of Engineering, Khon Kaen University, Khon Kaen, 40002, Thailand
[2]Department of Civil Engineering, Faculty of Engineering, Khon Kaen University, Khon Kaen, 40002, Thailand

Abstract. In the past, the railway network has not been developed, and new construction added has been mostly degraded caused by long useful life, and lack of maintenance due to a budget constraint. Currently, 93% of railway in Thailand is single track which results in delays due to low capacity and low average speeds, delayed train on a single track will also delay the train passing it. In some areas of the network, the amount of travel and transportation exceeds the capacity received, so delays in travel and transportation is quite high. Due to this above reason, the service of railway is not effective in terms of convenience and speed of service. Punctuality and cost is high compared with road transport. The biggest challenge that railways faces is the insufficient number of tracks and outdated or old trains. Most the trains cannot travel at their optimum speed because they do not get clearances on time. Thai government has a policy to develop railway track network and services to improve transportation efficiency and save fuel energy used in the transportation sector of the country. Thus, it is necessary to develop the rail transportation system from the existing network as a single track to the current double track network. This study focuses on investigating the degradation of the existing railway structure during the construction track doubling Project. Data were inspected by Track Inspection Car EM120N before the commencement of construction for analysing the condition with the Track Quality Index / TQI) in order to predict the degradation of the railway to plan maintenance and further renewal of the railway. The study found that the TQI value of the existing track structure during the two years of the doubling track construction (2017-2018) has changed significantly.

1 Introduction

1.1 Railway track structure

Railway track structure can be divided into two main parts i.e. superstructure and substructure. The superstructure consists of rails, fastening system, rail pads, and sleepers. The substructure consists of ballast, sub-ballast, and subgrade. Figure 1 shows track design and elements of ballast track structure.

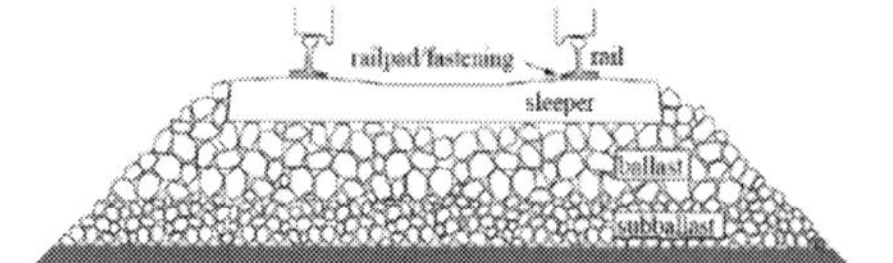

Fig. 1. Track design of a ballasted track. [Dahlberg, 2003]

1.2 Track quality index definition

TQI refers to figures of merit that objectively quantify the condition of the track. The index is computed from data collected by Track Inspection Car (TIC). Track geometry parameters consist of gage, cross level, longitudinal level, alignment, and twist. TQI is a statistical summary of a track geometry parameter measured over a prescribed length of track and effectively summarized a large number of measurements of each parameter for a given track segment.

1.3 Track Inspection car (TIC)

In prior maintenance works, the measurement of the track geometry parameters is normally carried out within a specific time interval. The results are recorded as numerical values, which can be used to indicate the level of track quality. In Thailand, track condition monitoring for Railway Track has inspected by the Track Inspection Car (EM120N) (Figure2&3).

* Corresponding author: thawatchai.enkku@gmail.com

1.4 Track irregularity measurement

Development of track degradation model has mentioned to the use of track geometry or shape data to classify the track geometry irregularities. Calculation of Track Quality Index (TQI) has been specifically developed to be more flexible to application to the degradation of parameters and to become more efficient and optimal to indication or description of various behaviors correctly. Some track parameters are identified as "track geometry" or track shape. The design of geometry usually has specific criteria to assure the safety and convenience in train operation. The standard of State Railway of Thailand (SRT) has specified the criteria of track irregularities required to be measured the deviation in following five types:

- Track gauge
- Cross level
- Longitudinal level
- Alignment

1.5 Track quality index tolerance

Track quality index tolerance is track irregularity index or acceptable track condition index. SRT defined the index properly with track maintenance work at present by considering significantly in terms of competency, economy, and riding comfort. The track quality Index tolerance is divided into 5 types, and each consists of index in each deviation list as follows:

Table 1. SRT standard TQI tolerance. [SRT, 2017]

Track Quality Condition	P Value or TQI Value each Item (%)					Accumulate P or TQI value (%)
	Grade	Profile	Cross Level	Alignment	Twist	
Very Good	A	0-13	0-5	0-9	0-14	0-10
Good	B	14-20	6-13	10-20	15-27	11-20
Fair	C	21-30	14-22	21-31	28-36	21-30
Not Fair	D	31-40	23-31	32-41	37-46	31-40
Not Acceptance	E	>41	>32	>42	>47	>41

Fig. 2. Track inspection car [EM120N] .(SRT, 2015)

Fig. 3. Monitoring TQI by EM 120N. [SRT, 2018]

1.6 Basics of factors influencing degradation

The ability to forecast geometry degradation is very necessary for deciding when maintenance and renewal (M&R) are required. In addition, the same train intervals may have a degradation rate, so it is necessary to gather geometry data to find the ideal value for most of the maintenance and improvement. To identify the influencing factors, the degradation of the quality index will be calculated for both the millimeter/MGT traffic and the millimeter/year if there is no subsidence and rapid degradation of the railway after the tamping. This paper presents about the factors influencing the construction of the double track railway at Thanon Chira Junction to Khon Kaen Station. However, to see the degradation of the railway structure in overview from the literature review, additional factors arising from the double rail construction are also presented as shown in figure 4.

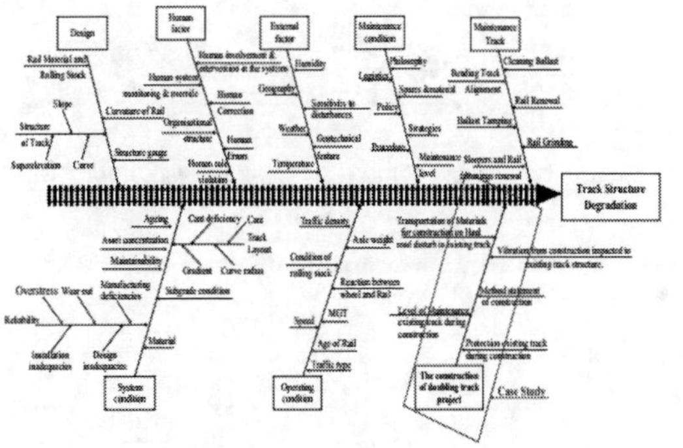

Fig. 4. Fishbone diagram showing factors influencing degradation.

Although track quality index (TQI) along the route has the same structural features such as rail, sleeper, ballast, and fastener installed at the same time and under same traffic, the degradation is significantly different that can be explained in Figure 7 and 8. The TQI values for each kilometer range under the same weight each year, but the TQI values are not equal because of natural factors. And other external factors are different, and the TQI value increases every year.

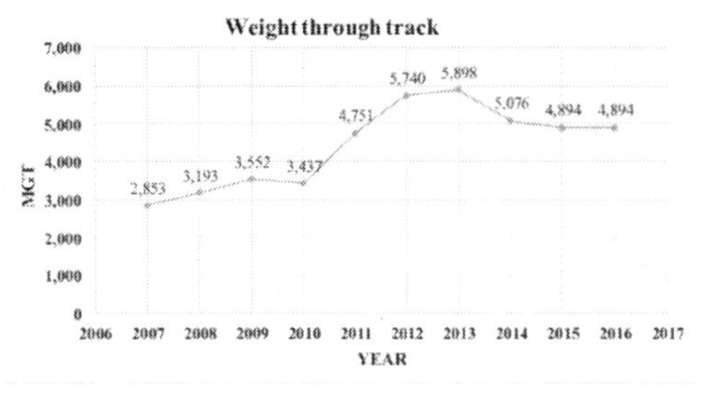

Fig. 5. Annual weight through track of NE line from year 2006-2016. [1]

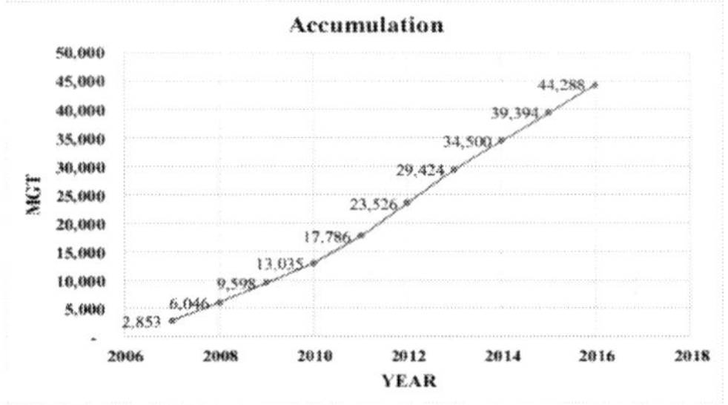

Fig. 6. Accumulated weight through track of NE line year 2006-2016. [1]

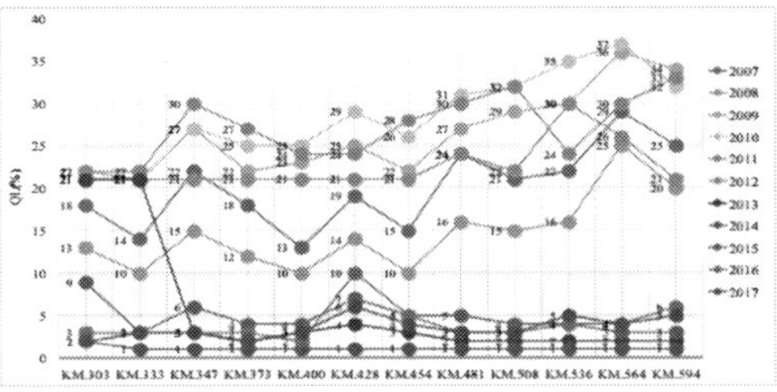

Fig. 7. TQI (%) of NE line (KM.303-KM. 594). [SRT, 2017]

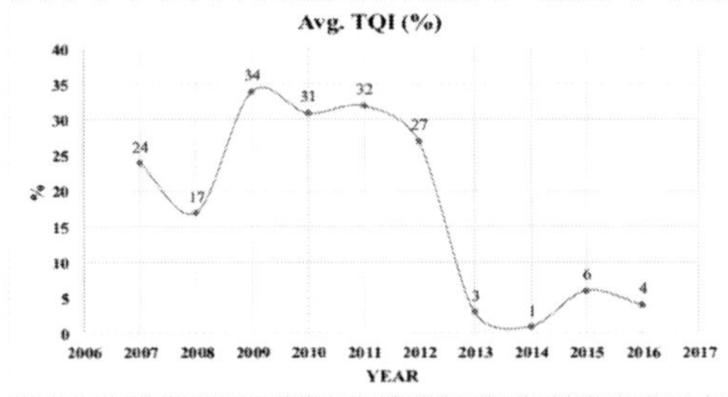

Fig. 8. Average TQI (%) of NE line (KM. 303-KM. 594). [SRT, 2017]

Figure 7 and 8 shown high degraded track condition during 2007-2012 due to the increase of cumulative weight through track as shown in Figure 6. The existing track structure in 2007-2011 used BS7or on timber sleeper found by record TQI in 2009 was high up to 34 ("Not Fair" level which exceeded the standard of requirement on proper track condition prescribed by the SRT, so track rehabilitation was needed to be conducted for this line in 2012.For safety, timber sleeper was replaced by concrete one, and BS70R rail is also replaced by BS100A one to support more weight through track. After the rehabilitation, TQI was better 3% - 6% ("Very Good" level), and the service was regularly provided.

2 Study Area

In 2016, the government made a policy for improving rail transport to shorten transport time and increase efficiency of rail logistic to reduce environmental pollution as well as road accidents, and the policy was to induce more people to use rail transport service. Based on the policy, SRT was assigned to construct 187- kilometer double track parallel with existing track. According to the contract, the construction must be completely done within 2019. In the construction management, the level crossing is disused and replaced by 76 points of box underpass to prevent the train- car crashes that often occurred at a previous time.

Fig. 9. Track doubling project. [SRT,2018]

Fig. 10. Examples of factors influenced by substructure track collapse.

3 Result

Data collected by the EM120N track inspection Car has revealed that during 2016-2018, the existing structure of 187-kilometer Chira Junction - Khon Kaen has gotten higher TQI i.e. at "Very Good" level. During 2013-2016,

TQI is 3%, 1%, 6%, and 4%, respectively while TQI during 2017-2018 is highly increased 15% and 25%, respectively as shown in Figure 11.

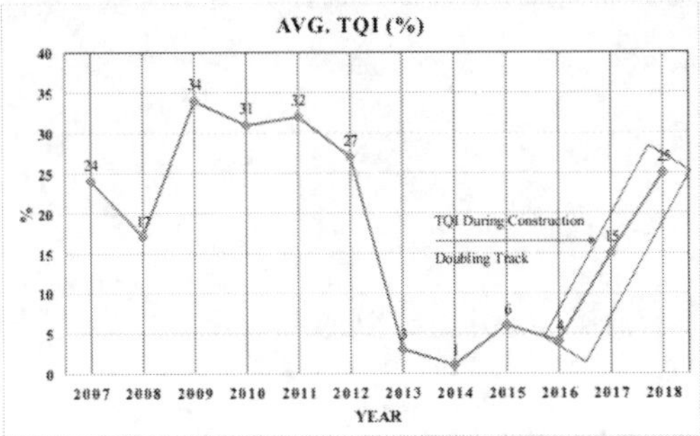

Fig. 11. TQI during 2017-2018 from EM 120N.

Table 2. Evaluation of the northeastern line 2007-2018.

Year	Annual Accumulated Weighing (Ton)	Avg. TQI. Value (%)	TQI Compared with SRT Standards	Remark
2007	2,853,537	24	Fair	
2008	3,193,792	17	Good	
2009	3,552,172	34	Not Fair	
2010	3,437,941	31	Not Fair	
2011	4,751,526	32	Not Fair	
2012	5,740,382	27	Fair	
2013	5,898,270	3	Very Good	Track Rehabilitated
2014	5,076,001	1	Very Good	Track Rehabilitated
2015	4,894,307	6	Very Good	Track Rehabilitated
2016	2,853,537	4	Very Good	Track Rehabilitated
2017	2,853,537 (N/A)	15	Good	Doubling Track
2018	2,853,537 (N/A)	25	Fair	Doubling Track

4 Conclusion and future study

The construction of doubling track project may have many parameters affecting the existing track according to TQI showing that the existing track structure has changed significantly. However, it is not over the point that can lead to the train derailment because SRT has strictly measured and controlled the operation of contractor. Thus, the study can predict the track degradation that it is caused by many factors from the construction of doubling track project. If we construct without any prevention of the existing track structure, and lack maintenance in time, it will be able to lead to the increase of degradation that exceeds a prescribed allowable value which produces transport insecurity of derailment. Thus, this study would like to recommend that degradation factors should be additionally studied after completed the track doubling project. Furthermore, when full service of double track is provided, it can possibly influence the degradation as well because weight through track of a train running on exiting track probably influences the degradation of a new track next to it. Benefits of the study can bring about to condition and behavior control of track infrastructure on the optimal areas in time that can improve reliability of service and safety. In addition, the factors that influence the existing track from the construction can be used for monitoring closely to maintenance in the future.

This study would not have been successful without the support of the Department of Civil Engineering, Faculty of Engineering, Khon Kaen University, Thailand. In addition, sincere gratitude goes to Maintenance Division Chief Khon Kaen Permanent Way Inspector of the State railway of Thailand (SRT), who is one part of this accomplishment supporting valuable data for the study.

References

1. Freight Service Department, State Railway of Thailand, Report on Weight and Train through the Railway 2007- 2016, the State Railway of Thailand, (2017)
2. Maintenance Department, Track Maintenance Manual, State Railway of Thailand, (1964)
3. Maintenance Department, Training Paper on Permanent Track, the State Railway of Thailand, (1994)
4. Stasha Jovanovic, Railway Track Quality Assessment and Related Decision Making, Delft University of Technology, Faculty of Civil Engineering & Geosciences, Road and Railway Engineering Group, The Netherlands.
5. H. Guo,W. Wang, W. Guo, X. Jiang, and H. Bubb, "Reliability analysis of pedestrian safety crossing in urban traffic environment," Safety Science, vol. 50, no. 4, 968–973 (2012)
6. RITA, "Table 1 U.S. Domestic Freight Ton-Miles by Mode (2007).
7. National Bureau of Statistics of China, "China Statistical Yearbook 2012 [WWW]," (2012), http://www.stats.gov.cn/tjsj/ndsj/2012/indexeh.htm.
8. Grimes, G. A., and Barkan, C. P. L. (2006). "Cost-effectiveness of railway infrastructure renewal maintenance." J. Transp. Eng., 132(8), 601–608.
9. Bocciolone, M., Caprioli, A., Cigada, A., and Collina, A. (2007). "A measurement system for quick rail inspection and effective track maintenance strategy." Mech. Syst. Sig. Process., 21(3), 1242–1254.

Plastic collapse load analysis for circumferential cracked cylinder under pure torsion

Sutham Arun[1*] and *Thongchai* Fongsamootr[2]

[1]School of Engineering, University of Phayao, Phayao, Thailand
[2]Department of Mechanical Engineering, Chaing Mai University, Chaing Mai, Thailand

Abstract. This paper aims to analyze the plastic collapse moment of circumferential cracked cylinder under pure torsion using the NSC approach and 3D FE model. The material considered in this work is assumed to be elastic-perfectly plastic. The influences of geometric parameters of crack and cylinder, such as R_m/t, a/t and θ/π on solution of plastic collapse load are also investigated. The analysis shows that for the case of $a/t < 0.75$, the values of limit torsion moment can be estimated by NSC analysis which provides conservative results. However, for the case of deeper crack, $a/t \geq 0.75$, the limit load solution predicted by NSC approach may not be safe, because the distribution of stress at yielding state does not correspond to the NSC assumption. Therefore, the approximated solution of collapse torsion moment for the case of deeper crack with $a/t \geq 0.75$ is proposed based on FE analysis.

1 Introduction

The cylinder is one of the most commonly used components in a wide range of industries. During operation, it may experience various load types which lead to in-service cracking. When cracks are discovered, the analysis of plastic limit load is often performed. This is because the plastic limit load is the important key in structural integrity assessment of the cylinder.

At present, analytical and finite element (FE) solutions of limit load have been well documented for circumferential cracked cylinders [1-4]. These solutions are available for Mode I type of loading of internal pressure, axial tension and bending. However, in addition to pressure, tension and bending, the cylinder is also subjected to torsion due to spatial arrangements. Therefore, the failure assessment method without taking the effect of torsion into account may be incomplete and may affect the operational safety.

Recently, there have been few studied about limit load solution for cracked cylinder subjected to combined loading which include the effect of torsion into their works. For example, Yinsheng Li et al. [5] studied the effect of torsion moment on the limit load solution of cylinder with circumferential part-through thickness crack using FE analysis. They proposed the method for estimating the collapse load for cracked cylinder with the ratios of crack depth to wall thickness ranging between 0.5 and 0.75. Kunio Hasegawa et al. [6] extended the work proposed by Yinsheng Li et al. They found that when the value of torsion moment is less than 20% of yield stress, the plastic collapse moment for circumferential cracked cylinders under combined bending and torsion moment can be estimated by the pure bending moment.

As mentioned above, the details about effects of torsion on the limit load solution is still inadequate. Therefore, the main aim of this paper is to provide the limit load solution for circumferential cracked cylinder subjected to pure torsion. The method used in the paper is based on NSC and FE analysis. The information obtained from this work can be used to support the structural integrity assessment of cracked cylinder.

2 Limit Load Solution Based on NSC Approach

Figure 1 shows the geometry of a cracked cylinder considered in this work. The outer diameter, mean radius and wall-thickness are denoted by D_0, R_m and t, respectively. The crack is assumed to be symmetric with respect to the y-axis where 2θ and $a(\xi)$ represent the total angle of the crack and the crack depth in a function of angular coordinate ξ.

Assuming that there is only torsion exists through all the cross section along the axial direction of the cylinder, there will be only shear stress, τ, acting on any cross section of cylinder. According to this, the applied torsional moment, T, can be computed from [7]:

$$T = \int \rho\,(\tau dA) \qquad (1)$$

where ρ is the radial distance from the axial of the cylinder to any small area dA.

According to the NSC approach, when the material has already yielded, the stresses will uniformly distributed over the net section and are equal to the value corresponding to the yielding condition [1]. From this, the

* Corresponding author: sutham.arun@gmail.com

plastic collapse moment, T_f, for the cracked cylinder defined in Figure 1 can be computed based on Eq.(1) by

$$T_f = \frac{2}{\sqrt{3}} t\sigma_f R_m^2 \left[\pi - \frac{1}{t}\int_0^\theta a(\xi)d\xi\right] \quad (2)$$

where σ_f is the flow stress of the cylinder material.
For a constant-depth crack which $a(\xi) = a_0$, Eq.(2) can be reduced to

$$\frac{T_f}{T_0} = 1 - \frac{a_0}{t}\frac{\theta}{\pi} \quad (3)$$

where the ratio T_f/T_0 is the normalized formulation of T_f and

$$T_0 = \frac{2}{\sqrt{3}}\pi R_m^2 t\sigma_f \quad (4)$$

represents the value of collapse torsion moment for uncracked cylinder. Eq.(3) clearly illustrates that the value of collapse torsional moment is directly proportional to the ligament area of the crack section.

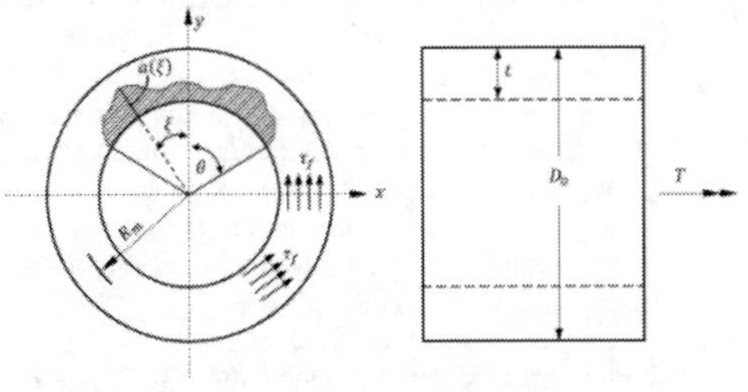

Fig. 1. Nomenclature and stress distribution of a cylinder with a circumferential crack under pure torsion.

3 Finite element model for a circumferential cracked cylinder

In order to investigate the correctness of Eq.(3), the FE models of cylinder with circumferential crack were established in ABAQUS [8]. The continuum element with 20 nodes, reduced integration (C3D20R) was used throughout the models. To improve the accuracy of the model, partitions were created on the cracked zone in order to form a dense mesh along the crack front. During analysis, the number of element in each model was increased ranging from 15,000 elements to 38,000 elements in order to confirm that the results obtained from FE analysis are mesh-independent. Details of the model construction are given in the following sections.

3.1 Model construction

The FE-model of cylinder employed in this work is illustrated in Figure 2. In all cases, the outer radius of the cylinder is 300 mm, and the length of the cylinders is set to be longer than $8D_0$ in order to avoid boundary effects on the crack ligament [5,6].

The circumferential crack is introduced into the middle section of the FE model. Geometric parameters of crack and cylinder are systematically varied to quantify their effects. These factors are the ratio of mean radius to wall thickness (R_m/t), the ratio of crack depth to wall thickness (a/t) and the half angle of crack (θ/π). The cases considered in this work are summarized in Table 1.

The material used for cylinder is a high-ductility stainless steel [5,6]. The Young's modulus is selected as 200 GPa with Poisson's ratio of 0.3. The stress-strain relationship is assumed to be elastic-perfectly plastic and the value of yield stress is equal to 3.72 MPa.

Two reference nodes are created at the center of cross section at end surfaces. The motion of these reference nodes is coupled with the motion of all nodes on each end surfaces by coupling constraints. All translation and rotation of left side reference node are prevented, and the torsional angle is applied to the right side reference node.

As the FE analysis does not directly provide the limit load, the twice-elastic-slope (TES) method is chosen to define the 'plastic collapse moment". This is because TES method is suitable for elastic-perfectly plastic materials which their load deformation curves show apparent collapse behavior [9].

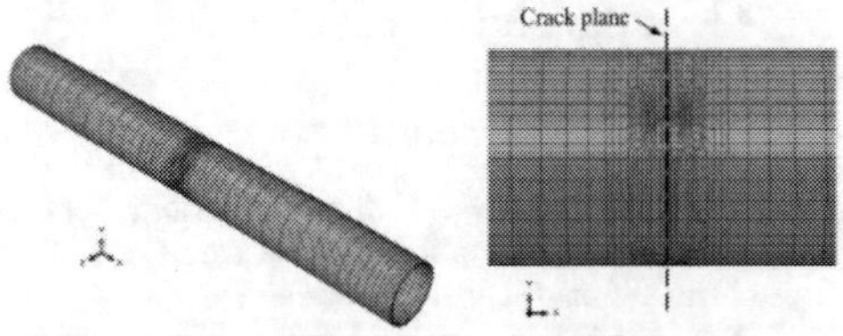

(a) (b)

Fig. 2. (a) The specimen mesh overview and (b) mesh near the crack in longitudinal direction.

Table 1. Cases considered in the present FE analysis.

Geometric parameter	Symbol	Values
The ratio of mean radius to wall thickness	R_m/t	5, 10, 20
The ratio of crack depth to wall thickness	a/t	0.25, 0.50, 0.75, 1.00
The half angle of crack	θ/π	0.0625, 0.125, 0.250, 0.375

3.2 Validation of the model

To gain confidence in the present FE analysis, the FE-model was first loaded by assuming that the cylinder has no crack. The collapse torsion moment obtained from the FE analysis is compared with the solution in Eq.(4). The results show that the present FE results agree very well with the analytical solution, Eq.(4), with the difference less than 0.1%. Further confidence is gained by comparing FE results with the results computed from the equation in elastic theory. This equation is [7]

$$\beta = \frac{L}{GJ}T = C\,T \quad (5)$$

 https://doi.org/10.1051/matecconf/201819202023

where β is the angle of twist, G is the modulus of elasticity in shear, L is the cylinder length and J is the polar moment of inertia of the cross-section area. The compliance, C, in Eq.(5) is the ratio of torsional moment to the angle of twist.

The relationship between T and β obtained from FE analysis and Eq.(5) are plotted together in Figures 3. It can be seen that in elastic region, the relationships obtained from both analysis are in good agreement.

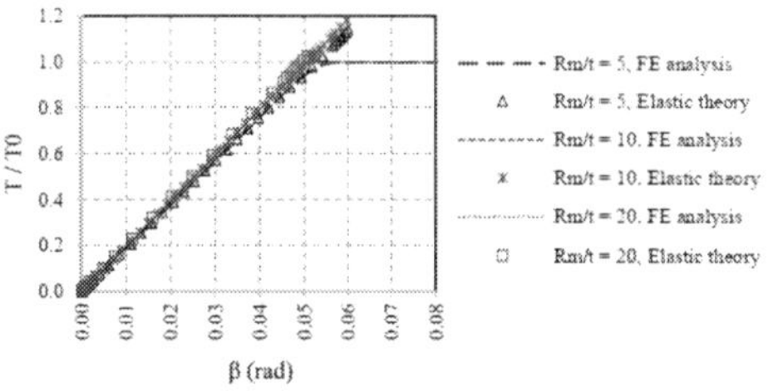

Fig. 3. The relationship between torsional moment and torsional angle (comparison between FE analysis and elastic theory).

4 Results and discussion

4.1 Global behaviour

Figure 4 presents the relationship between the normalization of applied torsional moment, T/T_0, and β obtained from FE analysis for the cases of $R_m/t = 5$ and $R_m/t = 10$. It is very clear that when the cylinder has a crack, the change of crack ligament area, a/t, does not affect the value of C in elastic region. In other words, the reduction of the ligament area in the crack section does not affect the value of J in Eq.(5). This means that when material behaves elastically, the elastic theory can still be used to predict the global behavior of the circumferential cracked cylinder.

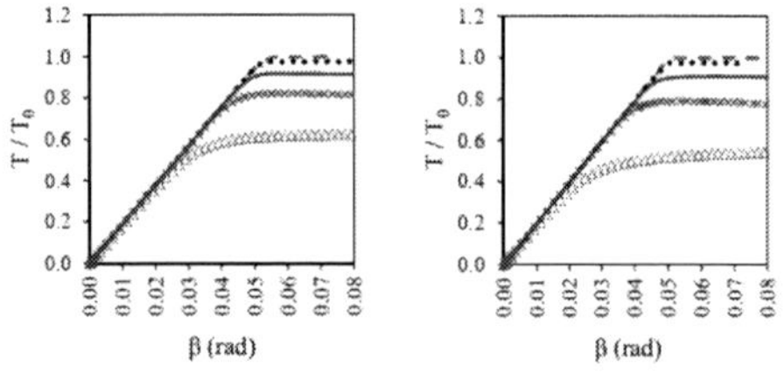

(a) (b)

Fig. 4. Relationship between T/T_0 and torsional angle : (a) $R_m/t = 5$ and (b) $R_m/t = 10$.

4.2 Effect of geometric factor of cracked cylinder on limit load

The influences of R_m/t, a/t and θ/π on the normalized limit torsional moment, T_f/T_0, are presented in Figures 5. It can be seen that when the value of a/t is less than 0.75, the ratio R_m/t affects slightly, not over 10 %, on the change of T_f/T_0. However, when the value of a/t is higher, the limit torsion moment changes significantly according to the value of R_m/t. This behavior is inconsistent with Eq.(3) which clearly indicates that the values of T_f/T_0 only depends on the ligament area of the crack section.

Figure 5 also shows that, at $a/t < 0.75$, the solutions of T_f/T_0 obtained from the FE-analysis are slightly higher than those predicted by Eq.(3). However, when a/t is higher, the values of T_f/T_0 provided by FE-analysis drop rapidly to lower than those predicted from Eq.(3). The reason of this behavior can be explained by the concept of equilibrium of torsion moment in Eq.(1) as follows;

Figure 6 presents the distribution of Von-Mises stress on the cracked plane at the state of yielding. It can be seen that when the cylinder has no crack, Figure 6a, the stress acting on this cross-section area is uniform and is equal to the value of yield stress (3.72 MPa). This corresponds to the assumption of NSC analysis. Therefore, for the case of uncracked cylinder, the ratio T_f/T_0 estimated by the FE-analysis is very close to the solution calculated from Eq.(3).

For the cracked cylinder with $a/t < 0.75$, the stresses in the region ahead the crack front become singular after subjecting to torsional moment, as shown in Figures 6b and 6c. These values are much higher than the value of yield stress. Therefore, the average value of shearing force, τdA, on cracked section is higher than the assumption of NSC method.

For the case of deep crack, $a/t \geq 0.75$, although there is stress singularity ahead the crack front, the value of average shearing force is still lower than the NSC assumption. This is because, when the cross section area is greatly reduced due to crack, the values of stresses acting on the area opposite to the crack location must decrease in order to maintain a self-equilibrated distribution of stress, as shown in Figures 6(d) and 6(e). According to this reason, the limit torsion moment obtained from the FE-analysis is lower than the one predicted from Eq.(3).

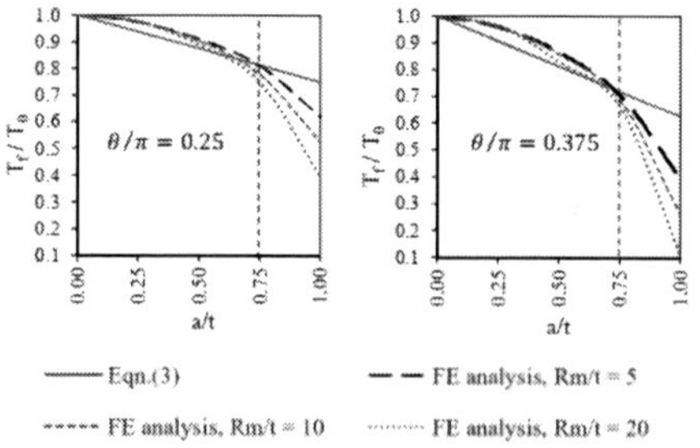

Fig. 5. Comparison of T_f/T_0 versus θ/π between NSC approach and FE analysis.

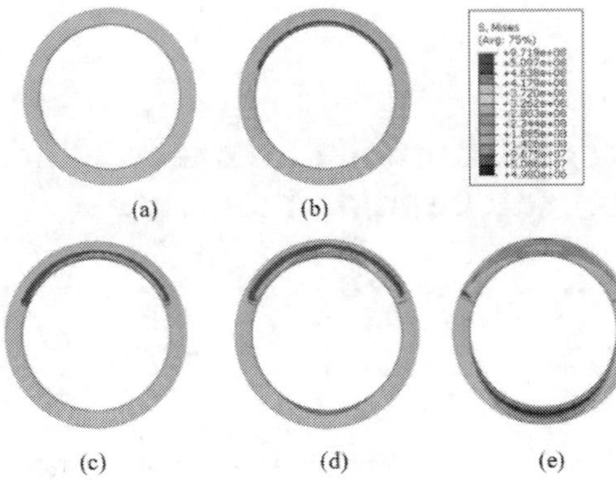

Fig. 6. The contour plot of von Mises stress at yielding state on the crack plane for the case of $\theta/\pi = 0.375$: (a) $a/t = 0.25$, (b) $a/t = 0.50$, (c) $a/t = 0.75$ and (d) $a/t = 1.00$.

4.3 The possible candidate for plastic collapse torsion solution

According to section 4.2, for the case of $a/t < 0.75$, the value of limit moment estimated by NSC approach, Eq.(3), is slightly lower than one obtained from FE-analysis. As, the difference is not over 5%, Eq.(3) can be used to obtain the conservative solution of the plastic collapse torsional moment.

However, for the case of deeper crack with $a/t \geq 0.75$, the limit load obtained from FE-analysis is much lower than the results from Eq.(3). Therefore, the estimation of collapse torsional moment by Eq.(3) may not be safe. According to this, the following linear approximation based on the FE analysis is proposed.

$$\frac{T}{T_0} = 0.5817 + 0.0258\left(\frac{R_m}{t}\right) + 3.1165\left(\frac{\theta}{\pi}\right) + 0.5468\left(\frac{a}{t}\right) - 0.0218\left(\frac{R_m}{t}\right)\left(\frac{\theta}{\pi}\right) - 4.9732\left(\frac{\theta}{\pi}\right)\left(\frac{a}{t}\right) - 0.0331\left(\frac{a}{t}\right)\left(\frac{R_m}{t}\right) \quad (6)$$

Eq.(6) provides the values of T_f/T_0 within 10% of FE results at $a/t \geq 0.75$, $2\theta \leq 135^o$ and $5 \leq R_m/t \leq 20$.

5 Conclusion

In this work, both the NSC approach and 3D FE model are used to analyze plastic collapse load for cylinder with circumferential crack subjected to torsional moment. The material considered in this work is assumed to be elastic-perfectly plastic. The influence of geometric parameters of crack and cylinder, such as R_m/t, a/t and θ/π on the solution of plastic collapse load are investigated. The following conclusions can be made :

1. As material behaves elastically, the change of crack ligament area does not affect the slope of the relationship between torsion moment and torsion angle.
2. The ratio R_m/t affects slightly on the change of T_f/T_0 when value of a/t is less than 0.75. However, when the values of a/t is higher, the value of limit torsion moment changes significantly according to the value of R_m/t.
3. For the case of $a/t < 0.75$, the values of limit torsion moment can be estimated by NSC analysis, Eq.(3). However, for the case of deeper crack, $a/t \geq 0.75$, the limit load solution predicted by NSC approach may not be safe as it provides very high value of T_f/T_0, comparing with FE analysis.
4. For the case of deeper crack with $a/t \geq 0.75$, the limit torsion moment solution can be estimated by linear approximation in Eq.(6), with the difference not over than 10% from FE analysis.

References

1. S.Rahmana and G.Wilkowskib, "Net-section-collapse analysis of circumferentially cracked cylinders-part I: arbitrary-shaped cracks and generalized equations," Engineering Fracture Mechanics, vol.61, pp.191-211, 1998.
2. Y.J.Kim, D.J.Shim, N.Kamran, Y.J.Kim , S.S.Hwang and J.S.Kim, "Finite element based plastic limit loads for cylinders with part-through surface cracks under combined loading," International Journal of Pressure Vessels and Piping, vol.80, pp.527-540, 2003.
3. Y.Lei and P.J.Budden, "Limit load solutions for thin-walled cylinders with circumferential cracks under combined internal pressure, axial tension and bending," The Journal of Strain Analysis for Engineering Design, vol.39(6), pp.673-683, 2004.
4. Y.Lei, Y.Li and Z.Gao, "Global limit load solutions for thick-walled cylinders with circumferential cracks under combined internal pressure, axial force and bending moment Part I: Theoretical solutions," International Journal of Pressure Vessels and Piping, vol.114-115, pp.23-40, 2014.
5. Y.Li, K.Hasegawa, P.H.Hoang and B.Bezensek, "Prediction Method for Plastic Collapse of Circumferentially Cracked Pipes Subjected to Combined Bending and Torsion Moments," 2012 ASME Pressure Vessels and Piping Division Conference, Toronto, Ontario, Canada, July 15–19, 2012.
6. K.Hasegawa, Y.Li and K.Osakabe, "Collapse loads for circumferentially through-wall cracked pipes subjected to combined torsion and bending moments," Engineering Fracture Mechanics 123 (2014) 77–85.
7. A.C.Ugural and S.K.Fenster, "Advanced Strength and Applied Elasticity," 3rd ed, PTR Prentice Hall, Englewood Cliffs, New Jersey
8. Hibbitt, Karlsson and Sorensen Inc, "ABAQUS/Standard User Manual," Rhode Island.
9. Y.Li, Y.Lei and Z.Gao, "Global limit load solutions for thick-walled cylinders with circumferential cracks under combined internal pressure, axial force and bending moment _ Part II: Finite element validation," International Journal of Pressure Vessels and Piping, vol.114-115, pp.41-60, 2014.

 https://doi.org/10.1051/matecconf/201819202024

Collapse load analysis for circumferentially cracked cylinder subjected to combined torsion and bending moments

Sutham Arun[1*] and *Thongchai* Fongsamootr[2]

[1]School of Engineering, University of Phayao, Phayao, Thailand
[2]Department of Mechanical Engineering, Chaing Mai University, Chaing Mai, Thailand

Abstract. Cylinder is one of the most commonly used components which has a risk of having circumferential cracks, especially in the welding zone. When cracks are discovered, it is necessary to perform the failure strength assessment of cracked cylinder and the limit load play an important part as the input of the assessment. At present, the limit load solution for circumferential cracked cylinder under combined bending and torsion can be estimated by using the methods of equivalent moment or biaxial failure parameter. However, these methods still have some limitations. The main aim of this paper is to propose the alternative method for predicting the failure moment of circumferential cracked cylinder under combined bending and torsion. The method used in this paper is based on the modification of biaxial failure parameter and the data from finite element analysis. Details of this method is presented in this paper.

1 Introduction

Cylinders used in power station, petroleum, steel and other industries may experience internal pressure, axial load, bending and torsion. This makes cylinder at the risk of having circumferential cracks, especially in the welding zone. When cracks are discovered, it is necessary to perform the failure strength assessment to prevent the severe accidents.

Plastic limit load is one of the most important factors when performing structurally integrity assessment of cracked structure. According to this, many analytical and finite element (FE) solutions have been developed for predicting the plastic limit load of circumferential cracked cylinder [1-4]. However, these solutions are usually developed for Mode I type loading (opening mode) of internal pressure, axial tension and bending. The effects of Mode III type of torsion moment are not included in the solutions.

As cylinder may be subjected to torsion during installation or piping arrangement, ignoring the effect of torsion when deriving the limit load solution may results in incomplete or overestimated solutions. At present, several methods for predicting the failure load of cracked cylinder under combined torsion and bending moments have been developed. These methods are based on the equivalent collapse moment and biaxial failure parameter [5,6]. However, these methods still have some limitations when they were used to estimate the failure load.

The main aim of this paper is to provide the alternative method to predict the failure load for circumferential cracked cylinder under combined bending and torsion. The method is based on the modification of biaxial failure parameter and the data from finite element analysis.

* Corresponding author: sutham.arun@gmail.com

2 Research background

2.1 Description of geometry

Figure 1 shows the geometry of circumferential cracked cylinder considered in this work. The dimension of the cylinder is described by its mean radius R_m and wall-thickness t. The crack is assumed to be symmetric with respect to the y-axis. The total angle of crack and the crack depth measured from cylinder inner surface are defined by 2θ and a. The cylinder is subjected to combined external bending moment M_B about x-axis and external torsion moment M_T about z-axis.

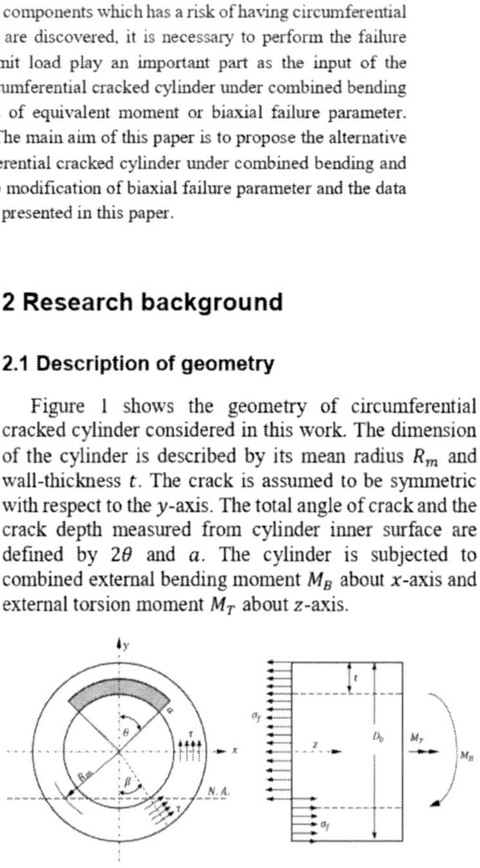

Fig. 1. Nomenclature and stress distribution of a cylinder with a circumferential crack.

According to the geometry defined in Figure 1, the plastic collapse moment for cylinder without crack under pure bending and pure torsion conditions can be given by;

$$M_{B0} = 4\,R_m^2 t\sigma_f \tag{1}$$

$$M_{T0} = \frac{2}{\sqrt{3}}\pi R_m^2 t\sigma_f \tag{2}$$

where M_{B0} and M_{T0} are the plastic collapse moment for uncracked cylinder subjected to bending and torsion, respectively. Eqs. (1) and (2) are developed based on an elastic-perfectly plastic material with flow stress σ_f.

2.2 Limit load solutions for cylinders with circumferential crack

For circumferential cracked cylinder under pure bending, the solution of collapse bending moment is provided in ASME B&PV Code Section XI [7]. This solution was developed based on net-section-collapse (NSC) principle which can be expressed as follows :

$$\beta = \frac{1}{2}\left(\pi - \frac{a}{t}\theta\right) \quad (3)$$

$$M_{BF} = 2\sigma_f R_m^2 t\left[2\sin(\beta) - \frac{a}{t}\sin(\theta)\right] \quad (4)$$

where σ_f is the flow stress, β is the neutral angle of the cylinder cross section, see Figure 1. Eqs. (3) and (4) are applicable to the case which whole crack is inside the tensile stress zone $(\theta + \beta \leq \pi)$.

The collapse moment solution for circumferential cracked cylinder under pure torsion has been mentioned in few literatures. This solution is derived based on the NSC approach. For the case of constant-depth crack, the solution is given as follows [6];

$$M_{TF} = \frac{2}{\sqrt{3}}\pi R_m^2 t\sigma_f\left(1 - \frac{\theta}{\pi}\frac{a}{t}\right) \quad (5)$$

However, there are some evidences showing that Eq.(5) can provide accurate prediction only for the case of short cylinder length or shallow surface crack [6,8].

For the case of cracked cylinder subjected to combined bending and torsion moments, Bezensek et al. [5] showed that the collapse moment can be described as equivalent moments M_{eq} defined by;

$$M_{eq} = \sqrt{M_B^2 + M_T^2} \quad (6)$$

where M_B and M_T are the external bending moment and the external torsion moment at collapse. Bezensek [5] also found that when the torsion stress τ acting on the crack section was less than 20% of flow stress $(\tau/\sigma_f < 0.2)$, the values of M_{eq} were almost constant. This implies that M_{eq} can be used to predict the collapse plastic failure for the case of small amount of torsion.

As M_{eq} is not valid for the case of large torsion moment, Li et al. [6] defined a non-dimensional failure parameter r which can be used to estimate the collapse failure of cylinder for general magnitude of torsion moment. This parameter is given by

$$r = \sqrt{\left(\frac{M_B}{M_{BF}}\right)^2 + \left(\frac{M_T}{M_{TF}}\right)^2} \quad (7)$$

where M_{BF} and M_{TF} are the plastic collapse moments of cracked cylinder subjected to pure bending moment and pure torsion moment, respectively. Li [6] also found that for any crack depth and crack angle, the value of r is almost equal to 1. However, there is some difficulties in using Eq.(7) in the prediction of failure moment. This is because there is no standard solution for computing the value of M_{TF} for general cases. Therefore, FE analysis may be the only way to obtain the value of M_{TF}.

3 Finite element analysis model

The FE models of cylinder with circumferential crack were generated in ABAQUS [9]. These models are used to analyze the correctness of the biaxial failure parameter defined in Eq.(7). Details of the models are given in the following sections.

3.1 Model construction

Figures 2 and 3 present the three-dimensional FE model of straight cylinder with circumferential crack. In all cases, the outer diameter of a cylinder, D_0 is 600 mm, and the length of the cylinder is eight times larger than the outer diameter in order to avoid boundary effects on the crack ligament [6,10]. To prevent problem associated with shear locking, 20-node quadratic brick element with reduced integration (C3D20R) is used to generate the models.

The material used for cylinder is a high-ductility stainless steel [6,10]. The Young's modulus is selected as 200 GPa with Poisson's ratio of 0.3. The stress-strain relationship is assumed to be elastic-perfectly plastic with the yield stress of 372 MPa.

The FE analysis contains 2 loading steps. The first step is to apply the specified value of torsion moment on the cylinder end. After the torsion moment is maintained at the given value, the bending moment is then applied to the cylinder in the second loading step. This process is performed until the maximum moment is achieved.

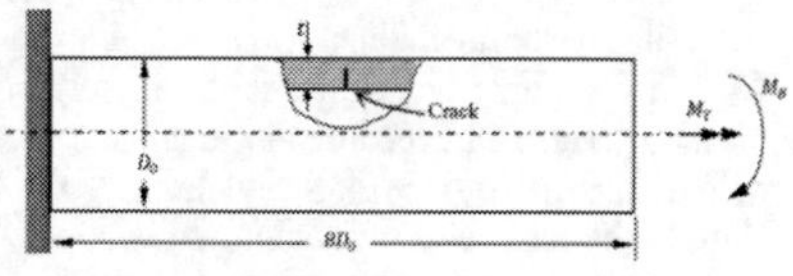

Fig. 2. Model of a cracked cylinder under bending moment and torsion moment.

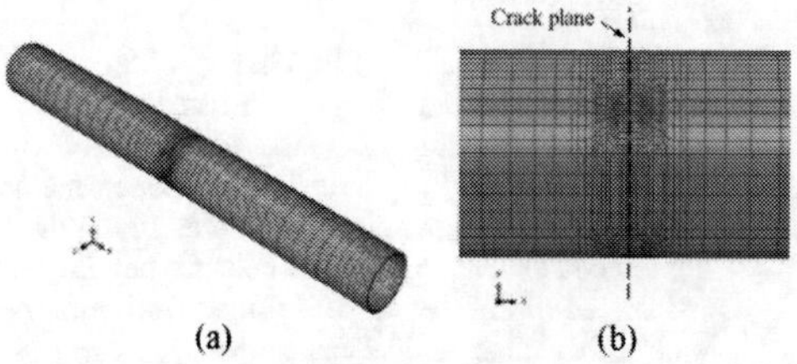

Fig. 3. (a) The specimen mesh overview and (b) mesh near the crack in longitudinal direction.

3.2 Limit load analysis

The models are analyzed using elastic-plastic analysis with large deformation formulation. The further post-processing is performed to obtain the value of limit load. For the case of cylinder subjected to a single load, e.g. pure bending or pure torsion, the collapse moment is defined using the twice-elastic-slope (TES) method [11]. For the case of cylinder subjected to combined bending and torsion moments, the plastic collapse load is defined from the value of the maximum bending moment which the model can carry. The reason in choosing this method is given in the next section.

3.3 Validation of the model

The confidence in the FE analysis is gained by comparing the current FE-results with the results from Hasegawa et al. [10]. Figure 4 shows that the present FE results agree very well with Hasegawa's work, with the difference less than 5%.

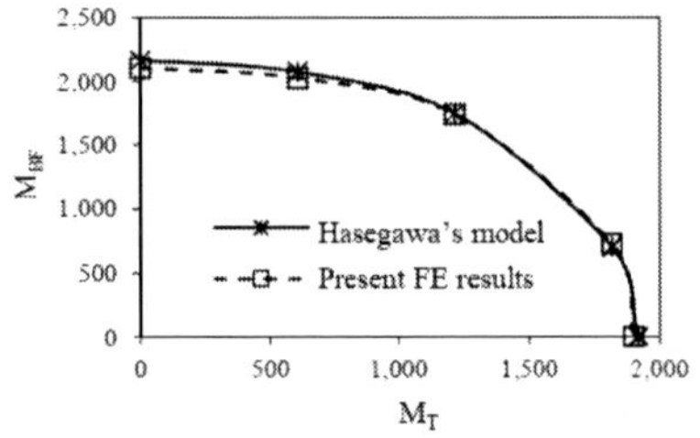

Fig. 4. Comparison of plastic collapse bending moment for cylinder with crack of $2\theta = 90$ and $a/t = 1.0$.

4 Results and discussion

4.1 Global behaviour

Figure 5 shows the relationship between the normalized bending moment (M/M_{B0}) and bending angle (θ_b) as a function of normalized torsion moment M_T/M_{T0}. This is the load-displacement curve for the case of $R_m/t = 10$, $a/t = 0.75$ and $2\theta = 90^o$. The figure shows that in elastic region, the slope of load-displacement curve is almost equal for any value of M_T/M_{T0}. This implies that when material behaves elastically, the elastic beam theory can still be used to predict the global behavior of the circumferential cracked cylinder subjected to combined torsion and bending moment.

From Figure 5, it can be observed that the bending moment decreases gradually with increasing of θ_b after the maximum bending moment is achieved. According to this, TES method may not be the proper method for estimating the collapse moment. Therefore, for the case of combined moments, the plastic collapse bending moment is defined from the value of the maximum bending moment which the model can carry.

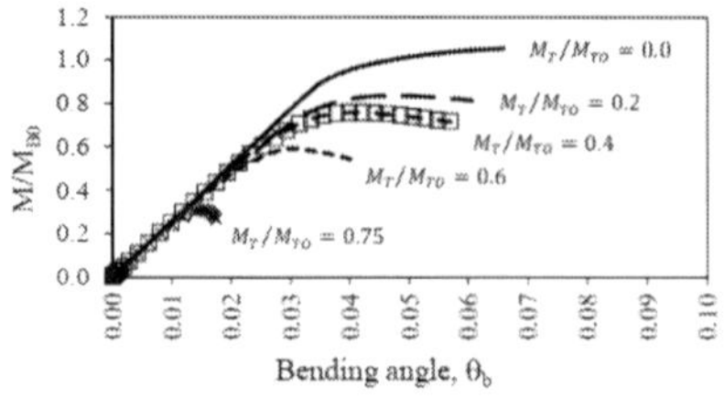

Fig. 5. Load – displacement curve for cylinder with crack of $2\theta = 90$ and $a/t = 0.75$.

4.2 Equivalent collapse moment (M_{eq})

The collapse bending moment and collapse torsion moment obtained from FE analysis are used to compute the values of M_{eq} by Eq.(6). Figure 6 illustrates that the values of M_{eq} is almost constant when the value of M_T/M_{T0} are less than 0.25. This information corresponds to the works presented in literatures [5] which indicates that M_{eq} can be used to predict the plastic collapse failure for the case of small amount of torsion.

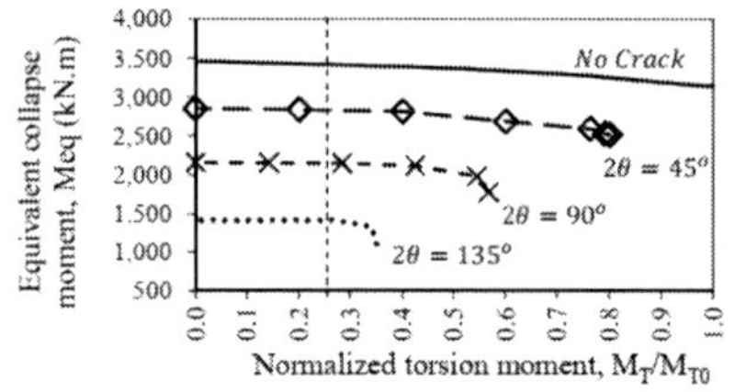

Fig. 6. Relationship between equivalent collapse moment and the collapse torsion moment for the case of $R_m/t = 10$ and $a/t = 0.75$.

4.3 Biaxial failure parameter (r)

Figure 7 illustrates the relationship between M_B/M_{BF} and M_T/M_{TF} for the case of $R_m/t = 10$. As presented in Eq.(7), this relationship is in the form of equation of circle centered at the origin with radius r. Figure 7 shows that the values of r is almost equal to 1 for any crack depth and crack angle. This result confirms that the parameter r can be used to predict the plastic collapse moment for a general range of torsion moment.

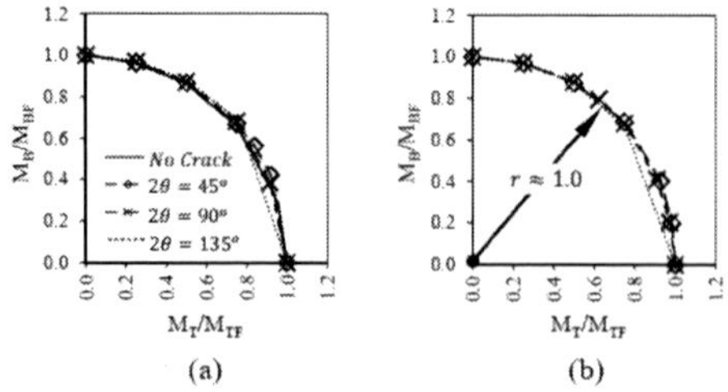

Fig. 7. Relationship between M_B/M_{BF} and M_T/M_{TF}: (a) $a/t = 0.25$ and (b) $a/t = 0.75$.

4.4 Modified biaxial failure parameter (r_0)

As mentioned before, it is necessary to assess the values of M_{TF} when Eq.(7) is employed to estimate the failure moment. The only way to obtain this value is from FE analysis. To overcome this issue, Eq.(7) is modified by replacing M_{BF} and M_{TF} with M_{B0} and M_{T0} as follows;

$$r_0 = \sqrt{\left(\frac{M_B}{M_{B0}}\right)^2 + \left(\frac{M_T}{M_{T0}}\right)^2} \quad (8)$$

Since M_{B0} and M_{T0} can be computed by Eqs.(1) and (2), the problem of computing M_{TF} is solved. Figure 8 presents the relationship between M_B/M_{B0} and M_T/M_{T0}. It can be seen that the relationship is still in the form of circle equation but the radius r_0 decreases with the increase of cracked area. In other words, r_0 is a function of a/t and θ/π. From this, the following quadratic approximation of r_0 is proposed based on the data from the FE analysis;

$$r_0 = 0.985 + 0.4\left(\frac{a}{t}\right) - 0.04\left(\frac{\theta}{\pi}\right) - 1.97\left(\frac{a}{t}\right)\left(\frac{\theta}{\pi}\right)$$

$$-0.33\left(\frac{a}{t}\right)^2 + 0.84\left(\frac{\theta}{\pi}\right)^2 \quad (9)$$

Eq.(9) provides the values of r_0 within 9% of FE results at crack angle $2\theta \leq 135^o$ and $5 \leq R_m/t \leq 20$.

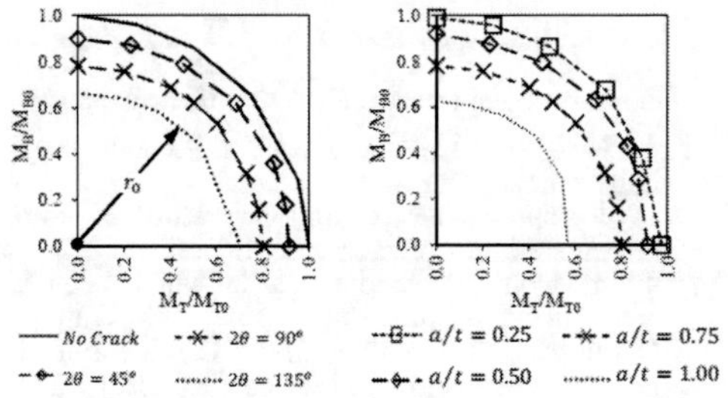

Fig. 8. Relationship between M_B/M_{B0} and M_T/M_{T0} for the case of (a) $a/t = 0.75$ and (b) $2\theta = 90^o$.

5 Conclusion

In this work, the 3D FE model is used to analyze plastic collapse load for circumferential cracked cylinder subjected to combined bending and torsion moments. The analysis is based on the methods of equilibrium collapse moment M_{eq} and biaxial failure parameter r. The following conclusions can be made :

1. As material behaves elastically, the global behavior of cracked cylinder can be predicted by the elastic beam theory.
2. The equivalent moment M_{eq} can be employed to predict the plastic collapse moment for the case of small amount of torsion, $M_T/M_{T0} \leq 0.25$.
3. The biaxial failure parameter r can be used to estimate the collapse moment of cylinder for general magnitude of torsion moment. However, it may not be convenient in using this parameter, since there is no standard solution of M_{TF} for general cases.
4. In order to avoid the calculation of M_{TF}, the parameter r is modified to be r_0 as shown in Eq.(8).
5. The values of r_0 can be estimated by Eq.(9) for the case of $2\theta \leq 135^o$ and $5 \leq R_m/t \leq 20$, with the difference less than 9% from the FE analysis.

References

1. S.Rahman, "Net-section-collapse analysis of circumferentially cracked cylinders Part II: idealized cracks and closed-form solutions, "Engineering Fracture Mechanics, vol.61, pp.213-230, 1998.
2. Y.J.Kim, D.J.Shim, N.Kamran, Y.J.Kim , S.S.Hwang and J.S.Kim, "Finite element based plastic limit loads for cylinders with part-through surface cracks under combined loading," International Journal of Pressure Vessels and Piping, vol.80, pp.527-540, 2003.
3. Y.Lei and P.J.Budden, "Limit load solutions for thin-walled cylinders with circumferential cracks under combined internal pressure, axial tension and bending," The Journal of Strain Analysis for Engineering Design, vol.39 (6), pp.673-683, 2004.
4. Y.Lei, Y.Li and Z.Gao, "Global limit load solutions for thick-walled cylinders with circumferential cracks under combined internal pressure, axial force and bending moment Part I: Theoretical solutions," International Journal of Pressure Vessels and Piping, vol.114-115, pp.23-40, 2014.
5. B.Bezensek, Y.Li, K.Hasegawa and P.H.Hoang, "Inclusion of torsion loads in Section XI flaw evaluation procedures for pipes containing circumferential planar surface crack-like flaw on the basis of Limit Load Analysis," Journal of Pressure Vessel Technology, vol.134(3), 2012.
6. Y.Li, K.Hasegawa, P.H.Hoang and B.Bezensek, "Prediction Method for Plastic Collapse of Pipes Subjected to Combined Bending and Torsion Moments," in Proceedings of ASME 2010 pressure vessel & piping conference, Bellevue, Washington, USA, July 18–22, 2010.
7. ASME Boiler & Pressure Vessel Code, Section XI. Rules for inservice inspection of nuclear power plant components, American Society of Mechanical Engineers, New York; 2010.
8. H.W.Ryu, J.J.Han and Y.J.Kim, "Limit loads for circumferentially cracked pipes under combined cending and torsion with tension," in Proceedings of ASME 2013 pressure vessel & piping conference, Paris, France, July 14–18, 2013.
9. Hibbitt, Karlsson and Sorensen Inc, "ABAQUS/Standard User Manual," Rhode Island.
10. K.Hasegawa Y.Li and K.Osakabe, "Collapse loads for circumferentially through-wall cracked pipes subjected to combined torsion and bending moments," Engineering Fracture Mechanics, vol.123, pp.77-85, 2014.
11. M.Muscat D.Mackenzie and R.Hamilton, "A work criterion for plastic collapse," International Journal of Pressure Vessels and Piping, vol.80(1), 2003.

Simplification of sound-pressure characteristics using narrow flow channels subject with temperature gradient

Tomoaki Kyoden[1*]

[1] National institute of Technology, Toyama College, Imizu, Toyama, Japan

Abstract. A thermoacoustic engine is one of promising system for utilization of unused energy. The object of this paper is to study the system of unused energy, such as heat energy and sound energy in noise using thermoacoustic engine. The thermoacoustic engine is composed of a stack sandwiched by two heat exchangers and tube. The stack has many narrow flow channels and its length for 50 mm. For example, in order to create narrow flow channels, the stack is comprised of wire-mesh screen. When the steep temperature gradient is made in the stack, a self-exited gas oscillation is generated. Thus, heat energy can be converted into sound energy. In this paper, thermoacoustic conversion device applied for noise energy generated from engine plants was presented. The noise has complicated frequency characteristics and low energy density. We demonstrated selective amplification of acoustic pressure only one-frequency in noise by utilization of the technique of thermoacoustic engine. In addition, the other frequencies were attenuated by viscosity dissipation in the stack. Summarizing these results, the simplification and amplified sound was created from complex noise.

1 Introduction

In order to realizing sustainable growth in a low-carbon society, utilization of unused energy is an important issue. The energy conversion techniques utilizing thermoacoustic phenomena, which are based on heat transport using compression and expansion of gas in many narrow flow channel called stack, is one of promising method, because of its availability for various energy sources, simple structure with non-moving parts and no harmful gas for refrigeration. The thermoacoustic energy conversion system have been propsed into two types: one is thermoacoustic engine [1-6], and other is thermoacoustic cooler. These thermoacousitc engine and cooler have been studied actively.

In the case of thermoacoustic engine, self-excited oscillation in pipe is caused by installing the stack which has steep temperature gradient [7,8]. The heat energy can be converted into sound energy due to the thermoacoustic effect. On the other hand, when the sound wave passes through the narrow flow channel, the heat transport due to compression and expansion of gas occurs. Thus, heat pump action is induced by sound wave. Thermoacoustic phenomena are attractive effect for development of new type of energy conversion devise.. Moreover, acoustic amplification by thermoacoustic effect was proposed by Ceperley [9], and it was realized by Biwa et al. [10,11]. These recent surge of research activity had brought significant findings. However, most of these research paid attention to waste heat or solar energy as input source energy. In order to extend the application of the thermoacoustic device, development of the system for various unused energy is important.

We propose a new energy harvesting technology using thermoacoustic phenomena. Specifically, by using a thermoacoustic phenomenon in which acoustic waves are induced by providing a rapid temperature gradient (about 200 ° C. to about 500 ° C.) in a fine flow path, noise of low energy density can be selectively The final goal is to propose a system that amplifies and leads to effective utilization. In this paper, we carried out the basic study of connecting method between thermoacoustic field and a noise. Especially, sound pressure amplified effect. The effect of thermoacoustic pressure amplification was measured experimentally. Summarizing this result, the effect of thermoacoustic amplification was discussed in relation to the stack structure and resonance state.

2 Experimental apparatus

Figure 1 illustrates a schematic diagram of the experimental apparatus. The resonance pipe was a circular pipe of 1,69 m in length. Its inner diameter was for 24 mm. The end of pipe was closed and the the speaker was set at the other end. The atmospheric air was filled in pipe. A sound wave was injected from the speaker. The coordinate *X* was defined along axis of the passage, and origin of *X* was set at the end of pipe. The pressure transducers were set on the wall and closed end. The thermoacoustic section was consisted of a stack at *X* = 507 ~ 557 mm and two heat exchangers (these thickness was for 10-mm). The top of the stack was set

* Corresponding author: kyouden@nc-toyama.ac.jp

at one-third length of the pipe. The one of heat exchanger was heated by heater. And the other was kept at room temperature. The temperature difference between top and bottom of the stack was kept to 300 K. When the sound wave passes through the stack, the sound energy increased by thermoacoustic effect.

In this study, the stack was composed of stainless-steel mesh screens. In the experiments, sound with selected frequency was supplied by using speaker driven by electric input power of 1 W.

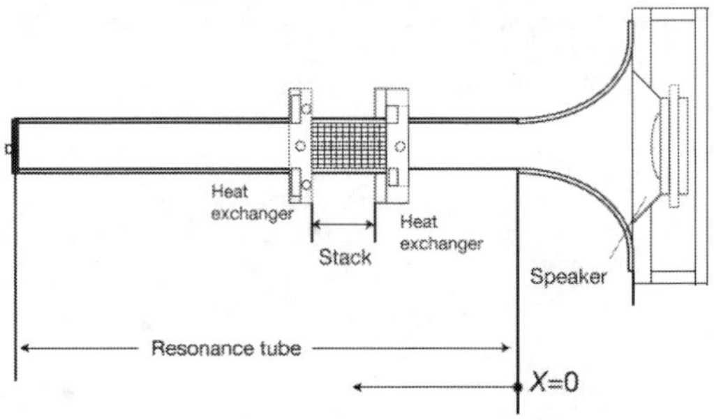

Fig. 1. Experimental apparatus.

3 Results and Discussion

3.1. Resonans state of circuler pipe with closed end

Firstly, we mesured the sound pressue at closed end for cralyfing resonance state. The resonance frequency are defined at maximum value of pressure. The consumption power of speaker was comstant for 1 W. Fig. 2 shows the resonance curves of pressure at closed end (X = 1.69 m). White circles were results without stack, and red circles were with stack and temperature gradient (300K), In the circular pipe with closed end, sound resonance with odd number times of a one-quarter wavelength is occurred, which is called the first harmonic, third harmonic, and fifth harmonic mode and more. In this paper, since the length of circular pipe L is for 1.69 m, the third harmonics and fifth harmonics mode were obtained when maximum pressure at the 50, 150 and 248 Hz, respectively. By comparing with case without stack and case with stack, it was found that the sound was attenuated by installing the stack, especially fifth harmonic frequency at 50 and 208 Hz. The number of stack which was used for stack was #30 and #60.

3.2. Applied to noise as sound source

In general noise has complex frequency characteristics. The noise energy could be simplified by using resonce effect and thermoacoustic device. Therefore, we choised the engine room noise in WAKASHIMARU. WAKASHIOMARU is trainning ship of Toyamanational institute of Technology, Toyama College.

Figure 2 shows the comparing between sound frequency characteristics, one is sound noise in the engine room in WAKASHIMARU and other one is measurement pressure amplitude at closed end under using the noise as sound sourse. In the case of the engine room noise, the dominant frequencies were spread out under 1 kHz. These frequencies were generated by main engine. Main engine is largest system in the engine room. The low frequency is generated large system. And higher value of frequencies were generated from dynamo generator. The dynamo generator has higher rotation speed comparing to the main engine.

The engine room noise was captured in a computer. By using the noise data in the computer, the engine room noise could be outputted from speaker using function generator. Therefore, we could use the engine room noise as sound source. The sound pressure measured at closed end is shown in Fig. 3 using red line. The highest value of pressure frequency was at 150 Hz. This frequency was third harmonic frequency in the experimental device. In addition, any frequencies were not appeared, especially high frequencies' region. These frequencies were difference at resonance frequencies, and the resonance frequencies except to third harmonic frequencies were damped by stack narrow flow channels.

Finally, the distribution of sound pressure in pipe is shown in Fig. 4. The engine room noise was used as sound source. The plotted value was sound pressure at component only 150 Hz. When input sound have the third harmonics mode frequency at 150 Hz, it was found that the stack position was located near the point of antinode in pressure amplitude. In the case of standing-wave, the position of antinode in pressure amplitude become the position of node in velocity amplitude.Thus, the maximum value of pressure amplitude and minimum value of velocity amplitudes were obtained at the end of stack. In addition, the acoustic loss is increased with velocity amplitude in the stack. By setting the stack at one-third position in the pipe, the acoustic loss of third harmonic frequency (it was for 150 Hz in this paper) was less than other frequency by low value of the velocity amplitude. The acsound pressure amplification by thermoacoustic effects was found by installing stack at pressure antiode.

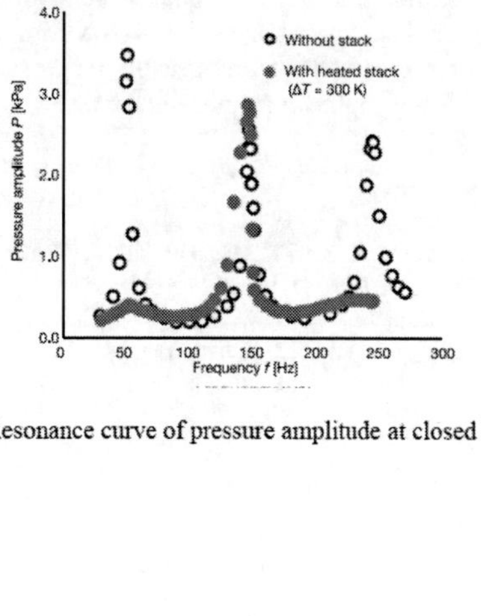

Fig. 2. Resonance curve of pressure amplitude at closed end.

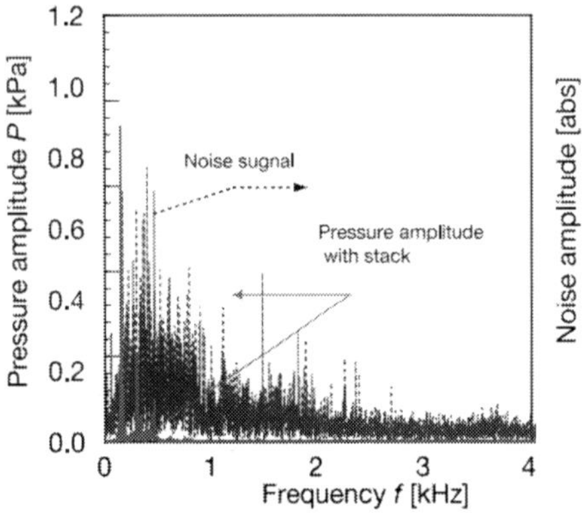

Fig. 3. Consists of sound noise in engine room in WAKASHIOMARU and pressure simplification utilizing stack.

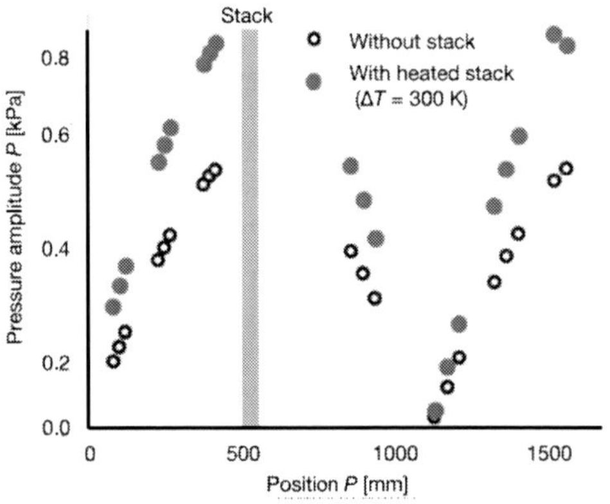

Fig. 4. Sound pressure distribution in pipe.

3 Conclusion

In order to utilizing noise energy for thermoacoustic cooler, we demonstrated the sound amplification in a pipe in which stack to temperature gradient was installed. The results showed the amplification of specified frequency and the attenuation of other frequency of sound could be achieved by installing the mesh stack at the pressure amplitude antinode in the pipe.

References

1. T. Yazaki, A. Iwat, T. Maekawa, and A. Tominaga, A., *Phys. Rev. Lett.*, **81**, pp. 3128, (1998).
2. S. Backhause, and G. W. Swift, G "A thermoacoustic Stirling heat engine," *Nature*, **399**, pp. 335-338, (1999).
3. Y. Ueda, T. Biwa, and U. Mizutani, *J. Acoust. Soc. Am.*, **115**, 1134, (2004).
4. S. Sakamoto, and Y. Watanabe, *Ultrasonics*, **42**, 53, (2004).
5. Y. Nakamura, and K. Kobayashi K. *Proc. of the 3rd International Forum on Heat Transfer*, paper No.IFHT2012-202, (2012).
6. S. Hasegawa, T. Yamaguchi, and Y. Oshinoya, *Applied Thermal Engineering.*, **58**, 394, (2013).
7. Y. Ueda, and C. Kato, *J. Acoust. Soc. Am.*, 124, 851, (2008).
8. S. Nagata, and Y. Ueda, *TEION KOGAKU*, 43, 561, (2008).
9. P. C. Ceperley, *J. Acoust. Soc. Am.*, **66**, 1508 (1979).
10. T. Biwa, M. Kozuka, and T. Yazaki, *Phys. Rev. E*, **69**, 066304, (2004).
11. T. Biwa, and R. Komatsu *J. Acoust. Soc. Am.*, **129**, 132, (2011).

The role of infrastructure sustainability and success of construction projects on owner benefits

Thanachon Krajangsri[1,*] and *Jakrapong* Pongpeng[1]

[1]Civil Engineering Department, Faculty of Engineering, King Mongkut's Institute of Technology Ladkrabang, Bangkok, Thailand

Abstract. The infrastructure sustainability and success of construction projects likely influence owner benefits. However, these influences have rarely been examined. Thus, the research aimed to examine such influences and intervening behavior of the success of construction projects between the sustainability of infrastructure and owner benefits using structural equation modeling (SEM). The study finds that there are a direct influence of the sustainability of infrastructure on owner benefits (regression weight: 0.63) and an indirect direct influence of that (regression weight: 0.22) through the intervening role of the success of construction projects. The findings ensure that implementation of infrastructure sustainability projects can enhance the success of construction projects and owner benefits.

1 Introduction

A successful construction project yields several benefits to an owner. The benefits should cover various terms to maintain an owner organization in business, society and community. In terms of economics, financial returns of the project should satisfy the owner company by supporting its financial viability. Therefore, accomplishing success in construction projects is crucial to the owner benefits [1]. Modern infrastructure projects need to concern about sustainable issues of sociality, environment and economics during their life cycle because achieving infrastructure sustainability can increase the owner competitiveness, reputation and business opportunities while conserving environmental characteristics of the projects [2].

Previous studies demonstrate possible relationships between sustainability of infrastructure and success of construction projects and owner benefits. However, these relationships have seldom been proved. This study was aimed to develop a model to examine (1) the influence between the infrastructure sustainability and owner benefits and (2) the intervening role of success of construction projects between the sustainability of infrastructure and owner benefits. The results from this study provide apparent understanding of the influence of the sustainability of infrastructure and success of construction projects on owner benefits.

2 Literature review

The literature on the three elements of owner benefits, sustainability of infrastructure and success of construction projects were reviewed to form research hypotheses and to develop conceptual models.

2.1 Owner benefits

To determine the owner benefits, the Triple Bottom Line concept of social, environment and economic aspects, was used as a framework for categorizing the benefits.

2.1.1 Social benefits

Institute for Infrastructure Sustainability [3] proposed the social benefits as follows: improved quality of life, encouragement of community growth and development, development of community workforce skills, increased public health, safety and well-being, increased collaboration and involvement of stakeholders, higher stakeholders satisfaction, and preservation of cultural heritage. Guarnieri [4] suggested social benefits as follows: improved community quality of life and increased project users health and comfort. Based on previous studies, the social benefits of sustainability in this study are described by: improved project owner's reputation, improved country's reputation, improved project users' quality of life, and improved community acceptance.

2.1.2 Environmental benefits

ISI [3] suggested the following environmental benefits: reduced energy and water consumption, reduced risks from natural disasters, preservation of ecological systems, reduced air pollutants and harmful emissions, less use of natural resources and impacts on ecological systems, and increased resilience to climate change and natural disasters. Olsen and Fenhann [5] identified various sustainable environmental benefits as follows: improved air quality, reduced harmful emissions, improved water quality, and protection of natural

* Corresponding author: thanachon@hotmail.com

resources. Based on previous studies, the environmental benefits of sustainability in this study are indicated by: improved project resilience to natural disasters, reduced environmental impacts from the project, protection of natural resources and ecological systems, and provision of education opportunities on sustainability.

2.1.3 Economic benefits

Dobson et al. [6] suggested the following economic benefits: reduced operation costs, reduced utilities costs, and reduced maintenance costs. Lee et al. [7] found economic benefits as follows: reduced energy costs, reduced water costs, economic growth improvement, increased income, and lower environmental impact costs. ISI [3] proposed the following economic benefits: reduced energy costs, reduced water costs, reduced maintenance costs, and reduced construction material costs. Based on previous studies, the economic benefits of sustainability in this study are determined by: reduced energy costs, reduced water costs, reduced operation and maintenance costs, and improved access to budget.

2.2 Sustainability of infrastructure

The infrastructure sustainability assessment systems and related studies are reviewed in this section to determine criteria and sub-criteria to obtain and evaluate infrastructure sustainability. CEEQUAL suggested 9 criteria in the latest version 5, including project strategy, project management, people and communities, land use and landscape, the historic environment, ecology and biodiversity, the water environment, physical resources-use and management, and transport [8]. Envision proposed 5 major criteria for assessment including quality of life, leadership, resource allocation, natural world, and climate and risk [3]. Assessment criteria for GreenLITES were proposed as follows: sustainable sites, water quality, materials and resources, energy and atmosphere, and innovation [9]. Greenroads used 7 assessment criteria: project requirements (mandatory), environment and water, access and equity, construction activities, materials and resources, pavement technologies and custom credits [10].

The sustainability of infrastructure for this study was described by 9 criteria: project management, location, energy, water, materials and resources, waste management, transportation, environmental impacts on surrounding areas, and community as found in the study of Krajangsri and Pongpeng [11].

2.3 Success of construction projects

A construction project is considered successful if it meets traditional indicators of time, cost and quality [12]. Alzahrani and Emsley [13] identified nine categories of success indicators: health, safety and quality, past performance, environment, management and technical aspects, resources, organization, experience, size/type of previous projects, and finance. Tabish and Jha [14] suggested 24 success indicators which were grouped into 5 categories comprising time, cost, quality, safety and no-dispute. Enshassi et al. [15] found ten success indicators as follows: cost, time, quality, productivity, client satisfaction, community acceptance, society, health and safety innovation, and environment. The success of construction projects for this research was described by 6 indicators: time, cost, quality, safety, client and community satisfaction, and environment as found in the study undertaken by Krajangsri and Pongpeng [11].

2.4 Hypothesis of the research

Improving infrastructure projects to be more sustainable have become an important agenda for the owner because ignoring social and environmental issues can negatively affect the owner reputation and long-term benefits. Then, implementing infrastructure sustainability assessment of projects can minimize these negative effects while enhancing the owner business opportunities, competitiveness and status [3]. Although previous studies displayed a possible relationship between sustainability of infrastructure and owner benefits, the relationship has never been examined. Therefore, the hypothesis was formed:

> H1: Sustainability of infrastructure directly influences owner benefits.

Krajangsri and Pongpeng [11] found that the sustainability of infrastructure directly influenced the success of construction projects. Based on these studies, the hypothesis was established:

> H2: Sustainability of infrastructure directly influences success of construction projects.

Herzig and Schaltegger [16] found that adopting sustainable success criteria can improve public acceptance, increase reputation, enhance competitiveness, increase transparency, motivate employee, act as a benchmark for a company, and determine economic benefits from sustainable activities. Previous studies show that there may be a relationship between success measures of construction project and owner benefits. The hypothesis was formulated to test the relationship:

> H3: Success of construction projects directly influences owner benefits.

Several studies suggested that successful sustainable construction creates both direct and indirect benefits to the owners. Direct benefits are minimized environmental impacts, lower or equal initial costs, reduced energy and water costs; while, indirect benefits are better health and safety of project users, improved comfort and satisfaction of project users, increased productivity, longer project lifespan, and improved project image [17-18]. From these studies, there may be an indirect relationship between sustainability of infrastructure and owner benefits through an intervening role of success of construction projects. The hypothesis was set to test this role:

H4: There is an indirect relationship between sustainability of infrastructure and owner benefits through an intervening role of success of construction projects.

3 Structural equation modelling and conceptual SEM model

Structural equation modeling (SEM) is a multivariate statistical tool combining confirmatory factor analysis, multiple regression analysis and path analysis to test a hypothesis on the (either direct and indirect or both) relationship of observed and latent (unobserved) variables. Observed variables carry the data that a researcher can directly measure. Whereas, latent variables carry the data that a researcher is interested but cannot directly measure. Variables in SEM can also be divided into exogenous and endogenous variables, depending on the direction of influence. Exogenous variables are independent variables that influence other variables. Endogenous are dependent variables that are influenced by other variables. In SEM, both exogenous and endogenous variables, which can be observed or latent, are considered [19].

4 Questionnaire for data collection

In order to test the hypotheses, 500 questionnaires were handed to samples in public and private sectors in Thai construction industry. 371 completed questionnaires were returned, translating to a response rate of 74 %, which is considered as excellent [20]. The samples were asked to rate the significance level of each criterion/indicator according to a Likert scale of 1 to 5.

5 Results and discussion

First, the measurement models were confirmed that they fit the sample data from the following goodness-of-fit criteria: p-value of 0.236 (more than the accpeted value of 0.05), relative chi-square (χ^2/df) of 1.029, goodness-of-fit index (GFI) of 0.891 (more than the accepted value of 0.09), and root mean square error of approximation (RMSEA) of 0.009 (more thant the minimum value of 0) [21]. To validate the conceptual SEM model, hypotheses H1, H2 and H3 were tested to examine direct influences among the latent variables as shown in Fig. 2. H1 tested that sustainability of infrastructure directly influences owner benefits. H1 was accepted because the p-value was less than 0.05 with a regression weight of 0.63. H2 and H3 tested that sustainability of infrastructure directly influences the success of construction projects and that the success of construction projects directly influences owner benefits.

H2 and H3 were accepted because the p-values were less than 0.05 with regression weights of 0.84 and 0.26, respectively. Since H2 and H3 were accepted, this implied an existence of the indirect influence between sustainability of infrastructure on owner benefits via the success of construction projects. The indirect influence was calculated by multiplying a regression weight of 0.84 (derived from the path of sustainability of infrastructure to the success of construction projects) by that of 0.26 (derived from the path of the success of construction projects to owner benefits), resulting in a regression weight of 0.22. Then, the regression weight (representing the indirect influence on the path of sustainability of infrastructure to the success of construction projects to owner benefits) of 0.22 was added to the regression weight (representing the direct influence on the path of sustainability of infrastructure to owner benefit) of 0.63, producing the total regression weight (representing the total influence) of 0.85. Obviously, the total influence was more than the direct influence; thus, H4 was accepted. The influences of sustainability of infrastructure on the success of construction projects and on the owner benefits are shown in Fig. 1. (Details of observed variables describing latent variables of criteria can be seen in [11])

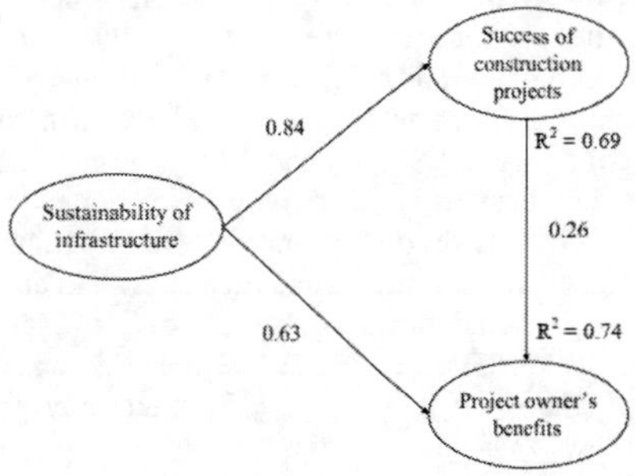

Fig. 1. Final SEM model

The results demonstrate that sustainability of infrastructure has higher direct influence on owner benefits (regression weight: 0.63) than that of the success of construction projects on owner benefits (regression weight: 0.26). This reinforces the belief that implementation of the infrastructure sustainability results in not only higher level of success of construction projects but also more owner benefits in terms of environment, economics and society. The owner, therefore, should incorporate sustainability concepts, such as preparation of flood risk assessment, waste management plan, and enhancement of community accessibility into the design of infrastructure. When the success of construction projects plays an intervening role between sustainability of infrastructure and owner benefits, the total influence was increased from 0.63 to 0.85 (a combination of direct influence (0.63) and indirect influence (0.22)). This demonstrates that successful construction projects, as an intervening variable, can generate long-term indirect benefits to the owner such as increased comfort and safety of project users, and improved reputation of the owner.

6 Conclusions

Infrastructure sustainability is one of key components in sustainable development strategy, which holistically consider social, environmental and economic dimensions. Although previous studies have suggested that sustainability of infrastructure enhances project success but the studies to examine the influence of the infrastructure sustainability on owner benefits were limited. Additionally, the intervening role of the success of construction projects between the sustainability of infrastructure and owner benefits has not been investigated. Therefore, the research was aimed to examine (1) the influence of sustainability of infrastructure on owner benefits and (2) whether the success of construction projects intervene between the sustainability of infrastructure and owner benefits. Then, hypotheses were formulated to test the influence and the intervening role. Questionnaires were used to collect the data from government and public agencies. After that, the data were used to validate the conceptual SEM model.

The SEM analysis results showed that all the hypotheses were accepted, meaning that the sustainability of infrastructure directly influenced the owner benefits. In addition, the total influence was significantly increased when the success of construction intervened between the sustainability of infrastructure and owner benefits, emphasizing its intervening behavior. The study originally contributes to examine the relationships between the infrastructure sustainability, success of construction projects and owner benefits. However, the relationships may alter depending on types and geographical locations of the infrastructure projects. Thus, further studies to examine the relationships based on the alternation are recommended.

7 References

1. Pankaj, A. J. J. and Bhangale, P., To Study Critical Factors Necessary for a Successful Construction Project, IJITEE , **2**, 5 (2013)
2. Levitt, R., CEM research for the next 50 years: Maximizing economic, environmental, and societal value of the built environment, J. Constr. Eng. M. ASCE **133**, 9 (2007)
3. ISI, *Envision: A Rating System for Infrastructure sustainability v.2.0.* Institute for Infrastructure Sustainability. Washington, DC, USA (2012)
4. Guarnieri, T.J., The real cost of sustainable development, *AACE International Transactions* (2008)
5. Olsen, K.H. and Fenhann, J., Sustainable development benefits of clean development mechanism projects. A new methodology for sustainability assessment based on text analysis of the project design documents submitted for validation, Energ. Policy. **36** (2008)
6. Dobson, D.W. Sourani, A., Sertyeilisik, B.,. and Tunstall, A., Sustainable construction: Analysis of its costs and benefits, Am. J. Civil. Eng., **1**, 2 (2013)
7. Lee, K.H., Ahn, Y.H., Jeon, M., and Suh, M.J., Organizational strategies to support sustainability in the construction company, *In the World SB14 Barcelona*, October 28-30th, 2014 (2014).
8. CEEQUAL, *Assessment Manual for Projects in the UK & Ireland v.4.1*, http://www.ceequal.com (2010)
9. NYSDOT, New York State Department of Transportation (2012) http://www.dot.ny.gov/programs/greenlites
10. Greenroads, *Greenroads Abridged Manual v1.5*, http://www.greenroads.org/366/download-the-manual.html (2011)
11. Krajangsri, T., Pongpeng, J.,. Effect of Sustainable Infrastructure Assessments on Construction Project Success Using Structural Equation Modeling. J. Manage. Eng. **33**, 3 (2017)
12. Atkinson, R., Project management: cost, time and quality, two best guesses and a phenomenon, its time to accept other success criteria, Int. J. Proj. Manag. **17** (1999)
13. Alzahrani, J. and Emsley, M. W., The impacts of contractors' attributes on success of construction projects: A post construction evaluation, Int. J. Proj. Manag. **31** (2013)
14. Tabish, S. Z. S. and Jha, K. N., "Important Factors for Success of Public Construction Projects." In *the2nd International Conference on Construction and Project Management, IPEDR*, **15** IACSIT Press, Singapore, (2011)
15. Enshassi, A., Mohamed, S., and Abushaban, S.,. Factors affecting the performance of construction projects in the Gaza Strip, J. Civ. Eng. Manag. **15**, 3 (2009)
16. Herzig, C. and Schaltegger, S., Corporate Sustainability Reporting. An Overview Sustain. Account. Rep. Dordrecht, Springer (2006)
17. Bordass, G., *The Economics Benefits of Sustainable Design* (2003) http://www1.eere.energy.gov.femp/pdfs/buscase_section2.pdf
18. Brightman, H. a. M., J., *The Social Benefits of Sustainable Design.* (2001) http://www1.eere.energy.gov?femp/pdfs/buscase_section3.pdf
19. Cho, K. M., Hong, T. H., and Hyun, C. T, Effect of project characteristics on project performance in construction projects based on structural equation model, Expert Syst. Appl. **36** (2009)
20. Babbie, E. R., *The practice of social research.* Belmont, CA: Wadsworth (1989)
21. Marsh, H. W., Balla, J. R., and McDonald, R. P. "Goodness-of-fit indices in confirmatory factor analysis: the effect of sample size." *Psychol. Bull.*, **103**, 3 (1988)

MATEC Web of Conferences **192**, 02027 (2018) https://doi.org/10.1051/matecconf/201819202027
ICEAST 2018

A comparison of green building assessment systems

Thanachon Krajangsri[1,*] and *Jakrapong* Pongpeng[1]

[1]Civil Engineering Department, Faculty of Engineering, King Mongkut's Institute of Technology Ladkrabang, Bangkok, Thailand

Abstract. Construction is one of industries that have negative impacts on the environment. Relevant organizations both in Thailand and other countries have been trying to minimize these impacts. Developing green building assessment systems is one of efforts in reducing environmental impacts and using natural resources efficiently. Currently, there are many green building assessment systems with different objectives and assessment scopes in some details. This could cause some confusion in selecting the suitable system for the project in accordance with its environmental, social and economical contexts. This study was aimed to compare and provide overview and components of these systems. The comparison results can be used as a guideline for further development of green building assessment systems. It is found that most major green building assessment systems still concern only environmental issues while newly developed systems have included social and economic issues in their criteria. Furthermore, the system which is developed in one environment could not be used with full capabilities in another. This will lead to more research in developing the assessment system that is more comprehensive and adaptable enough to be used effectively in various environment.

1 Introduction

Realizing environmental impacts from construction activities, relevant organizations in several countries developed building environmental assessment systems, also known as green building assessment systems, in early 1990s as guidelines for design, construction and performance assessment of green buildings. Leading systems are Building Research Establishment Environmental Assessment Method (BREEAM), developed by The Building Research Establishment Ltd. from the UK; Leadership in Energy and Environmental Design (LEED), developed by U.S. Green Building Council from the United States and Comprehensive Assessment System for Built Environment Efficiency (CASBEE), developed by Japan Greenbuild Council/Japan Sustainable Building Consortium from Japan. These systems are widely used both in their origin countries and abroad. Responding to this growing movement, Thai Green Building Council developed Thai's Rating of Energy and Environmental Sustainability and National Housing Agency developed Ecovillage Criteria in 2012 to encourage sustainable construction practices in Thailand.

With different systems and their different objectives, assessment scopes, building types, users, life cycle assessment, environmental, social and economic issues, hesitation could occur in selecting the best system to be used for the project effectively. Therefore, efforts have been made by academics and relevant organizations in energy and environmental sectors to classify these systems.

For instance, ATHENA Institute [1] classifies building environmental assessment systems into 3 levels, while International Energy Agency [2] classifies them into 5 levels according to their scopes of assessment, such as product comparison tools, life cycle assessment tools and whole building performance assessment tools. Reijnders and van Roekel [3] divided the assessment tools into 2 groups as qualitative tools and quantitative tools. The qualitative tools were based on scores and criteria. The quantitative tools considered life cycle assessment, input and output of material uses and energy consumption [4].

Currently, there are increasing numbers of building environmental assessment systems both local and international. Each system has different objectives and assessment scopes. International systems are used either directly or as guidelines for green building assessment system development in some countries. Haapio and Viitaniemi [5] studied 14 building environmental assessment tools including quantitative, qualitative and life cycle assessment tools to classify them according to their scopes. They were compared within the same level with comparison factors: assessed building types, users of the tools, life cycle phases, database of the tools, forms of assessment results.

* Corresponding author: thanachon@hotmail.com

The research was concluded that:

1) The assessment tools were much different depending on needs and purposes of development. The comparison is therefore very difficult.
2) The tools users were unable to assess reliability of the tools and their results.
3) The tools covered different life cycle depending on guidelines and database used.
4) There should be an examination on factors affecting tools selection of the users.

Forsberg and von Malmborg [4] studied 5 quantitative environmental assessment tools and compared them to present their overviews and current status. Some of these tools are already available and some are under development. The comparison factors are: users of the tools, overall purpose, specific purpose, assessed building types, considered issues, research dimensions, basis of comparison, scope of tools, result presentation.

The research was concluded that:

1) The tools studied were developed from life cycle analysis of buildings and materials used.
2) Life cycle assessment still had difficulties in environmental impact assessment.
3) Selecting scope of the assessment tools was important depending of required results.
4) Development of qualitative environmental assessment tools based on life cycle analysis to be sustainability assessment tools should consider social and economic issues as well.

Ding [6] studied 20 environmental assessment tools to present overviews and analyzed assessment methods used in several countries. Their limitations were examined and would be used as basis for improvement. Analysis results were as follows:

1) Using environmental assessment tools as design tools – The tools were useful when used during design phase but will be more useful if use during pre-design phase.
2) Selecting appropriate project – The tools were less useful in selecting the project options than assessing the project design.
3) Financial issues – Financial issues should be considered along with environmental issues.
4) Regional variations – Most assessment tools were developed according to their regional environment. Currently there is no single tool that can be used effectively worldwide.
5) Complexity – Environmental assessment is a complex task. Developing a tool that has complete coverage but still simple enough to use was challenging.
6) Quantitative and qualitative data evaluation – Most quantitative evaluation was performed in comparison with collected data and standards. However environmental issues were mostly qualitative data. Obtaining accurate evaluation results was more complex.
7) Weighting – Weighting should be adjusted on project-by-project basis to reflect development objective.
8) Measurement scales – Measurement scales use a scoring system but there was no common basis in determining the score level of each assessment criterion.

With these limitations, there should be a development of assessment tool that could evaluate data in various dimensions. Since construction was an activity involving many complex decisions in environmental, social and economic issues. Developing sustainable indexes was an approach to address multiple criteria in decision-makings of the project. Reijinders and van Roekel [3] studied comprehensiveness and adequacy of environmental building improvement tools. The tools were divided into 2 main groups:

1) Requirement type instruments – Requirement type instruments of the public and private sectors currently had inadequate coverage on environmental issues. They tend to focus on energy and water consumption.
2) Guidance type instruments – Five guidance type instruments studied had more coverage on environmental issues.

However, from literature review it is found that there are few researches comparing components of leading green building assessment systems, which could give the readers more understanding and could be used as guidelines for further system development and improvement on a whole building. This study was aimed to compare green building assessment systems which could be used on a whole building. Six systems were selected for comparison to present their principles and main components. This comparison was expected to assist stakeholders in choosing the best system for the design and construction of green projects effectively in accordance with their environmental, social and economic conditions. It could also be used as a guideline in developing and improving the local green building assessment system.

2 Methodology

This research used a qualitative method by examining components of green building assessment systems obtained from previous research works [3-6] and websites of relevant organizations (i.e, manuals of green building assessment systems). Then, the data were verified with data triangulation method by consulting experts and practioners in this area whether the obtained contents are accurate. By examining the different sources of data, the result of data triangulation test showed that all the data from the three different sources (manuals of green building assessment systems, experts, and practitioners) were consistent, which verified the obtained data. After that, the verified data were analyzed with component analysis to compare contextual and methodological aspects. The processes are as follows:

1) Determining framework in selecting systems: For this study, ATHENA Institute classification of green building assessment tools were used as follows:
   - Level 1: Product comparison tools and information sources

- Level 2: Whole building design or decision support tools
- Level 3: Whole building assessment frameworks or systems

The qualitative tools in Level 3 were tools were selected for this study according to Reijnders and van Roekel [3] concepts.

2) Selecting tools: Level 3 tools that could be used as whole building design and construction guidelines and assessment tools were selected.
3) Determining comparison aspects: From the assessment framework for environmental analysis proposed by Baumann and Cowell [7] as used by Forsberg and Malmborg [4] in their study covering contextual and methodological aspects were selected and improved for this study as follows:

Contextual Aspects

- Assessment system developers
- Assessment system users
- Assessment system categories
- Assessment building types
- Assessment system scopes

Methodological Aspects

- Assessment dimensions
- Main assessment criteria
- Assessment result ratings

4) Analyzing information: The related information and aspects were gathered and listed in comparison tables.
5) Summarizing the study.

3 Results

Green building assessment systems included in this study are used locally and internationally. BCA Green Mark, BEAM, and TREES are used at locally, mostly in Asian countries. BREEAM, CASBEE and LEED are used internationally as direct implements and guidelines for development of other systems. Results from research in their websites covering contextual and methodological aspects for comparison are shown in Tables 1.

4 Summary

From the study, it is found that green building concept has been practiced for more than 20 years to mitigate environmental problems from construction activities. The concept should be implemented in the project as early as possible, from planning to design, construction and operation phases to maximize its benefits. Green building assessment systems in this study have similar main assessment criteria, such as site, energy efficiency, water efficiency and indoor environmental quality. They are the core concepts of every system. The assessment procedures, users and types of building are also practically similar.

However, comparing these systems to determine which one is better is still difficult because their objectives, contextual and methodological contexts are different in details. The system that is developed to perform most effectively in one environment might not be as effective in another; due to environmental, social and economic differences.

Apparently, LEED is the most widely-used system in Thailand due to its popularity, coverage, proven environmental benefits and added marketability for the certified project. Selecting the right system for the project is quite complicated if the user starts to consider social and economic benefits in addition to environmental in his local condition. Currently, there is no single system that can be used effectively to assess sustainability of construction in all environmental, social and economic conditions. Further study is recommended to develop the system that is comprehensive and adaptable enough to be used with full capability in various environment as a solid guideline for sustainable construction.

5 References

1. Athena Sustainable Materials Institute. www.athenasmi.org
2. International Energy Agency, www.iea.org
3. L. Reijnders, A. van Roekel, Comprehensiveness and Adequacy of Tools for the Environmental Improvement of Buildings, J. Cleaner Prod. Elsevier, **7** (1999)
4. A. Forsberg, F. van Malmborg, Tools for Environmental Assessment of the Built Environment, Build. Environ. Elsevier, **39** (2004)
5. A. Haapio, P. Viitaniemi, A Critical Review of Building Environmental Assessment Tools, Environ. Impact Assess. Rev. Elsevier, **28** (2008)
6. G.K.C. Ding, Sustainable Construction – The Role of Environmental Assessment Tools, J. Environ, Manage. Elsevier, **86** (2007)
7. H. Baumann, S. Cowell, An Evaluative Framework for Conceptual and Analytical Approaches Used in Environment Management, Greener Management International, **26** (1999)
8. Building and Construction Authority, Singapore, www.bca.gov.sg
9. BEAM Society, Hong Kong, www.beamsociety.org.hk
10. BRE Global, UK, www.breeam.org
11. Japan GreenBuild Council (JaGBC)/Japan Sustainable Building Consortium (JSBC), Japan, www.ibec.or.jp
12. U.S. Green Building Council, USA, www.usgbc.org
13. Thai's Rating of Energy and Environmental Sustainability for New Construction and Major Renovation and Core and Shell Building, www.tgbi.or.th

Table 1: Comparison of Green Building Assessment Systems

Assessment Systems	Comparison Aspects							
	Contextual						Methodological	
	Developers	Users	System Categories	Building Types	Scope	Assessment Dimensions	Main Criteria	Ratings
BCA Green Mark	Building and Construction Authority, Singapore, 2005 [8]	Owner, Developer, Designer, Contractor, Private sector	New Buildings, Existing Buildings, Beyond Buildings, Occupant-Centrics	Residential, Non-residential, Landed house, Data centers, Healthcare, Schools, Offices, Restaurants, Retail, Super markets, Laboratories	Design, Construction, Operation	Environment	Climatic Responsive Design, Building Energy Performance, Resource Stewardship, Smart and Healthy Building, Advanced Green Efforts	Green Mark Platinum, Green Mark Gold^{Plus}, Green Mark Gold
BEAM	BEAM Society, Hong Kong, 1996 [9]	Owner, Developer, Designer, Contractor	New Buildings, Existing Buildings, Interior, Neighborhood, Bespoke	Residential, Commercial	Planning, Design, Construction, Management, Operation & Maintenance	Environment	Site Aspects, Materials Aspects, Energy Use, Water Use, Indoor Environmental Quality	Platinum, Gold, Silver, Bronze
BREEAM	BRE Global, UK, 1990 [10]	Owner, Developer, Designer, Contractor, Building Manager, Tenant, Researcher	Communities, Infrastructure, New Construction, In-use, Refurbishment & Fit-out	Office, Retail, Industrial, Data Centers, Education, Healthcare, Residential, Mixed Use, Other Buildings	Planning, Design, Construction, In-Use, Refurbishment	Environment	Management, Health and Wellbeing, Energy, Transport, Water, Waste, Materials, Land Use, Pollution, Innovation	Outstanding, Excellent, Very Good, Good, Pass, Acceptable
CASBEE	Japan GreenBuild Council (JaGBC)/Japan Sustainable Building Consortium (JSBC), Japan, 2001 [11]	Owner, Developer, Designer, Contractor, Private sector, Researcher	New Construction, Existing Buildings, Renovation, Market Promotion, Commercial Interiors, Temporary Construction, Heat Island, Urban Development, Cities, etc.	Residential, Commercial, Houses	Planning, Design, Construction, Management, Operation & Maintenance	Environment	Built Environment Quality: Indoor Environment, Quality of Service, Outdoor Environment Built Environment Load: Energy, Resources and Materials, Off-site Environment	Excellent, Very Good, Good, Fairly Poor, Poor
LEED	U.S. Green Building Council, USA, 1996 [12]	Owner, Developer, Designer, Contractor, Private sector, Researcher	Building Design & Construction, Interior Design & Construction, Building Operation & Maintenance, Neighborhood Development, Homes	Residential, Commercial, Schools, Retail, Data Center, Warehouse, Hotel, Healthcare, Home, Commercial Interior, Building Core & Shell	Design, Construction, Operation & Maintenance	Environment, Society, Economy	Location and Transportation, Sustainable Site, Water Efficiency, Energy & Atmosphere, Materials & Resources, Indoor Environmental Quality, Innovation, Regional Priority	Platinum, Gold, Silver, Certified
TREES	Thai Green Building Council, 2012 [13]	Owner, Developer, Designer, Contractor, Private sector, Tenant, Researcher	New Buildings, Existing Buildings	Residential, Commercial	Design, Construction	Environment	Building Management, Site and Landscape, Water Conservation, Energy and Atmosphere, Materials and Resources, Indoor Environmental Quality, Environmental Protection, Green Innovations	Platinum, Gold, Silver, Certified

MATEC Web of Conferences **192**, 02028 (2018) https://doi.org/10.1051/matecconf/201819202028
ICEAST 2018

Detonation effects of shallow buried explosive in sandy soil on target plate acceleration in a small-scale blast test

Zulkifli Abu *Hassan**, Aniza *Ibrahim and* Norazman *Mohamad Nor*

National Defence University of Malaysia, Sungai Besi Camp, 53100 Kuala Lumpur, Malaysia

Abstract. Small-scale blast tests were carried out to observe and measure the influence of sandy soil towards explosive blast intensity. The tests were to simulate blast impact imparted by anti-vehicular landmine to a lightweight armoured vehicle (LAV). Time of occurrence of the three phases of detonation phase in soil with respect to upward translation time of the test apparatus were recorded using high-speed video camera. At the same time the target plate acceleration was measured using shock accelerometer. It was observed that target plate deformation took place at early stage of the detonation phase before the apparatus moved vertically upwards. Previous data of acceleration-time history and velocity-time history from air blast detonation were compared. It was observed that effects of soil funnelling on blast wave together with the impact from soil ejecta may have contributed to higher blast intensity that characterized detonation in soil, where detonation in soil demonstrated higher plate velocity compared to what occurred in air blast detonation.

1 Introduction

Vehicles that operate in conflict zones are susceptible to landmine or IED threats. Explosive shock front from detonation of landmine could damage the vehicle and the impact could also cause bodily injuries due to vehicle floor board deformation and acceleration [1]. Preventing casualties or reducing severe injuries and ensuring the safety of the occupants require necessary measures which include developing protective systems for both the vehicle and personnel.

Physical damage and degree of impairment caused by detonations of buried mines vary when the explosion takes place in different type of soils. Depending on soil types and conditions, when shallow buried explosive is detonated, the impact delivered to the above ground object or structure impart distinct intensity of blast loading. This has been observed in numerous experimental works [2] [3] [4] [5] that were carried out to investigate the phenomenon. Properties and conditions of the soil that are crucial in causing variation of magnitude and intensities of blast loads have been identified and presented in many documented researches. Soil properties and conditions such as moisture content, particle size distribution, density and depth of burial (DoB) of the explosive charge are reported to be key dependent factors in causing variations in blast intensity of a landmine explosion [6] [7].

In order to develop optimal design of protective systems against landmine blast, accuracy of finite elements model simulations is crucial. Finite elements model validation, and improved data gathered from experimental tests based on specific type of soils and conditions would enhance simulation capability. In this study, small-scale blast tests have been carried out using portable tests apparatus to measure the effect of in-situ sandy soil on shallow buried explosive blast intensity. Target plate acceleration was measured using shock accelerometer and high-speed video camera was used to record the movements of the apparatus during the tests. This paper presents the finding of experimental tests of the blast loading from detonation of shallow buried explosive on target plate acceleration with correlation to detonation phase of explosion in soil.

2 Test apparatus

The apparatus is modelled at a 1/10th scale (scale factor 10) to represent the undercarriage dimension and weight of an armoured vehicle. The underside dimension of the apparatus however, is made into a square shape instead of rectangular, total weight of the apparatus is about 21.74 kg, it consists of a steel jig and a target plate.

As shown in Figure 1, the main structure of the test jig is made up of a C-section mild steel structural channel with 100 mm web x 56 mm flange x 5 mm thickness. Four pieces of the C-section mild steel members are welded and bolted together to form a 500 mm x 500 mm square frame that leaves a square opening of 390 mm side in its centre. A sacrificial steel target plate of 500 mm x 500 mm x 5 mm is attached at the bottom side of the frame.

* Corresponding author: zulabha@yahoo.com

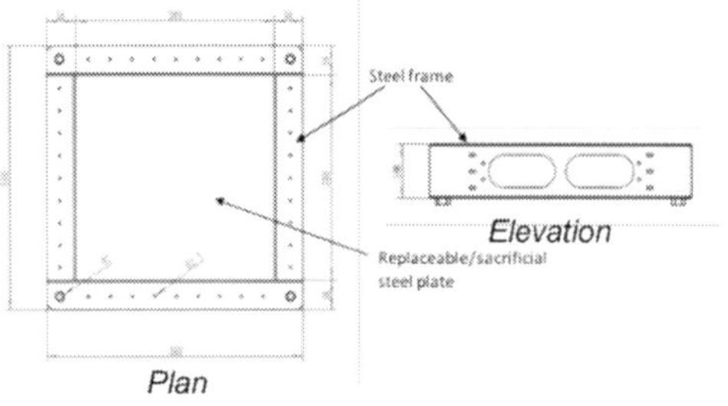

Fig. 1. Schematic drawing of a steel frame

3 Explosive charge

Each of the blast tests in this experiment was conducted by detonating the mass of 20 g AN Emulsion High Explosive commercial grade explosive charge. The charge assembly is shown in Figure 2.

Component	Specifications
Explosive charge	20 g AN Emulsion H E ρ = 1.13 g/cc - 1.24 g/cc VOD = 4500 - 5500 m/s Expl. energy = 2.85 MJ/kg
Detonator	containing secondary charge mass of 720 mg PETN

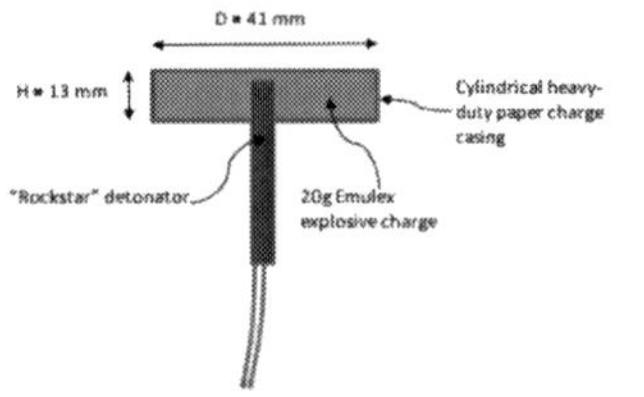

Fig. 2. Explosive charge assembly

4 Measurement methods

Observation on the movement of the apparatus and detonation phase during the test is by mean of optical method using Phantom V Series high-speed camera with maximum speed capability of 680000 fps. The camera was positioned approximately 5.0 m from the test apparatus during the blast test.

The plate acceleration is measured using PCB Piezotronic ICP shock accelerometer. The accelerometer has range up to 50000 G's peak (490000 m/s²) and frequency response of 10 kHz. Adapter for accelerometer was mounted on the top face of the target plate, positioned at 80 mm from the centre of the target plate. Plate acceleration measurements were carried out in both air-blast and buried-in-sand tests.

5 Test set-ups

Experimental tests were performed according to the following test set-up;

i) Air-blast test setup

The air-blast test setup is shown in Figure 3, the apparatus is placed on 450 mm height concrete stands to make free-space from the ground. A 60 mm height polystyrene spacer block is then positioned at the centre of the target plate. This block gauged the stand-off-distance (SOD) where explosive charge is mounted on the other end of the block.

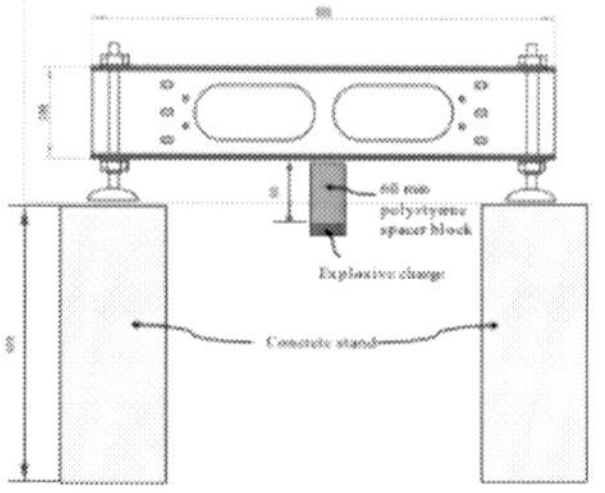

Fig. 3. Position of the apparatus and explosive charge for air-blast test setup

ii) Buried-in-sand test setup

Buried-in-sand test setup is done by placing the test apparatus on a sandy soil test bed. As shown in Figure 4, the explosive charge is buried 10 mm from the surface of the ground with SOD specified at 50 mm from the ground surface to the centre of the target plate.

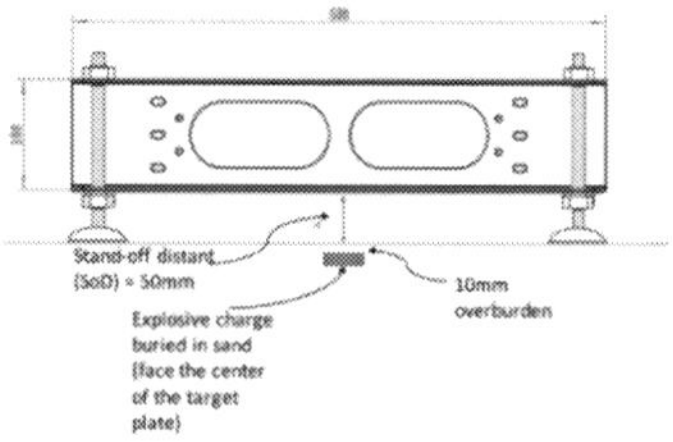

Fig. 4. Position of the apparatus and explosive charge in the buried explosive test setup

6 Results

6.1 Optical observations

High-speed camera was set to shoot at 30000 fps and interval at 33.31 μs pf. Optical observation presented in this paper is the observation that was carried out in buried-in-soil test. This is to demonstrate its correlation with detonation phases in soil which also see the effect and time of impact from soil ejecta.

Phase 1: Detonation and early interaction with the soil.

Fig. 5a

Fig. 5b

At the instance of detonation, from 0 m.s – 0.066 m.s was the point at which the explosive was totally consumed by detonation wave (Fig. 5a, 5b)

Fig. 5c

Fig. 5d

At this stage the initial expansion sent a pressure wave through the soil, at 0.133 m.s fireball started to shrink, however at 0.233 m.s the target plate started to deform (Fig. 5c, 5d).

Phase 2: Gas expansion.
This phase seemed to occur at 0.266 m.s, at which point wave reflected from the air-soil interface and created fissures, and target plate was seen to experience more deformation (Fig. 5c). As at 1.565 m.s, high-pressure gasses expanded out of the explosive and compressed the soil below it and forced the soil above it to evacuate at high speeds. Soil cap was ejected from the surface of the soil at supersonic velocity and at this stage the test apparatus started to translate upwards (Fig. 5f).

Fig. 5e

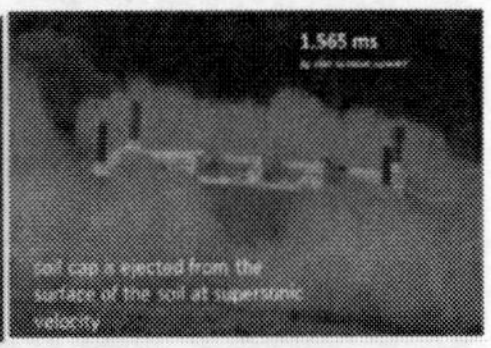

Fig. 5f

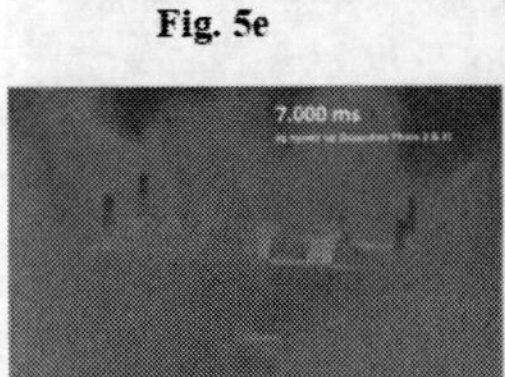

Fig. 5g

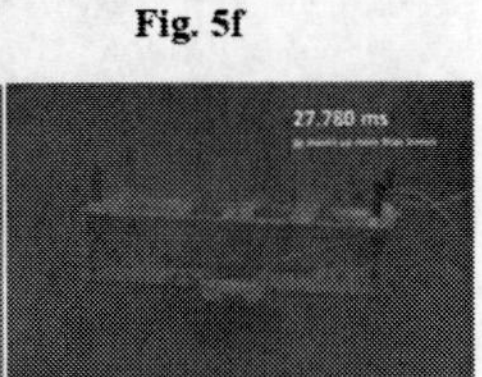

Fig. 5h

At 7.000 m.s the test jig continued moving vertically upwards, the high pressure gas of detonation products continued to shear the region of soil surrounding it and on the cavity walls (Fig 5g). This point also marked the boundary of Phase 2 and Phase 3. At 27.780 m.s the jig had moved up to about 50 mm height (Fig. 5h). This was the beginning of Phase 3, where the high-pressure gas of detonation products continued to shear the region of soil surrounding it and on the cavity walls [8].

Phase 3: Soil Ejecta
At 82.5 m.s the jig seemed to move further upwards reaching more than 100 mm in height, and soil ejecta could be seen spreading out from the bottom of the jig (Fig. 5i). At this stage, a complex interaction of compression waves, rarefaction waves and soil within the cavity to further erode and eject the surrounding soil upwards at high velocity. At 160 m.s, the jig moved up to more than 200 mm, and the spread of soil ejecta was more apparent (Fig. 5j).

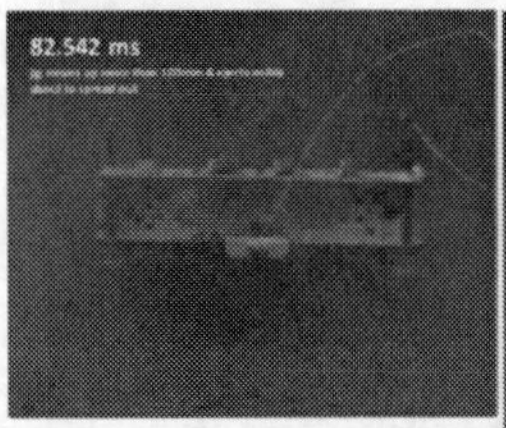

Fig. 5i

Fig. 5j

6.2 Plate Acceleration

The presented results are based on series of tests in air-blast and buried-in-sand tests. Figure 6 shows the comparison of average plate acceleration time history recorded in air-blast and buried-in-sand tests.

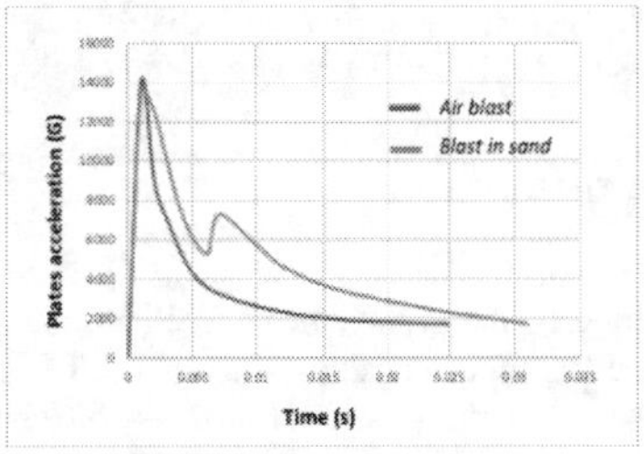

Fig. 6. Average plate acceleration – time history for air-blast and blast in sand tests

It is clear that the average peak plate acceleration in both air-blast and blast-in-sand tests reached almost 14000 G before gradually declining. However, in blast-in-sand test, another rapid surge occurred once again when the time passed 5 m.s, it surges about 1500 G before continuing to decrease.

When plate acceleration in the buried-in-sand test is compared to optical observation of the same test, the time between 5 to 10 m.s coincided with the boundary of gas expansion phase and ejecta phase, or phase 2 and 3 in phases of blast evolution in soil [9]. The plate although experiencing blast impact and acceleration, at the same time is moving upwards together with the apparatus. Although peak accelerations occurred at about 1 m.s, the optical observation showed that as the time passed 0.23 m.s, deformation of the target plate has already taken place wherein its surface was beginning to bulge, the apparatus then started to move upwards at about 1.56 m.s. In contrast with the recorded acceleration, the moment the apparatus was about to translate upwards, the plate acceleration has already begun to enter its receding phase.

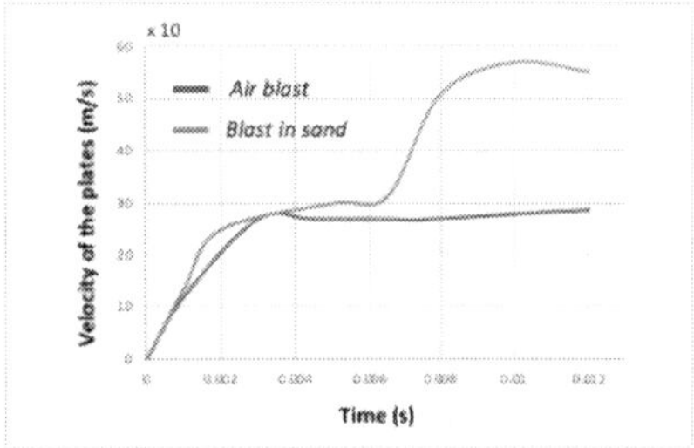

Fig. 7 Average plate velocity – time history for air-blast and blast in sand tests

Velocity of the target plate time history in air-blast tests and blast-in-sand tests are shown in Figure 7. Average velocity in air-blast tests peaked at around 3 m.s reaching as high as 280 m/s before it became constant. However average velocity in blast-in-sand tests show distinct pattern compared to air-blast tests, where there are two instances of velocity upsurge. The first 1 m.s looked similar to the velocity in air-blast tests, and reached in average of 285 m/s at 5 m.s. However, at between 6 - 7 m.s, another rapid rise in velocity occurred where it peaked at around 10 m.s with average velocity of 550 m/s.

7 Conclusion

Comparing peak acceleration in air-blast and blast-in-sand tests, the average peak acceleration in both tests did not differ much in value, wherein the average of both tests plate acceleration was around 14000 G. Nevertheless, data in terms of velocity showed better indication of the blast impact received by the target plate. Plate velocity was almost twice higher in blast-in-sand tests compared to in air-blast tests, although the first peak velocity may be slightly lower in blast-in-sand tests, the possible impact from ejecta and funnelling effect of the ground have resulted in higher peak velocity. There is a clear shoulder between two steep velocity upsurges occurring in buried-in-sand tests. The first sudden increase in velocity was similar to those of air-blast tests, which was caused by blast wave and detonation products. However, it was followed by another rapid rise after about 4 m.s of constant velocity. Correlation of optical observation with loading time of the second upsurge in plate velocity coincided with the boundary of phase 2 and phase 3 of blast evolution in soil, which was likely due to ground funnelling effect and soil ejecta impacted the target plate.

References

[1] A. Ramasamy, A. M. Hill, A. E. Hepper, A. M. Bull and J. C. Clasper, "Blast Mines: Physics, Injury Mechanisms and Vehicle Protection," *Journal of Royal Army Medical Corps 155(4)*, pp. 58-264, 2009.

[2] S. D. Clarke, S. D. Fay, J. A. Warren, A. Tyas, S. E. Rigby and I. Elgy, "A large scale experimental approach to the measurement of spatially and temporally localised loading from the detonation of shallow buried explosives," *Measurement Science and Technology 26(1)*, p. 26:015001, 2015.

[3] S. D. Clarke, S. D. Fay, J. A. Warren, A. Tyas, S. E. Rigby, J. J. Reay and R. Liversey, "Geotechnical causes for variation in output measured from shallow buried charges," *International Journal of Impact Engineering*, pp. 274-283, 2015.

[4] S. D. Clarke, J. Warren and A. Tyas, "The influence of soil density and moisture content on the impulse from shallow buried explosive charges," in *International Symposium on Interaction of the Effects of Munitions with Structures*, Seattle, US, 2011.

[5] S. L. Hlady, "Effect of soil parameters on land mine blast," in *18th Int. Sym. on Military Aspects of Blast and Shock (MABS18)*, Bad Reichenhall, Germany, 2004.

[6] J. Q. Ehrgott, S. A. Akers, J. E. Windham, D. D. Rickman and K. T. Danielson, "The influence of soil parameters on the impulse and airblast overpressure loading above surface-laid and shallow-buried explosives," *Shock and Vibration 18*, pp. 857-874, 2011.

[7] L. C. Taylor, R. R. Skaggs and W. Gault, "Vertical impulse measurement of mines buried in saturated sand," *International Journal for Blasting and Fragmentation, Fragblast*, vol. 9, pp. 19-28, 2005.

[8] V. S. Deshpande, R. M. McMeeking, H. N. Wadley and A. G. Evans, "Constitutive model for predicting dynamic interactions between soil ejecta and structural panels," *Journal of the Mechanics and Physics of Solids 57*, pp. 1139-1164, 2009.

[9] D. Bergeron, R. Walker and C. Coffey, "Detonation of 100-gram anti-personnel mine surrogate charges in sand: A test case for computer code validation. Technical Report 668," Defence Research Establishment Suffield, Ralston, Alberta, Canada, 1998.

https://doi.org/10.1051/matecconf/201819202029

Influence of load increment ratio on one-dimensional consolidation behaviour

Takahiro Yoshidomi[1], *Koichi* Iinuma[2], *Motohiro* Sugiyama[3*]

[1]Course of Architecture and Civil Engineering, School of Engineering, Tokai University, Kanagawa, Japan
[2]Design Department, Tokyo Branch, Ohba Co., Ltd., Tokyo, Japan
[3]Department of Civil Engineering, Faculty of Engineering, Tokai University, Kanagawa, Japan

Abstract. In this study, we conducted a stage loading consolidation test with varying load increment ratio and investigated its effect. By adjusting the primary consolidation ratio of one-dimensional consolidation analysis with secondary consolidation taken into account, it is possible to reproduce the consolidation amount-time curve affected by the load increment ratio.

1 Introduction

In the incremental loading consolidation test (JIS A 1210), load increment ratio $\Delta p/p$ is set to 1 and loaded at one-day intervals. $\Delta p/p$=1 is considerably excessive as compared with loading and construction conditions at the site. The incremental ratio for load distribution due to local load on thick clay layer and low loading such as low embankment and housing is much smaller than 1. It has been pointed out that the *e*-log*p* curve of the test with varying loading conditions (loading time interval, load increment ratio, and strain rate) varies greatly as shown in Figure 1 (a) and (b)[1-3]. On the other hand, some samples show different results from previous studies[4]. Showing the presence of such a resultant sample leads to settlement prediction accuracy. It is also a well-known fact that when the increment ratio is small, the consolidation-time curve becomes convex upwards and does not become an inverted S-shaped settlement curve[5]. As the secondary compression is dominant, the primary consolidation may be less than half of the total consolidation in some cases.

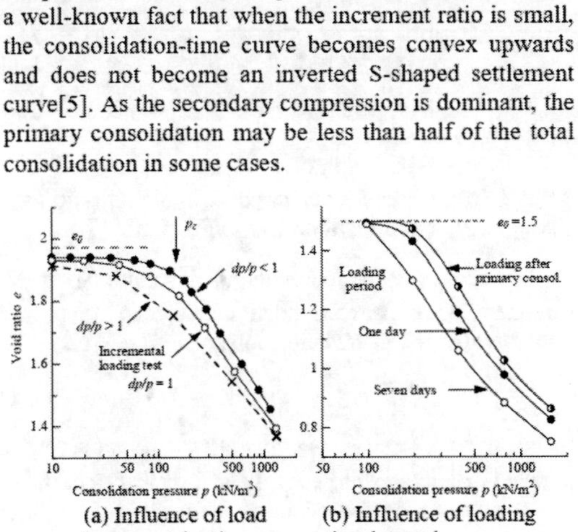

(a) Influence of load incremental ratio

(b) Influence of loading time interval

Figure 1. Influence of loading condition on the *e*-log*p* curve [2]

In this study, an incremental loading consolidation test was performed by varying the load incremental ratio to investigate its effects. It shows that adjusting the primary consolidation ratio of one-dimensional consolidation analysis considering secondary consolidation can reproduce the consolidation-time curve subject to the effect of the load incremental ratio.

2 Samples and test method

Four kinds of undisturbed clayey soil (U1 to U4) and two kinds of reconstituted clayey soil (R1, R2) were used as samples. Table 1 shows the physical properties of each sample. Samples U3 and R2 are high organic soils. Loading was carried out from p=9.81 kN/m^2 with a load increment ratio of 0.2 to 1.5, and the loading time of each load was adjusted so that the consolidation period was all 8 days. In addition, the loading Δp=39.2 to 235.5 kN/m^2 was loaded on six U4 samples consolidated at p=157 kN/m^2 for 2 days, respectively, and the settlement was measured for 24 hours.

Table 1. Physical properties of clayey soil

Sample	U1	U2	U3	U4	R1	R2
ρ_s	2.55	2.69	1.62	2.70	2.67	2.31
LL(%)	64.1	67.8	525	96	68.4	201.8
PI(%)	29.4	33.3	370	52	44.6	68.2
Sand(%)	5	0	-	11	5	-
Silt(%)	50	41	-	33	42	-
Clay(%)	45	59	-	56	53	-

3 Test results and consideration

3.1 Influence of load increment ratio

The relationships of volumetric strain (=axial strain) with log*p* of the undisturbed and reconstituted samples

* Corresponding author: sugi@keyaki.cc.u-tokai.ac.jp

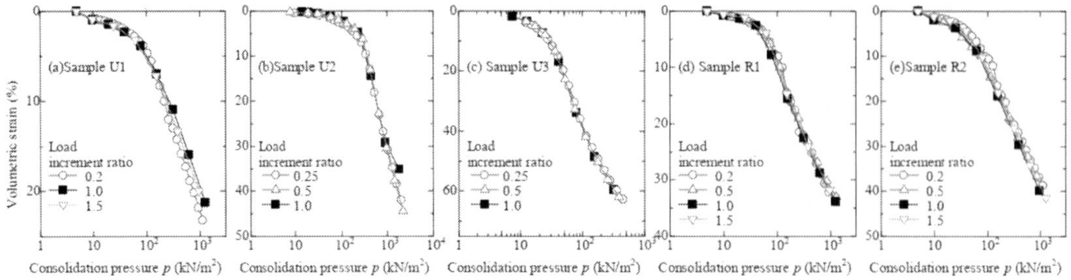

Figure 2. ε-log$p$ curve varying with a load incremental ratio

are shown in Figure 2(a) to (e). In the undisturbed samples, there was scattering of the initial void ratio e in particular, for this study, the volumetric strain was sorted out instead of the void ratio e.

Regardless of the sample and load increment ratio, there is no significant difference in each e–log$p$ relationship as in Figure 2. A similar result was observed in undisturbed undersea clay in Osaka Bay[4], which contrasts with past conspicuous results[1]. In the conceptual diagram relating to delayed compression of Bjerrum, the e-log$p$ curves with different consolidation times are parallel to each other as shown in Figure 3, the compression index Cc is equal, and the parallel lines are assumed to be strain rate constant lines[3]. Therefore, the strain rates in Figure 4 shows the strain rate in Figure 2 where the ε-log$p$ curves became almost equal despite the difference in the load increment ratio. Even though the ε-log$p$ curves are nearly equal, the strain rate changes in each test. This is contrary to the fact that the strain rate on the e-log$p$ curve assumed by Bjerrum is equal.

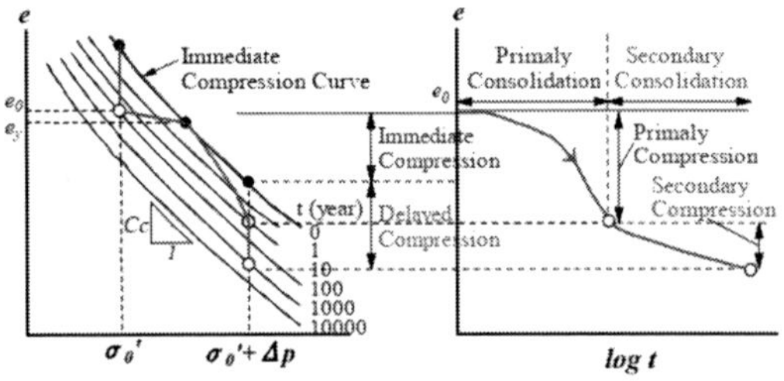

Figure 3. Conceptual figure of Bjerrum's delayed consolidation[3]

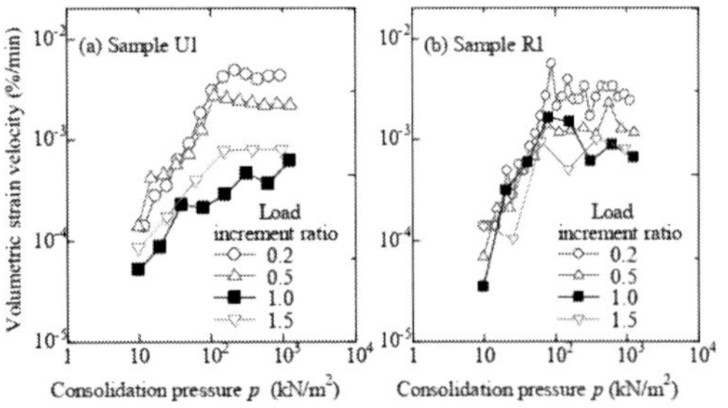

Figure 4. Load incremental ratio and volumetric strain rate

3.2 Load incremental ratio and consolidation–time curve

3.2.1 Change of consolidation–time curve

The symbol in Figure 5 is the measured value of the consolidation test with changing the load increment ratio. The figure shows that the shape of consolidation-time curve varies with decreasing load increment ratio as in previous studies, and it is not a typical inverted S shape. Because of the change in the shape of the curve, the end of primary consolidation time is unclear. However, the influence of the load increment ratio on the secondary consolidation is not significant, and the difference in the coefficient of secondary consolidation of each test is small.

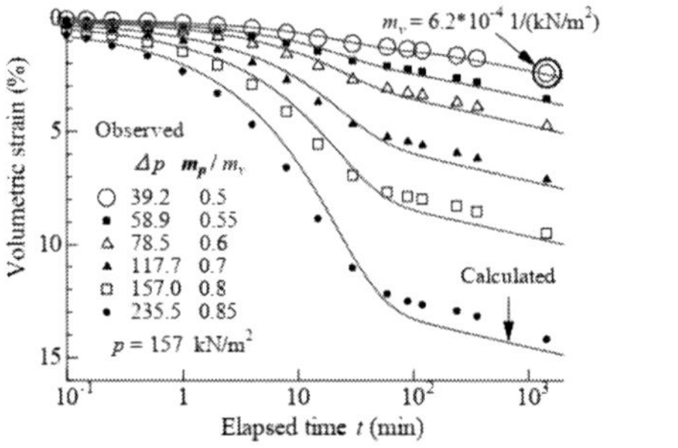

Figure 5. Consolidation–time curve: measured values and calculation results

3.2.2 One-dimensional consolidation equation and secondary consolidation model[6]

It is thought that the total volumetric strain rate $\dot{\varepsilon}(=\varepsilon_f/t_f)$ in one-dimensional consolidation is divided into primary and secondary consolidation components.

$$\dot{\varepsilon}\left(=\frac{\partial\varepsilon}{\partial t}\right)=\dot{\varepsilon}_p(=-m_p\cdot\dot{u})+\dot{\varepsilon}_s \qquad (1)$$

where $\dot{\varepsilon}_p$ and $\dot{\varepsilon}_s$ are the primary and secondary consolidation rates respectively, and m_p is the coefficient of volume compressibility defined by primary consolidation (volumetric strain). Substituting Eq. (1) in the continuous condition Eq. (1) that governs one-dimensional consolidation gives Eq. (2).

$$\frac{\partial u}{\partial t} = c_v^* \frac{\partial^2 u}{\partial y^2} + \dot{\varepsilon}_s / m_p \tag{2}$$

where $c_v^*(k/(\gamma_w * m_p))$ is the coefficient of consolidation that is defined using the primary consolidation, and the second term on the right side $\dot{\varepsilon}_s / m_p$ is the excess pore water pressure created by secondary consolidation. $\dot{\varepsilon}_s$ is obtained by Eq. (3) according to the secondary consolidation $\Delta\varepsilon_s$, the coefficient of secondary consolidation α defined by the volumetric strain and the initial rate of secondary consolidation $\dot{\varepsilon}_i$.

$$\dot{\varepsilon}_s = \dot{\varepsilon}_i \cdot \exp\left(\frac{-\Delta\varepsilon_s}{\alpha}\right) \tag{3}$$

Since the secondary consolidation behavior during primary consolidation is unknown and $\dot{\varepsilon}_i$ can not be experimentally investigated, $\dot{\varepsilon}_i$ is assumed in one-dimensional consolidation analysis. The authors propose to calculate the initial rate of the secondary consolidation from Eq.(4) using the secondary consolidation $\Delta\varepsilon_{sf}$ and its rate $\dot{\varepsilon}_{sf} = \alpha / 1\ (1/day)$ after one day of the incremental loading consolidation test [6].

$$\dot{\varepsilon}_i = \dot{\varepsilon}_{sf} \cdot \exp\left(\frac{\Delta\varepsilon_{sf}}{\alpha}\right) \tag{4}$$

With this proposed method, as the constant that represents the primary consolidation, the coefficient of volumetric consolidation m_p is defined and distinguished from ordinary m_v. Since m_p cannot be determined experimentally, so it need to assume. The consolidation–time curve is calculated based on the one-dimensional consolidation equation, Eq. (2), considering secondary consolidation, and by the explicit differential method according to the following boundary conditions and initial conditions.

$$\begin{array}{lll} t=0 \ \& \ H \geq y \geq 0\,; & u = u_0 \ (= \Delta\sigma_y) & \\ t \geq 0 \ \& \ y = H\,; & \partial u / \partial y = 0 & (5) \\ t > 0 \ \& \ y = 0\,; & u = 0 & \end{array}$$

where H is the maximum drainage distance and u_0 is the initial excess pore water pressure. The soil constants needed for the calculation are m_v, c_v, α, and m_p. With the proposed method, $\dot{\varepsilon}_i$ is calculated by assuming m_p or the primary consolidation ratio.

The solid line in Figure 5 shows the results based on the consolidation–time curve of Δp =39.2 kN/m^2 for which the load incremental ratio is smallest at 0.25, obtaining m_v=6.2×10^{-4} 1/(kN/m^2), c_v*(=c_v)=0.02 cm^2/min, and α=0.005, and applying this constant to other test results to reproduce the calculations of the tests. By changing the assumed value of the ratio of the primary consolidation m_p/m_v as shown in the figure, calculation results close to the measurement value can be obtained.

In the calculation shown in Figure 5, the assumed value of m_p/m_v that matches the test result of the load increment ratio Δp=39.2 kN/m^2 ($\Delta p/p$=0.25) is 0.5. Figure 6 shows the results of the calculation and comparison by revising the value of m_p/m_v to 0.7 and 0.9. In the case of m_p/m_v=0.9 where the proportion of the primary consolidation increases, the shape of the calculated consolidation-time curve is significantly different from the measured values. The difference in the calculation result in Figure 6 indicates the rationality of the assumed m_p/m_v value (=0.5). It is considered that the load increment ratio affects the ratio of the primary consolidation and the secondary consolidation of the total consolidation.

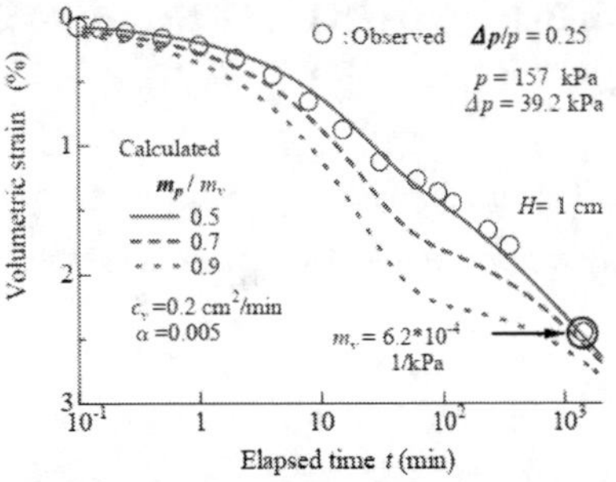

Figure 6. Settlement curves with differing primary consolidation ratios

4 Conclution

As a result of performing a consolidation test in which the load increment ratio was changed variously, test results different from the well-known past research results were obtained. In order to evaluate this result, it is necessary to examine whether physical properties etc. are indicators. The consolidation-time curve showing different shapes depending on the load increment ratio can be reproduced by changing the primary consolidation ratio. It is also considered that the smaller the load increment ratio, the smaller the proportion of primary consolidation.

References

1. G.A. Leonards, B.K. Ramiah, Time effects in the consolidation of clays, Am.Soc. Test., Matter. STP. **254**, (1959).
2. C.B. Crawford, Interpretation of the consolidation test, *Proc. ASCE*, **90**, SM5, (1964).
3. L. Bjerrum, Engineering geology of Norwegian normally consolidated marine clays as related to settlements of buildings, Geotechnique, **17(2)**, pp. 81-118, (1967).
4. T. Tsuchida et al., Comparison of results of consolidation testing using volumetric strain rate as the index, Thirty-fourth Japan National Conference on Geotechnical Engineering, pp. 487-488, (1999).
5. T. Mogami, Soil mechanics, *Gihodo*, pp.331-478, (1970).
6. T. Takeda, M. Sugiyama, M. Akaishi, H.W. Chang, Secondary compression behavior in one-dimensional consolidation test, Journal of GeoEngineering, TGS, **7(2)**, pp. 53-58, (2012).

Numerical simulation of two-dimensional laminar unsteady flow past a right trapezoidal cylinder at low Reynolds number: study of sharpening angle, time step, grid independence and domain size

Sodsai Lamtharn[1,*], *Monsak* Pimsarn[1]

[1]Mechanical Engineering Department, Faculty of Engineering, King Mongkut's Institute of Technology Ladkrabang, Bangkok, Thailand

Abstract. Numerical simulation of two-dimensional laminar unsteady flow past a right trapezoidal cylinder at low Reynolds number (Re = 100), zero of the flow approaching angle and sharpening angle of the right trapezoidal of 22.5° with a side ratio B/A = 1 are carried out to provide more applicable data for engineering design of barred tee in aspect of structural integrity. A finite volume method, non-uniform meshing with second-order implicit time discretization into eight-node quadratic quadrilateral finite elements is employed. An incompressible flow SIMPLEC code with constant fluid properties is used. The convective terms using a third-order QUICK scheme. The numerical simulation result is compared against the published results of flow past a square cylinder. The effect of sharpening angle on the response of the right trapezoidal cylinder is investigated. A special study of the effects of flow on significant factor for time step, grid independence, blockage ratio, domain size, upstream and downstream extents, size domain next to cylinder and size domain extent are performed systematically. The Strouhal number and RMS lift coefficients of fully saturated flow are calculated. The result shown that increasing of sharpening angle, the Strouhal number is negligible changed whilst the RMS lift coefficients significantly increased.

1 Introduction

In long piping or pipeline system normally used pigging for various internal activities such as separating fluids, cleaning and inspecting the pipeline. Pig is a spherically shaped device which can be propelled through a main pipeline by pressurized fluid flow from pig trap launcher through main line, branch line and together with all associated piping, valves, pipe supports and instruments to pig trap receiver. The main line is major portion of a pipeline, between pig traps which compose many branch connection line. When piping or pipeline system is liquid filled into system to operate, the flow past main line and flow into branch line and past barred tee. In order to prevent the passing of pigs/spheres through branch pipe as show on Fig. 1. and Fig. 2. the barred tees (guided bars) is provided, filled-welded attached cross inside branch at connection point. Typically barred tees be installed on all branches larger than 50% or 25% of the main pipeline diameter where unwrapped foam pigs or spheres are to be used.

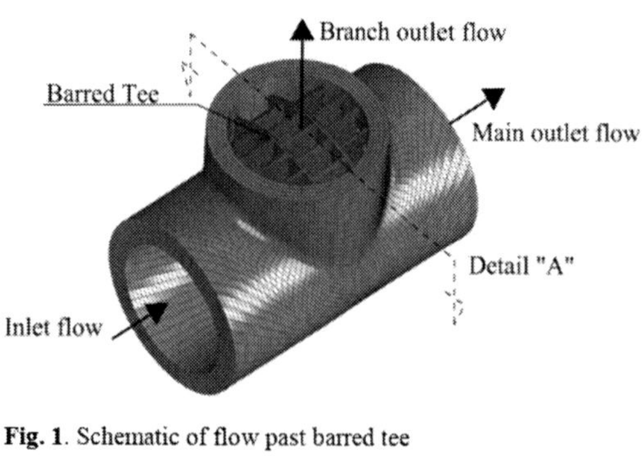

Fig. 1. Schematic of flow past barred tee

The barred tees have to integrity against to fluids flow induced vibration. In engineering design point of view, we concern to the synchronize between natural frequency of barred tee with vortex shedding frequency of fluid flow as past barred tee, the synchronize vibration of them cause barred tee experienced severe vibration then fatigue, detached from wall pipe into main pipe line which can be resulted pig stuck, valve face crash or scraper flow in to other inline equipment such as flow control valve and pump impeller. More additional cause

* Corresponding author: sodsai.l@hotmail.co.th

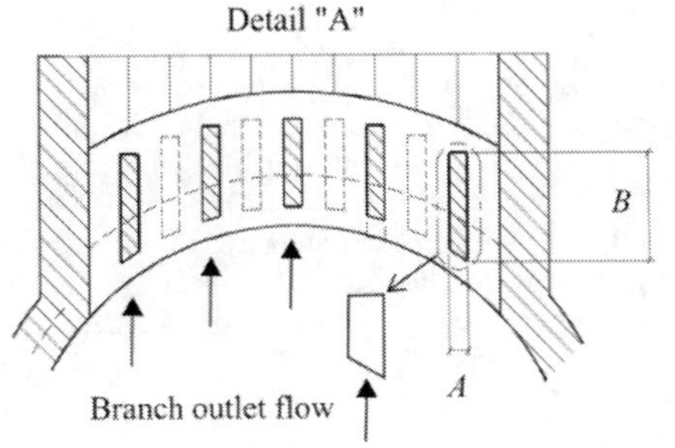

Fig. 2. Cross-section of barred tee

piping or pipeline stress exceed at branch connection. As shown on Fig. 2. the flow past a barred tee in which composed with several cross-section shape of a square, a rectangular and a right trapezoidal cylinder.

By numerical simulation, the enormous increase on the powerful processing of modern computers simulation has made it possible to identify and to study several mechanisms on flow past two dimensional cylindrical bluff body, many work used numerical simulation for flow past over cylindrical shape.

Sohankar et al. [1] carried out numerical simulations of two dimensional unsteady flow around a square cylinder at zero approaching angle $\alpha = 0°$, $Re = 100$ to study the influence of time step Δt, size domain next to cylinder ζ, grid size, extend of computation domain size Δ, upstream extends X_u and downstream extends X_d of the calculation domain, side wall height, respectively. The method for numerical simulation is an incompressible SIMPLEC finite volume code, non-staggered grid arrangement and Neumann outlet boundary condition. The convective terms are a third-order QUICK scheme and a second-order Van Leer scheme. Implicit time discretization and second-order Crank-Nicolson scheme is employed. For accurate calculation [1] finally used of $\Delta t = 0.025$ and $\zeta = 0.004$. The appropriate value of $X_u = 11.1$ and $X_d \geq 25$ units.

Sohankar et al. [2] performed and provide a number of quantities same as [1] with rectangular cylinders at $\alpha = 0°\text{-}90°$ with $Re \leq 200$, ratio of elongated side B per short side A = 1-4. The Strouhal number St and RMS lift coefficient Cl' at $B/A = 1$ are increase where Re increase up to angle of approaching about 45° then it decrease symmetrically.

Sohankar et al. [3] calculated and provide a number of quantities same as [1-2] a square cylinder at $\alpha = 0°\text{-}45°$ and Re = 45-200 included studied the outlet of the computational domain for a convective Sommerfeld boundary condition (CBC) and a traditional Neumann condition (NBC). For CBC is suitable for flow around a square cylinder with non-zero angle of approaching and the CBC more effective in reducing the CPU time, reducing the upstream influence from the outlet and reducing the necessary downstream extent. Within Re = 45-200, [3] recommended to use appropriate value $X_d > 15$ units for $Re < 100$ and $X_d > 10$ units for Re = 100-200 for CBC and the upstream influence of the outlet is effectively damped out on NBC at $X_d > 25$ units approximately. The blockage ratio γ reduced from 5% to 2.5% for Re = 100 - 200 at $\alpha = 0°$ resulted in a small decrease in Strouhal number and RMS lift coefficient.

Ming et al. [4] investigated effect of flow approaching angle at α = 0°, 45°, and 22.5° on the response of the square cylinder on vortex induced vibrations at $Re = 100$ and a low mass ratio of 3 with the Navier-Stokes equations using the finite element method. Rectangular computational domain in the flow direction of a length of 60 units and a height in the cross-flow direction of a length of 40 units are discretized into quadrilateral four-node linear finite elements. The center of the cylinder is distance of 20 units from the inlet boundary and equal distance to the up and down for lateral boundaries of 20 of the computational domain with $\gamma = 2.5\%$. Mesh dependence study is included in [4]. [4] also provide the Strouhal number and RMS lift coefficients at $\gamma = 2.5\%$ for α = 0°, 45°, and 22.5°.

Arnab et al. [5] performed numerical simulations of two-dimensional laminar flow past a triangular cylinder in which placed in free-stream at $10 \leq Re \leq 250$ by method a finite volume, second-order accurate in space and time, employing non-staggered arrangement of the variables with momentum interpolation for the pressure-velocity coupling. Result shown that vortex shedding phenomena of a triangular cylinder is same as to the square cylinder. Effect of grid refinement, X_u and X_d of the calculation domain are investigated and also study on the RMS lift coefficient and Strouhal number.

Mittal et al. [6] investigated effect of the blockage ratio on numerical simulation for flow past circular cylinder at $Re \leq 150$ with a stabilized space–time finite element formulation, γ = 1% and 5%. From literature as mention above and in the past, the most applicable engineering design data for flow past over circular, squre cylinder, rectangular, trangle and isosceles trapezoidal are emphasis intense investigated but more absence applicable engineering design data for right trapezoidal cylindrical cross section in which to used in barred tee engineering design.

In this study, numerical simulations are carried out to investigate Strouhal number, RMS lift coefficient, of a right trapezoidal cylinder in unsteady flow to provide further more applicable data for engineering design of barred tee in aspect of structural integrity. The effects of time step, grid independence, sharpening angle β, blockage ratio, domain size, upstream and downstream extents, size domain next to trapezoidal cylinder are studied systematically. Fig. 3. show the schematic of the computational geometry of flow in this study and the definition of α, β, B, A, X_u, X_d, vertical distance between the lower and upper walls of computational domain H and projected width of the cylinder in the streamwise direction d. The streamwise and tranverse direction is placed on the D and L axis, respectively. The center of cylinder is on origin. The constant flow inlet velocity U_∞ approached into side A and B at some degree of the α as depicted in Fig. 3. The Reynolds number is defined as $Re = U_\infty d/\nu$, where $d = A\cos(\alpha+\beta)/\cos\beta + B\sin\alpha$, $0° \leq \alpha \leq (90\text{-}\beta)°$ and $d = B\sin\alpha$, $(90\text{-}\beta)° < \alpha \leq 90°$ and $0° \leq \beta \leq 60°$ and ν is kinematic viscosity. All length dimensional are scaled to nondimensionless with d included Strouhal number $St = f_S d/U_\infty$, where f_S is the shedding frequency.

The confined boundary parameter which is blockage ratio is defined as $\gamma = d/H$, where velocities scaled with U_∞. The origin of force coordinates is placed at the center of right trapezoidal cylinder geometry with drag and lift force positive in the D and L-direction, respectively.

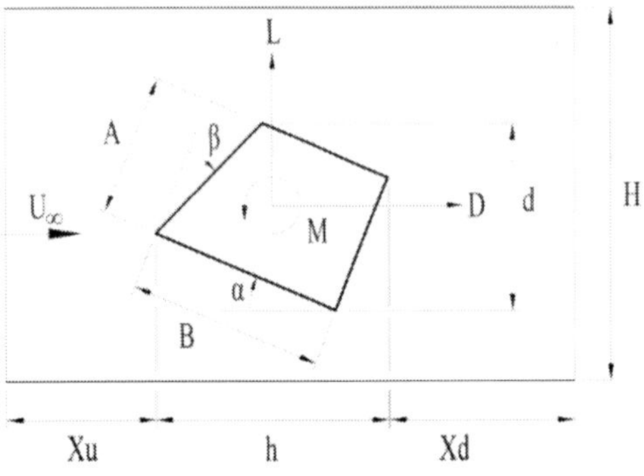

Fig. 3. Schematic of the computational geometry

2 Numerical details

The following numerical parameter were used in this study, unless otherwise specific, Re =100, α = 0°, β = 22.5°, B/A = 1, γ = 0.05, X_u = 10 and X_d = 15, ζ = 0.004 and extends all side of 5 units from cylinder, the mesh distribution was made non-uniform with a vary mesh size of 0.004-0.5 and outside a region from the cylinder which extends 5 units, the mesh distribution was made uniform with a constant mesh size Δ = 0.5, the number of nodes on each side of cylinder N_b = 20 nodes as shown in Fig. 4.

A finite volume method, non-uniform meshing with second-order implicit time discretization into eight-node quadratic quadrilateral elements is employed. An incompressible flow SIMPLEC code with constant fluid properties such as kinematic viscosity and density are used and the convective terms using a third-order QUICK scheme same as [1-3]. At the inlet which is located X_u units upstream from the cylinder, a uniform flow velocity was definded in streamwise direction U = 1, in tranverse streamwise direction V = 0. At the outlet which is located X_d units downstream from the cylinder,

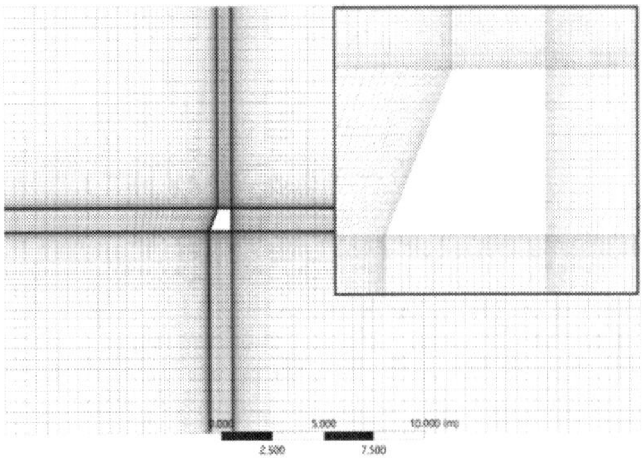

Fig. 4. Computational mesh for α = 0° and β = 22.5°, on the right top is near cylinder

the zero gage pressure is definded. The confined flow surface frictionless stationary wall with specific shear stress = 0 (V = 0) at upper and lower side of cylinder (H/2) are employed and the right trapezoidal cylinder surface is prescribed to No-slip stationary wall condition (U = 0, V = 0) with double precision. Initiated calculations with the fluid at rest (U = 0, V = 0) after that in few time step the flow velocity is increased smoothly to unity with constant Δt = 0.05, maximum iteration per time step of 40, the monitor convergence criteria in continuity and velocity are set to 0.001 to use compares with residual at the end of each iteration during iterative sequence.

3 Result and discussion

To avoid inaccuracy from numerical simulation, the study effect of modeling parameter are performed. By using the Strouhal number and RMS lift coefficients as indicator the strong sensitivity [2-3] represent to another sensitivity numerical parameters from fully saturated flow calculation.

3.1 Influence of time step

In order to capture the shedding phase and amplitude accuracy, the influence of the time step is investigated as presented in Table 1. These results indicated that with the decrease of time step the Strouhal number and RMS lift coefficient are increased. Decreasing it from 0.2 to 0.1, the Strouhal number and RMS lift coefficient are increased of 3.03% and 2.29% respectively. Decreasing it further to 0.05, the Strouhal number and RMS lift coefficient are increased of 0.76%, 0.87%, respectively. It negligible changed and next decreasing to 0.025, the Strouhal number and RMS lift coefficient are negligible changed. The time step of 0.05 and Courant number of 12.5 is chosen to used for grid independence study further from reason as mention above and minimize the Courant number. This value less than [1-3] which them used 0.025.

Table 1. Effect of time step.

Courant number	Δt	*St*	*Cl'*
50.0000	0.2000	0.1471	0.7405
25.0000	0.1000	0.1515	0.7575
12.5000	0.0500	0.1527	0.7641
6.2500	0.0250	0.1533	0.7683

3.2 Grid independence study

To avoid rapidly changing in high gradient area [1] the grid distribution is studied, the result is shown in Table 2. Decreasing the size domain next to cylinder from 0.004 to 0.002, consequently grid density changed 150x138 to 151x139. From the result both the Strouhal number and RMS lift coefficient are decreased of 0.76% and 0.24% respectively. More additional study, fixed the size domain next to cylinder of 0.002, increase the nodes on each side of cylinder to 40 nodes per cylinder side, the Strouhal number and RMS lift coefficient are again increased of 0.76% and 0.12% respectively. Finally it is decided to used the size domain next to cylinder by 0.004, grid density 150x138 and the nodes on each side of cylinder to 20 nodes for further investigate.

Table 2. Effect of grid at Δt = 0.05.

ζ	*Nb*	*Grid*	*St*	*Cl'*
0.004	20	150x138	0.1527	0.7641
0.002	20	151x139	0.1515	0.7622
0.002	40	171x159	0.1527	0.7631

3.3 Influence of domain size, upstream extents, downstream extents, convergence criteria and blockage ratio

To study influence of upstream and downstream extents distance on present flow past cylinder. From previous decided, increase X_u to 15 units, increase X_d to 20 units and reduced γ to 2.5% result as presented in Table 3. The results indicated that with the increase X_u from 10 to 15 unit the Strouhal numer and RMS lift coefficient are negligible decreased less than 1% and with the increase X_d from 15 to 20 unit, the Strouhal number not changed and RMS lift coefficient is slightly increased. Where reduced γ to 2.5% the result indicated that both the Strouhal number and RMS lift coefficient are slightly decrease of 0.76% and 1.24%, respectively. In contrast, refined convergence criteria to 0.0001, both parameter are negligible increased. It finally was decided that increasing X_u, X_d, domain size and convergence criteria not effected to previous decided. So it reasonable to final concluded numerical parameter to utility on this and further study of typical cylinder are following; γ = 0.05, X_u = 10 unit, X_d = 15 units, ζ = 0.004, ζ-Δ = 0.004-0.5, Δ = 0.5, N_b = 20 nodes almost aligned with [1-3]

Table 3. Effect of domain size at *Courant number* = 12.5, Δt = 0.05, ζ = 0.004, *Nb* = 20 and *Grid* = 150x138.

X_u	X_d	*Convergence criteria*	*γ*	*St*	*Cl'*
10	15	0.001	5.0%	0.1527	0.7641
15	15	0.001	5.0%	0.1515	0.7566
10	20	0.001	5.0%	0.1527	0.7646
10	15	0.001	2.5%	0.1515	0.7546
10	15	0.0001	5.0%	0.1538	0.7701

3.4 Effect of sharpening angle

Fig. 5. Shows the time histories of the RMS lift coefficient of β = 0° ,5°, 22.5° and 45°. It is indicated that for β = 0, which is square cylinder, symmetry. Where sharpening angle is increased, RMS lift coefficient evidently increase but the Strouhal number constant that conformable to square cylinder at some approaching angle [4]. From Table 4. It can see that flow past the right trapezoidal cylinder is quite similar to the square cylinder on phase of transverse oscillation although in RMS lift coefficient is more effect than square cylinder at zero flow approaching but tend to similar to the square cylinder with flow approaching rather than zero.

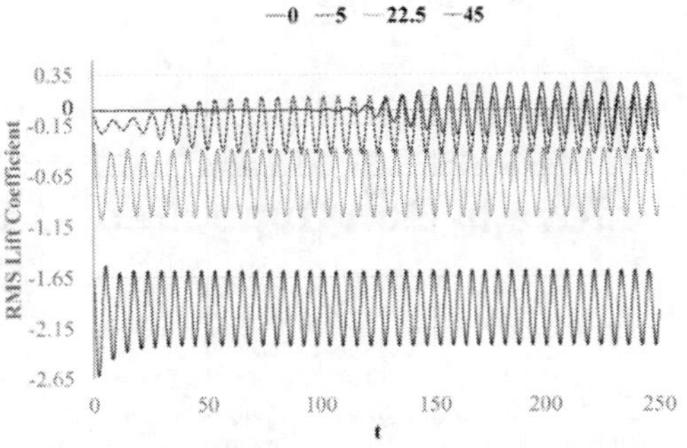

Fig. 5. Effect of sharpening angle on RMS lift coefficient

Table 4. Comparison *St* and *Cl'* at *Re* = 100 with other cylinder cross-section.

Approaching angle	*Type of cylinder*	*γ*	*St*	*Cl'*
α = 0°	Square [3]	0.05	0.1460	0.1560
	Square [4]	0.025	0.1447	0.1908
	Present	0.05	0.1527	0.7641
α = 22.5°	Square [4]	0.025	0.1408	0.5437
α = 45°	Square [4]	0.025	0.1388	0.6539

4 Conclusions

Numerical simulation of two-dimensional laminar flow past a right trapezoidal cylinder in unsteady flow at *Re* = 100, α = 0° and β = 22.5° with B/A = 1 have been carried out. The flow past the right trapezoidal cylinder is slightly difference to the square cylinder on phase of transverse oscillation at zero flow approaching. The RMS lift coefficient is increased follow sharpening angle increasing and to seem to similar square cylinder at flow approaching rather than zero.

The increasing of sharpening angle, the Strouhal number is negligible changed whilst the RMS lift coefficients significantly increased.

The Strouhal number of 0.1527 and RMS lift coefficient of 0.7641 in this study are provided for more applicable data for engineering design of barred tee in aspect of structural integrity further.

References

1. A. Sohankar, L. Davison, C. Norberg, Proc. Twelfth Australasian Fluid Mech. Conf., 517-520 (1995)
2. A. Sohankar, C. Norberg, L. Davison, J. Wind Eng. Ind. Aerodyn **69-71**, 189-201 (1997)
3. A. Sohankar, C. Norberg, L. Davison, Int. J. Numer. Meth. Fluids **26**, 39-56 (1998)
4. M. Zhao, L. Cheng, T. Zhou, Phys. Fluids **25**, 023603-25 (2013)
5. A. Kumar, A. Dalal, Int. J. Numer. Meth. Fluids **52**, 801–821 (2006)
6. T. Prasanth, S. Behara, S. Singh, R. Kumar, S. Mittal, J. Fluids and Structures **22**, 865–876 (2006)

Comparison of the heat for rectangular and circle salt boiling stove in Ban Dung area, Udon Thani Province.

Aphichat Srichat[1,*], *Ponthep* Vengsungnle[2], *Kaweepongt* Hongtong[1], *Weeraphon* Kaewka[1] and *Jarinee* Jongpluempiti[2]

[1]Department of Mechanical Engineering, Faculty of Technology, Udon Thani Rajabhat University, Udon Thani, Thailand 41000
[2]Department of Agricultural Machinery Engineering, Faculty of Engineering and Architecture,
Rajamangala University of Technology Isan, Nakhon Ratchasima, Thailand 30000

Abstract. This research aims a study to compare the thermal parameter of the original (rectangular) and circle salt boiling stove by computational fluid dynamics (CFD) of the boiling salt process in Ban Dung area, Udon Thani province. The first study, the rectangular stove has U-shaped channel for heating flow through chimney that the stove size is 2.4 m of width, 6.0 m of long, 0.8 m of high and cross-sectional area of heating flue was 0.8 x 0.8 m^2. The second study, the circle salt boiling stove has a similarly volume of the rectangular stove of 4.32 m^3 and the diameter of 4.58 m. From the results of mathematical models by CFD found that the rectangular stove have the thermal efficiency, the highest fired temperature and the velocity outlet of hot air of 17.78%, 537.77°C and 4.862 m/s, respectively. Then, the circle stove have 24.59%, 758.32°C and 3.1424 m/s, respectively. it can be calculated the thermal efficiency reduce of the boiling salt production is 27.69%. Comparison between the rectangular and the circle salt boiling stove found that the circle stove have the thermal efficiency and heating flue more than the rectangular stove. It can reduce the cost of production and reduce the fuel consumption in the salt boiling process.

1 Introduction

Northeastern region of Thailand is produced the salt area of 1,840 km^2 and about 30% of the agricultural area [1]. In Udon Thani province, Ban Dung district is mainly rock salt production region about 18% of total salt production area in Thailand. The amount of rock salt reserve is 18 trillion tons [2]. Now, Ban Dung has the salt production area about 4.99 km^2 and the amount of salt production is 194,500 tons/year [3]. The rock salt production, by pumping salt-water from the resource and then make the salt-water becomes crystalline (salt products) by boiling [4]. The salt boiler made from the steel plate in rectangular shape, size 2.4 x 6.0 x 0.3 m. The salt boiler is placed on a salt boiling stove that the original stove made from clay in U-shaped channel, size of 2.4 x 6.0 x 0.80 m., and then boiling salt by burning with firewood as a fuel until the salt-water becomes crystalline salt (salt products) [5]. For salt boiling process, there are the study to increasing the thermal efficiency of fuel and reducing the cost of salt boiling. If the salt boiling stove can be reduce the cost of firewood fuel that can increase the profitability of the salt production as well. If the salt boiling stove is more than higher thermal efficiency that can be reduced the amount of firewood in the salt boiling process and the conservation of forest resources in the future. From the study showed that the original salt stove was low efficiency, so it was used to amount of fuel was large quantities and the high cost of production.

Therefore, the mathematical model was used to analyze increasing the thermal efficiency of salt boiling stove. The design of the new salt boiling stove is different to the original which the new salt boiling stove is circle salt boiling stove. However, the same feature of both stoves is the volume of salt-water used for boiling is 4.32 m^3. Then, the stoves have been analyzed by computational fluid dynamics (CFD) program to compare thermal parameter and analysis to find ways to reduce the fuel cost of boiling salt process between the original (rectangular) and circle salt stove.

2 The combustion salt boiling equation

2.1 The computational fluid dynamics

Computational fluid dynamics (CFD) is analyzing fluid flow behavior [6]. Currently, CFD is critical to analyzing fluid flow behavior whether it is speed, pressure and temperature. There are three components as; partial differential equations, boundary conditions and geometry. The analysis is based on the equation of Reynolds-averaged Navier-Stokes equation can be expressed as (1).

$$\rho \bar{u}_j \frac{\partial \bar{u}_i}{\partial x_j} = \rho \bar{f}_i + \frac{\partial}{\partial x_j}\left[-\bar{p}\delta_{ij} + \mu\left(\frac{\partial \bar{u}_i}{\partial x_i} + \frac{\partial \bar{u}_j}{\partial x_i}\right) - \rho \overline{u_i' u_j'}\right] \quad (1)$$

* Corresponding author: saphichat@udru.ac.th

Eq. (1) is solved by the mass conservation equation, momentum conservation equation and energy conservation equation [7]. Eq. (1) is solved by the mass conservation equation, momentum conservation equation and energy conservation equation [7].

2.2 The combustion salt boiling process

The combustion salt boiling process equation that used to analyze the initial heat transfer is the internal energy can be expressed as Eq. (2). The heat flux for the system and the heat flux emitted from the system can be calculated from Eq. (3) [8].

$$\Delta Q = \frac{\Delta E}{V} \tag{2}$$

$$\Delta q = \frac{\Delta Q}{A} \tag{3}$$

The equation used to estimate the difference in pressure between inlet pressure and outlet pressure can be calculated from Eq. (4) [9].

$$\Delta P = Cah\left(\frac{1}{T_o} - \frac{1}{T_i}\right) \tag{4}$$

3 Model and hypothesis

3.1 The model of salt boiling stove

The original salt boiling stove in Ban Dung of salt production was U-shaped channel burner size of 2.4 x 6.0 x 0.8 m that made from clay and the firewood used in boiling salt process was a rubber wood from sawmills industrial. The model of rectangular salt stove is a U-shaped channel burner size of 2.4 x 6.0 x 0.8 m and cross-sectional area of heating flue was 0.8 x 0.8 m^2, as shown in Figure 1(a). The model of circle salt stove is developing for increase the thermal efficiency that the diameter of stove is 4.58 m. and the high of 0.3 m., as shown in Figure 1(b). A study to the thermal efficiency of salt boiling stove is calculating the related thermal energy by using the similarly conditions and mathematical model.

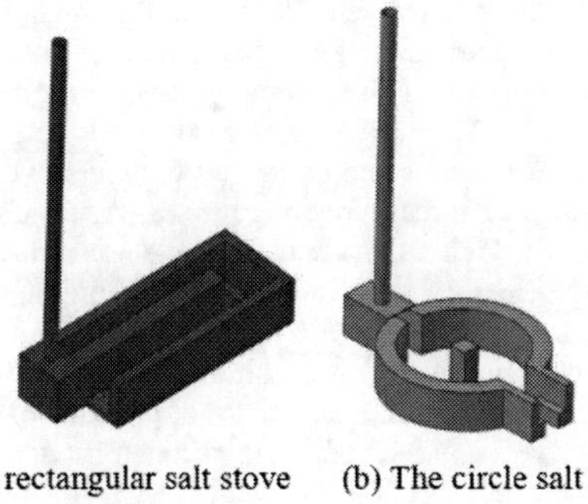

(a) The rectangular salt stove (b) The circle salt stove

Fig. 1. The model of salt boiling stove use for analyzing

3.2 The hypothesis of salt boiling stove

Mathematical model has been used to compare the thermal efficiency between the rectangular and the circle salt boiling stove using CFD analyze. The conditions of model consists of; the steady state condition, the fluid used in the analysis is air at ambient temperature, density of air is 1.225 kg/m^3, specific heat (Cp) of air is 1,006.43 J/kg-K, using standard k-epsilon equation to predict fluid turbulence of analysis, the heat source (calculated from the thermal energy of the fuel, the mass of the fuel per total time for test) of 22,500 W, the air inlet is exposed to atmospheric pressure, the air outlet is a difference in pressure of 30 psi with the air inlet, the pressure difference between the inlet pressure and the output pressure (Pa) is the constant a value of 0.0342 K/m [10], and the lose heat of the surface sink in salt boiling process (calculated from the rate of heat loss to salt in the area) of 2,500 W/m^3. Then, it can be simulated with CFD to compare the thermal parameter between the rectangular and the circle salt boiling stove.

4 Results

4.1 The temperature of salt boiling stove

From the models, it was found that the temperature of the circle salt stove has a higher average temperature than the rectangular salt stove. The heat dissipation of the rectangular salt stove is uniform that the inlet area has a higher temperature and reduced by the distance in the stove to the chimney, as shown in Figure 2. Similarly, the circle salt stove has a uniform heat dissipation but it is higher more than rectangular stove, as shown in Figure 3.

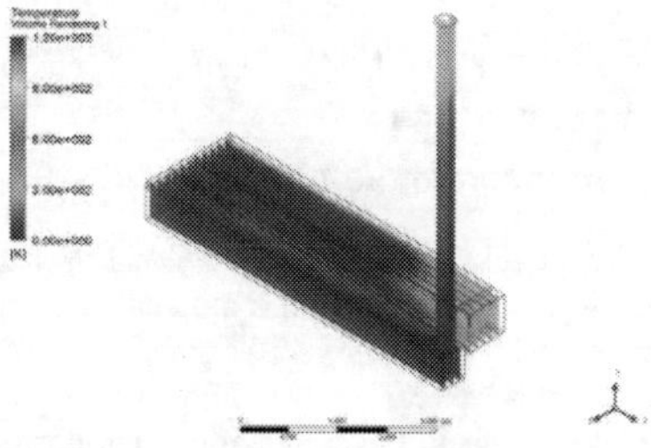

Fig. 2. The temperature of the rectangular salt boiling stove

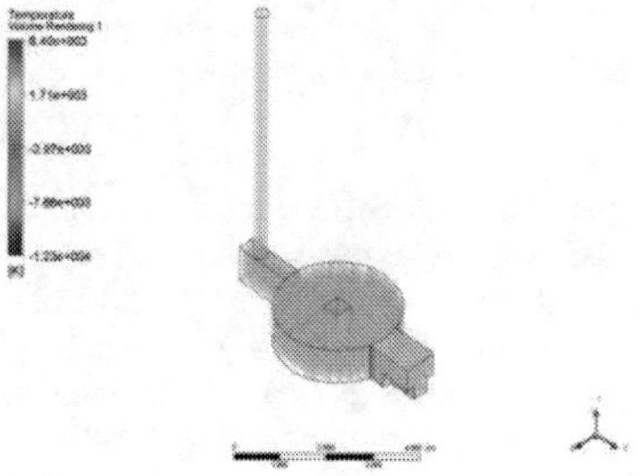

Fig. 3. The temperature of the circle salt boiling stove

4.2 The air flow velocity of salt boiling stove

From the models, it was found that the hot air flow velocity of the circle salt stove was a higher than the rectangular salt stove, but the hot air flow velocity values have a similar. The air flow velocity profiles were shown in Figure 4 and 5, respectively. However, the circle salt stove causes the hot air to circulate inside the stove rather than the rectangular stove. The hot air of rectangular stove will flow through the channel to the chimney. This is the result that the hot air flow velocity of the circle salt stove was a higher.

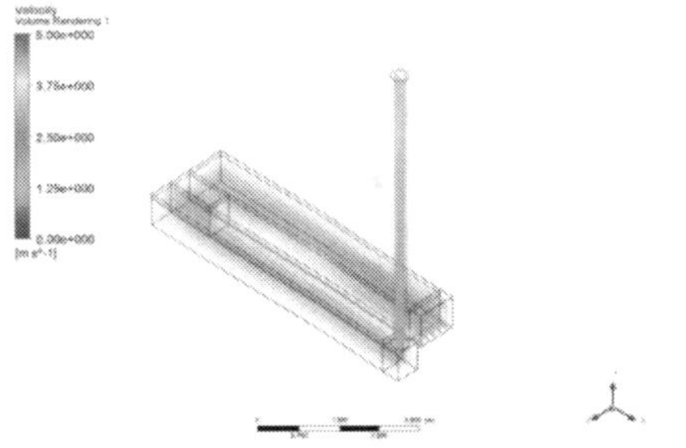

Fig. 4. The hot air flow velocity of the rectangular salt boiling stove

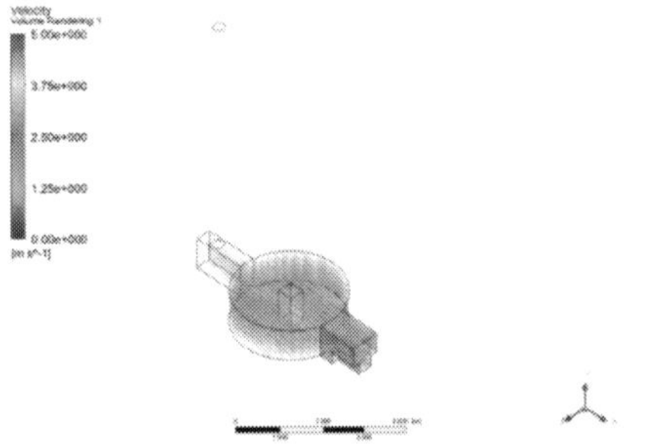

Fig. 5. The hot air flow velocity of the circle salt boiling stove

4.3 The pressure of salt boiling stove

When, the air flow of the circle stove is lower speed and the hot air circulate inside the stove that cause to the pressure of circle salt stove is higher than rectangular salt stove. The circle salt boiling stove has higher pressure more than the rectangular salt boiling stove. The pressure distribution profiles of the salt stoves were shown in Figure 6 and 7, respectively.

4.4 The comparison efficiency of salt boiling stove

From the thermal efficiency calculation of the rectangular and the circle salt boiling stove, it can be seen in Table 1. The rectangular salt stove has a thermal efficiency of 17.78%, but the circle salt stove has a thermal efficiency of 24.59%. It is evident that the thermal efficiency of the circle salt stove is greater than the rectangular salt stove. It can reduce amount of fuel consumption and heating energy consumption by 27.69%.

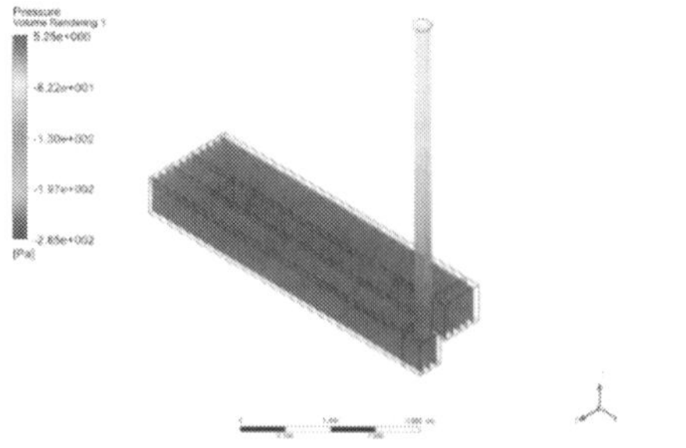

Fig. 6. The pressure distribution of the rectangular salt boiling stove

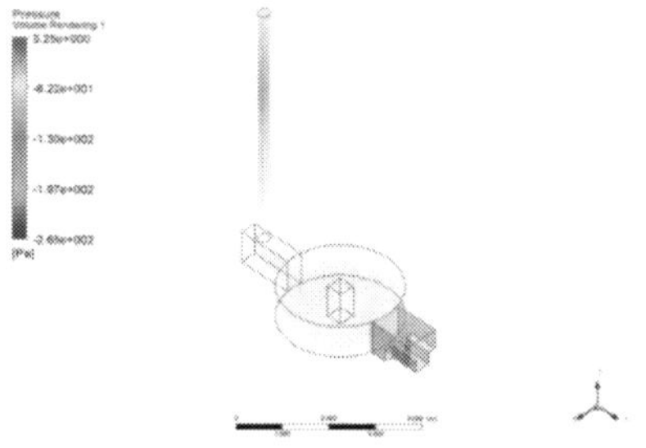

Fig. 7. The pressure distribution of the circle salt boiling stove

Table 1. Comparison of salt boiling stove efficiency (average).

Type of salt boiler	Heat input (MJ)	Heat Utilization (MJ)	η (%)
Rectangular salt stove	28.95	5.15	17.78
Circle salt stove	28.95	7.12	24.59

4.5 The parameter result for CFD

From the mathematical model results, it can be seen that the hot air temperature inlet of the rectangular salt stove and the circle salt stove were similar of 300 K (27°C) and the hot air temperature outlet of the rectangular salt stove and the circle salt stove were 1,239.16 K (966.16°C) and 1,432.34 K (1,159.34°C), respectively. Also, the results shown that the air chamber temperature of the circle salt stove is higher than the rectangular salt stove, the average temperature within the rectangular and circle salt boiling stove is 810.77 K (537.77°C) and 1,031.32 K (758.32°C), respectively. The inlet mass flow rate of the circle stove is lower than the rectangular stove of 0.432 and 0.535 kg/s, respectively. The air outlet velocity of the circle stove is lower than the rectangular stove is 3.124 and 4.86 m/s, respectively. The pressure of the rectangular and circle salt boiling stove is -0.372 and -0.232 Pa, respectively. From the result of the inside air flow velocity, the circle salt stove was reduced the air flow velocity and increased the pressure. When considering the value of temperature of pan area that transfer heat between the salt stove into salt boiler, the result shown that the temerature of pan area of the circle salt stove is higher than the rectangular salt stove of 1,254.32 K (981.32°C) and 739.69 K (466.69°C), respectively. From the result that it causes the heat for the circle salt stove is quickly generate and the heat time period is longer than the rectangular salt stove. From that reaon, it will reducing the fuel as firewood for boiling salt process. CFD analysis can be showed in Table 2.

Table 2. Analysis results by CFD.

Parameter [unit]	value	
	Rectangular stove	**Circle stove**
1. Air temperature inlet [K, °C]	300, 27	300, 27
2. Air temperature outlet [K, °C]	1,239.16, 966.16	1,432.34, 1,159.34
3. Air chamber temperature [K, °C]	810.77, 537.77	1,031.32, 758.32
4. mass flow rate of inlet [kg/s]	0.535	0.432
5. Velocity inlet [m/s]	0.780	0.654
6. Velocity outlet [m/s]	4.862	3.124
7. Furnace pressure [Pa]	-0.372	-0.232
8. Pressure of heat Source [Pa]	-0.374	-0.234
9. Temperature of pan area [K, °C]	739.69, 466.69	1,254.32, 981.32

5 Conclusion

This research is a mathematical model for the analysis of the thermal efficiency of salt boiling stove using firewood as fuel that is analyze by computational fluid dynamics (CFD) to compare thermal parameter and analysis to find ways to reduce the fuel cost of boiling salt process between the rectangular and circle salt stove. The salt boiling stove in Ban Dung, Udon thani Province is U-shaped channel salt boiling stove that size of 2.4 x 4.0 x 0.8 m and the volume of salt-water used for boiling is 4.32 m^3. The new salt boiling stove is the circle salt stove has the diameter of 4.58 m., the high of 0.3 m. and the volume is similar with rectangular stove. The experiment found that the output air temperature of the circle stove is higher than the rectangular stove. The chamber temperature of the salt stove is 810.77 K (537.77°C) and 1,031.32 K (758.32°C), respectively. The outflow velocity of the improvement stove is lower than the original stove of 3.124 and 4.86 m/s, respectively. The pressure in the salt stove is -0.372 and -0.232 Pa, respectively. For the circle salt stove found that the air flow velocity is decreases and the pressure is increases. The temperature of pan area of the circle salt stove is higher than the rectangular salt stove of 1,254.32 K (981.32°C) and 739.69 K (466.69°C), respectively. The rectangular salt stove has a heat input of 28.95 MJ, a heat utilization of 5.154 MJ and the thermal efficiency of 17.78%. And then, the circle salt stove has a heat input of 28.95 MJ, a heat utilization of 7.12 MJ and the thermal efficiency of 24.59%. It can be seen that the thermal efficiency of the circle salt stove is more valuable than the rectangular salt stove. As a result, it can be reduce the fuel as firewood and can be calculate the higher thermal energy of 27.69%, which can benefit the salt industry. If the salt producers want to reduce cost of production and reduce the fuel consumption in the salt boiling process, they should be use the circle stove for the salt production process.

Acknowledgements

The authors acknowledge the grant from Udon Thani Rajabhat University that financially supported in this paper. The authors are also grateful to the Pipatpong Boil Salt in Ban Dung for supporting the tools and equipment used in this research.

References

1. P. Buangnawa. *Report - Strategic planning for the development saline soil area at north-eastern of thailand.* (Land Development Department, Ministry of agriculture and Coorperatives, Bangkok, Thailand, 2015)
2. R. Phruchayawaranon. *Environmental Management in Salt Production Area, Ban Dung, Udon Thani.* (Thesis Master of Science, Department of Resource management, Kasetsart University, Thailand, 2012)
3. N. Sirithongkam. *Report - Project of Saline Soil Development in the North-Eastern Area (Phase 2: Expansion), Udon Thani Province.* (Department of Mineral Resources, Ministry of Industry, Thailand, 2012)
4. National Science and Technology Development Agency. *Salt more salty.* (National Science and Technology Development Agency, Ministry of Science and Technology, Thailand, 2012)
5. A.Srichat, P.Vengsungnle and A.Bootwong. *A Study to increasing the thermal efficiency of the improvements salt boiling stove which the fuel was firewood in Ban Dung area, Udon Thani province by CFD.* Energy Procedia **138**, (2017)
6. J. Jongpluempiti and K. Tangchaichit. *STUDY OF BEHAVIOUR THE HEAT AND FLOW IN CASSAVA RHIZOME FIRED DAN KWEAN KILNS.* (Thesis Ph.D. in Mechanical Engineering. Faculty of Engineering. Khonkaen University. Thailand, 2013)
7. Bichkar, P., Dandgaval O., Dalvi P., Godase R and Dey T. *Study of Shell and Tube Heat Exchanger with the Effect of Types of Baffles.* Procedia Manufacturing, **20**, (2018).
8. Cengel Y. A. and Ghajar A. J. *Heat and Mass Transfer (5th ed.).* (United States: McGraw-Hill Education, 2014)
9. Pei P., Ouyang M., Feng W., Lu L., Huang H., and Zhang, J. *Hydrogen pressure drop characteristics in a fuel cell stack.* International Journal of Hydrogen Energy. (2006)
10. Guo P., Li J., Wang Y. and Wang, Y. *Evaluation of the optimal turbine pressure drop ratio for a solar chimney power plant.* Energy Conversion and Management. (2016)

Real time monitoring of soil moisture content for landslide early warning: wn experimental study

Panupong Thumtuan[1], *Tanan* Chub-Uppakarn[1], and *Tanit* Chalermyanont[1,*]

[1]Department of Civil Engineering, Prince of Songkla University, Songkhla, Thailand

Abstract. Landslides occur commonly after heavy rainfall. More accurate and immediate prediction of landslides for early warning purpose can be achieved when real time water content of the soil slope is known. In this experimental study, the water content was measured using time domain reflectometers (TDR). Five TDRs were installed with equal vertical spacing in a test pit. The measured results were sent to and stored on a web server and real time monitoring was made online. All TDRs results showed a good and accurate water content response of the soil to the rainfall from top to the bottom of the test pit.

1 Introduction

Landslides are natural disaster events that always cause damage to surrounding areas and even loss of life. The landslides usually takes place in the rainy season when the soil is wet and weak [1]. Thus, the soil moisture content is the major factor that reduces soil shear strength and decreases accordingly the stability of the soil slope. Soil moisture content seasonally varies and depends significantly on rainfall. In summer, the soil is unsaturated and pore water pressure within the soil is negative, so-called "matric suction". The matric suction increases the shear strength of the soil. In contrast, in rainy season, with sufficient rainfall, the soil becomes saturated and the matric suction disappears which, in turn, decreases the shear strength of the soil and triggers the landslide. A landslide occurs when the weight of the wet soil is greater than the decreasing resisting force due to reduction of the soil shear strength [2, 3].

Stability of the slope can be assessed if the soil suction or moisture content is known. Correspondingly, early warning of a landslide can be achieved if its temporal moisture content data are available. The data can be made available via real time field monitoring. This paper describes a set of systems that we used in real time filed monitoring of soil moisture content conducted at a test site in Prince of Songkla University. A series of time domain reflectometers (TDR) was used for measuring the soil moisture constant. Sensor controller and communication units were used to sending the monitored data to a server at Southern Natural Disaster Research Center (NadRec). Real time soil moisture content data can be monitored online via NadRec's website (www.nadrec.psu.ac.th).

2 Site description

A test site was situated on a hill at the faculty of engineering, Prince of Songkla University. Instead of conducting the experiments on the real slope in the mountain, the site was selected because of its accessibility and experimental suitability. The hill is a part of the Korhong Mountain located in Hat Yai district, Songkhla province, southern Thailand (Fig. 1). Geologically, parent rocks of the mountain are carboniferous sandstone, shale, and conglomerate. While the residual soil above the parent rocks is mainly lateritic soil [4]. A slope angle of residual soil at the test site is about 22 degrees. The soil is relatively thin with a thickness of 2 – 3 m.

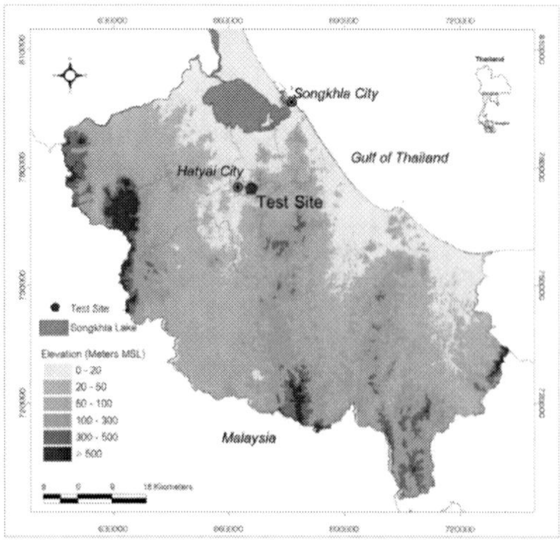

Fig. 1. A topographic map showing location of the test site.

* Corresponding author: tanit.c@psu.ac.th

In primary investigation program, field and laboratory tests were conducted to determine index and engineering properties of the soil including: natural water content (ASTM D 2216), soil unit weight (ASTM D 4718), specific gravity (ASTM D 854), sieve analysis (ASTM D422), Atterberg limits (ASTM D 4318), soil classification (ASTM D 2487). A boring log showing the testing results is shown in Fig. 2. The soil is classified into two layers. The upper layer is about 1.0 m-thick and classified as clayey gravel (GC), while the lower layer is about 1.2 m-thick and classified as high plasticity clay (CH).

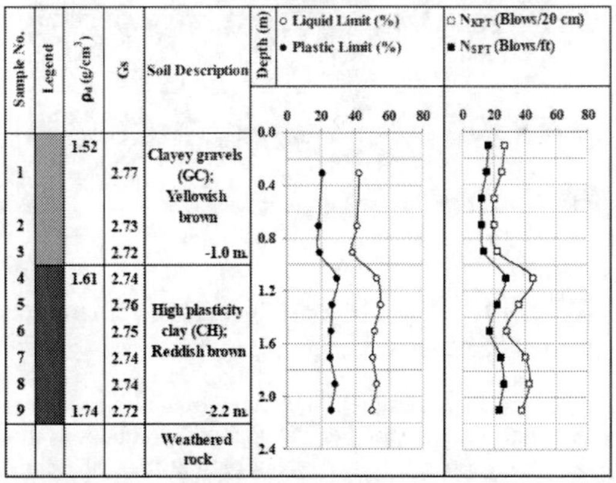

Fig. 2. Boring log of the test pit showing soil types and their properties.

3 Field instrumentations

Many instrumentations were installed at the test site in order to study the seepage characteristics and variation of soil moisture content with depth such that the reduction of shear strength of the soil intimately can be assessed real time. Fig. 3 shows a picture of the instrumentations installed at the test site. An automatic rain gauge was used to measure the rainfall. Water runoff on the slope surface was measured within the closed runoff testing area using flume and water depth probe. Double-ring infiltrometer and TDRs which were locally available and cheap, were used to determine infiltration.

Fig. 3. Picture of the test site showing instrumentations installed.

3.1 Time domain reflectometers

A TDR is used to determine soil moisture content via a dielectric constant of the soil from measurements of the propagation speed of an electromagnetic wave propagated through stainless steel rods as described in [5, 6]. A series of the TDR was installed vertically in a test pit (Fig. 3) to measure the soil moisture content (in term of volumetric water content) at various depths. Only TDR measurement results are shown and discussed in this paper. Fig. 4 shows locations of five TDRs (i.e., TDR 1 to TDR 5) installed in the test pit. The test pit was about 1.0 m in diameter and 2.4 m in depth. TDRs 2 to 5 were installed at 0.40 to 1.90 m. TDR 1 was installed perpendicular to the other TDRs at about 0.15 m from the slope surface.

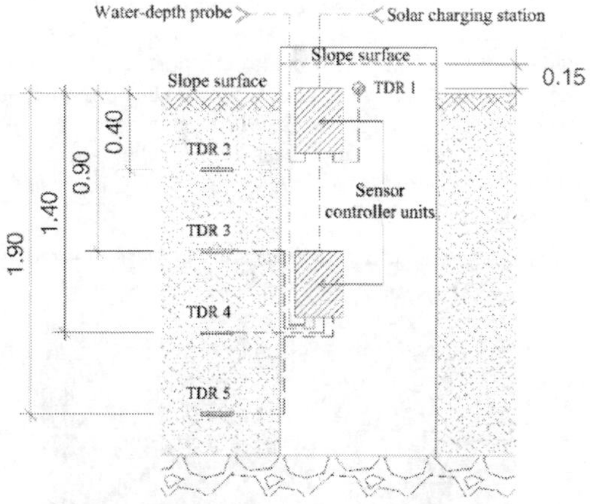

Fig. 4. A cross section of the test pit showing the locations of the TDRs installed (unit of depth is in meter).

The TDRs (Model TMEC-04 as shown in Fig. 5) used in this study were produced by the National Electronics and Computer Technology Center of Thailand (NECTEC). Their specifications include: range 0 to 90% (calibration dependent), accuracy 3%, resolution 0.5%, operation temperature 20 to 100 degrees Celsius, excitation voltage 5 volts.

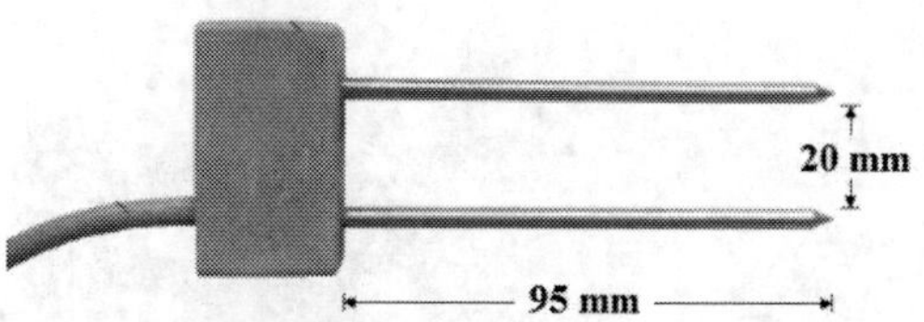

Fig. 5. A time domain reflectometer (TDR).

The TDRs were calibrated with soil samples collected from the test pit using a procedure described in Quinones et al. (2003) [5]. The soil samples were compacted to have a similar density as the intact soil in the test pit. Each TDR was inserted into a particular sample and water was added. The gravimetric and volumetric water content of the sample, thus, increased accordingly. As the water content increased, reading output voltage from the TDR decreased correspondingly. Relationships between the water content and the output voltage for TDR 1 are shown in Eqs. 1 and 2 for wetting and drying cases, respectively.

Wetting: $$\Theta = -20.96V + 60.890 \quad (1)$$

Drying: $$\Theta = -25.79V + 65.364 \quad (2)$$

where Θ is volumetric water content (dimensionless) and V is reading output voltage (volt). Goodness of fit values (R^2) are 0.9883 and 0.9976 for Eqs. 1 and 2, respectively.

3.2 Data acquisition system

Data acquisition system used in this study consisted of 3 main components: a sensor controller, a web server, and a solar charging station. The sensor controller (Fig. 6a) are wired with the TDRs so excitation voltage can be supplied and the output voltage can be acquired. The output voltage then was sent to and stored into the web server by 2.4 GHz antenna equipped within the sensor controller unit. Fig. 6b shows sensor controller units along with TDRs installed in the test pit.

(a) Sensor controller.

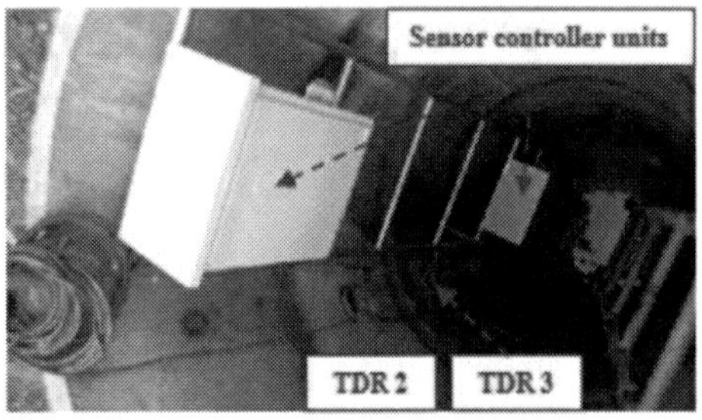

(b) Sensor controller units and TDRs in the test pit.

Fig. 6. Sensor controllers installed in the test pit; (a) sensor controller, (b) sensor controller units and TDRs in the test pit.

Electricity supplied to the sensor controllers was generated by a solar panel and stored in a battery installed in the solar charging station (Fig. 7). The station was located on the ground surface next to the test pit as shown in Fig. 3.

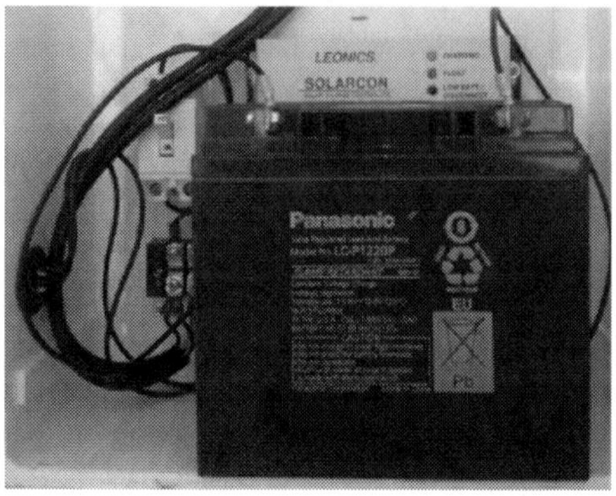

Fig. 7. Battery and charging unit inside the solar charging station.

4 Real time monitoring of the volumetric water content

Real time monitoring of the volumetric water content can be made online via the webpage of Southern Natural Disaster Research Center (NadRec), Prince of Songkla University (www.nadrec.psu.ac.th) as shown in Fig. 8. Information shown in Fig. 8 includes: (1) menu for selecting types of data, such as TDR output voltage, volumetric or gravimetric water contents and rainfall and (2) a graph showing the monitoring data with time.

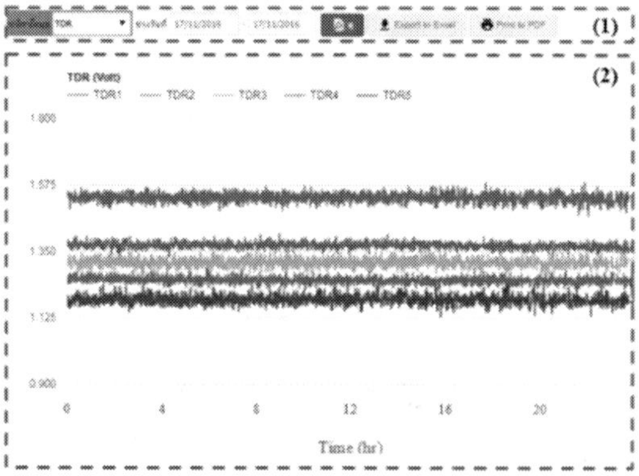

Fig. 8. NadRec's webpage displaying the monitoring data.

In this study, the volumetric water contents from five TDRs were measured from November 2015 to April 2017. For demonstration purpose, TDRs and rainfall data from October 17th to November 19th, 2016 are shown in Fig. 9. Before October 17th, the soil was dry as the volumetric water contents were 0.237, 0.215, 0.210, 0.223, and 0.221 for TDRs 1 to 5, respectively. After a couple days of rain, TDR 1 made contact with seeping water first as the volumetric water content increased on

October 20th whereas other TDRs still idled. TDR 2 then did make contact with the water on October 27th, and the other TDRs followed later on (see Fig. 9). This was because of the fact that the soil was moderately impervious and water required more time to seep through, the deeper TDRs. In additions, for TDR 1 which was located very close to ground surface (i.e., 0.15 m), its volumetric water content responded to the rain very quickly as its peaks synchronously appeared with the heavy rainfall events.

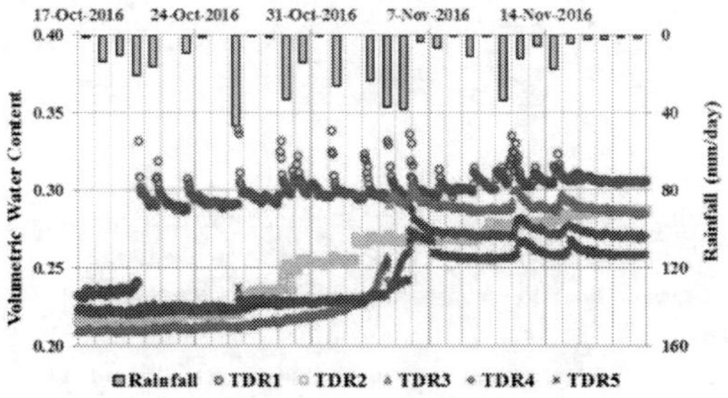

Fig. 9. TDR volumetric water contents and daily rainfall measured at the test site from October 17th to November 19th.

Another example of the variation of volumetric water content with depth is depicted in Fig. 10. The same set of data as Fig. 9 was used but only the volumetric water content data at particular times were selected. In Fig. 10, a clearer picture of how the volumetric water content changes with depth and time is shown. It can be seen that significant increase for TDR 3 was on November 3rd, or about 7 days after TDR 2. After that, on November 8th, TDRs 4 and 5 were found to have the significant increase. From November 8th to November 19th, no major change of the volumetric water content was observed while the rain kept falling. This indicated that the soil was nearly saturated throughout the soil layers. The volumetric water contents for the lower soil layer (i.e., TDRs 4 and 5) were less than that of other TDRs because its soil type was different from the upper one.

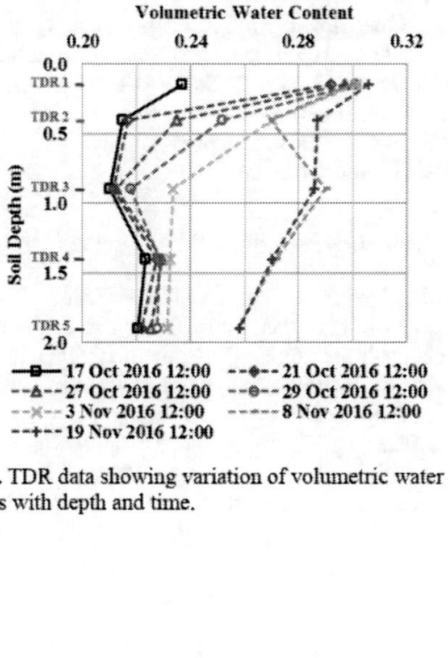

Fig. 10. TDR data showing variation of volumetric water contents with depth and time.

5 Conclusion

Real time monitoring of soil volumetric water content was made possible by a set of systems including time domain reflectometers (TDR), sensor controllers, a 2.4 GHz communication system, a solar power system, and a web server. A slope test site at Prince of Songkla University was selected and variation of the water content of the soil in the test pit was measured using TDRs. The measured data then were sent to and stored on the web server. Real time monitoring of the water content was made via the NadRec's webpage. All TDRs results showed a good and accurate response of the soil volumetric water content to the rainfall from top to the bottom of the test pit. In summary, these real time water content data are good for immediate predicting of the stability of slope for early warning purpose.

References

1. Zhang, L.L., Zhang, J., Zhang, L.M., and Tang, W.H. (2011), *Stability analysis of rainfall-induced slope failure: a review*. *Geotechnical Engineering*, **164**(GE5), p. 299-316.
2. Lim, T.T., Rahardjo, H., Chang, M.F., and Fredlund, D.G. (1996). *Effect of rainfall on matric suctions in a residual soil slope*. Can. Geotech. J., **33**, p. 618-628.
3. Rahardjo, H., Leong, E.C., and Rezaur, R.B. (2005). *Response of a Residual Soil Slope to Rainfall*, Can. Geotech. J., **42**, p. 340-351.
4. Department of Geology (2001), *Digital Geological Map of Songkhla Province*, in Thai.
5. Quinones, H., Ruelle, P., and Nemeth, I. (2003). *Comparison of three calibration procedures for TDR soil moisture sensors*, Irrig. and Drain. **52**, p. 203–217
6. Topp, G.C., Davis, J.L., and Annan, A.P. (1980). *Electromagnetic determination of soil water content: measurements in coaxial transmission lines*, Water Resources Research, **16**, p. 574–582.

The vulnerability analysis of catchment areas using rapid assessment method (case study in Universitas Indonesia, Depok)

Eva Rolia[1*], *Dwita* Sutjiningsih[1], and *Fitria* Fitria[1]

[1]Civil Engineering Department, Faculty of Engineering, Universitas Indonesia, Depok, Indonesia

Abstract. Catchment area of aquatic system in Universitas Indonesia (UI) campus has a function as groundwater recharge for Depok area, which is the buffer zone of Jakarta. This catchment area has high level of imperviousness as a consequence of increasing population. The high imperviousness implicates in the degradation of the catchment area health. For that reason, the vulnerability assessment of UI catchment area is needed to arrange of restoration recommendation. This research applies Rapid Assessment Method which implements eco-hydrological concepts including aquatic and riparian condition as well as imperviousness of catchment area. Locations of the research are Kenanga and Agathis catchments area. The result of initial classification based imperviousness both catchment areas are categorised as Non-Supporting, where the imperviousness of Kenanga and Agathis is 31,0% and 77,18% respectivelly. Based on the assessment on aquatic and riparian condition as well as future land use, the final classification of both catchment areas is Restorable Non-Supporting catchment area.

1 Introduction

The availability of water resources in terms of quantity and quality tends to be critical and become a worldwide problem [1]. Sustainability of water resources is determined by watersheds functioning as catchment areas. Since watersheds are formed by inseparately ecosystem interaction from upstream to downstream, the healthiness of catchment areas is important to be considered. It can be done by measuring the quality of the water located in water bodies which are parts of catchment areas such as lakes, rivers, and others. [2] stated that catchment areas damage indicators could be recognised by hydrological behaviour shifts such as high frequency of flood, more erosion and sedimentation, as well as decreasing water quality. Therefore, catchment area management should be done through sustainable natural resource utilisation [3].

Human activities and the needs of space implicate on land covering conversion from pervious to impervious cover. Increasing number of impervious land covering strongly affected on the change of water quality and ecosystem especially for small scale catchment areas within 13 – 77 km^2 [4,5]. Excessive impervious land covering brought bad impacts on the healthiness of catchment areas [6].

This paper will discuss the vulnerability assessment of catchment areas in UI campus namely Kenanga and Agathis using Rapid Assessment Method. Both catchment areas are characterised by education and residential uses. Estimates of impervious cover using GIS technology by digitizing land cover features, water quality data [7] and aquatic [8] used derived from previous studies [9]. The result of this study is expected to be a foundation to formulate necessary recommendations which finally aim to do conservation and increase environmental carrying capacity of UI campus and surrounding areas.

2 Methodology

2.1 General Description of the Study Area

Kenanga catchment area is astronomically located in 106°49'44,66"-106°49'51,51" East Longitude and 6°21'56,96"-6°22'04,64" South Latitude. Meanwhile, Agathis catchment area is located in 106°49'28,47"–106°49'29,69" East Longitude and 6°22'5,75" LS–6°22'9,18" South Latitude.

2.2 Analysis Procedure

Watershed vulnerability assessment can be done by Tools 12 of Watershed Vulnerability Analysis proposed by Zielinski in 2002. This method aims to classify watershed condition. Furthermore, watershed vulnerability analysis basically illustrates eight steps to create planning for watershed, to figure sub-watershed out, to forecast land covering based on existing condition, and to identify factors causing a change on initial sub-watershed classification. There are three

* Corresponding author: roliaeva@yahoo.com

basic steps which are initial classification, final classification, and priority scale formulation of restoration. In this research, initial classification was done by measuring impervious land covering using Geographical Information System (GIS). Then, final classification was done by assessing the existing condition of Kenanga pond and Agathis pond perimeters; and their projection as it was stated in the spatial planning of the city of Depok, West Java Province. To be more detail, the stages can be seen through this following figure.

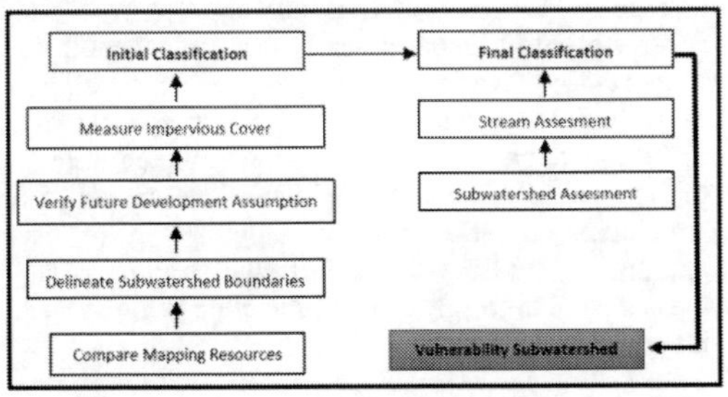

Figure 1. Steps of catchment area vulnerability assessment [4]

3 Result and Discussion

3.1 Initial Classification of Catchment Area

Initial classification of the catchment area was done by measuring land covering in the catchment area. This is important since impervious cover is one of dominant factors influencing the healthiness of catchment areas. Increasing number of impervious cover indicates development that occurred in the area. Direct measurement is considered as the most accurate method among various techniques. However, using Geographical Information System (GIS) is considered as the most efficient method [10,11]. These following figures illustrate impervious cover in Kenanga catchment area (A1) and Agathis catchment area (A2). This study uses ArcGIS 10.1 software by digitizing land cover features such as roof, asphalt road, concrete road, and then visual interpretation.

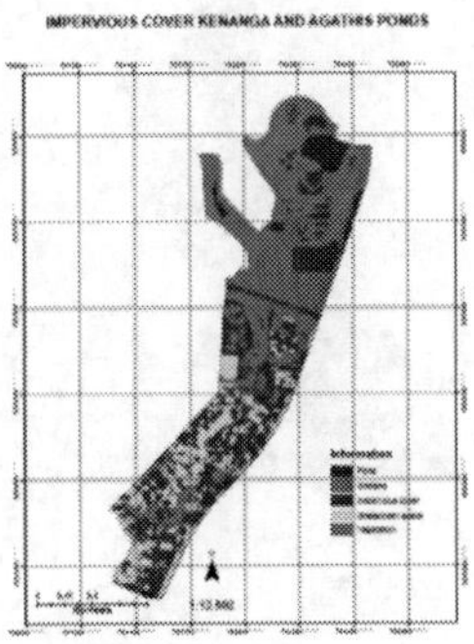

Figure 2. Existing condition of impervious cover of Kenanga (A1) and Agathis (A2) Catchment Areas

Furthermore, the percentage of impervious land covering was calculated as presented by this following table.

Table 1. The Percentage of Impervious Cover

No.	The Name of Catchment Area	Total Area of Impervious Cover (Ha)	Total Catchment Area (Ha)	Percentage (%)
1.	Kenanga	16,84	54,67	31,0
2.	Agathis	61,998	80,33	77,18

It is clearly seen that the percentage of impervious land covering reached 31,0% for Kenanga catchment area and 77,18% for Agathis catchment area. According to Impervious Cover Model developed by the Centre for Watershed Protection, Kenanga and Agathis can be categorised as "*Non-Supporting*" since the percentage exceeded 25%; the model can be seen in figure 3. It means that the water body cannot accommodate flood and the river undergo severe enlargement. In addition, the population of fish decrease significantly and the river cannot provide liveable habitat for insects. It also indicates bad quality of water and high level of bacteria. Biological water quality is also low and is dominated by insects and fish which are resistant to pollution. This following table shows the impervious cover model.

Table 2. The Impervious Cover Model [4]

No.	Impervious Cover (%)	Categories
1.	0 -10	Sensitive
2.	11 – 25	Impacted
3.	> 25	Non Supporting

3.2 Final Assessment of Catchment Areas

Final Assessment of catchment area was conducted in the water body around perimeter. Criteria that were considered in this assessment included water quality, the population of rare plant species, and land cover in riparian [7,8]. [7] conducted water quality test on cascade-pond in UI by using WQI. The result of her measurement can be seen in these following tables.

Table 3. The Result of Physical and Chemical Parameters Test on Kenanga Pond

No.	Parameter (Unit)	Result		Average	Standard
		Inlet	Outlet		
1.	Total Suspended Solid (mg/L)	67,5	19,20	43,5	50
2.	Turbidity (NTU)	5,11	2,66	3,88	25
3.	pH	6,19	6,37	6,28	6,5-9
4.	BOD (mg/L)	3,0	0,26	1,63	3
5.	DO (mg/L)	3,39	5,61	4,5	5
6.	Temperature	30,4	30,5	30,45	Deviation 3°C

Table 4. The Result of Physical and Chemical Parameters Test on Agathis Lake

No.	Parameter (Unit)	Result		Average	Standard
		Inlet	Outlet		
1.	Total Suspended Solid (mg/L)	6,10	5,50	5,80	50
2.	Turbidity (NTU)	1,68	0,86	1,27	25
3.	pH	6,14	6,20	6,17	6,5-9
4.	BOD (mg/L)	1,10	1,60	1,35	3
5.	DO (mg/L)	2,43	1,79	2,11	5
6.	Temperature	29,30	27,60	28,75	Deviation 3°C

Table 3 and 4 above show that water quality based on Physical and chemical parameters of Kenanga and Agathis ponds are still in the standard interval except pH. UI cascade-pond damage due to levels of excessive pollutant solute resulted in decreased water quality, reduced aquatic animals and border ecosystems [8]. Some parameters have been tested to investigate pollutant indicators in UI cascade-pond [9]. Those parameters are BOD, COD, Ammonium, and Phosphate which exceeded the standard interval. In addition, these results were strengthened by the result of Pinasti's research conducted in 2017 as it was shown in table 3 and 4, which explains that the Kenanga pond is heavily polluted and polluted on moderate in Agathis pond.

In addition to researching water quality, riparian zones were also observed about existing plants around it. The function of the riparian zone is the ecological function that is as a food source, as a nutrient filter, and habitat [12]. As a habitat, riparian zones affect the abundance of macroinvertebrates. Riparian buffer zones can significantly reduce the impact of deforestation on tropical flows. Macroinvertebrates are animals sensitive to existing pollution.

Furthermore, final classification of Kenanga and Agathis based on the assessment of river corridor and land use is categorised as *Restorable Non-Supporting* Watershed. This is because both lakes still have five assessment criteria of riverbank and two assessment criteria of sub-watershed scale. Some indicators show Kenanga and Agathis catchment areas can be classified as vulnerable catchment areas because there is an increasing trend of impervious cover, polluted water, and the extinction of some water plants. Moreover, future conditions of land cover were projected to decide appropriate rehabilitation efforts and priority scale of measures. The result of this projection is illustrated by these following figures.

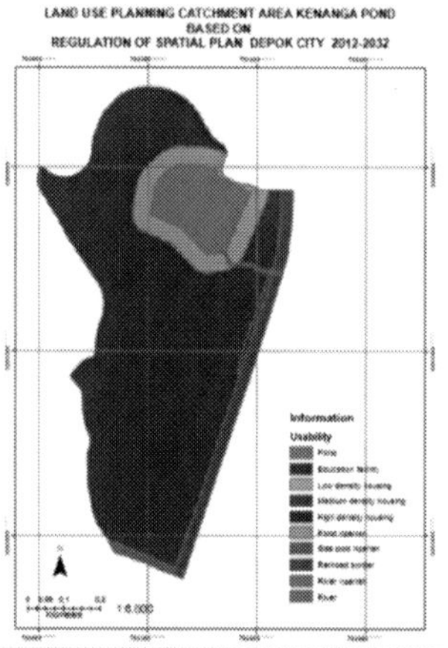

Figure 3. Spatial Planning Map of Kenanga Pond

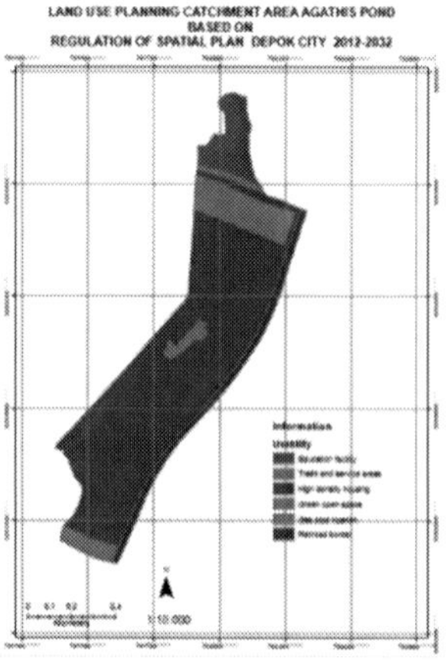

Figure 4. Spatial Planning Map of Agathis Pond

These two figures show that both catchment areas are dominated by educational areas. Land use Agathis catchment area is also dominated by residential and commercial uses. According to the Spatial Planning of Depok City 2012-2023, development planning proposed in Kenanga catchment area is to develop educational and governmental facilities in UI campus. However, there are still some areas dedicated to green open spaces. Therefore, impervious cover decreases to 29,57%. The same trend also occurs in Agathis catchment area. In this location, impervious cover decline to 73, 48%. To be more detail, the data is presented in Table 5.

Table 5. The percentage of Impervious Cover Based on the Spatial Planning of Depok City 2012-2032

No.	DTA	Impervious Cover (Ha)	The Large of DTA (Ha)	Percentage (%)
1.	Kenanga	16,16	54,67	29,57
2.	Agathis	59,02	80,33	73,48

From the table above, it is estimated that up to 2032 there is a percentage of land cover with a downward trend, although only less than 5%. It is possible there are restoration efforts made by the government based on Spatial Plans (RTRW) which has been made by the government of Depok City.

4 Conclusion

Based on rapid assessment conducted in this research, it can be concluded that Kananga and Agathis catchment areas is vulnerable. It is indicated by the percentage of both catchment areas, which are 31,0% in Kenanga and 78,18 in Agathis. Since the percentages are more than 25%, the vulnerability classification can be categorised as Non-Supporting. From the existing condition of water quality, water biota, riverbank and the projection of impervious land cover, both Kenanga and Agathis catchment areas can be classified as Restorable Non-Supporting. The estimated is a decrease in percentage of land cover (<5%) from the GIS calculation of 2017 condition compared with the condition according to RTRW Depok City 2012-2032.

Acknowlegment

I am very grateful to Universitas Indonesia for the International Publication Grant (PITTA 2018) funding No. 2422/UN.2.R3.1/HKP.05.00/2018, Universitas Muhammadiyah Metro for the cooperation. I would also like to thank to Water Resource Management Master Students, Mas Ardi, Mas Dwi, and Sugi for all your supports.

References

1. Asdak, *Hidrologi dan Pengelolaan Daerah Aliran Sungai*, Gadjah Mada University, (2002)
2. I. Mawardi, *Kerusakan Daerah Aliran Sungai dan Penurunan Daya Dukung Sumber Daya Air di Pulau Jawa serta Upaya Penanganannya*, Hidrosfir Indonesia Journal, **5**, No. 2, 1-11, (2010)
3. Marlena. B, *Kajian Pengelolaan sub-DAS Garang Hulu terhadap Kualitas Air Sungai*, National Conference on Water and Environmental Resource Management, (2012)
4. J. Zielinski, *Watershed Vulnerability Analysis*, Center for Watershed Protection, (2002)
5. Schiff, Benoit, *Effect of Impervious Cover at Multiple Spatial Scales on Coastal Watershed Streams*, Journal of The American Water Resources Association, **43**, 712-730, (2007)
6. Schueler, *Is Impervious Cover Still Important? Review of Recent Research*, Journal of Hydrologic Engineering, **14**, (2009)
7. MF. Prinasti, *Characteristic of Macroinvertebrates Abundance in Cascade-Pond System at Universitas Indonesia Campus, West Java, Indonesia*, The 6th International Conference of Euro Asia Civil Engineering Forum (EACEF), (2017)
8. F. Zulkarnain, *Cascade-pond System Health Assessment Based on Macroinvertebrate Indices and Its Relationship with Impervious Cover and Aquatic Buffer Zone in Urbanized Cathments*, The 6th International Conference of Euro Asia Civil Engineering Forum (EACEF), Matec Web of Conferences 138, (2017)
9. D. Sutjiningsih, *Water Quality Index for Determining The Development Treshold of Urbanized Catchment Area in Indonesia*, International Journal of Technology, **8**, No.1, 143-157, (2017)8
10. S. Hamidreza, *Spatiotemporal Vriation of Watershed Health Propensity through Reliability-Resielence-Vulnerability Based Drought Index (Case Study : Shazand Watershed in Iran)*, Science of The Total Environment, 587-588, 168-176, (2017)
11. Tiburan. Jr, *Geospatial Based Vulnerability Assesment of an Urban Watershed*, Procedia Environmental Science, (2013)
12. R. Naiman. H, *The Ecology of Interfaces : Riparian Zones*, Decamps, Annual Review of Ecology System, 28, 621 (1997)
13. *Rencana Tata Ruang Wilayah (RTRW) 2012-2032*, Depok City Government

The modelling of ground water quality in urban area based on demographics factor and building coverage ratio by using geographically weighted regression approach (case study in Jakarta, Indonesia)

Fitria Fitria[1*], *Dwita* Sutjiningsih[1], and *Titin* Siswantining[2]

[1]Civil Engineering Department, Faculty of Engineering, Universitas Indonesia, Depok, Indonesia
[2]Mathematics Department, Faculty of Mathematics and Natural Sciences, Universitas Indonesia, Depok, Indonesia

Abstract. The population density as consequences of urbanisation in certain area may cause the occurrence of land utility shifting from natural site to impervious. Thus, this imperivious cover can be used as indicator of environment quality, particularly the water quality, such being the case of Jakarta, that 80% of ground water sample in Jakarta has been contaminated by E.Coli. Due to this condition, there are indicators which might be correlated with the water quality, namely building coverage ratio (BCR) and demographic factors such as population and building density. This study argues the influence of those three factors toward the ground water quality. The aim of this study is to formulate a spatial regression-correlation model between those three factors with the main indicator of water quality, namely E.Coli. The sampling data were obtained from a densely populated areas in Duren Sawit sub-district, in Jakarta. In term of statistical process, Geographically Weighted Regression (GWR) was used to analyse the spatial data. The result suggested that the value of R^2 at 68.9% and the largest influence of E.Coli value in this models was more affected by the BCR rather than by demographic factors. This outcome would be an early recommendation for Jakarta.

1 Introduction

The land transfer function in urban areas tends to increase the surface area with the closure of cement, asphalt, and waterproof area (*impervious*) [1,2]. In the last few decades, a large number of big cities in Indonesia are faced by a similar problem, that is a high population growth rate. This condition is triggering an increasing demand of residential areas and city infrastructures. In the further development, it would impact on the shifting of land usage in urban area. The rise of impervious cover could lead to eco-hydrologist problems such as an increase of surface runoff, a declining water quality, the damage on habitat of water aquatic animal [3,4,5]. Nevertheless, the land covers (impervious cover) is the characteristic of urban area [2,6]. The impervious cover could be used as indicator of environmental quality mainly related to the water resources quality [2]. In addition, the previous studies discuss majority on how much the influence of population density toward the water quality in a certain watershed [4,5,6,7]. The data mapping has been conducted by employing GIS and statistical method includes anova, correlation, regression as well as spatial regression (Geographical Weighted Regression) [8,9]. Returning briefly to the environmental issues, DKI Jakarta become one of the largest metropolitan cities which experienced with the biggest impact of development. This condition was illustrated by its rate of city growth and development which is higher than other cities [10]. Regarding to the environmental quality monitoring report in DKI Jakarta in 2015, there has been a decreasing quality of river water, ground water and air, thereby the water pollution in Jakarta has reached a harmful level. This water quality monitoring was conducted toward well water in 197 urban village. The data sample was obtained twice during period September – October and November – December by employing few parameters such as physical, chemistry, and biology (microbiology). In term of quantity and quality aspects, the ground water notes better performance than the surface water. Although, 80% of ground water samples from 75 wells in 75 urban villages have contaminated with the Escherichia Coli (E.Coli) [11]. The spread of E. Coli bacteria in the ground is heavily influenced by the ground porosity. The horizontal movement is hard to be ascertained because it depends on several factors such as a type of soil, the height of ground water surface, the flow of ground water, the construction of hand pumps well as well the number of people who use it, also the number of people who throw feces [12]. Simultaneously, other impacts from the fast development in Jakarta is a quality reduction of space utilisation and the imbalance of space function and structure. According to the result

* Corresponding author: fitria65@@eng.ui.ac.id

of satellite images data processing lansat 8 and GIS technology in 2017, DKI Jakarta showed a high building density. Consequently, the building density and building coverage ratio (BCR) increased, which is caused by a higher population density in the middle of the constant variable of land areas. For those purpose, this study attempts to obtain a spatial regression correlation model among the variables of population density, the building density, and the building coverage ratio, toward one main indicator of water quality, that is E.Coli bacteria in urban area. The outcome of this study is expected would be able to deliver a positive contribution and the policy direction related to the law enforcement.

In the meantime, the research would be conducted in East Jakarta as one of municipalities in DKI Jakarta with densely residential. The location being precisely taken for research is Duren Sawit, as one of the most densely populated sub-districts in East Jakarta and considered to be housing zone based on the Local Regulation DKI Jakarta No.1 year 2014.

Subsequently, the modelling research of ground water quality was implemented through the spatial regression approach or Geographical Weighted Regression (GWR), in which the geographical location was applied as predictor variable that could influence the response variable[13].

2 Materials and Methods

2.1 Study Site

The location of this study was determined in urban area in DKI Jakarta, in Duren Sawit sub-district, East Jakarta municipality to be precisely. This sub-district consist of 7 urban villages.

2.2 Sampling Methods

The data sample was collected from Duren Sawit sub-district in 30 site of location based on the result of population density mapping. The ground water sampling was represented by the resident well water and the measurement of building coverage ratio in 30 houses. The proportional probability sampling (pps) method was applied by using the population density as guidance, thereby each urban village (study location) hold a different number of sample.

2.3 Data Collection

2.1.1 Population density (PD)

The population density data was obtained from Indonesia Central Bureau Statistic 2016, which includes the landmass of 7 urban villages, as well as the population number and density. Based on the population density as a guidance, the data sample was determined by using the proportional probability technique, namely the stratified random sampling. Thence, the number of sample was defined around 30 from total number of population density in entire urban villages.

2.1.2 Building density (NDBI)

The building density was estimated through the approach of NDBI (Normalized Difference Building Index). This index data was derived from the processing of satellite images data of landsat 8. Specially NDBI is a building density index in which the type of date shaped raster with pixel value between -1 until 1. The value of -1 determined the area with low building density, whereas the value of 1 for the high density.

2.1.3 Building Coverage Ratio (BCR)

The value of BCR is calculated by formula:

$$BCR = \frac{building\ area}{site\ area} \tag{1}$$

The value of BCR is presented in percentage (%).

2.1.4 Water Quality

The ground water quality sample was tested by using E Coli bacteria. The early interview conducted in 30 site was concluded that the ground water was mostly used as drinking water and sanitation. In term of microbiology, the sampling of ground water quality procedure has followed the SNI [14].

2.4 Modeling Methods

The Geographically weighted regression (GWR) method is a regression model for each site of location. The GWR method would generate a unique models because each site of location is different. This model involves a spatial factor (latitude and longitude coordinate) from each sample. The GWR model is developed by using the software of R 3.2.1 version. The model is presented by formula (2):

$$y_i = \beta_0(u_i, v_i) + \sum_{k=1}^{p} \beta_k(u_i, v_i) x_{ik} + \varepsilon_i \tag{2}$$

The weighted scheme in GWR could apply different method, such as Kernel Gaussian as the most commonly used techniques and the weighted function of bisquare [13] While, the Gaussian function to calculate the closest point –n is:

$$W_{ij} = \exp\left(-\frac{dij/b}{2}\right)^2 \tag{3}$$

Where:

j = the point - n, as the closest site to point - i

b = the distance of point - n as the closest (bandwith)
dij = the Euclidean distance between point - i to point - j

$$dij = \sqrt{(ui-uj)^2+(vi-vj)^2} \quad (4)$$

Bandwidth is the radius of a circle where the point in the radius of circle is still considered influential in forming parameter of location model. One method to determine the value of optimum bandwith is cross-validation (CV) method.

$$CV = \sum_{i=1}^{n}(y_i - \hat{y}_{\neq i}(b))^2 \quad (5)$$

The estimation of regression coefficient in Geographically Weighted Regression (GWR) was conducted by using the Weighted Least Square (WLS). This WLS method provides different weight to each location of observation. Nevertheless, the presence of spatial heterogeneity is one of the requirement of GWR could be applied. The spatial effect test (spatial heterogeneity) referse to the variation in each location. The heterogeneity data could be spatially tested by using statistical method of Breusch-Pagan (BP) with value:

$$BP = \left(\frac{1}{2}\right) f^T X(X^T X)^{-1} X^T f \sim x^2{}_k \quad (6)$$

While the vector element f is

$$f_i = \left(\frac{\varepsilon_i^2}{\sigma^2} - 1\right) \quad (7)$$

Where:
ε_i : least squares residual for observation - i
X : the matrix with size n x (k+1), which is consist of standardised vector (z) for every observation.
The decision making: Ho is rejected if BP $> x^2{}_k$, which means there is spatial heterogenity.

3 Result and Discussion

3.1 Data and research variable

The response variable (Y) is E.Coli bacteria, whereas the predictor variable (X) is the Population density (X1). While the building density is represented by NDBI (X2) and Building Coverage Ratio (X3).

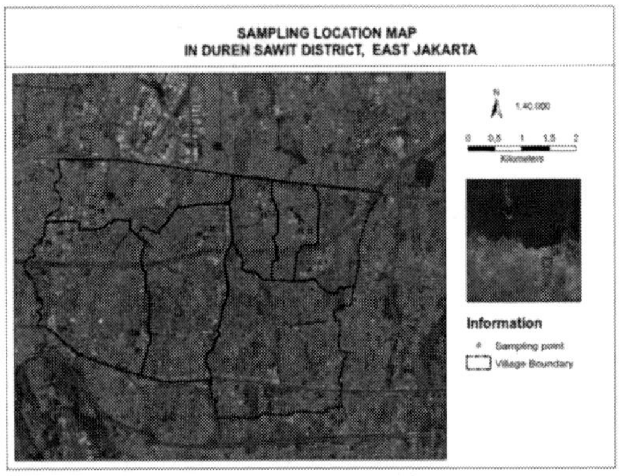

Fig. 1. Sampling Location

3.2 Regression Model

Prior to the analysis of GWR model, the OLS regression model analysis need to be employed by equation:

$$\hat{y} = -87.3594 + 0.1088X1 + 152.9582X2 + 162.6955X3 \quad (8)$$

GWR is frequently used in data analysis related with the spatial heterogeneity. In this study, the value of BP was obtained equal to =34.217>7.815= $X^2_{(0.05;3)}$ (Ho is rejected). It seem that there was a spatial heterogeneity in the model.

3.3 Model GWR

The result of GWR model by using software R 3.2.1 version is as follow:

Table 1. GWR Diagnostic information

Parameter	Values
bandwidth	0.05709193
CV score	22215.88
Number of data point	30
SS GWR residual	15582.71
SS OLS residual	15582.71
Quasi-global R^2	68.53%

Table 1. described the value of Quasi R^2 at 68.53 %, which means that the predictor variable in the model only able to explain the response variable at 68.53%, while 31.47% is explained by other variable outside the model. The GWR model generated the value of SS residual smaller than OLS model. This condition indicated that the GWR model in more appropriate to use than other. Also, the GWR model provided an equation in each sampling location.

Table 2. Estimation of local parameter variable GWR in every point

Location no	$\beta 0$	$\beta 1$	$\beta 2$	$\beta 3$	R^2
1	-80.12	0.0073	229.69	155.23	0.684
2	-80.30	0.0072	229.54	155.44	0.684
3	-80.78	0.0064	227.49	156.49	0.681
4	-79.34	0.0049	223.44	156.28	0.679
5	-78.53	0.0057	224.86	154.92	0.682
6	-78.48	0.0056	224.59	154.94	0.682
7	-82.06	0.0068	228.64	157.45	0.681
8	-81.97	0.0067	228.52	157.39	0.681
9	-81.94	0.0068	228.59	157.35	0.681
10	-81.90	0.0067	228.31	157.38	0.680
11	-81.70	0.0067	228.49	157.14	0.681
12	-81.36	0.0070	229.17	156.61	0.682
13	-81.23	0.0070	229.16	156.48	0.682
14	-81.28	0.0069	228.99	156.58	0.682
15	-80.86	0.0070	229.03	156.14	0.683
16	-80.81	0.0070	229.19	156.06	0.683
17	-80.82	0.0070	229.15	156.08	0.683
18	-83.98	0.0047	223.55	160.98	0.675
19	-83.76	0.0047	223.69	160.71	0.675

Location no	$\beta 0$	$\beta 1$	$\beta 2$	$\beta 3$	R^2
20	-83.77	0.0046	223.43	160.83	0.675
21	-83.52	0.0047	223.61	160.51	0.675
22	-83.62	0.0046	223.44	160.68	0.675
23	-83.46	0.0049	223.97	160.31	0.675
24	-83.27	0.0050	224.32	160.00	0.676
25	-81.91	0.0059	226.35	157.97	0.679
26	-81.25	0.0056	225.48	157.58	0.679
27	-80.78	0.0055	225.18	157.19	0.679
28	-83.71	0.0041	222.64	161.11	0.674
29	-83.31	0.0042	222.79	160.66	0.675
30	-82.94	0.0040	222.45	160.45	0.675

Table 2. showed the the GWR model introduced β for every point which eventually formed the model in every point, thereby it created 30 models. Based on the data above, the water quality variable and the parameter of E. Coli bacteria in entire site is affected by the building coverage ratio variable. Also, from the Table 2 , the GWR model in site-1 in Pondok Kopi village is:

$$\hat{y} = -80.12 + 0.0073X1 + 229.69X2 + 155.23X3 \quad (9)$$

From t test stated that the rising of 100% building coverage ratio would lead to the increase of E.Coli value.

4 CONCLUSION

This study was designed to determine the influence of building coverage ratio, population density and building density toward the ground water quality. The result of this research showed that the GWR approach is more appropriate to build a model of ground water quality in urban area, because this approach has covered the spatial element required in the research (sampling of geographic area). The GWR approach generated a local regression model in every site (in every neighbourhood association) with building coverage ratio as influential variable. In addition, a high level of building coverage ratio approaching 100% become one of the reasons why the septic tank in certain house is built nearby to the other house. This condition increase the possibility that the ground water may be polluted by E.Coli bacteria. Beside, a high level of building coverage ratio impact on the decreasing of rain-water infiltration to the ground, thereby it would not be able to dilute the pollutant E.Coli in the soil. In particular, the local model obtained from each neighbourhood association would provide a recommendation in enforcing the DKI Jakarta Local Regulation No.1 year 2014 related to the arrangement of building coverage ratio in each house in entire areas.

Acknowledgements

The author would like to thank the financial support provided by Universitas Indonesia through the PITTA 2018 funding scheme under grant No. 2422/UN.2.R3.1/HKP.05.00/2018, managed by the Directorate for Research and Public Service and Water resources Management Civil Engineering Universitas Indonesia. Special thanks to Duren Sawit Sub-District, and One-door integrated investment and public services agency Government DKI Jakarta for their permit to survey.

References

1. S. F. Indung, *System design of green open space for green city*. Dissertation IPB (2012)
2. C.L.Arnold and C.J.Gibbons, Impervious surface coverage: the emergence of a key environmental indicator, J. of the American planning Association, **62**, no. 2, 243-258 (1996)
3. C. Belluci, Stormwater and aquatic life: Making the connection between impervious cover and aquatic life impairments for TMDL development in Connecticut streams, in *Proceedings of the Water Environment Federation*(2007), **5**, 1003-1018 (2007)
4. E. Brabec, S. Schulte and P. L. Richards, Impervious surfaces and water quality: a review of current literature and its implications for watershed planning. CPL bibliography, **16**, no. 4, 499-514 (2002)
5. D. Sutjiningsih. Water quality index for determining the development threshold of urbanized catchment area in indonesia, Int. J. of Technology, **8**, no. 1, 143-157 (2017)
6. D. Lu and Q. Weng. Use of impervious surface in urban land-use classification, Remote Sensing of Environment, **102**, no. 1, 146-160 (2006)
7. T. Schueler, Impacts of impervious cover on aquatic systems,Watershed protection research Monograph, **1**, 142 (2003)
8. Q. Chen, K. Mei, R. A. Dahlgren, T. Wang, J. Gong et M. Zhang, Sc. Tot. Env, **572**. , 450-466 (2016)
9. J. Tu and Z.-G. Xia, Examining spatially varying relationships between land use and water quality using geographically weighted regression I: Model design and evaluation, Sc. Tot. Env, **407**, no. 1, 358-378 (2008)
10. B.P.L.H.D. Jakarta, *Laporan Status Lingkungan Hidup Daerah (SLHD) DKI Jakarta*, DKI Jakarta, (2015)
11. B. R. Kosasih, S. and N. I. Astuty, Kualitas Air Tanah di Kecamatan Tebet Jakarta Selatan Ditinjau dari Sebaran Escherichia Coli, J.Teknik Lingkungan Universitas Trisakti, **5**, No. 1, 1-7 (2009)
12. Marsono, *Faktor-faktor yang berhubungan dengan kualitas bakteriologis air sumur gali di permukiman (Studi di Desa Karanganom, Kecamatan Klaten utara, Kabupaten Klaten)*, (2009)
13. A. S. Fotheringham, C. Brunsdon and M. Charlton, *Geographically weighted regression: the analysis of spatially varying relationships*, John Wiley & Sons Ltd (2002)
14. SNI No.6989.58:2008, National Standarization Agency Indonesia (2008)

The design of the botanical indoor air biofilter system for the atmospheric particle removal

Izdihar Zahirah Ibrahim[1,*], *Wen-Tong* Chong[1], and *Sumiani* Yusoff[2]

[1]Department of Mechanical Engineering, Faculty of Engineering, University of Malaya, 50603 Kuala Lumpur, Malaysia
[2]Department of Civil Engineering, Faculty of Engineering, University of Malaya, 50603 Kuala Lumpur, Malaysia

Abstract. Indoor environmental quality (IEQ) objective generally focus on providing energizing and comfortable environments for occupants and minimizing the risk of building-related health problems. Living green walls are natural air-filters that creates a cleaner and revitalizing work environment that will lead to better IEQ. The research presented here describes the design (the new concept) of the botanical indoor air biofilter (BIAB) and modelling conducted to determine the effectiveness of the system in reducing the indoor airborne particulate matter levels. The BIAB was also evaluated for its single-pass filtration for particles ranging in diameter from 2.5 to 10 μm along with total suspended particles. The system is comprised of three functional components; a region of vertically grown plants as botanical section, an evaporative cooling pad as cooling section (additional section from a commercial BIAB), and a mechanical ventilation system that supply cool filtered air to surrounding. The complete system recorded highest removal efficiencies of 85% for TSP, 75.2% for $PM_{2.5}$, and 71.9% for PM_{10}. It indicated that with the additional component in the BIAB system (cooling component), it provides enhancement of the particulate removal due to the ability in absorbing the dust particles and filtration dynamics as the polluted air pass through the wetted cooling pad and the light shower of water.

1 Introduction

People are frequently concerned that they will have unhealthy symptoms or health conditions from exposures to the contaminants in the buildings where they spend time for a long period. People's satisfaction with the indoor environmental quality (IEQ) is important because it influences productivity and health [1]. Indoor air quality and thermal comfort are the important main subjects to consider of living indoor [2]. The combination of high temperature and high relative humidity serves to reduce thermal comfort and indoor air quality [3]. Most of the people who works in the office within 8 hours are more productive when they are satisfied with their indoor environment [4].

Thus, to maintain the comfort of the indoor environment, many buildings normally use an air conditioning and mechanical ventilation (ACMV) system. This solution requires high levels of energy consumption and also contribute to one of a major environmental issue, such as increasing emissions of greenhouse gases to the atmosphere that causes global warming. In addition, even in buildings with efficient ACMV systems, fine particulate matter (PM) might infiltrate into indoor environments as it cannot be effectively filtered out of the incoming air [5]. Indeed, the indoor air quality are affected from the concentration of outdoor pollutants and most of the large proportion of the indoor particles are from outdoor [6]. In addition, even with the filtration process from the portable indoor air filters, it cannot ensure the effectiveness of the particle removal will stand for a long period [7].

In this circumstance, the practice of the indoor greening system alone or in combination with conventional ACMV systems in buildings may provide environmental value and in addition contribute to reduce energy consumption [8]. By employing a greening system in the interior building, the effect on indoor air temperature and humidity are significant [9]. In the trend of indoor greening, green wall technology has led to the development of activated systems, termed botanical biofilters, that move air through the plant growth substrate to increase the rate at which the interior atmospheric environment is exposed to the components of the plant-substrate system that are active in air pollutant removal [10]. The majority of research that has been conducted on green wall had focused on defining it as botanical indoor air biofilter (BIAB) that can remove VOC, CO2 and particulate matter [10-13]. For example, Irga et al. [10] determined the effectiveness of BIAB in reducing $PM_{2.5}$ and PM10 that evaluated with single-pass filtration efficiency and compare with portable in-room air cleaner.

The research presented here describes the design (the new concept) of the botanical indoor air biofilter (BIAB) and modelling conducted to determine the effectiveness of the system in reducing particulate matter levels. The BIAB was also evaluated for its single-pass filtration for particles ranging in diameter from 2.5 to 10 μm along with total suspended particles.

2 BIAB system design

The study uses a new concept of a mini type active green wall (desktop size), see Fig. 1. Briefly, the system is comprised of three functional components; a region of vertically grown plants as botanical section, an evaporative cooling pad as cooling section (additional section from a commercial BIAB), and a mechanical ventilation system that supply cool filtered air to surrounding. The system consists of a 0.05 m^3 (38 x 38 x 34 cm) polyethylene module with 9 holes on the front face which plants grow horizontally from the circular compartment in the casing. Each hole in the module has 27.3% of the front face allowing air to pass. For these experiments, each circular compartment in the hole contains *Epipremnum aureum* (Golden photos) as the botanical component and Kenaf fibre as growth medium in the system. The cooling sections in these system uses a corrugated paper type of cooling pad (1.2 m^2 with 5 cm thickness) and a submersible mechanical water pump (8 watt) to wetted the cooling pad, at the same time function as water irrigation system in the module. At the back face of the module consists 8 no. of axial brushless fan (DC 12-volt, 1.44 watt in 80 x 80 x 15 mm casing size) with total airflow rate 150 $L.s^{-1}$ and function to draw air from the indoor space through the botanical component, within the growth medium, cross the moistened pad and back into indoor surrounding.

Fig. 1. (a) The front view of BIAB system module; (b) the side view.

3 Methodology

3.1 Single pass efficiency test of the BIAB

The study described by Irga et al. [14], sealed tight acrylic Perspex test chamber (0.6 x 0.6 x 0.6 m; 216 L with 10 mm thk.) was used in these experiments. At the one side of the chamber was designed to be a detachable door (sealed with adhesive foam rubber and adjustable metal clamps) to allow the placement of the module into the chamber and at the opposite side, flexible ducting (13 cm diameter) was fixed to the centre of the chamber (see Fig.2). The fitted ducting led to a combustion chamber in which PM was generated by burning an 80-gsm square paper (210 x 297 mm) with a 4 mL of filtered retail-grade diesel fuel (Petronas).

The generated PM flowed through the fitted ducting with active airflow provided by an axial brushless fan (DC 12-volt, 1.8 watt in 120 x 120 x 25 mm casing size) housed in the centre of the chamber (fitted with the ducting), before flowing through the BIAB module where pollutant air is distributed through the front face of the module (botanical section). A fan within the test chamber helped to spreading of the filtered airflow throughout the chamber to reduce precipitation of particles before exhaust into another ducting system (vacuum exhaust) that fixed to the detachable door of the chamber, which led to a tube of laser nephelometer (TSI Dustrak II 8531) to record average particle density and size distribution of the filtered airstream. After each sampling, the air was exhausted to spare via a vacuum exhaust pump.

To compare the efficiency of the system, the experiment was conducted in three different variations; firstly, to test a new concept of BIAB complete system (botanical and cooling component), secondly to test only cooling component without botanical component and lastly to test only botanical component without cooling component. For each variation, average PM concentration was recorded for three particle sizes; total suspended particle (TSP), $PM_{2.5}$ and PM_{10}. Duplicate tests were conducted for each experimental condition and all of the replicate trials were recorded for 10 min, which was sufficient time for the PM concentration to return to ambient levels for all treatments.

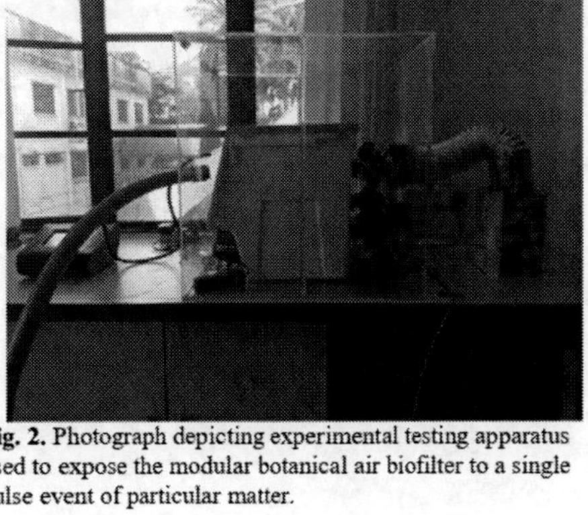

Fig. 2. Photograph depicting experimental testing apparatus used to expose the modular botanical air biofilter to a single pulse event of particular matter.

3.2 Data analysis

The analysis presented by Irga et al. [10], real-time concentration data from each single pass efficiency test was used to construct concentration versus time profiles. In order to account for the differences in pressure created by the different treatments, as well as the residence time distribution (RTD) differences (which is the amount of time a pollutant that is not filtered spends within the system until reaching the detector), the area under the concentration versus time data was calculated and subsequently used as the dependent variable.

Single pass filtration efficiency (η) was calculated using Equations (1) [15]:

$$\eta = \left(\frac{C_{in} - C_{out}}{C_{in}}\right) \times 100\% \quad (1)$$

where:

C_{in} = integration of the average particle concentration decay curve in the duct with no application of treatment

C_{out} = integration of the average particle concentration decay curve in the duct with application of treatment

Particle resolved CADR derived from the single pass efficiency data CADR was calculated utilizing Equation (2) [15]:

$$CADR = \eta \times G \quad (2)$$

where:

η = the single pass efficiency of the system
G = airflow rate through the system [$L.s^{-1}$]

4 Result and discussion

Single pass removal efficiencies for three particle sizes; total suspended particle (TSP), $PM_{2.5}$ and PM_{10} with across three different variances are shown in Fig. 3, and a comparative summary of treatment CADR values derived from η for the particle fractions measured is presented in Table 1.

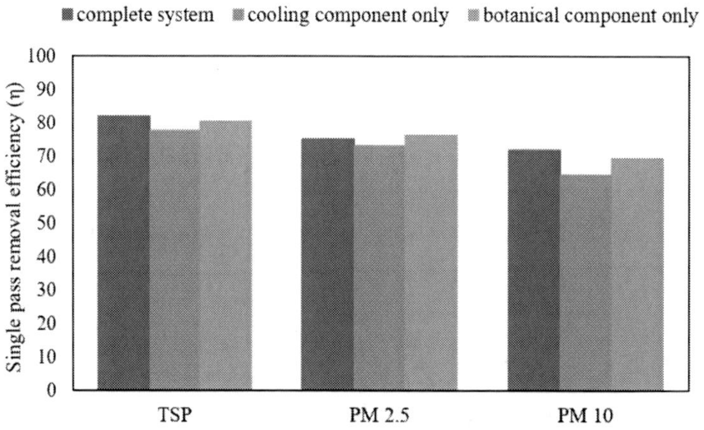

Fig. 3. Calculated single-pass removal efficiency (η) for TSP, $PM_{2.5}$ and PM_{10}.

Table 1. CADR ($L.s^{-1}$) calculated from single pass efficiency tests, when air flow is set to maximum.

Pollutant	Complete system	Cooling component only	Botanical component only
TSP	123.00	116.40	120.72
$PM_{2.5}$	112.80	109.66	114.52
PM_{10}	107.88	96.86	104.17

Removal was achieved in all treatments, and it can be seen that there were strong similarities in response across the three particle sizes tested. As expected, the new concept of BIAB complete system module treatment showed the highest removal rates, demonstrating removal efficiencies of 85% for TSP, 75.2%. for $PM_{2.5}$, and 71.9% for PM_{10}. Comparatively, the system that tested only with cooling section and without the botanical component present recorded removal efficiencies of 77.6% for TSP, 73.1% for $PM_{2.5}$, and 64.57% for PM_{10}, which were significantly lower than the removal efficiencies recorded for the complete system module. Additionally, when the system was tested with only botanical component present and without cooling component, removal efficiencies were 80.48% for TSP, 76.3% for $PM_{2.5}$, and 69.4% for PM_{10}.This study provides data for the characterisation of the removal efficiency of atmospheric particles by a new concept of botanical indoor air biofiltration (BIAB) system. We state that a higher removal of particulates from air can be achieved by ventilation of the polluted airstream through the two stages of the particle removal (botanical and cooling component) by the BIAB tested.

Recent study of active living wall focusing on the effect of PM removal efficiency, Irga et al. [10] is the first to document the CADR value for PM. The system is slightly different to the system presented here, as it consists of plant with packing media (plant-substrate system), where the air move through the plant growth substrate to increase the rate of PM removal efficiency. *Chlorophytum comosum* (Spider plant) being used as the botanical component in that system. Irga et al. [10] demonstrated that their system had removal efficiencies 53.35 ± 9.73% for TSP, 53.51 ± 15.99% for PM_{10}, and 48.21 ± 14.71% for $PM_{2.5}$, which were significantly lower than the removal efficiencies recorded for the HVAC filter at an air flow rate of 11.25 L s^{-1}. Additionally, Irga et al. [19] defined the CADR value of their ALW slightly less efficient than HVAC filter, with a total value of 28.70 m^3 h^{-1} for PM_{10} and 25.86 m^3 h^{-1} for $PM_{2.5}$.

Comparatively, the removal efficiencies recorded here are significantly higher than the results in Irga et al. [10], the experimented BIAB system have a higher airflow rate than their tested botanical biofilter module. Furthermore, this could be due to the additional cooling component inside the system and plant species differences. The PM real-time reduction appears likely that the additional cooling component in the system has an effect on absorbing the dust particles and filtration dynamics as the polluted air pass through the wetted cooling pad and the light shower of water.

5 Conclusion

The research presented here quantifies the additional functionality (new concept) of a botanical indoor air biofilter system, with regards to the removal of particulate matter from air. The complete system recorded highest removal efficiencies of 85% for TSP, 75.2%. for $PM_{2.5}$, and 71.9% for PM_{10}. It indicated that with the additional component in the BIAB system (cooling component), it provides enhancement of the particulate removal due to the ability in absorbing the dust particles and filtration dynamics as the polluted air pass through the wetted cooling pad and the light shower of water. The future studies are to focus on system's ability to remove VOCs and CO_2 and improve temperature and humidity of the indoor environment. Nonetheless, further controlled laboratory experiments are needed to investigate the long-term performance of the system, and to better describe the simultaneous removal of PMs, VOCs and CO_2. These investigations will provide empirical data on which to develop a simulation model that can be used to optimize the system's design, as well as to advance the implementation of the device.

The authors would like to thank the University of Malaya for the research grants allocated under the Sustainability Science (SuSci) Research Cluster, UM Living Lab Grant Programme (LL019-16-SUS).

References

[1] S. Kang, D. Ou, C.M. Mak, The impact of indoor environmental quality on work productivity in university open-plan research offices, Building and Environment (2017).
[2] J. Kim, T. Hong, J. Jeong, C. Koo, M. Kong, An integrated psychological response score of the occupants based on their activities and the indoor environmental quality condition changes, Building and Environment 123 (2017) 66-77.
[3] L. Fang, D.P. Wyon, G. Clausen, P.O. Fanger, Impact of indoor air temperature and humidity in an office on perceived air quality, SBS symptoms and performance, Indoor Air 14 Suppl 7 (2004) 74-81.
[4] C. Huizenga, S. Abbaszadeh, L. Zagreus, E.A. Arens, Air quality and thermal comfort in office buildings: Results of a large indoor environmental quality survey, 2006.
[5] R.B. Mosley, D.J. Greenwell, L.E. Sparks, Z. Guo, W.G. Tucker, R. Fortmann, C. Whitfield, Penetration of Ambient Fine Particles into the Indoor Environment, Aerosol Science and Technology 34(1) (2001) 127-136.
[6] R. Meier, C. Schindler, M. Eeftens, I. Aguilera, R.E. Ducret-Stich, A. Ineichen, M. Davey, H.C. Phuleria, N. Probst-Hensch, M.Y. Tsai, N. Kunzli, Modeling indoor air pollution of outdoor origin in homes of SAPALDIA subjects in Switzerland, Environ Int 82 (2015) 85-91.
[7] M.S. Zuraimi, M. Vuotari, G. Nilsson, R. Magee, B. Kemery, C. Alliston, Impact of dust loading on long term portable air cleaner performance, Building and Environment 112 (2017) 261-269.
[8] B. Raji, M.J. Tenpierik, A. van den Dobbelsteen, The impact of greening systems on building energy performance: A literature review, Renewable and Sustainable Energy Reviews 45 (2015) 610-623.
[9] D. Tudiwer, A. Korjenic, The effect of an indoor living wall system on humidity, mould spores and CO2-concentration, Energy and Buildings 146 (2017) 73-86.
[10] P.J. Irga, N.J. Paull, P. Abdo, F.R. Torpy, An assessment of the atmospheric particle removal efficiency of an in-room botanical biofilter system, Building and Environment 115 (2017) 281-290.
[11] Z. Wang, J.S. Zhang, Characterization and performance evaluation of a full-scale activated carbon-based dynamic botanical air filtration system for improving indoor air quality, Building and Environment 46(3) (2011) 758-768.
[12] A.B. Darlington, J.F. Dat, M.A. Dixon, The Biofiltration of Indoor Air: Air Flux and Temperature Influences the Removal of Toluene, Ethylbenzene, and Xylene, Environmental Science & Technology 35(1) (2001) 240-246.
[13] D. Llewellyn, M. Dixon, 4.26 - Can Plants Really Improve Indoor Air Quality?, in: M. Moo-Young (Ed.), Comprehensive Biotechnology (Second Edition), Academic Press, Burlington, 2011, pp. 331-338.
[14] T. Pettit, P.J. Irga, P. Abdo, F.R. Torpy, Do the plants in functional green walls contribute to their ability to filter particulate matter?, Building and Environment 125(Supplement C) (2017) 299-307.
[15] U.S. EPA, Evaluation of In-Room Particulate Matter Air Filtration Devices, U.S. Environmental Protection Agency, Washington, DC, 2008.

* Corresponding author: izdiharzahirah@gmail.com

Fabrication of polyaniline coated conductive cotton for ammonia gas detection

*M*on Myat Swe[1], *T*anthip Eamsa-ard[1], and *T*eerakiat Kerdcharoen[2,*]

[1] Materials Science and Engineering Programme and Center of Nanoscience and Nanotechnology,Faculty of Science. Mahidol University, Bangkok, Thailand
[2] Department of Physics, Faculty of science, Mahidol University and NANOTEC Center of Excellence at Mahidol University, National Nanotechnology Center, Bangkok, Thailand

Abstract—Nowadays, ammonia becomes major toxic pollutant. The industrial zones, agricultural zones and mining zones are facing this problem all over the world. Ammonia gives irritating strong smell in the environment and it can affect on the health of human with a long exposure time. Moreover, ammonia is also the biomarker which can indicate the health status of human. For these situations, many ammonia gas sensors are produced. In this work, flexible ammonia gas sensor is prepared based on polyaniline through in-situ polymerization process. We developed a wearable smart textile sensor for ammonia gas detection. 100% cotton which does not irritate the wearer was used for this sensor. The polyaniline-cotton composite was characterized by optical microscope, SEM and FTIR spectroscopy. The morphology of the composites reveal PANI was successfully coated on the cotton. The sensitivity of the gas sensors was monitored directly increasing the concentration of ammonia gas. The resistance changes of the sensors can be seen with increasing the concentration. It could also be proved that the sensor is flexible with bending test. In the future, this flexible PANI-cotton composite sensor can be used for ammonia detection in real applications.

1.Introduction

Ammonia (NH_3) is one of the major toxic pollutants while air pollution is the huge impact on the world today. Ammonia is widely used in chemicals, materials processing, refrigeration and fertilizers etc [1]. In chemical industry, ammonia is used as a main purpose for the production of nitric acid. Because of its ability to cool below 0° C, almost all refrigeration facilities use in mechanical system and food processing make use of ammonia [2,3]. In agriculture sector, farms become to use a lot of fertilizers and animals manures rich of nitrogen based on compounds like ammonia for plant growth and higher yield of fruit and seed production. When fertilizer or manure is spread over the farmland, the smell of ammonia causes the farm very unpleasant smell [4]. Ammonia is the source for ammonium nitrate based on explosives used in mining operations and sodium cyanide (NaCN) used in gold extraction from sulphide ores. On the other hand, creating a safe and cool environment for miners is one of the most important factors. For this reason, ice made by refrigeration is sent down the mine. In that way, the cold melt water is circulated through air cooler and provide cold environment [5].

The long terms exposure of ammonia can cause effect on our eyes, nose, mouth, lungs and throat. Our cardiac system and respiratory system can be damaged by the ammonia in the environment and it might lead to death [6-7]. For dangerous ammonia concentration, a detecting and warning alarm system should be applied in these industries [4].

Ammonia (NH_3) is considered to an important biomarker as it plays a significant role in human body [8,9].

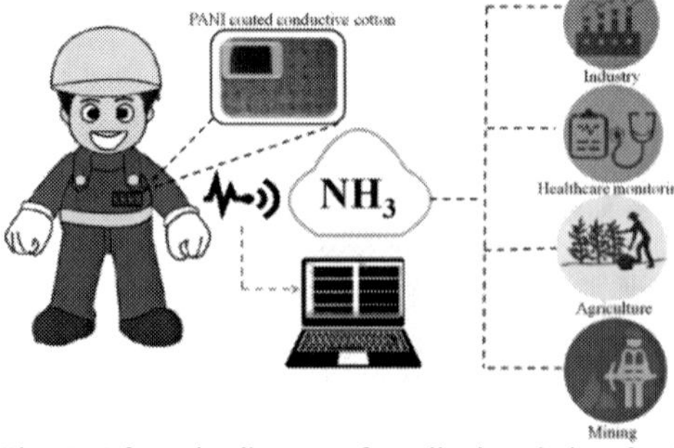

Fig. 1 Schematic diagram of application design for PANI coated conductive cotton as ammonia gas sensor

*Corresponding author: teerakiat@yahoo.com

Human body excretes ammonia in the form of urea and ammonium salt in urine, some ammonia through sweat glands and breath [10,11]. Patients with kidney failure and ulcers caused by Helicobacter pylori bacterial stomach infection have unbalanced ammonia in their urine and breath [12-14].

In order to detect ammonia (NH_3), sensors with great efficiency and quick response are required. Electronic textiles (e-textiles) can combine the functionality of smart electronic devices with the comfort and the flexibility of stylish clothing [15]. Polyaniline (PANI) is an extraordinary material to work as an active layer in gas sensing applications. Polyaniline (PANI) is the most appropriate conducting polymer for detecting ammonia (NH_3). Polyaniline (PANI) is the most appropriate conducting polymer for detecting ammonia (NH_3), due to its high conductivity, stability, easy synthesis, low cost and high reactivity with gases, such as ammonia [16]. In this work, it aims to fabricate conductive cotton based gas sensor for ammonia detection in industry, agriculture, mining and health care monitoring. Polyaniline (PANI) coated conductive cotton was prepared by in situ polymerization.

2. Experimental

2.1 Reagents and Materials

Aniline hydrochloride 97% was obtained from Aldrich, and used as received. Ammonium peroxy-disulfate, analytical grade, from Ajax Finechem Pty Ltd; was used. Sodium hydroxide (NaOH) was from RCL Labscan Limited. Undyed 100% cotton was employed as substrates in the experiments.

2.2 Preparation of PANI coated conductive cotton

Fig. 2 shows the fabrication of PANI coated conductive cotton. 100% cotton was prepared into five different sample sizes (1cmx1cm, 2cmx2cm, 3cmx3cm, 4cmx4cm and 5cmx5cm). The cotton samples were dipped in aqueous 5 M NaOH solution for 5 minutes for mercerizing, rinsed twice and dried before PANI coating.

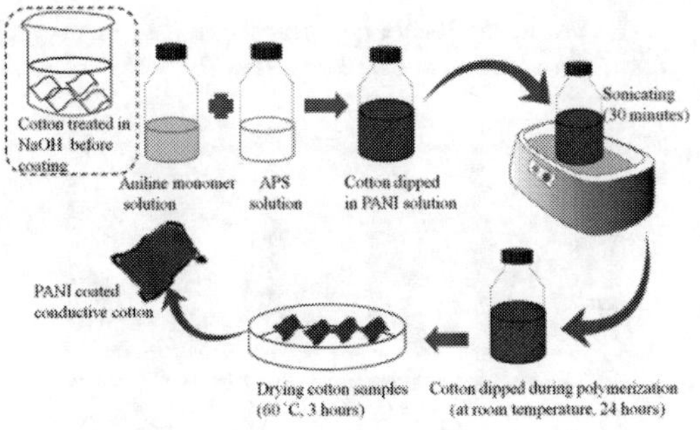

Fig. 2 Schematic diagram shows fabrication of PANI coated conductive cotton

PANI solution is made with monomer/oxidant molar ratio 1:2. Aniline hydrochloride (1.296 g) was dissolved in 50 mL DI water in a flask. Next, ammonium peroxydisulfate (APS) (2.738 g) was dissolved in 50 mL DI water. The solutions were mixed in a bottle and the cotton pieces were immersed in the mixed solution at that time. The mixture was sonicated for 30 minutes and left at room temperature for 24 hours to polymerize. After 24 hours, the cotton pieces were taken out of the solution and dried by heating in the oven at 60 C for 3 hours

3. Results and Discussions

3.1 Optical microscope

Fig. 3 shows optical microscope images of PANI coated cotton with different lengths (1cm-5cm). It can be observed the interlaced structure of warp and weft yarn clearly from the low magnification version of optical microscope images.

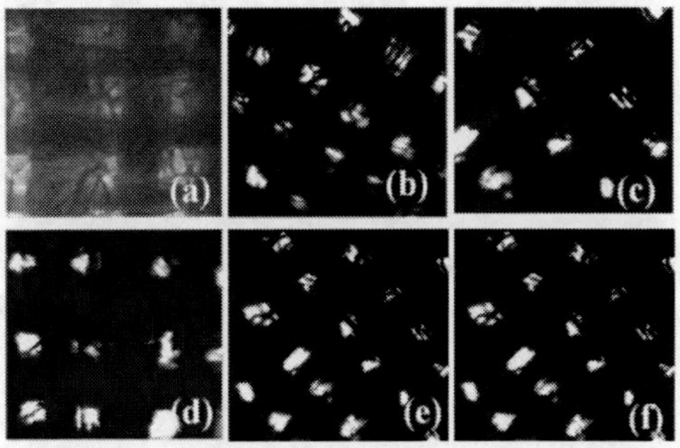

Fig. 3 Optical microscope images of (a) normal cotton and PANI coated conductive cotton (b) 1cmx1cm (c) 2cmx2cm (d) 3cmx3cm (e) 4cmx4cm (f) 5cmx5cm

The deposited PANI causes the white cotton surfaces turned to dark green. Therefore, the polyaniline coated cotton in figure (b) to (f) tends to dark green colour indicating the well dispersion of PANI. It was shown that PANI can be successfully deposited on the cotton surface by in-situ chemical polymerization of aniline. Furthermore, polyaniline deposition process is the convenient way to reduce the surface hairiness of cotton as shown in the optical microscope images of PANI coated conductive cotton.

3.2 Scanning Electron Microscope (SEM)

The SEM images of the PANI coated cotton samples were taken at different magnifications with 100 X (Fig. 4) and 1000 X (Fig. 5). In SEM image of untreated cotton, it shows a very smooth surface. The presence of PANI can clearly be seen on PANI cotton. The non-smooth surface in figures shows evidently that PANI did not cover uniformly. It can still be seen non-uniform PANI coated areas although some areas of the fabric were fully coated. However, PANI perforation in the fabric yarn provided the cotton to be conductive.

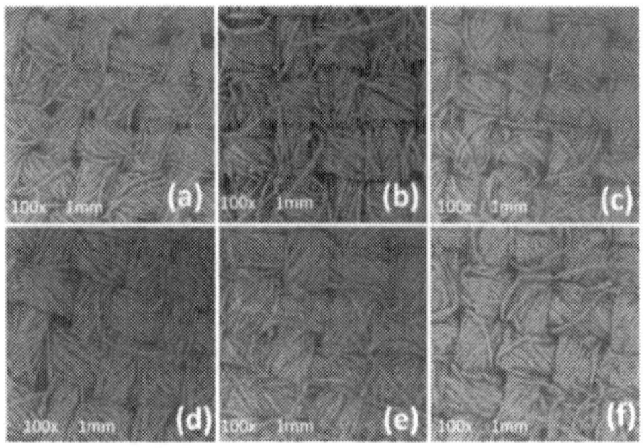

Fig. 4 Scanning Electron Microscope (SEM) images of (a) untreated cotton and PANI coated cotton (b) 1 cmx1cm (c) 2cmx2cm (d) 3cmx3cm (e) 4cmx4cm (f) 5cmx5cm

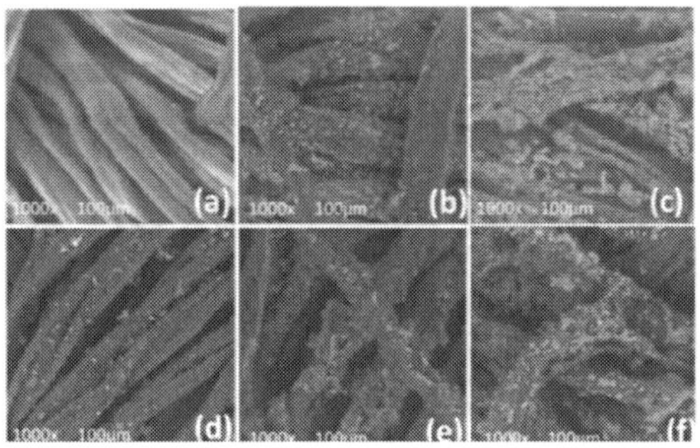

Fig. 5 Scanning Electron Microscope (SEM) images of (a) untreated cotton and PANI coated cotton (b) 1 cmx1cm (c) 2cmx2cm (d) 3cmx3cm (e) 4cmx4cm (f) 5cmx5cm

3.3 FTIR Spectroscopy

The FTIR spectra for untreated cotton and PANI coated cotton with different ranges which were recorded in the range of 400-4000 cm^{-1} are shown in Fig.6. The peak at 3337.4 cm^{-1} in the spectrum of untreated cotton was characteristics of CH_2 antisymmetric stretching vibrations of the secondary CH_2OH groups present in the glucose units of cellulose[17]. At this area of spectra of PANI coated cotton samples, the absence of this peak may be due to the decreasing of secondary CH2OH groups after PANI inclusion. At 2837.4 cm^{-1} of untreated cotton, the peak was CH_2 symmetric stretching band.

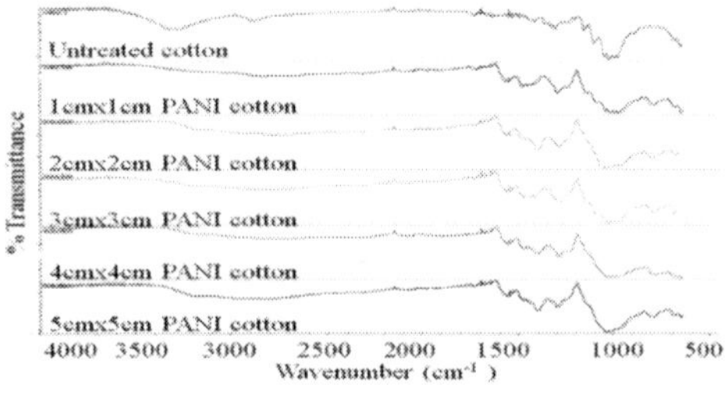

Fig. 6 FTIR spectra of untreated cotton and PANI treated cotton samples

The spectra of the PANI cotton with different ranges were the same as all of the samples were prepared under the same condition. The characteristic of PANI peaks were observed in the spectra of PANI coated cotton. The peak at 1403.9 cm^{-1} was attributed to the C=N stretching modes of quinoid rings and 1305.1 cm^{-1} peak represented to the C-N stretching of benzenoid rings. The N=Q=N stretching of the quinonoid units of PANI was observed at 1148cm^{-1}. This peak was due to electron delocalization. There was still small peaks in the PANI coated cotton spectra. It means that all of CH_2OH groups in the glucose units of cellulose coud not interact completely with conductive PANI.

3.4 Measuring the sensitivity of gas sensors

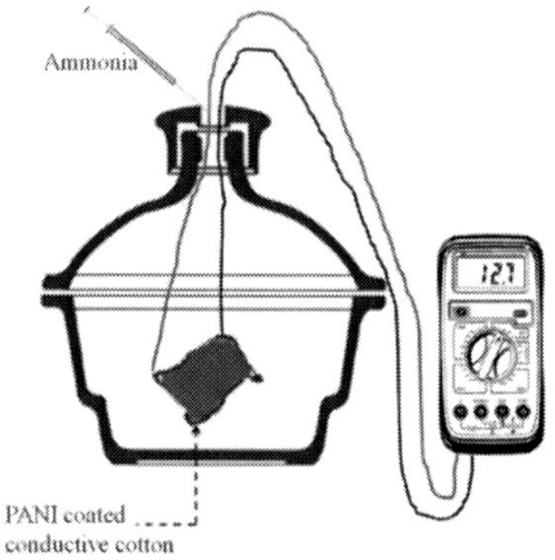

Fig. 7 Schematic measurement system for NH_3 detection

For measuring gas sensors performance, the measurement system for NH_3 gas detection is as shown in Fig. 7 The specially prepared gas chamber (24000 cm3) was used. The PANI coated cotton was put inside the gas chamber and the two edges of the cotton were attached with the clips of digital multimeter. In order to get the desired gas concentration for measuring, the measured amount of ammonia gas was injected with a syringe. The changing resistance values of the PANI coated cotton were monitored continuously using a digital multimeter.

Fig. 8 shows the sensitivity of PANI coated cotton gas sensor to ammonia gas from 25 ppm to 100 ppm. The sensitivity of the composite gas sensor is defined as

$$\% S = [(R - R_0) / R_0] \times 100 \quad (1)$$

where R and R_0 are the resistances of the sensor after and before VOCs exposure, respectively.

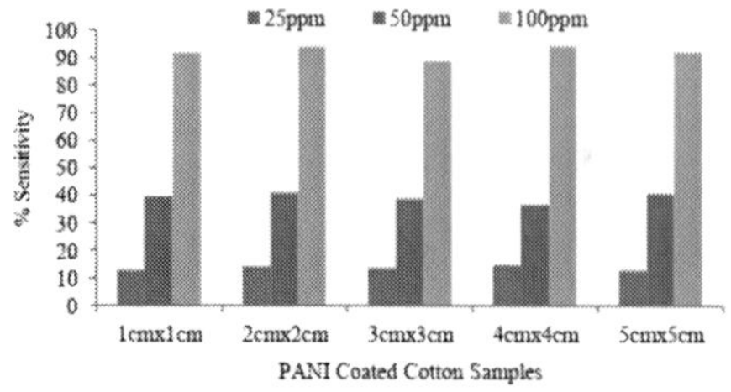

Fig. 8 The sensitivity of PANI coated cotton gas sensor

The results reveal that the resistance values of PANI coated cotton changed according to the concentrations of the ammonia gas. The resistance increased with the increasing concentration (25ppm-100 ppm). However, the length of the cotton cannot affect on the sensitivity of the sensors. There is not too much difference in sensitivity of the PANI cotton with various lengths.

When the ammonia gas passes through the sensor,the interactions between the reactive sites of PANI and ammonia molecules occur. As ammonia has electron donating nature, it accepts hydrogen from PANI backbone and form ammonium ion. So, PANI looses H^+ ion and its electron density increases. In that way, the resistance of the gas sensors become increased.

3.5 Flexibility Experiment

Table I. The resistance of PANI coated cotton gas sensor sample in various condition.

Cotton samples	Initial State	Bending	Normal
1cmx1cm	43 kΩ	40 kΩ	43 kΩ
2cmx2cm	48 kΩ	45 kΩ	48 kΩ
3cmx3cm	24 kΩ	20 kΩ	24 kΩ
4cmx4cm	19 kΩ	16 kΩ	19 kΩ
5cmx5cm	31 kΩ	28 kΩ	31 kΩ

4. Conclusions

PANI-cotton composite gas sensors have been successfully prepared with in-situ polymerization process. The morphology of the PANI-cotton showed that PANI can be successfully deposited on the cotton. The PANI based gas sensor can be operated efficiently at room temperature to detect ammonia with various concentrations. Varying the length of the cotton cannot affect on the sensing performance of the sensor. We hope that this wearable smart textile sensor will be very useful in real applications of ammonia gas detection in industry, agriculture and mining areas and for providing health status of the patients with kidney failure and stomach infection.

Acknowledgment

Mahidol University and National Nanotechnology Center are acknowledged.

References

[1] M. Das, D. Sarkar, "One-pot synthesis of zinc oxide - polyaniline nanocomposite for fabrication of efficient room temperature ammonia gas sensor," Ceramics International, **43**, pp. 11123–11131, (2017).

[2] A.T. Bulgan, "Use of low-temperature energy sources in aqua-ammonia absorption refrigeration systems," Energy Conversion and Management, **38**, pp. 1431–1438, (1997).

[3] P. Colonna, S. Gabrielli, "Industrial trigeneration using ammonia-water absorption refrigeration systems (AAR)," Applied Thermal Engineering, **23**, pp. 381–396, (2003).

[4] B. Timmer, W. Olthuis, A. van den Berg, "Ammonia sensors and their applications—a review," Sensors and Actuators B, **107**, pp. 666–677, (2005).

[5] J. Jermakka, L. Wendling, E. Sohlberg, H. Heinonen, E. Merta, J. Laine-Ylijoki, T. Kaartinen, and U.M. Mroueh, "Nitrogen compounds at mines and quarries", VTT Technology, **225**, (2012).

[6] T.R. Burns, S.D. Greenberg, M.L. Mace, J.A. Jachimczyk, "Pulmonary ultrastructure in acute ammonia toxicity in humans," Archives of Environmental & Occupational Health, pp. 127-180, (1983).

[7] M. Caplin, "Ammonia-gas poisoning: forty-seven cases in a London shelter," The Lancet, pp. 95-96, (1941).

[8] S. Van de Velde, F. Nevens, P. Van hee, D. Steenberghe, "GC–MS analysis of breath odor compounds in liver patients," Chromatography B, **875**, pp. 344-348, (2008).

[9] S.J. Davies, P. Spanel, D. Smith, "Breath analysis of ammonia, volatile organic compounds and deuterated water vapor in chronic kidney disease and during dialysis," Bioanalysis **6**, (2014).

[10] N.A. Campbell, J.B. Reece, "Biology," Pearson Education Inc., (2002).

[11] W. Ament, J.R. Huizenga, E. Kort, T.W.van der Mark, R.G. Grevink, G.J. Verkerke, "Respiratory ammonia output and blood ammonia concentration during incremental exercise," Sports Medicine, **20**, pp. 71–77, (1999).

[12] L.R. Narasimhan, W. Goodman, C. Kumar, N. Patel, "Correlation of breath ammonia with blood urea nitrogen and creatine duringhemodialysis," Proceedings of the National Academy of Sciences of the United States, **8**, pp. 4617–4621, (2001).

[13] B. Marshall, J.R. Warren, "Unidentified curved bacillus and gastric epithelium in active chronic gastritis," The Lancet, **1**, pp. 1273–1275, (1993).

[14] D.J. Kearney, T. Hubbard, D. Putnam, "Breath ammonia measure-ment inHelicobacter pyloriinfection," Digestive Diseases and Sciences, **47**, pp. 2523–2530, (2002).

[15] T. Seesaard, P. Lorwongtragool, S. Season, T. Kerdcharoen, "A novel creation of thread-based ammonia gas sensors for wearable wireless security system," International Conference on Electrical Engineering/Electronics, Computer, Telecommunications and Information Technology (ECTI-CON), IEEE, (2014).

[16] A. Kukla, Y. Shirshov, S. Piletsky, "Ammonia sensors based on sensitive polyaniline films," Sensors and Actuators B, **37** pp. 135–140, (1996).

[17] R.T. O'Connor, "Instrumental Analysis of Cotton Cellu-lose and Modified Cellulose," Marcel Dekker, Newyork, (1972).

MATEC Web of Conferences 192, 02037 (2018) https://doi.org/10.1051/matecconf/201819202037
ICEAST 2018

Prequalification criteria, according to the government procurement and supplies management Act, B.E. 2560 (2017): branch/type, class, right to bid, ability to carry out

Boonruk Vanborsel[1,*], and *Kumpon* Subsomboon[1]

[1]Civil Engineering Department, Faculty of Engineering, Naresuan University, Phitsanulok, Thailand

Abstract. The objective of this article is to study new bidding prequalification criteria of the Royal Thai government sectors according to the Government Procurement and Supplies Management Act, B.E. 2560 (2017) and compare with the seven government agencies according to Regulations on Procurement, 1992 by emphasizing branch/type of construction, class of the construction entrepreneur/contractor, right to bid per one contract, and number of project to carry out on concurrent time. Content analysis was used in this research, by studying and analysing the prequalification criteria of the Government Procurement and Supplies Management Act, B.E. 2560 (2017) comparing to Regulations on Procurement, 1992 The findings of prequalification criteria concluded that Act, 2017 defined 7 branches/types of construction for construction entrepreneur whereas building construction, road maintenance, and traffic safety were repealed. The right to bid per one contract volume lower than 5 million baht are allowed. There is no limit for the number of project and total project value on concurrent period comparing with Regulations, 1992.

1 Introduction

In the past decade, a contracting for works of Thai government generally, carried out according to Regulations of the Office of the Prime Minister on Procurement B.E. 2535 (Regulations, 1992) that applied to government agencies, while Regulations of the Ministry of Interior's on procurement of Local Government Administration B.E. 2535 (Regulations, 1992) applied only government body established under the law on local administration. The significant principal as regards the procurement, that is procurement officer of each step of procurement must perform by openness, transparency and giving an opportunity to have fair price competition. Consideration will be given to bidders' qualification and capability and specified that any government agency or local administration organization who requires making a pre-qualification for contracting for works can be performed only in the case which necessitates limiting for the qualified bidders. Contractor prequalification is the most effective way to support client's decision in order to select contractors from the short list of qualified contractors [1]. Selecting the right contractor for the right project is the most crucial challenge for any construction client [2], however contracting for works of government construction projects had many problems: like, co-interest bidder, action impeding fair and free competition, corruption, and, moreover the important agencies for example; local administration organization, office of the basic education commission, and so on did not make a list of pre-qualified bidder. As a result, bidding process occasionally got unqualified, lack of capacity, and availability contractor to carry out constructions. It leads to construction delay, low quality, and abandonment of construction projects. In 2017, procurement in Thai government had a big change: by repealing all of the traditional Regulations, 1992 and enforce "Government Procurement and Supplies Management Act, B.E 2560" (Act, 2017) [3] for all government sector using same standard framework, transparency/disclosure, accountability, fairness competition, and value for money. For this reason, this article aims to study prequalification criteria of Thai government sector according to the new Act, 2017 and compare with seven government agencies which make a list of pre-qualified bidder according to the traditional Regulations, 1992 by emphasizing branch/type of construction, /class of the construction entrepreneur/contractor, right to bid per one contract, and the number of projects to carry out on concurrent time.

2 Methodology

Contents analysis was used in this research to study and analyze the prequalification criteria of (1) Thai government sector which make a list of construction entrepreneur according to Act, 2017 and (2) the seven government agencies: Department of Public Works and Town Planning (DPW&TP), Department of Highways

* Corresponding author: boonruk vanborsel@yahoo.com

(DOH), Department of Rural Roads (DRR), Marine Department (MD), Bangkok Metropolitan Administration (BMA), Royal Irrigation Department (RID), and Department of Water Resources (DWR); which make a list of pre-qualified bidder according to the Regulations, 1992 by emphasizing branch/type of construction, class of the construction entrepreneur/contractor, right to bid per one contract, and the number of projects that ability to carry out on concurrent time. As a result of content analysis, the differences of prequalification criteria between Act, 2017 and Regulations, 1992 were compared.

3 Results

3.1 Branch/Type of construction

Table 1 shows Act, 2017 defined branch of construction that should make a list of construction entrepreneur for all government sector were specified into 7 branches: (1) road construction and special class, (2) bridge construction and tunnel/underpass construction and special class, (3) irrigation construction, (4) bank protection and coastal construction, (5) construction of structure in the sea, (6) dredging, and maintenance of coastal waterways construction, and (7) dredging, and maintenance of domestic water channels construction. While the seven government agencies which make a list of pre-qualified bidder according to the Regulations, 1992 defined branch of construction were (1) Department of Highways (DOH) divided construction into four types: road construction, bridge construction, roads maintenance and traffic safety, and special class construction, (2) Department of Rural Roads (DRR) divided construction into three types: road construction bridge construction and tunnel or underpass, road maintenance work, and traffic safety, (3) Marine Department (MD) divided construction into three types: bank protection construction and coastal protection wall construction, dredging and maintenance of domestic water channels, and dredging and maintenance of coastal waterways, (4) Royal Irrigation Department (RID) divided irrigation construction into four types: headwork structure and/or appurtenant structure, delivery system and/or appurtenant structure, drainage system and/or appurtenant structure, land consolidation and/or on-farm development, (5) Department of Water Resources (DWR) divided work of water resource conservation, rehabilitation and development into four types: headwork structure, delivery/ diversion/ distribution system, drainage system, work of water resource improvement rehabilitation, (6) Department of Public Works and Town Planning (DPW&TP) divided construction into 8 types: building, dam, drainage channel, wastewater treatment plant, flood prevention, road, bridge, and landscape improvement work, and (7) Bangkok Metropolitan Administration (BMA) divided construction into two types: buildings and road work.

Table 1. Comparison of branch/type of construction between all government sector which make a list of construction entrepreneur according to Act, 2017 and the seven government agencies which make a list of pre-qualified bidder according to Regulations, 1992

Branch/type of construction (Based on Regulations, 1992)	Act, 2017	Regulations, 1992						
	Government sector (all)	DPW &TP	DOH	MD	BMA	DRR	RID	DWR
1 Building construction								
1.1 Building construction, and other relations	No	Yes	No	No	Yes	No	No	No
2 Road, bridge, and box culvert construction								
2.1 Road construction	Yes (1)	Yes	Yes	No	Yes	Yes	No	No
2.2 Bridge construction	Yes (2)	Yes	Yes	No	Yes	Yes	No	No
2.3 Tunnel/underpass construction	Yes (2)	Yes	Yes	No	Yes	Yes	No	No
2.4 Road maintenance	No	Yes	Yes	No	Yes	Yes	No	No
2.5 Traffic safety	No	Yes	Yes	No	Yes	Yes	No	No
2.6 Road/bridge construction (special class)	Yes (1) (2)	No	Yes	No	No	No	No	No
3 Irrigation construction								
3.1 Headwork structure and/or appurtenant structure	Yes (3)	Yes	No	No	Yes	No	Yes	Yes
3.2 Delivery system and/or appurtenant structure	Yes (3)	No	No	No	No	No	Yes	Yes
3.3 Drainage system and/or appurtenant structure	Yes (3)	Yes	No	No	Yes	No	Yes	Yes
3.4 Land consolidation and/or on-farm development	Yes (3)	No	No	No	No	No	Yes	No
3.5 Water resource improvement rehabilitation	Yes (3)	No	No	No	No	No	No	Yes
3.6 Flood protection	Yes (3)	Yes	No	No	No	No	No	No
4 Other construction								
4.1 Construction of structure in the sea	Yes (5)	No	No	No	No	No	No	No
4.2 Bank protection construction, and coastal protection wall construction	Yes (4)	Yes	No	Yes	Yes	No	No	No
4.3 Dredging, and maintenance of coastal waterways	Yes (6)	No	No	Yes	No	No	No	No
4.4 Dredging, and maintenance of domestic water channels	Yes (7)	No	No	Yes	Yes	No	No	No

Table 2. Right to bid per one contract, (baht) subjected to class of construction entrepreneur and branch/type of construction according to Act, 2017

Class	Road construction	Bridge construction	Irrigation construction	Bank protection and coastal construction	Construction in the sea	Dredging, and maintenance of coastal waterways	Dredging, and maintenance of domestic water channels
Special	Unlimited	Unlimited	Unlimited	Unlimited	Unlimited	Unlimited	Unlimited
1	< 500 M	Type 1-4, unlimited	< 1,000 M	< 150 M	< 150 M	< 80 M	< 40 M
2	< 300 M	Type 2-4, unlimited	< 500 M	< 100 M	< 80 M	< 40 M	< 20 M
3	< 150 M	Type 3-4, unlimited	< 300 M	< 60 M	< 30 M	< 20 M	≥ 5 & < 10 M
4	< 60 M	Type 4, unlimited	< 100 M	< 30 M	< 15 M	≥ 5 & < 10 M	-
5	< 20 M	Type 4, ≥ 5 & < 10 M	< 25 M	≥ 5 & < 10 M	≥ 5 & < 10 M	-	-
6	≥ 5 & < 10 M	-	≥ 5 & < 10 M	-	-	-	-

Table 3. Right to bid per one contract (baht) subjected to class of contractor and branch/type of construction according to Regulations, 1992

Class	DPW&TP	DOH		MD		
	Construction	Road	Bridge	Bank protection and coastal protection wall	Dredging and maintenance	
					Coastal waterways	Domestic water channels
Special,A	Unlimited	-	-	-	-	-
Special	< 150	Unlimited	Unlimited	Unlimited	-	-
1	< 100 M	Standard 1-5, special; < 500 M	Type 1-4	< 100 M	Unlimited	Unlimited
2	< 60 M	Standard 1-5, special; < 300 M	Type 2-4	< 60 M	< 80 M	< 60 M
3	< 30 M	Standard 1-5; < 150 M	Type 3-4	< 30 M	< 40 M	< 30 M
4	< 10 M	Standard 3-5; < 6 M	Type 4	< 10 M	< 20 M	< 10 M

Table 3. (Cont.) Right to bid per one contract (baht) subjected to class of contractor and branch/type of construction according to Regulations, 1992

Class	BMA		DRR			RID	DWR
	Road	Buildings	Road	Bridge	tunnel or underpass	Irrigation	Water resource conservation, rehabilitation and development
Special	-	-	Unlimited	Unlimited	Unlimited	Unlimited	-
1	Unlimited	Unlimited	< 150 M	< 80 M	< 150 M	< 1,000 M	< 500 M
2	< 60 M	< 50 M	< 60 M	< 40 M	< 60 M	< 500 M	< 100 M
3	< 30 M	< 20 M	< 20 M	< 20 M	-	< 100 M	< 50 M
4	< 5 M	< 5 M	< 10 M	< 50 M	-	>10 < 25 M	< 10 M

Table 4. Comparison of number of projects (nos.) or total project value (baht) that ability to carry out on concurrent time between all government sector according to Act, 2017 and the seven agencies according to Regulations, 1992

Class	Government sector (all)	DPW&TP	DOH		MD		
					Bank protection and coastal protection wall	Dredging and maintenance	
	Construction (7 branches)	Construction (8 types)	Road	Bridge		Coastal waterways	Domestic water channels
Special, A	-	< 2,500	-	-	-	-	-
Special	Unspecified	< 1,500	6, (4)	unlimited	10 nos.	-	-
1	Unspecified	< 500	3	Type 1-4; < 3 nos.	5 nos.	3 nos.	3 nos.
2	Unspecified	< 300	2	Type 2-4; < 2 nos.	5 nos.	3 nos.	3 nos.
3	Unspecified	< 90	2	Type 3-4; < 2 nos.	5 nos.	3 nos.	3 nos.
4	Unspecified	< 50	2	Type 4; < 2 nos.	5 nos.	3 nos.	3 nos.
5	Unspecified	-	-	-	-	-	-
6	Unspecified	-	-	-	-	-	-

Table 4. (Cont.) Comparison of number of projects (nos.) or total project value (baht) that ability to carry out on concurrent time between all government sector according to Act, 2017 and seven agencies according to Regulations, 1992

Class	BMA		DRR			RID	DWR
	Road	Buildings	Road	Bridge	Tunnel or underpass	Irrigation	Water resource conservation, rehabilitation and development
Special	-	-	Unspecified	Unspecified	Unspecified	Unspecified	-
1	Depend on performance	Depend on performance	Unspecified	Unspecified	Unspecified	Unspecified	Unspecified
2			Unspecified	Unspecified	Unspecified	Unspecified	Unspecified
3			Unspecified	Unspecified	-	Unspecified	Unspecified
4			Unspecified	Unspecified	-	Unspecified	Unspecified

3.2 Class of construction entrepreneur /contractor and right to bid per one contract

Table 2 (right to bid per one contract subjected to class of construction entrepreneur and branch of construction follow Act, 2017) for (1) road construction, (2) bridge construction, (3) irrigation construction, (4) bank protection and coastal construction, (5) construction of structure in the sea, (6) dredging, and maintenance of coastal waterways construction, and (7) dredging, and maintenance of domestic water channels construction has divided contractors into 7, 6, 7, 5, 5, 4, and 3 classes respectively. All lowest class of construction entrepreneur have the right to bid per one contract ≥ 5 and < 10 million baht, while all special classes are unlimited for the right to bid.

Table 3 (right to bid per one contract subjected to class of contractor and branch of construction follow Regulations, 1992) for DOH, DRR (except tunnel or underpass construction), and MD (only bank protection construction, and coastal protection wall construction) has divided contractors into 5 classes. DPW&TP has divided contractors into 6 classes. The lowest class of all was BMA which has the right to bid per one contract not exceeding 5 million baht.

3.3 Number of project (nos.) or total project value (baht) to carry out on concurrent time

Table 4 showing the construction of 7 branches of all government sector according to Act, 2017 did not specify these issues. Whereas class 4 of DOH would be able to construct maximum 2 projects. All class of MD would be able to construct maximum 5 projects but depend on project value of each class. Class 4 of DPW&TP could carry out total project value not exceeding 50 million baht. BMA could carry out the projects depend on contractor performance. DRR, RID, and DWR did not specify these issues.

4 Conclusion

Prequalification criteria of Thai government sector which make a list of construction entrepreneur according to Act, 2017 are shown as follows:

4.1 Branch/Type of construction

Act, 2017 compiled definition of construction from seven government agencies which make a list of pre-qualified bidder according to Regulations, 1992, and defined the branch of construction into 7 branches: road construction, bridge construction, irrigation construction, bank protection and coastal construction, construction of structure in the sea, dredging, and maintenance of coastal waterways construction, and dredging, and maintenance of domestic water channels construction. The building construction, road maintenance, and traffic safety were repealed from the up-to-date Act, 2017.

4.2 Class of construction entrepreneur /contractor and right to bid per one contract

All lowest class according to Act, 2017 had the right to bid per one contract volume ≥ 5 and < 10 million baht. While all seven government agencies based on Regulations, 1992 except RID had the right to bid per one contract lower than 5 million baht. These are the interest data, why it did not define the right to bid per one contract lower than five million baht.

4.3 Number of project to carry out on concurrent time

The finding on these issues showed that all government sector did not specify the number of projects or total project value that construction entrepreneur has the ability to carry out on concurrent time according to Act, 2017.

References

1. F. Kog, H. Yaman, ECAM **23**, 6 (2016)
2. J.R.S. Cristóbal, J Constr Eng Manage. **138**, 6 (2012)
3. Royal Gazette, http://www.chiangmailocal.go.th/web master/20171207_0023-5-1349.pdf [In Th ai]. 23 November, URL (accessed: 24.12.2017) (2017)

Design analysis of a hot strip mill runout table bed under ANSYS workbench

Sudhansu Mohan Padhy[1], *Rudrajit* Sinha[1*], Ruby Mishra[1], *Purna* Chandra Mishra[1]

[1] School of Mechanical Engineering, Kalinga Institute of Industrial Technology Deemed to be University, Bhubaneswar-24, India

Abstract. A structural analysis has been performed on the setup of a spray cooling system for a model of a hot strip mill. The objective was to design a run out table (ROT) whose length is reduced for a spray cooling system. The setup consists of a bed (ROT) whose height can be adjusted as desired, and a carriage which would be carrying the hot plate, moving along the bed, for experimentation. The CAD modelling was done in CATIA and the analysis was performed in ANSYS Workbench. A prototype of the whole setup was built from the virtual model and was controlled using products from National Instrument Corporation (NI). The analysis is being conducted for the design optimization of the entire setup and its safety.

1 Introduction

The run-out table (ROT) in a hot strip mill is the place where the process of ultra-fast cooling takes place. [1] The hot strip manufactures coils of the required gauges from a slab of unknown dimensions and chemical composition. A ROT generally consists of a bed through which the hot strip moves, and an array of cooling nozzles that cools down the strip itself. Ultra-fast cooling is done by using a high flow of water and air. Cooling methods include curtains of jets, spray, circular jet, etc. A new design for the bed of the run-out table has been developed which has a degree of freedom in the Y-axis. A carriage has also been designed and developed to carry the plate/work piece [1, 2, 4,] at 850 to 900 °C moving along the bed, having and horizontal motion.

According to most researchers, [3] controlled cooling of the strips is required to attain a high grade of rolled steel which is important in industries such as automobile, construction and house-appliances. A. Suebsomran et al. [4] conducted a study to determine the effective cooling parameters for the run-out table (ROT) of strip steel in a hot rolling process by developing the 2-dimensional transient heat transfer of strip steel by using a mathematical model. R.-M. Guo [5] made a simple control algorithm for the run-out table cooling control based on a mathematical model which can deactivate and activate the cooling headers at any point that also uses a tracking system to monitor the temperature control. The length of the bed has been reduced to make the setup as compact as possible. In a traditional mill, [6] the values for strip thickness, temperature and strip thickness are given as input but in the ROT, [7] the material is cooled by arrays of the top and bottom cooling headers in a distance of about 70% of the run-out table length. [8] Usually, spray cooling is used for roll cooling or descaling of the plate's surface, and jet cooling is used for plate cooling since it can effectively transport its momentum to the plate's surface. [9] The cooling process contains a lot of complex physics which can be difficult to solve, due to which, studies of the problem have been largely restricted to experimental methods. After the cooling process, [10] the microstructure of the metal defines its mechanical physical properties. [11] NKK's Fukuyama Works completed installing a Super-OLAC H (on-line accelerated cooling for hot strip mill) system at the runout cooling zone of its No. 1 hot strip mill, which had a length of 30-45 m and was able to achieve a cooling rate of 700 °C/s for a sheet of thickness of 3mm.

This design is meant as a laboratory setup to perform experimental studies. The objective of this work is to analyse the effect of the load of the carriage and plates of two different thicknesses on the bed using ANSYS Workbench so as to find any flaw in the design and optimize it accordingly.

1.1 Modelling of the bed in runout table

The design model created in CATIA software environments consists of a horizontal bed and a carriage, as shown in Fig.1, which are the major components of the whole setup. The bed is made of flat plates 4 mm thick and is 1.7 m long in length. It is supported by two hollow cylindrical rods of 50 mm in diameter. The bed is supported by rollers so as to have a vertical motion. The created vertical motion of the whole bed enables experiments to be conducted with different settings.

* Rudrajit Sinha: rudrajit94@gmail.com

1.2 Methodology

1.2.1 Simulation in ANSYS Workbench

The structural analysis is done in ANSYS Workbench under static structural, where a force of 2 N, which simulates the maximum force that will be applied on the carriage by the spray and the force due to the weight of the carriage is applied on the plate. The analysis has been performed for three different positions of the carriage on the bed. Plates of two different thicknesses (4 mm and 10mm) have been used [12].

1.2.2 Development of Automated ROT

The prototype of the whole setup built from the virtual model was controlled using NI Instrumentation. A LabVIEW based programme was developed for the wireless control of carriage motion in the ROT. Under the LabVIEW programme, the speed of the carriage movement is controlled through MyRIO which in turn controls the stepper motor of the carriage. The Virtual Interface (VI) of the LabVIEW programme is depicted in Fig. 3. The carriage was designed and created to carry a hot metal plate, which would be cooled by spraying nozzles containing internally mix cooling fluids fitted onto an aluminium frame above.

2 Simulation parameters and results

2.1. The parameters for the analyses

The structural analysis of the setup was performed by conducting a series of 6 simulations with different parameters. The parameters consist of three different positions, 300 mm, 750 mm, 1200 mm, from the vertical rods and two different flat plates of thickness, 4 mm and 10 mm. The three positions were taken according to the movement range of the carriage.

The meshing for the whole setup was done with the auto-mesh feature of ANSYS Workbench. The mesh was a tetrahedral type of mesh for the whole setup as can be seen in Fig 2.

Table 1 shows the parameters taken for the analyses that have been simulated in ANSYS:

Table 1. Analyses Performed.

Sl. No.	Distance of the carriage from the vertical rods (D) (mm)	Plate thickness (t) (mm)
1	300	4
2	750	4
3	1200	4
4	300	10
5	750	10
6	1200	10

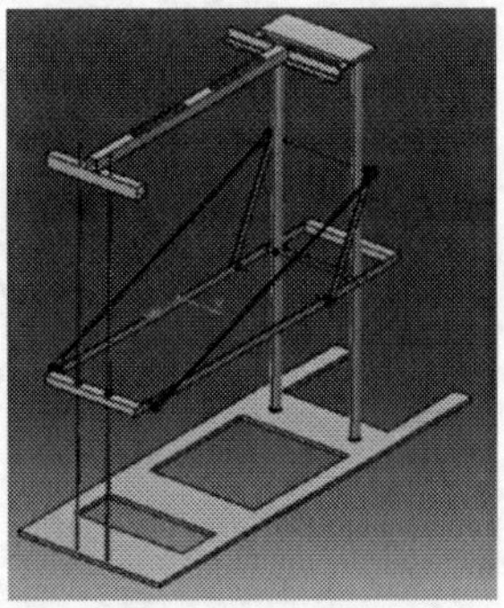

Fig. 1. The model of the whole setup consisting of the bed and the carriage modelled in CATIA.

Fig. 2. The tetrahedral mesh used for the simulations.

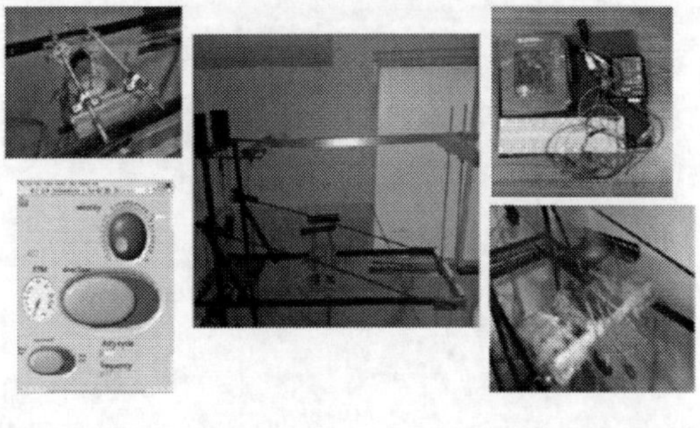

Fig. 3. The prototype controlled using MyRIO and LabVIEW.

2.2 Results and discussion

A force of 55N and 65N has been calculated and used for the simulation and the base is taken as the fixed support and for the solution, the maximum deformation, the maximum equivalent von-mises stress and the maximum principal stress was found.

From Figs. 4, 5 and 6 the maximum deformation was found at analysis no. 5 (Table 1) where the carriage is at a distance of 750 mm from the vertical rods, which is at the middle of the bed, carrying a 10 mm thick plate and a force of 65 N is used in the simulation.

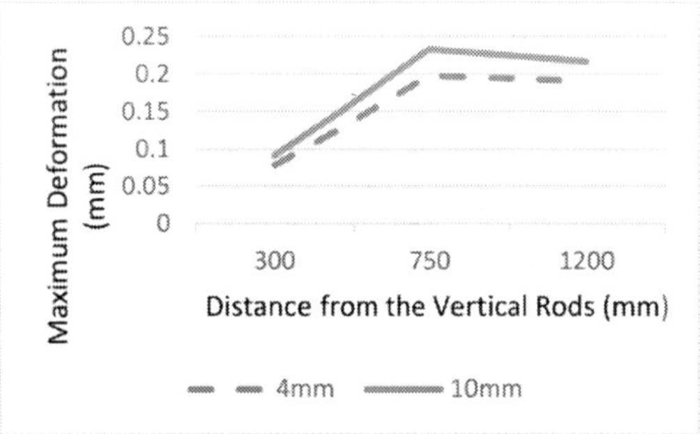

Fig. 4. The graphical representation of Maximum Deformation

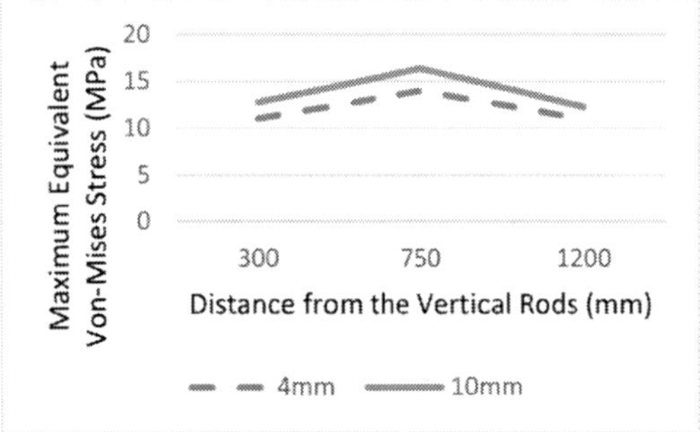

Fig. 5. The graphical representation of Maximum Equivalent Von-Mises Stress

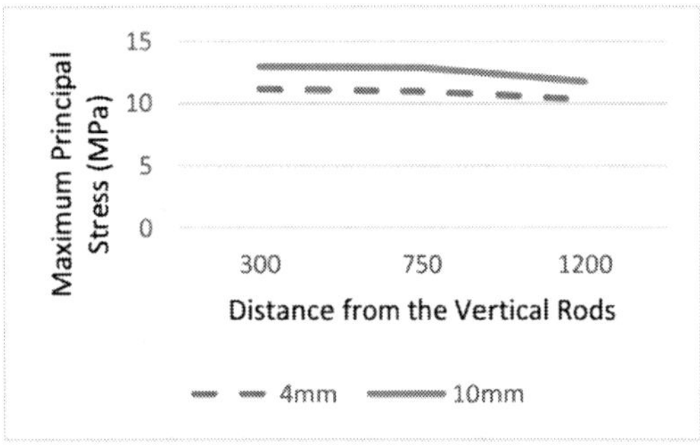

Fig. 6. The graphical representation of Maximum Principal Stress

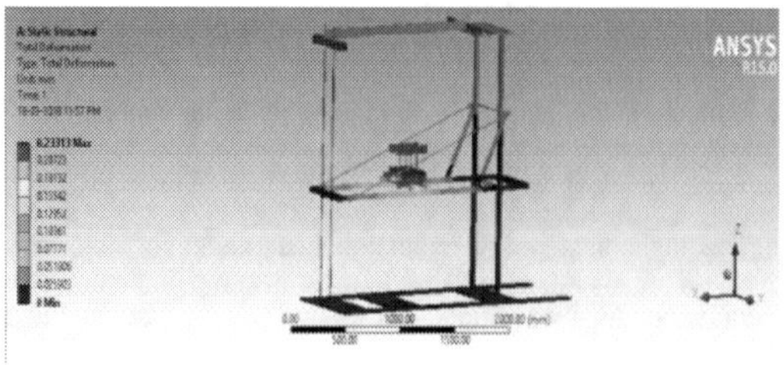

Fig. 7. The Maximum Deformation for the analysis no. 5

2.3 The effect of the force on the bed

The values for the maximum deformation were obtained using probes on the bed in the most deformed region as shown in Fig. 9 for each set analysis.

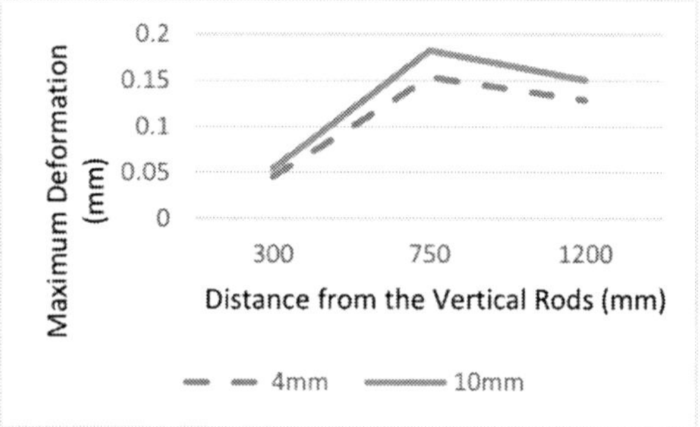

Fig. 8. The graphical representation of Maximum Deformation on the bed

Fig. 9. The Maximum Deformation on the Bed.

3 Calculations

The Forces used in the analyses were calculated using the weights of the carriage and the two different plates.

Weight of the Carriage = 4.515 kg

Weight of 4mm plate = 0.455 kg

Weight of 10 mm plate = 1.15 kg

Carriage + 4mm = 4.97 kg ≈ 5 kg

Force = 5 X 9.81 = 49.05 N

Carriage + 10mm = 5.665kg ≈ 5.7 kg

Force = 5.7 X 9.81 = 55.917 N

Since the plate is subjected to spray cooling, a force is also exerted on the plate from the spray. It is calculated using formulas for impact of jets on a surface normal to it and the Bernoulli's equation.

The force from jet:

Bernoulli's equation states that -

$$p + \frac{1}{2}\rho V^2 + \rho gh = constant \quad (1)$$

Where,

ρ = the density of the fluid,

p = the pressure,

V = the velocity,

h = elevation,

g = gravitational acceleration

Now,

$$\frac{p_1}{\rho} + \frac{V_1^2}{2} + gh_1 = \frac{p_2}{\rho} + \frac{V_2^2}{2} + gh_2 \quad (2)$$

$gh_1 = 0$ as the elevation is zero

$V_1 = 0\ (initial\ velocity)\ and\ V_2 = V$

$$\therefore V = \sqrt{2\left[\left(\frac{p_1 - p_2}{\rho}\right) - h_2 g\right]} \quad (3)$$

Taking the pressure as 0.5 MPa, density as 1000 kg/m^3 and height as 100 mm

We get $V = 31.65\ \mathrm{m/s}$

Solving the Bernoulli's equation we get the velocity of the jet and then we use it to find the force exerted by the jet.

Dynamic forces exerted by fluid on plate normal to jet:

$$F_x = \rho a V^2 \quad (4)$$

ρ = density of water = 1000 kg/m^3

a = area of the mouth of the nozzle

Diameter of the nozzle = 1.5 mm

$$= 10^3\ X\ \frac{\pi}{4}\ X\ (1.5\ X\ 10^{-3})^2\ X\ 31.65^2$$

$$= 1.77\ N\ \approx 2\ N$$

The above calculation is for a water jet on a plate normal to it and considering that the force from a water spray would be much less we take the value of the force as 2 N for the analysis.

Total force, F_1 (with 4 mm plate) = 51.05 N ≈ 55 N

Total force, F_2 (with 10 mm plate) = 61.91 N ≈ 65 N

4 Conclusion

A simple set of analyses were performed on the setup model using ANSYS Workbench. The advantage of this setup is that it requires lesser space due to the fact that the length of the bed has been reduced. Nozzles are placed on the aluminium frame on the top of the setup, and the hot plates are cooled using a spray mixture of water and air. The movement of the carriage in the prototype setup was controlled wirelessly with the help of MyRIO and LabView, which gives a greater advantage as one can observe and control the whole process of the cooling from a faraway distance. The results show that the maximum deformation formed on the bed is 0.182 mm, which can be considered as negligible for the different types of experiment being conducted on the whole setup. Therefore, the whole setup needs no optimization but if required, supports can be introduced on the bed at a distance of 750 mm from the vertical rods.

References

1. Z. Liu, Experiments and Mathematical Modelling of Controlled Runout table Cooling in a Hot Rolling Mill, The University of British Columbia, Department of Metals and Materials Engineering, (2001)
2. A. Suebsomran, S. Butdee, The Study of Cooling Process on Runout Table by Simulation Method, *ISRN Industrial Engineering*, **2013**, 7 (2013)
3. V. H. Hernandez- A Vila, Heat Transfer Model of the Hot Tolling Runout Table-Cooling and Coil Cooling of Steel, The University of British Columbia, Department of Metals and Materials Engineering, (1994)
4. A. Suebsomran, S. Butdee, Cooling process on a run-out table by the simulation method, *Case Studies in Thermal Engineering*, **1**, 51-56, (2003)
5. R.-M. Guo, Modeling and Simulation of Run-Out Table Cooling Control Using Feed forward-Feedback and Element Tracking System, *IEEE Transactions On Industry Applications*, **33**, (March/April 1997)
6. S. K. Yildiz, J. F. Forbes, B. Huang, Y. Zhang, F. Wang, V. Vaculik, M. Dudzic, Dynamic modelling and simulation of a hot strip finishing mill, *Applied Mathematical Modelling*, **33**, 3208-3225, (2009)
7. R.-M. Guo, Heat transfer of laminar flow cooling during strip acceleration on hot strip mill run-out tables, Trans. ISS-AIME, **8**, 49-59, (Aug. 1993)
8. M. J. Kwon, I. S. Park, Nozzle Arrangement for Optimal Performance of Cooling Process of Run Out Table, ISIJ International, **53**, 2192-2196, (2013)
9. I. S. Park, Numerical Analysis for Film Boiling Heat Transfer of a Moving Hot Steel Plate, ISIJ International, **51**, 743-747, (2011)
10. M. Hadjiysky, M. Petkova, Multimodel Based Constrained Predictive Control of Hot Strip Mill Runout Table, *IFAC Proceedings Volumes*, **31**, 817-822 (1998)
11. T. Ohashi, Production and Technology of Iron and Steel in Japan during 2002, ISIJ International, **43**, 957-974, (2003)
12. S. K. Nayak, P. C. Mishra, S. K. S. Parashar, Influence of spray characteristics on heat flux in dual phase spray impingement cooling of hot surface, Alexandria Engineering Journal, **55**, 1995-2004, (2016).

Blast mitigation by water in a bag on a tunnel floor

Tomotaka Homae[1,*], *Yuta* Sugiyama[2], *Kei* Shimura[3], *Kunihiko* Wakabayashi[2], *Tomoharu* Matsumura[2] and *Yoshio* Nakayama[4]

[1]Department of Maritime Technology, National Institute of Technology, Toyama College, Toyama, Japan
[2]Research Institute of Science for Safety and Sustainability, National Institute of Advanced Industrial Science and Technology, Tsukuba, Japan
[3]Department of Mechanical Engineering, Keio Univ, Yokohama, Japan.
[4]Planning Headquarters, National Institute of Advanced Industrial Science and Technology, Tsukuba, Japan

Abstract. Mitigation of blast wave, caused by explosion of explosives, from a straight tube using water in a bag (water bag) was evaluated. The length of the tube was 330 mm and the cross-section area was 30 x 30 mm^2. One end of the tube was closed. The water bag was placed on the floor or closed end wall of the tube near the explosive. The thickness of water was 3 mm. A specially designed small detonator, which contains lead azide of 100 mg, was ignited near the closed end wall of the tube. The blast pressure outside the tube was measured and examined. The blast wave was remarkably mitigated by the water bag. Equivalent ratio analysis revealed that the glass beads absorbed 33%-45% of explosion energy.

1 Introduction

Magazine used for storing explosives is generally installed in mountainous areas away from urban areas. The land suitable for residential area is limited in Japan, as Japan consists of islands and the ratio of flatlands is small. Thus, land development for housing near magazines is often taken place. As a result, residential buildings are increasingly located near magazines. In fact, maintaining such structures at a safety distance is becoming increasingly difficult. A subsurface magazine has an explosive storage chamber, a horizontal passageway, and a vertical shaft for vent. It was proposed and legislated in Japan. Subsurface magazines were proposed to reduce the effects of blast waves, fragments, and ground shock and to maintain a safety distance easily when storing explosives underground. Therefore, blast pressure evaluation of the area around a subsurface magazine is important.

Blast pressure around a subsurface magazine has been described in several reports [1-4]. The authors also carried out the experiments to evaluate the blast pressure. In a series of the experiments, the authors found that small amount of water on the floor of its storage chamber mitigated the blast pressure remarkably [5]. The authors have been studied the mitigation mechanism by experiments [6, 7] and numerical simulation [8, 9]. Our previous study [6] concluded that the high-speed camera images showed that the water in the chamber did not move apparently after the explosion, hence, the energy conversion from the blast wave to kinetic energy of water was not main origin of the mitigation. In addition, it was also concluded [7] that the water far from the explosive are not dominant for the blast-wave mitigation, because the blast pressure was not mitigated if the water was not put right under the explosive but put the other part of the chamber. The results conclude that the mitigation of blast wave by water is mainly caused by the interaction between the explosion and the water very near the explosion point. The numerical simulations [8, 9] of this system concluded that internal energy transfer from the blast wave to the water caused the mitigation. This result can apply to reduce the damage by explosion in closed places, such as subsurface magazines, underground magazines and tunnels.

The blast mitigation by little amount of water on the floor of the closed places shown above was obvious; however, it is not easy to apply this technique to actual subsurface magazines, underground magazines, and tunnels, as the magazines should keep away from humidity in general. Accordingly, the authors propose that water in a bag (hereinafter, water bag) can be placed on the floor/wall instead of a water pond. If the water bag is adopted, it is easier to maintain, while the blast wave cannot directly contact with or enter the water surface because of the bag. The behaviour should be different.

In this study, small-scale experiments for evaluating the mitigation effect by the water bag were carried out. The water bag was placed on the floor or closed end wall of the tube near the explosive. The mitigation effect of the water bag is discussed.

* Corresponding author: homae@nc-toyama.ac.jp

2 Experiment

2.1. Test Explosives

A specially designed electric detonator with 100 mg lead azide (Showa Kinzoku Kogyo Co. Ltd.) was used as a test explosive. 4 kV was applied to initiate the detonator (test explosive) using a firing system (FS-43; Teledyne RISI, Inc.).

2.2 Straight Tube Model

A straight tube was adopted as a model in this study, because examination of interaction between explosion and water bag was focused on. Although a subsurface magazine is L shape, as mentioned above, this technique can also apply to underground magazines and tunnels. The schematic diagram of the tube is shown in Fig. 1. The tube was made of transparent PMMA, thickness of 10 mm. A cross section of inside the tube was square, whose sides were 30 mm. The length was 300 mm. Considering the stress near the explosion point, the test explosive was placed inside a cap, made of aluminum alloy. The distance between the top of the test explosive and the bottom wall of the cap was set to be 10 mm. The cap was connected with the PMMA tube described above. Then, total length of the tube was 330 mm.

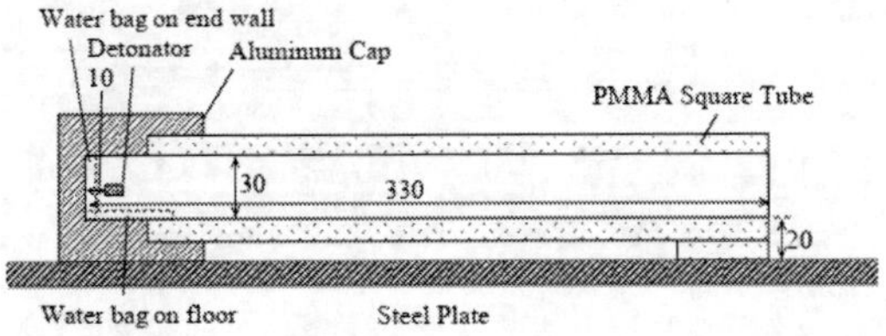

Fig. 1. Schematic diagram of the tube model used in this study.

2.3 Ground Surface Model and Outside Pressure Measurement

A steel plate, length of 3510 mm, width of 2200 mm, and thickness of 10 mm, was regarded as a ground surface. As the steel plate is rigid compared with soil, the reflection of blast wave is large. Hence, the worst disaster can be estimated using this model. The tube was put on the steel plate and initiated. The height of the lower edge of the tube exit above the ground surface was 20 mm (See Fig. 1).

The blast pressure outside the tube was evaluated using four pressure transducers (113B28; PCB Piezotoronics, Inc.). The transducers were on an extended line of the centerline of the tube. The distances from the exit of the tube to the transducers were 200 mm, 600 mm, 1000 mm, and 1400 mm, respectively. As the weight of the test explosive was 100 mg, corresponding Hopkinson scaled distance were from 4.3 m/kg$^{1/3}$ to 30.2 m/kg$^{1/3}$. This measurement was similar to that of the underground magazine or the earth-covered magazine model [10]. The pressure transducers were set with the vibration isolator (GEL Tape; Taica Corp.). The transducer diaphragm was flush with the steel plate surface. The pressure measurement outside the model is precisely described elsewhere [7].

2.4 Water Bag

A small polyethylene bag was prepared. The thickness of polyethylene was 0.04 mm. The placement of the water bag in the tube were varied for two ways. Firstly, the water bag was placed right under the explosive. Secondly, it was placed on the closed end wall (See Fig.1.). The size of the water bags depended on the way of placement. The water bag on the floor was 30 mm x 50 mm. The one on the closed end wall was 30 mm x 30 mm. Thickness of water layer was 3 mm in both cases. We presumed that the depth of water bag layer could be 10 % of the inner side of a tunnel. The amount of the water was 4.5 cm^3 and 2.7 cm^3, respectively. The water bag for floor is shown in Fig. 2. The test explosive did not touch the water bag.

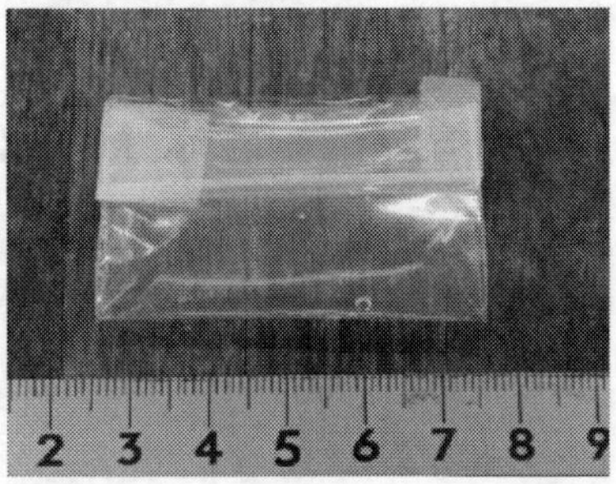

Fig. 2. Water bag for the floor.

2.5 Number of Experiments

The number of experiments are as follows. Without water bag was four times, the water bag on the floor (water bag on floor, hereinafter) was twice, and the water bag on the end wall (water bag on end wall, hereinafter) was once.

3 Results and Discussions

The obtained pressure histories were similar to those of surface explosion except for those at 200 mm from the tube exit. Consequently, blast parameters, such as peak overpressure and scaled impulse, were obtained and evaluated same as general blast waves caused by surface explosion. On the other hand, the pressure histories at 200 mm from the tube exit showed complex decay process because of the reflection of the blast wave in the tube. Then, comparison the data at 200 mm with the others was difficult. These data are omitted from the analysis, although these data are shown on the graphs.

The peak overpressure and the scaled impulse outside the tube were calculated from the pressure histories. The relation of those blast parameters and the scaled distance is shown in Figs. 3 and 4. The scaled distance is defined as the time integral of positive part of the pressure history divided by cubic root of explosive weight (100 mg in this study). The scaled distance is defined as the distance from the tube exit divided by cubic root of explosive weight.

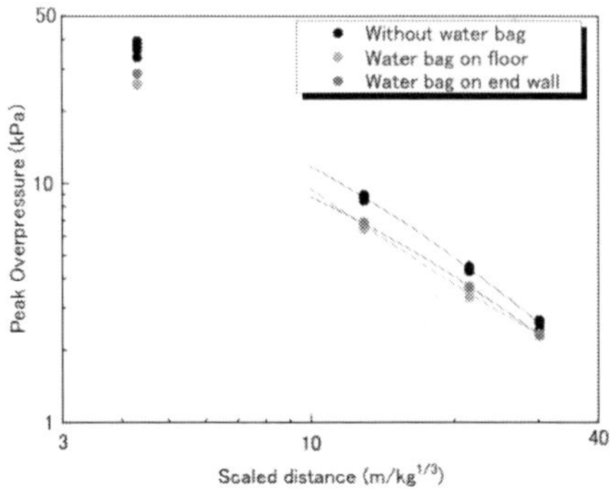

Fig. 3. Relation between the scaled distance and the peak overpressure outside the tube. The solid lines are the quadratic equation fitting for the data set obtained at the further three points for each experimental condition.

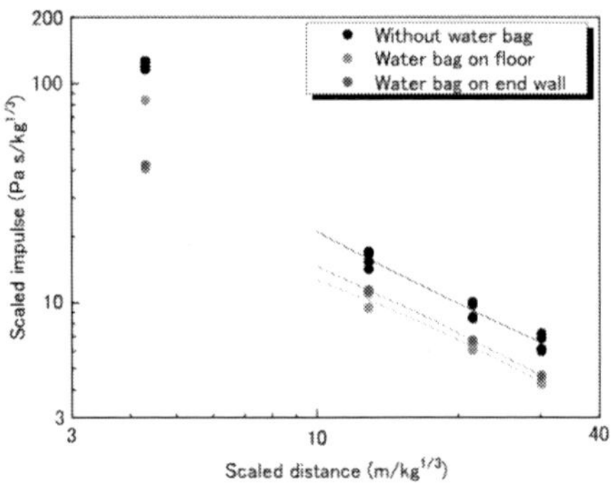

Fig. 4. Relation between the scaled distance and the scaled impulse outside the tube. The solid lines are the quadratic equation fitting for the data set obtained at the further three points for each experimental condition.

The peak overpressures and scaled impulses in the cases with water bags were mitigated. The results demonstrated the water bag mitigated the blast wave apparently.

Equivalent ratio was calculated to quantitatively evaluate the mitigation effect of the water bag. The equivalent ratio r is defined as follows [11, 12]

$$r = \frac{W_{\text{without waterbag}}}{W_{\text{with waterbag}}} \quad (1),$$

where $W_{\text{without waterbag}}$ is the weight of the explosive in explosion without water bag required to produce a selected blast wave parameter of equal magnitude to that produced by a weight of the explosive, $W_{\text{with waterbag}}$, in each experimental condition with the water bag. If $r = 1.0$, the weight of explosive was equal whether the water bag was placed or not. Namely, the water bag did not mitigate the blast wave at all. Small value of r corresponds to the weak blast wave. Namely, the water bag mitigate the blast wave. In this study, the equivalent ratio was calculated using the peak overpressure (r_P) and the scaled positive impulse (r_I). The equivalent ratio also depends on the scaled distance. The method of calculation is described elsewhere [11].

In order to obtain the equivalent ratio, the relations between peak overpressure, scaled impulse, and scaled distance were fitted using quadratics on log-log plane. The solid lines in Figs. 3 and 4 represent the fitted curves. As the three data at the scaled distance from 12.9 m/kg$^{1/3}$ to were analysed in this study, the equivalent ratio from 10 m/kg$^{1/3}$ to 30 m/kg$^{1/3}$ was calculated. The average values in this range for each experimental condition were calculated for comparison [12]. The results are represented in Figs 5 and 6.

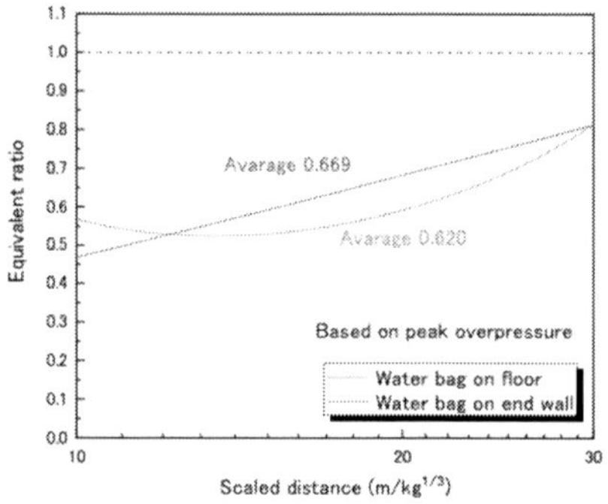

Fig. 5. Relation between equivalent ratio r_P, based on peak overpressure, and scaled distance.

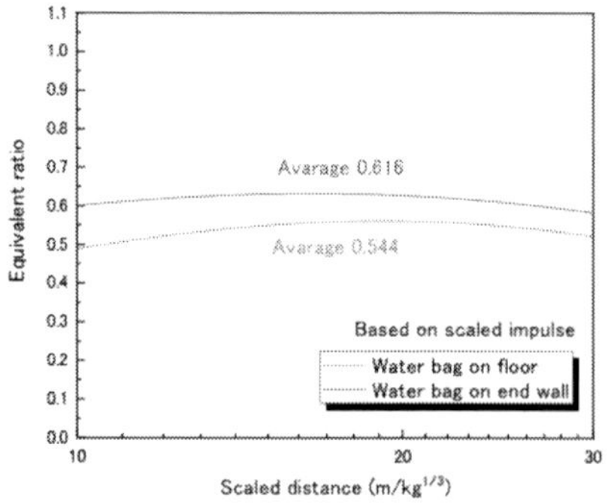

Fig. 6. Relation between equivalent ratio r_I, based on scaled impulse, and scaled distance.

The equivalent ratio was small, when the water bag was placed on the floor. The average value was 0.625 (based on the peak overpressure) or 0.549 (based on the scaled impulse). This result indicates that the explosive amount of only approximately 60% can generate the same blast parameter if the water bag is not placed. Namely, approximately 40% of explosion energy was absorbed by the water bag. The mitigation effect was smaller when the water bag placed on the end wall. However, note that the total amount of water is different. The dependencies of equivalent ratio, based on peak overpressure, on scaled distance are different in Fig. 5. It may be caused by the scatter of the data of the peak overpressure. The dependencies should be examined in next subject.

The results demonstrated that the peak overpressure was mitigated for 33-45 % by the water bag. Even though the blast wave did not contact with the water surface directly, the blast pressure was obviously mitigated. Energy transfer from blast wave to water by shock compression of water may be the origin of mitigation.

This new information can extensively apply to the explosion in closed places, such as subsurface magazines, underground magazines, and tunnel, for mitigation of blast wave.

This work was supported by JSPS KAKENHI Grant Number JP26350461.

References

1. T. Adachi,K. Munemasa, K. Hasue, and S. Nakahara, Propellants, Explosives Pyrotechnics, **16**, 1-5 (1991)
2. H. Ichino, T. Ohno, K. Hasue, S. Date, Sci. Tech. Energetic Materials, **70**, 38-42 (2009) (in Japanese)
3. H. Ichino, T. Ohno, K. Hasue, S. Date, Sci. Tech. Energetic Materials, **71**, 51-57 (2010) (in Japanese)
4. Y. Nakayama, K. Wakabayashi, T. Matsumura, M. Iida, Applied Mechanics and Materials, **82**, 663-668 (2011)
5. T. Homae, Y. Sugiyama, K. Wakabayashi, T. Matsumura, Y. Nakayama, Sci. Tech. Energetic Materials, **77**, 18-21 (2016).
6. T. Homae, Y. Sugiyama, K. Wakabayashi, T. Matsumura, Y. Nakayama, Proc. of SPIE, **10328**, 10328 1D-1-1D-6 (2017)
7. T. Homae, Y. Sugiyama, K. Wakabayashi, T. Matsumura, Y. Nakayama, Materials Science Forum, **910**, 149-154 (2018)
8. Y. Sugiyama, T. Homae, K. Wakabayashi, T. Matsumura, Y. Nakayama, Materials Science Forum, **910**, 78-83 (2018)
9. Y. Sugiyama, T. Homae, K. Wakabayashi, T. Matsumura, Y, Nakayama, Int. J. Multiphase Flow, **99**, 467-473 (2018)
10. D. Kim, T. Matsumura, Y. Nakayama, Sci. Tech. Energetic Materials, **74**, 100-105 (2013)
11. E.D. Esparza, Int. J. Impact Engng., **4**, 23 (1986)
12. M.M. Swisdak, Jr., *Explosion Effects and Properties Part I – Explosion Effects in Air*, NSWC/WOL/TR 75-116 (1975)

Estimating sediment delivery ratio by stream slope and relief ratio

Kieu Anh Nguyen[1] and *Walter* Chen[1,*]

[1]Department of Civil Engineering, National Taipei University of Technology, Taipei, Taiwan

Abstract. Nowadays, the storage capacity of a reservoir reduced by sediment deposition is a concern of many countries in the world. Therefore, understanding the soil erosion and transportation process is a significant matter, which helps to manage and prevent sediments entering the reservoir. The main objective of this study is to examine the sediments reaching the outlet of a basin by empirical sediment delivery ratio (SDR) equations and the gross soil erosion. The Shihmen reservoir watershed is used as the study area. Because steep terrain is a characteristic feature of the study area, two SDR models that depend on the slope of the mainstream channel and the relief-length ratio of the watershed are chosen. It is found that the Maner (1958) model, which uses the relief-length ratio, is the better model of the two. We believe that this empirical research improves our understanding of the sediment delivery process occurring in the study area.

1 Introduction

In recent years, soil erosion has become one of the most serious environmental issues in the world due to climate change. The determination of the amount of loose sediments in an area such as a watershed is an important first step to understand the problem.

Not all of the sediments will reach river channels or reservoirs. Sediment delivery involves soil erosion, transportation, and deposition. Sediment delivery ratio (SDR) is defined as the percentage of sediment yield to the total amount of soil erosion occurred in the erosion process. It can be written as follows:

$$SY = SDR \times SE \quad (1)$$

where

SDR = sediment delivery ratio

SY = sediment yield

SE = total soil erosion

SDR is controlled by many factors such as land use, catchment area, relief-length ratio, soil texture, transportation process, and slope gradient of the mainstream channel. Many studies have examined the relationships between the SDR and the factors above. However, most of the studies focused on the drainage area because the size of the area was easy to determine. We believe that there are other parameters more suitable for the empirical determination of SDR. Therefore, the purpose of this study is to use the slope of the mainstream channel and the relief-length ratio of the watershed to compute SDR, and to compare the results of the study area.

2 Study area and previous work

Shihmen reservoir watershed is located in the northern part of Taiwan and mostly in the Taoyuan City and the Hsinchu County (Fig. 1). The watershed is mostly covered by forest with an area of 759.53 km^2, and the elevation ranges from 220 m to 3527 m.

Shihmen reservoir watershed receives an average of 2200 mm to 2800 mm of precipitation annually, delivered by typhoons and rainstorms typically from May to August every year. Because of the location of Taiwan, tropical storms frequently cause flooding, landslides, and soil erosion.

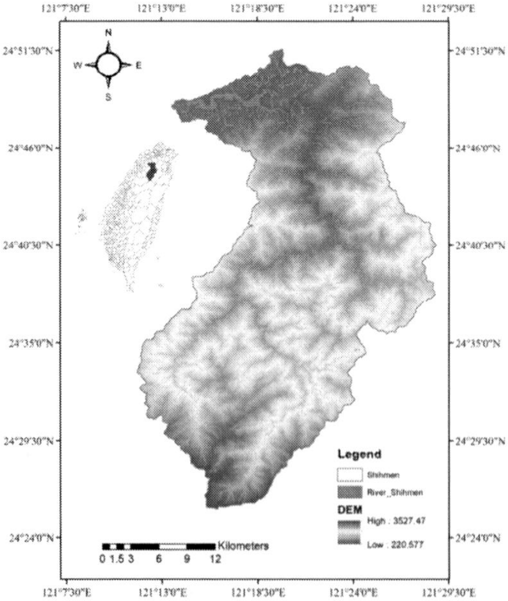

Fig. 1. Study area of the Shihmen reservoir watershed.

Previously, we have used eight empirical equations to evaluate SDR [1]. The independent variables include the

* Corresponding author: waltchen@ntut.edu.tw

drainage area, the runoff of the basin, the slope of the mainstream channel, and the relief-length ratio (Fig. 2). Among them, the relief-length ratio gives the best result, while the slope of the mainstream channel gives the worst estimate of SDR. However, since both models use similar topographic parameters, the question remains why such a big difference exists. In order to clarify this problem, we further compare the two models in this study.

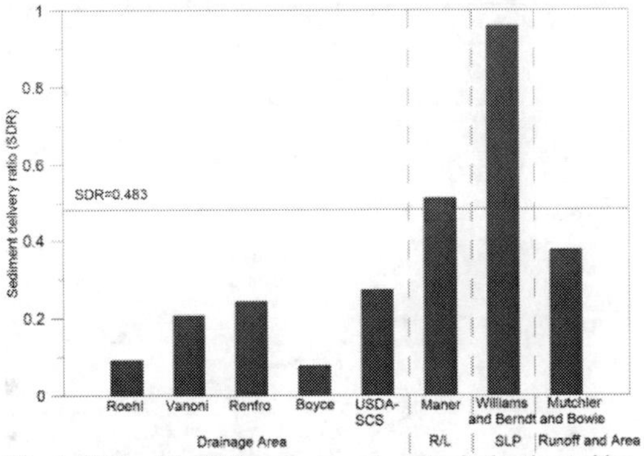

Fig. 2. SDRs of the Shihmen reservoir watershed estimated by eight empirical equations.

3 Research method

The two empirical models are described below.

3.1 SLP

Williams and Berndt (1977) derived the sediment delivery ratio for the catchment basin of Brushy creek of Texas [2] that connected SDR to the geomorphological and sediment-logical feature of the catchment. The equation is written as follows (Fig. 3a):

$$\mathrm{SDR} = 0.627 \times SLP^{0.403} \quad (2)$$

where

SLP = percent slope of mainstream channel

The mainstream is defined as the stream of the highest stream order. The slope of the mainstream depends on the stream profile, and is defined as the horizontal distance along the mainstream channel divided by the elevation difference between the two end points of the mainstream.

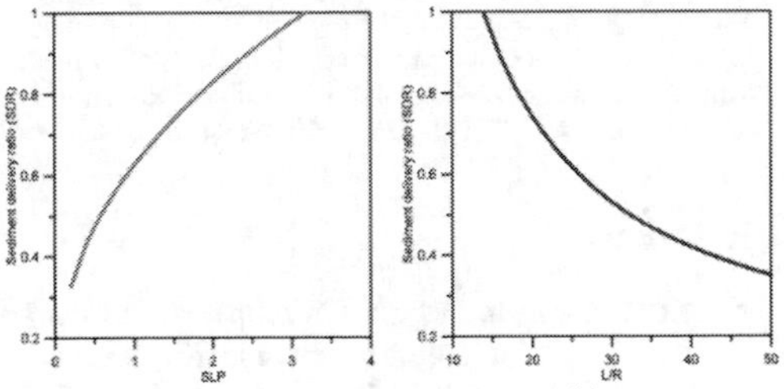

(a) (b)

Fig. 3. Relationships between SDR and (a) SLP, and (b) R/L

For the study area of the Shihmen reservoir watershed, the mainstream is shown in Fig. 4. We use the ArcMap software and the Flow functions to determine the length of the mainstream. The result shows that the mainstream length of the Shihmen reservoir watershed is 105.309 km, and the difference in elevations is 3,023 m.

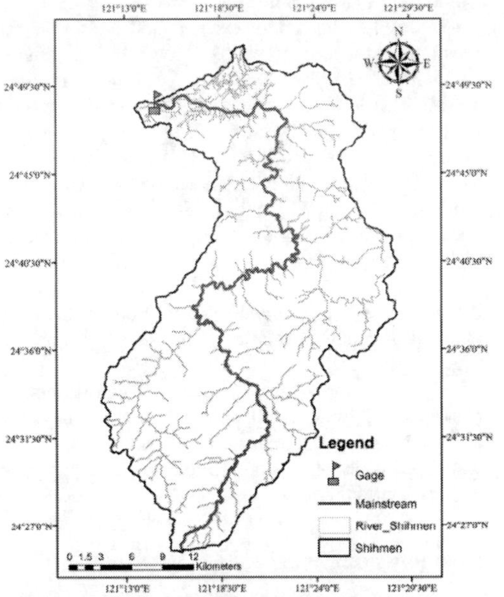

Fig. 4. Mainstream of the Shihmen reservoir watershed.

3.2 R/L

Relief-length ratio is the ratio between the relief of the basin (R) and the length of the basin (L). Contradictory to the SLP factor, the inverse of the relief-length ratio is inversely related to the sediment delivery ratio (Fig. 3b).

The relief-length ratio provides the general steepness of a drainage basin and the power of erosion process along the slopes of the basin. Schumm [3] illustrated that the soil loss per unit area of a basin was linked to the relief-length ratio, and recommended that a close relationship might exist between the relief-length ratio and the hydrological characteristics of the basin. In this paper, the Maner equation is used [4]:

$$\log \mathrm{SDR}(\%) = 2.943 - 0.82\,\mathrm{colog}\frac{R}{L} \quad (3)$$

where R is the relief of the watershed. It is defined as the difference between the average elevation at main headwaters and the elevation at basin outlet. L is the maximum length of the watershed parallel to the mainstream channel. The Shihmen reservoir watershed is divided into three main drainage sub-basins which contain three headwaters, as shown in Fig. 5 (points A, B, and C). The relief R is calculated to be 1513 m, and the length L is found to be 48304.87 m.

4 Results

As mentioned earlier, Fig. 2 showed the comparison of SDRs estimated by eight empirical equations for the

 https://doi.org/10.1051/matecconf/201819202040

Shihmen reservoir watershed, and the figure was used to examine the relative errors of different equations. Since the SDR of the Shihmen reservoir watershed is 0.483 (from dredging data), it can be seen from the figure that the relief-length ratio of Maner (1958) yields a much more accurate estimate of the SDR than the SLP of William and Berndt (1972) in the Shihmen reservoir watershed.

Table 1 further compares R/L and SLP. The table illustrates the similarity and difference between the two parameters. The R/L model reflects the variation of elevation with respect to distance along the length of the watershed. The elevation difference between the headwaters of the main tributaries and the outlet is used as the average elevation difference of the basin. On the other hand, in the SLP model, the elevation difference between the two end points of the mainstream divided by the stream length represents the slope of the mainstream, not necessarily the slope of the terrain of the watershed. To illustrate the difference, the profiles along the mainstream and the basin length are shown in Fig. 6 and Fig. 7, respectively.

The topography of the Shihmen reservoir watershed is characterized by steep slopes and fast moving flows. Although the mainstream of the Shihmen reservoir watershed is very long and tortuous, the SLP model still over estimates the SDR.

5 Conclusions

In this study, we compared and contrasted two very similar empirical equations used in the assessment of SDR. It is found that the relief-length ratio equation from Maner (1958) is better than the SLP equation. In fact, the Maner (1958) equation is the best equations among the eight empirical equations compared in a related study [1]. It is the best choice for the Shihmen reservoir watershed.

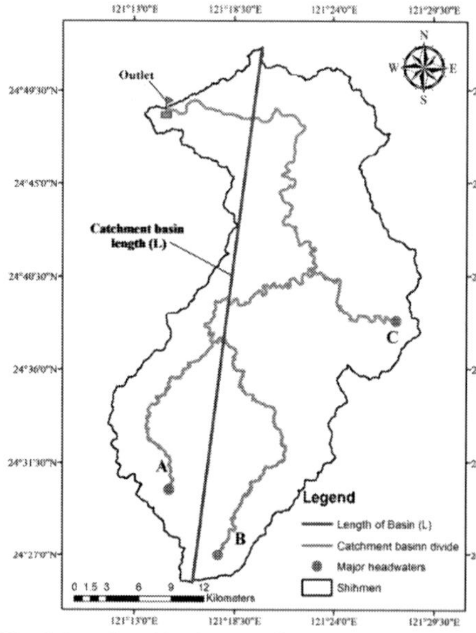

Fig. 5. Length and headwaters of the Shihmen reservoir watershed.

Table 1. Comparison between R/L and SLP.

	Relief-length ratio (R/L)	Stream slope (SLP)
Elevation parameters	Average elevation of headwaters Elevation of outlet	Maximum elevation of mainstream Minimum elevation of mainstream
Length parameter	Basin length	Mainstream length
Slope parameter	Slope of basin	Slope of mainstream

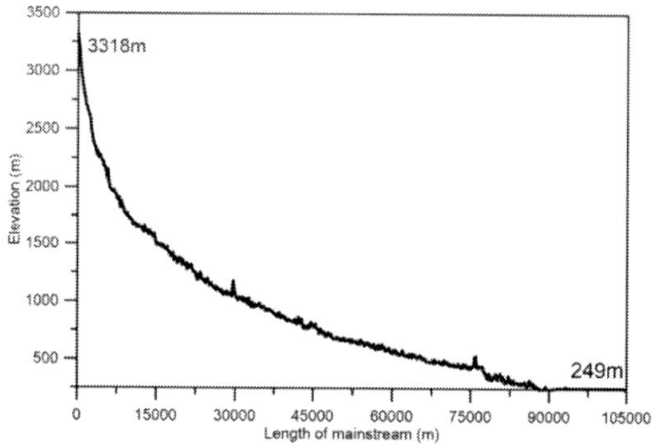

Fig. 6. Profile along the mainstream channel.

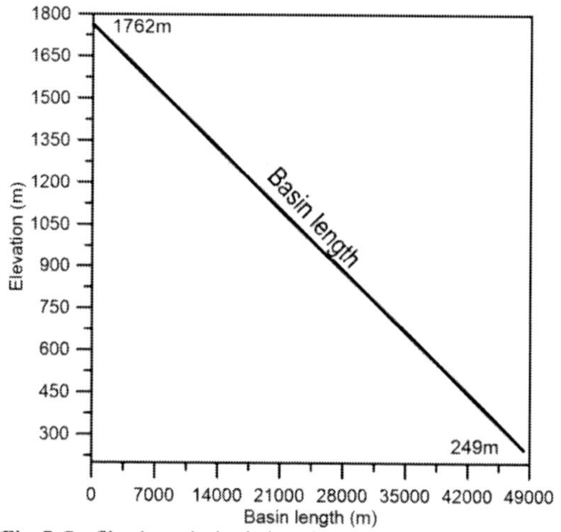

Fig. 7. Profile along the basin length.

This study was partially supported by the National Taipei University of Technology-King Mongkut's Institute of Technology Ladkrabang Joint Research Program (grant number NTUT-KMITL-106-01 and NTUT-KMITL-107-02) and the Ministry of Science and Technology (Taiwan) Research Project (grant number MOST 106-2119-M-027-004).

References

1. K. A. Nguyen, W. Chen, Comparison of empirical SDR equations. In: *International Symposium on Remote Sensing 2018*, Pyeongchang, Korea (2018)
2. J. R. Williams, H. D. Berndt, Sediment yield computed with universal equation. Journal

Hydraulics Division Proceedings, ASCE, 98(HY12), pp. 2087-2097 (1972)

3. S. A. Schumm, Evolution of drainage systems and slopes in badlands at Perth Amboy, New Jersey. Geological Society of America Bulletin, 67(5), pp. 597-646 (1956)
4. S. B. Maner, Factors affecting sediment delivery rates in the Red Hills physiographic area. Transactions, American Geophysical Union, 39(4), pp. 669-675 (1958)

Comparing watershed soil erosion of Taiwan and Thailand

Yi-Hsin Liu[1], *Kieu* Anh Nguyen[1], *Walter* Chen[1,*], *Jatuwat* Wattanasetpong[2], and *Uma* Seeboonruang[2]

[1]Department of Civil Engineering, National Taipei University of Technology, Taipei, Taiwan
[2]Department of Civil Engineering, King Mongkut's Institute of Technology Ladkrabang, Bangkok, Thailand

Abstract. Tropical watersheds in Taiwan and Thailand face the same severe soil erosion problem that is increasing at an alarming rate. In order to evaluate the severity of soil erosion, we quantitatively investigate the issue using a common soil erosion model (Universal Soil Loss Equation, USLE) on the Shihmen reservoir watershed of Taiwan and the Lam Phra Ploeng basin of Thailand, and compare their respective erosion factors. The results show an interesting contrast between the two watersheds. Some of the factors (rainfall factor, slope-steepness factor) are higher in the Shihmen reservoir watershed, while others (soil erodibility factor, cover and management factor) are higher in the Lam Phra Ploeng basin. The net result is that these factors cancel each other out, and the amount of soil erosion of the two watersheds are very similar at 68.03 t/ha/yr and 67.57 t/ha/yr, respectively.

1 Introduction

In contrast to cold regions where soil formation is slow, the tropical Taiwan and Thailand both have prevailing weathering environments and thick deposits of soils. As a result, soil erosion is taking place at an alarming rate. With the loss of precious topsoil, agricultural productivity is negatively impacted (in the case of Thailand) and reliable drinking water supply is jeopardized (in the case of Taiwan). To better identify and understand the different contributing factors of soil erosion in typical watersheds of Taiwan and Thailand, we continue a previous study [1] in this paper to quantitatively compare the characteristics of the watersheds of Taiwan and Thailand and their respective erosion factors.

2 Materials and methods

One watershed was selected from each country and the major characteristics of the watersheds are summarized in Table 1. In Taiwan, the Shihmen reservoir watershed is selected as the study area. It is geographically located from 121°10'15" E to 121°23'10" E longitude and from 24°25'45" N to 24°51'20" N latitude. The Shihmen reservoir is the third largest reservoir in Taiwan and has an effective storage capacity of 207,000,000 m^3 [2]. The reservoir was built in 1956. The catchment area to which the reservoir belongs is 75953 ha. The general topography of the watershed is steep and rugged. Because the reservoir supplies drinking water to major cities in northern Taiwan, most of the watershed was declared protected area from farming and cultivation activities. However, suspended sediments and mudflow are often caused by torrential rains in the monsoon seasons. The reservoir faces a crisis to continuously supply safe drink water to the public.

On the other hand, the Lam Phra Ploeng basin is chosen as the study area of Thailand. It is located between latitude 14°18'30" and 14°38'30" north and longitude 101°30'00" and 101°54'00" east. The Lam Phra Ploeng dam was constructed in 1963 for a basin mostly dedicated for agricultural uses nowadays. The size of the Lam Phra Ploeng basin is 81745 ha, slightly bigger than the Shihmen reservoir watershed. The average annual rainfall amount is 1070 mm, only equivalent to 42.8% of the rainfall received in the Shihmen reservoir watershed.

Table 1. Comparison of watersheds in Taiwan and Thailand.

	Shihmen reservoir watershed	Lam Phra Ploeng basin
Latitude and longitude	24°25'45" - 24°51'20" N 121°10'15" - 121°23'10" E	14°18'30" - 14°38'30" N 101°30'00" - 101°54'00" E
Area of watershed	75953 ha	81745 ha
River system	Dahan River	Mekong River - Lam Phra Ploeng River
Reservoir capacity	207,000,000 m^3	155,000,000 m^3
Annual average temperature	22 °C	27 °C
Annual rainfall	2500 mm	1070 mm
Land use type	Mostly protected area	Mostly farmland
Elevation	220-3527 m	256-1332 m
UTM coordinate system	UTM 51N	UTM 47N
Year of construction	1956	1963

* Corresponding author: waltchen@ntut.edu.tw

To compare the soil erosion in the watersheds of Taiwan and Thailand, the Universal Soil Loss Equation (USLE) is used. It "is an erosion model designed to compute longtime average soil losses from sheet and rill erosion" [3]. The equation is (in metric units):

$$A_m = R_m \times K_m \times \mathrm{L} \times \mathrm{S} \times \mathrm{C} \times \mathrm{P} \quad (1)$$

A_m = computed soil loss per unit area (t/ha/year)
R_m = rainfall and runoff factor (MJ-mm/ha/hr/year)
K_m = soil erodibility factor (t-hr/MJ/mm)
L = slope-length factor
S = slope-steepness factor
C = cover and management factor
P = support practice factor

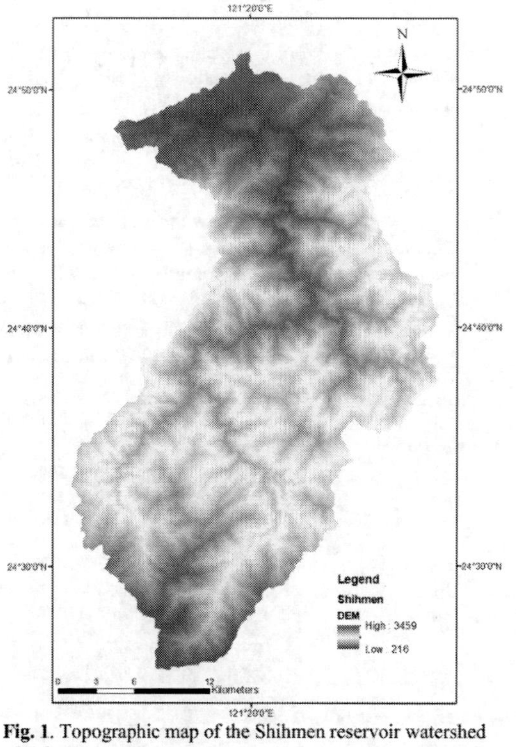

Fig. 1. Topographic map of the Shihmen reservoir watershed of Taiwan.

Due to the lack of concurrent data, erosion factors of different time periods are assumed to be representative of the study areas and used in this study. In order to apply Equation 1, Rm is calculated from readings of rainfall stations (Thailand) or obtained from Lu et al. (2005; Taiwan [4]). Km is referenced from the land use data issued by the Thai government (Thailand) or obtained from Wann and Hwang (1989; Taiwan [5]). Both the L and S factors are computed from the Global Digital Elevation Model (GDEM) produced by the Advanced Spaceborne Thermal Emission and Reflection Radiometer (ASTER), which was released in 2011 (version 2). The C factor of Thailand is obtained from the land use data issued by the Thai government, and the C factor of Taiwan is derived from the correlation table between the land use data and the C factor [6]. Finally, the P factor is assumed to be one for both study areas.

The topographic maps of the Shihmen reservoir watershed and the Lam Phra Ploeng basin are shown in Fig. 1 and Fig. 2. The Shihmen reservoir watershed is long in the north-south direction and narrow in the east-west direction, whereas the Lam Phra Ploeng basin is equally-sized in all directions. The analysis system is based on ArcGIS [7-9].

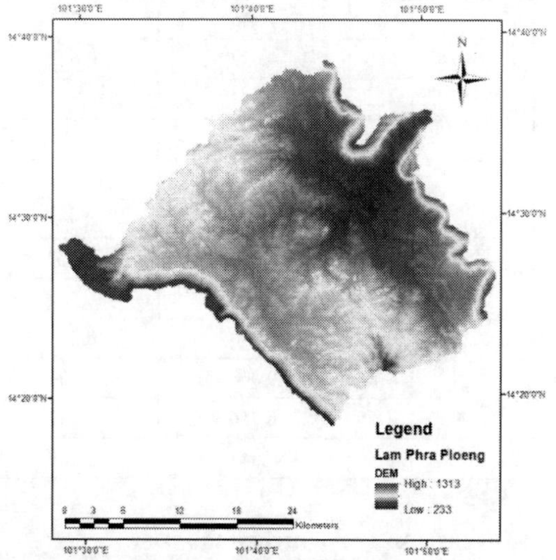

Fig. 2. Topographic map of the Lam Phra Ploeng basin of Thailand.

3 Results

The results of erosion factors computed from USLE are summarized in Table 2. As noted before, the Shihmen reservoir watershed has more than twice as much of rainfall as the Lam Phra Ploeng basin. This is reflected in the big difference of the R_m factor (which also depends on rainfall patterns). Because of the sharp terrain, the Shihmen reservoir watershed also has a higher S factor than the Lam Phra Ploeng basin (18.291 vs. 3.925). In contrast, the Lam Phra Ploeng basin has a much higher C factor than the Shihmen reservoir watershed due to its agricultural use of the basin. The soil of the Lam Phra Ploeng basin is also more erodible than the Shihmen reservoir watershed (perhaps due to cultivation). The average K_m factors are 0.261 (Lam Phra Ploeng basin) vs. 0.022 (Shihmen reservoir watershed).

Table 2. Comparison of soil erosion and erosion factors.

		Shihmen reservoir watershed	Lam Phra Ploeng basin
Soil erosion t/ha/yr	Max	11922.59	1816.65
	Min	0.00	0.00
	Mean	68.03	67.57
	Std	210.01	115.33

K_m t-hr/MJ/mm	Max	0.042	0.300
	Min	0.004	0.000
	Mean	0.022	0.261
	Std	0.006	0.044
R_m MJ-mm/ha/hr/yr	Max	15672.11	669.80
	Min	9897.87	484.56
	Mean	12797.92	547.19
	Std	998.47	26.07
L	Max	1.415	1.291
	Min	0.672	0.672
	Mean	0.795	0.725
	Std	0.082	0.076
S	Max	63.748	60.753
	Min	0.065	0.065
	Mean	18.291	3.925
	Std	10.222	5.582
C	Max	1.000	0.600
	Min	0.000	0.000
	Mean	0.016	0.245
	Std	0.043	0.241

4 Discussion and conclusions

Using erosion classes defined in Thailand, the maps of erosion classes of the Shihmen reservoir watershed and the Lam Phra Ploeng basin are shown in Fig. 3 and Fig. 4, respectively. Distinct patterns emerge from both figures. For the Shihmen reservoir watershed, high erosion classes exist mostly in the mountain range. However, for the Lam Phra Ploeng basin, high erosion classes are everywhere in the bottom of the basin. Interestingly, the erosion factors cancel each other out over the study areas. The end result is that both watersheds have very similar amounts of soil erosion, which are 68.03 t/ha/yr (Shihmen) and 67.57 t/ha/yr (Lam Phra Ploeng), respectively. The results suggest that the difference in soil erosion factors between the two watersheds can be quantified and is significant. They also suggest that more studies are required to elucidate these interesting findings.

This study was partially supported by the National Taipei University of Technology-King Mongkut's Institute of Technology Ladkrabang Joint Research Program (grant number NTUT-KMITL-106-01 and NTUT-KMITL-107-02) and the Ministry of Science and Technology (Taiwan) Research Project (grant number MOST 106-2119-M-027-004).

References

1. Y.-H. Liu, K. A. Nguyen, W. Chen, J. Wattanasetpong, U. Seeboonruang, Comparison of soil erosion between watersheds of Taiwan and Thailand. In: *International Symposium on Remote Sensing 2018*, Pyeongchang, Korea (2018)
2. Northern Region Water Resources Office, About Shihmen Reservoir, Retrieved March 25, 2018, from http://www.wranb.gov.tw/ct.asp?xItem=928&CtNode=424&mp=4 (2017; in Chinese)
3. W. H. Wischmeier, D. D. Smith, *Predicting rainfall erosion losses—a guide to conservation planning*. U.S. Department of Agriculture, Agriculture Handbook No. 537 (1978)
4. J.-Y. Lu, C.-C. Su, I.-Y. Wu, Revision of the isoerodent map for the Taiwan area. Journal of Chinese Soil and Water Conservation, 36(2), pp. 159-172 (2005; in Chinese)
5. S. S. Wann, J. I. Hwang, Soil erosion on hillslopes of Taiwan. Journal of Chinese Soil and Water Conservation, 20(2), pp. 17-45 (1989; in Chinese)
6. C. C. Wu, K.-F. Lo, L.-L. Lin, *Handbook of soil loss estimation*. National Pingtung University of Science and Technology (1996; in Chinese)
7. Y.-K. Jhan, *Analysis of soil erosion of Shihmen reservoir watershed*. Dissertation, National Taipei University of Technology (2014; in Chinese)
8. K.-J. Yang, *Terrain factor analysis of soil erosion in Shihmen reservoir watershed*. Dissertation, National Taipei University of Technology (2016; in Chinese)
9. D.-H. Li, *Analyzing soil erosion of Shihmen reservoir watershed using slope units*. Dissertation, National Taipei University of Technology (2017; in Chinese)

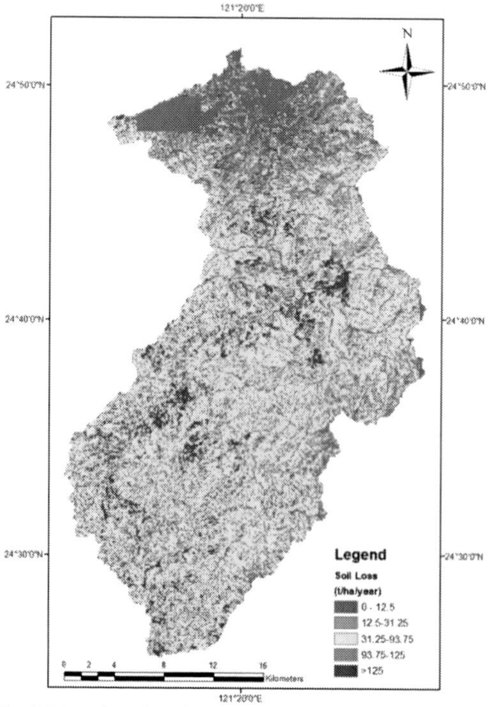

Fig. 3. Map of erosion classes of the Shihmen reservoir watershed.

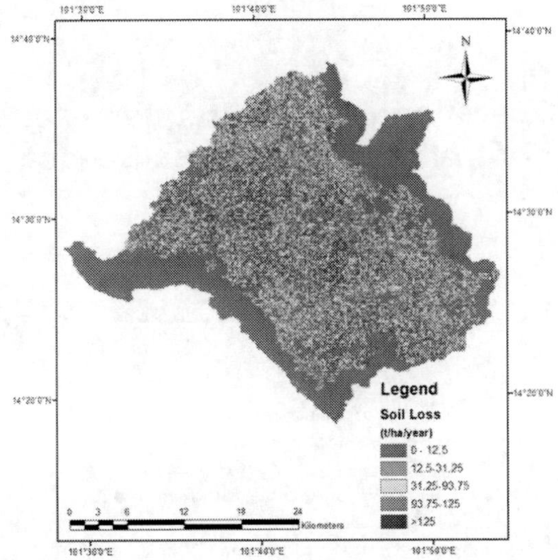

Fig. 4. Map of erosion classes of the Lam Phra Ploeng basin.

Study of Interaction Between Operational Clock-Frequency and Energy Consumption of Microcontrollers for Wireless Sensor Applications

Subhawat Jayasvasti [1], and *Don Isarakorn* [1*]

[1]Instrumentation and Control Engineering Department, Faculty of Engineering, King Mongkut's Institute of Technology Ladkrabang, Bangkok, Thailand

Abstract. This work focuses on the diagnosis of energy consumption of commonly used microcontrollers with various driving clock-frequency. The interaction among the driving clock-frequency, processing time and current consumption are determined to be a guidance for the design of low-power wireless sensor applications. In implementation, the microcontrollers are supposed to process the data following the operational procedures of wireless sensor application, which is programmed to solve mathematical equation with various driving clock-frequencies, and then sleep to reduce the power consumption. With a single circle procession, the current consumption at each driving clock-frequency are captured and analyzed. As the results, the current consumption is proportional to the driving clock-frequency, however energy consumption is minimized at highest clock-frequency.

1 Introduction

The wireless sensor applications have been ubiquitous in various field of works including sport, heath monitoring, transportation, agriculture and industrial [1-3]. With its considerable growth, the reduction in energy consumption of wireless application has been recently trendy, since the low-power device can operate with a few batteries during its lifespan. It can decrease the use of batteries and enables the realization of autonomous energy system powered by harvested ambient energy. Typically, the main strategy for decreasing the power consumption is to reduce the size of electronic components, and program the wireless devices to automatically wake up only for data measurement and then sleep after data was transmitted [4]. lower case. You should leave 35 mm of space above the title and 6 mm after the title.

Out of strategy mentioned above, one more main challenge is the appropriate speed of data procession of microcontroller composed in wireless sensor application because processing speed is proportional to the current consumption and clock-frequency driving the microcontroller. For example, Figure. 1 demonstrates the interaction between clock-frequency and current consumption of microcontroller (ATMega32) [5]. As can be seen, the high clock-frequency resulting high current consumption, while low current consumption is observed at low clock-frequency. However, these current consumptions

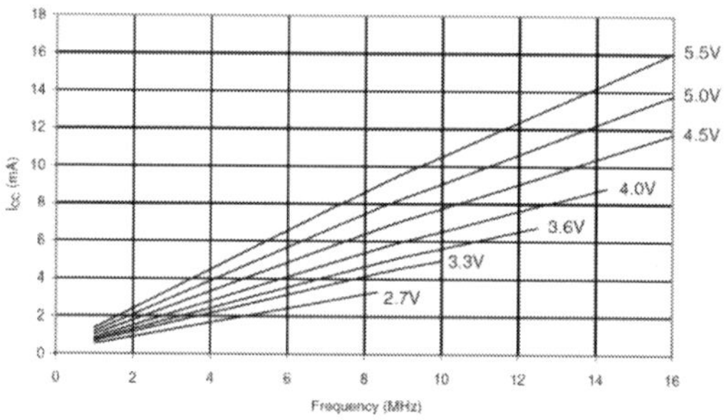

Fig. 1. The interaction between clock-frequency and current consumption.

The cannot obviously demonstrate the optimal clock-frequency resulting in low energy consumption, since the energy consumption derive from the power consumption and time interval of data procession of microcontroller. With a same dataset procession, the high processing speed take a short time for data procession, but the high-power consumption is required. On the other hand, the low processing speeds can operate with the low-power consumption, but it takes a long processing time [6].

Therefore, purpose of this paper is to find out the optimal clock-frequency, and microcontroller consuming the lowest energy among the commonly used microcontrollers for further design of low-power wireless sensor application. Firstly, the investigation method is designed following the operational feature of

* Don Isarakorn: don.is@kmitl.ac.th

wireless sensor application. Then, the experiment is implemented. After that, the experimental results are discussed. And, finally the conclusion is given.

2 Investigation Method

To achieve the reasonable results of investigation, the interaction between clock-frequency and energy consumption of microcontrollers is considered following the working sequence of wireless sensor application in a working circle: data measurement, procession, communication and sleep as shown in Fig. 2. The clock-frequency driving microcontrollers is increased from 1 MHz to the maximum frequency. The current consumptions at each driving clock frequency are captured by using digital multi-meter (Agilent 34411A) [7]. After that, the experimental results are analyzed and utilized to find out the microcontroller consuming the lowest-energy. The microcontrollers investigated in this work are shown in Fig.3 and listed in Table 1.

3 Experimental Design

Following the section 2, the experiment can be designed into five cases as below:

First, the microcontrollers are programmed to process the inconstant pointer changing the pointing location with a delay function (delay ()) in an infinite while loop (while (1)) for considering the current consumption at various driving clock frequency.

Second is the computation of a complex mathematical equation taking a bit long processing time, which supposes the procession of data derived from sensor. In the implementation, the microcontrollers are used to calculate the 2000 digits of Pi following a Nilakantha series as below:

$$\pi = 3 + 4\sum_{k=0}^{\infty} \frac{-1^k}{(2k+2)(2k+3)(2k+4)} \quad (1)$$

where, k is the number of digits.

And, the answer is stored in a long double variable (64 bits). Then, the processing time with each driving clock-frequency is observed.

Third, an equation (1) is processed in loop function to find out the average current consumption at each clock frequency.

Fourth is to examine the current consumption of microcontrollers in energy saving mode. This case assumes that the wireless sensor is stayed in sleeping or power-down state.

And finally, the microcontrollers are supposed to process in one operating circle of wireless sensor node as calculating the 2000 digits of Pi and sleeping for 5 seconds. Then, by adjusting the clock frequency, the current consumption and time interval of procession are analyzed for considering the energy consumption [8].

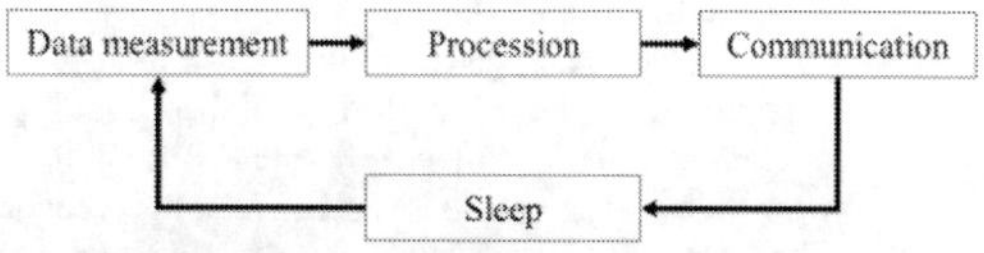

Fig. 2. Working circle of wireless sensor application.

Fig. 3. The complete microcontroller boards

Table 1. The investigated microcontrollers.

No.	Microcontrollers	Maximum clock frequency (MHz)
1	ATmega168 AVR/8bit	16
2	ATmega32 AVR/8bit	16
3	ATXmega128A1 AVR/16bit	32
4	ATSAMD21J ARM Cortex M0/32bit	48
5	ATSAMR21G ARM Cortex M0/32bit	48
6	ATSAM4S ARM Cortex M4/32bit	120
7	ATSAMG55 ARM Cortex M4/32bit	120
8	ATSAMV71 ARM Cortex M7/32bit	300

4 Results and Discussion

From the first experiment, the current consumption resulting from the procession of inconstant pointer was shown in Fig. 4. The current consumptions were in the same trend, which were maximized at highest driving clock frequency. The microcontroller consuming the lowest current at driving clock frequency of 1 MHz is SAMG55. The consumed current was 0.9 mA.

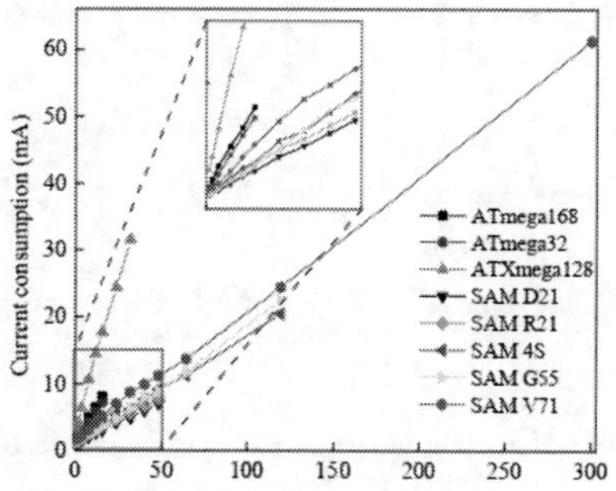

Fig. 4. Current consumption with increasing of clock frequency

With the second experiment, the processing times taken for calculating 2000 digits of Pi at each driving clock frequency were shown in Fig.5. The shortest processing time is noticed at highest clock frequency. The increasing clock frequency results the data processing speed increased. On the other hand, the results also presented the mathematic calculation performance of each microcontroller. The highest-performance microcontroller spends a shortest processing time. As the results, at driving clock frequency of 1 MHz, a SAMV71 present the best performance with processing time of 455.2 ms.

Figure. 6 show the results of third experiment. The current consumptions of microcontrollers are higher than the first experiment because the microcontrollers write and read the data into RAM. The lowest current consumption of 1.18 mA was found at SAMD21.

Following the fourth experiment, the current consumptions in the energy saving mode (sleeping mode) of microcontrollers were found as show in the Table II. As the results, the microcontrollers consume the different amount of energy due to their architectures. The ATSAMR21 and ATSAMV71 microcontrollers consume the much higher current than others, since SAMR21 composed of the wireless module inside which still consume the current even stayed in the energy saving mode, and SAMV71 is designed to consist of the complex integrated circuit for high-performance of data procession. With very high current consumption, the ATSAMR21 and ATSAMV71 microcontrollers are not appropriate for low-power wireless application. On the other hand, the xmega128 presented the very low current consumption which is 0.27 μA.

The trends of energy consumptions illustrated in Fig. 7 derived from last experiment. The energy consumptions decreased due to the increase of driving clock frequency. With maximum driving clock frequency, the lowest energy consumption of each microcontroller was found as shown in Fig. 8. As can be seen, the SAMG55 consume the lowest energy which is 1.22 mJ.

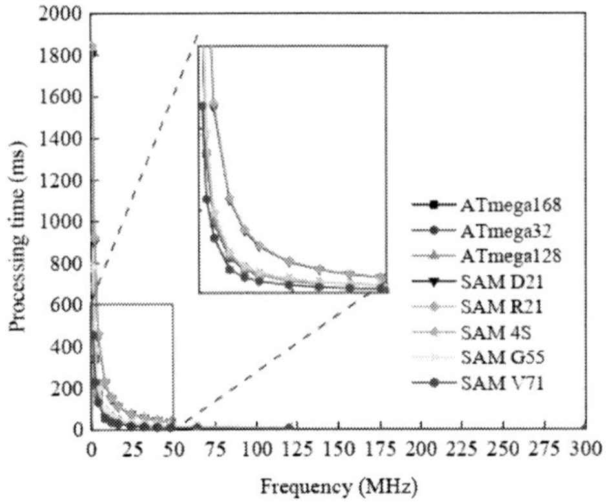

Fig. 5. Processing time for calculating the 2000 digits of Pi.

This work can be summarized that even through the high driving clock frequency is a cause of high current consumption, it results in the decrease of energy consumption. Therefore, the microcontrollers should be driven with their maximum clock frequency for wireless sensor application consuming low-energy.

Table 2. Current consumption for sleeping mode.

Current consumption (μA)	
ATMega168	32.00
ATMega32	41.59
ATXmega128	0.27
ATSAMD21	3.01
ATSAMR21	460.59
ATSAMG55	57.14
ATSAM4S	20.45
ATSAMV71	1344.18

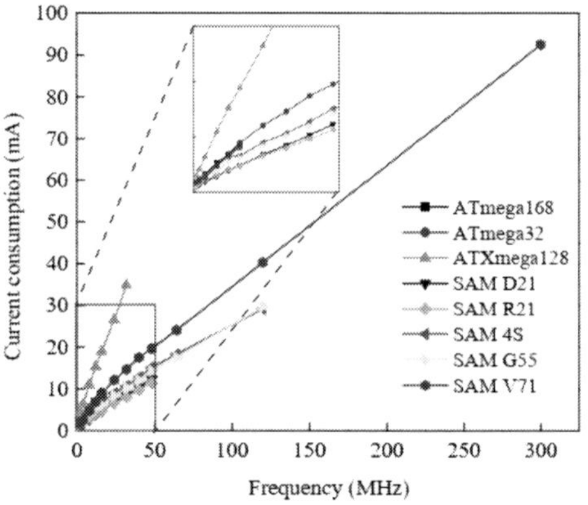

Fig. 6. Interaction between current consumption and driving clock frequency for calculating the 2000 digits of Pi.

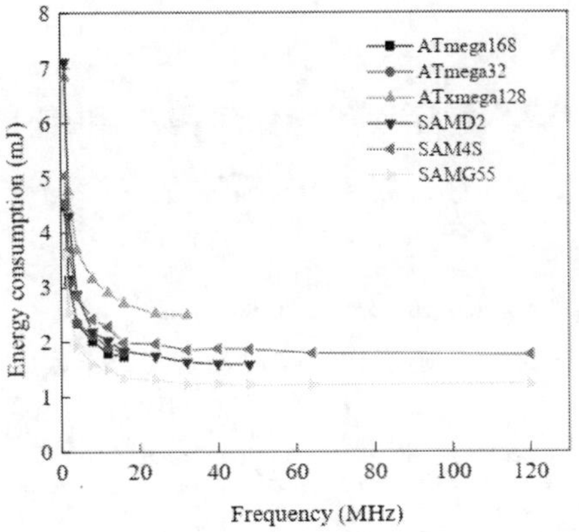

Fig. 7. Energy consumption with increasing of driving clock frequency.

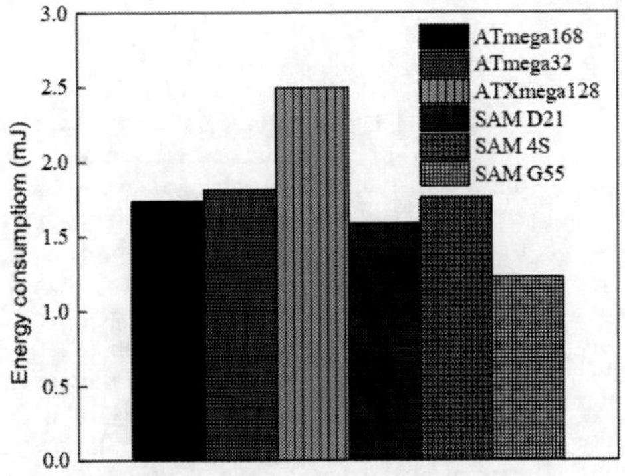

Fig. 8. The energy consumption of microcontrollers at maximum driving clock frequency.

5 Conclusion

This paper presented the experimental study of interaction between clock frequency and energy consumption of microcontrollers for wireless sensor application. The current energy consumptions resulted from the various driving clock frequencies were investigated with following the operational feature of wireless sensor application. As the results, the microcontrollers consumed the lowest energy at their maximum driving clock frequency.

Moreover, this work also demonstrated that the selection of microcontrollers for wireless application consuming the low energy can be considered by flowing the performances below:

- The mathematic data processing speed.
- The maximum driving clock frequency.
- The current consumption in the active and sleep modes.

Acknowledgement

This work was supported by King Mongkut's Institute of Technology Ladkrabang Research Fund [KREF016101].

References

1. T. Sudhawiyangkul, D. Isarakorn, *Design and optimization of a small-scale linear electromagnetic energy harvester*, pp. 1-6, (2015)
2. J. Kim ; C. Chu, *Analysis of energy consumption for wearable ECG devices*, pp. 962-965, (2014)
3. K. Masaba, A. Ntakirutimana, T. Ustun, *Design and implementation of a smart irrigation system for improved water-energy efficiency*, pp. 1-5, (2016)
4. T. Krongtripop, *A performance comparisong of energy conservation between idle mode and power down mode for microcontroller in wireless mobile node*, pp. 552 – 556, (2010)
5. *Atmel ATMega32 datasheet* "Active Supply Current vs. Frequency", pp 308.
6. I. Tsekoura et al, *An evaluation of energy efficient microcontrollers*, pp. 1–5, (2014)
7. S. Janković, V. Drndarević, *Microcontroller power consumption measurement based on PSoC*, pp. 673 – 676, (2015)
8. S. Suradhanwiar ; S. Durbha ; J. Adinarayana *Energy optimization in wireless sensor networks using duty-cycling approach*, pp. 1–6, (2017)
9. *Microchip ATSAMV71 datasheet* "Atmel-44003E-ATARM-SAM V71-Datasheet_12-Oct-16", (2016).

Peristaltic pumping through porous medium in presence of electric double layer

Ashu Yadav[1*], *Shashi* Bhushan[2], and *Dharmendra* Tripathi[2]

[1]Automobile Engineering Department, Faculty of Engineering, Manipal University, Jaipur-303007, India

[2]Mechanical Engineering Department, Faculty of Engineering, Manipal University, Jaipur-303007, India

Abstract. The present paper investigates the peristaltic flow of Newtonian fluids through the porous medium. The effects of electroosmosis mechanism on peristaltic pumping are also considered. An analytical solution is obtained under lubrication approach. Poisson Boltzmann equations are also simplified using Debye linearization. The effects of permeability parameter, electrical double layer thickness and electro-osmotic parameter on the flow characteristics, pressure distribution and shear stress distributions are computed. Numerical computations reveal that electroosmosis and permeability play vital role in peristaltic pumping. The findings of present study may be applicable in biomedical engineering and chemical engineering where peristaltic micropumps may be designed.

1 Introduction

Mathematical modeling of various real systems is governed by navier-stokes equations in various emerging field of Fluid dynamics. Understanding the basic transport mechanisms in biological systems are not as easy like engineering applications, because in-vivo experiments are not easy to perform. The concepts of modern bio-fluid mechanics can be applied to many systems in the human body. Gastrointestinal tract is an interesting part of the human physiological system that has many physical processes coupled with fluid dynamics. The physiology of the intestinal movements was investigated for the first time, confined to the small intestine of the dog, by Bayliss and Starling[1]. In the literature, the transport mechanism of chyme in intestines presets a special challenges to bio engineers Lew H. et al. [2] developed a mathematical model of peristalsis involved in small intestines based on low Reynolds number flow of a Newtonian fluid in a circular cylindrical tube. This idea was extended by many researchers to reflect the non-linear behavior of the chyme. Srivastava and Srivastava [3] studied fluid dynamic applications of vas deferens and small intestine in a uniform and non-uniform tube under the consideration of a non-Newtonian (Power law) fluid model.

Flow of fluids in channels under the effect of an electrical field is a key area in both medical engineering and energy sciences. Both electro-hydrodynamics and electro-kinetics (electro-osmotic transport) [4-6] are important aspects of this arena. The latter plays a prominent role in micro- and nano-channel systems in ionic flow delivery [7-9], petro-chemical sampling in low-permeability reservoirs [10], geotechnical remediation and field-based stabilization of geological masses [11]. Electrokinetics can be very successfully utilized to transform mechanical energy into electrical energy, to regulate flow delivery, achieve specified mass transfer rates etc. enhanced. Electro-osmotic flows in porous media are also of significance and arise in both biological systems and industrial systems and have been shown to markedly influence transport rates [12]. The classical approach to simulating transport in porous media is the Darcy model which is valid for viscous-dominated, low Reynolds number flows. It has been implemented extensively in electrokinetic modelling for porous media and such studied provide a very important compliment to laboratory-based investigations [13]. Gupta et al. [14] analysed the electrokinetic flow in porous media micro-channels with Stern layer effects, computing expressions for surface charge density, electrical conductivity, and electroosmotic coupling coefficient for various porous structures and physico-chemical boundary conditions

After reviewing the above literature, the present study aims to analyze the peristaltic flow of Newtonian fluids through porous medium in presence of electroosmosis. The influences of pertinent physical parameters on velocity, pressure and wall shear stress are examined. The present model can further be explored for various model of non- Newtonian fluids which are more relevant in biological environments.

2 Mathematical model

The geometric model for peristaltic flow through a finite length (L) porous medium is depicted in Fig.1, and mathematically expressed as:

* Corresponding author: ashu.yadav@jaipur.manipal.edu

$$\bar{h}(\bar{x},\bar{t}) = a + b\sin\frac{2\pi}{\lambda}(\bar{x} - c\bar{t}), \qquad (1)$$

where $a, b, \lambda, \bar{x}, c, \bar{t}, L$ are the half width of the channel, amplitude of wave, wavelength, axial coordinate, wave velocity, time and channel length.

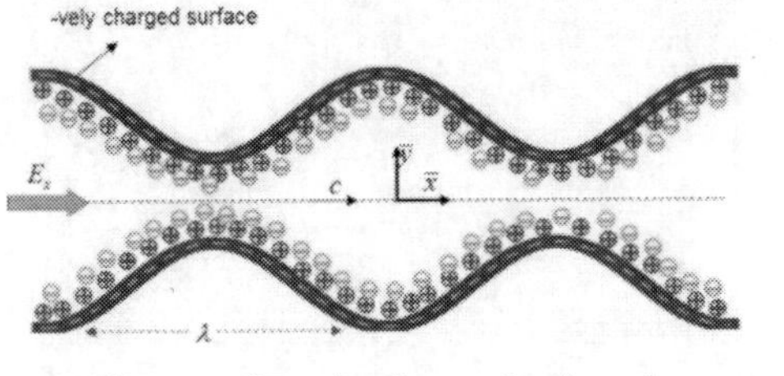

Fig. 1. Geometry for peristatic pumping through porous medium in presence of EDL.

The governing equations for two dimensional unsteady peristaltic flow of Newtonian fluids through porous medium in presence of EDL are expressed as:

$$\frac{\partial \bar{u}}{\partial \bar{x}} + \frac{\partial \bar{v}}{\partial \bar{y}} = 0, \qquad (2)$$

$$\rho\left(\frac{\partial \bar{u}}{\partial \bar{t}} + \bar{u}\frac{\partial \bar{u}}{\partial \bar{x}} + \bar{v}\frac{\partial \bar{u}}{\partial \bar{y}}\right) = -\frac{\partial \bar{p}}{\partial \bar{x}} + \mu\left(\frac{\partial^2 \bar{u}}{\partial \bar{x}^2} + \frac{\partial^2 \bar{u}}{\partial \bar{y}^2}\right) - \mu\frac{\bar{u}}{\bar{k}} + \rho_e E_x, \qquad (3)$$

$$\rho\left(\frac{\partial \bar{v}}{\partial \bar{t}} + \bar{u}\frac{\partial \bar{v}}{\partial \bar{x}} + \bar{v}\frac{\partial \bar{v}}{\partial \bar{y}}\right) = -\frac{\partial \bar{p}}{\partial \bar{y}} + \mu\left(\frac{\partial^2 \bar{v}}{\partial \bar{y}^2} + \frac{\partial^2 \bar{v}}{\partial \bar{y}^2}\right) - \mu\frac{\bar{v}}{\bar{k}}, \qquad (4)$$

where $\rho, \bar{u}, \bar{v}, \bar{p}, \mu, \bar{k}$ and E_x denote the fluid density, axial velocity, transverse velocity, pressure, fluid viscosity, the permeability of the porous medium and electrokinetic body force.

The Poisson's equation is defined as :

$$\nabla^2\phi = -\frac{\bar{\rho}_e}{\varepsilon}, \qquad (5)$$

in which ρ_e is the density of the total ionic charge and ε is the permittivity. The Boltzmann equation is expressed as :

$$n^{\pm} = n_0 Exp\left[\pm\frac{ez\bar{\Phi}}{K_B T}\right], \qquad (6)$$

where n_0 represents concentration of ions at the bulk, which is independent of surface electrochemistry, e is the electronic charge, is the charge balance, K_B is the Boltzmann constant, and T is the average temperature of the electrolytic solution.

Employing the following boundary conditions as ;

$$x = \frac{\bar{x}}{\lambda}, y = \frac{\bar{y}}{a}, t = \frac{c\bar{t}}{\lambda}, u = \frac{\bar{u}}{c}, v = \frac{\bar{v}}{c\delta}, \delta = \frac{a}{\lambda}, h = \frac{\bar{h}}{a},$$
$$\phi = \frac{b}{a}, p = \frac{\bar{p}a^2}{\mu c\lambda}, \mathrm{Re} = \frac{\rho c a}{\mu}, k = \frac{\bar{k}}{a^2}, \Phi = \frac{\bar{\Phi}}{\zeta}, \qquad (7)$$

where δ is wave number, Re is the Reynolds number, ζ is the zeta potential, Applying long wavelength and low Reynolds number approximations, the above Eqs. (2) - (4) reduce to

$$\frac{\partial u}{\partial x} + \frac{\partial v}{\partial y} = 0, \qquad (8)$$

$$\frac{\partial p}{\partial x} = \frac{\partial^2 u}{\partial y^2} - \frac{u}{k} + m^2 U_{HS}\phi, \qquad (9)$$

$$\frac{\partial p}{\partial y} = 0 \qquad (10)$$

where $m = aez\sqrt{\frac{2n_0}{\varepsilon K_B T}}$, is known as the electroosmotic parameter and $U_{HS} = -\frac{E_x\varepsilon\zeta}{\mu c}$ is the maximum electroosmotic velocity. Applying Debye-Hückel linearization approximation, Poisson-Boltzmann equation reduces to :

$$\frac{\partial^2\Phi}{\partial y^2} = m^2\Phi, \qquad (11)$$

The boundary conditions are :

$$\left.\frac{\partial\Phi}{\partial y}\right|_{y=0} = 0, \Phi|_{y=h} = 1, \left.\frac{\partial u}{\partial y}\right|_{y=0} = 0, u|_{y=h} = 0, v|_{y=0} = 0,$$
$$v|_{y=h} = \frac{\partial h}{\partial t}\; p|_{x=0} = p_0 \; and \; p|_{x=L} = p_L \qquad (12)$$

The solution of above governing equations subjected to boundary conditions are

$$\Phi = \frac{\cosh(my)}{\cosh(mh)}, \qquad (13)$$

$$u = k\frac{\partial p}{\partial x}\left\{\frac{\cosh(\frac{y}{\sqrt{k}})}{\cosh(\frac{h}{\sqrt{k}})} - 1\right\} + \frac{m^2 U_{HS}}{m^2 - \frac{1}{k}}\left\{\frac{\cosh(\frac{y}{\sqrt{k}})}{\cosh(\frac{h}{\sqrt{k}})} - \frac{\cosh(my)}{\cosh(mh)}\right\} \qquad (14)$$

Furthermore, utilizing Eq. (13) with continuity equation, and boundary condition (11), the *pressure gradient* emerges as :

$$\frac{\partial p}{\partial x} = \frac{1}{k^{3/2}\left(\frac{h}{\sqrt{k}} - \tanh(\frac{h}{\sqrt{k}})\right)}\left[G_0(t) + \int\frac{\partial h}{\partial t}dx - \frac{m^2 U_{HS}}{m^2 - \frac{1}{k}}\left\{\begin{matrix}\sqrt{k}\tanh(\frac{h}{\sqrt{k}}) \\ -\frac{\tanh(mh)}{m}\end{matrix}\right\}\right] \qquad (15)$$

The pressure difference can be obtained as :

$$\Delta p = p(\mathrm{x}, t) - p(0, t)\int_0^x \frac{\partial p}{\partial s}ds, \qquad (16)$$

where $G_0(t)$ is expressed as:

$$G_0(t) = \frac{(p_l - p_0) - \int_0^L \frac{1}{k^{3/2}\left(\frac{h}{\sqrt{k}} - \tanh(\frac{h}{\sqrt{k}})\right)}\left[\int\frac{\partial h}{\partial t}dx - \frac{m^2 U_{HS}}{m^2 - \frac{1}{k}}\left\{\begin{matrix}\sqrt{k}\tanh(\frac{h}{\sqrt{k}}) \\ -\frac{\tanh(mh)}{m}\end{matrix}\right\}\right]dx}{\int_0^L \frac{1}{k^{3/2}\left(\frac{h}{\sqrt{k}} - \tanh(\frac{h}{\sqrt{k}})\right)}dx} \qquad (17)$$

The local wall shear stress is derived as :

$$\tau_w = \left.\frac{\partial u}{\partial y}\right|_{y=h} = \sqrt{k}\tanh(\frac{h}{\sqrt{k}})\frac{\partial p}{\partial x} - \frac{m^2 U_{HS}}{m^2 - \frac{1}{k}}\left\{\frac{\tanh(\frac{h}{\sqrt{k}})}{\sqrt{k}} - m\tanh(mh)\right\} \qquad (18)$$

3 Results and Discussion

Numerical computations for axial velocity, pressure distribution and local wall shear stress have performed under the effects of debye length, electroosmotic velocity and permebility parameter. To visualize the effects of physical parameters, Figs. 2-4 are illustrated.

Figs. 2a-c illustrate the evolution of axial velocity (u) with EDL thickness (m), external electric field (U_{HS}) and permeability parameter (k) respectively. Fig. 2a shows the effect of EDL thickness on velocity profile, and it is noted that the axial velocity increases with increasing the EDL thickness. Fig. 2b depicts the effect of external electric field on velocity and it is noted that axial velocity increases with increasing external electric field. Fig. 2c depicts the effect of permeability parameter on velocity profile and it is noted that axial velocity increases with increasing permeability parameter.

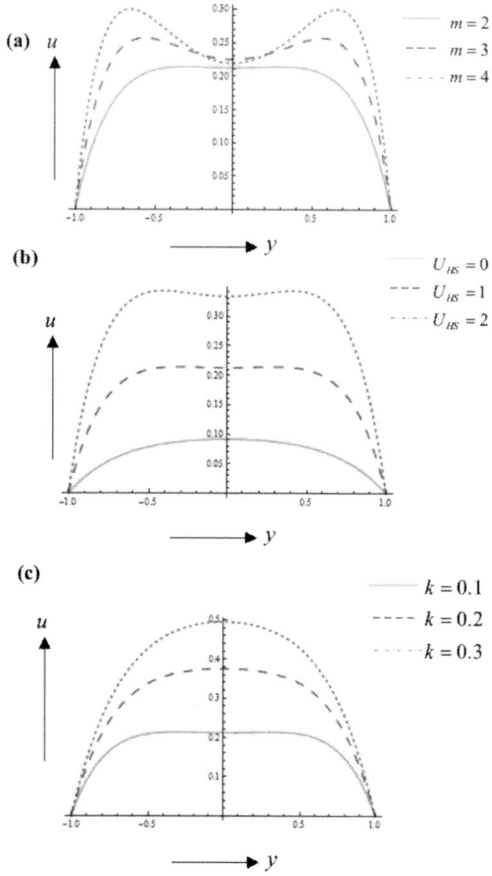

Fig. 2. Velocity profile (axial velocity vs. transverse coordinate) at $\varphi = 0.5, x = 0$ and **(a)** $U_{HS} = 1, k = 0.1$ **(b)** $m = 2, k = 0.1$ **(c)** $U_{HS} = 1, m = 2$.

Figs. 3 a-c present the pressure difference across the channel with EDL thickness (m), external electric field (U_{HS}) and permeability parameter (k) respectively. It is observed that pressure difference is highest at the wall contraction and its magnitude increases with increasing the value of EDL thickness and external electric field while it decreases with increasing permeability parameter. The effect of external electric field is quite significant as compared with effect of EDL thickness.

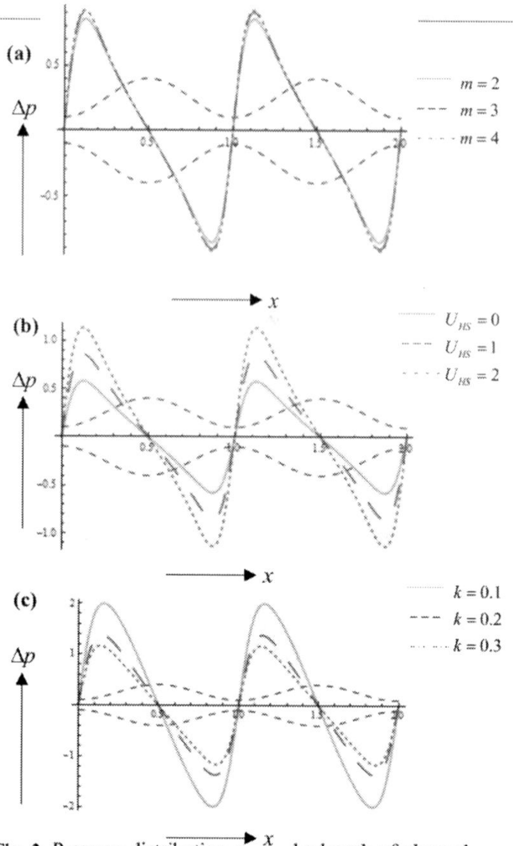

Fig. 3. Pressure distribution along the length of channel at $\varphi = 0.5, t = 0.25, p_L = p_0 = 0$ and **(a)** $U_{HS} = 1, k = 0.1$ **(b)** $m = 2, k = 0.1$ **(c)** $U_{HS} = 1, m = 2$.

Figs. 4 a-c depict the shear stress distribution along the channel length with EDL thickness (m), external electric field (U_{HS}) and permeability parameter (k) respectively. The magnitude of shear stress is hightest near the contraction and minimum where channel is releaxed. Further it is observed that local wall shear stress increses with increasing the value of EDL thickness and external electric field, but permeability parameter has reverse effect on wall shear stress.

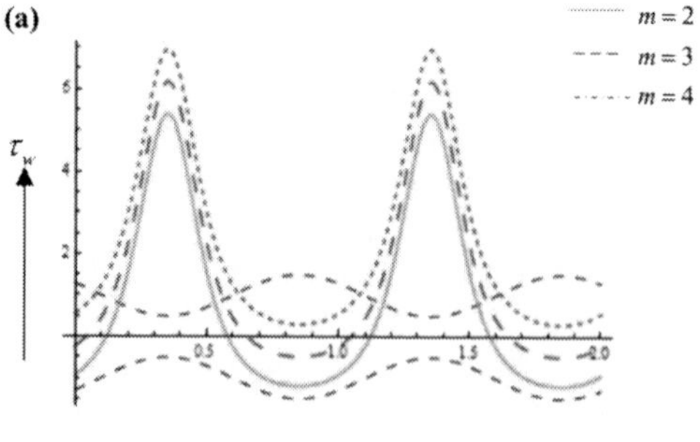

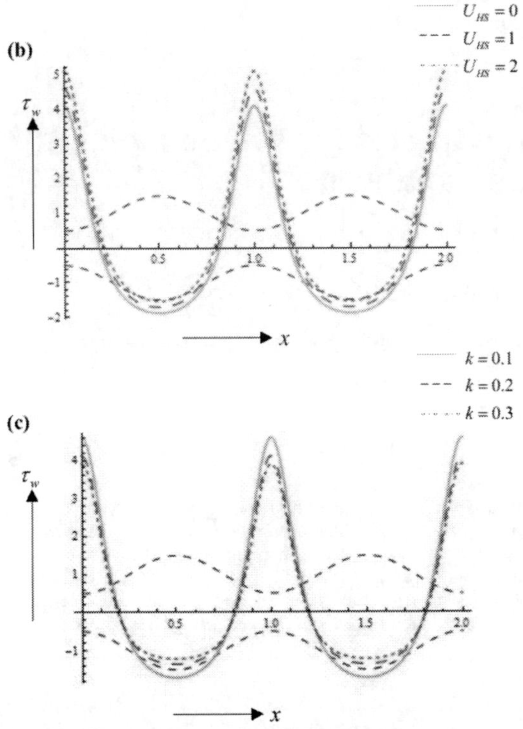

Fig. 4. Local wall shear stress along the length of channel at $t=0.25, \varphi=0.5$, (a) $U_{HS}=1, k=0.1$ (b) $m=1, k=0.1$ (c) $U_{HS}=1, m=1$

4 Conclusions

An analytical model is developed to study the peristaltic flow of Newtonian fluids through porous medium. The effects of permeability parameter, EDL thickness and external electric field on flow characteristics, pumping characteristics (i.e. pressure distribution) and wall shear distribution along the channel length are analyzed. It is noted that axial velocity strongly depends on the permeability, EDL thickness and external electric field. It is also concluded that peristaltic mechanisms can be controlled with applying the external electric field. The permeability of the medium resists the peristaltic flow process. That means if the medium is highly porous then peristaltic pumping can be maintained by external electric field. It is further noted that wall shear stress alters with electroosmosis and permeability. The finding of present model can be helpful in biomedical and industrial applications.

References

1 W. Bayliss and E. Starling, The Journal of physiology, **24,** pp. 99-143, (1899).
2 H. Lew, Y. Fung, and C. Lowenstein, Journal of Biomechanics, **4,** pp. 297-315, (1971).
3 L. Srivastava and V. Srivastava, Annals of biomedical engineering, **13,** pp. 137-153, (1985).
4 C.-O. Ng and Q. Zhou, Fluid Dynamics Research, **44,** p. 055507, (2012).
5 Y. Kang, K. Ooi, C. Yang, and T. Wong, Journal of Micromechanics and Microengineering, **15,** p. 301, (2004).
6 G. D. Ngoma and F. Erchiqui, Journal of Micromechanics and Microengineering, **16,** p. 83, (2005).
7 K. Kim, H. S. Kwak, and T.-H. Song, Fluid Dynamics Research, **43,** p. 041401, (2011).
8 K.-D. Huang and R.-J. Yang, Nanotechnology, **18,** p. 115701, (2007).
9 Y. Ren and D. Stein, Nanotechnology, **19,** p. 195707, (2008).
10 W. Zhang, J. Yao, Y. Gao, Q. Zhang, and H. Sun, Journal of Petroleum Science and Engineering, **134,** pp. 150-157, (2015).
11 R. Azzam and W. Oey, Transport in Porous Media, **42,** pp. 293-314, (2001).
12 D. B. Pengra, L. Shi, S. X. Li, and P.-Z. Wong, "*Experimental study of electrokinetics in porous media,*" in *MRS Proceedings*, p. 201, 1994.
13 R. S. Maier, E. Nybo, J. D. Seymour, and S. L. Codd, Transport in Porous Media, **113,** pp. 67-89, (2016).
14 A. Gupta, D. Coelho, and P. Adler, Journal of colloid and interface science, **316,** pp. 140-159, (2007).

Optimization of geometrical parameter for V-type alpha Stirling engine based on dimensionless analysis

Varit Senviboon[1], *Banterng* Silpsakoolsook[2], *Sutapat* Kwankaomeng[1,*]

[1] Department of Mechanical Engineering, Faculty of Engineering, King Mongkut's Institute of Technology Ladkrabang, Bangkok, Thailand

[2] Department of Chemistry, Faculty of Science, Siam University, Bangkok, Thailand

Abstract. This paper aims to study the effect of mechanism and geometrical parameters on the shaft work of alpha-type Stirling engine with V-cylinder arrangement and find the appropriate value of geometrical parameters to get the maximum value of shaft work output. The equation of motions of both expansion and compression pistons and the variations of volumes and pressures inside the cylinders are expressed in dimensionless forms. Schmidt analysis and Senft's shaft work theory are used to build dimensionless thermodynamic models in order to derive dimensionless shaft work and optimal geometrical parameters for achieving the maximum efficiency

1 Introduction

Stirling engine is a heat engine that works as a cycle of expansion and compression of air or other working gas by the effects of temperature differences. From these processes, the conversion of energy from heat to mechanical energy in addition, the Stirling engine is a closed-loop cycle engine that has a constant working gas is sealed without transferring to the environment. In practice, the Stirling engine are classified into three types, based on the configurations and varying displacements of piston and displacer, which are alpha, beta, and gamma. The alpha Stirling engine the simplest pattern form of the Stirling engine.

Karabulut H. [1] built and tested performance of the alpha Stirling engine. The engine provided the maximum power under the heater temperature of 1100 °C and 2.5 bar of engine's pressurization. Batmaz I, Ustun S [2] employed double heaters for solar power heating V-type Stirling engine which the unpressurized engine gave the best performance at 950 °C heater surface. A.A. El-Ehwany et al. [3] tested and tried to use elbow-bend as heater and cooler part of alpha Stirling engine. The result of the engine power delivers about 13% more power per cc per ΔT than the literature from this previous study. The optimization of the alpha Stirling engine were proposed by Homutescu and Balanescu [4]. The results show that the diameter of the expansion piston must be greater than the compression piston diameter. Lately, Scollo et al. [5] proposed the constructive and redesign of the prototype of the twin cylinder alpha Stirling engine. The engine performance was greatly increased. Recently, Martaj N., Rochelle P. [6] presented a 1D modeling of an alpha double-acting Stirling engine engines and carried out the optimal values of the engine geometry.

According to optimization of the Stirling engine, Senft [7] introduced the concept of effectiveness of mechanism and the relationship between force work and the shaft work of an engine. Later, the same author [8] the Schmidt theory was used to combined and find the optimal values for the phase angle and the swept volume ratio for gamma type Stirling engine. Cheng and Yang [9]. also used this method to evaluate and compare the power of Stirling engine between the alpha, beta, and gamma. The result shows that a beta type has the highest power, followed by the alpha and gamma. But the gamma Stirling engine is the most suitable for low temperature different applications. As the same author [10] the rhombic mechanism variable of the beta Stirling engine is also considered to be integrated with the method, in order to find the optimal mechanism variable and other variable. So that in this paper, propose the optimization by using the slider crank mechanism variable of an alpha Stirling engine to cooperate with this method. All variables considered were phase angle, area ratio between expansion side and compression side, temperature ratio, dead volume ratio and the ratio between connecting rod and crank radius at expansion side and compression side.

* Corresponding author: sutapat.kw@kmitl.ac.th

2. Dimensionless analytical models

2.1 Displacement of expansion and compression pistons

Since the changing of the displacement of expansion, $\bar{y}_e$ and compression, $\bar{y}_c$ pistons which is driven by slider crank mechanisms affect to the variation of volume inside the engine. The equations of displacement of slider crank mechanisms can be presented in dimensionless form as

$$\bar{y}_e(\theta) = \cos\theta + \sqrt{e_d^2 + (\sin\theta)^2} \quad \text{(1a)}$$

$$\bar{y}_c(\theta) = \cos(\theta-\alpha) + \sqrt{e_p^2 + (\sin(\theta-\alpha))^2} \quad \text{(1b)}$$

Where $e_d = l_e/r_e$ and $e_p = l_c/r_c$. Note that in practical, the variable e_p and e_d must be more than one. If it isn't, it cannot be a slider crank mechanism.

2.2 The variations of the volumes inside the cylinders

The volumes of the expansion side and compression side are changing because of movement of these two pistons in this case are assumed that no dead volume from clearance displacement between cylinder and piston. Let V_{se} is the swept volume of the expansion side $(= A_e r_e \bar{S}_e)$ so the equation of volume variation of expansion side, $\bar{V}_e$ and compression side, $\bar{V}_c$ can be written in dimensionless form as

$$\bar{V}_e(\theta) = [\bar{y}_{e,max} - \bar{y}_e(\theta)]/\bar{S}_e \quad \text{(2a)}$$

$$\bar{V}_c(\theta) = [\bar{y}_{c,max} - \bar{y}_c(\theta)]/a\bar{S}_c \quad \text{(2b)}$$

Where a is an area ratio that is $a = A_e/A_c$, $\bar{V}_e(\theta) = V_e/V_{se}$, $\bar{V}_c(\theta) = V_c/aV_{se}$, when $r_e = r_c$ so that the equation of total working gas volume variation can be written as

$$\bar{V}_t(\theta) = \bar{V}_h + \bar{V}_r + \bar{V}_k + \bar{V}_e(\theta) + \bar{V}_c(\theta) \quad \text{(3)}$$

At the first three terms are dimensionless volumes of heater $\bar{V}_h$, regenerator $\bar{V}_r$, and cooler $\bar{V}_k$, respectively The first three terms can be written as $\chi = \bar{V}_h + \bar{V}_r + \bar{V}_k$.

2.3 Pressure and back pressure

From the Schemidt theory, this theory is used to determine the pressure inside the working space. The temperature at total dead space is estimated as an average of T_e and T_c. The Schemidt theory can be rearranged and presented in dimensionless form as

$$\bar{p}(\theta,\alpha,a,\tau,\chi,e_p,e_d) = p(\theta)/\frac{mRT_e}{V_{se}}$$

$$= \tau/(2\tau\chi/(1+\tau) + \tau\bar{V}_e(\theta) + \bar{V}_c(\theta)) \quad \text{(4)}$$

Where $\tau = T_c/T_e$ is the temperature ratio, and $\bar{p}$ is dimensionless pressure

As discussed by Senft [7], the optimal value of the back pressure is the average cyclic pressure, which is

$$\bar{p}_b = \sqrt{\bar{p}_{max}\bar{p}_{min}} \quad \text{(5)}$$

2.4 Force, Indicated, and shaft works

The area under $\bar{p} - \bar{V}_t$ diagram is the dimensionless indicated work can be evaluated from $\oint \bar{p}d\bar{V}_t$.

Force work is work loss in the mechanisms. For the alpha Stirling engine force work will act at the expansion and compression pistons, the result of the pressure different on both sides of this two pistons. Based on the definition given by Senft [7]. Dimensionless force work can be calculated by

$$\bar{W}_f = \oint[(\bar{p}-\bar{p}_b)d\bar{V}_t]^- \quad \text{(6)}$$

Where the symbol $[]^-$ means

$$[n]^- = \begin{cases} -n, & n \le 0 \\ 0, & n > 0 \end{cases} \quad \text{(7)}$$

From equation (7) as it is described by Senft [7] can be determine dimensionless shaft work by

$$\bar{W}_s = E\bar{W}_i - (1/E - E)\bar{W}_f \quad \text{(8)}$$

Where E is the effectiveness of mechanism, depending on the mechanisms design, friction, speed revolution, mechanism parts manufacturing, lubrication and etc. In this study, value of E is defined to 0.8.

3. Numerical Results

The study effects of mechanism size and parameters were analyzed by theoretical analysis with search method. The results of the parameters are as follows: Fig. 1 shows the effect of temperature ratios on $\bar{W}_i$, $\bar{W}_s$ and $\bar{W}_f$ at E = 0.8, χ = 0.1, α = 90°, a = 1 and $e_p = e_d = 5$, then the values of $\bar{W}_i$ and $\bar{W}_s$ are continuously decreasing, but the value of $\bar{W}_f$ increases with the value of τ. From τ at 0.759, the value of $\bar{W}_s$ is zero. This indicates that the engine cannot work and it implies this engine type is not suitable for use at low temperature differences. The value of $\bar{W}_i$ is zero when the value of τ = 1, where this point means there is no difference in temperature between hot and cold side.

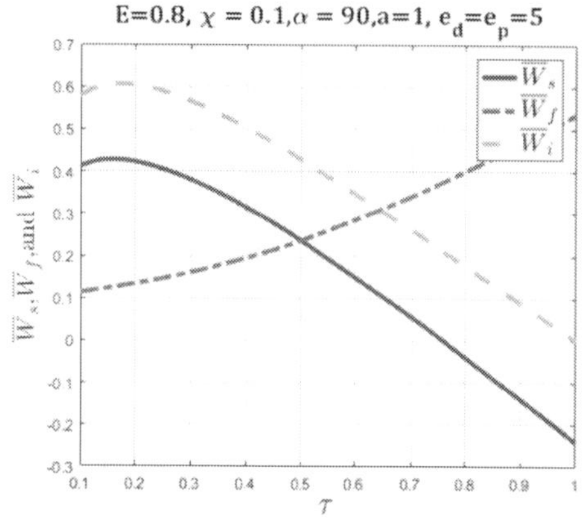

Fig. 1. Effects of temperature ratio on dimensionless works

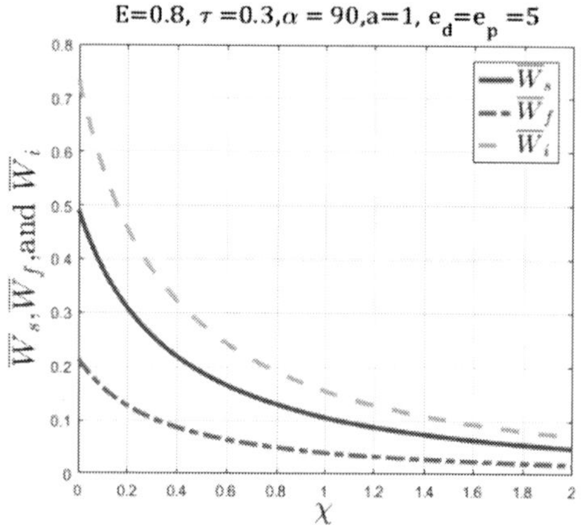

Fig. 2. Effect of total dead volume ratio on dimensionless works

The effect of dead volume ratio χ on $\overline{W}_i$, $\overline{W}_s$ and $\overline{W}_f$ is shown in Fig. 2 at $E = 0.8$, $\tau = 0.3$, $\alpha = 90°$, $a = 1$ and $e_p = e_d = 5$. The results clearly show that the increase in the dead volume ratio will decreases the value of the output. It is observed that the values of $\overline{W}_i$, $\overline{W}_s$ and $\overline{W}_f$ in the range χ are between 0 and 0.8, the value of the output decreases rapidly. But after this, the value starts to drop slowly. In theory, the value of the dead volume ratio should be as low as possible, but in practices, the design and construction of the engine, the heater, regenerator and the cooler part also have dead volume inside. So that this problem cannot be avoided. That why the study of the effect of the dead volume ratio is important.

Fig. 3 show the effect of e_p and e_d on dimensionless work at $E = 0.8$, $\tau = 0.3$, $\chi = 0.1$, $\alpha = 90°$ and $a = 1$. Let's $e_p = e_d$. The results of e_p and e_d are between 1.1 and 5, the value of the dimensionless work is high rate variable. And sooner or later, the work will increase slowly and little by little. As a result, the greater e_d and e_p will be better transmit work.

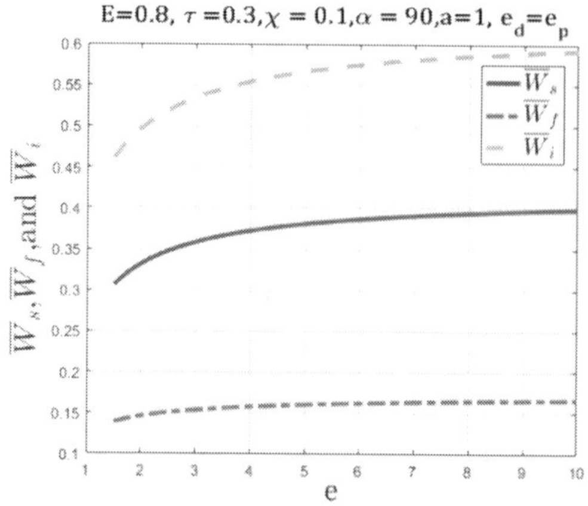

Fig. 3. Effect of e_p and e_d on dimensionless works

The effect of the phase angle on $\overline{W}_i$, $\overline{W}_s$ and $\overline{W}_f$ is shown in Fig. 4 at $E = 0.8$, $\tau = 0.3$, $\chi = 0.1$, $a = 1$ and $e_p = e_d = 5$. The result is that in the initial phase, the values of $\overline{W}_i$ and $\overline{W}_s$ are increased to the inverted curve by the maximum $\overline{W}_s$ is $\alpha = 77°$, and then it decreases.

Fig. 5 shows the effect of the area ratio of the expansion on the compression side in case $E = 0.8$, $\tau = 0.3$, $\chi = 0.1$, $\alpha = 90°$ and $e_p = e_d = 5$.From Fig. 5, it is notice that a must more than 0.22 so the engine can operate. In the range of $a = 0.1$ to 1. It is found that the values of $\overline{W}_i$ and $\overline{W}_s$ are increasing rapidly, while the value of $\overline{W}_f$ decreases rapidly. And then $\overline{W}_i$ and $\overline{W}_s$ will increase slowly and very little. $\overline{W}_s$ have the maximum value of $a = 1.79$, and then $\overline{W}_s$ will gradually decrease but very little. The result is that the value of the area ratio must be greater than 1, because if the area ratio is less than 1, then the area of the expansion side smaller than the compression area will be a mechanical disadvantage, resulting in the output power will be decrease or in practices, the engine may not be run.

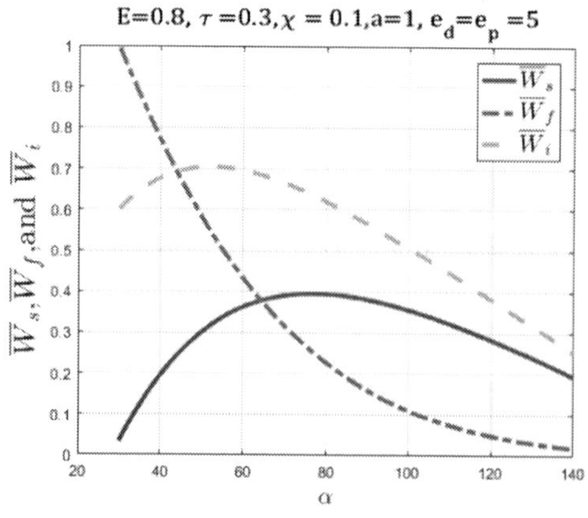

Fig. 4. Effect of phase angle on dimensionless works

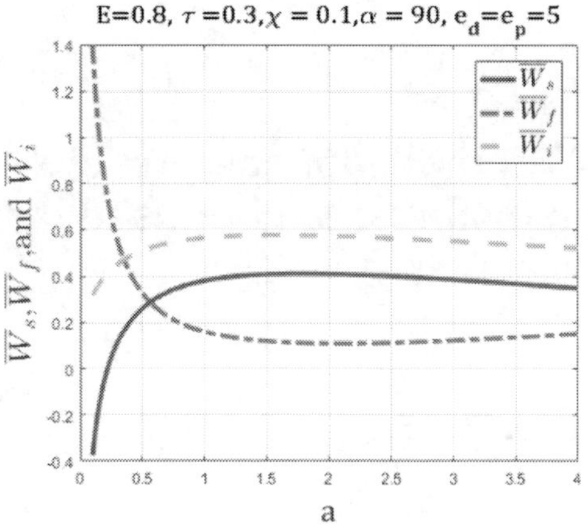

Fig. 5. Effect of area ratio on dimensionless works

The maximum value of the shaft work can be found from two variables that are area ratio and phase angle. So we use these two variables primarily to find the work from the maximum shaft work. Fig. 6 shows contour diagram as the parametric dimensionless shaft work in the function of the area ratio and the phase angle for the alpha V-type Stirling engines at. $E = 0.8$, $\tau = 0.3$, $\chi = 0.1$ and $e_p = e_d = 5$. We discover that dimensionless shaft work optimize value are 0.4292 at $a = 1.73$ and $\alpha = 76°$. However the output from variable of area ratio and the phase angle are also depends on the value of effectiveness of mechanism, temperature ratio, dead volume ratio and e_p, e_d. If one value changes, the optimize value will change.

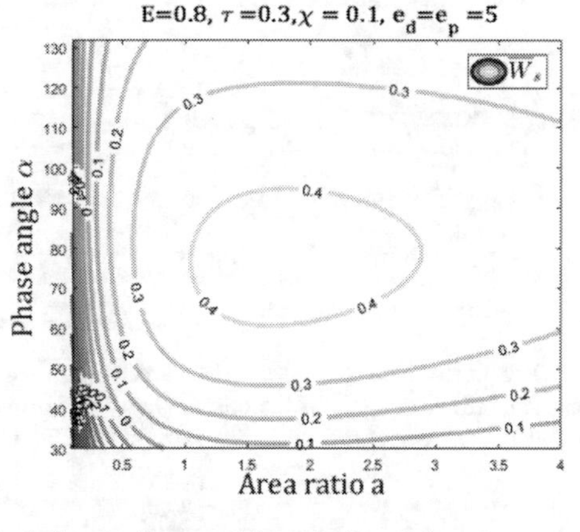

Fig. 6. Contour diagram of dimensionless shaft work as a function of phase angle and area ratio

4. Conclusion and Remarks

The result of the assumption at $E = 0.8$, $\chi = 0.1$, $\alpha = 90°$, $a = 1$ and $e_p = e_d = 5$ indicates that when the temperature ratio increases to 0.759, the value of dimensionless shaft work $\overline{W}_s$ is zero. It's implies that this engine is not suitable for use in low temperature different applications. In addition, the dead volume ratio too much affective to the work of the engine. It is notice that $\overline{W}_i$, $\overline{W}_s$ and $\overline{W}_f$ are rapidly decreasing as the volume ratio increases. Especially in the range $\chi = 0$ to $\chi = 0.8$. For the area ratio, it is found that should greater than 0.22 so the engine can operate. At $a = 0.1$ to $a = 1$ where $\overline{W}_i$ and $\overline{W}_s$ increase rapidly. It's mean the value of the area ratio must be greater than 1 and the expansion cylinder area must be greater than the compression side.

In terms of the variable of size of the slider crank mechanism, we discover that $e_d > 1$ and $e_p > 1$, but the results of these two variables don't have an optimize values. When these two values are higher, the mechanisms are greater transmission power and reduce the effect of side forces on the cylinder wall. Therefore, it is concluded that the variables e_p and e_d should be designed according to the appropriate proportions. In the other way, when the variable are tried to optimize at. $E = 0.8$, $\tau = 0.3$, $\chi = 0.1$ and $e_p = e_d = 5$. The maximum value of dimensionless shaft work is 0.4292 at $a = 1.73$ and $\alpha = 76°$

References

[1] H. Karabulut, Turk J Engin Environ Sci *Manufacturing and Testing of a V-Type Stirling Engine*. **24**, 71-80 (2000)

[2] I. Batmaz, S. Ustun, Appli. Ener. *Design and manufacturing of a V-type Stirling engine with double heaters*. **85**, 1041–1049 (2008).

[3] A.A. El-Ehwany, et al, Ener. Conver. & Mang. *Development of the performance of an alpha-type heat engine by using elbow-bend transposed-fluids heat exchanger as a heater and a cooler*. **52**, 1010-1019 (2011)

[4] V M. Homutescu, D-T. Bălănescu, *Optimization of diameter Ratio for Alpha-Type Stirling Engine* (2010) [Online]. Available at: https://www.agir.ro/buletine/808.pdf [Accessed on 6 March 2018]

[5] L.S. Scollo, et al, Inter. Jour. of Hydro. Ener. *Twin cylinder alpha stirling engine combined model and prototype redesign*, **38**, 1988-1996 (2013)

[6] N. Martaj, P. Rochelle, Inter. Jour. for Sim. of heat eng. *1D modeling of an alpha type Stirling engine*, **5**: A07 (2014)

[7] JR. Senft, Cambridge Univ Pr, *Mechanical efficiency of heat engines*, (2007)

[8] JR. Senft, Int. J. Energy Res. *Optimum Stirling engine geometry*, **26**, 1087-1101 (2002)

[9] CH. Cheng, HS. Yang, Appli. Ener. *Optimization of geometrical parameters for stirling engine based on theoretical analysis*, **92**, 395-405 (2012)

[10] CH. Cheng, HS. Yang, Ener. *Optimization of rhombic drive mechanism used in beta-type Stirling engine based on dimensionless analysis*, **64**, 970-978 (2013)

Influence of grid distribution on CFD model of compressible flow inside the primary nozzle and mixing chamber used in refrigeration application

Natthawut Ruangtrakoon[1,*], and *Eakarach* Bumrungthaichaichan[2]

[1]Department of Mechanical Engineering, Faculty of Engineering, King Mongkut's Institute of Technology Ladkrabang, Bangkok, Thailand
[2]Department of Chemical Engineering, Faculty of Engineering, King Mongkut's Institute of Technology Ladkrabang, Bangkok, Thailand

Abstract. In this study, the influence of grid distribution on CFD model of the primary nozzle and mixing chamber used in refrigeration application was primarily investigated. The only one geometry of primary nozzle and mixing chamber was modeled. The two different grid distributions, fine near-wall grid and regular grid with the identical total grid number, were simulated to investigate the flow phenomena inside the considered system. The appropriate boundary conditions and numerical methods were carefully employed. The simulated entrainment ratios obtained by two different grid arrangements were validated by comparing with the reliable experimental data. The results revealed that the Mach number distributions of these models were different. Further, the outlet total pressure predicted by fine near-wall grid was about 1.3% higher than that obtained by regular grid.

1 Introduction

A steam jet refrigeration system can convert waste heat (low-grade thermal energy), that is rejected from many industrial processes, to useful refrigeration. So, the electrical energy used in air-conditioning system can be reduced. Another advantageous point is that water, the most environmental friendly substance, can be used as the refrigerant in the system.

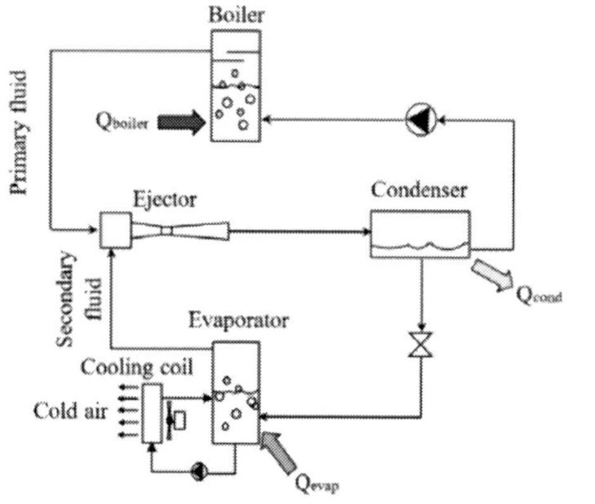

Fig. 1. Schematic view of a steam jet refrigeration cycle

Fig. 1 presents a schematic view of a steam ejector refrigeration system. The stream ejector, which is used as the thermal compressor for steam jet refrigeration system, provides four main parts, including primary nozzle, mixing chamber, throat, and subsonic diffuser, as shown in Fig. 2.

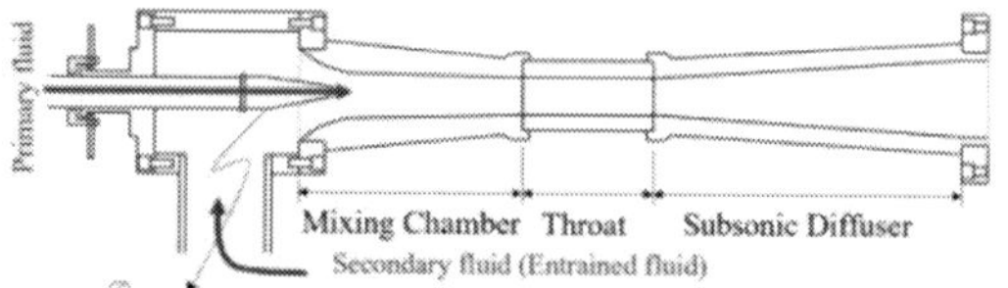

Fig. 2. Schematic of steam ejector

The performance of steam jet refrigeration system can be defined in term of the coefficient of performance (COP) for steam jet refrigeration as depicted in Eq. (1).

$$\mathrm{COP} = R_m \cdot \frac{h_{g-evap} - h_{f-con}}{h_{g-boiler} - h_{f-con}} \qquad (1)$$

where R_m is the entrainment ratio of the steam ejector, which can be defined as a ratio of mass flow rate of primary fluid to mass flow rate of secondary fluid and h_g and h_f respectively are enthalpy of vapor and liquid.

Since, the enthalpy change at the boiler is not much different from the enthalpy change at the evaporator. Then, it can be assumed that:

* Corresponding author: natthawut.ru@kmitl.ac.th

$$\text{COP} \approx R_m \tag{2}$$

So, the performance of jet refrigeration system depends almost on the performance of steam ejector. There are two parameters used to define the steam ejector performance, which are entrainment ratio as mentioned earlier and critical discharge pressure. In term of critical discharge pressure, it can be used to define the highest possible pressure of mixed fluid which the steam ejector can discharge to the condenser.

Over the past decades, many researchers have been attempted to improve the performance of steam jet refrigeration system by considering the experimental investigations [1,2]. Moreover, the computational fluid dynamics (CFD) is rapidly developed. It can be used to explain the mixing phenomenon inside the ejector. The predicted results of CFD were found to agree well with experimental values [3-5]. That is, CFD technique can be used to accurately predict the performance of the steam jet refrigeration system.

In order to improve the CFD results, the effect of grid distribution on CFD compressible flow model was studied. The two principle components of steam ejector (primary nozzle and mixing chamber) were primarily selected as the computational domain. The two different grid distributions were constructed and the simulated entrainment ratios obtained by these models were validated by comparing with the reliable experimental data. The boundary conditions as well as the numerical models were carefully employed. From this preliminary CFD study, the understand of the influence of grid distribution on flow behavior inside primary nozzle and mixing chamber, which directly affects on the performance of steam ejector jet refrigeration system, was obtained.

2 Description of CFD modelling

2.1. Computational domain

The dimensions of primary nozzle and mixing chamber used in the calculation domain as shown in Fig. 3 were similar to the steam ejector of Ruangtrakoon et al. [3], which based on Engineering Sciences Data Unit (ESDU) design. The diameters of primary nozzle throat (*d*) and exit (*D*) are 1.7 mm and 7.6 mm, respectively. The divergent angle of primary nozzle is 10°. This allows the nozzle to produce the Mach number of 4 at the primary nozzle exit plane.

2.2 Grid generation

The GAMBIT 2.4.6 program was employed to perform the two-dimensional axis-symmetrical solid model and grid generation of the domain. The quadrilateral grids were carefully generated inside the flow domain to achieve the accurate results. The grids inside primary nozzle of these models were different. Whereas, the axial grid sizes of mixing chamber were controlled to be identical. The total number of grids for two different grid distributions were identical (110,460 cells). The grid generation of these simulations are shown in Fig. 4.

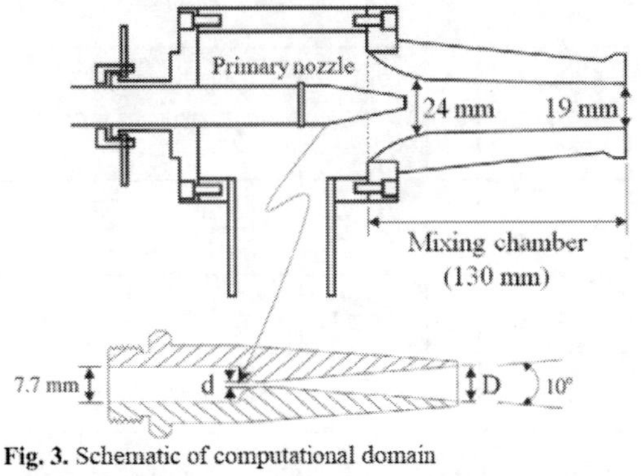

Fig. 3. Schematic of computational domain

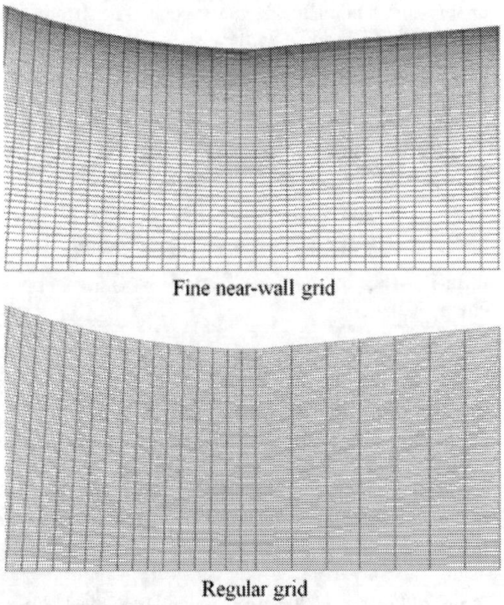

Fig. 4. Grid generation at nozzle throat

2.3 Governing equations

For CFD simulations of compressible flows, the steady state flow pattern and turbulence field were obtained by using the Reynolds average equations for conservation of mass and momentum together with the shear stress transport k-omega model (SST-k-omega). The compact form of steady state Reynolds average equation can be written as:

$$\frac{\partial(\rho U \phi)}{\partial x_j} = \frac{\partial}{\partial x_j}\left(\Gamma_\varphi \frac{\partial \phi}{\partial x_j}\right) + S_\varphi \tag{3}$$

where $\boldsymbol{U}$ is mean velocity vector, ϕ is a universal dependent variable, Γ_φ is the diffusivity, and S_φ is the source term. Further, the details of variables for continuity equation, momentum equation, and SST-k-omega are represented in Table 1.

Table 1. Variables for continuity equation, momentum equation, and SST-k-omega [6]

Equation	ϕ	Γ_φ	S_φ
Continuity	1	0	0
Momentum	U_i	μ	$-\frac{\partial P}{\partial x_i}+\frac{\partial}{\partial x_j}\left(\mu_t\frac{\partial U_i}{\partial x_j}\right)+S_{M,i}$
k-transport	k	$\mu+\frac{\mu_t}{\sigma_k}$	$G_k-Y_k+S_k$
ω-transport	ω	$\mu+\frac{\mu_t}{\sigma_\omega}$	$G_\omega-Y_\omega+D_\omega+S_\omega$

Further, the energy equation was used to achieve the energy distribution inside the system. The steady state energy equation can be written as:

$$\frac{\partial(U(\rho E+p))}{\partial x_j}=\frac{\partial}{\partial x_j}\left(k_{eff}\frac{\partial T}{\partial x_j}-\sum_j h_j\vec{J}_j+(\bar{\bar{\tau}}_{eff}\cdot U)\right)+S_h \quad (4)$$

where E is energy, k_{eff} is effective conductivity, $\vec{J}_j$ is the diffusion flux of species j, and S_h is source term for energy equation.

2.4. Boundary conditions

For CFD simulations of compressible flows, FLUENT [7] recommended to specify the zero operating pressure to minimize the errors due to pressure fluctuations. According to Fig. 3, the pressure-inlet and pressure-outlet boundary condition types were adopted at inlet of primary nozzle and mixing chamber and outlet of mixing chamber, respectively. Moreover, water vapor was used as the working fluid and was assumed to be ideal gas.

At primary nozzle inlet, the gauge total pressure and total temperature of primary fluid were 313,201 Pa and 408 K, respectively. Whereas, for mixing chamber inlet, the gauge total pressure of 873 Pa and total temperature of 280.5 K were imposed as the conditions of secondary fluid. At outlet section, the gauge pressure of 611.1 Pa and total temperature of 300 K were set. Further, the 1% turbulence intensity was adopted for all simulations. At the wall, the no-slip boundary condition and adiabatic condition were used.

2.5. Numerical methods and solution strategy

In this work, the double precision density-based solver of ANSYS FLUENT program was employed. The implicit formulation with Roe flux-difference splitting scheme was adopted. The spatial discretization scheme for gradient was least squares cell-based. Further, the spatial discretization schemes of flow and turbulence quantities were second order upwind.

Further, in order to obtain the converged solutions, the CFD models were computed until the residuals of all quantities were no longer change.

3 Results and discussion

Table 2 represents the model validation between the simulated entrainment ratios and experimental data of Ruangtrakoon et al. [1]. The results showed that the predicted results of these two different grid distributions were in good agreement with experiment. Meaning that, the present CFD models were properly developed.

Table 2. Model validation

Entrainment ratio (R_m)		
EXP [1]	**Fine near-wall grid**	**Regular grid**
0.358	0.364	0.362

Fig. 5 shows filled and line contours of Mach number for different grid distributions. According to this figure, the Mach number inside the primary nozzle and mixing chamber obtained by two different grid distributions are slightly different.

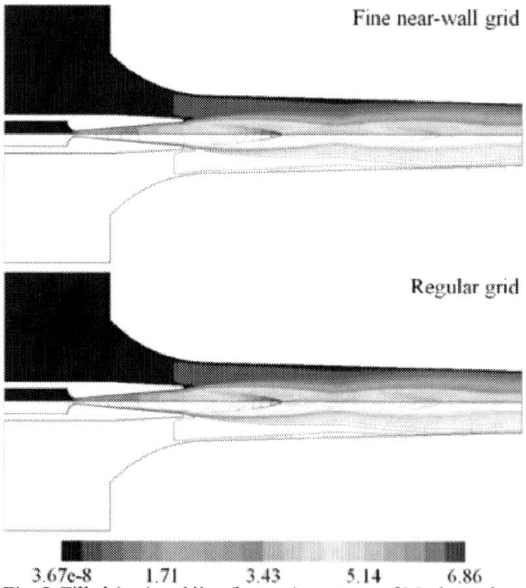

Fig. 5. Filled (top) and line (bottom) contours of Mach number

The difference in Mach number contour of these models is significantly observed at the region near primary nozzle exit as depicted in Fig. 6. From Fig. 6, the appearance of first shock obtained by fine near-wall grid is faster than that predicted by regular grid. This difference may be due to the growth rate of grid at the primary nozzle exit. That is, the growth rate of regular grid was higher than fine near-wall grid, which directly degrades the solution accuracy in the region of large gradient [8].

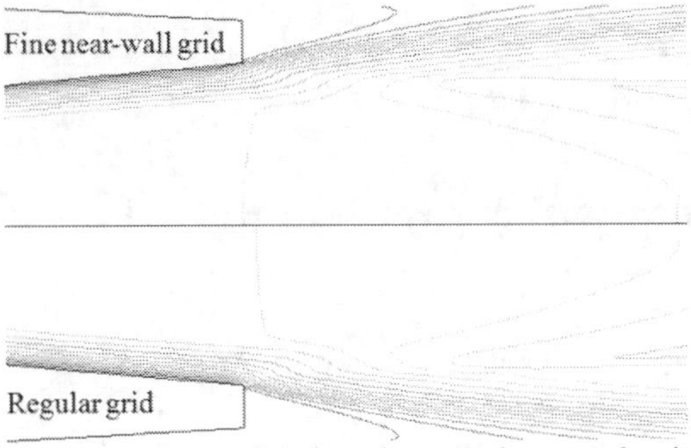

Fig. 6. Line contours of Mach number near primary nozzle exit

Further, the difference in flow phenomena inside this considered system can be illustrated by considering the centerline Mach number of two different grid distributions as shown in Fig. 7.

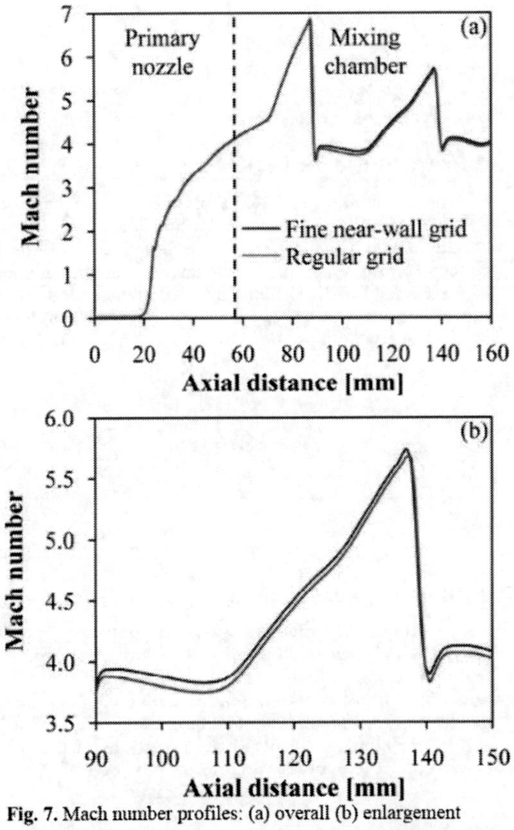

Fig. 7. Mach number profiles: (a) overall (b) enlargement

From Fig. 7, the simulated results show that the centerline Mach number inside the primary nozzle of these two different grids are similar. However, after the primary nozzle exit, the centerline Mach number of fine near-wall grid is higher than that predicted by regular grid, especially for axial distance between 90 - 160 mm, as shown in Fig. 7(b). Further, the predicted outlet total pressure of fine near-wall grid (89,640 Pa) is about 1.3% higher than that predicted by regular grid (88,507 Pa).

According to these preliminary simulated results, even when the axial grid size in mixing chamber of two different models are identical, the flow patterns inside the mixing chamber are different. Hence, it can be implied that the reason of this difference in flow pattern would be due to the grid distribution inside the primary nozzle. Meaning that, the grids inside the primary nozzle should be carefully generated to achieve the accurate downstream results, such as the location of first shock, Mach number, critical discharge pressure, etc.

4 Conclusion

In this work, the effect of grid distribution inside the primary nozzle on downstream phenomena was numerically studied by considering the fine near-wall grid and regular grid. The predicted entrainment ratios of these models were in good agreement with experimental data of Ruangtrakoon et al. [1]. The results revealed that even when the axial grid sizes of mixing chamber for two different models were identical, the Mach number distributions inside these models were different, especially the region near primary nozzle exit, e.g. the appearance of first shock obtained by fine near-wall grid was faster than that predicted by regular grid. The outlet total pressure of these models were also different. So, the grid generation inside primary nozzle should be carefully created. For future work, in order to achieve the comprehensive model for predicting the compressible flow inside the steam ejector, the three-dimensional model of actual configuration should be simulated.

The authors would like to thank College of Advanced Manufacturing Innovation, King Mongkut's Institute of Technology Ladkrabang, Thailand for supporting the ANSYS FLUENT software.

References

1. N. Ruangtrakoon, S. Aphornratana, T. Sriveerakul, Exp. Therm. Fluid Sci. **35**, 676 (2011)
2. F. Li, R. Li, X. Li, Q. Tian, Appl. Therm. Eng. **137**, 784 (2018)
3. N. Ruangtrakoon, T. Thongtip, S. Aphornratana, T. Sriveerakul, Int. J. Therm. Sci. **63**, 133 (2013)
4. S. Varga, A.C. Oliveira, B. Diaconu, Int. J. Refrig.-Rev. Int. Froid **32**, 1203 (2009)
5. G. Besagni, F. Inzoli, Appl. Therm. Eng. **117**, 122 (2017)
6. ANSYS Inc., *ANSYS Fluent Theory Guide: Release 15.0* (ANSYS Inc., USA, 2013)
7. FLUENT Inc., *Tutorial 7. Inviscid and Compressible Flow through a Converging-Diverging Nozzle* (FLUENT Inc., USA, 2006)
8. S.-E. Kim, T. Barbat, P. Spicka, *Meshing and CFD Accuracy* (FLUENT Inc., USA, 2005)

MATEC Web of Conferences **192**, 02046 (2018) https://doi.org/10.1051/matecconf/201819202046
ICEAST 2018

An evaluation on the performance of two simple triangular bending plate elements

Dian Rahmawati[1], *Imam* Jauhari Maknun[1], and *Irwan* Katili[1,*]

Universitas Indonesia, Civil Engineering Department, Kampus UI, Depok 16424, Indonesia.

Abstract. This paper will study and compare two different three-node triangular bending plate elements with three degree of freedom per node, i.e. MITC3 and DKMT. Both elements, which were developed based on Reissner-Mindlin plate theory and independent shear strain field, have simple formulation and have already been used widely. In this paper, numerical tests for circular plate case are conducted to verify the performance and show the convergence of these two triangular elements.

1 Introduction

In structural modelling, the use of triangular element is interesting due to its simplicity and flexibility. Three-node triangular element is mostly used for complex configuration. However, research on triangular is not as much as quadrilateral element. In finite element analysis, many researches show that quadrilateral element have a good performance. As a result, many analyses prefer to use the quadrilateral element [1-10]. This condition should encourage researchers to develop study about triangular element.

The challenge in finite element is how to generate a simple and applicable element formulation to reduce the computational cost, yet still has high accuracy and good convergence.

MITC (Mixed Interpolation of Tensorial Components) concept has been applied for triangular element. MITC3, proposed by Lee and Bathe [11], is one of the popular bending elements. There are quite a lot of studies and developments of MITC3 [12-15]. This three-node triangular element has a simple and general formulation. The behaviour of the element is isotropic, meaning that the stiffness matrix of the element does not depend on the sequence of node numbering.

Previously, Katili [16] proposed simple formulation for triangular bending plate element. The element, called DKMT, is an extension of DKT (Discrete Kirchhoff Triangular) element [17] which is a simple, efficient and reliable element for analysis of thin plates. In the case of thin plate problem, the solution obtained with DKMT element converged to that given by DKT element.

The purpose of this paper is to compare formulation and performance of two triangular elements, i.e. MITC3 and DKMT. Both of them have simple formulation. At the end of this paper, we conduct numerical studies for circular bending plate to compare and evaluate the convergence of both elements.

2 Formulation of DKMT

The formulation of DKMT plate element is based on Reissner-Mindlin hypothesis, which only require C^0 continuity, and generated by assumed natural strain concept. Transversal shear deformation for this element is expressed with special interpolation called shear projection method. The triangular element has three-degrees of freedom per node (Fig.1).

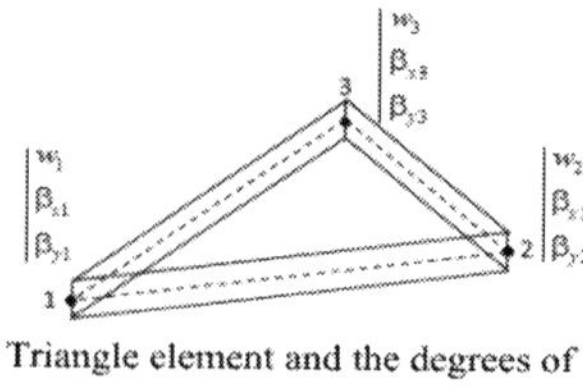

Fig. 1. Triangle element and the degrees of freedom

2.1 Bending strain matrix

The DKMT element was first published in 1993 [16]. DKMT element has 3 nodes with 3 degrees of freedom each, which are: w (translation in the z direction), β_x (rotation in the z-x plane) and β_y (rotation in the z-y plane). Incomplete quadratic rotation fields for β_x and β_y are considered in terms of rotations at the three corners and a temporary variable at mid-side i-j (Fig.2).

The displacement function is given as:

$$w = \sum_{i=1}^{3} N_i\, w_i$$

$$\beta_x = \sum_{i=1}^{3} N_i \beta_{x_i} + \sum_{k=4}^{6} P_k C_k \Delta\beta_{s_k} \qquad (1)$$

* Corresponding author: irwan.katili@eng.ui.ac.id

$$\beta_y = \sum_{i=1}^{3} N_i \beta_{y_i} + \sum_{k=4}^{6} P_k C_k \Delta\beta_{s_k}$$

where: w_i is the vertical displacement at node-i
N_i is the shape function at node-i
β_{x_i}, β_{y_i} are the rotation in plane of z-x and z-y

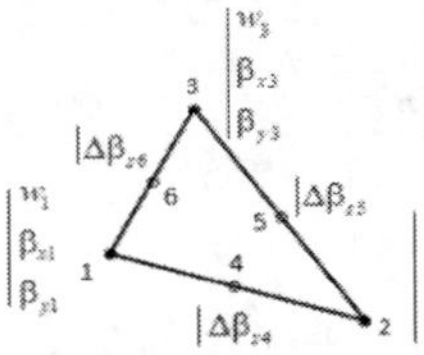

Fig. 2. Kinematic variable of corner node and temporary variable at mid side i-j.

N_i are the linear shape functions and P_k is the quadratic functions:

$$N_1 = \lambda \ ; \ N_2 = \xi \ ; \ N_3 = \eta \ ; \ \lambda = 1 - \xi - \eta$$
$$P_4 = 4\lambda\xi \ ; \ P_5 = 4\xi\eta \ ; \ P_6 = 4\lambda\eta \qquad (2)$$

C_k and S_k are the direction cosines of side i-j (Fig.2)

$$\{\chi\} = \begin{Bmatrix} \chi_x \\ \chi_y \\ \chi_{xy} \end{Bmatrix} = \begin{Bmatrix} \beta_{x,x} \\ \beta_{y,y} \\ \beta_{x,y} + \beta_{y,x} \end{Bmatrix} \qquad (3)$$

$\beta_{x,x}$ and $\beta_{x,y}$ denote the first derivatives of β_x with respect to x and y, respectively.

The relation between curvature and nodal displacement is expressed in the equation below.

$$\{\chi\} = \left[B_{b_\beta}\right]\{u_n\} + \left[B_{b_{\Delta\beta}}\right]\{\Delta\beta_{s_n}\} \qquad (4)$$

where: $\{u_n\} = \langle u_n \rangle^T = \langle \dots \ w_i \ \ \beta_{x_i} \ \ \beta_{y_i} \ \dots \rangle^T_{i=1,2,3}$

$$\{\Delta\beta_{s_n}\} = \langle \Delta\beta_{s_n} \rangle^T = \langle \Delta\beta_{s_4} \ \ \Delta\beta_{s_5} \ \ \Delta\beta_{s_6} \rangle^T$$

From eq. (1-4), we obtain the expression of bending strain matrix,

$$\left[B_{b_\beta}\right] = \begin{bmatrix} & 0 & N_{i,x} & 0 & \\ \dots & 0 & 0 & N_{i,y} & \dots \\ & 0 & N_{i,y} & N_{i,x} & \end{bmatrix}_{i=1,2,3} \qquad (5)$$

$$\left[B_{b_{\Delta\beta}}\right] = \begin{bmatrix} & P_{k,x} C_k & \\ \dots & P_{k,y} S_k & \dots \\ & P_{k,y} C_k + P_{k,x} S_k & \end{bmatrix}_{k=4,5,6} \qquad (6)$$

$N_{i,x}$, $P_{k,y}$, $N_{i,x}$ and $P_{k,y}$ denote the first derivatives of N_i and P_k with respect to x and y, respectively.

2.2 Shear strain interpolation

Shear deformation field is approximated linear in each element.

$$\{\underline{\gamma}\} = \begin{Bmatrix} \underline{\gamma}_x \\ \underline{\gamma}_y \end{Bmatrix} = \sum_{i=1}^{3} N_i \begin{Bmatrix} \underline{\gamma}_{x_i} \\ \underline{\gamma}_{y_i} \end{Bmatrix} \qquad (7)$$

where: $\underline{\gamma}_{x_i}$ and $\underline{\gamma}_{y_i}$ are the shear deformations at nodes-i

Shear deformation is assumed constant along the edge of element (Fig.3). Shear deformation at node–i is obtained from projection of constant shear deformation $\underline{\gamma}_{s_k}$ from the edge of element to element node.

$$\begin{Bmatrix} \underline{\gamma}_{s_4} \\ \underline{\gamma}_{s_6} \end{Bmatrix} = \begin{bmatrix} C_{12} & S_{12} \\ C_{31} & S_{31} \end{bmatrix} \begin{Bmatrix} \underline{\gamma}_{x_1} \\ \underline{\gamma}_{y_1} \end{Bmatrix} \ ; \ \begin{Bmatrix} \underline{\gamma}_{s_5} \\ \underline{\gamma}_{s_4} \end{Bmatrix} = \begin{bmatrix} C_{23} & S_{23} \\ C_{12} & S_{12} \end{bmatrix} \begin{Bmatrix} \underline{\gamma}_{x_2} \\ \underline{\gamma}_{y_2} \end{Bmatrix} \qquad (8)$$

$$\begin{Bmatrix} \underline{\gamma}_{s_6} \\ \underline{\gamma}_{s_5} \end{Bmatrix} = \begin{bmatrix} C_{31} & S_{31} \\ C_{23} & S_{23} \end{bmatrix} \begin{Bmatrix} \underline{\gamma}_{x_3} \\ \underline{\gamma}_{y_3} \end{Bmatrix}$$

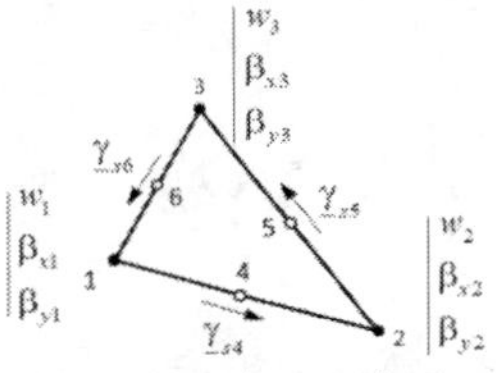

Fig. 3. Constant transverse shear strain along edge ij for DKMT

where:

$$C_{ij} = \frac{x_{ji}}{L_{ij}} \ ; \ S_{ij} = \frac{y_{ji}}{L_{ij}} \ ; \ L_{ij} = \sqrt{x_{ji}^2 + y_{ji}^2} \qquad (9)$$
$$x_{ji} = x_j - x_i \ ; \ y_{ji} = y_j - y_i$$

From eq. (7-9), we obtain:

$$\{\underline{\gamma}\} = \begin{Bmatrix} \underline{\gamma}_x \\ \underline{\gamma}_y \end{Bmatrix} = \left[B_{s_\gamma}\right]\{\underline{\gamma}_{s_n}\} \qquad (10)$$

with: $\{\underline{\gamma}_{s_n}\} = \langle \underline{\gamma}_{s_n} \rangle^T = \langle \underline{\gamma}_{s_4} \ \ \underline{\gamma}_{s_5} \ \ \underline{\gamma}_{s_6} \rangle^T$

$$\left[B_{s_\gamma}\right] = \begin{bmatrix} \left(\frac{S_{31}}{A_1}\lambda - \frac{S_{23}}{A_2}\xi\right) & \left(\frac{S_{12}}{A_2}\xi - \frac{S_{31}}{A_3}\eta\right) & \left(\frac{S_{23}}{A_3}\xi - \frac{S_{12}}{A_1}\lambda\right) \\ \left(\frac{C_{23}}{A_2}\xi - \frac{C_{31}}{A_1}\lambda\right) & \left(\frac{C_{31}}{A_3}\eta - \frac{C_{12}}{A_2}\xi\right) & \left(\frac{C_{12}}{A_1}\lambda - \frac{C_{23}}{A_3}\eta\right) \end{bmatrix} \qquad (11)$$

where:

$$A_1 = C_{12} S_{31} - C_{31} S_{12} \ ; \ A_2 = C_{23} S_{12} - C_{12} S_{23} \ ; \ A_3 = C_{31} S_{23} - C_{23} S_{31}$$

Katili [17] proposed the assumed independent transverse shear strain along the side i-j expressed as:

$$\underline{\gamma}_{s_k} = -\frac{2}{3}\phi_k \Delta\beta_{s_k} \ ; \ \phi_k = \frac{2}{\kappa(1-\upsilon)}\left(\frac{h^2}{L_{ij}^2}\right) \qquad (12)$$

Where h is the thickness of element and κ is the shear correction factor (usually $\kappa = 5/6$).

If (12) is applied on all sides of the element, the following matrix relation is obtained:

$$\{\underline{\gamma}_{s_n}\} = \left[A_\phi\right]\{\Delta\beta_{s_n}\} \ ; \ \left[A_\phi\right] = -\frac{2}{3}\begin{bmatrix} \phi_4 & 0 & 0 \\ 0 & \phi_5 & 0 \\ 0 & 0 & \phi_6 \end{bmatrix} \qquad (13)$$

Substituting (13) into (12) we get:

$$\{\underline{\gamma}\} = \begin{Bmatrix} \underline{\gamma}_x \\ \underline{\gamma}_y \end{Bmatrix} = \left[B_{s_\gamma}\right]\left[A_\phi\right]\{\Delta\beta_{s_n}\} \qquad (14)$$

If the assumed shear force and strains are constant on the side, then we obtain:

$$\underline{\gamma}_{s_k} = \frac{1}{L_{ij}}\int_0^{L_{ij}} \gamma_s \, ds = -\frac{2}{3}\phi_k \Delta\beta_{s_k} \ ; \ \gamma_s = w_{,s} + \beta_s \qquad (15)$$

Applying (15) on each side-ij, we obtain

$$\{\Delta\beta_{s_n}\} = [A_\Delta]^{-1}[A_u]\{u_n\} \quad \text{where} \qquad (16)$$

$$[A_\Delta] = -\frac{2}{3}\begin{bmatrix} (1+\phi_4) & 0 & 0 \\ 0 & (1+\phi_5) & 0 \\ 0 & 0 & (1+\phi_6) \end{bmatrix} \tag{17}$$

$$[A_u] = \frac{1}{2}\begin{bmatrix} -\frac{2}{L_{12}} & C_{12} & S_{12} & \frac{2}{L_{12}} & C_{12} & S_{12} & 0 & 0 & 0 \\ 0 & 0 & 0 & -\frac{2}{L_{23}} & C_{23} & S_{23} & \frac{2}{L_{23}} & C_{23} & S_{23} \\ \frac{2}{L_{31}} & C_{31} & S_{31} & 0 & 0 & 0 & -\frac{2}{L_{31}} & C_{31} & S_{31} \end{bmatrix} \tag{18}$$

Substituting (16) into (4), we define the bending curvatures as:

$$\{\chi\} = [B_b]\{u_n\}$$
$$[B_b] = \left[B_{b_\beta}\right] + \left[B_{b_{\Delta\beta}}\right][A_\Delta]^{-1}[A_u] \tag{19}$$

Introducing (16) into (14) gives:

$$\{\underline{\gamma}\} = [B_s]\{u_n\}$$
$$[B_s] = \left[B_{s_\gamma}\right]\left[A_{\phi\Delta}\right][A_u] \;;\; \left[A_{\phi\Delta}\right] = \left[A_\phi\right][A_\Delta]^{-1} \tag{20}$$

3 Formulation of MITC3

3.1 Bending strain matrix of MITC3

The displacement function is given as:

$$w = \sum_{i=1}^{3} N_i w_i \;;\; \beta_x = \sum_{i=1}^{3} N_i \beta_{x_i} \;;\; \beta_y = \sum_{i=1}^{3} N_i \beta_{y_i} \tag{21}$$

The relation between curvature and nodal displacement can be expressed in the equation below.

$$\{\chi\} = [B_b]\{u_n\} \;;\; [B_b] = \left[B_{b_\beta}\right] \tag{22}$$

3.2 Shear strain interpolation

The key of MITC3 [11] technique is the use of appropriate assumed strain interpolation and the proper choice of tying points to relate the displacement interpolation and strain interpolation, which is done separately.

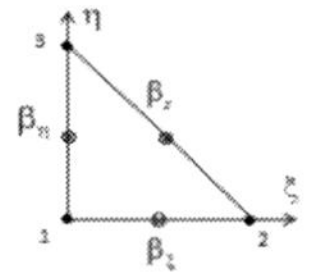

Fig. 4. Tying points

For MITC3, tying points [11-13] are chosen in the mid-points (Fig.4) of sides 1-2, 1-3, and 2-3. Distribution of β_ξ is assumed constant along ξ direction and β_η is assumed constant along η:

$$\beta_\xi = a_1 + a_2\,\eta \;;\; \beta_\eta = b_1 + b_2\,\xi \;;\; \beta_\lambda = \frac{1}{\sqrt{2}}\left(\beta_\xi - \beta_\eta\right) \tag{23}$$

We obtain the values of β_ξ, β_η, β_λ at the tying points *A*, *B*, and *C*.

– At point *A* (η=0), we obtain that $a_1 = \beta_{\xi(A)}$.

– At point *B* (ξ=0), we obtain that $b_1 = \beta_{\eta(B)}$.

– At node 2 (ξ=1,η=0), $\beta_\lambda(1,0) = \frac{1}{\sqrt{2}}\left(\beta_\xi(1,0) - \beta_\eta(1,0)\right)$

– At node 3 (ξ=0,η=1), $\beta_\lambda(0,1) = \frac{1}{\sqrt{2}}\left(\beta_\xi(0,1) - \beta_\eta(0,1)\right)$

– Along edge 3-2, $\beta_\lambda(1,0) = \beta_\lambda(0,1)$ leading to:

$$\frac{1}{\sqrt{2}}\left(\beta_\xi(1,0) - \beta_\eta(1,0)\right) = \frac{1}{\sqrt{2}}\left(\beta_\xi(0,1) - \beta_\eta(0,1)\right)$$

Which gives: $a_2 = -b_2 = c$

Then the eq. (22) becomes:

$$\beta_\xi = \beta_{\xi(A)} + c\,\eta \;;\; \beta_\eta = \beta_{\eta(B)} - c\,\xi \tag{24}$$

– At point *C* (ξ=1/2,η=1/2):

$$\beta_{\lambda(C)} = \frac{1}{\sqrt{2}}\left\{\beta_{\xi(C)} - \beta_{\eta(C)}\right\}$$ and

$$\beta_{\lambda(C)} = \frac{1}{\sqrt{2}}\left\{\left(\beta_{\xi(A)} + \frac{1}{2}c\right) - \left(\beta_{\eta(B)} - \frac{1}{2}c\right)\right\}$$

and we obtain the value of *c*,

$$c = \beta_{\eta(B)} - \beta_{\xi(A)} + \beta_{\xi(C)} - \beta_{\eta(C)} \tag{25}$$

Substuting eq. (25) into (24), we obtain the expression,

$$\beta_\xi = \beta_{\xi(A)} + \left(\beta_{\eta(B)} - \beta_{\xi(A)} + \beta_{\xi(C)} - \beta_{\eta(C)}\right)\eta$$
$$\beta_\eta = \beta_{\eta(B)} - \left(\beta_{\eta(B)} - \beta_{\xi(A)} + \beta_{\xi(C)} - \beta_{\eta(C)}\right)\xi \tag{26}$$

β_ξ and β_η at tying points are average value of two points at node.

$$\beta_{\xi(A)} = \frac{1}{2}\left(\beta_{\xi_1} + \beta_{\xi_2}\right) \;;\; \beta_{\eta(B)} = \frac{1}{2}\left(\beta_{\eta_1} + \beta_{\eta_3}\right)$$
$$\beta_{\xi(C)} = \frac{1}{2}\left(\beta_{\xi_2} + \beta_{\xi_3}\right) \;;\; \beta_{\eta(C)} = \frac{1}{2}\left(\beta_{\eta_2} + \beta_{\eta_3}\right) \tag{27}$$

Transverse shear strain field in parametric space:

$$\begin{Bmatrix} \underline{\gamma}_\xi \\ \underline{\gamma}_\eta \end{Bmatrix} = \begin{Bmatrix} w_{,\xi} + \beta_\xi \\ w_{,\eta} + \beta_\eta \end{Bmatrix} = \begin{Bmatrix} w_2 - w_1 + \beta_\xi \\ w_3 - w_1 + \beta_\eta \end{Bmatrix} \tag{28}$$

Then, from (26-28) we get the expression of shear strain in parametric space:

$$\begin{Bmatrix} \underline{\gamma}_\xi \\ \underline{\gamma}_\eta \end{Bmatrix} = \left[B_{s_\xi}\right]\{u_n\} \tag{29}$$

$$\left[B_{s_\xi}\right] = \frac{1}{2}\begin{bmatrix} -2 & (1-\eta) & \eta & 2 & 1 & -\eta & 0 & \eta & 0 \\ -2 & \xi & (1-\xi) & 0 & 0 & \xi & 2 & -\xi & 1 \end{bmatrix}$$

Transverse shear strain field in Cartesian system:

$$\{\underline{\gamma}\} = \begin{Bmatrix} \underline{\gamma}_x \\ \underline{\gamma}_y \end{Bmatrix} = [j]\begin{Bmatrix} \underline{\gamma}_\xi \\ \underline{\gamma}_\eta \end{Bmatrix} \tag{30}$$

Where [*j*] is the invers of Jacobian matrix.

Introducing (29) into (30), we obtain:

$$\{\underline{\gamma}\} = \begin{Bmatrix} \underline{\gamma}_x \\ \underline{\gamma}_y \end{Bmatrix} = [B_s]\{u_n\} \;;\; [B_s] = [j]\left[B_{s_\xi}\right] \tag{31}$$

Mentioned by Jeon *et al.* in the article [14], MITC3 uses three-points Hammer integration to compute shear stiffness.

4 Numerical Analysis

To study the convergence behaviour of MITC3 and DKMT elements, a circular plate problem is analysed. Because of symmetry, only a quarter of circular plate as presents in (Fig.5) is evaluated. Two different boundary conditions, i.e. a case of simply supported and a case of clamped supported in perimeter edge, are used with number of element (NELT) of 6, 24, 96, and 384.

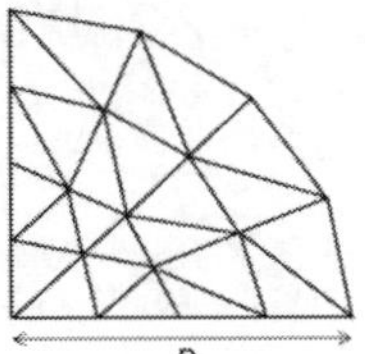

Fig. 5. Circular plate with number of element NELT=24. R = 5; h = 0.1, 1, 2; E = 10.92; Poisson's ratio = 0.3; and uniform load f_z = 1.

Table 1 presents the displacements at the centre of circle (w_c) for soft simply supported case, whereas Table 2 shows the results for clamped boundary condition. Fig. 6 and Fig. 7 show the convergence of w_c of both elements.

Table 1. w_c for simply supported boundary condition

R/h	NELT	MITC3	DKMT	Exact [16]
50	6	29056	37780	39831
	24	36374	39353	
	96	38986	39716	
	384	39690	39802	
5	6	35.019	39.404	41.599
	24	39.705	41.038	
	96	41.143	41.459	
	384	41.491	41.567	
2.5	6	5.044	5.554	5.870
	24	5.636	5.793	
	96	5.814	5.852	
	384	5.857	5.866	

Table 2. w_c for clamped supported boundary condition

R/h	NELT	MITC3	DKMT	Exact [16]
50	6	472.3	9975.2	9783.5
	24	4381.4	9991.4	
	96	8944.1	9841.2	
	384	9681.9	9796.0	
5	6	6.976	11.601	11.551
	24	10.341	11.694	
	96	11.271	11.587	
	384	11.484	11.560	
2.5	6	1.564	2.079	2.114
	24	1.969	2.127	
	96	2.080	2.118	
	384	2.106	2.115	

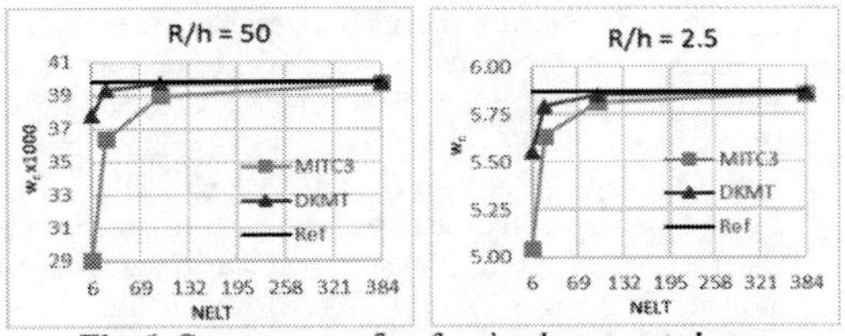

Fig. 6. Convergence of w_c for simply supported case

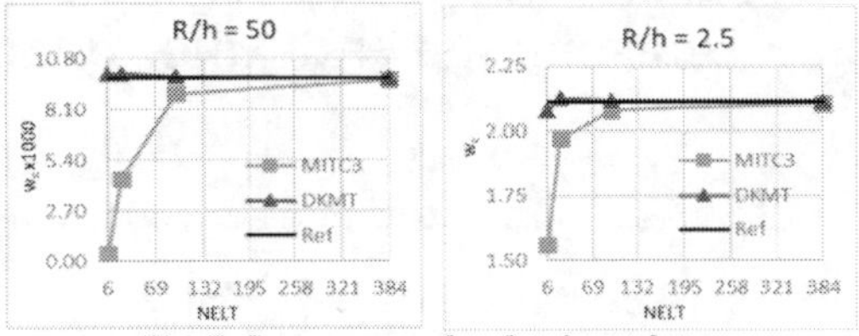

Fig. 7. Convergence of w_c for clamped case

5 Conclusion

We observe that both elements have good convergence towards the exact solution. DKMT element gives very good results and converges faster than MITC3 element. This is due to the quadratic interpolation functions for the rotations used in DKMT.

The authors gratefully acknowledge financial support from Indonesian Endowment Fund for Education (LPDP).

References

1. I. Katili, I.J. Maknun, O. Millet, A. Hamdouni, Compos Struct, **132**, 166-174 (2015)
2. M. Mahjudin, P. Lardeur, F. Druesne, I. Katili, Struct Saf, **61**, 12-21 (2016)
3. F.T. Wong, Erwin, A. Richard, I. Katili, Procedia Eng, **171**, 805 – 812 (2017)
4. I. Katili, J.-L. Batoz, I.J. Maknun, A. Hamdouni, O. Millet, Finite Elem Anal Des, **100**, 12-27 (2014)
5. I.J. Maknun, I. Katili, O. Millet, A. Hamdouni, Int J Comput Methods Eng Sci Mech, **17**, 391-400 (2016)
6. H. Irpanni, I. Katili, I.J. Maknun, Int J Mech Eng and Robot Res, **6**, 248-252 (2017)
7. I. Katili, I.J. Maknun, E. Tjahjono, I. Alisyahbana, Int J Tech, **6**, 1060-1069 (2017)
8. I. Katili, I.J. Maknun, J.-L. Batoz, A. Ibrahimbegović, https://doi.org/10.1016/j.compstruct.2018.01.043.
9. I. Katili, R. Aristio, Eur J Mech - A/Solids, **67**, 92-107 (2018)
10. I. Katili, J.-L. Batoz, I.J. Maknun, P. Lardeur, Comput Struct, **204**, 48-64 (2018)
11. P.S. Lee, K.J. Bathe, Comput Struct, **82**, 945-962 (2004)
12. P.S. Lee, H.C. Noh, K.J. Bathe, Comput Struct, **85**, 404–418 (2017)
13. O.P. Gupta (2015), www.researchgate.net/profile/DrOPGupta/contributions/ May 2015
14. H.M. Jeon, Y. Lee, P.S. Lee, K.J. Bathe, Comput Struct, **146**, 91-104 (2015).
15. Y. Lee, K. Yoon, P.S. Lee, Comput Struct, **110-111**, 93-106 (2012)
16. I. Katili, IJNME, **36**, 1859-1883 (1993)
17. J.-L. Batoz, K.J. Bathe, L.W. Ho, Int J Numer Meth Eng, **15**, 1771-1812 (1980)

Development of correlation-regression model between land use change and water quality indices in Ciliwung watershed

Yopy Arfan[1,*], *Dwita* Sutjiningsih[1]

[1] Department of Civil Engineering, Faculty of Engineering, Universitas Indonesia, Depok 16424, Indonesia

Abstract. Urbanization and industrialization lead to the change of land cover from pervious into impervious. This can impact environmental problems such as water quality degradation that affects human health and water ecosystems. The study aimed to develop a regression-correlation model between impervious cover in Ciliwung watershed and water quality indices in Ciliwung river. The correlation-regression model can be used to predict changes in the status of Ciliwung river water quality due to impervious cover changes. Methods of assessing the indices of water quality are CCME-WQI, NSF-WQI, and STORET within the period of 2005-2016. Monitoring locations from the most upstream to downstream are Atta'awun, Katulampa, Kedung Halang, Pondok Rajeg, Panus Bridge, Kelapa Dua, Condet, Kalibata, MT Haryono and Manggarai. Impervious cover data for each water quality monitoring location is processed using ArcGIS Software. Test of correlation significance between percentage of impervious cover and water quality indices using Pearson Correlation test method. The result of correlation test is significantly a strong inverse relationship between impervious cover and water quality indices. The result of regression test is trend line between impervious cover change and water quality indices that can be used to predict the change of water quality status in Ciliwung River.

1 Introduction

The development of urbanization and industrialization resulted in increased the land cover from pervious to impervious. During 2000-2010 there was an increase of 20% settlement land in Ciliwung watershed [1]. The rapid change of land into commercial building, residential, and agricultural areas that are not properly managed will produce waste water that impact on the decrease of river water quality. The increasing impervious cover resulted in environmental problems such as water quality degradation that affected human health and water ecosystem [2]. The impermeable surface precludes the infiltration of rainwater into the soil and ground water, which cause increased surface runoff water to the river, then leads to changes in water quality [3, 4].

Impervious cover (IC) is one of the important factors that affect the water quality status [5, 6]. Impervious cover model at first was developed by Schueler (1994), that aim to find out how the impact of impervious cover toward various water quality and biotic indicators. The Impervious cover model (ICM) defines four categories of urban streams based on how much IC exists in their contributing subwatershed, namely : sensitive, impacted, nonsupporting, and urban drainage [5]. Impervious cover model can be used to predict stream quality. Impervious cover model (ICM) was later modified by Schueler et al. (2009). This modification prevents the misperception that streams with low subwatershed IC will automatically possess good or excellent quality [5].

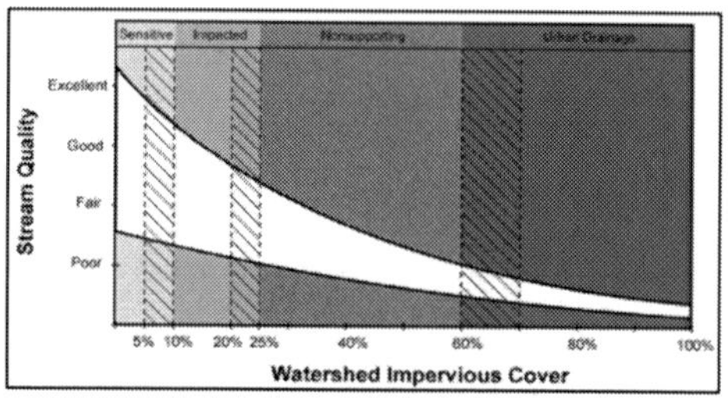

Fig. 1. Modified Impervious cover model [5].

Determination of river water quality status is very important to know the condition of river pollution level. Water Quality Index (WQI) is a method of simplification of the number of water quality parameters data into a number that can describe the status of water quality so that it can be understood by the public in general. There are various methods of Water Quality Index (WQI) such as The CCME-WQI Method [7], STORET method [8], NSF-WQI method [9], and other methods.

The study aimed to develop a correlation-regression model between impervious cover in Ciliwung watershed and water quality indices in Ciliwung river. Determination of Ciliwung river water quality status using CCME-WQI, NSF-WQI and STORET method. The correlation-regression model can be used to predict the water quality status in Ciliwung river due to the changes of impervious cover.

* Corresponding author: arfan.yopi@gmail.com

2 Research Methodology

2.1. Study Location

The study was conducted in Ciliwung watershed starting from the upper ciliwung in Pangrango and Gede Mountain flowing northward through Bogor, Depok, and Jakarta up to Manggarai. Ciliwung river located in West Java and DKI Jakarta Province. The area of Ciliwung watershed up to Manggarai is approximately ± 344.2 km^2 and the main river length approximately ± 114 km.

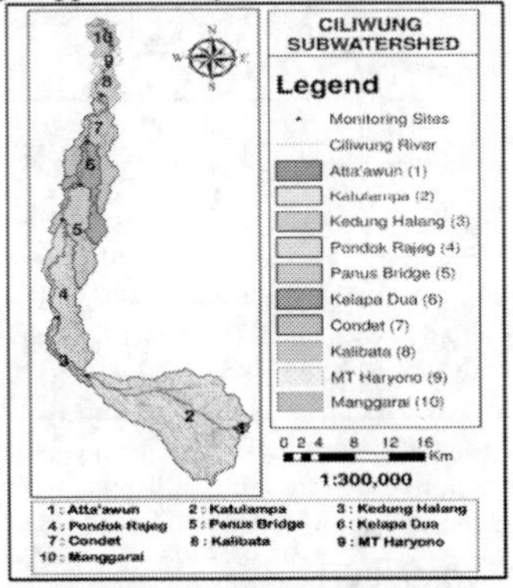

Fig. 2. Map of Ciliwung Watershed.

Table 1. Water Quality Monitoring Locations

No	Monitoring Locations	Coordinates		Total Area (km^2)*
		South Lat.	East Long.	
1	Atta'awun	06°41'56.2"	106°59'14.5"	1.42
2	Katulampa	06°37'59.7"	06°50'12.6"	150.75
3	Kedung Halang	06°33'45.4"	106°48'29.5"	159.70
4	Pondok Rajeg	06°26'43.7"	106°48'46.2"	214.75
5	Panus Bridge	06°24'02.4"	106°49'54.3"	250.11
6	Kelapa Dua	06°20'56.2"	106°50' 18.2"	275.32
7	Condet	06°17'34.2"	106°51'12.5"	312.78
8	Kalibata	06°15'28.5"	106°51'37.7"	324.91
9	MT Haryono	06°13'53.5"	106°51'48.6"	331.35
10	Manggarai	06°15'28.5"	106°51'37.7"	344.20

* Total Area up to monitoring location of ciliwung river

2.2 Framework of The Study

This study was conducted with several stages, namely data collection, land use data processing, determine water quality indices, and correlation-regression analysis.

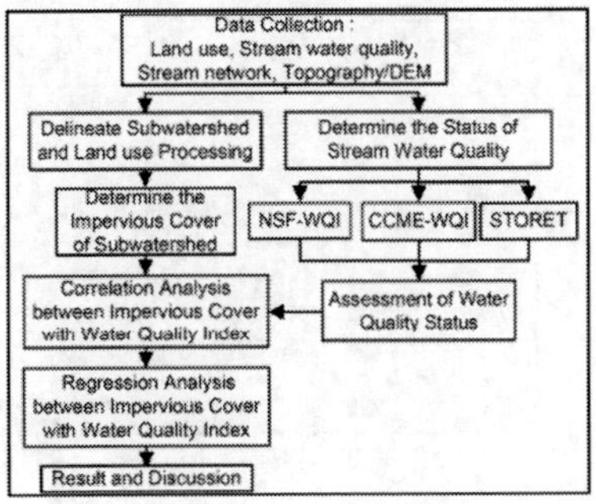

Fig. 3. Flow Diagram of The Study.

2.3 Water Quality Indices Methods

CCME-WQI Method

CCME-WQI method is the calculation to determine the status of the water quality by combining three vectors (F_1 *Scope*, F_2 *Frequency* and F_3 *Amplitude*). Water quality parameters assessed include DO, BOD, pH, temperature, Turbidity, TSS, Nitrate, Total phosphate, and Fecal Coliform in the period of 2005-2016. Calculation of CCME-WQI method using the following equation [7]:

$$F_1 = \left(\frac{Number\ of\ failed\ paramters}{Total\ number\ of\ parameters}\right) \times 100 \quad (1)$$

$$F_2 = \left(\frac{Number\ of\ failed\ tests}{Total\ number\ of\ tests}\right) \times 100 \quad (2)$$

$$F_3 = \left(\frac{nse}{0.01\ nse + 0.01}\right) \quad (3)$$

$$excursion_i = \left(\frac{Failed\ tests\ Value_i}{Objective_i}\right) - 1 \quad (4)$$

$$excursion_i = \left(\frac{Objective_i}{Failed\ tests\ Value_i}\right) - 1 \quad (5)$$

$$nse = \frac{\sum_{i=1}^{n} excursion_i}{\#\ of\ tests} \quad (6)$$

$$CCME\ WQI = 100 - \left(\frac{\sqrt{F_1 + F_2 + F_3}}{1.732}\right) \quad (7)$$

STORET Method

STORET is a method of determining the status of water quality based on government regulations (KepMen LH RI No. 115/2003 on Guidelines for Determination of Water Quality Status) by comparing water quality parameter data with the water quality standard. Water quality parameters assessed include DO, BOD, pH, temperature, Turbidity, TSS, Nitrate, Total phosphate, and Fecal Coliform in the period of 2005-2016. Calculation of STORET method using the following steps [8]: (1) Compare the value of the minimum, maximum, and average water quality data, (2) The result of water quality data complies with the water quality standard then given score 0, (3) The result of water quality data does not comply with the water quality standard then given the score according to table 2, (4) All negative values are summed to get water quality status according to criteria.

Table 2. Parameter Value of Water Quality STORET [8].

Number of Examples	Value	Parameters		
		Chemistry	Physics	Microbiology
< 10	Maximum	-1	-2	-3
	Minimum	-1	-2	-3
	Average	-3	-6	-9
≥10	Maximum	-2	-4	-6
	Minimum	-2	-4	-6
	Average	-6	-12	-18

NSF-WQI Method

The National Sanitation Foundation Water Quality Index (NSF-WQI) is one of the water quality analytical tools that commonly used to determined water quality status [9]. Calculation of NSF-WQI method using the following steps : (1) Determine the weight factor (Wi) and sub-index value (Li) for each parameter by using the sub-index function curve (Li). (2) Calculate the value of NSF-WQI using the following equation [9]:

$$NSF\ WQI = \sum_{i=0}^{n} Wi\ Li \quad (8)$$

(3) NSF-WQI values are classified according to NSF-WQI criteria.

2.4 Determination of Impervious Cover

Impervious cover is any surface in the landscape that cannot effectively absorb or infiltrate rainfall [10]. These include residential roofs, roads, buildings, and the other impervious cover. In this study the impervious cover is defined as the Total Impervious Area (TIA) determined by BIG land use map 2009 [11] and the conversion of digital land use maps in 2006 and 2016 based on satellite imagery of google earth history. Total impervious area in each sub-watershed is obtained by using residential rooftops, building rooftops and transport systems as the component of impervious cover [9]. Land use data is processed using ArcGIS Software.

2.6 Correlation and Regression

Pearson correlation analysis is used to find the relationship between two variables, the impervious cover as the independent variable and the water quality indices as the dependent variables. The Kolmogorov-Smirnov normality test is used to determine whether the data distribution is normal or not. Regression analysis is used to determine the trend line of relationship between impervious cover and water quality indices. The change of impervious cover is analyzed to predict the change of water quality status due to impervious cover change.

3 Result and Discussion

3.1 Impervious Cover Area

Land use of Ciliwung watershed in 2006, 2009 and 2016 are processed using ArcGIS as shown as in Figure 4. Total impervious cover areas are calculated on each subwatershed cumulatively up to monitoring locations [9].

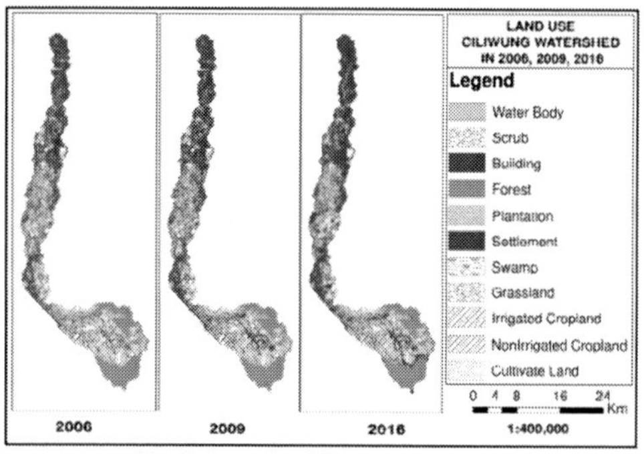

Fig. 4. Land Use Ciliwung Watershed

Figure 4 shows that impervious cover (residential rooftops, building rooftops and road) in Ciliwung watershed in 2006 to 2016 significantly increased from Atta'awun to Manggarai subwatershed. Percentage of impervious cover in Ciliwung watershed in 2016 includes Atta'awun 5%, katulampa 17%, Kedung Halang 21%, Pondok Rajeg 28%, Panus Bridge 30%, Kelapa Dua 34%, Condet 39%, Kalibata 41%, MT Haryono 42%, and Manggarai 44%. If calculated not cumulatively Kelapa Dua subwatershed up to Manggarai subwatershed have % IC greater than 60%. Impervious cover areas in other years (2005-2016) are obtained using linear regression based on the data 2006, 2009 and 2016 as shown as in Table 3.

Table 3. Total Area IC in Ciliwung Subwatershed

SW*	Total Area IC * (Km²)			Trend Line **	R^2
	2006	2009	2016		
1	0.01	0.02	0.07	y =0.57 x-1135.10	0.96
2	12.48	19.73	26.12	y =142.61 x-284794	0.94
3	17.66	26.15	32.96	y =159.98 x-319107	0.93
4	34.84	47.30	60.88	y =248.60 x-495009	0.96
5	43.89	59.75	74.99	y =294.45 x-585989	0.94
6	59.26	77.06	93.87	y =327.29 x-650292	0.94
7	82.92	101.98	121.69	y =368.87 x-731323	0.95
8	93.00	112.37	132.67	y =377.79 x-748220	0.95
9	97.91	117.47	138.62	y =388.45 x-769114	0.96
10	109.32	129.13	151.12	y =399.60 x-790336	0.96

* SW = Ciliwung Subwatershed ** y = Total Area IC, x = years

Table 3 shows that Total Area IC in 2006 to 2016 significantly increased in each subwatershed (Atta'awun 4%, Katulampa 9%, Kedung Halang 10%, Pondok Rajeg 12%, Panus Bridge 12%, Kelapa Dua 13%, Condet 12%, Kalibata 12%, MT Haryono 12%, and Manggarai 12%). Total Area IC up to Manggarai subwatershed in 2006 to 2016 significantly increased from 32% to 44%. Linear regressions value (R^2>0.9) indicated that trend line can be used to estimate IC areas in other years.

3.2 WQI Value

Water quality indices (NSF-WQI, CCME-WQI and STORET methods) using Water quality parameters namely DO, BOD, pH, temperature, Turbidity, TSS, Nitrate, Total phosphate, and Fecal Coliform in the period of 2005-2016. The results of quantification of water quality indices in Ciliwung River and the impervious cover percentage from 2005 to 2016 are plotted to Impervious Cover Model (ICM) chart. ICM classification results from 2005 to 2016 are as follows: 1 Sensitive stream with less than 10% IC (Attawun), 4 Impacted streams with 10-25% IC (Katulampa, Kedung Halang, Pondok Rajeg, and Panus Bridge) and 4 Nonsupporting streams with 25-60% IC (Kelapa Dua, Condet, Kalibata Bridge, MT Haryono, and Manggarai).

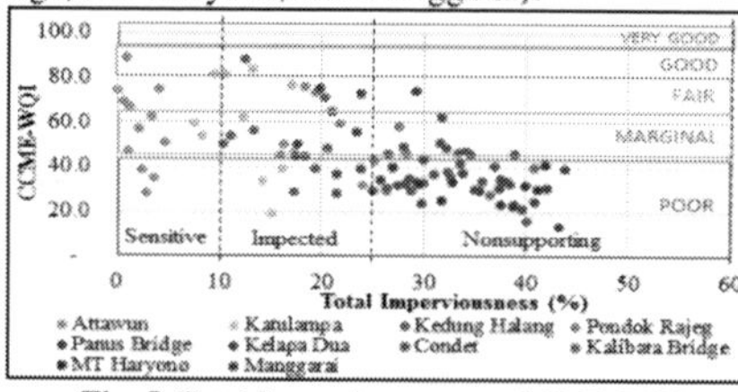

Fig. 5. Total Imperviousness vs CCME-WQI.

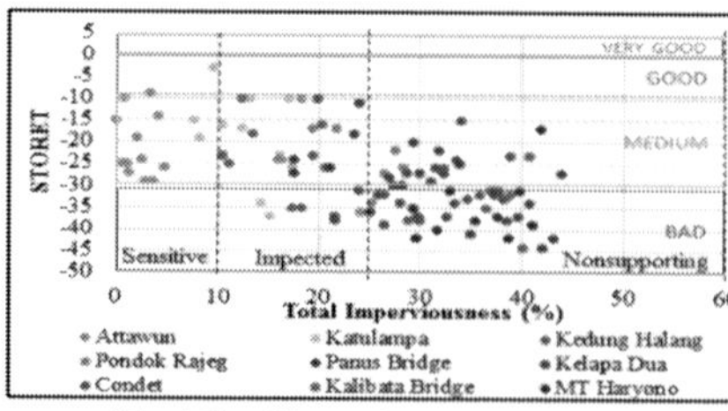

Fig. 6. Total Imperviousness vs STORET.

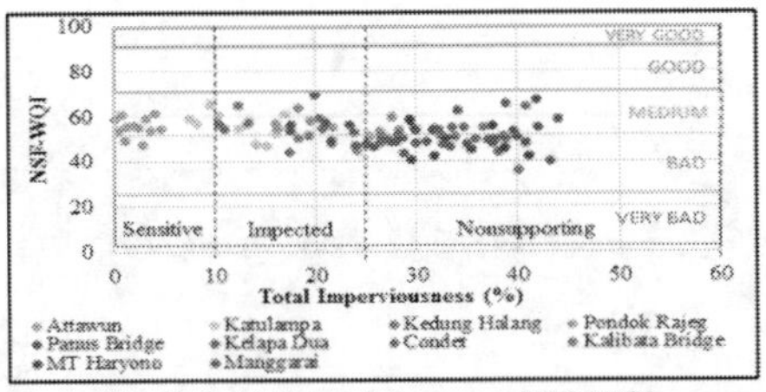

Fig. 7. Total Imperviousness vs NSF-WQI.

The ICM charts showed that the greater percentage of impervious cover, the water quality significantly decreased.

3.3 Correlation and Regression Test

The Kolmogorov-Smirnov normality and runs test showed that IC, NSF-WQI, CCME-WQI and STORET variables are normal distribution and not auto correlation with Asymptotic two-tailed significance/*p*-value of each variable > 0.05.

Table 4. Kolmogorov-Smirnov Normality and Runs Test

	IC	CCME-WQI	STORET	NSF-WQI
N	104	104	104	104
Normality Sig.	0.44	0.12	0.94	0.54
Runs Test Sig.		0.43	0.69	0.24

The Pearson correlation analyses showed that impervious cover was significantly strong negative correlated with Water Quality Index CCME ($r = 0.58$, $p < 0.01$) and STORET ($r = 0.54$, $p < 0.01$). While impervious cover was significantly negatively correlated with Water Quality Index NSF ($r = -0.32$, $p < 0.01$).

Table 5. Correlations & Regression Test Results

	CCME-WQI	STORET	NSF-WQI
Pearson (*r*)	-0.58**	-0.54**	-0.32**
Sig.	0.000	0.000	0.001
Regression WQI =	-0.83 x IC(%) + 65.52;Sig.=0.00 R^2= 0.34	-0.53 x IC(%)- 14.07;Sig.=0.00 R^2= 0.30	-0.16 x IC(%) + 56.93;Sig.=0.00 R^2= 0.10

** Correlation is significant at the 0.01 level (2-tailed).

The results of linear regression showed that Impervious cover significantly effecting Water Quality Index with CCME-WQI ($R^2 = 0,34$, $p < 0.01$), STORET ($R^2 = 0.30$, $p < 0.01$), and NSF-WQI ($R^2 = 0.10$, $p < 0.01$), which means that the R^2 values are variations of Impervious cover factor that can affect the water quality indices and the rest is influenced by other factors. Therefore, the linear regression model can be used to predict the change of water quality status in Ciliwung River in 2030.

Table 6. Results Predicted Water Quality Indices in 2030

SW*	% IC	NSF-WQI	Status	CCME-WQI	Status	STORET	Status
1	10	55	Medium	57	Marginal	-21	Medium
2	31	52	Medium	39	Poor	-30	Bad
3	35	51	Medium	36	Poor	-32	Bad
4	45	50	Medium	28	Poor	-36	Bad
5	47	49	Medium	26	Poor	-37	Bad
6	51	49	Bad	23	Poor	-38	Bad
7	56	48	Bad	19	Poor	-40	Bad
8	58	48	Bad	18	Poor	-41	Bad
9	59	47	Bad	17	Poor	-41	Bad
10	61	47	Bad	15	Poor	-42	Bad

* SW = Ciliwung Subwatershed

Overall predicted results in 2030 status of water quality in each subwatershed decreased, e.g. CCME-WQI in Katulampa and Kedung Halang decreased from "Marginal" to "Poor", NSF-WQI in Kelapa Dua decreased up to Manggarai from "Medium" to "Bad".

4 Conclusion

Impervious cover proved to have a strong inverse relationship with water quality indices. The results of the analysis show that Impervious cover factor can influence the water quality significantly and the rest is influenced by other factors. The correlation-regression models can be used to predict the water quality status of Ciliwung River. Therefore, the result can be a warning for all stakeholders that the management Ciliwung watershed is needed especially urban planning, monitoring and supervision to the waste management for residential, industrial and commercial areas in Ciliwung watershed.

Acknowledgment

The author would like to thank the financial support provided by Universitas Indonesia through the PITTA 2018 funding scheme under grant No.2422/UN2.R3.1/HKP.05.00/2018 managed by the Directorate for Research and Public Service and Special thanks to Evi Anggraheni and Environmental Agency Government DKI Jakarta and West Java Province.

References

[1] Kementerian Lingkungan Hidup RI, *Status Lingkungan Hidup Indonesia 2012*, (2012).

[2] R. Schiff and G. Benoit, J. Am. Water Resour. Assoc., **43**, 712–730 (2007).

[3] Z. Liu, Y. Wang, Z. Li, and J. Peng, Environ. Earth Sci., **68**, 2365–2373 (2013).

[4] J. B. Moring, U.S. Geol. Surv. Sci. Investig. Rep.,5101–C (2006).

[5] T. R. Schueler, L. Fraley-McNeal, and K. Cappiella, J. Hydrol. Eng., **14**, 309–315 (2009).

[6] C. Bellucci, Proc. Water Environ. Fed., 1003–1018 (2007).

[7] Canadian Council of Ministers of the Environment, *Canadian Water Quality Guidelines for the Protection of Aquatic Life : CCME Water Quality User's Manual Update 2017*, (2017).

[8] Kementerian Lingkungan Hidup RI, *KepMenLH RI No.115/2003 Tentang Pedoman Penentuan Status Mutu Air*, (2003).

[9] D. Sutjiningsih, Impervious Cover and Stream Quality Prediction Based on Impervious Cover Model and Water Quality Index at Ciliwung River, West Java, (2016).

[10] T. Schueler, Cent. Watershed Prot. Ellicott City, MD, 1 (2003).

[11] BIG, Data Penggunaan Lahan Se-Jawa. Badan Informasi Geospasial, Cibinong Bogor, (2009).

Development of ammonia gas sensor based on Ni-doped reduce graphene oxide

Thanva Tubthong[1], *Anurat* Wisitsoraat[3], *Chookiat* Tansarawiput[4], *Pakorn* Opaprakasit[2], and *Paiboon* Sreearunothai[2,*]

[1]Thailand Advanced Institute of Science and Technology (TAIST-Tokyo Tech), Sirindhorn International Institute of Technology (SIIT), Thammasat University, Phathum Thani 12121, Thailand

[2]School of Bio-Chemical Engineering and Technology, Sirindhorn International Institute of Technology (SIIT), Thammasat University, Phathum Thani 12121, Thailand

[3]National Electronics and Computer Technology Center (NECTEC), National Science and Technology Development Agency, Thailand Science Park, Klong Luang, Pathumthani 12120, Thailand

[4]National Nanotechnology Center (NANOTEC), National Science and Technology Development Agency, Thailand Science Park, Klong Luang, Pathumthani 12120, Thailand

Abstract. The work aims to develop a simple and low cost ammonia gas sensor based on reduced graphene oxide (rGO). Reduced graphene oxide doped with nickel sulfate ($NiSO_4$/rGO) was used as a sensing material. The sensor was fabricated by a simple drop-cast and spin-coat technique. The performance of the nickel-doped reduce graphene oxide were studied in terms of electrical changes as well as chemical interactions. It was found that after the fabricated sensor was exposed to ammonia vapour for 10 min, the average resistivity was increased to 43% from initial resistance and retained about 8% resistance change upon ammonia removal. The mechanism of the sensor reaction with the ammonia gas is also studied using Fourier Transform Infrared Spectroscopy (FTIR) and is discussed. This preliminary work may help develop the highly sensitive ammonia gas sensor.

1 Introduction

Ammonia is extensively used in several fields of industries including fertilizers, food processing, automotive, rubber, refrigerator and medical diagnostics[1]. High concentration of ammonia greater than 300 ppm can be harmful to human health[2]. It can cause suffocation, severe irritation to respiratory tract, eyes, and skin. Moreover, exposure to over 5,000 ppm can be lethal within 5-10 min [3]. Development of ammonia sensor which is inexpensive, robust and easy to fabricate is noteworthy.

Graphite oxide is the graphite that has been oxidized using strong oxidizers leading to various oxygenated functional groups in its structure. It can be obtained by treating graphite with strong oxidizers. The bulk material disperses in basic solutions to yield monomolecular sheet, known as graphene oxide (GO) by analogy to graphene, the single-layer form of graphite. Graphene oxide sheets have been used to prepare a strong paper-like material, membranes, thin films, and composite materials. Initially Graphene oxide attracted substantial interest as a possible intermediate for the manufacture of graphene. When the graphite stack exfoliated by oxidation[4], the oxidized graphene oxide can be reduced back using chemical or thermal treatment to remove oxygenated functional group, producing materials with large surface area. This state is called reduced graphene oxide (rGO). Transformation of GO to rGO can be performed either by heat or by chemical treatment. rGO is studied in many applications such as biological engineering[5, 6], optical electronics, absorbent material[7], ultrafiltration, energy storage, and gas sensor[8-10].

Nickel is a lustrous, silvery-white metallic element. Pure nickel, powdered to maximize the reactive surface area, shows a significant chemical activity, but larger pieces are slow to react with air under standard conditions because of an oxide layer formed on the surface preventing further reaction. Nickel(II) sulfate, or just nickel sulfate, usually refers to the inorganic compound with the formula $NiSO_4(H_2O)_6$. The yellow anhydrous form, $NiSO_4$, is a high melting solid that is rarely encountered in the laboratory. This material is produced by heating the hydrates above 330 °C. It decomposes at still higher temperatures to nickel oxide. Aqueous solutions of $NiSO_4\cdot 6H_2O$ and related hydrates can react with ammonia to give $[Ni(NH_3)_6]SO_4$. Addition of ammonium sulfate to concentrated aqueous solutions of nickel sulfate precipitates $Ni(NH_4)_2(SO_4)_2\cdot 6H_2O$. This blue-coloured solid is analogous to Mohr's salt, $Fe(NH_4)_2(SO_4)_2\cdot 6H_2O$.
In this study, nickel sulfate doped reduce graphene oxide is used as an active layer for detection of ammonia vapour by monitoring resistance changes. The mechanism of interaction is also investigated using Fourier-transform infrared spectroscopy (FTIR).

* Corresponding author: paiboon_sree@siit.tu.ac.th

2 Materials and Method

2.1 Materials

Graphene oxide (GO) used in this study was synthesized according to the Hummers' method [11]. In brief, 1.5 g of sodium nitrate and 3 g of graphite powder were mixed with 23 ml of sulphuric acid in a beaker in an ice bath. Potassium permanganate ($KMnO_4$, 9 g) was slowly added and heated to 35°C for 30 minutes. After that, 138 ml of deionize water was added and temperature maintained at 80°C for 30 minutes. Finally, 420 ml of the deionized water (DI) was added together with 50 ml of hydrogen peroxide (6%) in order to stop the reaction. The obtained products were washed with 5% hydrochloric acid and deionized water four and three times respectively, and centrifuged at 4000 rpm for 10 minutes, and dried in a vacuum oven at 40°C for 48 hr. The obtained graphene oxide was further ball-milled to reduce the particle size using ball milling (Retsch, Emax) at 800 rpm, 5 mm bead size and 30 minutes at 20 mg/mL concentration in DI water. Nickel sulfate (20 mg/ml) was added to the GO grinder solution ($NiSO_4$/GO) at the 1:1 (w/w) ratio, mixed together and used as active compound for fabricating $NiSO_4$/rGO sensor. In another set of active sensing compound, multi-wall carbon nanotube (MWNT, synthesized using arc discharge, from Nanogenstore [12]) was mixed with GO solution in the 1:1 ratio.

2.2 Preliminary selection of active compounds

Three active compounds have been initially tested for their responses to ammonia vapour which are reduced graphene oxide (rGO), carbon nanotube doped reduce graphene oxide (MWNT/rGO) and nickel sulfate doped reduce graphene oxide ($NiSO_4$/rGO). For the preliminary testing, drop cast technique have been used. The sensor is fabricated by dropping a solution of active compounds (10 mg/ml) onto a glass slide which has two aluminium electrodes separated by a gap of 0.5 mm. The sensor is then dried and GO thermally reduced using an oven at 180°C for 2 hr. The sensors are exposed directly to ammonia vapour from 30% ammonia solution (LOBA), for 75 minutes. This preliminary screening is to select the materials which show high response and quick recovery for further sensor fabrication.
Fig. 1. shows percent resistance change of the sensor upon exposure to ammonia atmosphere of:)a(reduce graphene oxide (rGO),)b(MWNT-doped rGO and)c(nickel sulfate ($NiSO_4$) doped rGO to ammonia vapour during 75 minutes)left(and after ammonia vapor removed)right(, nickel sulfate-doped reduce graphene oxide has almost a 100% change in sensing and also good recovery than the reduce graphene oxide alone which has some response and but cannot fully recovered to the initial resistance. The MWCT-doped reduced graphene oxide shows the highest response but cannot be recovered into the initial resistance too. As a result, nickel sulfate-doped reduce graphene oxide was chosen for further testing.

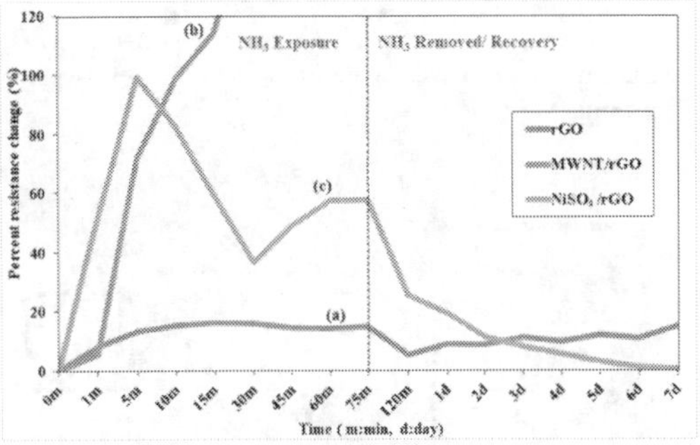

Fig. 1. Percent change of (a) reduce graphene oxide, (b) carbon nanotube doped reduce graphene oxide and (c) nickel sulfate doped reduce graphene oxide to ammonia vapor in the first 120 minutes (left), and after ammonia vapor removed (right).

2.3 Sensor fabrication

For fabrication of ammonia detection gas sensors, Fig. 2. shows the interdigitate electrodes with a 0.1 mm interdigitate spacing that was fabricated by sputtering an aluminium interdigitated pattern onto an alumina ceramic substrate.
The active sensor layer was prepared by spin casting a 50 μL of $NiSO_4$/GO solution onto the above interdigitated substrate at 5000 rpm for 30 seconds and dried in two steps: first on a hot plate at 50°C for 10 minutes, then in an oven at 200°C for 2 h to transform the $NiSO_4$/GO film into its reduced form ($NiSO_4$/rGO).

Fig. 2. (a) the interdigitated pattern with a 0.1 mm interdigitate spacing on an alumina substrate, and (b) $NiSO_4$/GO mixed solution for spin coating onto the electrode

2.4 Ammonia vapour testing

Fig. 3. shows diagram of a gas-test setup. Ammonia gas was generated by bubbling a 300 mL ammonia solution at various flow rates. This vapour is mixed with air zero for dilution at the ratio of 50% for the initial acclimatization, 50% for the 1st duty cycle, 25% for the 2nd duty cycle, and 12.5% for the 3rd duty cycle each with the interval of 10 min exposure to the ammonia gas and 30 min under air flow (air zero) for sensor recovery. The sensor response, or percent resistance change, may be calculated using equation (1),

$$\textit{Percent resistance change (\%)} = (R-R_0) \times R_0/100 \quad (1)$$

where R_0 is the sensor's initial resistance, and R is the sensor's resistance during gas exposure taken to be the value after the burn-in period (the 1st exposure period to ammonia vapor), just before the 2nd exposure to ammonia vapor.

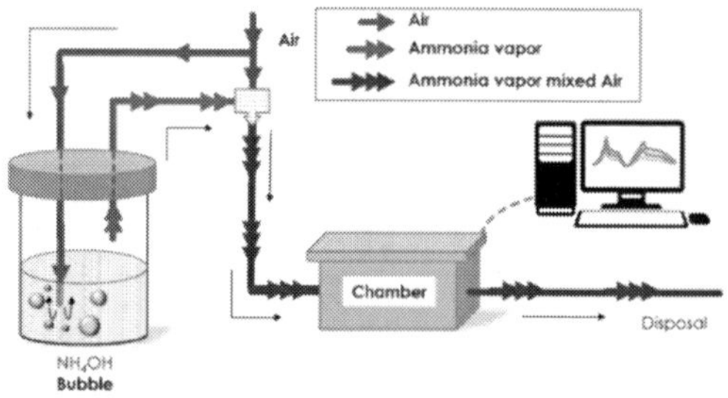

Fig. 3. Schematic diagram of the gas test set-up

3 Result and Discussion

3.1 Sensor's percent resistance change

Resistance of the nickel sulfate doped reduce graphene oxide ($NiSO_4$/rGO) sensor on an alumina substrate was measured in each sample for four duty cycles. Fig. 4. displays the percent change of electrical resistance during exposure to the ammonia vapour which is allowed to flow in consecutive stages as follows: the burn-in period or the first time the sensor is exposed to ammonia vapour at 50% ratio (mixing with air zero) for 10 min, followed by a recovery period by flowing air-zero for 30 min. During this period, the sensor resistance values change a lot and not completely recovered to the initial resistance value. The new resistance after 30 min recovery is taken as a baseline resistance of the sensor.

During the next duty cycle, taken as actual operating cycle, the sensor resistance change is increased to about 43%. In the following recovery period in air-zero flow, the resistance change of the sensor falls back to about 6% from the baseline value. In the 3rd duty cycle, ammonia vapour is reduced to 25%, the resistance change is increased to about 38%, and the sensor can recover to about 8% above the baseline value within 30 min after the ammonia gas turned off. In the 4th duty cycle, ammonia vapour concentration is reduced to 12.5% and the resistance change is increased to about 30%, the sensor can also recover very well with the resistance back to about 8% above the baseline value. From this result, it seemed that the sensors still change a little bit even after 4th cycles. The 8% retention of resistance may indicate some irreversible reaction of ammonia with sensor. The baseline resistance, or the resistance after ammonia removed, seemed to be quite stable after the 3rd cycle. The longer burn-in period may be needed to help stabilize the sensor performance.

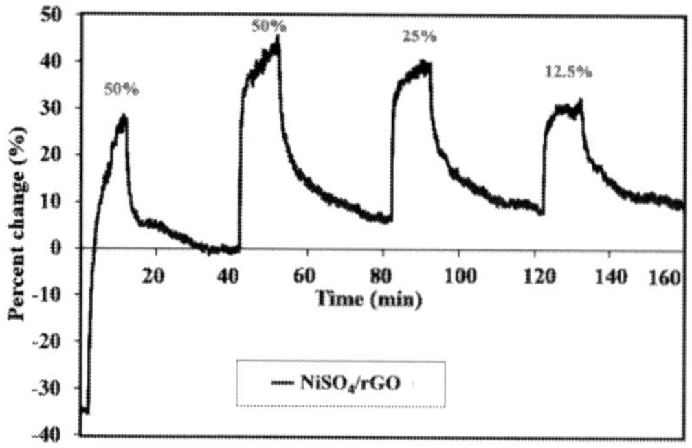

Fig. 4. Percent change response of $NiSO_4$/rGO exposure to ammonia bubble vapor for four duty cycles, the first cycle was taken as a burn-in period.

3.2 FTIR spectra

Fourier-Transform Infrared (FTIR) spectroscopy was used to examine the chemical characteristics as well as the interaction of $NiSO_4$/rGO with ammonia during exposure and recovery. The FTIR spectra of samples were measured in the attenuated total reflectance (ATR) mode using diamond head with the deuterated, L-alanine doped triglycine (DTGS) detector (Thermo Scientific, Nicolet iS5 FT-IR). FTIR spectra were obtained over the average of 32 scans from 3700 cm^{-1} to 570 cm^{-1}, with spectral resolution of 4 cm^{-1}.

Fig. 5. shows FTIR results compared to spectra of initial GO, $NiSO_4$, and $NiSO_4$/rGO sample used for gas ammonia sensing. Major peaks of GO spectrum were at 1718, 1618, 1210, and 1043 cm^{-1} which can be assigned to the C=O, C=C, and C-O of epoxy and ether vibrations, respectively [13]. For nickel sulfate, the major peaks were at 3409, 1600, 1070, 776 and 610 cm^{-1} which can be assigned to the O-H stretching, OH bending, SO_4 and Ni-O stretching, respectively [14]. After heat treatment, $NiSO_4$/GO is transformed into $NiSO_4$/rGO, the intensity of the bands representing graphene oxide decreases and peak of $NiSO_4$ still remained at 3405, 1070 and 610 cm^{-1}[13].

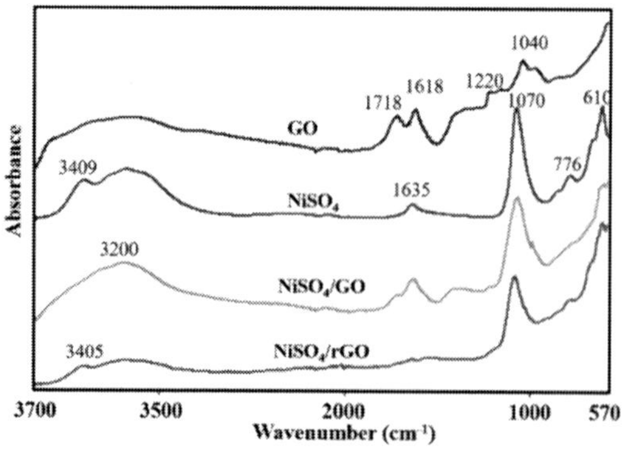

Fig. 5. FTIR results compared among spectra of initial GO, $NiSO_4$, $NiSO_4$/GO and $NiSO_4$/rGO sample used for gas ammonia sensing.

Fig. 6. shows the FTIR spectra of $NiSO_4$/rGO sample upon exposure to ammonia vapour, major peaks of acid-base reactions between ammonium and reduce graphene oxide were observed at 3300, 3200, and 1240 cm^{-1}. New emerging peak at 1440 cm^{-1} also shows up after sample's recovery under air flow. This bond is formed due likely to the ammonia desorption from the sensor's film. Upon re-exposure to ammonia vapor, this peak is suppressed again showing specific reversible interaction of ammonia gas with the sensor. The presence of the 3344 cm^{-1} upon ammonia exposure in both 1st and 2nd exposure is likely from the ammonium, which indicates that the ammonia gas may have been adsorbed onto the sensor's surface in the form of ammonium.

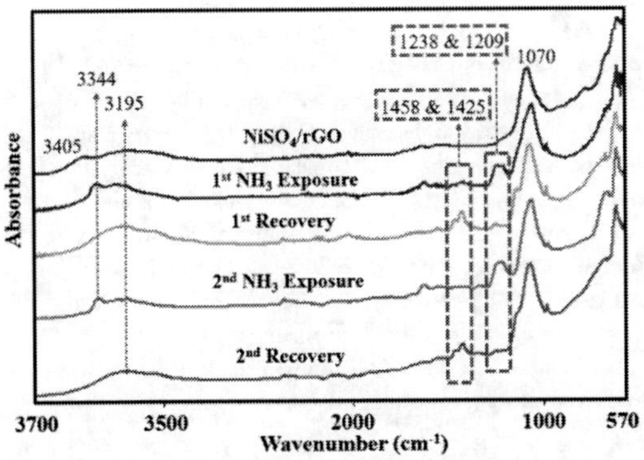

Fig. 6. FTIR of $NiSO_4$/rGO interaction with ammonia vapor for 5 minutes in two duty cycle.

4 Conclusion

Nickel-doped reduced graphene oxide ammonia sensor has been successfully fabricated using interdigitated sensors. The sensor can successfully detect ammonia gas generated by bubbling ammonia aqua solution with about 43% increase in the sensor's resistance during 10 min exposure (after initial burn-in period). Upon ammonia gas removal, the sensor's resistance can be almost fully recovered to about 6% above the baseline value in about 30 min. The sensors can be used for many cycles of ammonia gas exposure although the baseline shifted up slightly after each recovery indicating that some irreversible interactions also occurred. From the FTIR spectra, $NiSO_4$/rGO show interaction with ammonia vapor, causing the peak to appear at 3300, 3200, 1240 cm^{-1} indicating that the ammonia gas may have been adsorbed onto the film in the form of ammonium. The exact interaction mechanism remains to be further investigated.

This study was supported by the Center of Excellence in Materials and Plasma Technology (CoE M@P Tech), and Low Carbon Technology grant, Bangchak Petroleum Public Company. Thanva is grateful to TAIST-Tokyo Tech scholarship. Authors thank Thitima M. Daniels for helpful supports and guidance on equipment.

References

1. Mani, G.K. and J.B.B. Rayappan, Applied Surface Science. *Selective detection of ammonia using spray pyrolysis deposited pure and nickel doped ZnO thin films*. **311**: p. 405-412 (2014).
2. Zhang, D., C. Jiang, and Y.e. Sun, Journal of Alloys and Compounds. *Room-temperature high-performance ammonia gas sensor based on layer-by-layer self-assembled molybdenum disulfide/zinc oxide nanocomposite film*. **698**: p. 476-483 (2017).
3. Timmer, B., W. Olthuis, and A.v.d. Berg, Sensors and Actuators B: Chemical. *Ammonia sensors and their applications—a review*. **107**(2): p. 666-677 (2005).
4. Dimiev, A.M. and J.M. Tour, ACS Nano. *Mechanism of Graphene Oxide Formation*. **8**(3): p. 3060-3068 (2014).
5. Sun, P., et al., ACS Applied Materials & Interfaces. *Small Temperature Coefficient of Resistivity of Graphene/Graphene Oxide Hybrid Membranes*. **5**(19): p. 9563-9571 (2013).
6. Vuorinen, T., et al., Sci Rep. *Inkjet-Printed Graphene/PEDOT:PSS Temperature Sensors on a Skin-Conformable Polyurethane Substrate*. **6**: p. 35289 (2016).
7. Tippo, N., et al., *Slow Release of Menthol Using Sorbents Developed from Microwave Graphene Oxide*. 2017.
8. Bannov, A.G., et al., Sensors (Basel, Switzerland). *Investigation of Pristine Graphite Oxide as Room-Temperature Chemiresistive Ammonia Gas Sensing Material*. **17**(2): p. 320 (2017).
9. Seekaew, Y., et al., Organic Electronics. *Low-cost and flexible printed graphene–PEDOT:PSS gas sensor for ammonia detection*. **15**(11): p. 2971-2981 (2014).
10. Verma, S. and R.K. Dutta, RSC Advances. *A facile method of synthesizing ammonia modified graphene oxide for efficient removal of uranyl ions from aqueous medium*. **5**(94): p. 77192-77203 (2015).
11. Marcano, D.C., et al., ACS Nano. *Improved Synthesis of Graphene Oxide*. **4**(8): p. 4806-4814 (2010).
12. Singjai, P., S. Changsarn, and S. Thongtem, Materials Science and Engineering: A. *Electrical resistivity of bulk multi-walled carbon nanotubes synthesized by an infusion chemical vapor deposition method*. **443**(1): p. 42-46 (2007).
13. Onuki, K., et al., Bulletin of the Chemical Society of Japan. *Kinetics of the Thermal Decomposition of Nickel Sulfate*. **56**(11): p. 3294-3296 (1983).
14. Sharma, A.K., et al., *Extraction of Nickel Nanoparticles from Electroplating Waste and Their Application in Production of Bio-diesel from Biowaste*. Vol. 6. 2014.

A correlation between pavement skid resistance and wet-pavement related accidents in Thailand.

Jumrus Pitaksringkarn[1,*], *Ladda* Tanwanichkul[2], *Kwanruen* Yamthale[2]

[1]Civil Engineering Department, Faculty of Engineering, King Mongkut's Institute of Technology Ladkrabang, Bangkok, Thailand
[2]Civil Engineering Department, Faculty of Engineering, Khon Kaen University, Thailand

Abstract. This research aims to study a correlation between a pavement skid resistance and wet-pavement related accidents in order to determine the minimum friction threshold so called 'Investigatory Level (IL)' for the roads in Thailand. An accident database, a skid resistance database and a traffic volume of road network from the department of Highways, totalling 19 routes and 386kilometers, were used in the analysis of this study. In the analysis, 500-meter subsection intervals are used to determine a correlation between an average pavement skid resistance and wet-pavement accident rate using a non-linear regression analysis model. It was discovered that the pavement skid resistance has a major influence on the accident rate, depending on various types of road geometry. Moreover, the preliminary investigatory level is determined by using a past accident rate information as a reference for an investigatory level basis. We discover that the single carriageway (non-event) road is at highest threat due to low pavement skid resistance. The recommended investigatory level for each of the five road geometries all site, single carriageway, dual carriageway, horizontal alignment and curve road categories are 035, 0.50, 0.30, 0.30, and 0.40, respectively, which are in accordance to the investigatory level values in other countries.

1 Introduction

Three key elements of accident causes are road users, vehicles and road environments which could be one or a combination of those elements. From Department of Highway report, it was shown that 3 major causes of accident in the network including speeding, short cut-off driving and illegal overtaking [1]. The single vehicle accident as the type of run-off road crash is mostly caused by speeding on a low pavement skid resistance road. As a result, vehicle speeding is a potential factor in causing road accident especially on low friction pavement which could increase road crash numbers on the road network [2]. In fact, friction surface or skid resistance of road pavement could be one of the accident causing factors. This is because at the same level of speeding, the vehicle requires longer distance to stop on the road with low friction. The skid resistance of an asphalt pavement is of critical importance because it directly relates to the safety of motorists and their ability to stay on the road and be able to stop effectively. Skid resistance can be affected by several factors including the environment, roadway surface characteristics, and roadway geometry.

A correlation between traffic accidents and pavement surface characteristics has been researched and examined in many studies. In the study of [3], the number of wet-pavement skid crashes increases as the skid friction value decreases. Other studies focus on wet-pavement accidents by observing pavement skid resistance of road surface in different road geometries, such as single/dual carriageway, bending curve, and intersection area. Road pavements provide two main services including a functional service and a structural service. One of the important factors that determine a pavement functional service is the level of skid resistance due to the fact that safe driving depends on an adequate surface friction for vehicle manoeuvring, turning and braking. Skid resistance is therefore considered as one of the most important surface characteristics as it has a direct impact on traffic safety [4].

Therefore, the role of skid resistance standard in reducing accidents together with the role of surface texture as major factors in accident reduction is currently under reviewed by a number of highways agencies. Various studies employed using a relationship between accident rates and skid resistance in different road types and road geometries in order to establish an investigatory level of skid resistance [5-6]. This research thus aims to determine a correlation between wet-pavement accident rates and pavement skid resistance in order to initially establish the investigatory level of skid resistance for National Highway in Thailand adopted from the Department of Highway (DOH).

2 Road accident and skid resistance

Wet pavement accidents and a total number of accidents were used to study the relationships between

* Corresponding author: jumrus@gmail.com

https://doi.org/10.1051/matecconf/201819202049

pavement skid resistance and a total number of wet pavement accidents on major highways in the Kingdom of Saudi Arabia [7]. It was concluded that the number of accidents approaches almost a constant number at a skid number greater than 0.45. Moreover, the correlation between total traffic accident rates and the skid number (SN) was found to be statistically significant with high accident rates within and outside the capital city, Amman. The statistically significant negative correlation indicates that skid resistance is an important contributing factor in traffic accidents causation. This agrees with the findings of other research works stating that lower levels of Skid Number (SN) are likely to lead to a higher potential traffic accidents and also to a higher rate of fatal accidents [8].

The skid resistance offered by a wet road surface decreases with increasing travel speeds [9].This indicates that wet surface is likely to affect on the number of accidents, especially in speeding vehicles. [10] studied a relationship between skid resistance and wet pavement accidents on local roads in the Southwest of England aiming to determine the safe skid resistance threshold (according to IL HD28/04 standard) for local roads. And the results in both individual local counties and combined one showed strong relationships between wet accident rate and skid resistance which is similar to the study by [3, 11-12].

Road geometry type is the other factor that is usually concerned in an analysis of skid resistance and traffic accidents. In the studies by [2,13], both linear and non-linear correlation between skid resistance and accident indicating different road geometries especially sharp curves which affect skid resistance, resulting in a higher number of wet pavement accidents comparing to other types of road geometries. [2] provides an overview of the four locations the skid resistant material was placed and their respective performance and wear characteristics over one winter and after two years.

This research will explore and determine relationship between wet pavement accident rates and pavement skid resistance in order to estimate an Investigatory Level (IL) for Asphalt Surface Road considering in different accident locations and road types. Thus, relationship between wet accident and skid resistance will be determined as well as divided into different road types in order to preliminarily establish investigatory level for DOH network in Thailand.

3 Research methodology

3.1 Data collection and analysis

Skid resistance data, accident data, and traffic data are used for establishing skid resistance and traffic accident relationship in order to determine preliminary result of threshold investigatory level.

Accident and skid resistance databases during the year of 2011 to 2012 were acquired from the Department of Highway (DOH). Wet pavement accident locations and skid resistance data are encoded into a GIS map. Then the overlays of both layers, i.e. data wet accident and skid resistance data in the GIS map are used to count accident frequency and calculate an average friction of the 500-meter subsection interval of road section. Accordingly, 19 routes with the total distance of 386 kilometres covering 9 provinces in 9,000 kilometres in Thailand were selected for analysis in this study as shown in Figure 1. Skid resistance and accident data associated traffic volume and other data will be used to calculated following to selected study area.

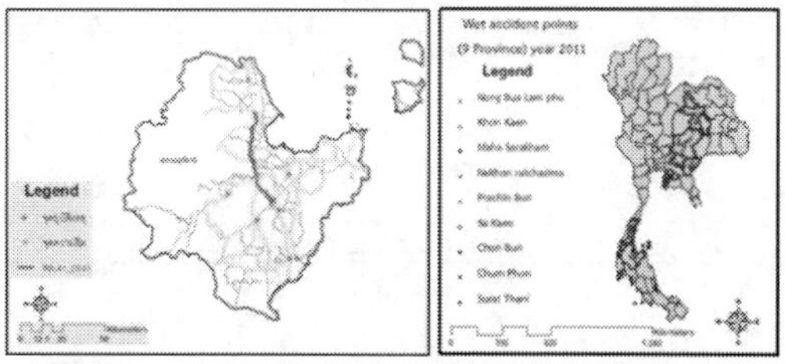

Fig. 1. Overlaid maps of skid resistance and wet accident location and selected wet accident Locations

3.2 Investigatory levels

The investigation level (IL) is a skid resistance maintenance priority level that has been set with the objective of equalizing the personal risk (crash rate rather than crash number) of a wet road skidding crash across the state highway network, while maintaining an economic balance between the cost of achieving the skid resistance and the crash savings [14], p 4. Moreover, investigatory levels for skid resistance are generally used as maintenance strategies in many road agencies [15]. It is therefore required to understand the measuring device outputs with respect to intuitive investigatory levels will need to be a prerequisite to such exercises. Many countries for example United Kingdom, Australia, and New Zealand consider to monitoring skid resistance level of pavement of road network [3]. In fact, along a highway with different geometry of the elements (gradient, bend radius), environmental conditions and circumstances, the threshold between 'safe' and 'potentially unsafe' are difficult defined by a unique level of skid resistance. Therefore, a set of skid resistance levels are considered for different situations and conditions.

Table 1. Minimum Skid Resistance Threshold in different countries

Country	Minimum Skid Resistance Threshold	
United Kingdom (HD 28/04)	Motorway	0.35
	Dual Carriageway non event	0.35
	Single Carriageway non even	0.40
	Dual Carriageway (All purposes)	0.40
	Bend – Dual carriageway	0.45
	Bend – Single carriageway	0.50
Australia (Long, Wu et al. 2013)	Intersection with controls	0.55
	Curves with tight radius	0.50
	Intersections	0.45
	Free areas of Undivided roads	0.40
	Free areas of Divided road	0.35
	Road Junction Approaches	0.35
	Curve < 250 m.	0.40
	Undivided Carriageway (non-event)	0.35
	Divided Carriageway (non-event)	0.25

3.3 Road category: accident and skid resistance models

Roadway geometry is one of important factors affecting skid resistance of road surface which is related to road accident [2] since skid resistance directly related to the ability of motorist to control the vehicle to be able to stop effectively. Moreover, in a larger-scale study of the link between skid resistance and personal injury accidents, based on 1,000km of road network [16], it has been confirmed that the different levels of accident risk for different types of road site and the increase in risk for sites with lower skid resistance. As a result, in this research, types of road geometry will be considered for using in categorize relationship between wet accident rate and skid resistance. The set of category will be also used for establishing the investigatory levels of minimum skid resistance number of road pavement surface. By different types of road geometry, there five models will be categorized using for analysis relationship between skid resistance and road accident in the research as details following:

Table 2. Model Road Types

Model	Description
M1	All site - Accident on two or more -lane roads
M21	Single carriageway - Accident on two-lane roads
M22	Dual carriageway - Accident on more than two-lane roads
M31	Horizontal Alignment - Accidents on all types of road
M32	Curve – Accident on two or more -lane roads

4 Analysis results

Non-linear regression analysis was used to establish relationship between skid resistance and wet accident for all five types of road geometry. Correlation coefficient, value of R^2 represents the friction of the road influence the accident rate whereas testing of the F-Test of the model indicates level of the road friction influencing the accident rate which is more than 95. In fact, the single carriageway and curve trends have higher decreasing trend comparing to the trends of dual carriageway and horizontal. In the conclusion, single carriageways and curve roadways are higher risk to wet accidents comparing with dual carriageways and horizontal roadways in term of skid resistance of roadway. Figure 4 to 6 present relationships and trends of 5 models of relationship between skid resistance and road friction.

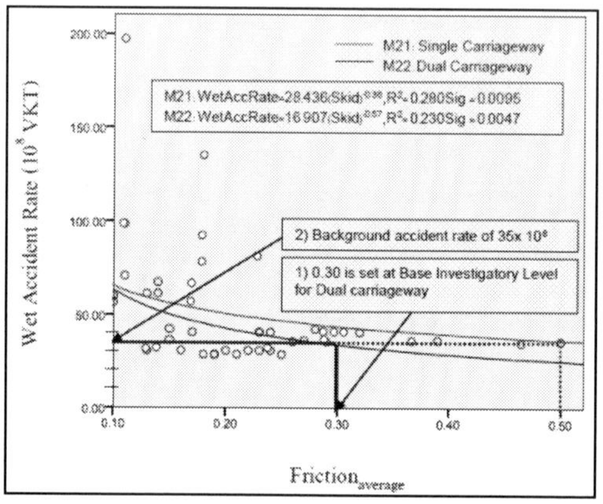

Fig. 5. M21-Single carriageway and M22-Dual carriageway non-events

Guideline from [17] to determining investigatory level is summarized from a work example in a guideline study. A 'base investigatory level' of different site categories will be set and then it will be used to determine an investigatory level for the other categories. Therefore, 4 steps taken from Austroads work will be used to determine IL for different site categories of road geometry in this research:

1. Establish trend of relationship between wet accident rate and skid resistance: Trends of relationship between wet accident rate and skid resistance for 5 categories.

2. Get the base investigatory level from existing guideline in different site categories: guidelines of investigatory level from United Kingdom, Australia, and New Zealand are summarised in Table 3. Base investigatory level is set at 0.30 for dual carriageway based on values from UK, Australia, and New Zealand at 0.35, 0.35 and 0.25 respectively.

3. Select a background accident rate for using in comparison of crash risk across different risk categories: in this research, wet accident rate from a low risk road category i.e. dual carriageway type will be used as back ground accident rate of 35 (108 VKT) for base investigatory level at 0.30.

4. Set investigatory level for each category: At 35 of wet accident rates, investigatory level for five categories can be set as detail in Table 4.The wet accident rate

slightly decreases or remains steady when average friction increase therefore investigatory level (IL) can be approximately established. The friction increments 0.05 are used for consideration in the trend shown in previous section. Summary of preliminary investigatory level (IL) for five categories are M1, M21, M22, M31, and M33 are 0.35, 0.50, 0.30, 0.30, and 0.40 respectively.

5 Conclusion and discussion

The relationships between wet accident rates and skid resistance of Highway Road in Thailand were examined for five models including all site, Single carriageway non-event, Dual carriageway non-event, Horizontal Alignment and Curve. It was found that wet accident rate is inversely proportional to skid resistance for all categories. As a result, wet accident rate decreased as and skid resistance increased. By using method recommended in a guideline from Austroads, dual carriageway road type was selected as the base investigatory as 0.3 and resulting of a background accident rate at 35 (108 VKT). As a result, preliminary investigatory Level (IL) can then be determined for three categories of five sub-models including All Site, Single carriageway non-event, Dual carriageway non-event, alignment and Bend for 0.35, 0.50, 0.30, 0.30 and 0.40 respectively. It was found that ILs from this study corresponding to ILs from other study such as study of [6, 10]. As a result, the preliminary investigatory level can be used as support tool in making decision for maintenance schedule of road surface monitory. Preliminary ILs could then be used as verify-tool to be more precise for road maintenance schedule for the road surface friction of Thailand. As a result, appropriate skid resistance would be maintained and wet accident could then be reduced.

References

1. Bureau of Highway Safety. Bangkok, Department of Highway, Ministry of Transport, Thailand. (2013)
2. M. Karim, J. Chyc-Cies, et al. *Conference of the Transportation Association of Canada Fredericton.* New Brunswick, Canada: 1-10. (2012)
3. B. Crisman and R. Roberti. Procedia - Social and Behavioral Science **53**: 1055-1068. (2012)
4. N. Piyatrapoomi, J. Weligamage, et al. *Proceedings of the International Conference on Managing Road and Runway Surfaces to Improve Safety,* Cheltenham, England. (2008)
5. J. Kudrna, A. Vojtelek, et al. *Road Skid Resistance Influence on the Number of Crash Accidents.* (2007)
6. A. Stevenson and M. Phillips. *International Safer Roads Conference,* May 2008. (2008)
7. A. I. Al-Mansour. The Journal of Engineering Research **3(1)**: 75-78. (2006)
8. K. Ramadan and I. M. Muslih. Jordan Journal of Applied Science 'Natural Sciences Series' **11(1)**: 11-18. (2013)
9. Road Safety Observatory. *Synthesis title: Surfaces, Observatory main category: Roads.* Retrieved 26 September 2015, from http://www.roadsafetyobservatory.com/Review/10051.
10. Kennedy, C. and J. Donbavan. *International Safer Roads Conference 2008.* (2008)
11. J. S. Kuttsech. *Master of Science In Civil and Environmental Engineering:* Blacksburg, Virginia, Virginia Polytechnic Institute and State University. 147. (2004)
12. H. E. Viner, R. Sinhal, et al. *Linking Road Traffic Accidents With Skid Resistance – Recent.* UK Developments. 1-13. (2005)
13. M. Saplioglu, E. Eriskin, et al. *10th International Congresson Advances in Civil Engineering.* Ankara, Turkey: 1-10. (2012)
14. W. Kotahi. *Notes to specification for state highway skid resistance managment.* NZ Transport Agency: 20. (2012)
15. C. Blanchard and D. Hollaway. *Road Surface Friction (Skid Resistance & Pavement Texture): Recommended Investigatory Levels.* Department of Planning, Transport and Infrastructure: 4. Adelaide. (2003)
16. M. P. Rogers and T. A. Gargett. Highways and Transportation, **38**, 10-16. (1991)
17. Austroads. *Development of Safety Related Investigatory Level Guidelines: A Worked Example of Methodology.* Sydney, Department of Planning, Infrastructure and Transport SA: 17. (2013)

The contribution agriculture waste reinforcement to the floor topping of renovation building architectures: compressive strength test by trial the percentage of the pseudostem of banana dried powder

Prang Subpa-asa[1,*], *Komson* Maleesee[1], Shigeyuki *DATE*[2], Tetsurou *KASAI*[2], *Nutchaphon* Phromsuwansiri[1], and *Chakorn* Duangpan[1]

[1]Civil Engineering Department, Faculty of Engineering, King Mongkut's Institute of Technology Ladkrabang, Bangkok, Thailand, 10520.

[2]Civil Engineering Department, Faculty of Engineering, Tokai University, Kanagawa-pref., Japan, 259-1292.

Abstract. The growing demand of construction around the world has led to an increased usage of concrete. However, convention concrete making materials are not entirely environmental friendly and this has enthused research on seeking greener alternative for concrete production. Agriculture industries are one of the major industries globally that harvested products such as food and biomass for organism around the world. After harvesting products, there are effluence of agriculture waste as left, straw and pseudostem. This research focusing on mechanical properties as compressive Strength Test of mortar cement adding with the pseudostem of banana dried powder (PBDP) by trial the percentage of PBDP and chemical properties as using scanning electron microscope (SEM). These findings indicate that 10% of PBDP replacement is the best ratio and element of PBDP which C was the most element in PBDP as 59.96% of mass and 71.49 % of atomic.

1 Introduction

Nowadays, to build new constructions in Thailand, approval drawing is required. There are also rules, standards and restrictions to follow. Meanwhile there are many constructions that have been built for more than 10 years in Thailand. As a result, building renovation is more popular. The renovation more concerns the theory of structure which in Thailand used American concrete institute (ACI) design standard for design the building. Consequently, superimposed dead load (SDL) like decorating, celling and system work be able to significant factor that affect to renovate construction. Thus, the research was focusing on renovate construction affect so the example was floor leveling method by pouring the concrete topping to searching for the materials reinforcement to compressive strength and lighten weight materials.

Agriculture farming is one of the major industries globally as most of the harvested agricultural products are sources of food of people around the world. Countries such as China, India, United States, Brazil and Nigeria are among the world's largest producer of agriculture products, which include cereal, vegetable, and fruits. [1]. However, after harvesting and consumption of the agricultural products, there are abundance of waste materials leftover, such as leaf, straw, stalk and ash. Most of these agriculture wastes are disposed to the surrounding and there is little effort in reusing these materials. In recent times, researchers have begun to utilize these wastes as partial replacement for conventional concrete-making materials and came up with interesting findings. While the use of agriculture wastes in concrete such as those from palm oil, coconut, sugarcane and paddy industry were well documented in the past. [2]

The agriculture in Thailand, among the total area of 520 million square kilometers, more than 65% is occupied by agriculture related activity. With the ever-increasing market demand for agricultural products, most of the agricultural residues often end up in the municipal waste streams and not efficiently. [2] The area for growing the banana is 760.72 square kilometers, the largest area of cultivated banana growing up to 461.72 square kilometers with 773,732 tons per year and increasing 13.4 percent a year. [1]

Banana fiber could play role in emerging bio-economy. Currently, millions of tons of banana pseudostem are dumped as waste and most of the

* Corresponding author: prangsub@gmail.com

farmers are facing huge problems in disposing the accumulated banana pseudostem.[3]

Hence, this paper will be researching on the properties of construction materials as cement and aggregate which renewable agriculture waste for sustainable environment friendly.

The banana plant is the largest herbaceous flower plant (2-16m in height), composed of long fibers strongly overlapping forming a pseudostem. Banana trees produce generally 30 large leaves (almost 2 m long and 30-60 cm wide). The pseudostem of banana was studied. [4] Banana grows in variety of soil and grows by corm structure. They are tall and fairly study, there are often mistaken for tree but it is actually a "false stem" or "pseudostem". After fruiting, the pseudostem dies, but offshoots will normally have developed from the base, so that the plant as a whole is perennial. As a result, the agriculture industries waste has been affected to environment.

According to [5] the banana leaf ash (BNNLA) consisted mainly of silica, which was about 49% and the loss on ignition was about 5%. The Blaine's specific surface area and specific gravity of the BNNLA was about 14,000 cm^2/g and 2.44, respectively.

Due to the higher fineness of the BNNLA compared to cement, [5] observed that the BNNLA had filler effect which contributed to the lower amount of entrained air in mortar specimens with up to 10% BNNLA cement replacement level. Similarly, the compressive and tensile strength of concrete specimens containing up to 20% BNNLA were approximately 12% and 20% higher respectively than the corresponding control concrete. In addition, the incorporation of BNNLA reduced the tendency of concrete specimens towards corrosion [5]

2 Methodology

2.1 Materials and Mix propoties

Banana fiber (Musa sapientum) obtained from the pseudostem of plant by production in Ladkrabang district, a suburb area in Bangkok, Thailand. Banana fiber was expected from the pseudo stem (average diameter of 200mm) of the banana plant. The pseudostem were cut in parts about 5 mm in length using long knife and then dried in a hot air oven at 60̊ C for 48 hours. [6] They were further cut in a knife mill to get banana powder of sieves NO.30 (600 microns) which show in fig 1. The material used and its quality was show in table 1. The pseudostem of banana dried powder used by the percentage of sand replacement that show in table 2

Table 1. Materials and properties

Materials	Symbol	Properties	ratio
Cement	C	Ordinary Portland cement	1
Fine aggregate	S	River sand	1
Pseudostem of banana	PBDP	Pseudostem of banana dried powder of sieves NO.30	sand replacement
Sikament FF	FF	A super plasticizing concrete admixture based on Melamine-Formaldehyde.	10% of water
Water	W	-	0.5

The mortar reference mix proportion were 1:1 (Portland cement type1: Sand pass sieve NO.30) by mass(w/w). PBDP were adding 10%, 20%, 30%, 40% and 50%by replacement sand. The compressive strength of the PBDP mortar were determine at age of 7 days using dimensional specimens of 50 mm x50 mm x50 mm.

Fig.1. The pseudostem of banana tree

Fig.1(a) Shows the pseudostem of banana tree used in the present study after cutting for 24 hours.

Fig.1(b) Shows the cutting of the pseudostem about 5 mm long by removing the cover of pseudostem and cleaning by water (soil).

Fig.1(c) Shows pseudostem that was dried in a hot air oven at 60̊ C for 48 hours.

Fig.1(d) Shows cellulose banana powder, cut in knife mill as 600 microns.

Table 2. Mix proportions

Sample	% of PB	C (g.)	PB (g.)	S (g.)	W/C	FF
A	0		0	1100		
B	2.5		27.5	1072.5		
C	5	1100	55	1045	0.5	10% of water
D	7.5		82.5	1017.5		
E	10		110	990		

2.2 Testing

The chemical phases of PBDP were obtained from scanning electron microscope (SEM)

A scanning electron microscope (SEM) is a type of electron microscope that produces images of a sample by scanning the surface with a focused beam of electrons. The electrons interact with atoms in the sample, producing various signal that contain information about the sample's surface topography and composition. The electron beam is scanned in a raster scan pattern, and the beam's position is combined with the detected signal to produce an image. SEM can achieve resolution better than 1 nanometer. Specimens can be observed in high vacuum in conventional SEM, or in low vacuum or wet conditions in variable pressure or environmental SEM, and at a wide range of cryogenic or elevated temperatures with specialized instruments [7]

The mechanical phases of PBDP were obtained from triaxle tester.

Compressive strength or compression strength is the capacity of a material or structure to withstand loads tending to reduce size. Measuring the compressive strength of a PBDP mortar cement by using Triaxial tester. Standard triaxial System with analogue measurement is the ideal basic solution for performing standard triaxial tests such as effective and total stress laboratories. Measurements of compressive strength are affected by the specific test method and conditions of measurement.

3 Result and Discussion

The micrographs of fibers surfaces by using SEM are presented in Fig 3. The surface of PBDP is relatively hairy. The representative element of PBDP are presented in the Fig 3(b) and Table 3 which C, O, K, Cl and Mg are 59.96%, 25.87%, 11.9%, 1.84% and 0.43% of mass respectively. The number of atomic of C, O, K, Cl and Mg are 71.49%, 23.15%, 4.36%, 0.74% and 0.25% respectively. The pozzolanic reactivity increased with mass of C element and affect to compressive strength load.

Table 3. Result of element of PBDP

Element	% Mass	%Atomic
C	59.96	71.49
O	25.87	23.15
K	11.90	4.36
Cl	1.84	0.74
Mg	0.43	0.25

(a)

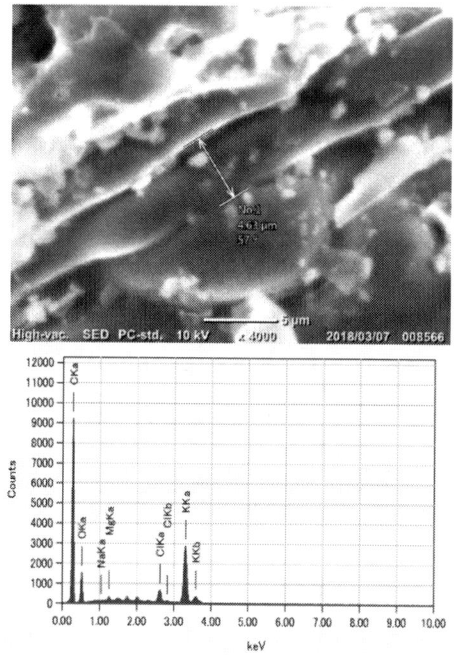

(b)

Fig.3. The result of PBDP by scanning electron microscope (SEM)

The representative of compressive strength of PBDP mortar were measured by triaxial tester in the Table 2 which mass of concrete and water cement ratio controled by mortar cement ratio 1 :1 (Portland cement type 1 : sand pass sieve NO.30). The general mortar (non-PBDP) on dimensinal specimens of 50x50x50 mm^3 were 227.9 ksc of average compressive strength and 1,928 kg/m^3 of cement mortar density. Table 4 was presented result of this study as PBDP sand replacement retio 2.5%, 5%, 7.5% and 10% were 1,788.59 kg/m^3, 1.616.59 kg/m^3, 1,466.26 kg/m^3, and 1,319.27 kg/m^3 of cement mortar density and 4.984 ksc, 3.205 ksc, 2.762 ksc and 3.768 ksc of compressive strength, respectively. Comperision of compressive strength and cement mortar density by ratio were 0.028, 0.0020, 0.0019, 0.0029 which are 2%, 2.5%, 7.5% and 10%, respectively.

Table 4. Result of mechanical

% of banana powder	Sample	W	L	H	Area	Volumn	Weidge	Density	Compressiven strength	Compressive strength
		cm	cm	cm	cm^2	cm^3	(Kg)	(kg/m^3)	(KN)	(ksc)
0.0%	A1	4.02	5.10	5.09	20.50	104.36	0.197	1887.784	42.250	210.069
	A2	3.95	5.11	5.05	20.18	101.93	0.196	1922.856	48.900	246.957
	A3	4.25	5.00	5.10	21.25	108.38	0.214	1974.625	47.310	226.947
2.5%	B1	4.61	5.15	5.07	23.74	120.37	0.212	1761.245	1.562	6.707
	B2	4.57	5.00	5.01	22.85	114.48	0.209	1825.670	1.000	4.461
	B3	4.35	5.01	5.03	21.79	109.62	0.195	1778.851	0.809	3.784
5.0%	C1	4.95	5.01	5.01	24.80	124.25	0.203	1633.862	0.911	3.745
	C2	4.96	5.06	5.03	25.10	126.24	0.204	1615.958	0.656	2.664
	C3	5.03	5.02	5.00	25.25	126.25	0.202	1599.962	0.794	3.205
7.5%	D1	5.07	5.03	4.99	25.50	127.26	0.184	1445.910	0.731	2.922
	D2	5.09	4.99	5.03	25.40	127.76	0.186	1455.884	0.550	2.207
	D3	5.02	5.01	5.02	25.15	126.25	0.189	1496.982	0.779	3.157
10.0%	E1	5.04	5.04	4.99	25.40	126.75	0.168	1325.402	1.085	4.354
	E2	5.00	5.07	5.01	25.35	127.00	0.169	1330.672	0.743	2.989
	E3	4.99	5.04	5.04	25.15	126.75	0.165	1301.734	0.977	3.960

4 Conclusion

This study investigated the mixing of PBDP sand replacement, compressive strength testing, scanning electron microscope testing were carried out in a ratio can be summarized that the optimal ratio in this study was 10% sand replacement as 3.768 ksc and 1,319.27 kg/m^3 of compressive strength and cement mortar density, respectively. As a result, the PBDP mortar cement ratio was 10% PBDP sand replacement which is the first phase of thesis, The Contribution Agriculture Waste Reinforcement to The Floor Topping of Renovation Building Architectures, that X-ray diffraction, Particle size distribution analyser, curing condition and shrinkage condition are further study.

References

1. Simpson, S.D., *Top Agricultural Producing Countries*, (2015) Retrieved from: www.investopedia.com/financial-edge/0712/top-agricultural-producing-countries.aspx.

2. C. Visvanathan and C. Chiemchaisri., *Management of Agricultural Wastes and Residues in Thailand: Wastes to Energy Approach*, International Conference on Agricultural Wastes, Putrajaya, Malaysia, (2006).

3. Mahapatra, D., Mishara, S. and Sutar, N, *Banana and its by-product utilization: an overview*. Journal Scientific & Industrial Research, 69: 323-329 (2010).

4. Bilba, K., Arsene, Marie-Ange. Ouensanga, A., *Study of Banana and Coconut fibers Botanical composition,thermal degradation and textural observations*, Bioresource technology, (2015).

5. Kanning, R.C., Portella, K.F., Braganca, M.O.G.P., Bonato, M.M., dos Santos, J.C.M., *Banana leaves ashes as pozzolan for concrete and mortar of Portland cement*, Construction and Building Material, 54, 460-465 (2014).

6. Ouensanga, A., *Variation of fiber composition in sugarcan stalks,* Wood and fiber science 21 (2), 105-111 (1989).

7. Stokes, Debbie J., *Principles and Practice of Variable Pressure/Environmental Scanning Electron Microscopy (VP-ESEM)*, Chichester: John Wiley & Sons. 234 pages (2008).

Investigation on pile-soil-tunnel interaction due to adjacent loaded pile row by 3D FEM

Narunat Heama[1], *Pornkasem* Jongpradist[2], *Prateep* Lueprasert[1], *and Suchatvee* Suwansawat[1]

[1]Faculty of Engineering, King Mongkut's Institute of Technology Ladkrabang, Bangkok, Thailand
[2]Faculty of Engineering, King Mongkut's University of Technology Thonburi, Bangkok, Thailand

Abstract. The effect of adjacent loaded pile row on the existing tunnel is analysed by 3D finite element method. The MRTA project is used as reference case in this study. The case study is divided into two cases, tunnel and pile tip located in soft clay and stiff clay, respectively. The pile diameter of 1 m and pile tip located at the elevation of tunnel crown with various number of piles within the row were considered. The analysis results are shown in terms of total soil displacement (without tunnel) and maximum tunnel deformations. Both total soil displacements and tunnel deformation increase with the number of piles and approach a maximum value when the number of piles is greater than 13 and 11 for cases tunnel in soft and stiff clay, respectively. This reveals that the tunnel deformation mechanism is primarily due to the soil displacement.

1 Introduction

The tunnel construction increases in the urban development to deal with the traffic congestion problems and resource supply. Typically, the tunnels are constructed underground along the major roads to avoid the building. In recent years, however, new buildings and infrastructures have been substantially constructed in the close proximity. Then, the existing tunnel is inevitably affected by the adjacent piles under loading as shown in Fig.1. An evaluation of the effect of loaded pile on the stability and integrity of tunnel is essential.

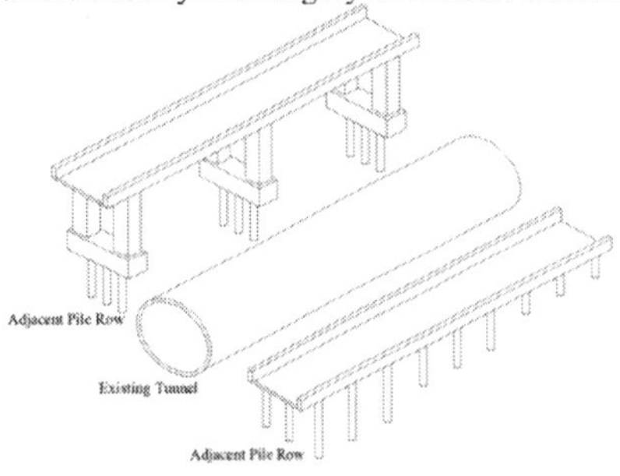

Fig. 1. New construction adjacent to existing tunnel in urban environment

Several researches have studied the mechanisms of soil–tunnel–pile interactions. Examples are the effect of nearby loaded bore pile on existing tunnel in London [1-2], the effect of multi-storey commercial building on two existing tunnels of Toronto subway in Canada [3]. In addition, the effect on existing tunnel due to adjacent bore pile under loading was analyzed by 3D finite element method (FEM) in Bangkok, Thailand [4-6]. Many researches focused on the effect on existing tunnel due to adjacent single pile under loading to study the response of tunnel. Only a few researches considered the effect of pile rows under loading on the existing tunnel. In previous research, the pile row and tunnel problem was analyzed assuming a 2D plane strain condition. Analysis under 2D plane strain condition using transformation technique for pile and 3D slice (using symmetry plane) were studied to evaluate and compare the effect of loaded pile row [7]. However, both conditions considered the ideal condition of which the number of pies in the row is great enough so that the plane strain condition can be considered at centre plane of the model. The results on that plane are thus in the conservative manner if the number of piles in the row is not that many. In engineering practice, the pile row condition is often encountered for bridge structures of which the number of piles in the row depend on the width. It is thus worth to know the minimum number of piles in the group that provides 2D plane strain condition at the centre plane. Besides, the 2D FEM cannot calculate the behaviour of soil between columns. Therefore, a 3D FEM analysis is required to analyse the effect of adjacent loaded pile rows on existing tunnel. Thus, this research focused on the number of piles within pile row in term of fully effect on existing tunnel by 3D FEM program. The data of MRTA was used in this research. The results are discussed in terms of total soil displacement and tunnel deformations.

* Corresponding author: narunat2835@gmail.com

2 Methodology

2.1. Geometrical parameters and a case study

The geometrical parameters considered in this study are shown in Fig2(a). The tunnel diameter (D_T) of 6.3 m with, the lining thickness of 0.3 m, are considered with the constant depth (L_T) of 20.0 m below the ground surface. The bored pile dimeter (D_P) of 1.0 m with the length of -13.7 m is defined. The pile spacing (S) of $3D_P$ and the clearance (C) of 0.5 m, which is the distance between the edge of bored pile to the edge of tunnel, were considered.

In this study, there are two case studies subsoil conditions. For the first case (Case 1), the tunnel and pile tip are located in soft clay. The second case (Case 2), the tunnel and pile tip are located in stiff clay. The number of piles within row is varied. The subsoil conditions and a piezometric drawdown of typical water pressure profile in Bangkok [8] are depicted in Fig 2(b).

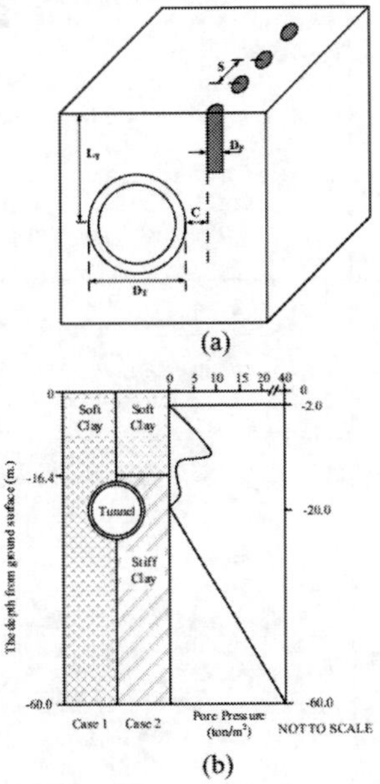

Fig. 2. Geometric parameters [4].

2.2 Numerical model and FEA

The static analysis of pile under loading was carried out by the FEM with implicit integration algorithm using PLAXIS 3D code. The mesh is shown in Fig 3. The 10-node tetrahedral elements or the volume elements. were used to model the soil and hardening grouting layers with a suitable aspect ratio. The tunnel lining and EPBs were discretized into the 6-node triangular plate elements. The bored pile was modelled by the embedded elements. The dimension of model is 50 m (≈8DT) in the transverse direction and 60 m (≈9.5DT) in the vertical direction. In the longitudinal direction is 54 m (≈8.5DT) ahead and behind the monitoring plane [10]. The simulation processes were divided into two steps. The first step, the tunnelling with shield method was simulated by following [9]. The second step, the impact of pile row on the tunnel deformation was simulated. The pile construction is not considered. The working load on plie rows is calculated by the α – method (P=416.54 kN) [11].

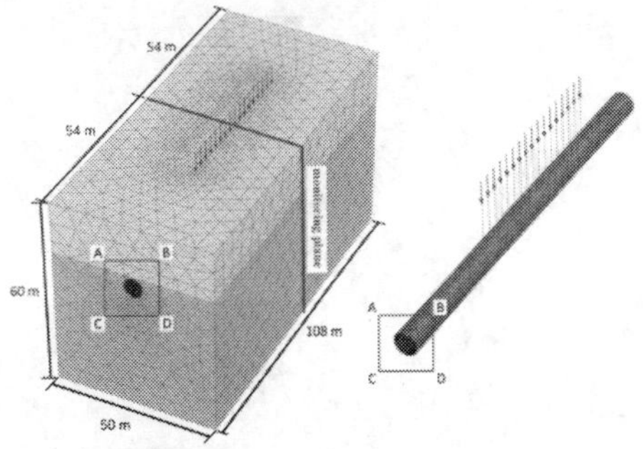

Fig. 3. Model diamantions and meshes.

2.3 Boundary Condition and Initial Conditions

2.3.1 Boundary conditions

Regarding the boundary conditions of numerical model, the movement at the sides of mesh was allowed in only vetical direction. At the bottom of mesh was fixed in all directions.

2.3.2 Initial conditions

The initial stress state is applied through the unit weight of soil layers and the coefficient of earth pressure at rest, K_0.

2.4 Material properties

Table 1. Soil model parameters [14]

Soil layer	Soft clay	Stiff clay
Material model	HS	HS
E_{oed}^{ref} (kPa)	5,000	60,000
E_{50}^{ref} (kPa)	5,000	60,000
E_{ur}^{ref} (kPa)	15,000	180,000
γ_{sat} (kN/m^3)	16	18
υ' (-)	0.33	0.33
ϕ' (°)	22	22
c (kPa)	5	18
m (-)	1	1
p_{ref} (kPa)	100	95

Table 1 depicts the soil parameters. Tunnel lining, pile and grouting layer uses the linear elastic model. The soft and stiff clay are assigned by hardening soil model (HS) [12]. The properties of soils and material parameters for the simulations in Bangkok subsoil are calibrated by [13-14] and has been used in previous research [15].

2.5 Measurement

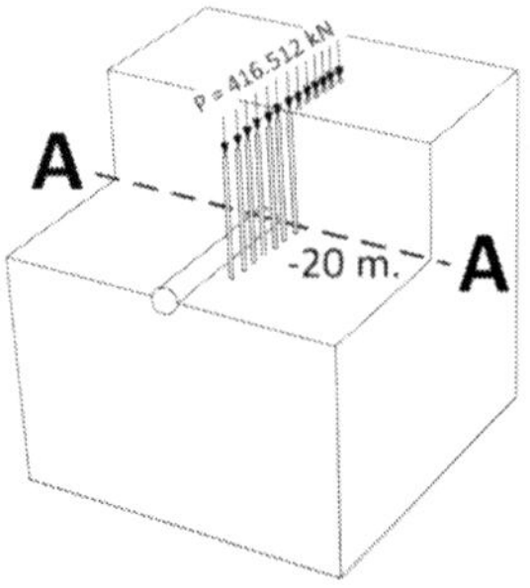

Fig.4. Monitoring section A-A.

The influence of number piles is evaluated in terms of the total displacement of soil at the tunnel axis (without tunnel) referred from Schroeder et al., 2004 [1] as shown in Fig.4. From study of Lueprasert et al., (2017) [6], the shape of tunnel deformation is ellipse and non-symmetric nature when affected by adjacent loaded pile located in single side. The proposed method of Lueprasert et al., (2017), which can appropriately capture the ellipse and non-symmetric shapes in terms of the maximum tunnel deformations, are used to assess the influence of number piles in this study.

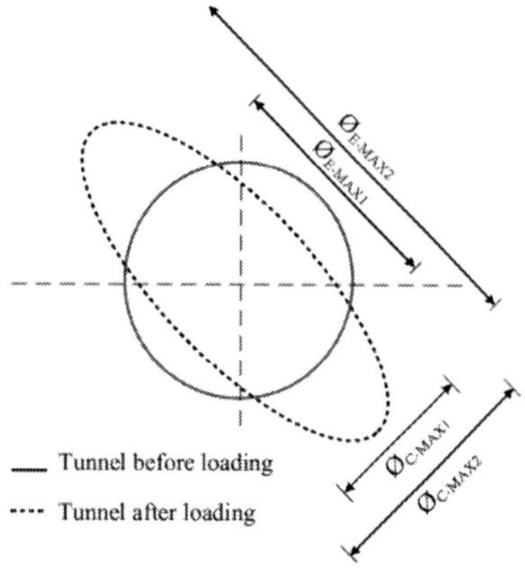

Fig. 5. The maximum changes of tunnel diameter

Figure 5 depicts the calculating method of maximum tunnel deformations. The calculation of the maximum changes of tunnel diameters are shown in Eq. (1) and (2). When ϕ_1 and ϕ_2 are tunnel diameters before and after the effect of pile rows under loading, respectively.

$\Delta\phi_E$ = maximum extension change of tunnel dimeter

$$\Delta\phi_E = \phi_{E2} - \phi_{E1} \quad (1)$$

$\Delta\phi_C$ = maximum contraction change of tunnel dimeter

$$\Delta\phi_C = \phi_{C2} - \phi_{C1} \quad (2)$$

3 Analysis Results

The total soil displacement and the change in tunnel diameter in terms of both maximum extension and maximum contraction deformations due to adjacent pile rows under loading are presented in this section.

3.1 Total displacement of soil at depth of 20 m.

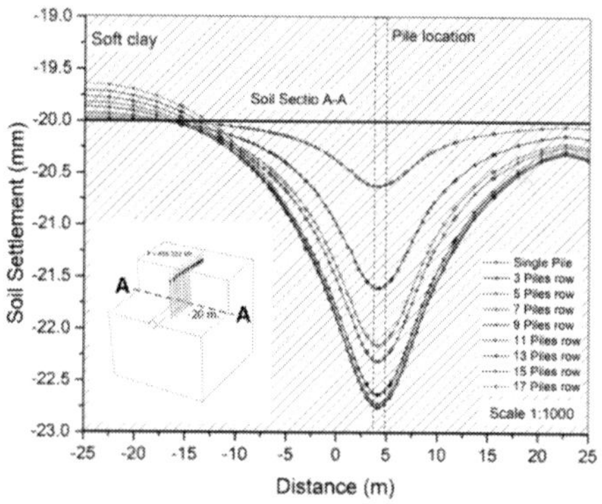

Fig. 6. soil displacement in case 1.

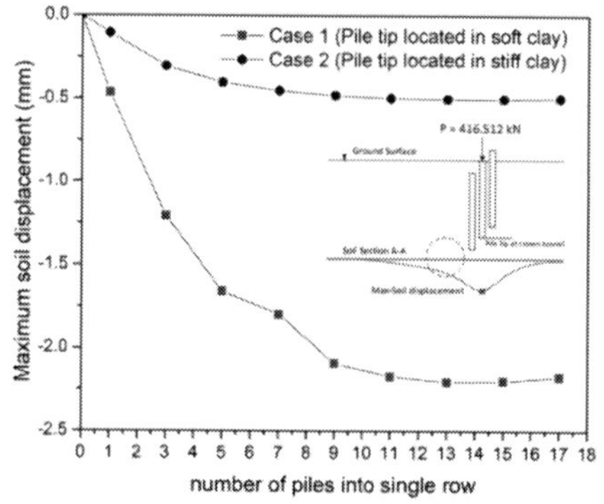

Fig. 7. the maximum soil displacement at depth of -20 m.

Figure 6 shows the total displacement profile of soil at the A-A section (without tunnel) for the case where pile tip is located in soft clay (Case 1). The total soil displacement increases with increasing number of piles. The maximum values with different pile numbers occur when the number of bored pile reaches 11. The maximum total soil displacement with number of piles within row are depicted in Fig.7. A similar tendency of both cases can be seen, but the magnitude of total soil displacement for Case 2 is smaller. The maximum total soil displacements are constant when the number of pile within row reaches 13 and 11 for case 1 and case 2 respectively.

As mentioned above, the soil displacements when the tunnel is located in soft clay (case 1) are larger than those in stiff clay (case 2). The movements of soil are principally due to pile load, load transfer behaviour from the pile and stiffness of the soil layer.

3.2 Tunnel deformation

The maximum extension change ($\Delta\phi_E$) and the maximum contraction change ($\Delta\phi_C$) in tunnel diameter due to the influence of the number piles in the row are shown in this section. $\Delta\phi_E$ and $\Delta\phi_C$ gradually increase with increasing pile number. However, when the 11 piles and 9 piles for tunnel located in soft and stiff clay respectively is modelled, $\Delta\phi_E$ and $\Delta\phi_C$ becomes maximum and constant even larger number of piles is modeled. Since, load transfer from the pile to surrounding soil of the tunnel induces the tunnel deformation and the tunnel lining is stiffer than the soil.

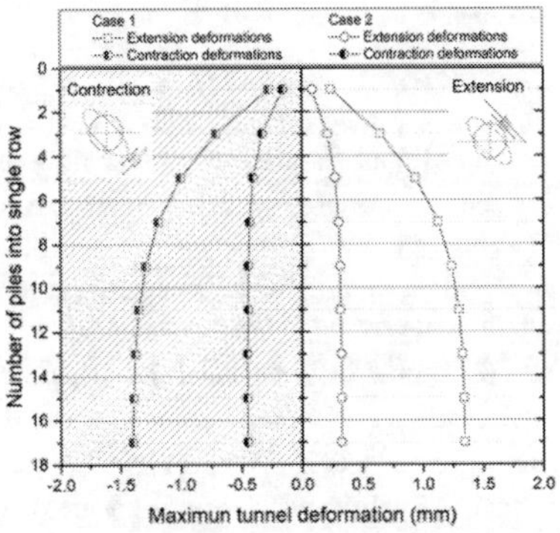

Fig. 8. the change of the tunnel diameters.

4 Conclusion

In this research, the effect of the number of piles in the row for piles under loading on the existing tunnel are investigated by 3D finite element method. The total soil displacements and the maximum change of tunnel diameters were observed. The conclusion of this study are as follows:

1. The total soil displacements are fully mobilized when number of piles within row reaches 13 and 11 for case 1 and case 2, respectively.
2. The maximum tunnel deformation in terms of both extension and contraction deformations are constant when the number of piles in row are 11 piles or Case 1 and 9 for case 2.
3. Both soil movement and tunnel deformation in case 2 are smaller than those of case 1inspite that the applied load of case 2 is higher. This implies that the tunnel deformation is primarily due to the surrounding soil movement.

5 Acknowledgment

The authors wish to express their gratitude to Thailand Research Fund (TRF) and 1D2 Group Co. Ltd. for the financial support through Grant PHD60I0032 and the Ladkrabang Underground and Tunneling Innovation Center (LUTIC) under Grant KREF035001.

References

1. F.C. Schroeder, D.M. Potts and T.I. addenbrooke, Geotechnique, **54**, 351-362 (2004)
2. L.J. Benton and A. Fhillips, in: Proc. 10th Eur. Conf. Soil Mech. Fdn Engng. Florence, **2**, 665-668 (1991)
3. Abdel M-Meguid, R.K. Rowe and K.Y. KY, Geomechanics, **2**, 447-469 (2002)
4. N. Heama, P. Jongpradist, P. Lueprasert and S. Suwansawat, Int. J. Geomate, **12**, 63-70 (2017)
5. P. Lueprasert, P. Jongpradist, K. Charoenpak, P. Chaipanna and S. Suwansawat, Maejo Int. J. sci. technol., **9**, 209-223 (2015)
6. P. Lueprasert, P. Jongpradist and P. Jongpradist, Tunn. Under. Space Technol., **70**, 166-181 (2017)
7. J.C. Chai, S. Shrestha, T. Hino, W.Q. Ding, Y. Kamo, J. Carter, Comput. Geotech., **68**, 28-37 (2015)
8. S. Suwansawat, *Ph.D. Thesis of Massachusetts Institute of Technology*, (2002)
9. P. Lueprasert, P. Jongpradist and S. Suwansawat, Int. J. Geomate, **12**, 51-57 (2017)
10. E.O. Mroueh and I. Shahrour Int. J. Numer. Anal. Methods Geomech, **26**, 217-230.
11. A.W. Skempton, Geotechnique, **2**, 153-173 (1959)
12. T. Schanz, P.A. Vermeer and P.G. Bonnier, In: Beyond 2000 Comput. Geotech. 10 Years PLAXIS Int. Proc. Int. Symp. Beyond 2000 Comput. Geotech. Amsterdum Netherlands 1820 March, 281-296 (1999)
13. R.E. Prust, J. Davies and S. Hu, Transport. Res. Rec. J. Transp. Res. Board, 207-217 (2005)
14. T. Rukdeechuai, P. Jongpradist, A. Wonglert and T. Kaewsri, EIT Res. Dev. J., **20**, 17-28 (2009)
15. P. Jongpradist, T. Kaewsri, A. Sawatparnich, A. Suwansawat, S. Youwat, W. kongkitkul and J. Sunitsakul, Tunn. Under. Space Technol., **34**, 96-109 (2013)

A nonlinear optimization model on the reinforcement length of a modular block wall by varying surcharge and soil strength parameters

Vuttichai Chatpattananan[1*], and *Thanadol* Kongsomboon[1]

[1]Department of Civil Engineering, Faculty of Engineering, King Mongkut's Institute of Technology Ladkrabang, Bangkok, Thailand

Abstract. This study proposes a nonlinear optimization approach in designing a modular block wall which is a type of the mechanical stabilized earth wall. A nonlinear optimization model is proposed based on minimizing the reinforcement length where the constraints considered are the external stability and the internal stability. The optimum reinforcement length can be determined based on available soil strength parameters and the maximum surcharge. This study also includes the parametric study of the reinforced soil, retained soil, and foundation soil by varying the ranges of the wall height, surcharge, and soil strength parameters in density and friction angle to see the behaviours of the aforementioned external stability and internal stability. This can be beneficial in designing this modular block wall encountering a poor soil condition or a large amount surcharge.

1 Introduction

Mechanically stabilized earth wall or MSE wall, is a retaining wall that uses soil with reinforcement in adding tension. With rigid facing units, MSE wall can set vertically not like reinforced soil slope (RSS) that can set slope up to 70 degrees. Two types of MSE walls are commonly used. One is segmental precast concrete panel (SPCP) that uses metal strips as reinforcements and precast concrete panel as the facing unit. The other is modular block wall (MBW) that uses geogrid as reinforcements and modular block as facing units as shown in Fig. 1.

SPRC MBW

Fig. 1. MSE Wall Types (FHWA-NHI-00-043) for

An MSE wall needs to resist the failure modes in external stability, internal stability, and global stability. External stability requires this rigid body MSE wall to pass the safety factor in sliding, overturning, and bearing capacity. Internal stability prevents the failure in inside the MSE wall by setting factors of safety in rupture and safety in pullout. Global stability considers all other failure that may occurs anywhere especially outside the MSE wall.

Designing an MSE mainly is to specify the reinforcement length, reinforcement spacing, and soil strength property. [1] gives a preliminary reinforcement length (L) should be $0.7H$ and 2.5 m where H is the height of the wall. [2, 3] recommends L/H at 0.7 where [4] recommends at $0.6L/H$. Also [1] recommends L/H to be less than 1.1. However, designing an MSE wall faces many varying variables, especially the soil strength parameters and the imposed surcharge load. Some studies introduce optimization based approach in designing MSE wall such as [5-7]. [5] uses optimization in finding minimum cost to design a segmental precast concrete with metal strip reinforcements. Simulation is also utilized in [5] to assist in approximating the nonlinear optimum solution. [8] uses an optimization approach on cost. [9] introduces an advanced technique in finding a nonlinear optimization model for MSE wall design parameters. This paper rather investigates the behaviour of varying parameters in designing an MBW first then utilizing a nonlinear optimization model via GAMS nonlinear software to minimize the geogrid reinforcements satisfying the constraints in external stability and internal stability.

2. Optimizing reinforcement length for a modular block wall

*Corresponding author: vuttich@hotmail.com

Designing a segmental precast concrete panel is different from designing a MBW only the differences in pullouts due to the difference of the potential failure surface in the internal stability as shown later in Fig. 3. For illustration, only MBW that uses geogrid as reinforcement is shown in this study. Also, the optimization will consider only the external stability and internal stability based on FHWA [1], the global stability that uses limit equilibrium method (LEM) is omitted here. LEM can be analyzed with many available software.

Based on [1], calculating factors of safety (FS) are shown in Fig. 2 as follows:

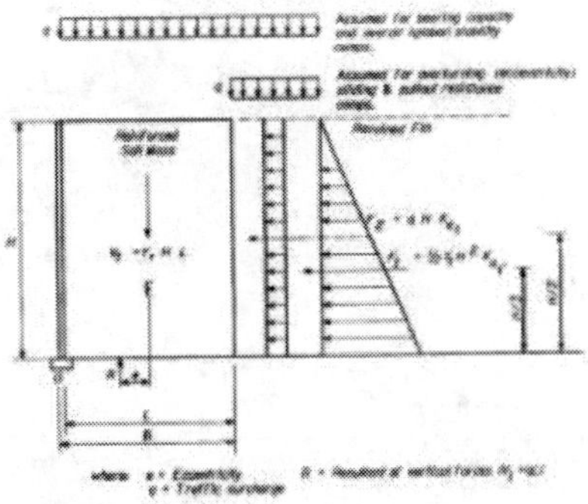

Fig. 2. MSE wall loads [1]

To formulate an optimizing model from Fig. 2, the decision variables can be classified as 3 groups which are as follows. Group 1 decision variables are wall height, surcharge loads, and soil strength parameters. Group 2 decision variables are external stability involving the factors of safety for sliding, overturning, eccentricity, and bearing. Group 3 decision variables are internal stability factors of safety for rupture and pullout.

Group 1 decision variables are denoted as wall height H m, geogrid length L m, surcharge load q kN/m^2, soil parameters are density γ kN/m^3, cohesion c kN/m^2, and internal friction angle ϕ degree. The active earth pressure coefficient is $K_a = tan^2(45 - \emptyset/2)$. Reinforced soil parameters are denoted as $\gamma_r, c_r, \emptyset_r, K_{ar}$. Retained (backfill) soil parameters are denoted as $\gamma_b, c_b, \emptyset_b$. Foundation soil are denoted as γ_f, c_f, K_{af}. For loads, vertical fill $V_1 = \gamma_r HL$ kN/m, vertical surcharge $V_2 = qL$ kN/m, horizontal retained soil $F_1 = 0.5\gamma_b H^2 K_{af}$ kN/m, horizontal surcharge $F_2 = qHK_{af}$ kN/m.

Group 2 decision variables are external stability factors of safety (FS) including sliding, overturning, eccentricity, and bearing. The sliding FS is

$$FS_S = V_1 \tan(\min\{\emptyset_r, \emptyset_f\})/(F_1 + F_2) \geq 1.5 \quad (1)$$

The overturning moment $M_O = F_1H/3 + F_2H/2$ kN-m, resisting moment $M_{RO} = V_1L/2$ kN-m, bearing moment $M_{RBP} = V_1L/2 + V_2L/2$ kN-m, then the overturning FS is

$$FS_O = M_{RO}/M_O \geq 2 \quad (2)$$

For eccentricity,

$$e = L/2 - (M_{RBP} - M_O)/(V_1 + V_2) \leq L/6 \quad (3)$$

The bearing capacity of foundation soil, where maximum bearing pressure is $\sigma_V = (V_1 + V_2)/(L - 2e)$ kN/m^2, ultimate bearing capacity, $\sigma_{ult} = c_f N_c + 0.5(L - 2e)\gamma_f N_\gamma$ kN/m^2 and the bearing FS is

$$FS_b = \sigma_{ult}/\sigma_v \geq 2.5 \quad (4)$$

The bearing capacity factors are assumed as follows. $N_q = e^{\pi tan\emptyset_f} tan^2(45 + \emptyset_f/2)$, $N_c = (N_q - 1)cot\emptyset_f$, $N_\gamma = (N_q - 1)\tan(1.4\emptyset_f)$.

Group 3 decision variables are internal stability factors of safety. With maximum tensile strength, T_{max} of geogrid at layer i is $T_{max-i} = \sigma_H V_i$, the horizontal stress $\sigma_H = K_{ar}(\gamma_b d_i + q)$, $T_{max-i} \leq T_{ai}$ where T_{ai} is the allowable tensile strength of geogrid at layer i. V_i m^2 is the grid area of geogrid at layer i at depth d_i m, $V_1 = d_1 + (d_2 - d_1)/2$ for the top first layer, $V_i = (d_i - d_{i-1})/2 + (d_{i+1} - d_i)/2$ for layer i to n-1. $V_n = (H - d_n) + (d_n - d_{n-1})/2$ for the bottom layer.

For pullout, the active zone length at layer i is $L_{ai} = (H - d_i)\tan(45 - \emptyset_r/2)$. The resistant zone length at layer i is $L_{ei} \geq 1.5T_{max-i}/(Ctan\emptyset_r C_R \gamma_r d_i R_c \alpha) \geq 1$ m $R_c = 1.0$ is % coverage of reinforcement ranging from 0.71 to 1.0. $C_R = 0.8$ is the interaction pullout coefficient derived a product, $C = 2$ for geogrid, $\alpha = 1.0$ is the scale effect correction factor , $L_T = L_a + L_e$ is the total length of geogrid at layer i where all at all layer $L_T \leq L$. The pullout factor of safety for layer i is

$$FS_{Pi} = L_e Ctan\emptyset_r C_R \gamma_r d_i R_c \alpha / T_{max-i} \geq 1.5 \quad (5)$$

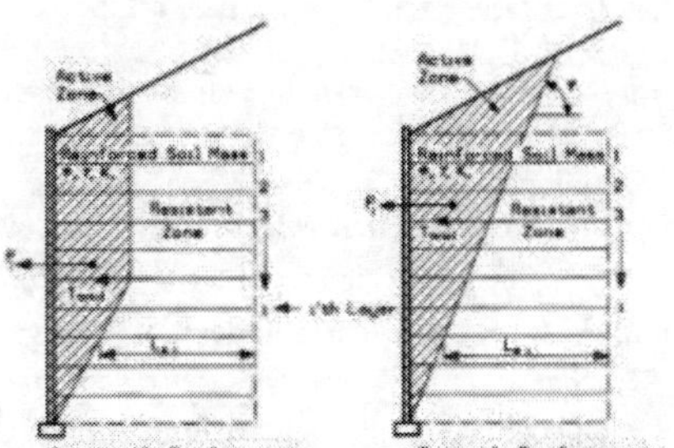

Fig. 3. Active zone and resistant zone of MBW wall [1]

In conclusion, the nonlinear optimization model is as follows.

$$\begin{array}{llll} Min & L & & \\ st. & FS_S & \geq & FS_{Sa} \\ & FS_O & \geq & FS_{Oa} \\ & e & \leq & L/6 \\ & FS_B & \geq & FS_{Ba} \\ & FS_{P1} & \geq & FS_{P1a} \end{array} \quad (6)$$

The constraint 1 to constraint 5 in the optimization model (6) are from the aforementioned (1) to (5). Following FHWA, allowable factors of safety for sliding (= 1.5), overturning (=2.0), bearing (=2.5), resistant zone pullout length (= 1.5). Also from Fig. 3, the pullout lengths are largely dominated by the top layer geogrid where the active zone length is largest at the top layer.

From the FHWA MBW example [1], where H = 9 m, L = 7.5 m, q = 11.97 kN/m, reinforced soil γ_r = 19.6 kN/m^3, c_r = 0 kN/m^2, $\emptyset_r$ = 34, retained soil γ_b = 19.6 kN/m^3, c_b = 0 kN/m^2, $\emptyset_b$ = 30, foundation soil γ_r = 19.6 kN/m^3, c_f = 0 kN/m^2, $\emptyset_f$ = 30. For the reinforcement layer depth, d_1 = 0.47 m, d_2 = 1.07 m. For this FHWA example, GAMS Software with NLP model is used to obtain the nonlinear optimization result for the reinforcement. If the minimum reinforcement constraint at 0.7L/H is relaxed, the optimum reinforcement length is shortened to 0.62L/H where the pullout is bounding constraint with the eccentricity as shown in Table I.

Table 1. Nonlinear optimization results on FHWA example.

Items	Allowable FHWA FS	FHWA Example	Nonlinear Optimization
H (m)	9	9	9
L (m)		7.5	5.54
L/H	0.7	0.83	0.62
Sliding FS	1.5	2.54	1.88
Overturning FS	2.5	5.19	2.83
Eccentricity (m)		0.68	0.92
Allowable Eccentricity (m)	L/6	1.25	0.92
Bearing FS	2.5	5.87	2.89
Tmax at top layer (kN/m)		4.61	4.61
Pullout Length at top layer(m)		5.54	5.54

3 Parametric study on an MBW

To see the effect on varying parameters effecting this MBW reinforcement lengths, [6] suggests the ranges of soil strength parameters in density, cohesion, and internal friction angle. For simplicity, the soil parameter varied in this study is only the soil internal friction angle (ϕ) ranging from 22 to 45 degree where soil density (γ) ranging from 17 - 22 kN/m^3 from [10] is set to be constant at 19.6 kN/m^3 since soil internal friction angle is normally implies its density. In addition, this study set soil cohesion to be zero since a soil type with cohesion and internal friction angle can be roughly converted to internal friction angle only.

Fig. 4 to Fig. 8 shows the effects on varying wall heights and surcharge on the aforementioned FHWA MBW wall example. Fig. 6 (left) shows that increasing wall height almost has no effect to L/H eccentricity at all, and has effect to L/H of the rest including sliding, overturning, bearing, and pullout if surcharge increased significantly to 50 kN/m^2, equivalent to 5 tons/m^2 or 2.5 m of wall weight height, as shown in the figures. The result is similar to weak soil strengths with low internal friction angles where not only eccentricity but overturning have no effect their L/H. Also as shown in Fig. 7 (right), wall height is limited by bearing L/H under weak soil condition.

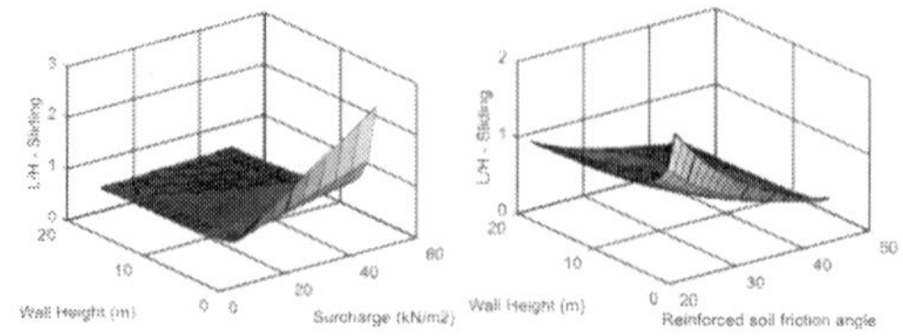

Fig. 4. L/H Sliding by varying H - q (left) and H - ϕ (right)

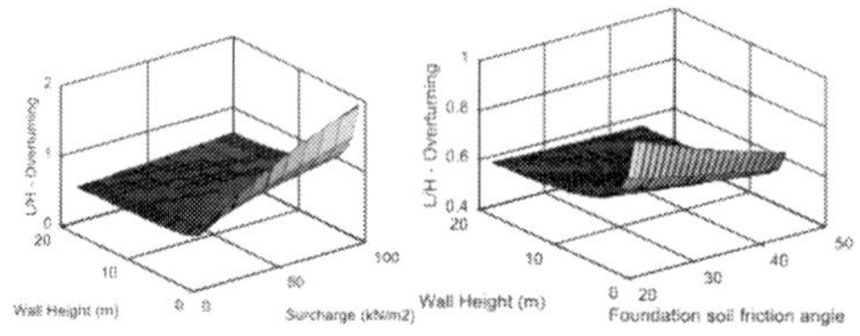

Fig. 5. L/H Overturn by varying H - q (left) and H - ϕ (right)

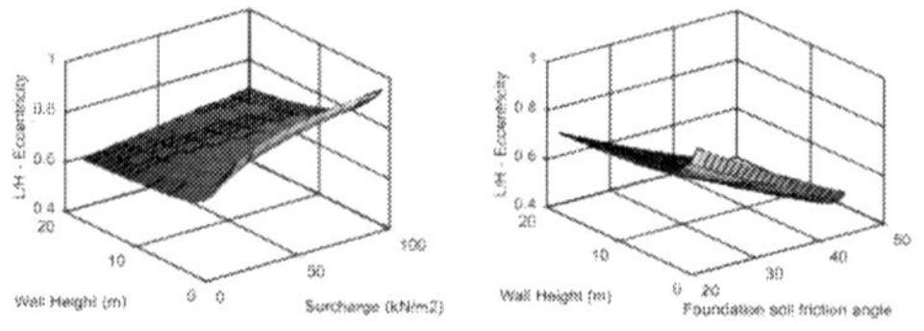

Fig. 6. L/H eccentricity by varying H - q (left) and H - ϕ (right)

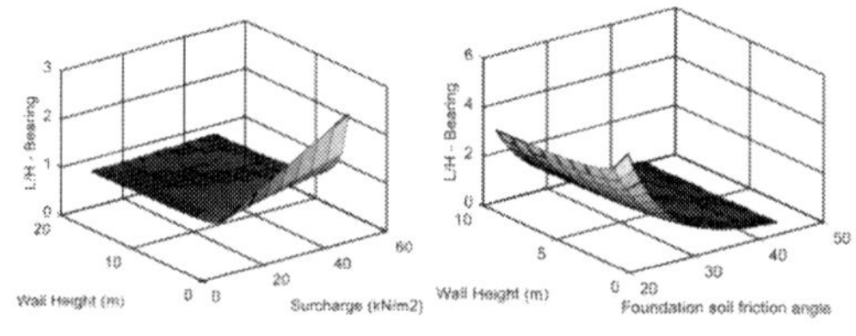

Fig. 7. L/H Bearing by varying H - q (left) and H - ϕ (right)

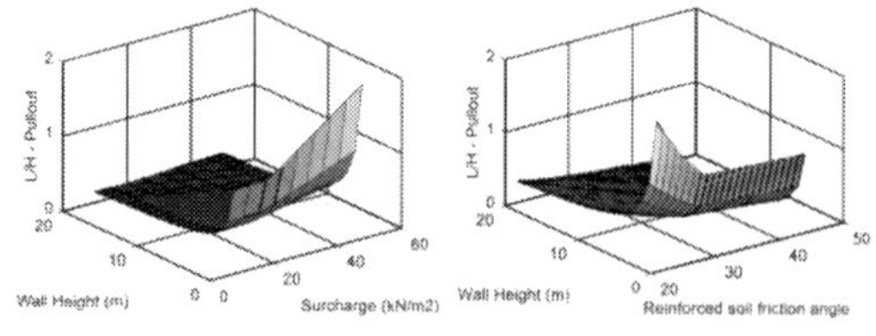

Fig. 8. L/H Pullout by varying H - q (left) and H - ϕ (right)

4 Optimizing based on site limitation

An example of using this nonlinear optimization is elucidated here. FHWA MBW example is used here for illustration that the wall height is 9 m and surcharge is q = 11.97 kN/m. Suppose that an MBW wall project facing a difficulty in the limitation under the weak soil strength condition. The soil strength properties for reinforce soil, retained soil, and foundation soil possess density from 17 - 18 kN/m^3 and internal friction angles are from 22 - 23.

Instead of limiting the geogrid length to 1.1*L/H*, this project is rather temporary and allow the factors of safety for sliding, overturning, and bearing to be just greater than 1.0 and eccentricity is less than *L/6*. The nonlinear optimization model is shown in (7) where its result is shown in Table 2. From Table 2, *L/H* is 0.76. Table 2 also shows the result that strictly design follows FHWA that yields *L/H* at 0.85. This result in reducing *L/H* is rather useful when some part of the constructed MSE wall facing weak soil condition the is much deviated from the soil investigation report.

$$
\begin{array}{llcl}
Min & L & & \\
st. & FS_S & \geq & FS_{Sa} \\
 & FS_O & \geq & FS_{Oa} \\
 & e & \leq & L/6 \\
 & FS_B & \geq & FS_{Ba} \\
 & FS_{P1} & \geq & FS_{P1a} \\
 & 17 \leq & \gamma_r, \gamma_b, \gamma_f & \leq 18 \\
 & 22 \leq & \emptyset_r, \emptyset_b, \emptyset_f & \leq 23 \\
 & FS_{Sa}, FS_{Oa}, FS_{Ba} & \geq & 1.0
\end{array}
\quad (7)
$$

The optimization model is shown in (7) where constraint 1 to 5 are from (1) to (5) same as the optimization model (6). The adding constraint 6 and constraint 7 are from the relaxing condition of soil strength condition from the site limitation. The constraint 8 is from the assumed relaxing design criteria.

Table 2. Relaxing optimization results on weak soil.

Items	Relaxing FS		FHWA FS	
	Actual	Minimum	Actual	Minimum
H (m)	9		9	
L (m)	6.82		7.61	
L/H	0.76		0.85	
Sliding FS	1.23	1.0	1.5	1.5
Overturning FS	3.25	1.0	4.20	2.0
Eccentricity (m)	0.98	L/6	0.84	L/6
Bearing FS	1.94	1.0	2.74	2.5
Tmax at top layer (kN/m)	6.73		7	
Pullout Length at top layer(m)	6.82	1.0	8	1.5

Another application is for a multi-tiered MBW project that the bottom wall needs to support the weight of the upper wall that is considered as a surcharge. In this case the surcharge is rather large. To illustrate, suppose that this FHWA MBW wall has a surcharge of 100 kN/m^2 which is roughly about 5 m of the upper wall height resting on the bottom wall. By using the nonlinear optimization model in (7), the result is shown in Table III. The solution is to use the maximum *L/H* at 1.1 with unlimited soil strength parameters with maximum soil density at 22 kN/m^3 and friction angle at 45. The solution in Table III on the left takes the maximum values of the density and friction angle for the reinforced soil which is practically impossible. Suppose at the construction site limits the soil strength parameters at the maximum density at 20 kN/m^3 and friction angle at 34, the solution in Table 3 in limiting soil strength parameters takes the maximum density and friction angle for the reinforced soil and strengthen the friction angle for the foundation soil.

Table 3. Optimization Soil Strength Parameters.

Items	Unlimited Soil Strengths		Limited Soil Strength	
	Actual FS	Minimum	Actual FS	Minimum
H (m)	9		9	
L (m)	9.9		9.9	
L/H	1.10		1.10	
Sliding FS	1.50	≥ 1.5	1.50	≥ 1.5
Overturning FS	4.77	≥ 2	4.77	≥ 2
Eccentricity (m)	0.69	≤ 1.65	0.67	≤ 1.65
Bearing FS	2.50	≥ 2.5	2.58	≥ 2.5
Tmax at top layer (kN/m)	14.27		23.51	
Pullout Length at top layer(m)	4.83		8.01	
Reinforced soil density (kN/m^3)	22	≤ 22	20	≤ 20
Retained soil density (kN/m^3)	17	≤ 22	17	≤ 20
Foundation soil density (kN/m^3)	17	≤ 22	17	≤ 20
Reinforced friction angle	45	≤ 45	34	≤ 34
Retained friction angle	30.07	≤ 45	32	≤ 34
Foundation friction angle	22	≤ 45	22	≤ 34

5 Conclusion

Employing the nonlinear optimization approach helps facilitating the design of a modular block wall that often facing difficulties such as varying soil strength parameters and maximum surcharge bearable for the bottom wall that needs to support the upper wall that is counted as a surcharge resting on the bottom wall. By limiting the ranges of soil strength parameters in reinforced soil, retained soil, and foundation soil, the minimum geogrid reinforcement length can be determined based on considering only the external stability and internal stability.

References

1. Federal Highway Administration, *FHWA-NHI-00-043*, (2001)
2. AASHTO, *Standard specifications for highway bridges*, (1997)
3. The British Standards Institution, BS8006, (2010)
4. National Concrete Masonry Association, *Design manual for segmental retaining walls. 2 ed*, (2002)
5. H. Ghiassian, K. Aladini, *Asian J. of Civil Engr.*, V. 10, 6, (2009)
6. D. Leshchinsky, D., V. Kaliakin, P. Bose, J. Collin, Soils and Foundations, Vol. 34, 4, (1994)
7. D. Leshchinsky, C. Vulova, *Geosynthetics International*, Vol. 8, 4, (2001)
8. P. K. Basudhar, A. Vashistha, K. Deb, A. Dey, Geotech Geol. Eng., 26, (2008)
9. K. N. Manahiloh, M. M. Nejad, M. S. Momeni, Int. J. of Geosynth. and Ground Eng., (2015)
10. R. Obrzud, A. Truty, The hardening soil model - a practical guidebook, (2010)

Influence of impervious cover determination method of upper Ciliwung watershed on flood warning system level change in Katulampa weir

Niswatin Farika[1,*], *Dwita* Sutjiningsih[1], and *Evi* Anggraheni[1]

[1]Civil Engineering Department, Faculty of Engineering, Universitas Indonesia, Depok 16424, Indonesia

Abstract. The Katulampa Weir is a part of Jakarta Flood Early Warning System for the Ciliwung River. The measured water level fluctuation in the Katulampa Weir is affected by the physical condition of the Upper Ciliwung Watershed. In this study, the definition and relevance of the impervious cover determination method based on the Total Impervious Area (TIA) and Effective Impervious Area (EIA) in generating design flood discharge and its effect on the change of flood early warning level in Katulampa Weir will be evaluated. Identification of land use distribution is based on digitized process used combined GIS maps using visual interpretation of high resolution satellite images 2017. The flood analysis for both methods is applied to the same rain conditions. Evaluation of flood early warning level changes are based on flood discharge simulated results and rating curve discharge in the Katulampa Weir. The simulation by WinTR-20 gives that maximum discharge using TIA method is 150 m^3/s and EIA method 139.5 m^3/s. There is no significant difference between and both are classified as the same level of flood early warning system level, which is on stage 3. However, for large watersheds, it takes much effort to identify and digitize an effective impervious area.

1 Introduction

The Katulampa Weir is part of the flood early warning system from Ciliwung River which will flow into Jakarta. Water level fluctuation measured is affected by physical characteristic of Upper Ciliwung watershed. Ecological condition in Ciliwung watershed is going worse, with decreasing forest land cover from 9.4% in 2000 to 2.3% in 2010, or environmental degradation rate of 7.14% in the last decade [1].

Impervious cover is defined as a land cover that prevents infiltration of water from the surface into the soil [2]. Some study suggest that impervious cover is one of the main indicators of water quality in a watershed. Water quality will start to decrease if more than 10% of watershed area is impervious [3]. Urbanization has caused in increasing number of impervious cover (road, parking lot, roof, pedestrian, etc.) which decreases the number of forest, wetland, and other open space parts which infiltrates rainfall [3].

Recent study shows that surface runoff in a watershed can be better described by Effective Impervious Area (EIA) than TIA [4]. The method usually used to determine the level of imperviousness is based on the total area of impervious area (TIA). Total Impervious Area (TIA) is an area measurement which resists infiltration of rainfall into the soil, while Effective Impervious Area (EIA) is part of TIA which has hydraulic connection straight into drainage system [5]. Another parameter related to impervious area is effective impervious area (EIA) which is the portion of TIA that is directly connected to the drainage system. Definition of hydraulically connected is water that falls on impervious cover will travel through impervious lane into the inlet of drainage channel [6].

Referring to the result of the study, this research will predict the flood discharge in Katulampa Weir using the information of imperviousness determined based on TIA and EIA method. Further predictions of the proposed hydrograph flood calculation results are used to identify differences in flood levels in the flood early warning system in the Katulampa Weir.

2 Research Methodology

Data analysis process in order to obtain research goals are as follows:

- Identification of land use distribution based on GIS map from visual interpretation and high-resolution satellite image Pleaides and World View 2
- Calculation of imperviousness using TIA and EIA
- Calculation of weighted Curve Number (CN) in sub-watersheds to be used to calculate flood discharge
- Evaluation of flood discharge with water level based on rating curve in Katulampa Weir to determine flood status

2.1. Study Area

The research was conducted in Upper Ciliwung Watershed. The upstream part of Ciliwung River is located at Gunung Gede-Pangrango which then flows through Bogor Regency, Bogor Municipality, Depok City and empties into Jakarta Bay. The water level fluctuation in the Ciliwung River is affected by the physical condition of the Upper Ciliwung Watershed.

* Corresponding author: niz_ika@yahoo.co.id

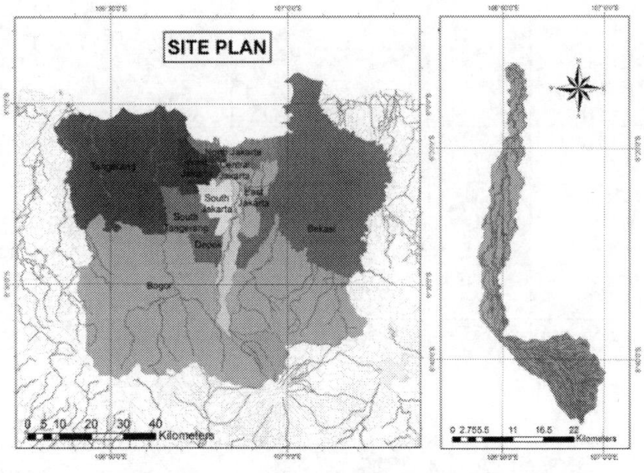

Fig. 1. Study Area

2.2 Data Requirement

Required data include rainfall data, water level data, soil type, DEM/topography, land use, building plot, road network and river network. Rainfall data is used as input model representing the research area of Gadog Station, Cilember Station and Gunung Mas Station. The runoff analysis of the spatial distribution difference of spatial impervious area is applied to the same rain conditions. The water level data in the Katulampa Dam is used for model calibration/validation purposes.

Delineated watershed of Upper Ciliwung includes the area of 15.075 hectare started from Tugu Village Puncak to Katulampa Weir located in 3 sub-district administrations of Cisarua, Megamendung and Ciawi.

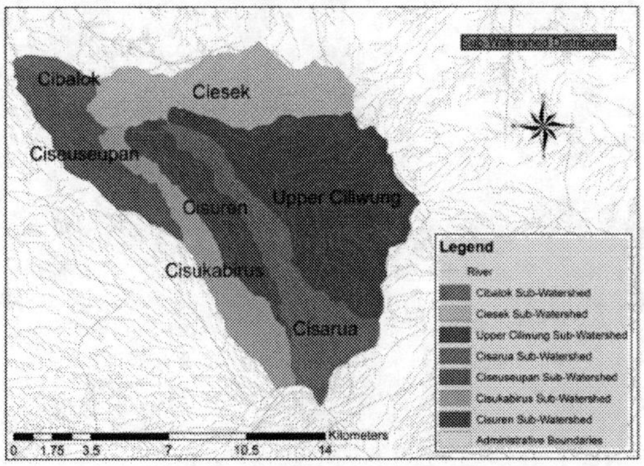

Fig. 2. Upper Ciliwung Watershed

Land use distribution in Upper Ciliwung Watershed using 2017 data obtained from BBWS Ciliwung Cisadane. The Ciliwung River Watershed delineation process into Upper Ciliwung Sub-Watershed uses ArcGis Application Version 10.1 with an Earth Range Map of Indonesia (RBI) data source containing contour, high point, river network, administration and road network from Geospatial Information Agency (BIG), Cibinong Bogor.

Table 1. Sub-watershed Division

No	*Sub-watershed*	*Area(hectare)*
1	Ciliwung Hulu	4.799,97
2	Cisarua	2.293,75
3	Cisuren	1.483
4	Cisukabirus	1.741,97
5	Ciesek	2.804
6	Ciseuseupan	1.523,61
7	Cibalok	429,66

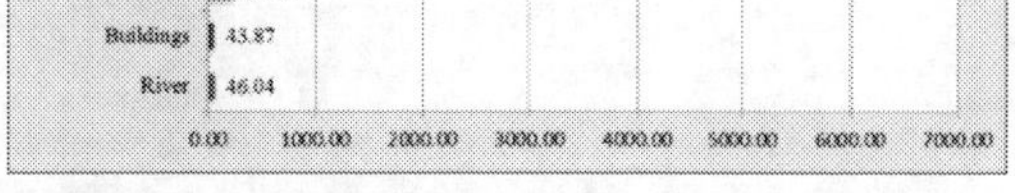

Fig. 3. Land Use Distribution of Upper Ciliwung Watershed

2.3 Katulampa Weir Technical Data

Geographically, Katulampa Weir is located in 6°37'07"LS dan 106°47'38"BT. Ciliwung discharge in Katulampa Weir be known with water level reading which will be converted using rating curve.

Table 2. Katulampa Weir Warning Level Criteria

Level	*Water Level Limit (cm)*		*Discharge Limit (cms)*	
	Upper	*Lower*	*Upper*	*Lower*
I	200	-	441.98	-
II	150	199	276.25	438.48
III	80	149	90.05	273.23
IV	0	79	0	87.85

Source : BPBD, Jakarta Province

* Corresponding author: niz_ika@yahoo.co.id

Table 3. Rooftops and Road Network Area Distribution

No	*Sub-Watershed*	*Area (Hektar)*	*Rooftops Area*		*Road Network Area*		*Total Rooftops & Road Network Area (EIA)*	
			(Hektar)	*(%)*	*(Hektar)*	*(%)*	*(Hektar)*	*(%)*
1	Ciliwung Hulu	479897	251.70	25.80%	109.56	32.84%	361.26	27.59%
2	Cisarua	229375	94.18	9.65%	43.61	13.07%	137.79	10.52%
3	Cisuren	280437	64.08	6.57%	34.86	10.45%	98.94	7.56%
4	Cisukabirus	148337	70.32	7.21%	16.07	4.82%	86.38	6.60%
5	Ciesek	174197	190.37	19.51%	60.40	18.11%	250.78	19.15%
6	Ciseuseupan	152361	277.24	28.42%	55.16	16.53%	332.40	25.39%
7	Cibalok	42966	27.78	2.85%	13.95	4.18%	41.73	3.19%
DAS Ciliwung Hulu		*1507570*	*975.67*		*333.60*		*1309.27*	

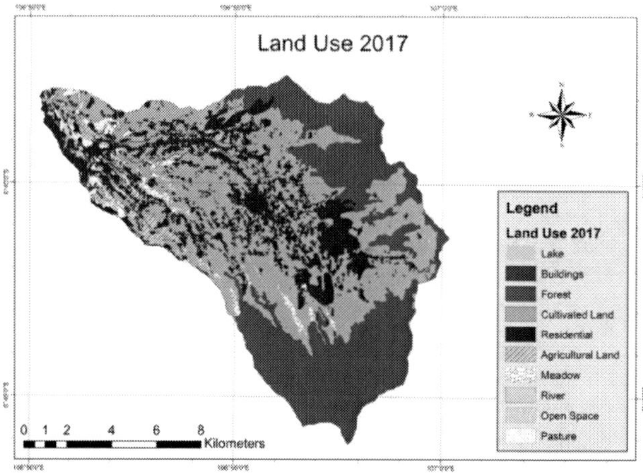

Fig. 4 Land Use of Upper Ciliwung Watershed 2017

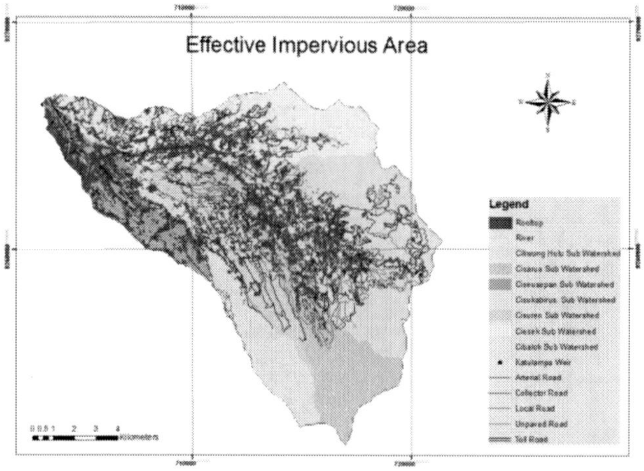

Fig. 5 Effective Impervious Area as the number of rooftops and road network

2.3 Impervious Area Determination

The determination of TIA in this study is based on land use or zoning mapping used by the Indonesian Geospatial Information Agency. Meanwhile, to obtain an effective impervious area based on digitized process used combined GIS maps and visual interpretation results from high resolution satellite images of World View 2 in 2017 with a scale of 1: 25.000 obtained from the Department of Geography Faculty of Mathematics and Natural Sciences University of Indonesia.

The effective impervious area determined based on recognized rooftops and road-network. Table 3 shows distribution of effective impervious area as the number of rooftops and road network areas in each sub-

* Corresponding author: niz_ika@yahoo.co.id

watershed. The Effective Impervious Area of Upper Ciliwung watershed is shown in Figure 5.

2.5. Model Simulation and Calibration

Hydrological modelling in this study is use WinTR-20 v. 3.20. WinTR-20 adopts work principal of SCS-CN model to calculate direct runoff. CN table refers to method used by United States Department of Agriculture (Chow, 1988). Flood discharge analysis with two scenarios: 1) TIA parameter, and 2) EIA parameter. Simulation of flood hydrograph based on rainfall event on April 13rd, 2017 and water level reading in Katulampa Weir at the same time. The method used for calibration and validation are determination coefficient (R^2) and Nash-Sutcliffe (NSE).

3 Result and Discussion

Rainfall intensity based on Thiessen method on 2017 was 45.4 mm/day with hypothetical distribution following Wanny Model distribution starts at noon and lasting in 4 hours. Table 4 summarizes information needed to simulate flood hydrograph using WinTR-20.

Tabel 4. Input Data for WinTr-20 Simulation Model

No	*Sub-watershed*	*L (km)*	*Tc (hr)*	*$CN_{weighted}$*	
				TIA	*EIA*
1	Ciliwung Hulu	15.49	1.56	73.84	75.30
2	Cisarua	17.45	1.32	75.32	76.86
3	Cisuren	15.01	1.42	71.93	72.64
4	Cisukabirus	15.75	1.33	80.29	80.71
5	Ciesek	14.53	1.36	71.87	72.35
6	Ciseuseupan	12.26	1.58	81.36	81.70
7	Cibalok	2.96	0.42	77.77	78.09

Simulation result provided in Figure 5 as flood hydrograph. Table 5 summarizes comparation of parameters, which consist of peak flow (QP), time to peak (TP), and flood duration (TB). It should be noted that $Q_{observed}$ is used as measured discharge data, while Q_1 is used for discharge approximation using TIA method, and Q_2 used using EIA method.

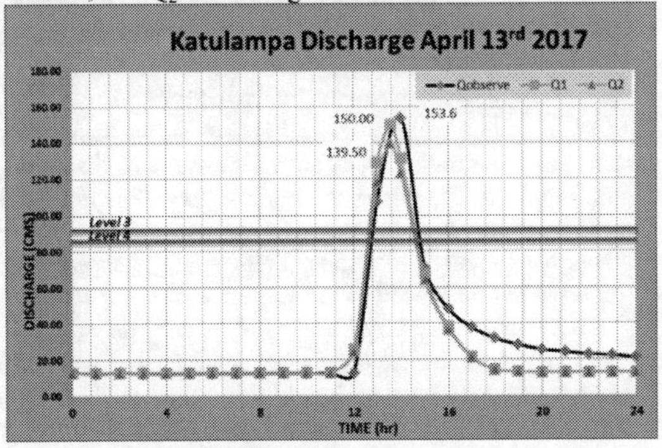

Fig. 6 Flood Hydrograph Simulation

Tabel 5. Summarized Flood Hydrograph Parameter

Observe Point	*Peak Flow Discharge (cms)*			*Time Peak (hr)*			*Flood Duration (hr)*		
	Q_{obs}	*Q_1*	*Q_2*	*Q_{obs}*	*Q_1*	*Q_2*	*Q_{obs}*	Q_1	Q_2
Katulampa Weir	153.6	150	139.5	2	1.56	1.54	11	8	8

The model hydrograph generally has a slightly different value than the observed hydrograph. The peak flow model between TIA parameters and measured discharge has a difference of 3.6 m^3/s. This condition is different from the result of an EIA parameter that is smaller than the measured discharge of 14.1 m^3/s. However both methods have peak time (Tp) which is smaller than observation.

The coefficient value of determination R^2 and NSE value for TIA method is 0.973 and 0.925 and for EIA method is 0.971 and 0.924. Thus the watershed simulation results can be said good results category.

4 Conclusion

Calculation using WinTR-20 gives the result that maximum discharge which is possible to obtained using TIA method is 150 m^3/s with water level reading of 196 cm. Compared with EIA method which produces the discharge of 139.5 m^3/s with water level reading of 182.5 cm. This study only performs simulation based on one rain event so that further study is needed to see the effect of both methods on design flood discharge.

The result shows that both methods are acceptable. There is no significant difference between and both are classified as the same level of flood early warning system level, which is on stage 3. However, for large watersheds, it takes much effort to identify and digitize an effective impervious area.

This work was supported by Universitas Indonesia through the PITTA 2018 under grant No.2422/UN2.R3.1/HKP.05.00/2018, which is greatly appreciated. The author would like to thank the anonymous reviewers for their valuable comments as well.

References

1. W.Y. Bhakti, *Prediksi Laju Erosi Potensial Dan Laju Timbulan Sampah Potensial Pada Luasan Penutup Lahan Kedap Air (Studi Kasus Das Ciliwung)*, Universitas Indonesia, Depok, 2014.
2. M. Zalewski, *Ecohydrology-The Use of Ecologycal and Hydrologycal Processes For Sustainable Management Of Water Resources*, Warsave, Hydrological Sciences Journal Vol. 47.
3. Chithra S.V., Dr. M.V. Harindranathan Nair, Amarnath A and Anjana N.S, *Impacts of Impervious Surfaces on the Environment,* International Journal of Engineering Science Invention, ISSN (Online): 2319 – 6734, ISSN (Print): 2319 – 6726 Volume 4 Issue 5, May 2015.
4. A. Ebrahimian [et al.], *Determination Of Effective Impervious Area In Urban Watershed*, Minnesota Department of Transportation, Minneapolis, 2015.
5. S.S Nalini and P. Sreeja, *Impact of Total and Effective Imperviousness on Runoff Prediction*, Springer International Publishing Switzerland, 2016.
6. W.S Han and S.J. Burian, *Determining Effective Impervious Area for Urban Hydrologic Modeling*, Journal of Hydrologic Engineering, Volume 14, Issue 2, February 2009.

A novel device for inclined compaction test on soils

Panu Promputthangkoon[1,*] and *Tavorn* Kuasakul[2]

[1]Natural Disaster Prevention and Control for Sustainable Environment Research Unit, Faculty of Engineering, Rajamangala University of Technology Srivijaya, Songkhla, Thailand
[2] Civil Engineering Department, Faculty of Engineering, Rajamangala University of Technology Srivijaya, Songkhla, Thailand

Abstract. It can be said that the soil compaction test is currently the standard method for obtaining the right amount of water to be added in order to achieve a maximum dry density. Then, the water content obtained from laboratory work, known as optimum moisture content, is utilised in the field for compacting the soil. It should be noted that the compaction test is carried out on a soil sample prepared in a mould horizontally laid. In the field, however, quite often the compaction is done on side embankments or sloping grounds. Hence, using the laboratory result to control the field density for such cases is problematic. Therefore, this study developed a device that could be used to conduct the compaction test concerning the following conditions: (1) compaction is vertically applied to a soil sample inclined at various angles (VC), and (2) compaction is normal to an inclined soil sample (IC). Some initial tests on lateritic soil using both methods developed showed that at the same energy applied the densities are quite different. These results confirm that, in the case of sloping ground, the standard compaction test may not be appropriate.

1 Introduction

Lateritic soil has continuously been employed for road and embankment constructions because of its abundant quantity around the world (see Fig 1 [1]). In addition, its basic and engineering properties are both desirable and suitable for construction projects. Holland [2] reported that the lateritic soil is created in tropical climate, as evident in Fig. 1. Note that the tropical climate occurs in regions within the equatorial belt, experiencing hot and humid weather with abundant rainfalls.

According to [3], lateritic soils are simply residual soils having Silicon dioxide (Silica) to Sesquioxide ratio between 1.33 to 2.00. One distinct aspect of the soil is its red colour. As such, Blight described the term lateritic soil as reddish tropically weathered geomaterials [4]. Its engineering properties are the mix of fine-grained sands, gravels, and soft rock. Even though it visually looks strong, but some particles tend to easily crush under impact. Besides, some particles left after crushing may exhibit plastic behaviour.

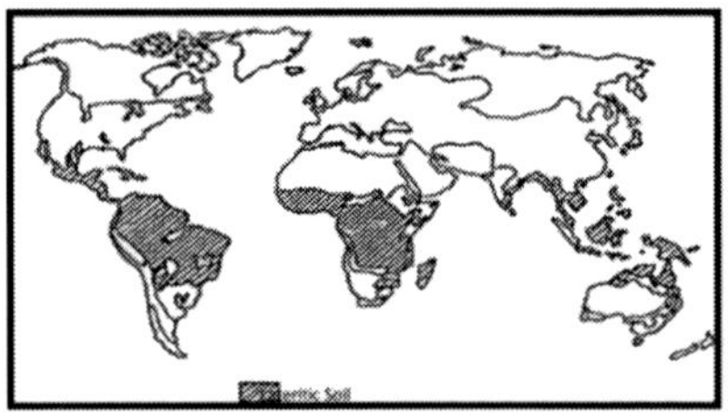

Fig. 1. Distribution of lateritic soil around the world [1]

The engineering properties of the laterite in Thailand were extensively studied in 1969 by U.S. Agency for International Development [5]. According to Brown et. al. [6], Thailand can be divided into five physiographic regions: (1) Northwest Highland, (2) Chao Phraya Plain, (3) Khorat Plateau, (4) Chanthaburi District, and (5) Peninsular Area. Of all of those five regions, lateritic soils are most found in the Khorat Plateau and Chanthaburi District.

The study concerning the quantity and sources of lateritic soil is important for developing a country. The soil is one of the main sources for road construction. The more the road is constructed the more economy is foreseen thereby increasing the wealth of people. Quite often a source of lateritic soil is far from a construction site, resulting in the need for transportation. This practice has led to the study for soil compaction. This is because it is well known that soil when excavated will be looser than that of its original state. Thus, when transported and filled in another location the soil must be compacted.

The maximum dry density of a lateritic soil when compacted in the field can be obtained with the right amount of added water. In addition, suitable compacting machines are also important. The standard practice concerning the soil compaction test currently carried out in the laboratory is performed on a soil being horizontal. In the field, however, in some situations the ground may be sloped thereby applying a laboratory result for a field operation is problematic. Therefore, this study attempted to overcome this problem by developing a device that can conduct the compaction test on a soil sample being non-horizontal.

* Corresponding author: panu pptk@hotmail.com

2 Development of the device

2.1. Design principles

The design of this device was based on the following assumptions: (1) applied energy can be generated from the standard impact hammer used in the compaction test, (2) a mould has a rectangular shape of 15 by 15 cm with around 4.10 cm high, (3) the mould must be able to support a collar in order to be able to apply the inclined compaction effort, and (4) comprises a base that can be adjusted to incline at various angles.

Fig 2 displays the schematic diagram of the mould developed. Basically, it comprises a steel plate supporting the mould having the dimensions of 15.00 by 15.00 by 4.10 cm. One important aspect of this design was that the top of the mould is recessed such that it can support a collar shown in Fig 3. The steel plate shown in Fig 2 was drilled such that it can be attached to the base show in Fig 4 using bolts and nuts. The two vertical frame components shown in Fig 4 were also drilled so that the base can be lifted at a specified angle thereby enabling the inclined compaction test carried out.

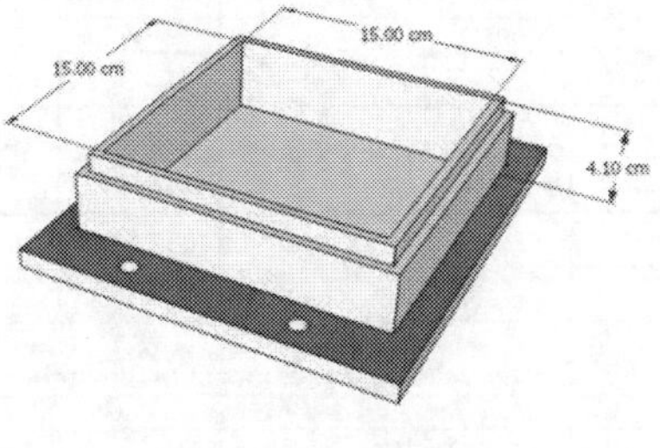

Fig. 2. Mould developed for inclined compaction test

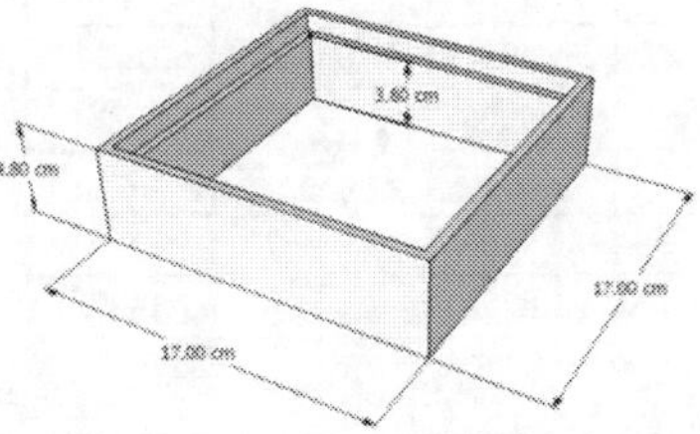

Fig. 3. Collar of the mould

Fig. 4. Picture of the base for supporting the mould

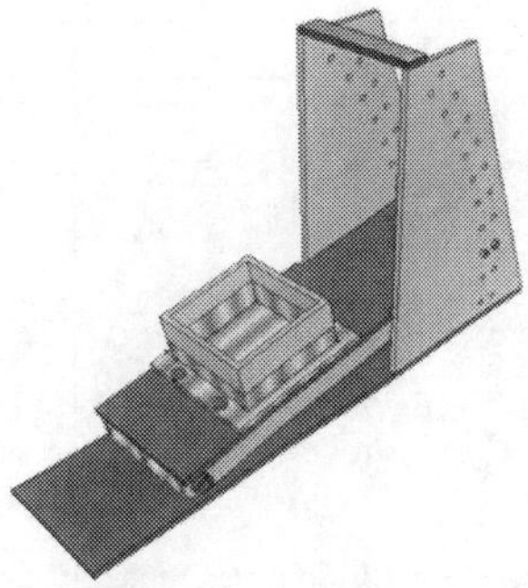

Fig. 5. Schematic diagram of the full system

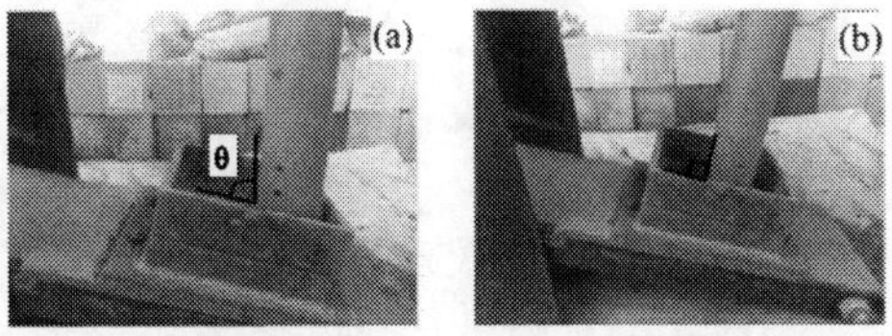

Fig. 6. Vertical compaction on inclined mould, VC (a) inclined compaction on inclined mould, IC (the energy applied is normal to the mould) (b)

Fig 5 illustrates the schematic diagram of the full system developed by the authors. The two rows of holes at the two vertical frames were drilled such that they can lift the base at the angles of 5, 10, 15, 20, 25, 30, 35, 40, and 45°, meaning the inclined compaction test could be carried out corresponding to these pre-set inclined angles.

2.2 Mechanism of inclined compaction

This study attempted to make the following tests applicable: (1) compaction energy is vertically applied to a soil sample inclined at various angles, and (2) compaction effort is normal to an inclined soil sample. The former is schematically depicted in Fig 6 (a); while the latter can be done according to the illustration in Fig 6 (b). It should be noted herein that as an inclined angle is greater the collar must be attached so that the compaction could be carried out.

As the mould used in this study was not the same as the standard mould normally employed in the compaction test according to the ASTM D1557-12 [7], the number of blows, *NL*, must be first determined. Because the mould developed was just 4.10 cm high, the number of layers for compaction was just two, instead of three as used in the standard compaction that uses the 4.584 in high mould (11.64 cm). This was accomplished by using the following equation

$$NL = \frac{Energy\ (from\ ASTM) \times Soil\ Volume}{Ram\ Wt.\ (kg) \times 9.81 \times Drop\ heigh \times Layers} \quad (1)$$

where energy = 2,700 $kN.m/m^3$, soil volume = $0.15 \times 0.15 \times 0.041\ m^3$, ram weight = 4.54 kg, drop height = 0.457 m, and layers of compaction = 2. This resulted in the *NL* of 63 blows per layer.

Table 1. Soil properties and classification results

Property	Value
Texture Composition:	
Gravel, > 75 – 4.75 mm, %	36
Sand, 4.75 – 0.075 mm, %	59
Silt and Clay, < 0.075 mm, %	5
Physical Properties:	
Liquid Limit (LL), %	31
Plastic Limit (PL), %	18
Plastic Index (PI), %	13
Specific Gravity (G_S)	2.70
Classification:	
AASHTO	A-2-6
Unified Soil Classification System	SC

3 Inclined compaction tests

3.1. Soil sample

The soil sample employed for this investigation was laterite of which is the most common material for the road construction in Thailand. It was obtained from Moo 8, Ban Bangonaetae, Korlortanyong Subdistrict, Nongjig District, Pattani Province, Southern Thailand.

The soil was first sieved to determine its size distribution. It was found that the coefficients of uniformity (C_u) and curvature (C_g) are 23 and 0.98, respectively, indicating that it was well-graded. In addition, the permeability test revealed that the coefficient is 0.4212 mm/s. The atterberg test showed that the plasticity index is 13%; and, the specific gravity is 2.70. According to the AASHTO standard, these properties indicate that the soil is A-2-6, meaning very good for road construction. For the case of the USCS, it was classified as SC, clayey sand. Table 1 displays the summary of the laboratory test results as well as soil classification by both AASHTO and USCS.

3.1. Test programmes

To assess the performance of the device developed, two compaction methods were conducted: (1) compaction effort was vertically applied on an inclined soil sample, namely VC, and (2) applied compaction effort was normal to an inclined soil sample, namely IC. For both testing methods, the soil samples were inclined, with respect to the horizontal, at the angles of 5, 15, and 30°.

In order to obtain a consistency result, each test set-up was carried out twice. If the results were not much different, they were hold and averaged; otherwise, a third or even fourth test was conducted. These resulted in a total of 12 tests experimented: VC5-1, VC5-2, VC15-1, VC15-2, VC30-1, VC30-1, IC5-1, IC5-2, IC15-1, IC15-2, IC30-1, and IC30-2. Please note that the number just after the alphabets VC and IC indicate an inclined angle of a soil sample (see Fig 6), and the number after the dash indicates the sample number (1 or 2).

4 Results and discussion

Test results obtained from this study were simply the maximum dry densities, $\rho_{d,max}$ and their corresponding optimum moisture contents, OMC. Table 2 shows the compaction test results for the VC. For the IC, the compaction test results were summarised and shown in Table 3. To visually observe the results obtained, all of the testing data were plotted and shown in Fig 7.

For the VC samples, it was found that as the inclined angle is gradually increased the $\rho_{d,max}$ is steadily decreased, i.e., the $\rho_{d,max}$ values are 1.883, 1.800, and 1.764 g/cm^3 with respect to the inclined angles of 5, 15, and 30°. Nonetheless, for the case of the OMC value, it was observed that the greater the inclined angle the higher the OMC value. Notice that these findings were also true for the case of the IC samples.

Even though the values of $\rho_{d,max}$ and OMC for both cases are very similar in terms of either increasing or decreasing, but their average values for the same inclined angle are quite different.

Table 2. Results for vertical compaction on inclined mould

Sample	$\rho_{d,max}$ (g/cm^3)	Avg. $\rho_{d,max}$ (g/cm^3)	OMC (%)	Avg. OMC (%)
VC5-1	1.870	1.883	15.20	15.00
VC5-2	1.895		14.80	
VC15-1	1.800	1.800	15.80	15.60
VC15-2	1.800		15.40	
VC30-1	1.765	1.764	17.20	16.90
VC30-1	1.762		16.60	

Table 3. Results for inclined compaction on inclined mould

Sample	$\rho_{d,max}$ (g/cm^3)	Avg. $\rho_{d,max}$ (g/cm^3)	OMC (%)	Avg. OMC (%)
IC5-1	1.890	1.890	14.30	14.15
IC5-2	1.890		14.00	
IC15-1	1.900	1.888	15.00	14.80
IC15-2	1.876		14.60	
IC30-1	1.810	1.819	17.40	16.00
IC30-1	1.828		14.60	

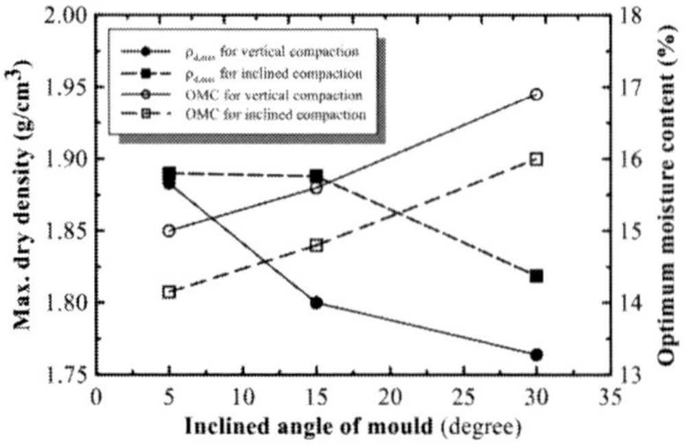

Fig. 7. $\rho_{d,max}$ and OMC of both vertical- and inclined compactions versus inclined angle of mould

Table 4. Differences of $\rho_{d,max}$ and OMC between inclined- and vertical compactions versus inclined angle of mould

Inclined angle (°)	Avg. $\rho_{d,max}$ (g/cm^3)		% Diff.	OMC (%)		% Diff.
	VC	IC		VC	IC	
5	1.883	1.890	0.37	15.00	14.15	-5.67
15	1.800	1.888	4.89	15.60	14.80	-5.13
30	1.764	1.819	3.12	16.90	16.00	-5.33

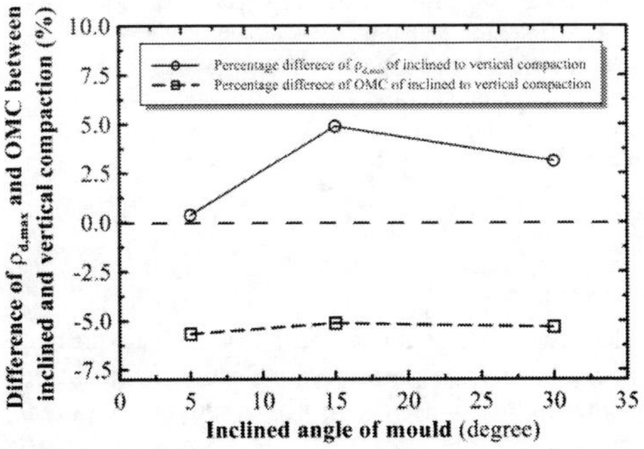

Fig. 8. Differences of $\rho_{d,max}$ and OMC between inclined- and vertical compactions versus inclined angle of mould

For instance, in the case of IC, when the inclined angled was changed from 5 to 15°, the average $\rho_{d,max}$ values were also changed, but just from 1.890 to 1.888 g/cm^3. Comparing this result with the case of VC, it can be seen that at the same inclined angle change the average values are changed from 1.883 to 1.800 g/cm^3, which is quite larger.

Furthermore, when the inclined angle was 5°, the difference of $\rho_{d,max}$ between the VC and IC samples is just 0.37%. This finding suggests that when a soil sample is almost horizontal the angle of the hammer relative to a soil sample has very little effect on the energy transferred to the soil being compacted.

Another point should be noted from this experiment is that the difference of the OMC values between the VC and IC samples when the inclined angles were gradually changed is almost constant: they are -5.67, -5.13, and -5.33% with respect to the inclined angles of 5, 15, and 30°. Thus, it may be concluded that the inclined angle of soil sample has no effect on the OMC value.

5 Conclusions

Lateritic soil is one of the most sought after material for road construction, especially in Thailand. The common practice is first looking for a material source. If the material is in accordance with the requirements, it is then excavated and transport to a construction site. To obtain a desired density, however, some kinds of compaction effort are essential.

Normally, the compaction test, either standard or modified, is carried out in a mould being horizontally laid. Then, the results were employed for quality control in the field. In the case of the ground is not horizontal this practice may raise a problem in terms of getting a correct result. As such, a device has been developed in order to observe the compaction behaviour of the ground not being horizontal. The compaction test then was carried out on a soil sample laid at various inclined angles with two conditions: (1) compacting effort was vertical to a soil sample (VC) and (2) compacting effort was normal to a soil sample (IC). From the experiment, the following points have been drawn:

1) for all inclined angles, the IC yields greater maximum density,
2) when the inclined angle is just 5° the difference of $\rho_{d,max}$ between VC and IC is minimal,
3) when the inclined angles are greater than 5°, the difference of $\rho_{d,max}$ between the two methods is more obvious, and
4) in the case of OMC, it may be concluded that the two testing methods have no effect on the result because the percentage differences of the OMCs between those two methods are relatively consistent.

6 Acknowledgements

The authors wish to thank RMUTSV for their financial aid and support. Laboratory work was carried out by Mr. N. Inchana, Mr. P. Mongkol, and Mr. A. Binborsor, of which is kindly acknowledged.

References

1. M.D. Gidigasu, *Laterite Soil Engineering*, Elsevier Scientific Publishing Company, (1976)
2. T.H. Holland, Geol. Mag., 4(10), (1903)
3. A. G/medhin, Compaction properties of lateritic soils (The Case of Assossa), Master Thesis, Addis Ababa Uuniversity, (2008)
4. G.E. Blight, *Mechanics of Residual soils*. A.A Balkema, the Netherlands, (1997)
5. B.A. Vallerga, J.A. Shuster, A.L. Love, C.J. Van Til, *Engineering Study of Laterite & Lateritic Soils in Connection with Construction of Roads, Highways, & Airfields*, U.S. Agency for International Development, (1969)
6. G.F. Brown, and others, *Geological Reconnaissance of the Mineral Deposits of Thailand*, U.S. Geological Survey Bulletin No. 894. Washington, (1951)
7. ASTM D1557-12, Standard test methods for laboratory compaction characteristics of soil using modified effort (56,000 ft-lbf/ft^3(2,700 kN-m/m^3))

Application of magnetorheological fluids in the design of a leg prosthesis with active damping

Oscar Arteaga[1,*], *Alex* Cevallos[1], *Maria I.* Erazo[1], *Klever D.* Morales[1], *Daniel* Tenezaca B.[1], and *Edison* Argüello M.[1]

[1]Departamento de Ciencias de la Energía y Mecánica, Universidad de las Fuerzas Armadas ESPE, Sangolquí, Ecuador.

Abstract. The article is about of research of the behavior of magnetorheological materials (MR), later it will be implemented in a prototype leg prosthesis with active damping for people who have suffered amputations in lower extremities, which considers the use of an actuator with Magnetorheological Fluids (MRF) *LORD MRF – 140 CG,* whose control is based on the adjustment of the magnetic field applied to the MRF using in its design the main parameters of anatomy and biomechanics of the foot-leg system, so that the force applied in the course of the gait cycle it can be absorbed by the prosthesis, reducing the impact on the user's column thus allowing the prosthesis to be subjected to tests for emulation in different positions according to the applied load.

1 Introduction

The amputation of a lower extremity is a problem that affects a specific number of people in the world, who when exposed to a totally different environment, can develop feelings such as distrust and insecurity until they adapt to the imminent need to use a prosthesis that allows them to balance their rhythm of life and re-establish the way they have lived up to that moment. [1]. Nowadays the technological advance of the prosthesis has allowed the patient to overcome the inconveniences presented in everyday life [2,3]

In order to minimize the impact caused by the absence of the lower extremity, Mechatronics as a synergistic effect of the integration of different disciplines in the design of a multidisciplinary system [4], it has developed a significant revolution in fields such as assistance robotics and rehabilitation of the lower extremities [5]. In Ecuador from 2007 to 2013 there are around 875 cases of traumatic amputations of the lower limb, and this number is expected to increase [6].

The willingness of patients to use a leg prosthesis with active damping allows drastically improve their mobility and enables them to participate in activities of daily living improving their quality of life [7,8,9], to this is added the natural movement and dexterity that a prosthesis provides. [3]

With the aim of avoiding the damages represented by the use of a prosthesis, this prosthesis is based on the principle of mass-spring-cushioned, the damping generated by a MRF which is a material that can change viscosity according to the applied magnetic field, can get to form a thick mass, all this without changing its volume. [10]

This article has 5 sections, including the Introduction. Section 2 describes the mechanical characteristics of the MRF. Section 3 describes the design of the shock absorber. Section 4 describes the tests and results of the prosthesis. Finally, the Conclusions are presented in Section 5. In future works it is expected to publish the results in the tests with real patients.

2 Fluid MR

The experimental procedure for the magnetorheological characterization of the fluid MRF-140CG of LORD is established. This procedure involves the use of the rotational rheometer MRC-501 Anton Paar Physica together with the cell MRD-70/1T coupled [11]. Together, a thermostatic bath Julabo F-25 is made to maintain a temperature of 20 ° C, the results obtained for a magnetic field of 0 T are shown in the figure 1.

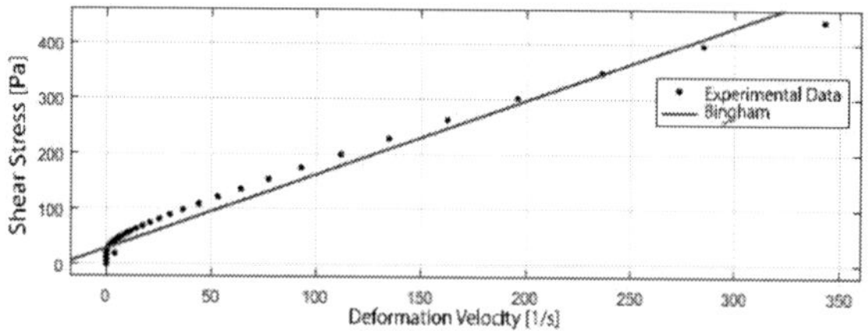

Fig. 1. Shear stress vs. strain rate at 20 °C

Applying the Bingham model for MRF, the shear stress equation is obtained (τ), depending of the strain rate ($\dot{\gamma}$), where the slope (1.338 Pa.s) represents the plastic viscosity (η) independent of the magnetic field.

$$\tau = 29.8 + 1.338\,\dot{\gamma} \qquad (1)$$

Similarly, the equation that predicts the MRF shear threshold for a magnetic field was obtained ($\dot{\gamma}$) of 600 mT, and strain rate of 350 $1/s$, shown in the figure 2, from which a value of $\tau_y(H)$ is obtained and it is 35117.639 Pa.

* Corresponding author: obarteaga@espe.edu.ec

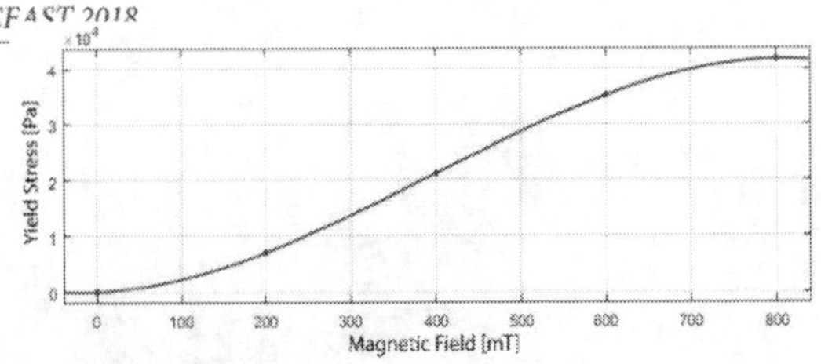

Fig. 2. Yield Stress Vs Magnetic Field at 20 °C

3 Shock absorber design

3.1 Design Parameters

Ecuador is a multi-ethnic and pluricultural country. The size and weight of its inhabitants vary according to genetics and nutrition, taking as initial parameters, an average height of the Ecuadorian man of $1.68\,m - 1.70\,m$ and a total body mass of $70\ kg$[12].

According to the parameters of the Dempster and Clauser model, the weight of the leg represents 4.3% of the total weight of the body [13]. Therefore, the critical force that the prosthesis must withstand in critical operating conditions is 3432.1 N generated by the mass of the entire body and supported by the legs. For study reasons, half of the impact force is considered to multiply it by a load factor of 1.2, resulting in 2059.26 N which represents the design strength (F_d).

It is considered that the MR damper works in valve mode [18], shown in the figure 3.

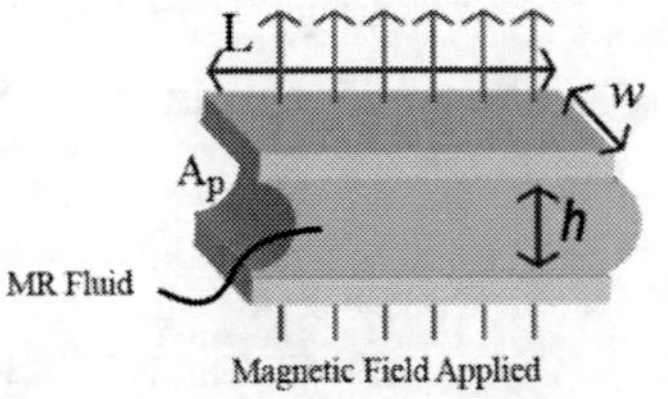

Fig. 3. Mode of operation of the MR fluid as valve.

Where, w is the width of the magnetic pole $w = \pi(D_p + h)$, L is a length of magnetic pole and h is the angular pipeline or gap.

In the design of the shock absorber is based on the following established parameters, whose nomenclature is shown in the figure 4.

Damping stroke	$c_a = 0.020\ m$
Outer diameter of the cylinder	$D_c = 0.0421\ m$
Thickness of the cylinder liner	$t_w = 0.006\ m$
External diameter	$D_e = 0.0301\ m$
Cylinder sleeve length	$L = 0.01\ m$
Maximum diameter of the magnetic pole	$D_n = 0.022\ m$
Coil length	$L_b = 0.01\ m$

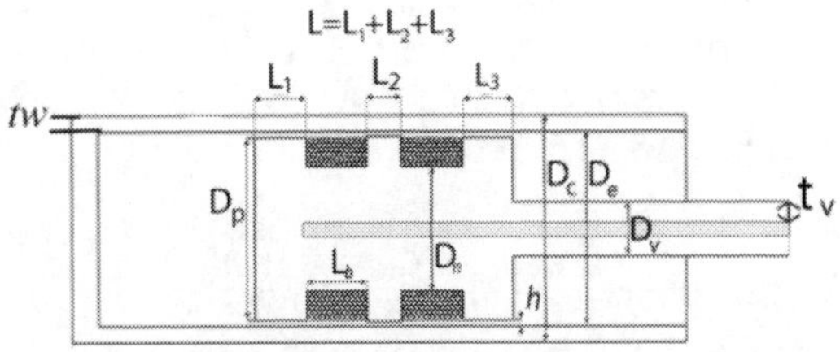

Fig. 4. Scheme MR damper.

To calculate the average speed of the piston requires the damping time that results 40% of the total duration of the human gait cycle [14], which is $\Delta t_c = 0.456\ s$, which is the time when all the weight of the body is supported by one leg.

Then, the average piston velocity that is calculated:

$$V_a = \frac{c_a}{\Delta t_c} = 0.044\ \frac{m}{s} \tag{2}$$

3.2 Dimensioning of the MR shock absorber

The controllable force and the dynamic range are two of the most important parameters in the evaluation of the overall performance of the MR damper, [15] and for their analysis are decomposed into:

$F_{\tau(H)}$: Controllable force, dependent on a shear stress, $\tau_y(H)$. [2]

$$F_{\tau(H)} = c\frac{\tau_y(H)LA_P}{h}sgn(V_a) \tag{3}$$

F_{uc}: Uncontrollable force, which is equal to the sum of the frictional force and the plastic viscosity force.

$$F_{uc} = 2F_n = 343.35\ N$$

F_n : Plastic viscosity force.

$$F_\eta = \left(1 + \frac{wh.V_a}{2.Q}\right)\frac{12.\eta.Q.L.A_P}{w.h^3} \tag{4}$$

$$c \approx 2.07 + \frac{1}{(1+0.4\tau)} \tag{5}$$

Friction force F_f, is of the same magnitude as the plastic viscosity force F_n. [18]. Where the value of c oscillates between 2.07 and 3.07 calculated by the equation 7, Q is the flow that is equal to $A_p.V_a = \pi(D_e - 2h)^2 * \frac{V_a}{4} = 2.8743\ x10^{-5}\ \frac{m^3}{s}$ y A_P is the cross section of the piston that is equal to $\pi(D_e - 2h)^2/4 = 6.5325 * 10^{-4}\ m^2$. [15] and with the previous equations the angular duct is calculated that is equal to 0.0063 m.

The dynamic range (D) is defined as the ratio between the strength of the shock absorber and the uncontrollable force F_{uc} [15].

$$D = \frac{F}{F_{uc}} = 1 + \frac{F_\tau}{F_n + F_f} \tag{6}$$

From where it is determined that the value of the dynamic range is 7, taking into account that this force is equal to the contact force of the ground when a person is standing upright, in the resting state is equal to half the body weight, [15] obtaining that the controllable force is:

$$F_\tau = F_d = 2059.26\ [N] \tag{7}$$

3.3 Shock absorber cylinder design

It is necessary to consider the pressure that is generated internally in the cylinder to determine the thickness of the cylinder liner, the same that will be made of steel AISI 1050 HR, the thickness of the cylinder liner is determined t [17], with the following equation:

$$\sigma = \frac{\Delta P * r}{2 * t} \tag{8}$$

Where, the inner radius of the cylinder liner $r = 0.01505\,m$ and maximum yield stress $\sigma = 338\,MPa$, resulting from the previous equation the thickness $t_w = 0.0361\,mm$, 112/5000

that is the minimum to withstand the pressure generated, so for the design is assigned a wall thickness of 2 mm.

For the design of the stem take into account that it must be able to transmit all the force without buckling or fracture, also that it must be hollow to give space to the cables that excite the coil, as shown in the figure 4.

t_v : wall thickness of the stem = $0.001\,m$

L_v : stem length= $0.06\,m$

D_v : diameter of the stem = $0.01\,m$

3.4 Design of the shock absorber stem

The stem is subjected to an analysis per column, in which the critical load is determined as a short column, this analysis is performed with the equation of J. B. Johnson [16], con $S_y = 276x10^6\,Pa$, $E = 207x10^9 Pa$ y $A = \pi({D_v}^2 - (D_v - 2t_v)^2)/4$.

$$P_{cr} = A * S_y \left[1 - \frac{S_y\left(\frac{Le}{r}\right)^2}{4\pi^2 E}\right] = 7744\ N \tag{9}$$

$$P_a = \frac{P_{cr}}{N} = 2581\ N \tag{10}$$

Where, P_{cr} is the critical load of the column, A the cross-sectional area of the hollow stem, P_a is the allowable load for the hollow stem and N is a safety factor that is equal to 3 for this type of design. [18]

The admissible load is 2581 N which is greater than 2059.26 N which is the maximum load produced in the cylinder, therefore, it is concluded that the design is safe.

3.5 Control Circuit Design

The design of the control circuit consists of a microcontroller with the function of operating the electromagnet, which in turn generates the magnetic field, with the help of a PWM signal, which will be modulated according to the information received by a load cell, shown in the figure 5.

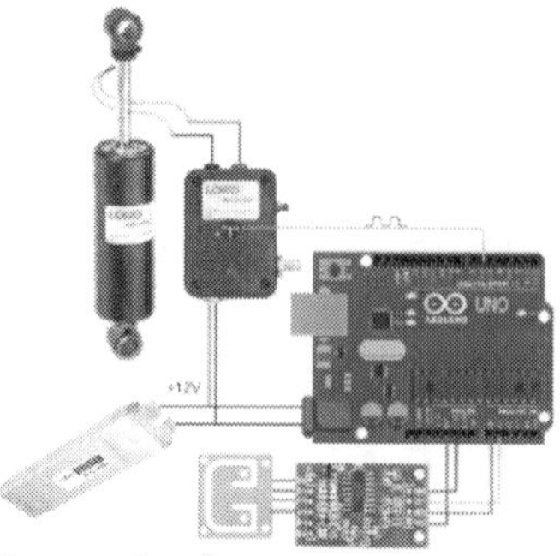

Fig. 5. Electronic connection diagram

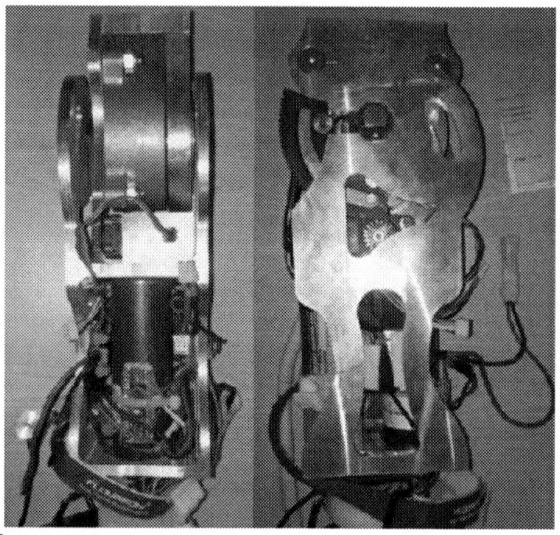

Fig. 6. Leg prototype

4 Test and Results

To validate the damper design several tests were performed, the first test was conducted by using load sensor, which records the load on the leg in function of time, this time is equal to the damping time during the gait cycle human, and its results are shown in figure 7.

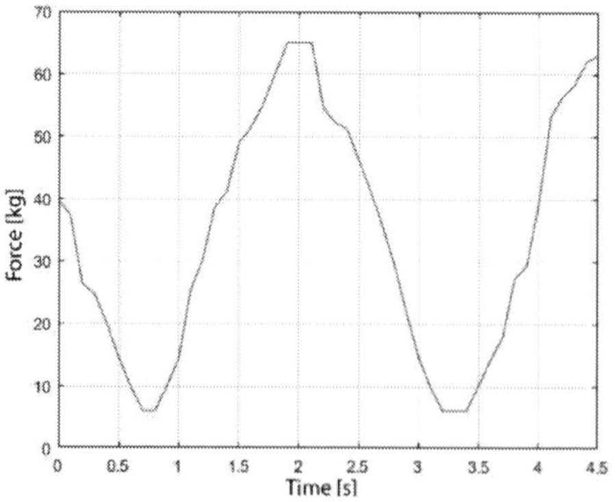

Fig. 7. Force vs. Damping time during the human gait cycle

The second test was performed in a "Dyno-Shock" system that subjects the MR damper to expansion and compression continuously, with a current of 0 A, magnetic field of 0 T and frequency of 1 Hz, the test results are shown in figure 8, from which it can be seen that the maximum load during the expansion is 102 kg.

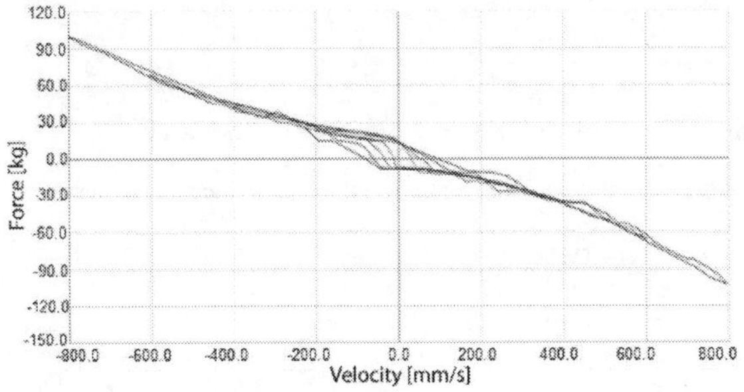

Fig.8. Force vs. Velocity at 0 A

In a third test, similar to the previous one, with a current of 1.8 A, magnetic field of 600 mT and frequency of 1 Hz, the results of the test are shown in Figure 9, from which it can be seen that the maximum load during the expansion is 250 kg which is equivalent to 2450 N.

Through the calculations, it was established that the necessary force for the damping is of 2059 N with a current of 1.8 A, and a magnetic field of 600 mT, but the tests showed that the shock absorber can reach higher values, so that the damper fulfills with the design requirements.

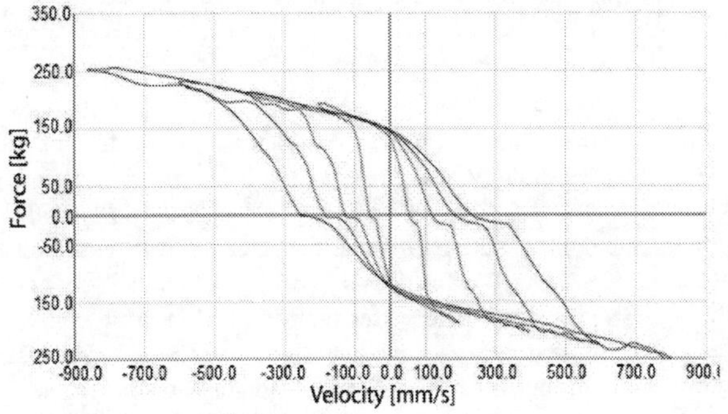

Fig.9. Force vs. Velocity at 1.8 A

5 Conclusions

- The maximum force required to obtain a suitable damping is 2059.26 N, which is achieved with an input current to the coil of 1.8 A creating a magnetic field of 600 mT.
- The minimum force provided by the prototype is 1000 N, equivalent to 102 kg, with a current magnetic field of 0T, this force is sufficient to maintain the weight of the user at rest, as well as providing energy savings that can lengthen the time of use of the prosthesis.
- In the results for expansion and compression tests, these show that as the current value increases, the damping force increases and the displacement of the piston decreases. The damping force is low for zero current and increases gradually as the current increases.

References

[1] Fragassa C., Berardi L., Balsamini G. Magnetorheological fluid devices: an advanced solution for active control. Faculty of Mechanical Engineering, (2016) 333-339.
[2] García J., Rodríguez C. Diseño de una Prótesis Transtibial Amortiguada. México DF. México: Instituto Politécnico Nacional, (2012).
[3] Gonzales M., & al, e. Amputación de extremidad inferior y discapacidad. Prótesis y Rehabilitación. Barcelona: Masson, (2005).
[4] Victor Ragusilaa, M. Reza Emami. Mechatronics by analogy and application to legged locomotion, Mechatronics, Inform. Process. 000(2016) 1–19.
[5] Michael R Tucker, Jeremy Olivier, Anna Pagel, Hannes Bleuler, Mohamed Bouri, Olivier Lambercy, José del R Millán, Robert Riener, Heike Vallery, Roger Gassert. Control strategies for active lower extremity prosthetics and orthotics: a review. Journal of NeuroEngineering and Rehabilitation, 12(1)(2015).
[6] Andrés G, Luis C, Diseño y construcción de protesis de miembro inferior monitoreado desde un computador personal, (2015).
[7] Au, Samuel K., Weber J., Herr H. Powered ankle--foot prosthesis improves walking metabolic economy. IEEE Transactions on Robotics, 25(1)(2009), 51-66.
[8] Martinez V., E. Mooney L., Elliott G., Herr H. Antagonistic active knee prosthesis. A metabolic cost of walking comparison with a variable-damping prosthetic knee. In Engineering in Medicine and Biology Society. (2011) 8519-8522.
[9] Farris R. J., Quintero H. A., Murray S. A., Ha K. H., Hartigan C., Goldfarb M. A preliminary assessment of legged mobility provided by a lower limb exoskeleton for persons with paraplegia. IEEE Transactions on neural systems and rehabilitation engineering, 22(3)(2014) 482-490.
[10] Zhan H, Liao C, Chen W, Huang S, A magnetic design method of MR fluid dampers an FEM analysis on magnetic saturation, (2006),
[11] H.M., L., G., S., C., G., & C., K. Reliable Plate-Plate MRF Magneto-Rheometry Based on Validated Radial Magnetic Flux Density Profile Simulations. Rheo, (2008).
[12] Lema, V. (Julio de 2013). Comparación estadística de medidas antropométricas entre mestizos,afroecuatorianos e indígenas del Ecuador.,http://repositorio.usfq.edu.ec/bitstream/23000/2631/1/107724.pdf
[13] Antonio, J., & Mas, D. (2015). Recuperado el 07 de Agosto de 2017,http://www.ergonautas.upv.es/metodos/biomecanica/biomecanica-ayuda.php
[14] Rajtukova, V, Michaliková, M, Bednarciková, L, Balogová, A, Zicak, Biomechanics of lower limb prostheses, 2014.
[15] Yang G., Spencer Jr., B., Carlson J., Sain M.. Large-scale MR fluid dampers: Modeling and dynamic performance considerations, ELSEVIER. (2002) 309-323.
[16] Richard G. Budynas, kate Gleason, Shigley´s mechanical engineering design- Tenth editionMexico: McGraw-Hill., 2014.
[17] R. Mott, Resistencia de materiales, Mexico: Pearson educación, S.A., 2009.
[18] F. D. Goncalves, M. Ahmadiam y J. D. Carlson, Investigating the magnetoreological effect at high Flow velocities, 2006.

 https://doi.org/10.1051/matecconf/201819202056

Design of natural-rubber panel railroad crossing using finite element method

Aunna Sukhom[1], *Ittipol* Jangchud[2], *Monsak* Pimsarn[3], *Jarruwat* Charoensuk[3], and *Veerachai* Treeporncharoen[4]

[1]Automotive Engineering, International College, King Mongkut's Institute of Technology Ladkrabang, Chalongkrung, Bangkok, Thailand

[2]Department of Chemistry,Faculty of Science, King Mongkut's Institute of Technology Ladkrabang, Bangkok, Thailand

[3]Mechanical Engineering Department, Faculty of Engineering, King Mongkut's Institute of Technology Ladkrabang, Bangkok, Thailand

[4]Extensive Research Polymers Co. Ltd., 26 Moo 4 Tamboon Nonghong, Panthong, Chonburi, Thailand

Abstract.
Thailand has a railway system that is available throughout the country, so there are several railroad crossings. These crossings are generally made of concrete or logs with multiple constraints. There are some disadvantages of concrete railroad crossing, such as, crack, noise during car passing over. To overcome these disadvantages, the softer materials should be used instead. Therefore, this research proposes the natural rubber, widely grown throughout Thailand, panel railroad crossing. However, the natural rubber alone is not enough to withstand the harsh condition. Thus, it is necessary to have some addition ingredients that will enhance the natural rubber properties. The material used in this research is a rubber compound between Chloroprene Rubber (CR) 75% and Natural rubber (NR) 25% blend with additives such as carbon black (CB), magnesium oxide (MgO) and sulfur (S8). The objectives of this article were to analyze the deformation of the natural rubber panel railroad crossing and to evaluate its safety factor, defined as the ratio of strain at break and the maximum equivalent strain, using finite element method. In the analysis, the applied loading of the model was obtained from the State Railway of Thailand. The analyzed results reveal that the deflection of rubber panels passes the standard from State Railway of Thailand. Safety factor of external rubber panel is 27.03 and for internal rubber panels are 9.12 and 15.29. The metal pads had elastically deformed and concrete railroad sleeper deformation was very small.

1 Introduction

Thailand in one of the countries that has the highest production capacity of rubber in the world. Although Thailand's rubber exports around 87%, export revenues are about 193,067 million Baht per year, but domestic processing about 13.4% of rubber can generate about 230,000 million Baht annually. Thailand is experiencing several problems with natural rubber, such as the volatility of raw rubber prices, the problem of natural rubber production that does not meet the market demand, the problem of the quality of raw rubber, the problems of the rubber market and competitors from the other countries. Therefore, Thailand focuses on the development of rubber processing industry in the country to increase the amount of rubber before selling, it will increase revenue for farmers. At present, the railroad industry is growing in the future due to investment in the development of the rail system in Thailand. The BTS network in Bangkok and the government's "nationwide double-track railway project ". According to the State Railway of Thailand. Currently, Thailand has a railroad track of approximately 4,346 kilometers. There are more than 2,500 railroad crossings across the country. Generally, these crossings are made of concrete or logs. The disadvantages of using these materials are lifetime limit, cracking problem, damaged concrete cushions due to impact with crossing panels. Additionally, there are problems in rainwater drainage and in noise generation during the cars passing through. The vehicle passing through must slow down, resulting in traffic problems at the intersection. Concrete and logs are brittle material that are not resistant to repetitive force. Breaking or cracking causes accidents and wastes the repair budget. The problem can be solved by finding materials that can withstand repeated repetitive forces. Rubber is a good choice. The advantages of using rubber to produce rubber crossing panels such as rubber are soft material, no rupture, good for vibration and silence while the car runs through. There are also promoting the use of rubber in Thailand, design and production of prototypes that are feasible to commercialize by Thai companies. The design of the workpiece should be appropriate for the actual use, including the finite element method for deformation and the estimated lifetime of the rubber panels. If the project is successful, it will be a key variable for the country and the government. As a result, the country's economic, social, agricultural and security development has increased. The Institute hopes that the project will be able to respond positively and significantly to the government.

2 Methodology

2.1 Material properties

Initially, the analysis needed to determine the material properties. The material used is the compound rubber that has been researched and develop a formula for support the impact between the rails and concrete pillars. The chemical composition is given in Table 1. The mechanical properties of the rubber are derived from the tensile test data, as shown in Fig 1. Table 2 shows the derived mechanical properties. Moreover, the rubber hyperelastic properties were also considered in this paper and Yeoh's model was employed.

Table 1. Chemical composition of rubber. [1]

Ingredient	CR/NR ratio	carbon black(CB)	magnesium oxide(MgO)	sulfur (S_8)
composition	75/25	60 phr	1 phr	1 phr

Table 2. Mechanical properties of rubber. [1]

Tensile Strength (MPa)	%Elongation at break	Modulus at 100 %	Poisson's ratio
19.718	513.916	3.8605	0.49

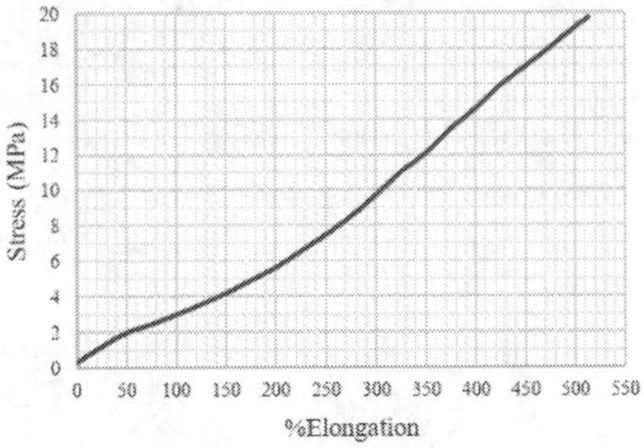

Fig. 1. Tensile test result of rubber.

Yeoh 3rd order model is the most suitable model that can predict behaviour of compound rubber and its constitutive model is given in equation 1.

$$W = C_{10}(I_1 - 3) + C_{20}(I_1 - 3)^2 + C_{30}(I_1 - 3)^3 \quad (1)$$

From curve fitting process using ANSYS, the Yeoh 3rd order model parameters are C_{10} = 82.595 x 10^5 Pa, C_{20}=15.303 x 10^4 Pa and C_{30} = -54.453 Pa

Table 3. Mechanical properties of used material. [2,3]

Material	Ultimate Tensile Strength (MPa)	Yield Strength (MPa)	Young's modulus (MPa)	Poisson's ratio
Structural Steel	460	250	200,000	0.3
Concrete	5	0	40,000	0.18

2.2 Loading and Boundary condition

Analysis of 3D problems by finite element method was done using ANSYS workbench program. The load is coming from vehicle wheel. The load value is 22. 5 kN[4] , which is the standard value set by the State Railway of Thailand calculate form real load 15 kN multiple with 1.5 safety factor. Press the load into 3 areas of the model to evaluate deformation. The components of simulation model shown in figure 2. Loading apply area and fixed condition shown in figure 3 4 and 5.

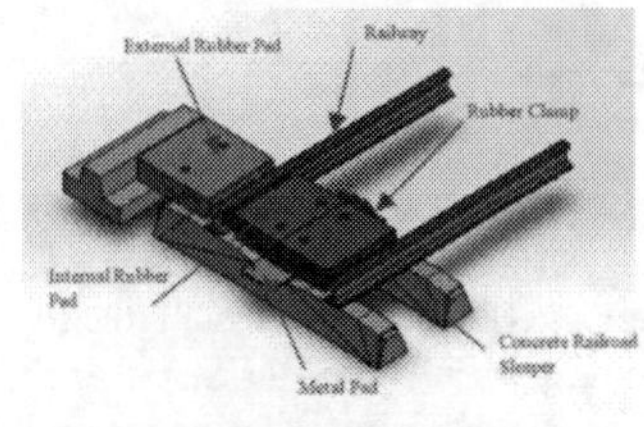

Fig. 2. Model components

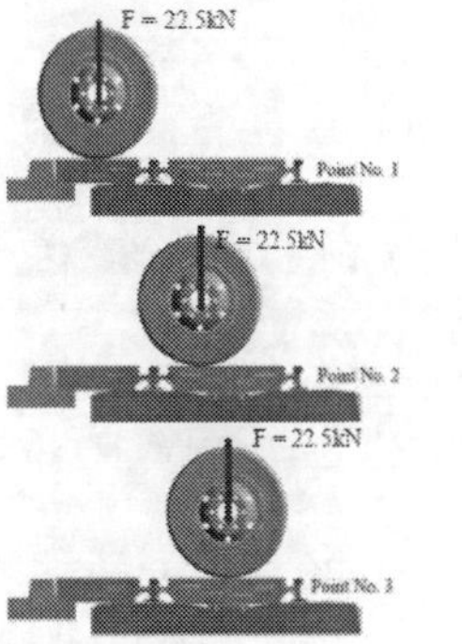

Fig. 3. Loading condition

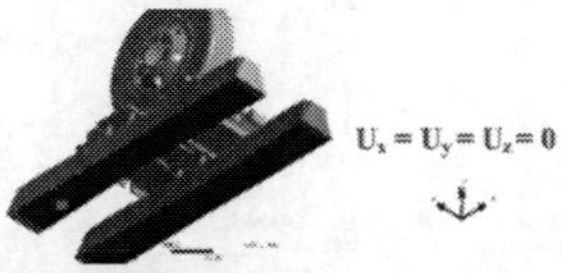

Fig. 4. Bottom fixed condition

Fig. 5. Rubber panels fixed condition

The contact relation between two internal rubbers are suppressed because the installation is done by inserting the two panels together without adhesive or fastener. The contact between rubber panels and metal pad was set as rough contact. For other contact relation set as bonded contact, they cannot separate from each other. The centre of contact between the two panels is wedge- shaped, which support assembly and allows the rubber to not separate from one another while the vehicle is passing. The wheel on this model is assumed to be a rigid body.

2.3 Safety factor

The rubber panels safety factor calculates by equation 2.

$$SF_{Rubber} = \frac{Fracture\ strain(elongaton\ at\ break)}{Maximum\ equivalent\ strain} \quad (2)$$

The metal pad safety factor calculates by equation 3.

$$SF_{Metal\ pad} = \frac{yiels\ stress\ of\ material}{Maximum\ von{-}mises\ stress} \quad (3)$$

The concrete safety factor calculates by equation 4.

$$SF_{Concrete} = \frac{Ultimate\ tensile\ strenght}{Maximum\ priciple\ stress} \quad (4)$$

3 Result and discussions

The Finite Element Model with the boundary conditions is submitted in Ansys for Static Analysis. FEM model results of rubbers are obtained which include deflection. The simulation result in millimetre unit shown in figure 6 7 and 8.

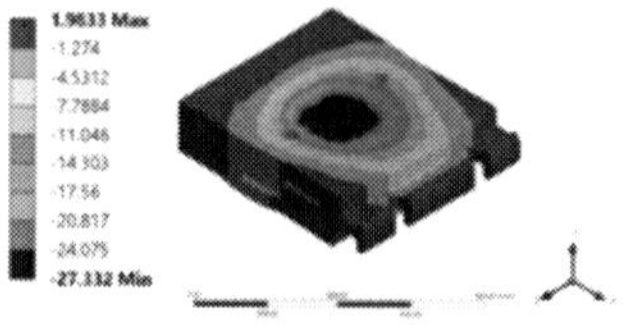

Fig. 6.Y axis deformation of External rubber point 1

Fig. 7.Y axis deformation of Internal rubber point 2

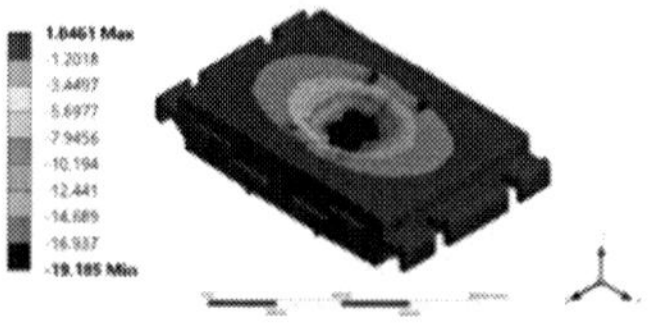

Fig. 8.Y axis deformation of Internal rubber point 3

The equivalent strain of external rubber panel shows as sample because this part is most deflated. Unit of equivalent strain is mm/mm.

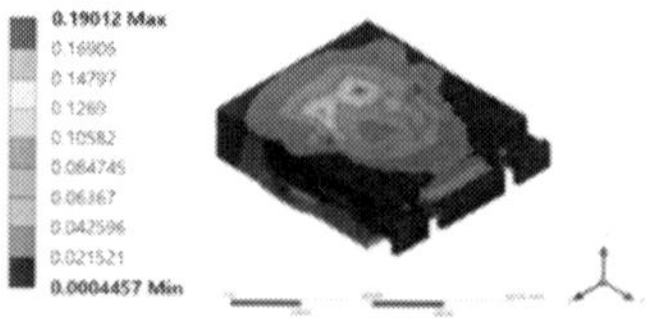

Fig.9. Maximum Equivalent Strain of External rubber point 1

When applying the force at each point, it was found that the centre of the external rubber panel had the greatest collapse. Cause this area is a wider than the internal rubber panel. To consider the internal rubber panel, the contact area between two rubber panels is less collapse than the centre of the rubber panel. Due to the metal pads under the rubber. If there are no metal pads, the collapse or separation of two rubber panels will increase. The collapse of the rubber panel occurs based on the accepted and can be restored to the original when the force removed. Because of the metal pad affects the deformation of the internal rubber panel, it must analyse the damage of the metal pads as well. The simulation result in millimetre unit shown in figure 10 and 11.

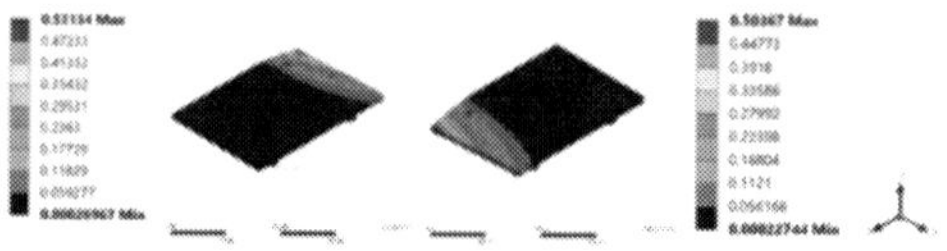

Fig. 10. Deformation of metal pad at point 2

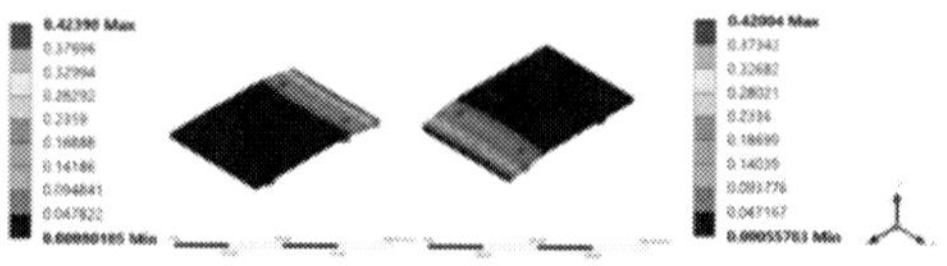

Fig. 11. Deformation of metal pad at point 3

Deformation and recovery of the metal pad can be determined from the highest stresses shown in figure 12 and 13 in MPa unit.

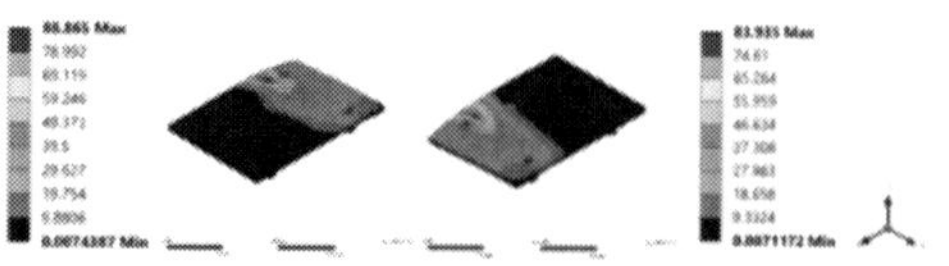

Fig. 12. Von-Mises Stress of metal pad at point 2

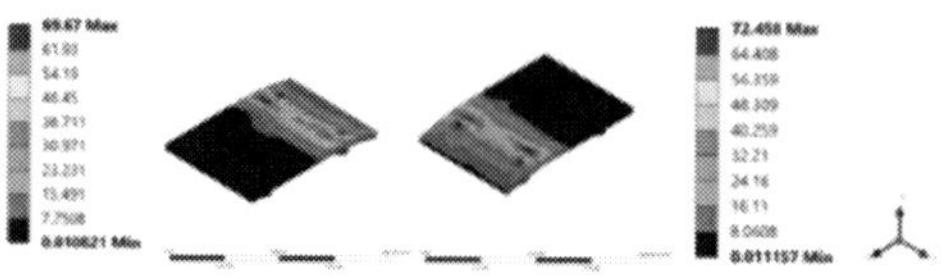

Fig. 13. Von-Mises Stress of metal pad at point 3

Maximum stress that occurs in metal pad is 83.943 MPa at point 2 loading. But the yield stress of this component is 250 MPa, according to mechanic properties of structural steel. That' s metal pad just occurs elastic deformation and can be recovered to the initial shape. In case of the concrete railroad sleeper, it also has deformation from contact with metal pad. The deformation occurs very small section in millimeter unit but permanent. The most deformed area is loaded at point 1. Sample picture of the simulation results from loading on the external rubber panel shown below.

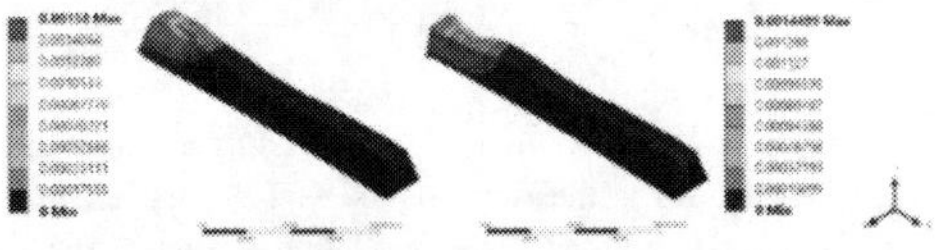

Fig. 14. Deformation of concrete railroad sleeper at point 1

For maximum principal stress of concrete railroad sleeper shown in figure 15.

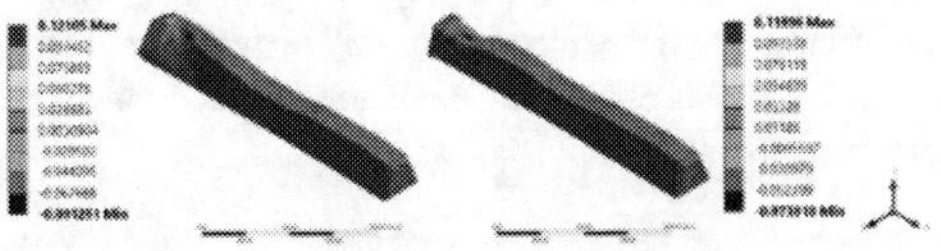

Fig. 15. Maximum principal stress of concrete railroad sleeper at point 1

According to FEM result can be following as shown in table 4.

Table 4. Total of model

Part	Result	Load point 1	Load point 2	Load point 3
Rubber	Y axis deflection (mm)	27.332	22.614	19.185
	Maximum equivalent strain (MPA)	0.19012	0.56379	0.33609
Metal pad	Deformation (mm)	-	0.53134	0.42398
		-	0.50367	0.42004
	Von-Mises Stress (Mpa)	-	88.865	72.458
		-	83.935	69.67
Concrete railroad sleeper	Deformation (mm)	0.00145	0.00859	0.00711
		0.0158	0.00943	0.00739
	Principle Stress (Mpa)	0.11896	0.9767	1.1749
		0.12105	1.62	1.34

From table 4 can use the maximum value to calculate safety factor.

Table 5. Safety factor of model

Part	Load point 1	Load point 2	Load point 3
Rubber	27.03	9.12	15.29
Metal pad	-	2.81	3.45
Concrete railroad sleeper	41.31	3.09	3.73

4 Conclusions

The rubber has suitable properties and can be used in real production the external rubber panel deflection is 27.332 mm. For internal rubber panels maximum deflection is 22.614 mm from loading at point 2. Safety factor of external rubber panel is 27.03. Safety factor of internal rubber panels are 9.12 and 15.29. The deformation according to metal pads that support below rubber panels. Metal pads also deflection from load, but the deformation of metal pad occurs in elasticity period. The concrete railroad sleeper very small section. From the analysis result can be estimated that the model can be used.

5 References

[1] S. Kavin, Development of rail pads from chloroprene rubber and natural rubber blend, 110 (2005)

[2] O. Uwe, P. Martin, S. Eck, W. Daves, Validation of a finite element crossing model using measurements at an instrumented turnout (2013)

[3] K. Yuttakit, Manufacturing and testing concrete railway sleeper manual (2014)

[4] TIS 2667-2015, Rail way rubber standard (2015)

[5] ISO 37 international standard "Rubber, vulcanized or thermoplastic - Determination of tensile stress- strain properties" (2011)

[6] ISO 2285 international standard "Rubber, vulcanized or thermoplastic - Determination of tension set under constant elongation, and of tension set, elongation and creep under constant tensile load" (2013)

[7] ISO 7743 international standard "Rubber, vulcanized or thermoplastic - Determination of compression stress-strain properties" (2011)

[8] The state railway of Thailand, NSTDA, KMUTT, NRCT, The Knowledge Accumulation of Rail Transportation: Case Study of Infrastructure- Basic railway (2015)

[9] The state railway of Thailand, Specification for the supply of rubber solepad. (1982)

An exploratory factor analysis of government construction procurement problems

Narong Leungbootnak[1,*], *Tanayut* Chaithongrat[2], and *Preenithi* Aksorn[3]

[1]Department of Civil Engineering, Faculty of Engineering, Khon Kaen University, Khon Kaen, Thailand
[2]Faculty of Architecture, Urban Design and Creative Arts, Mahasarakham University, Mahasarakham, Thailand
[3]Faculty of Architecture, Khon Kaen University, Khon Kaen, Thailand

Abstract. This study aims to apply Exploratory Factor Analysis in government construction procurement problems. The questionnaire used in this study to collect data with is the Cronbach's Alpha Coefficient equal to 0.986. The data was collected through Web Survey and 353 participants completed the questionnaires. The data was analysed with the use of percentages, mean, standard deviation, and Exploratory Factor Analysis. When ranking the effect of each component on government construction procurement problems, it was found that the most influential component is procurement process problems, followed by internal and external influence problems, and project management and technical problems, respectively. The top 3 detailed problems on government construction procurement were: 1) Problems arising from fix announcement period, 2) Restrictions according to the regulations that all agencies need to send the announcement details by the approval date of procurement, and 3) Problems arising from fix announcement periods, especially associated to mandatory plans.

1 Introduction

Government procurement is very important to a country's development because the government must spend the budget for materials to provide public goods such as education, security, facilities, basic infrastructure, safety, etc. It is argued that government procurement is essential to the delivery public services and the operation of government agencies [1]. Therefore, the governments spend budget, is limited solely by the effectiveness of the procurement analysis. However, effective procurement does not necessarily mean buying at the lowest price. Instead, the procurement depends primarily on the objective to help develop industry and technology in the country [2].

In many countries, procurement has developed several aspects rapidly to support the growth of domestic and global economy. Procurement is important as it aids basic and supporting factors of public and private sectors. As a result, the public procurement must adapt and develop rapidly to increase its flexibility. Due to the changes in missions among public sector, private sector, and civil society, the bureaucracy system becomes smaller, but more flexible and effective which makes the public and private processes quicker, more transparent, and more efficient [3]. Thus, government can strengthen the private sector so that both sectors can develop the country, and achieve sustainable growth [4]. Therefore, in the era of technology which plays a main role in every organization, both public and private sectors need to rely on information technology because ample, and rapid up-to-date information leads to more efficient work [5].

In each country, the expenditure related to government procurement accounts for 10-15% of Gross Domestic Product (GDP). Moreover, it is reported that the budget for construction projects is about 2,000 million dollars per year [6]. However, nowadays, the government procurement still cannot meet nor solve effectively the needs of stakeholders. It can be seen from the research and media as a reflection of several procurement problems such as corruption within the government construction procurement projects, procurement cost due to higher than actual cost, the cronyism resulting in poor quality of work, additional expenses or budget losses, and/or poor overpriced materials [5],[7-8].

Therefore, this study aims to examine the components of government construction procurement problems. The public organizations were selected to be samples and investigated using Exploratory Factor Analysis. Moreover, the relationship model was developed by using statistical software to analyse these factors.

2 Government Procurement Problems

From the literature reviewed, there were 4 main government procurement problems. The first problem is caused by government officials including both unintended and intended offenses acted by the officials seeking illegal benefits [9]. The second problem arises

* Corresponding author: Narongl.fec@gmail.com

from poor planning or not having a plan, unclear work specification details which were not defined in advance nor during the procurement process [10-13] identified the causes of the procurement problems such as execution plan, procurement, tender, contract audit and compliance. The third problem arises from complex rules, regulations, and official letters. The last problem is caused by the corruptions. Based on Tanayut C., Narong L., Preenithi A. & Patrick M. (2017) [14]., it was found that some government officials were purposely dishonest in seeking benefits from the procurement process, which included violating, distorting, or avoiding regulations and/or laws [15].

3 Methodology

This section presents the methodology used in this study as follows.

3.1 Population and Sample

The population in this study consisted of 3,025 government construction procurement projects, observed from January 2014 to January 2015. 353 projects were selected to be samples by adopting the sampling method introduced by Yamane at the significant level of 95%. The sample size was defined by considering the basic requirements of Factor Analysis which the acceptable sample size is not less than 300 [16]. Moreover, the samples were randomly selected by adopting Systematic Random Sampling approach [21].

3.2 Instrumentation and Validity Test

To create and validate the questionnaire, the researcher defined 45 questions from the review of literature and relevant research. The questionnaire was designed as Likert scale consisting 5 levels which are 5-Strongly agree, 4-Agree, 3-Undecided, 2-Disagree, and 1-Strongly disagree, respectively. After that, 5 experts considered the content validity of the questionnaire. According to the experts, only 40 questions were passed, the criteria since the mean scores were between 0.50 to 1.00. After that, the researcher improved the questionnaire and sampled out 30 participants who were not the selected as part of the test model. To test the accuracy of the questionnaire, Cronbach's Alpha Coefficient (α) of the questionnaire was determined. Consequently, the coefficient (α) was 0.986 indicating that the questionnaire was highly reliable.

3.3 Data Collection

The researcher collected the data through Web Survey [21] and cooperated with The Comptroller General's Department to disseminate the questionnaires on the website (www.gprocurement.go.th), and sent letters to government agencies selected as samples to participate in the online survey between April to May 2016. As a result, 353 participants completed and returned the questionnaires accounting for 100% response rate. After the results from the survey were analyzed using statistical technique.

3.4 Data Analysis

SPSS from Microsoft Windows, a statistical software application was used for statistical analysis. The analysis was divided into three levels Univariate analysis, Bivariate analysis, and Multivariate analysis. Univariate analysis consists of percentage, mean, standard deviation, skewness and kurtosis to examine the distribution of the government construction procurement problems, and to present the general characteristics of the samples. Bivariate analysis consists of correlation and Pearson Product Moment to examine correlation coefficient between the variables by conducting Bartlett's Test of Sphericity. It was found that the matrix was not an identity matrix which indicated the results could be used for further analysis, considering the variables with coefficient 0.3-0.9 [16]. Lastly, Multivariate analysis was conducted with the use of Exploratory Factor Analysis. The Principal Component Factor was selected to be use with Orthogonal rotation and Varimax method. Factor loadings higher than 0.40 were focused and labelled [16].

4 Results

This section presents the methodology used in this study as follows.

4.1 General Information of the Respondents

As presented in Table 1, most of the respondents were male (54.4%), aged between 41-50 years (33.7%), and have a Bachelor degree (63.74%). Furthermore, most of participants are the head of supplies or supplies officer (46.7%), followed by procurement auditor (22.9%). Most of the respondents have experienced in the procurement of construction projects for 9.5 years on average, with 7.6 standard deviation. Moreover, most of the projects belong to local government (26.5%), followed by general government (34.6%). 38% of the projects are in the central part of Thailand. It is also found that the special procurement is adopted most (34.6%), followed by tender (25.8%), and e-bidding (24.4%), respectively. Most of the projects are building construction (35.1%), followed by road construction (34%).

4.2 Results of Exploratory Factor Analysis

According to the results, it was found that KOM (Kaiser-Meyer-Olkin Measure of Sampling Adequacy) is equal to 0.971 which is very close to 1. This indicates that the data are suitable for Factor Analysis. From Bartlett's test of Sphericity, if found that Chi-square is equal to 15249.527 with p-value less than 0.01 meaning that the correlation matrix is not an identity matrix. Therefore, it can be concluded that all 38 variables are correlated to each other and suitable for Factor Analysis (Table 1).

Table 1. Validity test of correlation matrix between variables.

Variable	KMO		Bartlett's Test	
	Ideal Value	Actual Value	Ideal Value	Actual Value
40 variables	> 0.50 (> 0.9 = Very good)	0.971	p-value < 0.05	< 0.01

Table 2. The number of components, Eigenvalues, percent and cumulative percent of Eigenvalues of government construction procurement problems.

Component	No. of Items	Eigen values	%	Cumulative %
1	13	22.340	65.704	65.706
2	14	2.298	6.758	72.464
3	9	1.112	3.269	75.733

Table 2 presents the number of components, Eigenvalues, percent and cumulative percent of Eigenvalues of government construction procurement problems. All 3 components can explain the total variance which is equal to 75.733%. Each of the 3 components of government construction procurement problems consists of at least 3 variables, so the components satisfied the rule in which each component must contain at least 3 variables. Furthermore, each factor loading variable is greater than 0.4 [16] indicating that all 3 components with 40 variables satisfied the Factor Analysis rule.

Table 3-5, it was found that Component 1, consisting of 13 variables, has a factor loading between 0.548 – 0.802. Eigenvalue is 22.340, and the percent of variance is 65.706. Therefore, it is indicated that 13 variables are the best predictors from component 1. When comparing Eigenvalue with other components, this component is the most important since it has the highest Eigenvalue. Consequently, this component is named as 'Procurement process problems'.

Table 3. Component 1: Procurement process problems.

Variable	Text	Factor Loading
PP1	Officers lack the skills to draft TORs and contracts.	0.736
PP2	Officers lack knowledge of the supplies and other regulations related to procurement.	0.764
PP3	Some officers are worried that the information will leak.	0.707
PP4	There are no clear criteria for defining performance measures.	0.707
PP5	Officers lack precision and are afraid of rules and regulations.	0.749
PP6	Problems arising from fix announcement periods, especially associated to mandatory plans.	0.787
PP7	Problems arising from fix announcement period.	0.802

Table 3. (cont.)

Variable	Text	Factor Loading
PP8	Restrictions according to the regulations that all agencies need to send the announcement details by the approval date of procurement.	0.794
PP9	Problems arising from unclear clarification.	0.735
PP10	Problems arising from inadequate time for pricing.	0.669
PP11	Problems arising from defining the mid-price, incomplete pricing, no allowance for loss, and inappropriate mid-price.	0.623
PP13	Problems arising from referring to the mid-price defined by the Comptroller General's Department in which the price material is not updated.	0.548
PP15	Lack of accurate procurement planning	0.568
13 factors	**Eigenvalue**	**22.340**
	Percent of variance	**65.706**

Table 4. Component 2: Internal and External Influence Problems

Variable	Text	Factor Loading
PP26	Approved construction budget is not enough.	0.515
PP27	Changing executive affects, changes in utility and work progress.	0.617
PP28	Problems arising from complex workflows.	0.518
PP29	Officers have negative attitude toward their colleagues. Problems from different opinions and conflicts within the workplace.	0.615
PP31	Problems arising from political impact.	0.735
PP32	Problems arising from external audits such as audit agencies and NGOs.	0.672
PP33	Problems arising from contractors not participating in biding or leaving their jobs.	0.666
PP34	Problems arising from bidders' complaints in procurement such as unclear announcement.	0.816
PP35	Problems arising from community complaints.	0.875
PP36	Political and social problems such as protest.	0.887
PP37	Economic problems such as rising material prices.	0.715
PP38	Problems from poor contractors who bid lower than what the mid-price actually is.	0.646
14 factors	**Eigenvalue**	**2.298**
	Percent of variance	**6.758**

Table 5. Component 2: Internal and External Influence Problems

Variable	Text	Factor Loading
PP16	Problems arising from delayed work.	0.679
PP17	Problems arising from insufficient number of procurement officers.	0.664
PP18	Problems arising from officers' potential.	0.717
PP19	Problems arising from working with several difficult functions.	0.721
PP20	Problems arising from coordinated errors.	0.689
PP21	Problems of mismatch between construction models, and incomplete pricing.	0.633
PP23	Problems arising from time conflicts.	0.555
PP24	Problems arising from poor management.	0.629
PP25	Problems arising from different working quality on each function.	0.666
9 factors	**Eigenvalue**	**1.112**
	Percent of variance	**3.269**

These components of problems are in consistent with the government construction procurement problems proposed by Dzuke & Naude [1] stating that the components are procurement planning, bidding, and contract audit and compliance. Also, the concept of the procurement problem is in consistent with Gardenal [12] stating that the components of procurement problems include supply planning and procurement process.

5 Conclusions

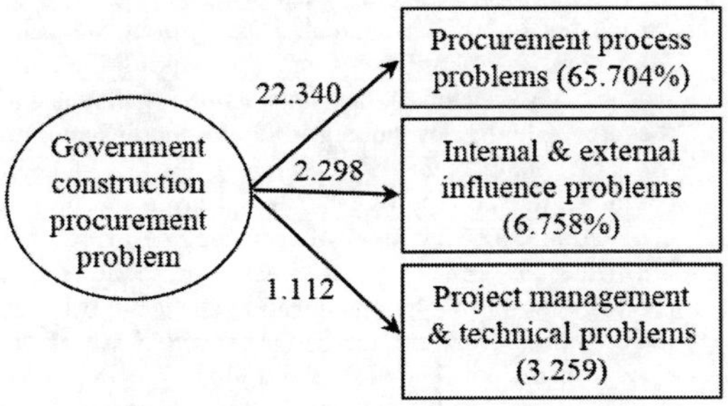

Fig. 1. Summary of component analysis of government construction procurement problems.

According to the factor analysis, there are 3 components affecting government construction procurement projects as presented in Figure 1. When ranking the effect of each component on government construction procurement problem, it is found that the most influential component is procurement process problems accounting for 65.704% in explaining the problem with 13 observed variables. Internal and external influence problems is the second influential component accounting for 6.758% in explaining the problem with 14 observed variables. Lastly, the third influential component is project management and technical problems accounting for 3.269% in explaining the problem with 9 observed variables.

References

1. Dzuke, A. & Naude, M.J.A., JTSCM, **9(1)**, 1–9 (2015)
2. Wathanam, T., *"WTO agreement on government procurement: GPA"*, TRF, Bangkok, Thailand (2007)
3. NESDB, *"The Twelfth National Economic and Social Development Plan, 2017–2021"*, Retrieved 10 October 2017, (2017)
4. Walker, H. & Brammer, S., Supply Chain Management: An International Journal, **14 (2)**, 128-137, (2017)
5. Du, J., Jiao,Y.Y., Jiao, R.J., Kumar, A. & Ma, M., Proc IEEE IEEM, 1337-1341, (2007)
6. Pimsi, A., PhD Thesis. King Mongkut's University of Technology North Bangkok, (2005)
7. Hazra, J., IJSTM, **7(5)**, 1364-1368, (2011)
8. McKevitt, D & Davis, P., *IJPSM,* **26(6)**, 469-480, (2013)
9. Licenji, A., ESJ, **11(25)**, 229–240, (2015)
10. Pesämaa, O., Eriksson, P.E. & Hair, J. F., IPMA, **27(6)**, 552-559, (2008)
11. Ameyaw, C., Mensah, S. & Osei-Tutu, E., IJCSCM, **2(2)**, 55–65, (2012)
12. Gardenal, F., JOPP, **13(2)**, 215-242, (2013)
13. Dzuke, A. & Naude, M.J.A., JTSCM, **9(1)**, 1–9, (2015).
14. Tanayut C., Narong L., Preenithi A. & Patrick M., ITJEMAST, **8(3)**, 221-231, (2017)
15. Dza, M., Fisher, R., & Gapp, R., PAR, **2(2)**, 49–57, (2013)
16. Field, A., *"Discovering statistics using IBM SPSS statistics"*, SAGE Publications, London, UK, (2013)
17. Chadcham, S. & Kornpeipanee, S., Journal of Educational Research and Measurement, **1(2)**, 1-24, (2003)
18. Kim, J.O., & Mueller, C.W., *"Factor analysis: Statistical Methods and Practical Issues"*, Beverly Hills, CA: Sage, (1978)
19. Hooper, D., Coughlan, J., Mullen, M., EJBRM, **6(1)**, 53-60, (2008)
20. Fabrigar, L.R., Wegener, D.T., MacCallum, R.C., & Strahan, E.J., Psychological Methods, **4 (3)**, 272-299, (1999)
21. Neuman, W.L. *"Social research methods: qualitative and quantitative approaches"*, Printed in United States of America, (2006)

The Karen villagers' knowledge development in road construction: a case study of Panoki –Kopadoh (PNK) route

Uba Sirikaew [1*], *Songklod* Sae-Ueng[1], *Somkiat* Khwanpruk[1], and *Thanom* Sriworasa[4]

[1]Civil Engineering Department, Faculty of Engineering, King Mongkut's Institute of Technology Ladkrabang, Bangkok, Thailand

Abstract. The Karen, of Panoki village agreed to upgrade the unpaved PNK route to a bamboo reinforced concrete road which used money from the donation. At the beginning of the road construction project, 2010, there were many problems concerned with construction technique and accidents. Civil Engineering Department, Engineering Faculty, King Mongkut's Institute of Technology Ladkrabang (KMITL) was invited to be a road construction consultant of Panoki-Kopadoh (PNK) route during 2012 to 2017. As the villages of Panoki in Tak province and Kopadoh in Chiang Mai province are in Thailand's northern remote mountainous areas connected by a 16.5 km unpaved narrow track of Panoki-Kopadoh (PNK) route. In fact, the majority of volunteer villagers understand only the Karen language, lack the basic construction skills, and show no regard for personal safety and hygiene. The consulting team thus provided a training program prior to construction and assisted with technical advice since the onset of the project. As an achievement of the project depends on people participation, this research is to examine the pre- and post-training local workers' knowledge development and attitudes toward the academic services provided by KMITL through direct observations, a Likert-scale-based questionnaire, and in-depth interviews. The findings reveal that the participating villagers have learned to construct a bamboo reinforced concrete road and are intent on applying the knowledge to similar projects in the future. They exhibit the higher level of knowledge development following the training and also hold a positive attitude toward KMITL's academic services.

1 Introduction

The 16.5-km Panoki-Kopadoh route is an unpaved narrow track that connects Thailand's mountainous village of Panoki in Tak province with another mountainous village of Kopadho in Chiang Mai. However, in the rainy season (May-October), the unpaved local route of high plastics fine grained soil becomes very difficult for motorized vehicles to travel on. Therefore, the villagers are required to travel the distance of 53 km (almost 5 hours) through Tak's Tha Songyang district to Omkoi district of Chiang Mai, Thailand. To shorten a travel time, a bamboo reinforced concrete road has thus been laid over the existing unpaved Panoki-Kopadoh (PNK) route where the construction of the PNK route is underway, whose elevations are in a range of 200-1300 meters above mean sea level. Approximately, 14 km of the PNK route has been completed as at the end of April 2017. Apart from the reduced travel time, other benefits gained are, for example, increased trade of local agriculture products, greater access to formal education, and transportation of patients in emergency. King Mongkut's Institute of Technology Ladkrabang (KMITL)'s Civil Engineering Department was invited to take part in the project as academic services provider and construction consultant. The construction began in December 2011 and could be carried out only 2 km annually during the dry season from January to April (i.e. 4 months). During the rainy season (May – October), the construction is halted due to the area's high elevations and thus the difficulty to transport equipment and machinery to the site. In addition, the shortage of labor has delayed the project completion. Since the local villagers are of Papa Kayour hill tribe descent, they understand only the Karen language, posing the communication challenge between the consulting team and local workers who also lack the basic construction knowledge and skills. To overcome the challenges, a 3-day practical training session was arranged during the early stage of the PNK project. The training covered the topics of ground assessment, concrete and construction materials, road construction techniques and personal safety and hygiene. Techniques of communication, signs for construction and interpreters, use for Thai and Karen. The aim of this research is to investigate the villagers' knowledge

* Corresponding author: ksuba@staff.kmitl.ac.th

development and attitudes toward the training and academic services provided by KMITL. KMITL's academic service provided a long term consulting for the Karen villagers, then their knowledge development was observed for 5 years. The knowledge development of the Karen villagers were qualitative description. The representative persons of KMITL comprised of the lecturers, staffs and students with the total number of 20 people. Since the construction has been started in Tak's Panoki village of Tha Songyang district of Tak province, the representative samples were the local residents of Panoki. The collection of data by was carried out from March to May 2013 through observations and deep interviews with the villagers with help from Thai-fluent Karen interpreters.

2 Literature Review

PNK route is the concrete road which used a bamboo as a wire mesh. Because it is a natural inexpensive material that can be sought locally. There are many bamboo reinforced concrete roads in use today and their advantages and disadvantages have been widely studied. In Thailand, bamboo reinforced concrete was used as a good alternative construction material for road [1]. To solve aboved-mentioned problems of the PNK project, the Voluntary Protection Program (VPP) was reviewed and has been adopted for the PNK project [2]. The VPP has been established to promote minimum safe practices in construction as well as personal health and hygiene. Since the PNK route is a chartable project, the volunteer local workers receive no daily wage. The communication at the construction site relies on simple drawings and signs to overcome the language barrier and guarantee the Karen workers' correct understanding. Due to the constraints of communication language and time, a preliminary survey was undertaken to measure the post-training attitudes and skill development of the villagers using a Likert-scale-based questionnaire and in-depth interviews. The Likert scale is a 5-point scale with "Strongly Disagree" on the left end and "Strongly Agree" on the opposite end. In the middle is "Neither Agree nor Disagree". In general, the scores on attitudinal items on a Likert scale are relatively reliable predictors of respondents' attitudes. For measuring the knowledge development of the Karen workers, the works of Swamy and Arends et al were considered. Swamy (2007) reported that questionnaires employed to evaluate respondents' attitudes toward an issue, product or service usually contain several biased questions that result in inflated responses [3]. Arends et al (1998) thus suggested that, in addition to the Liker scale, behavioral observations are an effective mean to determine the success of training [4].

3 Study Method

Prior to the construction, the 3-day practical-based training activities were provided. Table 1 presents the training activities and their respective objectives. The main activities entail site investigation, concrete work, safety and hygiene, service and maintenance, and public relations. The training purposes are to create a correct understanding and prevent human errors in the construction. The volunteers from the temple and KMITL's engineering students and academics, totaling 20 persons, were training instructors, and there were about 120 local Karen villagers participating in the training. The participants were equally divided into five small groups, each of which was assigned to attend two classes (i.e. main activities) on the first day, then another two classes on the second day, and the remaining class on the last day. The third day was also reserved for the post-training assessment of the participants' basic construction skills. Every participating villager was required to learn about the five main activities to complete the training. The content and workshop were simplified to enable ease of understanding. The success of a road construction project depends in large part on the careful investigation and selection of proper construction site and construction materials. The aim of site investigation is to determine the quality of the sub-grade natural ground of the route. Construction materials must be inspected to assure that they are suitable and satisfy the standards for road construction.

Table 1. Activities and objectives of the 3-day training

No.	Activity	Objective
1.	Site investigation	To assess the sub-grade natural ground, sand and gravel pits, slope failure, slope protection.
2.	Concrete work and construction technique	To prepare the construction materials, concrete mixing, curing and testing.
3.	Safety and hygiene	To understand and adopt traffic control, safety wares, e.g. mask, earplugs, helmet, safety boots.
4.	Service & maintenance	To learn about refreshment provision, transportation of construction materials, maintenance of equipment and tools, service during work.
5.	Public relations	To build a rapport with concerned individuals and organizations.

At the construction site, the drawings and signposts were improvised and emphasize quick and correct understanding of the local workers and other villagers. For instance, Fig 1 shows the sign of communication for construction which 1a is an "under construction" sign along the main road to warn motorists of danger and exercise caution. For 1B is a sign to remind the workers to perform the compaction of the sub-grade natural ground and inspection of the hill's slope stability. At the packing area, a drawing instructing of 1C sign reminds the local workers to bag 30-35 kg of aggregate with about five shovelfuls. Safety and hygiene signs show as 1D. Trainees studied and practiced the mix design of

 https://doi.org/10.1051/matecconf/201819202058

concrete. The compressive strength of a standard cylindrical concrete sample is 210 ksc with the density of 2,400 kg/cc. The required slump value of concrete shows in the range of 8 to 12 cm. For 1 cubic meter of concrete, the proportion of water, cement, coarse aggregate and fine aggregate presents as 143 kg, 257 kg, 1,240 kg and 752 kg, respectively.

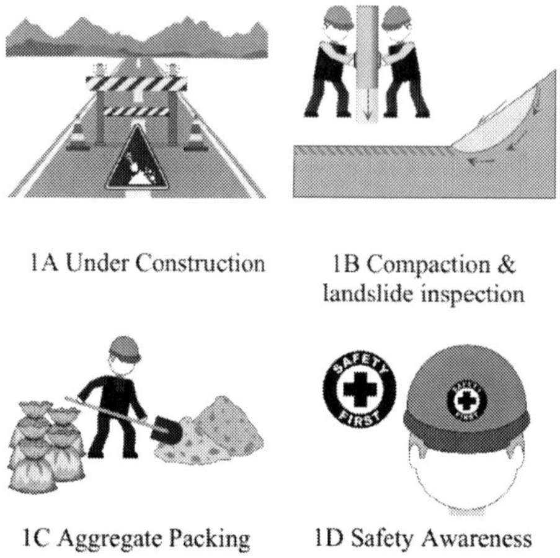

Fig.1 Drawing Signs at the Construction Site

Direct observations, in-depth interviews and the Likert-scale questionnaire were employed to the post-training collective skill development and attitudes of the villagers. The workers' skill development includes their understanding of work regulations and system, application of knowledge, participation and leadership, while the villagers' attitudes cover self-esteem and overall satisfaction toward the academic services offered by KMITL. The in-depth interview sessions were conducted with the help of Thai-fluent Karen interpreters. During March to May 2013, the direct observations were carried out during the training sessions and construction. The analysis of the data was carried out and the findings are of qualitative nature.

4 Result and Discussion

4.1 Description of the Sample

The population of Thailand's northern village of Panoki is around 300, constituting 80 households. The village has one day care center but no electricity and tap water. A total of 120 villagers were randomly sampled, equally split between male and female. The largest age group was 20-30 years old, accounting for 43% of the total. The interviews revealed that a majority of villagers did not seek employment in major cities outside their community because of the language barrier and parental responsibility. As shown in Fig.2, around 20% of the sample villagers are able to communicate in Thai, 11% of whom fluently and 9% somewhat fluently, whereas the rest (80%) barely understand the Thai language. Nearly 90% of the samples receive no formal schooling. However, some of the sample villagers could converse in Thai fluently as a result of their occupations, e.g. local civil servants, traders and students. Most of village people have no formal education and work as farmers. It was discovered during interviews that the village heads, medical service personnel and government officers are also regarded as villagers' opinion leaders. The villagers' belief in the fruits of good karma from good deeds has led them to offer manual labor to the construction project, anticipating no financial reward.

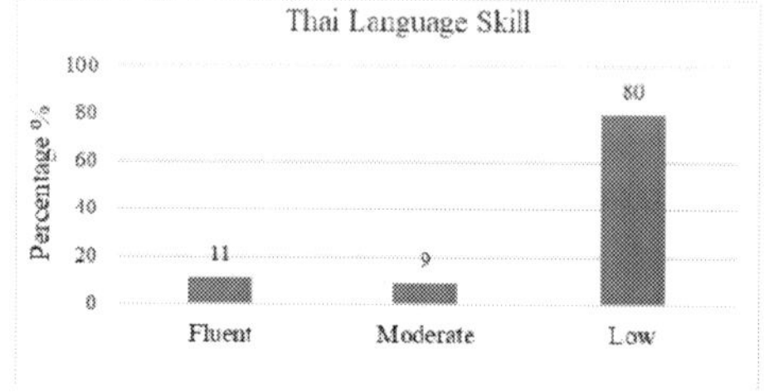

Fig.2 Thai language skill of Panoki villagers

4.2 Villagers' Knowledge Development

The villagers' post-training knowledge development was measured through direct observations and in-depth interviews to determine their attitudes. The measurements covered their understanding of the work process, application of knowledge, and leadership, all of which were rated "strongly agree" by the villagers. Fig. 3 to Fig.5 show more than 79% of the respondents gave a "strongly agree" rating with regards to their knowledge development and self-improvement after the training. By comparison, prior to the training, their attitudes were mostly either "undecided" or "disagree" on the Likert scale. The in-depth interviews revealed that most villagers gave a "strongly agree" rating for the 3-day training because of its hands-on nature whereby the villagers were encouraged to perform actual work. The hands-on training method was adopted to guarantee a correct theoretical and practical understanding of road construction. The villagers were very resourceful in locating the sources of raw materials and water. The villagers' post-training attitudes with regard to participation and overall satisfaction toward KMITL's.

However, prior to the training, the majority of villagers rated "undecided" and "disagree", the finding which was attributable to mutual distrust as the consulting team was perceived as strangers and the language barrier. It was discovered that local Karen villagers exhibit more trust in people of Karen lineage than in those of different descent. The training and collaboration in the PNK project helped bridge an understanding between the KMITL team and local villagers. The in-depth interviews revealed that those who volunteered in the PNK project hold a belief that, according to Buddhist canons, they were doing good deeds and accumulating good karma. A handful of villagers however were unable to contribute manual labor since they have to seek employment in other larger towns for economic reason. This minority group was found to give either an "undecided" or "disagree" rating to the project. Some of those who were against the project viewed the road would ruin the community's existing good cultures and replace them with materialism and indebtedness. The construction project was faced with insufficient funding, barring the hiring of skilled construction labor to complete the project. Thus, active participation and labor contribution from local villagers are crucial to the success of the project. This can only be realized through a good public relations campaign where word-of-mouth has been used to persuade the local residents to contribute to the project. The technical training and field practices given by KMITL for 5 years to the Karen trainees, brought their knowledge and community development. Concrete technology, road construction technique, management and safety and hygiene can be applied in their daily lives. Although, there was no budget for fiscal year 2017 to support PNK project.

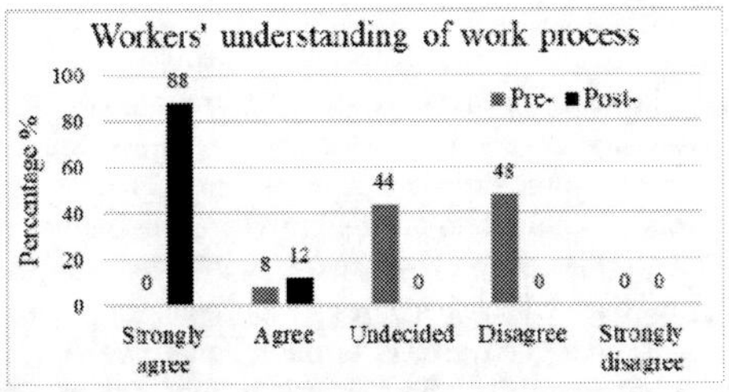

Fig.3 Pre- and Post-training Attitudes of Villagers For Understanding of Work Process

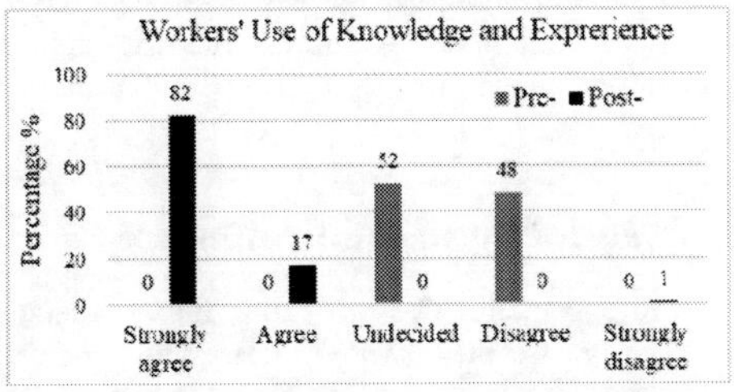

Fig. 4 Pre- and Post-training Attitudes of Villagers For Knowledge and Experiences

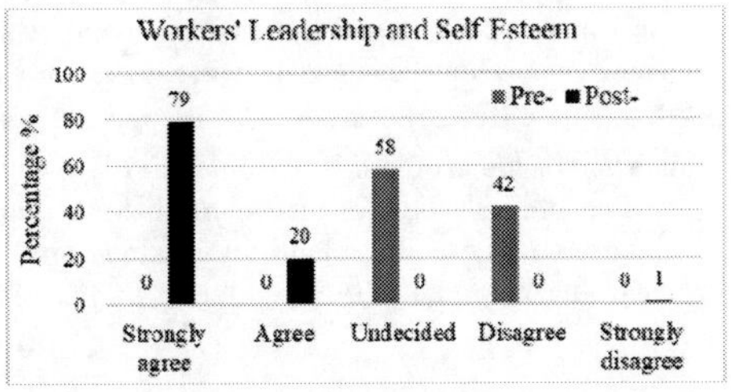

Fig.5 Pre- and Post-training Attitudes of Villagers For Leadership and Self-Esteem

5 Summary

In the construction of the PNK route, the 3-day training was one solution to overcome the obstacles, i.e. lack of skills, distrust and language barrier, and also crucial to the success of the project. According to the qualitative research findings, KMITL has been favored by the people of Panoki village to provide the training and support for the project. In addition, this research has revealed the following interesting facts with regard to the local residents of Panoki village:

(1) After a good rapport was established, the villagers were more open to share information about their life and livelihoods. They have also formed a positive attitude toward the academic services of KMITL and expressed their intention to seek consultation from the university in the future.

(2) The villagers have shown improvement in their work skills and exercised greater care of personal health and hygiene, especially their knowledge development on a construction of bamboo reinforced concrete road.

(3) With the future completion of this road, more economic development and thus materialism will find ways to their doorsteps, a phenomenon that would undesirably alter their life and livelihoods. Moreover, it is apparent that the cultivation of corn plants is increasing with the advance of the road. The more the corn cultivation areas, the severer the deforestation in the area. To preserve their cultures and ways of life as well as conserve the natural resources, future quantitative research has been planned for the village, e.g. quality of the PNK concrete road, road safety and maintenance.

(4) The observation data on behaviors and attitudes of the villagers could be used in redesigning the academic curricula to better suit the cultures and requirements of Karen children and youths.

The authors would like to express deep gratitude to Mr. Pawadee, Head of Panoki Village, and to students from Hui Nokkok School for their translation, without whom a good understanding and close rapport between the consulting team and local workers would not have established.

References

1. I. Yordyiam and N. Weeraphoprasit, *The study of Behavior of Bamboo-Reinforced Concrete Slab to flexural load.* Thesis of civil engineering, Burapha University, Chonburi, Thailand. (2007) (in Thai)

2. K. Maleesee, Concrete Technology for Villagers of Prabhudhabat Huai Tohm Temple, *document published in Prabhudhabat Huai Tohm Temple*. 46 p., (2008) (in Thai)

3. S. Swamy, How Should You Frame Questions to Measure User Attitudes Accurately? An Experimental Design Study. *UI-HCII 2007 Proc,II, China*, p. 496-505 (2007)

4. R. I. Arends, N. E. Winitzky and M.D.Tannenbaum, *Exploring Teaching*, McGraw-Hill, 443-451 p. (1998)

Hydromagnetic rotating flow of Casson fluid in Darcy-Forchheimer porous medium

Gauri Shanker Seth[1], and *Prashanta Kumar* Mandal[1,*]

[1] Department of Applied Mathematics, Indian Institute of Technology (Indian School of Mines), Dhanbad-826004, India

Abstract : Present study investigates three dimensional rotating flow of Casson fluid in the presence of magnetic field over a convectively heated linear stretching sheet. Concept of nonlinear radiative heat transfer is considered. The governing nonlinear partial differential equations are converted into ordinary differential equations with the help of similarity transformation and then solved by using shooting method along with Runge-Kutta-Fehlberg integration technique. The primary and secondary velocities and temperature profiles are plotted and analysed corresponding to various pertinent flow parameters. Also, the skin friction for both directions and rate of heat transfer at the surface are computed and explained.

1 Introduction

The effects of thermal radiation have massive impacts in industrial and engineering fields where devices are working at very high temperature under various non-isothermal conditions. Heat transfer via radiation is frequently applied in nuclear power plants, gas turbines, missiles and supersonic planes. Result for both small and large differences between fluid temperature at the sheet and free stream temperature can be obtained by considering nonlinear Rosseland diffusion approximation. The idea of nonlinear radiative heat transfer is presented by many researchers [1, 2].

Almost all industrial fluids disobey Newtonian law of viscosity. Therefore, study of non-Newtonian fluid flow is of supreme importance due to their inevitable presence in petrochemical industries, food and beverage processing and biotechnology. Casson fluid is one of the non-Newtonian fluid whose characteristics incluide human blood, jelly, honey, fruit juice with fibres etc. This type of fluid may find applications in medical and engineering sciences. Keeping in view the application of such flow, many researchers studied the flow of Casson fluid considering different aspects of the problem [3-6].

Rotating flow plays vital role in geophysics, astrophysics and fluid engineering. The engineering applications of rotating flows consist of filtration process based on centrifugal force, turbines, material processing and rotatary MHD generators. Wang [7] analysed viscous fluid flow near stretching surface in a rotating environment. Several researchers investigated the characteristics of flow behaviour considering rotation [8-10].

Flow of fluid through porous medium finds many applications in hydrology, agriculture, civil and petroleum engineering. In most of the cases, modeling and analysis of flow through porous medium are based on classical Darcy's theory, which is only valid for smaller porosity and lower velocity. Thus, at higher flow rates and higher permeability porous media, there is a departure from classical Darcy's law and in such a case inertial effect becomes important. Forchheimer [11] analysed such effects through square velocity vector in addition to Darcian velocity in momentum equation. A lot of researchers studied flow through porous medium considering Darcy-Forchheimer model [12-14].

Present investigation deals with the study of impact of rotation and non-linear thermal radiation over hydromagnetic flow of Casson fluid near a stretching sheet.

2 Modeling of the problem

Hydromagnetic Casson fluid flow generated by a linearly stretched sheet in Darcy-Forchheimer porous medium is considered. The sheet is aligned with the xy-plane. The stretching velocity of the surface in the x-direction is $u_w = ax$, where $a(>0)$ is the stretching rate. Fluid and sheet rotate in unison with uniform angular velocity Ω about z-axis in which direction magnetic field of uniform strength B_0 is applied. Schematic diagram of the problem is presented in Fig. 1.

The conservation of mass, momentum and energy equations (using Rosseland approximations for non-linear thermal radiation) in a rotating frame are given by

$$\frac{\partial u}{\partial x}+\frac{\partial v}{\partial y}+\frac{\partial w}{\partial z}=0, \qquad (1)$$

*Corresponding author: ism.prashanta@gmail.com

$$u\frac{\partial u}{\partial x}+v\frac{\partial u}{\partial y}+w\frac{\partial u}{\partial z}-2\Omega v=\upsilon\{1+(1/\beta)\}\frac{\partial^2 u}{\partial z^2}-\frac{\sigma}{\rho}B_0^2u$$
$$-\{1+(1/\beta)\}\frac{\upsilon}{K^*}u-\left(C_b/\sqrt{K^*}\right)u^2, \quad (2)$$

$$u\frac{\partial v}{\partial x}+v\frac{\partial v}{\partial y}+w\frac{\partial v}{\partial z}+2\Omega u=\upsilon\{1+(1/\beta)\}\frac{\partial^2 v}{\partial z^2}-\frac{\sigma}{\rho}B_0^2v$$
$$-\{1+(1/\beta)\}\frac{\upsilon}{K^*}v-\left(C_b/\sqrt{K^*}\right)v^2, \quad (3)$$

$$u\frac{\partial T}{\partial x}+v\frac{\partial T}{\partial y}+w\frac{\partial T}{\partial z}=\frac{\partial}{\partial z}\left[\left(\alpha+\frac{16\sigma^*T^3}{3(\rho C_p)k^*}\right)\frac{\partial T}{\partial z}\right]. \quad (4)$$

The booundary conditions to be satisfied are given by

$$u=u_w=ax,\ v=0,\ w=0,\ -k\frac{\partial T}{\partial z}=h_f\left(T_f-T\right)$$
$$\text{at} \quad z=0, \quad (5a)$$

$$u\to 0,\ v\to 0,\ T\to T_\infty \ \text{ as } \ z\to\infty, \quad (5b)$$

where,
$u,\ v,\ w,\ \rho,\ \upsilon,\ \sigma,\ \beta,\ C_b,\ K^*,\ T,\ T_f,\ T_\infty,\ \alpha,\ \sigma^*,\ k,\ C_p,\ k^*$ and h_f are, respectively, the components of velocity along the directions of x, y, z - axis, fluid density, kinematic coefficient of viscosity, electrical conductivity, Casson parameter, the drag coefficient, permeability of the porous medium, fluid temperature, temperature of the opposite side of the sheet, free stream temperature, thermal diffusivity, Stefan-Boltzmann constant, thermal conductivity of the fluid, specific heat at constant pressure, mean absorption coefficient and coefficient of heat transfer.

Now, we use the following similarity transformations :

$$\left.\begin{aligned}&\eta=\sqrt{(a/\upsilon)}\,z,\ \ u=axf'(\eta),\ \ w=-\sqrt{a\upsilon}\,f(\eta),\\&v=axg(\eta),\ \theta(\eta)=(T-T_\infty)/(T_f-T_\infty).\end{aligned}\right\}\ (6)$$

With the help of eq. (6), eqs. (2) to (4) become

$$\{1+(1/\beta)\}f'''+ff''+2\lambda g-\left[M^2+K\{1+(1/\beta)\}\right]f'$$
$$-(1+F_r)f'^2=0, \quad (7)$$

$$\{1+(1/\beta)\}g''-2\lambda f'-\left[M^2+K\{1+(1/\beta)\}\right]g$$
$$+(fg'-f'g)-F_rg^2=0, \quad (8)$$

$$\left(\left(1+Rd\left(1+(\theta_w-1)\theta\right)^3\right)\theta'\right)'+\Pr f\theta'=0, \quad (9)$$

Non-dimensional form of eqs. (5a) and (5b) are

$$\left.\begin{aligned}&f(\eta)=0,\ g(\eta)=0,\ f'(\eta)=1,\\&\theta'(\eta)=-\gamma\{1-\theta(\eta)\} \qquad \text{at } \eta=0,\\&g(\eta)\to 0,\ f'(\eta)\to 0,\ \theta(\eta)\to 0\ \ as\ \eta\to\infty,\end{aligned}\right\}(10)$$

where, $\lambda=\Omega/a$, $M^2=\left(\sigma B_0^2\right)/(\rho a)$, $K=\upsilon/(aK^*)$, $\Pr=(\rho\upsilon C_p)/k$, $F_r=(C_b x)/\left(\sqrt{K^*}\right)$, $\theta_w=T_f/T_\infty(>1)$, $Rd=\left(16\sigma^*T_\infty^3\right)/(3kk^*)$, $\gamma=(h_f/k)\sqrt{(\upsilon/a)}$ are rotation parameter, magnetic parameter, porosity parameter, Prandtl number, local inertia coefficient, temperature ratio parameter, radiation parameter and Biot number.

In view of engineering and practical importance, significant physical entities of flow-field and temperature field are presented in terms of skin friction coefficients C_{fx}, C_{fy} and Nusselt number Nu_x as

$$\sqrt{\mathrm{Re}_x}\,C_{fx}=\{1+(1/\beta)\}f''(0),$$
$$\sqrt{\mathrm{Re}_x}\,C_{fy}=\{1+(1/\beta)\}g'(0),$$
$$Nu_x/\sqrt{\mathrm{Re}_x}=-\left[1+Rd\{1+(\theta_w-1)\theta(0)\}^3\right]\theta'(0),$$

in which $\mathrm{Re}_x=u_w x/\upsilon$ is the local Reynolds number.

3 Validation of numerical solution

Eqs. (7) to (9) are coupled non-linear differential equations which are numerically solved by making use of shooting technique via Runge-Kutta-Fehlberg method. Suitable finite value of $\eta\to\infty$ say η_∞ is taken as $\eta_\infty\approx 10$. The step size $\Delta\eta=0.001$ is issued to obtain the numerical solution. Suitable guesses values of $f''(0)$, $g'(0)$ and $\theta'(0)$ are choosen and the integration is carried out. Our results for $-f''(0)$ and $-g'(0)$ agree excellently with the numerical results presented in [7, 8] (see Table 1).

4 Discussion of results

Obtained numerical solution of fluid velocities along x and y - directions and fluid temperature profiles are plotted in Figs. 2 to 6. Default values for parameters are $M^2=\lambda=\beta=\gamma=0.5, K=0.2, F_r=0.1, Rd=2, \theta_w=1.5$ and $\Pr=6.2$ unless otherwise stated.

It is evident from Figs. 2 and 3 that both primary velocity f' and secondary velocity g decrease on increasing M^2. f' decreases on increasing λ whereas g increases near the sheet with greater values of λ. This means that magnetic field has a propensity to reduce the velocity of the fluid in both directions. Also, it is concluded that rotation reduces the primary flow and accelerates the secondary flow near the sheet which is in agreement with the characteristics of Coriolis force [15]. It is evident from Figs. 4 and 5 that both f' and g decrease on incresing K and F_r. f' decreases whereas g increases as β enlarges. Thus, porosity of the medium including inertia force has a tendency to reduce the primary and secondary fluid velocities. Casson parameter has opposite effects on primary and secondary velocities. Fig. 6 revels that, by increasing the values of of Rd, θ_w and γ, an enhancement in the temperature field θ is observed. Also, the curves become S - shaped for large values of temperature ratio. Increasing values of θ_w indicates larger surface temperature on opposite side of sheet compared to the free stream temperature.

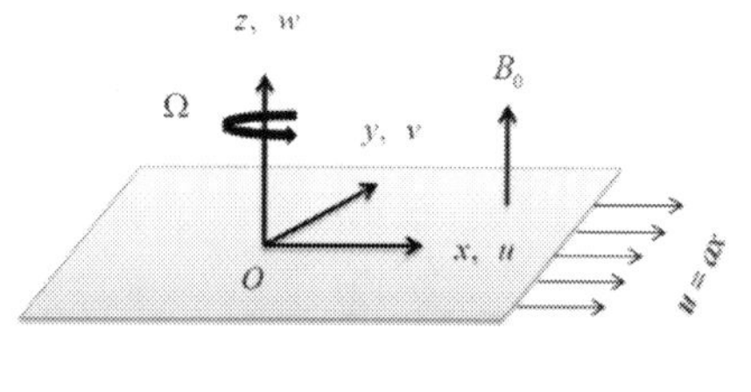

Fig. 1. Schematic diagram of the problem.

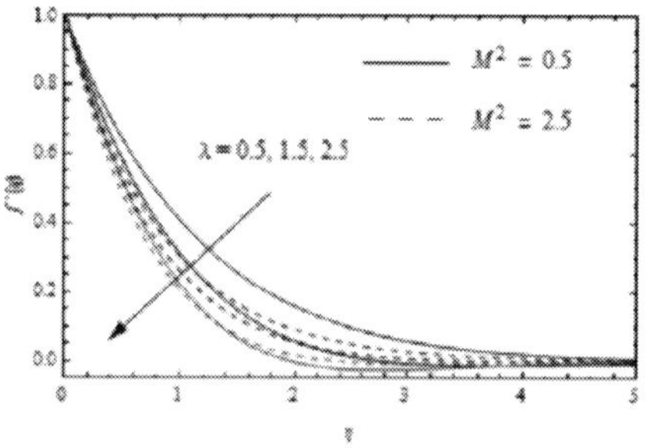

Fig. 2. Primary velocity profiles for different values of M^2 and λ.

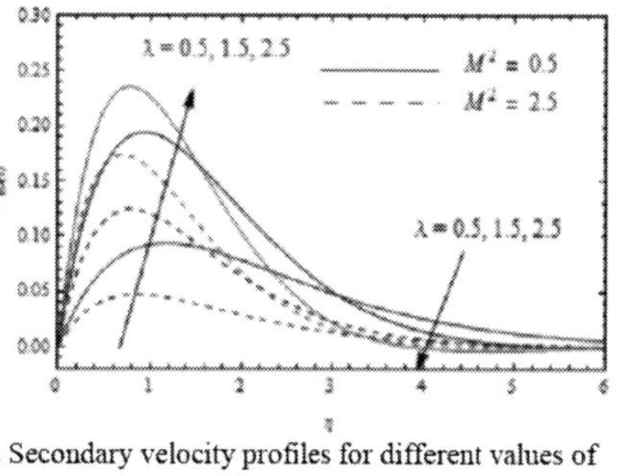

Fig. 3. Secondary velocity profiles for different values of M^2 and λ.

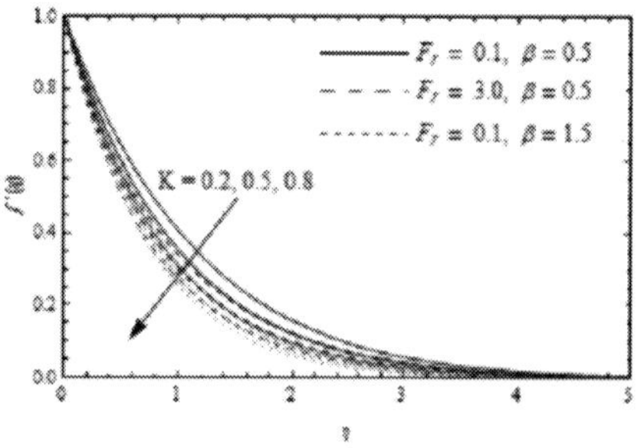

Fig. 4. Primary velocity profiles for different values of K, F_r and β.

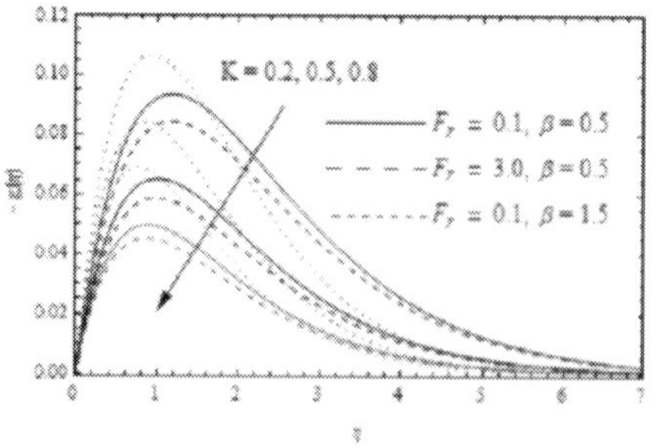

Fig. 5. Secondary velocity profiles for different values of K, F_r and β.

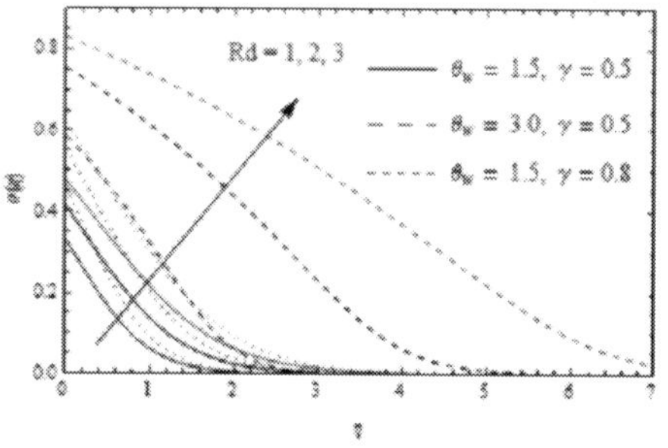

Fig. 6. Temperature profiles for different values of Rd, θ_w and γ.

Moreover, larger Biot number γ produces stronger convection which leads to higher temperature field and more thermal boundary layer thickness.

Numerical values of skin friction coefficients due to primary and secondary flows and local Nusselt number, are presented in Tables 2 and 3. It is noticed that, $\sqrt{\text{Re}_x}C_{fx}$ enhances with increasing M^2, K, λ and F_r and ruduces on increasing β, whereas there is an enhancement in $\sqrt{\text{Re}_x}C_{fy}$ due to increasing λ and reduction due to increament in β, M^2, K and F_r. Due to rising values of Rd, θ_w and γ, there is a marked enhancement in the values of $Nu_x/\sqrt{\text{Re}_x}$ whereas there is significant reduction in the values of $Nu_x/\sqrt{\text{Re}_x}$ due to the increment in λ, M^2, β, K and F_r.

5 Conclusion

The important findings of this investigation are summarised below :

- The impact of rotation and Casson parameter on primary velocity is opposite to that of secondary velocity near the stretching sheet.
- Thermal radiation, temperatute ratio and Biot number enhance the heat transfer rate whereas it attenuates when magnetic field, rotation, Casson parameter, porosity of the medium and local inertia coefficient are intesified.

Table 1. Comparison of present results with previous published results when $M^2 = K = F_r = 0$ and $\beta \to \infty$.

λ	Wang [7]		Nazar et al. [8]		Present Results	
	$-f''(0)$	$-g'(0)$	$-f''(0)$	$-g'(0)$	$-f''(0)$	$-g'(0)$
0	1	0	1	0	1	0
0.5	1.1384	0.5128	1.1384	0.5128	1.13838096	0.51276039
1	1.3250	0.8371	1.3250	0.8371	1.32502883	0.83709841
2	1.6523	1.2873	1.6523	1.2873	1.65235799	1.28725883

Table 2. Numerical data of skin friction coefficients and local Nusselt number.

M^2	λ	β	K	F_r	$-\sqrt{\mathrm{Re}_x}C_{fx}$	$-\sqrt{\mathrm{Re}_x}C_{fy}$	$Nu_x/\sqrt{\mathrm{Re}_x}$
0.5	0.5	0.5	0.2	0.1	2.62934	0.626940	1.32361
1.5					3.12564	0.511611	1.31136
2.5	0.5	0.5	0.2	0.1	3.56354	0.441663	1.29937
	1.5				3.02305	1.592230	1.30437
0.5	2.5	0.5	0.2	0.1	3.44740	2.286630	1.27785
		1.0			2.05930	0.538751	1.31107
0.5	0.5	1.5	0.2	0.1	1.85281	0.500770	1.30326
			0.5		3.07888	0.520507	1.31257
0.5	0.5	0.5	0.8	0.1	3.47993	0.453420	1.30174
				1	3.10462	0.602219	1.31801
0.5	0.5	0.5	0.2	3	3.54763	0.581886	1.30691

Table 3. Numerical data of local Nusselt number.

Rd	θ_w	γ	$Nu_x/\sqrt{\mathrm{Re}_x}$
1	1.5	0.5	0.86382
2			1.32361
3	1.5	0.5	1.75035
	2		1.92839
2	3	0.5	4.00365
		0.2	0.58673
2	1.5	0.8	1.84113

- The behaviour of magnetic field, porosity of the medium and local inertia coefficient on primary and secondary velocities are same.
- Rotation and porosity of the medium have same effect on skin friction coefficient in both the directions.
- For sufficiently large temperature ratio, the temperature profiles become S - shaped which indicates the occurrence of adiabatic case.

References

1. R. Cortell, J. King Saud Univ-Sci., **26**, 2, 161-167 (2014).
2. A. Mushtaq, M. Mustafa, T. Hayat, A. Alsaedi, J. Taiwan Institute of Chem. Eng., **45**, 1176-1183 (2014).
3. S. Nadeem, R. U. Haq, N. S. Akbar, IEEE Trans. Nanotech., **13**, 109-115 (2014).
4. S. A. Shehzad, T. Hayat, A. Alsaedi, J. Appl. Fluid Mech., **9**, 1, 215-223 (2016).
5. C. Sulachana, G. P. Aswinkumar, N. Sandeep, J. Niger. Math. Soc., **35**, 1. 128-141 (2016).
6. I. S. Oyelakin, S. Mondal, P. Sibanda, Alexandria Eng. J., **55**, 1025-1035 (2016).
7. C. Y. Wang, Z. Angew Math. Phys (ZAMP), **39**, 177-185 (1988).
8. R. Nazar, N. Amin, I. Pop, Mech. Res. Commun., **31**, 121-128 (2004).
9. M. Sheikholeslami, M. Hatami, D. D. Ganji, J. Mol. Liq., **190**, 112-120 (2014).
10. M. Mustafa, A. Mushtaq, T. Hayat, A. Alsaedi, Plos One, 11, 2, 1-16 (2016).
11. P. Forchheimer, Zeitschrift Ver D. Ing., **45**, 1782-1788 (1901).
12. M. A. Seddeek, J. Colloid Interface Sci., **293**, 137-142 (2006).
13. B. K. Jha, Appl. Math., **2**, 1432-1436 (2011).
14. M. A. Sadiq, T. Hayat, Results Phys., **6**, 884-890 (2016).
15. H. P. Greenspan, *The Theory of Rotating Fluids* (Cambridge Univ. Press, New York, 1968).

https://doi.org/10.1051/matecconf/201819202060

An evaluation of public truck terminal implementation and truck ban policy in a medium-size city of Thailand: a case study of Khon Kaen city

Vuchlung Chhorn[1], *Jaruwit* Prabnasak [1]*, and *Chow* Chompoo-Inwai[2]

[1] International College, King Mongkut's Institute of Technology Ladkrabang, Bangkok, Thailand
[2] Faculty of Engineering, King Mongkut's Institute of Technology Ladkrabang, Bangkok, Thailand

Abstract. This study aims at investigation the potential impacts of a public truck terminal implementation along with truck ban policy adaptation in Khon Kaen City, Thailand. A four-step model is developed and used for examining traffic condition changes due to the terminal relocation. Three different terminal locations are set. Truck traffic of each scenario are estimated and converted to total network cost, and then compared with the base-case scenario. The most suitable truck terminal location is suggested. Also, several concerns regarding the public truck terminal and truck ban policy in the study area are given.

1 Introduction

In large urban area, truck traffic is known as one of the major concern and one popular scheme to tackle is "truck ban policy". Although truck ban policy is well accepted by a large number of cities, banning large vehicles can reduce problems, such as parking and loading issue, traffic obstruction, vibration, air and noise pollution and road accident. However, some studies have been found that the policy can cause increase in direct cost to the truck operators [1, 2].

In 2015, the Department of Land Transport proposed a project to develop a number of truck terminals in major regional cities in the country [3]. This leads to a research question that: what would potentially be the result if an urban truck terminal is implemented along with truck ban policy. Therefore, this study aims at investigating the impact of truck terminal implementation along with truck ban policy using Khon Kaen as the principal study area.

2 Methodology

The four-step modeling technique [4] is employed. Data used for developing the model was collected in Khon Kaen in 2015. Three different truck terminal location scenarios are set in the study. The total of vehicle-kilometer-travelled (VKT) and vehicle-hour-travelled (VHT) of each scenario is calculated by the model on the daily basis. The vehicle-operating-cost (VOC) used in the study is referred to [5].

Fig. 1 presents the assumption of the truck terminal locations for the purpose of examining the impact of truck ban policy. Three test locations are defined by land-use, town master plan, and referred to a study on truck terminal development held by the Department of Land Transport recently [3].

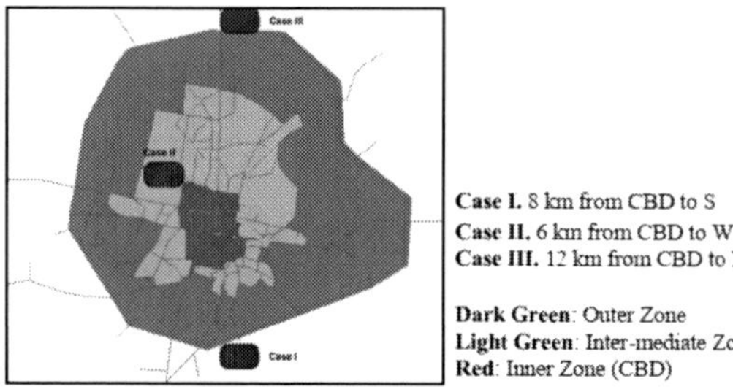

Fig. 1. Three truck terminal location scenarios

3. Model Results and Discussion

This section presents model estimated results and discussions for Base-Case and the three test scenarios.

3.1 Actual Truck Traffic Behaviors in the Study Area (Base-Case Scenario)

In this section, Base-Case scenario is developed in order to reveal the actual truck traffic behavior currently in the study area. The estimated result was shown in Table 1.

Table 1. Model estimation results for truck traffic behaviors in base case scenario

Daily truck demand (24 hours)	3,521	PCU
Total daily truck travel distance	58,504	P-kilometer
Total daily truck travel time	1,020.08	PCU-hour

* Corresponding author: jaruwit.pr@kmitl.ac.th

Table 1 reveals that the total demands of trucks for the normal operation with large trucks were approximately 3,521 PCU (about 16,671 tons of light trucks) moving inside city area in the road networks for 24 hours. In addition, it consisted of 58,504 vehicle-kilometers within the networks. This results in accounting for 1,020 hours with the average trip length of 0.7 kilometers as well as the average trip time around 0.81 hour with average trip speech 49.1 km/hour.

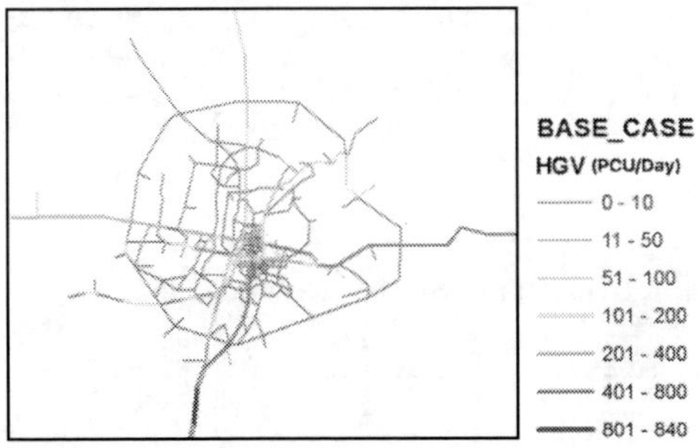

Fig. 2. HGV traffic flow in base-case scenario

In Fig. 2 demonstrates the truck traffic flow on the links for Base-Case scenario. It is showed that the traffic on the link arrives from all direction of the street within the study area. All trucks directly go inside the city area which lead to the heavy congestion problem on the link road as can be seen the yellow line and read line. The red line indicates the heavy traffic on the road. This result makes this assumption becoming more realizable due to the movement of traffic from everywhere crossing city center or coming from city center.

3.2 Case I Scenario

For Case I, the truck terminal is located 8 kilometers to the South of the CBD. In Table 2, it is found that the truck demands moving in outer zone are about 3,521 PCU which approximately around 8,802 Ton for large trucks which is moving outside to truck terminal Zone 71 and the truck demands which need to move inside city is about 8,337 PCU which roughly converted to 16,674 tons. This was double differences of the existing demands with the total vehicle hours around 2,152 hours with the same study area.

Table 2. Model estimated result of Case I

Daily truck demand HGV	3,521	PCU
Daily truck demand DGV	8,337	PCU
Total daily truck travel distance	118,843	P-kilometer
Total daily truck travel time	2,152.60	PCU-hour

Fig. 3 presents that, after having a truck terminal in the Case I location, the HGVs (Heavy Goods Vehicles) switch to use the outer ring road instead of inner city roads. Then all large trucks that move from outside city turning to use outer link road as shown as the yellow line and orange line through circle around map.

Fig. 3. HGV Traffic flow on links in Case I

Meanwhile, Fig. 4 indicates that the traffic flows on link road within CBD area with the DGV (Delivery Goods Vehicles – are assumed to be 2-Ton Pickup Trucks) moving within 24hrs operation. The result shows that traffic moving desire line is as per expected with 8,337 PCU (about 16,674 tons). The red color indicates the traffic along the small block from intermediate zone. The traffic is partly crowded until the truck terminal zone. However, as the truck movements are already converted from larger trucks to the small trucks, this is assumed that the traffic flow on the link is normal with small demand of large trucks on link.

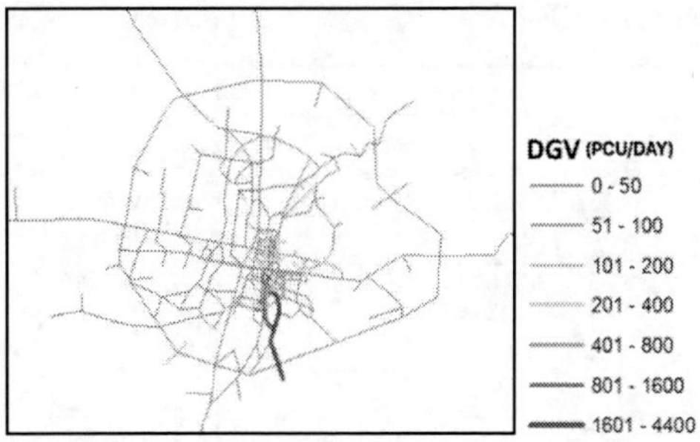

Fig. 4. Delivery Traffic flow on links in Case I

3.3 Case II Scenario

For Case II, the truck terminal is assumed to be located 6 kilometers to the West of the CBD (Inside the ring road). Results in Table 3 shows the same demand of HGV and DGV as Case I, however the differences are found between total truck traveling distances about 290,939 PCU-Kilometers which was 144 percent greater than Case I and truck traveling time about 4,841 PCU-hours which is about 125 percent greater than Case I. This obviously shows that Case II are not satisfied the study expectation.

Table 3. Model estimation results for truck traffic behaviors in Case II

Daily truck demand HGV	3,521	PCU
Daily truck demand DGV	8,337	PCU
Total daily truck travel distance	290,939	P-kilometer
Total daily truck travel time	4,841.17	PCU-hour

Fig. 5 presents the flow on the desire link within Case II as the result from GIS map. Result shows that the location condition in intermediate zone is quite far from the terminal location when compare with Case I. This location condition forces all the large vehicles to use outer link road instead causing the distance and traveling time increase as well.

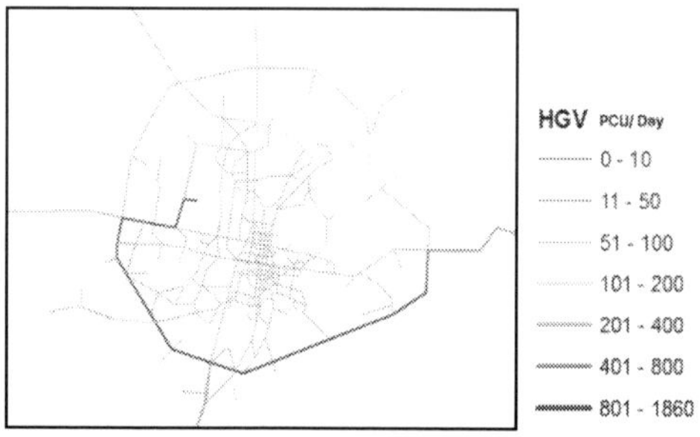

Fig. 5. HGV Traffic flow on links in Case II

Fig. 6, same method is applied on the truck movement so most goods turned to use inside city road and other roads along with intermediate zone. This demonstrates that the road along intermediate zone became significantly red. This indicated the heavy traffic on the link for those delivery trucks.

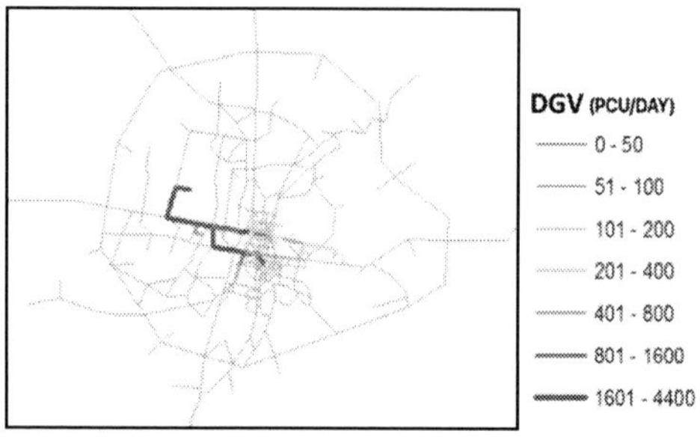

Fig. 6. Delivery Traffic flow on links in Case II

3.4 Case III Scenario

For the Case III which the truck terminal is assumed to be located 12 kilometers to the North of the CBD. The result is presented in Table 4. Total truck traveling distances are 148,507 PCU-kilometers, which is about 50 percent less than Case II. However, it is yet greater than Case I by 25 percent. Total traveling distance is 2,625 hours greater than Case I about 22 percent but it is 45 percent better than Case II.

Table 4. Model estimation result for truck traffic behaviors in Case III

Daily truck demand HGV	3,521	PCU
Daily truck demand DGV	8,337	PCU
Total daily truck travel distance	148,507	P-kilometer
Total daily truck travel time	2,625.74	PCU-hour

Fig. 7 and Fig. 8 demonstration the traffic flow of large trucks in the Case III. After the road that used for goods delivery is forced to close for truck, the ring road network is showed in the red color and orange color, while the DGVs use the main road from north into CBD.

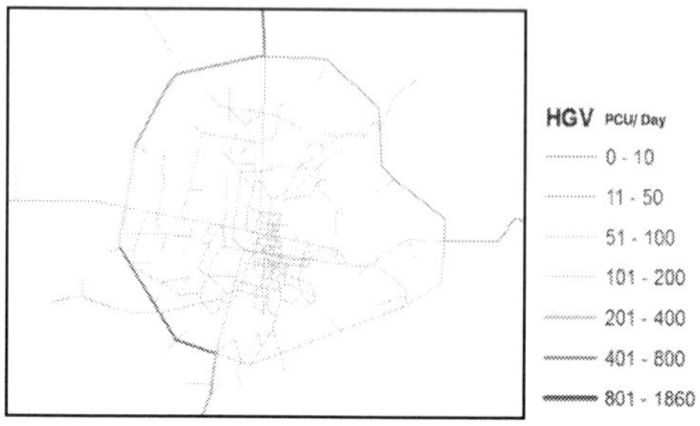

Fig. 7. HGV Traffic flow on links in Case III

Fig. 8. Delivery Traffic flow on links in Case III

3.5 Scenarios Comparison

In this section, the total kilometers and total hours of all three difference cases are evaluated as showed in Fig. 9.

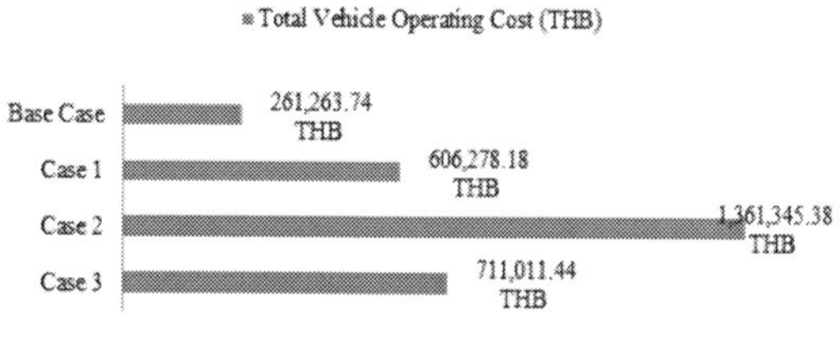

Fig. 9. Total vehicle-operating cost by cases

According to the figure, vehicle operation cost per day for each scenario is compared. Interestingly, Base-Case scenario provides the least cost, with the total cost of approximately 261,000 Thai Baht per day. In contrast, Case II is found to be the worst scenario with the total cost of over 1.36 million Thai Baht, 4 times greater than Base-Case. The truck terminal which gives the best result is Case I. Nevertheless, Case I generates the cost more than 2 times of Base-Case.

3.6 Discussion on Results

After the model results are compared, several interesting discussion can be given.

Firstly, it is found that the location of truck terminal is very crucial. Among the test scenarios, Case I is found to be the best. That is because the location set in the case matches to the actual fright behavior of the city where most of the CBD truck traffics come from or go to the south. For Case II and Case III, all CBD truck traffics are forced to travel along the ring road until reaching the terminal in the other side of the city before switching to small trucks and going back to the CBD Thus, in selecting a truck terminal location, clear understanding on current traffic pattern is necessary.

Secondly, it seems that the idea of implementing truck ban along with an urban truck terminal does not present a good result compared to actual situation. In fact, this finding is expected. In this study, the only direct costs generated by all freight activities in the CBD are taken into account. In the real situation, all trucks go straight from their origin to destination. In all test scenarios on the other hand, all truck traffics are forced to stop by the terminal which are located in suburban area, then all cargo are transshipped to small last-mile delivery trucks and vice versa. This additional activities lead to considerable extra travel distances. Besides, average cost per carrying capacity of delivery trucks, which are basically pick-up trucks, is significantly larger than that of those heavy trucks.

Thirdly, it is also important to note that in this study several costs from activities inside the terminal itself are not taken into account: those costs are, for example, handling, inventory, lost and damage and other terminal overhead costs. If these costs are concerned, it could extend the gap between Base-Case and all three test scenarios.

Although the study findings might lead to a conclusion that the implementation of urban truck terminal along with truck ban policy does not seem to benefit if focusing on direct transport cost. It cannot be said that the scheme is not valuable to the community. There are several benefits in which cannot be measured by the research methodology used in this study: for instance, benefits in term of reducing noise, vibration and air pollutions, traffic accidents, roadside parking issue and traffic obstruction while loading and unloading. These indirect benefits are crucial enough to induce a large number of cities and metropolitans to implement the scheme even if it cannot generate direct benefit in term of direct transport cost.

4. Conclusion

This study employs a four-step urban transport model to examine the potential benefits and costs of introducing the truck ban policy along with a development of an urban truck terminal in term of total direct transport cost of the network. Different terminal location scenarios are set to compare with Base-Case scenario. According to the study findings, it is interestingly found that the scheme does not improve or benefit the network in term of direct transport cost. In contrast, the scheme is fond to generate extra cost to the network. Additionally, it is found that the location of truck terminal plays a very important role and need to be matched to the actual behavior.

As discussed earlier, implementation of a truck ban policy along with an urban truck terminal development might not present benefits in term of direct transport cost. Nonetheless, there are several other benefits to the communities. This leaves a room for conducting future work in evaluating direct and indirect benefits of the scheme in other aspects.

Reference

1. C. Vuchlung & J. Prabnasak, An Evaluation of the Truck Ban Regulation and Suburban Truck Terminal Implementation Scheme in Regional Cities of Thailand: A Case Study of Khon Kaen City, Thailand., *The Sixth International Conference on Transportation and Logistics Proceedings*, Taipei (2016)
2. C. Vuchlung, *An Evaluation of the Truck Ban Regulation and Suburban Truck Terminal Implementation Scheme in Khon Kaen City of Thailand*, King Mongkut's Institute of Technology Ladkrabang, Bangkok (2017)
3. Departement of Land Transport, *A Feasibility Study on Regional Truck Terminal Development and Operation in Thailand: the Final Report*, Bangkok, Thailand. (2015)
4. M.G. McNally, *The Four Step Model*. Institute of Transportation Studies, University of California, Irvine, CA. (2007)
5. Expressway Authority of Thailand, *A Feasibility Study and Preliminary Design of Urban Expressway Network in Khon Kaen City: the Final Report,* Bangkok, Thailand. (2015)

Employee perception of work stress in a large construction company in Australia

Teck Heng Benson, Lim[1], and *Adam* Perica[1]

[1] Construction Management and Property, Faculty of Built Environment, University of New South Wales UNSW, Sydney, Australia

Abstract. Managing stress effectively is touted to be a key determinant of increased performance. It is however believed that individuals perceive and cope with stress differently. This thus makes stress management more complicated in construction. The aim of this research is to investigate the perceived work stress of construction employees by examining how perceived stress vary across employees and the perceived stress correlate with employee performance. In addressing these, an online questionnaire survey of 187 construction employees was undertaken, and data were analysed using the relative prevalence indexing method, ANOVA and Pearson correlation tests. The overall results show that poor work-life balance and work overloads are the major negative perceived stressors for construction professionals. Furthermore, there were differences in the perceived stress among cadets, operational staff, middle and top management. Lastly, the findings reveal the interpersonal and organisational climate have strong association with employee performance. In conclusion, the findings shed lights on the spectrum of work stress underling employee performance and how different cohorts of employees perceive and experience work stress.

1 INTRODUCTION

In construction, workplace stress is often reported as the key cause to adverse reactions that are detrimental to employees' health and performance (e.g., [1], [2]). Hitherto, a considerable amount of research has documented a list of factors that could attribute to workplace stress: a toxic work environment [2], negative workload [2], role ambiguity [1], lack of autonomy [1], bullying and harassment [3], career development barriers [4] and organisational climate [5]. As suggested by Lingard et al. [6], unlike other organisational assets, employees' needs and mental wellbeing cannot be overlooked as their capabilities cannot be easily replicated. It is critical for management to create a positive work environment, devoid of distress, for higher employee productivity and improved organisational performance.

The aim of this paper is to investigate the perceived work stress of construction professionals, by specifically examining how the perceived work stress varies across different roles; and how the perceived stress correlate with employee performance. By knowing these, company management could develop better targeted stress management strategies for different cohorts of employees.

2.0 LITERATURE REVIEW

According to Leung et al. [2], workplace stressors in construction could be classified into five categorises. *Personal stressors* are specific to individual employees' attitude toward work-life balance. *Interpersonal stressors* refer to the relationships with supervisors, colleagues, sub-contractors or clients. *Task stressors* relate the demands of a job. *Organisational stressors* relate to an organisation's decision-making processes. Lastly, *physical stressors* refer to the work environment that each employee is exposed to.

Over the past decades, a considerable amount of research has been done to explore how construction professionals and workers perceive their work stress across different countries. For example, Loosemore and Waters' [4] analysis of the stress perception of 190 male and female construction professionals in Australia have shown that excessive workload and low salary are the key stressors for men and women, respectively. Their findings further point out that male managers perceived 'implications of mistakes' and 'conflict or confrontational situations' as the primary workplace stressors, while their female counterparts pointed to the low job control and work overload. Similarly, Lingard and Francis [7] found that there exists a strong association between job scheduled irregularity, excessive working hours and emotional exhaustion in their sample of 231 male construction professionals, managers and administrators. Later, Lingard et al.'s [1] research further showecased that there is a strong negative association between employees' quality of life and their weekly hours worked, and that when employees worked consistently for more than 25-hour overtime per week, they were very likely to feel stressed, overloaded. More recently, Sunindijo and Kamardeen's [3] research on work stress in the Australian construction industry have

* Corresponding author: b.lim@unsw.edu.au

shown that both men and female construction professionals shared the top 10 stressors, with time pressure, excessive workload, long work hours and unpleasant work environment being the most critical issues.

3.0 RESEARCH DESIGN AND METHODS

In this study, an online survey questionnaire was designed based on the stressor- and performance- related measurement items proposed by Leung et al. [2, 5]. The questionnaire was pretested, validated and amended before being distributed to employees in an Australian construction and civil engineering company. In the first section, respondents were required to provide their background information (i.e., age, years of experience, average weekly hour worked and position). For subsequent sections, the respondents were requested to rate statements, regarding to: personal/ interpersonal stressors; task stressors; organizational stressors; physical stressors; and performance, based on a seven-point Likert Scale, ranging from 1 (strongly disagree) to 7 (strongly agree). Of these, some items had been worded reversely to increase the validity of questionnaire responses (e.g., TS2, TS3, TS4, OS1, OS2, PS1 and PS2). Table 1 summarises the measurement items used in this study.

For data analysis, the relative prevalence indexing (RPI) method, ANOVA and Pearson Correlation tests were conducted. The RPI method was preferred over the arithmetic average method because the former can derive relative indices within the range of 0-1 for each item and therefore enable researchers to undertake relative comparisons of items relating to the different types of stressors. This is an outcome that could be not achieved by directly comparing the arithmetic average of each items considering that items could have different maximum mean values (see [8]). The ANOVA test was conducted to determine if there are significant differences in the perceived stress among cadets, operational staff and middle and top management. Lastly, the Pearson Correlation test was undertaken to examine the relationship between perceived stress and performance of employees. For ease of interpretation, we had adopted Hinkle et al.'s [9] guideline to determine the magnitude of association.

Table 1. Measurement items

Code	Description
Personal Stressors	
PS1	My personal relationships are not affected by my work commitments
PS2	My role requirements allow me plenty of time to spend with my family
PS3	My role is flexible as long as I work my required hours
Interpersonal Stressors	
IP1	I have a good relationship with my colleagues
IP2	My colleagues are trustworthy and personable
IP3	My supervisor gives me recognition when I do a good job
IP4	I feel comfortable talking to my supervisor when I make a mistake
IP5	I feel comfortable requesting and using my leave entitlements
Task Stressors	
TS1	I am in control of my workload
TS2	I often need to shorten or skip my lunch breaks to stay on top of my work demands
TS3	Sometimes I must work from home or on my annual leave to stay on top of my work
TS4	I work more hours than most people I know
TS5	I rarely need to work overtime
TS6	I am happy with the hours I work each week
TS7	My job responsibilities are often vague or unclear
TS8	I am often caught between conflicting demands from different supervisors
Organisational Stressors	
OS1	Small matters often require referral to senior management for decision making
OS2	Most decisions I make need my supervisors' approval
OS3	I am often asked to complete a job differently to how I prefer
OS4	I am often required to carry out complicated procedures that do not make sense to me
OS5	I have control over when I start and finish work
OS6	I feel stable within my role in the company
OS7	I feel there are promotional opportunities available to me throughout my career with the company
OS8	My company provides mentoring, peer support and other training to support my professional development
Physical Stressors	
PS1	My office is overcrowded
PS2	My office is too noisy
PS3	The temperature of my office is uncomfortable
PS4	The safety equipment where I work is in good condition
PS5	The safety equipment where I work is regularly checked
PS6	I am satisfied with the overall levels of health and safety implemented by my company
Performance	
P1	I always meet my required deadlines
P2	I always meet the quality requirements of my tasks
P3	My planned schedules of programmes are effective
P4	I am a good communicator
P5	I feel confident in developing good relationships with others
P6	People enjoy working with me
P7	I believe in my company's goals and objectives
P8	I work hard to achieve the objectives of my role
P9	I have no intentions of leaving my company

For the data collection, 483 employees were invited to complete the online questionnaire survey. After three weeks, a total of 187 valid and complete responses were collected. Generally, about two-fifth of the 187 respondents were from middle management (38%), aged at least 40 years (47%) and had more than 16 years of working experience (46%) in the construction industry. More than 50% of them had worked about 50-60 hours per week on average.

4.0 RESULTS AND DISCUSSION

Table 2 summarises the perceived work stress among cadets, operational staff, middle and top management. The overall results reveal that most participants perceived the "Interpersonal Stressors (IS)" and "Physical Stressors (PYS) as quite positive, with the corresponding overall RPIs of 0.79 and 0.78. On the other hand, it appears that the "Task Stressors (TS)" and "Personal Stressors (PS)" had caused distress to employees, with the corresponding overall RPIs ranging from 0.45 to 0.52. These collectively add weight to the current debate that poor work life balance and work overload are the key negative stressors for construction professionals [1]. Our findings also tend to agree with Loosemore and Lim's [10] conclusion that middle management were the main victims of workplace injustice; as they were found to have suffered excessively poor work-life balance and task uncertainty comparing to other groups of employees.

Table 2 Perceived stress among different employees

CODE	Roles				
	Cadet	*O.S.*	*M. M.*	*T.M*	*F statistic*
PS1	4.03(0.57)	3.41(0.48)	3.02(0.43)	3.2(0.45)	2.528
PS2	4.15(0.59)	2.9(0.41)	2.59(0.37)	3.5(0.5)	6.999*
PS3	3.69(0.52)	3.07(0.43)	2.85(0.40)	3.5(0.5)	1.719
IP1	6.03(0.86)	5.94(0.84)	5.87(0.83)	6.05(0.86)	0.477
IP2	5.92(0.84)	5.72(0.81)	5.74(0.82)	6.05(0.86)	1.177
IP3	5(0.71)	4.6(0.65)	4.67(0.66)	4.6(0.65)	0.445
IP4	5.61(0.80)	5.28(0.75)	5.43(0.77)	5.55(0.79)	0.678
IP5	4.84(0.69)	5.3(0.75)	5.25(0.75)	5.5(0.78)	0.758
TS1	5.07(0.72)	4.57(0.65)	4.52(0.64)	4.55(0.65)	0.84
TS2	3.30(0.47)	2.87(0.41)	2.45(0.35)	2.2(0.31)	3.435**
TS3	4.84(0.69)	3.92(0.56)	3.07(0.43)	2.75(0.39)	9.428**
TS4	3.69(0.52)	2.27(0.32)	2.11(0.30)	2.75(0.39)	10.68**
TS5	3.30(0.47)	2.35(0.33)	1.94(0.27)	2.4(0.34)	6.535**
TS6	4(0.57)	3.48(0.49)	3.11(0.44)	4.3(0.61)	3.817*
TS7	4.65(0.66)	4.72(0.67)	4.95(0.70)	4.9(0.7)	0.379
TS8	5.15(0.73)	4.47(0.63)	4.25(0.60)	4.3(0.61)	1.896
OS1	4(0.57)	4.61(0.65)	5.05(0.72)	4.35(0.62)	3.124*
OS2	2.61(0.37)	4.07(0.58)	5.07(0.72)	4.75(0.67)	17.56**
OS3	4.73(0.67)	4.7(0.67)	4.76(0.68)	4.75(0.67)	0.022
OS4	4.76(0.68)	5.14(0.73)	5.11(0.73)	4.75(0.67)	0.771
OS5	3.5(0.5)	3.98(0.56)	3.56(0.50)	4.9(0.7)	3.596*
OS6	5.03(0.71)	5.11(0.73)	5.14(0.73)	5.55(0.79)	0.587
OS7	4.19(0.59)	4.11(0.58)	4.04(0.57)	4.8(0.68)	1.003
OS8	4.07(0.58)	4.28(0.61)	3.71(0.53)	4.15(0.59)	1.271
PYS1	5.84(0.83)	5.21(0.74)	5.25(0.75)	4.95(0.70)	2.045
PYS2	5.03(0.71)	4.77(0.68)	4.61(0.65)	4.9(0.7)	0.47
PYS3	4.84(0.69)	4.78(0.68)	4.85(0.69)	5.2(0.74)	0.361
PYS4	5.76(0.82)	5.85(0.83)	6.01(0.85)	5.8(0.82)	0.674
PYS5	5.65(0.80)	5.91(0.84)	6.04(0.86)	6.1(0.87)	1.242
PYS6	5.73(0.81)	5.91(0.84)	5.97(0.85)	6.1(0.87)	0.619

NB: O.S, M.M. and T.M. stand for operational staff, middle management and top management, respective. Mean value and RPI of respective items are reflected outside and inside the parenthesis respectively. * and ** denote p significant at < 0.05 and <0.001, respectively.

It is notable that there were significant differences in the perception of PS2, TS2, TS3, TS4, TS5, TS6, OS1, OS2 and OS5 among employees, with the F statistics ranging from 3.124 to 10.68. Of these, the post hoc test results reveal that cadet employees had significantly more time than their operational and middle management counterparts in spending time with their family (PS2).

Our findings further indicate that cadet employees were significantly less stressful than other employees in performing their tasks; considering that they were less likely to: shorten or skip their lunch to stay on top of their work (TS2); work from home or during annual leave (TS3); and work for a longer hour (TS4) and overtime (TS5). This could partially be explained to the supportive role of cadets to senior and more experienced employees. Furthermore, we found operational staff were significantly less likely to work from home or during annual leave than middle and top management did. This thus could indicate that the higher the position an employee hold, the more likely she or he would need to work from home or during annual leave. Interestingly, the post hoc test results further reveal that middle management were significantly less content than top management in terms of their working hour (TS6). An overall picture that emerges from here is that middle management could have spent more time than top management in managing subordinates and aligning project requirements and organisational priorities. This tends to agree with Caye et al.'s [11] assertions that middle management is the link between top management and team members, and that if top management set aspirations and strategy, they would then be responsible to translate those strategies into business plans and communicate them with other employees.

Adding to the above, the results suggest that cadet employees were significantly more inclined to experience organisational stress than the other employees; as they lacked autonomy and control over their work (OS5), and often required to seek approval from their supervisors (OS1 and OS2). Further to this, there seems to be a spill-over effect of deficiency of autonomy between operational staff, and middle and top management. Particularly, it is found that operational staff were more inclined to consult their supervisors for decision making than middle management required (OS2), while the latter had lesser control than top management in controlling work schedule. Interestingly, this further point to the "sandwiched" role of middle management.

Table 3 summarises the correlations between employees' perceived work stress and performance. It is notable that most of the stressor-related items have moderately strong positive correlation with employees' trust with their company (P7). Overall, the results reveal that employees are more likely to trust their employers, when there exists a positive, friendly and safe organisational climate where employees are trustworthy and encouraged to interact among each other, and have been given opportunities for career progression. These results tend to support McCarthy et al.'s [12] claim that employees, who experience high levels of organisational support, are more likely to portray high job satisfaction, and low turnover intentions. This further helps explaining the low positive correlation between P9 and OS8, pointing that employees are less likely to think of leaving the company when they had been given proper guidance and sufficient training for their professional development. This also tends to support Lim and Ling's [13] findings that effective competence development strategies could help improving employees' soft skills and gaining their behavioural commitment.

Turning to their task performance, we found that when employees can control their overload (TS1) and receive less conflicting but challenging work orders from their supervisors (TS8 and OS3), they are more likely to be efficient (P1) and effective (P3). This adds weight to Crane's [14] conclusion that task autonomy and interest play a key role in dictating an individual's goal setting behaviour and overall performance. Lastly, we found that peer relations (IP1) are positively correlated with individuals' ability and confidence of building up social relationships (P5) and developing a positive and interactive work environment (P6). This tends to support Lim and Ling's [13] assertion that effective intra-organisational relationship management is a key driver in promoting learning culture and a 'no-blame' attitude among employees.

Table 3 Correlations between perceived stress and performance

Items	P1	P2	P3	P4	P5	P6	P7	P8	P9
PS1		.219	.170						
PS2							.218		.203
IP1		.193		.259	.508	.429	.333	.160	.152
IP2					.295	.318	.342		.193
IP3					.152	.156	.440		.222
IP4	.208	.150	.285	.202	.254	.198	.454		.171
IP5			.244				.278		
TS1	.300	.286	.387	.152			.231		.215
TS2							.175	-.212	
TS3							.227	-.160	.162
TS4							.179		.161
TS5	.239	.178	.149				.237		.223
TS6			.159				.229		.209
TS7							.209		.192
TS8	.309	.148	.198	.162	.156		.202		
OS1							.198		
OS2		-.169							
OS3	.353		.224		.168	.156	.228		
OS4	.183						.204		
OS5							.164		.162
OS6			.235		.201	.156	.351		.221
OS7			.145				.344		.294
OS8			.214		.175	.210	.403		.352
PYS1							.172		
PYS2			.166				.248		.174
PYS3							.229		
PYS4			.152	.192	.164	.153	.273		
PYS5		.168	.223	.232	.196	.244	.335		
PYS6		.179	.224	.225	.177	.246	.333		.148

NB: Those insignificant relationships had been removed from the table. Moderately strong relationship (0.5 to 0.7) and low relationships 0.3 to 0.5) are highlighted in dark and light grey colours respectively.

5.0 CONCLUSION

This exploratory research aimed to investigate the perceived work stress of 187 employees in a large construction company. The overall findings show that employees were more disposed to be distressed by personal work-life imbalance and task uncertainty. Particularly, the findings highlight the plight of middle management, who appear to consistently suffer the highest levels of work stress. Furthermore, the findings shed light on the key function of interpersonal climate and organisational supports in stimulating or undermining employees' performance.

Lastly, it is acknowledged that there are limitations in this study and the findings are indicative and not conclusive. First, the sample size of 187 is small and not representative. Particularly, data was collected from a single company. However, the findings could further add to the existing knowledge of stress management for future studies. Further analyses (such as predictive modelling) should be conducted to examine the multilevel effect of perceived work stress on performance. A cross-cultural investigation of work stress should also be considered in future studies

6.0 REFERENCES

[1] Lingard H., Francis. V., and Turner. M., Work-life strategies in the Australian construction industry: Implementation issues in a dynamic project-based work environment. *Int. Journal of Project Management*, **30** (2012), 282-295.

[2] Leung, M., Chan, I., and Cooper, C., *Stress management in the construction industry*. John Wiley & Sons, (2015).

[3] Sunindijo, R. Y. and Kamardeen, I., Work stress is a threat to gender diversity in the construction industry. *Journal of Construction Engineering and Management*, **143** (2017).

[4] Loosemore, M. and Waters, T., Gender differences in occupational stress among professionals in the construction industry. *Journal of Management in Engineering*, **20** (2004), 126-132.

[5] Leung, M. Y., Skitmore, M. and Chan, Y. S., Subjective and objective stress in construction cost estimation. *Construction Management and Economics*, **25** (2007), 1063- 1075.

[6] Lingard. H. Brown. K. Bradley. L. and Bailey. C. Townsend. K., Improving employees work life balance in the construction industry: Project alliance case study. *Journal of Construction Engineering and management*, **133** (2007), 807-815.

[7] Lingard H. and Francis. V., *Managing work-life balance in construction*. Taylor and Francis. New York, (2009).

[8] Loosemore, M. and B.T.H Lim, Inter-organizational unfairness in the construction industry. *Construction Management and Economics*, **33** (2015), 310-326.

[9] Hinkle D.E, Wiersma, W. and Jurs S.G., *Applied Statistics for the Behavioral Sciences*, 5th ed. Boston: Houghton Mifflin, (2003).

[10] Loosemore, M. and Lim B.T.H., Intra-organisational injustice in the construction industry. *Engineering, Construction and Architectural Management*, **23** (2016), 428-447.

[11] Caye, J.M., Strack, R. Orlander, P., Kilmann, J., Espinosa, E.G., Francoeur, F. and Haen, P. *Creating a New Deal for Middle Management: Empowering a Neglected but Critical Group*. Available at https://www.bcg.com/documents/file52425.pdf, (2010).

[12] McCarthy, A., Darcy, C. and Grady, G., Work life balance policy and practice: understanding line Manager attitudes and behaviours. *Human Resources Management Review*, **20** (2010), 158-167.

[13] Lim B.T.H. and Ling F.Y.Y., Contractors' human resource development practices and their effects on employee soft skills. *Architectural Science Review*, **54** (2011), 232-24.

[14] Crane, M. E., *The effects of task autonomy and task interest on goal-setting behavior and task performance*. ETD Archive, Paper 476, (2012).

Thermal performance in a tubular heat exchanger with delta-winglets

Pattarapan Tongyote, *Pongjet* Promvonge*, *Nattawoot* Depaiwa and *Withada* Jedsadaratanachai

Mechanical Engineering Dept., Faculty of Engineering, King Mongkut's Institute of Technology Ladkrabang, Bangkok, Thailand

Abstract. The paper presents an experimental heat transfer enhancement study in a tubular heat exchanger fitted with delta-winglets. The experimental work was conducted by varying the airflow rate in the test tube having a constant wall heat-flux for turbulent regime, Reynolds number (Re) from 5200 to 23,000. Effects of three pitch ratios (PR=P/D=1.5, 2.0 and 3.0) and two attack angles, $\alpha = 45^o$ and 60°, of the winglets at a single blockage ratio (BR=b/D = 0.15) on thermal characteristics are examined. The experimental results show that the winglet-inserted tube yields, respectively, the heat transfer, friction factor and thermal performance in the form of TEF around 1.99–4.08, 4.9–14.3 times higher than the plain tube and 0.85–1.85, depending on the operating condition.

1 Introduction

Vortex flows have been extensively applied in thermal systems to increase their performance of such systems in industrial applications, for example, vortex tube [1], solar air heater [2,3], heat exchanger systems [4,5], etc. Vortex devices such as rib/baffle [6,7], coiled-wire [8,9], conical nozzles [10–13], twisted tape [14–16], fined tape [17,18], and winglet [19–21] are commonly employed for heat transfer enhancement in heating or cooling passages. Thermal improvement by those devices relies on induced vortex in the flow of fluid. The vortex generators (VGs) usually are inserted into the flow field to provide an interruption of boundary layer development and to accelerate the flow velocity, especially for tangential velocity, leading to increasing more turbulence intensity and then, causing the enhancement of the heat transfer rate. There have been many reports on the influences of insertion of vortex-flow devices in heating or cooling tubes/ducts on heat transfer, friction loss and thermal performance.

The application of rib/baffle and coiled-wire to promote heat transfer in a tube/duct regarded as a traditional style, is very plentiful in the literature. Promvonge and Eiamsa-ard [10–13] experimentally examined thermal characteristics in a tube with conical-nozzle inserts and pointed that the nozzles provided the considerable heat transfer over the plain tube around 250 to 350% depending on nozzle arrangements. Thermal behaviors in a tube contained with various twisted tapes were investigated and performance improvement by using those devices were reported in [14–16]. Promvonge et al. [17,18] experimentally and numerically studied on thermal characteristics in a square duct with angle-finned tape inserts and suggested that the smallest fin pitch ratio provides the maximum Nusselt number, friction factor and thermal performance. The turbulent convection heat transfer in a round tube with winglet insert was investigated as offered by [18,19]. Recently, Tamna et al. [21] numerically studied thermal performance in a round tube inserted with winglet-pairs placed on two tape sides before insertion. High thermal improvement including vortex-flow structure was reported.

In the mentioned literature review, a lot of vortex-flow devices or VGs have been offered to improve thermal performance of the heat exchanger tube. Among the mentioned devices, the winglets as reported in [21] is considered to be investigated due to its lowest pressure loss apart from easy manufacture. In the present work, delta winglet pairs are introduced. Due to curved surface of the round tube, the winglet pairs are placed repeatedly on double sides of a flat tape used as the support of winglets before inserting the tape into the test tube. The experimental results using air as the test fluid are presented for turbulent flow, Reynolds number ranging from 5200 to 23,000 in the current study.

2 Experimental procedure

Fig. 1 shows a schematic diagram of the experimental apparatus. In the figure, room air supplied by a 1.5-kW blower entered into an orifice flow-meter before passing through the test section. The volumetric rate of airflow in the system was measured by the flow-meter, built under ASME standard [22]. The pressure drop across the orifice was measured using an inclined manometer. The volumetric airflow rate was varied by adjusting the speed of the blower motor through an inverter.

The copper test tube with 50-mm inner diameter (D) was 3000-mm long, included the test section (L) of 1600-mm as demonstrated in Fig. 1. Using continually

* Corresponding author: kpongjet@gmail.com

winding flexible electrical wire around the tube was suggested to yield a uniform heat-flux condition. The outer surface of the test tube was well insulated to minimize convective heat loss to the surroundings. The inlet and outlet temperatures (T_i and T_o) of the bulk air were measured at certain points with a multi-channel temperature measurement unit in conjunction with the RTD PT-100 type temperature sensors. The wall temperatures (T_w) were measured by using sixteen T-type thermocouples mounted on the outer tube surface and located equally on the top wall along the test section. Two static pressure taps coupled with a digital manometer were used to measure pressure drops across the test tube. The measurement was done twice for all cases for examining its repeatability.

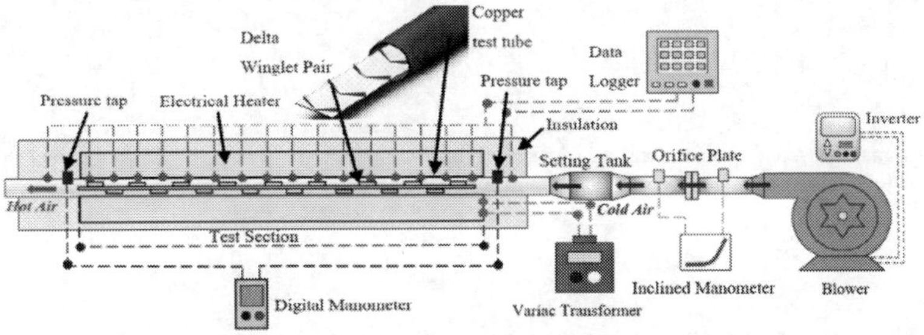

Fig. 1. Schematic diagram of experimental setup.

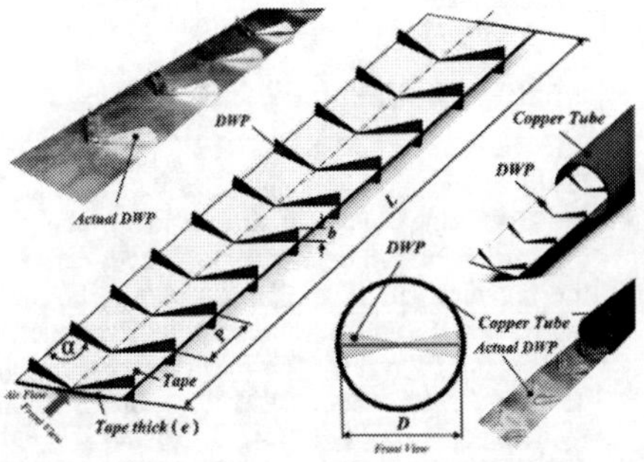

Fig. 2. Arrangements of delta winglet pairs (DWP) on a tape.

Fig. 2 portrays the arrangement of delta-winglet pairs (DWP) used in the present work. The winglet was made of a 0.5-mm aluminium strip and formed it to be delta winglet before attaching it on a tape by using hot glue as shown in Fig. 2. Each winglet was 19.8 mm in length (l) while the tape was 1600-mm long and 0.5-mm thick (e). The DWPs with a fixed blockage ratio (BR=b/D=0.15) were mounted repeatedly on both tape sides for two angles of attack (α=45° and 60°) and three relative pitches (PR=P/D = 1.5, 2.0 and 3.0) as displayed in Fig. 2.

3 Data reduction

Parameters of interest included Reynolds number (Re), heat transfer rate (Nusselt number, Nu), pressure drop (friction factor, f), and thermal enhancement factor (at the same pumping power), are described in this section. The Re is given by

$$\mathrm{Re} = UD/\nu \tag{1}$$

where U and ν are the mean air velocity in the test tube and kinematic viscosity of air, respectively.

The friction factor (f) computed by pressure drop across the length of the test section (L) is written as

$$f = \frac{\Delta p}{(L/D)\rho U^2/2} \tag{2}$$

In the experiment, air flowed into the test tube having a uniform heat-flux. In a steady state condition, the heat transfer in the test tube is assumed to be equal to the heat loss which is expressed as:

$$Q_{air} = Q_{conv} \tag{3}$$

where

$$Q_{air} = \dot{m}C_{p,air}(T_o - T_i) \tag{4}$$

The heat transfer by the test tube is written by

$$Q_{conv} = hA(\tilde{T}_w - T_b) \tag{5}$$

in which

$$\mathrm{T_b} = (\mathrm{T_o} + \mathrm{T_i})/2 \tag{6}$$

and

$$\tilde{T}_w = \sum T_w/16 \tag{7}$$

where T_w is local wall temperature along the tube. The average wall temperature, $\tilde{T}_w$ is evaluated from 16 local wall temperatures lined equally between the inlet and the exit of test tube. The average heat transfer coefficient (h) and mean Nusselt number (Nu) are estimated as follows:

$$h = \dot{m}C_{p,air}(T_o - T_i)/A\tilde{T}_w - T_b) \tag{8}$$

The heat transfer is computed in the form of Nu which can be obtained by

$$Nu = hD / k \tag{9}$$

All thermo-physical properties of air are determined at the overall bulk air temperature (T_b) from Eq. (6).

Thermal enhancement factor (TEF) defined as the ratio of the, h of an inserted tube to that of a smooth tube, h_0, at an identical pumping power (pp) is written as:

$$\mathrm{TEF} = \left.\frac{h}{h_0}\right|_{pp} = \left.\frac{\mathrm{Nu}}{\mathrm{Nu}_0}\right|_{pp} = \left(\frac{\mathrm{Nu}}{\mathrm{Nu}_0}\right)\left(\frac{f_0}{f}\right)^{1/3} \tag{10}$$

where Nu_0 and f_0 are Nusselt number and friction factor of the plain tube, respectively.

4 Results and discussion

4.1 Verification of plain tube

The validation of the present heat transfer (Nu) and friction loss (f) of the plain tube is made by comparing them with those from the published correlations [23] of Dittus-Boelter, Eq. (11) and Blasius, Eq. (12) as shown in Fig. 3. In the figure, both Nu and f are seen to agree very well within 6% with those from the correlations.

Dittus-Boelter correlation,

$$\mathrm{Nu} = 0.023\,\mathrm{Re}^{4/5}\,\mathrm{Pr}^{0.4} \tag{11}$$

Blasius correlation,

$$f = 0.316\,\mathrm{Re}^{-0.25} \tag{12}$$

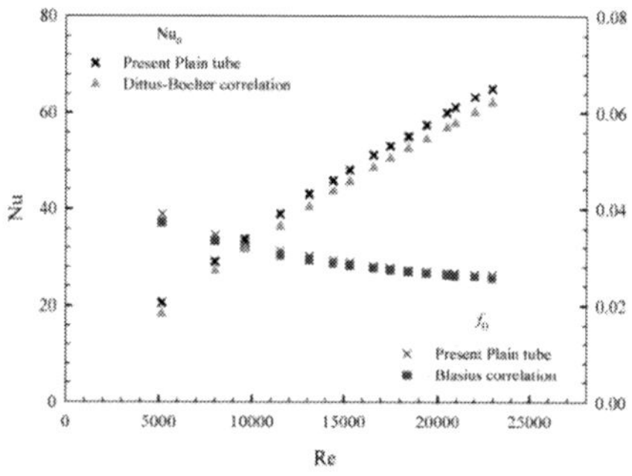

Fig. 3. Verification of Nu and f for plain tube.

4.2 Heat transfer result

The variation of Nusselt number ratio ($\mathrm{Nu/Nu_0}$) with Re for DWP tape inserts is displayed in Fig. 4. In the figure, the $\mathrm{Nu/Nu_0}$ shows the decreasing trend with the rise in Re for all cases. It is noted that the highest $\mathrm{Nu/Nu_0}$ around 4.08 times is seen for α=60° and PR=1.5 while the lowest one around 1.99 is at α=45° and PR=3.0. This is because of interrupting more often the flow at PR=1.5 and of stronger vortex strength at α=60° leading to more efficient heat transfer between the working fluid and the heated wall surface. The average increases in $\mathrm{Nu/Nu_0}$ for α = 45° and 60° are about 3.04, 2.48, and 2.33; and 3.22, 2.90 and 2.73 times at PR=1.5, 2.0 and 3.0, respectively. The increase in $\mathrm{Nu/Nu_0}$ for α=60° is about 12% in comparison with that for α = 45°.

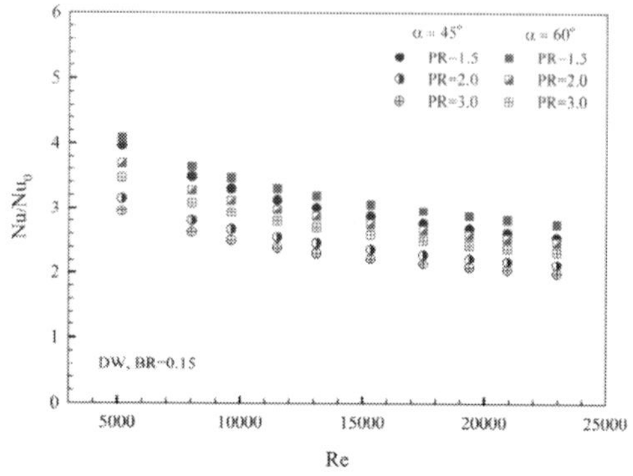

Fig. 4. Variation of $\mathrm{Nu/Nu_0}$ with Re for various DWP tapes

4.3 friction factor result

The variation of friction factor ratio (f/f_0) with Re for DWP tape inserts is depicted in Fig. 5. It is found that the DWP tape yields an extreme increase in f due to the dissipation of dynamic pressure, larger surface area and the reverse/swirl flow. The f/f_0 shows the uptrend with increasing Re and α values while gives the downtrend with rising PR value. The mean increases in f/f_0 for α = 45° and 60° are about 10.95, 7.67, and 5.99; and 13.01, 8.46 and 6.51 times at PR=1.5, 2.0 and 3.0, respectively. The DWP tape at α=60° and PR=1.5 gives the highest f/f_0 of about 14.3 times while the lowest of about 4.9 times is seen at α=45° and PR=3.0, similar to the $\mathrm{Nu/Nu_0}$ case. The mean f/f_0 for the DWP tape at PR=1.5 is seen to be higher than the one at PR=2.0 and 3.0 around 12 and 25%. However, the f/f_0 at α=60° is higher than that at α=45° around 20%.

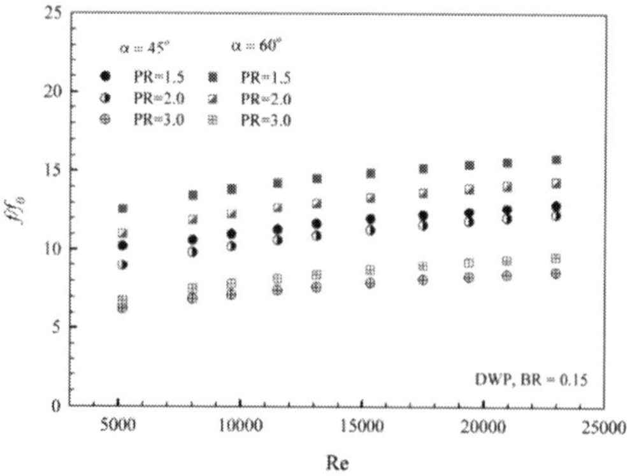

Fig. 5. Variation of f/f_0 with Re for various DWP tapes

4.4 Thermal performance

Figure 6 illustrates the variation of thermal enhancement factor (TEF) with Re for all inserts. The TEF is obtained by using Nu and *f* data of the inserted tube and the plain tube under a constant pumping power as shown in Eq. (10). In the figure, TEF displays the increasing trend with the rise in Re for all cases. The highest TEF around 1.85 is seen for α=45° and PR=1.5 at the lowest Re value while the minimum one around 0.88 is at α= 60° and PR=3.0. The best case is found at α=45° and PR=1.5 since TEF is higher than unity for all Re values.

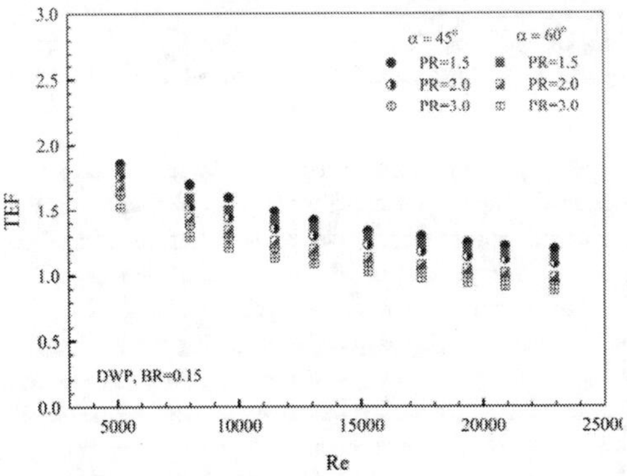

Fig. 6. Effect of DWP insert on TEF.

5 Conclusion

An experimental study on thermal behaviors in a circular tube contained with DWP tapes having three PR and two α values was conducted for turbulent flow, Re = 5200–23,000 under a constant heat-flux condition. The Nu/Nu_0 tends to decrease whereas f/f_0 increases with the rise in Re and PR values but both show the opposite trend with the increment in α value. The DWP tape insert provides the increases in Nu/Nu_0 and f/f_0 around 1.99 – 4.08 and 4.9 – 14.3 times. Also, the highest TEF of about 1.85 is found for α = 45° and PR=1.5 at the lowest Re.

References

1. S. Eiamsa-ard, K. Wongcharee, P. Promvonge, Int. Commun. Heat Mass Transf., **37**, 156-162 (2010)
2. P. Promvonge, S. Eiamsa-ard, Int. Commun. Heat Mass Transf., **34**, 838-848 (2007)
3. P. Promvonge, Int. Commun. Heat Mass Transf., **37**, 835-840 (2010)
4. S. Skullong, C. Thianpong, P. Promvonge, Heat Mass Transf., **51**, 1475-1485 (2015)
5. S. Liu, M. Sakr, Renew. Sustain. Energy Rev., **19**, 64-81(2013)
6. A. Tandiroglu, Int. J. Heat Mass Transf., **49**, 1559-1567 (2006)
7. S. Skullong, C. Thianpong, N. Jayranaiwachira, P. Promvonge, Chem. Eng. Process. Process Intensif., **99**, 58–71 (2016)
8. P. Promvonge, Int. Commun. Heat Mass Transf., **35**, 623-629 (2008)
9. P. Promvonge, Energy Convers. Manage., **49**, 980-987 (2008)
10. P. Promvonge, S. Eiamsa-ard, Int. Commun. Heat Mass Transf., **34**, 72-82 (2007)
11. P. Promvonge, S. Eiamsa-ard, Int. Commun. Heat Mass Transf., **34**, 838–848 (2007)
12. P. Promvonge, S. Eiamsa-ard, Exp. Therm. Fluid Sci., **32**, 332–40 (2007)
13. P. Promvonge, S. Eiamsa-ard, Energy Convers. Manage., **47**, 2867–82 (2006)
14. S. Eiamsa-ard, C. Thianpong, P. Promvonge, Int. Commun. Heat Mass Transf., **33**, 1225-1233 (2006)
15. P. Promvonge, S. Pethkool, M. Pimsarn, C. Thianpong, Int. Commun. Heat Mass Transf., **39**, 953-959 (2012)
16. C. Thianpong, P. Eiamsa-ard, P. Promvonge, S. Eiamsa-ard, Energy Procedia, **14**, 1117-1123 (2012)
17. P. Promvonge, S. Skullong, S. Kwankaomeng, C. Thianpong, Int. Commun. Heat Mass Transf., **39**, 617-624 (2012)
18. P. Promvonge, S. Skullong, S. Kwankaomeng, C. Thianpong, Int. Commun. Heat Mass Transf., **39**, 625-633 (2012)
19. S. Eiamsa-ard, K. Wongcharee, P. Promvonge, C. Thianpong, Appl. Therm. Eng., **30**, 310-318 (2010)
20. S. Chokphoemphun, M. Pimsarn, C. Thianpong, P. Promvonge, Chin. J. Chem. Eng., **23**, 605-614 (2015)
21. S. Tamna, S. Sripattanapipat, P. Promvonge, Energy Procedia, **100**, 518-521 (2016)
22. ASME, *Standard Measurement of fluid flow in pipes using orifice, nozzle and venture*, (United Engineering Center, 1984)
23. Incropera, F.P. Witt, P.D. Bergman, A.S. Lavine, *Fundamentals of heat and mass transfer*, (John-Wiley & Sons, 2006)

Development of a numerical simulation model for suction rate using CADMAS-SURF

Kornvisith Silarom[1,*], *Yoshimichi* Yamamoto[2], *Syota* Yoshizawa[3] and *Nunthawath* Charusrojthanadech[4]

[1]Graduate School of Science and Technology, Tokai University, Hiratsuka, Kanagawa, Japan
[2]Department of Civil Engineering, Tokai University, Hiratsuka, Kanagawa, Japan
[3]Graduate School of Engineering, Tokai University, Hiratsuka, Kanagawa, Japan
[4]Department of Civil Engineering, Faculty of Engineering, King Mongkut's Institute of Technology Ladkrabang, Bangkok, Thailand

Abstract. There are many cases where a coastal dike or a seawall constructed on a very shallow area was broken by a wave of a smaller height than the designed height. In many of these cases, the reason of the destruction was the suction phenomena. This phenomenon happens when waves reaches the front of the dike or the seawall, the wave pressure scours the front soil of the structure until it reaches the lowest edge of the structure followed by sucking of backfilling material of the structure. In this research, the authors proposed the numerical model which can calculate the suction rate with elapsed time by applying the pore water pressure and the flow velocity inside a dike or a seawall using "CADMAS-SURF". The authors proposed three coefficients for improving the calculated pressure from CADMAS-SURF. The reliability of the model was confirmed by reproducing the suction phenomena in Hirono Coast of Japan.

1 Introduction

There are many cases where a coastal dike or a seawall in a very shallow area was broken by repeating action of smaller waves than the designed wave. The main reasons of destructions are scouring in front of the structure, and suction from the inside of the structure. The suction phenomena occurred when the wave pressure scoured all the front soil layer of the structure, and then, the return flow sucks the backfilling materials out of the structure. If the structure loses its stability due to backfilling material loss, it will collapse. So, to design coastal facilities in the shallow area, it is important to consider the scouring and the suction phenomena.

Yamamoto et al. [1] found that the suction rate is proportional to the difference between shear force and shear resistance at the bottom of the structure (refer to Fig.1). Moreover, they proposed the empirical equation which can predict the failure of the structure.

Ioroi et al. [2, 3] proposed an empirical equation for calculating the suction rate using pore water pressure and velocity at the lowest edge of the structure. This equation can calculate the suction rate with sufficient accuracy. However, this equation is limited to the shape of the structure. They also evaluated the accuracy of their equation by using calculated pore water pressure and velocity from the numerical simulation model "CADMAS-SURF". However, the calculated results are not satisfying because this numerical model cannot consider the effect of porous media. To handle the above defect, Yoshizawa et al. [4] proposed an empirical equation for getting a modification coefficient. This coefficient can take the effect of median grain size into account.

In this research, the authors proposed a new suction rate formula which can calculate suction rates with elapsed time by using calculated pressure and velocity from CADMAS-SURF. Moreover, CADMAS-SURF cannot consider the effect of porous media consisting of fine grain such as sand or gravel, the authors proposed some empirical equations for calculating modification coefficient to improve the calculated pressure and velocity from CADMAS-SURF.

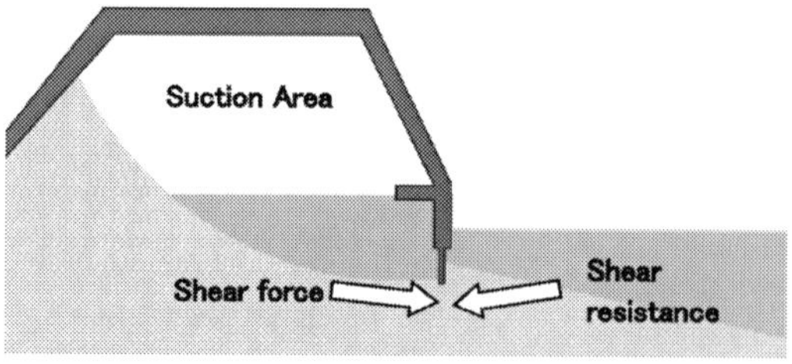

Fig.1 Illustration of suction in a coastal dike.

2. Related Literature

2.1 Original Suction Rate Formula

Ioroi et al. [2, 3] purposed a formula (Eq. (1)) for calculating the suction rate of the coastal dike or the seawall which the front side and the crown side are covered by concrete or asphalt. They proposed the

* Corresponding author: 7ltad005@mail.u-tokai.ac.jp

empirical equation, Eq. (3), for calculating the proportional coefficient (β) between the suction rate and the difference between shear force and shear resistance by using hydraulic experiments. This coefficient could convert the difference between the shear force and the shear resistance at the lowest edge of the structure to the suction rate with sufficient accuracy.

$$\frac{q}{w_s D_{50}} = \beta\left(\theta-\theta_c\right)\frac{1}{2}\left(1+\cos\left(\frac{\alpha T}{t}\right)\right), \tag{1}$$

$$\alpha = \begin{cases} -9.0\times10^{-5}\left(\dfrac{D_{50}}{0.2mm}\right)+0.0031, \left[1\le\dfrac{D_{50}}{0.2mm}\le 25\right] \\ -3.39\times10^{-5}\left(\dfrac{D_{50}}{0.2mm}\right)+0.0017, \left[25\le\dfrac{D_{50}}{0.2mm}\le 50\right] \end{cases} \tag{2}$$

$$\beta = 0.028e^{-0.35\left(\frac{D_{50}}{0.2mm}\right)}, \tag{3}$$

$$\theta_c = \tau_r / \left(\rho_s-\rho_w\right) g D_{50}, \tag{4}$$

$$\theta = \tau_f / \left(\rho_s-\rho_w\right) g D_{50}, \tag{5}$$

$$\tau_r = \left(\rho_s g d_t - \rho_w g d_t - P_{ob\,max}\right)\tan\phi, \tag{6}$$

$$\tau_f = \frac{1}{2} f \rho_w V_{max}^2, \tag{7}$$

$$V_{max} = \sqrt{2\left(\frac{P_{ob\,max}}{\rho_w}\right)\left(\frac{h}{H}+1.0\right)^{-1.11}} \tag{8}$$

where, q is the suction rate per unit width at the lowest edge of the structure, w_s is the settling velocity of Rubey's, D_{50} is the median grain size, θ_c is the dimensionless critical suction force, θ is the dimensionless suction force, T is the wave period, t is the time, α is the reduction coefficient, β is the proportional coefficient, τ_r is the shear resistance, τ_f is the suction force (shear force), ρ_s is the density of backfilling materials (= 1,800 kg/m^3), ρ_w is the density of water, g is the gravitational acceleration, d_t is the thickness of the sand layer in front of the structure, P_{obmax} is the maximum excess pore water pressure of the return flow, ϕ is the internal angle of the backfilling materials, f is the coefficient of the suction force at the lowest edge of the structure (≈1), and V_{max} is the maximum flow velocity of the return flow.

2.2 CADMAS-SURF

"CADMAS-SURF" [5] is a numerical model developed by Coastal Development Institute of Technology, Japan. This numerical model can simulate flow in a wave flume. This model based on 2D or 3D continuity equation (Eq. (9) for 2D) and 2D or 3D Navier Stroke's formulae (Eqs. (10)~(17) for 2D). The free surface is handled by Volume of Fluid (VOF) technique [6]. This model can calculate the water pressure and the velocity on every designated point of the simulation.

The 3D version of this model could consider the effect of the properties of some specified materials (e.g., big stone and concrete block). However, in the case of fine materials such as sand and gravel, this model cannot consider an effect of these materials.

$$\frac{\partial \gamma_x u}{\partial x}+\frac{\partial \gamma_z w}{\partial z}=0, \tag{9}$$

$$\lambda_v\frac{\partial u}{\partial t}+\frac{\partial \lambda_x uu}{\partial x}+\frac{\partial \lambda_z wu}{\partial z} = -\frac{\gamma_v}{\rho}\frac{\partial p}{\partial x}+\frac{\partial}{\partial x}\left\{\gamma_x\nu_e\left(2\frac{\partial u}{\partial x}\right)\right\} +\frac{\partial}{\partial z}\left\{\gamma_z\nu_e\left(\frac{\partial u}{\partial z}+\frac{\partial w}{\partial x}\right)\right\}-R_x, \tag{10}$$

$$\lambda_v\frac{\partial w}{\partial t}+\frac{\partial \lambda_x uw}{\partial x}+\frac{\partial \lambda_z ww}{\partial z} = -\frac{\gamma_v}{\rho}\frac{\partial p}{\partial z}+\frac{\partial}{\partial x}\left\{\gamma_x\nu_e\left(\frac{\partial w}{\partial x}+\frac{\partial u}{\partial z}\right)\right\} +\frac{\partial}{\partial z}\left\{\gamma_z\nu_e\left(2\frac{\partial w}{\partial z}\right)\right\}-R_z-\gamma_v g, \tag{11}$$

$$\lambda_v = \gamma_v+\left(1-\gamma_v\right)C_M, \tag{12}$$

$$\lambda_x = \gamma_x+\left(1-\gamma_x\right)C_M, \tag{13}$$

$$\lambda_z = \gamma_z+\left(1-\gamma_z\right)C_M, \tag{14}$$

$R_x = \frac{1}{2} C_D$ [redacted] $) u\sqrt{u^2+w^2}$, (15)

$R_z = \frac{1}{2} C_D$ [redacted] $) w\sqrt{u^2+w^2}$, (16)

$$\gamma_v\frac{\partial F}{\partial t}+\frac{\partial \gamma_x uF}{\partial x}+\frac{\partial \gamma_z wF}{\partial z}=0, \tag{17}$$

where, x and z are the horizontal and vertical coordinates, γ_x and γ_z are the sectional transform ratio in horizontal and vertical direction, respectively, γ_v is the porosity, u and w are the velocities in horizontal and vertical direction, respectively, ν_e is the kinematic viscosity, p is the water pressure, ρ is the water density, λ_v, λ_x, λ_z are coefficients expressed in Eqs. (12)~ (14), C_M is the coefficient of inertia, R_x and R_z are the drag force in the horizontal and vertical direction, respectively, C_D is the drag force coefficient, Δx, Δz are the mesh size in horizontal and vertical direction, respectively, and F is the VOF function.

3. Methodology

3.1 Wave flume experiment

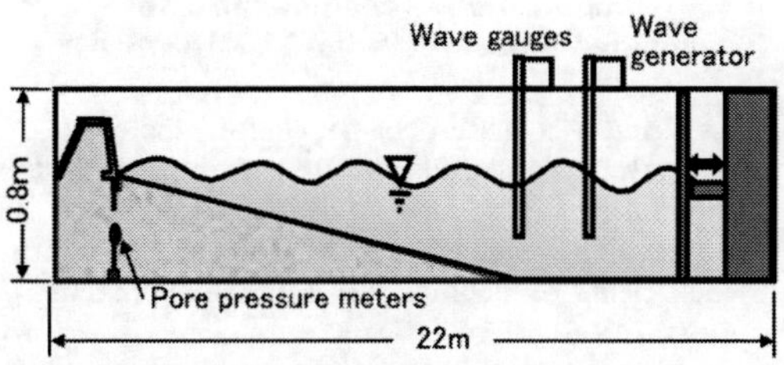

Fig.2 Wave flume experiment setup.

Some wave flume experiments were performed to get wave pressure distribution in a coastal dike and relation between suction rate and some concerned parameters. For these purposes, a dike model and a wave flume that has

0.5m in width, 0.8m in height, and 22.0m in length and the ball-screw-driven wave generator were setup as shown in Fig.2. To get pore water pressures and pressure distribution in the dike, pore water pressure gauges were equipped at the inner length inside the dike of 0cm, 5cm, 15cm, and 75cm. Moreover, to examine the effect of sand layer thickness affect the pore pressure, sand layer inside the dike was set to three cases: 1) before the suction, 2) in the middle of suction, and 3) at the end of suction.

The authors selected the case study of the Hirono coast, Shizuoka Prefecture, Japan as a prototype. This coast suffered suction damage due to Typhoon no.9 in 1997. At that time, the significant wave height was 6.91m and the significant wave period was 13.9s. So, in the experiments, the significant wave height and significant wave period were scaled to 0.23m and 2.65s by Froude law. Moreover, grain size was scaled from 0.4mm to 0.2mm by the similitude of beach profile given by Ito and Tsuchiya [7]. The results of the experiment are shown in Fig.3-5.

3.2 Development of a new suction formula

According to the Ioroi et al.'s formula (Eq. (1)) was designed for calculating suction rate by using maximum pore water pressure and maximum velocity of the single point at the lowest edge of the structure. So, to calculate suction rates on every point in the dike, the authors proposed a new suction formula, Eq. (18), which can use with calculated pressure and velocity from CADMAS-SURF.

$$\frac{q}{w_s D_{50}} = \beta(\theta - \theta_c) \quad (18)$$

And because of the empirical equation for calculating the proportional coefficient, Eq. (3), was derived for using with Eq. (1), the authors also proposed a new empirical equation for calculating the proportional coefficient, Eq. (18), which obtained by the relation between the suction rate and pore water pressure that can be measured in the experiments.

$$\beta = 0.142e^{-0.25(\frac{D50}{0.2mm})} \quad (19)$$

3.3 Modification coefficients to improve calculated pressures from CASMAS-SURF

As CADMAS-SURF does not consider the effect of the sand layer inside the dike. So, the authors performed some experiments mentioned above and proposed three empirical equations for calculating modification coefficients to improve the calculated pressure from CADMAS-SURF.

3.3.1 Modification coefficient for the inner length from the front of the dike, C_x.

The result from the experiments shows that pore water pressures decrease when the inner length of the dike increases, as shown in Fig.3. According to this result, the empirical equation to calculate a modification coefficient for the inner length from the front of the dike, Eq. (20), could be obtained.

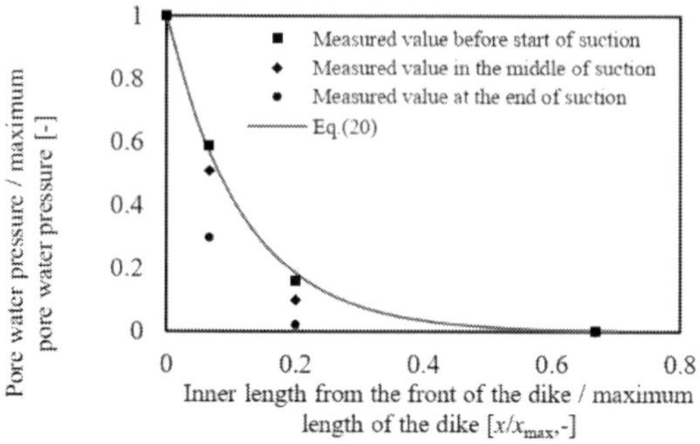

Fig. 3 Relation between pore pressure and inner length from the front of the dike.

$$C_x = 0.00022^{x/x_{max}} \quad (20)$$

3.3.2 Modification coefficient for the sand layer thickness, C_d.

From the above result, when we consider about the sand layer thickness affect pore water pressure, the result shows pore water pressures decrease when the sand layer thickness decrease. Fig. 4 shows the relation between modification coefficient and sand layer thickness. By this result, the authors obtained an empirical equation to calculate a modification coefficient for the sand layer thickness, Eq. (21).

$$C_d = e^{0.269 - \frac{0.263}{d/d_{max}} e^{44.996 - 23.715e^{\frac{-(x/x_{max})-3.265}{4.953}}} \ln(d/d_{max})} \quad (21)$$

where, d is the sand layer thickness, and d_{max} is the maximum sand layer thickness.

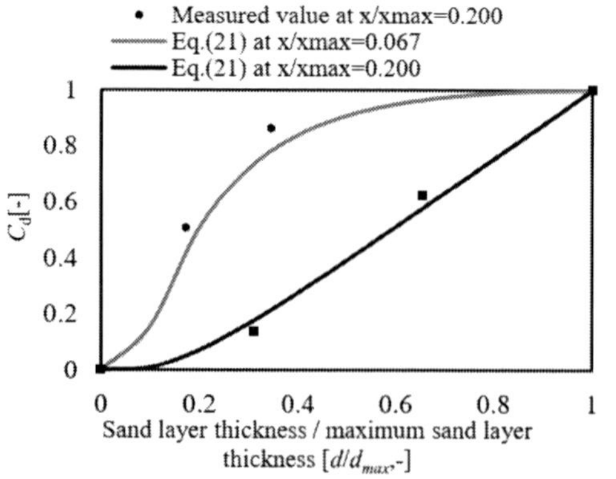

Fig. 4 Relation between C_d and sand layer thickness.

3.3.3 Modification coefficient for grain size, C_{D50}.

Yoshizawa et al. [4] presented the empirical equation to modify the calculated pressure from CADMAS-SURF, this equation can improve the effect of the median grain size into an account. However, by using this empirical equation to Eq. (19), the calculated result is overestimated. Therefore, the authors proposed a new empirical equation to calculate a modification coefficient for the grain size, Eq. (22).

$$C_{D_{50}} = 0.65(\frac{0.2}{D_{50}})^{0.85} + 0.35 \quad (22)$$

Finally, we can get the modified pressure by applying these three modification coefficients to the calculated pressure from CADMAS-SURF as expressed in Eq. (23). This modified pressure is used to calculate shear force instead of P_{obmax} in Eq. (6).

$$p = C_x C_d C_{D_{50}} p_{CAD} \quad (23)$$

where, p is the modified pressure, and p_{CAD} is the calculated pressure from CADMAS-SURF.

4. The reliability of the method

The reliability of the method was confirmed by a reproductive simulation of the experiment case. The calculated value from the simulation agreed with the measured value from the experiment as shown in Fig. 6.

Moreover, to confirm the applicability of the method, the suction phenomena on the Hirono coast was reproduced. Fig. 6 shows the simulation result on the Hirono coast. The simulated result agreed to the measured one, according to the field survey of Ioroi et al. that the accumulated suction rate was $28m^3/m$.

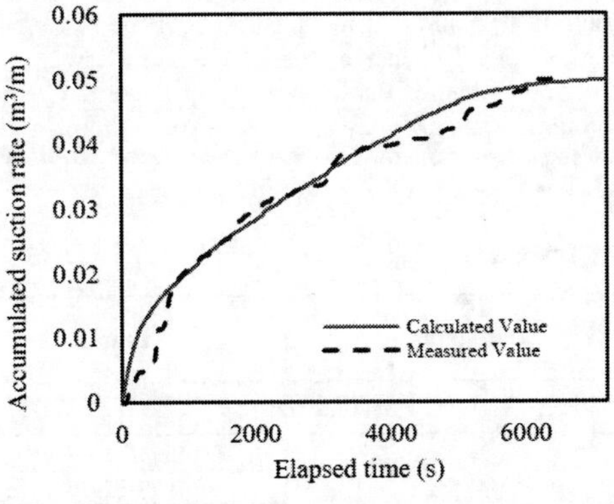

Fig. 5 Relation between accumulated suction rate and elapsed time.

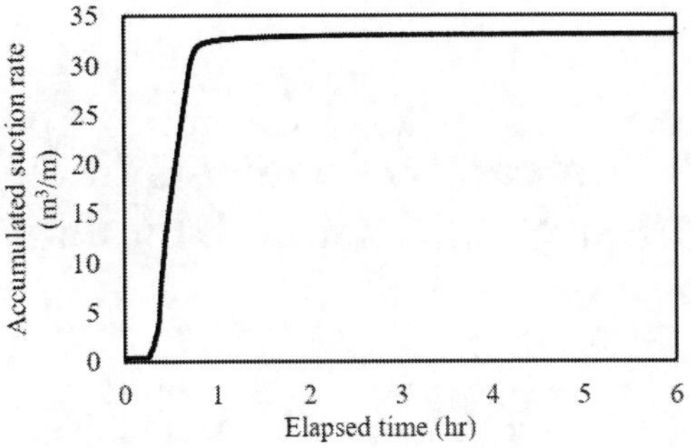

Fig. 6 Reproduction of suction phenomena in the Hirono coast.

5. Conclusion

The authors proposed Eqs.(18)~(23) that can be used in CADMAS-SURF to calculate the suction rates to elapsed time. The reliability and the applicability of the method were confirmed by reproducing simulations of the experiment and the Hirono coast.

References

1. Yamamoto, Y, Minanmi, N (2009). "Experimental Investigation on Destruction Mechanism of Coastal Dyke Due to Big Waves," Journal of Japan Society of Civil Eng, B2-Vol.**65**, No.1, 901-905 (in Japanese).
2. Ioroi, M, Yamamoto, Y (2013). "Formula for Evaluating Suction Rate of Backfilling Materials from a Coastal Dike by Big Waves," Proc 23th Int Offshore and Polar Eng Conf, Alaska, USA, ISOPE, 1210-1216.
3. Ioroi, M, Yamamoto, Y (2015). "Improvement of a Suction Rate Formula from the Bottom of the Sheet Pile of a Coastal Dike," Proc 25th Int Offshore and Polar Eng Conf, Hawaii, USA, ISOPE, **3**, 1483-1488.
4. Yoshizawa, S, Yamamoto, Y, and Kuisorn, W (2016). "Improvement of Methods for Calculating the Suction Rate from a Coastal Dike or a Seawall," Journal of Japan Society of Civil Eng, Vol. **72**, No. 2, 1147-1152 (in Japanese).
5. Coastal Development Institute of Technology (2008). "CADMAS-SURF," http://www.cdit.or.jp/program/cadmas-download.html (in Japanese).
6. Hirt, C, B. D. Nichols (1981). "Volume of Fluid (VOF) method for dynamics of boundaries," Journal of Computer Physics, **39**, 201-225.
7. Ito, M, Tsuchiya, Y (1984). "Scale-Model Relationship of Beach Profile," Coastal Eng Proc, **19**, 1386-1402.

Comparative analysis of city planning and land use change in Bangkok, Thailand, by using remote sensing and GIS

Suvimon Posuk[1], *Prof. Yoshitaka* Kajita[1], and *Arthit* Petchsasithon[2,*]

[1]Civil Engineering Department, Faculty of Engineering, Tokai University, Kanagawa, Japan
[2]Civil Engineering Department, Faculty of Engineering, King Mongkut's Institute of Technology Ladkrabang, Bangkok, Thailand

Abstract. This study proposed the comparative analysis of city planning and land use change by using remote sensing and GIS in Bangkok, Thailand. Bangkok had been developed and faced many land use problems. If the problems were not controlled, it will cause more problems in the future. Therefore, this study suggested the solution to solve occurring land use problems in Bangkok. Which, remote sensing was used to do the land survey and automatically map urban land cover from Landsat time-series satellite imagery. Then, the change of urban area and Bangkok Comprehensive Plan were compared by GIS. And, the results showed that urban area in Bangkok increased 403.99 km^2 over the past 21 years. While rural and agricultural zone in Bangkok Comprehensive Plan decreased due to residential area expansion. So, Area Division System and the district plan from urban planning system of Japan can solve the problems by developing the city and controlling urban areas expansion.

1 Introduction

Bangkok had been developed to be center of country and center of ASEAN Economic Community (AEC). Now, Bangkok has been faced unsuitable land use usage, flood prevention system, the disorder of urban area and environment problems [7]. If the land use change was not controlled, improved and planned, it will cause a lot of problems in future. Effective urban land use planning can help guide urban development. Also, Land survey is one of the processes that used for study land usage. Remote sensing can classify object from satellite imagery and can effective analyses time-series image. And, these techniques can be grouped into two types: (1) classification of input data and (2) directly segmenting the indices [2].

The objective of this study is to use remote sensing doing the land survey by applying Built-up Index (BUI), which has accuracy's result at 92.6% [8], to automatically map urban land cover from Landsat time-series satellite imagery. Then, to compare the change of urban area with established Bangkok Comprehensive Plan (land use zoning plan) by using Geographic Information Systems (GIS). Last, this study suggests the solution to solve occurring land use problems in Bangkok from the successful land use plan.

2 Study Area

The study area is Bangkok City, Thailand. Bangkok located in the central region of the country and the total area is 1,568.74 km^2, which is surrounding with six provinces and south part of the city is connected to the Gulf of Thailand. The approximate population is 5.69 million and population density is 3,629 people/km^2 (2014).

3 Data Resources

Time-series Landsat imagery was used in this study, which according to an established year or close to an established year of the Bangkok Comprehensive Plan. The Images have been downloaded from the U.S. Geological Survey. The images had area covering center region of Thailand that included all over the study area. Bangkok Comprehensive Plans have been downloaded from Department of Public Works and Town & Country Planning. Bangkok city and Bangkok administrative zone have been downloaded from BMA GIS center. The materials used in this study are listed in Table 1.

Table 1. Information of materials used in this study.

Data types	Year
Satellite imagery	
Landsat 5 (TM)	1993-11-07
	1999-12-26
	2006-10-26
	2006-12-29
Landsat 8 (OLI)	2014-11-17
Maps	
Bangkok Administrative Zone	2014
Bangkok Comprehensive Plan	1992 1999 2006 2013

* Corresponding author: arthit.pe@kmitl.ac.th

4 Methodology

4.1 The application of urban area extraction

The Built-up Index (BUI) was applied for urban area extraction, which used the difference between NDVI and NDBI [8]. However, the river needs to be excluded by using the stream space information of the area, because BUI is very irregular and has a negative effect [3]. Therefore, the Modified Normalized Difference Water Index (MNDWI) was used to extract open water area to reduce the negative effect from BUI. The equation of the applications is showed as follows:

$$\text{Urban area extraction} = BUI - MNDWI \quad (1)$$

$$\text{Built-up Index} = NDBI - NDVI \quad (2)$$

$$\text{NDVI} = (Near\ Infrared - Red)\ /\ (Near\ Infrared + Red) \quad (3)$$

$$\text{NDBI} = (Short\text{-}wave\ Infrared\ 1 - Near\ Infrared)\ /\ (Short\text{-}wave\ Infrared\ 1 + Near\ Infrared) \quad (4)$$

$$\text{MNDWI} = (Green - Short\text{-}wave\ Infrared\ 1)\ /\ (Green + Short\text{-}wave\ Infrared\ 1) \quad (5)$$

4.2 Analyzed established Bangkok Comprehensive Plan

Established Bangkok Comprehensive Plans Act. were analyzed to study the objectives, policies and promoted plans. Land Use Zoning Plan maps, which planned to control land use in Bangkok area, were digitized to GIS database map by using ArcMap 10.4.1 and were summarized to compare land use zoning plan change.

4.3 Urban area and Bangkok Comprehensive Plan change analysis and study city problems solving solutions

Urban area extraction results and urban growth causes [1] were summarized to define urban growth direction and were analyzed with Land Use Zoning Plan to do change analysis during an established period of Bangkok Comprehensive Plan, which compared the existing urban area with land use plan. Results of the analysis were considered the effect of city planning. Then, the solution of city planning problems was suggested by the study of concepts from successful land use plan.

5 Results

5.1 Urban area automatic extraction

Urban area in Bangkok increased in huge value over the past 21 years from 462.40 km^2 in 1993 to 866.39 km^2 in 2014 (403.99 km^2 increased). The most changed urban area was between 1999 and 2006, which total urban area changed from 579.55 km^2 to 764.50 km^2 (184.95 km^2 increased). Moreover, urban area in 2014 expanded to 54.72% or more than half of total area of city (Table 2).

Table 2. Comparison of urban area change

	1993	1999	2006	2014
Urban area (km^2)	462.40	579.55	764.50	866.39
Urban area (%)	29.21	36.60	48.29	54.72

Comparison of urban area in Bangkok Administrative zone (Table 3) shows that central zone had the largest urban area, which reached 89.89% of total zone area in 2014. Besides, the eastern zone had the less change in urban area with only 40.55% in 2014.

Table 3. Comparison of percentage of urban area in Bangkok zone

Bangkok zone	1993	1999	2006	2014
Central	72.33	80.12	86.19	89.89
Southern	65.92	72.36	79.31	83.70
Northern	42.14	51.94	66.06	74.86
Eastern	14.11	21.27	35.31	40.55
North Western	40.99	47.48	56.43	64.39
South Western	23.40	30.45	39.74	48.05

The comparison of urban area in Bangkok (Figure 1) displays the character of urban development of Bangkok city, which main part of urban development created along transportation. This type of development was difficult to control, manage and it made rapidly and limitless expansion [5].

5.2 Analysis of Bangkok Comprehensive Plan

The Bangkok Comprehensive Plans were formulated and were initiated as a guideline for developing Bangkok city. The plan was established four times and each plan had been in effect for seven years. The First Bangkok Comprehensive Plan was established in 1992 by Department of Public Works and Town & Country Planning. Thailand government announced a new regulation to give management power to local administration in 1999. Thus, Bangkok Administration decided to publish the First Revision of Bangkok Comprehensive Plan.

The Second Revision of Bangkok Comprehensive Plan (2006) was established to solve problems from the First Revision Plan. However, this plan had a problem in term of enforcement because of the mixed up with another law. In 2012, Thailand faced a great flood and it also affected to Bangkok city. Bangkok Metropolis Administration decided to publish the Third Revision Plan in 2013. The main objectives were improving plan's description, preventing disasters and solving global warming problem.

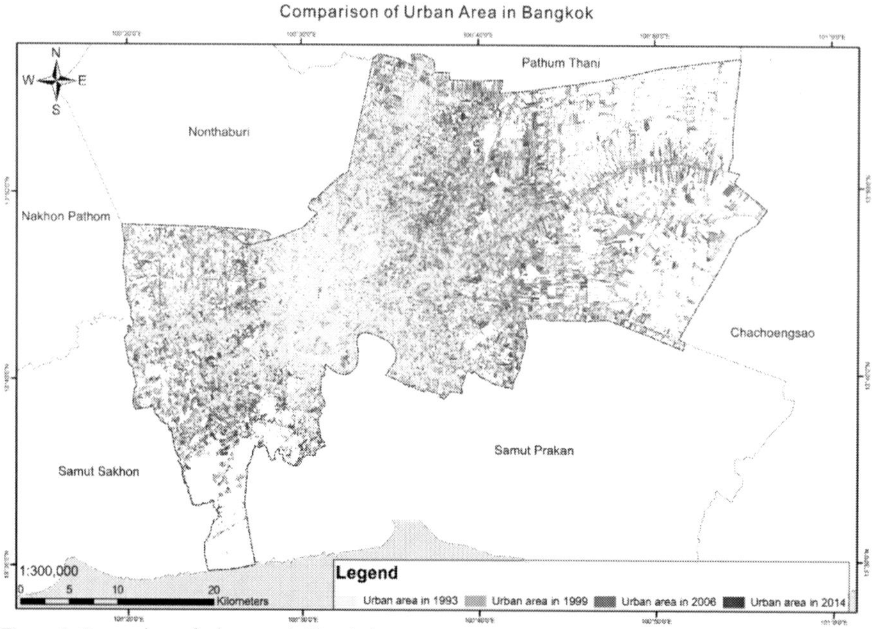

Figure 1. Comparison of urban area in Bangkok

Land Use Zoning Plan of Bangkok comprehensive plan was divided into four main groups: residential zone, commercial zone, industrial zone and rural and agricultural zone. The expected residential zone decreased from 760.92 km^2 (1992) to 695.57 km^2 (2013). Rural and agricultural zone decreased from 674.20 km^2 (1992) to 617.41 km^2 (2013). Industrial zone decreased from 19.06 km^2 (1992) to 14.08 km^2 (2013). While, commercial zone increased from 27.15 km^2 (1992) to 50.86 km^2 (2013) (Figure 2).

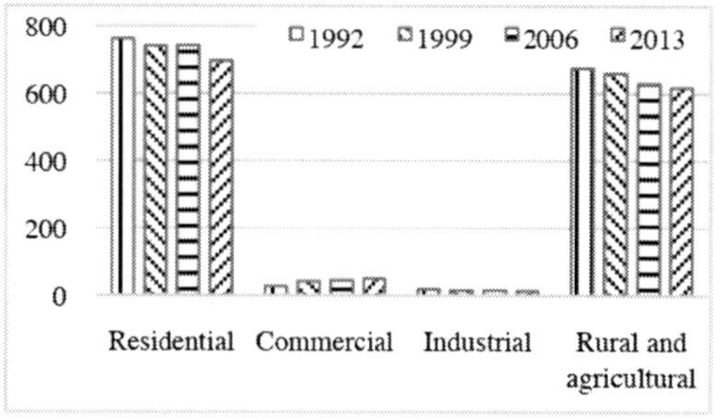

Figure 2. Comparison of land use zoning plan

5.3 Analysis of urban area and Bangkok Comprehensive Plan and suggest the solution to city planning problems solving

The population of Bangkok increased from 5.56 million in 1992 to 5.69 million in 2014. Gross provincial product (GPP) increased from 1.15 million (bath) in 1992 to 3.92 million (bath) in 2013. From the prediction of Department of city planning [6] show that in 2030 population will increase to 6.23 million and GPP will increase to 4.21 million (bath) (Figure 3).

Figure 3. Population and GPP of Bangkok

Urban area expanded 117.15 km^2 (19.53 km^2/year) from 1992 to 1999, 184.95 km^2 (26.42 km^2/year) from 1999 to 2006 and 101.89 km^2 (12.74 km^2/year) from 2006 to 2013 (Table 4).

Table 4. Average rate of urban growth in Bangkok

	1992-1999	**1999-2006**	**2006-2013**
Area change (km^2)	117.15	184.95	101.89
Average rate (km^2/y)	19.53	26.42	12.74

Comparison of existing urban area with the expectation area from land use zoning plan shows the increased of urban area in residential zone from 300.89 km^2 (1992) to 481.54 km^2 (2013). In the other hand, urban area in rural and agricultural zone also increased from 64.30 km^2 (1992) to 160.37 km^2 (2013) (Figure 4).

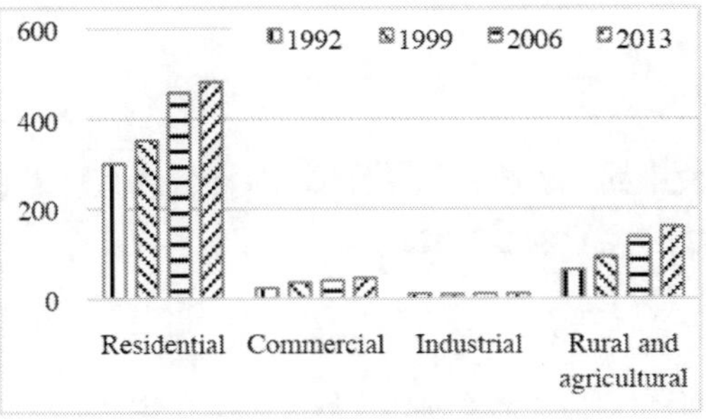

Figure 4. Comparison of urban area in land use zone

In 2015, ASEAN Economic Community (AEC) was formed and Bangkok aimed at developing city to be center of AEC region. Moreover, population and economics of Bangkok in 2030 was predicted that it will grow continuously. From the cause of urban growth showed that population growth and economic growth will create demand for new housing or urban expansion.

So, Bangkok has to develop the city to support the activities from AEC that will occur in the future. At the same time, Bangkok has to control the development of urban areas in residential zone and stop the expansion of urban areas in rural and agricultural zone. From the beginning, rural and agricultural zone was designed as green zones to control flooding. However, it had been rapidly changed into urban area.

From the study of urban planning system of Japan, Area Division System, which city planning area was classified into urbanization promotion area and urbanization control area. This system can prevent disorderly urbanization and control the expansion of urban area. And, the development of urban area, Bangkok should support districts to establish district plan. The district plan can develop and maintain a good urban area that matches the characteristics of different district, for example, to detailed district urbanization, to preserve the environment and to determine layout of public facilities [4].

6 Conclusion

This study applied urban area extraction to study urban area in Bangkok city by using BUI and MNDWI and analyzed Bangkok Comprehensive Plan with existing urban area to suggested solutions of city planning problems. The results of this study revealed three main results:

(1) Urban area in Bangkok increased from 462.40 km^2 in 1993 to 866.39 km^2 in 2014 or 403.99 km^2 increased over the past 21 years, which the central zone had the largest urban areas and the eastern zone had the smallest urban areas. The development character of Bangkok was the main part of urban area create along transportation, which it was difficult to manage development and it made rapidly and limitless expansion.

(2) The Bangkok Comprehensive Plans were established four times since 1992. Residential zone and rural and agricultural zone were the main zones of the Land Use Zoning Plan. Residential zone decreased from 760.92 km^2 (1992) to 695.57 km^2 (2013) because of commercial zone and transportation development. While rural and agricultural zone decreased from 674.20 km^2 (1992) to 617.41 km^2 (2013) due to residential area expansion.

(3) Bangkok aimed at developing city to be center of AEC region. Population growth and economic growth in Bangkok will create demand for new housing. Moreover, the comparison of extracted urban area with expected land use zoning showed that urban area in residential zone increased from 300.89 km^2 (1992) to 481.54 km^2 (2013). Urban area in rural and agricultural zone also increased from 64.30 km^2 (1992) to 160.37 km^2 (2013).

So, Bangkok has to develop the city to support the activities, control urban areas development and stop urban areas expansion. Area Division System from urban planning system of Japan can prevent disorderly urbanization and control the expansion of urban area. Furthermore, Bangkok should support district to plan the district plan, which the plan can control district urbanization, conserve the environment and develop public facilities.

References

1. B. Basudeb, Analysis of Urban Growth and Sprawl from Remote Sensing Data (Online) Available at: http://www.springer.com/978-3-642-05298-9

2. C. He, P. Shi, D. Xei, Y. Zhao, Remote Sensing Letters **1**, 213-221 (2010)

3. A. J. Lee, S. S. Lee, K.H. Chi, Korean Journal of Remote Sensing **27**, 637-651 (2011)

4. Ministry of Land, Infrastructure and Transport, Introduction of Urban Land Use Planning System in Japan (Online) Available at: http://www.mlit.go.jp/common/000234477.pdf

5. C. Nurat, Urban structure in Samut Prakarn province adjacent periphery area of Bangkok metropolitan, Thailand (2010)

6. Department of city planning, Bangkok Comprehensive Plan 2013 (Online) Available at: http://www.mlit.go.jp/common/000234477.pdf

7. N. Sumpaongern, The problems of acheiving the objectives of the Bangkok Comprehensive Plan (Online) Available at: http://203.131.219.242/cdm/compoundobject/collection/thesis/id/26200/rec/1

8. Y. Zha, J. Gao, S. Ni, International Journal of Remote Sensing **24**, 583-594 (2003)

MATEC Web of Conferences **192**, 02065 (2018)
ICEAST 2018
https://doi.org/10.1051/matecconf/201819202065

The stress analysis in electric poles with CFRP to prevent the consecutive demolishing from car accidents

Suchat Suvimolwan[1] and *Atavit* Sujaritpong[1]

[1]Civil Engineering Department, Faculty of Engineering, King Mongkut's Institute of Technology Ladkrabang, Bangkok, Thailand

Abstract. In statistically accidental reports, we could find progressive collapse of prestressed concrete electric poles from car accident, this is due to the poor properties of concrete which limits the impact resistance, the failure will therefore occur in the position of impact where maximum stress is presented near the base. The weight of broken concrete at above the failure position will cause the impact force to the adjacent electric poles, resulting in consecutive damage and increase the tensile force in electric cables. This research will emphasize repositioning of failure at pole base by application of CFRP material. Finite element analysis will be introduced to analyse the thickness of CFRP in order to control the desired point of failure. The result can be a guideline to prevent the consecutive demolishing for the existing electric poles later on.

1 Introduction

Generally, the manufacturing of electric pole according to the Metropolitan Electricity Authority's standard by concrete has been using as the main material for long, this is because concrete can be casted to shapes and sizes as desired including when considering cost and strength properties. However, due to the limitation of concrete itself especially flexural strength and brittle property of concrete are weak, we therefore can see many cases of consecutive broken down which occurred by the accidental from car crash.

Based on the incidents, this showed the strength of concrete is still not strong enough for the damage of electric poles. The consecutive broken down of electric poles mostly occurred at the position near by the pole base which highest moment was achieved. The tension force by weight of broken poles from the car crash was transferred thru the cable and to the next the poles at the top part, this caused the progressive broken to numbers of poles.

2 The failure of concrete poles, CFRP and related materials

.2.1. The failure of concrete poles

The failure of concrete pole will occur in the position of impact where maximum stress is presented near the base. The concrete weight above the failure position will cause the impact force to the adjacent electric poles, resulting in consecutive damage and increase the tensile force in electric cables. To minimize the weight beyond the failure position is therefore considered to decrease these impact forces to the adjacent poles and may present the consecutive damage consequently.

2.2 Resulting of bending moment from the tensile force thru the cable

The resistance to the bending moment of concrete pole which installed cables and various accessories can be explained in Eq.1.

$$B.M. = F \times S \qquad (1)$$

Where B.M. is bending moment, F is tensile force and S is the distance of tensile force to the ground base.

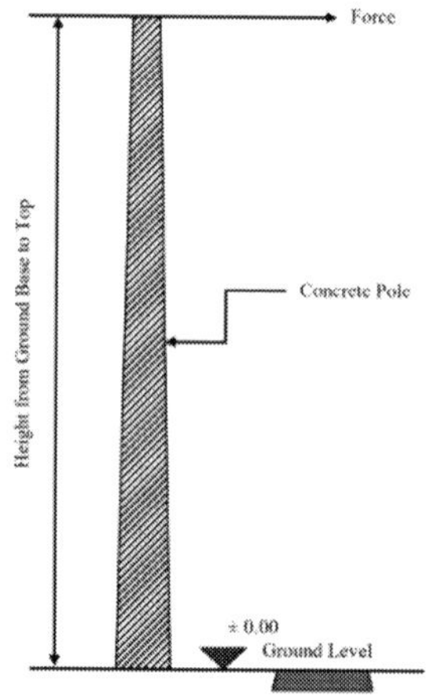

Fig. 1. Position of force from the cable

* Corresponding author: scsuchat2@gmail.com

2.3 Pre-stressed electric concrete poles

Based on the Metropolitan Electricity Authority's standard concrete electric poles has been manufactured in different heights with desired strength and bending moment for various functions. Table 1 is shown 4 main materials of concrete poles with the details of tensile strength as following.

Table 1. Strength of materials in concrete poles

Materials	Strength of Materials	Units
Concrete	450	kgf/cm^2
Pre-stressed Wires	12,500	kgf/cm^2
Stirrup Steel	2,400	kgf/cm^2
CFRP Fabric	35,500	kgf/cm^2

In this study, the concrete poles height of 12m. with the bending moment of 3.5T-m., 12m. with bending moment of 5T-m. and 12.35m. with bending moment of 6.5T-m. will be used as the models in Finite Element Analysis, these 3 models will be called 12GW, 12m and 12.35m respectively in this study.

2.4 CFRP materials and properties

The prevention of consecutive damage in electric poles can be achieved by the strengthening with the application of Carbon Fibre Reinforced Plastic (CFRP), unidirectional fibre fabric type is considered to use due to the reason of high strength of CFRP material, easily to wrap the shapes of concrete poles and the cost is also reasonable when compared to other types.

Table 2. Strength of Materials

Properties	CFRP 200g/m^2	CFRP 230g/m^2	CFRP 300g/m^2	CFRP 530g/m^2
Design Thickness	0.11	0.13	0.17	0.29
Ultimate Elongation	2.10%			
Tensile strength	35,500 kgf/cm^2 "Design Value"			
Tensile E-Modulus	2.35x10^6 kgf/cm^2 "Design Value"			

Wrapping the concrete poles with CFRP is to strengthen and to increase the ductility of the concrete poles where the maximum stress is expected. The height of wrapping will be desired as the position of cracking control for the selected concrete poles. Thickness of 0.13mm, 0.17mm, 0.26mm and 0.34mm by CFRP will be applied in each 3 models for FEM analysis, this is for the total of 12 models plus 3 reference models without CFRP wrapping. The properties and details of CFRP Materials are given in Table 2.

2.5 Concrete footing at the ground base

The analysis will also consider the practical of CFRP installation with the existing concrete poles, as a result the concrete footing will be applied and casted after CFRP wrapping, the desired position is at the ground level which is 30cm. deep down and upward from this ground base. The thickness of concrete footing will be 2 times of the section areas of pole at the base, this is also to avoid the stress occurred under the ground base and to be practical for on-site installation of CFRP.

3 Modelling and Analysis by FEM

In this paper the finite element analysis will be introduced to analyse which materials will take the maximum stress as well the safety factor ratio of compressive strength in concrete on the compression zone, including the analysis on the thickness of CFRP wrapping in order to control the desired position of failure.

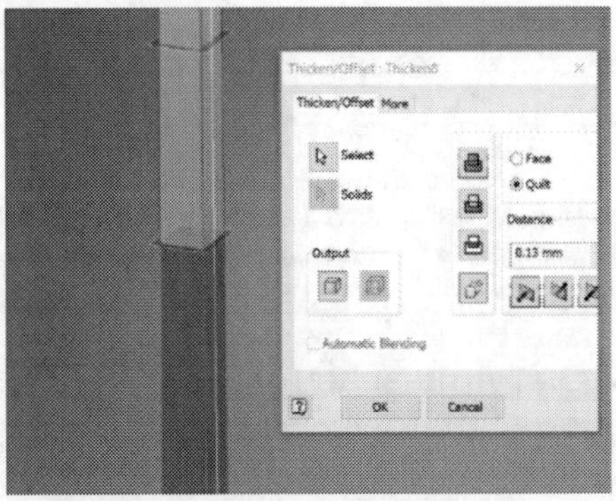

Fig. 2. Thickness of CFRP in FEM program

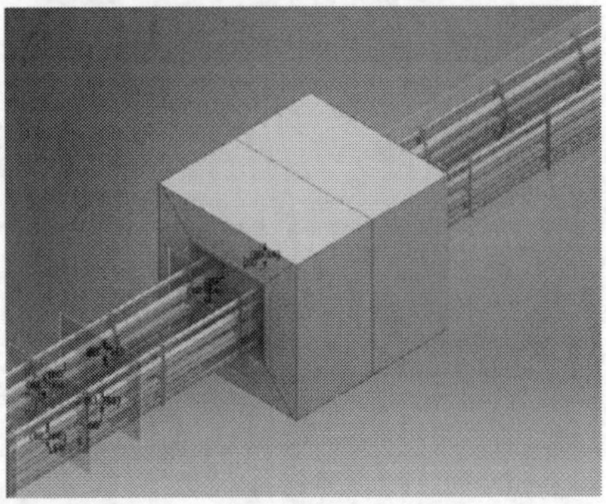

Fig. 3. Concrete Footing in the model

15 models will be created in Inventor program including reference models without CFRP, the solid body of models will consist of Concrete, Pre-stressed Wires, Stirrup Steel, CFRP Fabric and Concrete Footing at the ground level which require different menu in Inventor program, the thickness of CFRP were built in 4 types in between of 0.13mm until 2.8mm related to the result analysis required on each models as shown in Fig. 2 and 3.

Table 3 is shown the details of concrete poles, uniform loads, wrapping position, tensile forces and base level used with models in FEM analysis.

Table 3. Loads and tensile forces to use in FEM Analysis

Concrete Poles	Uniform Load 1 and 2 (Mpa)	CFRP Wrapping (m.)	Force (Wrap & Upwrapped) x Factor 2.0 (N)	Base Level (m.)
12GW (3.5T-m)	5.87	6.05	11,541.18	1.75
	14.12		6,699.51	
12m. (5T-m)	7.34	6.00	16,350.00	
	17.64		9,570.73	
12.35m. (6.5T-m)	9.27	5.30	18,089.36	2.10
	23.45		12,441.95	

The installation of CFRP are based on the standard procedure recommended by manufacturer with the special adhesives in reality, direction of fibre will be parallel to the height of poles in order to take on the action of tensile force transferred thru the cables from the broken poles.

5. Analysis results

The analysis was run by ANSYS program for all poles models, the deformation results of unwrapped are 27, 30, 20cm for 12GW, 12m and 12.35m respectively. Fig.4 is shown the contour of deformation of the FEM analysis.

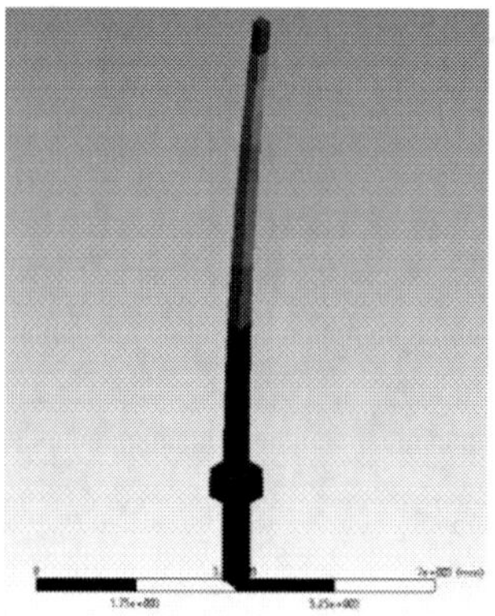

Fig. 4. Deformation contour from FEM Analysis of Models

The analysis result by FEM program on the wrapped model of 12m. (3.5T-m) gave the deformation value of 42, 41, 40 and 39cm which was slightly decreased when thickness was increasing but higher than unwrapped. While the maximum stress are 699, 675, 653 and 647 MPa and mainly occurred in CFRP materials, these stress were just 19% of CFRP's tensile strength, resulting by the thickness of 0.34mm, 0.51mm, 1.2mm and 1.4mm respectively. The details are shown in Fig. 5.

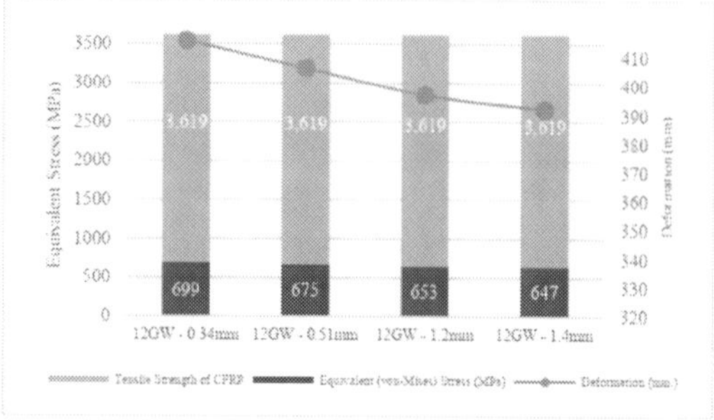

Fig. 5. Stress and Deformation Analysis of 12m. (BM 3.5T-m)

Considering the ratio of safety factor in compression zone, this analysis showed the potential of failure position on the height of poles, the result showed that increasing thickness of CFRP on the models could shift up the safety factor value at the poles base to be 1.05 while values at the position beyond the wrapping were shifted down below 1. This result showed that the thickness at 1.4mm is the most effective and safety for both deformation and prevention failure at the pole base. Fig. 6 is shown the relationship of safety factor ratio and the height of models.

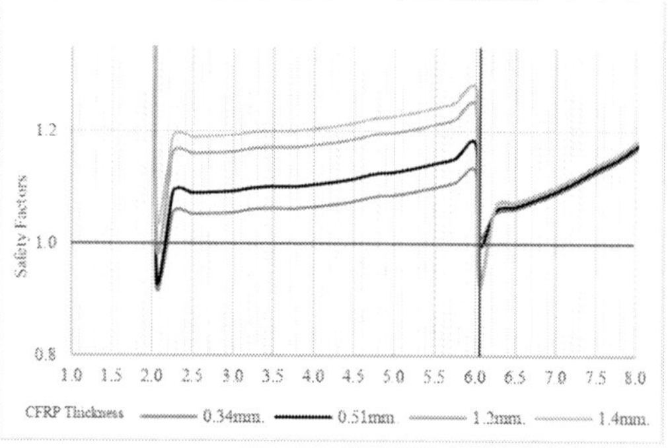

Fig. 6. Graph of Safety Factor Analysis of 12m. (3.5T-m)

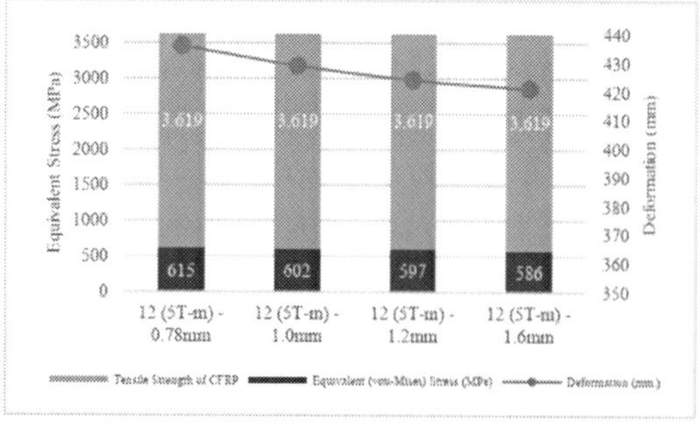

Fig. 7. Stress and Deformation Analysis of 12m. (5 T-m)

The analysis result of 12m. (5T-m) by FEM program is shown is Fig.7. The result gave the deformation value of 44, 43, 42 and 42 cm which were not much different when thickness was increasing. While the maximum stress showed some decrease from 615, 602, 597 to 586 MPa and mainly occurred in CFRP materials, these stress were just 17% of CFRP's tensile strength, resulting by the thickness of 0.78mm, 1.0mm, 1.2mm and 1.6mm respectively.

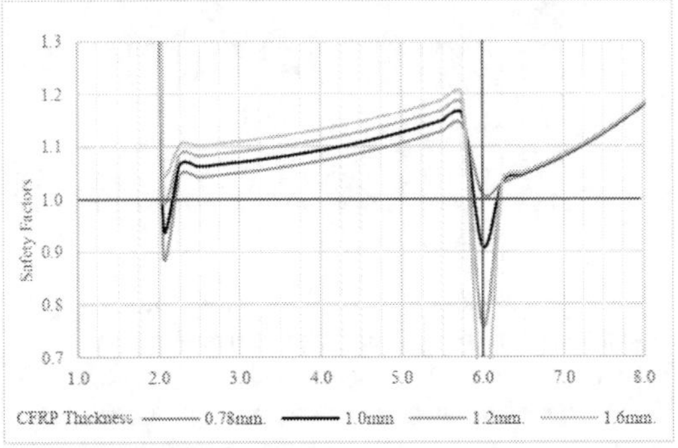

Fig. 8. Safety Factor Analysis of 12m. (BM 5T-m)

The analysis of safety factor in compression zone on 12m. (BM 5T-m) were started with CFRP at 0.78mm thickness which result showed the failure at the ground level, increasing thickness to 1.0mm, 1.2mm and 1.6mm were introduced with the pole models, and finally at 1.6mm could give the best result which shifted up the safety factor higher than 1 while the values at 6m were shifted down below 1. Fig. 8 is shown the safety factor ratio based on the height of models.

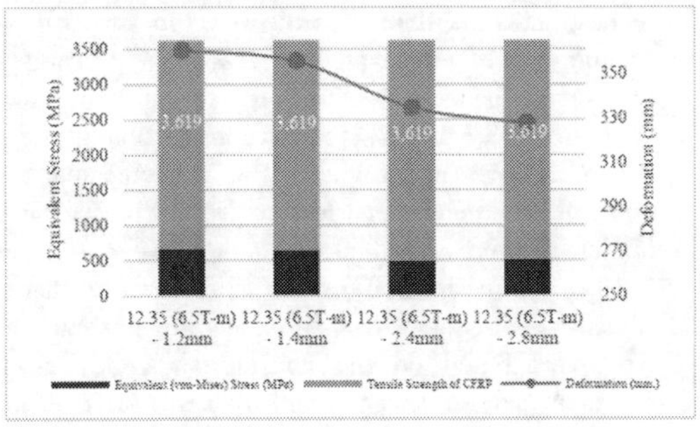

Fig. 9. Stress and Deformation Analysis of 12.35m.

The analysis result by FEM on model of 12.35m. (6.5T-m) gave the deformation value at 36, 36, 34 and 33cm for the different of CFRP thickness. While the maximum stress are 671, 655, 499 and 517 MPa and mainly occurred in CFRP, these stress were just 18.5% of CFRP's tensile strength resulting by the thickness of 1.2mm, 1.4mm, 2.4mm and 2.8mm respectively as shown in Fig. 9.
While the post-cracking by safety factor analysis of 12.35m pole model, the FEM analysis gave the similar result to the others. CFRP at 1.2mm thickness was introduced as the 1st model for the analysis, the result showed the failure at the ground base, therefore 3 more thickness of 1.4mm, 2.4mm and 2.8mm were built. CFRP at 2.4mm could give the safety factor higher than 1 at ground base while it shifted down the value below 1 at the position above the wrapping of CFRP. Fig. 10 is shown the relationship of safety factor and height of pole model.

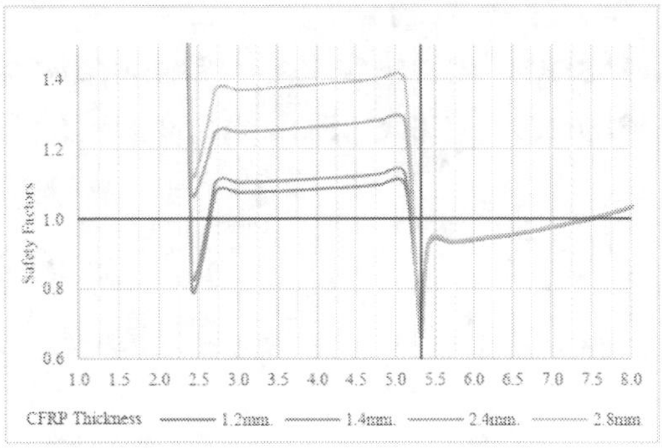

Fig. 10. Safety Factor Analysis of 12.35m. (BM 6.5 T-m)

5. Conclusion

The analysis result by FEM program from the application of CFRP material showed the prevention of failure at the poles base with the effective thickness of 1.4mm., 1.6mm. and 2.4mm, and the result of the maximum stress occurred at CFRP materials instead of rebar or concrete while the maximum stress values are at 19%, 17% and 18.5% for the models of 12GW, 12m. and 12.35m. respectively, these are still within the tensile strength of CFRP material. Considering the position of post-cracking in concrete poles which were analysed by the ratio of safety factor at the compression zones, the result of wrapped poles with CFRP could give the higher values than 1 at the ground base, while the results at above wrapping position was below 1, this could prove that the stress to break down the poles, moved up beyond the position of wrapping areas as per the desired position.
Therefore this study showed the improvement successfully to strengthen and prevent the consecutive demolishing for the electric concrete poles by CFRP material which were analysed by FEM.

Acknowledgement

This study was supported by Dr. Atavit Sujaritpong and Dr. Artit Petchsasithorn from the Faculty of Engineering at King Mongkut's Institute of Technology Ladkrabang for all the advices.

References

1. Metropolitan Electricity Authority, Products and Transportation Department, 2017, The Instruction of Standard Drawing of Electric Prestressed Concrete Poles 12m (GW), 12m (BM 3.5T-m.), 12.35m (BM 6.5T-m)
2. Ministry of Interior, Department of Public Works and Urban Planning, 2008, "MorYorPhor.1508-51 The Standard of Structure Strengthening Reinforced Concrete Structures with Composite Fibre Materials", Bangkok

Users' long-term satisfaction with post-disaster permanent housing: a case study of 2010 Merapi Eruption, Indonesia

Bee Lan Oo[1*], *Riza* Sunindijo[1], and *Fatma* Lestari[2]

[1] Faculty of Built Environment, University of New South Wales Sydney, Kensington, NSW 2052, Australia

[2] Occupational Health & Safety Department, Faculty of Public Health, Universitas Indonesia, Depok16424, West Java, Indonesia

Abstract. An objective evaluation of users' long-term satisfaction with the post-disaster permanent housing provided is important for judging the successes or failures of housing reconstruction programs. Focussing on four resettlement sites built via the REKOMPAK program after the 2010 Merapi eruption, this paper examines the users' long-term satisfaction with the provided housing, and explores how their satisfaction level relates to their demographic characteristics and level of participation in the reconstruction processes. The results show that the households are in overall satisfied with the present housing. Their overall satisfaction level is statistically positively associated with household income, but not their level of participation and other demographic characteristics. These findings provide evidence on a satisfactory outcome in the long-term for the REKOMPAK, a community-based reconstruction program that was implemented in Indonesia since 2004, and a useful insight for local authorities, humanitarian and other agencies involved in post-disaster reconstruction program.

1 Introduction

The Merapi volcano eruption in Oct 2010 was an unusually large magnitude explosive eruption that caused over 350 fatalities and more than 300,000 people evacuated [1]. Close to 3900 houses were damaged in the affected areas, and more than 75% of all housing damaged by the eruption were in the Slemen Regency in Yogyakarta Province [2]. Cangkringan district is one of the fourteen districts in the Slemen Regency, and was worst affected by the 2010 Merapi Eruption with over 2500 houses destroyed [3]. The post-disaster housing reconstruction was implemented through a scheme called REKOMPAK with funding from the Java Reconstruction Fund contributed by donors after Central Java earthquake in 2006 [4]. Under the REKOMPAK program, there are fifteen resettlement sites (called *huntap*, permanent settlements) with 1596 houses built in the five villages that make up the Cangkringan district. Focussing on four resettlement sites in Cangkringan district, this paper examines the users' long-term satisfaction with post-disaster permanent housing, and explores how their satisfaction level relates to their demographic characteristics and level of participation in the reconstruction processes.

2 The REKOMPAK

The REKOMPAK (community-based rehabilitation and reconstruction settlement project) is an owner-driven program that requires beneficiaries to make decision and to organize for their own recovery, giving them a sense of control of the future [4]. For reconstruction in Merapi, the REKOMPAK program was implemented with previous experiences and lessons learnt drawn from its implementation in natural disasters between 2004 and 2010 in Indonesia. Maly et al. [2] categorised that the fifteen resettlement sites in Cangkringan into five typologies based on the housing and infrastructure provision and sources of fund, namely: (a) collective resettlement sites; (b) independent collective resettlement, (c) combined support collective resettlement; (d) independent individual reconstruction; and (e) independently non-REKOMPAK supporter rebuilding. Type A was the typical type in most resettlement sites where the planning and construction of permanent housing were carried out by resident group representatives with the support from facilitators provided by REKOMPAK. The funding was in the amount of 30 million Rupiah for building a 6 x 6 meter reinforced concrete block house on 100m^2 lot of village-owned land for each household. While the resettlement sites vary in scale, characteristics, construction timelines, degree of housing expansion and inhabitation, and

* Corresponding author: b.oo@unsw.edu.au

livelihood opportunities, the REKOMPAK program had called for the community participation in the various reconstruction processes.

There has been much evidence in the literature on the importance of community participation in post-disaster reconstruction program for high level of beneficiary satisfaction (e.g., [5-7]). In addition, authors have emphasized on the importance of objective evaluations of permanent housing built for disaster victims in the long-term (e.g., [6-8]). Changes in resident satisfaction and permanency at a post-disaster housing sites in the long-term may lead to the home owners abandon the provided housing (e.g., [9-11]). Given that evaluation is difficult and may be omitted by donors, past evaluations often focus on easily enumerable aspects of reconstruction, such as the number of houses, schools and clinics built [6]. Literature on evaluation of users' long-term satisfaction remains scarce, this study aims to address this research gap by focussing on four resettlement areas built under the REKOMPAK program after 2010 Merapi Eruption. As it was a community-based reconstruction program, this study provides an insight into the relationship between the users' long-term satisfaction and their level of participation in the reconstruction. The empirical evidence would have implications on the development of the respective guidelines and policies for local authorities, humanitarian and other agencies.

3 Research method

A survey research design was adopted in this study for its abilities to provide a relatively quick and efficient method to (i) obtain information from the targeted sample, and (ii) generalize the research findings based on the sample involved. The data collection instrument is a structured survey questionnaire written in Bahasa Indonesia, and data was collected via a field survey between Dec 2017 and Feb 2018. As part of a larger project on testing a conceptual framework on users' long-term satisfaction with post-disaster permanent housing in Oo et al. [12], there are six sections in the questionnaire. The results reported in this present work are related to the users' overall satisfaction and how it relates to their demographic characteristics and level of participation in the reconstruction processes. Table 1 shows the number of households participated in the survey from four resettlement areas, namely: (i) Batur, (ii) Pagarjurang; (iii) Gondang 2; and (iv) Dongelsari, which are marked as Huntap 4, 5, 7 and 9 in Figure 1. These major resettlement areas are Type A collective resettlement areas built under the REKOMPAK on village-owned land. The total number of houses (738) built in these resettlement areas represents 46% of the total of 1596 houses built in Cangkringan District, Slemen Regency. For the data analysis focussing on long-term perspective, 285 (out of 297) sets of completed surveys were included based on a selection criterion that the households must have stayed in the post-disaster permanent housing for at least 5 years (i.e., moved in before year 2014). The response rates range between 31 and 57% for the four resettlement areas, which are considered high and representative.

* Corresponding author: b.oo@unsw.edu.au

Table 1. Number of households participated in the survey

Resettlement site (Huntap)	No. of houses built	No. of households moved in before 2014	Response rate (%)
Batur (4)	206	96	46.60
Pagerjurang (5)	301	96	31.89
Gondang 2 (7)	84	48	57.14
Dongelsari (9)	147	45	30.61
Total	738	285	38.62

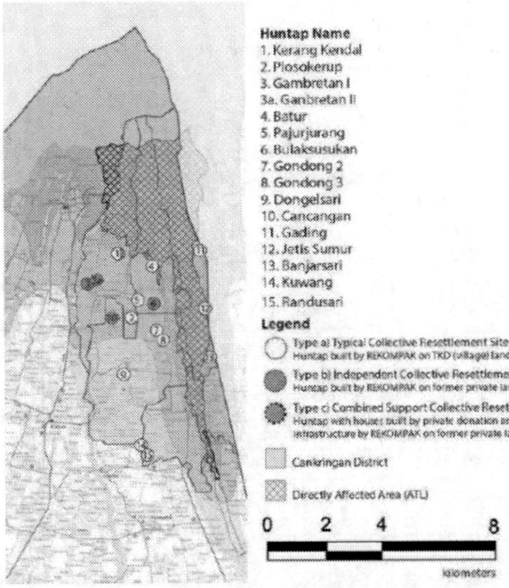

Fig. 1. Collective resettlement sites (Huntap) in Cangkringan District, Slemen Regency (Source: [2])

4 Results and discussion

Table 2 shows the characteristics of respondents and households participated in the study. The majority of them (72%) moved into the present house in year 2012, and the biggest age group (51%) is between 36 and 55 years old. The typical family size is three to four members with grown-up children as the majority (86%) of the households have no children younger than six years old. Almost all households (91%) have two to three economically active members who are either employed or self-employed. Sixty percent of the households with a monthly household income of less than 1 million Rupiah (around USD 70 at March 2018 exchange rates).

Table 2. Characteristics of respondents and households

Characteristics	Frequency	%
Gender of respondents		
Male	172	60.35
Female	113	39.65
Year moved in		
2010	5	1.75
2011	39	13.68
2012	204	71.58
2013	37	12.98
Age of respondents		
< 25	33	11.58
26 to 35	44	15.44
36 to 45	74	25.96
46 to 55	72	25.26
56 to 65	39	13.68
> 65	23	8.07
Employment status of respondents		
Employed/ self-employed	226	79.30
Housewife	42	14.74
Student	12	4.21
Unemployed	5	1.75
No. of economically active members in household		
1.0	9	3.16
2.0	168	58.95
3.0	92	32.28
4.0	16	5.61
Monthly household income		
< Rp 1 million	171	60.00
Rp 1-2 million	93	32.63
Rp 2-3 million	16	5.61
Rp 3-5 million	5	1.75
No. of people live permanently		
1	15	5.26
2	44	15.44
3-4	198	69.47
5-8	28	9.82
No. of elderly (56 and above)		
0	184	64.56
1	61	21.40
2	40	14.04
No. of children (younger than 6)		
0	245	85.96
1-2	38	13.33
3-4	1	0.35
5 and above	1	0.35

When the respondents were asked about their overall satisfaction with the present permanent housing, the mean scores of four resettlement areas are close to or slightly above 4 'satisfied' (Table 3). Although most of the respondents are in overall 'satisfied' with the present housing, it is noted that percentages of respondents from Batur (21%) and Dongelsari (11%) who were 'very satisfied' are higher than that of Pagerjurang (3%) and Gondang 2 (2%). Despite the lower perceptions from the two former resettlement areas, their rather high overall satisfaction provides indicative evidence that the permanent housing provided via the REKOMPAK resettlement program are well-received by the success of the REKOMPAK program in Merapi to past experiences and lessons learnt from its implementation in Aceh post-tsunami and Central Java earthquake in 2006 ([2, 3]).

Table 4 shows the respondents' (or households') level of participation in the permanent housing reconstruction processes. About one-third of them did not participate in any of the reconstruction processes. While the remaining 191 (67%) households had participated in different processes, only about 10% of the households were involved in more than three processes. The largest group (27%) had participated in at least two processes.

Table 4. The respondents' level of participation in the housing reconstruction processes

Level of participation	Frequency	%
No participation	94	32.98
Participated in different processes		
1 process	50	17.54
2 processes	76	26.67
3 processes	37	12.98
4 processes	14	4.91
5 processes	4	1.40
6 processes	3	1.05
7 processes	7	2.46

Table 5 shows the correlation test results between the respondents' overall satisfaction with the present housing and their demographic characteristics and level of participation in the reconstruction processes. The household monthly income is the only demographic factor that is statistically positively correlated to the respondents' overall satisfaction ($r_s = 0.179$, $p < 0.01$). This suggests that having higher household income is related to higher overall level of satisfaction. On the other hand, it is indeed surprising to find that there is no statistically significant positive correlation between the respondents' overall satisfaction and level of participation. This finding provides indicative evidence that the resident group representatives and the REKOMPAK facilitators were successfully in informing community during the reconstruction processes that leads to high satisfaction level, even not all the households had participated. A further examination shows that the mean overall satisfaction scores of those participated in the reconstruction process (3.95) is slightly lower than those who did not participate at all (4.02). These findings could possibly be explained because those involved were more critical of the housing conditions or had higher expectations that cannot be met during the reconstruction processes. Nonetheless, apart from household income, the overall satisfaction level of the respondents from all four resettlement areas are considerably high (mean scores of close to 4 out of 5) regardless of their age, education, family size, number of children and elderly in the family, number of years of living in, and level of participation in the reconstruction processes.

* Corresponding author: b.oo@unsw.edu.au

Table 3. The respondents' overall satisfaction with the present housing

Resettlement site (Huntap)	N	Percentage of respondents, scale 1 to 5					Mean score	Std. Dev.
		Very dissatisfied	Dissatisfied	Neither dissatisfied nor satisfied	Satisfied	Very satisfied		
Batur	96	0.00	0.00	14.58	64.58	20.83	4.06	0.59
Pagerjurang	96	0.00	3.13	3.13	90.63	3.13	3.94	0.43
Gondang 2	48	0.00	6.25	6.25	85.42	2.08	3.83	0.56
Dongelsari	45	0.00	0.00	8.89	80.00	11.11	4.02	0.45

Table 5. Correlation between the respondents' demographic characteristics, level of participation and overall satisfaction

	Spearman's rho correlation coefficient							
	Age	No. of people living in	No. of elderly	No. of children	Education	Household monthly income	No. of years living in	Level of participation
Overall satisfaction	-0.005	0.010	0.009	0.003	0.095	0.179**	-0.001	-0.081

** Correlation is significant at the 0.01 level (2-tailed)

5 Conclusion

As part of a larger project on testing a conceptual framework on users' long-term satisfaction with post-disaster permanent housing, this paper examined users' overall long-term satisfaction with post-disaster permanent housing in four resettlement areas following the 2010 Merapi eruption. Although the households have varying perceptions of the present housing, its location and neighbourhood, they are in overall satisfied with the present housing. Despite the high level of community participation (about two-third of the households had participated) in the reconstruction processes via the REKOMPAK program, the results show that their overall satisfaction level is statistically positively associated with household income, but not their level of participation and other demographic characteristics including age, education, family size, number of children and elderly in the family, and number of years of living in the present housing. Nonetheless, the evidence is suggestive that the REKOMPAK program were well-received by the households affected by 2010 Merapi eruption. For the next stage of this project, further statistical analysis will be conducted to examine the inter-relationships of users' demographic characteristics, reconstruction project delivery processes, attributes of permanent housing, and users' long-term satisfaction.

References

[1] Global Volcanism Program, Report on Merapi (Indonesia). In: Sennert, S K (ed.), Weekly Volcanic Activity Report, 26 October-1 November 2016. Smithsonian Institution and US Geological Survey (2016).

[2] Maly, E., Iuchi, K., Nareswari, A., Community-based Housing Reconstruction and Relocation: REKOMPAK program after the 2010 Eruption of Mt. *Merapi, Indonesia Institute of Social Safety Science Journal*, **27** (2015), 205-214.

[3] Samekto, C.B.D., Nuh, M., Community-based Housing Reconstruction and Relocation: REKOMPAK program after the 2010 Eruption of Mt. Merapi, *Journal of Public Administration Studies*, **1** (2017), 64-70.

[4] The World Bank, *REKOMPAK: Rebuilding Indonesia's Communities After Disasters. Jakarta, Indonesia.* The Secretariat of the Multi Donor Fund for Aceh and Nias and the Java Reconstruction Fund (2012).

[5] Davidson, C.H., Johnson, C., Lizarralde, G., Dikmen, N., Sliwinski A., Truths and myths about community participation in post-disaster housing projects. *Habitat International*, **31**(2007), 100-115.

[6] Fanany, I., Towards a model of constructive interaction between aid donors and recipients in a disaster context: the case of Lampuuk. In: Clarke M, Fanany I (eds), *Post-disaster reconstruction: Lessons from Aceh.* London: Taylor and Francis; (2010), 107-125.

[7] Dias, N.T,, Keraminiyage, K., De Silva, K.K., Long-term satisfaction of post disaster resettled communities: the case of post tsunami–Sri Lanka. *Disaster Prevention and Management*, **25**(2016), 581-594.

[8] Tas N, Cosgun N, Tas M., A qualitative evaluation of the after earthquake permanent housings in Turkey in terms of user satisfaction—Kocaeli, Gundogdu Permanent Housing model. *Building and Environment*, **42** (2007), 3418-3431.

[9] Barenstein, J.D., Housing reconstruction in post-earthquake Gujarat: a comparative analysis. Humanitarian Practice Network at Overseas Development Institute, (2006), network paper 54.

[10] Ahmed, I., McEvoy, D., Post-disaster housing reconstruction: Post-occupancy case studies from Sri Lanka. In Proceedings of ANZAScA 44th annual conference, Auckland (2010).

[11] Manatunge, J.M.A., Abeysinghe, U. Factors affecting the satisfaction of post-disaster resettlers in the long term: A case study on the resettlement sites of tsunami-affected communities in Sri Lanka. *Journal of Asian Development*, **3**(2017), 94-124.

[12] Oo, B.L., Sunindijo, R., Lestari, F. Users' long-term satisfaction with post-disaster permanent housing programs: A conceptual model, *International Journal of Innovation, Management and Technology*, **9**(2018), 28-32.

[13] Mei, E.T.W., Fajarwati, A., Hasanati, S., Sari, I.M., Resettlement following the 2010 Merapi Volcano eruption. *Procedia-Social and Behavioral Sciences*, **227** (2016), 361-369.

* Corresponding author: b.oo@unsw.edu.au

https://doi.org/10.1051/matecconf/201819202067

A design of LED driver circuit for reducing production cost in Thailand industry

Chantima Lapsongphon, and *Saroj* Pullteap*

Mechanical Engineering Department, Faculty of Engineering and Industrial Technology, Silpakorn University (Sanam Chandra Palace), Nakhon Pathom, Thailand.

Abstract. In this paper, the designing of a modified flyback converter circuit for 100 Watts LED lighting has been demonstrated. Some important factors, i.e. efficiency (ε), power factor (*PF*), and total harmonic current distortion (*THDi*), are considered for interpreting the performance of the proposed circuit. A commercial driver circuit which an efficiency of 94% and price of ~2,000 THB has, consequently, been employed as a reference circuit. Two experimental setups, performance investigation and cost analysis, have been demonstrated for confirming the concept designed. However, the experimental results shown that an efficiency of the developed driver is 93.93%, while the implementation cost is 1,257 THB respectively. Compared with the reference circuit, we found that the performance of modified flyback circuit is almost equivalent in any parameters, while the cost is cheaper than the reference of 37.15%. This implies that there is a possibility to transfer the knowledge of designing concept to mass production for Thailand industry. In addition, it would thus be reduced the investment cost by the importation driver circuits from other countries and also to support Thailand industry for producing the high quality goods for exporting aboard.

1 Introduction

Nowadays, the light emitting diodes (*LED*) lighting is well known as an *"energy saving lighting"* It's cause to there are many advantages over the bulb lighting such as lower energy consumption (90% more efficient), longer life spans (about 60,000 hours), higher safety, and lower maintenance requirements, etc [1]-[3]. Consequently, the proposed lighting is becoming popular equipment for using in the human life. However, the global lighting market is reported that the proportional usage of the LED lighting accounted for 30% of the total lighting [4]. From the purpose, the LED lighting can, increasingly, be extended to grow up in the future. However, the LED lighting production of Thailand industry has been imported many LED components such as chip LED, LED driver circuit from the foreign countries for a long time. It causes to insufficient of technology and less production volume to their produce. This effects to the industry of Thailand for losing some the budgets from the imported goods amounting in several million baht per year, and also affected on the consumers in terms of expensive goods. In recent year, several research works have been focused on the development of LED driver with different techniques for improving their efficiency such as A. Pollock was designed a high efficiency buck converter circuit for LED power supply in automotive system [5]. The results show that the proposed circuit offered efficiency of LED driver as 95%, which increased up to 3%. In addition, Y. Wang developed a LED driver based on the *BCM* boost circuit and *LLC* converter for a street lighting system. In the proposed circuit, two integrated switches were available for improving the reliability of the system [6]. It offered the efficiency of prototype as high as 91.1%, *THDi* of 20.2% and also a power factor (*PF*) of 0.98 under the full load condition. Furthermore, C. Cheng was presenting a novel LED driver circuit by combining a dual buck-boost *PFC* controller with the coupled inductors and half bridge type of the *LLC* resonant converter into a single stage power conversion circuit. However, the results demonstrated the high efficiency, high *PF*, and low *THD* of input utility line current which <5.5%, respectively [7]. Regarding to M. Arias work, he was studied the high efficiency LED driver without the electrolytic capacitor for a street lighting. The proposed circuit can be separated in three stage solutions i.e. *PFC* boost converter, galvanic isolation with transformer, and two input buck converter, respectively [8]. Experimental results show the efficiency as high as 93% for the whole topology and 95% for the cascade connection of the second and third stages. From the above mentioned, it was found that these drivers could not be able to sell into the market due to over qualify and high cost.

In this research presents the designing of a LED driver circuit for reducing a production cost by comparing with a commercial LED driver circuit. According to the designed circuit, a modified flyback converter has been considered. To study the experiment, the implementation process has been employed for investigating the results of the electrical performances and also cost analysis respectively. However, the

* Corresponding author: saroj@su.ac.th

experimental results are expected to become a major kind of trend for reducing the importation of driver circuits from the foreign suppliers. Moreover, the research output also make the LED operation cost in Thailand industry would be decreased.

2 Overviews of LED driver circuit

2.1. Principle of LED driver circuit

The LED driver circuit is an important part of LED lighting system. It is the electronic parts that converts the AC power into the proper DC voltage and regulates the current flowing through the LED during operation, like a ballast of fluorescent lamp [9]. Normally, the switching power supply can be applied for the LED driver circuit. It is containing of the converter circuits, which are an important factor to provide the efficient of whole LED lighting systems.

2.2 Mofidifed flyback converter circuit designing

Normally, the designing of modified flyback converter uses the multiphase of flyback converters that controlled circuit with interleaving control method. According to this technique, the two switch control (SW_1 and SW_2) would thus be operated in 180° out of phase as illustrated in Fig. 1.

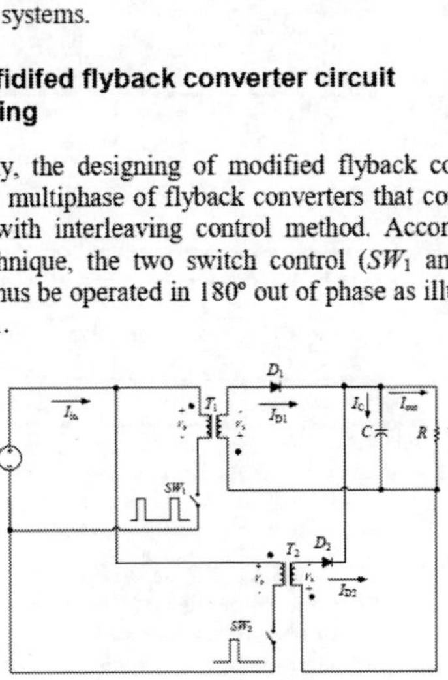

Fig. 1. Simplified schematic of modified flyback converter circuit

From Fig. 1, when SW_1, SW_2 are switched on, the energy is transferred from the input to the primary inductance of each transformer. In the other hand, when SW_1 and SW_2 are switched off, leads to all the energy stored in the flyback transformers are delivered to the secondary winding, capacitor and load, respectively [10]. It can be calulate the output voltage in (1). In the other way, the duty cycle (d) can be obtained from (2).

$$V_{out} = V_{in} \times \frac{d}{1-d} \times \frac{N_S}{N_P} \tag{1}$$

$$d = \frac{(1-d)V_{out}}{V_{in}} \times \frac{N_P}{N_S} \tag{2}$$

where: N_P is the primary winding, while N_S is corresponding to secondary winding, and also N_P / N_S is called *"transformation ratio (n)"* respectively. It is observed that the duty cycle of the signal is direct proportional to the transformation ratio. Moreover, the maximum duty cycle do not provide more than 44% at high frequencies [11].

2.3 Production cost analysis

Generally, the production cost is, certainly, the most important cost because it determines profit and selling prices [12]. It is referred to the cost which is, absolutely, need for producing the goods. According to the LED lighting cost, the industrial has been recognized the initial cost which consist of electronic component cost, maintenance cost, transportation cost, etc. In this paper, we might, however, be only focused on the electronic component cost for mass production. Normally, there are 3 common factors for considering such cost as followed by [13]:

- Book price is a price that can find on the distributors websites such as DigiKey, Mouser, Electronic source etc. It is separated into volume levels, i.e. 1-500, or 100-1,000 etc.
- Direct price is a price that directly quoted by the manufacturer to the original equipment manufacturer (*OEM*), where the manufacturer has agreed to sell direct to the *OEM*. It is, generally, buy in very large volumes.
- Broken price is the most important price type for production procurement. In competitive bids, the supplier may be agree to *"Break"* the book price that paid by the distributor due to they can produce by using the special lower cost. This system is called *"Ship from stock and debit,"* or *"Debit"*.

From the above mentioned, the book price might be used for reference to pricing of commercial LED drivers circuit. However, in this paper can be calculated the pricing of LED drivers production by using the cost-based pricing technique. [16] This method is setting prices (P) which based on the total cost (TC) and the percentage of cost (M) to desire profit margins of product. It is, normally, calculated by:

$$P = \frac{TC + M}{Q} + TC \tag{3}$$

where: M is called the markup percentage of cost, it is, normally, varies from 5% to more than 20% [14]

3 Experimental Setup

In this section, the 100 watts LED driver circuit has been designed under technique as mentioned above. However, it could be shown a block diagram in Fig.2

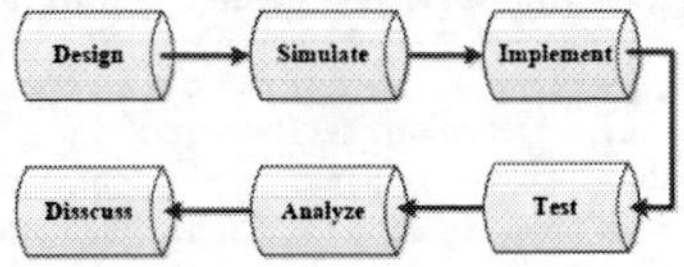

Fig. 2. Operational diagram of LED driver circuit

From the diagram above, there are 6 main steps to develop the driver circuit. According to the designing of LED driver circuits, it has been proposed for studying the possibility of the desired circuit by starting with simulation concept. The prototype of LED driver circuit has, next, been implemented for investigation the circuit performances and analyzing the operation cost. Consequently, the reference LED driver circuit has been compared with the designed circuit for summarizing of their performances and cost. The configuration of experimental circuit has been shown in Fig. 3.

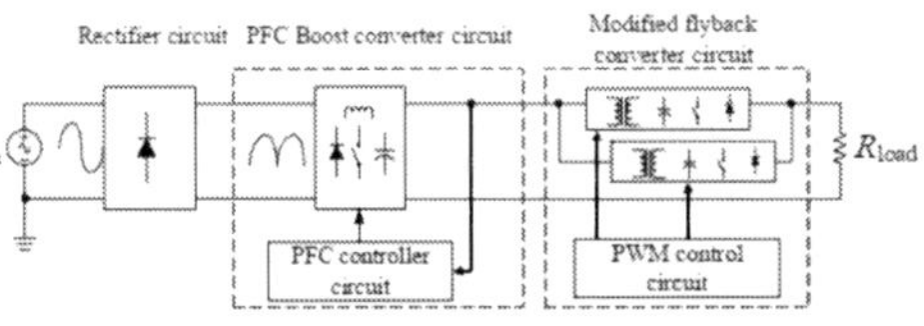

Fig. 3. Designing concept of proposed converter circuit.

According to Fig. 3, it has been divided into 2 major stages: the *PFC* boost converter circuit and proposed converter circuit. Consequently, the *PFC* boost converter circuit has been operated for reducing the harmonic currents and also improving the power factor of circuits. Moreover, the second stage shows the proposed circuits that has been utilized for obtaining the output power. In addition, the simulation program via *LTspiceXVII* programming has been employed for considering the possibility of designed circuit. After that, the prototype driver circuits have been demonstrated for characterized the performance and also production cost, respectively. Futhormore, the characterization of the driver circuit has been studied. The measurement setup has, however, been illustrated in Fig. 4

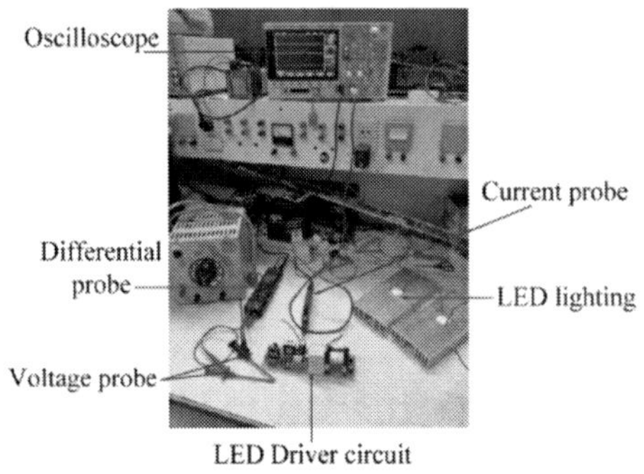

Fig. 4. Experimental setup for characterizing driver circuit

4 Results and Discussion

4.1. Results of driver circuit characterization

To ensure the designed circuit can be operated properly, the implementation has been chosen for studied the designed circuit as shown in Fig. 5(a). From the figure, we can observe that the actual designed circuit has small size with a diameter of 22 x 6.5 cm. Moreover, the driver circuit also has 2 small transformers for current dividing to parallel operation. This technique is an advantage over a classical flyback converter circuit.

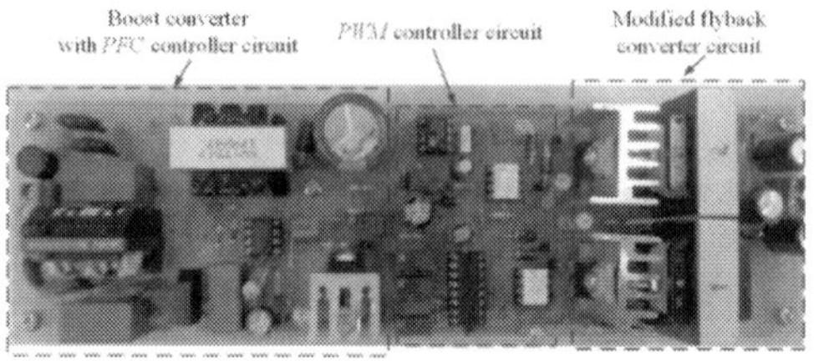

(a)

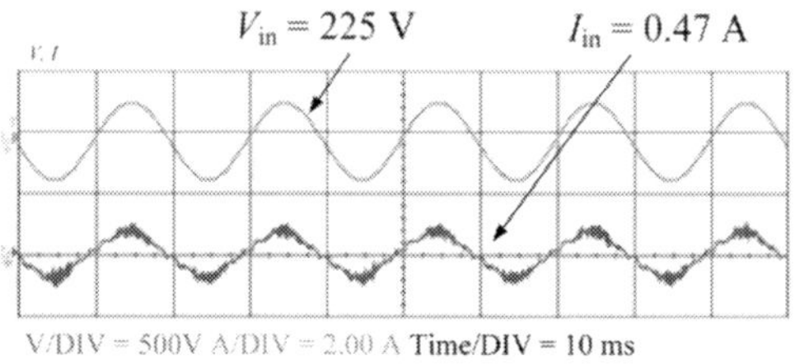

(b)

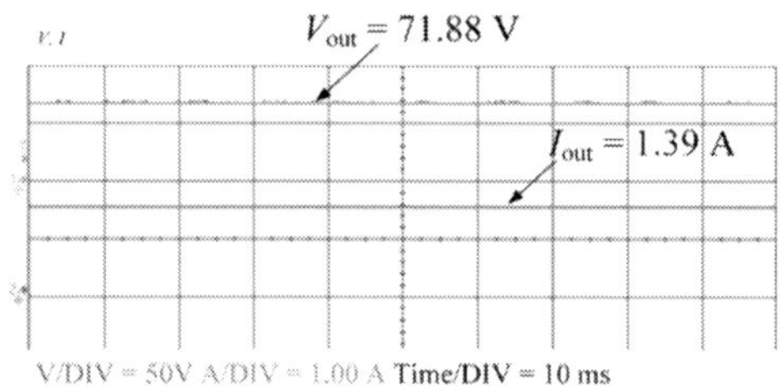

(c)

Fig. 5. Experimental result; (a) prototype circuit, (b) input voltage and input current waveforms, (c) output voltage and output current waveforms

Fig. 5(b) and 5(c) shows the input/output voltage and current which are obtained from the driver circuit. Moreover, we can observe that the input voltage (V_{in}) and input current (I_{in}) of 255 Volts and 0.47 Amp have been entered into the driver. Focusing on the input waveforms, we can summarize that the voltage and current waveforms are, similarly, indicated as sinusoidal waveform. In addition, it's offered approximately 0.95 of power factor (*PF*) and also indicating the *THDi* of 17.03% respctively. According to Fig. 5(c), the output voltage (V_{out}) of 71.88 Volts and the output current (I_{out}) of 1.39 Amp has been occurred. These values obtain the output power of 99.91 Watts by calculation. The results implied that the electrical parameters that exploited from the proposed circuit is closing to the desired value that exploited from the commercial driver in any parameters. However, a summarization of the experimental electrical characteristic of the developed LED driver circuit can thus be reported on Table 1.

Table 1. Summerization of electrical characteristics

Electrical characteristics	Reference driver circuit	Modified flyback converter circuit
Input voltage (V_{in})	225V	225 V
Input current (I_{in})	0.47 A	0.47 A
Input power (P_{in})	106.72 W	106.36W
Output voltage (V_{out})	71.66 V	71.88V
Output current (I_{out})	1.40 A	1.39A
Output power (P_{out})	100.32 W	99.91W
Efficiency (η)	94%	93.93 %
Power factor (*PF*)	0.96	0.95
Total harmonic current distortion (*THDi*)	17%	17.03%

According to data from the table, a summarization of electrical characteristics results from prototype circuits have been acquired. Compared with the commercial driver circuit, we conclude that the performance of the modified flyback converter circuit is very similarity to the commercial circuit in any criterions.

4.2. Results of production cost analysis

In this section explains the calculaiton of production cost for LED driver circuit implementation. Futhermore, we could estimate the price of prototype circuits by using the cost-based pricing method. The estimaiton results can thus be detailed in Table 2.

Table 2. Summerization of pricie of driver circuit

Price of products	Reference driver	Modified flyback converter circuit
Price / unit	~2,000	2,953.95
Price / 1,000 units		1,426.70
Price / 10,000 units		1,257.17

Table II shows the price of LED driver circuit in various quantities. It can be seen that the developed prototype can, probably, be produced in the industry because of lower cost than commercial circuit. In case of 10,000 pieces, it would be observed that the modified flyback converter has the lowest cost of 1,257.17 Bath

5 Conclusions

This paper was presented the designing of 100 Watts LED driver circuit by using the modified flyback converter circuit. Moreover, we were investigated the circuit performances and also analyzing the investment cost by comparing with the reference circuit. The experimental results were reported that the driver efficiency was 93.93%, while the implementation cost was 1,257 THB / 10,000 units. It was implied that the circuit performance was directly proportional to the driver circuit cost. However, the modified flyback converter was the most suitable for operation due to the all performances were similarity to the reference circuit. Moreover, it could be considered that the cost of the converter circuit was cheaper than the reference of 37.15%. Therefore, the usage of the designed circuits as demonstrated would thus be reduced the production cost and increasingly the large the market shared of lighting market in the future.

Acknowledgement

The authors wish to acknowledge the financial supported by the Thailand research fund (*TRF*) and Realtronics (Thailand) Co., Ltd. (Contract No. MSD59I0022) under the research and researchers for industries (*RRi*) project.

References

1. E. F. Schubert., *"Light-Emitting Diode,"* 2nd ed., Cambridge: Cambridge University Press. 2006.
2. P. D. Teodosescu, et al., "Resonant LED Driver with Inherent Constant Current and Power Factor Correction," *Electronics Letters*, vol. 50, no. 15, pp. 1086-1088, 2014.
3. M. H. Chang *et al*., "Light Emitting Diodes Reliability Review," *Microelectronics Reliability*, vol. 52, no. 5, pp. 762-782, 2012.
4. J. Wu, "LED inside 2016 Southeast Asian Lighting," *Market Report*, 2016.
5. A. Pollock, H. Pollock, and C. Pollock. "High Efficiency LED Power Supply," *IEEE Journal of Emerging and Selected Topics in Power Electronics*, vol. 3, no. 3, pp. 617-623, 2015.
6. Y. Wang, "A Single-stage LED Driver for The Street Lighting System," *IEEE Transl. Ind. Electron*., vol. 62, no. 9, pp. 546-5457, 2015.
7. C. Cheng, and T. Y. Chung, "A Novel Single-State High-Power-Factor LED Street-Lighting Driver," *IEEE Transl. Ind. Applications*, vol. 50, no. 5, pp. 3037-3045, 2014.
8. M. Arias, et al., "High-Efficiency LED Driver Without Electrolytic Capacitor for Street Lighting," *IEEE Transl. Ind. Applications*, vol. 49, no. 1, pp. 127-137, 2013.
9. Y. Wang, *et al*., "A Review of LED Drivers and Related Technologies," *IEEE Transl. Ind. Electron*., vol. 64, no. 7, pp. 5754-5765, 2017.
10. B. Tamyurek, "An Interleaved High-Power Flyback Inverter for Photovoltaic Applications," *IEEE Transl. Power Electronics*, vol. 30, no. 6, pp. 3228-3241, 2015.
11. S. Rahayu *et al*., "Interleaved Flyback DC-DC Converter Design with 350 W Power Output Using LT 3757 in LT Spice," *Journal of Physics: Conf. series*, vol. 824, pp. 1-6, 2017.
12. C. M. Drury, *"Management and Cost Accounting,"* Berlin: Springer, 2013.
13. D. J. Oyer, *"Cost-Based Pricing,"* United Kingdom: Management Concepts Press., 2012.
14. G. Eric, *"Sell What You Sow: the Grower's Guide to Successful Produce Marketing,"* 49308th ed., California: New World Publishing, 2005.

Microstructure evolution of Al-7wt%Si-2wt%Fe alloy processed by high-pressure torsion

Jittraporn Wongsa-Ngam[1,*], *Chakkrist* Phongphisutthinan[2], and *Terence* G. Langdon[3]

[1] Mechanical Engineering Department, Faculty of Engineering, King Mongkut's Institute of Technology Ladkrabang, Bangkok,10520, Thailand

[2] National Metal and Materials Technology Development Agency, National Science and Technology Development Agency, Bangkok, 12120, Thailand

[3] Materials Research Group, Faculty of Engineering and the Environment, University of Southampton, Southampton SO17 1BJ, United Kingdom

Abstract. The microstructure evolution of an aluminum silicon-based alloy after severe plastic deformation processing was examined. An aluminum silicon-based alloy; Al-7wt%Si-2wt%Fe, was processed by the severe plastic deformation technique called high-pressure torsion at room temperature under a high pressure of 6.0 GPa and rotational speed of 1.0 rpm with numbers of revolution up to 5 turns. Microstructure evolution, especially intermetallic phase, was observed using an optical microscope and a scanning electron microscope (SEM). The effects of high-pressure torsion on the Fe intermetallic compounds in Al-Si alloy were investigated. It was found that the intermetallic particles decreased in size with increasing imposed strains.

1 Introduction

Severe Plastic Deformation (SPD) is a well-established technique for producing bulk ultrafine-grained (UFG) metallic materials [1]. In SPD processing, materials are subjected to the imposition of very large strains without the introduction of any concomitant changes in the cross-sectional dimensions of the samples leading to superior properties [2]. In addition, the unusual phase transformations in SPD processing results in development of highly metastable states associated with the formation of supersaturated solid solutions, disordering or amorphization [3]

Aluminum alloys are widely used in modern engineering structures and components because of light weight and corrosion resistance. The most important cast aluminum alloy system is Al-Si, where the high levels of silicon contribute to give good casting characteristics. Al-Si-based casting alloys are interesting materials in the automotive and aerospace applications because of their low density, low thermal expansion coefficient and high wear resistance. However, Al-Si alloys exhibit low ductility and toughness that limit their applications. Previous studies demonstrated that severe plastic deformation processing applied to Al-Si alloy has a potential for producing a homogenous microstructure and refinement leading to improvement of the mechanical properties [4-6]. The influence of a rotary-die ECAP method on Al –Si alloys (11-23 wt.% Si) are possible to refine the size of the Si particles and this leads to a large improvement in ductility, toughness and wear resistance but little strengthening [7-10]. Remarkable refinement of coarse Si particles of Al-7 wt.% Si was achieved through five turns of HPT [11]. Furthermore, stable fine particles can also improve thermal stability of the UFG microstructure through inhibiting grain growth [12].

In addition, iron is a common impurity in aluminum alloys that arises from a number of possible sources, especially; in secondary aluminum i.e. aluminum recycling. This impurity causes both positive and negative influences. Fe alloying into aluminum alloys has a potential as an inexpensive alloying element for strengthening. However, due to its low solubility in Al, Fe may be present as secondary phases in the form of intermetallics. The iron-containing intermetallics participate directly in the fracture mechanism, the more intermetallics there are, the lower the ductility. In addition, as the iron level increases, porosity also increases, and this type of casting defect also has a significant impact to ductility [13]. In general, Fe contents are strictly controlled to be low in commercial alloys, depending on the required applications. In recent years, the capability of producing bulk Al-Fe alloys with supersaturation of Fe by HPT processing has been reported in a cast Al-11%Fe alloy [14]. The study on a cast Al-5 wt.% Fe subjected to ECAP processing reported grain refinement of the Al matrix and dispersion of second-phase particles so that the process led to the improvement of not only the microhardness but also the tensile strength and ductility [15].

The use of severe plastic deformation (SPD) to refine the secondary phase/intermetallic phase is interesting research, especially; the application of SPD to secondary

* Corresponding author: jittraporn.wo@kmitl.ac.th

aluminum alloys from recycle where coarse intermetallic particles occur. In this study, the high-pressure torsion (HPT) processing was applied for Al-Si based alloys with 2 wt.% Fe to examine the microstructural evolution.

2 Experimental materials and procedure

The materials used in this investigation were Al-7%Si-2%Fe (in wt.%) supplied by MTEC. They were prepared by casting of commercial A356 alloy billets and iron nail. A356 billets were melted at 800 °C for 2%Fe alloy. Then iron nails were added and isothermally held for 2 h. The molten aluminum was stirred every 30 min during casting after degasing by Nitrogen gas. Then, 1.3 kg molten aluminum ingot was cast into a steel mold and air cooled. The chemical compositions of A356 and the Al-7wt%Si-2wt%Fe are given in **Table 1.** Before HPT the ingots were machined into billets with diameters of 10 mm and lengths of 70 mm and cut into disks with diameters of 10 mm and polished to final thicknesses of ~0.83 mm.

Table 1. Chemical compositions of aluminum alloy

(wt%)	Si	Fe	Mg	Ti	Cr	Al
A356	7.035	0.096	0.442	0.103	0.018	Bal.
Al-7Si-2Fe	6.819	2.160	0.384	0.101	0.016	Bal.

The disks were processed using high-pressure torsion (HPT) [16]. The HPT processing was conducted under quasi-constrained conditions [17,18] using a facility consisting of upper and lower anvils having central depressions with diameters of 10 mm and depths of 0.25 mm. Each disk was placed in the central depression of the lower anvil and this anvil was then brought into position so that a pressure was applied on the disk. Torsional straining was achieved by rotation of the lower anvil at a constant speed of 1 rpm. In quasi-constrained HPT, there is a small outflow of material around the peripheries of the disks between the two anvils during the straining operation. All processing by HPT was conducted at room temperature using an applied force of 470 kN which corresponds to an imposed pressure, *P*, of 6.0 GPa with total straining through 1/4, 2 and 5 turns.
The samples were molded in plastic resins for grinding and polishing. The samples were ground with abrasive papers of silicon carbide (SiC) from 180 to 2500 grits subsequent to polishing with diamond powders of 3 and 1 μm on micro-cloths. The sample surface was finished with chemical etching with a suitable etching reagent. After sample preparation, microstructure evolutions were observed by a scanning electron microscope (SEM) operating at an accelerating voltage of 20 kV.

3 Results and discussion

3.1 Microstructure

A commercial A356 alloy belongs to a group of hypoeutectic aluminum-silicon alloys. This as-cast A356 exhibited a typical hypoeutectic solidification structure, consisting of a primary aluminum matrix surrounded by the eutectic network structure as shown in **Fig. 1**. Iron impurities in the alloy led to the formation of a small amount of iron intermetallic compounds beta phase, β-Al_5FeSi in a form of long platelet and alpha phase α-Al_8Fe_2Si in a form of Chinese script.

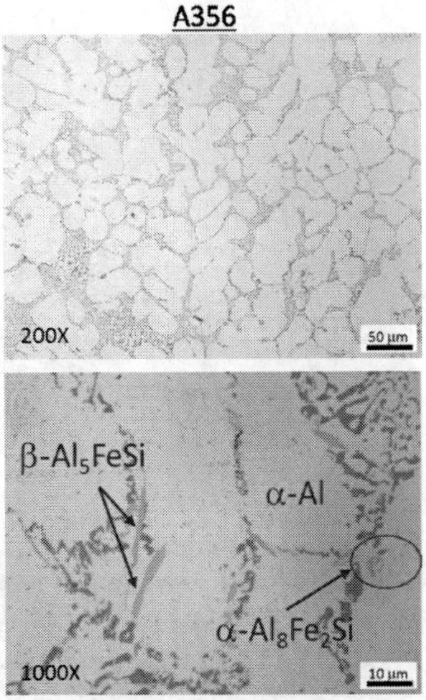

Fig.1. Microstructure of commercial A356 alloy.

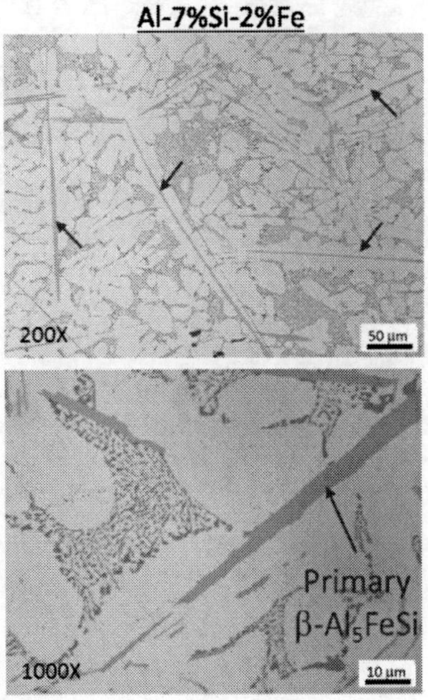

Fig. 2. The microstructure of Al-7wt%Si-2wt%Fe before HPT processing

 https://doi.org/10.1051/matecconf/201819202068

With increasing the Fe-content to 2wt%, Al-7wt%Si-2wt%Fe alloy, results in the earlier formation of the dominant intermetallic phases and more unconstrained growth of particles. As shown in **Fig. 2** the size of intermetallic β-phase is larger and longer and the fraction of intermetallic phase also increases in this material composition. It should be noted that the large and long particles cause negative effects to failure in which the failure propagation occurs easily.

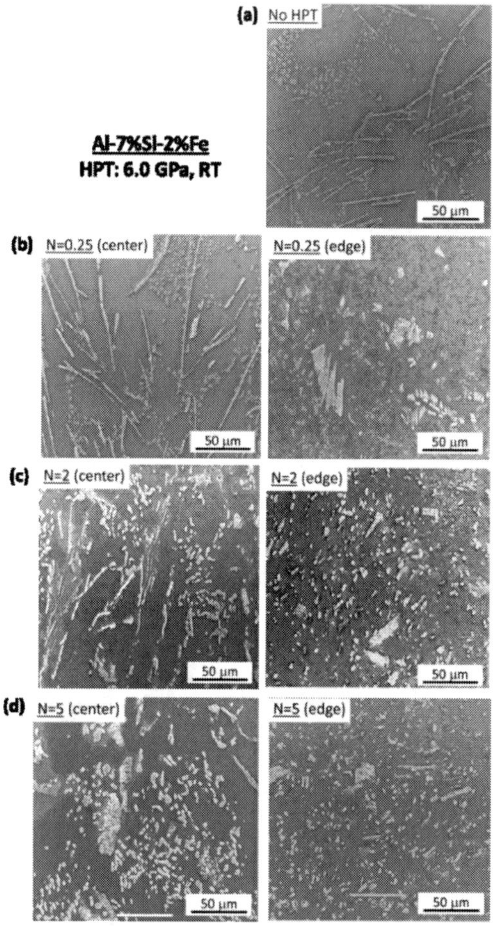

Fig.3. The microstructure of Al-7wt%Si-2wt%Fe in (a) as-cast sample (b) after HPT 1/4 turn (c) after HPT 2 turns and (d) after HPT 5 turns

The microstructures of samples after HPT were analysed in two regions; center and edge because shear strain introduced by HPT processing is a function of not only a number of turns *N*, but also the distance from the disk center *r* and the thickness *t* as following the relationship for equivalent strain [19]:

$$\varepsilon = \frac{2\pi N r}{\sqrt{3}t} \qquad (1)$$

After a quarter turn of HPT processing (**Fig. 3(b)**), there is a different microstructure at central and edge region because of the variation of imposed strain as a function of disk radius *r*, in Eq. (1). The microstructure on central area of the specimen closely resembles that in the as-cast specimen without HPT processing in **Fig. 3(a)**. However, there is a significant change near the edge of a disk because of the higher strains imposed that cause fragmentation of intermetallic particles.

After two turns (**Fig. 3(c)**), in the central area the long shape of β-phase become broken but the distribution remains inhomogeneous. Near the edge region of sample, intermetallic particles become fragmented to small size. The distribution of particles is quite homogeneous.

After five turns (**Fig.3(d)**), the intermetallic particles in the central region are continuously refined but the intermetallic particle are not homogeneously distributed. The microstructure near the edge after five turns is similar to the microstructure after two turns.

This results show that HPT processing has a potential to refine the intermetallic phases for the Al-7wt%Si-2wt%Fe alloy; which is representative of recycled aluminum alloys.

In future research, the effect of dispersion of second-phase particles on mechanical property improvement for Al-7%Si-2%Fe alloy after HPT processing will be studied.

4 Summary and conclusions

1. Al-Si-Fe alloy was subjected to large strain by high-pressure torsion (HPT) processing at room temperature under pressure of 6.0 GPa with different numbers of turns of 1/4, 2 and 5 turns
2. The effects of HPT on intermetallic particles of Aluminum-Silicon-Iron were investigated.
3. The microstructure evolution and intermetallic refinement during high-pressure torsion processing occur by introducing a large amount of plastic strain.
4. The large strains introduced by HPT at ambient temperature cause fragmentation of iron-intermetallic particles. The degree of fragmentation increases with increasing strains
5. The microstructure results demonstrate that the intermetallic particles decreased in size with increasing imposed strains.

Acknowledgement

We would like to acknowledge financial support of the KMITL Research Fund under project number KREF01590.

References

1. R.Z. Valiev, R.K. Islamgaliev, I.V. Alexandrov, Prog. Mater. Sci. **45** (2000) 103.

2. Y.T. Zhu, T.C. Lowe, T.G. Langdon, Scripta Materialia **51** (2004) 825.
3. R.Z. Valiev, Adv. Eng. Mater. **5** (2003) 296.
4. A. Ma, N. Saito, M. Takagi, Y. Nishida, H. Iwata, K. Suzuki, I. Shigematsu, A. Watazu, Mater. Sci. Eng. A **395** (2005) 70.
5. Y. Kume, M. Kobashi, N. Kanetake, Aluminium Alloys **519-521** (2006) 1441, Parts 1 and 2
6. Z.Y. Ma, S.R. Sharma, R.S. Mishra, Scripta Mater. **54** (2006) 1623.
7. A. Ma, K. Suzuki, Y. Nishida, N. Saito, I. Shigematsu, M. Takagi, et al. Acta Mater. **53** (2005) 211.
8. A. Ma, K. Suzuki, N. Saito, Y. Nishida, M. Takagi, I. Shigematsu, et. al. Mater. Sci. Eng. A **399** (2005) 181.
9. A. Ma, M. Takagi, N. Saito, H. Iwata, Y. Nishida, K. Suzuki, et. al. Mater. Sci. Eng. A **408** (2005) 147
10. Z. Zhang, S. Hosoda, I-S. Kim, Y. Watanabe, Mater. Sci. Eng. A **425** (2006) 55.
11. A.P. Zhilyaev, J.M. Garcia-Infanta, F. Carreno, T.G. Langdon, O.A. Ruano, Scripta Mater. **57** (2007) 763
12. K. Oh-ishi, Z. Horita, D.J. Smith, T.G. Langdon, J. Mater. Res. **16** (2001) 583.
13. J.A. Taylor, Procedia Mater. Sci. **1** (2012) 19-23.
14. O.N. Senkov, F.H. Froes, V.V. Stolyarov, R.Z. Valiev, J. Liu, Nanostruct. Mater.**10** (1998) 691.
15. V.V. Stolyarov, R. Lapovok, I.G. Brodova, P.F. Thomson, Mater. Sci. Eng. A **10** (2003) 159.
16. A.P. Zhilyaev, T.G. Langdon, Prog. Mater. Sci. **53** (2008) 893.
17. R.B. Figueiredo, P.R. Cetlin, T.G. Langdon, Mater. Sci. Eng. A 528 (2011) 8198.
18. R.B. Figueiredo, P.H.R. Pereira, M.T.P. Aguilar, P.R. Cetlin, T.G. Langdon, Acta Mater. 60 (2012) 3190.
19. R.Z. Valiev, Y. Estrin, Z. Horita, T.G. Langdon, M.J. Zehetbauer, Y.T. Zhu : JOM **58** (4) (2006) 33.

A Design and Construction of a Wind Tunnel for Engineering Laboratories

Chalida U-tapao [1]and *Seksun* Moryadee[2]

[1] Department of Civil Engineering, Faculty of Engineering, King Mongkut's Institute of Technology Ladkrabang, Bangkok, Thailand

[2] Department of Ordnance Engineering, Chulachomklao Royal Military Academy Nakhon Nayok, Thailand

Abstract. The propose of this research is to demonstrate a design and construction of a wind tunnel for engineering laboratories in order to study the principles and control wind speeds in the wind tunnel. In an experiment in aerodynamics and engineering, we found that diffuser must have a length equal to or more than twice the length of the test in order to prevent the turbulent flow in the test area. The wind speed control system uses Inverter to control a 3-phase frequency of electricity supplied to air blower. In the experiment, the frequency was adjusted in the range from 20.00 to 50.00 Hz. Experiment results show that wind speeds during the test area are in the range of 14.50 to 38.50 meters per second, and the relationship between frequency (Y) and wind speed (X) during the test is linear as follows: Y = (0.7945 × (X-20)) + 14.629. The maximum pressure is 90.31 kilograms per square meter. This wind tunnel can be used to design buildings with a maximum height of 20 meters according to the Bangkok Metropolis Building Control (2001).

1 Introduction

The damage in building structure mainly occurres at the lateral side. The design of equipment in engineering for the building structure in the civil engineering and aerospace engineering must have the standard and rule. The design of building structure for supporting the lateral force is related to the wind force. In Thailand, the Engineering Institute of Thailand (EIT) is an organization in charge of standard and rule of the building structure. Moreover, EIT identifies the standard of the building structure for supporting the lateral force related to the wind force. Therefore, one of the main tasks for EIT is to conduct a research on the lateral force. In order to do that, the realistic generator of the lateral force is design based on small scale for testing the material, equipment and structure. The strength per the lateral force is very important to use in the all engineering.

This research focuses on design and construction of the realistic generator of the lateral force, called the wind tunnel, which can conduct a measurement in wind speed material test and other tests in the laboratory.

2 Literature Review

Wilson et, al, (2008) conducted a study on the wind tunnel for testing the sensor in pressures force and wind speeds. The wind tunnel that is uniform laminar flow in the test section. This work applied the test section at diameter of 180 mm and 120 mmd with height (up to 60 meters per second). The model considers the flow characteristics at the flow entrance of the air by using Reynolds numbers. The result shows that the wind speeds is 3% within ± 0.3%. accuracy. These values are suitable for designing the wind tunnel.

Wittwer and Moller (2000) evaluated the wind tunnel which is low speed of National University of the Northeast (UNNE). This work analyzed the velocity flow and turbulence flow. The output of that research demonstates that velocity is constant and turbulence is low.

Huang et al., (2010) presented the Fuzzified Eigensystem Realization Algorithm (FERA) method. It is used to perform estimation in the set of aerodynamic parameters. The Initial of design fundamental and the result of design detail of Mini - UAV are based on LSWT which is a new method with combined theory fuzzy logic and Eigensystem Realization Algorithm (ERA). There are characteristics of aerodynamic parameters. For the FERA method obtains input variables which is better the testing tunnel realistic.

Boyle (2002) proposed the wind tunnel under uncertainty of data analysis. In the measuremed data an output is a dimensionless. It is an important of objective in design In case of laminar uniform for flowing into the test section impacts the factor. Moreover, the construction of the air flow is turbulence uniform. As the results, the factors affects wind velocity in a construct

* Corresponding author: kuchalid@kmitl.ac.th

ion ofwind tunnels with a cross-sections test of the wind tunnel.

3 Methodology

This paper proposed a design and construct the wind tunnel. In the first stage the wind tunnel is designed elements. Next, the wind tunnel is constructed with acrylic material. Then, the wind speed value is measured for testing the performance of the wind tunnel according to theory of the wind tunnel.

3.1 Design Elements

3.1.1 Contraction

The design contraction assigns the length of the contraction between 1 to 1.5 times of the diameter of the entrance as flows:

$$D_h \leq L_c \leq 1.5D_h \tag{1}$$

where
D_h is the diameter of hydraulic.
L_c is the length of contraction.

3.1.2 Contraction Ratio

Contraction Ratio (C) is the ratio of the cross-sectional area which is the entrance per the cross-sectional area of the exit. C is defined between 6 to 9.

$$6 \leq C \leq 9 \tag{2}$$

3.1.3 Test section

The test section is a ratio of height per the cross-sectional area that is in a range of 0.7 to 1 as follows

$$0.7 \leq \frac{H}{W} \leq 1 \tag{3}$$

where
H is the height of test section.
W is the width of test section.
Equation 4 is the length of the truncation.

$$L_T \geq 2D_h \tag{4}$$

where
L_T is the length of test section
D_h is the diameter of hydraulic

Fig. 1 shows the diffuser pile. The ratio of section area (A_R) is defined by

$$A_R = \frac{A_2}{A_1} \tag{5}$$

Then

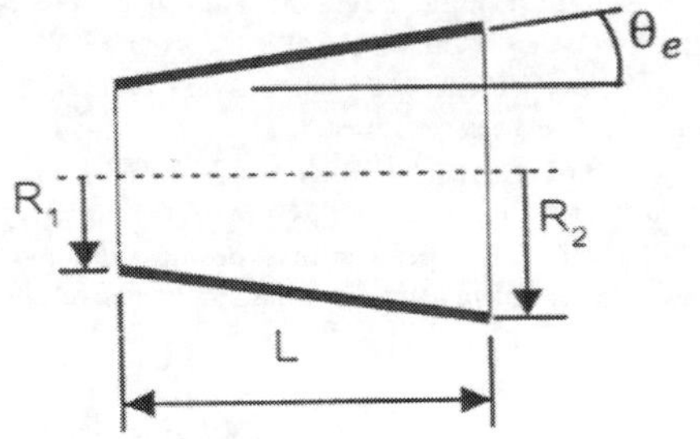

Fig 1. Diffuser pile.

$$2 \leq A_R \leq 3 \tag{6}$$

where
A_2 is the section area entrance of wind.
A_1 is the section area exists of wind.

3.1.4 Slope Angle

Generally, a slope angle θ_e is the range, we get

$$\theta_e = \tan^{-1}\left(\frac{R_2 - R_1}{L}\right) = \tan^{-1}\left(\frac{1}{2}\frac{\sqrt{A_R^{-1}}}{L/D_1}\right) \tag{7}$$

then $2^{o} \leq \theta_e \leq 3.5^{o}$.

3.1.5 Design the Grille

The width and the shape of the grille affected the reduction of the turbulent flow. Thus, we must select suitable for the wind tunnel.

3.2 Construction

In this sub-section presents the construction of the wind tunnel composed of acrylic material, the air duct, test section, diffuser, metal parts, the grate, and fan components as demonstrated in Fig.2. And Fig. 3 shows the completion of the wind tunnel used to test the material.

3.3 Testing to start a wind tunnel

3.3.1 Design and Construction

The open loop wind tunnel uses inverter to control a 3 – phase electricity supply. The performance of this open

 https://doi.org/10.1051/matecconf/201819202069

loop wind tunnel is measured for confirming competency before applying to any test. Hence, the testing a wind tunnel is started as follows:

1. Measuring the section test of the wind speed values. The adjusted frequency starts with 20 Hz increment by 2 Hzs to measure the wind speed values and record the results.
2. Repeat step 1. Bring the wind speed values at the same frequency estimate the average values.
3. We obtain the wind speed values in each frequency as the section test.
4. Plot graph the relationship between the wind speed values each as section test and the frequency.
5. The wind speed values of the section test are collected for calculating the wind force pressure unit.

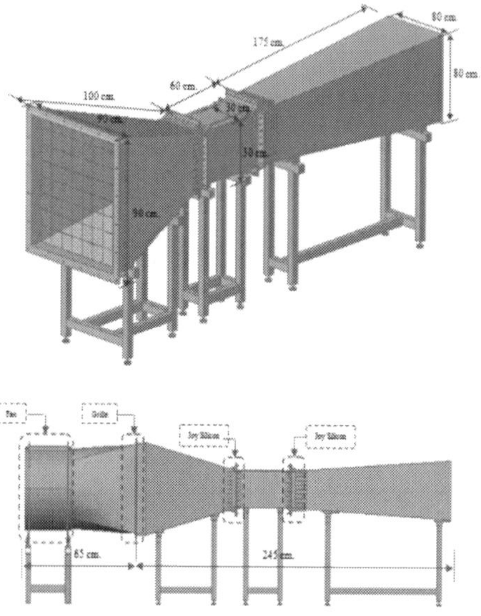

Fig 2. The dimension of the wind tunnel

Fig 3. The wind tunnel used to test the material.

3.3.2 Use of the wind tunnel

1. The open box test section is used to feed the material test. The materials can be placed at the center of the test section.
2. The material test must not exceed 30x30 cm for testing the lateral force of the wind tunnel as shown in Fig. 4.
3. Turn on the machine and the inverter fan. Adjusting the frequency between 20 Hz to 50 Hz to measure the wind speed of the material test.

Fig 4. The opened box test section for feeding material.

4 Results

In this framework, we present the results of the wind tunnel which can be utilized for using the material testing. This wind tunnel is designed and constructed. The experimental can be formed at the center of the test section, then the performance of the wind tunnel is measured as illustrated in Table 1.

Table1. The experimental results of the wind speed values.

Power supply (Hzs)	Wind Speed (m/s)		
	1st time	2nd time	Average
20	14.50	14.7	14.60
22	16.1	15.9	16.00
24	17.5	17.8	17.65
26	19.2	19	19.10
28	20.7	20.9	20.80
30	22.7	22.6	22.65
32	24.3	24.5	24.40
34	26.1	26.3	26.20
36	27.5	27.8	27.65
38	29.4	29.1	29.25
40	30.9	30.7	30.80
42	32.3	32.2	32.25
44	33.4	33.8	33.60
46	35.3	35.1	35.20
48	36.4	36.7	36.55
50	38.2	37.9	38.05

Table 1. shows the experiment results of the wind speed values. It can be concluded that this wind tunnel is the testing of wind speed during the test area at frequency in the range 20 to 50 Hz. Next, the results in Table 1. are

https://doi.org/10.1051/matecconf/201819202069

plotted the relationship between the wind speed values and frequency as shown in Fig. 5.

From Fig. 5, the relationship between wind speed values (Y) and frequency (X) is linear as follws:

$$Y = (0.7945 \times (X - 20)) + 14.629 \tag{8}$$

where

Y is the wind speed of the test section.

X is the electrical frequency power.

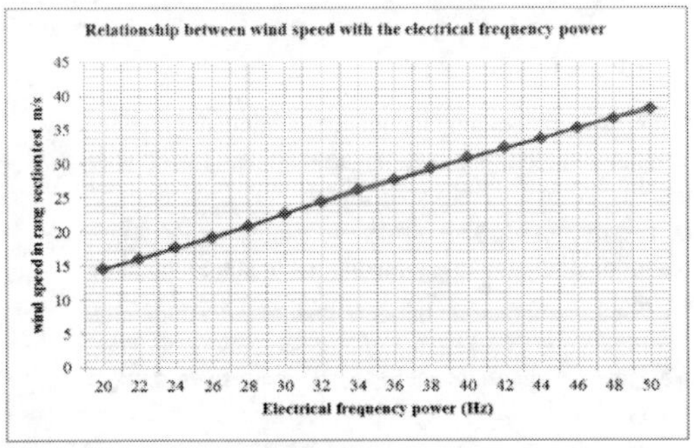

Fig 5. The relationship between wind speed values and frequency.

The wind speed values of the section test in Table 1. can be used to calculate the wind force pressure unit as follows

$$q = 0.004826v^2 \tag{9}$$

where

q is the wind force pressure unit (km/m^2).

v is the wind speeds km/hr.

From the experiment, asumming we select the maximum of the wind speeds of38 m/s or 136.8 km/hr in Table1. we will obtain the wind force pressure is 90.31 km/m^2.

5 Conclusion

In this paper, a design and construction of a wind tunnel for engineering Laboratories are presented. Also, we conduct a sample experimental test of the opened loop wind tunnel with varity of frequency in the range of 20 Hz to 50 Hz increment by 2 Hz. As the results, we obtain the wind speed values in the range of 14.50 meters per second to 38.50 meters per second. In addition, we found that the relationship between the wind speed values and freuqency is linear. This wind tunnel can be used for designing buildings with a maximum height 20 meters according to the Bangkok Regulations on Building Control 2001.

References

1. B. Jewel Barlow, H. William Rae, Jr. and A. Pope, Low speed wind tunnel Testing. New York : John Wiley, (1999)
2. C.F. Wilson, A.L. Camilletti, S.B. Calcutt and P.M. Ligrani. A wind tunnel for the calibration of Mars wind sensors. Planetary and Space Science.(2008)
3. C. Huei Huang, C.Liang Lin and Jy Maw Chao. FERA in parameter identification with application in low-speed wind tunnel test. 15 : 495-509 Aerospace Science and Technology (2010)
4. R. Adrian Wittwer and V. Sergio Moller. "Characteristics of the low-speed wind tunnel of the UNNE. 84 : 307-320. Journal of Wind Engineering and Industrial Aerodynamics. (2000)
5. M. T. Boyle. "Low speed wind tunnel testing. 31-39 Appears in Semiconductor Thermal and Temperature Measurement Symposium. SEMI-THERM IV Fourth Annual IEEE. (2002)

Zigbee wireless sensor network localization evaluation scheme with weighted centroid method

Lonesy Thammavong[1], *Khamphong* Khongsomboon[1], *Thanadol* Tiengthong[2] *and Sathaporn* Promwong[2,*]

[1] Department of Electronics and Telecommunication Engineering, Faculty of Engineering, National University of Laos Vientiane, LAOS

[2] Department of Telecommunications Engineering, Faculty of Engineering, King Mongkut's Institute of Technology Ladkrabang, Bangkok, Thailand

Abstract. Using wireless communication system, appropriate and correct indoor localization with Zigbee sensor network and could provide interesting services and applications. In this study the Zigbee transmission model with positioning method by using the relative-span exponentially weighted centroid method for the indoor localization. The experimental results and analyze results are evaluated a distance error. The ZigBee transmission model in measurement consists of 121 positions with distance between positions to positions is 0.3 meter. The experimental setup at every position operated at frequency band from 2.3 GHz to 2.5 GHz. The accuracy of estimated position is considered in the term of distance error with the cumulative distribution function (CDF) of distance error is shown. The result presents optimal value for REWL is 0.2 and mean of distance error is 0.65 m.

1 INTRODUCTION

The Zigbee wireless networks are widely used in several applications for indoor and outdoor localizations. The information of localization is a main criterion for the capability of wireless sensor network. The wireless network cannot work properly without position information. Moreover, the accurate location estimation is helpful in wireless sensor networks (WSNs) services such as routing, information processing [2], [3], and [5]. Thus, in WSNs, localization has become a very popular research orientation. Based on the requirement for measuring the received signal strength indicator (RSSI) between reference nodes and target nodes in the positioning process, WSNs localization algorithm can be split into the range-based localization algorithm and the range-free localization algorithm. A good localization, algorithm should be immune to ambient environmental impacts as well as inexplicit distances and should calculate a position as soon as possible. A promising algorithm of localization is the REWL algorithm with Zigbee technology.

The Zigbee is short-range wireless technology, low-cost and low power which satisfies IEEE 802.15.4 standard for both physical layer (PHY) and medium access control (MAC) [8]. The advantage of the Zigbee are time-saving and high reliability, distance ranging from 10 meter to 100 meter, suitable for construction, industrial control, work environment, Data rate ranging from 125 kbps to 250 kbps and medical

There are many techniques with positioning method. It's available either based on the hardware arrangement, or relied on simulation algorithm. In this research, the performance of WSNs depends on the density of nodes and the quality of data signals. The precision of position could be calculated from the co-ordinations of any four reference nodes.

In this paper we consider algorithm for indoor localization resolution. We evaluation the Weighted Centroid method based on RSSI for WSNs. The estimation of position is considered in term of distance error. The following part of this paper is arrange as follows. Section II explain of evaluation of indoor localization. Next, the measurement setup and results of this paper are explained in Section III and IV. Eventually, summary the conclusions are given in section V.

2 EVAULTION METHOD OF INDOOR LOCALIZATION

2.1. Receive Signal Strength Indicator (RSSI)

The RSSI (Receive Signal Strength Indicator) is a method to measure the strength of radio waves becomes different depending on the space between a reference node and a target node. The strength of received signals can compare by using probability distribution based on a statistical method and measures the location. To use this method, it should measure strengths of sample signals at predefined various locations in advance. The Zigbee used in sensor networks provides a REWL algorithm with RSSIs.

* Corresponding author: sathaporn.pr@kmitl.ac.th

2.2 Relative-Span Exponentially Weighted Localization Algorithm (REWL)

The REWL algorithm is based on Relative Weighted Localization (RWL) for wireless sensor networks proposed in [1] which the REWL is better than RWL. We show difference between two follow equations 1 and 2. This solution show how to approach the location p of the target node by calculating the RWL of the coordinate. More precisely, the position of the estimated target node is given by.

$$p = \frac{\sum_{i=1}^{n}\left[RSS_{\max} - RSS_i \times a_i\right]}{\sum_{i=1}^{n}\left[RSS_{\max} - RSS_i\right]} \tag{1}$$

While RWL performs solely average of the coordinates of reference node devices to localize blind folded devices, the REWL uses weights to guarantee an improvement of localization. establishing from the calculation of the RWL equation 1, the equation for determining the position with REWL is derived. Reveal the term as sum of ones and the multiplication of with ones, equation 1 is expanded to the REWL equation 2.

The REWL is an RSSI-based range-free localization algorithm. This algorithm is encouraged by the WCL method [4], [6], [7], [8] The weights are acquired by the respective positioning of the anchor RSSI value inside the span of all the RSS values measured by the target node. In the estimation of the target position, the REWL algorithm oblige the reference nodes, which present higher RSSI values and hence are probable to be closer to the target node. This is earning by used the weighting factor λ, as state by the exponentially moving average concept. The estimated target node position is given by

$$p = \frac{\sum_{i=1}^{n}\left[(1-\lambda)RSS_{\max} - RSS_i \times a_i\right]}{\sum_{i=1}^{n}\left[(1-\lambda)RSS_{\max} - RSS_i\right]} \tag{2}$$

Where n is number of reference node in range, p is the REWL approximated position of target node, $RSS_{\max}$ is the highest value in the span of the RSSI values measured by the target nodes, RSS_i is the RSSI value of the target node.

From equation 2 can be solution the equations 3 and 4

$$x = \frac{\left[(1-\lambda)^{RSS_{\max}-RSS_i} \times x_1\right]}{\left[(1-\lambda)^{RSS_{\max}-RSS_i}\right]} + \frac{\left[(1-\lambda)^{RSS_{\max}-RSS_i} \times x_2\right]}{\left[(1-\lambda)^{RSS_{\max}-RSS_i}\right]} + \frac{\left[(1-\lambda)^{RSS_{\max}-RSS_i} \times x_3\right]}{\left[(1-\lambda)^{RSS_{\max}-RSS_i}\right]} + \frac{\left[(1-\lambda)^{RSS_{\max}-RSS_i} \times x_4\right]}{\left[(1-\lambda)^{RSS_{\max}-RSS_i}\right]} \tag{3}$$

and

$$y = \frac{\left[(1-\lambda)^{RSS_{\max}-RSS_i} \times y_1\right]}{\left[(1-\lambda)^{RSS_{\max}-RSS_i}\right]} + \frac{\left[(1-\lambda)^{RSS_{\max}-RSS_i} \times y_2\right]}{\left[(1-\lambda)^{RSS_{\max}-RSS_i}\right]} + \frac{\left[(1-\lambda)^{RSS_{\max}-RSS_i} \times y_3\right]}{\left[(1-\lambda)^{RSS_{\max}-RSS_i}\right]} + \frac{\left[(1-\lambda)^{RSS_{\max}-RSS_i} \times y_4\right]}{\left[(1-\lambda)^{RSS_{\max}-RSS_i}\right]} \tag{4}$$

2.3 Performance Evaluation

The localization error is explain as a distance between the exact position and the approximated location of a target node.

3 ZIGBEE WIRELESS SENSOR NETWORK MEASUREMENT

In this research we have designed a model to measurement for location identifier of the target node in indoor environment. For more accurate localization, the experiment will be done without obstructions. RSSI measurement is carry on by using Zigbee in the term of transmission of the signal strength at frequency of 2.4 GHz. The model used for measurement consists of 121 positions with space between one position and next position is 0.3 meter. The model of measurement is depicted in figure 1. For x and y axes on the coordinates (0, A), (11, A), (0, K) and (11, K) are used as reference nodes. The measurement was done at the room 305 of 3th floor, E-Building, Faculty of Engineering, King Mongkut's Institute of Technology Ladkrabang.

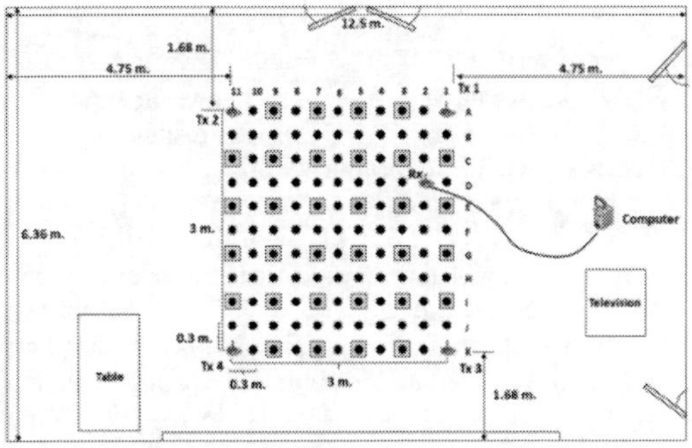

Fig. 1. The experimental setup.

4 EXPERIMENTAL RESULTS

In this section we proposed algorithm of performance analysis. By consider simulate the REWL based on WSNs for RSS. The accuracy calculates from reference node to target node follow equation 2 and the localization error calculates follow equation 5.

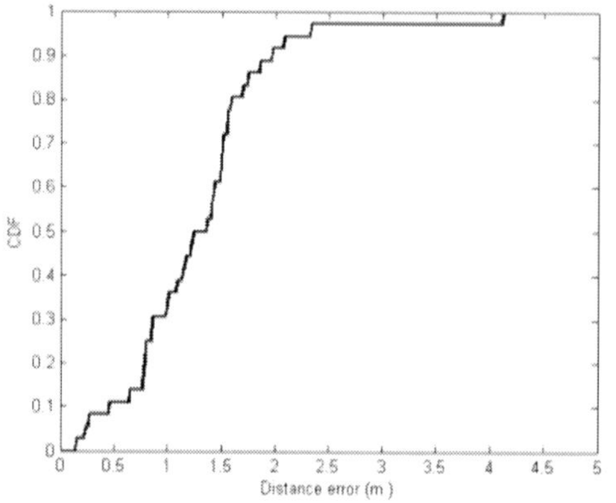

Fig. 2. The CDF of the distance error with range with different weighting factor at λ =0.10.

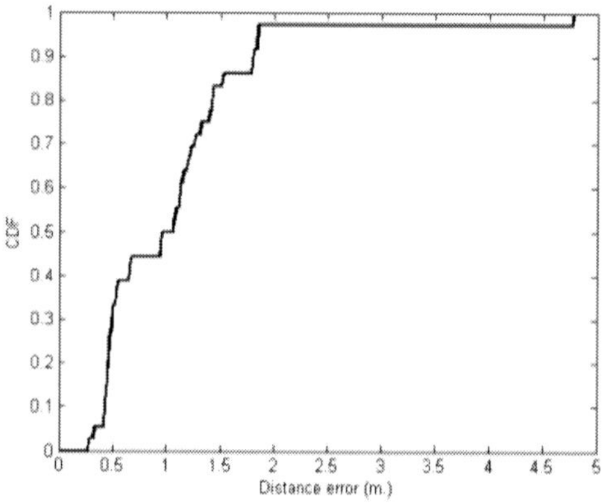

Fig. 3. The CDF of the distance error with range with different weighting factor at λ =0.15.

The distance error of each estimated distance is shown in term of CDF. The distance error in x, y axes of REWL algorithm with $\lambda = 0.1$ is shown in figure 2 the mean of distance error is 1.24 m. The distance error in x, y axes of REWL algorithm with $\lambda = 0.15$ is shown in figure 3 the mean of distance error is 0.9 m. The distance error in x, y axes of REWL algorithm with $\lambda = 0.2$ is shown in figure 4 the mean of distance error is 0.65 m. The distance error in x, y axes of REWL algorithm with $\lambda = 0.25$ is shown in figure 5 the mean of distance error is 0.75 m. To ascertain the optimal λ, we simulated a full equipped sensor network of the enclosed 2×2 reference node (4 nodes). In the figures 2, 3, 4, and 5 are demonstrating the CDF of the distance error with different weight factor (λ). The simulation result shows several minimum distance errors depending on the distance and the weight factor (λ). The smallest average of distance error exists at $\lambda = 0.20$, yields the best result as shown in figure 4. Consequently, before begin the localization process, a concentrated analysis of the transmission range, the dimensions of the network and distance are required to get the smallest distance error.

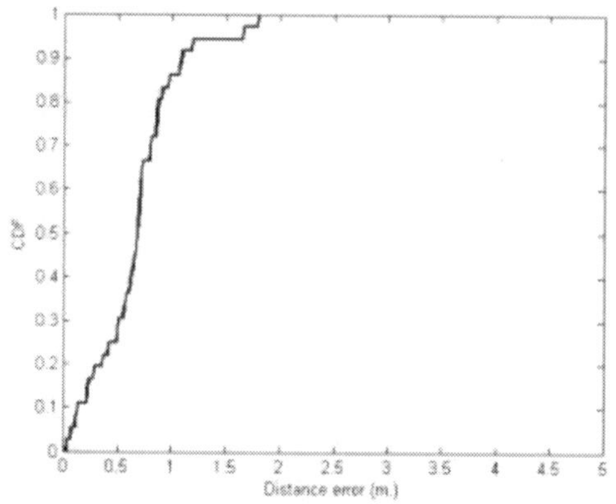

Fig. 4. The CDF of the distance error with range with different weighting factor at λ =0.20.

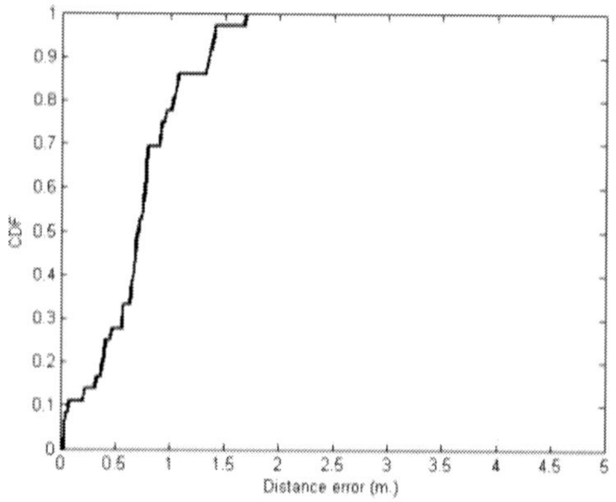

Fig. 5. The CDF of the distance error with range with different weighting factor at λ =0.25.

But in practice, interferences and obstacles as well as uncertain electronic transfer elements within the transceivers. Afterward, distances are estimated by using REWL algorithm. Consequently, the distance error of each position is evaluated by mesh contour and shown in figures 6, 7, 8 and 9.

5 CONCLUSION

This paper how to evaluate the Zigbee transmission model for indoor localization based on RSSI measurements collected by using REWL algorithm for WSNs. It provides the approximated position error of the target node in dissimilar test a points inside an inspection power transmission loss. This research we are summarizing the basic theoretical and practical facts impact the measured data exceedingly. Therefore, calculating distance using simple model will result in faulty distance with at least offset errors and scaling errors for concerning analysis of the distance error of REWL.

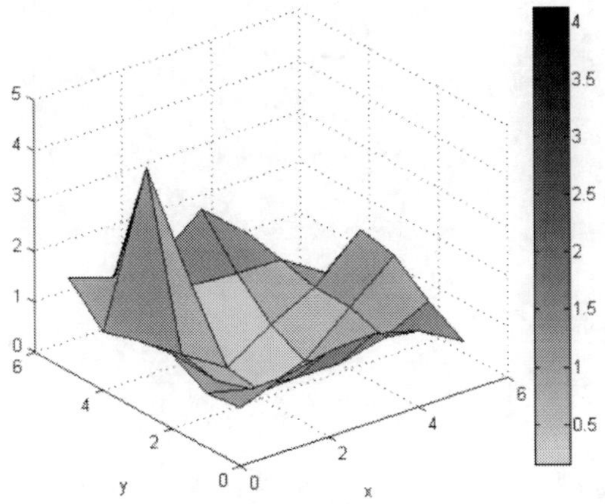

Fig. 6. The distance error of each position at λ =0.10.

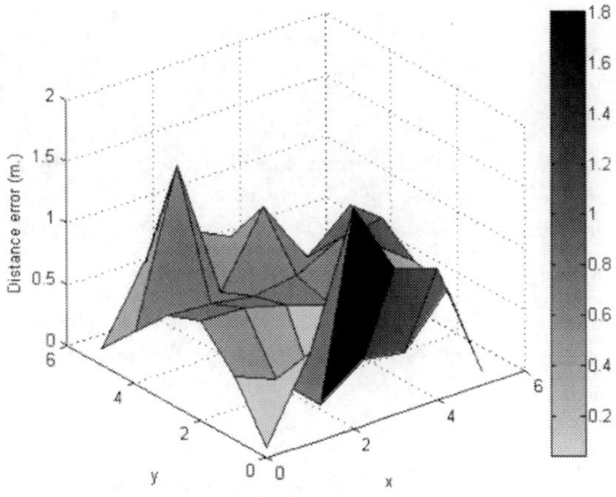

Fig. 8. The distance error of each position at λ =0.20.

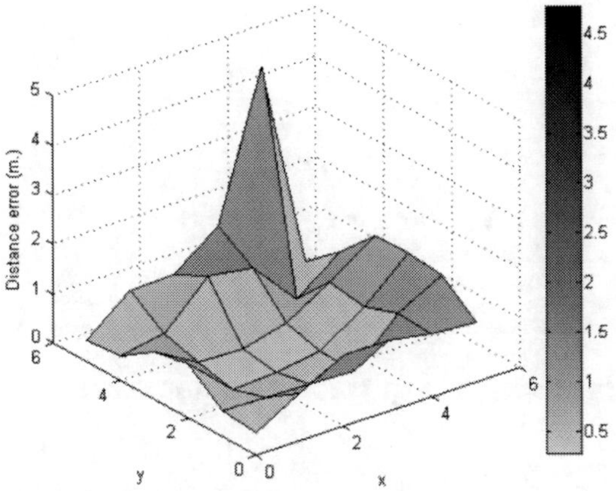

Fig. 7. The distance error of each position at λ =0.15.

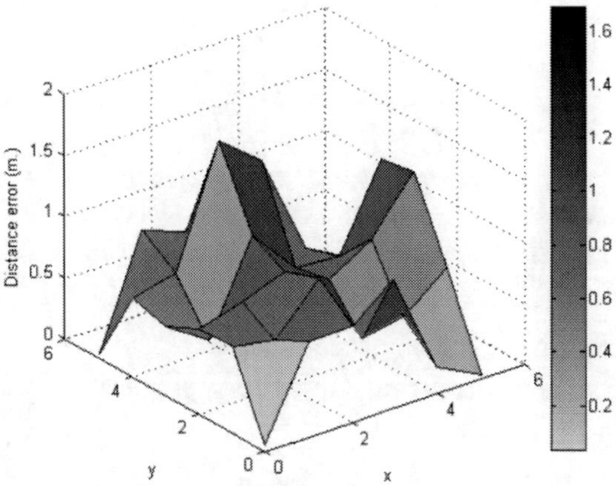

Fig. 9. The distance error of each position at λ =0.25.

Moreover, the REWL algorithm and against scaling error of distance measurement has been clarify. Localization results using Zigbee node indoor environment are described. Even though the positioning algorithm does not yet give the suitable results very precisely, the presented localization algorithm in merging with the Zigbee technology offer lots of advantages. It's very significant advantages are its minimum resource requirements, fast computation process and low complexity which advocate REWL as localization algorithm in Zigbee WSNs.

References

1. C. Laurendeau and M. Barbeau, "Relative span weighted localization ofuncooperative nodes in wireless networks," inProc. Int. Conf. WASA, pp. 358–367, August, 2009
2. F Ian. Akyildiz, W. Su, Y. Sankarasubramaniam, and E. Cayirci, "A survey on sensor network", IEEE Communication Magazine, pp102- 114, August 2002
3. H. Liu, H. Darabi, P. Banerjee, and J. Liu, "Survey of wireless indoor positioning techniques and systems," Systems, Man, and Cybernetics, Part C: Applications and Reviews, IEEE Transactions on, vol. 37, no. 6, pp. 1067–1080, November, 2007
4. J. Blumenthal, R. Grossmann, F. Golatowski, and D. Timmermann, "Weighted centroid localization in Zigbee-based sensor networks," in Proceedings of the IEEE International Symposium on Intelligent Signal Processing (WISP '07), Madrid, Spain, pp.1-6, October, 2007.
5. J. Blumenthal, R. Grossmann, F. Golatowski, and D. Timmermann, "Weighted centroid localization in Zigbee-based sensor networks," inProceedings of the IEEE International Symposium on Intelligent Signal Processing (WISP '07), Madrid, Spain, pp.1-6, October, 2007

6. J. Blumenthal, F. Reichenbach and D. Timmermann: Position Estimation in Ad hoc Wireless Sensor Networks with Low Complexity. In Joint 2 nd Workshop on Positioning, Navigation and Communication and 1st Ultra-Wideband Expert Talk, pp. 41–49, March, 2005.
7. Kristalina, P., Wirawan & Hendrantoro, G., "Improved Range-free Localization Methods for Wireless Sensor Networks", Proc. 2011 Int'l Conf. on Electrical Engineering and Informatics (ICEEI), pp. I5-1, July, 2011.
8. S. Schwarzer, M. Vossiek, M Pichler, A. Seltzer, "Precise Distance Measurement with IEEE 802.15.4 (Zigbee) Devices", Radio and Wireless Symposium IEEE, pp.779-782, January, 2008.
9. Y Chen, Q Pan, Y Liag, Z Hu, "Adaptive Weighted Centroid Target Localization Algorithm Based on RSSI in WSN", Computer Science and Information Technology (ICCSIT), 3rd IEEE International Conference, pp. 331-336, July, 2010.

Effect of freeze drying and hot air drying methods on quality of cordycepin production

Thitiphan Chimsook[1,*]

[1]Faculty of Science, Department of Chemistry and Applied Chemistry, Maejo University, Chiang Mai, Thailand

Abstract. This study determined the effect of drying methods including hot air drying and freeze drying on the quality of cordycepin production from *Cordyceps militaris*. The fruiting bodies and mycelium of *C. militaris* were used as the raw material. For hot air drying was performed at 55 °C for 24 hrs. Whereas the freeze drying was carried out under vacuum at 140 L min^{-1} for 48 hours. The bioactive compound extracted from dried powder of *C. militaris* from two drying methods was investigated. The results showed that both cordycepin and adenosine extracted from freezed drying sample had higher value than those of extracted from hot air dried sample. The bioactivities of *C. militaris* extract were investigated. The results revealed that the antioxidant activity and also total phenolic contents of *C. militaris* extract prepared from freeze drying had higher value than that of extracted from hot air drying. However, most of *C. militaris* production performed using hot air drying to dry sample because of its low cost technique.

Keywords: *C. militaris*, Cordycepin, Hot air drying, Freeze drying

1 Introduction

Drying of food is the method to remove the water in food. Drying is the application of warmth under controlled conditions to evaporation the water in liquid foods and produce the solid substance [1]. Because of the different products which have their own properties, there are developed different ways of drying. In general, the drying technologies compose of four principles including air drying, vacuum drying or puffing, spray drying and freeze drying, respectively. In this work, we evaluate two drying technologies of freeze drying and air drying (hot air drying) on quality of cordycepin production from *Cordyceps militaris*. Freeze-drying can preserve food and herb by extend and shelf life. With this technique it makes the material more convenient for transport. Freeze-drying is suitable for heat sensitive as it uses low processing temperature. The procedure of freeze-drying starts from freezing the material, then reducing the pressure and adding heat to allow the frozen water in the material to sublimate [2-4]. Compared to conventional food preservation technologies, the key benefits of freeze-drying include the following: retains original characteristics of the product such as colour, taste and size, reconstitutes to original state when placed in water, shelf stable at room temperature, reduces the weight of product, offers highest quality in a dry product compared to other drying methods [5, 6]. For hot air drying was invented to improve the efficient for drying herb and plant samples. The design of hot air drying was presented in Fig. 1 which showed the inside and outside figures of hot air drying.

Fig. 1. The design of hot air drying machine [7].

This invention was designed to save time and cost in drying process. From Fig. 1, the inside of oven composed of many shelves for the trays. For the operation, the automatic program was setted up [7]. This hot air drying machine can be enhanced the dehydration of various plants and also *C. militaris*. *C. militaris* is one of the most important traditional Chinese medicines. It has studied its pharmacological properties. The pharmacological properties are attributed to various active substances such as β-glucans, cordycepin, adenosine etc. [8,9]. In this work, cordycepin (3′-deoxyadenosine) was extracted from *C. militaris*. Cordycepin, a nucleoside analog, was first isolated from *C.militaris*. It has been regarded as a medicinal agent

* Corresponding author: thitiphan.cs@gmail.com

responsible for immunological regulation [10-12], anticancer [13], antifungus [14], antivirus [15], antihyperlipidemia [15] and antileukemia [16-18] activities. The quantity and quality of cordycepin production involved with the culture techniques and also drying methods. Therefore, this study chiefly highlights the drying methods on quality of cordycepin production from *C. militaris*.

2 Material and methods

2.1 Drying methods

Fresh of fruiting bodies and mycelium of *C. militaris* were a kind gift from SM Trading and Service Col, ltd. The factory was located in Maejo, Chiang Mai, Thailand. For hot air drying, the hygienic dryer was designed related to thermal engineer and food safety. All components of the dryer were suitable temperature distribution and accelerated moisture transfer out of system as shown in Fig. 1. 50 g of the fruiting bodies and mycelium of *C. militaris* were used as the material sample and dried at 55 °C for 24 hrs. For the freeze drying process, 50 g of *C. militaris* was frozen at -40 °C and freeze dried under vacuum at 140 L min^{-1} for 48 hours. The grinding machine was used to perform the powder freeze dried sample. All dried samples were prepared the cordycepin extract to investigate the quality of dried *C. militaris* sample after drying process.

2.2 Determinations of total phenolic content

The fruiting bodies and mycelium from freeze drying and hot air drying were dried to constant weight and homogenized to powder. About 20 g of dried sample in 500 ml ethanol was extracted by ultrasonic cell crushing, and then centrifuged at 10000 g for 15 min. The residue was re-extracted and then combined the supernatant before evaporated on a water bath at 55 °C until dry. The total phenolic content of extract was determined with Folin–Ciocalteu reagent according to the method of Khan et al. and Naczk M et al. [19, 20]. A 1.0 mL aliquot of sample was added to 1.5 mL deionized water and 0.5 mL of 0.1 mol/L. Folin–Ciocalteu reagent. The contents were mixed thoroughly immediately and added 1.0 mL of 20% sodium carbonate solution then mixed thoroughly. The controls contained all the reaction reagents except the sample. After 30 min, the sample was incubated at 37 °C and then measured the absorbance at 750 nm. The results were compared to a gallic acid calibration curve. The total phenolic content was expressed as mg gallic acid equivalent (GAE)/kg of dry weight (DW).

2.3 Determinations of antioxidant activities

The antioxidant activities of all substances were determined by DPPH assay. The methods was modified from that of Kim DO. et al. and Gulcin I. et al. [21, 22]. DPPH assay was based on the reduction of DPPH radical solution in the presence of hydrogen donating antioxidants. The DPPH radical solution (0.8 mM) in 95% ethanol was prepared. The extract (500 µl) was diluted to 5.0 ml using deionized water and 95% ethanol (1:1 V/V) and then 0.5 ml DPPH solution was added. The mixed solution was shaken vigorously. The sample was determined the absorbance after mixing. Trolox was used as chemical standard. The results were expressed in mg Trolox equivalents/kg of dry weight (DW). The assay was performed in triplicate.

2.4 Extraction of cordycepin and adenosine

The extraction of cordycepin and adenosine from dried *C. militaris* obtained from two drying methods was carried out prior to analysis the quantity of them by HPLC. All dried samples were extracted the bioactive compounds including cordycepin and adenosine. Dried samples were grinded into powder. Then, approximate 1.0 g of sample was precisely weighed and added into 10 ml 95 % ethanol and deionized water (1:1 V/V) in a 50 ml centrifuge tube. The mixed sample was subsequently placed in an ultrasonic machine for extracting cordycepin and adenosine at a power of 75 watt. After the centrifugation, the sample extraction procedure was repeated another twice. Supernatant obtained from the three times centrifugation was mixed and exactly measured of its volume. The sample was filtrated through a 0.45 µm filter prior to HPLC analysis [23].

2.5 Determination of the cordycepin and adenosine

All HPLC analysis work was carried out on Agillent model 1100 series, Agillent, Germany. Standards of cordycepin and adenosine were purchased from Sigma-Aldrich. The standard adenosine and cordycepin solvent was consecutively injected five times to draw calibration curves. The determination condition of the sample was set by using the ratio of mobile phase of water and methanol at 85:15(V/V), 90:10(V/V) and 92:8 (V/V), respectively. The separation was conducted in isocratic elution with a flow rate of 1.0 ml/min. The detection wavelength of photo-diode array was set at 210 - 400 nm and the column temperature was 30 °C. The injection volume was 10 µl [23].

2.6 Determinations of morphology of dried *C. militaris* powder

The morphology of dried *C. militaris* powder from two drying methods was investigated by Scanning Electron Microscopy (JSM 5410-LV, JEOL, Japan) with the tungsten filament K type, accelerate voltage of 10.0 kV and working distance of 15 mm. The *C. militaris* powder was milled sieved. Gold coating was performed by finer coater (JFC 1200, JEOL, Japan) for 150 sec.

3 Results and discussion

Cordycepin is one of the most important biologically active metabolites. In this work, cordycepin was extracted from dried powder of *C. militaris*. In general, drying of plant, food and active ingredients may be performed in different dryers and methods according to the initial properties, final desired characteristic of the products and economic factors. In this work, two drying methods of freeze drying and hot air drying were chosen to prepare the dried *C. militaris* before extracted the active compounds. The quantity and bioactivities of extracted active compounds may be revealed the efficiency of drying techniques to keep the precious substances from dried *C. militaris*. In recent years, researchers made remarkable progress in cordycepin production. They focused their attention on the three aspects to improve cordycepin production including strain improving, optimizing ingredients of medium and optimizing culture conditions, respectively, However, the drying process played an important role in the preservation of precious nutrients from *C. militaris*. Nowadays, the most drying method used to dry the fresh fruiting bodies and mycelium of *C. militaris* is hot air drying before grinded to fine powder. The hot air drying is chosen to dry the fruiting bodies and mycelium of *C. militaris* because it is the low cost technique, faster and provided higher yield than freeze drying. However, this study was conducted in order to evaluate the effect of different drying methods on quality of cordycepin production from *C. militaris*. The drying methods were the hot air drying at 55 °C for 24 hours and freeze drying for 48 hours. The active compound, bioactivities and also morphology of *C. militaris* extract from different drying methods were determined to clearify that drying methods had effected to the quality of cordycepin production. The results shown in Table 1 were the average yields, productivity and moisture contents of *C. militaris* powder obtained from hot air drying and freeze drying.

Table 1. Parameters of *C. militaris* powder obtained from the hot air drying and freeze drying.

Parameters	Hot air drying[1]	Freeze drying[1]
Yield (%)	65.00 ± 1.42^a	57.6 ± 1.78^b
Productivity (g h^{-1})	1.35a	0.6b
Moisture (%)	7.04 ± 0.19^a	4.15 ± 0.20^b

[1]Mean values and standard deviations. [a]Means in the same row with different superscripts differ significantly ($p < 0.05$) by the t-test.

The drying method performed the different yields of *C. militaris* powder. The higher yield of grinded powder after drying was obtained from hot air drying at 65.00 % whereas lower yield at 57.6% was found from the freeze drying. On the other hand, the moisture contents of *C. militaris* powder were 7.04 % and 4.15 % for hot air drying and freeze drying, respectively. These results revealed that the freeze drying can improve the efficient of dehydration process. The dehydration process can decrease the moisture contents of samples and it is very important on quality of *C. militaris* powder. The dried samples from two drying methods were prepared the *C. militaris* extracts by using ultrasonic cell crushing with 95% ethanol : deionized water at 1 :1 (V/V) as solvent in order to evaluate the bioactivities. including total phenolic contents and antioxidant activities using DPPH assay. Moreover, the bioactive substances of *C. militaris* including cordycepin and adenosine were carried out after extracting with 95% ethanol : deionized water at 1 :1 (V/V) and determined by HPLC analysis. All results were shown in Table 2. The antioxidant activity and also total phenolic contents of *C. militaris* extract from the freeze drying method were higher than those of the hot air drying method. In addition, both cordycepin and adenosine extracted from freezed drying sample had high value when compared to that from hot air drying sample.

Table 2. Bioactivities and bioactive compounds of *C. militaris* powder obtained from the hot air drying and freeze drying.

Parameters	Hot air drying	Freeze drying
Total phenolic content (mg GAE/kg (DW))	$3,197.50^a$	$3,597.20^b$
mg Trolox equivalents/kg sample (DW)	$24,012.15^a$	$28,816.00^b$
Cordycepin (mg/kg DW)	1950.50^a	$2,317.10^b$
Adenosine (mg/kg DW)	989.35^a	$1,097.05^b$

[a]Means in the same row with different superscripts differ significantly ($p < 0.05$) by the t-test.

The results revealed that different drying methods had significant effects on quantity and quality of cordycepin production from *C. militaris*. A literature search showed that drying methods had a significant effect on physical and chemical properties and also bioactive compounds. It can be concluded that the drying methods had effected on the proportion of the bioactive components in substances. Similar results of bioactive compounds were found in *C. militaris* and also *C. sinensis* which prepared with freeze drying methods by L. Huang et al. [23]. However, spray drying method was one of the best drying method which can help to remove the moisture in the sample. Due to its versatility and speed, spray drying became the most used drying technique for heat sensible substances such as foods and biological materials. For *C. militaris*, the spray drying method may not preserve the nutrient and bioactive compounds.

The surface morphology and size of *C. militaris* powder were analysed by SEM. All SEM images of *C. militaris* powder from the freeze drying and hot air drying methods were shown in Fig. 2a and Fig 2b, respectively. From Fig. 2a and Fig 2b, it can be seen in SEM images that *C. militaris* powder from both drying methods was similar which appeared as irregular shapes with variable sizes. However, the previous work have reported that the pure cordycepin crystals were in the form of rod-like crystals with a smooth surface [24]. For this work, it can be concluded that the drying methods have not affected the morphology of *C. militaris* powder.

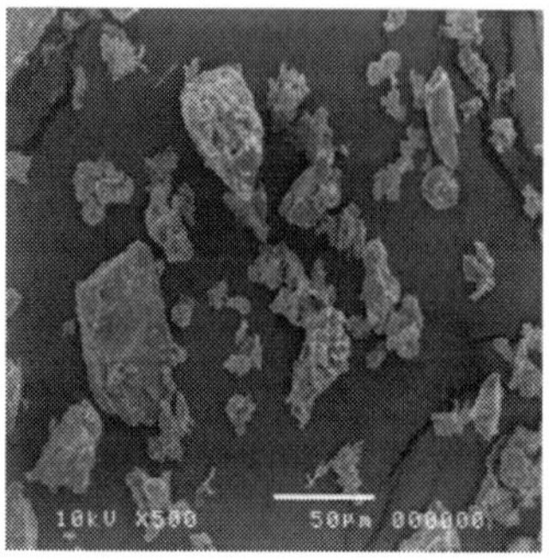

(a)

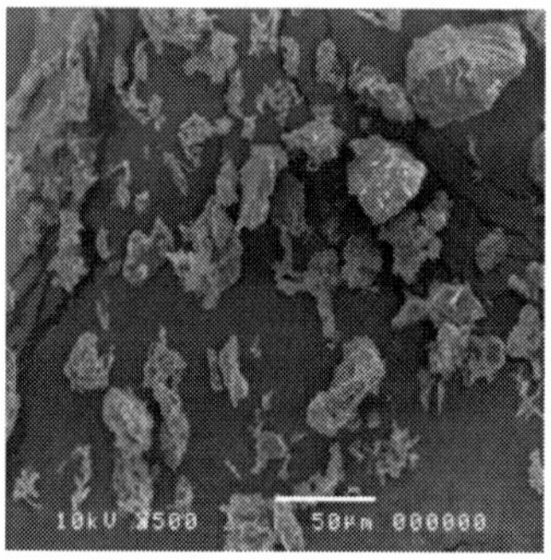

(b)

Fig. 2. SEM of *C. militaris* powder from (a) freeze drying method (b) hot air drying method.

4 Conclusion

Freeze drying and hot air drying methods have been successfully applied to dry *C. militaris* sample. All samples were extracted to evaluate the total phenolic contents and antioxidant activities. The bioactive compounds of *C. militaris* including cordycepin and adenosine were extracted and investigated. In term of the physical properties, hot air dried powders showed a promising result for easy handling of the dried sample when compared to the freeze drying process. The cost of hot air drying process was cheaper than that of freeze drying. However, the antioxidant activities and total phenolic contents of dried samples from freeze drying had higher value than that of hot air drying. Moreover, the cordycepin and adenosine which be extracted from freeze drying samples were higher than that of be extracted from hot air drying samples. The morphology of both drying methods was similar. All morphology of samples showed the irregular shapes with variable sizes and wrinkled shapes which were the result of low density particles. The XRD pattern and also morphology of cordycepin which extracted from each dried samples will be studied in further work.

5 Acknowledgement

This work was supported by NNR-National Science and Technology Development Agency and Faculty of Science, Department of Chemistry, Applied Chemistry, Maejo University, Thailand.

6 References

1. E.F.do E. Santo, L.K.F.de Lima, A.P.C. Torres, G.de Oliveiras and E.H.G. Ponsano, Food. Sci. Technol. **33**, 1 (2013)
2. A. Ciurzynska and A. Lenart, Pol. J. Food Nutr. Sci **61** (2011)
3. F. Franks, Eur. J. Pharm. Biopharm. **45** (1998)
4. F. Franks, Cryobiology. **40** (2000)
5. J. Gawalek, Inż. Roln. **71** (2005)
6. J.P. George and A.K. Datta, J. Food. Eng. **52** (2002)
7. T. Chimsook and R. Assawarachan, Key. Eng. Mater. **735** (2017)
8. Q. Zhang, Y. Liu, Z. Di, C.C. Han and Z. Liu, Fungal. Genom. Biol. **6**, 1 (2016)
9. S.K. Das, M. Masuda, A. Sakurai and M. Sakakibara, Fitoterapia. **81**, 8 (2010)
10. D.D. De Silva, S. Rapior, F. Fons, A.H. Bahkali and K.D. Hyde, Fungal. Divers. **55** (2012)
11. X. Zhou, C.U. Meyer, P. Schmidtke and F. Zepp, Eur. J. Pharmacol. **453** (2002)
12. E.M. Noh, J.S. Kim, H. Hur, B.H. Park and E.K. Song, Rheumatology. **48** (2009)
13. N. Yoshikawa, S. Yamada and C. Takeuchi, Naunyn. Schmiedebergs. Arch. Pharmacol. **77** (2008)
14. A.M. Sugar and R.P. McCaffrey, Antimicrob. Agents. Chemother. **42** (1998)
15. D.D. De Silva, S. Rapior, K.D. Hyde and H. Bahkali, Fungal. Divers. **56** (2012)
16. D.D. De Silva, S. Rapior and E. Sudarman, Fungal. Divers. **62** (2013)
17. E.N. Kodama, R.P. McCaffrey, K. Yusa and H. Mitsuya, Biochem. Pharmacol. **59** (2000)
18. H. Thomadaki, C.M. Tsiapalis and A. Scorilas, Cancer. Chemother. Pharmacol. **61** (2008)
19. R.A. Khan, M.R. Khan, S. Sahreen and M. Ahmed, Chem. Central. J. **6** (2012)
20. M. Naczk and F. Shahidi, J. Chromatogr. A. **1054** (2004)
21. D.O. Kim, S.W. Jeong and C.Y. Lee, Food. Chem. **81** (2003)
22. I. Gulcin, Arch. Toxicol. **86** (2012)
23. L. Huang, Q. Li, Y. Chen, X. Wang and X. Zhou, Afr. J. Microbiol. Res. **3**, 12 (2009)
24. J.-Q. Zhang, D. Wu, K.-M. Jiang, D. Zhang, X. Zheng, C.-P. Wan, H.-Y. Zhu, X.-G. Xie, Y. Jin, J. Lin, Carbohydr. Res. **406** (2015)

Polylactic acid (PLA) based green composites reinforced pineapple leaf fibres: evaluation of processing and tensile performance

Jedsada Chaishome[1,*] and *Suriyan* Supapvanich[2]

[1]Food Engineering Department, Faculty of Engineering, King Mongkut's Institute of Technology Ladkrabang, Bangkok, Thailand
[2]Department of Agricultural Education, Faculty of Industrial Education and Technology, King Mongkut's Institute of Technology Ladkrabang, Bangkok, Thailand

Abstract. This paper reports on a study of the compression moulding and the vacuum forming of unidirectional pineapple leaf fibres/polylactic acid composites and the influence of process variables on the tensile properties of the material. The characterisation of the micro and meso structures of the pineapple leaf fibres is reported. The effect of consolidation temperature on the fibre thermal stability and the tensile properties of the composites is investigated. The results show that vacuum forming was found to be preferable process with high stiffness modulus and UTS of the composites, compared to compression moulding. The insignificant detrimental effect of 165°C high consolidation temperature was observed. Finally, the fibre thermal degradation seems to dominate the composite tensile performance over its interfacial quality between the fibre and the matrix.

1 Introduction

There has been a sustainable agriculture itinerary toward short product life cycle made from renewable or biodegradable materials. For this season, academic research has been shifting from synthetic materials to natural materials, like fibres and bio-plastic composites [1-3]. Furthermore, the natural material offers several advantages i.e., light weighting, eco-friendliness and safety from hazards [4, 5].

Since 2006, environmental legislation in Europe and Japan has increased the pressure to reuse or to use renewable materials in the manufacture of vehicles [6]. For that reason, and the general drive toward sustainability, natural fibre reinforced polymer composite materials are of increasing interest in the automotive industry. They are also attracting interest for many other application areas including tribology, piping and electronic devices [7-9].

Natural fibres, such as pineapple leaf, ramie and hemp, are low density and can be used as the composite reinforcement, especially non-structural applications [6, 7]. Even though their tensile strength is low, compared to the E-glass fibres, it can compete in terms of specific properties [10-12].

Notwithstanding, their thermal stability of hemi-cellulose consisted of plant fibres limits composite properties after high temperature thermoplastic forming. The various natural constituents of plant fibre result in poor composite performance due to their low thermal degradation [13-15]. It depends on a number of factors namely heat rate, temperature and chemical composition of materials. Flax retains its strength at 170 °C for 2 hours while that of 200 °C for half an hour, for instance [14, 16]. In addition, polymers (as matrix materials) are limited from low thermal decomposition temperature of the plant fibres at high melted temperature processing. On the other hand, high viscosity melted thermoplastic property leads to low consolidation with high void content of composite, especially short dwelling period. Natural materials could generate a number of volatile products from their chemical constituents during high temperature and trap in the laminates as voidage[17].

2 Materials and experimental Method

2.1. Materials

Polylactic acid (PLA) and pine apple leaf fibrers are used in this investigation. PLA in a pellet form, supplied by Naturework Co., Ltd., is used as the matrix and the reinforcement is a unidirectional (UD) pineapple leaf fibres. Before the moulding studies, the fibres were fully characterised to determine, fibre diameter, fibre apparent density and fibre hydration levels.

2.2 Composite manufacturing

The UD pineapple leaf fibre/PLA composites were processed by isothermal film stacking of 40% fibre volume fraction (V_f) with compression moulding and vacuum forming process.

* Corresponding author: jedsada.ch@kmitl.ac.th

2.3 Thermogravimetric analysis

Thermal stability of pineapple leaf fibres was studied by thermogravimetric analysis. The samples were dried at 60 °C for 8 hours and were vacuum packed before performing. The fibres were weighted 10 mg and test using thermogravimetric (TG) technique to obtain a differential thermogravimetric (DTG) curves. The examination was conducted with a thermal analyzer setaram (Labsys evo TG/DTA, TG/DSC), under argon atmosphere. A heating process of room temperature to 600 °C at 10 °C/min of heating rate was used with an argon flow rate of 20 ml/min. Argon does not react with all materials as an inert gas.

2.4 Tensile test

Single fibres were performed tensile properties, according to ISO 11566: 1996. A TA XTplus of Texture Analyzer was used to examine 20 replicated samples with 2.5 mm/min of cross head speed. Each fibre cross-sectional area was calculated from each optical micrograph with image processing (ImageJ open source program).
The composite samples were examined their tensile perform according to ISO527-5:2009 (using an Instron universal testing machine). An extensometer gauge length of 50 mm at a cross head speed of 2 mm/min attached to ten replicated from each plaque test.

3 Results and discussion

3.1. Characterisation of the pineapple leaf fibres

The pineapple leaf fibre bundles were shown in Fig. 1 and single fibres were shown in Fig. 2. Each single fibre is not uniform shape. The mean equivalent diameter is 62.27 micrometres. A fibre apparent density of 1.32 g/cm^3 and the average moisture content at the initial condition of 11.24% for arbitrary storage RH were examined. Fig. 3 shows the stress-strain curves of pineapple leaf single fibre.

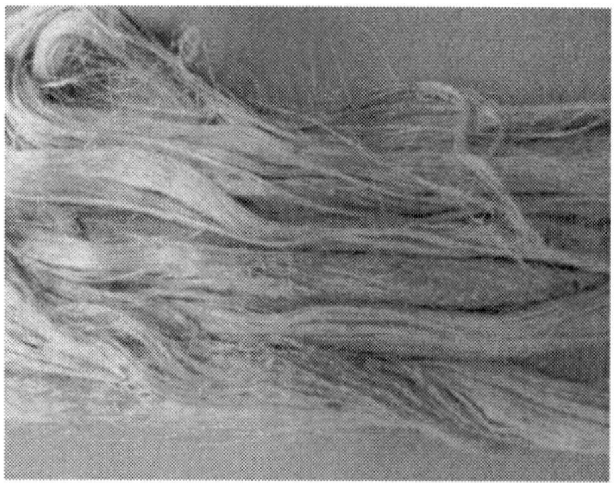

Fig. 1 The pineapple leaf fibre bundles.

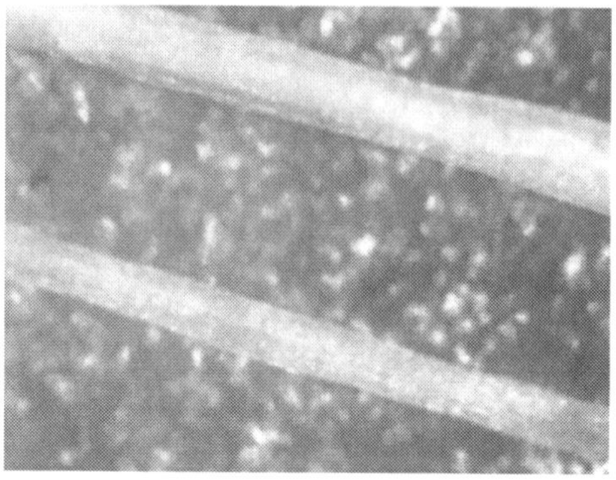

Fig. 2 The pineapple leaf single fibres.

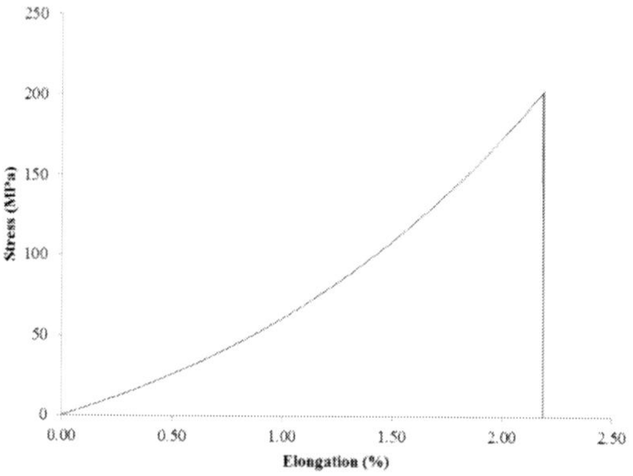

Fig. 3 Stress-strain curves of pineapple leaf single fibre.

3.2 Effect of temperature on the Thermal Stability of Fibres

Fig. 4 shows the differential thermogravimetric (DTG) peak between 30 °C and 600 °C. A small peak (between 45 °C and 120 °C) attributed to the water mass loss of the moist fibres. There is a shoulder peak (between 230 °C and 270 °C) attributed to hemi-cellulose and pectin degradation. A sharp peak at 350 °C attributed to cellulose degradation.
The DTG and the second time derivatives (D^2TG) curves were used to evaluate the initial decomposition temperature (T_0) of pineapple leaf fibres as described in Yao, et al.'s article[18]. The results from the thermal stability study reveal that the pineapple leaf fibre are stable until 250°C at a heating rate of 10°C/min. Nevertheless, TGA only shows the dynamic heating results (~40 to 200°C for 18 minutes). The result of quasi-static composite processing (~40 to 200°C for about 300 minutes) could show a much higher thermal degradation effect and the temperature limit of that is therefore likely to be below 250°C, resulting in a likely compromising of the mechanical performance of the fibres.

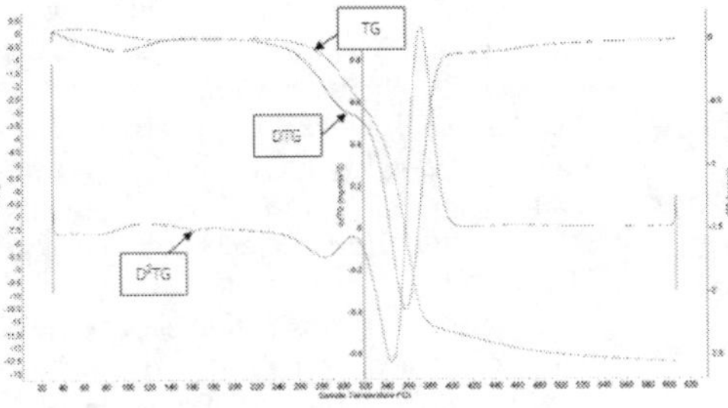

Fig. 4 The differential thermogravimetric (DTG) curves of the pineapple leaf fibre bundles.

3.4 Effect of processing on the tensile performance of composites

Tensile performance of 40%V_f UD pineapple leaf /PLA composites, are compared in Figs. 5. The materials processed with compression moulding of 50 kPa at 165, 180 and 190°C for 3, 5 and 7 mins were performed. The materials processed with vacuum forming at 165, 180 and 190°C for 30, 60 and 90 mins were examined.

Both modulus and UTS of the composites decrease with the dwelling time increase. The UTS of the compression moulding material is lower than that of the vacuum forming material. The weaker performance of the compression moulding material could be attributed to the higher void content level of the composite, caused by low consolidation time. Fig. 6 shows a micrograph of fibre-matrix interfaces.

Voidage presences in Fig. 6 expose their weak interfacial bonding between the fibres and the matrix due to inadequate consolidation time.

4 Conclusions

- Vacuum forming was found be preferable process with high stiffness modulus and UTS of the composites, compared to compression moulding.
- The insignificant detrimental effect of 165°C high consolidation temperature was observed.
- The fibre thermal degradation could dominate the composite tensile performance over their poor interfacial bonding effects.

5 Acknowledgements

The authors gratefully acknowledge the Biodiversity-Based Economy Development Office (Puplic Orgaization) (BEDO) for research funding and the department of Industrial Engineering, Faculty of Engineering, King Mongkut's Institute of Technology Ladkrabang, Bangkok, Thailand for the universal testing machine support.

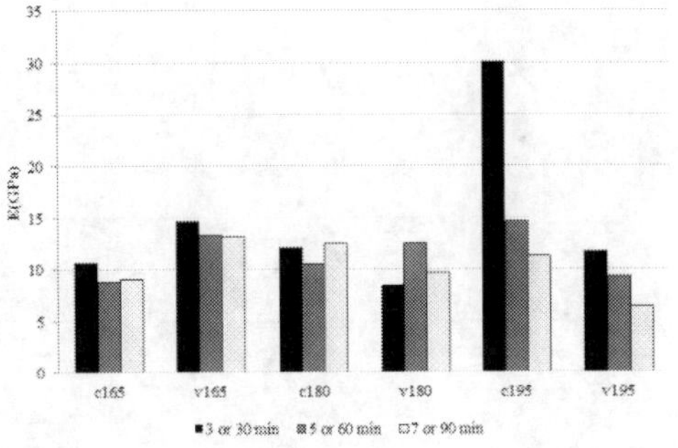

(a)

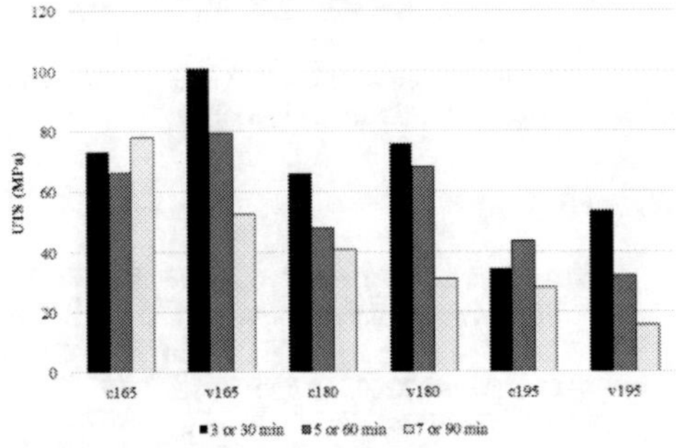

(b)

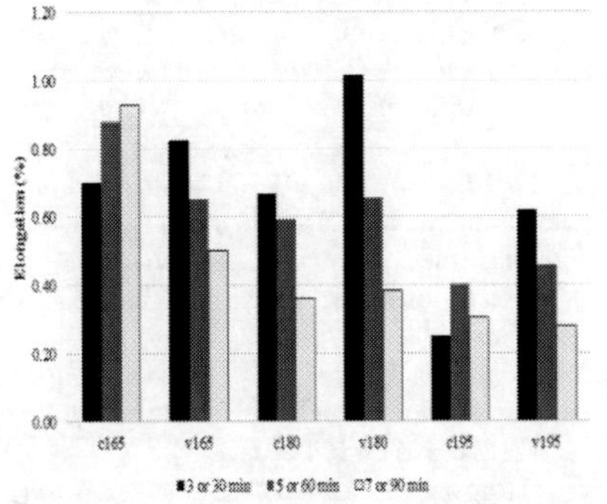

(c)

Fig. 5 (a) Stiffness modulus (E) of 40%V_f UD pineapple leaf /PLA composites (b) Ultimate tensile strength (UTS) of 40%V_f UD pineapple leaf /PLA composites
(c) Percentage of composite elongation at break of 40%V_f UD pineapple leaf /PLA composites
* c165 is compression moulding process of 50 MPa at 165°C of 40%V_f UD pineapple leaf /PLA composites
* v165 is vacuum forming process at 165°C of 40%V_f UD pineapple leaf /PLA composites.

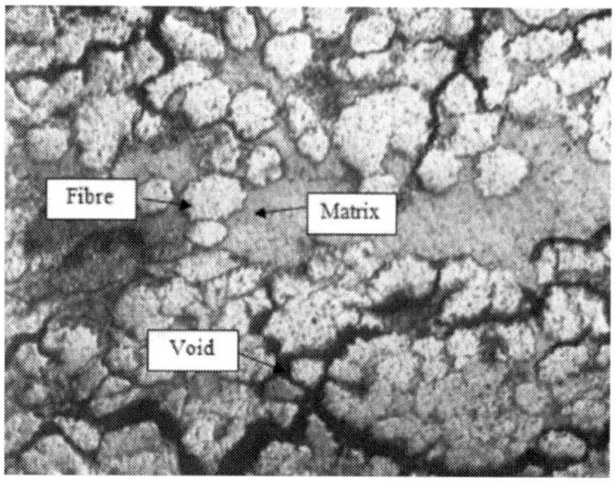

Fig. 6 poor-impregnated laminate with high void content of the pineapple leaf fibre bundles

References

1. Agopyan, V., et al., *Developments on vegetable fibre-cement based materials in Sao Paulo, Brazil: an overview.* Cement and Concrete Composites, 2005. **27**(5): p. 527-536.
2. Alavudeen, A., et al., *Review of natural fiber woven composite.* Reviews on Advanced Materials Science, 2011. **27**(2): p. 146-150.
3. Chaishome, J. and S. Rattanapaskorn. *The influence of alkaline treatment on thermal stability of flax fibres.* in *IOP Conference Series: Materials Science and Engineering.* 2017.
4. Bodig, J. and B.A. Jayne, *Mechanics of wood and wood composities.* Reprint ed. with corrections. ed. 1993, Malabar, Fla.: Krieger Pub.
5. John, M.J. and S. Thomas, *Biofibres and biocomposites.* Carbohydrate Polymers, 2008. **71**(3): p. 343-364.
6. Holbery, J. and D. Houston, *Natural-fiber-reinforced polymer composites in automotive applications.* JOM Journal of the Minerals, Metals and Materials Society, 2006. **58**(11): p. 80-86.
7. El-Tayeb, N.S.M., *A study on the potential of sugarcane fibers/polyester composite for tribological applications.* Wear, 2008. **265**(1-2): p. 223-235.
8. Rao, V.A., A. Satapathy, and S.C. Mishra. *Polymer composites reinforced with short fibers obtained from poultry feathers.* in *Proceeding of International and INCCOM-6 Future Trends in Composite Materials and Processing.* 2007. Kanpur.
9. Chang, K.H. and P.W. Richard, *Development of a bio-based composite material from soybean oil and keratin fibers.* Journal of Applied Polymer Science, 2005. **95**(6): p. 1524-1538.
10. Chaishome, J., et al., *Thermal Degradation of Flax Fibres as Potential Reinforcement in Thermoplastic Composites.* Advanced Materials Research, 2014. **894**: p. 32-36.
11. Chaishome, J. and S. Rakmae. *Thermal Stability of Bamboo Fibres.* in *7th International Conference on Agricultural engineering (TSAE2014).* 2014. Pranakhornsriayothanya, Thailand.
12. Chaishome, J. and S. Rattanapaskorn, *The Influence of Alkaline Treatment of Plant Fibres on Tensile Properties of Single Fibres and Composites.* Materials Science Forum, 2017. **894**: p. 46-49.
13. De Rosa, I.M., et al., *Morphological, thermal and mechanical characterization of okra (Abelmoschus esculentus) fibres as potential reinforcement in polymer composites.* Composites Science and Technology, 2010. **70**(1): p. 116-122.
14. Gassan, J. and A.K. Bledzki, *Thermal degradation of flax and jute fibers.* Journal of Applied Polymer Science, 2001. **82**(6): p. 1417-1422.
15. Ouagne, P., et al., *Analysis of the film stacking processing parameters for PLLA/flax fibre biocomposites.* Journal of Composite Materials, 2010. **44**(10): p. 1201-1215.
16. Wielage, B., et al., *Thermogravimetric and differential scanning calorimetric analysis of natural fibres and polypropylene.* Thermochimica Acta, 1999. **337**(1â€"2): p. 169-177.
17. Pielichowski, K.N., James Rapra Technology Limited Staff, *Thermal Degradation of Polymeric Materials.* 2005, Shrewsbury: Smithers Rapra. 320.
18. Yao, F., et al., *Thermal decomposition kinetics of natural fibers: Activation energy with dynamic thermogravimetric analysis.* Polymer Degradation and Stability, 2008. **93**(1): p. 90-98.

The effect of different additives on safety manipulation of cis-1,3,4,6-tetranitrooctahydroimidazo-[4,5-d]imidazole (BCHMX)

Ahmed K. HUSSEIN[1*], *Ahmed* ELBEIH[2], *Svatopluk* ZEMAN[1], *Marcela* JUNGOVA[1]

[1] Institute of Energetic Materials, Faculty of Chemical Technology, University of Pardubice, Czech Rep.
[2] Military Technical College, Kobry Elkobbah, Cairo, Egypt

Abstract. In this work, new cyclic nitramine named cis-1,3,4,6-tetranitrooctahydroimidazo-[4,5-d]imidazole (BCHMX) was premixed with insensitive explosives (3-nitro-1,2,4-triazol-5-one, NTO and 1,1-diamino-2,2-dinitroethene, FOX) and with two cyclic nitramines (1,3,5-trinitro-1,3,5-triazinane, RDX and, 1,3,5,7-tetranitro-1,3,5,7-tetrazocane, HMX) and bonded by polydimethyl-siloxane (PDMS) binder. Impact and friction sensitivities were evaluated. Detonation velocity was measured and the detonation characteristics were calculated by EXPLO 5 thermodynamic code. Relative explosive strength was determined by using the ballistic mortar test. The results showed a great positive influence on the sensitivity of BCHMX due to the addition of the insensitive explosive (NTO, FOX-7) however the performance was decreased. On the other side, mixing the nitramines improved the performance with reduction of BCHMX sensitivity. PDMS binder succeed to decrease the sensitivity of all the studied mixtures.

1. Introduction

An interesting explosive, *cis*-1,3,4,6-Tetranitrooctahydroimidazo-[4,5-d]imidazole (BCHMX) was prepared in the laboratory of the Institute of energetic materials as reported in Czech patent [1]. The performance of BCHMX was determined in comparison with several traditional explosives [2-4]. It has detonation performance in the level of HMX while it has high sensitivity similar to penterythritol tetranitrate (PETN) [5]. Reduction of its sensitivity is still considered a challenge. BCHMX was coated by different polymeric matrices in order to reduce the sensitivity without affecting its thermal stability [6-8]. polydimethyl-siloxane (PDMS) matrix seems to be an optimum binder for coating the explosive crystals: it has the ability to reduce the sensitivity of the explosives to acceptable level in addition to keep the detonation heat in the range of Plastic bonded explosives (PBXs) based on poly-fluorinated polymers [6]. After the addition of insensitive explosive such as 3-nitro-1,2,4-triazol-5-one, (NTO) or 1,1-diamino-2,2-dinitroethene (FOX-7), the sensitivity was improved significantly and reached the level of LOVA mixture (Ontallite; 50% TNT:50%NTO) [9, 10]. Consequently, in this paper and with the same vein, BCHMX was mixed with other nitramines explosive (1,3,5-trinitro-1,3,5-triazinane, RDX and, 1,3,5,7-tetranitro-1,3,5,7-tetrazocane, HMX) and bonded in PDMS. Their characteristics were studied and compared with the previous literature data.

2. Experimental work

2.1. Materials and preparation method

BCHMX was prepared at our institute where white crystals were obtained. NTO was prepared by adding formic acid to semicarbazide hydrochloride followed by the nitration using mixed acid as discussed in ref. [11]. FOX-7 is yellow crystals prepared by nitrating 2-methylpyrimidine-4,6-diol then the hydrolysis of intermediate 2-(dinitromethylidene)-5,5-dinitrodihydro pyrimidine-4,6(1*H*,5*H*)-dione to obtain FOX-7 [12]. Two types of polydimethylsiloxane (PDMS) were used (Wacker® AK 10 000) and (Wacker® AK 60 000). A ratio of 1:1 of these two kinds were used in order to produce the required average molecular mass and viscosity. The weight of energetic filler was 88% and in the mixture (44 % of BCHMX and 44% of other explosive) and the binder was 12%. Plastograph was used to prepare the samples and they were designed as BCHMX-Sil, BCHMX/NTO-Sil, BCHMX/FOX7-Sil, BCHMX/RDX-Sil, BCHMX/HMX-Sil.

2.2. Sensitivity to impact and friction

BAM impact sensitivity instrument [13] with exchangeable drop weight were used. Hammers with different weights were used from different heights to determine the impact energy required for initiation of the explosives. While BAM friction test apparatus was used to determine the friction sensitivity of the samples by spreading the material on a porcelain plate with rough surface. The normal force between the porcelain pistil and the plate was changed using different loads. The

* Corresponding author: ahmed92egypt@gmail.com

50% probability of initiation (H50) was determined for each sample based on the Probit analysis technique[14]. The results of the sensitivity measurements are reported in Table 1.

2.3. Detonation velocities measurements

The detonation velocitites of all the studied samples were obtained by using EXPLOMET-FO-2000 device (KONTINITRO AG). The tested samples were prepared on the form of cylindrical shape with 21 mm diameter and 20 cm length. Three optical probs were inserted inside each charge where the first prob was placed at 5 cm from the side with the detonator. The other two probs were inserted at a distance of 6 cm from the previous one. detonator no. 8 was used to initiate the charges. Three testes were pursuance for each sample. The mean value of each tested sample was reported in Table 2.

2.4. Detonation characteristics calculations

The main theoretical characteristics of detonation (velocity of detonation, D, pressure of detonation, P, and detonation heat, Q,) for all the studied samples were determined by EXPLO5 V 5.04 thermodynamic code [15]. BKWN set of parameters was applied in this study: $\alpha = 0.5$, $\beta = 0.176$, $\kappa = 14.71$, $\Theta = 6620$ [15]. The results of the theoretical calculation in comparison with the measured values are reported on Table 2.

2.5. Relative explosive strength measurements

Determination of the relative explosive strength for all the prepared samples, with respect to TNT, were done by using the ballistic mortar [13, 16]. 10 g of each sample was prepared and placed inside the ballistic mortar where the projectile locked the sample then fired by plain detonator. The results are reported in table 2 as the relative explosive strength compared to TNT [16].

3. Results and discussion

The performance of explosive material has a significant effect on its impact and friction sensitivity which was represent in recently accepted literatures [6, 17], at which, the high performance of explosive is accompanied by an increase of explosive sensitivity. However, modern techniques are used to discuss the relation between the performance and the reduction of sensitivity. Therefore, the plastic explosive mixture is one of the methods to combine between the performance and the sensitivity reduction. Moreover, presence of different explosive on the bimodal system in the PBXs has another effect on both sensitivity and performance. With a view to the sensitivity of the tested samples, a comparison between impact sensitivity (initiation due to uniaxial compression) and friction sensitivity (initiation due to shear slide of a fixed volume) of all the studied samples are shown in Fig. 1.

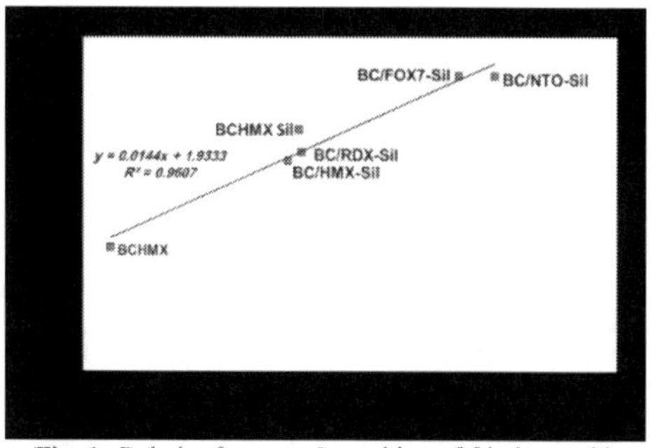

Fig. 1: Relation between Logarithm of friction and impact sensitivity

It was observed that the samples could be classified according to their sensitivity to two groups. The First group; mixture of BCHMX with nitramine and individual BCHMX in PDMS. The second group: mixture of BCHMX with FOX-7 and NTO.

The first group is more sensitive compared with second group. Regarding to the first group; BCHMX-Sil is the less sensitive to both impact and fiction compared with the mixtures with RDX and HMX, nevertheless, BC/RDX-Sil has the same value in impact sensitivity with BCHMX-Sil and still slightly sensitive to friction.

For the second group; the mixtures BC/FOX7-Sil and BC/NTO-Sil are the less sensitive PBXs among all the studied composition however BC/NTO-Sil is still less sensitive than BC/FOX7-Sil toward impact impulse.

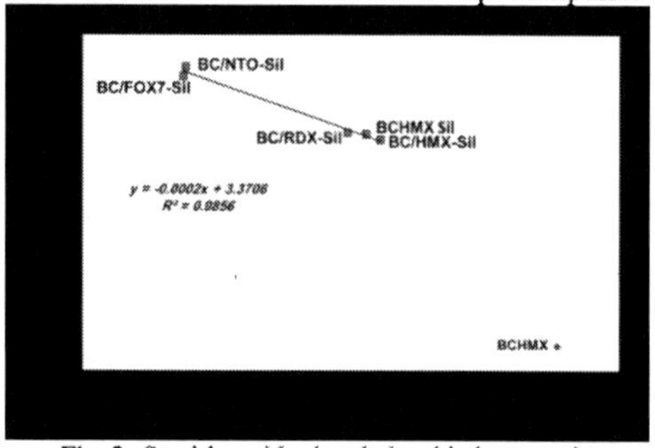

Fig. 2: Semi-logarithmic relationship between impact sensitivity and volume heat of explosion

An interesting relationship between volume heat of explosion (the product of loading density and the volume explosion heat $\rho.Q/\mathrm{MJ.m^{-3}}$) compared with the logarithm of impact sensitivity [18, 19] is presented in Fig. 2. Confirmation with what Licht [20] has pointed out, increasing the explosive performance is usually accompanied by an increase in the sensitivity [3, 6, 20] and cannot be proved by any theory. This relationship gave a clear approve with Licht and explain how less performance of BC/NTO-Sil and BC/FOX7-Sil are less sensitive than the other mixtures. However, It is also obvious here that BC/HMX-Sil the more sensitive and has the highest performance among the studied samples. It is also clear that the BCHMX is outside the line connected the studied PBXs, it is due to BCHMX is a pure explosive and has high performance and high sensitivity.

Table 1. Performance and detonation parameters of the samples studied

No	Code designation	Summary formula	Heat of combustion [J.g^{-1}]	Enthalpy of formation [kJ.mol^{-1}]	Impact sensitivity [J]	Friction sensitivity [N]
1	BCHMX	$C_4H_6N_8O_8$	9124	236.5	3.2 [a]	88[a]
2	BCHMX-Sil.	$C_{5.02}H_{9.18}N_8O_{8.47}Si_{0.54}$	11328	-27.16	24.3[a]	232[a]
3	BCHMX/NTO-Sil	$C_{2.42}H_{2.99}N_{4.84}O_{4.2}Si_{0.81}$	9512	-80.5	46.3[b]	>360[b]
4	BCHMX/FOX7-Sil	$C_{4.52}H_{9.82}N_6O_{7.36}Si_{0.446}$	11863	-286.5	42.3	>360
5	BCHMX/RDX-Sil	$C_{4.60}H_{9.60}N_{7.0}O_{4.56}Si_{0.53}$	11951	-93.8	24.6	192.7
6	BCHMX/HMX-Sil	$C_{5.50}H_{11.45}N_{8.0}O_{9.01}Si_{0.63}$	1.64	-135.7	23.1	179.7

a) [6]; b)[10]

Table 2. Results of the experimental measurements on the samples studied

Studied sample		Experimental			Calculated detonation parameters by Explo5				
No.	Code of samples	Density ρ [g.cm^{-3}]	RS %TNT	Ref.	Detonation velocity [m.s^{-1}]		Error %	Detonation pressure P [GPa]	Detonation heat Q [kJ.kg^{-1}]
					D_{exp}	$D_{cal.}$			
1.	BCHMX	1.79	-	[2]	8650	8840	+2.19	33.9	6447
2.	BCHMX-Sil.	1.62	137.5	[10]	7994	7742	-2.82	24.2	6245
3.	BCHMX/NTO-Sil	1.61	112.2	[10]	7864	7719	-1.84	23.7	5450
4.	BCHMX/FOX7-Sil	1.59	123.5	[9]	7889	7438	-5.71	21.3	5507
5.	BCHMX/RDX-Sil	1.61	137.8		7712	7633	-1.02	23.3	6197
6.	BCHMX/HMX-Sil	1.64	137.3		7834	7739	-1.21	24.3	6234

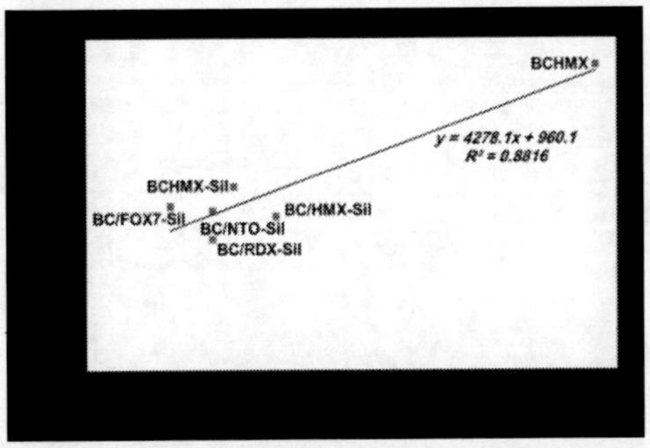

Fig. 3: A relationship between the density and the detonation velocity of all the studied explosives

The measured densities and detonation velocities were recorded in Table 2 and the results were presented on Fig. 3 where the well-known relationship between the density and the detonation velocity was confirmed. The BCHMX has the highest detonation velocity of the studied samples and as predicted the silicone matrix decreased the detonation velocity.

The detonation velocity of individual explosive BCHMX in PDMS has the higher detonation in spite of the lower density than the BC/HMX-Sil. This reason might be due to heterogeneity in the matrix. In the other way, the BC/FOX7-Sil has the lower density and has detonation velocity slightly lower than BCHMX-Sil (by 105 m s^{-1}). It is also obvious that addition of other explosive in the mixture with the BCHMX in the PBX matrix has reduced the detonation velocity.

Fig.4 presents a relationship between the volume heat of detonation, calculated by EXPLO5 (here Q in kJ kg^{-1}) and the relative explosive strength (% of TNT) measured by ballistic mortar. From Fig.4, the logical dependence of the relative explosive strength on the heat of detonation was confirmed.

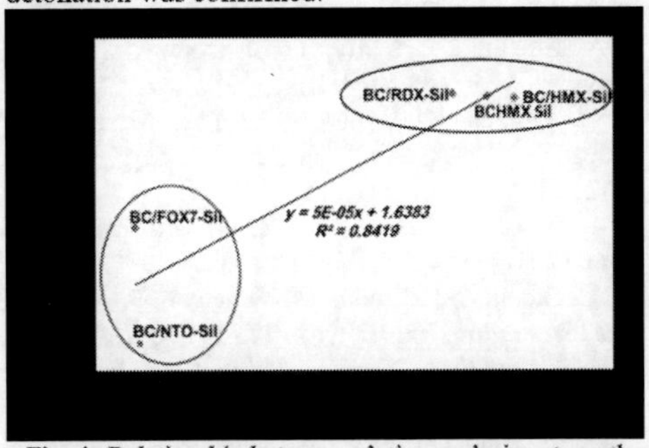

Fig. 4: Relationship between relative explosive strength and volume heat of detonation (Q in kJ kg^{-1} was calculated by the EXPLO5 code) of the studied PBXs.

BC/FOX7-Sil, BC/NTO-Sil have the relative strength of 123 and 112.2 % of TNT respectively while BC/HMX-Sil, BC/RDX-Sil and BCHMX-Sil have the relative strength of 137.3, 137.8 and 137.5 % of TNT respectively which confirm the high performance of the PBX however, BC/FOX7-Sil, BC/NTO-Sil have the lower performance among the studied samples, This is clear, as these two PBX have the heat of detonation of. 5507 and 5450 kJ kg^{-1} respectively. This behavior mainly comes from the FOX-7 and NTO which act as coolant for the gases generated and reduces the performance [10].

* Corresponding author: ahmed92egypt@gmail.com

4. Conclusion

A relative sensitive explosive *cis-1*,3,4,6-tetranitrooctahydroimidazo-[4,5-*d*]imidazole (BCHMX) is possible to be stabilized against mechanical impulses by incorporation into polydimethylsiloxane matrix (PDMS). However, using of the other explosive mixed with BCHMX in this matrix play a backbone in the sensitivity. The overall results confirm that mixing of FOX-7 and NTO with BCHMX in PBX bonded by Silicone matrix reduced the sensitivity of BCHMX significantly However, mixing of HMX with BCHMX slightly improve the performance but with increase of sensitivity in the same time. While mixing of RDX with BCHMX slightly decrease the performance and improve sensitivity. The calculated detonation characteristics of PBX by EXPLO5 thermodynamic code are compatible with the experimental values.

Acknowledgement

This paper was supported by means of the financial resources of Students Grant Projects No. SGSFCHT_2016002 of the Faculty of Chemical Technology at the University of Pardubice.

References

1. D. Klasovitý, S. Zeman, Czech Pat, **302068** (2010).
2. A. Elbeih, M. Mokhtar Mohamed, T. Wafy, Pro., Expl., Pyro., **41** (2016) 1044-1049.
3. A. Elbeih, T.Z. Wafy, T. Elshenawy, Cent. E. J. Energ. Mat., **14** (2017) 77-89.
4. W. Pelikán, S. Zeman, Q.L. Yan, M. Erben, A. Elbeih, Z. Akštein, Cent. E. J. Energ. Mat., **11** (2014).
5. A. Elbeih, T. Elshenawy, M. Gobara, Def. Sci. J., **66** (2016) 499.
6. A. Elbeih, S. Zeman, M. Jungova, P. Vávra, Z. Akstein, Pro., Expl., Pyro., **37** (2012) 676-684.
7. S. Zeman, Q.-L. Yan, A. Elbeih, Cent. E. J. Energ. Mat., **11** (2014) 395--404.
8. Q.-L. Yan, S. Zeman, P. Sánchez Jiménez, T.-L. Zhang, L. Pérez-Maqueda, A. Elbeih, J. Phy. Chem. C, **118** (2014) 22881-22895.
9. A.K. Hussein, A. Elbeih, M. Jungova, S. Zeman, Pro., Expl., Pyro.(2018)
10. A.K. Hussein, A. Elbeih, S. Zeman, Therm. Acta, **655** (2017) 292-301.
11. A.K. Nandi, S.K. Singh, G.M. Kunjir, J. Singh, A.K. Mandal, R.K. Pandey, Cent. E. J. Energ. Mat., **10** (2013).
12. J. Ottis, Z. Jalový, F. Liška, J. Energ. Mat., **26** (2008) 220-229.
13. M. Suceska, *Testing Methods of Explosives*, in, Springer, Heidelberg, (1995).
14. J. Šelešovský, J. Pachmáň, Cent. E. J. Energ. Mat., **7** (2010) 269-278.
15. M. Sućeska, Mat. Sci. Forum, Trans Tech Publ, 2004, pp. 325-330.
16. Notice of Czech Mining Authority No. 246/1996 of Law Collect, Establishing More Detailed Conditions for Allowing Explosives, Explosive Objects and Aids in Use, and their Testing, 1996.
17. A. Elbeih, S. Zeman, M. Jungova, Z. Akstein, Cent. E. J. Energ. Mat., **9** (2012) 131-138.
18. M. Jungová, S. Zeman, A. Husarová, Chiense J. Energ. Mat., **19** (2011).
19. A. Elbeih, S. Zeman, Cent. E. J. Energ. Mat., **11** (2014).
20. [20] H.-H. Licht, Pro., Expl., Pyro., **25** (2000) 126-132.

Efficiency of coconut water immersion inhibiting browning incidence on cut-surface of fresh-cut 'Gala' apples during storage

Suriyan Supapvanich[1,*], *Dararat* Anartnet[1], *and Chutima* Krungpree[1]

[1]Agricultural Education Department, Faculty of Industrial Education and Technology, King Mongkut's Institute of Technology Ladkrabang, Bangkok, Thailand

Abstract. The purpose of the work was to investigate the efficiency of coconut water preventing browning incidence of fresh-cut product using 'Gala' apple as the fruit model. The fresh-cut apples were dipped in coconut water at the concentration of 0, 50 and 100% for 2 min and then kept at 4 ± 1 °C for 7 days. Browning-related parameters such as browning index (BI), browning score (BS), lightness (L*), whiteness index (WI), total colour difference (ΔE*), total phenols (TP), and polyphenol oxidase (PPO) activity were determined. The result showed that visual appearance of the fresh-cut apple was maintained by coconut water dip. Coconut water dip apparently lowered BI, BS and ΔE* increases and maintained L* and WI throughout the storage. The increases in TP and PPO activity of the fresh-cut apples were inhibited by coconut water dips. No significant difference in browning prevention efficiency of the both 50 and 100% coconut water dips was found. Thus 50% coconut water dip is an effectively natural alternative preventing browning incidence of fresh-cut products.

1 Introduction

Browning on cut-surface is the main problem affecting the organoleptic properties including undesirable visual appearance of certain fresh-cut product. Regarding to the injury of tissue by minimal process, browning incidence is rapidly occurred on the cut-surfaces due to the reaction between PPO and phenolic compounds in the present of oxygen [1, 2]. To prevent browning incidence in fresh-cut products, sulphite has been industrially employed as antibrowning agent; however it has adverse effects on health aspects and is banned by government for fresh fruit and vegetables [3]. Recently, a range of natural chemicals including proteins, organic acids [4], carrageenan [5] and glutathione [6] were reported for browning prevention in fresh-cut fruits. Beside of these chemicals, the use of natural agents such as honey [7], pineapple juice [8] and Aloe vera gel [9, 10] preventing browning incidence have been considered. Browning inhibition property of natural agents is due to their antioxidant capacity [11]. Coconut water (CW) is recognised as a rich source of antioxidants especially free radical scavenging and superoxide radicals scavenging activities [12, 13]. In traditional Thai cuisine, CW dip could maintain visual appearance of craved fruit such marian plum. Thus, this work was focused on the browning inhibitory effects of CW using fresh-cut 'Gala' apple as the fresh-cut fruit model.

2 Materials and Methods

2.1 Coconut water preparation and experiment

CW from young coconut at the maturity of 190-200 day after full bloom was diluted with distilled water in the ratio of 1:1 (50%) and 1:0 (100%).

'Gala' apple (Malus demestica) fruit were purchased from a detail fruit market. The fruit were cleaned using 200 μL L-1 sodium hypochlorite. The fruit were vertically cut into 8 equal pieces and the core was removed using a knife. The cut apple pieces were dipped into prepared CW for 2 min. Six pieces of cut apple fruit were packed in a PE plastic box with lid and stored at 4±1 °C for 7 days. Four boxes of each treatment were sampled during storage. Factors related to browning were determined by comparing with untreated fresh-cut apple fruits (control).

2.2 Browning score (BS) index (BI) measurements

Three grams of the fresh-cut apple fruit flesh were extracted with 30 mL of 60% (v/v) ethanol. The absorbance of the extracted solution at the wavelength of 420 nm was measured. The unit of BI was expressed as OD420 per g fresh weight (OD420/g FW). The BS was estimated using 5 points scoring test. The appraisal was performed by 30 semi-trained panels. The fresh-cut fruit were evaluated using browning scores of 0 = no brown, 3 = moderate brown and 5 = extreme brown.

* Corresponding author: author@e-mail.org

2.3 Colour measurements

Colour attributes of the cut surface were measured using Minolta CR300 colorimeter (Minolta Co. Ltd. Japan). CIE L*, and ΔE* values of the cut surface of fresh-cut apple fruit were recorded. The whiteness index (WI) of the fresh-cut fruit was calculated using following equation (1).
WI = (100 – [(100 – L*)2 + a*2 + b*2]1/2).

2.4 Colour measurements

Five grams of fresh-cut apple fruit were homogenised with 60% (v/v) ethanol and then centrifuged at 6000 rpm for 20 min. The supernatant was used to determine total phenols content by using the method of SLINKARD and SINGLETON [12].

2.5 Total phenols assay

Five grams of fresh-cut apple fruit were homogenised with 60% (v/v) ethanol and then centrifuged at 6000 rpm for 20 min. The supernatant was used to determine total phenols content by using the method of SLINKARD and SINGLETON [12].

2.6 Polyphenol oxidase (PPO) activity assay

Two grams of the fresh-cut apple fruit were extracted with 0.1 M sodium phosphate [redacted] PPO activity was determined using the method described by GALEAZZI et al. [13].

2.7 Statistical Analysis

Statistical analysis was carried out using ANOVA and the means compared by the least significant difference test at a significance level of 0.05. The data are expressed as the mean of four replications and standard deviation (SD) bar.

3 Results

3.1 Coconut water preparation and experiment

The visual appearance and BI of fresh-cut apples during the storage are shown in Fig. 1 and 2, respectively. The visual appearance showed that browning incidence of the control was greater than the both CW dipped fresh-cut apples. The both 50% and 100% CW dips could inhibit cut-surface browning incidence of the fresh-cut apples. The visual browning incidence of the fresh-cut apples was concomitant with the BI value during the storage. The BI value of control was significantly higher than that of the both CW dipped fresh-cut apples throughout the storage ($P < 0.05$). The BI value of control was 1.76, 2.20 and 3.73 OD420nm g-1 during storage for 2, 5 and 7 d, consequently. The BI of both 50% and 100% CW were similar over the storage which it was approximately 0.47, 0.56 and 1.60 OD420nm g-1 during storage for 2, 5 and 7 d, consequently.

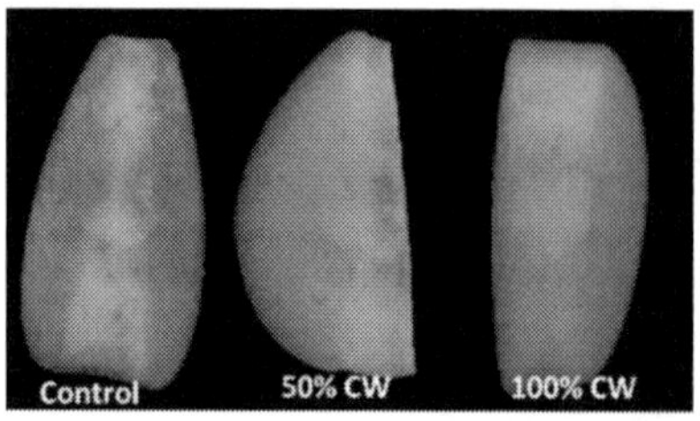

Fig. 1 Visual appearance of fresh-cut apple dipped in coconut water at various concentrations after storage for 7 d at 4±1°C.

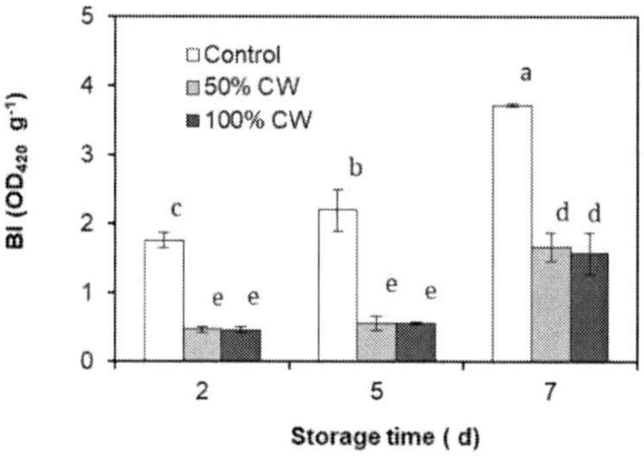

Fig. 2 BI of fresh-cut apple dipped in coconut water at various concentrations during storage for 7 d at 4±1°C.

3.2 Coconut water preparation and experiment

Fig. 3 shows BS and superficial colour attributes such as L* WI and ΔE* value of fresh-cut apples during the storage. BS of the control was markedly increased and significantly higher than those of both 50 % and 100 % CW dipped fresh-cut apples ($P < 0.05$) (Fig. 3A). The BS of both 50 % and 100 % CW dipped fresh-cut apples increased slightly during the storage and no significant difference between the both treatments was found throughout the storage. Both L* value and WI of the control decreased obviously and were significantly lower than those of the both 50 % and 100 % CW dipped fresh-cut apples throughout the storage ($P < 0.05$) (Fig. 3A and 3B). L* value of the both CW treatments and WI of 100 % CW dipped fresh-cut apples remained constant over the storage while WI of 50 % CW dipped fresh-cut apples was lower than that of 100 % CW dipped samples after storage for 5 and 7 days. The ΔE* value of the control was significantly higher than that of the both CW treated fresh-cut apples ($P < 0.05$) and increased continuously throughout the storage (Fig. 3D). Whereas, ΔE* value of the both CW treated fresh-cut apples seemed constant during storage. These indicate that CW immersion could maintain desirable appearance of fresh-cut apples due to inhibit browning incidence and maintain L* and WI values.

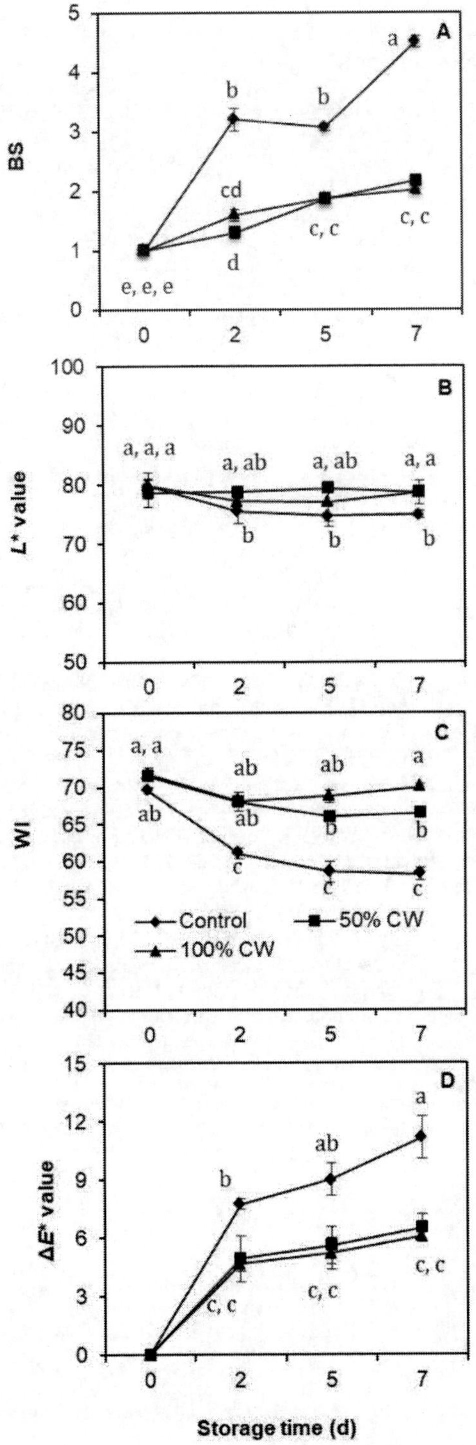

Fig. 3 BS (A), L* (B), WI (C) and ΔE* (D) values of fresh-cut apple dipped in coconut water at various concentrations during storage for 7 d at 4±1°C.

3.2 Total phenols and PPO activity

TP content and PPO activity in the cut-surface tissues of the fresh-cut apples are shown in Fig. 4. The both TP content and PPO activity of control were significantly higher than those of the both CW treated fresh-cut apples over the storage ($P < 0.05$). The increase in TP content during storage was found in the control (Fig. 4A). The TP content of control stored for 7 days was reached to 0.93 µg g-1 which was significantly higher than that on day 0. The TP content of both CW treated fresh-cut apples was significantly decreased after stored for 2 days ($P < 0.05$) and then seemed constant over the storage. In the similar vein, the PPO activity of control increased during storage (Fig. 4B). The PPO activity of control on day 7 was significantly higher than that of the control on day 0 and the both CW treated fresh-cut apples ($P < 0.05$). Whereas, the PPO activity of both CW treated fresh-cut apples seemed constant did not significantly differ and seemed constant over the storage.

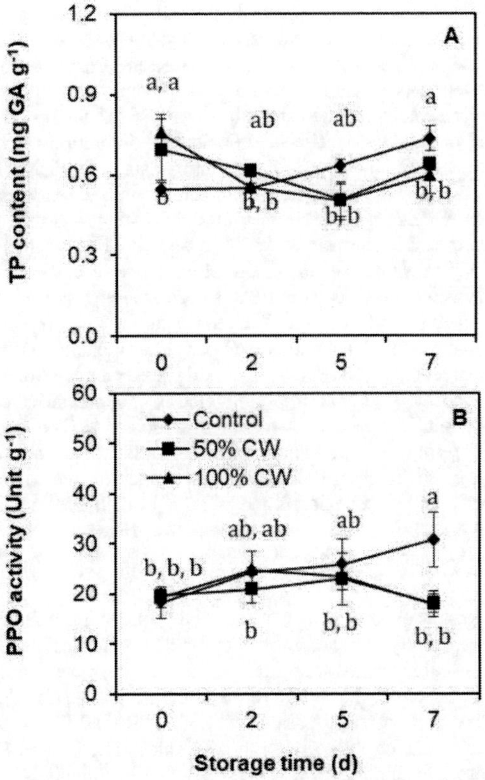

Fig. 4 TP content (A) and PPO activity (B) of fresh-cut apple dipped in coconut water at various concentrations during storage for 7 d at 4±1°C.

4 Discussion

The visual appearance shown in Fig. 1 indicated that CW dips obviously inhibited browning incidence on the cut-surface of the fresh-cut apples during the storage. The increases in both BI and BS of the CW treated fresh-cut apples were inhibited and apparently lower than those of the control. These were concomitant with the maintenance of L* and WI values. CW dips also lowered the increases in ΔE* values of the fresh-cut apples throughout the storage (Fig. 3 and 4). The changes in browning incidence and colour attributes on the cut-surface of fresh-cut apples were accompanied with the levels of TP concentration and PPO yield which were obviously high in the control compared to the both CW dipped fresh-cut apples. These suggest that CW has ability controlling enzymatic browning reaction due to retard the increase in PPO activity as well as TP concentration. It is commonly acknowledged that the cut-surface of fresh-cut products are directly presented to oxygen in atmosphere activating the reaction of released PPO and phenolic compounds resulting in complex polymers providing brown pigments [14, 11]. Kwak and Lim [15], Soysal [16] and Wessels et al. [11] suggested that antibrowning capacity of natural antibrowning agents might be associated with their antioxidant capacity, TP content and pH. In this present study, the DPPH free radical scavenging activity, TP content and pH of CW were determined which were 9.2 %, 2 g gallic acid L-1 and 6, respectively. Soysal [16] reported that the optimal pH for PPO extracted from 'Gloden Delicious' apple fruit was 5.5 and the enzyme activity is completely inhibited at pH below 3 [14]. Thus, the pH of CW might not affect PPO activity inhibition in the treated fresh-cut apples. The antioxidant activity and TP content of CW might play a role inhibiting enzymatic browning reaction. Wessels et al. [11] suggested that the reducing properties of polyphenols attributed antibrowning potential especially flavonoids and tannins which can inactivate PPO by complexing copper or reacting with proteins. Mantena et al. [17] had reported that CW had high antioxidant properties including DPPH scavenging, ABTS and superoxide scavenging as well as ascorbic acid concentration which is commercially used to inhibit browning in fresh-cut products [4]. Therefore, the antibrowning capacity of CW might be associated with the high level of antioxidant properties, the content of phenolic compounds and ascorbic acid concentration

5 Conclusion

Results show that CW at the both concentrations apparently inhibited browning incidence and maintained desirable visual appearance of the fresh-cut apples compared to the control throughout the storage. CW dips lowered browning parameters such as BI and BS values and delayed the increase in ΔE* value. The L* and WI were maintained by CW dips. CW dips also inhibited the increases in activity of PPO and TP content of cut-surface tissues of the fresh-cut apples. No significant difference in antibrowning capacity of the both 50% and 100% CW were found. We suggest that CW is an effective natural antibrowning alternative for fresh-cut fruits and 50% CW dip is sufficient to prevent browning incidence in fresh-cut 'Gala' apples during storage at 4 ± 1 °C for 7 days

References

1. A.E. Watada, L. Qi, Postharvest Biol. Technol. 15, 201-205 (1999)
2. R. Iyengar, A.J. McEvily, Trends Food Sci. Techol. 3, 60-64 (1992)
3. Food and Drug administration, Code of Federal Regulation (Washington, DC, US GPO, 1989)
4. S.M. Son, K.D. Moon, C.Y. Lee, Food Chem. 73, 23–30 (2001)
5. J.Y. Lee, H.J. Park, C.Y. Lee, W.Y. Choi, LWT-Food Sci. Technol. 36, 323-329 (2003)
6. S. Supapvanich , L. Samransuk, T. Somanusorn, N. Mesa, Acta Hortic. 1024, 339–346. (2014)
7. S. Supapvanich, P. Boonyaritthongchai, Inter. Food Res. J. 23(1), 389-394. (2016)
8. R. Srinath, C. Ramalingam, N. Nasimun Islam, Elixir Food Sci. 45, 7822-7826 (2012)
9. S. Supapvanich, P. Mitrsang, P. Srinorkham, P. Boonyaritthongchai , C. Wongs-Aree, J. Food Sci. Technol. 53(6), 2844–2850 (2016)
10. G.R.A. Alberio, G. Muratore, F. Licciardello, G. Giardina, G. Spagna, Food Sci. Technol. 35(2), 299-306 (2015)
11. B. Wessels, S. Damm, B. Kunz, N. Schulez-kaysers, J. Appl. Bot. Food Qual. 87, 16-23 (2014)
12. K. Slinkard, V.L. Singleton, Amer. J. Enol. Viti. 28, 49–55 (1997)
13. M.A.M. Galeazzi, V.C. Sgarbieri, S.M. Constantinides, J. Food Sci. 46, 150–155. (1981)
14. A.J. McEvily, R. Iyengar, W.S. Steven, Critical Rev. Food Sci. Nutr. 32, 253–273 (1992)
15. E.J. Kwak, S.I. Lim, J. Sci. Food Agric. 85(8), 1337-1342 (2005)
16. C. Soysal, J. Food Biochem. 33, 134-148 (2009)
17. S.K. Mantena, J. Sridhar, R. Badduri , K.B. Siripurapu, M.K. Unnikrishnan, Water. Nahrung/Food. 47, 126-131 (2003)

The study of the drying kinetic and the cutting performance in dried betel nut production machine

Naruebodee Srisang[1,*], *Siriwan* Srisang[1], and *Thatchapol* Chungcharoen[1]

[1]Department of Engineering, King Mongkut's Institute of Technology Ladkrabang, Prince of Chumphon Campus, Chumphon, Thailand

Abstract. Betel nut kernel is an important raw material in the leather and dye industries. The quality control after harvest is necessary, especially, the excess moisture content may cause the spoilage. Therefore, this research aims to study the betel nut fruit drying using infrared ray combined with the rotated trays. The betel nuts fruits were placed in rotated trays. These trays were installed within drying chamber which had the infrared rod in 2 positions namely the center and wall of drying chamber. The drying temperature used at 80, 100 and 120°C. The revolution speed of tray was set at 2, 4 and 6 rpm. The dried betel nut fruits were cut into 2 pieces at the blade speed of 25, 30 and 35 rpm. The experimental results showed the installation of infrared rods at the center could decrease the moisture content faster than at the wall. The betel nut fruit should use the drying temperature of 120°C for 10 h and the revolution speed of 2 rpm. The blade speed of 30 rpm gave the most cutting performance about 76%. The colour of dried betel nut kernel (L*, a* and b*) did not significantly difference with the commercial betel nut.

1 Introduction

Betel nut (*Areca Catechu Linn.*) also called as areca nut and widely plant for agricultural product in Chumphon province. Betel nut kernel has been widely used in medicine application including paint and leather manufacturing [1]. Moreover, betel nut kernel has been researched to extract the peroxidase as the major protein constituent in betel nut kernel to apply for useful in cancer therapy [2]. Generally, betel nut fruit is dried with sunlight around 3 weeks to reduce the moisture content. The dried betel nut fruit is cut with knife to easy separate kernel and husk. Then, the betel nut kernel is again dried with sunlight about 7 days to obtain the moisture content around 12-15% (d.b.). Kulanthaisami et al. [3] accounted the dried betel nut kernel should be reduced the moisture content to 13% (d.b.) before storage. The moisture content may be decreased to lower than 12% (d.b.) which depended on the husking or cutting process in next step. Jarimops et al. [1] reported the optimal moisture content of dried betel nut fruit about 6% (d.b.) before the husking using squeeze method. Subsequently, the requirement of low moisture content caused the increase of energy usage and production time. The long time in manufacturing process and the difficulty in quality control those were undesired for dried betel nut producer.

Hence, the objective of this work aims to study the betel nut fruit drying using the infrared ray combined with the revolved tray. The dried betel nut fruit after drying that was cut into 2 pieces using the cutting blade unit. The effect of drying temperature and the revolution speed of tray on the drying kinetic were investigated. The cutting performance for each the revolution speed of blade was reported. The quality of dried betel nut kernel was inspected in term of the colour.

2 Materials and methods

Betel nut fruits are mature which obtained from Chumphon province by they had the orange skin and brown kernel as shown in Fig. 1. Diameter of betel nut fruit was in the range of 3-5 cm and the average mass as 47.15 g. The betel nut fruits were dried using infrared drying combined with rotated trays. The betel nut fruit had the average of initial moisture content about 150% (Dry basis, d.b.). These betel nut fruits about 90 fruits were put in the cylindrical trays (8 trays) which were installed in drying chamber as represented in Fig. 2. These trays were driven by 1 HP motor and revolved around the revolution core at the revolution speed of 2, 4 and 6 rpm. The size of drying chamber was $0.7 \times 0.7 \times 0.98$ m^3. Inside drying chamber had the number of infrared lamp about 3 lamps by each infrared lamp about 1000 W. This research studied the infrared lamp installation in 2 places namely the wall and the center of drying chamber. For the installation on wall, a one infrared lamp was installed on each beside wall of drying chamber. On the other hand, in case the installation at center of drying chamber, three infrared lamps were set at the center chamber. The betel nut fruits in trays were dried at temperature of 80, 100 and 120°C simultaneously with the revolution of trays at predetermined time. The weight of betel nut fruit in each drying time was recorded to calculate the moisture content. The final moisture content of betel nut fruit desired about 15% (d.b.). The moisture determination was repeated and reported in the average value.

* Corresponding author: naruebodee.sr@kmitl.ac.th

The dried betel nut fruits were transported to the cutting blade unit that drove by 2 HP motor as displayed in Fig. 3. The revolution velocity of blade was 25, 30 and 35 rpm. The efficiency of betel nut fruit cutting was estimated from equation (1): [1]

$$\text{Cutting performance}_{(\%)} = [(F+0.5U) \times 100/N] \times 100 \quad (1)$$

Where F is the number of betel nut fruit which can be completely cut by blade unit, U is the number of betel nut fruit which can not completely cut by blade unit and N is the total number of betel nut fruit which is fed to blade unit.

Moreover, the colour of betel nut kernel was measured using Konica Minolta Charoma Meter CR-400. The colour value was reported in the form of L^*, a^* and b^*. Moreover, the change of betel nut kernel colour was displayed by the ΔE value as calculated in equation (2)

$$\Delta E = [(\Delta L^*)^2 + (\Delta a^*)^2 + (\Delta b^*)^2]^{1/2} \quad (2)$$

Where ΔL^* is the difference of the L^* value between sample and betel nut kernel before drying, Δa^* is the difference of the a^* value between sample and betel nut kernel before drying and Δb^* is the difference of the b^* value between sample and betel nut kernel before drying. All samples were duplicated and the average value from the repetition was reported.

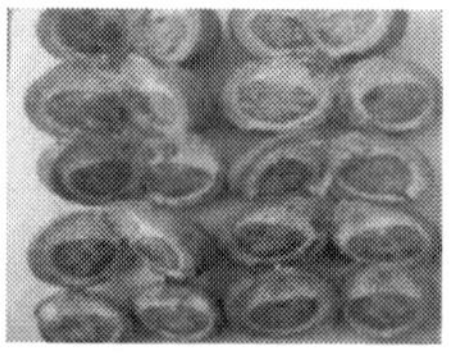

Fig. 1. Betel nut fruit.

Fig. 2. Betel nut fruits in rotated trays.

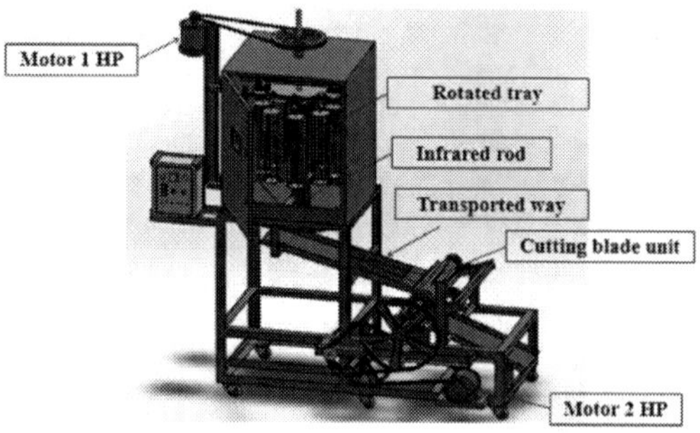

Fig. 3. Infrared dryer and cutting blade unit.

3 Results and discussion

3.1 Drying kinetic

The effect of the different infrared positions, i.e., the wall and the center of drying chamber on the drying kinetic of betel nut fruit as represented in Fig. 4, 5 and 6. The betel nut fruit was dried using the infrared drying at the temperature of 80, 100 and 120°C combined with the rotated trays. These trays revolved at the revolution velocity of 2, 4 and 6 rpm. Results exhibited that the initial moisture content about 150% (d.b.) rapidly decreased in the first period drying. Because of the infrared ray is a magnetic wave which does not use the medium for the heat transmission. Therefore, at the first period drying around 4 h, the betel nut fruit in rotated trays directly obtained the heat from the infrared ray and the partial ray penetrated and stimulated the diffusion of moisture from interior to exterior [4]. Then, the moisture evaporated from the surface of betel nut fruit. These incidents caused the rapid moisture removability.

When the further drying time, the moisture content gradually diminished to the final moisture content of 15% (d.b.). The moisture content slowly decremented due to the moisture movement in diffusion form slower than the evaporation form [5]. The drying time was in the range of 32-44, 18-24 and 10-14 h for the drying temperature of 80, 100 and 120°C, respectively. The reduction rate of moisture augmented with the enhanced drying temperature and resulted in the drying time shortened by it spent the shortest drying time for 10 h at temperature of 120°C and the revolution speed of tray about 2 rpm. The increased drying temperature that provoked the moisture diffusion within betel nut fruit. Jing et al. [6] and Saniso et al. [7] reported those the increment of diffusion coefficient with the increased drying temperature and resulted in the amelioration of moisture transfer.

The infrared position affected the drying kinetic of betel nut fruit by the infrared installation at the center of drying chamber gave the higher drying rate than the wall installation. This result may cause from the infrared installation at the center can considerably distribute the heat in around direction whereas the wall installation had the partial heat loss to the wall. The effect of infrared position did not clearly appear at the drying temperature of 80°C. However, this effect had obviously influenced when the increased drying temperature to 120°C as shown in Fig. 6.

The moisture reduction rate of betel nut fruit increased with the decrement of the revolution speed of tray. The tray speed of 2 rpm obtained the higher moisture reduction rate than the tray speed of 4 and 6 rpm as displayed in Fig. 4, 5 and 6. This result may be explained by the lower tray speed can regularly and continuously get the heat when collated with the higher tray speed.

 https://doi.org/10.1051/matecconf/201819203005

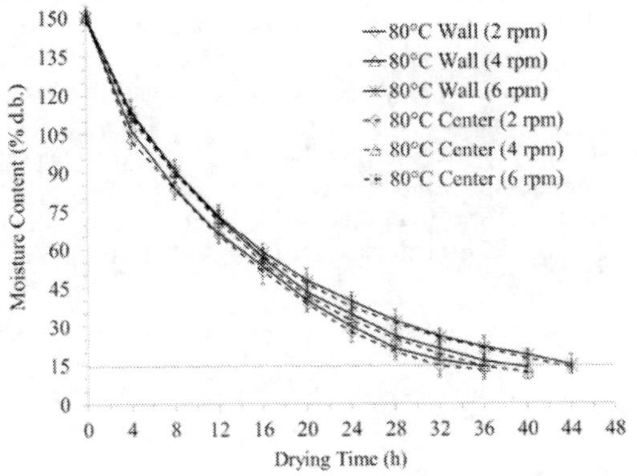

Fig. 4. Effect of different infrared positions (the wall and center of drying chamber) on the drying kinetic of betel nut fruit at drying temperature of 80°C.

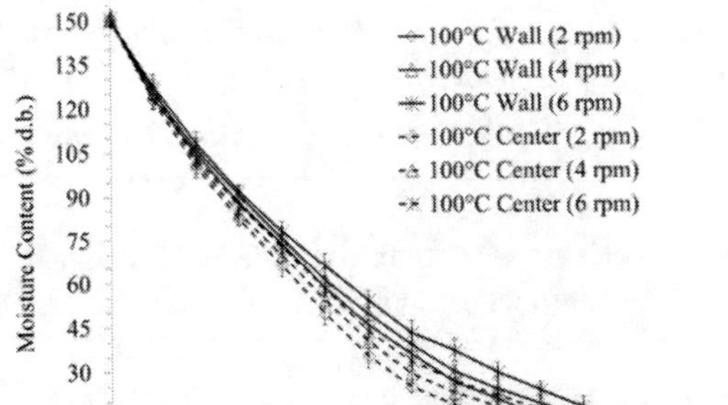

Fig. 5. Effect of different infrared positions (the wall and center of drying chamber) on the drying kinetic of betel nut fruit at drying temperature of 100°C.

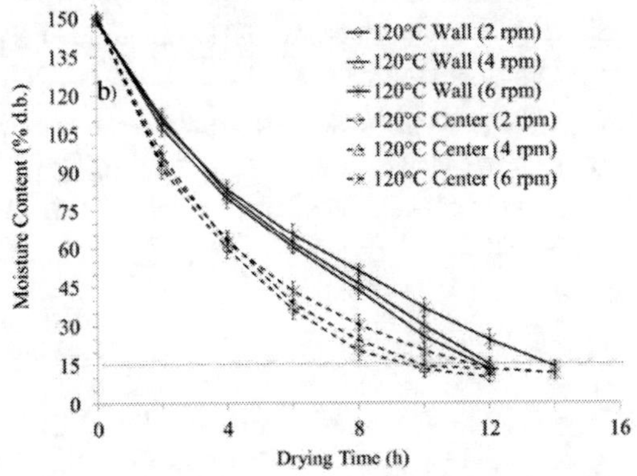

Fig. 6. Effect of different infrared positions (the wall and center of drying chamber) on the drying kinetic of betel nut fruit at drying temperature of 120°C.

3.1.1 Colour of betel nut kernel

Colour of betel nut kernel after drying that significantly changed as compared with the betel nut kernel before drying by the L* value essentially increased while the a* and b* value significantly decreased as presented in Table 1. Tirawanichakul et al. [10] reported the pennywort colour significantly changed after infrared drying.

The change of colour of betel nut kernel was indicated by the ΔE value. The betel nut kernel colour after drying contrasted with the betel nut kernel before drying as shown by the ΔE value significantly increased in the range of 18.79-19.99. However, the ΔE value of betel nut kernel colour after drying was insignificantly different as compared with commercial betel nut kernel. This result implied the betel nut kernel colour after infrared drying combined with rotated trays that was in the acceptable level in betel nut trade.

Table 1. Colour of betel nut kernel

Sample	Colour			
	L*	a*	b*	ΔE
Betel nut kernel before drying	39.13 ±0.98^{a}	9.35 ±0.64^{c}	17.69 ±0.49^{c}	0 ±0.00^{a}
Commercial dried betel nut kernel	57.70 ±1.69^{c}	5.16 ±1.66^{b}	12.82 ±2.05^{a}	19.81 ±2.32^{b}
80°C (2 rpm, 30 rpm)	57.58 ±0.97^{c}	4.95 ±0.26^{ab}	14.28 ±0.72^{b}	19.31 ±1.51^{b}
100°C (2 rpm, 30 rpm)	56.52 ±0.99^{b}	4.25 ±0.49^{a}	13.07 ±1.60^{a}	18.79 ±1.63^{b}
120°C (2 rpm, 30 rpm)	57.74 ±0.72^{c}	4.61 ±0.26^{ab}	12.26 ±0.72^{a}	19.99 ±1.13^{b}

a, b, c mean with different superscripts in the same column are significantly different ($p \leq 0.05$).

Above results were considered to use in the suitable condition selection for drying, that is, the revolution speed of tray and the installation position of infrared. To reduce the rapid moisture content and keep the colour quality, hence, the revolved tray speed used 2 rpm and the infrared rods were installed at the center of drying chamber.

3.2 Cutting performance

The cutting performance estimated from the betel nuts fruit got from the infrared drying at the tray speed of 2 rpm and the temperature of 120°C by the infrared rods were installed at the middle drying chamber. Then, the dried betel nut fruits at the moisture content around 15% (d.b.) were transported to the cutting blade unit and were cut at the blade velocity of 25, 30 and 35 rpm. Results showed the trend of cutting performance increased when the blade speed expanded from 25 to 30 rpm as presented in Table 2. On the other hand, the cutting performance diminished when the blade speed was excess 30 rpm. The above results indicated that the dried betel nut fruits should be cut at the blade speed of 30 rpm to obtain the most cutting performance about 76%. The excess blade speed may diminish the contact capability between blade and betel nut. However, the cutting performance in this research that corresponded with the result of Meein et al. [8] and Parathong [9]. Their results showed the cutting performance in the range of 75-90%.

The aspect of betel nut kernel after cutting as displayed in Fig. 7a that had a little crack and its aspect resembled with the commercial betel nut kernel as shown in Fig. 7b.

Table 2. Average cutting performance of blade unit.

Drying temperature (°C)	Tray speed (rpm)	Blade speed (rpm)	Cutting performance (%)
120	2	25	70.33
		30	76.67
		35	63

a)

b)

Fig. 7. Aspect of a) betel nut kernel after cutting using blade unit and b) commercial betel nut kernel.

4 Conclusions

The betel nut fruit drying using the infrared ray combined with the revolved tray should use the drying temperature of 120°C, the revolution speed of tray about 2 rpm and the rotation speed of cutting blade of 30 rpm. These conditions gave the high quantities of drying rate and cutting performance, including the colour quality of betel nut kernel was in the acceptable level as compared with commercial betel nut kernel.

The authors express the sincere appreciation to the King Mongkut's Institute of Technology Ladkrabang for supporting the research fund (income research fund 2558-0208009).

5 References

1. B. Jarimopas, S. Niamhom, A. Terdwongworakul, Biosystems Engineering **102**, 83-89 (2009)
2. Y.C. Liu, C.J. Chen, M.R. Lee, M. Li, W.T. Hsieh, J.G. Chung, H.C. Ho, *Evidence-Based Complementary and Alternative Medicine 2013*, 1-12 (2013)
3. S. Kulanthaisami, P. Subramanian, P. Venkatachalam, A. Sampathrajan, Madras Agric. J. **94**, 7-12 (2007)
4. N. Meeso, A. Nathakaranakule, T. Midhiyonon, S. Soponronnarit, Journal of Food Engineering **78**, 1248-1258 (2007)
5. S. Tirawanichakul, S. Lamaepae, Y. Tirawanichakul, Burapha Science Journal **17**, 117-129 (2012)
6. L. Jing, T. Zhaosheng, L. Haijun, W. He, Int. J. Agric. & Biol. Eng. **10**, 302-311 (2017)
7. E. Saniso, F. Saudi, R. Hanafee, Burapha Science Journal **17**, 142-149 (2012)
8. N. Meein, A. Kunchornrat, T. Tavattana, Journal of Industrial Education **8**, 13-23 (2014)
9. W. Parathong, *Master Thesis*, (2007)
10. S. Tirawanichakul, S. Chanchiew, Y. Tirawanichakul, KKU Research Journal **18**, 311-324 (2013)

Effect of barium loading on CaO derived from waste egg shell heterogeneous catalyst for canola oil biodiesel

Jakkrapong Jitjamnong[1,*]

[1]Petroleum Technology Department, Faculty of Industrial Education and Technology, Rajamangala University of Technology Srivijaya, Songkhla, Thailand

Abstract. The purpose of this research was to investigate the catalytic activity of Ba loading on calcium oxide (CaO) catalyst by varying the amount of barium added during the synthesis: 5-15 wt%. The waste egg shells were utilized as a CaO heterogeneous catalyst by calcined at 900 °C for 2 h. The Ba/CaO catalysts were prepared by impregnation method and were used as a catalyst in transesterification reaction of canola oil via microwave irradiation under microwave power 300 W. The characterization of catalyst and FAME composition of biodiesel were determined by X-ray fluorescence (XRF), scanning electron microscope (SEM), Fourier transform infrared spectrometer (FTIR), and gas chromatography (GC-FID). The conditions of biodiesel production were operated at 60 °C, 3 wt% of catalyst loading, 9:1 methanol-to-canola oil ratio, and microwave irradiation power was 300W for 2 min. The experimental results found that, the waste egg shells consist mainly of $CaCO_3$, which was decomposed to CaO more than 88 wt% after cacination step. The 15 wt% Ba/CaO catalysts exhibited the best catalytic performance with the FAME conversion higher than 97.68%.

1 Introduction

In resent year, the increasing exhaust of fossil based fuel and energy consumption due to rapidly increase of population and the expansion of economic. To solve these problems, biodiesel is an alternative energy becomes more significant due to biodegradable, renewable, low emission of SO_x and CO_x, non-toxic, and environmentally-friendly fuel, which can substitute petro-diesel. Biodiesel is a mixture of mono alkyl esters of long chain fatty acid methyl ester (FAME) can be synthesized from new and used plant oil or animal fats with an alcohol (methanol or ethanol) via transesterification [1-2]. In general, homogeneous base catalysts such as potassium hydroxide (KOH) or sodium hydroxide (NaOH) are some main disadvantages related to corrosion, non-reuse, and difficult to remove catalyst from the reactant mixture and consequently influences the abundant of waste cleaning water. However, the use of heterogeneous catalysts (solid based) has demonstrated higher catalysis, reused for the many production cycle time, easily separated out from the reaction system, less corrosive, and more environmentally friendly [3]. Calcium oxide (CaO) as heterogeneous catalyst has attracted much attention for transesterification of biodiesel production due to its high activity, low cost, and low solubility in biodiesel and methanol [4]. Several literature have studied the main resource of CaO from waste materials as heterogeneous catalyst such as crab shell [5], shrimp shell [6], river snail shell [7], animal bones [8], and egg shell [4,9] etc. Several studies have reported that the catalytic activity of the catalyst was strongly dependent on the basicity [10]. Notably, bi-functional heterogeneous base catalysts from alkaline-earth oxides had high basicity, which was improved the catalytic activity for biodiesel production. BaO had highly basic sites and seems to be appropriate base catalyst in the transesterification process [11].

The present framework was determined the effect of catalyst doped with barium on CaO derived from waste egg shells as environment-friendly heterogeneous catalyst. The synthesized catalysts were prepared by impregnation method. The biodiesel production was carried out with transesterification of canola oil as starting material and was applied microwave irradiation in the chemical reaction. The prepared catalysts were evaluated by X-ray fluorescence (XRF), scanning electron microscope (SEM) with an energy dispersive spectroscopy (EDS), and Fourier transform infrared spectrometer (FTIR). The biodiesel products were investigated by gas chromatography (GC-FID).

2 Experimental

2.1 Materials

The canola oil was used as feedstock for transesterification, procured from Lam Soon Co. Ltd.,

* Corresponding author: Alekjakkrapong@gmail.com

 https://doi.org/10.1051/matecconf/201819203006

Thailand. The waste egg shells as a source of CaO, were collected from local areas of Songkhla, Thailand. The powder of Barium chloride ($BaCl_2.2H_2O$, 99%) analytical grade reagent was bought from Ajax Finechem Pty Ltd. All of the chemicals were analytical grade.

2.2 Catalysts preparation

The waste egg shells were carefully rinsed with hot distillate water to remove the edible organic adhered to the egg shell, and was dried at 120 °C in an oven overnight. Then, the dried waste egg shells were grounded to powder. The powder egg shells were calcined in furnace at 900 °C for 2 h, which calcium carbonate was decomposed to calcium oxide, and are kept in dessicator. Synthesized the 5 wt% Ba/CaO, 2.66 g of barium chloride was mixed with distillate water and stirred for 15 min to become homogeneous. An aqueous solution containing barium chloride was impregned on 50 g of CaO and kept for 12 h at room temperature. Next, the catalyst was dried at 120 °C for 24 h to remove absorbed water, and last, calcined at 900 °C for 4 h. The same procedure is applied for preparation of 10 and 15 wt% Ba/CaO catalysts.

2.3 Catalyst characterization

The actual element compositions of calcined waste egg shell were measured by X-ray fluorescence spectrometer (XRF) (PW 2400, Philips, Netherlands). The chemical composition of neat calcium oxide derived from waste egg shells was analyzed using Thermo Nicolet Fourier Transform Infrared spectrometer (FTIR) Nexus 670 model. The sample was recorded in absorbance mode at 32 scans with a resolution of 4 cm^{-1} by KBr pressed disks. The frequency range was from 650 to 4000 cm^{-1}. The particle morphology and surface texture of Ba supported on CaO were studied using scanning electron microscopy (SEM). The SEM images of sample were recorded with an energy dispersive spectroscope (EDS) equipped (Oxford, Aztec).

2.4 Biodiesel production

100 g of canola oil was added to the glass equipped and plated at heater, which was operated at 60 °C. 3 g of calcined Ba/CaO (3 wt% compared to the starting vegetable oil) and methanol (9:1 methanol-to-canola oil ratio) was added into the glass equipped under agitation speed of 300 rpm with magnetic stirrer for 10 min. After that, the mixed solution was carried out under household microwave oven with a microwave irradiation power was 300W for 2 min. After transesterification, the catalyst was collected from mixture by filtration. The canola oil biodiesel and glycerol was separated by separatory funnel. Biodiesel was cleaned with hot distillate water at 65°C until pH was neutral in order to obtained pure biodiesel. Biodiesel was preserved in the glass bottle with sodium sulphate anhydrous to remove the absorbed excess water.

2.5 Canola oil-derived FAME analysis

The fatty acid methyl ester compositions of canola oil-derived biodiesel were analyzed by chromatograph (Hewlett Packard 5890 Series II) equipped with FID detector (GC-FID). The GC-FID conditions were carried out by helium (99.99%) as a carrier gas with a flow rate of 70 ml/min, 200°C of injector temperature with a split ratio at 75:1, and the temperature detector was set at 230 °C. 0.2 μL of fatty acids was injected into an oven at 130 °C. An isothermal time was held 2 min, and the oven was raised up to 220 °C with a rate of 2°C/min, and held for 15 min with analytical time was 62 min.

3 Results and discussion

3.1 Catalyst characterization

Table 1 shows the chemical compositions of calcined waste egg shells. Note that the major component is CaO approximately 88.29 wt%. This result clearly shows that the $CaCO_3$ from egg shells powder was decomposed to CaO after calcinations method. Nevertheless, the minor components of prepared catalyst are MgO, SiO_2, P_2O_5, SO_3, and SrO_3 was formed.

FTIR spectra of heterogeneous CaO catalyst as synthesized from waste egg shells which is illustrated in Figure 1. For the carbonyl groups, a board transmission in the carbonate region can be observed at 871 and 1444 cm^{-1}, respectively. This result exhibited the $CaCO_3$ phase in starting material (waste egg shells) and was transformed to CaO after calcination at 900 °C for 2 h. The intense band at 3641 cm^{-1} was assigned to the hydroxyl group stretching of $Ca(OH)_2$, which is identified to the absorption of atmospheric moisture on the surface of CaO catalyst [11,12].

The surface morphology of calcined catalysts were evaluated using SEM equipped (Figure 2). The neat CaO (Figure 2A) as heterogeneous catalyst display the dense surface with heterogeneous distribution of particle and irregular shapes of particle sizes. However, the SEM images of the CaO catalyst doped with 5, 10, and 15 wt% Ba and carried out by calcination step at 900 °C for 4 h are illustrated in Figure 2B-D. It was noticeably observed that the morphology of the catalyst was changed from irregular shapes into spherical shapes, which smaller particles size than calcium oxide. It possibly resulted from the deposition of barium species into CaO framework during the synthesis catalyst. Calcined catalysts from waste egg shells causes an increase surface area, leading to better catalytic performance.

The actual elements of catalyst (15 wt% Ba/CaO) were investigated with energy dispersive X-ray spectroscopy

Table 1. Chemical compositions of calcined waste egg shell.

Compound	Concentration (wt %)
CaO	88.29
MgO	0.95
SiO_2	0.05
P_2O_5	0.43
SO_3	0.23
SrO	0.07
Organic compound	9.98

Determined by XRF.

(SEM-EDS), and the result is shown in Figure 3. It was exhibited that, the content of Ca, O, and Ba were formed. Table 2 determined EDS analysis of barium doped on CaO. It was clearly seen that for 5 wt% Ba/CaO. The amount of Ca and Ba were 51.34 wt% and 1.78 wt%, respectively. For 10 wt% Ba supported on CaO, the actual elements of Ca and Ba were 54.77 wt% and 2.47 wt%, while 15 wt% Ba/CaO catalyst, the contents of Ca and Ba were 48.65 wt% and 2.80 wt%, respectively.

Table 2. EDS analysis of Ba loaded on CaO.

Catalyst	Element	Concentration (wt %)
5 wt% Ba/CaO	Ca	51.34
	O	46.89
	Ba	1.78
10 wt% Ba/CaO	Ca	54.77
	O	42.76
	Ba	2.47
15 wt% Ba/CaO	Ca	48.65
	O	48.55
	Ba	2.80

Measured by SEM-EDS.

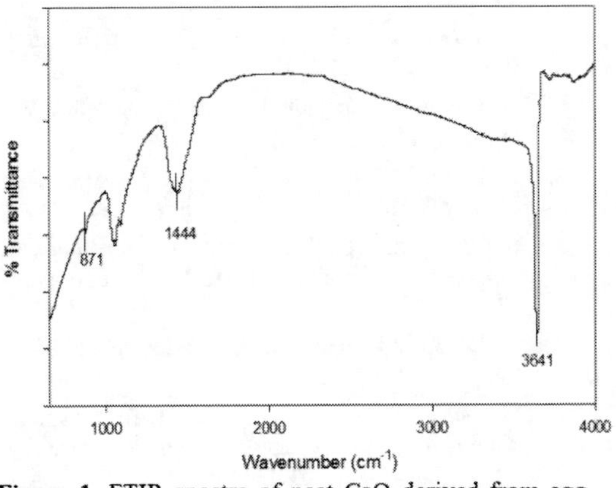

Figure 1. FTIR spectra of neat CaO derived from egg shells.

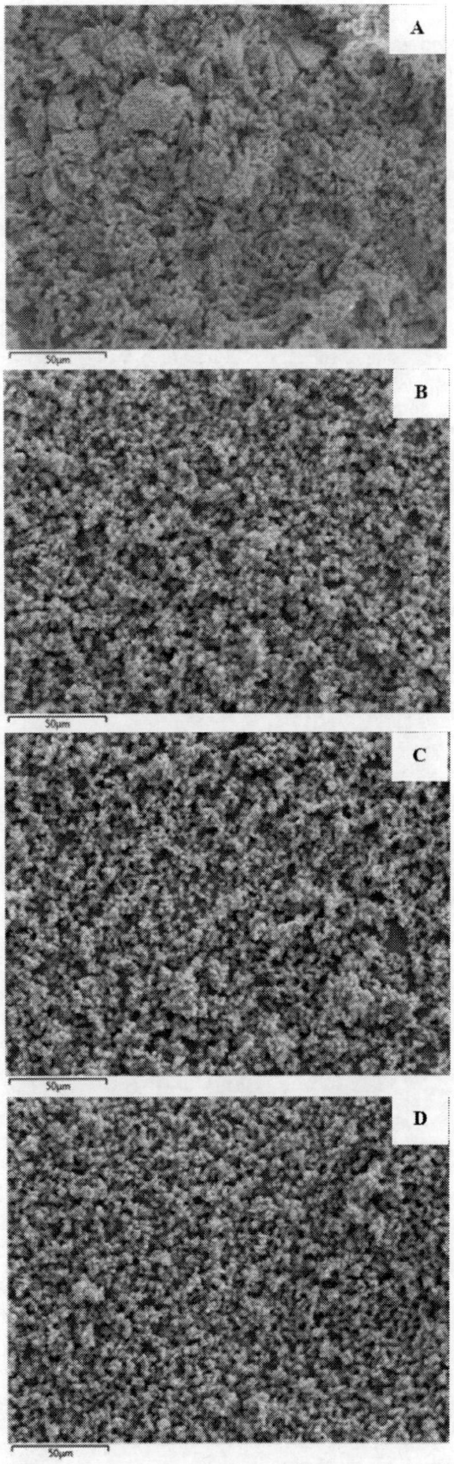

Figure 2. SEM photograph of (A) neat CaO, (B) 5 wt% Ba/CaO, (C) 10 wt% Ba/CaO, and (D) 15 wt% Ba/CaO. The scale bars for panels are 50 μm.

3.2 Effect of Barium loading

Figure 4 shows the FAME contents from the transesterification via microwave irradiation of canola oil with methanol and the results obtained from four different heterogeneous catalysts (3 wt% catalysts). As can be seen, the methyl ester compositions in the reaction using neat CaO catalyst exhibited 89.52% conversion since it has sufficiently strong basic strength. The FAME conversion of 5 wt% Ba loading on egg shells derived CaO was 90.74 wt%. However, when the barium doping is increased up to 10 and 15 wt%, the FAME conversion slightly increased were 94.33% and 97.68 wt%, respectively. According to the biodiesel yield comparison, it could be noted that Ba doped base catalyst well distributed on CaO phase which might have the total number of available active catalytic sites from Ba-CaO catalyst increased. It was reported that the FAME yield significantly increased when the barium loading was increased due to barium may have improved the amount of basic sites of CaO.

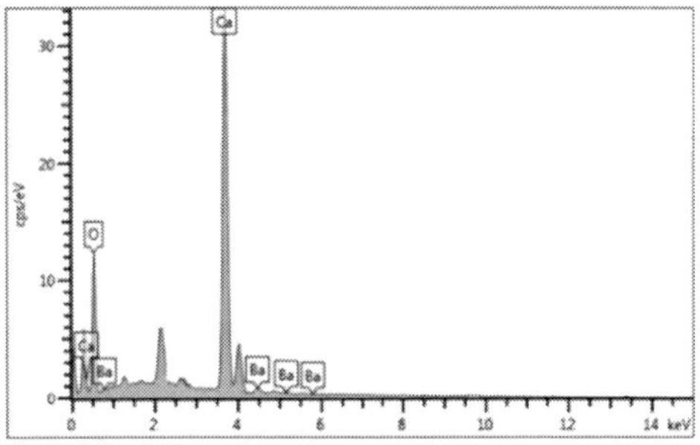

Figure 3. SEM-EDS photograph of 15 wt%Ba/CaO.

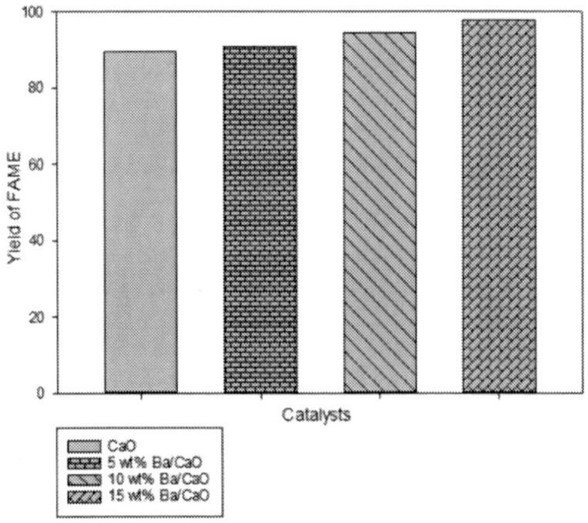

Figure 4. Effect of catalyst concentration (5, 10, and 15 wt% Ba/CaO) on yield of FAME.

4 Conclusions

The present work show that the calcined of waste egg shells at 900 °C for 2 h are suitable catalyst for production of biodiesel via transesterification with microwave irradiation of 9:1 methanol: oil molar ratio. The calcined of waste egg shells consist mainly of CaO more than 88 wt% conversion to FAME, which can be considered heterogeneous catalyst for biodiesel production. Moreover, the 15 wt% Ba/CaO catalyst gave the highest catalytic performance of biodiesel synthesis, which yielded of FAME conversion, was 99.68% due to the catalyst had excellent activity and stability during transesterification reaction. CaO derived from waste egg shells can be utilized as economical and reduce waste disposal problem.

Acknowledgments

The author is thankful for the financially supported by Rajamangala University of Technology Srivijaya, Thailand. And deeply thanks to Ms. Chachchaya Thunyaratchatanon and Ms. Chatrawee Direksilp, PhD student at The Petroleum and Petrochemical College, Chulalongkorn University, Thailand for their helps on instrumental analysis.

References

1. G. Baskar, I.A.E. Selvakumari, R. Aiswarya, Bioresour. Technol. **250**, 793 (2018)
2. K. Sudsakorn, S. Saiwuttikul, S. Palitsakun, A. Seubsai, J. Limtrakul, J Environ Chem Eng. **5**, 2845 (2017)
3. M. Alsharifi, H. Znad, S. Hena, M. Ang, Renew Energy. **114**, 1077 (2017)
4. P.R. Pandit, M.H. Fulekar, J. Environ. Manage. **198**, 319 (2017)
5. V. Shankar, R. Jambulingam, Sustainable Environ. Res. **27**, 273 (2017)
6. Y. Sun, V. Sage, Z. Sun, Chem. Eng. Res. Des. **126**, 142 (2017)
7. S. Kaewdaeng, P. Sintuya, R. Nirunsin, Energy Procedia. **138**, 937 (2017)
8. G. Corro, N. Sánchez, U. Pal, F. Bañuelos, Waste Manage. **47**, 105 (2016)
9. A. Piker, B. Tabah, N. Perkas, A. Gedanken, Fuel. **182**, 34 (2016)
10. S. Chaveanghong, S. M. Smith, C. Oopathum, C.B. Smith, A. Luengnaruemitchai, Renew Energy. **109**, 480 (2017)
11. J. Boro, L.J. Konwar, D. Deka, Fuel Process. Technol. **122**, 72 (2014)
12. H. Mazaheri, H. Chyuan Ong, H.H. Masjuki, Z. Amini, M.D. Harrison, C. Wang, F. Kusumo, A. Alwi, Energy. **144**, 10 (2018)

Ultrasonic-assisted extraction of gallic acid and isoquercetin from *aspergillus niger* fermented tri-phala waste

Teerin Chysirichote[1], and *Pattarabhorn* Pakaweerachat[1*]

[1]Department of Food Engineering, Faculty of Engineering, King Mongkut's Institute of Technology Ladkrabang, Bangkok, Thailand

Abstract. The *Aspergillus niger* fermented Tri-phala waste (FTW) was extracted with ultrasonic assisted extraction (UAE) using deionized water as an extraction medium at 30°C. The 40 kHz ultrasonic frequency was used for sonicate the FTW immerged in the water at the ratio of 1 : 100 for 10, 20, 30, 40, 50 and 60 min. The contents of gallic acid, isoquercetin obtained after extraction were measured by HPLC. The extraction yields of gallic acid and isoquercetin were compared with the yields from the water extraction without ultrasonic assistance (control condition). The results showed that using the ultrasonic assistance increased the extraction yield of gallic acid from 0.25±0.03 to 1.26±0.25 $g\,g^{-1}$ with the shorter extraction time from 60 min to 30 min. Moreover, isoquercetin extraction yield increased from 0.17±0.02 to 0.41±0.04 $g\,g^{-1}$ with the shorter extraction time from 60 min to 20 min.

1 Introduction

Tri-phala is a traditional Ayurvedic herbal medicine, consisted of equal parts of three herbal fruits: Indian gooseberry (*Emblica officinalis* Gaertn), Myrolan Wood (*Terminutesalia chebula* Retz.) and Belleric myrobalan (*Terminutesalia bellerica* (Gaertn) Roxb) [1-3]. Typically, Tri-Phala is boiled in water to extract the polyphenolic compounds which are gallic acid and rutin [4, 5]. Both of them have powerful antioxidant properties used in medical applications such as antimutagenic, lipid-lowering and anti-tumor agent [6-8]. Charoenchai et al. [4] measured gallic acid and rutin in Tri-phala powder using HPLC as 1.32-4.02% and 0.61-2.94% (w/w), respectively. However, Pawar and Alunkhe [5] measure gallic acid and rutin using UV Spectrophotometor as 0.67±0.01 and 0.76±0.01 in Tri-phala powder, respective.

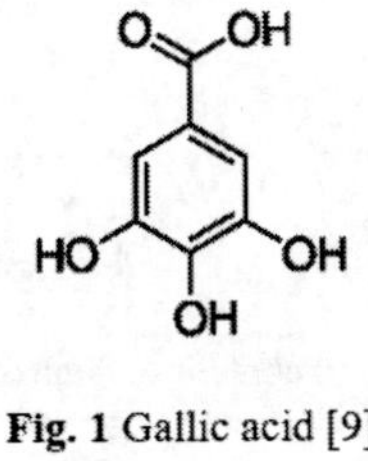

Fig. 1 Gallic acid [9]

Recently, Pattarabhorn and Teerin [2] studied the solid state fermentation (SSF) of *A. niger* on Triphala waste and found the gallic acid as the main product and a trace of isoquercetin (data not shown). Isoquercitrin is a kind of flavonoid which is derived from rutin [5]. Isoquercetin is an active ingredient that posesses many medical properties such as antioxidant, atheroprotective, allelopatic, plaque-stabilizing and anti-inflammatory [10-12] and has been in the GRAS (general regarded as safe) standard to be used as food additive.

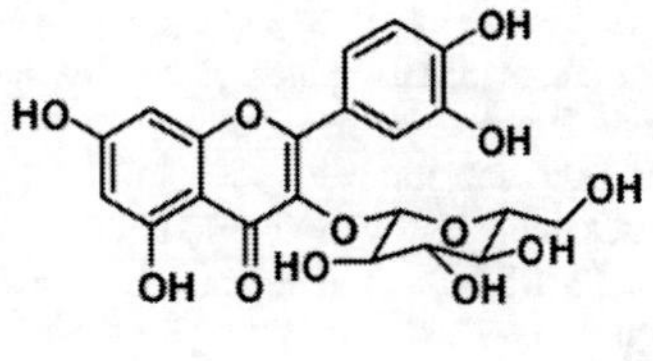

Fig. 2 Isoquercetin [13]

Gallic acid and isoquercetin are bioactive compounds that are soluble in water due to their strong electronegativity. The samples of bioactive compound extraction are shown in table 1.

Ultrasonic-assisted extraction has been well proven as a green technology and has enhanced the extraction process in different green aspects. Its advantages over other extraction methods are reducing extraction time, solvent usage, and energy consumption, increasing a yield of production, and enhancing bioactivity preservation of the separated biomolecules due to lower extraction temperatures [14]. Altemimi et al. [15] found that 50% ultrasonic power, frequency 37 kHz with extraction time 30 min at 40 °C increased the extraction yield, total phenols, flavonoids, DPPH-free radical

* Corresponding author: pattarabhorn.wee@gmail.com

scavenging activity and % ferric reducing antioxidant power in spinach extracts. Corrales et al. [16] also reported that ultrasonic enhanced anthocyanin extraction from grape by-products by disrupting its cell wall.

Table 1. Contents of isoquercetin and gallic acid extracted from different materials

Materials	Isoquercetin	Gallic acid	References
Apple juice	4.81±0.95 mg L^{-1}	-	[17]
Scutia buxifolia	6.66±0.04 mg L^{-1}	-	[18]
Leaves (*Morus nigra* L.)	1.84 mg g^{-1} dw	-	[19]
Ardisia japonica	0.09 mg g^{-1}	2.15 mg g^{-1}	[20]
Stem bark (*jatropha curcas*)	-	0.54 mg L^{-1}	[21]
Leaves (*Suaeda glauca* Bge.)	-	6.30 mg g^{-1}	[22]

Therefore, we were interested to increase the gallic acid and the isoquercetin yields during water extraction of the FTW by the ultrasonic assistance. The objective of this study was to compare the extraction rates and yields of gallic acid and isoquercetin released from the FTW by *A. niger* with the UAE and the water extraction.

2 Materials and methods

2.1 Materials preparation

TW was dried at 60 °C in a tray dryer for 24 h and ground using a hammer mill. The particles were screened using 30-mesh and 70-mesh sieves using a sieve shaker to obtain the particles size between 210 - 595 micron for the SSF [2]. The screened particles were adjusted their moisture content to 55 %(w/w) using 0.75 % sodium nitrate solution, which was used as a nitrogen source, and sterilized at 120°C for 20 min. Then, the spores of *A. niger* were inoculated at the concentration of $5x10^5$ spores·g^{-1} of dry substrate. The SSF were conducted in 250 mL Erlenmayer flask containing the substrate 30 g at 30 °C for 72 h. The fermented sample (FTW) was dried at 60 °C in a hot-air oven for 24 h and milled into a powder for the extraction.

2.2 Gallic acid and Isoquercetin Extraction

The FTW was sonicated for 10, 20, 30, 40, 50 and 60 min (40kHz) at 30 °C to extract the gallic acid and isoquercetin. Control sample was prepared by soaking sample into deionized water at 30 °C for 10, 20, 30, 40, 50 and 60 min. The temperature was controlled by a temperature controller and monitored all the extraction time. In case that the temperature was over than 30°C, cold water was added into the baht.The extraction rate was calculated from a slope of a time course curve of an extraction in a unit of g g^{-1}min^{-1}.

2.3 Gallic acid and isoquercetin determination

HPLC system with the Water 717 plus autosampler was used to carry out the analysis. The mobile phase was the gradient of acetonitrile and acetic acid (0.1%) in water. The flow rate was 1 mL·min^{-1} and a detection was at 280 nm. The column was phenomenex and the injection volume was 10 μL [2]. The gallic acid and isoquercetin contents were reported as a gram of gallic acid and isoquercetin per one gram of dry FTW. The chromatogram of the extracted FTW was shown in figure 3. The first and second peak represent gallic acid and isoquercetin, respectively.

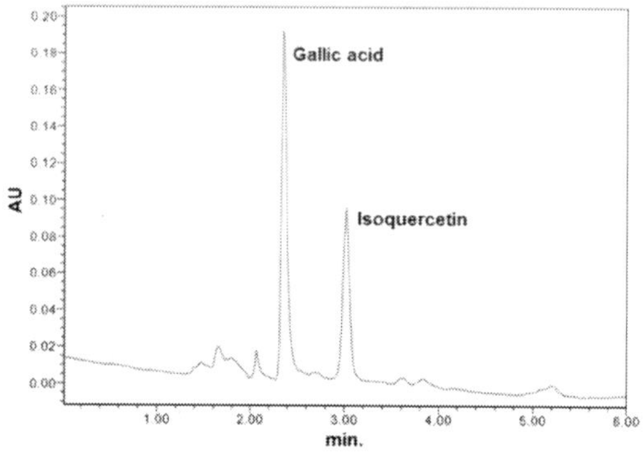

Fig. 3 Chromatogram of the FTW extracted.

2.4 Data analysis

The experiment was carried out in triplicate. An average values ± standard deviations were reported as results. Analysis of variance (ANOVA) at the 95 % significance level was used to analyse the significance of data.

3 Results and discussion

3.1 Gallic Acid Extraction

Figure 4 shows that the highest extraction rates obtained from both the UAE and the water extraction (control) were found from 0 to 30 min as 0.42±0.00 and 0.01±0.00 g g^{-1}min^{-1}, respectively.

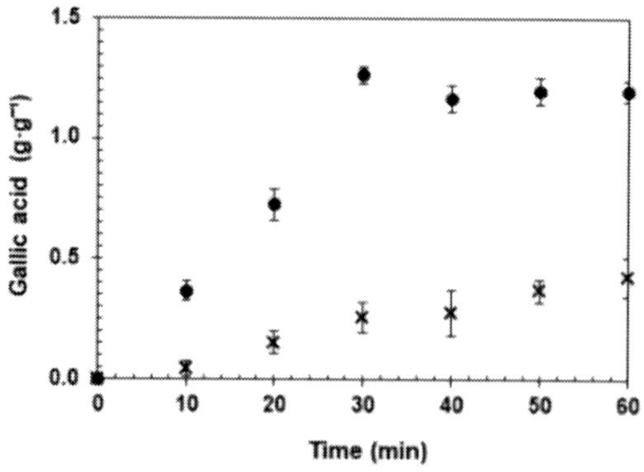

Fig. 4 Gallic acid content of control (×) and UAE (●) during the extraction

The maximum content of gallic acid released from the UAE was 1.26±0.25 g g^{-1} after 30 min. extraction, while the highest one from the water extraction (control)

was only 0.25±0.03 $g \cdot g^{-1}$ after 60 min extraction. It was believed that the releasing gallic acid in the UAE after 30 min. was occurred due to the equilibrium of the gallic acid concentration in the solution and the extracted FTW. The UAE helped shorten this extraction process since a violent shockwave from a sonication disrupted the substrate cell wall to allow solvent penetrating into the cell and extracting the desired component [23]. However, when using high frequencies and power of ultrasonic treatment caused a degradation or an oxidation of the phenolic compounds by the generation of highly reactive hydroxyl radicals [24].

3.2 Isoquercetin Extraction

The yield of isoquercetin from the FTW during the UAE and control water extraction was shown in figure 5. It is indicated that the isoquercetin releasing from the FTW in the UAE process highly increased from 0 to 20 min extraction. In contrast, that in the water extraction without sonicate-assisted (control) slowly increased until 60 min or the end of the study period. The highest yield of isoquercetin from the UAE was found at 20 min as 0.41±0.04 $g \cdot g^{-1}$ which could be calculated the extraction rate as 0.02±0.00 $g \cdot g^{-1} min^{-1}$. In comparison, the isoquercetin in the control extraction gradually released from the FTW until a terminal of study period (60 min). Its maximum yield measured at 60 min was only 0.17±0.02 $g \cdot g^{-1}$.

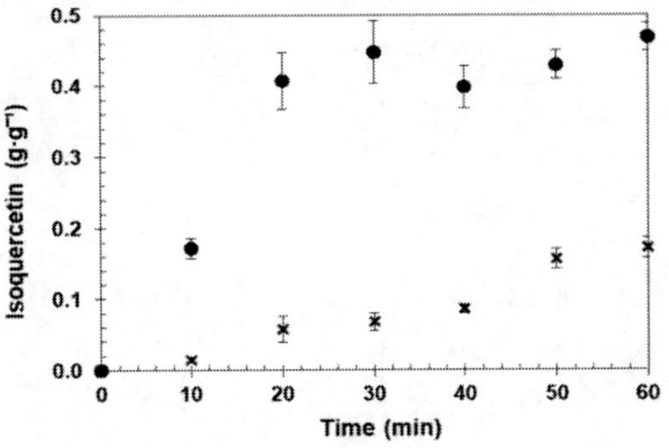

Fig. 5 Isoquercetin content of control (×) and UAE (●) during the extraction

Conclusion

The water extraction of *A. niger* fermented TW with ultrasonic-assisted increased the extraction yields of gallic acid and isoquercetin. It also reduced the extraction time down to 30 min for gallic acid and 20 min for isoquercetin to obtain their highest yields.

Acknowledgements

The authors would like to thank you for the financial support from the National Research Council of Thailand and the Institute of Thai Traditional Medicine, Department for Development of Thai Traditional Medicine and Alternative Medicine, Ministry of Public Health, Thailand for provided the facilities to carry out this research.

References

1. D. Frawley. *Ayurvedic healing, a comprehensive guide*, (Montilal Banarsidass, New-Delhi, 2003)
2. P. Pakaweerachat, T. Chysirichote, *Proceeding of The 9th TSAE International Conference*, Bankok 70 (2016)
3. N. Birla, K. Das Prabhat, Pharm. Biol. Eval. **3**, 366 (2016)
4. L. Charoenchai, P. Pathompak, F. Madaka, S. Settharaksa, W. Saingam, BHST. **14**, 57 (2016)
5. N. P. Pawar, V. R. Salunkhe, Int J Pharmtech Res. **5**, 724 (2013)
6. X. Q. Lin, F. Li, Y. Q. Pang, H. Cui, *Anal. Bioanal. Chem.* **378**, 2028 (2004)
7. X. W. Yu, Y. Q. Li, S. M. Zhou, Y. Y. Zheng, World J. Microbiol. Biotechnol. **23**, 1091 (2007)
8. R. Jasuja, F.H. Passam, D.R. Kennedy , S.H. Kim, L.V. Hessem, L. Lin , S.R. Bowley, S.S. Joshi, J.R. Dilks, B. Furie, B.C. Furie, R. Flaumenhaft, J. Clin. Invest. **6**, 2104 (2012)
9. B. Badhani, N. Sharma, R. Kakkar, RSC Adv. 35, 27540 (2015)
10. K. Motoyama, H. Koyama, M. Moriwaki, K. Emura, S. Okuyama, E. Sato, M. Inoue, A. Shioi, Y. Nishizawa, *Nutrition*, **4**, 421 (2009)
11. E. I. Salim, M. Kaneko, H. Wanibuchi, K. Morimura, S. Fukushima, Food Chem. Toxicol. **12**, 1949 (2004)
12. J. Fernandez, R. Reyes, H. Ponce, M. Oropeza, M. Vacalsteren, C. Jankowski, M. G. Campos, Eur. J. Pharmacol. **522**, 108 (2005)
13. J. C. R. Vellosa, L. O. Regasini, N. M. Khalil, V. D. S. B. Bolzani, O. A. K. Khalil, F. A. Manente, H. P. Netto, O. M. M. F. Oliveira, Eclet. Quím. **36** (2011) Doi : 10.1590/S0100-46702011000200001
14. I. Alzorqi, S. Sudheer, T. J. Lu, S. Manickam, Ultrason. Sonochem. **35**, 531 (2017)
15. A. Altemimi, R. Choudhary, D.G. Watson, Ultrason. Sonochem., **24**, 247 (2015)
16. M. Corrales, S.Toepfl, P. Butz, D. Knorr, B. Tauscher, Innov. Food Sci. Emerg. Technol. **9**, 85 (2008)
17. V. B. Suarez, J. Santamaria Victorero, J.J. Mangas Alonso, D. Blanco Gomis, J. Agric. Food Chem. **42**, 2732 (1994)
18. B. Aline Augusti, S. Michele Rorato, M. Luiz Filipe, F. Olmiro de Souza, M. Michel Mansur, M. C. Ivana Beatrice, A. Margareth Linde, *Molecules*. **17**, 5757 (2012)
19. M. M. Freitas, F. Pedro Ribeiro, S. Paula Monteiro, F. Christopher William, S. G. Eliete Neves, M. Yanna Karla, S. Damaris, F. B. Yris, S. Luiz Alberto, M. Mauri, M. Perola, PLOS ONE. **11** (2016). Doi:10.1371/journal.pone.0163130
20. X. Yu, J. Teye Azietaku, J. Li, H. Wang, F. Zheng, J. Hao,Y. Chang, , *Evid* Based Complement. Alternat. Med. (2018). Doi:10.1155/2018/4964291
21. D. M. Amirah Reddy Prasad, R. Maksudur Khan, J. Appl. Sci. **12**, 1106 (2012)

22. X. H. Wang, C. Cai, X. M. Li, Optimal Extraction of Gallic Acid from Suaeda glauca Bge. Leaves and Enhanced Efficiency by Ionic Liquids, Int. J. Chem. Eng., (2016) .Doi: 10.1155/2016/5217802
23. S. Hemwimol, P. Pavasant, A. Shotipruk, Ultrason. Sonochem. **13**, 543 (2006)
24. L. Galvan D'Alessandro, K. Kriaa, I. Nikov, K. Dimitrov, Ultrasound assisted extraction of polyphenols from black chokeberry, *Sep. Purif. Technol,* **93**, 42 (2012)

MATEC Web of Conferences **192**, 03008 (2018) https://doi.org/10.1051/matecconf/201819203008
ICEAST 2018

Formulation and preparation method of pine sawdust-based food for rumiants

Gerardo Espinosa-Garza[1], *Imelda* Loera-Hernandez[1] and *Natella* Antonyan[1*]

[1]Tecnologico de Monterrey, Escuela de Ingeniería y Ciencias, Eugenio Garza Sada 2501, Monterrey, N.L., 64849, México

Abstract. The waste products of the wood industries, particularly pine sawdust, represent a great potential source of dietary energy for ruminants. Even though the use of pine sawdust as part of a ruminant's diet represents an important environmental impact, studies on this subject are scarce. It is worth mentioning that pine sawdust is currently being used as a source of fibre in ovine diet. However, its digestion is limited due to its high lignin content. The objective of the present investigation consists in the design of a new treatment process for lignin degradation into pine sawdust through the use of the fungus *Pleurotus ostreatus*. From the results obtained, a new and sustainable food for the ruminants based on degraded pine sawdust is proposed. The experiments carried out showed significant elevation in the exploitation of the nutritious components of pine sawdust contained in the developed product due to the improvements in the digestive parameters of the ruminants. With the results obtained, besides the nutritional and environmental impact, a better control of the costs generated by the feeding of the ruminants is intended. This is due to the fact that pine sawdust does not suffer from production variation in comparison with other consumables.

1 Intoduction

In Mexico, the increased usage of forages has caused the supply of these agricultural by-products to be insufficient. The constant growth of the cost of stubbles of maize, straws of wheat barley, oats, etc. is being observed, causing a considerable increase in its cost. It is worth mentioning that these forages are not available all the year and their production usually lowers on rainy seasons. On the other hand, a little more than 8 million cubic meters of wood are processed annually in Mexico [1]. Sawdust is the main waste product in the wood industry, whose estimated production is 2.8 million cubic meters a year [2]. The natural degradation of sawdust is very slow and it is a waste of low use nowadays. Its storage is creating serious pollution problems in the soils in which it is settled [3-6]. Because it occupies a lot of space, sawdust causes considerable environmental problems. However, in the last years, the efforts to find alternatives for its utilization in order to obtain new high value products have increased. The incorporation of sawdust inside the nutrition of ruminants represents an economic and nutrimental advantage of its use [7]. One of its advantages as a raw material is that the sawdust does not compete directly with the food destined for human consumption. In the systems of food production for ruminants, the forages can constitute from 5 to 40% of the total of the diet. Usually, the forages are added into the ruminants food to have the necessary fiber to support the functions of the rumen and to avoid disorders of the digestive tract. Nevertheless, to be able to use lignocellulosic materials contained in sawdust in food for ruminants, the degradation of the lignin contained within is important [8]. The lignin, after the cellulose, is the principal component of wood. The importance of the degradation of the lignin is owed to the fact that this substance is completely indigestible for the ruminant bacteria and can cause an increase in the intestinal diseases, since its digestion requires the presence of oxygen [9-11]. The excess of fiber reduces the capacity of food ingestion, the synthesis of ruminal microbial protein, the digestibility of the ration and the energy supply [12]. Biological degradation of the lignin is very slow [13]. According to the existing research, there are several ways in which a treatment can be given to a substratum of sawdust to modify its composition. However, according to N. Mosier et al. (2005), the pre-treatment of the lignocellulosic biomass in order to raise the performance is one of the most expensive steps [14].

The objective of this research consisted in the design of a new treatment mechanism for lignin degradation into pine sawdust through a series of designed processes. The proposed development implied subjecting the sawdust to 4 different processes: physical, chemical, thermic and finally, a mixing process with the rest of the ingredients to reach the formulation of the final product. From the results obtained, a new and sustainable pine sawdust food for ruminants was proposed.

* Corresponding author: nantonya@itesm.mx

2 Processes

The development used in the present research implies subjecting the sawdust to 4 different processes:

- the first one is a physical process (mashing) with the objective of minimizing the size of the particle to achieve a better size for digestive use
- the second one is to reduce through a chemical process the percentage of lignin that the sawdust contains and to subsequently degrade the present cellulose in a sustainable manner.
- the third process refers to the thermic compression of the sawdust through the application of pressure and temperature to modify its properties.
- the fourth process is the interaction of the rest of the ingredients that will mix with the treated sawdust.

The particle size of the sawdust used in the present project was on average 2mm, with the purpose of increasing the contact area with the cellular and hemolytic microorganisms in the ruminants digestive tract. To obtain this particle size, a hammer mill with a screen size of 2 mm was used. Through the procurement of sawdust in 2 mm particle size, a chemical treatment using a solution of sodium hydroxide (NaOH) in a concentration range between 20-40% and a temperature between 80-90 °C during 45 minutes was used. The thermic process for lignin reduction was performed by heating up the sawdust in a microwave (MW). Nine different tests with 1 kg of sawdust each in which the temperature of the sawdust varied to achieve the desired properties was performed. The tests were performed so that the sawdust would reach a temperature of 40, 80 and 120 degrees through a heating process in the microwave with a power of 1100 W with an applied irradiation time of 70-80 seconds. The temperature that provided the best results was that off 80 °C. Under this processing, the lignin amount in the pine sawdust was reduced by 2-5%. Finally, the food for ruminants was formed adding the treated sawdust to wheat bran, soy, coconut flour, corn, oats, calcium supplement and salt with the proportions shown in Table 1.

Table 1: Food formulation for sheep using pine sawdust.

Ingredient	**Content (%)**
Treated sawdust	35
Wheat bran	14
Soy	10
Coconut flour	4
Corn	18
Oats	16
Calcium supplement	2
Salt	1

It is worth mentioning that the ingredients of the mix were of a size superior to 2 mm, except the coconut flour. The created food was characterized physiochemically in the manner presented in Table 2.

Table 2. Physicochemical formulation for sheep utilizing the nourishment with pine sawdust.

Analysis	**Content %**
Humidity	9.5
Protein	11.7
Fat	3.9
Ash	11.8
Calcium (mg per 100 g)	166.61
Total dietary fiber	20.48
Soluble fiber	11.5
Insoluble fiber	88.5

Posteriorly, 10 ruminants were fed with the food created for a period of 3 months. The analysis performed in the samples of food confirmed their nutritional value for a ruminant. The results from the Hematic Biometry Analysis performed on ovines are presented in Table 3.

Table 3. Hematic Biometry Analysis of the observed ovines. (1)

Sample number	Test	Result	Units	Reference value
1	Erythrocytes	5.8	$x10^6/mm^3$	8-16 x $106/mm^3$
1	Leukocytes	10.4	$x10^3/mm^3$	4-12 x $106/mm^3$
1	Hemoglobin	8.9	g/dl	8 – 15 g/dl
1	Hematocytes	28	%	24 – 50 %
1	M.C.V	48	fL	23 – 48 n
1	M.C.H	15.3	pg	8 – 12 pg
1	M.C.H.C	31.7	g/dl	31 – 38 g/dl
1	Plasma proteins	6.6	g/dl	6 – 7.5 g/dl
1	Fibrinogen	400	g/dl	100 – 500 g/dl
1	Plaquetes	527	mm3	250 – 750 mm
1	Band neutrophils	-	raros	0 raros
1	Segmented neutrophils	50	%	10 – 50 %

Table 3. Hematic Biometry Analysis of the observed ovines. (2)

Sample number	Test	Result	Units	Reference value
1	Lymphocytes	47	%	40 – 75 %
1	Monocytes	3	%	0 – 6 %
1	Eosinophils	-	%	0 – 10 %
1	Basophils	-	%	0 – 3 %
1	Absolute band neutrophils	-	raros	0 raros
1	Absolute segmented neutrophils	5200	mm3	400 – 6000 mm^3
1	Absolute monocytes	312	mm3	0 – 720 mm^3
1	Absolute eosinophils	-	mm3	0 – 1200 mm^3
1	Absolute basophils	-	mm3	0 – 360 mm^3
1	Absolute lymphocytes	4888	mm3	1600 – 9000 mm^3

The results of improvements in the observed analysis are presented in Table 4.

Table 4 Improvements that the sheep had by being fed with the created formulation

Test	Initial laboratory result	Final labortory result	Sheep number
M.C.H.C	High	Normalized	10
Fibrinogen	High	Normalized	10
Segmented neutrophiles	High	Normalized	10
Urea	High	Normalized	10
Urea nitrogen	High	Normalized	10
Phosphorus	High	Normalized	10
ASTTGO Aspartate aminotransferase	High	Improved	9
ALTTGP Alanine aminotransferase	High	Improved	9
Globulins	High	Improved	9
Lactate Dehydrogenase LDH	High	Improved	9
Glucose	High	Improved	8
Total cholesterol	Low	Improved	8

The sawdust treatment processes described above allowed to regulate in a better manner the ruminant conditions and therefore, propitiated a better exploitation of the food by the animal. Positive impact was generated and the digestive metabolism of the ovines was improved, which was reflected in the increase of consumption levels, weight increase and better food conversion. The interaction of the ingredients of the created formula propitiate digestion conditions with an increased microbial activity, especially due to the ruminant pH regulation and the energetic-proteic balance that the food consumption generates. This is reflected in an increased production of microbial protein, short length fatty acids (acetic, propionic and butyric) and primary metabolites (glucose, fatty acid and total protein), which were absorbed at the intestinal level. The serum concentrations of these metabolites and their proportions regulate the operation of certain organs and tissues. For the present investigation, the inclusion of pine sawdust in a diet formulated for ovines improved their hepatic physiology, which is evidenced through the measurement of diverse enzymes in the bloodstream of the observed ovines. An increase in the serum concentrations of these enzymes is positively correlated with a functional deterioration of the liver, this is to say that a reduction of these enzymes' concentrations in the blood serum indicated a better liver performance. In the same manner, the elaborated formula allowed to increase the levels of cholesterol in the blood serum, which can provoke a better energetic and protein use in the basal metabolism of the ovines.

3. Conclusions

The sawdust comes from pine wood, which is produced and processed year-round, no matter the season. The results of the present project impact the wood industries directly, giving them an opportunity to re-invest their money. Likewise, the present project offers ranchers the possibility of obtaining a food product for their cattle which is of high quality and of fixed price for any time of the year. The present project has a major impact on the wood industries, allowing money to re-enter the market to be able to sell the sawdust with a good profit margin. The processed sawdust comes from pine trees, which are produced and processed year-round, no matter the season. This product has the advantage of being produced all the time, as opposed to ruminant nourishments derived from other products, which can depend on the season and the availability of the product. The solution this project offers for the reuse of pine sawdust waste is the multiple possibility to obtain nourishment at any time of the year. The degradation of lignin from sawdust through the design presented in this research, generates a new form of nourishment for ruminants, which has the advantage of improving its digestibility. Furthermore, sawdust can be combined with different sources of protein and minerals to help generating the highest levels of microbial degradation in ruminants that is possible. This form of nourishment allows the increase of digestive parameters as well as liver performance, which translates into a higher biological efficiency and productivity of the animal. The invention presented in this research is currently under request for a patent.

References

1. Fregoso-Madueño, J. N., Goche-Télles, J. R., Rutiaga-Quiñones, J. G., González-Laredo, R. F., Bocanegra-Salazar, M., & Chavez-Simental, J. A. (2017). Alternative uses of sawmill industry waste. Revista Chapingo Serie Ciencias Forestales y del Ambiente, Volume XXIII, **(2)**, 2017, 243-260. http://dx.doi.org/10.5154/r.rchscfa.2016.06.040).
2. Pineda-Pineda, J., Sánchez del Castillo, F., Ramírez-Arias, A., Castillo-González, A., Valdés-Aguilar, L., Moreno-Pérez, E., "Aserrín de pino como sustrato hidropónico. I: Variación en características físicas durante cinco ciclos de cultivo", Revista Chapingo, Serie horticultura, 18**(1)**, pp. 95-111, 2012
3. Starbuck, C., " Producción y uso de composta de aserrín de avena y estiércol", III Simposium Internacional y IV Reunión Nacional de Agricultura Sostenible. Universidad de Guadalajara, 96 pp., 1997
4. Gonzalez, G. Gonzales, " Algunos residuos forestales y madereros en la alimentación del ganado", Investigación agraria. Sistemas y recursos forestales, Vol. 8**(1)**, January , 1999
5. Zheng W, Phoungthong K, Lü F, Shao LM, He PJ Evaluation of a classification method for biodegradable solid wastes using anaerobic degradation parameters. Waste Manag 33: 2632-2640, 2013
6. Anike FN, Yusuf M, Isikhuemhen OS (2016) Co-Substrating of Peanut Shells with Cornstalks Enhances Biodegradation by Pleurotus ostreatus. J Bioremed Biodeg 7:327. doi: 10.4172/2155-6199.1000327
7. A.L. Slyter, L.D. Kamstra "Utilization of pine sawdust as a roughage substitute in beef finishingrations", Journal of animal science, vol. 38 **(3)**, 1974
8. Espinosa Garza G., Loera-Hernández I., Antonyan N., Cárdenas Barrón L., Cano Méndez D., García Torres A., Robles Páez D., Troncoso Lozano R., Villarreal Garzao J . Pleurotus ostreatus in the decomposition of fibrous tissues for ruminants' feeding, Proceedings of 2018 IISE Annual Conference, (to be published)
9. Seung-Jin, L., Jin-Yoon, K., Na-Hae, P., Awji, E. G., Joo-Won, S., & Seung-Chun, P., "Effect of Dietary Fiber Source on the Growth Performance and Intestinal Microflora in Piglets. International Journal Of Applied Research In Veterinary Medicine", 14(2), pp.135-142, 2016
10. L. C. Rojas-Perez, Y. Pineros-Castro, M.E. Velasquez-Lozano, " Production of fermentable sugars from press fiber oil palm pre-treated biologically by Pleurotus ostreatus and Phanerochaete chrysosporium ", Revista ION, vol.24 **(2)**, Bucaramanga, 2011
11. Zheng W, Phoungthong K, Lü F, Shao LM, He PJ (2013) Evaluation of a classification method for biodegradable solid wastes using anaerobic degradation parameters. Waste Manag 33: 2632-2640.)(6. Anike FN, Yusuf M, Isikhuemhen OS (2016) Co-Substrating of Peanut Shells with Cornstalks Enhances Biodegradation by Pleurotus ostreatus. J Bioremed Biodeg 7:327. doi: 10.4172/2155-6199.1000327)
12. Bach, A., Calsamiglia, S., 2006, "La fibra en los rumiantes: ¿Química o física?", XXII curso de especialización Fedna. Barcelona: Universidad Autónoma de Barcelona., available in: http://www.produccion-animal.com.ar/informacion_tecnica/manejo_del_alimento/100-fibra_en_rumiantes.pdf).
13. O. Simon, B.K. Singh, M.J. Weil, Elaboración y caracterización de lignosulfonatos amonificados a partir de pinzote de banano y aserrín de laurel (Cordia alliodora) para utilizarse como fertilizante de liberación lenta ", Tierra Tropical, Vol.1 **(1)**: pp. 21-26 , 2005
14. Mosier N, Wyman C, Dale B, Elander R, Lee YY, Holtzapple M, Ladisch M., "Features of promising technologies for pretreatment of lignocellulosic biomass", Bioresour Technol. 96**(6)**, 673-86, 2005

Detection and enumeration of the dangerous food borne pathogens in cooked food that causes food poisoning and infectious diseases

Sumonthip KongtunJanphuk [1*]

[1]Department of Biotechnology, King Mongkut's University of Technology North Bangkok, Bangkok, Thailand

Abstract. This research was aim to analyzed the detection and enumeration of the dangerous food borne pathogens in cooked food that causes food poisoning and infectious diseases from the restaurants surround area of King Mongkut's University of Technology North Bangkok, Bangkok, THAILAND. Pot-stewed pork (Palow-Moo;PL) and Fried basil pork (Kapraw-Moo;PK) from five restaurants were collected and analyzed for food borne pathogens. The Analysis of bacterial contamination in that food by used to the methodology of Microbiology and diagnosis bacteria by Biochemistry methodology. The result were indicated that the cooked food from five restaurants contaminated with bacteria caused the risk of gastrointestinal disease as follows : PL dishes the microorganisms were founded between 1×10^5 to 3×10^8 colony/gram, MPN values in the ranges of *Escherichia coli* were infected < 3 to 64 and MPN values of *Staphylococcus aureus* < 3 to 43. There were founded *Salmonella* spp. 5 samples, *Pseudomonas aeruginosa* 4 samples, *Shigella* spp. 4 samples, *Bacillus cereus* 2 samples, *Proteus* spp. 2 sample and *Micrococcus* spp. 1 sample. In the PK dished, the total number of bacteria were founded between 5×10^3 to 2×10^8colony/gram, MPN values in the range of *E.coli* and *S.aureus* were infected < 3 to 39 and < 3 to 28. The results showed that the cooked food from five restaurants are contaminated with bacterial caused food poisoning and the risk of gastrointestinal diarrhea at the different levels.

1 Introduction

Food borne pathogens such as *Escherichia coli*, *Staphylococcus aureus*, *Pseudomonas aeruginosa*, *Salmonella* spp., *Shigella* spp., *Bacillus cereus*, *Proteus* spp., or *Micrococcus* spp. have been recognized for more than 100 years. Those pathogens can transmit through food and cause the food borne diseases which some of diseases are considered as a serious problem for public health. Symptoms of food borne diseases caused by pathogens depending on the type of infectious pathogen and the amount of food intake, which often include vomiting, fever, and diarrhea. Some pathogens can pass through the stomach into intestine and multiply, extended the food borne disease, and some can produce a toxin that can pass into the bloodstream, caused the more serious problems in the deeper body issues [1-3].

There are more than 30 restaurants around King Mongkut's University of Technology North Bangkok (KMUTNB) serving students, teachers, staff and people who live around this area. The pre-cooked food (ready-to-eat) restaurant is one of the most popular type because of the cheap price, large quantity, and various choices of foods. Two kinds of food that usually found in those restaurants are Pot-stewed pork (Palow-Moo;PL) and Fried basil pork (Kapraw-Moo;PK). In this study, PL and PK from 4 restaurants around KMUTNB and in the canteen of the university were collected and tested for food borne pathogens. The risky score of both type of food in all restaurants were evaluated according to the type of pathogens found in samples.

2 Materials and Methods

2.1. Sample preparation

Pot-stewed pork (Palow-Moo;PL) and Fried basil pork (Kapraw-Moo;PK) from four restaurants (South(S), North(N), West(W), East(E)) within the 1 km radius around the university and one restaurant in the campus(C) were collected 1 sample/week for 10 weeks consecutively. Pot-stewed pork samples were labeled as PLS, PLN, PLW, PLE and PLC, and Fried basil pork samples were labeled as PKS, PKN, PKW, PKE and PKC. Samples were kept in the plastic boxes with lids and immediately tested in the laboratory.

Samples were crushed by the food blender, and then squeezed until the juice was obtained. The pH and °Brix values of the samples were checked before the microbiological analysis.

* Corresponding author: skongtun@yahoo.com

2.2 Microbiological analysis [4-6]

All food samples were diluted using serial dilution technique. The isolation of the bacteria was performed using spread plate and streaking plate methods. The isolated bacteria were specified using Biochemical methods that described below;

2.2.1 Gram Stain Technique

The smear was prepared on a clean grease/oil-free slide. It allowed to air dry and then held the slide at one end and passed the entire slide through the flame of a Bunsen burner for a few times with the smear-side up. The crystal violet was gently drop over the smear and left to stain for 1 min. It was gently rinsed with distilled water and then flooded the smear with Gram's iodine, let it stand for 1 min. It was eluted using 95% ethyl alcohol until the eluted solution runs almost clear, and then immediately rinsed with water. The smear was continuously stain with safraninO for 1 min, then rinsed with distilled water, blot dry the slide with tissue paper before observed the smear under the microscope (100x).

2.2.2 Catalase test

Bacteria were grown on a Nutrient Broth and incubated at temperature of 37 °C for 18-24 hr. A few drops of H_2O_2 were applied onto a clean slide, followed by bacteria suspension. Gas wasusually evolved immediately.

2.2.3 Oxidase test

A filter paper which was saturated with 0.5% tetramethyl-*p*-phenylenediamine HCI was used then placed a small amount of bacteria to the paper. The positive reaction (bacteria change color to purple) would be observed immediately or within 10 seconds.

2.2.4 Methyl Red (MR) test

Bacteria were grown on a MR-VP liquid medium and incubated at temperature of 37 °C for 48 hr. 5 ml of Suspension was added into test tube, after that 5-6 drops of Methyl red were applied. The positive reaction was red color and negative color was yellow. If orange color at the top of the test sample was observed, let it stand for 48 hr and observed the result again.

2.2.5 Voges-Proskauer (VP) test

Bacteria were grown on a MR-VP liquid medium and incubated at temperature of 37 °C for 48 hr. 1 ml of Suspension was added into test tube, then added 0.5 ml of 5% naphthol solution and 0.2 ml of 40% potassium hydroxide. The test sample was shaken until it mixed well and observed within 5 min. The positive reaction was red color and negative color was yellow.

2.2.6 Indole test

Bacteria were grown on a Tryptone liquid medium and incubated at temperature of 37 °C for 24-48 hr. 1 ml of Suspension was added into test tube, then 5 ml of Kovacs' reagent was added. The positive reaction was red color on the top layer and negative color was no color.

2.2.7 Citrate utilization test

Bacteria were placed on Simmons' Citrate agar using a sterile loop and incubated at temperature of 37 °C for 24-48 hr (maximum 7 days). The positive reaction would turn the medium color from green to blue.

2.2.8 Gelatin liquefaction test

Bacteria were placed on the bottom of nutrient gelatin medium using a sterile needle and incubated at temperature of 37 °C for 48 hr. The control sample (no bacteria added) was prepared with the same procedure. After 48 hr, samples and control were placed into refrigerator for 2 hr, and then turbidity of samples was observed compared with the control.

2.2.9 Nitrate reduction test

Bacteria were grown on a trypticase-nitrate broth and incubated at temperature of 37 °C for 48 hr. 1 ml of 0.8 sulfanilic acid solution and 1 ml of 0.6% dimethyl-α-naphthylamine solution were added. Positive nitrite reduction would be red color. In the case of no color, zinc dust was added into the sample, the red color appeared in the Negative Nitrate Reduction.

2.2.10 Urease test

Bacteria were placed on the top of Christensen's urea agar slant using a sterile needle and incubated at temperature of 37 °C for 6-24 hr. The positive reaction would turn the medium color to pinkish-red.

2.2.11 Malonate test

Bacteria were placed on the top malonate broth and incubated at temperature of 37 °C for 24-48 hr. The positive reaction would turn the medium color to blue and the negative reaction was yellow, brown or no change.

2.2.12 Motility test

Bacteria were placed on the middle of motility test medium (semi-solid) with a depth of 1/2 inch using a sterile needle and incubated at temperature of 37 °C for 24-48 hr. The positive result would showed the turbidity spread around the growth area.

2.2.13 Triple Sugar Iron (TSI) agar test

Bacteria were inoculated into TSI using a sterile needle by stabbing through the center of the medium to the bottom of the tube and then streak the surface of the slant. The incubation was operated at temperature of 37 °C for 18-24 hr. The samples were observed for color change in the slant and the bottom and gas in the medium.

2.2.14 Carbohydrate Fermentation test

The bacteria were inoculated into Andrade's carbohydrate broth medium of lactose, sucrose, fructose, galactose, xylose, rhamnose, raffinose, trhalose, cellobiose, inulin, salicin, adonital, sorbitol and mannitol. The incubation was operated at temperature of 37 °C for 18-24 hr. The samples were observed for color change and gas in the Durham tube.

3 Results

The pH and Brix value of all 10 samples were shown in Table 1. There are 8 food borne pathogens which were *Bacillus cereus, Shigella* spp., *Staphylococcus aureus, Micrococcus* spp., *Salmonella* spp., *Proteus* spp., *Escherichia coli, Pseudomonas* spp., found in pre-cooked samples in this study compared with Bergey's manual of determinative bacteriology (Bergy's 1994) (Table 2). The microorganism contamination results of 10 samples were shown in Table 3, compared with the guidance for food safety in ready-to-eat food criteria (Department of Medical Science (DMSC), Ministry of Public Health, Thailand).

Table 1. pH and °Brix values of 10 pre-cooked samples.

Sample	pH	°Brix
PLS	5.24	18.1
PLN	6.58	15.6
PLW	6.94	16.9
PLE	6.21	16.3
PLC	5.79	17.5
PKS	5.90	32.0
PKN	6.03	24.3
PKW	6.73	19.7
PKE	5.43	22.9
PKC	6.61	26.1

PL = Pot-stewed pork, PK = Fried basil pork

Table 2. Food borne pathogenic bacteria found in pre-cooked samples.

Food borne pathogens	Pre-cooked samples									
	PLS	PLN	PLW	PLE	PLC	PKS	PKN	PKW	PKE	PKC
Bacillus cereus		+		+			+		+	
Shigella spp	+		+	+	+	+	+	+		+
Staphylococcus aureus	+	+	+	+	+		+	+	+	+
Micrococcus spp.			+			+		+	+	
Salmonella spp.	+	+	+	+	+		+	+	+	+
Proteus spp.		+		+		+	+		+	
Escherichia coli	+	+	+		+	+	+	+	+	+
Pseudomonas spp.	+		+	+	+			+	+	+

+ : Positive (found in samples)

Table 3. Evaluation of pre-cooked food sample (based on criteria of cooked food standard from Department of Medical Science, Ministry of Public Health, Thailand)

Criteria	**Pre-cooked food samples**										DMSC guidance*
	PLS	PLN	PLW	PLE	PLC	PKS	PKN	PKW	PKE	PKC	
Total Viable count	> $1x10^6$	> $1x10^6$	> $1x10^6$	> $1x10^6$	> $1x10^6$	> $1x10^6$	> $1x10^6$	> $1x10^6$	> $1x10^6$	> $1x10^6$	< $1x10^6$
MPN *E. coli*/gram	> 3	> 3	> 3	NF	< 3	< 3	> 3	> 3	> 3	< 3	< 3
S. aureus	< 100	< 100	< 100	< 100	< 100	NF	< 100	< 100	< 100	< 100	< 100
C. perfringens	NF	NF	NF	NF	NF	NF	NF	NF	NF	NF	< 100
B. cereus	NF	< 100	NF	> 100	NF	NF	< 100	NF	> 100	NF	< 100
Salmonella spp.	+	+	+	+	+	NF	+	+	+	+	NF
V. cholerae	NF	NF	NF	NF	NF	NF	NF	NF	NF	NF	NF
V. parahaemolyticus	NF	NF	NF	NF	NF	NF	NF	NF	NF	NF	NF
L. monocytogenes	NF	NF	NF	NF	NF	NF	NF	NF	NF	NF	NF

* based on DMSC food safety criteria guidance for pre-cooked food (Ready-to-eat) 11 January 2016.

4 Discussion and Conclusion

All pre-cooked foods from 5 restaurants were contaminated with the food borne pathogenic bacteria in the different levels and all of those samples did not passed the total viable count criteria of DMSC food safety guidance for pre-cooked food. According to Table 3, the test results PL and PK samples from the same restaurant were similar, which assumed as the contaminate of pathogens from the cooking process of each place.

Acknowledgement

This work was supported by grants from the King Mongkut's University of Technology North Bangkok Fund KMUTNB-GEN -59-44 and Faculty of Applied Science, King Mongkut's University of Technology North Bangkok. The author cordially thanks Department of Medical Science (DMSC), Ministry of Public Health, Thailand and Thai traditional medicine development center, Ministry of Public Health, Thailand, for the instrument support. Finally I would like to thank Ms. Chompoonuch Rassamee ,Ms. Patcharin Jittisak and Dr. Jutamas Sangsaiwho helped to complete the research.

References

1. R.G. Labbe, S. Garcia, editors, *Guide to Foodborne Pathogens* (John Wiley and Sons, Inc, New York, 2001)
2. M.D. Miliotis, J.W. Bier, editors, *International Handbook of Foodborne Pathogens* (Marcel Dekker, Inc., New York, 2003)
3. P.M. Fratamico, A.K. Bhunia, J.L. Smith, editors, *Food borne Pathogens Microbiology and Molecular Biology* (Caister Academic Press, Norfolk, 2005)
4. P.H. Clarke, S.T. Cowan, J. Gen. Micro. biol. 6, 187-197 (1952)
5. B. Priyanka, R.K. Patil, S. Dwarakanath, Indian J. Med. Res. **144**, 327-338 (2016)
6. D.H.Bergy, J.G Holt,editors, *Bergey's manual of determinative bacteriology* (Baltimore : Williams & Wilkins,1994)

A suitable k-epsilon model for CFD simulation of pump-around jet mixing tank with moderate jet reynolds number

Satapan Phapatarinan[1], *Eakarach* Bumrungthaichaichan[1], and *Santi* Wattananusorn[1,*]

[1]Department of Chemical Engineering, Faculty of Engineering,
King Mongkut's Institute of Technology Ladkrabang, Bangkok, Thailand

Abstract. This paper presents the appropriate turbulence model for predicting the overall mixing time inside an open 45° inclined side entry pump-around jet mixing tank with moderate jet Reynolds number of about 17,515. The model was carefully developed by using appropriate hexahedral grid arrangement and proper numerical methods. The two different k-epsilon turbulence models, including realizable k-epsilon model and low Reynolds number k-epsilon model, were simulated. The overall mixing times predicted by these turbulence models were compared with the previous data reported by Patwardhan (Chem. Eng. Sci. 57 (2002) 1307-1318). The results revealed that the low Reynolds number k-epsilon model was a suitable model for predicting the overall mixing time of jet mixing tank with moderate jet Reynolds number.

1 Introduction

Jet mixing is an important mixing device, which was firstly proposed by Fossett and Prosser [1]. It provides the various advantages, including simple design with non-moving part, inexpensive operating cost, and easy installation and maintenance. The liquid recirculation and entrainment inside the tank are driven by high velocity jet to mix the different components.

The jet mixing tanks were experimentally studied to achieve the mixing time correlations. However, these correlations were case specific [2]. So, the computational fluid dynamics (CFD) has been adopted to address this problem and to illustrate the details of fluid flow and mixing inside the vessel.

Over two decades or so, there are many different CFD models for pump-around jet mixing tanks. For high jet Reynolds number ($Re_j = \rho d_j U_j/\mu$, where ρ is fluid density, d_j is jet nozzle diameter, U_j is jet discharge velocity, and μ is fluid viscosity), most of these successful CFD models underpredicted the overall mixing times because of the overprediction in total momentum available for mixing due to the flat top liquid surface assumption as reported by Bumrungthaichaichan et al. [3]. However, Patwardhan [4] showed that the overall mixing time for moderate jet Reynolds number was overpredicted. Furthermore, Ghahremanian and Moshfegh [5] reported that the appropriate turbulence model should be selected for predicting the free turbulent round jet with low Reynolds number.

So, in this work, the two different k-epsilon turbulence models, including realizable k-epsilon model (RKE) and low Reynolds number k-epsilon model (LRKE), were simulated to investigate the proper turbulence model for predicting the overall mixing time inside the pump-around jet mixing tank with moderate jet Reynolds number because there is no turbulence model which is valid for all flow situations. Further, the previous CFD works only studied the suitable turbulence model for high Reynolds number jet mixing tank.

2 Description of CFD modelling

2.1. Jet mixing tank

The studied system was an open 45° inclined side entry pump-around jet mixing tank reported by Patwardhan [4]. The tank diameter and height of tap water were 0.5 m. The jet nozzle diameter and tank outlet pipe diameter were 0.008 m and 0.0381 m, respectively. The schematic of pump-around jet mixing tank is shown in Fig. 1.

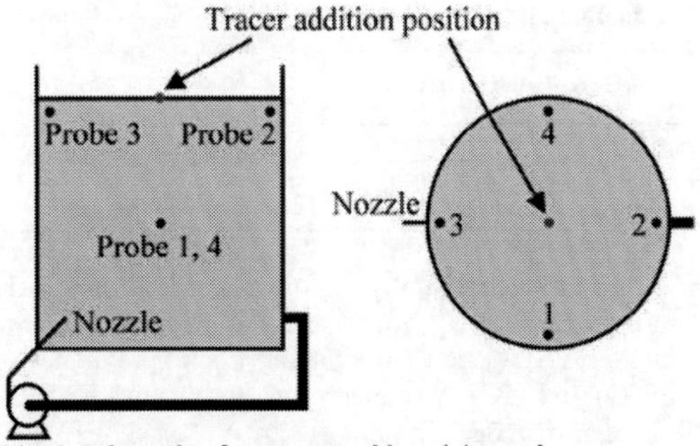

Fig. 1. Schematic of pump-around jet mixing tank

2.2 Grid generation

The three-dimensional solid model and grid generation of this jet mixing tank were carefully performed by using

* Corresponding author: cfdgroup_santi@hotmail.com

GAMBIT 2.4.6. The hexahedral grid topology was selected to obtain the accurate results. Further, the flow domain was manually split into the several blocks by using the specified faces to achieve the appropriate grid arrangement. That is, the grids were controlled to align with the flow direction to minimize the truncation error and numerical diffusion. The grid generation of this mixing tank is shown in Fig. 2.

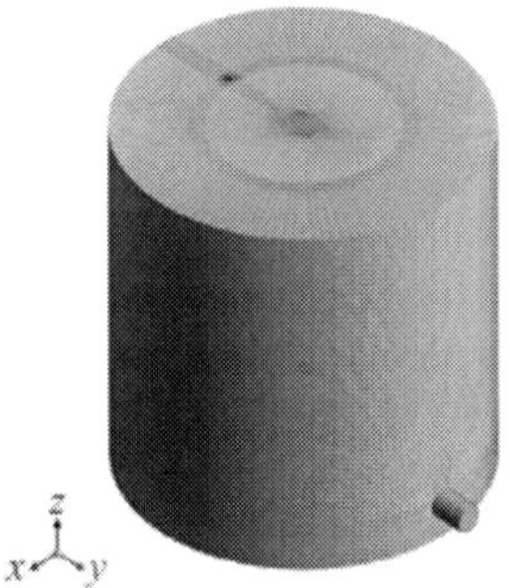

Fig. 2. Grid generation of pump-around jet mixing tank

2.3 Governing equations

For CFD simulation of jet mixing tank, the steady state jet flow and turbulence were achieved by using the Reynolds average equations for conservation of mass and momentum together with turbulence model. The steady state Reynolds average equation can be written in compact form as:

$$\frac{\partial(\rho \boldsymbol{U}\phi)}{\partial x_j} = \frac{\partial}{\partial x_j}\left(\Gamma_\varphi \frac{\partial \phi}{\partial x_j}\right) + S_\varphi \tag{1}$$

where $\boldsymbol{U}$ is mean velocity vector, ϕ is a universal dependent variable, Γ_φ is the diffusivity, and S_φ is the source term. Further, the details of variables for continuity and momentum equations are represented in Table 1.

Table 1. Variables for continuity and momentum equations

Equation	$\boldsymbol{\phi}$	$\boldsymbol{\Gamma_\varphi}$	$\boldsymbol{S_\varphi}$
Continuity	1	0	0
Momentum	U_i	μ	$-\frac{\partial P}{\partial x_i} + \frac{\partial}{\partial x_j}\left(\mu_t \frac{\partial U_i}{\partial x_j}\right) + S_{M,i}$

For turbulence models, the details of RKE and LRKE turbulence models were clearly described and reported by ANSYS Fluent Theory Guide [6] and previous work of Xin and Lei [7], respectively. By default, the LRKE turbulence model is hidden by ANSYS FLUENT. However, this model can be enabled by respectively typing the two additional commands in TUI, including define/models/viscous/turbulence-expert/low-re-ke? and yes [8].

In transient simulation of mixing time, the unsteady state species transport equations without chemical reaction [6] were solved and can be written as:

$$\frac{\partial(\rho Y_i)}{\partial t} + \frac{\partial(\rho \boldsymbol{U} Y_i)}{\partial x_j} = \frac{\partial}{\partial x_j}\left(\left(\rho D_{i,m} + \frac{\mu_t}{\mathrm{Sc}_t}\right)\frac{\partial Y_i}{\partial x_j}\right) + S_i \tag{2}$$

where Y_i is the local mass fraction of species i, $D_{i,m}$ is the mass diffusion coefficient for species i in the mixture, Sc_t is turbulent Schmidt number, μ_t is the eddy viscosity, and S_i is the source term for species transport equations.

Further, in this paper, the 95% mixing time was used to evaluate the mixing time. Generally, the 95% mixing time is the time required for concentration (c) to reach within 95% of the fully mixed value ($\bar{c}$) and can be written as:

$$t_{95\%} = \text{time for } \left|\frac{c-\bar{c}}{\bar{c}}\right| \le 0.05 \tag{3}$$

Finally, the overall mixing time of jet mixing tank was evaluated by using arithmetic average of mixing times obtained by four different probe locations as depicted in Fig. 1.

2.4. Boundary conditions

For steady state simulation, the velocity-inlet and pressure-outlet boundary condition types were adopted at inlet and outlet, respectively. Whereas, for transient simulation, the recirculation-inlet and recirculation-outlet were used [3]. At inlet, the uniform jet discharge velocity of 2.2 $\mathrm{m\cdot s^{-1}}$ was specified. Moreover, for turbulence quantities, the turbulence kinetic energy of 0.0726 $\mathrm{m^2\cdot s^{-2}}$ and its dissipation rate of 2.794518 $\mathrm{m^2\cdot s^{-3}}$ were directly imposed. At the tank wall and base, the no-slip boundary condition and standard wall functions were adopted. Further, the flat top water surface was assumed by using symmetry boundary condition, e.g. the normal velocity and normal gradient of all variables are zero.

2.5. Numerical methods

For this work, the double precision pressure-based solver of ANSYS FLUENT finite volume CFD code was used to simulate jet mixing tank. The pressure-velocity coupling scheme was SIMPLE. The spatial discretization scheme for all quantities and temporal discretization scheme were second order upwind and first order implicit, respectively.

2.6. Solution strategy

In order to identify the solution convergence for steady state simulation, the area weighted average of the velocity magnitude at plane $x = 0$ was monitored until it was constant. Further, the scaled residual of 10^{-5} for tracer was used as a convergence criterion for transient simulation. For unsteady state simulation of implicit solver, the time step size of 0.0025 s was selected [3, 9].

Moreover, in order to achieve the grid independent solutions, the velocity magnitude gradient adaption with the refine threshold of 10% of the maximum value [10] was used. From CFD work of Bumrungthaichaichan et al. [3], they suggested that the grids of CFD models should be adapted three times to obtain the accurate results. Hence, in this work, the final grid resolutions of the present CFD models were also achieved by adapting the CFD grids three times to obtain the grid independent solutions and to minimize or eliminate any uncertainties.

3 Results and discussion

As mentioned earlier, the two different k-epsilon models, including RKE and LRKE, were simulated to investigate the appropriate k-epsilon turbulence model for CFD simulation of open 45° inclined side entry pump-around jet mixing tank with moderate jet Reynolds number. The predicted overall mixing times were compared with the previous results of Patwardhan [4] as shown in Table 2. Furthermore, the profiles of normalized concentration, which is generally defined as a ratio of the tracer concentration to the fully mixed value, and jet axial velocity contours of these turbulence models are represented in Fig. 3 and Fig. 4, respectively.

Table 2. Comparison of overall mixing times

Model	Mixing time [s]	% Error [a]
EXP [4]	57.00	-
CFD [4] [b]	65.00	14.04
RKE	65.15	14.30
LRKE	54.59	4.23

[a] The percentage error is a ratio of absolute difference between the predicted overall mixing time and experimental mixing time to the experimental value.

[b] The overall mixing time was predicted by standard k-epsilon (SKE) turbulence model and power law discretization scheme.

From Table 2, the results show that the overall mixing times predicted by RKE and SKE reported by Patwardhan [4], are slightly different and are about 14% higher than that observed experimentally. Whereas, the LRKE underpredicts the overall mixing time by about 4%.

In Fig. 3, the normalized concentration profiles at probe 1 of RKE and LRKE models are different. The start of normalized concentration profile simulated by LRKE is faster than that predicted by RKE, which indicates that the predicted convective transport of LRKE is greater than that obtained by RKE. The RKE shows two small peaks of normalized concentration profile at time of about 35 s and 55 s. Whereas, the LRKE does not clearly represent the peak of normalized concentration profile. According to the absence in peak of normalized concentration profile, it can be implied that the dispersive transport of LRKE is higher than RKE. Furthermore, the normalized concentration profile of LRKE is faster to reach the 95% mixing time criterion as compared to RKE.

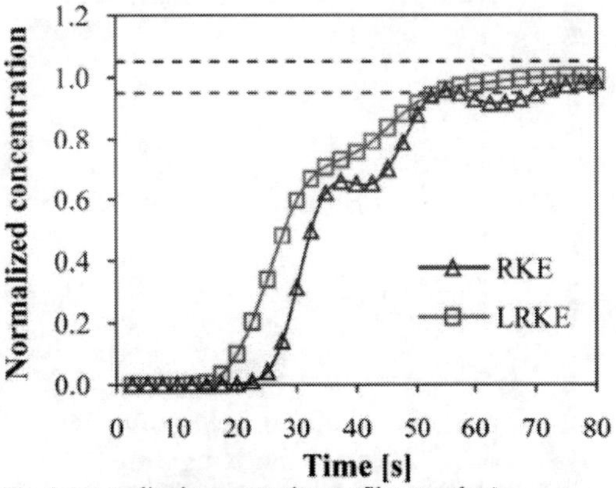

Fig. 3. Normalized concentration profiles at probe 1

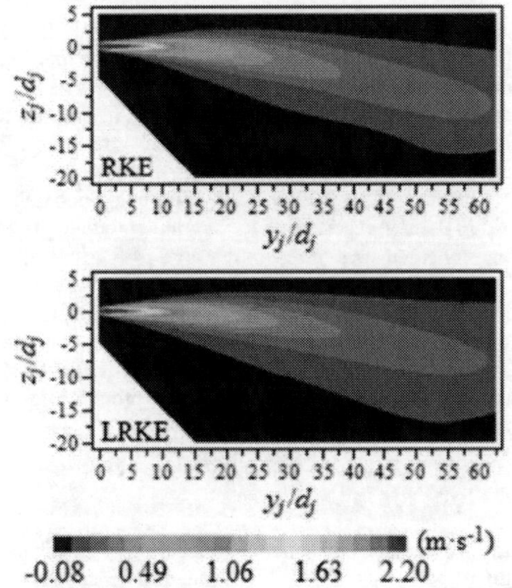

Fig. 4. Contours of jet axial velocity at plane $x = 0$ (y_j and z_j are jet longitudinal and jet radial distances.)

Fig. 4 represents the contours of jet axial velocity for two different k-epsilon models. The results reveal that the decay of centreline velocity for RKE is faster than that observed by LRKE. Moreover, the jet spreading of LRKE is wider than RKE. Hence, the total momentum available for mixing of LRKE would be higher than RKE, which directly results in lower predicted overall mixing time. These results confirm that the convective and dispersive transports of LRKE are greater than those obtained by RKE.

Due to the lack of experimental data of concentration and velocity distributions inside the pump-around jet mixing tank, the overall mixing time was only compared with measured data [4] and used to select the appropriate turbulence model for jet mixing tank with moderate jet Reynolds number. Hence, the present predicted overall mixing time of jet discharge velocity of 2.2 $m \cdot s^{-1}$ and previous CFD data of Bumrungthaichaichan et al. [3] for other jet discharge velocities were plotted and compared with the experimental data of Patwardhan [4] as shown in Fig. 5.

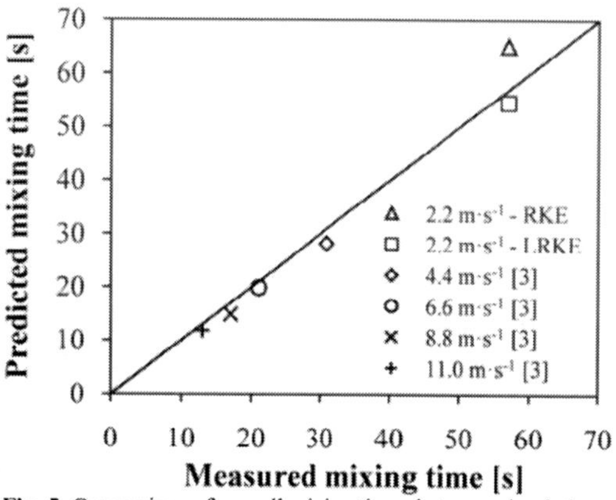

Fig. 5. Comparison of overall mixing times between simulations and experiments for different jet discharge velocities

From Fig. 5, the results show that the overall mixing times predicted by RKE are lower than experimental data for high jet Reynolds numbers ($U_j \geq 4.4$ $m \cdot s^{-1}$). For moderate jet Reynolds number ($U_j = 2.2$ $m \cdot s^{-1}$), the RKE overpredicts the overall mixing time, however, LRKE underpredicts the overall mixing time.

From the previous work of Bumrungthaichaichan et al. [3], they reported that the underpredicted overall mixing time was due to the overprediction in total momentum available for mixing obtained by flat top liquid surface assumption. So, it can be summarized that the RKE can only be used for predicting the overall mixing time at high jet Reynolds number. Further, for moderate jet Reynolds number, the proper turbulence model is LRKE.

4 Conclusion

In this work, a CFD model was carefully developed to investigate the suitable k-epsilon turbulence model for predicting the overall mixing time of pump-around jet mixing tank with moderate jet Reynolds number. The predicted overall mixing times of RKE and LRKE models were compared with the measured data of Patwardhan [4]. Due to the flat top liquid surface assumption, the proper k-epsilon turbulence model for predicting the overall mixing time inside this system was LRKE. For future work, in order to obtain the comprehensive CFD model for this system, the other low Reynolds number turbulence models should be employed and studied.

The authors would like to thank College of Advanced Manufacturing Innovation, King Mongkut's Institute of Technology Ladkrabang, Thailand for supporting the ANSYS FLUENT software.

References

1. H. Fossett, L.E. Prosser, Proc. I. Mech. E. **160**, 224 (1949)
2. E. Bumrungthaichaichan, Korean J. Chem. Eng. **33**(11), 3050 (2016)
3. E. Bumrungthaichaichan, A. Namkanisorn, S. Wattananusorn, J. Chin. Inst. Eng. (to be published)
4. A.W. Patwardhan, Chem. Eng. Sci. **57**, 1307 (2002)
5. S. Ghahremanian, B. Moshfegh, J. Fluids Eng.-Trans. ASME **136**, 011201-1 (2014)
6. ANSYS Inc., *ANSYS Fluent Theory Guide: Release 15.0* (ANSYS Inc., USA, 2013)
7. N. Xin, L. Lei, in 4th International Conference on Computer, Mechatronics, Control and Electronic Engineering, Hangzhou, China (2015)
8. ANSYS Inc., *ANSYS Fluent Text Command List: Release 15.0* (ANSYS Inc., USA, 2013)
9. E. Bumrungthaichaichan, N. Jaiklom, A. Namkanisorn, S. Wattananusorn, Sci. Res. Essays **11**(4), 42 (2016)
10. ANSYS Inc., *ANSYS Fluent Tutorial Guide: Release 15.0* (ANSYS Inc., USA, 2013)

Modified photoperiods and light intensities for grow-out production of nile tilapia *(O. niloticus)* under indoor tank culture system

Marvin Cinense[1,*], *Ireneo* Agulto[1], *Armando* Espino Jr. [1], *Romeo* Gavino[1], *Emmanuel* Sicat[1] and *Victorino* Taylan[1]

[1]Department of Agricultural and Biosystems Engineering Department, College of Engineering, Central Luzon State University, Science City of Munoz, Nueva Ecija, Philippines 3120

Abstract. The study aimed to evaluate the effect of modified photoperiods and light intensities on grow-out production of tilapia under indoor tank culture system; specifically on water quality and on growth performance of tilapia. The study was conducted inside the greenhouse and considered two factors: photoperiod (24L:0D, 20L:4D and 16L:8D); and light intensity (40 watts m^{-2}, 60 watts m^{-2} and 80 watts m^{-2}); also, a control treatment was set-up outside the greenhouse to receive the normal environmental condition. Light manipulations inside the greenhouse were able to sustain the water quality within the desirable level for tilapia cultured in tank; while control tanks shows high water quality variation. Growth performance of tilapia cultured in tank responds positively in prolonged photoperiod at different light intensities. However, light manipulation gave insignificant difference on feed conversion ratio and survival rate compared to normal condition. Results indicated that 20-hours photoperiod and 40 watts m^{-2} light intensity is just enough to sustain the recommended water quality and better growth performance of the cultured tilapia in indoor tank system. It can be concluded that longer photoperiod and restrained light intensity can improved the growth performance of tilapia under indoor tank culture.

1 Introduction

Aquaculture is affected by the impact of climate change such as increasing temperature and sudden changes in local weather condition that can cause low production or leading to fish kill. Also, increasing competition for freshwater may lead to the reduction of suitable areas for aquaculture. Indoor aquaculture has the potential and is a good alternative for outdoor production. However, it is necessary to determine its optimum environmental condition to maximize productivity. One of the major concerns in indoor production is the absence of sunlight; and currently, some information is available concerning the effects of artificial photoperiod and light intensity on the growth performance of tilapia for indoor tank culture.

This study generally aimed to devise modified photoperiods and light intensities for grow-out production of Nile tilapia (*Oreochromis niloticus*) under indoor tank culture system. Specifically, it aimed to: evaluate the effect of light manipulation on water quality and to analyze the effect of photoperiods and light intensities on the growth performance of tilapia under indoor tank culture system.

2 Materials and methods

The study considered two light manipulation factors arranged in split-plot: the photoperiod as main plot; P_1 (24L:0D), P_2 (20L:4D) and P_3 (16L:8D); and, light intensity as sub-plot; L_1 (40 watts m^{-2}), L_2 (60 watts m^{-2}) and L_3 (80 watts m^{-2}); also, a control treatment tanks were constructed outside to receive the normal outside condition. A total of 30 plastic tanks were used each having a 1000-liter or 1 cubic meter capacity to represent each treatment and was replicated three times.

The experiment was conducted inside the greenhouse structure. A net covering (80% shading) was used to minimize the effect of sunlight to light manipulation treatments. Light intensity in each photoperiod tank was provided by compact fluorescent lamps installed about 100 cm above the water surface. For light intensity treatment L_1, two 20-watts compact fluorescent bulbs were installed; for L_2, three 20-watts were installed; and for L_3, four 20-watts were installed. Each fish tank (except control) was covered with black cloth to minimize the effect of natural sunlight at the same time isolating them from the other treatments (Fig. 1).

All fish samples were acclimatized for three weeks under 12L:12D photoperiod [1]; to make sure that the effect of fish kill is due to the experimental treatments and not on fish stress or other outside factor. During this period, dead fish were removed. Prior to the start of the

* Corresponding author: mmcinense.dabe@gmail.com

experiment, the fish were starved for 24 hours, and their body weights measured. The tilapia stocked has an initial average weight of 25.0±0.5 grams. Feeding was done manually three times a day with the feeding rate of 4% biomass.

Fig. 1. The experimental set-up inside the greenhouse.

Ten fish samples per tank were used to determine the growth performance of tilapia in terms of weight gain (BWG), specific growth rate (SGR) and feed conversion ratio (FCR) every three weeks. Survival rate (SR) per tank were monitored daily. Also, uneaten feeds were measured and were deducted to the total feeds supplied [2]. These parameters were computed based on the following equations:

$$BWG = W_f - W_i \quad (1)$$

$$SGR = (Ln\ W_f - Ln\ W_i) * 100/t \quad (2)$$

$$FCR = D_f /(W_f - W_i) \quad (3)$$

$$SR = N_f * 100/N_i \quad (4)$$

where:

W_i = initial wet weight (grams)
W_f = final wet weight (grams)
t = time interval (days)
N_i = initial number of fishes stocked
N_f = final number of fishes stocked
D_f = dry feed intake

Water quality such as water temperature, dissolved oxygen (DO), pH and total ammonia were monitored daily using a multi-parameter water quality meter. Data on un-ionized ammonia were tabulated based on total ammonia readings using the temperature-pH table [3]. Water exchange and air blower were used to sustain the desirable water quality level for tilapia culture.

3 Results and discussion

3.1. Water quality

Good water quality is essential for growth and survival of the cultured species, and it must be monitored regularly to serve as guide for managing the aquaculture project. In tilapia production, water quality such as temperature, pH, ammonia and dissolved oxygen are some of the parameters to be considered to attain better growth performance and higher yield.

Water temperature in light manipulation (photoperiod and light intensity) treatment tanks ranged from 25-28 °C in the morning and 27-29 °C in the afternoon. It shows that even by exposing the treatment tanks into longer photoperiod from 16 hours up to 24 hours or by increasing the light intensity from 40 to 80 watts m^{-2}, it cannot significantly increase the water temperature in tanks. While, control tanks' temperature ranged from 24-28 °C in the morning and 28-31 °C in the afternoon. Record shows that water temperature in control tank is higher in the afternoon than the water temperature inside the greenhouse as it was influenced by the ambient air temperature (Fig. 2 and Fig. 3).

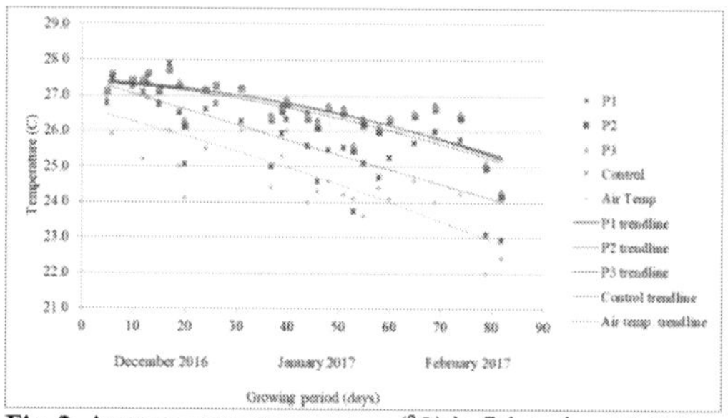

Fig. 2. Average water temperature (°C) in fish tanks at 8 AM for the different photoperiod treatments.

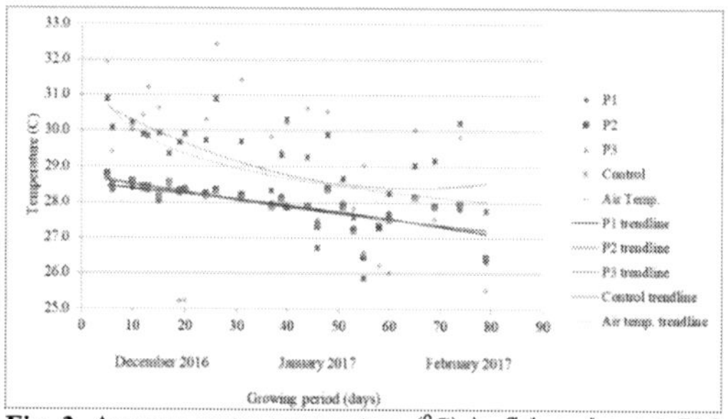

Fig. 3. Average water temperature (°C) in fish tanks at 2 PM for the different photoperiod treatments.

Water pH in light manipulation treatment tanks ranged from 7.2 to 7.6 in the morning and 7.1 to 7.5 in the afternoon. It shows that water pH is not directly influence by the increase or decrease of photoperiod and light intensity imposed in the study. While control tanks recorded 7.2 to 8.0 pH in the morning and 7.3 to 9.0 in the afternoon. Higher pH reading on control tanks in the afternoon can be attributed to the presence of plankton (Fig. 4 and Fig. 5).

Ammonia toxicity is dependent on both pH and temperature, wherein higher pH and water temperature make ammonia more toxic. Results showed that the average un-ionized ammonia during the initial stage of the growing period were above the desirable level in all light manipulation and control treatment tanks, but for the succeeding growing periods, it was within the acceptable level [4]. However, control tanks recorded high levels of un-ionized ammonia in the afternoon until the 43rd day of growing period, and during this period higher temperature and pH levels were also recorded which resulted to high mortality in these tanks (Fig. 6 and Fig. 7).

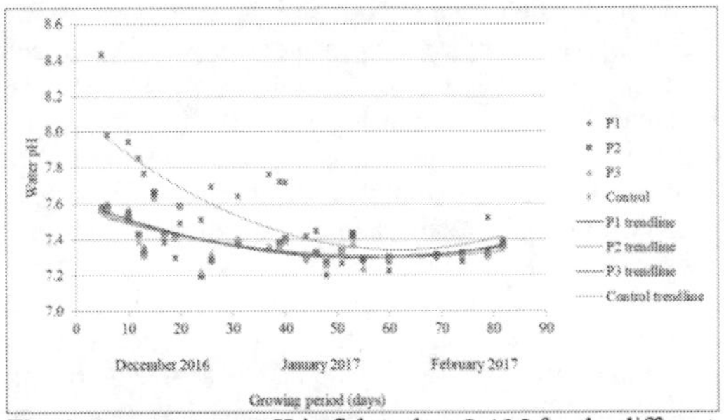

Fig. 4. Average water pH in fish tank at 8 AM for the different photoperiod treatments.

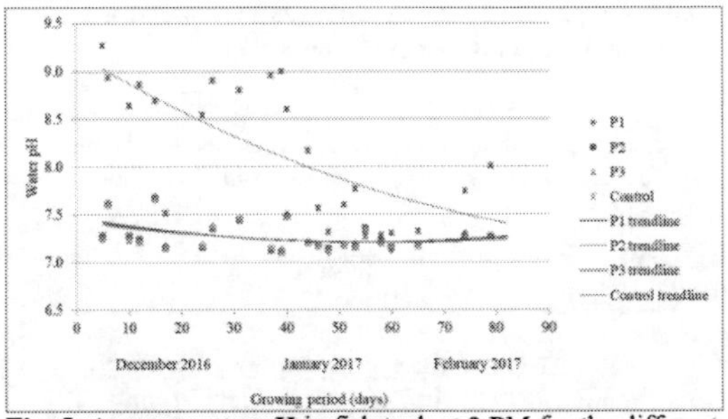

Fig. 5. Average water pH in fish tank at 2 PM for the different photoperiod treatments.

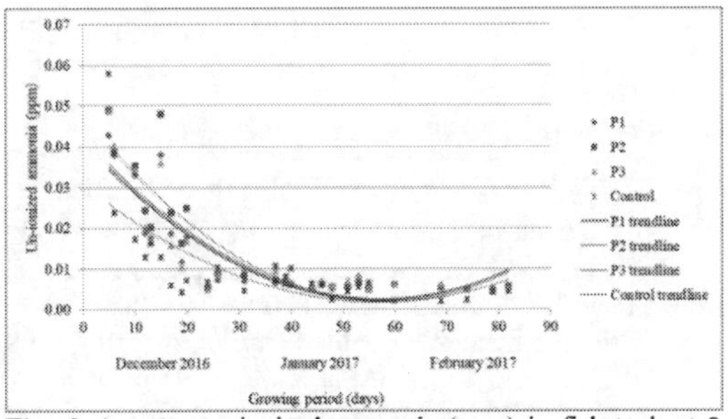

Fig. 6. Average un-ionized ammonia (ppm) in fish tank at 8 AM for the different photoperiod treatments.

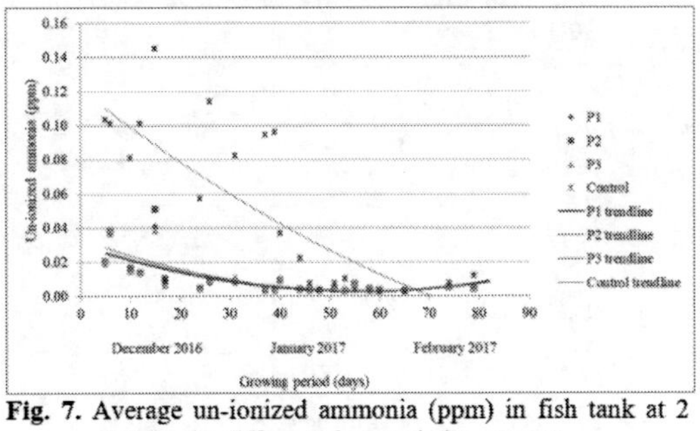

Fig. 7. Average un-ionized ammonia (ppm) in fish tank at 2 PM for the different photoperiod treatments.

Low DO was recorded for the different light manipulation treatment tanks throughout the growing period which ranged from 1.0 to 3.0 ppm. Similarly, the DO level recorded during the morning observations in the control tanks were even lower (< 2.0 ppm). However, it goes from normal to high levels in the afternoon which ranged from 4.0 to 14.0 ppm (Fig. 8 and Fig. 9).

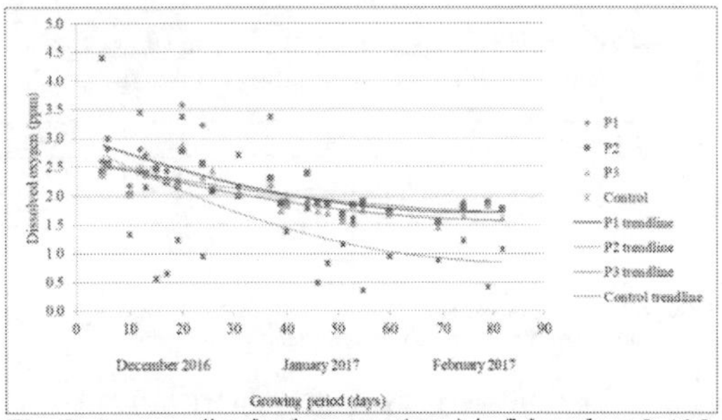

Fig. 8. Average dissolved oxygen (ppm) in fish tank at 8 AM for the different photoperiod treatments.

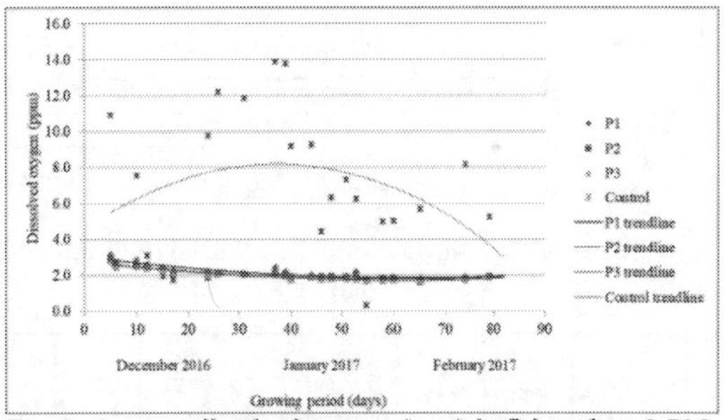

Fig. 9. Average dissolved oxygen (ppm) in fish tank at 2 PM for the different photoperiod treatments.

3.2 Growth performance

Tilapia stocked has an initial average weight of 25.0 ± 0.5 grams prior to the start of the experiment. Analysis on total body weight gain (BWG), specific growth rate (SGR), feed conversion ratio (FCR) and survival rate (SR) showed no significant differences among the means at different photoperiod and light intensity treatments and their interactions (Tables 1 to 4). However, comparing photoperiod treatments to control shows that the total body weight gain of tilapia under 20L:4D treatment was significantly higher by 52% than those under the control treatment (C); while means under 24L:0D and 16L:8D were not, with only 35% and 24% increase respectively (Table 5).

Table 1. Total body weight gain (grams) of fish subjected to different photoperiods and light intensities.

Treatment	L_1	L_2	L_3	Mean
P_1	55.60	49.43	49.83	51.62
P_2	56.17	56.83	60.70	57.90
P_3	40.52	53.13	48.03	47.23
Mean	50.76	53.13	52.86	

Means with the same subscript letter or without subscripts are not significantly different at 0.05 level of significance.

Table 2. Specific growth rate (%/day) of fish subjected to different photoperiods and light intensities.

Treatment	L_1	L_2	L_3	Mean
P_1	1.26	1.16	1.21	1.21
P_2	1.28	1.28	1.30	1.29
P_3	0.94	1.16	1.15	1.09
Mean	1.16	1.20	1.22	

Means with the same subscript letter or without subscripts are not significantly different at 0.05 level of significance.

Table 3. Feed conversion ratio of fish subjected to different photoperiods and light intensities.

Treatment	L_1	L_2	L_3	Mean
P_1	2.64	3.08	2.93	2.88
P_2	2.70	2.72	2.50	2.64
P_3	3.34	2.99	3.27	3.20
Mean	2.89	2.93	2.90	

Means with the same subscript letter or without subscripts are not significantly different at 0.05 level of significance.

Table 4. Total body weight gain (grams) of fish subjected to different photoperiods and light intensities.

Treatment	L_1	L_2	L_3	Mean
P_1	71.11	75.56	66.67	71.11
P_2	91.11	93.33	84.44	89.63
P_3	75.56	60.00	84.44	73.33
Mean	79.26	76.30	78.52	

Means with the same subscript letter or without subscripts are not significantly different at 0.05 level of significance.

Table 5. Performance of fish subjected to different photoperiods and light intensities.

Para	24L: 0D	20L: 4D	16L: 8D	40 w m^{-2}	60 w m^{-2}	80 w m^{-2}	C
BWG (g)	51_{ns}	57^{*}	47_{ns}	50_{ns}	53^{*}	52^{*}	38
SGR (%/day)	1.2^{*}	1.3^{*}	1.1_{ns}	1.2^{*}	1.2^{*}	1.2^{*}	0.9
FCR	2.8_{ns}	2.6_{ns}	3.2_{ns}	2.9_{ns}	2.9_{ns}	2.9_{ns}	3.8
SR (%)	71_{ns}	89_{ns}	73_{ns}	79_{ns}	76_{ns}	78_{ns}	57

Means in row with *ns* is not significant or * is significantly different compared to Control.

The BWG under light intensity 60 w m^{-2} and 80 w m^{-2} were significantly differ when compared to those under the control treatment, with an average increase of 39.0% and 38.5% respectively; however, 40 w m^{-2} was not, with only 33.0% increase. Similarly, SGR under light manipulation treatments shows significant effect as compared to control, except treatment 16L:8D. Data explained fish in the extended photoperiod group had faster growth rate [5]; and this can be attributed to a rise in appetite and feed intake, better feed efficiency and/or elevated digestibility [6].

Results showed that outside condition is not good enough for growing tilapia under tank culture system. Likewise, having a 24-hour photoperiod for tilapia did not give any significant increase in the growth performance of tilapia. It shows that a 20-hour photoperiod is good enough for the growth of tilapia inside the modified environment.

3 Conclusions

Based on the result of the study, it can be concluded that (a) light manipulation treatment in indoor tank culture were able to sustain the water quality within the desirable level; (b) extended photoperiod can improved the growth performance of tilapia cultured in tank regardless of extent and intensity of light. However, subjecting the tilapia to 20-hours photoperiod at 40 watts m^{-2} is good enough to improve the growth performance of tilapia in indoor tank culture compared to outside condition; (c) Feed conversion ratios of tilapia cultured in tank and treated with longer photoperiod at different light intensities showed better feed efficiency than those grown under normal daylight condition. However, exposing tilapia in tank for 20-hours photoperiod at 40 watts m^{-2} is worthy enough for better feed consumption as compared to normal daylight condition; and, (d) lower mortality rate were recorded on tanks treated with longer photoperiod at different light intensities. Indoor tank culture of tilapia treated with 20-hours photoperiod at 40 watts m^{-2} can increased the survival rate of tilapia cultured in tank up to 90% as compared to outside environment with only 58% survival rate.

Special thanks to Engineering Research for Development and Technology – Department of Science and Technology (ERDT-DOST).

References

1. A.K. Biswas, , M. Seoka, Y. Inoue, K. Takii, and H. Kumai. Aquaculture, **250**: 666– 673 (2005)
2. M.A.M. Ali and A.M.I. El-Feky. Egyptian J. Anim. Prod. **50**(3):186 -192 (2013)
3. K. Emerson, R.E. Lund, R.V. Thurston and R.C. Russo. J. Fish. Res. Board Can. **32**: 2379-2383 (1975)
4. V. Suresh and R.C. Bhujel. *Wiley-Blackwell Publishing Company*, U K, pp 338–364 (2012)
5. C. Fuentes-Silva, G.M. Soto-Zarazua, I. Torres-Pacheco, R.G. Guevarra-Gonzalez, J.F. Garcia-Trejo, A. Flores-Rangel, J. Caballero-Perez and A. Cruz-Hernandez. Int. J. of Agric. and Biol. ISSN Print: 1560-8530; ISSN Online: 1814-9596, 14-524/2015/17-4-785-790, DOI:10.17957/IJAB/14.0016 (2015).
6. E.M. Vera Cruz and C.L. Brown. J. Fish Biol., **75**: 130-141 (2009)

Experimental investigation of air characteristics during dehumidification in the multilayer desiccant bed column system

Prueksa Sawardsuk[1], *Jiraporn* Sripinyowanich Jongyingcharoen[2] and *Ekkapong* Cheevitsopon[1,*]

[1]Department of Food Engineering, Faculty of Engineering, King Mongkut's Institute of Technology Ladkrabang, Bangkok, Thailand

[2]Department of Agricultural Engineering, Faculty of Engineering, King Mongkut's Institute of Technology Ladkrabang, Bangkok, Thailand

Abstract. The gold of this research is to investigate air characteristics during dehumidification process using the multilayer desiccant bed. Silica gel packing in the multilayer column was used as a desiccant material. Airflow direction in the column was designed in a zigzag path passing each desiccant layer. The changes in temperature and relative humidity of exit air were recorded after dehumidification at different airflow rates of 18, 36, and 72 m^3/h. In the desiccant regeneration process, moisture in silica gel was removed by 85°C of hot air at varied airflow rates. The characteristics of exit air after regeneration were also monitored. The result revealed that air humidity ratio was significantly decreased using multilayer desiccant bed column. The highest rates of air dehumidification and desiccant regeneration were observed in the first 5 min operation. The highest air dehumidification rate was 12.82 g/min at the airflow rate of 72 m^3/h and the highest regeneration rate of desiccant was 6.70 g/min at the airflow rate of 72 m^3/h. In addition, the dual column of multilayer desiccant bed can be successfully applied to cyclic operation of dehumidification and regeneration when the cycle time were 5 min and airflow rate 36 and 72 m^3/h.

1 Introduction

The desiccant dehumidification of air is widely used in industrials and air conditioning. Either solid or liquid desiccant can be used. However, solid desiccant in forms of packed bed or rotary bed is preferable because of easy handling [1]. The packed bed system has less mechanical problems because its design is not complicate and higher amount of humidity is reduced as compared to the rotary bed system [2,3]. The commonly used solid desiccant is silica gel as it has high moisture adsorption capacity and great pore surface area to trap the water vapor. Adsorption and desorption of the packed bed of silica gel have been reported in several studies. Chang et al. [3] investigated the effect of regeneration conditions on the dehumidification process of modified silica gel and commercial silica gel packed bed. The adsorption of the silica gel was increased as the degree of regeneration increased. Awad et al. [4] fabricated a radial flow dehumidifier using silica gel as a desiccant. Hollow cylindrical bed patterns were designed and test at different diameter ratios. Low bed diameter ratio was appropriate for minimum pressure drop. On the other hand, higher bed diameter ratio caused greater air humidity drop. Dehumidifiers were applied in drying due to many benefits such as reducing drying time and retaining product qualities [5,6].

To date, limited reports are available in the literature on the performance of desiccant bed column. Therefore, the objective of this study was to design the zigzag airflow direction in desiccant bed column, and study the humidity ratio and temperature characteristics of the exit air after dehumidification and regeneration of the air in the multilayer desiccant bed column system. The system was composed of two columns performing the switch task of dehumidification and regeneration, i.e. cyclic operation. The cyclic operation of the system was also carried out at different air flow rates to determine the system performance.

2 Experiment study

2.1 Experimental setup

A schematic diagram of the multilayer desiccant bed column system is illustrated in Fig. 1. The system consisted of a dual column; column A and column B. Each column was 20 cm inner diameter and 100 cm height. Fifteen layers with sieve containers were inside the column. Each layer contained 150 g of silica gel. The sieve container had an air duct with a diameter of 2.54 cm to allow the air to pass through. The air duct was changed from left side of one layer to right side of the next layer in order that a zigzag pathway of the air was generated. Outside of the column was covered by rock wool insulation.

* Corresponding author: ekkapong.ch@kmitl.ac.th

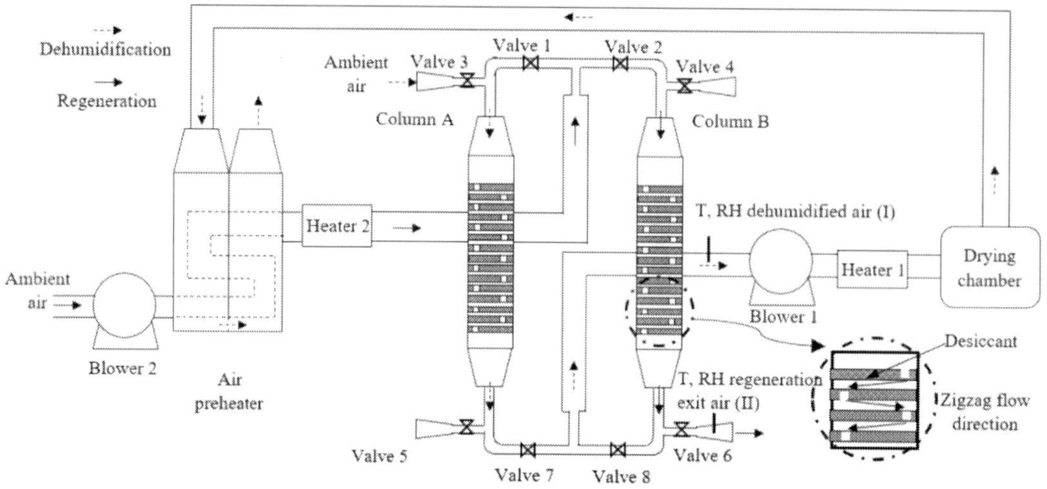

Fig. 1. A schematic diagram of the multilayer desiccant bed column system.

Dehumidification and regeneration operations of the system are shown in Fig. 1 as well. Column A and B were operated alternately between the process of dehumidification of air and regeneration of desiccant. Using column A for dehumidification, valves 3 and 7 were opened while valves 1 and 5 were closed. Ambient air was passed through valve 3, column A, and valve 7 respectively by blower 1. Dehumidified air was obtained prior to heating by heater 1 and supplying to a drying chamber. Simultaneously, the desiccant in column B was regenerated. Valves 2 and 6 were opened and valves 4 and 8 were closed. Ambient air was passed through blower 2, preheated by an air preheater, heated to the desirable temperature by heater 2, and used for desiccant regeneration in column B.

2.2 Instrumentation and experimental procedure

2.2.1 Dehumidification and regeneration process

To prepare fresh silica gel for the dehumidification process, commercial grade silica gel was dried in hot air oven at 90°C for 6 h and allowed to cool in a closed chamber. Saturated silica gel was required for the desiccant regeneration experiment. Silica gel was allowed to adsorb moisture in air at room temperature for 8 h until its maximum adsorption was reached.

The experiment on air dehumidification and desiccant regeneration were conducted individually. Air flow rates were set at 18, 36 and 72 m^3/h for both experiments. For the regeneration process, air temperature was 85°C. Temperature and relative humidity of ambient and exit air were recorded every 30 s for the whole period of 30 min by temperature hygrometer with the response time of 25 s (KT320, Kimo, France) at point I and II as show in Fig. 1. The dehumidification rate and regeneration rate were calculated by equation (1) and (2), respectively.

$$Dehumidification\ rate = \dot{m}_d \times (w_{i,d} - w_{o,d}) \quad (1)$$

$$Regeneration\ rate = \dot{m}_r \times (w_{o,r} - w_{i,r}) \quad (2)$$

Where $\dot{m}$ is mass flow rate of air (kg dry air/min), w_i and w_o were humidity ratio of inlet air and exit air (g water/kg dry air), respectively. The subscript letters of d and r are dehumidification and regeneration process, respectively.

Furthermore, specific dehumidification rate and specific regeneration rate were calculated and expressed as dehumidification rate per kilograms of silica gel and regeneration rate per kilograms of silica gel, respectively.

2.2.2 Cyclic operation process

To determine performance of the multilayer desiccant bed column system, cyclic operation was performed when alternating the air dehumidification and desiccant regeneration between columns A and B. Airflow rates varied at 18, 36 and 72 m^3/h for both processes. The regeneration air temperature was 85°C. One cyclic operation included 5 min dehumidification and 5 min regeneration. If column A was used for the dehumidification process, column B was simultaneously used for the regeneration process. The temperature and relative humidity of the exit air from both processes were recorded at the specified interval for 30 min.

3 Results and discussion

3.1 Dehumidification process: Effect of air flow rate on characteristics of the dehumidified air

Humidity ratio is defined as the proportion of the mass of water vapour to the mass of dry air. It is determined using a psychrometric chart given the data of dry bulb temperature and relative humidity of air. Fig. 2 presents the changes in humidity ratio and temperature of exit air (Point I) after the dehumidification process at the varied airflow rates 18, 36 and 72 m^3/h for 30 min. It can be seen that humidity ratio of the exit air sharply decreased at the initial period of dehumidification when compared to humidity ratio of the inlet air (12.09-13.94 g water/g dry air). Thereafter humidity ratio of the exit air increased with increasing time. This was because of lower adsorption capacity of silica gel with longer process time. Similar trend have been observed in a thin multilayer of activated alumina and packed bed of silica gel [7,8,9]. At first 5 min of the process, the highest dehumidification rates were 2.93, 6.52 and 12.82 g water/min (corresponding to the specific dehumidification rates of 1.30, 2.90 and 5.70 g water/(min·kg silica gel)) at the airflow rates of 18, 36 and 72 m^3/h, respectively. Regarding to temperature of the exit air, temperature slightly increased with increasing time because heat was released from the desiccant (heat of adsorption). It could be observed that in long term adsorption, humidity ratio and temperature of the exit air were the same as those of the inlet air. Fig. 3 shows a psychrometric process of air dehumidification at the represent airflow rate of 72 m^3/h. Humidity ratios of the inlet air reduced by about 64%, 50% and 29% when the process times were 1, 5 and 30 min, respectively. It is interesting that psychrometric path of the dehumidification process at 30 min conformed to the adiabatic path.

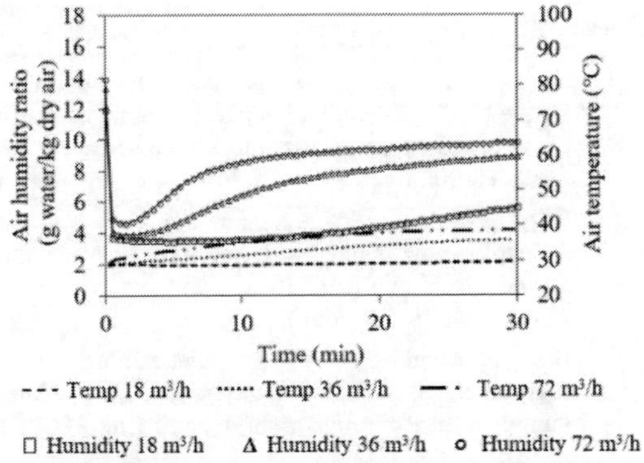

Fig. 2. Humidity ratio and temperature of exit air after the dehumidification process at various airflow rates.

3.2 Regeneration process: Effect of airflow rate on characteristics of the exit air

Saturated silica gel was regenerated using the air temperature of 85°C at the airflow rates 18-72 m^3/h. The changes in humidity ratio and temperature of the exit air after the regeneration process (Point II) for 30 min are illustrated in Fig. 4. Humidity ratio of the exit air increased continuously with increasing time. It indicated that moisture in the silica gel was evaporated by the regeneration process. Moreover, the change in humidity ratio of the exit air was dependent on the airflow rate. Humidity ratio of the exit air gradually increased with time at low airflow rate. And the exit air humidity ratio sharply increased with time at high airflow rate. Therefore, the highest regeneration rate was 6.70 g water/min (corresponding to the specific dehumidification rates of 2.98 g water/(min·kg silica gel))

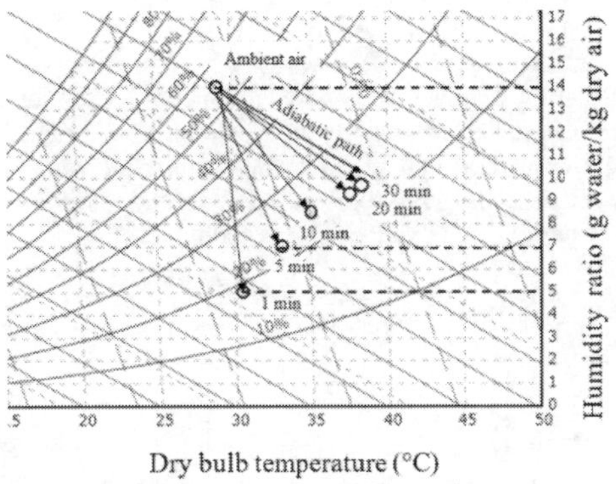

Fig. 3. Psychrometric paths of the dehumidification process with the airflow rate of 72 m^3/h at various time.

Fig. 4. Humidity ratio and temperature of exit air after the regeneration process at various air flow rates.

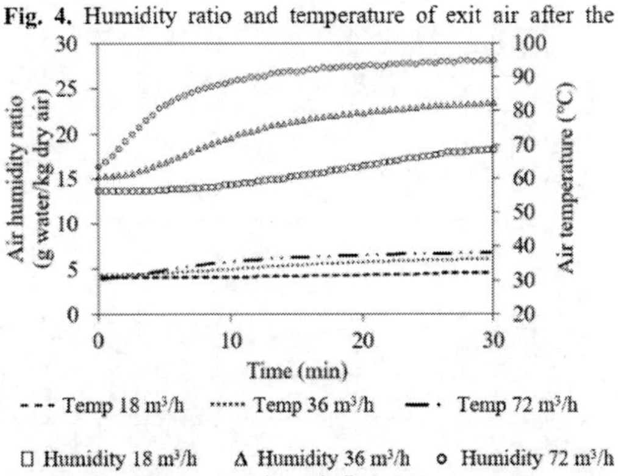

at the first 5 min when the airflow rate was 72 m^3/h. As can be seen in Fig. 4. temperature of the exit air increased with time. Abou-Ziyan et al. [7] also reported increasing temperature of the exit air for long process period until the temperature was equal to the inlet air temperature.

3.3 Cyclic operation of the system

Behaviors of air during the cyclic operation between the dehumidification and regeneration process at the airflow rate 18, 36 and 72 m^3/h are shown in Fig. 5, 6 and 7, respectively. Different characteristics of dehumidification process were found at airflow rate of 18 m^3/h. Humidity ratio of the dehumidification process increased in every cycle. It indicated that moisture was accumulated in the silica gel after the regeneration process. The lowest flow rate of hot air for the regeneration process cannot be applied in this system. For the cyclic operation at the airflow rate of 36 and 72 m^3/h, the humidity ratio of the exit air decreased during dehumidification and increased during regeneration. Similar characteristics were presented in previous research on cyclic operation [8,10]. At the airflow rate of 72 m^3/h, humidity ratio of the exit air was the lowest during dehumidification and the highest during regeneration. It should be concluded that the regeneration airflow rate was an important control parameter for a successful cyclic operation. In addition, performance of the cyclic operation was dependent on the cycle duration and air temperature used in the regeneration process [7]. When the airflow rate was 72 m^3/h, the multilayer desiccant bed column system could reduce air humidity ratio from the ambient condition (approximately 17 g water/kg dry air) to about 15 g water/kg dry air, yielding 12% reduction. Therefore, the results indicated that the dual column of the multilayer desiccant bed system could be successfully applied to provide continuous supplied air with low humidity ratio for various industrial processes such as drying.

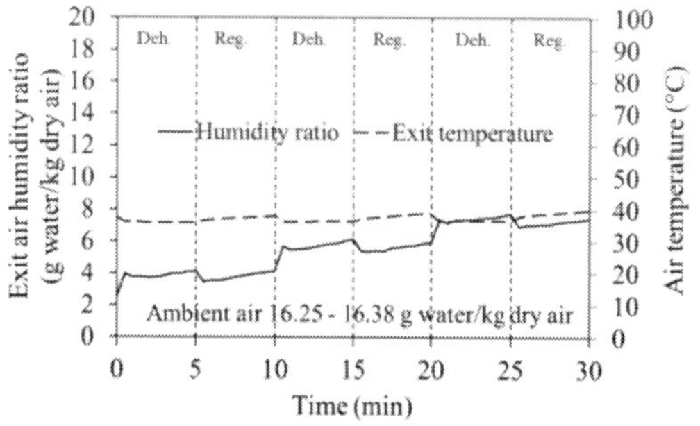

Fig. 5. Humidity ratio and temperature of the exit air during the cyclic operation at 18 m^3/h.

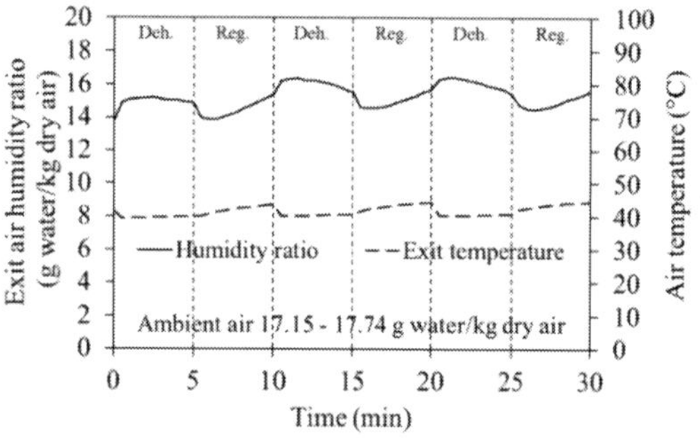

Fig. 6. Humidity ratio and temperature of the exit air during the cyclic operation at 36 m^3/h.

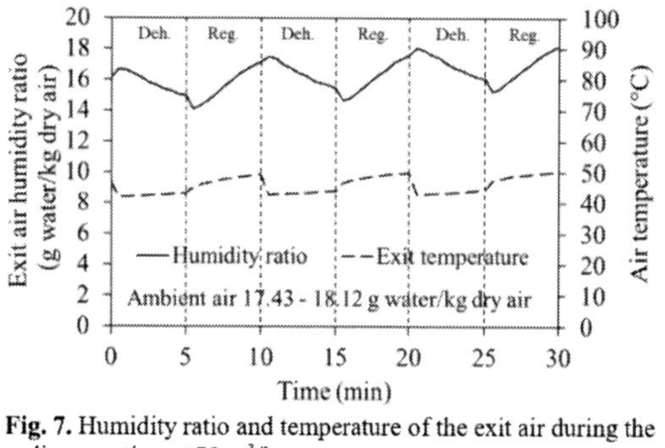

Fig. 7. Humidity ratio and temperature of the exit air during the cyclic operation at 72 m^3/h.

4 Conclusion

Air dehumidification and desiccant regeneration of the multilayer desiccant bed column system were studied by observing humidity ratio and temperature of the exit air. Cyclic operation between dehumidification and regeneration was run to test ability of the system. From the study, the following observations can be drawn.

1. The highest dehumidification rate was observed at the beginning of the dehumidification process, and then the dehumidification rate decreased due to lower adsorption capacity of desiccant.

2. The regeneration rate significantly increased with the airflow rate of hot air.

3. The cyclic operation at the airflow rates of 36 and 72 m^3/h presented continuous ability to conduct the dehumidification and regeneration process in the multilayer desiccant bed column system. Humidity ratio of the exit air reduced by about 12% (compared to humidity ratio of the inlet air) when the airflow rate was 72 m^3/h.

References

1. S. Misha, S. Mat, M.H. Rustan, K. Sopian, Review of solid/liquid desiccant in the drying applications and its regeneration methods, RENEW SUST ENERG REV **16**, 4686 (2001).
2. S.K. Yeboah, J. Darkwa, A critical review of thermal enhancement of packed beds for water vapour adsorption RENEW SUST ENERG REV **58**, 1500 (2016).
3. K.S. Chang, H.C. Wang, T.W. Chung, Effect of regeneration condition on the adsorption dehumidification process in packed silica gel beds, Appl. Therm. Eng **24**, 735 (2014).
4. M.M. Awad, A. Ramzy K, A.M. Hamed, H.M. Bekheit, Theoretical and experimental investigation on the radial flow desiccant dehumidification bed, Appl. Therm. Eng **28**, 75 (2018).
5. K. Nagaya, Y. Li, Z. Jin, M. Fukumuro, Y. Ando. A. Akaishi, Low-temperature desiccant-based food drying system with airflow and temperature control, J Food Eng **75**, 71 (2006).
6. G. O. Ondier, T. J. Siebenmorgen, A. Mauromoustakos, Low-temperature, low-relative humidity drying rough rice, J Food Eng **100**, 545 (2010).
7. H. Abou-Ziyan, D. Abd El-Raheim, O. Mahmoud, M. Fatouh, Performance characteristics of thin-multilayer activated alumina bed, APPL ENERG **190**, 29 (2017).
8. A.K. Ramzy, R. Kadoli, A. Babu T.P., Experimental and theoretical investigation on the cyclic operation of TSA cycle for air dehumidification using packed beds of silica gel, Energy **56**, 8 (2013).
9. A.E. Kabeel, Adsorption operations of multilayer desiccant packed bed for dehumidification applications, RENEW ENERG **34**, 255 (2009).
10. A.K. Ramzy, A.M. Hamed, M.M.Awad, M.M. Bekheit, Theoretical investigation on the cyclic operation of radial flow desiccant bed dehumidifier, JETR **2**, 6 (2010).

Effect of drying air condition and feed composition on the properties of orange juice spray dried powder

Kankanit Khwanpruk*, *Chanida* Akkaraphenphan, *Patipan* Wattananukit, *Worakarn* Kaewket, and *Suttiphat* Chusai

Food Engineering Department, Faculty of Engineering, King Mongkut's Institute of Technology Ladkrabang, Bangkok, Thailand

Abstract. This research aims to investigate the effect of inlet air humidity, drying temperature and feed composition on the properties of orange juice spray dried powder. Maltodextrin (DE10-12) was used as carrier material. Full factorial design was applied to this research. Inlet air with humidity of 10 g/kg dry air and 20 g/kg dry air, inlet drying air temperature of 140°C 160°C and 180°C and feed concentration of 40°Brix which varied the weight ratio of orange juice solid content to maltodextrin of 1:3 and 1:4.5 were examined. All spray drying conditions were performed in replicate. Ascorbic acid, moisture content water activity, pH and product yield were analysed. Analysis of variance revealed that humidity of inlet air influenced significantly on product yield, moisture content, water activity. Besides, the quality of product in term of ascorbic acid content was affected significantly by drying temperature and fraction of maltodextrin. Higher drying temperatures lead to lesser of ascorbic acid content and the more fraction of maltodextrin used in feed, the more ascorbic acid content retained in product.

1 Introduction

In food industry, drying process is widely used for preservation of foods and shelf life stability. Dried powder by spray drying method has the benefit of good reconstitution, low water activity and moisture content that good for transport and storage [1]. However, stickiness on the drying chamber wall is one of the problem occurred in spray drying step. Fruit juices such as orange juice contain highly hygroscopic ingredients such as fructose, glucose and organic acids which have the low glass transition temperature that lead to stickiness problem in spray drying chamber [2]. Feed solution can be successfully dried with improvement of spray drying condition. This research aims to investigate the effect of inlet air humidity, drying temperature and feed composition on the properties of orange juice spray dried powder. Full factorial design was set to find the optimal spray drying conditions for high quality powders and production yields.

2 Materials and Methods

2.1 Materials

In this study, Concentrated orange juice with a total solids mass concentrationof 65.0±0.2%, containing 3.0±0.2% Acid (as citric acid monohydrate), obtained from a local manufacturer, was used. maltodextrin with DE 10-12 (Maxway, Thailand)

2.2 Feed preparation

Orange juice concentrate of 65°Brix concentration was adjusted with maltodextrin (DE10-12) and water, in order to make the constant feed concentration of 40°Brix with the various mass ratio of Orange juice solid content to maltodextrin solution 1:3 and 1:4.5, respectively.

2.3 Spray drying and experiment design

2.3.1 Spray drying

Feed solution was dried by using a co-current spray dryer (JCM Minilab SDE-10, Thailand). The height of the spray dryer are 1.2 m. An atomiser is two fluid nozzle type. The atomiser air pressure was set at 0.1 MPa using a peristaltic pum for conveying the liquid feed to the chamber. Temperature of drying air was set by using electrical heater. Drying air flow rate was set constantly at 108 m^3/h. Humidity of inlet air was controlled by using a dehumidified unit which coupled with a conventional spray dryer as shown in Fig.1.

2.3.2 Experiment design

Experiments were planned applying a full factorial design.

* Corresponding author: kankanit.kh@kmitl.ac.th

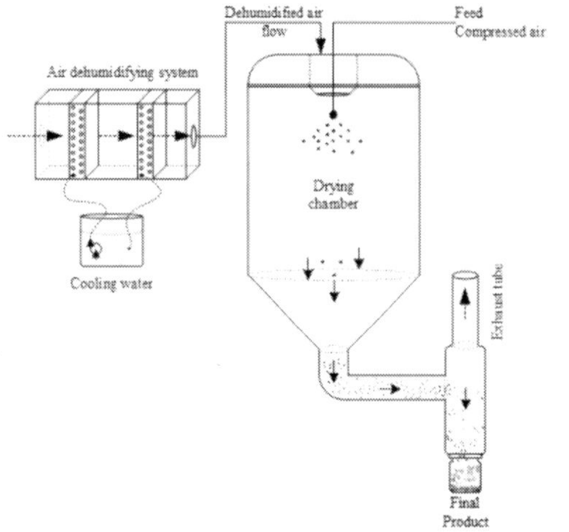

Fig. 1. Spray dryer with dehumidified system.

Inlet air with 2 levels of humidity of 10 g/kg dry air and 19 g/kg dry air, 3 levels of inlet drying air temperature of 140°C 160°C and 180°C and 2 levels of feed concentration of 40°Brix which varied the weight ratio of orange juice solid content to maltodextrin (O:M) of 1:3 and 1:4.5 were carried out.

2.4 Powder analysis

2.4.1 Production yield

The production yield was expressed as a percentage of the mass of orange juice spray dried powder gathered at the collection product bottle compare to the solid contents of orange juice solution and maltodextrin solution.

2.4.2 Moisture content and water activity (a_w)

The moisture content of orange juice powder was evaluated by vacuum oven method (MMM, VACUCELL model) [3]. Drying period was set at 70°C for 24 hours.

Water activity of the orange juice powder was determined using a water activity meter (an AquaLab 3TE Decagon, USA). The temperature was kept at 25.0±0.1°C during testing.

2.4.3 pH measurement

The pH values of orange juice solutions were analysed by using a pH metre (Consort C830, Belgium).

2.4.4 Ascorbic acid content

Ascorbic acid content was determined using the titration method modified from Kingwatee *et al.* (2015) [4]. In this context, spray dried orange powder was reconstitued to concentration of 40° Brix and then 10 ml aliquot was put into a volumetric flask and 90 ml of 0.4% (w/v) oxalic acid were added. Solution of 0.025%(w/v) 2,6-dichlorophenolindo phenol was used as a titrant. The end point of titration was a stable development of pink colour. L-ascorbic acid was used for construction a standard curve.

2.4.5 Particle morphology

Particle morphology was evaluated by scanning electron microscopy (SEM) ZEISS model EVO MA 10 at Instrument center of Advanced Manufacturing Innovation College, KMITL.

2.4.6 Statistical analysis

The experimental data were evaluated by using ANOVA (Minitab 16 software). The spray drying runs were carried out in duplicate and powder properties analysis were tested in triplicate. All data were reported as mean ± standard deviation.

3 Results and Discussions

Spray drying experiments were run by setting 2 levels of inlet air humidity, 2 levels of feed concentration with the different ration of orange solid content to maltodextrin at the constant feed concentrations of 40 °Brix and a range of inlet drying air temperature (140, 160 and 180°C). This experiments were used a constant
feed concentration of 40°Brix to control the amount of water evaporation. Humidity of drying air were set at 10 g/kg dry air and 20 g/kg dry air controlled by using the dehumidified system as shown in Fig. 1. The average humidity of drying air in this experiment was 10.42±0.24 g/kg dry air when using the modified spray dryer and was 19.07±1.55 g/kg dry air when using a conventional spray dryer. The effects of orange juice solid content ratio and different humidity and drying air temperature on the production yield, moisture content, water activity, pH and ascorbic acid content of spray dried product and ANOVA tests are shown in Table 1.

Production yield

The production yield from a range of humidity, mass content ratios and drying temperatures studied was shown in Fig. 2. The resulting yields from these experiments were about 24- 64%. In comparison with the yield from a small-scale spray dryer, this should be more than 60% to be acceptable [5]. Only products from using low humidity drying air showed an acceptable production yield. Fig.2 shows an increasing yield from spray drying as the humidity of drying air decrease and illustrates an increase of the yield as the drying temperatures increase. ANOVA tests suggested that both the humidity and drying temperature affected the yield from spray drying significantly (Table 1).

Table 1. Properties of orange juice powder under various conditions

Humidity (g/kg dry air)	Conc. (O : M)	Inlet temp. (°C)	Production yield (%)	Moisture content (%wb)	Water activity	pH*	Ascorbic acid (mg/g powder)
10	1:3	140	37.79±0.91^{cde}	1.59±0.21^{abc}	0.37±0.01^{ab}	3.05±0.02	2.94±0.04^{a}
		160	62.28±4.88^{ab}	0.81±0.40^{c}	0.33±0.00^{bc}	3.07±0.03	2.73±0.08^{ab}
		180	50.32±3.63^{bc}	0.55±0.20^{c}	0.31±0.01^{bc}	3.08±0.01	1.47±0.04^{h}
	1:4.5	140	46.63±3.34^{c}	2.83±0.04^{a}	0.40±0.00^{a}	3.05±0.00	2.50±0.19^{bcd}
		160	47.19±4.47^{c}	1.78±0.97^{abc}	0.32±0.01^{bc}	3.06±0.00	2.39±0.00^{bcde}
		180	63.98±1.83^{a}	0.98±0.19^{bc}	0.28±0.01^{c}	3.07±0.01	1.96±0.11^{efg}
20	1:3	140	24.95±2.41^{e}	1.89±0.28^{abc}	0.39±0.02^{a}	3.06±0.01	2.63±0.04^{abc}
		160	45.15±0.43^{c}	1.43±0.50^{abc}	0.33±0.02^{bc}	3.07±0.00	2.15±0.04^{defg}
		180	43.59±1.28^{c}	0.85±0.52^{c}	0.33±0.00^{bc}	3.10±0.00	1.82±0.08^{gh}
	1:4.5	140	25.17±3.36^{de}	2.93±0.29^{a}	0.40±0.00^{a}	3.06±0.03	2.39±0.11^{bcde}
		160	38.61±3.63^{cd}	2.62±0.56^{ab}	0.39±0.01^{a}	3.08±0.02	2.23±0.11^{cdef}
		180	49.61±5.84^{bc}	1.46±0.36^{abc}	0.28±0.04^{c}	3.09±0.01	1.99±0.19^{efg}

* no significant difference ($p>0.05$). The different of superscript in the same column means significant difference ($p<0.05$)

As the humidity of drying air decreased the yield was increased. This was caused by the drier air promoted a drying rate which improved the drying process by decreasing the stickiness problem [6]. As the drying temperature increased the yield was increased. This was caused by the lower moisture content of the powder at higher temperature, which reduced the high deposition of wet particles on the spray dryer walls [5]. However, for the sugar-rich feed spray drying, Goula and Adamopoulos (2010) reported that the production yield is largely affected by the inlet air temperature and the wall depositions increase by increasing the inlet air temperature [6].

Moisture content and water activity of orange juice powder

Spray dried Orange juice powders obtained from the experiments had moisture content in the range of 0.5-2.9% (wb) as shown in Table 1. Fig.3 shows a decrease of moisture content as the drying temperature was increased. High air inlet drying temperature increased the greater water evaporation, due to the higher rate of heat transfer to particles. ANOVA tests suggested that humidity and temperature of drying air temperature affect the moisture content and water activity of product significantly (Table 1). All of moisture content values obtained through this study were under the common observed value in industrial spray drying, which is lower than 5% [7]. Water activity is one of the most importance quality aspects for product storage and stabilityt. The average water activity values of between 0.28 and 0.40, which considered as microbiology safe and oxidative stable [8].

pH and Ascorbic acid content of orange juice powder

The values of pH of the orange juice powders in this study were significantly affected by humidity and drying air temperature included the fraction of orange juice solid content to maltodextrin content ($p>0.05$). The pH values of the orange juice feed solution were 2.79 ± 0.01for the O:M of 1:3 and the values of 2.81 ± 0.01 for the O:M of 1:4.5.

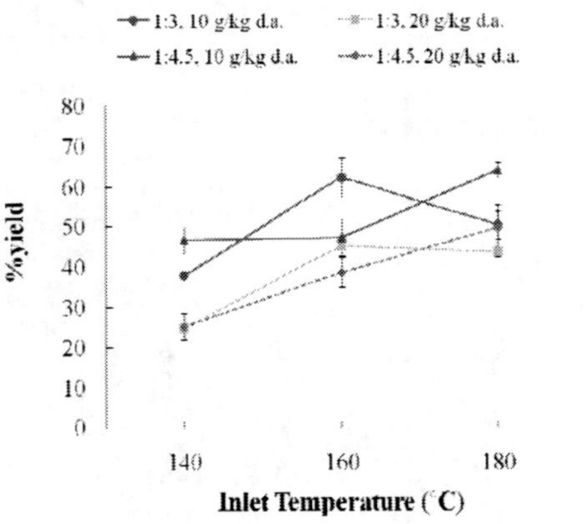

Fig. 2. Production yield from various spray drying conditions

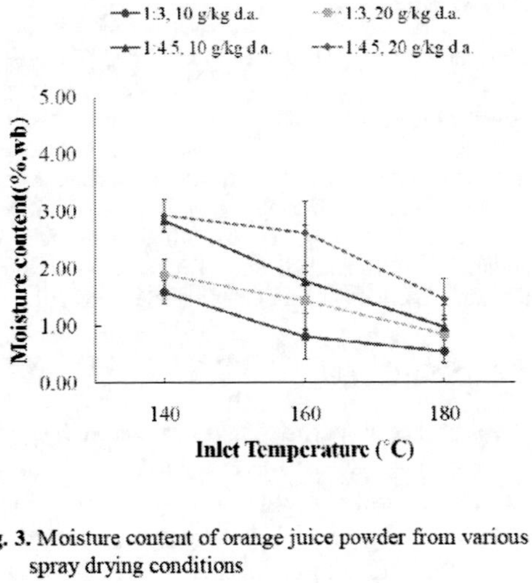

Fig. 3. Moisture content of orange juice powder from various spray drying conditions

The pH values of the spray dried powder (3.05-3.10) were slightly higher than pH values of feed which means that the acidity of orange juice product was decreased by the spray drying process. As well as the ascorbic acid content was affected significantly by drying temperature and the content of maltodextrin as drying aid. Ascorbic acid content in feed solution was 4.85 mg/ g feed solid for the O:M of 1:3 and was 3.39 mg /g feed solid for the O:M of 1:4.5. The contents of ascorbic acid in products are shown in Table 1. The results show that ascorbic acid contents are decreased as the temperature increased because of a degradation of ascorbic acid by temperature during spray drying process [9]. The comparison between ascorbic acid content before and after the drying process can be shown as percentage of retention displayed in Fig.4. The results show that maltodextrin plays the role of ascorbic acid protection as the more maltodextrin contained in feed, the more ascorbic acid content retained in products.

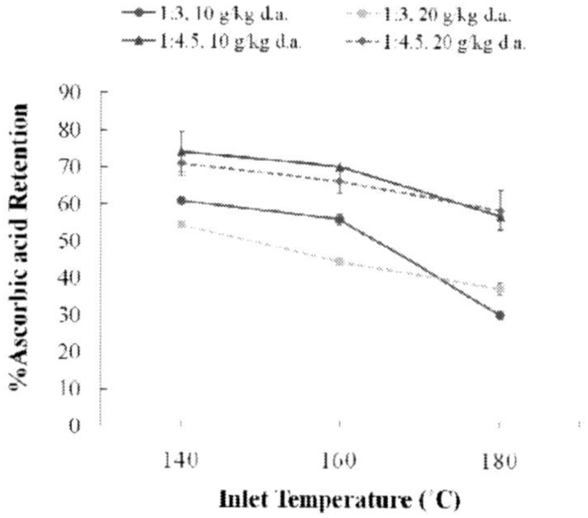

Fig. 4. Ascorbic acid retention of orange juice powder from various spray drying conditions

Powder morphology

The morphology of some spray dried orange juice powder was observed. Fig. 5 exhibits SEM image of powder illustrated various sizes of irregularly spherical particles with some smooth surface, some shrinkage and indentation which are one of the typical characteristics of particle formation during spray drying. Walton and Mumford (1999) reported that a hollow particle which collapsed and shrivelled, was formed during drying with the particle inflation from bubble nucleation [10].

4 Conclusions

The influence of spray drying conditions on the qualities of orange juice powder was examined. Production yield, moisture content and water activity were significantly affected by humidity and inlet drying temperature in this study. The using of low humidity drying air shows the greater production yield. In addition, the ascorbic acid content was significantly affected by maltodextrin concentration and drying temperature. Higher concentration of maltodextrin promotes the ascorbic acid retained in orange juice particles. However, higher drying temperatures lead to lesser of ascorbic acid content.

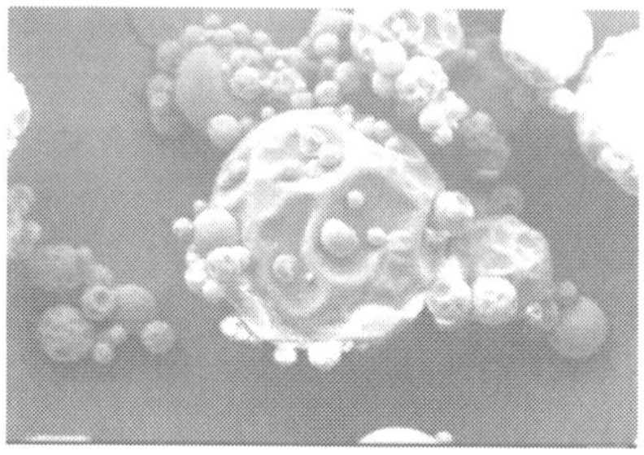

Fig. 5. SEM image (1000x) of spray dried orange juice powder at drying temperature of 160°C, humidity of 20 g/kg dry air and the ratio O:M of 1:4.5

References

1. K. Masters, *Spray drying handbook*, (Longman Scientific & Technical, New York, 1991)
2. A.K. Shrestha, T. Ua-arak, B.P. Adhikari, T. Howes and B.R. Bhandari, Int J Food Prop., 10, 661–673 (2007)
3. AOAC Method 979.12, *Official methods of analysis of AOAC International*, (AOAC Int., Arlington, 1995)
4. N. Kingwatee, A. Apichartsrangkoon, P. Chaikham, S. Worametrachanon, J. Techarung, T. Pankasemsuk, LWT-Food Sci Technol, **62**, 847-853 (2015)
5. H. Amagase, and N.R. Farnsworth, Food Res Int, **44**, 1702-1717 (2011)
6. A. M. Goula and K.G. Adamopoulos, Innov Food Sci Emerg Technol, **11**, 342-351 (2010)
7. H.A.C. Thijssen, J. Appl. Chem. Biotechnol., **21**, 372-377 (1971)
8. R.P.Sigh and B.A. Anderson, *The major types of food spoilage an overview. In: R. Steele (Ed.) Understanding and Measuring the Shelf-life of Food*, (CRC Press, Boca Raton, 2000)
9. J.R. Gregory III, *Vitamins. In: O.R. Fennema (Ed.) Food Chemistry*, (Marcel Dekker, Inc, New York, 1996)
10. D.E. Walton and C.J. Mumford, Transactions of IchemE, 77(A), 442-460 (1999)

Tensile properties of PLA/PBAT blends and PLA fibre-reinforced PBAT composite

Eakasit Sritham[1,*], *Phakaimat* Phunsombat[1] and *Jedsada* Chaishome[1]

[1]Food Engineering Department, Faculty of Engineering, King Mongkut's Institute of Technology Ladkrabang, Bangkok, Thailand

Abstract. The tensile properties of PLA/PBAT blends, PLA fibre reinforced PBAT composite (PLAF) at room temperature and -18°C were investigated. The concentrations of PLA in the blends were 10%, 20%, 30% and 40% (by volume). There was an improvement of elastic modulus (*E*) for PLA/PBAT blends when PLA was 40%. There was no significant difference of ultimate tensile strength (*UTS*) among the blends. For the same concentration of PLA (40%) in PLA-PBAT mixture, PLAF exhibited higher values of *E* and *UTS* than that of PLA/PBAT blends. Elongation of PLA/PBAT blends rapidly decreased upon the addition of PLA to the blends. The values of *E* and *UTS* for PLA/PBAT blends and composite, neat PLA, and PP increased with the decreasing of temperature from room temperature to -18°C. The effect of decreasing temperature was not observed on elongation. It was appeared from the results obtained for FTIR and DSC measurements that PLA and PBAT were immiscible, separating into two phases.

1 Introduction

Non-biodegradable plastics have posed an environment concern worldwide. The short time end-of-life products drastically affect global weather and landfill area hazards. The degree of concern has been raised along with the development in urbanization. [1-8] Accordingly, much of attention from a number of researchers has been devoted to replace synthetic plastics with bioplastics, such as poly(lactic acid) or (PLA), poly(3-hydroxy butyrate) or (PHB), and poly(butyl succinate) or (PBS) [1-5]. These bioplastics are biodegradable in nature.

Among bioplastics, PLA offers the highest tensile strength and being available at considerably low cost. PLA appears to be a high potential candidate to replace polypropylene (PP)—a thermoplastic polymer that has been commonly used in a broad range of applications including food packaging. However, PLA is typically brittle which limits its applications, particularly, at low temperatures. As to provide an alternative to synthetic plastics for eco-friendly food packaging industry, the impact resistance of PLA need to be enhanced [1, 3].

Poly(butylene adipate-co-terephthalate) or (PBAT) is biodegradable random copolymer which is flexible. By incorporating with PLA, the PBAT is expected to help improve the flexibility of the polymer blend/composite.

Though preparing PAL(fiber)/PBAT composite could be challenging, some mechanical properties might be positively improved making it suitable for using as food packaging material.

This study was primarily aimed to investigate the tensile properties of PLA/PBAT blends and PLA fiber-reinforced PBAT composite at room temperature and frozen temperature.

2 Materials and methods

2.1. Materials

PBAT (Ecoflex F Blend 1200) was purchased from BASF Company. PLA (grade 2003D) was purchased from Nature Work Company. PP (grade 1102k) was purchased from Global Connections Company. The polymers, as received, were dried in a hot air oven at 80°C for 6 hr before further preparation of experimental treatments.

2.2 Samples preparation

The neat polymers were manufactured by compression moulding for 5 min at 1 MPa and temperatures of 165°C, 130°C, and 200°C for PLA, PBAT and PP, respectively.

The formulations of PLA/PBAT blends were given in Table 1. The blends were mixed using an extruder at 165°C and 20 rpm, and were subjected to compression moulding at 165°C and 1 MPa for 5 min.

The treatment PLAF was prepared so that PLA remained in a form of fiber. PLA was first melted using an extruder at 170°C and 50 rpm. Then PLA fibers were pulled from the extruder to a spinner rotating at a speed of 150 rpm. Finally PLA fiber reinforced PBAT composite was manufactured by compression moulding at 130°C and 1 MPa for 5 min.

* Corresponding author: eakasit.sr@kmitl.ac.th

Table 1. Treatment formulations and code name of polymer blend/composite samples.

Treatments	PLA:PBAT ratio (by volume)	Code name
1	100:0	PLA
2	40:60	PLA40
3	30:70	PLA30
4	20:80	PLA20
5	10:90	PLA10
6	0:100	PBAT
7	40 (fiber):60	PLAF

2.3 Tensile property measurements

The measurements for tensile properties at room temperature were performed according to ISO 527:1996 using a computerized tensile testing machine model QC-536M1 (COMETECH). A total of five specimens for each treatment were tested at a rate of 200 mm/min. For the measurements at low temperature, sample specimens were infused in liquid nitrogen for 10 min before testing. The elastic modulus (E), elongation at break (E_b) and tensile strength (*UTS)* were obtained from a stress-strain curve.

2.4 Fourier transform infrared spectroscopy (FTIR)

FTIR measurements were carried out at room temperature using the IRPrestige-21 (Shimadzu) spectrometer in attenuated total reflectance (ATR) mode. Experimental control and spectral analyses were made with a Windows based IRsolution software bundled with the spectrometer. Samples were scan in a range of 4000 to 750 cm^{-1}, 20 scans per measurement, with a resolution of 8 cm^{-1}.

2.5 Differential scanning calorimetry (DSC)

Thermal property of samples was studied using the differential scanning calorimeter model DSC 204 *F1 Phoenix®* (NETZSCH, Germany). A 50-mg sample was weighed and tested under nitrogen atmosphere. Samples were subjected to a temperature scan where it was heated from -50°C to 200°C and subsequently cooled down to -50°C, both at the same rates of 10°C/min.

2.6 Statistical analysis

The experiments were laid out in a completely randomized design (CRD). Three replicates were made. The analysis of variance (ANOVA) was performed on experimental data at a significance level of 0.5. Sample means were compared using the Tukey's test when significant difference ($p<0.05$) was found.

3 Results and discussion

3.1 Tensile properties of polymer blends and polymer composite

Other than neat PBAT and neat PLA, for comparison purpose, PP sample was also tested along with the samples of PLA/PBAT blends and PLA(fiber)/PBAT composite. The results for elastic modulus (E), ultimate tensile strength (UTS), and elongation at break (E_b) are shown in Figs. 1-3, respectively.

3.1.1 Elastic modulus

From Fig.1, it could be clearly observed that the neat PLA exhibited much higher elastic modulus (around 567-974 MPa) than that of the neat PBAT (around 40-50 MPa), and even higher than that of PP (around 288-640 MPa). The modulus of neat PBAT, PLA10, PLA20 and PLA30 were not significant different. The improvement of modulus could only be seen when PLA concentration in the blend was increased to 40% (treatment PLA40). The results indicated poor interfacial adhesion between PBAT and PLA that led to the phase separation of PLA/PBAT blend. Ibrahim et al. reported that addition of more than 70% of PBAT deteriorated the properties of the PVC/PBAT blends [9].

It could be also seen that PLAF showed a higher modulus (around 213-501 MPa) as compared to that of PLA40. The finding suggests that for the same concentration of PLA in PLA-PBAT mixture, said 40%, preparing PLA in the form of fiber yields the higher elastic modulus than melting and blending with PBAT–treatment PLA40. The modulus of PLAF was close to that of PP, but statistically significantly lowers.

Statistical analysis indicated that temperature significantly affects elastic modulus of the samples. Lowering the temperature from room temperature down to -18°C caused the increase of elastic. These could be clearly seen in treatments PLA, PLAF and PP.

3.1.2 Ultimate tensile strength

It could be seen from Fig. 2. that the obtained values of ultimate tensile (UTS) strength of samples were of similar trend with that of the elastic modulus. PLA exhibited the highest value of ultimate tensile strength (UTS), around 52-67 MPa. The addition of PLA, up to 40%, to PLA/PBAT blends did not affect the UTS. The UTS values of these treatments were in a range of approximately 9 to 14 MPa. It was also found that, for the same concentration of PLA in PLA-PBAT mixture, at 40%, preparing PLA in the form of fiber yields the higher UTS value than melting and blending with PBAT–treatment PLA40. The UTS value of PLAF was close to that of PP, but statistically significantly lowers.

On the whole, lowering the temperature from room temperature down to -18°C significantly resulted in the increase of UTS. Like the case of elastic modulus, the finding could be clearly seen in treatments PLA, PLAF and PP.

3.1.3 Elongation at break

PBAT showed a distinctively high value of elongation at break (E_b), around 477-458%, as compared to that of PLA (around 8-12%) and PP (around 7-13%), indicating its highly flexible nature. The addition of PLA to PLA/PBAT blends gradually degraded the value of E_b, the flexibility. With the concentration of PLA in the blends from 30% upward, E_b dropped to the lowest values for PLA/PBAT mixture in this experiment, approximately 13% (Fig. 3).

In contrast to elastic modulus and the ultimate tensile strength, the decrease of temperature from room temperature to -18°C did not significantly affect the elongation at break.

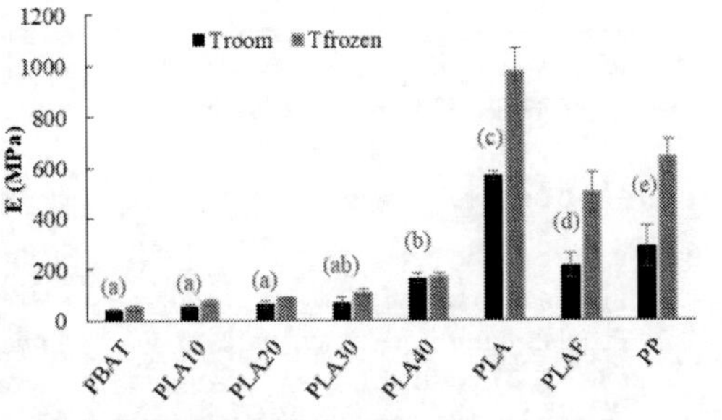

Fig. 1. Elastic modulus (E) of neat PLA, neat PBAT, neat PP, PLA/PBAT blends, and PLA fiber/PBAT composite.

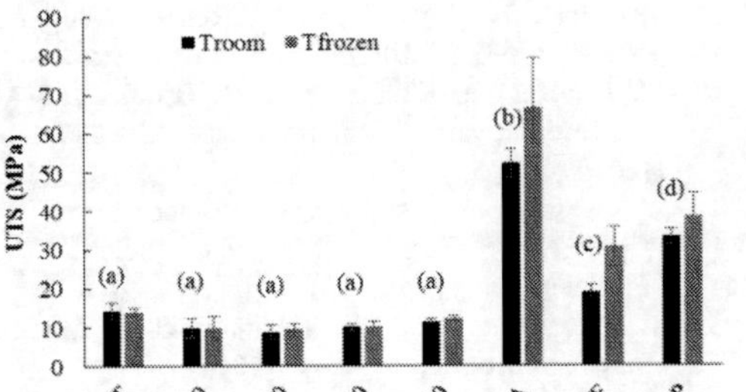

Fig. 2. Ultimate tensile strength (UTS) of of neat PLA, neat PBAT, neat PP, PLA/PBAT blends, and PLA fiber/PBAT composite.

3.2 FTIR absorption characteristics

FTIR absorption peaks for the neat PLA, PBAT and PLA/PBAT blends are given in Table 2.

For PLA, the peak at around 752 cm^{-1} associated with the rocking vibration of α-methyl; peak at around 864 cm^{-1} associated with the ester (O-CH-CH_3); the peak at around 1042 cm^{-1}, 1080 cm^{-1} and 1180 cm^{-1} associated

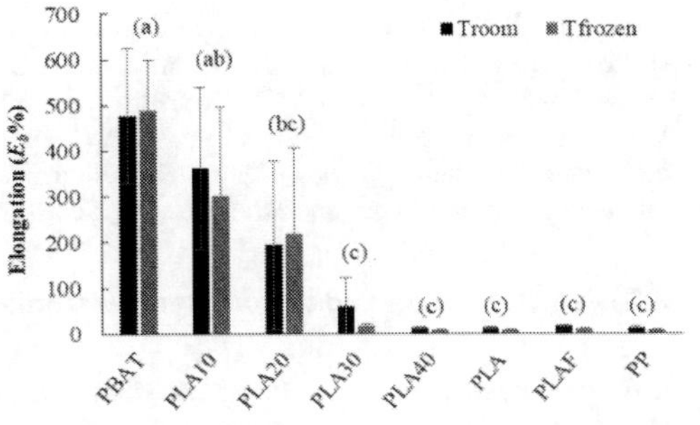

Fig. 3. Elongation at break (E_b) of of neat PLA, neat PBAT, neat PP, PLA/PBAT blends, and PLA fiber/PBAT composite.

with the stretching vibration of C-O-C; the peak at 1381 cm^{-1} associated with the CH symmetric bending vibration; the peak at around 1450 cm^{-1} associated with the CH_3 antisymmetric; the peak at 1748 cm^{-1} associated with the carbonyl C=O stretching vibration; and the symmetric and antisymmetric stretching vibration of CH_3 of saturated hydrocarbons were found at 2943 cm^{-1} and 2997 cm^{-1}, respectively [10, 11].

For PBAT, the peak at 725 cm^{-1} associated with the bending vibration of CH-plane of benzene ring; the symmetric stretching vibration of trans-C-O was found at 937 cm^{-1}; the peak at 1018 cm^{-1} associated with the bending vibration at the surface of adjacent hydrogen atoms on the phenyl ring; the peak at 1103 cm^{-1} associated with the left-right symmetric stretching vibration of C-O; the peak at 1265 cm^{-1} associated with the C-O symmetric stretching vibration; the peak at 1408 cm^{-1} associated with the trans-CH_2-plane bending vibration; the peak at 1504 cm^{-1} associated with the skeleton vibration of the benzene ring; the peak at 1713 cm^{-1} associated with the C-O stretching vibration; and the peak at 2959 cm^{-1} associated with the CH_2 asymmetric stretching vibration [10, 11].

Table 2. FTIR absorption peaks of neat PLA, neat PBAT and PLA/PBAT blends.

Wavenumber (cm^{-1})					
PBAT	PLA	PLA10	PLA20	PLA30	PLA40
725	-	729	729	729	729
-	752	-	-	-	-
-	864	872	872	872	872
937	-	937	937	937	937
1018	-	1018	1018	1018	1018
-	1042	-	-	-	-
-	1080	-	-	-	-
1103	-	1103	1103	1103	1013
-	1180	-	-	-	1180
1265	-	1265	1265	1265	1269
-	1381	1389	1389	1389	1389
1408	-	1408	1408	1408	1408
-	1450	1458	1458	1454	1454
1504	-	1504	1504	1504	1504
1713	-	1713	1713	1713	1713
-	1748	-	-	-	-
-	2943	-	-	-	-
2959	-	2959	2959	2955	2955
-	2997	-	-	-	-

Absorption spectral of PLA/PBAT blends showed the up-shift of CH-plane of the benzene ring vibration from 725 to 729 cm^{-1}. Ester vibration peak in PLA shifted from 864 to 872 cm^{-1}. Data obtained for PLA40 showed the up-shift of C-O symmetric stretching vibration from 1265 to 1269 cm^{-1}, also reported elsewhere [7]. There was however no clear evidence of interaction between PLA and PBAT in the blends.

3.3 Glass transition and melting temperatures

DSC thermogram of PLA/PBAT blends generally showed two T_g's and two T_m's corresponding to those for PLA and PBAT. The Fig.4 presents the typical thermogram of PLA40 showing the T_g values at -33.7°C and 56°C, for PBAT and PLA phases, respectively; and the T_m values at 129.9°C and 147.9°C, respectively for PBAT and PLA phases. Given in Table 3 are additional data for T_g and T_m values obtained for PLA/PBAT blends. The results suggested that PLA/PBAT blends were immiscible [2, 4].

Table 3. The thermal properties of neat PLA, neat PBAT and PLA/PBAT blends.

Polymers	T_{g1} (°C)	T_{g2} (°C)	T_{m1} (°C)	T_{m2} (°C)
PBAT	-33.8	-	122.9	-
PLA10	-36.2	55.3	126.3	148.6
PLA20	-34.9	55.5	128.0	148.1
PLA30	-36.3	55.6	128.3	147.5
PLA40	-34.0	55.7	128.6	147.6
PLA	-	56.1	-	152.9

However, T_m of PBAT phase in the blends slightly increased with increasing PLA as T_g of PLA phase in the blends were slightly decreased by the increase of PBAT according to the results obtained by Zhao et al. [5]. T_g and T_m of PLA were slightly changed by the addition of PBAT. The results accorded to the shifting of the FTIR spectrum.

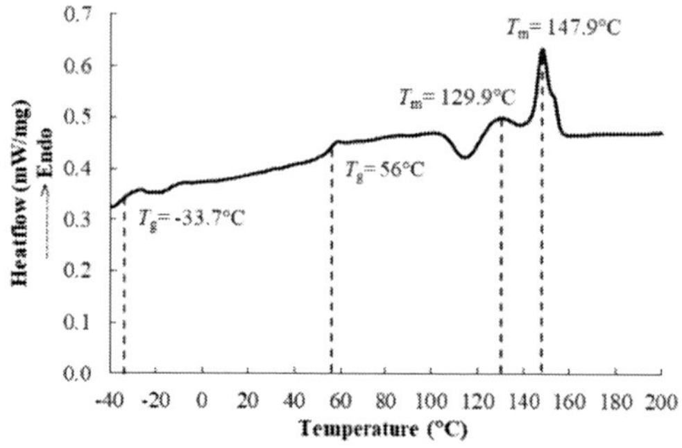

Fig. 4. Typical DSC thermogram of polymer blend sample: treatment PLA40

4 Conclusions

Elastic modulus of PLA/PBAT blends could be improved when the concentration of PLA was at least 40% (by volume). However, the addition of PLA to PLA/PBAT blends from 10% to 40% did not affect the ultimate tensile strength. For the same concentration of PLA (40%) in PLA-PBAT mixture, PLA(fiber)/PBAT composite exhibited the higher elastic modulus and ultimate tensile strength that that of PLA/PBAT blends. Elongation of PLA/PBAT blends rapidly decreased upon the addition of PLA to the blends. The decrease of temperature from room temperature to -18°C allowed elastic modulus and the ultimate tensile strength of PLA/PBAT blends and composite, neat PLA, and PP to increase significantly; but did not affect the elongation. Results from FTIR and DSC measurements suggested that PLA and PBAT were immiscible.

5 Acknowledgments

The authors would like to express thanks to the National Research Council of Thailand (NRCT) for its financial support–a graduate research scholarship 2017.

References

1. W. Jia, R.H. Gong, P.J. Hogg, Compos. Part B: Eng., **62**, 104-112 (2014).
2. E. Jalali Dil, P.J. Carreau, and B.D. Favis, Polymer **68**: 202-212 (2015).
3. W. Pivsa-Art, A. Chaiyasat, S. Pivsa-Art, H. Yamane, H. Ohara, Energy Procedia **34**, 549-554 (2013).
4. T. Yu, Y. Li, Compos. Part A: Appl. S., **58**, 24-29 (2014).
5. P. Zhao, W. Liu, Q. Wu, J. Ren, J. Nanomater. **2010**, 287082 (2010).
6. J. Chaishome, K.A. Brown, R. Brooks and M.J. Clifford, Advanced Materials Research., **894**, 32-36 (2014)
7. J. Chaishome and S. Rattanapaskorn. IOP Conference Series: Materials Science and Engineering., **191**, (2017).
8. J. Chaishome and S. Rattanapaskorn, Materials Science Forum., **894**, 46-49 (2017).
9. N.A. Ibrahim, N.M. Rahim, W.Z.W. Yunus, J. Sharif, J. Polym. Res. **18**(5), 891-896 (2011).
10. Y.X. Weng, Y.J. Jin, Q.Y. Meng, L. Wang, M. Zhang, Y.Z. Wang, Polym. Test. **32**(5), 918-926 (2013).
11. M. Xiuyu, W. Yufeng, W. Jianqing, X. Yaning, MATEC Web of Conferences, **88**, 02009 (2017).

MATEC Web of Conferences **192**, 03015 (2018) https://doi.org/10.1051/matecconf/201819203015
ICEAST 2018

Estimation of the organic film thickness on stainless steel using image intensity

Maethinee Songthai[1], *Taweepol* Suesut[2,*], and *Navaphattra* Nunak[1]

[1]Food Engineering Department, Faculty of Engineering, King Mongkut's Institute of Technology Ladkrabang, Bangkok, Thailand

[2]Instrumentation and Control Engineering Department, Faculty of Engineering, King Mongkut's Institute of Technology Ladkrabang, Bangkok, Thailand

Abstract. This research proposes an estimation of thickness of organic film coating on a Stainless Steel surface (SS) and a clear microscope slide using a visible optical method with the image intensity level analysis. Sweetened Condensed Milk (SCM) was used as a representative of the organic film. The uncertainty of inverse calibration equations was evaluated to determine the reliability of this technique. The results showed that the image intensity increased with the increasing of the film thickness in a linear relationship with a high coefficient of determination ($R^2 > 0.97$). Uncertainty of both inverse calibration equations was low and acceptable. It could be concluded that this technique could be used to estimate the thickness of the organic translucent film coating directly on a SS or a clear slide with the limitation of approximately 1.0 mm thickness using the reported linear inverse calibration equations.

1 Introduction

Cleaning of the food contact surface of processing equipment is the most important process in the food industry since it preserves the high quality of food products [1,2,3]. The cleaning method should be effective to ensure that all the food residues, dirt, lubricants and foulant on the food contact surface are eliminated [4]. The parameters of cleaning depend on soil types which can be classified into three types, i.e. viscoelastic film such as yogurt and toothpaste (type 1), microbial film (type 2) and solid/cohesive film such as pasteurized milk, sweetened condensed milk (type 3). [2,3] The most difficult soil type for cleaning is the group of cohesive film as it requires the contact time of water for removing the solid layers from the surface with the cohesion and adhesion forces and the chemical action for cleaning. The force that required to break soil from the internal surface of open equipment such as a large tank is a wall shear stress [5,6]. Each type of residues needs the different wall shear stress for soil removal causing to the different amount of water required in each cleaning step [5]. An understanding on soil behavior, especially soil film deposited on and removal from the surface during the process is a necessity of cleaning. There are several techniques to measure the film thickness; however, the optical method is the preferable technique which allows a real-time measurement of the organic film thickness [7]. In case of an optical UV-Vis method requires to maintain a constant temperature and pH of the liquid or cleaning detergent throughout the process [6], while the visible method can detect the fouling removal without any control of temperature and pH of chemicals [8].

To obtain the amount of soil residues on the surface can be evaluated from the thickness of soil film. Therefore, the thickness measurement should be reliable. As the calibration equation is a part of measurement, it should be taken into account. Several researchers have reported that the mean squared error of an inverse calibration equation is usually smaller than that of a classical calibration equation [9,10,11,12]. Besides, the inverse calibration equation is suitable for the small group of samples [13].

Therefore, the objectives of this research are to propose an optical method in the visible region for estimating the thickness of organic soil film on a Stainless Steel surface (SS) using the image intensity level and to evaluate the uncertainty of inverse calibration equation. The Sweetened Condensed Milk (SCM), a protein-sugar based food product, was chosen as the representative of the organic solid/cohesive film.

2 Materials and methods

2.1 Organic soil film preparation and experimental set up

SCM (Mali Co., Ltd, tube package) was coated as an organic film on a SS 316 having the average roughness value (R_a) of 0.4 µm and a clear microscope slide (Sail brand, 1.2 mm thickness) using the experimental unit (Fig. 1a). The film thickness was adjusted from 0.1 to 10 mm by adjusting the height of a wiper. The mass of SCM was weighted after coating on the surface for calculating the thickness of SCM film from its density.

* Corresponding author: taweepol.su@kmitl.ac.th

The experimental unit consisted of a CCD camera (Basler model acA2500-14gc, color 2590 x 1942 pixel) connected with the Gigabit Ethernet (Gig E), a tripod, two light sources (fluorescent lamp, LAMP-TAN, 50-60 Hz, 8 W), a control chamber with a black surface, and a computer set as shown in Fig 1b.

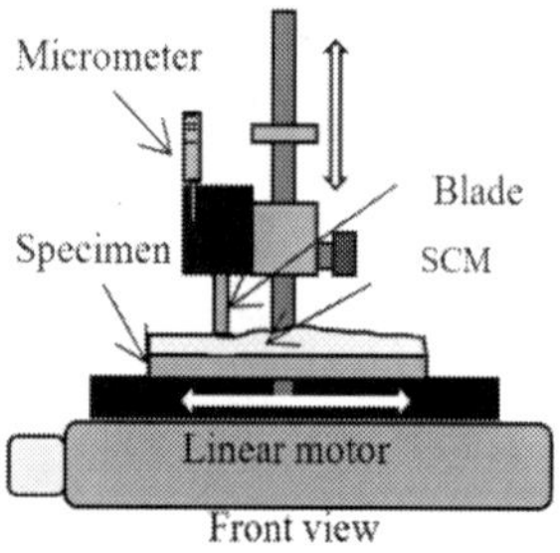

(a)

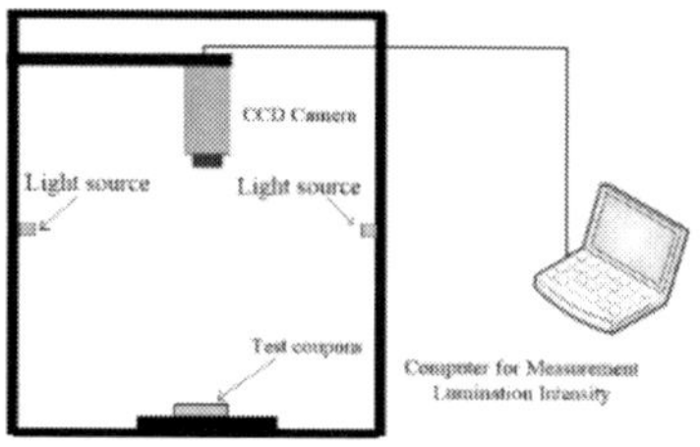

(b)

Fig. 1. Experimental unit

The inverse calibration equations were plotted between the intensity value obtained from an image analysis and the thickness of SCM film which directly coated on a SS and that coated on a clear slide putting on the SS. The examples of test specimens are presented in Fig. 2.

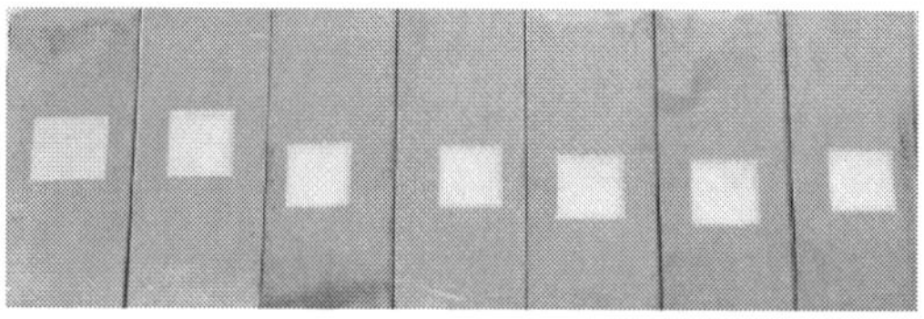

Fig. 2. The SCM film on SS with 0.2 to 0.8 mm of thickness

2.2 Determination of an intensity level of the SCM film

Intensity level of SCM film on the surface was determined by a computer vision system for real-time computing using the Vision Builder Software AI 2014 (National Instrument) by the following steps (Fig. 3):
Step 1: Images were acquired from the object via Gig E and CCD camera in real-time.
Step 2: Region of Interest (ROI) based processing was applied for segmentation of the SCM film area

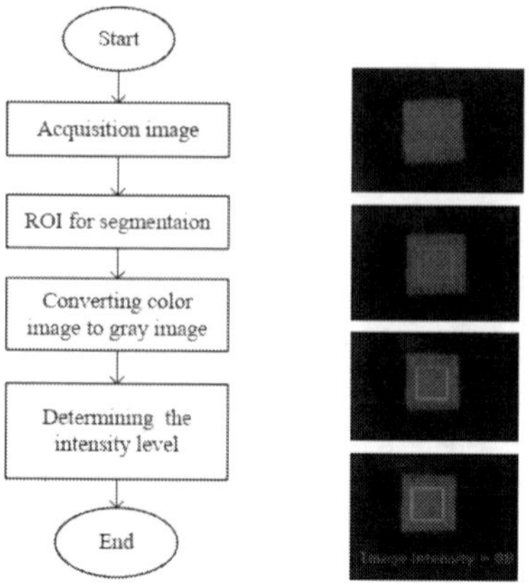

Fig. 3. Determination of image intensity level

Step 3: The color image was converted to grayscale by following steps; comparing the histogram of SCM film with different color model, finding the suitable color model, and extracting color image to gray image. It was found that the lightness from HSL model could identify the intensity level related to the film thickness better than RGB and HSV model.
Step 4: Intensity level was determined by using the measurement intensity, a function in Vision Builder Software.

2.3 The inverse calibration equation and uncertainty analysis

The estimation of SCM film thickness from an inverse calibration equation, presenting the relationship between reading values (image intensity) and thickness of SCM film, by the following steps: fitting the inversed calibration curve from the relationship between the image intensity level and the SCM film thickness, predicting the SCM film thickness, and calculating the uncertainty, respectively. The details of each step are described below:

To create an inverse calibration equation, x was fixed as a dependent variable (intensity level value) and y was a standard function or independent variable (film thickness). The calibration equation is

$$x = g(y) \qquad (1)$$

and the linear calibration equation is

$$x = b_0 + b_1 y_i \qquad (2)$$

where b_0 is an intercept and b_1 is a slope.

The residual standard deviation ($s(r)$) is calculated by

$$s(r) = \sqrt{\frac{\sum_{i=1}^{n}(y_i - \hat{y}_i)^2}{n-1}} \qquad (3)$$

where y_i is the observed value, $\hat{y}_i$ is the predicted value, and n is the number of calibration points. The uncertainty of predicted value from the inverse calibration ($u(x)$) is a type A uncertainty which calculates from the standard deviation of calibration equation ($s(x_c)$) in equation (4).

$$u(x) = s(x_c) = s\sqrt{\frac{1}{p} + \frac{1}{n} + \frac{(y_{pred}-\bar{y})^2}{\sum(y_i^2)-(\sum y_i)^2/n}} \quad (4)$$

where y_{pred} is a predicted value from equation (3), $\bar{y}$ is an average value of y, s is the standard deviation of calibration equation, p is the numbers of measurement for prediction, n is the number of measurement for calibration equation. The uncertainty of prediction values can be computed by equation (5).

$$u(y_{pred}) = \frac{s(x_c)}{b_1} \quad (5)$$

3 Results and discussion

Fig. 4 shows the relationship of intensity image of SCM film coated on a SS and a microscope slide at different thickness values. The image intensity increased with the increasing of the film thickness in a linear relationship. This could be explained by the principle of light which started from two lamps, light sources, generated the incident light on the film surface, and reflected to the camera. The light intensity which reflected from the SCM film to the camera depended on the thickness of film layer. In case of the thin film layer, it performed like a translucent surface which the absorption was much higher than the reflection and caused to the low image intensity level, and vice versa. However, the increasing of image intensity level was limited approximately at the thickness of 1 mm. It seems like at this thickness all incident light was fully reflected to a camera. Although the thickness increased, there was no difference of intensity values.

Linear equations were plotted as presented in equation (6) and (7) for SCM coated on a slide and that coated on a SS, respectively with a high coefficient of determination (R^2). They are called the inverse calibration equations. Slope of both calibration equations are in equal, 0.062.

$$y = 0.062x - 5.0377, R^2 = 0.98 \quad (6)$$

$$y = 0.062x - 3.8913, R^2 = 0.97 \quad (7)$$

where y is the SCM film thickness and x is an image intensity level. Besides, it was found that all image intensity analyzed from image of SCM coated on SS was lower than that from the one coated on a clear microscope slide for all measured thickness values since light could incident at the side of a clear microscope slide.

Fig. 5 (a) and (b) shows the residual plots of inverse calibration equations. It could be seen that the residuals of both equations randomly scattered around zero (calibration line). There was no relation between the spread of the residual and an average image intensity. This indicated that the calibration curve could be recognized as a linear model [10].

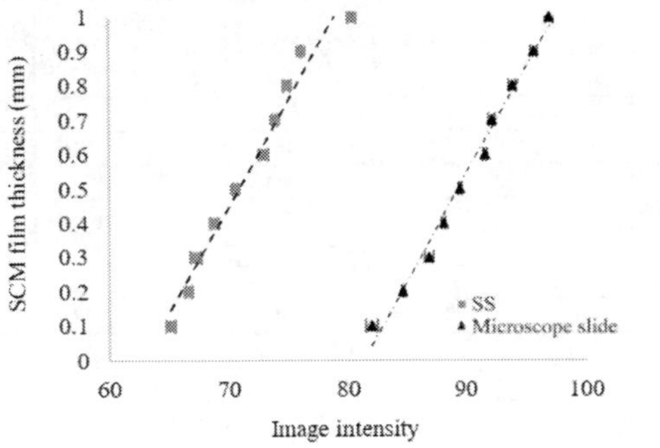

Fig. 4. Appling the inverse calibration to estimate the thickness of SCM film on microscope slides and SS by image intensity level

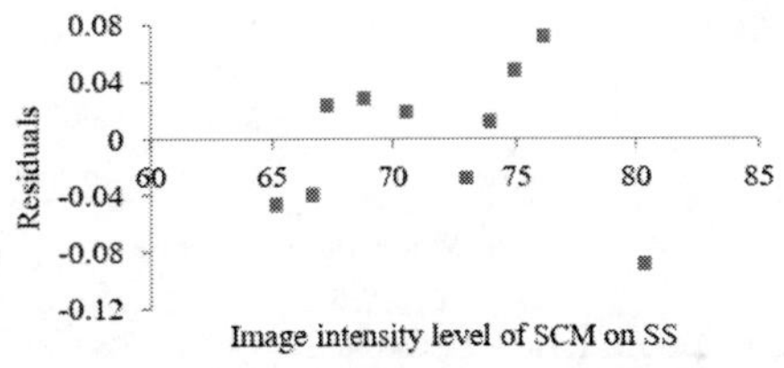

(a)

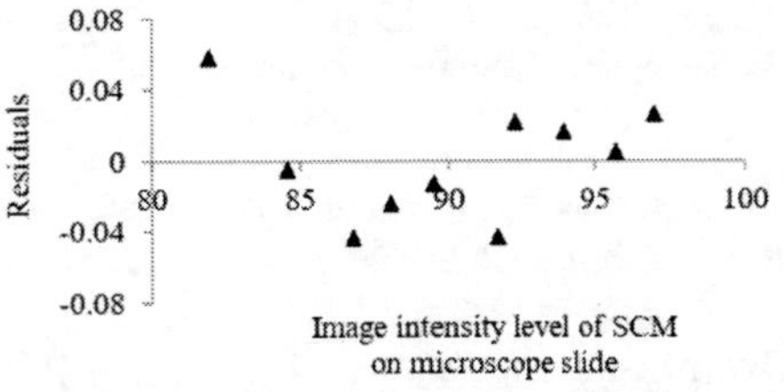

(b)

Fig. 5. Residual plots of inverse calibration equations

The thickness values of SCM film (0.1-1.0 mm) coated on a SS and a clear microscope slide were predicted using the inverse calibration equations as listed in table 1 and also their uncertainty values were evaluated as presented in Fig. 6. It could be observed that the thin layer of SCM coated on the surface (< 0.2 mm) had a high percentage of error. This might be affected from the human error in the procedure of SCM film preparation as the rheological characteristic of SCM is difficult to control in a constant thickness throughout the whole testing surface area. However, when considering uncertainty values, it was found that all evaluated values were acceptable as they were very low and less than the fixed resolutions (0.1 mm) as it was much enough to investigating and monitoring of the cleaning process.

Table 1. The prediction values, Standard Deviation (SD) and % Error of SCM film coated on a SS and a clear microscope slide at different thickness values

Measured thickness (mm)	SS			Microscope slide		
	Predicted thickness (mm)	SD	%Error	Predicted thickness (mm)	SD	%Error
0.1	0.11	0.03	9.77	0.10	0.03	1.27
0.2	0.23	0.02	13.78	0.23	0.01	13.67
0.3	0.28	0.03	6.92	0.27	0.01	8.44
0.4	0.37	0.02	6.43	0.41	0.03	1.78
0.5	0.48	0.03	4.89	0.54	0.01	5.71
0.6	0.59	0.02	1.79	0.62	0.02	3.33
0.7	0.69	0.01	1.36	0.70	0.02	0.09
0.8	0.76	0.01	5.16	0.78	0.02	3.13
0.9	0.96	0.00	3.11	0.86	0.02	4.24
1	1.09	0.02	4.14	1.00	0.02	0.39

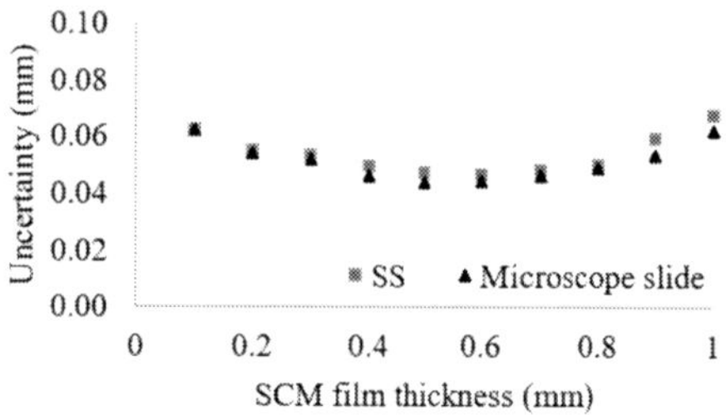

Fig. 6. The uncertainty of SCM film coated on a SS and a clear microscope slide at various thickness values

For these reasons, it could imply that the visible region could be applied to estimate the thickness of an organic film within the limitation of approximately 1.0 mm using the linear inverse calibration equation. The SCM film could be prepared by coating directly on a SS or coated on a clear slide putting on a SS.

4 Conclusions

An optical method in the visible region, an inexpensive and simple technique, proposed in this research with the inverse calibration equations obtained from performing the experiments by coating SCM film directly on a SS and on a clear slide was capable to estimate the thickness of organic film. Nevertheless, this technique was limited only for the translucent film which the absorption and reflection change according to changes of the thickness and the maximum thickness was approximately 1.0 mm. In addition, it was the technique having a great potential for improving to the real-time estimation. However, the further research is needed to improve the film preparation and investigate the other sources of uncertainty to calculate the combined uncertainty of this technique.

Reference

1. A.Y. Tamime, Cleaning-in-Place: Dairy, *Food and Beverage Operations Third Edition*. Blackwell Publishing, (2008)
2. K. R. Goode, K. Asteriadou, P. T. Robbins, P. J. Fryer, "Fouling and cleaning studies in the food and beverage industry classified by cleaning type," *Institute of food technologists*, **12**, 121-143, (2013)
3. P.J. Fryer, and K. Asteriadou, "A prototype cleaning map: A classification of industrial cleaning process." *Trends in food Science & Technology*, **20**, 255 – 262, (2009)
4. EHEDG. (2004). "Document No. 8: Hygienic Equipment Design Criteria (2nd Ed.)".
5. E. Fuchs, A. Boye, H. Stoye, M. Mauermann and J.-P. Majschak, "Influence of the film flow characteristic on the cleaning behavior." *International Conference on Heat Exchanger Fouling and Cleaning*. 411-420E. (2013)
6. E. Fuchs, R. Boye, J.-Murcek, P. Majschak, "An experimental comparison of film flow parameters and cleaning behavior of falling liquid film for different tilt angles," *Food and bioprocessing*, **93**, 318-326, (2015)
7. M. Mauermann, U. Eschenhagen, T. Wey-rauch, H. Köhler, Th. Bley, J. P. Majschak. "Monitoring the progress of cleaning using optical detection methods" *Fouling and Cleaning in Food Processing*, (2010)
8. N. I. Khalid, N. Nordin, N. Aziz, A. A. Ab. Nuraini, F.S. Taip, and A. Ahmedov, "Removal kinetic of pink guava puree fouling deposit from a stainless surface during alkaline cycle." *Proceedings of International on Heat Exchanger fouling and cleaning*, 396-404, (2015)
9. B. Klongratog, T. Suesut, and N. Nunak. "The uncertainty in sugar solution concentration measurement based on density approach" *Advanced Materials Research*, **811**, 358-364, (2013)
10. L. Tzehung, and C. Chaichung. "Uncertainty evaluation of humidity sensors calibrated by saturated salt solutions", *Measurement*, **40**, 591-599, (2007)
11. R. G. Krutchkoff, "Classical and inverse regression methods of calibration", *Techno-metrics*, **9 (3)**, 425-439, (1967)
12. V. Centner, D.L. Massart, S. de Jong, "Inverse calibration predicts better than classical calibration", *Fresenius' Journal of Analytical Chemistry*., **361**, 2-9, (1998)
13. J. Tellinghuisen, "Inverse vs. classical calibration for small data sets", *Fresenius J Anal Chem*. **368** :585–588, (2000)

Indoor production of loose-leaf lettuce (*Lactuca sativa L.)* using artificial lights and cooling system in tropical lowland

Mary Grace Semilla[1,*], *Ireneo* Agulto[2], *Armando* Espino Jr.[2], and *Emmanuel* Sicat[2]

[1]Graduate Student, and [2]Faculty, Department of Agricultural and Biosystems Engineering, College of Engineering, Central Luzon State University, Science City of Muñoz, Nueva Ecija, Philippines

Abstract. In the Philippines, crop production hindrances are climate change effects, typhoon occurrences, seasonal dependent cropping pattern and decreasing farm areas. There is a need to strategize farming technologies. Controlled environment agriculture (CEA) has potential in addressing these issues. A farming set-up in controlled environment was developed using locally available materials, light emitting diodes (LED) as sole-source of light and cooling system for temperature manipulation. This study was conducted to benchmark temperature and light intensity requirements for growing loose-leaf lettuce in lowland tropics. Light intensities of 50, 100 and 150 μmol m^{-2} s^{-1} were tested under temperature settings of 25 °C and 18 °C. Carbon dioxide and relative humidity were maintained at recommended levels. In actual chamber condition, average day/night temperatures inside were 25.3±0.4/25.5±0.2 °C and 18.9±0.6/18.7±0.3 °C, for the first and second cycles, respectively, whereas, day/night temperatures outside chambers were 29.6±2/25.9±0.5 and 26.2±1/23.6±0.6 during the first and second cycles, respectively. Under two temperature settings, best yield per unit area was observed at required light intensity of 150 μmol m^{-2} s^{-1}. No significant difference in productivity was observed under 25 °C and 18 °C. Also, no significant difference in productivity was observed between plants in two temperature settings and plants outside.

1 Introduction

Controlled environment agriculture (CEA), the cultivation of plants in enclosed environment in which growth factors are carefully controlled and programmed [1] has become a trend, especially in developed countries. CEA has been done both in completely indoor setting and greenhouse [2]. Commercial crop production and researches are conducted under environmentally controlled spaces [3] in which abiotic factors such as light, water carbon dioxide, temperature and management options like planting material and fertilizer application are manipulated [4].

In the Philippines, greenhouse has been adapted both in upland [5] and lowland [6]. However, initiatives are still done to encourage production in greenhouse than in open field [7]. The frequent occurrence of typhoon decreases production [8] and might as well threatens to destroy greenhouse structures. Moreover, production in the country is seasonal dependent. In a tropical lowland setting, greenhouse problems still arise due to temperature rise inside the structure [6, 9], thus the need to control environment favourable for crops. Indoor farming offers a lot of potentials in terms of food sustainability [10] and increased productivity, as crops can be planted all-year round. With this system, plants can be more resilient to typhoons and climate change effects such as flood or drought. This study was conducted to benchmark temperature and light intensity for loose-leaf lettuce as test crop.

2 Methodology

Three chambers were constructed using locally available materials with each having floor area of 2.5m x 6.0m and ceiling height of 2.5m. Each chamber was equipped with two 1hp window-type air-conditioning units (AC) operated alternately for 12 hours, two 70-watt circulating fans, and two sprinkler foggers (Figure 1). Plants were provided with 150, 100 and 50 μmol m^{-2} s^{-1} photosynthetic photon flux (PPF), for 14 hours. ECOLUM LED T8 tube lights, 6500K daylight were used such that 4, 3 and 2 tubes would provide PPF of 150 (L_3), 100 (L_2) and 50 (L_1), respectively. Also, planting bench with dimension of 1m x 5m x 0.8m was installed in each chamber. Planting pots (20cm x 20cm x 14cm) containing 1800g substrate mixture with 1:1:1 ratio by volume of garden compost, carbonized rice hull and vermicast, added with 2000 mL water was planted with two 15 days old fanfare (grand rapid or green ice) lettuce. Plants were grown in three rows along length of planting bench and harvested at 24 days after transplanting (Figure 2). Lettuce plants were also grown outside the chamber for comparison.

* Corresponding author: mgnsemilla@gmail.com

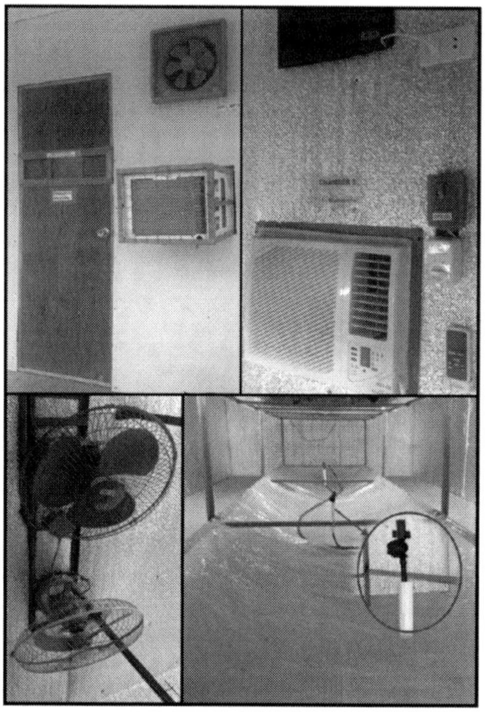

Figure 1. Components of the constructed CEA system

Figure 2. Plants grown in rows under LED T8 daylight

Temperatures of 25 °C (T_1, first cycle) and 18 °C (T_2, second cycle) were tested using the air-condition system. Carbon dioxide (CO_2), temperature and relative humidity (RH) were monitored using Desktop CO_2 Monitor 7788 Green Day Model by AZ Instruments Corp., Taiwan. PPF levels in each chamber were also monitored using Handheld Quantum Meter with integral sensor MQ-100 Model by Apogee Instruments Inc., USA.

2.1 Pre-test operation

Temperature and light distributions were evaluated prior to crop establishment. Without the plants inside the chambers, cooling system was operated to check the temperature distribution. Temperature was measured in 40 locations within the planting bench. On the other hand, light distribution was assessed by measuring the light intensity below the lighting fixture with the sensor at soil level approximately 13 cm from the LED tubes.

Three frequencies of monitoring environmental parameters such as temperature, CO_2 and RH were put to trial; hourly, every two hours and five measurements in a day or within 24 hours.

2.2 Monitoring of environmental parameters

2.2.1 Light intensity

Monitoring light intensities were done daily, one hour before lights on and one hour after lights off. Measurement was done by positioning the sensor at canopy level just below the lighting fixture [11-12]. As observed, the PPF value changed over time especially when the crop grows. Thus, to maintain PPF values of 50, 100 and 150, the height of the lighting fixtures were adjusted; a technique employed in other studies [13-15].

2.2.2 Temperature, CO_2 and RH

During crop establishment, temperature, CO_2 and RH were monitored daily at daytime (lights on) and night-time (lights off).

Hourly monitoring was necessary to monitor RH and to operate foggers, if its value reached 50%. During pre-test operation (hourly), no significant differences in temperature, CO_2 and RH were observed from 9:00 pm to 5:00 am. This was the night time condition. To acquire representative data for night condition, measurements were taken at 5:00am and at 9:00pm. Daytime measurements were taken hourly from 6:00 am to 8:00 pm. This monitoring schedule was based on the results of the pre-test operations.

3 Results and Discussion

3.1 Temperature and light Distribution

Temperature was found evenly distributed within the chamber. Also, light intensities of lighting fixture in each chamber were found with no significant differences between plant rows.

3.2 Environmental parameters in the chamber

3.2.1 Light intensity

Light intensities were within the range required for the treatments (Figures 3 and 4). There were fluctuations in average daily PPF because light intensity at canopy level was influenced by lettuce's growth. As leaves reached closer to the light source, PPF readings became higher. Adjustment of the height of lighting fixtures (about 5cm higher) then caused lower light intensity readings. PPF trend of L_3 varies from L_2 and L_1. For the first cycle, this occurred at 11-14 days, while for the second cycle at

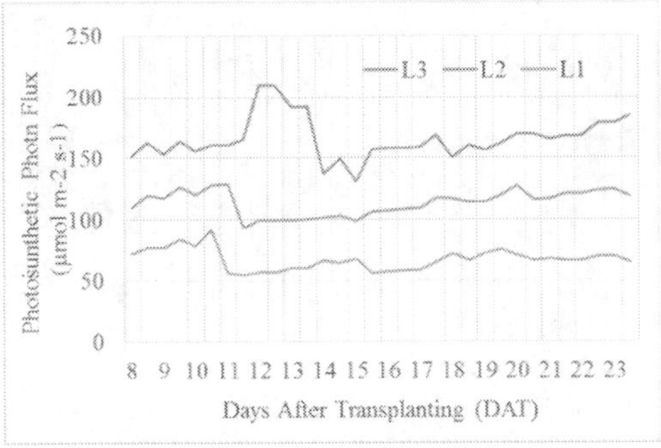

Figure 3. Average daily PPF measured at first cycle (25 °C)

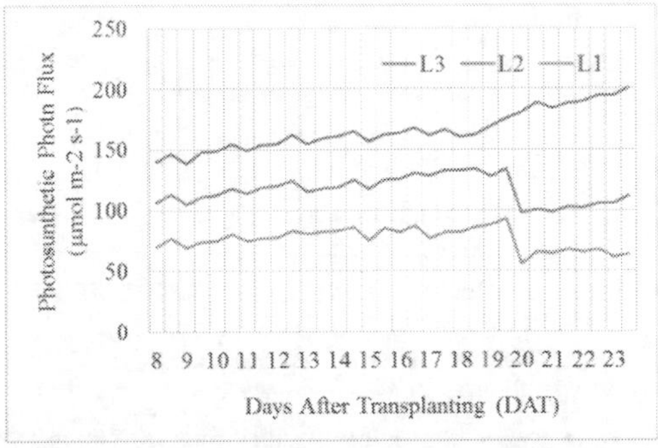

Figure 4. Average daily PPF measured at second cycle (18 °C)

21-23 days. The disparity was influenced by the difference in plant height response (data not shown) under 25 °C and 18 °C. Thus, the adjustment time of lighting fixtures also varies. Also, light intensity in chamber changes over time in a day even with the light sensor placed at same location [16].

3.2.2 Temperature, CO_2 and RH

Figures 5 and 6 show the actual day and night temperatures inside the chambers with the AC thermostat settings of 25 °C and 18 °C. In both settings, the current system achieved the required temperature treatments. Also, CO_2 and RH were maintained at levels required for normal plant growth.

Temperature is important since it controls the rate of plant growth [17], its physiological and chemical processes [17-18]. During first cycle, average day and night temperatures within chambers were 25.3±0.4 °C and 25.5±0.2 °C, respectively. Chamber temperatures during second cycle planting were 18.9±0.6 °C during day and 18.7±0.3 °C during night. Day and night temperatures outside the chambers were 29.6±2/25.9±0.5 and 26.2±1/23.6±0.6 during the first and second cycles, respectively.

Carbon dioxide (CO_2) levels were within acceptable range of 380 ppm [19] to 1500 ppm [20] and recommended levels of 800-1200 ppm [3]. Average day and night CO_2 inside the chambers were 496.1± 22 and

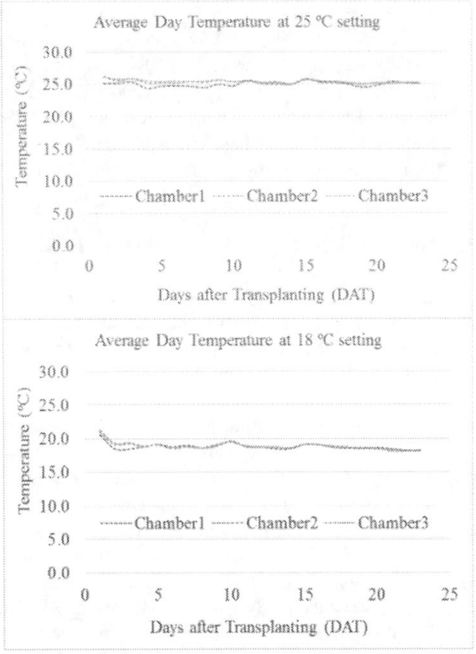

Figure 5. Day temperature at 25 °C and 18 °C setting

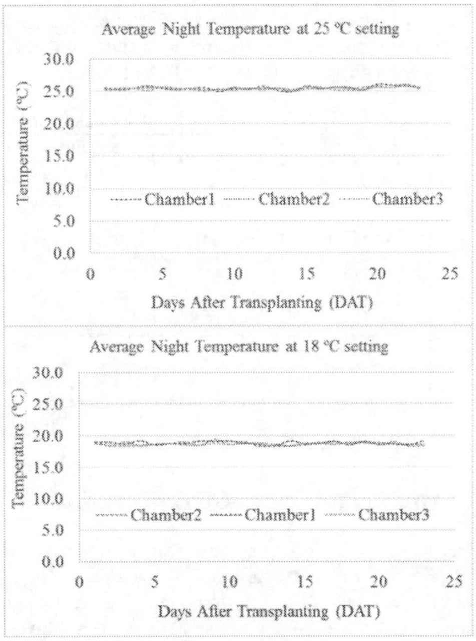

Figure 6. Night temperature at 25 °C and 18 °C setting

520.2± 37 ppm, respectively, during first cycle, and 463.7 ±35 ppm and 464.2±34 ppm, respectively during second cycle. Other researches had maintained CO_2 levels close to what were observed [21-23]. Thus, required CO_2 for plant's normal growth was achieved.

Relative humidity (RH) in the chambers was maintained according to acceptable RH of 40 - 80% [3] avoiding water stress, since RH affects plant water take up [3] and transpiration rate [17]. Average day and night RH within chambers were 65.8±4 and 80.3±3%, respectively, during first cycle, and 62.5±5 and 75.6±5%, respectively, during second cycle. Thus, relative humidity within chambers was sufficient enough for plant to perform normal physiological processes.

3.3 Yield performance of loose-leaf lettuce

On the average, yield of fanfare lettuce reached 29.6 grams per plant covering a planting density of about 14 plants per square meter. With the system, area productivity (grams of lettuce per square meter planted) was best under highest light intensity of 150 μmol m^{-2} s^{-1} at both temperature settings (Table 1). Analysis of variance in CRD showed that at 25 °C, L_3 with area productivity of 400.79 grams of lettuce harvested per square meter was significantly higher than those under treatments L_2 and L_1. Similarly, at 18 °C, L_3 with area productivity of 423.02 g m^{-2}, was significantly higher than those under L_2 and L_1. These results were comparable to other study where fresh weight of lettuce significantly increased at higher light intensity [24]. Using T-test (data not shown), no significant difference was found on the area productivity of lettuce at 18 °C and 25 °C temperature setting. Also, productivity of lettuce inside the chambers, at both temperature setting, were not significantly different from those outside the chambers with area productivity of 316.67 g m^{-2}.

Table 1. Area productivity (g m^{-2}) of loose-leaf lettuce

Light Intensity (μmol m^{-2} s^{-1})	First Cycle, T_1 (25 °C)	Second Cycle, T_2 (18 °C)
50 (L_1)	102.38±26 ^{c}	126.99±44 ^{c}
100 (L_2)	284.92±71 ^{b}	257.14±56 ^{b}
150 (L_3)	400.79±95 ^{a}	423.02±113 ^{a}

Means of the same column with same letter are not significantly different by LSD at 95% level of confidence

4 Conclusions and Recommendation

The constructed CEA system can maintain average temperature within the selected temperature settings of 18 °C and 25 °C. In conclusion, the system can produce loose-leaf lettuce at both temperature settings. However, it is highly recommended that the light intensity should be set at least 150 μmol m^{-2} s^{-1} for better crop performance. Also, improvement can be made on the lighting system. Although not included in the study, lettuce was found to be bitter at 25 °C. Thus, temperature settings between 18 °C and 25 °C can be tested. As such, the maximum temperature setting that will give the lowest power consumption which will not produce bitter lettuce can be determined. Further studies on quality of lettuce, other lettuce varieties and other cultivars as test crops, are recommended.

References

1. L. Fogg, K. Rauhala, E. Satterfield, US Patent 889-965 (1979)
2. L. Chin, K. Chong, Int. J. Phys. Sci. (2012)
3. ASHRAE, *ASHRAE Fundamentals Handbook*. Am. Soc. Agri. Eng. (2001)
4. T. Kozai, A. Sakaguchi, T. Akiyama, K. Oshima, K. Yamada. *Presentation, Assoc. for Vert. Farm* (2015)
5. J. Malamug, et.al., Tech. Rep. BSU, Phil (2011)
6. V. Ramos, *MS Thesis, CLSU, Phil.* (2008)
7. P. Armenia, K. Menz, G. Rogers, Z. Gonzaga, R. Gerona, E. Tausa, *ACIAR-PCARDD S. Phil. Fruits and Veg. Prog. Meeting*. Cebu, Phil. (2012)
8. F. Lansigan, W. de los Santos, J. Coladilla, Agric. Ecosys. & Environ. **82,** 1-3, 129-137 (2000)
9. M. Karlson, Univ. Alaska Fairbanks, U.S. Dept. of Agric. (2016)
10. K. Specht, R. Siebert, I. Hartmann, U. Freisinger, M. Sawicka, A. Werner, S. Thomaier, D. Henckel, H. Walk, A. Dierich, Agric. and Hum. Val. **31,** 1, 33-51 (2014)
11. R. Curry, T. Kozlowski, R. Prince, T. Tibbits. Cont. Envi. Guid. Acad. Press Inc. (1979)
12. Crop Science Society, https://dl.sciencesocieties. org/ files/publications/ces-guide.pdf. (2002)
13. R. Wojciechowska, A. Kolton, M. Zupnik, O. Dlugosz-Grochowska, W. Grzesiak. Folia Hort., **25**, 1 (2013)
14. G. Goins, N. Yorio, M. Sanwo, C. Brown, J. Exp. Bot. **48,** 312 (1997)
15. C. Brown, J. Amer. Soc. Hort. Sci., **120**, 5 (1995)
16. C. Cabacan, F. R. Cruz, I. Agulto, *IEEE 9th Int. Conf. on Hum. Nanotech. Info-Tech Com. and Contrl. Envi. and Mangmt.* (2017)
17. M. Brechner, A. Both, *Hydroponic Lettuce Handbook*, http://www.cornellcea. com/resources Publications/growersHandbooks/lettuce.html.
18. H. Lorenz, H. Wiebe, J. Scienta, **13,** (1980)
19. L. Xiaoying, G. Shirong, X. Zhigang, J. Xuelei, HortScience **46,** 2 (2011)
20. H. Shimizu, Y. Saito, H. Nakashima, J. Miyasaka, K. Ohdoi, *18th IFAC World Congress*, (2011)
21. K. Cope, B. Bugbee, HortScience, 504-509 (2013)
22. G. Massa, T. Graham, T. Haire, C. Flemming, G. Newsham, R. Wheeler, HortScience, **50**, 3, 501-506 (2015)
23. Z. Bian, R. Cheng, Q. Yang, J. Wang, J. Amer. Soc. Hort. Sci **141,** 2, 186-195 (2016)
24. W. Fu, P. Li, Y. Wu, J. Scienta, **135** (2012)

The study of quality attributes of parboiled rice during steaming process with revolved sieve

Thatchapol Chungcharoen[1,*], *Naruebodee* Srisang[1], and *Siriwan* Srisang[1]

[1]Engineering Department, King Mongkut's Institute of Technology Ladkrabang, Prince of Chumphon Campus, Chumphon, Thailand

Abstract. The steaming process with revolved sieve was studied as a function of steaming time and revolution velocity. The quality of parboiled rice in terms of thermal property, head rice yield and color was investigated and compared to the conventional steaming process such as fixed sieve. The experimental results showed that the steaming process with revolved sieve had advantage over the steaming process with fixed sieve. It led to a steamed sample with higher DG higher head rice yield and lower whiteness value. The required steaming time for a regularity quality by a steaming process with revolved sieve was also shorter. Moreover, the steaming time affected the qualities of parboiled rice. The long steaming time provided the higher DG and head rice yield and lower whiteness value of parboiled rice. Considering revolution velocity, it was found that the revolution velocity did not affect the qualities of parboiled rice. However, the revolution velocity of 5 rpm with steaming time of 5 min provided the difference of quality at each layer.

1 Introduction

Parboiled rice is the staple food in South Asian countries and it has a market share of 20-50% of world rice production [1]. Nowadays, parboiled rice is one of healthy foods which have increasingly been interested, especially in health conscious people because of the high nutrient value. Normally, rice parboiling process consists of three different operations, namely soaking, steaming and drying. First, rice was steeped in water at 70°C for 5 hr [2]. The main objective of rice soaking is to achieve uniform water absorption [3] and to provide the enough moisture content of rice about 44-45% (d.b.) [4] for the establishment of starch gelatinization in steaming process. Secondly, the soaked rice was steamed to achieve partial or complete gelatinization [5]. Finally, rice was dried to save the moisture content of 16% (d.b.) [6] for storage and milling.

Steaming is important process to provide the high quality of parboiled rice, especially head rice yield because this is the main process to create the gelatinization. This process, currently, is performed by the steaming with fixed sieve. The soaked paddy was placed on the fixed sieve and then, the saturated steam was blown throughout the paddy bed. This caused the incomplete gelatinization because the saturated steam was not completely contacted the paddy bed [7], resulting in the lower head rice yield. Moreover, the color of parboiled rice was not similar for all samples. To alleviate this problem, it should limit the amounts of soaked paddy which was placed on the sieve or increase the steaming temperature and steaming time to achieve complete gelatinization. However, their solving consumed more energy to produce the parboiled rice, leading to increase energy cost. One of possible solving which is the better way can be done by using the steaming with revolved sieve. It provided the moveable soaked paddy during steaming, leading to completely contact with the saturated steam, resulting in the easily complete gelatinization.

Therefore, the objectives of this work were to design and to create the steaming with revolved sieve and also to investigate the effects of revolution velocity and steaming time on the qualities of parboiled rice. The assessment of quality attributes of parboiled rice during steaming process was considered in terms of thermal property, head rice yield and color. They were compared with those of the steaming with fixed sieve.

2 Materials and methods

2.1. Materials

Suphanburi 1 paddy obtained from the Rice department, Suphanburi, Thailand was used as the sample rice in this study. Its moisture content and amylose content were about 12-13% dry basis (d.b.) and 25-30%, respectively. The paddy had already stored for a month prior to an experiment.

2.2 Preparation of sample

The 2.5 kg of paddy was steeped in controlled temperature water at 70°C for 5 hours. The hot water was then discarded and the paddy was stored in the

* Corresponding author: Thatchapol.ch@kmitl.ac.th

plastic box for 1 hour prior to steam. The soaked paddy was then steamed by a steaming process with revolved sieve shown in Fig. 1. The steaming system was comprised of stainless steel cylindrical revolved sieve of 30 cm diameter and 30 cm length, a pulley driven by a 1 HP motor, and a boiler with a capacity of 500 kg/h. The boiler produced the saturated steam at 106 kPa and flowed through steaming chamber. For stainless steel cylindrical revolved sieve, it had 4 fins inside the sieve and 2 holes on the side wall for separating sample layer by flat plate (shown in Fig. 2a). The steaming conditions were set at steaming times of 5, 10 and 15 min, revolution velocities of 5, 10, and 15 rpm, a bed height was fixed at 8 cm and a steaming temperature was set at 103°C. For steaming process with fixed sieve, the soaked paddy were taken to steam at the above steaming conditions but the sieve was not revolved. After finishing the steaming, the steam was stopped and suddenly blown out by bottom valve to avoid the over-heating and then the revolved sieve was taken out. The steamed paddy was divided into three layers such as top, middle and bottom (shown in Fig. 2b) by flat plate. In the last step, each layer was taken out for quality determination. The quality of sample obtained from the steaming process with revolved sieve was also compared with that of sample obtained from the steaming process with fixed sieve.

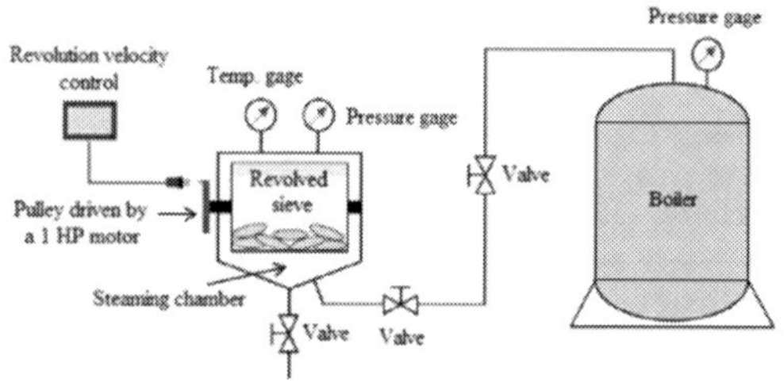

Fig. 1. Schematic diagram of steaming process with revolved sieve

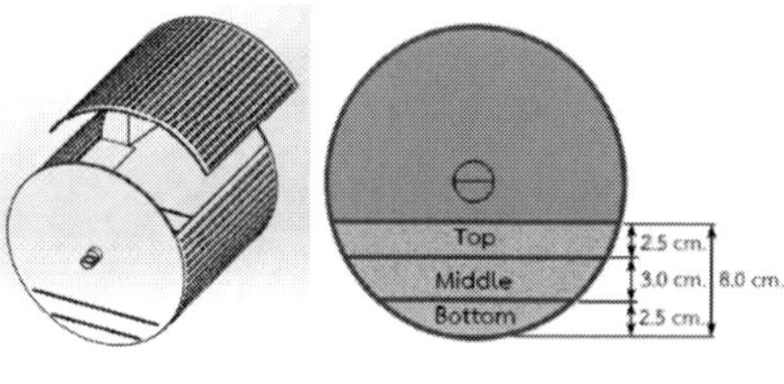

a) Sieve b) Layer of steamed paddy

Fig. 2. Details of cylindrical revolved sieve

2.3 Qualities determination

The degree of starch gelatinization (DG) was investigated using a differential scanning calorimeter (DSC; Perkin Elmer Co. Ltd., model DSC-7, Norwalk, USA). The head rice yield (HRY) was defined as a percentage of unbroken parboiled rice mass to paddy mass before milling. A digital whiteness meter (New Agronic, model W-600, Taiwan) was used for measuring the color of sample.

3 Results and discussion

3.1 Thermal property

Table 1 shows the enthalpy (ΔH) and degree of starch gelatinization (DG) of the parboiled rice steamed by fixed sieve and revolved sieve. The ΔH required for the complete gelatinization of reference rice was 9.28 J/g. When the rice sample was soaked and steamed by fixed sieve and revolved sieve, the ΔH decreased as compared to the reference rice, implying the occurrence of starch gelatinization [8]. The lower ΔH values for the treated samples were due to the modification of starch granule. During steaming, some starch granules lost their shape and were fused together. The ΔH required to melt starch was therefore lower. In the case of steaming process with revolved sieve at steaming time of 15 min, the ΔH values were 0 at every revolution velocity. This indicated that the destruction of starch granule was completed. The ΔH required to melt starch could not be found in these conditions. For steaming with fixed sieve, the DG at each layer was different when steaming at steaming time of 5 or 10 min. The DGs at middle layer were clearly lower than that at top and bottom layer at the same steaming time. This indicated that the saturated steam was not completely contacted the sample bed. When increasing the steaming time to 15 min, the DG was not different at each layer. From these data, it implied that the steaming time of 15 min was enough to obtain the complete of contact between sample and steam during steaming process with fixed sieve. However, it did not provide the complete gelatinization. For steaming with revolved sieve, the difference of DG at each layer was found specifically at steaming time of 5 min and revolution velocity of 5 rpm. When comparing between fixed sieve and revolved sieve, it can be seen that the DG was quite different although they were steamed at the same steaming time. The DG of sample with revolved sieve was relatively higher than that of sample with fixed sieve. Moreover, the gelatinization of rice starch was completed in the steaming process with revolved sieve at steaming time of 15 min. From these data, it implied that the steaming time of 10 min was enough to obtain the complete of contact between paddy and steam during steaming process with revolved sieve at every revolution velocity. In addition, The DG increased with an increase in steaming time while the revolution velocity did not affect the DG.

3.2 Head rice yield

Fig. 3 shows the head rice yield percentage of parboiled rice during steaming process with fixed sieve at steaming times of 5, 10 and 15 min. It was found that the steaming times of 5 and 10 min provided the different head rice yield percentage at each layer as shown in FIGURE 3 (a, b). The head rice yield percentages at middle layer were about 69.57% and 71.28% for steaming times of 5 and 10 min, respectively, while the head rice yield percentages at top and bottom layer were approximately 71.12 and 70.94% for steaming time of 5 min and 72.07 and 71.91% for steaming time of 10 min, respectively. These results corresponded to the thermal property result, showing the low DG at middle layer. However, the head rice yield percentage of each layer at steaming time of 15 min was not different.

Table 1. The degree of starch gelatinization (DG) of parboiled rice at various steaming conditions.

Sample	Steaming time (Min)	Revolution velocity (RPM)	Layer	ΔH (J/g)	DG (%)
Reference rice	-	-	-	9.28 ± 0.01^{a}	-
Parboiled rice (Fixed sieve)	5	-	Top	3.09 ± 0.02^{f}	66.69 ± 0.20^{fg}
		-	Middle	3.94 ± 0.02^{b}	57.60 ± 0.14^{i}
		-	Bottom	3.36 ± 0.07^{c}	63.75 ± 0.68^{h}
	10	-	Top	1.98 ± 0.05^{k}	78.63 ± 0.50^{c}
		-	Middle	2.33 ± 0.05^{i}	74.94 ± 0.53^{d}
		-	Bottom	2.13 ± 0.05^{j}	77.05 ± 0.53^{c}
	15	-	Top	0.18 ± 0.08^{m}	98.05 ± 0.88^{a}
		-	Middle	0.19 ± 0.03^{m}	98.01 ± 0.28^{a}
		-	Bottom	0.18 ± 0.01^{m}	98.10 ± 0.08^{a}
Parboiled rice (Revolved sieve)	5	5	Top	3.2 ± 0.01^{de}	65.56 ± 0.08^{gh}
			Middle	2.96 ± 0.02^{g}	68.06 ± 0.18^{ef}
			Bottom	2.86 ± 0.01^{h}	69.15 ± 0.07^{e}
		10	Top	3.19 ± 0.04^{de}	65.64 ± 0.40^{gh}
			Middle	3.18 ± 0.01^{e}	65.69 ± 0.13^{gh}
			Bottom	3.21 ± 0.03^{de}	65.44 ± 0.34^{gh}
		15	Top	3.22 ± 0.00^{de}	65.32 ± 0.03^{gh}
			Middle	3.25 ± 0.05^{d}	65.02 ± 0.50^{gh}
			Bottom	3.19 ± 0.01^{de}	65.54 ± 0.10^{gh}
	10	5	Top	0.701 ± 0.01^{l}	92.45 ± 0.10^{b}
			Middle	0.74 ± 0.02^{l}	92.08 ± 0.17^{b}
			Bottom	0.72 ± 0.00^{l}	92.29 ± 0.03^{b}
		10	Top	0.71 ± 0.01^{l}	92.34 ± 0.10^{b}
			Middle	0.72 ± 0.02^{l}	92.23 ± 0.16^{b}
			Bottom	0.73 ± 0.01^{l}	92.18 ± 0.12^{b}
		15	Top	0.71 ± 0.01^{l}	92.31 ± 0.08^{b}
			Middle	0.73 ± 0.03^{l}	92.17 ± 0.27^{b}
			Bottom	0.71 ± 0.01^{l}	92.36 ± 0.10^{b}
	15	5	Top	0	100.00
			Middle	0	100.00
			Bottom	0	100.00
		10	Top	0	100.00
			Middle	0	100.00
			Bottom	0	100.00
		15	Top	0	100.00
			Middle	0	100.00
			Bottom	0	100.00

a-m Means with different superscripts in the same column are significantly different ($p<0.05$)

Fig. 4 shows the effect of revolution velocity on the head rice yield percentage of parboiled rice during steaming process with revolved sieve at steaming time of 5 min. It was found that the revolution velocity affected the head rice yield percentage of parboiled rice. The difference of head rice yield percentage at each layer occurred when the sieve was revolved at 5 rpm. The head rice yield percentage at middle layer was also lower than that at top and bottom layer as shown in Fig 4(a). When the revolution velocity was increased to 10 or 15 rpm, the head rice yield percentage at each layer was not different. Moreover, it was found that the revolution velocity did not affect the head rice yield percentage of parboiled rice at each layer when increasing steaming time to 10 or 15 min (data not shown). It confirmed that the saturated steam was completely contacted the sample bed as already mentioned above. Furthermore, the head rice yield percentage increased with an increase of steaming time.

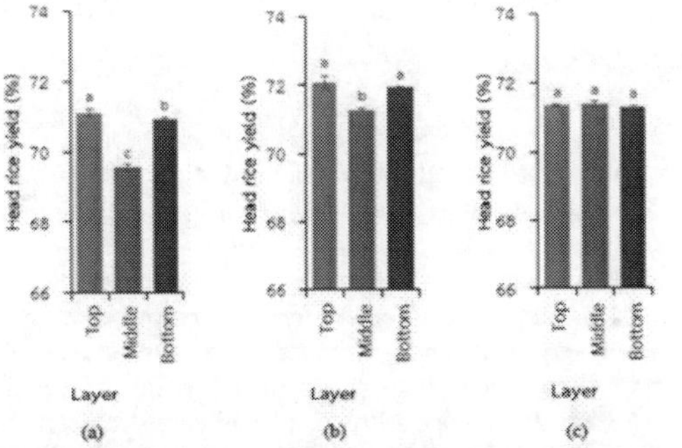

Fig. 3. The percentage of head rice yield of parboiled rice during steaming process with fixed sieve at (a) steaming time of 5 min, (b) steaming time of 10 min and (c) steaming time of 15 min.

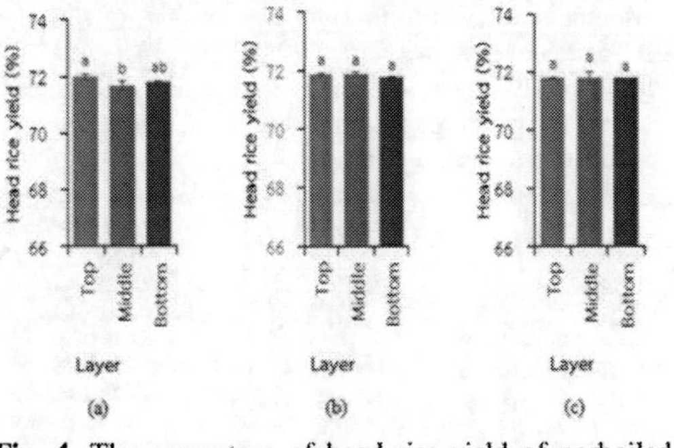

Fig. 4. The percentage of head rice yield of parboiled rice during steaming process with revolved sieve at steaming time of 5 min; (a) revolution velocity of 5 rpm, (b) revolution velocity of 10 rpm and (c) revolution velocity of 15 rpm

3.3 Color

The whiteness values of parboiled rice during steaming process with fixed sieve at various steaming times are presented in Fig. 5. The whiteness value at each layer was different when steaming at steaming time of 5 or 10 min. This result was expected since the DG and the head rice yield percentage at each layer was also different during these steaming times. The whiteness value of sample with middle layer at steaming time of 5 or 10 min was higher than that with top and bottom layer at the same steaming time (shown in Fig. 5 (a, b)). Moreover, it was found that the color of parboiled rice changed in a way that the whiteness value was decreased with increasing steaming time. This is because the long steaming time could produce more Maillard browning reactions [9].

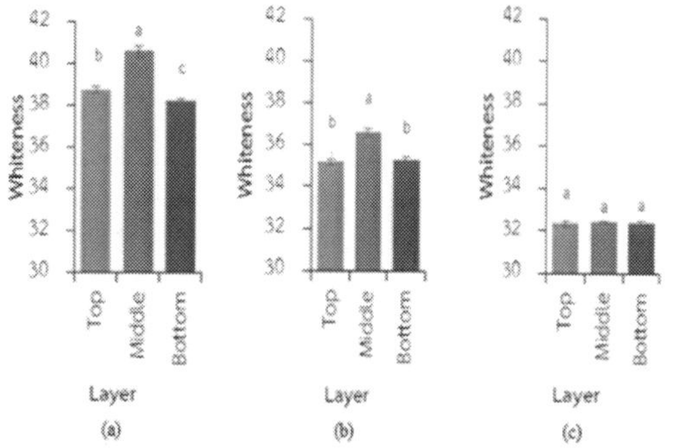

Fig. 5. The whiteness value of parboiled rice during steaming process with fixed sieve at (a) steaming time of 5 min, (b) steaming time of 10 min and (c) steaming time of 15 min

Fig. 6 shows the effect of revolution velocity on the whiteness value of parboiled rice during steaming process with revolved sieve at steaming time of 5 min. It was found that the difference of whiteness value at each layer did not take place during revolution velocities of 10 and 15 rpm the when steaming by revolved sieve at steaming time of 5 min (shown in Fig. 6 (b, c) as same as the results of DG and head rice yield percentage. As the steaming time was increased to 10 or 15 min, the whiteness value at each layer was similar. It was about 32 and 35% for steaming time of 10 min and 15 min, respectively (data not shown). It confirmed that the saturated steam was completely contacted the sample bed at these steaming times.

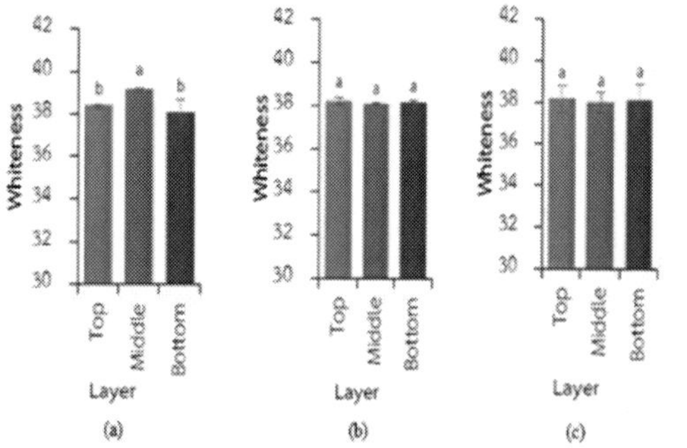

Fig. 6. The whiteness value of parboiled rice during process with revolved sieve at steaming time of 5 min; (a) revolution velocity of 5 rpm, (b) revolution velocity of 10 rpm and (c) revolution velocity of 15 rpm

4 Conclusions

The steaming process with revolved sieve could reduce the steaming time for a regularity quality. Its steaming time was 10 min while the time with regularity quality by steaming process with fixed bed was 15 min. The DG and the head rice yield were higher in the steaming process with revolved sieve than in the steaming process with fixed bed. Nevertheless, the whiteness value in the steaming process with fixed bed was higher. The qualities of parboiled rice in both steaming process changed with steaming time. The DG and the head rice yield were increased with increasing steaming time while the whiteness value was decreased. However, the revolution velocity did not affect the qualities of parboiled rice.

4 Acknowledgments

The authors express the sincere appreciation to the King Mongkut's Institute of Technology Ladkrabang (KREF046011) for supporting the study financially.

5 References

1. N.H. Choudhury, IRRI, 47-54 (1991)
2. S. Pruengam, S. Soponronnarit, S. Prachayawarakorn, S. Devahastin, Drying Technol. **32**, 1949-1955 (2014)
3. J.E. Wimberly, International Rice Research Institute, 101-116 (1983)
4. C. Taechapairoj, S. Prachayawarakorn, S. Soponronnarit, J. Food Eng. **76**, 411–419 (2006)
5. H. Dutta, C.L. Mahanta, Food Res. Int., 655-663 (2012).
6. S. Soponronnarit, S. Prachayawarakorn, Dry. Technol., 1667-1686 (1994).
7. T. Chungcharoen, S. Prachayawarakorn, S. Soponronnarit, G. Srzednicki, AIP Conference Proceeding, (International Symposium on Engineering, Energy and Environment, November 17-20, Bangkok, 2013).
8. S. Tirawanichakul, S. Prachayawarakorn, W. Varanyanond, P. Tungtrakul, S. Soponronnarit, Dry. Technol., 1731–1754 (2004).
9. L. Lamberts, I. Rombouts, K. Brijs, K. Gebruers, J. A. Delcour, Food Chem., 916–922 (2008).

Optimization of dimethyl ether production process synthesis using superstructure analysis

Teerapat Laiwatthanaphaisarn and *Amata* Anantpinijwatna*

Department of Chemical Engineering, Faculty of Engineering, King Mongkut's Institute of Technology Ladkrabang, Bangkok, Thailand

Abstract. Current world energy consumption is likely to increase over time. This is due to the growth of industry and transportation. The most important and most used energy sources are crude oil and natural gas. The consumption of energy is increasing continuously due to the economic expansion of the world fleet. At present, prices of primary energy sources such as oil and natural gas tend to increase. In addition, oil and gas are limited and likely to run out in the future. Currently, research and research on alternative energy is ongoing. To find the best alternative energy to replace in the future. Dimethyl ether is a substance that can be used as a substitute for liquefied petroleum gas (LPG) because of its similar physical properties. Most of them are used as fuel in vehicles. In addition, dimethyl ether is easier to liquefy than liquefied petroleum gas, giving advantages in terms of storage and transport, and a higher cetane value that can be used in the vehicle. Dimethyl ether is a substance that will burn completely. Dimethyl ether production has a wide variety of options. If the best option is difficult to analyse because of the complexity of the solution. Superstructure analysis will help to find alternatives for the production of dimethyl ether. Superstructure will identify the most economical alternative. The mathematical model is applied to the existing production process and new alternatives. In this work, the alternatives to produce dimethyl ether are displayed and the optimum alternative are chosen.

1 Introduction

Current world energy consumption is likely to increase over time due to the growth of industry and transportation. The most important energy sources are crude oil and natural gas which are globally consumed at approximately 5,700 million tons equivalent of crude oil per year [1]. At present, prices of primary energy sources such as oil and gas are in an increasing trend. Moreover, oil and gas are non-renewable and likely to run out in the near future. Current research on alternative energy to find the best alternative energy is ongoing.

Dimethyl ether (DME) is a substance that can be used as a liquefied petroleum gas (LPG) substitute because of their similar physical properties. DME is easier to liquefy than the LPG, giving advantages in terms of storage and transportation; moreover, it also has higher cetane value and smaller molecule which will be completely burnt benefitting the vehicle usage. It's also a nitrogen and sulphur lacked substrate; therefore, combustion of DME does not produce small particulates matters (PM), nitrogen oxides (NO_x) and sulphur oxide (SO_x). It can be said that DME is a clean alternative energy that is ideal for vehicle usage. The production of DME includes a wide variety of options ranging from simple reactor-distillations, reactor-distillation-pervaporation, single reactive distillation, or multiple reactive distillations. Due to the complexity of the synthesis and design of the process, the optimum processing pathway is hardly accomplished.

The process synthesis and design through superstructure optimization could help identifying the most economical alternative via mathematical method. The conventional production pathways, as well as, newly generated alternatives are combined into the form of superstructure, the generic mathematical model is applied and solved using mixed-integer linear programming (MILP) embed in the GAMs software.

2 Methodology

The integrated business and engineering framework for synthesis and design of processing networks [2,3] has been applied into this work. The potential alternatives for synthesizing and designing processes are presented as processes intervals; they are arranged in the form of superstructure as displayed in fig. 1. Raw materials are presented in the first column; while products are embodied in the last column. The processing steps sequentially proceed from the left- to the right-hand side of the superstructure, converting and/or separating the raw materials to the products.

Whereas, each process is presented through a generic process interval description displayed in fig. 2. Eqs. 1-7 are incorporated generic equations depicting the operation of a unit or a group of units.

* Corresponding author: amata.an@kmitl.ac.th

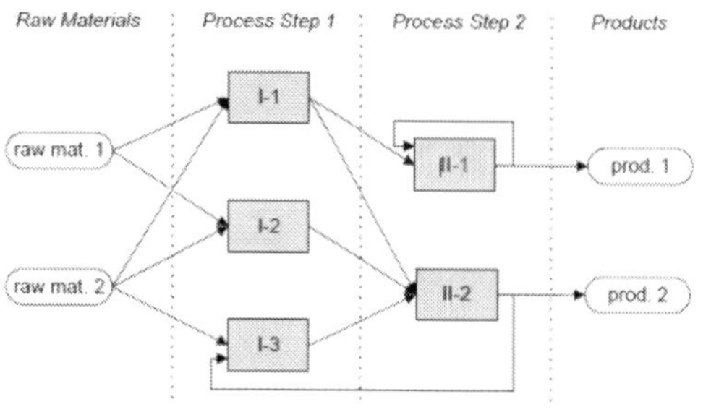

Fig. 1. Superstructure representation [3]

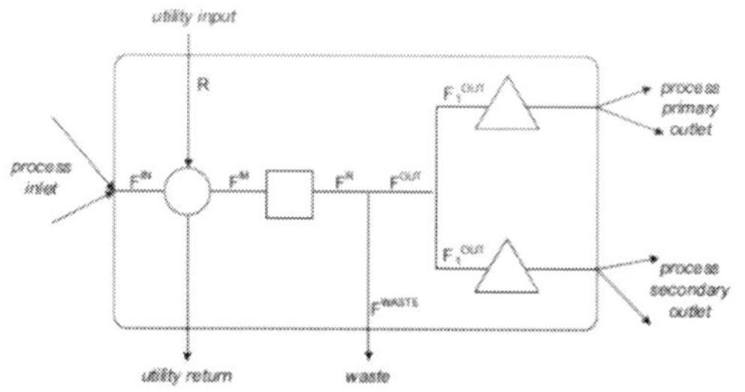

Fig. 2. Generic process interval [9]

$$F_{i,kk}^{M} = \sum_{k} F_{i,k,kk} + \alpha_{i,kk} \cdot R_{i,kk} \tag{1}$$

$$R_{i,kk} = \mu_{i,kk} \sum_{i,k} F_{i,k,kk} \tag{2}$$

$$F_{i,kk}^{R} = F_{i,kk}^{M} + \sum_{rr,rx} \left(\gamma_{i,kk,rr} \cdot \theta_{rx,kk,rr} \cdot F_{rx,kk}^{M} \right) \tag{3}$$

$$F_{i,kk}^{O} = F_{i,kk}^{R} \cdot \left(1 - SW_{i,kk}\right) \tag{4}$$

$$F_{i,kk}^{W} = F_{i,kk}^{R} - F_{i,kk}^{O} \tag{5}$$

$$F_{i,kk}^{OM} = F_{i,kk}^{O} \cdot SF_{i,kk} \tag{6}$$

$$F_{i,kk}^{OS} = F_{i,kk}^{O} - F_{i,kk}^{OM} \tag{7}$$

The mathematical model is formulated as mixed-integer linear programming problem with the objective function, the logical constraints, and the variable bounds defined as following by eqs. 8-12.

$$\max \quad f(x,y) \tag{8}$$

$$s.t. \quad g(x,y) \geq 0 \tag{9}$$

$$h(x,y) = 0 \tag{10}$$

$$x^{LO} \leq x \leq x^{UP} \tag{11}$$

$$y \in \{0;1\}^{n} \tag{12}$$

The objective function of this study is maximizing annual profit, in term of earnings before interest and taxes (EBITA), subject to logical decisions on choosing the unit operations and raw material flowrate.

3 DME Production and alternatives

3.1 Conventional DME production

The conventional production of DME by dehydration of methanol with the capacity 110 Mta has been reported by Luyben [4]. The process consists of an adiabatic reactor, two distillation columns, and feed preheater as displayed in fig. 3. Feed, 153 Mta of methanol, is preheated to 275 °C and then fed into the adiabatic reactor with solid catalyst. Methanol is undergoing the dehydration reaction; 82% of methanol is converted to DME with water as side product as displayed in eq. 13.

$$2CH_3OH \leftrightarrow CH_3OCH_3 + H_2O \tag{13}$$

The reactor product is used for preheating the feed, then cooled down before sending to distillation column. The first column separates DME from methanol and water; DME, as top product, comes out at 99.999% purity. The methanol and water mixture at the bottom of the first column is then sent to second column. The unreacted methanol is separated in this column and recycled back to mix with the feed.

The capital cost of the conventional process is approximately 3.4 MM$, with the net energy cost 2.6 MM$/y.

3.2 Processing alternatives

Pervaporation with alpha-alumina membrane has been reported employing for the separation between water and methanol by Chapman et al. [5]. Water-methanol are separated completely at 50 °C with 0.57 kg/m^2·h flux. This unit can be used as a substitution for the second distillation column.

Reactive distillation (RD) has also been reported for the production of DME, both from pure methanol [6] and methanol in glycerol [7]. The RD can be applied as a combined reactor-separation to enhance the methanol conversion; as well as, employed as product enhancer after the conventional reactor; moreover, it can also be used to convert the recycled methanol to reduce the main reactor load [8]. Products of RD could be methanol, water, DME, or mixtures of different compositions of these compounds depending on the whereabout of the column. In this study, the RD is employed as the single unit process, and as enhanced-substitution of both distillation columns.

The important mixing, reacting, and separating parameters of the conventional and alternatives processes are reported in tables 1a-c, and the DME production superstructure is displayed in fig. 4.

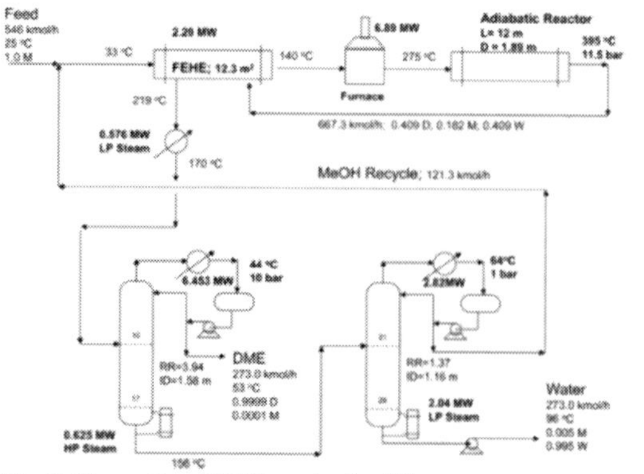

Fig. 3. Conventional DME production [4]

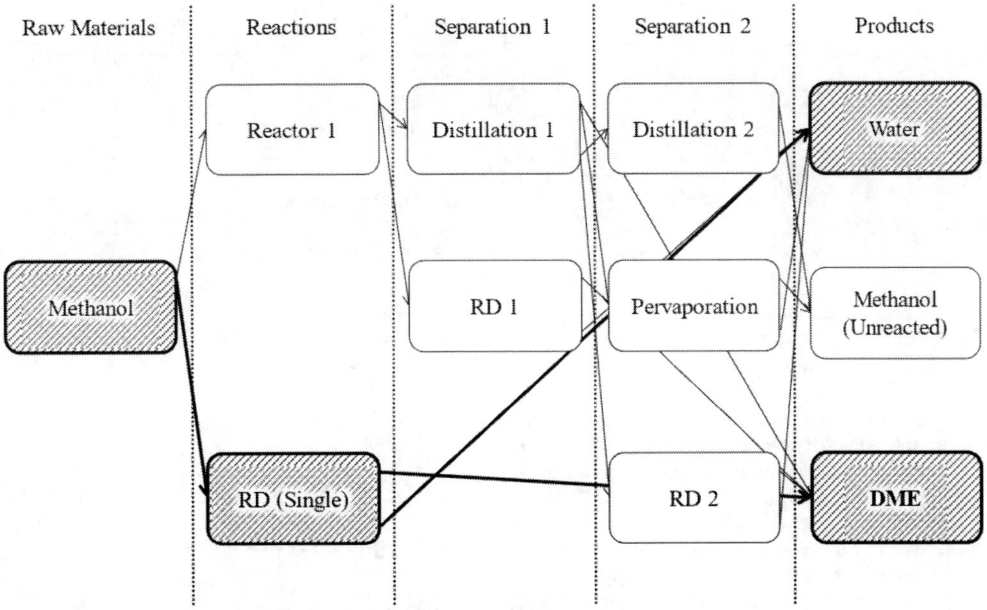

Fig. 4. DME production process superstructure and the optimum pathway

Table 1a. DME production processes reaction parameters

Intervals	θ_{MeOH}	γ_{MeOH}	γ_{DME}	γ_W
Rx 1	0.409	-2	1	1
RD (S.)	0.5	-2	1	1
RD 1	0.5	-2	1	1
RD 2	0.5	-2	1	1

Table 1b. DME production processes sepraration parameters

Intervals	SF_{MeOH}	SF_{DME}	SF_W
Dist. 1	1	0	1
Dist. 2	1	1	0
RD (S.)	0	1	0
RD 1	0	1	0
RD 2	0	1	0
Perva.	1	1	0

4 Results

Single reactive distillation operation has been chosen as the optimum production pathway for the dehydrogenation of DME, displayed as greyed box in fig. 4.

Comparing to the conventional process, the optimum pathway requires only one preheater and one column; moreover, due to the parallel reacting-separating operation of the RD, the reaction is shifted toward the product side, resulting in the increasing of the raw material conversion. An initial estimation is found that the capital investment of the optimum process is 2.3 MM$, 32% lowering from the conventional process; additionally, the total operating cost (including net energy cost) is approximately 2.2 MM$/y, 17% lower from the conventional operation.

Table 1c. DME production processes utilities parameters

Intervals	μ_{Pow}	μ_{NG}	μ_{HPS}	μ_{CW}
Rx 1	7.8	1.1		
Dist. 1			61.2	52.0
Dist. 2			337.6	38.4
RD (S.)	9.6		1017.2	90.0
RD 1			641.0	63.8
RD 2			641.0	63.8
Perva.	3.8			108.6

* μ_{Pow} ≡ kWh/tonne
μ_{NG} ≡ MMBTU/tonne
μ_{HPS} ≡ kg/tonne
μ_{CW} ≡ kg/tonne

The optimum process is then simulated in an open source process simulator DWSIM v.5.1. The result of the detailed simulation, as presented in fig. 5, indicates that the 54 stages-RD column can produce up to 110 Mtonne/y of above 99.99% purity of DME with the column operating pressure 11 atm., column reflux ratio 6.2 and the column diameter 2.1 m. Water, at above 99.99% purity, is also produced as a side product. The conversion of methanol reaches 100% with 34 reactive stages, and the methanol raw material is fed to the column at the 9th stage.

The detailed simulated result confirms the feasibility of the superstructure optimization methodology. The obtained design reduces capital and operating costs; moreover, the amount of waste generated is also reduced due to the enhanced conversion. On the other hand, there

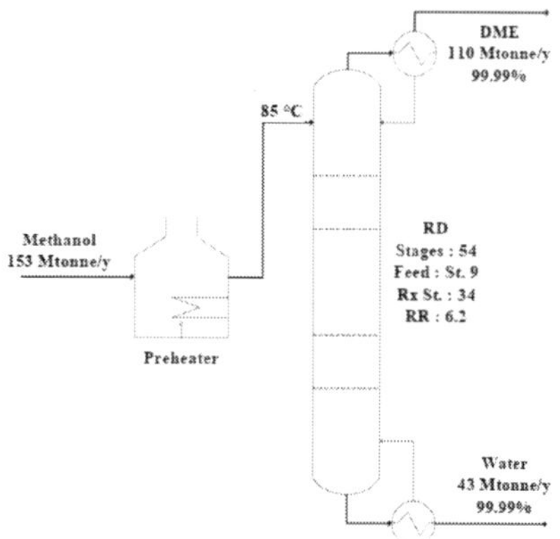

Fig. 5. Optimum DME production with single RD

is one particular disadvantage of the optimum design; since multiple units are integrated into one single unit, the degrees of freedom are reduced, and the controllability of the process is diminished.

Though, the result show promising design comparing to the conventional design, there are still room for the improvement regarding the energy integration such as the design displayed in fig. 6; more energy could be conserved by recycling the heat in the water from the column bottom.

5 Conclusion

In this work, the integrated business and engineering framework for synthesis and design of processing networks has been adopted for the synthesis and design of the DME production.

The conventional DME production process, as well as, processing alternatives are collected and organized in form of the superstructure. The mathematical model has been formulated as the mixed-integer linear programming problem following the generic structure-interval model.

The MILP problem has been solved with the objective to maximize the annual profit (in term of EBITA). The obtained result selects single RD unit as the optimum design for the production of DME. The design has also been validated with the detailed simulation. The result achieves 32% reduction of the capital investment and 17% reduction of the operating cost.

Despite achieving the optimum design with various improvements, the process can be further intensified for an even better result. The intensification step, specially, regarding energy intensification should also be included into the framework in the future.

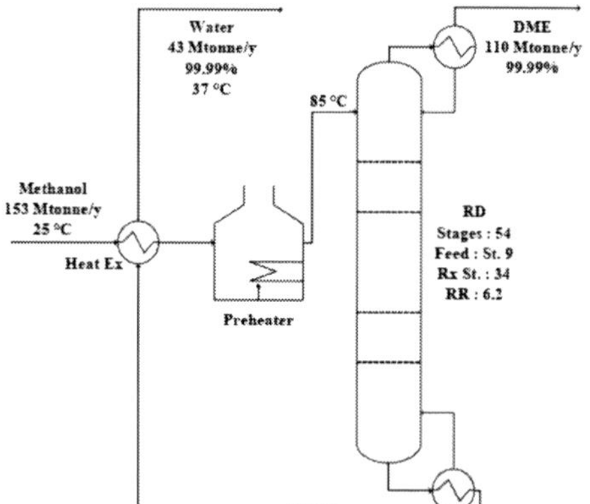

Fig. 6. Optimum DME production with single RD

Nomenclature

Subscripts/ Superscripts	
i	Component
k, kk	Process Interval
rr	Reaction
rx	Key Reactant
M	After Mixing
O	After Waste Splitter
OM	Primary Outlet
OS	Secondary Outlet
R	After Reaction
W	Waste Stream
Parameters/ Variables	
F	Stream Flow
R	Utilities Flow
SW	Waste Split Fraction
SF	Stream Split Fraction
α	Utilities Mixed Fraction
μ	Utilities/ Stream Ratio
θ	Conversion
γ	Stoichiometric Number
Compounds/ Utilities	
MeOH	Methanol
DME	Dimethyl Ether
W	Water
Pow	Power
NG	Natural Gas
HPS	High Pressure Steam
CW	Cooling Water

References

1. Enerdata, *Global Energy Statistical Yearbook 2017* (2017).
2. A. Quaglia, B. Sarup, G. Sin, and R. Gani, Comput. Aided Chem. Eng. **30**, 467 (2012).
3. A. Quaglia, B. Sarup, G. Sin, and R. Gani, Comput. Chem. Eng. **38**, 213 (2012).
4. W. L. Luyben, Comput. Chem. Eng. **106**, 17 (2017).
5. P. D. Chapman, T. Oliveira, A. G. Livingston, and K. Li, J. Memb. Sci. **318**, 5 (2008).
6. A. Kiss and D. J.-P. C. Suszwalak, Procedia Eng. **42**, 581 (2012).
7. P. Kaewwisetkul, L. Simasatitkul, and A. Arpornwichanop, Chem. Eng. Process. Process Intensif. **117**, 80 (2017).
8. Z. Lei, Z. Zou, C. Dai, Q. Li, and B. Chen, Chem. Eng. Sci. **66**, 3195 (2011).
9. A. Quaglia, B. Sarup, G. Sin, and R. Gani, Comput. Chem. Eng. **59**, 47 (2013).

Influence of jet discharge velocity profile on CFD simulation of pump-around jet mixing tank

Thanwa Jorakit[1], *Natthanon* Phaiboonsilpa[1], *Apinan* Namkanisorn[1], *Phisan* Ponpo[1], *Eakarach* Bumrungthaichaichan[1], and *Santi* Wattananusorn[1,*]

[1]Department of Chemical Engineering, Faculty of Engineering,
King Mongkut's Institute of Technology Ladkrabang, Bangkok, Thailand

Abstract. The present paper shows the effect of jet discharge velocity profile (or jet nozzle configuration) on CFD simulation of an open 45° inclined side entry pump-around jet mixing tank. The CFD model was carefully developed by using appropriate grid arrangement, boundary conditions, and numerical methods. The two different jet discharge velocity profiles, including top hat and fully developed profiles, were simulated by using the inlet mass flow rate of about 0.22 $kg{\cdot}s^{-1}$. The overall mixing times and normalized concentration profiles predicted by two different jet discharge velocity profiles were compared with the previous reliable experimental data. The results revealed that the different jet discharge velocity profiles resulted in different jet flow and mixing patterns inside the vessels.

1 Introduction

Jet mixing tank is a mixing device that uses high velocity liquid jet to mix the different components inside the vessel. It was firstly introduced by Fossett and Prosser [1] in 1949. Nowadays, it becomes an important mixing device because of various advantages, including simple design with non-moving part, inexpensive operating cost, and easy installation and maintenance.

Over 60 years or so, the different jet mixing time correlations were obtained by experiments. However, the universal mixing time correlation was unavailable. That is, these correlations predicted the overall mixing times well only for the range of studied parameters, i.e. these correlations were case specific [2]. Moreover, the details of fluid flow and mixing inside the vessel were not presented. So, the computational fluid dynamics (CFD) has been employed to address these problems.

One of the earliest CFD studies on jet mixing tank has been carried out by Brooker [3]. He revealed that his CFD model well predicted the overall mixing time with the maximum error of about 15% as compared to the experimental data. Later, the effects of jet mixing tank configurations and operating conditions, such as jet elevation, jet angle, tank shape, liquid height, jet injection rate, etc., on jet mixing behaviour were studied. Further, the effects of turbulence model and numerical methods on mixing behaviour were also investigated. However, the influence of jet discharge velocity profile (or jet nozzle configuration), which directly affects on jet flow pattern inside the vessels, was not represented.

Generally, for round jet, the different jet nozzle configurations provide the different jet discharge velocity profiles. The jet discharge velocity profiles of smooth contraction nozzle (SC) and long pipe nozzle (LP) respectively are top hat (TH) and fully developed (FD), which is generally described by one-seventh-power law. For constant inlet mass flow rate, the discharge velocity profiles of SC and LP are shown in Fig. 1. Further, these different nozzle geometries also represent the different jet flow patterns. For SC, the jet potential core length is about 4-6 nozzle diameter (d_j). Whereas, the potential core region of LP is not observed.

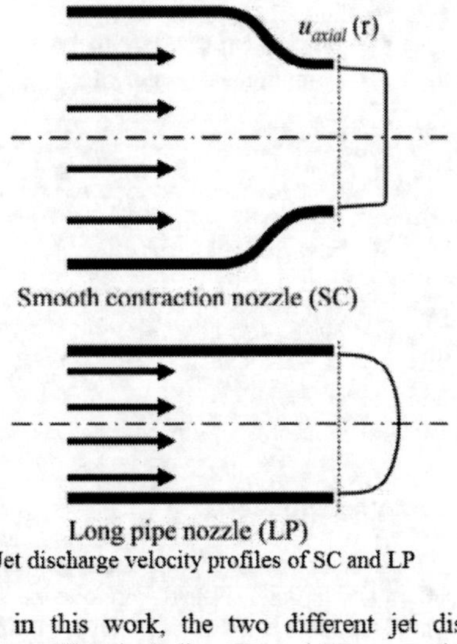

Fig. 1. Jet discharge velocity profiles of SC and LP

So, in this work, the two different jet discharge velocity profiles, including TH and FD, were simulated to investigate the jet flow pattern and mixing behaviour inside the open 45° inclined side entry pump-around jet mixing tank.

* Corresponding author: cfdgroup_santi@hotmail.com

2 Description of CFD modelling

2.1. Jet mixing tank

An open 45° inclined side entry pump-around jet mixing tank of Patwardhan [4] was employed to study the effect of jet discharge velocity profile. The diameter of flat bottom cylindrical tank was 0.5 m. The tap water height was 0.5 m. The diameter of jet nozzle and tank outlet pipe respectively were 0.008 m and 0.0381 m. The schematic of pump-around jet mixing tank is shown in Fig. 2.

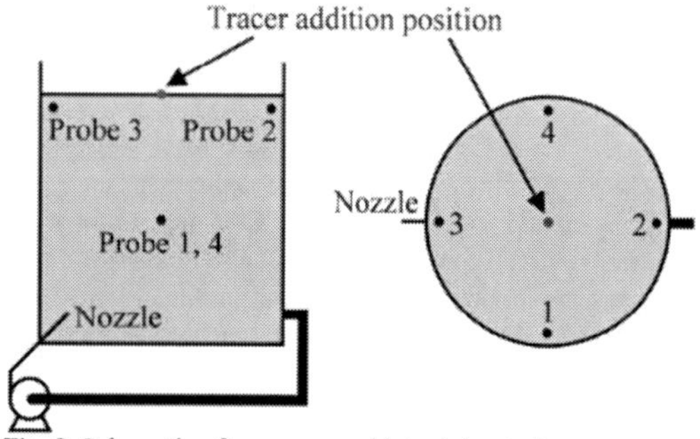

Fig. 2. Schematic of pump-around jet mixing tank

2.2 Grid generation

The GAMBIT 2.4.6 program was adopted to perform the three-dimensional solid model and grid generation of an open 45° inclined side entry pump-around jet mixing tank. The hexahedral grids were carefully generated inside the vessel to achieve the accurate results. That is, the grids were controlled to align with the flow direction to reduce the numerical errors by using the domain decomposition technique. Generally, in GAMBIT 2.4.6, the domain decomposition facility is unavailable. Thus, in order to use this technique, the jet mixing tank domain was manually split by using the specified faces. The grid generation of this mixing tank is shown in Fig. 3.

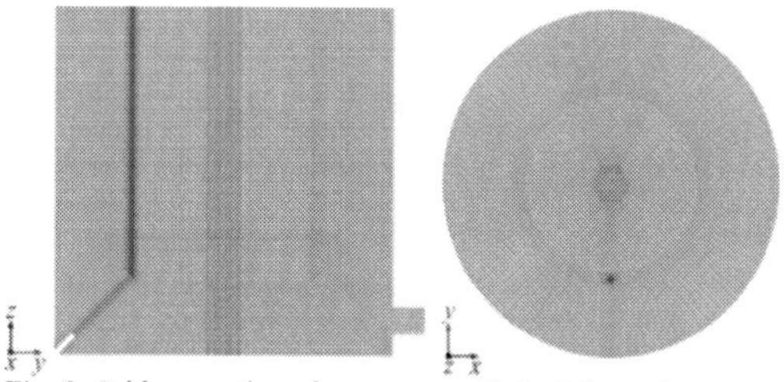

Fig. 3. Grid generation of pump-around jet mixing tank

2.3 Governing equations

For CFD simulation of jet mixing tank, the steady state flow pattern and turbulence field were obtained by using the Reynolds average equations for conservation of mass and momentum together with the realizable k-epsilon turbulence model (RKE). The compact form of steady state Reynolds average equation can be written as:

$$\frac{\partial(\rho U\phi)}{\partial x_j} = \frac{\partial}{\partial x_j}\left(\Gamma_\varphi \frac{\partial \phi}{\partial x_j}\right) + S_\varphi \quad (1)$$

where U is mean velocity vector, ϕ is a universal dependent variable, Γ_φ is the diffusivity, and S_φ is the source term. Further, the details of variables for continuity equation, momentum equation, and RKE are represented in Table 1.

Table 1. Variables for continuity equation, momentum equation and RKE [5]

Equation	ϕ	Γ_φ	S_φ
Continuity	1	0	0
Momentum	U_i	μ	$-\frac{\partial P}{\partial x_i} + \frac{\partial}{\partial x_j}\left(\mu_t \frac{\partial U_i}{\partial x_j}\right) + S_{M,i}$
k-transport	k	$\mu + \frac{\mu_t}{\sigma_k}$	$G_k + G_b - \rho\varepsilon - Y_M + S_k$
ε-transport	ε	$\mu + \frac{\mu_t}{\sigma_\varepsilon}$	$\rho C_1 S\varepsilon - \rho C_2 \frac{\varepsilon^2}{k + \sqrt{\nu\varepsilon}} + C_{1\varepsilon}\frac{\varepsilon}{k}C_{3\varepsilon}G_b + S_\varepsilon$

For jet mixing time investigation, the unsteady state species transport equations without chemical reaction [5] were iteratively solved and can be written as:

$$\frac{\partial(\rho Y_i)}{\partial t} + \frac{\partial(\rho U Y_i)}{\partial x_j} = \frac{\partial}{\partial x_j}\left(\left(\rho D_{i,m} + \frac{\mu_t}{\mathrm{Sc}_t}\right)\frac{\partial Y_i}{\partial x_j}\right) + S_i \quad (2)$$

where Y_i is the local mass fraction of species i, $D_{i,m}$ is the mass diffusion coefficient for species i in the mixture, Sc_t is turbulent Schmidt number, μ_t is the eddy viscosity, and S_i is the source term for species transport equations.

Further, in order to evaluate the mixing time, the 95% mixing time was used. Generally, the 95% mixing time is the time required for concentration (c) to reach within 95% of the fully mixed value ($\bar{c}$) and can be written as:

$$t_{95\%} = \text{time for } \left|\frac{c - \bar{c}}{\bar{c}}\right| \leq 0.05 \quad (3)$$

Finally, the arithmetic average of mixing times obtained by four different probe locations was adopted to achieve the overall mixing time.

2.4. Boundary conditions

For steady state simulation, the velocity-inlet and pressure-outlet boundary condition types were adopted at inlet and outlet, respectively. Whereas, for unsteady state simulation, the recirculation-inlet and recirculation-outlet were used [6]. At inlet, the two different jet discharge velocity profiles with constant inlet mass flow rate of about 0.22 $kg \cdot s^{-1}$ were specified. Moreover, for turbulence quantities, the turbulence kinetic energy of 0.2904 $m^2 \cdot s^{-2}$ and its dissipation rate of 22.35614 $m^2 \cdot s^{-3}$ were directly imposed. The no-slip boundary condition and standard wall functions were adopted at the tank wall. Further, the symmetry boundary condition was used at the top liquid surface.

Furthermore, the uniform inlet velocity of 4.4 $m \cdot s^{-1}$ was imposed at inlet for TH. For FD, the fully developed profile was obtained by preliminary CFD simulation of turbulent flow in circular pipe. The pipe diameter was 0.008 m. The pipe length was extended to 70 pipe diameter to ensure that the fully developed pipe flow was achieved. Then, the velocity profile at pipe outlet was extracted and written in C language as the user defined function (UDF). Finally, this UDF was specified at jet nozzle discharge section. The jet discharge velocity profiles for TH and FD are represented in Fig. 4.

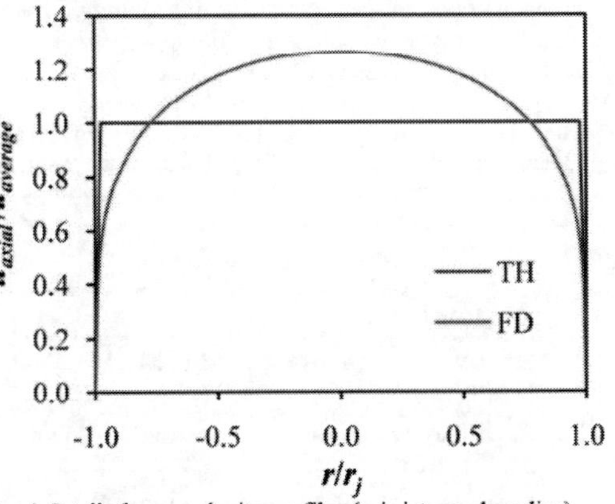

Fig. 4. Jet discharge velocity profiles (r_j is jet nozzle radius)

2.5. Numerical methods

For this work, the double precision pressure-based solver of ANSYS FLUENT program was employed to simulate pump-around jet mixing tank. The pressure-velocity coupling scheme, spatial discretization scheme, and temporal discretization scheme were SIMPLE, second order upwind, and first order implicit, respectively.

2.6. Solution strategy

The area weighted average of the velocity magnitude at plane $x = 0$ was monitored until it was constant to obtain the converged solutions of steady state simulation. For unsteady state simulation, the scaled residual of 10^{-5} for tracer was used as a convergence criterion. Further, the time step size of transient simulation was 0.0025 s [6, 7].

Moreover, the velocity magnitude gradient adaption with the refine threshold of 10% of the maximum value [8] was used to obtain the grid independent solutions. In order to obtain the grid independent solutions, the final grids of these models were achieved by adapting the grids three times as suggested by Bumrungthaichaichan et al. [6].

3 Results and discussion

In this work, the two different jet discharge velocity profiles, including TH and FD, were simulated to study the flow pattern and mixing behaviour inside an open 45° inclined side entry pump-around jet mixing tank. The predicted overall mixing times and normalized concentration profiles were compared with the results of Patwardhan [4] as represented in Table 2 and Fig. 5, respectively. Generally, the normalized concentration is defined as a ratio of the tracer concentration to the fully mixed value. Moreover, the jet axial velocity contours of two different jet discharge velocity profiles are shown in Fig. 6.

Table 2. Comparison of overall mixing times

Model	Mixing time [s]	% Error [a]
EXP [4]	31.00	-
TH	28.19	9.06
FD	25.27	18.48

[a] The percentage error is a ratio of absolute difference between the predicted overall mixing time and experimental mixing time to the experimental value.

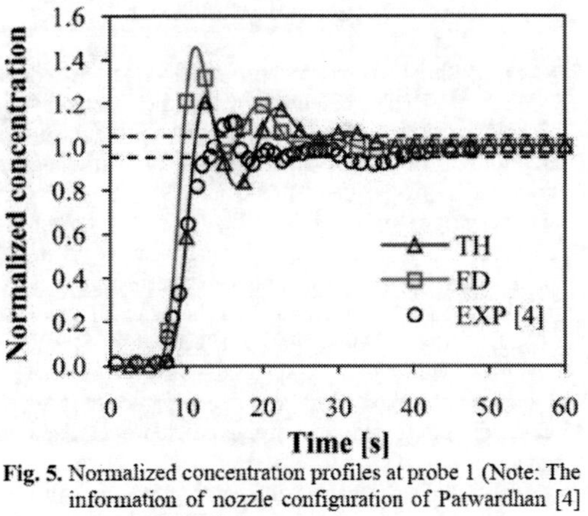

Fig. 5. Normalized concentration profiles at probe 1 (Note: The information of nozzle configuration of Patwardhan [4] was not reported)

From Table 2, the simulated results reveal that the predicted overall mixing times of TH and FD are lower than actual experiment. The percentage errors of TH and FD are about 9% and 18.5%, respectively. Further, the predicted overall mixing time of TH is higher than that predicted by FD, which indicates that the momentum available for mixing (convective transport) of FD is greater than that obtained by TH.

In Fig. 5, the results show that the normalized concentration profiles predicted by two different jet discharge velocity profiles are different. The start of normalized concentration profile obtained by FD is faster that that obtained by TH and the first peak value of normalized concentration profile for FD is higher than that simulated by TH. The start and first peak values of these predicted normalized concentration profiles are respectively faster and higher than experimental data because of the overprediction in convective transport [6]. However, these predicted normalized concentration profiles approach the experimental profile at the large times.

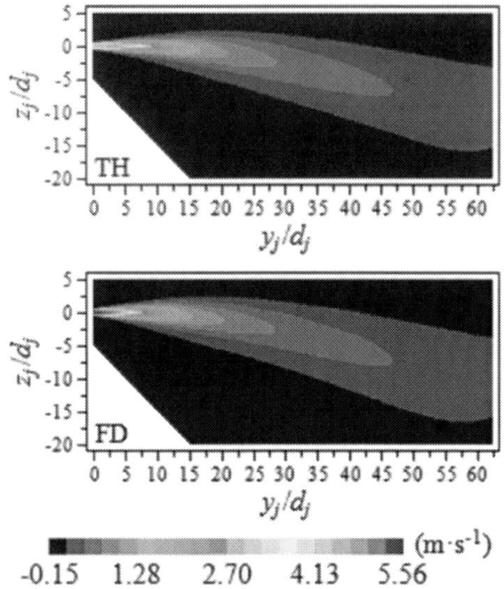

Fig. 6. Contours of jet axial velocity at plane $x = 0$ (y_j and z_j are jet longitudinal and jet radial distances.)

Fig. 6 shows the contours of axial velocity predicted by TH and FD. The results show that the axial velocities of TH and FD are found to decrease with increasing the longitudinal distance, which are similar to the general experiments of turbulent round jets [9]. For $y_j/d_j \leq 10$, the decay of centreline axial velocity of FD is faster than TH. For higher values of y_j/d_j, the decay of centreline axial velocity of two different jet discharge velocity profiles are slightly different. Further, the jet spreading of FD is slightly wider than TH, especially for y_j/d_j of about 12. This wider jet spreading of FD provides the higher convective transport as compared to the TH, which directly results in lower predicted overall mixing time.

According to these results, it can be seen that the overall mixing times predicted by TH and FD are lower than that obtained experimentally because these models overpredict the convective transport due to flat top liquid surface assumption [6]. Further, even when the jet inlet mass flow rate of TH and FD are identical, the different jet discharge velocity profiles represent the different jet flow and mixing patterns inside the vessels. Thus, the jet discharge velocity profile should be considered as another important parameter for developing the CFD model and mixing time correlation.

4 Conclusion

In this work, the effect of jet discharge velocity profile on jet flow pattern and mixing behaviour inside the pump-around jet mixing tank was numerically studied by considering the top hat (TH) and fully developed (FD) profiles. The overall mixing times and normalized concentration profiles predicted by two different jet discharge velocity profiles were compared with the measured data of Patwardhan [4]. The results revealed that even when the jet inlet mass flow rates of these CFD models were identical, the different jet discharge velocity profiles represented the different jet flow and mixing patterns inside the jet mixing vessels. That is, the different mixing patterns inside the tanks were directly affected by the different jet flow phenomena, such as decay of centreline velocity, jet spreading, etc. Further, the different jet flow characteristics of two discharge velocity profiles were caused by the differences in flow structures as reported by Xu and Antonia [9]. Hence, another important parameter for developing CFD model and mixing time correlation of pump-around jet mixing tank would be jet discharge velocity profile.

The authors would like to thank College of Advanced Manufacturing Innovation, King Mongkut's Institute of Technology Ladkrabang, Thailand for supporting the ANSYS FLUENT software.

References

1. H. Fossett, L.E. Prosser, Proc. I. Mech. E. **160**, 224 (1949)
2. E. Bumrungthaichaichan, Korean J. Chem. Eng. **33**(11), 3050 (2016)
3. L. Brooker, Chem. Eng. **30**, 16 (1993)
4. A.W. Patwardhan, Chem. Eng. Sci. **57**, 1307 (2002)
5. ANSYS Inc., *ANSYS Fluent Theory Guide: Release 15.0* (ANSYS Inc., USA, 2013)
6. E. Bumrungthaichaichan, A. Namkanisorn, S. Wattananusorn, J. Chin. Inst. Eng. (to be published)
7. E. Bumrungthaichaichan, N. Jaiklom, A. Namkanisorn, S. Wattananusorn, Sci. Res. Essays **11**(4), 42 (2016)
8. ANSYS Inc., *ANSYS Fluent Tutorial Guide: Release 15.0* (ANSYS Inc., USA, 2013)
9. G. Xu, R.A. Antonia, Exp. Fluids **33**(5), 677 (2002)

Rapid evaluation of moisture content in bamboo chips using diode array near infrared spectroscopy

Amrit Shrestha[1], *and Panmanas* Sirisomboon[1,*]

[1]Department of Agricultural Engineering, Faculty of Engineering, King Mongkut's Institute of Technology Ladkrabang, Bangkok 10520, Thailand

Abstract. Moisture in biomass plays important role during storage, combustion and pelletization. In order to measure moisture content in bamboo chips, two diode array near-infrared instruments, NIR-Gun (600-1100 nm at 2 nm intervals) and Micro-NIR (1150-2150 nm at 7 nm intervals), were used for scanning bamboo chips. Total number of samples used for developing model after removing outliers was 252. The circumference and moisture content of bamboos used were in the range between 16-39 cm and 39-86% wet basis (wb) respectively. Partial least squares regression technique was used to develop the model to predict the moisture content in bamboo chips. The R^2, SECV, SEP, bias and RPD of optimum model of NIR-Gun were found to be 0.924, 2.871% wb, 2.385% wb, -0.250% wb and 3.656, while for Micro-NIR model the values were found to be 0.743, 4.349% wb, 4.499% wb, 0.026% wb and 1.972 respectively. In prediction of moisture content in bamboo chip, both models show the effect of different constituents of bamboo more than moisture. This study indicates that the results are suitable for screening the moisture content in bamboo chips. This would be helpful for process controlling using the moisture parameter during drying, pelletization and thermochemical conversion.

1 Introduction

Biomass is multipurpose renewable energy which can be combusted directly for heat or convert into gas or liquid fuel. Bioenergy has always been one of the major sources of energy which contributes 10% in primary energy supply in 2012 where the share of renewable energy was 13.5% [1]. In the last few years, research on biomass has increased tremendously as an alternative energy because of increasing cost of fossil fuels, fossil fuel crisis and global warming problems originated mostly from fossil fuels combustion [2, 3]. The amount of greenhouse gases in atmosphere can be reduced by increasing the contribution of renewable energy for energy purpose [4]. A fast growing energy crops with high environmental benefits can commensurate the fossil fuels. In such case bamboos can be one of them.

Bamboos are a group of perennial evergreens vegetation which belongs to true grass family [5, 6] and are receiving great interest due to their high growth rate and better reduction of carbon footprint compared to an equivalent area of woody plants [6, 7]. Beside combustion and bioenergy application, bamboos can be used as a construction building material, household products, food, pulp and paper and composite boards [8, 9].

Moisture plays a vital role for determining the quality and price of fuel, storage, pellet production and combustion. High moisture results in swelling, disintegration and prone to mold growth. In other hand, moisture acts as a binder during pelletization. In addition, high moisture content will increase the cost of drying and makes the feedstock slippery producing lower quality pellets. Also, too dry material may plug the die holes during pelletization and increases the resistance [10, 11]. A research conducted on four different biomasses, oak, oak bark, pine, and cottonwood in the forms of sawdust, mulches, and chips, by Li and Liu [12] concluded that the necessary moisture for producing good-quality logs ranges from 5 to 12% for all the woody materials studied, and the optimum moisture content is in the neighborhood of 8%. Furthermore, Komilis *et al.* [13] concluded that both higher heating value and lower heating value decrease with increase in moisture content, whilst the calorific value depends on the individual organic matters. Moreover, Vamuka and Sfakiotakis [14] reported that the combustion rate decreases with increasing moisture content.

In diode array detector, several diodes are arranged in array in which each array measure the portion of incident light and hence increase the speed of measurement. In addition, diode array spectrometers are smaller in size which made possible to carry in a small box for field works. The ASTM hot air oven reference method for the moisture content measurement is time consuming. Near infrared (NIR) spectroscopy is a non-destructive, rapid, simple sample preparation, chemical free-analysis [15] and employs electromagnetic spectrum of wavelength ranging between 750 to 2600 nm which relates with vibration and combination overtones of the fundamental

* Corresponding author: panmanas.si@kmitl.ac.th

O—H, C—H and N—H bonds [16]. Thus, NIR spectroscopy has great potential to measure the moisture of the samples due to presence of O-H in water. NIR spectroscopy was applied for the evaluation of moisture content of rice straw by Jin and Chen [17], Miscanthus x giganteus and short rotational coppice willow by Fagan et al. [18], biomass pellet by Gillespie et al. [19]. NIR spectroscopy has been used not only to measure the moisture content of biomass but also the composition of biomass like cellulose, hemicellulose, lignin [20, 21], ash [18], heating value [22]. However, to this date no study has reported about potential of NIR spectroscopy to predict the moisture content of bamboo wood chips. The aim of this study was to measure the moisture content of bamboo chip by deploying NIR spectroscopy technique in conjugate with chemometric.

2 Experimental

2.1. Samples

The bamboo samples, *Dendrocalamus sericeus* cl. Phamon, were procured from Uttaradit, Thailand. The bamboo trees were cut about 10 cm from the ground level and about 1 m from the base of bamboos was selected for the study. The bamboo was then chopped by the chopping machine (P5508, Patipong, Thailand). The size of bamboo chips were less than 3 cm. The total number of bamboo trees were 90 consisting of various culm circumference ranging from 16-39 cm.

2.2 Near infrared scanning of bamboo chips, Dendrocalamus sericeus cl. Phamon

The chopped bamboo were transferred in aluminum cup (50 mm in diameter and 30 mm in height) and scanned by two NIR instrument: NIR-Gun (FQA, Fantec, Japan) in reflectance mode (over the short wavelength range from 600-1100 nm at 2 nm intervals and the reference material was polystyrene and Micro-NIR (JDSU, USA) in reflectance mode (operating in the long wavelength range between 1150-2150 nm) at 7 nm intervals and the reference material was spectralon. Each sample was divided into three sub-samples and each sub-sample was scanned two times. Hence, the total numbers of samples available for making NIR-model were 270. The samples were scanned on site in atmospheric condition of Thailand. The reflectance (R) spectra was transformed into absorbance, log (1/R), spectra using CA maker software (Fantec, Japan) and The Unscrambler X 10.3 (Camo, Norway) for spectra obtained by NIR-Gun and Micro-NIR respectively. The part of spectra that contain noise or unusual characteristics was erased and was not used for model development.

2.3 Determination of moisture content of bamboo chips, Dendrocalamus sericeus cl. Phamon

Moisture content on bamboo chips was determined according to the procedure described in ASTM International D4442 - 07 — Method A—Primary Oven-Drying Method. About 3g of the chopped bamboo chips, after NIR scanning, was taken in a dried aluminum cup and oven dried in a preheated oven for 3 hours at 103±2°C (ULM 500, Memmert, Germany) in a room relative humidity of less than 70%. The sensitivity of the balance (AR2140 Adventure, Ohauss,) was 0.1 mg. The temperature of the oven was calibrated by the thermocouple (51/52 II, Fluke, USA). The weight of the sample was measured in a 3 h drying interval until the weight was constant and moisture content, wet basis, in a sample was calculated as follows:

$$MC = \frac{m_i - m_f}{m_i} \times 100 \quad (1)$$

where, MC is moisture content in % wb, m_i and m_f are initial and final dried mass respectively.

2.4 Determination of outlier of reference data

The outliers of reference data were checked by standard normal distribution for the Z-score greater than as given by equation 2.

$$\frac{X_i - \overline{X}}{SD} \geq \pm 3 \quad (2)$$

where, X_i is the measured value of sample i $\overline{X}$ is the average value of all measured samples; SD is standard deviation of all measured samples.

2.5 Spectra pre-treatment and mathematical modelling

The spectra pre-treatments and model development were performed on The Unscramble X 10.3 software (Camo, Norway). After the wet-test, the reference data were merged corresponding spectral data. The data were then arranged in descending order, and data were separated into calibration and prediction set. The highest and lowest reference values were used in calibration set. Various pre-treatments on calibration data set were performed before model development which includes no pre-treatment, Savitzky-Golay (S-G) smoothing, area normalization, unit vector normalization, mean normalization, maximum normalization, range normalization, peak normalization, first derivative (second order polynomial with 11 and 21 points), second derivative (second order polynomial with 11 and 21 points), baseline offset, linear baseline correction, standard normal variate, de-trending, standard normal variate plus de-trending and multiple scatter correction. Partial least squares (PLS) regression technique was used to develop the NIR models and confirmed using full cross validation. The model was checked by test set, and the optimum model was selected for lowest number of factor, highest coefficient of determination of prediction, lower value of standard error of cross validation (SECV), standard error of prediction (SEP) and bias.

3 Results and discussion

The average pretreatment spectrum of bamboo chip scanned from two diode array instruments i.e. NIR-Gun and Micro-NIR of different moisture content is shown in Fig. 1 (a-b). On the spectrum of NIR-Gun, the peaks are seen around 657 nm and 959 nm. The bands vibration between 613-645 nm is associated with visible spectrum which represents the chlorophyll absorption [20, 23] and 950-1040 nm represents the second overtone of O-H stretch [16] For Micro-NIR spectrum, small rise is seen in the range of 1400-1500 nm and sharp peaks at 1876 nm which corresponds to first overtone of O-H bond (O-H stretch, internal OH bonds, single bridge and/or polymeric) and third overtone O-H deformation (primary and secondary alcohol) [16].

The circumference of bamboos used in this research to measure moisture content were in between 16-39 cm and the moisture content were in range of 39-86% wet basis. The descriptive statistics of calibration and prediction set for the measurement of moisture content are shown in Table 1. The outliers of reference data were removed by standard normal distribution for the Z-score greater than 3, while principal component analysis was used to find the spectral outlier using Hotelling's T-squared statistic which are isolated from the cluster. The total number of outlier was found to be eighteen, 7 from reference test and 11 from spectrum, in both cases. The model was created from remaining 252 sub-samples.

The selected wavelength range for creating the model was 615-1068 nm and 1158-2075 nm for short wavelength (NIR-Gun) and long wave length (Micro-NIR) respectively; however, the wavelength range of instruments are 600-1100 nm and 1150-2150 nm respectively. Because of noise at the beginning and end of spectra, some points were skipped. Firstly, the model was developed from the raw spectra without pre-treatment and later various pre-treatment on spectra were performed. The second derivative (second order polynomial and 5 points) was found to be the most optimum effective treatment with 7 PLS factors for the NIR-Gun spectra, while SNV+De-trending was found to be the most optimal pretreatment for the spectra scanned by Micro-NIR using 6 PLS factors. The second derivative helps to separate overlapping peaks and baseline shift on spectra caused by the light scattering and particle size into independent peaks of different concentration. After second derivative pretreatment of NIR-Gun spectra, the spectra are arranged in order of increasing moisture content and has common baseline. Higher moisture content depicts higher absorption peak and vice versa. On the other hand, the SNV+De-trending help to reduce multicollinearity, baseline shift and curvature in spectroscopic data.

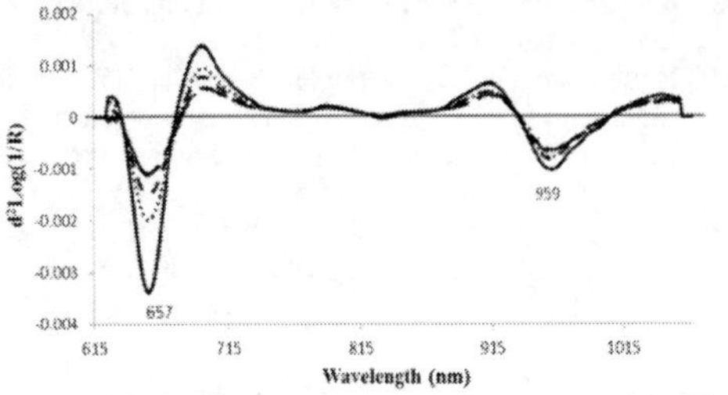

a)

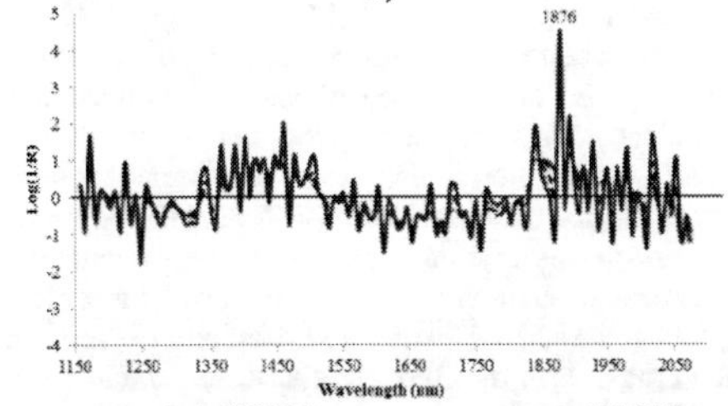

b)

Fig. 1. Spectra of bamboo chips of different moisture content (MC) on wet basis scanned by a) NIR-Gun spectrometer, pretreated by 2nd derivative and b) Micro-NIR spectrometer, pretreated by SNV + Detrending.

The R_p^2, SECV, SEP, bias and RPD of optimum model of NIR-Gun are 0.924, 2.871% wb, 2.385% wb, -0.250% wb and 3.656. In the case of Micro-Nir model, R_p^2, SECV, SEP, bias and RPD are found to be 0.743, 4.349% wb, 4.499% wb, 0.026% wb and 1.972 respectively. Williams [23], suggested that R^2 of 0.66-0.81 is ok for screening and some approximate calibration and 0.83-0.90 is usable with caution for most applications, including research. This shows that short wavelength was better than longer wavelength for the measurement of moisture content in bamboo.

Table 1. Statistical data for the measurement of moisture content (% wb) on bamboo chips sample used for developing PLS model.

Model	Calibration					Prediction				
	N	Max	Min	Mean	SD	N	Max	Min	Mean	SD
NIR-Gun	232	85.5853	39.0014	56.6868	9.0019	20	75.9423	41.3817	57.6277	8.7190
Micro-NIR	232	85.5853	39.0014	58.0856	9.5470	20	75.9423	41.3817	58.8670	8.8729

N is number of sample, Max is maximum, Min is minimum, SD is standard deviation

Regression coefficient plot of NIR-Gun model is shown in Fig. 2a. The most important variance used in computing the final calibration model can be described by the areas of spectrum for which the regression coefficients are biggest [23]. The highest peaks seen in regression coefficient plot were at 900, 942 and 977 nm which corresponds to the vibration of C—H str. 3rd overtone (CH_3) C—H str. third overtone (CH_2) and O—H str. second overtone (ROH and H_2O), CH_3 (C—H str. third overtone) respectively [24]. Minor peaks were also seen in the visible region between 600-700 nm which are related with chlorophylls absorption [20, 25].

Similarly, Fig. 2b depicts regression coefficient of the Micro-NIR model. The highest peaks seen in regression coefficient are 1366, 1809 nm. The peak at 1366 nm is assigned as the bond vibration of 2×C—O str. + C—H def. of CH_3 whereas the peak at 1809 nm is assigned as the bond vibration of O—H str. + 2×C—O str. of cellulose. In regression coefficient plot, the moisture peak at 1940 nm appears as a small shoulder peak with small peak at 1927 nm which is assign as O—H stretching and H—OH bending combination from water molecules [26].

The models, in both cases, were seen to be mostly affected by the different constituent of bamboo. The independent moisture peaks were not seen clearly, however, strong correlations can be seen between moisture and related absorbance peak. Shifting of bands was seen in more common due to the fluctuating environment temperature. The models mostly represent the bands associated with hydrocarbon present in bamboo rather than moisture in either case. Bamboo is a lignocellulose biomass which is mainly composes of cellulose, hemicellulose and lignin. The major constituent of cellulose is glucose units [27] while hemicellulosic is xylose [28], and lignin is a polymer of aromatic compounds [27].

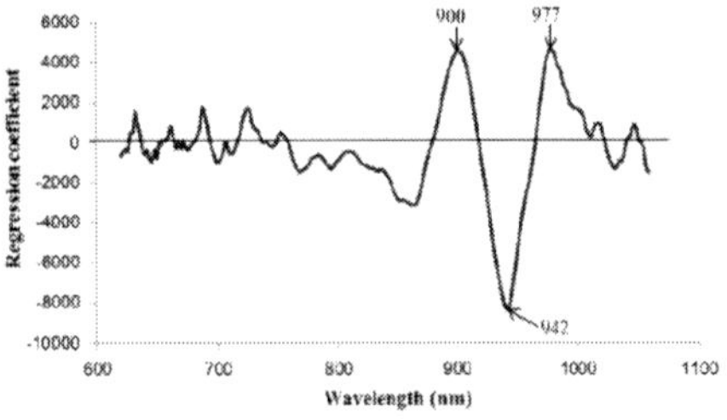

a)

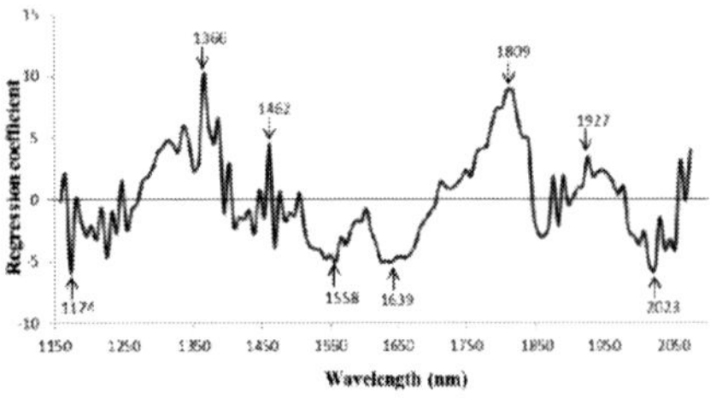

b)

Fig. 2. Regression coefficient plot of optimum PLS model for measurement of moisture content in bamboo chips developed by using spectra scanned from a) NIR-Gun spectrometer and b) Micro-NIR spectrometer.

Acknowledgements The authors would like to thank near infrared spectroscopy research center for agricultural product and food (www.nirsresearch.com), King Mongkut's Institute of Technology Ladkrabang for providing financial support and instruments. Also, the authors would like to acknowledge the plantation owner for providing samples, and all the people who helped to conduct this research.

References

1. K.B. Venkata, World Bioenergy Association (2015)
2. T.Y.Y.H. Sivasangar, Z. Zainal, K. Kitagawa, Int. J. Hydrog. Energy. **38,** 16011-16019 (2013)
3. S. Vassilev, D. Baxter, L. Andersen, C. Vassileva, Fuel. **89**, 913–933 (2010)
4. I.E.A. (IEA), *IEA Publications*
5. S. Biswas, Indian For. **114**, 514-531 (1988)
6. M. Kuttiraja, R. Sindhu, P.E. Varghese, S.V. Sandhya, P. Binod, S. Vani, A. Pandey, R.K. Sukumaran, Biomass Energy. **59**, 142-150 (2013)
7. A. Liebman, T. Einav, *Bamboo: an untapped and amazing resource*. Available from: http://www.unido.org/index.php? Id= 1000276.
8. P. Chaowan, J. Mater. Sci. Res. **2**, 90-102 (2013)
9. J.M.O. Scurlock, D.C. Dayton, B. Hames, Biomass Bioenergy. **19**, 229-244 (2000)
10. P. Lehtikangas, Biomass Bioenergy. **20**, 351–360 (2001)
11. H. Resch, *World Wood*. 14–21 (1989)
12. Y. Li, H. Liu, Biomass Bioenergy. **19**, 177-186 (2000)
13. D. Komilis, K. Kissas, A. Symeonidis, Waste Manage. **34**, 249-255 (2014)
14. D. Vamvuka, S. Sfakiotakis, Thermochim Acta. **526**, 192-199 (2011)
15. P.C. Williams, in P. Williams, K. Norris, Editors, *American Association of Cereal Chemists*. 107-142 (1987).
16. I. Murry, P.C. Williams, in P. Williams, K. Norris, Editors. *American Association of Cereal Chemists*. 17-34 (1987)
17. S. Jin, H. Chen, Ind Crop Prod. **26**, 207-211 (2007)
18. C.C. Fagan, C.D. Everard, K. McDonnell, Bioresource Technol **102**, 5200-5206 (2011)
19. G.D. Gillespie, C.D. Everard, K.P. McDonnell, Energy **80**, 582-588 (2015)
20. D. Sabatier, L. Thuries, D. Bastianelli, P. Dardenne J. Near Infrared Spectrosc. **20**(3), 371-385 (2012)
21. J. Xue, Z. Yang, L. Han, Y. Liu, Y. Liu C. Zhou, Appl Energy, **137**, 18-25 (2015)
22. J. Posom, P. Sirisomboon, Biosyst Eng, **130**, 52-59 (2015)
23. P. Williams, *PDK Grain*, (2007)
24. B.G. Osborn, T. Fearn, P.H. Hindle, *Longman Scientific & Technical* (1993)
25. L. Stryer, in K.B. Venkata, Editor. *World Bioenergy Association* (1995)
26. J. Workman, L.J. Weyer, *CRC press* (2008)
27. K. Othmer, *Concise Encyclopedia of chemical technology*, (2001)
28. J.L. Wen, L.P. Xiao, Y.C. Sun, S.N. Sun, F. Xu, R.C. Sun, X.L. Zhang, Carbohydr Res. **346**, 111-120 (2011)

* Corresponding author: panmanas.si@kmitl.ac.th

MATEC Web of Conferences **192**, 03021 (2018) https://doi.org/10.1051/matecconf/201819203021
ICEAST 2018

Gross calorific and ash content assessment of recycled sawdust from mushroom cultivation using near infrared spectroscopy

Jetsada Posom[1,3], *Jirawat* phuphanutada[2,4], and *Ravipat* Lapcharoensuk[2,4*]

[1]Department of Agricultural Engineering, Faculty of Engineering, Khon Kaen University, Khon Kaen, Thailand
[2]Department of Agricultural Engineering, Faculty of Engineering, King Mongkut's Institute of Technology Ladkrabang, Bangkok, Thailand
[3]Applied Engineering for Important Crops of the North East Research Group, Department of Agricultural Engineering, Faculty of Engineering, Khon Kaen University 40002, Thailand
[4]Post Harvest Innovation Research and Development Laboratory, Department of Agricultural Engineering, Faculty of Engineering, King Mongkut's Institute of Technology Ladkrabang, Bangkok, Thailand

Abstract. The aim of this study was to use the near infrared spectroscopy for predicting the gross calorific value (GCV) and ash content (AC) of recycled sawdust from mushroom cultivation. The wavenumber was in range of 12500-4000 cm^{-1} with the diffuse reflection mode was used. The NIR models was established using partial least square regression (PLSR) and was validated via using full cross validation. GCV model provided the coefficient of determination (R^2), root mean square error of cross validation (RMSECV), ratio of prediction to deviation (RPD), and bias of 0.90, 445 J/g, 3.19 and 4 J/g, respectively. The AC model gave the R^2, RMSECV, RPD and bias of 0.83, 1.7000 %wt, 2.44 and 0.0059 %wt, respectively. For prediction of unknow samples, GCV model provided the standard error of prediction (SEP) and bias of 670 J/g and -654 J/g, respectively. The AC model gave the SEP and bias of 1.84 %wt and 0.912 %wt, respectively. The result represented that the GCV and AC model probably used as the rapid method and non-destructive method.

1 Introduction

Mushroom, one of the most popular and a valuable edible fungus is highly appreciated in human diet because of its delicacy, texture and extreme amounts of proteins, minerals and bioactive compounds [1]. In Thailand, many cultivated mushrooms are well-established consumer acceptance such as *Auricularia spp*., *Volvariella volvacea*, *Pleurotus sajor-caju* [2-4]. Para rubber tree (*Hevea brasiliensis*) sawdust are used for mushroom cultivation popularly in Thailand due to the rubber tree sawdust is large waste from the local furniture industry [5]. Most of the time, the sawdust after mushroom cultivation is often reused to fertilizer, and it is either piled up in fields for a slow degradation. The recycle of these agricultural waste materials may help to improve the regional economies of the country.

Biomass are forms of renewable energy directly derived from organic materials such as plants, fast-growing trees and agricultural residues and waste. The most important characteristic of biomass is used to design planning and operation of thermal power plant is the energy content (heating value). The heating value defines how biomass is used to achieve process efficiency. Heating value can be divided into higher heating value (HHV) or gross calorific value (GCV); and lower heating value (LHV) or net calorific value (NHV). The bomb calorimeter is instrument for determination of the GCV. Ash (A) content is one of the largest problems of Biomass combustion. In combustion process, AC often called slagging and fouling deposit. These problems have negative impacts on the boiler, such as reducing the heat exchange and boiler efficiency [6]. The AC has a negative effect on the GCV, increasing 1% in A of wood fuels decrease a 0.2 MJ kg^{-1} in its GCV [7-8]. Nowadays, the AC can determine with dry ashing method. However, both standard methods for the determination of GCV and A require long analysis times and highly skilled technicians, and they are destructive to samples. Consequently, a novel analytical method that is fast, easy and non-destructive for evaluating GCV and AC of recycled sawdust is highly sought after within biomass processing.

Near infrared spectroscopy (NIRs) have been used to evaluate properties of biomass in many researches. GCV of various biomass was successfully predicted using NIRs such as millet [9], *leucaena leucocephala* [10], *jatropha curcas* kernels [11], and straw [12]. In addition, Fagen et al. [13] studied application of NIRs for measuring moisture content, gross calorific value (GCV), carbon (C) and ash content (AC) of two dedicated bioenergy crops which results of their showed performance of NIRs technique to evaluate these properties. Therefore, the NIRs is interesting technique to apply for GCV of recycled sawdust evaluation. The aim of this study was to the use of near infrared spectroscopy for predicting the

* Corresponding author: ravipat.la@kmitl.ac.th

GCV and AC of recycled sawdust from mushroom cultivation.

2 Material and method

2.1 Sample preparation

The recycled sawdust samples obtained from mushroom cultivation of 3 breeds which are Indian oyster (*Pleurotus pulmonarius*), Bhutan oyster (*Pleurotus pulmonarius*) and Poplar (*Agrocybe aegerita*). All samples were sawdust of Para rubber tree. The samples were collected from 16 mushroom farms in Nakhon Pratom province. Two samples were kept on each farm. Therefore 32 samples were for calibration set. These samples were used for modelling. After modelling, 10 samples of recycled sawdust were collected from Ratburi district which were defined to external validation set (unknow samples). Hence the total sample equal to 42 samples (32 for calibration set and 10 samples for external validation set). All samples were dried by a hot air oven (Memmert, model ULM 500, Germany) at 105 °C until got constant weight. After that, it was kept at the room temperature for 5 days to balance relative humidity and kept in aluminium bag for experiment.

2.2 NIR spectra scanning

Each sample was poured into sample cup, dimension was 35 mm of height and 50 of diameters. The samples were scanned using Fourier-Transform Near Infrared spectrometer (Bruker Ltd., Germany). The scanning condition was: diffuse reflectance mode, wavenumber range between 12500-4000 cm^{-1} (800-2500 nm), 32 scans per sample with a resolution of 8 cm^{-1}, at room temperature of 25±2 °C, and absorbance of log 1/R unit where R is reflection of NIR radian from the sample. The sample at the bottom of the cup, it was irradiated by NIR radiation, was obtained to determine the reference data i.e. GCV and AC since it contained more NIR spectral information. It was kept in plastic bag for experiment.

2.3 Gross calorific and ash content analysis

After scanning, about 0.5±0.1 g sawdust sample was compressed into pellet which used to determine GCV and carried on by bomb calorimeter (c200, IKA, Germany). Each sample was done duplicate and averaged to be one. After burning, AC was calculated by

$$\text{Ash}(\%\text{wt}) = \frac{w_a}{w_s} \times 100 \quad (1)$$

where w_a is weight of solid residue after burning (g) and w_s is weight of sample before burning (g).

2.4 Repeatability and maximum (R^2_{max})

Accuracy of reference method was determined as repeatability (Rep). Rep is the standard deviation between duplicate, it is the error obtained from man, machine, and method. Rep was also applied to calculate the maximum coefficient of determination (R^2_{max}), it was calculated as followed Dardenne [14].

$$R^2_{max} = \frac{SD_y^2 - Rep^2}{SD_y^2} \quad (2)$$

SD_y is standard deviation of reference data. As seen the equation, R^2_{max} is high when either high SD_y or low Rep. The R^2 will be highest if there is no spectral error [14].

2.5 NIR spectroscopy modelling

The total samples of 32 were used for model development. The PLS technique was used as mathematical algorithm and validated by full cross validation. The NIR spectra either non-preprocessed or preprocessed were applied. The following preprocessing technique were constant offset elimination, 1st derivative, 2nd derivative, multiplicative scatter correction (MSC) and standard normal variate (SNV), and scatter plot of regression coefficient and X-loading weight was represented. The performance of calibration model was then stated by statistical terms of coefficient of determination of calibration set (R^2) and validation set (r^2), root mean square error of estimation (RMSEE), root mean square error of cross validation (RMSECV), ratio of prediction to deviation (RPD) and bias. After model development, to ensure that whether model can predict the further sample. NIR spectroscopy models was tested again by the 10 samples of external validation set or unknow samples. The NIR spectroscopy models were performed using the multivariate statistical software of OPUS software V. 7.0.129 (Bruker Optik GmbH, German).

3 Results and discussion

3.1 GCV and ash content of sawdust

Fig. 1 displayed the relationship between GCV and AC of recycled sawdust, the GCV decreased with increasing AC. The coefficient of determination (R^2) equals to 0.785, the linear equation was:

$$Y = -0.0027X + 50.102 \quad (3)$$

when X is AC (%wt) and Y is GCV (J/g).

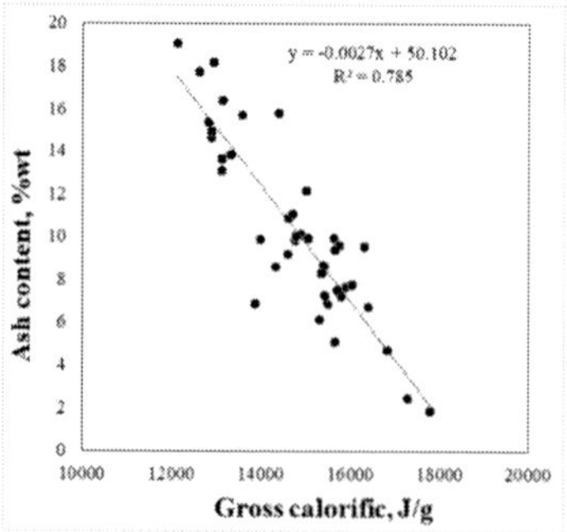

Fig. 1. The relationship between gross calorific and ash content of sawdust.

Table 1 demonstrated the statistical data i.e. maximum, minimum, mean, and standard deviation (SD) for calibration set and external validation set. The GCV and AC were between 12119-17777 J/g and 1.866-19.034 %wt, respectively.

Table 1. Statistical data used for model development and external validation set.

Data set	Parameters	Max	Min	Mean	SD
Calibration	GCV (J/g)	17777	12119	14590	1453
	AC (%wt)	19.034	1.866	10.619	4.297
External validation	GCV (J/g)	17282	14383	15353	867
	AC (%wt)	15.815	2.459	9.224	3.669

Repeatability and R^2_{max} including calibration set and unknown sample were illustrated in Table 2. It was found that the capacity of NIR spectroscopy can give the R^2_{max} of 0.98 and 0.90 for GCV and AC, respectively.

Table 2. Repeatability and maximum coefficient of determination (R^2_{max})

Parameters	Repeatability	R^2_{max}
GCV (J/g)	155	0.98
AC (%wt)	1.273	0.90

3.2 NIR spectra of sawdust sample

Fig. 2 demonstrated raw spectra of sawdust. The NIR spectra of sawdust show obvious absorption peaks at wave band for 5188 cm^{-1} (1928 nm), 4779 cm^{-1} (2092 nm) and 4409 cm^{-1} (2268 nm), they related to O-H stretching and HOH combination of water [15]; 2×O-H deformation + 2×C-O stretching of starch [16]; and lignin [15].

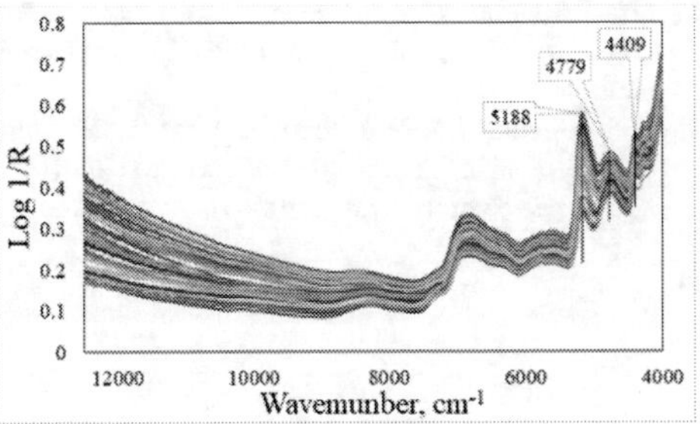

Fig. 2. NIR spectra of sawdust samples.

3.3 NIR spectroscopy modelling

Table 3 gave the result of PLS model predicted the GCV and AC. The optimal model for GCV was generated by wavenumber range of 7502.2-4597.7 cm^{-1}, using the preprocessing method of 1st derivative + MSC, and PLS factor of 7, the r^2 RMSECV, RPD and bias were 0.90, 445 J/g, 3.19, and 4 J/g, respectively. Meanwhile, AC model was developed using wavenumber range of 6102-5446.3 cm^{-1} without preprocessing, and PLS factor of 5, r^2, RMSECV, RPD and bias were 0.83, 1.7 %wt, 2.44 and 0.006 %wt, respectively. r^2 and RPD were recommended by Williams [17], r^2 was 0.83-0.90, the model could be used with caution for most application, and 0.66-0.81 could be used for screening and approximating, meanwhile, RPD between 3.1-4.9 used for any application and 2.4-3.0 for rough screening. Hence the GCV model could be used with caution for most application while AC model could be used for screening and approximating. However, the ratio between bias and its mean were approximately 0.025% (4 J/g to 14590 J/g) for GCV model and 0.055% (0.006%wt to 10.619 %wt) for AC model, their error was very small. Hence, the author recommended that it was acceptable for screening. Fig. 3 represents scatter plot between calibration set and validation set by full cross value validation of GCV and AC, respectively.

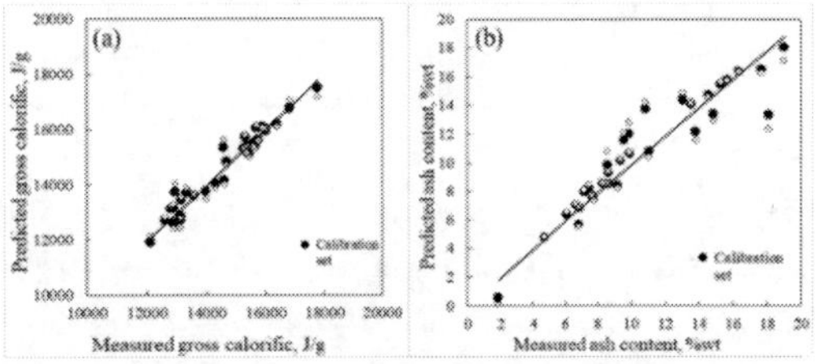

Fig. 3. Scatter plot between actual value and predicted value of a) gross calorific value and b) ash content both of calibration and validation set.

Fig. 4a and 4b showed the regression coefficient plot and X-loading of GCV model. The obvious peaks were at around 5419-4597 cm^{-1} (1845-2175 nm), related to starch (1960 and 2100 nm), C-O stretch combination (2100 nm); asym C−H stretch/C−H deformation combination of HC=CH (2170 nm) [18]. As we know, AC is the inorganic material, there is no interaction with NIR radian. Author opinions that NIR spectroscopy can predict AC because the inner relationship between AC and GCV was used to generate with mathematical algorithm. Fig. 4c and 4d illustrated the regression coefficient plot and X-loading of AC model. The wavenumber range between 6102-5402 cm^{-1} (1638-1851 nm) was optimal, this band covers the vibration band of aromatic (1685 nm), CH_3 (1695 nm and 1705 nm), CH_2 (1725 and 1765 nm), cellulose (17810 and 1820 nm), and water (1790 nm) [18]. The regression coefficient plot and X-loading of any wavenumber is high, means its vibrational band strongly effect to the prediction of predicted value.

Table 3 Result of PLS model for predicting gross calorific and ash content.

Parameters	Wavenumber (cm^{-1})	Pre-processing	Calibration set			Validation set			
			Factor	R^2	RMSEE	r^2	RMSECV	RPD	Bias
GCV (J/g)	7502.2-4597.7	1st derivative + MSC	7	0.95	354	0.90	445	3.19	3.69
AC (%wt)	6102-5446.3	Non- preprocessed	5	0.90	1.49	0.83	1.7	2.44	0.006

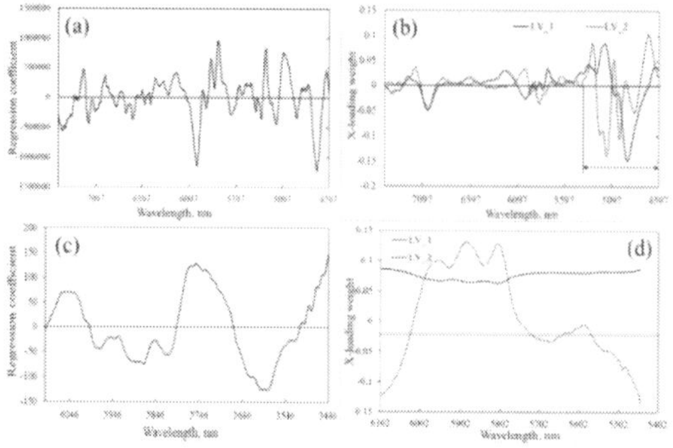

Fig. 4. regression coefficient (a and c) and X-loading (b and d) for gross calorific and ash content, respectively. LV_1 and LV_2 are latent variable for factor 1 and 2, respectively.

3.4 Prediction of unknow samples

Fig.5 shows the measured value, predicted value, and target line of unknown sample obtained from various area, the total sample was 10. The performance gave the SEP and bias of 671 J/g and -654 J/g for GCV model, respectively, while AC provided SEP and bias of 1.840 %wt and 0.912 %wt, respectively.

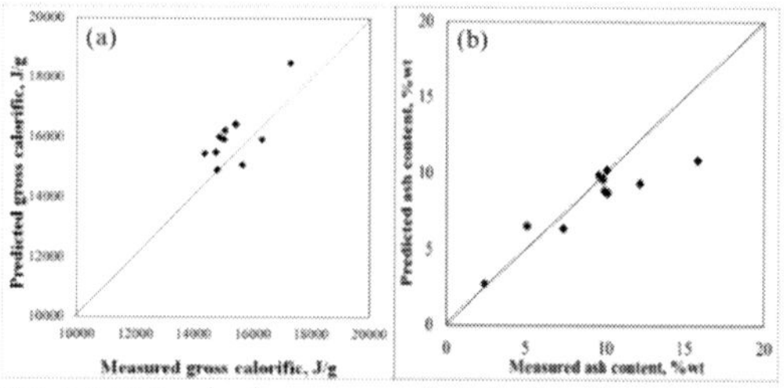

Fig. 5. Measured value versus predicted value of a(gross calorific value and b(ash content for unknown sample.

4 Conclusion

As seen the results, shows that NIR spectroscopy could be used as rapid and non-destructive for quality assurance and the prediction of GCV and AC content of sawdust. The model for GCV and AC gave a superior performance with validated by the full cross validation. To ensure the capacity of calibration model, the unknown sample collected from different sawdust milled plant were used to test. The performance for predicting unknown sample set was acceptable for screening. The meaning was that unknown sample is the representative population for future. This research was useful for qualitative screening the biomass to the thermal conversion plant.

Acknowledgment

The authors thank the Near Infrared Spectroscopy Research Center for Agricultural Product and Food (www.nirsresearch.com) at King Mongkut's Institute of Technology Ladkrabang, Bangkok, Thailand, for the use of their instruments.

References

1. B.A. Wani, R.H. Bodha, A.H. Wani, J. Med. Plant Res. **4**, 2598–2604, (2010)
2. R. Sanmee, B. Dell, P. Lumyong, K. Izumori, S. Lumyong, Food Chem. **82**, 527-532, (2003)
3. A. Kettawan, K. Chanlekha, R. Kongkachuichai, R. Charoensiri, Pak. J. Nutr. **10**, 1094-1103, (2011)
4. A. Srikram, S. Supapvanich, ANRES. **50(6)**, 432-436, (2016)
5. M. Thevasingh, W. Pale, W.C. Kyung, *Mushroom Growers Handbook 2 Shiitake Cultivation* (Mushworld, Seoul, 2005)
6. C. Yongtie, Y. Wenming, Z. Zhimin, X. Mingchen, P. Subbaiah, Energy Procedia **143**, 623-628 (2017)
7. A. Monti, N.D. Virgilio, G. Venturi, Biomass Bioenergy **32(3)**, 216-23, (2008)
8. G.D. Gillespie, C.D. Everard, K.P. McDonnell, Energy **80**, 582-588, (2015)
9. K. Zhang, L. Zhou, M. Brady, F. Xu, J. Yu, D. Wang, Energy **118**, 1353 – 1360, (2017)
10. J. Posom, A. Shrestha, W. Saechua, P. Sirisomboon, Energy **107**, 464 – 472, (2016)
11. J. Posom, P. Sirisomboon, Biosyst Eng., 130, 52 – 59. (2015)
12. C. Huang, L. Han, Z. Yang, X. Liu, Energy Convers. Manag. **49**, 3433 – 3438, (2008)
13. C.C. Fagan, C.D. Everard, K. McDonnell, Bioresour. Technol. **102(8)**, 5200-5206, (2011)
14. P. Dardenne, NIR News **21**, 8–14, (2010)
15. J. Workman, J.R.L. Weyer, *Practical Guide to Interpretive Near-infrared Spectroscopy* (Taylor and Francis, Boca Raton, 2007)
16. B.G. Osborne, T. Fearn, *Near Infrared Spectroscopy in Food Analysis* (Longman Science & Technical, London, 1986)
17. P. Williams, *Near-Infrared Technology Getting the Best Out of Light Edition 5.0.* (PDK Grain, Nanaimo, 2007)
18. J.S. Shenk, J.J. Workman, Jr., M.O. Westerhaus. *Hand book of Near-Infrared Analysis* (CRC press, Taylor & Francis Group, 2008)

Gross calorific value estimation for milled maize cob biomass using near infrared spectroscopy

Jetsada Posom[1, 2] and *Natrapee* Nakawajana[3,*]

[1]Department of Agricultural Engineering, Faculty of Engineering, Khon Kaen University, Khon Kaen, 40002, Thailand.
[2]Applied Engineering for Important Crops of the North East Research Group, Department of Agricultural Engineering, Faculty of Engineering, Khon Kaen University 40002, Thailand.
[3]Department of Agricultural Engineering, Faculty of Engineering, King Mongkut's Institute of Technology Ladkrabang. 1 Chalongkrung Road, Ladkrabang, Bangkok 10520, Thailand.

Abstract. The maize cob biomass is one of important biomass crops in Thailand. Nowadays, the use of the biomass as renewable resource is increasing, especially residue agriculture waste. As we know that the biomass properties impact combustion, in order to achieve boiler efficiency, its energy characteristics of biomass was required immediately before burning. This work uses the FT-near infrared spectroscopy to estimate gross calorific value (GCV) of maize cob as the rapid method. Each sample was scanned using diffuse reflectance mode at a wavenumber range between 12500-3600 cm^{-1}. The scanning was done with a resolution of 8 cm^{-1} and completed 32 scans per sample, then averaged to be one spectrum. The results showed that this technique could decrease a processing time to 1-2 minutes per sample to determine GCV whereas alternatively the current method used a processing time of 25-30 minutes per sample. The capacity of the model gave root mean square error of cross validation (RMSECV) of 91.1 Jg^{-1}, which was low. Hence, the model was acceptable and cloud be used for screening.

1 Introduction

Maize cob is a waste from agricultural process after the seed is removed from the cob. Maize cob can then become a habitat of bird and rat and moreover, bad smell could happen as well. This is the important problem of traders. Previously, the maize cob was used as a fertilizer but it was not popular for farmer because it takes a long time for decomposition. Hence, it has been initiated to use the maize cob as a fuel.

Thailand has imported maize cob. It's a waste of agricultural product from one of the economic crops of Thailand as a biomass fuel. When the maize cob is burned, it has calorific value approximately 17,000 kJ/kg, and it takes a long time to burn before becoming the ash. This implies that it has a characteristic of good fuel.

The major aspect of waste residue properties to be renewable energy is the gross calorific value (GCV) which is also known as the higher heating value. GCV is total energy in biomass released during burning process, which takes into account the latent heat of vaporization of water in the combustion products. Normally, GCV can be measured by bomb calorimeter but this method takes a processing time about 25-30 minutes per sample for measuring and is subject to high cost of analysis.

Near infrared (NIR) spectroscopy is a non-destructive and rapid technique for estimating a quality of food and agricultural products., In addition, there have been some reports that demonstrated the use of NIR spectroscopy for the evaluation of the GCV and other properties of biomass such as the evaluation of heating value of bamboo [1], the evaluation of moisture content and GCV of *Leucaena leucocephala* pellets [2], the prediction of GCV of *Miscanthus* [3], the GCV prediction of straw [4], the evaluation of moisture content and the GCV of oil-extracted residue of *Jatropha curcas* [9] and the analysis of GCV and elemental compositions of sorghum [10]. The results of their studies suggested that NIR spectroscopy had the potential to estimate the calorific value of biomass fuel. Hence, the objective of this research was to assess the potential of NIR spectroscopy to estimate GCV of milled maize cob biomass.

2 Materials and Methods

2.1 Sample

60 maize cob samples were collected from different area. The 60 samples were divided into two groups, 50 samples were used for utilizing calibration model and 10 samples were used as unknown sample for testing the calibration model. After harvest each sample from a plant, they were crushed into 5 cm and then dried to a constant weight by hot air oven (Memmert, model ULM 500, Germany) at 105 °C for 24 h, this process was

* Corresponding author: natrapee.na@kmitl.ac.th

preparing sample for easy milling in next process. Then, all sample were milled by milling machine (FRITSCH, 14.3000/10857, Germany) through a hole sizes of 2.00 mm and kept in an aluminium bag prior to the experiment. Next, each sample was measured the NIR absorbance using FT-NIR spectrometer and then some part of sample was measured the GCV by bomb calorimeter and other part was measured the moisture content for reference data by hot air oven at 105 °C, 24 h. The averaging moisture content of 60 samples were around 9.43% dry basis.

2.2 NIR spectral acquisition

Each milled maize cob sample was poured into a quartz bottom sample cup having dimension of 43 mm in diameter and 50 mm in height. For the reflectance mode, the thickness of sample was confirmed as infinity and must be ensured that there was no light leaked or transmitted through the sample. Fourier-transform infrared spectroscopy (FTNIR) (MPA, Bruker, Germany) was applied for scanning and absorbance was derived in log 1/R unit. Scanning conditions involved a resolution of 8 cm^{-1}, wavenumber range of 12500-3600 cm^{-1}, and the number of scanning of 32.

2.3 Reference data

After the milled maize cob samples had been scanned, the milled maize cob was sampled from the bottom of cup due to direct absorption by NIR radiation. About 0.5 g of sample was pelletized and the GCV was determined using a bomb calorimeter (C200, IKA, Germany) in isoperibol mode. The bomb calorimeter was calibrated by pelletized benzoic acid (IKA C 723, IKA, Germany).

After the GCV had been measured, the outliers were then checked by $((x_i-\bar{x}))/SD \geq \pm 3$, where x_i is the reference data of sample i. x and SD are the average and standard deviation of the reference data. If it was found, it was then rejected from the data set and was not used for modelling.

2.4 Data processing and NIR spectroscopy modelling

The NIR spectra and its corresponding reference data of GCV of fifty samples were used for modelling. The NIR models for predicting the GCV were constructed by PLS regression and validated by full cross validation. The spectral pre-treatment and modelling were carried on by OPUS software, version 7.0.129, Germany. Before the model development, the NIR spectra were pre-processed by constant offset elimination, straight line subtraction, vector normalization, min-max normalization, MSC (multiplicative scatter correction), first derivatives, second derivatives, the combination between first derivatives and straight line subtraction, the combination between first derivatives and vector normalization and the combination between first derivatives and MSC.

The effective model was obtained by lowest root mean squares error of cross validation (RMSECV), from which the number of PLS factor, spectral pre-treatment technique, and wavenumber range were listed. After that, the regression coefficient versus optimal wavenumber range was plotted. After modelling, the external sample of ten were used for tested the PLS model.

The performance of the GCV predicting model was stated by determination of the coefficient of determination (R^2), bias, root mean square error of cross validation (RMSECV) and ratio of standard error of cross-validation to deviation (RPD).

3 Results and discussions

3.1 Ground maize cob spectra

The raw spectra of 60 milled maize cob biomass samples in a range of 12500-3600 cm^{-1} were illustrated in Fig. 1. In the raw spectra of milled maize cob sample, the main absorption bands were observed at 5820 (1718 nm), 5180 (1930nm), 4408 (2268 nm), and 4266 cm-1 (2344 nm), corresponding to C-H stretching first overtone of hydrocarbon [5], O-H stretching/HOH deformation combination of starch [6], O-H stretching C-O stretch combination of cellulose [6], and C-H methylene of hydrocarbon [5].

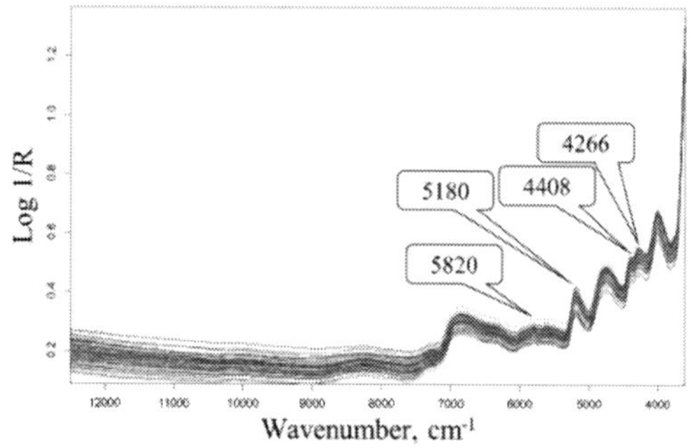

Fig. 1. The raw spectra of 60 milled maize cob biomass samples across the wavenumber of 12500 to 3600 cm^{-1}

3.2 Reference data of the GCV

Statistical data of calibration set, and external sample set for GCV of rice husk was demonstrated in Table 1 which shows the maximum, minimum, average, and standard deviation (SD). The maximum and minimum value of calibration set was higher and lower than external sample set, respectively, which covering the external sample. Then the calibration set can be representative sample for future prediction.

Table 1. Statistical values of gross calorific value data of maize cob milled samples used in model development.

	Calibration set	External sample
Max)Jg^{-1}(	17800.5	17431.0
Min (Jg^{-1})	17005.0	17253.0
Mean (Jg^{-1})	17405.0	17351.2
SD (Jg^{-1})	178.2	62.9

3.3 Near infrared spectroscopy models for GCV of milled maize cob sample

The effective calibration model was optimized using the wavenumber range of 5450.2-4246.7 cm^{-1}. The spectral pre-treatment of second derivative and the number of PLS factor listed from 1 to 3 was used. The PLS factor of optimal model was selected which gave the lowest RMSECV. Fig. 2a. displays the RMSEE versus PLS factor for calibration set whereas RMSECV versus PLS factor for validation set is shown in Fig. 2b. The PLS factor of three gave the lowest RMSECV which was then used for modelling.

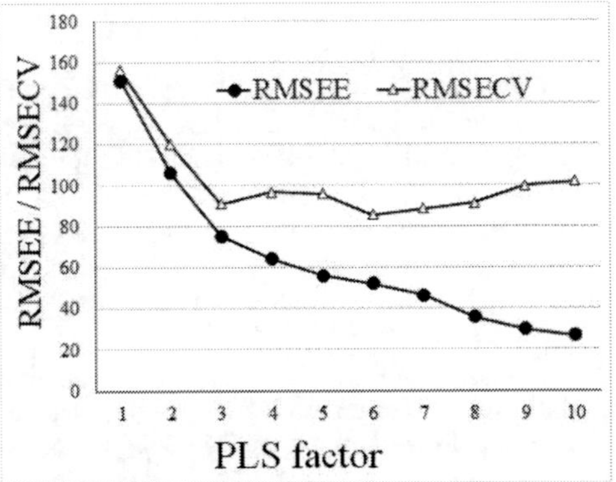

Fig. 2. RMSEE/RMSECV vs rank (PLS factor) for calibration set, and validation set

The PLS model was tested using leave-one-out cross-validation providing R^2, RMSEE, r^2, RMSECV, RPD, and bias of 0.83, 75 Jg^{-1}, 0.73, 91 Jg^{-1}, 1.94, and 0.293 Jg^{-1}, respectively. Zornoza et al. [7] recommended that if R^2 and RPD are between 0.66 to 0.80 and 2.0 to 2.5, respectively, it permitted only approximate prediction and could be used with screening and some other "approximate" calibrations [8]. The R^2 and RPD provided were less than 0.66 and 2, respectively, which meant the A content was poorly predicted [8]. William [8] suggested that R^2 between 0.50 and 0.64 could be used for rough screening. The ratio of bias to its mean of measured value were 0.0165 % (0.293 Jg^{-1}÷17405.0 Jg^{-1}) for GCV, which was very small. So that, we recommended that the models were suitable for estimation of the GCV of milled maize cob sample.

R^2 stated the percentage of proportion of the variance in GCV that can be explained by variance in absorption value [26]. For example, if R^2 was 0.90, means that 90% of variance in GCV was explained by variance in NIR spectra and 10% cannot be counted by NIR spectra (unexplained variance). Bias mentions the overall accuracy of the GCV model [26]. The low value of RPD means the measured value is not robust. If the model had a high RMSECV, then that modelling should need an increasing number of the sample set or sometimes it is seen that not necessary to develop model [26].

Fig. 3a and b illustrated comparison of the GCV of maize cob milled sample predicted by near infrared spectroscopy and measured by the bomb calorimeter of the calibration model and the validation model. The regression coefficient of GCV was shown in Fig. 4. Shown in Fig.5 was first derivative spectra of maize cob milled sample used for model development. The obvious peaks occurring on regression coefficient plot and second derivative spectra were at 5801 cm^{-1} (1723 nm), 4412 cm^{-1} (2266 nm), and 4293 cm^{-1} (2329 nm), related to C-H methylene (CH_2) of hydrocarbons, O-H stretching/C-O stretching combination of cellulose, and C-H stretching/CH_2 deformation combination of starch [6].

As reported by Posom and Sirisomboon [9], the band of cellulose and fiber had high impact to the forecast of the HV of the oil extracted residue of *Jatropha curcas* kernels. In addition, Zhang et al. [10] also reported that the important band for estimating HHV of sorghum was the C-H stretching and CH_3 structure.

To confirm whether calibration model can estimate the GCV of future sample, the 10 external samples were tested using calibration model, from which the outcome was shown in Table 2. The performance test had a low RMSEP and bias with 77.3 Jg^{-1} and -13.7 Jg^{-1}, respectively.

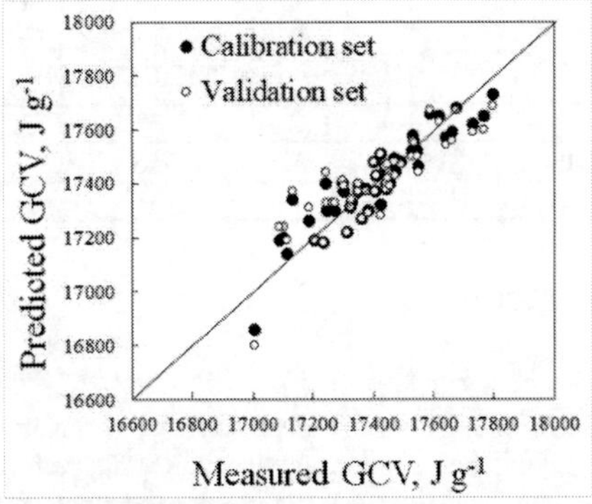

Fig. 3. Comparison of the gross calorific of maize cob milled sample predicted by near infrared spectroscopy and measured by the reference laboratory of the calibration model and the validation model.

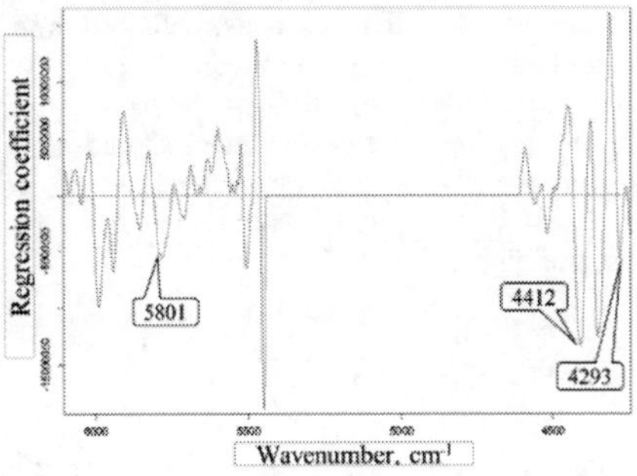

Fig. 4. PLS model regression coefficient plot for the gross calorific of maize cob milled sample.

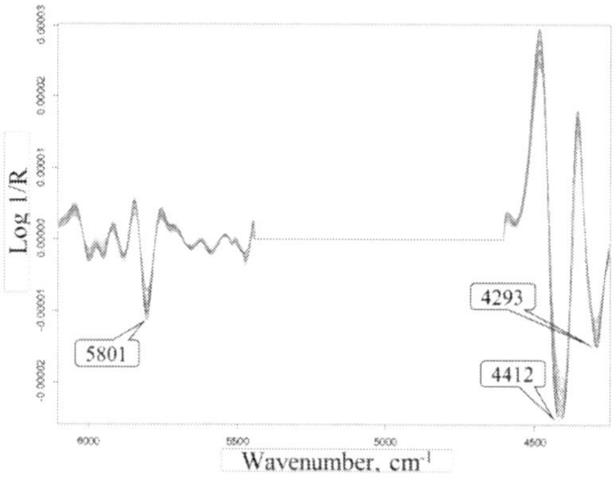

Fig. 5. First derivative spectra of maize cob milled sample used for model development.

Table 2. Measured value and predicted value of external sample predicted by calibration model

Sample	TRUE (Jg^{-1})	Prediction (Jg^{-1})
51	17431.0	17375
52	17305.5	17430
53	17404.0	17387
54	17327.5	17242
55	17387.5	17388
56	17348.5	17403
57	17368.0	17438
58	17422.0	17522
59	17253.0	17141
60	17264.5	17322
RMSEP (Jg^{-1})	77.3	
Bias (Jg^{-1})	-13.7	

4 Conclusions

The outcome showed that there is quite a possibility to the use of NIR spectroscopy as a rapid technique for classifying maize cob properties with a fair performance. Moreover, the time per sample can be reduced to 2-3 minutes where the current analysis took approximately 25-30 minutes per sample. For robustness, the model was tested by external sample (unknown sample) to confirm its ability again because it is the representative population for future. For model development, the vibrational band of hydrocarbons, cellulose, and starch strongly had an impact to the prediction of GCV. In the further study for improvement of the calibration model performance, the mode development should be done by collecting maize cob from various sources to get wider range of GCV, that way the calibration model may be a robust and global model.

Acknowledgments

The authors thank the Near Infrared Spectroscopy Research Center for Agricultural Product and Food at King Mongkut's Institute of Technology Ladkrabang, Bangkok, Thailand, for the use of their instruments. We also acknowledge the financial support from the King Mongkut's Institute of Technology Ladkrabang research fund.

References

1. Posom, J., and Sirisomboon, P. 2017. Evaluation of lower heating value and elemental composition of bamboo using near infrared spectroscopy. Energy 121, 147 – 158.
2. Posom, J., Shrestha, A., Saechua, W. and Sirisomboon, P. 2016. Rapid non-destructive evaluation of moisture content and higher heating value of *Leucaena leucocephala* pellets using near infrared spectroscopy. Energy 107, 464 – 472.
3. Everard, C. D., McDonnell, K. P. and Fagan, C. C. 2012. Prediction of biomass gross calorific values using visible and near infrared spectroscopy. Biomass and Bioenergy (45), 203 – 211.
4. Huang, C., Han, L., Yang, Z. and Liu, X. 2009. Ultimate analysis and heating value prediction of straw by near infrared spectroscopy. Waste Management (29), 1793 – 1797.
5. J. Workman, J.R.L. Weyer. 2007 Practical Guide to Interpretive Near-infrared Spectroscopy, Taylor and Francis, Boca Raton, FL, pp. 240-262.
6. John S. Shenk, Jerome J. Workman, Jr., and Mark O. Westerhaus. 2008. Application of NIR Spectroscopy to Agricultural Products. Text book: Editor: Donald A. Burns Emil W. Ciurczak. Hand book of Near-Infrared Analysis. CRC press. Taylor & Francis Group. pp. 356-357.
7. R. Zornoza, C. Guerrero, J. Mataix-Solera, KM. Scow, V. Arcenegui, J.M. Beneyto. 2008. Near infrared spectroscopy for determination of various physical, chemical and biochemical properties in Mediterranean soils, Soil Biol Biochem. 40 (7). 30-1923.
8. P. Williams. 2007. Near-Infrared Technology Getting the Best Out of Light Edition 5.0. A Short Course in the Practical Implementation of Near-infrared Spectroscopy for the User, PDK Grain, Nanaimo, Canada.
9. J. Posom, P. Sirisomboon. 2015. Evaluation of the moisture content of *Jatropha curcas* kernels and the heating value of the oil-extracted residue using near-infrared spectroscopy, Biosyst Eng. 130. 52-59.
10. K. Zhang, L. Zhou, M. Brady, F. Xu, J. Yu, D. Wang. 2017. Fast analysis of high heating value and elemental compositions of sorghum biomass using near-infrared spectroscopy, Energy. 118. 1353-1360.

Effect of drying methods on properties of green curry powder

Natthacha Chaloeichitratham[1,*], *Pornkanya* Mawilai[1], *Thadchapong* Pongsuttiyakorn[1] and *Pimpen* Pornchalermpong[1]

[1]Food Engineering Department, Faculty of Engineering, King Mongkut's Institute of Technology Ladkrabang, Bangkok, Thailand

Abstract. In this study, the effects of two drying methods: hot-air and freeze drying for Thai green curry paste in a terms of drying time and qualities have been investigated. The hot-air drying was carried out in tray dryer at temperature of 50, 60 and 70 °C. The freeze drying was carried out in freeze dryer at freezing temperature of -20°C, primary drying temperature of -10°C and secondary drying temperature of 50°C. Moisture content, water activity, colour, bulk density, and total phenolic content (TPC) were determined in samples. Freeze dried sample had significantly ($p<0.05$) lower moisture content, water activity, bulk density, total colour difference and browning index than hot air dried samples. For antioxidant activity, the results showed hot-air drying at 70°C effected highest TPC similar to freeze drying.

1. Introduction

Nowadays, most of people is paying an attention to healthy food. Especially, the food which have vegetables and herbs as main ingredient. Vegetables and herbs contain with many benefits such as antioxidant and phenolic compounds [1]. One of dishes used many herbs to be the main ingredient is Thai food named Thai green curry. Besides containing many benefits, Thai green curry paste is famous curry and accepted by people from the other country.

Thai green curry paste is kind of spicy paste included green chili, garlic, shallot, galangal, lemongrass and kaffir lime zest. These paste ingredients are the main power for antioxidant, anti-inflammatory and antimutation [2-3]. Short shelf-life of green curry paste is caused containing high moisture content over 70 percentage. There are several products was created for extending the paste shelf-life such as green curry paste in retort pouch or saterilized canned. However, the quality of ingredients is reduced by these process and cost of transportation is been higher from the higher weight [4]. With all of these reasons, drying a green curry paste was created for reducing the transportation cost and extended shelf-life of the paste.

To antimicrobial by reducing moisture content and water activity from a green curry paste is the process most selected by entrepreneurs. In present, there are many ways to drying a product but the easy and low-cost way, which is selected by entrepreneurs, is hot air drying; nevertheless, nutrition, flavor and colour of food were reduced by hot air-drying process [5-6]. The research of Huttakovit *et al.* in 2007 [7] found that high temperature was caused curry paste turns more browning as same as the research of Inchuen *et al.* [8] and Kausilya *et al.* [9] said the high temperature was caused the lower quality product, especially, total phenolic content and antioxidant activity were been lower by increasing temperature. In some case, it was found that total phenolic content of some herbs increased when it passed high temperature [1]. Meanwhile, total phenolic content and antioxidant activity were reduced by drying at the lower temperature but taking longer time because some type of foods are composed with sensitive substances. Then, drying at low temperature like freeze drying was selected to be the best process to maintain the quality of food and also maintain food property as a fresh food. [10-11].

In addition, drying process and condition which effect with product qualities, physical and chemical properties were studied by many researchers. For instance, Chee fah chili [12], oregano [13], rosemary, basil, thyme, marjoram and sage [5], olive [14], aerial and leaf of coriander [15], kimchi [16], onion [18], and Thai red curry paste [8]. Following the study of An *et al.* [17] has been reported that antioxidant activity total phenolic content and other important substances were obtained by freeze drying which is higher than hot air drying. On the other hand, time and energy were more spent with freeze drying process and it also caused higher cost than hot air drying. Moreover, it was found that high rehydration is affected by low moisture content of freeze drying process which is not proper with some products [19-20].

In conclusion, drying method and drying temperature are directly affected with food quality. However, the comparison between hot air and freeze drying have not been performed to dried Thai green curry paste. The aim of this study was to investigate drying which affects with

* Corresponding author: ch.natthacha@gmail.com

the qualities of green curry powder. Moisture content, water activity, colour, bulk density and total phenolic content were investigated in the study.

2. Materials and Methods

2.1 Materials and chemicals

Fresh herbs i.e. green chilli, garlic, shallot, lemon grass, galanga, kaffir lime zest and some spices (coriander seed, cumin and pepper) were purchased from local market in Khlong Luang district, Pathum Thani Province, Thailand. Folin Ciocalteu's reagent and gallic acid were sourced from Sigma-Aldrich (USA). All other chemicals and reagents were of analytical grade.

2.2 Sample preparation

Fresh green curry paste was prepared by blending sliced herbs, spices and seasoning together in milling machine (Ngowhuatyoo, Thailand). The 500 g of paste sample was packed in polyethylene bag, then frozen at -35°C for 90 min and stored at -18°C until used. The initial moisture content of the green curry paste was determined as 75.9% wet basis (w.b.).

2.3 Drying experiment

Frozen green curry paste samples were thawed at 4°C in refrigerator and spread on trays of each dryer, with 5 mm thickness. All samples were dried until final moisture content was below 10% w.b. [9]. Dried samples were ground into powder by using electric blender (HR2118, Philips, Nederland). The powdered was keep in aluminum foil bag for further studies.

2.3.1 Hot-air drying (AD)

Fresh green curry paste samples were dried using tray dryer (Convection Dryer, KluayNamThai - KarnChang, Thailand) at constant air velocity of 1 m/s and three different temperatures: 50, 60 and 70°C.

2.3.2 Freeze drying (FD)

Green curry paste samples were dehydrated in freeze dryer (Kryo 'D' Freezer, ITC, Thailand). The samples were frozen at -50°C until the core temperature of samples were -20°C and then were dried in primary drying at -20°C and 40 Pa vacuum pressure. After that, samples were dried in secondary drying at 50°C until core temperature of samples were 50°C, the process was completed.

2.4 Moisture content

Moisture content was determined using hot air oven method [21]. Powder sample was dried in the oven at 100°C until a constant weight was achieved. Then, the following relation was used:

$$\text{Moisture content} = \frac{(W_1 - W_2)}{W_1} \times 100 \quad (1)$$

Where W_1 is weight of powder before drying (g) and W_2 is weight of powder after drying (g). Moisture content was expressed as percentage (wet basis). The analysis was triplicated.

2.5 Water activity

Air dried and freeze dried paste samples were taken around 1-2 g to determine water activity using electronic water activity meter (Decagon Model, AquaLab, USA). Three replications of water activity were performed for each sample.

2.6 Bulk Density

Bulk density of samples were determined according to Seeranguryar *et al.* [20] with some modifications. Sample receiving cup was set up distance between bottom of loading funnel and top of sample receiving cup as 15 cm, using 60 cm^3 cylindrical cup. The powder was loaded to flow through the funnel into the cup until it overflows and carefully scraped excess powder from the top of the cup by smoothly moving spatula. The cup occupied by powder was then weighed. Bulk density was expressed using the equation:

$$\rho = \frac{m}{V} \quad (2)$$

Where ρ (g/ cm^3) is bulk density, m (g) is the mass of powder samples and V (cm^3) is the volume of receiving cup. The bulk density measurements were performed in triplicate.

2.7 Colour and browning index

The fresh and dry-rehydrated green curry paste samples were measured using spectrophotometer (ColorFlex, Hunter Lab, UK) in terms of CIE L^*, a^* and b^* values. L^* represents the range from lightness to darkness (0 - 100) of the object. a^* indicates redness (+) or greenness (−). b^* represents yellowness (+) or blueness (−). For each sample, three replications of the colour test were performed. total colour difference (ΔE) was also determined using the equation:

$$\Delta E = [(\Delta L^*)^2 + (\Delta a^*)^2 + (\Delta b^*)^2]^{1/2} \quad (3)$$

where $\Delta L^* = L^* - L_f^*$, $\Delta a^* = a^* - a_f^*$, $\Delta b^* = b^* - b_f^*$, and L^*, a^* and b^* are the colour values of sample and L_f^*, a_f^* and b_f^* are the colour values of fresh sample [22]

Browning index (BI) was determined using method of Park *et al.* [16] with slight modification. Fresh and powder samples about 1 g were extract with 50 mL of distilled water for 24 hours. The extract was filtered through Whatman No.1 filter paper (UK). The filtrate was diluted with an equal volume of 95% ethanol and then centrifuged at 4,000 rpm, 4°C for 15 min by centrifuge (MIKRO 220R, Hettich, Germany). The supernatant was measured based on its absorbance at 420 nm using spectrophotometer (GENESYS 10S, Thermo Scientific, UK). Measurements were performed in triplicate. The BI values were expressed as absorbance per g dry matter.

2.8 Total phenolic content

The sample extract was obtained from determination of Incheun *et al.* [8] with some modification. Powder (1 g) was extracted with 25 mL of 90% ethanol and incubated at 150 rpm, 30°C for 24 h. The extract was centrifuged at 6,000 rpm, 4°C for 20 min.

Total phenolic content (TPC) in the green red curry product extracts was determined using the Folin-Ciocalteu method described by Chan *et al.* [23] with some modifications. The diluted extract (2.25 mL) was mixed with 0.25 mL of Folin–Ciocalteau's reagent by vortex. After 5 min, 1.00 mL of 10% w/v Na_2CO_3 was added to the mixture. The absorbance was measured at 730 nm using a spectrophotometer after 10 min reaction. Distilled water was added instead of sample taken as blank. The amount of TPC was expressed as gallic acid equivalent per g dry matter.

3. Results and Discussion

3.1 Effect of drying methods on drying time and properties of Thai green curry paste powder.

Fresh Thai green curry paste with 5 mm thickness samples were dried with four drying methods: hot-air drying at 50, 60, and 70°C and freeze drying. The drying time are presented in Table 1. Freeze dry had the longest drying time with drying time of 36 hours and 40 min. Hot-air dried at 50°C samples went through the second longer drying time of 480 min, while dried at 70°C had shortest drying time.

The drying temperature had significant difference in water activity ($p < 0.05$) as shown in Table 1. Freeze dried sample had lowest water activity of 0.129, like the freeze drying can extremely reduce moisture content and water activity. For hot-air dried powder at 70°C resulted in lower water activity than dried at 50°C and 60°C due to increasing of temperature.

Bulk density is important in packaging, storage and transportation for powdered food [20]. Table 1 show the increasing of temperature caused reduction in bulk density of hot-air dried powder. The highest bulk density (0.57 g/cm^3) was at 50°C air drying and it was reduced by 8% at 70°C. Whereas, freeze dried sample had lowest bulk density as 0.30 g/cm^3 because lyophilization during freeze drying affected high porosity in sample [24].

3.2 Effect of drying methods on colour properties of Thai green curry paste.

From consumer acceptance viewpoint, colour properties are very important of dried product. A main colour of Thai green curry paste is green attribute to chlorophyll from the fresh green chilli [25].

The colour values in a term of CIE L^*, a^* and b^* system of products and total colour difference (ΔE) are presented in Table 2. There were significant differences of ΔE at the 95% confidence level between 4 products. Freeze dried sample had the lowest ΔE (4.13) and ΔE of 50, 60 and 70°C hot-air dried samples were increased with temperature increasing. Due to a^* values of dry-rehydrated green curry paste from hot-air drying increased when compared with fresh sample, whereas the value of freeze dried sample was more similar than them. That can indicate products from hot-air drying were less green. The discoloration of samples during drying was related to pigment destruction, enzymatic browning and non-enzymatic browning. Hence, results showed hot-air dried samples were more influenced by browning reaction than freeze dried sample, like BI value as shown in Table 2. There was no significant difference between BI value of fresh and freeze dried sample. But BI value of fresh product was significantly less than the 50, 60 and 70°C hot-air dried samples, which indicated that brown pigments formation occurred during processes. Similar to results were obtained from Park *et al.* [16] which drying of kimchi.

Table 1 Drying time, moisture content and water activity of Thai green curry paste at different drying methods.

Drying method	Drying time (min)	MC (%w.b.)	Water activity	Bulk density (g/cm^3)
AD50	480	8.71 ± 0.15	0.311 ± 0.004^d	0.5701 ± 0.0052^d
AD60	340	8.54 ± 0.20	0.247 ± 0.003^c	0.5476 ± 0.0022^c
AD70	250	8.48 ± 0.15	0.226 ± 0.005^b	0.5262 ± 0.0052^b
FD	2,200	3.84 ± 0.05	0.129 ± 0.008^a	0.3047 ± 0.0010^a

Means (± standard deviation) with different superscripts in each column significantly different ($p < 0.05$)

Table 2 Colour values, total colour difference and browning index of green curry powder at different drying methods.

Sample	L^*	a^*	b^*	ΔE	BI (abs/g dry matter)
Fresh	37.96 ± 0.43^a	-2.89 ± 0.27^a	30.14 ± 0.56^a	-	1.224 ± 0.021^a
AD50	38.73 ± 0.04^b	3.13 ± 0.16^c	31.11 ± 0.53^{ab}	6.16 ± 0.10^b	1.368 ± 0.007^b
AD60	38.28 ± 0.33^{ab}	3.99 ± 0.03^d	30.57 ± 0.54^a	6.92 ± 0.08^c	1.346 ± 0.003^b
AD70	38.04 ± 0.10^a	3.92 ± 0.30^d	32.09 ± 1.38^b	7.17 ± 0.18^d	1.409 ± 0.010^c
FD	41.34 ± 0.13^d	-1.04 ± 0.11^b	31.57 ± 0.35^{ab}	4.13 ± 0.05^a	1.226 ± 0.018^a

Means (± standard deviation) with different superscripts in each column significantly different ($p < 0.05$)

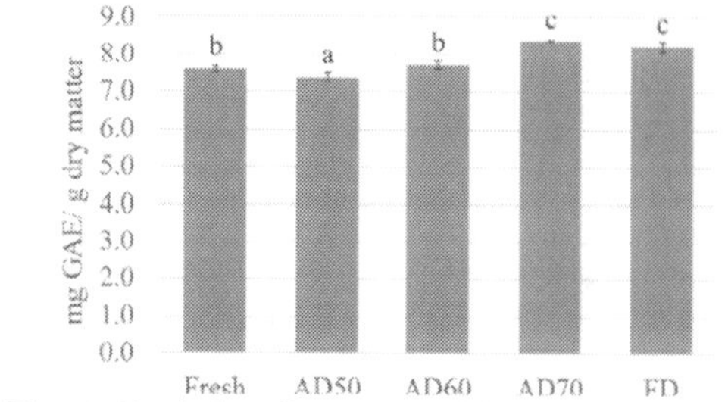

Fig. 1. Total phenolic content of fresh and dried green curry paste products.

3.3 Effect of drying methods on total phenolic content of Thai green curry paste powder.

In addition, TPC is one of the most important properties that indicates health benefit of the paste. Fig. 1 show TPC of four dried samples. It has been found that there was a significant difference at the 95% confidence level between hot-air and freeze dried sample at 50°C. Freeze dried powder was much higher than hot-air dried powder. Similar to results were obtained from Asami *et al.* [26] which reported that freeze dried marionberry, strawberry and corn had higher TPC hot-air dried samples. During hot-air dry process, the product is dried at atmospheric pressure whereas there is very little oxygen in the freeze drying chamber, which is the cause of oxidation reaction [27]. For effect of temperature of hot-air drying, it was found that TPC was increased when temperature was increased. Due to the higher temperature would promote decomposition of complex phenolic compounds affected higher TPC [17].

4. Conclusion

Based on the results of present investigation, properties of dried Thai green curry paste products were significantly affected by drying techniques. Freeze drying took longest drying time. However, physical properties of Thai green curry paste powder from freeze drying were lower than hot-air dried products. The lowest percentage of moisture content was freeze dried powder, whereas it had no significant difference among hot-air dried powder. Similar to water activity, freeze dried powder had lowest water activity. But water activity of hot-air dried samples were decreased when temperature was increased. And bulk density of freeze dried sample was lowest. For colour properties, it was found that freeze drying gave lowest ΔE and BI value. The results of TPC showed that hot-air drying at 70°C effected highest TPC similar to freeze drying.

Acknowledgements

The authors also thank to I.T.C (1993) CO., LTD Thailand for freeze dry machine (Kryo 'D' Freezer) supporting.

References

1. K. Ruanma, L. Shank, and G. Chairote, Maejo Int. J. Sci. Technol, **4**, 193 - 200 (2010)
2. B. Lee, J. Jung, and H. Kim, Food Chem Toxicol, **50**, 3912 – 3919 (2012)
3. N. Zobia, C. Zhihui, W. Cuinan, W. Yanbin, and D. Haiyan, Postharvest Biol Technol, **131**, 39–45 (2017)
4. K. Laohasongkram, S. Chaiwanichsiri, and O. Chanprasartsuk, *Proceedings of the 1st International Conference on Science and Applied Science*, **1**, 306-311 (2006)
5. M.B. Hossain, C. Barry-Ryan, A.B. Martin-Diana, and N.P. Brunton, Food Chem, **123**, 85 – 91. (2010)
6. C. Borompichaichartkul, KMUTT R&D J., **35**, 269 – 283. (2012)
7. W. Huttakovit, W. Khuikhoen, K. Mongkolwan, N. Suteebut, J. Boonyasawat, and N. Sakulyeunyongsuk, RMUTP Research J., **1**, 9 – 20 (2007)
8. S. Inchuen, W. Narkrugsa, and P. Pornchaloempong, Kasetsart J. (Nat. Sci.), **44**, 142 – 151 (2010)
9. S.R. Kausilya, S.T. Farah, M.Z.A. Mazidah, and R.I.S. Mohammad, J Saudi Society of Agri. Sci. (2017) (In Press)
10. L. Ma, H. Chen, W. Zhu, and Z. Wang, Food Res Int. **50**, 633 – 640 (2013)
11. T.L. Jeng, C.C. Lai, T.C. Liao, S.Y. Lin, and J.M. Sung, J Food Drug Anal, **23**, 701 – 708 (2015)
12. N. Toontom, M. Meenune, W. Posri, and S. Lertsiri, Int. Food Res J, **9**, 1023 – 1031 (2012)
13. K. Jałoszyński, A. Figiel, and A. Wojdyło, Acta Agrophysica, **11**, 81 – 90 (2008)
14. M.H. Ahmad-Qasem, E. Barrajón-Catalán, V. Micol, A. Mulet, and J.V. García-Pérez, Food Res Int, **50**, 189 – 196 (2013)
15. A.G. Pirbalouti, S. Salehia, and L. Craker, J Appl Res Med Aromat Plants, **4**, 35 – 40 (2017)
16. H.J. Park, Y. Lee, and J.B. Eun, Biocatal Agric Biotechnol, **5**, 193 – 198 (2016)
17. K. An, D. Zhao, Z. Wang, J. Wua, Y. Xu, and G. Xiao, Food Chem, **197**, 1292–1300 (2016)
18. F. Ren, C.A. Perussello, Z. Zhang, J.P. Kerry, and B.K. Tiwari, LWT - Food Sci. Technol., **87**, 102 – 111 (2018)
19. J.J. Ma, X.Y. Mao, Q. Wang, S. Yang, D. Zhang, S.W. Chen, and Y.H. Li, LWT - Food Sci. Technol. **56**, 296 - 302 (2014)
20. T. Seerangurayar, A. Manickavasagan, M.A. Abdulrahim, and A.A. Yaseen, J Food Eng. **215**, 33 – 43. (2017)
21. AOAC, *Official Methods of Analysis of AOAC INTERNATIONAL 17th ed.* (The Association of Official Analytical Chemists, Maryland, 2000)
22. S. Phoungchandang, and S. Saentaweesuk, Food Bioprod Process, **89**, 429–437 (2011)
23. E.W.C. Chan, Y.Y. Lim, S.K. Wong, K.K. Lim, S.P. Tan, F.S. Lianto, and M.Y. Yong, Food Chem, **113**, 166-172 (2009)
24. Y.H. Roos, *Phase Transitions in Foods* (Academic Press, San Diego, 1995)
25. S. Hemathulin, and S. Techawongstien, Khon Kaen Agr. J., **42**, 760 – 765 (2014)
26. D.K. Asami, Y.J. Hong, D.M. Barrett, and A.E. Mitchell, J. Agric. Food Chem, **51**, 1237-1241 (2003)
27. G.J. Leusink, D.D. Kitts, P. Yagnmaee, and T. Durance, J. Food Sci., **75**, 311-316 (2010)

* Corresponding author: ch.natthacha@gmail.com

Effects of low temperature drying processing on longan fruit

Thadchapong Pongsuttiyakorn[1,*], *Pachareeporn* trusphimai[1], *Pitikhate* Sooraksa[2], and *Pimpen* Pornchaloempong[1]

[1]Food Engineering Department, Faculty of Engineering, King Mongkut's Institute of Technology Ladkrabang, Bangkok, Thailand
[2]Computer Engineering Department, Faculty of Engineering, King Mongkut's Institute of Technology Ladkrabang, Bangkok, Thailand

Abstract. In this study, the single-stage drying in tray dryer at air temperatures of 40, 50, 60, 70 and 80°C is modelled and investigated. The longan fruits, E-dor variety, are peeled and seeded before testing. The drying rate is significantly influenced by the drying techniques and temperatures. Drying rats are initialized adjustment constant rate periods at 60 70 and 80°C. The rate of moisture removal is rapidly changed drastically during the falling rate period. The Midilli model with high R^2 and low χ^2 and RMSE is the most suitable model for predictability of longan drying. Variation rates of quality of the water activity, the shrinkage, and the browning index are also reported.

1 Introduction

Longan (*Dimocarpus Longan* Lour.) is a tropical fruit mainly produced in Thaïland. The fruit has a brown to light reddish-brown peel. The pulp is juicy and sweet, with a white, translucent flesh. Transformed longans are exported as fresh (76%), dried (22%), canned (2%) and frozen fruits (<0.5%) (Ministry of Agriculture and Cooperatives). That are mainly exported to China, Vietnam, Hong Kong, Indonesia and others. But longan is a seasonal fruit crop, it involves variations in production values. This sometimes leads to, over supply problems and low prices during high production periods. From the export value, dried longan is interested product. There are a lot of longan drying methods, such as solar drying of peeled longan [1].

Actually, rapid drying at high temperatures may reduce the cost of processing. However, the quality of the product may be reduced. Drying of high quality Longan have been proposed by various ways which are: a two–stage superheated steam drying followed by hot air drying process [2]. There are also forced convection and hot air recirculation in longan drying [3], multi-stage dying processes using heat pump for longan drying at temperature of 55°C, combined far-infrared-heat pump drying at temperature of 65°C, and combined far-infrared -hot air drying with 80% recycled air [4].

The qualities of dried longan such as moisture content, water activity and color effect on prices of product. The color characteristics of longan flesh are increased at the beginning and decreases as drying time and drying air temperature increase during air drying [5].

Prediction of longan drying by theoretical methods assumes that material moisture is conveyed by geometric diffusion. The answer to this model is effective diffusion coefficient. However, for empirical equations it is easy to use and has been widely used since it is based on experimental data. [6]

Development of a low temperature drying process for whole longan fruit is of great interest, in maintaining quality. The purpose of this research is to study the effects of low temperature drying on the drying kinetics.

2 Material and method

2.1 Material

Longans of E-dor variety (*Dimocarpus Longan* Lour.) were purchased from Talaadthai market in Pathum Thani province. The samples were peeled and seeded before experiment. The longan fleshes were approximate inner and outer diameters of 15 and 25 mm, respectively. The longans had average total soluble solid of 17.0-18.4·%brix and initial moisture content of 77.7-80.4 %(w.b.)

2.2 Drying of longans

2.2.1 Drying experiment

The drying experiments were performed in a laboratory scale hot air dryer. The experiments were conducted at drying temperatures of 40, 50, 60, 70 and 80°C with a constant air velocity of 1.08 m/s and recorded temperature and relative humidity every 1 minute and sampling of measurement every 1 hour.

2.2.2 Drying rate

In each drying experiment, 180 longan fruits were placed on three trays (average weight of 300 g). During drying, at temperatures of 40 and 50°C, the samples were weighed every half hour, and at temperatures of 60, 70

* Corresponding author: author@e-mail.org

and 80° C, the samples were weighed every 10 minutes until the weight of the sample was constant.

Rate of drying curves. Data acquired usually obtained from the total weight is W_t and the weight of the dry solid is W_s in kg then:

$$M_t = (W_t - W_s)/W_s \quad (1)$$

For the given constant drying conditions, the equilibrium moisture content is M_e ($kg_{equilibrium\ moisture}/kg_{dry\ solid}$), and the free moisture content is M ($kg_{free\ water}/kg_{dry\ solid}$), the water can be removed by mechanical methods and can be calculated for moisture content at different times over the drying period ($kg_{water}/kg_{dry\ solid}$) is M_t

$$M = M_t - M_e \quad (2)$$

The drying rate (R) is calculated for each drying time increment using:

$$R = (W_s / A) \text{ x } (dx/dt) \quad (3)$$

where A is the exposed surface area for drying (m^2), dx/dt is free moisture content of bulb at different times.

The moisture ratio (MR): In thin layer drying, the moisture ratio was calculated using equation:

$$MR = (M_t - M_e)/(M_i - M_e) \quad (4)$$

Where M_t is the moisture content at different times ($kg_{water}/kg_{dry\ solid}$), M_i is the initial moisture content ($kg_{water}/kg_{dry\ solid}$), and M_e is the equilibrium moisture content of the samples ($kg_{water}/kg_{dry\ solid}$) [7].

2.2.3 Mathematical modelling of drying curves

In the present study, the obtained drying data (time versus MR) are empirical modeled by 5 different models and equations of them are presented in Table 1.

Table 1. Empirical models used in the present study to fit data obtained from drying at different conditions.

No.	Model	Equation	Ref
1	Lewis	$MR = \exp(-kt)$	[8]
2	Page	$MR = \exp(-kt^n)$	[9]
3	Henderson and Pabis	$MR = a \text{ x } \exp(-kt)$	[10]
4	Wang and Singh	$MR = 1 + at + bt2$	[11]
5	Midilli	$MR = a \text{ x } \exp(-kt^n) + bt$	[12]

Drying data of the samples were analyzed using SPSS 2.0 software (International Business Machines Corp., United States). Nonlinear regression analysis was calculated corresponding parameters of the models mentioned in Table 1. Validation of the statistical models was determined using R^2, χ^2 and RMSE values calculated using the following equations:

$$R^2 = 1 - \sum(y_{exp} - y_{pred})^2 / \sum(y_{exp} - y_{ave})^2 \quad (5)$$

$$\chi^2 = (1/N) \text{ x } \sum(y_{exp} - y_{pred})^2 \quad (6)$$

$$RMSE = [(1/N) \text{ x } \sum(y_{exp} - y_{pred})^2]^{1/2} \quad (7)$$

where y_{exp} and y_{pred} are the experimental and predicted, and N are the number of observations [13].

2.3 Moisture content

Take a sample of longan fruit before and after drying about 3 – 5 g. and cut into small pieces. To measure the moisture content of the product before and after drying, longans dried measuring in the moisture analyzer with MA37-1 (Sartorius, German) at 90°C.

2.4 Water activity

Take a sample of longan fruit after drying to about 3 g. and cut into small pieces. To measure the water activity of the product after drying, the sample was measured by using Aqua lab LITE (Decagon, USA).

2.5 Shrinkage

Longans fruit 10 units, measured the volume of the product before and during drying with Hexane using the density kit analytical balance ydk03 (Sartorius, German). And then calculated the Shrinkage of the product using the following equations:

$$\%\text{Shrinkage} = [(V_0 - V)/V_0] \text{ x } 100 \quad (8)$$

where V_0 and V are the volume initiate and volume, respectively.

2.6 Browning index

Samples were analysed at different times of longan drying and for each temperature, to extract brown pigments. Take the same amount of solids (1.0 g). Samples were ground and mix with 25 ml of 20% acetic acid. After 24h, they were filtered through a funnel with Whatman No.2 filter paper. After that, the absorbance was measured in 10 mm cells against 20% acetic acid at 420 nm. Two measurements for each situation were carried out by using. THERMOSCIENTIFIC Genesys 10S UV-vis Absorbance [14-15].

3 Results and discussion

The resulting longan drying at 40, 50, 60, 70 and 80°C. Data was analyzed to determine the drying rate and the quality of the change over time during drying.

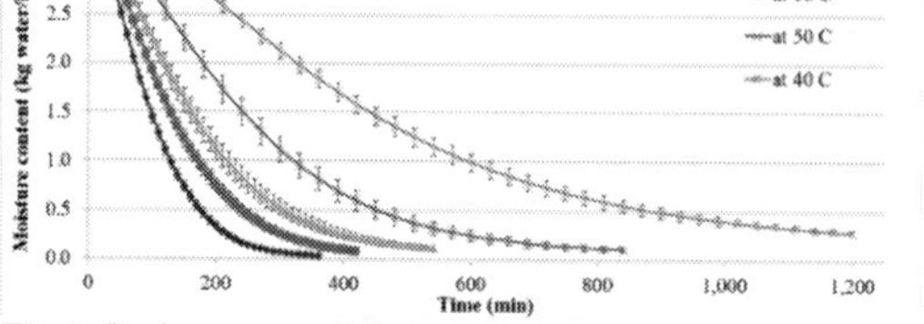

Fig. 1. Drying curves of the longan.

Drying curves and moisture content versus time were given in Figure 1. Our results were agreed with the other studies using different materials for drying [2, 6-7]. The drying time was shorter when the temperature rises. Drying time the longest in the shortest drying time at 80° C, it took around 360 min.

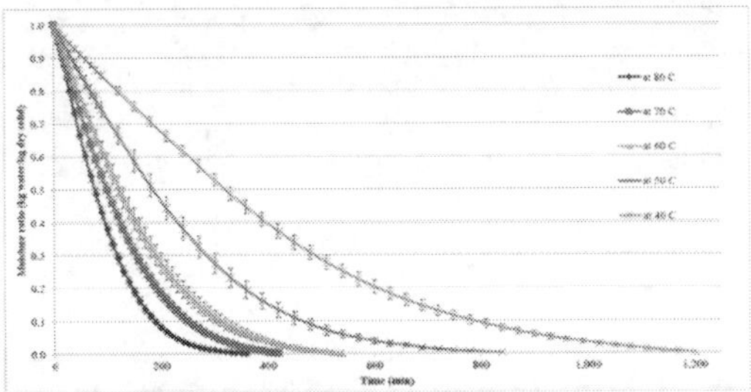

Fig. 2. Moisture ratio versus time for drying of the longan.

The moisture content for all drying at different temperatures was reduced. As shown in the Figure 2, the slope of the curve represents the change in drying rate during the drying period.

Figure 3 shows the drying rate of dried longans at various temperatures drying at 60, 70 and 80°C with an initial adjustment period. Due to the increase in surface temperature, the higher the temperature, the more steep, none for drying at 40 and 50°C. During the constant rate period, at constant rates, the surface moisture removal rate of the sample is saturated with water for only a short time [7]. Overall, drying in the falling rate period followed by a constant rate period.

Drying time versus moisture ratio with different temperatures. The corresponding parameters of the

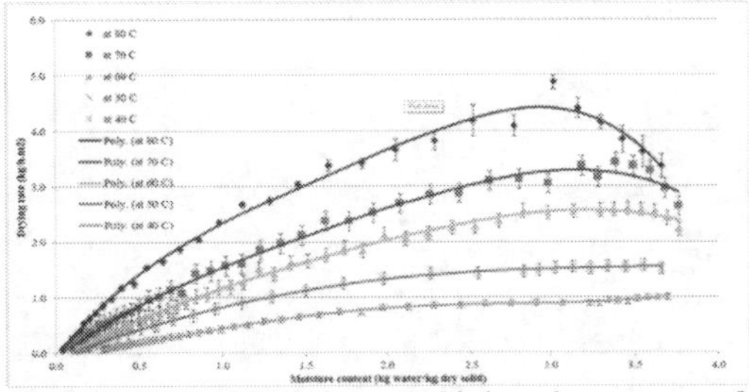

Fig. 3. Variation of the drying rate with the moisture content of longan.

empirical models and the R^2, $\chi 2$ and RMSE values for the accuracy of the models are presented in Table 2.

The R^2 values is between 0.984038 and 0.999915, which is more than 0.95, indicating that the predicted data is close to the experimental data. Therefore, all models can explain the relationship between time and relative humidity and the drying process [13]. The RMSE values represent the difference between the predicted value and the experimental value. Between 0.000008 and 0.072017, the most suitable models for longan were Midilli and Wang and Singh, respectively. Obtaining the drying kinetics data is essential for optimizing the drying process, The *k* coefficient represents the drying rate [16]. From tables 2, *k* increases as the drying temperature increases.

The water activity of the dried food product must be less than 0.6. However, the final drying product at 40 c is more than 0.6, is 0.7440, which was not met the criterion. For drying at 50, 60, 70 and 80°C, the water activity was less than 0.6, is 0.5447, 0.5220, 0.5040 and 0.4110, respectively as shown in Figure 4.

Table 2. Empirical model parameters and statistical parameters obtained from fitting of the drying models for longans.

No.	Model		40	50	60	70	80
1	Lewis $MR = \exp(-kt)$	k	0.002488	0.004212	0.006408	0.007912	0.010318
		R^2	0.984960	0.988093	0.986819	0.985279	0.984038
		χ^2	0.000241	0.000158	0.000990	0.000583	0.000662
		RMSE	0.015313	0.012408	0.031191	0.023889	0.025409
2	Page $MR = \exp(-kt^n)$	k	0.000448	0.001056	0.001651	0.001990	0.002429
		n	1.279378	1.248919	1.259123	1.275542	1.305251
		R^2	0.998623	0.999410	0.999448	0.999356	0.999569
		χ^2	0.001895	0.000552	0.000709	0.000660	0.000432
		RMSE	0.042983	0.023151	0.026390	0.025414	0.020525
3	Henderson and Pabis $MR = a \text{ x } \exp(-kt)$	k	0.002640	0.004515	0.006830	0.008472	0.011145
		a	1.056636	1.061607	1.066873	1.070407	1.079348
		R^2	0.989326	0.992672	0.991622	0.990789	0.990457
		χ^2	0.005167	0.003018	0.005277	0.004532	0.004633
		RMSE	0.070979	0.054147	0.072017	0.066584	0.067212
4	Wang and Singh $MR = 1 + at + bt^2$	a	-0.001816	-0.002994	-0.004533	-0.005652	-0.007174
		b	0.000001	0.000002	0.000005	0.000008	0.000013
		R^2	0.999246	0.996710	0.996518	0.997761	0.995546
		χ^2	0.000002	0.000159	0.000202	0.000026	0.000059
		RMSE	0.001366	0.012442	0.014078	0.005009	0.007582
5	Midilli $MR = a \text{ x } \exp(-kt^n) + bt$	k	0.000437	0.000935	0.001679	0.002117	0.002327
		n	1.271245	1.263392	1.248481	1.254274	1.308277
		a	0.980728	0.984676	0.988718	0.990528	0.989360
		b	-0.000025	-0.000016	-0.000030	-0.000049	-0.000029
		R^2	0.999807	0.999830	0.999913	0.999915	0.999844
		χ^2	0.000000	0.000001	0.000000	0.000000	0.000001
		RMSE	0.000636	0.000873	0.000015	0.000498	0.000776

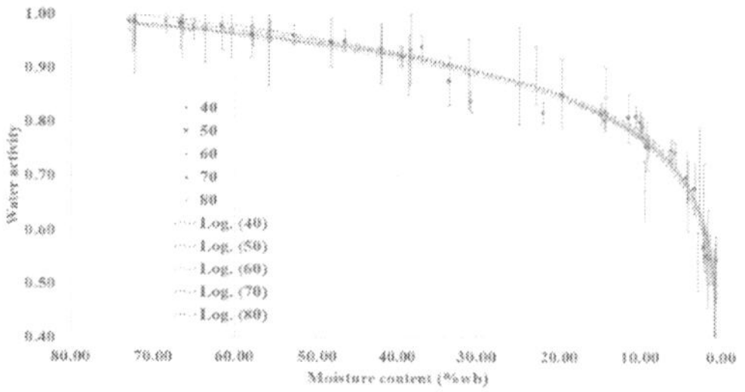

Fig. 4. Variation of the water activity with the moisture content of longans.

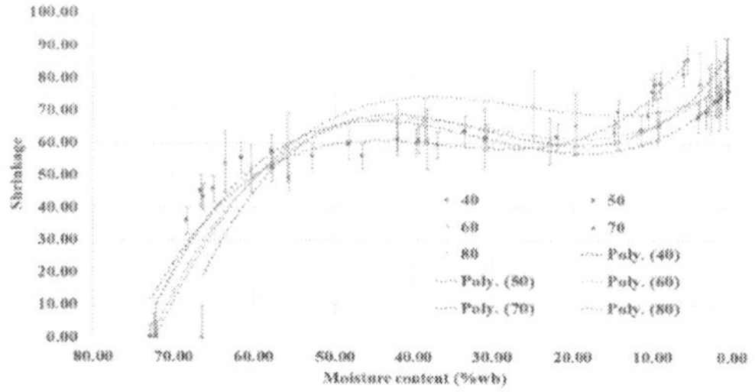

Fig. 5. Variation of the shrinkage with the moisture content of longan.

The rates of shrinkage of longan drying at the beginning of the experiment are shown in Figure 5. The shrinkage rate was high and decreased and the rate of contraction was even higher. The final shrinkage of longans for drying at temperatures of 40, 50, 60, 70 and 80 °C were 85.8014, 75.901, 80.165, 82.9295 and 81.6251, respectively.

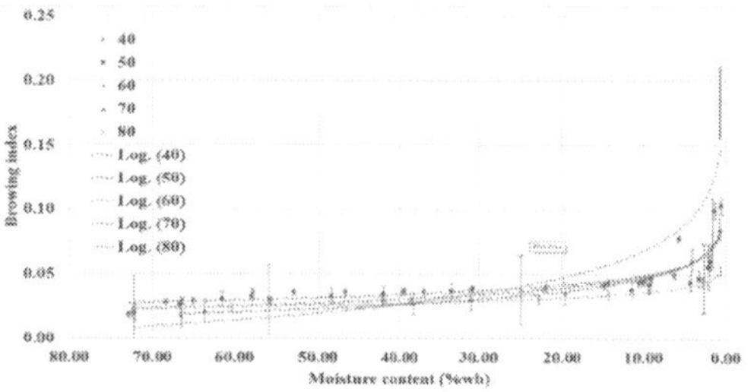

Fig. 6. Variation of the Browning index with the moisture content of longan.

Figure 6 shows the browning rate during drying increases due to increasing in solids content. The rate of browning in first period was slightly changed due to the high moisture content. When moisture content dropped below 20% wb, the brown index became more variation by the temperature affecting the rate of change. The Browning index of longans for drying processes at 40, 50, 60, 70 and 80°C were 0.077, 0.102, 0.052, 0.084 and 0.184, respectively

4 Conclusion

The study of product temperature and moisture content on the drying time reveals practical information for the final decision of drying processing with low temperature. Drying of longans showed initial adjustment periods at 60 70 and 80°C. At the constant rate period, the moisture removal was rapidly increased and was drastically removed at the falling rate period. The midilli model with the higher R^2 and lower χ^2 and RMSE values has the most suitable model for predictability of longan drying. The water activity of the dried longans was lower than 0.6 as desired, except for drying at 40°C. The shrinkage rate was high and decreased. The rate of browning index has increased, when the moisture content drops below 20% wb. For future research, the effect of moisture ratio of this drying on the quality will be observed.

References

1. S. Janjai, N. Lamler, P. Intawee, B. Mahayothee, B.K. Bala, M. Nagle, J. Müller. Elsevier Ltd., Solar Energy, **83**, 1550-1565 (2009)
2. T. Somjai , S. Achariyaviriya, A. Achariyaviriya, K. Namsanguan. Journal of Food Engineering, **95**, 313–321 (2009)
3. N. Tippayawong*, C. Tantakitti, S. Thavornun, V. Peerawanitkul. Biosystems engineering, **104**, 199–204 (2009)
4. A. Nathakaranakule, P. Jaiboon, S. Soponronnarit. Journal of Food Engineering, **100**, 662–668 (2010)
5. S. Chunthaworn, S. Achariyaviriya, A. Achariyaviriya, K. Namsanguan. Procedia Engineering, **32**, 104 – 111 (2012)
6. S. Surbkar, R. Assawarachan. Thai Society of Agricultural Engineering Journal, **17**, 1 59-66 (2011)
7. R.Ş. Çakmak, O. Tekeoğlu, H. Bozkır, A.R. Ergün, T. Baysal. Food Science and Technology, **69**, 197-202 (2016)
8. D.M.Bruce. Journal of Agricultural Engineering Research, **32**, 337–347 (1985)
9. P. S. Madamba, R. H. Driscoll, K. A. Buckle. Journal of Food Engineering, **29**, 75–97 (1996)
10. S. M. Henderson, S. Pabis. Journal of Agricultural Engineering Research, **6**, 169–174 (1961)
11. G.Y. Wang, R. P. Singh. American Society of Agricultural Engineers, **73**, 3001 (1978)
12. A. Midilli, H. Kucuk. Energy Conversion and Management, **44**, 1111-1122 (2003)
13. M. Başlar, M. Kılıçlı, O.S. Toker, O. Sağdıç, M. Arici. Innovative Food Science and Emerging Technologies, **26**, 182-190 (2014)
14. P. Udomkun, M. Nagle, B. Mahayothee, D. Nohr, A. Koza, J. Müller. Food Science and Technology, **60**, 914-922 (2015)
15. H. JinPark, Y. Lee, J. BangEun. Biocatalysis and Agricultural Biotechnology, **5**, 193-198 (2016)
16. M.J. Perea-Flores, V. Garibay-Febles, J.J. Chanona-Pérez, G. Calderón-Domínguez, J.V. Méndez-Méndez, E. Palacios-González, et al. Industrial Crops and Products, **38**, 64–71 (2012)

* Corresponding author: author@e-mail.org

MATEC Web of Conferences **192**, 03025 (2018) https://doi.org/10.1051/matecconf/201819203025
ICEAST 2018

Optimization study on Niosome Loaded *Ficus Deltodia* Extract.

Noorazwani Zainol[1], *Chen Siew Chin,*[2] *Nor Rashidah Ahmed*[1], *Noorazwani* Zainol[1], Azila Abdul Aziz[2], Rahimah Sabtu[1], *Harisun Yaakob* [2*]

[1]Food and Nutraceutical Unit, Institute of Bioproduct Development, Universiti Teknologi Malaysia, 81310, Johor Bahru, Johor, Malaysia
[2]Department of Bioprocess Engineering, Faculty of Chemical and Energy Engineering, Universiti Teknologi Malaysia, 81310, Johor, Malaysia.

Abstract. *Ficus deltoidea* (FD) plant extract is not able to cross the lipidic biomembranes of the skin as they are hydrophilic in nature. In order to overcome this problem FD were encapsulated with noisome, a novel lipid delivery system that potentially increases the absorption. This research study on optimum formulation and process condition of niosome loaded FD extract. The factors involved in formulation were amount of Span 60, amount of cholesterol and amount of Labrasol. Their corresponding responses for the design were entrapment efficiency (EE), particle size (PS) and zeta potential (ZP). The optimum formulation obtained were 0.6 mmol of Span 60, 0.4 of mmol Cholesterol, and 0.1 mmol of Labrasol. Results obtained showed improved activity on skin penetration compared to free FD.

INTRODUCTION

In recent years, natural herbs are widely use in cosmeceutical area due to their fewer side effects as compared to the synthetic or chemical based active ingredient products. Consumer awareness towards hazardous side effects of those chemical derived cosmetic products highly increase the demand of herbs based cosmetic. However, there are several common problems to have herbs applied in cosmeceutical field, as several phytoconstituents (such as flavonoids, terpenoids, tannins etc.) have poor absorption through skin. This is due to their poor lipid/water solubility and/or incompatible molecular size [1]. Benefits of *Ficus deltoidea* extract not only on promoting health effect but also giving a beneficial effect to the skin such as antiaging and whitening effect [2]. However *Ficus deltoidea* is mainly hydrophilic in nature, which may reduce their absorption through skin as they not able to cross the lipidic bio-membrane [1]. There are some methods were reported to overcome phytochemicals bioavailability problems such as purification and isolation, however this kind of method will reduce the efficacy of herbs due to the loss of synergy effect. Therefore, a good delivery vehicle is needed to deliver the extracts to targeted site and increase the using bioavailability. There were only few herbs being reported whereby niosome as a delivery vehicle such as *Ginkgo Biloba, Curcuma longa linn*, and *Silybum marianum*. Coacervation and phase separation, a method for proniosomes preparation is chosen for preparation of niosomal dispersion among other methods as this is a simple practical method for the routine and large scale production with the minimum use of organic solvent [3]

In this study, optimum formulation for DF loaded niosomes was investigated. Design of experiment (DOE) was carried out to deal with the parameters that affect niosomes production. The niosomes were characterized based on encapsulation efficiency, particle size, zeta potential and skin penetration.

MATERIALS AND METHODS

Tween 20, Tween 60, Span 20, Span 60, Span 80, Sephadex G-50 (medium), monobasic sodium phosphate, and dibasic sodium phosphate were purchased from Sigma Aldrich (St. Louis, MO, USA). Absolute ethanol (99.9 %) and 1-propanol was purchased from Qrec (Chonburi, Thailand). Labrasol was purchased from Gattefosse Corporation (Saint-Priest, France).

Dried leaves FD purchased from Malaysian Agriculture and Research Development Institute (MARDI), Muadzam, Pahang. The dried leaves were dried further for 2 days at 45 °C before being ground to a powder, which was extracted with boiling water for 1 h, after which the infusion was filtered and the filtrate was spray-dried to form a powder [4]. 1 g of obtained powdered FD extract was dissolved in 50 mL of water and centrifuged at 10000 g-force for 5min, the supernatant was then filtered with nylon membrane filter (Agilent techologies) with 0.2 μm pore size, the resultant filtrate was acts as nonsolvent for span 60 in niosomes preparation.

2.1. Preparation of Niosome

Niosome with FD extract encapsulated were prepared by coacervation phase separation method and followed by hydration as described by [5]. To study the effect of cholesterol, Labrasol, and surfactant on niosomes formation, first of all, Span 60 was mixed with an appropriate amount of lipid (Labrasol and cholesterol) in a 10 mL glass vial. 1 mL of solvent (absolute ethanol) was then added to the mixtures and the vial was tightly sealed before putting in a water bath at 60°C while shaking until

* Corresponding author: harisun@ibd.utm.my

complete dissolution of the mixture. The specified amount of FD extract (nonsolvent) was then added to the mixture while warming in the water bath for 3-5 min till clear or translucent solution obtained. Niosomal suspension was then obtained by hydration of 200 mL of distilled water for 10 min in water bath. The obtained niosomal suspensions were finally proceed with downsizing with a high-pressure homogenizer (Microfluidizer M-110P, Newton, MA, USA) with constants parameter (750 bar and 4 cycles). The encapsulation efficiency and niosome mean particle size were determined.

2.2 Characterization of Niosomes

Niosome were characterized based on their entrapment efficiency (EE), particle size (PS), zeta potential (ZP) and penetration study.

2.2.1 Entrapment efficiency (EE)

The EE was determined by minicolumn centrifugation method [6]. The column (EconoSpin size exclusion columns, Epoch life science, USA) were prepacked hydrated sephadex G-50 which come with inert support as shown in Figure 3.2. These columns were then placed in the microcentrifuge tubes and the whole assembly centrifuged to remove excess water. Niosomes suspension was then applied to the dried bed and the assembly centrifuged at 1700 rpm for 2 minutes. The niosomes were expelled in the void volume, on the other hand the excess unentrapped extract was trapped in the void volume. The expelled niosomes which collected in the microcentrifuge tubes were disrupted with 50 % n-propanol with ratio 1:1 in an ultrasonic bath for 20 minutes. The resultant mixture was filtered with 0.2 μm syringe nylon filter and the filtrate was subjected to total phenolic content assay. The entrapment efficiency can be represented as following Equation 1.0:

$$EE\% = \frac{\text{total phenolic content inside the vesicles} \times 100}{\text{total phenolic content added}} \quad (1.01)$$

2.2.2 Measurement of Particle Size

For the niosomal suspension, vesicular size and distribution will be determined by Zetasizer (Malvern Zetasizer Nano ZS, Malvern Instruments, Worcestershire, UK). The analyses were performed by dynamic light scattering (DLS) principle at constant temperature 25°C, material refractive index at 1.45 and absorption index 0.001. Each sample was measured three times and results are expressed as mean ± standard deviation. The polydispersity index (PDI) was used as a measure of unimodal size distribution. A small value of PDI (0.1) indicates a homogenous population, while a PDI. 0.3 indicates a higher heterogeneity.

2.2.3 Zeta potential determination

Zetasizer (Malvern Zetasizer Nano ZS, Malvern Instruments, Worcestershire, UK) was used to measure the zeta potential of niosomes with Electrophoretic Light Scattering principle. Zeta potential was determined 3 times for each sample, and results were automatically calculated by the analyzer using the Smoluchowski equation.

2.2.4 Skin penetration study

This study was performed by a transdermal Franz diffusion system (PermGear, Germany) that composed of six horizontal diffusion cells (effective area= 1.07 cm^2), a thermally control circulating water bath and a magnetic stirrer for each of cells. The experiments were conducted by using the abdnominal skin of 9 weeks old rats [7]. The shaved skin was cut into squares and the subcutaneous fat was carefully removed, and the skin was pre-equlibrated in Phosphate Buffer Solution (PBS, pH=6.5) at 4°C for 24 hour.

The skins were mounted on the receptor compartment with the stratum corneum side facing the donor cell. The receiver compartments (3.7 mL) were filled with PBS solution (pH 6.5) and the donor compartments are filled with 2.7 mL of the same PBS solution. The diffusion calls were equilibrated by the re-circulating water bath at 37 ± 1 °C, provides a skin surface temperature of 32 ± 1°C (Misik et al., 2011). The fluid in the donor and receptor compartment was stirred continuously by magnetic stirrer. Setup was run for half an hour to allow the equilibrium of temperature. 1 mL of samples was then pipette into the donor compartment.

Sampling was done at 0.5, 1, 1.5, 2.0, 3, 4, 5, 6, 8, 10, 14, 18 and 24 h. At each point, 1 mL of aliquots was withdrawn from the receiver compartment and thereafter, 1 mL of PBS solution was added as replacement to the receiver compartment. The collected samples were disrupted with 50 % n-propanol with ratio 1:1 in an ultrasonic bath for 20 minutes. The resultant mixture was filtered with 0.2 μm syringe nylon filter and the filtrate was subjected to total phenolic content assay in order to obtain the concentration of extract.

Flux ($\mu g/cm^2h$), were calculated using the linear regression interpolation of the experimental data at a steady state using Fick's Law [7][8]

Result and Discussion

3.1 Formulation Optimization

Formulation optimization involved three parameters which were the amount of cholesterol (A), Span 60 (B) and labrasol (C). The lower and upper limits of parameters were determined from the previous studies (unpublish data). Table 1 presents the DOE for FD niosomes where their responses were particle size (PS) (Y1), zeta potential (ZP)(Y2) and encapsulation eficiency (EE)(Y3).

https://doi.org/10.1051/matecconf/201819203025

Table 1. DOE for FD Niosomes

Run	A (mmol)	B (mmol)	C (mmol)	Y1 (nm)	Y2 (-mV)	Y3 (%)
1	0.29	0.62	0.09	203.267	45.2	67.3
2	0	1	0	232.9	50.13	76.9
3	0.18	0.64	0.18	176.2	45.267	84.63
4	0	0.8	0.2	269.2	47.7	51.8
5	0.09	0.82	0.09	283.5	50.3	84.13
6	0.4	0.6	0	198	49.233	78.57
7	0.09	0.62	0.29	300	40.76	61.1
8	0	0.6	0.4	335.967	33	35.8
9	0.25	0.5	0.25	180.13	42.4	75.3
10	0	0.6	0.4	335.8333	42.467	54.3
11	0	1	0	224.633	52.2	66.67
12	0.2	0.8	0	191.767	48.267	87.9
13	0.25	0.5	0.25	176.2	44.7	87.3
14	0.4	0.6	0	183.867	51.3	67.3

3.2 Effets of Formulation Ingrédients and their interactions on Réponses

Figure 1 shows the three-dimensional plot representing the effect of the interaction of variables on PS for FD niosomes. Based on the contour plot, particle size decreased while cholesterol (X1) increased, or when labrasol decreased. Therefore the minimum vesicle size can be achieved at a region where high amount of cholesterol and Span 60, low amount labrasol. Figure 2 shows the three-dimensional plot represents the effect of the interaction of variables on ZP. Based on the contour plot, the maximum particle size can be achieved at high amount of cholesterol and Span 60, low amount labrasol. Effect of mixture component towards EE could be observed through Figure 3, maximum EE could be obtained at a region where high cholesterol and low labrasol, however when increment of cholesterol exceed 30 mole %, EE started to decreased as existence of cholesterol may compete with the drug for packing space [9]. The response that significantly affected the mixture component for niosomes production was PS, ZP and EE. The optimum value for the parameter were shown in Table 2, with the difference is not more than 15%.

Table 2. Check point analysis

Check point	Processing parameters			Output		
	Cholesterol (mmol)	Span 60 (mmol)	Labrasol (mmol)	PS (nm)	ZP (-mV)	EE (%)
Optimum	0.37	0.57	0.07	176.2	47.8329	74.2205
Modified	0.40	0.60	0.10	151.20	49.70	78.43

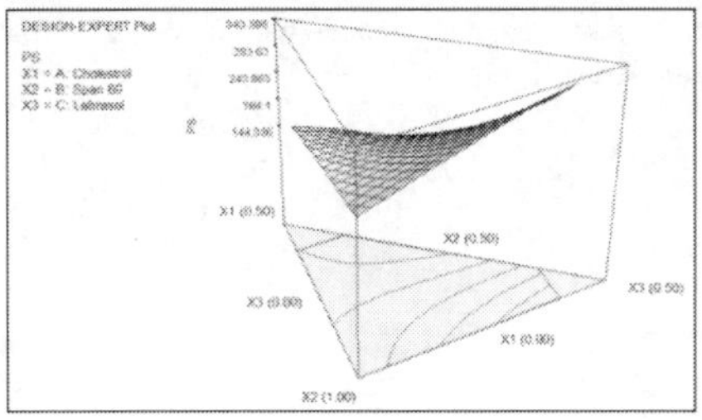

Figure 1. 3-dimensional response surface plot representing the effect of the interaction of variables on PS

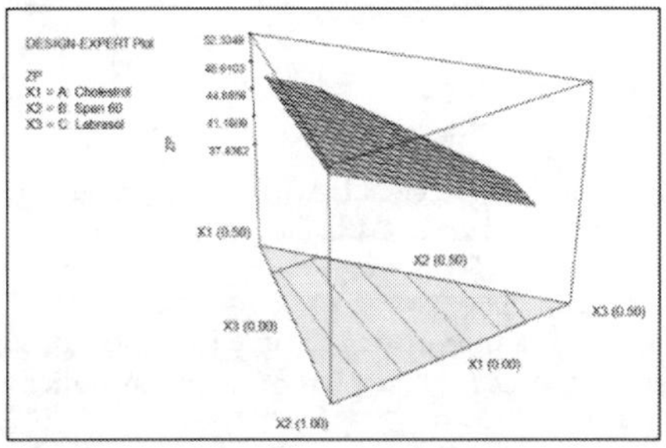

Figure 2. 3-dimensional response surface plot representing the effect of the interaction of variables on ZP

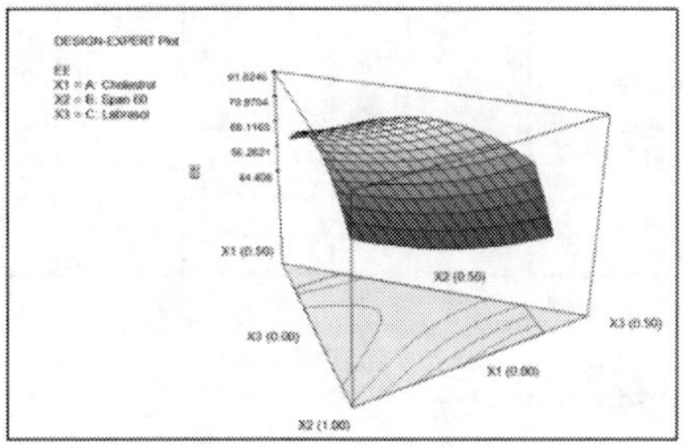

Figure 3. 3-dimensional response surface plot representing the effect of the interaction of variables on EE

3.3 Penetration of FD Niosomes in Rat Skin

Penetration study was done using Franz horizontal diffusion cells with rat skin used as the barier. The FD niosomes were obtained using the optimized formulation of 6 :1 :4 of Span 60 :labrasol :cholesterol. The permeations of free FD extracts and FD niosomes through rat skin were compared. Table 3 showed that the FD niosomes diffuse faster than FD extract without any encapsulation. Higher values of flux (1.56 times) obtained with FD niosomes 0.067 μg/cm²h than that obtained with FD extract only 0.043 μg/cm²h. Besides, deposition of extract by FD niosomes 0.157 μg also higher than FD extracts 0.124 μg. Fig 4 shows the trend of culmulative permeation of free FD extracts and FD niosomes through lab skin for the period of 24 hours and permeation coefficient through rat skin, which were calculated and tabulated in Table 3.

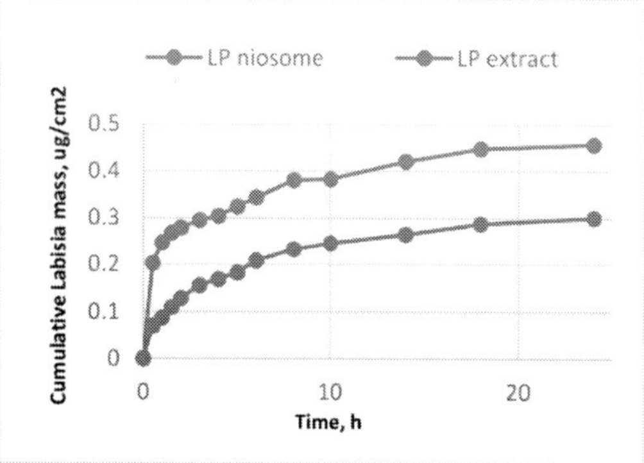

Figure 4. Trend of cumulative permeation through rat skin for the period of 24 hours.

Table 3. Comparison of transport parameters for penetration and the amount of active ingredients in the skin

Sample	Cumulative Permeation Q_t/S (μg/cm²)	J_{ss}, (μg/cm² h)	Kp (× 10^{-1} cm/h)	Amount of active ingredients in the skin, μg
FD extract without encapsulation (FDe)	0.212 ± 0.02	0.0430	0.72	0.124±0.027
FD niosomes	0.234 ± 0.34	0.0672	1.12	0.157±0.012

Conclusion

The optimum formulation obtained for FD loaded niosomes was Span 60, labrasol and cholesterol in the ratio of 6 :1 :4. The optimized niosomes showed the feasibility to be applied in cosmetic formulation.

Acknowledgement

The authors would like to thank the Ministry of Education Malaysia, Institute of Bioproduct Development, Universiti Teknologi Malaysia and research vote 4J271 for financial assistance to this research study.

Referrences

1. Manach, C., Scalbert, A., Morand, C., Rémésy, C. and Jiménez, L. (2004). Polyphenols: food sources and bioavailability. *The American Journal of Clinical Nutrition*. 79(5), 727–47.
2. Oh, M.-J., Abdul Hamid, M., Ngadiran, S. and Seo, Y.-K. (2010). Ficus deltoidea (Mas cotek) extract exerted anti-melanogenic activity by preventing tyrosinase activity in vitro and by suppressing tyrosinase gene expression in B16F1 melanoma cells. Arch Dermatol Res.
3. Fang, J., Hong, C., Chiu, W. and Wang, Y. (2001). Effect of liposomes and niosomes on skin permeation of enoxacin. International Journal of Pharmaceutics. 219, 61–2.
4. Sulaiman, M.R., Hussain, M.K., Zakaria, Z.A., Somchit, M.N. and Moin, S. (2008). Evaluation of the antinociceptive activity of Ficus deltoidea aqueous extract. Fitoterapia. 79, 557–561.
5. Vora, B., Khopade, A. and Jain, N. (1998). Proniosome based transdermal delivery of levonorgestrel for effective contraception. Journal of controlled release. 54, 149–165.
6. Fry, D., White, J. and Goldman, I. (1978). Rapid separation of low molecular weight solutes from liposomes without dilution. Analytical biochemistry. 90(2), 809–815.
7. Shah, K. a, Date, A. a, Joshi, M.D. and Patravale, V.B. (2007). Solid lipid nanoparticles (SLN) of tretinoin: potential in topical delivery. International journal of pharmaceutics. 345(1-2), 163–71.
8. Sintov, A.C. and Shapiro, L. (2004). New microemulsion vehicle facilitates percutaneous penetration in vitro and cutaneous drug bioavailability in vivo. *Journal of controlled release : official journal of the Controlled Release Society. 95(2), 173–183.*
9. Balakrishnan, P., Shanmugam, S., Lee, W.S. and Lee, W.M. (2009). Formulation and in vitro assessment of minoxidil niosomes for enhanced skin delivery. International Journal of Pharmaceutics. 377(1), 1–8.

Effects of pandan aqueous extract on physico-chemical properties of normal and waxy rice starches

Pawinee Deetae[1,2*], *Thungpon* Klumrat[2], *Warot* Nakakitwibool[2] and *Atikorn* Panya[3]

[1]Faculty of Agro-Industry, King Mongkut's Institute of Technology Ladkrabang, Bangkok,10520, Thailand.
[2]Food Technology Program, Mahidol University, Kanchanaburi Campus, Saiyok, Kanchanaburi,71150, Thailand.
[3]National Center for Genetic Engineering and Biotechnology, NSTDA, 113 Thailand Science Park, Paholyothin Road, Klong Luang, Pathumthani, 12120, Thailand.

Abstract. Pandan aqueous extract (PAE) has traditionally been used as a food additive in Southeast Asia to enhance their flavor, color and health benefits. The purpose of this research was to investigate the effects of PAE on physico-chemical properties of normal rice starch (NRS) and waxy rice starch (WRS). Results indicate that addition of PAE significantly ($p < 0.05$) affected the physico-chemical properties of rice starches. For pasting properties, PAE significantly increased peak viscosity, breakdown and pasting temperature of NRS, while PAE decreased peak viscosity, breakdown, final viscosity and setback values of WRS. The addition of PAE to NRS and WRS significantly ($p < 0.05$) reduced gel hardness and gel adhesiveness by more than 50% compare to native control gels. The gelatinization thermal properties of both NRS and WRS were significantly ($p < 0.05$) increased when PAE was included. Only the gelatinization enthalpy of WRS was influence by the PAE. It was concluded that the addition of PAE considerably improved the thermal stability of WRS.

1. Introduction

Pandan (*Pandanus amaryllifolius* Roxb.), a member of screw pine family Pandanaceae, widely distributes in various parts of South-East Asia such as Thailand, India, Malaysia and Indonesia. The plant bears no fruits as it is sterile and is to be propagated vegetatively. Pandan leaves have a desirable characteristic aroma, that has been described as popcorn-like as in jasmine and basmati rice (1-2). Various studies have demonstrated that pandan also has other properties including antioxidant and antimicrobial activities (3-4).

In some South-East Asian countries, pandan aqueous extract has traditionally been used as a drink and/or as an additive in food products as a flavoring, green coloring as well as for helping health benefits. Some component of herb extracts, such as phenolics, may interact with other food ingredients and Wu *et al.*, (5) and Xiao *et al.* (7) reported that physico-chemical properties of starch were affected when some herb extracts were added to food. Also Zhu *et al.* (6) showed that peak viscosity of wheat starch was increased in the presence of pomegranate extract, while, peak viscosity was not affect in the presence of Chinese hawthorn and gel hardness of wheat starch was reduced in the presence of green tea extract. Retrogradation of potato and maize starches was retarded more than 5 days of storage by the addition of green tea extract (7). Hence, inclusion an herb extract into a starch based food product could affect the properties during processing and end-use quality.

Non glutinous (normal) or glutinous (waxy) rice starch are important ingredients in various Thai desserts. Rice starch either from non-glutinous or glutinous varieties mainly composes of amylose and amylopectin, which its ratio is difference. The major constituent of glutinous rice starch is amylopectin and it possesses unique physical and chemical properties such as low pasting temperature and low final viscosity. In Thailand, addition of pandan aqueous extract prior to cooking rice starch based foods, especially desserts, is common, but interactions between rice starch and pandan aqueous extract have not been reported. Therefore, the aim was to investigate the effects of pandan aqueous extract on pasting, texture and gelatinization thermal properties of normal and waxy rice starches.

2. Materials and Methods

2.1 Material and pandan aqueous extract preparation: Normal rice starch (NRS) and waxy rice starch (WRS) were kindly supplied by Cho Heng Rice Vermicelli Factory Co. Ltd., Nakornpathom, Thailand. Pure dried pandan leaves, purchased from a local market in Kanchanaburi, Thailand, were grounded in a blender (Philips, Japan) then sieved through a No. 60 stainless steel sieve (60 meshes) and sealed in plastic bags and kept at -80 °C until use. For preparation of pandan aqueous extract (PAE), 10 grams of pandan powder were extracted in 100 ml of deionized water at 80 °C for 20 min according to the method described (6) then stored at 4 °C and used within 24 h.

2.2 Chemical compositions of starches and pandan aqueous extract (PAE): Proximate analysis of starches and pandan aqueous extract was done by using the standard methods AOAC (1990). Amylose content of rice starches was determined by Iodine Affinity method (8). Total phenolic content of PAE was measured by using the Folin-Ciocalteu assay (9). The pH of PAE was measured directly using pH meter (IQ, Scientific Instrument, USA).

2.3 Pasting properties: A Rapid Visco-Analyzer (RVA model 4, Newport Scientific Australia) was used to investigate the effect of PAE on both normal and waxy rice starch pasting properties. The PAE was added to rice starch (3 g dried basis) in a RVA canister to obtain a total constant sample weight of 28 g. The slurry was manually homogenized to prevent lump formation before the RVA run. A heating and cooling cycle was programmed in the following manner; the sample was heated from 50 to 95 °C for 7.50 min, held at 95 °C 5 min and cooled to 50 °C within 8.50 min, further held at 50 °C for 2 min. The pasting properties of the sample were then recorded and presented in rapid visco units (RVU). After RVA testing, the paddle was immediately taken off and the starch paste in the canister was covered by parafilm. It was kept at 25 °C for 48 h prior to texture analysis.

2.4 Texture analysis: A texture analyzer (TA-XT2i, Stable Micro Systems, Surrey, England) was used to investigate the force-time curve according to the method described (6) with some modifications. The gel was compressed by using a probe (0.5 HS Hemispherical plastic) at a speed of 1 mm/s to a distance of 15 mm. The maximum force peak in the TPA profile expressed the gel hardness and the negative area of the curve during the first retraction of the crosshead was termed adhesiveness. Springiness was measured by the distance of the detected height of the product on the second compression divided by the original compression distance. Cohesiveness was defined as the ratio of the positive force areas during the second compression to that during the first compression.

2.5 Thermal properties: Gelatinization thermal properties of samples were assessed by using a Differential Scanning Calorimeter (DSC), (Mettler Toledo, USA) as method described by Zhu *et al.* (6) with some modifications. Rice starch (based on dry weight) was dispersed in PAE to obtain the starch/ PAE ratio of 1:3. A 15 µg sample was then transferred to an aluminum pan DSC and hermetically sealed. After equilibration at room temperature (25◦C) for 1 h, the sample was heated in a DSC from 20°C to 100 °C at a rate of 10 °C/min. The onset (To), peak (Tp), conclusion (Tc) temperatures were recorded. Gelatinization enthalpy (ΔHg) was evaluated based on the area of the main endothermic peak. An empty pan was used as references for all measurements.

2.6 Statistical analysis: All experiments were carried out in triplicate. Results were reported as mean ± standard deviation. Data analyses were performed by using the XLSTAT PRO 2012 statistical package for Windows (New York, USA). Analysis of variance (ANOVA) was performed and Duncan's multiple range test was applied on the individual variables to compare the means and to assess their significant differences ($p < 0.05$).

3. Results and Discussion

3.1 Chemical compositions: The amylose contents in NRS was 31.52±0.5% and in WRS 4.85±0.1% respectively. All samples contained less than 1.25% protein, less than 1.25% fat and less than 0.3% ash. These results were consistent with the results reported by Puncha-arnon *et al.*, (10). The PAE contained 1.98% total solids with 1.13% carbohydrate, 0.23% protein and 0.62% ash. Total phenolics in the extract was found to be 0.11 g/100 ml with a pH of 5.74±0.04. Previously, it was reported that the carbohydrate in pandan leaves were mainly maltodextrin (11) and the polyphenolic in PAE were reported to be mainly gallic acid, catechin, kaempferol and naringin (12).

3.2 Pasting properties: Pasting properties can give information about starch suitability for industrial applications because it reflects changes in viscosity during processing when heated in excess water (13). Amylose content was reported to be the major factor controlling almost all physicochemical properties of rice starch including pasting and gelatinization (14). Since WRS had lower amylose content, had lower peak viscosity, peak time and pasting temperature than NRS, the physico-chemical characteristic of NRS and WRS are significantly difference (Table1). The addition of PAE significantly ($p < 0.05$) affected all pasting properties of both starches. During the heating and holding process, the mixture of NRS with PAE increased peak viscosity by up to 25 RVU), increased breakdown by up to 29 RVU and increased pasting temperature values by up to 3 ◦C compare to NRS alone, whereas, the mixture of WRS with PAE decreased peak viscosity by up to 71 RVU, decrease breakdown by up to 26 RVU but increase pasting temperature values by up to 11◦C compare to WRS alone. During cooling, PAE significantly ($p < 0.05$) reduced the final viscosity by approximately 50 RVU and reduced setback values by approximately 40 RVU for NRS and 5 RVU for WRS. The PAE used in this study contained 1.13% carbohydrate and carbohydrate such as maltodextrin may increase pasting temperature of starches in the system by reducing swelling capacity due to restriction amount of water available for starch granules (15). The PAE had a pH of 5.74 and acids, mainly phenolics like gallic acid in the extract, may reduce final viscosity and set back values of the starch pasting properties by destroying glyosidic bonds in starch granule (16).

3.3 Texture properties: WRS had lower hardness, adhesiveness, and springiness than NRS as shown in Table 2. Addition of PAE significantly ($p < 0.05$) reduced more than 50% gel hardness (up to 134 g for NRS, 22g for WRS) and gel adhesiveness (130 gs for NRS, 5 gs for WRS) but had little influence on gel springiness and gel cohesiveness. The texture properties of starch gels depend on the interaction between the disperse amylose and the continuous phase of starch (17). The presence of PAE may weaken the hydrogen boning and van der wale force of that interaction, thus altering the gel structure.

3.4 Thermal properties: WRS had lower gelatinization onset temperature (T_o), peak temperature (T_P), and conclusion temperature (T_C) than NRS as shown in Table 3. Addition of PAE significantly ($p < 0.05$) increased the gelatinization temperature of both NRS and WRS by up to approximately 4°C for T_o, T_P and T_C, while the addition of PAE significantly ($p < 0.05$) increase only the gelatinization enthalpy value (ΔH gel) in WRS (by up to 0.9 J/g). Temperature and enthalpy values were reported to be associated with the degree of crystallization of starch, amylose, amylopectin chain as well as other component present (18). The effect of PAE on the gelatinization properties of NRS and WRS might be explained by the components present in PAE such as carbohydrate, phenolic compounds could interact with water molecule, which restrict amount of water available for starch granule (7,15). This require higher amount of thermal energy for gelatinization.

4. Conclusions

The result of this work demonstrate that the addition of pandan aqueous extract significantly affect the pasting, textural and gelatinization thermal properties of both normal and waxy rice starches with the result that pandan extract considerably improved the thermal stability of waxy rice starch. This information may provide a basic knowledge for further pandan aqueous extract application in rice starch based product.

5. Acknowledgment

Professor Keith Thompson, a visiting professor at Faculty of Agro-Industry, King Mongkut's Institute of Technology Ladkrabang, is gratefully acknowledged for reviewing the English language.

6. References

1. V. Lakanalamai S., Ilangantileke, Cereal Chem, 70, 381-384 (1993).
2. RG. Buttery, BO. Juliano, LC. Ling, Chem Ind. 23, 478-479 (1983).
3. S.W. Yan, R. Asmah, Int Food Res J. 17, 417-423 (2010).
4. A.F. Faras, S.S. Wadkar, J.S. Ghosh. Int Food Res J. 21,421-423 (2014).
5. Y. Wu, Z. Chen, X. Li and Mei Li, Food Res Int., 42, 221-225 (2009).
6. F. Zhu, Y.Z. Cai, M. Sun, H. Corke, Food Chem., 112, 919-923 (2009).
7. H. Xiao, Q. Lin, G.Q. Liu, Y. Wu, W. Wu, X. Fu, Food Bioprocess Technol. Chem., 6, 2177-2183 (2013).
8. C.A. Knutson, Cereal Chem., 63, 89–92. (1986).
9. P. Deetae, P. Parichanon, P. Trakunleewatthana, P., C. Chanseetis, S. Lertsiri, Food Chem., 133, 953-959 (2012).
10. S. Puncha-arnon, D. Uttapap, Carbohyd Polym., 91, 85-91 (2013).
11. S. Chaiseri, V. Cheetangdee, Kaset J Nat Sci. 40, 67-74 (2006).
12. A. Ghasemzadeh, HZ. Jaafar, BMC Complement Altern Med., 13, 341, (2013).
13. J. Ahmed, L. Thomas, A. Taher, A. Joseph, Carbohydr. Polym. 152, 639–647 (2016).
14 HAM. Wickramasinghe, T. Noda, Food Sci Technol Res., 14, 49–54 (2008).
15. L. Juszczak, D. GaBkowska, T. Witczak, T. Fortuna, Int. J. Food Sci. 2013, (2013).
16. F. Zhang, M. Liu, F. Mo, M. Zhang, J. Zheng Polymers. 9, 695 (2017).
17. S.G., Ring. Starch/Stärke, 37, 80–83 (1985)
18. M.B. Cardoso, J. L. Putaux, D. Samios, D., Samios, D., N.P. Silveira. Carbohydr. Polym., 70, 160–165 (2007).

Table 1: Pasting properties of normal rice starch (NRS) and waxy rice starch (WRS) with and without pandan aqueous extract (PAE)

Samples	Pasting properties					
	Peak Time (Min)	Peak viscosity (RVU)	Breakdown (RVU)	Final Viscosity (RVU)	Setback (RVU)	Pasting Temperature (◦C)
NRS	9.60±0.00^{a}	357.67±0.50^{a}	133.30±1.37^{a}	419.25±2.61^{a}	194.89±1.54^{a}	79.40±0.23^{a}
NRS+ PAE	9.47±0.04^{b}	382.72±1.65^{b}	162.28±2.54^{b}	371.39±3.48^{b}	150.95±0.70^{b}	81.78±0.16^{b}
WRS	5.18±0.03^{c}	345.11±1.65^{c}	188.83±1.50^{c}	201.31±1.76^{c}	45.03±1.69^{c}	63.58±0.04^{c}
WRS+ PAE	5.87±0.04^{d}	273.47±0.54^{d}	162.97±0.57^{b}	151.31±0.96^{d}	40.80±0.48^{d}	75.20±0.17^{d}

Values are means ±SD (n=3). Mean (±SD) with different superscript letters (a, b, c d) in the same column indicate significant differences ($p < 0.05$).

Table 2: Texture properties of normal rice starch (NRS) and waxy rice starch (WRS) with and without pandan aqueous extract (PAE)

Samples	Texture properties			
	Hardness (g)	Adhesiveness (gs)	Springiness	Cohesiveness
NRS	240.24±6.53^{a}	143.00±29.48^{a}	1.56±0.15^{a}	0.41±0.03^{b}
NRS + PAE	106.12±16.70^{b}	12.71±8.03^{b}	2.06±0.43^{a}	0.46±0.07^{ab}
WRS	32.60±2.72^{c}	8.00±2.98^{b}	0.98±0.02^{c}	0.43±0.03^{ab}
WRS+PAE	10.26±0.23^{d}	2.80±0.47^{b}	1.15±0.09^{b}	0.48±0.01^{a}

Values are means ±SD (n=3). Mean (±SD) with different superscript letters (a, b, c d) in the same column indicate significant differences ($p < 0.05$).

Table 3: Gelatinization temperatures and enthalpy of normal rice starch (NRS) and waxy rice starch (WRS) with and without pandan aqueous extract (PAE)

Samples	To (◦C)	Tp (◦C)	Tc (◦C)	ΔH gel (J/g)
NRS	72.35±0.07^{a}	77.30±0.15^{a}	81.92±0.12^{a}	12.51±0.06^{a}
NRS + PAE	76.64±0.25^{b}	81.13±0.18^{b}	85.49±0.15^{b}	12.31±0.34^{a}
WRS	60.13±0.25^{c}	67.60±0.27^{c}	75.12±0.34^{c}	13.62±0.13^{a}
WRS+ PAE	63.88±0.10^{d}	71.27±0.18^{d}	78.48±0.08^{d}	14.54±0.25^{b}

To, Onset temperature; Tp, Peak temperature; Tc, Conclusion temperature; ΔH_{gel}, gelatinization enthalpy. Values are means ± SD (n = 3). Means (±SD) with different superscript letters (a, b, c and d) in the same column indicate significant differences ($p < 0.05$)

Comparison of the effect of freezing on the quality of 'nam dokmai' mango fruit

Suwicha Prakobsang[1,*], and *Pimpen* Pornchalermpong[1]

[1]Food Engineering Department, Faculty of Engineering, King Mongkut's Institute of Technology Ladkrabang, Bangkok, Thailand

Abstract. This study evaluated effect of freezing methods on freezing rate and physical qualities of mango flesh before and after thawing. The method of freezing included still air freezing (at -20°C), air blast freezing (at -35°C, velocity 3 m/sec) and Air Blast Combined Vacuum Freezing (at -40°C, were configured core temperature 2 level is -5°C and -10°C) until the temperature reaches -20°C. The frozen flesh were thawed and evaluated for drip loss, pH, total soluble solid, colour, texture (firmness and %firmness decrease). The results showed that Air Blast Combined Vacuum Freezing (-5°C) Quality of mango frozen after thawing did not show significant effect on pH and total soluble solid of fresh mango. The frozen mango had highest lightness (L*) and the total colour differences (ΔE) had lowest and can maintain the maximum. Therefore, it's possible to introduce air blast combined vacuum freezing techniques applied to frozen mango industry as a result of significant increases freezing rate.

1. Introduction

Mango (*Mangifera indica* Linn.) is an economic tropical fruit of Thailand. Ripe mango is high nutrient fruit, consisted of vitamin, carotenoids and beta carotene. As referred to year 2016, Both mango fruits and mango products were exported to other countries created value more than 521,472 million baht. Frozen mangos were exported around 2,409 tons which created value 365 million baht [8]. In present, air freight is selected to be the best way of ripe mango exportation because mango was categorized as short shelf life fruit. However, air freight caused higher cost than other shipping. Nowadays, freezing is the most selected process for ripe mango. The freezing is the method of reducing temperature until lower than -18°C which inhibits microorganism growth or deterioration of food. Water activity was reducing by freezing process to restrain the growth of bacteria and reduce enzyme activity. [1]; [2]; [5].

The quality of freezing fruit is depended on many components which are fruit types, variety, maturity stages or shelf life. However, the problem of producing freezing mangos were occurred from high water volume, changing texture, such as soften, changing of color and changing of important nutrient, such as vitamin C, carotenoids and microstructure, were occurred from freezing mango but all changing was high or low depended on the drip loss while freezing. [1]; [4-5])

Karl and Da-Wen (2000) studied about cooling by vacuum chilling which was used with product contained water volume than 90% such as mushroom and lettuce. Vacuum chilling can be used with freezing mango. Another advantage of vacuum chilling is to specify temperature and degree. Moreover, the study of vacuum chilling and mushroom browning by Polyphenol oxidase (PPO) were shown the mushrooms frosted by vacuum chilling were browned lower than other process. The study of Cheng and Lin (2007) showed the liquid state was changed faster when reducing pressure. Therefore, there are possibility to compile with reducing pressure and freezing to keep quality of mango.

In the past, the problems between different freezing was not studied and there was no the study about the using of vacuum system in freezing. Then, Still Air Freezing and Air Blast Freezing, which were most selected in freezing industry and Air Blast Combined Vacuum Freezing were compared for the advantages of freezing industry.

2. Materials and Methods

2.1 Materials

Mango (cv. Namdokmai) the variety of mango used in this study, No.4 were purchased from Chachoengsao province, Thailand, the fruits were selected for the pH value and total soluble solids were 4.12±0.22 and 14.78±0.98°Brix, amount 60 kg.

* Corresponding author: suvicha.pha@gmail.com

2.2 Sample preparation

The sample were washed in chlorine solution (concentration 200 ppm) for 2 minutes, air dried and peeled. Only the central segments of the fruit sample were cut into size 2 x 2 x 1.5 cm and soaked in solution (1% ascorbic acid :1% calcium chloride (w/w)) for 10 minutes. Place them apart in a stainless-steel tray.

2.3 Freezing experimental

Mango were divided into 4 sections for four freezing methods by Still Air Freezing (at -20°C) was carried out in a still air freezer (SF-PC997, Panasonic Co., Ltd., Japan), Air Blast Freezing (at -35°C velocity 3 m/sec) was performed in an air blast freezer (Compact freeze, PATKOL Co., Ltd., Thailand) and Air Blast Combined Vacuum Freezing (ABCVF) (at -40°C, were configured core temperature 2 level is -5°C and -10°C the system will stop the air blast freezing process and was used a vacuum system to reduce the pressure at 40 Pa until the core temperature of sample reaches -20°C) was executed in a Kryo 'D' Freezer (ITC Co., Ltd. Thailand), Four freezing process were used thermocouples K type, (TREX-8, LogTag Recorder, USA) insert into the core temperature of the sample, Record temperature every 30 seconds since the core temperature of 24°C, the temperature reached to -20°C. The freezing rate of the samples the freezing rate was calculated by modifying the equation: [4]

$$\text{freezing rate (cm/hr.)} = \text{Distantce}/\text{Freezing time} \quad (1)$$

Where Distance from the surface to the thermal center of fruits per freezing times of temperature decrease from the initial temperature (24 °C) to -20°C).

2.3.1 Thawing processes

The frozen sample were placed on napkin on plastic grate and thawed in control temperature and control relative humidity room (at temperature 25 °C, 50% relative humidity). Were used thermocouples K type insert into the core temperature of the sample until the temperature reaches 0 - 4°C and quality was then analysed.

2.4 Drip loss

Drip loss of the frozen samples were sampling ten sample drip loss was defined as the percentage and was calculated following equation:

$$\text{Drip loss (\%)} = (W_0 - W_1) \times 100 / W_0 \quad (2)$$

Were W_0 is weight of frozen mango (g) and W_1 is weight of tawed mango (g). Drip loss was expressed as percent (%).

2.5 pH and Total Soluble Solid (TSS)

Frozen mango sampled after thawed was blended with a blender (HR2115, Philips Electronics Co., Ltd., Nederland), measurement of pH Volume 10 ml with pH meter (Lab855,SI Analytics, Germany) and total soluble solid was determined by refractometer (HI96800, Hanna, Romania) and repeat three repetitions.

2.6 Colour and Colour difference

The colour of the fresh and frozen-thawed sample was measured through a Hunter Lab Colorimeter (ColorFlex EZ, Hunter Associates Laboratory, Inc., UK). In this system, L^* is presented to lightness (0-100 scale, black to white); a^* relate to red colour (+) and to green colour (-). and b^* corresponds to yellow colour (+) and blue colour (-). Compare colour difference (ΔE) of fresh and frozen was calculated following equation

$$\Delta E = [(L_0^* - L_1^*)^2 + (a_0^* - a_1^*)^2 + (b_0^* - b_1^*)^2]^{1/2} \quad (3)$$

Where L_0^*, a_0^* and b_0^* are the colour of fresh mango sample, and L_1^*, a_1^* and b_1^* are the colour of frozen-thawed mango sample.

2.7 Texture analysis

The texture of the sample both fresh and frozen samples was determined a Texture Analyzer (TA.XT2i Stable Micro Systems, Surrey, UK) with a 100 mm cylinder probe (P/100), using a compression of 50% strain and Pre-test speed : 1.0 mm/s. ; Test speed: 0.2 mm/s.; Post-test speed: 10.0 mm/s. The maximum peak force was represented a firmness value in Newton. five pieces of fruit were tested for each treatment. The results are used to calculate the percentage decrease in firmness (% Firmness decrease) by equation: [1]

$$\%\ \textit{Firmness decrease} = (Fi - F_f) \times 100 / Fi \quad (4)$$

Where Fi is the firmness of fresh sample, and F_f is the firmness of frozen-thawed sample.

2.8 Statistical analysis

The experiment was completely Randomize Design (CRD), the data were analysed by analysis of variance (ANOVA) according to the Duncan's Multiple Range Test (DMRT) at 95 percentage confidence intervals. Statistical analysis was presented by SPSS Statistics Software 20.0 for Windows (SPSS, Inc., Chicago, Illinois)

3. Results and Discussion

3.1 Effect of freezing time and freezing rate

The freezing times of each freezing methods were present in Table 1. The ABCVF(-5°C) method had the fastest time 30 minutes, freezing rate was 2.00 cm/hr. Each of these freezing methods has a different rate depending on the method of heat transfer, can be divided into freezing rates according to the efficiency of the equipment. The freezing rate it is classified the ABCVF and air blast freezing method was quick-freezing. In addition, found that still air freezing spends more time in the critical zone than other freezing (Fig 1.). The longer it takes for the critical zone, the greater the ice crystals [10]. On the other hand, the ABCVF and air blast freezing methods has the slope of the freezing temperature curve and passes through the critical zone. For this reason, the possibility of accelerating the freezing rate is significant of the vacuum combined air blast freezing.

3.2 Effect of freezing methods on Drip loss of mango.

Effect of difference of freezing method on drip loss of sample after frozen-thaw (Table 4.). It was found that the samples of frozen mangoes with the ABCVF (-5°C) has lowest value was 4.82%. The drip loss increases when the freezing rate decreases [10]. The results are consistent with Table 1. This results in a more complete microstructure of fruit. So, the loss of water after thawing is associated with reduced cell integrity.

Table 1. Freezing time and freezing rate of each freezing methods.

Freezing methods	Freezing time (minutes)	Freezing rate (cm/hr)
still air freezing	307.33±0.05^{d}	0.20±0.01^{d}
air blast freezing	33.33±0.03^{b}	1.80±0.09^{b}
ABCVF (-10°C)	36.50±0.01^{c}	1.64±0.03^{c}
ABCVF (-5°C)	30.00±0.02^{a}	2.00±0.03^{a}

Data are recorded as the mean ± standard deviation from 3 replications. And different letter in each column are significantly different (p□0.05)

3.3 Effect of freezing methods on pH and Total Soluble Solid (TSS) of mango.

In the Table 2. pH and total soluble solids content of the frozen-thawed samples was indicated air blast freezing and ABCVF (both -10°C and -5°C) there was no statistically significant difference in fresh mangoes. At the same time, the frozen sample of still air freezing observed increase pH value and decrease total soluble solids content statistical significance when compared to fresh mango. Due to during frozen, the water in the sample formed ice crystal which damaged the cellular. This causes the solids inside the cell to come out of the cell with the fluid lost [1].

Table 2. pH and Total Soluble Solid (TSS) content of fresh and frozen-thawed mangoes

sample	pH	TSS
fresh	4.12±0.22^{b}	14.78±0.98^{a}
still air freezing	4.50±0.03^{a}	12.37±0.06^{b}
air blast freezing	4.01±0.01^{b}	13.20±0.10^{b}
ABCVF (-10°C)	3.92±0.01^{b}	14.77±0.06^{a}
ABCVF (-5°C)	3.91±0.01^{b}	14.67±0.15^{a}

Data are recorded as the mean ± standard deviation from 3 replications. And different letter in each column are significantly different (p□0.05)

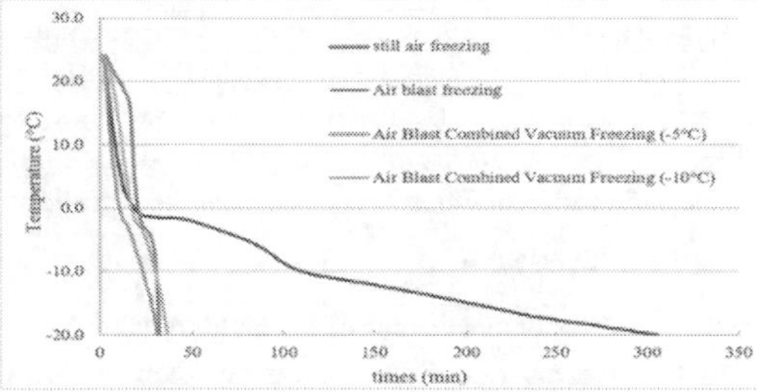

Fig. 1. Freezing curves obtained from four freezing method of mango.

3.4 Effect of freezing methods on Color and Color difference of mango.

The experiment data on the color parameters (L*, a* and b*) of mango frozen-thawed by four freezing methods, and color difference. In the Table 3 show that the brightness (L *) of the frozen mangoes ABCVF (-5°C) with the highest brightness. The red colour (a *) of frozen-thawed mangoes with still air freezing has highest value and more than fresh mango. Arasorn et al.,(2558) reported that color changes after freezing and defrosting were the main problem of enzymatic browning reactions. This is an oxidation reaction when cell injury enzyme and oxygen are in contact with monophenols, resulting in browning index in the mango. For this reason, freezing slows the freezing rate, the cells are highly destroyed, resulting in higher reaction rates. When all values were calculated for colour difference (ΔE), it was found that the samples of mangoes,

Table 3. Chang of color L* value, a* value, b* value and color difference color of fresh and frozen-thawed mangoes.

sample	*L**	*a**	*b**	ΔE
fresh	64.06±0.16^{a}	14.50±0.05^{b}	61.72±0.17^{a}	-
still air freezing	60.08±0.14^{f}	15.51±0.02^{a}	57.97±0.05^{c}	54.10±0.31^{d}
air blast freezing	60.43±0.07^{d}	10.80±0.02^{f}	57.53±0.11^{c}	45.81±0.16^{c}
ABCVF (-10°C)	61.34±0.12^{c}	12.35±0.09^{c}	61.56±0.55^{a}	30.88±0.26^{b}
ABCVF (-5°C)	62.23±0.10^{b}	11.22±0.06^{d}	60.26±0.11^{b}	24.75±0.20^{a}

frozen-thawed with ABCVF (-5°C) had the difference of mango colour least than compared to other frozen-thawed samples statistically significant. So, ABCVF (-5°C) method is possible can be maintain the quality of frozen mango. colour.

3.5 Effect of freezing methods on texture of mango.

Table 4. Firmness and drip loss or the both fresh and frozen-thawed mangos.

sample	Firmnes s (N)	Firmness decrease (%)	Drip loss (%)
fresh	8.31±0.83^{a}	-	-
still air freezing	2.62±0.33^{c}	68.22±4.95^{b}	15.99±0.66^{d}
air blast freezing	2.94±0.24^{c}	64.51±2.91^{b}	11.46±0.36^{c}
VCAF(-10°C)	3.67±0.19^{b}	55.38±5.04^{a}	9.53±0.59^{b}
VCAF (-5°C)	3.68±0.16^{b}	55.30±6.47^{a}	4.82±0.11^{a}

Data are recorded as the mean ± standard deviation from five replications for firmness and ten replications for drip loss. And different letter in each column are significantly different (p□0.05)

The texture change of frozen foods is mainly due to the freezing rate. The experiments showed that the firmness value of fresh was 8.31 N. Frozen samples with ABCVF (-5°C) where have significant of densely more than those at various levels was 3.68 N. (Table 4) The results of the experiment corresponded to the effect of freezing rate as shown in Table 1. In addition, different freezing methods also contributed to the firmness decrease (%) after thawed. It was found that ABCVF (-5°C) still had the lowest percentage reduction, and same trend as the firmness. Furthermore, the less drip loss effected less destroyed cell it possible to maintain many internal structures.

4. Conclusion

Among the four freezing methods studied in this research, the ABCVF (-5°C) method was found to be a greatly freezing method, with quality similar to fresh mango, whereas still air freezing method led to deterioration of fruits. The higher freezing rate and lowest drip loss of frozen-thawed mango of ABCVF (-5°C) method likely explained the higher preservation quality in terms of reduce pH, TSS, color and color different and texture quality. Among these treatment, the high freezing rate in ABCVF (-5°C) method the most effect to good quality during freezing because of the small size of ice crystal and reduced damage to the fruit cellular. However, since father studies the effect of long-term preservation on the quality, microstructure of frozen mango should be studied.

*Corresponding author: suvicha.pha@gmail.com

References

1. A. Sirijariyawat and S. Charoenrein, JST. **34** (5), 517-523. (2012)
2. H.-P. Cheng, C.-T Lin, J Food Eng. **78**, 569-576 (2007)
3. N. Crafot, CPFAC. p.F2.3.1-F2.3.15. (2001)
4. Official Method of Analysis of AOAC International. 1995.
5. D. Liang, F. Lin, G. Yang, X.Yue, Q. Zhang, Z. Zhang, H. Chen. LWT-Food Sci. Technol, **60**, 948-956. (2015)
6. V. Gustavo, B.-Cánovas, B. Altunakar and J. Danilo, M.-Lorío. Food and Agriculture Organization of the United Nations (FAO). **158**, 1010-1365. (2005)
7. IIR (International Institute of Refrigeration) . *Recommendations for the Processing and Handing of Frozen Foods*. International Institute of refrigeration, Paris. (1972)
8. K. McDonald and D.W. Sun, J Food Eng. **45**. p55-65. (2000)
9. Office of Agricultural Economics. *Mango*. Bankok : Thailand Ministry of Agriculture and Cooperatives. (2017)
10. G. Laura G. Mascaraque, R.P. Masia, R.G. Barrio, M.J. Periago, A.L. Rubio, Food Hydrocoll. **73**, 1-12. (2017)
11. L. Otero, M. Martino, N. Zaritzky, M. Solas, and P.D. Sanz, J Food Eng. **65**, No. 3. (2000)
12. L. Otero, MarõÂa T. Solas, Pedro D. Sanz Carlos de Elvira, Juan A. Carrasco, Z Lebensm Unters Forsch. **206**, 338-342. (1998)
13. S.C. Berces, F. Fonseca, M. Citeau and M. Marin, LWT-Food Sci. Technol, **32**(3) 1441-1449. (2010)
14. A.L. Vásquez, S. Neidhart, P. Pathomrungsiyounggul, P. Wiriyacharee, A. Chattrakul, P. Sruamsiri, et al., ESCAP. **8**(11), 1-13. (2002)
15. Y.H. Root, Phase Transitions in food (San Diego, CA; Academic Press, 1995)

Ammonia removal from giant fresh water prawn farm by using zeolite from oil palm ash

Rewadee Anuwattana[1,*], *Pattamaphorn* Phungngamphan[1], *Patthanant* Natpinit[1], *Narumon* Soparatana[1], *Worapong* Pattayawan[1], and *Supinya* Yindee[1]

[1]Expert Centre of Innovative Clean Energy and Environment, Thailand Institute of Scientific and Technological Research (TISTR) 35Mu 3, Technopolis, Khlong 5, Khlong Luang, Pathum Thani 12120, Thailand.

Abstract. The Empty Fruit Bunch (EFB) and Mesocarp fiber ash (FB) was obtained from the boiler at NAM HONG Palm Oil Company Limited, Thailand. The optimum condition was used in Na-A zeolite synthesis from oil palm ash. The obtain produce was $Na_{12}Al_{12}Si_{12}O_{48}\cdot 27H_2O$. The sample was dried in hot air oven. The calcium exchange capacity (CEC) of 386.39 mg $CaCO_3$/g of product, and the 86.05% of pure Na-A zeolite phase was formed. In batch experimental study, the synthetic wastewater was prepared. In order to find out the optimal zeolite dosage was added into 10 mg/L of total ammonium nitrogen solution. The results of unionized ammonia (NH_3) removal efficiency indicated that zeolite A standard and zeolite from oil palm ash are better than commercial zeolite. The optimum dosage of 0.2, 0.25 and 0.2 g of zeolite A standard, zeolite from oil palm ash and commercial zeolite, respectively. The NH_3 removal capacity from giant freshwater prawn farm, zeolite A standard is better than zeolite from oil palm ash and commercial zeolite, respectively. In the large volume of synthetic wastewater (1000 mL), the zeolite from oil palm ash is the best ammonia removal capacity when compared with zeolite A standard and commercial zeolite.

1 Introduction

Palm oil has been the leading vegetable oil in term of consumption and production. Approximately 4.5 million Rai of the total area planted with oil palm in the southern Provinces of Thailand is concentrated. [1] In Thailand, more than 100,000 families of farmers was involved in oil palm cultivation and achieve annual FFB yield transported to the crude palm oil industry. The extraction of 200 kg of crude palm oil from fresh fruit bunch are based on the processing of 1,000 kg of EFB. Thus, the amount of solid waste is 750 kg which disposed as waste to the environmental. The high nutrients of EFB used to produce organic fertilizer. The pressed EFB can be used as fuel and mesocrap in the boiler. The heating value of EFB and mesocrap are about of 3,700 and 4,420 Kcal/kg of dry fuel, respectively. [2] It is equal to 200 kw/hr and processed of EFB into pulp and paper process or fiberboard. [3] The ash of biomass is a mixture of 50-60 of SiO_2 and 2-4% of Al_2O_3 which is the alumino silicate compound. Another option for reuse of ash is silica gel and silicon carbide. On the other hand, the main composition of SiO_2 and Al_2O_3 in the palm ash allow the Na-A zeolite conversion which is the promising approach to improve the palm ash utilization. Zeolite are crystalline aluminum silicate which a framework of SiO_4 and AlO_4 tetrahedral linked to each other at the corner via sharing their oxygen atom. The structure of three-dimensional network is made up by tetrahedral and the negatively charged lattice accounting for the substitution of Si by Al define many special properties of zeolite, such as adsorption at molecule in huge internal channels. [4] These zeolite materials are used as adsorbents for the removal of NH_4^+ ion [5] and different heavy metal ions from wastewater.

Ammonia is toxic to giant fresh water prawn farm or fish if allowed to accumulate in giant fresh water prawn and fish production systems. If the ammonia concentration gets high enough (1-2 mg N/dm^3), the aquatic animal will be come lethargic and eventually fall and die. The relative proportion of the two forms present in water is mainly affected by pH and temperature following the equation (1). Unionized ammonia is the toxic form and predominates when pH is high. [6-7]

$$NH_4^+ \leftrightarrow NH_3 + H^+ \qquad (1)$$

Ammonium ion is relatively nontoxic and predominates when pH is low. The main source of ammonia in giant fresh water prawn is directly related to the feeding rate of protein level in feed. The main cause

*Corresponding author: rewadee_a@tistr.or.th

of nitrogen pollution in aquatic system is the disposal of nutrients directly from industrial effluents or indirectly from agriculture run off such as the giant prawn fresh water feeding or fish feeding farm. Various technology treatment such as chemical precipitation, biological treatment, ion exchange, adsorption, electrochemical technology, and membrane process have been used for the removal of ammonium ions from wastewater. [8] Zeolite Na-A is known as the most value form and high cation exchang capacity.[9] Zeolite Na-A, with the 4.1A° is used for removal of ammonium ion from waste water. [10] Therefore, natural zeolite for NH_4^+ removal is studied from wastewater having different composition.

In this work, we introduced mesocarp and empty fruit bunch ashes as a raw material to prepare Na-A zeolite and used as an adsorbent to removal unionized ammonia (NH_3) from synthetic wastewater and giant fresh water prawn feeding wastewater, respectively.

2 Methodology

The obtain product, zeolite from oil palm ash, was prepared. The chemical structure was $Na_8Al_6Si_6O_{24}SO_4$, it was dried at 105 ◦C for 24 hours.

For Cation exchange capacity (CEC) values were determined using the ammonium acetate method [11]. The results were expressed as millgrams $CaCO_3$ per gram of obtained product.

For the adsorption studies, the batch experiments for adsorption of NH_3were carried out to investigate the effect of zeolite dosage (0.05, 0.1, 0.15, 0.2 and 0.25 g/20 mL of synthetic waste water. (Total ammonia nitrogen: TAN in synthetic wastewater = 10 mg N/dm^3 and equal to unionized ammonia = 1.95 mg N/dm^3) at the pH value about 8.6 and 26 ◦C, each measurement was repeated 3 times to take the average value.

Then the samples were agitated on the shaker with the speed of 120 rpm at the room temperature and the ammonium solution was separated from the adsorbent under filtration with 0.45 μm membrane filters. The initial and final ammonium concentration remaining in solution were analyzed by the standard Nesslerization method by UV spectrophotometer at 425 nm.

The optimum for adsorbent dosage was used in the giant fresh water prawn farm to find the adsorption capacities of adsorbent. The optimum zeolite dosage was carried out to volumn of the synthetic wastewater by using zeolite A standard, zeolite from oil palm ash and commercial zeolite, respectively, for the continuous experiments.

The optimum zeolite dosage and volumn of synthetic wastewater were used in the giant fresh water prawn farm to find the adsoption capacity of various adsorbrents.

3 Results and Discussion

3.1 Characteristic of synthetic zeolite

The synthetic zeolite from oil palm ash (Krabi Province) was used in the experimental. The calcium exchange capacity (CEC) of zeolite A standard, zeolite from oil palm ash and commercial zeolite were approximately 420.8, 386.4 and 110 mg $CaCO_3$/g zeolite product, respectively. The characteristic of the structure was analyzed using X-ray diffraction as shown in Fig.1.

The XRD pattern of the raw oil palm ash and activated oil palm ash (Fig 1.) showed that the major crystalline phases of the activated oil palm ash was kaolinite (JCPDS NO. 06-0221). The Scanning Electron Microscope (SEM) confirmed that zeolite had the cubic shape morphology as shown in Fig.2.

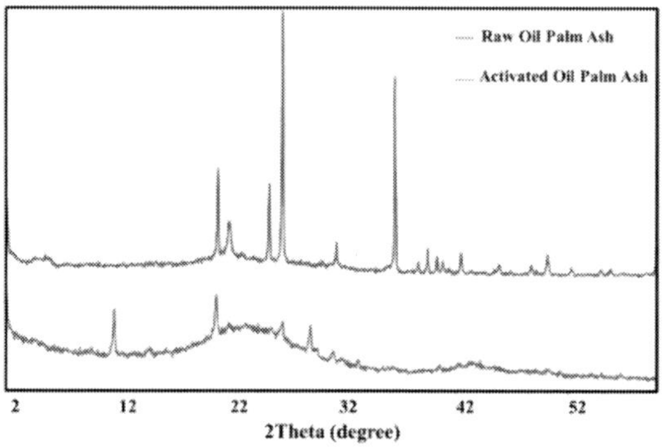

Fig1. XRD pattern of the raw oil palm ash and activated oil palm ash.

Fig 2. SEM images of (a) Raw oil palm ash and (b) Zeolite from oil palm ash.

The XRD pattern of the zeolite A standard, synthesized zeolite from oil palm ash and commercial zeolite are shown in Fig 3. It confirmed that the major crystalline phases of synthesized zeolite from oil palm ash were Sodium Aluminum Silicate Hydrate, (JCPDS NO. 11-0590).

3.2 Unionized ammonia removal capacity

In batch experimental study, the synthetic wastewater was prepared. The pH was adjusted to achieve the pH value of 8.6 at 26 °C. In order to find out the optimal dosage of various zeolites (0.05, 0.1, 0.15, 0.2 and 0.25 g) was added into synthetic wastewater of 20 mg/L of the

Total Ammonia Nitrogen, equal to unionized ammonia 1.95 mg N/dm^3, which were stirred for 60 minutes at 26°C with 120 rpm.

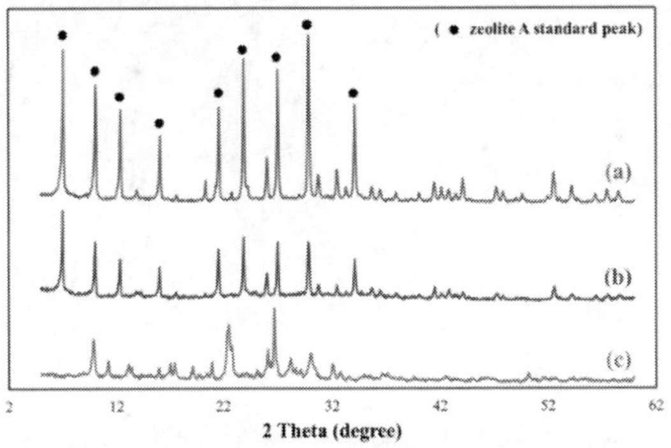

Fig 3. XRD pattern of the synthesized Na-A zeolite product from oil palm ash, (a) zeolite A standard, (b) zeolite from oil palm ash and (c) commercial zeolite

Table 1. Unionized ammonia (NH_3) removal capacity of various dosage of zeolites

dosage of zeolites (g)	NH_3 removal capacity (%)*		
	zeolite A standard	zeolite from oil palm ash	commercial zeolite
0.05	84.4 ± 0.2	83.2 ± 0.2	71.6 ± 0.1
0.1	88.7 ± 0.1	88.7 ± 0.1	81.4 ± 0.2
0.15	90.3 ± 0.1	90.5 ± 0.2	88.4 ± 0.1
0.2	92.6 ± 0.1	90.6 ± 0.1	91.4 ± 0.2
0.25	91.7 ± 0.2	91.4 ± 0.1	91.4 ± 0.2

* average value

The average unionized ammonia (NH_3) removal capacity of various dosage of zeolites in batch experimental has shown as Table 1.The result indicated that NH_3 removal capacity was not significantly different in zeolite A standard vs. zeolite from oil palm ash. The zeolite A standard and zeolite from oil palm ash are better than commercial zeolite.

For the NH_3 removal capacity from giant fresh water prawn farm as shown in Table 2. The results indicated that NH_3 removal capacity depended on cation exchange capacity value (CEC) because of the CEC value is representative as the amount of positive charge which can be exchanged. The result indicated that ammonia removal capacity of zeolite A standard is better than zeolite from oil palm ash and commercial zeolite respectively.

Table 2. Unionized ammonia (NH_3) removal capacity from Giant Fresh Water Prawn Farm (Initial concentration of NH_3 0.127 mgN/dm^3)

Zeolite	CEC* (mg $CaCO_3$/ g zeolite)	Unionized ammonia remain* (mgN/dm^3)	NH_3 removal capacity* (%)
zeolite A standard	420.8± 0.4	0.007 ± 0.0003	94.49 ± 0.09
zeolite from oil palm ash	386.4 ± 1.0	0.010 ± 0.0006	92.13 ± 0.17
commercial zeolite	110.0 ±0.8	0.027 ± 0.0015	78.74 ± 0.14

* average value

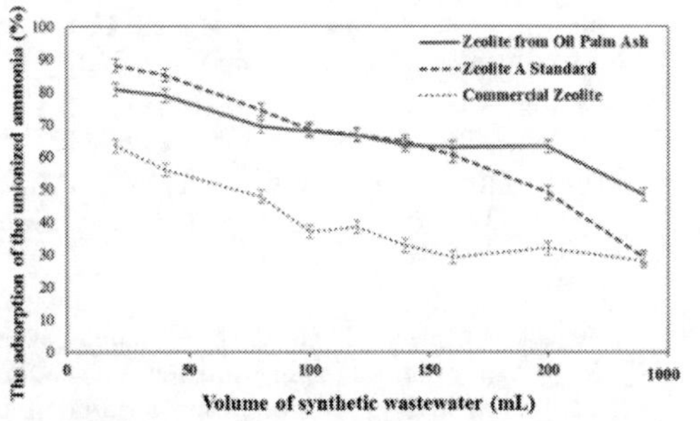

Fig 4. The adsorption of the unionized ammonia of the zeolite from oil palm ash, zeolite A standard and the commercial zeolite for the continuous experiments.

The adsorption of NH_3 in continuous experiments of the volume of synthetic wastewater (20, 40, 80, 100, 120, 140, 160, 200 and 1,000 mL) indicated that in the large volume, the zeolite from oil palm ash is better than zeolite A standard and commercial zeolite, respectively as shown in Fig 4.

The zeolite from oil palm ash is better removal for the large volume of synthetic wastewater because of the various cation in composition of oil palm ash. The presence of the cation can improve the high unionized ammonia (NH_3) removal efficiency. [12]

4 Conclusion

The Na-A zeolite synthesis from the oil palm ash was the calcium exchange capacity (CEC) of 386.39 mg $CaCO_3$/g of product, and the 86.05% of pure Na-A zeolite phase was formed. The CEC of zeolite Astandard and commercial zeolite was 420.8 and 110 mg $CaCO_3$/g Zeolite. The results of NH_3 removal efficiency indicated that zeolite A standard and zeolite from oil palm ash are better than commercial zeolite. The NH_3 removal capacity from giant freshwater prawn farm, zeolite A standard is better ammonia removal capacity than zeolite from oil palm ash and commercial zeolite, respectively. While, in the large volume (1000 mL), the zeolite from oil palm ash is the best ammonia removal capacity when compared with zeolite A standard and commercial zeolite.

5 Acknowledgment

The financial supports from Agricultural Research Development Agency (Public Organization) are greatly acknowledge.

6 References

1. Office of Agricultural Economics, *Agricultural Statistics Of Thailand 2016*, Thailand. (2016)
2. M. Faisal, Mahidin, Biomass Residue from Palm Oil Mills in Aceh Province: A Potential Usage for Sustainable Energy. *Adv. Eng. Inf. Tech*, **3** (2013)
3. M.Husin, A.Mokhtar, W. H. W.Hassan,, Oil palm plywood, *MPOB Information Series*, **206** (2003)
4. M. Ahmaruzzaman, A review on the utilization of fly ash, *Progress in Energy and Combustion Science*,
36 (2010)
5. Y. Zhao, B. Zhang, X. Zhang, J. Wang, J. Liu, R. Chen, Preparation of highly ordered cubic NaA zeolite from halloysite mineral for adsorption of ammonium ions, *Journal of Hazardous Materials,* **178** (2010)
6. J.A. Hargreaves, C.S. Tucker, Managing Ammonia in Fish Ponds. *SRAC Publication*,**4603**(2018)
7. D.J. Randall, T.K.N. Tsui, Ammonia toxicity in fish. *Marine Pollution Bulletin*,**45** (2002)
8. Y. Wimalasiri, M. Mossad, L. Zou, Thermodynamics andkinetics of adsorption of ammonium ions by graphene laminateelectrodes in capacitive deionization. *Desalination*,**357** (2015)
9. J. Izidoro, D. Fungaro, J. Abbott, S. Wang, Synthesis of zeolites X and A from fly ashes for cadmium and zinc removal from aqueous solutions in single and binary ion systems. *Fuel*,**103** (2013)
10. D.Y. Wu, Z.B. Hu, X.Z. Wang, S.B. He, H.N. Kong, Use of zeolitized coal fly ash in the simultaneous removal of ammonium and phosphate from aqueous solution. *Front Environ Sci Eng China*, **1** (2007)
11. A. Molina, C. Poole, A comparative study using two methods to produce zeolite from fly ash, *Miner. Eng.* **17** (2004).
12. A.B.Ariffin, Production of zeolites from oil palm ash. *Universiti Malaysia Pahang*, Malaysia. (2010)

Performance analysis of a pressure swing adsorption unit in removing biogas impurities using zeolite 13X

Magomnang Antonio-Abdu Sami[1], *Villanueva* Eliseo[2]

[1,2]Mechanical Engineering Department, College of Engineering and Architecture, University of Science and Technology of Southern Philippines – Cagayan de Oro Campus, Cagayan de Oro City, Philippines
[2]Mechanical Engineering Department, College of Engineering, MSU-Iligan Institute of Technology, Tibanga, Iligan City, Philippines

Abstract. This study aims to assess the performance of a Pressure Swing Adsorption (PSA) unit in removing the carbon dioxide from biogas by evaluating the breakthrough and adsorption capacity of the adsorption process as well as determining the effects of cyclic regeneration of the adsorbent. The PSA system developed and composed only a main vessel made up of 316 stainless steel components. It was then operated up to 10 bars pressure at ambient temperatures and gas flows at a rate from 0 to 15 L min^{-1}. Use of physical adsorbent (zeolite 13X) will consume the gaseous impurities such as CO_2. Product gas was collected into 1 L Tedlar bags and analyzed using SRI gas chromatograph with TCD and HID detector to validate the CO_2 and CH_4 composition. The results of the pressure swing adsorption (PSA) experiments showed an average increase of 160% in the net heating value over that of a certified gas standard. The amount of methane was also significantly higher although the amount of the other gasses (i.e. nitrogen) remained comparatively the same. The number of other gases was significantly lower and leaving no traces of carbon dioxide was observed in the PSA system product gas indicating that carbon dioxide had been completely adsorbed by the system. This study greatly helps to reduce CO_2 emitted to the atmosphere from the anaerobic co-digestion of biogas to produce high energy content bio-methane fuel.

1 Introduction

Biogas technology has been developed and widely used over the world because it has several advantages – the reduction of the dependence on non-renewable resources, high energy-efficiency, environmental benefits, available and cheap resources to feedstock, relatively easy and cheap technology for production, and additional values of digestate as a fertilizer.

Biogas as one of the sources of renewable energy significantly due to some ecological advantages, mainly being CO_2 as neutral; hence, it reduces greenhouse gas formation. Further, biogas production minimized the waste disposal as agricultural, commercial and municipal waste from anthropogenic sources.

Biogas mainly comprises methane (40-75%) and carbon dioxide (20-55%), and trace components of hydrogen sulfide (H_2S), ammonia (NH_3) and siloxanes. This trace components may destroy the engine, e.g., due to corrosion [1]-[2]. Removing the impurities leads to higher the amount of methane, the higher the output of energy from biogas.

There are technologies commercially available for removal of CO_2 from biogas are typically used for applications such as gas wells, sewage treatment plants, and landfills. Because of the different contaminants, there are processes that can be considered for CO_2 removal from biogas such as pressure swing adsorption, using of amines, vacuum swing adsorption, among others.

A Pressure Swing Adsorption (PSA) processes separate out CO_2, oxygen, and nitrogen by absorption and desorption on activated carbon at different pressures in four stages - adsorption, depressurizing, desorption and pressure build up. The principle of the pressurized process as described by Santos, et. al, 2010 [3] more readily dissolved in activated carbon than methane. Their proposed biogas upgrading plant presented in their research of Santos, et. al, 2010 [3] consists of a scrubber vessel for scrubbing (absorption of CO_2 into adsorbent like activated carbon), a flash tank for methane gas recovery, and a desorption tower for the regeneration.

In Pressure Swing Adsorption (PSA) processes, the absorbent is used as a medium to remove selectively the carbon dioxide gas. The adsorbent is a porous solid, normally with a high surface area.

Among those absorbents, zeolite, activated carbon, carbon molecular sieves (CMS) but also activated carbons, zeolites and other materials (titanosilicates) are used commercially. However, using these Zeolite has encountered difficulty to the adsorption properties to the chemical composition of the surfaces as identified by Siriwardane et.al, (2001) [4]. Moreover, Wang and Levan (2009) [5] measures the adsorption Isotherms for pure water vapor and carbon dioxide on zeolites over a wide of temperatures and the adsorption results to

* Corresponding author: a_magomnang@yahoo.com

confirmed their isotherm model used. Aside from that, Konduru, et. al (2007) [9] studied Zeolite 13X regeneration and recovery efficiency and suggests the determination of the optimum system parameters in order to establish that Zeolite 13X is the best absorbent. Hauchhum and Mahanta (2014) [6], also suggests that extensive study on the regeneration capability on Zeolite 4A and 13X.

From all the technologies presented, the pressure swing adsorption method of CO_2 removal was chosen because of the capability of its material presents stronger surface interactions to CO_2, to adsorb larger loadings as compared to methane [3]. This also supported by the study of Hauchhum & Mahanta, 2014 [6] showing a promising option to separate CO_2 due to its ease of applicability over a wide range of temperature and pressure conditions, also, it's low energy requirements and its low capital investment.

Based on the studies and gaps presented, this study contributes on assessing and determining the optimum system parameters required for zeolite 13X as an adsorbent in removing the carbon dioxide from biogas as a renewable source of fuel. After removal of the carbon dioxide, the methane gas then is used for possible engine testing in a stationary non-road engine. This technology overcomes the problem in increasing the methane recovery from biogas upgrading.

Thus, this study aims to evaluate the performance parameters of a Pressure Swing Adsorption (PSA) unit in removing the carbon dioxide from biogas by evaluating the breakthrough and adsorption capacity of the adsorption process as well as determining the effects of cyclic regeneration of the adsorbent from the anaerobic co-digestion of agricultural biomass waste utilizing dairy manure co-digested with various biomass feedstocks for power generation applications.

2 Materials and methods

2.1. Biogas upgrading experimental conditions

An experimental set-up is created only for the removing carbon dioxide (CO_2) from biogas. Wherein, a representative gas mixture, comprised of 63.18 %vol. CH_4 and 36.82 %vol. CO_2 will be passing through the adsorbent bed in an up-flow motion. After, pure methane in a down-flow motion in order to regenerate the adsorbent [7]. For this study, the researcher aims to compare the effects on the adsorption of using Zeolite 13X (Molecular Sieve) for the removing of carbon dioxide (CO_2) from a representative biogas mixture.

The experiments were conducted using a certified gas mixture standard prepared by Airgas (Airgas Southwest, Woodlands TX). This gas mixture is a good representation of the biogas produced from anaerobic co-digestion processes. For this experiment, the pressure was set at 400 kPa and the product gas outlet flow rate at 0.5-1.0 LPM. The product gas was also analyzed in real time using the Horiba NDIR gas analyzer and simultaneously analyzed using SRI gas chromatograph (SRI Instruments, Torrance, CA) with TCD and HID detector to validate the CO_2 and CH_4 composition.

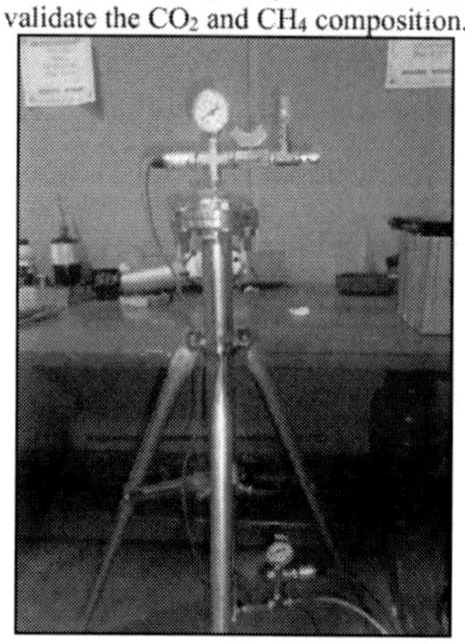

Fig. 1. Pressure Swing Adsorption Unit Schematic Diagram

2.2 CO_2 adsorption studies

Mérel, et.al, 2006 [8] described Pressure Swing Adsorption (PSA), working between high-pressure level during adsorption and a low-pressure level for desorption. Further, the study of Mérel, et.al, 2006 [8] employed the basic sequence of steps, referred to as the skarkstorm cycle, consist of pressurization, followed by adsorption and depressurization (blown down) followed by a purge or evacuation of the highly adsorbed component. On this study, the simple skarkstorm four-step cycle used and selected as the baseline because of the simplicity of the processes.

The removal of carbon dioxide on this study are operated with four steps: pressurization, adsorption, depressurization, and desorption. During the pressurization, gas representing the biogas composition consisting of 63.18% CH_4 and 36.82% CO_2 at ambient temperature is supplied to the bottom of the bed. During the adsorption, carbon dioxide (CO_2) is adsorbed on Zeolite 13X, where methane (CH_4) is obtained as product gas and the high-pressure feed gas enters in the pressurization step.

2.2.1 Breakthrough experiments

The Breakthrough curve will be determined and plotted in Fig. 2. For simplicity, a representative mixture of 63.18% CH_4 and 36.82% CO_2 represented in a typical biogas composition were assumed in this analysis.

2.2.2 Regeneration and reusability of the zeolite 13X at room temperature

Out of the adsorbent studied, zeolite 13X showed the optimal adsorption capacity of the zeolite 13X adsorbent is at 30°C. Therefore, Zeolite 13X were studied for reusability with free-flowing air desorption at room temperature in the same run followed by vacuum purge.

In the same manner, Jadhav, et.al, 2007 [10] reported that the adsorbent successfully retained adsorption capacity in three consecutive reuse cycles. The slight loss of capacity after the first cycle probably owing to some irreversible adsorption that could have been desorbed at ambient temperature.

2.2.3 Selectivity studies

The selectivity of the zeolite 13X adsorbents on the removal of carbon dioxide was studied using a mixture of 63.18 % vol. of CH_4 and 36.82 %vol. of CO_2 of biogas prepared by setting the flows around 0.63 l/min. prior to the adsorbent study, the adsorbent was pre-weighted and placed in the PSA unit. The mixture then flowed over the Zeolite 13X, and the product gas was gathered using the Horiba NDIR gas analyzer.

3 Results and discussions

3.1 CO_2 upgrading system evaluation

3.1.1 Breakthrough experiments

Hauchhum and Mahanta, 2014 [6] defined Adsorption as "a transient process and the amounts of material adsorbed within a fixed bed will depend on both position and time".

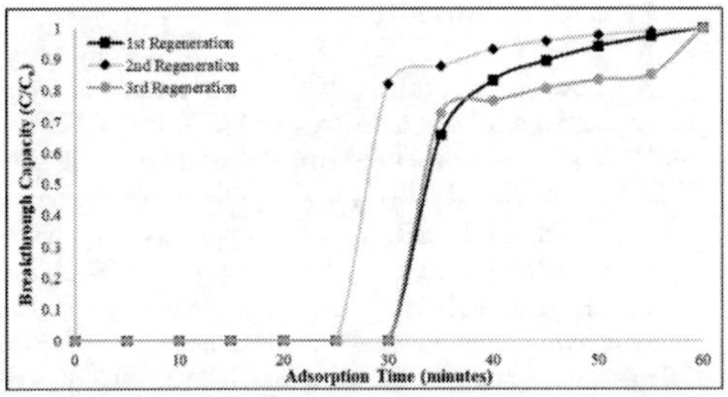

Fig. 2. Breakthrough curve for carbon dioxide (CO_2) removal using zeolite 13X at 400 kPa pressure

The CO_2 breakthrough curves for the zeolite 13X is shown in Fig. 2, and the experimental conditions are a binary gas mixture represented in a typical biogas composition is assumed in this analysis. Further, Hauchhum & Mahanta, 2014 [6] represents breakthrough curves as the ratio of the outlet concentration and the influent concentration against the contact time at an atmospheric pressure, and at some temperatures. For zeolite 13X, the adsorption experiencing breakthrough around 25 – 35 minutes depending on the efficiency of desorption (regeneration) of the adsorbent.

This implies that change in pore diameter of the adsorbent, zeolite 13X (10 Å) plays a vital role in entering the zeolite channels. Also, their cations, which is Na and K also helps in the adsorption of carbon dioxide. Due to its larger pore volume, zeolite 13X saturation is longer. In addition, the difference in chemical composition and nature of the surface and porosity was observed during the adsorption. Since the zeolite 13X has higher surface area can contribute to the higher adsorption capacity of zeolites, as compared with the study of Hauchhum and Mahanta, 2014 [6].

Furthermore, the reported values of carbon dioxide adsorption are in agreement with the published data of Hauchuum and Mahanta, 2014 [6], although the reported values of amount adsorbed are higher than all previous literature. In low-pressure range, the amount adsorbed compares very well with the data reported by [6].

3.1.2 Effects on reusability of adsorbent

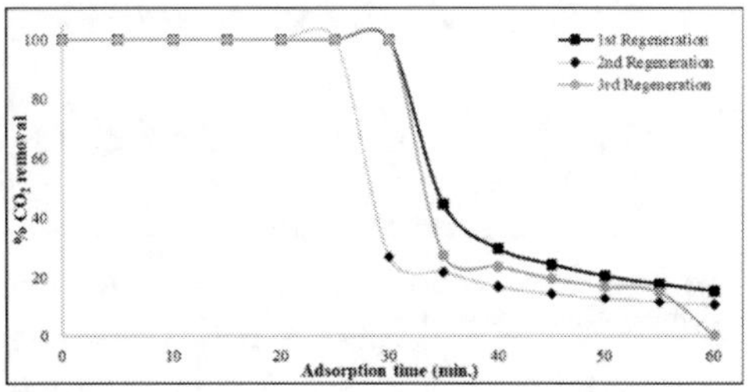

Fig. 3 Percentage removal of CO_2 sorption at different regeneration cycles

The adsorption capacity of the zeolite 13X indicates the reusability (regeneration capacity) of the adsorbent as presented in Fig. 3. Based on the experimental results, a 100% removal of carbon dioxide from the biogas is maintained after some time and the zeolite can be reused by passing through a countercurrent flow of air.

However, removal (desorption) efficiency of zeolite decreases every time it regenerates and it implied that the zeolite is not completely regenerated by air. This issue will be addressed by introducing hot air during the regeneration process.

3.1.3 Adsorption kinetics and contact time studies

The effluent concentration of the CH_4-CO_2 gas mixtures dropped to zero ppm$_v$ for approximately 26 - 33 minutes at initial capacity is 22.5 g_{CO2} /$kg_{Zeolite13X}$. These results are slightly lower than the reported values of by Konduru, et. al, 2007 [9], for a gas mixture containing N_2 (75%) and CO_2 (25%).

The experimental results were expected as dismal since the gas mixture is different (63.18% CH_4 and 36.82% CO_2) from related studies. Also, adsorption was terminated prior to the complete saturation due to instrumentation limitations. Moreover, Figure 4 shows the time required in reaching the saturation capacity decreases with each regeneration cycle. These results attributed to the residual (remaining) CH_4-CO_2 mixture were not removed during the regeneration cycle.

The adsorption isotherms of breakthrough and carbon dioxide adsorption were shown in Fig. 4. Based on the results, indicated in Figure 4 the adsorption of zeolite 13X is fully reversible and complete regeneration can have done and obtainable by evacuation of the adsorbent after adsorption process.

In addition, final adsorption curve shown with CO_2 was similar with Mérel, et. al, 2006 [8] and it indicates adsorbent was not affected by the adsorption of methane (CH_4) even after third regeneration. These results suggest that separation of carbon dioxide of carbon dioxide from a CH_4-CO_2 mixture is possible with Zeolite-13X adsorbents.

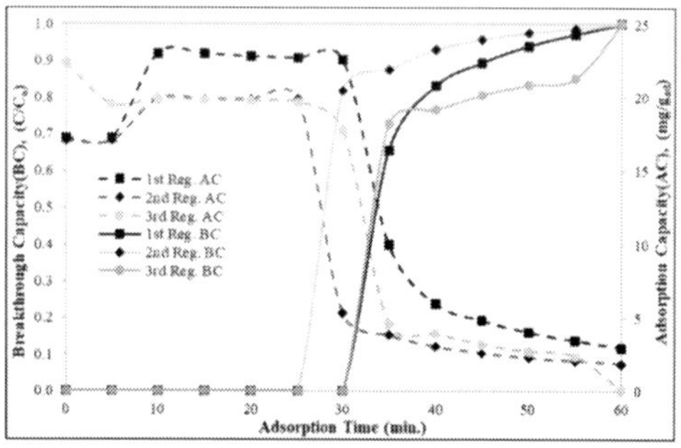

Fig. 4 CO_2 sorption isotherms of Breakthrough and Adsorption Capacity at different regeneration cycles

3.1.4 Selectivity studies

As shown in Figure 4, CO_2 adsorption breakthrough occurred around 25-30 minutes of adsorption, until the maximum adsorption capacity of the zeolite 13X was spent.

As reported by Jadhav, et.al (2007) [10], that the adsorption capacity of the zeolite 13X is found to be higher for CO_2 than for CH_4 [8]. As compared with the saturation adsorption capacities of Zeolite 13X for CO_2 and CH_4 are for this experiment are (22.50, 19.92 and 17.74 mg/g_{ads}) respectively.

Thus, CH_4 breakthrough shows a bulge or hump of ($C/Co > 1$) as reported by Jadhav, et. al, 2007 [10]. This implies that competitive adsorption of carbon dioxide from biogas was taking place [10]. Since CH_4 being in higher concentration occupies more sites and kinetic selectivity for CO_2 plays a vital role that the adsorbed CH_4 is displaced making space for CO_2. Thus, the product gas values of CH_4 shows an increase of values over than in the inlet.

4 Conclusion

The results of the PSA experiments showed an average increase of 160% in the net heating value over that of the certified gas standard. The amount of methane was also significantly higher although the amount of the other gasses (i.e. nitrogen) remained comparatively the same. The quantity of carbon monoxide (CO_2) in the product gas was significantly lower and no traces of carbon dioxide was observed in the PSA product gas indicating that carbon dioxide had been completely adsorbed by the system. The raw gas inlet flow rate started at 0.71 LPM and ended at around 0.80 LPM until the vessel was regenerated in a countercurrent of air. Based on the product gas composition during the adsorption and desorption cycles, the desired concentration of the product gas was achieved in a short span of 1 minute during adsorption. A longer stable gas concentration may be achieved by using more adsorbent.

The authors would like to thank the DOST-PCIEERD for providing financial assistance for the International Conference presentation. Also, the Bio-Energy Testing and Analysis Laboratory of Texas A&M University, College Station, Texas, USA for offering me the opportunity to conduct this research. Likewise, the Engineering Research and Development for Technology (ERDT) under the Science Education Institute (SEI) of Department of Science and Technology in the Philippines, and the College of Engineering and Architecture, USTSP-CDO for their support to this research presentation.

References

1. Biogaszusamensetzung (Composition of biogas), Available from https://bit.ly/2zblhl4. Date Accessed: 01/21/2018.
2. M. Harasimowicz, P. Orluk, G. Zakrzewska-Trznadel, AG Chmielewski. Application of polyimide membranes for biogas purification and enrichment. Journal of Hazardous Materials. 144 (3), (2007).
3. M.P. Santos, C.A. Grande, A.E. Rodrigues. Pressure swing adsorption for biogas upgrading. Effect of recycling streams in pressure swing adsorption design. Industrial & Engineering Chemistry Research. 50(2), (2010).
4. R.V. Siriwardane, M.S.Shen, E.P. Fisher, and J.A.Poston. Adsorption of CO2 on molecular sieves and activated carbon. Energy & Fuels, 15(2); (2001).
5. Y. Wang, and M.D.LeVan. Adsorption equilibrium of carbon dioxide and water vapor on zeolites 5A and 13X and silica gel: pure components. Journal of Chemical & Engineering Data, 54(10); (2009).
6. L. Hauchhum, P. Mahanta. Carbon dioxide adsorption on zeolites and activated carbon by pressure swing adsorption in a fixed bed. International Journal of Energy and Environmental Engineering. 5(4), (2014)
7. A.L. Maglinao, S.C. Capareda, H. Nam. Fluidized bed gasification of high tonnage sorghum, cotton gin trash and beef cattle manure: evaluation of synthesis gas production. Energy Conversion and Management. 105 (2015).
8. J. Mérel, M. Clausse, F. Meunier. Carbon dioxide capture by indirect thermal swing adsorption using 13X zeolite. Environmental Progress & Sustainable Energy. 25(4), (2006).
9. N. Konduru, P. Lindner, N.M. Assaf-Anid. Curbing the greenhouse effect by carbon dioxide adsorption with zeolite 13X. AIChE journal. 53(12). (2007).
10. P.D. Jadhav, R.V. Chatti, R.B. Biniwale, N.K. Labhsetwar, S. Devotta, S.S. Rayalu, Monoethanol amine modified zeolite 13X for CO_2 adsorption at different temperatures. Energy & Fuels. 21(6); (2007).

Kinetic modeling and simulation of bio-methanol process from biogas by using aspen plus

Rujiroj Tamnitra[1,*], *Rujira* Jitwung[2], *Tarawipa* Puangpetch[1], *Weerawat* Patthaveekongka[1], *and Kamonrat* Leeheng[2]

[1]Chemical Engineering Department, Faculty of Engineering and Industrial Technology, Silpakorn University, Nakorn pathom, Thailand
[2]Expert Center of Innovative Clean Energy and Environment, Thailand Institute of Scientific and Technological Research (TISTR), KhongLuang Patumthani Thailand

Abstract. A process of bio-methanol from biogas was studied by modifying kinetic model of reaction's Richardson and Paripatyadar comparing with laboratory data. Bio-methanol process consists of 2 steps: reforming reaction (at atmospheric pressure, temperature 500 - 750 °C) and methanol synthesis (at constant pressure 40 bar, temperature 140 - 280 °C). The reaction model of each step was individual simulated. Next both steps were integrated, then they were simulated using ASPEN PLUS software. This work investigated the optimum operating condition and predicted result of both reactions. The developing model was obtained, then it was applied for ten thousand liters per day of methanol. The simulation result received from reforming reaction showed increasing temperature effect to rising in CH_4 and CO_2 conversion and relating with laboratory result. The optimum condition of methanol synthesis is temperature 200 °C under constant pressure 40 bar.

1 INTRODUCTION

In most industry, the main target requires in increasing productivity and low operating cost. The modification or remodel of process is hardly adjusted without enough supporting data. The process simulation is one of alternative tools which it is useful to predict process and to support major result before the process is modified. The two main cases of simulation are i) an innovative process: studying variable parameters or condition effecting to process for prediction and comparison with alternative process and design process. ii) a conventional process: obtaining behaviour and mechanism of process. The advantages of simulation are not only saving cost, time and life but also providing fast and reliable result. For example, this tool can be applied for testing a critical condition (high pressure, high temperature, and limitation of chemical).[1]

Nowadays, renewable energy is interested for substituting fossil fuels such as coal, petroleum and etc. Biogas can be produced from manure, food, waste, and tapioca starch or sugarcane residue. Biogas process is a practical process to eliminate massive waste water from industries. Biogas contains mainly methane and carbon dioxide. Therefore, it is used as an alternative fuel for electricity generation. Only methane in biogas is combusted for producing electricity. However, carbon dioxide is wasted to environment. To subsidize biogas production, CO_2 utilization has been studied by many researchers. An attractive research is using CO_2 for bio-methanol production. The bio-methanol can be supplied as a raw material in biodiesel process. In addition, methanol requirement in Thailand was imported more than 500 million liters per year (data form department of industrial works, 2017). As a result, the production of bio-methanol is interested not only solving CO_2 problem form using biogas as an electricity but also alleviating amount of methanol importing from outside country.

Methanol production from biogas comprises of two main reactions: methane reforming and methanol synthesis. The methane reforming is a process which carbon dioxide and methane are reacted to form synthesis gas (syngas) consisted of carbon monoxide and hydrogen. For practical process water is added to the process. The methane reforming process consists of three main reactions: dry reforming of methane reaction (1), steam reforming of methane (2), and water gas shift reaction (3). The methanol synthesis occurs via three reactions: hydrogenation of carbon monoxide (4), hydrogenation of carbon dioxide (5), and water gas shift reaction (6).[2-4]

Methane Reforming

$$CH_4 + CO_2 \rightarrow 2CO + 2H_2 \quad (1)$$
$$CH_4 + H_2O \rightarrow CO + 3H_2 \quad (2)$$
$$CO + H_2O \rightarrow CO_2 + 2H_2 \quad (3)$$

Methanol Synthesis

$$CO + 2H_2 \rightarrow CH_3OH \quad (4)$$
$$CO_2 + 3H_2 \rightarrow CH_3OH + H_2O \quad (5)$$
$$CO + H_2O \rightarrow CH_3OH + H_2O \quad (6)$$

* Corresponding author: rujira_j@tistr.or.th

2 RESEARCH METHODOLOGY

This research focuses on the design of bio-methanol production process from biogas using process simulation by ASPEN PLUS software and applies for ten thousand liters per day capacity.

2.1 Reaction modeling

Reaction modeling can be divided in two reactions, including methane reforming reaction and methanol synthesis reaction. The previous work of kinetic studies was presented in this section and developed kinetic models by using laboratory operating condition.

Methane reforming reaction, the reaction model was first used by Richardson and Paripatyadar. The reaction rate equation was calculated using a Langmuir-Hinshelwood rate expression. This reaction was developed by Yacine Benguerba and Lila Dehimi adding a reverse reaction term in the previous model to explain the laboratory results. The developed kinetic reaction are shown in equation (7) and (8).[5, 6]

$$r_1 = \frac{k_1 K_{CO_2,1} K_{CH_4,1} P_{CH_4} P_{CO_2}}{(1+K_{CO_2,1}P_{CO_2}+K_{CH_4,1}P_{CH_4})^2}\left(1-\frac{(P_{CO}P_{H_2})^2}{K_{P_1}(P_{CH_4}P_{CO_2})}\right) \quad (7)$$

$$r_2 = \frac{k_2 K_{CO_2,2} K_{H_2,2} P_{CO_2} P_{H_2}}{(1+K_{CO_2,2}P_{CO_2}+K_{H_2,2}P_{H_2})^2}\left(1-\frac{(P_{CO}P_{H_2O})}{K_{P_2}(P_{CO_2}P_{H_2})}\right) \quad (8)$$

Methanol Synthesis Reaction, the kinetic was modeled and calculated using a Langmuir-Hinshelwood rate expression. This kinetic model of methanol synthesis is suitable in low pressure condition. However, it is limited in high pressure environment. Van den Bussche and Froment developed the model by adding equilibrium term to eliminate problem of high pressure condition. The model are shown in equation (9) and (10).

$$r_{MeOH} = k_{MeOH}\frac{(p_{CO_2}p_{H_2}) - (1/K_{pMeOH})(p_{CH_3OH}p_{H_2O}/p_{H_2^2})}{\left(1+K_a(p_{H_2O}/p_{H_2})+K_b\sqrt{p_{H_2}}+K_c p_{H_2O}\right)^3} \quad (9)$$

$$r_{RWGS} = k_{RWGS}\frac{p_{CO_2} - (1/K_{pRWGS})(p_{CO}p_{H_2O}/p_{H_2})}{\left(1+K_a(p_{H_2O}/p_{H_2})+K_b\sqrt{p_{H_2}}+K_c p_{H_2O}\right)} \quad (10)$$

In this present work, the modification of some constant values of (9) and (10) are adjusted to meet with the laboratory result by using linear regression method. The operating conditions were used: temperature range 140 - 280 °C and pressure 40 - 45 bar.[3, 7-9]

2.2 Process simulation

Many researchers attempt to acquire accurate modeling of the relevant kinetic processes because the kinetic modeling is a key of the methanol process indicating the behaviour of reactor. The expressions for reactor models are adapted from the work of Richardson et al. and Van den Bussche et al. Catalytic reaction in fixed-bed reactor was simulated in one dimensional heterogeneous model. Temperature and concentration gradients are accounted in axial direction. The flow diagram of the bio-methanol production process from biogas is shown in Fig. 1. ASPEN PLUS software is tool for this process simulation.

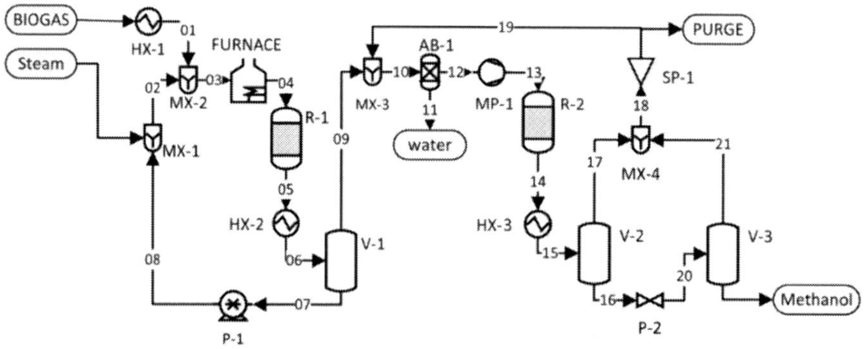

Fig. 1. Process flowsheet of methanol production

Biogas is preheated by low pressure steam (HX-1), mixed with steam in low pressure steam (HX-2) and then heated up in furnace to raise high temperature to meet up desired temperature in reactor (R-1). After reforming reaction, the fluid is cooled down in heat exchanger (HX-2) and trapped in vessel (V-1). The fluid is separated in two phases. Liquid is water which fed back by pump (P-1), mixing with steam in mixer (MX-1). Vapour or syngas was fed to absorption unit (AB-1) removing water before passing the syngas to next step. The methanol reaction is occurred in reactor (R-2) operating temperature (140 - 180 °C). The methanol product and residue gas are cooled down to 40 °C by cooling water in heat exchanger (HX-3). The fluid contains two phases: vapour stream and liquid stream, methanol in liquid stream is firstly trapped in vessel (V-2) then it depressurized to atmospheric pressure by relief valve (P-2) sending to vessel (V-3) for second trapped and finally obtaining methanol. The upper stream is residue gas passing and mixing with the vapour stream from vessel (V-2) in mixer (MX-4). This recycle gas is partially purged (10%) and returned (90%) to mixer (MX-3) for producing methanol production.

2.3 Experimental section

The reforming reactor was made from 304 stainless steel with the length, inside diameter and outside diameter of 1.6 cm. and 2.0 cm, respectively. The flow of mixed gas was controlled by mass-flow controller and the methanol synthesis reactor was fixed-bed continuous flow reactor made from 304 stainless steel belonging dimension of 30 cm length and 1.6 cm inside diameter. [10, 11]

3 RESULT AND DISCUSSION

3.1 reactor model

3.1.1 Methane Reforming Reactor

The results of modeling data were compared with laboratory data. The operating conditions were used by varying temperature range 500 - 750 °C under atmospheric pressure. In term of methane percentage conversion was shown in Fig. 2-3. The results were investigated that rising temperature effected to increasing methane conversion and relating to the developing model specification in high temperature range (500 - 750 °C). The error percentage is in range of 5.7 - 21.6%.

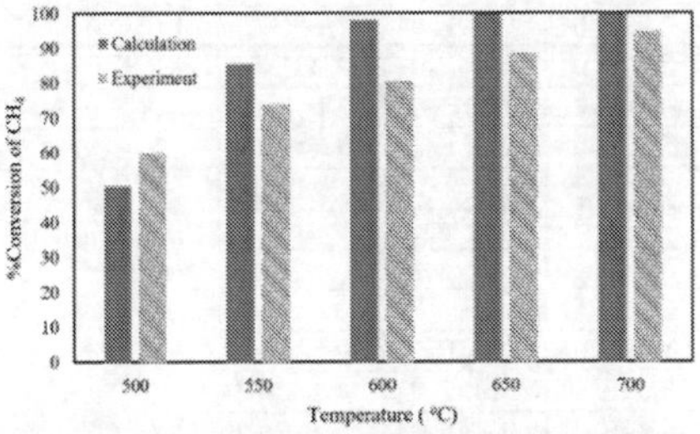

Fig. 2. Comparison of CH_4 developing model with laboratory result.

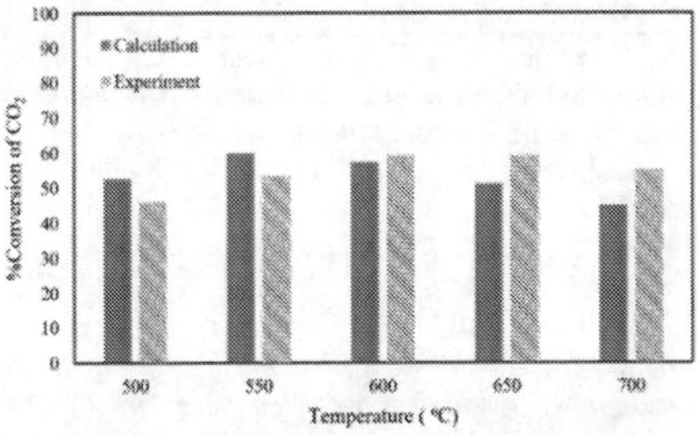

Fig. 3. Comparison of CO_2 developing model with laboratory result.

The simulation results were shown that CO_2 conversion was increased when increasing temperature in range of 500 - 600 °C and then it was decreased after temperature over 600 °C. Comparing the developing model with laboratory, the error percentage is 3.1 - 18.4% at 500 - 750 °C. Meanwhile, H_2/CO molar ratio was gradually increased when rising temperature. The percentage of H_2/CO is deviated of 9.6 - 29%.

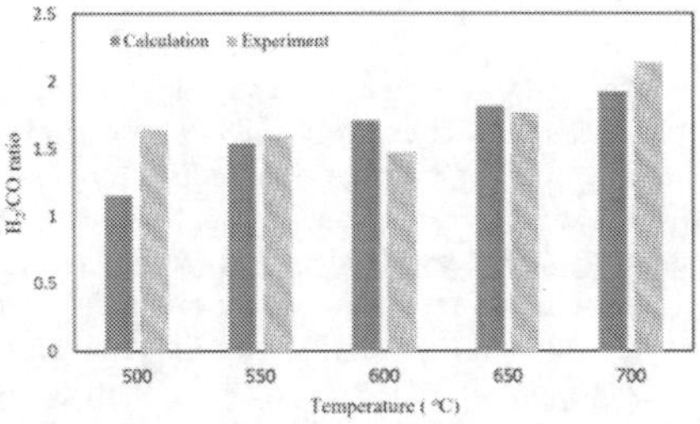

Fig. 4. Comparison of H_2/CO ratio developing model and laboratory result.

3.1.2 Methanol Synthesis Reaction

The simulation and laboratory results of methanol synthesis were depicted in Fig. 5-6. The CO and H_2 conversion generated from the reforming reaction are the main factor of methanol synthesis. The developing model is fitted with the CO and H_2 conversion obtaining from laboratory in temperature above 200 °C; however, it is widely differed under 200 °C (140 - 200 °C). This result can be explained that the developing model is out of the minimum range which the previous work [5, 12, 13] used (130 - 300 °C). The optimum conversion of CO and H_2 is temperature at 200 °C acquired from the developing simulation agreed with the laboratory.

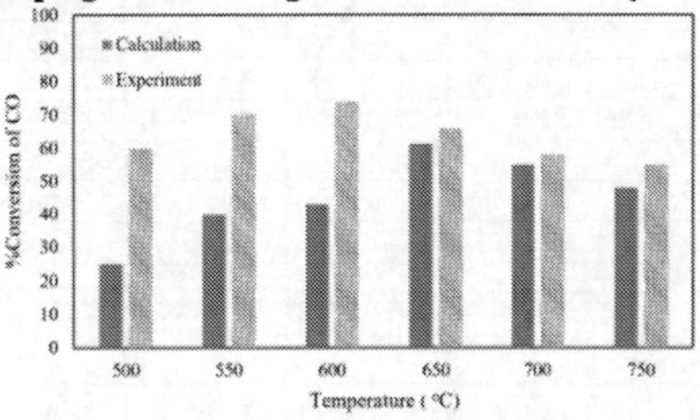

Fig. 5. Comparison of CO and H_2 developing model with laboratory result.

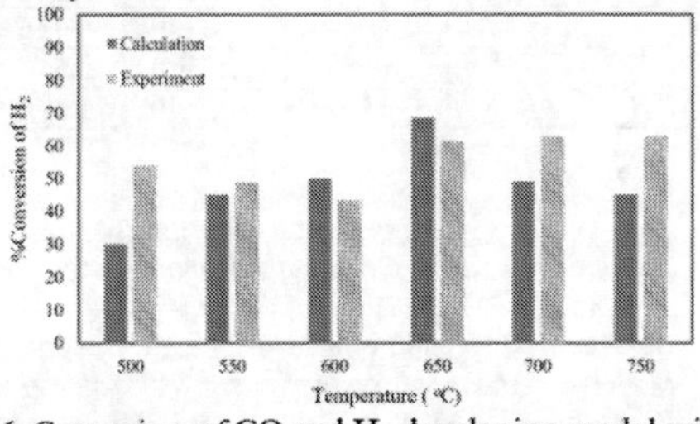

Fig. 6. Comparison of CO and H_2 developing model with laboratory result.

3.2 Process simulation

The study of methanol simulation process was applied for ten thousand liters/day. The result was shown in terms of material balances, chemical compositions, temperatures and pressure streams in table 1. It is found that 8483.65 liters/min (343.29 mol/min) of biogas and 3062.5 kg/min (170.00 mol/min) of water produce 10,000 liters/day (223.40 mol/min) of methanol. Some equipment units are installed to methanol process for specific reasons as following. Furnace was used for increasing temperature because reaction in reactor (R-1) is highly endothermic reaction that requires high heat source to increase CH_4 conversion. After reactor (R-1), water dissolved in gas stream can effect to the catalyst activity in methanol synthesis and can react with carbon monoxide reducing reactant for methanol synthesis and also generating CO_2 in the system as explained in equation (3). The vessel (V-1) and (V-2) were installed because the liquid methanol cannot be perfectly trapped in single stage and it is not liquefied under high pressure. The expansion valve (P-2) reduces the pressure of methanol stream to the atmosphere, then the liquid methanol is easily condensed into liquid phase.

4 CONCLUSION

Methanol is an important chemical for downstream chemicals. The process simulation of methanol produced from biogas is tooled by ASPEN PLUS using the developing kinetic data shown that the conversion of CH_4, CO_2 and H_2/CO ratio fitted in ranging of high temperature in both reactors. This work can apply for the optimum condition, scale up the process capacity and evaluate the feasibility economic.

Table 1. Stream summary of bio-methanol production using developing model in ASPEN PLUS□

Stream	BIOGAS	STEAM	1	2	3	4	5	6	7
Total Flow mol/min	343.29	170.00	343.29	301.13	644.42	644.42	1021.69	1021.69	131.13
Temperature C	25.00	125.00	105.00	99.46	72.09	750.00	750.00	35.00	35.00
Pressure bar	1	1	1	1	1	5	1	1	1
CH4	5.50E-01	0.00E+00	5.50E-01	1.38E-06	2.93E-01	2.93E-01	1.75E-04	1.75E-04	3.17E-06
CO2	4.50E-01	0.00E+00	4.50E-01	3.24E-03	2.41E-01	2.41E-01	8.50E-02	8.50E-02	7.44E-03
CO	0.00E+00	0.00E+00	0.00E+00	2.89E-04	1.35E-04	1.35E-04	2.52E-01	2.52E-01	6.65E-04
H2	0.00E+00	0.00E+00	0.00E+00	9.17E-10	4.29E-10	4.29E-10	4.87E-01	4.87E-01	2.11E-09
H2O	0.00E+00	1.00E+00	0.00E+00	9.96E-01	4.66E-01	4.66E-01	1.76E-01	1.76E-01	9.92E-01
CH3OH	0.00E+00	0.00E+00	0.00E+00	0.00E+00	0.00E+00	0.00E+00	0.00E+00	0.00E+00	0.00E+00
Stream	8	9	10	11	12	13	14	15	16
Total Flow mol/min	131.13	890.55	1314.16	50.00	1264.12	1264.12	828.55	828.55	313.98
Temperature C	35.00	35.00	30.58	30.58	30.58	200.00	200.00	35.00	35.00
Pressure bar	1	1	1	1	1	45	40	40	40
CH4	3.17E-06	2.00E-04	4.42E-04	0.00E+00	4.59E-04	4.59E-04	7.01E-04	7.01E-04	4.45E-04
CO2	7.44E-03	9.64E-02	2.03E-01	0.00E+00	2.11E-01	2.11E-01	3.20E-01	3.20E-01	2.71E-01
CO	6.65E-04	2.89E-01	2.69E-01	0.00E+00	2.79E-01	2.79E-01	1.65E-01	1.65E-01	2.01E-02
H2	2.11E-09	5.58E-01	4.85E-01	0.00E+00	5.04E-01	5.04E-01	2.42E-01	2.42E-01	5.70E-08
H2O	9.92E-01	5.61E-02	3.80E-02	1.00E+00	0.00E+00	0.00E+00	2.06E-03	2.06E-03	5.41E-03
CH3OH	0.00E+00	0.00E+00	5.05E-03	0.00E+00	5.25E-03	5.25E-03	2.71E-01	2.71E-01	7.03E-01
Stream	17	18	19	20	21	MeOH	PURGE		
Total Flow mol/min	514.57	605.15	423.61	313.98	90.58	223.40	181.55		
Temperature C	35.00	22.09	22.09	8.19	8.19	8.19	22.09		
Pressure bar	40	1	1	1	1	1	1		
CH4	8.57E-04	9.50E-04	9.50E-04	4.45E-04	1.48E-03	2.64E-05	9.50E-04		
CO2	3.49E-01	4.27E-01	4.27E-01	2.71E-01	8.65E-01	3.00E-02	4.27E-01		
CO	2.54E-01	2.26E-01	2.26E-01	2.01E-02	6.89E-02	2.84E-04	2.26E-01		
H2	3.89E-01	3.31E-01	3.31E-01	5.70E-08	1.98E-07	2.99E-16	3.31E-01		
H2O	1.05E-05	2.51E-05	2.51E-05	5.41E-03	1.08E-04	7.56E-03	2.51E-05		
CH3OH	7.11E-03	1.57E-02	1.57E-02	7.03E-01	6.43E-02	9.62E-01	1.57E-02		

REFERENCES

1. Foo D.C.Y. and Elyas R., *Introduction to Process Simulation*, in *Chemical Engineering Process Simulation*,3-21.(2017)
2. Bussche K.M.V. and Froment G.F., *A Steady-State Kinetic Model for Methanol Synthesis and the Water Gas Shift Reaction on a Commercial Cu/ZnO/Al2O3Catalyst*. JCat. **161** 1-10.(1996)
3. Skrzypek J., Lachowska M., and Moroz H., *Kinetics of methanol synthesis over commercial copper/zinc oxide/alumina catalysts*. ChEnS. **46** 2809-2813.(1991)
4. Snoeck J.W., Froment G.F., and Fowles M., *Kinetic Study of the Carbon Filament Formation by Methane Cracking on a Nickel Catalyst*. JCat. **169** 250-262.(1997)
5. Benguerba Y., et al., *Modelling of methane dry reforming over Ni/Al₂O₃ catalyst in a fixed-bed catalytic reactor*. Reaction Kinetics, Mechanisms and Catalysis. **114** 109-119.(2015)
6. Richardson J.T. and Paripatyadar S.A., *Carbon dioxide reforming of methane with supported rhodium*. Applied Catalysis. **61** 293-309.(1990)
7. Hansen John B. and Højlund Nielsen Poul E., *Methanol Synthesis*.
8. Peter M., et al., *Detailed kinetic modeling of methanol synthesis over a ternary copper catalyst*. Chem. Eng. J. **203** 480-491.(2012)
9. Šetinc M. and Levec J., *On the kinetics of liquid-phase methanol synthesis over commercial Cu/ZnO/Al₂O₃ catalyst*. ChEnS. **54** 3577-3586.(1999)
10. Rujira J. K.K., et al., *The effect of contaminated gases in syngas to biomethanol production*. PACCON.(2018)
11. Rujira J. K.K., et al., *Reforming of residue gas from CBG process to synthetic gas for bio-methanol production*. PACCON.(2018)
12. Graaf G.H., et al., *Chemical equilibria in methanol synthesis*. ChEnS. **41** 2883-2890.(1986)
13. Graaf G.H., Stamhuis E.J., and Beenackers A.A.C.M., *Kinetics of low-pressure methanol synthesis*. ChEnS. **43** 3185-3195.(1988)

Effect of postmortem storage on shear force value and calpain/calpastatin activity in longissimus dorsi muscle of rusa deer (*cervus timorensis*)

Orapin Jantasaeng[1], *Kanya* Jirajaroenrat[1,*]and *Kunya* Tuntivisoottikul[2]

[1]Department of Animal Production Technology and Fishery, Faculty of Agricultural Technology, King Mongkut's Institute of Technology Ladkrabang, Bangkok, Thailand
[2]Department of Agricultural Education, Faculty of Industrial Education and Technology, King Mongkut's Institute of Technology Ladkrabang, Bangkok, Thailand

Abstract. The objective of this study was to determine the effects of aging period on shear force values and activities of calpain/calpastatin enzymes of the longissimus dorsi (LD) muscle from rusa deer (*Cervus timorensis*). The shear force values were measured with instron materials testing machine. Results showed that postmortem storage influenced meat tenderness with a highly significant reduction of shear force value from 9.01±0.83 kg/cm^2 at day 1 to 4.34±0.10 kg/cm^2 at day 21 ($P<0.01$). Determination of the enzyme activities indicated that the μ-calpain activity decreased significantly from 1.50±0.42 to 0.19±0.28 units/g of meat ($P<0.01$), whereas m-calpain activity (22.88±9.64 to 16.95±8.34 units/g of meat) and calpastatin activity (9.93±2.37 to 6.82±2.96 units/g of meat) slightly decreased ($P>0.05$). Shear force values were significant correlated with higher levels of μ-calpain (r=0.915) and m-calpain (r=0.758), respectively. However, the values were not significant related to the calpastatin activity (r=0.462). Therefore, this study confirmed that in the case of rusa deer, μ-calpain is a key factor controlling postmortem meat tenderness compared to m-calpain and calpastatin. The results suggested that storage of the rusa deer meat for 7 days is enough to achieve aging of the LD muscle, which may help to reduce the cost of meat production.
Keywords: postmortem storage, shear force, calpain/calpastatin activity, rusa deer.

1 Introduction

Rusa deer (*cervus timorensis*) have been an alternative farm animal. Most important commercial product in the tropical countries is deer meat [1]. Deer meat contents a high level in protein and minerals, but low-fat content and cholesterol [2-3]. In Thailand, consumption of deer meat is increased. Improvement of farm management, feeding and reproduction were studied, but improvement of the deer meat quality is uninformative. In meat quality traits, tenderness is commonly known as one of the most important traits for consumer. Tenderness is controlled by the level of proteolysis of the endogenous proteolytic enzymes of the calpain protease system including calpains and calpastatin [4-7]. The calpains are cysteine peptidases present in different forms and tissues while the calpastatin acts as a specific inhibitor. Activities of the calpains is mainly controlled by Ca^{2+}, phospholipids and their specific inhibitor. Several studies reported that the calpain concentration in the muscle and the calpain/calpastatin ratio could be used as a good predictor of ultimate tenderness. The influences of the calpain enzyme system on meat tenderness can be expected to apply in deer meat [4-5, 8]. However, studies on the meat tenderization of rusa deer has been limited in Thailand. Here, the shear force values, and the μ-calpain, m-calpain and calpastatin in the longissimus dorsi muscle of the rusa deer associated with postmortem storage were revealed.

2 Material and Methods

2.1 Samples Preparation

The rusa deer meat was raised by Deer Co-operative of Thailand LTD., Kasetsart University, Kamphaeng Saen Campus Thailand. Ten female rusa deer (2 to 3 years of age, live weight 40 to 60 kg) were slaughtered and longissimus dorsi (LD) muscle were obtained within 45 min after exsanguination. The LD muscles were removed from the left side of the carcass, wrapped in plastic wrap, placed in cooling pack and transported immediately to meat science laboratory at Meat Technology Research Network Center, King Mongkut's Institute of Technology Ladkrabang, Bangkok, Thailand. The LD muscles were trimmed free from external fat and connective tissue. After that, LD muscles were divided into six pieces and vacuum packed. All was aged at 4 °C for 1, 3, 5, 7, 14 and 21 days postmortem.

2.2 Shear Force Values Measurement

The LD muscles were chopped into 2.5 cm thick, were boiled on water bath at 80°C until reached an internal

* Corresponding author: kanyatran@gmail.com

temperature of 70 to 75°C. The sample was cooled by rinsing with water until the internal temperature decreased to 25°C. They were chopped into 1 cm in diameter parallel to the longitudinal orientation of the muscle fibers. Theshear force values were measured by instron materials testing machine (Instron Model 1011, USA). The average maximum force (kg/cm^2) to cut across the fibers. Ten pieces was used for each sample [9].

2.3 Calpain/Calpastatin Activity Analysis

Ten grams of LD muscles was homogenized in 30 mL of extraction buffer (100 mM Tris/HCl, pH 8.3, 10 mM EDTA,.05% [vol/vol] MCE, 100 mg/L ovomucoid, 2 mM phenylmethylsulfonyl fluoride (PMSF), 6 mg/L leupeptin, 4°C) at 30 second cooling period between bursts. The homogenate was centrifuged at 30,000×g max for 1 hour at 4°C and filtered through glass wool. After clarification, the sample were dialyzed against 25 volumes of dialysis buffer (40 mM Tris/HCl, pH 7.35, 5 mM EDTA,.05% [vol/vol] MCE) for 24 hours at 4°C and centrifuged 30,000×g max for 30 minutes at 4°C. The extract applied to a DEAE-Sephacel by gravity. After loading, the column was washed with 5 column volumes of elution buffer (40 mM Tris/HCl, pH 7.35, 0.5 mM EDTA, 0.05% [v/v] MCE) to removed unbound proteins. The μ-calpain and calpastatin were eluted with 80 ml of 200 mM NaCl elution buffer, the m-calpain was subsequently eluted with 80 ml of 400 mM NaCl elution buffer. The eluted was collected 2 mL/fractions and measured at A280 nm and then the fraction consisted of protein were pooled for analysis of the calpains system activity. One unit of calpastatin activity was defined as the amount of the inhibitor that inhibits one unit of m-calpain activity. One unit of μ-calpain and m-calpain activity was defined as the amount of enzyme that catalysed an increase of 1.0 absorbance until at A280 nm [10].

2.4 Statistical Analysis

Completely Randomized Design (CRD) was used to design the experiment. Independent variable was aging period with separated into 6 levels (1, 3, 5, 7, 14, and 21 days). Shear force values, and μ-calpain, m-calpain and calpastatin activities were dependent variables. Ducan's New Multiple Range Test was used to compare the difference of means. The Pearson Correlation between the shear force value and the enzyme activities was also analysed.

3 Results and discussions

3.1 Shear Force Values

The shear force values decreased dramatically from day 1 to 7 postmortem aging from 9.01±0.83 kg/cm^2 to 6.47±1.12 kg/cm^2 ($p<0.01$) and then declined slowly to 4.34±0.10 kg/cm^2 after 21 days of aging ($P<0.01$), as shown in Table 1.

Table 1. Mean±SD of shear force value (kg/cm^2) during aging period

Aging period (day)	Shear force values (kg/cm^2)
1	9.01±0.83^{a}
3	8.20±0.63^{ab}
5	6.92±0.67^{bc}
7	6.47±1.12^{bc}
14	5.12±0.88^{dc}
21	4.34±0.10^{d}

^{a,b,c} Different superscripts showed highly significant differences ($P<0.01$)

After 3 days of aging, the shear force value was 8.20 kg/cm2, however, the values could be varied between 2.1 to 4.9 kg/cm2 [6]. In this study, the longer aging period decreased shear force values in agreement with the previous report found in the loin muscle of red deer (*Cervus elaphus*) ($P<0.001$) [11]. Likewise, the tenderness of loin muscle of red deer was improved after 21 days of aging [12].

3.2 Calpain/calpastatin activities

After calpain analysis, the results showed that the μ-calpain activity decreased significantly from 1.50±0.42 to 0.19±0.28 units/g of meat ($P<0.01$), whereas m-calpain activity deceased slightly from 22.88±9.64 to 16.95±8.34 units/g of meat and calpastatin activity decreased slightly from 9.93±2.37 to 6.82±2.96 units/g of meat ($P>0.05$) during aging period, as present in the Table 2.

Table 2. Mean ± SD of μ-calpain, m-calpain and calpastatin activities (units/g of meat) during aging period

Aging period (day)	μ-calpain	m-calpain	calpastatin
1	1.50±0.42^{a}	22.88±9.64	9.93±2.37
3	1.05±0.54^{ab}	21.10±8.57	9.29±2.10
5	0.74±0.47^{ab}	20.19±8.94	8.96±1.99
7	0.60±0.46^{bc}	20.02±8.95	7.59±2.60
14	0.46±0.33^{bc}	18.68±8.72	7.18±3.00
21	0.19±0.28^{c}	16.95±8.34	6.82±2.96

^{a,b,c} Different superscripts in column show highly significant differences ($P<0.01$)

In this study, μ-calpain activity decreased significantly during aging time. The result agreed with previous study which found that the μ-calpain activity was gradual decline ($P<0.05$) after 1 and 3 days postmortem, and after 15 days postmortem μ-calpain activity was not detected ($P>0.05$) [17]. At the same time, μ-calpain was activated and then they activity lost though an autolytic process [6]. Similarity to the change of calpain/calpastatin activity in beef, the activity of μ-calpain decreased rapidly during postmortem aging but activity of m-calpain activity changed slightly during the first 7 day, whereas calpastatin activity decreased substantially during this period [18].

3.3 Correlation of shear force values and Calpain/calpastatin activities

The correlation between shear force values and μ-calpain, m-calpain and calpastatin activities (Table 3.) shows some interesting among the analysis variables. Shear force value was significantly correlated with higher levels of μ-calpain activity (r =0.915) and m-calpains activity (r=0.758), respectively. However, the value was not significantly related to calpastatin activity (r=0.462). The activities of m-calpains and μ-calpain had highly significantly related (r = 0.632). Furthermore, both had significantly correlated with calpastatin activity (r=0.852). These results summarized that there was the correlation between shear force value, μ-calpain, and m-calpain activities (P<0.01), but not calpastatin activity. There were reported that the postmortem storage and processing affect the meat tenderness [13], and the most of postmortem tenderness improvement is attributed to μ-calpain [14], because after aging storage, the myofibrillar proteins were degraded by calpain system enzyme. They broke muscle protein by autolysis, thus, the resulting in the degradation of myofibrillar proteins and therefore improved tenderness [15-16]. These findings are consistent with previous reports as it showed that a strong correlation between the calpain and meat tenderization postmortem [19]. However, some reported that there was the good correlation between calpastatin activity and shear force in fallow deer meat [20]. The rate of tenderization indicates that the calpain and calpastatin system are closely linked to the proteolytic breakdown of myofibrillar proteins [21].

Table 3. Pearson's correlation coefficients between shear force and μ-calpain, m-calpan and calpastatin

	μ-calpain	m-calpain	calpastatin
Shear force	0.915^{**}	0.758^{**}	0.462^{ns}
μ-calpain		0.632^{**}	0.745^{**}
m-calpain			0.852^{**}

^{ns} non significant difference (P>0.05), and ** P<0.01

In conclusion, this study confirms that in the case of rusa deer, μ-calpain plays a key factor controlling postmortem meat tenderness compared to m-calpain and calpastatin. We suggest that postmortem storage should not exceed 7 days is enough to achieve aging of the LD muscle, which may help to reduce the cost of meat production.

Financial support for this work was provided by the Graduate Studies, King Mongkut's Institute of Technology Ladkrabang, Bangkok, Thailand and the Center of Excellence on Agricultural Biotechnology, Science and Technology Postgraduate Education and Research Development Office, Commission on Higher Education, Ministry of Education (AG-BIO/PERDO-CHE), Bangkok, Thailand.

References

1. R. Sookhareea, K.B. Woodford, G. McL. Dryden, Asian-Australas J Anim Sci, **14** (2001)
2. H.T. Shin, R.J. Hudson, X.H. Gao, J.M. Suttie, Asian-Aust. J Anim Sci, **13** (2000)
3. L.C. Hoffman, E., Wiklund. Meat Sci. **74**(2006)
4. M. Koohmaraie, Meat Sci, **43** (1996)
5. M.A. Sentandreu, G. Coulis, A. Ouali, Trends Food Sci. Technol, **13**(2002)
6. E. Wiklund, V.M.H. Barnier, F.J.M. Smulderd, K. Lundstriim, Meat Sci, **46** (1997)
7. E. Dransfield, Meat Sci, **36** (1994)
8. D.E. Goll, V.F Thompson, H. Li, W. Wei, J. Cong, Physiol. Rev, **83** (2003)
9. E.F. Delgado, G.H. Geesink, J.A. Marchello, D.E. Goll, M. Koohmaraie, J Anim Sci, **79** (2001)
10. G.H. Geesink, M. Koohmaraie, J. Anim Sci, **77** (1999)
11. E. Wiklund, J.M. Stevenson-Barry, S.J. Duncan, R.P. Littlejohn, Meat Sci, **59** (2001)
12. E. Wiklund, P. Dobbie, A. Stuart, R.P. Littlejohn, Meat Sci, **86** (2010)
13. N. Perry, Int. J. Gastron, Food Sci, **1** (2012)
14. M.J. Colle, J.A. Nasados, J.M. Rogers, D.M. Kerby, M.M. Colle, J.B. Van Buren,R.P. Richard, G.K. Murdoch, C.J. Williams, M.E. Doumit, Meat Sci, **135** (2018)
15. D. E. Goll, R. G. Taylor, J. A. Christiansen, V. F Thompson, Proceedings of the Reciprocal Meat Conference, **44** (1991)
16. M.I. Khan, S. Jung, K.C. Nam, C. Jo, Korean J. Food Sci An, **36** (2016)
17. E. Veiseth, S.D. Shackelford, T.L. Wheeler, M. Koohmaraie, J Anim Sci, **79** (2001)
18. M.L. Boehm, T.L. Kendall, V.F. Thompson, D.E. Gol, J Anim Sci, **76** (1998)
19. M. Koohmaraie, G.H. Geesink, Meat Sci, **74** (2006)
20. L.A. Volpelli, S. Failla, A. Sepulcri, E. Piasentier, Meat Sci, **69** (2005)
21. J.D. Morton, R. Bickerstaffe, M.P. Kent, E. Dransfield, G.M Keeley, Meat Sci, **52** (1999)

Response of black rice (Oryza sativa L Indica.) to organic fertilizer and seedling number on grain yield, amylum, antioxidant, and thiamin hydrochloride contents

Okti Herliana[1,*], *A. H. Saeful* Anwar[2], and *Ida* Widiyawati[1]

[1]Jenderal Soedirman University, Faculty of Agriculture, Departement of Agrotechnology, Purwokerto, Central Java, Indonesia

Abstract. This study aimed to investigate the influence of organic fertilizers and seedling numbers each holes of black rice on yield, antioxidant, amylum, and thiamin hydroclorine contents, and to observe the interaction effect between types of organic fertilizers and seedling numbers each hole planting. This study was conducted on rice field in Karanglewas Kidul Village, Karanglewas, Banyumas Regency, Central Java from April until September 2016. The location altitude in this study for about 93 meters above the sea level. The study was arranged by Split Plot Design by the main plot consist with three types of organic fertilizers that was chicken, goat, and cow manures, and the sub plot was consists by seedling number were, three, two, and one seedling by each holes, with three replicates of each combination treatments. The result showed that chicken manure treatment provide the best influences on grain weight per hectare by 5.154 tons, amylum content by 33.86%, anthocyanin content by 275.40 ppm, and Thiamin hydrocloride content by 0.056 mg/10g. The three seedling each hole provide the best influences on amylum content by 33.78%, anthocyanin content by 275.18 ppm, and thiamine hydrochloride content by 0.058 mg/10g.

1 Introduction

Rice is a staple food consumed by almost of all Indonesian people. Based on its color, there are three types of rice white, red and black rice. Black rice has higher anthocyanin content than the white one [1]. Black rice is also considered as a source of fibers, minerals and some amino acids required by the body [2]. Black rice contains anthocyanin and antioxidants which are beneficial to our health. Black rice also contains fibers and hemi-cellulose. The study conducted by Mangiri *et al.* [3] show that black rice has fiber content by 0.8%, vitamin C by 0.6 mg and vitamin E by 31.6 mg. The dominant components of black rice are carbohydrate by 64.98%, total protein content by 15.41%, fat content by 4.23%, minerals by 2.04%, and water content 13,34% [4].

Rice plants require the nourishment substance N, P and K to support the growth and quality of the rice produced. The nutrients serve as the nutrient sources for the plant growth, the formation of saplings as well as chlorophyll for assimilation processes, which eventually produce amylum and determine the quality of rice, such as amylum, anthocyanin and vitamin B contained in rice grains. Thus, to obtain grains with high quantity and good quality, then the plants are required to be given a complete nutrients [5]. Nutrient sources may be obtained from both anorganic and organic fertilizers. In organic farming system, the source of nutrients used may derived from the agricultural and livestock wastes, such as manure, green manure, agricultural waste, biological fertilizer, and household waste [6].

The application of organic fertilizers can maximally absorbed by the black rice plants, affected by several factors, such as plant population level. The plant population density is determined by the planting space and seedling number each holes. Indonesian Farmers usually grow their rice using a large number of seedlings. It is considered inefficient in terms of seedling availability and competition among plants that gave an influence on plant growth and production. The utilization of young seedling and one plant may support better for growth [7].

The number of seedlings per planting hole may influence the number of productive saplings and weight of 1000 grains, on the planting of 3 seedlings per planting hole on a plot of 3m^2 and with the urea dosage of 120 g/plot, SP-36 60 g/plot and KCL 60 g/plot shows the production of dry milled grains per plot by 2.18 kg or equivalent to GKG 7.2 tons/ha [8]. Christanto *et al.* [9] argue that the upland rice (*gogo* paddy) conducted with SRI method and then treated using a variation of planting spaces may influence the maximum sapling number and leaf area index. The treatment conducted using seedlings per planting hole may influence the weight of 1000 oven- dry grains, number of grain per panicle, oven-dry grain weight harvested per hectare, oven-dry grain weight per hectare and harvest index.

The utilization of organic fertilizers and the seedling number per planting hole has not been frequently applied

* Corresponding author: o.herliana@gmail.com

to the cultivation of black rice plants. Thus, it is considered necessary to conduct a study in order to investigate the best organic fertilizer and most optimum seedling number per planting hole for the black rice plant.

2 Material and methods

This study was conducted in the rice fields of Karanglewas Kidul Village, Karanglewas Sub-district, Banyumas Regency, Central Java, from April to September 2016. The average daily temperature of the research area is 29.3 °C with humidity between 60 to 99%. The research location altitude is 93 meters above the sea level. The materials used in this study include black rice seedlings and the manures of chicken, goat, and cow. Supporting tools used in this research include water gauge, soil moisture meter, lux meter, thermo hygrometer, seed separation and counter, altimeter, hoe, scissor, hand sprayer, hand counter, cutter, millimeters blocks, weigh balance and ovens.

This study used Split plot design with three replications. This is considered as a field study with 2 x 3 meter plot system with the main plot in the form of organic fertilizers, covering chicken manure (P1), goat manure (P2), cow manure (P3), while the sub-plot is the number of seedlings per planting hole, covering one seedling per planting hole (B1), two seedlings per planting hole (B2), and three seedlings per planting hole (B3).

The variables observed are plant growth rate, net assimilation rate, leaf area index, chlorophyll content of a and b, amylum, anthocyanin, thiamin hydrochloride, and grain weight per hectare. Chlorophyll content was analyzed using spectrofotometry on 665 nm spectrum. Amylum content of rice was analyzed using the method colorimetric iodide [10], antocyanin was analyzed by Shi method [11], and Thyamine hydrochloride was analyzed by Sudarmaji method [12]. Data analysis is conducted with analysis of varians, if the result is significant among the observed treatments, it will be followed by DMRT (Duncan Multiple Range Test) at the level 5%.

3 Result and discussion

The analysis of varians results of various treatments conducted with several fertilizer types and number of seedling per planting hole presented in Table 1 show that the treatments conducted with several types of organic fertilizers influence the variables of chlorophyll a, grain weight per hectare, amylum content, anthocyanin and thiamin hydrochloride. Meanwhile, the treatments conducted with several numbers of seedlings per planting hole influence the variables of plant growth rate, leaf area index, the content of amylum, anthocyanin and thiamin hydrochloride. However, there is no significant interaction between those two treatments to the observed variables.

Table 1. The results of various observations on the yield and quality of black rice

Variabel	Treatment		
	P	B	PxB
Plant Growth rate	ns	s	ns
Leaf area index	ns	s	ns
Clorofil b	ns	ns	ns
Clorofil a	s	ns	ns
Amilum	vs	vs	ns
Antosianin	vs	vs	ns
Thiamin Hidroklorida	vs	vs	ns
Grain weight (ton hectar $^{-1}$)	s	ns	ns

P : Organic fertilizer
B : Seeding number
PxB : Interaction between two treatments
ns : non significant
s : significant
vs : very significant

3.1. Influence of organic fertilizer to graind yield and black rice quality

The black rice plant has various responses to the application of organic fertilizers. The results of further Duncan test are shown in Table 2.

Table 2. The average data on the influence of treatments conducted with several organic fertilizer types on the black rice yield

Variable	Type of Organic Frertiliz		
	P1	P2	P3
KK a	2,11 a	2,11 a	1,90 b
Am	33,86 a	32,59 b	32,13 c
An	275,40 a	265,39 b	263,30 c
TH	0,056 a	0,054 b	0,050 c
WGH	5,154 a	4,271 b	3,933 b

P1 : Chiken manure
P2 : Goat Manure
P3 : Cow manure, Clorofil a on 665 nm spectrum,
Am : Amilum content (%)
An : Antocyanin (ppm)
TH : Thiamin Hidroklorida (mg/10 g sample)
WGH : Wight Grain hectar^{-1} (Tons). Numbers followed by different letter at each column are significantly different at P=0.05

The influence of treatments conducted with several organic fertilizers on black rice yield and quality in Table 2 shows that the treatments conducted with several types of manures influence the variables of chlorophyll, amylum, anthocyanin and thiamin hydrochloride content as well as the grain weight per hectare. The best treatment is characterized by a notation, the second best treatment is indicated by b notation, and that is not quite good is characterized by c notation. The variable of chlorophyll content has the best result at treatment P1 (chicken manure) and the second best at P2 (goat manure) by 2.11, while last one at P3 (cow manure) by 1.90. The variable of amylum content has the best result at treatment P1 by 33.86%, the second best at P2 by 32.59% and the last one is at P3 by 32.13%. In the variable of anthocyanin content, the best result is at treatment P1 by 275.40 ppm, the second best at P2 by 265.39 ppm and the last one at P3 by 263.30 ppm. The variable of thiamin hydrochloride content, the best result is at treatment P1 by 0.056 mg/10g material, the second best at treatment P2 by 0.054 mg/10g material and the last one at P3 by 0.050 mg/10g material. lastly, the variable of grain weight per hectare, the best result is at P1 by 5154.67 kg/ha, the second best is at P2 by 4271.33 and the last one at P3 by 3933.26 kg/ha.

The manures influence the rice plant chlorophyll content. Damanik *et al.* [13] explain that the administration of nitrogen directly influences the carbohydrate synthesis in the plant cells. Nitrogen has also the role as a chlorophyll composer causing the leaves have the green color. Hartanto and Maya [14] who conduct a study on Ciherang rice variety explain that time for the chicken manure application influences the length of the plant roots. However, it is unknown whether the types of manure may also influence the length of the rice plant roots or not.

Based on the results of study conducted by Purnomo *et al*, [15], the addition of vermicompost/ cow manure (5 tons/ha) + inorganic fertilizer (urea 226 kg/ha + phonska 360 kg/ha) with the planting of three seedlings per planting hole may produce the highest rice amylum content. Hardjowigeno [16], explains that the addition of organic material derived from vermicompost or cow manure on the rice field soil plays an important role in increasing the availability of N nutrients in the soil. The decomposition result of organic materials on water-flooded soil like the common rice fields is, one of them, in the form of NH_4 which may be directly utilized by the plants. Application of organic fertilizer also influenced antocyanin content on grape fruit, that the high doses 30 t/ha can increase the antocyanin content 12% higher than 20 t/ha [17].

The treatments of manures influence the components of the rice yield which is in accordance with research conducted by Bachtiar *et al.* [18] stating that the administration of SP-36 100% + manure 5 t/ha on the soil with the manure 20 t/ha may produce the dry grain weight by 648 g per the treatment plot which is almost twice bigger than that in control plot by only 344 grams. Organic fertilizer with inorganic fertilizer increased the fertilizer use efficiency and improved the physical and chemical properties of soil and it would be a reason towards increased yield [19].

3.2. Influence of seeding number to yield and quality of black rice

The influence of treatments conducted with the number of seedlings per planting hole on the black rice yield and quality is presented in Table 3. The treatments conducted with the number of seedlings per planting hole influence the variables of plant growth rate, leaf area index, the content of amylum, anthocyanin and thiamin hydrochloride.

Table 3. The average data on the influence of treatments conducted with number of seedlings per planting hole on yield and quality of black rice.

Variable	Seedling Number		
	B1	B2	B3
LPT	0,12 b	0,27 a	0,32 a
ILD	0,28 b	0,35 a	0,39 a
Am	31,94 c	32,86 b	33,78 a
An	260,02 c	268,89 b	275,18 a
TH	0,049 c	0,053 b	0,058 a

B1 : ne seedling
B2 : two seedling
B3 : three seedling
LPT : Plant Growt Rate (g/day), Numbers followed by different letter at each column are significantly different at P=0.05

The best result for the plant growth rate variable is at treatment B3 (3 seedlings per planting hole) and the second best at treatment B2 (2 seedlings per planting hole) respectively by 0.32 g/day and 0.27 g/day, while the last one at treatment B1 (1 seedling per planting hole) by 0.12 g/day. The best result for the leaf area index variable is at treatment B3 and the second best at treatment B2 respectively by 0.39 and 0.35, while the last one at treatment B1 by 0.28. The best result for amylum content variable is at treatment B3 by 33.78%, the second best at treatment B2 by 32.86% and the last one at B1 by 31.94%. The best result for anthocyanin content variable is at treatment B3 by 275.18 ppm, the second best at treatment B2 by 268.89 ppm and the last one at treatment B1 by 260.02 ppm. Lastly, The best result for thiamin hydrochloride content variable is at treatment B3 by 0.058 mg/10g material, followed by treatment B2 0.053 mg/10g material and the last one at treatment B1 by 0.049 mg/10g material.

Number of seedlings hill-1 is an important factor for successful rice production because it influences tiller population unit-1 area, which ultimately influence the yield [20]. Optimum number of seedlings hill-1 ensures proper crop establishment. Generally if single seedling is used than there is a chance for missing hills. On the other hand, if more than optimum seedlings hill-1 is used then there will be misuse of seedlings resulting excess use of seeds. But in these research the best result of all variable is three seedling hill^{-1} because used of single and second seedling is Seedling vigor contributes to successive tillering, yield and quality of transplanted rice. Growth and production of rice depend on timely cultivation and growth duration of cultivar which is affected by the number of seedlings hill-1 [21].

Hutagaol [22] explains that the treatments conducted with the number of seedlings per different planting hole for the upland rice plants produce vegetative variables with the best influence on the plant height and number of saplings at the administration of 4 seedlings. Wangiyana *et al.* [23] further explain that the planting of 3 seedlings per planting hole may result in more productive yield. The utilization of 3 seedlings per planting hole produces higher number of leaves, clumps and dry rice straw weight than the planting utilizing 1 and 2 seedlings per planting hole. Similarly, Zulhendi [24] also explains that the higher the number of seedlings per planting hole may result in higher number of saplings. Meanwhile, the higher the number of saplings may correlate more with the value of leaf area index that the value of ILD may also increase. The higher value of ILD, the wider the photosynthesis area and the photosynhtate produced. Local black rice varieties have a less adaptive ability to grow so that the use of the number of seeds 3 per hole better and support the development of tillers, panicles and grain. Grain quality is also determined by the resulting photosynthate.

According to Purnomo *et al.* [15], the process of photosynthesis in plants greatly influences the amylum content since the result of photosynthesis process is stored in the forms of carbohydrate and energy. The declining photosynthesis process may lead to the reduced photosynthesis result. This photosynthate is then stored in the forms of carbohydrate in stems and grains. Thus, the better the process of photosynthesis in plants the higher the formation of carbohydrates.

In this study, the highest thiamin content in black rice is 0.058 mg/10 g material or about 5.8 mg/kg material. This is in accordance with the research conducted by Laksmiwati *et al.* [25] stating that the thiamin content in black rice after cooking is about 5.3 mg/kg, extremely different from that of white rice which is only about 2.6 mg/kg.

4 Conclusions

The results showed that chicken manure treatment has gave the best influences of grain weight per hectare, amylum, anthocyanin and thiamin hydrochloride contents. In facts, 3 seedlings per planting hole treatment has gave the best influence of plant growth rate, leaf area index, anthocyanin, amylum, and thiamin hydrochloride contents.

Acknowledgements The authors would like to thank to the Research and Public Services Institution (LPPM) of Jenderal Soedirman University (Unsoed), Indonesia, for funding this study.

References

1. Q. Wang, P. Han, M. Zhang, M. Xia, H. Zhu, J. Ma, M. Hou, Z. Tang, W. Ling, Asia Pac J. Clin Nutr **16**, 295 (2007)
2. M. Hiomori, E. Koh, A.E Mitchell, J. Agric. Food Chem. **57**, 1908 (2009)
3. J. Mangiri, N. Maluyu, S.E.S Kawengian, e-Biomedik, **4** (2016)
4. S. Hartati, J. Food and Nutrient, **4**, 57 (2012)
5. Rasam, A. Rasna, *Pemupukan pada Tanaman Padi* (Balai Besar Penelitian Tanaman Padi, Indonesia, 2015)
6. E.A. Stockdale, N.H. Lampkin, R. Keatinge, E.K.M. Lennartsson, D.W. Macdonald, S. Padel, F.H. Tattersall, M.S Wolfe, C.A. Watson, Adv. Agron. **70**, 261 (2001)
7. A.H. Pinem, A. Barus, H. Chairani, Agroteknologi **1**, 921 (2013)
8. J. Susilo, Ardian, E. Ariani, Jom Faperta **2**, 1 (2015)
9. H. Christanto, I.G.M.S. Agung, J. Bumi Lestari **14**, 1 (2014)
10. B.O. Juliano, Cer. Sci. Today **16**, 334 (1971)
11. Z. Shi, M. Lin, F.J. Prancis., J Food Sci. **57**, 758 (1992)
12. S. Sudarmaji, B. Haryono, Suhardi, *Prosedur Analisis untuk Bahan Pangan dan Gizi* (Gadjah Mada University, Yogyakarta, Indonesia, 1997)
13. M. Damanik ,M.B. Bachtiar, E.H. Fauzi, *Kesuburan Tanah dan Pemupukan* (USU Press, Medan, Indonesia, 2011)
14. M. Hartanto, M. Melati, Bul. Agrohorti **1**, 9 (2013)
15. B.B. Purnomo, S. Jauhari, H. Sri, *Kajian Penambahan Vermikompos dan Pupuk Anorganik Terhadap Kualitas dan Hasil Tanaman Padi (Oryza sativa L.) di Lahan Sawah Palur Sukoharjo* (Sebelas Maret University, Surakarta, Indonesia, 2011)
16. S. Hardjowigeno, *Ilmu Tanah* (Akademika Presindo, Jakarta, Indonesia, 2005)
17. K. Seung-Heui, K. Seon-Kyu, J. Plant Biotech. **1**, 7 (2002)
18. T. Bachtiar, S.H. Waluyo, S.H. Syaukat, J. Ilmiah Aplikasi Isotop dan Radiasi **9**, 151 (2013)
19. T.P. Gangmei, P.J. George, J. Pharma. Phytochem. **6**, 550 (2017)
20. M.O. Farouk, M.A. Rahman, M.A. Hasan. Int. J. Suatain. Crop. Prod **4**, 58 (2009)
21. A. Mishra, V.M Salokhe, Exptl. Agric. **44,** 365 (2008)
22. H.P. Hutagaol, J. Ginting, S. Ginting, Agroteknologi **1**, 467 (2013)
23. W. Wangiyana, Z. Laiwan, Sanisah, Crop Agro **2**, 70 (2009)
24. Zulhendi, *Pengaruh Jumlah Bibit per Titik Tanam dan Umur Bibit Terhadap Pertumbuhan dan Hasil Padi Sawah Dalam Sistem Intensifikasi Padi (SRI)* (Andalas University, Sumatra Barat, Indonesia, 2005)
25. A.A.I.A.M. Laksmiwati, K. Ratnawati, N.W. Agustini, J. Chemistry **6,** 47 (2012)

Study of atrazine removal by an adsorbent synthesised by water hyacinth

Sitanan Nuntamongkol[1], *Pawinee* Deetae[2], and *Patthranit* Wongpromrat[1,*]

[1]Chemical Engineering Department, Faculty of Engineering, King Mongkut's Institute of Technology Ladkrabang, Bangkok, Thailand
[2]Food Fermentation Department, Faulty of Agro-Industry, King Mongkut's Institute of Technology Ladkrabang, Bangkok, Thailand

Abstract. In this work, water hyacinth was selected as a raw materials for synthesising the activated carbon used for removing atrazine, one of the most famous herbicides used in Thailand, from water. Three different treating methods were performed in order to attain the best adsorbent; untreated (AC), HCl-treated (HCl-AC) and NaOH-treated (NaOH-AC) activated carbons. After pyrolysis, NaOH-AC became ash, so it was not suitable for using as an adsorbent. Among these, HCl-AC was the best adsorbent for removing atrazine from water because of its highest surface areas. The adsorption experiments using AC and HCl-AC as the adsorbents confirmed that the adsorption efficiency of HCl-AC was higher than that of AC. In addition, the adsorption isotherm of HCl-AC was investigated by varying the atrazine initial concentration in the range of 3-15 ppm. The results best fitted with Langmuir isotherm that means this adsorption phenomenon was chemisorption with the maximum adsorbent capacity of 24.510 mg/g, Langmuir isotherm constant of 10.462 L/mg and the separation factor of 0.006 indicating that the equilibrium sorption of atrazine by HCl-AC was favourable.

1 Introduction

Thailand is an agricultural country where exports various kinds of agricultural products. Therefore, there is the using of some herbicides for producing high-quantity agricultural products. Thailand imports the large amount with high value of herbicides between 2010-2015 [1]. Atrazine, ($C_8H_{14}ClN_5$) is the Thailand top ten imported organochloride herbicide [2], is used to control weeds in farms and crop land sand for example corn, sugarcane, or millet farm. Due to their extensive usage and its long half life about 13-261 days and more than 200 days in soil and water resources, respectively [3]. This affects the contamination in surface water and ground water of Thailand [4]. Atrazine also affects to frog mutation [5], ecology [4] and human health [6]. Therefore, because of its hazard, it is necessary to remove atrazine from water resources.

The examples of atrazine removal techniques from water are incineration, reverse osmosis, electrodialysis, chemical degradation and etc. Nevertheless, all of these techniques are costly and produce corrosive and toxic gases [7]. Adsorption is a simply attractive method since the atrazine is adsorbed on the adsorbent's surface. The common adsorbent is an activated carbon because of its high surface area and porosity. However, activated carbon is an expensive material thus it is necessary to develop a low cost effective carbon that can be used to remove the contaminant from water. One property affecting the adsorption capacity is porosity which depends on nature of raw material and activation method. Since Thailand is an agricultural country, the large amount of agricultural wastes exist and be many studies for producing an effective activated carbon e.g. coconut shell, watermelon peel, rice straw and etc. One of the waste causing many problems in Thailand is a water hyacinth (Eichhornia crassipes) that is a free-floating perennial aquatic plant which can propagate and grow easily. The large amount of water hyacinth makes a rotten water because oxygen cannot dissolve into water, blocks waterways, affects both navigation and drainage, and increases siltation and flooding probability.

In the consequence, this study aims to prepare a low cost activated carbon from water hyacinth for atrazine adsorption. This means that this study will change waste material (water hyacinth) to be a valuable material. As complex structure of atrazine, chemical and physical treatments are needed to increase the porosity of water hyacinth. The effect of different kinds of activated carbon preparation methods on adsorption is investigated. In addition, the adsorption isotherm of the best adsorbent prepared in this work was determined in order to understand the phenomenon of atrazine adsorption.

2 Experiments

2.1. Dried water hyacinth preparation

* Corresponding author: patthranit.wo@gmail.com

The stem of water hyacinth collected from the river was peeled, cleaned with water, chopped into pieces with the thickness of 0.2 cm and dried under the sunlight for 3 days. After that, water in the stem pieces was removed by baking at 105°C until the weight was constant. Then, it was cooled down to room temperature and treated by 3 different methods as follow.

2.2 Treatment methods

1) Untreated method
The dried water hyacinth was pyrolysed at 500 °C under N_2 in the furnace for 1 h and, then the activated carbon was obtained. After cooling to room temperature, the activated carbon was ground and sieved to be 200-250 mesh-size. This activated carbon will be abbreviated as AC in the following.
2) HCl-treated method
The 10 g of dried water hyacinth was left in 1 L of 1 M HCl solution for 2 days. After that, water hyacinth was separated from solution by filtration and washed by distilled water until the pH of washed water was neutral. Then, it was baked at 105 °C for 2 days and pyrolysed at 500 °C under N_2 in the furnace for 1 h. After cooling to room temperature, the activated carbon was ground and sieved to be 200-250 mesh-size. The activated carbon prepared from this method is called HCl-AC.
3) NaOH-treated method
The activated carbon attained from this method is called NaOH-AC. Its preparation was similar to HCl-AC, but the chemical used for treatment was NaOH instead of HCl.

2.3 Material characterisations

The adsorbents prepared from 3 different treating methods mentioned above were characterised by Brunauer–Emmett–Teller (BET) method in order to observe the surface areas, pore volumes and pore diameters and CHNS Analysis for determining the compositions of the adsorbents.

2.3 Adsorption isotherm study

The adsorption isotherm determination was performed by batch test. The 200-mL of atrazine with the various initial concentrations in the range of 3-15 ppm was prepared and stirred at the agitation speed of 150 rpm. Then, 0.1 g of adsorbent was added into atrazine solution. Samples were taken at appropriate time intervals and atrazine concentration of the samples was determined by using Gas Chromatography-Mass Spectroscopy (GC-MS) with HP-5 method.

3 Results and discussion

3.1. Material characterisations

After pyrolysis of water hyacinth treated by NaOH, this material became ash. This was probably because NaOH could remove more lignin than HCl [8], so the amount of NaOH-treated water hyacinth pyrolysed was less than that of other treated methods. Due to high temperature of pyrolysis, most of NaOH-treated water hyacinth was therefore changed into ash. Thus, NaOH-AC was not further studied in this work.

The carbon percentages (%C), surface areas and pore diameters of AC and HCl-AC characterised by BET and CHNS Analysis are shown in Table 1. It can be seen that HCl treatment was able to improve the characteristics of the adsorbent, in terms of both %C, surface area and pore diameter, comparing with untreated method. Although, the pore diameter of HCl-AC was smaller than that of AC, but it was still bigger than the size of atrazine (10.36 Å) [9], so, among these, HCl-AC should be the best adsorbent for atrazine removal.

Table 1. Characteristics of the adsorbents

Adsorbents	%C	Surface area (m^2/g)	Pore diameter (Å)
AC	38.921	160.4	80.94
HCl-AC	71.192	436.0	29.01
NaOH-AC	Ash		

Batch adsorptions by AC and HCl-AC with the atrazine initial concentration of 12 ppm were performed in order to confirm the hypothesis that HCl-AC was the best adsorbent in this work. The results are shown in Fig. 1. It can be seen that the atrazine concentrations in solutions decreased with time as a result of the adsorption process. However, time to equilibrium of AC was about 4 h that was shorter than that of about 5 h of HCl-AC. Furthermore, the remaining atrazine concentration in solution at equilibrium of the batch used HCl-AC as an adsorbent was lower than that of AC. That means the adsorption capacity of HCl-AC was higher than that of AC. These, both time to equilibrium and capacity, confirmed that HCl-AC was the best adsorbent because HCl treatment was able to improve the characteristics of the activated carbon which directly affected its adsorption efficiency.

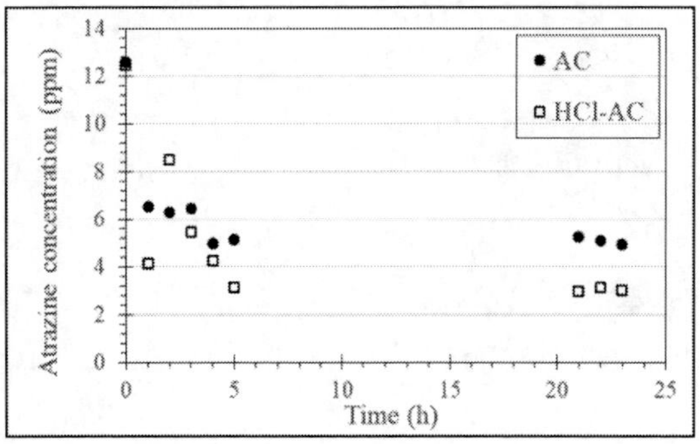

Fig. 1. The concentration of atrazine in solution with time during adsorption by AC and HCl-AC.

3.2. Adsorption isotherm

The atrazine adsorption isotherm by HCl-AC was investigated by varying the atrazine initial concentrations of 3, 6, 9, 12 and 15 ppm. For the adsorption of solution with the initial concentrations of 3 and 6 ppm, the adsorbent could adsorb all atrazine in solution, so only the results of 9, 12 and 15 ppm of initial concentrations were considered. The adsorption capacities at equilibrium were observed and fitted with 3 isotherm models [10-13] that were Langmuir isotherm (Eq. 1) assuming that the adsorption phenomenon is chemical adsorption and the maximum sorption corresponds to the saturated monolayer of adsorbate on adsorbent surface, Freundlich isotherm (Eq. 2) assuming that the adsorption is physical adsorption with multilayer of adsorbate on adsorbent surface and Temkin isotherm (Eq. 3) taking into account the effect of indirect interaction between adsorbate molecules and assuming that the heat of adsorption of all molecules in the layer decreases linearly due to the increasing of surface coverage. The linearised forms of Langmuir, Freundlich and Temkin isotherms were formatted as shown in Eqs. 4-6, respectively, and the fitted graphs are shown in Fig. 2.

Langmuir isotherm equation:

$$Q_e = \frac{Q_m K_L C_e}{1 + K_L C_e} \tag{1}$$

Freundlich isotherm equation:

$$Q_e = K_F C_e^{1/n} \tag{2}$$

Temkin isotherm equation:

$$Q_e = \frac{RT}{K_T} \ln(A_T C_e) \tag{3}$$

Linearised form of Langmuir isotherm equation:

$$\frac{C_e}{Q_e} = \frac{1}{K_L Q_m} + \frac{C_e}{Q_m} \tag{4}$$

Linearised form of Freundlich isotherm equation:

$$\log Q_e = \log K_F + \left(\frac{1}{n}\right) \log C_e \tag{5}$$

Linearised form of Temkin isotherm equation:

$$Q_e = \frac{RT}{K_T} \ln A_T + \frac{RT}{K_T} \ln C_e \tag{6}$$

where

Q_e = the amount of metal adsorbed per gram of the adsorbent at equilibrium (mg/g)
Q_m = the adsorption capacity (mg/g)
C_e = the equilibrium concentration of adsorbate (mg/L)
K_L = Langmuir isotherm constant (L/mg)
K_F = Freundlich isotherm constant (mg/g)
K_T = Temkin isotherm constant
n = adsorption intensity
A_T = Temkin isotherm equilibrium binding constant (L/g)
R = Gas constant (J/mol K)
T = Absolute temperature (K)

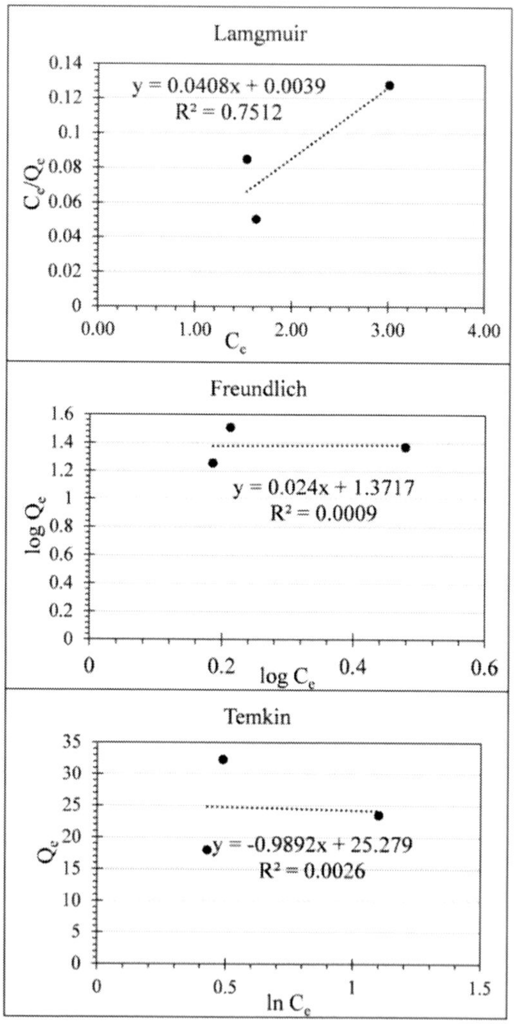

Fig. 2. (a) Langmuir, (b) Freundlich and (c) Temkin plotted of adsorption isotherms in the atrazine adsorption process using HCl-AC as an adsorbent.

Fig. 2. shows that Langmuir isotherm model best fitted with adsorption results with R^2 of 0.7512. Even the R^2 value was not high but it was the highest value comparing with those of other isotherm models. This means that the adsorption phenomenon of atrazine by HCl-AC was chemisorption revealing that there was a chemical interaction between atrazine and the homogeneous active site of HCl-AC.

The adsorption parameters in Langmuir equation which are Q_m and K_L can be calculated by fitting the equation in linearised form (Eq. 7) [10-11]. The slope and y-intercept of the graph represent $1/Q_m$ and $1/Q_m K_L$, respectively. It can be found that the adsorption capacity (Q_m) of HCl-AC was 24.510 mg/g and Langmuir isotherm constant (K_L) which is related to the energy of adsorption was 10.462 L/mg.

$$\frac{C_e}{Q_e} = \frac{1}{K_L Q_m} + \frac{C_e}{Q_m} \quad (7)$$

The essential characteristics of Langmuir isotherm can be expressed in terms of a separation factor (R_L), so-called equilibrium parameter, which is a dimensionless constant given by Eq. 8.

$$R_L = \frac{1}{1 + K_L C_0} \quad (8)$$

where C_0 is the highest initial concentration of adsorbate in ppm.

The value of R_L indicates the nature of adsorption to be either unfavourable ($R_L > 1$), linear ($R_L = 1$), favourable ($0 < R_L < 1$) or irreversible ($R_L = 0$) [11-12]. In this work, the value of R_L was 0.006 indicating that the adsorption of atrazine by HCl-AC was favourable.

4 Conclusions

1). The best treatment method for synthesising the activated carbon adsorbent was HCl treatment, since the characteristics, surface area and %C, of HCl-AC was improved comparing with those of AC. Although, the pore diameter of HCl-AC was finer than that of AC, but HCl-AC was still able to adsorbed atrazine because its pore diameter was larger than the size of atrazine molecule. For NaOH-AC, ash presented after pyrolysis that means this treatment method was not suitable for synthesising adsorbent.

2). The adsorption experiments using AC and HCl-AC as the adsorbents showed that the adsorption capacity of HCl-AC was higher than that of AC. This confirmed the characterisation results and indicated that HCl treatment can improve the characteristics of the adsorbent affecting the better performance of the material.

3). Langmuir, Freundlich, Temkin and Dubinin-Radushkevich isotherms were fitted with the experimental adsorption data of the batch using HCl-AC as an adsorbent to described the adsorption phenomenon. Among these, Langmuir isotherm best fitted with the R^2 of 0.7512. From the calculation, the adsorption capacity (Q_m) of HCl-AC was 24.510 mg/g, Langmuir isotherm constant (K_L) which is related to the energy of adsorption was 10.462 L/mg and the dimensionless separation factor (R_L) was 0.006 indicating that the adsorption of atrazine by HCl-AC was favourable.

References

1. Office of Agricultural Economoics. (2016). [Online]. Available: http://www.oae.go.th/ewt_news.php?nid=146. [Accessed: 18-Nov-2017].
2. A. Tawatsin, U. Thavara, P. Siriyasatien, Medical Research Archives, 3 (2015).
3. U.S.Department of Health and Human Service Agency for Toxic Substances and Disease Registry. (2003). [Online]. Available: https://www.atsdr.cdc.gov/toxguides/toxguide-153.pdf. [Accessed: 18-Nov-2017].
4. M. Graymore, F. Stagnitti, G. Allinson, Environmental International **26**, 7-8, 483-495 (2001).
5. T.B. Hayes, V. Khoury, A. Narayan, M. Nazir, A. Park, T. Brown, L. Adame, E. Chan, D. Buchholz, T. Stueve, S. Gallipeau, Ecology **107**, 10, 4612-4617 (2010).
6. Pesticide action network.). [Online]. Available: http://www.panna.org/resources/atrazine [Accessed: 18-Nov-2017].
7. R.K. Pathak, A.K. Dikshit, International Conference on Life Science and Technology IPCBEE **3**, 19-22 (2011).
8. M. Badiei, N. Asim, J.M. Jahim, K. Sopian, APCBEE Procedia **9**, 170-174 (2014).
9. S.S. Chen, J.S. Taylor, L.A. Mulford, C.D. Norris, Desalination **160**, 2, 103-111 (2004).
10. X. Chen, Information **6**, 14-22 (2015).
11. B.H. Hameed, J.M. Salman, A.L. Ahmad, Journal of Harzardous Materials **163**, 121-126 (2009).
12. A.O. Dada, A.P. Olalekan, A.M. Olatunya, O. Dada, IOSR Journal of Applied Chemistry **3**, 1, 38-45 (2012).
13. J.M. Salman, V.O. Njoku, B.H. Hameed, Chemical Engineering Journal **174**, 41-48 (2011).

Application of ozone ultrafine bubble water for cleaning and sterilizing the dialysis circuit

Yukiko Nakahara[1*], *Kazutoku* Miyazaki[2], *Junji* Nakao[3], and *Shijie* Zhu[4]

[1] Department of Medical Engineering, Junshin Gakuen University, Fukuoka, Japan
[2] Hakujinkai Hospital, Fukuoka, Japan
[3] Tosslec Co., Ltd., Kyoto, Japan
[4] Department of Intelligent Mechanical Engineering, Fukuoka Institute of Technology, Fukuoka, Japan

Abstract. The dialysis circuit was washed by the ultrafine bubble (UFB) water and ozone UFB water and sodium hypochlorite aqueous solution. The effect of washing time, ozone concentration of UFB water, influence of UFB water on artificial dialysis equipment, etc. were investigated. The cleaning and sterilizing effect of ozone UFB water passed through the filter for artificial dialysis was verified. It was confirmed that there were no colony bacteria in the circuit which was washed by the ozone UFB water for 1 hour. Ozone UFB water has the same bactericidal effect as sodium hypochlorite. As an alternative of sodium hypochlorite, we are expecting its application to the cleaning and sterilization of ozone UFB water for dialysis circuit.

1 Introduction

In Japan, the number of chronic dialysis patients are increasing in trend, and already over 320,000 [1]. The artificial dialysis equipment is used for treatment of chronic dialysis. As Detergents, sodium hypochlorite and peracetic acid are used for cleaning and sterilizing dialysis circuit. Sodium hypochlorite can remove and disinfect organic matter such as protein, fat, carbohydrate, and the peracetic acid can remove bacteria in dialysis circuit. However, conventionally used detergents have corrosiveness and irritating odor, and they have adverse effects such as detrimental effects when adhering to the skin [1]. In recent years, research on fine bubble technology was thriving and applied in industrial field, agriculture field, fishery field, daily life. The bubble diameter of micro bubbles is visible at 100 μm or less. It has a feature that it disappears in water after rising very slowly. Ultrafine bubble is a bubble smaller than micro bubbles and has a maximum peak at about 100 to 200 nm. It is not visible, has a characteristic that it remains in water for a long time (a few months) [2] and shows Brownian motion. Typical effects of fine bubbles include physical adsorption, cleaning sterilization effect, gas dissolution effect, gas inclusion effect, and physiological activity effect [2-10]. We focused on the gas inclusion effect and attempted to clean the dialysis circuit using ozone ultrafine bubble water.

The purpose of this study was to evaluate the cleaning and sterilizing effect of the ultrafine bubble (UFB) water with or without ozone in the artificial dialysis equipment.

----

*Corresponding author: nakahara@junshin-u.ac.jp

2 Materials and experimental methods

The experimental equipment included an RO equipment which manufactured by Nipro Co., Ltd., an ozone ultrafine bubble generator (HMB - 0 Z01) manufactured by Tosslec Co. Ltd., an endotoxin measuring instrument (Toxinometer Mini) manufactured by Wako Pure Chemical Industries. The verification sample was the used dialysis equipment circuit, provided by Hakujinkai Hospital. The experimental instruments were a 3.9 L polypropylene washing vessel. A bacterial culture was a 37 mm quality monitor (GN-6 Metricel) of Nippon Pole Co., Ltd., and a liquid culture medium of M-TGE broth was used. The soiled dialysis drain pipe (hose) was incorporated into a washing container, and the water around it was taken out and used as a test substance. The cleaning time was from 0.5 hour to 24 hours.

3 Results and Discussion

The soiled dialysis circuit was washed with ozone UFB water and tap water for 1 h, respectively. The ozone concentration of ozone UFB water was about 10.00 - 10.80 mg / L. 50 mL of the test substance was collected and cultured at 37 ° C, for 7 days using a 37 mm quality monitor. As shown in Fig. 1, when washed with ozone UFB water, colony bacteria were not observed. On the other hand, in the case of washing with general tap water, as shown in Fig. 2, colony bacteria were observed after washing for 1 hour.

3.1 Effect of ozone on cleaning effect

The cleaning effect of oxygen UFB water and ozone UFB water was compared. The concentration of oxygen was 99.999%. The ozone concentration was about 10.00 - 10.80 mg / L. Oxygen UFB water was placed in a washing container, soiled dialysis piping was placed in a washing container and washed for 1 hour. As shown in Fig. 3, colony bacteria were observed. Comparison of the number of colonies after washing for 1 hour by tap water, oxygen UFB water and ozone UFB water respectively was shown in Fig. 4. It was confirmed that there was a cleaning and sterilizing effect only by use of ozone UFB water.

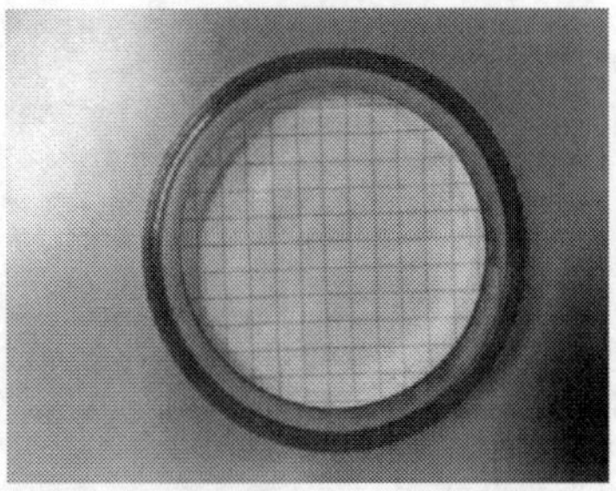

Fig. 1 Cleaning with Ozone UFB for 1 hour.

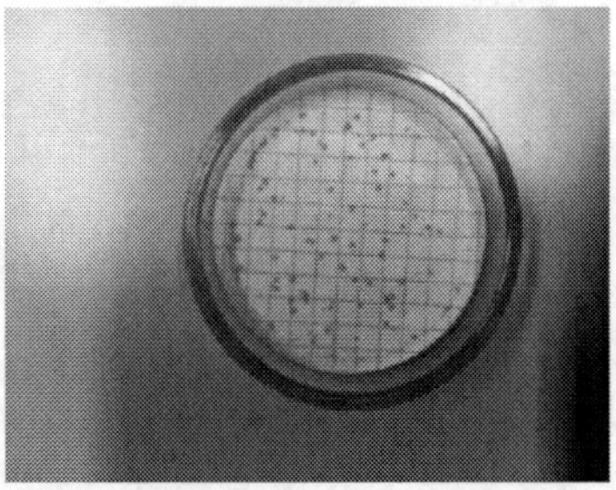

Fig. 2 Cleaning with tap water for 1 hour.

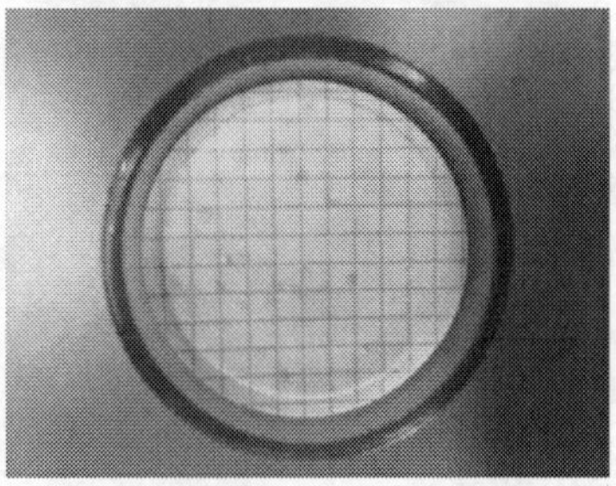

Fig. 3 Cleaning with UFB for 1 hour.

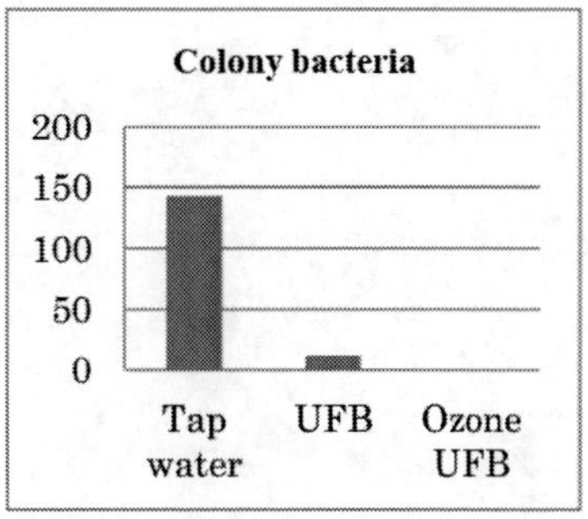

Fig. 4 Comparison of the number of colonies.

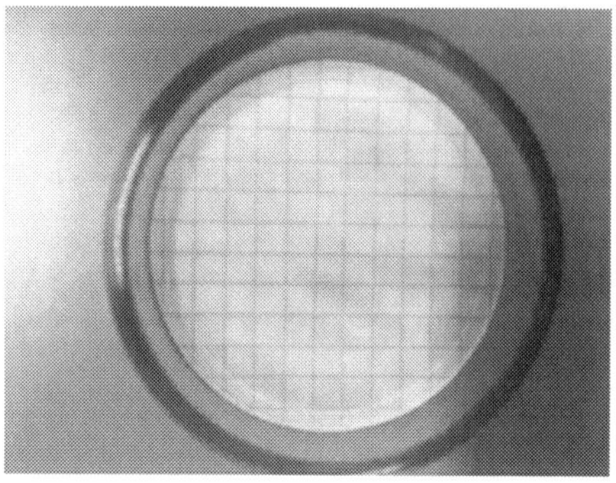

Fig. 5 Cleaning with 300 ppm sodium hypochlorite water for 1 hour.

3.2 Comparison of cleaning effect of sodium hypochlorite and ozone UFB water

The effect of washing time, ozone concentration of UFB, influence of UFB water on artificial dialysis equipment, etc. were investigated using sodium hypochlorite of ECO-200, diluted 200 times with general tap water, with the concentration of sodium hypochlorite aqueous solution of 300 ppm. The cleaning and sterilizing effect of ultrafine bubble water passed through a filter for artificial dialysis was verified.

4 Conclusion

The bactericidal effect of ozone ultrafine bubble water was obtained. As an alternative of sodium hypochlorite, we are expecting its application to the cleaning and sterilization of ozone UFB water for dialysis circuit and artificial dialyzer.

References

1. E. Masakane et. al., Journal of Japanese Society for Dialysis Therapy, Vol. 50, No. 1, 1-62 (2017)
2. Y. Kanai, K. Terasaka, S. Fujioka and K. Suzukawa, J. Chem. Eng. Japan, Vol.50 (2), 86-93 (2017)
3. K. Terasaka, J. Nitta, H. Fukuda, M. Iwahashi, T. Goshima, S. Fujioka, J. Chem. Eng. Japan, 46 (1), 33-39 (2013)
4. M. Takahashi, E. Izawa, J. Etou and T. Ohtani, Journal of the Physical Society of Japan, 107-10, 2171-2173 (2003)
5. K. Terasaka, Chemical Engineering of Japan, Vol.78, No.9, 580-584 (2014)
6. N. Sumida, 1st International Symposium of Fine Bubbles, (2012)
7. S. Maeda et al., 1st International Symposium on Multiscale Process Engineering (MMPE), Kanazawa, Japan, P. 24-1 (2011)
8. M. Takahashi et al., J. Phys. Chem. B 107, 2171, (2003)
9. N. Sumida, Chemical Engineering of Japan, Vol. 78, No. 9, 589-592 (2014)
10. T. Fujita, Chemical Engineering of Japan, Vol. 78, No. 9, 585-588 (2014)

A study of atrazine adsorption by using the rice straw synthesized adsorbent

Thachanan Samanmulya[1,*], *Pawinee* Deetae[2] and *Patthranit* Wongpromrat[1]

[1]Chemical Engineering Department, Faculty of Engineering, King Mongkut's Institute of Technology Ladkrabang, Bangkok, Thailand
[2]Food Fermentation Department, Faculty of Agro-Industry, King Mongkut's Institute of Technology Ladkrabang, Bangkok, Thailand

Abstract. This study focused on adsorption of atrazine in adsorbent synthesized from Rice Straw. Rice Straw biochar were used in the study as they exhibited significantly high adsorption capacity for pesticide. Briefly, rice straw was pyrolysis at 600 ˚C under nitrogen then biochar was modified using H_3PO_4. Biochar were characterized for their physico-chemical characteristics. The elemental analysis of biochar was performed using CHNS Analysis. The surface area and pore volume of adsorbents were estimated using the Brunauer, Emmett, and Teller (BET Analysis). In parts of Batch adsorption experiments was performed in range of room temperature and were divided to 2 parts. Part 1: Study of the kinetics adsorption was investigated using two models, namely, the pseudo-first-order and the pseudo-second-order models. Part 2: study of Adsorption isotherms. The results of CHNS analysis shows the elemental analysis that 51.795% of C, 0.763% of H, 1.634% of N and 0.654 % of S and the result of BET Analysis shows the surface area is 372.4 m^2/g and pore volume of adsorbents is 0.315 cm^3. From plotting between q_t(mg/g) and time (t) shows the dynamics of the adsorption of atrazine onto the biochar at 15 ppm of initial atrazine concentrations. In general, atrazine was adsorbed quickly around 1-30 min. This could be due to fast mass transfer of solute from the solution to surface of adsorbent s as a result of concentration gradient of the solute. The study of the kinetics adsorption showed that the pseudo-second-order model best described the adsorption which agrees with chemisorption as the rate controlling mechanism.

1 Introduction

Nowadays, an increasing worldwide consumption of fossil resources has not only led to a major shortage of energy resources but also accelerated the global warming and environmental problems. Many researchers have pointed out and made societal change towards more sustainable resources. One approach to reduce the use of fossil resource is the utilization of biomass materials, which is renewable and carbon neutral energy resource. Agricultural waste has contained a high percentage of carbon content that can be biomass materials candidate which is interested to produce energy, materials and chemicals instead of fossil resource. Almost of agricultural waste has brunt in the fields after harvest and its leads to produce emissions which can adverse to human respiratory and health.

Rice is a primary source of carbohydrates in many parts of the world, especially in Asia. Every kilogram of harvested rice gains a lot of straw wastes. Thailand is one of the biggest rice producers and the generated rice straw waste approximately 90% is abandoned. Due to a high availability, low or no cost and literal renewability, Biochar production would decrease the numbers of abundant rice straw towards the idea of "waste to wealth".

Currently, Thailand has a tendency of the use of pesticides in agricultural. 2010-2011, RASFF (Rapid Alert System for Food) of the EU had shown that atrazine, a kind of pesticide, was found in the agricultural products of Thailand about 55 times which were the highest number of detection comparing with the products from other countries. In addition, atrazine affects an environment and causes cancer and hormonal toxicity. As agriculture is the main industry in Thailand and there is a contamination of atrazine in water sources therefore the management of atrazine in water is necessary.

Atrazine molecule structure is shown in Fig. 1 and all above mentioned is reasons for our investigation of biochar production from rice straw to synthesize the adsorbent. The study of method for removing atrazine from water is also being offered.

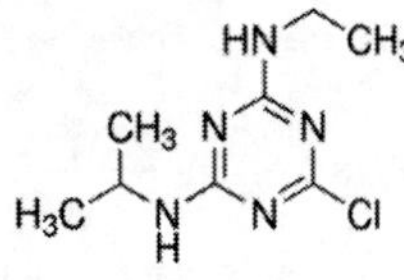

Fig. 1. Molecule structure of Atrazine.

* Corresponding author: thachanan.sa@kmitl.ac.th

2 Experimental

2.1. Biochar production procedure

Rice Straw raw materials are dried at 60 ˚C for 24 hrs. The dried rice straw was pyrolysis at 600 ˚C for 1 hr. under nitrogen gas and then ground to be a powder. Only 150 – 250 mesh size was used and heated in H_3PO_4 at 80 ˚C for 1 hr. After it cooled down for 24 hrs, biochar was filtrated, washed until neutral and dried before using.

2.2. Biochar characterization

The elemental analysis of biochar was performed using CHNS Analysis. The surface area and pore volume of adsorbents were estimated using the Brunauer, Emmett, and Teller (BET Analysis).

2.3. Adsorption experiments procedure

Atrazine solution was prepared at the desired concentration. A fixed amount of biochar was put in the atrazine solution and stirred. Each samples were took with the varied time in a range of 2 – 300 mins and filtrated. Gas Chromatography Mass Spectrometry quantified the amount of atrazine in sampling solutions.

3 Results and discussion

3.1. CHNS analysis

Elemental compositions of our synthesized adsorbent were qualified and quantified by CHNS analyser and shown in Fig. 2. Figure 2 is also shown a comparison with others synthesized adsorbents. Carbon content of our synthesized adsorbent is in a good range when compared with others works [1-3].

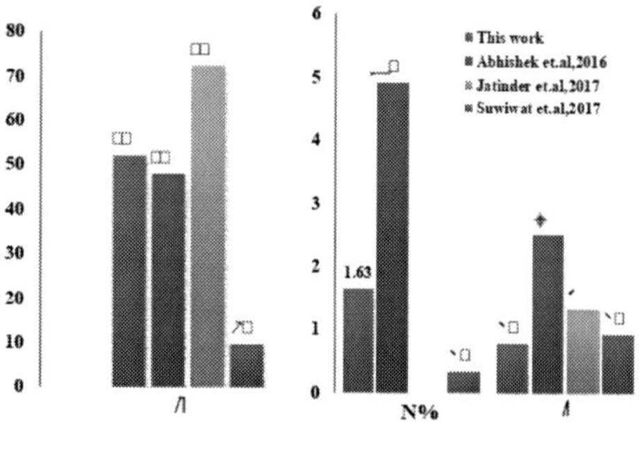

Fig. 2 Elements comparison of synthesized adsorbents.

3.2. BET analysis

BET analysis shows the surface area of 372.4 m^2/g and pore volume of 0.315 cm^3 for our synthesized adsorbent. Figure 3 is also shown a comparison of BET analysis with other adsorbents. Our synthesized adsorbent shows the highest values of surface area and pore volume when compared with other adsorbents [1-3]. It realized that our synthesized adsorbent from rice straw can be a good candidate for adsorption.

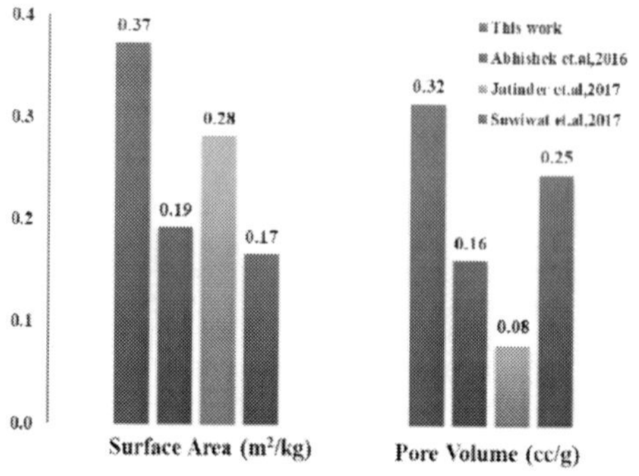

Fig. 3 Surface are and pore volume comparison of synthesized adsorbents.

3.3. Atrazine adsorption

The amount of atrazine adsorption, q_t (mg/g), at time t (hr), was calculated by:

$$q_t = \frac{(C_0 - C_t)V}{W} \tag{1}$$

Where C_0 (mg/L) is initial concentration. C_t (mg/L) is concentration at time t. V (L) is the volume of the solution and W (g) is the mass of dry adsorbent used.

Figure 4 shows the effect of contact time and initial concentration on atrazine adsorption and it can be realize that the time necessary to reach equilibrium is short. Even longer contract time shows swing line but it can be assumed that reached equilibrium rapidly.

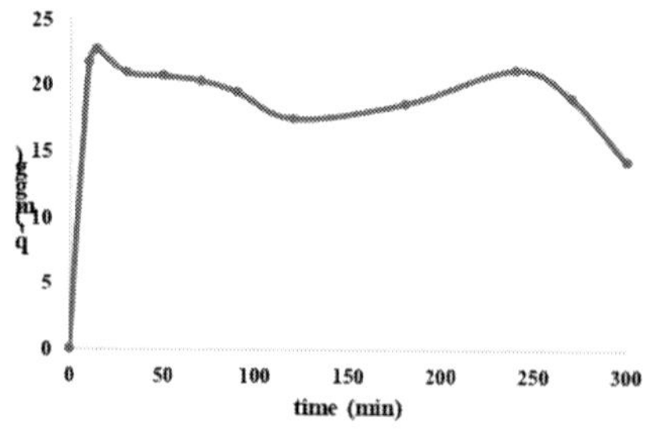

Fig. 4 Effect of contact time and initial concentration on atrazine adsorption.

3.4. Kinetics analysis

The modelling of the kinetics of atrazine adsorption was investigated by two common models, namely, the pseudo-first order model and pseudo-second-order model. We assumed Lagergren pseudo first order kinetics equation in the linear form as shown below:

$$log\,(q_e - q_t) = logq_e - \left(\frac{k_1}{2.303}\right)t \quad (2)$$

Where q_e and q_t (mg/g) are the amounts of atrazine adsorbed at equilibrium and at time *t* (hr.), respectively, and k_1 (1/hr) is the pseudo first order rate constant. If the plot was found to be linear with good correlation coefficient, indicating that Lagergren's equation is appropriate to atrazine adsorption. But the plot, which is shown in Fig. 5, is not express well by Lagergren pseudo first order kinetics equation and correlation coefficient is 0.59 which is very low.

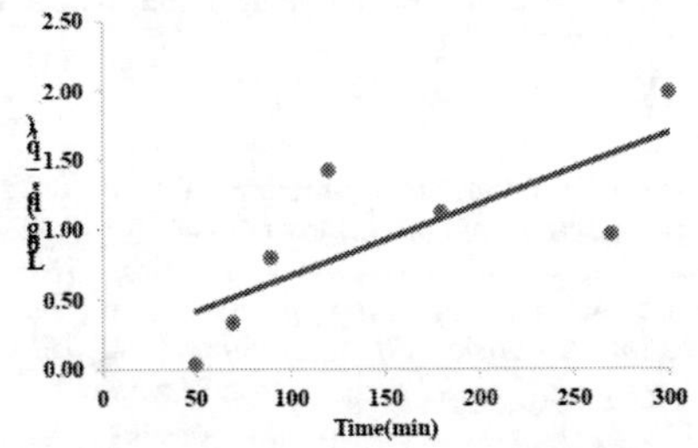

Fig. 5 Pseudo-first-order kinetic for adsorption of atrazine.

Since pseudo first order model is not appropriate for our atrazine adsorption. We also assumed pseudo second order model for this investigation. The pseudo second order kinetics equation is expressed as below:

$$\frac{t}{q_t} = \frac{1}{k_2 q_e^2} + \left(\frac{1}{q_e}\right)t \quad (3)$$

Where k_2 (g/mg h) is the rate constant of adsorption, q_e (mg/g) is the amount of atrazine adsorbed at equilibrium and q_t (mg/g) is the amount of atrazine adsorbed at time *t*.

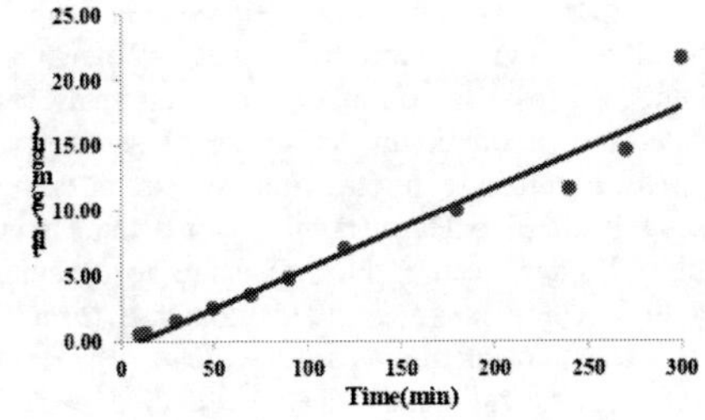

Fig. 6 Pseudo-second-order kinetic for adsorption of atrazine.

Figure 6 shows that the pseudo second order kinetics equation can express well on our atrazine adsorptions with correlation coefficient of 0.95. It can be indicated that our atrazine adsorption by using synthesized adsorbent from rice straw seem to be chemisorption.

4 Conclusion

In comparison with another 3 works [1-3] elemental analyses, %C of this work was closed to the data of Abhishek et al. [1], due probably to the similar preparation of rice straw. The result of Suwiwat et al. [3] which chemically activated the raw material by KOH before pyrolysis and it may affect carbon content of the adsorbent. Considering BET analysis, the adsorbent of this work contained the highest surface area and pore volume revealing that pyrolysis affected to pore volume. Biochar preparation by pyrolysis followed by chemical activation using H_3PO_4 were a good precursor for the adsorbent preparation.

Atrazine adsorption was expressed well by the pseudo-second-order kinetics model and agreed with chemisorption. Atrazine was adsorbed rapidly in first 30 min.

References

1. A. Mandal, N. Singh, T.J. Purakayastha, Sci. Total Environ. **577**, 376–385, (2017).
2. J. K. Ratana, M. Kaura, B. Adirajua, Materials Today Proceedings **5**, 3334–3345, (2018).
3. S. Sangon, A. J. Hunt, T. M. Attard, P. Mengchang, Y. Ngernyen, N. Supanchaiyamat, Journal of Cleaner Production **172**, 1128-1139 (2018).

Detection of arsenite-oxidizing bacteria in groundwater with low arsenic concentration in Rayong province, Thailand

Phurinat Pipattanajaroenkul[1], *Prinpida* Sonthiphand[2], *Supeerapat* Kraidech[1], *Satika* Boonkaewwan[1], and *Srilert* Chotpantarat[3, 4, 5*]

[1]International Programs in Hazardous Substance and Environmental Management, Graduate School, Chulalongkorn University. Center of Excellence on Hazardous Substance Management (HSM), Chulalongkorn University, Thailand
[2]Department of Biology, Faculty of Science, Mahidol University, Thailand
[3]Department of Geology, Faculty of Science, Chulalongkorn University, Thailand
[4]Research Program on Controls of Hazardous Contaminants in Raw Water Resources for Water Scarcity Resilience, Center of Excellence on Hazardous Substance Management (HSM), Chulalongkorn University, Thailand
[5]Research Unit Control of Emerging Micropollutants in Environment, Chulalongkorn University, Thailand

Abstract. As contamination in groundwater has become one of the global problems. It has been reported that million people around the world get adverse effects on their health by direct and indirect As-contaminated groundwater consumption. In groundwater, the most abundant As species are arsenite (As^{3+}) and arsenate (As^{5+}). Arsenite is more toxic and mobile than arsenate. Consequently, arsenite oxidation is considered as an important process in groundwater As bioremediation. It has been reported that arsenite-oxidizing bacteria play an important role in reducing As toxicity in contaminated groundwater environment. A functional gene involved in arsenite oxidation is *aioA* gene. The key objective of this study was to investigate the arsenite-oxidizing bacteria in groundwater with low arsenic concentration in Rayong province, Thailand. The results showed that the arsenite-oxidizing bacteria were detected in groundwater with low arsenic concentration. Phylogenetic analysis revealed that they were closely related to *Rhizobium* sp., *Bradyrhizobiaceae* sp., *Hydrogenophaga* sp., and *Stenotrophomonas* sp. The knowledge gained from this study will help better understand the distribution of arsenite-oxidizing bacteria in groundwater with low As concentration.

1 Introduction

Arsenic (As) is considered as toxic metalloids that widely distributes on the earth's surface. It can cause both acute and chronic effects to human health [1]. It has been reported that there are several countries around the world such as Vietnam, China, Taiwan, Mexico, Argentina, India, Bangladesh and Thailand have As contamination in groundwater problems and millions of people get effects from As by the consumption of this As-contaminated groundwater. Due to As toxicity, the World Health Organization has set the standard of As concentration in drinking water to be <10 μg/l. In groundwater, the most abundant As species are arsenite and arsenate. It is found that arsenite has more toxicity and mobility than arsenate [2]. Both geogenic and anthropogenic origins are considered as sources of As in environment. According to the geological information, As can be released to an environment by the weathering of As-bearing rocks such as granite or the oxidation and dissolution of As-bearing minerals for instance, sulfide minerals and oxide minerals. Furthermore, As can be discharged to an environment by human activities such as agriculture, industry and mining operations which can generate high As concentration. [2, 3]. There are many techniques which are used to remove As from groundwater for example, physiochemical techniques such as chemical oxidation, coagulation, reverse osmosis and ion-exchange. It has been reported that these techniques are supposed to be very expensive, inefficient, particularly when As concentration in groundwater is very low [4]. In the present, As bioremediation is considered as an alternative choice to treat As contaminated groundwater. It is found that several microorganisms can survive in high As concentration environments by relying on their cellular mechanisms [5]. In oxidizing environments which arsenate is a predominant As specie. Arsenate can be converted to arsenite by the mechanism of arsenate-reducing bacteria. Heterotrophic cytoplasmic arsenate-reducing bacteria can reduce arsenate to arsenite by using their cytoplasmic arsenate reductase (*arcC*) and this activity is considered as a detoxification process [6]. Conversely, dissimilatory arsenate reducing bacteria is considered as an arsenate respirer. These microbes enable to transform arsenate to arsenite by their

* Corresponding author: csrilert@gmail.com

respiration process and gain energy for growth [7]. It has been reported that this mechanism is driven by their respiratory arsenate reductases (*arr*) [8]. In reducing environments, arsenite is dominant. Arsenite-oxidizing bacteria have an important role in controlling this toxic arsenite [9]. Arsenite is oxidized to arsenate by mechanism of heterotrophic and chemoautotrophic arsenite-oxidizing bacteria by using their specific enzyme called respiratory arsenite oxidase (*aio/aox*) [10]. The arsenite oxidation of heterotrophic arsenite-oxidizing bacteria (HAOs) is considered as a detoxification process because these microbes unable to gain any energy from this activity to produce their cell materials [11]. While chemoautotrophic arsenite oxidizing bacteria (CAOs) are regarded as an arsenite eater. They can conserve energy for growth through their oxidation process [12]. It has been reported that Bordetella sp. strain SPB-24 and Achromobacter xyloxosidans strain SPB-31, could survive in high As contaminated garden soil by their As detoxification process [13]. It has also been reported that Herminiimonas arsenicoxydans strain UPLAs could chemoautotrophically grow in arsenite concentration up to 500 mg/l [14]. Since several arsenite oxidizing bacteria were found in various rich As concentration environments, the purpose of this study was to investigate the communities of arsenite-oxidizing bacteria in groundwater with low As concentration to better understand their inherent distributions in this natural environment.

2 Material and methods

2.1 Site description and sample collection

The groundwater sampling point (GW1) located in Muang district, Rayong province, Thailand with the coordinate of 12° 44' 2" N and 101° 9' 51" E. This area was considered as a dense residential area. The groundwater samples were collected from the groundwater well in the volume approximately 3 liters and stored on ice during transportation.

2.2 Geochemical analysis

Geochemical parameters such as pH, electrical conductivity (EC), oxidation-reduction potential (ORP), dissolved oxygen (DO) and temperature were measured on site. Total organic carbon was analyzed using a total organic carbon analyser (TOC-VCPH). Total As was analyzed using inductively coupled plasma mass spectrometry (ICP-MS) at the Central Laboratory (Thailand) Co., Ltd. Chachoengsao province.

2.3 DNA extraction

The genomic DNA was extracted by using the FastDNATM SPIN Kit for Soil and the FastPreo® instrument (MP Biomedicals, Santa Ana, CA). DNA quality was determined by using agarose gel electrophoresis, while DNA quantity was evaluated by using spectrophotometer nanodrop, respectively. Then, extracted DNA was diluted for the final concentration of 5 ng/µl for using as a DNA template in PCR reaction.

2.4 Polymerase chain reaction (PCR)

A DNA template was used to amplify an arsenite oxidase gene (*aioA* gene) by following conditions of the manufacturer protocol of BioLabs (Taq DNA Polymerase with Thermopol® Buffer: M0267S). The primers: Aox BM1-2F ND and Aox BM2-1R ND were used to amplified *aio* gene [15]. PCR profile was performed as follow: initial denaturation at 95°C for 30 seconds followed by 35 cycles of denaturation 95°C for 30 seconds, annealing 55°C for 30 seconds, extension 68°C for 30 seconds, and final extension at 68°C for 5 minutes.

2.5 DNA purification

DNA purification was performed by following NucleoSpin® Gel and PCR Clean-up [16]. The PCR product was mixed with Buffer NT1 in NucleoSpin® Gel and PCR Clean-up column within 2ml. collection tube, after that the centrifugation was done for 30 seconds at 11,000 x g. Then, 700 µl. of Buffer NT3 was added, and the centrifugation was performed again at the same condition. After that, the column was put into 1.5 ml. Microcentrifuge-tube. The 30 of Buffer NE was added, then the microcentrifuge tube was centrifuged for 1 minute at 11,000 x g. Then, incubation at room temperature was done for 1 minute. Finally, purified DNA in the microcentrifuge tube was kept for cloning.

2.6 Clone library

The pGEM®-T and pGEM®-T Easy Vector Systems' protocol was used as a ligation and transformation guideline. The ligation product was incubated overnight at 4°C. After that, transformation was conducted by using XL1-blue super competent cells. The heat shocking was performed at 42°C for 45 seconds following by incubation in shaker at 37°C for 1 hour with shaking at 250 rpm. After that, the transformed product was spread on an LB agar plate which contained 100 mg/ml. of Ampicillin, 50mg/ml. of X-gal and 1mM. of IPTG. The incubation was performed overnight at 37°C. Finally, 20-25 white colonies which contain arsenite oxidase gene were selected for sequencing.

2.7 DNA sequencing and phylogenic analysis

The cloning samples were sent to Macrogen company in South Korea for bacterial DNA sequences analysis. The obtained sequences were compared with database of Genbank. After that the neighbor-joining tree was constructed with 1,000 bootstrap tests by using MEGA 7.021 [17].

3 Result and discussion

3.1 Geochemical properties of groundwater

The geochemical properties of groundwater sample could be briefly described. The pH of the groundwater sample was 7.1, whereas the dissolved oxygen (DO) of the groundwater was unexpectedly high showing 8.90 mg/l. Total dissolved solid (TDS) and total organic carbon (TOC) were found at 306.56 mg/l and 5.10 mg/l, respectively. Furthermore, the value of oxidation-reduction potential was 182.80 mV and total As concentration in groundwater was 0.64 μg/l. From the geochemical properties, it could be described that the groundwater was relatively neutral and moderately oxidizing condition. With this condition, As could be appeared in both arsenite and arsenate species [18]. The TDS was fairly high representing the salinity of groundwater. Moreover, this elevated TDS corresponds to the location of the study area which located near the Gulf of Thailand. The high TDS might result from the influence of saltwater intrusion [19]. A very high DO might be affected from the shallow depth of groundwater level and the continuous utilization by dense population. In the part of As concentration in groundwater, it was still in the range of the standard of 10 μg/l, reflecting the high-quality management in protecting the toxic substance releasing to groundwater resources. However, the existence of As and other geochemical physiochemical parameters could be supportive factors for the arsenite-oxidizing bacteria to survive.

3.2 Detection of arsenite-oxidizing bacteria

The arsenite-oxidizing bacteria in groundwater with low As concentration was detected by specific primers targeting *aioA* gene using the polymerase chain reaction (PCR). The size of *aioA* gene PCR product is 550 base pairs which was verified using an agarose gel electrophoresis. The groundwater sample analyzed in this study showed a positive result. This result implied that arsenite-oxidizing bacteria could be found in the environment which low As concentration [20]. Then this sample was analyzed to find the diversity of arsenite-oxidizing bacterial species by performing the clone library. Five (GW1-1, GW1-2, GW1-3, GW1-4 and GW1-5) analyzed sequences were blasted against the Genbank database using the BLASTn program. The results demonstrated that all analyzed sequences showed 76 – 84 % identity to previously reported arsenite-oxidizing bacteria (Table 1).

Table.1 The identification of *aioA* gene sequences analyzed by the BLAST^n program.

Sequence ID	Bacterial species	Genbank acc. No.
GW1-1	*Stenotrophomonas* sp.	JQ739209.1
GW1-2	*Hydrogenophaga* sp.	KM884951.1
GW1-3	*Rhizobium* sp.	KT992344.1
GW1-4	*Stenotrophomonas* sp.	JQ739209.1
GW1-5	*Bradyrhizobiaceae* sp	AB974343.1

Phylogenetic analysis showed that all arsenite-oxidizing bacteria detected in this study were closely related to α, β and γ-Proteobacteria (Fig 1). The bacterial clone GW1-3 and GW1-5 were closely related to *Rhizobium* sp. and *Bradyrhizobiaceae* sp., respectively. These two bactererial species were a member of α-Proteobacteria. The bacterial clone GW1-2 was highly similar to *Hydrogenophaga* sp. which belongs to β-Proteobacteria. Furthermore, the result also demonstrated that the bacterial clone GW1-1 and GW1-4 were closely related to *Stenotrophomonas* sp. which were a member of γ-Proteobacteria (Fig. 1). It has been reported that the closely related bacterial species such as *Rhizobium* sp., *Bradyrhizobiaceae* sp., *Hydrogenophaga* sp. and *Stenotrophomonas* sp. carried the *aioA* (*aoxB*) gene which was considered as a genetic marker of arsenite-oxidizing bacteria [15]. *Rhizobium* sp. NT-26 was found on arsenopyrite rocks in gold mining where As concentration was in a range of 2,000-13,000 μg/l [21]. This bacteria showed chemolithoautotrophic growth which used arsenite as an electron donor, oxygen as an electron acceptor and carbondioxide or bicarbonate as carbon source [21, 22]. *Hydrogenophaga* sp. was detected in the shallow groundwater aquifer with various As concentrations (1-763 μg/l.) [23]. In addition, *Stenotrophomonas* sp. isolated from soil with low As concentration of 8.8 mg./kg. demonstrated the heterotrophic arsenite-oxidation by using organic substances such as glucose and maltose as a carbon source, but this bacteria cannot gain energy for its growth [24]. This information indicated that arsenite-oxidizing bacteria could be found in several environments for example, mineralized area, soil, and groundwater which various As concentrations. Furthermore, all of these bacteria for instance, *Rhizobium* sp., *Bradyrhizobiaceae* sp., *Hydrogenophaga* sp. and *Stenotrophomonas* sp. could relieve the toxicity of As by transforming to its less toxic form. The knowledge from this study will be useful for better understanding the distribution of arsnite-oxidizing bacteria in groundwater found in low As concentration. The knowledge gain from this study is the first important step which can be built on for further development of As bioremediation.

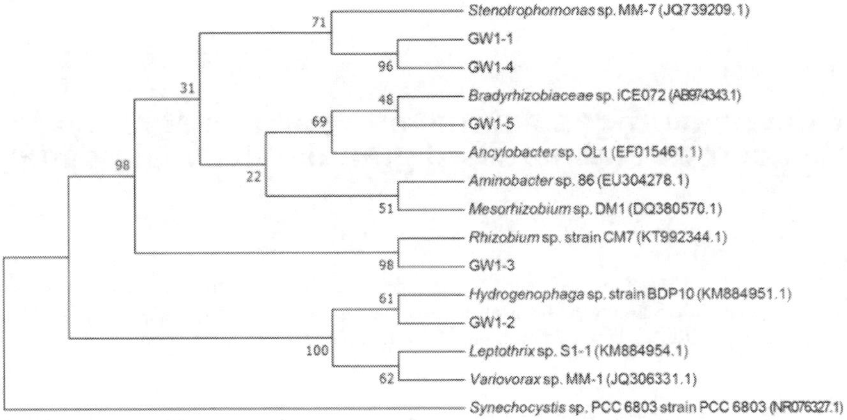

Fig. 1. *aioA* gene based phylogenetic tree constructed by using the neighbor-joining model with 1000 bootstrap value.

Acknowledgement

We are thankful to the International Postgraduate Program in Hazardous Substance and Environmental Management (HSM), Chulalongkorn University for financial support. We also would like to thank a New Researcher Grant from Mahidol University for financial support. We express our sincere thanks to the Grant for International Research Integration: Chula Research Scholar, Ratchadaphiseksomphot Endowment Fund (GCURS-59-06-79-01), the Office of Higher Education Commission (OHEC) and the S&T Postgraduate Education and Research Development Office (PERDO) for the financial support of the Research Program.

References

1. R. T. Lugtu, S. C. Choi and Y.S. Oh, J Microbiol, **47**, 6, pp. 686-692 (2009)
2. P. L. Smedley and D. G. Kinniburgh, Applied Geochemistry, **17**, 5, pp. 517-568 (2002)
3. D. Lievremont, P. N. Bertin and M. C. Lett, Biochimie, **91**, 10, pp. 1229-1 (2009)
4. A. C. Kao, Y. J. Chu, F. L. Hsu and V. H. C. Liao, Journal of Contaminant Hydrology, **155** (supplement C), pp. 1-8 (2013)
5. S. Z. Abbas, M. Riaz, N. Ramzan, M. T. Zahid, F. R. Shakoori and M. Rafatullah, Brazilian Journal of Microbiology, **45**, 4, pp. 1309-1315 (2014)
6. J. R. Lloyd and R. S. Oremland, Elements, **2**, 2, pp. 85, (2006)
7. R. S. Oremland and J. F. Stolz, Trends Microbiol, **13**, 2, pp.45-49, (2005)
8. M. Roy, A. K. Giri, S. Dutta, and P. Mukherjee, Environ Int, **75**, pp. 180-198 (2015)
9. J. F. Ferguson and J.Gavis, Water Research, **6**, 11, pp. 1259-1274 (1972)
10. S. Silver and L. T. Phung, Appl Environ Microbiol, **71**, 2, pp. 599-608 (2005)
11. L. Cai, G. Liu, C. Rensing and G. Wang, *BMC Microbiol*, **9**, 4, (2009)
12. C. Valenzuela, V. L. Campos, J. Yanez, C. A. Zaror and M. A. Mondaca, *Bull Environ Contam Toxicol*, **82**, 5, pp. 593-596, (2009)
13. S. P. Bachate, R. M. Khapare and K. M. Kodam, *Applied Microbiology and Biotechnology*, **93**, 5, pp. 2135-2145, (2012)
14. W. Weeger, D. Lièvremont, M. Perret, F. Lagarde, J. C. Hubert, M. Leroy and M. C. Lett, *Biometals*, **12**, 2, pp. 141-149 (1999)
15. M. Quéméneur, A. Cébron, P. Billard, F. Battaglia-Brunet, F. Garrido, C. Leyval and C. Joulian, *Applied and Environmental Microbiology*, **76**, 13, pp. 4566-4570 (2010)
16. B. Vogelstein and D. Gillespie, *Proceedings of the National Academy of Sciences of the United States of America*, **76**, 2, pp. 615-619 (1979)
17. S. Kumar, G. Stecher and K. Tamura, *Mol Biol Evol*, **33**, 7, pp. 1870-1874 (2016)
18. K. F. Akter, G. Owens, D. E. Davey and R Naidu, *Reviews of Environmental Contamination and Toxicology*, **184**, pp. 97-149 (2005)
19. S. Sadeg and N. Karahanoðlu, *Environmental Geology*, **40**, 9, pp. 1151-1168 (2001)
20. K. Q. Xiao, L. G. Li, L. P. Ma, S. Y. Zhang, P. Bao, T. Zhang and Y. G. Zhu, *Environmental Pollution*, **211**, pp.1-8 (2016)
21. J. M.Santini, L. I. Sly, R. D. Schnagl and J. M. Macy, *Applied and Environmental Microbiology*, **66**, 1, pp. 92-97 (2000)
22. S. Tamaki and W. T. Frankenberger, *Reviews of Environmental Contamination and Toxicology: Continuation of Residue Reviews*, **124**, pp. 79-110 (1992)
23. P. Li, Y. Wang, X. Dai, R Zhang, Z Jiang, D. Jiang, S.Wang, H. jiang, Y. Wang and H. Dong, *PLoS ONE*, **10**, 5 (2015)
24. M. M. Bahar, M. Megharaj and R. Naidu, *Biodegradation*, **23**, 6, pp. 803-812 (2012)

Behaviors of hydrogen sulfide removal using granular activated carbon and modified granular activated carbon

Songkiat Roddaeng[1], *Pongjet* Promvonge[1,*]and *Rewadee* Anuwattana[2,*]

[1]Mechanical Engineering Dept., Faculty of Engineering, King Mongkut's Institute of Technology Ladkrabang, Bangkok, Thailand
[2]Expert Center of Innovative Clean Energy and Environment, Thailand Institute of Scientific and Technological Research, 35 Mu 3 Tambon Khlong-Ha, Amphoe Khlong-Luang, Pathumthani 12120, Thailand.

Abstract. An experimental study on hydrogen sulphide (H_2S) adsorption behaviours using amine-impregnated solid adsorbent (GAC/PEI) was carried out for H_2S concentration in the range of 200 and 400 ppm. The dynamic adsorptions of GAC and modified GAC (2.0 and 1000 g/L PEI) in a fixed-bed column were investigated by determining the breakthrough curves and adsorption capacities of various adsorbents. The adsorbent exhibits an excellent adsorption capacity of 106.87 and 231.45 mgH_2S/g-adsorbent for 200 ppm and 400 ppm H_2S, respectively. The H_2S breakthrough capacity is found to relate to the surface adsorption and chemical adsorption.

1 Introduction

Biogas known as a renewable energy has been produced by anaerobic digestion of animal manure. The biogas can be converted directly into heat energy whereas the by-product can be used for agricultural fertilizer. The main composition of biogas is a mixture composed of approximately 50-70% CH_4, 30-50% CO_2, smaller amounts of NH_3 (80-100ppm), H_2S (500-1,000 ppm) and trace of hydrocarbon (<100 ppm) [1]. The merit of biogas is that it can be used directly to generate electricity and heat energy for engineering applications. However, the disadvantage for use in a gas engine is the reduction in recovery of the energy content in the biogas. Therefore, biogas purification is required by H_2S removal technique to upgrade its property to be similar to natural gas for employing as car fuel. Many technologies have been studied for the H_2S removal such as liquid absorption by amine and solid adsorbent. However, amine aqueous solution adsorption is found to unfit for low concentration of H_2S removal because of expensive cost and corrosion. Activated carbon is a highly microporous material with larger surface area. It is utilized because of their wide availability, low cost and high thermal stability. Schrier et al. [2] investigated the H_2S removal using nanoporous graphene material for gas separation. Mabayoje et al. [3] studied the improvement of H_2S adsorption on the compositions of graphene oxide with copper. Plaza et al. [4] reported CO_2 capture using activated carbon and alumina impregnated with various amine such as DETA, pentaethylenehexamine (PEHA) and polyethyleneimine (PEI). Their result indicated that DETA impregnated on alumina has a high CO_2 adsorption capacity. The carbon-base support derived from sewage sludge impregnated with PEI can increase the CO_2 adsorption capacity. Gutie'rrez et al. [5] examined the biogas desulfurization by thermally threaded sewage-sludge. Liu and Wang [6] studied the H_2S removal using 4A molecular sieve zeolite synthesis from attapulgite and showed that the sulphur adsorption capacitor was up to 10 mg/g-sorbent. Wang et al. [7] developed a nanoporous composite sorbent by loading PEI on SBA-15 and MCM-41 and indicated that a high sorption capacity of 87-mg H_2S/g-sorbent is obtained. Chen Q et al. [8] introduced PEI-loaded hierarchical porous silica monolith developed as a high stability, recyclable H_2S sorbent at low temperatures having a large H_2S breakthrough capacity of 1.27-mmol of H_2S/g-sorbent at 22°C. This could easily be regenerated at 75 °C. Boonyawan et al. [9] studied the effect of amine addition on low cost fumed silica (fSi). Their result indicated that the fSi impregnated with PEI at 40 wt.% has the highest H_2S adsorption capacity among amine addition. Wang et al. [10] used the nanoporous composite sorbent PEI/SBA-15 having a high sorption capacity and showed that the moisture had a promoting effect on the removal of H_2S from gas streams. However, few researches have focused on the development of sorbents with the ability to selectively remove H_2S.

The main objective of the present work is to prepare Granular Activated Carbon (GAC) and a PEI–modified GAC on H_2S adsorption performance. The suitable PEI loading concentration (2.0 and 1,000 g/L) on GAC using various concentration of (200 and 400 ppm) H_2S was also prepared.

*Corresponding author: Tel: +66-8-1633-2856, E-mail: kppongje@kmitl.ac.th, +66-8-6972-3560, E-mail: rewadee_a@tistr.or.th

2 Materials and Methods

2.1 Preparation and characterize for adsorbents

The commercial Granular Activated Carbon (GAC) was obtained from METRACo., Ltd. It was sieved to sizes ranging from 20–40 mesh and dried at 105°C for 6 hours before impregnation process. Polyethyleneimine (PEI) with average molecular weight 25,000 g/mol obtained from Sima-Aldrich. 15-g of PEI was dissolved in 30-g of 99.9% ethanol solution under stirring for 20 minutes and dilution for various concentrations (2.0%w/w and 1000 g/L). In the wet impregnation process, the 10-g adsorbents (GAC) were placed in a vial containing 20 mL of various PEI concentrations (2.0%w/w and 1000 g/L). The GAC with PEI loading was stirred at 120 rpm speed for 3 hours. Finally, the adsorbent was dried at 80 °C for 10 hours under vacuum conditions. The obtained GAC was impregnated with two PEI concentrations denoted as GAC/PEI-2.0 and GAC/PEI-1000. The characterization analysis of the adsorbent was conducted by scanning electron microscopy analysis for morphological structure based on JSM-6400 (JEOL, Japan) and surface area estimation was performed by N_2 adsorption.

2.2 H_2S adsorption performance testing method

The H_2S adsorption tests were carried out using a stainless-steel tube with 12-mm inner diameter and 200-mm in length at which 1 g of sorbent was packed in the middle of the column support with quartz wool. Before the test, the adsorbent was treated under a N_2 flow of 50 mL/min. Each concentration of H_2S (200 and 400 ppm) was passed through the adsorbent at a flow rate of 5 mL/sec. The H_2S outlet concentration was detected by a biogas 5000-analyser. The adsorption experiment was shown in Fig. 1. In the present work, the breakthrough concentration of H_2S was defined as 25% of the initial concentration. The test was stopped when the outlet H_2S concentration reaches the breakthrough concentration.

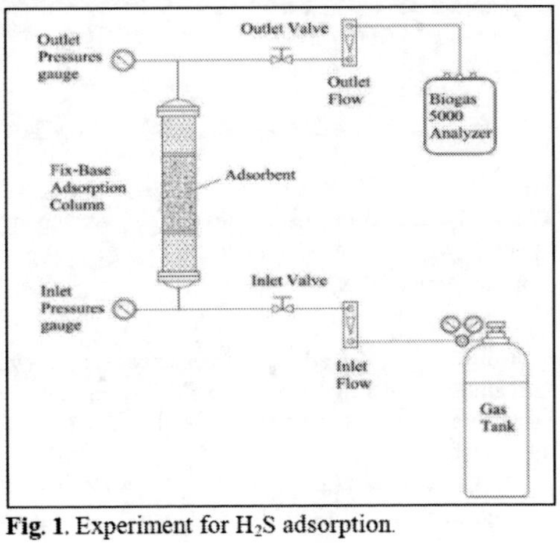

Fig. 1. Experiment for H_2S adsorption.

3 Results and Discussion

3.1 Characterization of Adsorbent

The scanning electron microscope (SEM) image of GAC and modified granular activated are shown in Fig. 2 and 3, respectively.

Fig. 2. Scanning electron microscope (SEM) images of GAC at magnification scale (x1500 magnification).

Fig. 3. Scanning electron microscope (SEM) images of GAC/PEI-2.0 at magnification scale (x1500 magnification).

Figs. 2 and 3 show the surface morphology of the amine modified adsorbents, as determined by SEM analysis. The adsorbent showed the agglomeration of the primary particles GAC to form larger particles, comparing the SEM image of the GAC/PEI-2.0 with granular activated carbon (GAC) adsorbent.

Table 1. Structure properties of granular activated carbon (GAC) before and after wet impregnation with PEI.

Sample	S_{BET} (m^2/g)	V_t (cm^3/g)	Pore size (Å)
GAC	981.6	0.5413	22.06
GAC/PEI-2.0	925.8	0.5194	22.44
GAC/PEI-1000	14.6	0.0168	45.94

Table 1 displays the surface area and the pore volume of the adsorbent. The GAC displays a specific surface area of 981.6 m^2/g and a pore volume of 0.5413 cm^3/g. The modified GAC with PEI is also displayed by the surface area and pore volume obtained. The GAC/PEI-2.0 displays a specific surface area of 925.8 m^2/g and a pore volume of 0.5194 cm^3/g and GAC/PEI-1000 displays a specific surface area of 14.6 m^2/g and a pore volume of 0.0168 cm^3/g.This result indicates that the amount of PEI on the GAC has an effect on the GAC properties. The high loading PEI impregnated on GAC will decrease the surface area and pore volume as shown in Table 1. The surface area of GAC/PEI is decreased because of the filled PEI impregnation on granular activated carbon (GAC).

3.2 H_2S Adsorption performance

The result indicates that the lower amount of PEI impregnated on GAC (GAC/PEI-2.0) provides good H_2S adsorption capacity at low initial concentration of H_2S (200 ppm). The increasing amount of PEI impregnated on GAC (1000 ppm) gives poor performance for H_2S adsorption. For high initial concentration of H_2S (400 ppm), GAC shows the best adsorbent. The results can be concluded in Table 2.

Table 2. H_2S adsorption capacity using GAC, GAC/PEI-2.0 and GAC/PEI-1000.

Sample	H_2S Adsorption Capacity (mgH_2S/g adsorbent)	
	200 ppm	400 ppm
GAC	18.36	231.45
GAC/PEI-2.0	106.87	62.22
GAC/PEI-1000	6.50	11.93

Figs. 4 and 5 show H_2S adsorption performance on various adsorbent (GAC, GAC/PEI-2.0 and GAC/PEI-1000) to find the breakthrough time at low and high H_2S concentration, the breakthrough concentration of H_2S is as 25% of the initial concentration. The breakthrough time at low H_2S concentration, GAC/PEI-2.0, yields higher performance for H_2S adsorption. The result indicates that the GAC/PEI-2.0 give a high H_2S adsorption capacity for low H_2S concentration as shown in Fig. 4 while the GAC shows high H_2S adsorption capacity for high H_2S concentration as shown in Fig. 5.

Figs. 4 and 5 also indicated that the adsorption capacity of adsorbents that strongly depends on the amount of micropores of a specific diameter presented in GAC as well as their volume as given in Table 3.

Table 3. Adsorption time of GAC, GAC/PEI-2.0, and GAC/PEI-1000)

Sample	Adsorption time (Breakthrough 25%)	
	200 ppm	400 ppm
GAC	60.25	375
GAC/PEI-2.0	354.37	102.11
GAC/PEI-1000	18.90	18.99

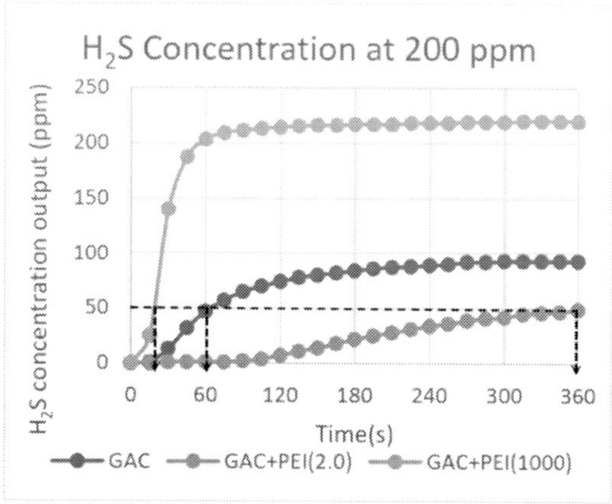

Fig. 4. H_2S adsorptionisotherm of GAC and modified GAC (GAC/PEI-2.0 and GAC/PEI-1000) using low concentration of H_2S (200 ppm)

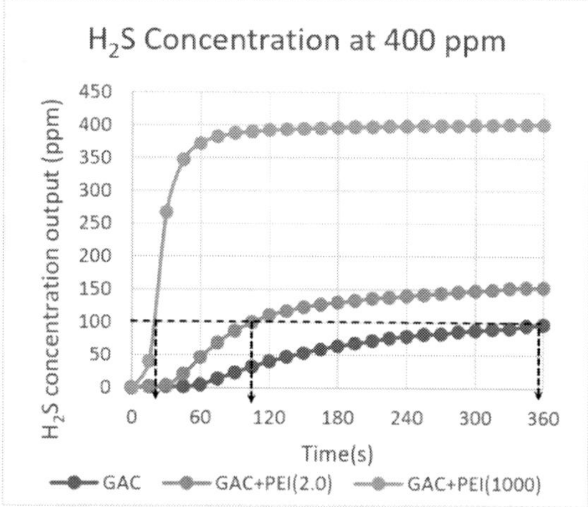

Fig. 5. H_2S adsorption isotherm of GAC and modified GAC (GAC/PEI-2.0 and GAC/PEI-1000) using high concentration of H_2S (400 ppm)

However, the H_2S adsorption capacity decreases when the amount of PEI is increased. Accordingly, the adsorption process takes place only at the surface of the adsorbent.

It is shown as Figs. 4 and 5 that the merit of PEI-impregnated on modified GAC is due to their larger pore diameter and the interconnected porous channel networks that prevent clogging of the pores and maintain the porosity of the adsorbents after PEI impregnation. The amine composition from PEI interacts strongly with low concentration of H_2S. However, it may inhibit the adsorption with high concentration of H_2S because of the lower surface area.

The reaction of H_2S on PEI impregnated activated carbon at the normal temperature could be chemical adsorption as the main reaction and physical adsorption. For chemical, H_2S reacts with amine group in PEI structure at carbon surface. The reaction of H_2S amine is shown as:

$$H_2S + 2RNH_2 - C \leftrightarrow RNH_3HS - C \quad (1)$$

$$H_2S + 2R_2NH - C \leftrightarrow (R_2NH)_2S - C \quad (2)$$

For H_2S adsorption, the two H atoms in the H_2S molecule interact with a N atom in the amine group in the first type of sorption site, whereas each of the two H atoms in the H_2S molecule interact with a N atom from one of two amine groups in the second type as shown in equation (1) and (2). [11].

4. Conclusion

PEI impregnated on GAC sorbents was developed and tested for H_2S removal. For the breakthrough concentration of H_2S defined as 25% of initial concentration, Granular Activated Carbon (GAC) is the best adsorbent for high concentration of H_2S (400 ppm) while GAC/PEI-2.0 is the best adsorbent for low concentration of H_2S (200 ppm). The chemical adsorption as the main reaction via the amine composition from PEI interacts strongly with low concentration of H_2S. However, it may inhibit the adsorption with high concentration of H_2S because of the lower surface area.

5 Acknowledgments

The financial support for this work is provided by the Thailand Institute of Scientific and Technological Research.

References

1. Q. J. Chen, Z. Wang, D. H. Long, X.J. Liu, L. Zhan, X.Y. Lian, W.M. Qiao, L.C. Ling, Ind. Eng. Chem. Res. **49**, 3152 (2010).
2. J. Schrier, ACS appl. Mater. & interface, **3**, 4451 (2011).
3. O. Mabayoje, M. Seredych, T.J. Bandosz, ACS appl. Mater. & interface, **4**, 3316 (2012)
4. M.G. Plaza, K.J. Thurecht, C. Pevida, F. Rubiera, J.J. Pis, C.E. Snape, T.C. Drage, Fuel Process Tech., **110**, 53 (2013)
5. F. J. Gutie·rrez, P. G. Aguilera, P. Ollero, Sepa. Purif. Tech., **123**, 200 (2014)
6. X.Liu, R.Wang, J.Haz. Mat., **326**, 157 (2017)
7. X.Wang, C.Song, Energy Fuels, **28**, 7742 (2017)
8. Q. Chen, F. Fan, D. Long, X. Liu, X. Liang, W. Qiao, Ind Eng Chem Res, **49**, 11408-14, (2010)
9. B. Yoosuk, T. Wongsanga, P. Prasassarakich, fuel, **168**, 47-53, (2016)
10. X. Wang, X. Ma, L. Sun, C.A. Song, Green Chem, **9**, 695–702, (2007)
11. X. Ma, X. Wang, C. Song, J. Am, Chem Soc, **131**, 5777, (2009)

Effect of parameters on the morphology and fibre diameters of edible electrospun chitosan-cellulose acetate-gelatin hybrid nanofibres

Jaruayporn Somsap[1,*], *Kobsak* Kanjanapongkul[2], and *Racha* Tepsorn[1]

[1]Department of Food Science and Technology, Faculty of Science and Technology, Thammasasat University (Rangsit Center), Phatum Thani, Thailand
[2]Department of Food Engineering, Faculty of Engineering at Kamphaeng Saen, Kasetsart University, Kamphaeng Saen Campus, Nakhon Pathom, Thailand

Abstract: Electrospinning is the favorite process to fabricate fibres with diameter in the range nanoscale through the action of electric field. In this study, 3-7% chitosan, 18.0% cellulose acetate and 30.0% gelatin solution in aqueous 80% acetic acid solution were blended at the volume ratio of 4:1:5 have been successfully electrospun. The effect of processing parameters and the concentration of the polymer solution on the morphology and diameter of electrospun were investigated. The morphology and diameter of electrospun fibres were observed by scanning electron microscope. The diameters of chitosan-cellulose acetate-gelatin nanofibres ranging from 78.94 to 421.05 nm. The results showed that the fibre diameters increased when the solution concentration and flow rate were increased, whereas the fibre diameters decreased when the applied voltage and distance between tip to collector were increased. The conditions of the solution concentration 18.8 %wt, applied voltage at 23 kV, flow rate at 11.67 μL/min and collector distance at 10 cm were selected to prepares the desirable electrospun nanofibres for applications and the further research.

Keyword: Electrospinning, electrospun nanofibre, edible electrospun

1 Introduction

Nanofibres was a nanomaterial that was tiny fibre with diameter in the range of 1-100 nanometre. It has a variety technique to prepared nanofibres such as drawing, template synthesis, phase separation, self-assembly and electrospinning [1-3]. However, electrospinning was the most attractive process to fabricate nanofibres due to that the apparatus was uncomplicate and flexible for the raw material selection. The basis apparatus of electrospinning consists of (i) high voltage power supply, (ii) syringe with a blunt ended stainless needle, (iii) syringe pump and (iv) ground collector [4,5]. In the electrospinning process, polymer solution is taken in the syringe with needle then the positive electrode which has a high voltage power supply was connected to a needle while the negative electrode was connected to a ground collector. When the high voltage was applied, the droplet shape at the end of needle will be changed from a semicircle into a conical shape as known "Taylor's cone" due to the charge - charge repulsion [5-8]. During the jet transportation cross the high voltage electric field, the whipping instability motion and stretching of fibre will be observed. The solvent will be evaporated. Thin and dry fibres on the collector are obtained [8,9]. These fibres were called "electrospun nanofibres".

Electrospun nanofibres presented the outstanding properties such as high surface area to volume or mass ratio, light weight, small inter-fibrous pore size with high porosity [5,7,10]. They were used for several applications such as filtration membranes, nanofiber composites, wound dressing, tissue scaffolding, drug delivery systems and food packaging [7,11]. Morphology and fibre diameters of electrospun depended on various parameters that including solution properties, process parameters, and ambient parameters [2,5,12]. Understanding the parameters will help to fabricate the desirable electrospun nanofibers.

Chitosan, cellulose acetate and gelatin are the natural polymers that were selected in this study due to there are edible, biocompatible, biodegradable, eco-friendly and low toxicity. Dhandayuthapani et al. [1] and Haider et al. [13] successfully fabricated electrospun nanofibre mats from chitosan-gelatin blend for applications in the biomedical field. In addition, several reports demonstrated that cellulose acetate -gelatin blend could use for electrospun nanofibre mats preparation as well [14,15]. However, as far as we have known that the availability data for the electrospun nanofibre from the blend of chitosan-cellulose acetate-gelatin hybrid was

* Corresponding author: jaruayporn12@gmail.com

never presented. The aim of this study was to investigate the effect parameter including of concentration of polymer solution, applied voltage, flow rate and distance between tip to collector on the electrospinning process. The morphologies of electrospun including diameters and diameter distribution were determined. The outcomes of this study may give the optimum condition for fabricate uniform electrospun fibres with desirable properties for applications in food sector.

2 Materials and methods

Chitosan powder (degree of deacetylation 95.04%) and gelatin powder (250 bloom) were purchased from Thai Food and Chemical Co.Ltd (Thailand). Cellulose Acetate (CA, average Mn ~ 30,000) was purchased from Aldrich Co. Ltd (USA). Glacial acetic acid (>99%) was purchased from QRec (New Zealand). Chitosan, CA and gelatin were separately dissolved in 80%acetic acid (v/v) under continuous stirring at room temperature until completely dissolved. Then, the Chitosan-Gelatin-Cellulose acetate were prepared at the volume ratio of 4:1:5, respectively. The concentration of Chitosan-CA-Gelatin in mixed polymer were 18.0, 18.8 and 19.6% wt with the constant concentration of Gelatin and CA. The morphology and diameters of the electrospun nanofibres was investigated using scanning electron microscopy (SEM JSM-7800F, JEOL, MA, USA). Statistical analyses were performed by ANOVA with 3 replicates. Differentiate analysis among the data was analyzed using Duncan at *p-value* <0.05.

Electrospinning set up and process

The mixed Chitosan-CA-Gelatin solution was loaded in 10 mL plastic syringe connected with 18 gauges metal needle. The flow rate (11.67-25.00 μL/min.) was controlled by a syringe pump (New Era NE-300, USA). A high voltage DC power supply (20-26 kV) was applied at distance of 8-12 cm. The electrospinning was performed at room temperature (27±2°C) and all electrospun nanofibres were dried overnight to remove residual solvent.

3 Results and discussion

Effect of solution concentration

The solution of Chitosan in acetic acid was varied from 8.0%, 9.0%,10.0%, 11.0% and 12.0% wt at 25kV, the distance collector was 8 cm and flow rate was 25.00 μL/min. It was found that no fibre formation and only droplets were collected as shown in Fig. 1. Average droplet's size was 1.042 - 43.235 μm. The diameter and roughness were increased by increasing Chitosan concentration. Gelatin electrospun fibre mats were obtained by the 30.0%wt gelatin at 23kV, flow rate was 11.67 μL/min. The electrospun fibre mats were only obtained by 18.0%wt CA at 25kV, flow rate was 50.00 μL/min. and the distance was 10 cm (Fig. 2A and 2B). The diameter of Gelatin fibres ranged from 384.61-2564.10 nm and average diameter was 1061.53±500.25 nm. On the other hand, Oraby et al. [16] reported that no fibres at 30.0% wt Gelatin concentration was investigated, the fibres were first appeared at 40.0% wt Gelatin (69-138 nm) and increased to when Gelatin concentration was increased.

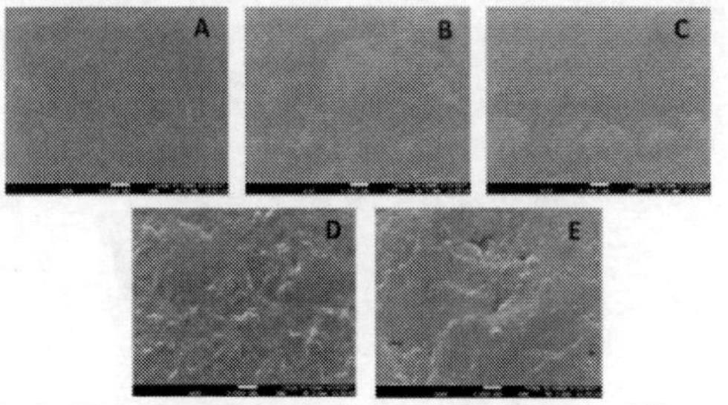

Fig 1. SEM images (100x) of electrospun drops at different Chitosan concentration (A) 8.0%, (B) 9.0%, (C) 10.0%, (D) 11.0% and (E) 12.0% wt

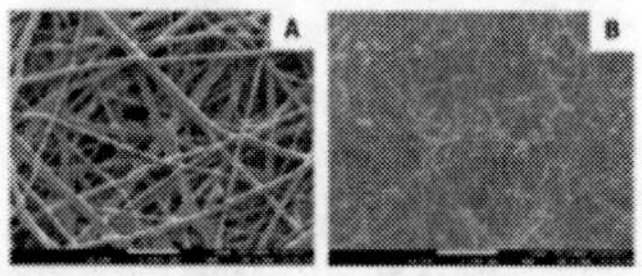

Fig. 2. SEM images of electrospun fibre mats (A) Gelatin electrospun; (B) CA electrospun (2,000x (A) and 2,500x (B))

Whereas Haider et al. [13] showed that at below 15.0%wt Gelatin solution, the fibres were not detected and the smooth fibres electrospun nanofibres was firstly detected at 15.0%wt with average diameter of 149 nm at 20 kV and the electrical field distance at 10 cm. The increasing of the solution concentration could be resulted to the viscosity properties, especially the entanglement properties of polymer. At the high concentration, above entanglement concentration, the nanoscaled fibres would be no optained. However, the flat fibres and the ribbon-type fibre were achived [12,17]. For the diameter of CA fibres ranged from 47.61-142.85 nm and the average diameter was 85.142±24.05 nm. Moreover, beads were observed on the fibres. On the contrary, Han et al. [18] have reported that CA at higher concentrations than 17.0 wt% in acetic acid could be electrospuned, long uniform nanofibres without beads were observed.

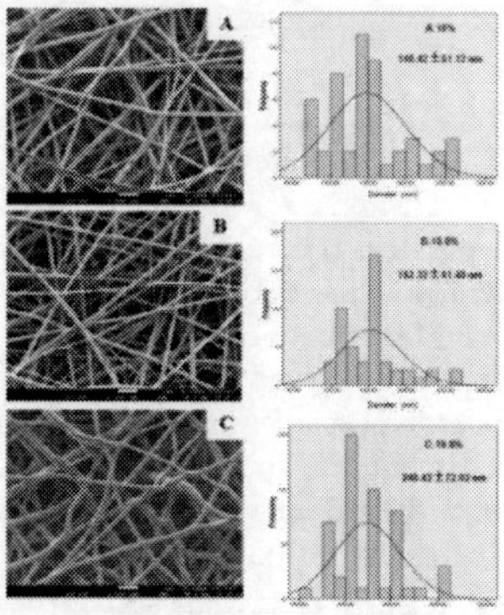

Fig. 3 SEM images (10,000x) of electrospun nanofibres and diameters distribution for the various solution (A) 18.0wt% (B) 18.8wt% and (C) 19.6wt%

Fig. 3 showed the Chitosan-CA-Gelatin nanofibres and diameter distribution obtained at concentration of 18.0, 18.8 and 19.6 %wt in 80% acetic acid. All of solution concentrations of Chitosan-CA-Gelatin could be electrospuned. The diameter of Chitosan-CA-Gelatin nanofibres ranged from 78.94-421.05 nm and the average diameter were 148.42±51.12, 152.32±41.48 and 248.42±72.02 nm, respectively. The fibres diameter were increased continuously by increasing polymer concentration from 18wt% to 19.6wt% (Fig. 4).

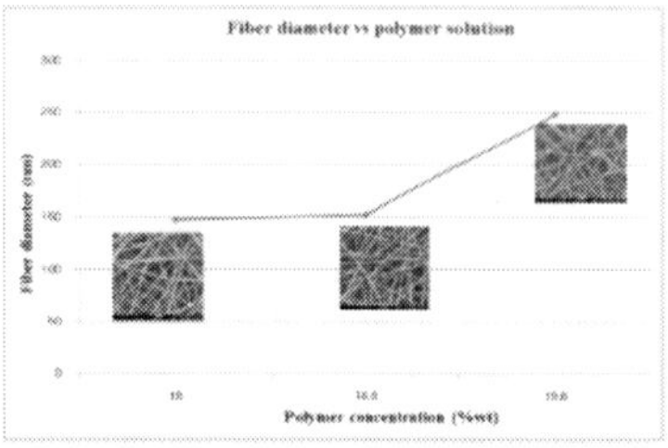

Fig. 4 The relation between the fibre diameters and polymer concentration

However, the diameters at 18.0 and 18.8wt% solution concentration were different without significant (p>0.05). The smooth, non-bead and continuously fibres were obtained. At 19.6 wt% solution concentration, the junction and sticky were observed due to the higher viscosity and surface tension of polymer jets [19]. In contrast, the diameter fibres were significant increased (p<0.05). Polymer concentration and viscosity are related factor, when one increased, the other would be increased that resulted to the size of diameter fibres. At low concentration, polymeric micro or nano-particle will be obtained, the suitable concentration, smooth, uniform and continuous nanofibres was presented [5,10,12,16].

Effect of applied voltage

Voltage involved the force in the electrical field. Applied voltage resulted on the formation of Taylor cone according to the repulsive force in an opposite direction of the surface tension. The overcome state resulted on the initiation of fine jet from the cone and flying to collector [5,6,12]. In this experiment, effect of applied voltage on fibre morphology at 20, 23 and 26 kV were investigated. The average fibres diameters at 20, 23 and 26 kV were 170.21±42.15, 152.32±41.48 and 106.95±35.39 nm respectively. Smooth, non-beads and continuous fibres were obtained from all treatments. In addition, fibres diameters were decreased while applied voltage was increased (Fig.5 and 6), it was in accordance with the previous reports [12,18]. However, the beads of PVA/alginate was also detected [7]. Moreover, the electrospun fibres with bigger beads were obtained at 45 kV. It was investigated that when applied voltage was increased, number of junctions and fibres diameters decreased while the size of beads increased. Effect of increasing voltage could be that was increased the density of charge in polymer solution, then the Taylor cone was formed, stretched and formation of smaller fibres due to the higher repulsive forces. Moreover, the more increasing of voltage, the more electric field strengths between needle and ground collector also increased. The polymer jet could fly to collector in short time that results in increasing of bead formation and/or larger fiber diameters [5,7,12,19].

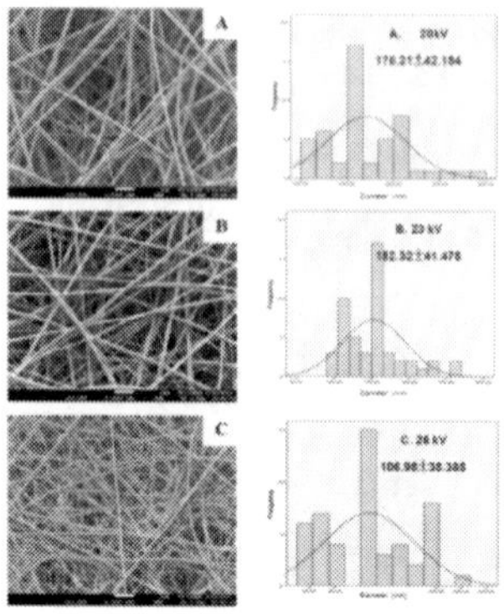

Fig. 5 SEM images (10,000x) of electrospun nanofibres and diameters distribution at different applied voltage (A) 20kV (B) 23kV and (C) 26kV

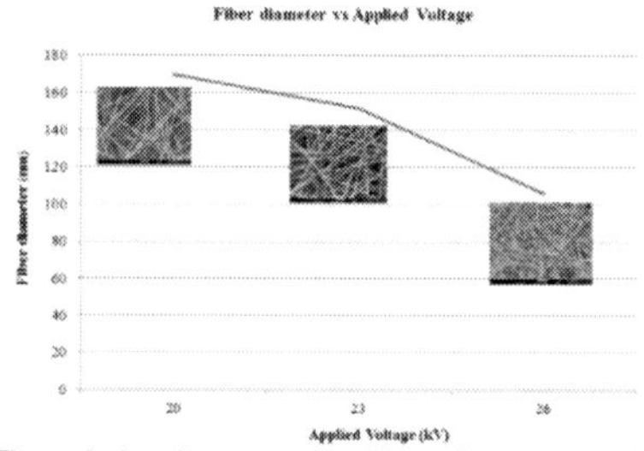

Fig.6 The relation between the fibre diameters and applied voltage

Effect of flow rate

The average fibres diameter at flow rates of 11.67 µL/min, 16.67 µL/min, and 25.00 µL/min were 152.32±41.48, 154.21±45.40 and 165.68±54.75 nm respectively. As show as Fig. 7 and 8. Smooth, non-bead and continuous fibres were obtained.

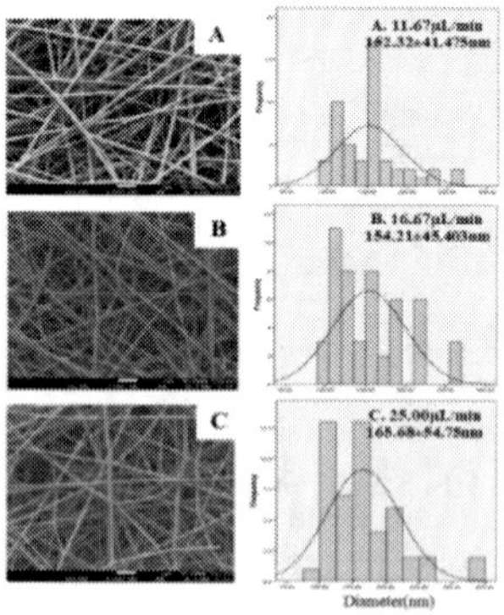

Fig. 7 SEM images (10,000x) of electrospun nanofibers and diameter distribution at the different flow rate ;(A) 11.67µL/min (B) 16.67 µL/min and (C) 25.00 µL/min

The flow rate influence to the initiate droplet shape. At high flow rates, bigger droplet was formed, and the flying ability across electrical field was decreased. The thick fibres were formed on the collector. Therefore, the lower flow rate was more recommend [8,12,21].

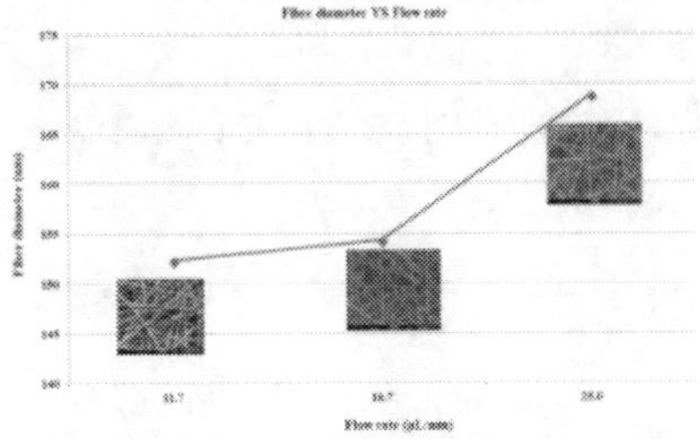

Fig. 8 Relation between the fibres diameters and flow rate

In this research, diameters at flow rates 11.67 μL/min, and 16.67 μL/min were different without significant (p>0.05).

Effect of distance

The distance showed a direct effect on the jet flying time and solvent evaporation time. At the short distance, bead formation, thick and wet fibres could be obtained, whereas at the longer, small bead, smooth and continuous fibres with fibres diameters decreasing could be obtained owing to the flying time of jet and completely evaporated [2,12,19]. The average diameter of electrospun Chitosan-CA-Gelatin fibres at distance 8 cm, 10 cm, and 12 cm were 158.32±48.84, 152.32±41.48 and 146.74±40.22 nm respectively.

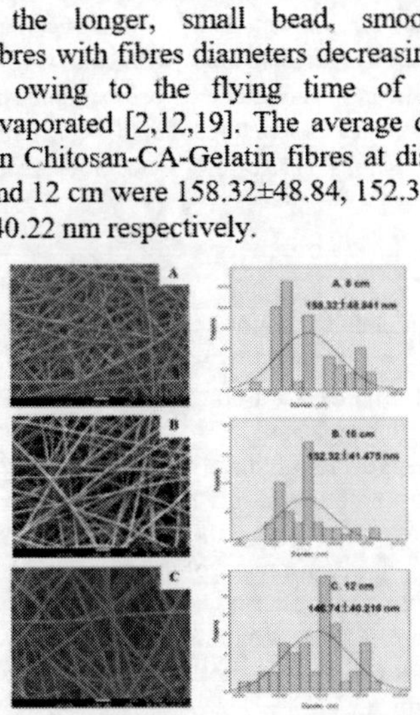

Fig.9 SEM images (10,000x) of electrospun nanofibers and diameter distribution at the different the collector distance (A) 8 cm (B) 10cm and (C) 12 cm

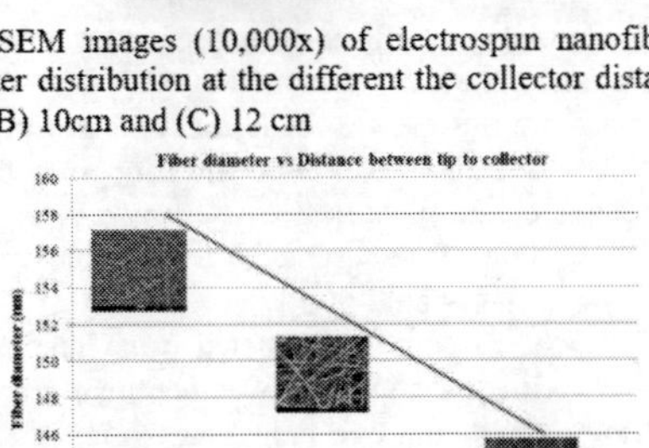

Fig. 10 Relation between the fibre diameters and collector distance

It could be indicated that fibres diameters decreases when distance was increased (Fig. 9 and 10). Increasing the distance could give time for the fluid jet to stretch and for the solvent to evaporate completely [19,20].

4 Conclusion

Formation of Chitosan-CA-Gelatin was successfully obtained from the mixed of those at the ratio of 4: 1: 5. According to the study, it can be indicated that the morphology of Chitosan-CA-Gelatin was depended on polymer concentration, applied voltage, polymer's flow rate and the distance between tip to collector. The optimum conditions were 18. 8 % wt of polymer concentration, 23 kV, 11.67 μL/ min and 10 cm in distance.

This study was provided by the Thammasat University scholarship and National Research Council of Thailand. Scanning Electron Microscope was supported by TU-ARC.

References

1. B. Dhandayuthapani, U.M. Krishnan, S. Sethuraman, J. Biomed Mater Res B Appl Biomater, **94** (2010)
2. R. Jalili, A. Hosseini, M. Morshed, Iran Polym J, **14** (2005)
3. R.Y. Tsai, S.C. Hung, J.Y. Lai, D.M.Wang , H.J. Hsieh, J Taiwan Inst Chem E, **45** (2014)
4. J. Anu Bhushani, C. Anandharamkrishnan, Trends Food Sci Technol **38**, (2014)
5. A. Rogina, Appl. Surf. Sci. **296** (2014)
6. A. Baji, Y.W. Mai, S.C. Wong, M. Abtahi, P. Chen, Compos. Sci. Technol. 70 (2010)
7. A.G. Sener, A.S. Altay, F. Altay, 7th IEEE, pp. I-324-I-328M. (2011)
8. S. Zargham, S. Bazgir, A. Tavakoli, A.S. Rashidi, R. Damerchely, J. Eng. Fiber. Fabr. **7**, 42- 49 (2012)
9. R. Erdem, I. Usta, M. Akalin, O. Atak, M. Yuksek, A. Pars, Appl. Surf. Sci. **334**, 227-230 (2014)
10. C. Santos, C.J. Silva, Z. Büttel, R. Guimarães, S.B. Pereira, P. Tamagnini, A. Zille, Carbohydr Polym. **99**, 584-592 (2014)
11. Z. Li, C. Wang, Springer Briefs in Materials. pp.15-28 (2013)
12. S. Haider, W.A. Al-Masry, N. Bukhari, M. Javid, Polym Eng Sci **50**, 1887-1893 (2010)
13. S. Khalili, S. Khorasani, N. Saadatkish, K. Khoshakhlagh, Polym Sci **58**, 399-408 (2016)
14. E. Vatankhah, M.P. Prabhakaran, G. Jin, L.G. Mobarakeh, S. Ramakrishna, J Biomater Appl. **28(6)**, 909-921 (2014)
15. D. Hussain, F. Loyal, A. Greiner, J.H. Wendorff, Polym J. **51**, 3989-3997 (2010)
16. M.A. Oraby, A.I. Waley, A.I El-Dewany, E.A. Saad, B.M. Abd El-Hady, J Appl Sci Res 9, 534-540. (2013).
17. S.O.Han, J.H. Youk, K.D. Min, Y.K. Kang, W.H. Park, Mater Lett **62**, 759-762. (2008)
18. M. Chowdhury, G. Stylios, Int J Basic Appl Sci **10** (2010)
19. X. Yuan, Y. Zhang, C. Dong, J. Sheng, Polym Int **53,** 1704-1710. (2004)
20. D. Rodoplu, M. Mutlu, J Eng Fiber Fabr 7, 118-123. (2012)

Atmometer – based irrigation scheduling system for drip – irrigated onion (*Allium cepa L.)*

Cresan Joy Villaroman[1+], *Armando* Espino Jr[2]., *Jeffrey* Lavarias[2], and *Victorino* Taylan[2]

[1] Masteral Student of College of Engineering, Central Luzon State University and Faculty of Institute of Engineering and Applied Technology, Bulacan Agricultural State College, San Ildefonso, Bulacan, Philippines [2]Department of Agricultural and Biosystems Engineering, Faculty of College of Engineering, Central Luzon State University, Science City of Muñoz, Nueva Ecija, Philippines

Abstract. This study was conducted to develop an atmometer-based irrigation scheduling system for drip-irrigated onion production. The study was conducted at San Agustin, San Jose City Nueva Ecija from November 2016 – March 2017. Three treatments composing of three replicates were considered in the research. Treatments 1 and 2 were based on the recorded atmometer readings with an irrigation interval of two days and five days respectively. Treatment 3 is a soil moisture-based irrigation scheduling with a management allowed deficit of 50 %. Calibration curved was obtained by comparing the atmometer readings with the estimated evapotranspiration using Modified Penman-Monteith equation. It was used in computing the crop water requirement for Treatments 1 and 2. The important parameters that used to answer the objective of the study such as plant height, crop yield, bulb weight, bulb diameter, water use and water productivity, were acquired during and after crop production. The statistical analysis used in the study was Analysis of Variance for Complete Randomized Design and paired T-test. Based on the result, Treatment 1 was highly useful in increasing water productivity without sacrificing the crop qualities.

1 Introduction

Bulb onion (*Allium cepa*) belongs to allium family which includes garlic and shallots. It is a popular vegetable widely grown throughout the world for its pungent bulbs and flavorful leaves. It was originated from Southeast Asia, but the wild progenitor has yet to be found [1]. The variety mostly used in the Philippines is the red "pinoy" and yellow onion. In year 2012, the total world production of dry bulb onion was 82,851,732 MT and 4,342,135 MT for shallot and green onions. The largest producer of onion includes China, India, USA, Pakistan, and Turkey, wherein China and India accounted about 47 % [2].

Water application affects the crop yield and production. Improper practice may lead to yield and quality reduction of the crop. Water management can be executed through irrigation scheduling. It does not only answer the question when to irrigate but also includes the estimation of how much water to be applied to achieve the maximum potential of crops [5]. Irrigation scheduling can be based on soil moisture and crop evapotranspiration (ET). Soil moisture - based irrigation scheduling was not applicable for large scale irrigation and requires additional work during installation.

ET-based irrigation scheme are based on empirical equation and weather data. The most common and recommended method by Food and Agriculture Organization (FAO) in estimating the crop ET was based on empirical equation which is the Modified Penman-Monteith equation. However, this method requires knowledge about the meteorological data which is indeed not common in all farmers.

The availability of the data is another problem in this method. Agrometeorological data are not always available in every local municipalities and the acquisition and installation of weather instruments in every locality was not possible because it was not economical.

An alternative tool in estimating the crop ET is through the use of atmometer or the electronic evapotranspiration gauge. This instrument was easy to install even in remote areas and the data collection does not requires complex work because it can be easily read in the sight tube. It also reduced the complexity of ET-based irrigation system.

The cost of one atmometer was also cheaper compared with the complete weather instruments. The technology could improve irrigation management while simplifying ET-based irrigation scheduling and it could give close estimate with the ET computed from weather station and is suitable for automated irrigation management in quantifying crop water needs [3]. It could also be used in area where there is no standard meteorological data available, it could be used in

+ Corresponding author: cresanjoyvv@gmail.com

irrigation scheduling of both the farm and extension level [4].

2 Materials and method

The experimental design was illustrated in Figure 1. Treatments 1 and 2 represents the two – day and five – day irrigation interval based on atmometer readings. Treatment 3 is a soil moisture – based irrigation scheme with 50 % management allowed deficit. The irrigation for the control is the farmers' practice which is flooding. Drip irrigation system was used to irrigate the experimental treatments. Buffer crops were used to lessen the error of the treatments.

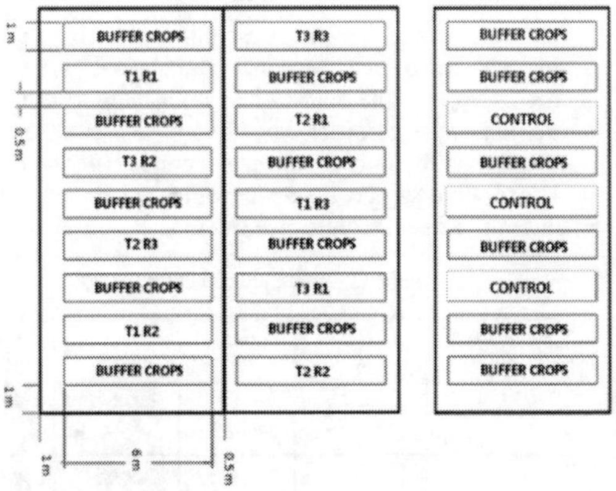

Fig. 1. Experimental layout

2.1 Calibration of atmometer

Atmometer is an instrument that estimates the crop Et_o. The data can be obtained by simply looking in the sight tube. Other model of the instrument has a data logger wherein the data can be directly downloaded to a computer.

The estimated Et_o from the weather parameters were calculated using the Modified Penman-Monteith Equation[6] (Eq. 1).

$$ET_o = \frac{0.048\Delta\,(R_n - G) + \gamma \frac{900}{T + 273} u_2 (e_s - e_a)}{\Delta + \gamma\,(1 + 0.34\,u_2)} \qquad (1)$$

The data from atmometer (Figure 2) was downloaded from the data logger attached to the instrument. It was compared and calibrated with the calculated Et_o using the correlation and regression analysis to come up with a calibration curve and equation. It was used to compute the crop irrigation requirements for Treatment 1 and 2.

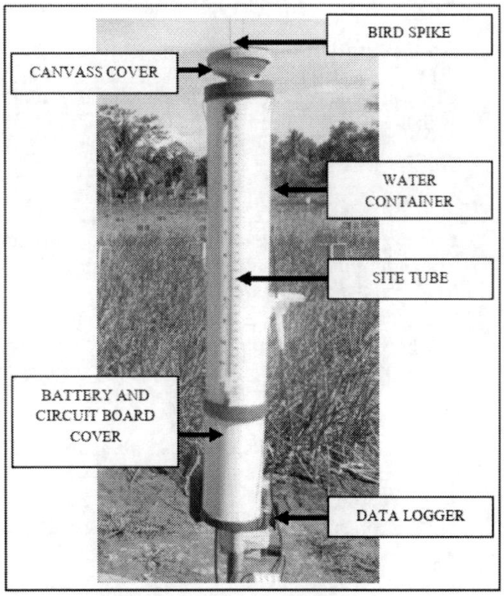

Fig. 2. Atmometer

2.2 Crop Production

The onion variety used for crop production is the "Red Pinoy" which is a standard short-day onion variety commonly used in the local municipality. The cultural practices followed in the study were seedlings preparation, land preparation, transplanting, fertilizer application, weeds and pests control, and harvesting. The study used three irrigation treatments as mentioned before.

2.3 Parameters of the study

In order to achieve the objectives of the study, the following data was gathered: atmometer reading, agrometeorological data, soil moisture behavior, plant height, total volume of water, crop yield, bulb diameter, bulb weight and other observations.

2.4 Statistical analysis

The experimental design of the study was Complete Randomized Design (CRD). It was used to compare the three experimental treatments to identify the significant difference within the treatments with a 0.05 probability levels. The experimental treatments were also compared with the control using the paired T-test to see the response of onion in different irrigation scheme.

NS – no significant difference with the control
* - significant difference

3 Results and discussion

3.1 Calibration of atmometer

The generated equation for the calibration of atmometer with the Modified Penman-Monteith equation was:

$$y = 0.688x + 1.522 \quad (2)$$

The independent variable was the atmometer readings and the dependent variable was the computed evapotranspiration (ET) from the weather data using Modified Penman-Monteith equation.

Figure 3 shows the computed ET and atmometer readings during crop production. A correlation value of 0.89 that was calculated was substantial considering that the distance between the experimental site and adjacent weather station was about 15 km.

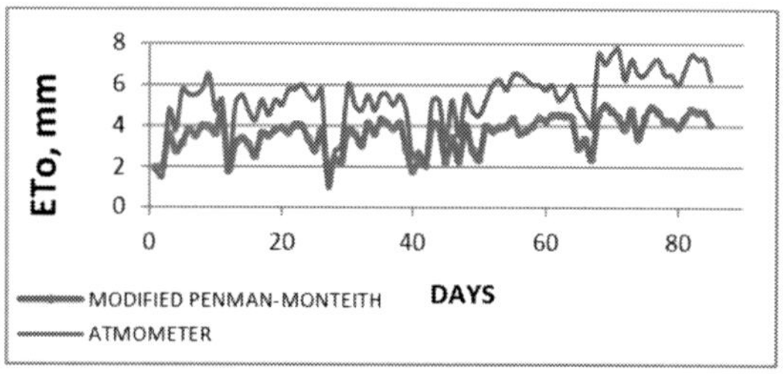

Fig 3. ETo estimated from atmometer and weather data

3.2 Plant height

The generated Weekly plant height was monitored starting from 30 days after transplanting until approximately one week before the harvest. Highest growth rate was observed in farmers' practice compared with other treatments which portrayed almost uniform growth rate. In general, all treatments showed the same growth behaviour as shown in figure 4.

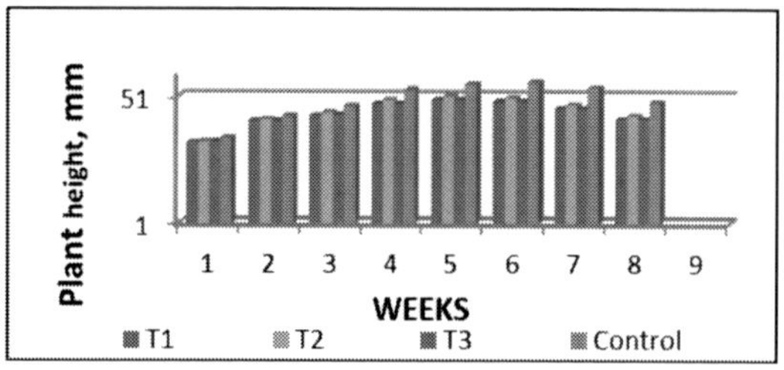

Fig 4. Weekly plant height of onion

3.3 Crop yield

The tabulation of average crop yield per hectare was represented by table 1. The crop yield data gathered in the study undergo statistical analysis and the result shows no significant difference between the treatments. However, the result of the comparison of the treatments with the control reveals that Treatment 1 does not affect the yield.

NS – no significant difference with the control
* - significant difference

Table 1. Computed yield, tons per hectare.

Treatment	R_1	R_2	R_3	MEAN
T_1	30.75	32.50	37.50	33.58 ^{NS}
T_2	29.50	26.50	30.50	28.83 *
T_3	32.50	34.00	27.00	31.17 *
Control	37.50	38.00	29.00	34.83

3.4 Bulb diameter

The onion bulbs were classified into three groups based on the size which is small, medium and large according to the Philippine National Standard (PNS). The measured average diameter of bulb onion was listed in Table 2. The comparison of treatments with the control shows that the bulb diameter in control was significantly larger than the Treatment 2 while Treatments 1 and 3 does not affect the bulb size of onion.

Table 2. Average bulb diameter, cm

Treatment	R_1	R_2	R_3	MEAN
T_1	4.31	4.35	4.65	4.44^{NS}
T_2	4.40	4.33	4.34	4.36 *
T_3	4.30	4.42	4.57	4.43^{NS}
Control	4.55	4.57	4.59	4.57

3.5 Bulb weight

The individual bulb weight of the onion was also identified in the study with the use of digital weighing scale (1 g sensitivity). The average bulb weight of the treatments was shown in Table 3. The result of paired T-test reveals that control was significantly heavier in terms of average bulb weight than Treatment 2 while the other treatments yield no significant difference with the control.

Table 3. Average bulb weight, g

Treatment	R_1	R_2	R_3	MEAN
T_1	37.60	42.23	48.63	42.82 ^{NS}
T_2	40.00	39.60	41.13	40.24*
T_3	38.53	43.30	45.90	42.58 ^{NS}
Control	50.10	45.27	50.97	48.78

3.6 Water productivity

Table 4 is the tabulation of the average water productivity. It showed that Treatments 1, 2 and 3 have considerable increase in water savings as compared with the control unit.

3.7 Uniformity distribution

The uniformity distribution in every treatment was tabulated in Table 5. Based on the statistical analysis there is no variation between the Treatment 1 and control in terms of bulb uniformity. On the other hand, Treatments 2 and 3 produced more uniform onion bulbs than control.

Table 4. Water productivity, kg/m^3

Treatment	R_1	R_2	R_3	MEAN
T_1	7.91	8.35	9.64	8.63 *
T_2	7.19	6.46	7.44	7.03 *
T_3	11.89	12.44	9.88	11.40 *
Control	1.90	1.93	1.47	1.77

Table 5. Water productivity, kg/m^3

Treatment	R_1	R_2	R_3	MEAN
T_1	90.42	86.77	87.83	88.34 ^{NS}
T_2	92.22	93.16	91.79	92.39 *
T_3	89.66	90.00	89.33	89.66 *
Control	86.17	88.26	85.76	86.73

3.8 Other observations

Pungency of the onion determines it's sharpness of smell and taste. The higher the value means the stronger the taste and smell of the onion. Based on ANOVA results, it revealed that there is no significant difference within the treatments.

The color of onion bulb was determined through physical observations. It was divided into different categories (red, pink, red with pink) depending on variation of colors. The most common color of onion was red, followed by red with pink and the least was pink.

The bulb shape was categorized into five types (round, elongated, deformed, split, and, round and elongated) by means of physical observation. The least frequent is the split and the most frequent was circle.

4 Conclusion

Based on the results of the study, the following conclusions were drawn: 1) more uniform - sized bulb can be produced under drip irrigation system, 2) the use of treatment 1 increase water savings without sacrificing the crop quality; and, 3) atmometer can be a substitute for automatic weather station in managing irrigation scheme based on crop evapotranspiration and it can be used in large scale irrigation system.

This study will not be possible without the financial support from the Department of Science and Technology – Science Education Institute (DOST – SEI), through the Engineering Research and Development for Technology (ERDT) scholarship grant.

5 References

1. K. Adam, Organic allium prod., Natl sustainable agric. info. service (2006)
2. Department of Agriculture Luzon A Cluster, Val. chain analysis and competitiveness strategy: Bulb onion, Phil. rural devt proj. (2014)
3. S. Irmak, J.O. Payero, D.L. Martin, Using modified atmometers (ET gage) for irri. mngt. (2005)
4. V. Magluilo, R. d'Andria, G. Rana, The modified atmometer: a useful tool to estimate Eto in mediterranean environment, (2002)
5. E. Martin, Methods of measuring for irri. scheduling-when, **30**, (2009)
6. L. Zotarelli, M. Dukes, C. Romero, K. Migliaccio, K. Morgan, Step by step calc. of the Penman-Monteith Evapotranspiration (FAO – 56 Method), (2015)

NS – no significant difference with the control
* - significant difference

Effect of hot air assisted infrared drying on drying characteristics and quality of rice bran pellets

Sahachart Sansak and *Jiraporn* Sripinyowanich Jongyingcharoen*

Department of Agricultural Engineering, Faculty of Engineering, King Mongkut's Institute of Technology Ladkrabang, Bangkok, Thailand

Abstract. The objective of this study was to determine drying characteristics and quality of rice bran pellet subjected to hot air assisted infrared (HA-IR) drying at different levels of infrared (IR) intensity (750 – 3750 W/m^2) and air temperature (40 – 80°C). The rice bran pellets were dried from 0.18 to 0.08 g water/g dry matter. The maximum drying rate (DR) and drying time were in the ranges of 0.0030 – 0.0165 g water/g dry matter·min and 8 – 54 min, respectively. Higher IR intensity and air temperature resulted in greater maximum DR and shorter drying time. The same trend was also observed for the effective moisture diffusivity (D_{eff}) values which were between 0.3103×10^{-7} and 2.7603×10^{-7} m^2/s. As compared to the commercial reference sample of dried rice bran pellets, the products of this study had higher oil content and lower FFA content. The oil content was affected by IR intensity only while the FFA content was affected both by IR intensity and air temperature. HA-IR drying could improve drying characteristics of the rice bran pellets and produce better quality of the dried products.

1 Introduction

Rice bran is a by-product obtained during the process of rice milling. It is 8 – 10% of the whole grain weight and generally used for rice bran oil production. In the industrial level of rice bran oil production, the rice bran is subjected to steam cooking and extruding to be rice bran pellets. During this step, enzymatic degradation is delayed and porous structure of the rice bran is formed. However, the pellets are high in moisture content, at risk of lipid oxidation, and unsuitable for oil extraction. Typically, the pellets must be dried prior to oil extraction. Conventional drying method used for rice bran pellets is hot air drying, which takes long time and needs high temperature.

This research was focused on investigating an alternative drying technique to reduce moisture from the rice bran pellets. IR drying has been reviewed as a potential method for drying various agricultural products. Penetration of IR radiation provides more uniform heating than hot air as the radiation impinges and penetrates on material and then its energy is converted into heat [1]. In the field of paddy drying and rice bran stabilization, many researchers have reported a promising potential of IR heating in inactivating the lipase enzymes in rice bran [1, 2]. The lipases are the main cause of the hydrolysis of neutral bran oil to free fatty acids (FFA), leading to the development of hydrolytic rancidity [3]. Application of HA-IR drying is reported to be advantage as it had best synergistic effect and saved drying time as compared to the hot air or IR drying alone [4, 5].

In the prior studies, advantages of IR heating have been clearly established for treatment of rice bran. However, the research on drying of rice bran pellets is lacked. Therefore, the objective of this study was to determine drying characteristics and quality of rice bran pellet undergoing HA-IR drying. Effects of IR intensity (750 – 3750 W/m^2) and hot air temperature (40 – 80°C) on rice bran pellet drying were also investigated.

2 Materials and methods

2.1 Sample preparation

The rice bran pellets, prepared by steam cooking and extruding, were provided by CEO AGRI FOOD Co. Ltd., Singburi, Thailand. Its moisture content was 0.18 g water/g dry matter. It was packed in polypropylene zip-locked bags and kept in a refrigerator at 4 – 5°C prior to use. The sample was left at room temperature before running the experiments.

2.2 Architecture of the hot air assisted infrared dryer

The HA-IR dryer is presented in Fig. 1. A double-walled stainless steel drying chamber was cylindrical with a diameter and length of 30 and 60 cm. Two infrared heaters of 800 W each were installed inside the chamber at the top. The heaters generated a surface temperature of 750°C with the peak wavelength of 3.1 mm. An aluminum cover was placed over the heaters as an infrared reflector. The perforated stainless steel drying tray (0.20×0.40×0.05 m^3) was fitted in the chamber with the distance from the heaters at 20 cm. Fin heaters of 4

* Corresponding author: jiraporn.jo@kmitl.ac.th

kW and a blower of 0.5 HP were used for producing hot air. The hot air temperature was measured by thermocouples (type K, Lega model AK-01, Lega Coperation Co., Ltd., Bangkok, Thailand) and controlled by PID (Proportional–Integral–Derivative Controller) with an accuracy of ±1°C.

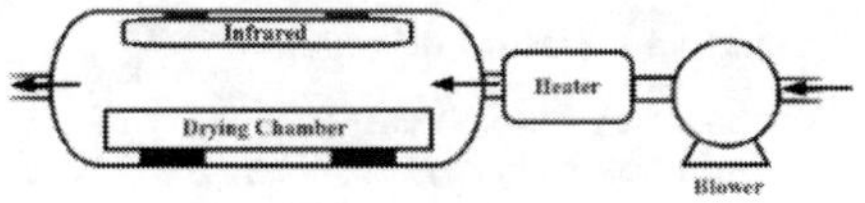

Fig. 1 A HA-IR dryer

2.3 Experimental plan

Full factorial experiment in complete randomized design was conducted for HA-IR drying of the rice bran pellets with the independent variables of infrared intensity (750, 2250, and 3750 W/m^2) and hot air temperatures (40, 60, 80°C). From the preliminary experiment of this study, the levels of infrared intensity and hot air temperature higher than 3750 W/m^2 and 80°C caused the rice bran pellets to be burnt. Air velocity was fixed at 1 m/s. The rice bran pellets were placed in single layer on the drying tray and dried to the final moisture content of 0.08 g water/g dry matter. During drying, the sample was weighed at the predetermined intervals to measure its moisture content. Drying characteristics and D_{eff} were determined as shown in 2.4 and 2.5. Quality of the dried product was also evaluated as shown in 2.6. Reference sample, the hot air dried rice bran pellets provided by CEO AGRI FOOD Co. Ltd., Singburi, Thailand, was also used for comparison purpose. The reference was the rice bran pellets dried using the commercial scale hot air dryer at the temperature of 140°C for 30 min, which is the typical condition used by the company.

2.4 Drying characteristic determination

Moisture content was determined in a hot air oven at 105°C for 24 h [6]. The equation for calculation of the moisture content in dry basis is as follows:

$$M = \frac{W_w - W_d}{W_d} \quad (1)$$

where M, W_w, and W_d are moisture content (g water/g dry matter), wet weight of the material (g), and dry weight of the material (g), respectively.

Moisture ratio (MR) and DR were calculated as below:

$$MR = \frac{M_t - M_e}{M_i - M_e} \quad (2)$$

$$DR = \frac{M_{t+dt} - M_t}{dt} \quad (3)$$

where M_i, M_t, M_e, and M_{t+dt} are moisture content (g water/g dry matter) at initial, specific time, equilibrium, and t+dt, respectively; and t is drying time (min). M_e can be assumed to be zero when drying at high temperature and for radiation drying [7].

2.5 Calculation of effective moisture diffusivity

To present the mechanism of moisture movement during drying, D_{eff} was calculated. The diffusion mechanism can be explained by Fick's second law based on shape of the material. Rice bran pellets are finite cylindrical. The general series of Fick's second law for a finite cylindrical object is given in Eq. (4) [8]. The first term equation was used for the solution of D_{eff}.

$$MR = \frac{8}{\pi^2} \sum_{m=1}^{\infty} \frac{4}{\lambda_m^2} \exp\left(\frac{\lambda_m^2 D_{eff} t}{r_0^2}\right) \sum_{n=1}^{\infty} \frac{1}{(2n-1)^2} exp\left(\frac{-\pi^2 (2n+1)^2 D_{eff} t}{4l}\right) \quad (4)$$

where D_{eff} is the effective diffusivity (m^2/s); λ_m is the root of the Bessel function of the first kind of zero order; R0 is initial radius of the cylinder (m); and l is half length of the cylinder (m).

2.6 Quailty evaluation

2.6.1 Oil content

Oil content was determined using the solvent extraction, AACC method 30-10 [9]. *n*-hexane was used as the extraction solvent and after the extraction process, was removed using a rotary evaporator. The oil content was expressed as g oil/g dry matter.

2.6.2 Free fatty acid

FFA content was determined by AOCS official method Ca 5a-40 [10]. The rice bran oil extracted from 2.6.1 was dissolved in a mixture of ethanol and diethyl ether and titrated with a KOH/ethanol solution. The FFA content was calculated as oleic acid equivalent.

2.7 Statistical analysis

The experiments were triplicated. Mean and standard deviation were used to present the experimental results. Two-way ANOVA was used to analyse the data statistically. Duncan's multiple range test was used to determine the significance of differences between treatments at 95% confident interval.

3 Results and discussion

3.1 Drying characteristics

Representative drying curves and DR curves of HA-IR drying of rice bran pellets are shown in Figs. 2-5. Effects of air temperature and infrared intensity were considerable and only two DR periods i.e. the heating up and falling rate period were observed. Increased values of these two parameters resulted in enhanced DR and shortened drying time. As presented in Table 1, the highest maximum DR and the shortest drying time were obtained when using the highest IR intensity and air temperature of 3750 W/m^2 and 80°C, respectively. It was also observed that the varied IR intensity used in this study had more crucial effect than the varied air temperature. The greater effect of the IR intensity over the air temperature was also observed by Ponkham [5] for HA-IR drying of ring shaped-pineapple at 1000-5000 W/m^2 and 40-60°C. Particularly, when increasing the level of IR intensity from 2250 to 3750 W/m^2, the DR dramatically increased by about 0.009 g water/g dry matter·min at each constant level of the air temperature.

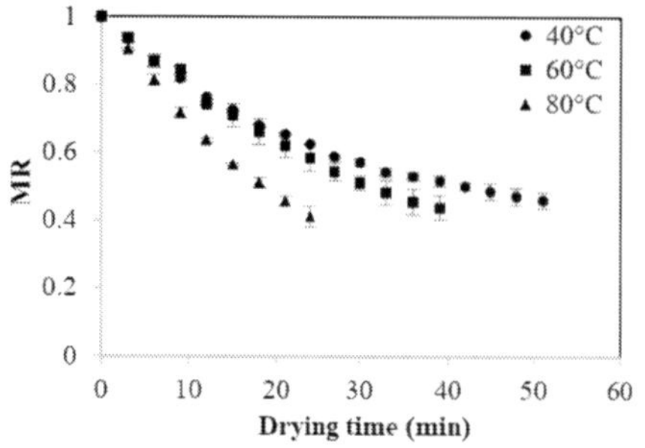

Fig. 2 Drying curves of HA-IR drying of the rice bran pellets at the IR intensity of 750 W/m².

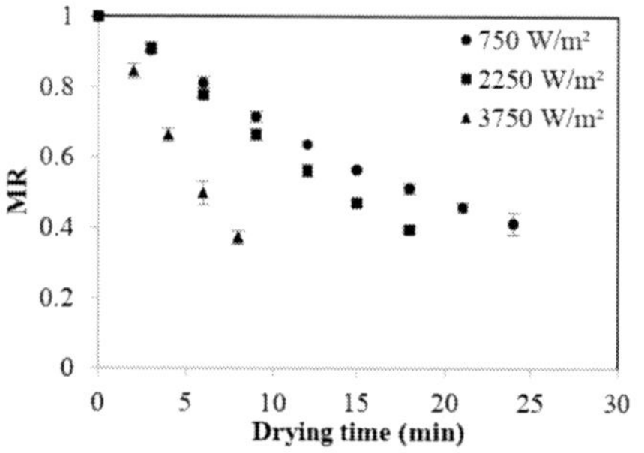

Fig. 3 Drying curves of HA-IR drying of the rice bran pellets at the air temperature of 80°C.

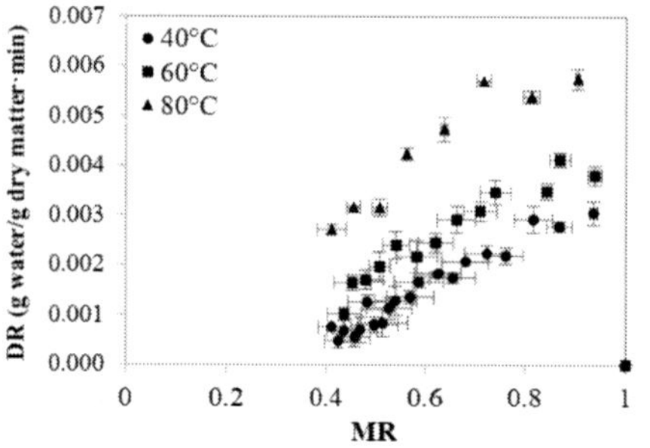

Fig. 4 DR curves of HA-IR drying of the rice bran pellets at the IR intensity of 750 W/m².

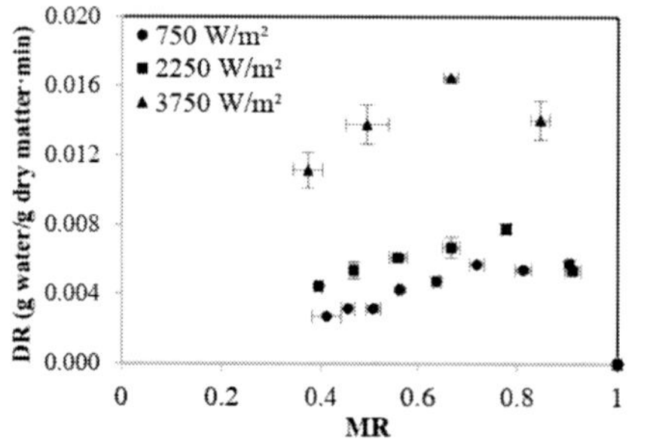

Fig. 5 DR curves of HA-IR drying of the rice bran pellets at the air temperature of 80°C.

Table 1 Maximum DR (g water/g dry matter·min) and drying time (min) of HA-IR drying of rice bran pellets.

IR intensity (W/m²)	Air temperature (°C) 40	60	80
750	0.0030 (54)	0.0041 (39)	0.0057 (24)
2250	0.0056 (24)	0.0061 (21)	0.0078 (18)
3750	0.0133 (10)	0.0156 (8)	0.0165 (8)

Drying time is presented in parenthesis.

3.2 Effective moisture diffusivity

D_{eff} presents the moisture transport property of rice bran pellets during drying in a HA-IR dryer. The effects of air temperature and infrared intensity were also obvious as the D_{eff} increased with increasing air temperature and infrared intensity (Table 2). It was notable that the D_{eff} obtained in this study were much greater than the typical range of D_{eff} of agricultural materials (10^{-11} – 10^{-9} m²/s) [11]. This could be due to the porous structure of the rice bran pellets and the drying technique used in this study which improved the moisture transport property of the material.

Table 2 $D_{eff} \times 10^{-7}$ (m²/s) of HA-IR drying of rice bran pellets.

IR intensity (W/m²)	Air temperature (°C) 40	60	80
750	0.3103	0.5289	0.9203
2250	0.9240	0.9244	1.2293
3750	1.9094	2.0146	2.7603

3.3 Quailty evaluation

The statistical results show that there is no interaction effect between air temperature and IR intensity on oil content and the significant effect on this quality was provided only by the IR intensity as shown in Table 3 (p<0.05). The highest IR intensity resulted in the significant lowest value of oil content. This result conforms to that reported by Trisnawati et al. [12], hot fat due to drying caused reduction of the oil content as the fat components may split into volatile products. However, as compared to the reference, all of the products of this study had higher oil content (Fig. 6).

On the other hand, the interaction effect between air temperature and IR intensity was found on FFA content of the HA-IR dried rice bran pellets (p<0.05). At the fixed levels of IR intensity, increasing air temperature caused reduced formation of FFA in the materials, except at the lowest IR intensity of 750 W/m². When keeping the air temperature constant, higher IR intensity seemed to provide significant lower value of FFA, except at the lowest air temperature of 40°C. As a result, the lowest FFA content was given by HA-IR drying of the rice bran pellets at the highest IR intensity and air temperature of 3750 W/m² and 80°C, respectively. It should be noted that all of the products of this study had lower FFA content than the reference. HA-IR could improve the quality of the dried rice bran pellets.

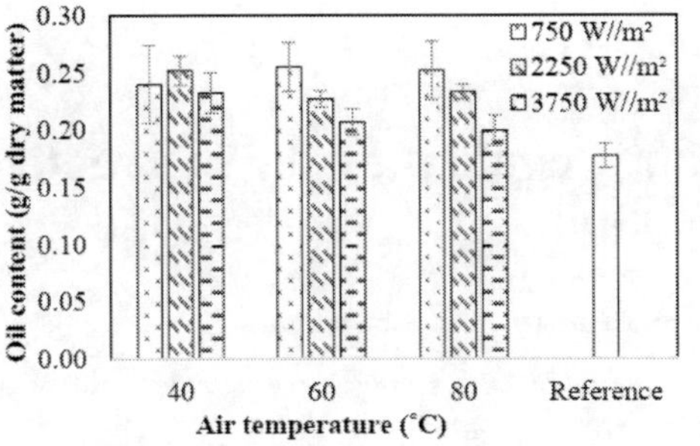

Fig. 6 Oil content of the dried rice bran pellets as compared to the reference.

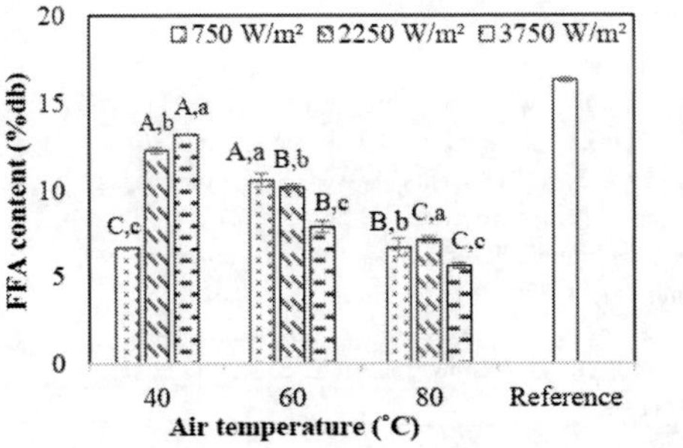

Fig. 7 FFA content of the dried rice bran pellets as compared to the reference.
Uppercases represent the main effect of air temperature at the same IR intensity.
Lowercases represent the main effect of IR intensity at the same air temperature.

Table 3 Oil content of the dried rice bran pellets.

IR intensity (W/m^2)	Oil content (g/g dry matter)
750	0.249±0.022a
2250	0.238±0.014a
3750	0.213±0.019b

4 Conclusions

Drying characteristics and quality of rice bran pellets undergoing HA-IR drying were determined in this study. IR intensity and air temperature had considerable effects on these parameters. The higher the IR intensity and air temperature were applied, the greater the maximum DR and D_{eff} and the shorter drying time were obtained. The greatest maximum DR of 0.0165 g water/g dry matter and D_{eff} of 2.7603×10^{-7} m^2/s and the shortest drying time of 8 min were provided by HA-IR of the rice bran pellets at the highest IR intensity and air temperature of 3750 W/m^2 and 80°C. This drying condition also caused the lowest formation of FFA content although the highest IR intensity produced the dried rice bran pellets with the lowest oil content.

References

1. T. Wang, R. Khir, Z. Pan, Q. Yuan, LWT-Food Science and Technology **78**, 281-288 (2017)
2. N. Yilmaz, Food Chemistry **190**, 179-185 (2016)
3. S.M. K, H.J. Chung, S.T. Lim, Journal of Cereal Science **60**, 243-248 (2014)
4. D.G.P. Kumar, H.U. Hebbar, M.N. Ramesh, LWT **39**, 700-705 (2006)
5. K. Ponkham, N. Meeso, S. Soponronnarit, S. Siriamornpun, Food and Bioproducts Processing **90**, 155-164 (2012)
6. AOAC, *Official methods of analysis* (Association of Official Agricultural Chemists, Washington, DC., 1995)
7. C. Contreras, M.E. Martin-Esparza, A. Chiralt, N. Martinez-Navarrete, Journal of Food Engineering **88**, 55-64 (2008)
8. A. Ayala-Aponte, L. Serna-Cock, J. Libreros-Triana, C. Prieto, K.D. Scala, DYNA **81**, 145-151 (2014)
9. AACC, *Approved methods of American Association of Cereal Chemists* (The Am. Assoc. Cereal Chem. Inc., St. Paul, MN., 2000)
10. AOCS, *Official methods and recommended practices of the American Oil Chemists Society* (American Oil Chemists Society, Champaign, 2004)
11. J. Sripinyowanich, A. Noomhorm, Drying Technology **29**, 735-748 (2011)
12. W. Trisnawati, K. Suter, K. Suastika, D.N.K. Putra, Food Science and Quality Management **31**, 28-36 (2014)

Drying characteristics and quality evaluation in convective drying of *Cissus quadrangularis* Linn.

Setthawat Thanimkarn[1], *Ekkapong* Cheevitsopon[2], *Jiraporn* Sripinyowanich Jongyingcharoen[1,*]

[1]Department of Agricultural Engineering, Faculty of Engineering, King Mongkut's Institute of Technology Ladkrabang, Bangkok, Thailand

[2]Department of Food Engineering, Faculty of Engineering, King Mongkut's Institute of Technology Ladkrabang, Bangkok, Thailand

Abstract. This study aimed to investigate the effect of drying temperature (40, 60, 80, and 100°C) on drying characteristics of *Cissus quadrangularis* Linn. (CQ) undergoing convective drying. Physical properties and phytochemicals of the dried CQ were also evaluated. CQ with the thickness of 5 mm was dried from about 10 to 0.1 g water/g dry matter. The results showed that increasing drying temperature increased drying rate (DR) and effective moisture diffusivity (D_{eff}) and consequently decreased drying time. The drying time, maximum DR, and D_{eff} were in the ranges of 85-1920 min, 0.0059-0.0248 g water/g dry matter·min, and $0.7302\text{-}9.1281\times10^{-9}$ m^2/s, respectively. Lower drying temperature could preserve quality of the dried CQ. Decreasing drying temperature resulted in greener and lower bulk density and shrinkage. The greatest total phenolic content (TPC) and quercetin content were obtained by drying the CQ at 60°C.

1 Introduction

CQ, a conventional medicinal plant, is used for treatment of various conditions such as hemorrhoid and irregular menstruation [1]. Normally, fresh CQ cannot be consumed as it contains calcium oxalate which leads to throat irritation. As same as most herbal plants, it must be dried, ground and capsulated for consumption [2]. Convective drying has been the most common way to produce the dried CQ.

Drying temperature is a crucial parameter required for dryer operation. This parameter can affect the moisture transport phenomena and stability of phytochemicals in material. Evaluation of the moisture diffusion mechanism and quality degradation as affected by different drying temperatures can help making decision in drying process design of convective drying of CQ and hence optimizing quality of the dried CQ.

However, little information is available in the literature on drying characteristics and product quality of CQ undergoing convective drying. Therefore, the aim of this study was to investigate the effect of drying temperatures on drying characteristics and moisture transport property of CQ during drying in a convective dryer. Quality of the dried CQ (greenness, shrinkage, bulk density, total phenolic content (TPC), and quercetin content) affected by drying temperature was evaluated.

2 Materials and methods

2.1 Sample and sample preparation

Fresh CQ was purchased from a herbal farm in Prachin Buri Province, Thailand. It was washed and cut at the thickness of 5 mm, of which the top and bottom portions were excluded. The initial moisture content was 10.10±0.15 g water/g dry matter.

2.2 Architecture of experimental apparatus

The convective drying system was designed and fabricated at the Agricultural Engineering Workshop, King Mongkut's Institute of Technology Ladkrabang, Thailand. A schematic diagram of the overall experimental setup is illustrated in Fig. 1. The dryer consisted of three main components, including heaters, a drying chamber, and a control system. A double-walled stainless steel drying chamber was cylindrical with a diameter and length of 30 and 60 cm. The perforated stainless steel drying tray ($0.20\times0.40\times0.05$ m^3) was fitted in the chamber. Fin heaters of 4 kW and a blower of 0.5 HP were used for producing hot air. The hot air temperature was measured by thermocouples (type K, Lega model AK-01, Lega Coperation Co., Ltd., Bangkok, Thailand) and controlled by PID (Proportional–Integral–Derivative Controller) with an accuracy of ±1°C.

* Corresponding author: jiraporn.jo@kmitl.ac.th

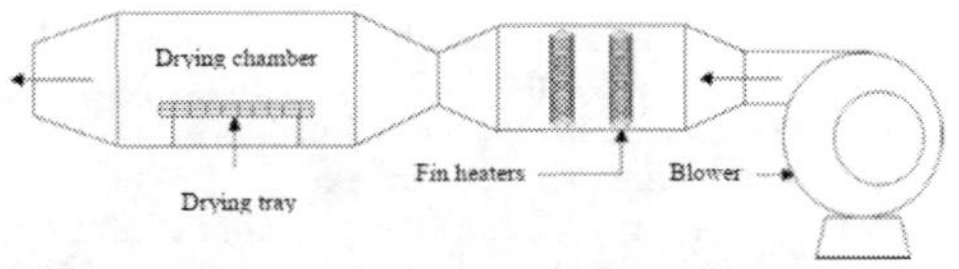

Fig. 1 Schematic view of the convective dryer.

2.3 Experimental Procedure

The experiments of convective drying of CQ were performed at the temperatures of 40, 60, 80, and 100°C. A fixed air velocity of 1 m/s was used. CQ (260 g) was placed on the drying tray in single layer. During the process of drying, the sample was weighed at each predetermined interval for determination of drying characteristics and D_{eff} as shown in 2.4 and 2.5. Drying was performed until the moisture content of QC reached 0.10 g water/g dry matter. Quality of the dried CQ was evaluated as listed in 2.6.

2.4 Drying characteristic determination

Moisture content was determined in a hot air oven at 105°C for 24 h [3]. The equation for calculation of the moisture content in dry basis is as follows:

$$M = \frac{W_w - W_d}{W_d} \quad (1)$$

where M, W_w, and W_d are moisture content (g water/g dry matter), wet weight of the material (g), and dry weight of the material (g), respectively.

Moisture ratio (MR) and DR were calculated as below:

$$MR = \frac{M_t - M_e}{M_i - M_e} \quad (2)$$

$$DR = \frac{M_{t+dt} - M_t}{dt} \quad (3)$$

where M_i, M_t, M_e, and M_{t+dt} are moisture content (g water/g dry matter) at initial, specific time, equilibrium, and t+dt, respectively; and t is drying time (min).

2.5 Calculation of effective moisture diffusivity

To evaluate the moisture transport property, that is, diffusion mechanism of CQ during drying in a convective dryer, values of D_{eff} at different drying temperatures were determined. The diffusion mechanism can be explained by the Fick's second law. The solution of this equation developed by Crank [4], and the form of Eq. (4) is applicable for material with slab geometry by assuming uniform initial moisture distribution:

$$\mathrm{MR} = \frac{8}{\pi^2} \sum_{n=0}^{\infty} \frac{1}{(2n+1)^2} \exp\left(- \frac{(2n+1)^2 \pi^2 D_{\mathrm{eff}} t}{4L^2} \right) \quad (4)$$

where D_{eff} is the effective diffusivity (m^2/s), L is the half-thickness of slab (m), and n is the number of terms taken into consideration.

2.6 Quality evaluation

2.6.1 Greenness (negative a)*

Color in CIE L*, a*, and b* system of the dried CQ was measured by a spectrocolorimeter (ColorFlex, Hunter Associates Laboratory, Inc., Virginia, USA). As CQ is a plant, its main color was expressed by CIE $a*$ (redness/greenness) coordinate.

2.6.2 Shrinkage

Calculation of shrinkage was based on the concept of fluid replacement. In this work, n heptane was used as a fluid. Shrinkage was expressed in terms of the percentage of change in the volume of the dried CQ over the fresh CQ.

$$\%\,Shrinkage = \left(\frac{V_i - V}{V_i} \right) \times 100 \quad (5)$$

where V_i and V are respectively the volumes of material at the beginning and dried product.

2.6.3 Bulk Density

A 25-mL cylinder was filled with the dried CQ and tapped 5 times. Weight of the CQ inside the cylinder was measured. Bulk density was computed by dividing the weight of CQ by its volume.

2.6.4 Total phenolic content

TPC of the dried CQ was determined by Folin-Ciocalteu method [5]. Dried CQ was ground and sieved through a 180 mesh screen. Five millimeters of 70% methanol was used to extract a sample of 0.2 g. The mixture was extracted under reflux condenser at 70°C for 2 h. The extract was vigorously agitated using a vortex mixer for 5 min and centrifuged at 3500 rpm for 10 min. UV-vis spectrophotometer (UV-1800, Shimadzu, Kyoto, Japan) was used for TPC analysis with an absorbance of 765 nm. A calibrating agent was pure ethanol. The content was expressed as gallic acid equivalents.

2.5.5 Quercetin content

The quantitative HPLC method for determining quercetin in the dried CQ was modified from Thiangtham [6]. The HPLC system (SPD-10A, Shimadzu Co., Kyoto, Japan) with C-18 reverse phase was used in this analysis. A mobile phase was 65:35 v/v of 0.05% ortho-phosphoric acid: acetonitrile. A standard agent was trans-resveratrol. The retention time of quercetin was 8.78.

2.6 Statistical analysis

The experimental data were analyzed using one-way ANOVA with the experimental design of a completely randomized design. Mean and standard deviation were used to present the experimental data. Duncan's multiple range test was used to determine the significance of differences between treatments at 95% confidence level.

3 Results and discussion

3.1 Drying characteristics

Drying curves of convective drying of CQ are shown in Fig. 2. It can be observed that the MR decreased with drying time following an exponential decay. The drying results showed that increasing drying temperature led to increase in DR and consequently decrease in drying time. It took only 85 min to dry CQ at 100°C while the drying times were 1920, 230 and 130 min when drying at 40, 60 and 80°C, respectively.

As can be seen from Fig. 3, the constant rate drying period was not observed in convective drying of CQ; the drying process took place in the falling rate period except a very short accelerating period at the beginning. The moisture loss rate of CQ was faster at the beginning than at the end. This observation is in agreement with previous results on thin-layer drying of biological products [7]. The maximum DR obtained from each drying condition was determined. It was found that the maximum drying rates were 0.0059, 0.0107, 0.0173, and 0.0248 g water/g dry matter·min when drying CQ at 40, 60, 80 and 100°C, respectively. As compared to the lowest drying temperature of 40°C, the maximum drying rate increased by 81, 193 and 320% when drying at 60, 80, 100°C, respectively.

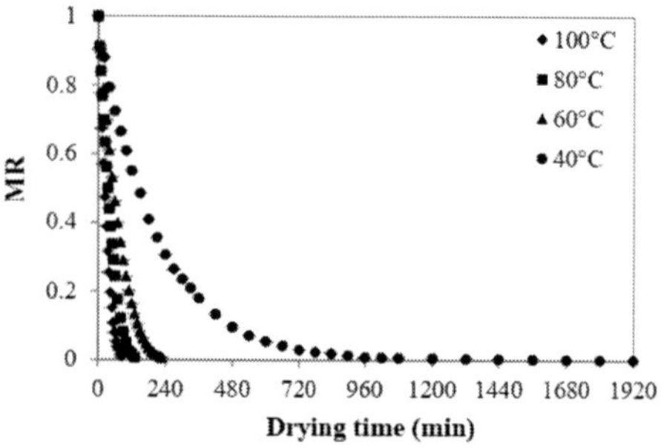

Fig. 2 Drying curves of convective drying of CQ at different drying temperatures.

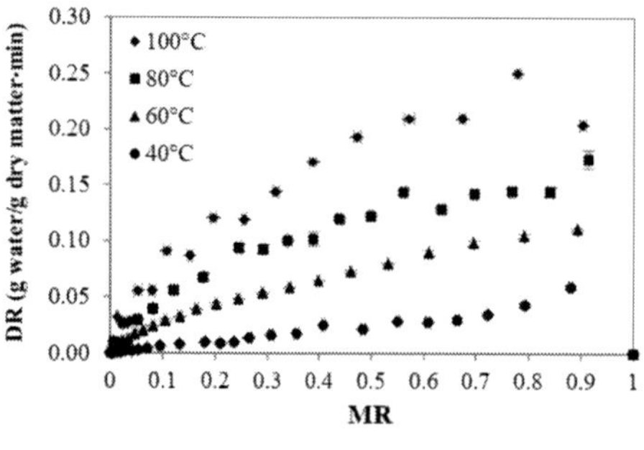

Fig. 3. DR curves of convective drying of CQ at different drying temperatures.

3.2 Effective moisture diffusivity

Values of D_{eff} of CQ dried at different drying temperatures were determined by plotting experimental data in terms of ln MR versus drying time. It is clear that increasing drying temperature resulted in higher D_{eff} (Table 2). These values met the standard range of D_{eff} of food and agricultural products subjected to drying (from 10^{-10} to 10^{-9} m^2/s) [8].

Table 1. D_{eff} $\times 10^{-9}$ (m^2/s) of convective drying of CQ at different drying temperatures.

Temperature (°C)	D_{eff} $\times 10^{-9}$ (m^2/s)	R^2
40	0.7302	0.9780
60	3.6918	0.9630
80	6.0043	0.9582
100	9.1281	0.9591

3.3 Quality of the dried CQ

Table 2 shows the results of physical property determination of the CQ subjected to convective drying at different drying temperatures. Significant differences of the dried CQ were observed for their greenness (a*), bulk density, and shrinkage ($P < 0.05$). When drying temperature increased, greenness value decreased. This could be due to a non-enzymatic browning reaction occurring at high drying temperature [9]. On the other hand, increasing drying temperature resulted in increased bulk density and shrinkage of the dried CQ. High drying temperature can cause high stress development in material during drying and result in shrinkage of the material [10]. From these properties, it could be concluded that better physical properties was provided by lower drying temperature.

Significant differences in TPC and quercetin content of the dried CQ were also observed ($p < 0.05$) as shown in Fig. 4. It is interesting that the peak of TPC and quercetin content were given by convective drying of CQ at the drying temperature of 60°C. Chan et al. [11] reported that browning reaction products formed during drying have higher antioxidant activity. This should be the reason of the lowest values of TPC and quercetin content in the CQ dried at 40°C. However, cell walls of CQ may be disrupted at high drying temperature. As a result, its oxidative and hydrolytic enzymes are released and its phytochemicals are degraded [5].

Table 2. Physical properties of the dried CQ.

Temp. (°C)	Greenness (a*)	Bulk density (g/cm³)	Shrinkage (%)
40	-2.26±0.13^d	0.182±0.002^c	88.47±0.51^c
60	-1.12±0.01^c	0.184± 0.010^c	89.62±2.94^b
80	4.49±0.08^b	0.211±0.001^b	89.63±0.55^b
100	4.75±0.08^a	0.233±0.002^a	92.47±0.32^a

The same letter in the same column is not significantly different at p<0.05.

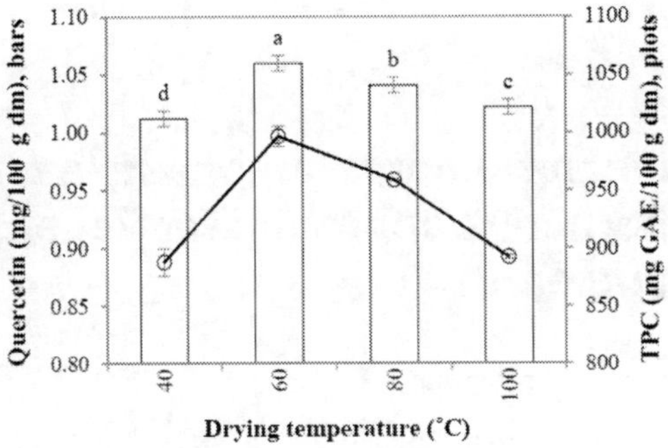

Fig. 4. Quercetin contents (bars) and TPC (plots) of the dried CQ.

4 Conclusions

The effect of drying temperatures on convective drying of CQ was analysed in the present work. Increased drying temperature resulted in increases in DR and D_{eff} and decrease in drying time. The highest values of maximum DR of 0.0248 g water/g dry matter·min and D_{eff} of 9.1281×10^{-9} m^2/s and the lowest value of drying time of 85 min were obtained by drying CQ at the highest temperature of 100°C. However, lower drying temperature trended to provide better quality of the dried CQ. The highest TPC and quercetin content were provided by convective drying of CQ at the drying temperature of 60°C.

Acknowledgement

This work was supported by a research grant of the academic year 2015 offered by King Mongkut's Institute of Technology Ladkrabang. The authors would like to express their gratefulness to the Institute.

References

1. A. Vijayalakshmi, P.R. Kumar, S.S. Priyadarsini, C. Meenaxshi, C., Journal of Chemistry doi:10.1155/2013/150675, 1-9, 2013.
2. S. Panpimanmas, C. Sithipongsri, C. Sukdanon, C. Manmee, J Med Assoc Thai **93**, 12 (2010)
3. AOAC, *Official methods of analysis* (Association of Official Agricultural Chemists, Washington, DC., 1995)
4. J. Crank, *The mathematics of diffusion* (Clarendon Press Oxford, UK, 1975)
5. Q. Chen, Z. Li, J. Bi, L. Zhou, J. Yi, X. Wu, LWT-Food Science and Technology **80**, 178-184 (2017).
6. J. Thiangtham, Determination of antioxidant constituents in the medicinal plant *Cissus quadrangularis* Linn. (*Master's thesis, Chulalongkorn University*). Available from the Chulalongkorn University Intellectual Repository. (ISBN No. 9741739184) (2003).
7. I. Doymaz, M. Pala, Journal of food engineering **60**, 125-130 (2003).
8. J. Sripinyowanich, A. Noomhorm, Drying Technology **29**, 735-748 (2011)
9. S. Honma, M. Murata, Food preservation and processing study, *Tokyo kagaku dojin*, Tokyo, 168 (2011).
10. Z. Pan, C. Shih, T. H. McHugh, E. Hirschberg, Food Science and Technology **41**, 1944-1951 (2008).
11. E.W.C. Chan, Y.Y. Lim, S.K. Wong, K.K. Lim, S.P. Tan, F.S. Lianto, et al., Food Chemistry **113**, 166-172 (2009).

The effect of ethoxylate nonionic surfactants on phase inversion temperature and salinity: an alternative approach for vegetable oil recovery from spent bleaching earth

Gitsada Panumonwatee[1], *Ampira* Charoensaeng[2] and *Noulkamol* Arpornpong[1,*]

[1]Faculty of Agriculture, Natural Resources and Environment, Naresuan University, Thailand.
[2]Petroleum and Petrochemical College, Chulalongkorn University, Thailand.

Abstract. An accurate determination of the hydrophilic-lipophilic nature of surfactants plays an important role in guiding microemulsion formation. The objective of this study is to determine the effect of ethoxylate numbers (EONs) (3, 5, and 7 moles) of nonionic surfactants on a phase inversion temperature (PIT) and optimum salinity based on the equivalent alkane carbon numbers (ACNs) of vegetable oils. Three vegetable oils, soybean oil, crude rice bran oil and crude palm oil, were selected for use as a surrogate oil to represent the residual oils found in spent bleaching earth. In this study, the hydrophilic-lipophilic deviation (HLD) was used to predict the optimum salinity (0-20 %wt.) at various temperatures (25-55°C). The results showed that the ACNs of crude rice bran oil, crude palm oil, and soybean oil were 15.41±0.35, 13.71±0.41, and 17.60±0.28, respectively. In comparison, these predictions with the experimental results, the data showed slight deviations in the optimum salinity with the specific temperature. Finally, the ACN and the surfactant characteristics obtained in this study were combined with the HLD equation and used to validate its practically and utility for guiding the optimum microemulsion formulation.

1 Introduction

The Bleaching step in the refinery edible oil process operates using bleaching earth as an adsorbent for removing undesirable matters (e.g., colour pigments, phosphatides free fatty acids, or gums). Spent bleaching earth (SBE) containing a high fraction of residual vegetable oil approximately 20-40 wt.% is discarded waste [1]. It is classified as industrial waste by the federal regulation (DIW Waste Code, 02 03 99, Thailand) and required to have proper treatments prior its disposal [2].

One of the alternative approaches for enhancing vegetable oil's recovery from spent bleaching earth through the application of microemulsification technology using microemulsion-based washing agent. Microemulsions (ME) consisting of four types (Winsor Type I, II, III, and IV) are isotropic liquid mixtures of surfactant-water-oil [3]. At low salinity concentration or low temperature, hydrophilic surfactants can self-aggregate to form normal micelles, also called Winsor Type I microemulsion. As the salinity and temperature are increased (more hydrophobicity), phase transition can shift from Type I to Type III and then to Type II [3, 4]. Regarding the Winsor Type III microemulsion, bi-continuous micelles can be obtained when the interfacial tension (IFT) between water and oil is balanced, resulting in the lowest IFT [4].

Recently, the hydrophilic-lipophilic deviation (HLD) concept has been applied to design optimum formulation of microemulsion-based washing agents. It also includes the parameters of the oil polarity, temperature ranges, salinity concentrations, alcohol levels, and surfactant hydrophobicity. The detail of HLD can be seen elsewhere [5-7]. For nonionic surfactants, the temperature is a key parameter for determining microemulsion phase transition. The microemulsion phase transition from Type I to III can occur once the system reaches the phase inversion temperature (PIT).

Arpornpong et al. [8] formulated microemulsion-based washing agent to enhance rice bran oil recovery from SBE using HLD concept. The HLD equation predicted a Winsor type III microemulsion (HLD =0) at an optimum condition of 10% v/v Dehydol LS3TH/crude rice bran oil system with 8.5% wt. NaCl at 25°C. As compared to the experimental phase behavior, the result showed that an optimum condition of 10% v/v Dehydol LS3TH/crude rice bran oil system with 15% wt. NaCl at 25°C showed the highest oil extraction efficiency for up to 71.38±1.75%.

However, there are several types of crude vegetable oils (e.g., palm oil, rice bran oil, soybean oil) remaining in SBE. The residual oils in SBE depends on the raw materials used to make purified oil in vegetable oil industries. Each oil has its own hydrophobicity which is similar to the equivalent alkane carbon number (ACN). The ACN was initially introduced by Cash et al. [9]. Even though the ACN of general oils such as *n*-alkanes have been reported, the regional or specific oils for industrial application still requires further detailed

* Corresponding author: noulkamola@nu.ac.th (N. Arpornpong).

research. The ACN of each oil has often been determined by studying the microemulsion phase behaviour. This procedure seems to be far more complex for the operator who oversees several types of oils, surfactants, or other additives. As a result, the HLD concept could potentially be used as a guide for selecting the optimum formulation of microemulsion-based washing agent for removing vegetable oil in SBE.

The aims of this study are (1) to determine the alkane carbon number (ACNs) of three vegetable oils, (2) to study the effect of the ethoxylate oxide groups on fatty alcohol ethoxylate surfactant (EONs) on phase inversion temperature (PIT) and salinity, and (3) to compare the phase behaviour study results with the HLD predicted data.

2 Materials and Methods

2.1 Materials

Three fatty alcohol ethoxylate nonionic surfactants which share the same carbon chain length (C12-14) with 3, 5, and 7 ethylene oxide (EO) groups were used as in this research and received from the Thai ethoxylate Co., Ltd (Thailand). The surfactant properties are shown in Table 1.

Vegetable oils used as a surrogate residual oil in SBE were crude rice bran oil, crude palm oil and soybean oil. The crude rice bran oil and crude palm oil that were used was receive from the Surin Bran oil Co., Ltd. and the Suksomboon palm oil Co., Ltd., respectively. The soybean oil was purchased from the Morakot Industries.

Four alkane hydrocarbons, hexane (C6, 99%), decane (C10, 99%), dodecane (C12, 99%) and hexadecane (C16, 99%) were purchased from the Acros Organics. Their ACN values are 6, 10, 12 and 16, respectively [9].

Table 1. The properties of the surfactants.

Name	Formula	MW[a]	HLB[b]	Cloud point(°C)
Dehydol LS3 TH	$C_{12\text{-}14}(C_2H_4O)_3OH$	318	7.9	51-53
Dehydol LS5 TH	$C_{12\text{-}14}(C_2H_4O)_5OH$	406	10.3	68-73
Dehydol LS7 TH	$C_{12\text{-}14}(C_2H_4O)_7OH$	494	12.1	52-58

[a] MW is the molecular weight (g/mol) providing by the manufacturer.

[b] HLB is the hydrophilic-lipophilic balance providing by the manufacturer.

2.2 Methods

2.2.1 Determination of the vegetable oil ACN

The alkane carbon number of oil (ACN) is a key parameter for determining the hydrophilic-lipophilic balance of surfactants. In order to determine the ACN value, the HLD equation for nonionic surfactant system must be simplified by way of the operating microemulsion system without NaCl and alcohol while adjusting the temperature up until the phase inversion temperature point (PIT) has been reached. The HLD equation for nonionic surfactant systems was shown in Eq.1 [6, 10].

$$HLD = \beta + b(S) - k(ACN) - f(A) + c T\Delta T \quad (1)$$

Where β is the surfactant characteristic value, b is the electrolyte constant (0.13 for NaCl), S is the salinity concentration (g/100 ml), k is the constant of the surfactant (0.17 is normal), f(A) is the concentration of alcohol, cT which is a temperature coefficient is 0.06 K-1 and ΔT is a different temperature between the studied and the reference temperature (25 °C). At equilibrium of the hydrophilic potential and the lipophilic potential, the middle phase will appear (known as Winsor type III) and result in a minimum interfacial tension between water and oil in the system and HLD being equal zero.

Phase inversion temperatures was studied in flat bottom test tubes with screw caps. An equal volume of the water phase and oil phase (5 ml) were added into a test tube and kept in a thermostatic water bath. The surfactant concentration used was 10% v/v in the distilled water (a non-salinity system). The PITs of known ACNs of oils including hexane (C6), decane (C10), dodecane (C12) and hexadecane (C16) were measured and used to determine the unknown ACNs of vegetable oils.

2.2.2 Effect of EON on PIT study

The EON in surfactant showed a hydrophilic potential. In this study, the effect of EON on PIT of individual oils studied in this experiment was determined by plotting a linear regression between EON and PIT. Three surfactants which have different numbers of polyethoxylated group (EON = 3, 5 and 7) were mixed with each hydrocarbon oil.

2.2.3 Determination of optimum salinity using HLD

The theoretical PIT point from HLD equation was obtained by using the Eq.2. The surfactant characteristic parameter (*β*) can be estimated by calculating the lipophilic potential of surfactant (*α*) minus a number of polyethylene mole (EON): α-EON. The α parameter was evaluated by assessing the alkyl carbon chain of surfactant, namely SACN. Salager et al. [11] found a linear regression between α and SACN for alcohol ethoxylate surfactant as shown is Eq.3.

$$\alpha = 0.34SACN + 2.00 \quad (3)$$

After the ACN of vegetable oil was studied (section 2.2.1), the *β* value of each surfactant was determined by using the Eq.2. The *β* value was used for prediction of optimum salinity (S*). The S* was found by microemulsion phase behaviour method. A 1:1 ratio of water phase (surfactant-water-NaCl) and crude palm oil was mixed to reached equilibrium. The experimental-based S* and theoretical-based S* were compared for an estimation errors.

3 Results and discussions

3.1 Determination of the ACN of vegetable oils

The ACN of vegetable oils were determined by measuring the temperature (°C) at an optimum phase transition (Winsor Type I to III). All surfactant-oil systems were conducted at the system of 10% v/v surfactant/vegetable oil. The PIT was plotted as a function of the alkane carbon number (ACN), as shown in Fig. 1.

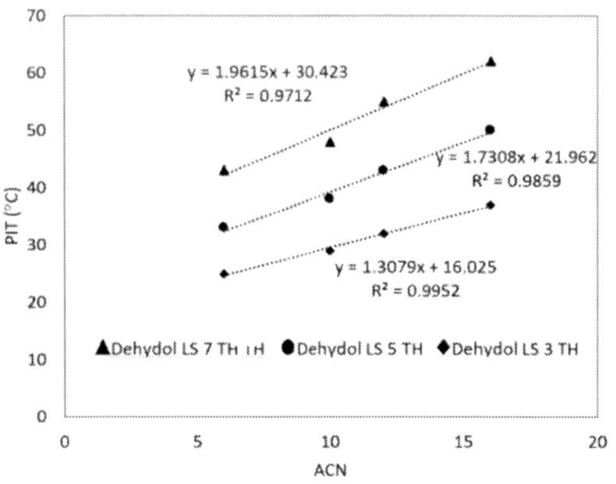

Fig.1 A plot of the phase inversion temperature (PIT) vs the ACN of the oil in various surfactant systems, Dehydol LS3 TH, Dehydol LS5 TH and Dehydol LS7 TH (10% v/v) without salinity

According to Fig. 1, the results showed that the PIT of each surfactant increased with the increasing of the alkane carbon number of oil (*n*-alkane, C6-C16). The increment of the ACN required higher temperature for the microemulsion phase inversion from Winsor type I to Winsor type III.

The relative linear graph was used for calculation the ACN regarding the three vegetable oils (soybean oil, crude rice bran oil and crude palm oil). Table 2 shows the PIT of all surfactant-oil systems. According to the linear regression, the ACNs of vegetable oils were 17.60±0.28, 15.41±0.35 and 13.71±0.41 for soybean oil, crude rice bran oil and crude palm oil, respectively. This was a result of the structure of each oil and the differences in their main fatty acid composition [12].

Table 2. Effect of surfactant types on the phase inversion temperature with three different vegetable oils

Surfactants	PIT (°C)		
	Soybean oil	Crude rice bran oil	Crude palm oil
Dehydol LS3 TH	39	36	34
Dehydol LS5 TH	53	48	45
Dehydol LS7 TH	65	61	57

Moreover, at high ACN of oils, the system requires more free energy (high temperature) for transferring oil molecule from oil phase to microemulsion phase. The effect of ACN on the microemulsion formations can be seen elsewhere [4].

3.2 Effect of EON on PIT study

The effect of EON on PIT was investigated by using a 10% v/v Dehydol LS series with four hydrocarbons. A plot of the phase inversion temperature (PIT) versus the ethylene oxide groups (EON) of the surfactants is shown in Fig. 2.

From Fig.2, the results showed that the PIT of all systems increased with the increasing of the EON of surfactant. The slope of the plot shows a relation between the optimum phase transition temperature and numbers of ethoxylate group. The results showed that the PIT increased, 5.31±0.72°C per one unit of ethylene oxide group. Furthermore, these results are consistent with other researchers [4, 8].

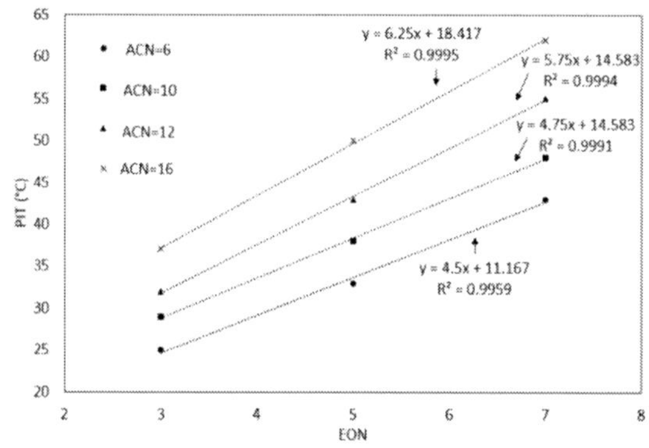

Fig.2 A plot of the phase inversion temperature (PIT) vs the ethylene oxide groups of the surfactant (EON)

3.3 Determination of optimum salinity using HLD

From the phase inversion temperature (PIT) study, the β value of Dehydol LS3 TH, Dehydol LS5 TH and Dehydol LS7 TH were determined to be 1.80, 1.09 and 0.31, respectively. The β values of these surfactants were similar with the reported values of Arpornpong et al. [8].

Fig.3 illustrates the relationship between temperature and optimum salinity. Regarding all the surfactants, the predicted optimum salinities (b(S*)) from HLD equation (dash trend lines) were found in agreement with the experimental data (solid trend lines) at elevated temperatures.

The decreasing of temperature affects the increment of optimum salinity in the system. The phase transitions were observed from Winsor Type I to III microemulsion (at HLD=0) with an increasing temperature parameter (ΔT) and salinity concentration (S). Arpornpong et al. [8] reported that this slight differences may come from the calculated ACN, β, and other environmental parameters in the experiment. Thus, it could be summarized that the HLD equation has the ability for guiding the optimum salinity concentration and temperature for a microemulsion formation.

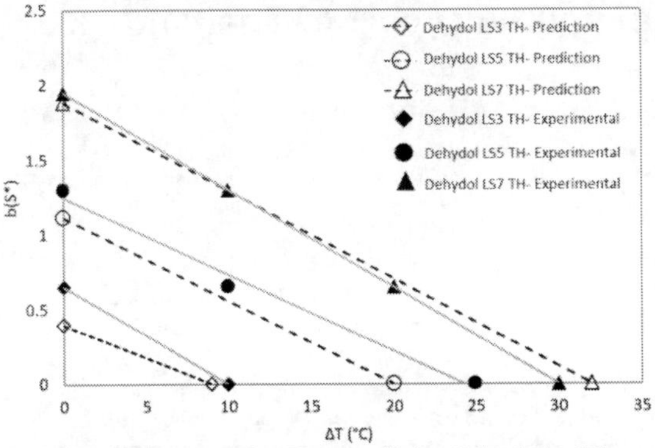

Fig.3 The comparison of optimum salinity (S*) obtained by prediction using HLD and the experimental phase behavior study of the system of 2% v/v surfactant solution with crude palm oil

4 Conclusion

This study investigated the effect of ethoxylate nonionic surfactants on the phase inversion temperature of a surfactant-vegetable oil system. The ACN of vegetable oil is one of the key parameters in the microemulsion formation, and was thus investigated in this study. The ACNs of soybean oil, crude rice bran oil and crude palm oil were 17.60, 15.41 and 13.71, respectively. The ACN and β data calculated in this work were then used in the HLD equation for predicting the optimum salinity and temperature. The result showed that the increase of one EON group of the ethoxylate surfactant increased the phase inversion temperature by 5.31°C. In comparison, the data regarding these predictions along with the experimental results showed a slight deviation in the optimum salinity. Therefore, the HLD equation has the ability to be used as a guidance for formulating microemulsion-based washing agent, in accordance with the enhancing vegetable oil recovery from spent bleaching earth.

5 Acknowledgement

Financial support for this work was provide by the National Research Council of Thailand (NRCT2561004) and the Thailand research fund (MRG6080015). Moreover, the researchers would like to thanks the Surin Bran Oil Co., Ltd., the Suksomboon palm oil Co., Ltd., and the Thai ethoxylate Co., Ltd. for providing us with the materials and chemical compounds to successfully accomplish this research.

6 References

1. K. Loh, F. Cheng, M.Choo, N. Ma., J American Applied Sci **3**, 2063-2067 (2006)
2. DIW., *3Rs for industrial wastes management.* (Ministry of Industry, Bangkok, 2012)
3. M.J. Rosen, *Micelle Formation by Surfactants, In Surfactants and Interfacial Phenomena* (John Wiley & Sons Inc., New Jersey, 2004)
4. J.L. Salager, A.M. Forgiarini, B. Johnny, J. Surfact. Deterg. **16**, 449-472 (2013)
5. W.H. Wade, J.C. Morgan, R.S. Schechter, J.K. Jacobson, J.L. Salager, Soc. Pet. Eng. **18**, 242-252 (1978)
6. J.L. Salager, N. Marquezm A. Graciaa, J. Lacgaise, Langmuir **16**,5534-5539 (2000)
7. J.L. Salager, J.C. Morgan, R.S. Schechter, W.H. Wade, E. Vasquez, soc. Pet. Eng. J. **19**, 107-115 (1979)
8. N. Arpornpong, A. Charoensaeng, S. Khaodhiar, D. Sabatini, Colloids Surf. A. **541**, 87–96 (2018)
9. L. Cash, J.L Cayias, G. Fournier, D. Macaister, T. Schares, E.H. Wade, J. Colloid Interface Sci. **59**, 39–44 (1977)
10. M. Bourrel, J.L. Salager, R.S. Schechter, W.H. Wade, J. Colloid Interface Sci. **75**, 451–461 (1980)
11. J.L. Salager, L. Manchego, L. Marquez, J. Bullo, A. Forgiarini, J. Surfact. Deterg. **17**, 199–213 (2014)
12. A. Witthayapanyanona, E.J. Acosta, J.H. Harwell, D.A. Sabatini, J. Surfact. Deterg. **9**, 331-339 (2006)

Effects of climate change on agriculture water demand in lower Pak Phanang river basin, southern part of Thailand

Natapon Kaewthong[1], *and Pakorn* Ditthakit[2,*]

[1] Graduate Student, School of Engineering and Resources, Walailak University, Nakhon Si Thammarat, Thailand
[2] Assistant Professor, School of Engineering and Resources, Walailak University, Nakhon Si Thammarat, Thailand

Abstract. The aim of the research is to analyse the effects on agricultural water demand in the Lower Pak Phanang River Basin area due to climate change. The climate data used in the analysis were rainfall, maximum, minimum, and average temperatures. The climate datasets were obtained from statistical downscaling of global circulation model under the CMIP5 project by means of bias correction with Optimizing Quantile Mapping implemented by the Hydro and Agro Informatics Institute. To determine agriculture water demand, reference evapotranspiration (ETo) based on Hargreaves method was calculated for both baseline climate data (1987-2015) and forecasted climate data in 2038. For agriculture water demand in the Pak Phanang river basin, we considered paddy field, palm oil, rubber, grapefruit, orchard, vegetable, ruzy and biennial crop, based on land use data of the Land Development Department of Thailand in 2012. The results showed that forecasted agriculture water demand in 2038 with existing land use data in 2012 will be increased with the average of 18.9% or 61.78 MCM as compared to baseline climate condition. Both water demand and supply management measures would be suitably prepared before facing unexpected situation.

1 Introduction

Climate change affects several aspects of global community, especially food and livelihood security. It is therefore inevitable to comprehend such phenomena and to prepare for facing negative effect occurrences. In recent years, many studies have shown that Thailand has the potential to be affected by climate change. According to Thailand's Department of Meteorology, from 1981 to 2007, Thailand had an average temperature increase of about 1 degree Celsius. The Intergovernmental Panel on Climate Change (IPCC) predicted that in 2050 Thailand will have an average temperature rise of 1.2-1.9 degrees Celsius, and there is likely to increase sea surface temperatures. The moment 10 years ago, the water temperature in the sea area of South East Asian with a temperature increase of 0.1-0.3 ° C (1951 to 2000). The results from several researches show that Thailand will have a lower number of rainy days and lower rainfall, especially during the dry season, on the other hand, there is a high rainfall intensity during the rainy season [1-6]. It has been found that this change in agricultural production has a significant effect on agricultural output, especially rice, which is likely to decrease [7-9]. For this research, we analyzed and compared the amount of water demand of each plant species based on the Hargreaves method [10] using climate data including highest and lowest temperatures, average temperatures, and daily rainfall for baseline and future events.

1.1 Study Area.

The Pak Phanang River Basin (PPRB) is located on the east coast of southern Thailand. Pak Phanang River is the main waterway which has a length of about 180 km covers Pak Phanang District, Hua Sai District, Ron Phibun District, Chian Yai District, Chulabhorn District, Phra Phrom District, Chalerm Phrakiat District, Cha-am District and some parts of Lan Saka District and Nakhon Si Thammarat District. The terrain can be divided into three parts: the upper watershed is a high mountain and parallel to the coast. The middle part of watershed is alternating area between high and low, and the lower part of watershed is adjacent to the Gulf of Thailand, where the flood plain and some lowland areas are below normal sea level. In the past, PPRB had abundance of a swampy forest area in the middle part of watershed. Such area was a significant natural water source for agriculture. However, when the swamp area was destroyed, the Lower PPRB has been facing water shortage as well as sea water intrusion problems. Additionally, the Lower Lower PPRB, which is a coastal lowland, has been affected by sea level rise and fall and deposited by sediment at the river mouth estuary. Such situations affects to deficiency of cultivation, economic downturns, and people emigration. In the year 1992, His Majesty the late King Bhumibol Adulyadej gave an order problem-solving measure for the Pak Panang watershed,

* Corresponding author:

including three main tasks: 1) constructing Pak Phanang floodgate to solve the water shortage problem for agriculture and household use, 2) dredging the canal for the purpose of drainage during rainy season and storing water during dry season, 3) dividing land use areas into freshwater and saltwater areas by construction of flood gates and water control structure along the tributaries. The command areas and water control structures of the Lower PPRB are shown in Figure 1. It shows components of the system including main river (Pak Phanang river), canals, command area, locations of pumping station, agricultural water demand in each command area, water flow direction, return flow, and water storage or community pond in each command area. Main Canal (MC) command area is the area which irrigation water is obtained from pumping water from river or canals into gravitational irrigation canal system, whilst Main Drainage (MD) command area which irrigation water is obtained from pumping water from river or canals into drainage canal system (water flowing with adverse slope). River and canals storage system of approximately 70 million cubic meters are the main water sources of the study area, especially during dry season. It is insufficient for more growing water demand. The schematic diagram of the Lower PPRB in Figure 2 shows the linkage components of the system, including Pak Phanang River, canals, command areas, locations of pumping station, initial water flow direction, and return flow.

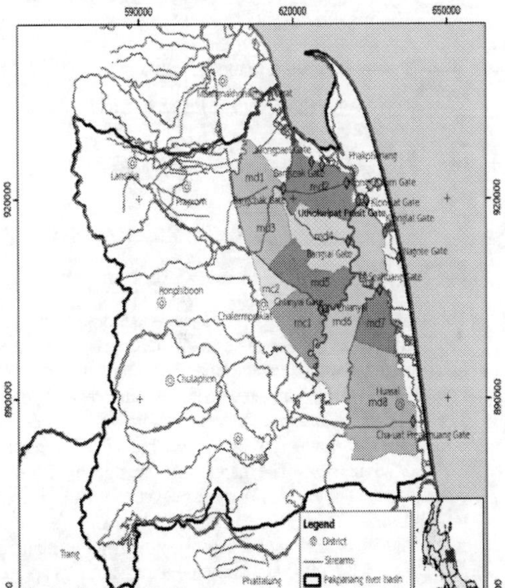

Fig. 1. The command areas and water control structures of the Lower PPRB

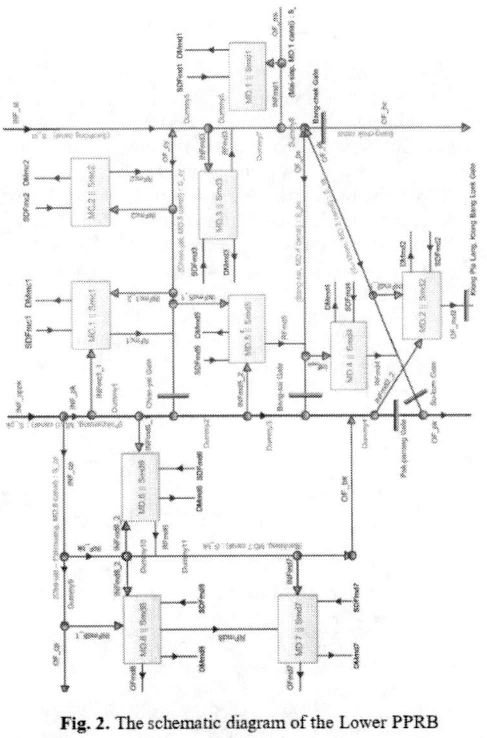

Fig. 2. The schematic diagram of the Lower PPRB

1.2 Climate Change Data.

This research used the rainfall data of the rainfall stations in the Pak Phanang watershed, consisting of 552004, 552007, 552010, 552011, 552023, 552025 and 552027 stations between 1987 and 2015. Agricultural water demand was estimated for both recorded climate data and forecasted climate data under climate change condition. The climate datasets were obtained from statistical downscaling of global circulation model under the CMIP5 (the world climate research program coupled model inter-comparison phase5) project. For downscaling of global circulation model, the Hydro and Agro Informatics Institute developed bias correction with Optimizing Quantile Mapping method. More accuracy was obtained when compared to using traditional bias correction method [11]. The climate conditions include 4-5 scenarios depending on the data portal contribution such as historical, RCP2.6, RCP4.5, RCP6.0 and RCP8.5. The concerning period can be separated into 4 periods such as historical scenario for 1 period (1979-2006) and RCP scenarios for 3 periods (2015-2039, 2045-2069 and 2075-2099). Furthermore, the bias corrected GCM data were conversed to the 0.25° x 0.25° gird format (885 grids) [12]. This study used the CNRM-CM5 model using the Scenario RCP4.5 and Scenario RCP8.5.

2 Methodology

2.1 Land Use Analysis

Land use data in 2012 obtained from the Land Development Department of Thailand was used for determination of agricultural areas of the Lower Pak Phanang Basin. It was the third level of land use. Land use data consisted of paddy field, deserted paddy field, palm, tropical rain forest, swamp forest, mangrove forest, vegetable crop, biennial crop, rubber, deserted orchard, perennial, industry land, aquaculture land, river/canal/irrigation canal, soil puddle/sand/lateritic/old mine, farm pond, and others. It indicated that the Lower PPRB had an area of about 66,772.32 acres and the most used areas were paddy field with total area of 39,272 hectares, equivalent to 42.82 percent of total area. The second is the total area of 12,101 hectares of palm plantation, which is 13.20 percent of the total area.

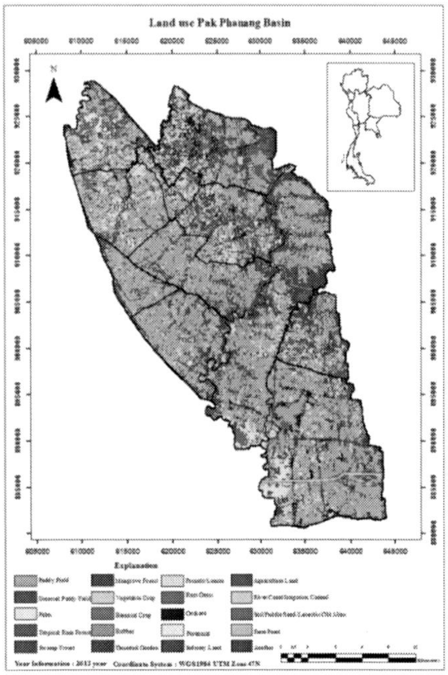

Fig. 3. Map showing land use classification in 2012.

2.2 Water Demand Analysis

The agricultural water demand analysis in the Lower PPRB was based on the calculation of the water demand for 8 plant species, including paddy field, palm oil, rubber, grapefruit, orchard, vegetable, ruzy and biennial crop. It was analysed for both baseline climate data (1987-2015) and forecasted climate data in 2038. Hargreaves equation [10] as shown in equation 1 was applied herein for determination of reference evapotranspiration (ETo) since it required only four climate data, including rainfall, maximum, minimum, and average temperatures. It is coincident with climate data obtained from the Hydro and Agro Informatics Institute. The water equivalent of the extraterrestrial radiation can be derived from locations of climate stations. And water requirements at farm level were analysed using equation 2. Due to number of pages limitation, the more detail information about Kc values for each plant can be found in Plant Coefficient (kc) of 40 plants [13] from Office of Water Management and Hydrology, Royal Irrigation Department, Thailand.

The Hargreaves equation [10] can be written as:

$$\boldsymbol{ETo = 0.0023Ra\,(T + 17.8)\,\sqrt{Tmax - Tmin}} \quad (1)$$

where

ETo is reference evapotranspiration (mm d^{-1});

Ra is the water equivalent of the extraterrestrial radiation (mm d^{-1});

Tmax, Tmin and T are the daily maximum, minimum and mean air temperature (°C).

Water demand of plants at farm level can be calculated as:

$$W_{farm} = ET_c + W_{LP} + W_{PS} + W_L - P_e \quad (2)$$

Where

W_{farm} is water demand of plants at farm level;

ET_c is crop evapotranspiration ($ET_c = K_c$ x ET_o);

K_c is crop coeffiecient;

W_{LP} is water required for land preparation;

W_{PS} is percolation and seepage losses of water from paddy field;

W_L is water to establish standing water layer;

P_e is effective rainfall.

3 The results

Table 1 shows water demand of plants per area. It indicated that from forecasting climate change in the year 2038 resulted in increasing average water demand of plants per area in the Lower PPRB during dry season of approximately 22 percent. The highest percentage change of increasing water demand of plants per area were ruzi grass and biennial crop of about 27, and 26 percent. The percentage change of increasing water demand of rubber, orchard, vegetable, and grapefruit was 22, 22, 22, and 23, respectively. And the lowest percentage change of increasing water demand of plants per area were palm oil and paddy field of about 17 percent. Additionally, more agricultural water demand for all command areas in the Lower PPRB Basin was approximately 61.78 million cubic meters. It required 326.57 million cubic meters for current climate condition and 388.35 million cubic meters in the year 2038.

Table 1. Water demand of plants per area.

Plant Types	Water demand of plants per area in cubic meter per hectares (Jan-Apr)		
	Current climate	Climate change	Percentage change
Paddy field	4,881.25	5,749.60	+17.79
Palm oil	6,488.75	7,625.62	+17.94
Rubber	1,016.88	1,241.80	+22.12
Grapefruit	4,331.25	5,361.49	+23.78
Orchard	5,084.38	6,208.99	+22.12
Vegetable	2,890.00	3,531.09	+22.17
Ruzy	3,290.63	4,189.52	+27.32
Biennial crop	4,887.50	6,166.13	+26.16

4 Discussion and conclusions

This research aimed to study the impacts on the amount of water demand of plants in response to climate change in the year 2038 to be ready to deal with the situation in the future. The results showed that in the future, especially during the dry season, the agricultural water demand in the Lower PPRB Basin, a complicated irrigation area, will be higher with the average water demand of plants per area of approximately 22 percent due to higher temperature and lower rainfall. These coincided with the previous research [14, 15]. The preparedness measures for facing an unexpected event could be significantly considered for all concerning agencies as well as local farmers, for example, optimal cropping pattern, water infrastructure development projects, community or farm pond promoting rainwater harvesting.

Acknowledgment

This research is based on the cooperation of all sectors in the area. The research results will lead to the utilization for water management in the area in the future. The authors thank to irrigation officers and local stakeholders for their valuable and beneficial recommendation to complete the research study.

References

1. W. Chaowiwat, O. Kamnoet and S. Weesakul (2016) *Future Drought Characteristics over Thailand by Using Bias Corrected Multi CMIP5 General Circulation Model*, International Conference of Multidisciplinary Approaches on UN Sustainable Development Goals (UNSDGs), December 28th – 29th, 2016.
2. Denny Marks (2011). Climate Change and Thailand : *Impact and Response, Contemporary Southeast Asia* Vol.33, No.2 (2011), pp.229-58
3. W.Chaowiwat, P.Danusatianpong, K.Sarinnapakorn and S.Boonya-aroonnet (2017). *Extreme Climate Prediction for Water Management Community Network under Changing Climate*, The 22nd Nation Convention on Civil Engineering July 18-22, 2017, Nakhon Ratchasima, THAILAND
4. IPCC, 2007 : *Summary for Policymakers. In : Climate Change 2007 : The Physical Science Basic.* Contribution of Working Group I to the Forth Assessment Report of the Intergovernmental Panel on Climate Change [Solomon, S.D. Qin, M. Manning, Z. Chen, M. Marquis, K.B. Averyt, M.Tignor and H.L. Miller (eds.)] Cambridge University Press, Cambridge, United Kingdom and New York, NY, USA.
5. Thanapakpawin, P.,Boonya-aroonnet, S., Chankarn, A., Chitradon, R., & Snidvongs, A. (2011). Thailand *Drought Risk Management : Macro and Micro Strategies*. In R.Shaw, & H.Nguyen, Droughts in Asian Monsoon Region (Vol.8, pp. 121-140). Emerald Books.
6. Royol Chitradon, Surajate Boonya-aroonnet and Porranee Thanapakpawin (2009). *Risk Management of Water Resources in Thailand in the Face of Climate Change*. Sasin Journal of Management 2009, pp.64-73
7. Jintana Kawasaki and Srikantha Hereth (2011). *Impact Assessment of Climate Change on Rice Production in Khon Kaen Province, Thailand.* J.ISSAAS Vol.17, No.2 :14-28 (2011)
8. John Felkner, Kamilya Tazhibayeva and Robert Townsend (2009). *Impact of Climate Change on Rice Product in Thailand.* Am Econ Rev.2009 May ; 99(2) :205-210.
9. Mukand S. Babel, Anshul Agarwal, Dillip Kumar Swain and Srikantha Herath (2011). *Evaluation of climate change impacts and adaptation measures for rice cultivation in Northeast Thailand.* Climate Research Vol.46 :137-146,2011.
10. Hargreaves, G.H., Samani, Z.A., 1985. *Reference crop evapotranspiration from temperature.* Appl. Eng. Agric. 1(2), 96–99.
11. W. Chaowiwat, T. Charoensuk, P. Srisompron and S.Boonya-aroonnet (2015). *Development of Statistical Downscaling of Global Circulation Model Data by Optimizing Quantile Mapping Method.* The 20th National Convention on Civil Engineering 8-10 July 2015, Chonburi, Thailand
12. Taylor KE, Stouffer RJ, Meehl GA (2012). *An Over of CMIP5 and the Experiment Design.* Bull Am Meteorol Soc 93(4) :485-498.
13. Office of Water Management and Hydrology, Royal Irrigation Department, Thailand (2003). *Plant Coefficient (kc) of 40 plants.* Retrieved from http://water.rid.go.th/hwm/cropwater/CWRdata/Kc/kc_th.pdf
14. Y. Wada, D.Wisser, S.Eisner, M.Florke, D.Gerten, I.Haddeland, N.Hanasaki, Y.Masaki, F.Portmann, T.Stacke, Z.Tessler and J.Schewe (2013). Multimodel projection and uncertainties of irrigation water demand under climate change. Geophysical Research Letters, Vol. 40, 4626-4632.
15. Shamsuddin Shahid (2011). *Impact of climate change on irrigation water demand of dry season Boro rice in northwest Bangladesh.* Climate Change (2011) 105 : 433-453

Production and analysis of volatile flavor compounds in sweet fermented rice (Khao Mak)

Jittimon Wongsa[1,*], *Vilai* Rungsardthong[2], and *Tamaki* Yasutomo[3]

[1]Department of Agricultural Engineering for Industry, Faculty of Industrial Technology and Management, King Mongkut's University of Technology North Bangkok Prachinburi Campus, Prachinburi, Thailand

[2]Department of Agro-Industrial, Food and Environmental Technology, Faculty of Applied Science, Food and Agro-Industry Research Center, King Mongkut's University of Technology North Bangkok, Bangkok, Thailand

[3]Department of Bioresource Technology, National Institute of Technology, Okinawa National College of Technology, Okinawa, Japan

Abstract. Khao Mak is a sweet fermented rice-based dessert with a unique flavor profile commonly found throughout Thailand. The traditional starter culture (Look Pang) contains yeast, mold and herbs, which is used to ferment cooked glutinous rice. This research studied production of Khao Mak which resulted in volatile flavor compounds that were affected by rice varieties, including white glutinous rice (Kor Khor 6), Japanese rice (Hitomebore) and black glutinous rice (Kam Doi and Leum Phua). Total soluble solids (TSS) as degree Brix, pH, and alcohol concentrations were measured daily during the fermentation period. Volatile flavor compounds were separated and identified by gas chromatography mass spectrometry (GC-MS). At the end of the fermentation, samples had pH ranging from 3.91±0.16 to 4.30±0.09, total soluble solids of 32.65±1.65 to 44.02±1.72°Brix, and alcohol concentrations between 0.33±0.03 and 0.38±0.03% (v/v). The potent odors associated with Khao Mak were alcohol, wine-like, whiskey-like, solvent-like, sweet and fruity. The major volatile compounds, which have stimulant flavor in Khao Mak, included 1-propanal, 1-propanol-2-methyl, 1-butanal-3-methyl and acetic acid ethyl ester. Among all rice varieties tested, the white glutinous rice produced Khao Mak containing the highest TSS, alcohol concentrations followed by Hitomebore rice and black glutinous rice, respectively.

1 Introduction

Fermented sweet rice called Khao Mak is a traditional sweet, low-alcoholic fermented product from Thai folk wisdom made mainly in the north-eastern and eastern provinces of Thailand. Apart from lactic acid and alcohol that give special flavours to Khao Mak, the fermented rice is also known for its health properties [1]. To produce Khao Mak, cooked glutinous rice and traditional fermentation starter were used in the process. The starter or Look Pang is a semi-circular starch ball that incorporates various fermentative biological species such as molds (*Aspergillus* species, *Amylomyces* species, *Rhizopus* species and *Mucor* species) and yeast (*Saccharomyces cerevisiae, Issatchenkia* species, *Candida* species, *Saccharomycopsis* species and *Pichia* species) [2]. In the production process of Look Pang, Thai herbs such as pepper, garlic, and galangal are added as antibacterial agents [3, 4].

After the fermentation, texture of the cooked rice became soft, agglutinated clumps that give mild alcoholic taste with lactic acid flavour [5]. Molds produce enzymes (e.g. α-amylase, glucoamylase), which can hydrolyze raw starch in the glutinous rice and convert it into sugar. The process also yields other organic acids such as lactic acid [6]. After fermentation, Khao Mak is consumed as a sweet dessert dish without cooking.

Rice (*Oryza sativa* L.) is a predominant food and agricultural product of Thailand [7] which includes white rice varieties and pigmented rice varieties such as purple and black rice (e.g. Hom Nil, Rice Berry, Kam Doi, Leum Phua) and red rice (Sung Yod) [8].

Traditionally, Khoa Mak has been prepared from white glutinous rice. Pigmented rice which has been used as an ingredient for deserts and snacks is gaining more popularity due to its nutritional values [9]. When compared to normal white rice, black rice is found to harbour higher level of vitamin E and B, food fibres, and phytic acid. The rice also contains higher level of antioxidant activity when compared to red rice [10].

The production of Khao Mak with pigmented rice will make the product more nutritious. Also, the sweet aroma flavor of the product is also important for consumer's satisfaction.

In this study, Khao Mak will be produced from 4 varieties of rice. Highland black glutinous rice (Kam Doi), one of the aromatic and indigenous black sticky rice which is enriched with flavonoids, especially anthocyanins. Leum Phua is also a special variety of

* Corresponding author: jittimon.w@fitm.kmutnb.ac.th

aromatic glutinous rice variety which has been reported to have very high antioxidant property. Thai white glutinous rice (Kor Khor 6) is a popular choice for producing Khao Mak because of its fragrance. Japanese rice (Hitomebore) is a special rice with a gentle flavor and extremely smooth texture. This research aims are to investigate volatile flavour compounds and its physical properties in Khao Mak that are affected by rice varieties.

2 Experimental procedures

2.1 Materials

Glutinous white rice (Kor Khor 6) and glutinous black rice (Kam Doi and Leum Phua) were all purchased from local market in Lampang, Thailand. Japanese rice (Hitomebore) was obtained from supermarket in Okinawa, Japan. Look Pang was supplied by Tao-namchoke company which is located in Sisaket province, Thailand.

Fig. 1. Look Pang.

2.2 Preparation of sweet fermented rice

Sweet fermented rice was produced following the traditional process. Rice samples were submerged in water for 2 hours prior to being steamed in an autoclave at 121°C for 20 minutes. Cooked rice was washed with water and drained for 15 minutes to cool down at room temperature. Look Pang starter were crushed into a powder. The cooked rice (300g) was treated with fermentative starter at 0.2 g per 100 g of rice in the glass container, then closed tightly and kept at 30°C for 3 days. The process flow chart of Khao Mak is shown in Fig. 2.

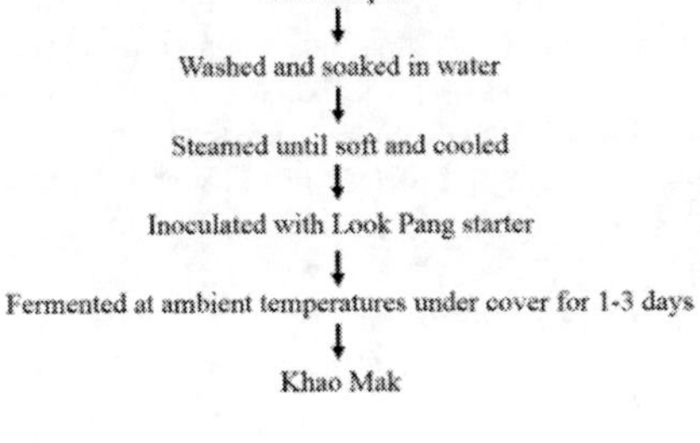

Fig. 2. Process for the production of Khao Mak.

2.3 Total soluble solids, pH and alcohol concentration analysis

Total soluble solid (°Brix) was measured in a small extracted solution sample of Khao Mak using a digital hand-held refractometer. The pH values were measured by using a pH meter. Alcohol content of samples was examined using a density/specific gravity meter.

2.4 Assessment of DPPH radical scavenging's activity

To assess DPPH radical scavenging's activity, samples were treated with 2,2-diphenyl-1-picrylhydrazyl at 0.1 mM concentration and kept at room temperature, in the absence of light for 30 minutes. Spectrophotometer was used to measure level of absorbance at 517 nm [6] and percentage of the above said activities were measured and calculated by using the following formula;

$$\text{Scavenging activity (\%)} = [(A_0 - A_1)/A_0] \times 100 \quad (1)$$

A_0 = the absorbance of control reaction
A_1 = the absorbance of test compound

Interpolation of linear regression analysis was used to attain a concentration that inhibit half of the activity (IC_{50})

2.5 Volatile Compound Analysis

Khao Mak sample was extracted and then centrifuged at 3000 rpm for 10 min. Once the phases had been separated, the supernatant liquid phase was prepared for analysis of volatile compounds and n-amyl alcohol was used directly as internal standard. In brief, 1.0 mL of the sample and 1.0 mL of n-amyl alcohol were pipetted into a 20 mL chromatography headspace vial. The vial was sealed with aluminium crimp cap PTFE/ Butyl rubber septum, vortexed and placed in the headspace autosampler system (TurboMatrix 110, PerkinElmer, USA). The headspace gas was analyzed by Shimadzu GCMS QP2010 Ultra (Shimadzu Corporation, Tokyo, Japan) with a Stabilwax column; 60 m × 0.25 mm i.d. × 0.50 µm film thickness (Restek, USA). Temperatures program used were as following: 60°C hold for 1 min, 60 to 200°C hold for 7 min and 200°C hold for 1 min.

3 Results and discussion

The characteristics of Khao Mak samples depended on the varieties of rice. Figure 3 illustrates changing pH level of each rice samples which were used in this 7 days experiments. The samples from white glutinous rice (Kor Khor 6) on the 3rd day of the fermentation were sweet acidic taste at the pH of 4.30±0.12. Similar results were observed from the experiments which indicated that, after the first day of fermentation, Khao Mak's pH was 6.45±0.20 before abruptly dropped to 3.17±0.16 to 4.10±0.09 after day 1 - 3 of fermentations.

Khao Mak fermentation system converted monosaccharide in starch to ethyl alcohol and acids by using yeast and mold activities [4].

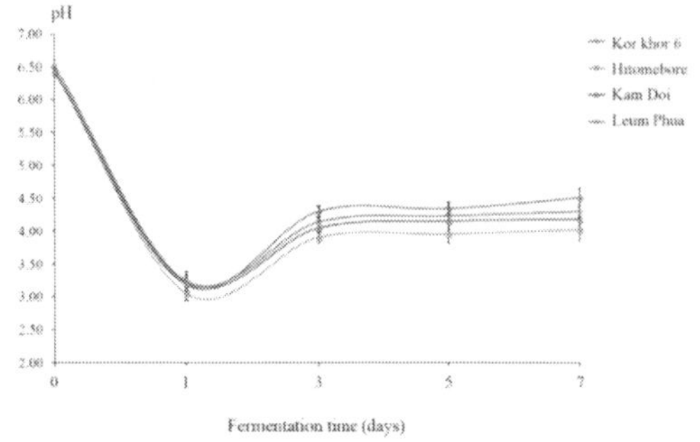

Fig. 3. Change of pH values in Khao Mak during fermentation.

Figure 4 shows levels of alcohol contents produced by the fermentation for which white glutinous rice exhibited high alcohol content. Khao Mak's alcohol level was at 0% on the first day of fermentation before sharply rose to approximately 0.35±0.03% on the 3rd day and continued to increase until the end of the fermentation period. All treatments were exhibited similar trend, but with higher contents found in glutinous white rice more than that of black rice.

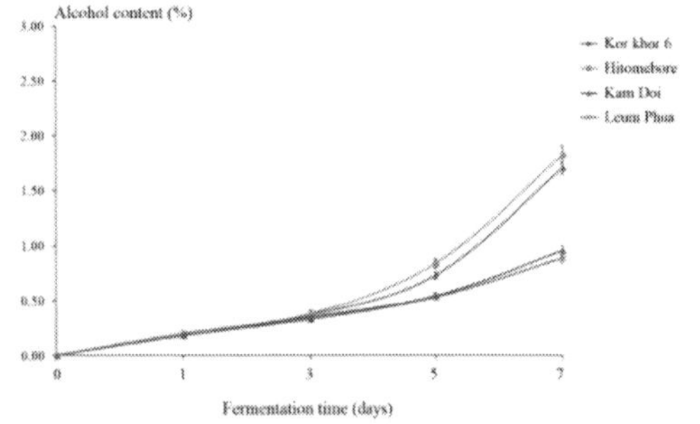

Fig. 4. Alcohol content of Khao Mak during fermentation.

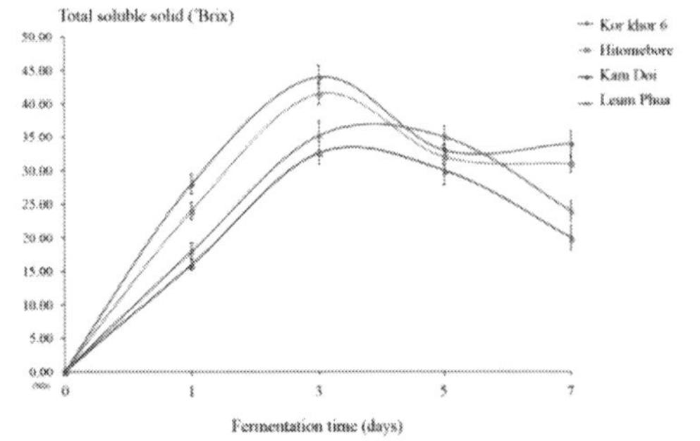

Fig. 5. Total soluble solid of Khao Mak during fermentation.

Figure 5 showed changes of Khao Mak's soluble solid during 7 days of fermentation. Total soluble solids content of all samples increased rapidly around 32.60±2.14 to 44.05±1.72°Brix after the fermentation started. Starch hydrolysis by mold resulted in increased sugar [11]. However, after the 4th day, Khao Mak's total soluble solid were fluctuated and decreased.

In Figure 6, Khao Mak produced from black glutinous rice (Kam Doi and Leum Phua) gave the strongest free radical scavenging activity with the IC_{50} value of 43.40±2.42 mg/mL and 32.26±1.65 mg/mL, respectively. Sangkitikomon et al. [10] suggested that black rice's anthocyanin performs higher antioxidant activity than other rice varieties.

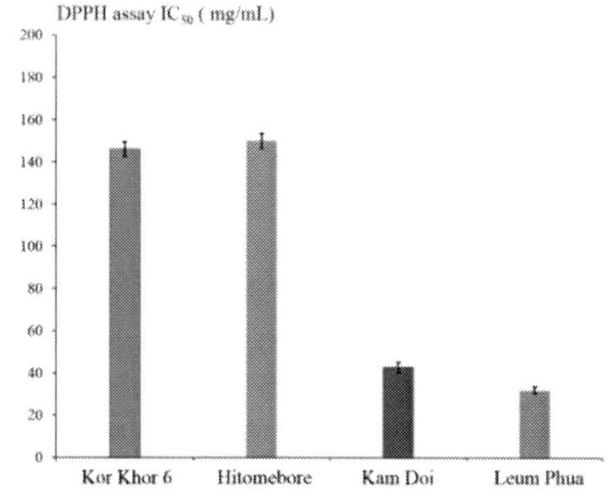

Fig. 6. DPPH radical scavenging activity IC_{50} of Khao Mak.

Volatile compounds in Khao Mak samples with various rice varieties were studied using gas chromatography mass spectrometry. Volatile compounds which shown as compound identification was attained from analysis of the provided qualitative data.

Four volatile compounds were commonly found in all samples (Table 1); acetic acid ethyl ester, 1-propanol, 1-propanol-2-methyl and 1-butanal-3-methyl. These compounds (1-propanol, 1-propanol-2-methyl and 1-butanal-3-methyl) also found in Thai rice wine (Sato and Ou product) [12, 13]. The concentration of volatile compounds in black glutinous rice were higher than white rice fermented samples. This might be due to the difference of their initial nitrogen concentration (correlation with protein content) of raw material. Amino acid metabolism serves as an important factor that facilitates volatile compound productions in food and alcoholic beverages [12]. Khao Mak fermented with Luem Phua rice contained higher concentration of 1-propanol (118.851±0.263 μg/mL) and 1-propanol-2-methyl (19.005±0.095μg/mL) compared with other samples.

Table 1. Volatile flavor compounds in Khao Mak analysed by GC-MS

Compounds	RT	Concentration (μg/mL)				Odor Descriptors [Ref.]
		Kor khor 6	Hitome bore	Kam Doi	Luem Phua	
Acetic acid, ethyl ester	6.88	0.650 (0.020)	4.050 (0.025)	0.897 (0.012)	0.924 (0.008)	Pungent, Sour, Vinegar [13]
1-propanol	10.07	10.122 (0.135)	14.176 (0.118)	63.966 (0.205)	118.851 (0.263)	Solvent [13] Alcohol, Fruity [14]
1-propanol-2-methyl	11.48	6.278 (0.091)	7.193 (0.046)	10.810 (0.150)	19.005 (0.095)	Alcohol, Wine [14] Solvent [15]
1-butanol-3-methyl-acetate	13.12	0.005 (0.002)	0.036 (0.007)	ND	0.010 (0.004)	Solvent [12] Fruity, Malty [16]
1-butanal-3-methyl	15.06	4.709 (0.015)	4.695 (0.060)	5.719 (0.066)	8.121 (0.085)	Solvent [12] Wine-like, Sweet [17]
Hexanoic acid, ethyl ester	16.38	0.023 (0.008)	0.007 (0.003)	0.024 (0.010)	ND	Sour [16] Sweat [14]
1-pentanal	16.37	1.050 (0.000)	1.050 (0.000)	1.050 (0.000)	1.050 (0.000)	Whiskey-like, Solvent [13]
Octanoic acid, ethyl ester	21.39	0.005 (0.004)	0.014 (0.009)	ND	ND	Sweat, Cheese [14]

ND: Not detected
The numbers in brackets indicates the standard deviations.

4 Conclusion

Various rice, including glutinous white rice (Kor Khor 6), glutinous black rice (Kam Doi and Leum Phua) and Japanese white rice (Hitomebore), were fermented with Look Pang for volatile flavor compounds analysis. Highest content of TSS and alcohol were found in Khao Mak that was made from white rice. Aroma flavour of Khao Mak provided alcohol, solvent-like, wine-like, whiskey-like, sweet and fruity aromas. The volatile compound of 1-propanal, 1-propanol-2-methyl, 1-butanal-3-methyl and acetic acid ethyl ester was detected in the Khao Mak products.

Acknowledgment. This research was supported by National Institute of Technology, Okinawa National College of Technology, Japan and King Mongkut's University of Technology North Bangkok, Thailand under the Research Internship Program (June-July 2017).

References

1. S. Tanasupawat, K. Komagata, World J. Microbiol. Biotechnol. **11** p.253-256 (1995)
2. A. Taechavasonyoo, J. Thaniyavarn, C. Yompakdee, As. J. Food Ag-Ind. **6** p.24-38 (2013)
3. W. Tochampa, N. Jittrepotch, T. Kongbangkerd, K. Kraboun, K. Rojsuntornkitti, Int. Food Res. J. **18** p.239-248 (2011)
4. A. Manosroi, W. Ruksiriwanich, B. Kietthankorn, W. Manosroi, J. Manosroi, Food Res. Int. **44** p.2757-2765 (2011)
5. Y. Plaitho, K. Kangsadalampai, M. Sukprasansap, J. Med. Plants Res. **7**,2 p.91-98 (2013)
6. N. Mongkontanawat, W. Lertnimitmongkol, J. Agri. Technol. **11**,2 p.501-515 (2015)
7. S. Chanta, P. Prathepha, B. Jongdee, Indian J. Tradit. Know. **13**,3 p.473-483 (2014)
8. P. Pramai, S.Jiamyangyuen, Songklanakarin J. Sci. Technol. **38**,5 p.463-472 (2016)
9. K. Tananuwong, W. Tewaruth, LWT-Food Sci. Technol. **43** p.476-481 (2010)
10. V. Sangkitikomon, T. Tentumnou, A. Rodchanasasod, J. Nutr. **43**,2 p.16-21 (2008)
11. A. Tongyai, S. Riebroy, S. Maneerat, N. Siriwong, S. Chulakarungka, *Proceedings of the 50th Kasetsart University Annual Conference* (2012)
12. T. Amatayakul, N. Somsap, P. Rotsatchakul, KKU Res. J. **17**,6 p.939-949 (2012)
13. P. Chuenchomrat, A. Assavanig, S. Lertsiri, Sci. Asia **34** p.199-206 (2008)
14. S. Cortés-Diéguez, R. Rodriguez-Solana, J. Manuel Domínguez, D. Emilia, J. Inst. Brew. 121 p. 628-635 (2015)
15. T. E. Siebert, H. E. Smyth, D. L. Capone, C. Neuwöhner, C. H. Pardon, G. K. Skouroumounis, M. J. Herderich, M. A. Sefton, A. P. Pollnitz, Anal. Bioanal. Chem. **381** p.937-947 (2005)
16. S. Kaneko, K. Kumazawa, Biosci. Biotechnol. Biochem. **79**,3 p.484-487 (2015)
17. E. Falqué, E. Fernández, D. Dubourdieu, Talanta. **54**, p.271-281 (2001)

Synthesis of glycerol carbonate from dimethyl carbonate and glycerol using CaO derived from eggshells

Wanichaya Praikaew[1], *Worapon* Kiatkittipong[1,*], *Kunlanan* Kiatkittipong[2], *Navadol* Laosiripojana[3], *Navin* Viriya-empikul[3], *Sunya* Boonyasuwat [4], *Farid* Aiouache[5], *Vesna* Najdanovic[5] *and Suttichai* Assabumrungrat[6]

[1] Department of Chemical Engineering, Faculty of Engineering and Industrial Technology, Silpakorn University, Nakhon Pathom 73000, Thailand

[2] Department of Chemical Engineering, Faculty of Engineering, King Mongkut's Institute of Technology Ladkrabang, Bangkok 10520, Thailand

[3] The Joint Graduate School of Energy and Environment, King Mongkut's University of Technology Thonburi, Bangkok 10140, Thailand

[4] Verasuwan CO., LTD Setthakij 1 Road, Nadi, Muang, Samut Sakhon 74000, Thailand

[5] Department of Engineering, Lancaster University, Lancaster LA1 4YW, United Kingdom

[6] Center of Excellence in Catalysis and Catalytic Reaction Engineering, Department of Chemical Engineering, Faculty of Engineering, Chulalongkorn University, Bangkok 10330, Thailand

Abstract. Waste eggshell is proposed as a highly active catalyst for glycerol carbonate production from dimethyl carbonate (DMC) and glycerol. The effect of reaction temperature, reaction time and catalyst loading on the reaction performance were investigated in order to find a suitable operating condition. CaO derived from waste eggshell exhibits catalytic activity comparable to commercial CaO. By using CaO eggshell, glycerol conversion of 96% can be achieved within 90 min of reaction time under 2.5:1 feed molar ratio of DMC to glycerol, 0.08 mole ratio of CaO to glycerol and reaction temperature of 60°C. The catalyst was examined by XRD, TGA/DSC, SEM, N_2 adsorption-desorption and Hammett indicators method. Utilization of eggshell as a catalyst for glycerol carbonate production not only provides a cost-effective and value-added of waste eggshell as a green catalyst, but also decrease amount of waste and its treatment cost which is ecologically friendly.

1. Introduction

Biodiesel is a world-recognized renewable fuel for fossil diesel substitution. It is generally produced from transesterification of vegetable oils with alcohols in the presence of an alkaline-based catalyst which gives glycerol as a by-product. However, for every 9 kg of biodiesel production, about 1 kg of crude glycerol is formed. Consequently, the price of glycerol has fallen significantly since 2006 [1]. This has prompted to convert low cost glycerol to value-added products [2]. Moreover, converting of glycerol to value added product is crucial to secure the sustainability of biodiesel production in term of cost competitive [3]. In addition, glycerol now becomes one of the top twelve major building blocks suggested by researchers of National Renewable Energy Laboratory (NREL).

Glycerol carbonate is one of the glycerol derivatives that capture more scientific and industrial attention at present. By preliminary economic screening of NREL researchers, glycerol carbonate was proposed as one of promising glycerol utilizations for value added chemicals. Glycerol carbonate has a potential benefit as bio-lubricant, due to its adhesion to metallic surfaces and resistance to oxidation, hydrolysis and pressure. In addition to surfactants and pharmaceuticals, glycerol carbonate can be used as a building block to synthesize polymers, such as polycarbonates, polyglycerol esters, hyperbranched polyols, and non-isocyanate polyurethanes [4]. Thus far, several routes have been explored to synthesize glycerol carbonate from glycerol [5,6] However, with industrial feasibility consideration, transesterification (or transcarbonation) of glycerol with organic carbonate sources such as dimethyl carbonate, ethylene carbonate or propylene carbonate using basic catalysts becomes the most of interest [7, 8]. CaO, one of

* Worapon Kiatkittipong: kiatkittipong_w@su.ac.th

the reported catalysts, is very interesting due to high activity, very cheap and easily available. Complete glycerol conversion and 95.3% yield could be achieved at 1.5 h reaction time, reaction temperature = 95°C, and dimethyl carbonate/glycerol molar ratio = 3.5. However, lower activity is found when CaO is exposed to the air and moisture environment, thus converted into $Ca(OH)_2$ and $CaCO_3$ during the reaction[9]. CaO can be derived from natural calcium carbonate ($CaCO_3$) source for example dolomite, mollusk shell, crab shell and egg shell by calcination. Egg shells are excellent source due to which composed a high content of calcium carbonate (96% of shell weight) [10]. The waste eggshell as catalyst was used in biodiesel production [11-15], synthesis of DMC [16], synthesis of H_2/syngas [17-19], etc.

In this work, we investigated the possibility of application of eggshell as a catalyst for glycerol carbonate production from glycerol and dimethyl carbonate. Suitable operating condition is investigated.

2. Materials and methods

2.1 Chemicals and preparation of catalysts

Glycerol (99.5 v/v%, Ajax Finechem Ltd) and dimethyl carbonate (≥ 99 wt%, Merck) were employed as a reactant. Butanol (99.8 v/v%, Sigma Aldrich) and and glycerol carbonate (99 wt% Sigma Aldrich) were used as the internal standard was used and calibration standard, respectively, for gas chromatography analysis. Commercial CaO was purchased from Lab reagent & Fine Chemicals Ltd. Waste eggshell contain mainly $CaCO_3$ which converted to CaO phase by calcinations at temperature 900°C for 2.5 h. The CaO was kept in desiccators to avoid contacting with CO_2 and humidity in the air.

2.2 Characterization of catalysts

The catalyst was characterized by thermo gravimetric analysis under air flow condition with a heating rate 10°C/min from 25°C to 1000°C to examine decomposition of catalyst. The morphology of catalysts were determined using scanning electron microscope (SEM). The structure and crystalline phases of calcined eggshell were analyzed by X-ray diffraction and basic strength was determined by using Hammett Indicator (H_). For basic strength determination, 25 mg of catalyst was shaken with 5 ml of a solution of Hammett indicator diluted in methanol and left to equilibrate for 2 h. The following Hammett indicators were used: phenolphthalein (H_ = 9.8), 2,4-dinitroaniline (H_ = 15.0) and 4-nitroaniline (H_ = 18.4). N_2 physisorption was performed to measure the surface area and pore volume of catalysts.

2.3. Reaction procedure

Transesterification of glycerol with DMC was performed in 250 ml three-necked flask equipped with condenser and water bath. The reaction conditions are as follows: 2.5:1 molar ratio of DMC to glycerol, 4% mole ratio of initial amount of CaO to glycerol, reaction temperature from 40°C to 60°C for 3 h.

2.4. Product analysis

The samples were analyzed by gas chromatography(Shimadzu GC-14B) with flame ionization detection (FID) using an ZB-WAX (30 m long, 0.32 mm, 0.5 μm) capillary column. Injector and detector temperatures were kept 300°C. Oven temperature started at 40°C and was increased at 15 °C/min to 240 °C which was held for 30 min.

3. Result and discussion

3.1 Characterization of CaO egg shell

According to the thermogravimetric analysis (TGA) results reported in Fig 1, the first weight loss of eggshells at approximately 300-600°C is assigned to decomposition of moisture and organic compounds, whereas the final weight loss between 600°C and 750°C is due to loss of CO_2 from carbonate converting to CaO phase. In addition, the DSC curve, attributed to an endothermic peak at around 600-700°C, confirming the decomposition of $CaCO_3$ to CaO also implies that the calcination temperature of eggshells should be above 800°C to completely CaO forming. The surface area of CaO derived from eggshell was 4.9 m^2/g. which is slightly lower than the surface area of commercial CaO (5.5 m^2/g). The strength of the basic sites was 15.0 <H_<18.4 for the fresh CaO eggshell which was similar to commercial CaO catalyst (Table 1). Both catalysts were mesopores (from BET technique) with generally irregular crystal structure as aggregate shown in SEM images (Fig.2).

Table1. BET surface area and basic strength of catalysts

Catalysts	BET surface area(m^2/g)	Basic strength (H_)
Commercial CaO	5.5	15.0< H_<18.4
CaO derived from eggshell	4.9	15.0< H_<18.4

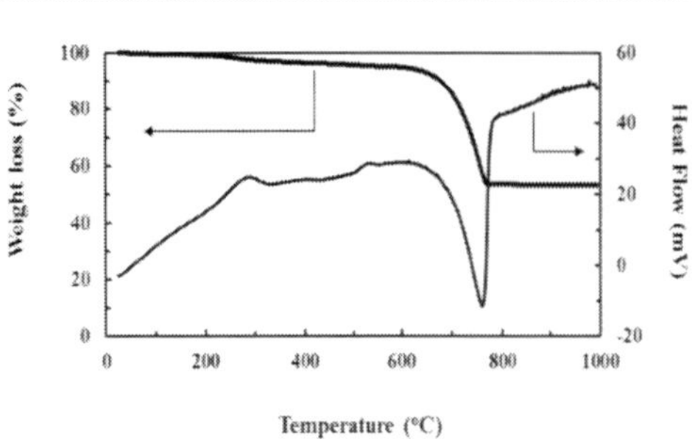

Fig. 1. TGA-DSC curves of eggshells.

Fig.2. Morphology of (A) commercial CaO and (B) CaO eggshell.

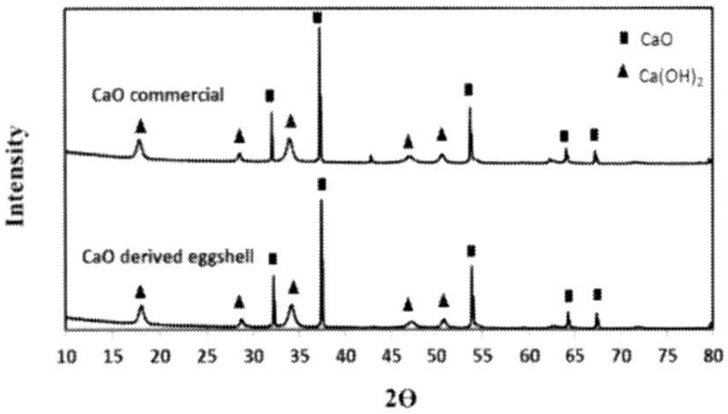

Fig. 3. XRD pattern of commercial CaO vs CaO eggshell.

The crystal structure of catalysts were obtained using an X-ray diffraction (XRD). The main composition of the calcined eggshell was CaO species while $Ca(OH)_2$ was found as a minor (Fig 3). The observation of $Ca(OH)_2$ phase might be because the sample was exposed to humidity in the air.

3.2 Catalytic activity of transesterification

Comparison between CaO derived from eggshell and commercial CaO for transesterification of glycerol with DMC are shown in Fig.4. The result shows that commercial CaO obtained complete conversion and CaO eggshell was nearly as same as one to approach glycerol conversion 96%.

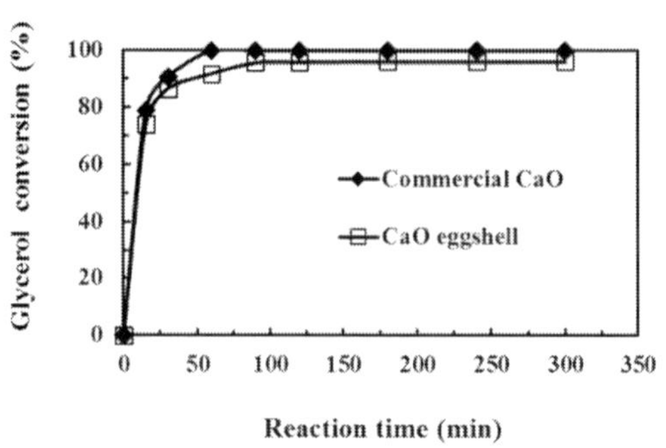

Fig. 4. Catalytic activity of commercial CaO vs. CaO eggshell.

3.3 Effect of reaction temperature

Transesterification of glycerol and DMC was investigated at varying temperature range between 40°C and 60°C under atmosphere pressure. As shown in Fig.5, initial conversion of glycerol increase with increasing reaction temperature due to the dependence of the kinetic rate with the temperature. Moreover, the equilibrium conversion increased from 83% to 91% and 96% with increasing reaction temperature from 40°C to 50°C and 60°C, respectively. Higher operating temperatures is favor due to a nature of endothermic reaction. In addition, increasing operating temperature enhance the miscibility of reactant mixture.The highest glycerol conversion of 96% can be achieved at 60°C within 90 min of reaction time.

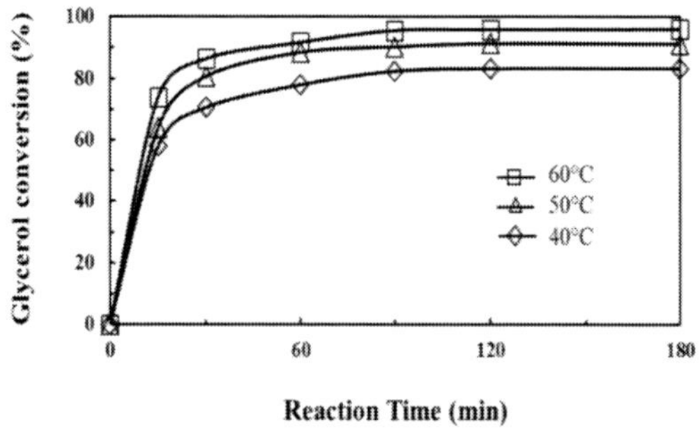

Fig. 5. Effect of reaction temperature on glycerol conversion. Reaction condition: molar ratio of DMC: glycerol = 2.5, mole ratio of catalyst to glycerol = 0.08, speed stirrer = 350 rpm.

3.4 Effect of catalyst loading

Fig.6. indicate that the amount of catalyst influence on conversion of glycerol when increasing mole ratio of CaO to glycerol from 0.02 to 0.08 the conversion was raised from 76% to 96%. This catalytic activity is enhanced due to the increase in the concentration of the active basic sites that are crucial to catalyze the transesterification reaction. Therefore mole ratio of CaO to glycerol of 0.08 was obtained the highest conversion of glycerol.

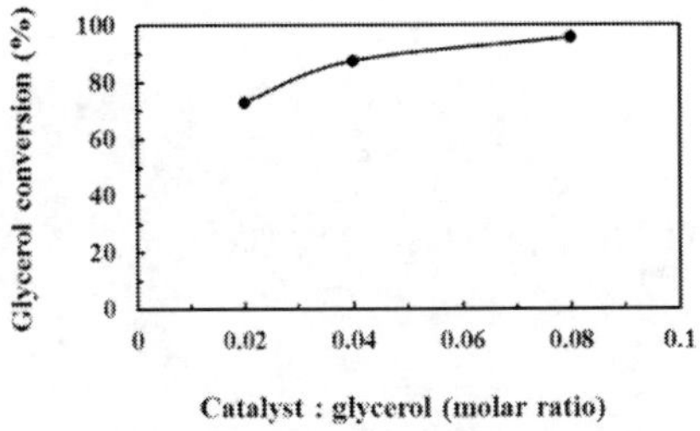

Fig. 6. Effect of catalyst loading on glycerol conversion. Reaction condition: molar ratio of DMC: glycerol = 2.5, reaction temperature = 60°C, reaction time = 90 min, speed stirrer = 350 rpm.

4. Conclusion

CaO derived from waste eggshell has strong basic sites and exhibits high catalytic activity of transesterification between DMC and glycerol. It was found CaO eggshell showed similar performance to the commercial CaO. Three reaction temperature levels of 40, 50, and 60 were investigated with different reaction time and catalyst loading. Glycerol conversion of 96% can be achieved within 90 min under suitable condition (reaction temperature of 60°C and catalyst loading of 0.08 mol/mol glycerol).

Acknowledgements

The authors also would like to acknowledge Research and Researchers for Industries (RRi) PhD program from the Thailand Research Fund (TRF) and Verasuwan CO., LTD.

References

1. Quispe, C.A.G., C.J.R Coronado and J.A. Carvalho Jr., *Glycerol: production, consumption, prices, characterization and new trends in combustion*. Renew Sust Energy Rev, (2013). **27**: p. 475-493
2. Teng, W.K. et al., *A review on the performance of glycerol carbonate production via catalytic transesterification: Effects of influencing parameters*. Energy Conversion and Management, (2014). **88**: p. 484-497
3. McNutt, J. and J. Yang, *Utilization of the residual glycerol from biodiesel production for renewable energy generation*. Renewable and Sustainable Energy Reviews, (2017). **71**: p. 63-76.
4. Dibenedetto, A., et al., *Converting wastes into added value products: from glycerol to glycerol carbonate, glycidol and epichlorohydrin using environmentally friendly synthetic routes*. Tetrahedron, (2011). **67**: p. 1308-1313.
5. Jung, K.S., et al., *Method of manufacturing glycerol carbonate*. (2012), U.S. 8,314,259 B2.
6. Rousseau, J., et al., *Tosylated glycerol carbonate, a versatile bis-electrophile to access new functionalized glycidol derivatives*. Tetrahedron, (2009). **65**: p. 8571-8581.
7. Ochoa-Gómez, J.R., et al., *A Brief Review on Industrial Alternatives for the Manufacturing of Glycerol Carbonate, a Green Chemical*. Organic Process Research & Development, (2012). **16**: p. 389-399.
8. Sonnati, M.O., et al., *Glycerol carbonate as a versatile building block for tomorrow: synthesis, reactivity, properties and applications*. Green Chemistry, (2013). **15**: p. 283-306.
9. Ochoa-Gómez, J.R., et al., *Synthesis of glycerol carbonate from glycerol and dimethyl carbonate by transesterification: Catalyst screening and reaction optimization*. Applied Catalysis A: General, (2009). **366**: p. 315-324.
10. Sun, Z., H. Chi, and L.-S. Fan, *Physical and Chemical Mechanism for Increased Surface Area and Pore Volume of CaO in Water Hydration*. Industrial & Engineering Chemistry Research, (2012). **51**: p. 10793-10799.
11. Oliveira, D.A., P. Benelli, and E.R. Amante, *A literature review on adding value to solid residues: egg shells*. Journal of Cleaner Production, (2013). **46**: p. 42-47.
12. Wei, Z., C. Xu, and B. Li, *Application of waste eggshell as low-cost solid catalyst for biodiesel production*. Bioresource technology, (2009). **100**: p. 2883-2885.
13. Chavan, S.B., et al., *Synthesis of biodiesel from Jatropha curcas oil using waste eggshell and study of its fuel properties*. RSC Advances, (2015). **5**: p. 63596-63604.
14. Sharma, Y.C., B. Singh, and J. Korstad, *Application of an Efficient Nonconventional Heterogeneous Catalyst for Biodiesel Synthesis from Pongamia pinnata Oil*. Energy & Fuels, (2010). **24**: p. 3223-3231.
15. Viriya-empikul, N., et al., *Biodiesel production over Ca-based solid catalysts derived from industrial wastes*. Fuel, (2012). **92**: p. 239-244.
16. Gao, Y. and C. Xu, *Synthesis of dimethyl carbonate over waste eggshell catalyst*. Catalysis today, (2012). **190**: p. 107-111.
17. Fan, S., et al., *Experimental and kinetic study of catalytic steam gasification of low rank coal with an environmentally friendly, inexpensive composite K 2 CO 3–eggshell derived CaO catalyst*. Fuel, (2016). **165**: p. 397-404.
18. Taufiq-Yap, Y.H., et al., *Hydrogen production from wood gasification promoted by waste eggshell catalyst*. International Journal of Energy Research, (2013). **37**: p. 1866-1871.
19. Karoshi, G., et al., *Calcined eggshell as an inexpensive catalyst for partial oxidation of methane*. Journal of the Taiwan Institute of Chemical Engineers, (2015). **57**: p. 123-128.

Optimization of drum drying parameters for gac powder

Pornthip Suriyajunhom[1], and *Maradee* Phongpipatpong[1]

[1]Food Engineering Department, Faculty of Engineering, King Mongkut's Institute of Technology Ladkrabang, Bangkok, Thailand

Abstract. Gac (Momordica cochinchinensis Spreng.) is a tropical fruit which contains a rich nutrient source, especially lycopene. However, fresh gac fruit can spoil rapidly due to its high moisture content. Drum drying is a process to remove moisture from food in order to preserve nutrients and extend shelf life. In this study, gac fruit powder was produced by using drum drying technique. The present research was aimed to determine the optimum drum drying condition for gac powder. Response Surface Methodology was employed with three processing variables, including drum surface drying temperature (115, 125, 135°C) maltodextrin content (100, 150, 200%) and gum arabic content (10, 20, 30%). The response variables were moisture content, solubility, hygroscopicity, total phenolic content and lycopene content of gac fruit powder. Regression analysis was then performed to obtain the mathematical relation between the processing variables and the responses in the form of second-order polynomial equations.. The results showed that optimum conditions for drum drying of gac fruit was at 135°C of drum surface drying temperature, 138% of added maltodextrin and 10% of added gum arabic, which provided the highest amount of total phenolic and lycopene content.

1 Introduction

Gac is a tropical fruit which classified in the cucurbitaceae family. It grows in many countries in Asia including China, Malaysia, Vietnam and Thailand. This plant can grow fast and has a low price, but it is a good source of nutrients such as vitamin E, fatty acid, phenolic compound, carotenoid – Lycopene and beta-carotene [1,2]. However, fresh gac fruit has a short shelf life due to its high moisture content that susceptible to deterioration. Thus, food processing can be applied in order to prevent spoilage, develop and add value to the product. Drum Drying is one of popular processing technique to convert food into a dried flake or powder which can help to keep nutrition and to prolong shelf-life of fruit product. The advantage of drum drying is that the simpler structure of the dryer, less operating costs, less drying time, higher drying capacity, and can be applied to wider ranges of liquid food or viscous food [3]. A typical drum dryer may have either a single drum with smaller applicator rolls or double drums. In this dryer, pureed food is spread uniformly over the surface of a heated drum. While the drum rotates, drying takes place and dried product is scraped off as flake from the drum using a knife system. The product can further grind into powder form [3,4].

Several studies have reported the problem of the development of stickiness and caking during fruit drying due to the presence of low molecular weight sugars and acids which have low glass transition temperature, results in lower product yield, operation shut-downs, equipment wear, difficulties in material storage and handling. This causes a considerable economic loss and consumers' discrimination [5]. Caking and stickiness are enhanced by numerous factors such as glass transition temperature (Tg) and hygroscopicity of the food materials, and environmental temperature and relative humidity etc. Therefore carrier agents e.g. maltodextrins, gum arabic, waxy starch, and microcrystalline cellulose are usually added into the feed prior to drying step, influence the properties and stability of the powder. Maltodextrin is the most common drying aids which helps increase in Tg of the food product. The addition of maltodextrin can be applied to reduce stickiness of the powder during drying especially for sugar rich foods [6]. Gum arabic is mixture of polysaccharides and glycoproteins, it is used primarily in the food industry as a stabilizer and emulsifier to enhance the favour and stability of food product. However, the addition level of carrier agents is important to obtain the desired food powder characteristics. [7]

The optimization of processing parameters is essential in food engineering and product development. Response surface methodology (RSM) has been successfully applied to optimize food processing operations including drying. This method requires identifying and fitting experimental data and appropriate response surface model with the use of statistical experimental design, regression modeling techniques, and optimization methods.[8] The most common forms of response variables and design variables are low-order polynomials (first or second-order).

Since drum drying is an energy-efficient and low production cost process, it is interesting to apply this technique for gac fruit. Therefore, this research is aimed to study drum drying process for gac powder production and

* Corresponding author: maradee.ph@kmitl.ac.th

to optimize the process conditions to obtain the desirable characteristics of drum-dried gac fruit powder.

2 Materials and methods

2.1 Sample Preparation

Fresh gac fruits were collected from plantation area in Kanchanaburi province after approx. 150 days after flowering. Only mesocarp and endocarp were manually separated from the fruit, blended and pasteurized at 72°C for 15 sec. before packed into polyethylene bags and kept in a freezer at -18oC until be analysed and processed.

2.2 Drum Drying Operation

For drum drying experiments, the frozen gac pureed samples were thawed overnight at room temperature then mixed with Maltodextrin DE-10 and gum arabic at different amounts based on experiment design. A double-drum dryer was used for drying trials. Steam is used as heating medium and fed into the drum. Drum surface temperature can be adjusted by controlling steam pressure (at 10, 20, 30 psig). The dried gac samples were scraped out from the drums and kept in aluminum foil bag until analysis.

2.3 Experimental design

Response surface methodology was used as an optimization approach. Experimental data of gac fruits drying tests carried out based on Box-Behken design as shown in Table 1. The effect of drum surface drying temperature, maltodextrin concentration and gum Arabic concentration, selected as independent variables, on physicochemical properties of gac powder including moisture content, solubility hygroscopicity, total phenolic content and lycopene content was described by polynomial models. The experiment results were analysed and expressed in the form of Eq. (1). Optimization procedure was finally performed using graphical analysis with superimposed contour plot.

$$Y = \beta_0 + \sum_{i=1}^{n} \beta_i X_i + \sum_{i=1}^{n} \beta_{ii} X_i^2 + \sum_{i=1}^{n-1} \sum_{j=i+1}^{n} \beta_{ij} X_i X_j \quad (1)$$

Where Y = Physicochemical properties of gac fruit powder

X_i = Variable such as drying temperature (°C) maltodextrin (%) and gum arabic (%)

β_0 = Regression coefficient at Y-intercept

β_i = Regression coefficient in term of linear, quadratic and interaction

Table 1 Independent variables and their levels in the Box-Behnken design.

Independent variables	symbol	Coded values		
		-1	0	+1
Drying temperature (°C)	X_1	115	125	135
Maltodextrin content (%)	X_2	100	150	200
Gum Arabic content (%)	X_3	10	20	30

2.4 Physicochemical Properties Analysis

2.4.1 Moisture content

The gac powder sample was dried in the oven at 105°C until the weight of sample remained unchanged. The results were computed as shown in Eq. (2).

$$\text{Moisture content (\%)} = \frac{W_1 - W_2}{W_1} \times 100 \quad (2)$$

When W_1 = initial sample weight before oven drying (g)

W_2 = dried sample weight after oven drying. (g)

2.4.2 Solubility

25 gms of dried gac powder was dissolved in 30 ml of water for 10 min. and centrifuged at 3,000 rpm for 15 mins. Supernatant liquid was poured and dried in hot air oven at 105°C until the weight remained unchanged. The results were computed as shown in Eq. (3). [9]

$$\text{Solubility(\%)} = \frac{\text{Weight of solids in the dried supernatant}}{\text{Weight of initial solids sample}} \times 100 \quad (3)$$

2.4.3 Hygroscopicity

Hygroscopicity were determined according from Bhusari et al (2014) by placing 1 g of dried gac powder in saturated solution of sodium chloride for 14 days. Hygroscopicity was then calculated based on Eq. (4) and expressed in $g_{moisture}\, g^{-1}_{dry\ solid}$. [10]

$$\text{Hygroscopicity}\left(\frac{g_{moisture}}{100\, g_{dry\ solid}}\right) = \frac{\frac{\Delta m}{(m_i + m)}}{\frac{\Delta m}{m} + 1} \quad (4)$$

When m_i = The free water contents of the powder before placing into the saturated environment. (g)

m = The initial mass of powder. (g)

Δm = The increase in the weight of powder after equilibrium. (g)

2.4.4 Total phenolic content [11]

Total phenolic content was determined using the Folin–Ciocalteau reagent method. The absorbance was measured using UV-vis spectrophotometer at 765 nm wave length. Total phenolic contents of gac powder were calculated using a standard curve of Gallic acid and report as mg GAE g/100 mL

2.4.5 Lycopene content

Lycopene content were determined according to the Kimura method [12]. The absorbance was measured with UV-vis spectrophotometer at different wave lengths including 453, 505, 645 and 663 nm. Lycopene content in gac juice powder were calculated based on Eq.5 and reported unit in (mg/100g).

$$\text{Lycopene content} = -0.0458A_{663} + 0.204A_{645} + .372A_{505} - 0.0806A_{453} \quad (5)$$

When A663, A645, A505 and A453 is absorbance at 663, 645, 505 and 453 nm respectively

3 Result and discussion

The relationships between the response variables and three independent variables were investigated and correlated using a 2nd order polynomial equation, Eq. 1. ANOVA evaluations of these models are presented in Table 2.

3.1 Moisture content

In this study, the experiments show that the increase in drying temperature, maltodextrin content and gum arabic content result in a negative effect on gac powder moisture content. The most significant factor on product moisture was drying temperature ($p<0.1$). Similar results were also observed by Galaz et. al. [4], they reported that the effect of steam pressure or drying temperature on product moisture content is highest, followed by drum speed. Higher temperature reduces feed viscosity, reduces the entrainment flow of liquid on drum surface and thereby reduces dried product moisture. The minimum moisture content of gac powder was 1.5 % when drying at 135°C of drum surface temperature, 200% of maltodextrin addition and 30% of gum arabic addition to the gac fruit pureed before fed to the drum dryer. The contour plot of drum-dried gac powder is presented in Fig. 1.

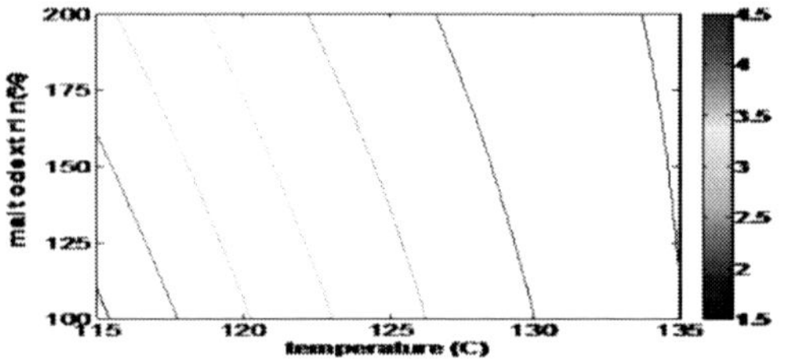

Fig.1 .Moisture content of drum-dried gac powder affected by drying températu re and maltodextrin content at 10%. gum arabic

3.2 Solubility

Solubility of drum-dried gac powder was increased with drying temperature and maltodextrin content. While the addition of gum arabic had little effect on product solubility. Increasing the concentration of maltodextrin or gum arabic result in the decrease intermolecular forces by interacting with the functional groups of the polysaccharide, providing an increase in solubility. The contour plots of product solubility is present in Fig. 2 and it shows that the highest solubility (76%) of drum-dried gac powder was at the high temperature and high amount of maltodextrin addition. Higher drying temperature increases the volume of gac powder and make it more porous. This cause product powder can dissolve faster. It was also observed that maltodextrin and gum arabic are a good dissolvable substance. [11].

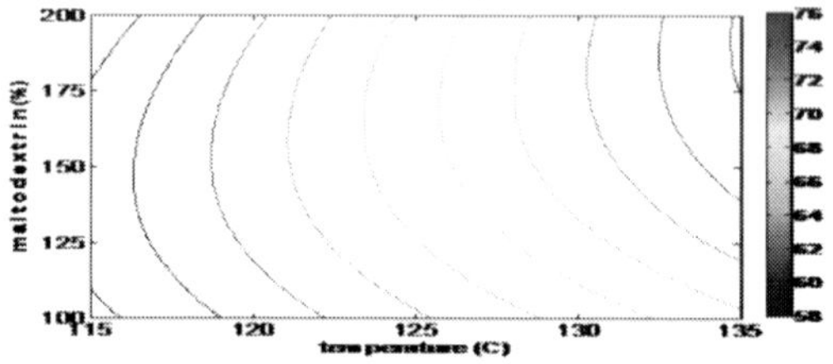

Fig. 2. Solubility of gac powder as affected by drying temperature and maltodextrin content at 10% gum arabic

3.3 Hygroscopicity

In this study it was found that the interaction of maltodextrin and gum arabic content was the most significant positive effect on hygroscopicity of gac fruit powder ($p<0.01$). While drying temperature has little effect on hygroscopicity. The highest value of hygroscopicity of gac fruit powder also occurred at high drying temperature and higher amount of carrier agents (maltodextrin and gum Arabic) as shown in Fig 3

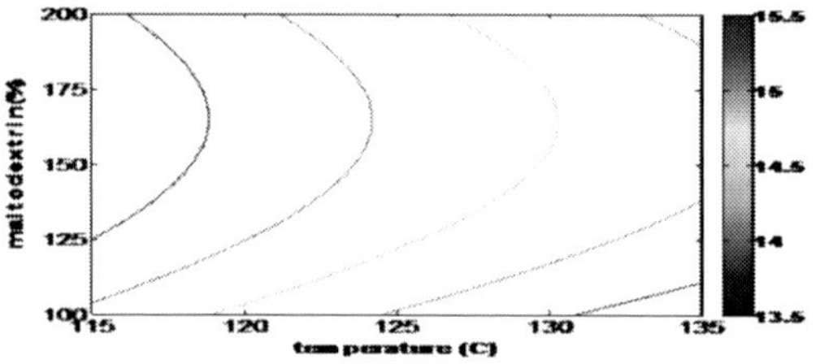

Fig. 3. Hygroscopicity of gac powder as affected by drying temperature and maltodextrin content at 10% of gum Arabic

Table 2 Regression coefficients of the second-order polynomial response models.

Variables	Regression coefficient	Moisture content (%)	Solubility (%)	Hygroscopicity ($g_{water}/100g_{powder}$)	Total phenolic content	Lycopene content
intercept	β_0	87.84399	-36.4979	0.740489	-64.6843	-0.09806
x_1	β_1	-1.13828*	0.989454	0.288347	0.91774	0.001247
x_2	β_2	-0.06309	-0.22095	0.050087*	0.025576	-0.00088
x_3	β_3	-0.18276	1.312556	0.208455	0.508405	0.01567
$x_1 \cdot x_1$	β_{11}	0.003685	-0.00257	0.435144	-0.00249	1.21E-05
$x_2 \cdot x_2$	β_{22}	1.17E-06	-0.00079**	0.002973***	9.41E-05	3.44E-06
$x_3 \cdot x_3$	β_{33}	-0.00163	0.007121	0.060509	-0.00024	2.87E-05
$x_1 \cdot x_2$	β_{12}	0.00046	0.003851**	0.855142	-0.00061	-2.8E-06
$x_2 \cdot x_3$	β_{13}	0.00024	0.00052*	0.026412	0.000131	-2.1E-06
$x_1 \cdot x_3$	β_{23}	0.001667	-0.01266	0.214111**	-0.0042	-0.00013
R Square (R^2)		0.9476	0.9883	0.9794	0.8207	0.6381
Standard Error (S.E.)		0.3573	1.0032	0.2064	0.9613	0.0194

Note * Significant at $p<0.1$, ** Significant at $p<0.05$ and *** Significant at $p<0.01$

3.4 Total phenolic content

Total phenolic content of gac powder is shown in Fig 4. The study shows that the maximum total phenolic content of gac powder was 8.5 mg GAE g/100 mL and occured at high drying temperature (135°C) and less addition of maltodextrin (100%) and gum arabic (10%). Increasing drying surface temperature provided higher amount of total phenolic content due to the synthesis of polyphenols occur at high temperature, while the increase in maltodextrin and gum arabic decrease the total phenolic content of the dried gac powder.

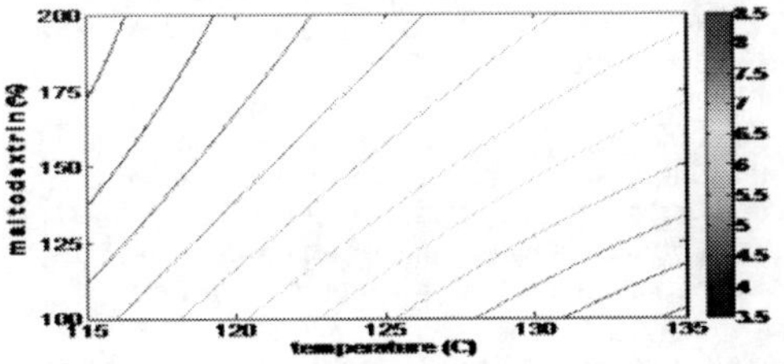

Fig. 4 . The contour plot of total phenolic content of gac powder as affected by drying temperature and maltodextrin content at 10% gum Arabic

3.5 Lycopene content

Fig 5. shows the variation of drying temperature and maltodextrin content at 10% of gum arabic. It can be seen that drum surface drying temperature has a positive effect on lycopene content, while the addition of maltodextrin results in decreasing lycopene content in gac powder. The maximum lycopene content was 0.18 mg/100g and occurred at high drying temperature (135°C) with less addition of maltodextrin (100%) due to heat induces isomerization of the all-trans to cis forms which increase with temperature. While the addition of gum arabic had little positive effect on lycopene content.

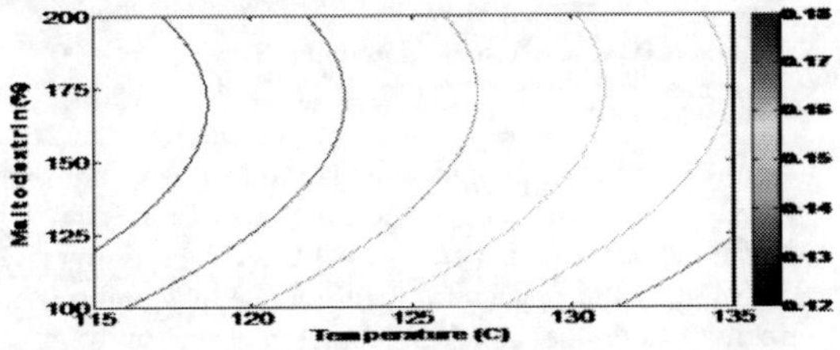

Fig. 5. The contour plot of lycopene content of gac powder as affected by drying temperature and maltodextrin content at 10% of gum arabic

3.6 Optimization of drum drying processing parameter to produce gac fruit powder

Optimization of process conditions for drum drying of gac fruits was done by using response surface methodology and superimposed of contour plots of response variables at different processing conditions. The decision criteria of drum dried gac powder properties consist of moisture content < 5%, Solubility > 70%, Hygroscopicity < 15 $g_{water}/100g_{powder}$, total phenolic content > 6.5 mg/g and lycopene content > 0.15 mg/100 g_{powder}. Fig. 6 shows the graphical optimization. It was found that the optimum condition was at 135°C drying temperature, maltodextrin and gum arabic at 138% and 10% respectively. Physicochemical properties of gac powder composed with 1.48% of moisture content, 73.91% of solubility, 14.99 g /100g of hygroscopicity, 7.35 mg/g of phenolic content and 0.16 mg/100g of lycopene content.

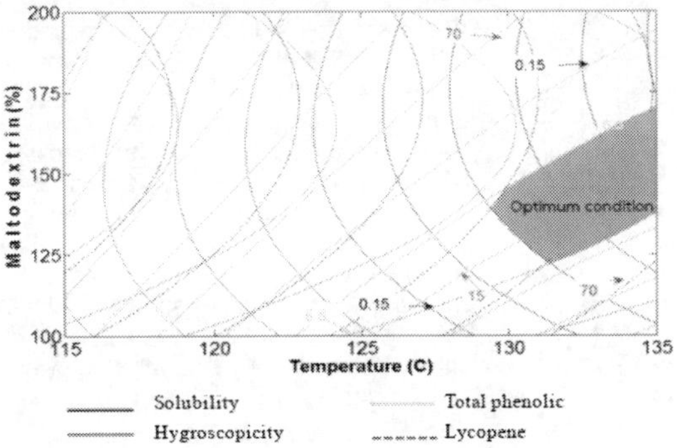

Fig. 6. Superimposed contour plot of response variables.

4 Conclusion

Drum drying was used for gac fruit powder production. Response surface optimization was carried out to determine the optimum processing condition of drum surface drying temperature, maltodextrin content and gum arabic content, that provide the desired properties of dried gac powder. The second order polynomial models were successfully developed to describe the response variables at different processing condition.

References

1. T.C. Kha, M. H. Nguyen, P.D. Roach, S.E. Parks, C. Stathopoulos, Food Reviews International, 29(1), 92-106 (2013)
2. J. Kubola, S. Siriamornpun, Food Chemistry, Volume 127 (3), 1138-1145 (2011)
3. A,S, Mujumdar, Handbook of Industrial Drying, CRC Press, Taylor & Francis Group (2014)
4. P. Galaz, M. Valdenegro, C. Ramírez, H. Nuñez, S. Almonacid, R. Simpson, Food Engineering, 208, 19-27 (2017)
5. B. Adhikari, T. Howes, D. Lecomte, B.R. Bhandari, Powder Technology, 149 (2–3), 168-179 (2005)
6. D. P. S. Oberoi, D. S. Sogi, Food Engineering, 165, 172-178 (2015)
7. C. Agustina, D. Battista, D. Constenla, M. V. Ramírez-Rigo, J.Piña., Powder Technology, 286, 193-201, (2015)
8. I. Atalar, M. Dervisoglu, LWT - Food Science and Technology, 60 (2), 751-757, (2015)
9. S. Vidović, J.Z.Vladić, Ž.G. Vaštag, Z.P. Zeković, L.M. Popović, Powder Technology, 258, 209-215 (2014)
10. S.N Bhusari, K. Muzaffar, P. Kumar, Powder Technology, 266, 354-364 (2014)
11. J.Yang, R.Gadi, R. Paulino, T.Thomson, Food Chemistry, 122(3), 627-632 (2010)
12. M. Nagata, I. Yamashita, J. Japan. Soc. Food sci. Technol, 39(10), 925-928 (1992)

* Corresponding author: maradee.ph@kmitl.ac.th

Photoactivity of Fe and Zn-doped TiO_2 in phenol degradation under visible light

Panadda Yodsomnuk[1], *Kanya* Junjeam[1], *Mutsee* Termtanun[1*]

[1]Department of Chemical Engineering, Faculty of Engineering and Industrial Technology, Silpakorn University, Nakhon Pathom, 73000

Abstract. This study developed bandgap energy of TiO_2 into the visible region for the photo-activation. The photocatalytic performance of TiO_2 was improved when Fe-doping at 2% and 4% by weight, Zn-doping at 2% and 4% by weight, and doping Fe/Zn 1+1% and 2+2% by weight using flame spray pyrolysis as a synthesis method. Brunauer-Emmett-Teller (BET), ultraviolet-visible spectroscopy (UV-Vis), and x-ray diffraction (XRD) were used to characterize the properties of synthesized and all doped TiO_2 photocatalysts. To prove the effectiveness, all the modified TiO_2 were also compared with the pure synthesized TiO_2 by performing photodegradation of phenol under 1hr visible light. The remaining phenol concentration was estimated using gas chromatography-flame ionization detector (GC-FID). Doped TiO_2 enhances smaller grain size, larger specific surface area and narrower the bandgap energy after comparing with the pure synthesized TiO_2. Zn-doped leads to large surface area that improved the photocatalytic activity of catalyst. However, there was no significant synergistic effect between Fe and Zn doping. The optimum condition for phenol photodegradation was at about 4%Fe doped TiO_2.

1 Introduction

Advanced Oxidation Process (AOP), the application of photocatalysis for detoxification of wastewater, has been discovered as one of clean technology to remove organic and inorganic compounds, pesticides, eradication toxins, and destruction contaminants. This technique uses semiconductors in photocatalysis since they display a great ability with a low cost, and environment friendly. Various semi-conductors, for example, TiO_2, ZnO, ZrO_2, CdS, MoS_2, Fe_2O_3, and WO_3 have been studied and modified to use in photocatalysis degradation of organic and inorganic contaminants. Many researchers have highlighted on TiO_2 as a potential photocatalyst in wastewater treatment due to several factors [1]. TiO_2 is non-toxic, highly chemical and mechanical stability. It has outstanding potential in degradation of contaminants in wastewater, applied in completely oxidization of the organic pollutants to environmentally friendly products, which are H_2O and CO_2. However, titanium dioxide has major drawback in that its bandgap energy is equal to 3.2 eV, which is not only required UV to activate but also can be absorbs only 5% of visible light. Therefore, the improvement of the light absorption wavelength of TiO_2 into visible light region has been widely interested among researchers, in order to use under sunlight. To achieve this goal, the narrowing bandgap energy of TiO_2 by using various metals loading has been established as a possible option for TiO_2 modification. It has been noticed that doping metals such as Fe, Cu, V, Pt, Au, Ag, Mn, Mo, Cr, and Ni could improve the photocatalytic activity [2-13]. Among these metals, Fe and Zn seem to be considerable potential photocatalyst promoters[14-19]. However, until now there was no research studying the effect of co-doping of Fe and Zn as promoter on TiO_2. In this study, flame spray pyrolysis was selected as a method to synthesize doped TiO_2. Flame spray pyrolysis (FSP) is an effective technique to synthesize high performance nanoparticle catalysts, which produces high purity nanoparticles with controlled particle size by adjusting the process conditions such as the concentration and flow rate of precursors, and fuel flow rate. Therefore, the purposes of this study are to synthesize Fe/Zn-doped TiO_2 catalysts using FSP, to characterize their physical and chemical properties, to find out any synergistic effect between Fe and Zn, and to measure their photocatalytic activity in phenol degradation under visible light.

2 Materials and method

2.1 Preparation of the Fe/Zn doped TiO_2

Titanium (IV) tetra isopropoxide (TTIP) (Sigma-Aldrich, United States 97%), ferrous nitrate (III) nonahydrate (Sigma-Aldrich, United States 99.95%) and zinc chloride (Ajax Finechem, Australia 95%) were dissolved in ethanol (CT Chemical, Thailand 99.9%) to a 0.4 M solution and used as a precursor feed for flame spray pyrolysis. The mixture was injected through the center capillary of the FSP nozzle with oxidizing gas into flame zone by syringe pump at the flow rate of 5 ml/min. The liquid spray was ignited by methane (at the flow rate of 1.5 l/min) and oxygen (at the flow rate of 3 l/min). After combustion, the metal precursor sprayed in FSP was then converted into nanoparticles. The product nanoparticles were collected on a glass fiber filter (Whatman GF/D, 15 cm diameter).

2.2 Characterization of catalysts

Pure synthesized TiO_2 and Fe/Zn doped TiO_2 were analyzed with a number of characterization techniques. XRD using CuK_α radiation with Ni filter in the 2θ range of 20-80 degree with resolution 2°/min was carried out to detect the phase identification and to measure the phase composition. The light absorption property of each

* Corresponding author: Termtanun_m@su.ac.th

catalyst sample was measured using UV-Vis under an ultra violet and visible wavelength (200-800 nm). BET surface area analysis (N_2 adsorption) was carried out with firstly pretreated in helium gas flow of 50 ml/min at 180°C for 4 hr. SEM at 5000X magnification, and EDS were performed to study the variation of the morphology and to analyze elemental compositions of doped photocatalysts.

2.3 Photocatalytic activity measurement

100 ppm of phenol aqueous solution was used as a stock solution of organic compound. The photocatalytic degradation was carried out with 150 ml of phenol solution using 0.6 g of catalyst. 30 minutes starting up procedure under dark adsorption (no irradiation source) of TiO_2 was carried out before running through the photocatalytic degradation of phenol. The mixture was stirred at 300 rpm for an hour under visible light irradiation (Phillips, mercury lamp 16 watt) with the distance of 100 mm above the solution surface. Then, the tested solution after using photocatalyst sample was collected at the end of an hour to monitor the remaining concentration of phenol using GC-FID. The product distributions from photocatalytic degradation of phenol were observed using GC-MS. The type of GC column was reverse phase fused silica capillary using helium as a carrier gas, at the column temperature of 220 °C with column flow rate of 0.81 ml/min.

3 Results and discussion

3.1 X-Ray diffraction

XRD spectra of the pure synthesized TiO_2 and doped TiO_2 are shown in the Figure 1. In case of the pure synthesized TiO_2, the strongest peak representing anatase TiO_2 occurred at $2\theta = 25.2°$. On the other hand, XRD patterns exhibited the strongest diffraction peaks of the rutile phase at $2\theta = 27.4°$. However, the results indicated no any characteristic peaks of Fe and Zn which implied that they were well dispersed on the surface or because of low doping amount.

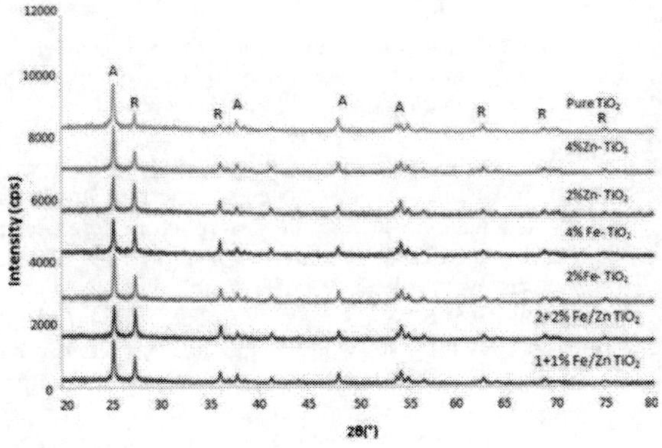

Figure 1 XRD patterns of various doped-TiO_2 comparing with the pure synthesized TiO_2

In addition, Table 1 lists the crystalline size associated with $2\theta = 25.2°$ and $2\theta = 27.4°$ which were the sharpest peak that indicated anatase and rutile phases in all samples, respectively. By using Debye-Scherrer equation [13], the crystallite size of anatase was 41.1 nm, and 39 nm for 2% Fe-TiO_2 and 4% Fe-TiO_2, respectively. Therefore, the crystalline size of the anatase TiO_2 diminished while increasing doping amount of Fe. Similarly, in case of Zn doping, it reduced the grain size of anatase TiO_2. The anatase crystallite size of 2%Zn-TiO_2 (41.1 nm) and 4% Zn- TiO_2 (41nm) were smaller than that of the pure synthesized TiO_2 (45.4 nm). The results of anatase phase show the identical trend with rutile phase in that Fe and/or Zn doping on TiO_2 lead to smaller crystalline size.

The percentages of rutile and anatase phase in synthesized and doped TiO_2 were calculated using the Spurr-Myers equation [20]:

$$\%Anatase = \frac{1}{\left(1+1.265\frac{Ir}{Ia}\right)} x\ 100 \quad (1)$$

$$\%Rutile = 100 - \%Anatase \quad (2)$$

Where I_r = intensity of the diffraction peak of rutile, and I_a = intensity of the diffraction peak of anatase.

Table 1 The crystalline size of anatase and rutile phases of all synthesized TiO_2 determined from XRD patterns in Figure1.

Catalyst samples	Crystalline size (nm)	
	Anatase	Rutile
Pure TiO_2	45.4	45.9
2%Fe-TiO_2	41.1	35.8
4%Fe-TiO_2	39	37.3
2%Zn-TiO_2	41.1	35.8
4%Zn-TiO_2	41	38.9
1+1%Fe/Zn-TiO_2	41	36.4
2+2%Fe/Zn-TiO_2	41.1	35.7

After this calculation, 2% Fe-doped TiO_2 has approximately 67.02% anatase and 32.98% rutile which is more anatase-to-rutile than pure synthesized TiO_2 (65.2% anatase and 34.8% rutile). This revealed the fact that adding Fe and Zn into TiO_2 photocatalyst inhibits the phase transformation from anatase to rutile which results in more anatase content, corresponding to the prior study [21].

3.2 BET surface area and pore distributions

In Table 2, BET surface area of Fe and Zn doped TiO_2 are larger than that of the pure synthesized TiO_2 and range between 34.86 m^2/g and 64 m^2/g. After comparison among doped TiO_2, 4% Fe-doped TiO_2 has the largest surface area which is equal to 64 m^2/g.

Table 2 BET surface area measurement of Fe and Zn doped TiO_2 catalysts.

Catalyst samples	BET surface area (m^2/g)
P-25	20.4
pure TiO_2	50.5
2%Fe-TiO_2	52.2
4%Fe-TiO_2	64
2%Zn-TiO_2	37.7
4%Zn-TiO_2	48.5
1+1%Fe/Zn-TiO_2	34.9
2+2%Fe/Zn-TiO_2	41.7

3.3 UV-Vis spectroscopy

The bandgap energy calculated from Kubelka-Munk relationship [22] using UV-Vis results shown in Table 3, is the sufficient required energy of photons that is used to active electron and hole in the photocatalytic process. In this theory, the graph was plot between transform reflectance spectra $(F(R_\infty)h\nu)^{1/2}$ versus hv.

According to Kubelka-Munk equation;

$$F(R_\infty) = \frac{(1-R_\infty)^2}{2R_\infty} \quad (1)$$

Transformation of absorbent unit into reflectance

$$A = log10(1/R) \quad (2)$$

$$R_\infty = 10^{-A} \quad (3)$$

Where A = absorbent unit

h = Plank constant (6.624 x 10^{-34} Joules sec)

ν = velocity of light (3.0 x 10^8 meter/sec)

R_∞ = reflectance unit

Table 3 Calculated bandgap energy (E_g) of various doped TiO_2 compared with the pure synthesized TiO_2

Catalyst samples	E_g(eV)
P-25	3.42
Pure-TiO2	3.39
2%Fe-TiO_2	3.2
4%Fe-TiO_2	2.87
2%Zn-TiO_2	3.26
4%Zn-TiO_2	3.16
1+1%Fe/Zn-TiO_2	3.08
2+2%Fe/Zn-TiO_2	3.05

The less bandgap energy of the photocatalyst, the greater probability for photocatalytic reaction. Doping Fe and/or Zn on TiO_2 are able to narrow the bandgap energy and shift light absorption region to longer wavelength which lead to higher photocatalytic efficiency for organic removal application under visible light when compared to the pure synthesized TiO_2. Among all doped TiO_2, 4%Fe-doped TiO_2 achieves the maximum amount of light adsorption at the visible region. This was related to the calculation of bandgap energy from Kubelka-Munk relationship in that 4%Fe-doped TiO_2 has lower bandgap energy (2.87 eV) than that of 4%Zn-doped TiO_2 (3.16 eV) and 2+2% Fe/Zn co-doped TiO_2 (3.05 eV).

3.4 Photocatalytic activity

The photocatalytic degradation activities of phenol of the various catalysts under visible light irradiation are shown in Figure 2. According to the percentage remaining of phenol from Gas Chromatography (GC), the percentage of phenol conversion (X) can be determined from the following equation [23];

$$X(\%) = \left(1 - \frac{CA}{CA0}\right) x\ 100 \quad (3)$$

While C_A is equal to remaining concentration of phenol in an aqueous solution measured by GC-FID, and C_{A0} is equal to an initial concentration of phenol in an aqueous solution. With the dark adsorption experiment, the adsorbed concentration of phenol was 8.2% using pure synthesized TiO_2. However, after 1 hr photocatalytic degradation under visible light irradiation, the percentage conversion of phenol was 14.3% for the pure synthesized TiO_2. With Fe and/or Zn doping, the photocatalytic activities increased according to calculated phenol conversion. After consideration all doped TiO_2, both metal types and metal loading show remarkable influence on photocatalytic activity. The results showed that 4%Fe doping on TiO_2 has the most efficiency on phenol degradation with 95.72% of phenol conversion, which is higher than that of 4% Zn doping and 2+2% Fe/Zn co-doping (73.36% and 79.34% phenol conversion, respectively).

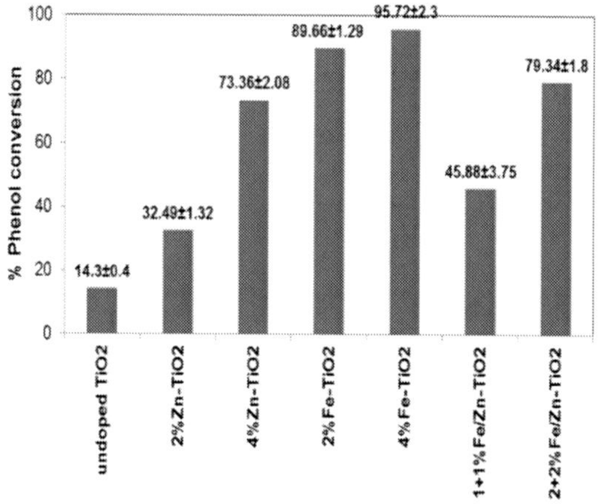

Figure 2 The comparison of photo-degradation activity of different photocatalysts measured by %phenol conversion after 1 hr degradation under visible light irradiation

These results are corresponding with UV-Vis spectra that Fe-loading has lower bandgap energy than that of Zn- loading and shows better adsorption performance at longer wavelength which leads to the higher photocatalytic efficiency under visible light irradiation. The influence of Fe-doping on the photocatalytic activity reflects on the narrowing of the bandgap energy and leading to exhibit superior visible-light-driven photocatalytic activity and increasing BET surface area (S_{BET}). The reasons of why Zn improves photocatalytic activity of TiO_2 was that it decreases the grain size, shifts the light absorption to higher wavelength which was confirmed by UV-Vis and XRD results in the former section. This was corresponding to the prior study that it enhances charge separation of electron-hole pairs which results in increase in oxidation and reduction reaction [19,24]. However, co-doping of Fe and Zn on TiO_2 leads to less conversion than Fe single metal-doping. Though Fe and Zn have prominently potential in improving TiO_2 photocatalytic activity, it was presumably surprising that there was rarely synergistic effect between them. Moreover, it can also be noticed that there are two factors which influence on photocatalytic activity: the bandgap energy and the specific surface area. The most potential photocatalyst should have combination of two characteristics as narrower the bandgap energy and larger the specific surface area. For example, 4% mol Zn loading provides the wider the bandgap energy than co-doped 1%Fe and 1%Zn but the larger the specific surface area, leading to the better photocatalytic performance in phenol photodegradation. Considering at all different doping amounts of Fe or Zn onto TiO_2, using 4%mol Fe loading is accomplish the highest of photocatalytic activity in phenol degradation.

4. Conclusions

This research proved the influences of doping level and doping type on photocatalytic activity, which was corresponding with some previous studies. Based on Fe and Zn modification, they help improving TiO_2 for photocatalytic degradation under visible light. There was apparent evidence that Fe and Zn doping on TiO_2 can narrow the bandgap energy. Hence, doping metal on TiO_2 supports the photocatalytic activity in the range of visible light which can be activated by using solar energy. Among the different doping levels and between Zn or Fe doping metals on TiO_2, Fe-doped TiO_2 shows the best photocatalytic performance, and the optimal amount of Fe doping was 4% mol. Unfortunately, there was no sign of synergistic effect between Fe and Zn because co-doping of Fe and Zn on TiO_2 had lower photo-activity than Fe single-doped TiO_2.

Abbreviations

AOP: Advanced Oxidation Process
FSP: Flame Spray Pyrolysis
XRD: X-Ray Diffraction
UV-Vis: Ultraviolet Visible Spectroscopy
BET: Brunauer-Emmett-Teller
SEM: Scanning Electron Microscopy
EDS: Energy Dispersive X-Ray Spectroscopy
GC-FID: Gas Chromatography-Flame Ionization Detector
GC-MS: Gas Chromatography-Mass Spectroscopy
X: percentage of phenol conversion
C_A: remaining concentration of phenol in aqueous solution
C_{A0}: initial concentration of phenol in aqueous solution
S_{BET}: Surface area from BET analysis (m^2/g)
E_g: bandgap energy (eV)
A.U.: Arbitrary Unit
cps: counts per second

This research was financially supported by Faculty of Engineering and Industrial Technology, Silpakorn University, Nakhon Pathom, Thailand.

References

[1] Konstantinou L. K., and Albanis. T. A., J. App. Cat. B: Environ., 49, 1-14(2004)
[2]Colón G., Maicu M., Hidalgo M.C., Navío J.A., Appl. Cat. B: Environ., 67 (1–2), 41–51(2006)
[3]Ananpattarachai J., Boonto Y., and Kajitvichyanukul P., Environ. Sci. Pollut. Res., 1-9(2015)
[4]Binas V. D., Sambini K., Maggos T., Katsanaki A., and Kiriakidis G., Appl. Cat. B: Environ., 113–114, 79–86.(2012)
[5] Hajjaji A., Trabelsi K., Atyaoui A., Gaidi M., Bousselmi L., Bessais B., and Khakani M. A. E., Nanoscale Res. Lett. 9: 543(2014)
[6] Maeda M. and Yamada T., J. Phys. Conf. Series, 61, 755–759(2007)
[7] Mohamed R. M., Mkhalid I. A., Al-Thabaiti S. A., and Mokhtar M., J. Nanosci. Nanotech. 13(7), 4975-4980(2013)
[8]Zhou W., Liu Q., Zhu Z. and Zhang J, *J. Phys. D: Appl. Phys.*, 43 (3)(2010)
[9] Karunakaran C., Abiramasundari G., Gomathisankar P., Manikandan G., and Anandi V., J. Coll. Int. Sci., 352(1), 68-74(2010)
[10] Nainani R., Pragati T., and Chaskar M., 2012 J. Mat.Sci. Eng., 2 (1), 52-58.
[11] Paulauskas I. E. and Modeshia D. R. Platinum Metals Rev., 57, (1), 32–43(2013)
[12] Loganathan K., Bommusamy P., Muthaiahpillai P., and Velayutham M., J. Environ., 16(2), 81-90(2011)
[13]Luu C. L., Nguyen Q. T. and Ho S. T., Nanosci. Nanotech., 1(1), 1-10(2010)
[14] Kokila P., Senthilkumar V., and Nazeer K. P., Arch. Phy. Res., 2(1), 246-253(2011)
[15] Hung W. C., Chen Y. C., Chu H., and Tseng T. K., Appl. Surf. Sci., 255, 2205-2213(2008)
[16]Zhao Y., Li C., Liu X., Gu F., Du H. L., and Shi L., Appl. Cat. B: Environ., 79, 208-215(2008)
[17] Zhu J., Chen F., Zhang J., Chen H., and Anpo M., J. Photochem. Photobio. A: Chem., 180, 196-204(2006)
[18] Wei Y.L., Chen K.W., and Wang H.P., J. Vac. Sci. Technol. B, 27(3), 1385-1387(2009)
[19] Aware V.D., and Jadhav S.S., Appl. Nanosci., 1-8(2015)
[20] Spurrs R. and Myers H. Quantitative, Anal. Chem. 29(5) (1957)
[21] Chen C., Wang Z., Ruan S., Zou B., Zhao M., and Wu F., Dyes and Pigments 77, 204-209(2008)
[22]Lavand A. B. and Malghe Y. S., J. Saudi Chem. Soc. 19(5) 471-478(2015)
[23] Cheng C. K., Deraman M. R., Ng K. H., and Khan M. R., J. Cleaner Product., 112, 1128-1135(2016)
[24]Singla P., Sharma M., Pandey O.P., and Singh K., Appl. Phys. A, 116, 371-378(2014)

Effect of doping Fe/Cu/Ti on WO_3 on furfural degradation

Guntapon Pitak[1], *Suphitchaya* Srisodsai[1], *Mutsee* Termtanun[1*]

[1]Department of Chemical Engineering, Faculty of Engineering and Industrial Technology, Silpakorn University, Nakhon Pathom, 73000

Abstract. This research improved tungsten oxide catalysts to increase efficiency in photocatalytic degradation of furfural under visible light. The aim of this research was to compare the efficiency of modified tungsten oxide with undoped and commercial tungsten oxide. Tungsten oxide nanoparticles were doped with 3 single metals, which were Fe/Cu/Ti at 1%wt, 2%wt, and 3%wt, synthesized by flame spray pyrolysis technique (FSP) and then characterization by X-Ray Diffraction (XRD), N_2 adsorption/desorption (BET surface area analysis), UV-Vis Spectroscopy (UV-Vis). Photocatalytic degradation experiments using doped WO_3 were carried out with 5 ppm initial concentration of furfural solution using 0.6 M catalyst concentration under visible light. From the results, FSP-synthesized WO_3 has better efficiency in furfural degradation than the commercial WO_3. All catalysts have mesoporous structure because an average pore size is in the range of 6-10 nm. Among all synthesized and doped WO_3, it can be concluded that 3%wt Fe-doped tungsten oxide provides the highest acceleration rate in photocatalytic degradation of furfural.

1 Introduction

Tungsten oxide is one of the most promising inorganic materials which exhibit excellent electrochromic, photochromic and gasochromic properties and it has been widely investigated to be used in gas sensor application. However, tungsten oxide (WO_3) also has been developed to be a potential photocatalysts [1] because of its outstanding characteristics, highly stable physicochemical properties, good electron transport properties, nontoxicity and no-photo-corrosion. WO_3, with a band gap between 2.2 and 2.8 eV [2], is a visible light response catalyst, make it as an alternative choice for the photocatalytic oxidation of organic pollutants under solar irradiation instead of TiO_2. WO_3 can be synthesized by a variety of techniques such as precipitation [3], sol-gel[4], hydrothermal method[5], etc. However, the one of the disadvantages is that it has high possibility in occurrence of electrons-holes recombination. Thus, in order to increase the photocatalytic performance of WO_3, the addition of transition metals could either be alleviating of electrons-holes recombination process or supporting electrons to go to conduction band easier. Recently, there have been many studies and developments of tungsten oxide by doping of some transition metals, for example Fe and Mo [6-7] or making composite with other metal oxide such as TiO_2, CuO, $AgIO_3$ [8-12]. In this research, WO_3 samples were synthesized using flame spray pyrolysis, doped with Fe/Cu/Ti at different amount by wt%, characterized and tested the photocatalytic performance with 3 hours furfural degradation reaction under visible light.

2 Materials and method

2.1 Preparation of doped WO_3

Precursor solutions prepared with tungsten hexachloride used in WO_3 synthesis were fed into flame spray pyrolysis machine. Firstly, the precursor was prepared by dilution of tungsten hexachloride in ethanol and added 0.1M of Iron(III) nitrate nonahydrate, Copper(II) chloride and Titanium(IV) tetra isopropoxide with a calculation by wt% doping of single metal Fe/Cu or Ti equal to 1, 2, 3, respectively. The mixed precursor was injected through the center capillary of the flame spray pyrolysis nozzle with oxidizing gas into flame zone by syringe pump at flow rate was 5 ml/minute. Feed solution was dispersed with oxygen gas at flow rate of 5 l/min. Fianlly, after combustion, the precursor sprayed in FSP was then converted into WO_3 nanoparticles. WO_3 and doped WO_3 nanoparticles were collected on a glass fiber filter (Whatman GF/D, 15 cm diameter).

2.2 Characterization of WO_3 catalysts

X-ray diffraction was used to report phase composition of catalyst samples using CuK_α radiation with Ni filter in the 2 theta range of 10-80 degrees with a resolution of 2° per min. UV-VIS spectroscopy was used to measure the amount of light and intensity in the UV range and visible light (in the range of 190 nm to 900 nm) that absorbed by the sample to define bandgap energy of doped tungsten oxide samples by calculation through Kubelka-Munk equation. Brunauer-Emmett-Teller (BET) was used for analyzing the surface area of doped tungsten oxide samples by replacing the surface or porous surface with nitrogen gas. The pre-treatment using a helium gas flow of 50 ml / min at 180 °C for 3 hours was performed to remove the water contained in the catalyst and then tested by nitrogen gas substitution.

2.3 Photocatalytic activity measurement

The photoactivity measurements of prepared samples in furfural photodegradation were performed at an ambient

* Corresponding author: Termtanun_m@su.ac.th

temperature. The small scale-batch reactor is comprised of beaker within the black box with dimension length 30 cm, width 21.5 cm, and height 23 cm with 2 ventilation fan speed 1675 rpm inside. 4 compact 35 W Philips lamps with 2450 lumen/lamp (wavelength around 620 nm) were used as light source located about 10 cm far from beaker. In a typical process, aqueous suspensions of furfural 500 ml at 5 ppm initial concentration and 300 mg of nanocatalyst were placed in a vessel. Prior to irradiation, the suspension were magnetically stirred under dark for 30 min. 3 ml sample was drawn from furfural suspension every 10 min and removal of catalyst by centrifugation. The light absorption intensity at 273 nm was measured by UV-Vis spectrophotometer to identify furfural remaining concentration. The percentage of degradation was recorded as C/C_0 versus with time, while C is concentration of furfural for each irradiated time interval at wavelength 273 nm and C_0 is an initial concentration of furfural.

3 Results and discussion

3.1 X-Ray diffraction

In Figure 1, according to the JCPDS card no. 01-085-0950, all WO_3 catalysts were in monoclinic phase. Furthermore, there was also a peak corresponding to Cu at 2 theta equals to 41.85, 45.0, 47.42. After Fe/Cu/Ti metals were doped, the peak-height of metals slightly increased and the peak-height of tungsten somewhat decreased due to small level of doping. Doping metals would account for forming some oxide compositions of Fe/Cu/Ti and combining into the WO_3 crystalline structure. According to the 2 peaks of (002) and (200) WO_3 almost disappear to merely the rest of (020), it seems to be inferred that doping Fe possesses the most modified tungsten oxide phase, while the insignificant results appeared in case of Cu/Ti doping.

i)

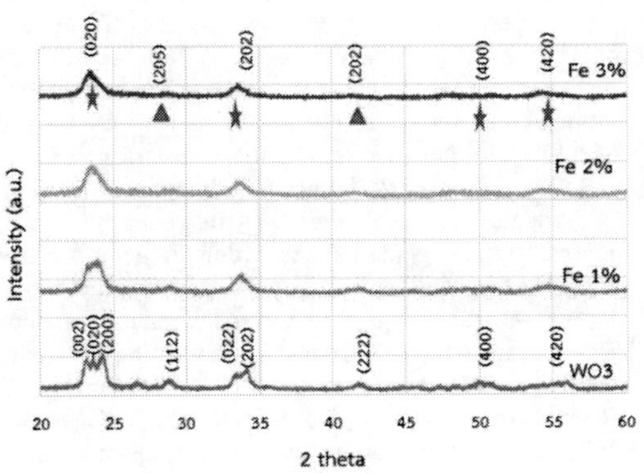

ii)

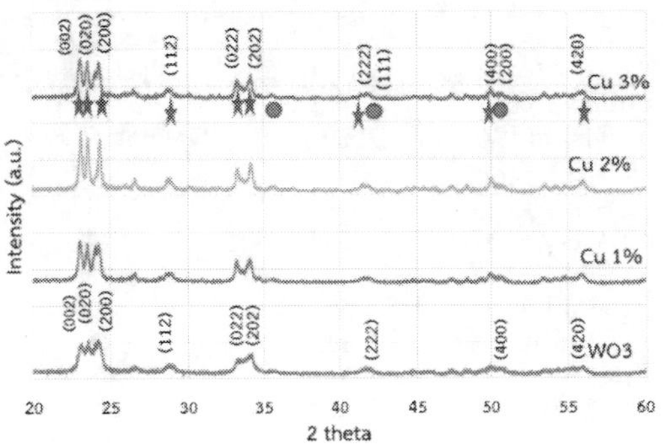

iii)

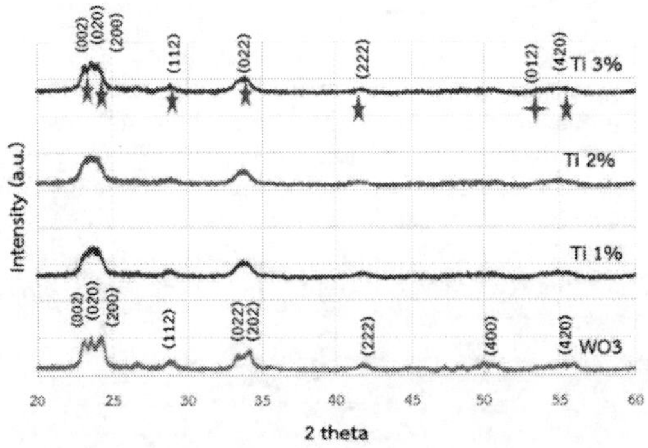

Figure 1 i), ii), iii) represent XRD patternsof Fe/Cu/Ti doped compared with undoped WO_3 (★, ▲, ●, ✦ represented WO_3, Fe, Cu, Ti, respectively)

3.2 BET surface area and pore size distributions

From Table 1, BET surface area (S_{BET}) of synthesized WO_3 using flame spray pyrolysis has larger surface area than that of the commercial WO_3. Flame spray pyrolysis method, FSP produces nanopowder of 20–100 nm average particle sizes (APS) with specific surface areas of 30-60 m^2/g whereas the commercial WO_3 is prepared by sol-gel technique [4]. When doping amount increased, the surface area was also increased. 1wt% Cu-doped WO_3 has the largest specific surface area at 59 m^2/g. However after increasing Cu amount to 3 wt%, particles started to be agglomerate particles which lead to minimal surface area. From the results of an average pore size, it can be concluded that all catalysts have mesoporous structure.

Table 1 BET surface area and average pore size in nanometre of Fe/Cu/Ti-doped WO_3 compared to commercial and undoped WO_3

Catalyst samples	S_{BET} (m^2/g)	Average pore diameter (nm)
Commercial	8.26	6.69
Undoped-WO_3	20.84	11.11
1%Fe-WO_3	35.75	8.34
2%Fe-WO_3	48.82	7.56
3%Fe-WO_3	53.73	8.02
1%Cu-WO_3	59.30	6.98
2%Cu-WO_3	59.50	6.49
3%Cu--WO_3	15.97	8.34
1%Ti-WO_3	39.10	8.05
2%Ti-WO_3	41.70	8.07
3%Ti-WO_3	30.80	7.62

3.3 UV-Vis spectroscopy

After UV-Vis analysis, the ability in light absorbance of photocatalyst is also the significant characteristic that the promising photocatalyst should massive absorb and able to activate under visible wavelength [13-15]. The bandgap energy of all WO_3 samples was calculated by Kubelka-Munk relationship [16], shown in Table 2. In Kubelka-Munk relationship, equation (1), the graph is plot between transform reflectance spectra $(F(R_\infty)h\nu)^{1/2}$ and $h\nu$.

$$F(R\infty) = \frac{(1-R\infty)^2}{2R\infty} \quad (1)$$

Transformation of absorbent unit into reflectance

$$A = log10\ (1/R) \quad (2)$$

$$R_\infty = 10^{-A} \quad (3)$$

Where A = absorbent unit
h = Plank constant (6.624 x 10^{-34} Joules sec)
ν = velocity of light (3.0 x 10^8 meter/sec)
R_∞ = reflectance unit

In Table 2, bandgap energy is the one of the key factors that directly impacts on the photocatalysis mechanism [17]. The narrower the bandgap energy, the lower the energy required on photocatalytic reaction. The commercial WO_3 has nearly equal bandgap energy and light adsorption under visible region when compared with an undoped WO_3. From this Table, it was obvious that 3%Fe- WO_3 has the smallest bandgap energy when compared with other catalysts. In case of Cu and Fe-doping, an increase in the doping amount leads to smaller the bandgap energy. Doping titanium enhances larger bandgap energy when compared with undoped and commercial WO_3 due to its wider bandgap energy, which would act as the pathway of electrons transition. However, there should be aware of doping amount as if too minimal bandgap energy might bring about electrons-holes recombination [19-20].

Table 2 Calculated bandgap energy of doped WO_3 compared with commercial and undoped WO_3

Catalyst samples	**E_g (eV)**
Commercial	2.68
Undoped-WO_3	2.65
1%Fe-WO_3	2.53
2%Fe-WO_3	2.47
3%Fe-WO_3	2.42
1%Cu-WO_3	2.63
2%Cu-WO_3	2.57
3%Cu-WO_3	2.48
1%Ti-WO_3	2.64
2%Ti-WO_3	2.70
3%Ti-WO_3	2.70

3.4 Photocatalytic activity

Photoactivity of synthesized and doped WO_3 was evaluated by furfural degradation under visible light for 3 hours. From the results, doping Fe/Cu/Ti on WO_3 shows difference influence on photocatalytic activity. Table 3 shows the total conversion of furfural using synthesized and various doped WO_3 after 3 hours. The percentage of furfural conversion (X) can be determined from the following equation [16];

$$X(\%) = \left(1 - \frac{CA}{CA0}\right) x\ 100 \quad (4)$$

While C_A is equal to remaining concentration of furfural measured by UV-Vis, and C_{A0} is equal to an initial concentration of furfural. The photocatalytic degradation of furfural using WO_3 is the pseudo first-order linear reaction [21], the relationship between $-ln(C_A/C_{A0})$ versus time is appeared to be linear, therefore the reaction rate constant (k value) can be calculated from equation (6), shown in Table 3.

Degradation reaction rate of furfural is assumed to be equal to R_A,

$$R_A = -dC_A/dt \quad (5)$$

$$-ln(C_A/C_{A0}) = kt \quad (6)$$

The catalyst that has lower C/C_0 shows the better the phocatalytic performance. From Table 3, it was found that FSP-synthesized WO_3 has better efficiency in furfural degradation than commercial WO_3 because FSP-synthesized WO_3 has larger surface area than commercial WO_3, resulted in more active sites. According to the equation (6), the highest k is belonged to 3 wt% Fe-doped WO_3, meaning that doping Fe at 3 % by weight is the most facilitated in furfural photocatalytic degradation. According to the results from Table 2, the bandgap energy of catalyst samples after doping with Fe or Cu were lower than that of undoped WO_3 because Fe and Cu assist in shifting to smaller bandgap energy. However, there is an optimum limit of Fe level since 2 wt% Fe-doped WO_3 and 3 wt%Fe-doped WO_3 obtain rather similar reaction rate, meaning nearly equivalent in enhancing catalytic ability. In case of Ti, it was observed from the results that low amount of doping Ti onto WO_3 accelerates furfural degradation since the conversion of furfural from undoped WO_3 is lower than that from Ti- doped WO_3. However, the large amount of Ti doping showed detrimental influence on furfural photodegradation rate.

Table 3. The percentage of furfural conversion and reaction rate constant (k) after 3 hours photodegradation under visible light using different WO_3 samples.

Catalyst samples	Conversion(%)	k value (min^{-1})
Commercial	25±0.13	0.0015
Undoped-WO_3	43±0.18	0.0027
1%Fe-WO_3	36±0.2	0.0036
2%Fe-WO_3	65±0.31	0.0083
3%Fe-WO_3	68±0.28	0.0090
1%Cu-WO_3	30±0.15	0.0021
2%Cu-WO_3	32±0.24	0.0021
3%Cu-WO_3	40±0.37	0.0028
1%Ti-WO_3	67±0.4	0.0077
2%Ti-WO_3	49±0.3	0.0044
3%Ti-WO_3	38±0.08	0.0034

4. Conclusions

Tungsten oxide was successfully synthesized by Flame spray pyrolysis to use as a photocatalyst in furfural degradation. It was modified and developed photocatalytic activity by doping metal of Fe, Cu, and Ti at different mass fraction. From XRD, all the catalysts were appeared in the monoclinic phase. According to the light absorption results, it was found that after three metals-doping, the light absorption intensity increased and the bandgap energy changed. The effect on light absorption depends on type of metals. The addition of Fe helps in reducing the bandgap energy. The smallest bandgap energy is about 2.42 eV, as observed from 3 wt% Fe-doped WO_3. During photocatalytic degradation of furfural, among various catalysts, 3 wt% Fe-doped WO_3 shows the most potential in furfural degradation with the conversion of furfural equal to 68% and the highest reaction rate constant, k calculating based on the pseudo first order assumption.

Abbreviations

FSP: Flame Spray Pyrolysis
XRD: X-Ray Diffraction
UV-Vis: Ultraviolet Visible Spectroscopy
BET: Brunauer-Emmett-Teller
TEM: Transmission Electron Microscopy
EDS: Energy Dispersive X-Ray Spectroscopy
APS: Average Particles Size
X: The percentage of furfural conversion
C_A: The remaining concentration of furfural in aqueous solution
C_{A0}: The initial concentration of furfural in aqueous solution
S_{BET}: Surface area from BET analysis (m^2/g)
E_g: bandgap energy (eV)
A.U.: Arbitrary Unit
cps: counts per second

This research was financially supported by National Science and Technology Development Agency (NSTDA), Bangkok, Thailand and instrumental supported by Department of Chemical Engineering, Faculty of Engineering and Industrial Technology, Silpakorn University, Nakhon Pathom, Thailand.

References

[1] G. Xin, W. Guo, T. Ma., App. Surf. Sci. **256** (2009) 165–169
[2] L. E. Frage, J. H. Franco, M.O. Orlandi, M. Valnice, B. Zanoni, J. Environ. Chem. Eng. **1** (2013) 194–199
[3] D. S. Martinez, A. M. L. Cruz, E. L.Cuellar, App. Catal. A: Gen. **398** (2011) 179-186
[4] H. Yang, R. Shi, K. Zhang, Y. Hu, A. Tang, X Li, J. Alloy Comp. **398** (1-2) (2005) 200-202
[5] D. J. Ham, A. Phuruangrat, S. Thongtem, J. S. Lee, Chem. Eng. J. **165** (1) (2010) 365-369
[6] H. Song, Y. Li, Z. Lou, M. Xiao, L. Hu, Z. Ye, L. Zhu, App. Catal. B: Environ. **166–167** (2015) 112–120
[7] X. C. Song, E. Yang, G. Liu, Y. Zhang, Z. S. Liu, H. F. Chen, Y. Wang, Res. Paper **12** (2010) 2813-2819
[8] H. Widiyandari, A. Purwanto, R. Balgis, T. Ogi, K. Okuyama, Chem. Eng. J. **180** (2012) 323– 329
[9] O. Arutanti, A. Bayu D. Nandiyanto, T. Ogi, F. Iskandar, T. O. Kim, K. Okuyama, J. Alloys Comp. **591** (2014) 121-126
[10] Q. W. Cao, Y. F. Zheng , X. C. Songa, J. Taiwan Inst. Chem. Eng. **70** (2017) 359–365
[11] P. N. Gaikwad, T. M. Wandre, K. M. Garadkar, P. P. Hankare, J. R. Sasikala, Col. Surf. A: Physicochem. Eng. Asp. **506** (2016) 804-811
[12] J. A. Mendoza, D. H. Lee, J.-H. Kang, Chemosphere **182** (2017) 539-546
[13] C.-F. Lin, C.-H. Wu, Z.-N. Onn, J. Hazard. Mat. **154** (2008) 1033-1039
[14] J. Zhu, W. Zheng, B. He, J. Zheng, M. Anpo, J. Mol. Catal. A: Chem. **216** (2004) 35-43
[15] N. R. Khalid, Z. Hong, E. Ahmed, Y. Zhang, H. Chan, M. Ahmad, App. Surf. Sci. **258**(2012) 5827-5834
[16] A. B. Lavand and Y. S. Malghe, J. Saudi Chem. Soc. **19** (5) (2015) 471-478
[17] F. Wong, C.D. Valentin, G. Pacchioni, J. Phys. Chem. C. **116** (2012) 8901-8909
[18] C. K. Cheng, M. R. Deraman, K. H. Ng, M. R. Khan, J. Clean. Product. **112** (2016), 1128-1135
[19] V.D. Aware, and S.S. Jadhav, Appl. Nanosci., 1-8(2015)
[20] P. Singla, M. Sharma, O.P. Pandey, and K. Singh, Appl. Phys. A, **116**, 371-378(2014)
[21] Z.-D. Meng, L. Zhu, J.-G. Choi, C.-Y. Park, W.-C. Oh, Nanoscales Res. Lett. **6** (459) (2011) 1-11

https://doi.org/10.1051/matecconf/201819203049

Precision test for spectral characteristic of NIR spectroscopy and bomb calorimeter for measuring higher heating value of Leucaena Leucocephala

Jetsada Posom[1,2], *Panmanas* Sirisomboon[3], and *Wanphut* Saechua[3,*]

[1]Department of Agricultural Engineering, Faculty of Engineering, Khon Kaen University, Khon Kaen, 40002, Thailand
[2]Applied Engineering for Important Crops of the North East Research Group, Department of Agricultural Engineering, Faculty of Engineering, Khon Kaen University 40002, Thailand
[3]Agricultural Engineering Department, Faculty of Engineering, King Mongkut's Institute of Technology Ladkrabang, Bangkok 10520, Thailand

Abstract. The higher heating value (HHV) plays a significant role in determining of the energy potential of biomass. The method to receive this value using a bomb calorimeter takes a long time approximately of 15 minutes. The implementation of near infrared spectroscopy (NIR) technique by generating a prediction model can be replaced the former method in order to reduce time into 2 minutes. Therefore, the precision of both techniques have to be concerned in order to get the reliable values. The repeatability and reproducibility processes were carried out to evaluate the precision of both methods. It was found that the percentage difference of standard deviation (SD) from the mean of repeatability for the NIR was less than 5 % and for reproducibility was slightly over 5 % (but not over than 10%). The SD of repeatability and reproducibility of the bomb calorimeter were 54.6 J/g (0.3 % difference of average HHV value) and 81.3 J/g (0.5 % difference of average HHV value) respectively. The average HHV from the bomb calorimeter was 18,166.77±236.29 J/g. Both techniques showed very low percentage difference of the SD from the mean that revealed very high reliability and precision of measurements.

1 Introduction

The continuously decreasing in fossil fuel leads to finding of alternative energy which can be used instead of conventional energy. The renewable energy such as biomass, wind, geothermal, solar, waves, tides, and hydroelectric power recently plays important role in the alternate energies. Thailand, one of the agricultural countries has also desired in those energy resources. The department of alternative energy development and efficiency (DEDE) has now given an impetus to grow the energy plants in order to use as biomass energy. Leucaena leucocephala is one of the energy plants that have been promoted due to fast growing, good disease resistance, and high potential energy in combustion process. Not only it can be used as an alternative energy, but also can be utilised as wood furniture work. Traditionally, the trade of biomass as wood chip or saw dust used moisture content in materials for indicating selling price. However, higher heating value (HHV) of the biomass better represents potential and capacity of energy contained in the biomass. The agricultural wastes such as corn straw, rice straw, rice husks, and palm kernel also can be used as biomass fuel. Their HHV were 18,195, 15,511, 15,986, and 19,405 J/g respectively[1]. The conventional method (bomb calorimeter) to determine this value takes a long time approximately of 15 minutes per sample excluding preparing process. Using the NIR technique to evaluate the HHV may be a better way to evaluate the HHV of biomass due to shorter time approximately of 2 minutes and easier procedure. However, the NIR technique also requires reference data used for generating the model to predict value based on those references. It was clearly seen that the conventional method took longer time than the NIR method. However, the accuracy of the model created by the NIR scanning requires piles of data and wide and variety range of them. This may take times and attempts to collect data in order to get the robust model but it subsequently saves time and money in long term. There were many researchers done experiments with the NIR scanning to generate model for prediction some specific values or properties of organic substances such as foods, fruits, beverages, and biomass[2-6]. The NIR technique has advantages as non-destructive method, easy procedure, short time process, low cost, environmental friendly and provides precision prediction[7]. Therefore, the procedure before getting the reference data has to be more concerned. The repeatability and reproducibility processes have been claimed and accepted to use to confirm the precision and reliable of the test [8-10]. In this study aimed to reveal

[3*] Corresponding author: wanphut.sa@kmitl.ac.th

 https://doi.org/10.1051/matecconf/201819203049

the precision test of both instruments (the NIR spectrometer and the bomb calorimeter) in order to diagnose possible error and to confirm the reliability of the measurements. Additionally, the HHV of Leucaena leucocephala was measured in order to create the model to predict the HHV in future.

2 Materials and Methods

The experiments implemented of two instruments which were the NIR spectrometer and the bomb calorimeter with using Leucaena Leucocephala as a sample of biomass. The repeatability and reproducibility were carried out on both techniques.

2.1. Samples

Leucaena Leucocephala ranged of 8±1 cm dimeter used in this study was 120 samples which were 40 samples from each three parts of Leucaena Leucocephala (lower trunk, middle trunk and upper trunk). Next, they were cut and chopped into small pieces as wood chips by using a chopper machine (P5508, Patipong, Thailand). All samples were dried in an oven at 105 °C for 24 hours and left in room temperature 24 hours for moisture equilibrium. The repeatability and reproducibility samples were randomly selected of 10 sets of samples for the precision test of bomb calorimeter experiments with making in a pellet shape approximately range of 0.5-1.0 gram. The other random of two sample sets for the precision of the NIR scanning used dried Leucaena Leucocephala chips. The feature of samples was shown in Figure 1.

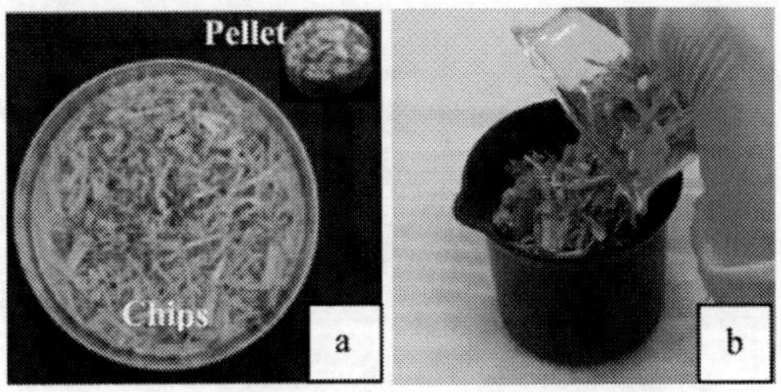

Fig. 1. a) A sample used for the bomb calorimeter
b) A sample used for the NIR spectrometer

2.2 Instruments

The NIR spectrometer used for scanning was FT-NIR (MPA, Bruker, Germany) in wavenumber range of 12,500–3,600 cm^{-1} at 8 cm^{-1} resolution. The bomb calorimeter (C200, IKA, Germany) was used to measure of higher heating values which was divided into two parts each measurement done with 2 replicates per a sample set. The first part was carried out for repeatability and reproducibility, and the second part was determining the average HHV. The calibration process before measuring was carried out in the sample vessel using pelletized benzoic acid (IKA C 723, IKA, Germany).

2.3 Repeatability and reproducibility

Repeatability is a method to show the precision of the instrument when the measurement was replicated many times at the same condition .The repeatability process of the NIR was done by scanning the same sample in a quartz cup for 10 times at the same position. Then absorbance values of the same of three wavenumbers (4011.4, 5191.7 and 6800.2 cm^{-1}) of each spectrum were used to determine the average absorbance and the standard deviation values. Reproducibility of the NIR presents the homogenous of the sample. The method was carried out by scanning the same sample and re-loading this sample for 10 times. The three wavenumbers were chosen to evaluate the average absorbance and the standard deviation values.

The repeatability of the bomb calorimeter also presented the precision of the HHV measurements. It was done by combustion of the 10 sample sets with two replications for each set. The differences of higher heating value (HHV) of each sample set were determined and the standard deviation of the differences was calculated. The reproducibility of the bomb calorimeter shows the precision of the process whether different researchers did the experiment the results still accuracy. It was carried out by using of 10 blinded samples compared with their reference values. The different HHV between blinded samples and reference values and the standard deviation of the differences were determined.

The interpretation of the repeatability and reproducibility was consideration of the value of the SD. Typically the accepted SD value depends on the acceptance level for each experiment and normally using not much than 5 % difference from the mean [11, 12]. The lower SD presented higher precision of the measurement.

3 Results and Discussions

The average HHV of Leucaena Leucocephala from the bomb calorimeter showed in Table 1. The average HHV was 18,166.07±236.29 J/g which showed a good energy potential of biomass compared with other agricultural wastes[13] and has an excellent potential for using as combustion fuel as well as saw dust or wood pellet[14]. Additionally, it was found that the different position of Leucaena Leucocephala had no effect on the HHV by analysing with statistical method one way ANOVA at confident level of 95 percentages (Table 1). Figure 2 showed average raw spectra of NIR scanning for 120 samples of Leucaena Leucocephala which consisted peaks of cellulose at wavenumber 6,808 cm^{-1}, water at wavenumber 5,188 cm^{-1}, starch at wavenumber 4,740 cm^{-1} , water at wavenumber 4,277 cm^{-1}, and benzene at wavenumber 3,980 cm^{-1}.

The experiments for repeatability and reproducibility of the NIR scanning and the bomb calorimeter were shown in Table 2 and Table 3 respectively. The average HHV of Leucaena leucocephala in repeatability and reproducibility processes from the bomb calorimeter were 17,943.7 J/g and 18,208.0 J/g respectively. It was found that the NIR spectrometer had an excellent capability for scanning which could notice from very low standard deviation values of repeatability and reproducibility procedures.

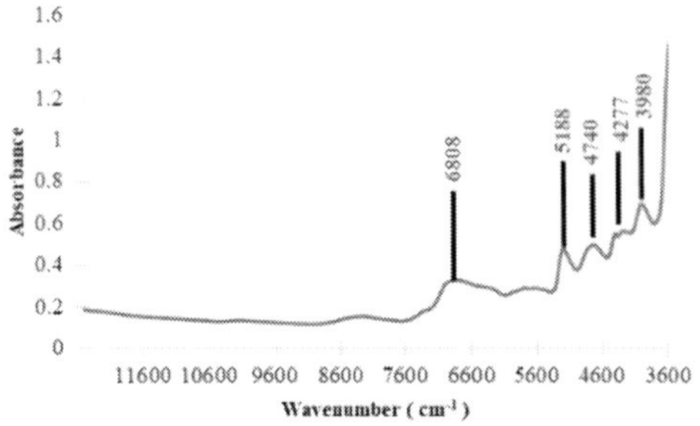

Fig. 2. The average of raw spectra of Leucaena leucocephala

The percentage difference of the SD of the NIR in repeatability showed lower than 5% compared with the mean of each wavenumber which revealed of precision and reliability of the NIR spectrometer. It was noticed that the percentage difference of the SD of the NIR in reproducibility showed a little higher than 5% in wavenumber of 5191.7 cm^{-1} and 6800.2 cm^{-1} which were 6% and 7% respectively. This showed that the sample may be nonhomogeneous if it relied on the accepted level of 5% difference form the mean. However, this research could imply that the sample had homogenous due to acceptance of the percentage difference of SD value which was not higher than 10 %. The reason was the samples could not be sieved because it had to be used like receiving from the chopper process as the commercial trade. This meant that the NIR Spectrometer had ability to replicate the measurement and the samples used in scanning were homogenous.

Table 1. The HHV of Leucaena Leucocephala

Position	Avg. HHV ±SD (J/g)
Lower trunk	18,168.28±243.93^{NS}
Middle trunk	18,182.15±226.67 ^{NS}
Upper trunk	18,147.78±238.28 ^{NS}
Overall Avg. HHV±SD	18,166.07±236.29

*NS : Non Significant

Consistently, the repeatability and reproducibility of the bomb calorimeter showed very low of the standard deviation. The SD of repeatability was 54.6 J/g which was approximately of 0.3 percentage of the average HHV from repeatability procedure. The SD of reproducibility was 81.3 J/g which was approximately of 0.5 percentage of the average HHV of reproducibility process. This meant that the bomb calorimeter also had ability to replicate measuring and different performers had no effect on measurements. These results of repeatability and reproducibility for both instruments showed very low standard deviations could be helped in analysis process of the prediction in term of the variance of data or the out liner of some spectra could not come from the error of instruments. If there will some errors occur in experiments, the performer have to focus on other causes of which do not from instruments.

Table 2. The repeatability and reproducibility of the NIR scanning

Wave-number (cm^{-1})	Repeatability		Reproducibility	
	Avg. Abs. ± SD	% different from the mean	Avg. Abs. ± SD	% different from the mean
4011.4	0.6705± 0.0008	0.1	0.7256± 0.0318	4.4
5191.7	0.4586± 0.0010	0.2	0.5153± 0.0299	5.8
6800.2	0.3114± 0.0012	0.4	0.3626± 0.0262	7.2

Table 3. The repeatability and reproducibility of the bomb calorimeter.

Samples	Repeatability	Reproducibility
	The difference of HHV (J/g)	The difference of HHV (J/g)
1	173.0	184.0
2	127.0	288.0
3	35.0	71.0
4	14.0	100.5
5	54.0	241.5
6	149.0	187.5
7	70.0	42.5
8	54.0	95.0
9	138.0	83.5
10	58.0	94.5
SD of differences	54.6	81.3

. 4 Conclusions

It can be concluded that the NIR spectrometer and bomb calorimeter have an excellent precision and reliability for using to evaluate the HHV of Leucaena Leucocephala. This could help the experiments to be achieved in generating of reliable prediction model. However, the success model generated from the NIR in order to reduce the process time for determining the HHV with the bomb calorimeter requiring more robust model which should be developed with more samples and more variation samples.

5 Acknowledgements

Thanks to Near Infrared Spectroscopy Research Centre for Agricultural Product and Food (www.nirsresearch.com), the Department of Agricultural Engineering, Faculty of Engineering, King Mongkut's Institute of Technology Ladkrabang, Thailand.

References

1. H. Quian, X. Guo, S. Fan, K. Hagos, X. Lu, C. Liu, D. Huang, J. Chem. Eng. **61**,12(2016)
2. P. Jetsada, S. Panmanas, Energy.**121**(2017)
3. P. Jetsada, S. Panmanas, S. Wanphut, Energy.**103**(2017)
4. P. Jetsada, S. Panmanas, Bio. Eng.**130**(2015)
5. B.M. Nicolai, B. Katrien, B. Els, P. Ann, S. Wouter, I.T. Karen, L. Jeroen, Postharvest Bio. and Tech.**46**,2(2007)
6. J.M. Triolo, A.J. Ward, P. Lene, M.L. Mette, Q. Haiyan, G.S. Sven, Appl. Energy. **116**(2014)
7. G. Ainara, F.O.V.Silvia, E. Nerea, D. Alberto, M.M. Juan, Trends in Analytical. Chem.**76**(2016)
8. S. Panmanas, K. Apidul, W. Phil, J. NIR. **21**,2(2013)
9. A. Ferragina, C. Cipolat-Gotet, A. Cecchinato, M. Pazzola, M.L. Dettori, G.M. Vacca, G. Bittante, J. Dairy Sci. **100**,5(2017)
10. D.M. Musingarabwi, H.H. Nieuwoudt, P.R. Young, H.A. Eyeghe-Bickong, M.A. Vivier, Food Chem.**190**(2016)
11. J.W. Bartlett, C. Frost, Ultrasound Obstet Gynecol. **31**,4(2008)
12. J.M. Bland, D.G. Altman, Lancet. **1**,8476(1986)
13. Z. Ke, Z. Ling, B. Michael, X. Feng, Y. Jianming, W. Donghai, Energy. **118**(2017)
14. Z. Yaning, G, Xiaoyan, L. Bingxi, Z. Haochun, Q. Baoheng, W. Yifan, Fuel. **159**(2015)

Effects of gender of dairy or beef cattle and the number of pair of permanent incisors on carcass traits

Kunya Tuntivisoottikul[1*] and *Rutcharin* Limsupavanich[2]

[1]Department of Agricultural Education, Faculty of Industrial Education and Technology, King Mongkut's Institute of Technology Ladkrabang, Bangkok, Thailand
[2]Department of Animal Production Technology and Fisheries, Faculty of Agricultural Technology, King Mongkut's Institute of Technology Ladkrabang, Bangkok, Thailand

Abstract. Effects of group of cattle (gender of dairy or beef cattle) and number of pair of permanent incisors (0-4 pairs) on carcass traits were investigated. Data collected from 2,059 carcasses were obtained from beef cattle cooperative in 2016. Carcass traits included slaughter weight (wt.), hot carcass wt., percentage of hot or chilled carcass wt., chilling loss, skin wt., and marbling score. Factors included group of cattle and number of pair of permanent incisors observed in each cattle group, while slaughter wt. was a co-variable in the statistical model. Descriptive statistics were used to analyze marbling score. Results showed that group of cattle affected slaughter wt., hot carcass wt., percentage of hot or chilled carcass wt., skin wt. ($P<0.0001$), and chilling loss ($P<0.05$). Number of pair of permanent incisors within each cattle group affected all carcass traits ($p<0.05$), except chilling loss ($p>0.05$). Female and male beef cattle in all ages had higher hot carcass wt. and percentage of hot or chilled carcass wt. than female and male dairy cattle. The highest marbling score of 5 was found in female and male beef and female dairy cattle with 4 pairs of permanent incisors. However, majority of cattle had marbling score 2.

1 Introduction

In 2006, the beef cattle population in Thailand was approximately 8.04 million heads, but it decreased to 4.41 million heads in 2015, of which 3.90 % were finishing beef cattle [1]. Most of the finishing beef cattle are crosses of *Bos indicus* (Native or Brahman or their crosses) and *Bos taurus* (Charolais, Simental, Limousin, etc.). Currently, beef production is not enough for local consumption. An alternative was done by insemination of dairy cows to produce crossbreds for meat production. In addition, culled dairy cows or male dairy cattle are raised as beef cattle [2]. Most dairy cattle are at least 75% Holstein Friesian (HF) blood. Only females are needed for milk production. Culled dairy cattle, especially male, are sold at a cheaper price. Some dairy farmers, therefore, continue raising them for meat production. At present, about 20% of dairy meat is shared in beef market. Important intrinsic factors influencing carcass traits are breed, gender, and age or maturity of the animals. Carcass wt. and gender of cattle are common criteria for carcass classification [3]. Dentition is generally used to evaluate cattle age, as actual birthdates are not usually available [3-4]. South African carcass classification scheme and the Australian Meat system use dentition to estimate physiological age or maturity of cattle for beef carcass classification [3]. Dentition classification and its criteria for classification can be different among countries. In Thailand, the dentition at slaughter is recorded in number of pair of incisors (0, 1, 2, 3, and 4 mean none, the 1st, 2nd, 3rd, and 4th pairs of permanent incisors were observed). For commercial beef production, physiological age is as important factor influencing carcass traits as slaughter wt. In general, cattle with 0, the 1st, 2nd, 3rd, and 4th permanent incisors are <18, 18–24, 24–30, 36, and 42–48 months of age, respectively. Reports comparing carcass traits based on gender of dairy or beef cattle and their physiological ages, especially under commercial production in Thailand, are a few [5-7], comparing to those in other countries [8-12]. Therefore, we investigated carcass traits as affected by gender of dairy or beef cattle and their physiological age, according to the number of pair of permanent incisors.

2 Material and methods

2.1 Data collection

Data were collected from 2,059 cattle obtained from the Beef Cluster Cooperative Limited located in Nakhon Pathom Province during 2016. Information on animal identification and farm origin was recorded. Cattle were

* Corresponding author: kunya.tu@kmitl.ac.th

separated into 4 groups including 398 female and 173 male dairy cattle and 354 female and 1,134 male beef cattle (FD, MD, FB, and MB, respectively). The beef cattle are minimum 50%Charolais and 50%Brahman or Native or their crosses, while the dairy cattle are at least 75 % HF. Cattle were raised intensively by feeding with fresh grasses and concentrate until body wt. reached 500 to 550 kg. The cattle were then transported to a slaughterhouse. They were rested for 12 hrs prior to slaughtering. After bleeding, images of their teeth were taken using a digital camera. The number of pair of permanent incisors was observed from the image. The skin was removed and weighed. After evisceration, carcass splitting, and slaughtering process, hot carcass wt. was obtained from both sides of each carcass. After 7 days of aging (4°C), chilled carcass wts. were obtained. Carcass was then quartered between the 12th to 13th ribs. Intramuscular fat (marbling) on the surface of *M. longissimus dorsi* on the 12th ribs of the right carcass side was evaluated (score 1 to 5, where 1 = devoid, 2 = slight, 3 = small, 4 = moderate, and 5 = abundant) [13]. Percentage of hot or chilled carcass wt. was calculated.

2.2 Data analysis

Effects of cattle group and the numbers of pair of permanent incisors on carcass traits were analyzed using two general linear models. The difference in slaughter wts. was tested by model (1), while other carcass traits, except marbling score, were analyzed using model (2). Slaughter wt. was used as covariate in model (2). Since the categories of independent variable, the number of pair of permanent incisors of animal within each cattle group, were unequal, nested model was performed as follows:

$$y_{ijk} = \mu + GC_i + Pt(GC)_{ij} + \varepsilon_{ijk} \quad (1)$$

$$y_{ijk} = \mu + GC_i + Pt(GC)_{ij} + \beta(SW_{ijk} - \overline{SW}) + \varepsilon_{ijk} \quad (2)$$

Where: y_{ij} = Observation of slaughter wt. of animal k in each gender i; y_{ijk} = Observation of the studied traits of animal k in each gender i; μ = Overall mean; GC_i = Group of gender as fixed effect (i=1, 2, 3 and 4, where 1=female beef cattle, 2=female dairy cattle, 3=male beef cattle, and 4= male dairy cattle); $Pt(GC)_{ij}$ = Pairs of permanent incisors of animal within each gender of beef or dairy cattle as co-factors (j=0, 1, 2, 3, and 4, where 0 = none, 1, 2, 3, and 4 = 1st, 2nd, 3rd, and 4th pair of permanent incisors observed); β = Coefficient regression of the studied traits to the slaughter wt.; SW_{ijk} = Slaughter wt. of the animal k in each gender group i and in each group of observed pairs of permanent incisors j; $\overline{SW}$ = Mean of slaughter wt.; ε_{ijk} = Error of experiment of the observation y_{ijk}

Differences among means were compared with pdiff option in GLM procedure using statistical software [14].

Descriptive statistic in term of measured frequency was used for marbling score data analysis.

The distribution of all carcass trait data was as fallows. Average slaughter wt., hot carcass wt., chilling loss and skin wt. were 589.48, 342.30, 7.73, and 44.75 kg, respectively. Dressing percentage, both hot carcass and chilled carcass wts, were 58.03 and 56.71 %, respectively. Animals in this population obtained marbling in mode with 2.

3 Results and discussions

3.1 Factors affecting carcass traits

The effects of group of cattle and the number of pair of permanent incisors of animals within group of cattle on carcass traits are shown in Table 1. Group of cattle and the number of pair of permanent incisors significantly influenced slaughter wt. ($P<0.0001$), but with low R^2 values (0.06). Because there were many factors which could affect slaughter wt., for example, feeding, animal handling at the farms or during transportation to the slaughterhouse. All carcass traits were affected by group of cattle and by number of pair of permanent incisors ($P<0.05$). But chilling loss was affected only by group of cattle ($P<0.05$). Our results agreed with previous study [5], who reported that group of cattle, including beef steers, dairy steers, and culled dairy cows, and the number of pair of permanent incisors had an influence on cold carcass wt., dressing percentage, and skin wt. ($P<0.05$).

Table 1. Effects of slaughter weight, group of cattle, and number of pair of permanent incisors within group of cattle on carcass traits

Carcass traits/ Factor	Slaughter weight	Group of cattle (GC)	No.pair of permanent incisors within GC (Pt(GC))	R^2
Slaughter weight, kg	-	<0.0001	<0.0001	0.06
Hot carcass weight, kg	<0.0001	<0.0001	<0.0001	0.91
Hot carcass weight, %	<0.0001	<0.0001	0.0001	0.42
Chilled carcass weight, %	<0.0001	<0.0001	0.0001	0.43
Chilling loss,kg	<0.0001	0.0156	0.1930	0.25
Skin weight,kg	<0.0001	<0.0001	<0.0001	0.56

3.2 The effect of gender of dairy or beef cattle (group of cattle)

Least squares means and standard errors of carcass traits as affected by group of cattle were as follows: FB had the lowest slaughter wt. (546.42±11.04 kg). MB showed the highest slaughter wt. (603.70±3.74 kg), but it did not differ ($P>0.05$) from MD (589.79±10.26 kg). FB and

MB had higher (P<0.05) hot carcass wts., and percentages of hot and chilled carcass wts. than FD and MD. Previous study reported that carcass wt. obtained from live wt. or dressing percentage for HF steers was typically less than that of beef steers [15]. The reduced yields were affected by increased proportion of gut [16-17], reduced subcutaneous fat [18], increased liver size, and increased proportion of abdominal fats as mesenteric and omental fat [17]. It is interested that although FB had the lowest slaughter wt., they had higher hot carcass wt. and percentages of hot and chilled carcass wts. than FD and MD. The hot carcass wt. and percentages of hot and chilled carcass wts. of FD in the present study were similar to those from dairy cows with slaughter wt. of 550 to 649 kg, previously studied in Thailand [19], which had hot carcass wt. of 318.05±1.84 kg, percentage of hot carcass wt. of 53.87±0.23, and percentage of chilled carcass wt. of 52.51±0.24, respectively.

FB and MB carcasses had lower chilling loss than MD carcasses (P<0.05), but did not differ from FD carcasses. Skin wt. of MB was the greatest (48.11±0.32 kg), while that of FD was the least (38.89±0.60 kg). There was no difference (P>0.05) in skin wt. of FB and MD (44.66±0.94 and 42.87±0.87 kg, respectively). Previous studies reported skin wts. of 35.95±0.53 kg for culled dairy cows with slaughter wt. of 550 to 649 kg and 43.50±2.55 kg for dairy steers with 550 to 599 kg slaughter wt. [20-21].

For percentage distribution of animals in each group of cattle graded into each marbling score, results showed that total percentages of cattle in score 1 to 5 were 39.58, 40.26, 15.06, 3.98, and 1.12 %, respectively. All groups of cattle were graded into score 1 to 5, except for MD which had score 1 to 4. The majority of cattle (40.26%) had marbling score 2. Interestingly, 9.28 % of FD had score 2 and 6.36 % was graded score 1. For female, 6.95 % FB had score 1, which was similar to FD (6.36%). However, the percentage of FD graded score 2 was more than that of FB (9.28 and 6.70 %, respectively). For male cattle, the percentages of MB and MD with marbling score 1 were 21.81 and 4.47 %, respectively, which was slightly more than those with score 2 (21.03 and 3.25 % for MB and MD, respectively).

3.3 Effects of number of pair of permanent incisors on carcass traits

Pairs of permanent incisors of animal within gender of cattle breeds affected the carcass traits (Table 2). No FD with zero pair of permanent incisors was found. Because culled dairy cows were generally older than one year. MD and FD with the 4th pair of incisors had the least hot carcass wts., but were not different from those of FD with the 1st and 3rd pairs of incisors. Similar results were also found in percentages of hot and chilled carcass wts. FB and MB in all maturity categories, except the most mature groups (4th pair of incisors), had more hot carcass wt. and dressing percentages than FD and MD at all ages. Interestingly, carcass trait values of the most mature group (4th pairs of permanent incisors) were lower than those of the younger (1st, 2nd, and 3rd pairs).

Table 2. Least squares means and standard errors (LSM±SE) of studied traits affected by number of pair of permanent incisors of each cattle group.

Pt	GC	n	HCW LSM ±SE	PHCW LSM ±SE	PCCW LSM ±SE	SKW LSM ±SE
0	FB	2	354.95 ±10.63^{f}	60.59 ±1.79^{f}	59.13 ±1.79^{f}	50.08 ±4.44^{de}
1	FB	30	347.74 ±2.75^{def}	58.96 ±0.46^{def}	57.68 ±0.46^{def}	45.80 ±1.15^{cd}
2	FB	85	349.89 ±1.63^{ef}	59.39 ±0.27^{ef}	58.12 ±0.27^{f}	44.52 ±0.68^{bc}
3	FB	104	347.84 ±1.48^{ef}	59.03 ±0.25^{def}	57.79 ±0.25^{ef}	42.61 ±0.62^{b}
4	FB	133	342.86 ±1.31^{d}	58.17 ±0.22^{d}	56.86 ±0.22^{d}	40.27 ±0.55^{b}
1	FD	9	324.28 ±5.01^{abc}	54.84 ±0.84^{abc}	53.44 ±0.84^{abc}	41.84 ±2.09^{b}
2	FD	46	325.31 ±2.22^{bc}	55.04 ±0.37^{abc}	53.64 ±0.37^{abc}	38.46 ±0.92^{a}
3	FD	96	321.54 ±1.53^{ab}	54.56 ±0.26^{ab}	53.21 ±0.26^{ab}	38.26 ±0.64^{a}
4	FD	247	317.55 ±0.96^{a}	53.85 ±0.16^{a}	52.52 ±0.16^{a}	37.01 ±0.39^{a}
0	MB	29	351.68 ±2.79^{ef}	59.57 ±0.47^{f}	58.26 ±0.47^{f}	52.22 ±1.16^{e}
1	MB	380	350.52 ±0.77^{ef}	59.43 ±0.13^{f}	58.12 ±0.13^{f}	48.87 ±0.32^{d}
2	MB	517	351.79 ±0.66^{f}	59.61 ±0.11^{f}	58.31 ±0.11^{f}	48.15 ±0.28^{d}
3	MB	155	349.74 ±1.22^{ef}	59.21 ±0.20^{ef}	57.91 ±0.20^{ef}	47.35 ±0.51^{d}
4	MB	53	346.55 ±2.07^{de}	58.70 ±0.35^{de}	57.35 ±0.35^{de}	43.97 ±0.86^{bc}
0	MD	6	330.13 ±6.14^{c}	55.68 ±1.03^{bc}	54.41 ±1.03^{bc}	45.09 ±2.56^{bcd}
1	MD	76	329.66 ±1.72^{c}	55.91 ±0.29^{c}	54.51 ±0.29^{c}	42.24 ±0.72^{b}
2	MD	62	332.77 ±1.91^{c}	56.40 ±0.32^{c}	54.94 ±0.32^{c}	44.64 ±0.80^{bc}
3	MD	25	327.85 ±3.00^{bc}	55.62 ±0.51^{bc}	54.22 ±0.51^{bc}	43.59 ±1.25^{bc}
4	MD	4	307.38 ±7.52^{a}	52.79 ±1.26^{a}	51.33 ±1.27^{a}	38.80 ±3.14^{ab}

^{abc}: Different letters in the same column differed (P<0.05). HCW=Hot carcass weight, PHCW = Percentage of hot carcass weight, PCCW= Percentage of chilled carcass weight, SKW= Skin weight.

Skin wt. of MB with zero pair of permanent incisors was the highest (52.22±1.16 kg), but did not differ from FB in the same age group (50.08±4.44 kg). The lowest skin wt. was found in FD with the 4th, 3rd, and 2nd pairs of permanent incisors (37.01±0.39, 38.26±0.64, and 38.46±0.92 kg, respectively) and in MD with the 4th pairs of permanent incisors (38.80±3.14 kg)

Figure 1 shows percentages of animals at each age category within each cattle group obtained marbling scores 1 to 5. Marbling score is an important trait for adding value to the chilled carcass weight. The most mature (4th pair) FB, MB, and FD had the highest marbling score of 5 (30.43, 52.17, and 17.39 %, respectively). Marbling score 5 was not observed in any age categories of MD. MB with the 2nd pairs of permanent incisors were graded marbling score 4 (39.02%).

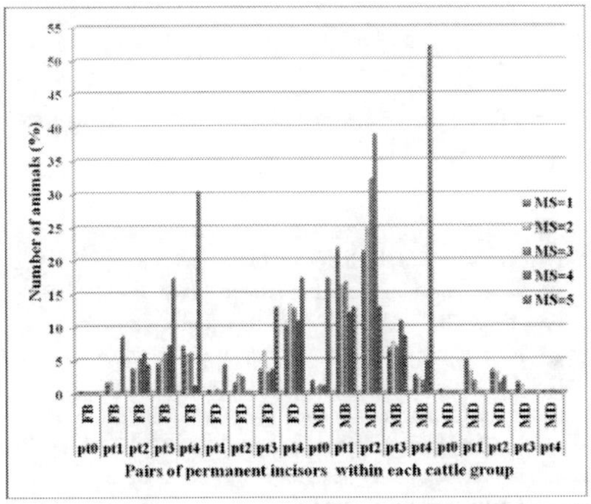

Fig. 1. Percentages of cattle with 0, 1st, 2nd, 3rd, and 4th pairs of permanent incisors (pt0 to pt4) within each group of cattle, which were graded 1 to 5 marbling score (MS).

In conclusion, hot carcass wt., percentage of hot or chilled carcass wt., and skin wt. of FB and MB from all ages were greater than those of the dairy cattle. MD cattle with 0-3 pairs of permanent incisors had greater hot carcass wt., percentage of hot or chilled carcass wt., and skin wt. than FD cattle, except for marbling score. Therefore, MD cattle with less than 4 pairs of permanent incisors can be raised as finished cattle. However, profit optimization from finishing FD and MD cattle should be further studied.

We would like to thank Mr.Sitthiporn Boorananath, Beef Cluster Cooperative Limited (Maxbeef), Kamphaengsaen, Nakhon Pathom Province, for data supporting.

References

1. *Number of Livestock in Thailand in the Year 2015*. Department of Livestock Development, 193 pp. (2015)
2. *Report Operating Result: Strategy for Promotion and Development of Thai Beef Cattle Industrial System*. Sub-committee on Agricultural Productivity Examination, 139 pp. (2017)
3. P.E. Strydom. S. Afr. j. anim. sci. **41**, 3 (2011)
4. T. E. Lawrence, J. D. Whatley, T. H. Montgomery, and L. J. Perino. J. Anim. Sci. **79** (2001)
5. K. Tuntivisoottikul and K. Worawong. Agricultural Sci. J. Vol. **48**, 2 (2017)
6. S. Sawanon. Kasetsart Livestock Magazine. **39**, 153 (2012)
7. P. Tavitchasri, T. Kanloung, T. Phonmum, N. Prachachet, and N. Ngamyeesoon. *63rd International Congress of Meat Science and Technology*, Cork, Ireland (2017)
8. Crawford, G. *Minnesota Farm Guide*, University of Minnesota Beef Team. USA. (2010)
9. Fiems, L.O., Campeneere, S. De, Caelenbergh, W. Van, De Boever, J.L. and Vanacker J.M. Meat Sci. **63** (2003)
10. Minchin, W., Buckley, F., Kenny, D.A., Monahan, F.J., Shalloo, L., and O'Donovan, M. Meat Sci. **81**, 1 (2009)
11. Schnell, T. D. K. E. Belk, J. D. Tatum, R. K. Miller, and G. C. Smith. J. Anim. Sci. **75** (1997)
12. Wardynski, F. 2012. Available source: http://msue.anr.msu.edu/experts/frank wardynski
13. *TACFS 6001: Beef.* 2004. Available source: http://www.acfs.go.th/eng/commodity_standard.php
14. SAS. 1999. *SAS/STAT User's Guide: Statistic*, SAS Institute Inc. North Carolina.
15. D. R. Buege. Meat Facts and Analysis **88**, 1 Univ. WI-Extension (1988)
16. A. Y. M. Nour, M. L. Thonney, J. R. Stouffer, and W. R. C. White. J. Anim. Sci. **57** (1983)
17. St. C. S Taylor, and J. I. Murray. Anim. Prod. **53** (1991)
18. A. Y. M. Nour, M. L. Thonney, J. R. Stouffer, and W. R. C. White. J. Anim. Sci. **57** (1983)
19. W. Sukjai, Y. Opatpatanakit and T. Supphakitchanon. KHON KAEN AGR. J. **40** SUPPLMENT (2012)
20. S. Noidad, R. Lertpatarakomol,and M. Rakthong. *Proceeding of the 5th Meat Science and Technology*, KMITL. (2014a).
21. S. Noidad, R. Lertpatarakomol,and M. Rakthong. *Proceeding of the 5th Meat Science and Technology*, KMITL. (2014b).

The fatty acid profiles in *M. longissimus dorsi* and *M. semimembranosus* of short-haired and long-haired crossbred sheep meat

Pattraphorn Pattharararangsarith[1] *and Kunya* Tuntivisoottikul[1,*]

[1]Agricultural Education Department, Faculty of Industrial Education and Technology, King Mongkut's Institute of Technology Ladkrabang, Bangkok, Thailand

Abstract. This study was conducted to investigate the effects of crossbred groups and anatomical muscle positions on the chemical composition and fatty acid profiles in the loin (*M. longissimus dorsi)* and chump (*M. semimembranosus)* muscles of the short-haired and long-haired crossbred sheep. Five sheep per group with the 25 kg initial body weight were fed with 2 kg/d TMR and free choice grass (*Brachiaria mutica*), and those were raised 84 days. After slaughter, loin and chump were used to examine the chemical composition and fatty acid profiles. The results showed that the loin from the short-haired and long-haired crossbreds had Palmitic acid at the highest level among the other saturated fatty acid (SFA), whereas this fatty acid was not detected in the chump. However, the highest SFA in the chump of both crossbreds was the Stearic acid. Furthermore, the fatty acids in both groups were detected in the chump more than in the loin. There was found that the nutritional values were not affected by sheep crossbred groups ($P>0.05$). The loin and chump had affected the percentage of crude protein ($P<0.01$), the amount of Lauric Acid (C12:0), and the total fatty acids ($P<0.01$). Interaction factor between the crossbred groups and the anatomical muscle positions was unaffected all the traits studied ($P>0.05$).

1 Introduction

Sheep (*Ovis* spp.) is the small ruminant which is widely propagated all over Thailand, especially in the southern part where has many Muslim populations. The characteristics of sheep production system in Thailand are the native breed, a small body size with 8-12 kg body weight, a small flock size, the low fertility traits, and the low technology used; therefore, the animal performance indices are below the biological potential, resulting in the low quality of final products. However, the advantages of these sheep are the disease resistance and the high potential for the environmental resource utilization. To enhance the productivity of the native sheep, crossbreeding with the foreign breeds is the common practice. Recently, Department of Livestock Development (DLD) Thailand imported the plenty of pure breeds for crossbreeding with the Thai native breed. Therefore, many crossbred lines were breed from Thai native and various foreign breeds such as Katahdin, Santa Inês, Dorper, Barbados Blackbelly, South African Mutton Merino, and Corriedale. Thus, these crossbreds are broadly contributed to the local farmers and the plenty sheep in Thailand were the undefined crossbred. Recently, the Santa Inês and Dorper breed that are recognized for their outstanding meat production were used as the terminal crossing between native breeds. Therefore, those exotic breeds widely emerged at the local farms [1]. In order to classify the undefined crossbred sheep, the appearance characteristics were described. In this experiment, the long-haired crossbred was expected that it has been dominated with the Dorper breed; in the other hand, the short-haired crossbred was illustrated the characteristics of the Santa Inês breed.

Adipose fat deposits have been emphasized in assessment studied of meat quality, as these deposits are increasingly being used in the improvement of texture and palatability of meat. The fatty acid profiles are the major contributor to meat quality in small ruminants. Moreover, they play the important role in coping with the health issue in the view of the high incidence of diseases. The pertinent attention was toward decreasing the percentage of saturated fatty acid (SFA) and increasing the percentage of unsaturated fatty acid (USFA); in addition, the poly-unsaturated fatty acid (PUFA) was desired. The fatty acid composition of meat might be influenced by a variety of factors, including animal breeding, feeding, and rearing system [2]. However, there is no documented result on the chemical composition and fatty acid profiles of the crossbred sheep in Thailand. Therefore, the objective of the present study was to investigate the effects of crossbreed groups and anatomical muscle positions on the chemical composition and the fatty acid profiles in the loin and chump of short-haired and long-haired crossbred sheep.

* Corresponding author: kunya.tu@kmitl.ac

2 Materials and Methods

2.1 Animals and production system

The research was carried out at Suwanvajokkasikit Animal R&D Institute, Kasetsart University, Kamphaeng Saen Campus, Nakorn Pathom Province, Thailand. The local altitude is 6 m above sea level at 13°59′2 N and 99°59′38″ E. The temperature in the region varies between the minimum 18 °C to the maximum 40°C, with an annual rainfall of 500-1,650 mm. Ten male sheep, one-year-old with 25 kg initial body weight, equally represented the two crossbred groups; short-haired and long-haired crossbred sheep. Animals were raised in the confinement with the slatted floor. The animals were fed with 2 kg/d TMR and free choice grass (*Brachiaria mutica*). The feed samples were analysed as the proximate analysis [3].

Table 1. Chemical composition of TMR and grass (*Brachiaria mutica*)

Chemical composition (%)	Feed	
	TMR	***Brachiaria mutica***
Dry matter	88.83	21.97
Crude Protein	13.65	6.20
Crude Fat	4.70	0.15
Crude Fiber	4.94	21.11

2.2 Chemical composition and fatty acid profiles

At slaughter, sheep were weighted, stunned and killed by exsanquination. The gastrointestinal tract and skin were removed. The carcasses were split into two identical longitudinal halves. Tested muscle samples from *M. longissimus dorsi* and *M. semimembranosus* were taken for laboratory analysis component as the following: the percentages of moisture and dry matter– by oven drying 105 °C, crude protein– by the Kjeldahl method, crude fat– by Soxhlet extractor and crude ash – by 550 °C heat [3] and fatty acid profiles– by gas chromatography [4].

2.3 Statistical analysis

Descriptive statistics were performed to calculate mean values and standard deviations for all variables. In addition, least squares procedures were used to analyse data for the chemical composition and the fatty acid profiles. The General Linear Model (GLM) were used in the analyses of these characteristics included the fixed effects of crossbred groups and the muscle anatomical positions. The option PDIFF was used for calculating significant differences between least squares means.

$$Y_{ijk} = \mu + B_i + M_j + e_{ijk}$$

where:

Y the dependent variable
μ the overall mean
B_i fixed crossbred group effect, (i = short-haired, long-haired crossbred)
M_j fixed anatomical muscle position effect (j = *M. longissimus dorsi*, *M. semimembranosus*)
e_{ijk} residual error

A pre-analysis showed that the interaction was not significant, so it was excluded out of the model.

3 Results and discussions

3.1 The chemical composition in loin and chump of short-haired and long-haired crossbred sheep meat

The chemical composition in loin and chump of short-haired and long-haired crossbred sheep obtained in the experiment are described in Table 2. It was evidence that the percentage of moisture, crude protein, and crude fat of long-haired crossbred loin are greater than those short-haired crossbred loin (75.83±1.19, 23.37±1.77, and 8.29±1.29 vs 75.27±1.23, 23.02±2.13, and 8.11±1.10, respectively). On the other hand, the percentage of dry matter and crude ash of short-haired crossbred loin are higher than those of long-haired crossbred loin (24.73±1.23 and 4.01±0.22 vs 24.17±1.19 and 3.68±0.36). It had been shown that the percentage of dry matter, crude fat and crude ash of long-haired crossbred chump are greater than those of short-haired crossbred chump. (24.31±1.77, 9.86±2.80, and 4.17±0.36 vs 24.09±0.94, 8.20±1.66, and 3.90±0.33, respectively). Nonetheless, the percentage of moisture and crude protein of short-haired crossbred chump are higher than those of long-haired crossbred chump (75.91±0.94, 19.91±1.09 vs 75.69±1.77, 19.80±0.84).

Table 2. Means (± standard deviation), minimum and maximum values of the chemical composition in Loin and chump of the long-haired and short-haired crossbred sheep

Chemical composition	Long haired crossbred			Short haired crossbred		
	Mean±SD	**Min**	**Max**	**Mean±SD**	**Min**	**Max**
Loin (*M. longissimus dorsi*)						
Moisture	75.83±1.19	73.89	76.91	75.27±1.23	73.09	75.92
Dry matter	24.17±1.19	23.09	26.11	24.73±1.23	24.08	26.91
Crude protein	23.37±1.77	20.65	25.04	23.02±2.13	19.79	25.40
Crude fat	8.29±1.29	6.34	9.55	8.11±1.10	6.84	9.54
Crude ash	3.68±0.36	3.30	4.08	4.01±0.22	3.63	4.18
Chump (*M. semimembranosus*)						
Moisture	75.69±1.77	74.26	78.54	75.91±0.94	74.89	77.13
Dry matter	24.31±1.77	21.46	25.74	24.09±0.94	22.87	25.11
Crude protein	19.80±0.84	19.04	20.71	19.91±1.09	18.72	21.11
Crude fat	9.86±2.80	7.10	13.85	8.20±1.66	6.55	10.04
Crude ash	4.17±0.36	3.91	4.80	3.90±0.33	3.55	4.30

The data from Table 2 have been shown that the percentage of crude fat of both mention muscles (loin and chump) of long-haired crossbred sheep were higher than those of short-haired crossbred sheep. The long-haired crossbred sheep which was expected to be the crossbred between the native and Dorper breed has the greater average body weight than those of the short-haired crossbred sheep which was anticipated to be the crossbred between the native and Santa Inês breed (the live body weight at slaughter 35.26±3.24 and 34.00±2.22 kg, respectively). There was the substantial increase in the energy and fat contents of the weight when the body weight was increased from 15-30 kg [5]. It had been known that fat depot in the muscle as the intramuscular fat is play the important role for the meat quality. The intramuscular fat was significantly associated with the sensory of meat such as the tenderness, juiciness, and flavour. Therefore, the Dorper breed has been suggested to be the terminal line of crossing breed for improving the great potential for growth and the ability for accumulate intramuscular fat in muscle. However, many previous studies showed that the Santa Inês breed, a naturalized hair breed, is known for its low-fat meat production and good maternal traits providing an opportunity for crossbred with specialized meat breeds [6], [7]. This breed was suggested to be the terminal line for producing the crossbred as well.

3.2 The fatty acid profiles in loin and chump of short-haired and long-haired crossbred sheep meat

Table 3. Means and standard deviation values of the fatty acid profiles (mg/100 g) in the loin and chump muscle of the long-haired crossbred sheep meat

Fatty acid	Loin		Chump	
	Mean	SD	Mean	SD
Lauric Acid (C12:0)	22.47	9.03	13.27	4.90
Myristic Acid (C14:0)	212.82	136.26	nd	-
Pentadecanoic Acid (C15:0)	16.33	12.10	nd	-
Palmitic Acid (C16:0)	1494.77	912.00	nd	-
Heptadecanoic Acid (C17:0)	51.47	40.24	38.53	24.11
Stearic Acid (C18:0)	933.86	774.81	635.65	427.18
Myristoleic Acid (C14:1)	9.97	2.63	6.18	0.65
Palmitoleic Acid (C16:1)	143.58	79.47	99.32	29.15
Elaidic Acid (C18:1n9t)	11.80*	-	16.26	6.02
Cis-11-Eicosenoic Acid (C20:1)	7.15	4.75	9.06*	-
Linoleic Acid (C18:2n6c)	77.78	47.60	70.26	48.60
Linolenic Acid (C18:3n3)	14.06	9.09	12.00	8.73
Cis-11,14,17-Eicosadienoic Acid (C20:3n3)	15.62	6.11	15.29	3.35
Total Fatty Acid	**2990.09**	**1995.64**	**865.65**	**545.61**

* found from one sample, nd = not found

The fatty acid profiles in loin and chump of long-haired and short-haired crossbred sheep meat were presented in Table 3 and 4. The number of fatty acid profiles in loin are greater than those in chump from both crossbreds. It was not found the Myristic acid (C14:0), the Pentadecanoic acid (C15:0), and the Palmitic acid (C16:0) in chump; nevertheless, those were found in loin. For the saturated fatty acid (SFA), the Palmitic acid (C16:0) was the most found in loin of both crossbred sheep (long-haired vs short-haired crossbred sheep; 1494.77 and 1138.60 mg/100g), the Steric acid (C17:0) was found in both loin and chump of long-haired crossbred (933.86 and 544.48 mg/100g); while, they were 635.65 and 544.48 mg/100g in both loin and chump of short-haired crossbred. It had been known that fatty acid composition in ruminant has abundant of the saturation fatty acid as the result of biohydrogenation in rumen [2]. From the study of [4], the result showed that the Myristic acid (C14:0) and Palmitic acid (C16:0) are the two main fatty acids involved in increasing low-density lipoprotein (LDL), a risk factor for heart disease; therefore, reducing the C14:0 content in the muscle of sheep is an interesting and desiring. The presence of high levels of Palmitic Acid in different muscles was affected to the total fatty acid composition (TFA) in the muscles. Fatness level was related to the ratio of polyunsaturated fatty acid (PUFA) to saturated fatty acid (SFA), which is inversely correlated with intramuscular fat [2].

Table 4. Means and standard deviation values of the fatty acid profiles (mg/100 g) in the loin and chump muscle of short-haired crossbred sheep meat

Fatty acid	Loin		Chump	
	Mean	SD	Mean	SD
Lauric Acid (C12:0)	21.17	7.82	11.42	5.14
Myristic Acid (C14:0)	190.67	61.53	nd	-
Pentadecanoic Acid (C15:0)	13.54	3.05	nd	-
Palmitic Acid (C16:0)	1138.60	251.41	nd	-
Heptadecanoic Acid (C17:0)	33.72	3.81	23.62	8.12
Stearic Acid (C18:0)	544.48	79.93	338.43	90.57
Myristoleic Acid (C14:1)	11.80	5.15	8.31	4.27
Palmitoleic Acid (C16:1)	138.63	35.81	102.43	29.09
Elaidic Acid (C18:1n9t)	16.72	3.28	10.38	3.63
Cis-11-Eicosenoic Acid (C20:1)	3.37	0.20	nd	-
Linoleic Acid (C18:2n6c)	75.92	13.04	70.81	11.23
Linolenic Acid (C18:3n3)	15.02	0.38	13.34	3.33
Cis-11,14,17-Eicosadienoic Acid (C20:3n3)	17.80	1.37	18.64	1.95
Total Fatty Acid (TFA)	**2210.72**	**446.05**	**564.33**	**152.99**

3.3 The effects of crossbred groups and anatomical muscle positions on the chemical composition and fatty acid profiles in meat of short-haired and long-haired crossbred sheep

Least squares means (±standard error) of the chemical composition and fatty acid profiles effected by the short-haired and long-haired crossbred sheep are presented in Table 5. There was no different (P>0.05) of the chemical composition, the saturated fatty acid (SFA), the monounsaturated fatty acid (MUFA), and the polyunsaturated fatty acid (PUFA) profiles between the crossbred groups. The experimental data indicate that the long-haired crossbred had the trend to accumulate the fat in muscle more than those the short-haired crossbred. Furthermore, the long-haired crossbred meat presented higher SFA, MUFA, and lower PUFA than those the short-haired crossbred. Fatty acid content has become very important as it implicates the health issue of human, the high ration of PUFA : SFA is required [4]. When the diet has the high unsaturated fatty acid (UFA), the use of ionophores may alter the meat lipid composition by affecting the ruminal dehydrogenation. In order to enhance the polyunsaturated fatty acid ω3 deposition in the muscle tissue, the higher forage levels in sheep diet is desired. It showed that Linoleic acid (C18:2n6c) are the essential fatty acid which cannot synthesize in sheep body. Therefore, it must be get from the diet of sheep [6].

Table 5. Least squares means (± standard error) of the chemical composition and the fatty acid profiles (mg/100 g) effected by the long-haired and short-haired crossbred sheep muscle

Traits	Crossbred sheep	
	Long hair	**Short hair**
Moisture (%)	75.76±0.42	75.59±0.42
Dry matter (%)	24.24±0.42	24.41±0.42
Crude protein (%)	21.58±0.53	21.46±0.50
Crude fat (%)	9.07±0.58	8.16±0.58
Ash (%)	3.93±0.10	3.96±0.10
Saturated Fatty Acids (SFA)		
Lauric acid (C12:0)	17.87±2.28	16.30±2.14
Margaric acid (C17:0)	45.00±7.55	28.67±7.55
Stearic acid (C17:0)	784.76±141.19	441.45±141.19
Monounsaturated Fatty Acids (MUFA)		
Myristoleic acid (C14:1)	8.07±1.29	10.05±1.23
Palmitoic acid (C16:1)	121.45±15.24	120.53±15.24
Elaidic acid (C18:1n9t)	14.03±2.42	13.55±1.40
Polyunsaturated Fatty Acids (PUFA)		
Linoleic acid (C18:2n6c)	74.02±11.76	73.36±11.76
Alpha-Linolenic acid (C18:3n3)	13.03±2.33	14.18±2.04
cis-11,14,17-Eicosatrienoic acid (C20:3n3)	15.46±1.31	18.22±1.43
Total fatty acid	**1927.87±335.51**	**1387.53±335.51**

Table 6. Least squares means (± standard error) of the chemical composition and the fatty acid profiles (mg/100 g) effected by the loin and chump muscle

Traits	Muscle Type	
	Loin	**Chump**
Moisture (%)	75.55±0.42	75.80±0.42
Dry matter (%)	24.45±0.42	24.20±0.42
Crude protein (%)	23.20±0.50[a]	19.85±0.53[b]
Crude fat (%)	8.20±0.58	9.03±0.58
Ash (%)	3.84±0.10	4.04±0.10
Saturated Fatty Acids (SFA)		
Lauric acid (C12:0)	21.82±2.28[a]	12.34±2.15[b]
Margaric acid (C17:0)	42.60±7.55	31.08±7.55
Stearic acid (C17:0)	739.17±141.19	487.04±141.19
Monounsaturated Fatty Acids (MUFA)		
Myristoleic acid (C14:1)	10.88±1.29	7.24±1.23
Palmitoic acid (C16:1)	141.11±15.24	100.87±15.24
Elaidic acid (C18:1n9t)	14.26±2.34	13.32±1.53
Polyunsaturated Fatty Acids (PUFA)		
Linoleic acid (C18:2n6c)	76.85±11.09	70.53±12.39
Alpha-Linolenic acid (C18:3n3)	14.54±2.15	12.67±2.22
cis-11,14,17-Eicosatrienoic acid (C20:3n3)	16.71±1.23	16.97±1.49
Total fatty acid	**2600.40±335.51[a]**	**714.99±335.51[b]**

[ab] Different letters within the same row differ significantly (P<0.01)

Least squares means (±standard error) of the chemical composition and fatty acid profiles effected by the lion and chump muscle are presented in Table 6. It had been shown that the parameter traits; the crude protein (CP), the Lauric acid (C12:0), and the total fatty acid of the loins were highly significant greater than those chumps (P<0.1). The outcome of the plenty deposition of protein in the muscle was related to the number of muscle fiber, the muscle fiber type, and the protein type. The loin which is the sides between the lower ribs and pelvis has been realized as the high quality of muscle because it contains the plenty numbers of myofibrillar protein [3]. Nevertheless, the chump which is the muscle where the loin section meets the top of the hind legs is greater contraction than loin. This is the reason that chump has less myofibrillar protein and less fat deposit.

Lauric acid (C12:0), the 12-C saturated fatty acid (SFA) of intramuscular fat, is greater deposited in loin than in chump. It has been known that the chump has more muscle contraction than loin due to the enormous activity for moving at that muscle. This is the reason that the less muscle contract, the grater fat deposit. It has been shown that the interaction factor between the crossbred groups and the anatomical muscle position was unaffected all the traits studied (P>0.05).

4 Conclusion

The nutritional values (chemical composition and fatty acids profiles) were not affected by sheep crossbreds groups (P>0.05) in this study. Whereas, the anatomical muscle positions were the influence of some value traits; crude protein (CP), SFA (Lauric acid: C12:0), and total fatty acid (TFA). Interaction factor between the crossbred groups and the anatomical muscle positions was unaffected all the traits studied (P>0.05).

References

1 Department of Livestock Development. [online] http://www.dld.go.th/service/sheep/type.html (2018)
2 S. De Smet, K. Rase, and D. Demeyer, Animal Research, 53 (2004)
3 AOAC, *Official methods of analysis* (Association of official Chemist, Washington DC, 1992)
4 M.M.Y. Carneiro, R.H.de T.e B. de Goes, L.H.X. da Silva, A.R.M. Fernandes, R.T. de Oliveira, C.A.L. Cardoso, A.S.O. Hirata, Zootec 45, 6 (2016)
5 J.G.L. Regadas Filho, E.S. Pereira, P.G. Pimentel, A.B.S. Villarroel, A.N. Medeiros, Small Ruminant Research 109, 2-3 (2013)
6 T.P. Paim, A.F. Silva, R.F.S. Martins, B.O. Borges, P.M.T. Lima, C.C. Cardoso, G.I.F. Esteves, H. Louvandini and C. McManus, Small Ruminant Research 112, 1-3 (2013)
7 A.V. Fisher, M. Enser, R.I. Richardson, J.D. Wood, G.R. Nute, E. Kurt, L.A. Sinclair, R.G. Wilkinson, Meat Science 55 (2000)

2 Materials and Methods

2.1 Animals and production system

The research was carried out at Suwanvajokkasikit Animal R&D Institute, Kasetsart University, Kamphaeng Saen Campus, Nakorn Pathom Province, Thailand. The local altitude is 6 m above sea level at 13°59'2 N and 99°59'38″ E. The temperature in the region varies between the minimum 18 °C to the maximum 40°C, with an annual rainfall of 500-1,650 mm. Ten male sheep, one-year-old with 25 kg initial body weight, equally represented the two crossbred groups; short-haired and long-haired crossbred sheep. Animals were raised in the confinement with the slatted floor. The animals were fed with 2 kg/d TMR and free choice grass (*Brachiaria mutica*). The feed samples were analysed as the proximate analysis [3].

Table 1. Chemical composition of TMR and grass (*Brachiaria mutica*)

Chemical composition (%)	Feed	
	TMR	*Brachiaria mutica*
Dry matter	88.83	21.97
Crude Protein	13.65	6.20
Crude Fat	4.70	0.15
Crude Fiber	4.94	21.11

2.2 Chemical composition and fatty acid profiles

At slaughter, sheep were weighted, stunned and killed by exsanquination. The gastrointestinal tract and skin were removed. The carcasses were split into two identical longitudinal halves. Tested muscle samples from *M. longissimus dorsi* and *M. semimembranosus* were taken for laboratory analysis component as the following: the percentages of moisture and dry matter– by oven drying 105 °C, crude protein– by the Kjeldahl method, crude fat– by Soxhlet extractor and crude ash – by 550 °C heat [3] and fatty acid profiles– by gas chromatography [4].

2.3 Statistical analysis

Descriptive statistics were performed to calculate mean values and standard deviations for all variables. In addition, least squares procedures were used to analyse data for the chemical composition and the fatty acid profiles. The General Linear Model (GLM) were used in the analyses of these characteristics included the fixed effects of crossbred groups and the muscle anatomical positions. The option PDIFF was used for calculating significant differences between least squares means.

$$Y_{ijk} = \mu + B_i + M_j + e_{ijk}$$

where:

Y	the dependent variable
μ	the overall mean
B_i	fixed crossbred group effect, (i = short-haired, long-haired crossbred)
M_j	fixed anatomical muscle position effect (j = *M. longissimus dorsi*, *M. semimembranosus*)
e_{ijk}	residual error

A pre-analysis showed that the interaction was not significant, so it was excluded out of the model.

3 Results and discussions

3.1 The chemical composition in loin and chump of short-haired and long-haired crossbred sheep meat

The chemical composition in loin and chump of short-haired and long-haired crossbred sheep obtained in the experiment are described in Table 2. It was evidence that the percentage of moisture, crude protein, and crude fat of long-haired crossbred loin are greater than those short-haired crossbred loin (75.83±1.19, 23.37±1.77, and 8.29±1.29 vs 75.27±1.23, 23.02±2.13, and 8.11±1.10, respectively). On the other hand, the percentage of dry matter and crude ash of short-haired crossbred loin are higher than those of long-haired crossbred loin (24.73±1.23 and 4.01±0.22 vs 24.17±1.19 and 3.68±0.36). It had been shown that the percentage of dry matter, crude fat and crude ash of long-haired crossbred chump are greater than those of short-haired crossbred chump. (24.31±1.77, 9.86±2.80, and 4.17±0.36 vs 24.09±0.94, 8.20±1.66, and 3.90±0.33, respectively). Nonetheless, the percentage of moisture and crude protein of short-haired crossbred chump are higher than those of long-haired crossbred chump (75.91±0.94, 19.91±1.09 vs 75.69±1.77, 19.80±0.84).

Table 2. Means (± standard deviation), minimum and maximum values of the chemical composition in Loin and chump of the long-haired and short-haired crossbred sheep

Chemical composition	Long haired crossbred			Short haired crossbred		
	Mean±SD	Min	Max	Mean±SD	Min	Max
Loin (*M. longissimus dorsi*)						
Moisture	75.83±1.19	73.89	76.91	75.27±1.23	73.09	75.92
Dry matter	24.17±1.19	23.09	26.11	24.73±1.23	24.08	26.91
Crude protein	23.37±1.77	20.65	25.04	23.02±2.13	19.79	25.40
Crude fat	8.29±1.29	6.34	9.55	8.11±1.10	6.84	9.54
Crude ash	3.68±0.36	3.30	4.08	4.01±0.22	3.63	4.18
Chump (*M. semimembranosus*)						
Moisture	75.69±1.77	74.26	78.54	75.91±0.94	74.89	77.13
Dry matter	24.31±1.77	21.46	25.74	24.09±0.94	22.87	25.11
Crude protein	19.80±0.84	19.04	20.71	19.91±1.09	18.72	21.11
Crude fat	9.86±2.80	7.10	13.85	8.20±1.66	6.55	10.04
Crude ash	4.17±0.36	3.91	4.80	3.90±0.33	3.55	4.30

The data from Table 2 have been shown that the percentage of crude fat of both mention muscles (loin and chump) of long-haired crossbred sheep were higher than those of short-haired crossbred sheep. The long-haired crossbred sheep which was expected to be the crossbred between the native and Dorper breed has the greater average body weight than those of the short-haired crossbred sheep which was anticipated to be the crossbred between the native and Santa Inês breed (the live body weight at slaughter 35.26±3.24 and 34.00±2.22 kg, respectively). There was the substantial increase in the energy and fat contents of the weight when the body weight was increased from 15-30 kg [5]. It had been known that fat depot in the muscle as the intramuscular fat is play the important role for the meat quality. The intramuscular fat was significantly associated with the sensory of meat such as the tenderness, juiciness, and flavour. Therefore, the Dorper breed has been suggested to be the terminal line of crossing breed for improving the great potential for growth and the ability for accumulate intramuscular fat in muscle. However, many previous studies showed that the Santa Inês breed, a naturalized hair breed, is known for its low-fat meat production and good maternal traits providing an opportunity for crossbred with specialized meat breeds [6], [7]. This breed was suggested to be the terminal line for producing the crossbred as well.

3.2 The fatty acid profiles in loin and chump of short-haired and long-haired crossbred sheep meat

Table 3. Means and standard deviation values of the fatty acid profiles (mg/100 g) in the loin and chump muscle of the long-haired crossbred sheep meat

Fatty acid	Loin		Chump	
	Mean	SD	Mean	SD
Lauric Acid (C12:0)	22.47	9.03	13.27	4.90
Myristic Acid (C14:0)	212.82	136.26	nd	-
Pentadecanoic Acid (C15:0)	16.33	12.10	nd	-
Palmitic Acid (C16:0)	1494.77	912.00	nd	-
Heptadecanoic Acid (C17:0)	51.47	40.24	38.53	24.11
Stearic Acid (C18:0)	933.86	774.81	635.65	427.18
Myristoleic Acid (C14:1)	9.97	2.63	6.18	0.65
Palmitoleic Acid (C16:1)	143.58	79.47	99.32	29.15
Elaidic Acid (C18:1n9t)	11.80*	-	16.26	6.02
Cis-11-Eicosenoic Acid (C20:1)	7.15	4.75	9.06*	-
Linoleic Acid (C18:2n6c)	77.78	47.60	70.26	48.60
Linolenic Acid (C18:3n3)	14.06	9.09	12.00	8.73
Cis-11,14,17-Eicosadienoic Acid (C20:3n3)	15.62	6.11	15.29	3.35
Total Fatty Acid	**2990.09**	**1995.64**	**865.65**	**545.61**

* found from one sample, nd = not found

The fatty acid profiles in loin and chump of long-haired and short-haired crossbred sheep meat were presented in Table 3 and 4. The number of fatty acid profiles in loin are greater than those in chump from both crossbreds. It was not found the Myristic acid (C14:0), the Pentadecanoic acid (C15:0), and the Palmitic acid (C16:0) in chump; nevertheless, those were found in loin. For the saturated fatty acid (SFA), the Palmitic acid (C16:0) was the most found in loin of both crossbred sheep (long-haired vs short-haired crossbred sheep; 1494.77 and 1138.60 mg/100g), the Steric acid (C17:0) was found in both loin and chump of long-haired crossbred (933.86 and 544.48 mg/100g); while, they were 635.65 and 544.48 mg/100g in both loin and chump of short-haired crossbred. It had been known that fatty acid composition in ruminant has abundant of the saturation fatty acid as the result of biohydrogenation in rumen [2]. From the study of [4], the result showed that the Myristic acid (C14:0) and Palmitic acid (C16:0) are the two main fatty acids involved in increasing low-density lipoprotein (LDL), a risk factor for heart disease; therefore, reducing the C14:0 content in the muscle of sheep is an interesting and desiring. The presence of high levels of Palmitic Acid in different muscles was affected to the total fatty acid composition (TFA) in the muscles. Fatness level was related to the ratio of polyunsaturated fatty acid (PUFA) to saturated fatty acid (SFA), which is inversely correlated with intramuscular fat [2].

Table 4. Means and standard deviation values of the fatty acid profiles (mg/100 g) in the loin and chump muscle of short-haired crossbred sheep meat

Fatty acid	Loin		Chump	
	Mean	SD	Mean	SD
Lauric Acid (C12:0)	21.17	7.82	11.42	5.14
Myristic Acid (C14:0)	190.67	61.53	nd	-
Pentadecanoic Acid (C15:0)	13.54	3.05	nd	-
Palmitic Acid (C16:0)	1138.60	251.41	nd	-
Heptadecanoic Acid (C17:0)	33.72	3.81	23.62	8.12
Stearic Acid (C18:0)	544.48	79.93	338.43	90.57
Myristoleic Acid (C14:1)	11.80	5.15	8.31	4.27
Palmitoleic Acid (C16:1)	138.63	35.81	102.43	29.09
Elaidic Acid (C18:1n9t)	16.72	3.28	10.38	3.63
Cis-11-Eicosenoic Acid (C20:1)	3.37	0.20	nd	-
Linoleic Acid (C18:2n6c)	75.92	13.04	70.81	11.23
Linolenic Acid (C18:3n3)	15.02	0.38	13.34	3.33
Cis-11,14,17-Eicosadienoic Acid (C20:3n3)	17.80	1.37	18.64	1.95
Total Fatty Acid (TFA)	**2210.72**	**446.05**	**564.33**	**152.99**

3.3 The effects of crossbred groups and anatomical muscle positions on the chemical composition and fatty acid profiles in meat of short-haired and long-haired crossbred sheep

Least squares means (±standard error) of the chemical composition and fatty acid profiles effected by the short-haired and long-haired crossbred sheep are presented in Table 5. There was no different (P>0.05) of the chemical composition, the saturated fatty acid (SFA), the monounsaturated fatty acid (MUFA), and the polyunsaturated fatty acid (PUFA) profiles between the crossbred groups. The experimental data indicate that the long-haired crossbred had the trend to accumulate the fat in muscle more than those the short-haired crossbred. Furthermore, the long-haired crossbred meat presented higher SFA, MUFA, and lower PUFA than those the short-haired crossbred. Fatty acid content has become very important as it implicates the health issue of human, the high ration of PUFA : SFA is required [4]. When the diet has the high unsaturated fatty acid (UFA), the use of ionophores may alter the meat lipid composition by affecting the ruminal dehydrogenation. In order to enhance the polyunsaturated fatty acid ω3 deposition in the muscle tissue, the higher forage levels in sheep diet is desired. It showed that Linoleic acid (C18:2n6c) are the essential fatty acid which cannot synthesize in sheep body. Therefore, it must be get from the diet of sheep [6].

Table 5. Least squares means (± standard error) of the chemical composition and the fatty acid profiles (mg/100 g) effected by the long-haired and short-haired crossbred sheep muscle

Traits	Crossbred sheep	
	Long hair	**Short hair**
Moisture (%)	75.76±0.42	75.59±0.42
Dry matter (%)	24.24±0.42	24.41±0.42
Crude protein (%)	21.58±0.53	21.46±0.50
Crude fat (%)	9.07±0.58	8.16±0.58
Ash (%)	3.93±0.10	3.96±0.10
Saturated Fatty Acids (SFA)		
Lauric acid (C12:0)	17.87±2.28	16.30±2.14
Margaric acid (C17:0)	45.00±7.55	28.67±7.55
Stearic acid (C17:0)	784.76±141.19	441.45±141.19
Monounsaturated Fatty Acids (MUFA)		
Myristoleic acid (C14:1)	8.07±1.29	10.05±1.23
Palmitoic acid (C16:1)	121.45±15.24	120.53±15.24
Elaidic acid (C18:1n9t)	14.03±2.42	13.55±1.40
Polyunsaturated Fatty Acids (PUFA)		
Linoleic acid (C18:2n6c)	74.02±11.76	73.36±11.76
Alpha-Linolenic acid (C18:3n3)	13.03±2.33	14.18±2.04
cis-11,14,17-Eicosatrienoic acid (C20:3n3)	15.46±1.31	18.22±1.43
Total fatty acid	**1927.87±335.51**	**1387.53±335.51**

Table 6. Least squares means (± standard error) of the chemical composition and the fatty acid profiles (mg/100 g) effected by the loin and chump muscle

Traits	Muscle Type	
	Loin	**Chump**
Moisture (%)	75.55±0.42	75.80±0.42
Dry matter (%)	24.45±0.42	24.20±0.42
Crude protein (%)	23.20±0.50[a]	19.85±0.53[b]
Crude fat (%)	8.20±0.58	9.03±0.58
Ash (%)	3.84±0.10	4.04±0.10
Saturated Fatty Acids (SFA)		
Lauric acid (C12:0)	21.82±2.28[a]	12.34±2.15[b]
Margaric acid (C17:0)	42.60±7.55	31.08±7.55
Stearic acid (C17:0)	739.17±141.19	487.04±141.19
Monounsaturated Fatty Acids (MUFA)		
Myristoleic acid (C14:1)	10.88±1.29	7.24±1.23
Palmitoic acid (C16:1)	141.11±15.24	100.87±15.24
Elaidic acid (C18:1n9t)	14.26±2.34	13.32±1.53
Polyunsaturated Fatty Acids (PUFA)		
Linoleic acid (C18:2n6c)	76.85±11.09	70.53±12.39
Alpha-Linolenic acid (C18:3n3)	14.54±2.15	12.67±2.22
cis-11,14,17-Eicosatrienoic acid (C20:3n3)	16.71±1.23	16.97±1.49
Total fatty acid	**2600.40±335.51[a]**	**714.99±335.51[b]**

[ab] Different letters within the same row differ significantly (P<0.01)

Least squares means (±standard error) of the chemical composition and fatty acid profiles effected by the lion and chump muscle are presented in Table 6. It had been shown that the parameter traits; the crude protein (CP), the Lauric acid (C12:0), and the total fatty acid of the loins were highly significant greater than those chumps (P<0.1). The outcome of the plenty deposition of protein in the muscle was related to the number of muscle fiber, the muscle fiber type, and the protein type. The loin which is the sides between the lower ribs and pelvis has been realized as the high quality of muscle because it contains the plenty numbers of myofibrillar protein [3]. Nevertheless, the chump which is the muscle where the loin section meets the top of the hind legs is greater contraction than loin. This is the reason that chump has less myofibrillar protein and less fat deposit.

Lauric acid (C12:0), the 12-C saturated fatty acid (SFA) of intramuscular fat, is greater deposited in loin than in chump. It has been known that the chump has more muscle contraction than loin due to the enormous activity for moving at that muscle. This is the reason that the less muscle contract, the grater fat deposit. It has been shown that the interaction factor between the crossbred groups and the anatomical muscle position was unaffected all the traits studied (P>0.05).

4 Conclusion

The nutritional values (chemical composition and fatty acids profiles) were not affected by sheep crossbreds groups (P>0.05) in this study. Whereas, the anatomical muscle positions were the influence of some value traits; crude protein (CP), SFA (Lauric acid: C12:0), and total fatty acid (TFA). Interaction factor between the crossbred groups and the anatomical muscle positions was unaffected all the traits studied (P>0.05).

References

1 Department of Livestock Development. [online] http://www.dld.go.th/service/sheep/type.html (2018)
2 S. De Smet, K. Rase, and D. Demeyer, Animal Research, 53 (2004)
3 AOAC, *Official methods of analysis* (Association of official Chemist, Washington DC, 1992)
4 M.M.Y. Carneiro, R.H.de T.e B. de Goes, L.H.X. da Silva, A.R.M. Fernandes, R.T. de Oliveira, C.A.L. Cardoso, A.S.O. Hirata, Zootec 45, 6 (2016)
5 J.G.L. Regadas Filho, E.S. Pereira, P.G. Pimentel, A.B.S. Villarroel, A.N. Medeiros, Small Ruminant Research 109, 2-3 (2013)
6 T.P. Paim, A.F. Silva, R.F.S. Martins, B.O. Borges, P.M.T. Lima, C.C. Cardoso, G.I.F. Esteves, H. Louvandini and C. McManus, Small Ruminant Research 112, 1-3 (2013)
7 A.V. Fisher, M. Enser, R.I. Richardson, J.D. Wood, G.R. Nute, E. Kurt, L.A. Sinclair, R.G. Wilkinson, Meat Science 55 (2000)

A CFD study on wall shear stress of falling water film on food contact surface

Nawaporn Suthanarak[1], and *Navaphattra* Nunak[1,*]

[1]Food Engineering Department, Faculty of Engineering, King Mongkut's Institute of Technology Ladkrabang, Bangkok, Thailand

Abstract. This paper proposes a CFD model to simulate the wall shear stress of the falling water film on the hydrophilic surface (aluminum) of the inclined flat plate in the range of 30°, 60° and 90° at a flow rate of 100, 200 and 300 L/h. Wall shear stress simulated from the CFD using SST k-ω with the VOF model was compared with the real data obtained from the experiment. It was found that both inclined angle and water flow rate had an effect to the wall shear stress and the CFD with the proposed model had the potential to predict it with the maximum averaged error of 2.9%.

1 Introduction

In food industry, the removal of soil and microorganism from food processing equipment and machines in wet-cleaning with water is the most important process indicating the level of food safety. If inadequate or improper cleaning occurs, soil residues may remain on the equipment surface causing to product contamination and deterioration and foodborne disease in the consumers.

Generally, cleaning of a large tank with the principle of falling film using a spray ball is preferable. A series of jets of water spraying injected through spray balls spread out on the interior surface of a tank and flowed down by gravity force creating the wall shear stress for removing soil from the surface. From a theory of fluid mechanics, the wall shear stress is an important factor to indicate the cleaning performance. It depends on many parameters such as liquid properties, flow characteristics and inclination of the surface [1-4]. Fuchs et al. [3] and Lerch et al. [5] have studied the cleaning process with falling water film and found that the turbulent flow of water can enhance the cleaning performance. This finding also is in the agreement with Tamime [2]. Therefore, an understanding on flow characteristics of cleaning fluid effecting to cleaning performance in term of wall shear stress is importance.

To obtain the wall shear stress, it can be achieved by directly calculating from the measured thickness of liquid film. However, it spends a lot of time and costs a lot. Therefore, Computational Fluid Dynamics (CFD) is introduced to simulate the wall shear stress [6-9]. Moreover, it is also widely used to simulate the flow velocity of fluid and other applications about a decade [10-11]. In this study, the CFD model was proposed to estimate the wall shear stress of falling water film created on the inclined surface at various flow rates.

2. Theory of falling liquid film

Falling liquid film is a film flow downward under gravity, the characteristics of film flow can be indicated by film Reynolds number (Re_f) as illustrated in equation (1) [3, 12]:

$$Re_f = \Gamma/\eta = \Gamma_V/\upsilon = \dot{V}/B\,\upsilon \qquad (1)$$

where Γ is the mass flow rate per wetted width (kg/h·m), η is the dynamic viscosity of the liquid (kg/m·s), Γ_V is the volumetric wetting rate (m^3/h·m), υ is the kinematic viscosity of the liquid (m^2/s), $\dot{V}$ is the volume flow rate (m^3/h) and B is the wetted width (m).

According to equation (1), the film flow characteristic can be classified into three types as follow: $Re_f < 4$ is laminar flow or smooth laminar film, $4 < Re_f < 400$ is transient flow which is a laminar film with a small wave on the interface and $Re_f > 400$ is a turbulent flow or wavy film having a large wave on the interface (fig. 1) [12].

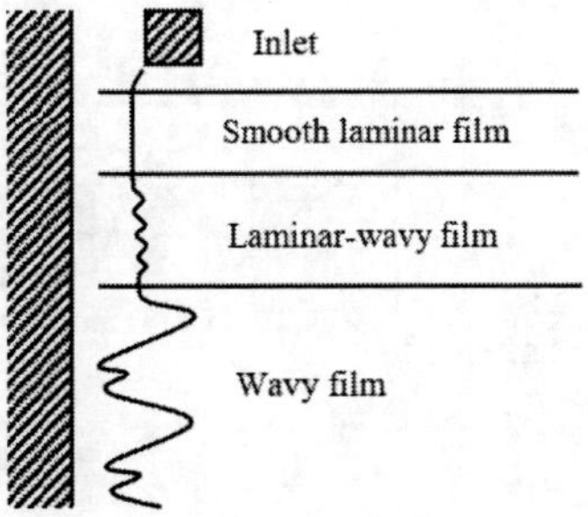

Fig. 1. Falling film characteristic [12]

The thickness of liquid film can be evaluated from the theory of Nusselt film which assumes that the film flow characteristic is laminar under the action of gravity,

* Corresponding author: Navaphattra@gmail.com

the shear stress at interface is neglected, and fluid properties are constant [3, 4, 13]. The film thickness (δ) can be calculated by equation (2)

$$\delta = (3\upsilon^2 Re_f / g\sin\theta)^{1/3} \quad (2)$$

where **g** is the gravitational acceleration (m/s^2) and θ is the angle of inclination of the surface with respect to a horizontal.

During the film flows down, there is a force acting upwards created between the surface and the liquid film that is called the wall shear stress, influenced by the liquid film thickness. It is an important parameter indicating the performance of soil removal from the surface. It can be calculated by equation (3) with the several assumptions, i.e. the fluid flow is uniform and incompressible, the net pressure force is zero, and the momentum flux through the control volume is zero [14].

$$\tau_w = \rho g \delta \sin\theta \quad (3)$$

where τ_w is the wall shear stress (Pa), ρ is density (kg/m^3).

3 Materials and methods

3.1 The experimental setup

The experimental unit (see fig. 2) consisted of a centrifugal pump, a tube, a ball valve, a water container, and a testing surface. For the testing surface, an 10x50 cm aluminum plate used as representative of food contact surface, can be adjusted for any angle. Water was circulated by the centrifugal pump (Sonic AP1600 20-23 watt), installed in a water container with a tube and a ball valve connecting, from a water container to overflow on a testing surface.

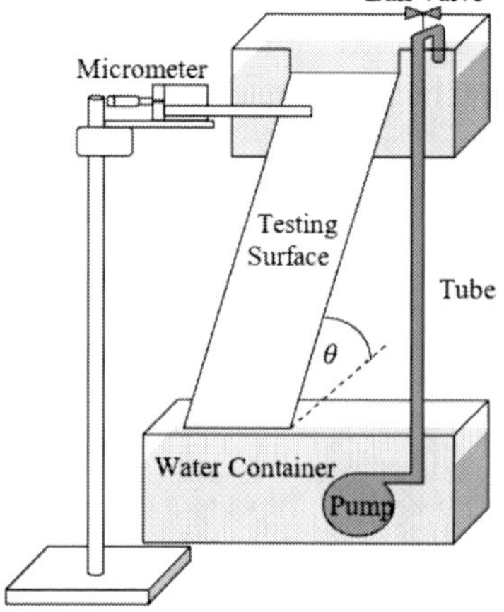

Fig. 2. Experimental Unit

3.2 Experimental method

The experiments were carried out under gravity driven falling water film on a vertical and inclined testing surface at various angles (30°, 60° and 90°). Water was circulated at least 5 minutes before the data were recorded. The flow rate of water was measured using weighing method (SF-400A resolution 0.1g) and the temperature of water was measured by thermocouple type K (Fluke 54IIB Thermometer S2). Experiments were performed at the flow rate of 100, 200 and 300 L/h with Reynolds number of 311, 622 and 932, respectively. The micrometer with resolution of 0.01 mm was installed on the tripod to measure the thickness of water film at every 5 cm long from the origin point (0 cm) (fig. 3), then compared with the theoretical results calculated from the equation (2) to consider the real data of water film thickness for calculating wall shear stress created on the surface by equation (3) [3, 4, 13]. They were referred as the real data for validating the results simulated from CFD.

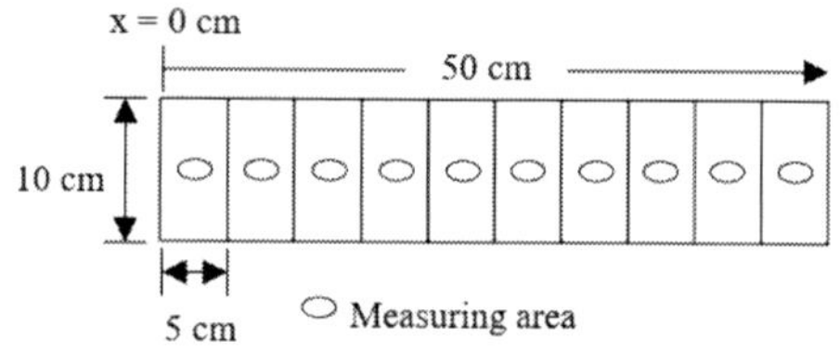

Fig. 3. Film thickness measuring position

3.3 CFD Modeling

In this study, the falling film flow was investigated by the three-dimensional CFD simulation (ANSYS FLUENT 16.0) with Navier-Stokes equation, which assumes that fluid is incompressible flow and constant viscosity. The shear stress transport k-ω (SST k-ω), a turbulent model, was used to simulate film flow with the sensitivity inside and outside shear layer [15]. From the preliminary experiments, it was found that the SST k-ω model gave the minimum error comparing with the wall shear stress simulated from other turbulent models, i.e. STD k-ε, RNG k-ε and Realizable k-ε models (results not shown).

In addition, the multiphase model, Volume of Fluid (VOF) was also applied with SST k-ω model and hexagonal mesh for capturing the interface between two phases since the film flow in this study was two-phase flow (gas and liquid) [8, 15].

The wall shear stress of falling water film was simulated with the CFD model by separating the water film into many small elements, then it was considered under the conditions as shown in Table 1.

To validate the CFD model, the simulated wall shear stress was compared with the real data or theoretical data obtained from section 3.2 in term of a relative error as presented in equation (4)

$$\%error = |\,(x_{sim} - x_t) / x_t\,|\, x100 \quad (4)$$

where $\%error$ is a relative error, x_{sim} is a simulation data and x_t is a real data.

https://doi.org/10.1051/matecconf/201819203052

Table 1. Simulation conditions fixed in the CFD

Parameter	Detail
Operating Conditions	
Gravity (m/s^2)	9.81
Pressure (kPa)	101.325
Surface Tension (N/m)	0.072
Model	
Viscous Model	SST k-ω
Multiphase Model	VOF
Phase	
Primary Phase	Air (25°C)
- Density (kg/m^3)	1.184[a]
- Viscosity (kg/m-s)	1.849×10^{-5} [a]
Secondary Phase	Water (25°C)
- Density (kg/m^3)	997[b]
- Viscosity (kg/m-s)	0.000891[b]
Boundary Conditions	
Inlet water flow rate (L/h)	100, 200 and 300

[a] Physical properties of air [16]
[b] Physical properties of water [16]

4 Results and discussion

4.1 Wall shear stress validation

Fig. 4 shows the thickness of water film calculated from the Nusselt's film theory and measured by a micrometer. The thickness measured from the experimental unit are in a good agreement with the findings of Fuchs et al. [3]. The thickness values obtained from both methods were very close for flow characteristic in the laminar or transient (100 L/h), whereas the values obtained from the measurement were higher than that from the theory at higher flow rate (200 and 300 L/h) or in the turbulent flow. Since the Nusselt's film theory is scoped only for the laminar film flow, the calculated results are lower than the actual film thickness. Besides, cleaning the internal surface of a large tank should be processed with the turbulent falling film flow. Therefore, in this study, the wall shear stress considered to be the real data was calculated from the measured film thickness and used to validate the CFD model.

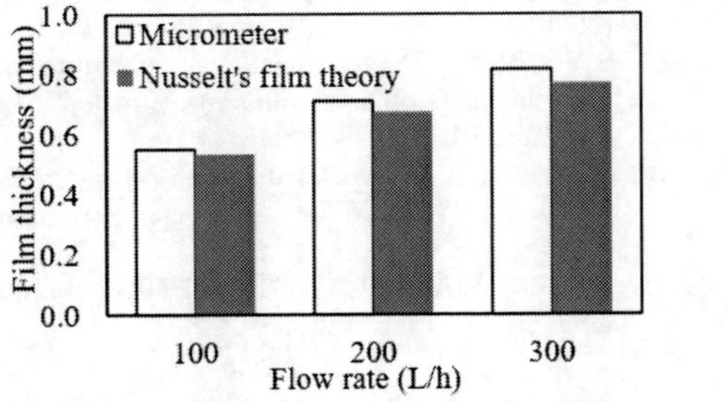

Fig. 4. Water film thickness on the inclined surface at 30° calculated from Nusselt's film theory and measured by micrometer

Fig. 5 shows an example of comparing the wall shear stress obtained from the CFD simulation and the experiment at every 5 cm long from the origin point (0 cm). It was found that the CFD simulation had much relative error at the first one fifth of the whole distance due to the influence of the water inlet of the experimental unit. For this reason, the proposed CFD model could be used to simulate the film flow at the distance of 10-50 cm from the origin point. The contour of wall shear stress created on the inclined surface as shown in fig. 6. Therefore, the wall shear stress hereafter was the CFD averaged from the distance of 10 to 50 cm as presented in Table 2.

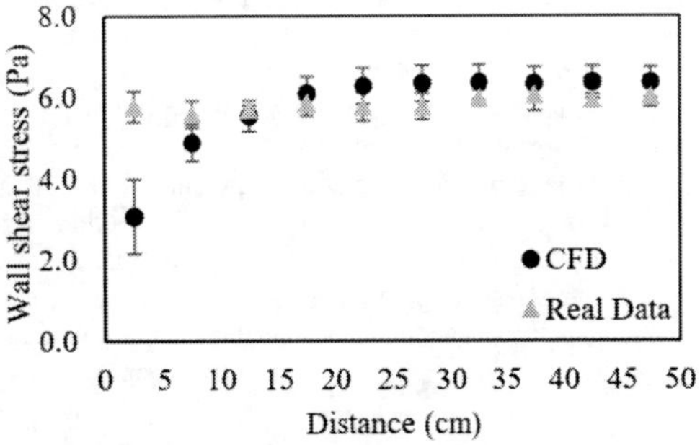

Fig. 5. Wall shear stress simulated from CFD and the real data obtained from the experiment on the inclined surface at 60° with a flow rate of 300 L/h

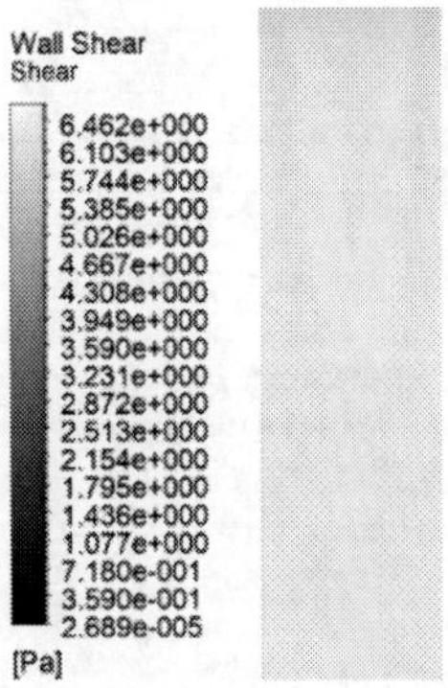

Fig. 6. Wall shear stress contour of water film flow on the inclined surface at 90° with a flow rate of 200 L/h

Considering Table 2, It was found that the thickness of water film simulated from CFD had a good agreement with Fuchs et al. [3] and represented the low relative error. The maximum relative error under all experimental conditions was less than 3%. This probably due to different assumptions of the CFD model and the equation (3). Since aluminum plate is a hydrophilic surface having the surface tension close to other types of food contact materials such as stainless steel, this model could also be applied to simulate the wall shear stress created on the food contact surface. The higher surface tension is, the higher wall shear stress is [6]. However, for more accuracy the further work on the correction factor for any types of materials would be needed.

https://doi.org/10.1051/matecconf/201819203052

Table 2. Comparison of experimental calculation and CFD simulation with relative error

Flow rate (L/h)	Wall shear stress (Pa) with different inclined angles and relative error (%)								
	30°			60°			90°		
	Experiment	CFD	Error	Experiment	CFD	Error	Experiment	CFD	Error
100	2.689	2.624	2.437	3.879	3.767	2.895	4.323	4.198	2.900
200	3.472	3.464	0.247	5.082	5.061	0.407	5.634	5.592	0.747
300	3.986	3.941	1.112	5.844	5.761	1.432	6.367	6.385	0.283

4.2 Effect of inclined angle and flow rate

From Table 2, wall shear stress of water film on the surface increased with the increasing of surface inclination and water flow rate as water film was influenced by gravity at a more inclined angle. When considering the water flow rate, the wall shear stress occurred form water film flow at 100 liters per hour, non-turbulent flow, was lower than that of at 200 and 300 liters per hour, turbulent flow. The higher water flow rate caused the film flow more turbulent, and more wall shear stress. However, the film wall shear stress is influenced by the water flow rate more than an inclined angle.

5 Conclusions

Wall shear stress is an important factor for cleaning by falling liquid film. It was influenced by water flow rate and inclined angle. The increasing of inclination angle and water flow rate caused the greater wall shear stress. When comparing wall shear stress simulated from the CFD using SST k-ω with VOF model and obtained from the experiments, it was found that the CFD was a potential tool to predict the wall shear stress at different conditions with the maximum error of 2.9%. Nevertheless, further research would be needed in case of changing the material type. Finally, this study would be more benefits if this concept was applied to the cleaning behavior of any types of soils using falling water film in the future work.

The authors would like to thanks Mr. Kittisak Srichaitamrong and Ms.Natnaree Phromma for performing film thickness measurement.

Reference

1. K. R. Goode, K. Asteriadou, P. T. Robbins, P. J. Fryer, "Fouling and cleaning studies in the food and beverage industry classified by cleaning type," *IFT*, **12**, 121-143, (2013)
2. Y. Tamime, *Cleaning-in-place: Dairy, Food and beverage operations Third Edition*, Blackwell Publishing, Oxford, (2008)
3. E. Fuchs, A. Boye, R. Murcek, J.-P. Majschak, "An experimental comparison of film flow parameters and cleaning behavior of falling liquid film for different tilt angles," *Food Bioprod. Process.*, **93**, 318-326, (2014)
4. A. To, S. R. Annapragada, "Experimental parameters for small-scale cleaning characterization. Part II: effect of fluid velocity on the kinetics of cleaning, *Insitute of validation technology*, (2015)
5. K. Lerch, J. Hinrichs, P. Dittmer, J. Rauschnabel, "Cleanability of surfaces from active pharmaceutical ingredient surrogate riboflavin by falling film," *Chemie Ingenieur Technik*, **85**, 323-332, (2013)
6. H. Lan, J. L. Wegener, B. F. Armaly, J. A. Drallmeier, "Developing laminar gravity-driven thin liquid film flow down an inclined plane," *J. Fluids Eng.*, **132**, (2010)
7. B. B. B. Jensen, A. Friis, Th. Benezech, P. Legentilhomme, C. Lelievre, "Local wall shear stress variations predicted by computational fluid dynamics for hygienic design," *Food Bioprod. Process.*, **83(C1)**, 53-60, (2005)
8. Y. Xu, J. Yuan, J. U. Repke, G. Wozny, "CFD study on liquid flow behavior on inclined flat plate focusing on effect of flow rate," *Eng. Appl. Comp. Fluid*, **6**, 186-194, (2012)
9. Y. Yu, "A computational fluid study of falling film behavior on flat plate," *FEDSM2014*, (2014)
10. T. Norton, D.-W. Sun, "Computational fluid dynamics (CFD) – an effective and efficient design and analysis tool for the food industry: A review," *Trends Food Sci. Technol.*, **17**, 600-620, (2006)
11. S. Salo, A. Friis, G. Wirtanen, "Cleaning validation of fermentation tanks," *Food Bioprod. Process.*, **86**, 204-210, (2008)
12. F. Weise, S. Scholl, "Evaporation of pure liquids with increased viscosity in falling film evaporator," *Heat Mass Transfer*,**45**, 1037-1046, (2009)
13. S. Portalski, "Studies of falling liquid film flow: Film thickness on a smooth vertical plate," *Chem Eng. Sci.*, **18**, 787-804, (1963)
14. W. R. Fox, T. A. McDonald, *Introduction to fluid mechanics*, John Wiley & Sons, New York, (1994)
15. ANSYS FLUENT, *ANSYS FLUENT Theory guide*, ANSYS Inc., (2011)
16. Y. A. Cengel, J. M. Cimbala, *Fluid mechanics: fundamentals and applications*, McGraw-Hill, New York, (2006)

Antimicrobial Potential of Vapour Phase Acetic Acid in Combination with Ethyl Alcohol against *Salmonella* Typhimurium Contaminated on Bird Eye Chili (*Capsicum frutescens* L.)

Pasit Nichayapun[1], *Theeraporn* Khawsa-ard[1], *Todsaphon* Pimrat[1], *Pokpong* Lilitsajja[1], and *Racha* Tepsorn[1,*]

[1]Food Science and Technology Division, Faculty of Science and Technology, Thammasat University (Rangsit Centre), Phathum Thani, Thailand

Abstract. Bacterial contamination is the main cause of food poisoning. The currently decontamination methods arise to find novel, effective and safe approach. Acetic acid (AC) in both aqueous and vapour forms have been conducted on its effectiveness against several bacterial contaminations. The use in high concentration can cause strong smell and corrosives that have an effect on the qualities of food. This study aim to assess the potential of Mechanically Vapourized AC solution in combination with ethyl alcohol (EA) (MVA-E) on the reduction of *Salmonella enterica* serotype Typhimurium ATCC 13311 contaminated on both microbiological media and on fresh Bird Eye Chili (*Capsicum frutescens* L.). In vitro surface inhibitions of *S.* Typhimurium at low population, ca. 1.0 CFU/cm^2, and high population, ca. 10.0 CFU/cm^2 were examined on Tryptic Soya Agar (TSA). MVA-E at the concentration of 10:95 (AC:EA) presented the absolutely inhibited S. Typhimurium within 20 min at 27±2°C. At the concentration of 10:75 (AC:EA), the absolutely inactivated S. Typhimurium was observed within 10 min at 50±2°C. For the evaluation of antimicrobial activity of MVA-E over time, the results indicated that ca. 8.00 Log_{10} reductions were observed within 20, 25 and 25 min at the concentration of AC:EA ratio as 10:95, 10:75, and 10:45, respectively at 27±2°C. The effectiveness of MVA-E increased when the temperature of MVA-E process increased. The reduction of *S.* Typhimurium contaminated on fresh Bird Eye Chili by MVA-E was also determined at 4°C, 27±2°C and 50°C. The efficiency of MVA-E on the reduction of S. Typhimurium depended on the concentration of EA in MVA-E process, the fumigation time and also the temperature. To the best of knowledge this is the first time a combination of AC and EA in vapour phase has been tested as a preservative method prevent microorganism proliferation.

1 Introduction

Chilies (*Capsicum* sp.) have been cultivated in many regions of the world [1]. One of the most consumed chili in Thailand is Brid Eye Chili, (*Capsicum frutescens* L.), a wide range of chili fruits is available in ASEAN's markets [2]. It was used as pungent ingredients in several culinary practices [3]. Also it is green ingredient used to prepare several meals, some people consume as fresh fruit. Chilies, such as Bird Eye Chili are often contaminated by microorganisms on the surface and can cause food poisoning illness [4-7]. The outbreaks caused by consumptions of contaminated those with foodborne pathogens as *Salmonella* sp., *Listeria monocytogenes*, *Vibrio* sp. or others [8-10]. As a consequence of such situation, contaminations of *Salmonella* sp. in vegetables become the greatest for public health concerns. Postharvest handling such as washing or spraying with sanitizer is most important procedure in view of the fact that removed, eliminated or reduced the surface microbiological contaminant. The US-FDA suggested the use of 200 mgL^{-1} chlorine, at a pH of 6.0-7.5, for 1-2 min as disinfection processes for fresh produce. Several reports had indicated that chlorine at that level lacks the effectiveness to eliminate spoilage microorganisms. In addition, chlorine presented the potential to form carcinogenic compounds [11]. Several researchers have projected other sanitization processes [12-13]. Among of those, AC presented the special interest [14-15]. Volatile antimicrobial have become popular in research and have also been exhibited antimicrobial properties in many food product [14, 16-18]. The use of acetic AC in vapour phase was reported about their potential to control the growth of microorganism. However, the use of AC alone in vapour phase caused the strong vinegar odour and affected to the appearance quality of vapourized fresh produces. The use of volatile antimicrobial substances such as EA or in combination with another should be the good alternative technique. Thus, in this study the efficacy of mechanically vapourized AC (MVA) in combination with mechanically vapourized EA (MVE) as mechanically vapouized AC-EA (MVA-E) on the

* Corresponding author: rtepsorn@tu.ac.th

reduction of *S.* Typhimurium in both *in vitro* study and contaminated on model fresh Bird Eye Chili was investigated.

2 Materials and Methods

2.1 Chemical and microbiological media

Tryptic Soya Agar (TSA), Tryptic Soya Broth (TSB) and Peptone were purchased from Difco (Dico, USA). AC was purchased from QRëC (QRëC, New Zealand). 95% EA was kindly provided by The Liquor Distillery Organization of Thailand.

2.2 Preparation of Test Organism

S. Typhimurium, was kindly provided by Department of Food Science and Technology, Thammasat University (Rangsit Centre). Period of expose, *S.* Typhimurium was sub-culture twice in TSB at 37°C for 18 h before use as inocula.

2.3 *In vitro* susceptibility of *S.* Typhimurium to the MVA-E

2.3.1 Agar overlay method

The susceptibility of *S.* Typhimurium to MVA-E was determined *in vitro* using modified agar overlay method. Briefly, The *S.* Typhimurium at *ca.* 7.0 Log_{10} CFU/mL was used as high inoculums and at *ca.*4.00 Log_{10} CFU/mL as low inoculum. Contaminated Petri dishes without cover were aseptically placed in fumigation chamber as presented in Fig. 1. AC solution at the concentration of 10.0% in combination with EA at the concentration of 0%, 30.0%, 45.0%, 75.0% and 95.0% was individually placed and directly connected with air pump.

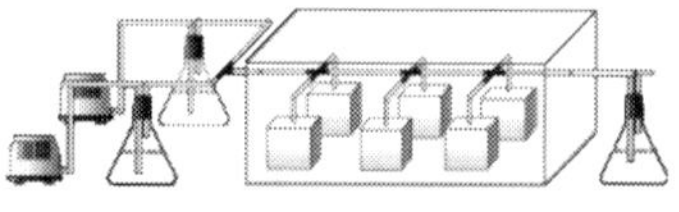

Fig. 1. Schematic illustration represented the fumigation chamber

Plates were contacted with MVA-E and withdrew at the interval time as 0, 5, 10, 15, 20 and 10 min. The reduction ratio was calculated after incubation. The impact of temperature was determined at 4°C and 50°C. The rate of vapour production was calculated.

2.3.2 Time killing analysis

For time killing analysis, the susceptibility of *S.* Typhimurium to MVA-E was determined *in vitro*. *S.* Typhimurium at *ca.* 8.00 Log_{10} CFU/mL was prepared in 50 mL TSB. Fumigation tube connected with MVA-E was aseptically placed into suspension. MVA-E was generated. At the interval time as 0, 5, 10, 15, 20 and 25 min the microbial populations were enumerated by spread plate technique on TSA. As described above, the impact of temperature was also determined.

2.4 MVA-E effects on fresh Bird Eye Chili inoculated with *S.* Typhimurium

2.4.1 S. Typhimurium suspension preparation

S. Typhimurium was sub-cultured twice in TSB. Cells were harvested by centrifugation at 1000x*g* for 15 min at 4°C. The inocula was adjusted to the final concentration at *ca* 7.00-8.00 Log_{10} CFU/mL

2.4.2 Challenging Test

Bird Eye Chilies were purchased from local markets near Thammasat University. Each fruit was sanitized in 1000 ppm chlorinated water. The fruits were dry under UV lamp. 0.1 mL of prepared *S.* Typhimurium was contaminated on each to obtain 6.00 Log_{10} CFU/g. The contaminated those were then exposed to MVA-E in the fumigation chamber at 65±2% RH. Each fumigated fruit was then examined for evidence of survival of *S.* Typhimurium by spread plate technique. The population of organism was calculated as Log_{10} CFU/g. The impact of temperature was also determined.

3 Results and Discussions

3.1 Susceptibility of *S.* Typhimurium to MVA-E

3.1.1 In vitro agar overlay method

Efficacies of MVA-E against *S.* Typhimurium on the surface of TSA are demonstrated in Fig. 2. The results presented the inhibitory effect of MVA-E against *S.* Typhimurium at 95.0% and 75.0% of EA in combination with 10.0% AC at 27±2°C within 20 and 25 min, respectively. At the low concentration, the complete destructive effect was not detected. These inhibitory properties were detected in both low and high inoculums level. As shown, the temperature affected the antimicrobial properties of MVA-E. At low temperature (4°C), the efficacy of MVA-E was decreased compared with room temperature (27±2°C), and increased when exposed with high temperature as 50°C. According to Fig 3, the increasing rate of pH and weight loss of both AC and EA under vaporized process were the main factors affecting the antimicrobial properties. These increasing were obviously presented when those were exposed to high temperature. Hence, pH of test solution and weight loss of AC and EA increased along with the rising of temperature, the antimicrobial properties of MVA-E also increased when the temperature increased. As mentioned in the previous research, the inhibition potential of AC was acted upon the pKa and pH [20,21]. The bacteriostatic and bactericidal of AC and the other derivatives have been attributed to the lower pH below t' . . . 1 [...]

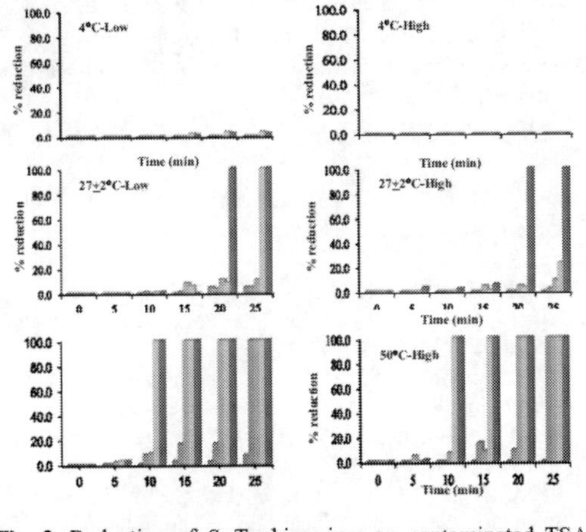

Fig. 2. Reduction of *S.* Typhimurium on contaminated TSA during the fumigation with MVA-E and different concentration of EA (▪; 0%, ▪; 30%, ▪, 45%, ▪; 75%, ▪; 95%)

The pH affected the survival of *Salmonella* sp. was 4.9 [23]. As a result, the pH of MVA-E was nearly3.0 it was in range of inhibition. Therefore, the confirmation of *S.* Typhimurium inhibition by MVA-E was subjected to the decreasing of pH.

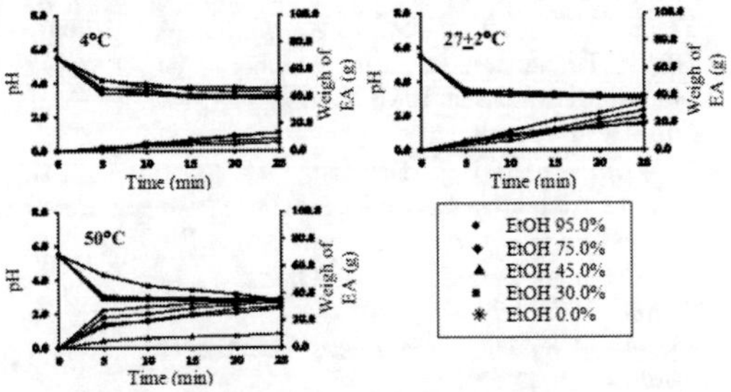

Fig. 3. The change in weight of EA and pH during fumigation.

The other possibility of the impact of temperature on the antimicrobial properties of MVA-E is that the upsurge in dissociated molecules. The proton motive activity of the outer membrane might be interfered, and then resulted to penetration of the undissociated form of AC. For this reason, intracellular pH was reduced and inducing cell death. EA demonstrated the assist properties. It changed the properties of outer membrane. Because of the fact that, Phospholipid, the main component of cell membrane, was dissolved by EA and caused the damage, AC in both undissociated and dissociated form could simply penetrate into cell and caused the death of cell. Glycoprotein, the protein component of cell membrane, was denaturized by EA and also caused the nonsurvivable of cell.

3.1.2 Time killing analysis

In Fig. 4 the Time Killed-Curve, represented the antimicrobial activity of MVA-E were shown. The MVA-E at the concentration of EA at 45.0, 75.0 and 95.0% had no ability to provide the lethal effect within 15 min. However, *S.* Typhimurium was reduced to undetectable level within more than 20 min when the above mentioned concentrations of EA in MVA-E were used.

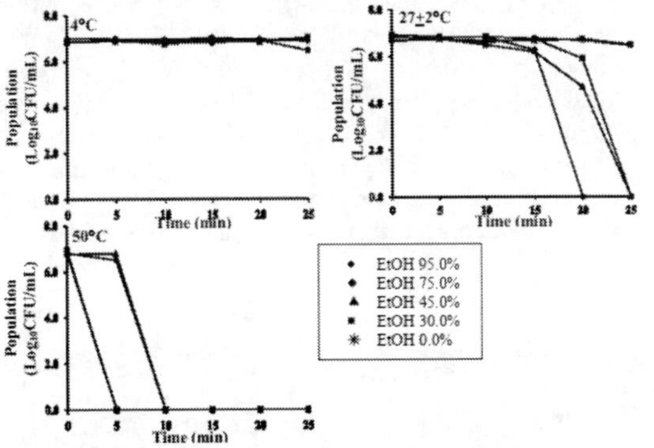

Fig. 4. The population of *S.* Typhimurium during contact with MVA-E at different concentration of EA

The temperature demonstrated the effect on antimicrobial properties of MVA-E. At the higher temperature, the antimicrobial activity was greater than at the low temperature. The time killing analysis is equal to the inhibition curve, known as the 'killing curve'. In generally, this studies showed that the survivors of those was decreased by concentration of EA was increased, and completely inhibited at the concentration more than 45.0%. Moreover, it could be indicated that the longer the fumigation time the lower the necessary concentration of the MVA-E. The Time-Dependent characteristic was observed according to this study.

3.2 MVA-E effects on fresh Bird Eye Chili inoculated with *S.* Typhimurium

3.2.1 Challenging Test

The initial amount of contaminated *S.* Typhimurium on Bird Eye Chili at *ca.* 6.00 Log_{10} CFU/g were exposed to MVA-E at the concentration of AC at 10.0% in combination with the different concentration of EA along with different fumigation time.

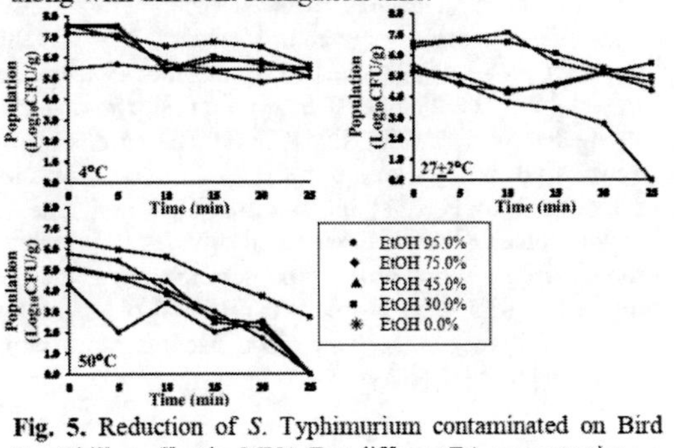

Fig. 5. Reduction of *S.* Typhimurium contaminated on Bird Eye Chili as affect by MVA-E at different EA concentration.

The effect of operating temperature was also determined. The results were shown in Fig. 5. The natural contaminated Bird Eye Chilies were also tested. The numbers of Total Bacteria Count (TBC) were determined as describe above. The results of TBC were presented in Fig 6. Fumigation with MAV-E at room temperature (27±2°C) and 50°C at all concentration of EA demonstrated the lethal effect. At the concentration of AC at 10.0% in combination with 95.0% EA the complete destruction was presented within 25 min. When the temperature decreased to 4°C the complete elimination was not detected along the period of fumigation. The downward trend of fumigation time was observed when the temperature was rising. Weak acid always demonstrate the microbial inhibition properties [19-20].

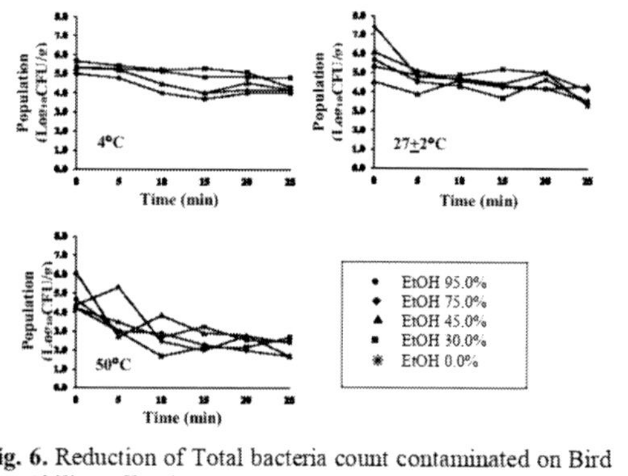

Fig. 6. Reduction of Total bacteria count contaminated on Bird Eye Chili as affect by MVA-E at different EA concentration.

Consideration on the effective of MVA-E against natural contaminated microorganism on Bird Eye Chili, it could be indicated that the complete elimination was not detected within the period of exposure.

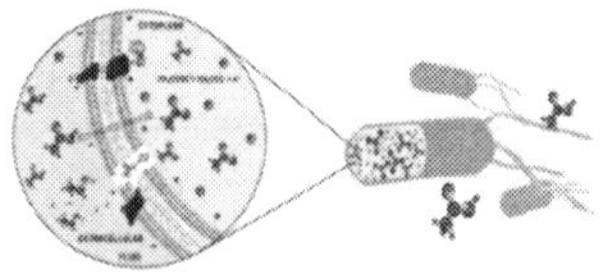

Fig. 7. Schematic illustration of the possible mechanism of MVA-E against *S.* Typhimurium

As presented in Fig. 7, the possible antimicrobial mechanism of MVA-E was demonstrated. It seems that MVA-E decontaminates surfaceborne microorganism and thus sterilizes vegetable surfaces [24]. In this research, the results demonstrated that MVA-E presented the inhibitory effect on both *S.* Typhimurium *in vitro* and in model food as Bird Eye Chilis. The mechanically vapourized process allowed the AC in the form of dissociated to become the undissociated molecule in vapour phase. For this reason, the MVA-E presented more inhibitory properties. Another advantage, the gas phase can diffuse much more than liquid phase so the rate of cell membrane penetration become more rapid compared to liquid phase.

4 Conclusions

Mechanically vapourized AC in combination with EA (MVA-E) presented the inhibitory potential to eliminate and decontaminate the number of *S.* Typhimurium in both *in vitro* and model food as contaminated on fresh Bird Eye Chilies. It is free of carcinogenic risk. The antimicrobial properties depended on the concentration, fumigation time and the temperature. When these key factors increase, the antimicrobial properties are also increased.

References

1. K.H., Kraft, C.H., Brown, G.P., Nabhan, E., Luedeling, J.J., Luna-Ruiz, G.C., d'Eeckenbrugge, R.J., Hijmans, P., Gepts. PNAS, **111** (2014)

2. Servicio de Información Agroalimentaria y Pesquera. Secretaría de Agricultura, Ganadería, Desarrollo rural, Pesca y Alimentación (SIAP). *Cierre de la producción agrícola por cultivo*, (2015)

3. NC., Elviss, C.L., Little, L., Hucklesby, S., Sagoo, S., Surman-Lee, E. de Pinna, E.J., Threlfall. *Int J Food Microbiol.*, **134**, (2009)

4. J.M., Farber. *J Food Prot.*, **60**, (1997)

5. M.M.I., Osornio, A.R., Chaves., *J Food Prot.*, **60**, (1997)

6. G.A., Francis, C., Thomas, D., O'Breine., *Int J Food Sci Technol.*, **34**, (1999).

7. A.M., Piagentini, E.M., Pirovani, D.R., Güemes, J.H., di Pentima, M.A., Tesisi., *J. Food Sci.*, **62**, (2006)

8. M., Friedman, P.R., Henika, R.E., Mandrell., *J Food Prot.*, **65**, (2002).

9. L., Pezzoli, R., Elson, C.L. Little, H., Yip, I., Fisher, R., Yishai, J., Threfall., *Foodborne Pathog Dis.*, **5**, (2008)

10. J., Pakalniskiene, G., Falkenhorst, M., Lisby, S.B., Madsen, K.E., Olsen, E.M. Nielsen, K., Mølbak., *Epidemiol Infect*, **137**, (2009)

11. M.E., Parish, L.R., Beuchat, T.V., Suslow, L.J., Harris, E.H., Garrett, J.N., Farber, F.F., Busta., *Compr. Rev. Food Sci. and Food Saf.* **2**, (2003)

12. S., Sanz, M., Giménez, C., Olarte, C., Lomas, J., Portu., *J. Appl. Microbiol.* **93**, (2002)

13. A., Allende, J., McEvoy, Y., Tao, Y., Luo., *Food Control.*, **20**, (2009)

14. P.L., Sholberg, P., Haag, R., Hocking, K., Bedford., *J. Hortic. Sci.*, **35**, (2000)

15. B.L., Meatherall, D., Gregson, T., Ross, J.D., Pitout, K.B., Laupland, *Am J Med*, **122**, (2009)

16. N.G., Tzortzakis., *Int. J. Food Microbiol.*, **142**, (2010)

17. M.A., Gatto, A., Ippolito, V., Linsalata, N., Cascarano, F., Nigro, S., Vanadia., *Postharvest Biol. Technol.*, **61**, (2011)

18. W., Krusong, P., Dansai, A., Itharat., *KMITL Science and Technology.*, **12**, (2012)

19. F.N., Arroyo-López, J., Bautista-Gallego, M.C., Durán-Quintana, A., Garrido-Fernández., *Food Microbiol.*, **25**, (2008)

20. R.G., Bell, K.M., de Lacy., *Food Microbiol.*, **4**, (1987).

21. C., Bell, A., Kyriakides., *Salmonella*: A praticalapproch to the organism and its control in food, (2002)

22. J., Yeesibsan, W., Krusong, *Asian Journal of Food and Agro-Industry*, **2**, (2009)

23. A.L., Branen, P.M., Davidson, S., Salminen., *Food Additive.*, (1989)

24. P.L., Sholberg, A.P., Gaunce., *Crop Prot.*, **15**, (1996)

Effect of ractopamine on physicochemical characteristics and collagen solubility in finishing pigs

Chanporn Chaosap[1*], *Krongkaew* Kaewthaworn[2], *Ronachai* Sitthigripong[2], *Panneepa* Sivapirunthep[1], *and Kamon* Chaweewan[3]

[1] Department of Agricultural Education, Faculty of Industrial Education and Technology, King Mongkut's Institute of Technology Ladkrabang, Bangkok 10520, Thailand

[2] Department of Animal Production Technology and Fisheries, Faculty of Agricultural Technology, King Mongkut's Institute of Technology Ladkrabang, Bangkok 10520, Thailand

[3]Nakornratchasima Livestock Research and Breeding Center, Bureau of Animal Husbandry and Genetic Improvement, Department of Livestock Development (DLD), Khanongphra, Pakchong, Nakornratchasima 30130, Thailand

Abstract. The objective of this study was to investigate physicochemical characteristics and collagen solubility of pigs fed with dietary ractopamine (RAC) during finishing period. Thirty of three-way crossbred pigs (Large White × Landrace × Duroc) with an initial bodyweight (BW) of 18.0 ± 0.5 kg were randomly assigned to one to three dietary RAC inclusions (0, 20, or 40 mg/kg) fed during finishing period (60-100 kg BW). Each group consisted of five gilts and five barrows. After 109 day of raising period, all pigs were slaughtered at the average weight of 100 kg. *Longissimus dorsi* muscle (LD) was taken from each carcass after overnight chill and analyzed for physical characteristics, chemical composition, and collagen solubility. The LD from pigs fed 40 mg/kg of RAC had lower a* value ($p < 0.01$) than others. Pigs fed with 40 mg/kg of RAC had higher Warner-Bratzler shear force ($p < 0.05$) than pigs fed without RAC. Pigs fed with dietary RAC had lower glycogen content than control ($p < 0.01$). While other chemical compositions and collagen solubility of LD were not affected by dietary RAC ($p > 0.05$). Gilts had larger muscle fibers ($p < 0.01$), lower glycogen content ($p < 0.05$), higher content of insoluble collagen ($p < 0.05$), total collagen ($p = 0.054$), and collagen solubility ($p = 0.054$) than barrows.

1. Introduction

Using ractopamine as growth-promoting in pig industry, it is approved that the feed efficiency and carcass quality have been improved while it has affected meat quality. Such as the studies of [1] showed that RAC increased carcass lean content and acted directly to nutrients from fat deposition toward protein deposition in finishing pigs. RAC has been used to improve feed efficiency and increase leanness [2]. However, the study of [3] and [2] reported that dietary RAC supplemented increased toughness of pork. The addition of RAC at the level of 5 ppm in diet slightly reduced tenderness of pork [4]. As well as, the study of [5] reported that RAC supplemented at 10 mg/kg in pig diet had negatively impacted on tenderness.

Color of fresh meat is one of the most important traits influencing consumers' purchasing decisions [6]. The amount of glycogen at slaughter affects meat quality by influencing post mortem lactic acid production and pH decline [7]. The effect of pH on water holding capacity and color of meat in pork has been widely documented [8, 9]. The other important attribute in terms of meat quality is tenderness [10]. Collagen content and its solubility had an impact on meat tenderness [11, 12]. Due to glycogen content and collagen solubility had significant effects on meat quality, but the information of these traits in pig fed with RAC is still limit. Therefore, the objective of this study was to investigate the effect of RAC supplement diet on physicochemical characteristics and collagen solubility in finishing barrows and gilts.

2. Materials and Methods

2.1 Animals

The experimental procedure on animal care and use was approved by the Institutional Animal Care and Use Committee of King Mongkut's Institute of Technology Ladkrabang (AGRI-KMITL003/2015).

A total of thirty 3-way crossbred pigs (Large White X Landrace X Duroc) with an average initial bodyweight of 18.0±0.5 kg were randomly assigned to 1 to 3 dietary inclusions of RAC (0, 20, or 40 mg/kg; Paylean™). Each group consisted of five gilts and five barrows. Animals were fed *ad libitum* with 18, 16, and 14% protein diet during starting (20 to 40 kg), growing (40 to 60 kg), and

* Corresponding author: chanporn.ch@kmitl.ac.th

finishing period (60 to 100 kg), respectively. RAC was supplemented to diet during finishing period. Pigs were raised under the same condition in individual pens for 109 days. After the raising period, all pigs were slaughtered at the average weight of 100 kg.

2.2 Physical characteristics

The pH at 45 min and 24 h post mortem were measured in duplicate directly at the 10 to 11th ribs of the left side carcass using pH meter equipped with a spear tip glass electrode (Model SG2 - ELK Seven Go™, Mettler Toledo International Inc., China). After 24 hours chilling, *Longissimus dorsi* (LD) was removed from the right side of each carcass for further analysis.

Drip loss was determined in duplicate as the weight loss during suspension of meat samples (weighing approximately 100 g) in a closed plastic bag over 48 hour at 4 °C [13]. Drip loss was expressed as a percentage relative to the initial weight. For meat color measurement, LD muscle was cut into 3-cm-thick slice and allowed to bloom for 30 min before measuring CIE L*, a*, and b* color values using a Minolta Chromameter CR-300 (Minolta Camera Co., Japan; Illuminant D65). For measuring cooking loss, two pieces of 3-cm-thick slices obtained from each LD muscle were weighed, placed into high-density polyethylene bag, heat sealed, and then cooked in a water bath set at 80°C for 30 min or until internal temperature of meat sample reached 70°C. After cooked samples were cooled down by running tap water to room temperature before weighing then cooking loss percentage was estimated by calculating the difference between before and after cooking weight. Cooked samples were cut along the muscle fiber into 10 cubes (1 x 1 x 3 cm^3). Shear force values were determined using a Warner–Bratzler shear attachment on a texture (model 2519-104, Instron®, Norwood, MA, USA) equipped with a 50 kg load cell using 200 mm/min crosshead speed.

Muscle fiber diameter was determined from 200 muscle fibers of 2 cubes (2 × 2 × 2 CM^3) of each muscle sample using 4X compound microscope equipped with Dino-Eye eyepiece camera to capture images and subsequently evaluated with Dino Capture version 2.0 software (AnMo Electronics Corporation, New Taipei City, Taiwan) [14]. Sarcomere length data were determined from 2 cubes (2 × 2 × 2 CM^3) of muscle samples according to [15]. The average of 30 sarcomere lengths was acquired from each sample.

2.3 Chemical characteristics

After 24 hours chilling, 100 g of LD muscle from the right side of each carcass was collected and kept in vacuum bag at -20°C until evaluating moisture, fat, and collagen content. Moisture and crude fat content of meat samples were performed following the procedures of the Official Association of Official Analytical Chemists (AOAC) method [16]. For measuring glycogen content, 50 g of LD muscle from the left side of each carcass was collected within 1 hour post mortem then immediately frozen in liquid nitrogen and stored at -40°C until measuring glycogen content following the method of [17]. Soluble, insoluble, and total collagens were determined by the method of [16, 18].

2.4 Statistical analysis

Analyses of variance was generated by using the GLM procedure (SAS Inst. Inc., Cary, NC) with RAC and sex as the main effects with interaction. Least squares means were separated using the probability of difference option (PDIFF), and the results were considered significant difference when $p<0.05$.

3. Results and Dicussion

3.1 Physical characteristics

The pH at 45 min and 24 h postmortem were not different among RAC supplemented levels and between sexes in the present study (Table 1). The LD muscles from pigs fed with 0 and 20 mg/kg of RAC had higher a* value than pigs fed with 40 mg/kg of RAC ($p < 0.01$) while L* and b* values were not affected by dietary RAC ($p > 0.05$). Level of RAC supplemented and sex had no affected on drip loss and cooking loss percentage ($p > 0.05$). Pigs fed with 0 mg/kg of RAC had lower WBSF than pigs fed with 40 mg/kg of RAC ($p < 0.05$) at day 1 and day 5 post mortem. While WBSF of pigs fed with 20 mg/kg of RAC showed no significant difference from others ($p > 0.05$). The level of supplemented RAC and sex had no significant effect on sarcomere length and muscle fiber diameter. However, barrows had shorter ($p < 0.05$) muscle fiber diameter than gilts. There were no interaction between RAC level in diet and sex on physical characteristics in LD muscles in the present study ($p > 0.05$).

As the pH at 45 min and 24 hour postmortem were not different among RAC supplemented levels and between sexes in the present study. Several studies showed that there was no significant difference in muscle pH in RAC treated pigs [2, 3, 4, 5]. Also, the LD muscles of pigs fed with 0 and 20 mg/kg RAC had higher a* value than pigs fed with 40 mg/kg of RAC while L* and b* values were not affected by dietary RAC in the present study. The study of [1, 2, 3, 5] demonstrated no effect of RAC feeding on L* value which was similar to this study. In contrast to this study, [1, 3, 4] reported the lower b* value of LD in RAC treated pigs. While [5] did not detect an effect of feeding RAC on b* values of the LD. Similar to this study, a persistent reduction in a* values has been reported in RAC fed pigs [1, 2, 4, 5]. The lower a* values indicate less oxymyoglobin in the muscle [1]. As described by [19], RAC caused an increase in the proportion of white fibers. A decrease in myoglobin content in muscle is suggested by this shift; because white fibers have less myoglobin than red and intermediate fibers. Several studies reported that RAC affected muscle fiber distribution. As in the study of [20], RAC supplemented at 20 or 60 mg/kg did not affect the percentage of type I and IIA fibers but the percentage of type IIAX fibers were decreased while type IIB fibers were increased. Furthermore, the researchers reported a strong significant negative correlation (R = -0.768) between type IIAX fibers and type IIB fibers. [21] also demonstrated that RAC supplemented at 20 mg/kg caused an increase in the mRNA expression of type IIB fibers at the expense of type IIX fibers while the expression of type I fibers was unaffected.

Table 1 Physical characteristics of pork from barrows and gilts fed with and without RAC

Traits	RAC (R)			Sex (S)		RMSE	P-value		
	0	20	40	Gilt	Barrow		R	S	RXS
pH_{45}	6.15	5.97	6.14	6.02	6.15	0.37	0.485	0.361	0.986
pH_{24}	5.49	5.47	5.52	5.46	5.52	0.10	0.568	0.166	0.785
Color									
L*	59.66	58.91	58.18	57.80	60.04	4.70	0.818	0.238	0.181
a*	8.26^{a}	7.14^{a}	5.84^{b}	7.48	6.67	1.21	0.002	0.101	0.401
b*	3.35	2.04	1.58	2.26	2.39	1.68	0.122	0.843	0.138
DL[1](%)	6.68	6.54	6.61	7.39	5.83	2.38	0.991	0.086	0.091
CL[2](%)									
Day 1	20.05	19.41	19.83	19.95	19.58	2.81	0.876	0.719	0.198
Day 5	20.39	20.12	19.40	20.42	19.51	2.42	0.646	0.317	0.137
SF[3] (kg)									
Day 1	3.73^{X}	4.17^{XY}	4.71^{Y}	4.27	4.14	0.78	0.032	0.654	0.168
Day 5	3.95^{X}	4.24^{XY}	4.79^{Y}	4.41	4.24	0.70	0.039	0.524	0.206
SL[4] (μ)	1.70	1.71	1.71	1.71	1.71	0.06	0.934	0.866	0.225
MD[5](μ)	79.76	86.04	80.43	87.20^{a}	76.96^{b}	7.02	0.111	0.0005	0.276

[a,b] within a row and main effect, least squares means with different superscripts differ ($p < 0.01$)
[X,Y] within a row and main effect, least squares means with different superscripts differ ($p < 0.05$)
[1-5] DL=Drip loss, CL=Cooking loss, SF=Shear force, SL=Sarcomere length, MD=Muscle fiber diameter

Level of RAC supplemented and sex had no affected on drip loss and cooking loss percentage in the current study. However, the study of [4] reported that barrow had higher drip loss at 24 and 48 hour post mortem than gilt. The results from [3, 4, 5] showed that there were no difference of drip loss and cooking loss between pigs fed with and without RAC supplemented. The study of [1] found the higher drip loss percentage of 20 mg/kg of RAC than control but there was not different in cooking loss among treated RAC. In this study, pigs fed with 40 mg/kg of RAC had higher WBSF than pigs fed without RAC while pigs fed with 20 mg/kg of RAC showed no significant difference compared to pigs fed with 0 and 40 mg/kg of RAC. From the study of [2, 3, 5], pigs fed with more than 10 mg/kg RAC had higher WBSF and more toughness. Moreover, [2] reported that pigs fed with RAC showed slower protein degradation rate than control pigs. LD muscle from pigs fed with RAC tended to have a slower protein degradation rate than control which indicated by the delayed emerging of 14 – 45 kDa polypeptides. The less postmortem degradation of myofibrillar protein may be responsible for increasing in toughness of pork from pigs fed with RAC.

There were no significant difference on sarcomere length and muscle fiber diameter due to RAC supplemented level and sex in this study. However, barrows had shorter muscle fiber diameter than gilts. This could be due to gilts had higher proportion of type IIb fiber than barrows. [22] reported that gilts had a larger muscle fiber especially in type IIb fibers than barrows. According to the length of sarcomere, when the muscle contracted it would have shorter sarcomere length then meat would likely to be tougher. ß-agronist did not affect the range of sarcomere length in Lambs [23] and in cattle [24]. In the present study, sarcomere length of pig fed with different levels of RAC were not significantly different thus RAC did not affected muscle contraction. There is controversial for the effect of gender on muscle fiber size. The studies of [25] reported larger fibers in gilts than in barrows, whereas [26] reported there was no effect of gender on muscle fiber size of Longissimus muscle.

3.2 Chemical composition and collagen solubility

The percentage of moisture and crude fat from LD of pigs fed with different RAC levels and different sexes were not significant different (Table 2). Pigs fed without RAC had higher ($p<0.05$) glycogen content than pigs fed with 20 and 40 mg/kg of RAC. The effect of sex on glycogen content was found in this study as barrows had higher glycogen content than gilts ($p < 0.05$). RAC levels had no effect on collagen content and collagen solubility ($p > 0.05$). Gilts had higher insoluble collagen ($p < 0.05$), total collagen ($p = 0.054$), and percentage of collagen solubility ($p = 0.054$) than barrows. There were no interaction between RAC level in diet and sex on chemical composition and collagen solubility in LD muscles in the present study ($p > 0.05$).

Table 2 Chemical composition and collagen characteristics of pork from barrows and gilts fed with and without RAC

Traits[1]	RAC (R)			Sex (S)		RMSE	P-value		
	0	20	40	Gilt	Barrow		R	S	RXS
M	73.09	71.96	73.58	72.80	72.95	1.95	0.182	0.832	0.073
EE	2.13	2.41	2.23	2.00	2.51	0.70	0.689	0.059	0.531
Gly	40.10^{a}	22.56^{b}	16.36^{b}	22.10^{X}	30.94^{Y}	9.26	<0.0001	0.022	0.081
SC	0.42	0.36	0.35	0.38	0.37	0.11	0.357	0.800	0.731
IC	2.66	2.95	3.08	3.27^{X}	2.53^{Y}	0.94	0.598	0.042	0.360
TC	3.08	3.32	3.43	3.65	2.90	1.02	0.731	0.054	0.375
CS	13.82	11.25	12.86	10.69	14.56	5.23	0.550	0.054	0.356

[a,b] within a row and main effect, least squares means with different superscripts differ ($p < 0.01$)
[X,Y] within a row and main effect, least squares means with different superscripts differ ($p < 0.05$)
[1] M = % Moisture ; EE = % Ether extract ; Gly = Glycogen content (μmol/g) ; SC = Soluble collagen (mg/g) ; IC = Insoluble collagen (mg/g) ; TC = Total collagen (mg/g) ; CS = % Collagen solubility

Crude fat percentage of LD was not affected by dietary RAC in the present study. In agreement with the results reported by [3, 5]. The study of [5] also reported that there was no effect of sex on fat content which was consistent with this study. Pigs fed without RAC had higher glycogen content than pigs fed with dietary RAC in the present study. In agreement with [27] who reported that the amount of glycogen stored in skeletal muscle tissue of the pigs fed without RAC higher than pigs fed with RAC. [21] explained the possibility that chronic supplementation of RAC leads to the decline of glycogen levels, it may be due to the needs for energy increased associated with increased muscle growth. Barrow had higher glycogen content than gilt in this study. In contrast to [28] reported that gender had no effect on glycogen content neither at 45 min nor at 24 hour postmortem.

Dietary RAC levels did not impact collagen content and collagen solubility in this study. Similar to the current study, there were no effect of RAC on collagen content and collagen solubility in heifer [29] and in steer [30]. In this study, gilts had significant higher content of insoluble collagen and tended to have higher content of total collagen and percentage of collagen solubility than barrows. In contrast to [30] who reported that gender had no significant effect on total collagen content and collagen solubility. The average of total collagen 4 mg/g in Longissimus muscle reported by [31] was slightly higher than the results in this study which were 2.90 and 3.65 mg/g for barrow and gilt, respectively. While the range of 10.69 – 14.56 % of collagen solubility in the present study was in the similar range, 11 – 13 %, as reported by [31].

Conclusion

The LD from pigs fed with 40 mg/kg of RAC was less redness than LD from others. Pigs fed with 40 mg/kg of RAC had tougher LD than control. Dietary RAC decreased glycogen content in LD muscle but not affected other chemical composition and collagen solubility. Compared to barrows, gilts had lager muscle fiber, lower glycogen content, and higher insoluble collagen content, also tended to have higher total collagen content and collagen solubility.

Acknowledgments

This research was supported by a grant of the National Bureau of Agricultural Commodity and Food Standards, Thailand.

References

1. S.N. Carr, D.J. Ivers, D.B. Anderson, D.J. Jones, D.H. Mowrey, M.B. England, J. Killefer, P.J. Rincker, F.K. McKeith, J Anim Sci **83**, 2886-93 (2005a)
2. Y.L. Xiong, M.J. Gower, C. Li, C.A. Elmore, G.L. Cromwell, M.D. Lindemann, Meat Sci **73**, 600-4 (2006)
3. S.N. Carr, P.J. Rincker, J. Killefer, D.H. Baker, M. Ellis, F.K. McKeith, J Anim Sci **83**, 223-30 (2005b)
4. J.F. Patience, P. Shand, Z. Pietrasik, J. Merrill, G. Vessie, K.A. Ross, A.D. Beaulieu, Can J Anim Sci **89**, 53-66 (2009)
5. N.B. Athayde, O.A. Dalla Costa, R.O. Roca, A.L. Guidoni, C.B. Ludtke, G.J.M.M. Lima, J Anim Sci **90**, 4604-10 (2012)
6. S.J. Moeller, R.K. Miller, K.K. Edwards, H.N. Zerby, K.E. Logan, T.L. Aldredge, C.A. Stahl, M. Boggess, J.M. Box-Steffensmeier, Meat Sci **84**, 14–22 (2010)
7. G. Monin, P. Sellier, Meat Sci **13**, 49-63 (1985)
8. J.L. Melody, S.M. Lonergan, L.J. Rowe, T.W. Huiatt, M.S. Mayes, E. Huff-Lonergan, J Anim Sci **82**, 1195-1205 (2004)
9. E. Huff-Lonergan, S.M. Lonergan, Meat Sci **71**, 194–204 (2005)
10. M. Maltin, D. Balcerzak, R. Tilley, M. Delday, Proc Nutr Soc **62**, 337–47 (2003)
11. S.H. Fang, T. Nishimura, K. Takahashi, J Anim Sci **77(1)**, 120-30 (1999)
12. T.L. Wheeler, S.D. Shackelford, M. Koohmaraie, J Anim Sci **78(4)**, 958-65 (2000)
13. K. Honikel, Meat Sci **49**, 447-57 (1998)
14. H.J. Tuma, J.H. Venable, P.R. Wuthier, R.L. Henrickson, J Anim Sci **21**, 33-6 (1962)
15. H.R. Cross, R.L. West, T.R. Dutson, Meat Sci **5(4)**, 261-266 (1981)
16. W. Horwitz, *Official methods of analysis of AOAC International 17th ed.* (AOAC International, 2002)
17. C.E. Dreiling, D.E. Brown, L. Casale, L. Kelly, Meat Sci **20**,167-77 (1987)
18. F. Hill, J Food Sci **31**,161-66 (1966)
19. J.L. Aalhaus, A.L. Schaefer, A.C. Murray, S.D.M. Jones, Meat Sci **31(4)**, 397-409 (1992)
20. F.F.S. Depreux, A.L. Grant, D.B. Anderson, D.E. Gerrard, J Anim Sci **80**, 1888-94 (2002)
21. A.M. Gunawan, B.T. Richert, A.P. Schinckel, A.L. Grant, D.E. Gerrard, J Anim Sci **85**, 2115-24 (2007)
22. C. Larzul, L. Lefaucher, P. Ecolan, J. Gogue, A. Talmant, P. Sellier, P. Le Roy, G. Monin, J Anim Sci 75 (12), 3126-37 (1997)
23. Y.B. Lee, H. Jung, Y.S. Kim, R.H. Dalrymple, J Anim Sci **66** (Suppl. 1), 279 (1988)
24. I. Girard, J.L. Aalhus, J.A. Basarab, I.L. Larsen, H.L. Bruce, Can J Anim Sci **92**, 175-88 (2012)
25. C. Larzul, L. Lefaucher, P. Ecolan, J. Gogue, A. Talmant, P. Sellier, P. Le Roy, G. Monin, J Anim Sci **75 (12)**, 3126-37 (1997)
26. G. Bee, J Anim Sci **82**, 826-36 (2004)
27. T.S. Araujo, L.C.J. Porto, E.G. Mario, L.J. Pereira, M.S.D.S. Ferreira, M.G. Zangeronimo, M.H. Napimoga, L.M. Botion, R.V. De Sousa, Turk J Vet Anim Sci **38**, 54-62 (2004)
28. C. Terlouw, A. Berne, T. Astruc, Livest Sci **122**, 199–213 (2009)
29. S.M. Ebarb, K.J. Phelps, J.S. Drouillard, K.R. Maddock-Carlin, M.A. Vaughn, D.D. Burnett, J.A. Noel, C.L. Van Bibber-Krueger, C.B. Paulk, D.M. Grieger, J.M. Gonzalez, J Anim Sci **95(3)**,1219-31 (2017)
30. P.E. Strydom, L. Frylinck, J.L. Montgomery, M.F. Smith, Meat Sci **81**, 557–64 (2009)
31. J.A. Correa, L. Faucitano, J.P. Laforest, J. Rivest, M. Marcoux, C. Gariepy, Meat Sci **72(1)**, 91-9 (2006)

Feasibility study on using evaporator condensate to subcool refrigerant in a vapor-compression refrigeration system

Kiattisak Roonprasang[1], and *Eakasit* Sritham[1,*]

[1]Food Engineering Department, Faculty of Engineering, King Mongkut's Institute of Technology Ladkrabang, Bangkok, Thailand

Abstract. A computer simulation based on MATLAB was developed to investigate the effectiveness of using evaporator condensate to subcool refrigerants in a vapor-compression refrigeration system. Three different types of refrigerant including R22, R410a, and R134a were studied. The simulation clearly showed the effectiveness of the technique to subcool refrigerant using evaporator condensate. Throughout the RH range of surrounding in this study, 40%-100%RH, the vapor-compression model system using R22 as a refrigerant provided the highest COP linearly ranging from 3.40 to 3.76. The greatest improvement in COP was found in the model system with R410a where the COP values increased by 3.33% to 15.36.

1 Introduction

Refrigeration plays an essentially important role in food industry. It involves in transportation and storage of both raw materials and final products, and in a number of processing lines. Hence, any improvement in the efficiency of a refrigeration system—known as the coefficient of performance (COP), would help reducing the production cost of food products and consequently increasing the market potential of the products.

Refrigeration systems widely adopted in food industry are of a vapor-compression type which mainly includes a compressor, a condenser, an expansion valve, and an evaporator. The COP is defined as a ratio between evaporator heat load and compressor work [1-2]. A common technique to improve COP is done by subcooling liquid refrigerant after exiting a condenser using the refrigerant from evaporator outlet, theoretically in the state of saturated vapor. A subcool unit is basically sort of a heat exchanger. Once absorbed heat from liquid refrigerant during passing through a subcool unit, the refrigerant would turn into superheat causing the increase of compressor work–thing to be sacrificed when doing subcool this way.

Under tropical climate like in Thailand, refrigeration systems release considerable amount of condensate from evaporator. The part of evaporator workload spent to condensate moisture in the air is quite a large portion. Accordingly, there seems to be a room to improve the efficiency of a vapor-compression refrigeration system if we could gain back the energy by using evaporator condensate to subcool liquid refrigerant after exiting a condenser. Once subcooled, the enthalpy of liquid refrigerant would decrease, and, theoretically, the vapor quality of the refrigerant after exiting an expansion valve would lower. These two effects would yield a result in the same direction–increasing the ability of evaporator to absorb heat, and the higher COP could be expected.

This study was aimed at proposing a technique to improve the efficiency of a vapor-compression refrigeration system by using an evaporator condensate to subcool the refrigerant. A schematic diagram of the proposed systems is given in Fig. 1 along with the P-H diagram in Fig. 2.

2 Methodology

2.1 Computer simulation

Computer simulation was developed on MATLAB®. Three different refrigerants including DuPont™ Freon® 22 (Chlorodifluoromethane), DuPont™ Suva® 410A (R-410-A), and HFC-134A (1,1,1,2-tetrafluoroethane) were considered. Thermal properties data base of these refrigerants [3-5] were also created in MATLAB. A computer function module that was written is capable of fetching the refrigerant properties from data base and performing interpolation. The surrounding humidity shall directly affect the amount of condensate being released from an evaporator, and hence the effectiveness of the proposed technique–subcooling. Hence, the range of a surrounding humidity was set to cover that of the year-round climate condition in Thailand, approximately 40% – 100%RH [6].

2.2 Parameters for a refrigeration model system in the simulation

The parameters for a vapor-compression refrigeration model in this simulation were set similar to those found in common residential air conditioning systems as follows.

* Corresponding author: eakasit.sr@kmitl.ac.th

The evaporator heat load (Q_{ev}) was 26,000 BTU/hr (7.6198 kW), the inlet and outlet air temperatures at the evaporator were, respectively, 27°C and 4°C (T_{ev}). The surface area and the operating temperature of the condenser (T_{cond}) were 0.114 m^2 and 50°C, respectively. The velocity of air flowing through a condenser was assumed to be 2 m/s. The ambient temperature was fixed at 35°C.

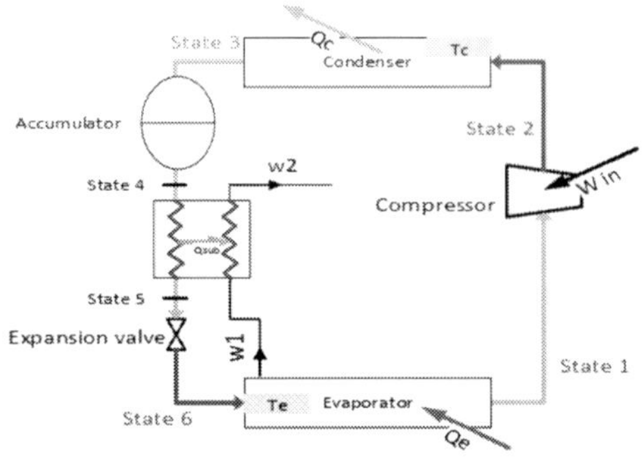

Fig. 1. A schematic diagram of vapor-compression refrigeration system with a subcool unit. Adapted from [1-2].

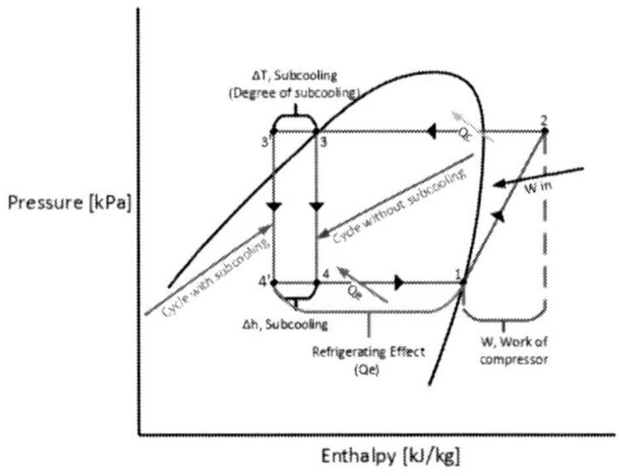

Fig. 2. P-H diagram for a vapor-compression refrigeration system: with and without subcooling. Adapted from [1-2].

2.3 Calculation

2.3.1 Refrigerant flow rate

The total flow rate of refrigerant (m_R) was calculated as follows:

$$m_R = m_v + m_l \quad (1)$$

$$m_l = Q_{ev}/q_{ev} \quad (2)$$

$$m_v = (m_l * x)/(1-x) \quad (3)$$

where m_v and m_l are flow rates of vapor and liquid refrigerant (kg/s), respectively, and x is a vapor quality.

2.3.2 Properties of refrigerant in a vapor compression cycle

State 1: Saturated vapor

$$P_1 = P \text{ at } T_{ev}=4°C \quad (4)$$

$$h_1 = h_g \text{ at } P_1 \quad (5)$$

$$s_1 = s_g \text{ at } P_1 \quad (6)$$

State 2: Superheated vapor

$$P_2 = P \text{ at } T_{cond}=50°C \quad (7)$$

$$h_2 = h \text{ at } P_2 \text{ and } s_1 \quad (8)$$

State 3: Saturated liquid

$$P_3 = P_2 \quad (9)$$

$$h_3 = h_f \text{ at } P_3 \quad (10)$$

State 4: Totally saturated liquid

$$h_4 = h_f \text{ at } P_3 \quad (11)$$

where h is the enthalpy of refrigerant (kJ/kg) and s is the entropy of refrigerant (kJ/kg.K) [2]. Subscripts g and f represent the states of saturated liquid and saturated vapor, respectively.

2.3.3 Subcooling process

Given q as the amount of heat being exchanged at a subcool unit and ε is the effectiveness of heat exchanger/subcool unit [7]:

$$q = \varepsilon q_{max} \quad (12)$$

$$q_{max} = C_R(T_4 - T_{W1}), \text{ when } C_R < C_W \quad (13)$$

or $$q_{max} = C_W(T_4 - T_{W1}), \text{ when } C_W < C_R \quad (14)$$

C_R is the specific heat of refrigerant; $C_R = m_R c_{p,R}$, where $c_{p,R} = h_4/T_{cond}$. C_W is the specific heat of condensate $C_W = m_W c_{p,w}$ where $c_{p,w}$ = 4.2166 kJ/kg°C. The ε was assumed to be 0.8 in this study.

Then, from Fig. 1., the enthalpy of refrigerant after subcooling (state 5) could be calculated using Eq.15.

$$q = m_R(h_4 - h_5) \quad (15)$$

2.3.4 COP of the system with subcooling

In order to calculate the COP of a system with subcooling, a new state of refrigerant before entering an evaporator needs to be determined. Details of the calculation were given below;

$$x_{subc} = (h_6 - h_{f@P6}) / h_{fg@P6} \quad (16)$$

$$m_{v,subc} = x_{subc}(m_R) \quad (17)$$

$$m_{l,subc} = m_R - m_{v,subc} \quad (18)$$

$$q_{ev} = h_1 - h_6 \quad (19)$$

$$w_{comp} = h_2 - h_1 \quad (20)$$

$$Q = m_l(q_{ev}) \quad (21)$$

$$W = m_R(w_{comp}) \quad (22)$$

$$COP = Q/W \quad (23)$$

3 Results and Discussion

The results obtained with computer simulation are as follows.

3.1 The dependence of condensate flow rate on relative humidity of surrounding air (RH)

The condensate flow rate at evaporator strongly related to the humidity of surrounding air. The rate linearly increased with RH. The calculated flow rate was in a range of 0.001 – 0.0046 kg/s when the RH was set to 40% – 100% (Fig. 3). It was found that condensate flow rate increased for $6x10^{-5}$ kg/s at every percentage increase of RH.

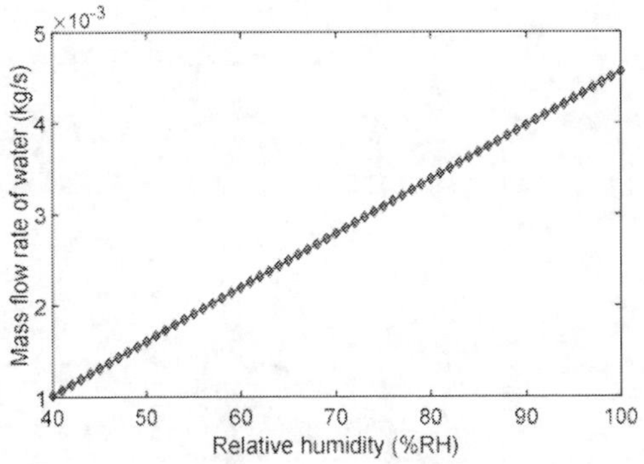

Fig. 3. The relationship between mass flow rate of a condensate and relative humidity of the surrounding

3.2 Effects of RH on vapor quality of refrigerants

The simulation showed that vapor quality of refrigerant before entering an evaporator was linearly dependent on RH for every type of refrigerants. This was as expected since the greater flow rate of condensate would be obtained when the system is operated under higher RH atmosphere (see also Sec.3.1 above). The higher amount of condensate would exhibit the greater subcooling effect.

The vapor quality before entering an evaporator gradually decreased with the increasing RH, for every refrigerant. The lowest vapor quality was found in the model system with R410a, while the system with R134a showed the highest vapor quality. The vapor quality fell in the ranges of 0.1617-0.2075, 0.2430-0.2794, and 0.2993-0.3308, for the systems with R410a, R22, and R134a, respectively (Fig. 4).

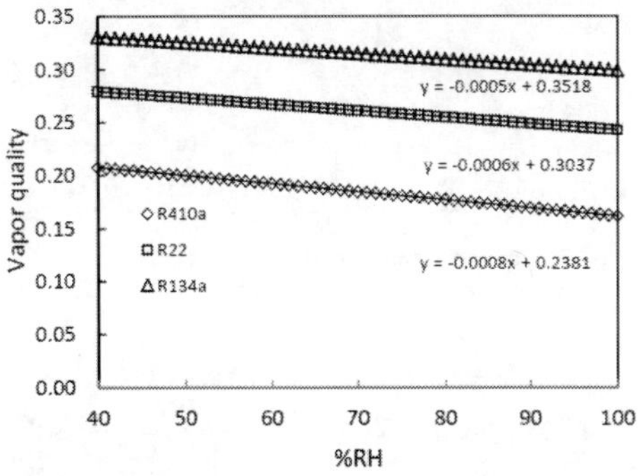

Fig. 4. The vapor quality of refrigerants before entering the evaporator and relative humidity of the surrounding

3.3 Performance of vapor-compression refrigeration system with subcool unit

As for comparison, the simulated COP values for typical (without subcooling) vapor-compression refrigeration systems were 3.31, 3.22, and 3.01, respectively, when using R22, R410a and R134a as refrigerants.

It was found that the COP obtained for the model systems with subcooling linearly increased with the increasing RH. The model system with R22 and R410a refrigerants provided comparable COP throughout the range of RH considered in this study. The COP ranged from 3.40 to 3.76, and from 3.33 to 3.72, for the system with R22 and R410a, respectively. The model system with R134a showed lower COP, in a range of 3.09–3.39. In general, the rates of increase in COP values with RH appeared to be similar for all three different types of refrigerant (Fig. 5).

Though the model system with R22 could provide the highest COP value, the greatest improvement in COP, as compared to that of the system without subcooling, was obtained from the model system using R410a as a refrigerant. The COP of R401a model system increased by 3.33% to 15.36%. It could be also seen that, the higher the RH, the greater rate of improvement in COP the system exhibited. The rate of change of percentage increase of COP was 0.201% per %RH (Fig. 6).

The improvement of COP for the model systems with R22 and R134a were rather similar, with the values falled in a ranges of 2.96%–13.64%, and 2.75%–12.65%, respectively. The rates of change of percentage increase of COP were 0.178% and 0.165% per %RH, respectively, for the system with R22 and R134a (Fig. 6).

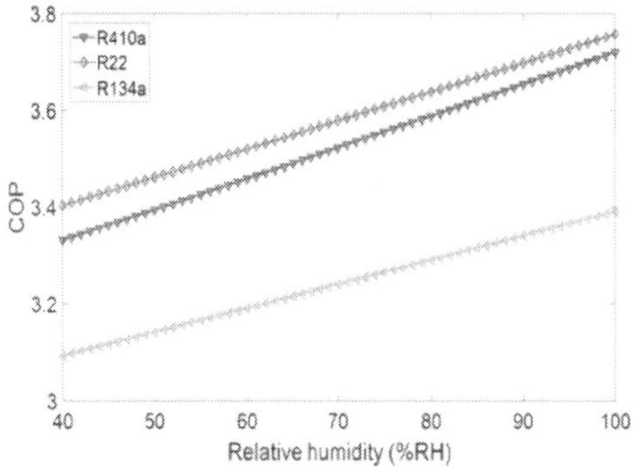

Fig. 5. The dependence of COP for a vapor-compression refrigeration model system (with subcool unit) on RH

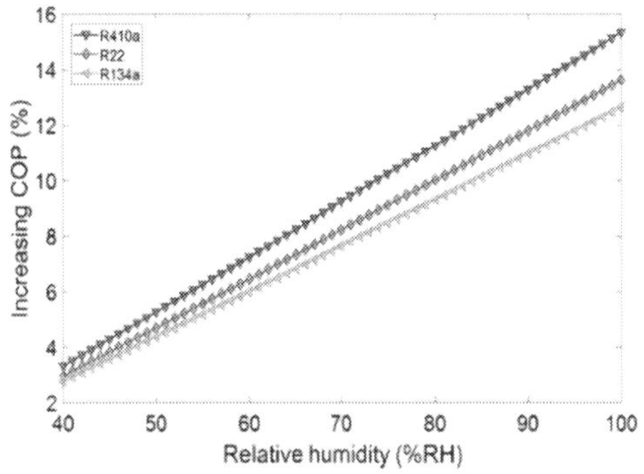

Fig. 6. The percentage increase in COP as a function of RH for a vapor-compression refrigeration system with three different refrigerants

4 Conclusions

A computer simulation developed in this study clearly showed the effectiveness of a technique to subcool refrigerant using the condensate from an evaporator. Throughout the RH range of surrounding considered in this study, 40%–100%RH, the vapor-compression model system using R22 as a refrigerant provided the highest COP values which linearly ranged from 3.40 to 3.76. The greatest improvement in COP was found in the model system with R410a as a refrigerant, where the COP values increased by 3.33% to 15.36.

References

1. C.P. Arora, *Refrigeration and Air Conditioning*, 3rd ed., New York: McGraw-Hill (2009)
2. Y.A. Cengel, M.A. Boles, *Thermodynamics : An Engineering Approach*, 8th Ed., New York: McGraw-Hill (2015)
3. DuPont Fluorochemicals, *Thermodynamic Properties of DuPont Suva 410A refrigerant*, (2004)
4. DuPont Fluorochemicals, *Thermodynamic Properties of HFC-134a* (2004)
5. DuPont Fluorochemicals, *Thermodynamic Properties of DuPont Freon 22 refrigerant* (2004)
6. The Metheorological Department, Ministry of Information and Communication Technology, Thailand, *Relative humidity by metheorology station: 2014*, [Available online] https://www.m-society.go.th/ewt_news.php?nid=14571 (Accessed : Jan. 2016)
7. F.P. Incropera, D.P. DeWitt, T.L. Bergman, A.S. Lavine, *Fundamentals of heat and mass transfer*, 6th ed. John Wiley & Sons, p.686-687 (2007)

Meat characteristics from four different cutting parts of Cherry Valley duck

Chanporn Chaosap*, *and Panneepa* Sivapirunthep

Department of Agricultural Education, Faculty of Industrial Education and Technology, King Mongkut's Institute of Technology Ladkrabang, Bangkok 10520, Thailand

Abstract. The objective of this study was to investigate the duck meat characteristics in 4 main parts : breast, fillet, leg, and thigh. A total of 30 Cherry Valley ducks were used in this study. They were slaughtered at the age of 42 days. Carcasses were cut and collected the boneless and skinless of breast, fillet, leg, and thigh for measuring meat characteristics. The results showed that fillet and leg had the longest sarcomere length at 6, 12, and 24 hour post mortem (P<0.05). The highest pH was from leg and the lowest pH was from breast and fillet (P<0.01). The cooking loss percentage of breast was the highest while the lowest was from leg (P<0.01). Thigh and leg were tougher than breast and fillet. Shear force value negatively correlated with sarcomere length at 24 hour post mortem and with pH but positively correlated with L* value.

1. Introduction

Cherry Valley duck is a commercial crosses of Pekin ducks and it is one of a major duck crosses that has been used for commercial duck meat production in Thailand. It has high growth rate and reaches a market live weight of 3.45 kg at 42 days of age with a FCR of 1.92 for a medium-sized commercial duck [1]. For supporting duck production, not only the information of growth performance, carcass yield, and cuttability but also the information of duck meat characteristics is required because it is involved in cooking and processing process. In duck meat production industry, the duck carcass is sale both in a whole carcass and in different cutting parts such as breast, thigh, leg, and fillet. Each part of carcass has different meat characteristics in terms of tenderness, color, water holding capacity, and etc. According to different meat characteristics, each duck cutting part will be cooked in the various methods to meet consumer taste choices. Some scientific studies regarding each part of carcass are needed in order to provide reasons for using in how to cook and process which temperature and time are mainly involved. However, the scientific information of duck meat characteristics in each duck carcass cut up are limited. So the objective of this study was to investigate the duck meat characteristics in 4 main parts : breast, fillet, leg, and thigh.

2. Materials and Methods

2.1 Animals

A total of 30 Cherry Velley ducks were used in this study. They were slaughtered at the age of 42 days . All carcasses were kept in ice box during transferring around 2 hours from Duck King slaughter house in Chachangsao province, Thailand to the laboratory at the Department of Agricultural Education, Faculty of Industrial Education and Technology, King Mongkut's Institute of Technology Ladkrabang. Carcasses were cut and collected the boneless and skinless of breast, fillet, leg, and thigh at approximately 6 hour post mortem for measuring sarcomere length, pH, color, cooking loss, and shear force.

2.2 Sarcomere length measurement

Sarcomere length was measured according to the method of [2]. At different post mortem time, 10-15 g of fresh muscle sample from each cutting part of duck carcass were immersed in solution A (0.1 M potassium chloride, 0.39 M boric acid, and 5 mM ethylenediaminetetraacetic acid in 2.5% glutaraldehyde) for 2 hours. The samples were then transferred to new vials containing solution B (0.25 M KCl, 0.29 M boric acid, and 5 mM ethylenediamine-tetraacetic acid in 2.5% glutaraldehyde) for 17 to 19 hours. Sarcomere length was measured by light diffraction using a 0.5-mW helium-neon laser (Model No. 31004, REO, USA). The lengths of 30 sarcomeres were measured in 30 myofibrils from each sample.

2.3 pH measurement

The pH at 6 hour post mortem were measured in duplicate directly at each cutting part of the carcass using pH meter equipped with a spear tip glass electrode (Model SG2 - ELK Seven Go™, Mettler Toledo International Inc., China).

2.4 Color measurement

The boneless and skinless of 4 main cutting parts of the carcass were allowed to bloom for 30 min before measuring CIE L*, a*, and b* color values using a Minolta Chromameter CR-300 (Minolta Camera Co., Japan; Illuminant D65).

2.5 Cooking loss and Shear force measurement

Two pieces of approximately 3-cm-thick slices obtained from each cutting part were weighed, placed into high-density polyethylene bag, heat sealed, and then cooked

* Corresponding author: chanporn.ch@kmitl.ac.th

in a water bath set at 80°C for 30 min or until internal temperature of meat sample reached 70°C. After cooked samples were cooled down by running tap water to room temperature before weighing then cooking loss percentage was estimated by calculating the difference between before and after cooking weight. As many as possible of 1 x 1 x 3 cm^3 cuts of each cooking loss sample were removed from across the slice parallel to the muscle fiber orientation. Each cut was sheared once perpendicular to the muscle fiber orientation using a Warner-Bratzler shear head attached to a single column materials testing machine (model H1KS, Hounsfield, England) equipped with a 50 kg load cell using 50 mm/min crosshead speed.

2.6 Statistical analysis

Analyses of variance was generated by using the GLM procedure (SAS Inst. Inc., Cary, NC) with cutting parts as the main effect. Least squares means were separated using the probability of difference option (PDIFF), and the results were considered significant difference when $P<0.05$. The relationships between shear force value and other traits from four parts of duck carcasses were evaluated by Pearson correlation coefficients.

3. Results and Discussion

3.1 Sarcomere length

In this study, sarcomere length from fillet and leg were longer than from breast and thigh at 6, 12, and 24 hour post mortem (Table 1). However, the study of [3] showed that the sarcomere length from breast and leg of Chungdong ori duck were not significant different. [4] reported that the sarcomere length of age 6 week old Cherry Valley duck measured immediately after slaughter was 1.38 µm which was shorter than in this study. Leg and fillet may contain red or slow muscle fiber having slow contraction which causes longer sarcomere length as explained by [5].

Sarcomere is the contractile unit of myofibrils and sarcomere lengths shorten during muscle contraction [6]. The sarcomere length is shorthen by the overlapping of actin filaments in the center of the I-band and myosin filaments close to the Z-disk causing muscle shrinkage which is reducing meat tenderness [7]. The degree of muscle shrinkage depended on the contractile stage during post mortem storage. The highest degree of muscle shrinkage is during rigor mortis which has a crucial influence on meat tenderness. [8] reported that duck pectoralis major muscles showed a resting sarcomere length of 1.92 ± 0.12 µm at 15 min post mortem during storage at 0°C and shrank continuously until 24 hour post mortem. These authors reported that sarcomere lengths for meat stored at 0°C for 2 and 24 hour post mortem were, respectively, 24 and 37% shorter than at 15 min post mortem. As stated by [9] leg muscles of chicken such as soleus muscle is red muscle which contents more red type I fiber and breast muscle or pectoralis muscle is white muscle which contents more white type IIB fiber. [4] stated that duck leg has a large ratio of red muscle fiber than breast. Red muscle has more muscle contraction than white muscle [10]. In this study, sarcomere length of leg from 6 hour post mortem to 24 hour post mortem showed more contraction (3.3%) than sarcomere length contraction in breast (2.1%) that was in agreement with [9, 10].

Table 1 Sarcomere length (µm) at different time point of 4 major cutting parts of Cherry Valley ducks

Trait[1]	n	LSMeans				RMSE	P value
		Breast	Fillet	Leg	Thigh		
SL_{6h}	30	1.90^b	2.07^a	2.09^a	1.92^b	0.28	0.022
SL_{12h}	10	1.73^b	2.00^a	1.96^a	1.85^b	0.19	0.011
SL_{24h}	30	1.86^b	2.04^a	2.02^a	1.94^b	0.28	0.046

^{a,b} within a row, least squares means with different superscripts differ ($p < 0.05$)
[1] SL_{6h}, SL_{12h}, SL_{24h} = sarcomere length at 6, 12, and 24 hour post mortem

3.2 pH

The pH of leg was the highest following by thigh, and the lowest pH was from breast and fillet ($P<0.01$). The locomotive muscles ; leg and thigh had higher pH than supportive muscles; breast and fillet (Table 2). In agreement with [3] that reported duck leg had higher pH than breast, 6.52 and 5.95, respectively. [11] reported that Cherry Berry duck breast has pH values ranging from 6.6 to 6.0 at 15 min post mortem and at 24 hour post mortem, respectively. Type IIB white breast muscle has higher glycogen content thus its pH post mortem declines more than in type I red leg muscle [9]. [12] stated that pH during post mortem storage decreased faster in fast glycolytic muscle (breast) than in slow oxidative muscle (leg).

3.3 Color

The most redness muscle indicated by the highest a* value was fillet, followed by leg, thigh, and breast, respectively ($P < 0.01$) as shown in Table 2. Breast and thigh had the same b* value which was higher than from leg and fillet ($P < 0.01$). L* value from fillet and leg were higher than from breast and thigh ($P < 0.01$). Breast had lower a* value than leg might be explained by breast had less myoglobin than leg so breast color was less redness [9]. However, from [13] showed that breast and leg had the same a* value approximately 13.5 and [14] also reported that a* value of breast and leg were 17.8 and 17.7, respectively. Breast and thigh had the same b* value and were higher than from leg and fillet in the present study. In agreement with [13] reported higher b* value from breast than from leg (9.15 VS 6.86). In addition [14] reported that breast had b* value 1.89 and leg had b* value 1.50. Breast had significantly lower L* value than leg (28.99 VS 30.13) in the current study. In agreement with [14] found lower L* value from breast than from leg, 40.3 and 41.5, respectively. These authors explained that breast had lower L* value than leg which was probably due to the higher fat content in leg than in breast. In contrast, [13] reported that breast had higher L* value than leg (49.53 VS 47.93).

3.4 Cooking loss

Cooking loss of breast was the highest followed by from fillet and thigh (P<0.01) and the lowest cooking loss was from leg (Table 2). Because the lower pH post mortem causes denaturation of muscle protein in breast so breast has lower water holding capacity and increased exudation and cooking loss of meat [12].

3.5 Shear force

In this study, leg and thigh had the highest shear force value followed by breast and the lowest was fillet (P< 0.01). Fillet was more tender than others this might be associated with its longer sarcomere length. [3] reported that leg had higher shear force value than breast which was similar to this study. [12] informed that post mortem acidification kinetic of muscle had strongly influence on meat texture. In the present study, breast and fillet had lower pH so it might be decresed shear force value.

Table 2 pH, color, cooking loss, and shear force of 4 major cutting parts of Cherry Valley ducks (n=30)

Trait	LSMeans				RMSE	P value
	Breast	Fillet	Leg	Thigh		
pH	5.84^c	5.85^c	6.28^a	6.16^b	0.23	<.0001
color						
a*	22.13^c	25.52^a	23.79^b	22.39^{bc}	2.72	<.0001
b*	9.85^a	7.59^c	8.37^b	9.27^a	1.08	<.0001
L*	28.99^b	30.75^a	30.13^a	28.71^b	1.49	<.0001
CL^1 (%)	29.19^a	26.69^b	23.59^c	25.69^b	2.98	<.0001
SF^2 (kg)	4.84^b	2.57^c	5.62^a	6.02^a	1.18	<.0001

[a,b,c] within a row, least squares means with different superscripts differ (p < 0.01)
[1]CL = cooking loss, SF^2 = shear force

3.6 Correlation

The correlation between shear force value and other study traits as shown in Table 3.

Table 3 Correlation coefficient of parameter studies and shear force value

Trait	Shear force value	
	r	P value
$SL_{6h}{}^1$	-0.17	0.063
$SL_{12h}{}^1$	-0.24	0.138
$SL_{24h}{}^1$	-0.22	0.015
pH	-0.46	<0.0001
a*	-0.12	0.210
b*	0.17	0.059
L*	0.41	<0.0001
% Cooking loss	-0.05	0.556

[1]SL_{6h}, SL_{12h}, SL_{24h} = sarcomere length at 6, 12, and 24 hour post mortem

Shear force value negatively correlated with sarcomere length at 24 hour post mortem (r=-0.22, P=0.015) and had a trend of negative significant correlation with sarcomere length at 6 hour post mortem (r = -0.17, P =0.063). The pH had a negatively significant correlation with shear force value (r=-0.46, P<0.01). For meat color, b* value tend to have a positive correlation with shear force value (r=0.17, P=0.059) while L* positively correlated with shear force value (r=0.41, P<0.01). The shortening of sarcomere was a main contributor to toughess as indicated by a strong negative correlation between shear force and sarcomere length in chicken breast during chilled at −12 °C and 0°C [15].

4. Conclusion

Different cutting parts from duck carcass had different meat characteristics. Sarcomere length from leg and fillet were longer than from breast and thigh. Leg had the highest pH followed by thigh and the lowest pH were from breast and fillet. The most redness part was fillet followed by leg and thigh and the lowest redness part was breast. Breast and thigh were more yellow color than leg and fillet, respectively. Leg and fillet were lighter color than breast and thigh. Breast had the highest cooking loss and the lowest was from leg. Thigh and leg were tougher than breast and fillet, respectively. In conclusion, fillet was the most tender due to its lowest shear force value which might be associated with the longest sarcomere length. In addition, shear force value negatively correlated with sarcomere length at 24 hour post mortem and with pH at 6 hour post mortem while it positively correlated with L* value. From the results, leg and thigh were tougher than breast and fillet so the appropriate method to cook them should be slow cook or long time low temperature cooking.

Acknowledgement

The research was financially supported by Faculty of Industrial Education and Technology, King Mongkut's Institute of Technology Ladkrabang, Thailand and Duck King Co. Ltd.

References

1. Cherry Valley Farms. [online] https://www.cherryvalley.co.uk. Available at: https://docs.wixstatic.com/ugd /949670_45e8451249894c97ac1fae0879aaf6dd.pdf [Accessed 21 March 2018].
2. H.R. Cross, R.L.West, T.R. Dutson . Meat Sci **5**(4), 261-266 (1981)
3. M.S. Ali, H. Yang, J. Jeong, S. Moon, H. Hwang G. Park, S. Joo. Asian-Aust. J. Anim. Sci. **20**(12),1895-1900 (2007)
4. L. Jing, C. Guo-bin, X. Qi, Z. Yi-yu, Z. Hai-bo, Z. Qiong, C. Guo-hong. Agri. Sci. Tech. **10** (5), 107-109 (2009)
5. Y. Hwang, G. Kim, J. Jeong, S. Hur, S. Joo. Meat Sci. 86, 456-61.
6. K.O. Honikel, C.J. Kim, R. Hamm, P. Roncales. Meat Sci.**16**(4), 267-282 (1986)
7. B.B. Marsh and W.A. Carse. J. Food Tech. **9**, 129-139 (1974)
8. Y.S. Choi, D.J. Han, J.H. Choi, K.E. Hwang, D.H. Song, H.W. Kim, Y.B. Kim, C.J. Kim. Poult. Sci. **95**, 1198-1204 (2016)

9. S. McKee (2003)
[online] https://www.poultryscience.org
Available at:
https://www.poultryscience.org/docs/pba/1952-2003/2003/2003%20McKee.pdf
[Accessed 20 March 2018].
10. J. R. Bendall. J. Sci. Food Agric. **26**, 55–61 (1975)
11. M.S. Ali, G.H. Kang, H.S. Yang, J.Y. Jeong, Y.H. Hwang, G.B. Park, S.T. Joo. Asian-Aust. J. Anim. Sci. **20**(6), 1002-1006 (2007)
12. A. Listrat, B. Lebret, I. Louveau, T. Astruc, M. Bonnet, L. Lefaucheur, B. Picard, J. Bugeon. Sci. World J. (2016)
13. Muhlisin, D.S. Kim, Y.R. Song, H.R. Kim, H.J. Kwon, B.K. An, C. W. Kang, H.K. Kim, S.K. Lee. Korean J. Food Sci.An. **33**(1), 89-95 (2013)
14. D. Kokoszyński, R. Wasilewski, K. Stęczny, Z. Bernacki, K. Kaczmarek, M. Saleh, P.D. Wasilewski, M. Biegniewska. Europ.Poult.Sci. **79** (2015) DOI: 10.1399/eps.2015.110
15. A.A. Dunn, D. J. Kilpatrick, N.F.S. Gault. Br. Poult. Sci. **36**, 401–413 (1995)

Effect of calcium precursors on pelletized property and cyclic CO_2 capture performance

Suwimol Wongsakulphasatch [1], *Krasanai* Sukchoknamchai [2], *Natthapong* Suthapot [2], *Wanichaya* Praikaew [2], *Choowong* Chaisuk [2], *Worapon* Kiatkittipong [2,*], *Navadol* Laosiripojana [3] *and Suttichai* Assabumrungrat [4]

[1] *Department of Chemical Engineering, Faculty of Engineering, King Mongkut's University of Technology North Bangkok, Bangkok 10800, Thailand*
[2] *Department of Chemical Engineering, Faculty of Engineering and Industrial Technology, Silpakorn University, Nakhon Pathom 73000, Thailand*
[3] *The Joint Graduate School of Energy and Environment, King Mongkut's University of Technology Thonburi, Bangkok 10140, Thailand*
[4] *Center of Excellence in Catalysis and Catalytic Reaction Engineering, Department of Chemical Engineering, Faculty of Engineering, Chulalongkorn University, Bangkok10330, Thailand*

Abstract. CaO is widely used for capturing CO_2 in large scale application however the work on the manufacture of CaO based pellets for the application of CO_2 capture is relatively limited. In this research investigated the effect of calcium precursors ($Ca(OH)_2$, $Ca(NO_3)_2$ and $Ca(CH_3COO)_2$) and their synthesis method (sol-mixing and wet-mixing) on pelletization. Adsorbent synthesized from $Ca(CH_3COO)_2$ as a precursor using sol-mixing pelletize to porous spherical provide the best mechanical strength and CO_2 adsorption capacity 0.51 and 0.41 g CO_2/ g CaO at first cycle and tenth cycle respectively. The outstanding cyclic performance of $Ca(CH_3COO)_2$ sol-mixing can be ascribed to the presence of inert support material i.e. $Ca_9Al_6O_{18}$ and $Ca_{12}Al_{14}O_{33}$ with suitable pore volume to prevent sintering of CaO and structure collapsing at high temperature.

1 Introduction

A dramatic increase of carbon dioxide, a known greenhouse gas, in atmosphere is a worldwide environmentally concern. Calcium looping is one of potential technologies for the separation of CO_2 from flue gas mixture. Moreover, it can be applied as a sorbent in sorption enhanced reforming for H_2 production [1]. The CO_2 capture reaction for CaO adsorbent is occurred by the reversible reaction between CaO and CO_2, named carbonation-calcination reaction [2]

$$CaO(s) + CO_2(g) \leftrightarrow CaCO_3(s) \quad \Delta H^\circ = \text{-178 kJ mol}^{-1}$$

Such a calcium looping process involves in multi-cycles carbonation-calcination; however, the makeup flow of fresh sorbent in calcium looping is still required due to the deactivation of it lead to higher operating cost. Consequently, several technologies have been explored for enhancing the CO_2 capture activity of the sorbents, such as steam or water hydration [3-4], incorporation of metal ion using wet mixing method , sol–gel process [5], flame spray pyrolysis technology [6], co-precipitation[7,8]

It is worth to note that, no matter what method was employed, the type of calcium precursor used plays a significant role in the carbonation-calcination performance. The dependent on the type of calcium precursor used is one of interested in this research.

On the other side, because CaO particles are soft and are able to erode quickly in fluidized bed, mass loss is therefore another problem when using a calcination reactor operating in fluidization mode [9]. Also the pelletization has outgrowth in reducing pressure drop more than the powder sorbent when applied in fixed bed adsorpter, which is of more interest for the use in industry; however the work on the manufacture of CaO based pellets for the application of high-temperature CO_2 capture is relatively limited.

It has been reported that the pellet sorbent can increase the mechanical strength for sorbent [10-11]. Moreover, Wu et al. [12] evaluated the use of $Ca(OH)_2$ as a precursor of Ca-base pellet sorbent for the application of CO_2 capture. Experimental results show that $Ca(OH)_2$ precursor provides better adsorption capacity than CaO precursor and $CaCO_3$ precursor, of which 10.7 mol CO_2/kg of adsorbent can be obtained at temperature of 650°C and at CO_2 partial pressure of 0.4 atm. This is due to the decomposition of $Ca(OH)_2$ which release water creates more surface area. Ridha et al

* Worapon Kiatkittipong: kiatkittipong_w@su.ac.th

studied the effect of pellet size on CO_2 uptake capacity. Natural kaolin and $Al(OH)_3$ obtained from acid leaching of kaolin were employed as binders. They found the smaller pellets have higher ability to capture about 20% more CO_2 than the larger pellets in the first cycles. However, after a large number of cycles, the effect becomes insignificant. In addition, the pellet with $Al(OH)_3$ binder has higher thermal stability than that without binder [13].

In this study, a dependent of calcium precursor on pelletization and their carbonation/calcination performance are investigated.

2. Materials and Methods

2.1 Chemicals

Calcium hydroxide ($Ca(OH)_2$), calcium acetate ($Ca(CH_3COO)_2$), calcium nitrate ($Ca(NO_3)_2$), and calcium oxide (CaO) were used as calcium precursors. Aluminium nitrate ($Al(NO_3)_3$) was used as aluminium precursor. Deionized water and 2-propanol were used as solvent. All chemicals were used as received.

2.2 Preparation of sorbents

Two techniques of mixing method were used to prepare sorbents: sol-mixing and wet-mixing. The synthesis procedure for sol-mixing technique, briefly, calcium precursor was firstly decomposed to pure CaO by calcination in the furnace; at 850°C for calcium acetate precursor and at 900°C for calcium hydroxide precursor. Then CaO was mixed with aluminium precursor in the solution of propanol and stirred for 1 h at 75°C. The reaction was aged for overnight and then recovered by drying in an oven at 120°C for 12 h followed by calcination at 900°C for 1.5 h. In the case of wet-mixing, the solution of calcium and aluminium were mixed and stirred at 75°C for 1 h. The solution was then dried in an oven at 120°C for 12 h, dry white powder was obtained. The obtained material was calcined at 900°C for 1.5 h to form CaO. The ratio of calcium to aluminium was calculated to be 75/25 wt% for all samples. In this work, we denoted the synthetic materials prepared from wet-mixing as –w and from sol-mixing as –s, for example, the adsorbent prepared from wet-mixing method using calcium hydroxide as a precursor was denoted as $Ca(OH)_2$-w. Pellet-type adsorbents were prepared by mixing water to the powder of CaO-based sorbent with the ratio of 1.5 g/g of sorbent powder. Pellet was prepared by gradually dropping water in powder sorbent and stirred until a gel was obtained. Then the gel was subjected to molding and pressed until a stiff consistency is reached. After shaping, the samples were dried at room temperature for 24 h.

2.3 Adsorbent characterization

Crystallinity of synthetic adsorbents were characterized by powder X-ray diffraction (XRD). The spectra were recorded on a Rigaku, Miniflex II desktop X-ray diffractometer using Cu Kα radiation in the 2θ range of 10-80°C at room temperature. Structure and morphologies of the adsorbents were imaged by scanning electron microscopy (SEM, JEOL JSM-5800 LV). The surface properties, including BET surface area, average pore size, and pore size distribution, were determined by N_2 adsorption/desorption at 77 K, provided by BEL SORP mini II.

2.4 CO_2 adsorption test

CO_2 adsorption capacity was performed by introducing the CaO pellets in a vertical alumina tube placed in the furnace at constant carbonation temperature of 450°C. CO_2 gas was flowed into the reactor at 10 bar, the weight and pressure changes were monitored until it became constant. Regeneration of CaO was performed by heating the sorbent to calcination temperature of 900°C for 1.5 h.

3. Results and Discussion

3.1 Effect of calcium precursor on sorbent characteristic

The synthetic adsorbents were characterized by XRD as shown in Fig.1. All the samples are composed of CaO, $Ca_{12}Al_{14}O_{33}$, and $Ca_9Al_6O_{18}$ with different compositions depend upon calcium precursor and synthesis method.

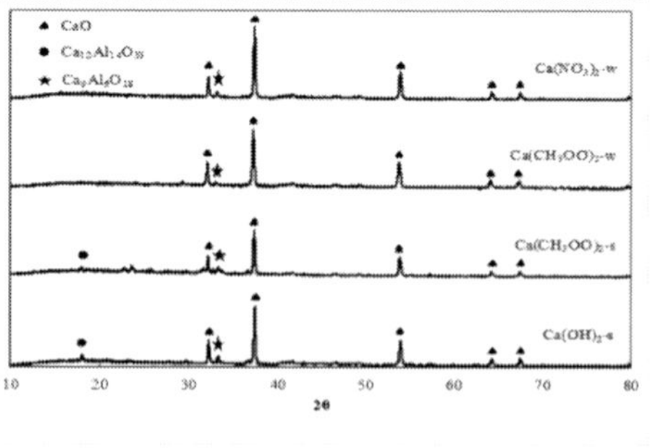

Fig.1: XRD pattern of synthetic CaO-based sorbents.

Fig.2 presents SEM images of the adsorbents CaO-based alumina. The adsorbents obtained by the use of $Ca(CH_3COO)_2$ as calcium precursor has the smallest size of crystal where the others aggregated to larger particle. Moreover comparing between wet-mixing and sol-mixing of $Ca(CH_3COO)_2$ precursor, the wet-mixing exhibit comparatively lower particle size. N_2 adsorption/desorption isotherm measurement was conducted to determine surface properties texture of the sorbents as summarized in Table 1. Two sorbents, $Ca(OH)_2$-s and $Ca(NO_3)_2$-w have low surface area and are similar to that of commercial CaO, whereas, $Ca(CH_3COO)_2$-w and $Ca(CH_3COO)_2$-s offer distinctive high surface area than others. The significantly increased of specific surface area is in consistent with the results of SEM as smallest particle of the sorbent obtained from calcium acetate precursor via wet-mixing method (Fig. 2b). In addition, using wet-mixing in $Ca(CH_3COO)_2$-w increased their pore volume remarkably comparing to the sol-mixing of $Ca(CH_3COO)_2$-s (Table 1).

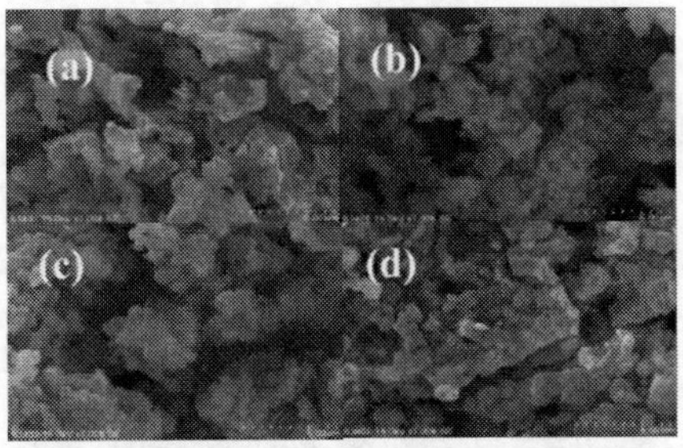

Fig.2: SEM images of fresh sorbent sample (a) $Ca(NO_3)_2$-w (b) $Ca(CH_3COO)_2$-w, (c) $Ca(CH_3COO)_2$-s, (d) $Ca(OH)_2$-s

Table 1: Surface properties of the synthesized materials

Adsorbent	S_{BET} ($m^2\,g^{-1}$)	Total pore volume ($cm^3\,g^{-1}$)
Commercial CaO	5.6	0.0384
$Ca(NO_3)_2$ wet-mixing	4.5	0.0018
$Ca(CH_3COO)_2$ wet-mixing	16.6	0.2284
$Ca(CH_3COO)_2$ sol-mixing	13.1	0.0525
$Ca(OH)_2$ sol-mixing	7.4	0.0463

3.2 Pelletization of adsorbent

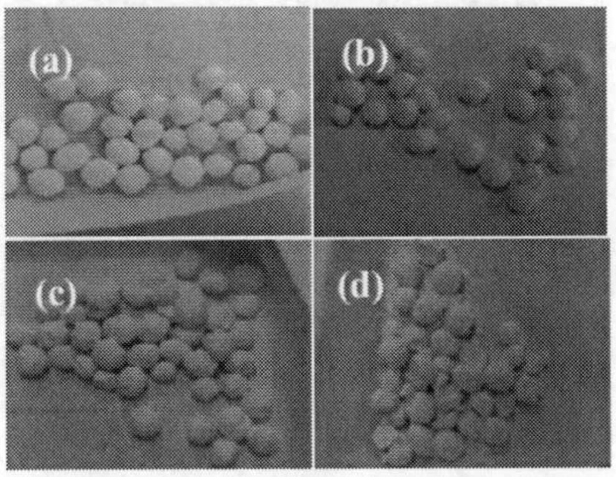

Fig. 3: Photographs of spherical pellet after exposed at 900°C for 1 hr: (a) $Ca(NO_3)_2$-w, (b) $Ca(CH_3COO)_2$-w, (c) $Ca(CH_3COO)_2$-s, (d) $Ca(OH)_2$-s

Thermal stability of the sorbents in pellet form has been examined before performing high temperature CO_2 adsorption test. Thermal stability test has been conducted at 900°C similar to the temperature where calcination reaction takes place. Fig. 3 shows thermal resistance of spherical pellets of different preparation techniques. The results show that CaO-based alumina in spherical shape, which was derived from $Ca(OH)_2$ cracked (Fig. 3d) while others can retain their shape and structure (Fig. 3a-c)

3.3 CO_2 adsorption performance

Since the mechanical property of $Ca(OH)_2$-s was unsuitable, the CO_2 adsorption performance test of $Ca(OH)_2$-s was neglected in this section. The value of CO_2 sorption capacity of different pellet-type sorbent are shown in Fig. 4. $Ca(CH_3COO)_2$-w and $Ca(CH_3COO)_2$-s exhibit higher CO_2 sorption capacity than that of commercial CaO (the values are 0.58, 0.51 and 0.42 g CO_2/g CaO, rescpectively), while $Ca(NO_3)_2$ had the lowest sorption capacity (0.22 g CO_2 / g CaO). Although the understanding is not straightforward, the much lower in total pore volume of $Ca(NO_3)_2$-w might be a reason of a much lower CO_2 sorption capacity while the surface area also a dominant parameters of the sorption capacity as observed from the tendency of others sorbent. The cyclic carbonation/calcination performance are further tested for three sorbents except $Ca(NO_3)_2$-w.

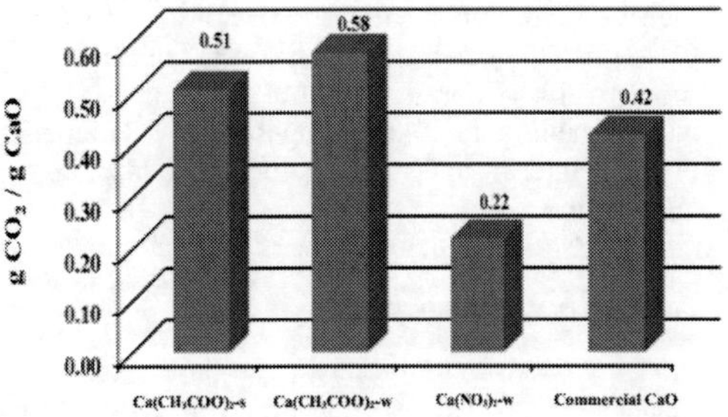

As shown in Fig. 5, although $Ca(CH_3COO)_2$-w can exhibit the highest sorption capacity in the first cycle, the capacity drastically decreased in the 2nd and 3rd cycle. Generally, the presence of $Ca_9Al_6O_{18}$ and $Ca_{12}Al_{14}O_{33}$ inert support material effectively prevents sintering of CaO to some extent during regeneration of sorbent at high temperature. Although, $Ca(CH_3COO)_2$-w consists of $Ca_9Al_6O_{18}$ inert materials acting as structural supports, a very large pore volume of $Ca(CH_3COO)_2$-w are hypothesized as a reason for structure collapse after only three cycle carbonation/calcination. This undesirable loss of pellet strength during carbonation/calcination cycles can be explained due to the presence of cracks and/or structure collapsing in the $Ca(CH_3COO)_2$-w and commercial CaO after three cycles.

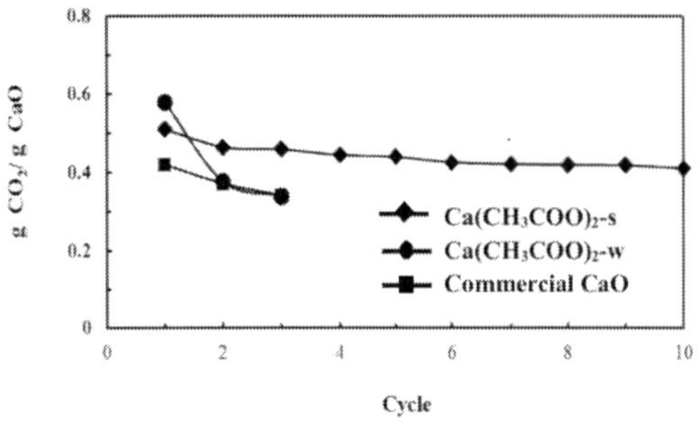

Fig 5: Show adsorption capacities of pellet sphere sorbents during carbonation/calcination cycles

4. Conclusion

A screening of potential calcium precursors and their synthesis method for pelletization of CaO-based sorbents for CO_2 capture was conducted in this work. Among calcium precursors including $Ca(OH)_2$, $Ca(NO_3)_2$ and $Ca(CH_3COO)_2$), CaO derived from $Ca(CH_3COO)_2$ precursors using both wet- and sol-mixing methods exhibit higher CO_2 sorption capacity than that of commercial CaO. However, the pelletized CaO from $Ca(CH_3COO)_2$ wet-mixing and commercial CaO cannot retain their structure after carbonation/calcination for 3 cycles lead to a drastic loss in sorption capacity. $Ca(CH_3COO)_2$ sol-mixing exhibit the most carbonation/calcination stability with 0.51 and 0.41 g CO_2/ g CaO at first cycle and tenth cycle respectively. The presence of $Ca_9A_{16}O_{18}$ and $Ca_{12}Al_{14}O_{33}$ with suitable pore volume might be key parameters to prevent sintering of CaO and structure collapsing at high temperature.

Acknowledgment

Financial supported by Silpakorn University Research and Development Institute (SURDI) is gratefully acknowledged.

References

1. Udomchoke, T., et al., *Performance evaluation of sorption enhanced chemical-looping reforming for hydrogen production from biomass with modification of catalyst and sorbent regeneration.* Chemical Engineering Journal, (2016). **303**: p. 338-347.
2. Van Selow, E., et al., *Carbon capture by sorption-enhanced water– gas shift reaction process using hydrotalcite-based material.* Industrial & Engineering Chemistry Research, (2009). **48**: p. 4184-4193.
3. Phromprasit, J., et al., *Activity and stability performance of multifunctional catalyst(Ni/CaOandNi/$Ca_{12}Al_{14}O_{33}$CaO) for bio-hydrogen production from sorption enhanced biogas steam reforming.* International Journal of Hydrogen Energy, (2016). **41**: p. 7318-7331.
4. Phromprasit, J., et al., *H_2 production from sorption enhanced steam reforming of biogas using multifunctional catalysts of Ni over Zr-, Ce-and La-modified CaO sorbents.* Chemical Engineering Journal, (2017). **313**: p. 1415-1425.
5. Akgsornpeak, A., et al., *Development of synthetic CaO sorbents via CTAB-assisted sol–gel method for CO_2 capture at high temperature.* Chemical Engineering Journal, (2014). **237**: p. 189-198.
6. Lu, H., et al., *Flame-made durable doped-CaO nanosorbents for CO_2 capture.* Energy & Fuels, (2008). **23**: p. 1093-1100.
7. Kierzkowska, A.M. and C.R. Müller, *Development of calcium-based, copper-functionalised CO_2 sorbents to integrate chemical looping combustion into calcium looping.* Energy & Environmental Science, (2012). **5**: p. 6061-6065.
8. López-Periago, A., et al., *CO_2 capture efficiency and carbonation/calcination kinetics of micro and nanosized particles of supercritically precipitated calcium carbonate.* Chemical engineering journal, (2013). **226**: p. 357-366.
9. Anthony, E., *Ca looping technology: current status, developments and future directions.* Greenhouse Gases: Science and Technology, (2011). **1**: p. 36-47.
10. Manovic, V., E.J. Anthony, and D. Loncarevic, *CO_2 looping cycles with CaO-based sorbent pretreated in CO_2 at high temperature.* Chemical Engineering Science, (2009). **64**: p. 3236-3245.
11. Manovic, V. and E.J. Anthony, *CaO-based pellets with oxygen carriers and catalysts.* Energy & Fuels, (2011). **25**: p. 4846-4853.
12. Wu, S., et al., *Properties of Ca-base CO_2 sorbent using $Ca(OH)_2$ as precursor.* Industrial & engineering chemistry research, (2007). **46**: p. 7896-7899.
13. Ridha, F.N., et al., *High-temperature CO_2 capture cycles for CaO-based pellets with kaolin-based binders.* International Journal of Greenhouse Gas Control, (2012). **6**: p. 164-170.

Study on wind energy potential for agricultural water pumping system in the middle part of Thailand

Chamlong Prabkeao and *Akapot* Tantrapiwat*

Department of Mechanical Engineering, Faculty of Engineering, King Mongkut's Institute of Technology Ladkrabang, Bangkok, Thailand

Abstract. A study on wind powered water pumping system aimed for agriculture was carried out in the middle part of Thailand. In this alluvial plain, wind energy potential was determined by making a survey on 21 observation sites. The survey was made in a period of one year, and it has shown that this region locates in a clam climate zone with average wind speeds at about 2 m/s. A wind turbine-water pumping system was installed and evaluated for its performance and efficiency. The result has shown a linear relationship between water discharge capability and the wind speeds. Due to the type of turbine and low wind speed in this region, the system efficiency turned out to be minimal, yet it was practical because the wind power was free. A simple cost analysis from the survey data also has shown that using a wind turbine in this region will be worthwhile when it can be operated for about two decades.

1 Introduction

Among all renewable energy resources, the wind energy is clean and simple to be harvested. It has been used for many centuries, from mainly agricultural applications in the past to the industrial and household usage as in today. A report in 2011 from the Global Wind Energy Council [1] has told that the electricity generated by wind energy in 2010 had increased 35,802 MW which was 22.5% of the total 194,390 MW electricity made by wind energy in the world. Half of these were produced in China where the economy was blooming as well as the need of energy resources. In other countries such as India, Japan, South Korea and Taiwan, the use of wind power was also increasing every year. These made Asia becoming one of the growing regions in wind energy usage with the capacity of 19,022MW in 2010.

Besides harvesting it to produce electricity, wind energy was often gained directly for specific tasks such as irrigations in farming, salt pond, and livestock. Couple decades ago Smulders [2] had presented water pumping system for agriculture using wind energy in developing countries. By designing a wind turbine incorporated with water pump configuration, wind energy can be used efficiently and affordable for cultivation. Several years later Mohsen and Akash [3] studied the possibility of using such systems in different areas of Jordan. The study could pinpoint 11 potential regions for using wind energy irrigation. However the availability of wind energy did not tell that it was worthwhile to be harvested. Al-Suleimani and Rao [4] did the efficiency test for wind turbine-water pump at different levels of wind speeds and the amount of water obtained. For certain region such as Adrar in Algeria, Bouzidi [5] had done a test to show that water pumping by wind energy was practical and cheaper in comparison to the use of solar cells. Although several researches had confirmed the potential use of wind energy in irrigation, it was important to evaluate the availability of wind energy for any specific region.

Despite all of its advantages, the biggest limitation of wind energy is its inconsistency which is also varied between regions and seasons. However the variation of the wind power is not so crucial in some applications such as agricultural irrigation for certain types of farm. Because some plants and crops do not require frequent irrigation, water pumping via available wind power can be sufficient. In 2007 Saudi Arabia, a team of researchers led by Rehman [7] was able to utilize wind power for irrigation in a remote area using small wind turbines. Another fact about farming is that a good cultivation area usually locates near natural water resources, and should be a suitable living habitat. The location is often in a calm climate region which is not windy, and yet the low speed wind turbine can be used.

This research was carried out with the aim to evaluate the wind power energy for water pumping system in geological lowland region in the middle part of Thailand. This large alluvial plain is surrounded by four main rivers i.e., Chao Phraya, Tha Chin, Mae Klong, and Bang Pakong rivers. It is known as one of the best place for cultivation in South East Asia and locates the largest rice farming area in Thailand. In 2011, a similar work was done at northeast of Thailand. Poonnoi and Tangchaichit have conducted a wind energy survey [6] at high altitude in a period of one year. They found that the average wind speed in that region was 4.0-4.5 m/s

* Corresponding author: akapot.ta@kmitl.ac.th

 https://doi.org/10.1051/matecconf/201819203058

2 The wind energy survey

This wind energy survey was simple and straight forward. By installing anemometers with data-loggers on the target sites, the wind speeds can be recorded. Due to the large size of target area and the long period of survey, this research turned out to be a tedious work and costly. Fortunately it was supported by the Department of Alternative Energy Development and Efficiency, under the Ministry of Energy, Thailand. With the aim to collect wind energy potential data in the prominent cultivation area of Thailand, the outcome can be useful for the future development in both agriculture and energy conservation.

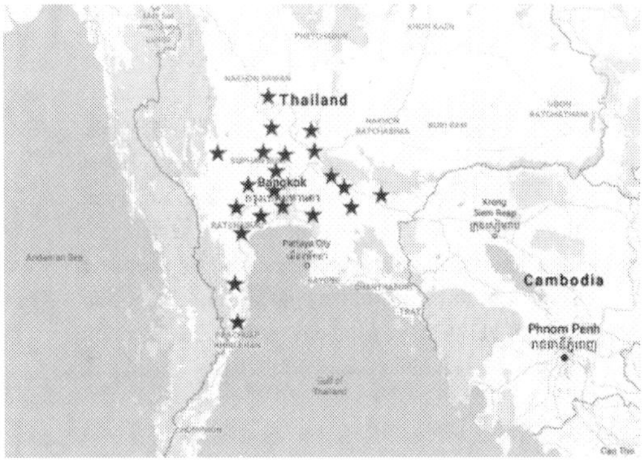

Fig. 1. The 21 survey sites in the middle part of Thailand

2.1. Locations and data record

The win energy survey was conducted in the middle part of Thailand. There were 21 survey sites that cover 21 provinces where agriculture is the main occupation of the residents. Figure 1 shows the distribution of the observation sites. These locations were selected based on several factors which include district zoning, accessible road, and facilities. Also the availability of nearby water reservoirs was taken into the considerations. After all sites were selected, the wind speed measurement and record equipment were installed in every location.

In each observation site, wind speed and direction of wind were measured using anemometer, Winlog from Rainwise, with the wind speed range between 0-60 m/s and the accuracy within ±2%. It was mounted on a tall column and installed at the height of 12 meters. Theoretically the taller wind turbine the more power can be obtained. However large and tall wind turbine tower would require too expensive construction, and in this case, it is not practical. At this height, the wind speed and direction can be observed with insignificant disruption from nearby trees and resident houses. The anemometer can rotate freely for 360 degree, and it was equipped with a digital data-logger with the capability to record with 10 minutes logging intervals for one year on three AA-batteries. The data can be downloaded via USB port into a computer for analysis.

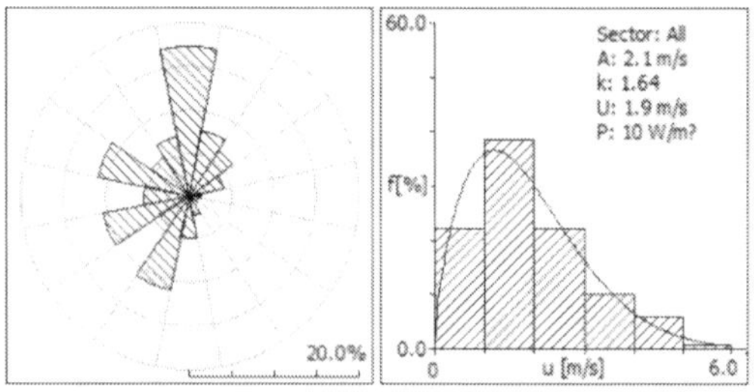

Fig. 2. The sector-wise and Weibull distributions of wind directions and speeds in Chachoengsao observation site.

2.2 Wind power statistic

The measurement of wind speeds and directions were collected for a period of one year from September 2015 to August 2016. This duration began at the end of raining season and was followed by a short period of winter. After several months of summer in the first half of 2016, rains had started, and the data record was stop at the end of that raining season. Although there might be a slight shift of seasons, during the survey in 2015-2016 it was considered a typical climate in that period.

Once the data were collected from all 21 sites, they were analysed into Observed Wind Climate Report (OWC) using a program called WAsP version 8.1 which is developed by Riso National Laboratory, Denmark. The software can present data in different forms. The formats used in this research include:

- Weibull distribution of wind speeds in the processed time-series.
- Sector-wise distribution of winds in the processed time-series
- Diurnal variability of wind speed for 24 hours
- Annual variability of wind speed

The Weibull distribution tells the most often wind speed can be observed and the variation of other less often data set. The sector-wise distribution or wind rose diagram tells the distribution of wind frequent directions. Figure 2 has shown an example of both wind rose and Weibull distribution in one year data which belong to the Chachoengsao observation site.

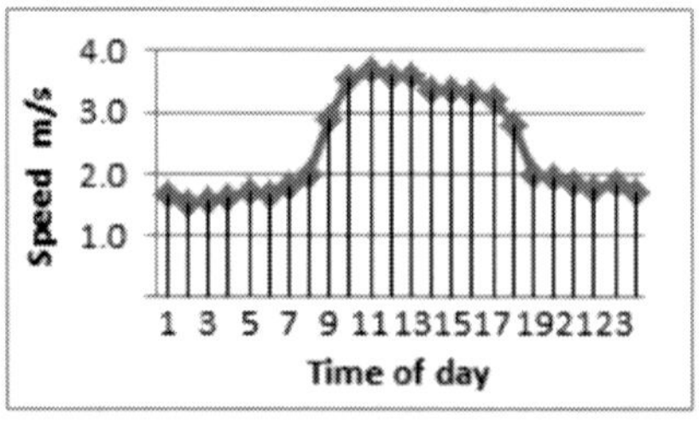

Fig. 3. The example of diurnal variability of wind speed in April 2016 taken from Chachoengsao observation site

In this survey, there were huge amounts of data to be analysed. It was more convenient to average these overwhelm information in order to represent a data set collaborated with a certain domain. For example, the wind speeds for each hour of any day within a month were averaged and represented a wind speed of that particular hour for the entire month. When combining all data from all hours, a diurnal variability chart can be presented as shown in figure 3, an example wind speed in April 2016 from Chachoengsao observation site. This chart can tell the distribution of average wind speed within each day of a month. Still, there were many observation sites, and the data cannot be presented all in here.

To look at the overall wind energy potential for the entire region, Figure 4 presents the statistical mode of wind speeds that were recorded in the duration of one year for the 21 observation sites. It can be seen that the most often wind speeds for all sites were not significantly different. The average most often wind speed for the entire region was 1.93 m/s. This value was considered small but not too much out of our expectation. As mentioned early, most of plentiful cultivation areas usually locate at a calm climate zone. Fortunately low speed wind turbine was practical for the agriculture water pumping application.

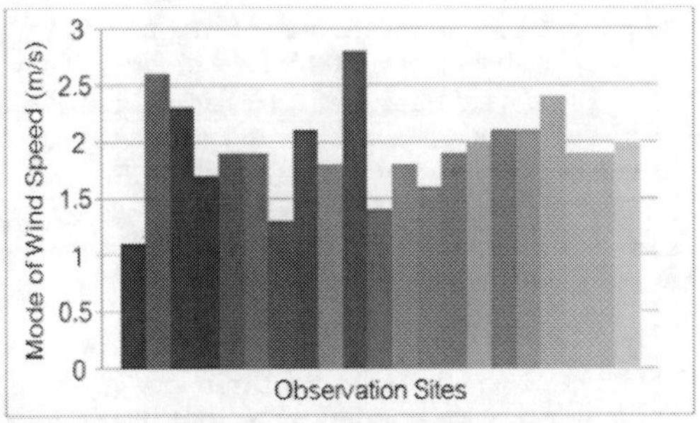

Fig. 4. Statistical mode of wind speeds for all observation sites

3 The water pumping system

Because the Based on the wind energy survey data, it was clear that the target region has an average low speed wind throughout the year. Nonetheless the research was to study the potential of using wind power for irrigation in typical cultivation region. At this minimal wind speed, Multi-Blade wind turbine was selected to be installed on a pilot site. There were two main reasons i.e., the Multi-Blade wind turbine is capable of working at low speed wind, and lower cost in comparison to other types of wind turbine. The wind turbine has 30 blades with the diameter 4.2 meters. It was installed at the height of 12 meter same as the anemometer height during the survey. The wind turbine was directly connected with a piston pump via a set of gears, crank shaft, and rods. The piston pump has the diameter at 4.5 inches and the maximum stroke of 7 inches.

Fig. 5. The wind turbine-water pumping system at the pilot site in Ayutthaya province

Using data from the survey, the site in Ayutthaya province was selected to be the pilot site used in testing the wind turbine-water pumping system. At this location, the wind energy survey has shown the record in average value for almost all aspects such as Weibull distribution, and the statistical mode of wind speed. Also Ayutthaya is considered a popular cultivation area with a lot of natural water reservoirs. Figure 5 shows the wind turbine-water pumping system at this location.

The wind turbine at this pilot site was able to run from completely stop at about 2.5 m/s wind speed, and it can maintain the operation with system inertia at the wind speeds between 1.5-2 meters depending on the load. The water pump, which was attached to this wind turbine, was able to operate at the suction head about 6 meters, and discharge head approximately 25 meters. Because the water pump may operate at 24 hours, the system had a water tank with an overflow valve as the reservoir. The pumping system was also equipped with flow meter for telling the amount of water obtained in relation to the wind speed.

To determine the system performance, wind turbine and water pump were considered as one device. All the frictions between these two devices were counted as the system loss. Therefor the input power for the system was the wind power which can be determined by:

$$P_i = \frac{1}{2}\rho A V^3 \tag{1}$$

where ρ is the air density at 1.2 kg/m^3, A is the projection area of the wind turbine perpendicular to the turbine spindle axis, and V is the wind speed. For the system output which is the water discharged by the pump, the power can be written as

$$P_o = \frac{\rho_w g H Q * 10^{-3}}{60} \tag{2}$$

where ρ_w is the water density at 1000 kg/m^3, g is the acceleration of gravity, H is the head of pump, and Q is the water flow rate. With both power input and output, the efficiency of the wind turbine-water pumping system can be obtained via:

$$\eta = \frac{\rho_w g H Q * 10^{-3}}{\rho A V^3 * 30} \quad (3)$$

During the experiment, the amounts of water discharged from the pump were recorded in relation to the magnitude of wind speed. This relationship is presented in figure 6, and shown a linear response with the coefficient of determination, R^2 =96.9%. When applying this relationship to the equation stated above, the variation of system efficiency at different wind speed can be obtained as shown in figure 7. It should be noted that the overall efficiency was very low because the input power was determined by the wind force entering the turbine without deducting the wind momentum that leave the system. It is impossible to absorb all kinetic energy from the wind. Also, because the Multi-Blade turbine was designed to work at low speed, this makes the efficiency better at slower wind speed.

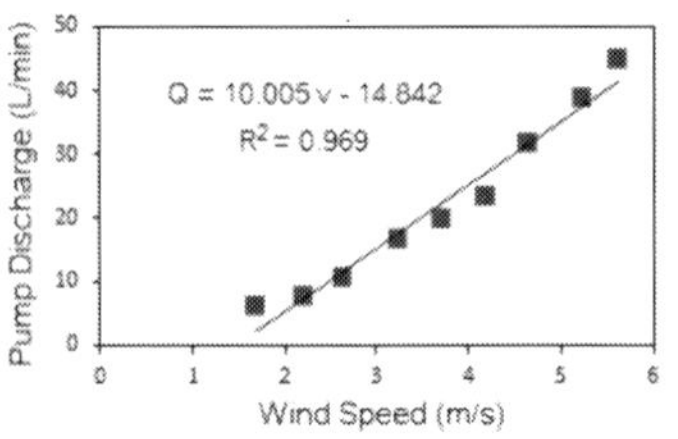

Fig. 6. Relationship between wind speed and pump discharge performance

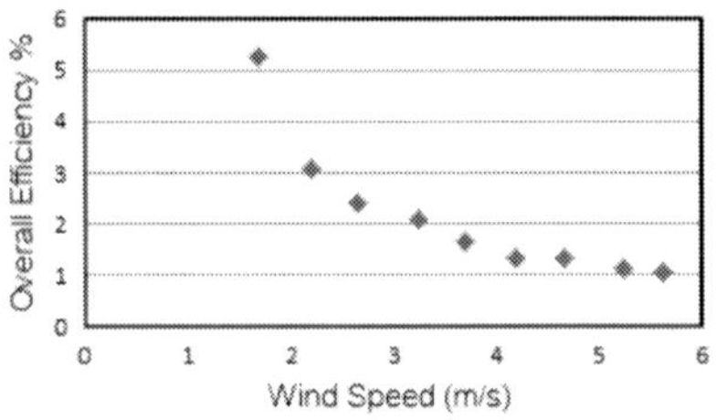

Fig. 7. The overall efficiency of the wind powered water pumping system at different wind speeds

Although the system efficiency was relatively low, the input energy was free. This system can be made economically worthwhile. By making a simple analysis on the investment and operation cost in comparison to other types of power sources which include electrical motor and diesel engine, the result can be seen in table 1. The estimation was made according to the wind data from the survey. At the average wind speed 2 m/s, the wind turbine can pump water approximately 432,000 liter per month. The initial investment for wind turbine is relatively high, and it costs about 200,000 Bahts for the size used in the pilot test site. Fortunately it requires no fuel, and only small annual maintenance is needed. With the same amount of water pumped, the electrical motor and diesel engine only work for 2 hours per day. The engine has lowest initial price because it can be operated everywhere. On the other hand, the electric motor can operate by cheaper power source, but the initial cost is relatively higher than the engine due to the electricity may not be available and needed to be established in the rural area. This estimation was made based on today average equipment and energy price. The investment of wind turbine can be returned sooner if the future energy price goes up as it tends to be.

Table 1. Cost comparison of the water pumping system with different power sources

Power sources	Price (Baht)	Fuel cost per hour (Baht)	Total cost 10 years (Baht)	Total cost 20 years (Baht)
Wind turbine	200,000	-	212,000	224,000
Electrical motor (with wiring)	62,000	8	125,400	208,800
Diesel engine	33,000	20	189,000	345,000

4 Conclusion

The wind energy survey in the middle part of Thailand was successfully carried out. The result has shown that this cultivation region has calm climate throughout the year, and the most often wind speed can be observed at about 2 m/s. At this level of low speed wind, Multi-Blade wind turbine was selected and installed at a pilot site. It was tested and able to evaluate the potential of wind powered water pumping system. Based on the turbine performance and the data from the survey, a cost analysis was created, and it has indicated that the use of wind turbine is worthwhile only when it can operate for about two decades.

This research was funded by the Department of Alternative Energy Development and Efficiency, under the Ministry of Energy, Thailand

References

1. GWEC, *Global Wind Report 2011, the Annual market update,* (2011)
2. P. Smulders, *6th EWEC conf*, 99-105, (1986)
3. M. Mohsen, Ba. Akash, Renew Energ, **14**, 441-446, (1998)
4. Z. Al-Suleimani, Nr. Rao, Appl Energ, **65**, 339-347, (2000)
5. B. Bouzidi, Renew Sust Energ Rev, **15**, 4436-4442, (2011)
6. M. Poomnoi, K. Tangchaichit, KKu Res.J, **18(5)**, 803-810, (2013)
7. S. Rehman, Im. El-Amin, F. Ahmad, Sm. Shaahid, Am. Al-Shehri, Jm. Bakhashwain. Renew Sust Energ Rev, **11(5)**, 937-950, (2007)

Development of an automatic tracking system to determine field efficiency of agricultural machines

Natussapol Tamrai, *Boripun* Masarn, *Apidul* Kaewkabthong, *Ravipat* Lapcharoensuk, *Songvoot* Sangchan, and *Vasu* Udompetaikul*

Department of Agricultural Engineering, Faculty of Engineering, King Mongkut's Institute of Technology Ladkrabang, Bangkok, Thailand

Abstract. Field efficiency of machines tells how efficient the farm machines are operating in the field. Measuring of the field efficiency used to be a tedious and laborious work which is not worth to collect for further operational optimization. The objective of this study was to develop an automatic system for monitoring the field activities and then evaluation of the field efficiency of farm machines. The system consisted of a microcontroller to collect working data including position, speed heading, and working status of the machine. The system was installed on a farm tractor with plowing disc to test on two fields with the same size, but in different traveling directions, i.e., lengthwise and crosswise. The results showed lengthwise operation yielded a higher field efficiency due to less number of turning at headlands. The proposed system allowed to collect necessary information for detailed efficiency evaluation of farm machines. This technique enables further utilization of the operational information and benefit to use in the optimization of the farm works.

1 Introduction

Field efficiency of agricultural machines is the ratio of actual theoretical field time to the total time spent in the field [1]. It tells how efficient the system is due to operating pattern and field physical conditions. In ideal operation, the machine spends all time (theoretical field time) to operate the major task at its optimum forward speed and perform over its full width of action. However, in actual, the machine also spends time in other unproductive time elements such as turning or stopping for removing the obstructions causing to decrease the field efficiency.

Factors affecting field efficiency are the theoretical capacity of the machine, machine maneuverability, field patterns, field shape, field size, crop yield (in harvesting operation), soil and crop conditions, and other system limitations [1]. Optimizing of the field efficiency must consider these parameters. However, in field works, many parameters are not easy to collect accurately due to their high spatial variation.

Traditionally, field efficiency was determined in manual basis. Time used for all turns and all other activities during the operation were manually recorded. Optimum operating speed was measured during the machine working over the tracks that represent the major field condition. Measurement of field efficiency involves tedious and laborious field works. It is not easy to test under all field and crop conditions to get enough information for precise optimization of the operation.

Advances in computing and information technology allow ease of data collection. GNSS (Global Navigation Satellite System) provides real-time location and traveling speed information. It is possible to collect all operating information while the machine is working without any interruption from the researchers. This allows us to get big data from numerous field conditions that might cover various real situations.

In 2015, Vasu et al. installed a GNSS tracking system to a sugarcane harvester to evaluate the time efficiency in mechanized harvesting operation [2]. Later in 2017, Apidul and Vasu used a similar tracking system installed on three cane harvesters with different horsepower to study the effect of field accessibility to the field efficiency of the harvesting operation [3]. Data could be easily obtained with a lot of operational details. However, it was difficult to install sensors to detect the working status (cutting or no-cutting). Without, the cutting status sensor, the turning time could be obtained with less accurate.

The objective of this study was to develop an automatic system, including a working status sensor for the field efficiency determination. The system was deployed in a plowing operation of a tractor with plowing disc installed. This is a first step to collect complete working information to discover the nature of farm machine operations.

* Corresponding author: vasu.ud@kmitl.ac.th

2 Material and Methods

2.1. Field Efficiency Monitoring System

The system used an Arduino MEGA R3 microcontroller system (Italy) as the central unit to collect and record the machine's spatial and operational information. A GNSS Module (U-blox M8N, Switzerland) was used to acquire positioning, speed, and heading information of the machine's travel at 5 Hz acquisition rate (200 ms interval). Also, a digital compass module (Honeywell HMC5883L, USA) was also used to assist the system to recognize traveling direction (forward/ backward) of the machine.

The system was installed on a tractor with plowing discs. An accelerometer (InvenSense MPU-6050, USA) was used as an inclination sensor attaching to the raising arm between the tractor and the implement. This sensor was used to measure the raising angle for indicating whether the arm was at working depth or not. This indicator represents the working status of the operation. The major components of the agricultural machine monitoring system are shown as a diagram in figure 1. Figure 2 shows locations of each component installed on an agricultural tractor.

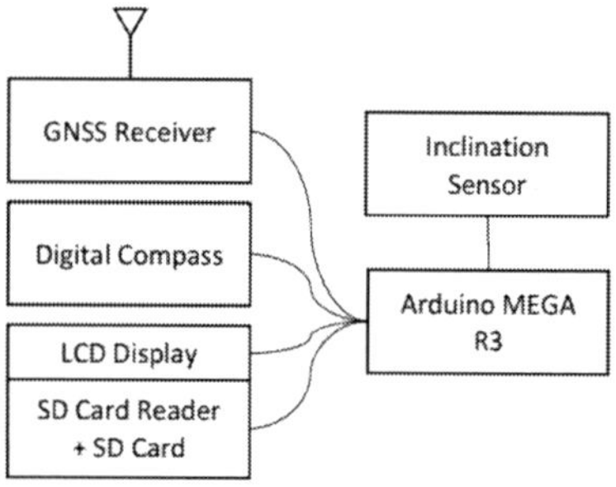

Fig. 1. Major components of the agricultural machine monitoring system.

Fig. 2. The test tractor with the instruments installed.

2.2 Field Test

The system was tested on two plots located in the faculty of Agricultural Technology, KMITL, Bangkok (13°43' 52.1"N 100°47'26.8"E). Both plots had approximately the same size of 0.425 ha (85 m × 50 m). The different operation between 2 plots was the plowing direction, i.e., lengthwise and the crosswise directions of the plots. Theoretically, the first one should result in a better field efficiency due to it has a fewer number of turns.

For each test, the operator performed plowing throughout the field at a depth of approximately 30 cm. Since the headlands were flat, the tractor could make the turns easily outside the field in the fish-tail shape and continue the adjacent rows consecutively.

Total operating time was manually recorded for determining the actual capacity of the operation. Each turning time was also recorded as well. Other related information such as the factory working width of the plowing width (2.2 m) and actual working widths for each pass were also measured and noted.

2.2 Data Analysis

Positioning and moving information from the start to the end of the operation were first plotted using QGIS software to show the trace of the machines in the field. Next, unproductive points were then excluded by following techniques. The backward movement could be detected by comparing the direction of GNSS points and the heading angles from the electronic compass. If they were in opposite directions, the machine was moving backward and must be excluded. The working status recorded by the inclination sensor was used to classify whether it was in productive status or not. An additional step was to check if the machine was not moving by verifying if the speed (from the GNSS) was slower than 0.5 km/h (in stationary, GNSS generated a small speed value due to its positioning calculation).

Mode or the most frequent value of the speed from all productive points was used as the representation of the optimum speed. The reason was that it represented the most desirable operating speed under the corresponding field and crop conditions.

The field efficiency could be calculated from the ratio of actual field capacity and the theoretical field capacity. The actual capacity requires field size and the total operation time to calculate. The theoretical capacity is the multiplication of rated width and the optimum operating speed. (Eqns. 1-3)

$$\eta = \frac{C_{act}}{C_{theo}} \quad (1)$$

$$C_{act} = \frac{A}{t_{tot}} \quad (2)$$

$$C_{theo} = \frac{w_r v_{opt}}{10} \quad (3)$$

where η = field efficiency *[in decimal]*,
C_{act} = actual field capacity *[ha/h]*,
C_{theo} = theoretical field efficiency *[ha/h]*,

 https://doi.org/10.1051/matecconf/201819203059

A = working area *[ha]*,
t_{tot} = total working time *[h]*,
w_r = rated width of the implement *[m]*, and
v_{opt} = optimum operating speed *[km/h]*.

3 Results and Discussion

Field parameters and working information for both lengthwise and crosswise directions are described and noted in table 1. Tracing points of the operations in both directions recorded from the system are plotted in figure 3. Light green dots show operation in active status (performed plowing), and red dots show inactive status (turning and backing). The working paths shown in the maps are not equally distributed due to the accuracy of the GNSS receiver. However this misalignment is not affect the determination of field efficiency.

Table 1. Field and working parameters including the calculated capacity and efficiency of the operations in both lengthwise and crosswise directions.

Parameters	Lengthwise	Crosswise	
Area	4250		*sq.m.*
	0.425		*ha*
Plot Length	85	50	*m*
Total Time	21.1	23.3	*min*
	0.351	0.389	*h*
Actual Capacity	1.21	1.09	*ha/h*
Rated Width	2.20		*m*
Actual Width	2.08	2.13	*m*
Optimum Speed	6.84	6.92	*km/h*
Theoretical Capacity	1.51	1.52	*ha/h*
Field Efficiency	80.4	71.8	*%*

Theoretical capacities for both plots were similar due to the ignorance of turning operations. It considers working at the optimum speed with the rated implement width (which is the same). The optimum speed of the operation in the crosswise direction (6.92 km/h) was a little bit higher than that of the lengthwise direction (6.76 km/h). Typically, the machine is set to working speed. If the soil conditions are the same and the field allows the adequate length of operation, the speed should not be significantly different. Noted that the plots were already plowed two weeks earlier, the plowing test could be easily performed with a higher optimum speed comparing to the typical operation. Theoretical field capacities were 1.51 and 1.52 ha/h for lengthwise and crosswise operation, respectively.

In actual, the total time for the crosswise operation (23.3 min) was slightly higher than that of the lengthwise operation (21.1 min). This due to more number of turns to operate in the transverse direction. The actual capacity or the working area (same for both plot) and the actual operation time are 1.21 and 1.09 ha/h for the lengthwise and crosswise operations, respectively.

Field efficiency determined from the recorded data were 80.4% and 71.8% for the longitudinal and transverse operations, respectively. There was no doubt that working in lengthwise would yield higher field efficiency. However, it is not necessarily true. There are many factors affecting the efficiency such as soil condition, field shape, machine maneuverability, etc.

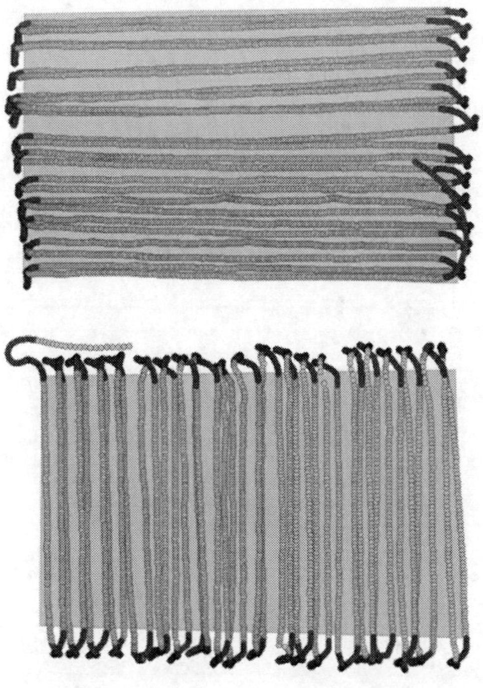

Fig. 3. Tracing points of the operations in lengthwise (upper) and crosswise (lower) directions recorded from the system. Light green dots show operation in active status (performed plowing), and red dots show inactive status (turning and backing).

4 Conclusion

The monitoring system was developed. The center controller recorded the necessary information such as position, speed, heading, and working status of the operations. The tests were performed on two plots with the same size of 0.425 ha. The difference of 2 plots was the working direction, one on lengthwise and another on crosswise directions. Information collected from the system was used to determine the field efficiency. The operation in the longitudinal direction yielded a higher field efficiency due to it contained fewer time losses for turning.

The proposed system shows a possibility of collecting necessary information for detailed efficiency evaluation of farm machines. This technique enables further utilization of the operational information and benefit to use in the optimization of the farm works.

References

1. D. Hunt. *Farm Power and Machinery Management, 10th ed.* Waveland Press, Inc. (2001)
2. V. Udompetaikul, L. Luenam, and C. Chiawchanwattana. Determination of Time Efficiency of Agricultural Machines using GNSS. *The 16th TSAE National Conference, March 17-19, 2015. Bangkok, Thailand.* (2015)
3. A. Kaewkabthong and V. Udompetaikul. Influences of the Field Accessibility to the Performance of Mechanical Sugarcane Harvesting. *The 10th TSAE International Conference, September 7-9, 2017. Bangkok, Thailand.* (2017)

Vibratory bed assisted infrared drying of parboiled rice and its rice bran stability

Pattarachai Vichaiya, *Jiraporn* Sripinyowanich Jongyingcharoen*

Department of Agricultural Engineering, Faculty of Engineering, King Mongkut's Institute of Technology Ladkrabang, Bangkok, Thailand

Abstract. The purposes of this research were divided into two parts: (1) to determine the drying characteristics and quality of parboiled paddy undergoing vibratory bed assisted infrared (VIR) drying, and (2) to study the stability of rice bran from the VIR-dried parboiled paddy. Infrared (IR) drying and hot air (HA) drying were applied for comparison purpose. For the experiment 1, VIR drying of parboiled paddy provided the best drying characteristics with the shortest drying time of 50 min and the maximum drying rate of 0.00146 g water/g dry matter·min. VIR drying also produced the greatest head rice yield of 61.3%. The experiment 2 presented that rice bran from VIR-dried parboiled paddy had the highest oil content of 0.256 ± 0.041 g/g dry matter and the most stable in term of minimum change in free fatty acid (FFA) content. The increase in FFA during storage for 14 days was 0.36%. However, the increases in FFA of rice bran from IR- and HA-dried parboiled rice and normal rice were 0.38, 1.36, and 4.10%, respectively.

1 Introduction

Parboiled rice is ranged in the top five exported rice products of Thailand. Three basic steps of parboiling include soaking, steaming, and drying. Hot air drying is the common drying technique used for parboiling. To improve drying in the parboiling process, VIR drying is considered as a promising technique. IR radiation provides more uniform heating than convective heating. It can reduce moisture gradient and stress development in rice kernels during drying [1]. However, to overcome short penetration depth of IR, mechanical vibration would be useful. Therefore, VIR drying was applied for parboiled paddy drying in this study.

After drying, the parboiled paddy must be subjected to rice milling process to produce parboiled rice. During the milling process, the valuable by-product, i.e. rice bran, is obtained and can be used as a raw material of rice bran oil production. IR radiation has been reported to be useful for rice bran stabilization as well. It can inhibit the reaction of FFA formation by inactivating the lipase enzymes in rice bran and then prolong the shelf life of rice bran oil [2, 3]. However, researches mostly focus on direct IR treatment of rice bran. Report on effect of IR drying of paddy on its rice bran stability is very rare. Short cut the IR treatment of rice bran by applying IR radiation at the step of drying of paddy is of interest.

Therefore, the main aims of this study were to determine drying characteristics of parboiled paddy using VIR drying and to determine head rice yield of parboiled rice and stability of parboiled rice bran as affected by VIR drying of parboiled paddy. For comparison, IR drying and HA drying of parboiled paddy were conducted. Fresh paddy was also employed as a control as well.

2 Materials and methods

2.1 Sample preparation

Paddy of Khao Dawk Mali 105 was from Surin province, Thailand. It was vacuum packed with 5 kg per bag and kept in a refrigerator at 4-5°C. The paddy was allowed to equilibrate to the ambient condition before using in all experiments. The initial moisture content of the paddy was 0.13±0.01 g water/g dry matter. This untreated paddy sample was refered to fresh paddy in this study.

2.2 Preparation of moist parboiled paddy

The paddy was cleaned and soaked in water at a temperature of 55°C for 260 min so that its moisture content was increased to 45 g water/g dry matter [4]. The soaked paddy was then steamed at 100°C for 20 min and tempered at an ambient condition for 10 min before drying.

2.3 Architecture of the dryer used in this study

The dryer used for IR, VIR, and HA drying is presented in Fig. 1. A double-walled stainless steel drying chamber was cylindrical with a diameter and length of 30 and 60 cm. Two infrared heaters of 800 W each were installed inside the chamber at the top. The heaters generated a surface temperature of 750°C with the peak wavelength of 3.1 mm. An aluminum cover was placed over the heaters as an infrared reflector. The perforated stainless

*Corresponding author: jiraporn.jo@kmitl.ac.th

steel drying tray (0.20×0.40×0.05 m^3) was fitted in the chamber with the distance from the heaters at 20 cm. The drying tray was mounted over a vibratory bed arranged to be vibrated vertically by a variable speed motor attached to an eccentric joint. Fin heaters of 4 kW and a blower of 0.5 HP were used for producing hot air. The hot air temperature was measured by thermocouples (type K, Lega model AK-01, Lega Coperation Co., Ltd., Bangkok, Thailand) and controlled by PID (Proportional–Integral–Derivative Controller) with an accuracy of ±1°C.

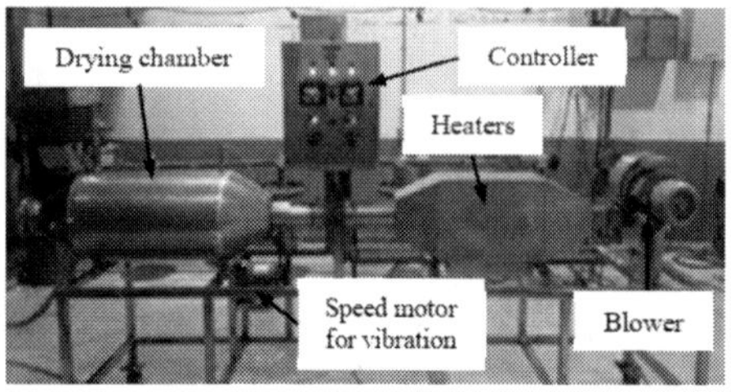

Fig. 1 A dryer used for IR, VIR, and HA drying.

2.4 Experimental procedure

The moist parboiled paddy (200 g) was dried by three drying techniques including IR, VIR, and HA drying. The infrared intensity used in this study was 750 W/m^2 for IR and VIR drying. To supply vibrational mechanism to the drying tray for VIR drying, the vibrational system was run at 500 rpm. HA drying was conducted at 100°C. Air velocity was set at 1 m/s. The parboiled paddy was dried to the final moisture content of 0.13 g water/g dry matter. During the drying process, the paddy was weighed at the predetermined intervals for moisture content determination. Drying characteristics were determined in terms of moisture ratio (MR) and DR as shown in Section 2.5. The dried parboiled paddy was milled and determined for its HRY as presented in Section 2.6. Rice brans obtained after milling these samples were stored and determined for its stability in terms of FFA (see Section 2.7).

2.5 Drying characteristic determination

Moisture content was determined in a hot air oven at 105°C for 24 h [5]. The equation for calculation of the moisture content in dry basis is as follows:

$$M = \frac{W_w - W_d}{W_d} \quad (1)$$

where M, W_w, and W_d are moisture content (g water/g dry matter), wet weight of the material (g), and dry weight of the material (g), respectively.

MR and DR were calculated as below:

$$MR = \frac{M_t - M_e}{M_i - M_e} \quad (2)$$

$$DR = \frac{M_{t+dt} - M_t}{dt} \quad (3)$$

where M_i, M_t, M_e, and M_{t+dt} are moisture content (g water/g dry matter) at initial, specific time, equilibrium, and t+dt, respectively; and t is drying time (min). M_e can be assumed to be zero when drying at high temperature and for radiation drying [6].

2.6 Milling process and head rice yield determination

The dried parboiled paddy of 200 g were dehusked using the testing husker (THU35B, Satake Corporation, Hiroshima, Japan). The resulted brown rice was polished for 3 min using the test mill (TM05, Satake Corporation, Hiroshima, Japan) to obtain milled rice and rice bran. The rice bran was used for the study in Section 2.7. Head rice and broken rice were separated using the rice length grader (TRG, Satake Corporation, Hiroshima, Japan). Head rice was defined as the rice kernel having a length longer than 6/8 of the whole kernel length [7]. HRY was calculated by dividing the mass of head rice over the mass of paddy before milling.

2.7 Storage of rice bran and determination of its stability

Rice bran, the by-product obtained after milling, was packed in a 180 ml polypropylene container and stored at the temperature of 25°C and the relative humidity of 50% approximately. The total storage time was 14 days. The stored rice bran was randomly sampled every two days for oil content and FFA evaluation. Oil contents of the rice bran samples were determined using the solvent extraction, AACC method 30-10 [8]. The extraction solvent was *n*-hexane. The values of oil content were expressed as g oil/g dry matter and reported as grand means. FFA content was determined by AOCS official method Ca 5a-40 [9]. The resulted rice bran oil after *n*-hexane extraction was dissolved in a mixture of ethanol and diethyl ether. The solution was titrated with a KOH/ethanol solution. The values of FFA content were calculated as oleic acid equivalent.

2.8 Statistical analysis

The experiments were triplicated. Error bars in the plots present standard deviation. One-way ANOVA was used for statistical hypothesis testing. Duncan's multiple range test was used to determine the significance of differences between treatments at 95% confident interval.

3 Results and discussion

3.1 Drying characteristics of parboiled paddy

Drying characteristics of parboiled paddy undergoing different drying methods including IR, VIR, and HA drying are presented in terms of drying curves and DR curves as shown in Figs. 2 and 3. In Fig. 2, the drying curve of VIR drying exhibits the most rapid decrease in MR over drying time. To obtain the final moisture content of 0.13 g water/g dry matter, VIR drying required the shortest drying time of 50 min, followed by 60 min and 80 min for IR and HA drying, respectively. DR curves (Fig. 3) also presents that VIR drying resulted in the greatest maximum DR of 0.0146 g water/g dry matter·min while the maximum DR obtained by IR and HA drying were 0.0109 and 0.0078 g water/g dry matter·min. These curves also shows that there were two

DR periods, i.e. heating up and falling rate periods for IR, VIR, and HA drying of parboiled paddy.

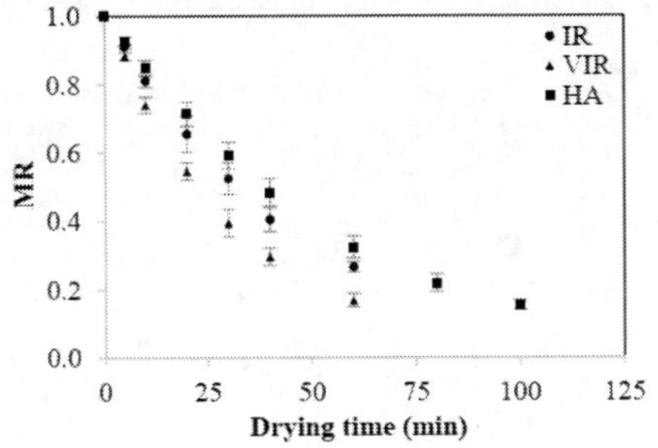

Fig. 2 Drying curves of IR, VIR, and HA drying of parboiled paddy.

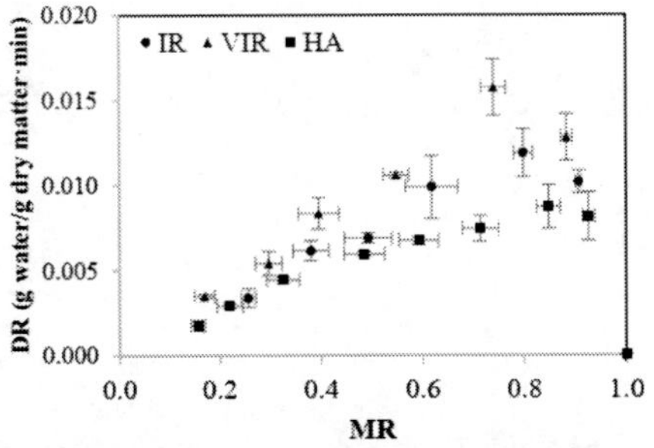

Fig. 3 DR curves of IR, VIR, and HA drying of parboiled paddy.

3.2 Head rice yield of milled parboiled rice

Head rice yield is the important parameter defining the quality of parboiled rice. During parboiling, rice starch is gelatinized, hence milling yield can be increased [4]. Fig. 4 shows the results of HRY of three parboiled milled rice and one untreated milled rice. Statistical difference of HRY among these sample was observed (p<0.05). Athough all of the parboiled mill rice had higher HRY than the fresh rice, non-significant difference was obtained between the HA-dried rice and fresh rice. It should be noted that VIR drying of parboiled paddy resulted in the greatest HRY of the milled sample of about 65%. The second highest HRY was obtained by IR drying of parboiled paddy as well. These results confirmed the benefit of IR radiation which can reduce moisture gradient and stress formation in rice kernels during drying and lead to higher HRY [1, 4].

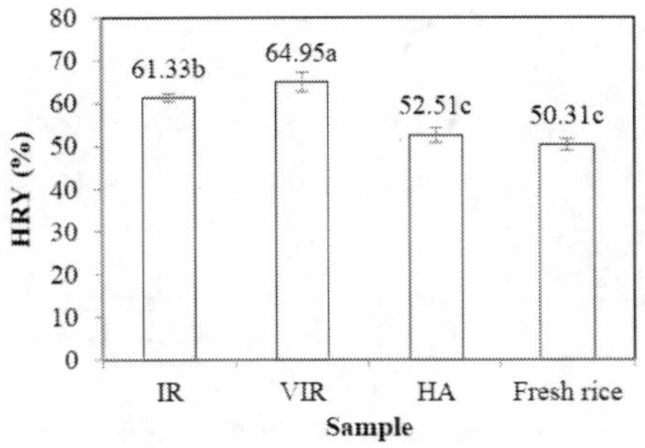

Fig. 4 HRY of milled rice from parboiled paddy undergoing IR, VIR, and HA drying and fresh paddy.

3.3 Oil content and stability of rice bran

Stabilization of rice bran (also known as enzyme inactivation process) is the process widely used to extend the shelf life of rice bran. Rancidity caused by presence of lipases is the key parameter indicating the shelf life of rice bran as it contributes to the development of off-flavor and taste. During polishing, bran layers of rice are removed and individual cells are disrupted. Endogenous lipase enzymes hence cause the hydrolysis neutral bran oil to FFA and glycerol, which lead to the development of hydrolytic rancidity [10, 11].

In this study, rice bran samples obtained from polishing the parboiled brown rice were stored, oil extracted and determined for its stability. It is clear from Fig. 5 that oil content of the rice bran from VIR dried parboiled paddy was the highest. However, the values of oil content from IR, VIR, and HA drying treatments were not significantly different ($p \geq 0.05$). Fig 6 presents the changes in FFA content of rice bran during storage for 14 days. At day 0, FFA content of each rice bran sample was about 0.7% db. FFA content of the rice bran from untreated paddy increased dramatically for the whole period of storage and reached about 5% db at day 14. However, FFA contents of the rice bran from parboiled paddy were relatively stable during storage. HA drying of parboiled paddy led to slowly increase in FFA content of the rice bran and was about 1.7% db at day 14. This Fig obviously presents that FFA contents of the rice bran from IR and VIR dried parboiled rice increased slightly and were about 0.9% db at the last day of storage. Both IR and VIR drying gave the similar result of the best drying techniques applied to parboiled paddy for rice bran stabilization. In addition, these results confirm that we can short cut the IR treatment of rice bran by applying IR radiation at the step of drying of paddy.

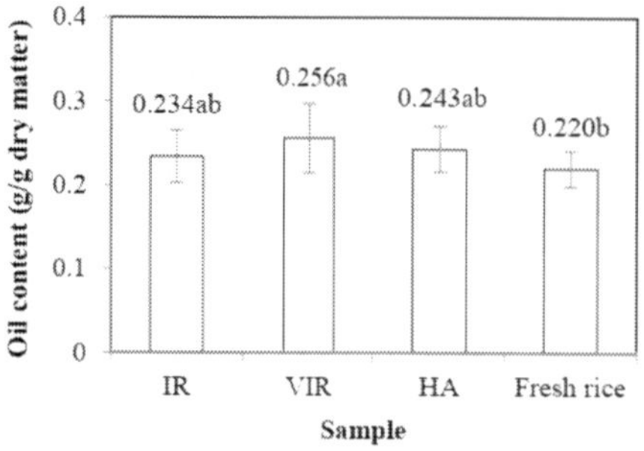

Fig. 5 Oil contents of rice bran from parboiled paddy undergoing IR, VIR, and HA drying and fresh paddy.

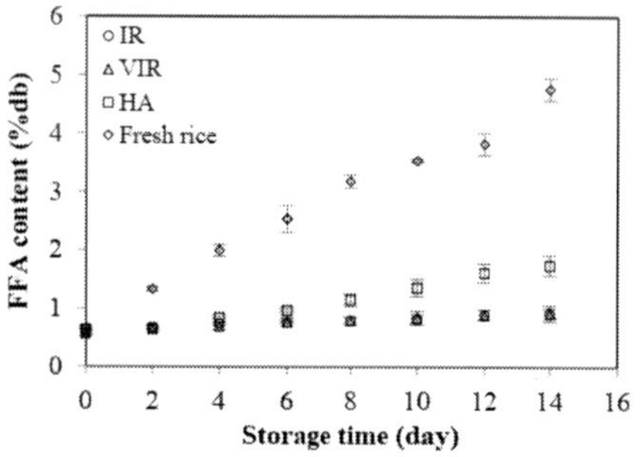

Fig. 6 Changes in FFA content of rice bran from parboiled paddy undergoing IR, VIR, and HA drying and fresh paddy during storage.

4 Conclusions

VIR drying was the most suitable drying technique providing the best drying characteristics of parboiled paddy drying and the best quality of parboiled rice as compared to IR drying and HA drying.

However, rice bran from the VIR, IR, and HA dried parboiled paddy was not significantly different in oil content. Both IR and VIR drying of parboiled paddy gave the similar result of good rice bran oil stability with their FFA consideration.

References

1. R. Khir, Z. Pan, A. Salim, B.R. Hartsough, S. Mohamed, LWT-Food Science and Technology **44**, 1126-1132 (2011)
2. N. Yilmaz, Food Chemistry **190**, 179-185 (2016)
3. T. Wang, R. Khir, Z. Pan, Q. Yuan, LWT-Food Science and Technology **78**, 281-288 (2017)
4. P. Bootkote, S. Soponronnarit, S. Prachayawarakorn, Food Bioprocess Technol **9**, 1574-1586 (2016)
5. AOAC, *Official methods of analysis* (Association of Official Agricultural Chemists, Washington, DC., 1995)
6. C. Contreras, M.E. Martin-Esparza, A. Chiralt, N. Martinez-Navarrete, Journal of Food Engineering **88**, 55-64 (2008)
7. FAO, http://www.fao.org/docrep/x5048e/x5048E1o.gif
8. AACC, *Approved methods of American Association of Cereal Chemists* (The Am. Assoc. Cereal Chem. Inc., St. Paul, MN., 2000)
9. AOCS, *Official methods and recommended practices of the American Oil Chemists Society* (American Oil Chemists Society, Champaign, 2004)
10. K. Gul, B. Yousuf, A.K. Singh, P. Singh, A.A. Wani, Bioactive Carbohydrates and Dietary Fibre **6**, 24-30 (2015)
11. S.M. Kim, H.J. Chung, S.T. Lim, Journal of Cereal Science **60**, 243-248 (2014)

Modeling of hot air drying of coconut residue

Pattawee Wutthigarn, *Jeerayut* Hongwiangjan, *Jiraporn* Sripinyowanich Jongyingcharoen*

Department of Agricultural Engineering, Faculty of Engineering, King Mongkut's Institute of Technology Ladkrabang, Bangkok, Thailand

Abstract. In this study, the effect of drying temperature (50-110°C) on hot air drying characteristics of coconut residue was investigated. The drying time and drying rate (DR) were in the ranges of 540-100 min and 0.0048-0.0182 g water/g dry matter·min at the drying temperature of 50-110°C, respectively. Six drying models (Lewis, Page, Henderson and Pabis, Logarithmic, Midilli et al, and linear-plus-exponential model) were used to determine the change in moisture ratio (MR) with drying time. The linear-plus-exponential model provided best fitting of the predicted MR to the experimental MR with the highest average R^2 of 0.9985 and the lowest *RMSE* of 0.01463. The variation of drying temperature with the constants and coefficient of the model was polynomial type. The generalized linear-plus-exponential model as a function of drying temperature gave best result of prediction of MR with the R^2 of 0.9709.

1 Introduction

Coconut residue is the by-product of coconut milk extraction. It is high in fiber as almost double as the amount found in wheat bran [1]. To date, use of ground coconut residue as a wheat alternative is of interest. In addition to be renowned as high in fiber, coconut residue also considered as a good source of protein with gluten free. Ramaswamy [1] reported that flour from coconut residue can improve digestion, help regulate blood sugar, protect against diabetes, help prevent heart disease and cancer, and aid in weight loss.

Due to the extremely high amount of coconut residue leaving from the coconut milk extraction process of about 40 tons/day and the low drying capacity of the industries, a lot of coconut residue are spoiled each day. To solve the above problem, study on drying of coconut residue should be useful.

In the literature, several drying models are available for many fruits and vegetables. The models can be used to calculate the change in MR with time. It provides important parameters of drying optimization and is good for selection of the ideal drying conditions [2]. This study therefore focused on determining the best drying models for hot air drying of coconut residue.

As drying temperature is considered as the most important parameter of hot air drying, defining the constants or coefficients of the best models as a function of drying temperature is also important. Therefore, the aims of this study were to investigate the effect of drying temperature on drying characteristics of coconut residue undergoing hot air drying and to create the generalized drying model of coconut residue as a function of drying temperature.

2 Materials and methods

2.1 Material

Coconut residue with the initial moisture content of 1.84 g water/g dry matter·min was supplied by the Perfect Companion Group Co., Ltd., Samut Prakarn province, Thailand. The coconut residue (5 kg each) was packed in polyethylene bags and kept in a refrigerator at 4-5°C. It was allowed to equilibrate at room temperature before using in all experiments.

2.2 Experimental procedure

Coconut residue was weighed 200 g and placed as single layer with the thickness of 1 cm on a drying tray. The experiment was carried out using a hot air oven (UF260, Memmert GmbH+Co.KG, Schwabach, Germany) at the temperatures of 50, 70, 90, and 110°C. Air velocity was 1 m/s for all treatments. The final moisture content required for the dried product was 0.03 g water/g dry matter. During drying, the coconut residue was weighed at the predetermined intervals for moisture content determination. Drying experiment at each drying temperature was triplicated and mean with standard deviation was reported herein.

2.3 Drying characteristic determination

Moisture content was determined in a hot air oven at 105°C for 24 h [3] and calculated using the following equation:

$$M = \frac{W_w - W_d}{W_d} \tag{1}$$

where M, W_w, and W_d are moisture content (g water/g dry matter), wet weight of the material (g), and dry weight of the material (g), respectively.

*Corresponding author: jiraporn.jo@kmitl.ac.th

Followings are the questions for calculation of MR and DR:

$$MR = \frac{M_t - M_e}{M_i - M_e} \quad (2)$$

$$DR = \frac{M_{t+dt} - M_t}{dt} \quad (3)$$

where M_i, M_t, M_e, and M_{t+dt} are moisture content (g water/g dry matter) at initial, specific time, equilibrium, and t+dt, respectively; and t is drying time (min).

2.4 Mathematical modeling of drying curves

Five semi-theoretical and empirical drying models (Table 1) were used for fitting with the drying curves of hot air drying of coconut residue. The nonlinear regression analysis for curve fitting was performed and goodness of fit between the models and the experimental data was evaluated using two criteria, including coefficient of determination (R^2) and root mean square error (*RMSE*) as shown in eqs. (4) and (5).

$$R^2 = 1 - \frac{\sum_{i=1}^{N}(MR_{exp,i} - MR_{pre,i})^2}{\sum_{i=1}^{N}(MR_{exp,i} - \overline{MR})^2} \quad (4)$$

$$RMSE = \left[\frac{1}{N}\sum_{i=1}^{N}\left(MR_{pre,i} - MR_{exp,i}\right)^2\right]^{1/2} \quad (5)$$

where $MR_{exp,i}$ and $MR_{pre,i}$ are the ith experimental and predicted MR, respectively; and N is the number of observations.

Table 1 Mathematical drying models applied for hot air drying of coconut residue [4].

Model No.	Model name	Model
1	Lewis	$MR = \exp(-kt)$
2	Page	$MR = \exp(-kt^n)$
3	Henderson and Pabis	$MR = a\exp(-kt)$
4	Logarithmic	$MR = a\exp(-kt) + c$
5	Midilli et al.	$MR = a\exp(-kt^n) + bt$
6	Linear-plus-exponential	$MR = \exp(-kt^n) + bt + c$

To create the generalized drying model, the relationship between constants and coefficients (*coc*) in the best drying model with drying temperature (*T*) was determined by multiple regression analysis using the following equations:

Linear type $\quad coc = a + bT \quad$ (6)

Power type $\quad coc = aT^b \quad$ (7)

Exponential type $\quad coc = a\exp(bT) \quad$ (8)

Arrhenius type $\quad coc = a\exp\left(\frac{-b}{8.314T}\right) \quad$ (9)

Logarithmic type $\quad coc = a + b\ln(T) \quad$ (10)

Polynomial type $\quad coc = a + bT + cT^2 \quad$ (11)

The equation with the greatest R^2 was selected as the best equation for describing the relationship of *coc* and *T*.

The performance of the generalized model was also evaluated using R^2. Drying curves showing comparison between experimental MR and predicted MR calculated from the generalized model were presented as well.

3 Results and discussion

3.1 Drying characteristics

DR curves of hot air drying of coconut residue are illustrated in Figs. 1 and 2. DR of every treatments started to decrease since the first 5 min of drying. The suitable drying times to reach the final moisture content of dried coconut residue (0.03 g water/g dry matter) were 540, 240, 150, and 100 min at the drying temperatures of 50, 70, 90, and 110°C, respectively. Although no constant rate period was observed in these experiments, DR during the falling rate period gradually decreased with MR. The average DR at the drying temperatures of 50, 70, 90, and 110°C were 0.0048, 0.0088, 0.0136, and 0.0182 g water/g dry matter·min, respectively. It is clear from the above results that drying temperature had a considerable effect on drying time and DR of coconut residue. The drying time decreased substantially with increasing drying temperature. The values of DR almost doubled when the drying temperature was increased by 20°C.

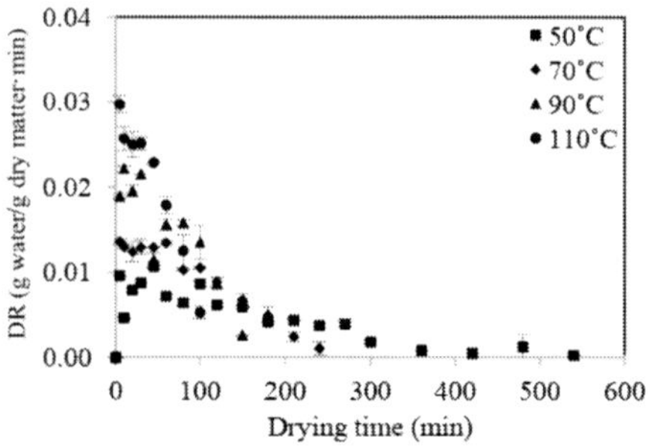

Fig. 1 DR versus drying time of hot air drying of coconut residue at different drying temperatures.

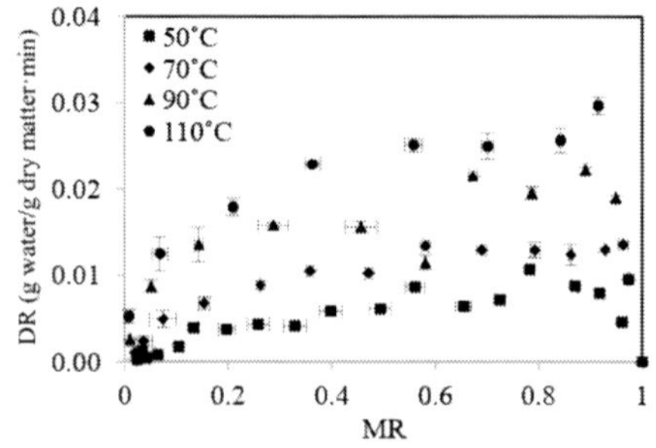

Fig. 2 DR versus MR of hot air drying of coconut residue at different drying temperatures.

Table 2 Constants and coefficients and statistical analysis for the drying models of coconut residue.

Model No.	T(°C)	k	n	c	a	b	R^2	RMSE
1	50	0.0062					0.9928	0.03063
	70	0.0060					0.9927	0.02393
	90	0.0153					0.9696	0.06356
	110	0.0233					0.9737	0.06044
						ave.	**0.9822**	**0.04464**
2	50	0.0024	1.186				0.9993	0.01006
	70	0.0028	1.156				0.9994	0.00697
	90	0.0035	1.356				0.9912	0.03609
	110	0.0060	1.364				0.9960	0.02527
						ave.	**0.9965**	**0.01960**
3	50	0.0065			1.037		0.9952	0.02578
	70	0.0062			1.026		0.9955	0.01959
	90	0.0163			1.049		0.9745	0.06140
	110	0.0247			1.051		0.9788	0.05808
						ave.	**0.9860**	**0.04121**
4	50	0.0058		-0.0537	1.080		0.9970	0.02070
	70	0.0043		-0.2654	1.276		0.9991	0.00927
	90	0.0087		-0.4159	1.427		0.9953	0.02784
	110	0.0143		-0.3435	1.359		0.9974	0.02188
						ave.	**0.9972**	**0.01992**
5	50	0.0021	1.213		0.992	0.000010	0.9993	0.01018
	70	0.0031	1.121		1.000	-0.000146	0.9994	0.00774
	90	0.0048	1.227		0.986	-0.000798	0.9962	0.02695
	110	0.0073	1.269		0.987	-0.000823	0.9991	0.01394
						ave.	**0.9985**	**0.01470**
6	50	0.0021	1.212	-0.0085		0.000026	0.9993	0.01006
	70	0.0031	1.121	-0.0001		-0.000146	0.9994	0.00774
	90	0.0048	1.228	-0.0145		-0.000724	0.9962	0.02688
	110	0.0072	1.268	-0.0137		-0.000719	0.9991	0.01385
						ave.	**0.9985**	**0.01463**

3.2 Modeling of drying curves

Drying curves presenting the changes in MR with time during hot air drying of coconut residue are shown in Fig. 3. As seen in these curves, MR decreased exponentially with drying time. In this study, 6 drying models with exponential terms were used to describe the drying curves. The constants and coefficients of these models for coconut residue undergoing hot air drying at the temperatures of 50, 70, 90, and 110°C are presented in Table 2. The statistical criterion used for model selection, i.e. R^2 and *RMSE*, are also given in the Table with their average values in bold characters. The results show that average R^2 of four models (Page, Logarithmic, Midili et al., and linear-plus-exponential models) were higher than 0.99. The findings are in accordance with the review of Onwude et al. [5], the Lewis, Midilli et al., Page, 2-term, logarithmic, Midified Page, and the approximation of diffusion models were the most suitable models in describing the drying behavior of fruits and vegetables. In this study, Model 6, the linear-plus-exponential model, had the highest average R^2 of 0.9985 and the lowest *RMSE* of 0.01463. Therefore, the linear-plus-exponent model was proposed to be the best model for describing the MR of coconut residue undergoing hot air drying for interval of drying temperature in this study. The linear-plus-exponential model was first introduced for modeling of drying curves by Sripinyowanich and Noomhorm [4]. They reported that this model can provide reasonable trend of moisture loss with time. The typical exponential model like the Page model ultimately decreases the MR value to zero while the linear-plus-exponential model adjusts well to the characteristic of moisture loss during the last falling rate period, the remaining moisture is strongly bound and cannot totally disappear at the end point of drying.

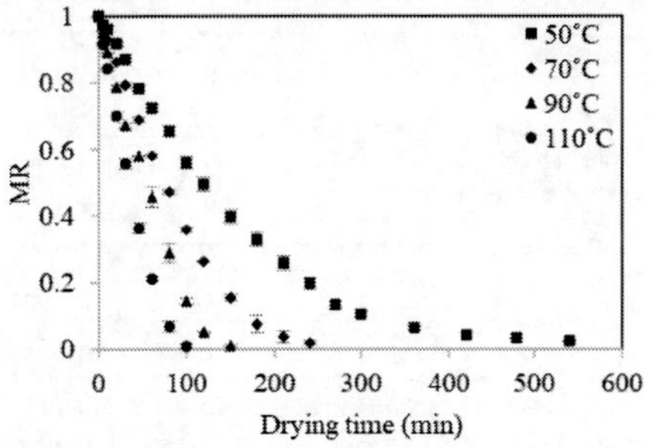

Fig. 3 Drying curves of hot air drying of coconut residue at different drying temperatures.

To combine the effect of drying temperature in the selected models, the constants and coefficients of the models were derived as a function of drying temperature. The polynomial type equation was the best in describing the relationship between the constants and coefficients of the models and the drying temperature. The generalized model with the regression equations of the constants and coefficients of the linear-plus-exponential model against drying temperature is expressed as:

* Corresponding author: jiraporn.jo@kmitl.ac.th

$$MR(b, c, k, n, t) = \exp(-kt^n) + bt + c \quad (12)$$

where

$$k = 0.0030 - 6.4 \times 10^{-4}T + 9.3 \times 10^{-7}T^2 \quad (13)$$

$$n = 1.58 - 0.0116T + 8.2 \times 10^{-5}T^2 \quad (14)$$

$$c = -0.0257 + 6.2 \times 10^{-4}T - 4.8 \times 10^{-6}T^2 \quad (15)$$

$$b = 0.0014 - 3.2 \times 10^{-5}T + 1.1 \times 10^{-7}T^2 \quad (16)$$

This generalized linear-plus-exponential model can be used to estimate the MR of coconut residue subjected to hot air drying at any drying temperature and time in the interval of study with the average R^2 of 0.9709. The drying curves of experimental and predicted MR using the generalized model are illustrated in Fig. 4. Residual plots were also examined to measure the model adequacy (Fig. 5). The residual patterns mostly depict overpredicted values. This means that the predicted MR values were mostly higher than the corresponding experimental MR values. Fig. 6 also presents validation of the generalized model using plots of the experimental MR versus the predicted MR. It is observed that the experimental MR versus the predicted MR were in consistency as the data points approximately banded around the reference 1:1 line, especially for the 50°C and 110°C treatments.

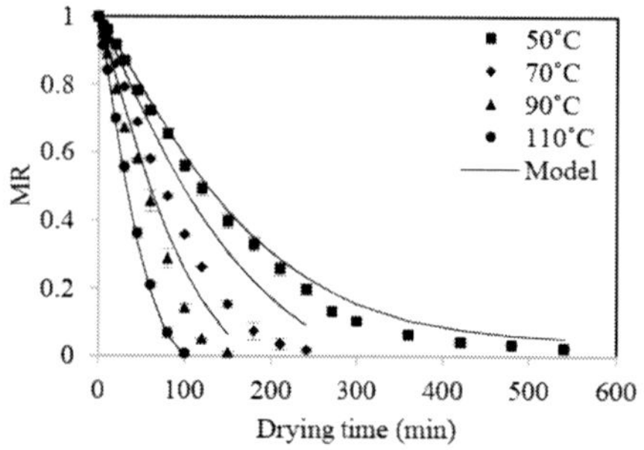

Fig. 4 Experimental and predicted drying curves of coconut residue at different drying temperatures by the generalized linear-plus-exponential model.

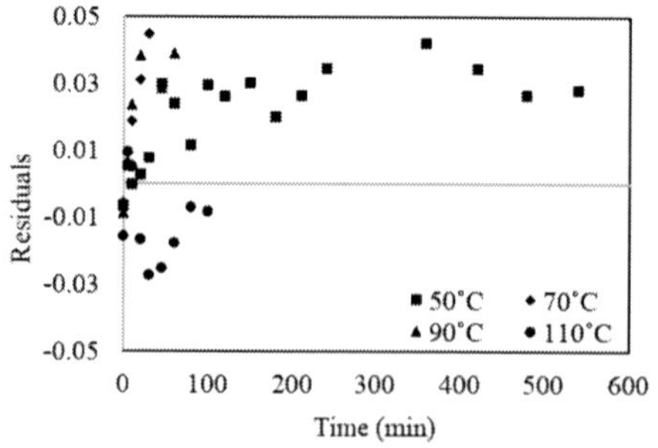

Fig. 5 Residual plot of the generalized linear-plus-exponential model for coconut residue undergoing hot air drying at different drying temperatures.

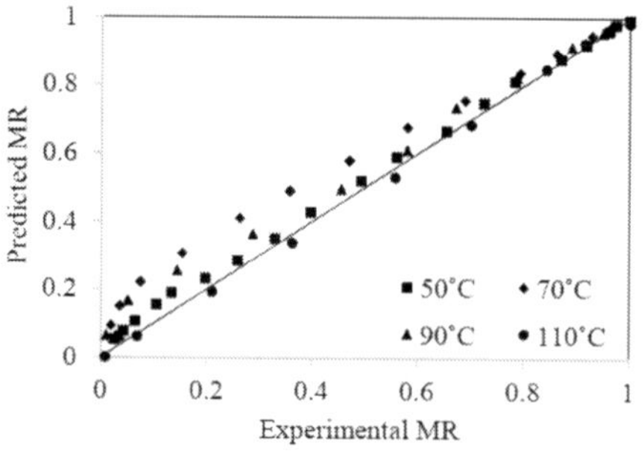

Fig. 6 Comparisons of experimental and predicted MR of coconut residue at different drying temperatures by the generalized linear-plus-exponential model.

4 Conclusion

Drying temperature had a considerable effect on drying characteristics of coconut residue subjected to hot air drying. Increasing drying temperature by 20°C resulted in shortening drying time by 50 min or higher and improving DR as almost double.

The hot air drying of coconut residue occurred in the falling rate period. Among the six drying model, the linear-plus-exponential model was the best in describing the change in MR of coconut residue with time in the temperature range of 50-110°C.

The generalized linear-plus-exponential model taking into account the effect of drying temperature provided good fitting of the predicted MR to the experimental MR with the R^2 of 0.9709. It would be useful engineering application for hot air drying of coconut residue.

References

1. L. Ramaswamy, International Journal of Ayurvedic and Herbal Medicine **4**, 1426-1436 (2014)
2. P. Suvanvisan, E. Cheevitsopon, J.S. Jongyingcharoen, the 10th TSAE International Conference, 36-41 (2017)
3. AOAC, *Official methods of analysis* (Association of Official Agricultural Chemists, Washington, DC., 1995)
4. J. Sripinyowanich, A. Noomhorm, Drying technology **29**, 735-748 (2011)
5. D.I. Onwude, N. Hashim, R.B. Janius, N.M. Nawi, K. Abdan, Comprehensive Reviews in Food Science and Food Safety, doi: 10.1111/1541-4337.12196 (2016)

Potential biodiesel production from palm oil, coconut oil and soybean oil for Thailand

Somkiat Khwanpruk and *Chalida* U-tapao *

Department of Civil Engineering, Faculty of Engineering, King Mongkut's Institute of Technology Ladkrabang, Bangkok, Thailand

Abstract. The government began subsidizing the use of B20 in large trucks on a voluntary basis beginning in 2016 and intends to implement the B10 requirement in 2018 for all diesel sales. However, policy makers in both the MOE and the Ministry of Agriculture and Cooperatives (MOAC) recently agreed that the mandatory biodiesel consumption plan for 2036 may be unattainable (given the strategy does not permit reliance on imports) and is therefore being reexamined. Despite an increase in harvested area, crude palm oil (CPO) production, the only feedstock used for biodiesel in Thailand, stagnated at 1.8-2.0 million from 2014-2016 due to unfavourable weather conditions. Thailand experiences a great economic and industrial development and is the second largest energy consumer in South East Asia. Being a net oil importer, Thai government has declared a renewable energy development programme in order to secure sustainable development and energy security. Thailand spends more than 10% of GDP for energy imports and transport sector accounts for 36% of total final energy consumption of which 50% is diesel. Diesel marks a huge impact on Thai economy.

1. Introduction

Biodiesel is one of the major biofuels for alternative energy sources of Thailand now. In the process of biodiesel producing, one of the main raw materials for biodiesel producing is oil palm, (scientific name of oil palm is Elaeis Guineensis); however, there is some limitation on the amount of oil palm feedstock ready for being raw material for biodiesel processing. The Ministry of Energy of Thailand has released the strategy for biodiesel policy of the year 2012; defining that the B10 biodiesel shall be used nationwide. [1] From the study on the Thailand's biodiesel report, Thailand's biodiesel production capacity for the year of 2015 appeared totally at 2,866.72 million liters. From the total of such 2,866.72 million liters, 2,230.33 million liters are processed from palm oil feedstock, 473.25 million liters name; jatropha, common name; physic nut, ettlespurge), and 163.14 million liters are processed from used oil [2]. are processed from jatropha oil feedstock (scientific Such the capacity should respond optional diesel replacement capacity demand at the percentage of 9.95 for the year of 2015, and it could be representing the renewable energy cost of diesel imports up to 74,964.73 million baht at current prices as shown in Table 1. In addition, the national increasement of bio – yield will be resulting in massive biodiesel production capacity in Thailand and also will be increasing the yield of biodiesel if the government focuses on agricultural capability of palm oil producing to higher yield and on utilization of modern agricultural technology able to practically raise the yield to that of 2-3 times of the actual one, and also on supporting of regionwide palm plantation to all over agricultural areas of Thailand. [1]

Table 1. Diesel replacement capacity of biodiesel.

Year	Diesel Demand	Bio - oil yield for feedstock of biodiesel producing (million liter)			Diesel replacement capacity of biodiesel (percent)	Value (million baht)
		Palm oil	Coconut oil	Total		
2012	26,191.30	2,163.21	0	2,163.21	10.35	70,863.10
2013	27,069.20	2,160.87	0	2,160.87	10.11	71,584.06
2014	27,947.20	2,186.57	0	2,186.57	9.99	73,038.26
2015	28,825.10	2,230.33	0	2,230.33	9.95	74,964.73

Note * Calculated from average diesel price, January – September, 2016, at 26.15 baht per liter.

Table 2. Raw palm oil of Thailand Market, 2014 – 2017 (ton)

* Corresponding author: chalida.ut@kmitl.ac.th

Years	Early feedstock	Yield	import	Export	Domestic consumption	
					consumption	Biodiesel processed
2014	204,000	2,000,610	-	255,252	939,369	842,398
2015	167,591	2,068,475	53,279	68,102	1,053,329	833,223
2016	334,691	1,814,496	-	56,638	970,945	890,585
2017	230,929	2,162,496	-	70,000	1,050,000	1,000,000

Source: Department of Internal Trade

Table 3. Yearly regional coconut product of Thailand

Region / Province	Product (Ton) Coconut		
	2014	2015	2016
Nationwide	970,730	904,025	857,710
Northern	6,821	5,000	3,520
Northeastern	3,598	2,890	2,123
Central	419,789	419,789	400,745
Southern	516,595	476,346	451,322

2. Property requirements for materials in biodiesel

2.1 Oil palm

For the year of 2016, it had the effective plantation area of 4.56 million rai, up from 4.30 million rai in 2015, at 6.21 percent with the yield of 11.00 million tons, down from 12.05 million tons in 2015, at 8.72 percent at the due to continual drought disaster from 2014 to early yield rate of 2,409 kg per rai, down from 2,803 kilograms per rai in 2015, at the percentage of 14.06, 2019; having been fed with lower rainfall than that of normal one, the palm bunch yield became hardly grewas shown in Table 2. However, it always remains the major bio-oil plant favourable for being used as the main raw material for biodiesel production in Thailand because it is the more potential plant for biofuel producing than others; it processes low production cost with high yield production rate. [3]

2.2 Soybean

For the years of 2012/13 – 2016/17, soybean plantation areas and its yields were likely to annually decrease by 1.97 percent and 3.42 percent, respectively. In 2016/17, the planting areas were at 0.212 million rai and its yielded at 55,979 tons, reduced from 0.217 million rai and from 56,963 tons for the years of 2015/16 at 2.30% and 1.73%, respectively [3]. The decrease in planting areas and yields was mainly due to lower benefits than that of other competitive alternative crops and due to being lack of productive soybean breeding seed.

The soybean productive yield per rai in 2012/13 - 2016/17 annually decreased at 0.47 percent in 2016/17, and that of 2016/17 appeared at 264 kilograms per rai increasing from 262 kilograms per rai for that of 2015/16 at 0.76 percent.

2.3 Coconut

Coconut plants have long been economical plant both for consumption and for industries' processes. It has been planted all over regions of Thailand as shown in the table 3.

That is the major cause to make it lower price compared to other bio-oil products such as sesame oil, soybean oil, rice bran oil, peanut oil, and sunflower oil. Due to the cost of biodiesel processing at 80% is depending on that of raw material cost, the costs of bio - oils being processed for a raw material of biodiesel processing could be found that price of the raw material, sesame oil, soybean oil, rice bran oil, peanut oil, and sunflower oil were not suitable for being used as bio-diesel raw materials, because of the high costs of such raw materials compared to that of coconut [4]. When having calculated the cost of processed biodiesel, it was found that at the investment of biodiesel production of 6.42 baht per liter (Walailak University, refer to the Department of Naval Shipyard, 2005), It could be seen that the cost of biodiesel from sesame oil, soybean oil, rice bran oil, peanut oil, and from sunflower oil were very high, costing from 33.70 - 63.52 baht. [3]

3 Analysis of results

B100, or biodiesel, in Thailand is currently produced from palm oil-derived feedstock such as crude palm oil (CPO), refined bleached deodorized palm oil (RBDPO), palm stearin, and free fatty acids of palm oil (FFA). Thailand's campaign to utilize used cooking oil for biodiesel production exists among one or two biodiesel companies as a corporate social responsibility campaign; however, the use is limited to 5-6 million liters of unused cooking oil per annum. Biodiesel production is driven by government mandates, mainly aimed to help palm farmers. All palm oil feedstocks used for biodiesel are domestic since the government strictly controls the imports of palm oil [5]. Meanwhile, the blending of biodiesel among petroleum refineries is strictly controlled and monitored to comply with the mandatory biodiesel blending requirements. All domestic diesel is required to meet these blending requirements including diesel for on-road inland shipping, trains, agriculture, and industrial. [2]

3.1 Marketing

The demand for raw palm oil in Thailand was likely to increase both for consumption and for biodiesel producing at 1.78% and at 8.11%, respectively. For 2016, the demand for crude palm oil for consumption was at 970,945 tons, decreased from 1,053,329 tons in early 2015, at 7.82%. Raw palm oil demand for biodiesel production is at 890,858 tons, raised from 833,223 tons in 2015, at 6.88%, due to the demand for actual diesel increased from 58.52 million liters in 2015 to 61.22 million liters in 2016. While, for the years of 2013 – 2016, the demand for soybean breeding seeds has annually increased by 8.03 percent [3]. For 2016, the demand was at 2.66 million tons, increasing from 2.62 million tons in 2015, at 1.53 percent. There are many purposive usages of soybean such as for oil extraction, breeding, and for processing of food products; it appears at 67.3 percent, 0.15 percent, and 32.05 percent of the overall soybean seed demand, respectively. [3]

3.2 Prices

In case of oil palm, for the years of 2012 – 2015, oil palm and palm oil prices in Thailand are likely to increase following the crude palm oil price in the global market. Additionally, there were more palm oil factories than the marketed palm oil products and resulted in competition of raw palm oil purchasing. The prices of fresh palm always were under movements. The price of fresh palm fruit sold by the farmers over the past 5 years was expected to annually increase by 4.23 percent. In 2016, the average price per kilogram was at 5.50 baht, up from 4.12 baht in 2015, at the percentage of 33.50.

In case of soybean, soybean seed and soybean oil domestic prices identically varied to that of the world market and higher global soybean yields. The negotiated price of mix-graded soybean seeds for the year of 2016 could be at 14.47 baht per kilogram, decreased from that of 15.46 baht per kilogram of the year 2015 at the percentage of 6.40 percent. The import price of soybean seeds for 2016 is at 14.43 baht, decreased from that of 2015 from 15.00 baht per kilogram and at the percentage of 3.80. The wholesale price of purified soybean oil for the year of 2016 was at 38.98 baht per kilogram, decreased from 47.48 baht per kilogram of that of 2015, at the percentage of 18.55.

In case of coconut, it has been forecasted for the prices as shown in figure 1.

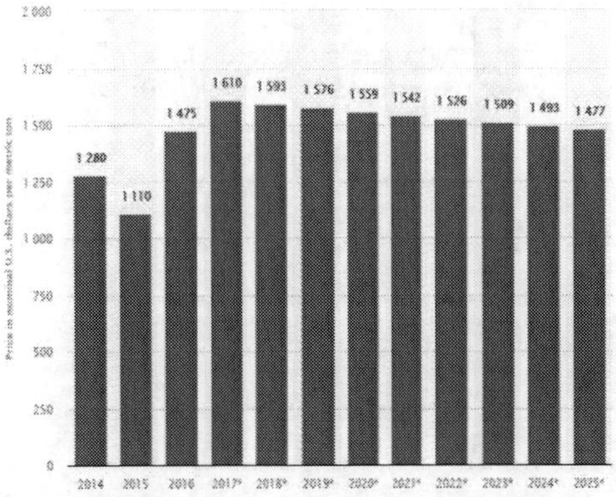

Fig 1. Statistic depicts the average annual prices for coconut.

(Source:https://www.statista.com/statistics/675808/average-prices-coconut-oil-worldwide/)

This statistic depicts the average annual prices for coconut oil from 2014 through 2025. In 2016, the average price for coconut oil stood at 1,475 nominal U.S. dollars per metric ton.

4 Conclusions

Based on the results, Thailand is currently implementing their Alternative Energy Development Plan 2015 (AEDP 2015). However, the Ministry of Energy is reportedly revising the plan's targets downward due to lower petroleum price expectations and limited feedstock supplies of ethanol and biodiesel. While the new AEDP plan may not be officially endorsed until late 2017, early revisions reduce the final target for both gasohol and biodiesel consumption in 2036 to 2.6 billion liters each. Thailand experiences a great economic and industrial development and is the second largest energy consumer in South East Asia. Being a net oil importer, Thai government has declared a renewable energy development programme in order to secure sustainable development and energy security. Thailand spends more than 10% of GDP.

References

1. T. Silalertruksa, S. H. Gheewala, Environmental sustainability assessment of palm biodiesel production in Thailand. Energy 2012, 43, 306–314.
2. BIODIESEL, THAILAND INDUSTRY OUTLOOK 2017,Available online: https://www.krungsri.com/bank/getmedia/2a1ede06-1b73-4282-99d9 3ef7c787714a/IO_Biodiesel_201705_EN.aspx
3. FAO. 2017. Agriculture Statistic(online). Available: http://www.fao.org [2017,September 20]
4. S. Patumsawad, 2nd Generation Biofuels: Technical Challenge and R&D Opportunity in Thailand. JSEE. Special Issue, 47–50 (2011)
5. FAOStat. Food and Agricultural Organization Statistics. 2012. Available online:http://faostat.fao.org/site/567/default.aspx#ancor (Accessed on 1 December 2012)

A complementarity-based equilibrium model of biodiesel market, in Thailand

Somkiat Khwanpruk and *Chalida* U-tapao*

Department of Civil Engineering, Faculty of Engineering, King Mongkut's Institute of Technology Ladkrabang, Bangkok, Thailand

Abstract. This paper presents a complementarity-based model making a decision for energy production in biodiesel market. This model has three strategic players complete against each other on the biodiesel productions. All three players make up biodiesel from agricultural products and they are price-takers. The first player represents producers who produce and sell biodiesel from palm oil. The second and third players produce and sell biodiesel from coconut oil and soybean oil, respectively. The strategic player's decisions involve quantities of palm, coconut and soybean oil related with the market price. This paper uses operation research to optimize a decision maker's objectives within the limits of available agricultural products and also provide results for sustainable energy production.

1 Introduction

Biodiesel is a significant alternative energy. Ministry of Energy. In Thailand, biodiesel is an alternative renewable energy producing from agricultural products such as palm oil, coconut oil, soybean oil and jatropha oil. Biodiesel produced from agricultural products was about 2,700 million liters (ML) which they were calculated almost 100 percentage of the potential of biodiesel production of Thailand in 2015 [1]. Moreover, Thai population used biodiesel from agricultural product to reduce import 74,964 Million Baht (at present price) Consumption of biodiesel is increased in transportation sector due to demand and supply. Thailand is not a produced petroleum country. As such, Thai government pushes many policies to increase palm, coconut and soybean production and getting oil from them. However, palm, coconut and soybean are also important raw material for various food products. For example, palms are used to produce cooking oil, soybeans are raw materials for human and animal food and coconut is general food. As mention before, palm, coconut and soybean productions are divided into two groups of producers, which are energy and food producers. Therefore, not only prices of raw material for food products prices attract palm, coconut and soybean producers but also biodiesel price is another benefit from these agricultural products.

This paper presents and analyzes a competitive market equilibrium model that can be applied for biodiesel energy market. The decision can help palm oil, coconut oil and soybean oil producers or decision-makers sell their products to either energy or food producers.

2 An equilibrium model for biodiesel market

In what follow, we describe all producer participants including palm oil, coconut oil and soybean oil producers, as well as a detail description of biodiesel market. We also present about decision variables, objective function, constraints, and their associated optimization problem [2].

2.1 Producer participants and biodiesel market

Producer participants in this equilibrium model for biodiesel market include palm, coconut and soybean oil producers. Three producers can make a decision on selling palm (q_p), coconut (q_c) and soybean (q_s) either to energy producers (q_{pE}, q_{cE}, q_{sE}) or to food producers (q_{pF}, q_{cF}, q_{sF}), respectively. The growers can decide to sell all their products to energy producers, sell some of them, or don't sell any of them depending on the demand on biodiesel market. Fig. 1 shows flow diagram of palm, coconut and soybean oil to biodiesel market.

2.2 Decision variables and parameters

For our work decision variables and parameters as follows [1], [3], [4] and notations below;

q_{pE} = palm oil production (liter)

q_{cE} = coconut oil production (liter)

q_{sE} = soybean oil production (liter)

q_{bp} = biodiesel is produced from palm oil (liter)

q_{bc} = biodiesel is produced from coconut oil (liter)

q_{bs} = biodiesel is produced from soybean oil (liter)

π_b = biodiesel prices (\$/liter)

γ_{bp} = production costs of biodiesel from palm oil (baht/liter)

γ_{bc} = production costs of biodiesel from coconut oil (baht/liter)

γ_{bs} = production costs of biodiesel from soybean oil (baht/liter)

* Corresponding author: chalida.ut@kmitl.ac.th

 https://doi.org/10.1051/matecconf/201819203063

$\bar{q}_{EP}$ = maximum amount of biodiesel is produced from palm oil

$\bar{q}_{EC}$ = maximum amount of biodiesel is produced from coconut oil (liter)

$\bar{q}_{ES}$ = maximum amount of biodiesel is produced from soybean oil (liter)

b = -6.6823

a = 49.97

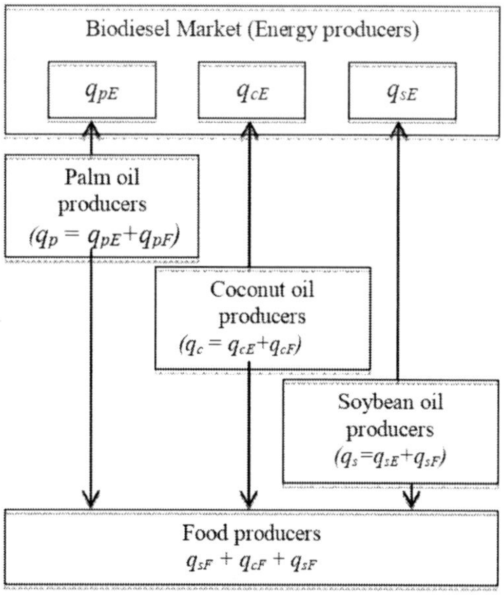

Fig. 1 Flow diagram of palm, coconut and soybean oil to biodiesel market

2.3 Objective function and constraints

Objective function of this equilibrium model is to maximize energy producers' profits (revenues minus cots). Three players are involved including producers who produce biodiesel from palm oil (described in (1) and (2)), from coconut oil (described in (3) and (4)) and from soybean oil (described in (5) and (6)). The capacity constraints describe the maximum amount of palm, coconut and soybean oil used to produce biodiesel [5].

$$Max_{q_{pE}} \pi_E q_{pE} - \gamma_{pE} q_{pE} \quad (1)$$

$$\text{s.t.} \quad q_{pE} \leq \bar{q}_{pE}(\lambda_{pE}) \quad (2)$$

$$Max_{q_{cE}} \pi_E q_{cE} - \gamma_{cE} q_{cE} \quad (3)$$

$$\text{s.t.} \quad q_{cE} \leq \bar{q}_{cE}(\lambda_{cE}) \quad (4)$$

$$Max_{q_{sE}} \pi_E q_{sE} - \gamma_{sE} q_{sE} \quad (5)$$

$$\text{s.t.} \quad q_{sE} \leq \bar{q}_{sE}(\lambda_{sE}) \quad (6)$$

Karush-Kuhn-Tucker (KKT) conditions

$$0 \leq -\pi_E + \gamma_{pE} + \lambda_{pE} \perp q_{pE} \geq 0 \quad (7)$$

$$0 \leq \bar{q}_{pE} - q_{pE} \perp \lambda_{pE} \geq 0 \quad (8)$$

$$0 \leq -\pi_E + \gamma_{cE} + \lambda_{cE} \perp q_{cE} \geq 0 \quad (9)$$

$$0 \leq \bar{q}_{cE} - q_{cE} \perp \lambda_{cE} \geq 0 \quad (10)$$

$$0 \leq -\pi_E + \gamma_{sE} + \lambda_{sE} \perp q_{sE} \geq 0 \quad (11)$$

$$0 \leq \bar{q}_{sE} - q_{sE} \perp \lambda_{sE} \geq 0 \quad (12)$$

Market-clearing condition (Supply=Demand)

$$\pi_E = a - b(q_{pE} + q_{cE} + q_{sE}) \quad (13)$$

2.4 Scenarios consideration

Each scenario considers a combination of three agricultural products. Palm, coconut and soybean productions are separated into three segments. For example, palm production is used to produce biodiesel 100%, 50% and 0%. Energy producers can also produce biodiesel from coconut 100%, 50% and 0% and soybean are used 100%, 50% and 0% to produce biodiesel. Therefore, this optimization model considers totally 27 scenarios related with the quantities of each agricultural products. For example, scenario 1 considered 100% of palm oil (5,169 million liters), 100% of coconut oil (0.584 million liters) and 100% of soybean oil (0.024 million liters), and scenario 2 considered 100% of palm oil, 100% of coconut oil and 50% of soybean oil, respectively. Fig 2 shows the detail of each scenario.

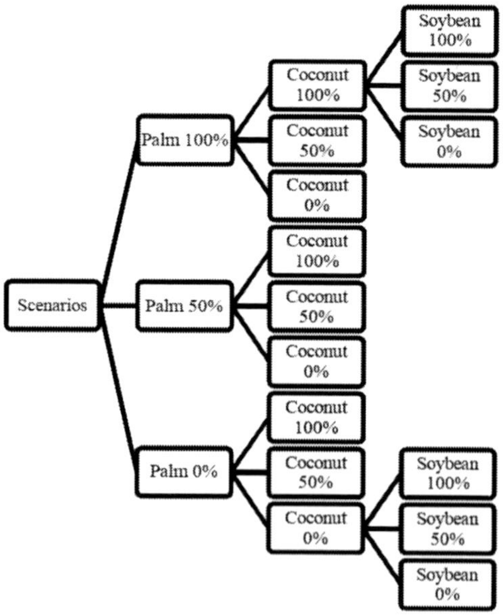

Fig. 2 Detail of consideration scenario

3 Result and discussions

Energy market equilibrium model is shown in this results. The palm oil is always selected to produce biodiesel. The quantity of palm oil was 4.96 million liters and the biodiesel prices was 17.262 baht per liter. Nevertheless, if the quantity of palm oil in the energy market were decreased, coconut oil and soybean oil would affect energy market prices. For example, the energy producers should produce biodiesel from coconut oil and soybean oil if the quantity of palm oil were decreased to 50% and zero (scenario 10 to 27) and coconut oil would affect energy market if the quantity of soybean oil and palm oil decrease to 50% and zero. The biodiesel prices would increase from 31.926 to 49.890 baht per liter if energy producers had to use coconut oil and/or soybean oil. The maximum amount of coconnut oil and soybean oil were produced biodiesel in order to reduce the demand and the prices for this energy market.

Table 1. Biodiesel prices and quantities of biodiesel form palm oil, coconut oil and soybean oil

Scenario	Quantities Palm, Coconut, Soybean oil	*qbP* (10^6 liters)	*qbC* (10^6 liters)	*qbS* (10^6 liters)	Biodiesel prices (baht/liter)
1	5,169,435, 584,220, 24,239	4.9558	0	0	17.262
2	5,169,435, 584,220, 12,120	4.9558	0	0	17.262
3	5,169,435, 584,220, 0	4.9558	0	0	17.262
4	5,169,435, 292,110, 24,239	4.9558	0	0	17.262
5	5,169,435, 292,110, 12,120	4.9558	0	0	17.262
6	5,169,435, 292,110, 0	4.9558	0	0	17.262
7	5,169,435, 0, 24,239	4.9558	0	0	17.262
8	5,169,435, 0, 12,120	4.9558	0	0	17.262
9	5,169,435, 0, 0	4.9558	0	0	17.262
10	2,584,718, 584,220, 24,239	2.5847	0.1250	0.0242	31.926
11	2,584,718, 584,220, 12,120	2.5847	0.1371	0.0122	31.926
12	2,584,718, 584,220, 0	2.5847	0.1492	0	31.926
13	2,584,718, 292,110, 24,239	2.5847	0.1250	0.0242	31.926
14	2,584,718, 292,110, 12,120	2.5847	0.1371	0.0122	31.926
15	2,584,718, 292,110, 0	2.5847	0.1492	0	31.926
16	2,584,718, 0, 24,239	2.5847	0	0.0242	32.751
17	2,584,718, 0, 12,120	2.5847	0	0.0122	32.751
18	2,584,718, 0, 0	2.5847	0	0	32.751
19	0, 584,220, 24,239	0	0.5842	0.0242	45.954
20	0, 584,220, 12,120	0	0.5842	0.0122	46.034
21	0, 584,220, 0	0	0.5842	0	46.114
22	0, 292,110, 24,239	0	0.2921	0.0242	47.882
23	0, 292,110, 12,120	0	0.2921	0.0122	47.962
24	0, 292,110, 0	0	0.2921	0	48.042
25	0, 0, 24,239	0	0	0.0242	49.810
26	0, 0, 12,120	0	0	0.0122	49.890
27	0, 0, 0	0	0	0	49.970

Considering at scenario 27 (none of palm oil, coconut oil and soybean oil), the market price was increased to 49.970 baht per liter and alternative sources might be needed to deliver to energy market. All results for each scenario shows in table 1.

4 Conclusion

This equilibrium model presentes energy producers and market. We consider the competitive model for biodiesel energy market in this paper. Three producers complete against each other in energy market. Each player is a price-takers, so the market structure is considered as a perfect competition. This equilibrium problem results when considering the Karush-Kuhn-Tucker (KKT) conditions for optimization problems (maximizing profits of energy producers) faced by energy market participants along with market-clearing condition constraint [5], [6]. The results show that the biodiesel price is reduce (17.262 baht per liter) if farmers sold palm to energy market (5.17 million liters). However, if the quantity of palm used to produce biodiesel is decrease to half and zero, then the market price will be increase. The biodiesel prices can be increased to 31.926, 32.751 and 49.970 baht per liter of biodiesel depended on the amount of palm, coconut and soybean oil in energy market.

References

1. Development and Efficiency, Ministry of Energy, 2013, *Annual report* (2012)
2. L. Laokhongthavorn, C. U-tapao, and W. Dejtisakdi, *Seventh Annual International Business Heath and Engineering Conference*, (Las Vegas, Nevada, USA, 2014)
3. Economic and social development department, *Proceedings of the Fiji/FAO Asia Pacific Sugar Conference*, (1997)
4. Bank of Thailand, 2012 "Ethanol yearly report 2012", Websit http://www.bot.or.th/Thai/EconomicConditions/Thai/Northeast/commodities/Doclib_CommodityYearly/Ethanol%20Report_2555%20-last.pdf>.
5. S. A Gabriel, S. Kiet, and J. Zhuang., Oper. Res. **53**, 799-818 (2005)
6. S. A Gabriel, A. J. Conejo, J. D. Fuller, B. F. Hobbs, and C. Ruiz, *Complementarity modeling in energy markets*, (Springer, 2010)

AUTHOR INDEX

AUTHOR INDEX

AUTHOR INDEX

AUTHOR INDEX

AUTHOR INDEX